Christopher Bailey.
~~8~~ June 1995

GW00702978

SIMULATION OF MATERIALS PROCESSING:
THEORY, METHODS AND APPLICATIONS

PROCEEDINGS OF THE FIFTH INTERNATIONAL CONFERENCE ON NUMERICAL METHODS IN INDUSTRIAL FORMING PROCESSES – NUMIFORM '95
ITHACA / NEW YORK / USA / 18-21 JUNE 1995

Simulation of Materials Processing: Theory, Methods and Applications

Edited by
SHAN-FU SHEN & PAUL R. DAWSON
Cornell University, Ithaca, New York, USA

A.A. BALKEMA / ROTTERDAM / BROOKFIELD / 1995

The texts of the various papers in this volume were set individually by typists under the supervision of each of the authors concerned.

Published by

A.A.Balkema, P.O. Box 1675, 3000 BR Rotterdam, Netherlands (Fax: +31.10.413.5947)

A.A.Balkema Publishers, Old Post Road, Brookfield, VT 05036, USA (Fax: 802.276.3837)

ISBN 90 5410 553 4

Printed in the Netherlands

Table of contents

Material behaviour

Simulation methods and large scale computation

Design and inverse methods

Forming of metal sheet

Metals – General topics

Rolling of metals

Polymer – General topics

Simulation of Materials Processing: Theory, Methods and Applications, Shen & Dawson (eds)
© 1995 Balkema, Rotterdam. ISBN 90 5410 553 4

Preface

The present volume contains the papers presented at the 5th International Conference on Numerical Methods of Industrial Forming Processes (NUMIFORM '95), held at Cornell University, Ithaca NY, June 18-21, 1995. This series of conferences was launched in 1982 at the University of Swansea, under the pioneering efforts of Prof. O.C. Zienkiewicz. Its emphasis of generating useful numerical answers from first principles underlying the highly complicated industrial forming processes immediately roused considerable interest. The number of participants maintained a steady increase in the succeeding ones, which took place in Gothenburg (1986), Fort Collins (1989), and Sophia-Antipolis (1992) prior to Ithaca.

Rapid advances of computing, meanwhile, have enabled the coverage of the studies to expand into new frontiers. Efforts extend beyond numerical '*methods*' in the strict sense of techniques of producing numerical solutions of given sets of idealized mathematical equations with specified initial and boundary conditions. More and more the task becomes the numerical '*simulation*' of real-life industrial processes. To cope with tough practical questions, better modelling of the processes in key aspects of the mechanics and materials behavior is required; with these new mathematical challenges arise. Answering the questions of how to best design or control a process demands new simulation tools as seen in the recent attention to inverse methods. Effective uses of large-scale computation and modern visualization techniques obviously are important to practical success.

Papers presented at the 5th NUMIFORM Conference have been collected in this volume. They serve to convey a general view of the current problems of industrial need, and the status quo of capabilities as well as limitations in providing quantitative answers to them. To those who have followed these conferences since the early days, the growing community and the tremendous achievement cannot fail to impress. Slowly but surely, numerical simulation is making impacts of real value to various highly complicated industrial forming processes.

Simulation of Materials Processing: Theory, Methods and Applications, Shen & Dawson (eds)
© 1995 Balkema, Rotterdam. ISBN 90 5410 553 4

Organization

CHAIRS

Shan-Fu Shen
Paul R. Dawson

STEERING COMMITTEE

J.-L. Chenot, France
A. Samuelsson, Sweden
R. I. Tanner, Australia
E. G. Thompson, USA
K. K. Wang, USA
R. D. Wood, United Kingdom

TECHNICAL & SCIENTIFIC COMMITTEE

T. Beltyschko, USA
M. J. Crochet, Belgium
J. C. Gelin, France
J. Huétink, Netherlands
A. Isayev, USA
J. Kihara, Japan
S. MacEwen, Canada
H. A. Nied, USA
M. Predeleanu, France
O. Richmond, USA
M. Saran, USA
C. Tucker, USA
R. Wagoner, USA

LOCAL ORGANIZING COMMITTEE

N. Zabaras
M. Miller
S. Mukherjee
S. L. Phoenix
N. Hieber

SPONSORS

AC Tech
ALCOA
Ford Motor Company
Reynolds Metals Company

Keynote lectures

Simulation of Materials Processing: Theory, Methods and Applications, Shen & Dawson (eds)
© 1995 Balkema, Rotterdam. ISBN 90 5410 553 4

Some simulation issues in polymer processing

Morton M. Denn
Department of Chemical Engineering, University of California, Berkeley & Materials Science Division, Lawrence Berkeley Laboratory, Calif., USA

ABSTRACT: There are a number of unresolved physical problems which have a bearing on the simulation of polymer melt processes. The nature of the stress field near geometric singularities is unknown. There is evidence for wall slip at high stresses, but the form and generality of slip remains unclear. Solidification under stress cannot be predicted adequately. Liquid-liquid phase transitions may occur for some molecular architectures at high stresses.

INTRODUCTION

Most objects fabricated from polymers are processed in the melt, where a substantial amount of property development results from macromolecular orientation which is induced during liquid-phase shaping. The morphology resulting from the solidification step is also important in determining properties, as is subsequent solid-phase thermal and mechanical treatment. In this paper we will identify major issues associated with the simulation of melt processing and solidification.

Polymer melts are made up of entangled chains. The relaxation processes associated with the response of the chains to macroscopic deformations induce time scales for the stress which are often of the order of seconds; hence, polymer melts are viscoelastic, in that the present stress state depends on the history of deformation. Stress constitutive relations adequate to describe melt behavior are typically families of nonlinear differential equations of high order, or nonlinear integral equations. Furthermore, processing operations may be much faster than characteristic relaxation times, so equilibrium material properties may not be the relevant variables. The most comprehensive analytical analysis of polymer processing is probably to be found in Pearson (1985). Detailed discussions of the computational approaches to simulation of polymer processing operations are in Pearson and Richardson (1983) and Tucker (1989), as well as more specialized texts. A discussion of unresolved fluid-mechanical issues is given by Denn (1990). An overview of extrusion instabilities and related issues concerning polymer/nonpolymer interactions can be found in Denn (1994).

NUMERICS

Considerable attention has been devoted to the numerical simulation of polymer melt flow, and commercial software applicable to a wide range of continuous and batch processes is available. The complexity of stress constitutive equations and the three-dimensional nature of many processes have been important issues in the development of practical algorithms, but there is an underlying theoretical issue of overriding concern. Computational algorithms for many constitutive equations fail to converge in flows with geometric singularities (corners, die lips, etc.) when the dimensionless product of a characteristic relaxation time and the deformation rate (the "Weissenberg number" or, sometimes, the "Deborah number") grows large, and the range of processing interest is often unattainable. The extensive literature on this subject is mostly to be found in *J. Non-Newtonian Fluid Mechanics*, including publication of special refereed issues based on continuing biannual workshops. It is believed that most viscoelastic constitutive equations lead to stresses near geometric singularities and stagnation points which grow much more rapidly than those in Newtonian liquids, and the numerical problem is associated with resolution of these large gradients. The critical Weissenberg number depends on the flow problem, the particular constitutive equation, and the details of the algorithm, and some procedures are in place which allow computation to very large Weissenberg numbers. It is not obvious, however, that existing models adequately represent the true physical situation near the singularity, and the real problem may be one of inappropriate constitutive theory or improper boundary conditions (cf. Lipscomb *et al.*, 1987; El-Kareh & Leal, 1989; Bhave *et al.*, 1991; Coates *et al.*, 1992). Local solutions in the neighborhood of boundary singularities have not been found for constitutive equations of interest, nor have existence theorems been established. These are challenging mathematical problems which bear directly on practical processing computations.

POLYMER/NONPOLYMER INTERFACES

The "no-slip" boundary condition, in which fluid in contact with a solid surface is assumed to have the same velocity as the surface, has long been a fundamental principle of fluid mechanics, despite occasional suggestions that the condition is approximate (see a brief review in Schowalter, 1988). On the other hand, wall slip has been accepted as normal in rubber processing. At the high stress levels encountered in many processing operations (in excess of 0.1 MPa) it would not be surprising if polymer melts behaved more like rubber than the low-molecular-weight liquids upon which classical fluid mechanics is based. Nevertheless, reports of wall slip in polymer melts have generally been ignored or discredited. During the past decade the situation has changed somewhat, largely as a result of reports from Union Carbide (Ramamurthy, 1986) that the onset of an extrusion instability in linear polyethylenes known as "sharkskin" can be dependent on the material of construction of the extrusion die. (This observation had been made previously but had not attracted the same attention.) The implication of this claim is profound, for it makes the die surface an active participant in the mechanics of forming, rather than a passive element whose sole purpose is to provide anchoring for no-slip.

The picture of wall slip which has developed draws heavily on analogies to solid-phase adhesion and dynamic fracture. The formulation by Hill and coworkers (1990) relates the relative interfacial velocity between melt and metal (or metal oxide) to the local shear stress, temperature, pressure, and melt shear modulus, as well as to the thermodynamic work of adhesion between polymer and wall and to a power-law exponent measured in a peel test on the solid polymer. Several theories are compared in Denn (1992), where it is noted that similar functional forms emerge despite different starting assumptions. A paper by Hatzikiriakos (1995), using molecular dynamics to calculate a slip function with memory, is the most recent contribution to a growing literature.

Whether wall slip really affects extrusion is controversial. El Kissi and Piau (1994) argue strongly against slip as a factor in the sharkskin instability, for example. Convincing direct measurements of slip at low stresses (but not in polyethylene) have been reported by Migler and coworkers (1993, 1994). Person and Denn (unpublished) have carried out extrusion experiments with linear low-density polyethylene in a slit die in which the die faces can be changed; they have shown that within the sharkskin regime different heat treatments of brass die faces can affect the throughput by ten percent for a given pressure drop. Die dimensions are unchanged, but spectroscopy shows differences in surface alloy compositions. Though this effect is not large, it is difficult to find any explanation other than different degrees of wall slip. In a less-convincing experiment, Halley and Mackay (1994) showed that the exit pressure, which is associated with flow rearrangement, changes for linear low-density polyethylene with the use of different metal inserts just prior to the die exit.

The consequences for process simulation are significant if the generality of wall slip at high stress levels is established. The stress singularity near boundary discontinuities cannot develop with slip, and the numerical problem described above is moderated. Wall stress levels beyond the onset of slip will be lower than otherwise computed, resulting in less orientation. Slip combined with the small degree of compressibility characteristic of polymer melts can result in unsteady flow under constant extrusion conditions (Georgiou and Crochet, 1994a,b), and dynamical simulations may sometimes be necessary.

SOLIDIFICATION

Many engineering polymers are semicrystalline, and properties depend on the degree of crystallinity and crystal morphology. It is known that crystallization rates can be enhanced significantly by stress, but the details of the process are not well-understood and useful kinetic expressions are not available. Quiescent crystallization kinetics are usually employed in simulations, and these are sometimes adequate for estimating the spatial distribution of the degree of crystallinity. Extensional processes, like the formation of fibers from melts, result in high degrees of stress-induced orientation, and it is likely that crystallization rates are greatly enhanced.

The formation of fibers from polymer melts has been simulated extensively, and at least for polyesters at moderate takeup speeds (up to 3,500 m/min), where the solidification appears to be mostly amorphous and to take place near the glass-transition temperature, the subject is viewed as mature; most fiber producers are known to use steady-state simulations for guidance in process operation, and dynamic analyses seem to capture major features and to explain instabilities. An overview is given in the chapter by Denn in Pearson and Richardson (1983). The situation is quite different when the polymer is crystalline. High-speed spinning of polyester (above 5,000 m/min) leads to a highly-crystalline fiber, and available models have been unable to describe the observed formation of a "neck" which forms (apparently) just prior to solidification (Ziabicki and Kawai, 1985). Devereux and Denn (1994) recently found that a dynamic simulation of disturbance propagation in melt spinning of polypropylene, using the "standard" model and the best available rheological information, gave amplitudes that were in error by orders of magnitude, failed to predict the location of the dominant resonant peak properly, and even predicted dynamic instability under conditions where the system was operating stably. The model works well for polyester dynamics; the most likely cause of the problem is the absence of an adequate description of the solidification process, and this is where attention is needed. Solidification under stress is also likely to be an important factor in the properties of products formed from the blown-film process.

MESOPHASE FORMATION

An unexplored area which may have significant processing significance is stress-induced phase change to form a polymer nematic liquid crystal, or *mesophase*. Nematic mesophases are highly-oriented and have viscosities which are considerably lower than amorphous systems of the same molecular weight. It has been suggested that necking in high-speed polyester spinning is the result of a stress-induced mesophase transition.

Waddon and Keller (1990, 1992) have reported a significant decrease in extrusion pressure for high-molecular-weight polyethylenes in a narrow temperature range, which they ascribe to a stress-induced mesophase transition. Pudjijanto and Denn (1994) found a stable "island" and a large drop in extrusion pressure in the interior of the "slip-stick" regime, an unstable region of (apparently) alternating slip and stick which is always observed for linear polyethylenes following sharkskin. The island was observed for a linear low-density polyethylene of considerably lower molecular weight than the polymers used by Waddon and Keller, but it was not found with a linear low-density polyethylene of slightly different backbone structure (different short-chain branching) but with the same molecular weight and viscosity function. It is likely that the observations of reduced extrusion pressure by both groups are of the same phenomenon, although Pudjijanto and Denn could find no evidence of a liquid-crystalline phase.

The possibility of a phase transition at high stresses to a liquid with significantly-reduced viscosity raises interesting processing possibilities, but a direct experimental test is unlikely. A stress-induced liquid-crystalline layer at the wall in extrusion would be difficult to distinguish from true slip flow. One provocative observation is that whereas linear polyethylenes exhibit apparent slip with extrusion instabilities (i.e., the flow is less dissipative), polystyrene becomes more dissipative with the onset of unstable flow (Shidara and Denn, 1993); polystyrene has bulky side groups on the chain which would reduce the probability that the molecule could be pulled out by a stress field into the extended-chain conformation needed for a mesophase. Hence, stress-dependent liquid-liquid phase transitions are likely to be very sensitive to molecular architecture.

ACKNOWLEDGMENT

Our work on extrusion instabilities and polymer/nonpolymer interactions is supported by the Director, Office of Energy Research, Office of Basic Energy Sciences, Materials Science Division of the U. S. Department of Energy under Contract No. DE-AC03-76SF00098.

REFERENCES

Bhave, A. V., R. C. Armstrong & R. A. Brown 1991. Kinetic theory and rheology of dilute, nonhomogeneous polymer solutions. *J. Chem. Phys.* 95:2988-3000.

Coates, P. J., R. C. Armstrong, & R. A. Brown 1992. Calculation of steady-state viscoelastic flow through axisymmetric contractions with the EEME formulation. *J. Non-Newtonian Fluid Mech.* 42:141-188.

Denn, M. M. 1990. Issues in viscoelastic fluid mechanics. *Ann. Rev. Fluid Mech.* 22:13-34.

Denn, M. M. 1992. Surface-induced effects in polymer melt flow. In P. Moldenaers & R. Keunings (eds.), *Theoretical and Applied Rheology*: 45-49. Amsterdam: Elsevier.

Denn, M. M. 1994. Polymer Flow Instabilities: A Picaresque Tale , *Chem. Engng. Education* 28:162-166.

Devereux, B. M. & M. M. Denn 1994. Frequency response analysis of polymer melt spinning. *Ind. Eng. Chem. Research* 33:2384-2390.

El-Kareh, A. W. & L. G. Leal 1989. Existence of solutions for all Deborah numbers for a non-newtonian model modified to include diffusion. *J. Non-Newtonian Fluid Mech.* 33:257-287.

El Kissi, N. & J. M. Piau 1994. Adhesion of linear low density polyethylene for flow regimes with sharkskin. *J. Rheology* 38:1447-1463.

Georgiou, G. C. & M. J. Crochet 1994a. Compressible viscous flow in slits with slip at the wall. *J. Rheology* 38:639-654.

Georgiou, G. C. & M. J. Crochet 1994b. Time-dependent extrudate-swell problem with slip at the wall. *J. Rheology* 38:1745-1755.

Halley, P. J. & M. E. Mackay 1994. The effect of metals on the processing of LLDPE through a slit die. *J. Rheology* 38:41-51.

Hatzikiriakos, S. G. 1995. A multimode interfacial constitutive equation for molten polymers. *J. Rheology* 39:61-71.

Hill, D. A., T. Hasegawa & M. M. Denn 1990. On the apparent relation between adhesive failure and melt fracture. *J. Rheology* 34:497-523.

Lipscomb, G. G., R. Keunings & M. M. Denn 1987. Implications of boundary singularities in complex geometries. *J. Non-Newtonian Fluid Mech.* 24:85-96.

Migler, K. B, H. Hervet & L. Leger 1993. Slip transition of a polymer melt under shear stress. *Phys. Rev. Letters* 70:287-290.

Migler, K. B, G. Massey, H. Hervet & L. Leger 1994. The slip transition at the polymer solid interface. *J. Physics-Condensed Matter* 6:A301-A304.

Pearson, J. R. A. 1985. *Mechanics of polymer processing*. London: Elsevier Applied Science.

Pearson, J. R. A. & S. Richardson (eds.) 1983. *Computational analysis of polymer processing*. London: Applied Science.

Pudjijanto, S. & M. M. Denn 1994. A stable "island" in the slip-stick region of linear low-density polyethylene. *J. Rheology* 38:1735-1744.

Ramamurthy, A..V. 1986. Wall slip in viscous fluids and influence of material of construction. *J. Rheology* 30:337-357.

Schowalter, W. R. 1988. The behavior of complex fluids at solid boundaries. *J. Non-Newtonian Fluid Mech.* 29:25-36.

Shidara, H. & M. M. Denn 1993. Polymer melt flow in very thin slits, *J. Non-Newtonian Fluid Mech.* 48:101-110.

Tucker, C. L. III (ed.) 1989. *Fundamentals of computer modeling for polymer processing*. Munich: Hanser.

Waddon, A. J. & A. Keller 1990. A temperature window of extrudability and reduced flow resistance in high-molecular weight polyethylene: interpretation in terms of a mobile hexagonal phase. *J. Polym. Sci.* B28:1063-1073.

Waddon, A. J. & A. Keller 1992. The temperature window of minimum flow resistance in melt flow of polyethylene: further studies on the effect of strain rate and branching. *J. Polym. Sci.* B30:923-929.

Ziabicki, A. & H. Kawai (eds.) 1985. *High-speed fiber spinning: science and engineering aspects*. New York: Wiley.

Simulation of Materials Processing: Theory, Methods and Applications, Shen & Dawson (eds)
© 1995 Balkema, Rotterdam. ISBN 90 5410 553 4

Advanced computational strategies for 3-D large scale metal forming simulations

D.R.J.Owen, D.Perić, E.A.de Souza Neto, Jianguo Yu & M.Dutko
Department of Civil Engineering, University of Wales, Swansea, UK

A.J.L.Crook
Rockfield Software Ltd, The Innovation Centre, University of Wales, Swansea, UK

ABSTRACT — The paper discusses some of the relevant computational advances which permit the simulation of industrial metal forming problems within realistic time frames and computational resources. The need for rigorous consideration of both theoretical and algorithmic issues is emphasised, particularly in relation to the computational treatment of finite strain elasto-plastic (viscoplastic) deformation, the modelling of frictional contact conditions and element technology capable of dealing with plastic incompressibility. Commercially important aspects such as adaptive mesh refinement procedures are discussed and attention is given to choice of appropriate error estimators for elasto-plastic materials and the transfer of solution parameters between succesive meshes. The role of explicit solution techniques in the simulation of metal forming problems is also discussed. The benefits of recent computational developments are illustrated by the solution of a range of industrially relevant problems.

1. INTRODUCTION

Computational procedures, which are almost exclusively based on the finite element method, are now routinely employed in the modelling of metal forming operations and the acceptance of such approaches to industrial design is continually increasing, due to improved awareness, availability of appropriate software and reducing computational costs.

Probably the most striking example of developments in this field is the strides recently made in the numerical solution of finite strain plasticity problems. The formulation of rigorous solution procedures has been the subject of intense debate over the last decade and only recently has some consensus been reached on an appropriate constitutive theory based on tensorial state variables to provide a theoretical framework for the macroscopic description of a general elasto-plastic material at finite strains. In computational circles, effort has been directed at the formulation of algorithms for integration of the constitutive equations relying on operator split methodology. In particular the concept of consistent linearisation has been introduced to provide quadratically convergent solution procedures. By employing logarithmic stretches as strain measures a particularly simple model for large inelastic deformations at finite strains is recovered. In particular, the effects of finite strains appear only at the kinematic level and the integration algorithms and corresponding consistent tangent operators for small strain situations can be directly employed.

A further class of nonlinear problem for which considerable advances in numerical modelling have been made in recent years is that of contact-friction behaviour. Contact-friction phenomena arise in many important areas of forming and numerical treatment in the past has, of necessity, relied on temperamental and poorly convergent algorithms. This situation has changed markedly with recognition of the complete analogy that exists between contact-friction behaviour and the classical theory of elasto-plasticity. Hence, the operator split algorithms and consistent linearisation procedures developed for the latter case translate directly to contact-friction models to provide robust and rapidly convergent numerical solutions.

A general feature encountered in the FE simulation of finite strain plasticity problems is that the optimal mesh configuration changes continually throughout the deformation process requiring mesh derefinement as well as mesh refinement during the adaptive process. Considerable benefits may accrue for such problems in terms of robustness and efficiency, realizing that the requirements of computational efficiency are ever increasing. At the same time, error estimation procedures will play a crucial role in quality assurance by providing reliable finite element solutions.

A current area of crucial debate in the computational modelling of metal forming operations is the relative merits of explicit and implicit solution strategies. Issues of particular concern are the accuracy of explicit approaches in relation to implicit solutions and the relative computational efficiency of both approaches.

The remainder of this paper discusses the above (and other) issues in more detail and provides numerical examples illustrating some of the recent advances made in the finite element analysis of forming problems.

2. FINITE STRAIN ELASTO-PLASTICITY

2.1 Constitutive equations for large strain elasto-plasticity

It is only recently that the constitutive theory for finite strain plasticity, commonly based on the multiplicative decomposition of the deformation gradient, and the associated numerical integration algorithms, have received a complete rational basis. Notable contributions in this area are presented in [4,7,18-19,25,28-29,31]. It is generally accepted that a sufficiently general constitutive model with possible wide application, including most of metal plasticity, together with considerable simplifications in the numerical treatment of finite strain elasto-plasticity, may be defined by employing a quadratic strain energy function in the form of a scalar symmetric function of its principal stretches $\lambda^e_{(i)}, (i = 1,2,3)$

$$W(\lambda^e_{(1)}, \lambda^e_{(2)}, \lambda^e_{(3)}) = \mu \left((\ln \lambda^e_{(1)})^2 + (\ln \lambda^e_{(2)})^2 + (\ln \lambda^e_{(3)})^2 \right) + \tfrac{1}{2}\lambda \, (\ln J^e)^2 \quad (2.1)$$

where μ, λ are non-zero material constants and $J^e = \lambda^e_{(1)}\lambda^e_{(2)}\lambda^e_{(3)}$ is the Jacobian. By denoting a logarithmic strain as $\boldsymbol{E} := \ln \boldsymbol{U}$ and its work conjugate stress as $\boldsymbol{T} := \boldsymbol{R}^T \boldsymbol{\tau} \boldsymbol{R}$ the constitutive equations for large deformations of elasto-plastic solids at finite strains, with attention restricted to isotropic elasto-plasticity, can be represented in a form similar to the standard small strain plasticity equations as (see [4,7,18-19,25,28-29,31] for details)

$$\boldsymbol{F} = \boldsymbol{F}^e \boldsymbol{F}^p \quad (2.2a)$$
$$\boldsymbol{T} = \partial_{\boldsymbol{E}^e} W = 2\mu \boldsymbol{E}^e + \kappa \operatorname{tr}[\boldsymbol{E}^e] \boldsymbol{1} \quad (2.2b)$$
$$\operatorname{sym}\left[\dot{\boldsymbol{F}}^p \boldsymbol{F}^{p\,-1}\right] = \dot{\gamma}\, \partial_{\boldsymbol{T}} \Phi(\boldsymbol{T}, K) \quad (2.2c)$$
$$\dot{K} = \dot{\gamma} H(\boldsymbol{E}^p, K) \quad (2.2d)$$

Here $\Phi(\boldsymbol{T}, K)$ is the yield surface, K is the scalar variable describing isotropic hardening while function $H(\boldsymbol{E}^p, K)$ defines the hardening law. Finally, loading/unloading conditions may be formulated in the standard Kuhn-Tucker form

$$\Phi \le 0, \qquad \dot{\gamma} \ge 0, \qquad \dot{\gamma}\Phi = 0. \quad (2.3)$$

2.2 Numerical integration of the large strain elasto-plastic constitutive model

Procedures for numerical integration of the rate type constitutive equations which are particularly suitable for implementation are connected with general operator split methodology, where the original problem of evolution is solved through composition, applying first the elastic and then the plastic algorithm.

As a crucial point in the derivation of the algorithm, an *exponential approximation* is used in the discretisation of the plastic flow rule in the plastic corrector stage first employed in its original form in the computational literature by Weber and Anand [31] (see also [4,7,25,29]). It leads to the following incremental evolution equation

$$\boldsymbol{F}^p_{n+1} = \exp\left[\Delta\gamma \partial_{\boldsymbol{T}} \Phi_{n+1}\right] \boldsymbol{F}^p_n \quad (2.4)$$

It can be easily shown that the above approximation results in

$$\boldsymbol{E}^e_{n+1} = \boldsymbol{E}^{e\,\mathrm{trial}}_{n+1} - \Delta\gamma \partial_{\boldsymbol{T}} \Phi_{n+1} \quad (2.5)$$

which is valid whenever $\boldsymbol{C}^{e\,\mathrm{trial}}_{n+1} := (\boldsymbol{F}^{e\,\mathrm{trial}}_{n+1})^T \boldsymbol{F}^{e\,\mathrm{trial}}_{n+1}$ and $\partial_{\boldsymbol{T}}\Phi_{n+1}$ commute. In conjunction with the quadratic strain energy function expressed in terms of logarithmic stretches (2.1), equations (2.2-5) lead to a particularly simple format of the integration algorithms for multiplicative finite strain plasticity which is equivalent to the small strain plasticity algorithms.

3. CONTACT-FRICTION MODELLING

At the stage of the deformation process corresponding to the deformation mappings $\boldsymbol{\chi}^s$ and $\boldsymbol{\chi}^m$ of the slave and master bodies respectively, the *gap* separating a material point $\boldsymbol{x}$ on Γ_s from the master boundary Γ_m is defined by

$$g_N(\boldsymbol{x}) := [\boldsymbol{\chi}^s(\boldsymbol{x}) - \boldsymbol{\chi}^m(\boldsymbol{y})] \cdot \boldsymbol{N} \quad (3.1)$$

where $\boldsymbol{y}$ is the material point on Γ_s currently defining the closest distance between $\boldsymbol{x}$ and the master body.

Assume that contact has been established between $\boldsymbol{x}$ and $\boldsymbol{y}$, i.e., $g_N = 0$. The subsequent relative displacement (under contact conditions) between the two material points will be denoted $\boldsymbol{u}$. For convenience, it will be decomposed as

$$\boldsymbol{u} = \boldsymbol{u}_T + u_N \boldsymbol{N} \quad (3.2)$$

where $\boldsymbol{u}_T := (\boldsymbol{I} - \boldsymbol{N} \otimes \boldsymbol{N})\boldsymbol{u}$ and $u_N := \boldsymbol{u} \cdot \boldsymbol{N}$ are, respectively, the tangential and normal components of $\boldsymbol{u}$.

The same additive decomposition will be considered for the contact pressure, $\boldsymbol{p}$, acting on the slave body, i.e.,

$$\boldsymbol{p} = \boldsymbol{p}_T + p_N \boldsymbol{N}. \quad (3.3)$$

3.1 Normal behaviour

In the definition of the normal contact behaviour, it is assumed that penetration between the two bodies is admissible. In addition, a linear relation between the normal reaction and the penetration, $g_N < 0$, is postulated resulting in the following constitutive function for p_N:

$$p_N = \begin{cases} -k_N g_N & \text{if} \quad g_N < 0 \\ 0 & \text{otherwise} \end{cases} \tag{3.4}$$

where k_N is the *normal stiffness* or *penalty factor.*

3.2 Tangential behaviour

Following standard arguments of the elastoplasticity theory of friction [5,14,20,22,30], the decomposition of the tangential relative velocity $\dot{\boldsymbol{u}}_T$ into an *adherence* and a *slip* component is adopted,

$$\dot{\boldsymbol{u}}_T = \dot{\boldsymbol{u}}_T^a + \dot{\boldsymbol{u}}_T^s \tag{3.5}$$

along with the following constitutive law for the tangent reaction on the slave boundary

$$\boldsymbol{p}_T = -k_T\, \boldsymbol{u}_T^a \tag{3.6}$$

where k_T is the *tangential contact stiffness.* The evolution law for the slip component is defined by

$$\dot{\boldsymbol{u}}_T^s = -\dot{\gamma}\, \frac{\partial \Psi}{\partial \boldsymbol{p}_T} \tag{3.7}$$

where the *slip potential* Ψ defines the direction of frictional sliding and $\dot{\gamma}$ is consistent with the loading/unloading condition:

$$\Phi(\boldsymbol{p}, A) \leq 0 \qquad \dot{\gamma} \geq 0 \qquad \dot{\gamma}\, \Phi(\boldsymbol{p}, A) = 0. \tag{3.8}$$

The *slip criterion* Φ above, is assumed to be a function of the contact reaction $\boldsymbol{p}$ and (possibly) a set A of internal variables taking into account the history dependence of the friction phenomenon.

For *isotropic* frictional contact, frictional sliding may occur only in the direction opposite to the tangential reaction. Hence, the slip potential Ψ is given by

$$\Psi(\boldsymbol{p}_T) = ||\, \boldsymbol{p}_T\, ||. \tag{3.9}$$

As a central hypothesis of the present model for friction with hardening, the following particular form for the slip criterion is postulated:

$$\Phi(\boldsymbol{p}, w) = ||\, \boldsymbol{p}_T\, || - \mu(w)\, p_N \tag{3.10}$$

where *the friction coefficient, μ, is assumed to be a function of the single scalar internal variable w* defined on the *slave* boundary.

In addition, the internal variable w is chosen as the density of *frictional work* expended on the point considered. Hence, its evolution equation is defined by

$$\dot{w} = -\boldsymbol{p}_T \cdot \dot{\boldsymbol{u}}_T^s. \tag{3.11}$$

The resulting constitutive model for frictional contact with hardening is analogous to classical work hardening plasticity.

3.3 Numerical integration of the frictional contact model

The similarity between rate independent elastoplasticity and the contact problem with friction makes techniques such as the operator split methodology, employed in Section 2.2 in the context of large strain plasticity, particularly suitable for the numerical integration of the constitutive equations for frictional contact presented above.

Firstly, at a given configuration χ_{n+1}, defined by an incremental displacement $\Delta \boldsymbol{u}$, the normal contact reaction is updated as

$$\boldsymbol{p}_{N\,n+1} = g_N(\chi_{n+1}) \tag{3.12}$$

Then, the corresponding tangential contact reaction, $\boldsymbol{p}_{T\,n+1}$, is computed by means of the procedure described below.

3.3.1 Elastic predictor With the fixed contact pressure determined above, the elastic trial state is evaluated:

$$\boldsymbol{p}_{T\,n+1}^{\text{trial}} = \boldsymbol{p}_{T\,n} - k_N\, \Delta \boldsymbol{u}_T \tag{3.13}$$

If the consistency condition

$$\Phi_{n+1}^{\text{trial}} = ||\boldsymbol{p}_{T\,n+1}^{\text{trial}}|| - \mu(w_n)\, p_{N\,n+1} \leq 0 \tag{3.14}$$

is violated, then frictional sliding occurs and the corrector procedure must be employed to compute the frictional force:

3.3.2 Frictional slip corrector In this stage, the rate evolution equation for w, discretized by a one-step backward Euler scheme, is solved in conjuction with the requirement that the slip function Φ vanishes during frictional sliding, i.e., the following system is solved for the variables $\Delta\gamma$ and w_{n+1}:

$$||\boldsymbol{p}_{T\,n+1}^{\text{trial}}|| - k_T\, \Delta\gamma - \mu(w_{n+1})\, p_{N\,n+1} = 0 \tag{3.15}$$

$$w_{n+1} - w_n - \mu(w_{n+1})\, p_{N\,n+1}\, \Delta\gamma = 0 \tag{3.16}$$

and the frictional force is updated as

$$\boldsymbol{p}_{T\,n+1} = \mu(w_{n+1})\, p_{N\,n+1}\, \frac{\boldsymbol{p}_{T\,n+1}}{||\boldsymbol{p}_{T\,n+1}||} \tag{3.17}$$

EXAMPLE 3.1 Friction modelling for coated steel sheet material. The experimental identification of the frictional hardening curve $\mu(w)$ was carried out by means of sliding tests for Electrogalvanised steel sheet, typically employed in the manufacture of automotive body shells, resulting in the following polynomial relationship:

$$\begin{aligned} \mu(w) = &- 0.4096{\times}10^{-6}\, w^5 + 0.2890{\times}10^{-4}\, w^4 - \\ &- 0.8212{\times}10^{-3}\, w^3 + 0.1035{\times}10^{-1}\, w^2 - \\ &- 0.3148{\times}10^{-1}\, w + 0.1568 \end{aligned} \tag{3.18}$$

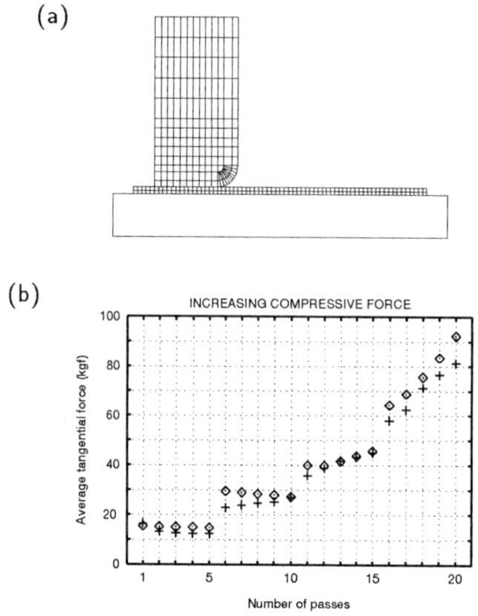

Figure 3.1: Sliding tests with Electrogalvanised steel sheet: Numerical simulation and experimental results. (a) Finite element model; (b) Comparison of numerical predictions ($\Diamond$) with experimental results ($+$).

To illustrate the predictive capability of the proposed model, the numerical simulation of a series of sliding tests is presented and the results are compared with the corresponding experiments. The above functional relationship is implemented as a sequence of linear segments which allows for a general frictional law of the type $\mu = \mu(w)$ to be easily included for simulation.

The finite element model employed is illustrated in Figure 3.1(a). The numerical simulation is compared with the experimental results in Figure 3.1(b) where the tangential reaction on the tip is plotted against the sliding distance for the situation where the normal force is increased incrementally from 100 to 400 kgf.

4. ELEMENT TECHNOLOGY

A crucial task for forming simulation, particularly bulk forming operations, is the development of appropriate elements which can model the incompressible nature of plastic flow without exhibiting locking behaviour. In this respect, the geometrically nonlinear *assumed strain* method, proposed by Simo and Armero [27] is particularly relevant and the essential steps in the formulation and discretisation of the problem are presented in Box 4.1.

The crucial idea is to assume the displacement gradient $\boldsymbol{H}$ to be a sum of the compatible part $\mathrm{Grad}\,\boldsymbol{u}$ and the enhanced part $\widetilde{\boldsymbol{H}}$, which also implies multiplicative decomposition $\boldsymbol{F} = \widetilde{\boldsymbol{F}}\boldsymbol{F_x}$ where $\widetilde{\boldsymbol{F}} = \boldsymbol{1} + \widetilde{\boldsymbol{H}}\boldsymbol{F_x}^{-1}$ and $\boldsymbol{F_x} = \partial \boldsymbol{x}/\partial \boldsymbol{X}$ represent, respectively, the enhanced and compatible contributions. It is employed in a suitable form of the 3-field Hu-Washizu variational principle, which reduces to a 2-field expansion ($\boldsymbol{u}$ and $\widetilde{\boldsymbol{h}}$) after the stress field $\boldsymbol{\tau}$ is effectively eliminated by enforcing L_2-orthogonality between the approximated spaces of stress and enhanced displacement gradient. The two remaining equations are numerically discretized in the usual manner of the finite element method. Standard material and geometrical parts of the element stiffness matrix are recovered consistent with the finite strain model employed and a radial return algorithm is used in the integration of the constitutive equations. In that context, the consistent elasto-(visco)plastic tangent operator for small strain situations, is directly employed. Interelement discontinuity of the enhanced part of the displacement gradient enables static condensation on the element level, hence global displacement-type equations are obtained. These are solved in the usual incremental-iterative manner where quadratic convergence of the algorithm is ensured with consistent linearization.

It should be noted that although locking is avoided in the simulations using the resulting formulation, the opposite effect of hourglassing may be encountered, depending on a particular choice of the reference configuration.

4.1 Remark on practical experiences with the enhanced strain method

It is emphasized here that, in contrast with the more elaborate hybrid finite elements, also employed in the solution of problems where the enforcement of incompressibility is crucial, assumed enhanced strain methods preserve the strain driven format of algorithms for integration of inelastic constitutive equations which accompany the purely kinematical formulation. Due to the simplicity, robustness and reliability of such integration schemes, the use of the assumed enhanced strain methodology is particularly convenient in the development of finite element codes for general inelastic analysis.

Some significant benefits are derived from this class of assumed enhanced strain methods. As compared to the standard isoparametric formulation, a substantial diminution of incompressibility locking and accuracy gain in bending dominated problems is observed when enhanced assumed strain elements are used. Nevertheless, drawbacks still exist and need to be carefully addressed. Particularly in highly constrained compressive regimes, the presence of undesirable hourglass modes may be detected. This fact is dramatically illustrated in Figure 4.1 which shows deformed meshes during the simulation of the upsetting of an elastoplastic cylindrical billet with the assumed enhanced strain axisymmetric element named EAS-5

(i) Assumed deformation gradient:

$$\boldsymbol{F} = \mathbf{1} + \operatorname{Grad}\boldsymbol{u} + \widetilde{\boldsymbol{H}}$$

(ii) Enhanced spatial variational equations:

$$\int_{\Omega} \widetilde{\nabla}(\delta\boldsymbol{u}) : \widetilde{\boldsymbol{\tau}}(\boldsymbol{u},\widetilde{\boldsymbol{h}})\,\mathrm{d}\,x - {}^{\mathrm{EXT}}G(\delta\boldsymbol{u}) = 0$$

$$\int_{\Omega} \delta\widetilde{\boldsymbol{h}} : [\widetilde{\boldsymbol{\tau}}(\boldsymbol{u},\widetilde{\boldsymbol{h}}) - \boldsymbol{\tau}]\,\mathrm{d}\,x = 0$$

$$\int_{\Omega} \delta\boldsymbol{\tau} : \widetilde{\boldsymbol{h}}\,\mathrm{d}\,x = 0$$

$$\widetilde{\boldsymbol{h}} = \widetilde{\boldsymbol{H}}\boldsymbol{F}^{-1}$$

$$\widetilde{\boldsymbol{\tau}}(\boldsymbol{u},\widetilde{\boldsymbol{h}}) = \widetilde{\boldsymbol{R}}^{e}\,\partial_{\widetilde{E}^{e}}\widehat{W}(\widetilde{\boldsymbol{E}}^{e})\,\widetilde{\boldsymbol{R}}^{e\,T}$$

(iii) Finite element interpolation of $\widetilde{\boldsymbol{H}}$:

$$\widetilde{\boldsymbol{H}}_{e} = \sum_{I=1}^{\mathrm{NENH}} \boldsymbol{\alpha}_{I}^{e} \otimes \boldsymbol{Q}_{I}^{e} \quad ; \quad \langle \boldsymbol{P}_{e}, \widetilde{\boldsymbol{H}}_{e}\rangle_{\mathrm{L}_2} = 0$$

(iv) Element linearized operators:

$$\widetilde{\nabla}^{s}\boldsymbol{u}_{e}^{h} = \boldsymbol{b}_{d}(\widetilde{\operatorname{grad}}\,N_{e}^{I})\boldsymbol{d}_{e}\ \widetilde{\boldsymbol{h}}^{s} = \boldsymbol{b}_{a}(\boldsymbol{q}_{e}^{I})\boldsymbol{\alpha}_{e}$$

$$\widetilde{\nabla}\boldsymbol{u}_{e}^{h} = \boldsymbol{g}_{d}(\widetilde{\operatorname{grad}}\,N_{e}^{I})\boldsymbol{d}_{e}\ \widetilde{\boldsymbol{h}} = \boldsymbol{g}_{a}(\boldsymbol{q}_{e}^{I})\boldsymbol{\alpha}_{e}$$

$$\widetilde{\operatorname{grad}}\,N_{e}^{I} = \boldsymbol{F}_{e}^{-T}\operatorname{Grad}N_{e}^{I}\ \boldsymbol{q}_{e}^{I} = \boldsymbol{F}_{e}^{-T}\boldsymbol{Q}_{I}^{e}$$

(v) Element residual equations:

$$\boldsymbol{r}_{d} = \int_{\Omega_e} \boldsymbol{b}_{d}^{T} : \widetilde{\boldsymbol{\tau}}\,\mathrm{d}\,x - {}^{\mathrm{EXT}}\boldsymbol{f}$$

$$\boldsymbol{r}_{a} = \int_{\Omega_e} \boldsymbol{b}_{a}^{T} : \widetilde{\boldsymbol{\tau}}\,\mathrm{d}\,x$$

(vi) Newton's method. Linearization:

$$\boldsymbol{k}_{dd}\Delta\boldsymbol{d}_{e} + \boldsymbol{k}_{ad}^{T}\Delta\boldsymbol{\alpha}_{e} = -\boldsymbol{r}_{d}$$

$$\boldsymbol{k}_{ad}\Delta\boldsymbol{d}_{e} + \boldsymbol{k}_{aa}\Delta\boldsymbol{\alpha}_{e} = -\boldsymbol{r}_{a}$$

$$\boldsymbol{k}_{dd} = \int_{\Omega_e} \boldsymbol{b}_{d}^{T}\,\mathbf{c}\,\boldsymbol{b}_{d}\,\mathrm{d}\,x + \int_{\Omega_e} \boldsymbol{g}_{d}^{T}\widehat{\boldsymbol{\sigma}}\boldsymbol{g}_{d}\,\mathrm{d}\,x$$

$$\boldsymbol{k}_{ad} = \int_{\Omega_e} \boldsymbol{b}_{a}^{T}\,\mathbf{c}\,\boldsymbol{b}_{d}\,\mathrm{d}\,x + \int_{\Omega_e} \boldsymbol{g}_{a}^{T}\widehat{\boldsymbol{\sigma}}\boldsymbol{g}_{d}\,\mathrm{d}\,x$$

$$\boldsymbol{k}_{aa} = \int_{\Omega_e} \boldsymbol{b}_{a}^{T}\,\mathbf{c}\,\boldsymbol{b}_{a}\,\mathrm{d}\,x + \int_{\Omega_e} \boldsymbol{g}_{a}^{T}\widehat{\boldsymbol{\sigma}}\boldsymbol{g}_{a}\,\mathrm{d}\,x$$

(vii) Static condensation on the element level:

$$\widetilde{\boldsymbol{k}}_{b}^{e}\Delta\boldsymbol{d}_{e} = \widetilde{\boldsymbol{r}}^{e}$$

$$\Delta\boldsymbol{\alpha}_{e} = -\boldsymbol{k}_{aa}^{-1}(\boldsymbol{r}_{a} + \boldsymbol{k}_{ad}\Delta\boldsymbol{d}_{e})$$

$$\widetilde{\boldsymbol{k}}^{e} = \boldsymbol{k}_{dd} - \boldsymbol{k}_{ad}^{T}\boldsymbol{k}_{aa}^{-1}\boldsymbol{k}_{ad}$$

$$\widetilde{\boldsymbol{r}}^{e} = \boldsymbol{r}_{d} - \boldsymbol{k}_{ad}^{T}\boldsymbol{k}_{aa}^{-1}\boldsymbol{r}_{a}$$

(viii) Global displacement-type equation:

$$\widetilde{\boldsymbol{K}}\Delta\boldsymbol{d} = -\widetilde{\boldsymbol{G}}$$

$$\widetilde{\boldsymbol{K}} = \mathop{\mathbf{A}}_{e=1}^{\mathrm{NEL}} \widetilde{\boldsymbol{k}}^{e} \quad ; \quad \widetilde{\boldsymbol{G}} = \mathop{\mathbf{A}}_{e=1}^{\mathrm{NEL}} \widetilde{\boldsymbol{r}}^{e}$$

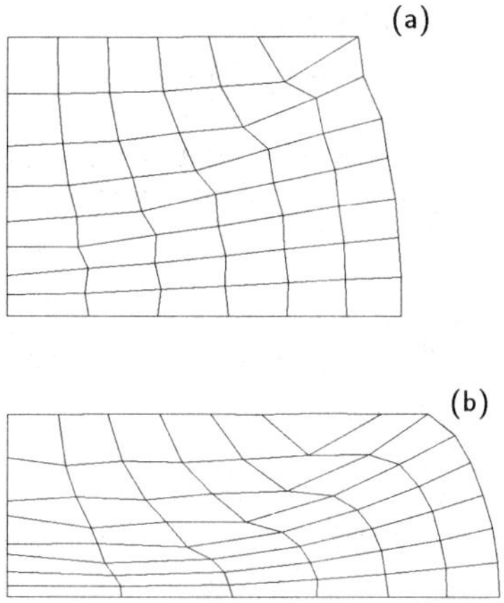

Figure 4.1: Upsetting of a cylindrical billet. Hourglass patterns. (a) 4 Gauss point quadrature rule and (b) 9 Gauss point quadrature rule.

by Simo and Armero [27]. Figures 4.1.(a)-(b) correspond, respectively, to 4 and 9 gauss point quadrature integration rules. Hourglass patterns are clearly visible and indicate that further research is required for the design of enhanced assumed strain finite elements with optimal performance in a wider range of situations.

5. IMPLICIT SOLUTION OF THE DISCRETISED NONLINEAR MODEL: LINEARISATION ASPECTS

5.1 Consistent linearisation of the discrete problem

In order to achieve the assymptotically quadratic rates of convergence which characterize the full Newton-Raphson algorithm, it is crucial that continuum-numerical models for the description of general nonlinear behaviour of solids be consistently linearized. One of the most striking features of the above large strain elasto-plastic formulation, is that, as far as consistent linearization is concerned, the results of small strain theories can be extended to account for geometrically non-linear effects in a rather simple manner.
In its spatial version, the weak form of the incremental momentum balance is expressed as

$$G(\chi_{n+1},\boldsymbol{\eta}) := \int_{\chi_{n+1}(\Omega)} (\boldsymbol{\sigma}_{n+1} : \nabla\boldsymbol{\eta} - \boldsymbol{b}_{n+1}\cdot\boldsymbol{\eta})\ \mathrm{d}\,v - \int_{\chi_{n+1}(\partial\Omega_\sigma)} \boldsymbol{t}_{n+1}\cdot\boldsymbol{\eta}\ \mathrm{d}\,a = 0, \qquad \forall\,\boldsymbol{\eta}\in\mathcal{V} \tag{5.1}$$

where Cauchy stress, $\boldsymbol{\sigma}_{n+1}$, is the outcome of the numerical integration algorithm described above, $\boldsymbol{b}_{n+1}$ is the body force and $\boldsymbol{t}_{n+1}$ the external surface tractions. The linearization of (5.1) provides the basis of the Newton-type iterative schemes for solution of the momentum balance equations. In this context, the directional derivative of the internal virtual work plays an essential role. The virtual work of the internal forces,

$$G^{\text{int}}(\chi_{n+1}, \boldsymbol{\eta}) := \int_{\chi_{n+1}(\Omega)} \boldsymbol{\sigma}_{n+1} : \nabla \boldsymbol{\eta} \, \mathrm{d}\, v \tag{5.2}$$

has its derivative in the direction of an incremental displacement $\Delta \boldsymbol{u}$ given by the formula

$$\mathrm{D}\, G^{\text{int}}(\chi_{n+1}, \boldsymbol{\eta}) \cdot \Delta \boldsymbol{u} = \int_{\chi_{n+1}(\Omega)} \nabla \boldsymbol{\eta} : \mathsf{a} : \nabla(\Delta \boldsymbol{u}) \, \mathrm{d}\, v \tag{5.3}$$

where a is the standard *spatial tangent modulus.*

After some lengthy but straightforward algebra, the following explicit expression for the cartesian components of a is obtained:

$$\mathsf{a}_{ijkl} = \frac{1}{J_{n+1}} \left[(\mathsf{R} : \widehat{\mathsf{H}} : \mathsf{L} + \mathsf{K} : \mathsf{M}) : \mathsf{N} \right]_{ijkl} + \sigma_{lj} \delta_{ik} \tag{5.4}$$

where $\widehat{\mathsf{H}}$ denotes the *small strain algorithmic consistent tangent operator* and is the only term taking part on the assemblage of a that depends on the particular material model (and integration algorithm) adopted. The remaining terms are related *exclusively to the kinematics of large strains*. The cartesian components of the fourth order tensors K, N and R are given by

$$\begin{aligned} \mathsf{K}_{ijlk} &:= \left[\boldsymbol{R}^e_{n+1} \boldsymbol{T}_{n+1} \right]_{ik} (\boldsymbol{F}^{e\,\text{trial}}_{n+1})_{jl} + \\ &\quad + \left[\boldsymbol{R}^e_{n+1} \boldsymbol{T}_{n+1} \right]_{jk} (\boldsymbol{F}^{e\,\text{trial}}_{n+1})_{il} && (5.5a) \\ \mathsf{N}_{ijkl} &:= (\boldsymbol{F}^{e\,\text{trial}}_{n+1})_{li} (\boldsymbol{F}^{e\,\text{trial}}_{n+1})_{kj} + \\ &\quad + (\boldsymbol{F}^{e\,\text{trial}}_{n+1})_{lj} (\boldsymbol{F}^{e\,\text{trial}}_{n+1})_{ki} && (5.5b) \\ \mathsf{R}_{ijkl} &:= (\boldsymbol{R}^e_{n+1})_{ik} (\boldsymbol{R}^e_{n+1})_{jl} && (5.5c) \end{aligned}$$

and L and M have been defined as

$$\mathsf{L} := \frac{\partial \boldsymbol{E}^{e\,\text{trial}}_{n+1}}{\partial \boldsymbol{C}^{e\,\text{trial}}_{n+1}}, \quad \text{and} \quad \mathsf{M} := \frac{\partial (\boldsymbol{U}^{e\,\text{trial}}_{n+1})^{-1}}{\partial \boldsymbol{C}^{e\,\text{trial}}_{n+1}} \tag{5.6}$$

5.2 Consistent linearisation of the discrete frictional contact problem

If contact occurs on a certain portion $\partial \Omega_c$ of the body analysed, then the contact contribution to the virtual work of the external forces:

$$G^c := \int_{\chi_{n+1}(\partial \Omega_c)} \boldsymbol{p}_{n+1} \cdot \boldsymbol{\eta} \tag{5.7}$$

must be included in the weak statement of the momentum balance (5.1). Thus, in order to preserve the quadratic rates of convergence of the Newton-Raphson scheme for solution of the incremental equilibrium problem, the consistent linearization of this extra term must be carefully addressed.

In this context, the appropriate linearization of the frictional contact algorithm plays an essential role.

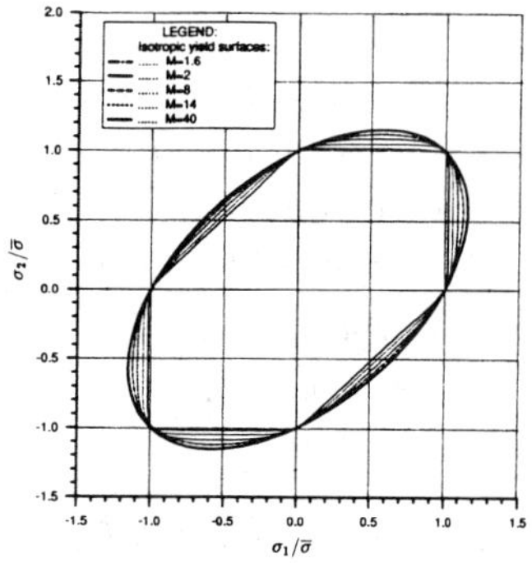

Figure 6.1 Isotropic yield surfaces for several values of material constant M and for shear stress σ_{xy}=0.

Since for given values of $\boldsymbol{p}_n$ and w_n and an incremental displacement $\Delta \boldsymbol{u}$, the numerical integration algorithm described above determines uniquely the reaction $\boldsymbol{p}_{n+1}$, an *algorithmic* constitutive function $\widehat{\boldsymbol{p}}$ can be defined such that the frictional force $\boldsymbol{p}_{n+1}$ is delivered as

$$\boldsymbol{p}_{n+1} = \widehat{\boldsymbol{p}}(\boldsymbol{p}_n, w_n, \Delta \boldsymbol{u}) \tag{5.8}$$

Following a standard procedure of classical elastoplasticity, the differentiation of the algorithmic function at converged values for t_{n+1} provides the tangent relation

$$\widehat{\boldsymbol{D}} := \frac{\partial \boldsymbol{p}_{n+1}}{\partial \Delta \boldsymbol{u}} \tag{5.9}$$

where the second order tensor $\boldsymbol{D}$ is the so called *consistent tangent operator* associated with the present model (and integration algorithm).

6. FURTHER ASPECTS OF THE DEFORMATION OF INELASTIC SOLIDS AT FINITE STRAINS

6.1 Yield surface representation

For thin sheet metal forming operations, due to the processing of the material, the plastic behaviour of textured polycrystalline sheet is predominantly anisotropic. Use of the Hill anisotropic criterion, which contains no shear stress, is restricted to a planar isotropy or for the cases where the principal stress axes coincide with the anisotropy axes.

Full planar anisotropy is described by the yield function introduced by Barlat and Lian [1] which for the plane stress state is of the superquadric form (computational aspects of this criterion are discussed in [6])

$$\begin{aligned} f(\boldsymbol{\sigma}, \overline{\sigma}) = & a|K_1 + K_2|^M + a|K_1 - K_2|^M + \\ & + (2-a)|2K_2|^M - 2\overline{\sigma}^M \end{aligned} \tag{6.1}$$

in which

$$K_1 = \frac{\sigma_{xx} + h\sigma_{yy}}{2} \tag{6.2a}$$

and

$$K_2 = \sqrt{\left(\frac{\sigma_{xx} - h\sigma_{yy}}{2}\right)^2 + p^2\sigma_{xy}^2} \qquad (6.2b)$$

where a, h, p and M are material constants and $\bar{\sigma}$ is the yield stress from a uniaxial tension test. For a given value of the exponent M material constants a, h, p can be evaluated using R values, i.e. plastic strain ratios of the in-plane strains to the thickness strain obtained from uniaxial tension tests in three different directions. The role of the material constant M is illustrated in Figure 6.1 obtained by plotting the function (6.1) for various M values in the normalized σ_{xx} and σ_{yy} plane for the isotropic case (i.e. for a=h=p=1) and taking σ_{xy}=0. The resulting set of functions span the set of yield surfaces which include the standard von Mises and Tresca yield surfaces for M=2 and $M \to \infty$, respectively.

An important property of the yield function described by equations (6.1-2) is its convexity when constants a, h, p are positive and $M > 1$, as proved by Barlat and Lian [1].

EXAMPLE 6.2 Stretching of a circular thin sheet by a hemispherical punch : elasto-plastic material. The geometry, material characteristics and other parameters for this problem are given in Ref. [16].

The punch force versus punch displacement diagram, presented in Figure 6.2(a), gives comparison between results obtained for various M-values. The maximum punch force decreases with increase of the M-value, which indicates a strong influence of the curvature of the yield surface on the initiation of strain localisation and its development.

The distribution of true strain in the radial direction is shown in Figure 6.2(b) for various M-values and for punch displacements D_p = 35 mm. A typical localisation behaviour may be observed where strain accumulates in a narrow zone, reaching high levels and leading to failure. The appearance of localisation and associated failure is less pronounced with decrease of the curvature on the yield surface, specified by decrease of M-value, and is clearly delayed for a quadratic yield surface corresponding to the standard von Mises yield criterion.

6.2 Elasto-viscoplastic solids

The constitutive model for J_2 elasto-viscoplasticity is represented as

$$f(\boldsymbol{\sigma}, \bar{\sigma}) = F(\boldsymbol{\sigma}) - \bar{\sigma} \qquad (6.3a)$$

$$F(\boldsymbol{\sigma}) = \sqrt{3J_2} \qquad (6.3b)$$

$$\dot{\varepsilon}^{vp} = \gamma\langle\Phi(\bar{\sigma})\rangle\sqrt{\frac{3}{2}}\frac{\text{dev}\,[\boldsymbol{\sigma}]}{||\,\text{dev}\,[\boldsymbol{\sigma}]\,||} \qquad (6.3c)$$

in which Φ is the *viscoplastic flow potential.*

For metal forming operations under high temperature conditions, the effective stress (usually termed

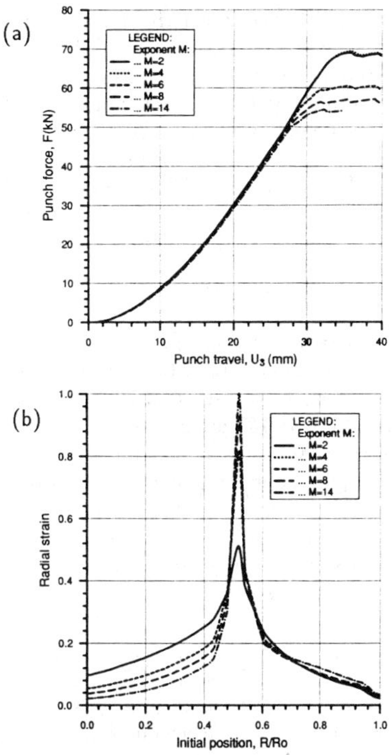

Figure 6.2 Stretching of a circular thin sheet by a hemispherical punch: (a) Punch force versus punch displacement curves for various M-values. (b) Distribution of true strain in the radial direction plotted over the initial configuration for various M-values at punch displacement D_p = 35.0 mm.

the flow stress) is rate-dependent making viscoplastic approaches suitable for simulation. Several empirical relations exist for flow stress prediction, which are based on experimental tests, and are applicable to particular materials under specific conditions (see Rodič [22] for details). Three of the most commonly employed expressions are summarised below:

(1) *Hajduk expression.* The flow stress $\bar{\sigma}$ is assumed to be expressed in the following form:

$$\bar{\sigma} = K_{fo}K_T K_{\bar{\varepsilon}} K_{\dot{\bar{\varepsilon}}} \qquad (6.4)$$

where the three coefficients are functions of the form:

$$K_T = A_1 \exp\left[-m_1 T\right]\,;\; K_{\bar{\varepsilon}} = A_2\bar{\varepsilon}^{m_2}\,;\; K_{\dot{\bar{\varepsilon}}} = A_3\dot{\bar{\varepsilon}}^{m_3} \qquad (6.5)$$

where A_i and m_i are material constants determined from tests.

(2) *Sellars-Tegart expression.* This is based on the following interpolation equation:

$$Z = \dot{\bar{\varepsilon}}\,\exp\left[\frac{Q}{R(T+273)}\right] = C[\sinh\left[\alpha\bar{\sigma}\right)]^n \qquad (6.6)$$

where Z is the Zener-Hollomon parameter, Q is an activation energy usually independent of temperature and in many cases also independent of strain, R is the gas constant 8.31 J/molK, T is the temperature in degrees Celsius, while C, α and n are material constants.

(3) *ALSPEN expression*: This expression is found to cover closely the properties of some aluminium alloys by fitting experimental curves in the form:

$$\bar{\sigma} = c(T)\,(\alpha + \alpha_0)^{n(T)}\,\bar{\varepsilon}^{m(T)} \tag{6.7}$$

Coefficients $c(T)$, $n(T)$ and $m(T)$ are described in the references [22,23], α_0 is a constant, and $\mathrm{d}\,\alpha = \mathrm{d}\,\bar{\varepsilon}^{vp}$ for temperatures below the onset limit $T_0 \approx 700K$, otherwise $\mathrm{d}\,\alpha = 0$.

Since practically $\bar{\varepsilon} = \bar{\varepsilon}^{vp}$ and $\dot{\bar{\varepsilon}}^{vp} = \gamma\Phi$ for J_2 plasticity, parameters γ and Φ can be obtained, as shown in Box 6.1, for all three interpolation functions. In all three cases, the yield stress is taken to be zero, thus the assumption is made that some part of the strain is always inelastic.

BOX 6.1 Elasto-viscoplastic material model

(i) Additive strain rate decomposition:

$$\varepsilon = \varepsilon^e + \varepsilon^{vp}$$

(ii) Elastic response:

$$\dot{\varepsilon}^e = \mathbf{C}^{-1} : \dot{\sigma}$$

(iii) Flow rule:

$$\dot{\bar{\varepsilon}}^{vp} = \gamma\langle\Phi(\bar{\sigma})\rangle\,\sqrt{3/2}\,\frac{\operatorname{dev}[\sigma]}{||\operatorname{dev}[\sigma]||}$$

(iv) γ and Φ from interpolation functions:

(1) Hajduk

$$\gamma = \left[\frac{1}{A_1 A_2 A_3}\exp\left[m_1 T\right]\bar{\varepsilon}^{vp(-m_2)}\right]^{(1/m_3)}$$

$$\Phi = \left(\frac{\bar{\sigma}}{K_{fo}}\right)^{(1/m_3)}$$

(2) Sellars-Tegart

$$\gamma = C\,\exp\left[\frac{-Q}{R(T+273)}\right]$$

$$\Phi = \left[\sinh[\alpha\bar{\sigma}]\right]^n$$

(3) ALSPEN

$$\gamma = [c(T)]^{(-1/m(T))}\,[\alpha(\bar{\varepsilon}^{vp}) + \alpha_0]^{(-n(T)/m(T))}$$

$$\Phi = (\bar{\sigma})^{(1/m(T))}$$

7. THERMO-MECHANICAL COUPLING

For some classes of forming problems consideration of the coupling between mechanical and thermal phenomena is essential for realistic simulation, since temperature changes can induce thermal stresses and change the material properties for mechanical analysis and mechanical deformations can modify the thermal boundary conditions and generate heat by frictional sliding or dissipation of plastic work.

A staggered solution approach is commonly adopted for such problems in which separate analyses are undertaken for each phenomenon with data exchange performed at the end of each time step or increment. In particular, the nodal temperatures are transferred to the mechanical analysis, while the displacements, plastic work, frictional heat flux and contact data are communicated to the thermal solution. Such a solution procedure is convenient for implementation in a two processor environment and represents the simplest form of parallel processing. On the simplest level, identical meshes can be employed for analysis of the two phenomena. However, when mesh adaption procedures are to be introduced for both phenomena different meshes will result, requiring data mapping between the two solution phases. Details on the formulation and numerical analysis of thermo-mechanical coupled processes may be found in a recent comprehensive review by Simo and Miehe [28].

7.1 Metal cutting processes

EXAMPLE 7.1 Metal cutting process simulation. Sequences of the simulation of a metal cutting process are shown in Fig. 7.1. The problem involves the removal of a continuous chip by a cutting tool and modelling requires, in addition to finite strain elasto-plasticity and frictional contact representation, the introduction of an appropriate material separation criterion. Realistic modelling of this class of problem necessitates the introduction of thermo-mechanical coupling effects, since the generation of heat by, principally, frictional sliding between the chip and cutting tool and also the dissipation of plastic work can result in large temperature increases in the tool.

The present work adopts a fracture, or cleavage, condition based on a softening plasticity model in a narrow band of elements situated along the cutting line. In particular, the element currently ahead of the cutting tool is assumed to separate when the work associated with plastic softening reaches the critical energy release rate of the material. To avoid mesh size dependency, a regularisation technique is adopted in which the slope of the softening branch of the material within this band is made a function of the element width.

Fig. 7.1 shows the development of effective stress during the cutting process. Steady state conditions have been achieved by the time the tool has reached the position shown in Fig. 7.1(c). It is interesting to note that the residual effective stress near the newly formed surface is almost 0.7 the initial yield stress of the material. Development of the reactive force on the tool is illustrated in Fig. 7.2.

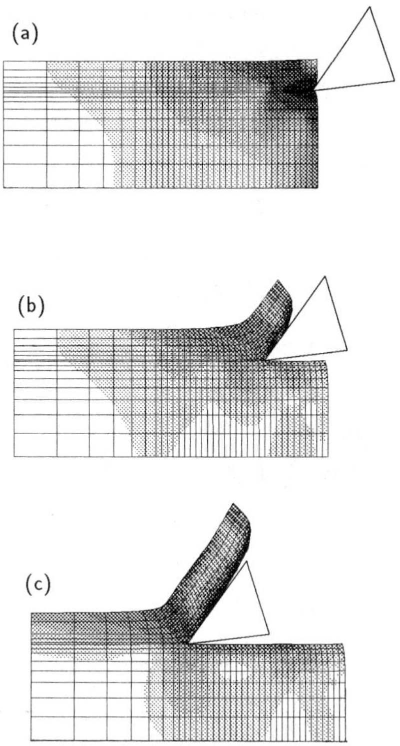

Figure 7.1: Metal cutting process simulation: (a)-(c) Deformed finite element meshes at various stages of the cutting process with effective stress distribution.

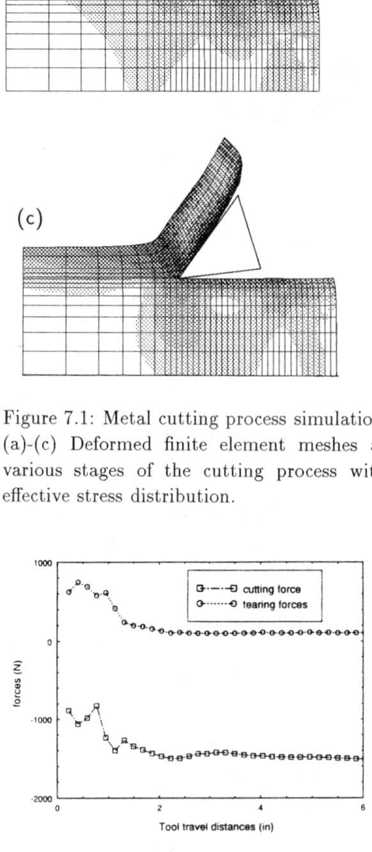

Figure 7.2 Metal cutting process simulation: Evolution of the force on the tool during the operation.

7.2 Prediction of tool wear for forging operations

A statistical investigation covering more than 100 forging geometries has categorised the principal causes of tool failure. Different types of wear (mainly at corners and roundings) is the cause for scrapping of some 60 % of tools, various types of crack formation account for approximately 25 % and local plastic deformation is responsible for 5 %. Consequently, the prediction of tool wear is of primary concern for the implementation of a preventive maintenance strategy within the forging industry

Current approaches adopt a semi-empirical wear model which is incorporated into a finite element model to predict the 'instantaneous' tool wear, which then must be integrated to estimate the accumulated wear occurring over a larger number of forming cycles. Typical adhesive and abrasive wear models take the form

$$Z = K\,\frac{q \times s}{H} \tag{7.1}$$

where Z is the wear volume per unit area, q is the normal pressure, s the sliding length, H the surface hardness and K is a wear constant.

The local hardness is a strong function of temperature which necessitates a fully coupled thermo-mechanical approach to finite element modelling.

The sliding length, which is the amount of material passing a specific point on the die surface, is considered to be the most important parameter in the wear calculation, due to its strong influence on the temperature distribution generated through frictional heat.

An important aspect is the variation of the friction stress with the normal pressure. Normally the law of constant friction is used, but it is often advantageous to take the variation of the friction stress with the normal pressure into account by using the general friction model formulated by Bay and Wanheim [2].

EXAMPLE 7.2 A plane strain bulk forming of a crane hook. Figures 7.3-4 show the predicton of wear for the forging of a crane hook. Figures 7.3(b)-(d) show the temperature development in the tool over one forging cycle, which is primarily due to frictional sliding between workpiece and tool. It is seen that the areas of greatest temperature increase coincide with regions of high curvature of the tool profile, which results in large normal pressures and consequently relatively large amounts of frictional work dissipated as heat. Figure 7.4 illustrates the wear profile prediction, presented in the scaled form of Z/K, where the coincidence of high wear with regions of high temperature (low hardness) and normal pressure is immediately apparent.

8. ADAPTIVE STRATEGIES FOR FORMING PROBLEMS

Adaptive stategies for linear elliptic problems have achieved a high level of mathematical understanding and are already routinely performed within finite element computations [32]. In contrast, due to its complexity adaptive strategies for nonlinear problems have received little attention. Notable contributions in this area are presented in [12-13,15,21,33]. In this

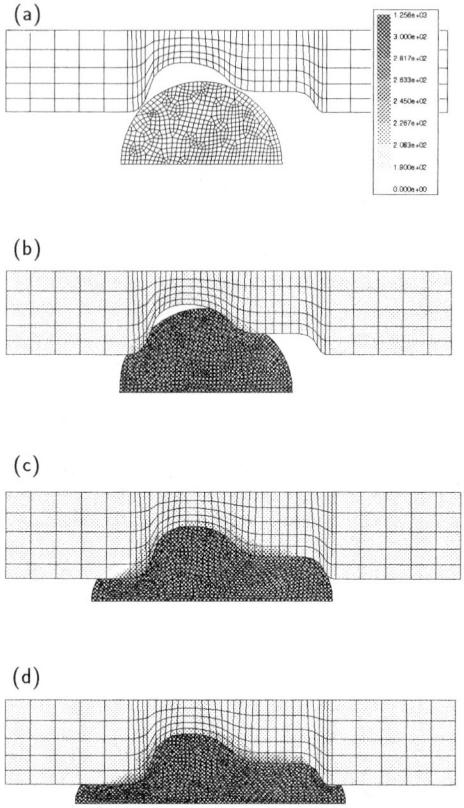

Figure 7.3: A plane strain bulk forming of an industrial component: (a) Initial finite element meshes for tool and workpiece, (b)-(d) Deformed finite element meshes with temperature distributions at various stages of punch displacement.

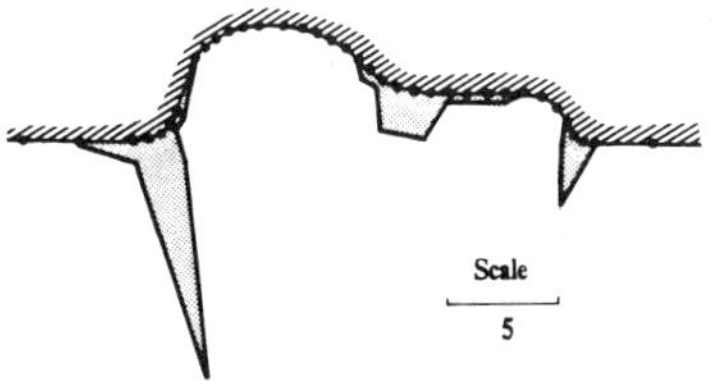

Figure 7.4 A plane strain bulk forming of an industrial component: The wear profile prediction presented in the scaled form of Z/K.

section, some basic concepts already in standard usage for the linear problems are generalized in order to develop an adaptive strategy for elasto-plastic problems of evolution. In particular, *a posteriori* error estimates

TABLE 8.1 Error estimation procedure – Energy norm error

(i) Energy functional

$$\Psi(\boldsymbol{\Sigma}) = \frac{1}{2}\int_{\Omega} \boldsymbol{\Sigma} : \mathbb{G}^{-1}\,\boldsymbol{\Sigma}\,\mathrm{d}x = |||\boldsymbol{\Sigma}|||^2$$

(ii) Stress error in energy norm

$$|||e_{\Sigma}|||_K^2 = \int_K (\boldsymbol{\Sigma} - \boldsymbol{\Sigma}^h) : \mathbb{G}^{-1}(\boldsymbol{\Sigma} - \boldsymbol{\Sigma}^h)\,\mathrm{d}x = |||\boldsymbol{\Sigma} - \boldsymbol{\Sigma}^h|||_K^2$$

$$|||e_{\Sigma}|||^2 = \sum_K |||e_{\Sigma}|||_K^2$$

(iii) Stress projection

$$\int_{\Omega} \boldsymbol{\Pi}(\boldsymbol{\Sigma}^* - \boldsymbol{\Sigma}^h)\,\mathrm{d}x = \mathbf{0}$$

(iv) Error estimate

$$\varepsilon_{\Sigma,K}^2 = \int_K (\boldsymbol{\Sigma}^* - \boldsymbol{\Sigma}^h)^T : \mathbb{G}^{-1}(\boldsymbol{\Sigma}^* - \boldsymbol{\Sigma}^h)\,\mathrm{d}x = |||\boldsymbol{\Sigma}^* - \boldsymbol{\Sigma}^h|||_K^2\,\mathrm{d}x \approx |||e_{\Sigma}|||_K^2,$$

$$|||\varepsilon_{\Sigma}|||^2 = \sum_K |||\varepsilon_{\Sigma}|||_K^2,$$

(v) Relative error

$$\eta_{\Sigma} = \frac{|||e_{\Sigma}|||}{\Psi^{1/2}} \approx \frac{\varepsilon_{\Sigma}}{(\Psi^h)^{1/2}}$$

based on the Zienkiewicz-Zhu adaptive strategy and the energy norm (see [32-33] and references therein) are appropriately modified to account for the elasto-plastic deformation at small and finite strains. We emphasize that, in this context, the generalized energy norm appears as a natural metric for this class of problems. In addition, the intrinsic dissipation functional, associated with the second law of thermodynamics, is by heuristic arguments, utilized as a basis for the development of a complementary *a posteriori* error estimator.

8.1 Basic error indicators and adaptive refinement

8.1.1 Error indicators. Denoting by $\boldsymbol{\Sigma}^h := (\boldsymbol{\sigma}^h, \boldsymbol{q}^h)$ the approximate generalized stresses obtained as the finite element solution, where $\boldsymbol{\sigma}^h$ and $\boldsymbol{q}^h$ denote the stress tensor and set of internal variables, respectively, the corresponding error may be defined as

$$\boldsymbol{e}_{\boldsymbol{\Sigma}} := \boldsymbol{\Sigma} - \boldsymbol{\Sigma}^h \tag{8.1}$$

Following standard usage for linear elliptic problems we introduce the generalized stress error in the energy norm for a generic element K, and a corresponding global error, respectively, as

$$|||\boldsymbol{e}_{\boldsymbol{\Sigma}}|||_K^2 = \int_K (\boldsymbol{\Sigma} - \boldsymbol{\Sigma}^h) : \mathbb{G}^{-1}(\boldsymbol{\Sigma} - \boldsymbol{\Sigma}^h)\,\mathrm{d}x \tag{8.2a}$$

$$|||\boldsymbol{e}_{\boldsymbol{\Sigma}}|||^2 = \sum_K |||\boldsymbol{e}_{\boldsymbol{\Sigma}}|||_K^2 \tag{8.2b}$$

where $\mathbb{G} = \text{diag}\,[\mathbf{C}, \mathbf{D}]$ is the generalised modulus with $\mathbf{C}$ being the standard elastic modulus and $\mathbf{D}$ denoting the generalised hardening modulus.

An elementary procedure for the error estimation may be defined by the replacement of the exact values of variables and relevant derivatives of the problem by some post-processed values obtained from the available finite element solution and the problem data. The post-processed solution is expected to have superior accuracy compared to the original finite element solution. This characteristic of the post-processed solution is attributed to the so-called *superconvergence* properties, which are at present proved for certain regular meshes. In particular, the *a posteriori* error estimation procedure originally proposed and used by Zienkiewicz and Zhu [32] for linear elliptic problems is based on the observation that exact stresses $\boldsymbol{\sigma}$ may be represented accurately by smoothed stresses $\boldsymbol{\sigma}^*$ obtained by a suitable projection of approximate stresses $\boldsymbol{\sigma}^h$ which satisfies

$$\int_{\Omega} \boldsymbol{\Pi}(\boldsymbol{\sigma}^* - \boldsymbol{\sigma}^h) = \mathbf{0} \qquad (8.3)$$

Some possible choices for the projection matrix $\boldsymbol{\Pi}$ are listed in Zienkiewicz and Zhu [32].

These ideas can be easily extended to the elasto-plastic problem of evolution (see Perić *et al.* [21] for details). There exist several possible options for the choice of error indicators for elasto-plastic problems. Although much of the research remains to be completed, our initial experience suggests that: (i) the error indicator based on the generalised energy and (ii) the error indicator based on the plastic dissipation (work), have provided good error indicators for most metal forming simulations undertaken. For convenience, the error estimation procedure based on energy error indicator is summarised in TABLE 8.1.

8.1.2 Refinement strategy. The mesh refinement procedure is constructed in a standard way (see e.g. Zienkiewicz *et al.* [32-33]) with the objective of achieving a uniform distribution of local error. In addition, the relative error η_{Σ} is required to be within specified limits. In our numerical examples, the element size is predicted according to the asymptotic rate convergence criteria of the linear model, although this is not realistic for an elasto-plastic material with the presence of the short wave-length deformation pattern induced by strain localisation.

8.2 Mesh generation/adaption techniques

Structured and unstructured meshes can be used in the discretisation of the domain. There exists little doubt that there is a role for both categories of meshes in the finite element simulation of industrial forming applications.

Mesh adaption is generally required in the finite element simulation of industrial forming operations as the geometry and physical characteristics of the material are evolving drastically in the course of the forming process. Taking the above constraints into account a two-stage Delaunay unstructured approach in mesh generation/adaption is adopted in this work.

The mesh adaption scheme implemented is capable of performing the mesh adaption according to the mesh prediction data. This mesh prediction data, usually in a form of mesh density variation or mesh refinement indices, is interpreted from the error data created by some error estimator as was discussed in Section 8.1.

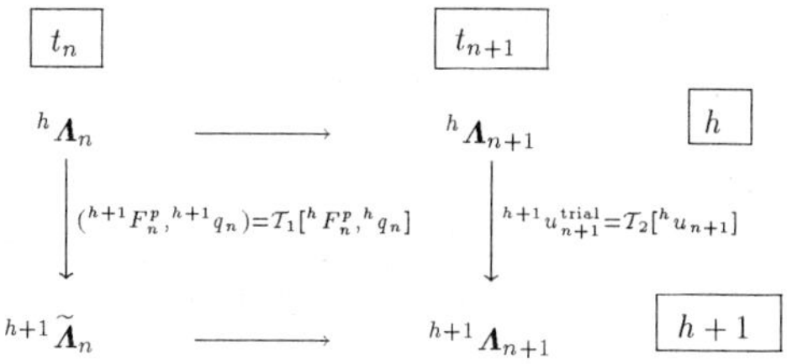

Figure 8.1: Transfer operator diagram.

8.3 Transfer operators

The issue of transfer of variables to new meshes must be properly addressed as it is expected to be critical in FE simulations based on the implicit approach [11]. Important aspects which must be considered include:

(i) consistency with the constitutive equations,
(ii) requirements of equilibrium,
(iii) compatibility of the state transfer with the displacement field on the new mesh,
(iv) compatibility with evolving boundary conditions,
(v) minimisation of the numerical diffusion of the state fields.

Let ${}^h\boldsymbol{u}_n, {}^h\boldsymbol{F}_n, {}^h\boldsymbol{F}^p_n, {}^h\boldsymbol{\sigma}_n, {}^h\boldsymbol{q}_n$ denote values of the displacement, deformation gradient, plastic part of deformation gradient tensor, stress tensor and a vector of internal variables at time t_n for the mesh h (see Figure 8.1). For simplicity of notation, we define a state array ${}^h\boldsymbol{\Lambda}_n = ({}^h\boldsymbol{u}_n, {}^h\boldsymbol{F}_n, {}^h\boldsymbol{F}^p_n, {}^h\boldsymbol{\sigma}_n, {}^h\boldsymbol{q}_n)$. Assume, furthermore, that the estimated error of the solution ${}^h\boldsymbol{\Lambda}_n$ respects the precribed criteria, while these are violated by the solution ${}^h\boldsymbol{\Lambda}_{n+1}$. In this case a new mesh $h+1$ is generated and a new solution ${}^{h+1}\boldsymbol{\Lambda}_{n+1}$ needs to be computed.

As the backward Euler scheme is adopted the plastic strain ${}^{h+1}\boldsymbol{F}^p_n$ and the internal variables ${}^{h+1}\boldsymbol{q}_n$ for a new mesh $h+1$ at time t_n need to be evaluated. In this way the state ${}^{h+1}\widetilde{\boldsymbol{\Lambda}}_n = (\bullet, \bullet, {}^{h+1}\boldsymbol{F}^p_n, \bullet, {}^{h+1}\boldsymbol{q}_n)$ is constructed. It should be noted that this state characterises the history of the material and, in the case of a fully implicit scheme, provides sufficient information for computation of a new solution ${}^{h+1}\boldsymbol{\Lambda}_{n+1}$.

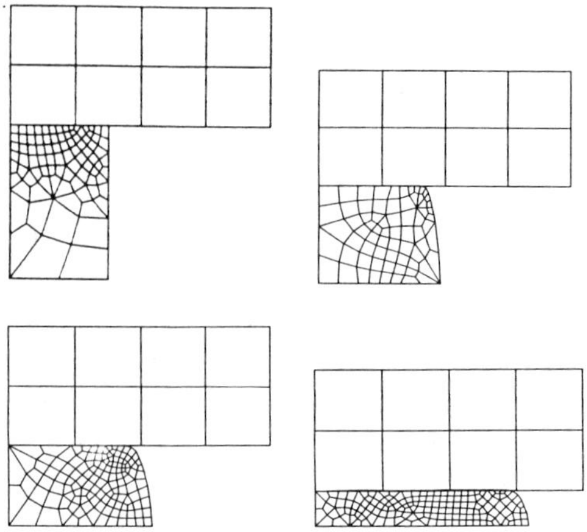

Figure 8.2: Upsetting of a cylindrical billet: Adapted finite element meshes at various stages of the process.

- Transfer operator $\mathcal{T}_1$

Let $\mathcal{T}_1$ be the *transfer operator* between meshes h and $h+1$ defined by

$$({}^{h+1}\boldsymbol{F}_n^p, {}^{h+1}\boldsymbol{q}_n) = \mathcal{T}_1[{}^h\boldsymbol{F}_n^p, {}^h\boldsymbol{q}_n] \tag{8.4}$$

The variables $({}^h\boldsymbol{F}_n^p, {}^h\boldsymbol{q}_n)$ specified at quadrature points of the mesh h, are transferred by the operator $\mathcal{T}_1$ to every point of the domain Ω, in order to specify the variables $({}^{h+1}\boldsymbol{F}_n^p, {}^{h+1}\boldsymbol{q}_n)$ at the quadrature points of the new mesh $h+1$. The operator $\mathcal{T}_1$ can be constructed in different ways. Some possible choices are discussed in [11].

- Transfer operator $\mathcal{T}_2$

As a trial guess for the computation of solution ${}^{h+1}\boldsymbol{\Lambda}_{n+1}$ which is required to respect equilibrium and constitutive relations, the state ${}^{h+1}\boldsymbol{\Lambda}_n = \mathcal{T}_1[{}^h\boldsymbol{\Lambda}_n]$ may be employed. It should be noted that the solution ${}^{h+1}\boldsymbol{\Lambda}_n$ does not satisfy the equilibrium and constitutive relations at time t_n.

An improved initial guess may be defined by the transfer

$${}^{h+1}\boldsymbol{u}_{n+1}^{\text{trial}} = \mathcal{T}_2[{}^h\boldsymbol{u}_{n+1}] \tag{8.5}$$

of the displacement field ${}^h\boldsymbol{u}_{n+1}$ obtained with the mesh h at time t_{n+1}. It is important to observe that the transfer operator $\mathcal{T}_2$ can be easily constructed knowing that the displacement field ${}^h\boldsymbol{u}_{n+1}$ is defined at each point of the domain Ω.

EXAMPLE 8.1 Upsetting of a cylindrical billet. To illustrate the above concepts, finite element simulation of a cylindrical billet composed of an elasto-plastic material is presented. Frictional contact conditions are imposed between the billet and the tool. For this simulation four noded enhanced strain elements are employed. Due to symmetry, only one half of the billet is considered.

Figure 8.2 depicts several finite element meshes adapted during the upsetting process.

EXAMPLE 8.2 Stretching of a strip with a hole. In this example, we present the finite element simulation of a problem with a shear band formation that causes localization of deformation. The specimen is assumed to be made of an elastic perfectly–plastic material with Young's modulus $E = 100$ [GPa], Poisson's ratio $\nu = 0.3$ and yield stress $\sigma = 100$ [MPa]. Due to symmetry, only one quarter of the strip is considered. In analysis an error indicator based on inelastic energy is used and the maximum size of elements is restricted to $h_{\max} = 0.7$ and the minimum size is $h_{\min} = 0.15$.

Meshes obtained after adaptive remeshing are shown in Figures 8.3(b)-(d).

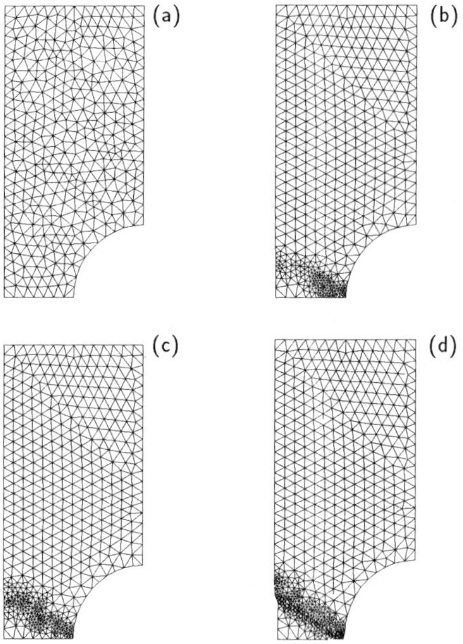

Figure 8.3: Stretching of an elastic-plastic strip with a hole.: (a) Initial mesh, (b) $U_2 = 0.04$, (c) $U_2 = 0.06$, (d) $U_2 = 0.415$.

9. EXPLICIT SOLUTION STRATEGIES

The solution strategies for the explicit dynamic analysis of large strain plasticity problems with contact are well established and their essential features are summarised below.

9.1 The discretised dynamic equations

Commencing from the linear momentum equation, the weak form of the equilibrium equations can be derived, which when discretised in the normal finite element manner leads to

$$M\ddot{u}_n + C\dot{u}_n + P(u_n) = F(t_n) \qquad (9.1)$$

In which u_n represents the displacement vector at time t_n, M and C are respectively the mass and damping matrices and $P(u_n)$ represents the internal force contribution from the element stress field which satisfies the (non-linear) constitutive relations. The term $F(t_n)$ represents the external forces arising from applied tractions and contact conditions.

The introduction of central difference approximations for the velocity and acceleration in terms of displacements, together with the use of mass lumping procedures and the assumption of mass proportional damping, leads to the following uncoupled recurrance relation from which the nodal displacements u^I_{n+1} at time t_{n+1} can be evaluated in terms of the corresponding quantities at time stations t_n and t_{n-1} as

$$u^I_{n+1} = \left[M^I(1+\alpha\Delta t/2)\right]^{-1}\{-\Delta t^2(P^I_n + F^I_n) + \\ +2M^I u^I_n - M^I(1-\alpha\Delta t/2)u^I_{n-1}\} \qquad (9.2)$$

in which α is the mass proportional damping coefficient. This expression permits the evaluation of displacement on an individual nodal basis with internodal coupling occurring only through the calculation of the internal forces P_n.

Using co-rotational measures of stress and strain the incremental large strain constitutive relation can be written

$$\Delta\sigma^c = \mathbf{C}[\Delta\varepsilon^c] \qquad (9.3)$$

Particular forms of the constitutive tensor $\mathbf{C}$ follow from standard elasto-plastic or elasto-viscoplastic descriptions. Different strain measures may be employed, but considerable computational benefits arise from use of the logarithmic stretch so that

$$\Delta\varepsilon^c = \ln[U] \qquad (9.4)$$

The total stress at any time t_{n+1} is obtained from

$$\sigma^c_{n+1} = \sigma^c_n + \Delta\sigma^c \qquad (9.5)$$

and the Cauchy stress is then given by

$$\sigma_{n+1} = R_{n+1}\sigma^c_{n+1}R^T_{n+1} \qquad (9.6)$$

where R_{n+1} is an orthogonal rotation tensor defined by polar decomposition at time instant t_{n+1}.

9.2 Element methodology

The finite element models employed for the simulation of industrial forming processes are invariably large and therefore in analysis a compromise between accuracy and computational efficiency has to be reached. This has lead to the use of reduced integration procedures to primarily reduce computation time, but which also counters the overly stiff behaviour associated with full integration. For linear elements this implies single point integration which, however, can result in spurious zero energy (hourglass) modes of deformation [3,9,10,24].

In order to obtain reliable results, various control methods have been proposed to eliminate hourglassing by providing restraint which the element lacks under single point integration, but without stiffening the element's adequate response to other modes [3,8]. Two principal ways of resisting hourglassing are with viscous damping and by introduction of artificial stiffness, but it should be stressed that such control techniques do not fully remove the kinematic modes and, in particular, coarse meshes and meshes loaded with large nodal forces, resulting either from boundary conditions or from contact, are still susceptible to hourglassing.

9.3 Explicit solution of forming problems

In practice, the majority of forming operations are sufficiently slow to be classified as quasi-static, with the material response being rate independent. Therefore, the sole justification of using explicit transient dynamic solution procedures for metal forming simulations is the much reduced computational times required, in comparison to quasi-static implicit analysis, for large scale industrial problems.

However, in order to achieve significant computational advantage several numerical artefacts have to be introduced into the explicit solution procedure. Whereas for quasi-static implicit analysis the process and material parameters utilised are the physical ones, some parameters are given artificial values in explicit analysis in order to provide acceptable CPU times. In particular, the following parameters are invariably modified:

Material density. Since the maximum permissible time step length, as defined by the Courant stability limit, is directly proportional to the square root of the material density, this parameter is increased; usually by one order of magnitude at least.

Punch velocity. In order to reduce the total number of time steps necessary to model the forming process, the punch velocity is increased; again by at least an order of magnitude.

Loading history. Since increasing both the material density and punch velocity results in increased inertia forces, the punch travel must be suitably controlled so as to minimise the inertia effects.

The amounts by which the material density and punch speed can be increased is limited, as the inertia forces may become unacceptably large; even with a judiciously designed loading history. For example, in deep drawing operations the predicted material

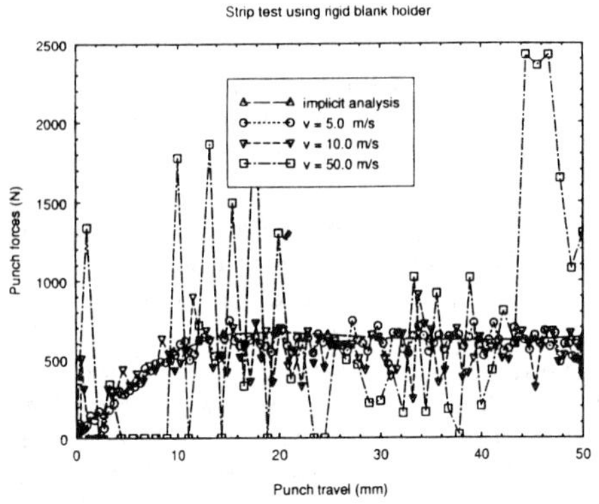

Figure 9.1 Drawing of a U-strip specimen: Punch force with increasing punch velocity.

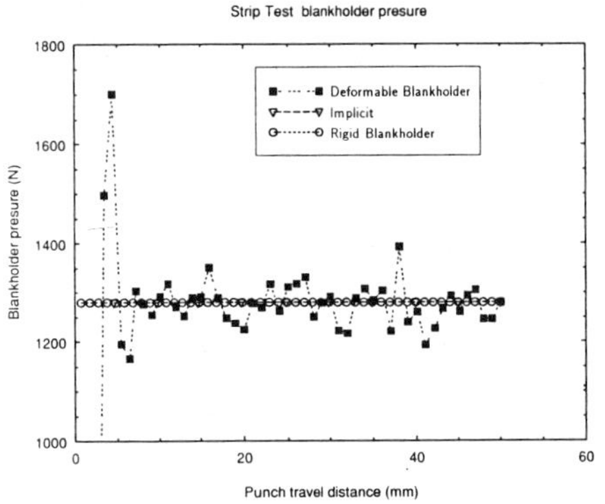

Figure 9.2 Drawing of a U-strip specimen: Comparison of rigid and deformable blankholder algorithms.

stretching near the punch is often larger than seen in practice whilst the predicted draw is too small.

Other issues which must be given attention in explicit analysis include contact modelling (particularly blankholder simulation) and element mesh orientation. The two main options open for modelling blankholder contact conditions are:

(i) *Deformable blankholder* in which the blankholder is represented by solid continuum elements with a pressure load applied to the upper surface.

(ii) *Rigid model* where a rigid contact surface is used in conjunction with a constraint equation to ensure that the total normal interface pressure is equal to the blankholder pressure.

Explicit analysis of sheet forming problems also necessitates careful design of the finite element mesh. Care must be taken that elements are not skewed, particularly in relation to any die radius over which they must pass. This places considerable restriction on mesh design for non simple configurations, which

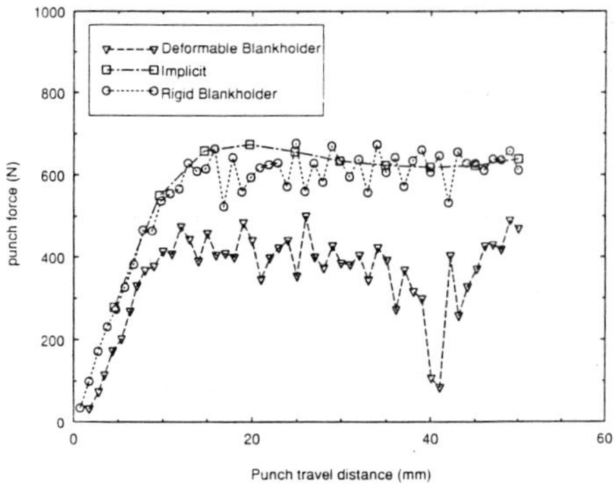

Figure 9.3 Drawing of a U-strip specimen: Punch force against the punch travel.

has clear implications for automatic mesh generation and adaptive strategies. The poor performance of skewed elements occurs even when specific warping terms are included in the element formulation [3]. Indeed, without the additional warping terms artificial localisation may occur at the die radius.

EXAMPLE 9.1 Drawing of a U-strip specimen. To illustrate the effects of the above parameters the drawing of a U-strip specimen is modelled both by a full 3-D explicit analysis and also as a 2-D implicit quasi-static problem. For the 3-D simulation 4 node shell elements are employed whilst the blankholder is modelled as deformable, using 3-D elements, and rigid, using a force controlled rigid contact surface.

Figure 9.1 illustrates the effect of punch speed on the prediction of the forming force. It is seen that the solution is significantly affected by the punch speed, with increasing velocities resulting in greater oscillations. The results of the explicit analysis could be considered acceptable up to a punch speed of 5 [m/s] provided that the loading is suitably ramped. The material density has been scaled by 10 for solution.

The history of the contact force between sheet and blankholder with a very low punch speed is shown in Figure 9.2. The rigid contact surface algorithm provides an almost exact blankholder pressure, due to the constraint equation and consequently the total friction force on the blankholder is also accurate. The results for the deformable blankholder show large oscillations in the blankholder pressure, primarily due to points making and losing contact due to inertia effects. This results in a friction force which is on average 25 % lower than the expected value. The effect of this under-prediction of the blankholder friction force may be seen in Figure 9.3. The punch force history for the rigid blankholder case agrees well with implicit solution, whereas the deformable blankholder prediction is some 30 % too low.

EXAMPLE 9.2 Deep drawn automotive headlamp panel. Finally, the explicit simulation of an industrial

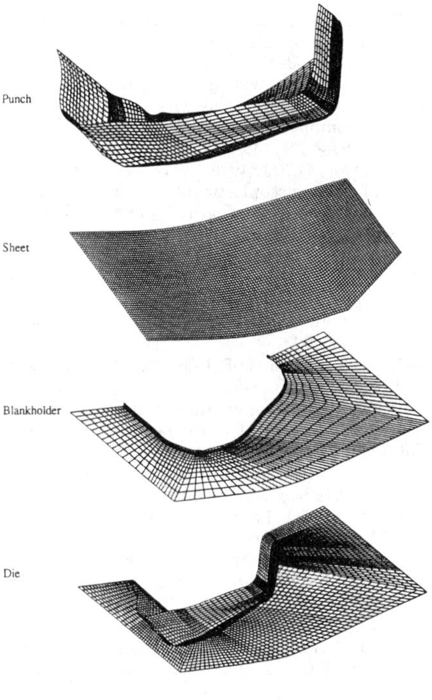

Figure 9.4 Deep drawn automotive headlamp panel: Definition of finite element model.

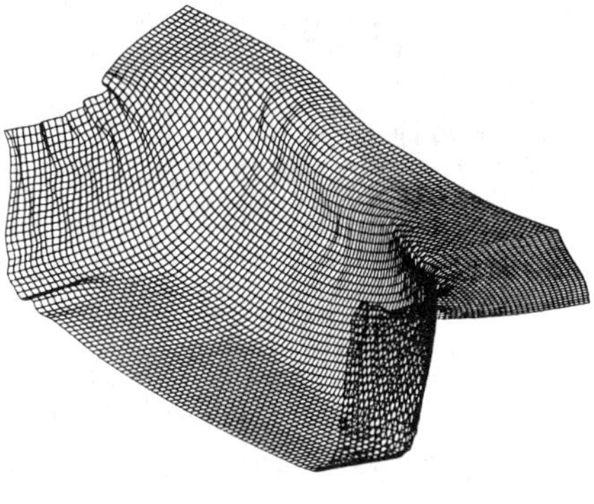

Figure 9.5 Deep drawn automotive headlamp panel: Location of wrinkles in simulation.

component is illustrated in Figure 9.4 which defines the finite element description of the sheet and tooling for the pressing of a deep drawn automotive headlamp panel. Features of the problem include a non-planar blankholder, a deep draw resulting in relatively large sliding distances and wrinkling in certain regions of the finished part. The sheet is modelled by 5500 four node shell elements and rigid tools are assumed, with the blankholder being pressure controlled. The final pressed shape is shown in Figure 9.5 where it is seen that the wrinkling that occurs in the prototype is reproduced by the finite element simulation. This wrinkling is caused by both lack of blankholder restraint at one end and a mismatch between the punch and die.

10. CONCLUDING REMARKS

Some recent advances in the finite element analysis of metal forming problems have been reviewed, indicating the progress that has been made both in the theoretical understanding of inelastic material behaviour under finite strains and the associated numerical implementation. Whilst the state of knowledge in some areas is relatively mature, considerable further understanding and development is required in others. For example, issues related to the modelling of complex contact-friction phenomena are far from settled and a more comprehensive treatment of friction may necessitate the integration of micromechanical studies with computational approaches.

Although adaptive strategies are, at present, routinely performed for linear elliptic problems, their extension to nonlinear elliptic problems – in particular to metal forming problems where, typically, large inelastic deformations at finite strains are standard working conditions – is by no means trivial. Apart from the issues briefly mentioned in this paper, several important aspects of adaptive strategies related to nonlinear industrial applications which need further attention include the introduction of various types of error estimators and their comparative analysis, and the consideration of data transfer strategies.

The adequacy of explicit dynamic transient solutions for metal forming simulations is still a subject of debate. The primary justification of using such techniques is the much reduced computational effort involved; for large scale industrial problems at least. However the quality of the solutions obtained is invariably inferior to the corresponding implicit quasi-static solution, which is brought about by inertia effects and by the scaling of process and material parameters necessary to provide acceptable CPU times.

Adequate metal forming simulations can be provided by explicit methods, but considerable experience is required on the part of the user for optimal choice of the crucial solution parameters.

In future the competitiveness of implicit quasi-static solution methods may be improved by developments in sparse matrix techniques for equation solving, which when accompanied by increased computer memory availability, may make the solution of large scale industrial problems by such approaches a realistic proposition.

ACKNOWLEDGEMENT

This work was partially supported by the U.K.

EPSRC through Grants GR/H45858, GR/J13779 and GR/J89644 awarded to the University of Wales Swansea, and this support is gratefully acknowledged.

REFERENCES

[1] F.Barlat and J.Lian, 'Plastic behavior and stretchability of sheet metals. Part I: A yield function for orthotropic sheets under plane stress condition', *Int. J. Plasticity,* 5, 51–66 (1989)

[2] N.Bay and T.Wanheim, 'Contact phenomena under bulk plastic deformation conditions', In *Advanced Technology of Plasticity 1990 – Proceedings of the Third International Conference on Technology of Plasticity,* pp. 1677–1691, The Japan Society for Technology of Plasticity, (1990)

[3] T.Belytschko and I.Leviathan 'Projection schemes for one-point quadrature shell elements', *Comp. Meth. Appl. Mech. Engng.,* 115, 277–286 (1994)

[4] A.Cuitiño and M.Ortiz, 'A material-independent method for extending stress update algorithms from small-strain plasticity to finite plasticity with multiplicative kinematics', *Engng. Comp.,* 9, 437–451 (1992)

[5] A.Curnier, 'A theory of friction', *Int. J. Solids Struct.,* 20, 637–647 (1984)

[6] M.Dutko, D.Perić and D.R.J.Owen, 'Universal anisotropic yield criterion based on superquadric functional representation. Part 1: Algorithmic issues and accuracy analysis', *Comp. Meth. Appl. Mech. Engng.,* 109, 73–93 (1993)

[7] A.L.Eterović and K.-J.Bathe, 'A hyperelastic-based large strain elasto-plastic constitutive formulation with combined isotropic-kinematic hardening using logarithmic stress and strain measures', *Int. J. Num. Meth. Engng.,* 30, 1099–1114 (1990)

[8] D. P. Flanagan and T.Belytschko, 'A uniform strain haxahedron and quadrilateral with orthogonal hourglass control', *Int. J. Num. Meth. Engng.,* 17, 679–706 (1981)

[9] J. O. Hallquist, *LS-DYNA3D Theoretical Manual,* (1991)

[10] E.Haug, J.Clinckemaillie and F.Aberlenc, 'Computational Mechanics in Crashworthiness Analysis', In Post Symposium Course of the *Second Internatioanl Symposium on Plasticity,* Nagoya, 4-5 August (1989)

[11] Ch. Hochard, D. Perić, D.R.J.Owen, M.Dutko, 'Transfer operators for evolving meshes in elastoplasticity', In *Computational Plasticity IV: Fundamentals and Applications,* (edited by D.R.J.Owen et al.), Pineridge Press, Swansea (1995)

[12] P.Ladeveze, G. Cofignal and J.P. Pelle 'Accuracy of elastoplastic and dynamic analysis', In *Accuracy Estimates and Adaptive Refinements in Finite Element Computations,* (edited by I.Babuška et al.), 181–203, John Wiley, New York (1986)

[13] N.-S.Lee and K.-J.Bathe, 'Error indicators and adaptive remeshing in large deformation finite element analysis', *Fin. Elem. Anal. Des.,* 16, 99–139 (1994)

[14] J.T.Oden and J.A.C.Martins, 'Models and computational methods for dynamic friction phenomena', *Comp. Meth. Appl. Mech. Engng.,* 52, 527–634 (1985)

[15] M.Ortiz and J.J.Quigley IV, 'Adaptive mesh refinement in strain localization problems', *Comp. Meth. Appl. Mech. Eng.,* 90, 781–804 (1991)

[16] D.R.J.Owen and D.Perić, 'Recent developments in the application of finite element methods to nonlinear problems', *Fin. Elem. Anal. Des.,* 18, 1–15 (1994)

[17] D.Perić, 'On a class of constitutive equations in viscoplasticity: Formulation and computational issues', *Int. J. Num. Meth. Engng.,* 36, 1365–1393 (1993)

[18] D.Perić, D.R.J.Owen and M.E.Honnor, 'A model for large strain elastoplasticity based on logarithmic strain: computational issues', *Comp. Meth. Appl. Mech. Engng.,* 94, 35–61 (1992)

[19] D.Perić and D.R.J.Owen, 'A model for large deformations of elasto-viscoplastic solids at finite strains: computational issues', In *Finite Inelastic Deformations – Theory and Applications,* (edited by D. Besdo and E. Stein), 299–312, Springer-Verlag, Berlin (1992)

[20] D.Perić and D.R.J.Owen, 'Computational model for 3-D contact problems with friction based on the penalty method', *Int. J. Num. Meth. Engng.,* 35, 1289–1309 (1992)

[21] D.Perić J.Yu and D.R.J.Owen, 'On error estimates and adaptivity in elasto-plastic solids: applications to the numerical simulation of strain localisation in classical and Cosserat continua', *Int. J. Num. Meth. Engng.,* 37, 1351–1379 (1994)

[22] T.Rodič, *Numerical Analysis of Thermomechanical Processes During Deformations of Metals at High Temperatures,* Ph.D Thesis, University of Wales, Dept. Civil Eng., Swansea (1989)

[23] M.Schönauer, T.Rodič, and D.R.J.Owen, 'Numerical modelling of thermomechanical processes related to bulk forming operations', *J. Physique IV,* suppl. *J. Physique III,* 3, 1199–1209 (1993)

[24] K. Schweizerhof, L.Nilsson, and J.O.Hallquist, 'Crashworthiness analysis in the automotive industry', *Int. J. Comp. Appl. Techn.,* 5, 134–156 (1992)

[25] J.C.Simo, 'Algorithms for static and dynamic multiplicative plasticity that preserve the classical return mapping schemes of the infinitesimal theory', *Comp. Meth. Appl. Mech. Engng.,* 99, 61–112 (1992)

[26] J.C.Simo and T.J.R.Hughes, 'General return mapping algorithms for rate-independent plasticity', In *Constitutive Laws for Engineering Materials: Theory and Applications,* (Edited by C.S.Desai et al.), 221–231, Elsevier Science Publ., New York (1987)

[27] J.C.Simo and F.Armero, 'Geometrically non-linear enhanced strain mixed methods and the method of incompatible modes', *Int. J. Num. Meth. Engng.,* 33, 1413–1449 (1992)

[28] J.C.Simo and C.Miehe, 'Associative coupled thermoplasticity at finite strains: formulation, numerical analysis and implementation', *Comp. Meth. Appl. Mech. Engng.,* 98, 41–104 (1992)

[29] E.A. de Souza Neto, D.Perić and D.R.J.Owen, 'A computational model for ductile damage at finite strains: Algorithmic issues and applications', *Engng. Comput.,* 11, 257–281 (1994)

[30] E.A. de Souza Neto, K.Hashimoto, D.Perić and D.R.J.Owen, 'A phenomenological model for frictional contact of coated steel sheets', In Proc. 2nd Int. Conf. *Numerical Simulation of 3-D Sheet Metal Forming Processes - Verification of Simulation with Experiment,* 31 August - 2 September, 1993, Tokyo, Japan

[31] G.Weber and L.Anand, 'Finite deformation constitutive equations and a time integration procedure for isotropic, hyperelastic-viscoplastic solids', *Comp. Meth. Appl. Mech. Engng.,* 79, 173–202 (1990)

[32] O.C.Zienkiewicz and J.Z.Zhu, 'A simple error estimator and adaptive procedure for practical engineering analysis', *Int. J. Num. Meth. Engng.,* 24, 337–357 (1987)

[33] O.C.Zienkiewicz, G.C.Huang and Y.C.Liu, 'Adaptive FEM computation of forming processes – application to porous and non-porous materials', *Int. J. Num. Meth. Engng.,* 30, 1527–1553 (1990)

Simulation of Materials Processing: Theory, Methods and Applications, Shen & Dawson (eds)
© 1995 Balkema, Rotterdam. ISBN 90 5410 553 4

Evolution of hybrid stress finite element method and an example analysis of finite strain deformation of rigid-viscoplastic solid

Theodore H. H. Pian
Department of Aeronautics and Astronautics, Massachusetts Institute of Technology, Mass., USA

ABSTRACT The evolution of the assumed stress hybrid element method is reviewed to indicate how the element stiffness matrices can now be formulated to achieve its robustness for the finite element solutions. For the metal forming problems it is essential that the accuracy of such solutions can be maintained when the element geometry is severely distorted. As an illustration of the assumed stress hybrid finite element method the formulation of finite strain deformation of rigid-viscoplastic solid is included.

1. INTRODUCTION

In using finite element methods for the analysis of metal forming problems it is important to adopt solid elements which can handle thin plates and shells and nearly incompressible materials without locking difficulties. They should also be insensitive to the distortion of the element geometry. Because the formulation of hybrid finite element methods are based on multi-field variables, there are many effective ways to improve the performance of the resulting elements. In recent years there have been very significant advances in the technology of assumed stress hybrid finite element methods that elements with robust properties essential to the analysis of forming process can now be formulated.

The main purpose of this paper is a review of the evolution of the assumed stress hybrid finite element methods. The discussion is concentrated in the formulation by variational methods. One special problem that is related to metal forming problems is the analysis of flow problems for rigid-viscoplastic material. This is included as an illustration of the assumed stress hybrid finite element method. Here again, the discussion is concentrated on the variational formulation of the corresponding finite element method.

2. EVOLUTION OF HYBRID STRESS FINITE ELEMENT METHODS

2.1 Definition of hybrid stress finite element

The term *hybrid* was initially coined to designated finite elements which are formulated based on the approximation of one field variable within the element, and another one along the interelement boundaries (Pian and Tong 1968) The original hybrid element (Pian 1864) was based on the principle of minimum complementary energy with equilibrating stresses in the interior of the element and compatible displacements along the interelement boundary. In the finite element formulation, the assumed element stresses are equilibrating and are approximated by finite number of terms with unknown parameters represented by stress vector $\boldsymbol{\beta}$, and the boundary displacements are expressed in terms of nodal displacements $\mathbf{q}$. The variational principle will yield $\boldsymbol{\beta}$ in terms of $\mathbf{q}$. The element strain energy can then be expressed in terms of $\mathbf{q}$, and the expression of the element stiffness matrix can be determined.

It was realized later that element stiffness matrices can also be formulated by using another two-variable variational functional, the Hellinger-Reissner functional, which contains both stresses and displacements in the interior of the elements. In fact, when the same equilibrating stresses are used and the assumed displacements are compatible along the inter-element boundary, the resulting element stiffness matrix is the same as that formulated by the corresponding complementary energy principle.

In this situation the original meaning of *hybrid* no longer holds., and the terms *hybrid/mixed* and *mixed/hybrid* have often been used in the literature. It is felt that it is more clear-cut when the term *hybrid finite element* for solid mechanics is redefined as *an element that can be formulated based on a multi-field variational functional, but only the nodal displacements are left as unknowns in the final matrix equations*. Thus, *Hybrid stress elements* are those elements which are formulated by using both displacements and stresses as the field variables.

2.2 Formulation of element stiffness matrix by Hellinger-Reissner variational functional

The Hellinger-Reissner variational functional, which for a single element of volume V_n, is also the element strain energy, is,

$$\Pi_{HR} = \int_{V_n} [-\frac{1}{2}\boldsymbol{\sigma}^T \mathbf{S}\boldsymbol{\sigma} + \boldsymbol{\sigma}^T(\mathbf{Du})]dV \quad (1)$$

where $\boldsymbol{\sigma}$ = stress vector, $\mathbf{S}$ = elastic compliance matrix, and the strain displacement relation is given by $\boldsymbol{\varepsilon} = \mathbf{Du}$. In the finite element formulation, the element displacements $\mathbf{u}$ are interpolated in terms of nodal displacements $\mathbf{q}$ by $\mathbf{u} = \mathbf{Nq}$. The stresses are approximated by $\boldsymbol{\sigma} = \mathbf{P}\boldsymbol{\beta}$. Then the element strain energy is given by

$$U_n = -(1/2)\,\boldsymbol{\beta}^T\mathbf{H}\boldsymbol{\beta} + \boldsymbol{\beta}^T\mathbf{G}\mathbf{q} \quad (2)$$

where

$$\mathbf{H} = \int_{V_n} \mathbf{P}^T\mathbf{S}\mathbf{P}dV;\ and\ \ \mathbf{G} = \int_{V_n} \mathbf{P}^T(\mathbf{DN})dV \quad (3)$$

are, respectively, the flexibility and leverage matrices. By setting the variation with respective to $\boldsymbol{\beta}$ to zero, $\boldsymbol{\beta}$ can be expresses as $\boldsymbol{\beta} = \mathbf{H}^{-1}\mathbf{Gq}$, and the element stiffness matrix is given by

$$\mathbf{k} = \mathbf{G}^T\mathbf{H}^{-1}\mathbf{G} \quad (4)$$

2.3 Construction of robust finite elements

A robust finite element must fulfill the following requirements:

(1) It does not have any kinematic deformation modes.

(2) It can pass the constant strain patch test.

(3) It is invariant with respect to the choice of reference coordinates for the assumed stresses.

(4) It will not have locking difficulties in handing nearly incompressible material.

(5) It is insensitive to the distortion of element geometry.

Obviously the last two requirements are essential to the finite element solutions of metal forming problems which involve elements of severe geometric distortions and plastic strains which are incompressible.

It was recognized very early that in the formulation of a hybrid stress finite element, the assumed stresses and displacements should be properly balanced (Pian and Tong 1969). Rational methods for determining the stress terms for obtaining an element of optimal performance were later established based on the idea that the polynomial expansions for stresses and displacements should be both complete and balanced (Pian 1985). Here the displacements $\mathbf{u}$ are separated into compatible part $\mathbf{u}_q$ which are interpolated in terms of nodal displacements $\mathbf{q}$, and incompatible part $\mathbf{u}_\lambda$ which are expressed in terms of parameters $\boldsymbol{\lambda}$. The static condensation of $\boldsymbol{\lambda}$ in the element level will lead to constraint conditions to the initially assumed stresses. The incompatible displacements are introduced to maintain the completeness. The assumed stresses are initially unconstrained in complete polynomial expansion, and are of the same order as the strains corresponding to $\mathbf{u}$. But, on account of the introduction of incompatible displacements, a boundary integral term should be added to enforce the compatibility. This result is to introduce the following weak equilibrium constraint on the element stresses,

$$\int_{V_n} (\mathbf{D}^T\boldsymbol{\sigma})^T\mathbf{u}_\lambda = 0 \quad (5)$$

This is the satisfaction of the homogeneous equilibrium equations in variational or integral sense (Pian and Chen 1982). However, for many cases, an inclusion of certain higher order perturbation of the element geometry is required to obtain sufficient constraining equations (Pian and Sumihara 1984).

For 4-node quadrilateral plane elements, the incompatible displacements to be used is the Wilson displacements with four λ parameters. The resulting optimal stress pattern for a 4-node quadrilateral plane element of arbitrary geometry, when expressed in isoparametric coordinates, contains 5 β-parameters of the following form:

$$\boldsymbol{\sigma} = \{\ \sigma_x\ \sigma_y\ \tau_{xy}\ \} = [\ \mathbf{I}\ \ \mathbf{P}_h\ ]\ \boldsymbol{\beta} \quad (6)$$

where $\mathbf{I}$ is a 3x3 identity matrix and $\mathbf{P}_h$ is given by,

$$\mathbf{P}_h = \begin{bmatrix} a_3^2\xi & a_1^2\eta \\ b_3^2\xi & b_1^2\eta \\ a_3b_3\xi & a_1b_1\eta \end{bmatrix} \tag{7}$$

where a_1, a_3, b_1 *and* b_3 are related to the coordinates (x_i, y_i) of the corner nodes by,

$$\begin{bmatrix} a_1 & b_1 \\ a_3 & b_3 \end{bmatrix} = \frac{1}{4}\begin{bmatrix} -1 & +1 & +1 & -1 \\ -1 & -1 & +1 & +1 \end{bmatrix}\begin{bmatrix} x_1 & y_1 \\ \vdots & \vdots \\ x_4 & y_4 \end{bmatrix} \tag{8}$$

It was also found that the above stress pattern can be obtained by the following alternative procedure:

For a square 4-node plane element of a dimension 2x2, the isoparametric coordinates (ξ,η) and the Cartesian coordinates (x,y) coincide. It has been determined from the consideration of minimum number of required stress modes to suppress the spurious kinematic deformation modes and of the resulting stresses being uncoupled (Pian and Chen 1983) that an optimal stress pattern is given by Eq. (6) with

$$\mathbf{P}_h = \begin{bmatrix} 0 & \eta \\ \xi & 0 \\ 0 & 0 \end{bmatrix} \tag{9}$$

It is noted that the five stress modes corresponds to the five straining modes due to the displacements **u**.

It can be argued that for an irregular shaped element the optimal stresses, if expressed in terms of tensor stress components τ^{ij} in the local isoparametric coordinates, should be the same as the above. But the corresponding physical components σ_{ij} in global Cartesian coordinates will need the following tensor transformation:

$$\sigma_{ij} = \sigma^{ij} = J_k^i J_l^j \tau^{kl} \tag{10}$$

where $J_k^i = \partial x^i / \partial \xi^k$ are the values of the Jacobian.

In the finite element finite element formulation, the values of the Jacobian are replaced by their values at the origin of the local system ($\xi = 0$, $\eta = 0$), so that the resulting constant stresses will not couple with higher order terms. This is a requirement for the convergence of the finite element solutions. For the 4-node plane elements the resulting stress field turns out to be that same as that given by Eq. (7). The element which is constructed by using this stress expansion is often referred as the Pian-Sumihara element. It is labeled here as PS5β element. Equation (7) shows that for irregular shaped elements the assumed stresses are coupled.

An alternative approach for choosing the stress terms has been suggested based on a different measure that should be taken on account of the introduction of incompatible displacements (Pian and Wu 1988). The corresponding constraint condition for the stresses is that the virtual work due to the boundary tractions resulting from the stresses of order higher than constant and the incompatible boundary displacements should vanish. It turns out that when the incompatible displacements are the same Wilson displacement field the resulting stress pattern is the same as that of the PS5β element. In this case, however, the process of perturbation of element geometry is no longer required.

The stress pattern for an 8-node hexahedral element can also be determined from similar consideration (Pian and Chen 1983). For a hexahedral element of dimensions 2x2x2, the stresses $\boldsymbol{\sigma}\left(=\left\{\sigma_x\,\sigma_y\,\sigma_z\,\tau_{yz}\,\tau_{zx}\,\tau_{xy}\right\}\right)$ have eighteen β–parameters as shown in the following:

$$\boldsymbol{\sigma} = \mathbf{I}_c\boldsymbol{\beta}_c + \mathbf{P}_h\boldsymbol{\beta}_h \tag{11}$$

where

$$\mathbf{P}_h = \begin{bmatrix} 0 & 0 & 0 & \eta & 0 & 0 & \varsigma & 0 & 0 & \eta\varsigma & 0 & 0 \\ \xi & 0 & 0 & 0 & 0 & 0 & 0 & \varsigma & 0 & 0 & \varsigma\xi & 0 \\ 0 & \xi & 0 & 0 & \eta & 0 & 0 & 0 & 0 & 0 & 0 & \xi\eta \\ 0 & 0 & \xi & 0 & 0 & 0 & 0 & 0 & 0 & 0 & 0 & 0 \\ 0 & 0 & 0 & 0 & 0 & \eta & 0 & 0 & 0 & 0 & 0 & 0 \\ 0 & 0 & 0 & 0 & 0 & 0 & 0 & 0 & \varsigma & 0 & 0 & 0 \end{bmatrix} \tag{12}$$

For regular brick-shaped elements the 18 stress modes correspond to 18 straining modes due to the element displacements. Here again, for irregular elements a transformation is taken and the higher order stress terms are all coupled. This element is labeled as PT18β element (Pian and Tong 1986).

Di and Ramm (1994) have presented a number of quadrilateral plane and hexahedral solid elements constructed by hybrid stress approach using different constrained stresses. Both PS5β and PT18β elements are included. These elements all satisfy the first four requirements listed earlier. For problems modeled by elements with geometric distortions, these elements all yield much more accurate results than those formulated by expressing stresses in Cartesian coordinates.

2.4 Relief of locking difficulty for analyzing thin plates and shells using 8-node solid elements

Metal forming process may often involve structure part which consists both thick and thin sections. In that case, it is more convenient to model the thin sections also by solid elements instead of plate or shell elements. Thus, for such applications, the solid element should be specially designed so that it can also handle the analysis of plates and shells without locking difficulty

As it will be indicated later, the PT18β element does lead to locking when it is used to analyze thin plates and shells using elements of distorted plane geometry. The reason for the locking phenomenon is that for thin plates and shells the transverse normal stress and the two transverse shearing stresses should approach zero. As shown in Eqs. 11 and 12, there are 8 stress modes for these three stresses. This may lead to 8 constraint conditions.

A convenient way to examine whether constraint conditions will lead to locking phenomenon is to determine the value of *constraint index* (CI) which is define as the difference between NK, the number of kinematic degrees of freedom brought by an element, when added to the existing finite element mesh, and NC, the number of independent constraints per element (Malkus and Hughes 1978). For the present 8-node solid element, NK = 6, and NC = 8. Thus, CI is equal to -2. This an indication of the existence of locking difficulty. In the case of the use of rectangular PS5β elements to model the bending of slender beams, NC = 3, NK = 4 and CI = 1. Hence, the element is locking free.

Sze and Ghali (1993) have suggested a selective scaling procedure to relieve the locking difficulty of the PT18β solid element. The scheme is to reduce the magnitudes of the transverse normal strain modes and the two transverse shearing strain modes by multiplying those columns of the leverage matrix **G** that are associated with these three transverse stresses by a factor equal to t/L, where t is thickness of the plate or shell and L is the average dimension the element. It is seen from Eq. (12) that by applying the scaling factor only to the 2nd. 3rd, 5th, 6th and 12th columns of the leverage matrix there will be enough reduction of constraints to make the resulting element locking free. This element is labeled as SS18β element. A study of the performances of these two 8-node solid elements has been made by these authors. For the solution of simply supported and thin square plate of $L/t = 10^4$, subjected to uniform loading using regular mesh, there is no evidence of locking when square shaped elements are used. However, for the solutions of clamped circular plate of radius R loaded by central load using irregular shaped elements, while PT18β element is able to yield good accuracy when one quarter of the plate is modeled by 48 elements and for R/t = 50, the normalized central displacement is only 0.129 for R/t = 5000. On the other hand, the accuracy of the SS18β element is not affected by the decreasing in plate thickness.

2.5 Modification of stress interpolations for enhancing computational efficiency

In the construction of hybrid stress finite elements the most costly operation is the inversion of the flexibility matrix **H.** One way to reduce the computational effort is to decouple the constant and higher order stress modes such that the **H** matrix will contain only diagonal sub-matrices, hence the inversion of **H** is reduced to that of matrices of much smaller order. It can be shown that such a process involves only a rescaling of the parameters defining the constant stress terms. For example, Zienkiewicz and Taylor (1989) have shown that the element stiffness matrix for a 4-node quadrilateral plane element is reduced to the sum of two terms. The first one is identical to a one-point quadrature evaluation of the displacement-based element. The second term is a rank 2 stabilization matrix based on assumed stresses. That matrix can be computed analytically. Thus, with such an implement, the Pian-Sumihara element can be made also competitive with displacement based finite elements from the point of view of computational efficiency.

Sze et al.(1991) suggested an alternative approach for stress formulations to reduce the **H** matrix to only diagonal submatrices. For example, for the 4-node quadrilateral element, it involves, simply the division of $\mathbf{P}_h$ of Eq.(7) by the Jacobian determinant J. Sze (1992) has shown that further reduction of the computing cost is justified by keeping only the diagonal terms of the $\mathbf{H}_{hh}$ matrix. A significant recent advance is a procedure suggested by Sze (1993, 1994) to decouple the lower and higher order stress terms, for higher order elements such as 9-node membrane and shell elements and 20-node hexahedral elements. The resulting scheme is that the tiffness matrix can all be reduced to two terms, one of which can be obtained by displacement based formulation with reduced integration, while the other is to stabilize the element by appropriate assumed stresses.

It can be concluded that the assumed stress hybrid finite element method has now reached its maturity both from the robustness of the resulting elements and from the point of view of computing efficiency.

3. FLOW OF RIGID-PLASTIC MATERIAL BY HYBRID STRESS FINITE ELEMENT METHOD

Finite deformation analyses of metal forming process have been formulated by finite element methods based on rigid-viscoplastic constitutive models (Kobayashi et al 1989). The use of hybrid finite element method for such analysis is most attractive because of its capacity of alleviating the incompressibility "locking" difficulty. The following presentation is based on the work of Chen and Cui (1993).

In their transient analysis of the forming process the Updated Lagrangian formulation is employed. In addition, an implicit time integration procedure is used.

3.1 Rigid-viscoplastic constitutive equations

In contrast to the development in Section 2, tensors are used for the stress and rate of deformation in the present section. Here, $\boldsymbol{\sigma}$ and $\boldsymbol{\sigma}'$ are respectively the Cauchy stress tensor and the deviatoric stress tensor, and $\mathbf{V}$ is used to represent the velocity field. The rate-of-deformation is, then, given by

$$\boldsymbol{\varepsilon} = \frac{1}{2}(\nabla\mathbf{V} + \mathbf{V}\nabla) \tag{13}$$

where ∇ is the gradient operator.

Under rigid-viscoplastic hypothesis the stress and rate-of-deformation relation is given by

$$\boldsymbol{\varepsilon} = \boldsymbol{\varepsilon}^{e} + \boldsymbol{\varepsilon}^{vp} \approx \boldsymbol{\varepsilon}^{vp}\,; \quad \boldsymbol{\varepsilon}^{vp} = \gamma\langle\Phi(\mathrm{F})\rangle\frac{\partial \mathrm{F}}{\partial\boldsymbol{\sigma}} \tag{14}$$

For isotropic-hardening von-Mises material, $\mathrm{F} = \sqrt{J_2}$, the stress and rate-of-deformation relation is given by

$$\boldsymbol{\varepsilon} = (1/2\mu)\boldsymbol{\sigma}' \tag{15}$$

where μ is viscosity and

$$\boldsymbol{\sigma}' = \boldsymbol{\sigma} + \mathrm{p}\mathbf{I} \tag{16}$$

Here, $\mathbf{I}$ is identity tensor, p is the hydrostatic pressure and the viscosity is given by $\mu = \frac{\bar{\sigma}}{3\bar{\varepsilon}}$, $\bar{\sigma}$ and $\bar{\varepsilon}$ being the effective stress and effective strain rate respectively. For a given material the effective stress is a function of the effective strain, the temperature and an internal variable. The other condition to be considered is the incompressibility of plastic strain, i.e.,

$$\nabla\cdot\mathbf{V} = 0 \tag{17}$$

3.2 Variational functional for hybrid stress finite element method

In the variational principle used in the present formulation, the introduction of the incompressibility constraint is based on a penalty method. This idea is similar to that for the solution of Stokes flow problem by the penalty-hybrid finite element method (Chen and Zhao 1990). For the case of prescribed tractions along the boundary S_T the variational functional takes the form,

$$\Pi = \int_{\Omega}[-\frac{1}{4\mu}\boldsymbol{\sigma}':\boldsymbol{\sigma}' + \boldsymbol{\varepsilon}:\boldsymbol{\sigma}' - \mathrm{p}(\nabla\cdot\mathbf{V}) - \frac{\varepsilon}{2}\mathrm{p}^2]d\Omega$$
$$- \int_{S_V}\bar{\mathbf{T}}\cdot\mathbf{V}dS \tag{18}$$

3.3 Finite element formulation

For the finite element formulation, $\boldsymbol{\sigma}'$, p and $\mathbf{V}$ are expressed in terms of shape functions $\mathbf{P}$, $\mathbf{A}$ and $\mathbf{N}$,

$$\boldsymbol{\sigma}' = \mathbf{P}\boldsymbol{\beta}\,; \quad \mathrm{p} = \mathbf{A}\boldsymbol{\alpha}\,; \quad \mathbf{V} = \mathbf{N}\mathbf{q} \tag{19}$$

where $\mathbf{q}$ is nodal velocities, $\boldsymbol{\beta}$ and $\boldsymbol{\alpha}$ are internal parameters which are independent from one element to the other. By defining the expressions of $\mathbf{H}$, $\mathbf{G}$, $\bar{\mathbf{H}}$ and $\bar{\mathbf{G}}$ in the following manner:

$$\begin{aligned}
\boldsymbol{\beta}^T\mathbf{H}\boldsymbol{\beta} &= \int_{\Omega_n}(\frac{1}{2\mu})\boldsymbol{\sigma}':\boldsymbol{\sigma}'d\Omega \\
\boldsymbol{\beta}^T\mathbf{G}\mathbf{q} &= \int_{\Omega_n}\boldsymbol{\varepsilon}:\boldsymbol{\sigma}'d\Omega \\
\boldsymbol{\alpha}^T\bar{\mathbf{G}}\mathbf{q} &= \int_{\Omega_n}\mathrm{p}(\nabla\cdot\mathbf{V})d\Omega \\
\boldsymbol{\alpha}^T\bar{\mathbf{H}}\boldsymbol{\alpha} &= \int_{\Omega_n}\mathrm{p}^2 d\Omega
\end{aligned} \tag{20}$$

one obtains for the assembladge of N elements,

$$\Pi = \sum_{n}^{N}(-\frac{1}{2}\boldsymbol{\beta}^T\mathbf{H}\boldsymbol{\beta} + \boldsymbol{\beta}^T\mathbf{G}\mathbf{q} - \boldsymbol{\alpha}^T\bar{\mathbf{G}}\mathbf{q}$$
$$-\frac{\varepsilon}{2}\boldsymbol{\alpha}^T\bar{\mathbf{H}}\boldsymbol{\alpha} - \mathbf{Q}_n^T\mathbf{q}) \tag{21}$$

The stationary condition of the functional then leads to,

$$\boldsymbol{\beta} = \mathbf{H}^{-1}\mathbf{G}\mathbf{q}, \qquad \boldsymbol{\alpha} = -(\frac{1}{\varepsilon})\overline{\mathbf{H}}^{-1}\overline{\mathbf{G}}\mathbf{q}, \tag{22}$$

and the element stiffness matrix **k** is given by

$$\mathbf{k} = \mathbf{G}^T\mathbf{H}^{-1}\mathbf{G} + \frac{1}{\varepsilon}\overline{\mathbf{G}}^T\overline{\mathbf{H}}^{-1}\overline{\mathbf{G}} \tag{23}$$

It can be seen that by using this penalty-hybrid method, the internal parameters $\boldsymbol{\beta}$ and $\boldsymbol{\alpha}$ can all be eliminated from the element level and the resulting matrix equation is of the following conventional form for finite element method for structural analysis.

$$\mathbf{Kq} = \mathbf{Q} \tag{24}$$

In the analysis of Stokes flow by Bratianu and Atluri (1983), the formulation is also based on a hybrid finite element method using deviatoric stress $\boldsymbol{\sigma}'$, pressure p and velocity $\mathbf{V}$ as variables. However, without the present treatment of incompressibility constraint by penalty method it is not possible to eliminate the parameters $\boldsymbol{\alpha}$ in the element level. The final matrix to be solved contains the nodal displacements $\mathbf{q}$ and the $\boldsymbol{\alpha}$ parameters of all elements as unknowns.

For the metal forming problem the configuration of the solid continuum, the viscosity μ and hence also the element stiffness matrix **k** all vary with respective to time. In the updated Lagrangian formulation of this nonlinear problem, the above finite element solution is performed on a unknown configuration C^{N+1} at the termination point t^{N+1} of the current time increment step. For the configuration updating, an implicit time integration scheme is used i.e.,

$$\mathbf{X}^{N+1} = \mathbf{X}^{N} + \mathbf{V}^{M}\Delta t \tag{25}$$

where $\mathbf{V}^M$ represents the velocity field at an interim configuration between C^N and C^{N+1}.

3.4 Example solutions for compression of ring of rectangular cross-section

The example problem analyzed by the present method is the same one analyzed by Chen and Kobayashi (1978). It consists of a ring of rectangular cross-section and of annealed aluminum 1100, compressed by two rigid plates. Four-node quadrilateral elements are used for analyzing the deformation of this axisymmetric problem. Here, the global coordinates are r, z and θ, and the deviatoric stresses $\sigma'_r, \sigma'_z, \tau_{rz}\, and\, \sigma'_\theta$ are designated, respectively, as $\sigma'_{11}, \sigma'_{22}, \sigma'_{12}\, and\, \sigma'_{33}$. For a square axisymmetric solid element of cross section dimension 2x2, and with the isoparametric coordinates ξ and η in parallel, respectively, with the r and z axes, the optimal stress modes has been determined as (Wu et al 1987),

$$\begin{Bmatrix} s^{11} \\ s^{22} \\ s^{12} \\ s^{33} \end{Bmatrix} = \begin{bmatrix} 1 & 0 & 0 & 0 & \eta & 0 & 0 & 0 \\ 0 & 1 & 0 & 0 & 0 & \xi & 0 & 0 \\ 0 & 0 & 1 & 0 & 0 & 0 & 0 & 0 \\ 0 & 0 & 0 & 1 & 0 & 0 & \xi & \eta \end{bmatrix} \begin{Bmatrix} \beta_1 \\ : \\ : \\ \beta_8 \end{Bmatrix} \tag{26}$$

At the initial stage of the metal forming analysis all the elements are rectangular in shape, therefore, the about stress pattern may be used. But in order to account for severe distortion of the elements, the first three stress components can be modified by the tensor transformation given by Eq, (10).

The initial mesh pattern for the finite element analysis is 8x12 for the half of the ring cross section. For the case that the coefficient of friction between the plate and the ring equal to 0.25, the plot, obtained by the present method, for the change in inner equatorial diameter of the ring vs. the reduction of its height up to 50%, agrees quite well with that obtained by Chen and Kobayashi. The reduction in height for each increment step adopted in the present analysis is 5%, while the increment step required for the latter is 1%. Thus, the advantage of the present hybrid finite element method can be clearly demonstrated.

REFERENCES

Bratianu, C. & S.N. Atluri, 1983. A hybrid finite element method for Stokes flow: Part 1--Formulation and numerical studies. *Computer methods in Appl. Mech. and Engng.* 36: 23-37.

Chen, C.C. & S.Kobayashi. 1978. Application of numerical methods to forming process. ASME, AMD--Vol. 26, New York

Chen, D.P. & H.J. Cui 1993. Hybrid/mixed finite element method for the analysis of metal forming processes. *Computational Mechanics* -- Proc. of the 2nd Asian-Pacific Conference on Computational Mechanics/Sydney/Australia. Eds. S. Valliappan et al. :55-60. Rotterdam. Balkema.

Chen, D.P. & Z. Zhao. 1990. A penalty-hybrid finite element analysis of Stokes flow. *Applied Mathematics and Mechanics.*11:501-

511. Shanghai University of Technology.

Di, S.L. & E. Ramm. 1994. On alternative hybrid stress 2D and 3D elements, *Engineering Computations*. 11:49-68.

Kobayashi, S., S.N. Oh & T. Altan. 1989. *Metal Forming and the Finite Element Method.* Oxford University Press. Oxford

Malkus, D.S. & T.J.R. Hughes. 1978. Mixed finite element methods--reduced and selective integration techniques: An unification of concepts. *Comput. Meth. Appl. Mech. Engng.* 15:63-81

Pian, T.H.H. 1964. Derivation of element stiffness matrices by assumed stress distributions, *AIAA J.* 2:1333-1336

Pian, T.H,H, & P. Tong 1968. Basis of finite element methods for solid continua. *Int. J. Numer. Meth. Engng.* 1: 3-28.

Pian, T.H.H. & P. Tong 1969. Rationalization in deriving element stiffness matrix by assumed stress approach. *Proc. 2nd Conf.on Matrix Methods in Structural M*echanics: 441-469. Air Force Flight Dynamics Lab. Wright Patterson Air Force Base.

Pian, T.H.H. & D.P. Chen. 1982. Alternative ways for formulation of hybrid stress elements. *Int. J. Numer. Meth. Engng.* 18:1679-1684.

Pian, T.H.H. & D.P. Chen.1983.On the suppression of zero energy deformation modes. *Int. J. Numer. Meth. Engng.* 19:1741-1752.

Pian, T.H.H. & P. Tong. 1986. Relations between incompatible displacement model and hybrid stress model. *Int. J. Numer . Meth Engng.* 22:173-181.

Pian, T.H.H. & C.C.Wu.1988.A rational approach for choosing stress terms for hybrid finite element formulations. *Int. J. Numer. Meth. Engng. 26:2331-2343.* .

Sze, K.Y., C.L. Chow & W.-J. Chen, 1991. Hybrid element for enhancing computational efficiency. *Int. J. Numer.Meth. Engng.* 31:999-1008.

Sze, K.Y.. 1992. Efficient formulation of robust hybrid elements using orthogonal stress/strain interpolants and admissible matrix formulation. *Int. J. Numer.. Meth. Engng.* 35:1-20

Sze. K.Y. 1993. A novel approach for devising higher order hybrid elements, *Int. J. Numer. Meth. Engng.* 36:3303-3316.

Sze, K.Y. & A. Ghali. 1993. Hybrid hexahedral element for solids, plates, shells and beams by selective scaling. *Int. J. Numer. Meth. Engng,* 36 :1519-1940.

Sze, K.Y. 1994. Control of spurious mechanism for 20-node and transition sub-integrated hexahedral elements. *Int. J. Numer. Meth. Engng.* 37:2235-2250.

Wu, C.C., S.L. Di & T.H.H. Pian.1987.Optimizing formulation of axisymmetric hybrid stress elements. *Acta Aeronautica et Astronautica Sinica.* 8:639-644.

Zienkiewicz, O.C. & R.L. Taylor, 1989. *The Finite Element Method-- 4th Edition.* Vol. 1. Chapter 13 . McGraw-Hill, London

Invited lectures

Simulation of Materials Processing: Theory, Methods and Applications, Shen & Dawson (eds)
© 1995 Balkema, Rotterdam. ISBN 90 5410 553 4

The consistent tangent operator and BEM formulations for usual and sensitivity problems in elasto-plasticity

Marc Bonnet
Laboratoire de Mécanique des Solides, CNRS URA, Centre Commun Polytechnique, Mines, Ponts et Chaussées, École Polytechnique, Palaiseau, France

Subrata Mukherjee*
Department of Theoretical and Applied Mechanics, Cornell University, Ithaca, N.Y., USA

ABSTRACT: This paper presents boundary element method (BEM) formulations for usual and sensitivity problems in (small strain) elasto-plasticity using the concept of the local consistent tangent operator (CTO). 'Usual' problems here refer to analysis of nonlinear problems in structural and solid continua, for which Simo and Taylor first proposed the use of the CTO within the context of the finite element method (FEM). A new implicit BEM scheme for such problems, using the CTO, is presented first. A formulation for sensitivity analysis follows. It is shown that the sensitivity of the strain increment, associated with an infinitesimal variation of some design parameter, solves a linear problem which is governed by the (converged value of the) same global CTO as the one that appears in the usual problem.

1 *Introduction*

In this paper, we address two important topics within the context of boundary element method (BEM) analysis of (small strain) elasto-plastic problems. The first is an implicit BEM formulation for usual elasto-plastic analysis and the second is a sensitivity formulation for such problems. Both these formulations involve the consistent tangent operator (CTO – see Simo and Taylor 1985).

Sensitivity analysis of nonlinear (material and/or geometrical) problems in solid mechanics is an active research area at present. In this context, design sensitivity coefficients (DSCs) are rates of change of response quantities, such as stresses or displacements in a loaded body, with respect to design variables. These design variables could be shape parameters, sizing parameters, boundary conditions, material parameters etc. DSCs are useful in diverse applications, a very important one being optimal design using gradient based optimization algorithms. Such analyses can be applied, for example, to optimal design of certain manufacturing processes.

Currently, the direct differentiation approach (DDA) or the adjoint structure approach (ADA) are popular for accurate sensitivity analysis. Either of these can be applied in conjunction with general purpose numerical methods such as the finite element method (FEM) or the BEM. The FEM has been used for sensitivity analysis of nonlinear problems by, among others, Arora and his co-workers (Arora and Cardoso 1992, Yao and Arora 1992a,b), Choi and his co-workers (Choi and Santos 1987, Santos and Choi 1988), Haber and his co-workers (Vidal et al. 1991, Vidal and Haber 1993), Kleiber and his co-workers (Kleiber 1991, Kleiber et al. 1994, Kleiber et al.1995), Michaleris et al. (1994) and Badrinarayanan and Zabaras (1995). Haber, Kleiber and their associates were the first to point out that the consistent (or algorithmic) tangent operator (CTO) (as opposed to the continuum tangent operator) plays a key role in nonlinear sensitivity analysis. The CTO was originally proposed by Simo and Taylor (1985) for FEM analysis of (usual) nonlinear problems. The sensitivity problem is always linear (even if the usual problem is not) and the global or system matrix related to the CTO is precisely the stiffness matrix for these problems. Use of the CTO,

†Most of the research presented was performed when S. Mukherjee was on sabbatical leave at the Laboratoire de Mécanique des Solides, Ecole Polytechnique, Palaiseau, France. S. Mukherjee also acknowledges partial financial support from a research contract from the General Motors Research Laboratories, Warren, Michigan, USA, with Cornell University.

as pointed out by Kleiber, Haber and their associates, provides very accurate numerical results for sensitivities, while other approaches (e.g. using the continuum tangent) might lead to significant errors. These researchers present numerical results for materially nonlinear problems. Michaleris et al. (1994), in a recent paper, present sensitivity formulations for general transient nonlinear coupled problems, together with an accurate numerical procedure for the calculation of the CTO. Badrinarayanan and Zabaras (1995) present a consistent scheme for sensitivity analysis of nonlinear (both material and geometric) problems in solid mechanics. This paper presents very accurate numerical results for sensitivities at the end of a large deformation (extrusion) process.

All the researchers cited above have employed the FEM in order to obtain their numerical results. Mukherjee and his co-workers have been active in solving nonlinear (both material and geometric) sensitivity problems by the BEM. Examples of the work of this group, using the explicit BEM, are Zhang et al. (1992a,b) and Leu and Mukherjee (1993) (see also the forthcoming book: Chandra and Mukherjee 1995). Wei et al. (1994) have used sensitivities to carry out shape optimal design of an elasto-plastic problem.

Most of the publications on BEM analysis of (usual) nonlinear problems in solid mechanics report on the use of the explicit approach for time integration of the appropriate rate equations. Banerjee and his co-workers (Banerjee 1994) have presented variable stiffness formulations for such problems. Implicit BEM formulations have been presented by Jin et al. (1989) and Telles and Carrer (1991,1994). Mukherjee and his co-workers have been interested in implicit sensitivity calculations, using the BEM, during the last few years. Leu and Mukherjee (1994a,b) have presented implicit objective integration schems for recovery of stress sensitivities at a material point. This work addresses large strain viscoplastic problems but only considers integration of the algorithmic constitutive model (analogous to the radial return algorithm) at a material point. They have coupled this analysis with the BEM (Leu and Mukherjee 1995) to solve general boundary value problems. The CTO, however, has not been employed in the work by Leu and Mukherjee cited above. It is observed (Leu and Mukherjee 1995) that stress sensitivities at some material points, at the end of a large deformation process, can exhibit significant numerical errors.

The remedy appears to be an implicit BEM formulation that employs the consistent tangent operator. Within the context of the BEM, this paper presents, for the first time, an implicit scheme that explicitly utilizes the CTO. Small-strain elastoplastic problems, with isotropic and kinematic hardening behavior, are considered in this paper, but further generalizations present no conceptual difficulties. Next, the corresponding sensitivity formulation is derived. It is shown that the (converged value of) the "global" CTO appears, as expected, as the stiffness matrix for the linear system of equations that govern the sensitivity of the strain increment over a time step. It is expected that numerical results for sensitivities, using the formulation presented in this paper, will be very accurate.

2 *Constitutive law*

Following Simo and Taylor (1985), considering the evolution problem from a discrete incremental standpoint for a finite time step Δt (as opposed to continuous time), the elastoplastic constitutive law reduces to giving a rule which outputs, $\boldsymbol{\sigma}_{n+1}$ consistent with the yield criterion, for any given strain increment $\Delta\boldsymbol{\varepsilon}_n = \boldsymbol{\varepsilon}_{n+1} - \boldsymbol{\varepsilon}_n$ (input):

$$\boldsymbol{\sigma}_{n+1} = \bar{\boldsymbol{\sigma}}(\boldsymbol{\varepsilon}_n, \boldsymbol{\sigma}_n, \bar{e}^p_n, \Delta\boldsymbol{\varepsilon}_n) \qquad (1)$$

Here, the notation $\bar{\boldsymbol{\sigma}}$ symbolically denotes the action of the *radial return algorithm* of Simo and Taylor (1985). Also, the subscript n refers to time (or pseudo-time) t_n.

3 *Equilibrium (FEM version)*

In a FEM framework, the supplementary constraint provided by the equilibrium equation (e.g. virtual work theorem) is then used to determine which $\Delta\boldsymbol{\varepsilon}$ exactly should be input in (1). The necessary condition for equilibrium at step $n+1$ is, using (1):

$$\begin{aligned} G(\boldsymbol{u},\boldsymbol{v}) &\equiv \int_\Omega \bar{\boldsymbol{\sigma}}(\boldsymbol{\varepsilon}_n, \boldsymbol{\sigma}_n, \bar{e}^p_n, \Delta\boldsymbol{\varepsilon}_n) : \boldsymbol{\nabla}\boldsymbol{v}\,\mathrm{d}V \\ &\quad - \int_\Omega \boldsymbol{b}^{n+1}.\boldsymbol{v}\,\mathrm{d}V - \int_{S_T} \boldsymbol{t}^{n+1}.\boldsymbol{v}\,\mathrm{d}S \\ &= 0 \qquad (\forall \boldsymbol{v} \in \mathcal{V}) \end{aligned} \qquad (2)$$

with $2\Delta\boldsymbol{\varepsilon}_n = (\boldsymbol{\nabla} + \boldsymbol{\nabla}^T)(\boldsymbol{u} - \boldsymbol{u}_n)$. Then, the displacement increment Δu_n, such that $\boldsymbol{u}_{n+1} = \boldsymbol{u}_n + \Delta u_n$ solves the nonlinear equation (2), is sought for iteratively using a Newton method: the additive correction $\delta u^i_n = \Delta u^{i+1}_n - \Delta u^i_n$ to Δu^i_n solves

$$
\begin{aligned}
& G(\boldsymbol{u}_{n+1}^i, \boldsymbol{v}) \\
& \quad + \int_\Omega \frac{\partial \bar{\boldsymbol{\sigma}}}{\partial \Delta \boldsymbol{\varepsilon}_n}(\boldsymbol{\varepsilon}_n, \boldsymbol{\sigma}_n, \bar{e}_n^p, \Delta \boldsymbol{\varepsilon}_n^i) : \boldsymbol{\nabla}(\delta u_n^i) : \boldsymbol{\nabla} \boldsymbol{v} \, \mathrm{d}V \\
& \quad = 0 \qquad (\forall \boldsymbol{v} \in \mathcal{V}) \qquad (3)
\end{aligned}
$$

with $2\Delta \boldsymbol{\varepsilon}_n^i = (\boldsymbol{\nabla} + \boldsymbol{\nabla}^T)\Delta u_n^i$. The *consistent tangent operator* (CTO) is the fourth-order tensor $\boldsymbol{c}_{n+1} = \partial \bar{\boldsymbol{\sigma}} / \partial \Delta \boldsymbol{\varepsilon}_n$. It depends on the particular algorithm $\Delta \boldsymbol{\varepsilon}_n \rightarrow \boldsymbol{\sigma}_{n+1}$ chosen. The expression of $\boldsymbol{c}_{n+1}$ is given in Simo and Taylor (1985).

4 *Equilibrium (BEM version)*

BEM FORMULATION FOR ELASTIC PROBLEMS WITH INITIAL STRAIN. We use notation similar to those in Telles and Carrer (1991) but consider an initial strain $\boldsymbol{\varepsilon}^p$ instead of an initial stress.

Elastic constitutive law ($2\boldsymbol{\varepsilon} = (\boldsymbol{\nabla} + \boldsymbol{\nabla}^T)\boldsymbol{u}$: total strain)

$$
\boldsymbol{\sigma} = \boldsymbol{C} : (\boldsymbol{\varepsilon} - \boldsymbol{\varepsilon}^p) \qquad \boldsymbol{\varepsilon}^p = \boldsymbol{\varepsilon} - \boldsymbol{C}^{-1} : \boldsymbol{\sigma} \qquad (4)
$$

BEM equations on the boundary (in regularized form):

$$
\begin{aligned}
& \int_{\partial\Omega} [u_i(\boldsymbol{y}) - u_i(\boldsymbol{x})] P_i^k(\boldsymbol{x}, \boldsymbol{y}) dS_y \\
& \qquad - \int_{\partial\Omega} p_i(\boldsymbol{y}) U_i^k(\boldsymbol{x}, \boldsymbol{y}) dS_y \\
& \quad = \int_\Omega U_{i,j}^k(\boldsymbol{x}, \boldsymbol{y}) C_{ijab} \varepsilon_{ab}^p(\boldsymbol{y}) dV_y
\end{aligned}
$$

where U_i^k, P_i^k denote the components of the Kelvin displacement and traction associated with a unit point force applied at $\boldsymbol{x}$ along the k-direction and p_i is the traction vector. The above BIE symbolically reads:

$$
\begin{aligned}
& [\boldsymbol{H}]\{\boldsymbol{u}\} - [\boldsymbol{G}]\{\boldsymbol{p}\} = [\boldsymbol{Q}]\{\boldsymbol{C} : \boldsymbol{\varepsilon}^p\} \\
& \quad \Rightarrow \; [\boldsymbol{A}]\{\boldsymbol{y}\} = \{\boldsymbol{f}\} + [\boldsymbol{Q}]\{\boldsymbol{C} : \boldsymbol{\varepsilon}^p\} \qquad (5)
\end{aligned}
$$

where $\boldsymbol{y}$ collects the boundary unknowns.

BEM representation at internal points. The displacement gradient is given, in regularized form, by:

$$
\begin{aligned}
u_{k,\ell}(\boldsymbol{x}) &= \int_{\partial\Omega} u_i(\boldsymbol{y}) D_{i\ell}^k(\boldsymbol{x}, \boldsymbol{y}) dS_y \\
&- \int_{\partial\Omega} p_i(\boldsymbol{y}) U_{i,\ell}^k(\boldsymbol{x}, \boldsymbol{y}) dS_y \\
&- C_{ijab} \varepsilon_{ij}^p(\boldsymbol{x}) \int_{\partial\Omega} n_\ell(\boldsymbol{y}) U_{a,b}^k(\boldsymbol{x}, \boldsymbol{y}) dS_y \\
&- \int_\Omega U_{i,j\ell}^k(\boldsymbol{x}, \boldsymbol{y}) C_{ijab} [\varepsilon_{ab}^p(\boldsymbol{y}) - \varepsilon_{ab}^p(\boldsymbol{x})] dV_y
\end{aligned}
$$

using the notation $D_{i\ell}^k = C_{ijab} n_j U_{a,b\ell}^k$. The total strain at $\boldsymbol{x}$ is then readily obtained from the above equation. In symbolic form, one has, with $\{\boldsymbol{\varepsilon}\}$: "vector" of strains at all internal points:

$$
\begin{aligned}
\{\boldsymbol{\varepsilon}\} &= [\boldsymbol{G}']\{\boldsymbol{p}\} - [\boldsymbol{H}']\{\boldsymbol{u}\} + [\boldsymbol{Q}']\{\boldsymbol{C} : \boldsymbol{\varepsilon}^p\} \\
&= -[\boldsymbol{A}']\{\boldsymbol{y}\} + \{\boldsymbol{f}'\} + [\boldsymbol{Q}']\{\boldsymbol{C} : \boldsymbol{\varepsilon}^p\}
\end{aligned}
$$

Substituting for $\{\boldsymbol{y}\}$ from (5) into the above equation, we have:

$$
\{\boldsymbol{\varepsilon}\} = \{\boldsymbol{n}\} + [\boldsymbol{S}]\{\boldsymbol{C} : \boldsymbol{\varepsilon}^p\} \qquad (6)
$$

where

$$
\begin{aligned}
\{\boldsymbol{n}\} &= \{\boldsymbol{f}'\} - [\boldsymbol{A}'][\boldsymbol{A}]^{-1}\{\boldsymbol{f}\} \\
[\boldsymbol{S}] &= [\boldsymbol{Q}'] - [\boldsymbol{A}'][\boldsymbol{A}]^{-1}[\boldsymbol{Q}]
\end{aligned}
$$

Note that $\{\boldsymbol{n}\}$ denotes the purely elastic solution, i.e. the one obtained for the same loading but in the absence of plastic strain. Then, $(4)_2$ is incorporated (in the form $\{\boldsymbol{C} : \boldsymbol{\varepsilon}^p\} = \{\boldsymbol{C} : \boldsymbol{\varepsilon}\} - \{\boldsymbol{\sigma}\}$) into (6), giving:

$$
\{\boldsymbol{\varepsilon}\} = \{\boldsymbol{n}\} + [\boldsymbol{S}](\{\boldsymbol{C} : \boldsymbol{\varepsilon}\} - \{\boldsymbol{\sigma}\})
$$

Finally, the strain and the total stress are related through:

$$
[\boldsymbol{S}]\{\boldsymbol{\sigma}\} = \{\boldsymbol{n}\} + ([\boldsymbol{S}][\boldsymbol{C}] - [\boldsymbol{I}])\{\boldsymbol{\varepsilon}\} \qquad (7)
$$

Following Telles and Carrer (1991), the above formulae for elastic problems with "initial" strain are given in accumulated form (as opposed to rate form).

5 *The consistent tangent operator and the BEM elasto-plastic formulation*

Now, $\boldsymbol{\varepsilon}^p$ is the plastic strain, and we consider the evolution of the structure between time t_n and t_{n+1}. One has, from (7) and using the notation $\Delta()_n = ()_{n+1} - ()_n$:

$$
[\boldsymbol{S}]\{\Delta \boldsymbol{\sigma}_n\} = \{\Delta \boldsymbol{n}_n\} + ([\boldsymbol{S}][\boldsymbol{C}] - [\boldsymbol{I}])\{\Delta \boldsymbol{\varepsilon}_n\} \qquad (8)
$$

which includes the equilibrium constraint.

On the other hand, the radial return algorithm (1) relates $\bar{\boldsymbol{\sigma}} = \boldsymbol{\sigma}_{n+1} = \boldsymbol{\sigma}_n + \Delta \boldsymbol{\sigma}_n$ to $\Delta \boldsymbol{\varepsilon}_n$. Combining the constitutive and equilibrium equations in the form

$$
\{\boldsymbol{\sigma}_n\} + \{\Delta \boldsymbol{\sigma}_n\} = \{\bar{\boldsymbol{\sigma}}\}
$$

we obtain a nonlinear equation for $\Delta \boldsymbol{\varepsilon}_n$ of the form:

$$
\begin{aligned}
& G(\boldsymbol{\varepsilon}_n, \boldsymbol{\sigma}_n, \bar{e}_n^p, \Delta \boldsymbol{\varepsilon}_n) \\
& \quad \equiv [\boldsymbol{S}]\{\boldsymbol{\sigma}_n\} + \{\Delta \boldsymbol{n}_n\} + ([\boldsymbol{S}][\boldsymbol{C}] - [\boldsymbol{I}])\{\Delta \boldsymbol{\varepsilon}_n\} \\
& \qquad - [\boldsymbol{S}]\{\bar{\boldsymbol{\sigma}}(\boldsymbol{\varepsilon}_n, \boldsymbol{\sigma}_n, \bar{e}_n^p, \Delta \boldsymbol{\varepsilon}_n)\} \\
& \quad = 0 \qquad (9)
\end{aligned}
$$

The Newton method can also be applied in this case, and it is readily seen that the consistent tangent operator $\boldsymbol{c}_{n+1}$ appears here as well. The additive correction $\delta\boldsymbol{\varepsilon}_n^i = \Delta\boldsymbol{\varepsilon}_n^{i+1} - \Delta\boldsymbol{\varepsilon}_n^i$ to $\Delta\boldsymbol{\varepsilon}_n^i$ solves:

$$
\begin{aligned}
& G(\boldsymbol{\varepsilon}_n, \boldsymbol{\sigma}_n, \bar{e}_n^p, \Delta\boldsymbol{\varepsilon}_n^i) \\
& \quad + ([\boldsymbol{S}][\boldsymbol{C}] - [\boldsymbol{S}][\boldsymbol{c}_{n+1}^i] - [\boldsymbol{I}])\{\delta\boldsymbol{\varepsilon}_n^i\} = 0 \qquad (10)
\end{aligned}
$$

The quantity $([\boldsymbol{S}][\boldsymbol{C}] - [\boldsymbol{S}][\boldsymbol{c}_{n+1}^i] - [\boldsymbol{I}])$ is hereafter called the global CTO (see Kleiber et al. 1994 for the FEM version). Once the nonlinear equation (9) is solved for $\Delta\boldsymbol{\varepsilon}_n$, all the variables at time t_{n+1} are readily computed.

6 *Sensitivity analysis*

Now we are interested in computing the sensitivity of the mechanical variables associated with an infinitesimal perturbation of some design parameter b of an unspecified nature. Since this means that we are comparing two history-dependent mechanical processes, the sensitivity computation should also proceed in an iterative way, resulting in an accumulation of sensitivity increments.

Differentiation of the boundary integral equation (5) and internal strain representation (6) with respect to b for given $\boldsymbol{\varepsilon}^p$ is a well understood process (Chandra and Mukherjee 1995). One readily sees that the evaluation of the derivative $\overset{\star p}{\boldsymbol{\varepsilon}}$ plays a key role. We now concentrate on this particular task.

Let us differentiate eq. (9) w.r.t. b, so that:

$$
\begin{aligned}
& [\overset{\star}{\boldsymbol{S}}]\{\boldsymbol{\sigma}_n\} + [\boldsymbol{S}]\{\overset{\star}{\boldsymbol{\sigma}}_n\} + \{\Delta \overset{\star}{\boldsymbol{n}}_n\} \\
& \quad + [\overset{\star}{\boldsymbol{S}}][\boldsymbol{C}]\{\Delta\boldsymbol{\varepsilon}_n\} + ([\boldsymbol{S}][\boldsymbol{C}] - [\boldsymbol{I}])\{\overset{\star}{\Delta\boldsymbol{\varepsilon}}_n\} \\
& \quad - [\overset{\star}{\boldsymbol{S}}]\{\bar{\boldsymbol{\sigma}}(\boldsymbol{\varepsilon}_n, \boldsymbol{\sigma}_n, \bar{e}_n^p, \Delta\boldsymbol{\varepsilon}_n)\} \\
& \quad - [\boldsymbol{S}]\{\overset{\star}{\bar{\boldsymbol{\sigma}}}(\boldsymbol{\varepsilon}_n, \boldsymbol{\sigma}_n, \bar{e}_n^p, \Delta\boldsymbol{\varepsilon}_n)\} = 0 \qquad (11)
\end{aligned}
$$

(assuming that b is such that the elastic coefficients do not depend on it). This derivation aims at finding a governing equation for $\overset{\star}{\Delta\boldsymbol{\varepsilon}}_n$ in terms of known variables at times t_n and t_{n+1} and sensitivities at time t_n. Obtaining the derivatives $[\overset{\star}{\boldsymbol{S}}]$ and $\{\overset{\star}{\boldsymbol{n}}_n\}$ requires quite lengthy derivations and is thus by no means trivial. However, this aspect has been discussed, at length, in previous works on sensitivity analysis of elastic and elasto-plastic BEM formulations (e.g. in Chandra and Mukherjee, 1995). The focus in this paper is on obtaining $\{\overset{\star}{\bar{\boldsymbol{\sigma}}}(\boldsymbol{\varepsilon}_n, \boldsymbol{\sigma}_n, \bar{e}_n^p, \Delta\boldsymbol{\varepsilon}_n)\}$ from the appropriate constitutive algorithm, and group the terms multiplying $\overset{\star}{\Delta\boldsymbol{\varepsilon}}_n$ and the rest separately. Of course, one expects

$$
\begin{aligned}
\{\overset{\star}{\bar{\boldsymbol{\sigma}}}(\boldsymbol{\varepsilon}_n, \boldsymbol{\sigma}_n, \bar{e}_n^p, \Delta\boldsymbol{\varepsilon}_n)\} &= [\frac{\partial\bar{\boldsymbol{\sigma}}}{\partial\Delta\boldsymbol{\varepsilon}_n}]\{\overset{\star}{\Delta\boldsymbol{\varepsilon}}_n\} \\
& \quad + \mathcal{F}(\mathcal{S}_n, \overset{\star}{\mathcal{S}}_n, \mathcal{S}_{n+1})
\end{aligned}
$$

where $\mathcal{S}_n$ denotes the state at time t_n. Note that the derivative of (9) w.r.t. b must be taken for the converged state $\mathcal{S}_{n+1}$, i.e. with the value of $\Delta\boldsymbol{\varepsilon}_n$ such that $G(\boldsymbol{\varepsilon}_n, \boldsymbol{\sigma}_n, \bar{e}_n^p, \Delta\boldsymbol{\varepsilon}_n) = 0$.

RADIAL RETURN ALGORITHM (RRA). The RRA is summarized by the following equations (notation as in Simo and Taylor (1985) unless otherwise indicated):

$$
\begin{aligned}
\bar{\boldsymbol{\sigma}} &= K\Delta\boldsymbol{\varepsilon}_n : (\mathbf{1} \otimes \mathbf{1}) + \boldsymbol{s}_{n+1} && (12) \\
\boldsymbol{s}_{n+1} &= \boldsymbol{\alpha}_n + \sqrt{\tfrac{2}{3}}(\kappa_{n+1} + \Delta H_n)\hat{\boldsymbol{n}} && (13) \\
\bar{e}_{n+1}^p &= \bar{e}_n^p + \sqrt{\tfrac{2}{3}}[\gamma\Delta t] && \\
\hat{\boldsymbol{n}} &= \frac{1}{\left|\boldsymbol{\xi}_{n+1}^T\right|}\boldsymbol{\xi}_{n+1}^T && \\
\boldsymbol{\xi}_{n+1}^T &= \boldsymbol{s}_n + 2G\Delta\boldsymbol{e}_n - \boldsymbol{\alpha}_n && (14)
\end{aligned}
$$

where $[\gamma\Delta t]$ solves the nonlinear *consistency equation*

$$
\begin{aligned}
& \left|\boldsymbol{\xi}_{n+1}^T\right| - \sqrt{\tfrac{2}{3}}\kappa(\bar{e}_n^p + \sqrt{\tfrac{2}{3}}[\gamma\Delta t]) \\
& \quad - 2G[\gamma\Delta t] - \sqrt{\tfrac{2}{3}}\Delta H_n = 0 \qquad (15)
\end{aligned}
$$

with $\Delta H_n = H_{n+1} - H_n$, $H_n = H(\bar{e}_n^p)$, $\kappa_n = \kappa(\bar{e}_n^p)$. The evolution of the back-stress $\boldsymbol{\alpha}_n$ is governed by

$$
\boldsymbol{\alpha}_{n+1} = \boldsymbol{\alpha}_n + \sqrt{\tfrac{2}{3}}\Delta H_n\hat{\boldsymbol{n}} \qquad (16)
$$

SENSITIVITY OF $[\gamma\Delta t]$. Let us differentiate (15) w.r.t. b and with the converged value of $[\gamma\Delta t]$. Noting the following formulae:

$$
\begin{aligned}
\overset{\star T}{\boldsymbol{\xi}}_{n+1} &= (\boldsymbol{I} - \frac{1}{3}\mathbf{1} \otimes \mathbf{1}) : (\overset{\star}{\boldsymbol{\sigma}}_n + 2G\,\overset{\star}{\Delta\boldsymbol{\varepsilon}}_n) - \overset{\star}{\boldsymbol{\alpha}}_n \\
\left|\boldsymbol{\xi}_{n+1}^T\right|^{\star} &= \hat{\boldsymbol{n}} : \overset{\star}{\boldsymbol{\xi}}{}_{n+1}^T \\
&= \hat{\boldsymbol{n}} : (\overset{\star}{\boldsymbol{\sigma}}_n + 2G\,\overset{\star}{\Delta\boldsymbol{\varepsilon}}_n - \overset{\star}{\boldsymbol{\alpha}}_n) \qquad (17) \\
\overset{\star}{\kappa}_{n+1} &= (\overset{\star p}{\bar{e}}_n + \sqrt{\tfrac{2}{3}}[\overset{\star}{\gamma}\,\Delta t])\kappa_{n+1}' \\
\overset{\star}{\Delta H}_n &= \Delta H_n'\,\overset{\star p}{\bar{e}}_n + \sqrt{\tfrac{2}{3}}[\overset{\star}{\gamma}\,\Delta t]H_{n+1}'
\end{aligned}
$$

one gets, after grouping terms, the following scalar linear equation for $[\overset{\star}{\gamma}\,\Delta t]$

$$
\begin{aligned}
& \hat{\boldsymbol{n}} : (\overset{\star}{\boldsymbol{\sigma}}_n + 2G\,\overset{\star}{\Delta\boldsymbol{\varepsilon}}_n - \overset{\star}{\boldsymbol{\alpha}}_n) - \sqrt{\tfrac{2}{3}}(\Delta H_n' + \kappa_{n+1}')\,\overset{\star p}{\bar{e}}_n \\
& \quad = \frac{2}{3}(\kappa_{n+1}' + H_{n+1}' + 3G)[\overset{\star}{\gamma}\,\Delta t]
\end{aligned}
$$

which readily gives

$$2G[\overset{\star}{\gamma}\,\Delta t] = (1-\delta)\hat{\boldsymbol{n}} : (\overset{\star}{\boldsymbol{\sigma}}_n + 2G\,\overset{\star}{\Delta\boldsymbol{\varepsilon}}_n - \overset{\star}{\boldsymbol{\alpha}}_n) - \theta(1-\delta)\left|\boldsymbol{\xi}^T_{n+1}\right| \overset{\star p}{\bar{e}}_n \quad (18)$$

where the abbreviated notation

$$\theta = \frac{1}{\left|\boldsymbol{\xi}^T_{n+1}\right|}\sqrt{\tfrac{2}{3}}(\Delta H'_n + \kappa'_{n+1})$$

$$\delta = \frac{\kappa'_{n+1} + H'_{n+1}}{\kappa'_{n+1} + H'_{n+1} + 3G}$$

has been used.

SENSITIVITY OF THE RRA. Differentiation of (13) with respect to b gives

$$\overset{\star}{\boldsymbol{s}}_{n+1} = \overset{\star}{\boldsymbol{\alpha}}_n + \sqrt{\tfrac{2}{3}}(\kappa_{n+1} + \Delta H_n)\,\overset{\star}{\hat{\boldsymbol{n}}} + \sqrt{\tfrac{2}{3}}(\overset{\star}{\kappa}_{n+1} + \overset{\star}{\Delta H}_n)\hat{\boldsymbol{n}} \quad (19)$$

The sensitivity of $\hat{\boldsymbol{n}}$ (see Simo and Taylor 1985) is

$$\begin{aligned}\overset{\star}{\hat{\boldsymbol{n}}} &= \frac{1}{\left|\boldsymbol{\xi}^T_{n+1}\right|}(\boldsymbol{I} - \hat{\boldsymbol{n}}\otimes\hat{\boldsymbol{n}}) : \overset{\star T}{\boldsymbol{\xi}}_{n+1} \\ &= \frac{1}{\left|\boldsymbol{\xi}^T_{n+1}\right|}(\boldsymbol{I} - \frac{1}{3}\mathbf{1}\otimes\mathbf{1} - \hat{\boldsymbol{n}}\otimes\hat{\boldsymbol{n}}) : \\ &\qquad (\overset{\star}{\boldsymbol{\sigma}}_n + 2G\,\overset{\star}{\Delta\boldsymbol{\varepsilon}}_n) \\ &\quad - \frac{1}{\left|\boldsymbol{\xi}^T_{n+1}\right|}(\boldsymbol{I} - \hat{\boldsymbol{n}}\otimes\hat{\boldsymbol{n}}) : \overset{\star}{\boldsymbol{\alpha}}_n \end{aligned} \quad (20)$$

Using equations $(17)_3$, $(17)_4$, (18) and (20) in equation (19) and grouping terms appropriately, one gets

$$\begin{aligned}\overset{\star}{\boldsymbol{s}}_{n+1} &= [(1-\beta)\boldsymbol{I} + (\beta-\delta)\hat{\boldsymbol{n}}\otimes\hat{\boldsymbol{n}}] : \overset{\star}{\boldsymbol{\alpha}}_n \\ &\quad + [\beta(\boldsymbol{I} - \frac{1}{3}\mathbf{1}\otimes\mathbf{1}) - (\beta-\delta)\hat{\boldsymbol{n}}\otimes\hat{\boldsymbol{n}}] \\ &\qquad : (\overset{\star}{\boldsymbol{\sigma}}_n + 2G\,\overset{\star}{\Delta\boldsymbol{\varepsilon}}_n) \\ &\quad + \theta(1-\delta)\boldsymbol{\xi}^T_{n+1}\,\overset{\star p}{\bar{e}}_n \end{aligned} \quad (21)$$

where the new parameter β (as in Simo and Taylor 1985) is:

$$\beta = \frac{1}{\left|\boldsymbol{\xi}^T_{n+1}\right|}\sqrt{\tfrac{2}{3}}(\Delta H_n + \kappa_{n+1})$$

Also, $\beta - \delta$ is called $\bar{\gamma}$ in Simo and Taylor (1985). Combining the above equation with the sensitivity version of (12), we obtain the sensitivity of the radial return algorithm as

$$\begin{aligned}\overset{\star}{\boldsymbol{\sigma}} &= [\beta(\boldsymbol{I} - \frac{1}{3}\mathbf{1}\otimes\mathbf{1}) - (\beta-\delta)\hat{\boldsymbol{n}}\otimes\hat{\boldsymbol{n}}] : \overset{\star}{\boldsymbol{\sigma}}_n \\ &\quad + \Big\{K\mathbf{1}\otimes\mathbf{1} + 2G\beta(\boldsymbol{I} - \frac{1}{3}\mathbf{1}\otimes\mathbf{1}) \\ &\qquad - 2G(\beta-\delta)\hat{\boldsymbol{n}}\otimes\hat{\boldsymbol{n}}\Big\} : \overset{\star}{\Delta\boldsymbol{\varepsilon}}_n \\ &\quad + \theta(1-\delta)\boldsymbol{\xi}^T_{n+1}\,\overset{\star p}{\bar{e}}_n \\ &\quad + [(1-\beta)\boldsymbol{I} + (\beta-\delta)\hat{\boldsymbol{n}}\otimes\hat{\boldsymbol{n}}] : \overset{\star}{\boldsymbol{\alpha}}_n \end{aligned} \quad (22)$$

Note that $\overset{\star}{\boldsymbol{\alpha}}_n$ is updated using:

$$\overset{\star}{\boldsymbol{\alpha}}_{n+1} = \overset{\star}{\boldsymbol{\alpha}}_n + \sqrt{\tfrac{2}{3}}\left\{\overset{\star}{\Delta H}_n\,\hat{\boldsymbol{n}} + \Delta H_n\,\overset{\star}{\hat{\boldsymbol{n}}}\right\} \quad (23)$$

which results from a differentiation of (16) w.r.t. b. It is very important to note that the factor of $\overset{\star}{\Delta\boldsymbol{\varepsilon}}_n$ in (22) is equal to the converged value of the consistent tangent operator $\boldsymbol{c}_{n+1}$ as given in Simo and Taylor (1985).

SENSITIVITY ANALYSIS. Finally, substitution of (22) into (11) gives a *linear* equation for $\{\overset{\star}{\Delta\boldsymbol{\varepsilon}}_n\}$ of the form

$$([\boldsymbol{S}][\boldsymbol{C}] - [\boldsymbol{S}][\boldsymbol{c}_{n+1}] - [\boldsymbol{I}])\{\overset{\star}{\Delta\boldsymbol{\varepsilon}}_n\} = \{\boldsymbol{F}\} \quad (24)$$

The function $\{\boldsymbol{F}\}$ is completely known at this stage. It depends on the converged values of the variables at states n and $n+1$ and known sensitivities – those of the stress etc. at state n and the known loading sensitivity $\{\overset{\star}{\boldsymbol{n}}_{n+1}\}$. The matrix multiplying $\overset{\star}{\Delta\boldsymbol{\varepsilon}}_n$ is the converged value of the "global (or system) consistent tangent matrix" (see equation (10)).

REFERENCES

[1] ARORA J.S., CARDOSO J.B. – Variational principle for shape sensitivity analysis. *AIAA Journal,* ***30****, pp. 538-547, 1992.*

[2] BADRINARAYANAN S., ZABARAS N. – A sensitivity analysis for the optimal design of metal forming processes. *Submitted for publiction.*

[3] BANERJEE P.K. – The boundary element method in engineering (2nd. edition). *McGraw Hill, London, 1994.*

[4] CHANDRA A., MUKHERJEE S. – Boundary element methods in manufacturing. *Oxford University Press, Oxford, UK, 1995. (expected)*

[5] CHOI K.K., SANTOS J.L.T. – Variational methods for design sensitivity analysis in nonlinear structural systems; part. I: theory. *Int. J. Num. Meth. in Eng.* **24**, *pp. 2039-2055, 1987.*

[6] JAO S.Y., ARORA S.J. – Design sensitivity analysis of nonlinear structures using endochronic constitutive model. Part 1: general theory. *Comp. Mech.,* **10**, *pp. 39-57, 1992.*

[7] JAO S.Y., ARORA S.J. – Design sensitivity analysis of nonlinear structures using endochronic constitutive model. Part 2: discretization and applications. *Comp. Mech.,* **10**, *pp. 59-72, 1992.*

[8] JIN H., RUNESSON K., MATIASSON K. – Boundary element formulation in finite deformation plasticity using implicit integration. *Comp. & Struct.,* **31**, *pp. 25-34, 1989.*

[9] KLEIBER M. – Computational coupled nonassociative thermoplasticity. *Comp. Meth. in Appl. Mech. & Engng.,* **90**, *pp. 943-967, 1991.*

[10] KLEIBER M., HIEN T.D., POSTEK E. – Incremental finite-element sensitivity analysis for non-linear mechanics applications. *Int. J. Num. Meth. in Eng.* **37**, *pp. 3291-3308, 1994.*

[11] KLEIBER M., HIEN T.D., ANTÚNEZ H., KOWALCZYK P. – Parameter sensitivity of elasto-plastic response. *Comput. Eng.* (in press).

[12] LEU L.J., MUKHERJEE S. – Sensitivity analysis and shape optimization in nonlinear solid mechanics. *Engng. Anal. with Bound. Elem.* , **12**, *pp. 251-260, 1993.*

[13] LEU L.J., MUKHERJEE S. – Implicit objective integration for sensitivity analysis in nonlinear solid mechanics. *Int. J. Num. Meth. in Eng.* **37**, *pp. 3843-3868, 1994a.*

[14] LEU L.J., MUKHERJEE S. – Sensitivity analysis of hyperelastic-viscoplastic solids undergoing large deformations. *Comp. Mech.,* **15**, *pp. 101-116, 1994b.*

[15] LEU L.J., MUKHERJEE S. – Sensitivity analysis in nonlinear solid mechanics by the boundary element method with an implicit scheme. *J. Engng. Anal. & Design,* in press, 1995.

[16] MICHALERIS P., TORTORELLI D.A., VIDAL C.A. – Tangent operators and design sensitivity formulations for transient non-linear coupled problems with applications to elastoplasticity. *Int. J. Num. Meth. in Eng.* **37**, *pp. 2471-2499, 1994.*

[17] SANTOS J.L.T., CHOI K.K. – Sizing design sensitivity analysis of nonlinear structural systems, part II: numerical method. *Int. J. Num. Meth. in Eng.* **26**, *pp. 2097-2114, 1988.*

[18] SIMO J.C., TAYLOR R.L. – Consistent tangent operators for rate-independent elastoplasticity. *Comp. Meth. in Appl. Mech. & Engng.,* **48**, *pp. 101-118, 1985.*

[19] TELLES J.C.F., CARRER J.A.M. – Implicit procedures for the solution of elastoplastic problems by the boundary element method. *Mathl. Comput. Modelling.,* **15**, *pp. 303-311, 1991.*

[20] TELLES J.C.F., CARRER J.A.M. – Static and transient dynamic nonlinear stress analysis by the boundary element method with implicit techniques. *Engng. Anal. with Bound. Elem.* , **14**, *pp. 65-74, 1994.*

[21] VIDAL C.A., LEE H.S., HABER R.B. – The consistent tangent operator for design sensitivity of history-dependent response. *Comp. Syst. in Engng.,* **2**, *pp. 509-523, 1991.*

[22] VIDAL C.A., HABER R.B. – Design sensitivity analysis for rate-independent elastoplasticity. *Comp. Meth. in Appl. Mech. & Engng.,* **107**, *pp. 393-431, 1993.*

[23] WEI X., LEU L.J., CHANDRA A., MUKHERJEE S. – Shape optimization in elasticity and elastoplasticity. *Int. J. Solids Struct.,* **31**, *pp. 533-550, 1994.*

[24] ZHANG Q., MUKHERJEE S., CHANDRA A. – Design sensitivity coefficients for elasto-viscoplastic problems by boundary element methods. *Int. J. Num. Meth. in Eng.* **34**, *pp. 947-966, 1992a.*

[25] ZHANG Q., MUKHERJEE S., CHANDRA A. – Shape design sensitivity analysis for geometrically and materially nonlinear problems by the boundary element method. *Int. J. Solids Struct.,* **29**, *pp. 2503-2525, 1992b.*

Simulation of Materials Processing: Theory, Methods and Applications, Shen & Dawson (eds)
© 1995 Balkema, Rotterdam. ISBN 90 5410 553 4

The ALE method for the numerical simulation of material forming processes

J.-L.Chenot & M.Bellet
CEMEF, École des Mines de Paris, URA CNRS, Sophia-Antipolis, France

ABSTRACT: The standard flow formulation for quasi-static incompressible deformation of viscoplastic materials is briefly recalled. The penalty method and the mixed formulation are used to write the integral equation of the problem. Explicit and implicit time integration schemes are discussed and thermal coupling is considered. The ALE method with laplacien smoothing is presented in the rate form, that allows to propose first or second order, explicit or purely implicit time integration formulas. Then the dynamic viscoplastic problem is recalled with a penalty approach either on the velocity or the acceleration fields. The corresponding ALE formulations must take into account the convective contribution of the evolution of the velocity field. Again the rate form of our equations allows to propose first, second (or higher) order schemes.

1. INTRODUCTION:

For classical metal forming processes involving very large deformation, and even more for mold filling in casting, the same initial finite element mesh cannot be used during the whole process. Remeshing is therefore compulsory (see [Coupez,1992], [Coupez,1994]) but when the updated lagrangian formulation is used, the number of remeshing steps can be very high and results in a significant increase of computational time with a loss of accuracy due to interpolations from one mesh to the next one, which is difficult to control accurately.

Generally the elements distortion, due to internal shear, is much more important than the evolution of the boundary of the computational domain. This is the reason why the ALE formulation (Arbitrary Lagrange Euler) was proposed and is supposed to represent a computationaly efficient compromise between the Lagrange formlulation which leads to extremely distorted meshes, and the Eulerian one which does not allow any change in the domain.

The early works on the ALE method were devoted to fluid mechanics and the first F.E. application of the method is given in [Donea, 1977]. A great effort has been devoted to the development fo the method for fluid dynamics see e.g. [Ramaswany, 1987], [Huerta, 1988], [Baaijens,1993] or fluid-structure interaction [Nomura,1992]. Generalization to solid mechanics [Ghosh,1991] or solid-solid interaction [Liu,1988] are now available. The application of the ALE method to impact problems is also vely popular, see for example [Huerta, 1994].

In metal forming simulation, we can find a number of works where the ALE method is claimed to improve the computational accuracy, we shall quote for example:
- in rolling [Huisman, 1985], [Van der Lugt, 1986], [Liu, 1991], [Liu, 1992], [Huetink, 1993];
- in forging [Cescutti, 1988], [Ponthot, 1992];;
- in extrusion [Ghosh, 1986], [Huetink, 1993], [Ghosh, 1993a];
- in machining [Rakotomalala, 1993];
- in the mold filling stage of casting [Gaston, 1994];
- in coutinuous casting [Ghosh, 1993b];
- in wire drawing [Van der Lugt, 1986];
- in ring rolling [Ghosh, 1986], [Hu, 1992].

In polymer forming the finite element simulation with the ALE method was applied with success for die swell [Hurez, 1991] or mold filling in injection [Hayes, 1991], [Magnin, 1994].

However most of the published works utilize relatively simple time integration schemes which do not guarantee a second order approximation with respect to time. The purpose of this work is to investigate first and second order schemes in order to allow relatively large time increments or/and a great number of time increments without an excessive loss of accuracy, due to the convective terms.

2. THE FLOW FORMULATION FOR QUASI-STATIC PROCESSES

For more details the interested reader can refer to [Chenot, 1992].

2.1 *Constitutive equation and integral formulation*

We shall consider mainly hot forming so that we are allowed to neglect the elasticity effect and discuss only purely viscoplastic behavior. The 3-D isotropic viscoplastic Norton-Hoff law is written as follows:

$$\mathbf{s} = 2K\left(\sqrt{3}\,\dot{\bar{\varepsilon}}\right)^{m-1}\dot{\boldsymbol{\varepsilon}} \tag{2.1}$$

where $\mathbf{s}$ is the deviatoric stress tensor,

$\dot{\bar{\varepsilon}}$ is the effective strain rate: $\dot{\bar{\varepsilon}} = \sqrt{\frac{2}{3}\dot{\boldsymbol{\varepsilon}}:\dot{\boldsymbol{\varepsilon}}}$,

K is the material consistency, which can depend on strain hardening as:

$$K = K_0\left(\bar{\varepsilon}_0 + \bar{\varepsilon}\right)^n \tag{2.2}$$

m is the strain-rate sensitivity index.

For simplicity, we shall consider only dense materials for which the associated incompressibility condition is written:

$$\mathrm{div}(\mathbf{v}) = 0 \tag{2.3}$$

A viscoplastic friction behavior can be used, it is described by a non-linear relation between the shear stress $\boldsymbol{\tau}$ and the tangential velocity $\mathbf{v}_s$ at the tool-part interface $\partial\Omega_c$:

$$\boldsymbol{\tau} = -\alpha_f K \left|\mathbf{v}_s\right|^{q-1}\mathbf{v}_s \tag{2.4}$$

where $\mathbf{v}_s = \mathbf{v} - \mathbf{v}_{tool}$ is the relative sliding velocity (provided the non penetration condition is imposed: see eq. (2.7)),

α_f is the friction coefficient,

q is the sensitivity to the sliding velocity.

In the part domain Ω, when inertia and body forces are neglected, the equilibrium equation is:

$$\mathrm{div}(\boldsymbol{\sigma}) = 0 \tag{2.5}$$

with boundary conditions:

- on the free surface $\partial\Omega_s$:

$$\mathbf{T} = \boldsymbol{\sigma}.\mathbf{n} = 0 \tag{2.6}$$

- on tool-part interface $\partial\Omega_c$, in addition to the friction law, the non-penetration constraint is imposed:

$$(\mathbf{v}-\mathbf{v}_{tool})\cdot\mathbf{n} = \mathbf{v}_s.\mathbf{n} = 0 \text{ if } (\boldsymbol{\sigma}.\mathbf{n})\cdot\mathbf{n} < 0 \tag{2.7}$$

The virtual work principle gives the integral form, which holds for any virtual velocity field $\mathbf{v}^*$:

$$\int_\Omega \boldsymbol{\sigma}:\dot{\boldsymbol{\varepsilon}}^* d\mathcal{V} - \int_{\partial\Omega_c} \boldsymbol{\tau}.\mathbf{v}^* \, d\mathcal{S} = 0 \tag{2.8}$$

with the conditions:

$$\mathrm{div}(\mathbf{v}) = 0 \text{ in } \Omega \tag{2.9}$$

$$(\mathbf{v}-\mathbf{v}_{tool})\cdot\mathbf{n} = 0 \text{ and } \mathbf{v}^*.\mathbf{n} = 0 \text{ on } \partial\Omega_c \tag{2.10}$$

To impose easily the incompressibility condition, the penalty formulation is often used. For any virtual velocity field $\mathbf{v}^*$:

$$\int_\Omega \mathbf{s}:\dot{\boldsymbol{\varepsilon}}^* d\mathcal{V} - \int_{\partial\Omega_c} \boldsymbol{\tau}.\mathbf{v}^* d\mathcal{S} + \rho_p \int_\Omega K\mathrm{div}(\mathbf{v})\,\mathrm{div}(\mathbf{v}^*) d\mathcal{V} = 0 \tag{2.11}$$

where $\mathbf{s}$ is the deviatoric stress tensor, and with a penalty coefficient $\rho_p = 10^5$ to 10^7.

In order to impose more precisely the incompressibility condition and obtain better conditioned systems, we may prefer the mixed formulation. For any $\mathbf{v}^*$ and any p^* (belonging to appropriate functional spaces):

$$\int_\Omega \mathbf{s}:\dot{\boldsymbol{\varepsilon}}^* d\mathcal{V} - \int_{\partial\Omega_c} \boldsymbol{\tau}.\mathbf{v}^* \, d\mathcal{S} - \int_\Omega p\,\mathrm{div}(\mathbf{v}^*) d\mathcal{V} = 0 \tag{2.12a}$$

and

$$\int_\Omega p^*\mathrm{div}(\mathbf{v}) d\mathcal{V} = 0 \tag{2.12b}$$

Considering now heat transfer, the classical heat equation contains a source term $\dot{q}_V$ due to viscoplastic heat dissipation:

$$\rho\, c \frac{dT}{dt} = \mathrm{div}(k\,\mathbf{grad}(T)) + \dot{q}_V \tag{2.13}$$

The viscoplastic heat dissipation is equal to:

$$\dot{q}_V = f\,\boldsymbol{\sigma}:\dot{\boldsymbol{\varepsilon}} = fK\left(\sqrt{3}\,\dot{\bar{\varepsilon}}\right)^{m+1} \tag{2.14}$$

the f factor ranges generally between 0.9 and 1.

The constitutive law depends on temperature, for example equation (2.2) is replaced by:

$$K = K_0\left(\bar{\varepsilon}_0 + \bar{\varepsilon}\right)^n \exp\left(\beta/T\right) \tag{2.15}$$

with the rate sensitivity index:

$$m = m_0 + m_1 T \tag{2.16}$$

The mass conservation condition must take into account the thermal dilatation:

$$\text{div}(\mathbf{v}) - 3\,\alpha_d\,\dot{T} = 0 \tag{2.17}$$

where α_d is the linear dilatation coefficient. The left hand side of equation (2.17) is substituted to div(**v**) in equation (2.11) or (2.12a,b) when thermal dilatation is taken into account.

In most hot forming processes, the boundary conditions are generally expressed as follows:

- on the free surface $\partial\Omega_s$, radiation is approximated by:

$$-k\frac{\partial T}{\partial n} = \varepsilon_r\,\sigma_r\,(T^4 - T_0^4) \tag{2.18}$$

ε_r is the emissivity parameter, σ_r the Stefan constant, T_0 the outside temperature.

- on the surface of the tools $\partial\Omega_c$ we have conduction with the tool, and surface energy due to friction so that:

$$-k\frac{\partial T}{\partial n} = h_{cd}\,(T - T_{tool}) + \frac{b}{b+b_{tool}}\,\alpha\,K\,|\mathbf{v}_s|^{q+1} \tag{2.19}$$

where h_{cd} is the heat exchang coefficient with the tool with temperature T_{tool}, b and b_{tool} are the effusivity of the part and of the tool respectively.

2.2. *Space and time discretization*

Using isoparametric elements, the velocity field is expressed as:

$$\mathbf{v} = \sum_n \mathbf{V}_n\,N_n(\xi) = \mathbf{V}.\mathbf{N} \tag{2.20a}$$

with the parametric relation:

$$\mathbf{x} = \sum_n \mathbf{X}_n\,N_n(\xi) = \mathbf{X}.\mathbf{N} \tag{2.20b}$$

so that we can write the strain rate tensor with the usual **B** operator:

$$\dot{\boldsymbol{\varepsilon}} = \sum_n \mathbf{V}_n\,\mathbf{B}_n \tag{2.21}$$

For the penalty approach, the discretized form of the mechanical equation is then:

$$\mathbf{R}_n(\mathbf{V}) = \int_\Omega 2K\,(\sqrt{3}\dot{\bar{\varepsilon}})^{m-1}\,\dot{\boldsymbol{\varepsilon}}:\mathbf{B}_n d\mathcal{V} + \int_{\partial\Omega_C} \alpha_f K|\mathbf{v}_s|^{q-1}\mathbf{v}_s N_n d\mathcal{S} + \rho_p \int_\Omega K\text{div}(\mathbf{v})\,\text{tr}(\mathbf{B}_n) d\mathcal{V} = 0 \tag{2.22a}$$

which will be denoted for short:

$$\mathbf{R}(\mathbf{X}, \mathbf{V}) = 0 \tag{2.22b}$$

We must use an appropriate reduced integration scheme for the divergence integral in the penalty formulation. Similarly, for the mixed approach, compatible elements for the pressure field must be selected. Eqs. (2.12a,b) become:

$$\mathbf{R}_n^V(\mathbf{V},\mathbf{P}) = \int_\Omega 2K(\sqrt{3}\dot{\bar{\varepsilon}})^{m-1}\,\dot{\boldsymbol{\varepsilon}}:\mathbf{B}_n d\mathcal{V} + \int_{\partial\Omega_c} \alpha_f K|\mathbf{v}_s|^{q-1}\mathbf{v}_s N_n\,d\mathcal{S} - \int_\Omega p\,\text{tr}(\mathbf{B}_n)\,d\mathcal{V} = 0 \tag{2.23a}$$

$$\mathbf{R}_m^P(\mathbf{V}) = 0 = \int_\Omega M_m\,\text{div}(\mathbf{v})\,d\mathcal{V} \tag{2.23b}$$

Where M_m denotes the pressure interpolation function. As previously for the penalty formulation, eqs. (2.24a) and (2.24b) will also be denoted by the compact form:

$$\mathbf{R}(\mathbf{X}, \mathbf{V}, \mathbf{P}) = 0 \tag{2.24}$$

The nodal update for the simpler explicit Euler scheme is:

$$\mathbf{X}_n^{t+\Delta t} = \mathbf{X}_n^t + \Delta t\,\mathbf{V}_n^t \tag{2.25}$$

An implicit second order scheme (trapezoidal rule) improves the precision:

$$\mathbf{X}_n^{t+\Delta t} = \mathbf{X}_n^t + \frac{1}{2}\Delta t\,(\mathbf{V}_n^t + \mathbf{V}_n^{t+\Delta t}) \tag{2.26}$$

but equations (2.21) or (2.22a,b) must be written on the unknown configuration at time t+Δt, so that the resolution procedure is more complicated (see [Bohatier, 1985]).

The incremental incompressibility method and a discussion of various contact algorithms are given in [Chenot, 1995].

In our approach, the discretization of the temperature field uses the same elements as for the mechanical problem:

$$T = \sum_n T_n N_n(\xi) \tag{2.27}$$

With the Galerkin method, we obtain the following semi discretized form:

$$\mathbf{C}\frac{d\mathbf{T}}{dt} + \mathbf{H}\,\mathbf{T} + \mathbf{F} = 0 \tag{2.28}$$

with the heat capacity matrix **C** given by:

$$C_{ij} = \int_\Omega \rho\, c\, N_i N_j\, d\mathcal{V} \tag{2.29}$$

while the conductivity matrix **H** is defined as:

$$H_{ij} = \int_\Omega k\, \mathbf{grad}(N_i).\mathbf{grad}(N_j)\, d\mathcal{V} \tag{2.30}$$

The discretized viscoplastic heat generation contribution is expressed as:

$$\dot{q}_{v\,i} = \int_\Omega K(T)\, (\sqrt{3}\,\dot{\bar{\varepsilon}})^{m+1} N_i\, d\mathcal{V} \tag{2.31}$$

The boundary conditions contribute to the **F** vector and introduce some additional terms in the **H** matrix.

For time discretization, a second order three level scheme is generally chosen, where the following values are substituted in (2.28):

$$\mathbf{T} = a\,\mathbf{T}^{t-\Delta t} - (1.5 - 2a - g)\,\mathbf{T}^{t} + (a - 0.5 + g)\,\mathbf{T}^{t+\Delta t} \tag{2.32}$$

$$\frac{d\mathbf{T}}{dt} = (1\text{-}g)\,\frac{\mathbf{T}^{t} - \mathbf{T}^{t-\Delta t}}{\Delta t} + g\,\frac{\mathbf{T}^{t+\Delta t} - \mathbf{T}^{t}}{\Delta t} \tag{2.33}$$

Accordingly, matrices **C** and **H** and vector **F** which are affected by non-linear terms (due to temperature dependency) are expressed as follows in eq. (2.34):

$$\mathbf{C} = (0.5 - g)\,\mathbf{C}^{t-\Delta t} + (0.5 + g)\,\mathbf{C}^{t} \tag{2.34}$$

With $g = 1$ and $a = 0$ the scheme is similar to Crank-Nicholson, while the Dupont scheme corresponds to the values $a = .25$ and $g = 1$.

2.3 - The A.L.E. method for quasi static viscoplastic flow

2.3.1 - The mesh velocity

The mesh velocity **w** is defined mostly by a mathematical condition which is supposed to preserve an optimal shape of the mesh, the structure of the elements and their connectivity remaining the same. As an example the laplacian operator is often used as it preserves a satisfactory smoothness of the mesh and is easy to implement. The condition is written :

$$\Delta\mathbf{w} = \mathrm{div}(\mathbf{grad}(\mathbf{w})) = 0 \tag{2.35}$$

Obviously, the boundary of the domain must remain the same as the physical one, so that we must impose at least:

$$(\mathbf{v} - \mathbf{w}).\mathbf{n} = 0 \tag{2.36}$$

However if we consider that on the boundary $\partial\Omega$ the mesh velocity is equal to the material (i. e. kinematic) velocity, the problem can be decoupled component by component, so that several scalar field equations must be solved with the same linear operator:

$$\Delta w_i = 0 \;\text{ in } \Omega \tag{2.37a}$$

with prescribed boundary condition:

$$w_i = v_i \;\text{ on } \partial\Omega \tag{2.37b}$$

If we consider an internal variable such as the equivalent strain $\bar{\varepsilon}$, for which we do not introduce any diffusion, the time derivative for a geometrical point following the mesh will be:

$$\frac{d_g\bar{\varepsilon}}{dt} = \dot{\bar{\varepsilon}} - (\mathbf{v} - \mathbf{w}).\mathbf{grad}\,(\bar{\varepsilon}) \tag{2.38}$$

A global formulation can also be introduced for the calculation of $\bar{\varepsilon}$ using the Galerkin method if the convective term is small enough, or a SUPG (Stream Upwind Petrov Galerkin) method [Brooks, 1982]. However a nodal, or integration point, treatment is much more computationally effective.

For the temperature evolution we must transform the heat equation (2.13) into:

$$\rho c\,\frac{d_g T}{dt} = \mathrm{div}(k\,\mathbf{grad}(T)) + \dot{q}_V - \rho c\,(\mathbf{v} - \mathbf{w}).\mathbf{grad}(T) \tag{2.39}$$

The grid velocity is interpolated by:

$$\mathbf{w} = \sum_m \mathbf{W}_m N_m \tag{2.40}$$

Equation (2.345 is replaced by the classical integral formulation, so that we obtain the linear system:

$$\mathbf{D}(\mathbf{X}, \mathbf{V}, \mathbf{W}) = 0 \tag{2.41}$$

the components of which are:

$$D_{im} = \sum_n \int_{\Omega} \mathbf{grad}N_m.\mathbf{grad}N_n\, W_{in}\, d\mathcal{V} = 0 \quad (2.42a)$$

with the boundary condition:

$$W_{im} = V_{im} \text{ for } \mathbf{X}_m \in \partial\Omega \quad (2.42b)$$

Equation (2.38) can be written locally at each node provided a smoothing procedure allows to obtain a nodal mean value of $\dot{\bar{\varepsilon}}$ (initially discontinuous), and possibly of $\mathbf{grad}(\bar{\varepsilon})$. An alternative way is to compute $\bar{\varepsilon}$ only at integration points. On each element number e we put:

$$\int_{\Omega^e} \frac{d_g\bar{\varepsilon}}{dt} N_n d\mathcal{V} = - \int_{\partial\Omega^e} \bar{\varepsilon}(\mathbf{v} - \mathbf{w})\, .\mathbf{n}N_n\, d\mathcal{S}$$
$$+ \int_{\Omega^e} \dot{\bar{\varepsilon}} N_n d\mathcal{V} + \int_{\Omega} \bar{\varepsilon}\mathrm{div}(N_n(\mathbf{v}\text{-}\mathbf{w}))d\mathcal{V} \quad (2.43)$$

with eq (2.43) $\frac{d_g\bar{\varepsilon}}{dt}$ can be calculated at each node of a given element e and interpolated further at integration points.
Remarks: this procedure does not involve any global smoothing and avoids the computation of the derivatives of $\bar{\varepsilon}$ which can be discontinuous. The same procedure can be applied to discontinuous stress fields.

Using eq. (2.39), the semi discretized heat equation is transformed into:

$$\mathbf{C}.\frac{d_g\mathbf{T}}{dt} + \mathbf{H}.\mathbf{T} + \mathbf{K}^g.\mathbf{T} + \mathbf{F} + \;\; 0 \quad (2.44)$$

where the additional $\mathbf{K}^g$ matrix is defined by:

$$K^g_{mn} = - \int_{\Omega} \rho c(\mathbf{v} - \mathbf{w}).\mathbf{grad}(N_m)N_n d\mathcal{V} \quad (2.45)$$

For the temperature field, a local treatment is no more possible when the diffusion contribution cannot be neglected. For incremental forming processes (a typical example is forging) the convective term due to the ALE formulation is not predominant. When stationary or quasi stationary process (rolling, extrusion, metal cutting), or locally stationary processes (die filling in casting) are investigated, the convective therm can be very important so that the SUPG method must be used.

2.4 - Time integration schemes for the ALE method

2.4.1 - The pseudo explicit approach

At time t, the set of mechanical equations, corresponding to the mixed velocity and pressure formulation, is written in the compact form:

$$\mathbf{R}(\mathbf{X}^t,\ \mathbf{V}^t,\ \mathbf{P}^t) = 0 \quad (2.46)$$

$$\mathbf{D}(\mathbf{X}^t,\ \mathbf{V}^t,\ \mathbf{W}^t) = 0 \quad (2.47)$$

When $\mathbf{X}^t$ is known, it is possible to solve first (2.46) to obtain $\mathbf{V}^t$ and $\mathbf{P}^t$. Then (2.47) is solved in order to obtain the grid velocity $\mathbf{W}^t$, and the mesh is updated according to:

$$\mathbf{X}^{t+\Delta t} = \mathbf{X}^t + \Delta t\, \mathbf{W}^t \quad (2.48)$$

The equivalent strain is computed using the equation of evolution (2.38) for pure convection:

$$\bar{\varepsilon}^{t+\Delta t} = \bar{\varepsilon}^t + \Delta t \frac{d_g\bar{\varepsilon}^t}{dt} \quad (2.49)$$

For the temperature distribution it is also possible to compute the time grid derivative by solving eq. (2.44), and to update the nodal temperatures by a one step Euler explicit formula analogous to (2.49), but the stability condition will impose severe restrictions on the time increment size.

Moreover, an explicit second order Runger-Kutta scheme can be proposed, using the standard procedure. First, eqs. (2.46) and (2.47) are solved in the same way, but the time increment is $\Delta t/2$ so that eq. (2.48) is replaced by:

$$\mathbf{X}^{t+\Delta t/2} = \mathbf{X}^t + \Delta t/2\, \mathbf{W}^t \quad (2.50)$$

and equations similar to (2.49) for internal parameters and temperature. Then we can write the equilibrum equation at time $t + \Delta t/2$:

$$\mathbf{R}(\mathbf{X}^{t+\Delta t/2},\ \mathbf{V}^{t+\Delta t/2},\ \mathbf{P}^{t+\Delta t/2}) = 0 \quad (2.51)$$

and solve it for the velocity and pressure fields at time $t + \Delta t/2$. Then the grid smoothing equation can be also written:

$$\mathbf{D}(\mathbf{X}^{t+\Delta t/2},\ \mathbf{V}^{t+\Delta t/2},\ \mathbf{W}^{t+\Delta t/2}) = 0 \quad (2.52)$$

Eq. (2.52) allows to compute the grid velocity at time $t + \Delta t/2$. Finally the Runge and Kutta scheme provides a second order approximation by putting:

$$\mathbf{X}^{t+\Delta t} = \mathbf{X}^t + \Delta t\, \mathbf{W}^{t+\Delta t/2} \quad (2.53)$$

$$\bar{\varepsilon}^{t+\Delta t} = \bar{\varepsilon}^t + \Delta t \frac{d_g\bar{\varepsilon}^{t+\Delta t/2}}{dt} \quad (2.54)$$

Remark: an implicit scheme is often prefered for the temperature calculations; it is possible to combine it with and explicit scheme for the mechanical variable

2.4.2 - Implicit methods

Let us first consider a one step implicit method for a purely mechanical problem, i.e. when the effect of internal parameter and temperature are neglected. We write, using the same notations as previously:

$$\mathbf{R}(\mathbf{X}^{t+\Delta t},\ \mathbf{V}^{t+\Delta t},\ \mathbf{P}^{t+\Delta t}) = 0 \tag{2.55}$$

$$\mathbf{D}(\mathbf{X}^{t+\Delta t},\ \mathbf{V}^{t+\Delta t},\ \mathbf{W}^{t+\Delta t}) = 0 \tag{2.56}$$

$$\mathbf{X}^{t+\Delta t} = \mathbf{X}^{t} + \Delta t\ \mathbf{W}^{t+\Delta t} \tag{2.57}$$

We see immediately that eqs. (2.55) and (2.56) are coupled by eq. (2.57) as we cannot obtain separatly $\mathbf{V}^{t+\Delta t}$ and $\mathbf{W}^{t+\Delta t}$. Moreover the domain $\Omega^{t+\Delta t}$ on which the computation is made is an unknown. An iterative alternate scheme can be proposed, where we start with $\mathbf{X}^{(o)} = \mathbf{X}^{t}$ and use repeatedly the following steps, while all the variables undergo no more change:

i) for $\mathbf{V}^{(k)}$ and $\mathbf{P}^{(k)}$ solve:

$$\mathbf{R}(\mathbf{X}^{(k-1)},\ \mathbf{V}^{(k)},\ \mathbf{P}^{(k)}) = 0$$

ii) for $\mathbf{W}^{(k)}$ solve:

$$\mathbf{D}(\mathbf{X}^{(k-1)},\ \mathbf{V}^{(k)},\ \mathbf{W}^{(k)}) = 0$$

iii) update $\mathbf{X}$ by:

$$\mathbf{X}^{(k)} = \mathbf{X}^{t} + \Delta t\ \mathbf{W}^{(k)}$$

The problem is more complicated when the constitutive equation depends on the equivalent strain which must be updated according to:

$$\bar{\varepsilon}^{t+\Delta t} = \bar{\varepsilon}^{t} + \Delta t \frac{d_g \bar{\varepsilon}^{t+\Delta t}}{dt} \tag{2.58}$$

and possibly on the temperature field, the time derivative of which necessitates the solution of a global problem. In this case the above alternative scheme can be extended with two more steps:

iv) update of $\bar{\varepsilon}$,
v) update of temperature.

A second order scheme can also be proposed with $\alpha = 0.5$ in the generalized trapezoidal integration formula. We keep eqs. (2.55) and (2.56), but eq. (2.57) is replaced by:

$$\mathbf{X}^{t+\Delta t} = \mathbf{X}^{t} + \Delta t\left((1-\alpha)\mathbf{W}^{t} + \alpha \mathbf{W}^{t+\Delta t}\right) \tag{2.59}$$

The same generalized trapezoidal rule can be applied to the integration of the equivalent strain and of the temperature.

3 - THE VISCOPLASTIC DYNAMIC PROBLEM

3.1 - Mechanical formulation

The constitutive law is the same as that described in section 2.1 but the dynamic equation takes into account the acceleration $\boldsymbol{\gamma}$ and possibly the body forces $\mathbf{g}$:

$$\rho\, \boldsymbol{\gamma} - \rho\, \mathbf{g} - \mathrm{div}(\boldsymbol{\sigma}) = 0 \tag{3.1}$$

The integral form is written first:

$$\int_\Omega \rho\, \boldsymbol{\gamma}.\mathbf{v}^* d\mathcal{V} + \int_\Omega \boldsymbol{\sigma}:\dot{\boldsymbol{\varepsilon}}^* d\mathcal{V} - \int_\Omega \rho\, \mathbf{g}.\mathbf{v}^* d\mathcal{V} - \int_{\partial\Omega} \mathbf{T}.\mathbf{v}^*\ d\mathcal{S} = 0 \tag{3.2}$$

with the incompressibility constraint $\mathrm{div}(\mathbf{v}) = 0$. The Lagrange multiplier method allows to rewrite eq. (3.2) as:

$$\int_\Omega \rho\, \boldsymbol{\gamma}.\mathbf{v}^* d\mathcal{V} + \int_\Omega \mathbf{s}:\dot{\boldsymbol{\varepsilon}}^* d\mathcal{V} - \int_\Omega \rho\, \mathbf{g}.\mathbf{v}^* d\mathcal{V} - \int_\Omega p\, \mathrm{div}(\mathbf{v}^*)\ d\mathcal{V} - \int_{\partial\Omega} \mathbf{T}.\mathbf{v}^*\ d\mathcal{S} = 0 \tag{3.3a}$$

$$\int_\Omega \mathrm{div}(\mathbf{v})\ p^*\ d\mathcal{V} = 0 \tag{3.3b}$$

This form can be used only with implicit formulations where the incompressibility constraint is imposed at the end of the increment on the final velocity.

Another form can also be derived with the Lagrange multiplier method, when the incompressibility constraint is differentiated with respect to time, so that:

$$\frac{d}{dt}(\mathrm{div}(\mathbf{v})) = \mathrm{div}(\boldsymbol{\gamma}) - \mathrm{tr}\left((\mathbf{grad}(\mathbf{v}))^2\right) \tag{3.4a}$$

or with the component form:

$$\frac{d}{dt}(\mathrm{div}(\mathbf{v})) = \sum_i \frac{\partial \gamma_i}{\partial x_i} - \sum_{i,j} \left(\frac{\partial v_i}{\partial x_j}\right)^2 \tag{3.4b}$$

With the mixed formulation, we obtain a formula similar to equation (3.3a), while (3.3b) is transformed into:

$$\int_{\Omega} \left(\text{div}\boldsymbol{\gamma} - \text{tr}\left((\mathbf{grad}(\mathbf{v}))^2\right)\right) p^* \, d\mathcal{V} = 0 \qquad (3.5)$$

3.2 - Space and time discretization

The finite element discretization is made with the hypothesis of section 2, and the acceleration field is interpolated according to:

$$\boldsymbol{\gamma} = \sum_n \boldsymbol{\Gamma}_n N_n \qquad (3.6)$$

For simplicity the set of eqs. (3.3a,b), or (3.3a) and (3.4) will be denoted here:

$$\mathbf{R}(\mathbf{X}, \mathbf{V}, \ \boldsymbol{\Gamma}, \mathbf{P}) = 0 \qquad (3.7)$$

The explicit scheme is built in the following way:
i) at time t we suppose $\mathbf{X}^t$ and $\mathbf{V}^t$ are known;
ii) compute $\boldsymbol{\Gamma}^t$ and $\mathbf{P}^t$ from:

$$\mathbf{R}(\mathbf{X}^t, \ \mathbf{V}^t, \boldsymbol{\Gamma}^t, \mathbf{P}^t) = 0 \ ;$$

iii) update coordinate and velocity vectors respectively according to:

$$\mathbf{X}^{t+\Delta t} = \mathbf{X}^t + \Delta t \, \mathbf{V}^t + \frac{\Delta t^2}{2} \boldsymbol{\Gamma}^t \qquad (3.8a)$$

$$\mathbf{V}^{t+\Delta t} = \mathbf{V}^t + \Delta t \, \boldsymbol{\Gamma}^t \qquad (3.8b)$$

This scheme is often used for impact problems with very high velocities. For metal forming the contribution of inertia forces is generally much smaller, so that this scheme imposes to use very small time increments.

A simple implicit method can be built when the dynamic equation is solved at time t + Δt/2. We solve for $\boldsymbol{\Gamma}^{t+\Delta t/2}$, $\mathbf{P}^{t+\Delta t/2}$ the mechanical equation:

$$\mathbf{R}(\mathbf{X}^{t+\Delta t/2}, \ \mathbf{V}^{t+\Delta t/2}, \ \boldsymbol{\Gamma}^{t+\Delta t/2}, \mathbf{P}^{t+\Delta t/2}) = 0 \qquad (3.9a)$$

with:

$$\mathbf{V}^{t+\Delta t/2} = \mathbf{V}^t + \Delta t/2 \, \boldsymbol{\Gamma}^{t+\Delta t/2} \qquad (3.9b)$$

$$\mathbf{X}^{t+\Delta t/2} = \mathbf{X}^t + \Delta t/2 \, \mathbf{V}^t \qquad (3.9c)$$

At the end of the time increment, **V** and **X** are updated according to:

$$\mathbf{V}^{t+\Delta t} = \mathbf{V}^t + \Delta t \, \boldsymbol{\Gamma}^{t+\Delta t/2} \qquad (3.10a)$$

$$\mathbf{X}^{t+\Delta t} = \mathbf{X}^t + \Delta t/2 \, (\mathbf{V}^t + \mathbf{V}^{t+\Delta t}) \qquad (3.10b)$$

In this scheme, we have assumed implicitely that the acceleration field is constant during each time increment.
Remark: a higher order scheme can also be introduced if the acceleration is considered as linear over each time step (see [Arminjon, 1987]).

3.3 - The ALE formulation for dynamic problem

To the previous equations we must add the transport equation for the velocity field:

$$\frac{d_g \mathbf{v}}{dt} = \boldsymbol{\gamma} - (\mathbf{v} - \mathbf{w}).\mathbf{grad v} \qquad (3.11)$$

The discretized form of eq. (3.11) can be written on each element:

$$\frac{d_g \mathbf{v}}{dt} = \boldsymbol{\Gamma}.\mathbf{N} - (\mathbf{V} - \mathbf{W}).\mathbf{L}.\mathbf{V} \qquad (3.12)$$

where **L** is the discretized transport operator so that:

$$L_{kmn} = N_m \frac{\partial N_n}{\partial x_k} \qquad (3.13)$$

and the component form of (3.12) is:

$$\frac{d_g v_i}{dt} = \sum_n \Gamma_{in} N_n - \sum_{k,m,n} (V_{km} - W_{km}) L_{kmn} V_{in} \qquad (3.14)$$

We shall assume that this element equation will allow to compute the nodal derivative $\frac{d_g \mathbf{V}}{dt}$ by a smoothing procedure, which can be global or local.

3.3.1 The ALE explicit scheme for a dynamic problem

We suppose that $\mathbf{X}^t$ and $\mathbf{V}^t$ are known at time t, then the following steps are performed:

i) $\boldsymbol{\Gamma}^t$ and $\mathbf{P}^t$ are computed from:

$$\mathbf{R}(\mathbf{X}^t, \ \mathbf{V}^t, \boldsymbol{\Gamma}^t, \mathbf{P}^t) = 0 \ ;$$

ii) $\mathbf{W}^t$ is calculated from:

$$\mathbf{D}(\mathbf{X}^t, \ \mathbf{V}^t, \mathbf{W}^t) = 0;$$

iii) the nodal coordinates vector and the velocity field become respectively:

$$\mathbf{X}^{t+\Delta t} = \mathbf{X}^t + \Delta t \, \mathbf{W}^t \qquad (3.15)$$

$$\mathbf{V}^{t+\Delta t} = \mathbf{V}^t + \Delta t \frac{d_g \mathbf{V}^t}{dt} \qquad (3.16)$$

With the mesh derivative of the velocity which is expressed as:

$$\frac{d_g \mathbf{v}^t}{dt} = \mathbf{\Gamma}^t.\mathbf{N}^t - (\mathbf{V}^t - \mathbf{W}^t).\mathbf{L}^t.\mathbf{V}^t \,. \tag{3.17}$$

Remark: this formulation can be improved with respect to accuracy (not stability), by introducing a second (or higher) order Runge-Kutta method.

3.3.2 An ALE implicit formulation for dynamic problems

We suppose that $\mathbf{X}^t$ and $\mathbf{V}^t$ are known at time t. At time t+Δt/2 we write the following set of equations is written:

$$\mathbf{R}(\mathbf{X}^{t+\Delta t/2}, \mathbf{V}^{t+\Delta t/2}, \mathbf{\Gamma}^{t+\Delta t/2}, \mathbf{P}^{t+\Delta t/2}) = 0 \tag{3.18a}$$

$$\mathbf{D}(\mathbf{X}^{t+\Delta t/2}, \mathbf{V}^{t+\Delta t/2}, \mathbf{W}^{t+\Delta t/2}) = 0 \tag{3.18b}$$

$$\mathbf{X}^{t+\Delta t/2} = \mathbf{X}^t + \Delta t/2\, \mathbf{W}^{t+\Delta t/2} \tag{3.18c}$$

$$\mathbf{V}^{t+\Delta t/2} = \mathbf{V}^t + \Delta t/2 \frac{d_g \mathbf{V}^{t+\Delta t/2}}{dt} \tag{3.18d}$$

where $\frac{d_g \mathbf{V}^{t+\Delta t/2}}{dt}$ is again computed from a smoothing of:

$$\frac{d_g \mathbf{v}^{t+\Delta t/2}}{dt} = \mathbf{\Gamma}^{t+\Delta t/2}.\mathbf{N}^{t+\Delta t/2} - (\mathbf{V}^{t+\Delta t/2} - \mathbf{W}^{t+\Delta t/2}).\mathbf{L}^{t+\Delta t/2}.\mathbf{V}^{t+\Delta t/2} \tag{3.18e}$$

This set of equations can be solved iteratively by an iterative alternate procedure. First eq. (3.18a) is solved with respect to $\mathbf{\Gamma}^{t+\Delta t/2}$, then eq. (3.18b) for $\mathbf{W}^{t+\Delta t/2}$, and coordinates are updated according to (3.18c) and velocities according to eqs. (3.18d,e). The process is repeated until all the variables remain constant from an iteration to the next one.
Finally at t+Δt we have:

$$\mathbf{X}^{t+\Delta t} = \mathbf{X}^t + \Delta t\, \mathbf{W}^{t+\Delta t/2} \tag{3.19a}$$

$$\mathbf{V}^{t+\Delta t} = \mathbf{V}^t + \Delta t \frac{d_g \mathbf{V}^{t+\Delta t/2}}{dt} \tag{3.19b}$$

Any internal variable, e.g. $\bar{\varepsilon}$, can be updated according to:

$$\bar{\varepsilon}^{t+\Delta t} = \bar{\varepsilon}^t + \Delta t \frac{d_g \bar{\varepsilon}^{t+\Delta t/2}}{dt} \tag{3.20}$$

3.3.3 A partly implicit explicit ALE method

First we can use separately a purely implicit method for the kinematic variables $\mathbf{\Gamma}^{t+\Delta t/2}$ and $\mathbf{P}^{t+\Delta t/2}$, corresponding to eqs. (3.9a-c). Then a second order explicit scheme for the ALE procedure is used: at time t we compute $\mathbf{W}^t$ from eq. (2.47), so that we put:

$$\mathbf{X}^{t+\Delta t/2} = \mathbf{X}^t + \Delta t/2\, \mathbf{W}^t \tag{3.21a}$$

$$\mathbf{V}^{t+\Delta t/2} = \mathbf{V}^t + \Delta t/2 \frac{d_g \mathbf{V}^t}{dt} \tag{3.22b}$$

with the mesh derivative of the velocity given by eq. (3.18e). Then eq. (2.52) is solved with respect to $\mathbf{W}^{t+\Delta t/2}$ and the node coordinates and velocity are updated according to eqs. (3.18c) to (3.18d). But the mesh derivative of the velocity requires a slightly different treatment, we have:

$$\frac{d_g \mathbf{v}^{t+\Delta t/2}}{dt} = \left(\mathbf{\Gamma}^{t+\Delta t/2}\right)^* .\mathbf{N}^{t+\Delta t/2} - (\mathbf{V}^{t+\Delta t/2} - \mathbf{W}^{t+\Delta t/2}).\mathbf{L}^{t+\Delta t/2}.\mathbf{V}^{t+\Delta t/2} \tag{3.23}$$

where the acceleration $\left(\mathbf{\Gamma}_n^{t+\Delta t/2}\right)^*$ itself does not correspond to the same material point with nodal vector $\mathbf{X}_n^t$ at time t, so that:

$$\left(\mathbf{\Gamma}^{t+\Delta t/2}\right)^* = \mathbf{\Gamma}^{t+\Delta t/2} - \frac{\Delta t}{2} (\mathbf{V}^{t+\Delta t/2} - \mathbf{W}^{t+\Delta t/2}).\mathbf{L}^{t+\Delta t/2}.\Gamma\mathbf{\Gamma}^{t+\Delta t/2} \tag{3.24}$$

The advantage of this formulation is that it does not introduce a significant change in the mechanical resolution modules.
The procedure is schematically represented in Figure 1 where the necessity of a convection term at time t+Δt/2 is clearly shown.

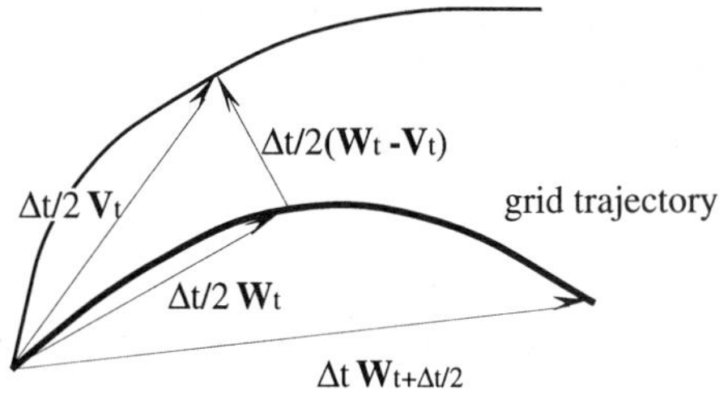

Figure 1: the grid explicit second order sheme.

4. CONCLUSIONS

We have recalled the finite element formulation for viscoplastic material flow and quasi static or dynamic problems, in view of hot metal forming and mold filling with liquid metal in casting. The ALE method is recalled with a grid velocity formulations and a laplacian type smoothing procedure. General notations are proposed, with the velocity or acceleration formulation, which allow to investigate easily first and second order explicit or implicit schemes.

In the future three important issues will be investigated:
- adaptive smoothing for the ALE method in order to achieve continuous adaptive remeshing as long as possible in order to delay complete remeshing;
- smooth transition when remeshing with an important topology change is necessary,
- generalization of the present approach to more complicated constitutive laws.

REFERENCES

Arminjon M. and Chenot J.-L., An implicit formulation for finite element analysis of dynamic plastic deformation in 2D-geometry, Second International Conference on Advances in Numerical Methods in Engineering : Theory and Applications NUMETA'87, Swansea, U. K. (1987) T2/1-10.

Baaijens F.P.T., An U-ALE formulation of 3-D unsteady viscoelastic flow, Int. J. numer. methods Engrg., 36 (1993) 1115-43.

Bohatier and Chenot J.-L., Finite element formulation for non-steady state large viscoplastic deformation. Int. J. Numer. Method. Eng., 21 (1985) 1697.

Brooks A. N. and Hughes T. J. R., Streamline upwind Petrov Galerkin formulations for convection dominated flows with particular emphasis on the incompressible Navier-Stokes equations, Comp. Meth. Appl. mech. engrg., 32 (1982) 199-259.

Cescutti J.-P., Wey E. Chenot J.-L. and Mosser P.-E., Finite element calculations of hot forging with continuous remeshing, Modelling of Metal forming processes, ed. by Chenot J.-L. and Onate E., Kluwer Academic Press, Dordrecht (1988) 207-18.

Chenot J.-L. and Bellet M., The viscoplastic approach for the finite element modelling of metal forming processes, in Numerical Modelling of Material Deformation processes. Research, Developments and Applications, ed. by Hartley P., Pillinger I. and. Sturgess C., Springer Verlag, London (1992) 179-224.

Chenot J.-L., Recent progresses in finite element simulation of the forging process, 4th Int. Conf. in Computational Palsticity. Fundamentals and Application, Barcelona, Spain (1995).

Coupez T. and Chenot J.-L., Mesh topology for mesh generation problems - Application to three-dimensional remeshing, Numerical Methods in Industrial Forming Processes, Ed. by Chenot J.-L. et al, A.A. Balkema, Rotterdam, (1992) 237-42.

Coupez T., A mesh improvement method for 3-D automatic remeshing, Numerical grid generation in computational fluid dynamics and related fields, ed. by Weatherill N. P. et al, The Cromwell Press Ltd, Melksham (1994) 615-26.

Donea J., Fasoli-Stella P. and Giuliani S., Lagrangian and Eulerian finite element techniques for transient fluid-structure interaction problems, Transaction of the 4th SMIRT Conf., San Francisco, CA (1977) Paper B 1/2 .

Gaston L., Glut B., Bellet M. and Chenot J.-L., An arbitrary lagrangian eulerian finite element approach to non-steady state fluid flows. Application to mold filling, Proc. 7th Int. Conf of modelling of casting, welding and advanced solidification processes, M. Goss (ed.), TMS, London (1995).

Ghosh S., Finite element simulation of some extrusion processes using the arbitrary lagrangian-eulerian description, J. Mater. Shaping Technol. 8, (1986) 53-64.

Ghosh S. and Kikuchi N., An arbitrary lagrangian-eulerian finite element method for large deformation analysis of elastic-viscoplastic solids, Comp. Methods Appl. Mech. Engrg, 86 (1991) 127-88.

Ghosh S. and Manna S.K., R-Adapted arbitrary lagrangian-eulerian finite element method in metal forming simulation, JMEPEG, 2 (1993a) 271-82.

Ghosh s and Moorthy S., An arbitrary lagrangian-eulerian finite element model for heat transfer analysis of solidification processes, Numer. Heat transfer, 23 (1993b) 327-50.

Hayes R.E., Dannelongue H.H. and Tanguy P.A., Numerical simulation of mold filling in reaction injection molding, Polym. Eng. Sci., 31 (1991) 842-8.

Huetink J. and Mooi H.G., Application of numerical simulation in design and control of forming processes, Journal de Physique IV, 3 (1993) 1129-34.

Huerta A. and Liu W.K., Viscous flow with large free surface motion, Comput. Methods. Appl. Mech. Engrg 69 (1988) 277-324.

Huerta A. and Casadei F., New ALE applications in non-linear fast-transient solid dynamics, Eng. Comput., 11 (1994) 317-45.

Hu Y.K. and Liu W.K., ALE finite element formulation forming rolling analysis, 33 (1992) 1217-36.

Hurez P. Tanguy P.A. and Bertrand F.H., A finite element analysis of die swell with pseudoplastic and viscoplastic fluids, Comput. Methods in Appl. Mech. Engrg, 86 (1991) 87-103.

Huisman H.J. and Huetink J., A combined eulerian-lagrangian three dimensional finite element analysis of edge rolling, J. of Mech. Working Tech., 11 (1985) 333-53.

Liu W.K., Chang H., Chen J.S. and Belytschko T., Arbitrary lagrangian-eulerian Petro-Galerkin finite elements for non linear continua, Comput. Methods Appl. Mech. Engrg, 68 (1988) 259-310.

Liu W.K., Chen J.S., Belytschko T. and Zhang Y.F., ALE finite elements with particular reference to external work rate on frictional surface, Comput. Methods Appl. Mech. Engrg, 93 (1991) 189-216.

Liu W.K., Hu Y.K. and Belytschko T., ALE finite elements with hydodynamic lubrication for metal forming, Nuclear Eng. Design, 138 (1992) 1-10.

Magnin B., Coupez T;,Vincent M. and Agassant J.-F., Numerical modelling of injection mold-filling with an accurate description of the front flow, Proc. of Numerical methods in industrial forming processes, ed. by Chenot J;-L. et al;, Balkema, Rotterdam (1992) 365-70.

Nomura T. and Hughes T.J.R., An arbitrary lagrangian-eulerian finite element body, Comput. Methods Appl. Mech. Engrg., 95 (1992) 115-38.

Ponthot J.-P., The use of the eulerian-lagrangian formulation including contact: applications to forming simulation via FEM., Proc. of Numerical methods in industrial forming processes, ed. by Chenot J;-L. et al;, Balkema, Rotterdam (1992) 293-300.

Rakotomalala R., Joyot P. and Tourratier M., Arbitrary lagrangian-eulerian thermomechanical finite element model of material cutting, Com. Numer. Meth. Engrg, 9 (1993) 975-87.

Ramaswany B. and Kawahara M., Arbitrary lagrangian-

eulerian finite element method for unsteady, convective, incompressible viscous free surface fluid flow, Int. J. Numer. Methods Fluids, 7 (1987) 1053-75.
Van der Lugt J. and Hueting J., Thermal mechanically coupled finite element analysis in metal forming processes, Comput. Methods. Appl. Mech. Engrg, 54 (1986) 145-60.

Simulation of Materials Processing: Theory, Methods and Applications, Shen & Dawson (eds)
© 1995 Balkema, Rotterdam. ISBN 90 5410 553 4

Modelling the glass fiber fabrics deformation processes

J.C.Gelin & A.Cherouat
Laboratoire de Mécanique Appliquée R. Chaléat, URA CNRS, Université de Franche-Comté, Besançon, France

ABSTRACT: The overall analysis presented in the paper concerns the modelling of glass fiber preforming before resin transfer moulding. First the mechanical behaviour of glass fiber fabrics is described as the discrete summation of the behaviour of each thread in the fabric. Then a finite element modelling is presented giving directly access to the tensions in threads and strongly differing from geometric approaches classically used for such problems. Finally numerical simulations of simple and complex parts, compared with experiments, prove the efficiency of the proposed approach.

1. INTRODUCTION

The RTM (Resin Transfer Moulding) manufacturing process for thin composite structures is composed of two main stages. In a first step the glass fiber fabrics are shaped by a drawing operation. Then a resin is injected at high temperature. After cooling, the obtained thin composite possesses high mechanical strength combined with low weight (Gay, 1987). They can substitute for metal components with some advantages such as gain of mass, facility to obtain complex shapes with a single operation or noise decreasing. These fabric reinforced composite components are used for instance in aeronautical and automotive industries. Nevertheless, the fabrication process is not totally controlled and studies are necessary to know if the manufacturing of a given shape is possible or not, and how it can be carried out. In a related area, a study concerning the forming of aligned fiber composites into complex shapes has been proposed (Gutowski et al., 1991).

The study proposed in the present work concerns the modelling and simulation of the first stage (glass fiber fabric drawing) to determine if the shaping is possible for a given geometry of the component and the related die and blank.

The behaviour of the fabric during the shaping process is very different of those of a metal sheet. During the sheet metal forming process the blank is usually subjected to large membrane extension and one of the main problem of the operation is the thickness variation. As opposed, the extension of fabric threads is small, but the angular variations between weft and warp threads can be very large and are the principal deformation mode of the fabric (sometimes called trellising). The fabrics used for the RTM process are built in order to render this angular variation possible (and so the shaping process).

However different problems can occur during the glass fiber shaping process: i) local folding due to overlapping of threads in case of too large in plane angle variations, ii) fractures of fibers due to excessive tensions in the threads, iii) curls of the threads due to large compression strains.

These problems (or at least their appearance) can be detected from the strain and tension state of the fabric during the drawing operation.

Some models are proposed in the literature for draping of a fabric on a given 3D surface, mainly based on geometrical approaches (Bergsma and Huisman, 1988). The problem is treated while considering only the trellising deformation mode of the fabric. Other algorithms for the fabric draping operation have been presented (Van der Ween, 1991). They are based on a structured mesh oriented by thread directions and an inextensibility assumption.

The presented study consists of a mechanical formulation of the shaping process simulation based on a simple behaviour of the fabric and the potential energy minimisation. The deformation energy of the fabric is calculated as the sum of the tension energy in the different threads. On the basis of these assumptions a numerical simulation is built. It uses 3D surface membrane finite elements composed of threads. A set of shaping simulations is performed and compared to the related experiments. Two industrial simulations are presented: i) the shaping of a square box that shows the possibility to optimize the tool's geometry to provide a feasible glass fiber

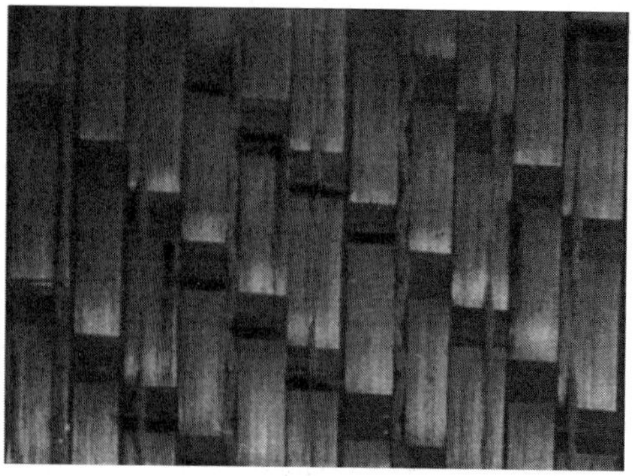

Figure 1 Photograph of the glass fiber fabric (S4x3) used in RTM

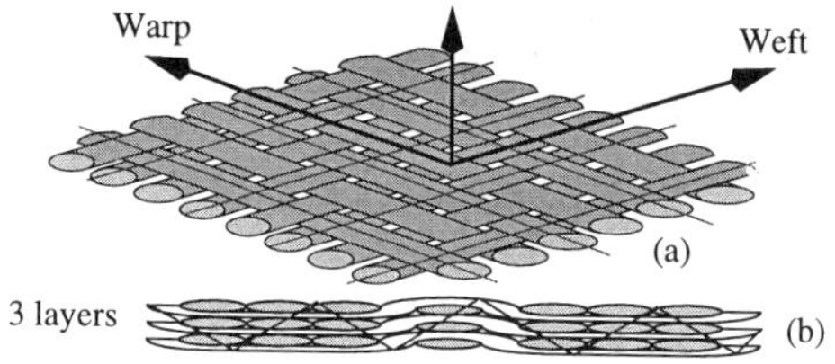

Figure 2 (a) Schematic representation of the glass fiber fabric, (b) Arrangement of threads in each layer.

Table 1. Physical and geometrical characteristics of threads.

Industrial Reference		580S4x3
Fiber diameter (μm)		17
Mass per unit length(g/Km)	Warp	1200
	Weft	1200
Number of threads per unit length (thread/cm)	Warp	2.2
	Weft	2.5

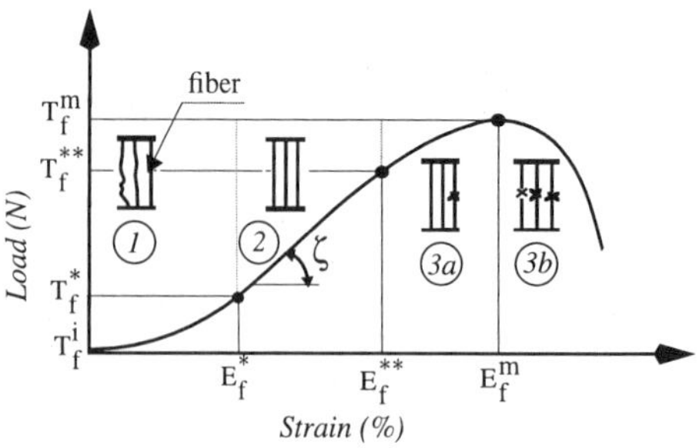

Figure 3 Tensile test of a single glass thread (2000 fibers).

component ii) the shaping of a complex automotive part. The main interest of the proposed software is the knowledge of the thread tensions and thread directions during and after the shaping process.

2. MODELLING THE MECHANICAL BEHAVIOUR OF GLASS FIBERS FABRICS

The present section relates experiments and models developed for the identification of the constitutive behaviour of glass fiber fabrics. First an experimental investigation is carried out on single glass thread specimens to identify the basic material behaviour of glass thread in tensile direction. Then tensile and shear tests carried out on glass fiber fabric specimens show that the extension of the proposed model for a single thread fit well, in numerous cases, the experimental results. Finally the interaction effects due to the undulations of the fabric are discussed from results obtained by biaxial tensile experiments.

2.1 *Description of the fabric*

The experiments and models presented are for the fabric shown in figure 1. It consists of three layers of orthogonal threads arranged as shown in figure 2. The physical and geometrical properties of each thread are given in table 1. The initial orientation between both thread groups is set to 90°.

2.2 *Characterization of the tensile behaviour of a single thread*

Three zones can be distinguished (Figure 3) in the tensile test of a single glass thread (Gelin and Sabhi, 1991): non-linear behaviour in zone I due to the progressive tensile straining of filaments, linear behaviour in zone II, and progressive failure of the filaments accompanied by load increase to a maximum followed by a decrease in zone III. A model for a single thread, based on a statistical approach, has been elaborated considering respectively progressive tensile and progressive damage (Boisse, Gelin and Sabhi, 1992). Such a model leads to the following behavior, in terms of thread tension vs. thread strain:

$$T_f^{11} = \begin{cases} \zeta\left[\left(1 - \dfrac{T_f^*}{\zeta E_f^*}\right)E_{11}^f + \dfrac{2T_f^*}{\zeta} - E_f^*\right]\left(\dfrac{E_{11}^f}{E_f^*}\right) + T_f^i \\ \zeta\,(E_{11}^f - E_f^*) + T_f^* \\ \zeta E_{11}^f \exp\left[\alpha\,(E_{11}^f - E_f^{**})^{\beta}\right] + T_f^* - \zeta E_f^* \end{cases} \quad (1)$$

where T_f^{11} stands for the tension in the thread's direction, E_{11}^f for the Green-Lagrange strain component in the thread direction, ζ the slope of the linear part of curve, (α, β) are parameters to identify and T_f^i, (E_f^*, T_f^*), (E_f^{**}, T_f^{**}), (E_f^m, T_f^m)

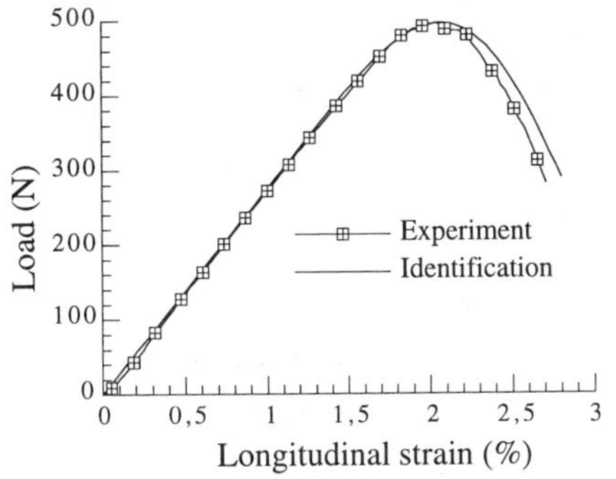

Figure 4 Tension vs. strain curve for the tensile test of a single glass fiber thread, (Experiment and model).

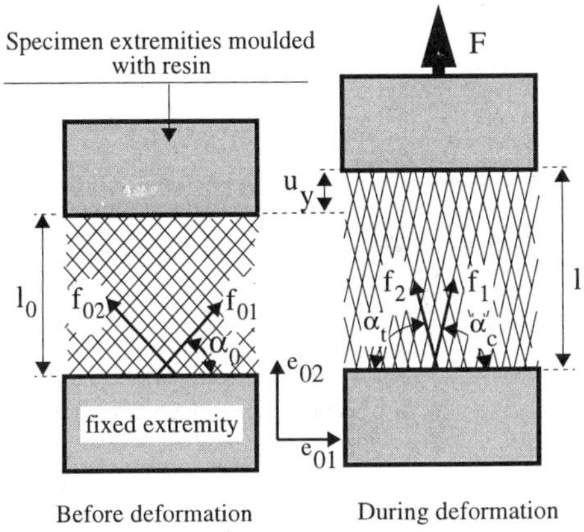

Figure 5 Unidirectional displacement experiment on a fabric specimen.

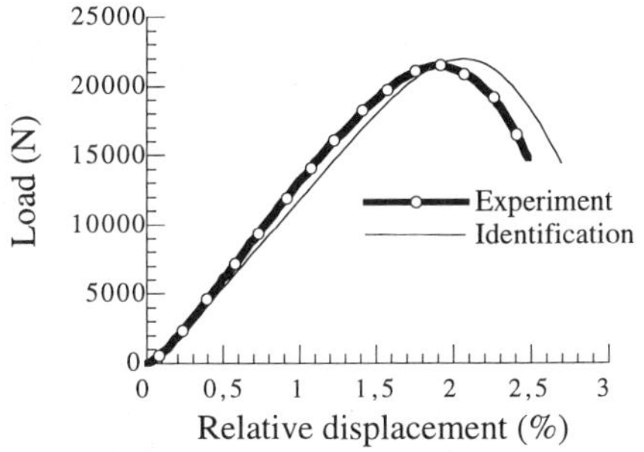

Figure 6 Load vs. displacement curves for a unidirectional displacement test (n_c= 48, α_{c0}= 0°), (n_t= 0, α_{t0} =90°).

are defined in figure 3. The comparison between this model and experimental results on a single thread is shown in figure 4. The agreement is good.

From the observations and assumptions made in that section, the mechanical model of the whole fabric is based on the behaviour of a single thread and the current fabric geometry, i.e., the directions of the threads at time t. Two experiments to validate this model performed on complete fabric specimens are described in sections 2.3. and 2.4. In these tests, the global experimental measures are compared to the theoretical values obtained from the fabric model.

2.3 *Unidirectional displacement test*

In this experiment (Figure 5), a displacement in the vertical direction is imposed at the moving extremity of the fabric specimen. This test is carried out for different thread orientations. When the warp and weft directions do not coincide with the laboratory orientations ($\mathbf{e}_{10}$ horizontal and $\mathbf{e}_{20}$ vertical), the vertical displacement induces some extra loads in the horizontal direction. If a conventional tensile machine is used, these induced loads generate horizontal displacements and rotations of the specimen extremities. In order to perform a strictly unidirectional displacement test, a special mechanical system has been defined (Sabhi, 1993). The extremities of the tensile specimen are moulded with resin and prepreg (Boisse, Gelin and Sabhi, 1992) to assure a good prehension of threads and to fix an extensometer that measures the true vertical relative displacement (Figure 5). Denoting by u_y the vertical displacement of the moving extremity of the specimen, by l_o the vertical initial length, and by $\mathbf{f}_{10}$ and $\mathbf{f}_{20}$ the unit vector in the initial thread directions, the Green-Lagrange strain component in the thread direction is given as:

$$E^f_{11} = \left(\frac{u_y}{l_0} + \frac{u_y^2}{2l_0^2}\right)\sin^2\alpha_0 \qquad (2)$$

Using the constitutive equation for a single thread defined in 2.2, equation (2) gives the tension in each thread. The load F on the moving part of the machine is obtained by adding the vertical projections of all the thread tensions

$$F = n_c T^{11}_{fc} \sin\alpha_c + n_t T^{11}_{ft} \sin\alpha_t \qquad (3)$$

where n_c and n_t are the number of threads in the warp and weft directions, respectively, T^{11}_{fc} and T^{11}_{ft} are the tensions in these threads, and α_c, α_t are the current orientation of these threads.

In Figure 6 and Figure 7 the loads obtained from equation (3) are compared with the experimental loads measured for a given vertical displacement for different initial orientation angles. The loads are plotted versus the relative displacement u_y/l_0.

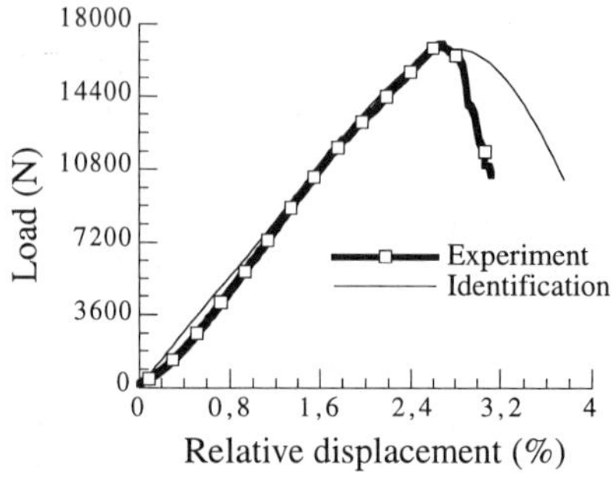

Figure 7 Load vs displacement curves for a unidirectional displacement test (n_c=39, α_{c0} = 30°), (n_t=9, α_{t0} = 60°).

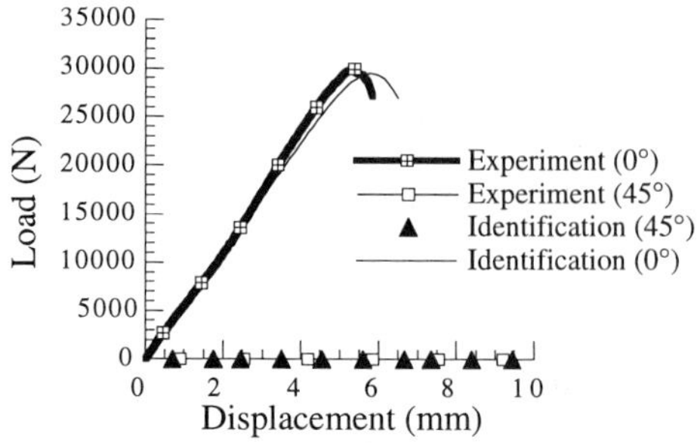

Figure 9 Load vs displacement curve for the shear test, with $\alpha_0 = 0^\circ$ and with $\alpha_0 = 45^\circ$.

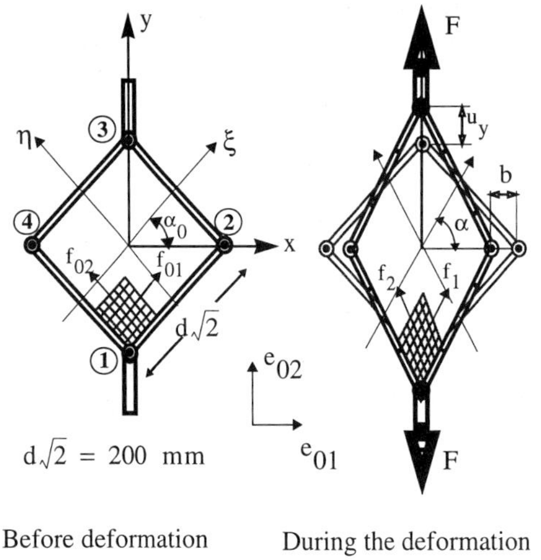

Figure 8 Shear test on a glass fibers fabric specimen.

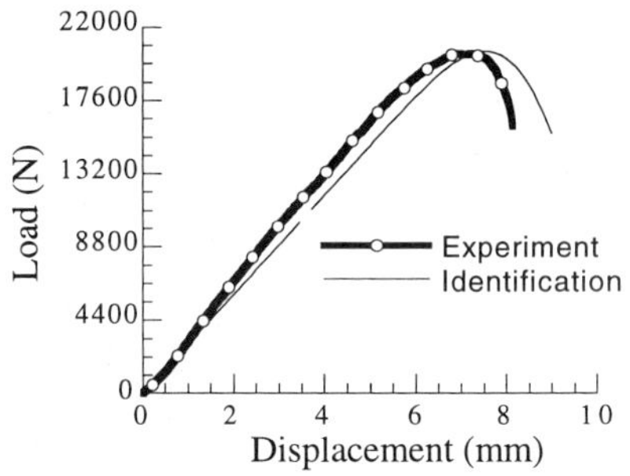

Figure 10 Load vs displacement curve for the shear test with $\alpha_0 = 20^\circ$

2.4 *Shear test*

A glass fibers fabric specimen is clamped on a hinged parallelogram (Figure 8). Each edge of the parallelogram is assumed to be non-deformable. Applying a vertical load at points 1 and 3, the parallelogram is deformed and the fabric specimen is subjected to a shear strain state.

The initial and current geometries of the parallelogram in this experiment (Figure 8) yield the longitudinal strain component in a given thread direction.

This longitudinal strain is constant for all warp and weft threads. When this strain is positive for one of the two thread sets, it is negative for the other and the tension T_f^{11} is equal to zero for these fibers. The positive component E_{11}^f determines the tension T_f^{11} in the thread through the constitutive equation (1) of the single thread. The equilibrium of the rigid bars comprising the parallelogram gives a relationship between the loads of the threads in tension to the vertical effort F at point 1 and 3 in the form

$$F = n_f T_f^{11} \frac{\cos\alpha_0}{\cos\alpha_0 + \sin\alpha_0} \left(\cos\alpha - \sin\alpha \frac{d + u_y}{d - b}\right) \quad (4)$$

where n_f is the number of threads in tension (warp or weft direction), and α is their current orientation angle, such that

$$\tan\alpha = \frac{d - b}{d + u_y} \tan\alpha_0 \quad (5)$$

The experimental load measured for a given displacement is compared to the theoretical value obtained by (4), see Figure 9 and Figure 10. The experiments are conducted for different initial fiber orientations.

From both classes of experiments previously described, a theoretical value of the load on the fabric specimen is established for a given displacement field, by accounting for each thread with the assumptions made in section 2. Both classes of tests were performed for different initial thread orientations. The agreement between theoretical and experimental values is good (Figures 6,7,9 and 10). For the shear test, the load displacement curves have been plotted with the same scale. The curves clearly show that the experimental in-plane stiffness of the fabric is very small when threads are oriented with α_0=45° (i.e. the in-plane stiffness of the fabric is very small). The maximum value of the load on the tensile machine is 120 N. This phenomenon is well described by the model that only accounts for thread tensions in the strain energy of the fabric, i.e. neglects the in-plane fabric shear stiffness. The in-plane shear strain can be large but does not contribute to the strain energy. Although some phenomena such as the inter-fiber frictions are not accounted for, the global behaviour of the fabric derived from the behaviour of single threads and their current positions is in a satisfactory agreement with these experiments. This model will be used for a numerical modelling of the drawing process, but some investigations are in progress to characterize the thread interactions and consequence on the mechanical behaviour of glass fabrics.

2.5 *Biaxial tensile tests and interaction effects*

The studies concerning the mechanical behaviour of woven composites are numerous (Ishikawa and Chou, 1982; Ishikawa and Chou, 1983; Ishikawa, Sushima, Hayashi and Chou, 1985). Those concerning the fabric without resin are less numerous. A reference work has been made in (Kawabata, Niwa and Kawai, 1973) and extended in (Realf, Boyce, Backer, 1993). The global behaviour of a fabric is defined by the tension behaviour of the threads in one hand and the relative position evolution of the thread within the fabric in the other hand.

The total deformation of the fabric is given by the extension of the threads and geometrical non-linearities due to the thread trend to become straight. The last aspect is a two-dimensional phenomenon. The strain in a direction (warp or weft) leads to a strain in the other direction. Several types of approaches can be used in order to determine the global behaviour of the fabric, i.e. to relate the warp and weft tensions to the strain state. A micro mechanical approach (or geometrical approach) is possible (Kawabata, Niwa and Kawai, 1973; Realf, Boyce, Backer, 1993). A phenomenological modelling concerning a large class of fabric, based on experimental studies using a bi-axial tension machine (Figure 11) is in progress (Borr, Cherouat, Gelin, 1995) and Figure 12 gives first results concerning interactions effects. Nevertheless, the glass fiber fabrics used in the R.T.M. process (to the least those on interest in this study) have a very flat geometry. Actually they have to be able to be exhibit very large in plane shear strains in order to render the shaping process possible.

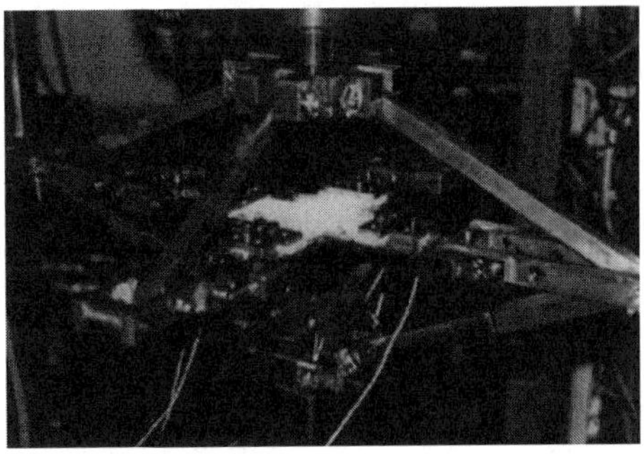

Figure 11 The biaxial tensile apparatus for fabrics

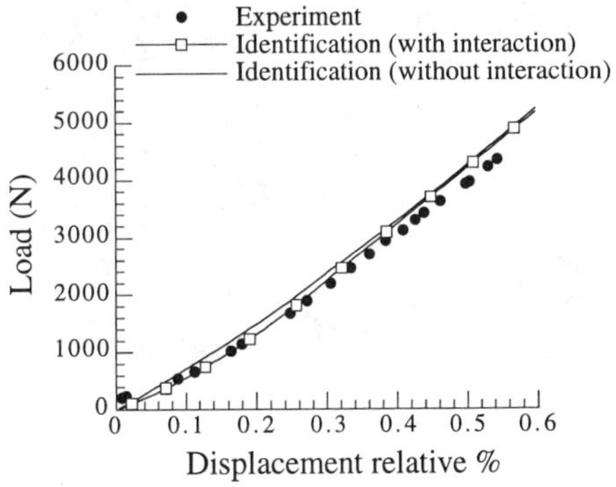

Figure 12 Load vs displacement curve for the uniform biaxial extension. Comparison between the experimental and the identification.

3. MODELING THE GLASS FIBER FABRICS DEFORMATION

The glass fabric deformation is described with membrane assumptions. The material coordinates system associated to each thread is use to describe deformation of the structure and a nonlinear solution procedure allows to take into account overall non-linearities.

Let a doubly curved surface (Figure 13) parametrized as:

$$x = x^0(\xi_1, \xi_2) + u(\xi_1, \xi_2) \tag{6}$$

and let the initial and current covariant vectors defined as:

$$g_i^0 = \frac{\partial x^0}{\partial \xi_i} \quad ; \quad g_i = \frac{\partial x}{\partial \xi_i} \tag{7}$$

The strain tensor can be expressed in the covariant frame as:

$$E = E_{ij} g^{0i} \otimes g^{0j} \quad E_{ij} = \frac{1}{2}(g_i \cdot g_j - g_i^0 \cdot g_j^0) \tag{8}$$

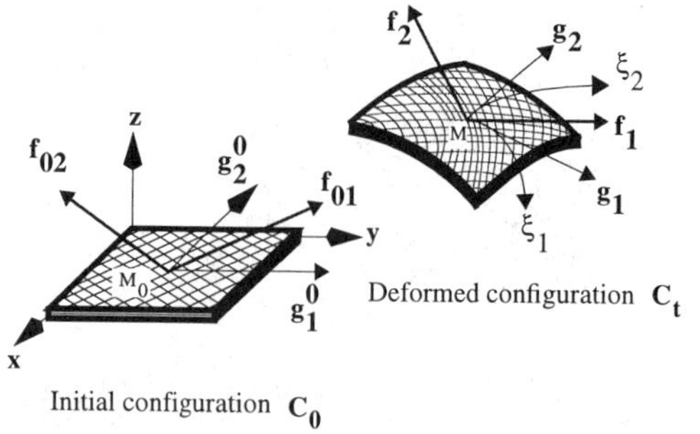

Figure 13 Schematic representation of glass fiber deformations

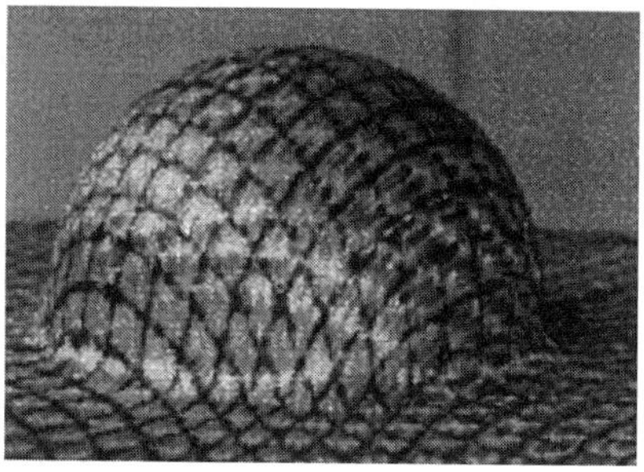

Figure 14 Final shape obtained with an hemispherical punch

From classical geometry it is possible to calculate E^f components from E components. The strain tensor can be also expressed in the thread direction as:

$$E^f = E^f_{11} f_{01} \otimes f_{01}$$
$$E^f_{11} = E_{ij} (g^0_i \cdot f_{01}) (g^0_j \cdot f_{01}) \tag{9}$$

where f_{01} is the material vector in the thread directions.

The tensile tensor in the thread directions is expressed as:

$$T_f = T_f^{11} f_{01} \otimes f_{01} \quad (T_f^{11} \geq 0) \tag{10}$$

The total potential energy over the glass fabric is obtained as a discrete summation

$$V(u) = \sum_{\text{thread}} \int_{l_0} W(E^f_{11})\, ds_0 - J^e_{ext}(u) \tag{11}$$

where $W(E^f_{11})$ is the deformation energy per unit length of a thread and $J^e_{ext}(u)$ the potential associated to the external loads. The constitutive equation for a thread is expressed as:

$$T_f^{11}(E^f_{11}) = \frac{\partial W(E^f_{11})}{\partial E^f_{11}} \tag{12}$$

and the equilibrium equation (11) can be written:

$$G(u, \eta) = \sum_{\text{threads}} \int_{l_0} T_f^{11} D_u [E^f_{11}]\, \eta\, ds_0 - \int_{\Gamma_{t0}} \bar{t} \eta\, ds_0 \tag{13}$$
$$\forall \eta / \eta = 0 \text{ on } \Gamma_u$$

4. FINITE ELEMENT SIMULATION

The finite element simulation of the deformation of glass fiber fabrics is based on the following remark: if we examine the deformation of a glass fiber fabric after deep drawing with an hemispherical punch we observe that straight lines drawn on the fabric before preforming remain continuous after deformation (Figure 14). This means that a continuous membrane assumption can be used.

The finite element approximation is based on such a remark and the position and displacements are interpolated with P1 or Q4 functions as:

$$x = \sum_{k=1}^{n} N^k(\xi_1, \xi_2)\, x^{0k} + \sum_{k=1}^{n} N^k(\xi_1, \xi_2)\, u^k \tag{14}$$

where N^k are the classical shape functions of P1 (n=3) or Q4 (n=4) membrane elements. The linear derivative of strain components is easily obtained from (14).

$$D_u(E_{ij})\,\eta = \frac{1}{2}\left(\frac{\partial \eta_m}{\partial \xi_i}(g^0_j)_m + \frac{\partial \eta_m}{\partial \xi_j}(g^0_i)_m + \frac{\partial \eta_m}{\partial \xi_i}\frac{\partial u_m}{\partial \xi_j} + \frac{\partial u_m}{\partial \xi_j}\frac{\partial \eta_m}{\partial \xi_i}\right) \tag{15}$$
$$m \in [1, 3] \qquad (i, j) \in [1, 2]^2$$

Equation (12) is evaluated over each element that leads to a non linear system solved with a classical Newton method. The displacement field u^{i+1} calculated at iteration i+1 is the solution of the linear system:

$$D_u G(u, \eta)^i \Delta u^i = -G(u, \eta)^i \tag{16}$$

with:

$$D_u G[u, \eta]^i \Delta u^i = \sum_{\text{thread}} \int_{l_0} D_u [E^f_{11}]^i \Delta u^i \frac{\partial T_f^{11}}{\partial E^f_{11}} D_u [E^f_{11}]^i \eta\, ds_0$$
$$+ \sum_{\text{thread}} \int_{l_0} D_u [D_u [E^f_{11}]^i \eta]\, \Delta u^i [T_f^{11}]^i ds_0 \tag{17}$$
$$= \eta_s F^e_{ext} - \sum_{\text{thread}} \int_{l_0} [D_u [E^f_{11}]]^i \eta\, [T_f^{11}]^i \Delta u^i ds_0$$

The linearized derivative of strain E^f_{11} is related to the nodal values of the homogeneous displacement η

field from the linear ${}^{L}B$ and non-linear ${}^{NL}B(u)$ strain interpolation matrix:

$$D_u(E^f_{11})\,\eta = \begin{cases} \dfrac{1}{\|g^0_1\|^2}\,[{}^{L}B_{1s} + {}^{NL}B_{1s}]\,\eta_s & \text{Warp} \\ \dfrac{1}{\|g^0_2\|^2}\,[{}^{L}B_{2s} + {}^{NL}B_{2s}]\,\eta_s & \text{Weft} \end{cases} \qquad (18)$$

$$s \in (1, nxddl)$$

Taking into account of finite element interpolation (17) this expression can be written:

$$\underset{elt}{\mathrm{A}}\,[K^e + K^e_T]^i\,\Delta u^i = \underset{elt}{\mathrm{A}}\,[F^e_{ext} - (F^e_{int})^i] \qquad (19)$$

In the case of the linear triangle directed by the thread directions, Figure 15, the strains in warp or weft threads are constant. A discrete quadrature permits to reduce the sum on any thread of the element to a sum on two threads. If the threads are regularly distributed between $0 \le 1 - \xi_1 \le 1$ and $0 \le 1 - \xi_2 \le 1$:

$$\xi^k = \frac{2k-1}{n} \qquad \sum_{k=1}^{n}(1-\xi^k) = \frac{n}{2} \qquad (20)$$

the stiffness matrix, the geometrical stiffness matrix and interior load vector are given by the sum of two scalars. Accounting for this, the stiffness matrices components and internal nodal load components are given by the following explicit equations (Cherouat, Gelin, Boisse, Sabhi, 1995):

$$K^e = \frac{\partial T^{11}_f}{\partial E^f_{11}} B_{1s} B_{1r} \frac{n_1/2}{\|g^0_1\|^3} + \frac{\partial T^{11}_f}{\partial E^f_{11}} B_{2s} B_{2r} \frac{n_2/2}{\|g^0_2\|^3} \qquad (21)$$

$$K^e{}_T = \frac{T^{11}_f}{2}\frac{\partial N^k}{\partial \xi_1}\frac{\partial N^l}{\partial \xi_1}\frac{n_1}{\|g^0_1\|} + \frac{T^{11}_f}{2}\frac{\partial N^k}{\partial \xi_2}\frac{\partial N^l}{\partial \xi_2}\frac{n_2}{\|g^0_2\|} \qquad (22)$$

$$F^e{}_{int} = \frac{T^{11}_f}{2} B_{1r}\frac{n_1}{\|g^0_1\|} + \frac{T^{11}_f}{2} B_{2r}\frac{n_2}{\|g^0_2\|} \qquad (23)$$

In the case of the quadrangular four node element a discrete quadrature is used and the different matrices are obtained as the sum of four scalars.

5. EXPERIMENTS AND COMPUTATIONAL RESULTS

5.1 *Shaping process with hemispherical punch and die*

A set of shaping experiments has been carried out in order to validate the proposed computational procedure. A 75 mm radius hemispherical punch is used combined with a 78 mm hemispherical die. A blank-holder pressure equal 0.6 MPa restrains the fabric in order to avoid wrinkling, Figure 16.

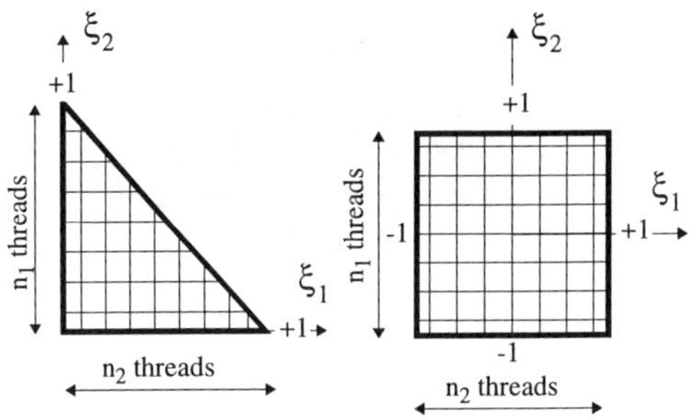

Figure 15 P1 and Q4 membrane elements

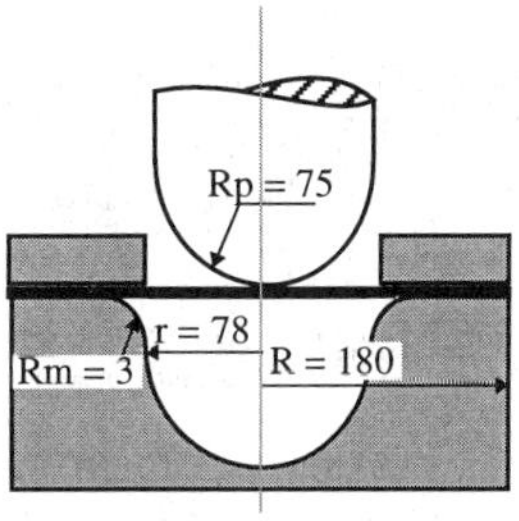

Figure 16 Geometry of the tools

The final computed and experimental shapes are shown in Figure 17 and Figure 18. Figure 17 corresponds to an initially square fabric with (0,90°) thread orientation, and Figure 18 corresponds to an initially square fabric with (-45°,+45°) thread orientation. Comparing Figure 17a and 17b and 18a and 18b reveals that the computed results are in very good agreement with the experimental results.
Furthermore it can be remarked that the final shape obtained with both thread orientations are very different revealing strong anisotropy effects. In figure 19 the computed and experimental values of the angle variations within the fabric along the line where these variations are the largest, i.e. 45° for (0,90°) orientation and 0° for (-45°,+45°) orientation, are shown to be in very good agreement.

5.2 *Shaping of a square cup with square punch and die*

A set of computations shows the possibility of tool optimization in order a glass fibers preform without defects (Cherouat, 1995). Figure 20a shows the geometry of a square cup shaping process, with an initial rectangular fabric part with (0,90°) thread orientation. Figure 20b give the final result in terms of angular distortion within the final part, the angular variations computed reach 80° near the corners of the square cup and are larger than the maximum value permitted by the fabric under consideration that appears to be experimentally inferior to 60°. A modification of the tool geometry is proposed in

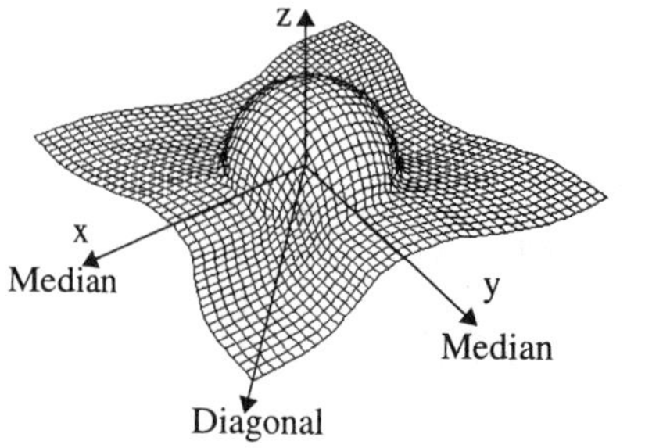

Simulation (a)

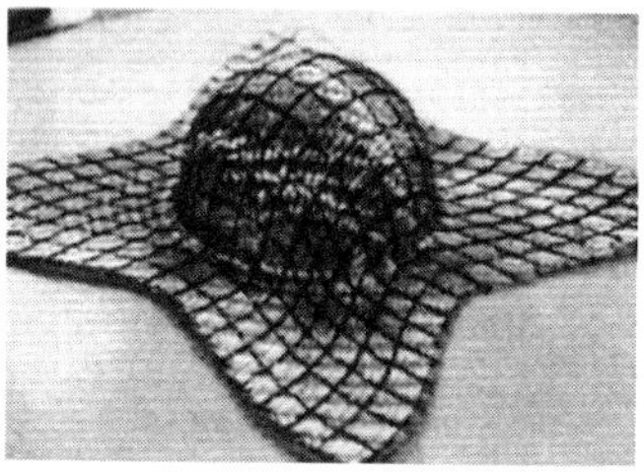

Experimental (b)

Figure 17 Experimental and computed shapes for the drawing with an hemispherical punch (0°,90°)

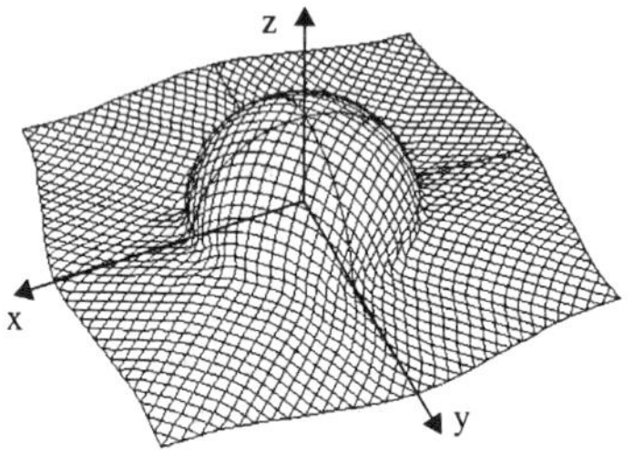

Simulation (a)

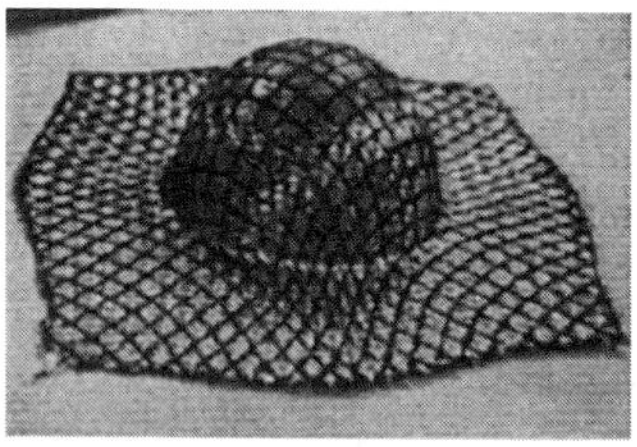

Experimental (b)

Figure 18 Experimental and computed shapes for the drawing with an hemispherical punch (45°)

Figure 21a and the computed result for angular variations is represented in Figure 21b. In that case the maximum angular distortion permitted is not reached and the shape can be used without problems.

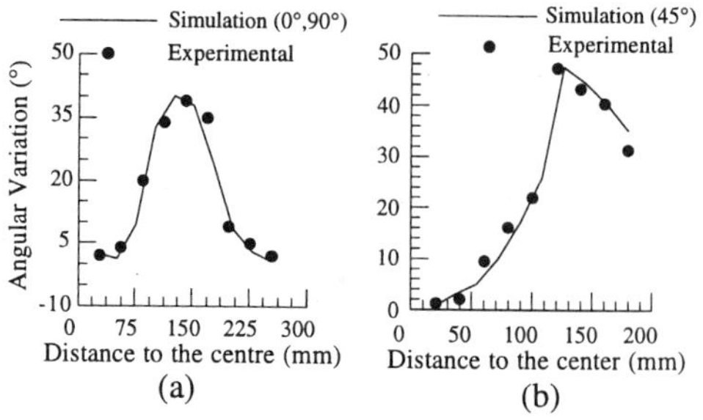

Figure 19 Comparison of shear angle along diagonal (a) and median (b) directions

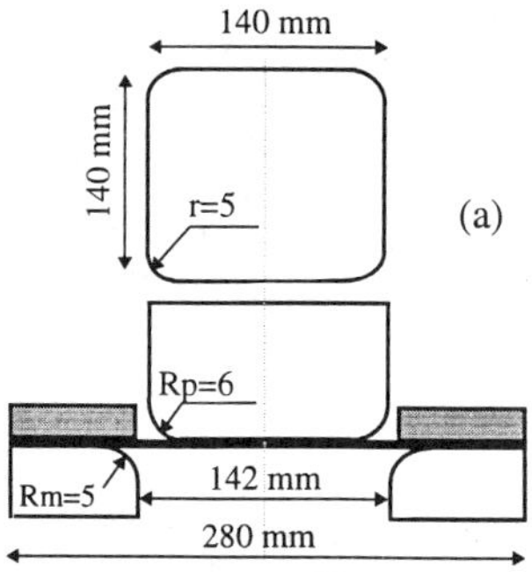

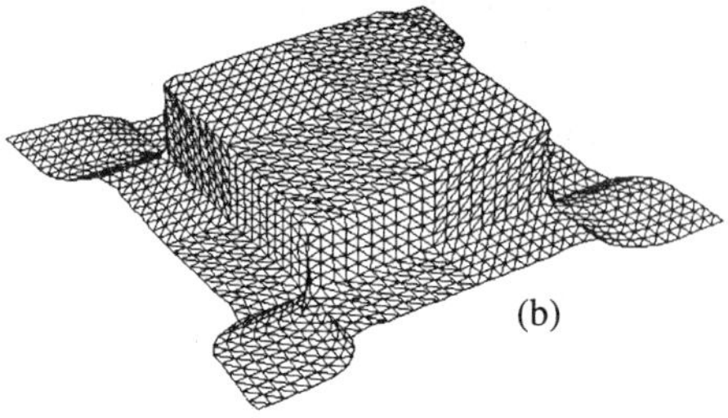

Figure 20 Deformed mesh (square cup shaping process)

5.3 *Shaping of an automotive component*

Components used in automotive industry are generally complex and Figure 22 illustrates a typical shape used in automotive industry. The punch and die are defined using Bezier patches with the indicated dimensions. The computed software developed take directly into account such tools definition and the final glass fiber shape obtained is illustrated in Figure 23. Figure 24 shows the angular variation contours and for the fabric under consideration there is no thread overlapping.

6. CONCLUSION

A general approach for shaping of glass fibers fabric is proposed. The proposed approach is based an a constitutive model for glass fiber deformation that includes the non-linear elastic behaviour of glass

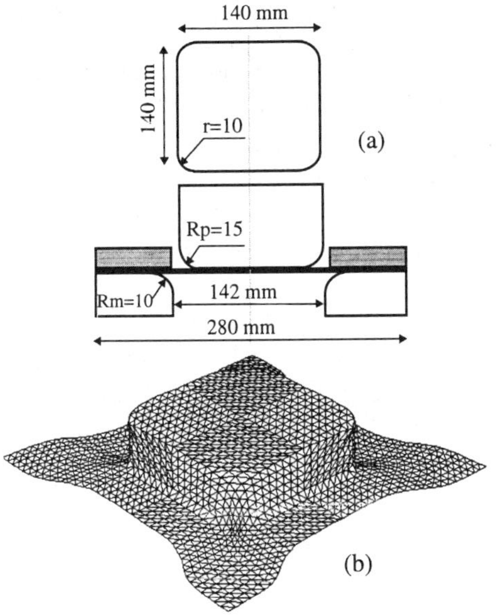

Figure 21 Deformed mesh (square cup with a modification of the tool geometry)

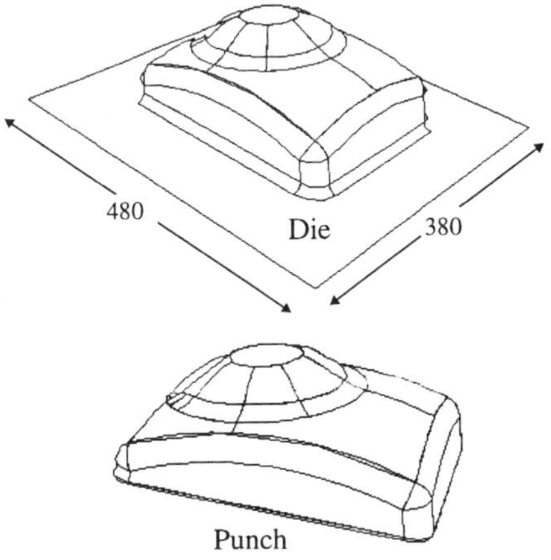

Figure 22 Geometry of the tools

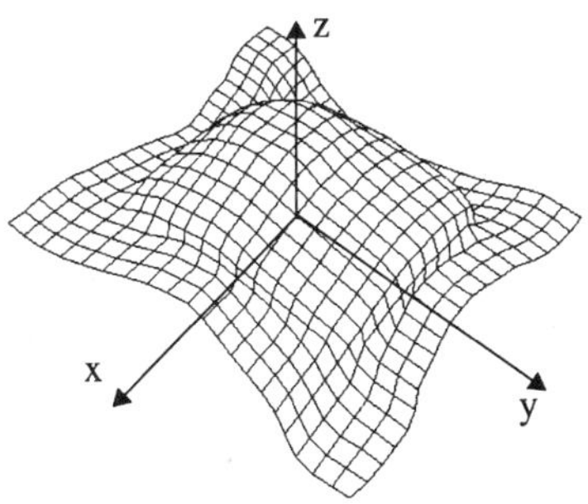

Figure 23 Deformed mesh

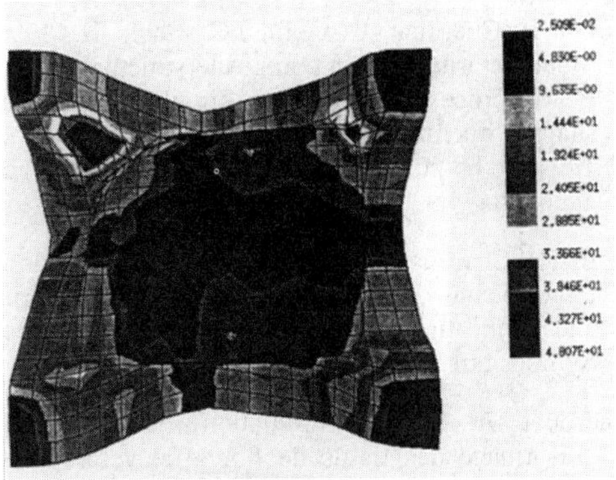

Figure 24 Shear angles between warp and weft threads (Degrees)

threads and takes into account the progressive damage. The constitutive fabric behaviour is obtained from the discrete summation of the behaviour of each threads, taking into account material strain component in the thread direction. It has been shown that the interaction effects due to the undulation of threads are not so important for the glass fabric generally used for RTM, but that point is under development by modelling the interaction effects with an inverse approach.

The finite element solution procedure is based on discrete membrane elements and leads in the case of meshes structured by the threads to a powerful computational tool with explicit expression of stiffness matrices and nodal loads.

Numerical examples presented show the agreement of the computed shapes and distortions with the experimental ones. The numerical results show also the possibility to optimize tool geometry to render feasible the glass preform, and the possibilities for the simulation of very complex shaping processes.

REFERENCES

Bergsma O.K., Huisman J., 1988, Deep drawing of fabric reinforced thermoplastics, Proceedings of the 2nd Int. Conf. on Computer Aided Design in Composite Material Technology., Ed. by CA Brebbia and all, Springer Verlag, 323-333.

Boisse P., Gelin J.C., Sabhi H., 1992, Forming of glass fiber fabrics into complex shapes: Experimental and computational aspects, Annals of the CIRP, Vol 41/1, 245-248.

Boisse P., Gelin J.C., Sabhi H., 1993, Experimental study and F.E. Simulation of glass fiber shaping processes, In Use of Plastics and Composites, ASME-MD 46, 125-147, To be published in Polymer Composites (February 1995).

Boisse P., Cherouat A., Gelin J.C., Sabhi H., 1994, Manufacturing of thin composite structures by the RTM process, numerical simulation of the shaping operation, JNC-9, St-Etienne, November 1994, To be published in Composites Science and Technology.

Borr M., Cherouat A., Gelin J.C., 1995, Modelling large biaxial deformations of glass fibers fabrics and application to shaping processes, Euromech Symposium 334, Lyon - May 1995, France.

Cherouat A., 1994, Simulation numérique du préformage des tissus de fibres de verre par la méthode des éléments finis, PhD Thesis University of Besançon, France.

Cherouat A., Gelin J.C., Boisse P., Sabhi H., 1995, Modélisation de l' emboutissage des tissus de fibres de verre par la méthode des éléments finis, To be published in the European Journal of Finite Elements, March 1995.

Gay D., 1993, Matériaux composites, Hermès, Paris.

Gelin J.C., Sabhi H., 1991, Constitutive equations for glass fiber networks and consequences on resin flow during processes, Seventh Annual Meeting of the Polymer Processing Society, PPS7, Ed by G Vlachapoulos, 119-121.

Gutowski T., Tam A., Dillon D., Stoller S., 1991, Forming aligned fiber Composites into complex shapes, Annals of the CIRP, Vol 40/1, 291-294.

Ishikawa T., Chou T., 1982, Elastic Behavior of Woven Hybrid Composites, Journal of Composites Materials, Vol 16, 2-19.

Ishikawa T, Chou T, 1983, Non Linear Behavior of Woven Fabric Composites, Journal of Composites Materials, Vol 17, 399-413.

Ishikawa T., Sushima M., Hayashi Y., Chou T., 1985, Experimental Confirmation of the Theory of Elastic of Fabric Composites, Journal of Composites Materials, Vol 19, 3443-458.

Kawabata S., Niwa M.and Kawai H., 1973, The finite deformation theory of plain weave fabrics, Part I: The biaxial-deformation Theory, Part II: The uniaxial-deformation Theory, Part III: The shear-deformation Theory, Journal of the Textile Institute, 64, 21-83.

Realf M.L., Boyce M C., Backer S., 1993, A micromechanical approach to modelling tensile behavior of woven fabrics, Use of Plastic and Plastic Composites: Material and Mechanics Issue, ASME New-York, MD-Vol 46.,285-293

Sabhi H., 1993, Etude expérimentale et modélisation mécanique et numérique du comportement des tissus de fibres de verre lors de leur préformage, PhD Thesis University of Besançon.

Van der Ween F., 1991, Algorithms for draping fabrics on doubly curved surfaces, Int. J. Num. Meth. Engng., 31, 1415-1426.

Simulation of Materials Processing: Theory, Methods and Applications, Shen & Dawson (eds)
© 1995 Balkema, Rotterdam. ISBN 90 5410 553 4

Computation of profile dies for thermoplastic polymers using anisotropic meshing

J. F. Gobeau, T. Coupez, B. Vergnes & J. F. Agassant
CEMEF, École Nationale Supérieure des Mines de Paris, URA CNRS, Sophia-Antipolis, France

ABSTRACT :
One of the main difficulties in modelling thermoplastic polymers flow in profile dies arises from the complex geometries, which lead to time and memory consuming computations if accurate velocity and pressure fields are expected. We propose a 3D finite element method including unstructured anisotropic meshes. It allows to make calculations in complex industrial die geometries with reasonable time and memory size using a free mesh generation.

1-INTRODUCTION

Die designing is a major challenge in profile extrusion of polymers. In particular, the flow at the die exit must be well balanced, meaning that all parts of the profile must exit the die with the same average velocity. Even a slight imbalance can cause the profile to bend. Generally, due to the complexity of the problem, dies are designed empirically : the correct shape is reached by trial and error, an expensive and time consuming approach.

Therefore, prediction of polymer flow through dies is of great interest. Numerous studies have been dedicated to that problem. Some of them are based on approximate 2D analytical calculations [1, 2, 3]. The basic idea is to divide the complex 3D flow into well-known 2D or axisymmetrical simple flows : flow in a thin slit, flow through a circular tube, converging flow in a dihedron or in a cone. Interconnection between these simple flows can be taken into account, which seems to provide good prediction of velocities at the exit of simple dies [3]. Other publications are based on 2D finite element computations. They consist in calculating the velocity field in complex cross-sections [4], supposing that, within the cross section, pressure gradient is negligible. 3D calculations using finite element method [5, 6, 7] are often restricted to simple die geometries : square dies, hole key dies,...

In fact, the main obstacle in modelling the polymer flow comes from the complex die geometry: generally, the polymer is distributed among several flow paths of different widths and lengths, each path feeding one part of the profile. Transversal flows occur between these paths, which are obviously interconnected. Hence, for a correct calculation, it is necessary to account for a three dimensional flow. Moreover, the polymer undergoes a severe contraction along the die : the thicknesses of the different parts of the profile at the exit are very low compared to the entry section and to the die length. The rate of reduction is about one per cent. A refined mesh must be chosen in order to get enough nodes at the die exit. With a regular mesh, the memory size and the C.P.U. time required to make the calculation in a complex die geometry would be prohibitive.

A 3D finite element computation using tetrahedric $P1^{+}/P1$ elements [8] and an iterative scheme [9] are used. An unstructured anisotropic meshing method of the profile die is proposed. It consists in choosing three different characteristic mesh sizes in three main directions of space. In the case of profile extrusion, the aim is to mesh the thin channels with elements stretched along the principal directions of the flow. This leads to reduce drastically the number of unknowns, and is proven to be very efficient.

2-THREE-DIMENSIONAL FINITE ELEMENT MODELLING

Two different geometries were studied : in the first one, the die evolves regularly from a circular section to the desired profile section (figure 1). In the second one, the polymer flows around a cylindrical mandrel and then undergoes a radial flow before reaching the specified shape (figure 2).

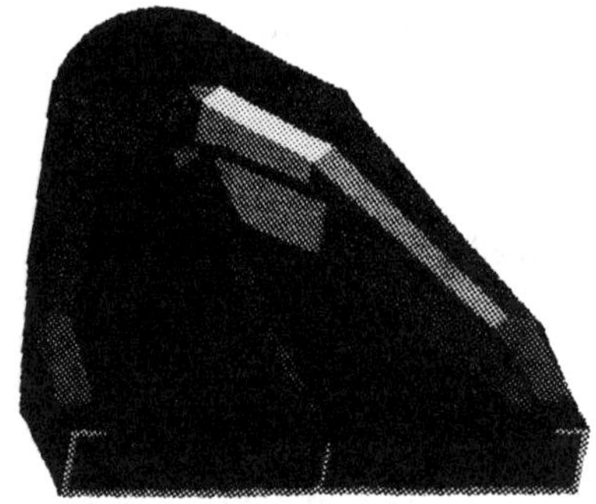

figure 1 : streamlined die

figure 2 : plate die

Steady-state isothermal finite element calculations were carried out in both geometries for a power-law fluid. A 3D finite element solver for incompressible materials was used. It is based on a mixed velocity/pressure formulation and uses the tetrahedric linear P1+/P1 element (mini-element) [8], which satisfies the Brezzi-Babuska compatibility condition.

The problem consists in solving the well known non-linear Stokes equations, where gravity and inertia forces are neglected :

$$\mathrm{div}\left(K\dot{\gamma}^{m-1}\dot{\varepsilon}(v)\right) - \nabla p = 0$$

$$\mathrm{div}(v) = 0$$

with : $\dot{\varepsilon}(v)$: strain rate tensor

$\dot{\gamma} = \sqrt{2\,\dot{\varepsilon}:\dot{\varepsilon}}$: equivalent strain rate

K : consistency of the molten polymer

m : shear-thinning power-law index

The boundary conditions are the following :

- sticking contact at the die walls.
- imposed pressures at entry and exit sections.
- no transversal velocities at entry and exit sections.

The corresponding variationnal formulation is as follows :

Find $(v,p) \in VxP$:

$$\begin{cases} \int_\Omega 2\eta\dot{\varepsilon}(v):\dot{\varepsilon}(v^*)\,d\Omega - \int_\Omega p\,\nabla.v^*\,d\Omega = 0 \\ \int_\Omega p^*\nabla.v\,d\Omega = 0 \end{cases}$$

$$\forall\ (v^*, p^*) \in VxP$$

(where generally V and P stand respectively for $\left(H_0^1(\Omega)\right)^3$ and $L^2(\Omega)$)

Then, the classical Galerkin finite element method leads to the resolution of non-linear equations, which are solved by the Newton-Raphson method.

At each step of the Newton-Raphson iterations, we must solve a linear system of the following kind :

$$Ax = b \qquad (1)$$

A is symmetric but not definite positive. In fact, the incompressibility leads to a saddle point which can be only solved iteratively by a Conjugate Residual Method [9]. This technic is used here with a simple diagonal preconditioning. Indeed, due to the static condensation of the P1+/P1 bubble, diagonal terms appear on the line corresponding to the pressure.

The computation cost is governed by the matrix vector products Aq (q is the unknown vector at one iteration step). Therefore, an element by element storage method [10] is used, which consists in calculating each element matrix contribution A_e, of size 16x16 (three velocity components and the pressure at each node of the tetrahedron). Then, the product Aq is obtained by :

$$Aq = \left(\sum_{e=1}^{NbElt} \widehat{A_e}\right) q = \sum_{e=1}^{NbElt} \left(\widehat{A_e q_e}\right)$$

with :

A_e element matrix contribution

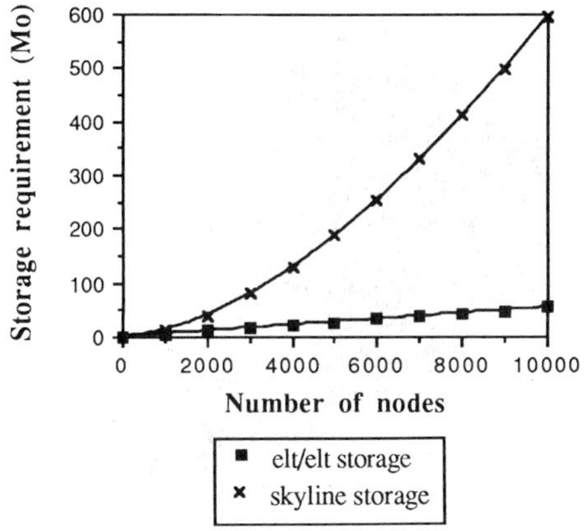

figure 3 : theoretical storage requirements with respect of the number of nodes.

$\widehat{A_e}$	element matrix contribution expanded to the full global system size.
NbElt	number of elements

The advantage of the method lies in the linear storage requirement : it is estimated to 680 N for element by element storage instead of 16 $N^{5/3}$ for the direct Choleski method with a skyline storage, N being the total number of nodes (figure 3).

3-MESHING OF THE PROFILE DIES

3-1 The existing automatic mesh generator

The meshing of our profile die is based on the automatic mesh generator developped in FORGE3© [11]. The existing version is able to generate isotropic tetrahedra in a volumic domain from a given triangular mesh of its boundary. The result is a tetrahedric mesh, which fits exactly the initial boundary mesh. It must be pointed out that it is possible to modify the surfacic mesh by adding or removing nodes, according to several chosen control parameters such as homogeneity or surfacic quality.

The originality of the mesh generator of FORGE3© has already been developped in details in previous publications [11, 12]. The basic idea is to separate the mesh topology, defined as a set of tetrahedra verifying specific conditions, and the geometry of the domain to be meshed. Then, different algorithms carry out topological changes including node creation, so that the final mesh topology fits the specified geometry.

Furthermore, while changing the current topology, the mesh generator aims at optimizing both average and minimal quality of tetrahedra, a tetrahedron being considered as perfect when it is equilateral. Therefore, the mesh generator needs a quantification of the quality of a tetrahedron. It uses the following criterion Cr :

$$C_r = \frac{(\text{volume})^2}{(\text{area})^3}$$

Cr is maximal for an equilateral tetrahedron.

3-2 Generation of anisotropic meshes

3-2-1 Change of metric

As explained in the introduction, one of our goals is to mesh the thin paths of the die with anisotropic tetrahedra stretched along the main directions of the flow, in order to improve the accuracy of the computed velocity field.

Different methods exist in the literature in order to obtain such anisotropic meshes [13, 14]. In this paper, a method based on non-euclidian metrics [13] is presented. This technic has already been used in 2D geometries within the frame of the Delaunay/Voronoï method [13, 15]. It is introduced here for 3D extrusion flow within the frame of the mesh-topology improvement method [12].

A metric can be defined as a 3x3 symmetric definite positive matrix M, the identity matrix corresponding to the euclidian case. The length of a vector u in the metric M is calculated as follows :

$$N_M(u) = ({}^t u M u)^{1/2}$$

The diagonalization of M provides a geometrical interpretation of the change of metric. In fact, λ_i being the eigenvalue associated with the eigenvector v_i, the unitary length l_i along v_i calculated in the euclidian geometry is given by :

$$l_i = \frac{1}{\sqrt{\lambda_i}}$$

Hence, the unity-circle in the non-euclidian metric M can be considered in the classical euclidian geometry as an ellipsoid (E), whose principal axes (Δ_i) are parallel to the eigenvectors v_i (figure 4). The intersection of (E) with (Δ_i) occurs respectively at the lengths l_i, i = 1, 3.

In the case of profile extrusion, we will typically choose a very high unitary length along the main directions of the flow compared to unitary length in the thickness.

In order to generate anisotropic meshes, the algorithms of the existing mesh generator are not modified, but lengths, areas and volumes are no more calculated in the euclidian metric but in the non-euclidian metric. In particular, the quality criterion of the tetrahedron is calculated as follows :

$$C_r = \frac{(V_M)^2}{(S_M)^3}$$

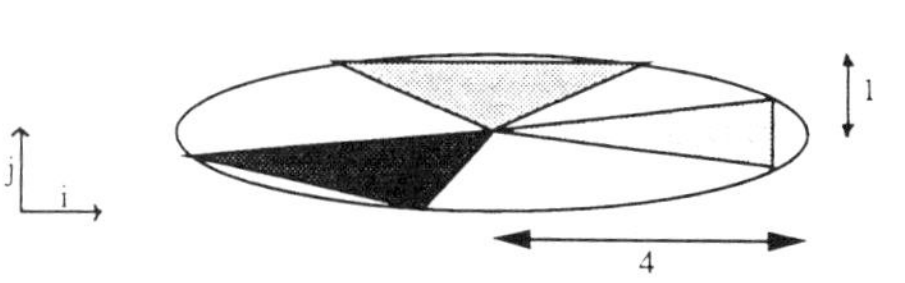

figure 4 : unitary-circle and unitary equilateral triangles in the metric $M = \begin{pmatrix} 1/16 & 0 \\ 0 & 1 \end{pmatrix}$

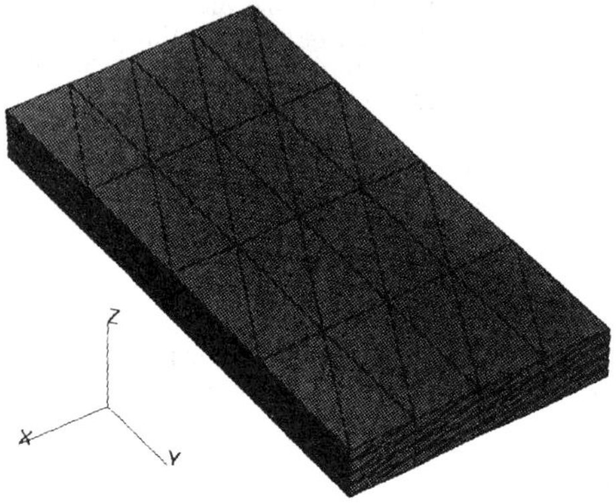

figure 5 : parallelepipedic box

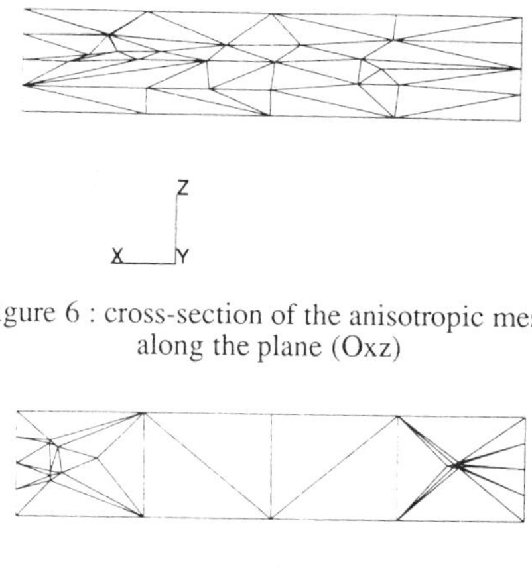

figure 6 : cross-section of the anisotropic mesh along the plane (Oxz)

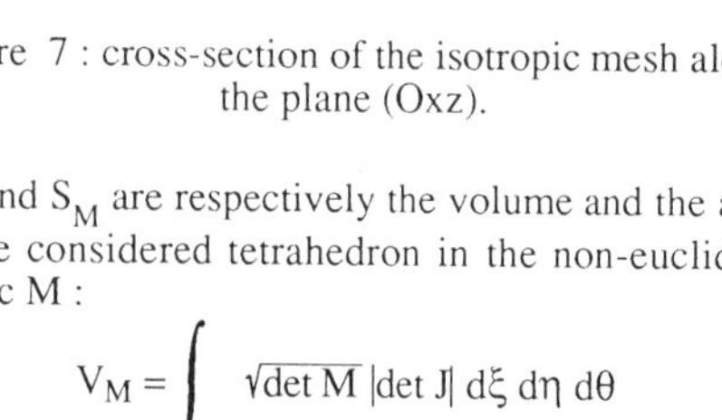

figure 7 : cross-section of the isotropic mesh along the plane (Oxz).

V_M and S_M are respectively the volume and the area of the considered tetrahedron in the non-euclidian metric M :

$$V_M = \int_{\Omega_{ref}} \sqrt{\det M}\, |\det J|\, d\xi\, d\eta\, d\theta$$

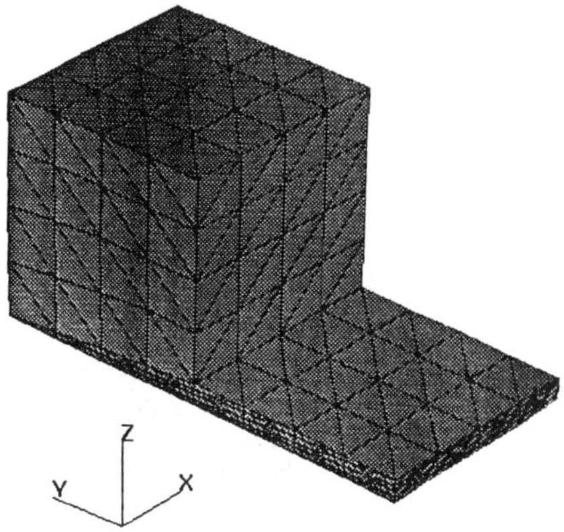

figure 8 : simple contraction die

$$S_M^2 = \frac{1}{4}\int_{T_{ref}} \left[N_M^2\left(\frac{\partial P}{\partial \xi}\right) N_M^2\left(\frac{\partial P}{\partial \eta}\right) - \left(\frac{\partial P}{\partial \xi},\frac{\partial P}{\partial \eta}\right)_M^2 \right] d\xi\, d\eta$$

J is the jacobian of the geometric transformation from the reference tetrahedron to the real one. P is a point of the reference tetrahedron.

Since the initial topological algorithms are not modified, the mesh generator still aims at improving the average quality of the tetrahedra, that is to say generating new tetrahedra as equilateral as possible in the sense of the specified metric. In an euclidian point of view, we are able to get in this way highly stretched elements.

An example is given in figure 5, which represents a parallelepipedic box of 10 x 20 x 2. The anisotropic mesh (figure 6) was generated in order to get at least four elements in the thickness, as well as in width and length. The following metric M was used :

$$M = \begin{pmatrix} 4 & 0 & 0 \\ 0 & 0.04 & 0 \\ 0 & 0 & 0.04 \end{pmatrix}$$

Nodes have been inserted in the volume insuring that there is no edge linking two opposite walls of the box. For comparison purpose, an isotropic mesh of that box was generated from the same triangular surfacic mesh (figure 7). It provides better shaped elements but poor discretization in the thickness.

3-2-2 Anisotropic meshes of profile dies

The geometry of profile extrusion dies evolves drastically from the entry section to the exit section (figures 1 and 2). The reduction of section along the flow depends strongly on the part of the profile considered. Therefore, it is not possible to impose an uniform metric in the whole die because a reasonable choice of the metric in a localized

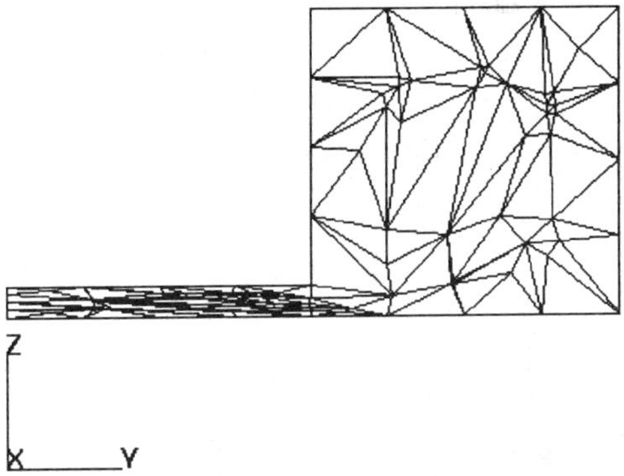

figure 9 : cross-section of the anisotropic mesh of the simple contraction die

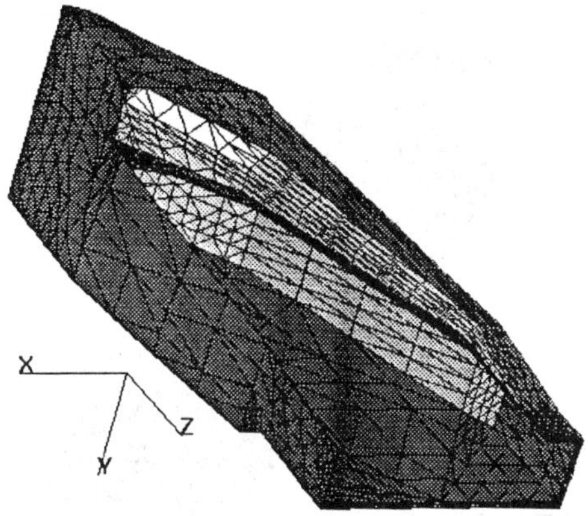

figure 11 : anisotropic mesh of the streamlined die

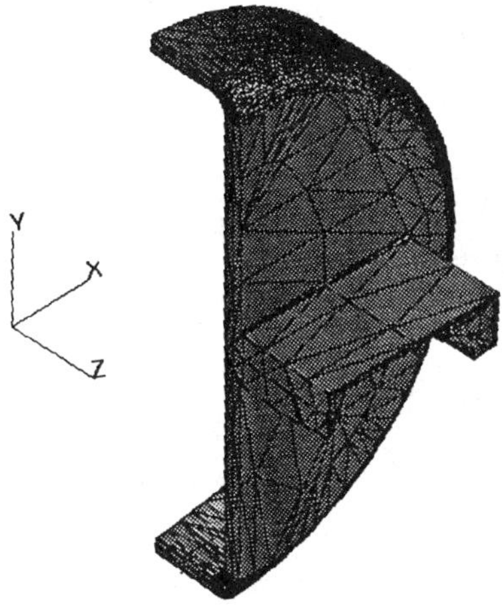

figure 10 : anisotropic mesh of the plate die

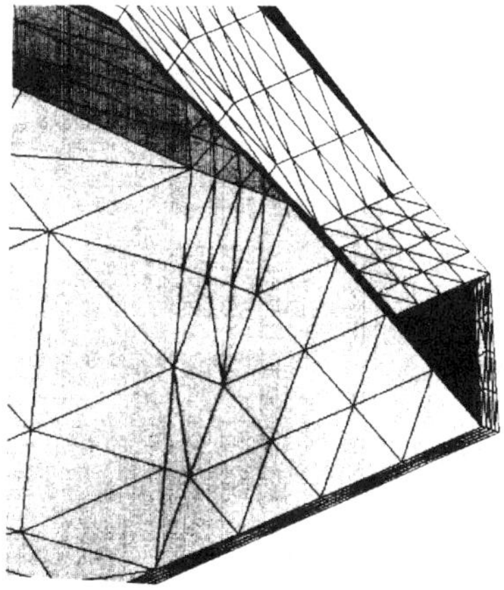

figure 12 : anisotropic mesh of the streamlined die (zoom on die exit)

domain might be not convenient at all in another one.

In order to illustrate this problem, we have considered a very simple die made of two juxtaposed boxes : a cube of 10 x 10 x 10 and a parallelepiped of 10 x 10 x 2 (figure 8). The aim is to get an anisotropic mesh of the thin parallelepiped: we keep the same metric as previously. On the other hand, large isotropic elements are expected in the cube : we choose there an euclidian metric.

A difficulty appears when one element overlaps the two boxes of different metrics : the problem is to choose which metric to use for the calculation of the criterion. Therefore, the volumes and areas of each element are calculated in non-euclidian metrics with a Gauss method using five integration points $(\overset{v}{\xi_k},\overset{v}{\eta_k},\overset{v}{\theta_k})$ for the volumes and four integration points $(\overset{s}{\xi_k},\overset{s}{\eta_k})$ for the surfaces. $\overset{v}{w_k}$ and $\overset{s}{w_k}$ are the corresponding weighting factors. The formulae are :

$$S_M = \sum_{k=1}^{4} \frac{1}{2}\left(f(\overset{s}{\xi_k},\overset{s}{\eta_k})\right)^{1/2} \overset{s}{w_k}$$

with :

$$f(\overset{s}{\xi_k},\overset{s}{\eta_k}) = N_M^2\left(\frac{\partial P}{\partial \xi}(\overset{s}{\xi_k},\overset{s}{\eta_k})\right) N_M^2\left(\frac{\partial P}{\partial \eta}(\overset{s}{\xi_k},\overset{s}{\eta_k})\right) - \left(\frac{\partial P}{\partial \xi}(\overset{s}{\xi_k},\overset{s}{\eta_k}),\frac{\partial P}{\partial \eta}(\overset{s}{\xi_k},\overset{s}{\eta_k})\right)_M^2$$

$$V_M = \sum_{k=1}^{5} \sqrt{\det M(\overset{v}{\xi_k},\overset{v}{\eta_k},\overset{v}{\theta_k})}\ |\det J|\ \overset{v}{w_k}$$

In the case of one element overlapping the two boxes, the Gauss integration points localized in the thin parallelepipeds will be associated with the anisotropic metric and the other points with the euclidian metric. The mesh obtained is given on figure 9.

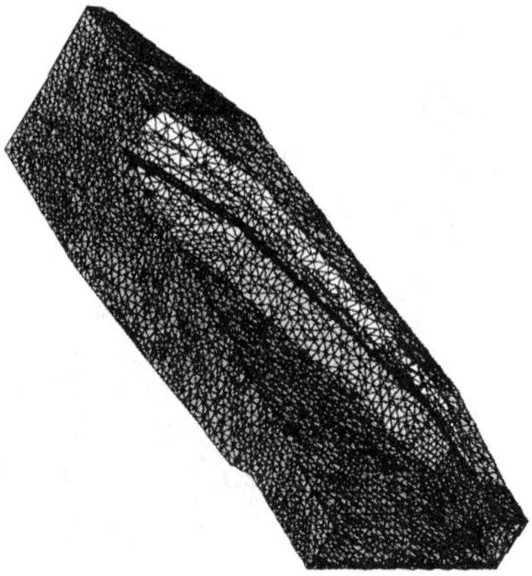

figure 13 : regular surfacic mesh of the streamlined die

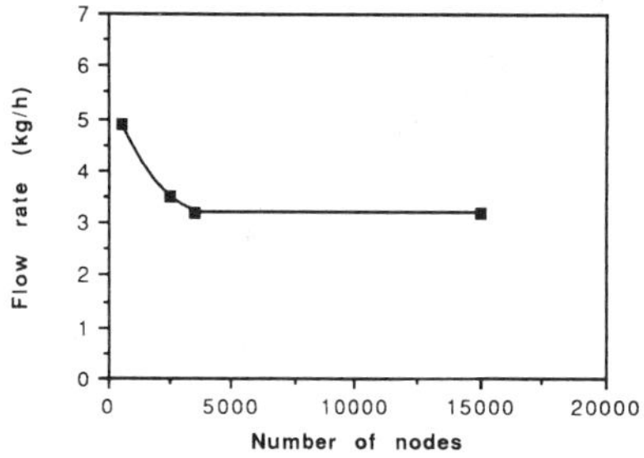

figure 14 : evolution of the estimated flow rate with the mesh size

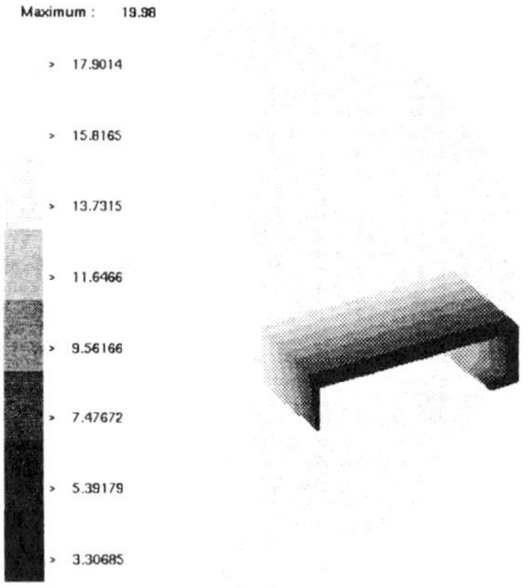

figure 15 : pressure in the plate die

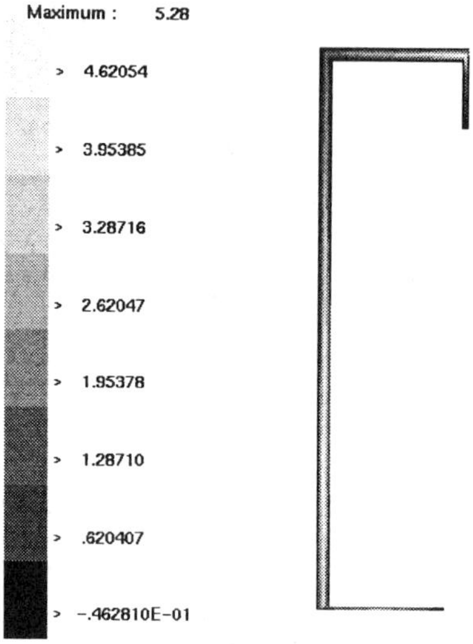

figure 16 : velocity field at the exit section of the plate die

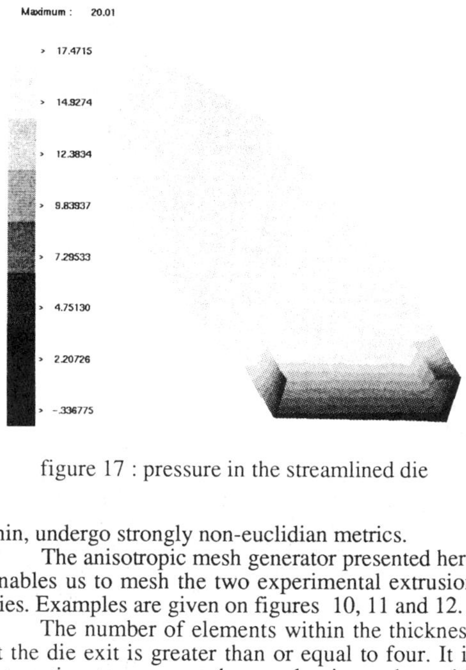

figure 17 : pressure in the streamlined die

Typically, a real extrusion die will be cut into three or four juxtaposed boxes along the flow : the nearest boxes from the entry section are associated with quasi-euclidian metrics because the shape-factor is quasi-isotropic. On the contrary, the boxes at the end of the die, where the cross section is very thin, undergo strongly non-euclidian metrics.

The anisotropic mesh generator presented here enables us to mesh the two experimental extrusion dies. Examples are given on figures 10, 11 and 12.

The number of elements within the thickness at the die exit is greater than or equal to four. It is interesting to compare these volumic meshes with the surfacic regular mesh of the streamlined die on figure 13 : it has been obtained by calculating a mesh size in order to get at least two nodes in the thickness at the exit. It clearly demonstrates the interest of our anisotropic mesh.

4-NUMERICAL CALCULATIONS

For each die, several anisotropic meshes were

generated with a growing number of nodes.
In order to study the influence of the mesh size, velocity and pressure fields were calculated with the same specified entry pressure for each mesh. The influence of the number of nodes on the numerical results is illustrated on figure 14 in the case of the streamlined die and a Newtonian viscosity : the flow rate is represented as a function of the number of nodes. It can be seen that the flow rate is stabilized over 5000 nodes.
Some finite element results are presented on figures 15, 16, 17 and 18.

These results show that the pressure gradient is maximal at the end of the die. As expected, the pressure decreases linearly before the exit section, where the polymer flows through an uniform thin cross section. The pressure there is quite uniform in a cross section, which justifies the assumption commonly used in the literature [3].

As for the velocity field, the calculation shows an imbalance between the different parts constituting the profile. The location of the fastest and slowest parts fits well with experiments carried out on both extrusion dies.

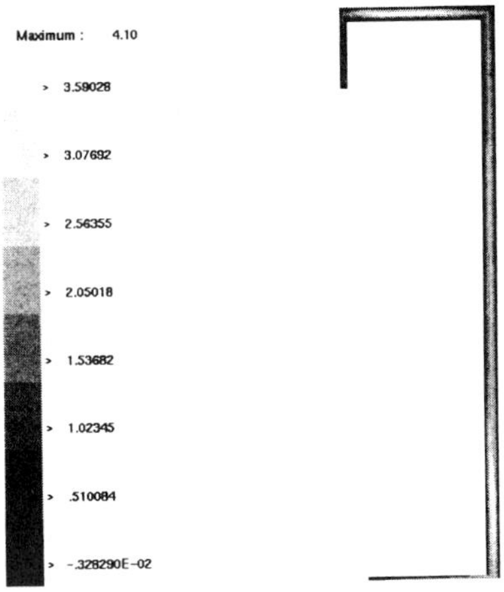

figure 18 : velocity field at the exit section of the streamlined die

In the case of large meshes, it is interesting to point out the storage requirement of the element by element storage compared to the skyline storage. It is presented in the following table for a 14611 nodes case (72248 finite elements, maximum width band of 1568 and average width band of 327):

skyline storage	elt/elt storage
583 Mo	75 Mo

It demonstrates the interest of the element by element storage.

5-CONCLUSION

A 3D finite element method for the computation of profile dies is proposed in this paper. The aim is to provide a helpfull tool for the prediction of the velocity field in and at the exit of the die, in order to reduce the cost and the delay for die design.

The originality of the study is based on an anisotropic meshing of the die obtained with a non-euclidian metric. In this way, the number of nodes is drastically reduced while preserving a good accuracy of the predicted velocities and pressures.

Moreover, the Element by Element Conjugate Minimal Residual Method used in the finite element calculations contributes to the reduction of the computational cost in terms of CPU time and storage requirements.

6-REFERENCES

[1] SUMMERS J.W., BROWN R.J., "Practical Pinciples of Die Design - A Simplified Procedure, in Table Form for Rigid PVC", *J. of Vinyl Technol.*, vol. 3, no. 4, p. 215-218, (1981)

[2] JARZEBSKI A.B., WILKINSON W.L., "Non-Isothermal Developing Flow of a Power-Law Fluid in a Converging Slit", *J. of Non-Newton. Fluid Mech.*, vol. 12, p. 1-11, (1983)

[3] HUNEAULT M.A., LAFLEUR P.G., CARREAU P.J., "Design of Profile Extrusion dies : a Systematic Approach", *SPE ANTEC Proc.*, vol. 38, p.431-435, (1992)

[4] HUREZ P., TANGUY P., BLOUIN D., "A Finite Element Based Design of Profile Extrusion Dies", *SPE ANTEC Proc.*, vol. 37, p.1028-1031, (1991)

[5] D'HALEWHYN, BOUBE M.F., VERGNES B., AGASSANT J.F., "Extrusion of Elastomers in Profile Dies : 3D Computations and Experiments", in P. Moldenaers and R.Keunings , Eds, *Theoritical and Applied Rheology*, Elsevier, Amsterdam, vol. 1, p.336-338, (1992)

[6] MARCHAL J.M., LEGAT V., "Extrusion and Co-Extrusion in Three-Dimensional Geometries", *SPE ANTEC Proc.*, vol. 37, p.819-821, (1991)

[7] SCHWENZER C., MENGES G., "Megapus - A Program Package for the Layout of Extrusion Lines", *SPE ANTEC Proc.*, p.109-111, (1987)

[8] ARNOLD D.N., BREZZI F., FORTIN M., "Stable Finite Element for Stokes Equations", *Calcolo*, 21, p.337-344, (1984)

[9] WATHEN W., SILVESTER D., "Fast Iterative Solution of Stabilized Stokes Systems, PART I : Using Simple Diagonal Preconditioners", *SIAM J. Numer. Anal.*, vol. 30, no. 3, p 630-649, (1993)

[10] CAREY G.F., "Parallelism in Finite Element Modelling", *APNUM 2*, vol. 3, p. 281-288, (1986)

[11] COUPEZ T., CHENOT J.L., " Mesh Topology for Mesh Generation Problems - Application to three-dimensional remeshing", in *NUMIFORM 1992*, Chenot, Wood and Zienkiewicz, Eds., p. 237-242, (1992)

[12] COUPEZ T., in J.Hauser, N.P. Weatherhill, P.R. Eiseman and J.F. Thompson, Eds., *Numerical Grid Generation in Computational Fluid Dynamics and Related Fields,* Pineridge Press, (1994)

[13] GEORGE P.L., HECHT F., VALLET M.G., "Creation of Internal Points in Voronoi's Type Method - Control Adaptation", *Adv. Eng. Software,* vol. 13, no 5/6 combined, p. 303-312, (1991)

[14] SIMPSON R.B., "Anisotropic Mesh Transformations and Optimal Error Control", *Applied Numerical Mathematics*, vol. 14, p. 183-198, (1994)

[15] MAGNIN B., "Modélisation du remplissage des moules d'injection pour polymères thermoplastiques par une méthode eulérienne-lagrangienne arbitraire", Ecole Nationale Supérieure des Mines de Paris, *PhD Thesis*, (1994)

Simulation of Materials Processing: Theory, Methods and Applications, Shen & Dawson (eds)
 ISBN 90 5410 553 4

Aspect of Lagrangian-Eulerian formulation

J. Huétink, R. Akkerman, H.G. Mooi & G. Rekers
University of Twente, Department of Mechanical Engineering, Enschede, Netherlands

ABSTRACT: In Euler-Lagrange formulations, commonly denoted as "Arbitrary" Lagrangian-Eulerian (ALE) formulations, the mesh and material deformations are uncoupled. These formulations can handle path-dependent material behaviour and free surfaces while keeping the mesh regular. The convective part of a time-stepping Euler-Lagrange finite element formulation can be formulated in various ways. A finite element interpolation is compared to a finite volume formulation. The first algorithm is based on an averaging procedure used in postprocessing of finite element calculations. The second algorithm describes the fluxes through the element sides and stems from a finite difference method for compressible fluid dynamics. Both approaches have complementary characteristics with respect to accuracy and implementation. The method can handle typically transient processes, but also stationary processes can be simulated as a limiting case of a transient simulation. Some applications to problems in forming processes are shown.

1. INTRODUCTION

In the simulation of forming processes large material deformations, history dependent material behaviour and contact phenomena play an important role. Lagrangian formulations are well suited for problems concerning path-dependent material properties and free surfaces but suffer from numerical problems when the mesh is distorted heavily. Eulerian formulations are able to cope with large material deformations but are less suited for the description of history dependent material and free surfaces in transient problems. In "Arbitrary" Lagrangian-Eulerian (ALE) formulations, the mesh and material deformations are uncoupled. These formulations can handle path-dependent material behaviour and free surfaces while keeping the mesh regular.

The method can handle typically transient processes, such as upsetting of a metal billet (Fig. 1) as well as stationary processes, such as polymer extrusion (Fig. 2)(Akerman 1993). The first example has mainly Lagrangian characteristics but keeps the mesh regular, the second example has mainly Eulerian characteristics but follows the free surface movements.

A number of Euler-Lagrange formulations for the simulation of forming processes are reported in literature. The methods can be divided into two groups, respectively coupled and split formulations. In the first formulation the coupled Euler-Lagrange equations are solved, see for instance the work of Liu et al (1988). The tangent stiffness matrix consists of an ordinary Lagrangian and an additional Eulerian part.

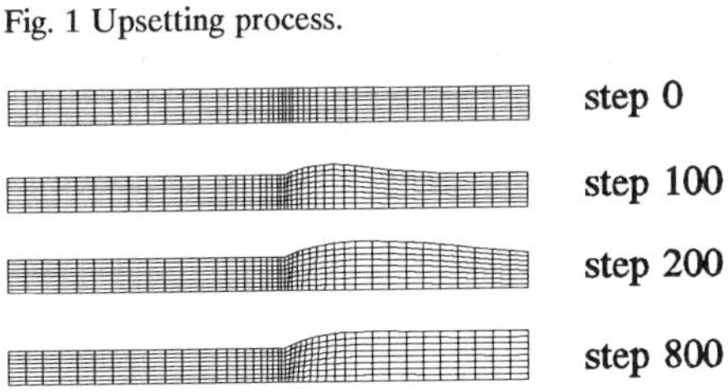

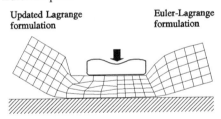

Fig. 1 Upsetting process.

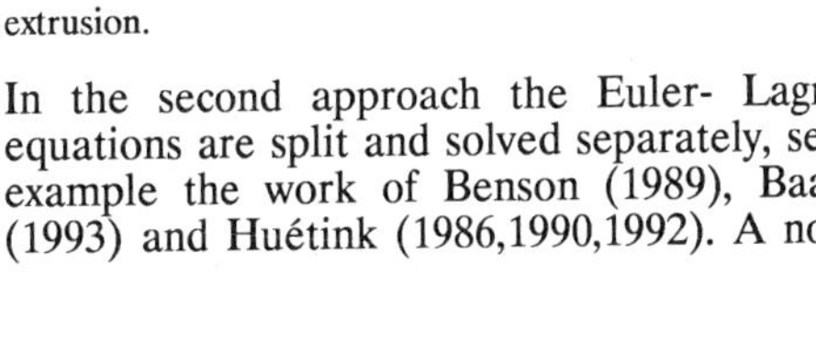

Fig. 2 Transient simulation of steady state polymer extrusion.

In the second approach the Euler- Lagrange equations are split and solved separately, see for example the work of Benson (1989), Baaijens (1993) and Huétink (1986,1990,1992). A normal

Lagrangian step is performed, followed by an explicit (purely convective) Eulerian step. In this work the split Euler-Lagrange formulation is used.

2. ARBITRARY LAGRANGIAN EULERIAN METHOD (ALE)

As the name indicates, in the Arbitrary Lagrangian-Eulerian (ALE) method the user can make his own choice whether a grid point is completely Lagrangian or Eulerian or arbitrary. This means that the material displacements are disconnected form the grid point displacements. So, also convective changes have to be calculated and both the material rate of change $\dot{f}^m$ of a magnitude f, as well as the rate of change in a grid point $\dot{f}^g$ have to be examined. These rates read, respectively:

$$\dot{f}^m = \left(\frac{df}{dt}\right)^m = \frac{\partial f}{\partial t} + \dot{x}^m \cdot \vec{\nabla} f \qquad (1)$$

$$\dot{f}^g = \left(\frac{df}{dt}\right)^g = \frac{\partial f}{\partial t} + \dot{x}^g \cdot \vec{\nabla} f \qquad (2)$$

where $\dot{x}^m$ and $\dot{x}^g$ represent the material and grid velocity, respectively. The magnitude f may represent any material associated quantity such as strains, stresses and temperatures. Subtracting eq.(2) from eq.(1), and integrating the result in time from t_0 to $t_0 + \Delta t$ a time incremental procedure is obtained:

$$f(x^g + \Delta x^g, t + \Delta t) = f(x^g, t) + \Delta f^m + (\Delta x^g - \Delta x^m) \cdot \vec{\nabla} f \qquad (3)$$

This can be seen as a Taylor expansion of:

$$f(x^g + \Delta x^g, t + \Delta t) = f(x^g + \Delta x^g - \Delta x^m, t) + \Delta f^m \qquad (4)$$

Mark that if the material displacement is equal to the grid displacement ($\Delta x^g = \Delta x^m$) the updated Lagrange method is obtained. Equation (3) and equation (4) coincide in that case. If the grid displacement equals zero ($\Delta x^g = 0$) the Eulerian method is obtained. Note that the first term at the right hand side of eq. (4) is unknown if the grid movement does not coincide with the material movement. Hence the main problem in addition to a Lagrangian formulation is to obtain an approximation of the value in an upstream point which does not coincide with a grid point. Consequently as a natural extension of a Lagrangian formulation a split ALE formulation is used.

In the split algorithm the (Lagrangian) material increment (Δf^m) is firstly calculated and added.

The asterisk indicates a fictive intermediate

$$f^*(x^g, t^*) = f(x^g, t) + \Delta f^m \qquad (5)$$

state and is temporary stored as an element (integration point) value.

In the second part convection is added by a map of the intermediate state to the new grid location. This is accomplished by a weighed sum (interpolation) of the nearest surrounding known grid point values

$$f^*(x^g + \Delta x^g, t^* + \Delta t) = f^*(x^g + \Delta x^g - \Delta x^m, t^*) = \sum_K W_K(x^g, \Delta x^g, \Delta x^m) f_K^*(x_K^g, t^*) \qquad (6)$$

Formally followed by replacing

$$f(x^g + \Delta x^g, t + \Delta t) = f^*(x^g + \Delta x^g, t^* + \Delta t)$$

N.B. The sequence of the split may be reversed.

If in an updated Lagrange finite element calculation the displacement increments are interpolated with C_0 continuity, then a C_{-1} continuity of stresses and strains is obtained. However, if the calculation is not Lagrangian, a C_{-1} continuity is not sufficient. The stresses and strains generally are discontinuous over the element boundaries and the gradient at element level is not reliable. Furthermore, if linear elements are used the stress and strain field calculated by means of this displacement field is constant on element level. In this case the gradient completely vanishes. Hence, equation (3) or equation (6) cannot directly be applied on stress and strain calculations within an element. A solution for these problems is the construction of an additional C_0 continuous approximation field for f by averaging the nodal point values of f of adjacent elements (Huétink 1986, 1992). If these nodal average values are calculated according to the smoothing procedure of Hinton (1974) it can be shown to be numerical diffusive (Huétink 1992). Numerical diffusion stabilizes the solution but replacing the initial C_{-1} continuous approximation by the smoothed approximation result into high overdiffusion.

In the next section we concentrate on the convective part of the step, which we call the Euler step of the split algorithm. Two different Euler algorithms are described: an artificial dissipation and a limited flux scheme.

3. CONVECTION/EULER FORMULATION

3.1 General remarks

A graphical interpretation of the Euler formulation is found by considering a mesh which is a result of a previous Lagrangian step and a neighbouring mesh where the simulation has to be continued. The integration point values (stress, strain, etc.) of the Lagrangian mesh have to be remapped to the new mesh, see figure 3.

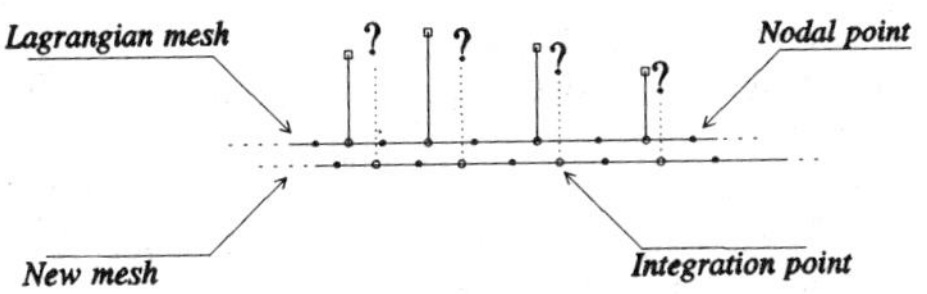

Fig. 3 Remapping integration point values.

(The map represented by the right hand side of eq. (6).)
The remapping is established by constructing a function based on the integration point data of the original Lagrangian mesh. This approximation is used to evaluate the integration point values of the new mesh.
The mapping procedure can be regarded as a discretized form of the advection equation.

$$\frac{\partial f}{\partial t} + V \cdot \frac{\partial f}{\partial x} = 0 \tag{7}$$

The relative velocity between the material and mesh is given by $V = \dot{x}^m - \dot{x}^g = V^m - V^g$. The coordinate x moves locally with the grid $x = x^g + V^g(t - t_0)$. The remapping can be established by solving the advection problem. In literature various schemes can be found which deal with this problem.

The Euler formulation has to satisfy some requirements (Benson 1989). Firstly the approximation should be accurate: error analysis can be performed to determine the order of accuracy. Secondly, the estimation should be stable. This requirement is important when discontinuous functions, which frequently occur in the numerical simulation of plasticity, have to be remapped. Furthermore, the Euler step should be consistent: when the Lagrangian mesh and the Arbitrary mesh coincide the integration point values of both meshes should be the same. Subsequently, the remapping should be efficient since each independent variable has to be remapped in every iteration or increment.

3.2 One-dimensional grids

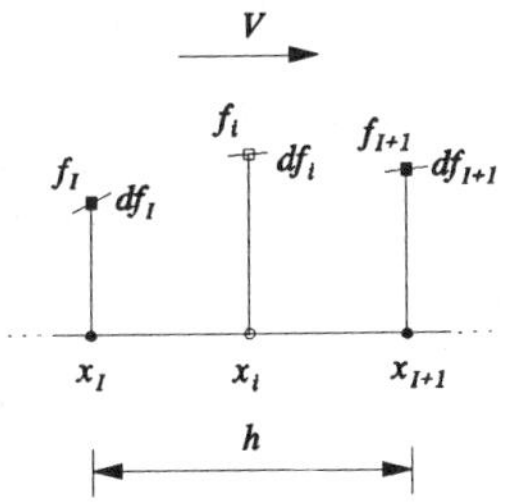

Fig. 4 1D staggered grid

The convective schemes will be illustrated using a regular, one-dimensional, staggered grid with one integration point and two nodes per element, see figure 4. Integration and nodal point values carry respectively lowercase and uppercase indices i and I. The values of the Lagrangian mesh at time t and the arbitrary mesh at time $t+\Delta t$ are respectively denoted by subscript and superscript indices. The relative velocity V is assumed to be positive for the time being. The extension of the following derivations to negative values of V is straightforward.

Both the finite element interpolation and the finite volume formulation are described as an adaptation of the Lax-Wendroff finite difference scheme:

$$\begin{aligned} f^i &= f_i - C(f_i - f_{i-1}) \\ &\quad - \frac{1}{2}C(1-C)(f_{i+1} - 2f_i + f_{i-1}) \end{aligned} \tag{8}$$

or, equivalently:

$$\begin{aligned} f^i &= \frac{1}{4}(1+2C)f_{i-1} + \frac{1}{2}f_i + \frac{1}{4}(1-2C)f_{i+1} \\ &\quad - \frac{1}{4}(1-2C^2)(f_{i+1} - 2f_i + f_{i-1}) \end{aligned} \tag{9}$$

where C is the Courant number $V\Delta t/h$ which is a measure for the relative displacement between the material and mesh.

The Lax-Wendroff scheme is second order accurate in space and time but exhibits spurious oscillations around discontinuities of the solution. To remedy these instabilities artificial dissipation can be added to the Lax-Wendroff scheme or the antidiffusive flux in the Lax-Wendroff scheme can be limited. Hence, one has to compromise between the accuracy and stability requirements. The Lax-Wendroff scheme is consistent since $C=0$ results in $f^i = f_i$.

Finite element interpolation scheme 1D
The starting point for the one-dimensional finite element scheme is equation (9). The term $2C^2$ is replaced by an artificial dissipation variable α:

$$\begin{aligned} f^i &= \frac{1}{4}(1+2C)f_{i-1} + \frac{1}{2}f_i + \frac{1}{4}(1-2C)f_{i+1} \\ &\quad - \frac{1}{4}(1-\alpha)(f_{i+1} - 2f_i + f_{i-1}) \end{aligned} \tag{10}$$

The variable α results in artificial dissipation when $\alpha > 2C^2$ and should vanish for zero Courant number C to satisfy the consistency requirement mentioned above. A suitable function for the artificial dissipation variable is $\alpha = A|C|$ where A is an artificial dissipation parameter which is normally set to $A=1.5$. ($A=2$ results into an up wind scheme which is to much overdiffussive)
The finite element form of equation (10) is given by:

$$f^i = \sum_{K=I}^{I+1} N_K(-C)f_K + (1-\alpha)(f_i - \sum_{K=I}^{I+1} N_K(0)f_K) \tag{11}$$

Thus, the new integration point value consists of

the locally smoothed function as proposed by Hinton (1974) (for postprocessing purposes) with an addition as proposed by Van Leer (1979). Note that all the information necessary for the update of the integration point value is kept within the element.

Finite volume scheme 1D
For this one-dimensional limited flux scheme we start from equation (8). The antidiffusive flux is restricted by introducing a limiter function ϕ in the second order term:

$$f^i = f_i - C(f_i - f_{i-1}) - \frac{1}{2}C(1-C) * \left(\phi(r_{i+\frac{1}{2}})(f_{i+1}-f_i) - \phi(r_{i-\frac{1}{2}})(f_i - f_{i-1})\right) \tag{12}$$

The ratios of successive gradients are given by:

$$r_{i+\frac{1}{2}} = \frac{f_i - f_{i-1}}{f_{i+1} - f_i}, \quad r_{i-\frac{1}{2}} = \frac{f_{i-1} - f_{i-2}}{f_i - f_{i-1}} \tag{13}$$

The limiter function can be constructed using the concept of Total Variation Diminishing (TVD) schemes, resulting in a solution without wiggles. A number of limiter functions have been proposed, see for instance the work of Van Leer (1979), Sweby (1984) and Hirsch (1988). Here, the Van Leer limiter function is used:

$$\phi(r) = \frac{r + |r|}{1 + |r|} \tag{14}$$

Equation (12) can be expressed in terms of integration point values, averaged nodal values and gradients of the considered element:

$$f^i = f_i - 2C(f_i - f_I) + - C(1-C)\left(\phi(r_{i+\frac{1}{2}})(f_{I+1} - f_i) - \phi(r_{i-\frac{1}{2}})(f_i - f_I)\right) \tag{15}$$

where the ratios of successive gradients are given by:

$$r_{i+\frac{1}{2}} = \frac{f_i - f_I}{f_{I+1} - f_i}, \quad r_{i-\frac{1}{2}} = \frac{f_I - f_i - df_i + 2df_I}{f_i - f_I} \tag{16}$$

with the element gradient and averaged nodal gradient respectively:

$$df_i = f_{I+1} - f_I, \quad df_I = \frac{1}{2}(df_{i-1} + df_i) \tag{17}$$

N.B. Uppercase subscribts denote the usual averaged nodal values.

Note that all the data necessary for the calculation of the new integration point value are available within the element when also the gradients are stored at the nodal points and the integration points.

3.3. Two-dimensional grids
We now proceed to regular, two-dimensional, staggered grids with one integration point and four nodes per element. The relative velocity vector $\boldsymbol{V}$ is assumed to have positive components. The extension of the following to an arbitrary direction of $\boldsymbol{V}$ is straightforward. For the artificial dissipation (based on finite element interpolation) and limited flux (based on finite volumes) schemes, a different location of the averaged nodes is chosen, see figure 5.

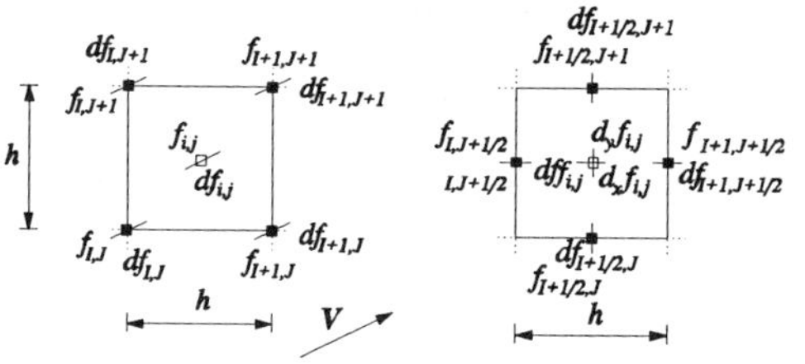

Fig. 5 2D grids for finite element (left) and finite volume scheme (right)

Finite element interpolation 2D
In multiple dimensions again the FE-discretization can be applied, giving the extension of equation (9):

$$f^{i,j} = \sum_{K=I}^{K=I+1} \sum_{L=J}^{L=J+1} N_{K,L}(-C_x, -C_y) f_{K,L} + +(1-\alpha)\left(f_{i,j} - \sum_{K=I}^{K=I+1} \sum_{L=J}^{L=J+1} N_{K,L}(0,0) f_{K,L}\right) \tag{18}$$

with Courant numbers $C_x = V_x\ \Delta t / \mathrm{h}$ and $C_y = V_y\ \Delta t / \mathrm{h}$.

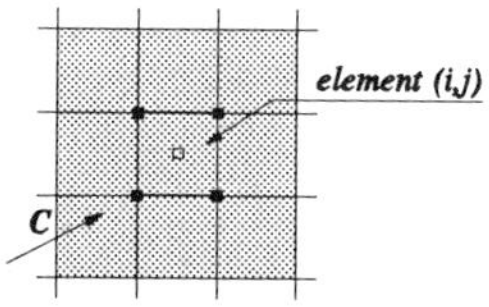

Fig. 6 Corner coupling

A graphical representation of the 2D artificial dissipation scheme (18) is given in figure 6. The shaded elements contribute indirectly by means of the averaged corner nodes to the update of the integration point value of element *(i,j)*.

The artificial dissipation variable for the 2D case is chosen $\alpha = A|C|$ similar to the 1D situation: the one parameter artificial dissipation scheme. The discretization error introduced by equation (18) is (Van der Helm 1992):
The accuracy of the artificial dissipation scheme (18) depends on the flow direction. When $|C_x| = |C_y|$ and the one parameter artificial dissipation scheme is adapted to $\alpha = 2C^2$ with

$$r^{i,j} = \tfrac{1}{4}h^2(\alpha - 2C_x^2)\left(\frac{\partial^2 f}{\partial x^2}\right)_{i,j} + \tfrac{1}{4}h^2(\alpha - 2C_y^2)\left(\frac{\partial^2 f}{\partial y^2}\right)_{i,j} + O(h^3) \qquad (19)$$

$C = \max(|C_x|,|C_y|)$, both terms on the right hand side of equation (19) vanish, resulting in a second order accurate scheme. For flow in non diagonal directions a term of order h^2 remains unequal to zero on the right hand side, leading to crosswind diffusion.

The function for the artificial dissipation variable can be refined by taking into account the curvature of the field f in the flow direction. Similar procedures have been reported in literature (Brooks et.al. 1982, Donea 1983). Suppose that a dimensionless measure for the curvature $\kappa \epsilon [0,1]$ in element (i,j) is available: κ is small when the curvature is small and equal to one when the curvature is large. The artificial dissipation variable can then be chosen as $\alpha = \kappa A|C|$: the two parameter artificial dissipation scheme. A possible measure for the curvature is given by:

$$\kappa = \begin{cases} K, & K \le 1 \\ 1, & K > 1 \end{cases} \qquad K = B\frac{|d_0|}{\min(|d_1|,|d_2|)} \qquad (20)$$

The nominator and denominator parameters are:

$$d_0 = \sum_{K=I}^{K=I+1}\sum_{L=J}^{L=J+1} \frac{N_{K,L}(-C_x,-C_y) - N_{K,L}(0,0)}{|C|} df_{K,L} \approx h^2\left(\frac{\partial^2 f}{\partial n^2}\right)_{i,j}$$

$$d_1 = \sum_{K=I}^{K=I+1}\sum_{L=J}^{L=J+1} \frac{N_{K,L}(-C_x,-C_y) - N_{K,L}(0,0)}{|C|} f_{K,L} \approx h\left(\frac{\partial f}{\partial n}\right)_{i,j}$$

$$d_2 = f_{i,j} \qquad (21)$$

where $\boldsymbol{n} = \boldsymbol{V}/|\boldsymbol{V}|$ is the unit vector in flow direction and the first denominator term d_1 is equal to the element gradient $df_{i,j}$. The derivative with respect to $\boldsymbol{n}$ represents the component of the gradient in the flow direction: $\partial f/\partial \boldsymbol{n} = \nabla f \cdot \boldsymbol{n}$. The gradients in the nodal points are determined with the usual averaging procedure

Appropriate values for the dissipation parameters A and B depend to a certain degree on the type of problem and can be found by performing numerical experiments on suitable test cases, see the next section. A proper choice for the parameter A appears to be $A=1.5$. The choice for B depends on the curvature of the field. For small curvatures $B=0.05$ is sufficient, for larger curvatures $B=0.5$ will do the job. Note that for large values of B the two parameter model reduces to the one parameter model since in that case κ equals unity.

Finite volume scheme 2D

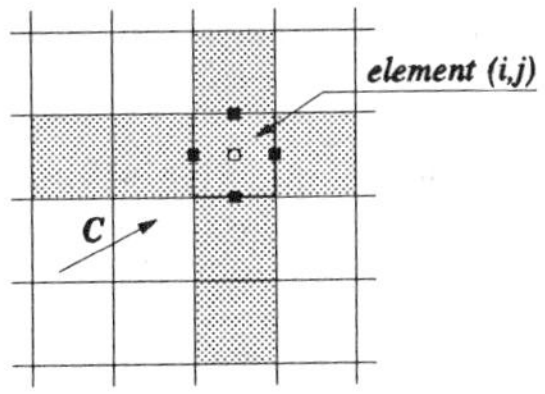

Fig. 7 Side coupling of the 2D limited flux scheme

The multi-dimensional finite volume scheme is found by addition of the fluxes through the element sides in both coordinate directions. This is a straightforward generalization of equation (15).

The scheme is represented in figure 7. The shaded elements add indirectly via the averaged mid-side nodes to the update of the integration point values of element (i,j). Note the side coupling of this element to the adjacent elements.

The accuracy of the limited flux scheme varies with the flow direction. When assuming that $|C_x| \gg |C_y|$, the scheme reduces to a set of 1D limited flux schemes (9) with nearly second order accuracy. The accuracy decreases when material is advected between elements that have only one corner node in common, for example when $|C_x| = |C_y|$.

When comparing the different Euler formulations it is clear that the one parameter artificial dissipation scheme is the most simple scheme and the cheapest with respect to the number of operations and data storage per element. The two parameter artificial dissipation scheme and the limited flux scheme are more complicated and require more data manipulation and storage.

At the end of each time step the averaged nodal values are determined. In the next step these nodal values can be elaborated to element gradients, which in turn can be elaborated to nodal averaged gradients. It will be clear that these nodal averaged gradients lag one extra step behind. For the artificial dissipation scheme this appears to give no problems, since the nodal gradients have only an indirect influence through the curvature parameter. For the limited flux scheme, however, the gradients have a direct influence on the accuracy and have to be updated in an extra loop at the end of each time step. A more extensive description of the numerical procedure is given in (Van der Helm 1995).

4. APPLICATIONS

Linear quadrilateral, constant dilatation elements for plane strain and axisymmetric analysis have been used here. The artificial dissipation scheme is evaluated at the four integration point locations, the limited flux scheme at the element center. The midpoint values are found by averaging the four integration point values.

The behaviour of the two convective schemes in different flow directions is illustrated by a pure advection problem on a course grid.

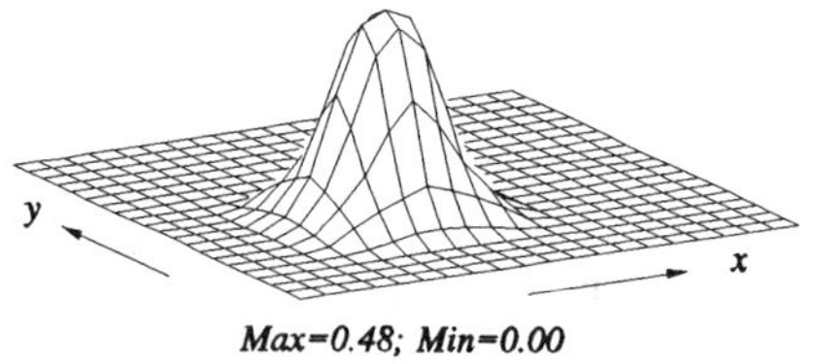

Fig. 8 Initial plastic strain distribution.

Consider a patch of *20×20 mm* rigid-ideally plastic material. The patch is modelled using *20×20* constant dilatation plane strain elements. The patch is prestrained, the initial plastic strain distribution is shown in figure 8.

The plastic strain pulse is advected by translating the mesh in the (ΔU,ΔV) direction and remapping the integration point values to the initial mesh. Two advection cases are considered: in the first case the displacement increments are $\Delta U = 0.5$ *mm* and $\Delta V = 0.0$ *mm* (advection parallel to mesh), in the second case $\Delta U = 0.5$ *mm* and $\Delta V = 0.5$ *mm* (advection diagonal to mesh). The total number of increments is *10*. Two Euler formulations are used: an adaptation of the one parameter artificial dissipation scheme with $\alpha = 2C^2$ where $C = \max(|C_x|, |C_y|)$ and the limited flux scheme. The results of the artificial dissipation and limited flux schemes for parallel and diagonal advection are shown below.It is clear that the accuracy of the Euler formulation depends on the flow direction. For this particular testcase the limited flux scheme gives the best results when the advection is parallel to the mesh, the artificial dissipation scheme gives the best

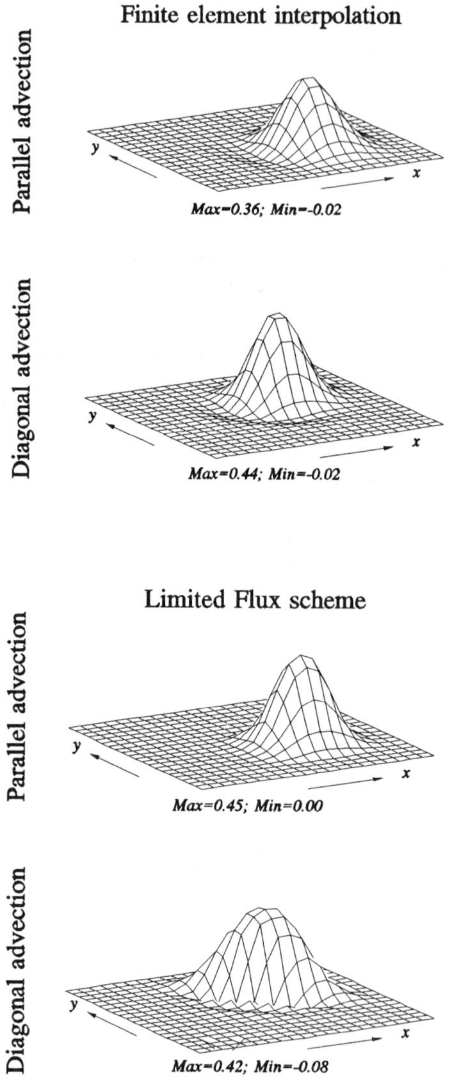

Fig. 9 Resulting plastic strain distributions.

outcome when the advection is diagonal to the mesh.

These results agree well with the discussion on the accuracy of the artificial dissipation and limited flux schemes in section 3. Note that the limited flux scheme is monotone when the advection is parallel to the mesh.

An application of the method is a wire drawing problem, with interaction between the Cauchy stress tensor and the plastic strain. When the constraining forces of the die are removed in the outlet zone, the elastoplastic behaviour of the material will lead to a redistribution of the internal stresses. All plots are based on corner node averaged values.

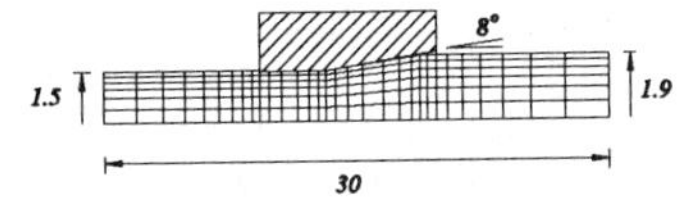

Fig. 10 Wire drawing mesh.

Consider a wire drawn from the right to the left through a die. One half of the wire is modelled using *6×31* constant dilatation axisymmetric elements. The elastic material behaviour of the wire is defined by the Young's modulus $E=200000\ N/mm^2$ and Poisson's ratio $\nu=0.3$. The initial yield stress is $\sigma_{y0}=250\ N/mm^2$. The plastic material behaviour is represented by a Von Mises model. Hardening is described according to the relation between the equivalent plastic strain ϵ_p and yield stress σ_y is given by a Voce equation:

$$\sigma_y = \sigma_{y0} + D\left(1 - e^{-\frac{\epsilon_p}{\epsilon_{p0}}}\right) \tag{22}$$

where $D=200\ N/mm^2$ and $\epsilon_{p0}=0.3$.

The wire is drawn in *320* increments to the total displacement $U=15\ mm$. The solution is continuously remapped to the initial mesh. The resulting solution is stationary. Three Euler-Lagrange simulations are carried out using respectively the one parameter artificial dissipation scheme with $A=1.5$, the two parameter artificial dissipation scheme with $A=1.5$, $B=0.05$ and the limited flux scheme. The resulting plastic strains and axial stresses are depicted below.

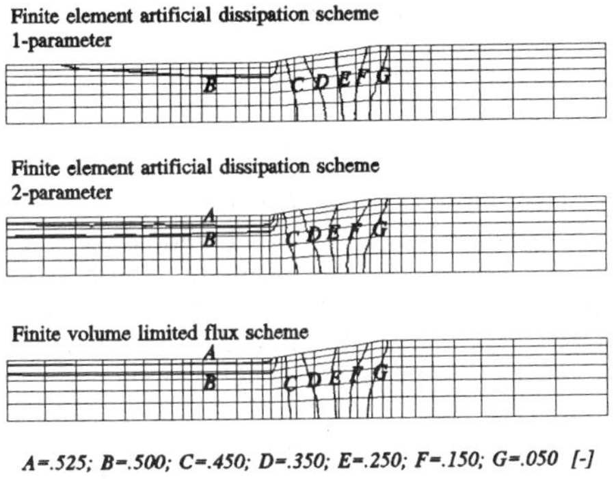

Fig. 11 Resulting plastic strain distributions

From these figures it is clear that all formulations give acceptable results. Differences between the three schemes are observed in the outlet zone left from the die. In that region the isolines of the equivalent plastic strain diverge for the one parameter artificial dissipation

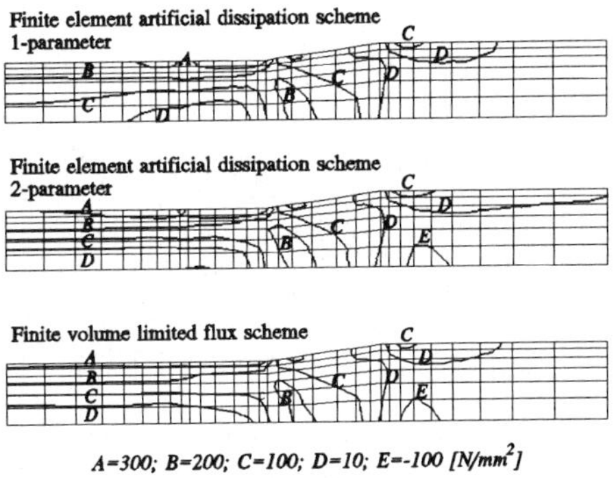

Fig. 12 Resulting axial stress distributions.

model due to crosswind diffusion.

For the two parameter model some small wiggles in the isolines are noticed. The limited flux scheme displays nearly no crosswind diffusion in the outlet zone and is stable.

Particle tracing

Another feature of the numerical procedure is tracing of the material movement. This is accomplished by treating the material displacement in a similar way as the stresses and strains. Notice that the algorithm is formulated for an arbitrary material associated quantity *f*. Hence it may also be applied to the material displacement. By subtracting the material displacement from the current position, a trace back of the material is found. A graphical presentation of the particle movement is given by isolines of (one of) the initial coordinates. The material tracing feature is shown for two examples in which large shear occurs due to sticking friction at the tool interface. The first example concerns direct extrusion of hollow aluminum profiles. These are extruded by means of a hollow (port hole) die: the inner surface is formed by a core which is suspended by three or more legs in the die. So, the aluminium flow is split by legs, and after the legs the flow has to weld together again (in so-called welding chambers, see Fig. 13). More details about the process are given in Mooi (1995).

A plane strain FEM approximation of the flow around a connection (leg) between the inner

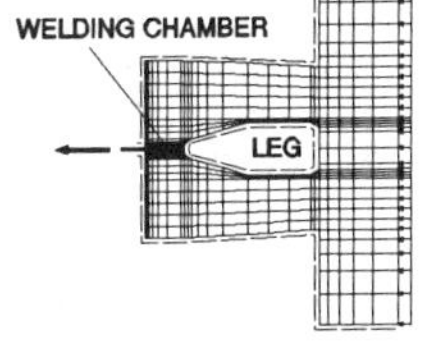

Fig. 13 Plane flow area

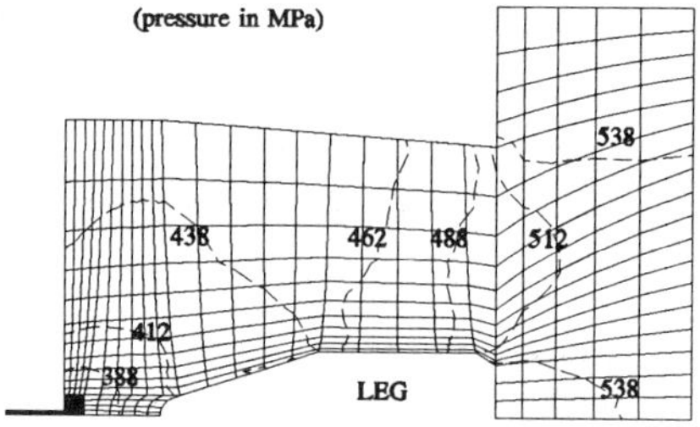

Fig. 14 Mesh with calculated pressure distribution

and outer part of a port hole die is used. The dotted lines represent the area with so-called contact elements, modelling the contact and Coulomb friction. Displacements were prescribed in the inlet.

The material particle tracing is shown by Fig. 15 for two different stages of the transient flow calculation. The numbers indicate the initial coordinate as shown on top of the figures.

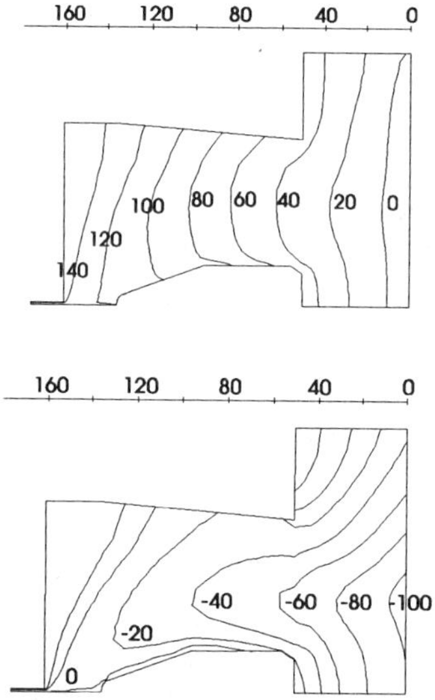

Fig. 15 particle tracing in aluminium extrusion

Around the leg a very high shear deformation can be observed. In this area the elements are sufficiently small to trace the high shearing. At the outer boundary (top of the Figs.) this effect is not found due to the large elements.

The next application is a finite element analysis is used to calculate the isothermal inertialess flow of a viscoelastic fluid around a sphere falling in a cylindrical tube as depicted in Fig. 16 .

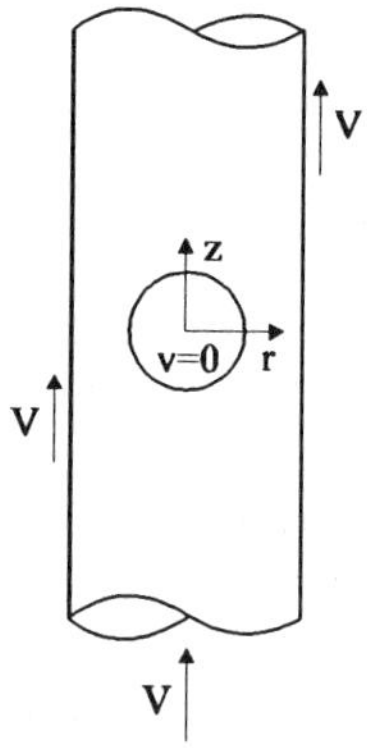

Fig. 16 Rigid sphere falling through a fluid in a tube, as seen from the reference frame of the sphere.

The flow is assumed to be axisymmetric and the reference frame is assumed to be moving with the sphere. The radius of the cylinder is denoted with R_c,the radius of the sphere is denoted with R_s and the velocity of the cylinder is denoted with V. The ratio of R_c and R_s is denoted with $X \equiv R_c/R_s$.

Furthermore, the calculations are performed using the Upper Convected Maxwell model (UCM) as is done by Lunsmann et al, (1993) and Crochet et al. (1992). To incorporate this nonlinear viscoelastic model in our finite element formulation use is made of the elastic Finger strain as described by Rekers et al., (1995). Using the Cauchy stress this model can be written as:

$$\boldsymbol{\sigma} = -p\boldsymbol{1} + \boldsymbol{\tau}, \tag{23}$$

$$\overset{\nabla}{\boldsymbol{\tau}} + \frac{1}{\theta}\boldsymbol{\tau} = 2\mu \boldsymbol{D}, \qquad \theta = \frac{\eta}{\mu}, \tag{24}$$

where μ is the Hookean shear modulus, θ is the relaxation time and η is the viscosity.

The simulations were carried out for X=2 and for a wide range of Weissenberg numbers, being defined as

$$We = \frac{\theta V}{R_s} \tag{25}$$

The mesh used for the calculations is depicted in Fig. 17 . The particle lines presented in Fig. 18 and Fig (25) are calculated for a Weissenberg number We=0.6 (V=1mm/s, R_s=1mm, θ=0.6s). Examining this figure, what strikes one most is the high elongation of the particle lines downstream from the sphere near the centerline. In Fig. 18 some irregularities are observed at the front surface of the sphere. This can be attributed to the small boundary layer which is more obvious from Fig (25) . Due to the sticking boundary conditions the material in turn inside out; layers at different constant r-value upstream are mixed downstream.

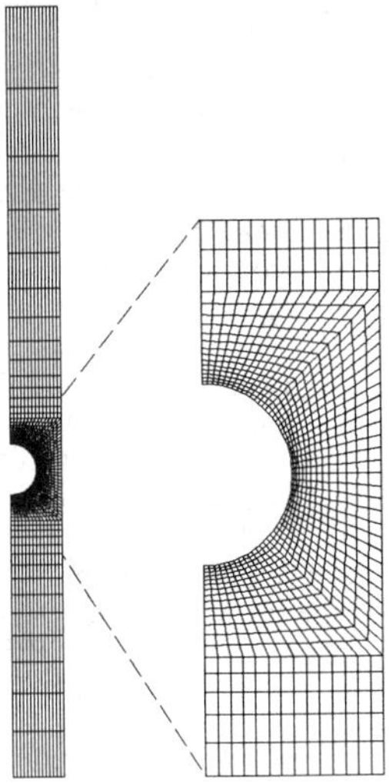

Fig. 17 Mesh used to calculate the steady-state motion of a sphere moving along the centerline of a cylindrical tube.

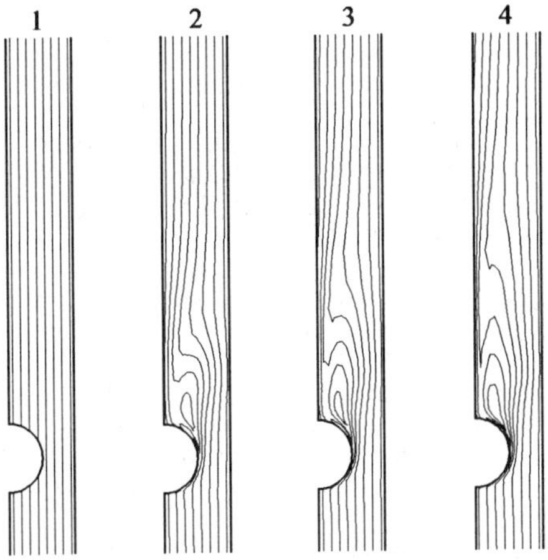

Fig. 19 Particle lines with the same initial r-value for different stages: definition of particle lines (1), particle lines after respectively 2s. (2), 4s. (3) and 6 s. (4)

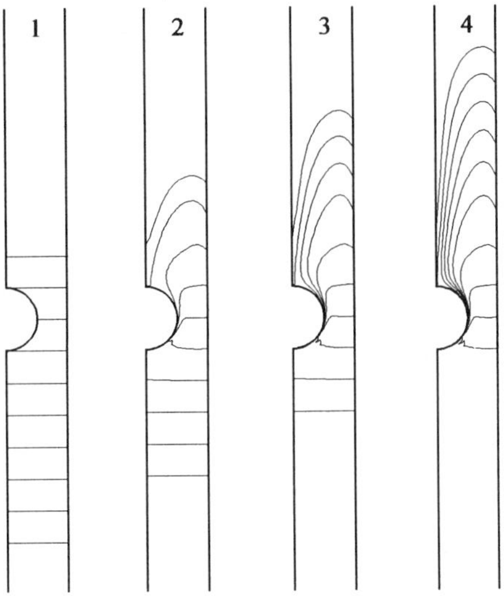

Fig. 18 Particle lines with the same initial z-value for different stages: definition of particle lines (1), particle lines after respectively 2s. (2), 4s. (3) and 6 s. (4)

5. CONCLUSIONS

In this paper two split Euler-Lagrange formulations are discussed. The Euler step of the first Euler-Lagrange formulation is based on an artificial dissipation scheme, the second on a limited flux scheme. For the one-dimensional case it can be shown that both schemes are basically a Lax-Wendroff finite difference scheme. For the two-dimensional case they differ considerably. The implementation of both methods is different: the artificial dissipation scheme uses averaged corner nodes, the limited flux scheme uses averaged mid-side nodes. Furthermore, the accuracy of both schemes is different: the artificial dissipation scheme is more accurate when the relative displacements are skew to the mesh, the limited flux scheme is more precise when the relative displacements are parallel to the mesh. We can conclude that both schemes have complementary virtues.

However, when the Euler-Lagrange simulation has mainly Lagrangian characteristics, which is the case for the upsetting process mentioned in the introduction, the accuracy of the Euler algorithm is of minor importance. When the Euler-Lagrange simulation has mainly Eulerian characteristics, which is the case for the extrusion and wire drawing process mentioned before, the accuracy of the Euler algorithm is of more importance.

8. REFERENCES

Akkerman, R. 1993. *Euler-Lagrange simulations of nonisothermal viscoelastic flows* (thesis), University of Twente, Enschede, Netherlands

Baaijens, F.P.T.1993, An U-ALE formulation of 3-D unsteady viscoelastic flow, *Int. J. Num. Meth. Eng.*, 36, 1115-1143.

Benson, D.J. 1989. An efficient, accurate, simple ALE method for nonlinear finite element programs, *Comp. Meth. Appl. Mech. Eng.*, 72, 305-350.

Brooks, A.N. and Hughes, T.J.R., 1982, 'SUPG formulations for Convection dominated flows with particular emphasis on the incompressible NS equations', *Comp. Meth. in Appl. Mech. and Eng.*, vol.32, pp 199-259.

Crochet, M.J. & V. Legat 1992, The consistent streamline-upwind/Petrov-Galerkin method for viscoelastic flow revisited, *Journal of non-Newtonian fluid mechanics*, 42, 283-299, Elsevier Science Publishers, Amsterdam.

Donea, J. 1983 Arbitrary Lagrangian-Eulerian finite element methods, In: T. Belytchko and T.J.R. Hughes, editors, *Comp. Meth. in Mech. 1*, Elsevier Science Publishers B.V., Amsterdam.

Helm, P.N. van der 1992 Investigation and implementation of an ALE algorithm, *TNO report* BI-92-112.

Helm,P.N. van der, J.Huétink & R.Akkerman, 1995 Comparison between Artificial dissipation and limitd flux Schemes in ALE Formulation submitted for publ. *Int. J. Num. Meth. Eng.*

Hinton, E.and Gampbell,J.S, 1974, 'Local and global smoothing of discontinuous finite element functions using least square method', *Int.J.Num.Meth. Eng.*, Vol.8, pp. 461-480.

Hirsch,C.1988, *Numerical computation of internal and external flows 2*, John Wiley & Sons, New York.

Huétink, J. 1986,*'On the simulation of thermo-mechanical forming processes'*, Dissertation, University of Twente, The Netherlands.

Huétink, J., Vreede,P.T. and Van der Lugt,J. 1990 Progress in mixed Eulerian-Lagrangian finite element simulation of forming processes, *Int. J. Num. Meth. Eng.*, 30, 1441-1457.

Huétink, J. and Helm, P.N. van der 1992, 'On Euler-Lagrange finite element formulation in forming and fluid problems', In: Chenot J.-L. e.a., *'Num.Meth.in Industr. Forming Proc.'*, Balkema Rotterdam, pp. 45-54.

Leer, B. Van. 1979, Towards the ultimate conservative difference scheme. V. A second-order sequel to Godunov's method, *J. Comp. Phys.*, 32, 101-136,.

Liu, W.K., Chang, H., Chen J.S. and Belytschko, T. 1988, Arbitrary Lagrangian-Eulerian Petrov-Galerkin finite elements for nonlinear continua, *Comp. Meth. Appl. Mech. Eng.*, 68, 259-310.

Lunsmann, W.J., L. Genieser, R.C. Armstrong & R.A. Brown, 1993, Finite element analysis of steady viscoelastic flow around a sphere in a tube: calculations with constant viscosity models, *Journal of non-Newtonian fluid mechanics*, 48, 63-99,Elsevier Science Publishers, Amsterdam, 1993

Mooi, H. G. & J. Huétink, 1995 Simulation of complex Aluminium extrusion using an ALE formulation, *Proceedings of the 5th Numerical Methods in Forming Processes* conference, Ithaca, New-York

Rekers, G., R. Akkerman & J. Huétink, 1995 Finite element simulation of viscoelastic flow through a converging channel, *Proceedings of the 5th Numerical Methods in Forming Processes* conference, Ithaca, New-York, 1995

Sweby, P.K. 1984, High resolution schemes using flux limiters for hyperbolic conservation laws, *Siam J. Num. Anal.*, 21, 995-1010, 1984.

Simulation of Materials Processing: Theory, Methods and Applications, Shen & Dawson (eds)
© 1995 Balkema, Rotterdam. ISBN 90 5410 553 4

Ultrasonic devulcanization of waste rubbers: Experimentation and modeling

A. I. Isayev, J. Chen & S. P. Yushanov
Institute of Polymer Engineering, The University of Akron, Ohio, USA

ABSTRACT: Tire and rubber waste recycling is an important issue facing the rubber industry. In addressing this issue, the present paper describes the first attempt to formulate a model and to simulate a novel continuous ultrasonic devulcanization process. The proposed model is based upon a mechanism of rubber network break up caused by cavitation, which is created by high intensity ultrasonic waves in the presence of pressure and heat. The break up of a three-dimensional network in crosslinked rubbers is combined with flow modeling. Experimental data obtained in several devulcanization runs using SBR rubber are compared with the results predicted by the simulation.

1 INTRODUCTION

During the last decade, Isayev and co-workers (1987,1990) have carried out extensive studies in an attempt to develop a polymer processing technology which utilizes high power ultrasonics. It has been shown that during extrusion high intensity ultrasonic waves affect the die characteristics by reducing the pressure, die swell and also postpone melt fracture. The ultrasonic waves can also breakdown the molecular chains which permanently reduces the viscosity of the original polymer melt. Breakdown of molecular chains occurs not only in polymer melts but also in polymer solutions. Degradation of polymer solutions has a long history (Jellinek 1966, El'piner 1964, Casale and Porter 1988). There is now overwhelming evidence that degradation in polymer solutions occur as a result of the cavitation process associated with the stresses generated by the ultrasonic waves and their rate of change. The effect of cavitation is due to the presence of voids or density fluctuation in a liquid. The mechanism of the ultrasonic effect on fluids was extensively studied by Suslick and co-workers (1989,1990). Acoustic cavitation is also observed in polymer melts (Peshkovskii et al 1983). It has also been found that high intensity ultrasonics imposed during foam formation improve the uniformity, reduce the size of the cell structure and enhance the mechanical properties of the foam (Isayev and Mandelbaum 1991).

Recently, it was discovered that ultrasonic waves of certain levels, in the presence of pressure and heat, rapidly break up the three-dimensional network in vulcanized rubbers (Isayev 1993, Isayev and Chen 1994). The devulcanized rubber becomes soft. It can be reprocessed, shaped and revulcanized in much the same way as a virgin rubber (Isayev et al 1995). The process of ultrasonic devulcanization is very fast and occurs on the order of a second or less. The latter allowed us to develop a continuous process of devulcanization. Isayev et al (1995) also describes some recent preliminary experiments and efforts to understand a possible mechanism of the devulcanization and attempt to scale up the process. In particular, the performed measurements indicate that the rubbers are partially devulcanized and the devulcanization process is accompanied by some degradation.

The present study describes the attempt to build a simplified theoretical model for the devulcanization process, including simulation, and comparison between predicted and experimental data.

1.1 *Materials and methods of investigation*

A solution polymerized styrene-butadiene rubber (SBR) Duradene 706 manufactured by the Firestone Company is used in the present experiments. It contains 23.5% bound styrene, 76.5% butadiene, and a nonstaining antioxidant stabilizer system. The weight average molecular weight and polydispersity index of the SBR measured by gel permation chromatography (GPC) are found to be 2.86×10^5 (g/mole) and 2.3, respectively. A recipe of the prepared compound includes 1.5 and 1.0 part respectively of sulfur and santocure, per 100 parts of SBR. The rubber and curing ingredients are mixed using a two-roll mill (Albert) by passing the material through the nip of the rolls approximately thirty times. The prepared compounds are vulcanized in a compression molding press at 171°C for 30 min. The vulcanizates are chopped and charged into an ultrasonic reactor. The reactor consists of a 1 inch Killion laboratory plastic extruder with an L/D = 24 and an ultrasonic crosshead die attachment, see Figure 1. A 900 W ultrasonic power supply, a converter, and a booster are used to provide the longitudinal vibrations of the horn with a frequency of 20 kHz. The horn diameter is 12.7 mm, the diameter of the die opening is 6.3mm, the gap between the flat face of the horn and the chamber bottom is 0.5 mm. The temperature of the extruder barrel, T_E, is fixed at 120°C. The amplitude of the horn oscillations, A, and the screw rotation speed, N, are varied.

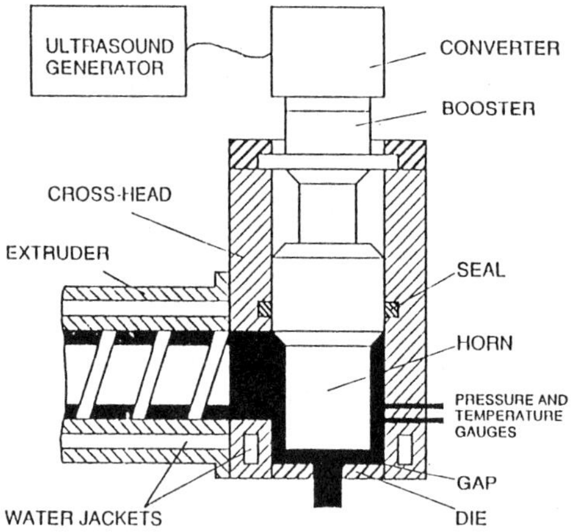

Fig. 1. Schematic of the devulcanization reactor.

The devulcanized samples are obtained at various processing conditions. The flow rate, pressure at the die entry, and the material temperature are measured. In the present setup, it is difficult to control the melt temperature in the die due to significant heat generation by the ultrasonics. Thus, various die temperatures are achieved for each processing condition depending on the amplitude and the residence time of the material in the die.

A Monsanto Processibility Tester is utilized for the viscosity measurements of virgin and devulcanized SBR. The viscosity is measured in the shear rate range of 1 to 1,000 s^{-1} at temperatures of 70, 100, and 130°C.

A Monsanto Tensiometer (Model 10) is used to obtain the stress-strain characteristics of cured virgin compound. The dumbbell samples for the testing are punched from the cured rubber sheets using the ASTM D412, type C, die. The measurements are carried out at room temperature at an elongation rate of 25 mm/min.

The gel fraction of the devulcanized SBR is measured by a BFGoodrich device according to ASTM D3616 using benzene as the solvent. The crosslink density of the vulcanized and devulcanized rubbers is measured using a swelling technique proposed by Flory (1950). Benzene is used in the swelling experiments.

The GPC technique is used to measure the molecular weight and the polydispersity index of the soluble component in the devulcanized SBR. Polystyrene samples with narrow molecular weight distribution are used for the calibration.

2 THEORETICAL MODELING

The proposed model is the first attempt to simulate the ultrasonic devulcanization process for rubber-like materials. Theoretical modeling of the devulcanization process includes two main aspects: 1) on a micro-level, network degradation in ultrasonic fields, and 2) on a macro-level, material behavior (temperature, pressure, velocity and shear rate) during flow in the devulcanization process. Evidently, the temperature, pressure, etc. affect the rate of the network degradation and vice versa. These two problems are inherently coupled. They cannot be solved separately. In addition, there are a lot of particular sub-problems such as molecular structure, structural behavior of network, etc. which have to be studied theoretically in more detail. Thus, some additional assumptions should be made to handle this task. The proposed model inevitably includes fitting parameters which are specified based on comparison between theoretical modeling and experimental data. The number of the fitting parameters is kept to a minimum.

2.1 *Devulcanization modeling*

In formulating the devulcanization model, the following assumptions are made: 1) the break up of the network chains and crosslinks is an independent process; 2) the over-all rate of molecular break up is inversely proportional to the relative strength of the bond, and proportional to the number of locally over stressed molecular bonds; 3) that cavitation is a dominant mechanism governing the devulcanization process. Collapsing cavities are considered to be non-interactive; 4) the rate of devulcanization is governed by an average residence time. This assumption leads to a uniform rate of devulcanization at any particular cross section. Based on assumptions 1 and 2, the rate of breakup for various bonds can be written as:

$$-\frac{\partial N_i(t)}{\partial t} = k_1 N_{ia}(t)\frac{E_o}{E_i}; \quad N_i(0) = N_{i0} \tag{1}$$

where $N_i(t)$ is the number of the molecular bonds present in the system at an average residence time t, with N_{i0} being their number at some initial time, $N_{ia}(t)$ is the current number of "active" molecular bonds which are subjected to a local over stressing, E_i is the bond strength, E_o is the reference strength (carbon-carbon bond strength can be chosen as E_o, i.e.: $E_o = E_c$), and k_1 is a rate constant. In Eq. (1), the subscript i denotes the specific type of the molecular bonds under consideration: i = C refers to the C - C bonds of the main chain, i = c_k refers to the crosslinks of k-th type, for example, monosulfidic C - S - C and polysulfidic C - $\underbrace{S...S}_{x}$ - C crosslinks.

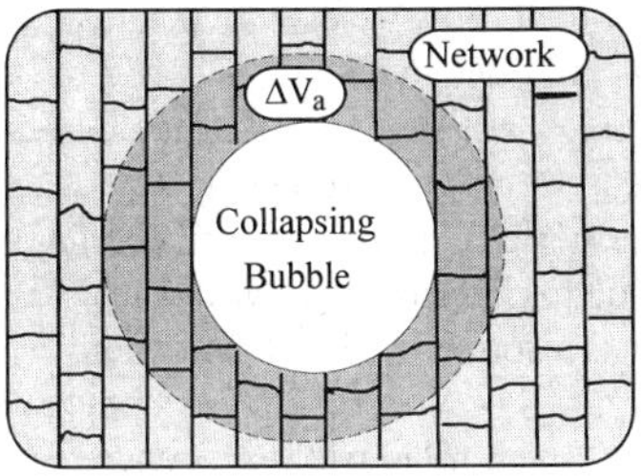

Fig. 2. Schematic of the overstressed network fragment around the collapsing bubble.

Based on assumption 3, the number of "active" bonds per unit volume can be expressed as:

$$N_{ia} = \Delta V_a N_b N_i \tag{2}$$

where ΔV_a is the "active" volume around a single collapsing bubble, and N_b is the concentration of the collapsing bubbles. Thus, the product of $\Delta V_a N_b$ is the fraction of the total existing bonds which are affected by local over stressing around a collapsing bubble. In turn, the "active" volume ΔV_a is the volume of the spherical layer bounded by R_{max} and R_{max} $+\Delta L_a$, where R_{max} is the bubble radius just before collapsing, and ΔL_a is a radial distance from the bubble surface at which the network is locally over stressed during bubble collapse. It is reasonable to assume that bond ruptures are localized near the surface of the collapsing bubble: $\Delta L_a << R_{max}$. Hence:

$$\Delta V_a = \frac{4}{3}\pi\left[(R_{max} + \Delta L_a)^3 - R_{max}^3\right] \approx 4\pi R_{max}^2 \Delta L_a \tag{3}$$

The magnitude of R_{max} is determined from the simulation of the cavitation process in crosslinked rubber based on the incompressibility approximation. It is a function of the intensity of the ultrasonic waves, the initial radius of the bubble nuclei, the current hydrostatic pressure, the frequency and amplitude of the ultrasonic waves, and the elastic characteristics of the material (Chen 1995). Mooney potential function, which is found to satisfy the experimental stress-strain behavior of the SBR vulcanizate, is utilized in the calculation of R_{max}. For brevity, a description of modeling of the cavitation process is omitted here.

Nucleation and growth of gas bubbles in crosslinked elastomers have been studied experimentally by Gent and Tompkins (1969). It was found that the number of primary bubbles per unit volume depends exponentially on pressure P. Taking into account this experimental result one can write:

$$N_b = k_2 \exp\left(\alpha \frac{P}{\sigma_o}\right) \tag{4}$$

where σ_o is the tensile strength of the material, α and k_2, are some constants which depend on the conditions of bubble formation and material properties.

Integration of Eq. (1) and making use of Eqs. (2)-(4) leads to:

$$N_i(t) = N_{i0} \exp\left[-4\pi K \frac{E_C}{E_i} \int_0^t R_{max}^2 \exp\left(\alpha \frac{P}{\sigma_o}\right) dt\right] \tag{5}$$

where $K = k_1 k_2 \Delta L_a$ is a constant. In Eq. (5), the values P and R_{max} are time dependent since during flow the pressure varies along the die length, and R_{max}, which in turn, depends on hydrostatic pressure. The constants K and α are two fitting parameters which are to be specified for comparison of the theoretical and experimental results on gel fraction of the material after devulcanization.

Now one needs to specify initial conditions for Eq. (5). Suppose that the network was produced from originally "infinite" linear molecules by the addition of different kinds of crosslinks: $c = \sum_k c_k$, where c_k is the number of moles of the k-th crosslinks in a unit volume, c is the total number of moles of various crosslinks in a unit volume. The number of moles of the network chains (the main chain segments between two subsequent crosslinks) in a unit volume is equal to n = 2c. Hence, the number of the segments, N_C and the number of k-th crosslinks, N_{c_k} in a unit volume are given by:

$$N_C = nN_A = \frac{\rho N_A}{M_c}, \quad N_{c_k} = c_k N_A \tag{6}$$

where ρ is material density, M_c is molecular weight of the network chain, and N_A is Avogadro number. Relations (6) are the initial conditions for the calculation of the kinetics of network chains and crosslink degradation.

With increasing amount of main chain and crosslink scissions the gel fraction decreases. Modeling the gel fraction in relation to the number of broken main chains and crosslinks is a complex one and needs special consideration. One can use, for example, the Monte-Carlo technique to solve this problem. It is clear that if there are no broken crosslinks the network gel fraction is equal to one. If all crosslinks are broken the gel fraction is equal to zero. As a first approximation satisfying this condition we will use the following equation to estimate the network gel fraction ξ(t) at time t:

$$\xi(t) = \xi_0 - \frac{2\sum N_{c_k}(t)}{N_C(0)} \tag{7}$$

where ξ_0 is a gel fraction of the original material.

2.2 *Die filling modeling*

The flow in a disk die with outer radius R_1 and inner radius R_2 is modeled as a one-dimensional flow in a strip of length $L = R_1 - R_2$ and varying width w. The simulation of a one-dimensional flow in a cavity of simple geometry can be summarized in terms of the set of transport equations (see, for example, Hieber 1987). For the case of a strip of width w and half-gap thickness h these equations are:

$$Q = w \int_{-h}^{h} u dz \tag{8}$$

$$0 = \frac{\partial}{\partial z}\left(h \frac{\partial u}{\partial z} \right) - \frac{\partial P}{\partial x} \tag{9}$$

$$\rho C_p u \frac{\partial T}{\partial x} = k_{th} \frac{\partial^2 T}{\partial z^2} + \Phi + \dot{Q}_s + \dot{Q}_d \tag{10}$$

where Q denotes the volumetric flow rate, x the streamwise direction, z the gapwise (transverse) direction. Further, u denotes the velocity in the x direction, η the shear viscosity, P the pressure, T the temperature, with ρ, c_p, and k_{th} denoting the material density, specific heat and thermal conductivity. $\Phi = \eta\dot{\gamma}^2$ and $\dot{Q}_s = 2\pi^2\varepsilon^2 fE''$ are rates of viscous and ultrasound dissipation, respectively, where $\dot{\gamma} = |\partial u/\partial z|$ is the shear rate, ε = A/2h is the ultrasound strain amplitude with A being the ultrasonic amplitude, f is the ultrasound frequency, and E" is the loss modulus at frequency f. $\dot{Q}_d = \left(E_C \dot{N}_C + \sum_k E_{c_k} \dot{N}_{c_k} \right) / N_A$ is the rate of energy consumption due to devulcanization. Here E_C is the energy to rupture a C - C bond and E_{c_k} is the energy of the scission of c_k-type crosslinks.

Eq. (8) represents a balance of mass, whereas Eq. (9) gives the force balance in the streamwise direction between the viscous shear stress and the pressure gradient. Inertial effects have been omitted from the left side of Eq. (9) due to the high viscosities typical of polymer melts. Eq. (10) is the quasi-stationary energy equation. It indicates that the change in temperature as one follows a material particle is due to the net effect arising from the gapwise thermal conduction, viscous and ultrasound heating, and energy consumption by devulcanization.

Equations (8) - (10) have to be supplemented with a viscosity function as well as boundary conditions. In the present case, the viscosity is a function of a temperature, shear rate, and a gel fraction. This function is represented by a power-law model:

$$\eta = \eta_o m(T,\xi)\dot{\gamma}^{n-1} \tag{11}$$

The parameters of Eq. (11) are based upon fitting experimental data obtained for dependence of viscosity vs. shear rate, temperature and gel fraction of devulcanized rubber. The following function is used to describe temperature and gel fraction dependence of viscosity of devulcanized rubber obtained in a wide range of processing conditions:

$$m(T,\xi) = \exp\left[\frac{b(\xi)}{RT}\right] \quad (12)$$

where $b(\xi)$ is a polynomial function.

The boundary conditions on $u(x,z)$, $T(x,z)$, and $P(x)$ are:

$$u(x,h) = 0; \quad \frac{\partial u(x,0)}{\partial z} = 0; \quad 0 \le x \le L; \quad (13a)$$

$$T(0,z) = T_o, \; 0 \le z \le h; \; T(x,h) = T_w, \; 0 < x \le L; \quad (13b)$$

$$\frac{\partial T(x,0)}{\partial z} = 0, \; 0 < x \le L; \quad P(L) = 0 \quad (13c)$$

The present simulation is carried out for a constant flow rate. The pressure distribution in the die is an unknown function. It has to be calculated using an iterative procedure.

2.3 *Simulation algorithm*

The simulation procedure of the flow in the die is simplified by introducing the function $S = \int_0^h \frac{z^2}{\eta} dz$ which is a measure of the fluidity of the polymer melt. Then, the pressure gradient $\Lambda_x = -dP/dx$, velocity, and shear rate are expressed from Eqs. (8) and (9) taking into account the boundary conditions (13). Resulting equations are:

$$\Lambda_x = \left(\frac{Q}{2wS}\right)^n; u = \Lambda_x^{1/n} \int_z^h \left[\frac{z}{\eta_o m(T,\xi)}\right]^{1/n} dz; \dot{\gamma} = -\Lambda_x \frac{z}{\eta} \quad (14)$$

The energy equation(10) is approximated by a finite-difference scheme. For stability reasons, an upwind difference is used to express the streamwise convection term. An implicit representation in the gapwise direction is used for the temperature in the conduction term.

A uniform mesh (x_i, z_j) is introduced. The simulation algorithm is as follows. When the melt front moves from a position x_i to x_{i+1}, the temperature distribution along the new melt front is initially calculated using the old values of $u, \dot{\gamma}, \xi$, and T at x_i. Then the pressure gradient, velocity, and shear rate are calculated by Eq. (14) based upon the updated temperature field. The pressure is determined by numerical quadrature of Λ_x, starting with zero at the melt front and integrating back towards the die entrance. During this procedure the value of the pressure is updated first. With this updated pressure the bubble collapse process is simulated and R_{max} is evaluated. Then making use of Eq. (5) the number of broken bonds is calculated and material gel fraction is determined by Eq. (7). After that, the temperature T, pressure gradient Λ_x, and kinematic characteristics $u, \dot{\gamma}$ are re-evaluated using the updated values of the gel fraction and pressure. This cycle is then repeated. The iterative process is continued until a prescribed convergence criterion is satisfied between successive cycles. The calculation is continued until the die is filled. The final distribution of various quantities corresponds to the stationary solution of the devulcanization process.

2.4 *Model parameters*

A power-law model is employed to describe flow behavior of devulcanized rubber obtained under various processing conditions. The flow curve of each devulcanized rubber is measured at various temperatures. As indicated in Eq. (11), there are two parameters η_o and n and a function $m(T,\xi)$ which specifies shift factor of the function, η vs. $\dot{\gamma}$, with respect to temperature and gel fraction variation. Flow curves of the devulcanized rubber at a constant gel fraction and various temperatures are found to superimpose to a flow curve at some reference temperature, T_r, by shifting along the shear rate axis. By carrying out this procedure, the shift factor as a function of temperature is determined at a constant gel fraction. Then, the second shift is performed in which the reduced flow curves at the reference temperature and various gel fractions are superimposed to a master curve corresponding to some reference gel fraction, ξ_r. The shift factors as a function of gel fraction at various temperatures are shown in Fig. 3. The master curve for reduced viscosity vs. reduced shear rate is given in Fig. 4. Some scattering of the data is seen, especially in the region of high shear rates. This scattering is due to the use of a constant power-law index. In fact, some deviation of power-law index n is found as gel fraction in devulcanized rubber is varied (Isayev et al 1995). Apparently, a modified Cross model would be a better model for describing the viscosity function of devulcanized rubber. However, a power-law model used here adds a simplicity to the present treatment of the devulcanization process. The parameters in Eqs. (11) and (12) are determined by a curve fitting of the data depicted in Figs. 3 and 4 using the least square method. The fitted curves are shown by the solid lines in these figures. The values of these parameters are η_o = 1.5

Table 1. *Initial data for simulation*

C_p, J/kg·K	1884[1)]	E_C, kJ/mole	352[3)]
k_{th}, J/m·s·K	0.22[1)]	E_S, kJ/mole	285[3)]
ρ, kg/m³	930[1)]	E_{SS}, kJ/mole	268[3)]
f, kHz	20	R_1, mm	6.35
E", MPa	1[2)]	R_2, mm	3.175
c, No/m³	$4.1\cdot10^{25}$	h, mm	0.25
ξ_o	0.958	σ_o, MPa	1.25

[1)] Roff and Scott (1971)
[2)] Mark and Gaylord (1971)
[3)] Kircher (1987)

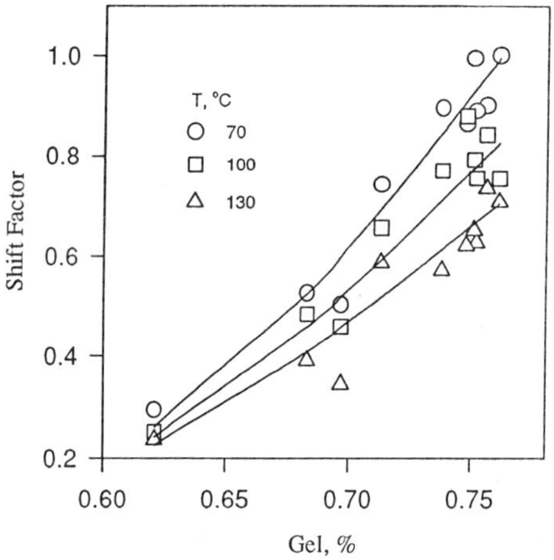

Fig. 3 Fitted (curves) and experimental (symbols) shift factors versus gel fraction at various temperatures.

x 10^4 Pa·s^{2-n}, $b(\xi) = 859 - 950\xi + 9162\xi^2$ (J·K/mole) and n = 0.28. The other parameters used in the simulation are listed in Table 1. Dependences of R_{max} on the ambient pressure P, and gas pressure, P_g in the bubble are derived from the simulation of the cavitation process for SBR rubber.

2.5 *Simulation results*

Simulation is carried out for the various devulcanization conditions. An example of these results is presented for the following conditions: Q = 0.257 cm³/s, T_w = 441°K, A=38.5 μm. Calculations are carried out for uniform meshes 20x20. Fig. 5 shows the gapwise distribution of velocity at various cross sections in the die. The velocity increases as the rubber moves from the die entrance towards the die exit. This increase in the velocity is due to a decrease in

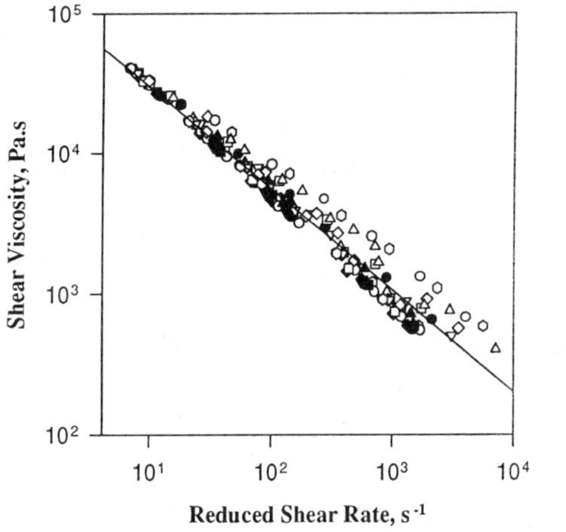

Fig. 4 Fitted (curves) and experimental (symbols) master curve of viscosity versus shear rate at T_r = 70°C and ξ_r = 0.761.

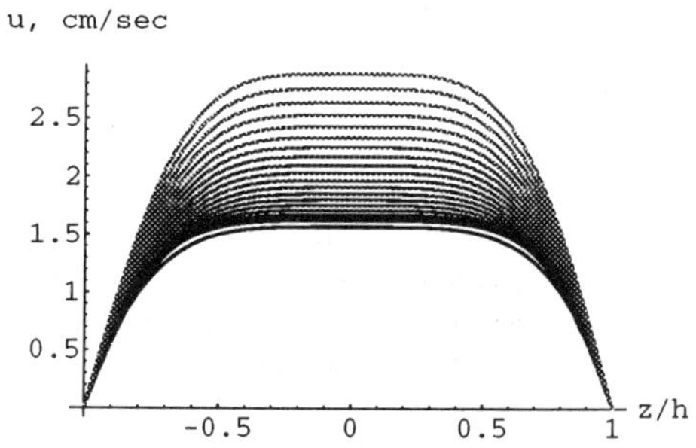

Fig. 5 Velocity profiles at various cross sections in the die.

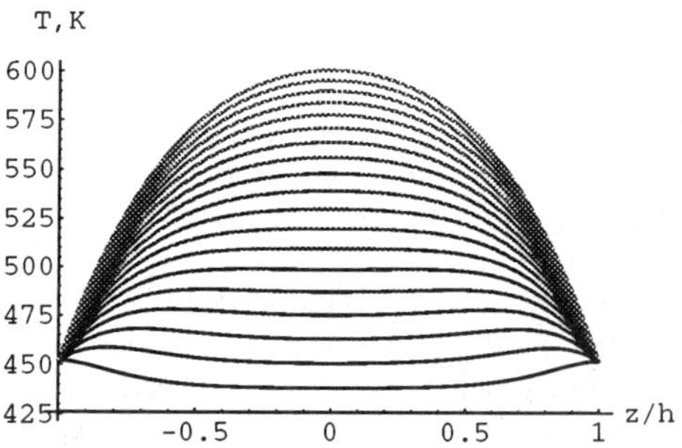

Fig. 6 Temperature profiles at various cross sections in the die.

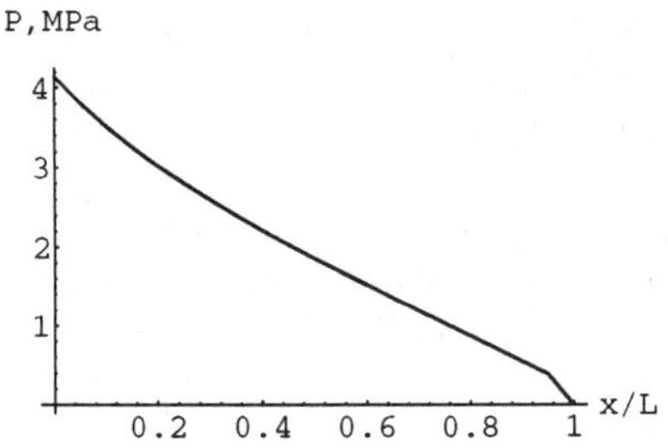

Fig. 7 Pressure distribution along die length.

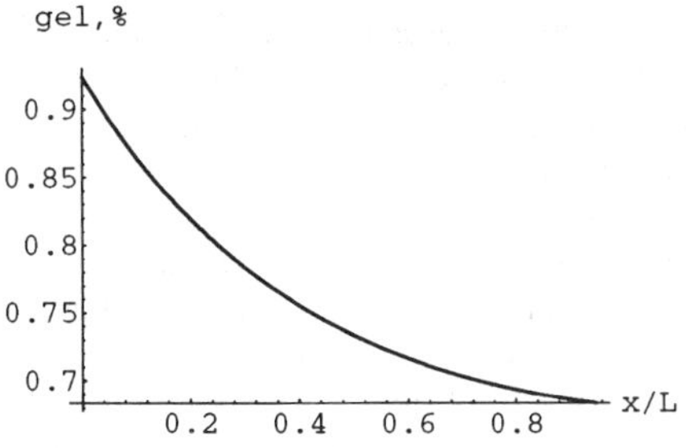

Fig. 8 Gel fraction variation along die length.

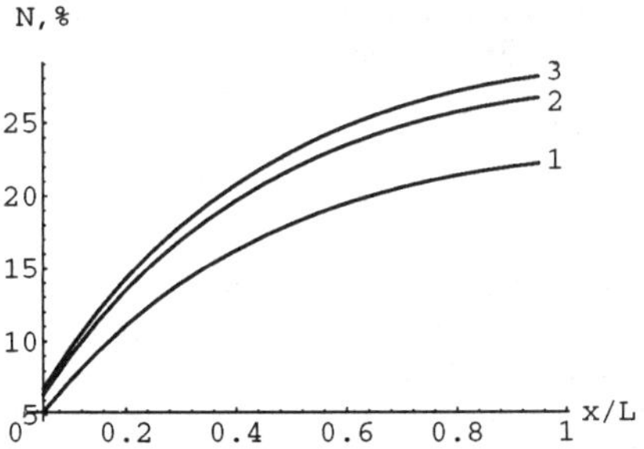

Fig. 9 Fraction of various bond ruptures along die length. (1) C - C; (2) C - S - C; (3) C - (S...S) - C.

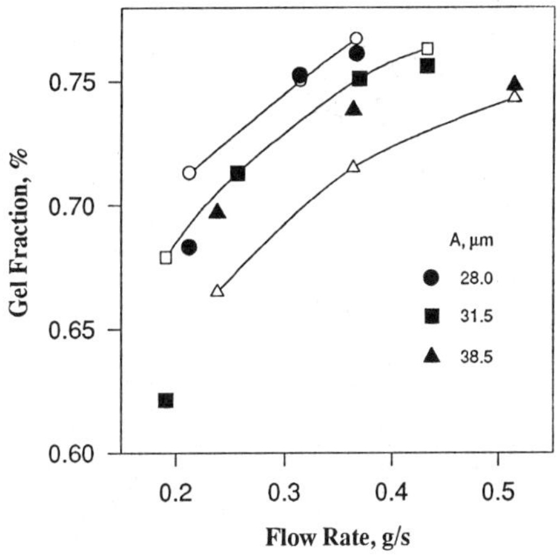

Fig. 10 Theoretical (lines with open symbols) and experimental (closed symbols) gel fraction of the devulcanized SBR vs. flow rate.

cross-sectional area of the disk die. Shape of the velocity profile also changes. At the die entrance the velocity profile is more flat than at the die exit. The latter is caused by the temperature profile variation as rubber is heated up due to the ultrasonic dissipation, as indicated in Fig. 6. In particular, rubber enters the die under uniform temperature. It is continuously heated up as the distances away from the wall increase and as the rubber moves toward the die exit. This increase in temperature leads to a decrease of the viscosity in the core region. In turn, the viscosity reduction in the core affects the velocity profile.

Fig. 7 shows variations of the pressure along the die length. It is seen that the pressure gradient at the die entrance is high despite the fact that the cross-sectional area of the die is highest in that region. This high pressure gradient at the entrance is caused by the significant viscosity reduction due to the high rate of devulcanization. This conclusion can be made based on Fig. 8, which depicts gel fraction as a function of distance from the entrance. In particular, gel fraction decreases monotonically. The rate of the gel fraction reduction is highest in the vicinity of the die entrance. This conclusion is plausible since it shows that the rate of various bond ruptures is extremely high at the die entrance, as indicated by Fig. 9. Here the fraction of various bonds broken is given as a function of the distance from the die entrance. Due to the absence of information concerning the concentration of various crosslinks in the network, for the present calculations, the number of monosulfidic and polysulfidic crosslinks is considered to be equal. It is seen from Fig. 9 that the bonds are broken up as soon as the vulcanized rubber contacts the edge of the ultrasonic horn. The rate of bond rupture is highest at this location. As the rubber moves away from the horn edge, the portion of broken bonds increases monotonically with continuous decay of the rate of their rupture. From Fig. 9 it is also seen that at the die exit about 22% of carbon-carbon, 26% of monosulfidic, and 27% of polysulfidic bonds are broken. In fact, this observation is consistent with the strength of respective bonds being highest for carbon-carbon and lowest for polysulfidic bonds. It is clear that the ultrasonic devulcanization process does not lead to complete devulcanization. Namely, a partial devulcanization takes place accompanied by degradation of the main chain.

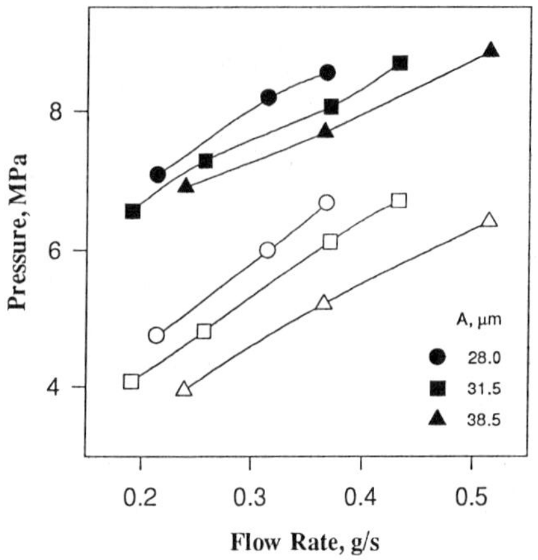

Fig. 11 Theoretical (open symbols) and experimental (closed symbols) pressure vs. flow rate.

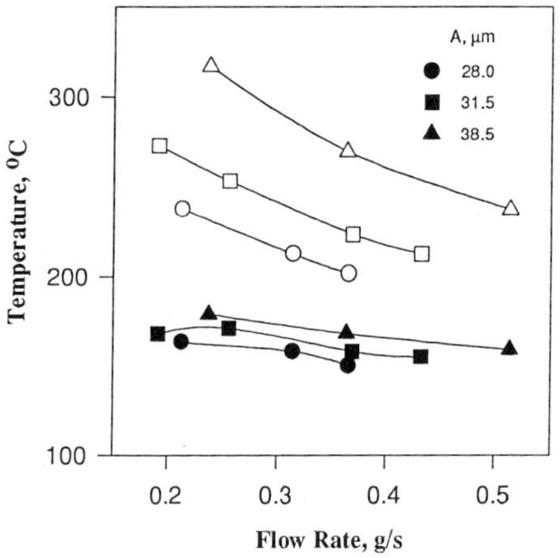

Fig. 12 Theoretical "mixing cup temperatures" (open symbols) at the die exit and experimental (closed symbols) as a function of flow rate.

2.6 *Comparison of the Theory with Experiment*

The predicted and experimental results for the gel fraction are presented in Fig. 10 for various conditions of devulcanization. It should be noted that adjustable parameters K and α are used to fit the experimental data. Comparison indicates that the theory is qualitatively correct in capturing a tendency of gel fraction variation of devulcanized rubbers obtained at various processing conditions.

Fig. 11 shows the predicted and experimental die characteristics in the devulcanization process. The predicted pressures are lower than the experimental ones. This discrepancy is due to the fact that the pressure transducer is located at some distance from the die entrance (see Fig. 1), leading to an extra pressure drop. Moreover, an extra pressure drop arises due to the entry and exit pressure losses which are not taken into account in the present simulation.

Predicted and experimental temperature dependence on the flow rate are shown in Fig. 12. Predicted temperatures are calculated at the die exit as "mixing cup temperatures," i.e. velocity weighted average temperatures. It can be seen that the predicted temperatures are very sensitive to ultrasonic amplitude and flow rate variation. Namely, as the amplitude increases the temperature also increases due to higher dissipation of the ultrasonic energy. On the other hand, as flow rate increases the temperature decreases due to the decrease of the residence time of the rubber in the die. Qualitatively, a similar tendency is observed in the measured temperatures. However, the measured temperatures are significantly lower than the predicted ones. The highest difference between experimental and calculated temperatures is about 140°C and is observed at the lowest flow rate and highest amplitude. This difference is due to the difficulties in temperature measurements. The thermocouple is placed in the bottom plate of the die where temperature gradients caused by conduction could be quite substantial.

CONCLUSIONS

The present study describes theoretical and experimental results related to a novel ultrasonic devulcanization technology for recycling used tires and other rubber waste. Experimental data indicate that ultrasonic devulcanization leads to a partial devulcanization along with main chain degradation. A simple model of devulcanization is proposed. The model is based upon the ability of high intensity ultrasonic waves to introduce cavitation in a vulcanized rubber. Numerical simulation of the process is carried out. Velocity, temperature, pressure fields, and structural characteristics such as gel fraction and fraction of broken bonds are calculated. The predicted results are found to be in qualitative agreement with experimental data. Future efforts should be directed toward improvement of the proposed model and accumulation of more experimental data related to physical and structural properties of devulcanized rubber.

ACKNOWLEDGMENTS

This work is supported by a grant DMI-9312249 from the National Science Foundation, Division of Engineering.

REFERENCES

Casale, A. and R.S. Porter 1977. Polymer Stress Reactions. New York: Academic Press.

Chen, J., 1995. Ph.D. Dissertation, The University of Akron, Akron, Ohio.

El'piner, I.E. 1964. Ultrasound: Physical, Chemical and Biological Effects. New York: Consultants Bureau.

Ensminger, D., 1988. Ultrasonics: Fundamentals, Technology and Applications. New York: Marcel Dekker

Flory, P.J. 1950. Statistical Mechanics of Swelling Network Structure. *J. Chem. Phys.*, 18:108-111.

Gent, A.N., and D.A. Tompkins, 1967. Nucleation and Growth of Gas Bubbles in Elastomers. *J. Appl. Phys*. 20: 2520-2525.

Hieber, C.A., 1987. Melt-Viscosity Characterization, in A.I. Isayev (ed.) Injection and Compression Molding Fundamentals. 1-136. New York: Marcel Dekker.

Isayev, A.I., C.M. Wong and X. Zeng, 1987. Flow of Thermoplastics in Dies with Oscillatory Boundary. *SPE ANTEC Tech. Papers*, 33: 207-210.

Isayev, A.I. 1990. Effect of Sound and Ultrasonic Waves on Polymer Extrusion. *Proceed. of the 23rd Israel Conference of Mechanical Engineering*, paper #5.2.3, 1-5.

Isayev, A.I., C.M. Wong, and Z. Zeng 1990. Effect of Oscillations During Extrusion on Rheology and Mechanical Properties of Polymers. *Adv. Polym. Technol.*, 10: 31-45.

Isayev, A.I. and S. Mandelbaum 1991. Effect of Ultrasonic Waves on Foam Extrusion. *Polym. Eng. Sci.*, 31: 1051-1056.

Isayev, A.I. 1993. U.S. Patent 5,258,413.

Isayev, A.I., and J. Chen 1994. U.S. Patent 5,284,625.

Isayev, A.I., J. Chen and A. Tukachinsky 1995. Novel Ultrasonic Technology for Devulcanization of Waste Rubbers. *Rubber Chem. Technol.*, (in press).

Jellinek, H.H.G. 1966. Depolymerization. In Encyclopedia of Polymer Science and Technology, H.F. Mark, N.G. Gaylord and N.M. Bikales, eds., Vol. 4, New York: Wiley Interscience.

Kircher, K., 1987. Chemical Reactions in Plastics Processing. New York: Hanser Publishers.

Mark, H.F., and N.G. Gaylord, 1971. Encyclopedia of Polymer Science and Technology. Vol. 14, New York: John Wiley & Sons.

Peshkovskii, S.L., M.L. Friedman, A.I. Tukachinsky, G.V. Vinogradov and N.S. Enikolopian 1983. Acoustic Cavitation and Its Effect on Flow in Polymers and Filled Systems. *Polym. Compos.*, 4: 126-134.

Roff, W.J., and J.R. Scott, 1971. Handbook of Common Polymers. London: Butterworth.

Suslick, K.S. 1989. The Chemical Effects of Ultrasound. *Scientific American*, 260: 80-86.

Suslick, K.S., S.J. Doktycz and E.B. Flint 1990. On the Origin of Sonoluminescence and Sonochemistry. *Ultrasonics*, 28: 280-290.

Simulation of Materials Processing: Theory, Methods and Applications, Shen & Dawson (eds)
© 1995 Balkema, Rotterdam. ISBN 90 5410 553 4

Some experiences and perspective of inverse analysis of the problem in metal forming engineering

J. Kihara
The University of Tokyo, Japan

ABSTRACT: In metal forming engineering, the residual stress distribution of product and the tractions acting across the contact between tool and work are the important variables in order to control the engineering system and investigate to improve the technology, and rquested to be evaluated. For the purpose, some trials using the boundary element technipue have been performed by the present author and his colleagues. The possibility and subjects to be solved in future will be discussed.

1 INTRODUCTION

In the last decade the bounary element method has been much developed to be a powerful engine in numerical analysis. Other than the application of this method to modeling and simulation, some trials have been done to apply the method to the inverse problems. In this report, the author intends to summarize his experiences in development of such system to analyze inversely the process parameters in metal working using the boundary element method, and clarify the subjects to solve in order to make the trial successful.

The calculàtion engine to be tried may be either finite element method or boundary element method, but the author chose in actual the latter. He thought that the metal forming was performed under the action of the external tractive forces, and therefore, the action through the boundary including the contact surface made the work deform to be the part and the relations among the surface tractions and displacements played the main role in the process. This is the reason that he chose boundary element method.

As for the subjects which the author investigated, there were the estimation of the residual stresses in rolled sheet, the pressure and frictional stress distribution on the contact surface between the work and tool in free forging, the evaluation of tribological situation with respect to the surface roughness of work and viscosity of liquid lubricant in the ring compression test and the observation of change in thermal property of steel during cooling from its austenitic temperature, ie. 1273 K.

In this report the author will review these works with a discussion of criticism and, at the same time, some suggestions to overcome or improve difficulties included in the work respectively.

2 FORMULATION OF INVERSE ANALYSIS

The way of formulation is different in detail depending on the choice of the computational method, but in principle there are two ways of formulation for the inverse analysis. One is a usual and traditional way, which can be called a curve fitting method. In this method, we suppose some governing parameters of the phenomenon, obtain a plenty of the calculated results for changing the parameters in wide range and determine the value of the controlling parameters by optimization of them to minimize the dicrepancy between the observed results and the calculated ones. The author applied this way to the observation of change in the thermal property of steel during cooling.

The other way is one to be applied to the case in which the number of unknown boundary parameters is in excess of the number of equations and the boundary element equation cannot be solved. In this case we find some subsidial relations among the unknown boundary parameters and some informations possible to be obtained experimentaly and can determine the whole unknowns by solving the synthesized equation composed of main boundary equation and the adopted subsidial equation. Such situation may occur in analysis of the contact problems. Unless we assume the friction law nor kinematical condition, we cannot obtain the frictional stress distribution on the contact surface. However, either the friction law or the kinematical condition are also unknown condition in real, and must be investigated.

3 THE ESTIMATION OF RESIDUAL STRESS DISTRIBUTION IN ROLLED SHEET

3.1 The residual stress

The residual stress is present in the body which is heat treated and/or plasticaly deformed, corresponding to the permanent strain gradient generated during the process. In the case of rolled sheet distortion and warping may occur in blanks when it is cut from the sheet, in which residual stress is induced during rolling process.

In order to control the rolling process to produce the sheet of well flatness, the rolling conditions must be controlled to reduce the sheet guage evenly reduced in the width direction. Though the flatness does not seems to be harmed in rolling, some magnitude of residual stress distribution may occur and it reveals itself as distortion or warping in a small pieces when it is cut from the sheet.

If it is necessary to refine the rolling conditions for the sheet to keep its flatness strictly, the monitoring system of the conditions must be set up on the basis of the residual stress investigation even when the residual stress is not so high as buckles the sheet to lose its flatness.

3.2 The architecture of the inverse analysis of the residual stress in rolled sheet

The experimental information is supposed to be obtained by so-called "Sach's" method, in which a part of the inspected body is cut from the rest, the deformation or strain introduced to the rest at the instant is measured and the residual stress distribution is evaluated by the computational method based on some feasible simulation model.

In the present case, a rolled sheet is cut along the width direction, and the introduced strains are measured by the strain guages pasted on the sheet near by the cutting line.

The elastic boundary elemnt modeling is applied to the whole boundary including the cutting line. The stress state is assumed to be 2 dimensional. In this case the bounary parameters on the cutting line are unknown as for both displacements and tractions. Therefore, the usual boundary element equation has excess unknown parameters, and is impossible to be solved.

As the subsidual relationship, here is adopted the bounary element modelling equation of the relation between the measured strain data and the whole boundary parameters. A linear simultanious equation is constructed to estimate the tractions distributed along the cutting line, and the values of the inverse sign are the estimated residual stress component.

3.3 Feasible study with a practical example

An example of the feasible study was reported by the author and his colleagues (Kihara, J. et al, 1983). They first set the subject that was how much the method was sensitive. And then they analysed the residual stress distribution in a rolled sheet which was geometrically equivalent to the modelled sheet in the sensitivity study.

In this problem, on the boundary

except for the cutting line the tractions are known parameters, and on the cutting line both tractions and displacements are unknown parameters. First they set a proper supposed traction distribution on the supposed cutting line. The strain corresponding to the traction distribution is calculated at the chosen point considering the distance from the cutting line.

The calculated strain data is utilized as the virtual measured data input to the synthesized symultanious equation. Some trials reveal that the well performed results can be obtained if the strain data in the area at distance of about 1 element length from the line are utilized. If the distance is larger than 1 element length, the calculation tends to be unstable, and unless the displacement at a far-from end point is fixed, any feasible solution cannot be obtained when the distance is 3 times of the element length. Such situation is due to St′ Venant principle in the static elasticity.

The principle gives the present system a restriction in area where the subsidual experimental informations are sampled, and is one of the largest difficulties in application of the static and elastic bounary element method to estimate the unknown bounary parameters with help of measured strain data in the body region.

Unless the singularity of the kernel function is not considered in calculation of the coefficints and the calculated data of strains at the distance nearer than 1 element length are used as the virtual measurement, the reproducibility of the virtual boundary data is much poor. Such difficulty may be overcome by improvement of accuracy employing the technique of singularity treatment in calculation, and is not substantial one.

A sample sheet is rolled to the same geometry as that in the virtual sample supposed in feasible study. At the middle the sheet is cut, and the strains at the distance of about 1 element length are measured by the rozett strain-guage method. The distribution of both normal tractions and shear traction are well evaluated.

3.4 Further development

The sheet has much smaller dimension in thickness in comparison with the other directions and its stress state is considered to be plane stress as mentioned previously. However, if we strip off even thin layer from the one side surface, the sheet often warps up or down. That shows there exists at least the distribution of normal stress acting in plane, and locally in thickness direction the stress condition is not plane one.

The author has interest in evaluation of such stress distribution, and the problem is thought of as an important subject, although he has not yet organized the investigation. It is hopeful that such inverse analysis system may be developed to estimate 3 dimensional distribution of stresses in the thin sheet. One of the feasible solution is that the mean stress through the thickness is evaluated by the previously mentioned method, and the distribution of stresses to thickness direction 0is assumed to be quadratic and is estimated by the calculation based on the warping measurement with stripping of surface layer and the mean stress evaluation.

4. ESTIMATION OF THE SURFACE TRACTIONS ON THE CONTACT PLANE BETWEEN THE TOOL AND THE WORK IN COMPRESSION OF A CYLINDRICAL SPECIMEN WITH THE PLANE DIES

4.1 Non-destructive evaluation of the tractions acting on the contact surface between the work and the tool

The metal forming technology has evaluated or modeled from the beginning era of engineering mathmatics by the anlytical approach which ranges from obtaining the strict mathematical function to the construction of numerical calculating system. However as for the mechanical situation of the contact surface the boundary conditions are usually assumed and treated as prescribed conditions. There are set some assumptions that the shape of the contact surface is not changed during the process, the Coulomb frictional law is valid and the value of frictional coefficient is sometimes a prescribed parameter. The mathematical models to calculate the load or the torque necessary to deform the work have been constructed on the basis of such treatment of the contact boundary conditions. The measured load or torque is compared with the calculated one to determine the parameters concerning with the boundary parameters on the contact surface.

The direct or experimental determination of the boundary tractions has been tried by many investigators, and almost all of them has adopted the so-called pin-contact method. Pins are embedded in the tool, and the contact end surface of the pins transduces the tractions supposed to act on the contact surface. The load applied to the end of the pin is measured by small load cell set in the body of the tool. These investigations have revealed many important behaviors of the tractions acting on the contact surface. However, the tooling on the contact surface to set the pin affects the boundary condition to change the values from the natural ones.

As the other non-destructive method than the classical curve-fitting mentioned at the first section of this statement, the author and his colleagues suggested a numerical method using the boundary element method (Kihara, J. et al. 1984). The surface of the tool which keep itself elastic is discretized to the boundary elements and the boundary element equation is constructed. The parameters on the element bolonging to the contact surface are unknown as for both tractions and diplacements, and therefore, the equation is impossible to be solved. Same as in the case of the residual stress analysis in rolled sheet, the subsidual informations and those relationships with the boundary parameters are necessary to obtain the contact surface parameters.

4.2 The architecture of the evaluation system of the contact surface tractions

The surface strain measurement of the tool, the compression load and the shape of crown of the compressed cylinder are utilized as the subsidual informations. These informations are related with the boundary parameters, and the relating equations are merged to the boundary element equation. Here we obtain a linear simultanious equation to solve the boundary parameters including those on the contact surface.

4.3 A practical example

The material of the work is a plain carbon machine structural steel, Several specimens of different ratio of height to diameter are prepared. The two lubrication conditions are applied and one is dry, and the other is in use of molybdenum di-sulphide powder dispersed in grease. In the latter condition the lubrication is expected to be well performed.

First it must be noticed that the distribution of the surface traction cannot be finely defined. The first reason is due to St′ Venant principle which degrades the utility of the strain data measured at a little far point from the contact area, and the second is related to the orthogonality of the subsidual informations with the evaluated bounary parameters. Therefore, the tractions at either the outer boundary of the contact or the center are only determined.

When the ratio of height to diameter is less than unity, the pressure is higher at the center than at the outer boundary, and the result is reversed when the ratio is more than unity. The result is parallel with the observation using the pin-contact method (Takahashi, S. 1965, Nagamatsu, A. et al., 1970).

The distribution of pressure is the more serious, the poorer the lubrication is. The estimated frictional stresses at the outer boundary are ranging from 30~60 MPa which corresponds to 5~10% of the normal pressure in well lubricated condition, and are 280MPa in dry lubrication which is 40~45% of the normal pressure with its change.

For other than finess of estimation, the present method gives a reasonable results.

5. THE SURFACE TRACTION ESTIMATION IN THE CASE OF RING COMPRESSION TEST WITH A SPECIAL RESPECT TO TRIBOLOGY IN METAL WORKING

5.1 Ring compression test

The ring compression test was originated by Kunoki in order to evaluate the frictional traction or frictional coeffcient on the contact surface during compression (Kunoki, M., 1954) Some classical modeling of the effect of friction on the deformation behavior was established, and the frictional traction on the contact surface was evaluated by the comparison of experimental displacement of the inner radii with the calculated one of the suitable frictional condition.

The inner radii expands in the well lubricated condition and shrinks in poorly lubricated condition, because the frictional work depends on the product of frictional stress by the slip displacement of the work against the tool, and the total displacement is saved when the slip movement of the work surface is locked in the middle of the contact radii due to the enough magnitude of the resistance against slip that is frictional sticking strength.

This method has been utilized as one of the orthogonal test method to evaluate the frictional stress or the frictional coefficient in metal working science.

5.2 Improvement of the previous method mentioned in 4

The inverse analysis of the contact surface traction is reported in the previous section, but here the author and his colleagues reconstructed the system to be improved (Aizawa, T. et al. 1991). The main improvement is the usage of only surface strains measured on the surface of the tool but the other informations of the shape of crown of the deformed work and the total load. In the old technique, the coefficient matrix has the zero diagonal coefficients relating the subsidual informations with the boundary parameters, which is a serious defect from the point of view of accuracy, and the homogeneity of variables is also defected.

For the 3 dimensional boundary element calculation, the surface strain cannot in general be determined by the ordinery numerical integration, but by analytical integration or numerical way associated with the kinematical principle, which means that the rigid displacement does not produce any strain. Here is no zero diagonal coefficients in the matrix and the homogeneity among the boundary parameters and the subsidual informations is also certifyed.

5.3 The experimentals of the ring compression test

The tool on which the surface strains are measured by rozett strain guages is a cylindrical die with a hole in the center. On the surface of the inner hole the strain guages are also pasted and the surface strains are measured. The die is made of the tool steel, and the work is made of copper. The surface of the work contacting with the tool is ground to some classes of roughness in about cocircular pattern.

Several synthetic oils of different viscosity are used as the lubricant, and of course the expepiment without any lubricant is also performed.

The tractions at the outer radii, inner radii and the middle of them are evaluated by solving the synthetic bounary element equation.

The inner radii change corresponds well to the magnitude of frictional stress as expected by the rough thinking. The important information of this analysis is concerning with the nature of distribution of frictional stress and also the relation between of the direction of the stresses and the behavior of the inner radii displacement.

The frictional stresses for almost all experiments are directed to outer radial direction, and in the case of fine surface with no lubrication the stress is the highest, and in the case of rough surface with the least viscous lubricant that is the lowest. The stress distributes linearly in the radial direction as highest at the outer and lowest at the inner.

The frictional stresses at the inner radii are ranging only from

20MPa to 42MPa for all experimental conditions, but on the other hand the frictional stresses at the outer radii are ranging from 38MPa to 155MPa depending seriously on the lubricating conditions. The normal pressures are ranging within only 20MPa about 470MPa as the middle value. Here it is clear that neither the constant frictional coefficient model nor the constant frictional stress model may be valid. This is an important discovery.

The paradoxical results are following. The direction of the frictional stress is outer radial in all case of analysis of the experiments, but in the case of poor lubrication the inner radii shrinks. At first sight it seems to be much ridiculous. The microscope observation of the surface of such work specimen deformed under the severe lubricating condition reveals that the folding of the inner wall has occured during deformation and the initial inner radii displaces to the outer direction. The new surface generates from the center by the folding behavior of the inner wall surface. The slip direction is always outer radial and the shrinkage of the inner hole occurs due to the folding behavior of the inner wall surface. It is one of the most important discoveries in this field.

Fortunately the stress distrubution almost linear and in this case the linear element model is adopted. Therefore the much reliable results may be obtained in this investigation, although only a small number of freedom can be given for the contact boundary informations.

6. THE THERMAL PROPERTY OF THE PLAIN CARBON STEEL DURING COOLING FROM ITS AUSTENITIC TEMPERATURE

6.1 The determination of thermal diffusivity of steels during cooling

This subject has in usual investigated using the finite differential method, and the established data base is easily accessed. Therefore this subject is a handy and proper bench mark test of the reliability of the numerical method. The reason why this subject has been deeply investigated is that it is of importance to catch accurate hold of the thermal transients of steel billet or slab during heating and cooling in the hot working processes and heat treatment engineerings.

When the boundary element modelling is applied to the non-steady heat transfer problem, the problem must be a boundary non-linear one. In the case of cooling steel ingot in the air from its austenitic temperature, the boundary is conditioned by heat flux determined by temperature. Though the boundary element method is applied, the inner thermal parameters are necessary in such non-steady heat transfer problem. The boundary element method is classified as one of the implicit methods, and therefore, in architecturing of the numerical method, the relation between the spatial discretization and the time increment interval governs strictly the accuracy of calculation. The author and his colleagues investigate the affectants on the accuracy and construct an accurate calculation program (Kihara, J. et al. 1988).

Other than resistance against heat flux on the boundary, the thermal diffusivity itself depends also on temperature. The temperature dependence of the analysed apparent thermal diffusivity reveals the latent heat at the transformation of steel from austenite to ferrite with pearlite. The temperature dependence of the thermal diffusivity increses nonlinearlity of the subject. As time is marched, temperature changes, and at every 25K, the boundary condition is up-dated and the proper value of the thermal diffusivity is researched to fit the calculated temperature with the measured one treating the discrepancy of the calculation from the measured result as a virtual heat source.

6.2 An experimental example

In order to test the method, a lectangular steel block which is plain carbon steel containing about 0.15% carbon is chosen as a experimental specimen. The length of the specimen is 300mm, and the cross section is square of which edge length is 50mm. At the center of the specimen and on the middle of the side surface the thermocouples are set to observe the temperature and its difference between the measuring points.

The specimen is heated in the electric furnace kept at 1273K for 2hrs. thereafter that is extracted in the air on the isolite blick block in the erectic pause.

At the short period after the extraction a small heating up is observed at the center of specimen. The author and his colleagues think it of as the heat input at the surface due to oxidation of the surface ferrous element of 10000 atomic depth based on the calculation. Such oxidation is feasible, because the content of oxygen is lowered in the semi-closed atmosphere in the electric furnace, and the extraction of the specimen from the furnace gives the surface oxygen enough to oxidize.

The resisitance on the surface agaist the heat flux at every 25K is reffered to the function of the radiation emissivity and the data base of the convective heat flow, and the bounary conditions are updated.

The virtual heat generation is evaluated to fit the calculation of the difference in temperature between that at the center and that on the surface with the observed one at every 25K during cooling. The real thermal diffusivity is evaluated based on the conversion of the virtual heat generation to the specific heat and the latent heat of phase transformation.

The analysis is in good agreement with thh traditional results in the literatures.

7. SUMMARY

Based on the review of the investigations worked by the author and his colleagues, who have tried to apply the boundary element technique to the inverse analysis of mechanics in metal working engineering, the inverse analystis using the boundary element technique is thought of as an effective and useful method. The utility is found to be due to the nature of the boundary element technique.

In the application of the static elastic boundary element, there comes difficulty due to St′ Venent principle in the linear elasticity. This interferes from the fine analysis on the contact boundary because of lack of the degree of freedom. Some improvement is possible, if we choose the proper element model corresponding to the real distribution of the surface variables. However it is only a sort of good luck. It is not the case in application of acoustic bounary element technique. This is a problem of this subject.

Other than this just mentione dabove, the thermal inverse problem can be fully and at the sametime satisfactorily analysed usiong the boundary element technique. The concept of the virtual heat generation is very powerful to make the subject easy to be solved as in the case of the application of the finite different technique.

REFFERENCES

KIHARA, J., SHEN, G., Yamauchi, T., Mimura, H., Makino, H. & Liu, B.. 1983. Boundary Elements, Proceedings of the Fifth International Conference, Hiroshima, Japan, November 1983, In Brebbia, C. A., Futagami, T. & Tanaka, M. (eds), :393-406. Berlin Heidelberg New York Tokyo: Springer-Verlag.

Kihara, J., Makino, H. & Shen, G.. 1984. Jounal ofthe Japan Society of Technology of Plasticity, vol. 25:806-812.

Takahashi, S.. 1965. Jounal of Japan Society of Technology of Plasticity, vol. 1:271

Nagamatsu, A., Murota, T. & Jimma, T.. 1970. Jounal of Japan Society of Technology of Plasticity, vol. 6:1276

Kunoki, M.. 1954. Determination of Friction coefficint by ring test. Reports of Scientific Research, : 63-76.

Aizawa, T., Sakamoto, H., Chihara, T. & Kihara, J.. 1991. Journal of the Fuculty of Engineering, The University of Tokyo (B) vol. 41- 2:351-367.

Kihara, J., Aizawa, T. and Taneda, K.. 1988. Boundary Element Methods in Applied Mechanics. In Tanaka, M & Cruse, T. A. (eds):463-472. Pergamon Press.

Simulation of Materials Processing: Theory, Methods and Applications, Shen & Dawson (eds)
© 1995 Balkema, Rotterdam. ISBN 90 5410 553 4

Characterization, validation and finite element modelling of extrusion

S.R. MacEwen & A. Langille
Alcan International Limited, Kingston Research and Development Centre, Ont., Canada

J. Savoie
Department of Mining and Metallurgical Engineering, McGill University, Montreal, Que., Canada

J. Root
AECL Research, Chalk River Laboratories, Ont., Canada

M.J. Stout, S.R. Chen & U.F. Kocks
Material Science Division, Los Alamos National Laboratory, N.Mex., USA

ABSTRACT: Characterization of material and interface properties, choice of appropriate numerical methods and validation of model results against experimental data are the three critical ingredients needed in order to develop a finite element model of extrusion that has true predictive capability and that is of use to industry. This paper focuses on the constitutive equations for the high temperature deformation of AA6063 extrusion billet and on the use of plant trials and neutron diffraction techniques to measure texture in the extrudate in order to validate model predictions.

1 INTRODUCTION

The use of finite element models to describe the performance of products and the processes of metal forming is finding ever-increasing use in industry. Mathematical models are augmenting or replacing traditional laboratory testing and plant-trial approaches to solving problems.

There are three critical aspects in developing a finite element model that will have true predictive capability. One must describe appropriately the constitutive behaviour of the metal being formed and the nature of the interface between the metal and the tooling; one must choose the appropriate mesh, boundary conditions and method of solution and one must demonstrate, by comparison with experiment, that the predictions of the model are realistic. Thus, characterization, numerical methods and validation are the three key ingredients in developing a finite element model for use in industry.

This paper will describe the use of finite element models to simulate the process of extrusion, with particular focus on the aspects of characterization and validation. Extrusion of aluminum alloy billet into a variety of products ranging from simple rod to exceedingly complex shapes, for applications ranging from aerospace to architecture, is a major concern for the aluminum producers. Annually, more than a million metric tonnes of aluminum billet in North America ends up in extruded products. Extrusion is second only to rolled products for aluminum usage.

2 CHARACTERIZATION

By far the most common approach to modelling extrusion is to use an Eulerian fluid flow code, such as POLYFLOW [1], FIDAP [2] or POLYCAD [3] to emulate the deformation of the billet that occurs as it is forced by the ram through the container, feeder and die. Thus, characterization of the plasticity of the metal must be expressed in terms of a shear-thinning viscosity. The most commonly used approach is to assume that a steady state strain rate equation, in the form:

$$\dot{\varepsilon} = A\sigma^{m} \exp(-Q / kT) \qquad [1]$$

describes the deformation process. Here, $\dot{\varepsilon}$ and σ are the axial strain rate and stress components, respectively and the stress exponent, m, and activation energy, Q are assumed independent of stress. The viscosity is then expressed as:

$$\mu = \sigma' / 2\dot{\varepsilon} \qquad [2]$$

where σ' is the deviatoric part of the axial stress component, and a representation of viscosity appropriate for the finite element code of choice can be defined. For aluminum alloy AA6063, the values of m and Q are generally taken to be about 6 and 1.6 ev, respectively.

The assumptions of constant m and Q are open to question. Clearly, if the rate controlling mechanism of deformation is dislocation glide, not diffusion as is assumed in equation [1], the activation energy will de-

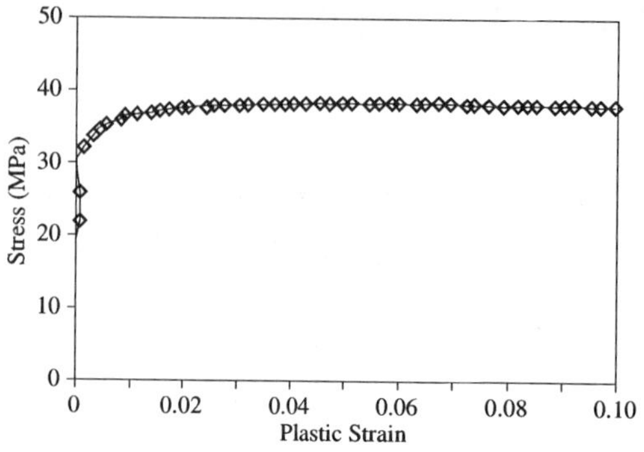

Figure 1: Stress–strain curve for AA6063 at 673K and 7 x $10^{-3}s^{-1}$.

pend on stress and the stress exponent will not be constant. A typical stress strain curve for aluminum alloys for temperatures above 400°C exhibits yield, followed by a relatively small transient leading to a saturation flow stress, as shown in Figure 1. Figure 2 shows the saturation flow stress data for AA6063 determined by compression testing for temperatures in the range 350°C to 550°C and strain rates in the range 10^{-4} s^{-1} to 2 x 10^3 s^{-1}. The data are plotted according to a procedure suggested by Kocks [4]. The abscissa is the activation energy, normalized by Gb^3, where G is the shear modulus and b the Burgers' vector. It is evident that all of the data, for the complete range of temperatures and strain rates, can be described by a single equation:

$$\ln(\sigma_S / G(T)) = \ln(\sigma_{SO} / G_0) - (kT / W_0 Gb^3)\ln(\dot{\varepsilon}_{SO} / \dot{\varepsilon}) \qquad [3]$$

Here, $\dot{\varepsilon}_{SO}$ is a scaling factor (actually the pre-exponential of the rate equation) that unifies the temperature and stress dependencies of the saturation stress; $\dot{\varepsilon}_{SO} = 10^7$ proves satisfactory. W_0 and σ_{SO} / G_0 are determined from the slope and intercept of Figure 2, respectively; the subscript 0 refers to 0K. With the formulation of equation [3],

$$\text{“}m\text{”} \equiv \frac{\partial \ln \dot{\varepsilon}}{\partial \ln \sigma} = W_0 \frac{Gb^3}{kT} \qquad [4]$$

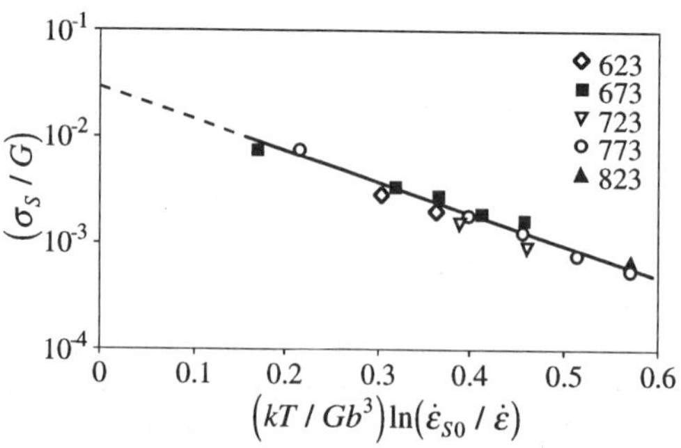

Figure 2: Saturation flow stress for AA6063.

Thus *"m"* depends on temperature, varying from 7 to 4.5 over the temperature range of interest for extrusion, and *"Q"* depends on temperature, through the temperature dependence of the shear modulus, and on stress. Clearly, neither are constant as required by equation [1]. The strain rate predicted for a given stress depends sensitively on the form of the constitutive law chosen. The significance of this difference would appear directly in predictions of the ram pressure and indirectly by changing the amount of dissipative heating predicted by an FE model.

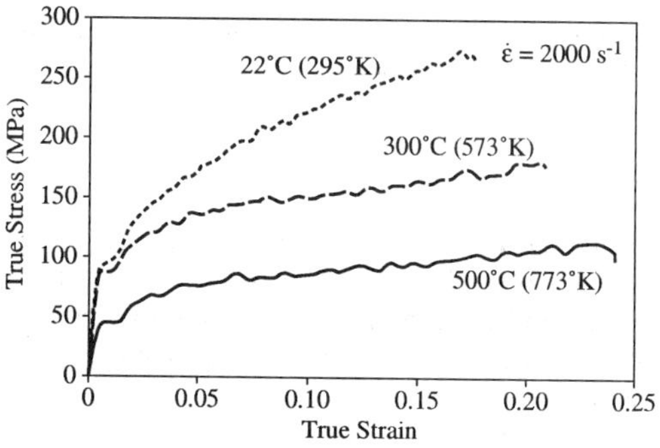

Figure 3: Hopkinson bar experiments; stress–strain curves for strain rates in excess of 2 x 10^3 s^{-1}.

A second concern relates to the assumption that the workhardening transient may be ignored for high temperature deformation. While the workhardening transient may be negligible at low to intermediate strain rates, Figure 3 shows that at high strain rates, in this case about 2 x 10^3 s^{-1} achieved by the use of Hopkinson bar experiments, the workhardening transient, even at 500°C, is significant. Incorporation of all aspects of high temperature deformation, including the strain rate dependencies of both yield and saturation and a description of the workhardening transient could be based on the "MTS" approach of Kocks and co-workers [5–7] and will be described in a future publication.

3 NUMERICAL METHODS

There are two basic issues in modelling extrusion: the choice of type of code and the choice of appropriate boundary conditions. Code options are Eulerian or Lagrangian and for each, transient or steady state analysis. Eulerian codes are by far the most widely used, as they do not require the extensive rezoning that would be the case if a Lagrangian code were to be used. Nevertheless, the Eulerian codes provide special difficulties in incorporating moving boundaries (i.e. the ram), in calculating stresses and deflections of the tooling and, most importantly, in calculating residual stresses in the extrudate. For real industrial applications, a transient analysis is almost always required, in order to capture the transient temperature distributions that occur in actual press tooling as billets are sequentially extruded and dies are changed. Definition of heat transfer coefficients, HTC's, and the conditions of friction (slip/stick) along the container, feeder and die are the primary requisite boundary conditions. Thermal boundary conditions become progressively more simple as the extent of the tooling being modelled increases. For example, if one includes all of the tooling one needs only to define heat transfer to the air surrounding the press. However, if one models only to the inside boundary of the container, feeder and die, the problem of assigning an appropriate thermal boundary condition becomes formidable, if not intractable. Clearly the solution is to model as much of the complete tooling set as is possible, given the limitations of available computing power. Friction at the billet-tooling interface remains an unsolved problem. Generally, full stick is assumed, but this may be questionable. For both heat transfer and friction, the best (and possibly only) solution is to use instrumented plant trials to "calibrate" both the heat transfer and friction conditions of the press.

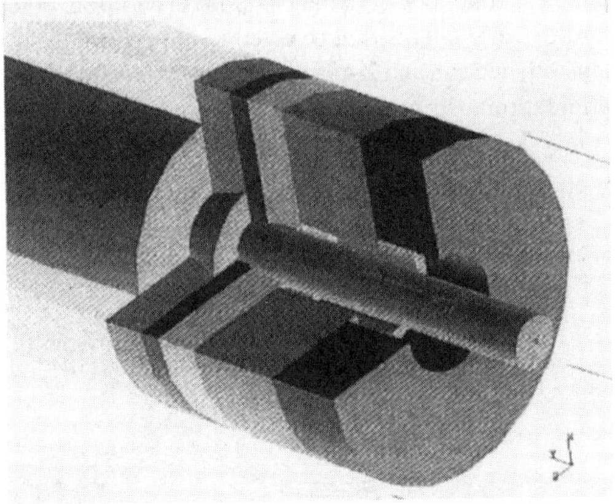

Figure 4: Solid model of a typical extrusion press tool stack.

A typical press geometry is shown in Figure 4. Note that all of the important components, from a metal flow and heat transfer point of view, are present in the model; these include the container, feeder, die, backer, bolster and sub-bolster. The partially extruded billet is also shown. During operation of the press all of the tooling components are in intimate contact due to the high force of the ram. However, during the time taken to extract the ram and to insert the next billet, the tooling components are under a much lower contact pressure. Consequently, in defining the HTC's between the tooling components one must be cognisant of the effect of interfacial pressure. Generally, the HTC's are high during extrusion and lower during the time when the ram pressure in not applied.

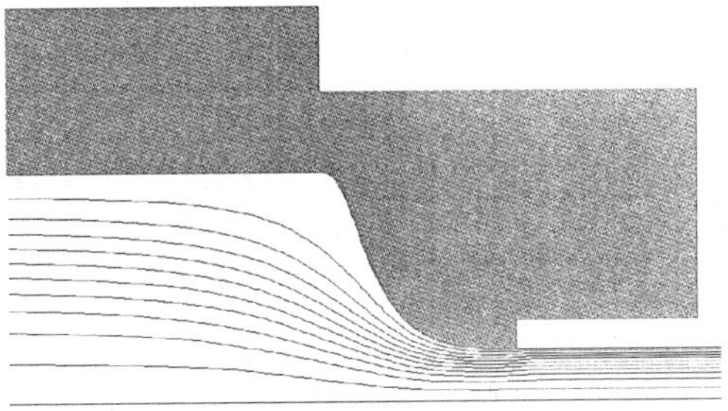

Figure 5: Streamline pattern for a second design of tool stack.

Analysis of the press configuration shown in Figure 4 has been done using FIDAP. The ram was not modelled and an "inlet velocity" was defined corresponding to the ram velocity imposed during the trials. Transient analysis followed the extrusion of three billets. Initially, it was assumed that the tooling and billet were at uniform pre-heat temperatures. Thereafter, temperatures were calculated in the billet and all of the tooling components. The coupled problem of "fluid flow" with dissipitative heating and heat transfer was solved. The analysis predicts strain rates varying from a low of about 10^{-3} s^{-1} at the inlet to the feeder to a maximum of about 10^{2} s^{-1} at the exit of the die bearing, and that the temperature increase due to the shear-rate heating is about 260°C after 20 seconds of extrusion. A second model was developed using the feeder and die as shown in Figure 5, which plots the streamline distribution after 20 seconds. Note that the inflection point angle of each streamline increases with increasing displacement from the axis of the rod.

4 VALIDATION: TEMPERATURE COMPARISONS

The tooling shown in Figure 4 was instrumented with 32 thermocouples placed at the die bearing and other strategic positions throughout the tool stack. Trials were done on a commercial press using an 0.25 metre diameter billet and ram speeds varying from about 0.005 to 0.01 m/s. Temperatures were recorded continuously as a series of billets was extruded. Initial estimates of the tooling interfacial HTC's gave reasonable, but not acceptable, agreement with the experiments. Several iterations of the model, using different estimates of the HTC's, were required before the agreement shown in Figures 6 and 7 were obtained. The effect of dissipative heating is clearly evident in Figure 6, which plots data for the die bearing. The effect of the gap conditions changing when the tool stack is not under the pressure of the ram is clearly evident in Figure 7, which plots data for the bolster. It is concluded that the coefficients so obtained are appropriate for the particular press of Figure 4 and could be used with reliable predictive capability for other feeder/die designs and press operating conditions.

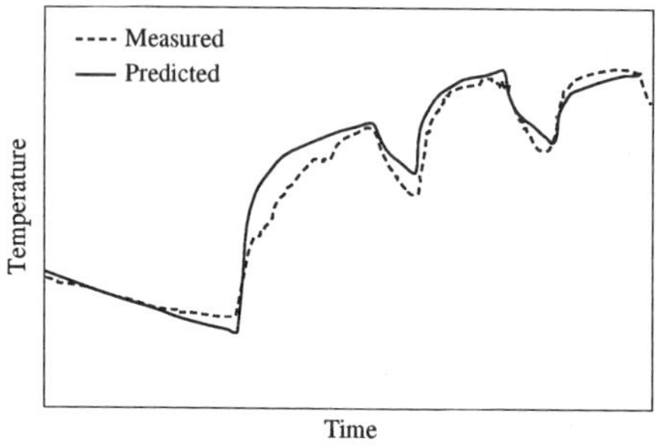

Figure 6: Temperature transients near the die bearing of Figure 4.

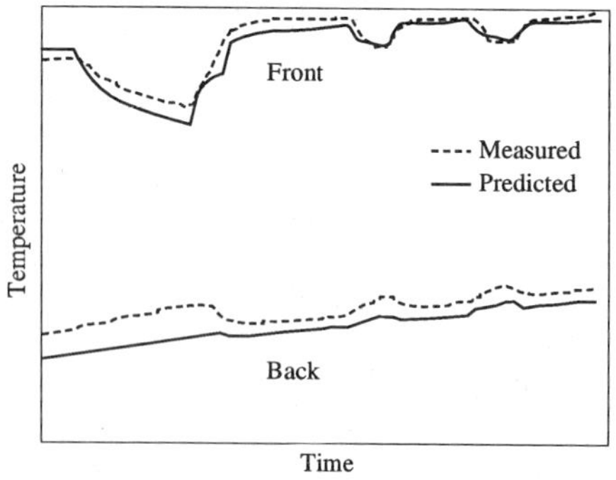

Figure 7: Temperature transients in the bolster of Figure 4.

5 VALIDATION: EXAMINATION OF THE EXTRUDATE

Although plant trials can provide valuable information with which to fine-tune and validate models, they are expensive and time consuming. Ideally, it would be possible to use some attribute of the extrudate itself as a tool for validation. The strong shear rate gradients seen in the models suggests a possibility. The evolution of crystallographic texture is known from decades of work to evolve according to strain path and the amount of imposed deformation. Thus, it is reasonable to hypothesize that the gradient in plasticity produced by extrusion should produce a corresponding gradient in crystallographic texture; essentially the fingerprint of extrusion.

Neutron diffraction experiments were done at the Chalk River Laboratories using a rod extruded using the feeder and die of Figure 5. Texture gradients were mapped out non-destructively using Cd masks with slits 1 mm wide x 9 mm high to define the incident and diffracted neutron beams. A computer-controlled X-Y table was used to translate the sample in a radial direction, thereby tracking the small diffracting volume across a diameter of the rod. For one series of measurements, the rod was oriented such that the scattering vector, q, was at a fixed angle ψ to the axial direction of the rod. In a second series of measurements, a fixed value of the diametral position was chosen and the angle ψ was scanned to map out the distribution of grain orientations at that location. This distribution was analyzed by fitting a Gaussian function to the distribution of intensity with ψ. In this way, the angular positions of the maxima, ψ_{max}, of the {111} and {200} diffraction peaks were determined as functions of radial position. In addition, pole figures were obtained at selected radial positions using both neutron and X-ray diffraction, and from these, orientation distribution functions were calculated.

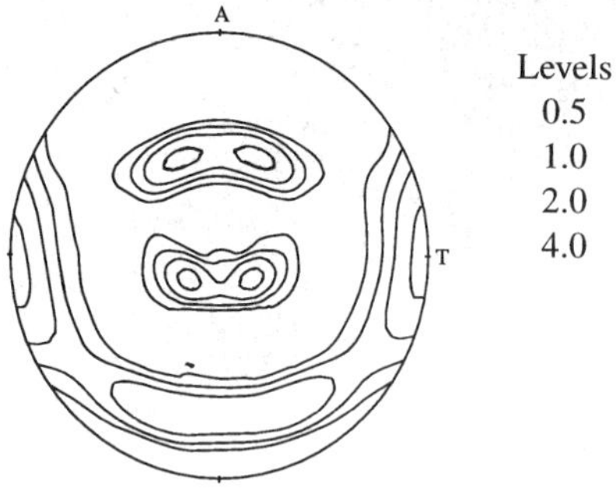

Figure 8: {111} pole figure 1 mm from the surface of the extruded rod.

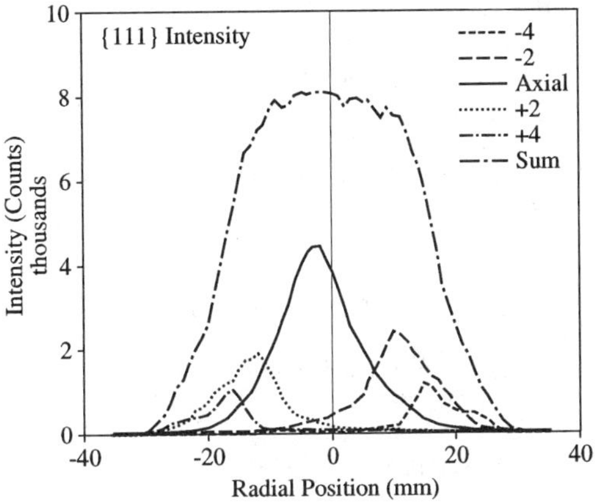

Figure 9: Variation of {111} diffraction intensity as a function of radial position.

{111} and {200} pole figures show that the centre-line of the rod exhibits a very strong "wire-draw" texture, with maximum pole densities 130 and 20 times random for the {111} and {200} fibres, respectively. There is little or no difference between the centre and mid-radius textures. The situation changes drastically at the surface where the {111} texture, Figure 8, is reminiscent of a brass rolling texture, but rotated by about 18 degrees about the circumferential (tangential) direction. Figure 9 shows the variation in intensity of the {111} diffraction line with radial position. When $\psi = 0$, the intensity peak has its maximum near the centre of the rod. As ψ is rocked to the left or right, the position of maximum intensity shifts radially, in one direction for positive ψ and in the opposite for negative ψ. The effect is most striking for the {111} diffraction line, but is also present for the {200}. Figure 10 plots ψ_{max} as a function of radial position for the {111} data. The shape of this plot, showing a symmetric pattern, indicates that all {111} poles lie within ±15° of axial, and is surely a most sensitive indicator of the way in which the aluminum flowed through the die.

6 DISCUSSION

The data of Figures 9 and 10 present a compelling correlation between the ψ rotation of the fibre texture and the nature of the streamlines of flow. At the centreline the texture is exactly as expected and well documented for wire-drawing operations, which impose the same strain path as extrusion. At a particular radial position, the plane normals of the crystals correctly oriented to diffract at a given ψ are rotated into qualitative alignment with the streamlines. Figure 11 plots the inflection point angle, θ, determined from the finite element analysis. Although Figures 10 and 11 differ in magnitude, their resemblance is remarkable.

The prediction of texture evolution for non-homogeneous processing routes, such as extrusion and rolling, has received considerable attention over the past five years. Work by Mathur and Dawson [8] in 1990 presented a methodology for analyzing the development of texture during wire drawing using models of polycrystalline plasticity. Van Houtte et al in 1993 [9] attempted to predict the "cyclic" surface texture of the ferritic phase of drawn, pearlitic wires using a Taylor model and, most recently, Aukrust et al [10] presented simulations and measurements of texture in an extruded AA6068 plate.

In order to predict extrusion textures according to crystal plasticity theory, an appropriate form of the velocity gradient L_{ij} must be found. At the center of the rod, its diagonal takes the following obvious form: (-0.5,-0.5,1.0). The procedure consisting of finding the velocity gradient acting at an arbitrary, radial position of the rod using integration along the stream-

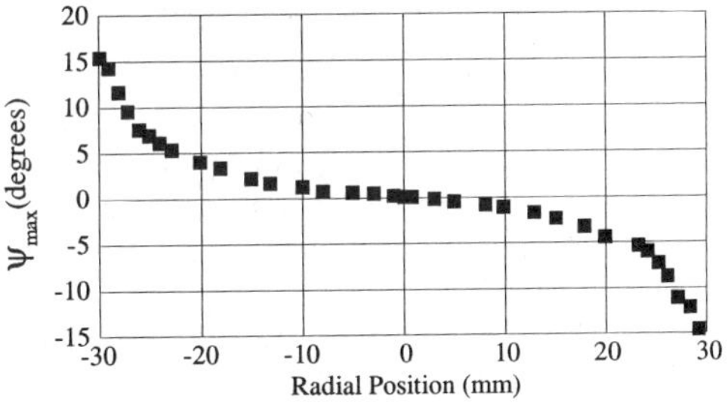

Figure 10: ψ_{max}, determined experimentally from the {111} neutron diffraction data as a function of radial position in the billet.

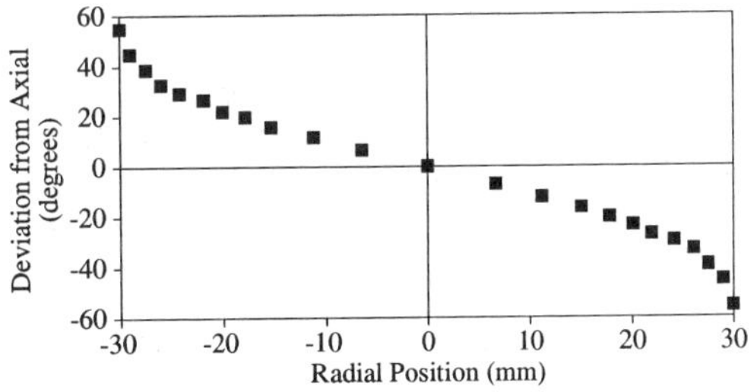

Figure 11: θ, determined from the inflection point angle of the streamlines in the finite element model, as a function of radial position in the billet.

lines calculated by finite element analysis is beyond the scope of this paper. However, as the measured surface texture has a strong resemblance to a rolling texture, it is reasonable to assume that the diagonal of L_{ij} at the surface should take the form: (-1,0,1). Assuming that the changes in the velocity gradient evolve gradually between the center and the surface, L_{ij} (without yet taking into account any shear) takes the following form:

$$L = \begin{bmatrix} -.5-\beta & 0 & 0 \\ 0 & -.5+\beta & 0 \\ 0 & 0 & 1.0 \end{bmatrix} \qquad [5]$$

where ß=0 at the center and ß=0.5 at the surface or any value between these limits at intermediate radial positions.

Experimental textures are characterized by a rotation of the {111} and {200} fibres about the tangential direction. This can be reproduced by simply rotating the velocity gradient, L_{ij} by an angle κ about the tangential direction. The velocity gradient in the new, local coordinates, L_{ij}^S, takes the final form:

$$L^S = \begin{bmatrix} L_{11}^S & 0 & 0 \\ 0 & -.5+\beta & 0 \\ L_{31}^S & 0 & L_{33}^S \end{bmatrix} \qquad [6]$$

where the values of L_{11}^S, L_{33}^S and L_{31}^S depend on the rotation angle, κ. Thus the two parameters ß and κ allow a definition of the velocity gradient at any radial position.

It has been found that by appropriate choice of ß in the range 0 to 0.5 and κ in the range 0 to -24° (centre to surface) the observed textures, varying from pure wire-draw at the centre to a rotated "brass" rolling texture at the surface, could be predicted acceptably well.

7 CONCLUSION

Finite element simulations can be valuable tools for industrial application provided that the appropriate care is taken in characterization of the material behaviour, choice of analysis code and boundary conditions and validation of the model.

ACKNOWLEDGEMENT

SRM would like to thank Dr. U.F. Kocks for many helpful discussions on the analysis of data and the application of the MTS model to aluminum alloys.

REFERENCES

1. POLYFLOW s.a. Place de l'Universite 16, Louvain-la-Neuve, Belgium.

2. FIDAP Fluid Dynamics International, 500 Davis St., Evanston, Il. 60201 USA.

3. POLYCAD Polydynamics Inc., 1685 Main St. W., Hamilton, Ont. Canada L8S 1G5

4. Kocks, U.F., private communication

5. Mecking, H. and Kocks, U.F., *Acta. Metall.,* vol. 29, 1981, p. 1865

6. Follensbee, P.S. and Kocks, U.F., "A Constitutive Description of Copper Based on the Use of the Mechanical Threshold Stress as an Internal State Variable," *Acta. Metall.,* vol. 36, 1988, pp. 81-93.

7. Kocks, U.F. and Chen, S-R., "Constitutive Laws for Deformation and Dynamic Recrystallization in Cubic Metals," *Aspects of High Temperature Deformation and Fracture in Crystalline Materials*, edited by Hosoi, Y., Yoshinaga, H., Oikawa, H., and Maruyama, K., Proceedings of the 7th JIM International Symposium on Aspects of High Temperature Deformation and Fracture in Crystalline Metals, Nagoya, Japan, pp. 593-600, (1993).

8. Mathur, K.K. and Dawson, P.R., "Texture Development During Wire Drawing" *J. Eng. Mater. Technol.,* vol. 112, pp. 292-297, July 1990

9. Van Houtte, P., Watté, P., Aernoudt, E., Sevillano, J. Gil, Lefever, I., and Van Raemdonck, W. "Taylor Simulation of Cyclic Textures at the Surface of Drawn Wires using a Simple Flow Field Model", Proceedings of the 10th International Conference on Textures of Materials, *Materials Science Forum,* vol. 157-6, pp. 1881-1886, 1994

10. Aukrust, T., Tjøtta, S., Skauvik, I., Vatne, H.E., and Van Houtte, P. "Modelling of Texture Development in Aluminium Extrusion", Proceedings of the 15th Riso International Symposium on Materials Science, 5-9 September 1994

Simulation of Materials Processing: Theory, Methods and Applications, Shen & Dawson (eds)
© 1995 Balkema, Rotterdam. ISBN 90 5410 553 4

Finite element simulation of high-speed machining

T.D. Marusich & M. Ortiz
Brown University, Providence, R.I., USA

ABSTRACT: A Lagrangian finite element model of orthogonal high-speed machining is developed. The model utilizes continuous remeshing to alleviate element distortion problems associated with the finite deformations encountered. The thermo-mechanically coupled model accounts for dynamic response of multiple discretized-bodies in contact, as well as interfacial heat transfer. Both ductile and brittle fracture initiation and propagation are accounted for during processing; an essential feature in determining shear-localized chip morphologies. No predetermined "line of separation" or slide-line contact has been assumed at the depth of cut near the tool tip. Rather, the workpiece material is allowed to soften and flow around the finite radius of the discretized cutting tool, fracturing only when the criterion is satisfied. Examples of continuous and shear-localized chip morphologies are given in the high-speed range.

1 Introduction

The development of a new machining process requires considerable investment of time and resources (Vasilash, 1992). Precise knowledge of the optimal range of the cutting parameters is essential for a timely startup. Process features such as tool geometry and cutting speed directly influence chip morphology, cutting forces, final product dimensionality and tool life. Computer simulation of the cutting process can potentially reduce the number of design iterations and result in a substantial cost savings. Considerable effort has therefore been devoted to the development of computational models of high-speed machining. Of primary concern is the determination of the steady-state temperature distribution in the workpiece, tool and chip (e. g., Tay *et al.*, 1976; Tay *et al.*, 1974; Lin *et al.*, 1992). Attempts to model the process of chip formation have for the most part been based on a predetermined line of separation between the workpiece and chip (e. g., Strenkowski and Carroll, 1985; Komvopoulos and Erpenbeck, 1991). Nodes on this line are separated when the tool tip is sufficiently close, or when a certain level of plastic strain is attained. Evidently, this
simple approach is not capable of predicting surface roughness and chip morphology, nor is it able to predict the extremely large strains encountered near the tool-chip interface. Sekhon and Chenot (1993), by contrast, have used mesh adaptivity to allow for an arbitrary surface of separation. However, elastic strains are not accounted for, which precludes the computation of residual stresses in the workpiece.

In addition, these models do not account for fracture, which severely limits the types of chip morphologies which can be predicted. It should also be noted that the majority of simulations conducted in the past idealize the tool as rigid. The tool stiffness, however, is known to directly influence surface roughness and tool chatter (Kalpakjian, 1984). Also, the lack of a discretized tool eliminates the capability to make a complete evaluation of the cutting tool performance since neither stresses nor tool temperatures can be determined. In this paper, we present a 2-D model of orthogonal high-speed machining, based on the developments of Marusich and Ortiz (1994), which overcomes some of the previously mentioned limitations. During the machining event, extremely large deformations occur which invariably result in distortion of the elements within a Lagrangian type setting. To alleviate this problem we employ continuous remeshing and adaptive refinement to maintain good element aspect ratios and an appropriate level of accuracy in the heavily deforming regions. Remeshing is also a tool used for the initiation and propagation of cracks in the workpiece; crucial to predicting shear-localized chip morphologies. In the following sections we will outline the finite element model used in the machining application. For the sake of brevity, we intentionally leave out some fine points of the procedure and implicitly refer to Marusich and Ortiz (1994) for a more in-depth discussion.

2 Numerical Procedure

2.1 *Equations of Motion*

Consider a solid initially occupying a reference configuration B_0, and a process of incremental loading whereby the deformation mapping over B_0 changes from $\boldsymbol{\phi}_n$, at time t_n, to $\boldsymbol{\phi}_{n+1} = \boldsymbol{\phi}_n + \mathrm{u}$, at time $t_{n+1} = t_n + \Delta t$. Dynamic equilibrium is enforced at time t_{n+1} weakly by recourse to the virtual work principle

$$\int_{B_0} \mathrm{P}_{n+1} : \nabla_0 \boldsymbol{\eta} \, dV_0 - \int_{B_0} (\mathrm{f}_{n+1} - \rho_0 \mathrm{a}_{n+1}) \cdot \boldsymbol{\eta} \, dV_0 - \int_{\partial B_{0\tau}} \mathrm{t}_{n+1} \cdot \boldsymbol{\eta} \, dS_0 = 0 \quad (1)$$

where P_{n+1} denotes the first Piola-Kirchhoff stress field at time t_{n+1}, f_{n+1}, a_{n+1} and t_{n+1} are the corresponding body forces, accelerations and boundary tractions, respectively, ρ_0 is the mass density on the reference configuration, $\boldsymbol{\eta}$ is an admissible virtual displacement field, and ∇_0 denotes the material gradient. Upon discretization of (1) with finite elements the governing equations become

$$\mathrm{M}\mathrm{a}_{n+1} + \mathrm{F}^{int}_{n+1} = \mathrm{F}^{ext}_{n+1} \quad (2)$$

where M is a standard mass matrix, lumped in our cased, F^{ext} is the external force array including body forces and surface tractions and F^{int} is the internal force array arising from the current state of stress.

The second-order accurate central difference algorithm is used to discretize (2) in time (Hughes and Belytschko, 1983; Belytschko, 1983), with the result

$$\mathrm{d}_{n+1} = \mathrm{d}_n + \Delta t \mathrm{v}_n + \frac{1}{2}\Delta t^2 \mathrm{a}_n \quad (3)$$

$$\mathrm{a}_{n+1} = \mathrm{M}^{-1}\left(\mathrm{F}^{ext}_{n+1} - \mathrm{F}^{int}_{n+1}\right) \quad (4)$$

$$\mathrm{v}_{n+1} = \mathrm{v}_n + \frac{1}{2}\Delta t(\mathrm{a}_{n+1} + \mathrm{a}_n) \quad (5)$$

where d, v and a denote the displacement, velocity and acceleration arrays, respectively. Although the minimum time step used in explicit dynamics is bounded by stability (Hughes, 1983), contact algorithms available for explicit dynamics are more robust and straightforward than their implicit counterparts. Explicit schemes are therefore more attractive for problems such as machining which involve complicated contact situations. Explicit integration is particularly attractive in three-dimensional calculations, where implicit schemes lead to system matrices which often exceed the available in-core storage capacity.

2.2 *Thermal Effects*

In applications such as machining, large amounts of heat may be generated due to the plastic working of the solid and friction at the tool-chip interface. The temperatures attained can be quite high and have a considerable influence on the mechanical response. The relevant balance law, in this case, is the first law, which can be expressed in weak form as

$$\int_{B_t} \rho c \dot{T} \eta \, dV + \int_{\partial B_{tq}} h \eta \, dS = \int_{B_t} q \cdot \nabla \eta \, dV + \int_{B_t} s \eta \, dV \quad (6)$$

where ρ is the current mass density, c the heat capacity, T the spatial temperature field, η an admissible virtual temperature field, h the outward heat flux through the surface, q is the heat flux, s is the distributed heat source density, and B_{tq} the current Neumann boundary. In machining applications, the main sources of heat are plastic deformation in the bulk and frictional sliding at the tool-workpiece interface. The rate of heat supply due to the first is estimated as

$$s = \beta \dot{W}^p \quad (7)$$

where $\dot{W}^p$ is the plastic power per unit deformed volume and the coefficient β is of the order of 0.9 (e. g. Kobayashi *et al.*, 1989). The rate at which heat is generated at the frictional contact, on the other hand is

$$h = -t \cdot [\![v]\!] \quad (8)$$

where t is the contact traction and $[\![v]\!]$ is the jump in velocity across the contact. This heat must be apportioned between the tool and chip. Using transient half-space solutions, the ratio of the heat supply to the chip, h_1, and the tool, h_2, can be computed as (*cf.* Grigull and Sandner, 1984)

$$\frac{h_1}{h_2} = \frac{\sqrt{k_1 \rho_1 c_1}}{\sqrt{k_2 \rho_2 c_2}} \tag{9}$$

where k_α, ρ_α and c_α, $\alpha = 1, 2$, are the thermal conductivity, mass density and heat capacity of the workpiece and the tool, respectively.

Inserting the finite element interpolation into (6) results in the semi-discrete system of equations (Belytschko, 1983)

$$C\dot{T} + KT = Q \tag{10}$$

where T is the array of nodal temperatures, C is the heat capacity matrix, K is the conductivity matrix, and Q is the heat source array. In the applications of interest here, the mechanical equations always set the critical time step for stability. It therefore suffices to lump the capacitence matrix and integrate the energy equation (10) explicitly by the forward Euler algorithm (Hughes and Belytschko 1983; Belytschko, 1983), with the result

$$T_{n+1} = T_n + \Delta t \dot{T}_n \tag{11}$$

$$C\dot{T}_n + K_n T_n = Q_n \tag{12}$$

2.3 *Thermo-Mechanical coupling*

A staggered procedure (Park and Felippa, 1983) is adopted for the purpose of coupling the thermal and mechanical equations. Mechanical and thermal computations are staggered assuming constant temperature during the mechanical step and constant heat generation during the thermal step. Following Lemonds and Needleman (1986), A mechanical step is taken first based on the current distribution of temperatures, and the heat generated is computed from (7) and (8). The heat thus computed is used in the thermal analysis where temperatures are recomputed by recourse to the forward Euler algorithm (11) and (12). The resulting temperatures are then used in the mechanical step and incorporated into the thermal-softening model described in Section 2.5, which completes one time-stepping cycle. A schematic flowchart of the staggered procedure is given in Box 1.

i) Initialize $T_1 = T_0 + \Delta t \dot{T}_0$, $n = 0$.
ii) Isothermal mechanical step:
$\{\Omega_n, T_{n+1}\} \rightarrow \{\Omega_{n+1}, T_{n+1}\}$
iii) Heat generation (bulk + contact)
iv) Rigid conductor step:
$\{\Omega_{n+1}, T_{n+1}\} \rightarrow \{\Omega_{n+1}, T_{n+2}\}$
v) $n \rightarrow n + 1$, GOTO (ii)
where $\Omega_{n+1} = \{x_{n+1}, v_{n+1}, a_{n+1}\}$

Box 1. Thermo-mechanical coupling procedure.

2.4 *Thermo-Mechanical Contact*

Machining involves contact between deformable bodies, e. g., the tool and workpiece or the faces of a crack propagating through the chip. Heat is generated at the contact site by friction and which conducts into the different bodies. A contact routine was developed based on the method by Taylor and Flanagan (1987) to handle deformable and rigid contact. This method was then extended to include heat transfer across an interface as well as the inclusion of frictional heat generation. The deformable contact algorithm consists of computing a predicted configuration for each time step assuming no contact has occurred. The boundary is checked for nodes which have penetrated the surface. A corrective force is applied to the penetrated nodes to enforce the contact conditions, taking into account the Coulomb friction law used. After the contact has been enforced the frictional heat is computed and apportioned to the nodes on the contacting surfaces by way of (9).

The method of interface heat transfer developed is based on the assumption of temperature continuity across contact boundary. The contacting nodes are predicted in the mechanical time step from the contact algorithm. These nodes then provide a thermal constraint on (12) at the contact boundary where temperature and rate continuity are enforced. For contacting nodes not lying on top of one another, the temperature continuity is enforced to be the linear interpolant along the boundary. Once continuity is enforced a reduced set of equations (12) is formed and the temperatures are computed in the normal fashion.

2.5 *Constitutive Model*

We adopt a standard formulation of finite deformation plasticity based on a multiplicative decomposition of the deformation gradient into its elastic and plastic components. From this an extension from small strain plasticity into the finite deformation range is made through purely kinematic manipulations, independent of the material model. In addition, we effect the requisite constitutive updates by a fully implicit update algorithm. The interested reader should consult Cuitiño and Ortiz (1992) for further elaboration.

In a typical high-speed machining event, very high strain rates in excess of 10^5 s^{-1} may be attained within the primary and secondary shear zones, while the remainder of the workpiece deforms at moderate or low strain rates. Under these conditions, a power viscosity law with constant rate sensitivity m is not adequate. Indeed, the experimental stress-strain rate curves (Klopp *et al.*, 1985; Clifton and Klopp, 1985; Tong *et al.*, 1992) for metals exhibit a transition at strain rates of the order of 10^5-10^6 s^{-1} from low to high rate sensitivity. At low strain rates, a rate sensitivity exponent in the range 100-200 adequately fits the data, while in the high strain rate regime a much lower rate sensitivity exponent in the range 5-20 applies. While the experimental observations indicate that the effect of stress elevation at high strain rates is due to contributions of increased flow stress rate sensitivity, contributions by the rate sensitivity of hardening are also found to be significant for some metals. However, it is felt that for structural steels of the type in our simulations the dominant behavior is that of flow stress rate sensitivity. Modelling this type of behavior is important, since a high rate sensitivity in the primary shear zone may lead to an elevation in stress which in turn can promote brittle fracture. A simple model which accounts for this behavior consists of assuming a stepwise variation of the rate sensitivity exponent m while maintaining continuity of stress. This leads to the relation

$$\left(1+\frac{\dot{\epsilon}^{\mathrm{p}}}{\dot{\epsilon}_0^{\mathrm{p}}}\right) = \left(\frac{\bar{\sigma}}{\mathrm{g}(\epsilon^{\mathrm{p}})}\right)^{\mathrm{m}_1} \quad (13)$$

$$\left(1+\frac{\dot{\epsilon}^{p}}{\dot{\epsilon}_0^{p}}\right)\left(1+\frac{\dot{\epsilon}_t}{\dot{\epsilon}_0^{p}}\right)^{\mathrm{m}_2/\mathrm{m}_1-1} = \left(\frac{\bar{\sigma}}{\mathrm{g}(\epsilon^{\mathrm{p}})}\right)^{\mathrm{m}_2} \quad (14)$$

where $\bar{\sigma}$ is the effective Mises stress, g the flow stress, ϵ^{p} the accumulated plastic strain, $\dot{\epsilon}_0^{\mathrm{p}}$ a reference plastic strain rate, m_1 and m_2 are low and high strain rate sensitivity exponents, respectively, and $\dot{\epsilon}_t$ is the threshold strain rate which separates the two regimes. For the low strain rate regime (13) applies if $\dot{\epsilon}^{\mathrm{p}} \leq \dot{\epsilon}_{\mathrm{t}}$, and (14) applies if $\dot{\epsilon}^{\mathrm{p}} > \dot{\epsilon}_{\mathrm{t}}$. In calculations, we begin by computing $\dot{\epsilon}^{\mathrm{p}}$ according to (13), and switch to (14) if the result lies above $\dot{\epsilon}_{\mathrm{t}}$.

Following Lemonds and Needleman (1986), we also adopt a power hardening law with linear thermal softening. This gives

$$\mathrm{g} = [1-\alpha(\mathrm{T}-\mathrm{T}_0)]\sigma_0\left(1+\frac{\epsilon^{\mathrm{P}}}{\epsilon_0^{\mathrm{P}}}\right)^{\frac{1}{\mathrm{n}}} \quad (15)$$

where n is the hardening exponent, T the current temperature, T_0 a reference temperature, α a softening coefficient, and σ_0 is the yield stress at T_0. It should be noted that, owing to the staggered integration of the coupled thermal-mechanical equations, the temperature T remains fixed during a mechanical step and, therefore, plays the role of a known parameter during a stress update.

2.6 *Fracture criteria*

The process of shear-localized and discontinuous chip formation involves the propagation of fractures through the deforming chip. The simulation of these chip morphologies therefore requires the formulation of suitable fracture criteria, in conjunction with numerical procedures for nucleating and propagating a crack through the mesh (see Marusich and Ortiz (1994) for an elaboration).

Structural steels can fracture in a brittle or ductile manner (see Ritchie and Thompson, 1985 for a review) based on the state of stress and deformation. Mixed mode fracture conditions, as expected in machining, will result in a brittle fracture when the hoop stress, $\sigma_{\theta\theta}$, on a plane reaches a critical value (Ritchie *et al.*, 1973) in accordance with the material fracture toughness, $\mathrm{K_{IC}}$, through

$$\sigma_f = \frac{\mathrm{K_{IC}}}{\sqrt{2\pi l}} \quad (16)$$

and

$$\max_{\theta} \sigma_{\theta\theta}(l,\theta) = \sigma_{\mathrm{f}} \quad (17)$$

Void growth and coalescence is known to be a principal mechanism of ductile fracture in structural steels (Cox and Low, 1974). Voids are nucleated as a result of fracture or decohesion of carbides and subsequently grow as the surrounding material strains plastically. The experimental evidence shows that the fracture toughness of metals

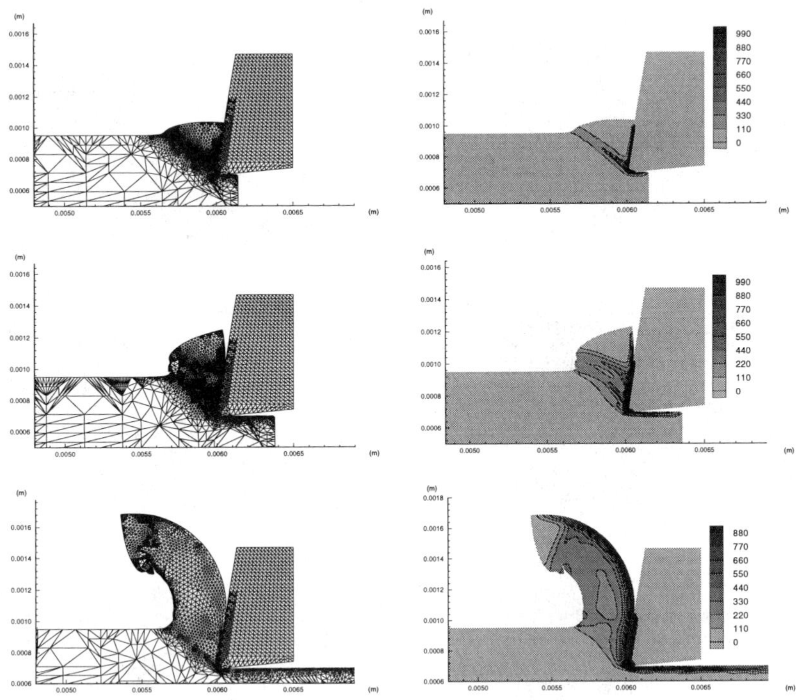

Mesh (a) Temperature contours (b)

Figure 1: Continuous chip formation. V=30m/s, Rake=10°

depends on the size and spacing of void nucleating second phase particles (Cox and Low, 1974; Rawal and Gurland, 1977). Ritchie *et al.* (1979) were able to cast the ductile fracture criterion in terms of an effective plastic strain ϵ_f^p at a distance l directly ahead of the crack tip with stress. Proceeding as before, we express this criterion in the form

$$\max_{\theta} \epsilon^p(l, \theta) = \epsilon_f^p \tag{18}$$

with the understanding that the crack propagates at the angle θ for which the criterion is met. Based on Rice and Tracey's solution (1969), the critical effective plastic strain can be estimated as (Ritchie and Thompson, 1985)

$$\epsilon_{\rm f}^{\rm p} \approx 2.48{\rm e}^{-1.5{\rm p}/\bar{\sigma}} \tag{19}$$

where $\rm p = \sigma_{kk}/3$ is the hydrostatic pressure ($\rm p > 0$ for hydrostatic tension). The strong dependence of the critical effective plastic strain $\epsilon_{\rm f}^{\rm p}$ on the triaxiality ratio $\rm p/\bar{\sigma}$ is apparent from (19).

2.7 *Adaptive meshing*

The primary difficulty with Lagrangian formulations applied to large plastic flows is the mesh distortion incurred. One remedy for this ill is continuous remeshing as the solution evolves. We use Delaunay triangulation for generating six noded quadratic elements (Marusich and Ortiz, 1994). Refinement of the mesh is executed by adding more nodes to elements which satisfy the criterion of maintaining an equidistribution of plastic work rate in each element

$$\int_{\Omega_{\rm h}^{\rm e}} \dot{\rm W}^{\rm p} {\rm d}\Omega > {\rm TOL} \tag{20}$$

where TOL is some suitable problem-dependent value. Here, $\Omega_{\rm h}^{\rm e}$ denotes the domain of element e, and, for the Mises solid, the plastic power density is given by

$$\dot{\rm W}^{\rm p} = \bar{\sigma}\dot{\epsilon}^{\rm p} \tag{21}$$

Triangulation alone does not entirely eliminate elements with poor aspect ratios, since the triangulation is based on nodal positioning. In this event a geometric smoothing is performed to im-

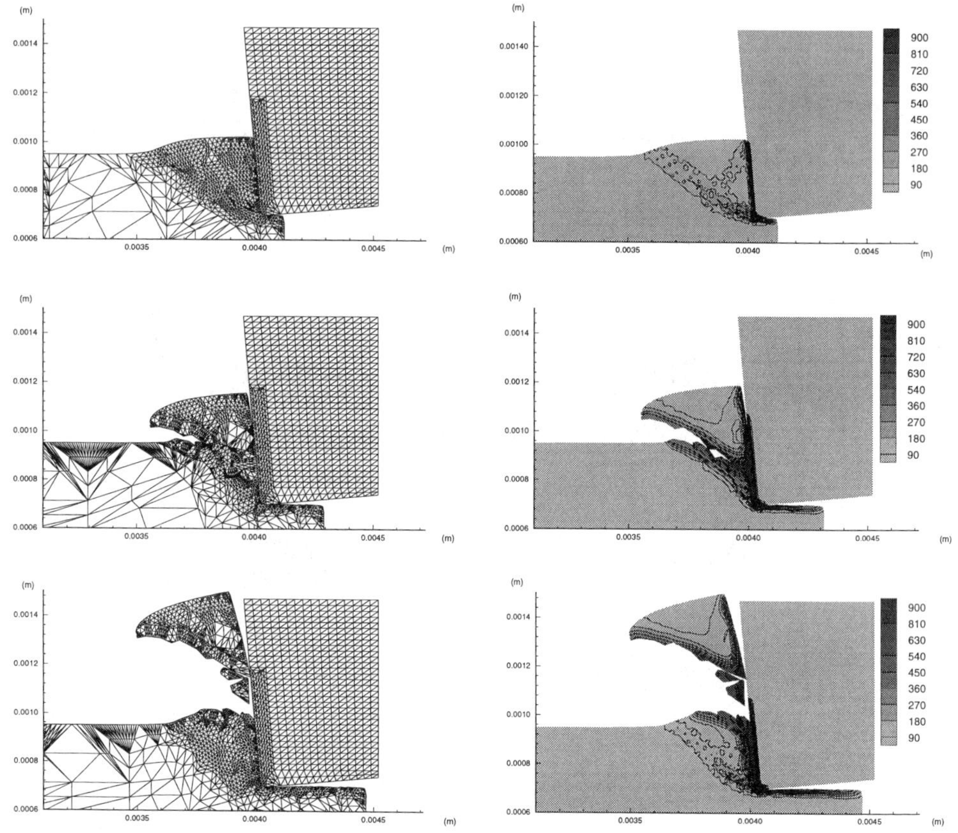

Mesh (a) Temperature contours (b)

Figure 2: Discontinuous chip formation. V=30m/s, Rake=-5°

prove aspect ratios in the mesh. The sum of these techniques allows continuation of simulation indefinitely.

3 Application

The following applications concern the cutting of a rectangular block of high strength AISI 4340 steel at a velocity of 30 m/s and feed of 250 μm/rev, with a cutting tool material of tungsten carbide. The tool is fixed along the back and top surfaces as to minimize the effect of the stiffness while preserving the stress and temperature analysis. The tool is allowed to deform in the elastic range only.

At positive rake angles, the deformation is largely confined to the primary shear zone and to a boundary layer adjacent to the tool, as expected, Fig. 1. A crack initiates on the free surface of the chip and propagates into the interior before becoming arrested. After this transient situation dissipates a continuous chip formation is realized in which no shear localization occurs. Temperatures in the wake of the tool average 900°C while interior chip temperatures are as low as 200°C. Interface temperatures increase from 1200°C at the tool tip to nearly 1400°C along the rake face. The very high temperatures can be partly attributed to the absence of a coolant, which will result in a noticeably shorter tool life.

The second simulation is an identical situation except that we have used a negative cutting angle, Fig. 2. A shear localization develops early in the cutting process as the material softens thermally. A crack initiates on the free surface in the root of a notch formed at the localization. The crack propagates the entire distance through the chip, severing the segment. A new localization occurs and the process repeats itself. Temperatures in the wake of the tool again average near

900°C. However, near the tool tip the temperatures increase from about 1000°C to 1400°C along the rake face. Perhaps the heating due to deformation is reduced near the tool tip in the case of a shear localized chip, yet remains high on the tool face due to frictional sliding.

4 Conclusions

We have assembled several numerical techniques which, in combination, enable the simulation of orthogonal high-speed machining. Finite deformation plasticity with thermal softening, explicit dynamics, mesh-on-mesh contact with friction, fully coupled heat conduction, continuous remeshing and fracture are the main elements of the model.

References

Belytschko, T., "An Overview of Semidiscretization and Time Integration Procedures," in: T. Belytschko and T. J. R. Hughes (eds.), *Computational Methods for Transient Analysis*, North-Holland (1983) 1-65.

Clifton, R. J. and Klopp, R. W. "Pressure-Shear Plate Impact Testing," *Metals Handbook Ninth Edition*, 8 (1985) 230-239.

Cox, T. B. and Low, J. R., "Investigation of the Plastic Fracture of AISI 4340 and 18 nickel-200 Grade Maraging Steels," *Metall. Trans.*, 5 (1974) 1457-1470.

Cuitiño, A. M. and Ortiz, M., "A Material-Independent Method for Extending Stress Update Algorithms from Small-Strain Plasticity to Finite Plasticity with Multiplicative Kinematics," *Engineering Computations*, 9 (1992) 437-451.

Grigull, U. and Sandner, H. "Heat Conduction," Springer-Verlag (1984).

Hughes, T. J. R., "Analysis of Transient Algorithms with Particular Reference to Stability Behavior," in: Belytschko, T. and Hughes, T. J. R. (eds.), *Computational Methods for Transient Analysis*, North-Holland (1983) 67-155.

Hughes, T. J. R. and Belytschko, T., "A Précis of Developments in Computational Methods for Transient Analysis," *J. Appl. Mech.*, 50 (1983) 1033-1041.

Kalpakjian, S., "Manufacturing Processes for Engineering Materials" (1984) Addison-Wesley.

Klopp, R. W., Clifton, R. J. and Shawki, T. G., "Pressure-Shear Impact and the Dynamic Viscoplastic Response of Metals," *Mechanics of Materials*, 4 (1985) 375-385.

Kobayashi, S., Oh, S.-I. and Altan, T., "Metal Forming and the Finite Element Method," *Oxford University Press* (1989).

Komvopoulos, K. and Erpenbeck, S. A., "Finite Element Modeling of Orthogonal Metal Cutting," *J. Engrg. Ind.*, 113 (1991) 253-267.

Lemonds, J. and Needleman, A. "Finite Element Analysis of Shear Localization in Rate and Temperature Dependent Solids," *Mechanics of Materials*, 5 (1986) 339-361.

Lin, J., Lee, S. L. and Weng, C. I., "Estimation of Cutting Temperature in High Speed Machining," *J. Engrg. Mater. Tech.*, 114 (1992) 289-295.

Marusich, T. D. and Ortiz, M. "Modelling and Simulation of High-Speed Machining," *Internat. J. Num. Methods Engrg.* Submitted for Publication (1994).

Park, K. C. and Felippa, C. A., "Partitioned Analysis of Coupled Systems," in: T. Belytschko and T. J. R. Hughes (eds.), *Computational Methods for Transient Analysis*, North-Holland (1983) 157-219.

Rawal, S. P. and Gurland, J., "Observations on the Effect of Cementite Particles on the Fracture Toughness of Spheroidized Carbon Steels," *Metall. Trans. A*, 8A (1977) 691-698.

Ritchie, R. O., Knott, J. F. and Rice, J. R., "On the Relationship Between Critical Tensile Stress and Fracture Toughness in Mild Steel," *J. Mech. Phys. Solids*, 21 (1973) 395-410.

Ritchie, R. O. and Thompson, A. W., "On Macroscopic and Microscopic Analyses for Crack Initiation and Crack Growth Toughness in Ductile Alloys," *Metall. Trans.*, 16A (1985) 233-247.

Rice, J. R. and Tracey, D. M., "On the Ductile Enlargement of Voids in Triaxial Stress Fields," *J. Mech. Phys. Solids*, 17 (1969) 201-217.

Sekhon, G. S. and Chenot, J. L., "Numerical Simulation of Continuous Chip Formation During Non-Steady Orthogonal Cutting," *Engineering Computations*, 10 (1993) 31-48.

Strenkowski, J. S. and Carroll, J. T., III, "A Finite Element Model of Orthogonal Metal Cutting," *J. Engrg. Ind.*, 107 (1985) 349-354.

Tay, A. O., Stevenson, M. G., de Vahl Davis, G. and Oxley, P. L., "A Numerical Method for Calculating Temperature Distributions in Machining from Force and Shear Angle Measurements," *Int. J. Mach. Tool Des. Res.*, 16 (1976) 335-349.

Tay, A. O., Stevenson, M. G., de Vahl Davis, G., "Using the Finite Element Method to Determine Temperature Distributions in Orthogonal Machining," *Proc. Instn. Mech. Engrs.*, 188 (1974) 627-638.

Taylor, L. M. and Flanagan, D. P., "PRONTO 2D: A Two-Dimensional Transient Solid Dynamics Program," Sandia National Laboratories, SAND86-0594, 1987.

Tong, W., Clifton, R. J. and Huang, S., "Pressure Shear Impact Investigation of Strain rate History Effects in Oxygen-Free High-Conductivity Copper," *J. Mech. Phys. Solids*, 40 (1992) 1251-1294.

Vasilash, G. S. "Cutting Tools," *Production Engineering Magazine*, (1992) 40-44.

Simulation of Materials Processing: Theory, Methods and Applications, Shen & Dawson (eds)
© 1995 Balkema, Rotterdam. ISBN 90 5410 553 4

Parallel algorithms for large scale simulations in materials processing

K.K. Mathur
D.E. Shaw & Co., New York, N.Y., USA

ABSTRACT: Accurate material description in forming processes involving polycrystals introduces new demands for computational resources. Parallel computing platforms and several novel algorithms have enabled the modeling of such complex forming processes at a reasonable cost. This article discusses some of the algorithmic techniques used to achieve the goals of performance and scalability for materials processing simulations.

INTRODUCTION

Most modern high performance computing systems have memory distributed among the processors. These processors along with their memory and communication hardware are often referred to as processing nodes. In turn, the processing nodes are interconnected by a network such as a mesh, a binary cube, or a tree. These computing systems hold a promise for extreme performance. However, careful attention to data allocation, data motion in the distributed data structures, memory hierarchies, and load balancing is required to achieve such extreme performance. This article summarizes important algorithmic issues related to an efficient parallel implementation of polycrystal plasticity in large scale finite element simulations where crystallographic texture is a critical feature of the material structure.

VISCOPLASTIC FORMULATION

A viscoplastic formulation is used for the three dimensional finite element analysis of the deformation process simulation. The reader is referred to Beaudoin et al. [1994a] or Dawson et al. [1994] for a detailed description of the formulation. Briefly, the finite element formulation is based on the weak forms of the balance laws. It is assumed that the elastic effects are negligible. The incompressibility constraint is applied via the consistent penalty method. Of particular interest here is the explicit use of polycrystalline plasticity to model the flow properties of the material during the transient Lagrangian analyses. The implementation of the formulation is aimed at computing systems which support programming languages with an array syntax. Some examples of such programming languages are High Performance Fortran, Fortran 90, Connection Machine Fortran, and C*. The current implementation has used the Connection Machine system, CM-5, and Connection Machine Fortran (CMF) as the target platform for parallel computing. Scalability has been demonstrated by maintaining constant throughput time for problems of increasing size that are given proportionally larger numbers of processing nodes [CMSSL 1994]. Further, the same implementation has successfully used an IBM RS6000 workstation with F90 as the platform for simulating applications that do not have very large computational and memory requirements [Beaudoin, Dawson, and Mathur, 1995].

PARALLEL IMPLEMENTATION

The entire data set used by the model is divided into three groups – a group of unassembled finite elements, a group of assembled nodal points, and a group of crystals. These three groups are mapped on to the same set of processing nodes. In general, any interaction between the three groups results in data motion. The mapping of the three groups to the processing nodes is such that the locality of reference is maximized and the contention for the

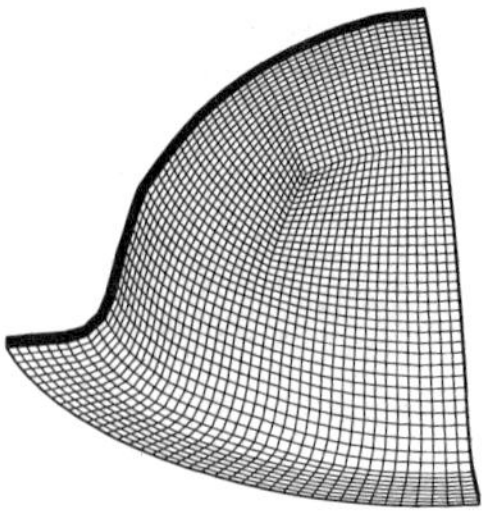

Figure 1. Finite element mesh for forming application

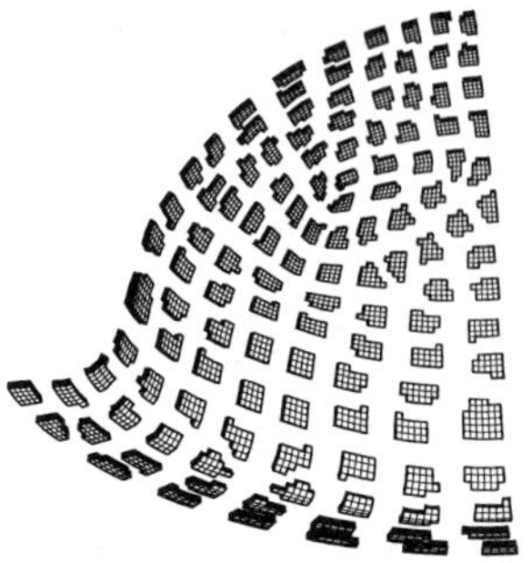

Figure 2. Partitioning of the finite element mesh onto 128 node configuration

communication channels in the network interconnecting the processing nodes is kept to a minimum.

The unassembled finite elements are mapped on to the processing nodes of the computing system using a parallel implementation of the recursive spectral algorithm [Pothen et al. 1990, Simon 1991, Johan et al. 1993 and CMSSL 1993]. No assumption is made about the topology of the finite element mesh; all the results reported here are based on meshes consisting of one element type only. The primary intent of the mapping is to assign the assembled nodal points and unassembled finite elements of the unstructured mesh to the processing nodes based on a partitioning of the graph representation of the mesh topology. The recursive spectral partitioning scheme is based on the eigenvector of the smallest nonzero eigenvalue of the Laplacian matrix associated with the graph (also called the Fiedler vector [Fiedler 1975]). In the context of finite element meshes, the graph representation of the mesh topology is also called the dual mesh. The dual mesh represents the finite elements in terms of their adjacencies rather than the spatial positions of nodal points. This is because computations are most naturally performed in the group of unassembled finite elements. Data motion is required when passing data between the global representation (group of assembled nodal points) and the representation of the collection of local finite elements. Partitioning is performed with the intent of minimizing this communication. In a finite element mesh, data transfer is necessary between finite elements that share a face, an edge, or a corner. An approximation for the dual mesh is simply a list of elements that share a face with a given finite element. Figures (1) and (2) show a typical finite element mesh used in the simulations reported here mapped on to a 128 processing node configuration.

The group of assembled nodal points is mapped to the processing nodes based on the partitioning described above. Nodal points that are internal to a partition are mapped to the processing node to which the partition is assigned. Boundary nodes must be assigned to one of the partitions with which they are associated, or replicated on all of the partitions in which they appear. Only boundary nodes require communication. The implementation reported here assigns the boundary nodes to one of the associated partitions in a random manner. Randomization minimizes contention for the communication channels that interconnect the processing nodes [Valiant 1982].

The group of crystals are assigned to the processing nodes in a consecutive (block) order. It is convenient to view the group of crystals as a two dimensional array where row i of the array represents the aggregate underlying finite element labeled i. The group of crystals reside in an array of shape $N_e \times N_c$ where N_e is the number of finite elements (aggregates) and N_c is the number of crystals in each aggregate. The local subarray on each processing node is of the shape $\lceil \frac{N_e}{N_1} \rceil \times \lceil \frac{N_c}{N_2} \rceil$, where N_1 and N_2 are the number of processing nodes along the two axes ($N_1 \times N_2 = N_p$ and N_p is the number of processing nodes in the computing system). The even distribution of the group of crystals described here is sufficient for achieving arithmetic load balance without sophisticated data dependence. Further, this data aggregation scheme is well suited for the data motion that results in off processing node memory references.

COMMUNICATION PRIMITIVES

Any interaction between the three data groups comprising the entire data set required for the simulation results in data motion among the processing nodes of the computing system. This section describes the communication primitives that are necessary for the data motion steps.

A sparse matrix vector multiply forms one computational kernel of the conjugate gradient method. Two different implementations of the sparse matrix vector multiply are used in the data parallel formulation. The first implementation is based on an element by element approach [Johnsson and Mathur, 1990, Johan *et al.*, 1992] where no explicit assembly of the global stiffness matrix is required. In this case, the sparse matrix vector multiply ($y = A \times x$) can be expressed as:

t = gather(x, P);
$s = A \oplus t$; for all elements
y = scatter(s, P);

where the global vectors x and y are defined in the group of assembled nodal points and the unassembled stiffness matrices A are defined in the group of unassembled finite elements. P is the nodal connectivity. The symbol $\oplus$ is used to denote a matrix vector multiplication operation. First, local copies of the the appropriate vector elements of x are made for each finite element by a "gather" operation. Next, a local matrix vector multiply, performed for each finite element, results in a unassembled product vector, s, which is assembled into the group of nodal points by means of a "scatter" operation. In this representation, the explicit assembly of the global stiffness matrix is not required. Instead, all computations are performed on the unassembled element–wise stiffness matrices.

For general mesh topologies, the gather and scatter operations result in global unstructured communication in the distributed data structure. This unstructured communication is also referred to as a global indirection step [Mathur and Johnsson, 1992]. For the operations described above, this indirection is done via the nodal connectivity of the finite element mesh. The performance of these operations is critically dependent on the data layout of the group of unassembled finite elements and the group of nodal points. The greater the extent of locality of reference, the better will be the net data motion rate. Further, unlike meshes that have a regular topology, address computations required to perform the global gather and scatter operations for arbitrary topologies are quite complex. Address computation in distributed data structures involves the evaluation of the addresses on each processing node and the routing information as well. When the mesh connectivity remains fixed, the address computation can be performed once and the preprocessing cost amortized over several calls to the gather and scatter functions. The performance numbers reported in this article are based on gather and scatter functions available in the scientific software library of the model computing architecture [CMSSL, 1994]. These library functions provide a mechanism to cache the address computation and reuse it as needed at a later stage of the computation.

The second algorithm that is used for the sparse matrix vector multiply is based on the assembled global stiffness matrix. The storage representation for the assembled global stiffness matrix is identical to the Harwell–Boeing scheme [Duff and Reid, 1982]. In this storage scheme, an arbitrary sparse matrix is represented as three one dimensional vectors r, c, and $G_k (= e_{rc})$. The extent of these vectors is equal to the number of nonzero matrix elements in the sparse matrix. Given a mesh nodal connectivity, a preprocessing step computes a pointer, Q, that maps the unassembled element stiffness matrix entries a_{ij} to an assembled global stiffness matrix G_k. The vectors r and c representing the row and column numbers of the nonzero entries of the global stiffness matrix can also be evaluated at the same time. The assembly of the global stiffness matrix is expressed as:

G = scatter(A, Q);

where A are the unassembled element–wise stiffness matrices. The sparse matrix vector multiply can be expressed as:

t = gather(x, c);
$s = G \oplus t$;
y = scatter(s, r);

where the gather operation is based on the column numbers, c, of the global stiffness matrix and the scatter operation is based on the row numbers, r, of the global stiffness matrix.

Consider a simple 4×4 matrix G with eight non–zero matrix elements. Further assume that the contents of the vectors r, c, and G are

$$r = \{1 \quad 1 \quad 2 \quad 2 \quad 3 \quad 3 \quad 4 \quad 4\}$$

$$c = \{1 \quad 3 \quad 2 \quad 4 \quad 1 \quad 3 \quad 2 \quad 4\}$$

$$G = \{k_{11} \quad k_{13} \quad k_{22} \quad k_{24} \quad k_{31} \quad k_{33} \quad k_{42} \quad k_{44}\}$$

After the gather step the contents of t are

$$t = \{x_1 \quad x_3 \quad x_2 \quad x_4 \quad x_1 \quad x_3 \quad x_2 \quad x_4\}$$

The multiplication step yields a vector s

$$s = \{k_{11}x_1 \quad k_{13}x_3 \quad k_{22}x_2 \quad k_{24}x_4 \quad k_{31}x_1 \quad k_{33}x_3 \quad k_{42}x_2 \quad k_{44}x_4\}$$

Finally the scatter step produces the result of the matrix vector multiplication in y

$$y = \{k_{11}x_1 + k_{13}x_3 \quad k_{22}x_2 + k_{24}x_4 \quad k_{31}x_1 + k_{33}x_3 \quad k_{42}x_2 + k_{44}x_4\}$$

As before, the gather and scatter operations are performed by calls to the functions available in the scientific software library on the model architecture. The assembly of the global stiffness matrix is performed only when the unassembled element–wise stiffness matrices are recomputed.

For the same mesh, the total number of floating point operations performed the two algorithms outlined above differ significantly. For three dimensional meshes, the element by element algorithm requires approximately $f_1 = 18n^2N_e$ floating point operations, where n is the number of nodal points per element. Neglecting the floating point operations that are performed during the assembly of the global stiffness matrix, the assembled global stiffness matrix algorithm requires approximately $f_2 = 2N_z$ floating point operations, where N_z is the number of nonzero matrix elements in the global stiffness matrix. For all realistic meshes, $f_2 < f_1$. However, the element by element algorithm has the potential of performing the arithmetic by dense matrix vector multiply which can be coded for very high efficiencies. In contrast, the arithmetic for the assembled stiffness matrix algorithm must be done by scalar operations. For the finite element meshes reported here, the assembled global stiffness matrix vector multiply takes less execution time when the number of elements on each processing node is small. When the element to processing node ratio is large, the execution time of the element by element algorithm is less even though this algorithm is performing extra arithmetic operations, primarily because of significantly higher execution rate of the dense matrix vector multiply. For the results presented in the later sections, the break even point for the two sparse matrix vector multiply algorithms is when the element to processing node ratio is about sixteen.

The preconditioning step forms the second computational kernel of the sparse linear system solver. During the course of this investigation, three different preconditioning strategies have been investigated. The communication primitives required for the three strategies are described here. The three preconditioning schemes used are:

1. element–wise (pointwise) diagonal scaling,
2. block diagonal scaling, and
3. substructuring.

The preconditioning step for the element–wise diagonal scaling is

$$r = D^{-1}z \tag{0.1}$$

where D is the diagonal of the assembled global stiffness matrix. Once the unassembled element–wise stiffness matrices have been computed, the evaluation of D requires the assembly of the diagonal elements of the unassembled matrices. This can be accomplished by a scatter operation which is based on the mesh connectivity, P,:

$$D = \mathsf{scatter}(A,\ P);$$

This assembly step is performed once whenever the unassembled element–wise stiffness matrices are computed. The preconditioning step in the inner loop of the conjugate gradient algorithm is a simple scalar multiplication operation. No additional data motion steps are required for preconditioning.

The block diagonal scaling is a simple extension of the pointwise diagonal scaling. The 3×3 block (for three dimensional simulations with three degrees of freedom per node) associated with a nodal point in the mesh is used in the preconditioning step. After the evaluation of the unassembled element–wise stiffness matrices, six scatter operations that are identical to the one used for the pointwise diagonal preconditioning are required to compute the block diagonal matrix D[1]. The inversion of the 3×3 blocks in D and the application of D^{-1} to z in the inner loop of the conjugate gradient algorithm involve local operations only.

Each processing node of the computing architecture is assigned a submesh based on the recursive spectral partitioning algorithm described above. A global stiffness matrix is evaluated for each submesh. The collection of the N_p different stiffness matrices can then be viewed as a block diagonal matrix D. This block diagonal matrix is used as a preconditioner for the conjugate gradient iterations. The evaluation of the preconditioner involves the following operations:

$$G = \mathsf{scatter}(A,\ Q);$$
$$D = \mathsf{gather}(G,\ Q');$$

where Q' is a pointer that maps nodal points in each partition to the location of the nonzero matrix element in the assembled global stiffness matrix. The application of D^{-1} to z can be expressed as:

$$t = \mathsf{gather}(z,\ P');$$
$$s = D^{-1} \oplus t;\ \text{for all partitions (submeshes)}$$
$$r = \mathsf{scatter}(s,\ P');$$

where P' maps the nodal points in each submesh with the nodal points of the mesh. In summary, the data motion primitives required by the three preconditioning schemes reported here are the same as the primitives that are required by the sparse matrix vector multiply.

[1] six scatters are required instead of nine because of symmetry.

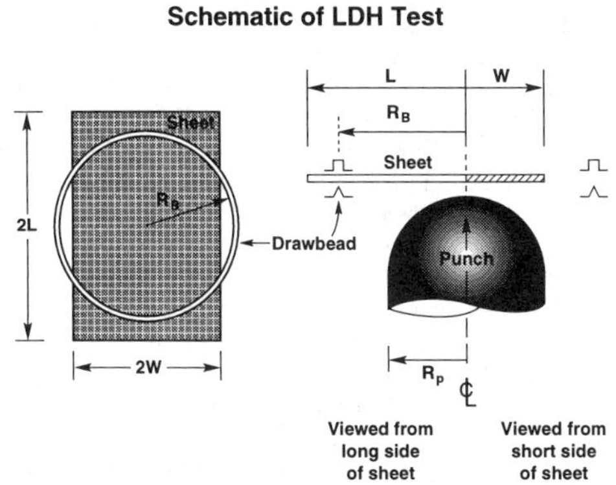

Figure 3. Limiting dome height schematic diagram

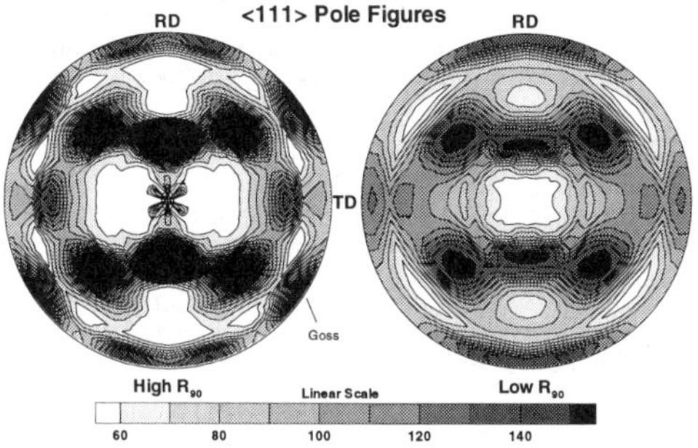

Figure 4. Pole figures for two processing practices

LIMITING DOME HEIGHT SIMULATION

As an example serving to exploit the computational and memory resources of a distributed memory, massively parallel computing platform, limiting dome height tests were simulated. The limiting dome height test provides a measure of a material's formability through the value of the dome height when the material fails. The dome is formed by stretching a flat sheet over a hemispherical punch. A restraining draw bead controls the degree of draw-in. By manipulating the width of the blank, strain states ranging from drawing to stretching are achieved. Figure (3) shows the experimental setup for this test.

The performance of an aluminum alloy (2036–T4) was examined for two conditions arising from differences in the processing practice. Crystallographic textures for the two practices are labeled as 'High r_{90}' and 'Low r_{90}' and are shown as $< 111 >$ pole figures in Figure (4). Each texture was numerically discretized using an aggregate of 256 weighted orientations. The finite element mesh of the LDH specimen was composed of

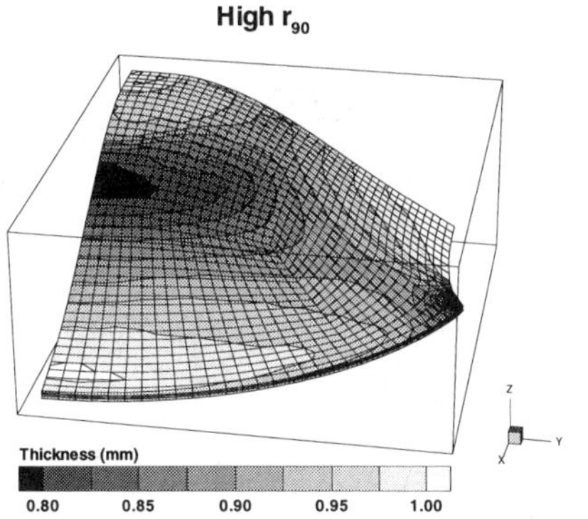

Figure 5. Thickness contours for high r_{90} texture

3600 eight–node brick elements. Identical aggregates were assigned to each element at the start of a simulation, giving a total of 921600 crystals and 3686400 microstructural state variables. On a 64 processing node Connection Machine CM–5 configuration, the simulations take approximately 40 minutes per time increment. On this configuration, 63 processing nodes store a sub–mesh with 57 finite elements each and one processing node stores the remaining nine finite elements. As the simulation proceeds the crystal orientations in every element are updated according to the deformation experienced by the element. The updating of crystal orientations quantifies the evolution of the flow properties during forming.

Figures (5) and (6) show the sheet thickness at failure for the two initial textures. 'High r_{90}' gives poorer LDH – there is a greater degree of thinning in the part. The part fails near the centerline ("X"–axis; darker contours). The reason for this is the presence of a "Goss" component in 'High r_{90}'. This is the dark region at position 45°, 135°, 225°, and 315° at the rim of pole figure corresponding to 'High r_{90}' (Figure 4). Comparison to experiment is reported by Bryant et al. [1994]. Through the use of parallel algorithms, the inclusion of a micromechanical description of the crystallographic texture in large scale finite element simulations becomes a practical avenue for studying processing alternatives to produce materials with superior properties.

Acknowledgements

This work is supported by the Office of Naval Research contract N00014-90-J-1810. Computing resources were provided by the Advanced Computing Laboratory at Los Alamos National Labora-

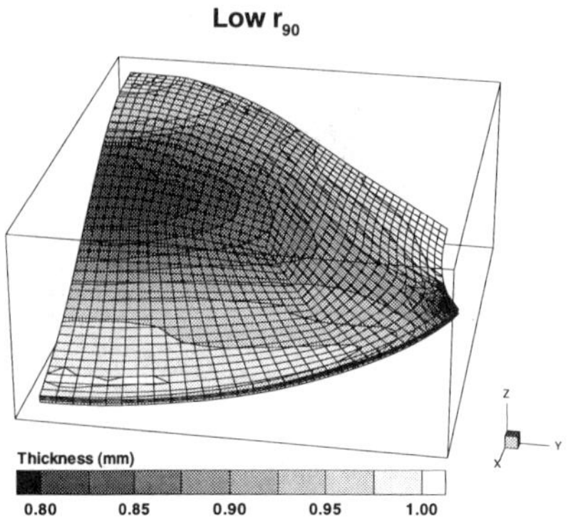

Figure 6. Thickness contours for low r_{90} texture

tory and by the National Center for Supercomputing Applications at University of Illinois Urbana–Champaign.

References

A. J. Beaudoin, P. R. Dawson, K. K. Mathur, U.F. Kocks and D. A. Korzekwa, "Application of polycrystal plasticity to sheet forming," Comp. Meth. Appl. Mech. and Engr., 117 (1994) 49–70.

P. R. Dawson, A. J. Beaudoin, K. K. Mathur and G. B. Sarma, " Finite Element Modeling of Polycrystalline Solids," European Review of Finite Elements, 3, no 4, (1994) 543–571.

I. S. Duff and J. K. Reid, "A set of fortran subroutines for solving sparse symmetric sets of linear equations," AERE R10533, HMSO, Harwell, London 1982.

M. Fiedler, "A property of eigenvectors of nonnegative symmetric matrices and its application to graph theory," Czech. Math. J, 25 (1975) 619–633.

Z. Johan, K. K. Mathur, T. J. R. Hughes and S. L. Johnsson, "A data parallel finite element method for computational fluid dynamics on the connection machine system," Comp. Meth. Appl. Mech. and Engr., 99 (1992) 113–134.

Z. Johan, K. K. Mathur, S. L. Johnsson, and T. J. R. Hughes, "An efficient communication strategy for finite element methods on the connection machine CM–5 system," Comp. Meth. Appl. Mech. and Engr., 113 (1994) 363–387.

S. L. Johnsson and K. K. Mathur, "Data structures and algorithms for the finite element method on a data parallel supercomputer," Int. J. Num. Meth. Engr., 29 (1990) 881–908.

K. K. Mathur and S. L. Johnsson, "Communication primitives for unstructured finite element simulations on data parallel architectures," Comp. Syst. Engr., 3 (1992) 63–71.

A. Pothen, H. D. Simon, and K.-P. Liou, "Partitioning sparse matrices with eigenvectors of graphs," SIAM J. Matrix Anal. Appl. 11 (1990) 430–452.

H. D. Simon, "Partitioning of unstructured problems for parallel processing," Comp. Systems Engr., 2 (1991) 135–148.

L. Valiant, "A scheme for fast parallel communication," SIAM J. Comp. 11 (1982) 350–361.

CMSSL for CM Fortran: CM–5 edition, Volume 3.2, Thinking Machines Corporation, Cambridge MA, 1994.

J. D. Bryant, A. J. Beaudoin and R. T. VanDyke, " The effect of crystallographic texture on the formability of AA2036 autobody sheet", SAE Technical Paper Series Number 940161, (1994), 1–8.

A. J. Beaudoin, P. R. Dawson, and K. K. Mathur, "Analysis of Anisotropy in Sheet Forming Using Polycrystal Plasticity", Proceedings of ASCE Symposium on Texture Development and Anisotropy, May 1995, in press.

Simulation of Materials Processing: Theory, Methods and Applications, Shen & Dawson (eds)
© 1995 Balkema, Rotterdam. ISBN 90 5410 553 4

Simulation of springback in sheet metal forming

Kjell Mattiasson
Volvo Car Corporation & Chalmers University of Technology, Göteborg, Sweden

Anders Strange
Volvo Car Corporation, Olofström Plant, Sweden

Per Thilderkvist
IUC, Olofström, Sweden

Alf Samuelsson
Chalmers University of Technology, Göteborg, Sweden

ABSTRACT: Modern Finite Element codes for sheet forming simulation have been shown to be able to produce excellent results regarding draw-in, strain distribution, punch force etc. Springback is, however, a phenomenon which so far has been proven surprisingly difficult to simulate accurately. In the present report an elaborate examination of the primary factors that influence the results of springback simulations is presented. As a basis of the discussion the 2D draw bending benchmark of the NUMISHEET'93 conference is used.

1 INTRODUCTION

1.1 *Background*

Springback is a phenomenon, which takes place when the workpiece is removed from the tools after completed forming, or, for instance, when the flanges are cut away in subsequent operations. The resulting stress release give rise to, normally, elastic deformations. Depending the shape of the part these elastic deformations result in more or less severe global shape changes.

In today's engineering practice the tools are modified in order to compensate for expected springback. These modifications are based on experience and empirical formulas regarding magnitude and appearance of the shape deviations. Since it is very difficult to make accurate judgements concerning the magnitude of the springback, even for highly experienced engineers, the design and tryout of the pressing tools often have to be done in several iteration loops, a process which tends to become both costly and time consuming.

Today the world's automobile makers are struggling hard to reduce the weight of the cars. An effort in this direction is to introduce new car body materials, like high strength steels and aluminium alloys. The strain release at unloading at a point in a structure is roughly proportional to the relation σ/E, where σ is the stress before unloading and E is Young's modulus of elasticity. Considering this fact, it is evident that both of the previously mentioned classes of materials have a much higher tendency for springback than traditional deep drawing steels. Also the fact that the engineers' experience of these materials is much less than for traditional ones, has highlighted the need for a simulation tool that can accurately predict springback.

The springback problem has been the subject of numerous investigations during the years. In a recent paper by Huang and Gerdeen (1994), a bibliography on springback in sheet metal forming is appended, including as many as 90 references.

1.2 *Finite Element simulation of sheet forming processes - current status*

In recent years sheet forming simulation by means of the Finite Element Method has become generally used in the tool design offices at most automobile makers as a means of evaluating different process layouts already at the design stage. Excellent predictive capabilities regarding draw-in, strain distribution, punch force etc. have been proved. The most commonly used codes for large scale simulations are based on a dynamic approach and an explicit time integration scheme, but also an implicit, special purpose code, like AUTOFORM (Kubli et.al. 1993, Kubli et.al. 1991, Anderheggen 1991), has attracted many users in the industry.

The simulation of the springback phase is seemingly a trivial problem compared to the simulation of the forming phase, since it involves no contacts, and the deformations are small or of moderate size and are usually elastic. Despite this fact, results from springback simulations have many times shown results, which are far from the level of accuracy obtained for other parameters, such as the ones mentioned above.

In connection to the NUMISHEET'93 conference (Makinouchi et.al. 1993) a few sheet forming benchmark tests were set up. One of these was a springback study of a to U-shape deep-drawn sheet strip. The participants were invited to provide numerical solutions, as well as experimental results. The predicted values exhibited, however, a large scatter, and many results were far beyond what can be considered to be acceptable ones.

It must be considered to be a highly serious matter that so many experts, using the best codes available, fail in solving a, as it looks like, very simple springback problem. In the authors' opinion it is of utmost importance that there is a complete understanding of all the factors that are influencing the results of numerical springback analyses, in order to avoid those mistakes that evidently must have been made in, for instance, the solutions of the above related benchmark problem.

In the present study a thorough investigation with this objective has been performed. As a basis for the investigation the NUMISHEET'93 benchmark has been used, since this problem is very well documented, both with respect to experimental as well as to numerical results.

1.3 *The dynamic, explicit method and springback simulation*

Springback analyses by means of dynamic, explicit codes must, in contrast to forming simulations, be performed in a real time scale. The normal procedure is to apply a suitable amount of damping and let the workpiece vibrate freely until a static equilibrium is reached. The drawbacks of this procedure is, in the first place, that it usually is very difficult to estimate what amount of damping to apply without doing a separate

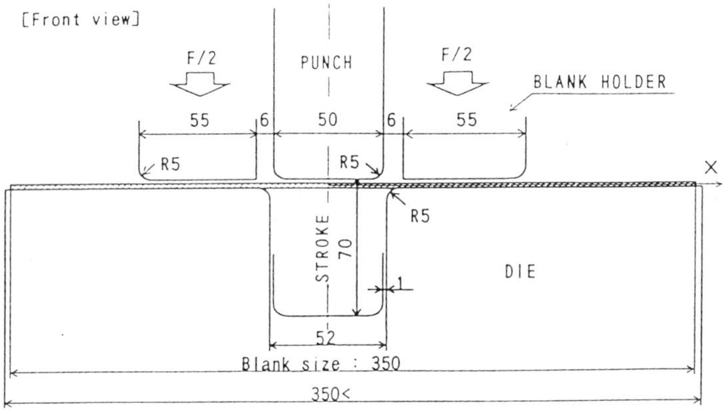

Fig.1 Geometry description of the NUMISHEET'93 2D draw bending benchmark problem (sheet width = 35 mm)

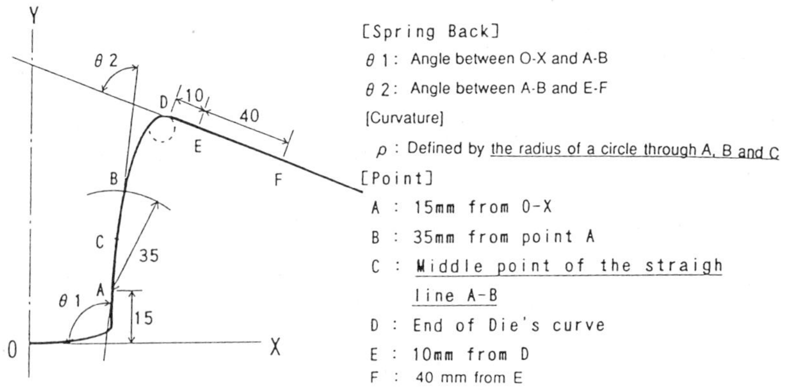

Fig. 2 Definition of springback parameters for the NUMISHEET'93 2D draw bending benchmark problem

eigenfrequency analysis, and, secondly, that the time for reaching a static equilibrium in many cases can be several times longer than the time needed for the actual forming operation.

In an implicit method, on the other hand, the springback simulation is a trivial task once the forming simulation has been carried through. The latter one is, however, in contrast usually a both difficult and very time consuming simulation.

The situation is, thus, a bit paradoxical. The dynamic, explicit method is very well suited for performing forming simulations, but is unsuitable for springback analyses. With the traditional implicit method the situation is exactly the opposite.

As a consequence of this the dynamic, explicit code LS-DYNA3D has been provided with a coupling to the implicit code LS-NIKE3D, so that the springback analysis can be performed by the implicit code, starting from the stress state at the end of the forming step, provided by the dynamic, explicit code. These two codes have been used in the simulations reported herein.

2 THE NUMISHEET'93 2D DRAW BENDING BENCHMARK PROBLEM

2.1 *Problem description*

The geometry of the benchmark problem is described in Fig. 1. The problem was intended to be analysed for three different sheet materials: mild steel, high strength steel, and aluminium, and for two different blankholder forces (F in Fig. 1): 2.45 and 19.60 kN, respectively. Thus, altogether there were six variants to be analysed.

The effective stress - plastic strain relation was given in the form

$$\sigma = K\,(\varepsilon_0 + \varepsilon_p)^n \qquad (1)$$

The material parameters, together with sheet thicknesses and friction coefficients for the various materials are given in Table 1.

In Fig. 2 the parameters used for characterising the springback are defined. These are the angles θ_1 and θ_2, and radius of curvature ρ. In the further discussion we have, however, found it more convenient to use the deviations from the right angles: $\Delta\theta_1=\theta_1-90^o$ and $\Delta\theta_2=90^o-\theta_2$, respectively, together with the curvature $K=1/\rho$.

2.2 *Results from the benchmark test*

The trends were the same for the results from all the six variants of the problem. Just the results from one of them will therefore be reproduced here: the high strength steel with high blankholder force. It was this variant that showed the largest deviations between numerical and experimental results.

We have chosen to divide the numerical results into two groups depending on which type of method that was used in the simulation. In the first group there are results obtained with traditional implicit methods and with the so called "static, explicit" method. In the second one there are results obtained with the dynamic, explicit method.

The numerical results are compared to the average of the experimental values. Altogether there were experimental results supplied from 11

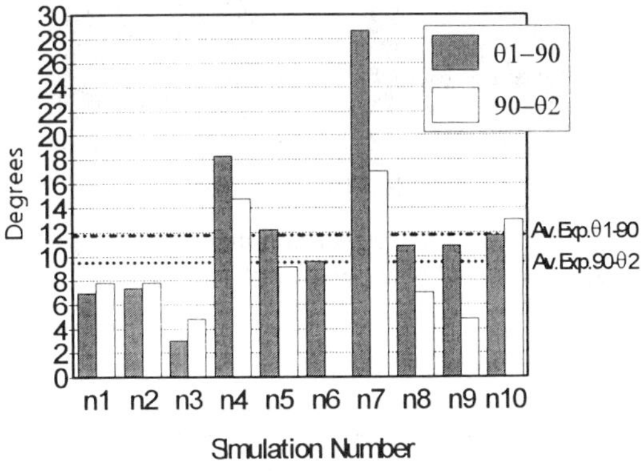

Fig. 3 Benchmark results from predictions of spring-back angles obtained with implicit codes

Table 1. Material parameters for the NUMISHEET'93 2D draw bending benchmark problem

MATERIAL	t (mm)	μ	E (MPa)	ν	R	σ_y (MPa)	ε_0	n	K (MPa)
Mild Steel	0.78	0.144	206	0.3	1.77	173.1	0.00712	0.2589	565.32
High Strength Steel	0.74	0.129	206	0.3	1.66	277.1	0.0125	0.2182	680.61
Aluminium	0.81	0.162	71	0.33	0.64	135.3	0.0166	0.3593	576.79

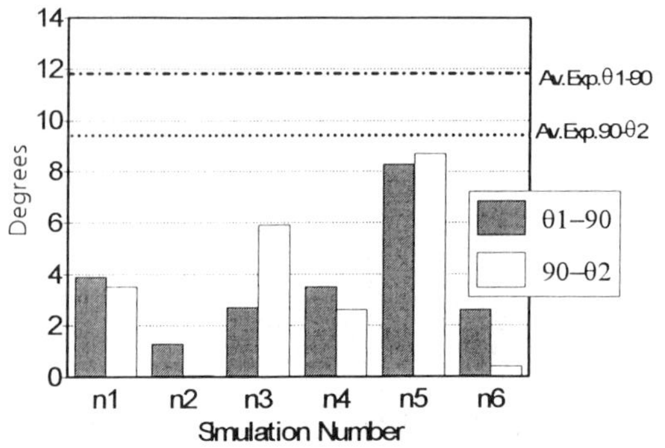

Fig. 4 Benchmark results from predictions of spring-back angles by means of dynamic, explicit codes

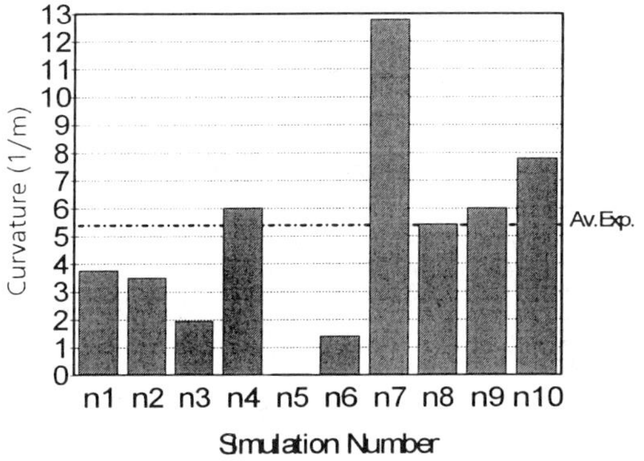

Fig. 5 Benchmark results from predictions of spring-back curvature by means of implicit codes

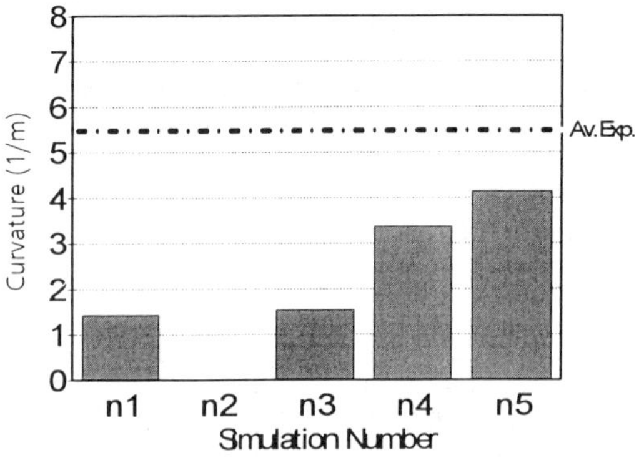

Fig. 6 Benchmark results from predictions of spring-back curvature by means of dynamic, explicit codes

different groups. When calculating the average of the experimental values, the largest and smallest value in each series have been omitted.

From the results displayed in Figs. 3-5 it can be observed that that the results obtained by implicit codes exhibit a rather big scatter. However, no general trend can be seen whether this type of codes tend to over- or underestimate the springback.

Regarding the dynamic, explicit codes, on the other hand, there is a clear tendency to strongly underestimate the springback. The errors are in most cases far too big for the results to be judged as acceptable. The average error in curvature, for instance, is as big as 62 %. This in spite of the fact that most of the results were obtained by means of well-known softwares such as LS-DYNA3D, PAMSTAMP, and RADIOSS (and that one of the solutions was submitted by two of the authors).

Based on these observations we were convinced of that there was some systematic error involved in the dynamic, explicit approach, maybe caused by dynamic effects, that could explain the characteristic underestimation of springback.

It should also be mentioned that a group from Chrysler, Changqing Du et.al.(1994), after the NUMISHEET'93 conference has re-analysed the current benchmark problems with the LS-DYNA3D/NIKE3D codes, and with an in-house, implicit code. Their results are in general in better agreement with experimental values than the ones presented at the conference, but the trend is still the same that springback is underestimated.

2.3 *Possible causes of inaccurate springback predictions*

In this study we will systematically investigate possible reasons for the in general inaccurate springback predictions reported in connection to the NUMISHEET'93 conference. The following possible causes will be studied:

- Element size in the sheet strip
- Hardening law (isotropic, kinematic)
- Finite element modelling of the draw radius
- Dynamic effects (in the dynamic, explicit approach)

3 INFLUENCE OF ELEMENT SIZE

3.1 *Introduction*

A previous study, Mattiasson et.al.(1992), has indicated that the mesh size in the sheet must be put in relation to the magnitude of the draw radius. When studying parameters such as strain distribution and punch force, an element size equal to half the draw radius was found to be near optimum. Many times in practice element sizes in the range of 0.75-1.0 times the draw radius must be chosen in order to keep the size of the critical time step as well as the finite element model within reasonable bounds. This should be kept in mind during the following discussion.

As previously pointed out the springback is primarily dependent on the magnitude of the stresses in the sheet after the completion of the forming phase, and of the modulus of elasticity E. In order to get an understanding of how the stresses depend upon the element size, the stress states in two points on the outer surface of the sheet strip will be traced during the forming operation. These points are situated approximately 40 (point A) and 80 mm (point B), respectively, from the centre line of the strip.

In Fig. 7 the expected stress history in a point on the outer strip surface is displayed. In region A-B the point slips on the horizontal part of the die, in B-C the point enters the die corner and the strip bends, in C-D the point exits the die corner and the strip is straightened , and, finally, in region D-E the point slips along the vertical die wall. Depending on what hardening law (isotropic or kinematic) the material is assumed to obey, the path C-D-E will have different appearances.

3.2 *Current simulations*

Simulations have been performed with four different element sizes: 3.0, 2.0, 1.0, and 0.5 mm side length, respectively, which in relation to the draw radius (5 mm) corresponds to values in the range 0.1-0.6. The sheet strip has been modelled as a plane strain problem. The punch velocity-time relation has been prescribed according to a sine curve with zero velocity at the start and at the end of the stroke, and with a total process time of 20 ms. The forming simulations have been performed with LS-DYNA3D and the springback simulations with LS-NIKE3D. The tool surfaces have been modelled by means of VDA CAD-surfaces. This implies that the die corner is completely smooth and no errors due to discretization of the radius enter the solution. The material in this study is assumed to be isotropic

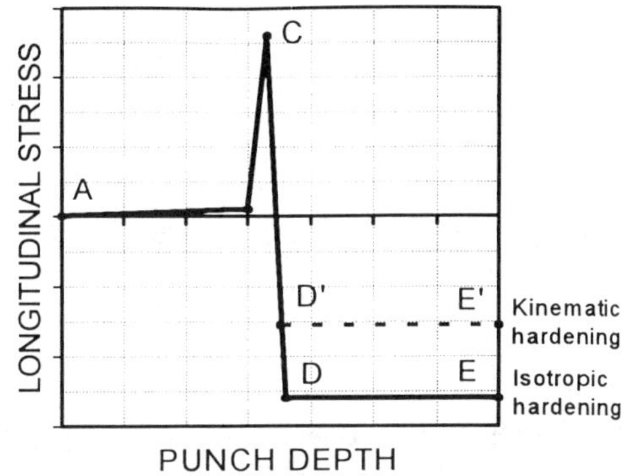

Fig. 7 Expected appearance of the stress history in a point on the outer surface of the sheet strip

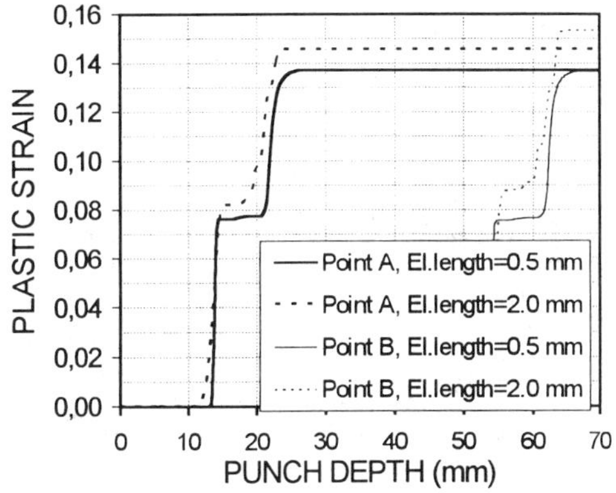

Fig. 8 Plastic strain histories, for two different element sizes, in two points on the outer surface of the sheet strip

hardening, as was the case in the vast majority of the supplied results to the benchmark tests.

In Fig. 8 the calculated plastic strains in point A and point B, respectively, are shown for two different element sizes. As can be seen the differences between the strain curves for the two element sizes are not particularly large. It should also be observed that the plastic strain seems to remain constant after the material point has left the draw radius (corresponding to point D in Fig. 7).

The longitudinal stresses in point A and B, respectively, for different element sizes are shown in Fig. 9. It is remarkable to note that for most element sizes there is a pronounced relaxation of stress after it having reached its peak point at the exit of the

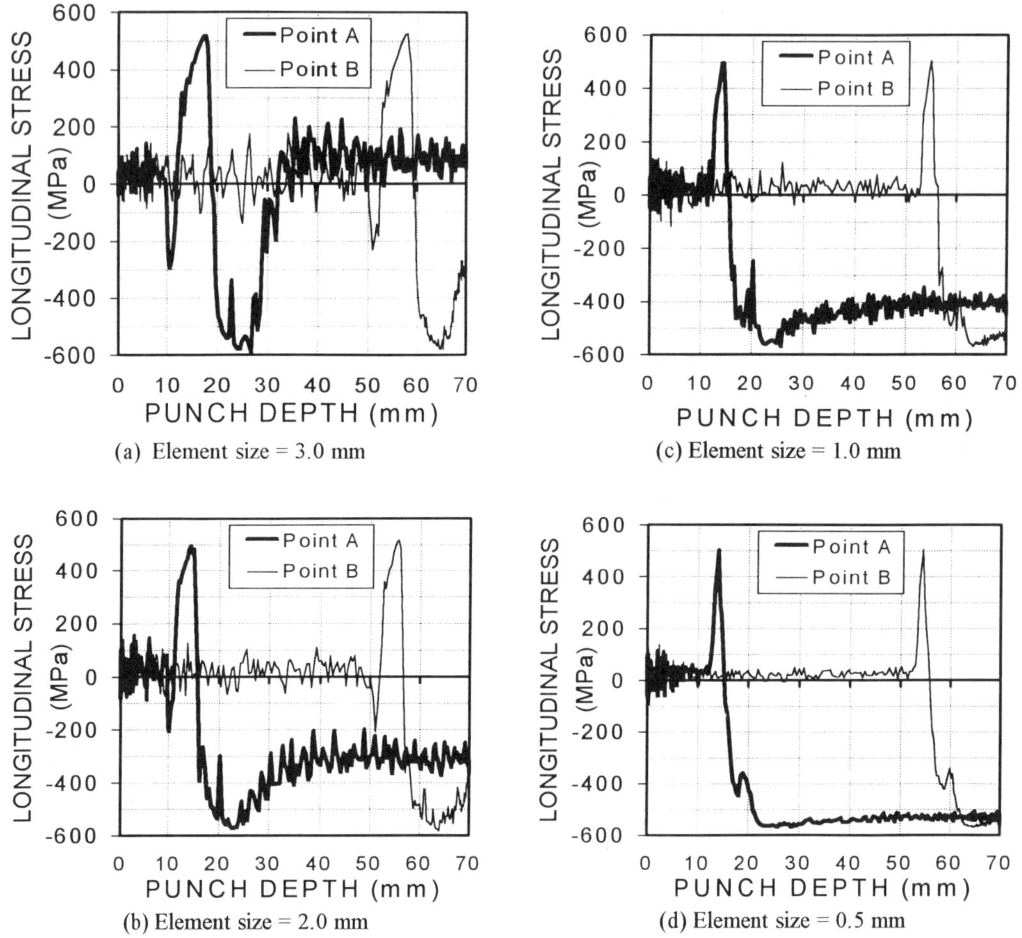

(a) Element size = 3.0 mm

(c) Element size = 1.0 mm

(b) Element size = 2.0 mm

(d) Element size = 0.5 mm

Fig. 9 Stress histories in two points on the outer surface of the sheet strip

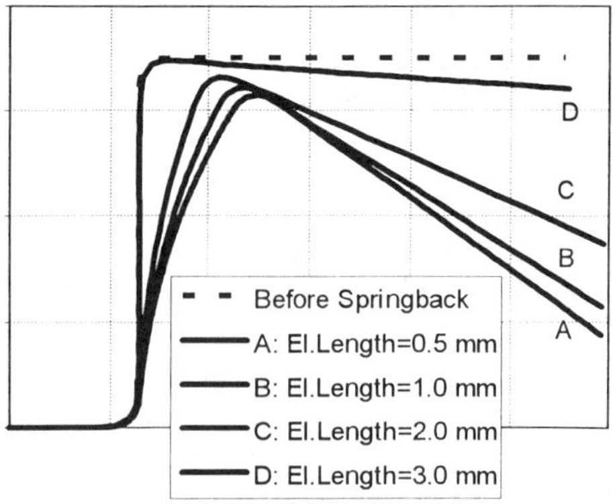

Fig. 10 Geometry after springback for different element sizes

draw radius. It should also be observed that this stress relaxation is not accompanied by any change of plastic strain according to Fig. 8.

The magnitude of the stress relaxation seems to be directly related to the element size. Just for the smallest element size, 0.5 mm, the relaxation is negligible, and the stress history curve resembles the expected one according to Fig. 7.

Since the magnitude of the springback is directly proportional to the magnitude of the stress at the end of the forming process, there should, in view of the above observations, exist an inverse relationship between magnitude of springback and element size. This presumption is confirmed in Fig. 10, in which the calculated geometries of the strip

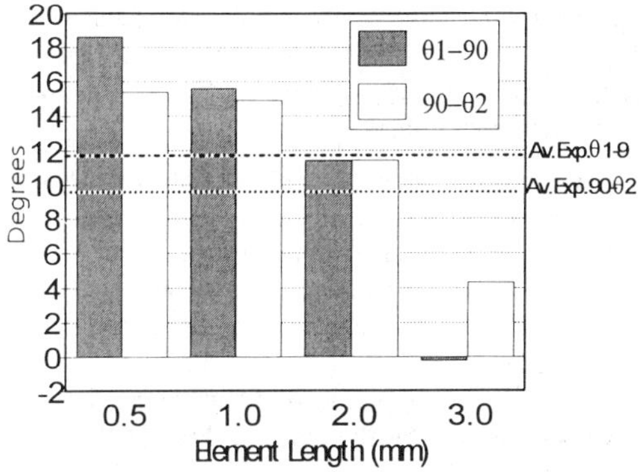

Fig. 11 Springback angles obtained with different element sizes

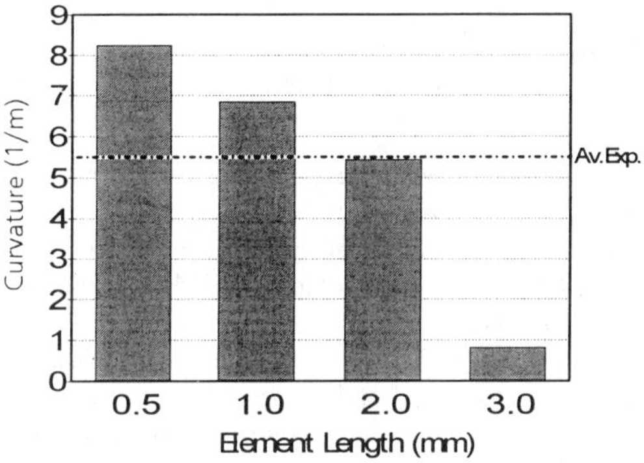

Fig. 12 Springback curvature obtained with different element sizes

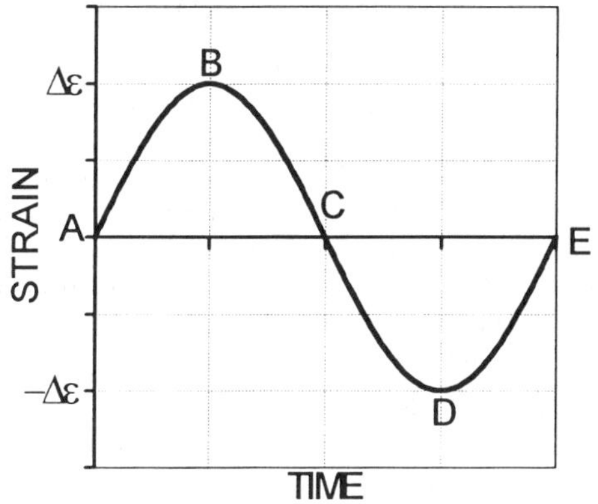

Fig. 13 Strain variation due to a bending wave

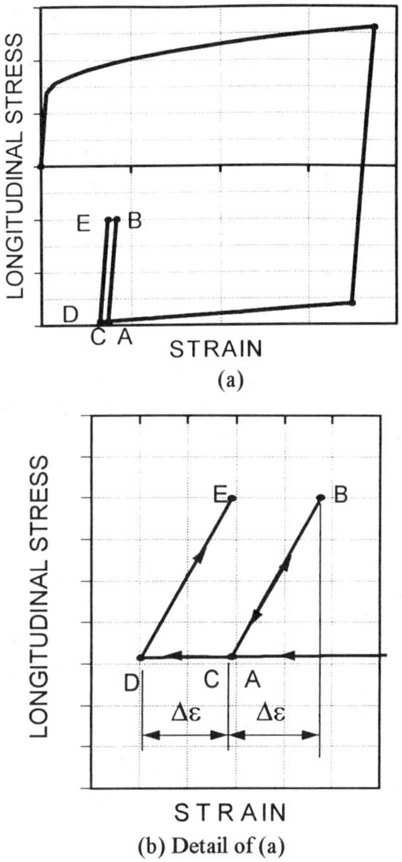

Fig. 14 Stress-strain history for a point on the surface of the strip

after springback, for various element sizes, are displayed. For the largest element size, 3.0 mm, the springback is very small, but it could be interesting to learn that for larger element sizes the springback will even take place in the wrong direction.

In Figs. 11 and 12 calculated springback angles and curvatures are compared to experimental values. It can be observed that for element sizes smaller than 2.0 mm the springback is overestimated. This is in contradiction to all previously reported results obtained by means of dynamic, explicit codes. We will in the next chapter discuss the reason for this overestimation of the springback.

3.3 *An explanation of the stress relaxation phenomenon*

The stress relaxation phenomenon demonstrated in Fig. 9, may at first seem a bit odd,

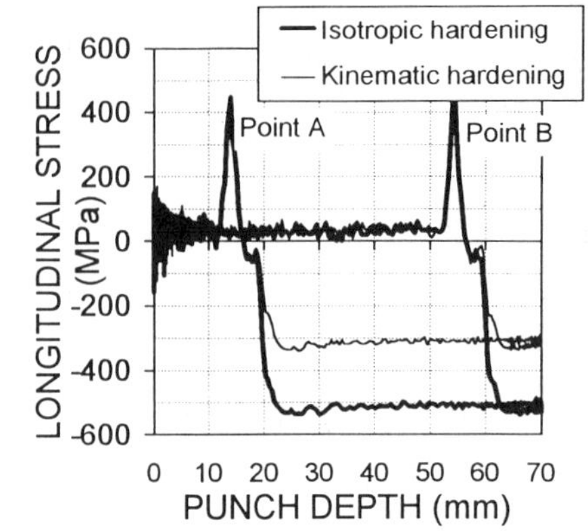

Fig. 15 Stress histories for isotropic and kinematic hardening models, respectively. Element length=0.5 mm

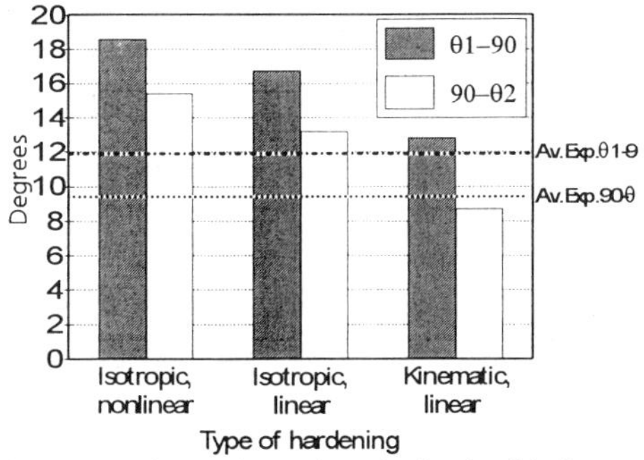

Fig. 16 Springback angles obtained with isotropic and kinematic hardening models, respectively

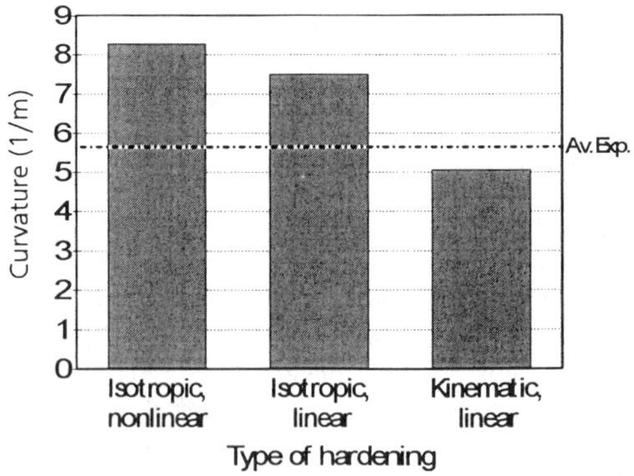

Fig. 17 Springback curvature obtained with isotropic and kinematic hardening models, respectively

especially since it seemingly is not accompanied by a change in strain. This is, however, not completely true. There is in fact a change of strain involved, although it is very small.

When the discrete finite elements of the strip slip over the draw radius, bending waves with small amplitudes are induced in the vertical part of the strip. The amplitudes of these waves are depending on size of the elements in the strip: the bigger elements, the bigger amplitude. In Fig. 13 the change in strain due to one period of an imaginary bending wave is displayed. Furthermore, in Fig. 14 the stress - strain relationship in a point on the outer surface of the strip is shown. At A the material point has left the draw radius and is exposed to a bending wave according to Fig. 13. The first half period, A-B-C, just gives rise to an elastic unloading-loading path, while the next quarter period, C-D, results in an elastic-plastic strain increment $\Delta\varepsilon$. The last quarter period, D-E, give finally rise to a new elastic unloading path.

Thus, the final stress state is represented by point E in the stress-strain diagram, and the stress relaxation is roughly given by the expression $E*\Delta\varepsilon$, where E is Young's modulus. This indicates that the stress relaxation would be less pronounced for aluminium than for steel, a statement which is supported by the NUMISHEET'93 benchmark results, where the numerical results for aluminium generally were in better agreement with the experimental values than were the cases for the steel materials.

4 ISOTROPIC AND KINEMATIC HARDENING

The sheet material that is sliding over the draw radius is subjected to bending and unbending, which in turn give rise to plastic loading and unloading. This is, thus, a case where the Bauschinger effect would play a significant rule. Although the kinematic hardening model is a better approximation of the Bauschinger effect than the isotropic one, most of the participants in the benchmark tests were using the isotropic hardening model.

One of the objects of the present investigation has therefore been to study the consequences of using a kinematic instead of an isotropic hardening model. Unfortunately LS-DYNA3D does not contain any nonlinear, kinematic hardening model in its material library, so we have had to content ourselves with a linear hardening one. The nonlinear stress-strain relation has been approximated by the following linear expression

$$\sigma = \sigma_y + H' \varepsilon_p \qquad (2)$$

A least squares fit in the range $0<\varepsilon_p<0.15$ gives $\sigma_y = 302$ MPa and $H' = 1149$ MPa. In order to be able to do a fair comparison, the above linear relationship has been used both in conjunction with a kinematic as well as with an isotropic hardening model. The computed stresses with the two hardening models are displayed in Fig. 15. As expected the kinematic model gives rise to considerably lower stress levels.

Results from springback analyses are shown in Figs. 16 and 17. A comparison between the first two columns gives an idea of the magnitude of the error introduced by doing a linear approximation of the stress-strain relationship. A comparison between the second and third columns reveals the difference between the two hardening models. It is interesting to note that, taking the error due to the approximate stress-strain relation into account, the kinematic model gives an almost perfect fit to the experimental results.

5 MODELLING OF THE DRAW RADIUS

Normally the pressing tools are modelled by means of finite elements. Smooth, curved surfaces are thus approximated by facet surfaces. Especially the modelling of the draw radius can influence the outcome of the whole simulation, see e.g. Gröber et.al. (1993).

Two simulations have been performed with element lengths 1.0 and 2.0 mm, respectively. In both cases the draw radius was modelled by means of three flat element, which must be considered to be a very rough approximation. The influence on the stress histories is demonstrated in Figs. 18 and 19. These figures should be compared to the corresponding stress histories in Fig. 9, where a completely smooth description of the draw radius has been used.

As can be seen there is only a minor influence on the stresses for the case with element size 1.0 mm, while there is a considerable influence for the element size 2.0 mm. The conclusion is, thus, that the effects of coarse meshes in the sheet and on the draw radius, respectively, seem to amplify each other.

6 DYNAMIC EFFECTS IN DYNAMIC, EXPLICIT CODES

In all the simulations presented so far, a process time of 20 ms has been used. This time has been found to

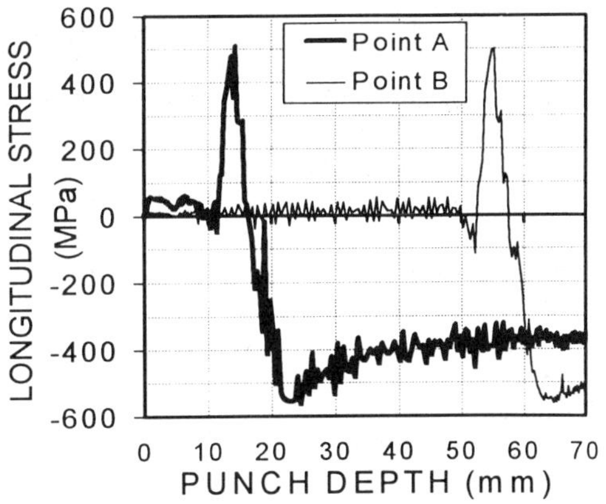

Fig. 18 Stress histories. Draw radius modelled by three facet elements. Element length=1.0 mm

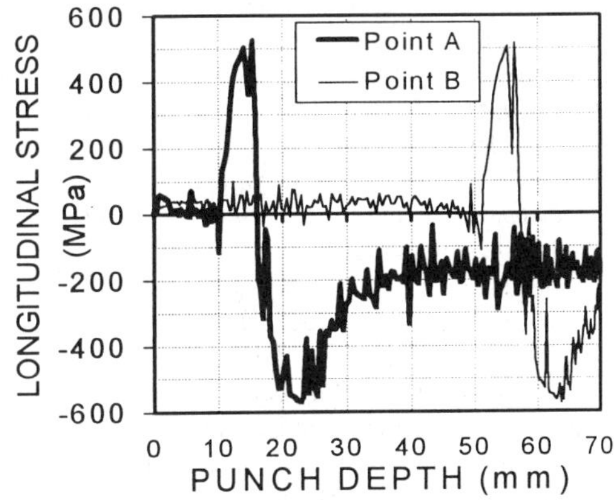

Fig. 19 Stress histories. Draw radius modelled by three facet elements. Element length=2.0 mm

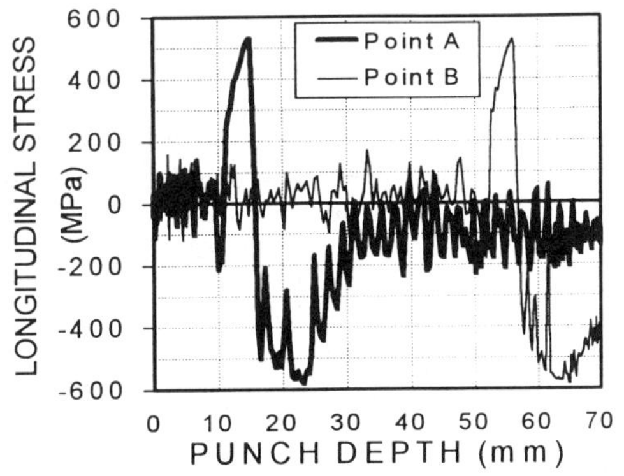

Fig. 20 Stress histories. Process time 10 ms. Element length=2.0 mm

be optimal for the current problem. Tests have also been performed with process times up to 100 ms with only negligible differences in results. However, for shorter times the dynamic effects will be noticeable. In Fig. 20 the stress history for a case with a process time of 10 ms and an element length of 2.0 mm is shown. Compared to the case with the process time 20 ms in Fig. 9, a marked influence of the dynamic effects can be observed. It should, however be pointed out that no differences between the plastic strain history curves could be observed. This indicates that stresses are much more sensitive to the imposed velocity than what strains are.

7 ADAPTIVE MESH REFINEMENT

As previously has been shown, it takes extremely small elements in the sheet to avoid the stress relaxation when the sheet passes the draw radius. In fact, the desired element size is five to ten times smaller than what is normally used in a finite element model of the sheet. A way to circumvent this problem could be the use of adaptive mesh refinement techniques, in which the mesh is automatically refined in areas with large curvatures and/or large strain gradients.

Such methods are now becoming available in commercial softwares. For instance, the code AUTOFORM is based on this technique, and also LS-DYNA3D offers this capability in its latest version. In a dynamic, explicit method techniques such as sub-cycling or mass scaling should be used together with adaptive mesh refinement in order to avoid problems with too small critical time steps.

8 SUMMARY AND CONCLUSIONS

Based on the results of the NUMISHEET'93 2D draw bending benchmark problem, our starting point for the present study was that there was some inherent, systematic error in the dynamic, explicit method, resulting in a strong underestimation of the springback. The investigation has, however, indicated that this basic presumption was wrong, and that most of the problems associated with a springback analysis are present, whether or not a static, implicit or a dynamic, explicit method is used.

The main problem in a springback simulation is the unphysical stress relaxation that takes place when the sheet is drawn over the draw radius. Due to the finite element discretization of the sheet a bending wave is induced, which in turn causes this stress relaxation.

The most important, single factor controlling this relaxation is the discretization of the sheet. It has been shown that it takes extremely small elements to completely avoid this phenomenon. Other factors that are influencing the stress relaxation are the discretization of the draw radius, and dynamic effects (in a dynamic, explicit method). It has also been demonstrated that these factors have a tendency of amplifying each other. Adaptive mesh refinement techniques is indicated as a way of circumventing the problem.

The importance of considering the Bauschinger effect by using a kinematic hardening law has been demonstrated.

REFERENCES

Anderheggen, E 1991. On the design of a new program to simulate thin sheet metal forming processes. *VDI Conference, Zurich (Switzerland): 231-245*

Du, C., Zhang, L. & Wang, N-M 1994. *1994 SAE Conference, Detroit, Michigan (USA)*

Gröber, M., Schweizerhof, K & Hallquist, J.O. 1993. Comments to the 2D draw bending benchmark problem. *NUMISHEET'93, Tokyo (Japan):* 663

Huang, M & Gerdeen, J.C. 1994. Springback of doubly curved sheet metal surface - an overview. *1994 SAE Conference, Detroit, Michigan (USA):* 125-138

Kubli, W., Anderheggen, E. & Reissner, J. 1991. Nonlinear solver with uncoupled bending and stretching deformation for simulating thin sheet metal forming. *VDI Conference, Zurich (Switzerland): 325-341*

Kubli, W. & Reissner, J. 1993. Optimization of sheet metal forming processes using the special purpose program AUTOFORM. *NUMISHEET'93, Tokyo (Japan): 271-280*

Makinouchi, M., Nakamachi, E., Onate, E. & Wagoner, R.H. (Editors) 1993. *NUMISHEET'93 - Second International Conference on Numerical Simulation of 3D Sheet Metal Forming Processes - Verification of Simulation with Experiment, Tokyo (Japan)*

Mattiasson, K., Bernspång, L., Hammam, T., Schedin, E., Melander, A. & Samuelsson, A. 1992. Evaluation of a dynamic approach, using explicit integration, in 3D sheet forming simulation. *NUMIFORM'92, Sophia-Antipolis (France)*

Simulation of Materials Processing: Theory, Methods and Applications, Shen & Dawson (eds)
 ISBN 90 5410 553 4

Evaluation of integral constitutive equations in viscoelastic modelling of polymer melts

E. Mitsoulis & G. Barakos
Department of Chemical Engineering, University of Ottawa, Ont., Canada

ABSTRACT: Integral constitutive equations of the K-BKZ type have been used recently in numerical simulations by a number of groups in a world-wide effort to model polymer melts in complex flow systems. It has been argued that due to a better fit of rheological data in simple flows and their intrinsic value in implementing suitable time- and strain-memory functions, these models better represent the behaviour of polymeric liquids in flow situations. To test this, three different models have been used to characterize a standard low-density polyethylene melt (IUPAC-LDPE) and predict its behaviour in flows through contractions. The models used are the Currie, Wagner, and PSM integral equations, which factorize the time- and strain-memory functions. The results for key flow features, such as *extrudate swell* and *excess pressure losses (end or Bagley correction)*, clearly show that good agreement can be achieved between simulations and experiments, especially with the PSM model, for low and moderate flow rates. The Wagner model is better for high flow rates, while the Currie model is inappropriate. It appears that the strain-memory function is most crucial for the correct behaviour, and further efforts have to be directed towards devising more appropriate dumping functions.

1 INTRODUCTION

Early attempts at modelling polymer melt flows employed relatively simple *differential* constitutive equations, such as the second-order, upper-convected Maxwell (UCM) and Oldroyd-B fluids. For the most part and for quite a number of years, these attempts failed to reach convergent solutions at relatively low values of a dimensionless number, such as the Weissenberg (We or Wi or Ws) or Deborah (De) numbers, characteristic of the viscoelastic nature of the fluids (Crochet *et al.*, 1984). Recently, good progress has been made on the numerical aspect by using the Streamline-Upwind/Petrov-Galerkin (SU/PG) scheme, which has overcome problems of convergence and has made solutions possible at very high elasticity levels (Marchal and Crochet, 1987). However, the numerical results in most cases have not produced a quantitative agreement on some of the key flow features observed experimentally (Debbaut *et al.*, 1988).

It became, therefore, apparent that it was necessary to search for and employ more realistic models, which would be able to simulate accurately the behaviour of polymer solutions and melts. It has been argued that *integral* constitutive equations with a spectrum of relaxation times are more capable of adequately describing the behaviour of viscoelastic fluids in simple shear, simple extension and any combination of mixed flows (Papanastasiou *et al.*, 1983).

Considerable success in modelling viscoelastic flows has been achieved recently with an integral constitutive equation based on the ideas of rubber elasticity and put forward by Kaye (1962), and independently by Bernstein *et al.* (1963). The original, so-called, K-BKZ model has been modified by various workers, notably by Wagner (1978), who proposed the separability (or factorization) of the integrand into a *linear time-memory* function to account for linear viscoelastic behaviour, and a *non-linear strain-memory* function to account for viscoelastic behaviour in strong deformations. While the time-memory function is pretty well established as a sum of exponentials involving a relaxation spectrum with multiple relaxation times, the strain-memory function can take different forms.

A particular form of the strain-memory function was proposed by Papanastasiou *et al.* (1983) and it involves a sigmoidal behaviour based on two parameters for strong (non-linear) shear and elongational effects. In this form, the K-BKZ model was proved capable of fitting adequately experimental data obtained from simple shear and elongational flows, and therefore, it was argued that it is a good candidate for more general and complex flows, such as those encountered in typical polymer processing operations. This equation has been used in several numerical simulations of polymer melts, notably by Dupont and Crochet (1988), Luo and Tanner (1986, 1988), Luo and Mitsoulis (1989, 1990a,b), Bernstein *et al.* (1994), etc.

The present work employs the K-BKZ integral constitutive equation in the simulation of a typical polymer melt flowing through a non-trivial contraction geometry. Three different types of memory functions are used in an attempt to provide an evaluation of the results and search for possible improvements in future viscoelastic simulations of polymer melts.

2 MATHEMATICAL MODELLING

The integral constitutive equation of the K-BKZ type is based on the potential theory of rubber elasticity and provides the stress tensor $\tau(t)$ as a time integral of the form:

$$\tau(t)=\int_{-\infty}^{t} m(t-t')\left[\varphi_1 C_t^{-1}(t')+\varphi_2 C_t(t')\right]dt' \quad (2.1)$$

where m(t-t') is the time-dependent term, $\varphi_1(I_1,I_2)$ and $\varphi_2(I_1,I_2)$ are strain-dependent terms, I_1 and I_2 are the first invariants of the Cauchy-Green tensor C_t and its inverse, the Finger tensor C_t^{-1}.

The time-dependent term is well-established as being a sum of relaxation modes corresponding to a discreet relaxation spectrum, i.e.,

$$m(t-t')=\sum_k \frac{a_k}{\lambda_k}\exp\left(-\frac{t-t'}{\lambda_k}\right) \quad (2.2)$$

where a_k are the relaxation moduli and λ_k are the relaxation times, respectively, for k relaxation modes. Note that the zero-shear-rate viscosity $\eta_0=\sum_k a_k\lambda_k$. For the Currie model (Currie, 1982), we have:

$$m(t-t')=\frac{96\eta_0}{\pi^4\lambda^2}\sum_{k=0}^{\infty}\exp\left[-(2k+1)^2\left(-\frac{t-t'}{\lambda}\right)\right] \quad (2.3)$$

The strain-memory term may take different forms that give rise to specific types of different constitutive models. For example, the Currie model has the following form:

$$\varphi_1=5\left[I_{C^{-1}}+2\left(I_C+3.25\right)^{1/2}-1\right]^{-1} \quad (2.4a)$$

$$\varphi_2=-\varphi_1\left(I_C+3.25\right)^{1/2} \quad (2.4b)$$

The Wagner model (Wagner, 1978) has the following form:

$$\varphi_1=\frac{h\left(I_{C^{-1}},I_C\right)}{1-\theta} \quad (2.5a)$$

$$\varphi_2=\frac{\theta h\left(I_{C^{-1}},I_C\right)}{1-\theta} \quad (2.5b)$$

where the *dumping function* or *kernel* $h(I_{C^{-1}},I_C)$ is given by:

$$h\left(I_{C^{-1}},I_C\right)=\exp\left[-n\sqrt{\beta I_{C^{-1}}+(1-\beta)I_C}\right] \quad (2.6)$$

Note that in the above n, β and θ are material constants. In particular, n relates to strong shear deformations, β relates to elongational deformations, while θ relates to the existence of second normal stress difference effects, and is defined by $N_2/N_1=\theta/(1-\theta)$, where N_1 and N_2 are the first and second normal stress differences, respectively.

The PSM model (Papanastasiou *et al.*, 1983) as modified by Luo and Tanner (1988) has the same form for φ_1 and φ_2 as the Wagner model, but the kernel $h(I_{C^{-1}},I_C)$ is given by:

$$h\left(I_{C^{-1}},I_C\right)=\frac{\alpha}{(\alpha-3)+\beta I_{C^{-1}}+(1-\beta)I_C} \quad (2.7)$$

In the above, α is a material parameter determined from strong shear flows, while β relates to elongational deformations as in the Wagner model.

In the above constitutive models, it is first necessary to determine the parameters by performing regression analysis to experimental data obtained from simple shear and elongational tests with standard instruments. Typically, a non-linear regression analysis is carried out, as explained by Papanastasiou *et al.* (1983) and more recently by Kajiwara *et al.* (1995).

The analysis provides the number of relaxation modes, the values of the relaxation times and moduli, and the other parameters n, β and θ appearing in the above models. The constitutive equations can thus be used to predict steady and time-dependent shear and extensional viscosities, normal stresses, and storage and loss moduli. The evaluation of the stress integral (2.1) in the case of such simple flows is straightforward.

As an example, we provide here the best-fitting values and curves for a number of experimental data given for a commercial low-density polyethylene melt, so-called IUPAC-LDPE, which has been the subject of an international experimental study (Meissner, 1975). Figure 1 provides the best-fit for shear and elongational viscosities, and N_1 data, while Table 1 gives the values for the relaxation spectrum and the other parameters. Figure 2 provides the dumping function h in simple shear and elongational flows, respectively, for the 3 special types of models, i.e., the Currie, Wagner and PSM. It is seen that in all cases of experimental data, the fit of the models is best for the PSM, followed by the Wagner, and then the Currie model, i.e., PSM>Wagner>Currie, in a qualitative sense.

3 METHOD OF SOLUTION

The above constitutive equations are solved together with the conservation equations of mass, momentum and energy (in the case of non-isothermal flows) using the Finite Element Method (FEM). A special numerical scheme is needed to calculate the viscoelastic stresses for the general case of flows with and without recirculation (Luo and Mitsoulis, 1990a,b). Galerkin discretization is maintained and the numerical algorithm for convergence is Picard iteration as described by Luo and Mitsoulis (1990b).

For constrained flows *without free surfaces*, convergence has always been good even for very high flow rates. Convergent solutions have been obtained independent of mesh size, provided enough elements are used, and the solution procedure advances slowly from low flow rates (Newtonian behaviour) to higher ones by using a *flow rate increment scheme*.

For flows *with free surfaces*, convergence has been more difficult to achieve, due to the extra non-linearity introduced by the presence of the unknown *a priori* free surface. Convergent solutions can be obtained independent of mesh size, provided enough elements are used especially in the die exit, and the solution procedure uses severe under-relaxation for the free-surface movement, so that the flow domain is not altered drastically during iterations for a fixed flow rate (see Barakos and Mitsoulis, 1995). In general, many iterations are performed to achieve convergence at a fixed flow rate for both the system of equations and the free surface movement, and then the apparent shear rate is increased to obtain results at higher flow rate values.

4 RESULTS AND DISCUSSION

Detailed results for the flow of a well-characterized low-density polyethylene melt (the IUPAC-LDPE Sample A) through capillary dies of varying L/R dimensions (see a schematic representation of the flow system in Fig. 3) have just appeared in the literature (Barakos and Mitsoulis, 1995). The working integral constitutive equation was the PSM model with a dumping function given by eq. (2.7). These results shown also here in Figs. 4 and 5, made clear that the simulations predict very well experimental findings for extrudate swell from orifice dies (L/R=0), while they tend to overestimate the corresponding values from very long dies (L/R=∞).

A careful examination has shown that this is due mainly to the inability of the strain-memory function of the PSM model (eq. 2.7) to capture well the fading memory of the polymer melt. In particular, when the die is very short, the melt has only to remember its slow flow in the preceding reservoir, and the strain-memory function does not contribute very much in the extrudate swell mechanism. However, as the die gets longer and for high flow rates, the melt travels fast in a longer domain and it should forget faster its previous stress state to correspond to the relatively moderate swell. This is not though true for the strain-memory function given by eq. (2.7). The model appears to be *too elastic*, giving much higher swelling ratios than the experimental ones.

The memory function given by Wagner (1978) appears to be more suitable but only at high flow rates, as evidenced in Figs. 4a and 5a. The same conclusions have been reached by Goublomme *et al.* (1992), who have examined a high-density

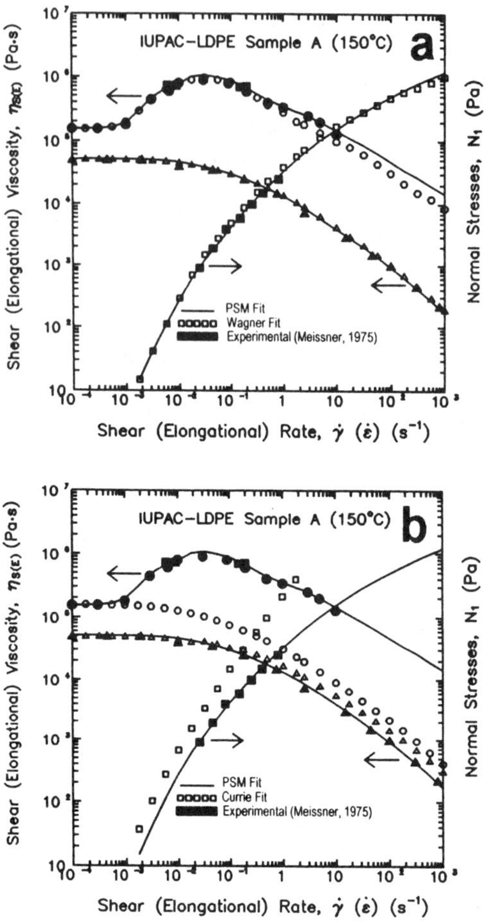

Fig. 1. Model predictions for the shear (η_S) and elongational (η_E) viscosities, and the first normal stress difference N_1 for the IUPAC-LDPE melt as obtained from different integral models: (a) PSM and Wagner, (b) PSM and Currie.

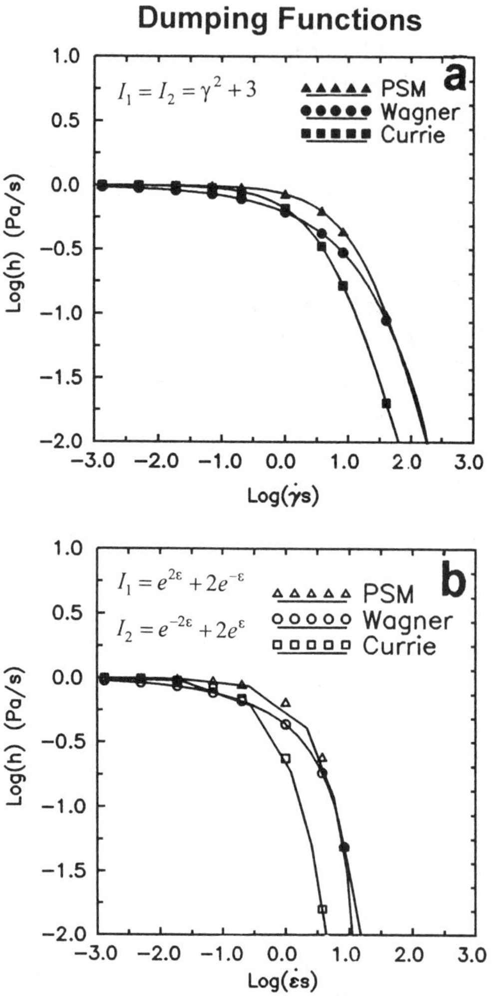

Fig. 2. Model predictions for the dumping function h for the IUPAC-LDPE melt as obtained from different integral models: (a) simple shear flow, (b) simple elongational flow.

polyethylene (HDPE) melt at very high flow rates with both the PSM and Wagner models. By contrast, the Currie model with a single relaxation time performs much worse, giving a reduction in swelling as the L/R ratio approaches zero (see Figs. 4b and 5b). This is contrary to the fading memory phenomena exhibited by polymeric liquids, and the model behaves similarly to the Maxwell and Oldroyd-B fluids, which also show a reduction in swelling at low L/R ratios (unpublished data from our group). Obviously, the Currie model is not capable of capturing well the behaviour of LDPE, as also shown in the fitting of rheological data earlier (see Fig. 1b).

The results for the excess pressure losses from the contraction and orifice flows can be used to calculate the *end (or Bagley) correction* according to Bagley's formula (Bagley, 1957). These results are shown in Fig. 6 for the three models studied. It is seen that again the PSM model is better for low to moderate flow rates, while the Wagner model is more appropriate for high flow rates. The Currie model is again the least performing.

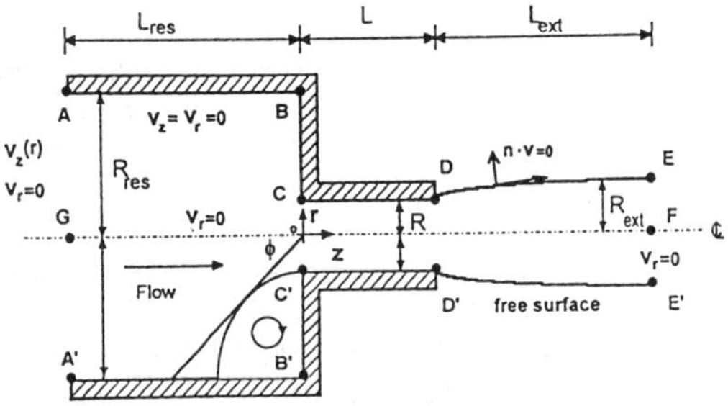

Fig. 3. Schematic diagram of entry flow in a contraction and exit flow with the accompanying phenomenon of extrudate swell.

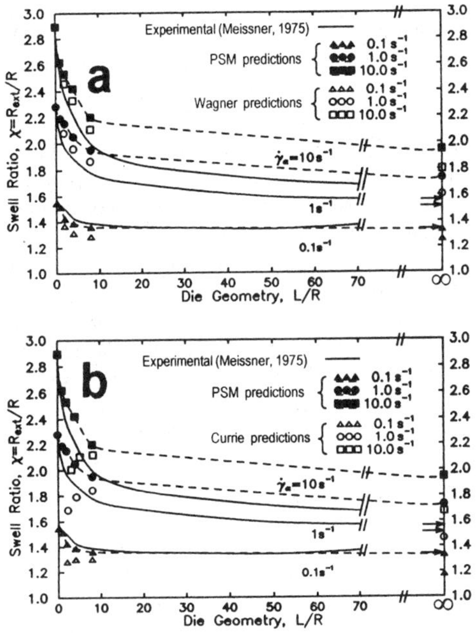

Fig. 4. Model predictions for the effect of L/R on extrudate swell for the IUPAC-LDPE melt as obtained from different integral models: (a) PSM and Wagner, (b) PSM and Currie.

5 CONCLUSIONS

The conclusions of this work can be summarized as follows:

- The Currie model without non-linear parameters in the strain-memory function gave unrealistic predictions for the extrudate swell and it underpredicted the Bagley correction.
- The Wagner model underpredicted swelling ratio and Bagley correction at low flow rates, but it seems more suitable at high flow rates.
- The PSM model gave good predictions for the swelling ratio and Bagley correction at low and moderate flow rates, but it exhibited a more elastic character than the Wagner model, overpredicting at high flow rates.

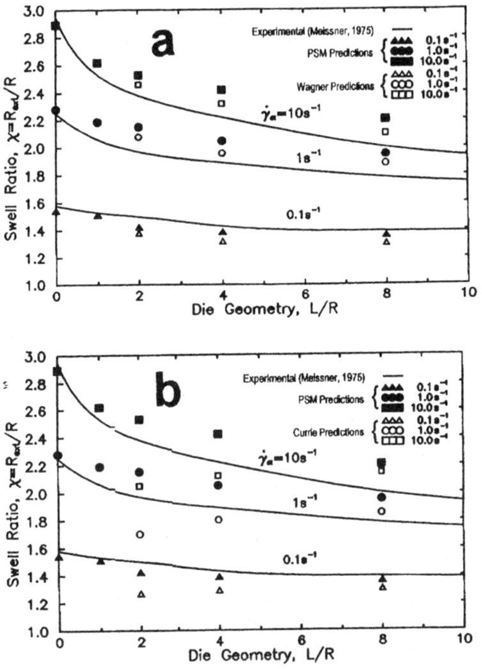

Fig. 5. Blown-up section of model predictions for the effect of L/R on extrudate swell for the IUPAC-LDPE melt as obtained from different integral models: (a) PSM and Wagner, (b) PSM and Currie.

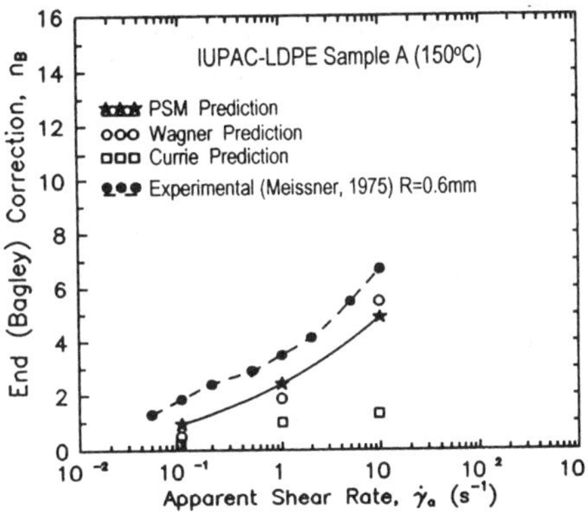

Fig. 6.

TABLE 1 Material parameter values used in eq. (2.2) for fitting data of the IUPAC-LDPE (sample A) at 150°C according to Luo and Tanner (1988).

k	λ_k (s)	a_k (Pa)	β_k (-)
1	10^{-4}	$1.29x10^5$	0.018
2	10^{-3}	$9.48x10^4$	0.018
3	10^{-2}	$5.86x10^4$	0.08
4	10^{-1}	$2.67x10^4$	0.12
5	10^0	$9.80x10^3$	0.12
6	10^1	$1.89x10^3$	0.16
7	10^2	$1.80x10^2$	0.03
8	10^3	$1.00x10^0$	0.002

Currie model: η_0 = 51,000 Pa.s, λ = 59 s
Wagner model: n = 0.184, θ = -1/9
PSM model: α = 14.38, θ = -1/9

It has been thus shown that integral constitutive equations of the K-BKZ type with a suitable dumping function are capable of fitting rheological data for polymer melts in simple flows, and therefore, they constitute good candidates for simulating complex flows as well. It is, however, of primary importance to employ an appropriate form for the strain-memory function, in particular the kernel $h(I_{C^{-1}},I_C)$, as this contributes most to the memory phenomena exhibited by polymeric liquids in strong flows.

More work is, therefore, needed to establish better strain-memory functions for the flow of polymer melts, especially in fast and fast-changing flows encountered in polymer processing.

ACKNOWLEDGMENTS

Financial assistance from the Natural Sciences and Engineering Research Council (NSERC) of Canada and the Ontario Centre for Materials Research (OCMR) is gratefully acknowledged.

REFERENCES

Bagley, E.B. 1957. End corrections in the capillary flow of polyethylene. *J. Appl. Phys.* 28:624-627.

Barakos, G. and E. Mitsoulis 1995. Numerical simulation of extrusion through orifice dies and prediction of Bagley correction for an IUPAC-LDPE melt. *J. Rheol.* 39.

Bernstein, B., E.A. Kearsley, and L. Zapas 1963. A study of stress relaxations with finite strain. *Trans. Soc. Rheol.* 7:391-410.

Bernstein, B., K.A. Feigl, and E.T. Olsen 1994. Steady flows of viscoelastic fluids in axisymmetric abrupt contraction geometry: a comparison of numerical results. *J. Rheol.* 38:53-71.

Crochet, M.J., A.R. Davies, and K. Walters 1984. *Numerical simulation of non-Newtonian flow.* Amsterdam: Elsevier.

Debbaut, B., J.M. Marchal, and M.J. Crochet 1988. Numerical simulation of highly viscoelastic flows through an abrupt contraction. *J. Non-Newtonian Fluid Mech.* 29:119-146.

Dupont, S. and M.J. Crochet 1988. The vortex growth of a KBKZ fluid in an abrupt contraction. *J. Non-Newtonian Fluid Mech.* 29:81-91.

Goublomme, A., B. Draily, and M.J. Crochet 1992. Numerical prediction of extrudate swell of a high-density polyethylene. *J. Non-Newtonian Fluid Mech.* 44:171-195.

Kajiwara, T., G. Barakos, and E. Mitsoulis 1995. Rheological characterization of polymer solutions and melts with an integral constitutive equation. *Int. J. Polym. Anal. Char.* in press.

Kaye, A. 1962. Note No. 134. College of Aeronautics, Cranford, UK.

Luo, X.-L. and R.I. Tanner 1986. A streamline element scheme for solving viscoelastic flow problems, part II. Integral constitutive models. *J. Non-Newtonian Fluid Mech.* 22:61-89.

Luo, X.-L. and R.I. Tanner 1988. Finite element simulation of long and short circular die extrusion experiments using integral models. *Int. J. Num. Meth. Eng.* 25:9-22.

Luo, X.-L. and E. Mitsoulis 1989. Memory phenomena in extrudate swell simulations from annular dies. *J. Rheol.* 33:1307-1327.

Luo, X.-L. and E. Mitsoulis 1990a. A numerical study of the effect of elongational viscosity on vortex growth in contraction flows of polyethylene melts. *J. Rheol.* 34:309-342.

Luo, X.-L. and E. Mitsoulis 1990b. An efficient algorithm for strain history tracking in finite element computations of non-Newtonian fluids with integral constitutive equations. *Int. J. Num. Meth. Fluids* 11:1015-1031.

Marchal, J.M. and M.J. Crochet 1987. A new mixed finite element for calculating viscoelastic flow. *J. Non-Newtonian Fluid Mech.* 26:77-114.

Meissner, J. 1975. Basic parameters, melt rheology, processing and end-use properties of three similar low density polyethylene samples. *Pure Appl. Chem.* 42:551-612.

Papanastasiou, A.C., L.E. Scriven and C.W. Macosko 1983. An integral constitutive equation for mixed flows: viscoelastic characterization. *J. Rheol.* 27: 387-410.

Wagner, M.H. 1978. A constitutive analysis of uniaxial elongational flow data of a low-density polyethylene melt. *J. Non-Newtonian Fluid Mech.* 4:39-55.

© 1995 Balkema, Rotterdam. ISBN 90 5410 553 4

Boundary element modeling of the electrochemical machining process

H.A. Nied
GE Corporate Research and Development, Schenectady, N.Y., USA

M.S. Lamphere
GE Aircraft Engines, Hooksett, N.H., USA

ABSTRACT: Since the electrochemical machining (ECM) process is essentially a surface phenomenon, a solution is readily obtained by employing classical potential theory. By using a 2D fundamental singular solution, the boundary value problem for the ECM process can be reduced to the solution of a boundary integral equation. When the surfaces are discretized, a numerical solution for the potential and the derivatives at the surface can be determined. The boundary element method (BEM) developed by Brebbia was used to solve the field equations for this non-conventional machining process. The electrochemical anodic reaction was furnished by Faraday's Law, which provided the relationship for the rate of dissolution of the workpiece. By this approach, a 2D computer process model was developed for the purpose of simulating the ECM process. As an example, application of this process model for determining the machined shape of an airfoil compressor blade is described.

1. INTRODUCTION

Electrochemical machining is a process for the removal of an electrically conductive metal by anodic dissolution using large surface current densities when the anode and cathode are separated by a narrow gap containing a rapidly flowing electrolyte. ECM offers many advantages over traditional machining processes like milling and turning. ECM is a non-contact machining process that can quickly shape any electrically conductive material regardless of the material's hardness or toughness. Since the workpiece is shaped through anodic dissolution, the process does not produce residual stresses in the parent material, resulting in a stress free component. Additionally, due to the nature of the process, ECM produces a smooth, burr free surface that usually does not require post machining operations.

The major characteristics of the ECM process are illustrated in Fig. 1. In this figure, showing two flat electrodes for simplification, metal is dissolved from the anodic workpiece due to the electric current flow between the cathodic tool and the anode by ion transport. The rate of material dissolution depends entirely on the total charge that is passed to the workpiece. Different alloys have different rates of material removal based on the dissolution valency of their constituents. It is worth mentioning that the theoretical material removal rate usually does not correspond to the experimental rate. This difference is called the current efficiency and is usually expressed as a percentage of the theoretical amount predicted by Faraday's law.

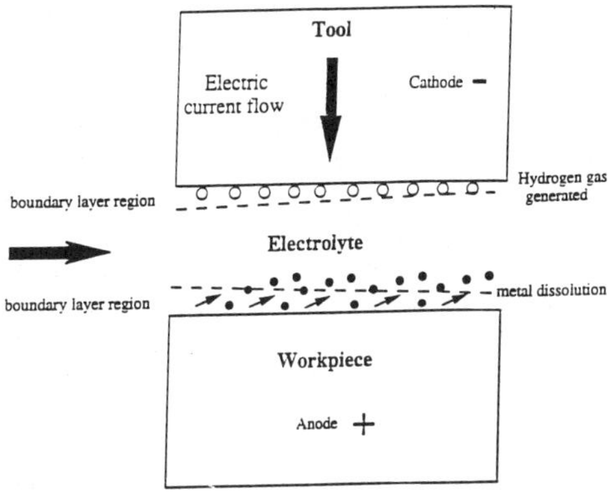

Fig. 1. Schematic of the electrochemical machining (ECM) Process.

The major functions of the electrolyte are to provide a conductive media to carry the current, remove heat generated from the tool and workpiece,

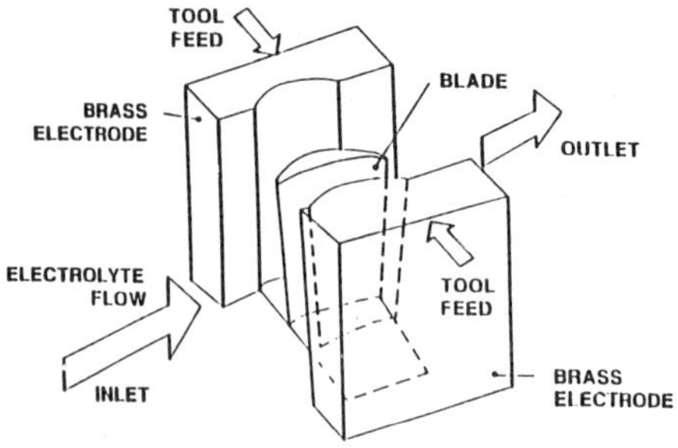

Fig. 2. Airfoil ECM process tooling configuration.

and flush away the dissolution products and gases generated downstream. Hydrogen gas is one of the primary by-products generated which tends to collect at the cathode. Presence of hydrogen in the interelectrode gap has the effect of reducing the electrolytes' bulk conductivity. While the current density controls the amount of hydrogen that is generated, maintaining high back pressure in the gap can minimize the effect by reducing the size of hydrogen bubbles that are generated. The basics of electrochemistry that are applied to the ECM process including equilibrium electrode potentials, irreversible electrode reactions, polarization and overpotential are described in detail by both De Barr [1] and McGeough [2].

The ECM process has been used to fabricate difficult to machine mechanical components such as airfoils made of super alloys. Fig. 2 shows schematically the typical tooling arrangement used for machining airfoils and blades in the aircraft engine industry using cross flowing electrolyte. Electrolyte under high pressure flows in the gap between the airfoil and the electrodes from the leading to trailing edge of the airfoil, or vice versa. ECM process parameters such as temperature, voltage, tool feedrate, and pressure are tightly controlled during the ECM process. Other parameters such as electrolyte conductivity, pH, and turbidity are monitored and adjusted as necessary. Cathodes for the convex and concave shapes are usually fed from opposite sides at equal feedrates during the process to maintain a prescribed rate of material removal.

The design of ECM tooling and process parameters is currently a costly and lengthy iteration process. For these reasons, a 2D ECM process model was developed for the purposes of improving general process understanding, predicting machined shapes, aid in the design of cathode tools, and process optimization.

2. ECM MODELING DESCRIPTION

2.1 Process Modeling Background:

Early ECM simulation was focused on the solution of the Laplace equation for single electrode geometries and mathematical methods for cathode design. It was recognized by these researchers Refs. [3]-[7], that the accuracy for the prediction of anode material dissolution was highly dependent on the current density distribution. The approach used by these authors was to use analytic mathematics rather than numerical methods to obtain a fundamental understanding of the current density distribution in the ECM process. For example, Nilson and Tsuei [4] used complex variables for obtaining solutions to simple single tool geometries. None of these highly mathematical methods could be practically used for simulating the entire complicated ECM process. A numerical method was introduced by Jain and Pandey [8]-[12] using the finite element method (FEM) which would allow handling more complicated geometries. This approach, although superior to previous efforts, required discretizing the interelectrode region which in most ECM processes is extremely small, in the order of 0.0254 cm. Narayanan et al. [13]-[14] modeled the ECM single tool process by using the boundary element method (BEM). This approach provides a more efficient numerical method for updating only the workpiece surface rather than remeshing the entire model, as would be required in finite elements.

All of the mentioned models of the ECM process were primarily directed to the solution of the Laplace equation for the determination of the current density distribution without detailed consideration of the effects of electrolyte flow on the machining process.

2.2 BEM Process Model Development:

Many problems of engineering can be classified as mixed boundary value problems when the function and its derivative are specified on sections of the boundary surrounding the interior domain. Such is the case for the ECM process when voltage or currents are specified on surfaces for single or multiply connected domains. The numerical method usually used for solving such mixed boundary value problem is by applying the finite element method (FEM). Using the FEM approach, one discretizes the material interior with a mesh that includes the boundaries. If gradients of the potential near the surface are sought, the interior mesh must be accordingly refined. The FEM formulation can then be reduced to a set of linear equations for the solution of the potential at each interior node when boundary condition nodal values are specified on the boundary. This method provides a solution at each interior point whether it is needed or not. The resulting linear matrix equations are banded and can be easily solved on a high speed digital computer. The major difficulty with this method is the complexity required during the remeshing of the model with each time step after the material removal from the surface of the anode.

On the other hand, the BEM formulation applied to ECM does not require an internal mesh, Narayanan et al.[13]. It requires only a mesh on the boundary, which is an advantage when applied to the ECM process since we are chiefly interested in the solutions on the surface. In addition, since the ECM process model requires a rigid body movement of the cathode and a moving boundary on the anode, the application of the BEM method provides a superior alternative since remeshing is only required on surfaces to simulate gap closure and workpiece dissolution on a surface. In the case of using FEM, remeshing at each time step would require updating the mesh in the interior as well as on the boundary.

Solution of the integral equations for the BEM formulation is accomplished by using a fundamental singular solution and by applying Gauss's theorems of potential theory that allow converting the problem from the solution of any interior domain to the distributed contribution of the potential and its normal derivative on the bounding surface. The application of the singular solution and the derivation of the boundary integral equations used in this formulation are contained in the appendix. The numerical solution of these boundary integral equations was developed using the method developed by Brebbia. For a more thorough exposition on this topic, the reader is referred to chapter 2 of Brebbia [15]. His formulation and approach were used as the basis for developing the electrochemical machining process model for simulating the electrochemical machining of gas turbine airfoils and blades. The most important variable that must be determined in the ECM process is the current density distribution on the workpiece with the subsequent application of Faraday's laws of electrochemistry. Consider the airfoil ECM application shown in Fig. 2. If a cutting plane is passed through the tooling and the blade normal to the airfoil longitudinal axis, a 2D model at that section is obtained. Fig. 3 shows a CAD/CAM section drawing of an airfoil blade with the dual opposed cathode electrodes (top) and the equivalent BEM model (bottom).

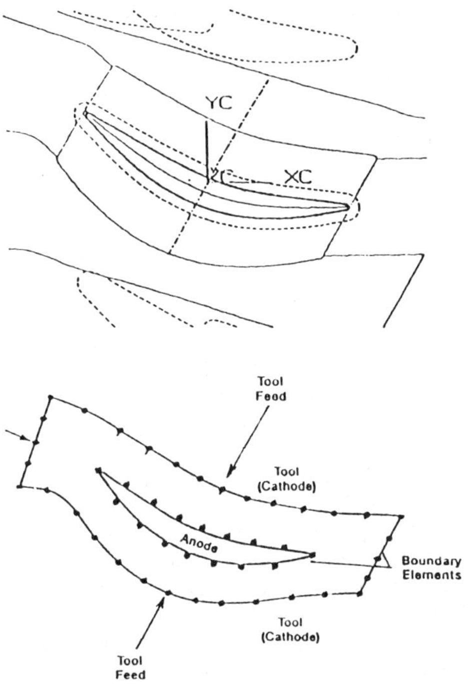

Fig. 3. CAD/CAM drawing of airfoil and tooling and representative BEM model.

The electrolyte conducts the electric current between the cathode and anode by an ionic transport phenomenon which provides the mechanism for surface dissolution of the airfoil (anode). This dissolution of the anode surface produces a slowly moving boundary condition in the mathematical formulation. Crank [16] rigorously derives the moving boundary conditions which must be incorporated into a precise formulation of the ECM process. In the case of a 2D problem, the current density distribution produced on the anode will not be uniform in this process. The purpose of this model is to accurately predict the current density distribution by the BEM method during the entire machining simulation to obtain the final predicted shape of the airfoil.

The electrolyte flow in the gaps between the cathode and anode is a high velocity turbulent flow. The electrolyte flow will split at the inlet for airfoil machining and merge at the outlet. In the present development, the universally accepted velocity profile for turbulent channel flow is assumed for the electrolyte fluid mechanics. The basic equations used to model the electrolyte flow characteristics

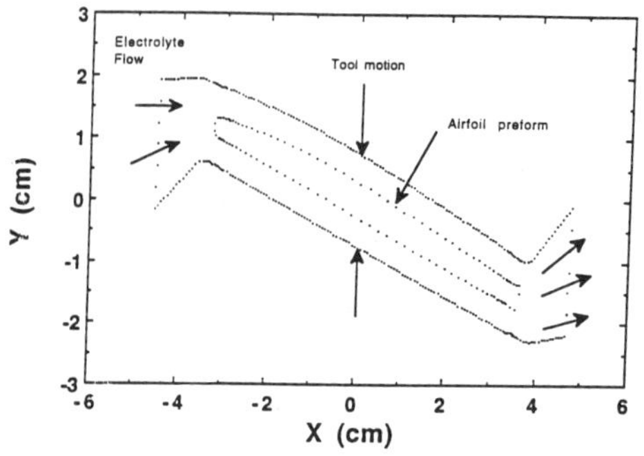

Fig. 4. BEM representation of tooling and airfoil preform at t=0.0 sec.

are contained in the appendix. The most important consideration is the temperature rise that occurs in the machining gaps during the machining process from inlet to outlet due to its effect on the electrolyte properties. The temperature rise in the electrolyte gap is due to Joule heating and is proportional to the local current density. The turbulence in the flow produces essentially a uniform temperature across the gap which varies inversely as the gap size. The wall shear stress resistance due to the viscosity of the electrolyte follows the widely accepted Darcy friction factor method which is dependent on the Reynolds number of the flow. The electrolyte properties such as electrical conductivity, viscosity, density, and specific heat are all a function of temperature and must be accounted for in order to have a realistic simulation.

Using the boundary integral equation method, many complicated geometries can be conveniently represented by a simple 2D contour of the ECM process model consisting of the boundaries enclosing the tool, electrolyte gap, insulation, and workpiece. The advantage of this formulation is that only the boundary composed of the tool, the workpiece, and insulators have to be discretized to construct a 2D model without the necessity of discretizing the interior of the electrolyte between the tool and workpiece as would be the case if the finite element method were employed. This analytical modeling method is most appealing since in the electrochemical machining process we are only interested in the reactions and dissolution of the anode at the surface. Another reason for using the boundary element method is that greater numerical accuracy can be achieved on the surfaces in the model since the functions and their derivatives are calculated directly at the surfaces. In the finite element formulation, the derivatives of the function are calculated at interior Gauss points and the values at surfaces are extrapolated from these interior points. The kinematics of the BEM model is also easier to implement for both the velocity of the surface mesh of the tools and the surface removal of the workpiece. It turns out that an initial coarse surface mesh on the airfoil becomes a finer mesh as material is removed since the distance between element nodes at the surface is reduced. At each time step, the current density distribution and the nodal displacement due to machining the workpiece, the total accumulated weight removed, and the tool displacement are calculated in sequence until the final process time is reached.

3. ELECTROLYTE PROPERTIES

For an effective process model, it is necessary to incorporate electrolyte properties into the ECM computer model to be able to simulate the electrochemical machining behavior at the surface. The electrochemical surface dissolution of the anode is a complicated phenomenon and is different for various cell combinations of metal and electrolyte. The properties of the electrolyte are chiefly dependent on the concentration of the solute and temperature.

The 2D ECM computer code developed has a materials properties library for user selected combinations of electrolyte type and workpiece materials. These properties were determined in a sub-scale laboratory facility especially built for that purpose. The material library requires electrolyte data as a function of temperature, concentration, and current density to determine the electrolyte's electrical and thermal conductivity, viscosity, density, overpotential and cutting efficiency. It is also essential to include the overvoltage in the model for the precise determination of the local current density distribution on the surface of the workpiece.

4. APPLICATIONS

As an example, the results of modeling a typical compressor fan blade in the aircraft engine industry are described. The process simulation for the electrochemical machining of a titanium alloy (Ti 6-4) compressor blade was conducted using a sodium chloride (NaCl) electrolyte. Actual machine and process parameter settings were used in the computer program to analyze the machined shapes, weight of material removed, current density distribution around the perimeter of the airfoil, interelectrode temperature increase in the electrolyte, and the current history used in the process. Since this code is a 2D computer program, the full height of the blade was modeled by a series of cutting planes normal to the stacking axis. However, due to space limitation, the results for only only one section will be described.

Fig. 4 is an illustration of a BEM model that was constructed to simulate the 2D ECM process for machining a typical cross section of fan blade near the tip. This figure shows the nodal points for the cathode tooling surrounding a forged airfoil preform at t = 0.0 sec. Note that only a surface mesh is required to define the entire model. Fig. 4 also shows the velocity direction of the dual cathodes that move in opposite directions. This model was oriented such that the trailing edge was at the electrolyte inlet on the left and the electrolyte exit at the leading edge on the right. The contour of the model is composed of series of boundary elements on the surface. A BEM solution for this model is obtained by treating this geometry as a multiply connected region using the conventional directions for traversing the outer and inner contours defining the tooling and airfoil, respectively.

Boundary conditions are specified on the entire model. For example, voltage potentials for each element are specified on both the cathode and anode. Elements used to represent insulation or the inlet and outlet channels of the electrolyte are specified by setting their normal derivatives to zero, thus indicating no flow of current. During the machining process, the cathode tools have a prescribed velocity and voltage. The velocity and voltage are set initially high to close the machining gap quickly, but are stepped down over the process cycle to obtain better conformance of the workpiece shape to that of the cathode.

Fig. 5 shows a computer graphics line plot of the entire ECM model with the tooling at the intial position and the initial shape of the workpiece.

As the machining process progresses in time, only the defining mesh for the surface of the tool and insulation remain the same. The tooling mesh is translated along a vector at each time step to simulate tool motion. Remeshing occurs over the entire surface of the workpiece at each time step due to anodic dissolution. Workpiece remeshing is required during the simulation to maintain optimal nodal spacing over the surface to prevent nodal crossover and improve the leading and trailing edge definition. Fig. 6 shows the location of the cathodes and the final machined shape after 215 seconds of simulated process time. Note that the interelectrode gap at the end of the process is very small and is approximately 0.0254 cm.

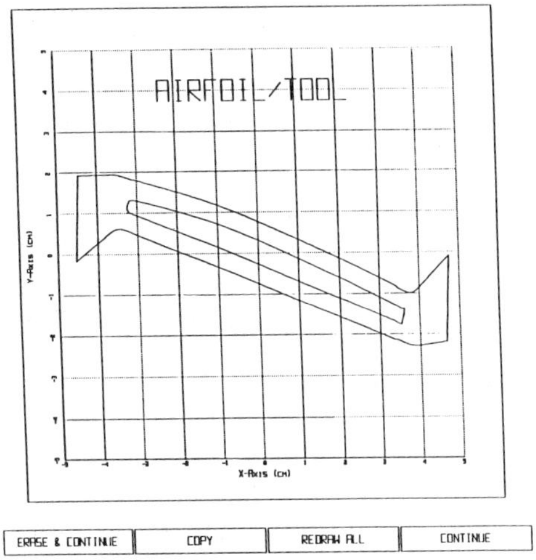

Fig. 5. Computer graphics plot of the workpiece and tooling at t=0.0 sec.

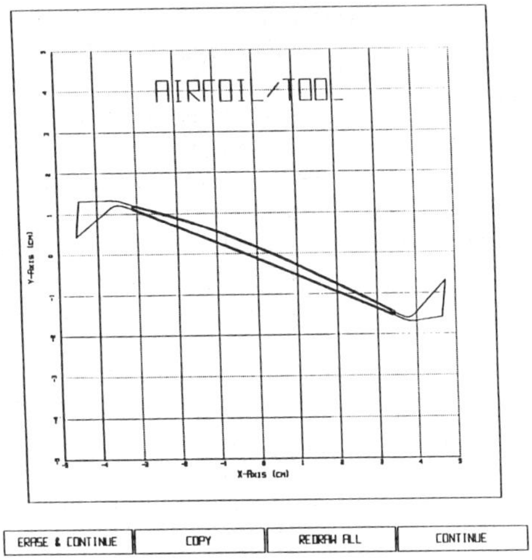

Fig. 6. Computer graphics plot of the workpiece and tooling at t= 215.0 secs.

With the zoom capability of the computer program, the analyst is able to investigate the geometry of the airfoil at the trailing and leading edges. As an illustration, Fig. 7 shows the shape of the airfoil at the trailing edge. The definition and accuracy of the simulated profile depend on the number of elements in the model, with a greater number yielding superior results. Using the computer graphics capability,

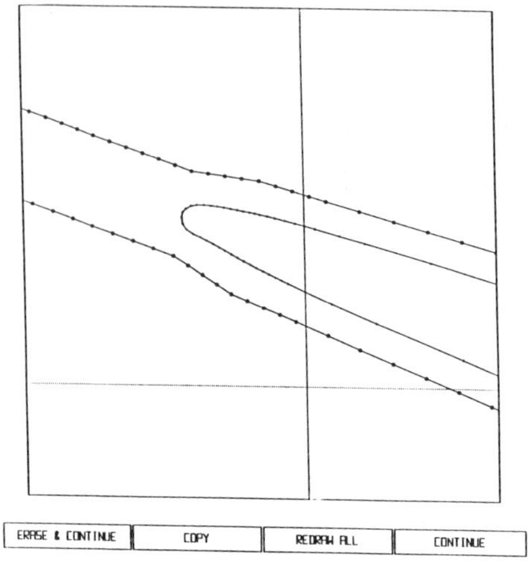

Fig. 7. Zoom view of trailing edge at t= 215.0 sec.

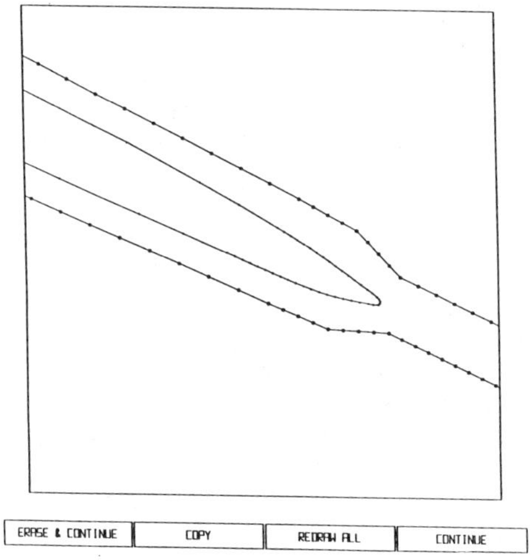

Fig. 8. Zoom view of leading edge at t= 215.0 sec.

one can determine the geometric details at both the leading and trailing edges. Fig. 8 shows the machined profile at the leading edge when the zoom feature is used.

The rate of dissolution at the airfoils leading and trailing edges limits the number of nodes that can be applied in practice. Higher nodal density requires smaller time steps to prevent nodal crossover on the workpiece surface. The analyst must make the final determination of ultimate model resolution verses the cost of running a larger model with smaller time steps.

The predicted dimensions and final shape of the airfoil are important for the process engineer to evaluate. The radii at both the leading and trailing edges and the chordal dimension from leading to trailing edge can easily be evaluated using the features of the computer graphics included in the program. Fig. 9 shows the prediction of the entire shape of the airfoil.

An accurate determination of the current density distribution around the outer contour of the airfoil is extremely important to precisely simulate the ECM process. Fig. 10 is a plot of the current density distribution around the airfoil at the termination of the ECM process. At t = 215.0 seconds in the process, the cathodes are close to the airfoil and produce the final shape. The current densities are largest at both the leading and trailing edges, which is due solely to geometric effects. The current density on the convex and concave sides of the airfoil are essentially uniform.

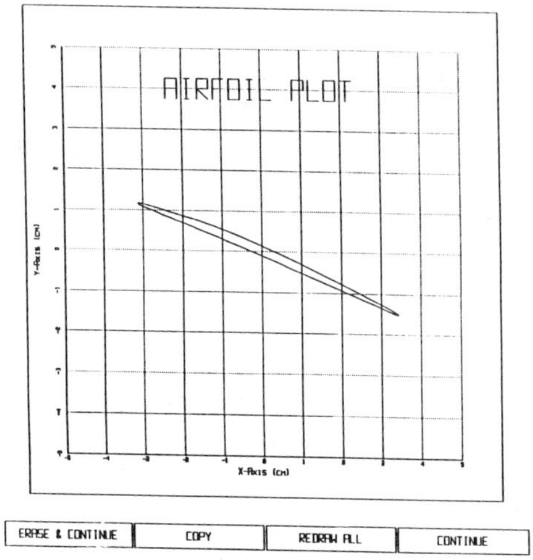

Fig. 9. Airfoil machined profile at t= 215.0 sec.

Once the predicted dimensions and profile have been determined, they can be compared to the nominal values of the airfoil for that section. Fig. 11 shows the overlay of the entire predicted airfoil on the plotted points of the drawing coordinates. This case shows that the predicted airfoil shape is close to the target profile. Further evaluation can be carried out by obtaining the zoomed plots at both the trailing and leading edges. Fig. 12 and Fig. 13 show the nominal dimension overlay on the predicted profile for the trailing and leading edges, respec-

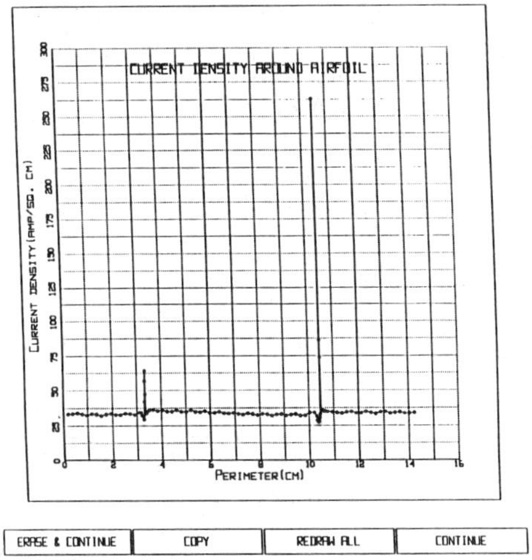

Fig. 10. Current density distribution over airfoil.

tively. With suitable computer graphics, the process engineer can quickly deterimine if the tooling or the process parameters should be altered.

One of the key results of the analysis is the determination of the total temperature rise in the gaps. While the maximum temperature rise near the root was 10.6 °C, near the tip the temperature rise was slightly greater, being equal to 11.2 °C. This temperature rise is directly proportional to the square of the current density and the flow rate of the electrolyte in the gap. The reduced flow rate at the root can be attributed to a narrower interelectrode gap. The temperature rise at the point in the outlet gap is dependent on the turbulent mixing when the flows merge. There is also a slight difference in the electrolyte flow rates on the convex and concave sides due to the split flow that occurs at the inlet and the initial position of the preform. If there is a significant difference in the balance of the flow through the two gaps, there can be a significant overall effect on the workpiece shape.

5. CONCLUSIONS

The boundary element method (BEM) is an effective analytic tool that can be used as an ECM process simulator for predicting the final machined shape of 2D workpiece geometries based on specified materials and process parameters. It has also been shown that the BEM technique can be used to successfully predict the shape of an aircraft engine compressor airfoil or blade, considered to be one of the most difficult surfaces to model due to the

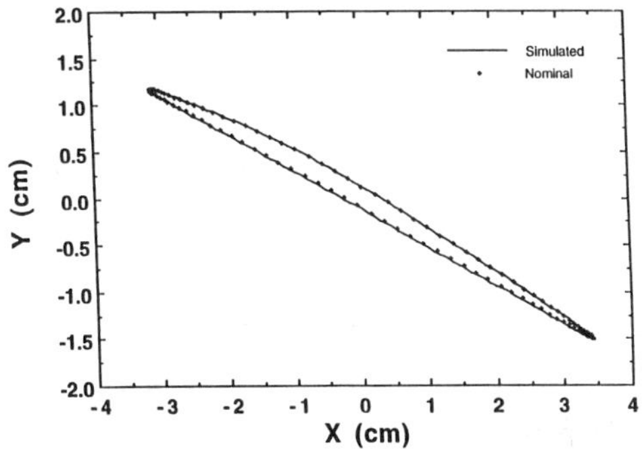

Fig. 11. Comparison of airfoil prediction with nominal dimensions.

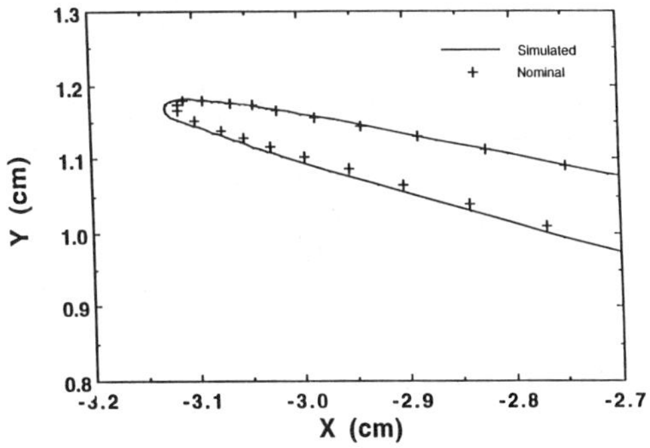

Fig. 12. Overlay view of trailing edge.

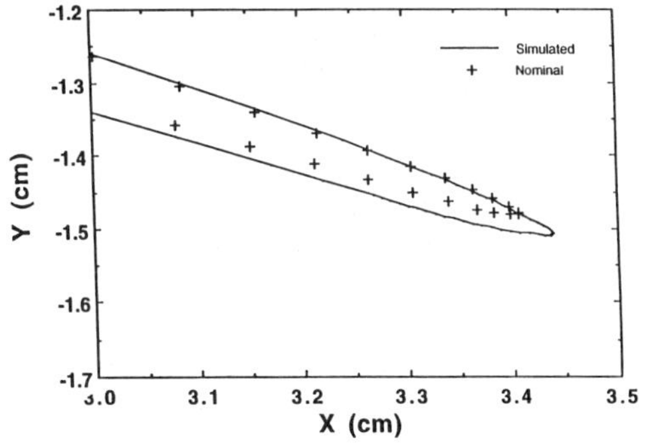

Fig. 13. Overlay view of leading edge.

complex surface geometry and sharp radii at both leading and trailing edges.

ECM simulation offers unique insights into the process that can only be effectively obtained from modeling. This includes such items as the interelectrode gap size, current density distribution, and the electrolyte pressure, temperature and velocity distribution. Models can be used to conduct parametric studies of future tool designs and to aid with the interpretation of experimental results. In addition, the ECM simulation furnishes an alternative rational way to establish the initial process parameters during the process development stage.

In practice, during electrochemical machining only external parameters such as inlet and outlet pressures, applied voltage and total electric current can be readily measured and monitored. A process model provides a tool to determine the state of machining in the interelectrode gap which is of major interest. This allows the user to investigate the electrolyte temperature, velocity, current density on the workpiece and cathode tooling, pressure distributions and gap size along the flow path. These variables are virtually impossible to measure except by laboratory methods. For example, consider the current density distribution on the anode workpiece. This single variable has a significant effect on the local machining mechanism by virtue of Faraday's laws of electrolysis. Being able to predict the current density distribution would allow the user to make the necessary changes in tool design and to adjust the tool velocities for better performance and surface finish.

6. REFERENCES

1. De Barr,A.E., and Oliver, D.A., Electrochemical Machining, American Elsevier Publishing Co. Inc., New York, (1968).
2. McGeough, J.A., Principles of Electrochemical Machining, William Clowes & Sons, London, (1974).
3. Nilson, R.H., and Tsuei,Y.G., Inverted Cauchy problem for the Laplace equation in engineering design, *Jour of Eng Math*, Vol.**8,** No. **4,** 329, October (1974).
4. Nilson, R.H., and Tsuei,Y.G., Free Boundary Problem for the Laplace Equation with Application to ECM Tool Design, *Transactions of the ASME*, 54, March (1976) .
5. Tsuei, Y.G., Yen, C.H., and Nilson, R.H., Theoretical and Experimental Study of Workpiece Geometry in Electrochemical Maching, *ASME publication*, **76-WA/Prod-6**, 1, (1976).
6. Nilson, R.H., and Tsuei, Y.G., Free Boundary Problem of ECM by Alternating-Field Technique on Inverted Plane, *Computer Methods in Applied Mechanics and Eng*, **6** , 265, (1976).
7. Forsyth, P., Jr. and Rasmussen, H., A Kantorvich Method of Solution of Time Dependetn Electrochemical Machining Problems, *Computer Methods in Applied Mechanics and Eng.* **23**, 129, (1980).
8. Jain, V.K., and Pandey, P.C., Design and analysis of ECM Toolings, *Precision Engineering*, Vol. **1**, No. **4**, 199, (1979).
9. Jain, V.K., and Pandey, P.C., Finite element approach to the two dimensional analysis of electrochemical machining, *Precision Engineering*, Vol. **2,** No. **1**, 23, Jan. (1980).
10. Jain, V.K., and Pandey, P.C., Tooling design for ecm, *Precision Engineering*, Vol. **2**, No. **4,** 195, Oct. (1980).
11. Jain, V.K., and Pandey, P.C., Tooling Design for ECM- A Finite Element Approach, *ASME Jour. of Eng. for Industry* Vol.**103**, 183, May (1981).
12. Jain, V.K., and Pandey, P.C., Tooling design for Electrochemical Machining of Complex Shaped Workpieces, *J. Inst. Eng. India*, Vol. **62,** 95, Nov. (1981).
13. Narayanan, O.H., Hindua, S., and Noble, C.F., The Prediction of Workpiece Shape During Electrochemical Machining by the Boundary Element Method, *Int. J. Mach. Tool Des. Res.*, Vol. **26**, No. **3**, 323,(1986).
14. Narayanan, O.H., Hindua, S., and Noble, C.F., Design of tools for electrochemical machining by the boundary element method, *Proc. Instn Mech Engrs*, Vol. **200** No. **C3**, 195, (1986).
15. Brebbia, C.A., and Walker,S., Boundary Element Techniques in Engineering, Newnes-Butterworths, London, (1980).
16. Crank, J., *Free and moving boundary problems*, Clarendon Press, Oxford, (1984).
17. Hewitt, G.F., and Hall-Taylor, N.S., Annular Two-Phase Flow, Pergamon Press, Oxford, (1970).
18. Nied, H.A., and Perry, E.M., Finite Element Simulation of the Electrochemical Machining Process, *ASME, MD-Vol.* **20**, *Computer Modeling and Simulation of Manufacturing Processes*, 37-57, Nov. (1990).

APPENDICES

(1) Integral Equation Formulation of Boundary Value Problem:

The basic concept of boundary integral equations has its roots in potential theory. Consider a material bounded by a surface. On this surface either the potential or its normal derivative must be specified. The solution objectives for such boundary value problems are to seek the value of the potential and its first derivative everywhere in the interior and on the bounding surface. On the surface where a potential has been specified as a boundary condition, the first derivative is sought. On the boundary where the derivative or flux has been specified, the value of the potential is sought.

The direct method for the formulation of the mixed boundary value problem is based on Green's identity. The potential problem can be written in terms of the Laplace equation:

$$\nabla^2\Phi=0 \text{ in domain } \Omega \tag{1.1}$$

where Φ is the potential. The mixed boundary conditions for the specification of the potential and the flux can be written as:

$$\Phi = \overline{\Phi} \text{ on the surface } \Gamma_1$$

$$q = \frac{\partial\Phi}{\partial n} = \overline{q} \text{ on the surface } \Gamma_2 \tag{1.2}$$

where the total boundary is given by: $\Gamma = \Gamma_1 + \Gamma_2$. The notation of the bar denotes the known boundary condition specified. Consider a weighting function Φ^* which has continuous first derivatives. This function will be required to satisfy the governing equation. Using the weighted residual method, the integral equation based on Greens's theorems can be written as:

$$\int_\Omega (\nabla^2\Phi)\Phi^* d\Omega = \int_{\Gamma_2} (q - \overline{q})\Phi^* d\Gamma - \int_{\Gamma_1} (\Phi - \overline{\Phi})q^* d\Gamma \tag{1.3}$$

where $q = \frac{\partial\Phi}{\partial n}$ and $q^* = \frac{\partial\Phi^*}{\partial n}$. The first integral is a volume interal, whereas, the two integrals on the right hand side are surface integrals. When the integral on the left side is integrated by parts, an integral equation can be obtained which will be the starting point for the boundary element method. Accordingly, (1.3) can be written as:

$$\int_\Omega \Phi(\nabla^2\Phi^*)d\Omega = -\int_{\Gamma_2} \overline{q}\Phi^* d\Gamma - \int_{\Gamma_1} q\Phi^* d\Gamma + \int_{\Gamma_2} \Phi q^* d\Gamma + \int_{\Gamma_1} \overline{\Phi} q^* d\Gamma \tag{1.4}$$

(2) Boundary Element Formulation of Boundary Value Problem:

The integral equation in Appendix (1.4) can be recast into a numerical form by introducing a fundamental solution which satisfies the governing equation and the associated boundary conditions. In the case of an electrostatic problem, if we assume a concentrated charge is acting at point "i," the original governing equation can be written as:

$$\nabla^2\Phi^* + \Delta^i = 0 \tag{2.1}$$

where Δ^i is a Dirac delta function. The solution of this equation is the fundamental solution. If eqn. (2.1) is satisfied by a fundamental solutions, then

$$\int_\Omega \Phi(\nabla^2\Phi^*)d\Omega = -\Phi^i \tag{2.2}$$

and Φ^i represents the unknown function at point "i." When this relationship is applied to eqn (1.4), the following equation will be obtained.

$$\Phi^i + \int_{\Gamma_2} \Phi q^* d\Gamma + \int_{\Gamma_1} \overline{\Phi} q^* d\Gamma = \int_{\Gamma_2} \overline{q}\Phi^* d\Gamma + \int_{\Gamma_1} q\Phi^* d\Gamma \tag{2.3}$$

where the fluxes are given by $q^* = \frac{\partial\Phi^*}{\partial n}$ and $q = \frac{\partial\Phi}{\partial n}$.

The integral equation solution can be obtained by using the fundamental singular solution for the 2D domain.

$$\Phi^* = \frac{1}{2\pi}\ln\left(\frac{1}{r}\right) \tag{2.4}$$

(3) Integral Equation Formulation of Boundary Value Problem:

Using Brebbia's method [15], the integral eqation can be rewritten in the following form by applying the singular solution and taking limits. Accordingly, the integral equation for the potential at any point on the surface can be written as:

$$\frac{1}{2}\Phi^i + \int_{\Gamma_2} \Phi \frac{\partial\Phi^*}{\partial n} d\Gamma + \int_{\Gamma_1} \overline{\Phi} \frac{\partial\Phi^*}{\partial n} d\Gamma = \int_{\Gamma_2} \overline{q}\Phi^* d\Gamma + \int_{\Gamma_1} q\Phi^* d\Gamma \tag{3.1}$$

Equation (3.1) can be generallized in a compact form for a later numerical format.

$$\frac{1}{2}\Phi^{i} + \int_{\Gamma} \Phi\, q^{*} d\Gamma = \int_{\Gamma} q\Phi^{*} d\Gamma \qquad (3.2)$$

where $\Gamma = \Gamma_1 + \Gamma_2$ and $\Phi = \overline{\Phi}$ on Γ_1 and $q = \frac{\partial \Phi}{\partial n} = \overline{q}$ on Γ_2.

The final desired numerical form for application in the ECM code is obtained as:

$$\frac{1}{2}\Phi^{i} + \sum_{j=1}^{n} \Phi_j \int_{\Gamma_j} q^{*} d\Gamma = \sum_{j=1}^{n} q_j \int_{\Gamma_j} \Phi^{*} d\Gamma \qquad (3.3)$$

The integrals in (3.3) are evaluated using the Gauss quadrature method.

(4) Basic Equations for Electrolyte Flow Formulation and Temperature Rise:

The derivation of the integral form for the momentum equation is derived in Ref. [17] which is essentially a statement of the balance of the pressure, wall friction, and momentum forces along a control volume on a streamline assuming that the gravitational effects are negligible from inlet to outlet. This equation is written as:

$$-\int_{A} \frac{dP}{ds} \Delta s dA = \int_{S} \tau_w \Delta s dS + \int_{A} \frac{d}{ds}(\dot{m}v)\Delta s dA \qquad (4.1)$$

where s is the distance along the streamline, τ_w is the wall shear stress, $\dot{m}$ is the mass flux rate, **v** is the velocity along the streamline, **A** is the cross-sectional area and **S** is the surface area of the stream tube.

This relationship can be reduced to a differential form that can be employed in the computer code:

$$-\Delta P = \frac{\tau_w}{H_{Rad}} \Delta s + \rho v \Delta v \qquad (4.2)$$

units where H_{Rad} is the hydraulic radius and ρ is the mass density of the electrolyte.

The other relationships necessary for the flow formulation are the wall shear stress τ_w, the dimensionless wall friction factor for turbulent flow λ and the Reynolds number R_n, respectively. Accordingly, they can be written as:

$$\tau_w = \frac{1}{8}\lambda \rho v_{av}^2 \qquad (4.3)$$

where λ is the widely known Darcy friction factor. These relationships have not been derived from first principles, but have been validated by extensive experimentation. For a highly turbulent flow this friction factor becomes:

$$\lambda = 0.495\left(Log_{10} R_n\right)^{-2.2} \qquad (4.4)$$

which is dependent on the Reynold's number of the flow.

$$R_n = \frac{\rho v H_d}{\mu} \qquad (4.5)$$

with μ being the electrolyte viscosity. In this formulation all the electrolyte properties are temperature dependent.

The temperature rise in the electrolyte can be determined from the basic energy equation. In Ref. [18], it was shown that the heat conduction term for the high Reynolds number application plays an insignificant role compared to the convective term. The following equation results:

$$\rho_e c_e v \frac{\partial T}{\partial s} = J^2 \sigma_e \qquad (4.6)$$

where the subscripts refer to the electrolyte property. The density, specific heat and specific resistivity are given by the symbols ρ_e , c_e and σ_e , respectively. The temperature rise in the electrolyte is essentially due to the local Joule heating where J is the current density. This equation can be rewritten in a numerical form for programming as:

$$\Delta T = \left(\frac{\sigma_e}{\rho_e c_e}\right) \frac{J^2(s)}{v(s)} \Delta s \qquad (4.7)$$

Simulation of Materials Processing: Theory, Methods and Applications, Shen & Dawson (eds)
© 1995 Balkema, Rotterdam. ISBN 90 5410 553 4

Thermomechanical analysis of porous media

M. Predeleanu
Laboratoire de Mécanique et Technologie, ENS Cachan, CNRS, Université Paris 6, France

ABSTRACT : The rheological models taking into account the porosity of materials are, at present, extensively used in forming process analysis, as well as in describing the behavior of initially void-containing materials in the evaluation of degradation (damage) due to large plastic deformations.
This paper reviews the main results concerning thermal aspects in modeling porous media behavior. Two approaches are considered : on the microscopic level, attention is focused on the evolution laws of porosity ; on the macroscopic level, coupled thermomechanical models, including isotropic damage, are used to describe the influence of thermal effects in deformation processes.

1. INTRODUCTION

The modeling of the behavior of porous media has received considerable attention in recent years due not only to its theoretical interest but also to broad technological applications concerning materials such as soils and rocks, sintered metals, ceramics, composite materials, bone and other biological tissues. In this paper, we are concerned with damaged materials, which, due to straining, are submitted to modifying microstructure processes, characterized essentially by nucleation, growth and coalescence of voids, and microcracks. The damage phenomenon leads to a progressive deterioration of the material in the sense that its resisting capacity to subsequent loading is diminished. An advanced evolution of damage induces the development of plastic instabilities and fractures. In material manufacturing processes, the main consequences of the development of these straining effects are unsatisfactory limitations in working operations and/or unacceptable products. Apart from the determination of forming limits, another equally important interest in the damage analysis is the evaluation of the "soundness" of the material undergoing a forming operation (Predeleanu, 1987, Gelin and Predeleanu, 1992).

On the microscale, a porous material is considered as a biphasic material composed of a solid skeleton (matrix) and voids, randomly distributed. During the straining process, this microstructure is changing : the number of voids may increase by nucleation of new voids or may diminish by the closing of other ones ; the geometrical form and the volume of voids are evolving, and this may lead to the coalescence of some cavities and therefore to the appearance of macrocracks.

Description of the behavior of a porous solid considered in this paper is based on the introduction of an internal variable taking into account the porosity or void volume fraction f (ratio of the void volume over the total volume of the matrix and voids) in classical models of continuous media. For isotropic damage evolution, f is a scalar-valued function defined at every point of the body and is dependent on the mechanical and thermal fields involved in the straining process. The presence of porosity affects all constitutive functions describing the elasticity, plasticity, viscosity or thermal properties of the material. Several theoretical models have been conceived for deducing, by homogeneisation proceduces, the macroscopic values of some material functions in terms of porosity, such as : elasticity moduli, viscosity coefficients, plastic yield locus, viscoplastic potentials, thermal expansion coefficients, specific heat, thermal conductivity, etc. Nevertheless, many other material functions, present in constitutive models, are defined heuristically and determined experimentally.

The second important step in the construction of the constitutive model for a porous solid is the determination of the evolution law of the porosity f in terms of the mechanical and thermal fields. This could be determined by analyzing, on the microscale, the deformation of a representative volume element (r.v.e.) under general mechanical and thermal loadings. The treatment of the general case, including the nucleation, growth and coalescence of voids under complex loadings and any cavity geometry, remains an open problem. In addition, for straightforward utilization in analysis, the porosity evolution law must be expressed in a relatively simple form. This has led to the consideration of simple geometrical models for the representative volume element (hollow sphere or hollow ellipsoid, cylindrical cavity) and specific loading programs. Simple analytical results are available only for the cavity growth/collapse phase and some particular matrix behavior. Quasi-static and isothermal regimes have been favored in the literature and some of them will be reviewed shortly. In this paper, attention is focused mainly on the thermal effects and also on the dynamic effects on porosity evolution due to high rate loadings.

2. MACROSCOPIC GOVERNING EQUATIONS

2.1. *Fully-coupled thermoelastic/viscoplastic constitutive law*

Thermoelastic response. Let us consider a porous solid $\mathfrak{B}$ in the current configuration $\mathcal{C}$ at t, occupying an open region of the three-dimensional Euclidian space $\mathbb{R}^3$. Let us denote **F**, the deformation gradient from a reference configuration $\mathcal{C}_0$ to the current configuration $\mathcal{C}$, with det $\mathbf{F} > 0$. The total deformation is composed of thermal, elastic and inelastic components. In the case of large deformations and large changes of temperature from a reference value T_0, a multiplicative decomposition of **F** can be considered as

$$\mathbf{F} = \mathbf{F}^e\,\mathbf{F}^p\,\mathbf{F}^T \tag{2.1}$$

where the indices e,p and T refer to the elastic, plastic and purely thermal parts of **F**, respectively. The decomposition (2.1) generalizes the Duhamel-Neumann postulate of addition of thermal and mechanical strains employed in linear thermoelasticity. For thermally isotropic materials, the thermal deformation gradient is $\mathbf{F}^T = \Gamma(T)\mathbb{1}$ where Γ is a scalar-valued function of temperature and $\mathbb{1}$ is the identity tensor. Following Lu and Pister (1975), we can define :

$$\Gamma(T) = \exp\,[\int_{T_0}^{T} \alpha(s)ds \tag{2.2}$$

where $\alpha(T)$ is the usual coefficient of linear thermal expansion, and T_0 and T are absolute temperatures in $\mathcal{C}_0$ and $\mathcal{C}$, respectively. If, in particular, α is independent of temperature, then $\Gamma(\theta) = \exp(\alpha\theta)$ where $\theta = T - T_0$.

For small thermal strains, $\Gamma(\theta) \cong 1 + \alpha\theta$.

By using the decomposition (2.1), the finite thermoelastic response can be defined by means of a free-energy potential expressed in terms of classical elastic tensors corresponding to the $\mathbf{F}^e$ component, temperature and the porosity.

Assuming that the thermoelastic strains are small with respect to the inelastic ones, the thermoelastic response can be defined by a hypoelastic law, generalizing Hooke's law for small strains. It can therefore be assumed that the Eulerian rate deformation tensor **D**, defined as the symmetric part of the velocity gradient with respect to the current configuration, is decomposed as :

$$\mathbf{D} = \mathbf{D}^e + \mathbf{D}^p \tag{2.3}$$

where superscripts e et p refer to the elastic (recoverable) and inelastic (non-recoverable and rate-dependent) parts, respectively.

Then, the thermoelastic response is defined by :

$$\mathbf{D}^e = \mathbf{L}^{-1}[\overset{\nabla}{\sigma}] + \frac{\alpha}{3}\dot{T}\,\mathbb{1} \tag{2.4}$$

where $\overset{\nabla}{\sigma}$ is the corotational (Jaumann) time rate of the Cauchy stress σ and $\mathbf{L} = 2G\mathbb{1}_4 + (K - \frac{2}{3}G)\,\mathbb{1}_2$ is the Hooke's isotropic elasticity tensor. $G = \bar{G}(T, f)$, $K = \bar{K}(T, f)$, $\mathbb{1}_4$ and $\mathbb{1}_2$ are the shear modulus, bulk modulus, four-order identity tensor and second-order identity tensor, respectively.

The porosity dependence of elastic moduli has been estimated by using various material models. For instance, the Mac Kenzie (1950) relations, modified by Johnson (1981) are, the following :

$$\bar{G} = G_m(1 - f)\,(1 - \frac{6K_m + 12G_m}{9K_m + 8G_m}) \tag{2.5}$$

$$\bar{K} = \frac{4G_mK_m\,(1 - f)}{4G_m + 3G_mK_m} \tag{2.6}$$

where G_m and K_m are the elastic moduli of the matrix. The temperature dependence of G and K is estimated as being the same as that for the elastic moduli of the matrix. Generally, a linear variation is assumed, with the following form :

$$G_m(T) = G_m(T_0)\,[1 - c_1\frac{T - T_0}{T_f}] \tag{2.7}$$

$$K_m(T) = K_m(T_0)\,[1 - c_2\frac{T - T_0}{T_f}] \tag{2.8}$$

where $G_m(T_0)$ and $K_m(T_0)$ denote the elastic moduli of the matrix at the reference temperature T_0, T_f is the melting temperature, and c_1 and c_2 are material constants(see e.g., Budianski, 1970, Frost and Ashby, 1982).

Ineslastic response. The inelastic rate of deformation can be defined by using different models of viscoplasticity theory, as a function of stress, absolute temperature T, porosity f and a set of internal variables ζ describing the plastic flow.

$$\mathbf{D}^p = \hat{\mathbf{D}}^p\,(\sigma, T, f, \zeta) \tag{2.9}$$

Generally, the function $\hat{\mathbf{D}}^p$ is derived from a convex viscoplastic potential by using the normality flow rule :

$$\hat{\mathbf{D}}^p = \frac{\partial \Psi}{\partial \sigma} \qquad (2.10)$$

where Ψ is an isotropic function.

If it is assumed that Ψ, as an isotropic function of σ depends only on the first and second invariants of σ, Ψ (J_1, J_2, T, f, ζ), the relation (2.10) can be written as :

$$\mathbf{D}^p = \frac{\partial \Psi}{\partial J_2(\sigma')} \frac{\partial J_2(\sigma')}{\partial \sigma} \text{ , tr } \mathbf{D}^p = 3\frac{\partial \Psi}{\partial J_1(\sigma)} \qquad (2.11)$$

where $J_1(\sigma) = \text{tr } \sigma$, $J_2(\sigma) = \frac{1}{2} \text{tr } \sigma^2$, and the prime superscript represents the deviatoric part.

Approximate expressions for the macroscopic potentials Ψ for porous materials have been derived using the micro-mechanical analysis of the representative volume element for various types of specific constitutive behavior of the matrix.

Mainly, three constitutive models have been used for local analysis. The rigid-perfectly plastic matrix behavior has been considered by Gurson (1977) in his deduction of the well-known yield criterion, which was slightly modified by Tvergaard (1981) to account for void interaction effects

$$\Psi = \frac{3J'_2}{\sigma_M^2} + 2 f q_1 \cosh\left(\frac{q_2}{2} \frac{J_1}{\sigma_M}\right) - 1 - q_3 f^2 \qquad (2.12)$$

where σ_M is the flow stress of the matrix. Gurson's model is obtained for $q_1 = q_3 = 1$, $q_2 = \sqrt{3}$ or $q_2 = 3$ corresponding to the cylindrical or spherical void model, respectively.

The Gurson-Tvergaard yield criterion was used and adapted to describe rate-dependent behavior by Pam et al (1983), Needleman and Tvergaard (1985), Gelin (1986) through the introduction of supplementary assumptions on matrix behavior.

The power law viscous matrix behavior (Norton law of exponent n, $1 \leq n < \infty$) has been considered in many recent studies for deducing macroscopic viscoplastic potentials : Duva and Hutschinson (1984), Duva (1986), Cocks (1989), Ponte-Castaneda (1991), Suquet (1992), Michel and Suquet (1992), J.B. Leblond and al. (1994) and Briottet and al (1995).

Haghi and Anand (1992) have proposed a new macroscopic viscoplastic potential using Duva and Hutschinson (1984) results obtained for a porous solid containing a dilute concentration of spherical voids. It was shown by finite-element calculations, for a large range of porosities and stress triaxiality ratios, that this potential gives better results compared with those predicted by Needleman-Tvergaard or Duva-Crow potentials. The Haghi-Anand model is expected to be applied in the study of hot working metals and also in the study of late stages of densification of metallic powders by hot viscostatic pressing.

The overstress model of Hohenemser and Prager (1932) generalized by Perzyna (1963) has been used for matrix behavior by Sun and Wang (1992). Their approximate macroscopic viscoplastic potential reduces to Gurson's yield function when the rate of the deformation-sensitive parameter tends to zero. It is shown that the normality rule no longer exists in general.

It is worth noting that all studies cited above, in order to obtain macroscopic plastic potentials for porous solids, have been based on quasi-static solutions of local micromechanical problems for the r.v.e. However, under high stress rate and/or when the value of the porosity becomes high, the inertial effects could be very significant.

The phenomenological viscoplastic model in the overstress form, widely used in dynamic problems, was generalized by Perzyna (1986) for porous, damaged materials. The viscoplastic strain rate is written as :

$$\mathbf{D}^p = \frac{\gamma}{\psi} < \phi\left(\frac{F}{k} - 1\right) > \frac{\partial F}{\partial \sigma} \qquad (2.13)$$

where γ is a temperature-dependent viscosity function, ψ a control function dependent on $I_2/I_2^s - 1$, I_2 the second univariant of the rate of the deformation, I_2^s its static value, ϕ the viscoplastic over stress function, F the quasi-static yield function, and k the material isotropic hardening-softening function that is also dependent on temperature.

The symbol <x> means :

$$<x> = \begin{cases} 0, \text{ if } x \leq 0 \\ x, \text{ ig } x > 0 \end{cases}$$

Usually, the functions ϕ and ψ are assumed to have a power form : $\phi(x) = x^m$, $m \geq 1$, $\psi(x) = x^n$ and must be deduced from tests under dynamic loading. The yield function F is defined either by heuristic considerations or by a micromechanical analysis. From the former, we cite the following, which is widely used : (cf : Shima and Oyane 1976, Hancock 1983, Doraivelu et al. 1984, Perzyna 1986) :

$$F = J'_2 + \nu f J_1^2 \qquad (2.14)$$

where ν is a material parameter, $J'_2 = J_2(\sigma')$

2.2. *Evolution laws for internal porosity variable*

To complete the mathematical constitutive model, the evolution law of the void volume fraction must be formulated in terms of mechanical and thermal fields. The determination of the corresponding time differential equation for the scalar-valued function f (porosity), reducing to in a suitable mathematical or for simple use in numerical computations, taking into account the three stages of the evolution process (nucleation, growth and coalescence of voids) and being valid for low and/or high rate loadings, represents a difficult and open problem. Approximate formulations have been proposed, many of which are based on heuristic considerations, with others being suggested by the solution of the void growth problem for the r.v.e. There is a paucity of comparative and critical results to establish the influence of different factors concerning inertial terms, heat conduction, constitutive models, rate sensitivity, etc...

Generally, an additive decomposition of the porosity rate corresponding to nucleation $\dot{f}_n$, growth $\dot{f}_g$ and coalescence $\dot{f}_c$ phases is introduced. Sometimes, a term corresponding to diffusion phenomena is added when the temperature is high $\dot{f}_d$,

$$\dot{f} = \dot{f}_n + \dot{f}_g + \dot{f}_c + \dot{f}_d.$$

Concerning the void nucleation phase, many experimental studies have emphasized various mechanisms of nucleation, and, as a consequence, corresponding global or local criteria have been proposed (cf. Montheillet and Moussy, 1986).

For high rate processes, the rate of void nucleation, as described by Seaman et al. (1976), Curran et al. (1977) and Perzyna (1986), is given by :

$$\dot{f} = \frac{h(f)}{1 - f} [\exp (\frac{\delta \, |\sigma_m - \sigma_{mn}|}{\beta T}) - 1] \qquad (2.15)$$

where $\sigma_m = \frac{1}{3} \operatorname{tr} \sigma$ is the mean stress, σ_{mn} the threshold mean stress for void nucleation, β the Boltzmann constant, δ a material coefficient and h(f) a function introduced to take into account the effects of void interaction on the void nucleation process.

For the void growth phase, the following relation, deduced from the macroscopic mass conservation equation, is generally used :

$$\dot{f} = (1 - f) \operatorname{tr} \mathbf{D} \qquad (2.16)$$

However, the micromechanical analysis of the r.v.e. suggests different void growth relations for specific loading programs.

So, for dynamic processes, an approximate relation for the void growth rate, as deduced from analyzing the hollow sphere unit cell (cf. Eftis and Nemes 1991, Caroll and Holt 1972, Johnson 1981, Perzyna 1986), is employed.

$$\dot{f} = \frac{1}{\eta} g(f) \, \varphi_1(f, f_0) \, |\sigma_m - \sigma_{mg}|$$

where :

$$\varphi_1 (f, f_0) = \frac{\sqrt{3}}{2} f (\frac{1 - f}{1 - f_0})^{2/3} ((\frac{f}{f_0})^{2/3} - f)^{-1} \qquad (2.17)$$

The function g(f) is associated to void interaction, $\sigma_{mg}(f)$ is the threshold mean stress for void growth and $\bar{\eta}$ is a material constant.

It is important to note that the simplified form (2.17) was obtained by neglecting the inertial effects in order to obtain a simple relation for computation. Also, the thermal effects are not generally included in void growth rate relations. Recently, Lee and Dawson (1993), based on numerical simulations of the behavior of a spherical unit cell under various conditions of stress triaxiality, proposed a new relation for void growth linearly dependent on equivalent strain rate. The analysis is performed under quasi-static loading conditions for an incompressible strain-hardening, viscoplastic matrix, described by Hart-type constitutive model. It is shown that over a temperature range of 373°K to 473°K, the mean stress required to produce a given void rate is approximately the same.

3. MICROMECHANICAL ANALYSIS : THERMAL EFFECTS

3.1. Hollow sphere model : basic equations

The study of the void evolution laws and the determination of the macroscopic strain rate potentials from the microscopic ones are based on the solution of the local response of the representative volume element (r.v.e.) isolated from the porous material and loaded as in remote state fields. The macroscopic fields are deduced from the microscopic ones by means of the Hill's well-known average relations (1967).

Considerable research has been devoted to both problems, and most results are concerned with quasi-static analysis for isothermal regime, including various specific constitutive assumptions for the intervoid matrix.

To highlight the influence of the thermal effects on void growth evolution, let us consider, for simplicity, that the representative volume element of the porous material in the reference undeformed configuration is an isotropic hollow sphere Ω_0 of outer radius b_0 and internal cavity radius a_0, subjected to a remote radial thermomechanical loading. The porosity or void volume fraction is defined as $f = (\frac{a}{b})^3$, where a and b are the current values at instant t of the internal and external radius of the hollow sphere, respectively.

Let $(r_0, \theta_0, \varphi_0)$ denote the spherical coordinates of a point in the reference undeformed configuration and (r, θ, φ) the corresponding coordinates in the current deformed configuration. The spherically symmetric motions are defined by :

$$r = \hat{r} (r_0, t), \theta = \theta_0, \varphi = \varphi_0 \qquad (3.1)$$

on the interior of the hollow sphere $\Omega_0 = \{(r_0, \theta_0, \varphi_0) \mid a_0 < r_0 < \varphi_0, 0 < \theta_0 \leq 2\pi, 0 \leq \varphi_0 \leq \pi\}$ and on the time interval $0 \leq t < \infty$.

The velocity vector is $v = v_r \otimes \mathbf{e}_r$, where $v_r = \frac{\partial \hat{r}}{\partial t}$, and the eulerian strain rate tensor has the form :

$$\mathbf{D} = \frac{\partial v_r}{\partial r} \mathbf{e}_r \otimes \mathbf{e}_r + \frac{v_r}{r} (\mathbf{e}_\theta \otimes \mathbf{e}_\theta + \mathbf{e}_\varphi \otimes \mathbf{e}_\varphi) \qquad (3.2)$$

where $(\mathbf{e}_e, \mathbf{e}_\theta, \mathbf{e}_\varphi)$ denotes an orthonormal basis for a spherical coordinate system (r, θ, φ).
The Cauchy stress tensor is :

$$\sigma = \sigma_{rr} \mathbf{e}_r \otimes \mathbf{e}_r + \sigma_{\theta\theta} (\mathbf{e}_\theta \otimes \mathbf{e}_\theta + \mathbf{e}_\varphi \otimes \mathbf{e}_\varphi) \qquad (3.3)$$

N.B. : In this section, we shall use the same symbols for microscopic fields as were used for macroscopic fields in Section 2.

The following local boundary value problem will be considered :

$$\text{for} \quad r = a(t), \sigma_{rr} = 0, T = T_a, t \geq t_0$$
$$\text{for} \quad r = b(t), \sigma_{rr} = p_0 \left(\frac{b_0}{b}\right)^2, T = T_b, \ t \geq t_0 \qquad (3.4)$$

where p_0 is a uniform nominal radial tensile stress (dead load).
The initial conditions are :

$$\sigma_{rr}(M, t_0) = \sigma_{\theta\theta}(M, t_0) = v_r(M, t_0) = 0, M \in \Omega^0$$
$$T(M, t_0) = T^0(M), M \in \Omega^0 \qquad (3.5)$$

Due to the spherical symmetry, the governing equations of the problem may be defined in terms of : $\sigma_{rr}, \sigma_{\theta\theta}, T, v_r$ and f.
So, if a constitutive model in the overstress form (2.13) is assumed, the temperature-dependent stress-strain relations become :

$$\frac{\partial v_r}{\partial r} + \frac{2v_r}{r} - 3\alpha\dot{T} = \frac{1}{3K}[\dot{\sigma}_{rr} + 2\,\dot{\sigma}_{\theta\theta}]$$
$$\frac{\partial v_r}{\partial r} - \frac{v_r}{r} = \frac{1}{2G}(\dot{\sigma}_{rr} - \dot{\sigma}_{\theta\theta})$$
$$+ \frac{\gamma}{\psi} < \phi\left[\frac{F}{k} - 1\right] > \left(\frac{\partial F}{\partial \sigma_{rr}} - \frac{\partial F}{\partial \sigma_{\theta\theta}}\right) \qquad (3.6)$$

where the dot denotes the material time derivative.
The motion equations are reduced to a single one :

$$\frac{\partial \sigma_{rr}}{\partial r} + \frac{2}{r}(\sigma_{rr} - \sigma_{\theta\theta}) = \rho\,\dot{v}_r \qquad (3.7)$$

The simplified heat conduction equation may be written as :

$$\rho C_v \dot{T} = -\operatorname{div} q + \chi\, \sigma : \mathbf{D}^p + S \qquad (3.8)$$

where C_v is the specific heat, q the heat flux, S(r, t) the external heat supply and χ the Taylor -Quinnet coefficient that takes into account the stored elastic energy ($\chi \sim 0.9$). The thermoelastic coupling was neglected. We note that in the case of adiabatic processes, the system of equations (3.6) - (3.8) is a first-order quasi-linear system of partial differential equations, written in the following matrix form :

$$A_1(W, r, t)W_t + A_2(W, r, t)W_r + A_3(W, r, t) = 0 \qquad (3.9)$$

where W(r,t) is a column vector grouping all unknowns of the problem, and W_t and W_r are the temporal and spatial derivatives of W, respectively. If the thermomechanical coupling is neglected, it has been proved that the Cauchy problem has a unique solution if Lipschitz-type regularity conditions are satisfied (cf. Jeffrey 1976, Ionescu and Sofonea 1993). Nevertheless, these conditions do not always ensure a global solution. For the dynamic coupled thermomechanical problem, the question of the existence and unicity of the solution is still open.Though the problem of the void growth has concerned many authors in the past, the study of the coupled thermomechanical problem described above has only been marginally treated for particular cases. Some important points concerning the thermal effects will be analyzed, namely : the deformation-induced temperature changes and their effects on the porosity evolution, the influence of the temperature-dependent material properties, the stress-focusing effect caused by thermal stress propagation.

3.2. Deformation -induced temperature effects

The limited results existing on this topic have been obtained by neglecting the elastic strains and by assuming the incompressibility of the inter-void matrix. In this case, the radially symmetric problem of the hollow sphere is greatly simplified ; for some specific constitutive models, the mechanical fields can be expressed in terms of a single variable, namely the void radius a(t).
Indeed, the incompressibility hypothesis requires that the motion be isochoric, so that det $\mathbf{F} = 1, t \geq 0$, i.e. from (3.1).

$$\frac{\hat{r}^2}{r_0^2} \frac{\partial \hat{r}}{\partial r_0} = 1 \qquad (3.10)$$

or, after integration :

$$r = \hat{r}(r_0, t) = [r_0^3 + a^3(t) - a_0^3]^{1/3},$$
$$\text{for } a_0 \leq r_0 \leq b_0, t \geq 0 \qquad (3.11)$$

The velocity field can be expressed by :

$$v_r = \dot{r} = a^2\dot{a}\, r^{-2} \qquad (3.12)$$

and the acceleration field by :

$$\dot{v}_r = \ddot{r} = 2ar^{-5}(r^3 - a^3)\,\dot{a}^2 + a^2\,\ddot{a}\,r^{-2} = Q(r, t) \quad (3.13)$$

The strain rate tensor is :

$$\mathbf{D} = -2a^2\,\dot{a}\,r^{-3}\,\mathbf{e}_r \otimes \mathbf{e}_r$$

$$+ a^2\,\dot{a}\,r^{-3}\,(\mathbf{e}_\theta \otimes \mathbf{e}_\theta + \mathbf{e}_\varphi \otimes \mathbf{e}_\varphi) \quad (3.14)$$

Let us consider that the behavior of the matrix is described by a viscoplastic overstress model (2.13) with $F = \sqrt{J'_2}$, (von Mises material) and $\psi(x) = 1$. As a result :

$$\sqrt{J'_2} = k[1 + \phi^{-1}\,(\frac{\sqrt{I_2}}{\gamma})] \quad (3.15)$$

or equivalently :

$$\bar{\sigma} = \sigma_0\,[1 + \phi^{-1}\,(\eta\,\bar{D})] \quad (3.16)$$

where ϕ^{-1} denotes the inverse of ϕ, $\bar{\sigma} = (\frac{3}{2}\mathrm{tr}\,\sigma'^2)^{1/2}$ the equivalent stress, $\bar{D} = (\frac{2}{3}\mathrm{tr}\,\mathbf{D}^2)^{1/2}$ the equivalent strain rate, $\sigma_0 = \sqrt{3}\,k$ the tensile yield stress and $\eta = \frac{\sqrt{3}}{2\gamma}$ a material function.

For spherical symmetric deformation of a hollow sphere :

$$J'_2 = \frac{1}{3}(\sigma_{rr} - \sigma_{\theta\theta})^2 \text{ and } \sqrt{I_2} = \sqrt{3}\,a^2\,\dot{a}\,r^{-3} \quad (3.17)$$

The motion equation (3.7) becomes :

$$\frac{\partial\sigma_{rr}}{\partial_r} - \frac{2\sqrt{3}k}{r}[1 + \phi^{-1}\,(\frac{\sqrt{I_2}}{\gamma}\,\frac{a^2\,\dot{a}}{r^3})] = \rho Q(r, t) \quad (3.18)$$

where Q(r, t) is given by (3.13).

By integrating with respect to r and taking into account the boundary conditions (3.4), we obtain :

$$p_0(t)\,(\frac{b_0}{b})^2 - \int_{a(t)}^{b(t)} \frac{2\sqrt{3}k}{r}\,[1 + \phi^{-1}(\frac{\sqrt{3}}{\gamma}\,\frac{a^2\,\dot{a}}{r^3})]dr = \rho\,P(r,t) \quad (3.19)$$

where :

$$P(r,t) = \frac{a^4\,\dot{a}^2}{2}\,(\frac{1}{b^4} - \frac{1}{a^4}) - (2a\,\dot{a}^2 + a^2\,\ddot{a})\,(\frac{1}{b} - \frac{1}{a})$$

where k and γ are temperature-dependent material functions and $b(t) = [b_0^3 - a_0^3 + a^3(t)]^{1/3}$.

Uncoupled thermomechanical problem. In this case, with the temperature field T(r,t) being known for $a \leq r \leq b$, $t \geq 0$, the solution of the second-order differential equation (3.19) gives the dynamic evolution of the internal void radius a(t) ; therefore, all mechanical fields can be determined.

For the particular case $\phi(x) = x$ and supposing that the temperature field is uniform, we obtain from (3.19) :

$$p_0(t)\,(\frac{b_0}{b})^2 - 2\sqrt{3}k\,\ln\frac{b}{a} + \frac{2k}{\gamma}\,a^2\,\dot{a}\,(\frac{1}{b^3} - \frac{1}{a^3}) = \rho P(r,t) \quad (3.20)$$

The study of the isothermal problem, that also uses an affine relation between the equivalent stress and equivalent rate strain, has been carried out by Johnson (1981), Perzyna (1986), Eftis and Nemes (1991) and Cortez (1992). For simplicity, the inertial terms in (3.20) have been neglected in formulating the viscoplastic-damage constitutive model for dynamic loading, as in the simplified formula (2.17), for instance. However, the viscosity effects don't always dominate the void growth process with respect to the inertia effects, as proved recently by Wang and Jian (1994).

Coupled thermomechanical problem. The heat conduction equation must be added to equation (3.18). For constitutive relation (3.16) with $\phi(x) = x$ and assuming an adiabatic regime, the heat conduction equation can be written :

$$\rho C_v\,\dot{T} = 2\chi\sigma_0\,[1 + 2\eta\,\frac{\dot{r}}{r}]\,\frac{\dot{r}}{r} \quad (3.21)$$

Carroll et al (1986) have studied the influence of the temperature-dependent coefficient changes on porosity in compaction, by considering for a cooper-like material :

$$\sigma_0 = \begin{cases} -\sigma_1\,(1 - \frac{T}{T_m}), & T \leq T_m \\ 0, & T \geq T_m \end{cases} \quad (3.22)$$

$$\eta = -\frac{3\eta_0}{\sigma_0}\,\exp(B/T)$$

where T_m is the melting temperature and σ_0, η_0 and B are material constants. The coupled differential equation system (3.19), (3.21) has been solved numerically in the case $\phi(x) = x$, for the pressure profile p_0 :

$$p_0(t) = \begin{cases} p_1 + (p_2 - p_1)t_2, & 0 \leq t \leq t_2 \\ p_2, & t \geq t_2 \end{cases} \quad (3.23)$$

and calculations were performed at room temperature 293°K and at 900°K as well as for a peak pressure of 2.4 and 14.3 GPa. The results show that the porosity evolution over time is quite different whether the model is temperature-independent or temperature-dependent. The temperature-independent model predicts an unrealistically long compaction time. The relative importance of inertial effects and viscoplastic effects is also shown. Thus, in the case of lower pressure, the inertial effect is shown to be dominant (for preheating to 900°K), and the viscoplastic effect is dominant if temperature dependence is ignored. It can be concluded that the model would predict a balance of the two effects with mild preheating. For the invicid case ($\eta = 0$) and with the temperature-dependent yield stress given by (3.22), a closedform solution can be obtained for the conduction equation (3.21), namely :

$$T = T_m - (T_m - T_0)\left(\frac{r_0}{r}\right)^s \qquad (3.24)$$

where $T_0 = T(r_0, 0)$ is the initial temperature and

$$s = \frac{2\chi\sigma_1}{\rho C_v T_m} .$$

From (3.24), we deduce that the maximum of the deformation-induced temperature is localized on the void boundary and tends to the melting temperature when the void radius decreases to zero. The temperature at the outer boundary of the hollow sphere will be close to the initial value T_0. For the same consitutive assumptions, Cortez (1992) has deduced that the linear thermal softening, defined by (3.22) for aluminium and copper-like materials, has a negligible influence on the dynamic tensile strength at high strain rates, due to excessively localized heat generation near the surface of the void. It is to be expected that the thermal conductibility effects in the analysis modify the above conclusion. Also, it was pointed out that the material viscosity, in addition to hardening, has a great influence on the porosity curves.
Obviously, many other numerical solutions of the coupled thermo-mechanical problem based on the above formulation are needed in order to more thoroughly evaluate the importance of various aspects impacting void growth evolution, such as : were propagation effects with or without elastic strains, strain rate sensitivity, inertial term influence, constitutive parametric changes, etc...

3.3 *Stress-focusing effect*

The accumulation of the stresses in certain zones of a solid, due to the interfering effects of the reflected waves produced by rapid heating, is known as the stress-focusing phenomenon. It was identified by Ho (1976) for cylindrical bodies and by Hata (1991) for spherical bodies. The stress localization zones are the center of the cross section of a cylindrical rod, the center of a solid sphere and the internal surface of a hollow sphere. In the case of a hollow sphere, it was proved that the stress accumulation increases as the porosity decreases.
To better understand the physical mechanisms of this phenomenon, let us consider the problem of a hollow sphere submitted to a thermal shock caused by a sudden rise in temperature uniformly throughout the body :

$$T - T_0 = (T_1 - T_0)\,[tH(t) - H(t - t_1)]/t_1,\ t \geq 0 \qquad (3.25)$$

where t_1 is a finite heating time, $T_1 > T_0$ a positive constant and H(t) the Heaviside step function. For elastic materials, this problem has been considererd by Tsui and Kraus (1965), Zaker (1968), Ho (1976) and Hata (1991). The latest author has obtained a closed form solution by using the Laplace transform and Goodier's potentials which enable decomposing the solution into one corresponding to the thermal contribution and one to the homogeneous dynamic elastic problem. Thus, it was possible to emphasize the stress waves occurring at the void boundary that propagate radially outward and the stress waves occurring at the external boundary that proceed radially inward. The interference of the two waves induces an accumulation of hoop stress at the void boundary. The numerical evaluations show that as the porosity decreases, the pick value of the hoop stress increases. Considering that the solid sphere is the limit case when the ratio a/b tends to zero, it can be deduced that very high stress occurs at the center of the solid sphere. The same stress wave accumulation effect was observed by Ho (1965), who analyzed the problem of an elastic cylindrical rod subjected to sudden heating uniformly over its cross section. It was observed that even though the initial thermal radial stresses are relatively small, the induced waves reflected from the boundary surfaces of the rod may accumulate at the center and give rise to very large stresses. For non-zero heating duration, the peak stress is finite, whereas this peak becomes infinite for instantaneous heating ($t_1 = 0$). The magnitude of the peak stress depends upon the magnitude of the temperature rise and the effective heating duration. Initially, the radial stresses are compressive during the heating period and, after reflection from the boundary, accumulate and produce a high peaking tensile stress.
The time duration t_1 is considered of the order $\underline{d/c}$ where d and c are the radius of the rod and the elastic wave speed, respectively. The thermal conductivity has been neglected because, for most materials, thermal diffusion time is much longer than d/c. Therefore, the temperature field has been considered as quasi-static subsequent to the end of the temperature rise.
In addition to the implication regarding the void growth problem, the results cited above have an intrinsic value in the prediction of the thermal shock endurance of various special ceramics used in the manufacturing of electronic devices. The type of temperature field defined by (3.25) may be developed by the absorption of infrared-ray radiation or electromagnetic radiant energy for pulses that are typically of a duration of much less than 1μsec.
The implications of the stress-focusing phenomenon on the nucleation and growth of voids need to be better evaluated quantitatively. Some recent studies concerning the cavitation in solids have predicted for some material models the appearance of new voids and the development of the existing, infinitesimal

voids, for very high loading intensity (cf. Chung et al. 1987, Horgan and Pence 1989). How is this prediction modified if an anisothermal approach is adopted that takes into account the stress-focusing effect ?

4. CONCLUDING REMARKS

Modeling the deformation behavior of porous solids for the non-isothermal regime requires including in the basic formulation: thermal expansion strains, a temperature-dependent constitutive model of the material and thermomechanical coupling effects in the energy equation. These ingredients have been taken into account in the treatment of the local problem on the microscopic level in order to evaluate the thermal effects on the porosity evolution. For simplicity, as the representative volume element has been considered a hollow sphere unit cell subjected to a radially symmetric loading.

Two important concluding remarks can be deduced from analytical and numerical evaluations. The first concerns the thermomechanical coupling. The deformation-induced heat localizes on the boundary of the void. In compaction process of an elastoplastic material characterized by a linear temperature dependence of the flow stress, the melting temperature is attained at the void boundary at void closure. For void expansion, the heat is also accumulated at the void boundary, but no general conclusions can be deduced concerning the influence of the temperature on the void growth evolution.

The second remark concerns the wave propagation efffects inducing the accumulation of high hoop stresses on the void boundary when the material is suddenly heated (stress focusing effect). The stress concentration increases as the porosity decreases.

These thermal effects can certainly impact the required conditions for the occurrence of the cavitation phenomenon in solids under high stressing.

Further studies concerning thermomechanical analysis of the nucleation, growth and coalescence of cavities are needed to formulate realistic evolution laws applicable to high-energy rate processes.

REFERENCES

Briottet L., Klocker H., Montheillet 1995. Damage prediction and associated overall stress-strain rate relationship of a voided metal matrix. Personal communication.

Budianski B. 1970. Thermal and thermoelastic properties of isotropic composites, J. Composites Materials 4: 286-295.

Carroll M.M. and Holt A.C. 1972. Static and dynamic pore-collapse relations for ductile porous materials, J. Appl. Phys. 43, 4:1626-1636.

Carroll M.M., Kim K.T., Nesterenko V.F. 1986. The effect of temperature on viscoplastic pore collapse, J. Appl. Phys. 59(6):1962-1967.

Chung D.T., Horgan C.O. and Abeyaratne R. 1987. A note on a bifurcation problem in finite plasticity related to void nucleation, Int. J. Solids Structures 23: 983-988.

Cocks A.C.F. 1989. Inelastic deformation of porous materials, J. Mech. Phys. Solids 37, 6: 693-715.

Cortez Raul 1992. The growth of microvoids under intense dynamic loading, Int. J. Solids Structures 29, 11:1339-1350.

Curran D.R., Seaman L., Shockey D.A. 1977. Dynamic failure in Solids Physics Today, January: 46-55.

Duva J.M. 1986. A constitutive description of materials containing voids, Mech. Materials 5: 137-144.

Duva J.M., Hutchinson J.M. 1984. Constitutive potentials for dilutely voided nonlinear materials, Mech. Materials 3: 41-54.

Eftis J., Nemes J.A. 1991. Evolution equation for the void volum growth rate in a viscoplastic-damage constitutive model, Int. J. Plasticity 7:275-293.

Frost H.J. and Ashby M.F. 1982. Deformation-mechanism maps : the plasticity and creep of metals and ceramics, Pergamon Press, Oxford.

Gelin J.C. 1986. Application of a thermo-viscoplastic model to the analysis of defects in warm forming conditions, Annales of C.I.R.P. 35, 1: 157-160.

Gelin J.C. and Predeleanu M. 1992. Recent advances in damage mechanics : modelling and computational aspects, In Numerical Methods in Industrial Forming Processes edited by J.L. Chenot et al., A.A. Balkema/Rotterdam: 89-97.

Gurson A.L. 1977. Continuum theory of ductile rupture by void-nucleation and growth : Part I - Yield criteria and flow rules for porous ductile media, ASME J. Eng. Mat. Tech., 99:2-15.

Haghi M. and Anand L. 1992. A constitutive model for isotropic, porous, elastic-viscoplastic metals, Mech. Materials 13: 37-53

Hancock J.W. 1983. Plasticity of porous metals in Yield, Flow and Fracture of Polycristals, edited by Baker T.M., Applied Science Pub. Essex, U.K.

Hata T. 1991. Thermal shock in a hollow sphere caused by rapid uniform heating, ASME J. of Applied Mechanics 58: 64-69.

Hill R. 1967. The essential structure of constitutive laws for metal composites and polycrystals, J. Mech. Phys. Solids 15: 79-95.

Ho C.H. 1976. Stress-focusing effect in a uniformly heated cylindrical rod, ASME J. of Applied Mechanics 43: 464-468.

Hohenemser K. and Prager W. 1932. Über die Ansatze der Mechanik isotroper Kontinua, ZAMM, 12: 216-226.

Horgan C.O., Pence T.J. 1989. Cavity formation at the center of a composite incompressible nonliearly elastic sphere, ASME J. Applied Mechanics, 56: 302-308.

Ionescu I.R. and Sofonea M. 1993. Functional and numerical methods in viscoplasticity, Oxford Science Publication.

Jeffrey A. 1976. Quasi-linear hyperbolic systems and waves, Pitman Publishing, London.

Johnson J.N. 1981. Dynamic fracture and spallation in ductile solids, J. Appl. Phys. 52(4): 2812-2825.

Leblond J.B., Perrin G. and Suquet P. 1994. Exact results and approximate models for porous viscoplastic solids, Int. J. Plasticity 10, 3: 213-235.

Lee Y.S. and Dawson P.R. 1993. Modeling ductile void growth in viscoplastic materials - Part I : void growth model mech. Materials 15: 21-34. Part II : Application to metal forming mech. Materials 15: 35-52.

Lu S.C.H. and Pister K.S. 1975. Decomposition of deformation and representation of the free energy function for isotropic thermoelastic solids, Int. J. Solids Structures, 11: 927-934.

Mackenzie J.H. 1950. The Elastic Constants of a solid containing spherical holes, Proc. Phys. Soc., 63B(2): 2-11.

Michel J.C. and Suquet P. The constitutive law of nonlinear viscous and porous materials, J. Mech. Phys. Solids 10: 783.

Montheillet F. et Moussy F. 1986. (eds) Physique et mécanique de l'Endommagement. Editions de Physique.

Needleman A. and Tvergaard V. 1985. Material strain-rate sensitivity in the round tensile bar, in Salençon J. (ed) Proceedings Considere Memorial Symposium, Presses de l'Ecole Nationale des Ponts et Chaussées: 251-262.

Pan J. Saje M and Needleman A. 1983. Localization of deformation in rate sensitive porous plastic solids, Int. J. Frac.: 261.

Perzyna P. 1963. The constitutive equations for rate sensitive plastic materials, Quart Applied Math. XX, 4: 321-332.

Perzyna P. 1986. Constitutive modeling of dissipative solids for postecritical behavior and fracture, ASME J. Eng. Mat. Technol.; 106: 410-419.

Ponte-Castaneda P. 1991. The effective mechanical properties of nonlinear isotropic materials, J. Mech. Phys. Solids 39: 45.

Predeleanu M. 1987. Finite strain plasticity analysis of damage effects in metal forming processes, In "Computational methods for Predicting Material Processing Defects", edited by M. Predeleanu Elsevier Sciences Publishers, B.V. Amsterdam: 295-308.

Seaman L., Curran D.R. and Shockey D.A. 1976. Dynamic failure in Solids Physics Today, January: 46-55.

Shima S. and Oyane M. 1976. Plasticity theory for porous solids, Int. J. Mech. Sci. 18: 285-291.

Sun Li-Zhi and Huang Zhu-Ping 1992. Dynamic void growth in rate-sensitive plastic solids, Inter J. of Plasticity 8: 903-924.

Suquet P. 1992. Bounds for the overall potential of power-law materials containing voids with an ? shape, Mech. Res. Com., 19: 51.

Tsui T. and Krausch 1965. Thermal stress-wave propagation in hollow elastic spheres, J. Acoust. Soc. America 37: 730-737.

Tvergaard V. 1981. Influence of voids on shear hand instabilities under plane strain conditions; Int. J. Fracture 17: 389.

Wang Z.E. Ping and Zeng Jian 1994. Dynamic failure in ductile porous materials, Engineering Fracture Mechanics 49, 1: 61-74.

Zaker T.A. 1968. Dynamic thermal shock in hollow spheres, Quart Applied Mathematics 26: 503-520.

Simulation of Materials Processing: Theory, Methods and Applications, Shen & Dawson (eds)
© 1995 Balkema, Rotterdam. ISBN 90 5410 553 4

Concurrent design of products and their manufacturing processes based upon models of evolving physicoeconomic state

O. Richmond
Alcoa Technical Center, Pa., USA

ABSTRACT: Until now NUMIFORM Conferences have been focused primarily on the simulation of unit manufacturing processes, with each succeeding Conference demonstrating an improving capability for realistic simulation of complex detail. To make connection with larger design questions, however, like design of factories containing chains of unit processes or life-cycle design of products where manufacture is just one phase, a different viewpoint is required. For that purpose it is suggested here that the life of a product should be viewed as a succession of physicoeconomic states. Physicoeconomic models then are aimed at describing the evolution of physical state parameters, like geometry and temperature and stress fields, together with an economic state parameter, cost. It is believed that such models can provide a much stronger basis for large-scale design decisions than current cost models which have little physical content. In order for detailed models of unit processes to be utilized in these physicoeconomic models, however, more attention will need to be given to capturing the essence of the detailed models in simpler forms.

1 INTRODUCTION

The NUMIFORM Conferences have from the beginning been focused primarily on the numerical analysis and *design of unit forming processes*. Over the years since the first conference tremendous progress has been made in the capability to simulate such processes including especially the effects of complex three-dimensional geometry and the effects of evolving microstructure along with evolving temperature and stress fields. Progress has also been made in the development of design methods for optimizing unit processes.

Perhaps it is time now to focus some attention on larger-scale design questions such as *design of processing factories* (or systems) containing chains of unit forming processes. The design challenge then is to optimize the entire processing system to achieve desired final product attributes at minimum manufacturing cost. Perhaps it is time also to focus attention on *design for product life cycle* where life in a unit process is only one component of manufacturing life, and manufacturing life is only one component of a life that also includes feed stock genesis, consumer use and disposal. The design challenge in this case is to achieve desired product attributes at minimum total life cycle cost.

The purpose of this brief article is to outline a framework for addressing these larger-scale design questions. The principal concept is to view the life of a product as an evolution of physicoeconomic states where cost is included as a state parameter along with such physical entities as geometry and temperature and stress fields. A very simple example is given for the case of homogeneous evolving states. The broader question for the NUMIFORM community is: how can we best capture the essential results of detailed inhomogeneous unit process models in simpler forms that are suitable as part of physicoeconomic models?

2 PRODUCT LIFE AS A SUCCESSION OF PHYSICOECONOMIC STATES

As used here, product life includes the genesis of feed stock materials (birth), manufacturing processes (nurture), product use (maturity) and disposal or recycling (death or reincarnation). We shall be concerned primarily with the manufacturing and use phases.

Manufacturing of sophisticated products generally involves the conjunction of many material streams brought together by processes like coating and joining. Manufacturing can also generate new material streams by processes like cutting and machining which must be recycled or disposed. For purposes of illustration, however, the present discussion is limited to the manufacture and use of a single product component generated from a single material stream.

The physicoeconomic state of this product component at a single moment of its manufacturing or use stages of life is characterized by its overall geometry, its surface structure, its interior temperature and stress fields, its interior structure fields and its accumulated cost. Geometry and internal stress and temperature fields have been included in unit process and component models for a long while. Internal structure fields are of more recent origin, although now fairly commonplace, especially in the form of internal variables representing certain features of the evolving microstructure and damage. The desirability of including surface structure fields as a separate entity has been recognized for some time but is still rare in practice. Such fields would represent evolving surface phenomena like surface topography (roughness), oxides, corrosion and fatigue damage. Finally, in the present work, it has been recognized that accumulating cost can be represented as another evolving state parameter.

3 EXAMPLE INVOLVING HOMOGENEOUS PROCESSING STATES

To illustrate the concept of using a physicoeconomic model for life cycle design, consider the design of a truck wheel which is approximated as a cylindrical disc that is cold formed from an aluminum alloy casting. The diameter of the wheel, the load it must carry and its desired use life are specified. The use and disposal costs can be reduced by decreasing the size, that is thickness, of the wheel. This can be achieved by increasing the strength of the wheel material, which in turn can be achieved either by increasing the alloy content or by increasing the cold work. The first method increases the raw material and disposal costs; the second increases the manufacturing costs. The question is: what alloy composition, casting geometry and final wheel thickness minimize the total cost of initial material, manufacture, use and disposal.

The physicoeconomic state of the wheel is assumed to be given by five parameters: alloy content, a; diameter, d; thickness, h; strength, s; and cost, c. The diameter, thickness, strength and cost all evolve during manufacture, but in the present model only cost continues to evolve during use. One can easily conceive of a more sophisticated model where strength continues to change (degrades) during use due to such phenomena as fatigue and corrosion, but such factors are not included here.

The volume of material, V, is assumed to be conserved during manufacture and use. Hence,

$$V = \frac{\pi}{4} d_o^2 h_o = \frac{\pi}{4} d_1^2 h_1 \quad , \tag{1}$$

where the subscript zero refers to the cast state and the subscript one to the forged state. The strength of the forging is assumed to be related to the imposed strain, ε, by the relation,

$$s = s_M \left(1 - e^{-A\varepsilon}\right) \tag{2}$$

where

$$\varepsilon = \ln\left(h_o / h\right) \tag{3}$$

and

$$s_M = C\left(1 + Ba_o\right) \tag{4}$$

where A, B and C are material constants.

The work of forging per unit volume of material is given by,

$$w = \int_0^{\varepsilon_1} s d\varepsilon = C\left(1 + Ba_o\right)\left[\varepsilon_1 - \frac{1}{A}\left(1 - e^{-A\varepsilon_1}\right)\right] \ . \tag{5}$$

The condition for meeting load requirements on the wheel is,

$$s_1 h_1 = D \quad , \tag{6}$$

where D is a constant determined by the diameter and elastic constants of the wheel and the design load.

The life cycle cost is assumed to be given by,

$$C_L = V\left[\alpha\left(1 + \beta a_o\right) + \gamma w + \delta + \varepsilon\left(1 + \mu a_o^2\right)\right] \ , \tag{7}$$

where the first term is the cost of the casting; the second, the cost of forging; the third, the cost of use; and the fourth, the cost of disposal. The coefficients α, β, γ, δ, ε and μ are estimated cost factors. Utilizing equations (1), (2), (4), (5) and (6) in equation (7) then gives,

$$C_L = \frac{\pi d_1{}^2 D}{4C(1+Ba_o)\left(1-e^{-A\varepsilon_1}\right)}[\alpha+\delta+\varepsilon+\alpha\beta a_o+$$

$$+\frac{\gamma C(1+Ba_o)}{A}\left(A\varepsilon_1 - 1 + e^{-A\varepsilon_1}\right)] \tag{8}$$

This relation gives the life cycle cost as a function of the two design variables: alloy fraction, a_o, and forging strain, ε_1.

The minimum life cycle costs are achieved by setting the partial derivatives of equation (8) with respect to a_o and ε_1 equal to zero. Thus,

$$e^{A\varepsilon_1} - 1 - A\varepsilon_1 = \frac{A\left(\alpha+\delta+\varepsilon+\alpha\beta a_o+\varepsilon\mu a_o^2\right)}{\gamma C(1+Ba_o)} \tag{9}$$

and

$$Ba_o{}^2 + 2a_o + \frac{[\beta\alpha \; - B(\alpha+\delta+\varepsilon)]}{\varepsilon\mu} = 0 \tag{10}$$

Solving (10) for the optimum value of composition, a_o, and then (9) for the optimum value of forging strain, ε_1, the optimum value of forging thickness can be obtained from (2), (4) and (6) which gives,

$$h_1 = \frac{D}{C(1+Ba_o)\left(1-e^{-A\varepsilon_1}\right)} \tag{11}$$

This completes the concurrent design for wheel thickness, forging strain and alloy composition; i.e., for the desired product together with its manufacturing process and material composition.

4 CLOSING REMARKS

In this brief article, I have defined the concept of evolving physicoeconomic states as a possible framework for models that can be used to design manufacturing sequences that achieve desired product attributes at minimum cost, or even more broadly, that can be used for the concurrent design of products and their manufacturing processes to achieve desired lifetime attributes at a minimum total life cycle cost. I have then illustrated this concept with a very simple example using only homogeneous processing states. In order to utilize detailed inhomogeneous unit process or product component models in these larger-scale physicoeconomic models, it is evident that means must be found for capturing the essential features of the detailed models in simpler model forms.

Simulation of Materials Processing: Theory, Methods and Applications, Shen & Dawson (eds)
© 1995 Balkema, Rotterdam. ISBN 90 5410 553 4

The use of tailor-welded blanks in automotive applications

F.I. Saunders & R.H. Wagoner
Department of Materials Science and Engineering, The Ohio State University, Columbus, Ohio, USA

ABSTRACT: The use of tailor-welded blanks for automotive applications to consolidate parts, improve tolerances, save weight, and increase stiffness is expanding greatly, starting from virtually no applications in 1990. While welds in steel are generally stronger than the base material, there is a need to understand the forming characteristics of tailor welded blanks in order to design and produce high-quality parts of cost. Three formability issues of tailor-welded blanks were addressed: the relative formability of three weld types (CO_2 and YAG laser welds and mash-seam welds with and without additional processing), the usefulness of various mechanical tests, and the prediction of the forming behavior using the finite element method. No significant difference in formability was measured among the mash seam (as welded) and laser welding processes. Finite element modeling was shown to predict well the weld-line motion for dome tests and a more complex scale-fender geometry. The role of various variables on weld motion was determined. The local weld properties were found to be insignificant, but the results are very sensitive to draw restraining forces.

1 INTRODUCTION

A tailor-welded blank is comprised of two or more sheets that have been welded together in a single plane previous to forming. The sheets can be identical, or they can have different thickness, mechanical properties or surface coatings. They can be joined by various welding processes, i.e. laser welding, mash seam welding, electron-beam welding and induction welding.

The advantages of such a process are numerous. 30 to 50 % of the sheet metal purchased by some stamping plants ends up as scrap; scrap which can be used for new blanks with tailor-welded blank technology [1]. Alternatively, tailor-welded blanks can be constructed leaving unused area open, thus minimizing offal directly. Part consolidation, made possible by distributing material thickness and properties, allows for reduced costs and better quality, stiffness and tolerances. Tailor-welded blanks provide greater flexibility for component designers. Instead of being forced to work with the same gage, strength or coating throughout an entire part, different properties can be selected for different locations on the blank. In some cases, for example, differential coating thickness are employed.

There have been only a few published results on the formability of tailor-welded blanks. Azuma et. al. [2] studied the behavior of tailor-welded blanks in three standard forming operations, while Nakagawa et. al. [3] and Iwata et. al. [4] used the finite element method (FEM) to analyze similar geometries. Radlymayr and Szinyur [5] evaluated the formability of tailor-welded blanks as a function of weld speed using the Nakazima Test (100 mm hemispherical dome stretch) and the uniaxial tensile test, with similar results to Azuma et. al.

Shi et. al. [6] enumerated formability issues in tailor welded blanks, especially finding that the weld bead should not be oriented parallel to the major stretch axis for optimum performance. When the weld is oriented normal to the major stretch axis, the difference in load bearing capacities of the base materials will result in the weld moving toward the stronger material, thus localizing the strain in the weaker base material.

Wang et. al. [7] determined the stress-strain behavior in the actual weld bead for a number of combinations of thickness and strength for both laser and mash seam welds, by adjusting tensile results for non-weld material The laser welds were significantly stronger than the base materials, while the mash seam weld bead had similar properties to the higher strength materials.

These studies present a wide range of information about welded sheet materials, but there is not a clear understanding of the factors which influence the overall press formability of tailor-welded blanks. A better understanding is needed of the failure criterion for tailor welded blanks, the influence of the welding process on the overall formability, and the influence of weld location on formability, and weld motion.

In the current study, laser and mash seam welded blanks with different strengths and thicknesses were tested to compare the formability of different weld types and to investigate the role of deformation mode and loading condition. An experimental analysis of a scale automotive fender was conducted using the *Hydraulic Forming Simulator* [8,9] to investigate the forming behavior of a tailor-welded blank under realistic conditions. The forming tests and the scale fender were simulated with the finite element method (FEM) in order to assess problems with such simulation, and to identify the important factors in overall press formability.

2 EXPERIMENTAL PROCEDURES

2.1 *Base Materials*

Two combinations of materials were used. A 1.8 mm aluminum killed drawing quality (AKDQ) steel (K = 500 - 530 MPa, n = 0.19) was joined to a 2.1 mm high strength low alloy (HSLA) steel (K = 740 MPa, n = 0.17), and a 0.8 mm AKDQ was joined to 1.8 mm AKDQ. The HS2.1/AK1.8 combination is a proposed combination for an automotive engine compartment rail, while the AK0.8/AK1.8 is being used for an automotive door inner.

2.2 *Welding Processes*

Two types of laser welding processes were used: CO_2 gas and solid state Nd:YAG laser. The CO_2 laser welding was performed by Armco Research and Technology [10] and the Nd:YAG laser welding was performed by Hobart Laser Products [11]. Mash seam welding was performed by Newcor Bay City [12], and optional post-welding hot or cold planishing was also evaluated.

2.3 *Experimental Forming Analysis*

The experimental forming analysis consisted of four different testing methods: standard and subsize uniaxial tensile testing [13], *OSU Formability Test* [14], a full-dome test based on LDH geometry [15], and the scale fender forming operation [8,9].

For uniaxial tensile tests, the standard 50.8 mm gage length tensile specimens had a deforming width of 12.8 mm, and subsize specimens had a 25.4 mm gage length and 6.4 mm deforming width. The weld line was oriented parallel to the tensile axis; a nominal strain rate of 5.0 x 10^{-4} s^{-1} and an extensometer with 25.4 mm gage length were employed.

The optimized *OSU Formability Test* blanks were 114 mm by 127 mm, where the 114 mm dimension was parallel to the major stretch axis and the weld line. The LDH test blanks were originally 133 mm by 178 mm, with the 178 mm dimension lying parallel to the weld line and the major stretch axis. However, the displacement-free boundary conditions for the LDH test could not be enforced (especially for mash seam blanks), so 178 mm by 178 mm blanks were employed for a full dome stretch. The weld line was parallel to one of the edges.

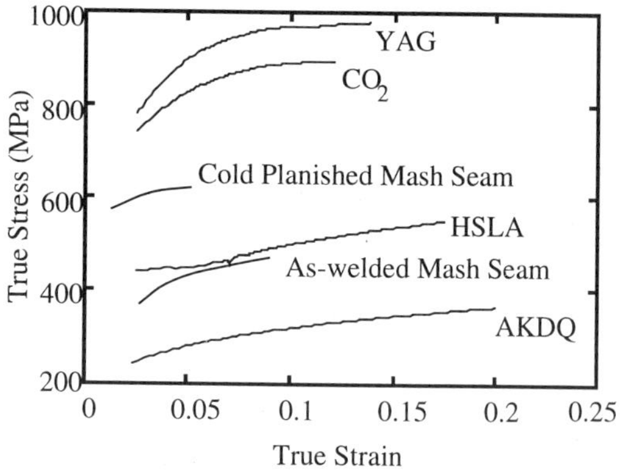

Figure 1. True stress-strain curves for welds compared to base material.

For dome and fender testing, the AK0.8/AK1.8 blanks, the dies were shimmed using 1 mm sheet deformed to the shape of the die. The shims were oriented such that the punch always contacted the flush side of the sheet.

The test specimens for the scale fender die were 356 mm by 432 mm, with the weld oriented parallel to the 356 mm direction at various locations. A constant blank holder force (BHF) of 10 tons was applied.

3 EXPERIMENTAL RESULTS AND DISCUSSION

The hardening laws for base materials were determined using a simplex curve-fitting program [16]. The HLSA material exhibited a yield point elongation of nearly 6% outside of this region, otherwise both base materials were well-represented by the Hollomon [17] hardening law,

$$\bar{\sigma} = K\bar{\varepsilon}^n \tag{1}$$

where $\bar{\sigma}$ is the effective stress, $\bar{\varepsilon}$ is the effective strain, n is the work hardening exponent and K is the strength coefficient.

An analytical procedure was developed to deconvolute the properties of the base and weld bead materials. The load supported by the base material is subtracted from the total load to obtain the load supported by the weld bead. As shown in

Table 1. Formability measures for welded HS2.1/AK1.8 specimens.

	Total Elong. (sub.)	Total Elong (stand.)	OSU Test (mm)	Full Dome (mm)
2.1 mm HSLA (Base)	40 %	41 %	18.8	35.0
1.8 mm AKDQ (Base)	41 %	44 %	19.0	35.8
CO_2 Laser	34 %	35 %	19.2	19.8
YAG Laser	27 %	27 %	18.7	19.6
Unplanished Mash Seam	23 %	35 %	18.8	21.5
Cold Planished Mash Seam	15 %	21 %	17.3	21.0
Hot Planished Mash Seam	22 %	30 %	18.1	20.9

Table 2. Formability measures for welded AK0.8/AK1.8 specimens.

	Total Elong. (sub.)	Total Elong. (stand.)	OSU Test (mm)	Full Dome (mm)
0.8 mm AKDQ (Base)	40 %	44 %	18.8	35.0
1.8 mm AKDQ (Base)	41 %	44 %	19.0	35.8
CO_2 Laser	34 %	35 %	19.2	19.8
YAG Laser	27 %	27 %	18.7	19.6
Unplanished Mash Seam	23 %	35 %	18.8	21.5
Cold Planished Mash Seam	15 %	21 %	17.3	21.0
Hot Planished Mash Seam	22 %	30 %	18.1	20.9

Figure 1, the calculated flow stress of the weld bead for laser welding processes is nearly twice that of the base materials, while mash seam welds have strengths similar to the HSLA steel.

The laser welds more closely followed a saturation type hardening law as described by Voce [18]:

$$\bar{\sigma} = \sigma_0[1 - A\exp(-B\bar{\varepsilon})] \qquad (2)$$

where σ_0 is the saturation stress, and A and B are other curve fit parameters.

The formability test results are summarized in Table 1 (HS2.1/AK1.8) and Table 2 (AK0.8/AK1.8). Except for two outliers, there is no significant difference in any of the formability measures among CO_2 and Nd:YAG laser welds and unplanished mash seam welds for the HS2.1/AK1.8 materials. With the AK0.8/AK1.8 blanks, the formability varies somewhat with weld type, in the order CO_2 (highest), YAG, mash, although the differences are fairly small and there are again two outliers.

The uniform and total elongations for all as-welded samples are roughly 2/3 of the base materials, with a range of roughly 1/2 to 1, depending on the size of the specimen (larger specimens closer to one), which base material is chosen for comparison, and which weld pair is compared (AK0.8/AK1.8 closer to one). Cold planishing greatly reduces formability of mash seam welds while hot planishing restores most of the ductility.

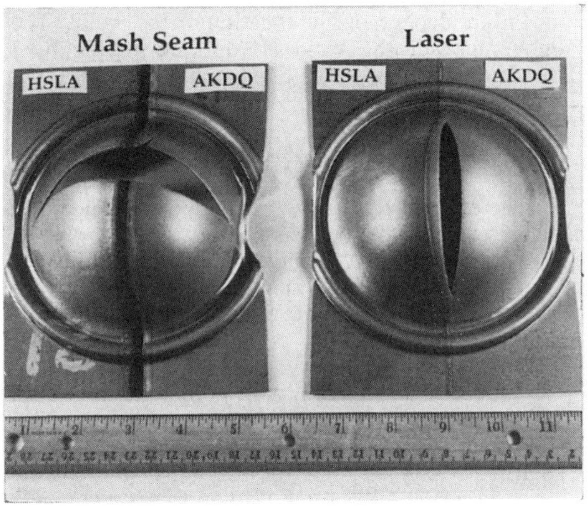

Figure 2. LDH failure modes for welded HS2.1/AK1.8 blanks.

The failures in the tensile specimens and *OSU Formability Tests* are similar: a crack nucleates in the weld, normal to the weld direction and then propagates into surrounding material. In LDH tests (Fig. 2) the failure pattern and orientation depended on weld type. Because of the difficulty in enforcing stretch conditions (note the draw-in near blank edges in Fig. 2), subsequent dome experiments were done with full blank widths. The failure pattern in all dome tests is similar to the laser weld shown in Fig. 2.

The OSU Test correlates closely with the tensile test, Tables 1 and 2. This is expected; the deformation modes of the OSU Test and the uniaxial tension test are similar, with principal loading parallel to the weld. The plane strain vs. uniaxial deformation state apparently creates little difference in this case. There is no correlation between these tests and the full dome results. Not only is the failure mode quite different, but the differential material strength and thickness causes significant motion of the weld line, thus changing the deformation pattern in complex ways.

The two failure modes observed near the weld in tailor-welded blanks under a variety of forming conditions are shown schematically in Fig. 3. The Type 1 failure is caused by exhaustion of ductility in the weld as straining along the weld direction proceeds. The formability in this mode presumably depends on the weld properties and dimensions.

The Type 2 failure is nearly a plane-strain one related to the material properties in the weaker section. The weld in this case serves primarily to inhibit lateral contraction because of the presence of the stronger material. Of course, the principal effect of differential strength may be to induce failure well away from the weld by altering deformation patterns. Trial forming of our scale fender illustrates this, Fig. 4. The failure location and punch-height-to-failure depend on the location of the weld. The movement of the weld is reported in the next section, with comparable FEM simulations.

4 FINITE ELEMENT MODELING METHODS

The experiments show that the performance of a tailor welded blank depends intimately on the amount of weld movement. Prediction of this aspect is thus critical to any evaluation of die performance of formability. Two finite element codes were used. SHEET-3 [19] is a 3-D program developed at the Ohio State University. It features a rigid viscoplastic formulation and membrane elements, which reduces the CPU time considerably, but precludes analysis of bending-dominated operations. SHEET-3 can account for the anisotropy of the material using Hill's 1948 [20] and 1979 [21] yield criteria, as well as Hosford's [22] yield function. (It has been shown in previous work that the yield criterion used can have a large effect on the resulting strain distribution of a simulated part [23, 24].) In this study, the code was internally modified to account for the different properties or thicknesses of the blanks.

ABAQUS is a well-known program [25] based on an elastic-plastic formulation and featuring a finite strain shell element. The plasticity theory is the classic von Mises yield function, without material anisotropy.

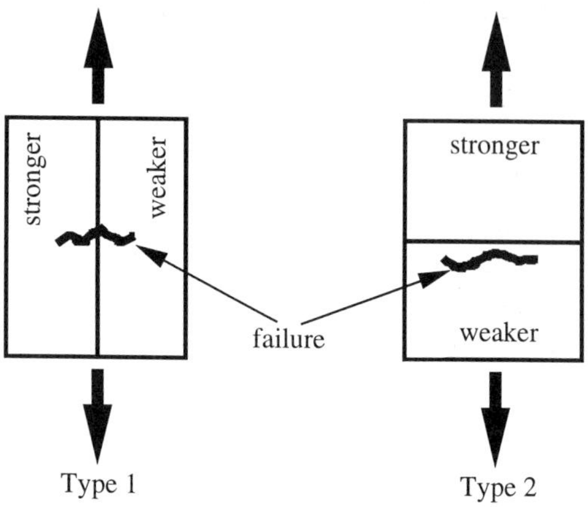

Figure 3. Failure modes for tailor-welded blanks.

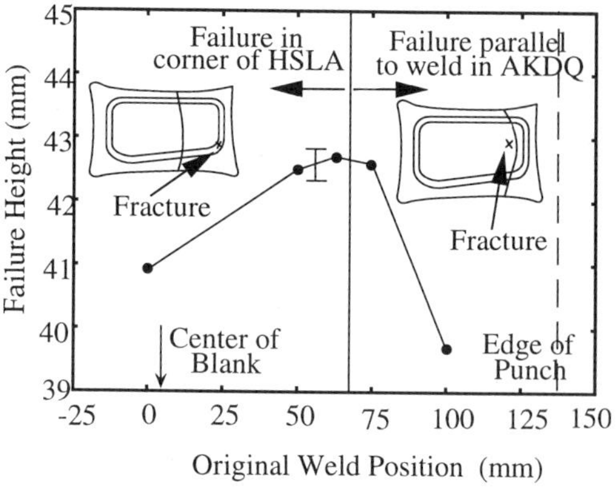

Figure 4. Failure height for scale fender forming as a function of the original position of the weld from the center of the blank.

4.1 *The Dome Test*

In order to simulate the dome test using ABAQUS, nodes around a 60 mm radius were held fixed, to represent the stretch boundary condition. For modeling narrow widths, the blank size and mesh were adjusted accordingly. For modeling a laser welded blank, a row of elements 1 mm in width contained the laser bead mechanical properties, while a row of elements 7 mm in width contained the mechanical properties of the mash seam beads. Overall mesh size was 19 x 30 elements.

In the simulations for dry conditions, the friction coefficient used was 0.25. For the lubricated conditions, the friction coefficient was 0.10. These numbers are typical of measurements in friction studies at the Ohio State University [26,27].

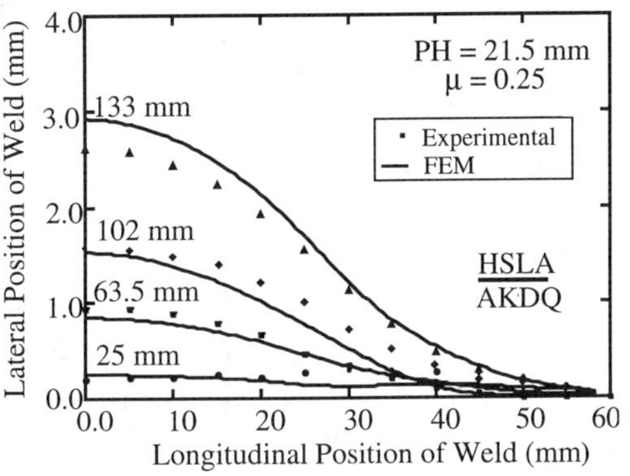

Figure 5. ABAQUS simulated weld movement compared to experimentally measured values for a number of different width blanks of the dome test for HS2.1/AK1.8 blanks.

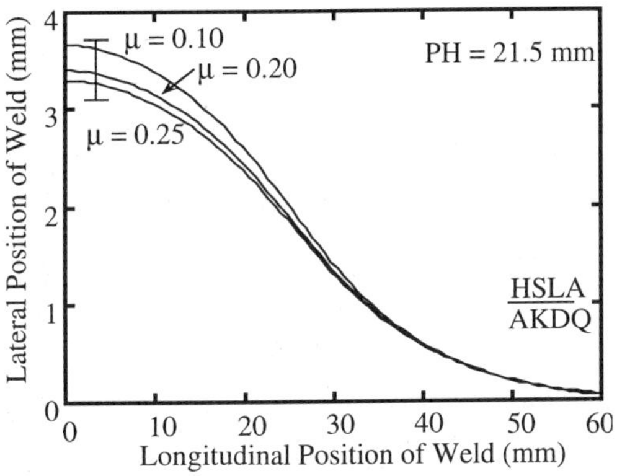

Figure 6. Influence of the global friction condition on simulated weld movement. Bar shows experimental scatter.

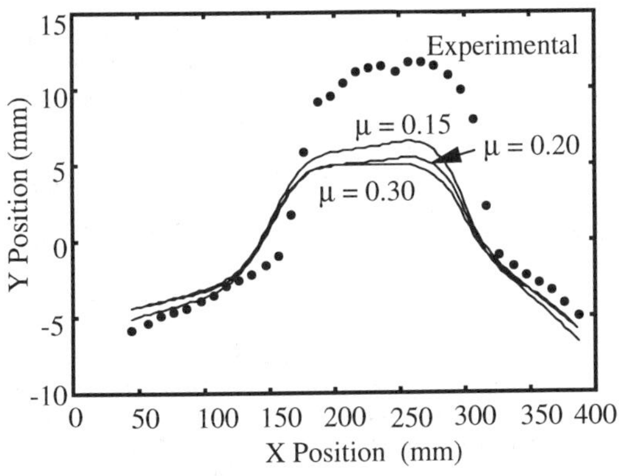

Figure 7. Simulated weld movement compared to measurements for a HS2.1/AK1.8 blank with the weld 102 mm from center for constant boundary forces.

4.2 *The Scale Fender Die*

SHEET-3 was used with a mesh of 800 elements and 441 nodes for the scale fender geometry. Only laser-welded blanks were modeled, and the mechanical properties of the weld bead were neglected. The boundary forces were initially assumed to be uniform, as determined by the blank holder pressure, friction and blank contact area. These conditions were later modified to represent the physical restraint more realistically.

5 FEM RESULTS AND DISCUSSION

5.1 *The Dome Test*

Figure 5 shows typical measured plain-view weld positions compared to the ABAQUS predictions for dome test specimens of 25 mm, 63.5 mm, 102 mm, and 133 mm. As the specimen width increases, more boundary region is clamped and the weld movement becomes more severe. For tailor-welded blanks comprised of base materials with different load bearing capacities, the amount of weld motion is related to the lateral deformation (and force). The model shows reasonable correlation to the experiment for each of the boundary conditions.

As control, this set of experiments and simulations was repeated for welded couples of the same material on both sides of the weld. The experiments showed a maximum weld movement of ± 0.25 mm, which is the uncertainty of the combined measurement and positioning error. This uncertainty illustrates that the differences are statistically insignificant, and in all cases the simulations agree.

The remainder of the dome simulations were conducted for the purpose of understanding what variables contribute significantly to determining the weld movement. In particular, the role of friction (μ = 0.1 - 0.3), weld properties and width (as shown in Fig. 1, w = 1 - 7 mm), and weld type were investigated with the dome geometry. None of these variables had a significant effect on weld movement. A typical result is shown in Fig. 6, where the role of overall friction coefficient was simulated.

5.2 *Scale Fender Die*

The scale fender forming was first simulated using SHEET-3 by applying uniform restraining forces at the periphery of the blank (although drawbeads are employed in the actual dies.) This distributed force is computed by converting the binder load to a friction force (with μ = 0.25) acting on both surfaces of the blank and then distributing it over all boundary nodes. A blank holder force of 20,000 lbf was thus converted to nodal forces of 494 N per node along the 356 mm side and 619 N per node for the 432 mm side. This simulation showed marked discrepancy in simulated and measured weld position, Fig. 7. Previous simulations [28] of this

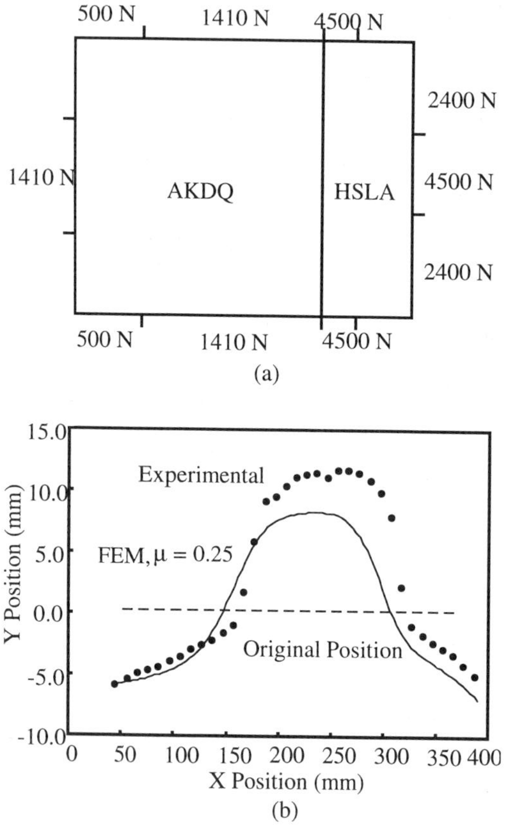

Figure 8. (a) Restraining forces used for a blank with the weld 102 mm from center, calculated using Stoughton's drawbead model [29]. (b) SHEET-3 predicted weld movement in the scale fender die compared to the experimental values for a HS2.1/AK1.8 blank with the weld 102 mm from center, no lubrication, using Stoughton's model.

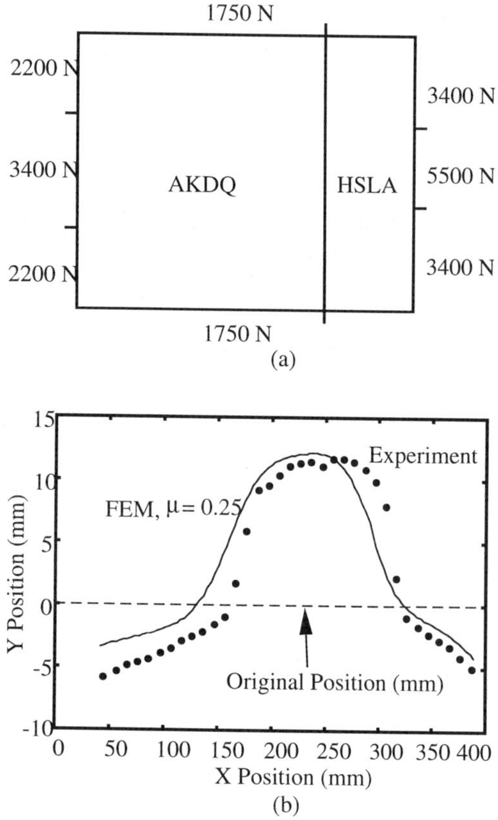

Figure 9. (a) Arbitrary restraining forces for the blank with the weld located 102 mm from the center, dry. (b) SHEET-3 predicted weld movement using fit boundary conditions compared to the experimental values for a HS2.1/AK1.8 blank with the weld 102 mm from the center, dry.

forming operation with standard blanks showed good agreement with experiment, so we suspected boundary conditions.

Clearly the use of a uniform restraining force is unreasonable in view of the presence of drawbeads and differential material properties and thickness. In an attempt to provide more justifiable boundary conditions, a drawbead restraining force model proposed by Stoughton [29] was employed. The drawbead restraining forces calculated by the model were used at the nodes corresponding to the drawbeads in the AKDQ. In the regions between the drawbeads a minimal restraining force of 500 N was applied to maintain numerical stability. In the HSLA region, a force corresponding to the blankholder force on the blank was applied, with a corresponding drawbead restraining force added where drawbeads were located. The resulting restraining forces and simulated weld movement are shown in Figure 8 for a blank with the weld located at 102 mm from the center under dry conditions. This modification allows much better, but still imprecise simulation.

We performed a similar simulation with arbitrary restraining forces (Fig. 9a) to see if this one variable can bring the results into good registry. The results, Fig. 9b, show very good agreement, suggesting that accurate knowledge of boundary conditions in the blankholder is critical for accurate simulation of forming with tailor-welded blanks.

While the boundary restraining forces used in the last simulation are arbitrary, it is possible to check whether they are reasonable independently. Comparison of total draw-in at four cross sections of the blank showed results within 3 mm of measured values.

6 CONCLUSIONS

1.) There is little difference in the intrinsic formability of laser and mash seam welds, based on several measures of formability.

2.) Weld line motion during forming operations is critically important to the development of strain patterns and failure sites. It depends principally on the differing material strengths, and very little on other factors such as friction or weld properties.

3.) Finite element simulation of weld-line motion can be quite accurate, depending on the accuracy of the boundary conditions at the periphery of the blank.

4.) Two kinds of failure patterns specific to tailor-welded blanks occur. When the principal stretch axis is parallel to the weld, the limited weld ductility allows cracks to nucleate and propagate normal to the weld line. When the principal stretch axis is perpendicular to the weld, the stronger material constrains the weaker one to fail near plane-strain tension with the crack in the weaker material, parallel to the crack.

ACKNOWLEDGMENTS

We would like to gratefully thank the Edison Materials Technology Center for funding project CT-41, the Center for Net Shape Manufacturing for providing a fellowship to the first author, and the Ohio Supercomputer Center for providing the computer time for the simulation work. The following industrial partners are deeply thanked for their assistance either through consultancy or performing tasks: Kumar Bhatt, Gary Neiheisel, Tim Webber, Bob van Otteren, Ming Shi and Tom Stoughton. Also, Dajun Zhou is thanked for his assistance in the finite element modeling. Finally, we are indebted to Mark C. Stasik and J. Michael Garrett for performing the final editing of this manuscript.

REFERENCES

1. B. Irving; *Blank Welding Forces Automakers to Sit Up and Take Notice* , Welding Journal, Sept. 1991

2. K. Azuma, K. Ikemoto, K. Arima, H. Sugiura and T. Takasago: *Proc. of 16th Biennial IDDRG Congress*, ASM International, 1990

3. N. Nakagawa, S. Ikura, and F. Natsumi: *SAE Paper #930522*, Society of Automotive Engineers, Inc. 400 Commonwealth Drive, Warrendale, PA 15096-0001, 1993

4. N. Iwata, M. Matsui, N. Nakagawa, and S. Ikura, *NUMISHEET '93, Proceedings of 2nd International Conference on Numerical Simulation of 3-D Sheet Metal Forming Processes,* 1993, Eds. A. Makinouchi, E. Nakamachi, E. Onate, R.H. Wagoner

5. K. M. Radlymayr, J. Szinyur, *IDDRG Working Groups Meeting*, Associazione Italiana Di Metallurgia, I-20121 MILANO, Piazzale Rodolfo Morandi, 2, Italy

6. M. F. Shi, K. M. Pickett and K. K. Bhatt: *SAE Paper #930278*, Society of Automotive Engineers, Inc. 400 Commonwealth Drive, Warrendale, PA 15096-0001, 1993

7. B. Y. Wang, M. F. Shi, H. Sadrina and F. Lin: Unpublished Research

8. Yuji Hishida and Robert H. Wagoner: *Experimental Analysis of Blank Holding Force Control in Sheet Forming* (SAE Paper Number 930285), Sheet Metal and Stamping Symposium, SAE SP-944, Warrendale, PA, 1993, pp. 93-100.

9. Yuji Hishida and Robert H. Wagoner: *Analysis of Blank Holding Force Control Forming using Hydraulic Forming Simulator*, Advanced Technology of Plasticity 1993, 1993, pp. 1740-1746.

10. Dr. Gary Neiheisel, Armco Research and Technology, 705 Curtis St., Middletown, Oh 45043

11. Tim Webber, Hobart Laser Products, 332 Earthart Way, Livermore, CA 94550

12. Bob van Otteren, Newcor Bay City, 1846 Trumbull Dr., Bay City, Mi 48707-0918

13. ASTM Standard E-8, American Standard of Testing of Materials

14. M.P. Miles, N. Krishnaiyengar, R.H. Wagoner, and J.L. Siles: *Metall. Trans. A,* May 1993

15. R. A. Ayres, W. G. Brazier and V. F. Sajewski: *Evaluating the GMR-Limiting Dome Height Test as a New Measure of Press Formability Near Plane Strain*, J. Appl. Metalworking, vol. 1, no. 1, pp. 41, 1979

16. M.S. Caceci, W.P. Cacheris: *Byte*, 1984, p. 340

17. J.H. Holloman: Trans. AIME, 1945, vol. 162, p. 268

18. E. Voce, *J. Inst. Met.*, 1948, vol. 74, pp. 537, 562 and 760; *The Engineer*, 1953, vol. 195, p.23; *Metallurgica*, 1953, vol. 51, p. 219

19. Y. Germain, K. Chung and R.H. Wagoner: *Int. J. Mech. Sci.*, 1989, Vol. 31, No. 1, pp 1-24

20. R. Hill: *Proc. Roy. Soc.*, 1949, vol. 198, p. 428

21. A. Parmer and P.B. Mellor: *Int. J. Mech. Sci.*, 1978, vol. 20, pp. 385-392

22. W.F. Hosford, *J. Appl. Mech.*, 1972, E39, pp. 607-609

23. J.R. Knibloe and R.H. Wagoner: *Metall. Trans. A*, 1989, vol. 20A, pp. 113-123

24. F. I. Saunders and R. H. Wagoner, *Computer Applications in Shaping and Forming of Materials,* eds. M. Demeri, 1993

25. ABAQUS, v.5.2 at the Ohio Supercomputer Center, under academic license from Hibbitt, Karlsson & Sorensen, Inc., Pawtucket, R. I., 1993

26. W. Wang and R. H. Wagoner, *SAE Paper #930807*, Society of Automotive Engineers, Inc. 400 Commonwealth Drive, Warrendale, PA 15096-0001, 1993

27. F. I. Saunders, J. M. Garrett, R. H. Wagoner, The Ohio State University, Columbus, Ohio, 43210, Unpublished research

28. R. H. Wagoner and D. Zhou: *NUMIFORM '92 - Proceedings on the 4th International Conference on Numerical Methods in Industrial Forming Processes,* A. AA. Balkema Publishers, eds. J.-L. Chenot, R. D. Wood and O.C. Zienkiewicz, p. 123-132, 1992

29. T. B. Stoughton: *Proc. of 15th Biennial IDDRG Congress*, ASM International, 1988, pp.205-215

Simulation of Materials Processing: Theory, Methods and Applications, Shen & Dawson (eds)
© 1995 Balkema, Rotterdam. ISBN 90 5410 553 4

Macromolecules and megabytes

Roger I. Tanner, S.C. Xue & N. Phan-Thien
Department of Mechanical and Mechatronic Engineering, The University of Sydney, N.S.W, Australia

ABSTRACT: Over a period of almost 40 years computers have been used to solve rheological and polymer processing flow problems. The rapid improvement in computer performance in this period has been accompanied by improved understanding of polymer rheology, but further progress in both areas is needed, especially in cases of three-dimensional flow. To this end some new three-dimensional finite volume computations are presented.

1 INTRODUCTION

One of us first used a digital computer in 1956, but the machines then available were very limited in speed and memory. By 1960 one could easily solve several ordinary non-linear differential equations and some progress in rheology was made at this time; for example, the solution of a complex (helical) viscometric flow was made possible by computation. At about this time the finite element method of computation was invented by Clough and others, so that an alternative to the existing finite difference methods was presented. However, finite elements were not used routinely in viscous fluid mechanics until about 1970. In rheology, the impact of Lodge's book, Elastic Liquids (1964), directed attention to the importance of elongational flows and got us away from thinking only about shear flows. Perhaps more importantly the work of M Yamamoto and Lodge on the 1946 network theory of Green and Tobolsky also directed thinking towards the necessity of using macromolecular ideas in formulating constitutive equations in rheology.

While network theories have in turn been challenged by reptation as a fundamental explanation of polymer physics, they continue to be useful in formulating rheological equations. In a similar way, finite elements, boundary elements and difference-type methods co-exist as computational methods for rheology and polymer processing.

As far as processing is concerned, the application of finite element methods to extrusion die problems and generally for free-surface flows began in 1974, and progress has been continuous since then. By about 1980 it was possible to obtain sufficient computing power to solve visco-elastic flow problems in plane or axisymmetric flows using finite element methods. As time went on attempts at extensions to three-dimensional flows were explored, but even today there are severe difficulties in solving three-dimensional problems with more realistic constitutive equations using finite element methods. Some recent work by us has instead used a staggered-grid difference type method with the PTT rheological model. This makes far less demands in terms of machine speed and memory, especially for 3-D flows, and some results are reported below.

2 GOVERNING EQUATIONS

We are concerned with three-dimensional incompressible, isothermal viscoelastic flows. The relevant equations are:

$$\nabla . u = 0 \tag{1}$$

$$\rho \left(\frac{\partial u}{\partial t} + u . \nabla u \right) = \nabla . \sigma , \tag{2}$$

where ρ is density of the fluid, t is time and u = (u, v, w) the velocity vector, and the total stress tensor σ is given by

$$\sigma = -p1 + 2\eta_N D + \tau , \qquad (3)$$

where p the hydrostatic pressure, 1 the unit tensor, $2\eta_N D$ is the Newtonian contribution to the stress tensor with η_N being the solvent viscosity and D the strain rate tensor, and τ is the non-Newtonian stress tensor which can be connected with kinematic quantities by the constitutive equations.

2.1 *Constitutive equations*

For viscoelastic fluids, the stress τ is assumed to satisfy the MPTT model (Tanner 1988):

$$g\tau + \lambda\left[\frac{\partial\tau}{\partial t} + u\,.\,\nabla\tau - \mathrm{L}\tau - \tau\mathrm{L}^{\mathrm{T}}\right] = 2\eta_m D , \qquad (4)$$

with

$$g = 1 + \frac{\lambda\epsilon}{\eta_{m0}}\, tr\,\tau; \quad \mathrm{L} = L - \xi D, \qquad (5)$$

where $L = \nabla u^{\mathrm{T}}$ is the velocity gradient tensor, T denotes the transpose operation, λ the relaxation time, ξ and ϵ are material parameters, η_{m0} is the zero shear rate molecular-contributed viscosity, and

$$\eta_m = \eta_{m0}\frac{1 + \xi(2-\xi)\lambda^2\dot{\gamma}^2}{(1 + \Gamma^2\dot{\gamma}^2)^{(1-n)/2}} , \qquad (6)$$

where $\dot{\gamma} = \sqrt{2trD^2}$ is the generalized shear rate, n is the power-law index and Γ is a time parameter. In this paper, we assume that $\Gamma = \lambda$ but this is not essential.

The zero-shear rate viscosity of the fluid is $\eta_0 = \eta_N + \eta_{m0}$. By defining $\beta = \eta_{m0}/\eta_0$ as the retardation ratio, we have $\eta_N = (1 - \beta)\eta_0$, and

$$\sigma = -p1 + 2(1-\beta)\eta_0 D + \tau , \qquad (7)$$

$$\lambda\left[\frac{\partial\tau}{\partial t} + \nabla.(u\tau)\right] = 2\beta\eta_0\mu D + \lambda\,(\mathrm{L}\tau + \tau\mathrm{L}^{\mathrm{T}}) - g\tau , \qquad (8)$$

$$\mu = \frac{1 + \xi(2-\xi)\lambda^2\dot{\gamma}^2}{(1 + \Gamma^2\dot{\gamma}^2)^{(1-n)/2}} , \qquad (9)$$

This covers the following six kinds of constitutive models:

(i) the Newtonian fluid model ($\lambda=0$, $\epsilon=0$, $\xi=0$, $\beta=0$ and $\eta_0=\eta_N$);
(ii) the upper-convected Maxwell (UCM) model ($\epsilon=0$, $\xi=0$, $\beta=1$ and $\eta_0=\eta_m=\eta_{m0}$);
(iii) the Oldroyd-B model ($\epsilon=0$, $\xi=0$, $0<\beta<1$ and $\eta_0=\eta_m=\eta_{m0}$);
(iv) the simplified PTT (SPTT) model ($\xi=0$, $\beta=1$ and $\eta_0=\eta_m=\eta_{m0}$);
(v) the PTT model ($\beta=1$ and $\eta_0=\eta_m=\eta_{m0}$);
(vi) the MPTT model ($\beta=1$).

Our goal is to solve the above governing equations with appropriate boundary conditions to predict the velocity and stress fields of incompressible viscoelastic fluid flow. We shall consider some problems of the flow in non-circular ducts, with and without singular stress points. For example, consider an incompressible, isothermal and steady-state fully developed flow in a pipe with z axis as the axis of the pipe in Cartesian coordinates (x,y,z). The flow is not generally rectilinear due to elastic effects. The general condition for steady rectilinear flow deduced by Oldroyd (1965) implies that there are four cases in which rectilinear flow can occur:

1. in an axisymmetric flow or in a plane channel flow;
2. if the second normal stress difference is zero (as for the UCM, Oldroyd-B, SPTT models);
3. if both the second normal stress coefficient $\psi_2(\gamma)$ and $\eta(\gamma)$ are constant (for the MPTT model with g=1 and n=1);
4. if $\psi_2(\gamma) = k\eta(\gamma)$, in which k is a constant (for the PTT or MPTT model with g=1).

For the model adopted, the viscometric functions (in a simple shearing flow) are determined from the following algebraic equations (note that g involves the stress components - we only wish to emphasize on the functional forms of the viscometric functions here)

$$\psi_1(\dot\gamma) = \frac{2\lambda}{g}\,(\eta(\dot\gamma) - \eta_N(\dot\gamma)), \tag{10}$$

$$\eta(\dot\gamma) = \eta_N(\dot\gamma) + \frac{g\eta_m(\dot\gamma)}{g^2 + \xi(2-\xi)\lambda^2\dot\gamma^2}, \tag{11}$$

and

$$\psi_2(\dot\gamma) = -\frac{\xi\lambda}{g}\,(\eta(\dot\gamma) - \eta_N(\dot\gamma)).$$

Therefore, for the PTT and the MPTT models, ψ_2 is not proportional to the viscosity function, and the flow cannot be rectilinear in pipes with non-circular cross sections, and some secondary flow is to be expected.

For the second problem, we shall set $\psi_2 = 0$, so no secondary circulations occur, but a region of "stress singularity" occurs.

3 NUMERICAL METHOD

To solve the above nonlinearly coupled governing equations economically, recourse to an iterative numerical method was made. Previous work on viscoelastic flows is recorded by Na and Yoo (1991) and Darwish et al. (1992). It may be noticed that all of the governing equations can be written in the form of the general transport equation as follows:

$$\frac{\partial}{\partial t}(\Lambda\Phi) + \frac{\partial}{\partial x_j}(\Lambda u_j\Phi) = \frac{\partial}{\partial x_j}\left(\Gamma\frac{\partial\Phi}{\partial x_j}\right) + S_\Phi, \tag{12}$$

where Φ is the working variable which can be a component of vector or tensor or even a constant. The coefficients Λ, Γ have different meanings for different working variables, and S_Φ is called the 'source term' which includes all the terms that cannot be accommodated in the convective and diffusion terms, and it has a different definition for different equations.

For the momentum equations, $\Phi = u_i$, $\Lambda = \rho$, $\Gamma = \eta_0$, and

$$S_\Phi = -\frac{\partial p}{\partial x_i} + \frac{\partial}{\partial x_j}\left(\eta_0\frac{\partial u_j}{\partial x_i}\right) + \frac{\partial}{\partial x_j}(\tau_{ij} - 2\beta\eta_0 D_{ij}). \tag{13}$$

For the continuity equation, $\Phi = 1$, $\Lambda = \rho$, $S_\Phi = 0$.

For the constitutive equations, $\Phi = \tau_{ij}$, $\Lambda = \lambda$, $\Gamma = 0$, and

$$S_\Phi = 2\beta\eta_0\mu D_{ij} + \lambda\,(L_{ik}\tau_{kj} + L_{jk}\tau_{ki}) - g\tau_{ij}. \tag{14}$$

3.1 *Numerical formulation*

A simple finite volume formulation is used for the spatial discretization and a first-order implicit formula is used for temporal differences because of its simplicity for implementation and unconditional stability for numerical calculations. The flow domain is divided into a set of non-overlapping control volumes ΔV with bounding surface area A as shown in Fig. 1 where a control volume for grid point P is given. Integrating (12) over the control volume and time step δt, and using the divergence theorem, we have

$$\Lambda\frac{\Delta V}{\delta t}(\Phi_p - \Phi_p^n) + \int_A\left(\Lambda u_j\Phi - \Gamma\frac{\partial\Phi}{\partial x_j}\right)n_j dA = \bar S_\Phi, \tag{15}$$

where the subscript p refers to the grid point, superscript n denotes the value evaluated at time-step n, and $\bar S_\Phi$ is the volume integral of the source term S_Φ. This can be linearized as

$$\bar S_\Phi = \int_{\Delta V} S_\Phi dV = \bar S_c + \bar S_p\Phi_p, \tag{16}$$

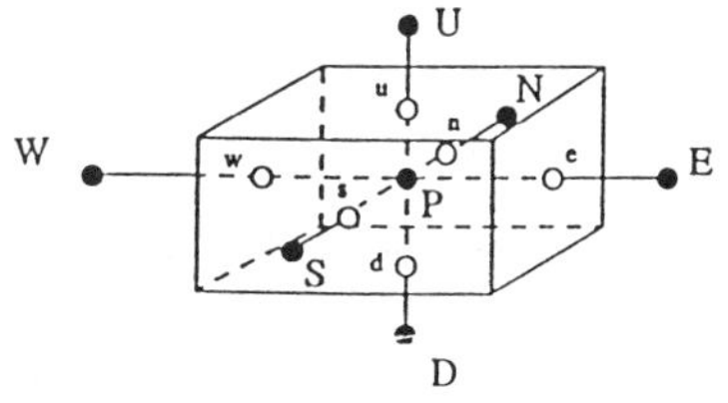

Figure 1. The control volume for grid point P.

in which $\bar{S}_c$ is the part of $\bar{S}_\Phi$ that does not explicitly depend on Φ and $\bar{S}_p$ is the coefficient of Φ_p, which has a negative value to enhance the numerical stability of the discretized equation system. Here, an overbar means the applied values are evaluated using the known fields for time level n. By using a proper spatial variation approximation scheme, the final discretized equation relating the Φ_p to its neighbouring grid point values can be symbolically expressed in a general form for each control volume:

$$a_p \Phi_p = \sum_{nb} a_{nb} \Phi_{nb} + \bar{S}_c + a_p^0 \Phi_p^n , \qquad (17)$$

where

$$a_p^0 = \Lambda \frac{\Delta V}{\delta t} , \; a_p = \sum_{nb} a_{nb} + a_p^0 - \bar{S}_p , \qquad (18)$$

and the summation is to be taken over all of the neighbour grid points nb of the central point p, and the coefficients a_{nb} are the functions of the working variable, and their structures depend on both the approximation scheme used and the form of the cell chosen. It is these coefficients that determine the spatial accuracy of the final solution. In our calculations, the PL (power law scheme) proposed by Patankar (1980) is employed for the formulation because it covers CD (central difference) and UD (upwind difference), and gives an excellent approximation to the exact exponential solution for the one-dimensional convective-diffusive equation. In addition, the FVM with PL, instead of CD or UD, has better conservational properties and thus produces physically realistic solutions even for coarse meshes. In PL, the form of the coefficients is chosen as follows:

$$a_{nb} = D_{nb} f(\mid P_{nb} \mid) + [sign(nb) F_{nb}, 0] , \qquad (19)$$

where $D_{nb} = (\Gamma A/\delta x_i)_{nb}$ is the local diffusion conductance; $F_{nb} = (\Lambda u_i A)_{nb}$ the "mass" flux passing through the corresponding face A normal to i direction of the control volume, sign (nb) is +1 for upstream face and -1 for downstream face, and $f(\mid P_{nb} \mid)$ is the function of the local Peclet number defined by $P_{nb} = F_{nb}/D_{nb}$, which is given by

$$f(\mid P_{nb} \mid) = [0, (1 - 0.1 \mid P_{nb} \mid)^5] , \qquad (20)$$

where the symbol [a,b] means the greater of a and b.

The constitutive equations adopted are hyperbolic, and the simplest stable scheme is the UD, which only ensures first-order spatial accuracy for stress. In order to attain unconditional stability, an artificial diffusion term $\nabla.(\alpha\nabla\tau)$ with α being an artificial diffusion coefficient is introduced on both sides of the constitutive equation, and discretized in the usual way; but the current value is taken for τ on the left hand side; while the known value from previous time level for τ on the right hand side. In this way, the discretized constitutive equations take the same form as (17):

$$a_p \tau_p^{ij} = \sum_{nb} a_{nb} \tau_{nb}^{ij} + \bar{S}_c^{ij} + a_p^0 \bar{\tau}_p^{ij} , \qquad (21)$$

where the superscripts ij refer to tensor components while reserving the subscripts for the grid points and the overbar for the values from time level n. In the constant part of the source term the stress is approximated piecewise-constantly in each control volume. The coefficients a_{nb} take the forms given by (19) with $\Gamma = \alpha$, $\Lambda = \lambda$; the corresponding P_{nb} is called the local artificial Peclet number.

To avoid violating the Scarborough criterion, it is necessary to ensure $f(\mid P_{nb} \mid) > 0$, which can be used for determining the necessary values of α. If UD is adopted, then $f(\mid P_{nb} \mid) > 1$, and there is no restriction to α, but the terms D_{nb} in the equation for a_p stabilize the numerical calculation if α is properly chosen, especially in the case when a_p is dominated by a very small value of g, which may appear during the iteration process; however, the stress diffusion terms do slow down the convergence rate.

3.2 *Solution method*

All of the discretized equations for each control volume in the computational domain consist of a set of nominally linear algebraic equations which is easily solved by means of the Line-by-Line technique based on the TDMA (Thomas algorithm or the tridiagonal matrix algorithm), which has been detailed by Patankar (1980) and is not repeated here.

For viscoelastic fluid flow computations, the extra stress is nonlinearly coupled via the source term of the momentum equations. Here, decoupled techniques can be adopted in such a way that the source term which contains the extra stress in the momentum equations is treated as a pseudo-body force with the known dynamics field obtained from the previous time (or iteration) level, and the stress is updated by solving the discretized constitutive equations after obtaining the kinematics field from the momentum equations.

To obtain the kinematics field, an equation for pressure field is obviously necessary because it is also an unknown. Here, the strategy of pressure correction based on splitting operator techniques is utilized to produce the pressure equation, in which the continuity of the flow field is enforced via a pressure correction so that the resulting pressure relation, which couples the pressure and the velocities, replaces the continuity relation; while the momentum equations retain their role for determining the velocity field. To avoid physically unrealistic fields, such as checker-board velocity and pressure distributions, staggered control volumes have to be used. For the P-centered control volume for scalar fields, the velocities u_i are discretized using their values on the faces normal to the x_i direction, thus, for example, the location of the control volume for u_i in the momentum equations is staggered only in x_i direction relative to the control volume for scalar fields. Therefore, the momentum equation for components u_i on the face f normal to x_i direction in the P-centered control volume takes the discretized form as follows (no sum in i, f is fixed):

$$a_{i,f}u_{i,f} = \sum_{nb} a_{i,nb}u_{i,nb} - \Delta_i(pA) + \bar{S}_{i,f} + a^0_{i,f}u^n_{i,f}\,, \quad (22)$$

in which the pressure difference term has been separated from the source term with the operator Δ_i standing for the difference of the applied values, and A refers to the area of the corresponding face normal to x_i direction (we use the subscript i' here). The continuity in this control volume, thus, can be expressed as the following discretized form

$$\Delta_k(u_{k,f}A) = 0\,. \quad (23)$$

The most commonly used algorithm developed based on this pressure correction strategy is the SIMPLER algorithm, which consists of two main steps; the pressure field computation step, and the velocity field computation step.

We have named the algorithm actually used here as the SIMPLEST algorithm (SIMPLE with Splitting Terms), partly because it is still based on the pressure correction strategy and an explicit step in the neighbour velocities in the momentum equations is used to deduce the pressure corrector equation; and partly because the techniques of splitting operators are utilized.

For steady-state calculations, unlike the SIMPLE and the PISO in which an initial guessed pressure field strongly affects the rate of convergence to the correct solution, the SIMPLEST takes advantage of the property of the SIMPLER that a pressure field compatible with the velocity field is generated in the first step. Hence, if the guessed velocity field happened to be the correct one, a correct pressure field would be obtained in this step and no more iteration would be necessary. Moreover, the p' field for which the equation is derived by taking an explicit step in the velocities of neighbour points, is used to correct the velocity field only; while the p equation for pressure, which prevents an overestimation of the pressure field, and the pressure field obtained from the p equation is closer to the correct solution because no approximation is used. By virtue of this, the need for under-relaxation will be greatly reduced, thus the number of iterations for convergence will be much less.

For transient calculations, like the PISO scheme which ensures that the solution obtained differs from the exact solution of the discretized equations by terms proportional to the powers of the time-step size, the pressure and velocity field obtained at the end of each time step are made to satisfy one and the same momentum equation so that it is not necessary to effect iterations at each time step as the SIMPLER does, and it is also stable even for large time steps and a good temporal accuracy can also be ensured.

In addition, in the SIMPLER scheme, the fields are updated only once at each iteration. While in the SIMPLEST, like the PISO, the fields are updated twice. Therefore, at the end of each time (iteration) step, the fields calculated in the SIMPLEST are much closer to the exact solution.

Because the formulation method adopted here is developed in terms of the primitive variables (u,v,w,p) with stresses as additional variables, the boundary conditions can be directly introduced by properly arranging the meshes so that the grid points for all variables with known values coincide on the boundary.

4 NUMERICAL RESULTS AND DISCUSSION

The algorithm developed here was first tested on a Poiseuille flow and compared with other algorithms using a simple case where an analytic solution is available to make sure that it works well and efficiently. Then, it was used for the flow studies briefly reported below.

In the viscoelastic fluid case, the UCM constitutive equation has no diffusion term, thus the central-difference scheme will result in a discretization equation that violates the Scarborough criterion and may cause the numerical solution to diverge. Several techniques can be used to stabilize the equation, as described in Patankar (1980). In this work, as mentioned above, we have introduced on both sides of the constitutive equation an artificial diffusion term $\alpha\nabla^2\tau$, where τ in one side represents the current value to be found, and in the other side it takes the known value from the previous iteration. The values of α (of dimension length^2) do not affect the final converged solution, but they do affect the numerical stability and the convergence speed. Generally, we require $\alpha/h^2 > 0.5\lambda u/h$ to ensure stability, where λ is the relaxation time given in Eqn. (8), u is the magnitude of the velocity in the convection term of (8), and h the mesh size. Too large a value of α will slow the convergence; too small a value may lead to divergence.

4.1 *Secondary flows*

To investigate the pattern and strength of secondary flows, computations have been carried out for the MPTT fluid flowing at a mean primary velocity $w_m = 8.0$ to 9.0 in straight pipes with rectangular cross section area (W×H) being kept at unity. The aspect ratios (r = W/H) chosen for calculations include 1, 1.56, 4, 6.25, 9 and 16, of which the square pipe (r = 1) is used to study the influence of the material parameters on the secondary flows. The basic material parameters chosen are shown in Table 1. Here, although the range of values of λ may appear rather low, due to the maximum shear rates chosen being of order 10^2, the relevant second normal stress differences are high enough to affect the flow behaviour of the fluid.

In all the calculations, the under-relaxation factors for u, v, w are 0.85, and no relaxation is needed for the pressure. The mesh sizes used in both x and y directions for different ratios are kept at the same level with Δx and $\Delta y \leq 0.025$.

The calculations were first carried out for the following cases in which it has been verified by theoretical analysis that there is no secondary flow:

1. Both ψ_2 and η are constant (when $\epsilon = 0$ and n = 1);

2. $\psi_2(\dot\gamma) = k\eta(\dot\gamma)$ in which $k = -\xi\lambda$ is a constant (when $\epsilon = 0$).

The numerical results show that the pattern of the transverse velocity is still in some regular form, but the maximum magnitude is of the order 10^{-6} which can be attributed to truncation error of the numerical method in comparison to the order 10^{-2} to 10^{-4} in the presence of secondary flows.

The secondary flows in straight pipes at four different aspect ratios for the MPTT fluids with above material parameters are shown in Fig. 2. Here, the velocity vector field and the streamlines of the secondary flows in one quadrant of each pipe are shown in view of the symmetry, in which the upper right corner is the centre of the pipe. It is clearly demonstrated that in all cases, there are two vortices in each quadrant as expected (Tanner 1988), but the pattern and strength of the two vortices depend on the aspect ratio r for a given mean primary velocity. They are completely symmetrical about the diagonal of the quadrant in square pipes; while with the increasing aspect ratio r, the vortex near the long wall expands and its centre gradually moves closer to the short wall squeezing the vortex near the short wall to vanishing.

Table 1. Basic material parameters

Parameters	ρ	η_{m0}	λ	ϵ	ξ	n	β
Values	1	1	0.01-0.015	0.1	0.2	0.65	1

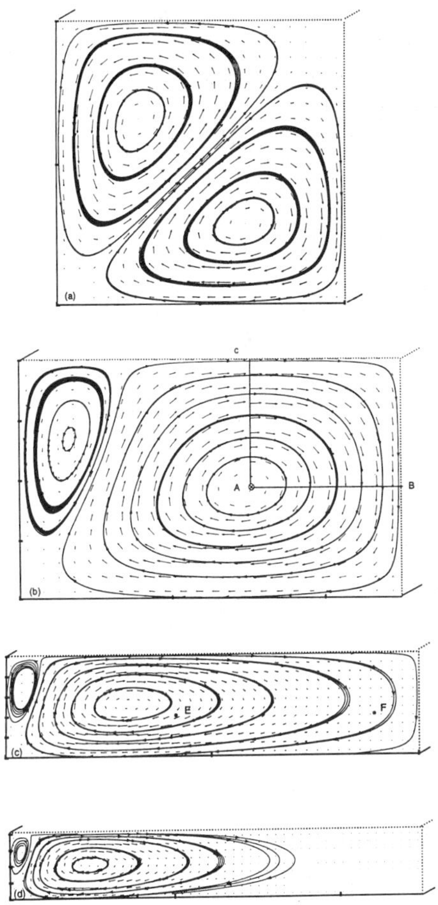

Figure 2. Velocity vector field and streamlines of secondary flow in one quarter of the cross section of pipe with aspect ratio r: (a) 1, (b) 1.56, (c) 4, and (d) 6.25.

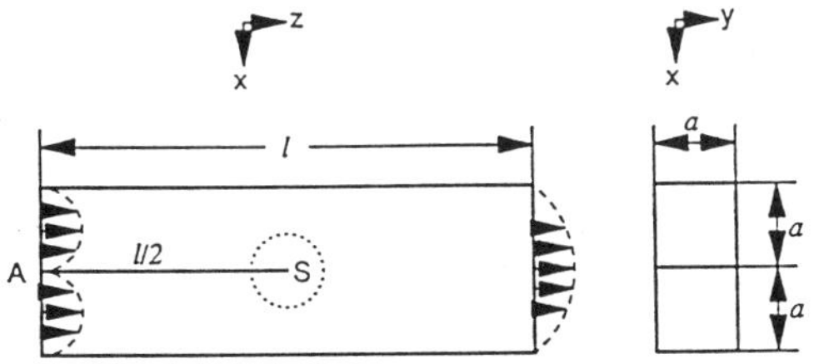

Figure 3. Schematic view of stick-slip flow geometry.

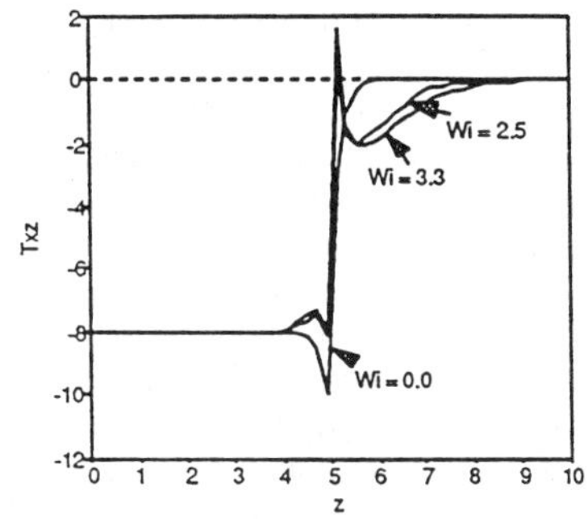

Figure 4. The shear stress τ_{xz} as a function of z.

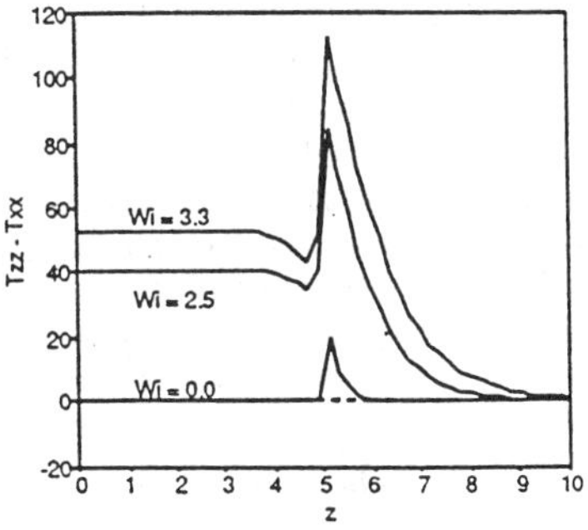

Figure 5. The first normal stress difference as a function of z.

4.2 *Stick-slip flow in a rectangular pipe*

The flow geometry is shown in Fig. 3, where AS is a thin blade, with no-slip boundary conditions, placed in a rectangular channel. This problem is challenging in view of the stress singularity at S.

We scale length with α, velocity with $Q/(2a^2)$, where Q is the flow rate, and pressure and stress with $\eta Q/(2a^3)$. In the calculations, we chose $1 = 6a$. Flows at different Weissenberg numbers ranging from 0 to 3.3 are modelled. Here the Weissenberg number (a dimensionless measure of the fluid's elasticity) is defined by $Wi = \lambda\dot{\gamma}_w$, where $\dot{\gamma}_w$ is the maximum wall shear rate in the fully developed upstream region. Computations for higher Wi numbers are still possible, but the length of the downstream domain needs to be extended as judged by velocity and stress contours.

No-slip boundary conditions are applied on walls and the AS plane. At the inlet and outlet we impose a pressure difference ΔP and set $u = v = 0$. By assuming that the flow in upstream region is fully developed, the stress components at the inlet can be expressed in terms of velocity gradients only, thus the stress boundary conditions can be calculated in each iteration.

Figures 4 and 5 show the shear stresses τ_{xx} and the first normal stress differences $N_1 = \tau_{zz} - \tau_{xx}$ along the centreline of the rectangular conduct respectively. In Fig. 4, both Newtonian and UCM fluids show a sudden change in shear stress in the vicinity of the point S. In the case of the UCM fluid, the shear stress is also found to change its sign in a small region just downstream of the singular point. The normal stress differences (Fig. 5) are seen to rise dramatically near the singular point, with the peak values growing with the Wi number. The maximum value of N_1 is 20.07 for the Newtonian case, 84.15 for Wi = 2.5 and 112.11 for Wi = 3.3. Local asymptotic analysis for the UCM fluid in a 2-D stick-slip flow given by Tanner and Huang (1993) shows that the normal stress near the singularity point behaves almost like $\tau \sim f(\theta)r^{-1}$, where (r,θ) represent a polar coordinate system with its origin at the singularity. Similar behaviour is found in the 3-D flow case.

5 CONCLUSION

A finite volume method for solving flows of viscoelastic fluids has been implemented for two- and three-dimensional problems. It gives promising solutions of complicated problems with singularities. Since the method does not require large amounts of memory, it is possible to solve large problems on small computer systems.

REFERENCES

Darwish, M.S., J.R. Whiteman and M.J. Bevis 1992. *J.non-Newt.Fluid Mech.* 45:311.

Lodge, A.S. 1964. *Elastic liquids.* New York:Academic Press.

Na, Y. & J.Y. Yoo 1991. *Comp.Mech.* 8:43.

Oldroyd, J.G. 1965. *Proc.Roy.Soc.A.* 283:115.

Patankar, S.V. 1980. *Numerical heat transfer and fluid flow.* New York:McGraw-Hill.

Tanner, R.I. 1988. *Engineering rheology.* revised ed. Oxford.

Tanner, R.I. & X. Huang 1993. *J.non-Newt.Fluid Mech.*50:135.

Simulation of Materials Processing: Theory, Methods and Applications, Shen & Dawson (eds)
© 1995 Balkema, Rotterdam. ISBN 90 5410 553 4

Evolution of the intragranular microstructure at moderate and large strains: Modelling and computational significance

Cristian Teodosiu & Zaiqian Hu
LPMTM-CNRS, University Paris Nord, Villetaneuse, France

ABSTRACT: The paper is devoted to the anisotropic work-hardening behaviour under strain-path changes at large strains. After reviewing some microscopic and macroscopic experimental evidence, an elastic-plastic model with three internal variables is presented. It is shown that the model gives a satisfactory description of the influence of the predeformation (amount of prestrain, strain-path changes, etc.) on the yield loci and on the subsequent work-hardening behaviour of a mild steel. Finally, the computational significance of the model for state-update algorithms is analysed within the context of an explicit time-marching scheme.

1 INTRODUCTION

In recent years, considerable attention has been paid to the anisotropic work-hardening of polycrystalline metals under strain-path changes at large strains. This development was mainly due to the importance of this topics, not only for the understanding of the underlying physical mechanisms (e.g. the evolution of dislocation structures and of texture), but also for the modelling of metal forming processes. Indeed, such processes may impose very intense forming sequences, leading to severe strain-path changes and accumulated strains up to several hundreds per cent. In this field, experimental investigation, especially two-stage strain-path tests, has revealed the existence of a complex work-hardening behaviour.

Several purely phenomenological models have been proposed to describe the deformation-induced evolution of the yield surface. For example, Mazilu and Meyer (1985), Cordebois and Boucher (1990) modelled the influence of the predeformation on the yield loci of a prestrained material, giving a successful description of cross-hardening and Bauschinger effects. However, their models cannot describe the influence of the predeformation on the subsequent work-hardening behaviour beyond a microplastic stage, such as the observed work-hardening stagnation under reversed deformation at large strains, and the work-softening during a subsequent orthogonal deformation (Rauch and Schmitt 1989).

The objective of the present paper is to propose a physically-based, phenomenological model of the evolution of the anisotropic work-hardening under arbitrary strain-path changes at moderately large strains. Microscopic evidence is used as a hint for choosing adequate internal variables and for postulating their evolution equations. In a previous paper (Hu et al. 1992), it has been possible, by using a one-dimensional model, to explain the work-hardening stagnation and the resumption of work-hardening during stress reversal at large strains.

More recently (Teodosiu & Hu 1993; Hu & Teodosiu 1995), the model has been extended to three-dimensional, continuous or sharp, strain-path changes and has been identified and validated for a mild steel, by using a rigid-plastic approximation. In the present paper, we propose an extension of this model to finite elastic-plastic deformations and focus our attention on the computational aspects related to the implementation of the model in finite element codes.

2 EXPERIMENTAL EVIDENCE

As stressed by Kuhlmann-Wilsdorf (1989), one of the most important deformation-induced microstructural organization is the formation and evolution of some low-energy dislocation structures (LEDS). Here we limit ourselves to presenting some aspects of this microstructural evolution, mainly related with strain-path changes.

2.1 *Monotonic deformation*

Whenever a sufficient amount of monotonic deformation is allowed for along the same deformation path, some LEDS gradually form. For a polycrystalline solid under monotonic deformation along the same strain path, several slip systems are generally activated in most of the grains.

In stage II of work-hardening, the dominant LEDS are planar persistent dislocation structures (Steeds 1966), which will be called dislocation sheets. Such sheets are more or less parallel to the main slip plane. The lattice misorientation across a sheet is closely connected with the polarity of the sheet. Here the term 'polarity' means that on each side of the sheet there exists an excess of dislocations of the same sign, the sign being different on the opposite sides of the sheet (Kocks et al. 1980). In fact, as shown by Rauch et al. (1990), the activation of non-coplanar slip systems is a prerequisite for the formation of dislocation sheets.

2.2 *Reversed deformation*

The simplest strain-path change is the stress reversal. During a reversed deformation after a sufficiently large amount of preshear, there exists a microplastic regime of rapid work-hardening, followed by a work-hardening stagnation, and then by a resumption of work-hardening. The extent of the stage of strain-hardening stagnation increases with the amount of preshear.

According to microscopic evidence (see, e.g. Hasegawa et al. 1975), the first stage of reversed deformation corresponds to the coarsening of preformed dislocation cell structures, the extended plateau in the stress-strain curve to the partial disintegration of the preformed cell structures, and the subsequent resumption of work-hardening to the formation of new cell structures.

2.3 *Two-stage strain-path changes*

A real metal forming process may generally involve quite complex strain-path changes. After the pioneering work of Ghosh and Backofen (1973), two-stage strain-path tests have been widely used to investigate the influence of stain-path changes on the work-hardening transients. In order to characterize such strain-path changes, Schmitt et al. (1985) have proposed the scalar parameter $\beta = \mathbf{N}_1 : \mathbf{N}_2$, where $\mathbf{N}_1$ and $\mathbf{N}_2$ are, respectively, the directions of the strain rate tensors during the first and second deformation stages, and a colon denotes the double-contracted tensor product. It is evident from the above definition that β varies from -1 (Bauschinger test) to 1 (monotonic test). A deformation sequence is called orthogonal when $\beta = 0$.

According to Schmitt et al. (1985), the parameter β is suitable for characterizing the yield loci of a severely prestrained material. Denoting by σ_S and σ_P, respectively, the yield stress after reloading and the flow stress before unloading, the ratio σ_S/σ_P equals one for $\beta = 1$, is smaller than one for $\beta = -1$, and reaches in general a maximum value for $\beta = 0$. Experimental data obtained from different strain-path changes with the same β-value show a non-identical but similar tendency, confirming that β is an adequate parameter to characterize strain-path changes (Bacroix et al. 1994).

The orthogonal sequence is particularly interesting, since it involves new slip systems which were latent during the predeformation. During a subsequent orthogonal deformation, the stress is always higher than that during the monotonic deformation at the same accumulated strain. When the amount of prestrain e_f is small, the work-hardening rate is always positive, whereas when e_f is large, first a work-softening occurs, followed by a resumption of work-hardening.

During the work-softening, while the macroscopic deformation can be still more or less homogeneous, the subsequent deformation is essentially localized at the intragranular scale. In fact, the shear rate is then mainly carried by parallel microbands along the newly active slip planes, whereas outside the microbands there is almost no shear. Moreover, the preformed dislocation sheets are sheared by the microbands, and this can be used as a marker of their presence.

2.4 *Microstructure vs. texture*

Juul Jensen and Hansen (1987) have measured the 0.2%-offset tensile yield stress on specimens cut at various angles α to the rolling direction from a cold-rolled sheet of aluminium. The amount of prestrain was up to 200%. The texture of the rolled sheet has been determined experimentally and the corresponding Taylor M-factor was calculated by a series expansion method. Then the variation with α of the ratio of the tensile yield stress to the corresponding M-factor has been investigated. At very large strains, this ratio is almost constant, indicating that the plastic anisotropy can be almost attributed to the crystallographic texture. On the contrary, after moderately large, monotonic strains, the ratio increases with α, showing a strong influence of the intragranular microstructure on the plastic anisotropy.

3 MODELLING

The plastic behaviour of metals depends not only on the current state of deformation, but also on the deformation history. This influence of the deformation history can be reasonably represented by the current values of a sufficient number of microstructural parameters, called internal variables. Whereas the mathematical structure of constitutive equations with internal variables has been repeatedly investigated, the real challenge consists in the appropriate choice of a reasonably small number of physically-significant parameters, describing the dominant microstructural evolution.

3.1 *Simplifying hypotheses*

We restrict ourselves to the cold deformation of metals and neglect any viscous effects on the work-hardening. Furthermore, the amount of elastic strains is considered negligibly small as compared to that of plastic strains.

As mentioned in the preceding section, the contribution of the microstructural evolution to the work-hardening is predominant at moderately large strains, hence only this contribution and the influence of the initial texture will be considered here, whereas the influence of the texture evolution on the work-hardening will be neglected.

Moreover, we shall assume that the material is initially stress-free and homogeneous, i.e. its initial plastic and elastic anisotropy are the same for all material particles.

3.2 *Kinematics*

In what follows, bold letters represent second- or fourth-order tensors. Components, whenever used, are referred to a Cartesian orthogonal frame. The summation convention over repeated indices of such components is used throughout the paper. Let $\mathbf{A}$, $\mathbf{B}$ denote second-order tensors and $\mathbf{S}$ a fourth-order tensor. We define the double-contracted tensor product between such tensors as $\mathbf{A}:\mathbf{B} = A_{ij}B_{ij}$, $(\mathbf{S}:\mathbf{A})_{ij} = S_{ijkl}A_{kl}$, $\mathbf{A}:\mathbf{S}:\mathbf{B} = A_{ij}S_{ijkl}B_{kl}$. We also define the norm of $\mathbf{A}$ as $|\mathbf{A}| = \sqrt{\mathbf{A}:\mathbf{A}}$ and its direction, if $\mathbf{A}$ is non-zero, as $\mathbf{A}/|\mathbf{A}|$. Finally, the norm of $\mathbf{S}$ is defined by $|\mathbf{S}| = \sqrt{S_{ijkl}S_{ijkl}}$.

The kinematics of large, elastoplastic deformations is based on the multiplicative decomposition of the deformation gradient $\mathbf{F}$ into a plastic part $\mathbf{F}^p$ and an elastic part $\mathbf{F}^e$, i.e.

$$\mathbf{F} = \mathbf{F}^e\mathbf{F}^p. \tag{1}$$

By time-differentiating Eq. (1), it follows that the gradient $\mathbf{L}$ of the velocity field can be written as

$$\mathbf{L} = \dot{\mathbf{F}}\mathbf{F}^{-1} = \dot{\mathbf{F}}^e(\mathbf{F}^e)^{-1} + \mathbf{F}^e\dot{\mathbf{F}}^p(\mathbf{F}^p)^{-1}(\mathbf{F}^e)^{-1}. \tag{2}$$

For cold deformation of metals, the elastic part of $\mathbf{F}$ involves strains that are minute fractions of unity, but possibly large rotations. That is why, by making use of the polar decomposition of $\mathbf{F}^e$, we shall write

$$\mathbf{F}^e = (\mathbf{1} + \mathbf{e})\mathbf{R}, \tag{3}$$

where $\mathbf{1}$ is the unit second-order tensor, $\mathbf{e}$ is a symmetric tensor of small elastic strains ($|\mathbf{e}| \ll 1$), and $\mathbf{R}$ is the rotation tensor ($\mathbf{R}^{-1} = \mathbf{R}^T$, hence $\mathbf{R}\mathbf{R}^T = \mathbf{1}$, where a superscript T denotes the transpose of a second-order tensor).

By introducing (3) into (2) and neglecting terms of second order in $|\mathbf{e}|$, we obtain

$$\mathbf{L} = \dot{\mathbf{R}}\mathbf{R}^T + \overset{\circ}{\mathbf{e}} + \mathbf{R}\dot{\mathbf{F}}^p(\mathbf{F}^p)^{-1}\mathbf{R}^T, \tag{4}$$

where

$$\overset{\circ}{\mathbf{e}} = \dot{\mathbf{e}} + \mathbf{e}\,\dot{\mathbf{R}}\mathbf{R}^T - \dot{\mathbf{R}}\mathbf{R}^T\mathbf{e} \tag{5}$$

denotes the objective time-derivative of $\mathbf{e}$, calculated with the spin associated with the 'elastic' rotation $\mathbf{R}$. The symmetric and antisymmetric parts of (5) give the strain rate tensor $\mathbf{D}$ and the spin $\mathbf{W}$ as

$$\mathbf{D} = \overset{\circ}{\mathbf{e}} + \mathbf{D}^p, \quad \mathbf{W} = \dot{\mathbf{R}}\mathbf{R}^T + \mathbf{W}^p, \tag{6}$$

where

$$\mathbf{D}^p = \mathbf{R}\widehat{\mathbf{D}}^p\mathbf{R}^T, \quad \mathbf{W}^p = \mathbf{R}\widehat{\mathbf{W}}^p\mathbf{R}^T \tag{7}$$

are, respectively, the plastic strain rate and the plastic spin, while $\widehat{\mathbf{D}}^p$ and $\widehat{\mathbf{W}}^p$ are the symmetric and antisymmetric parts of $\dot{\mathbf{F}}^p(\mathbf{F}^p)^{-1}$.

As a consequence of the non-uniqueness of the multiplicative decomposition (1), there exists a certain ambiguity in the choice of the local natural configuration. At the scale of a grain, this ambiguity disappears, since all dislocation structures are corotational with the atomic lattice of the crystal, and hence the corresponding internal variables should be corotational with the lattice, too. For a polycrystal, Mandel (1982) proposed a privileged frame, whose spin equals the mean lattice spin of all grains of the polycrystal, and which proves to be particularly suitable for the micro-macro transition (see, e.g., Teodosiu 1989). Clearly, the evolution of Mandel's frame is connected to that of the orientation distribution function, and hence it is outside the aim of the present paper. Still, we will retain one of the basic ideas of Mandel's analysis: all tensorial, structural variables are supposed to turn with the same spin $\dot{\mathbf{R}}\mathbf{R}^{-1}$. Furthermore, since the work-hardening behaviour of polycrystalline materials at moderately large strains is not very sensitive to the choice of this spin, we shall simply assume $\mathbf{W}^p = \mathbf{0}$ in Eq. (6). Consequently, all tensorial variables will be considered corotational with the material, while their objective rates, denoted by a small superposed circle, like in Eq. (5), will be of Jaumann type.

The remaining part of this section is essentially based on the rigid-plastic model previously developed by the present authors (Hu and Teodosiu 1995).

3.3 *Elastic behaviour*

Assuming again that the elastic strains are small against unity and neglecting the influence of the plastic deformation on the elastic constants, it may be shown (see, e.g., Teodosiu 1989) that the time-differentiation of the hyperelastic constitutive equation leads to the hypoelastic form

$$\overset{\circ}{\mathbf{T}} = \mathbf{c} : \overset{\circ}{\mathbf{e}} = \mathbf{c} : (\mathbf{D} - \mathbf{D}^p), \tag{8}$$

where

$$c_{ijkm} = R_{ip}R_{jq}R_{kr}R_{ms}\widehat{c}_{pqrs}, \tag{9}$$

whereas $\widehat{\mathbf{c}}$ is the the tensor of the initial elastic constants.

3.4 *Internal state variables*

The complete set of internal state variables of our model is denoted by $(\mathbf{S}, \mathbf{P}, \mathbf{X})$. $\mathbf{S}$ and $\mathbf{X}$ have the dimension of stress, $\mathbf{P}$ has no dimension. $\mathbf{S}$ is a fourth-order tensor, $\mathbf{P}$ and $\mathbf{X}$ are second-order tensors. For a well-annealled material, all their initial values are taken equal to zero.

The first tensorial variable $\mathbf{S}$ describes the directional strength of planar persistent dislocation structures.

The second tensorial variable $\mathbf{P}$ is associated with the polarity of the persistent dislocation structures, which is due to the excess of dislocations of the same sign on each side of a dislocation sheet (Kocks et al. 1980). When a microstructure is not at all polarized, $\mathbf{P} = \mathbf{0}$. On the other hand, when a microstructure is completely polarized after a monotonic deformation, then $\mathbf{P} = \mathbf{N}$, where $\mathbf{N}$ is the direction of the plastic strain rate tensor $\mathbf{D}^p$ during the predeformation. We may, therefore, assume, without loss of generality, that $0 \leq |\mathbf{P}| \leq 1$.

The third tensorial variable $\mathbf{X}$ is intended to describe the rapid changes in stress under reversed deformation, e.g. those produced by dislocation pile-ups.

3.5 *Yield condition and flow rule*

The yield condition is supposed of the form

$$F = \bar{\sigma}_e - R_0 - f\,|\mathbf{S}| = 0, \tag{10}$$

where R_0 is the initial yield stress, and $f\,|\mathbf{S}|$ denotes the contribution of the persistent dislocation structures to the isotropic hardening, $0 \leq f \leq 1$. $\bar{\sigma}_e$ is the equivalent effective stress, defined by

$$\bar{\sigma}_e = \sqrt{(\mathbf{T} - \mathbf{X}) : \mathbf{M} : (\mathbf{T} - \mathbf{X})}, \tag{11}$$

where $\mathbf{T}$ is the Cauchy stress tensor, and $\mathbf{M}$ is a fourth-order tensor that represents the texture anisotropy and has the properties $M_{ijkl} = M_{jikl} = M_{klji}$, $M_{iikl} = 0$. In fact, $\mathbf{M}$ should also be added to the set of internal variables. However, since we neglect the texture evolution, we will omit $\mathbf{M}$ from this set, and simply assume that the components of $\mathbf{M}$ in the corotational frame are constant (i.e. $\overset{\circ}{\mathbf{M}} = \mathbf{0}$) and describe the initial texture anisotropy.

The plastic deformation rate is given by the associated flow rule

$$\mathbf{D}^p = \dot{\lambda}\frac{\partial F}{\partial \mathbf{T}} = \dot{\lambda}\frac{\mathbf{M} : (\mathbf{T} - \mathbf{X})}{R_0 + f\,|\mathbf{S}|}, \tag{12}$$

where $\dot{\lambda}$ is the plastic multiplier, and a superposed dot denotes time differentiation. $\dot{\lambda}$ is zero within the elastic region ($F < 0$) or under elastic unloading ($F = 0$ and $\frac{\partial F}{\partial \mathbf{T}} : \overset{\circ}{\mathbf{T}} \leq 0$). During a plastic loading process ($F = 0$ and $\frac{\partial F}{\partial \mathbf{T}} : \overset{\circ}{\mathbf{T}} > 0$), $\dot{\lambda}$ is positive and determined by the consistency condition $\dot{F} = 0$.

3.6 *Evolution of the internal variables*

We propose for $\mathbf{P}$ the evolution equation

$$\overset{\circ}{\mathbf{P}} = C_P(\mathbf{N} - \mathbf{P})\dot{p}, \tag{13}$$

where C_P characterizes the polarization rate of the persistent dislocation structures, $\mathbf{N}$ represents the current direction of the strain rate tensor, and $\dot{p}$ is the power conjugate of $\bar{\sigma}_e$, i.e.

$$\bar{\sigma}_e \dot{p} = (\mathbf{T} - \mathbf{X}) : \mathbf{D}^p. \tag{14}$$

In particular, it follows from Eqs (11), (12) and (14) that $\dot{p} = \dot{\lambda}$.

According to Eq. (13), whatever the initial value of $\mathbf{P}$, $|\mathbf{P}|$ will approach unity and $\mathbf{P}$ will tend to $\mathbf{N}$, if the strain rate direction $\mathbf{N}$ remains unchanged for an amount of deformation that is sufficiently large with respect to $1/C_P$.

Next we assume that the evolution of $\mathbf{X}$ is given by the equation

$$\overset{\circ}{\mathbf{X}} = C_X (X_{\text{sat}} \mathbf{N} - \mathbf{X}) \dot{p}, \tag{15}$$

where C_X characterizes the saturation rate of $\mathbf{X}$, and X_{sat} is the saturation value of $|\mathbf{X}|$.

In general, C_X has a relatively high value, e.g. $C_X = 50$ for an AKDQ mild steel (Hu et al. 1992). Thus, according to Eq. (15), $\mathbf{X}$ approaches very rapidly its saturation value, and hence it may be assumed that $\mathbf{X} = X_{\text{sat}} \mathbf{N}$ after a few per cent of strain increment, wherever the strain rate tensor remains self-coaxial.

The dependence of $\mathbf{X}$ on the persistent dislocation structures is included in the scalar function $X_{\text{sat}}(\mathbf{S}, \mathbf{N})$. From a mathematical point of view, X_{sat} should be a function of the scalar invariants generated by $\mathbf{S}$ and $\mathbf{N}$. Since the main effect we want to take into account is the ralative orientation of the current strain rate to the persistent dislocation structures, we assume that

$$X_{\text{sat}} = \widehat{X} + (1 - f) \sqrt{r\, |\mathbf{S}|^2 + (1 - r) S_D^2}, \tag{16}$$

where $\widehat{X}$ is the initial value of X_{sat}, r is a material parameter, and

$$S_D = \mathbf{N} : \mathbf{S} : \mathbf{N}. \tag{17}$$

By introducing the parameter $\beta_S = S_D / |\mathbf{S}|$, Eq. (16) can be rewritten as

$$X_{\text{sat}} = \widehat{X} + (1 - f)\, |\mathbf{S}| \sqrt{r + (1 - r) \beta_S^2}. \tag{18}$$

When $r > 1$, X_{sat} has the highest value for $\beta_S = 0$, and the lowest one for $\beta_S = 1$. This is convenient for describing the yield loci of prestrained materials, since β_S is a measure of the change in orientation of the current strain rate tensor with respect to the persistent dislocation structures, and may be shown to be a generalization of the parameter β defined by Schmitt et al. (1985).

As mentioned in Section 2, one of the most striking features of the dislocation structures is their directionality. Thus, for a material deformed from a well-annealed initial state, dislocation sheets or cells develop roughly parallel to the active slip planes. On the other hand, for a severely cold-deformed material subjected to a subsequent orthogonal deformation, the strain rate is highly localized. The microbands are parallel to the newly active slip planes, and between them new dislocation sheets are gradually formed. This experimental evidence strongly suggests that dislocation structures associated with the current direction of the strain rate evolve quite differently from the rest of the persistent dislocation structures.

In order to describe such evolution processes, and taking into account the definition of S_D, we decompose $\mathbf{S}$ as

$$\mathbf{S} = S_D \mathbf{N} \otimes \mathbf{N} + \mathbf{S}_L, \tag{19}$$

where S_D represents the strength associated with dislocations of the currently active slip systems, whereas $\mathbf{S}_L$ is associated with the latent part of the persistent dislocation structures.

In order to describe the evolution of S_D, we set

$$\dot{S}_D = C_S \left[g \left(S_{\text{sat}} - S_D \right) - h S_D \right] \dot{p}, \tag{20}$$

where C_S characterizes the saturation rate of S_D, S_{sat} denotes the the saturation value of S_D, g is a function of S_D and $\mathbf{P} : \mathbf{N}$ that describes the influence of the polarity, and h is a function of $\mathbf{X} : \mathbf{N}$.

Neglecting the influence of g and h, i.e. setting $g = 1$, $h = 0$, the above equation describes a gradual saturation of S_D towards S_{sat}, corresponding to the formation and saturation of planar persistent dislocation structures associated with $\mathbf{N}$. In order to form dislocation sheets or cell walls, the amount of deformation along which $\mathbf{N}$ keeps unchanged should be larger than $1/C_S$.

The function h is defined as

$$h = \frac{1}{2} \left(1 - \frac{\mathbf{X} : \mathbf{N}}{\mathbf{X}_{\text{sat}}} \right), \tag{21}$$

and has a non-negligible value only during a microplastic stage. In fact, as has been shown by Hu (1993), the presence of h enables to evaluate the slight loss ΔS_D of S_D as $\mathbf{X} : \mathbf{N}$ approaches its saturation value X_{sat}. The result reads

$$\Delta S_D = -\frac{C_S}{C_X} S_D. \tag{22}$$

Experimental evidence shows that, for a severely prestrained material under a subsequent reversed deformation, there exists a work-hardening stagnation, followed by a resumption of work-hardening. This phenomenon has been already modelled in our previous work (Hu et al.

1992; Hu 1993). Here we give a more general description of this phenomenon, via the function g, which takes into consideration the present three-dimensional context. Specifically, by denoting

$$P_D = \mathbf{P} : \mathbf{N}, \tag{23}$$

we assume that

$$g = \begin{cases} 1 - \frac{C_P}{C_S + C_P} \left| \frac{S_D}{S_{\text{sat}}} - P_D \right| & \text{if } P_D \geq 0, \\ (1 + P_D)^{n_P} (1 - \frac{C_P}{C_S + C_P} \frac{S_D}{S_{\text{sat}}}) & \text{otherwise.} \end{cases} \tag{24}$$

Clearly, g is continuous with respect to P_D. Moreover, assume that a material is first severely deformed at a constant strain rate of direction $\mathbf{N}_1$. Then, according to Eq. (13), at the end of this deformation the polarity tensor $\mathbf{P}$ will be practically equal to $\mathbf{N}_1$. If the material is subsequently subjected to a reversed deformation, i.e. $\mathbf{N} = -\mathbf{N}_1$, then, by Eqs (23) and (24), $P_D = -1$ and $g = 0$. Considering also Eq. (20), it may be shown that the last condition corresponds to a stagnation of the work-hardening.

We finally discuss the evolution of $\mathbf{S}_L$, which results from the interaction between microbands and the preformed microstructures. Two physical mechanisms are possible: the annihilation of dislocations in the preformed structure, as proposed by Thuillier and Rauch (1994), and the softening of the preformed structures after being sheared by microbands, as suggested implicitly by Bay et al. (1992). Both mechanisms reduce the strength of the preformed structures, represented by $|\mathbf{S}_L|$. Hence we propose the following evolution equation to describe these phenomena

$$\overset{\circ}{\mathbf{S}}_L = -C_S \left(\frac{Z}{S_{\text{sat}}} \right)^n \mathbf{S}_L \dot{p}, \tag{25}$$

where n is a positive material parameter and

$$Z = |\mathbf{S}_L| = \sqrt{|\mathbf{S}|^2 - S_D^2}. \tag{26}$$

The factor $(Z/S_{\text{sat}})^n$ is introduced in order to explain the influence of the amount of prestrain ϵ_f. According to Rauch and Schmitt (1989), for a severely deformed material under a subsequent orthogonal deformation, the percentage of grains containing microbands increases with the prestrain ϵ_f. Since the diminution of $|\mathbf{S}_L|$ is mainly due to the interaction between microbands and preformed microstructures, the decreasing rate of $|\mathbf{S}_L|$ should increase with ϵ_f. When ϵ_f is very small, $Z \ll S_{\text{sat}}$, and hence the evolution of $\mathbf{S}_L$ is negligible, whereas when ϵ_f is large, Z approaches S_{sat}, and the evolution of $\mathbf{S}_L$ is speeded up.

3.7 *Discussion*

In the present model, the most important internal variable is the fourth-order tensor $\mathbf{S}$. The choice of its order is due to the necessity to describe the anisotropic contribution of persistent dislocation structures to the flow stress. Except for a slight loss during microplastic stage under stress reversal, it has an equal value in the forward and reversed deformation direction, but some higher values in other directions, especially in an orthogonal one. Furthermore, it does not decrease under stress reversal, unlike the back-stress.

Besides the three-dimensional framework, the present model differs from our previous models (Hu et al. 1992; Hu 1993) in the choice of internal variables. In previous models, two scalar internal variables P and R have been introduced, which correspond respectively to dislocations within the sheets and in the interior of cells. In the present work, $\mathbf{S}$ describes the total anisotropic strength of persistent dislocation structures, while $\mathbf{P}$ is redefined as the polarity of such persistent dislocation structures.

The model has been identified for rolled sheets of aluminium-killed mild steel, by using sequences of simple shear with intermediate strain-path changes (Hu and Teodosiu 1995). The values obtaind for the 10 material parameters occurring in the yield condition and in the evolution equations are $R_0 = 51$MPa, $\widehat{X} = 64$MPa, $S_{\text{sat}} = 145$MPa, $C_P = C_S = 2.7$, $f = 0.59$, $n_p = 0.2$, $n = 3.0$, $r = 2.4$, $C_X = 50$. It should be noted that R_0 represents now the initial shear yield stress in the rolling direction. The offset used to define yield loci in shear experiments has been chosen equal to 0.34%, corresponding to the conventional traction offset of 0.2% by a von Mises equivalence.

In order to test the validity of the model, its predictions have been compared with experimental results under additional independent tests. Figures 1 and 2 show two such validation experiments: a reversed deformation and, respectively, an orthogonal deformation, both of them following a preshear in the rolling direction. Clearly, the model predictions are in good qualitative and quantitative agreement with experimental results.

4 COMPUTATIONAL SIGNIFICANCE OF THE MODEL

We will examine now the state-update algorithm, which is the essential ingredient of the coupling between a finite element code and a constitutive model involving internal state variables.

4.1 *Rotation-compensated tensor quantities and equations*

As already mentioned in Sect. 3.2, we assume for simplicity that all tensorial, structural variables turn with the same spin $\dot{\mathbf{R}}\mathbf{R}^{-1} = \mathbf{W}$. It then proves convenient to reformulate the constitutive and evolution equations in terms of 'rotation-compensated' quantities, which will be denoted by a superposed hat. More precisely, if $\mathbf{A}$ and $\mathbf{S}$ denote as before a second-order and fourth-order tensor, respectively, then the corresponding rotation-compensated tensors, $\widehat{\mathbf{A}}$ and $\widehat{\mathbf{S}}$ will be defined by

$$A_{ij} = R_{ip}R_{jq}\widehat{A}_{pq}, S_{ijkm} = R_{ip}R_{jq}R_{kr}R_{ms}\widehat{S}_{pqrs}. \quad (27)$$

The main advantage of this transformation is that the Jauman-type derivatives of the initial tensors are related to the material time derivatives of the rotation-compensated tensors by relations similar to (27), i.e.

$$\overset{\circ}{A}_{ij}= R_{ip}R_{jq}\dot{\widehat{A}}_{pq}, \overset{\circ}{S}_{ijkm}= R_{ip}R_{jq}R_{kr}R_{ms}\dot{\widehat{S}}_{pqrs}. \quad (28)$$

It is also noteworthy that the transformation (27) preserves the norms, i.e. $|\widehat{\mathbf{A}}| = |\mathbf{A}|$, $|\widehat{\mathbf{S}}| = |\mathbf{S}|$. With this notation, the main equations presented in Sect. 3 may be rewritten as follows.
Hypoelastic equation:

$$\dot{\widehat{\mathbf{T}}} = \widehat{\mathbf{c}} : (\widehat{\mathbf{D}} - \widehat{\mathbf{D}}^p). \quad (29)$$

Yield condition:

$$F = \bar{\sigma}_e - R_0 - f|\widehat{\mathbf{S}}| = 0, \quad (30)$$

where

$$\bar{\sigma}_e = \sqrt{(\widehat{\mathbf{T}} - \widehat{\mathbf{X}}) : \widehat{\mathbf{M}} : (\widehat{\mathbf{T}} - \widehat{\mathbf{X}})}. \quad (31)$$

Associated flow rule:

$$\widehat{\mathbf{D}}^p = \dot{\lambda}\widehat{\mathbf{V}}, \quad (32)$$

with

$$\widehat{\mathbf{V}} = \frac{1}{\bar{\sigma}_e}\widehat{\mathbf{M}} : (\widehat{\mathbf{T}} - \widehat{\mathbf{X}}). \quad (33)$$

Evolution equations of the internal state variables (with $\dot{p}$ replaced by $\dot{\lambda}$):

$$\begin{aligned}
\dot{\widehat{\mathbf{M}}} &= \mathbf{0}, && (34)\\
\dot{\widehat{\mathbf{P}}} &= \dot{\lambda}C_P(\widehat{\mathbf{N}} - \widehat{\mathbf{P}}), && (35)\\
\dot{\widehat{\mathbf{X}}} &= \dot{\lambda}C_X(X_{\text{sat}}\widehat{\mathbf{N}} - \widehat{\mathbf{X}}), && (36)\\
\dot{S}_D &= \dot{\lambda}C_S[g(S_{\text{sat}} - S_D) - hS_D], && (37)\\
\dot{\widehat{\mathbf{S}}}_L &= -\dot{\lambda}C_S\left(\frac{Z}{S_{\text{sat}}}\right)^n \widehat{\mathbf{S}}_L, && (38)
\end{aligned}$$

where

$$\begin{aligned}
\widehat{\mathbf{N}} &= \frac{\widehat{\mathbf{D}}^p}{|\widehat{\mathbf{D}}^p|} = \frac{\widehat{\mathbf{V}}}{|\widehat{\mathbf{V}}|}, && (39)\\
\widehat{\mathbf{S}} &= S_D\widehat{\mathbf{N}} \otimes \widehat{\mathbf{N}} + \widehat{\mathbf{S}}_L, && (40)\\
Z &= |\widehat{\mathbf{S}}_L| = \sqrt{|\widehat{\mathbf{S}}|^2 - S_D^2}. && (41)
\end{aligned}$$

4.2 *Determination of the plastic multiplier*

For plastic loading, the plastic multiplier results from the consistency condition $\dot{F} = 0$, i.e.

$$\dot{\bar{\sigma}}_e - f|\dot{\widehat{\mathbf{S}}}| = 0. \quad (42)$$

A lengthy but straightforward calculation using Eqs (37), (38) and (41), gives

$$f|\dot{\widehat{\mathbf{S}}}| = H\dot{\lambda}, \quad (43)$$

where

$$H = \frac{fC_S}{\sqrt{Z^2 + S_D^2}}\left\{-\left(\frac{Z}{S_{\text{sat}}}\right)^n Z^2 + [g(S_{\text{sat}} - S_D) - hS_D]S_D\right\}$$

is the work-hardening modulus. Introducing now Eqs (31) and (43) into (42), and taking into account (33) and (34), yields

$$\widehat{\mathbf{V}} : (\dot{\widehat{\mathbf{T}}} - \dot{\widehat{\mathbf{X}}}) - H\dot{\lambda} = 0. \quad (44)$$

Next, by replacing (29) and (36) into (44) and solving for $\dot{\lambda}$, we deduce the expression of the plastic multiplier for plastic loading

$$\dot{\lambda} = \frac{1}{f_0}\widehat{\mathbf{V}} : \widehat{\mathbf{c}} : \widehat{\mathbf{D}}, \quad (45)$$

where

$$f_0 = \widehat{\mathbf{V}} : \widehat{\mathbf{c}} : \widehat{\mathbf{V}} + C_X(X_{\text{sat}}|\widehat{\mathbf{V}}| - \widehat{\mathbf{V}} : \widehat{\mathbf{X}}) + H, \quad (46)$$

while $\dot{\lambda} = 0$ for neutral loading, unloading, or in elastic state.

4.3 *Elastoplastic moduli*

Putting now (45) into (32) and the result obtained into (29) gives

$$\dot{\widehat{\mathbf{T}}} = \widehat{\mathbf{c}}^{ep} : \widehat{\mathbf{D}}, \quad (47)$$

where

$$\widehat{\mathbf{c}}^{ep} = \widehat{\mathbf{c}} - \frac{\alpha}{f_0}(\widehat{\mathbf{c}} : \widehat{\mathbf{V}}) \otimes (\widehat{\mathbf{c}} : \widehat{\mathbf{V}}) \quad (48)$$

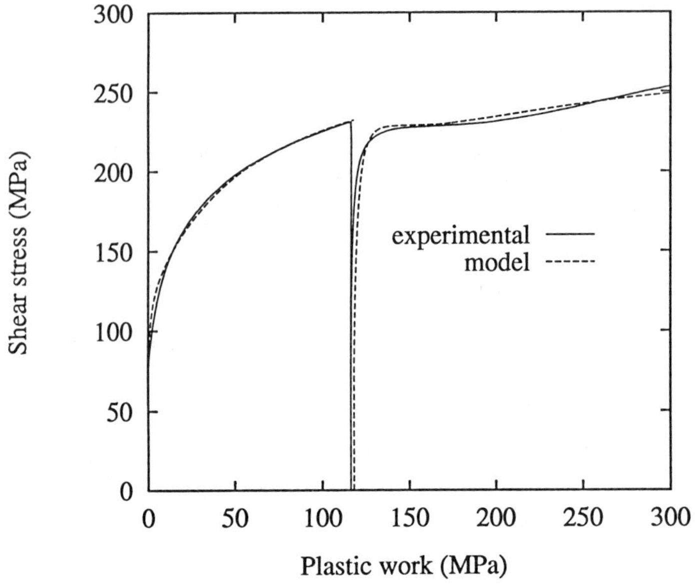

Fig. 1 Comparison of the experimental result with the theoretical prediction for the work-hardening behaviour of a prestrained steel sheet under a reversed shear. The amount of preshear is 53%.

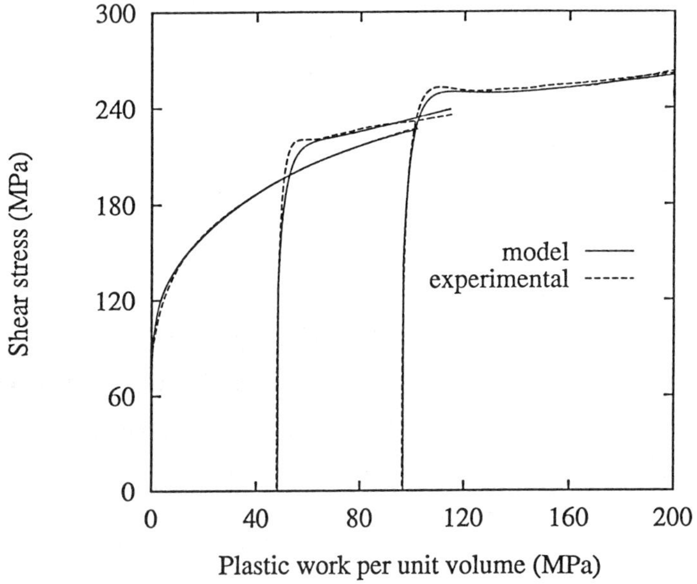

Fig. 2 Comparison of the experimental result with the theoretical prediction for the work-hardening behaviour of a prestrained steel sheet under an orthogonal shear. The amount of preshear is 31%.

are the (tangent) elastoplastic moduli. Here $\alpha = 1$ for the plastic loading and $\alpha = 0$ otherwise.

Eq. (47) can be used for calculating the (tangent) elastoplastic stiffness matrix occurring, e.g. in the rate form of the principle of virtual power. To this end, it can be also written in the more familiar form

$$\overset{\circ}{\mathbf{T}} = \mathbf{c}^{ep} : \mathbf{D}, \tag{49}$$

where now

$$\mathbf{c}^{ep} = \mathbf{c} - \frac{\alpha}{f_0}(\mathbf{c} : \mathbf{V}) \otimes (\mathbf{c} : \mathbf{V}). \tag{50}$$

4.4 *State-update algorithms*

The above equations allow a convenient update of the state variables whenever an explicit time marching scheme is being used. Indeed, suppose that the velocity field at time t has been determined, e.g. by using the rate form of the principle of virtual power. All tensor state variables are first rotation-compensated by using $\mathbf{R}(t)$. Then, Eqs (35) to (38) and (47) can be integrated by an explicit scheme to give the rotation-compensated state variables at time t. Next, by integrating the tensor equation $\dot{\mathbf{R}} = \mathbf{WR}$, we may obtain the rotation at time t. Finally $\mathbf{R}(t+\Delta t)$ is used to back-rotate the tensor state variables at time $t + \Delta t$.

Alternatively, the principle of virtual power can be written in incremental form and associated to implicit algorithms for the integration of the kinematic, constitutive, and evolution equations.

5 CONCLUSION

In this work, the anisotropic work-hardening behaviour under strain-path changes at moderately large strains has been modelled by using an internal-variable approach. Several intragranular deformation mechanisms, such as the formation of persistent dislocation structures under monotonic deformation, depolarization of preformed microstructure under stress reversal, and the interaction between preformed microstructure and microbands, have been taken into account in choosing the internal variables and their evolution equations.

Finally, the computational significance of the constitutive and evolution equations of the model has been investigated within the context of an explicit time-marching scheme.

It should be noted that in the present work only the contribution of dislocation structures and initial texture to the induced anisotropic work-hardening have been taken into account. At very large strains, however, the influence of texture evolution could be predominant. Moreover, additional physical mechanisms, e.g. recovery of dislocation structures, should be also taken into account.

REFERENCES

Bacroix, B., P. Genevois & C. Teodosiu 1994. Plastic anisotropy in low carbon steels subjected to simple shear with strain path changes. *European J.Mech.Ser.A/Solids* 13:661-675.

Bay, B., N. Hansen, D.A. Hughes & D. Kuhlmann-Wilsdorf 1992. Evolution of f.c.c. deformation structures in polyslips. *Acta Metall.*40:205-219.

Cordebois, J.P. & M. Boucher 1990. A new approach of elastic domain evolution. Theoretical and applied aspects. In S.K. Ghosh(ed), *Proc.Int.Conf. Development in Forming Technology:*1.

Ghosh, A.K. & W.A. Backofen 1973. Strain-hardening and instability in biaxially stretched sheets. *Metall.Trans.*4:1113-1123.

Hasegawa, T., T. Yakou & S. Karashima 1975. Deformation behaviour and dislocation structures upon stress reversal in polycrystalline aluminium. *Mater.Sci.Engng.*A20:267-276.

Hu, Z., E. Rauch & C. Teodosiu 1992. Work-hardening behaviour of mild steel under stress at large strains. *Int.J.Plasticity* 8:839-856.

Hu, Z. 1994. Work-hardening behaviour of mild steel under cyclic deformation at finite strains. *Acta metall.mater.*42:3481-3491.

Hu, Z. & C. Teodosiu 1995. Anisotropic work-hardening induced by microstructural evolution under strain-path changes at large strains (to be published).

Juul-Jensen, D. & D. Hansen 1987. Relations between texture and flow stress in commercially pure aluminium. In A.I. Andersen(ed), *8th Risø Int.Symp.Metall.Mat.Sci.*:353-360. Roskilde: Risø Nat.Lab.

Kocks, U.F., T. Hasegawa & R.O. Scattergood 1980. On the origin of cell walls of lattice misorentations during deformation. *Scripta Metall.*14:449.

Kuhlmann-Wilsorf, K. 1989. Theory of plastic deformation: properties of low energy dislocation structures. *Mater.Sci.Engng.*A113:1-41.

Mandel, J. 1982. Définition d'un repère privilégié pour l'étude des transformations anélastiques du polycristal. *J.Méca.Théor.Appl.*1:7-23.

Mazilu, P. & A. Meyers 1985. Yield surface description of isotropic materials after cold prestrain. *Ingenieur Archiv.*55:213-220.

Rauch, E.F., S. Hashimoto & B. Baudelet 1990. Simple shear deformation of iron-silicon single crystals. *Scripta Metall.*24:1081-1086.
Rauch, E.F. & J. Schmitt 1989. Dislocation substructures in mild steel deformed in simple shear. *Mater.Sci.Engng.*A113:441-448
Schmitt, J.H., E. Aernoutd & B. Baudelet 1985. Yield loci for polycristalline metals without texture. *Mater.Sci.Engng.*75:13-20.
Steeds, J. W. 1966. Dislocation arrangement in copper single crystals as a function of strain. *Proc.Roy.Soc.*292:343-372.
Teodosiu, C. 1989. The plastic spin: microstructural origin and computational significance. In D.R.J. Oden, E. Hinton & E. Oñate(eds), *Computational Plasticity: Models, Software and Applications:* vol.1,163-175. Swansea U.K.: Pineridge Press.
Teodosiu, C., & Z. Hu 1993. Modelling the work-hardening behaviour of mild steel under complex deformation at large strains. In A.S. Khan(ed), *Proc.Int.Symp. Plasticity and its Current Applications.* Baltimore, Maryland, USA (to be published).
Thuillier, S. & E. F. Rauch 1994. Development of microbands in mild steel during cross loading. *Acta Metall.* (to be published).

Simulation of Materials Processing: Theory, Methods and Applications, Shen & Dawson (eds)
© 1995 Balkema, Rotterdam. ISBN 90 5410 553 4

Profile extrusion and die design for viscoelastic fluids

Thanh Tran-Cong
Faculty of Engineering and Surveying, University of Southern Queensland, Toowoomba, Qld, Australia
Nhan Phan-Thien
Department of Mechanical and Mechatronic Engineering, The University of Sydney, N.S.W., Australia

ABSTRACT: A simple technique based on Boundary Element Methods (BEM) is developed to design profile extrusion dies. The technique is applied to obtain the die profiles corresponding to a square viscoelastic extrudate under different processing conditions. The results obtained compare well with available data, including our own experimental data using a commercial grade LDPE. Furthermore, the required die profiles change more rapidly in size while their relative shapes vary slowly with increasing Weissenberg number. Thus, an extrusion profile designed for a Newtonian fluid can be an excellent starting point for a viscoelastic extrusion profile.

1 INTRODUCTION

Die design and development is inherently an empirical and trial and error procedure associated with high cost, which could be considerably reduced by a careful computational design programme. One of the difficulties encountered in engineering computations for extrusion dies (Michaeli 1984) is the complex phenomenon of extrudate swell (Tran-Cong and Phan-Thien 1988a). Here, the distribution of the viscosity and the elasticity of the melt play a vital role in determining the extrudate shape (Tanner 1985). Although the study of the extrudate behaviour corresponding to a fixed die geometry (direct problem) would give an indication of the kind of die profile needed for experimental refinement, the solution to the inverse problems, i.e., the design of extrusion dies for a given extrudate profile, is highly desirable. Three-dimensional numerical simulation of direct problems is gaining attention recently (Bush and Phan-Thien 1985, Tran-Cong and Phan-Thien 1988b, 1988c, Shiojima and Shimazaki 1987, 1990, Karagiannis *et al* 1988). However, only few reported results are for viscoelastic fluids (Shiojima and Shimazaki 1987, 1990, Tran-Cong and Phan-Thien 1988c). In the case of extrusion die design (inverse problem), some results are first reported by Tran-Cong and Phan-Thien (1988a), using a Boundary Element Method. These results concern only Newtonian fluids; non-Newtonian results for the inverse problem are first obtained by Tran-Cong (1989). Further numerical and experimental results are obtained by Tran-Cong *et al* (1993). In this paper, we review the formulation and the implementation of a Boundary Element Method for the inverse die design problem for a viscoelastic fluid of the Oldroyd variety. The die design for a square extrudate is reported, with special emphasis on the convergence characteristics, with regard to increasing Weissenberg number and relaxation methods. We also report some preliminary experimental data, supporting the results of numerical simulations.

2 GOVERNING EQUATIONS AND BEM FORMULATION

The steady state, isothermal and creeping flow of incompressible viscoelastic fluids is considered. The momentum balance and continuity equations are

$$\nabla \cdot \boldsymbol{\sigma} = 0, \quad (1)$$

$$\nabla \cdot \mathbf{u} = 0, \quad (2)$$

where $\boldsymbol{\sigma}$ is the total stress tensor and $\mathbf{u}$ is the velocity vector.

In addition to the above field equations, the constitutive equation of the particular fluid is needed.

This work is concerned with viscoelastic fluids with a single relaxation time of the Oldroyd variety where the stress tensor can be written as

$$\boldsymbol{\sigma} = -p\mathbf{1} + 2\eta_s\mathbf{D} + \boldsymbol{\tau}, \qquad (3)$$

where p is the hydrostatic pressure which arises due to the incompresibility constraint (2), $\mathbf{1}$ is the unit tensor, η_s is the solvent viscosity, $\mathbf{D}$ is the rate-of-strain tensor, $\boldsymbol{\tau}$ is the extra stress tensor which is governed by constitutive equations of the Maxwell type

$$\lambda\frac{\Delta\boldsymbol{\tau}}{\Delta t} + \mathbf{R} = 0, \qquad (4)$$

in which λ is the relaxation time, $\mathbf{R}$ is model dependent, and

$$\frac{\Delta\boldsymbol{\tau}}{\Delta t} \equiv \frac{\partial\boldsymbol{\tau}}{\partial t} + \mathbf{u}\cdot\boldsymbol{\nabla}\boldsymbol{\tau} - \mathbf{L}\boldsymbol{\tau} - \boldsymbol{\tau}\mathbf{L}^T \qquad (5)$$

is the upper-convected derivative of the extra stress tensor, where $\mathbf{L}$ is the velocity gradient tensor and $\mathbf{L}^T$ denotes its transpose.

In BEM applications, Eq. (3) in conjunction with Eq. (4) is rewritten as

$$\boldsymbol{\sigma} = -p\mathbf{1} + 2\eta_p\mathbf{D} + \boldsymbol{\varepsilon}, \qquad (6)$$

where $2\eta_p\mathbf{D}$ represents a linear part of the stress tensor and $\boldsymbol{\varepsilon}$ the remaining. The linear part can be chosen arbitrarily, but a good choice would be either the solvent stress, or the total Newtonian stress (the extra stress in the limit of slow flow). Then the set of governing equations (1), (2) and (6) is recast in integral form (Bush and Tanner 1983)

$$\begin{aligned} C_{ij}(\mathbf{x})u_j(\mathbf{x}) &= \int_{\partial\mathbf{D}} u^*_{ij}(\mathbf{x},\mathbf{y})t_j(\mathbf{y})d\Gamma(\mathbf{y}) \\ &\quad - \int_{\partial\mathbf{D}} t^*_{ij}(\mathbf{x},\mathbf{y})u_j(\mathbf{y})d\Gamma(\mathbf{y}) \\ &\quad - \int_{\mathbf{D}} \varepsilon_{jk}(\mathbf{y})\frac{\partial u^*_{ij}(\mathbf{x},\mathbf{y})}{\partial x_k}d\Omega(\mathbf{y}) \qquad (7) \end{aligned}$$

where $\mathbf{D}$ is the open domain with connected bounding surface $\partial\mathbf{D}$, $\mathbf{x},\mathbf{y} \in \mathbf{D}$, $u_j(\mathbf{y})$ is the j-velocity component at $\mathbf{y}$, $t_j(\mathbf{y})$ is the j-component of boundary traction at $\mathbf{y}$, $\varepsilon_{jk}(\mathbf{y})$ is the jk-component of $\boldsymbol{\varepsilon}$ at $\mathbf{y}$, $u^*_{ij}(\mathbf{x},\mathbf{y})$ is the i-component of velocity field at $\mathbf{x}$ due to a "Stokeslet" in j-direction at $\mathbf{y}$ and $t^*_{ij}(\mathbf{x},\mathbf{y})$ is its associated traction. $C_{ij}(\mathbf{x})$ depends on local geometry, $C_{ij}(\mathbf{x}) = \delta_{ij}$ if $\mathbf{x} \in \mathbf{D}$ and $C_{ij}(\mathbf{x}) = \frac{1}{2}\delta_{ij}$ if $\mathbf{x} \in \partial\mathbf{D}$ and $\partial\mathbf{D}$ is a smooth surface. Details of $u^*_{ij}(\mathbf{x},\mathbf{y})$ and $t^*_{ij}(\mathbf{x},\mathbf{y})$ for three dimensional problems are given elsewhere (e.g., Tran-Cong and Phan-Thien 1988b).

3 METHOD OF NUMERICAL SOLUTION

The application of equation (7) to solve a number of direct problems is described in detail previously (Tran-Cong and Phan-Thien 1988b, 1988c). Here the procedure is briefly recaptured and extended for the inverse problem. Essentially, the problem is solved numerically by decoupling the non-linear effects, which are treated as small perturbations in an otherwise linear solution. In this method, the last integral on the right hand side of Eq. (7), i.e.,

$$\int_{\mathbf{D}} \varepsilon_{jk}(\mathbf{y})\frac{\partial u^*_{ij}(\mathbf{x},\mathbf{y})}{\partial x_k}d\Omega(\mathbf{y}) \qquad (8)$$

is considered as a pseudo-body-force in a linear problem. Hence for a given stress field, Eq. (8) is evaluated and Equation (7) is discretised over the boundary $\partial\mathbf{D}$ to yield a system of linear algebraic equation of the form

$$[\mathbf{G}]\{\mathbf{t}\} = [\mathbf{H}]\{\mathbf{u}\} + \{\mathbf{b}\}, \qquad (9)$$

where $\{\mathbf{t}\}$, $\{\mathbf{u}\}$, $\{\mathbf{b}\}$ are the global nodal traction, velocity and body force vectors respectively. The unknown boundary tractions $\mathbf{t}$ and velocity $\mathbf{u}$ in Eq. (9) are rearranged and the system is solved by standard Gauss elimination. Equation (7) is then reapplied to evaluate the velocity field in $\mathbf{D}$. The last step in the iterative procedure is to compute the stress field and the whole procedure described above is repeated until a convergence (or the lack of) is obtained. A pseudo-time "marching" technique is employed to obtain the extra stress field. The constitutive Eq. (4) governing the extra stress is written as

$$\mathbf{f} \equiv \frac{\partial\boldsymbol{\tau}}{\partial t} = -\mathbf{u}\cdot\boldsymbol{\nabla}\boldsymbol{\tau} + \mathbf{L}\boldsymbol{\tau} + \boldsymbol{\tau}\mathbf{L}^T - \frac{1}{\lambda}\mathbf{R}, \qquad (10)$$

Given an initial guess of the solution, the iterative procedure consists of the following steps:

1. For a given stress field, Eq. (8) is evaluated, and Eq. (9) is assembled and solved for the unknown boundary tractions and velocities;

2. The free surface is computed by a path-line method (Tran-Cong and Phan-Thien 1988b) according to the boundary velocities obtained;

3. The difference in the cross section of the extrudate found in step 2 and the required one is then computed and all the pathlines are translated by the corresponding amount to make the final cross section of the extrudate the required one (Tran-Cong and Phan-Thien 1988a), and the die profile is modified correspondingly;

4. Eq. (7) is then reapplied to compute the velocity field in the domain **D**;

5. The kinematics are kept constant while the stresses are marched using Eq. (10) according to a chosen scheme. In this study a first order Euler scheme is used, i.e.,

$$\boldsymbol{\tau}^{n+1} = \boldsymbol{\tau}^n + \Delta t\, \mathbf{f}^n, \qquad (11)$$

where $\boldsymbol{\tau}^{n+1}$ is the extra stress at step $(n+1)$, $\boldsymbol{\tau}^n$, $\mathbf{f}^n$ is the extra stress and its partial time derivative given by Eq. (10) at step n, Δt is the pseudo-time step. n is increased until little change is observed (in this study, $\Delta t = 0.01\lambda$, where λ is the relaxation time of the fluid, and the tolerance for the extra stresses is 10^{-5}). Other higher-order Runge-Kutta schemes, including some fully implicit schemes have been used (Zheng 1992), yielding essentially the same results, but with larger time steps.

A global convergence measure (CM) in the kinematics is defined by

$$CM = \frac{\{\sum_1^N \sum_{i=1}^3 (u_i^n - u_i^{n-1})^2\}^{\frac{1}{2}}}{\{\sum_1^N \sum_{i=1}^3 (u_i^n)^2\}^{\frac{1}{2}}}, \qquad (12)$$

where u_i is the i-velocity component at a node, N is the total number of nodes, n is the iteration number. The computation is stopped when the convergence measure is less than 10^{-4}.

The above technique gives a solution to the problem with a given set of parameters and boundary conditions (viscosity, elasticity, processing speed ...). In order to obtain a solution for arbitrary set of parameters, it is necessary to have a reasonably close guess to avoid divergence. For example, to investigate the viscoelastic effect, which is the case in this paper, the Weissenberg number (Wi, defined in a later section) is increased at small discrete steps such that the Newtonian solution can be used as the first guess for a solution corresponding to the smallest Wi. Subsequent solutions use the previous solution corresponding to the previously smaller Wi as the first guess at the beginning of the iterative cycle.

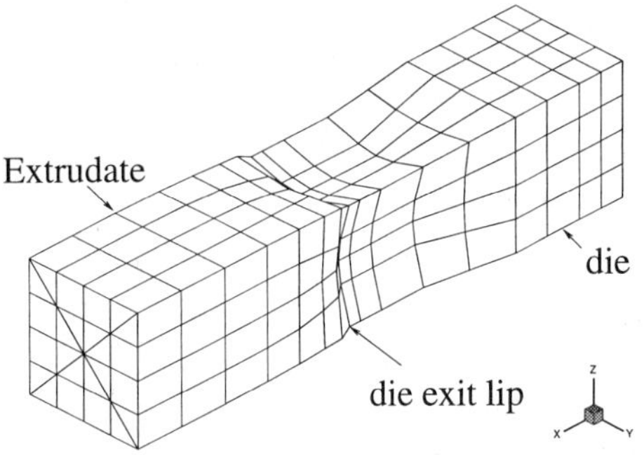

Figure 1: The die and extrudate shape for Wi=0.24

4 BOUNDARY CONDITIONS

The experience with Newtonian profile extrusion (Tran-Cong and Phan-Thien 1988a) suggests that a "taper" design is more appropriate. In this approach, the entry profile is chosen such that the best possible approximation to the velocity boundary condition can be found. The entire length of the die is obtained by blending the entry section with the required exit section. Thus only the transitional and the exit lip region are changing in the design process. The customary no slip boundary conditions at the wall and the traction free extrudate surface are assumed in all cases. The choice of Newtonian inlet velocity profile is justified *a posteriori* (Tran-Cong and Phan-Thien 1988c) when fully developed flow is observed at some distance downstream of the inlet. At the downstream section of the discretized flow domain, the plug flow condition is specified, i.e., zero traction in the flow direction and zero radial velocity components.

5 NUMERICAL RESULTS FOR A SQUARE VISCOELASTIC EXTRUDATE

5.1 *A Design Problem*

We now consider an illustrative problem of designing a die to produce a square extrudate. Owing to symmetry, only one eighth of the flow domain need be discretised. Then the discretised mesh contains a total of 156 nodes and 38 boundary ele-

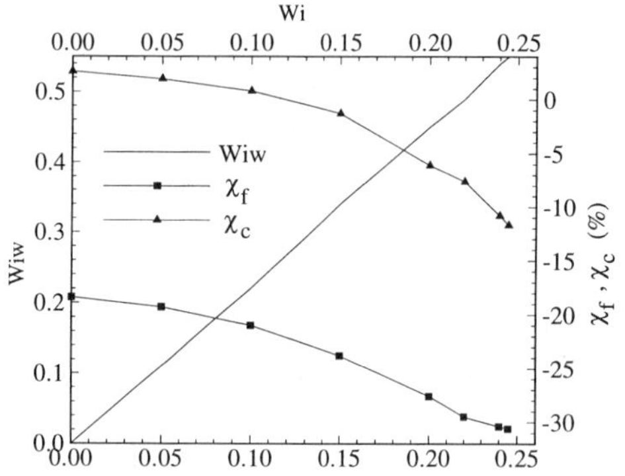

Figure 2: Die geometry as a function of Wi number. χ_f is the percentage reduction in the size of the die relative to the required extrudate size at the middle of the flat face of the extrudate; χ_c is the corresponding value at the corner.

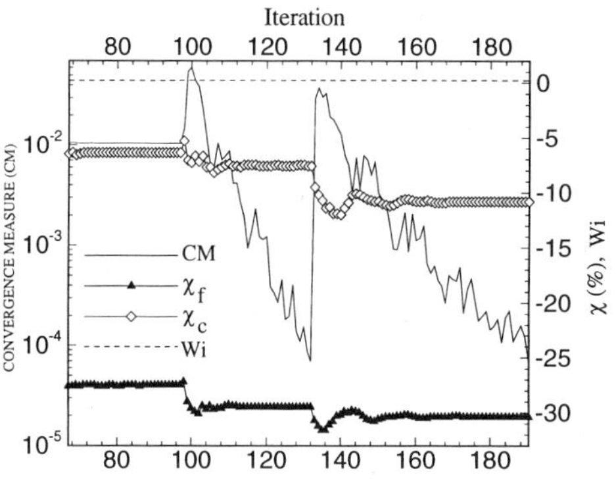

Figure 3: Convergence characteristics for Wi = 0.22 to 0.24.

ments (8-node quadratic quadrilaterals and 6-node quadratic trilaterals). The die length is 5 and the extrudate length is 4. Figure 1 shows an isometric view of a final mesh. On a DEC3000 M800, one iteration takes approximately 30 minutes and convergence at $Wi = 0.22$ requires about 40 iterations.

5.2 *A Model Viscoelatic Material*

The method discussed above is implemented and tested for a square extrudate profile using a model viscoelastic material. In the following discussion, the characteristic length, a, is half the side of the die entry cross section; the characteristic speed, U, is the centreline (maximum) speed at the inlet; the characteristic time, λ, is the constant relaxation time; η_0 is the zero shear rate viscosity; η_m is the polymer contributed viscosity. For the MPTT model, $\mathbf{R}$ (in Eq. (4)) is given by (Phan-Thien and Tanner 1977, Phan-Thien 1978, 1984)

$$\mathbf{R} = g\boldsymbol{\tau} + \lambda\xi(\mathbf{D}\boldsymbol{\tau} + \boldsymbol{\tau}\mathbf{D}^T) - 2\eta_m(\dot{\gamma})\mathbf{D}, \quad (13)$$

where

$$g = \left[1 + \epsilon\frac{\lambda}{\eta_0}\mathrm{tr}(\boldsymbol{\tau})\right], \quad (14)$$

$$\dot{\gamma} = \sqrt{2\mathrm{tr}(\mathbf{D}^2)}, \quad (15)$$

$$\eta_m(\dot{\gamma}) = \eta_0\frac{1 + \xi(2-\xi)\lambda^2\dot{\gamma}^2}{(1 + \Gamma^2\dot{\gamma}^2)^{\frac{1-n}{2}}}, \quad (16)$$

in which ϵ, ξ, n are dimensionless parameters and Γ is a time constant. The dimensionless variables are given by

$$t' = \frac{t}{\lambda},\ \mathbf{x}' = \frac{\mathbf{x}}{\mathbf{a}},\ \mathbf{u}' = \frac{\mathbf{u}}{\mathbf{U}},\ \boldsymbol{\tau}' = \frac{\boldsymbol{\tau}}{\eta_0\frac{U}{a}},\ \eta' = \frac{\eta_m}{\eta_0 + \eta_s},$$

where t is time, $\mathbf{x}$ is the position vector, $\mathbf{u}$ is the velocity vector, $\boldsymbol{\tau}$ is the extra stress tensor and η is the viscosity.

In this study, $\xi = 0$ and $n = 1$ were chosen. Then the constitutive Eq. (4) is given in dimensionless form by (dropping the primes)

$$\frac{\partial\boldsymbol{\tau}}{\partial t} = (2\eta\mathbf{D} - g\boldsymbol{\tau}) - Wi(\mathbf{u}\cdot\boldsymbol{\nabla}\boldsymbol{\tau} - \mathbf{L}\boldsymbol{\tau} - \boldsymbol{\tau}\mathbf{L}^T), \quad (17)$$

where $g = 1 + \epsilon\, Wi\, \mathrm{tr}(\boldsymbol{\tau})$, and $Wi = \lambda U/a$ is the Weissenberg number.

5.3 *Discussion of Results*

The only dimensionless parameter to be varied in this study is the Weissenberg number, Wi, defined above. In this study the inlet velocity is kept constant and Wi is increased by increasing the relaxation time λ. However, if the Weissenberg number was based on a wall shear rate at a point far upstream, Wi_w say, as is often the case reported in the literature, the computed wall shear rate $\dot{\gamma}_w = \sqrt{2\mathrm{tr}(\mathbf{D}^2)}$ at the mid-point of a flat face of the die at the inlet shows that $Wi_w \approx 2.25\, Wi$. Both Wi and Wi_w are presented for comparison. Table 1 and Figure 2 show two representative measures of the final die profile for the indicated Wi

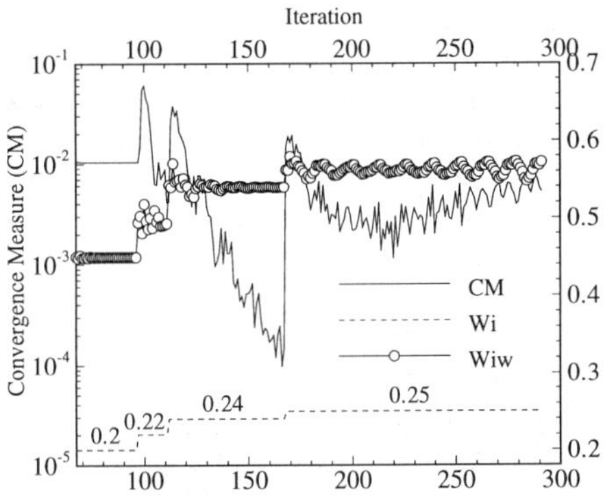

Figure 4: Convergence characteristics for Wi = 0.22 to 0.25. Convergence cannot be achieved if step size in Wi is large.

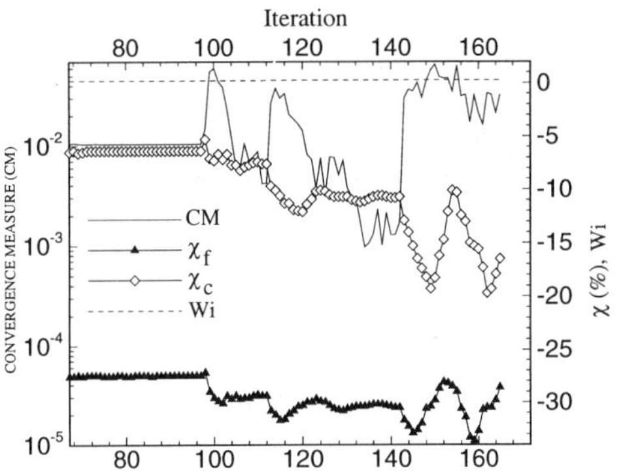

Figure 5: Convergence cannot be achieved if the quality of the previous solution is poor, by setting a low tolerance value.

number. Note that χ's are defined as the percentage difference, in the radial direction, between the die and the extrudate geometry relative to the extrudate geometry (a positive value of χ represents a swelling, and negative value, a shrinkage). Here χ_f is the maximum value of χ measured at the middle of the flat face of the extrudate (always negative at all Weissenberg numbers), and χ_c is the minimum value of χ measured at the corner of the extrudate (which can be positive at low Wi). Figure 2 indicates that the rate at which the die geometry changes in response to increasing in Wi. Initially, at low values of Wi, the magnitudes of the χ_f and χ_c increase in about the same proportion. Thus the shape of the die profile is only weakly dependent on the Weissenberg number. At larger values of Wi, greater than about 0.24, however, χ_f starts to decrease more slowly than χ_c, leading to a different shape of the die profile.

In this study the effects of the step size used to increase Wi, the degree of convergence in the preceeding iteration and free surface relaxation on the convergence are examined.

First we examine the effect of the free surface relaxation. For the case $Wi = 0.2$ no relaxation is applied (i.e., the free surface is updated using the currently obtained velocity field). It is found that convergence is extremely difficult as shown in Figure 3 (iteration 67 to iteration 96). It can be seen there that small oscillations persist, and the convergence measure cannot be reduced below 10^{-2}. Thus in subsequent iterations a relaxation is applied to the updating of the free surface. This is done simply by averaging the velocity fields between the current iteration and the preceeding one.

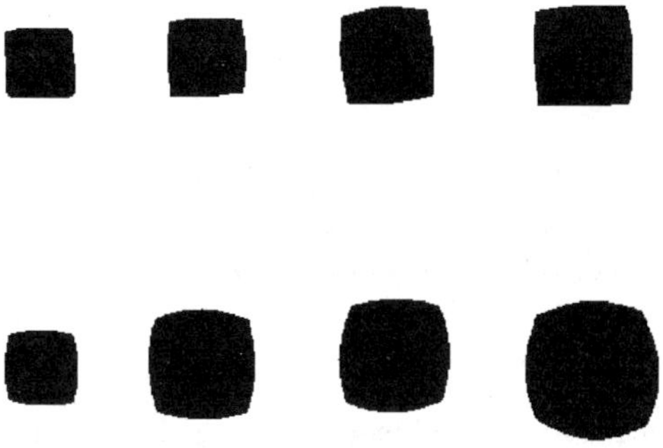

Figure 6: Some experimental extrudate shapes for a commercial grade LDPE obtained with a die specially designed by BEM (for a Newtonian fluid), and for a square die of the same size.

Table 1: Die shape relative to a square extrudate as a function of Wi: MPTT fluid ($\epsilon = 0.01$; $\xi = 0$; $n = 1$; $\lambda = 1$). χ_f is percentage reduction in the size of the die relative to the required extrudate at the middle of the flat face; χ_c is the corresponding value at the corner.

Wi	Wi_w	$\chi_f(\%)$	$\chi_c(\%)$
0	0	-18.4	+2.6
0.05	0.11	-19.3	+1.9
0.10	0.22	-21.0	+0.8
0.15	0.34	-23.8	-1.3
0.2	0.45	-27.6	-6.1
0.22	0.495	-29.5	-7.6
0.24	0.539	-30.4	-10.8
0.245	0.551	-30.6	-11.7

When an increment in the Weissenberg number is applied, the convergence first increases, then decreases to a value below $O(10^{-4})$ after some 30 iterations (Figure 3).

Secondly, we examine the effect of Wi step size. Initially, Wi was increased from 0 to 0.2, in steps of 0.05. In this range of Wi the convergence of the present method is reasonable with a step increase in Wi of 0.05. However, for Wi above 0.2, it is necessary to reduce the step size to 0.02 in order to achieve convergence, see Figure 4. This is fine until $Wi = 0.24$ above which a step size of 0.005 is necessary for convergence.

Finally, we attempt to economize by imposing a less stringent convergence tolerance, of $O(10^{-2})$, at intermediate Wi numbers (considered to be of no interest) in order to proceed quickly to the final value of Wi. This brings only limited success at low Wi numbers and the failure is clear in Figure 5, where a divergent trend is already seen.

6 SOME EXPERIMENTAL RESULTS

It is desirable to verify qualitatively the above numerical results at least at low Weissenberg numbers. For this purpose some preliminary experiments were carried out on a commercial grade LDPE using die profiles which are designed for square and triangular Newtonian extrudate. Given the numerical results of Table 1, which are also depicted in figure 2, it is expected within a small range of Wi's, less than 0.24, the "Newtonian" die profile should be adequate. This is confirmed by the experimental results for both square and triangular extrudate profile (Tran-Cong 1989, Tran-Cong *et al* 1993). Figure 6 shows the actual profiles of an LDPE extrudate obtained using a die specially designed for a Newtonian fluid (top row), and those obtained with a square die of the same cross dimensions (bottom row). The extruder was operated at a minimum speed to achieve a condition as close to Newtonian as possible. It is clear that the extrudates from the specially designed dies are close to being square; the asymmetry in the extrudate shape is due to an imperfect temperature and gravity control.

7 CONCLUSION

The die design method developed here is potentially a useful tool in extrusion industries. The prediction of the die profile, even if it is only qualitative, could reveal important trends in the behaviour of the material being processed. The fact that the shape of the extrudate profile only weakly depends on the Weissenberg number suggests that a die design using the Newtonian fluid model would form a good starting point of the design process. Further scaling of the die geometry is necessary, by an experimental process at a given Weissenberg number, to control the overall size of the extrudate. Thus the die design and development cost could be significantly reduced. Further work on die design for complex extrudate shape is being carried out.

ACKNOWLEDGEMENT

This work is supported by Australian Research Council Grants.

REFERENCES

Bush, M.B. and Phan-Thien, N. 1985. Three dimensional viscous flow with a free surface: flow out of a long square die. *J. Non-Newt. Fluid Mech.* 18:211-218.

Bush, M.B. and Tanner, R.I. 1983. Numerical solution of viscous flows using integral equation method. *Intl. J. Num. Meth. Fluids* 3:71-92.

Karagiannis, A., Hrymak, A.N. and Vlachopoulos, J. 1988. Three-dimensional extrudate swell of creeping Newtonian jets. *AIChE Journal* 34:2088-2094.

Michaeli, W. 1984. *Extrusion Dies Design and Engineering Computations.* Munich:Hanser.

Phan-Thien, N. 1978. A non-linear network viscoelastic model. *J. Rheology* 22:259-283.

Phan-Thien, N. 1984. Squeezing a viscoelastic liquid from a wedge: an exact solution. *J. Non-Newt. Fluid Mech.* 16:329-345.

Phan-Thien, N. and Tanner, R.I. 1977. A new constitutive equation derived from network theory. *J. Non-Newt. Fluid Mech.* 2:353-365.

Shiojima, T. and Shimazaki, Y. 1987. Three dimensional finite element analyses for a Maxwell fluid using the penalty function method. In G.N. Pande and J. Middleton (eds), *NUMETA'87*, Swansea, Martinus Nijhoff. 121-126.

Shiojima, T. and Shimazaki, Y. 1990. Three dimensional finite element method for extrudate swells of a Maxwell fluid. *J. Non-Newt. Fluid Mech.* 34:269-288.

Tanner, R.I. 1985. *Engineering Rheology.* Oxford:Claredon Press.

Tran-Cong, T. 1989. Boundary element method for some three-dimensional problems in continuum mechanics, PhD thesis, The University of Sydney, Sydney.

Tran-Cong, T. and Phan-Thien, N. 1988a. Die design by a Boundary Element method. *J. Non-Newt. Fluid Mech.* 30:37-46.

Tran-Cong, T. and Phan-Thien, N. 1988b. Three dimensional study of extrusion processes by Boundary Element Method. Part 1: an implementation of high order elements and some Newtonian results. *Rheologica Acta* 27:21-30.

Tran-Cong, T. and Phan-Thien, N. 1988c. Three dimensional study of extrusion processes by Boundary Element Method. Part 2: extrusion of viscoelastic fluid. *Rheologica Acta* 27:639-648.

Tran-Cong, T., Phan-Thien, N. and Paffard, P. 1993. Profile extrusion of viscoelastic fluids: Numerical and experimental results. In D.A. Siginer *et al* (eds), *Developments in Non-Newtonian Flows* AMD-Vol.175:5-10. New York:ASME.

Zheng, R. 1991. BEM for some problems in Fluid Mechanics and Rheology. PhD Thesis, The Univesity of Sydney, Sydney.

Simulation of Materials Processing: Theory, Methods and Applications, Shen & Dawson (eds)
© 1995 Balkema, Rotterdam. ISBN 90 5410 553 4

Advances in fiber orientation prediction for injection-molded composites

Charles L. Tucker III & Brent E. VerWeyst
Department of Mechanical and Industrial Engineering, University of Illinois, Urbana, Ill., USA

ABSTRACT: Injection-molded composites contain complex patterns of fiber orientation, which are caused by flow during mold filling. Fiber orientation affects the mechanical properties and dimensional stability of the part. Numerical simulations of mold filling now include prediction of the fiber orientation state, so that designers can anticipate the effect of orientation on part performance. This paper surveys the prediction methods for fiber orientation, with emphasis on the use of a second-order tensor as a microstructural state variable. A recent advance in this field, the orthotropic closure approximation of Cintra and Tucker (1995), is included in an injection-molding calculation and improves the agreement between predicted and measured fiber orientation in injection-molded parts.

1 INTRODUCTION

Injection-molded composites are made by combining short reinforcing fibers with a thermoplastic and injecting the mixture into a closed mold. The flow pattern during mold filling creates preferential orientation of the fibers, which become fixed feature in the part when the polymer solidifies. This fiber orientation pattern represents a type of non-homogeneous microstructure that significantly influences the mechanical properties and performance of the molded part. Of particular interest is non-uniform shrinkage and warping, which can result from the anisotropic stiffness and thermal expansion of the material.

Recent years have seen much effort devoted to the prediction of flow-induced orientational microstructure in injection-molded composites. Several comprehensive software packages for injection molding now offer modules to predict fiber orientation, and other modules to translate the final orientation pattern into thermo-mechanical properties (Friedl and Brouwer, 1991; Gupta and Wang, 1993; Henry, Kjeldsen and Kennedy, 1994; Randall and Chiang, 1994; Crochet, Dupret and Verleye, 1994). This paper summarizes some of the important aspects of microstructure prediction in the context of injection molding and presents recent advances in the field. Readers interested in a detailed review of the subject are referred to Tucker and Advani (1994).

2 THEORY

2.1 Fiber orientation

The orientational behavior of a single, rigid, axisymmetric particle in a Newtonian fluid at low Reynolds number was first solved by Jeffery (1922). The fibers in injection-molded composites are typically circular cylinders with an average length-to-diameter ratio between 10 and 50, but the typical volume fractions of 10 to 20% make the suspension far from dilute. Folgar and Tucker (1984) proposed modifying Jeffery's equation with an additional term to represent the effect of fiber-fiber interactions on the orientational motion. This term has the form of an isotropic rotary diffusion; that is, interactions between fibers cause randomization and prevent the fibers from becoming perfectly aligned. The rotary diffusivity is assumed to be proportional to the scalar strain rate $\dot{\gamma}$, with a dimensionless proportionality constant C_I called the interaction coefficient. The resulting equation for the motion of a single fiber is

$$\dot{p}_i = -\frac{1}{2}\omega_{ij}p_j + \frac{1}{2}\lambda\left(\dot{\gamma}_{ij}p_j - \dot{\gamma}_{kl}\,p_k p_l p_i\right) - \frac{C_I\dot{\gamma}}{\psi}\frac{\partial\psi}{\partial p_i} \tag{2.1}$$

Here p_i are the Cartesian components of a unit vector aligned with the fiber symmetry axis, $\dot{\gamma}_{ij}$ and ω_{ij} are the rate-of-deformation and vorticity tensors using the definitions of Bird, et al. (1987), λ is a constant that depends on the length to diameter ratio of the fibers, and $\psi(p_i, t)$ is a probability density function for fiber orientation.

The real composite has many different fiber orientations p_i in any given region, so one can adopt a continuum model and treat the probability density function ψ as a smooth function of position. Solutions for ψ at a single spatial point can be computed with some effort, but solutions for the many points required to model a typical injection-molded part would be prohibitively expensive.

To make the computation practical, the density function ψ is replaced by a second-order tensor that represents the second moments of ψ (Advani and Tucker, 1987). This is defined as

$$a_{ij} = \int p_i p_j \, \psi(p_k) \, dp_k \tag{2.2}$$

Called an orientation tensor, structure tensor, or conformation tensor, a_{ij} contains partial but useful information about the fiber orientation state.

Combining eqns. (1) and (2) with a conservation statement about the motion of fibers between different orientations produces an evolution equation for the orientation tensor:

$$\begin{aligned} \frac{Da_{ij}}{Dt} &= -\frac{1}{2}(\omega_{ik} a_{kj} - a_{ik} \omega_{kj}) \\ &+ \frac{1}{2}\lambda (\dot{\gamma}_{ik} a_{kj} + a_{ik} \dot{\gamma}_{kj} - 2\dot{\gamma}_{kl} a_{ijkl}) \\ &+ 2C_I \, \dot{\gamma} \, (\delta_{ij} - 3a_{ij}) \end{aligned} \tag{2.3}$$

where δ_{ij} is the unit tensor and D/Dt represents the material derivative. Note that a_{ij} plays the role of a microstructural state variable and the model has a rate form: the fiber orientation state a_{ij} convects with the fluid and evolves according to the current state and the instantaneous strain rate and vorticity.

2.2 *Injection Mold Filling*

Equation (2.3) has become the standard model for predicting fiber orientation in injection-molded parts. The velocity field during filling is calculated using a generalized Hele-Shaw formulation (Hieber and Shen, 1980), which includes shear thinning and temperature-dependent viscosity, but not melt elasticity. The extended lubrication approximation is accurate for the thin-cavity geometries typical of injection-molded parts.

Most models assume that the fiber orientation pattern does not alter the velocity field, though in all cases the velocity field drives fiber orientation. This "decoupled" assumption works well in many cases and can be supported as the proper limit of the coupled theory for flow in a sufficiently narrow gap (Tucker, 1991). Calculations with two-way coupling between flow and orientation are beginning to appear (Ranganathan and Advani, 1993), and they show that coupling effects are mainly important near the gate and at other locations where the cross-section changes rapidly.

A variety of numerical methods have been used to solve the flow and orientation equations. Bay and Tucker (1992a) wrote a 2-D finite difference code to model two simple geometries: a long rectangular strip and a center-gated disk. Finite element-based codes have been developed for more general part shapes by Friedl and Brouwer (1991), Gupta and Wang (1993), and Crochet, Dupret and Verleye (1994). A variety of schemes are used to handle the movement of the melt front as it progresses across the initially-empty cavity.

2.3 *Status of Current Predictions*

Fiber orientation can be measured in solid parts by cutting and polishing cross-sections and measuring the orientations of many fibers on the cross-section (Fischer and Eyerer, 1988; Bay and Tucker, 1992c; Hine et al., 1993). We adopt here the convention that the 1 axis is along the flow direction, the 2 axis is across the flow in the plane of the part, and the 3 axis is across the thickness of the part.

Predictions based on eqn. (2.3) show all of the right qualitative behavior and much of the quantitative behavior of the real parts. First, the fibers all lie nearly parallel to the 1-2 plane, so the tensor component a_{33} is usually very small.

Second, a traverse across the thickness of the part shows considerable variation in the nature and direction of fiber orientation. Near the surface of the part there is always a thick shell of fibers aligned mostly in the flow direction. In this layer one typically sees a_{11} equal to 0.8 to 0.9; $a_{11} = 1.0$ would represent perfect alignment in the flow direction, but fiber-fiber interactions keep this from occurring.

Third, as one continues across the thickness there is a transition region, leading to a core of fibers near the midplane that can have a very different orientation. If the in-plane flow is diverging, as in a center-gated disk, then the core fibers will align transverse to the flow (a_{11} close to zero). This is a response to the strong cross-flow stretching deformation of a radially diverging flow. If the in-plane flow is parallel, as in a rectangular strip, then the core will have a random-in-plane orientation (a_{11} close to 0.5). The thickness of the core depends on the gapwise velocity profile, which is determined by heat transfer effects and the dependence of polymer viscosity on temperature.

Fourth, some parts have a thin skin of more random orientation very close to the surface. This skin develops only for slow filling speeds.

The interaction coefficient C_I has a strong

influence on the predictions, but there is no theory to predict its value. C_I is known to vary with fiber volume fraction and aspect ratio (Bay, 1991). Usually C_I is adjusted to match the experimental results for the shell orientation in a long strip, this region having essentially the steady-state orientation for simple shear flow. This value is used to predict orientation in any other flow for the given material.

A careful comparison between predictions and experiments (Bay and Tucker, 1992b) shows that the calculated patterns have all of the desired trends, but the quantitative agreement is only fair. A major shortcoming is that the model predicts a far thinner core in the center-gated disk than is observed in experiments. Also the shell in the disk is not as highly aligned as in the strip, though the calculations say the two parts should have the same degree of alignment in the shell. In all cases the model over-predicts the value of a_{13}, a component that indicates an out-of-plane tilt of the principal axes of orientation, though even the predicted value is small.

While the model described so far has shown enough promise to be incorporated into commercial software, there is still room to improve its quantitative accuracy. Recent work has concentrated on improving the model.

3 RECENT ADVANCES

3.1 Closure Approximations

A key issue with the use of the second-order tensor a_{ij} as a microstructural state variable is the closure problem. Note that the fourth-order orientation tensor, defined as

$$a_{ijkl} = \int p_i p_j p_k p_l \, \psi(p_m) \, dp_m \qquad (3.1)$$

appears in eqn. (2.3). One can develop a similar equation of change for a_{ijkl}, but the sixth-order tensor appears there. The solution is to approximate the fourth-order tensor in terms of the second-order tensor. Any such formula is called a closure approximation, and has the form

$$a_{ijkl} = f(a_{mn}) \qquad (3.2)$$

This closes the equation set with eqn. (2.3).

A variety of closure approximations have been tested by Advani and Tucker (1990), but none were found to be highly accurate for all flows. Bay and Tucker (1992b) suggested that the closure approximation was the major source of error in predictions of fiber orientation in injection moldings.

A new generation of closure approximations has recently been developed, including the orthotropic closures of Cintra and Tucker (1995) and the natural closure of Verleye and Dupret (1995). These formulae are much more accurate in simple, homogeneous flows than any previous closure, and perform well over a wide variety of flow fields and orientation states.

The orthotropic closures (Cintra and Tucker, 1995) are based on three ideas. First, in any closure the approximate fourth-order tensor must be orthotropic, with the same principal axes as the second-order tensor. Here "orthotropic" has the same meaning as in the mechanics of composite materials: the material possesses three mutually perpendicular planes of symmetry. In the orthotropic closure method one finds the principal axes and principal values of a_{ij}, forms the corresponding approximate a_{ijkl} in those axes, and then transforms back to the problem axes. Orthotropy requires that many of the components of a_{ijkl} in the principal axes be zero.

The second insight for orthotropic closures is that the remaining non-zero components of a_{ijkl} must be functions of the eigenvalues of a_{ij}. Using additional symmetry and trace requirements, the closure problem reduces to approximating three scalar coefficients in terms of two eigenvalues. These are approximated as low-order polynomials.

The third insight is that a variety of exact values for these coefficients can be computed by solving for the complete probability density function $\psi(p_i, t)$. Such solutions are possible for single material points, but too expensive for modeling complete part shapes. Instead, these calculations are done off line, one time only. This data is then input to a least-squared fitting procedure, which generates the polynomial coefficients for the orthotropic closure.

The natural closure of Verleye and Dupret (1995) is based on similar insights, but their closure is fitted to an exact solution for $C_I = 0$ and $\lambda = 1$, whereas the orthotropic closures are matched to numerical solutions for typical values of C_I.

Figure 1 compares solutions of eqn. (2.3) using different closures to a numerical solution for the distribution function ψ. Two components are shown, a_{11} and a_{13}, and the flow field is simple shear flow with a velocity distribution $v_1 = Gx_3$, where G is the shear rate.

The initial condition for this figure is random-in-space, so that $a_{11} = a_{22} = a_{33} = 1/3$. As the suspension is sheared the fibers tend to align in the direction of shearing, and a_{11} increases. There is also a rise in a_{13}, as the fibers collect near an axis at 45° to the shearing, followed by a fall as the fibers lie down in the 1-2 plane. Note that the orthotropic closure does an excellent job of reproducing the actual behavior (the distribution function values). The other curves use the hybrid closure (Advani and Tucker, 1990).

Many such tests have been conducted using different flows and different closures (Cintra and Tucker, 1995). In all cases the orthotropic closure is very accurate and represents a dramatic

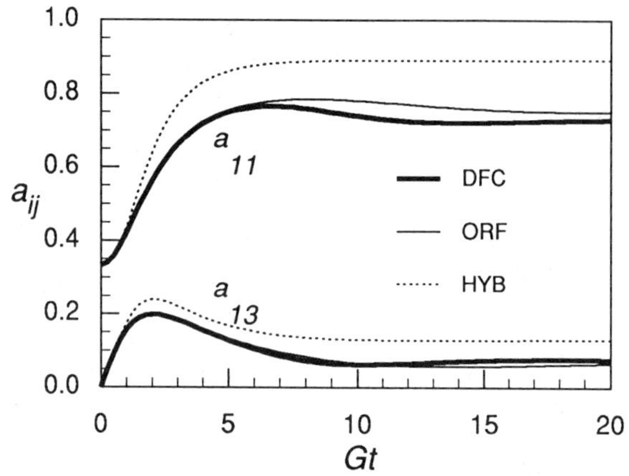

Figure 1. Orientation tensor components in simple shear flow with orthotropic (ORF) and hybrid (HYB) closures, compared to direct integration of distribution function (DFC).

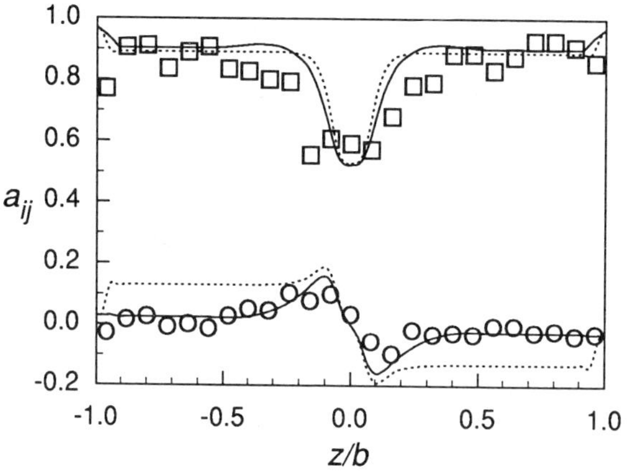

Figure 2. Predicted and measured orientation in an end-gated strip at $x/b = 34$.
Symbols: □ = a_{11} ;O = a_{13}; ——— orthotropic closure; .... hybrid closure.

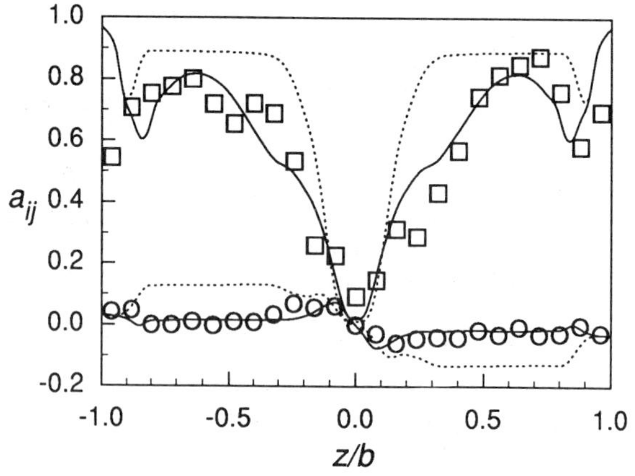

Figure 3. Predicted and measured orientation in a center-gated disk at $r/b = 22.8$.
Symbols as in fig. 2.

improvement over earlier forms. The natural closure is similarly much better than previous forms, but is not quite as accurate as the orthotropic closure when an interaction coefficient is used.

3.2 Injection Molding Predictions

To test the influence of the closure on fiber orientation predictions, we have incorporated the orthotropic closure into an injection mold filling simulation. We used the ORIENT software (Bay and Tucker, 1992a), a finite difference code based on a standard Hele-Shaw formulation. All details of the software are as described by Bay and Tucker except for the closure, which is the ORL version discussed by Cintra and Tucker (1995). For comparison we also include the same calculations using the hybrid closure, as originally used by Bay and Tucker.

The predictions were tested against the experimental data of Bay and Tucker (1992b). Nylon 6/6 with 43 wt% glass fibers was used to fill two molds: a film-gated strip 203.2 mm long and 3.18 mm thick, and a center-gated disk 76.2 mm in radius and 3.18 mm thick. Orientation was measured on polished cross-sections. Columns of micrographs were taken across the thickness at several locations and divided into slices, so that orientation could be reported as a function of the gapwise coordinate z. Each experimental point represents measurements on 30 to 100 fibers.

Figure 2 shows the orientation profile across the thickness of the end-gated strip, at a distance $x/b = 34$ from the gate. (b is the half-thickness of the part). For each closure the value of C_I is adjusted to match a_{11} in the shell. This requires quite different values: 0.010 for the hybrid closure and 0.001 for the orthotropic closure. With these values the orthotropic closure gives a much better match to the a_{13} component.

Figure 3 shows the orientation profile across the thickness in a center-gated disk at $r/b = 22.8$. Both closures reproduce all of the right qualitative features, but the new orthotropic closure has much better quantitative accuracy. It matches the transition from core to shell very closely, a result that has not been possible before. The orthotropic closure gives a better match to a_{13}, similar to the strip data in fig. 3.

A very interesting result is the improvement in fit to the shell-layer orientation in the disk. Comparing figs. 2 and 3 shows that the shell in the disk is less oriented (has lower a_{11}) than the strip. The new orthotropic closure reproduces this result, while the old hybrid closure does not. The same level of improvement is also obtained for the remaining data sets of Bay and Tucker.

4 DISCUSSION AND CONCLUSIONS

This problem provides some interesting lessons about microstructure prediction. The first is the power and utility of a good microstructure state variable. The tensor a_{ij}, because it is compact, makes the prediction of non-homogeneous structure in commercial parts a practical possibility. The orientation tensor also has all the required properties of a microstructural state variable: it is indifferent to the choice of coordinate system, it can be predicted once the initial state and deformation are known, and it can be used to predict the final mechanical properties of the material.

A second lesson is the importance of being careful in developing models. The use of the interaction coefficient C_I provides much more accurate predictions of fiber orientation than the classical Jeffery equation alone, despite our limited understanding of the basis for this model. However, the use of C_I as a fitting parameter disguised some of the weaknesses of the early closure formulas. In earlier calculations C_I was adjusted to match the dominant feature of fiber orientation, a_{11} in the shell layer. But the value required to do this was artificially high, owing to deficiencies in the closure approximation. In turn this combination gave rise to other errors in the predictions, such as the overly narrow core predicted for a center-gated disk. Improving the closure approximation immediately reduced these errors.

Third is the importance of experimental verification for microstructure predictions. Fiber orientation can be measured directly, so verification does not have to rely on indirect experiments, such as the measurement of a physical or mechanical property that depends on orientation. The presence of careful experimental measurements revealed the weak aspects of earlier models, and lead to the current level of accuracy. It now appears that much more orientation data, taken by methods in which we have a high degree of confidence, will be needed before the orientation models can be refined again.

A fourth significant point is that current models produce results of useful accuracy, even though they omit many well-known physical phenomena. Elasticity of the polymer melt and the influence of the fiber orientation on rheology are a few of the omitted items. Apparently these factors are not very important in the cases of practical interest.

In summary, improvements in closure approximations have greatly improved the accuracy of fiber orientation predictions for injection molding. The differences between experiments and predictions is now comparable to the experimental errors in measuring fiber orientation.

ACKNOWLEDGMENTS

This work was supported by General Electric Co. and General Motors Corp. through the Thermoplastic Engineering Design Venture. Brent VerWeyst is supported as an Office of Naval Research Graduate Fellow.

REFERENCES

Advani, S.G. & C.L. Tucker 1987. The Use of Tensors to Describe and Predict Fiber Orientation in Short Fiber Composites. *J. Rheology* 31:751-784.

Advani, S.G. & C.L. Tucker 1990. Closure Approximations for Three-Dimensional Structure Tensors. *J. Rheology* 34:367-386.

Bay, R. S. & C. L. Tucker 1992a. Fiber Orientation in Simple Injection Moldings, Part 1: Theory and Numerical Methods. *Polymer Composites* 13:317-321.

Bay, R. S. & C. L. Tucker 1992b. Fiber Orientation in Simple Injection Moldings, Part 2: Experimental Results. *Polymer Composites* 13:322-342.

Bay, R. S. & C. L. Tucker 1992c. Stereological Measurement and Error Estimates for Three-Dimensional Fiber Orientation. *Polym. Eng. Sci.* 32:240-253.

Bay, R. S. 1991. *Fiber Orientation in Injection Molded Composites: A Comparison of Theory and Experiment.* Ph.D. Thesis, University of Illinois.

Bird, R. B., R. C. Armstrong & O. Hassager 1987. *Dynamics of Polymeric Liquids, Vol. 1: Fluid Mechanics, 2nd ed.* New York: Wiley.

Cintra, J. S. & C. L. Tucker 1995. Orthotropic Closure Approximations for Flow-Induced Fiber Orientation. submitted to *J. Rheology.*

Crochet, M. J., F. Dupret & V. Verleye 1994. Injection Molding. In S. G. Advani (ed), *Flow and Rheology in Composites Manufacturing.* 415-463. Amsterdam: Elsevier.

Dupret. F. & V. Verleye 1995. to appear. See also Verleye, V. & F. Dupret 1993. Prediction of Fiber Orientation in Complex Injection Molded Parts. *Proc. ASME Winter Annual Mttg.* New York: ASME.

Fischer, G. & P. Eyerer 1988. Measuring Spatial Orientation of Short Fiber Reinforced Thermoplastics by Image Analysis. *Polymer Composites* 9:297-304.

Folgar, F. & C. L. Tucker 1984. Orientation Behavior of Fibers in Concentrated Suspensions. *J. Reinf. Plast. Compos.* 3:98-119.

Friedl, C. & R. Brouwer 1991. Fibre Orientation Prediction. *SPE Tech. Papers* 37:326-329.

Gupta, M. & K. K. Wang 1993. Fiber Orientation and Mechanical Properties of Short-Fiber-Reinforced Injection-Molded Composites: Simulated and Experimental Results. *Polymer Composites* 14:367-382.

Henry, E., S. Kjeldsen & P. Kennedy 1994. Fiber Orientation and the Mechanical Properties of SFRP Parts. *SPE Tech. Papers* 40:374-377.

Hieber, C. A. & S. F. Shen 1980. A Finite Element/Finite Difference Simulation of the Injection Mold Filling Process. *J. Non-Newtonian Fluid Mech.* 7:1-31.

Hine, P. J., R. A. Duckett, N. Davidson & A. R. Clarke 1993. Modelling of the Elastic Properties of Fibre Reinforced Composites. I: Orientation Measurement. *Compos. Sci. Tech.* 47:65-73.

Jeffery, G. B. 1922. The Motion of Ellipsoidal Particles Immersed in a Viscous Fluid. *Proc. Roy. Soc.* A-102:161-179.

Randall, C. & H. H. Chiang 1994. Applications of Fiber Orientation Analysis in Injection Molding of Fiber-Filled Composites. *SPE Tech. Papers* 40:397-401.

Ranganathan, S. & S. G. Advani 1993. A simultaneous solution for flow and fiber orientation in axi-symmetric diverging radial flow. *J. Non-Newtonian Fluid Mech.* 47:107-136.

Tucker, C. L. & S. G. Advani 1994. Processing of Short-Fiber Systems. In S. G. Advani (ed) *Flow and Rheology in Polymer Composites Manufacturing* 147-202. Amsterdam: Elsevier.

Tucker, C. L. 1991. Flow Regimes for Fiber Suspensions in Narrow Gaps. *J. Non-Newtonian Fluid Mech.* 39:239-268.

Simulation of Materials Processing: Theory, Methods and Applications, Shen & Dawson (eds)
© 1995 Balkema, Rotterdam. ISBN 90 5410 553 4

Simulation of polymer reactive molding

Cheng-Hsien Wu, L. James Lee & S. Nakamura
Engineering Research Center for Net Shape Manufacturing, The Ohio State University, Columbus, Ohio, USA

ABSTRACT: Polymer reactive molding covers many manufacturing processes in which reactive resins are injected into a closed mold cavity to produce polymer or composite products. During processing, the flow pattern, the temperature field, and the conversion distribution can be very complicated because of resin reaction. Many computer codes for the simulation of polymer processing are based on the control-volume finite element method (CVFEM) because it is more user friendly and more robust than the conventional finite element methods. However, using lower order interpolation functions in CVFEM may create significant numerical errors. This is especially true for convection - diffusion problems with a changeable source term. In this study, an Eulerian - Lagrangian approach is proposed where an Eulerian coordinate system is used to solve the velocity field, while the chemical species balance and the energy balance are based on a Lagrangian coordinate system. This approach is used to simulate the non-isothermal resin transfer molding (RTM) process. The simulation results from this approach and a CVFEM method are compared.

INTRODUCTION

A large number of polymer products involve polymerization in fabricating the final shape. Examples include compression, transfer and injection molding of rubbers; compression and injection molding of unsaturated polyester resins; reaction injection molding of polyurethanes and nylons; autoclave curing of epoxy/graphite prepregs; and encapsulation of electronic parts using epoxy or polyimide resins. A common feature in processing these materials is that polymerization (or a portion of polymerization) and processing take place at the same time. Accordingly, the term 'reactive processing' has been used by many researchers to describe these processes.

The basic techniques of process modeling and numerical analysis of thermoplastic polymer processing have been explained in detail in a recent book [1]. These include common assumptions and simplifications, and various numerical methods. Most of them can be directly applied to reactive processing. The presence of chemical reaction, a non-linear source term, and the irreversible property changes such as resin conversion and resin viscosity, however, make the simulation of polymer reactive molding a much more challenging task than that of thermoplastic polymer molding. In this study, a reactive injection molding process, resin transfer molding (RTM), is used as an example to elucidate the effect of chemical reaction on the molding process simulation.

MATHEMATICAL MODELS

In the RTM process, a reactive liquid resin is injected into a mold cavity preplaced with a fiber reinforcement. Mold filling is often analyzed by Darcy's law. If the x and y coordinates are defined along the planar directions of a three-dimensional thin cavity and the z coordinate is defined along the gapwise direction, the governing equations of momentum balance can be expressed as follows:

(a) momentum balance

$$\begin{bmatrix} \bar{u} \\ \bar{v} \end{bmatrix} = -\begin{bmatrix} s_{xx} & s_{xy} \\ s_{yx} & s_{yy} \end{bmatrix} \begin{bmatrix} \dfrac{\partial p}{\partial x} \\ \dfrac{\partial p}{\partial y} \end{bmatrix} \tag{1}$$

and

$$\begin{bmatrix} s_{xx} & s_{xy} \\ s_{yx} & s_{yy} \end{bmatrix} = \frac{1}{h_z} \int_{-\frac{h_z}{2}}^{\frac{h_z}{2}} \frac{1}{\mu(x,y,z)} \begin{bmatrix} S_{xx} & S_{xy} \\ S_{yx} & S_{yy} \end{bmatrix} dz \ . \tag{2}$$

Here h_z is the thickness of the three-dimensional mold cavity, μ is viscosity, and S_{ij} is the component of the permeability tensor of the fiber preform defined in the local coordinate system. Equation 1 follows Darcy's law for flow through porous media with negligible inertia effects.

(b) continuity

$$\frac{\partial \bar{u}}{\partial x}+\frac{\partial \bar{v}}{\partial y}=0 \tag{3}$$

Equation 3 is a consequence of the incompressible fluid assumption, where the superficial velocity components $\bar{u}$ and $\bar{v}$ are the gapwise average values in the planar directions.

Viscous dissipation is usually negligible compared with other contributions such as heat transfer between the fluid resin and the fiber reinforcement or reaction exotherm. Mass diffusion is also insignificant because the chemical reaction rate is often much higher than the mass diffusion rate. When the heat transfer coefficient between the fiber mat and the resin fluid is very large and the resin flow is very slow, local thermal equilibrium of the resin and the fiber can be assumed. The energy equation for such a lumped system can be written as:

(c) energy balance

$$\rho c_p \frac{\partial T}{\partial t}+\rho_r c_{pr}\left(\bar{u}\frac{\partial T}{\partial x}+\bar{v}\frac{\partial T}{\partial y}\right)$$
$$=k\left(\frac{\partial^2 T}{\partial x^2}+\frac{\partial^2 T}{\partial y^2}+\frac{\partial^2 T}{\partial z^2}\right)+\phi \Delta H \dot{m} \tag{4}$$

and

$$c_p = c_{pr}w_r + c_{pf}w_f$$
$$\rho = (\rho_r \rho_f)/(\rho_f w_r + \rho_r w_f)$$
$$k = (k_r k_f)/(k_f w_r + k_r w_f) \tag{5}$$
$$w_r = \frac{\phi}{\rho_f}\Bigg/\left(\frac{\phi}{\rho_f}+\frac{1-\phi}{\rho_r}\right)$$
$$w_f = 1 - w_r.$$

Here T is temperature, ρ is density, c_p is specific heat, k is thermal conductivity, ϕ is the fiber preform porosity, ΔH is the heat of reaction, $\dot{m}$ is the reaction rate, and w is the weight fraction. The subscript "f" stands for the fiber preform and "r" for the resin.

When the resin flow rate is high, the local thermal equilibrium of the resin and the fiber can not be assumed, a two-temperature model should be used [2]. Recently, Tucker and co-workers [3] derived a one-temperature model based on an early work of heat transfer through porous media [4]. The energy equation is

$$\{\phi(\rho c_p)_r + (1-\phi)(\rho c_p)_f\}\frac{\partial \langle T\rangle}{\partial t}+(\rho c_p)_r\langle v_r\rangle \vec{\nabla}\langle T\rangle$$
$$=\vec{\nabla}\cdot\{(\bar{\bar{k}}_e+\bar{\bar{k}}_D)\cdot\vec{\nabla}\langle T\rangle\} \tag{6}$$
$$+\phi\rho_r H_R f_c\{\langle\alpha\rangle^f,\langle T_r\rangle^f\}+\mu\langle\vec{v}_r\rangle\cdot\bar{\bar{S}}^{-1}\cdot\langle\vec{v}_r\rangle$$

where $\bar{\bar{k}}_e$ is an effective thermal conductivity tensor, $\bar{\bar{k}}_D$ is a tensor which counts the dispersion effect, H_R is the heat of reaction per unit mass and $\bar{\bar{S}}$ is the permeability tensor. In this study, the resin flow rate is assumed to be low. Therefore, Equation 4 is used as the energy balance equation.

(d) species balance

$$\frac{\partial \alpha}{\partial t}+\frac{1}{\phi}\left(\bar{u}\frac{\partial \alpha}{\partial x}+\bar{v}\frac{\partial \alpha}{\partial y}\right)=\dot{m} \tag{7}$$

Here, α is resin conversion.

NUMERICAL APPROACH

The above mentioned mathematical models have been solved numerically by either the control volume finite element method (CVFEM), the finite difference method with domain mapping approach, the boundary element method or a combination of them. Because of its robustness and the user friendly nature, CVFEM has been widely used in commercial software to simulate various manufacturing processes [5-7]. This method is based on the physical interpretation of mass conservation within a control volume as shown in Figure 1. The pressure, temperature, conversion or other physical quantities are defined at each node point.

Equations 1 and 3 can be manipulated and integrated over a control volume, which leads to the following equation, using the divergence theorem:

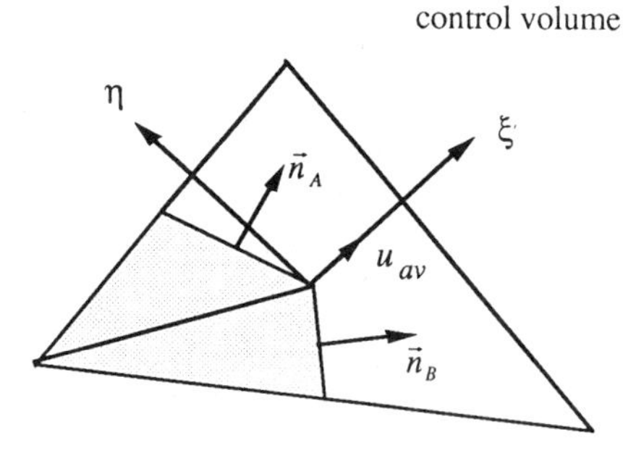

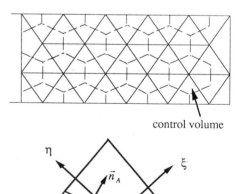

Figure 1. Typical control volume distribution and a triangular element.

$$\int_{\Gamma} -\frac{1}{\mu}\begin{bmatrix} n_x & n_y \end{bmatrix} \begin{bmatrix} S_{xx} & S_{xy} \\ S_{yx} & S_{yy} \end{bmatrix} \begin{bmatrix} \partial p/\partial x \\ \partial p/\partial y \end{bmatrix} d\Gamma = 0 \qquad (8)$$

where n_x and n_y are the components of the vector normal to the surface of the control volume, Γ. Equation 8 is the working equation for solving the problems of flow through anisotropic porous media and is basically a mass balance equation. The mold filling can be regarded as a quasi-steady state process by assuming a steady state condition at each time step. With the applied boundary conditions, the set of linear algebraic equations developed in the two-dimensional formulation can be solved to determine the pressure field during mold filling. The velocity field can then be computed using Darcy's law.

For non-isothermal cases, the heat conduction in the thickness needs to be considered. To discretize this term, the finite difference method, collocation method or other interpolation functions have been used [2]. Because of the nature of the mesh generated in the control volume method, the grids are fixed. Consequently, the physical flow front may be markedly different from the 'numerical' flow front. The 'numerical' flow front is the boundary between the filled control volumes and the partially filled control volumes. A simple way to solve the governing equations is to neglect the flow front region and to handle 'numerical' flow front as an interface condition. This is adequate for the flow equation since the pressure at the flow front region vanishes and does not play a significant role in pressure calculation. However, this method cannot precisely account for the energy balance in the flow front region where heat exchange between the resin fluid and the fiber is a dominant factor in the heat transfer calculation. An improved way to handle the heat transfer in the flow front region is to represent the energy equation or species mass balance equation by a balance of energy or mass in a Lagrangian form [2].

To consider the convection terms in a control volume, it is necessary to consider each individual element in a control volume because the velocity varies from element to element. The upwind scheme is the most widely used method. A better method to reduce the diffusion error is to introduce a special interpolation function which is exponential in the flow direction, but linear in the direction normal to the flow [8-10]. Lin et al. [2] applied this interpolation function to solve the energy balance equation and the species balance equation, i.e. Equations 4 and 7.

EFFECT OF REACTION ON PROCESS SIMULATION

To study the accuracy of CVFEM in simulating the reactive mold filling process, we start from checking the species balance equation. An isothermal mold filling process with reaction is tested. It is assumed that there is no reaction exotherm and the resin viscosity remains constant during mold filling.

The kinetic model used in this test is based on an epoxy resin [11], which can be described as

$$\frac{d\alpha}{dt} = \left[C_{H31_0} A_1 e^{(-E_1/RT)} + C_{H31_0}^2 A_2 e^{(-E_2/RT)} \alpha\right](1-\alpha)^2 \qquad (9)$$

where
C_{H31_0}=5.607x10^{-3} (mole/cm^3),
A_1=1.629x10^8 (cm^3/mole-s),
A_2=7.681x10^{11} (cm^6/mole^2-s),
E_1=6.205x10^4 (J/mole),
E_2=6.576x10^4 (J/mole).

If the molding temperature is kept constant (66°C in this case), the kinetic model can be simplified as

$$\frac{d\alpha}{dt} = (K_1 + K_2\alpha)(1-\alpha)^2 \qquad (10)$$

where $K_1 = 2.51\times10^{-4}$ (1/s) and $K_2 = 1.78\times10^{-3}$ (1/s). Because the ratio of K_1 to K_2 strongly affects the shape of the reaction rate vs. time curve, three different values of K_1, i.e. $K_1 = 2.51\times10^{-2}$ (1/s), $K_1 = 2.51\times10^{-4}$ (1/s), and $K_1 = 2.51\times10^{-6}$ (1/s), are used to analyze the numerical error. The conversion and reaction rate vs. time curves are shown in Figure 2. The kinetic model with the largest K_1 (2.51×10^{-2} 1/s) has a monotonously decreasing

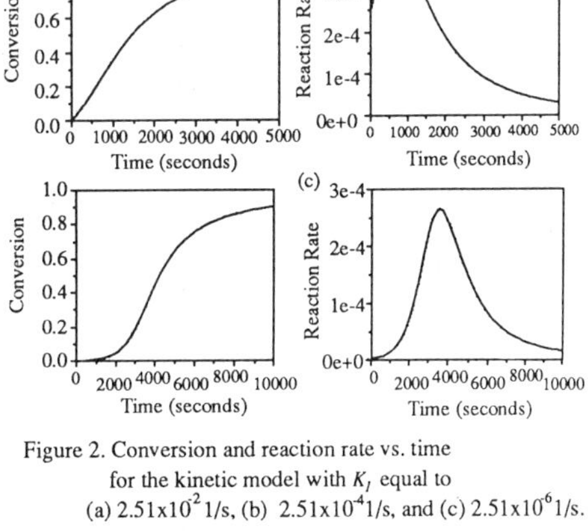

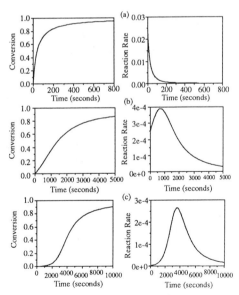

Figure 2. Conversion and reaction rate vs. time for the kinetic model with K_1 equal to (a) 2.51x10^{-2} 1/s, (b) 2.51x10^{-4} 1/s, and (c) 2.51x10^{-6} 1/s.

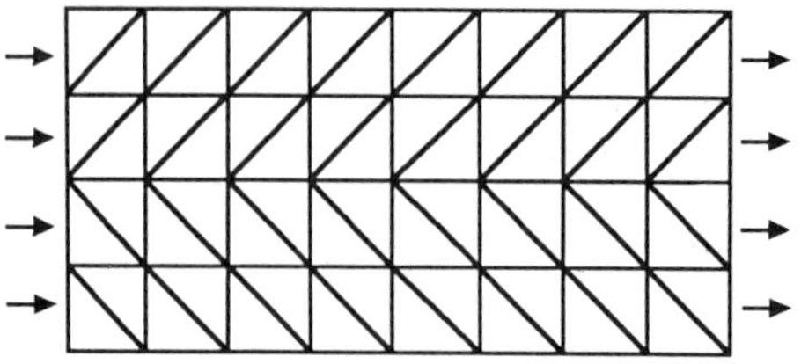

Figure 3. The finite element mesh used in the simulation.

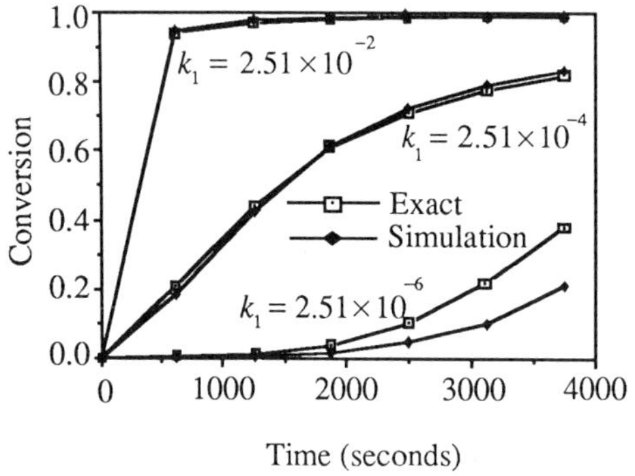

Figure 4. The comparison between CVFEM results and exact solutions of flow front conversion.

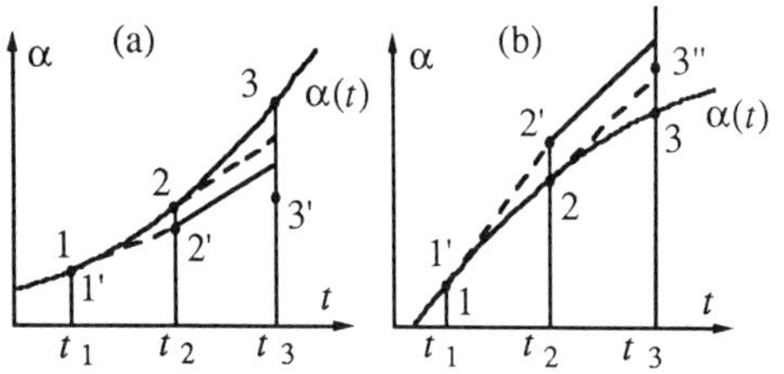

Figure 5. Schematic of numerical error for different types of kinetic models: (a) type I (b) type II.

reaction rate. The medium K_1 $(2.51 \times 10^{-4}\ 1/s)$ produces a rate profile with an increasing reaction rate at the beginning and a decreasing reaction rate after 700 seconds. The smallest K_1 $(2.51 \times 10^{-6}\ 1/s)$ produces a bell-shaped rate profile with the highest reaction rate at 3500 seconds.

A symmetric mesh is generated to produce a one dimensional flow as shown in Figure 3. A line source is assumed on the left side of the flow domain.

Errors in the flow front region

The flow front conversion vs. time relationships for different K_1 are compared with the exact solutions and the results are shown in Figure 4. The difference between the exact solution and the numerical results depends on the type of kinetic model. The numerical result agrees with the exact solution for $K_1 = 2.51 \times 10^{-2}$ (1/s). For $K_1 = 2.51 \times 10^{-4}$ (1/s), the simulated conversion is slightly under-estimated at the beginning, while over-estimated after the reaction rate reaches its peak value. As for $K_1 = 2.51 \times 10^{-6}$ (1/s), the simulated conversion is under-estimated and the discrepancy grows with the filling time in the first 3500 seconds during which the reaction rate increases with time.

These differences can be explained by the change of reaction rate and conversion vs. time. The conversion profiles shown in Figure 2 can be divided into two types. Type I is that the reaction rate increases with increasing conversion (or time), while type II is that the reaction rate decreases with increasing conversion (or time) as shown in Figure 5. If the simulation starts from $t = t_1$ (i.e. $\alpha_1 = \alpha_{1'}$), at next time step $t = t_2$, the exact conversion is at point 2, while the numerical solution is at point 2' due to the numerical error. At time step $t = t_3$, whether the discrepancy diverges or converges depends on the relationship between reaction rate and conversion. If the reaction rate increases with increasing conversion as shown in Figure 5(a), then $(d\alpha/dt)_{2'} < (d\alpha/dt)_2$ because $\alpha_{2'} < \alpha_2$. The numerical conversion at $t = t_3$ will be at point 3'. On the other hand, if the reaction rate decreases when the conversion increases as shown in Figure 5(b), then $(d\alpha/dt)_{2'} > (d\alpha/dt)_{2}$. The numerical conversion at $t = t_3$ will be at point 3''.

From the above discussion, the following conclusions can be obtained:

(a) If the reaction rate is an increasing function vs. time, then the conversion will be under-estimated and the numerical error will grow with time.

(b) If the reaction rate is a decreasing function vs. time, then the conversion will be over-estimated and the numerical error will decrease with time.

To improve the simulation accuracy one can use either smaller time steps or a higher order integration scheme.

Errors in the main flow region

The conversion distribution is shown in Figure 6. There is slight discrepancy between the simulated and the exact conversion profiles. The highest K_1 has less accuracy than the other two cases. This can be explained as follows:

Although the exponential function is the exact solution of a one-dimensional convection-diffusion equation [10], it does not satisfy the species balance equation because the species balance equation has the convection term and the source term as,

$$\frac{\bar{u}}{\phi}\frac{d\alpha}{dx} = \dot{m} = \frac{d\alpha}{dt} \qquad (11)$$

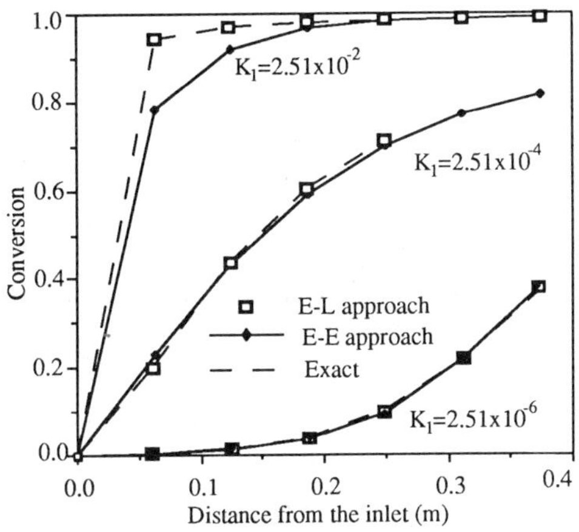

Figure 6. Conversion distribution along the flow direction for different K_1 at 3750 seconds (when the mold cavity is 3/4 filled).

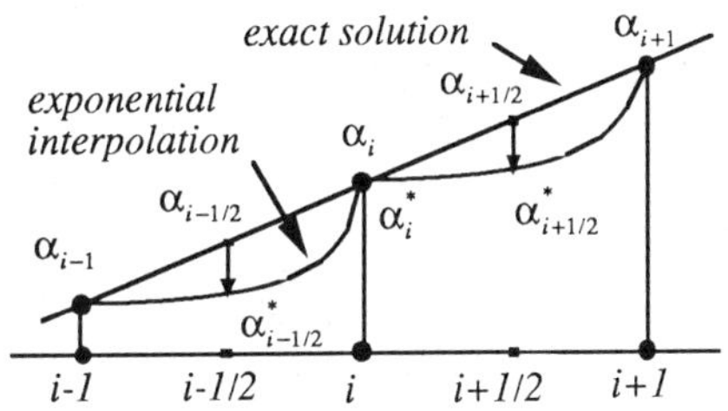

Figure 7. Species balance for a kinetic model with constant reaction rate.

with the boundary condition of $\alpha = 0$ at $x = 0$.

The exact solution of Equation 11 is

$$\frac{\alpha}{(k_1+k_2)(1-\alpha)} + \frac{k_2}{(k_1+k_2)^2} ln\left|\frac{(k_1+k_2\alpha)}{k_1(1-\alpha)}\right| = x \qquad (12)$$

This is obviously not an exponential function as used by Patankar [10] and Lin et al. [2].

The mass flux across the sub-surface of a control volume is the velocity multiplied by the conversion gradient. Therefore, an accurate interpolation at the sub-surface is necessary to produce a good simulation. For a one-dimensional mold filling process, a control volume around node *i* with two neighboring nodes, *i-1 and i+1* is shown in Figure 7. If the kinetic model has a constant reaction rate in this interval, the conversion distribution is linear as shown in Figure 7. The species balance for the control volume around node *i* is

$$\bar{u}\alpha_{i-1/2} + \int_{x_{i-1/2}}^{x_{i+1/2}} \phi \dot{m}_i dx = \bar{u}\alpha_{i+1/2} \qquad (13)$$

If exponential interpolation function is used, then,

$$\bar{u}\alpha^*_{i-1/2} + \int_{x_{i-1/2}}^{x_{i+1/2}} \phi \dot{m}^*_i dx = \bar{u}\alpha^*_{i+1/2} \qquad (14)$$

where $\alpha^*_{i-1/2}$ and $\alpha^*_{i+1/2}$ are the interpolated conversion based on the exponential form and $\dot{m}^*_i$ is the corresponding reaction rate of α^*_i.

Because of the linear conversion distribution, subtracting Equation 13 from Equation 14 gives

$$m^*_i = m_i \qquad (15)$$

Since the reaction rate is independent of the conversion, it leads to

$$\alpha^*_i = \alpha_i \qquad (16)$$

This means that any interpolation function may result in the exact conversion distribution if the reaction rate is constant. For kinetic models with varying reaction rate, however, an inappropriate interpolation function may create significant numerical errors. As mentioned in Figure 5, the reaction rate type I has an increasing function as shown in Figure 8(a). The dashed curve is the interpolation between the exact conversions α_i and α_{i+1}. Because of the increasing reaction rate, $\alpha^*_{i+1/2}$ is more under-estimated than $\alpha^*_{i-1/2}$. In order to balance the species, the predicted conversion at node *i*, α^*_i, is higher than α_i, and the conversion is over-estimated. The two errors tend to compensate with each other. Therefore, the overall error is smaller. On the other hand, the reaction rate type II has a decreasing function as shown in Figure 8(b). The dashed curve is the interpolation between the exact conversions α_i and α_{i+1}. Because of the decreasing reaction rate, $\alpha^*_{i+1/2}$ is less under-estimated than $\alpha^*_{i-1/2}$. In order to balance the species, the predicted conversion at node *i*, α^*_i, is lower than α_i, and the conversion is under-estimated. The overall error is larger when the two errors are added.

EULERIAN - LAGRANGIAN APPROACH

Several approaches have been proposed to improve the simulation of reactive flows. Prakash [9] proposed an improved shape function:

$$\Phi = A \cdot exp\left(\frac{\rho U_{avg} X}{\Gamma}\right) + BY + C + \frac{SX}{\rho U_{avg}} \qquad (17)$$

which is the exact solution of a one-dimensional steady-state convection-diffusion equation with a uniform source term, S. Prakash tested this shape function with three different problems and showed that this shape function worked better than the one without the last term in Equation 17. Another approach was proposed by Schneider and Raw [12]. A numerical method to relate the control volume surface values to the nodal point values was introduced by forming a governing differential equation discretization at the control volume surfaces. From this discrete equation, the flux across the control volume surface is determined in terms of the nodal values. Using this approach, the exponential type interpolation function is not needed in convection-diffusion computation.

In this study, an Eulerian - Lagrangian approach is used to improve the simulation of reactive flows. Among the governing equations of the mold filling process, the energy and the species balance equations have the convection terms. In order to avoid the numerical diffusion, the temperature and the conversion are solved in a Lagrangian frame of reference, whereas, the pressure and the velocity can be solved in an Eulerian frame of reference. This approach has been used to simulate the combustion process [13-15] and the thermoplastic injection molding process [16].

The numerical procedure is to insert a set of fluid particles, which behaves like tracers, into the flow from the inlet. The tracers move with the fluid flow. Tracers are released set by set until the mold filling ends as shown in Figure 9. It is necessary to track the fluid particles as they move. The trajectory is given by

$$\frac{d\vec{x}_{ij}}{dt} = \vec{u}_{ij} \tag{18}$$

where $\vec{x}_{ij}$ is the position vector of the center of the fluid particle which is the ith tracer of the jth set. The species balance equation can be calculated via

$$\frac{d\alpha_{ij}}{dt} = \left(K_1 + K_2\alpha_{ij}\right)\left(1-\alpha_{ij}\right)^2 \tag{19}$$

where K_1 and K_2 are functions of temperature, in general. The integration of these ordinary differential equations along the trajectories avoids the numerical diffusion that would result from the interpolation of the spatial derivatives in Equation 7.

One set of tracers is inserted into the flow near the inlet at the beginning of mold filling. At each time step, one can send another set of tracers from the inlet. A major advantage of this approach is that the number of tracers and their distribution can be flexibly chosen by the user. For example, in most cases, the conversion at flow front is more important than that at other locations due to its higher value, so one can send more tracers at the beginning and less tracers at the later time. For each time step, the flow equation is solved and the pressure and velocity distributions are obtained based on an initial guess of the viscosity distribution. The new flow front and

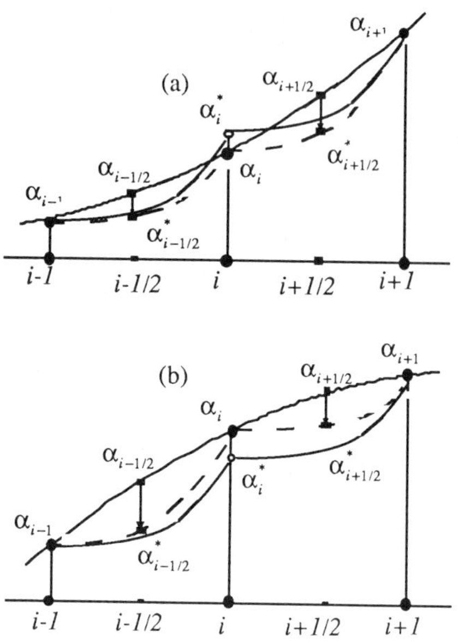

Figure 8. Species balance for kinetic models of (a) type I and (b) type II.

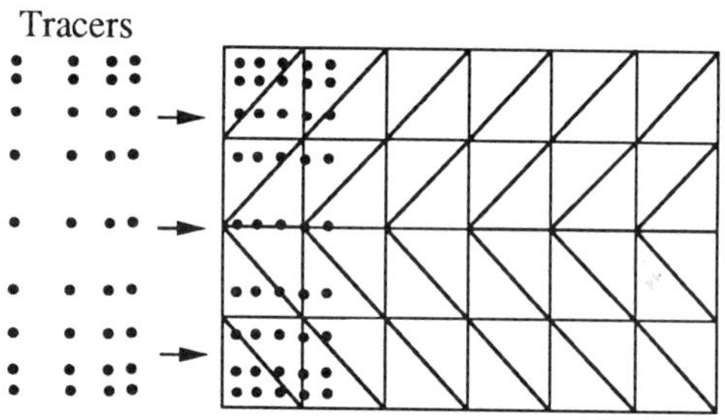

Figure 9. Schematic of the tracers with the Lagragian Approach.

the new time interval Δt are determined from the previous velocity field at the previous flow front.

For resins with high reaction rates, the time step may not be small enough to accurately calculate the conversion. Therefore, each time step can be divided into smaller sub-time intervals. The sub-time interval for each tracer is determined from the following criterion:

$$\Delta t \le \Delta\alpha_{max} / (d\alpha/dt) \tag{20}$$

where Δt is the time step, $\Delta\alpha_{max}$ is the specified allowable conversion change in one sub-time interval, and $d\alpha/dt$ is the reaction rate. Since the species balance equation is solved independently, each tracer has its own sub-time interval which depends on its reaction rate.

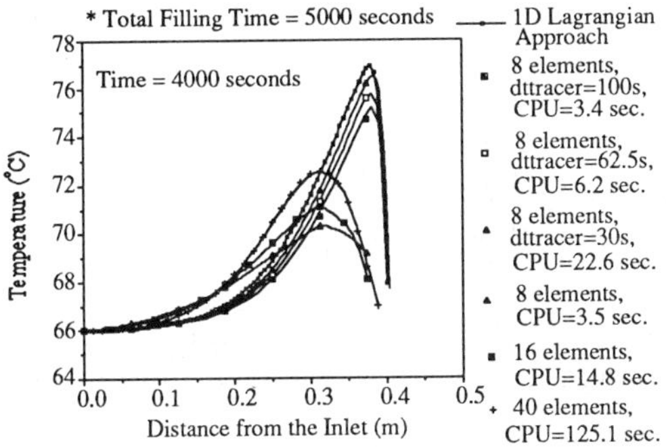

Figure 10. Temperature profiles for one dimensional non-isothermal molding processes.

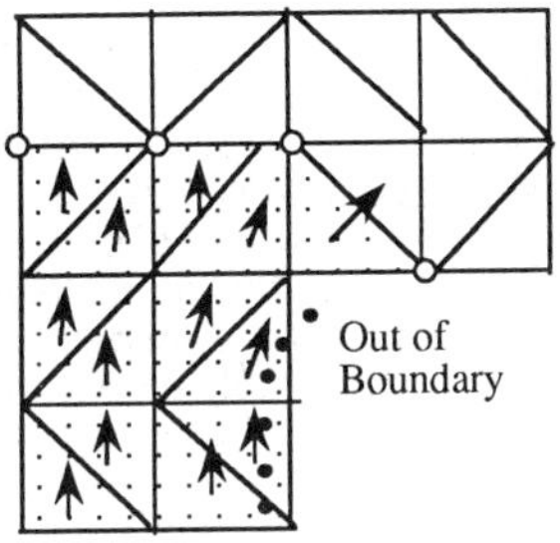

Figure 11(a). Velocity profile for an L-shaped part with Darcy's law.

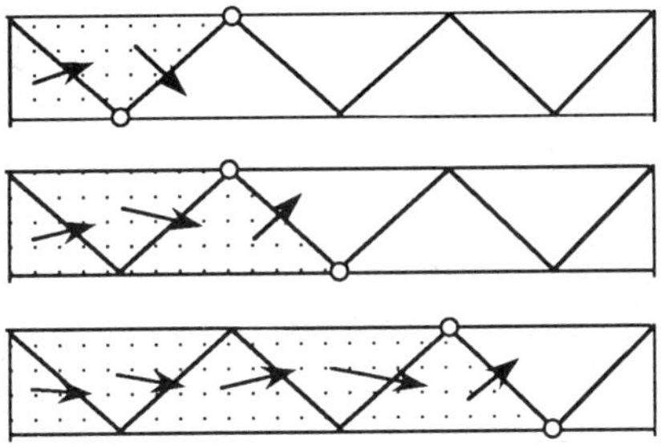

Figure 11(b). Velocity profiles during mold filling with Darcy's law.

For each sub-time interval within a time step Δt, the velocity field and the pressure distribution are assumed to remain unchanged. The conversion of each tracer is updated until a time step ends. Then the new pressure and velocity profiles are calculated again. All the procedures are repeated until the mold is filled.

The conversions of the tracers can be connected to plot the conversion profile. If the conversions at specific locations are needed, an interpolation may be necessary. There are several ways to do this. In this study, if the tracer distribution is very dense, the conversion at a location can be roughly represented by that of the closest tracer. If the tracer distribution is not dense enough, a bilinear interpolation is used.

One-dimensional flow simulation

The mesh of the one-dimensional flow simulation shown in Figure 3 is used here. First, an isothermal process is simulated. The conversion distribution along the flow direction is shown in Figure 6. The results show that with the Eulerian - Lagrangian approach, the numerical diffusion effect is eliminated in the conversion calculation. This approach only needs 1/5 of the CPU time than that based on the Eulerian - Eulerian approach [2]. This is because that the new approach does not need to compute the convection term and a flexible sub-time interval can be used for each tracer. As a result, the Eulerian - Lagrangian approach is able to provide better numerical accuracy with less computation time.

Next, a non-isothermal flow simulation is conducted by assuming that the reaction exotherm is not zero. The initial resin, fiber, and mold temperatures remain the same (i.e. 66°C). There is no exact solution for this non-isothermal reactive flow. However, for a one-dimensional flow with constant flow rate, a Lagrangian approach can be used to solve the energy balance equation and the result can serve as the 'correct' temperature distribution.

Figure 10 compares three cases based on the Eulerian - Lagrangian approach and three cases based on the Eulerian - Eulerian approach (i.e. Lin's approach [2]). The results show that the accuracy of using the Eulerian - Eulerian approach depends strongly on the mesh size. Using more elements provides a more accurate temperature distribution, however, more CPU time is needed. The Eulerian - Lagrangian approach provides not only better results but also with less CPU time. The simulation results are better with more tracers.

Errors in flow Calculation

In the Eulerian - Lagrangian approach, the temperature and conversion are calculated based on the tracers. Therefore, the accuracy of the tracer locations is very important. Darcy's law is a first-order momentum balance equation with only pressure being specified at the boundary (i.e. mold wall and flow front). Consequently, the simulated velocity distribution near the boundary could be inaccurate. For example, when flow is through an L-shaped mold cavity, the velocity distribution calculated based on Darcy's law may look like the one shown in Figure 11(a). Near the corner, the velocity vectors are not parallel to the boundary, which may cause the tracers flowing out of the boundary. Even in a simple one-dimensional flow as shown in Figure 11(b), the velocity distribution near the flow front can also be inaccurate. This error is caused by the angle of the flow front element and the pressure at the flow front nodes being assigned to be zero. After the flow front passes the corresponding elements, the velocity error gradually decreases.

An incorrect velocity distribution certainly affects the accuracy of the temperature and conversion distributions in the Eulerian - Lagrangian approach, because it would result in an incorrect tracer distribution. For the Eulerian - Eulerian approach, such error may also be significant.

BRINKMAN'S EXTENSION OF DARCY'S LAW

An approach to correct the error in velocity calculation is to use a second-order momentum balance equation such as Brinkman's extension of Darcy's law [17]. It can be written in the form:

$$\vec{v} = \frac{\bar{\bar{S}}}{\mu}\left(-\vec{\nabla}p + \tilde{\mu}\nabla^2\vec{v}\right) \quad (21)$$

where $\vec{v}$ is the velocity vector, μ is the viscosity, $\tilde{\mu}$ is the effective viscosity, and $\bar{\bar{S}}$ is the permeability tensor. The boundary conditions are

$$\left.\frac{\partial p}{\partial n}\right|_{wall} = 0,$$

$$p|_{front} = 0,$$

$$p|_{gate} = p_0,$$

and

$$\vec{v}|_{wall} = 0$$

$$\left.\frac{\partial \vec{v}}{\partial \vec{n}}\right|_{front} = 0 \text{ and } \left.\frac{\partial \vec{v}}{\partial \vec{t}}\right|_{front} = 0 \quad (22)$$

where $\vec{n}$ and $\vec{t}$ are the normal vector and tangential vector, respectively.

By assuming that pressure is a function of x and y only in the Cartesian coordinate, Equation 21 can be written as:

$$\begin{bmatrix} u(x,y) \\ v(x,y) \end{bmatrix} = \frac{1}{\bar{\mu}(x,y)}\begin{bmatrix} S_{xx} & S_{xy} \\ S_{yx} & S_{yy} \end{bmatrix}\begin{bmatrix} -\frac{\partial p(x,y)}{\partial x} + \tilde{\mu}\nabla^2 u(x,y) \\ -\frac{\partial p(x,y)}{\partial y} + \tilde{\mu}\nabla^2 v(x,y) \end{bmatrix} \quad (23)$$

$\bar{\mu}(x,y)$ is the averaged viscosity along the gapwise direction and can be expressed by

$$\frac{1}{\bar{\mu}(x,y)} = \frac{1}{h_z}\int_{-h_z/2}^{h_z/2}\frac{1}{\mu(x,y,z)}dz \quad (24)$$

where h_z is the part thickness. By substituting Equation 23 into the continuity equation and integrating over a control volume, we get

$$\int_C \frac{h_z}{\bar{\mu}(x,y)}\begin{bmatrix} n_x & n_y \end{bmatrix}\begin{bmatrix} S_{xx} & S_{xy} \\ S_{yx} & S_{yy} \end{bmatrix}\begin{bmatrix} -\frac{\partial p(x,y)}{\partial x} + \tilde{\mu}\nabla^2 u(x,y) \\ -\frac{\partial p(x,y)}{\partial y} + \tilde{\mu}\nabla^2 v(x,y) \end{bmatrix} d\Gamma = 0 \quad (25)$$

For isotropic fiber reinforcements, using the Divergence theorem, Equation 25 can be reduced to

$$\int_C \frac{h_z}{\bar{\mu}}\begin{bmatrix} n_x & n_y \end{bmatrix}\begin{bmatrix} S_{xx} & S_{xy} \\ S_{yx} & S_{yy} \end{bmatrix}\begin{bmatrix} -\frac{\partial p}{\partial x} \\ -\frac{\partial p}{\partial y} \end{bmatrix} d\Gamma = 0 \quad (26)$$

which is the same as solving Equation 8, i.e. the pressure field can be directly calculated without knowing the velocity field. This is the major advantage of using Brinkman's extension of Darcy's law, instead of the Navier-Stokes equation, to simulate the fluid flow. The results of using Brinkman's extension of Darcy's law in the Eulerian - Lagrangian approach to simulate the polymer reactive molding processes will be presented in the conference.

REFERENCES

1. C.L. Tucker III (ed.), "Fundamentals of Computer Modeling for Polymer Processing", Hanser, New York (1989).
2. R.J. Lin, L.J. Lee and M.J. Liou, Intl. Polym. Proc., 6(4), pp. 356 (1991).
3. C.L. Tucker III and R. B. Dessenberg, 'Governing Equations for Flow and Heat Transfer in Stationary Fiber Beds', in "Flow and Rheology in Polymeric Composites Manufacturing", edited by S.G. Advani, pp. 257, Elsevier, Amsterdam (1994).
4. J.C. Slattery, "Momentum, Energy and Mass Transfer in Continua", second edition, Krieger Publishing, New York (1981).
5. C.A. Hieber and S.F. Shen, J. Non-Newtonian Fluid Mech., 7, pp. 1 (1980).
6. V.W. Wang, C.A. Hieber and K.K. Wang, J. Polymer Eng., 7, pp. 21 (1986).
7. T.A. Osswald and C.L. Tucker, J. Int. Polym. Proc., 5, pp. 79 (1990).
8. B.R. Baliga and S.V. Patankar, Numerical Heat Transfer, 3, pp. 393 (1980).
9. C. Prakash, Numerical Heat Transfer, 9, pp. 253 (1986).
10. S.V. Patankar, "Numerical Heat Transfer and Fluid Flow", Hemisphere, Washington, D.C. (1980).
11. M. Perry, Ph.D. Dissertation, The Ohio State University (1993).
12. G.E. Schneider and M.J. Raw, Numerical Heat

Transfer, 9, pp. 1 (1986).
13. B. Seth, S.K. Aggarwal and W.A. Sirignano, Combustion and Flame, 39, pp. 149 (1980).
14. S.K. Aggarwal and W.A. Sirignano, Comuters and Fluids, 12, pp. 145 (1984).
15. S.K. Aggarwal and W.A. Sirignano, Combustion and Flame, 62, pp. 69 (1985).
16. C.M. Hsiung and M. Cakmak, Tenth Annual Meeting of the Polymer Processing Society, Akron, OH, (1994).
17. H.C. Brinkman, App. Sci. Res., A1, pp. 27 (1947).

Material behaviour

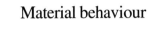

Simulation of Materials Processing: Theory, Methods and Applications, Shen & Dawson (eds)
© 1995 Balkema, Rotterdam. ISBN 90 5410 553 4

Experimental results of the friction stress distribution for the upsetting of cylinders

L. Baillet & J.C. Boyer
Laboratoire de Mécanique des Solides de l'INSA de Lyon, France

ABSTRACT : A measuring tool and a data processing procedure related to an inverse method developed for the experimental determination of local friction parameters during slow forming processes are presented. Experiments deal with different lubricating conditions and deliver the distribution of normal stress and friction stress at different interface radii for different height reductions.

1 INTRODUCTION.

The identification of realistic friction laws suited for forming processes still remains an experimental challenge:

- some in situ trials with special equipment as sensitive pins [2], [3] or sensitive balls [8] show how to measure the contact stress vector at one point of the workpiece-tool interface ;
- some tests developed for particular processes are proposed in order to determined global friction force versus global normal force for different tribological conditions. This approach is used for global friction parameter determination for drawing [1] or deep drawing [7].

As pointed out by [6] who also presented an elastic measuring tool, friction law identification has to be enhanced for numerical modelling of forming processes and it becomes necessary to observe experimentally the friction stress distribution at the workpiece-tool interface.

New theoretical developments [4], [5] predict the influence of :

- the tool roughness
- the mechanical and thermal properties of the tool material and of the workpiece material
- the thickness and the viscosity of the fluid lubricant
- the normal stress, the interface temperature and the contact duration
- the bulk plastic strain, the stress state of the workpiece and its relative motion inducing wear.

Some of these different coupled effects take place at some microscopic scale but as macroscopic friction laws are needed, an attempt of normal stress and friction stress measurements during a simple forming process with varying lubrication condition is presented for axisymetric geometries.

2 DESCRIPTION OF THE MEASURING TOOL.

The flat die developed for the 10000.kN hydraulic press is formed of two half cylinders tighten together with a thick hoop and bolts (Fig. I). Strain gage rosettes are bounded on the meridian plane of one half-cylinder near the die surface in front of the thin horizontal groove milled in the second half cylinder. The positions of the sensors have been predetermined by a numerical extension of the CERRUTI 's problem for the axisymetric geometry and then verified with the 3D finite element modelling of the die in order to consider the groove effects.

The sixty strain gages and 10 temperature sensors arranged along the diameter of the meridian plane of the two half cylinders at 7.mm depth under the die surface are connected with their associate amplifiers to 12 bits acquisition boards monitored by a personal computer. The force transducers of the hydraulic press, a vertical displacement transducer measuring punch stroke and two radial displacements transducers measuring two outer diameters of the workpiece fill up the 80 channels data acquisition system which collects for each upsetting 3000 sets of 60 strains, 10 temperatures, 2 to 5 forces and 3 displacements.

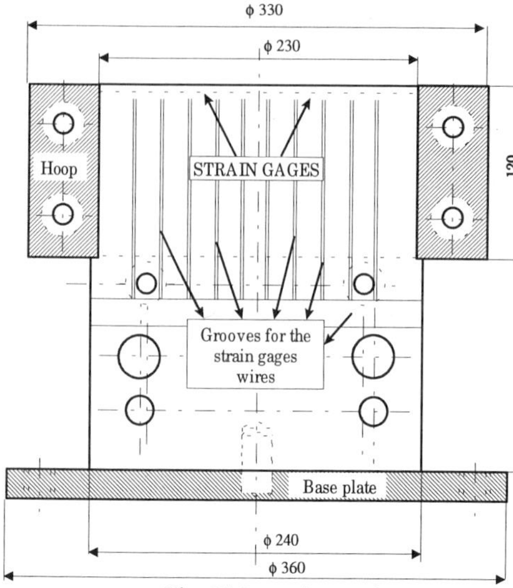

Fig. I Measuring tool.

The proprietary data acquisition software is tuned with the data acquisition board for delivering in R.A.M each set of 80 measurements every 1.5 ms.

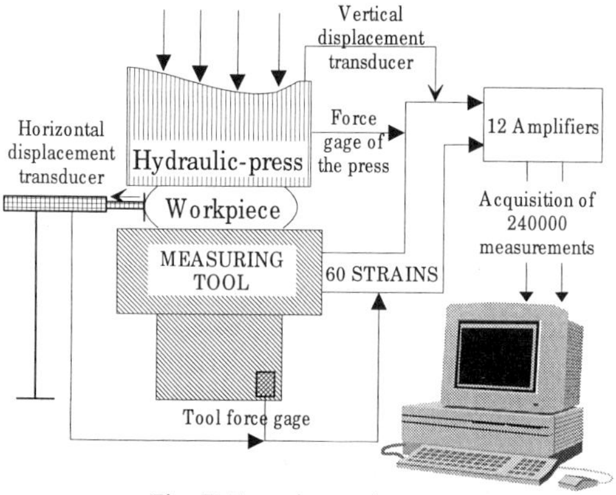

Fig. II Experimental set-up.

3 INVERSE MECHANICAL PROBLEM.

3.1 Introduction.

From the strain measurements at some points under the surface, the stress distribution acting on the workpiece-tool interface is deduced with some degree of approximation with an inverse method based upon an isothermal linear elastic behaviour of the measuring tool.

3.2 Equations of direct and inverse problem.

If $\{S\}$ is a vector constituted with the normal and tangential components of the stress vector at several points distributed along a radius of the workpiece-tool interface, the vector $\{\varepsilon\}$ of the radial, axial and tangential strains at points under the die surface is related to vector $\{S\}$ by a linear elastic operator [M] that can be evaluated analytically for simple geometries or numerically for real geometries.

The solution of the so-called direct problem is straightforward as :

$$\{\varepsilon\} = [M]\{S\} \qquad (1)$$

The associated inverse problem seems to be easily solved if the inverted matrix exists as :

$$\{S\} = [M]^{-1}\{\varepsilon\} \qquad (2)$$

Fig. III summarised the direct path and the inverse path for the upsetting of a cylindrical workpiece and the corresponding mechanical quantities involved for the interface balance.

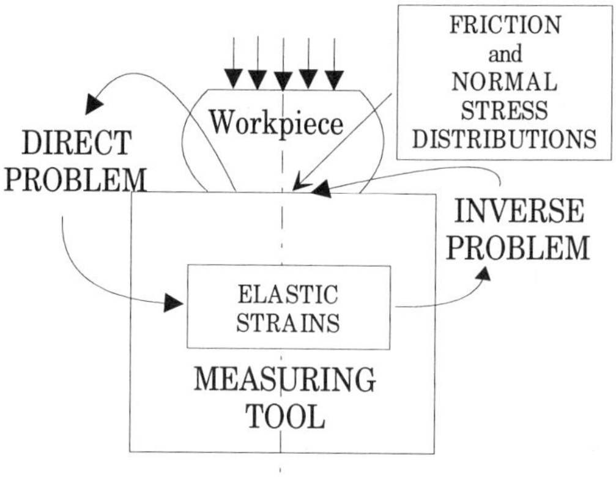

Fig. III. Direct and inverse problem.

Some difficulties arise with equation (2) because small discrepancies of strain values involve large errors on stress magnitudes and each measured strain value have to be considered as the sum of an unknown exact value and some experimental error.

In order to estimate the influence of the experimental errors, the case of the upsetting of a cylindrical workpiece is studied when the contact area diameter is 50.mm with 10 strain gage rosettes equally distributed along a radius of a flat die. The [M] matrix elements are calculated with a finite element software.

3.3. Strain sensitivity to stress errors.

Two linear distributions of normal and friction stresses are applied at the workpiece-tool interface. The second one (σ_{n2}, τ_2) exhibiting a small deviation of 5% with respect to the first one (σ_{n1}, τ_1) as represented on fig. IV.

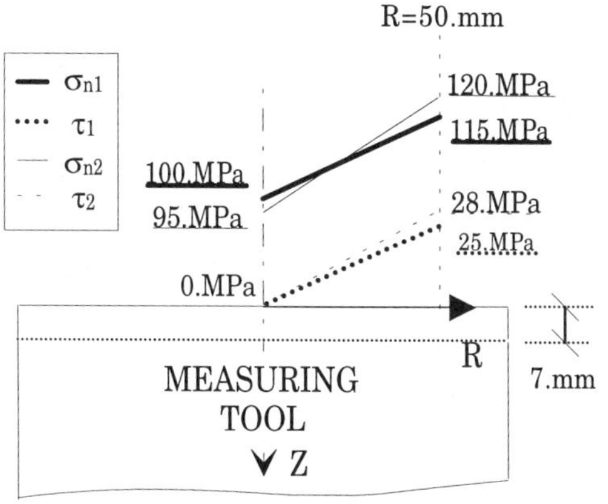

Fig. IV. Load of the measuring tool.

The corresponding strain distributions along the interface radius are presented on fig. V for the vertical, the 135° and the -45° gages of each rosette. The maximum strain deviation of 8% is observed for the inner -45° gage.

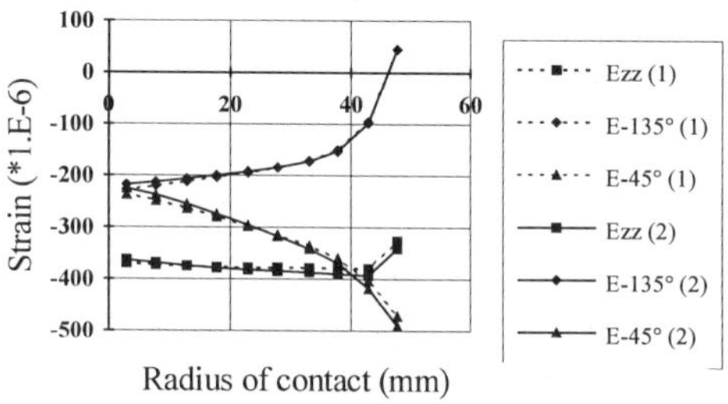

Fig. V. : 1st load = (σ_{n1}, τ_1)
2nd load = (σ_{n2}, τ_2).

The two sets of strain curves look quite similar, small stress deviation does not give rise to large strain variation.

3.4. Stress sensitivity to strain errors.

The strain values of the first loading (σ_{n1}, τ_1)

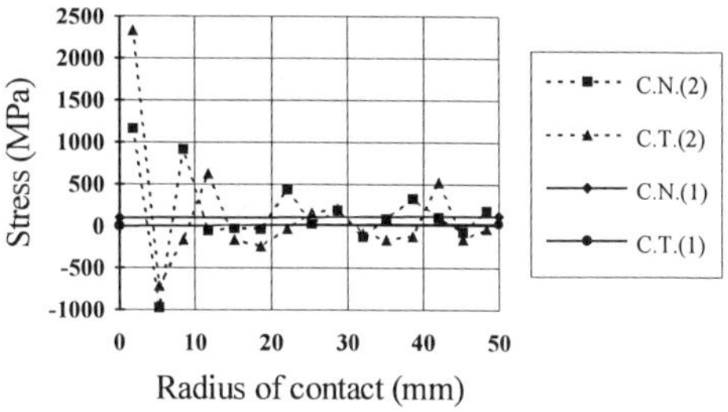

Fig. VI. : C.N. = normal stress
C.T. = friction stress.

described in the previous section is chosen as the unknown exact values, solving equation (2) for this strain vector $\{\varepsilon\}_t$ gives the right stress values. A second set of artificial experimental strain values is created by adding randomly ± 20.µm/m to the "exact values" $\{\varepsilon\}_t$ and then introduced in equation (2). The stress solution vector $\{S\}_e$ for this strain vector $\{\varepsilon\}_e = \{\varepsilon\}_t \pm 20\mu m$ is represented on Fig. VI.

The stress distribution has no physical meaning, small strain errors have tremendous effects on stress values.

3.5 Conclusion.

The determination of the interface stress distribution from internal strain measurements requires :

- an accurate data acquisition for the strains
- some improvements of the standard inverse method for minimising the stress sensitivity to strain deviation.

4 SPATIAL REGULARIZATION OF THE INVERSE METHOD.

4.1 Position of the problem.

The calculation of the [M] matrix coefficients comes from an elastic finite element modelling of the measuring tool with a fine mesh. The circular workpiece-tool contact area is divided in N_c rings of equal width. As the components of the vector $\{\varepsilon\}$ are the strain values induced on the strain gages by the stress distribution, column j of the [M] matrix is the set of coefficient relating the strain on each gage to a single constant normal stress applied only upon the ring number j.

The influence of the tangential stress on the strain is taken into account in the same way from column

N_c+1 to $2N_c$ of the [M] matrix for the same radius R_f of the contact area.

As the rate of representation of the stress components rise with the number of sensors, the accuracy of the measuring tool can be improved with more strain gages [9].

4.2 Application of the least square method.

The matrix [M] evaluated for each radius R_f which varies during the upsetting, has a number of lines equal to the number of strain gages N_m and a number of column equal to twice the number of ring N_c, considered on the contact area. The $2N_c$ stress values are deduced with the least square method from the strain measurements $\{\varepsilon\}_e$.

$$\sum_{i=1}^{N_m}\left(\varepsilon_{e_i}-\varepsilon_{t_i}\right)^2=\sum_{i=1}^{N_m}\left(\varepsilon_e-([M]\{S\}_e)_i\right)^2=F \quad (3)$$

with F minimal.

The minimisation of equation (3) is equivalent to solve the following linear system :

$$\begin{cases}\displaystyle\sum_{k=1}^{2.N_c}K_{jk}S_{e_k}=E_j \\ j=1,\dots,2.N_c\end{cases} \quad \text{equivalent to } \{E\}=[K]\{S\}_e$$

where $K_{jk}=\sum_{i=1}^{N_m}M_{ji}^TM_{ik}$

and $E_j=\sum_{i=1}^{N_m}\varepsilon_{e_i}M_{ij}$ (4)

4.3 Regularization of the inverse method.

In order to avoid the unrealistic stress variation involved by strain deviation, some continuity has to be introduced in the stress equation (4). Weighted conditions on the first and second derivative of the stress components are added to the basic least square equation (3).

$$\sum_{i=1}^{N_m}\left(\varepsilon_{e_i}-\varepsilon_{t_i}\right)^2+R_{Np}\sum_{i=1}^{N_c}\left(\frac{\partial S_{e_i}}{\partial r}\right)^2+R_{Tp}\sum_{i=N_c+1}^{2.N_c}\left(\frac{\partial S_{e_i}}{\partial r}\right)^2=F$$

$$\sum_{i=1}^{N_m}\left(\varepsilon_{e_i}-\varepsilon_{t_i}\right)^2+R_{Ns}\sum_{i=1}^{N_c}\left(\frac{\partial^2 S_{e_i}}{\partial^2 r}\right)^2+R_{Ts}\sum_{i=N_c+1}^{2.N_c}\left(\frac{\partial^2 S_{e_i}}{\partial^2 r}\right)^2=F$$

(5) and (6)

The R_{Np}, R_{Tp} ,R_{Ns} and R_{Ts} coefficients are respectively the regularization factors of the first and second derivative of the normal and tangential stress. The final equation set can be expressed as :

for j=1,..,2.N_c :

$$\sum_{k=1}^{2.N_c}\left(K_{jk}+R_{jk}\right)S_{e_k}=\sum_{k=1}^{2.N_c}T_{jk}S_{e_k}=E_j \quad (7)$$

where R=[R_{jk}] is the smoothing matrix and [T] the solving stress matrix.

4.4 Choice of the regularization coefficients.

The values of the "smoothing" coefficients R_{Np}, R_{Tp} ,R_{Ns} and R_{Ts} must be selected so that the [T] matrix turns toward its optimal conditioning defined with the factor :

$$C_{ond.}=\sqrt{\frac{\lambda_1}{\lambda_2}}\geq 1 \quad (8)$$

where :

V_p is the set of the eigen values of $[T][T]^T$

$\lambda_1=\text{Max}\{|\lambda|;\lambda\in V_p\}$

$\lambda_2=\text{Min}\{|\lambda|;\lambda\in V_p\}$ et $\lambda_2\neq 0$.

The [T] matrix is said ill-conditioned when $C_{ond.}>>1$.

The influence of the regularization coefficients upon the condition number of the matrix [T] is presented on figure VIII for 30 strains gages, 15 rings in the contact area and with R_f=50.mm. The minima of the different curves representing the condition number versus the first and second derivative regularization coefficients of the normal and tangential stress lie in the $\left[10^{-14};10^{-10}\right]$ range, the [T] matrix condition number increases exponentially for values of the regularization coefficients greater than 10^{-10} and lower than 10^{-14}.

An appropriate value of the regularization coefficients leads to a [T] matrix well conditioned and less sensitive to the strain experimental deviation.

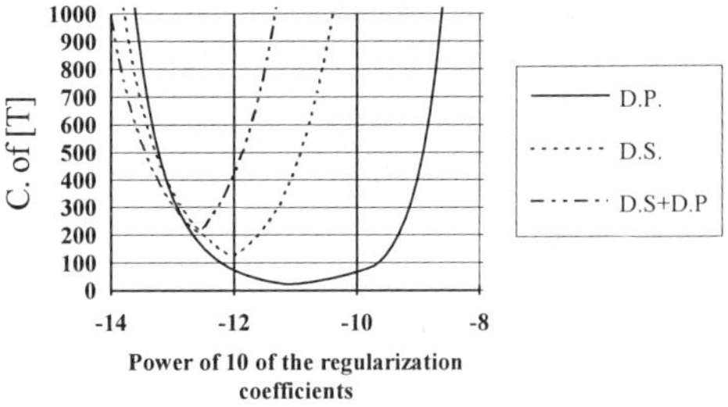

Fig. VII. : D.P.= first derivative
D.S.= second derivative.

5 EXPERIMENTAL RESULTS.

5.1 Introduction

The experimental material was Aluminium of commercial purity. The specimen were cylinders of 60.mm diameter and 80.mm height. The final tool surface preparation for the first tests presented in this paper consisted of turning followed by polishing with 400 grit emery cloth and wiping with clean dry cloth before each experiment. Three tests were carried out for each lubricating conditions : dry friction, mixed lubrication with vaseline, glycerine or graphite grease. All the experiments were conducted in a hydraulic press with punch velocity control in the 0.2-10 mm/s range. For the lowest punch velocity, a photo camera was used to record the shape of the workpiece for the different punch strokes.

At the end of each experiment, the data collected were stored on hard disk for further processing like the filtering of electrical noises and record translation of the strains versus time for each strain gage in strain variations versus interface radius for each time before the inverse method procedure previously described was applied.

The presented experimental results are obtained with the first and the second derivative spatial regularization and the additional condition of a zero tangential stress at the centre of the interface. For each determination of the interface stress distribution, the regularization coefficients are set in order to obtain the optimal condition number of the [T] matrix (8)

5.2 Normal stress and friction stress distribution for unlubricated aluminium.

The variation with height reduction from 6.mm to 52.9 mm at different interface radii are respectively presented at fig. VIII and IX.

The experimental curves show that :

- the normal stress is very low around the centre of the workpiece tool interface and maximum near the edge of the workpiece until a 50% height reduction is reached ;
- the friction hill appears only for height reduction greater than 60% ;
- the friction stress is zero under the mean radius of the workpiece where sticking condition and low normal stress level are present and then linearly increases still the outer radius for height reduction lower than 50%.

For higher height reduction, the friction stress varies quite linearly with the interface radii.

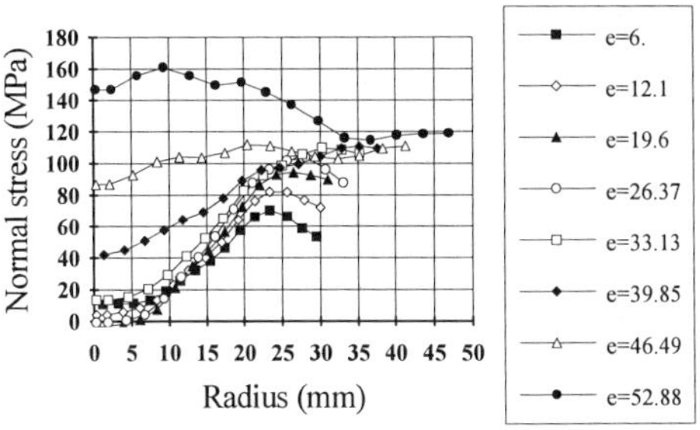

Fig. VIII. Normal stress for dry friction.

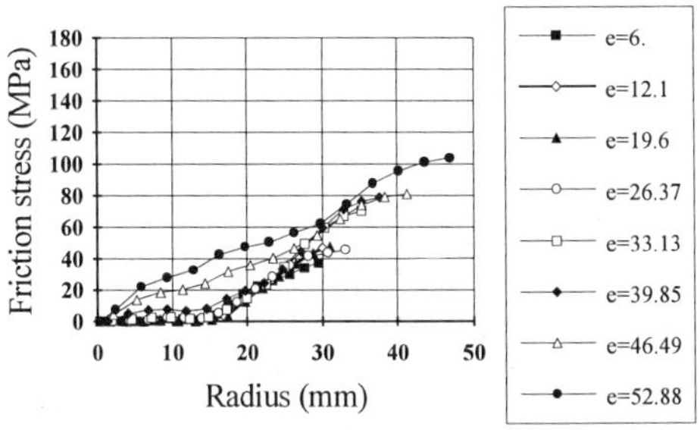

Fig. IX. Friction stress for dry friction.

5.3 Stress distribution for mixed lubrication with graphite grease.

With graphite grease sprayed on the platen surfaces, the barrelling of the workpiece occurs after a 50% height reduction. Before this transition value, the normal stress distribution along the workpiece radii exhibits a smooth "valley" shape unlike the dry friction case. For higher height reduction, the friction hill is flattened but exits as for the dry friction case.

The friction stress variations are different for height reductions lower and greater than 50.%, the transition value corresponds to the lubricant ejection.

In the first case as the highest normal stress takes place near the workpiece outer diameter, the lubricant is trapped in the interface, low values of the friction stress are found around the interface centre and the maximum friction stress at the outer edge could be representative of a mixed lubrication regime with a relative sliding of the workpiece.

When the lubricant is pushed out by the increasing normal stress that changes the "valley" shape distribution to the "hill" shape distribution, friction suddenly rises as the local interface properties are modified.

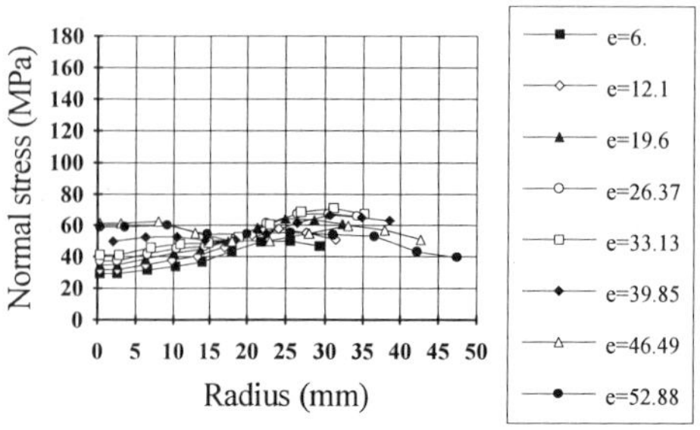

Fig. X. Normal stress for mixed lubrication with graphite grease.

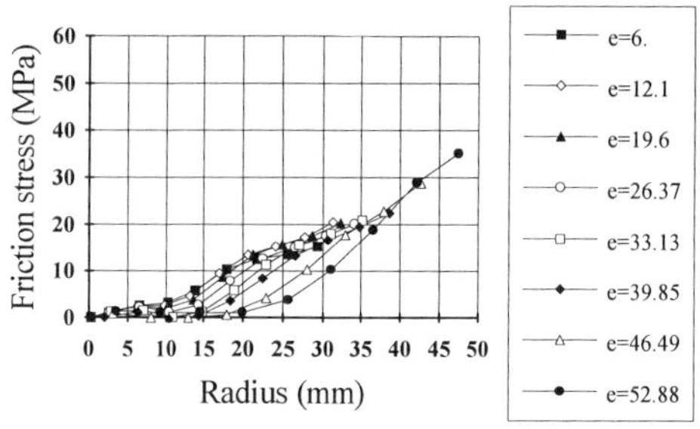

Fig. XI. Friction stress for mixed lubrication with graphite grease.

6. CONCLUSION.

The measuring tool and the mechanical inverse method allow the transformation of strain measurements under the die surface to interface stress distribution which is a key point for the analysis of interface behaviour under varying friction conditions.

This experimental technique will be used to validate the classical friction laws used as interface model behaviour in numerical modelling of forming processes and to compare numerically predicted interface stress distribution for simple forming operation with a flat die.

ACKNOWLEDGEMENTS.

The authors wish to thank the French Industrial Group for Research Advances in Forging for the financial support of this work and more specially the Gennevilliers' SNECMA plant and Pamiers' FORTECH plant.

REFERENCES

[1] RIGAUT J.M, OUDIN J.,BRICOUT J.P., CABEZON J., RAVALARD Y. A new friction test procedure for the improvement of drawing and similar processes. Journal of Materials Processing Technology, 1990, Vol. 3, p.3-28.

[2] PEARSALL, G.W. and BACKOFEN, W.A. Frictional boundary conditions in plastic compression. Journal of Engineering for Industry (ASME), 1963, Vol.85, p. 68-76.

[3] VAN ROOYEN, G.T and BACKOFEN, W.A. A study of interface friction in plastic compression. Int. J. Mech. Sci., 1960, Vol. 1, p.1-27.

[4] AVITZUR, B. Boundary and hydrodynamic lubrication. Wear, 1990, Vol. 139, p. 49-76.

[5] CHALLEN, J.M. ,McLEAN, L.J. and OXLEY, P.L.B. Plastic deformation of a metal surface in sliding contact with a hard wedge : its relation to friction and wear. Proc. R. Soc. Lond., 1984, Vol. 394, p. 161-181.

[6] DOEGE E., BEDERNA Ch. Indirect analysis of boundary stresses. Journal of Materials Processing Technology, 1994, Vol. 45, p.57-62.

[7] HAUG E., DI PASQUALE E., PICKETT A.K. Industrial sheet metal forming simulation using explicit finite elelment methods. VDI BERICHTE NR. 894, 1991, p. 259-291.

[8] LEHMANN G., BERNHARDT R. Messung von randspannungen und temperaturen in der wirkfuge von gesenkgravuren. International Conference and Workshop. Baden-Baden, Germany. 28-30 Sept 1994,p. 118-130

[9] BLANC G., RAYNAUD M., CHAU T.H. Detremination of the sensor location for 2D inverse heat conduction problems. Inverse problems in engineering mechanics. H.D. BUI, M. TANAKA & AL. BALKEMA, 1994, p179-184.

Simulation of Materials Processing: Theory, Methods and Applications, Shen & Dawson (eds)
© 1995 Balkema, Rotterdam. ISBN 90 5410 553 4

Modeling 304L stainless steel high temperature forgings using an internal state variable model

D.J. Bammann, P.S. Jin & B.C. Odegard
Sandia National Laboratories, Livermore, Calif., USA
G.C. Johnson
University of California, Berkeley, Calif., USA

ABSTRACT: A constitutive, or material response model is introduced which predicts the effects of deformation and temperature on the mechanical and microstructural properties of 304L stainless steel forgings. In this work, the hardening and recovery effects are tracked in a 304L stainless steel forging, forged at 850°C. Results are validated experimentally using compressive reload tests. The results shows good correlation between predictions and experimental results.

Introduction

Finite element analysis provides a tool in the design of many manufacturing processes. In particular, FEA allows the optimization of die design in multi-stage forgings which when coupled with an accurate description of the material response, provides predictive capabilities relating to the mechanical and microstructural properties of the forging. FEA will allow the designer to tailor the state of the deformed material. Further, these analyses will increase forging process efficiency, reduce unnecessary tool development, and avoid strain localization in the final forging stage. In this paper we examine the capabilities of internal state variable models to predict the material properties of the final state of a forging.

Constitutive Model

The model employed in these forging analyses is a strain rate and temperature dependent elastic-plastic model [1,2]. The kinematics associated with the model introduce a multiplicative decomposition of the deformation gradient into elastic and inelastic parts [3,4]. Unloading elasticity defines a zero stress state or natural configuration. It is with respect to this natural configuration, prescribed by the inelastic part of the deformation gradient, that the constitutive model is defined. Internal variables are introduced in this frame to describe the state of the material. The free energy is assumed to be a function of the elastic strain and the current values of the internal variables. We consider a hyper elastic material whereby the stress is defined as the derivative of the free energy with respect to the elastic strain and then specialize the model by assuming linear isotropic elasticity with respect to the natural configuration. Upon taking the material derivative and mapping to the current configuration, we obtain,

$$\overset{\circ}{\sigma} = \lambda \mathrm{tr}\boldsymbol{D}_e \boldsymbol{I} + 2\mu \boldsymbol{D}_e \tag{1}$$

where, σ is the Cauchy stress, $\boldsymbol{D}_e$ is the elastic symmetric part of the velocity gradient, defined as the difference between the total minus the thermal part, $\boldsymbol{D}_{th}$, and the inelastic part $\boldsymbol{D}_p$,

$$\boldsymbol{D}_e = \boldsymbol{D} - \boldsymbol{D}_p - \boldsymbol{D}_{th}, \tag{2}$$

λ and μ are the temperature dependent Lame constants, and $\circ$ denotes the convective derivative of the Cauchy stress defined by,

$$\overset{\circ}{\sigma} = \dot{\sigma} - \boldsymbol{W}_e \sigma + \sigma \boldsymbol{W}_e \tag{3}$$

where, $\boldsymbol{W}_e$ is the skew part of the elastic part of the velocity gradient given by,

$$\boldsymbol{W}_e = \boldsymbol{W} - \boldsymbol{W}_p. \tag{4}$$

The plastic flow rule is chosen to have a strong nonlinear dependence upon the deviatoric stress σ',

$$\boldsymbol{D}_p = f(\theta)\sinh\left\{\frac{|\sigma' - \alpha| - \kappa - Y(\theta)}{V(\theta)}\right\}\frac{\sigma' - \alpha}{|\sigma' - \alpha|} \tag{5}$$

where $f(\theta)$ and $V(\theta)$ describe a rate dependence of the yield stress at constant temperature θ. The deviatoric stress σ', is defined by,

$$\sigma' = \sigma - \frac{1}{3}\mathrm{tr}(\sigma)1. \qquad (6)$$

A tensor variable α and a scalar variable κ have been introduced to describe the deformed state of the material. The evolution of both state variables is cast into a familiar hardening minus recovery format. Both dynamic and thermal recovery terms were proposed for the variables. The dynamic recovery was motivated from dislocation cross slip that operates on the same time scale as dislocation glide. For this reason, no additional rate dependence results from this recovery term. The thermal recovery term is related to the diffusional process of vacancy assisted climb. Because this process operates on a much slower time scale, a strong rate dependence is predicted at higher temperatures where this term becomes dominant.

The evolution for these variables is defined by,

$$\dot{\kappa} = H(\theta)\boldsymbol{D}_p - \left\{R_s(\theta) + R_d(\theta)|\boldsymbol{D}_p|\right\}|\kappa|\kappa \qquad (7)$$

$$\overset{\circ}{\alpha} = h(\theta)\boldsymbol{D}_p - \left\{r_s(\theta) + r_d(\theta)|\boldsymbol{D}_p|\right\}|\alpha|\alpha \qquad (8)$$

The tensor variable, α, represents a short transient and results in a smoother "knee" in the transition from elastic to elastic-plastic response in a uniaxial stress-strain curve. What is more important, this variable controls the unloading response and is critical in welding or quenching problems during the cooling cycle of the problem. It is termed a short transient in that the variable hardens rapidly and saturates to a constant steady state value over a very short period of time during a monotonic loading at constant temperature and strain rate. This saturation value is maintained until the rate, temperature or loading path changes and the process repeats. This variable is responsible for the apparent material softening upon unloading termed the Bauschinger effect. The importance of the variable α in the prediction of localization in large deformation problems was detailed in [5]. The scalar variable, κ, is an isotropic hardening variable that predicts no change in flow stress upon reverse loading. This variable captures long transients and is responsible for the prediction of continued hardening at large strains. Unlike α, once steady state has been reached under constant conditions, this variable is not affected by a change in loading. The model prediction for various strain rates and temperatures

The response is dominated by dynamic recovery at lower temperatures where the rate dependence is weak. The effects of thermal recovery become significant at higher temperatures where the rate dependence is strong.

In Figure 2 the model prediction is compared with uniaxial compression tests that involve a significant change in temperature.

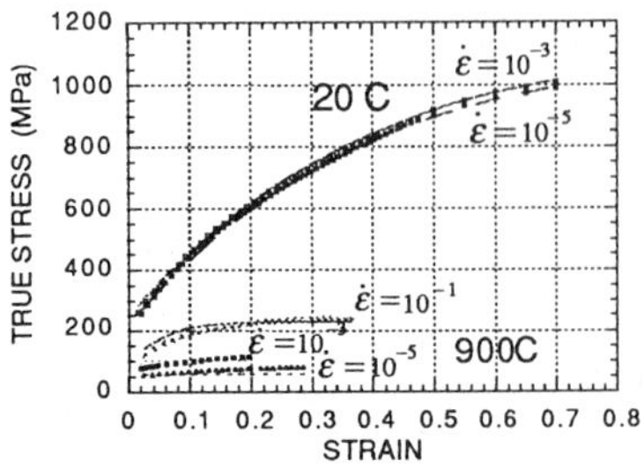

Figure 1 - Model prediction for 304L stainless steel tension tests or 304L stainless steel is depicted in Figure 1.

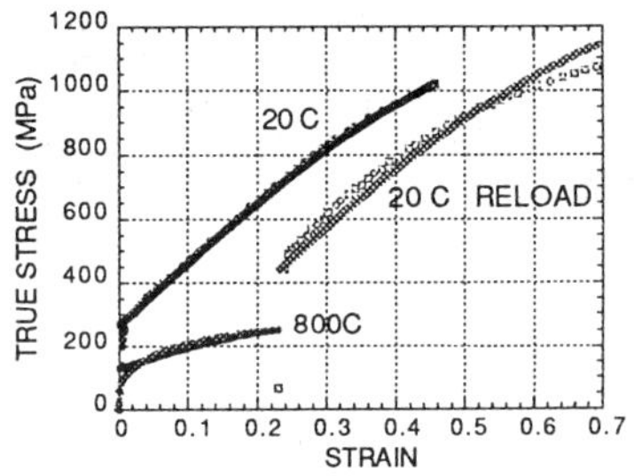

Figure 2 - Model prediction of compression tests and compression reload demonstrating temperature and history effects.

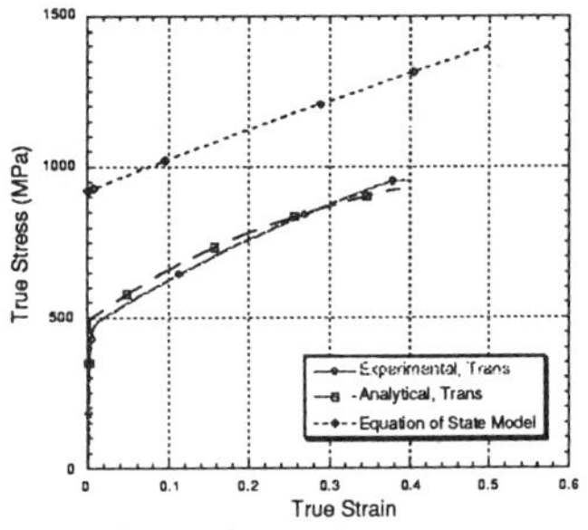

Figure 3 - Comparison of the state variable model prediction of the reload from the upset forge and the comparison with the equation of state model prediction.

Two specimens were loaded, one at 20C to a strain of nearly 0.5 and the other at 800C to a strain of 0.23. This specimen was unloaded and quenched rapidly. Once the specimen had cooled, it was reloaded at 20C. If the stress was a unique function of temperature and plastic strain, the flow stress upon reload would have been identical to the 20C specimen at that strain. As shown in Figure 2 [6], this is not the case. Rather, there is a strong temperature history effect upon reload that is

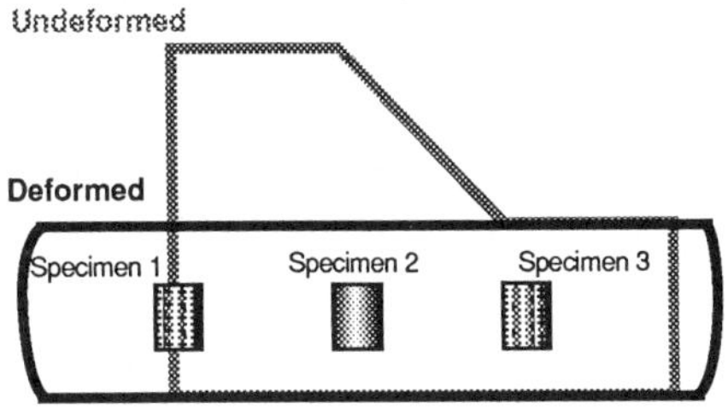

Figure 4 - Schematic of the preforged wedge and the locations of the reload specimens.

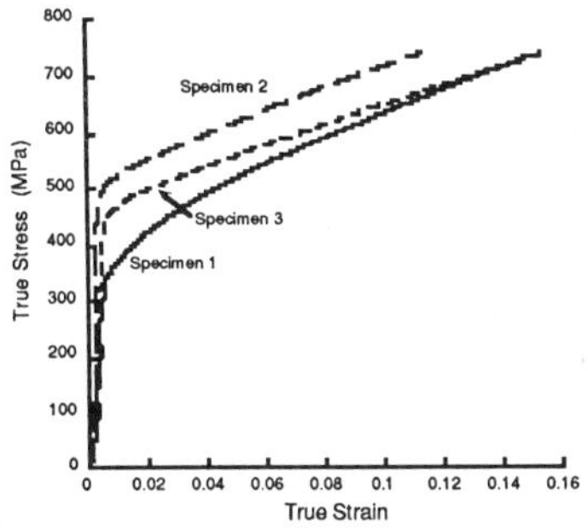

Figure 5 - Experimental results of the compression reload tests. Specimen 1, Specimen 2, and Specimen 3 correspond to highest, intermediate, and lowest deformations, respectively.

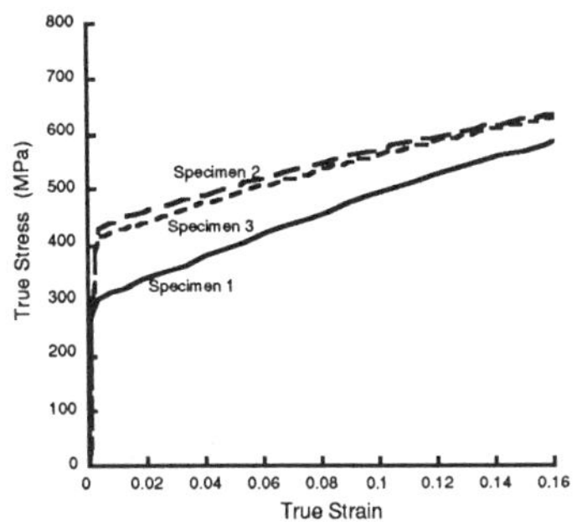

Figure 6 - Predicted results of the compression reload tests.

adequately captured by the state variable model.

Forging Analyses

The model has been implemented in to the finite element code PRONTO [7] and used in the analysis of both an axisymmetric upset forging and the forging of a three dimensional wedge into a plate. In the first problem a cylinder of 304L SS was deformed at high rate at a temperature of 850C in the axial direction to a deformation of 50%. Tensile specimens were then extracted from the cylinder and reloaded at 20C at a rate of 10^{-3} s^{-1}. The model has accurately captured the change of state that occurred during the high rate loading as seen by the prediction of the reload in Figure 3.

In the same figure, the prediction of a typical equation of state model is shown. These models assume that the stress is a unique function of strain, strain rate, and temperature independent of the loading path. Therefore, since the strain in the specimen from the forging process is 50%, the prediction of the stress on the reload at 20C, is the same as if the total deformation had occurred at 20C. Figure 4 shows an illustration of the wedge and the location of the compression reload specimens.

Reload specimens were extracted at each end where the most and least deformation has occurred as well as the center. The reload data is shown in Figure 5.

Notice that the specimen from the area of most deformation has the lowest yield stress upon reloading, while the specimen extracted from the center region is the strongest. Figure 6 depicts the model prediction from the finite element analysis of the forging process and the reloads.

The temperature dependence of the thermal recovery term was modified to accurately span the temperature range, and the model is seen to predict the observed response. This lends some insight into the results. Since the forging process is conducted at high rates, the material undergoes an adiabatic temperature change with the areas of highest deformation realizing the highest temperature change.. This change is sufficient to bring the temperature of the area of highest deformation to the recrystallization temperature, whereby the material begins to soften. This recovery process continues after the deformation until the material is quenched. This response is characterized by the variable, κ, which describes the hardened state of the material and can be related to the dislocation substructure size. As is seen in Figure 7, κ at the severely deformed end, initially increases until the temperature at which the recovery term, R_s, becomes large enough to dominate.

At this time recovery is greater that the hardening, and κ begins to decrease with increasing strain. By the end of the deformation associated with the forging process, κ, at the initially thick and thin ends is virtually identical., but the temperature is significantly higher in the end having undergone the greatest deformation. Therefore as the recovery process continues, the temperature is decreased in the quench, κ, at the higher temperature decreases more rapidly, resulting in the reload response observed and predicted in figures 5 and 6 respectively.

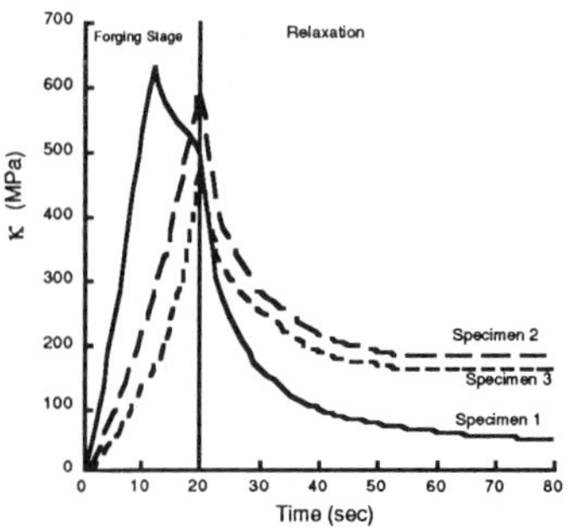

Figure 7 - Kappa (κ) as a function of time for the three deformation levels.

References

[1]. D. J. Bammann, "An Internal Variable Model of Viscoplasticity," Int. J. Eng. Sci., 22 (1984), pp. 1041-1053.

[2] D. J. Bammann, "Modeling Temperature and Strain Rate Dependent Large Deformations of Metals," Appl. Mech. Rev., 1 (1990), 312-318.

[3] D. J. Bammann and G. C. Johnson, "On the Kinematics of Finite Deformation Plasticity", Acta Mechanica, 70 (1987), pp. 1-13.

[4] D. J. Bammann and E. C. Aifantis, "A Model for Finite Deformation Plasticity", Acta Mechanica, 69 (1987), pp. 97-117.

[5] W. A . Kawahara, Unpublished data.

[6] L. M. Taylor, D. P. Flanagan, "PRONTO 2D, A Two-Dimensional Transient Solid Dynamics Program", Sandia Report, SAND86-0594 • UC-32, Unlimited Release, May 1992.

Simulation of Materials Processing: Theory, Methods and Applications, Shen & Dawson (eds)
© 1995 Balkema, Rotterdam. ISBN 90 5410 553 4

A plasticity model for materials undergoing phase transformations

D.J. Bammann, V.C. Prantil & J.F. Lathrop
Sandia National Laboratories, Livermore, Calif., USA

ABSTRACT: We propose a model based in mixture theory for metal alloys that undergo a phase transformation over broad temperature ranges. Single phase plastic flow is described by a state variable constitutive model. The transformation induced plasticity (TRIP) is driven by internal stresses whose form is based on micromechanical simulations of the response of a transforming representative volume. This work was supported by U.S. DOE under contract No. DE-AC04-94AL85000.

INTRODUCTION

For many applications in materials processing, the thermal and mechanical response are coupled with evolution of the underlying material microstructure. This coupling is evident in heat treatment, quenching and welding of a large class of metal alloys which undergo phase transformations that alter the constitutive response of the material. When modeling this coupled material response, calculation of the residual stresses and subsequent distortion are often of primary importance. Microstructural phase transformations often have substantial effects on these global facets of the material behavior. In particular, two features on the microscopic level have direct effects on the macroscopic response. The volume difference associated with the phase change imparts a purely dilatational deformation. In the presence of a deviatoric stress field, however, the weaker phase undergoes an additional deviatoric straining imposed upon it by the stronger phase. This transformation induced plasticity (TRIP) can have substantial effects on residual stresses and distortions.

Previous attempts to model the effects of phase transformations have met with mixed success. Empirical methods of incorporating a temperature dependent yield strength to simulate a phase change does not account for the volume change or any additional plasticity [1,2]. Mimicking the phase change with large changes in the thermal expansion coefficient can capture the effects of the spherical volume change, but not the deviatoric TRIP strains that accompany them [1,3,4]. Calculations which account for the dilatational volume change explicitly, but not the additional TRIP strains exhibit discrepancies with measured residual stresses [5,6]. Simulations both with and without account for this augmented plasticity indicate that neglecting this effect can result in residual stresses that are incorrect in sign and magnitude [1,7,8]. Because TRIP strains are accounted for locally, their effect on modeling global distortions and residual stresses will, in general, be problem dependent [3,7]. Such factors as levels of imposed restraint on the global level or inhomogeneity of the transformation locally may determine, in part, the effect of the TRIP phenomenon in any particular application. The thermal and mechanical boundary conditions can, therefore, play a significant role in determining the relative importance of incorporating TRIP in the analysis of a global boundary value problem.

We begin with a brief description of TRIP and the phenomenology of its dependence on stress and transformation kinetics. Based on these phenomenological descriptions, we motivate a formalism for incorporating the effects of TRIP by explicitly defining the microscopic stress field that drives this additional plastic straining. To illustrate this approach, we consider a two-phase system and comment on the role of this internal stress field in dual phase alloys. We restrict our attention to high strength, low alloy steels in which high temperature austenite transforms to martensite at low temperatures. We outline a multiphase state variable constitutive formulation and show how the flow law naturally accounts for the TRIP dependence on stress. Discrete micromechanical simulations of the phase transformation have been performed by Saeedvafa and Asaro [9]. Comparisons with the discrete model indicate that the dependence of TRIP strain on both the stress and volume fraction can be captured well by the multiphase state variable framework. Finally, we use this model to analyze the stress response during quenching of an idealized, long annular cylinder and comment on the effects of incorporating the TRIP phenomenon for global boundary value problems.

THE TRIP PHENOMENON

When alloys undergo a solid state phase change, the volume difference between the crystal structures induces a microplastic region in the weaker phase as the transformation proceeds. In the absence of a far field stress, this plastic field has no deviatoric part on average. There is, thus, no net contribution to the macroscopic plastic flow. In the presence of a deviatoric stress field, however, the straining from the volume misfit exhibits a deviatoric part that increases with the magnitude of the local stress deviator. This effect was described by Greenwood and Johnson [10] for applied stresses well below the yield strength of the weaker phase. There is evidence to indicate that under extreme cooling rates where a purely martensitic structure forms, the resulting platelet microstructure induces an additional shear deformation in the presence of an applied stress [11]. This imparts an inherent directionality to the response. In an effort to focus on the general features of TRIP, we consider only the Greenwood-Johnson mechanism here. An extensive theoretical treatment of TRIP mechanisms has been described by Leblond, Mottet and Devaux [12,13]. Leblond, Devaux and Devaux [14,15] provide a mathematical overview of TRIP wherein the additional strain rate in the austenite is given by

$$\dot{\varepsilon}^{TRIP} = \frac{K}{\Sigma_y}\left(\frac{\Delta V}{V}\right)\Psi(\Phi)\,\dot{\Phi}\sigma' \tag{1}$$

where Φ is the volume fraction of martensite.The transformation plastic strain rate is proportional to the applied stress deviator, σ', the rate of change of volume fraction, $\dot{\Phi}$, the volume misfit, $\Delta V/V$, and inversely proportional to the yield strength of the austenite, Σ_y. While the variations on equation (1) have met with some success, recent discrete micromechanical simulations of the transformation under applied stress indicate that the proportionality of TRIP strain rate on stress is not strictly linear over a broad range of stresses [9]. In addition, the discrete dependence of TRIP strains on volume fraction reported by Saeedvafa and Asaro [9] has shed light on the particular form of the function $\Psi(\Phi)$.

What is interesting is that the transformation plastic strain rate is proportional to the deviatoric stress, not the deviatoric stress rate [1]. As such, it closely resembles the phenomenological form of flow laws for state variable descriptions of viscoplasticity. This dependence has interesting implications because it indicates that the additional plasticity associated with the TRIP mechanism can be modeled by a state variable formulation. In the present work, we do this by defining a macroscopic stress which is a local average of the microscopic stress field acting between particles of different phases. To motivate this interaction stress, we appeal to the notion of an internal stress as a measure of material structure.

AN INTERNAL STATE VARIABLE MODEL

Internal state variable models have been utilized previously to describe the response of single phase metals over large strain rate and temperature ranges. These models have been formulated in a manner consistent with the kinematics of large deformation elastic-plastic response and are discussed elsewhere in this volume [16]. For multiphase materials, additional complications arise in predicting the material response. In particular, the effects of phase transformation induced plastic strain must be included. As a steel is cooled from above the austenization temperature, a solid state phase transformation occurs. The detailed nature of which phase forms depends upon the temperature and the cooling rate. Although we consider only martensite here, the model is easily generalized to consider additional product phases. The martensite occupies approximately 4% more volume than the austenite from which it transforms. The yield strength of martensite can also be an order of magnitude higher than that for austenite. As the austenite transforms to martensite, a volumetric strain develops in the austenite owing to the volume mismatch and difference in strength. In the presence of any deviatoric far field stress, this will be accompanied by a deviatoric component of strain. Since the martensite is much stronger, this results in a forward stress, $\pi^{(1)}$, in the austenite, resulting in an apparent softening of the material. Similarly, a backward stress, $\pi^{(2)}$, is created in the martensite. This stress field is a true, tensorial backstress that must be subtracted from the applied stress to get the glide resistance to macroscopic plastic flow. To first order, it is proportional to the volume fraction of hard particles and has its microscopic origins in the TRIP mechanism [19]. Introducing such a backstress in the yield condition and flow rule provides a natural way to model the apparent yield drop and additional plasticity characteristic of TRIP.

To generalize the state variable model for multiphase materials, we assume that each point of the continuum can be occupied simultaneously by all phases. Each phase is modeled by the appropriate state variable model for that phase with rate and temperature dependence as described in [16]. Then the material response can be modeled with as much complexity or simplicity as required, since the model reduces to rate independent bilinear response with an appropriate choice of parameters. We denote the current configuration deviatoric Cauchy stress in the austenite and martensite by $\sigma'^{(1)}$ and $\sigma'^{(2)}$ respectively. We assume a classic volume fraction weighted rule of mixtures, such that the total Cauchy stress, σ', is given by

$$\sigma' = \Phi\sigma'^{(2)} + (1-\Phi)\,\sigma'^{(1)}. \tag{2}$$

We assume a hypoelastic relation for each phase that

is consistent with an assumption of linear elasticity giving

$$\overset{\circ}{\sigma}'^{(i)} = 2\mu D'^{(i)}_e + \frac{\dot{\mu}}{\mu}\sigma'^{(i)} \qquad (3)$$

where $D'^{(i)}_e$ is the elastic symmetric part of the deviatoric velocity gradient, and μ is the temperature dependent shear modulus, assumed to be the same for both phases. Elastic deformation rates are defined as the difference between the total deformation rate and the sum of the thermal, $D'^{(i)}_{th}$, and the inelastic, $D'^{(i)}_p$, contributions in each phase

$$D'^{(i)}_e = D' - D'^{(i)}_p - D'^{(i)}_{th} \ . \qquad (4)$$

Here, $\circ$ denotes the convective derivative of the Cauchy stress defined by

$$\overset{\circ}{\sigma}{}^M = \dot{\sigma}^{(i)} - W^{(i)}_e \sigma^{(i)} + \sigma^{(i)} W^{(i)}_e \qquad (5)$$

where $W^{(i)}_e$ is the skew part of the elastic velocity gradient for each respective phase given by

$$W^{(i)}_e = W - W^{(i)}_p \quad . \qquad (6)$$

For the present purposes we choose a Jaumann derivative and all $W_p = 0$. Defining, $\pi^{(2)}$ and $\pi^{(1)}$ as the backward and forward long range stresses in the martensite and austenite respectively,

$$\pi^2 = \frac{-(1-\Phi)}{\Phi}\pi^1 \qquad (7)$$

This type of approach was taken by Freed, Raj and Walker [18] in modeling the hard and soft regions of a polycrystalline material as first proposed by Kocks [19]. Approximating compatibility of the hard and soft regions using the self-consistent scheme of Budiansky and Wu [20], they found that the total and local deviatoric strain fields must satisfy

$$\varepsilon'^{(i)} = \varepsilon' + \left(\frac{\beta}{1-\beta}\right)\frac{\pi^{(j)}}{2\mu} \qquad (8)$$

where i,j=1,2 and i is not equal to j. Here, β is the Eshelby shape factor for a spherical inclusion. We will make a simplifying Taylor-like assumption, and assume that the strains in each field are identical. In an effort to relieve this assumption, a numerical micromechanical simulation of the deformation partitioning will be the subject of a future work.

In addition to the long range forward and backward internal stress fields which act between the phases, we assume the existence of two short range internal stress fields $\alpha^{(i)}$ and $\kappa^{(i)}$, which act locally within each phase. We then define the net stress acting in each phase, as,

$$\xi^{(i)} = \sigma'^{(i)} - \alpha^{(i)} - \kappa^{(i)} - \pi^{(i)} \ . \qquad (9)$$

Now, we impose specific assumptions concerning the directionality of the stresses $\pi^{(i)}$ and $\kappa^{(i)}$. In particular we postulate that they act in the direction of the Cauchy stress minus the short range stress, $\alpha^{(i)}$, in each phase,

$$\pi^{(i)} = \pi^{(i)}\frac{\sigma'^{(i)} - \alpha^{(i)}}{\left|\sigma'^{(i)} - \alpha^{(i)}\right|} \qquad \kappa^{(i)} = \kappa^{(i)}\frac{\sigma'^{(i)} - \alpha^{(i)}}{\left|\sigma'^{(i)} - \alpha^{(i)}\right|} \qquad (10)$$

The effective stresses acting to cause plastic flow in each phase are then given by

$$\left|\xi^{(i)}\right| = \left|\sigma'^{(i)} - \alpha^{(i)}\right| - \kappa^{(i)} - \pi^{(i)}, \qquad (11)$$

where $\kappa^{(i)}$ are the scalar internal variables acting in each phase as discussed above. The plastic flow rule is chosen to have a strong nonlinear dependence upon the deviatoric stress

$$D^{(i)}_P = f^{(i)}(\theta)\sinh\left\{\frac{\xi^{(i)} - Y^{(i)}(\theta)}{V^{(i)}(\theta)(1+N(c))}\right\}\frac{\sigma'^{(i)} - \alpha^{(i)}}{|\sigma'^{(i)} - \alpha^{(i)}|}, \qquad (12)$$

where $f(\theta)$ and $V(\theta)$ describe a rate dependence of the yield stress at constant temperature and N(c) describes the yield increase with increasing carbon content, c. Tensor variables α^M, α^A and scalar variables κ^M, κ^A have been introduced to describe the deformed state of each phase as described previously for a single phase material [16]. The evolution of these variables is defined for phase (i) by

$$\begin{aligned}\dot{\kappa}^{(i)} &= H^{(i)}(\theta,c)\left|D^{(i)}_P\right| - \left\{R^{(i)}_S(\theta) + R^{(i)}_D(\theta)\left|D^{(i)}_P\right|\right\}\left|\kappa^{(i)}\right|\kappa^{(i)} \\ \overset{\circ}{\alpha}{}^{(i)} &= h^{(i)}(\theta,c)D^{(i)}_P - \left\{r^{(i)}_S(\theta) + r^{(i)}_D(\theta)\left|D^{(i)}_P\right|\right\}\left|\alpha^{(i)}\right|\alpha^{(i)} \ .\end{aligned} \qquad (13)$$

The fit to the carbon dependent data reported by Sjöström [17] is shown in Figure 1. Finally, we adopt the simplest evolution equation for the long range internal stress that is, to first order, proportional to volume fraction, but vanishes as the austenite transforms entirely to martensite

$$\dot{\pi} = C\,\frac{\Delta V}{V}\left\{\dot{\Phi} - 2\Phi\dot{\Phi}\right\} \qquad (14)$$

Here a simple exponential kinetics are used as described in [22]

$$\dot{\Phi} = \tilde{k}\exp(\lambda_j\{\eta - \eta_M\}^2)\,\dot{\eta} \quad ; j = 1,2 \qquad (15)$$

where η is the driving force for transformation,

$$\eta = k(M_s - \theta) \qquad (16)$$

and M_s is the martensite start temperature. Here, $\lambda_{1,2}$ and η_M are prescribed by the rate kinetics [22].

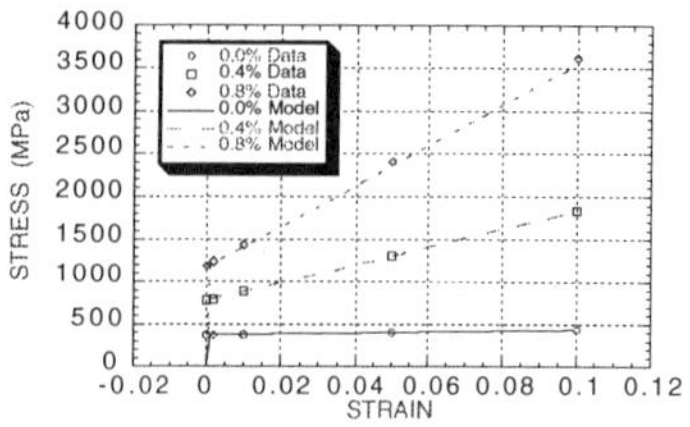

Figure 1. Fit of state variable model to hardening as a function of carbon content (from Sjostrom).

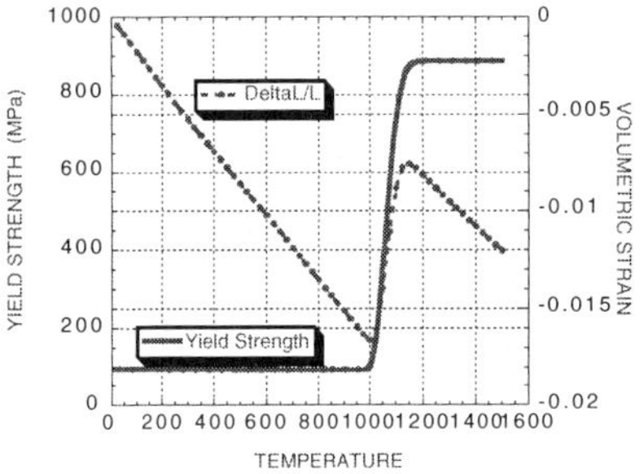

Figure 2. Yield drop and volumetric dilatation upon phase transformation.

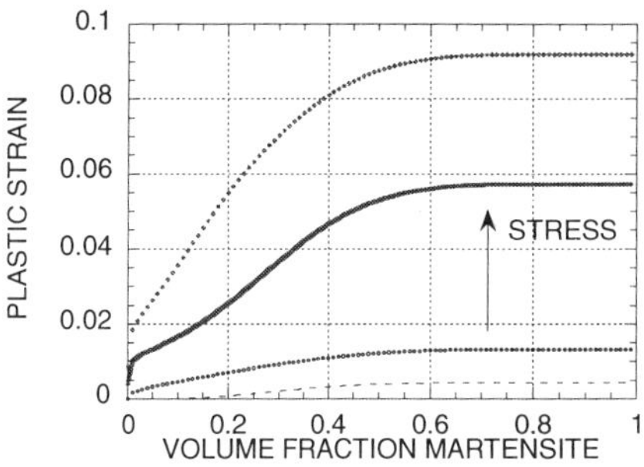

Figure 3. Dependence of TRIP strain on volume fraction as a function of stress.

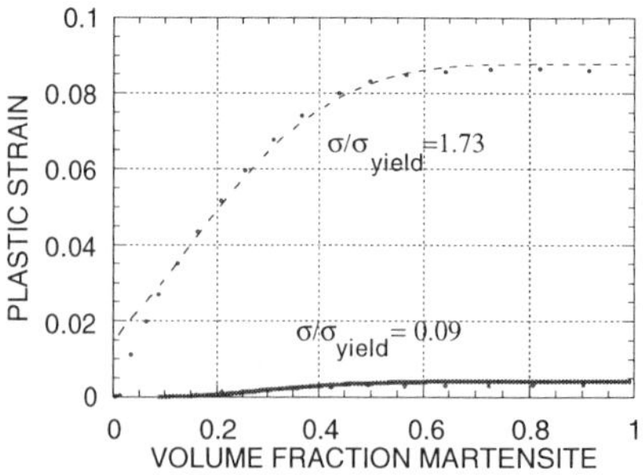

Figure 4. State variable fit to results of micromechanical simulations on a transforming representative volume.

MODEL SIMULATIONS

Local TRIP Behavior

We now investigate the predictions of this model with respect to previous models of phase transformation plasticity. The yield drop phenomenon and the dilatational volume change of a material point for constant cooling rate are depicted in Figure 2. These are well represented. One issue of concern in previous formulations for TRIP is the stress dependence of the additional strain rate. In our model, this choice is naturally determined from the assumption of the stress dependence of the flow in the austenite. Figure 3 illustrates the dependence of the plastic strain rate as a function of volume fraction of martensite. These curves were obtained by integrating equations (3) and (13-17) for various applied stresses. These curves illustrate the effects of the nonlinearity of the model. In addition to the obvious nonlinear stress dependence of the flow rule, these curves also reflect the nonlinear response of the state variables during the transformation. For high applied stress, the short transient $\alpha^{(i)}$, has saturated for very small values of Φ since $\alpha^{(i)}$ grows initially as the hyperbolic sine of the stress. Saeedvafa and Asaro [9] have recently investigated the stress dependence of the TRIP strain using a numerical micromechanical approach. A comparison of the state variable prediction with the discrete simulation is depicted in Figure 4 over a broad range of applied stresses. These results are encouraging, but further experimental investigations are currently being conducted to better understand the stress dependence of this complex phenomena.

Global TRIP Behavior

The multiphase state variable model has been implemented in the finite element computer program JAC3D [22] and used to simulate the quenching of a long annular rod. The rod is initially at uniform temperature, $T_a = 1073\,K$. The quenchant temperature, T_q, is $400\,K$ and the heat flux differs on inner and outer boundaries. The problem geometry is depicted in Figure 5. For long aspect ratio cylinders, the thermal response is one-dimensional. It can be computed analytically [23] and used to drive the transformation in equation (15).

The corresponding radial and hoop stresses are depicted in Figure 6. Including transformation plasticity has noticeable effects on both components. The tangential stresses are increased marginally at the inner surface, but relieved just beyond the transformation front. Near the outer surface, the tangential stress changes sign while maintaining nearly the same magnitude. Radial stresses are larger, in general, and, again, change sign at the outer surface. These results indicate that local TRIP behavior is manifest on the global macroscopic scale in a way that can be cast in a state variable framework. A corresponding experimental program is underway to characterize a class of low alloy steels and verify the simulations over a broad range of conditions. While the qualitative trends are reasonable, the quantitative predictions will need to be revisited when sufficient experimental data are available.

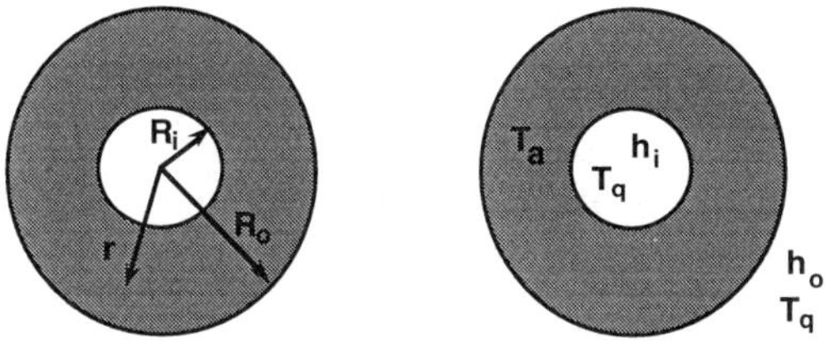

R_i = 18 mm T_a = 1073 K
R_o = 54 mm T_q = 473 K

Figure 5. One-dimensional, annular rod or disk.

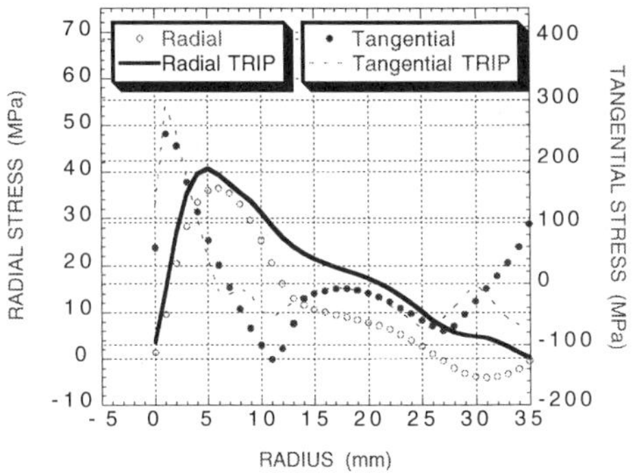

Figure 6. Radial and tangential stresses calculated with (TRIP) and without the effects of phase transformation.

CONCLUSIONS

An internal state variable formulation for phase transforming alloy steels is presented. We have illustrated how local TRIP strains can be accommodated by an appropriate choice for the corresponding internal stress field acting between the phases. The state variable framework compares well with a numerical micromechanical calculation providing a discrete dependence of TRIP strain on stress and volume fraction. The multiphase model is used to simulate the stress state of a quenched bar and show qualitative trends in the response when the TRIP phenomenon is incorporated on the length scale of a global boundary value problem.

REFERENCES

[1] J. A. Goldak, "Modeling Thermal Stresses and Distortions in Welds," *Recent Trends in Welding Science and Technology,* eds. S.A. David and M. Vitek, (Gatlinburg, TN: ASM International, 1990), 71-82.

[2] D. Dubois, J. Devaux, and J. B. Leblond, "Numerical Simulation of a Welding Operation: Calculation of Residual Stresses and Hydrogen Diffusion, " *Proceedings of the Fifth International Conference on Pressure Vessel Technology,* Vol. 2, Materials and Manufacturing (San Francisco, CA: American Society of Mechanical Engineers, 1984), 1210-1239.

[3] S. Sjöström, "Interactions and Constitutive Models for Calculating Quench Stresses in Steel," *Materials Science and Technology,* 1, (1985), 823-829.

[4] J. A. Goldak et al., "Computational Weld Mechanics," *Proceedings of the International Symposium on Computer Modeling of Fabrication Processes and Constitutive Behavior of Metals,* (Ottawa, Canada 1986).

[5] A. S. Oddy, J. A. Goldak, and J. M. J. McDill, "Transformation Plasticity and Residual Stresses in Single-Pass Repair Welds," *ASME Journal of Pressure Vessel Technology,* 114, (1992), 33-38.

[6] B. L. Josefson and C. T. Karlsson, "Transformation Plasticity Effects on Residual Stresses in a Butt-Welded Pipe," *ASME Journal of Pressure Vessel Technology*, 114, (1992), 376-378.

[7] A.S. Oddy, J.A. Goldak, and J.M.J. McDill, "Numerical Analysis of Transformation Plasticity in 3D Finite Element Analysis of Welds," *Eur. J. Mech., A/Solids,* 9 (3) (1990), 253-263.

[8] J. A. Goldak et al., "Coupling Heat Transfer, Microstructure Evolution and Thermal Stress Analysis in Weld Mechanics," *Proceedings of the IUTAM Symposium,* (Lulea, Sweden 1991), 1-31.

[9] M. Saeedvafa and R. J. Asaro, "Transformation Induced Plasticity," (LAUR-95-482, Los Alamos National Laboratory, 1995).

[10] G. W. Greenwood and R. H. Johnson, "The Deformation of Metals Under Small Stresses During Phase Transformation," *Proc. Roy. Soc.*, A 283 (1965), 403-422.

[11] C. L. Magee, "Transformation Kinetics, Microplasticity and Aging of Martensite in Fe-31 Ni," (Ph.D. Thesis, Carnegie Institute of Technology, 1966).

[12] J. B. Leblond, G. Mottet, and J. C. Devaux, "A Theoretical and Numerical Approach to the Plastic Behavior of Steels During Phase Transformations - I. Derivation of General Relations," *J. Mech. Phys. Solids*, 34 (4) (1986), 395-409.

[13] J. B. Leblond, G. Mottet, and J. C. Devaux, "A Theoretical and Numerical Approach to the Plastic Behavior of Steels During Phase Transformations - II. Study of Classical Plasticity for Ideal-Plastic Phases," *J. Mech. Phys. Solids*, 34 (4) (1986), 411-432.

[14] J. B. Leblond, J. Devaux, and J. C. Devaux, "Mathematical Modeling of Transformation Plasticity in Steels I: Case of Ideal-Plastic Phases, " *International Journal of Plasticity*, 5 (1989), 551-572.

[15] J. B. Leblond, J. Devaux, and J. C. Devaux, "Mathematical Modeling of Transformation Plasticity in Steels II: Coupling With Strain Hardening Phenomena," *International Journal of Plasticity*, 5 (1989), 573-591.

[16] D. J. Bammann et al,"Modeling A High Temperature Forging of 304L Stainless Steel," *NUMIFORM'95: The Fifth Intertnational Conference on Numerical Methods in Industrial Forming Processes*, this volume.

[17] S. Sjöström, "The Calculation of Quench Stresses in Steel," *(Ph.D. Thesis, Linkoping Univeristy, Sweden, 1982).*

[18] A. D. Freed, S. V. Raj, and K. P. Walker, "Three-Dimensional Deformation Analysis of Two-Phase Dislocation Substructures," *Metallurgica et. Materialia*, 27 (1992), 233-238.

[19] U. F. Kocks, "Constitutive Behavior Based on Crystal Plasticity," *Unified Constitutive Equations for Creep and Plasticity*, ed. A. Miller (London: Elsevier Applied Science, 1987), 1-88.

[20] B. Budiansky and T. T. Wu, "Theoretical Prediction of Plastic Strains of Polycrystals," *Proc. Fourth U. S. Cong. Appl. Mech.*, Vol. 2, (New York, NY: American Society of Mechanical Engineers, 1962), 1175-1185

[21] "Prediction and Control of Heat Treat Distortion of Helicopter Gears," (Interim Report to the U. S. Army #41686, Arthur D. Little, Inc., 1993).

[22] J. H. Biffle, "JAC3D - A Three-Dimensional Finite Element Computer Program for the Nonlinear Quasi-Static Response of Solids With the Conjugate Gradient Method," (SAND87-1305 UC-814, Sandia National Laboratories, 1993).

[23] V. C. Prantil, "Thermal Response During Quench of a Long, High Aspect Ratio Cylinder," (SAND Report in preparation, Sandia National Laboratories, 1995).

Simulation of Materials Processing: Theory, Methods and Applications, Shen & Dawson (eds)
© 1995 Balkema, Rotterdam. ISBN 90 5410 553 4

Development of local shear bands and orientation gradients in fcc polycrystals

A.J. Beaudoin, Jr
Reynolds Metals Company, Richmond, Va., USA

H. Mecking
Technical University Hamburg-Harburg, Germany

U.F. Kocks
Los Alamos National Laboratory, N.Mex., USA

ABSTRACT: A finite element formulation which derives constitutive response from crystal plasticity theory is used to examine localized deformation in fcc polycrystals. The polycrystals are simple, idealized arrangements of grains. Localized deformations within individual grains lead to the development of domains that are separated by boundaries of high misorientation. Shear banding is seen to occur on a microscopic scale of grain dimensions. The important consequences of these simulations are that the predicted local inhomogeneities are meeting various requirements which make them possible nucleation sites for recrystallization.

1 INTRODUCTION

Efforts that combine the finite element method with crystal plasticity models are often applied to problems of macroscale deformations. In such applications, polycrystal plasticity theories provide a useful means of introducing a description of material anisotropy which is both intializable and evolvable. There is however another research path that is providing valuable results: the study of material microstructure. The ability of finite element formulations to address gradients in the deformation field enables the researcher to probe grain-to-grain interactions in the microstructure.

With regard to fcc metals with high stacking fault energy, a question that has been the subject of much labor concerns the origin of recrystallization nuclei. It is well established that aluminum and copper, when rolled to sufficient reduction and then subsequently heat treated, will develop a strong recrystallization texture. Upon heat treatment, strain free grains with orientations different from adjacent grains appear. Growth of these recrystallized grains is aided by the motion of high-angle boundaries (Vandermeer & Juul Jensen 1994). In fcc metals, this recrystallization texture is often dominated by the cube orientation - an orientation that is not predicted when the usual crystal plasticity models are applied to homogeneous plane strain compression.

Looking back (before any heat treatment) to the prior cold working of the metal, it is the intent of this work to assess the potential of a finite element model to develop the mis-orientations, then - possibly - the specific cube orientations which have been the focus of so much experimental work.

1.1 *Development of misorientation in the deformation microstructure of fcc metals*

Generation of large-angle subboundaries may possibly procede from a variation of slip system activity throughout a grain (Kocks *et al.* 1994). Microscopic variations in slip lead to local domains with differences in the plastic rotation. Subdivision occurs within a grain, with some misorientation existing between adjacent domains. Several names have been given to microstructural features of this general kind: dislocation boundaries, dense dislocation walls, microbands, lamellar boundaries (Hansen & Juul Jensen 1992, Hughes & Hansen 1994). From the viewpoint of achieving a nucleus for recrystallization, we will restrict our consideration to "domains", separated by relatively high misorientation, developing within a grain.

1.2 *Generation of cube orientations in plane strain compression*

Numerous researchers have offered experimental evidence showing the presence of cube orientations in deformed microstructures. Inhomogeneity of the imposed strain field due to friction effects and tool geometry leads to mesoscale shearing which produces the rotated cube component (100)[011] (Hansen & Mecking 1975, Fedosseev & Gottstein 1993). For moderate reductions of 80-90%, cube orientations rotated about the sheet normal N have been observed in polycrystals (Doherty *et al.* 1993) and single crystals (Kamijo *et al.* 1993). At reductions of 95% or more, true cube orientations begin to appear from recrystallization (Nes & Dons 1986, Doherty *et al.* 1993)

Kamijo *et al.* (1993) have noted the importance of aspect ratio on the formation of cube nuclei. In experimental studies involving rolling of a single S-orientation, the generation of cube nuclei is enhanced by increasing the crystal length-to-thickness ratio. In addition, the cited work has pointed to the formation of (001)[uv0] (N-rotated cube) due to deformation inhomogeneity required to maintain compatibility amongst neighboring grains.

The Dillamore and Katoh (1974) mechanism provides for divergent rotation within a grain. As the orientation gradient within the grain increases, a cube orientation is formed in a transition band inside of the grain. Recently, Lee *et al.* (1994) have proposed a deformation banding model which produces cube orientations at reductions of 80%.

1.3 *Objective*

The objective of this work is to address two questions:

- can regions of high misorientation be developed due to deformation heterogeneities arising from grain-to-grain interaction, and
- are cube (or near cube) orientations developed in regions of deformation heterogeneity?

Two model arrangements of crystals are developed to study the development of deformation heterogeneity.

2 METHOD

2.1 *Description of the Hybrid Finite Element code*

The hybrid element formulation is a viscoplastic formulation centered on two residuals (treatment of the incompressibility constraint is detailed in Beaudoin *et al.* (1994). A statement of equilibrium is derived from the traction balance between elements, as

$$0 = \sum_e \left[\int_{\mathcal{B}_e} tr(\boldsymbol{\sigma}\text{grad}(\Phi))\, dV + \int_{\mathcal{B}_e} \Phi \cdot \text{div}(\boldsymbol{\sigma}) dV - \int_{\partial\mathcal{B}_t} \Phi \cdot \mathbf{t}\, dA \right] \quad (2.1)$$

where $\boldsymbol{\sigma}$ is the Cauchy stress, $\mathbf{t}$ are tractions, $\mathcal{B}_e$ is the element volume, $\partial\mathcal{B}_t$ is the element surface, and Φ are weighting functions. A second residual is formed for the deviatoric stress derived from the crystal consitutive response

$$\int_{\mathcal{B}_e} \mathbf{T} \cdot \mathcal{C}[\boldsymbol{\sigma}'] dV = \int_{\mathcal{B}_e} \mathbf{T} \cdot \mathbf{D}' dV \quad (2.2)$$

where $\boldsymbol{\sigma}'$ and $\mathbf{D}'$ are the deviatoric portions of the stress and deformation rate, $\mathcal{C}$ describes the constitutive response, and $\mathbf{T}$ are weighting functions. The combined use of a single orientation (or single polycrystal aggregate, though not utilized here) in each element with piecewise discontinuous shape functions for the deviatoric stress facilitates concurrent (parallel) computations. Hence, these simulations are particularly well-suited to massive parallel architectures. Spectral decomposition of the mesh is used to partition the unassembled global system of equations. Adherence to Fortran-90, with communications performed through library routines allows for porting of the code (even to workstations).

2.2 *A hypothetical bicrystal*

To explore the ability of a particular orientation to accommodate its surroundings in plane - strain compression, a hypothetical bicrystal structure was investigated (Figure 1). A single orientation, representing a grain, was imbedded in an environment consisting of a single S-orientation, $\{123\} < 634 >$. The entire assembly was then subjected to a plane strain compression.[1] Nodes on the two free faces (faces with normal to R) were constrained such that the structure retained a brick shape.

Three interior grains were examined using (ψ,θ,ϕ) in the convention of Kocks: (10,12,22), (35,45,30), and (55,12,22). Mesh configurations of $12 \times 12 \times 12$, $16 \times 16 \times 16$, and $32 \times 32 \times 32$ were used.

[1] To simplify discussion of the resulting texture, the axes of the sample coordinate system for the plane strain compression will be referred to using their rolling counterparts: R (rolling), T (transverse), and N (normal) directions.

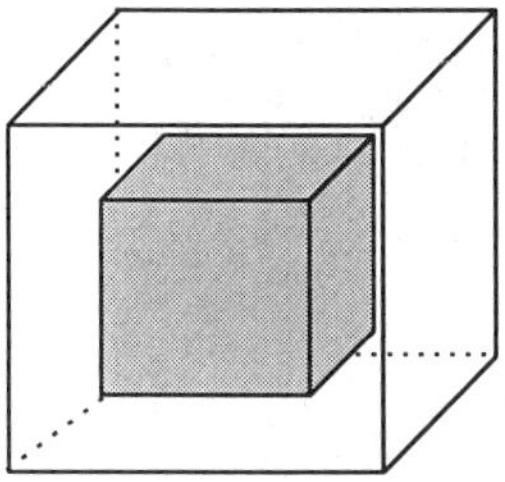

Figure 1. The bicrystal.

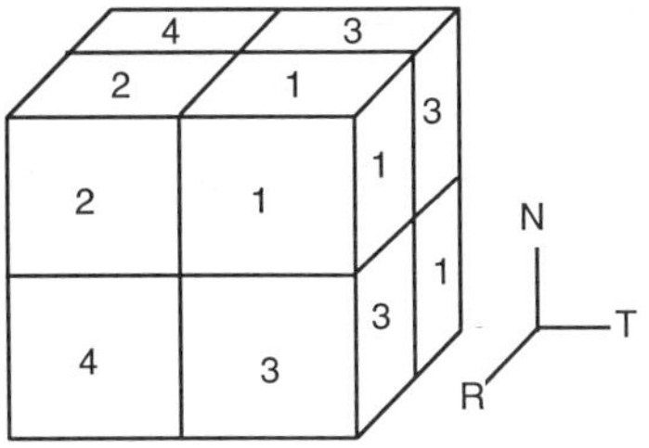

Figure 2. The S-array.

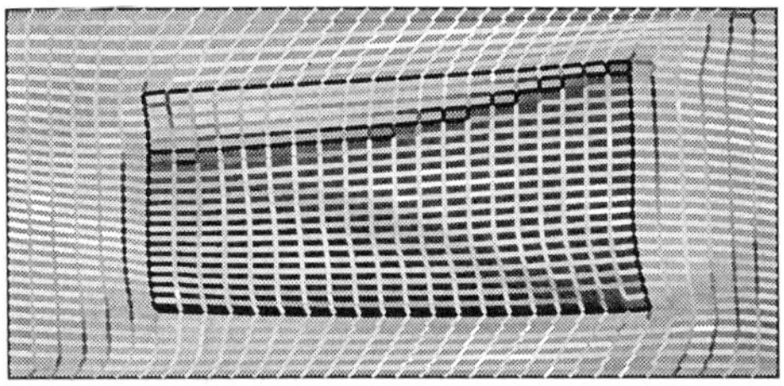

Figure 3. Bicrystal with (35,45,30) interior grain.

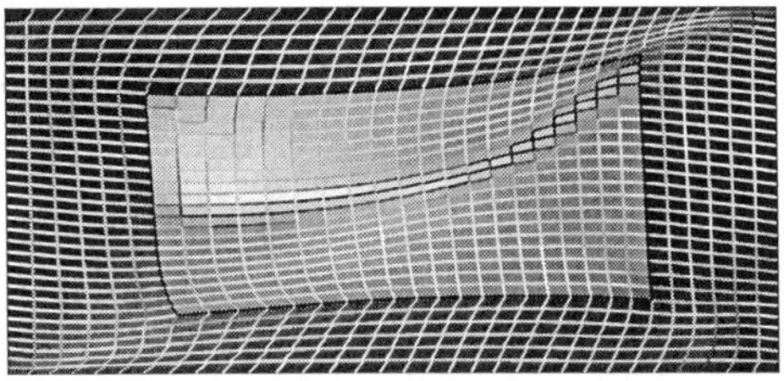

Figure 4. Bicrystal with (55,12,22) interior grain.

2.3 *Array of S-orientations*

This model was developed with the intention of producing a periodic structure which maintained orthotropic sample symmetry during plane strain compression (at least in the interior, on a local scale). In rolling of a single crystal with an S-orientation, there are two shears produced: ϵ_{RN} and ϵ_{TN}. The positive and negative sense of each of the two shears combine in four combinations for the four respective symmetrical orientations. In our model, the four symmetrical S-orientations were placed in a $2 \times 2 \times 2$ array (Figure 2). The symmetrical orientations were arranged such that the shears balanced along center planes with normals in the R and T directions. This structure will be referred to as the S-array.

Mesh densities of $16 \times 16 \times 16$ and $32 \times 32 \times 32$ were used. For reasons to be described subsequently, simulations were run with two initial geometries. One geometry was initially cube-shaped, the other was brick-shaped with relative proportions in directions of R:T:N as $2{:}1{:}\frac{1}{2}$.

3 RESULTS

3.1 *The bicrystal*

Results for the (35,45,30) and (55,12,22) interior grains are shown in Figures 3 and 4. In these plots, the inside of each element is shaded according to the middle Euler angle, θ. The boundary is shaded in proportion to the misorientation of an element with its neighbor; a solid black border indicates a misorientation of 10° or more.

Distinct structures were observed for each of the three trial orientations. For the orientation (10,12,22), a gradient in orientation was present across the grain, but not so great that any significant misorientation was developed. In contrast, a slice with normal in the T direction for the (35,45,30) orientation shows the development of a distinct boundary near the edge of the grain (Figure 3). The most interesting result was provided by the orientation (55,12,22). Here, a band courses through the grain which is sharply misoriented from the grain bulk (Figure 4). It was the resolution of this band that prompted refinement of the mesh outlined above; the band appeared in the same location for all mesh configurations.

In an effort to rule out hardening as an initiator of the localization, the simulation of the (55,12,22) orientation was re-run with a high hardening rate so as to prevent saturation of the flow stress. Changes in hardening rate did not alter the texture development.

For the bicrystal shown in Figure 4, orientations within the band have the (001) crystal axis quite close to the sheet normal direction: they are close to "cube". These orientations were rotated cube; band orientations with closest coincidence between the crystal (001) axis and the sheet

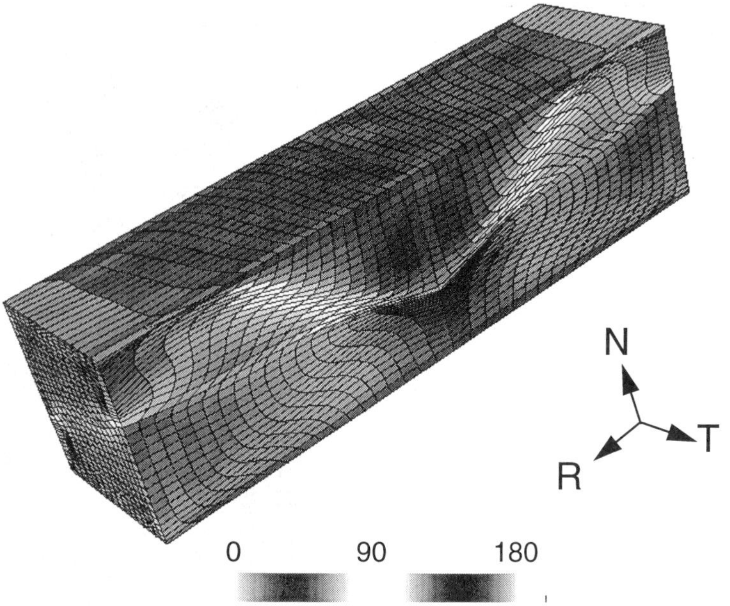

Figure 5. Array of S-orientations shown after 50% reduction. N-rotated cubes lie on shear shear band.

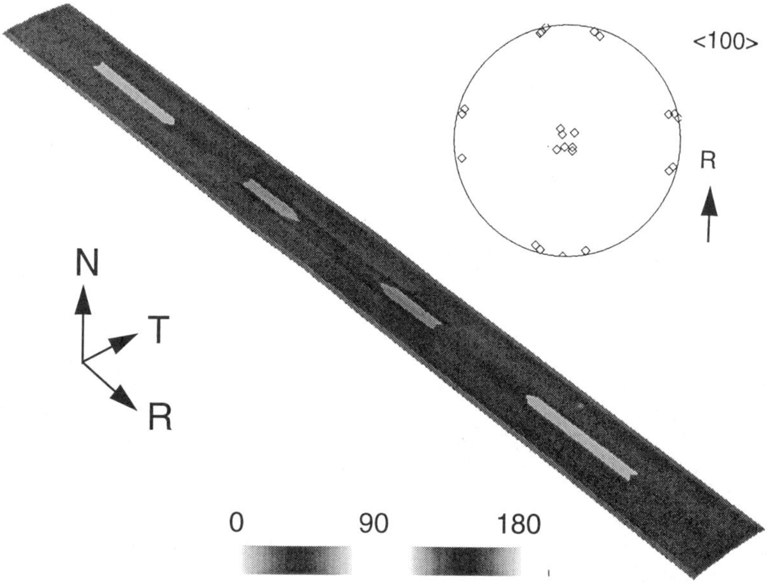

Figure 6. S-orientation with intial $2{:}1{:}\frac{1}{2}$ aspect ration taken to 80% reduction. The mid-plane with normal N is shown.

normal were rotated roughly 30° degrees from the sheet rolling direction.

3.2 *S-array*

Initial trials were carried out with a 1:1:1 aspect ratio. Figure 5 shows the mesh after a 50% reduction. The view shown in Figure 5 is a cut through the mid-plane with normal in the T direction. The symmetry designed into the problem, that center-planes with normals T and R would remain plane, was clearly maintained. Shearing deformations are quite severe. Similar to the bicrystal simulation of Figure 4, bands develop within the mesh. Situated in the bands are cube orientations - again, of the N-rotated cube variety.

Kamijo et al. (1993) have suggested that the grain aspect ratio plays a role in the development of cube orientations from an initial *S*-orientation. Prompted by this suggestion, simulations were re-run with the flattened aspect ratio of $2{:}1{:}\frac{1}{2}$ (R:T:N). Results after a reduction of 80% are shown in Figure 6. Orientations from the light-shaded regions of Figure 6 are also shown in the accompanying $<100>$ pole figure. There are clusters of cube orientations which are now within 20° of "true" cube.

The rotations of a single orientation - marked in Figure 6 - from S towards (001)[100] is shown in Figure 7. The rotation of an adjacent orientation is shown in Figure 8. Here, it is seen that the neighboring region re-orients towards a symmetrical orientation.

4 DISCUSSION

It is known that inhomogeneities in deformation may be introduced on a mesoscopic scale in the sense that they are not confined to a single grain as a consequence of the imposed strain field (due to friction effects and tool geometry) or by strain localizations due to instabilities of flow (caused by work hardening characteristics and rate sensitivity). The present paper shows that shear banding occurs also on a microscopic scale (of grain dimensions); it is presumably caused by the reaction forces set up between grains with no regard to the inhomogeneities of either the external strain field or by strain localization. These interactions again produce shear zones with a similar strain geometry to the mesoscopic shear bands and similarly, the rotated cube appears.

Several observations of structure development in these idealized models concur with descriptions made in experimental studies.

- Clearly, the present work points to the possibility of developing domains within a grain that are separated by boundaries of high (> 10°) misorientation. Two distinct structures were observed: a grain breaking up into domains and bands of orientations which are mis-oriented (on both sides of the band) from the grain bulk. These domains follow from slip variations within the grain leading to differential rotation between domains.
- At intermediate reductions, N-rotated cube orientations appear in bands. In the bicrystal and *S*-array models, the bands are produce by shear deformations initiated through grain interaction.
- To effect a rotation of an element towards true cube requires quite large deformations. Combining the aspect ratio of the initial configuration with reduction of the flattened *S*-array, a 95 % reduction from an equiaxed substructure was required to produce the near cube sites.

The important consequences of these simulations are that the predicted local inhomogeneities are meeting various requirements which make them possible nucleation sites for recrystallization. First, they are sufficiently narrow so that a nucleus picks up quickly a strong orientation difference with its neighborhood as it expands into its environment; a high-angle, and therefore highly mobile, boundary is formed quickly. Secondly, they are everywhere in the microstructure since they form in every grain at large enough strains and are dispersed throughout the whole volume.

Though the precise (001)[100] cube orientation was not produced in this work, with high grain aspect ratio and reduction orientations with closer rotation towards true cube were developed. In the Dillamore and Katoh (1974) theory, cube orientations form in transition bands generated by the deformation of (001)[uv0] orientated material. The suggestion by Kamijo *et al.* (1993) that a slip rotation from S towards an intermediate (001)[uv0] orientation may arise from to local inhomogeneities in deformation is supported by the present work.

Finally, a few comments must be made with regard to the numerical aspects of the work. Much effort was placed in refining the mesh to resolve gradients in the texture. In all of the simulations, the pattern of texture development was unchanged with the mesh discretization. For the bands containing the N-rotated cube shown in Figures 4 and 5, it was possible to capture two or more elements at or near rotated cube within the band. This was not the case for the *S*-array with flattened aspect that produced the truer cubes. Here, the texture

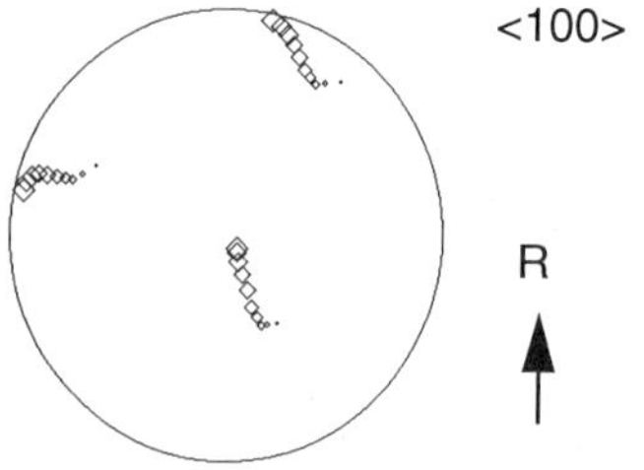

Figure 7. Rotation of an element orientation from "*S*" to cube.

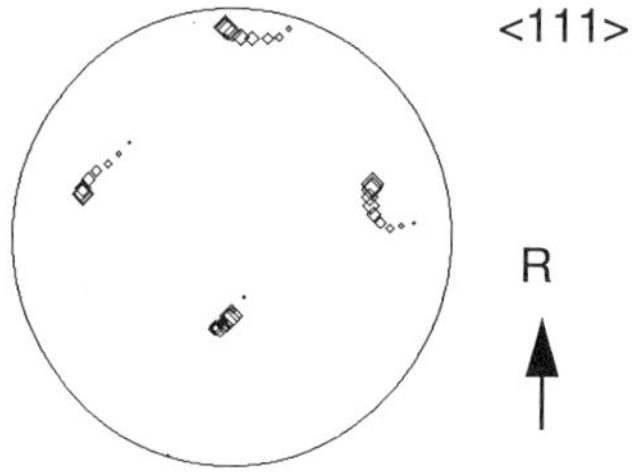

Figure 8. Rotation of an orientation from an element adjacent to that shown in Figure 7. The rotation is toward a symmetrical position.

gradients appear to be quite severe, with a cube orientation lying in the center of typical rolling orientations. Future efforts will focus on mesh refinement necessary to better capture this gradient and enable studies of higher reduction.

ACKNOWLEDGEMENTS

Dr. Stuart Wright provided software subroutines for the analysis of misorientation. The simulation software was developed under Office of Naval Research contract N00014-90-J-1810. Computing resources were provided through the Advanced Computing Laboratory at Los Alamos National Laboratory, sponsored by the U.S. Dept. of Energy.

REFERENCES

Beaudoin, A.J., Dawson, P.R., Mathur, K.K., Kocks, U.F., and D.A. Korzekwa 1994. Application of Polycrystal Plasticity to Sheet Forming. *Comput. Methods Appl. Mech. Engrg.* **117**:49-70.

Dillamore, I.L. and H. Katoh 1974. The Mechanisms of Recrystallization in Cubic Metals with Particular Reference to Their Orientation-Dependence. *Met. Sci.* **8**:73-83.

Doherty, R.D., Kashyap, K., and S. Panchadeeswaren 1993. Direct Observation of Recrystallization Texture inCommercial Purity Aluminum. *Acta metall. mater.* **41**:3029-3053.

Fedosseev and Gottstein, 1993. Inhomogeneous Deformation and Texture during Rolling. *Modelling of Rolling Processes*:296-308. London: The Institute of Metals.

Hansen, N. and D. Juul Jensen 1992. Flow Stress Anisotropy Caused by Geometrically Necessary Boundaries. *Acta metall. mater.* **40**:3265-3275.

Hansen, J. and H. Mecking 1975. Influence of the Geometry of the Deformation on the Rolling Texture of F.C.C. Metals. In G.J. Davies, I.L. Dillamore, R.C. Hudd and J.S. Kallend (eds.) *Texture and Properties of Materials*, Proceedings of ICOTOM IV, Metals Society, London, p. 34.

Hughes, D.A. and N. Hansen 1994. A Comparison of the Evolution of Cold and Hot Deformation Microstructures in FCC Metals. In J.J. Jonas, T.R. Bieler, and K.J. Bowman (eds.), *Advances in Hot Deformation Textures and Microstructures*:427-444. Warrendale, PA:TMS.

Kamijo, T., Adachihara, H., and H. Fukutomi, H. 1993. Formation of a (001)[100] Deformation Structure in Aluminum Single Crystals of an *S*-orientation. *Acta metall. mater.* **41**:975-985.

Kocks, U.F., Embury, J.D., Cotton, J.D., Chen, S.R., Beaudoin, A.J., Wright, S.I., and A.D. Rollett 1994. Attempts to Model the Generation of New Grain Boundaries During the Deformation of Polycrystals. In J.J. Jonas, T.R. Bieler, and K.J. Bowman (eds.), *Advances in Hot Deformation Textures and Microstructures*:459-468. Warrendale, PA: TMS.

Lee, C.S, Smallman, R.E., and B.J. Duggan 1994. Deformation Banding and Formation of Cube Volumes in Cold Rolled FCC Metals. *Mater. Sci. Technol.* **10**:862-868.

Nes, E., and A.L. Dons 1986. Nucleation of Cube Texture in Aluminium. *Mater. Sci. Tech.* **2**:8

Vandermeer, R.A. and D. Juul Jensen 1994. Modeling Microstructural Evolution of Multiple Texture Components During Recrystallization. *Acta metall. mater.* **42**:2427-2436.

Simulation of Materials Processing: Theory, Methods and Applications, Shen & Dawson (eds)
© 1995 Balkema, Rotterdam. ISBN 90 5410 553 4

Linking microscopic and macroscopic effects during solidification of semicrystalline polymers

A. Benard & S.G. Advani
Department of Mechanical Engineering, University of Delaware, Newark, Del., USA

ABSTRACT: Texture and crystallinity of semi-crystalline polymers are strong functions of crystallization kinetics, concentration of impurities and the rate of cooling during solidification. To establish a link between these phenomena, a mathematical framework recently introduced for metals was adopted to model the evolution of various crystalline morphologies at the microscopic scale by coupling heat and mass diffusion with the Lauritzen-Hoffman theory of crystal growth. This approach allows us to link the temperature of the crystalline structure with the heat evolved. A volume averaging process is used that allows the interactions between the solid and liquid phases at the microscopic level to be accounted for in the macroscopic heat conduction equations. A finite element formulation of the heat conduction equations in three dimensions, coupled with the averaged microscopic equations at the element level, permits us to predict the microstructure in addition to the temperature distribution for an entire part.

1 INTRODUCTION

Thermoplastic semicrystalline polymers constitute one of the most important family of polymers used in manufacturing products ranging from household appliances to high-performance composite structures. These polymers are generally more sensitive than other materials to the processing conditions and large microstructure variations can occur within the same finished product. This results in local variations of the mechanical properties. It is therefore necessary to understand the role of processing conditions on the heterogeneous microstructure of such materials.

The development of the microstructure in quiescent semicrystalline polymers and composites is directly dependent on the nucleation rate, the crystals growth rate and the diffusion of the non-crystallizable material. These phenomena are strongly linked to the evolution of the heat flow in a part. To model the coupling between the development of the microstructure with the macroscopic heat flow, it is possible to use a numerical scheme in which computations are performed at both the macroscopic and the microscopic levels. Microscopic models are used to describe the evolution of the microstructure in a small control volume, and these models are coupled with the computation of the overall heat flow over an entire part. It is the aim of this work to describe and model the relationship between the evolution of the microstructure to the macroscopic heat flow in semicrystalline polymeric materials.

The crystallization of small volume of polymer can be computed with the help of models that represents the evolution of the average microstructure. A wide variety of entities are encountered after the solidification of quiescent semicrystalline polymers and composites, and corresponding models have been developed (Benard & Advani 1995[a]). These models simulate the growth of entities with planar, cylindrical and spherical symmetries. The evolution of such entities in a small control volume is coupled through the latent heat release with the temperature evolution of an entire part. Some of the coupling schemes used for metals were recently reviewed by several authors (Beckermann & Viskanta 1993, Tseng et al 1990, Rappaz 1989) and the same techniques can be used with semicrystalline polymers. To allow interactions between the solid and liquid phases at the microscopic level to be accounted for in the macroscopic equations, a volume averaging procedure is used to describe the microstructure. A three dimensional finite element formulation of the heat conduction equation, coupled with the microscopic equations at the element level, permits the prediction of the microstructure and temperature distributions during solidification of an entire part.

2 MICROSCOPIC MODELS

Three basic microscopic models can be used to represent the evolution of the microstructure in semicrystalline polymers. These models are represented in Figure 1. The first model corresponds to the solidification of a planar front growing from a cold wall. The second model is associated with cylindrical crystals often encountered around carbon fibers. The third model represents the evolution of spherical entities called spherulites. The most common entities encountered during the solidification of quiescent semicrystalline polymers are spherulites. Spherulites are polycrystalline aggregates formed from a radiating array of crystalline fibers that branch regularly to create a three-dimensional structure of approximate radial symmetry. The physical and chemical properties of a part are directly linked to the evolution of such entities i.e. the degree of crystallinity, the size of the spherulites, the distribution of the non-crystallizing species, etc.

The growth of a spherulite can be represented with concepts similar to those introduced by Rappaz and Thevoz (1987) in modeling the solidification of metallic alloys. Three concentric zones can be defined within a unit cell : a liquid zone, an interdendritic zone and a solid zone. The equations that must be solved within the cell for the the crystallization process are crystal growth rate, heat balance, mass diffusion in the liquid domain, and a mass conservation for the entire cell. The growth rate of the crystals can be represented in a general manner by the Lauritzen-Hoffman theory of crystal growth (Hoffman et al 1976)

$$\frac{\partial R_g}{\partial t} = G_o \exp\left[\frac{-U^*}{R(T-T_\infty)}\right] \exp\left[\frac{-K_g}{T\Delta T f}\right] \quad (2.1)$$

where the first term represents the temperature dependence of the segmental jump rate in the polymer while the second term is the contribution from the net rate of nuclei formation. The description of each parameter is given elsewhere (Benard & Advani 1995^b). A lumped heat balance can be written a the level of the cell and is as follows

$$4q_{ext}\pi R_{tot}^2 = \left(-H_f\frac{\partial f_s}{\partial t} + \rho C_p\frac{\partial T}{\partial t}\right)\frac{4}{3}\pi R_{tot}^3 \quad (2.2)$$

where q_{ext} is the heat flow extracted from the cell, H_f is the volume heat of fusion, f_s is the solid fraction and C_p the heat capacity. Finally a mass balance of the non-crystallizable material can be written for the cell and it must be solved along with the diffusion equation in the liquid domain. The mass balance is given by

$$\int_o^{R_s} kc^* 4\pi d^2 dr + \frac{4}{3}\pi c^*(R_g^3 - R_s^3) + \int_{R_g}^{R_{tot}} c(r,t)4\pi r^2 dr = \frac{4}{3}\pi R_{tot}^3 c_o \quad (2.3)$$

where R_g and R_s are respectively the average spherulite and the solid radii. They are related to f_g and f_s with $f_g = R_g^3/R_{tot}^3$ and $f_s = R_s^3/R_{tot}^3$. c^* is the concentration of non-crystallizable material in the the interdendritic zone, k is a partitioning factor and R_{tot} gives the maximum size that a spherulite can reach. The diffusion equation in spherical coordinates is

$$\frac{1}{r}\frac{\partial^2}{\partial r^2}(cr) = \frac{1}{D}\frac{\partial c}{\partial t}, \quad (2.4)$$

where D is the mass diffusion coefficient.

Equations (1) through (4) are coupled and need to be solved simultaneously. The growth process is described with f_s and f_g, the solid and spherulite volume fractions respectively. These two parameters allow a detailed description of the evolution of the average microstructure within a small control volume. Other parameters such as the average spherulite size and the number of nuclei must also be tracked at each node. The governing equations and a solution scheme are presented by Benard & Advani (1995^b) for the spherulitic growth process. The macroscopic heat conduction equation, solved on the system level, provides q_{ext} required by the microscopic models for computing the evolution of the microstructure.

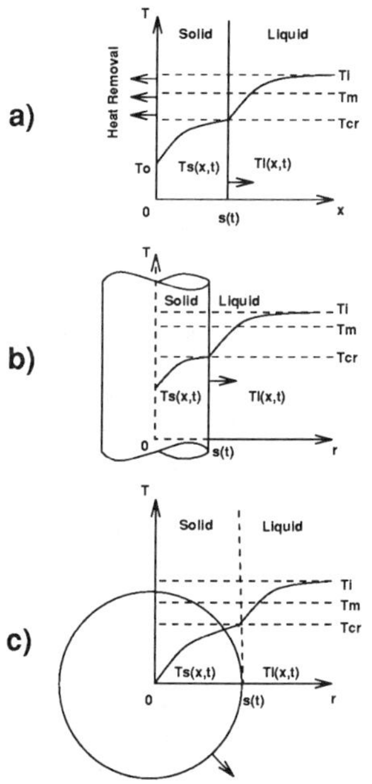

Figure 1. Schematic of three basic microstructures encountered in semicrystalline polymers and composites.

3 MICRO/MACRO COUPLING

The solidification process can be described with an enthalpy formulation of the energy equation, which is written for the entire physical domain considered. This equation encompasses the solid, liquid and mixed phases. The enthalpy formulation presents several advantages in modeling the cooling of semicrystalline polymers, the most important advantage being that such materials often solidify with a substantial solid/liquid zone. The nucleation and growth processes of the spherulites can sometimes occur over the entire thickness of a part.

3.1 *Governing Equations*

For quiescent melts, the energy equation, written with the temperature and the enthalpy as dependent variables, is given by

$$\frac{\partial H}{\partial t} = \frac{\partial}{\partial x_j}\left[k\frac{\partial T}{\partial x_j}\right] \tag{3.5}$$

where k is the thermal conductivity and T is the temperature. H is the volume enthalpy defined as

$$H = \int_{T_o}^{T} \rho C_p(\theta)d\theta + \rho L(1 - f_s) \tag{3.6}$$

where C_p is the heat capacity, f_s is the volume solid fraction and L is the latent heat of fusion. With this definition, we have

$$\frac{\partial H}{\partial t} = \rho C_p\frac{\partial T}{\partial t} - \rho L\frac{\partial f_s}{\partial t}. \tag{3.7}$$

As a first approximation, a linear relationship can be used to represent variations of the heat conductivity with temperature

$$\frac{k - k_m}{k_g - k_m} = \frac{T_m - T}{T_m - T_g} \tag{3.8}$$

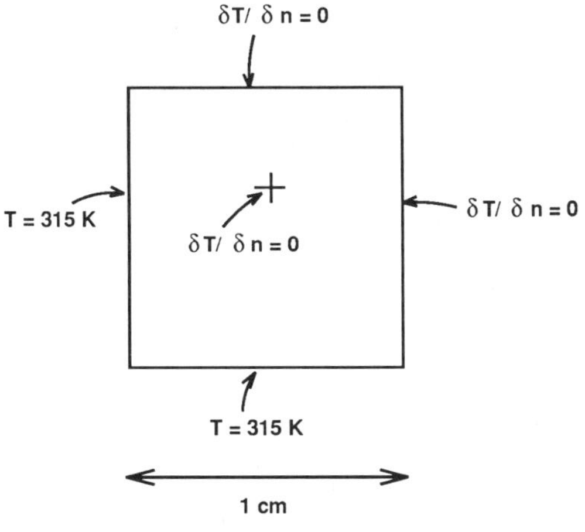

Figure 2. Schematic of the problem solved.

where k_m and k_g are the conductivities at the melting point (T_m) and at the glass transition (T_g) temperatures respectively.

3.2 *Numerical Solution of the Heat Conduction Equation*

It is possible to discretize the heat conduction equation (1) and obtain a system of nonlinear equations that can be written as

$$[K(T)]\{T\} + \{bc\} = [M]\frac{\Delta\{H\}}{\Delta t} \tag{3.9}$$

for the so-called latent heat method (Rappaz 1989). $[K(T)]$ is the temperature dependent conductivity matrix, $[M]$ is the mass matrix, $\{H\}$ is the vector of enthalpy at all nodes and $\{bc\}$ is the vector corresponding to the boundary conditions. The standard Galerkin finite element method was used in discretizing the system (Reddy and Gartling 1994). The possible boundary conditions that may be applied are constant temperature, heat flow and convection.

The temperature can be linearized by

$$\{T\}^{t+\Delta t} = \{T\}^t + \left[\frac{dT}{dH}\right]\Delta\{H\} \tag{3.10}$$

which simplifies the solution of the discretized equations. The change of enthalpy at each node can be found and it provides the condition required for computing the evolution of the microstructure.

This approach was implemented in a computer code written with the C programming language. The program allows the modelization of three dimensional structures.

4 RESULTS AND DISCUSSION

To illustrate the model, the solidification of a rectangular channel of 2 cm × 2 cm was computed. Only one fourth of the channel was enmeshed with symmetry boundary conditions on two of the four channel walls surfaces. Symmetry was also applied on the front and back surfaces. A constant wall temperature of 315 K was applied on the other walls of the channel. A grid of 20 × 20 was chosen with a macroscopic time-step of 10^{-3} sec. The problem is schematically shown in Figure 2 and material properties of polyethylene were used (Gaur & Wundelich 1981).

The evolution of the temperature, the solid fraction and the grain size distribution can be computed during the solidification process. The results for the computations are presented in Figures 3, 4 and 5 respectively.

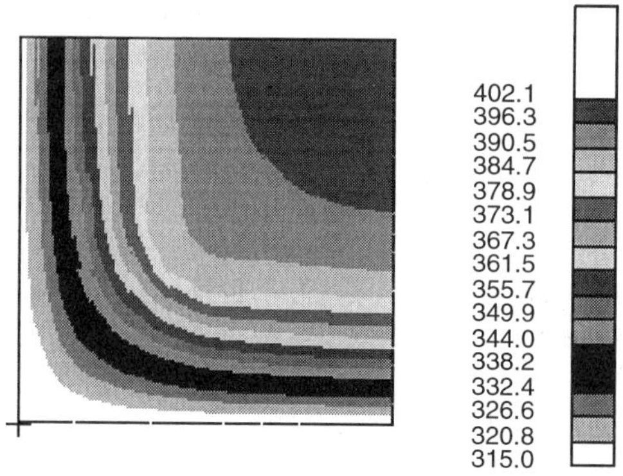

Figure 3. Isotherms in the channel after 50 seconds into the solidification process.

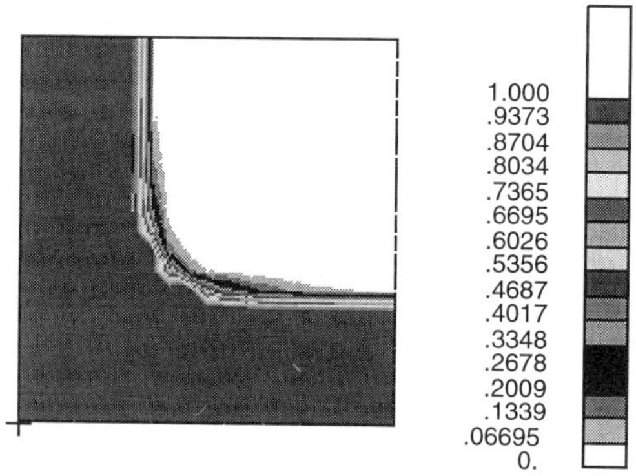

Figure 4. Solid fraction distribution after 50 seconds into the solidification process.

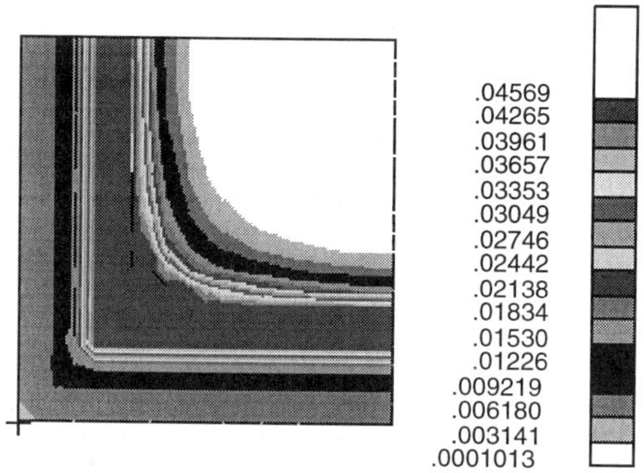

Figure 5. Spherulites size distribution after 50 seconds into the solidification process.

The temperature distribution can be observed in Figure 3, 50 seconds after the solidification process started. A scaled color bar shows the corresponding values of the isotherms in the domains (in Kelvin). It can be observed that the melt is undercooled at the center by about 15 K ($T_{melt} = 417.8K$) and that the isotherms follow an expected pattern.

The solid fraction f_s can be observed, in Figure 4, to form a rather narrow zone in which its value changes rapidly from 1 (solid) to 0 (liquid). The solid fraction is an indication of the of the amount of material that can be solidified under given heat transfer conditions. When the values of f_s and f_g (related to the spherulite size) are far apart, it indicates that the spherulites are not well formed and that secondary crystallization will occur.

The spherulites size distribution, which is one of the most interesting feature of these calculations, can be observed in Figure 5. It is shown that small spherulites are formed close to the wall, where the polymer is cooled rapidly. The size of the spherulites increases gradually as we move away from the cold walls. The maximum size is reached toward 1/4 of the channel half-thickness. As the polymer is still molten at the center the part, the spherulites size then decreases to zero, in the molten domain. After completion of the solidification process, a non-uniform spherulites size can be observed, especially close to the wall, as it can be observed in practice (Katti & Schultz 1982). Coupling the microscopic models with macroscopic phenomena allows one to predict this spherulitic distribution.

5 CONCLUSION

A micro-macro coupling scheme was used for modeling the solidification of semicrystalline polymers and the evolution of the microstructure in such materials. An enthalpy formulation of the energy equation was used and coupled with various microscopic models. It is then possible to obtain a detailed description of the microstructural changes occurring during solidification. A computer code has been developed to simulate the solidification process of semicrystalline polymers and results are presented for the isotherms, the spherulites size distribution and the solid fraction distribution.

REFERENCES

Benard, A. & S.G. Advani 1995[a]. Microscopic Models for the Spherulitic Growth in Polymers and Composites. in preparation

Benard, A. & S.G. Advani 1995[b]. A Cell Model to Describe the Spherulitic Growth in Semicrystalline Polymers. *Polym. Eng. Sci.* to appear.

Beckermann, C. & R. Viskanta 1993. Mathematical Modeling of Transport Phenomena During Alloy Solidification. *Appl. Mech. Rev.* 46:1-27.

Gaur U. & B. Wunderlich 1981. Heat Capacity and Other thermodynamic Properties of Linear Macromolecules. II. Polyethylene. *J. Phys. Chem. Ref. Data* 10:119-152.

Hoffman, J.D., G.T. Davis and J.I. Lauritzen 1976. The Rate of Crystallization of Linear Polymers With Chain Folding. In N.B. Hannay (ed), *Treatise on Solid State Chemistry* 3:497-614. New-York, NY:Plenum.

Katti, S.S. and J.M. Schultz 1982. The Microstructure of Injection-Molded Semicrystalline Polymers : A Review. *Pol. Eng. Sci.* 22:1001-1017.

Rappaz, M. 1989. Modelling of Microstructure Formation in Solidification Processes. *Int. Mater. Rev.* 34:93-123.

Rappaz, M. & Ph. Thevoz 1987. Solute Diffusion Model for Equiaxed Dendritic Growth : Analytical Solution. *Acta Metall.* 35:2929-2933.

Reddy, J.N. & D.K. Gartling 1994. *The Finite Element Method in Heat Transfer and Fluid Dynamics.* Ann Arbor : CRC Press.

Tseng, A.A., J. Zou, H.P. Wang and S.R.H. Hoole 1990. Numerical Modeling of Macro- and Micro-Behaviors of Materials in Processing. In ASME(ed),*Processing of Polymers and Polymeric Composites* MD V19:135-159. New-York, NY:ASME.

Simulation of Materials Processing: Theory, Methods and Applications, Shen & Dawson (eds)
© 1995 Balkema, Rotterdam. ISBN 90 5410 553 4

Stress concentrations around aligned fiber breaks in a unidirectional composite with an elastic-plastic matrix

I.J. Beyerlein & S.L. Phoenix
Theoretical and Applied Mechanics, Cornell University, Ithaca, N.Y., USA

A.M. Sastry
Sandia National Laboratories, Albuquerque, N.Mex., USA

ABSTRACT: This study develops a new computational technique for studying the crack-tip stress environment in a unidirectional fiber-reinforced composite, consisting of parallel equally-spaced elastic fibers embedded in an elastic-perfectly plastic matrix and under a remote tensile force. The method involves a recently developed Break-Influence Superposition (BIS) technique, within the framework of the shear-lag model first introduced by Hedgepeth. For a range of remote loads, the analysis examines the effects of plastic yielding on the fiber and matrix stresses around a "center crack," modeled by one to several contiguous fiber breaks. The model considers the case where the matrix yield region develops only in the matrix bay flanked between the last broken fiber and the first surviving fiber. Results show that the extent of the yield region increases linearly with applied load, and the growth rate of the yield zone increases with the number of breaks. Also, the remote stress required to initiate yielding at the crack tip decreases with increasing crack length, and once yielding develops, the stress concentrations and overload lengths in nearby unbroken fibers are substantially altered with magnitudes reduced but length scale increased.

1. INTRODUCTION

1.1 *Fracture Behavior of Composites*

The fracture of fiber-reinforced composites involves a complex progression of damage, characterized by random fiber breaks, stress transfer from broken to surviving fibers, interfacial debonding, and matrix yielding culminating in crack like structures [1]. This complexity in failure leads to the ability of fiber-reinforced composites to sustain high loads before catastrophic failure. In most fibrous composites, the fiber extensional stiffness is much greater than that of the matrix, and the matrix and interfacial shear strength are relatively weak. As a result, the matrix can plastically yield or debond from the fiber. Crack propagation begins when stresses in the fibers close to a "crack tip" exceed their strength; however, subsequent matrix yielding or interfacial debonding can greatly reduce these fiber stress concentrations. Therefore, a fracture criterion based on fiber fracture alone is questionable.

1.2 *Stress Concentrations and Matrix Yielding*

The local stress distributions in the fiber and matrix resulting from initial fiber breaks must be analyzed on a discrete microstructural level to gain understanding of the various failure mechanisms. In this work, damage due to initial fiber breaks and matrix yielding are described, where two important considerations are fiber stress concentrations and extent of the yielded region. Both correspond to achieving critical strain levels in the fibers and matrix ahead of the crack tip.

In composites, localized stress concentrations resulting from clusters of fiber breaks are the dominant driving force for crack propagation. For a transverse crack, represented by a row of fiber breaks, the maximum fiber stress concentrations lie along the crack plane; however, fibers do not necessarily fail in-line with the previous fiber break, since fiber strengths vary along the fiber length. Experimental observations on metal matrix composites show that the crack extends in a rough, non-planar fashion [2]. Therefore, both the fiber length where the fiber is overloaded and the magnitude of the peak stress concentration are important. It is essential to also consider matrix yielding as the load to cause yielding can be much lower than that to initiate fiber breaks. This affects how load is redistributed onto broken fibers. The present model assumes that a row of fiber breaks exists prior to application of the design load, matrix yielding is the only additional damage which occurs after application of the load, and yielding develops between the last broken and first intact fibers (See Fig. 1). This behavior is typically observed in ductile matrix composites. The results of our analysis will

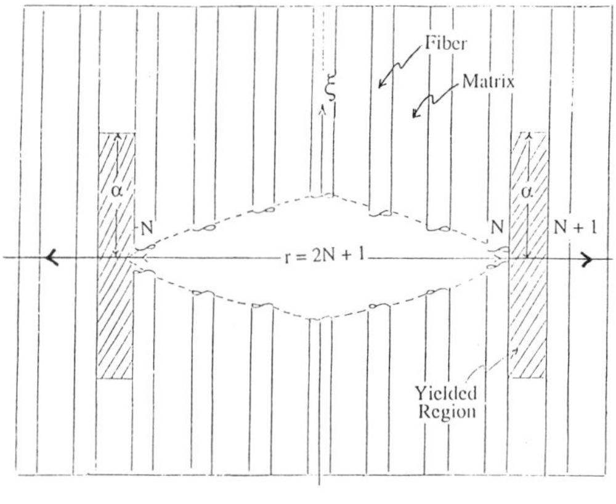

Fig. 1. Discretized two-dimensional unidirectional fiber composite lamina.

provide a relationship between the surrounding fiber and matrix stress concentrations, the plastic yield zone, number of fiber breaks, r, and the applied load.

2. PREVIOUS WORK

Predicting the fracture behavior of unidirectional fibrous composites has been tackled in a number of ways, including models extending the classical shear lag analysis, full elasticity solutions, linear elastic fracture mechanics (LEFM), and finite element methods [3, 4, 5, 6, 7]. We focus on the first. For a two-dimensional array of elastic matrix and fiber elements, the stress concentrations in the matrix and fiber elements immediately adjacent to the last broken fiber of a group of r breaks have been determined by Hedgepeth [6], using a shear lag analysis which assumes that the matrix transmits only shear between the fibers, and the fibers carry all the tensile load.

Recently, the BIS [8] technique was developed in Hedgepeth's framework to practically determine the stress concentrations *everywhere* due to *arbitrarily* located fiber breaks. The technique introduced in this study, further extends the BIS technique to handle matrix yielding. Many other models have been developed to extend Hedgepeth's model by eliminating restrictions on the number and arrangement of fiber breaks and on the size of the stress field which can be evaluated, and by accounting for matrix yielding or a debonding fiber-matrix interface, or by making it applicable to tensile-carrying matrix composites [3, 9, 10]. However, these improvements also increase the complexity of the computation. In several of these studies, yielding was assumed to occur only in the matrix bay between the last broken and first surviving fiber. Experiments on notched unidirectional fibrous sheets under tension show that the yield zone develops first at the crack tip and then extends in a matrix region parallel to the fibers [11]. Hedgepeth and Van Dyke [12] also developed a two-dimensional elastic-perfectly plastic matrix model and obtained an analytic solution for the stress concentration in fiber elements immediately adjacent to an single, isolated fiber break. His results indicated that stresses much above the stress required to initiate yielding are necessary to significantly reduce the stress concentrations.

Goree and Gross [13] and Dharani et al [10] variously developed the shear lag model to predict the length of the plastic zone or debond (split) length, the matrix shear and fiber stress at the end of a transverse notch, and the effect of transverse fiber damage ahead of the crack tip. Either type of damage reduced the maximum stress concentration on the fibers ahead of the crack tip.

One advantage of the BIS based model is that the extent of the yield zone and stress distributions everywhere can be computed given any arbitrary configuration of broken fibers, and applied stress. Thus, stress concentrations are estimated on a microstructural level. Also the solution requires discretization only of an area surrounding the fiber breaks and yielded regions, and therefore can analyze much larger composites with many randomly located fiber breaks, than using a full discretization scheme.

3. ANALYSIS

3.1 *2-D Lamina*

Determining the fiber and matrix loads and displacements for arbitrarily located fiber breaks in a large lamina involves first discretizing the two-dimensional composite lamina near fiber breaks. Each fiber and matrix element behaves elastically and is of equal length 2δ, where δ is much less than the *characteristic load transfer length*--the fiber length required for an isolated broken fiber to recover its load. The center fiber is numbered n = 0, and the fibers to the right are numbered n = 1, 2, . . . ∞ and those to the left, n = -1, -2, . . .-∞. The matrix bay to the right of fiber n is matrix bay n. Likewise, along the fiber direction, the respective elements are numbered in the + x direction from (n, 0) to (n, ∞) and in the -x direction, from (n, -1) to (n, -∞).

3.2 *The Analysis*

The shear lag assumptions are incorporated when deriving the force equilibrium equations on the fiber element along the fiber axis, x. Let $p_n(x)$ and $u_n(x)$ be the force and displacement, respectively, in fiber n at location x, and p is the load applied per fiber at x = $\pm\infty$. For those elements outside of the yielded region, the equilibrium of forces in the x-direction results in the following equilibrium condition [1],

$$\frac{EAd^2u_n(x)}{dx^2} = \qquad (1)$$

$$\frac{-Gh(u_{n+1}(x) - u_n(x)) + Gh(u_n(x) - u_{n-1}(x))}{w}$$

For convenience, non-dimensional loads and displacements are defined to simplify Eqn 1. However, unlike the classical shear lag model, all quantities in this study are normalized with respect to p*, the remote tensile force per fiber normalized by the matrix yield stress in shear, τ_y. P_n and U_n, the load and displacement in fiber n, and ξ the axial coordinate are,

$$P_n = p_n/p^*, \qquad U_n = \frac{u_n}{p^*\sqrt{\frac{w}{EAGh}}} \qquad (2a)$$

$$p^* = \tau_y\sqrt{\frac{wEAh}{G}} \qquad (2b)$$

$$\xi = \frac{x}{\sqrt{\frac{wEA}{Gh}}} \qquad (2c)$$

The non-dimensional shear stress T_n and strain Γ_n are

$$T_n = \frac{\tau_n}{p^*}\sqrt{\frac{wEAh}{G}} \qquad (3a)$$

$$\Gamma_n = U_{n+1} - U_n = \frac{\gamma_n}{p^*}\sqrt{wEAGh} \qquad (3b)$$

E is Young's modulus of the fibers, G is the effective shear modulus of the matrix, w is the effective fiber spacing, and h is the fiber width and laminate thickness. Also note that when the matrix is deforming elastically, $\tau = G\gamma_n$, and therefore, T_n and Γ_n are equivalent. According to Eqns 2b and 3, the non-dimensional yield stress and strain criteria are $T_y = \Gamma_y = 1$. By introducing these non-dimensional parameters, the equilibrium equation (Eqn 1) simplifies to,

$$P_n(\xi) = \frac{dU_n(\xi)}{d\xi} \qquad (4)$$

and

$$\frac{d^2U_n(\xi)}{d\xi^2} + U_{n+1}(\xi) - 2U_n(\xi) + U_{n-1}(\xi) = 0$$

3.3 *Two Unit Elastic Problems*

The problem of determining the fiber loads and displacements everywhere is reduced to solving for the stress redistribution due to two basic forms of damage: an isolated fiber break in fiber element (0,0) and a shear load couple applied to matrix element (0,0). These solutions are derived using the shear lag analysis and therefore, assume the composite contains fully elastic elements. The first one is the *isolated break problem*--the stress redistribution due to a single break located at fiber element (0,0) [14]. In this problem, a unit compressive force is applied to each end of the break and zero tensile force at infinity. The second elastic problem is the *load couple problem*--the stress distribution due to a shear force couple [14]. In a laminate containing no breaks, a positive unit load couple is applied at ξ=0 between fibers n = 0 and n = 1, and a zero edge force is applied at infinity. This load couple induces negative and positive unit load jumps in fibers n = 0 and n = 1, respectively, and negative shear in the matrix (0,0).

The displacement solutions to the isolated break $V^*_{b,n}$ and the isolated shear couple problem $V^*_{c,n}$ are called the virtual quantities, and are developed in [14]. The corresponding values for fiber and matrix loads $L^*_n(\xi)$ and $\Gamma_n(\xi)$ are determined using Eqns 4 and 3b with V in replace of U. The results for the isolated break and load couple problem are,

$$V^*_{b,n}(\xi) = \mathrm{sgn}(\xi)\frac{1}{4}\int_0^{\pi} [\cos(n\theta) \cdot e^{-2|\xi|\sin(q/2)}]\, d\theta \qquad (5a)$$

$$V^*_{c,n}(\xi) = \frac{1}{4\pi}\int_0^{\pi} \{\cos(n\theta)-\cos[(n-1)\theta]\} \cdot \{\sin(\theta/2)\}^{-1} e^{-2|\xi|\sin(\theta/2)}\, d\theta \qquad (5b)$$

respectively, where sgn(ξ) = 1 when $\xi \geq 0$ and -1 when $\xi < 0$.

Since these solutions are translation invariant, the solution for an arbitrarily located break or shear load couple at (n_i, ξ_i) is simply obtained by shifting (n, ξ) in Eqn 5 by (n_i, ξ_i). In this way, these shifted solutions become *load transmission factors*, representing the effects that any arbitrary fiber break and yielded matrix element have on the displacements and loads of all the other elements in the laminate, damaged or not.

This study covers conditions (i.e. the applied load and number of breaks), such that yielding does not occur in neighboring matrix regions. In this analysis, the locations of initial fiber breaks or discontinuities are known and restricted to occur only at the center of the fiber elements. Likewise, in matrix elements, shear deformation beyond the elastic regime, is represented by corrective point shear force couples applied only at the center of the matrix elements. Every yielded matrix element has a point force shear couple applied to its center. Based on the boundary conditions for both elastic unit problems, an auxiliary problem is solved where the load on the ends of each fiber break is -P (where P is the applied load), and the shear load in all yielded matrix elements is $2\delta T_y$. Each one of these loads must equal a weighted sum of the loads transmitted from all the other breaks and damaged matrix elements. Therefore, the first step is to determine the load transmission factors between each break and each shear couple. From the elastic load-break problem, we define $\Lambda_{ji} = L^*_{b,nj-ni}(\xi_j - \xi_i)$, the proportion of the load transmitted onto fiber break j due to a compressive load $-P_i$ on fiber break i [8]. Similarly, we define $\Omega_{ki} = 2\delta\Gamma_{b,nk-ni}(\xi_k - \xi_i)$ as the effective force transmitted onto matrix element k due to fiber break i. From the load couple problem, $\Phi_{ik} = L^*_{c,ni-nk}(\xi_i - \xi_k)$ is defined as the proportion of load transmitted onto fiber break i due a unit force couple imposed on matrix element k. Similarly, $\Psi_{lk} = 2\delta\Gamma_{c,nl-nk}(\xi_l - \xi_k)$ is defined as the effective force on matrix element l due to a unit force couple imposed on matrix element k.

The proper weights characteristic of each break and yielded element are solved, such that there exists a compressive load -P on all fiber breaks and a force couple $2\delta T_y$ on all damaged matrix elements and no load at $\pm\infty$. So, for r breaks and s damaged matrix elements, the following system of r + s equations is used to solve for r K_b's and s K_c's, the weighting factors for a break and damaged matrix element, respectively.

$$\begin{Bmatrix} -P_r \\ 2\delta T_{r+s} \end{Bmatrix} = \begin{Bmatrix} \Lambda_{r.r} & \Phi_{r.s} \\ \Omega_{r+s.r} & \Psi_{r+s.r+s} \end{Bmatrix} \begin{Bmatrix} K_{b.r} \\ K_{c.r+s} \end{Bmatrix} \qquad (6)$$

Note that the load couple weight factor, $K_{c,i}$ represents the differential shear force exerted by yielded matrix element i on its adjacent flanking fibers, thus compensating for the inelastic behavior of a damaged matrix element.

Finally, the fiber and matrix loads and displacements for any n and at any axial distance ξ are the weighted sums of the influences of all the breaks and yielded matrix elements in the lamina. Recalling the load transmission factors and the weights solved previously, the virtual fiber loads L^*_n and displacements V^*_n, and matrix shear load $\Gamma_n(\xi)$ everywhere are a weighted sum of the loads transmitted from all fiber breaks and shear load couples. For instance, the virtual fiber loads are [14],

$$L^*_n(\xi) = \sum_{i=1}^{r} K_{b,i} L_{b,n-n_i}(\xi - \xi_i) + \sum_{j=r+1}^{r+s} K_{c,j} L_{c,n-n_j}(\xi - \xi_j) \qquad (7)$$

On the other hand, the matrix shear T_n in the yield zone equals the shear strain in element i plus the corresponding uniform corrective shear loads $K_{c,i}/2\delta$.

To calculate the exact solution for a laminate loaded at $x = \pm\infty$ by a tensile force P (resulting in 0 load at fiber breaks), a tensile load P is superimposed to the solutions. The shear strain $\Gamma_n(\xi)$ and stress $T_n(\xi)$ remain the same, but the fiber loads $P_n(\xi)$ and displacements $U_n(\xi)$ become

$$P_n(\xi) = L^*_n(\xi) + P \qquad (8a)$$

$$U_n(\xi) = V^*_n(\xi) + P\xi \qquad (8b)$$

Of course identifying the matrix elements that undergo yielding at any stage must be done incrementally. The total number of yielded elements in a quarter plane, v, depends on the applied load P and r and indicates the amount of yielding (i.e. if v is 15, then 4 x 15 yielded elements surround the crack).

4. RESULTS

This section presents some typical fiber stresses, matrix shear stresses, and the extent of the yield zone for one to fifteen aligned, contiguous fiber breaks and various amounts of plastic yielding.

4.1 *Extent of the Yielded Region*

The results show that for all crack lengths or numbers of breaks r, the yield zone $\alpha = 2\delta v$ increases almost linearly with the applied load. This linear relationship agrees with results from other shear lag based models which account for elastic-perfectly plastic matrix behavior [12, 13]. Also, the slope of the α vs. P curve or the rate of plastic growth increases with r, which is expected since more fiber breaks lead to a higher maximum stress concentration, and P_y, the load which initiates yielding, decreases with r.

4.2 *Load Profiles on First Surviving Fiber*

Figure 2 compares the SCF along the axis of the first surviving fiber between the elastic and plastic cases for various amounts of yielding and r = 15. In both the purely elastic and plastic yielding cases, there is a definite decrease in the fiber load ahead of the crack tip as the load rises above the applied load. Including matrix yielding definitely decreases the stress concentration on the fiber immediately adjacent to the crack, but also decreases the rate of decay of the stress away from the crack tip. Thus, SCF is lower over a wider fiber length region.

4.3 *SCF_{max} versus P*

Figure 3 shows the effects of the applied load P on the SCF_{max} of the first surviving fiber ahead of r breaks for both the elastic and plastic case. As in the elastic case, the SCF_{max} in the first surviving fiber next to the yield zone increases with the number of contiguous breaks. Unlike the elastic case, in the plastic case, the SCF_{max} in the first surviving fiber strongly depends on and decreases with applied load. The applied load P where the plastic curve deviates from the elastic line is $P_{y,r}$, the applied load which initiates yielding for r contiguous breaks. As shown in Figure 3, P_y strongly depends on r, decreasing as r increases. Especially for a large r, this SCF_{max} curve drops quite rapidly with increased load.

4.4 *SCF_{max} versus Surviving Fiber N:*

Figure 4 shows how the SCF_{max} in the unbroken fibers ahead of the fiber breaks decays along the crack plane for both the elastic and plastic cases and for r = 15. When compared to the elastic case, the first intact fibers flanking the yield zone experiences a reduction in SCF, whereas sub-adjacent and further surviving fibers experience an increase in SCF. Note that although the yield zone consists of a portion of the outer matrix bays, plastic yielding significantly alters the stress distribution surrounding the fiber breaks. In all cases, only the first or first few surviving fiber experience a reduction in SCF_{max}, and both the number of fibers experiencing a reduced SCF (adjacent) and a higher SCF (subadjacent) increases with number of broken fibers and amount of yielding. Compared to the elastic case, the crack-tip fiber stress concentrations are significantly reduced, and compensating for this reduction, the SCF in the neighboring fibers is increased. Nonetheless, the substantial reduction in SCF_{max} near the crack tip typically lowers the likelihood of fiber failure in the vicinity of the crack tip; therefore, improving composite toughness and possibly strength. On the other hand, for the case of several fiber breaks, the crack plane normal stresses obtained from the elastic shear lag model closely follows a classical $r^{-1/2}$ distribution [15].

4.5 *Shear Strain*

Examining Γ_n in the yield zone shows that the maximum shear strain in both the plastic and elastic case occurs an infinitesimal distance away from the crack plane and is zero at $\xi = 0$. Typically, the maximum Γ_n in the yielded matrix bay is more than twice as high and the decay rate significantly lower than that in the elastic matrix. In addition, the recovery rate decreases for an increased number of breaks.

In the next matrix region, the dimensionless shear stress and strain are equivalent, since the matrix is behaving elastically. Shear strain profiles in this region show that the maximum shear strain actually occurs at some distance away from the tip, a distance greater than the length of the yielded region. Most importantly, for this range of r (1, 15) and P (0, 20), Γ_n does not exceed unity, and therefore, yielding only occurs in the assumed yielded region. If yielding

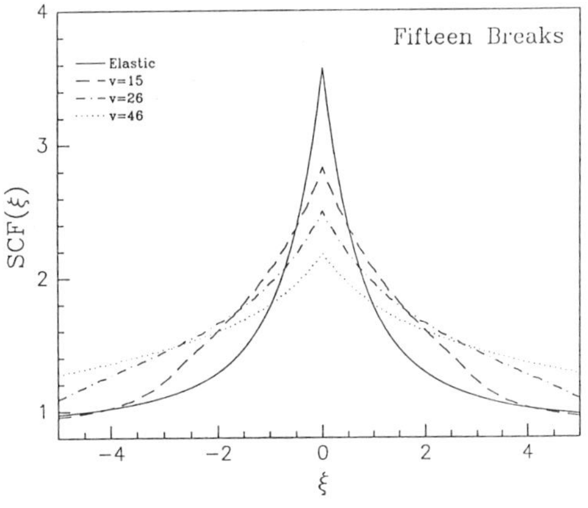

Fig. 2. The effects of matrix yielding on the SCF along the fiber axis of the first intact fiber next to 15 fiber breaks.

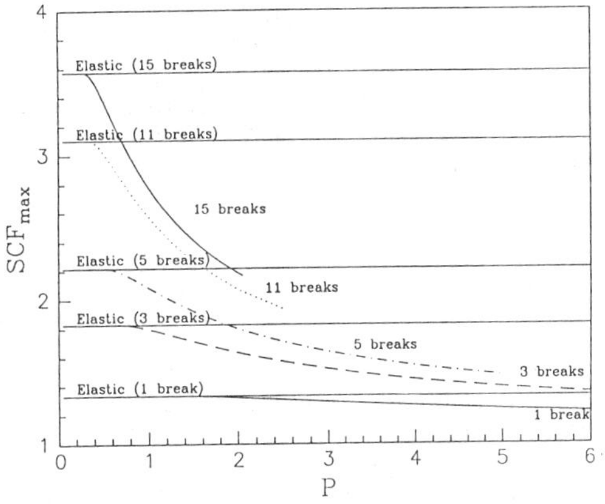

Fig. 3. The relationship between P and the SCF_{max} on the first intact fiber for different number of fiber breaks.

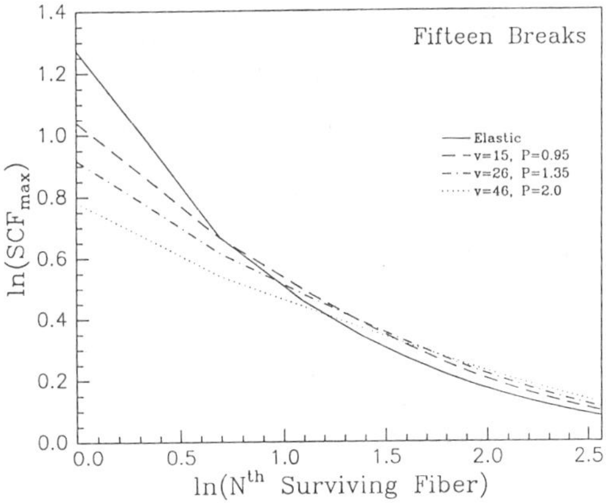

Fig. 4. The effects of matrix yielding on intact fibers in line and ahead of 15 fiber breaks.

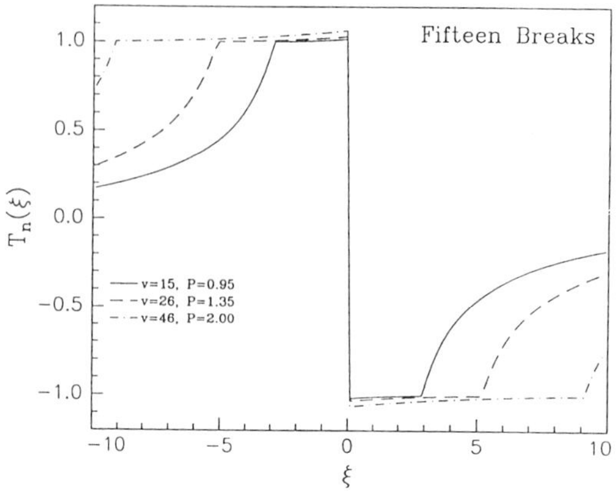

Fig. 5. The $T_n(\xi)$ in the matrix region containing varying amounts of yielding and adjacent to 15 fiber breaks.

develops in neighboring matrix regions, the model can easily be adapted to handle this case.

Compared to the elastic case, the region of maximum Γ_n is broader and located further away from the crack plane, and therefore, the plastic shear strain is higher over a larger part of the region.

4.6 *Shear stress in Yielded Region*

To calculate the shear stress of a yielded matrix element i, the corrective shear load $(K_{c,i}/2\delta)$ is added to Γ_n at all points within the yielded matrix element (i.e. the center and left and right boundaries). After applying a uniform corrective shear load, the average shear stress over the element length 2δ is calculated and assigned to its center. Figure 5 is the average shear stress at the *center* of matrix elements for 15 breaks and for various degrees of yielding. As expected, when T_n the shear stress is averaged locally, T_n is approximately unity, $T_y = 1$, along the plastic yield region, 2α, and outside this region, it quickly decays to zero. Only very small errors due to discretization are noticeable for ξ near zero.

5. DISCUSSION AND CONCLUSIONS

In analyzing the fracture behavior of composites, the model needs to account for the dominant and complex fracture mechanisms in the crack tip environment. This requires understanding and incorporating into the model, the effects of matrix plasticity, fiber-matrix debonding, statistical fiber fracture, matrix cracking, and fiber pull-out. The shear lag model presented in this study is a structural based, simplified model which is computationally capable of accounting for these various forms of damage for large numbers of arbitrarily placed breaks.

The main result is that for higher applied loads and more plastic deformation the effects of fiber breaks become more wide spread; overloading a larger area around the crack tip than that in the elastic case. Our results suggest that in analyzing composite materials which exhibit plastic behavior or ductile-type fracture, the model must account for the plastic behavior before ultimate failure.

REFERENCES

1. T. W. Chou. *Microstructural Design of Fiber Composites.* NY : Cambridge University Press, 1992.
2. J. G. Goree, L. R. Dharani, and W. F. Jones. *Metal Matrix Composites: Testing, Analysis, and Failure Modes*, ASTM STP 1032. (1989) 251-269.
3. K. Goda and S. L. Phoenix. *Comp. Sci. Tech.* **50** (1994) 457-468.
4. J. G. Goree and R. S. Gross. *Eng. Frac. Mech.*, **13** (1980) 395-405.
5. F. Hikami and T. W. Chou. *AIAA Journal.* **28** (1990) 499-505.
6. J. M. Hedgepeth. NASA TN D-882 (1961).
7. J. M. Hedgepeth and P. Van Dyke. *Textile Research Journal.* **39** (1969) 618-626.
8. A. M. Sastry and S. L. Phoenix. *J. Mat. Sci. Let.* **12** (1993) 1596-1599.
9. J. N. Rossettos and M. Shishesaz. *J. App. Mech.* **54** (1987) 723-724.
10. L. R. Dharani, W. F. Jones, and J. G. Goree.
11. W. F. Jones and J. G. Goree. *Mechanics of Composite Materials*, ASME AMD. **58** (1983) 171-178.
12. J. M. Hedgepeth and P. Van Dyke. *J. Comp. Mat.* **1** (1967) 294-309.
13. J. G. Goree and R. S. Gross. *Eng. Frac. Mech.* **13** (1979) 563-578.
14. A. M. Sastry and S. L. Phoenix, Calculation of stress concentrations around arbitrary multiple fiber breaks in a unidirectional composite with an elastic-plastic-debonding matrix, Submitted for publication.
15. I. J. Beyerlein and S. L. Phoenix, Comparison of Shear Lag Theory and Continuum Fracture Mechanics for Modelling Fiber and Matrix Stresses in a Cracked Composite Lamina, Submitted for publication.

Simulation of Materials Processing: Theory, Methods and Applications, Shen & Dawson (eds)
© 1995 Balkema, Rotterdam. ISBN 90 5410 553 4

Interactions among microfeatures in composites: A hybrid micro-macro beam approach

A.Chandra, Z.-Q.Jiang & Y.Huang
Department of Aerospace and Mechanical Engineering, The University of Arizona, Tucson, Ariz., USA

ABSTRACT: Local analysis schemes capable of detailed representations of micro-features of a problem are integrated with a macro-scale BEM technique capable of handling finite geometries and realistic boundary conditions. This paper focuses on micro-scale interactions among cracks and inclusions as well as their ramifications on macro-scale damage evaluations. The micro-scale effects are introduced into the macro-scale BEM computations through an augmented fundamental solution obtained from an integral equation representation of the micro-scale features. The proposed hybrid micro-macro BEM formulation allows complete decomposition of the real problem into two sub-problems, one residing entirely at the micro-level while the other resides at the macro-level. This allows for investigations of the micro-structural attributes while retaining the macro-scale geometric features and actual boundary conditions for the structural component under consideration. As a first attempt, dilute inclusion densities with strong inclusion-crack and crack-crack interactions are considered. The numerical results obtained from the hybrid BEM analysis establish the accuracy and effectiveness of the proposed micro-macro computational scheme for this class of problems. The proposed micro-macro BEM formulation can be easily extended to investigate the effects of other micro-features (e.g., interfaces, short or continuous fibers, in the context of linear elasticity) on macro-scale failure modes observed in structural components.

INTRODUCTION: Many engineering materials contain defects in the form of cracks, voids and inclusions that can significantly affect their load carrying capabilities. Damage is easily initiated as micro-cracks around an inclusion due to the residual stresses and thermal mismatch between inclusion and matrix properties. Accordingly, interactions among these micro-scale defects play crucial roles in determining the strength and life of components under service conditions. This is particularly true for relatively brittle materials such as ceramics, intermetallics or ceramic matrix composites.

In recent years, performance and weight goals for the next generation of high-temperature applications have spurred the development of improved intermetallics and ceramic matrix composites capable of sustained operations in the 1000-1500°C range. These materials contain multiple phases and secondary phase particles. To model real-life applications of these high-temperature components, one must incorporate general loading situations, finite and often complex geometries of particular components, and detailed representations of interacting inhomogeneities, along with their associated damage evolutions. And therein lies the fundamental difficulty in the analysis of these problems. Typically, the micro-defects and their spacings are of the order of a few micrometers while the overall dimensions of a component may range from a few centimeters to even a meter. Thus, the computational scheme is required, simultaneously, to provide a detailed representation of the underlying mechanics at two widely different scales: a local micro-scale ranging from 10 to 100 μm and a global macro-scale that may range from 10 to 1000 mm.

The computational techniques in existence today, such as the finite element method (e.g., Oden 1972, Gallagher 1975, Bathe 1982) and the boundary element method (e.g., Mukherjee 1982, Cruse 1988, Lutz et al. 1992, and Huang and Cruse 1994), are ideally suited for macro-scale analysis and can easily handle complex geometries along with general loading conditions. Thus, the defects are normally introduced as geometric entities in such macro-scale computational schemes. The resulting mixed boundary value problems essentially assume that the defect sizes are of the same order as the geometric dimensions of the body (e.g., Cruse and Polch 1986). In many cases, special quarter-point finite elements (e.g., Barsoum 1976, Yahia and Shephard 1985) or boundary elements (e.g., Crouch 1976, and Raveendra and Banerjee 1992) are introduced to capture the $\sqrt{r}$ singularity at the tip of an elastic crack. Snyder and Cruse (1975) have developed numerical fundamental solutions for fracture problems. Cruse and Novati (1992) have also utilized a displacement discontinuity approach to formulate a traction BIE formulation for nonplanar and multiple crack problems. In such cases, the details of micro-scale features are limited by the level of discretization, and the technique becomes prohibitively expensive if a large number of micro-features need to be represented. Moreover, such a technique is applicable only to isolated elastic cracks. These special elements cannot represent any effects due to crack interactions, and one must rely on numerical discretization to capture such effects. This poses significant difficulties whenever the cracks are closely packed. Besides, these special elements cannot be directly extended to model other types of defects, such as voids or inclusions (e.g., secondary phases and short or continuous fibers) that are commonly present in many real materials. And, an understanding of the evolution of such defects in a damage cluster is critically important for estimating the characteristics (e.g., strength and fracture toughness) of modern intermetallics and ceramic materials.

Thus, the computational techniques available today are capable of analyzing a problem at a macro-scale or at a micro-scale. They, however, cannot bridge these two widely different scales in a single analysis. Recently, Banerjee and Henry (1992) introduced special boundary elements and Nakamura and Suresh (1993) extended the unit cell analysis using FEM for modeling the effects of fiber packing in composite materials. However, much work is needed before one can directly investigate the effects of macro-scale design (geometric, loading, and boundary condition) considerations upon the evolutions of micro-scale defects, which essentially govern the strength and life of individual component in service.

Over the past few decades, analytical techniques have been used extensively to investigate various micro-mechanical phenomena. Several micro-mechanical models (e.g., Budiansky 1965, Weng 1984, and Huang et al 1994) have been developed to study the behavior of materials containing various distributions of inhomogeneities. Micro-mechanical models are particularly suited for the prediction of overall properties of composites, but they cannot açcurately represent the local stress and deformation fields around each inhomogeneity. These local fields, however, have been demonstrated to be of extreme importance for defect initiation, growth, and coalescence (Becker et al. 1988).

Various researchers have attempted to investigate the interactions among micro-defects in a damage cluster (e.g., Erdogan et al 1973, Horii and Nemat-Nasser 1986, Hu and Chandra 1993a,b)

The analytical and semi-analytical investigations cited above can provide crucial insights into the behaviors of interacting micro-defects in a damage cluster. However, these analyses are mostly carried out under assumptions of infinite bodies or extremely simplified geometry and loading conditions, which severely restricts their applicability to real-life situations involving complex finite geometries and general loading conditions. Thus, on one hand, there are analytical models capable of yielding very accurate results for various micro-scale phenomena involving evolutions of micro-defects in a damage cluster, but for very simple geometries and loading situations. On the other hand, very powerful computational techniques have been developed to handle real-life-macro-scale problems involving complex finite geometries and general loading conditions, yet it is very difficult to relate the effects of macro-scale parameters on the interactions and evaluations of micro-features present in a real material.

HYBRID MICRO-MACRO BEM APPROACH:

To address the above issues, a hybrid micro-macro BEM formulation is developed here. The fundamental solution is utilized as a bridge for transition from the micro- to the macro-scale.

As a first attempt, circular inclusions with dilute densities are assumed and particular attention is paid to crack-crack as well as crack-inclusion interactions. However, it should be emphasized that, following the work of Hu and Chandra (1993a, b), fundamental solutions can also be obtained for problems involving other micro-features, e.g., fibers and interfaces and elliptic inclusions. An augmented fundamental solution is obtained as,

$$\hat{U}_{ij}(p,q) = U_{ij}(p,q) + \overline{U}_{ij}(p,q) + \overline{\overline{U}}_{ij}(p,q) \qquad (1a)$$

and

$$\hat{T}_{ij}(p,q) = T_{ij}(p,q) + \overline{T}_{ij}(p,q) + \overline{\overline{T}}_{ij}(p,q) \qquad (1b)$$

where U_{ij} and T_{ij} are the conventional kernels obtained from the Kelvin solution, $\overline{U}_{ij}$ and $\overline{T}_{ij}$ are the kernels representing perturbations due to elastic inclusions; $\overline{\overline{U}}_{ij}$ and $\overline{\overline{T}}_{ij}$ are the kernels capturing the effects of the micro-cracks on the particle reinforced composite. The augmented kernels $\hat{U}_{ij}$ and $\hat{T}_{ij}$, capturing the micro-scale effects, may now be utilized directly in a macro-scale BEM analysis of general elastic structures subject to realistic loadings and boundary conditions.

Using Betti's reciprocal theorem or a weighed residual approach, a hybrid BEM formulation requiring only macro-scale discretization may be developed as (Cruse 1988):

$$u_j(P) = \int_{\partial B}\left[\hat{U}_{ij}(P,Q)\tau_i(Q) - \hat{T}_{ij}(P,Q)u_i(Q)\right]ds_Q \qquad (2)$$

Equation (2) captures the effects of inclusions and micro-cracks through the augmented kernels. A boundary integral equation for the unknown components of displacements and tractions in terms of the prescribed ones can be obtained by taking the limit as the internal source point p approaches a boundary point P. This leads to the boundary equation

$$C_{ij}u_i(P) = \int_{\partial B}\left[\hat{U}_{ij}(P,Q)\tau_i(Q) - \hat{T}_{ij}(P,Q)u_i(Q)\right]ds_Q \qquad (3)$$

Using standard procedures for elasticity, displacement gradients and stresses may now be obtained.

NUMERICAL RESULTS: In the present work, straight boundary elements are used with linear shape functions for both displacements and tractions. No other discretization is necessary for the macro-scale problem, and the micro-scale effects of the inclusion and micro-cracks are introduced through the augmented fundamental solutions. For the micro-scale problem, 9 integration points are used on each crack. A Gauss-Chebychev polynomial scheme (requiring only n quadrature points for accurate representations of polynomial functions of order 2n) is used. This makes it very effective for interacting inclusions and cracks at close spacings. The issues regarding the effectiveness of such a scheme are available in various references (e.g., Erdogan 1973, Theocaris and Ioakimidis 1977, Melin 1983, Li and Hills 1990, Rubinstein 1990, Hu and Chandra 1993a,b, and Chandra et al. 1994).

The proposed hybrid micro-macro BEM scheme is first verified against known solutions involving inclusion-crack interactions. For all cases, the stress intensity factors are normalized with respect to that for a single crack in an infinite plane ($\sigma_0\sqrt{\pi a}$ for uniaxial and biaxial tension, $\tau_0\sqrt{\pi a}$ for pure shear loading), where σ_0, τ_0 are the remote stresses and "a" is the half length of the crack. In this section, the radius of the inclusion is denoted as 'R". The proposed BEM scheme can handle plane stress or plane strain problems, however, only plane strain problems are considered in this section.

As shown in the inset of Figure 1, a problem with a circular inclusion of radius R at the center of a square (of dimension 2h x 2h) finite domain and accompanied by two collinear symmetric cracks of length 2a each is considered first. The square domain is subjected to uniaxial tension. The distance of the center of the crack from the center of the includion is denoted by "d". Accordingly, the non-dimensional parameter δ = (d-R)/a represents the proximity of the cracks to the inclusion or the boundary of the domain. The ratio h/R represents the size of the square domain relative to the inclusion. The normalized mode-I SIFs at the inner crack tips obtained from the hybrid micro-macro BEM computations at different h/R values of 5.0, 10.0 and 20.0 are plotted in Fig. 1 with respect to δ. Harder inclusions with G_2/G_1=4.0 and $\upsilon_2=\upsilon_1$=0.25 are used in these computations. It may be observed from Fig. 1 that the results compare very well to those obtained by Hu et al. (1993b) for h/R=∞, when the boundary of the domain remains far away from the micro-features. Significant deviations in mode-I SIFs are observed as the micro-features approach and interact with the boundary. Fig. 2 depicts the same situation as that in Fig. 1, but with a softer inclusion (G_2/G_1=0.25). It is observed that the crack-inclusion interactions are very strong at close spacings. When the micro-features remain far away from the edge of the body, the mode-I SIFs for softer inclusions also agree very well with those obtained by Hu et al (1993b) for h/R=∞. As the cracks move toward the edges of the square body, inclusion-crack interactions become weaker while the edge-crack interactions become stronger.

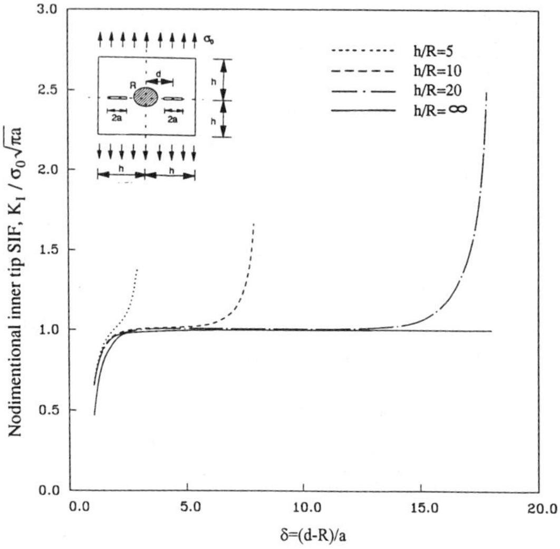

Fig 1. Variation of the normalized inner tip SIF vs. parameter δ ($N=2$, $G_2/G_1=4.0$, $R=1.0$)

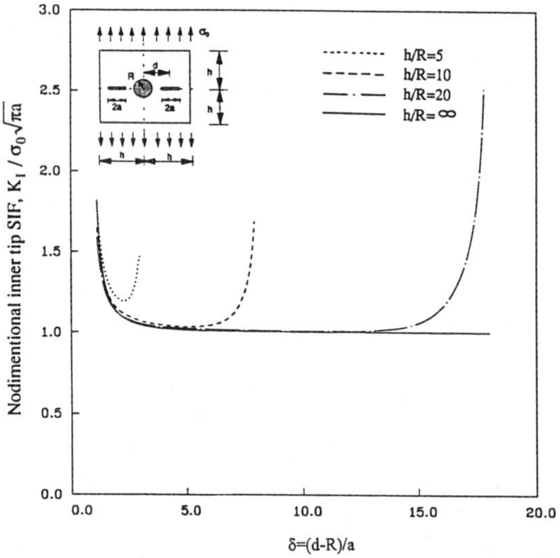

Fig 2. Variation of the normalized inner tip SIF vs. parameter δ ($N=2$, $G_2/G_1=0.25$, $R=1.0$)

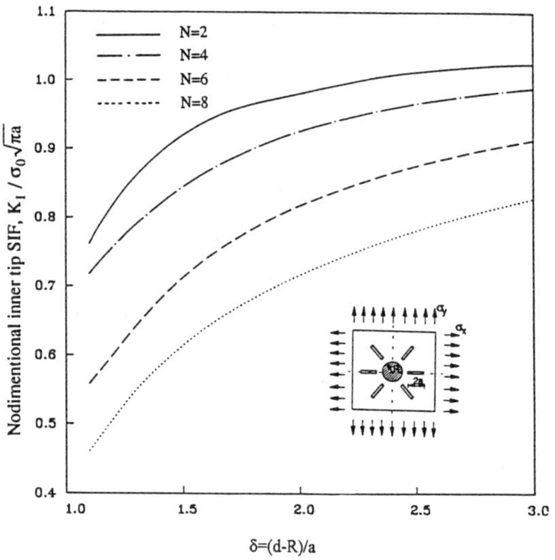

Fig 3. Variation of the normalized inner tip SIF vs. parameter δ ($G_2/G_1=4.0$, $\nu_1=\nu_2$, $R=1.0$, $h/R=20$)

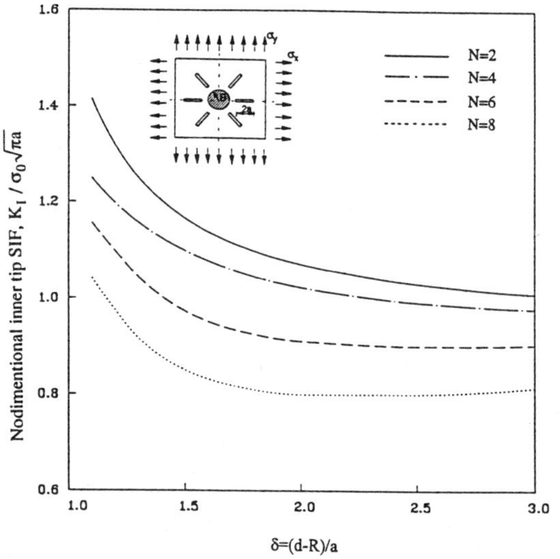

Fig 4. Variation of the normalized inner tip SIF vs. parameter δ ($G_2/G_1=0.25$, $\nu_1=\nu_2$, $R=1.0$, $h/R=20$)

A square domain containing an inclusion surrounded by a varying number of radial cracks and subjected to biaxial tension is investigated next. Such radial cracking may commonly occur from secondary phases due to mismatch in coefficients of thermal expansion (Huang et al., 1992) and associated transformation loadings. The number of radial cracks is varied from 2 to 8 with equal spacing. As observed in Fig. 3, a hard inclusion ($G2/G1 = 4.0$, $\nu_1=\nu_2=0.25$) provides shielding in general. This shielding, however, is tempered by crack-crack interactions that become stronger as the number of radial cracks increases and their spacing decreases. It is interesting to note here, that at closer spacings, the radial cracks also provide shielding due to crack-crack interactions. Accordingly, the overall shielding is stronger at higher number of radial cracks. At $\delta=1.1$, the normalized SIF drops from 0.76 at N=2 to 0.46 for N=8. Fig 4 shows the SIFs for the same problem but with a softer inclusion ($G_2/G_1 = 0.25$). A softer inclusion causes stress amplification, in general, due to inclusion-crack interactions. However, as the number of radial cracks increases (>2), this amplification is modulated by the shielding effects due to crack-crack interactions among radial cracks. It may be observed from Fig 4, that at δ=1.1, the normalized inner-tip SIF drops from 1.42 for N=2 to 1.04 for N=8. Thus, depending on the number of radial cracks, the softness of the inclusion and their spacings, there exists a critical number of radial cracks above which the overall behavior manifested by the normalized SIFs may change from amplification to shielding.

Discussion and Summary

The problem of interacting micro-cracks around an inclusion in a system involving complex finite geometries and general boundary conditions is considered in this paper. A hybrid micro-macro BEM formulation capable of handling interactions among the inclusion, arbitrarily distributed cracks and the boundaries of the system is developed. The effects of micro-scale features are introduced into the macro-scale BEM analysis through an augmented fundamental solution. Thus, the hybrid scheme retains the boundary nature of the problem and associated advantages of a BEM formulation.

Several numerical examples have been shown in order to verify and validate the capability of the proposed BEM formulation for analyzing problems stated above. It is observed that the proposed scheme can solve these problems accurately and effectively. The numerical results reveal that the inclusion-crack and crack-crack interactions can greatly affect the nature of local stress distributions.

Many real-life materials are reinforced as matrix materials by mixing some fibers to form composite materials. During the processing and later routine use of these composite materials, it is quite likely that some micro-cracks would nucleate around the fibers. The proposed hybrid micro-macro BEM formulation can capture the effects of such micro-scale phenomena within the context of macro-scale structural computations. The proposed technique can also be generalized through a unit cell approach to obtain effective macro-scale material properties based on its micro-scale features(Wei et al 1994).

Acknowledgment

The authors gratefully acknowledge the financial support furnished by grant No. DMC 8657345 from the U.S National Science Foundation.

References

Banerjee, P. K. and Henry, D. P., 1992, "Nonlinear Micro and Macromechanical Analyses of Composites by BEM," *Proceedings, HITEMP Conference,* Vol. 2, pp. 42.1-42.11, NASA Lewis Res. Center, Cleveland.

Barsoum, R. S., 1976, "On the use of Isoparametric Finite Elements in Linear Fracture Mechanics," *Int. J. Num. Meth. Eng.,* Vol. 10, pp. 25-27.

Bathe, K. J., 1982, *Finite Element Procedures in Engineering Analysis*, Prentice-Hall, Englewood Cliffs, N.J.

Becker, P., Hsueh, C.-H., Angelini, P., and Tiegs, T. N., 1988, "Toughening Behavior in Whisker-Reinforced Ceramic Matrix Composites," *J. Am. Ceramic. Soc.,* Vol. 71, pp. 11056-1061..

Budiansky, B., 1965, "On the Elastic Moduli of Some Heterogeneous material," *J Mech. Phys. Solids*, Vol. 13, pp. 223-227.

Chandra, A. and Mukherjee, S., 1994, *Boundary Element Methods in Manufacturing*, Oxgord University Press, New York (in press).

Chandra, A. and Tvergaard, V., 1993, "Void Nucleation and Growth During Plane Strain Exclusion," *Int. J. Damage Mech.*, Vol.2, pp. 330-348.

Chandra, A., Y. Huang, X. Wei and K.X. Hu 1994, "A Hybrid Micro-Macro BEM Formulation for Micro-Crack Clusters in Elastic Components", submitted to *Int.J.Num. Meth. Eng.*

Crouch, S. L., 1976, "Solution of Plane Elasticity Problems by the Displacement Discontinuity Method," *Int. J. Num. Meth. Eng.*, Vol. 10, pp. 301-343.

Cruse, T. A., 1988, *Boundary Element Analysis in Computational Fracture Mechanics*, Kluwer Academic Publishers, Dordrecht, The Netherlands.

Cruse, T. A. and Polch, E. Z., 1986, "Application of an Elastoplastic Boundary Element Method to Some Fracture Mechanics Problems," *Eng. Frac. Mech.*, Vol. 23, pp. 1085-1096.

Erdogan, F., Gupta, G. D., and Cook, T. S., 1973, "Numerical Solution of Singular Integral Equations," *Methods of Analysis and Solutions of Crack Problems* (G. C. Sih, ed.), pp. 386-425, Noordhoff, Leyden, The Netherlands.

Gallagher, R. H., 1975, *Finite Element Analysis: Fundamentals*, Prentice-hall, Englewood Cliffs, N.J.

Horii, H. and Nemat-Nasser, S., 1986, 'Brittle Failure in Compression: Splitting, Faulting and Brittle-Ductile Transition," *Phil. Trans. Royal Soc. Lond.*, Vol. A319, pp. 337-374.

Hu, K. X. and Chandra, A., 1993a, 'Interactions Among General Systems of Cracks and Anticracks: An Integral Equation Approach,". ASME *J. Appl. Mech.*, Vol. 60, pp. 920-928.

Hu, K. X., and Chandra, A., and Huang, Y., 1993b, 'Fundamental Solutions for Dilute Distributions of Inclusions Embedded in Microcracked Solids," *Mech. mater.*, vol. 16, pp. 281-294.

Huang, Q. and Cruse, T. A., 1994, 'On the Nonsingular Traction-BIE in Elasticity," *Int. J. Num. Meth. Eng.* (in press).

Huang, Y., Hu, K. X., Wei, X., and Chandra, A., 1994, "A Generalized Self-Consistent Mechanics Method for Composite Materials With Multiphase Inclusions," J. *Mech. Phys. Sol.*

Huang, Y., K.X. Hu and A. Chandra, 1992 'Damage evaluation for solids containing dilute distributions of inclusions and microcracks", J. Appl. Mech.

Li, Y. and Hills, D. A., 1990, "Stress Intensity Factor Solutions for Kinked Surface Cracks," *J. strain Analysis,* Vol. 25, pp. 21-27.

Lutz, E., Ingraffea, A. R., and Gray, L. J., 1992, 'Use of Simple Solutions for Boundary Integral Methods in Elasticity and Fracture Analysis," *Int. J. Num. Meth. Eng.*, vol. 35, pp. 1737-1752.

Melin, S., 1983, 'Why Do Cracks Avoid Each Other," *Int. J. Fracture*, Vol. 23, pp. 37-45.

Mukherjee, S., 1982, *Boundary Element Methods in Creep and Fracture*, Elsevier Applied Science, London.

Nakamura, T. and Suresh, S., 1993, 'Effects of Thermal Residual Stresses and Fiber Packing on Deformation of Metal-matrix Composites," *Acta Metall. Mater.*, Vol. 41, pp. 1665-1681.

Oden, J. T., 1972, *Finite Elements of Nonlinear Continua*, McGraw-Hill Pub., New York.

Raveendra, S. T. and Banerjee, P. K., 1992, 'Boundary Element Analysis of Cracks in Thermally Stressed Planar Structures," *Int. J. Sol. Struc.,* Vol. 29, pp. 2301-2317.

Rubinstein, A. A., 1990, 'Crack-Path Effect on Material Toughness," *J. Appl. Mech.,* Vol. 57, pp. 97-103.

Theocaris, P. S. and Ioakimidis, N. I., 1977, " Numerical Integration Methods for the Solution of Singular Integral Equations, " *Quart. Appl. Math.*, Vol. 35, pp. 173-182.

Wei, X., Chandra, A. and Huang, Y., 1994 "A Unit Cell Analysis with Hybrid Micro-Macro BEM", in preparation.

Weng, G. J., 1984, 'Some Elastic Properties of Reinforced Solids, with Special Reference to Isotropic Containing Spherical Inclusions," *Int. J. Eng. Sci.*, Vol. 22, pp. 845-856.

Yahia, N. A. B. and Shephard, M. S., 1985, 'On the Effect of Quarter-Point Element Size on Fracture Criteria," *Int. J. Num. Meth. Eng.,* Vol. 20, pp. 1629-1641.

Simulation of Materials Processing: Theory, Methods and Applications, Shen & Dawson (eds)
© 1995 Balkema, Rotterdam. ISBN 90 5410 553 4

Modelling failure of materials with random microstructure

P.M. Duxbury
Department of Physics and Astronomy, Michigan State University, Mich., USA
P.L. Leath
Department of Physics and Astronomy, Rutgers University, USA

ABSTRACT: A brief overview and summary is given of numerical and analytic methods currently being used to predict the effect of random microstructure on material failure.

1. INTRODUCTION

Advances in computational facilities in combination with new algorithms and analytical methods is enabling a deeper analysis of the interplay between random material microstructures and material failure. Although traditional finite element methods can be used to simulate failure of random microstructures, the majority of work has used finite difference or "lattice" models. This is primarily due to the fact that the new modelling methods initiated in the statistical physics community (see e.g. Herrmann and Roux 1990). Most of the failure simulations have been in quasistatic loading, although some work using more complex constitutive laws is being done. There has also been considerable progress in the calculation of the ideal strength of materials using "ab initio" electronic structure methods and model potentials. The model potentials have also been used to simulate the dynamics of fracture using molecular dynamics (MD) simulations (see Kimura et al. 1994). The latter simulations have been done for a large number of atoms (up to $\sim 10^8$ atoms) over time scales of up to a microsecond in real time. The MD simulations are thus relevant for fast fracture or impact, in contract to the quasistatic simulations mentioned above. The quasistatic method will be outlined in Section 2.

Traditional fracture testing involves machining a crack into a sample and the study of a crack which grows from this imposed initiation site. Analysis of this experiment is often handled adequately by fracture mechanics. This analysis tells us the critical crack size for a given load level and we can monitor the sample in service to test whether cracks of that size exist. It is also possible to monitor the rate of crack growth to determine the time taken for a crack to reach the critical (unstable) crack length. Although this procedure is powerful, it imposes a failure mechanism based on the existence of a dominant crack, even prior to the fracture instability. In many composites and other materials with many phase microstructures, the damage is distributed, and instability occurs when damage"localises" on a plane (see Figs. 1 & 2 - from Duxbury et al. 1995). This process is different than the failure mechanism probed using standard fracture testing. The time taken for an unstable crack to develop is also not correctly given by the result for the growth of a single crack. The statistical variability observed in failure is then caused primarily by the nucleation process which causes an unstable crack to develop. Naturally imposing a machined crack cannot probe this variability correctly. To go beyond traditional fracture mechanics thus requires serious consideration of crack nucleation, a process which is highly microstructure dependent.

2. METHOD FOR QUASISTATIC FAILURE SIMULATIONS

To simulate the failure of a material with random microstructure, it is necessary to first model and digitise the microstructure. The microstructure can either be generated computationally, or by using a digital image of experimental microstructures. Given the microstructure, the system is discretised (finite difference) or meshed (finite element). These procedures are relatively standard, although good results rely on performing these steps carefully. Each element or bond in the system is assigned a constitutive law. In the case of brittle materials the local constitute law is defined by the elastic moduli and threshold stresses. In the simplest case only the Young's modulus and tensile failure stress of each element or bond need be specified. The fracture algorithm is then as follows (e.g. in tensile loading). (i) Increase the applied strain by a small increment and calculate the local stresses everywhere. If any element or bond exeeds its threshold, it irreversibly fails. (ii) After each local failure, the stress is relaxed again and any element or bond which exceeds its threshold is again removed. This process is iterated until either failure occurs in which case stop OR the system arrives at a stable damage pattern. In the latter case we return to step (i). It is possible to carry out this procedure in a way which enables the reconstruction of the the stress-strain curve in any loading. An example of such a simulation for an electrical composite is shown in Figs. 1,2. These results can be roughly taken over to the mechanical problem by associating stress with current and strain with voltage.

This failure algorithm is quite general and can be applied to a variety of failure problems such as

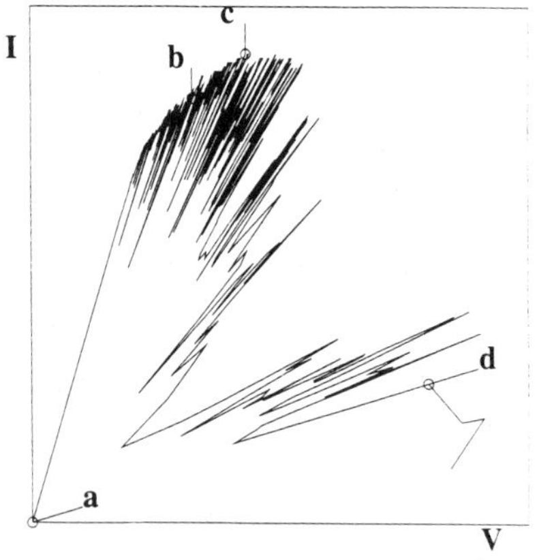

Figure 1. The I-V characteristic of a random composite undergoing electrical failure. The composite initially was composed of a 50/50 mixture of two resistors with conductance ratio $g = 8$. The resistors are brittle and fail when the current through them reaches a threshold. The ratio of thresholds $i = 1/4$. The damage patterns at, c and d are shown in Fig. 2.

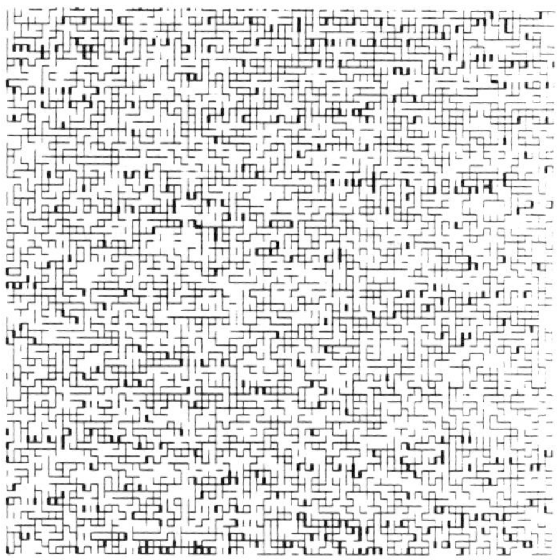

a)

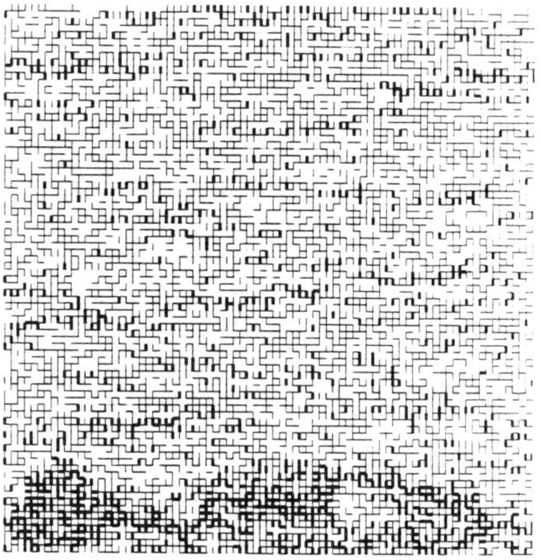

b)

Figure 2. Damage accumulation in a random composite undergoing electrical failure. *a)* The damage pattern at point, c, in the I-V curve of Fig. 1. *b)* The damage pattern at point, d, of the I-V curve of Fig. 1. The bonds which are not present in the graph are highly conducting and weaker. The darkest lines indicate failed bonds. It is clear that the damage accumulation is quite random prior to instability.

electrical, dielectric and superconducting failure as well as to mechanical failure. Many of the generic effects of microstructure on failure are common to all of these systems, although there are important differences as well.

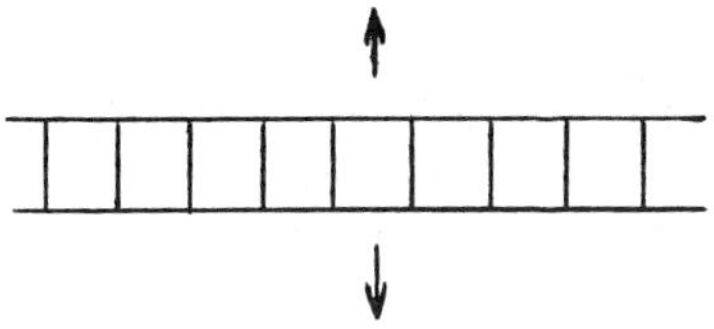

Figure 3. The parallel bar model.

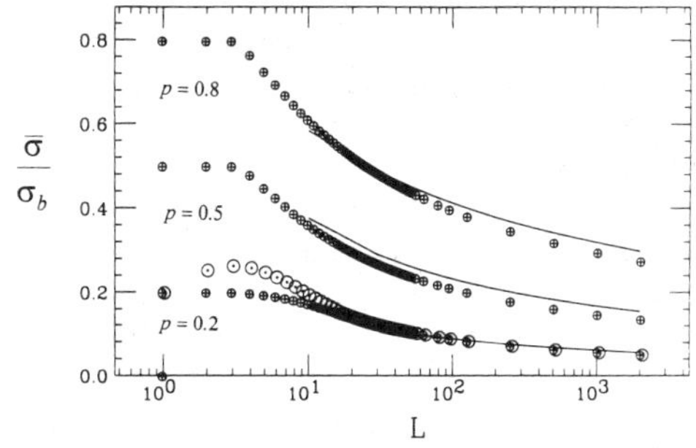

Figure 4. The average strength of a parallel bar model with volume fraction $f = 1 - p$ of random pores as a function of sample size L from Duxbury and Leath (1994a). The solid line is Eq. (1)

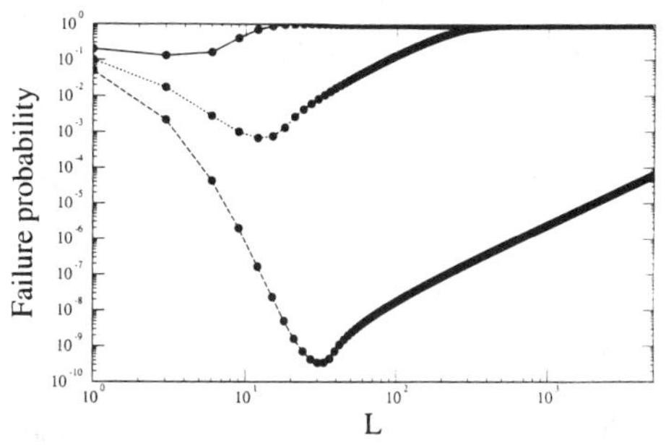

Figure 5. The failure probability of a parallel bar model as a function of sample size. The bonds of the parallel bar model are drawn from a uniform distribution of width W, and σ is the applied stress. The curves are for, starting at the top of the figure, $\sigma/W = 0.2$, $\sigma/W = 0.1$ and $\sigma/W = 0.05$ (Duxbury and Leath 1994b).

3. SIMPLE MODELS AND PRECISE RESULTS

Simple models for crack nucleation have been used for some time. However progress has been relatively slow due to the difficulty in doing a precise analytic analysis of these models. Even the simple "parallel bar" models like that shown in Fig. 3 are very complex mathematically.

One exception is the "democratic" or "equal-load sharing" model in which all surviving elements of the system carry equal load. This is a sort of mean-field theory for fracture and ignores local stress enhancements which are important in many materials. In this case the results are quite general and well known (Daniels 1945, Coleman 1958, Smith and Phoenix 1981). For fibers which are drawn from a "well behaved" probability distribution, for example a Wiebull or a Gaussian, the following properties hold.
(i) The system has a finite average strength in the large lattice limit and the stress-strain curve is parabolic near the critical stress.
(ii) The statistics of static failure are asymptotically Gaussian, although approach to the asymptotic limit is often quite slow.
Many other features and generalisation of these models are also analytically tractable, provided they preserve the mean-field load sharing property.

More recently, there has also been considerable progress in the analysis of models in which there are local stress enhancements after local fiber failures, the "local-load-sharing models". Two powerful methods have been developed. Firstly a transition matrix method developed by Harlow and Phoenix (1981, 1991) and secondly a recurrence-relation for the failure probability developed by Duxbury and Leath (Duxbury and Leath 1994b, Leath and Duxbury 1994). It has been shown that for "well behaved" probability distributions:
(i) The average strength decreases as the inverse of the logarithm of the sample size. Thus infinite size samples have zero strength. However this weak size effect is very difficult to see in bulk materials (see e.g. Fig. 4).

For large sample sizes, the average strength of the parallel bar model with local-load-sharing and random pores is well appoximated by (Duxbury and Leath 1994a),

$$\frac{\bar{\sigma}}{\sigma_b} = \frac{1}{1 - \frac{ln(pL)}{2ln(1-p)}}, \tag{1}$$

where σ_b is the strength when $f = 0$. The size effect in more general models is predicted to be quite similar to that of Eq. (1) (Duxbury and Leath 1994b).

(ii) At low applied stress levels, there is an optimal sample size at which the failure probability is a minimum (see Fig. 5).

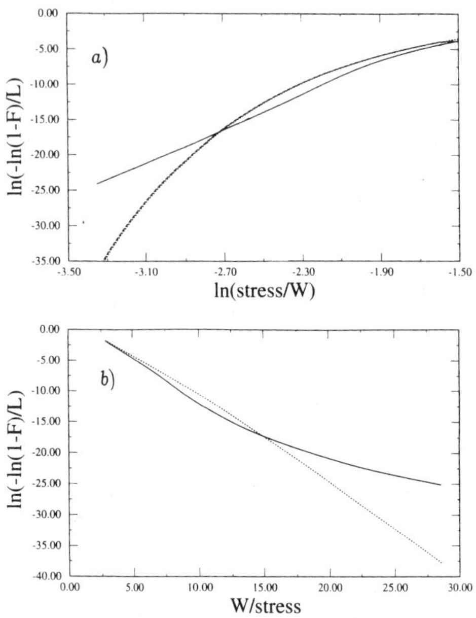

Figure 6. The statistics of static failure calculated for a parallel bar model with a uniform distribution of local failure thresholds (from Leath and Duxbury 1994)). *a)* A test of Weibull statistics and *b)* a test of modified Gumbel statistics. The dotted lines are for samples of size $L = 150$, while the solid line is for a small sample $L = 15$.

The probability minimum illustrated in Fig. 5 is exponentially deep and the sample size at which it occurs increases inversely as the applied stress. It occurs for a wide variety of bond failure thresholds (e.g. for the Weibull case) and it would be of technological advantage if materials and structures could be designed to operate near this minimum.

(iii) The statistics of static failure are usually of a modified Gumbel form, although Weibull statistics are a reasonable approximation over a range of sample sizes and stress levels (see e.g. Fig. 6). Typically the Weibull distribution provides a reasonable fit for smaller samples, while the modified Gumbel is more appropriate for large samples and especially in the high reliability tail of the distribution.

The simplest asymptotic form of the new failure distribution suggested by the analytic and numerical results is (Duxbury et al. 1987; Duxbury et al. 1994, Leath and Duxbury 1994)

$$F(\sigma) = 1 - exp(-kV exp(-c/\sigma)) \qquad (2)$$

Where k and c are microstructure dependent, but stress and sample size independent, quantities and V is the volume. However one of a family of related distributions may arise, depending on the details of the local stress concentrations (Duxbury et al. 1994).

Analytic methods based on various "critical configurations" are also very well developed. For example two crack configurations, wedges and funnels, needles and fibers have been analysed in many contexts. These results can be used in a statistical analysis and in some cases give a good estimate of failure properties of brittle systems (see e.g. Duxbury et al. 1987, Li and Duxbury 1987, Machta and Guyer 1987, Duxbury et al 1995).

REFERENCES

Coleman B.D. 1958. *J. Appl. Phys. 29:* 968-983.
Daniels H.E. 1945. *Proc. Roy. Soc. A183:* 405-435.
Duxbury P.M., Beale P.D. & Moukarzel C.M. 1995. Phys. Rev. B: in press.
Duxbury P.M., S.G. Kim & P.L. Leath 1994. *Mater. Sci. and Eng. A176:* 25-31.
Duxbury P.M. & Leath P.L. 1994a. *Phys. Rev. B49:* 12676-12687.
Duxbury P.M. & Leath P.L. 1994b. *Phys. Rev. Lett. 72:* 2805-2808.
Duxbury P.M., Leath P.L. & Beale P.D. 1987. *Phys. Rev. B36:* 367-380.
Harlow D.G. & Phoenix S.L. 1981. *Int. J. Fracture 17:* 601-630.
Harlow D.G. & Phoenix S.L. 1991. *J. Mech. Phys. Sol. 39:* 173-200.
Herrmann H.J. & Roux S eds. 1990. *Statistical models for the fracture of disordered media*, Amsterdam: North Holland.
Kimura H., Narita N., Suzuki T., Suzuki T. & Watanabe T. eds. 1994. *Proceedings of the fourth international conference on the fundamentals of fracture, Mater. Sci. and Eng. A176.*
Leath P.L. & Duxbury P.M. 1994. *Phys. Rev. B49:* 14905-14917.
Li Y.S. & Duxbury P.M. 1987. *Phys. Rev. B36:* 5411-5419.
Machta J. & Guyer R.A. 1987. *Phys. Rev. B36:* 2142-2190.
Smith R.L. & Phoenix S.L. 1981, *J. Appl. Mech. 48:* 75-82.

Simulation of Materials Processing: Theory, Methods and Applications, Shen & Dawson (eds)
© 1995 Balkema, Rotterdam. ISBN 90 5410 553 4

Requirements of material modeling for hot rolling

Jonas Edberg
MEFOS Metal Working Research Plant, Luleå, Sweden

Pekka Mäntylä
Rautaruukki Oy, Raahe, Finland

ABSTRACT: Different material models used in finite element simulations of the hot rolling process are compared. The results from the simulations are compared with experiments. The advantages, limitations and the relevance of the different material models are discussed. Improvements needed in the material modeling are outlined.

1 INTRODUCTION

Simulation of flat rolling is a process that has been simulated using many different methods. Most of them are empirical or semi empirical. Nowadays, the Finite Element Method (FEM) is frequently used for simulation of rolling. Using this method it is possible to include effects that had to be neglected in the empirical and semi empirical methods.

Accurate friction models and material models are very important when simulating the hot rolling process and other processes where metals are deformed at high temperatures. The relative importance of these parameters depend on the rolling conditions. Many friction models and material models common in finite element codes are not accurate enough, because they do not take into account all the important phenomena that occur when a solid metal is subjected to large strains, high strains rates and high temperatures.

The effect of different friction models and the associated parameters have been closely studied in order to improve the accuracy of the hot rolling simulations (Edberg 1992). One conclusion from this work is that for many rolling operations, the friction models are less important in comparison with the material models. This leads to the fact that the best way to improve the accuracy of the simulations is to improve the material models. The material model used should take strain, strain rate effects and temperature effects into account in order to be reasonable accurate. In addition, recovery, recrystallization, grain growth, microstructure and phase transformations should be taken into account to get a complete understanding of the problem. The later phenomena are missing in many material models.

Finite element simulations of hot plate rolling is taken as an example. During rolling there may be large temperature gradients in the rolled plate that are unevenly distributed, leading to asymmetrical rolling conditions. These temperature gradients develop in the reheating furnace, remain in the rolled material during the rolling process and lead to asymmetrical material flow in the roll gap. Besides evolution of residual stresses and flatness defects such as cross waves, ski-end formation, the asymmetrical metal flow also causes the nonuniform microstructure and mechanical properties in the final product. These types of defects result in relatively large and expensive rejections and in reduced productivity. It is evidently of great economic interest to minimize these type of defects, since they lead to problems and rejections during the manufacturing process and in poorly-functioning products.

Simulations are made using a few material models and the results are compared with experiments made in the Rautaruukki hot strip mill. Limitations and differences between these material models are pointed out. A material model that overcomes some of the limitations of the models used is outlined. It will be able to handle metallurgical effects such as recrystallization, recovery, grain growth, evolution of micro structure and phase transformations.

The two-dimensional finite element programs PALM and DYNA2D was used to evaluate the different material models.

2 THEORETICAL ANALYSES

2.1 Finite element program

The finite element programs PALM and DYNA2D (Hallquist 1988) obtained from Lawrence Livermore National Laboratory, USA, are used in the simulations. DYNA2D is an explicit code for the nonlinear transient dynamic response of two-dimensional solids and structures. It does not form and solve a large system of coupled equations typical of implicit codes and does not require iterations within each time step. The explicit central difference method is used to integrate the equations of motion in time. This method is only conditionally stable and stability is governed by a limit on the time step size. For solid elements this limit is essentially the time required for an elastic stress wave to propagate across the shortest dimension of the smallest element in the mesh. The acoustic wave speed is

$$c = \sqrt{\frac{E}{\rho}}$$

where E is Young's modulus and ρ is the density. The rolling velocity is fairly low in the present simulations. Inertial forces are small compared with the deformational forces. Therefore it is possible to reduce the computation time by using a density that is higher than the real density of the simulated material. This technique is described by Lindgren (1990) and Edberg (1992). PALM is an implicit thermal-mechanical coupled code for static and dynamic problems. It lacks limits on the time step size other than accuracy requirements as the code is based on an unconditionally stable time integrator. PALM is based on TOPAZ2D (Shapiro 1986) for thermal analysis and NIKE2D (Hallquist 1986) for mechanical analysis.

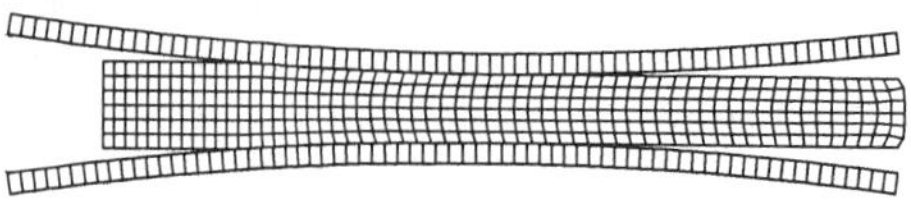

Figure 1. Finite element model at time 0.12 s.

2.2 Finite element model

The same finite element mesh is used in all simulations. Due to symmetry only the upper half of the plate and the upper roll is analyzed. The model can be seen in Figure 1. Plane strain conditions are assumed in the direction transverse to the rolling direction. Thus the simulations are representative of the mid section along the rolling direction. The roll is rotating with a peripheral velocity of 800 mm/s at the outer radius. The plate in the model is 102.5 mm long and 12.8 mm thick before rolling. The outer diameter of the work rolls is 1014 mm and the gap between the work rolls is 10 mm before rolling. The distance between the inner and the outer radii of the work rolls is 3 mm. The density of the materials are 1000 times larger than the real density for steel to speed up the calculation in DYNA2D. In the simulations using PALM, the plate is pulled into the roll gap during the first 0.001 seconds and no inertia forces are accounted for.

The rolling takes about 0.22 seconds. There is a steady state condition relative to the rolling section prevailing from about 0.06 to 0.15 seconds. The friction along the slideline between the rolls and the plate is modelled using Coulomb friction. The friction coefficient is assumed to be 0.3.

2.3 Material models

Finite element simulations are performed using three different material models for the rolled plate. The work rolls are assumed to be elastic in all the simulations. The material of the outer layer of the roll, which is included in the finite element model, is given a Young's modulus of 215 MPa and a Poisson's ratio of 0.3. Three different material models are used for the rolled plate. The first material model is an elastoplastic model with variable hardening. The hardening is assumed to be isotropic in these simulations but it can also be kinematic or a combination of isotropic and kinematic hardening. This model has no temperature or strain rate dependence.The Young's modulus for the plate material is assumed to be 100 GPa at 1000 C. The temperature is assumed to be constant. Poisson's ratio is set to 0.35. The yield condition according to von Mises and the associated flow rule is used. The effective stress-effective plastic strain relation is given from linear interpolation between the values in Table 1.

The data for plastic yielding are based on a strain rate of 2/s in Suzuki (1969:201). The simulations show that the strain rate is approximately 4/s. The material is not very strain rate sensitive. A more accurate material model will give slightly higher stress levels.

The second material model is a thermo-elastoplastic model with linear hardening. The hardening is assumed to be isotropic. This model has no strain rate dependence.The Young's modulus, Poisson's ratio, thermal expansion α, yield stress σ_y and the plastic hardening modulus E_p as a function of temperature are given from linear interpolation between the values in Table 2. The yield condition according to von Mises and the associated flow rule is used. Again the data for plastic yielding are based on a strain rate of 2/s in Suzuki (1969:201). Temperatures are calculated simultaneously with the mechanical analysis.

The third model is based on the constitutive relationship by Johnson and Cook (1983) where the model for the von Mises flow stress, σ, is expressed as

$$\sigma = [A + B\varepsilon^n]\,[1 + C\ln(\varepsilon_d')]\,[1 - T_h^m]$$

where ε is the effective plastic strain, $\varepsilon_d' = \varepsilon'/\varepsilon_0'$ is the dimensionless plastic strain rate for $\varepsilon_0' = 1.0\ s^{-1}$ and T_h = is the homologous temperature expressed as

$$T_h = (T - T_{room}) / (T_{melt} - T_{room})$$

The expression in the second set of brackets is never less than 1.0. The five material constants A, B, n, C, and m are optimized to fit experimental data and are shown in Table 3. In this case the material parameters are optimized to fit experimental data in the temperature interval 850-1100 °C only. Therefore it is dangerous to use these parameters outside this interval.Temperatures are the same as for the thermo-elastoplastic material model. The Young's modulus for the plate material is assumed to be 100 GPa and Poisson's ratio is set to 0.35. There is no thermal expansion in this model.

Table 1. Variable hardening model.

Effective plastic strain	Effective stress [MPa]
0.000	60.0
0.025	74.0
0.059	84.0
0.110	98.0
0.177	113.0
0.240	122.0
0.295	127.0
0.371	128.8

Table 2. Thermo-elastoplastic model.

Temp [°C]	800	1000	1100
Young [GPa]	120	100	95
Poisson ratio [-]	0.35	0.35	0.35
α [10^{-5}/°C]	1.235	1.292	1.320
σ_y [MPa]	80	60	45
E_p [MPa]	290	270	220

Table 3. Material parameters, model 3.

A [MPa]	4.085873
B [MPa]	205.817248
n [-]	0.22106369
C [-]	0.1702872
m [-]	0.585483
ε_0' [s^{-1}]	0.005680355
T_{room} [°C]	650.0
T_{melt} [°C]	1347.563

Table 4. Thermal material parameters.

Temperature [°C]	750	1250
Heat capacity [J/kgK]	650.0	720.0
Thermal cond [W/mK]	25.0	31.0

In the thermal simulation an isotropic temperature dependent material model is used. The thermal conductivity and heat capacity are shown in Table 4.

The density of the roll is assumed to be 7750.0 kg/m^3 and the plate density is 7860 kg/m^3 in all simulations. The thermal conductivity between the plate and the roll is assumed to be 300 W/m^2°C when there is no contact and 40 000 W/m^2°C during contact. The latter value is based on the heat conductivity of the oxide layer on the plate surface.

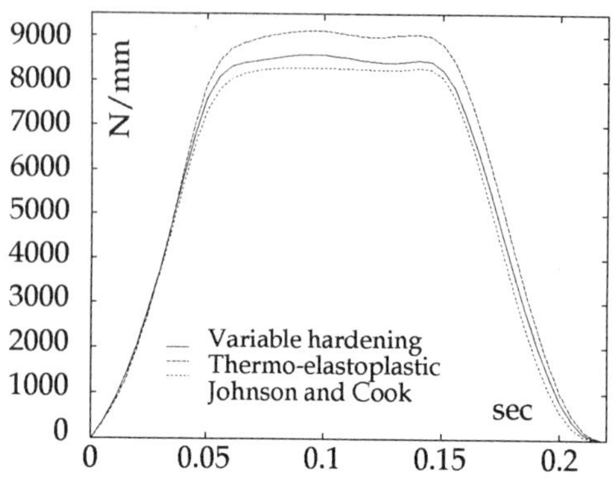

Figure 2. Vertical contact force.

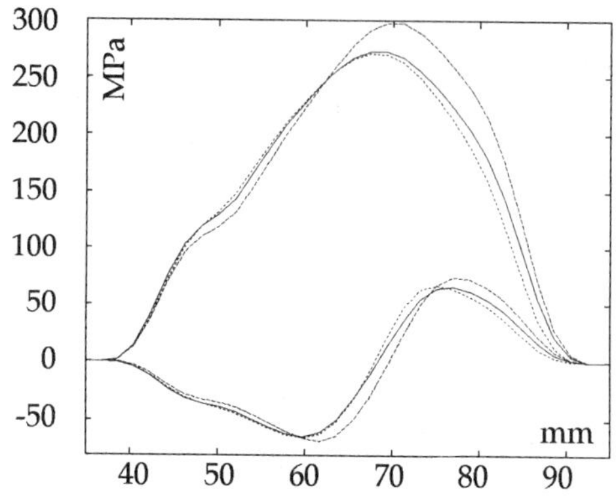

Figure 3. Pressure and shear stress distribution.

3 EXPERIMENTS

An experiment was performed at the Rautaruukki plate mill in Raahe, Finland. Using a Davy single-stand four-high plate mill. The width is 3600 mm and the maximum work roll diameter is 1045 mm and the maximum backup roll diameter is 1825 mm. The diameters in the experiment were 1014 mm and 1800 mm, respectively. A plate with a length of about 13 m was rolled to finishing dimensions of 2687 X 10.1 mm in the experiment. In the last pass the gauge adjustment was changed when three quarters of the plate had been rolled, so that the last quarter was not reduced in thickness at all. Therefore it was possible to measure the thickness before the last pass, which was 12.7 mm. The total rolling force, temperature, screw position and rolling velocity was recorded on a pen-recorder. The total rolling force was measured to be 22680 kN. The temperature before rolling was about 1000 C. In order to calculate the roll force distribution across the width of the plate, the CROWN software package for off-line calculation of plate profile and flatness in 4-high rolling mills was used (Wiklund 1987). The calculated rolling forces per unit length (i.e. per unit length transverse to the rolling direction) in the mid section of the plate is 8.55 kN/mm.

4 CALCULATED RESULTS

4.1 Contact forces

The calculated contact force is about 8.6, 9.1 and 8.5 kN/mm for the three material models, respectively. The time history for the vertical contact force can be seen in Figure 2for the models. The pressure distribution and the shear stress distribution for the three material models are shown in Figure 3. The shear stress distribution corresponds to the friction force per unit area. All three models calculate quite similar contact forces.

4.2 Stresses and deformation

The thicknesses after rolling are 10.10 mm in all material models. The thickness will decrease about 0.1 mm during cooling if the plate. The effective stress according to von Mises is shown in Figure 4 where part of the plate is shown magnified. The effective stress for the thermo-elastoplastic model is higher than for the variable hardening model. Using these models the effective stresses increase with deformation and the highest effective stress levels are found close to the surface on the exit side where the plastic deformation reaches its maximum value. The thermo-elastoplastic model has higher effective stress levels due to its temperature sensitivity. The Johnson and Cook model show high effective stress levels in regions with high deformation rates. Therefore there is a decrease in the effective stress levels in the middle of the contact area due to a hydrostatic situation with low deformation rates in this region.

The longitudinal stress state after rolling is shown in Figure 5,
where part of the plate is shown magnified. The stress is 15 MPa at the surface and -10 MPa in the middle of the plate for the variable hardening model, -60 MPa at the surface,

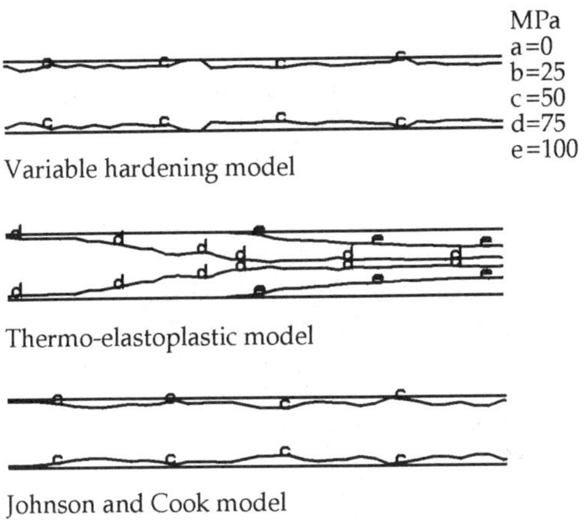

Figure 4. Effective stress after rolling.

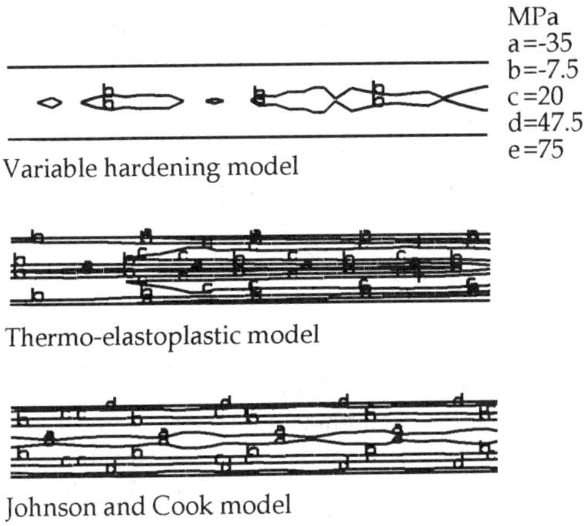

Figure 5. Longitudinal stress after rolling.

-35 MPa in the middle of the plate and 40 MPa at about 1/3 of the thickness for the thermo-elastoplastic model and 60 MPa at the surface and -35 MPa in the middle of the plate for the Johnson and Cook model. The reason for the differences in the longitudinal stress state after rolling for the thermo-elastoplastic model is that the surface of the plate is cooled during rolling due to roll contact and then heated after rolling by the hotter center parts of the plate. Then the surface thermally expands creating compressive stresses in the surface.

The effective stress after rolling is 50 MPa at the surface and 40 MPa in the middle of the plate for the variable hardening model, 130 MPa at the surface and 70 MPa in the middle of the plate for the thermo-elastoplastic model and 60 MPa at the surface and 40 MPa in the middle of the plate for the Johnson and Cook model.

The maximum effective plastic strain is about 0.35 for all the three material models. This value is found at the surface of the plate.

5 COMPARISONS

It is possible to obtain accurate results for the contact forces and plastic deformation using any of the three material models. The stress distribution during and after rolling differs between the models and it is easy to find unrealistic behavior in all models used.

Perhaps the most important remark is that none of the material models accurately describe the effect of changing strain rate and temperature during deformation. Most laboratory tests for studying flow stress under hot working conditions are performed at constant strain rates and temperature. Material parameters and models are then optimized to describe these laboratory tests accurately. In industrial hot working operations the strain rate and temperature changes continuously. In this paper it is shown that the strain rate varies from the entry to the exit side and that very steep temperature gradients are created during rolling. An accurate material model therefore has to describe the effect of changing the strain rate and temperature during deformation. A material model that is able to do this is proposed by Bergström (1982). In this model, the true flow stress is related to the total dislocation density, ρ, as

$$\sigma = \sigma_0 + \alpha G b \sqrt{\rho}$$

where σ_0 is the strain independent friction stress, α is a constant, G is the shear modulus and b is Burgers vector of the dislocation. Bergström shows how to calculate ρ as a function of strain, strain rate and temperature and the model is capable of describing the effect of changing the strain rate and temperature during deformation. Recovery effects are also included in the model. A drawback is that ρ is described by a differential equation that has to be integrated numerically. This is a minor problem in a finite element code. Static and dynamic recrystallization is not included in the model, but this phenomena has been studied by many researchers and there are many publications on that subject. If a recrystallization model is combined with Bergströms model, then this model would be able to describe many hot working operations more accurately. There is also the possibility in the future to include the effect of microstructure and phase transformations into the model since these phenomena affects the dislocation density.

6 CONCLUSIONS

The results show that a reasonable accuracy can be obtained in many cases with the material models used, but it also shows that a better material model is needed in order to get a complete understanding of the problem. The present material models works quite well for the simulation of the rolling operation, but are not capable of describing the evolution of stresses between the rolling passes. A material model based on the dislocation density according to Bergström combined with a recrystallization model would be a great step forward in the simulation of hot working operations.

REFERENCES

Bergström, Y. 1982. The plastic deformation of metals, A dislocation model and its applicability. Report of the Royal Institute of Technology, Stockholm.

Edberg, J. 1992. Three-dimensional simulation of plate rolling using different friction models, NUMIFORM 92. *Proc. of the fourth International Conference on Numerical Methods in Industrial Forming Processes.* Valbonne. France. September 14-18. A.A. Balkema. Rotterdam. pp. 713-718.

Edberg, J & L-E. Lindgren 1993. Efficient three-dimensional model of rolling using an explicit finite-element formulation. *Communication in Applied Numerical Methods in Engineering*. Vol. 9. pp.613-627.

Hallquist, J.O. 1986. NIKE2D-A vectorized implicit, finite deformation finite element code for analyzing the static and dynamic response of 2-D solids with interactive rezoning and graphics. Report UCID-19677. Rev 1. Lawrence Livermore National Laboratory. USA.

Hallquist, J.O. 1988. Users manual for DYNA2D - An explicit two-dimensional hydrodynamic finite element code with interactive rezoning and graphical display. Report UCID-18756. Rev 3. Lawrence Livermore National Laboratory. USA.

Johnson, G R. & W H. Cook 1983. A constitutive model and data for metals subjected to large strains, high strain rates and high temperatures. Presented at. *The seventh international symposium on ballistics*, Hague. Netherlands, April 1983.

Lindgren, L-E. & J. Edberg 1990. Explicit versus implicit finite element formulation in simulation of rolling. *Journal of Material Processing*. Vol. 24, pp. 85-94.

Shapiro, A B 1986. TOPAZ2D - A two-dimensional finite element code for heat transfer analysis, electrostatic, and magnetostatic problems. Report UCID-20824. Lawrence Livermore National Laboratory. USA.

Suzuki, H., S. Hashizume, Y. Yabuki, Y. Ichihara, S. Nakajima & K. Kenmochi 1969. Studies of flow stress of metals and alloys. Report of Institute of Science, University of Tokyo 18:3-101

Wiklund, O. N-G Johnsson & J Levén 1987. Simulation of crown, profile and flatness of cold rolled strip by merging severally physically based computer models. *4th Int Steel rolling conference.*

Simulation of Materials Processing: Theory, Methods and Applications, Shen & Dawson (eds)
© 1995 Balkema, Rotterdam. ISBN 90 5410 553 4

Finite element analysis of transformation hardening

H.J.M.Geijselaers & J.Huétink
University of Twente, Department of Mechanical Engineering, Enschede, Netherlands

ABSTRACT: A model is set up to determine temperatures, phase compositions and stresses during a thermal hardening cycle.
To model the different phases, a parallel fraction model is used, with a variable fraction for each phase. In this way phenomena like latent heat, transformation dilatation and transformation plasticity can easily be described.
Constitutive relations are derived, which are based on an independent stress history per phase.
Most phase transformations depend on diffusion of carbon and as such require some time to materialize. During a rapid thermal cycle large deviations from the equilibrium phase composition (superheating and undercooling) may occur. For the description of these transformations an incremental formulation of the Avrami equation for isothermal transformation is used.
The constitutive relations are implemented in a finite element model.
Examples are shown of simulations of laser transformation hardening, where superheating of the original structure plays an important role.

1. INTRODUCTION

A common and age-old heat treatment of steels is transformation hardening.
At room temperature an unalloyed hypo-eutectic carbon steel typically consists of two phases: ferrite, iron with practically 0% carbon and pearlite, a mixture of ferrite and cementite (Fe_3C, which contains 6.7% C).
The object to be hardened is heated to a temperature in excess of the Ac3 temperature (typically 800°- 1000° C) to get a homogeneous solution of the carbon in an austenitic iron matrix. Then it is quenched in water or oil and cooled sufficiently fast to obtain a metastable structure of carbon solved in a martensite matrix. The resulting structure has high hardness and good wear resistance.
A modern method of transformation hardening is laserhardening. Heat is applied locally, by a laser beam, passing over the surface of the object. The temperature in the laserspot rises rapidly to almost melting temperature. Due to the high thermal gradients, conduction to the cold bulk material will cause sufficiently rapid cooling of the heated part for most of the austenite to transform to martensite.
Simulation of transformation hardening involves a highly nonlinear thermal analysis which includes effects of latent heat of phase transformation.
This is complemented by a simulation of phase transformations, taking into account undercooling of austenite and, in the case of laserhardening, superheating of ferrite and pearlite.
When residual stresses are important, also a structural analysis, taking into account the dilatation due to phase transformations, is required.
All physical phenomena occurring during transformation hardening mutually interact, so only a fully coupled thermo-structural analysis will give satisfactory results under all circumstances.

2. PHASE TRANSFORMATIONS

Phase transformations of interest for hardening of steel are the so called austenite transformation and its reverse the pearlite transformation, both of which are diffusion controlled.
The martensite transformation involves no diffusion, but rather an instantaneous change of crystal structure.
The kinetics of diffusion controlled transformations is described by the Avrami equation.

$$\varphi = \varphi_0 + (\bar{\varphi}-\varphi_0)(1-e^{-(t/\tau)^n}) \qquad (1)$$

Here $\tau(T,\boldsymbol{\sigma})$ and $n(T,\boldsymbol{\sigma})$ are material parameters, which depend on the temperature T and the stresses $\boldsymbol{\sigma}$, $\bar{\varphi}(T,\boldsymbol{\sigma})$ is the volume fraction of the considered phase, that should be present in an equilibrium situation and φ_0 is the fraction at the start of the transformation. The Avrami equation is valid for isothermal transformations, it accounts for the time required for nucleation and growth of the new phase. Generalization to the case of non-isothermal transformation is done under the assumption, that the rule of additivity is valid (Cahn 1956, Leblond 1984).

$$\dot{\varphi} = (\bar{\varphi}-\varphi)\frac{n}{\tau}\left(\ln\left(\frac{\varphi_0-\bar{\varphi}}{\varphi-\bar{\varphi}}\right)\right)^{\frac{n-1}{n}} \qquad (2)$$

Superheating and undercooling is simulated by accounting for a carbon balance (Huétink 1990). The martensite transformation is described by the Koistinen-Marburger relation:

$$\varphi_m = \varphi_{\gamma Ms}(1-e^{-\beta(T_{Ms}-T)}) \qquad (3)$$

Here the martensite start temperature $T_{Ms}(\boldsymbol{\sigma})$, and the factor $\beta(\boldsymbol{\sigma})$ depend on the stresses, $\varphi_{\gamma Ms}$ is the amount of austenite, that is still present when upon cooling the T_{Ms} temperature is reached.
During any transformation the relations $\sum\varphi^i=1$ and $\sum\dot{\varphi}^i=0$ must be obeyed.

3. COUPLED THERMAL-STRUCTURAL ANALYSIS

3.1 Thermal Analysis

The energy conservation equation reads:

$$\lambda\boldsymbol{\nabla}.\boldsymbol{\nabla}T + \boldsymbol{\sigma}:\dot{\boldsymbol{\varepsilon}} - \rho\dot{e} = 0 \qquad (4)$$

λ Is the thermal conductivity, ρ is the mass density and e is the specific internal energy. The internal energy is a summation of the energies per fraction.

$$\rho e = \sum \varphi^i\rho^i e^i \qquad (5)$$

Per fraction the internal energy is a function of temperature.

$$e^i(T) = \int_{T_0}^{T} c_v^i(T)dT + e_0^i \qquad (6)$$

where c_v^i is the specific heat of the ith phase. So the rate of internal energy is:

$$\rho\dot{e} = \sum \varphi^i\rho^i c_v^i\dot{T} + \sum \dot{\varphi}^i\rho^i e^i \qquad (7)$$

The first term is the regular specific heat, the second term is the latent heat of phase transformation. The resulting rate equation is:

$$-\boldsymbol{\sigma}:\dot{\boldsymbol{\varepsilon}} + \sum \varphi^i\rho^i c_v^i\dot{T} = -\sum \dot{\varphi}^i\rho^i e^i + \lambda\boldsymbol{\nabla}.\boldsymbol{\nabla}T \qquad (8)$$

3.2 Structural Analysis

The equilibrium equation in the absence of body forces is

$$\boldsymbol{\sigma}.\boldsymbol{\nabla} = \mathbf{0} \qquad (9)$$

Here $\boldsymbol{\sigma}$ is the Cauchy stress tensor. Under the assumption of small deformation gradients, the rate form of this equation is written as

$$\dot{\boldsymbol{\sigma}}.\boldsymbol{\nabla} = \mathbf{0} \qquad (10)$$

A small strain approximation is adopted:

$$\dot{\boldsymbol{\varepsilon}} = 1/2\ (\boldsymbol{v}\boldsymbol{\nabla}+\boldsymbol{\nabla}\boldsymbol{v}) \qquad (11)$$

The strain rate is assumed to consist of a number of independent contributions:

$$\dot{\boldsymbol{\varepsilon}} = \dot{\boldsymbol{\varepsilon}}^{el} + \dot{\boldsymbol{\varepsilon}}^{pl} + \dot{\boldsymbol{\varepsilon}}^{th} + \dot{\boldsymbol{\varepsilon}}^{tr} + \dot{\boldsymbol{\varepsilon}}^{tp} \qquad (12)$$

where $\dot{\boldsymbol{\varepsilon}}^{el}$ is the elastic part, $\dot{\boldsymbol{\varepsilon}}^{pl}$ the plastic part, $\dot{\boldsymbol{\varepsilon}}^{th}$ the thermal dilatation, $\dot{\boldsymbol{\varepsilon}}^{tr}$ is the strain due to phase transformation and $\dot{\boldsymbol{\varepsilon}}^{tp}$ is due to transformation plasticity.
More than one phase may be present simultaneously. In that case it is not justified to use a macroscopic expression for the stress rate. Each phase has different elasticity coefficients, yield stress and hardening modulus. A different stress for each phase is assumed and the macroscopic stress is written as a weighted sum of the stress per phase:

$$\boldsymbol{\sigma} = \sum\varphi^i\boldsymbol{\sigma}^i \qquad (13)$$

So, the stress rate is:

$$\dot{\boldsymbol{\sigma}} = \sum\dot{\varphi}^i\boldsymbol{\sigma}^i + \sum\varphi^i\dot{\boldsymbol{\sigma}}^i \qquad (14)$$

A material model is constructed, where the thermal expansion and the transformation induced strain are macroscopic and the elastic strain, the plastic strain and the transformation plasticity are evaluated per phase (figure 1.).

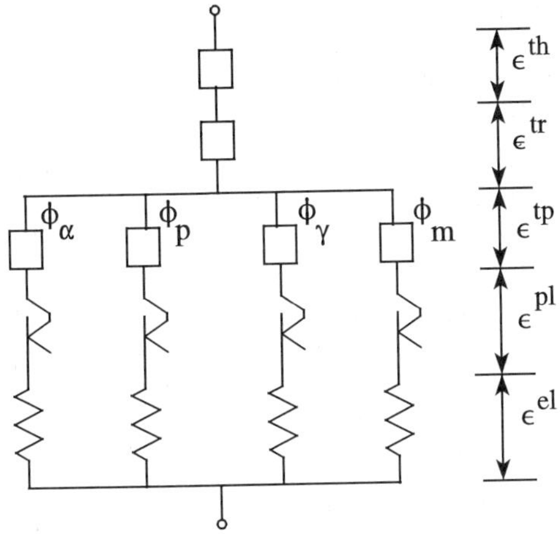

Figure 1: Parallel fraction model for stresses

The constitutive equations for the different phases have to be modified accordingly. The stress in the ith phase is assumed proportional to the elastic strain in the phase.

$$\boldsymbol{\sigma}^i = \boldsymbol{E}^i : \boldsymbol{\varepsilon}^{el(i)} \tag{15}$$

So for the stress rate we find:

$$\dot{\boldsymbol{\sigma}}^i = \boldsymbol{E}^i : (\dot{\boldsymbol{\varepsilon}} - \dot{\boldsymbol{\varepsilon}}^{pl(i)} - \dot{\boldsymbol{\varepsilon}}^{tp(i)} - \dot{\boldsymbol{\varepsilon}}^{th} - \dot{\boldsymbol{\varepsilon}}^{tr}) + \boldsymbol{\sigma}^i / G^i \, dG^i / dT \, \dot{T} \tag{16}$$

3.2.1 Transformation and thermal strain

The mass density can be written as a weighted sum of the phase densities.

$$\rho = \sum \varphi^i \rho^i \tag{17}$$

The mass density ρ^i of each phase is a function of temperature. The rate of the mass density is written as:

$$\dot{\rho} = \sum \dot{\varphi}^i \rho^i + \sum \varphi^i \, d\rho^i / dT \, \dot{T} \tag{18}$$

The first term on the right hand side is the density change due to phase transformation, the second term, due to thermal expansion. For isotropic materials density change and strain ε are related by:

$$\dot{\varepsilon} = -\dot{\rho} / 3\rho \tag{19}$$

The strain rates due to phase transformation and thermal expansion are

$$\dot{\boldsymbol{\varepsilon}}^{th} + \dot{\boldsymbol{\varepsilon}}^{tr} = -\sum (\varphi^i / 3\rho)(d\rho^i / dT) \dot{T} \, \boldsymbol{I} - \sum \dot{\varphi}^i (\rho^i / 3\rho) \boldsymbol{I} \tag{20}$$

Here $\boldsymbol{I}$ is the second order unit tensor.

3.2.2 Transformation plasticity

Experiments show that applying a stress, while phase transformation occurs, results in a permanent strain, which cannot be explained from yielding of one the phases involved (de Jong 1959). For low stress levels, this strain is found to be proportional to the applied stress and to the amount of phase transformed. The usual generalization to a multi axial stress state expresses the transformation plasticity proportional to the deviatoric stress $\boldsymbol{s}$ (Denis 1986). In our implementation however we assume independent stress and strain components for each phase. One of the explanations for the phenomenon of transformation plasticity is a stress relaxation in the newly formed phase, due to alignment of the crystal planes. It therefore seems justified to assume a transformation plastic strain in the forming phase as:

$$\dot{\boldsymbol{\varepsilon}}^{tp(i)} = 3/2 \, K^i \dot{\varphi}^i \boldsymbol{s}^i \tag{21}$$

The constants K^i depend on the chemical composition of the steel and on the type and sense of transformation. They have to be obtained from tests.

3.3 CONSTITUTIVE STRESS-STRAIN EQUATIONS

3.3.1. Elastic deformation

Substitution of (16), (20) and (21) into (14), with $\dot{\boldsymbol{\varepsilon}}^{pl(i)} = \boldsymbol{0}$, yields:

$$\dot{\boldsymbol{\sigma}}^i = \boldsymbol{E}^i : \dot{\boldsymbol{\varepsilon}} + \left(\sum_k \frac{\varphi^k}{3\rho} \frac{d\rho^k}{dT} 2G^i \frac{1+\nu}{1-2\nu} \boldsymbol{I} + \frac{\boldsymbol{\sigma}^i}{G^i} \frac{dG^i}{dT} \right) \dot{T} + \sum_k \frac{\rho^k}{3\rho} 2G^i \frac{1+\nu}{1-2\nu} \dot{\varphi}^k \boldsymbol{I} - 3G^i K^i \boldsymbol{s}^i \dot{\varphi}^i \tag{22}$$

So the stress rate is composed of three terms, a strain rate dependent part, a temperature rate dependent part and a phase transformation dependent part.

3.3.2 Plastic deformation

The description of plastic deformation is based on the Von Mises yield criterion, with isotropic hardening. Plastic deformation of a phase occurs,

when the equivalent stress exceeds the one dimensional yield stress $\sigma_y^i(\varepsilon^{p(i)},T)$, applicable to that phase.
It is common to define a yield tensor $\boldsymbol{Y}$ and a hardening parameter h:

$$\boldsymbol{Y}^i = \frac{3G^i \boldsymbol{s}^i \boldsymbol{s}^i}{(\sigma_y^i)^2} \; ; \; h^i = \frac{\partial\sigma_y^i/\partial\varepsilon^{p(i)}}{3G^i + \partial\sigma_y^i/\partial\varepsilon^{p(i)}} \qquad (23)$$

Using this notation we find an expression for the phase stress rate:

$$\begin{aligned}\dot{\boldsymbol{\sigma}}^i &= (\boldsymbol{E}^i - (1-h^i)\boldsymbol{Y}^i):\dot{\boldsymbol{\varepsilon}} + \\ &+ (\sum_k \frac{\varphi^k}{3\rho}\frac{d\rho^k}{dT} 2G^i \frac{1+\nu}{1-2\nu}\boldsymbol{I} + \frac{\boldsymbol{\sigma}^i}{G^i}\frac{dG^i}{dT} \\ &\quad -(1-h^i)(\frac{1}{G^i}\frac{dG^i}{dT} - \frac{1}{\sigma_y^i}\frac{\partial\sigma_y^i}{\partial T})\boldsymbol{s}^i)\dot{T} + \qquad (24) \\ &+ \sum_k \frac{\rho^k}{3\rho}\dot{\varphi}^k 2G^i \frac{1+\nu}{1-2\nu}\boldsymbol{I} - 3G^i h^i K^i \boldsymbol{s}^i \dot{\varphi}^i\end{aligned}$$

Like in the elastic case, the plastic stress rate is also composed of three terms, a strain rate dependent part, a temperature rate dependent part and a phase transformation dependent part.

4. INCREMENTAL EXPRESSIONS AND CONSISTENT TANGENTS

The rate equations obtained are not used as such, but rather, incremental expressions are derived, which are linearized to obtain consistent tangent equations.
A linearized prediction of e.g. the phase fraction increment is then found as:

$$\begin{aligned}\Delta\varphi^i &= \int_{t_0}^{t_0+\Delta t} \dot{\varphi}^i dt \qquad (25) \\ &\approx \Delta\varphi^i(T_0,\boldsymbol{\sigma}_0) + \frac{1}{2}\frac{\partial\Delta\varphi^i}{\partial T}\Delta T + \frac{1}{2}\frac{\partial\Delta\varphi^i}{\partial\boldsymbol{\sigma}}:\Delta\boldsymbol{\sigma}\end{aligned}$$

The stress dependence of the transformation kinetics has not been implemented yet. The temperature dependence gives rise to extra terms, similar to the phase transformation dependent part of the constitutive equations, but with $1/2\ \partial\Delta\varphi^i/\partial T\ \Delta T$ instead of $\Delta\varphi^i$.

5. FINITE ELEMENT DISCRETISATION

The incremental version of the rate form of the equilibrium equation is discretized following the standard Galerkin procedure and using the constitutive equations derived above. Bilinear plain strain, generalised plain strain and axi-symmetric isoparametric elements with constant dilatation are used. Primary unknowns are displacements and temperature increments. In general a small strain approach is sufficient.

6. EXAMPLES

The examples presented here consist of two series of simulations of laser hardening. Laser hardening is a rather modern technique (Chatterjee-Fischer 1984), where the surface to be hardened is heated by high density laser light. The temperature in the laser spot will rise to nearly (sometimes in excess of) melting temperature at rates of 1000 - 10000 °C/s.
Due to the high temperature gradients, cooling can be accomplished by conduction to the cold bulk material, so called self-quenching. The cooling rates are of the same order as the heating rates. However, for self-quenching it is necessary, that the object to be hardened possesses sufficient bulk in relation to the thermal penetration depth. This is demonstrated in the first series of simulations, where we show, that for a given energy density, a minimum thickness is required to obtain any hardening. The second set of simulations will demonstrate, that the hardening depth and the residual stresses may be manipulated by varying the energy density and the interaction time.
The data used in the simulations are representative for steel Ck45 and are summarised in table 1. Data for transformation kinetics of this steel are amply available (e.g. Orlich 1972). A clear description of the processes involved in the various transformations of this type of steels is given by Ashby e.a. (1984).
The model used for the simulations consists of a row of 20 axi-symmetric elements arranged to represent a cylindrical section through a uniformly heated infinite slab. The cylindrical sides are constrained to remain straight.
The heating is done on one side. To allow for total cooling down, conduction to ambient temperature is provided on both top and bottom, with a low heat transfer coefficient.

6.1 Influence of the Thickness.

In the first series of simulations four different slab thicknesses are analyzed: 4, 6, 8 and 10 mm. The heat input is kept constant at 25 MW/m^2. The interaction times, the times of the heating phase, are moderated such, that in all cases the surface temperature just reaches 1500 °C. The interaction time is approximately .5 seconds for all thicknesses.

Tabel I: Thermal and Mechanical Properties.

		ferrite/ pearlite	austen- ite	martens- ite
Youngs Modulus	20 °C	210	200	200
[GPa]	600 °C			158
	1500 °C	90	50	
Yield Stress	20 °C	360	190	1600
[MPa]	600 °C	40	30	1250
	1500 °C	10	10	
Hardening Modulus	20 °C	12.6	4	20
[GPa]	600 °C			15.8
	1500 °C	7.2	2.5	
Mass Density	20 °C	7850	8000	7760
[kg m^{-3}]	600 °C			7550
	1500 °C	7300	7300	
Conduction Coeff.	20 °C	15	49	43.1
[W m^{-1}°C^{-1}]	600 °C			30
	1500 °C	25	27	
Specific Heat	20 °C	480	520	485
[J kg^{-1}°C^{-1}]	600 °C			670
	1500 °C	900	680	
Internal Energy	20 °C		60000	40000
[J kg^{-1}]				

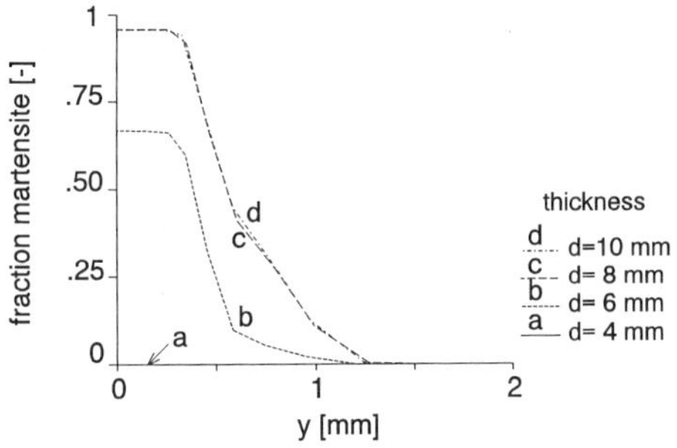

Figure 2: Influence of thickness on hardening depth

The results as presented in figure 2 show, that with thicknesses of 8 and 10 mm an identical hardening depth of approximately .5 mm is obtained.

With 6 mm thickness the cooling rate is already too low to preserve all austenite down to the T_{Ms} temperature. Back transformation to ferrite/pearlite has already partly taken place. With 4 mm no martensite forms at all.

6.2 Influence of the Energy Density.

Differences in the energy density may have a substantial effect on the hardening depths obtained, as well as on the residual stresses. To demonstrate this we redid the simulation on the 8 mm thick slab with energy densities of 8.3, 16.6 and 75 MW/m^2. In all four cases heating is sustained until a surface temperature of 1500 °C is reached.

With a low energy density no hardening is obtained at all. With an energy density of 16.6 MW/m^2 back transformation to ferrite/pearlite has started due to a too low cooling rate (figure 3).

Both the energy densities of 25 and 75 MW/m^2 yield full martensite formation at the surface. The hardened layer in the high energy density case is however an order of magnitude thinner. The magnitude of the residual stresses is of the same order, but the high compression stress peek at the surface is again much more localised in the high energy density case (figure 4).

The residual stresses give rise to a slight distortion of the slab. In both cases the heated surface comes out convex. In the 25 MW/m^2

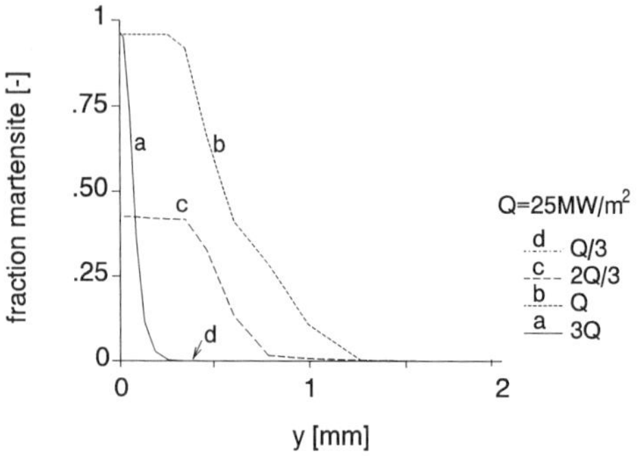

Figure 3: Influence of energy density on hardening depth

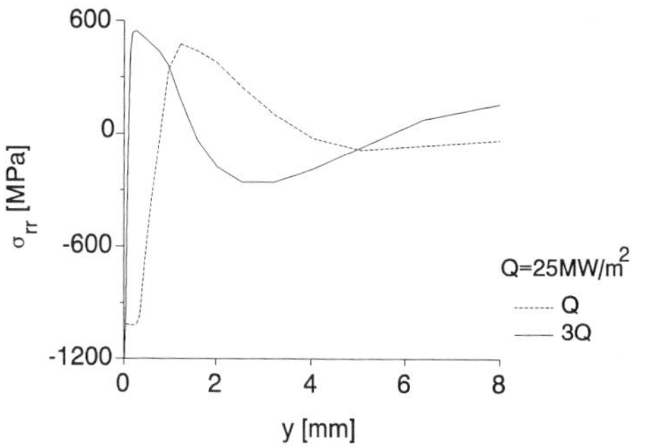

Figure 4: Influence of energy density on residual stresses

case, the overall curvature radius is approximately 8.5 m, whereas with the high energy density a curvature of 3 m is obtained.

7. CONCLUSIONS

A parallel fraction model allows for an elegant description of the phenomena occurring during phase transformations. The implementation in a finite element code, that is suited for coupled thermo-structural calculations, is straight forward. In most cases a small deformation approximation is sufficient.
Simulations, using a simple model, of laser hardening demonstrate some of the complexities of this process.

8. ACKNOWLEDGEMENT

The authors like to thank mr. E. Sloot for performing the analyses and for his assistance with the preparation of this paper.

9. REFERENCES

Ashby M.F., Easterling K.E.:'The Transformation Hardening of Steel Surfaces by Laser Beams-1. Hypo-Eutectoid Steels', *Acta Metall.*, Vol. 32 (1984) 1935-1948.

Cahn J.W.:'Transformation Kinetics during Continuous Cooling', *Acta Metall.*, Vol. 7 (1959) 572-575.

Chatterjee-Fischer R., Rothe R., Becker R.:'Überblick über das Härten mit dem Laserstrahl', *Härterei Techn. Mitt.* Vol. 39 (1984) 91-98.

De Jong M., Rathenau G.W.:'Mechanical properties of Iron and some Iron Alloys while undergoing Allotropic Transformation', *Acta Metall.*, Vol. 7 (1959) 246-253.

Denis S., Simon A.:'The Role of Transformation Plasticity in the Calculation of Quench Stresses in Steel', *Residual Stresses in Science and Technology*, ed. Macherauch E., Hauk V., (1986) 565-572.

Huétink J., Beckmann L.H.J.F., Geijselaers H.J.M.:'Finite Element Analysis of Laser Transformation Hardening', *Proceedings ECO3, SPIE Vol. 1276 CO_2 Lasers and Applications II*, (1990) 426-436.

Leblond J.B., Devaux J.:'A new Kinetic model for Anisothermal Metallurgical Transformations in Steels including effect of Austenite Grain Size', *Acta Metall.*, Vol 32 (1984) 137-146.

Orlich J. e.a.: 'Atlas zur Wärmebehandlung von Stähle', Verlag Stahleisen M.B.H., Düsseldorf, (1972).

Simulation of Materials Processing: Theory, Methods and Applications, Shen & Dawson (eds)
 ISBN 90 5410 553 4

Numerical simulation and experimental observations of initial friction transients

D.A. Hughes, L.I. Weingarten & D.B. Dawson
Sandia National Laboratories, Livermore, Calif., USA

ABSTRACT: Experiments were performed to better understand the sliding frictional behavior between metals under relatively high shear and normal forces. Microstructrual analyses were done to estimate local near-surface stress and strain gradients. The numerical simulation of the observed frictional behavior was based on a constitutive model that uses a state variable approach.
This work was supported by U.S. DOE under contract No. DE-AC04-94AL85000.

INTRODUCTION

Many manufacturing operations such as metal forming are highly influenced by the frictional state between metals. Improvements in product design efficiency through the use of numerical simulation of the manufacturing process will thus depend on how well friction models duplicate actual behavior. Friction behavior during metal forming operations is, however, difficult to characterize and model because friction transients are common over the short sliding distances encountered during forming. This study combines experiment, metallurgical analysis and a finite element method (FEM) simulation to look at friction behavior within this transient regime.

In the first part of this study, experiments were performed to better understand the dry sliding frictional behavior between metals under relatively high shear and normal forces and for short sliding distances. The friction tests were performed on a large area flat plate friction apparatus with copper samples and a steel platen. A micromechanical and metallurgical analysis of the deformation within near-surface layers was then performed on the samples to provide estimates of the steep local stress and strain gradients as a function of subsurface depth for comparison with friction models. These tests and experimental analyses also provide data on the repeatability of friction transients and the relevant size scales for modeling.

In the second part, a numerical methodology was implemented to simulate the frictional behavior seen in this study's sliding tests. Common relations that define sliding resistance as being proportional to normal pressure are attributable to Coulomb (1785). Since those simple relations do not describe what is observed experimentally, other relationships were sought for the simulation. An evolving constitutive model for friction recently developed by Anand (1993) was used to simulate the observed frictional behavior. As originally proposed, this model was rate-independent and assumed isothermal and isotropic behavior. It uses a state variable that accounts for the resistance of the surface layers to slip. The parameters in this model can be directly measured from tests.

FRICTION EXPERIMENTS AND RESULTS

The method of friction testing and metallurgical analysis was described in detail in Hughes et al. (1994,1995). For the friction tests, moderately high sliding speeds (0.25 and 25 mm/s) were imposed on high purity (99.99%) OFE copper samples at normal pressures near the yield of the annealed samples. Results showed that the evolution and magnitude of the coefficient of friction appears to be a function of load, rate, surface cleanliness, and roughness. The observed coefficients of friction ranged from approximately 0.4 to 1.4. Plots of either the shear force or coefficient of friction as a function of sliding distance show an initial increase in the coefficient or shear force to a relative maximum followed by a decrease to an asymptotic value (Figures 1,2). These figures show that the evolution of sliding resistance is a function of the relative motion of the Cu friction sample and the steel platen. Figure 1 shows that the sliding resistance is a function of the normal load. The effect of normal force is further illustrated by plotting the average shear force and peak shear force as a function of normal pressure for several tests (Figure 3). The trend for increasing shear forces with increasing normal pressure is clear, even though Figure 3 also illustrates the scatter encountered in friction testing. Other experiments, Figure 4, show the effect of rate (sliding speed).

The effect of these variables on the microstructure and on the large near-surface stress and strain

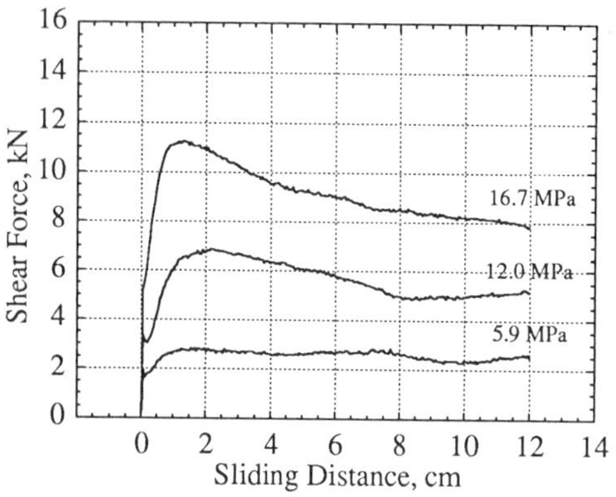

Figure 1. Slip resistance as a function of normal load at 25 mm/s

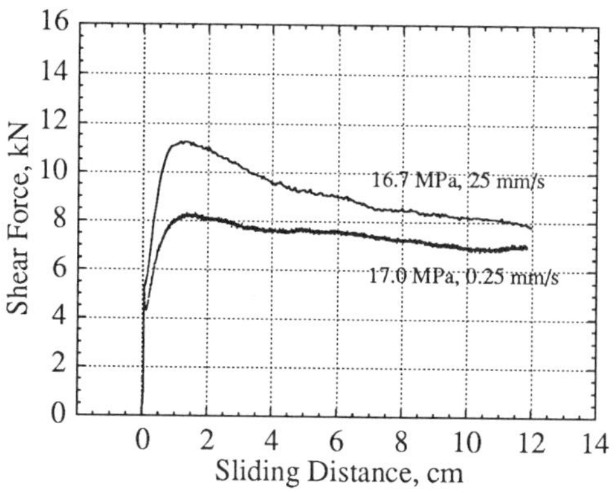

Figure 4. Slip resistance as a function of rate

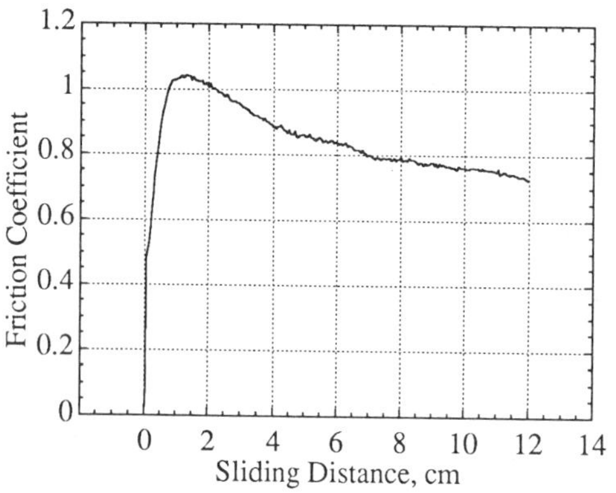

Figure 2. Coefficient of friction vs. Sliding distance (12.0 MPa, 25 mm/s)

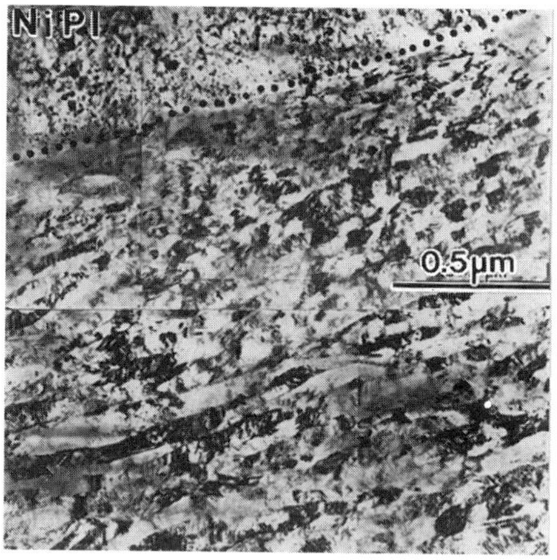

Figure 5. Near-surface friction microstructures

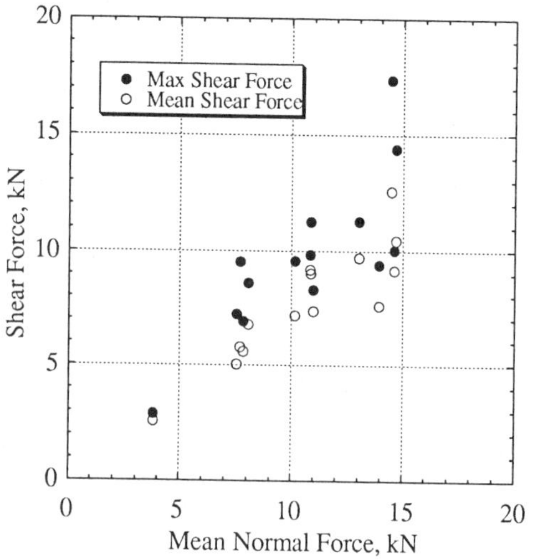

Figure 3. Mean and maximum shear force vs. Mean normal force

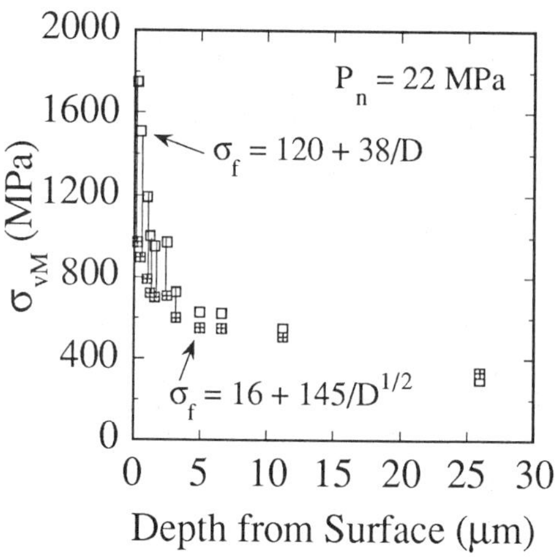

Figure 6. Estimate of local stress gradient (Hughes et al 1995).

gradients has been analyzed and reported previously (Hughes et al. 1994,1995). An example of those results is given here to show the relevant size scales of these gradients with depth below the surface. Figure 5 shows the gradient in the dislocation microstructure with depth below the surface following sliding. The dislocation boundaries have a spacing on a nanometer size scale, which indicates that the material has been deformed to very large strains ($\varepsilon_{vM} > 5$) at very high stresses. Figure 6 shows estimates of the stress gradients which develop during sliding. Very high stresses are encountered at the surface and decrease with depth. High stresses persist at deeper depths with higher normal pressures. Strain gradients were shown by Hughes et al. (1994,1995). Comparison of these results with micromechanical friction models shows that asperity wave models (Avitzur et al 1984 and Challen et al 1984) can account for some of the local gradients. However, additional friction mechanisms such as adhesion are also necessary.

These fine-scale steep gradients in stress, strain, and microstructure show that the evolution of the frictional behavior is clearly a function of subsurface deformation and the near-surface state of the material. Conventional finite element modeling techniques can not sufficiently discretize the geometry to simulate the gradients in the structure, even if the gradients could be modeled easily. However the evidence that the friction evolution is based on deformation processes suggests that friction may be treated within conventional FEM using ideas from plasticity.

The purpose of the next phase of the study was to propose a methodology that can produce a simulation of the frictional behavior seen in this study's sliding tests. In recent publications, Anand and his colleagues (Anand and Tong 1993, Anand 1993, Pisoni 1992) have proposed a "first-order" friction constitutive model that can simulate the interface sliding behavior at metal-to-metal contacting surfaces. Experiments by Anand and Tong (1993) and Courtney-Pratt and Eisner (1957) have shown a correlation in the type of behavior seen at contacting interfaces with the elastic-plastic behavior of ductile metals. In essence, the "stick-slip" behavior seen in curves of shear stress vs. sliding distance is quite similar to that seen by ductile metals that experience large deformations (J_2 flow theory). The "sticking" or "adhering" behavior is similar in nature to the elastic portion of the stress-strain curve, while the "slipping" behavior correlates to the plastic portion. This suggests that a friction constitutive model can be developed that has similar features to those developed for metal plasticity.

MODEL BACKGROUND

A complete friction constitutive model should be able to correlate what is occurring at the microstructural level (e.g., near-surface plasticity gradients) with observable macroscopic behavior (e.g., slip resistance). This model should be easily implemented in a finite element code and could be modified if necessary.

In the papers by Anand and his colleagues a friction constitutive model using a state variable approach is developed. Though they may only account for first order effects, these efforts are representative of the state-of-the-art in the field. They have modeled a complex surface state by algorithms that simulate one class of frictional behavior well. At the same time the model is fairly straightforward to implement in finite element codes. In this work we will apply their approach, modify as necessary and suggest improvements. The details of this development are not repeated here, but certain important features are discussed. A state parameter, α, is introduced which ideally takes into account the microstructural phenomena at or near the contact interface. Evolution equations are written for the traction at the interface and a variable, s, defined as the slip resistance of the contact surface. As shown below, s is a function of the contact pressure, p, and the state parameter α.

$$s = s(p, \alpha) \tag{1}$$

The state parameter α is replaced by a macroscopic variable u_s. The variable u_s is defined as the "accumulated or equivalent relative tangential slip" (Anand and Tong 1993). This could be interpreted as an observable metric of the changes taking place at the microstructural level. Given the functional form in Eqn. (1) for s, an evolution equation (ds/dt) can be written in terms of relationships that describe the change in slip resistance as a function of normal pressure and relative slip. These relationships are commonly called hardening functions.

Critical to the development of the constitutive model, is the notion of a *slip surface* which is similar to the concept of a *yield surface* in metal plasticity theory. In the flow theory of plasticity, the yield surface is the surface in stress space, within which there is no change in plasticity and incremental plasticity changes can only occur on the exterior. In the present context, the slip surface is similar in that adhering between surfaces is enforced internal to the slip surface. Slipping is analogous to the incremental plastic strain. The function

$$f(t, s) = 0 \tag{2}$$

represents the *slip surface*. Here, t is the traction vector and s is the slip resistance defined in (1). If, we assume

$$f(t, s) = \tau - s, \tag{3}$$

where $\tau = \sqrt{t_T \cdot t_T}$ is the absolute value of the tangential traction (t_T), then if $\tau < s$, we will have adhering between the surfaces. If this is not the case, we will have a slipping action.

CHOICE OF SLIP RESISTANCE FUNCTION

For over two centuries engineers have utilized what has generally come to be known as Coulomb's law to describe the frictional behavior at contact surfaces. This states that the slip resistance is proportional to the normal pressure. The proportional factor, μ, is commonly called the coefficient of friction. Anand (1993) reviews previous studies that show that the slip resistance tends to saturate (reach a constant value) for given values of relative slip. Experimental results by Anand and Tong (1993) confirm this result. The experiments were performed in a testing machine that could apply simultaneous compression and torsion to an annular workpiece made from OFE copper. The tool material in contact with the workpiece was AISI A2 tool steel. The workpiece was initially compressed and then torqued to produce a relative tangential slip. In Anand's experiments, the normal pressures varied from 40 MPa to 540 MPa with sliding velocities that ranged from 0.0288 to 2.88 mm/s. In the paper by Anand and Tong (1993) it is indicated that the slip-rate sensitivity was small. These normal pressures were considerably higher than the normal pressures used in the present study. As presented previously in this report, the maximum normal pressure used here was approximately 30 MPa, with sliding velocities that were either 0.25 or 25 mm/s. The materials used in our experiments were different, high purity OFE copper and hardened 4340 steel, but to a first order they could be considered to be similar. The results of our experiments were discussed earlier in this paper (Figures 1-4). Our experiments and Anand's both show a saturation of shear force and the coefficient of friction with sliding distance. Our data, unlike Anand's, show a transient in the shear force and coefficient of friction to a relative maxima before decreasing to an asymptotic value. An important differentiation between our experiments and Anand's, in addition to the distinct rate and normal pressure ranges, is the deformed copper in our experiments is continually coming into contact with portions of the tool (steel platen bar) that have not previously been in sliding contact with the workpiece material.

In spite of the differences in the experiments, it is possible to use the same relationship used by Anand and Tong (1993) that describes the slip resistance as a function of normal pressure, p, and the relative tangential slip, u_s. The assumed slip resistance function is

$$s = s^* \left(\tanh \left(\frac{\mu p}{s^*}\right) \right), \qquad (4)$$

where, $\mu = \mu(u_s)$ and $s^* = s^*(u_s)$. An examination of Eqn. 4 show that at low pressures, the slip resistance reduces to Coulomb's law and that at high pressures, produces a saturated value which is only a function of the relative tangential slip, u_s.

The functional forms of μ and s^* can be determined by fitting the experimental data to mathematical functions. We now deviate from Anand's approach in that the current experimental data shows a slightly different functional relationship with respect to u_s. In the paper by Anand and Tong (1993), s^* is determined by a fit to high pressure (540 MPa) data, while μ is obtained from a fit to low pressure (40 MPa) data. In this manner, μ and s^* were described by

$$\mu = \mu_o + (\mu_s - \mu_o)\left(1 - \exp(-u_s/u_1) \right), \qquad (5)$$

$$s^* = s^*_o + (s^*_s - s^*_o)\left(1 - \exp(-u_s/u_2) \right). \qquad (6)$$

The constants μ_o, μ_s, u_1, s^*_o, s^*_s, and u_2 were determined by Anand and Tong (1993) using a non-linear least squares fit of the low and high pressure data to be:

$$\mu_o = 0.37,\ \mu_s = 0.75,\ u_1 = 2.5 \text{ mm},$$
$$s^*_o = 105 \text{ MPa},\ s^*_s = 215 \text{ MPa},\ u_2 = 3.5 \text{ mm}. \qquad (7)$$

In the fit to the current data, due to the relatively low pressures as compared to Anand's experiments, we choose to fit μ to a representative low pressure experiment (Figure 2). We must select a functional relationship for μ that still saturates at high values of u_s, but has a functional relationship that can produce the transient peak seen in the present experiments. A relationship for μ that meets this criteria and is a multiplicative modification of Eqn. 5 is

$$\mu = \mu_o + (\mu_s - \mu_o)\left(1 - \exp(-u_s/u_1) \right)\left(1 + a_1 \operatorname{sech}(-u_s/a_2) \right). \qquad (8)$$

The behavior of the function in the last set of parentheses in Eqn. (8), containing the hyperbolic secant function, approaches unity for low and high values of u_s with an intermediate maximum. There are an additional two constants (a_1, a_2) that must be obtained in fitting the experimental data. As discussed above the constants in Eqn. (8) are obtained by a fit to low pressure data in the present experiments. Using the "Nonlinear Fit" package within Mathematica (Wolfram 1994) on the experimental data for a normal pressure of 12.0 MPa and a sliding velocity of 25 mm/s (Figure 2), the constants in Eqn. (8) are determined to be:

$$\mu_o = 0.375,\ \mu_s = 0.68,\ u_1 = 8.09 \text{ mm},\ a_1 = 0.914,$$
$$a_2 = 33.01 \text{ mm}. \qquad (9)$$

The concept of a slip surface, slip resistance function (Eqn (4)), and the fit of constants to experimental data fully describes a friction constitutive model for the interface between the workpiece (OFE copper) and the hardened steel platen. In the following section, we will show results of calculations that use the fits to Anand's data (Eqn. (7)) as well as the current data (Eqn. (7) as modified by (9)).

NUMERICAL RESULTS

The numerical implementation of the friction model was first written by Pisoni (1992). A subroutine was written which put a new friction interface model in the implicit version of the ABAQUS (Hibbitt, et al 1993) finite element computer code. Originally written in double precision for computer workstations, the subroutine was modified to run on the Cray YMP computer in single precision.

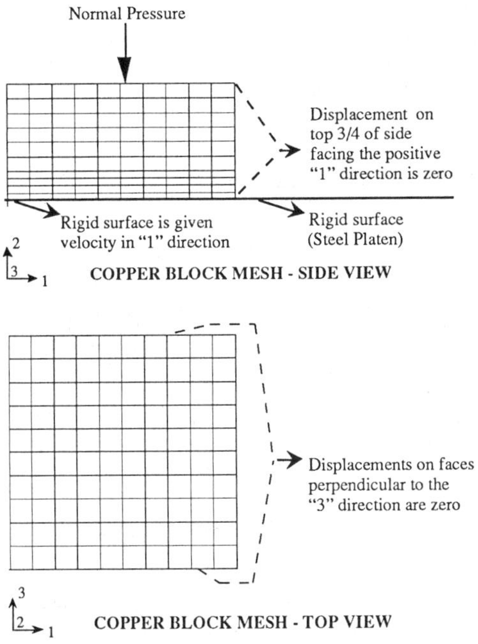

Figure 7. Finite element mesh

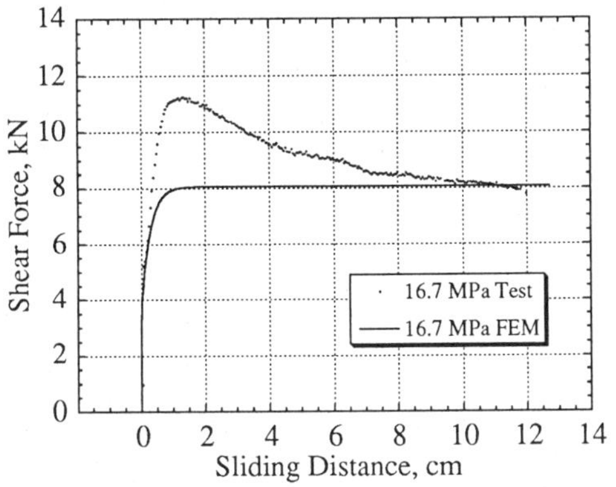

Figure 8. Shear force vs. Sliding distance (25 mm/s)

Utilizing the friction constitutive model user subroutine in ABAQUS, the OFE copper test block was modeled with appropriate boundary conditions that simulate the action of the Sandia flat plate friction tester (Hughes et al 1995). The copper was assumed to yield at 30 MPa and was assumed to have linear strain hardening simulating initial material plasticity (Hughes et al 1994). The hardened 4340 steel platen was assumed to be an infinite rigid surface. The mesh of the copper block is shown in Figure 7. The boundary conditions noted on Figure 7 were chosen to simulate the effects of the fixture containing the copper block.

At first, the properties of the friction interface were those developed by Anand as described above. The

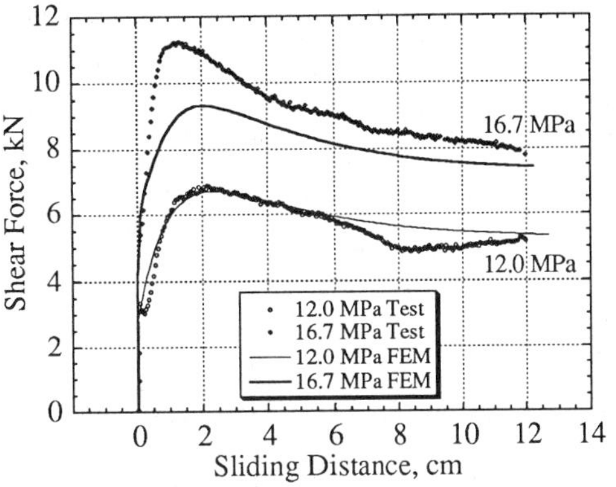

Figure 9. Shear force vs. Sliding distance (12.0 and 16.7 MPa, 25 mm/s)

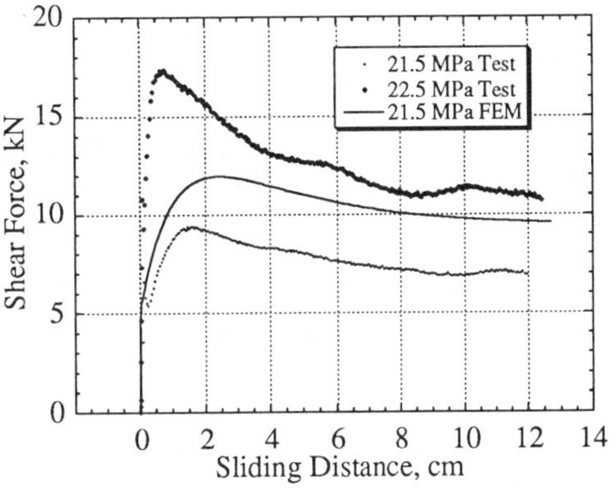

Figure 10. Shear force vs. Sliding distance (21.5 MPa, 25 mm/s)

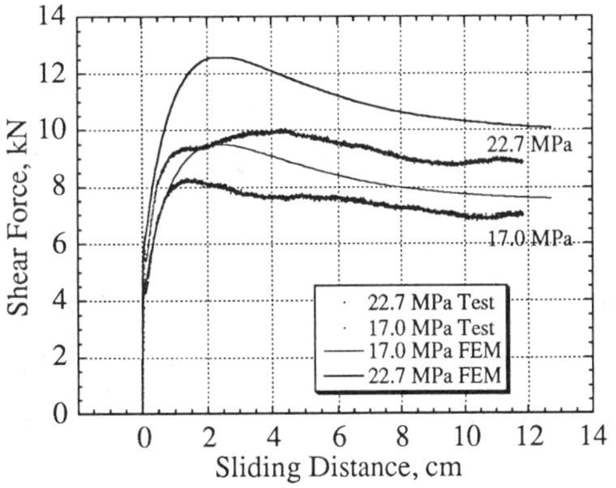

Figure 11. Shear force vs. Sliding distance (.25 mm/s)

result for the axial force on the steel platen (rigid surface) is shown in Figure 8 and compared to the test results for a sliding speed of 25 mm/s and a normal pressure of 16.7 MPa. As seen in this Figure, the numerical simulation accurately predicts the axial force for relatively large sliding distances, but significantly under predicts the force during the initial motion. In addition, there is no transient behavior with a relative maxima in the code calculation.

In the next series of calculations, Eqn (8) was used for μ in the model with constants fitted to the present test series, Eqn. (9). The results shown in Figure 9 are for a sliding speed of 25 mm/s and normal pressures of 12.0 and 16.7 MPa. The agreement between the experiment and the numerical analyses is good especially for the 12.0 MPa normal pressure. This is not surprising since the constants in the μ relation were fit for that experiment. Both sets of numerical simulations in Figure 9 show the relative maxima at early times and both then decrease asymptotically to values similar to values seen in the experiments. Figure 10 shows the comparison of a numerical simulation using a normal pressure of 21.5 MPa with two tests which nominally used the same normal pressure and the same sliding speed. These two tests represent the maximum range of scatter observed (see Figure 3). The simulation falls in the middle of the data range. A slow speed (0.25 mm/s) comparison is shown in Figure 11 for normal pressures of 17.0 and 22.7 MPa. There is good agreement been test and the numerical result for 17.0 MPa as in the corresponding normal pressure case (16.7 MPa) at the higher sliding speed in Figure 9.

SUMMARY AND CLOSING REMARKS

In this study, experiments, metallurgical analysis and numerical simulations were performed to better understand the sliding frictional behavior between metals under relatively high shear and normal forces. The numerical simulation of the frictional behavior is based on a constitutive model recently developed by Anand (1993). It uses a state variable that accounts for the resistance of the surface layers to slip. The implementation of the friction constitutive model was in the implicit version of the ABAQUS finite element code. There are no inherent limitations in the model that would preclude its inclusion in other implicit or explicit time integration codes. Experimental results showing a rise to a maximum in shear force followed by a decrease as sliding continues may show the importance of two other parameters that were neglected: rate and temperature. Of these two, the rate effects are straightforward to implement if one assumes a slip resistance function similar to Eqn. 4. An outline of a procedure that is an extension of the model used in this paper's calculations is documented in the Appendix of Pisoni's work (1992). The effect of temperature at the interface should be addressed by formulations in a finite element code that allow temperature changes due to material deformations. In turn, these thermal effects can cause property changes within the workpiece, including damage. If one includes the effect of temperature in the friction constitutive model, this model could conceivably be linked to material damage models such as those of Bammann, et al (1993).

REFERENCES

Anand, L., "A Constitutive Model for Interface Friction", Computational Mechanics, 12/4, 197-213, 1993.

Anand, L., Tong, W., "A Constitutive Model for Friction in Forming", Annals of the CIRP, 42/1, 361-366, 1993.

Avitzur, B., Huang, C.K. and Zhu, Y.D., "A friction model based on the upper-bound approach to the ridge and sublayer deformations", Wear 95, 59-77 1984.

Challen, J.M., McLean, L.J. and Oxley, P.L.B., "Plastic deformation of a metal surface in sliding contact with a hard wedge: its relation to friction and wear", Proc. Royal Society of London A 394, 161-181, 1984.

Coulomb, C.A., "Memoires de Mathematique et de Physique de l'Academie Royale des Sciences," 1785.

Courtney-Pratt, J.S., Eisner, E., "The effect of a tangential force on the contact of metallic bodies", Proc. of The Royal Soc. A 238, 529-550, 1957.

Bammann, D.J., Chiesa, M.L., Horstemeyer, M.F., Weingarten, L.I., "Failure in Ductile Materials Using Finite Element Methods", in *Structural Crashworthiness and Failure,* Edited by Jones, N., Wierzbicki, T., Elsevier Applied Science, 1993.

Hibbitt, Karlsson, and Sorensen, Inc., "ABAQUS Users Manual", Medway, RI, 1993.

Hughes, D.A., Dawson, D.B., Korellis, J.S. and Weingarten, L.I. "Near surface microstructures developing under large sliding loads", J. Mater., Eng. and Perf. 3, 459-475, 1994.

Hughes, D.A., Dawson, D.B., Korellis, J.S. and Weingarten, L.I., "A microstrucutrally based method for stress estimates", Wear, in press 1995.

Pisoni, A., "A Constitutive Model for Friction in Metal Working", MS Thesis, MIT, October 1992.

Wolfram Research, Inc., "Mathematica, Version 2.2", Champaign, Illinois, 1994.

Simulation of Materials Processing: Theory, Methods and Applications, Shen & Dawson (eds)
© 1995 Balkema, Rotterdam. ISBN 90 5410 553 4

Numerical implementation of a state variable model for friction

D.A. Korzekwa
Los Alamos National Laboratory, N.Mex., USA

D.E. Boyce
Cornell University, Ithaca, N.Y., USA

ABSTRACT: A general state variable model for friction has been incorporated into a finite element code for viscoplasticity. A contact area evolution model is used in a finite element model of a sheet forming friction test. The results show that a state variable model can be used to capture complex friction behavior in metal forming simulations. It is proposed that simulations can play an important role in the analysis of friction experiments and the development of friction models.

1 INTRODUCTION

Control of friction and lubrication is important for the success of many material processing and fabrication operations. Although the tribology of metal working has been studied extensively[1-4], there is no concise theory that can be easily implemented into numerical simulations of metal forming processes. The primary reason for this is that a wide range of phenomena may be responsible for friction and wear at the tool–workpiece interface. Consequently, for a given metal forming operation, a friction model should be tailored to reflect the appropriate physical mechanisms.

It is desirable to express models for friction and lubrication in terms of physically identifiable quantities that describe the state of the interface[5]. Candidates for these state variables are numerous, including measures of surface roughness, lubricant film thickness, fractional coverage of boundary lubricant, and the material strength at the surface. In some cases it is useful to pose a model in terms of the fraction of the interface area that is in boundary contact, which combines the effects of other state parameters. It is very important to recognize that these state variables can have different values at different points along the interface and that they evolve with time.

In this article a numerical framework for state variable friction models is presented with examples appropriate for metal forming. A general scheme for such models has been implemented in the viscoplastic finite element code HICKORY[6]. A simple model that uses the fractional area in boundary contact as a state variable is compared to Coulomb friction.

2 NUMERICAL IMPLEMENTATION

The implementation of a state variable friction model in a numerical code requires a data structure appropriate for the state variables at the interface. A routine to compute the friction tractions is executed using the current values of the state variables and other external variables from the solution. Routines to update the interface state variables are called at appropriate intervals. The state variables are defined at surface elements as a single value per element. For two-dimensional applications the bulk elements are either nine node quadrilaterals or six node triangles, and the surface elements are defined on the three nodes of an element boundary that lie on the surface of the mesh, as illustrated in Figure 1.

A relatively simple state variable model has been implemented with the contact area fraction and asperity lay vector as state variables. The contact area evolution model is taken from Korzekwa et al.[7]. This model calculates the fraction of the area that is in boundary contact from a rate of flattening of the surface asperities. The flattening rate $(1/E)$ is a function of the normal traction, or pressure, on the asperity, the effective deformation rate in the bulk, and the orientation of the

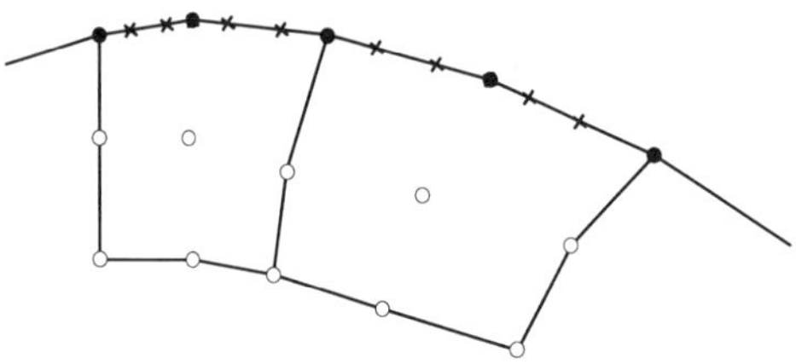

Figure 1: Surface and bulk elements with surface quadrature points.

long axis of the asperity, or asperity lay vector, relative to the deformation rate components. The model is based on work by Wilson and Sheu[8], who showed that the asperity flattening rate is a strong function of the strain rate. This results in much larger contact area fractions for a deforming workpiece than for non-deforming conditions. The contact area can vary substantially, depending on the strain and pressure history at a given point on the interface.

The friction traction τ_f is calculated from the simple relationship

$$\tau_f = cA \tag{1}$$

where A is the contact area fraction and c is a constant. This model is appropriate if it is assumed that the variation in contact area is the most important feature to describe the friction behavior, and a simple boundary friction coefficient c is adequate for the local friction traction on a contacting asperity.

The inputs to the model are the initial contact area fraction and asperity lay vector, the parameter c, and a table that describes the initial topography of the surface. Currently, the table gives the asperity spacing, l as a function of surface profile height, and the bearing area curve. The contact area evolution routines are accessed through a single call, and can be modified without changing the main program. The initialization of the state variables and input of surface roughness data and initialization of other data are in an initialization routine that is called once at the beginning of the problem.

The evolution law for A is expressed in terms of a non-dimensional flattening rate, $1/E$. The evaluation of E uses several external variables that must be obtained at each point on the surface where the friction traction is required. The function

$$E = f(H, A, \phi) \tag{2}$$

is evaluated from a lookup table, where

$$H = \frac{\tau_n}{k} \tag{3}$$

and

$$\phi = \tan^{-1}(d_{yy}/d_{xx}) \tag{4}$$

The normal traction applied by the tool to the surface of the workpiece is τ_n, and k is the flow strength in shear of the workpiece at the surface. The strain rate components d_{xx} and d_{yy} are components of the deformation rate $\bar{d}$ aligned transverse and parallel, respectively, to the asperity lay[7]. Therefore the deformation rate of the bulk material at the surface and the asperity lay vector are required to calculate ϕ. The rate of change of the contact area fraction is calculated from

$$\frac{dA}{dt} = \frac{dA}{dz}\frac{\bar{d}l}{E} \tag{5}$$

where dA/dz and l are taken from the surface roughness data table, $\bar{d}$ is the deformation rate, and z is height coordinate of the profile.

HICKORY uses a state variable formulation for the constitutive law for the bulk material, and the workpiece flow strength is calculated from the current effective stress at a point near the surface. In the current implementation the evolution of the constitutive state variable is not affected by the interface conditions, but in principal the constitutive state variable at the surface could also evolve as a state variable in the friction model, and the flow strength at the surface could be different than that of the bulk.

The velocity gradient components at the surface are needed to calculate E and ϕ in equation 2. The velocities of the nodes of the surface elements are passed from the velocity field solution to the routines that update the state variables. For an incompressible material the surface node velocities are sufficient. This scheme will also work for a displacement formulation, where the displacement gradient and time step can be used.

The current asperity lay vector is calculated from the initial lay vector and the current coordinates of the surface element. The initial asperity lay vector is expressed as a linear combination of the surface tangent vectors obtained by mapping a

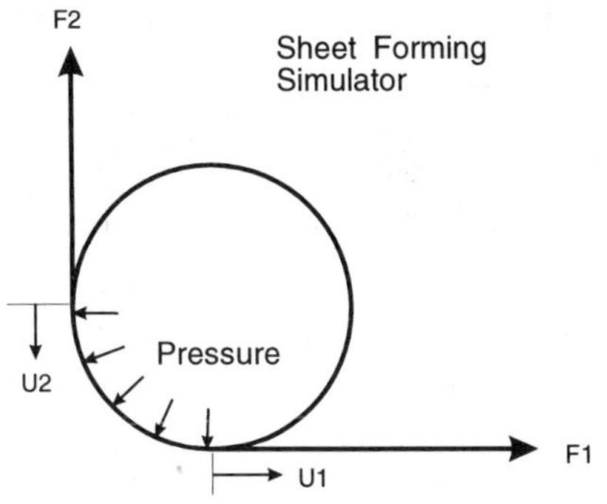

Figure 2: Sheet forming friction test.

reference element to the surface element. The lay vector is updated by forming the same linear combination on the new tangent vectors of the surface element.

3 FRICTION TEST EXAMPLE

The contact area evolution model described above can be an important component of a realistic friction model, but alone it is not adequate to describe the friction behavior in most cases. However, this model can capture the main features of some friction data obtained from a test by Wilson, et al.[9] The sheet forming friction test is shown schematically in Figure 2. A sheet metal sample is stretched and pulled around a pin by two independent actuators. The strain rate in the strip and the sliding velocity at the interface between the strip and pin can be controlled independently.

This experiment can separate the effects of normal pressure, sliding velocity and plastic strain by varying the angle of wrap and actuator velocities. Saha and Wilson[10] showed that for some materials the apparent coefficient of friction at the interface increases with increasing strain in the sheet during the test. It is assumed that this is caused by the increased contact area fraction due to asperity flattening by the mechanism described in the previous section. The fractional contact area at each point on the workpiece surface in contact with the tool will depend on the strain and normal traction history of that point.

The sheet forming friction test was modeled as a plane strain problem with either the contact area evolution model for friction or with Coulomb friction. The Coulomb model used the relationship

$$\tau_f = \mu \tau_n \tag{6}$$

where τ_n is the normal traction. Figure 3 shows a region of the mesh for a test with a 90 degree wrap angle. The results presented here are for an average strain rate of either 0.07/s or 0.105/s, and an average sliding velocity of $\approx$ 11mm/s. The material and surface properties were for a type 304 stainless steel sheet. The asperity lay was defined as perpendicular to the sliding and stretching direction, and therefore it did not evolve for the examples shown here. This results in a constant value of ϕ for the flattening rate calculation.

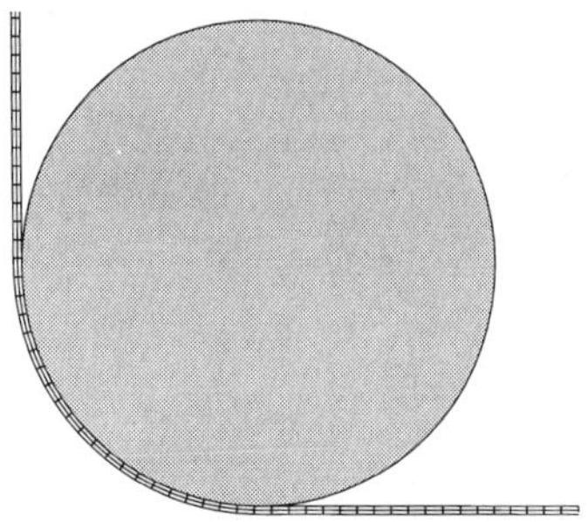

Figure 3: Sheet forming friction test finite element model.

For the state variable model, the contact area fraction is initialized to 0.04, and at the beginning of the test the friction traction is very low. As the sheet sample is stretched while in contact with the tool, the contact area increases, and the friction traction increases. Figure 4 shows the contact area state variable at various times during the test. The position coordinate is the distance along the strip surface from one end of the sample, and only a portion of the surface is in contact with the tool at any time.

As the test proceeds, points on the workpiece surface that were initially in contact with the tool will have the contact area increase until they leave contact. When the problem has progressed to the point where the strip surface at the beginning of the contact region has traveled along the complete tool-workpiece interface and come out of contact, the state variable distribution reaches an approximate steady state. If the workpiece is hardening

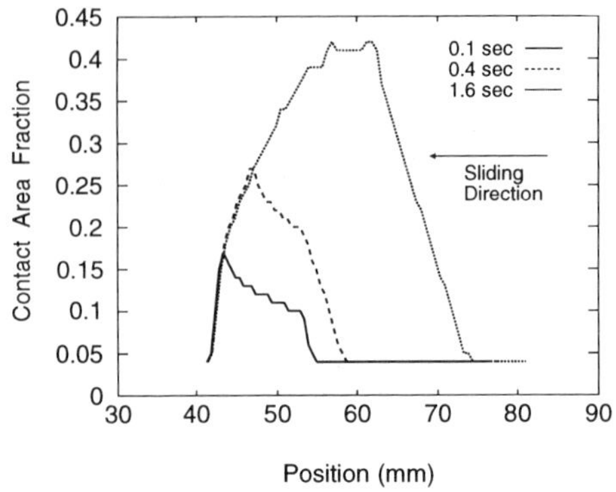

Figure 4: Contact area fraction evolution.

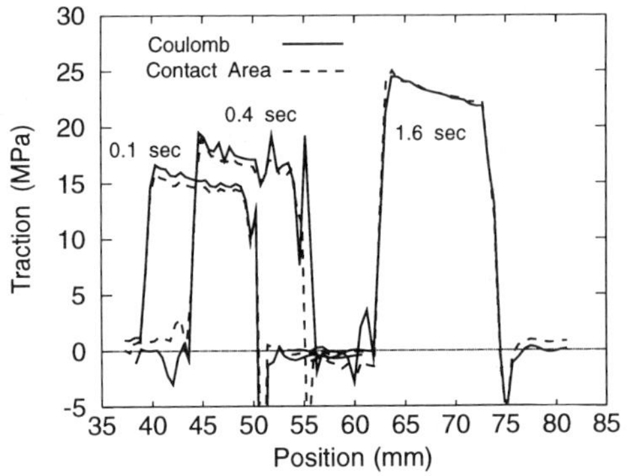

Figure 5: Normal traction at the interface for Coulomb and contact area models.

with strain this will not be strictly true, because of the influence of the workpiece flow strength on the normal traction.

At a time of 1.6 sec it can be seen in Figure 4 that the contact area distribution has evolved to this steady state condition at the interface. The region of contact is to the right of the peak. The contact area at a point on the strip increases monotonically as the point proceeds along the contact region, and stops increasing after it leaves the tool. The flat region on the peak is the portion of the strip that has traveled the full length of the contact region and developed the maximum contact area fraction.

For relatively low friction tractions, the contact area model and the Coulomb model give very similar normal traction distributions at the interface. Figure 5 shows the normal tractions for both models at the same three times as Figure 4. The Coulomb coefficient was 0.1, and c was 10.25 MPa for the contact area model. The tangential traction distributions, however, are quite different, as shown in Figure 6. The friction traction for the Coulomb model is nearly constant as would be expected based on the nearly constant normal traction. The results for the contact area model reflect the contact area evolution along the interface, giving an increasing friction traction as a material point proceeds along the interface.

The increase in friction traction with increasing strain rate is shown in Figure 7. The contact area fraction increases for the higher strain rate because the accumulated strain while in contact with the tool is higher for a given point on the surface of the sheet. When the results are expressed as an average friction coefficient, defined as the

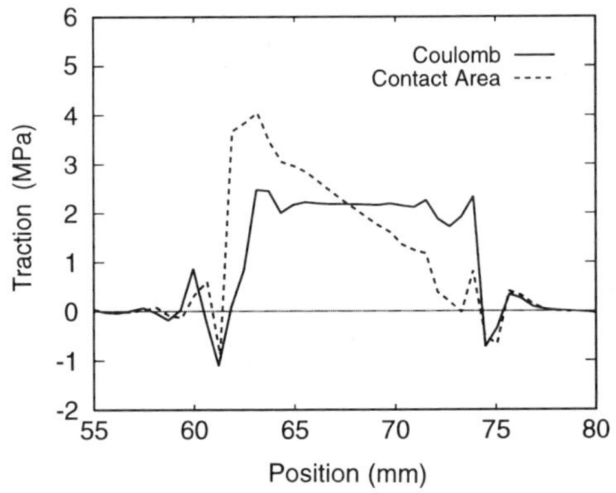

Figure 6: Comparison of tangential (friction) tractions for Coulomb and contact area models.

normal load divided by tangential load, the apparent coefficient of friction increases with strain rate, as observed in the experiment. The Coulomb model does not capture this effect since, by definition, the ratio of normal to tangential tractions is constant.

4 DISCUSSION

Some of the results from the finite element model of the sheet forming friction test are given in Table 1 for an average sliding velocity of $\approx$ 11mm/s and average strain rate of 0.07/s. The loads on the ends of the strip and the geometry were used to calculate the average normal traction P and average friction coefficient $\bar{\mu}$, after the method that was used to analyze the experiments in [10].

In order to match a typical experimental re-

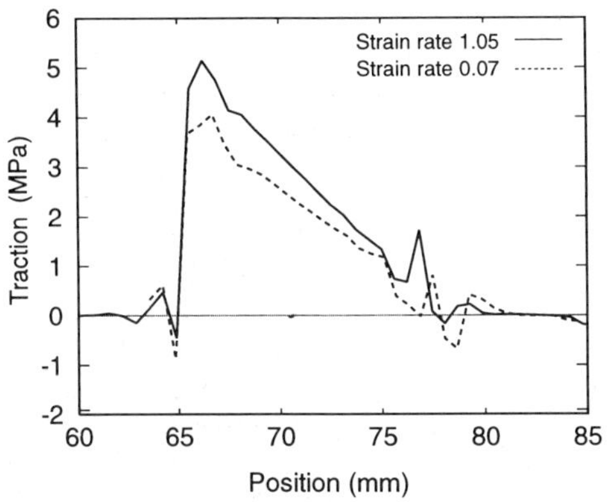

Figure 7: Friction tractions for two strain rates using the contact area model.

Table 1: Average Friction Coefficients for Coulomb and Contact Area Friction Models

Model	μ	c (MPa)	$\bar{\mu}$	P (MPa)
Coulomb	0.1	–	0.097	22.9
Contact	–	10.25	0.098	23.2
Contact†	–	10.25	0.109	26.0
Coulomb	0.2	–	0.195	22.2
Contact	–	20.0	0.192	22.5
Coulomb	0.4	–	0.450	20.0
Contact	–	41.0	0.410	21.1

† Strain rate = 0.105.

sult of $\bar{\mu} = 0.1$ values of c and μ were chosen to be 10.2 MPa and 0.1, respectively, for the two friction models in the finite element simulations. Doubling both μ and c gives $\bar{\mu} \approx 0.2$ for both models, and one can argue that the two models cannot be distinguished because $\bar{\mu}$ varies linearly with both μ and c over this range of observed values of $\bar{\mu}$. However, the value of $\bar{\mu}$ increases by $\approx$10% when the strain rate is increased from 0.07/s to 0.105/s when the contact area model is used. This is approximately the same increase that was experimentally observed in [10] for a galvanized sheet steel. At higher friction levels the two models do not agree as closely. When the Coulomb coefficient is set to 0.4, the value of $\bar{\mu}$ is calculated to be 0.45, while the comparable value for the contact area model is 0.41.

These results illustrate the limitations of using either experiments or simulations alone. Experiments usually only measure average tractions, and do not resolve the spatial variation in conditions at the interface. The sheet forming friction test cannot distinguish between the two results in Figure 6 with a single test, because the average friction coefficients are the same. On the other hand, a numerical simulation can generate very detailed results, and numerical experiments are possible that cannot be conducted in the physical world. However, the numerous limitations of the simulation tools and their underlying physical models make it impossible to independently verify the correctness of a given friction model using simulations alone.

A combined approach seems best, where the candidate models are chosen based on experimental results, and simulations are used to validate them. Simulations can then be used to predict the results of subsequent experiments, and the results will provide further validation or expose the limitations of the model. Experimental characterization to try to identify friction and wear mechanisms must be used to ensure that the various friction models are used appropriately. If a complex model for friction is to be incorporated into a finite element code and used to model complex forming operations, it must be validated for simple problems before it can be useful.

The contact area evolution model shown here will be used as part of more sophisticated models currently under development. In most cases a liquid lubricant is present, and the hydrodynamics of the lubricant film will have a very strong effect on the actual contact area[11]. The flattening model used is probably reasonable for some materials, but in other cases the surface will roughen as it is strained, and this will have a very strong effect on the actual contact area. Also, the coefficient c encompasses complex boundary friction behavior that depends on the composition of the lubricant, the tool and the workpiece, as well as roughness features. In some cases a model for boundary friction must be added to capture the important effects.

The general implementation method described here for state variable friction models is flexible enough to accommodate a wide range of physically-based models. Even if the mechanisms are not understood completely, empirical models can be implemented in terms of physically meaningful quantities.

The state variable structure presented here may

not be adequate for the implementation of some models involving hydrodynamic lubrication theory. The method described here allows evolution of state variables based on local variables on a point-by-point basis. Much of the current theory of hydrodynamic effects involves integration of the Reynolds equation along the interface to find the lubricant film thickness and pressure distribution. The theory is very attractive and can provide valuable predictions of friction behavior, but a different approach to implementation is required. An example of this type of lubrication model in a finite element code has been demonstrated by Liu et al.[12]

5 CONCLUSIONS

A state variable friction model has been implemented into the finite element code HICKORY. This implementation is sufficiently general to allow a wide range of physical phenomena at the tool–workpiece interface to be incorporated into metal deformation process simulations. A contact area evolution model for friction was employed in a simulation of a friction test for sheet forming. It was demonstrated that the contact area model can capture experimental effects that cannot be modeled by simple Coulomb friction.

It is proposed that numerical simulations be used to help analyze friction experiments and evaluate friction models. It seems likely that a combination of experiments and simulation will further the understanding of the basic friction mechanisms, as well as validate the friction model for use in complex metal forming simulations. The ability to use more realistic, physically based friction models promises to make numerical simulations of metal forming processes more accurate and useful.

6 ACKNOWLEDGMENTS

This work was supported by the Department of Energy under contract W-7405-ENG-36. Thanks go to to Paul Dawson at Cornell University and Bill Wilson at Northwestern University for their helpful discussions and encouragement. The authors also wish to acknowledge Stuart MacEwen at Alcan International and Roland Timsit, formerly at Alcan, for their collaboration in this work.

REFERENCES

[1] J.A. Schey, *Tribology in Metalworking.* American Society for Metals, 1983.

[2] W.R.D. Wilson, "Friction and lubrication in bulk metal forming processes," *Journal of Applied Metalworking*, vol. 1, pp. 7–19, 1979.

[3] N. Bay and T. Wanheim, "Real area of contact and friction stress at high pressure sliding contact," *Wear*, vol. 38, pp. 201–209, 1976.

[4] W.R.D. Wilson, "Friction models for metal forming in the boundary lubrication regime," *Trans. ASME Journal of Engineering Materials and Technology*, vol. 113, pp. 60–68, 1991.

[5] W.R.D. Wilson, "Strategy for friction modeling in computer simulations of metalforming," in *Proc. NAMRC XVI*, pp. 45–84, SME, 1988.

[6] G.M. Eggert and P.R. Dawson, "Assessment of a thermoviscoplastic model of upset welding by comparison to experiment," *International Journal of Mechanical Sciences*, vol. 28, pp. 563–589, 1986.

[7] D.A. Korzekwa, P.R. Dawson, and W.R.D. Wilson, "Surface asperity deformation during sheet forming," *International Journal of Mechanical Sciences*, vol. 34, pp. 521–539, 1992.

[8] W.R.D. Wilson and S. Sheu, "Real area of contact and boundary friction in metal forming," *International Journal of Mechanical Sciences*, vol. 30, pp. 475–489, 1988.

[9] W.R.D. Wilson, H.G. Malkani, and P.K. Saha, "Boundary friction measurements using a new sheet metal forming simulator," in *Proceedings of NAMRC XIX*, pp. 37–42, SME, SME, 1991.

[10] P.K. Saha and W.R.D. Wilson, "Influence of plastic strain on friction in sheet metal forming," *Wear*, vol. 172, pp. 167–173, 1994.

[11] W.R.D. Wilson and D. Chang, "Low speed mixed lubrication of bulk metal forming processes," in *Tribology in Manufacturing Processes* (K. Dohda, S. Jahanmir, and W.R.D. Wilson, eds.), TRIB-Vol.5, (New York, NY), pp. 159–167, ASME, ASME, 1994.

[12] W.K. Liu, Y-K Hu, and T. Belytschko, "ALE finite elements with hydrodynamic lubrication for metal forming," *Nuclear Engineering and Design*, vol. 138, pp. 1–10, 1992.

Simulation of Materials Processing: Theory, Methods and Applications, Shen & Dawson (eds)
© 1995 Balkema, Rotterdam. ISBN 90 5410 553 4

Application of FE simulation of the compression test to the evaluation of constitutive equations for steels at elevated temperatures

J. Kusiak & M. Pietrzyk
Akademia Gorniczo-Hutnicza, Krakow, Poland

J.G. Lenard
University of Waterloo, Ont., Canada

ABSTRACT: The objective of the paper is the application of the finite element technique to the interpretation of the plastometric tests and evaluation of the stress-strain equations independently on the conditions of the tests. The inverse technique combined with the rigid-plastic finite element model and with an optimization method is used in the analysis. Hot compression of the axisymmetrical steel samples is considered. The results for the C-Mn steel and niobium steel are presented.

1 INTRODUCTION

The behaviour of the materials during hot working is of interest from both scientific and industrial points of view. The accuracy of the commonly used finite-element models depends strongly on the accuracy of the description of the material properties. Stress-strain curves generated through tension, torsion and compression tests are routinely used in the finite-element programs. It is observed, however, that the flow curves obtained from various tests for one material are not identical (Lenard & Karagiozis 1987). The differences are due to the different distributions of stresses, strains and temperatures in the deforming sample. Various tests involve various frictional and thermal conditions. Heat generation and heat transfer to the tool and to the surrounding air result in inhomogeneity of temperatures. Friction leads to the inhomogeneity of strains. Finite-element modelling allows the prediction of these inhomogeneities and account of them is taken in the analysis of the results of a test. The objective of the present work is to apply the finite-element program developed for axisymmetric compression (Pietrzyk 1993) to the evaluation of the constitutive equations for the hot deformation of steel. An inverse technique combined with various optimization methods is used to determine the coefficients in the stress-strain relationships.

2 MATHEMATICAL MODEL

The mathematical model used in the work is composed of two parts. The first is the mechanical component computing the stresses, strains and strain rates during the compression. This is then coupled to a model of the heat transfer during the experiment. Detailed description of the method is given by Pietrzyk & Lenard (1991), and its application to the axisymmetric compression is described by Pietrzyk (1993).

The finite element code is used to simulate metal flow and heat transfer during hot compression of axisymmetrical samples. Simulation of the plastometric tests for various materials allows the determination of the stress-strain relationships independently of the condition of the test. The analysis is based on an inverse technique combined with the optimization of the objective function defined as the difference between the measured and calculated loads during the test:

$$\Phi = \sqrt{\frac{1}{N}\sum_{i-1}^{N}\left(F_{ci}-F_{mi}\right)^2} \tag{1}$$

where: N - number of experimental points on all curves, F_{ci}, F_{mi} - calculated and measured loads, respectively.

The analysis leads to the coefficients in various types of the constitutive equations which, when introduced into the finite-element programs, give the best agreement between the measured and calculated loads. It is concluded that these equations describe properly the local stress-strain behaviour of the material, independently of the conditions of a test. The method has been successfully applied by Pietrzyk et al. (1994) to the cold compression of various metals. Further applications include plane strain compression of aluminum at elevated temperatures (Gelin & Ghouati 1994, Pietrzyk & Tibballs 1994).

3 STRESS-STRAIN RELATIONSHIPS

Proper evaluation of the function describing the relationship between the strains and the stresses is essential for the accuracy of the finite-element simulation of hot forming processes. As shown by Pietrzyk & Tibballs (1994), simple functions commonly used as the strain hardening curves which give good results for a narrow range of the temperatures and/or strain rates, usually fail to predict proper values of the yield stress in the full range of process conditions.

The behaviour of metals during hot plastic deformation is determined by the joint influence of athermal strain hardening, thermally activated recovery and thermally activated recrystallization (Mecking & Kocks 1981). These three phenomena lead to a characteristic shape of the stress-strain curves with a peak (Laasraoui & Jonas 1991). Depending on the temperature and the strain rate, the curves may exhibit a state of saturation or one peak or oscillations. The type of the curve depends on the history of deformation conditions and it is very difficult to find a reasonably simple function which may describe all possible types of the behaviour.

Various researchers have tried to model the flow curves incorporating the effects of strain rate and temperature to develop constitutive equations applicable to hot metal forming processes. Among a number of relationships, the hyperbolic sine Arrhenius type rate equation requires particular attention:

$$\sigma_p = A_1 \sinh^{-1}(A_2 Z)^p \qquad (2)$$

where: Z - Zener Hollomon parameter, σ_p - yield stress, A_1, A_2, p - material constants.

This equation has been used in a number of publications (Ryan & McQueen 1990, Rao et al. 1993) to the simulation of the strain hardening of various metals. However, it describes only the relation between the steady state yield stress and the strain rate and temperature. Hodgson and Collinson (1990) suggested a description of the strain hardening using the power term:

$$\sigma_p = A_1 \varepsilon^n \sinh^{-1}(A_2 Z)^p \qquad (3)$$

where: ε - true strain, n - material constant, Equation (3) is used in the analysis that follows.

4 EXPERIMENT

Two sets of the experimental data are considered. The first, published by Laasraoui & Jonas 1991, includes hot compression of axisymmetrical low carbon steel (0.03%C, 1.54%Mn) samples. The second was performed in the University of Waterloo. It involved hot compression of axially symmetrical samples, made of a microalloyed steel, containing 0.043% C, 1.43%Mn, 0.312%Si and 0.075%Nb. The 8 mm diameter specimens of 12 mm height were compressed to a true strain of 1.1, at a temperature range of 860 - 1000°C and at constant rates of strain, varying from 0.0001 to 0.1 s^{-1}. Both experimental results show that, at some test conditions, the dynamic recrystallization leads to a characteristic shape of the stress-strain curve with a stress achieving a peak at some strain and decreasing after that to some steady state value.

5 RESULTS

Application of the inverse technique to the experimental data published by Laasraoui & Jonas (1991) leads to the activation energy in Zener Hollomon parameter Q = 329600 J/mol and constants given in Table 1. The results are presented in Fig.1. It is obvious that equation (3) cannot describe properly the stress strain relationship when dynamic recrystallization is involved. In order to find the equation which describes accurately the strain hardening itself, the compression to the strain below the peak value was considered next. An inverse technique resulted in the values of the material constants given in the second row of Table 1. Comparison of the measured and calculated stress strain curves below the strain of 0.25 is shown in Fig.2. Similar analysis was performed for the niobium steel and the optimum

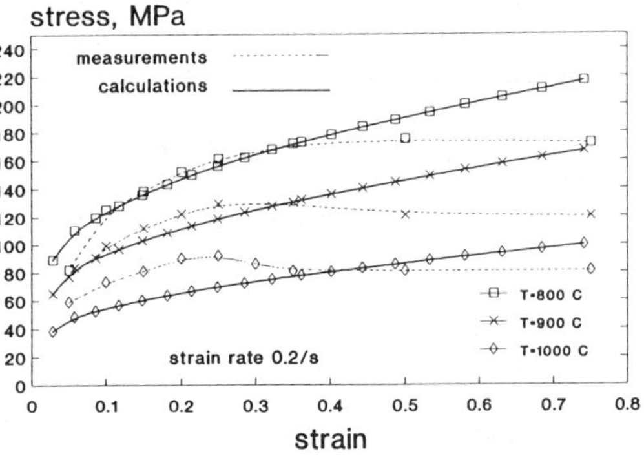

Fig.1. Stress-strain curves obtained from the inverse analysis for function (3) compared with the measurements by Laasraoui and Jonas (1991), C-Mn steel.

parameters are given in the third row of Table 1. Fig.3 shows resulting comparison of the measured and calculated compression loads. It can be concluded at this stage that equation (3) describes the strain hardening of the steel properly when the influence of dynamic recrystallization is negligible. It is adequate for the simulation of a single passes in hot rolling. Multipass rolling, however, often leads to the accumulation of strains and to dynamic recrystallization. Moreover, other processes such as hot forging may involve large local deformation, much exceeding the strain at the peak stress. In these cases, simulation of metal flow based on equation (3) may lead to erroneous results. Therefore, an attempt was made to introduce a stress strain function which describes the yield stress properly over a wide range of strains.

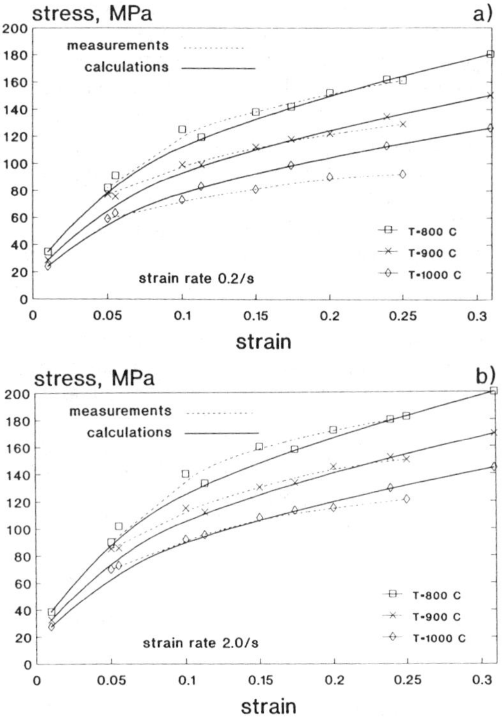

Fig.2. Stress strain curves obtained from the inverse analysis for function (3) below the strain of 0.25 compared with the measurements by Laasraoui and Jonas (1991); C-Mn steel, strain rate 0.2 s^{-1} (a) and 2.0 s^{-1} (b).

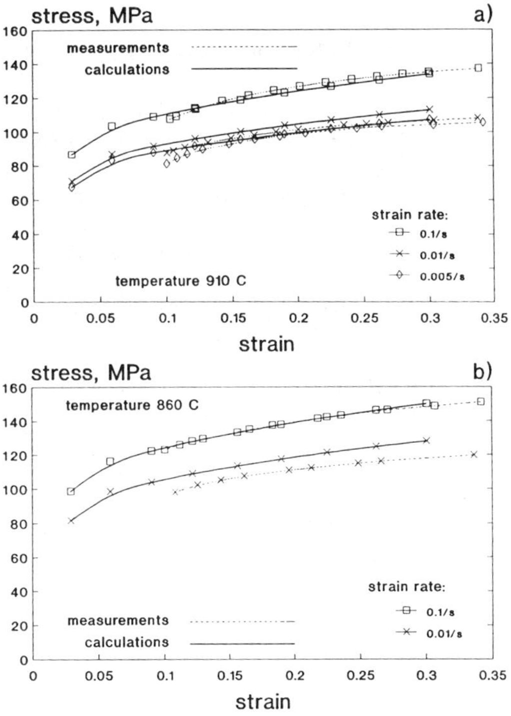

Fig.3. Stress-strain curves obtained from the inverse analysis for function (3) below the strain of 0.35 compared with the measurements; niobium steel, temperature 910°C (a) and 860°C (b).

Table 1. Coefficients in stress-strain equations obtained for various conditions

Fig.	steel	eq.	A_1 MPa	A_2x10^{-13} s	A_3	A_4	n	p	Q kJ/mol
1	C-Mn	(3)	126	1.38	-	-	0.235	0.137	329.6
2	C-Mn	(3)	171	1.87	-	-	0.393	0.086	355.3
3	niobium	(3)	156	2.62	-	-	0.138	0.108	306.8
4	C-Mn	(4)	167	2.0	1.858	-	0.573	0.218	374.3
5a	C-Mn	(5)	114	0.212	3.46	1.25	0.26	0.171	372.7
5b	niobium	(5)	233	3.13	0.729	0.089	0.081	0.159	275.8

This aim can be achieved by an introduction of the strain dependence of Q and p (Rao et al. 1993), requiring two additional parameters in the optimization procedure. Similar problem is connected with the Voce approach, described by Hodgson & Collinson (1990). The latter equation has the form of the sum of two terms, which might be inconvenient. Simpler approach introduces an additional exponential term with one constant. The following equation was suggested:

$$\sigma_p = A_1 \varepsilon^n \sinh^{-1}(A_2 Z)^p \exp(-A_3 \varepsilon) \qquad (4)$$

One exponential term leads to the situation presented in Fig.4 which is not correct qualitatively. For larger strains the theoretical curve tends to zero instead to a saturation steady value. Proper description of the dynamic recrystallization behaviour can be obtained by an introduction of two exponential terms, as suggested by Pozdeyev et al. (1973) and by Beynon et al. (1993). The influence of these terms starts above the peak strain:

$$\sigma_p = A_1 \varepsilon^n \sinh^{-1}(A_2 Z)^p \{1 + \exp[-A_3 R(\varepsilon - \varepsilon_p)] - \exp[-A_4 R(\varepsilon - \varepsilon_p)]\} \qquad (5)$$

where:

$$R(\varepsilon - \varepsilon_p) = 0 \quad \textit{for} \quad \varepsilon < \varepsilon_p$$
$$R(\varepsilon - \varepsilon_p) = \varepsilon - \varepsilon_p \quad \textit{for} \quad \varepsilon > \varepsilon_p \qquad (6)$$

In equation (6) ε_p is the strain corresponding to the peak stress.

Calculations based on equation (5) lead to the coefficients given in the last two rows of Table 1. Comparison of the measured and calculated stress-strain curves for these coefficients is presented in Fig.5. It can be concluded here that the conventional hyperbolic sine equation multiplied by an expression with two exponential terms allows the simulation of the behaviour of the materials beyond the start of dynamic recrystallization. It has to be pointed out, however, that closed form relationships of the type of equation (6) cannot describe stress-strain conditions in general. Since they do not account for the history of deformation, they often fail to predict the yield stress properly when softening phenomena play an important role. They also do not perform well in multistage processes with the interpass times too short for the recrystallization to be completed. In order to adjust better the equation (5) to the conditions of dynamic recrystallization, the temperature and strain rate dependence of coefficients A_3 and A_4 has to be introduced. This step would lead to a significant increase of unknown parameters in the optimization procedure. Models based on the internal variable method would be more realistic in this situation.

Recapitulating the work it should be emphasized that an analysis of various forms of stress-strain equations was not Authors' main intention. The

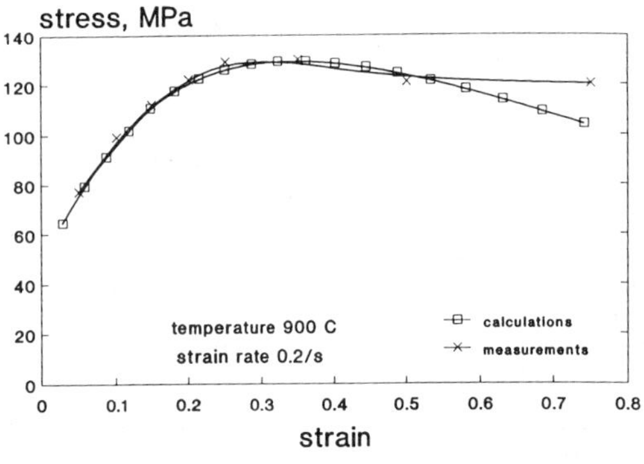

Fig.4. Stress-strain curves obtained from the inverse analysis for function (4) compared with the measurements; C-Mn steel, temperature 900°C, strain rate 0.2 s^{-1}.

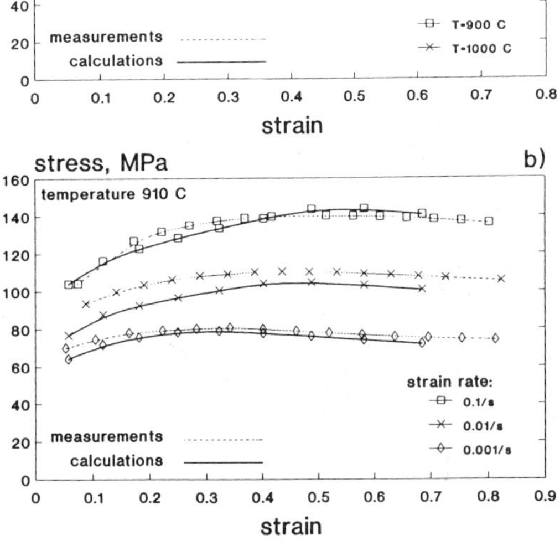

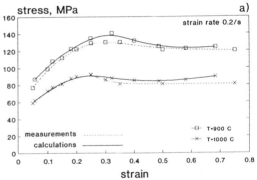

Fig.5. Stress-strain curves obtained from the inverse analysis for function (5) compared with the measurements; a) C-Mn steel, strain rate 0.2 s^{-1}; b) niobium steel, temperature 910°C.

objective of the work was to show the abilities of the inverse technique as far as an interpretation of the plastometric tests is considered. In conclusion it was shown that combining the inverse technique with the optimization procedures allows the determination of the coefficients in the stress-strain equation which give the best correlation between measured and calculated loads during the compression. The method accounts for the influence of the deformation heating and friction on the test. The importance of these phenomena is well seen in Fig.6 where the variations of the temperature and the strain rate at four locations in the deformation zone during the compression are shown.

The choice of the optimization technique is important. Since they do not require the calculation of derivatives of the objective function, the non-gradient methods are preferable. In the present work Monte Carlo method was used at the beginning of search to localize the minimum. The purpose was to avoid finding the local minima which may appear, in particular when functions with larger numbers of coefficients are used. The simplex method was used at the final stage of search to find the precise solution.

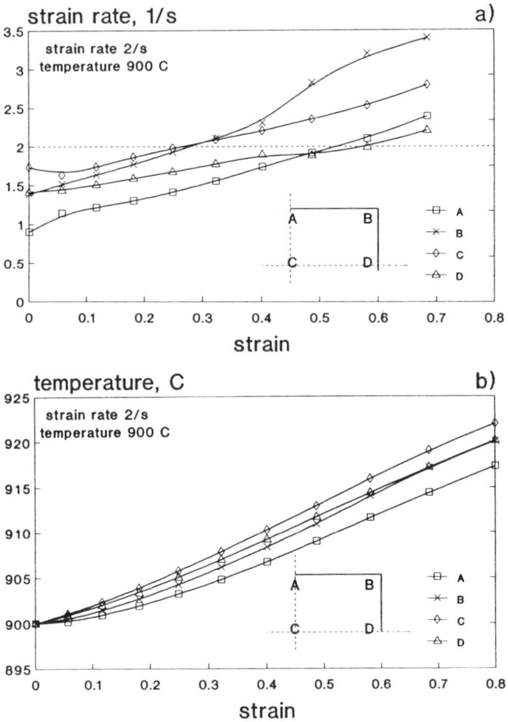

Fig.6. Variation of the strain rate (a) and temperature (b) during hot compression, at four locations in the deformation zone.

6 CONCLUSIONS

The work shows an application of the finite-element technique to the interpretation of the compression test. The finite-element analysis predicts the distributions of strain rates, strains and temperatures in the deformation zone and allows to account for the influence of deformation heating and friction on the stress-strain relationship. Optimization technique with the square of difference between the measured and calculated loads during compression leads to the real, local strain-hardening relationship. The accuracy of the method depends on the selection of the function describing the stress-strain curve. It is shown that commonly used sine hyperbolic function (equation (3)) describes properly the strain hardening below the peak strain. The function fails, however, when the dynamic recrystallization becomes a dominant factor affecting the flow stress. Introduction of the additional exponential terms (equation (5)) allows to account for the dynamic recrystallization.

The efficiency of the technique can be looked at in two ways. Since the method requires running of the finite element code for each experiment available in each iteration, it is time consuming. On the other hand, the starting point for optimization is taken from direct approximation of the experimental curves and it is usually close to the solution. The aim of the technique is only to introduce changes in the coefficients which account for the influence of friction and heat generation. These changes are small and often only few iterations are required to find the solution.

ACKNOWLEDGEMENTS

The financial assistance of the Polish Committee for Scientific Research (grant no. PB 0820/S2/94/06) and NATO is gratefully acknowledged.

REFERENCES

Beynon, J.H., X.K. Li & Ponter A.R.S. 1993. Inclusion of Dynamic Recrystallization in Process Models for Hot Working. *Mat. Sci. Forum.* 113-115: 293-298.

Gelin, J.C. & O. Ghouati 1994. An Inverse Method for Determining Viscoplastic Properties of Aluminum Alloys. *J. Mat. Proc. Techn.* 45: 435-440.

Hodgson, P.D. & D.C. Collinson 1990. The Calculation of Hot Strength in Plate and Strip Rolling of Niobium Microalloyed Steels. In: S. Yue, (ed.), Proc. Symp. Mathematical Modelling of Hot Rolling of Steel: 239-250. Hamilton.

Laasraoui, A & J.J. Jonas 1991. Prediction of Steel Flow Stress at High Temperatures and Strain Rates, *Metall Trans. A*, 22A: 1545-1558.
Lenard, J.G. & A.N. Karagiozis 1987. Accuracy of High Temperature, Constant Rate of Strain Flow Curves. In R. Papirno & C.H. Weiss (eds), *Factors that Affect the Precision of Mechanical Tests, STP 1025, ASTM*: 206-216, Bal Harbour.
Mecking, H. & U.F. Kocks 1981. Kinetics of Flow and Strain-Hardening. *Acta Metall.* 29: 1865-1875.
Pietrzyk, M. 1993. Comp_axi - komputerowy program do symulacji plastometrycznej proby speczania probek osiowosymetrycznych. *Hutnik* 60: 190-197.
Pietrzyk, M. & J.E. Tibballs 1995. Application of the Finite Element Technique to the Interpretation of the Plane-Strain Compression Test for Aluminum. In: *Proc. Conf. Complas 4*, Barcelona, (in press).
Pietrzyk, M. & J.G. Lenard 1991. *Thermal Mechanical Modelling of the Flat Rolling Process*, Berlin: Springer-Verlag.
Pietrzyk, M., J. Kusiak & M. Paćko 1994. Application of the Finite Element Technique to the Interpretation of the Axisymmetrical Compression test. In J. Bartecek (ed.), *Conf. Formability'94*: 339-346, Ostrava.
Pozdeyev, A., V.I. Tarnovskiy, V.I. Yeremyeyev & V.S. Vakaashvili 1973. *Primienienie teorii polzuchesti pri obrabotke metallov davleniem*. Moskva: Metallurgiya.
Rao, K.P., E.B. Hawbolt, H.J. McQueen & D.B. Baragar 1993. Constitutive Relationships for Hot Deformation of a Carbon Steel: a Comparison Study of Compression Test and Torsion Test, *Can. Metall. Quart.* 32: 165-175.
Ryan, N.D. & H.J. McQueen 1990. Work Hardening, Strength and Ductility in the Hot Working of 304 Austenitic Stainless Steel. *High Temp. Technol.* 8: 27-44.

Simulation of Materials Processing: Theory, Methods and Applications, Shen & Dawson (eds)
© 1995 Balkema, Rotterdam. ISBN 90 5410 553 4

Numerical analyses for material bifurcation and localization in plane sheet

Guo-Chen Li, Tao Huang & Chen Zhu
Institute of Mechanics, Academia Sinica, Beijing, People's Republic of China

ABSTRACT: Numerical method is utilized to well approximate the varieties of velocity forms variated at material bifurcation and provides better ground for searching the lowest critical load. It is also an exclusive approach to the analysis of the effects caused by the inhomogeneity of void damage and the nonuniform distributions of stress across sheet thickness. A pair of crossed and necked bands in the plane sheet under tension is numerically simulated as that observed, the success in simulation sheds light on revealing the post-bifurcation nature of the localization phenomenon.

1 THEORETICAL FORMULATIONS

Ensuring that the functional

$$\Pi = \frac{1}{2}\int_V \left[\frac{\mathscr{D}\tau_{ij}}{\mathscr{D}t}D_{ij} - \sigma_{ij}(2D_{ik}D_{jk} - V_{k,i}V_{k,j})\right]dv - \int_S \dot{F}_i V_i ds \quad (1.1)$$

takes an extremum is equivalent to satisfying the equilibrium equations together with the boundary conditions for any incremental solution of the boundary-value problem in the practical up-dated Lagrangian formulation (McMeeking & Rice 1975). In which, $\mathscr{D}\tau_{ij}/\mathscr{D}t$ is the Jaumann rate of Kirchhoff stress, σ_{ij} is the Cauchy stress, $\dot{F}_i$ denotes the rate of external load and the deformation rate D_{ij} is related to the velocity components V_i as

$$D_{ij} = \frac{1}{2}(V_{i,j} + V_{j,i}) \quad (1.2)$$

Under the condition of dead (conservative) loading, it has been demonstrated (Burke & Nix 1979 and Li 1991) that bifurcation occurs at the point when the second variation

$$Q = \delta^2 \Pi = \frac{1}{2}\int_V \left[\frac{\mathscr{D}\delta\tau_{ij}}{\mathscr{D}t}\delta D_{ij} - \sigma_{ij}(2\delta D_{ik}\delta D_{jk} - \delta V_{k,i}\delta V_{k,j})\right]dv \quad (1.3)$$

becomes stationary.

In (1.1) and (1.3) we have

$$\frac{\mathscr{D}\tau_{ij}}{\mathscr{D}t} = L_{ijkl}D_{kl} \quad (1.4)$$

and

$$\frac{\mathscr{D}\delta\tau_{ij}}{\mathscr{D}t} = L^*_{ijkl}\delta D_{kl} \quad (1.5)$$

In case the "linear comparison solid" model can be applied, then $L^*_{ijkl} = L_{ijkl}$. In view of the voiding damage in ductile materials, a dilatational plastic constitutive model was lately developed and justified (Li etal 1992), which gives the stiffness tensor as

$$L_{ijkl} = \frac{E}{1+\upsilon}\left[\frac{1}{2}(\delta_{ik}\delta_{jl} + \delta_{il}\delta_{jk}) + \delta_{ij}\delta_{kl}\frac{\upsilon - E/3E_{tm}}{1-2\upsilon+E/E_{tm}} - \frac{3}{2\sigma_e^2}\frac{S_{ij}S_{kl}}{1+2(1+\upsilon)E_{te}/3E}\right] \quad (1.6)$$

where E and υ are the Young's modulus and Poisson's ratio respectively. σ_e is the equivalent stress, which is related to the deviatoric stress S_{ij} as $\sigma_e = \left(\frac{3}{2}S_{ij}S_{ji}\right)^{1/2}$, δ_{ij} represents Kronecker delta. E_{te} ($= d\sigma_e/d\varepsilon_e$) and E_{tm} ($= d\sigma_m/d\varepsilon_m$) are the plastic tangent moduli along the equivalent stress-strain curve ($\sigma_e - \varepsilon_e$) and the mean stress-strain curve ($\sigma_m - \varepsilon_m$) respectively, with the mean stress and mean

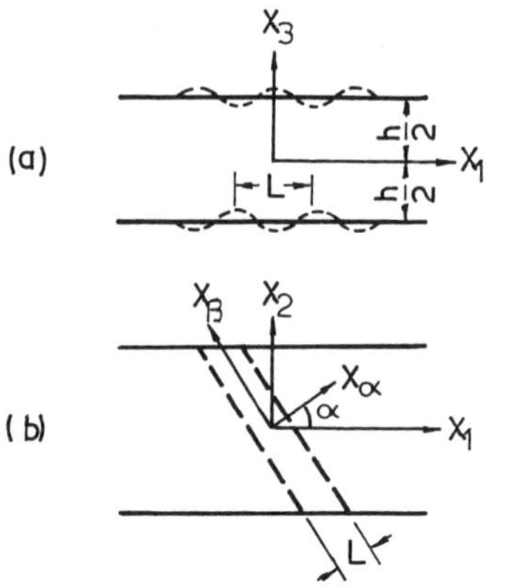

Fig. 1 Schematic demonstration of (a) surface wrinkling and (b) shear banding

strain defined as $\sigma_m = \frac{1}{3}(\sigma_{kk})$ and $\varepsilon_m = \frac{1}{3}(\varepsilon_{kk})$.

When finite-difference method is used for numerical solution, we need to normalize the second variation Q and the components of generalized velocity variation $V_{(m)}^{(n)}$ by taking

$$\overline{Q} = \frac{Q}{EL^2} \text{ and } \overline{V}_{(m)} = \frac{V_{(m)}^{(n)}}{L} \tag{1.7}$$

where L stands for the characteristic length parameter in the bifurcation pattern, (m) (=1,2⋯m) is the $(m)^{th}$ component at each nodal point and (n) (=1,2⋯n) denotes the sequence number of the discretized nodes with a total number of n. The stationary condition at bifurcation can be implemented by vanishing the partial derivatives of $\overline{Q}$ with respect to $\overline{V}_{(m)}$. Hence, at each nodal point we have m equations in the form of

$$\partial\overline{Q}/\partial\overline{V}_{(m)} = 0 \qquad (m)=1,2\cdots m \tag{1.8}$$

2 MATERIAL BIFURCATION IN PLANE SHEET

Figures 1 (a) and (b) schematically demonstrate the surface wrinkling and shear banding respectively. Their velocity variations can be represented approximately by taking

$$\begin{aligned} \delta V_1 &= \sin\frac{2\pi x_1}{L} V_1(x_3) \\ \delta V_3 &= \cos\frac{2\pi x_1}{L} V_2(x_3) \end{aligned} \tag{2.1}$$

with $\delta V_2 = 0$, $\delta V_{1,2} = \delta V_{3,2} = 0$ for surface wrinkling and

$$\begin{aligned} \delta V_\alpha &= \sin\frac{\pi x_\alpha}{L} V_1(x_3) \\ \delta V_\beta &= \sin\frac{\pi x_\alpha}{L} V_2(x_3) \end{aligned} \tag{2.2}$$

with $\delta V_3 = 0$, $\delta V_{\alpha,\beta} = \delta V_{\beta,\beta} = 0$ for shear banding. $V_1(x_3)$ and $V_2(x_3)$ are the two components that should be discretized along the sheet thickness h, by using finite-difference method.

In the case of surface wrinkling, the boundary conditions that should be satisfied are

$$\left.\begin{array}{lll} \delta V_1 = 0, & \delta V_{3,1} = 0 & \text{at } x_1 = 0 \\ \delta V_3 = 0, & \delta V_{1,3} = 0 & \text{at } x_3 = 0 \\ \int_{-h/2}^{h/2} \delta \dot{T}_{11} dx_3 = 0, & \int_{-h/2}^{h/2} \delta \dot{T}_{13} dx_3 = 0 & \text{at } x_1 = L/2 \\ \delta \dot{T}_{33} = 0, & \delta \dot{T}_{31} = 0 & \text{at } x_3 = h/2 \end{array}\right\} \tag{2.3}$$

here, $\delta\dot{T}_{11}$, $\delta\dot{T}_{13}$ and $\delta\dot{T}_{33}$ are the components of the variation of nominal stress rate. The boundary conditions for shear banding should change into

$$\left.\begin{array}{lll} \delta V_\alpha = 0, & \delta V_\beta = 0 & \text{at } x_\alpha = 0 \\ \delta V_{\alpha,3} = 0, & \delta V_{\beta,3} = 0 & \text{at } x_3 = 0 \\ \int_{-h/2}^{h/2} \delta \dot{T}_{\alpha\alpha} dx_3 = 0, & \int_{-h/2}^{h/2} \delta \dot{T}_{\alpha 3} dx_3 = 0 & \text{at } x_\alpha = L/2 \\ \delta \dot{T}_{33} = 0, & \delta \dot{T}_{3\alpha} = 0 & \text{at } x_3 = h/2 \end{array}\right\} \tag{2.4}$$

with regard to the (x_α, x_β, x_3) coordinates. For both cases, the pre-bifurcation stress state is taken to have the plane sheet stressed uniaxially, that is to say, only σ_{11} is finite while the other stress components are nil.

Table 1 The critical stresses σ_{cr}/σ_y ($\upsilon = 0.3$, $\varepsilon_y = 0.002$, $E_{tc}/E = 0$)

σ_{cr}/σ_y	surface wrinkling		shear banding ($\frac{h}{L} = 1000$)	
E/E_{tm}	numerical	analytical	numerical	analytical
60	2.733 ($\frac{h}{L}$=4—10)	2.750 (25—35)	2.750	2.750
80	2.058 (4—12)	2.065 (25—35)	2.068	2.068
100	1.651 (3—6)	1.657 (25—28)	1.657	1.657

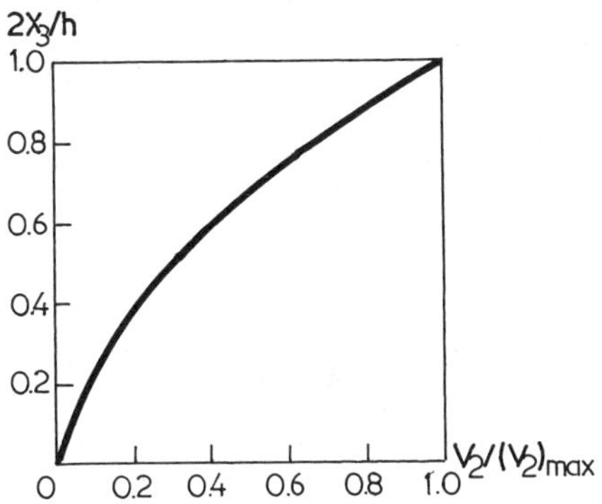

(a) surface wrinkling

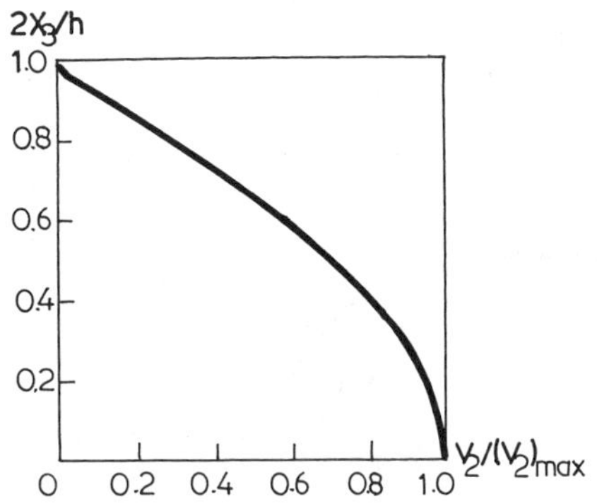

(b) shear banding

Fig. 2 The distribution of V_2 across the half thickness of the sheet

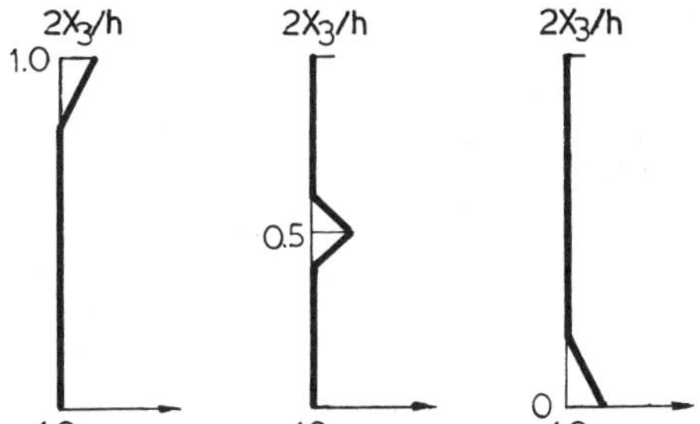

(a) (b) (c)

Fig. 3 Three patterns of inhomogeneity/nonuniformity distribution

Taking the material parameters at bifurcation to have $\upsilon=0.3$, $E_{te}/E=0$, ε_y (yield strain) $=0.002$ with various values for E_{tm}/E ($=60$, 80 and 100), we can use the numerical procedure stated previously to calculate the critical stresses, using approximately 50 finite-difference sections along the half thickness. The results of both the surface wrinkling and shear banding are listed in Table 1. Each stress reaches its lowest value (to three places of decimals) at the value of h/L shown in the circular parenthesis. The analytical results are based on some previous works (Li & Zhang 1990 and Li & Zhu 1995), using trigonometric series (two to three terms) for the functions V_1 and V_2 in (2.1) and (2.2).

Figures 2 (a) and (b) show the normalized functions of V_2 across the half thickness of the sheet. For surface wrinkling, the maximum V_2, i.e. $(V_2)_{max}$, occurs at the surface plane while conversely the $(V_2)_{max}$ of shear banding locates at the middle plane. Since the numerical method has better flexibility in simulating the varieties of velocity forms variated at material bifurcation, it is able to provide lower critical stress as shown in Table 1 for surface wrinkling. Secondly, the results indicated that the different patterns of bifurcation can have critical stresses close with each other. This fact implies that they might be sensitive to the inhomogeneity of materials and the nonuniformity of stress distribution.

3 THE EFFECTS OF INHOMOGENEITY AND NONUNIFORMITY

As the inhomogeneity of material damage and the nonuniformity of stress distrbution are existing inevitably in real sheet plate, investigation on these effects can help understand their deterioration on material bifurcation and set up theoretical basis for finding means to improve the stiffness of materials. Different patterns of inhomogeneity/non-uniformity distribution are given in the following to study the difference of their effects on critical stress.

Three patterns of distribution across the half thickness of sheet are demonstrated in Fig. 3 (a), (b) and (c). They are symmetric with respect to the middle plane at $x_3=0$. In which, we define

$$D=\frac{(E/E_{tm})_m}{(E/E_{tm})_a} \tag{3.1}$$

where, the subscript *m* stands for the maximum (if $D>1$) or the minimum (if $D<1$) and the subscript *a* denotes the average value of the parameter enclosed in the circular parenthesis. When $D=1$, the distribution is uniform. If $D>1$ or $D<1$, it means that there is a local bulge or indent with regard to the uniform distribution of E/E_{tm} within the region marked across the half thickness of sheet. Since the Young's modulus E is a constant, $D>1$ implies the material within the local region suffers more voiding damage so that the value of E_{tm} is smaller than that in the major part of distribution. Conversely, $D<1$ indicates that voiding is locally alleviated and results in larger value of E_{tm}.

Finite-difference method is again applied to the present analysis to calculate the critical stresses F_i for the cases of having inhomogeneity in voiding damage. The critical stress for the homogeneous case ($D=1$) is given as F_0. Figs 4(a) and (b) demonstrate the variation of the critical stresses with respect to varying the values of D. It is obvious that, bifurcations into either surface wrinkles or shear bands can be stabilized by the depression of voiding. Even a local alleviation can stiffen the materials to have critical stresses equal or slightly larger than the theoretical prediction for ideally homogeneous material (i. e. $D=1$). On the other hand, catastrophic decrease can happen if the situation is inversed when voiding is promoted locally.

Secondly, if the promotion of voiding is at the site where the disturbance at bifurcation is largest, then the case becomes more devastative. Consequently, the pattern (a) of inhomogeneity distribution for surface wrinkling but the pattern (c) for shear banding most impair the critical stresses. These are in accordance with the facts that the largest disturbance occurs at the top surface ($x_3=h/2$) in surface wrinkling but moves to the middle plane ($x_3=0$) in the case of shear banding, as shown in Figs 2 (a) and (b).

Similar process of calculation can be repeated for the case when the pre-bifurcation stress σ_{11} is not distributed uniformly along the thickness. Let the voiding damage be homogeneous, the effects of nonuniformity in stress distribution on critical stresses are calculated and show similar trend as that illustrated in Figs 4 (a) and (b). Then, local depression of stressing is stabilizing, but any flourish of stress amplitude is not favourable for stabilization, especially if the stress peak is located at the place where the velocity disturbance is large.

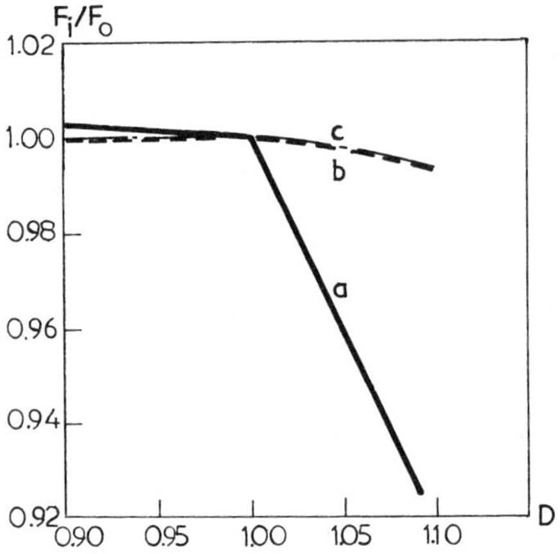

(a) surface wrinkling

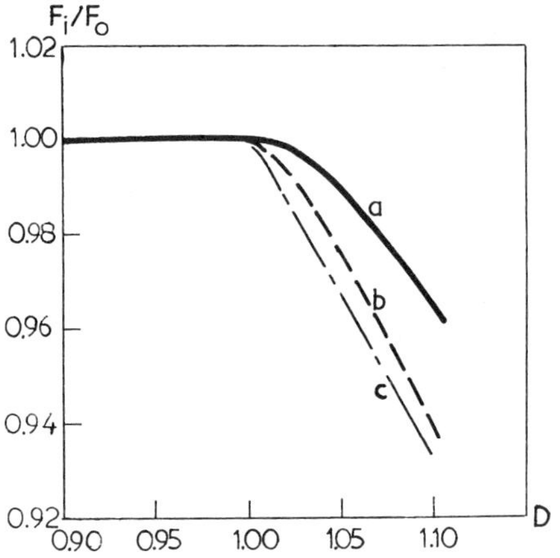

(b) shear banding

Fig. 4 The variation of the critical stress F_i (normalized by F_0) with respect to varying the values of D

4 LOCALIZATION AFTER BIFURCATION

It has been proved (Li & Zhu 1995) that shear-band bifurcation triggers high triaxiality in the central zone of the plate thickness and consequently induces much higher void volume. The initial band lies along a direction almost perpendicular to the uniaxial loading axis. Localization after this bifurcation is simulated by using the empty-element method and a relaxation procedure for the residual stresses.

When constant-strain triangular elements are employed for numerical computations, once the stress/strain in any element attains a critical value, the element is then retracted from providing further contribution to the global stiffness, henceforth it is emptied. Furthermore, as the surrounding elements are still subjected to the stresses retained by this element during its loading history before being emptied, the residual stresses should be relaxed in a proportional manner controlled by a relaxation rate parameter defined as

$$\dot{S}=\delta S/\delta t \tag{4.1}$$

where δS is the common percentage assigned for reducing stresses in each computation step follow-

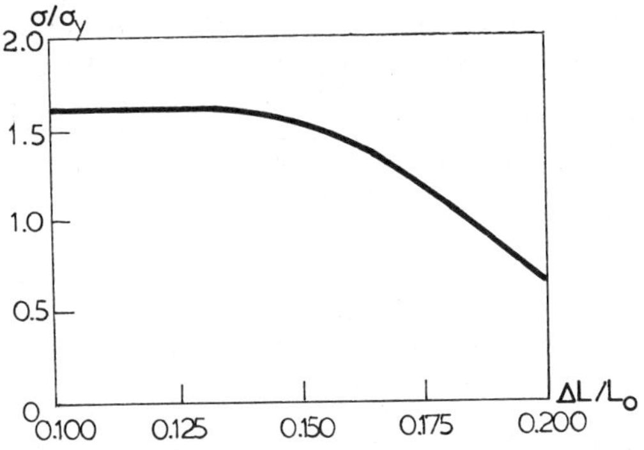

Fig. 5 Loading stress σ/σ_y versus elongation $\Delta L/L_0(\dot{S}=-25)$

ing afterwards and δt is the increase of the elongation $\Delta L/L_0$ or other monotonic parameter that can be chosen as generalized time. Here, L_0 is the initial length for measurement and ΔL is the total increase of this length, therefore $\Delta L/L_0 = t$.

Let a long, rectangular sheet be pulled uniaxially and incrementally by uniform displacements at its two short ends. It has an aspect ratio as 3. 25 : 1. Owing to the symmetry of shape and loading, only a quarter section of his sheet is needed for discretization into finite elements. A large strain computation was done based on minimizing the functional given in (1. 1).

The material used is assumed to follow Prandtl-Reuss law, which hardens according to a power law relation with a hardening exponent as n $=0.14$. It is pulled uniformly until the elongation reaches $\Delta L/L_0 = 0.1$. Then, let two elements be emptied (see Fig. 6) in the central zone along the direction perpendicular to the loading axis. This is to simulate the voiding condition stimulated by the high triaxial stress triggered during the shear-band type of bifurcation. Voids are nucleated within the initial band. The emptied elements link up (in Fig. 6) as a band which has an aspect ratio of 4 : 1. It is narrow enough to be taken as a long band.

New elements can also be emptied in ensuing elongating, in case the equivalent strains reach a value of $\varepsilon_c = 0.8$. The residual stresses in each emptied element are relaxed with a rate defined in (4. 1) and given here as $\dot{S}=-25$ (it means that a 0. 5% of stress reduction within each elongation increment of 2×10^{-4}). The average tensile stress σ loaded at the two ends can be calculated and normalized by the yield stress σ_y. Its relationship with the elongation $\Delta L/L_0$ is shown in Fig. 5. Giving different values to the relaxation rate $\dot{S}$, we can obtain different amount of extension for the loading

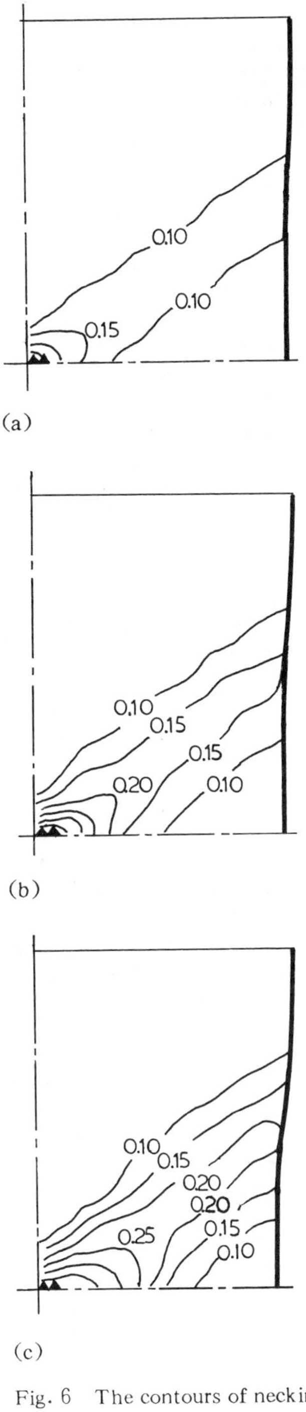

Fig. 6 The contours of necking $\Delta h/h_0$ at $\Delta L/L_0$ =(a) 0. 153, (b) 0. 164 and (c) 0. 176 (the elements emptied at $\Delta L/L_0 = 0.1$ are marked by solid triangles ▲)

stress-elengation curve. Smaller absolute value of rate parameter yields larger ductility. The curve in Fig. 5 is the representative of a middle case.

For the same example given in Fig. 5, we can illustrate the contours of necking at different stages of elongation. We use a parameter $\Delta h/h_0$ to characterize the extent of necking. Δh is the contraction of sheet thickness and is normalized by the initial thickness h_0. In Fig. 6 is shown the development of necked band during the final stages of elongation. Localized band is obviously seen lying approximately 45° with respect to the loading axis. It radiates from the elements emptied initially at $\Delta L/L_0 = 0.1$. Since the incompressibility assumption is used for plasticity in this case (it means that in (1.6) the tangent modulus $E_{tm} \to \infty$), the localization is purely caused by the initial voiding occured at the global elongation of $\Delta L/L_0 = 0.1$ and the ensuing voiding when the local equivalent strain in finite element reaches $\varepsilon_C = 0.8$. As seen from the three stages shown in Fig. 6, the necking band widens and deepens with respect to the extension of elongation. This phenomenon is in qualitative accordance with the observations in tests (Zhu, Hong & Li 1993).

5 CONCLUDING REMARKS

(a) Numerical analyses can provide good ground for searching the lowest critical loads in different patterns of material bifurcation, as numerical discretization can offer better representation for the distributions of various patterns.

(b) Investigation on the effects of inhomogeneity of material damage and the nonuniformity of stress distribution can be easily handled by using numerical method. Any means of alleviating void damaging or stressing, even locally, can promote material stability. However, it is most harmful to have more porosity or to enforce larger stress in the region where stronger velocity disturbance is predicted to occur at bifurcation.

(c) The success in simulating the crossed and necked bands seen in the plane sheet under tension sheds light on revealing the post-bifurcation nature of this localization phenomenon, instead of taking them as the formation at bifurcation.

REFERENCES

McMeeking, R. M. & J. R. Rice 1975. Finite element formulations for problems of large elastic-plastic deformation. *Int. J. Solids Structures.* 11: 601—616.

Burke, M. A. & W. D. Nix 1979. A numerical study of necking in plane tension test. *Int. J. Solids Structures.* 15: 379—393.

Li, G. C. & Y. Z. Zhang 1990. Bifurcation and fracture in sheet-metal forming with void growth effect (in Chinese). *Acta Mechanica Sinica.* 22: 302—310.

Li, G. C. 1991. On dilatational plastic constitutive equation of ductile meterials and plastic loading paths at bifurcation. *Science in China* (Ser. A). 34: 825—834.

Li, G. C., Liu, H. Q., Du, M. L., Hong, Y. S. & X. Zhang 1992. Crack tip behaviour and crack propagation in ductile materials. *Fatigue Fract. Engng, Mater. Struct.* 15: 187—202.

Zhu, C., Hong, Y. S. & G. C. Li 1993. Characteristics of shear banding in dual phase steel. *Materials Science & Technology.* 9: 1037—1043.

Li, G. C. & C. Zhu 1995. Formation of shear bands in plane sheet. *Int. J. Plasticity* (in press).

Simulation of Materials Processing: Theory, Methods and Applications, Shen & Dawson (eds)
© 1995 Balkema, Rotterdam. ISBN 90 5410 553 4

A general framework for porous viscoplasticity

Esteban B. Marin
Sibley School of Mechanical and Aerospace Engineering, Cornell University, Ithaca, N.Y., USA
David L. McDowell
George W. Woodruff School of Mechanical Engineering, Georgia Institute of Technology, Atlanta, Ga., USA

ABSTRACT: A general constitutive framework for porous viscoplasticity is used to study the role of specific void growth models in terms of both associative and non-associative viscoplastic flow rules. Three particular model frameworks for porous viscoplasticity are identified based on this study. These are denoted as associative, non-associative and partially coupled frameworks. Some applications are examined using Bammann's internal state variable viscoplastic model in the context of the three model frameworks.

1 INTRODUCTION

Damage in polycrystalline ductile metals subjected to large deformation occurs primarily by the nucleation and growth of voids (porosity) [1]. A number of explicit expressions for the rate of change of void volume fraction, ϑ, have been proposed [2,3], typically of the form

$$\dot{\vartheta} = \hat{\dot{\vartheta}}(\vartheta, \frac{\sigma_h}{\bar{\sigma}}, m, \dot{\bar{\epsilon}}^p) \tag{1}$$

where σ_h and $\bar{\sigma}$ are the hydrostatic stress and the von Mises equivalent stress, respectively, m is a strain rate sensitivity parameter, and $\dot{\bar{\epsilon}}^p$ is the inelastic equivalent strain rate of the surrounding matrix. Such models typically assume matrix strain-controlled cavity growth, neglecting diffusional growth.

Implicit void growth models, on the other hand, couple damage with the mechanical response of ductile metals by specifying a yield function which depends on pressure and porosity [4-7]. For an initially isotropic material, the yield function can be represented as

$$F = \hat{F}(I_1, J_2^{\circ}, R, \vartheta) \tag{2}$$

where I_1 is the first invariant of the overstress tensor $\boldsymbol{\sigma}$-$\mathbf{A}$, J_2° is the second invariant of the deviatoric overstress tensor ($\mathbf{s}$-$\boldsymbol{\alpha}$), and R is an isotropic scalar hardening parameter representing the static yield strength [8] of the matrix material; $\boldsymbol{\sigma}$ and $\mathbf{A}$ are the Cauchy stress and the backstress (kinematic hardening) tensors, respectively, and $\mathbf{s}$ and $\boldsymbol{\alpha}$ their deviatoric components. The invariants I_1 and J_2° are defined as

$$I_1 = \mathrm{tr}(\boldsymbol{\sigma} - \mathbf{A}) \quad , \quad J_2^{\circ} = \frac{1}{2}\mathrm{tr}(\mathbf{s} - \boldsymbol{\alpha})^2 \tag{3}$$

where $\mathbf{s}=\boldsymbol{\sigma}-1/3\mathrm{tr}(\boldsymbol{\sigma})\mathbf{I}$ and $\boldsymbol{\alpha}=\mathbf{A}-1/3\mathrm{tr}(\mathbf{A})\mathbf{I}$; the symbol tr( ) denotes the trace of a tensor and $\mathbf{I}$ is the second rank identity tensor. A common form for F which will be used in this work is the 'elliptic' model

$$F = \sqrt{3h_1 J_2^{\circ} + h_2 I_1^2} - h_3 R \tag{4}$$

where $h_i = \hat{h}_i(\vartheta, m)$, i=1,2,3. Specific expressions for the coefficients h_i (see Table I) have been derived using analytical (bound estimates) [6] and numerical [7] unit cell methods.

Specifically, the general internal state variable framework considered in this study incorporates the following basic elements [9]:

(1) A viscoplastic flow potential.
(2) Combined nonlinear isotropic-kinematic hardening with recovery.
(3) Evolution of isotropic damage (porosity).
(4) Elasto-viscoplastic damage coupling.

A rate formulation of the constitutive equations is employed, with the current configuration taken as the reference state for the next increment of deformation. We assume small elastic strains and the additive decomposition of the rate of deformation tensor **D** into elastic, $\mathbf{D}^e$, and inelastic components, $\mathbf{D}^p$, i.e. $\mathbf{D}=\mathbf{D}^e+\mathbf{D}^p$. Also, the Jaummann stress rate is used as the (objective) corotational rate for the tensorial variables $\boldsymbol{\sigma}$ and **A**, e.g. $\overset{\nabla}{\boldsymbol{\sigma}}=\dot{\boldsymbol{\sigma}}-\mathbf{W}\cdot\boldsymbol{\sigma}+\boldsymbol{\sigma}\cdot\mathbf{W}$, where **W** is the continuum spin tensor. Void nucleation effects are not considered in this paper.

2 A CLASS OF POROUS INELASTIC INTERNAL STATE VARIABLE MODELS

A class of porous viscoplastic state variable models can be characterized by the following constitutive equations [9]:

$$
\begin{aligned}
F &= \hat{F}(\boldsymbol{\sigma}-\mathbf{A},R,\vartheta)\\
\mathbf{D}^p &= \|\mathbf{D}^p\|\,\mathbf{n}_p = g(\langle F\rangle)\,\mathbf{n}_p\\
\overset{\nabla}{\boldsymbol{\sigma}} &= \underset{\sim}{\mathbf{C}}(\vartheta):(\mathbf{D}-\|\mathbf{D}^p\|\,\mathbf{n}_p) - \frac{\dot{\vartheta}}{1-\vartheta}\boldsymbol{\sigma}\\
\overset{\nabla}{\mathbf{A}} &= \sqrt{\frac{3}{2}}\,C\left[(1-\vartheta)b\,\mathbf{n}_A-\mathbf{A}\right]\dot{\bar{\epsilon}}^p-\Omega^A(\mathbf{A})\,\mathbf{A}\\
\dot{R} &= \sqrt{\frac{3}{2}}\,\mu^R(R^s-R)\,\dot{\bar{\epsilon}}^p-\Omega^R(R)\\
\dot{\vartheta} &= (1-\vartheta)\,\mathrm{tr}(\mathbf{n}_p)\,\|\mathbf{D}^p\|
\end{aligned}
\qquad (5)
$$

where $\|\mathbf{D}^p\|=(\mathbf{D}^p:\mathbf{D}^p)^{1/2}$ and **A**, R and ϑ are internal state variables.

For purposes of this paper, the inclusion or exclusion of I_1 (i.e. a pressure-dependent or pressure-independent flow potential), will define a damage-coupled ($F=\hat{F}(I_1,J_2^o,R,\vartheta)$) or partially damage-coupled ($F=\hat{F}(J_2^o,(1-\vartheta)R)$) inelastic model, respectively.

During inelastic deformation, the direction of evolution of inelastic straining is prescribed by the unit vector $\mathbf{n}_p$ which is given by

$$
\mathbf{n}_p=\frac{\beta}{\sqrt{1+\beta^2}}\mathbf{e}+\frac{1}{\sqrt{1+\beta^2}}\mathbf{n}_s\ ,\quad \beta=\frac{1}{\sqrt{3}}\frac{\mathrm{tr}(\mathbf{D}^p)}{\|\mathbf{D}^p_d\|} \qquad (6)
$$

Typically, $\beta=\hat{\beta}(\vartheta,I_1,(J_2^o)^{1/2},m)$ in most models. In general, $\mathbf{n}_p$ is not parallel to the outward unit vector normal to the flow potential surface, $\mathbf{n}_\sigma$, where

$$
\mathbf{n}_\sigma=\frac{\partial F/\partial\boldsymbol{\sigma}}{\|\partial F/\partial\boldsymbol{\sigma}\|}\ ,\quad \frac{\partial F}{\partial\boldsymbol{\sigma}}=\frac{\partial F}{\partial I_1}\mathbf{I}+\frac{\partial F}{\partial J_2^o}(\mathbf{s}-\boldsymbol{\alpha}) \qquad (7)
$$

In these general terms, the model can represent either an associative ($\mathbf{n}_p=\mathbf{n}_\sigma$) or a non-associative ($\mathbf{n}_p\neq\mathbf{n}_\sigma$) flow rule.

In state variable viscoplasticity, $\|\mathbf{D}^p\|$ is commonly specified by the kinetic equation

$$
\|\mathbf{D}^p\| = g\left(\left\langle\frac{F}{h_3 V}\right\rangle\right) \qquad (8)
$$

where g is the viscosity function of the porous material, which is assumed to be an homogeneous function of F, and V is the drag strength. For viscoplastic deformation, $F\geq 0$.

For small elastic stretch, the hypoelastic elastic response of the porous material is given by Eq. $(5)_3$. For isotropic elasticity, the damaged-coupled elastic stiffness tensor $\mathbf{C}(\vartheta)$ is

$$
\underset{\sim}{\mathbf{C}}(\vartheta) = 2\mu(\vartheta)\,\underset{\sim}{\mathfrak{I}} + \lambda(\vartheta)\,\mathbf{I}\otimes\mathbf{I} \qquad (9)
$$

where $\mu(\vartheta)=(1-\vartheta)\mu^m$ and $\lambda(\vartheta)=(1-\vartheta)\lambda^m$ are the damaged-coupled Lamé elastic constants of the porous material, $\underset{\sim}{\mathfrak{I}}$ is the fourth rank identity tensor; here, superscripted "m" quantities are associated with the matrix.

The hardening rules for the backstress of the aggregate **A**, Eq. $(5)_4$, and the static yield strength of the matrix R, Eq. $(5)_5$, follow a hardening minus recovery format, which has proven very successful to model complex loading histories [10]. The static thermal recovery functions Ω^A and Ω^R introduce rate dependence in the evolution of **A** and R. In these equations, C, b, μ^R and R^s are material constants. The directional index $\mathbf{n}_A$ gives the direction of the kinematic translation of the yield surface in stress space. It is obtained by assuming an equivalence in material response (evolution of porosity) for both pure isotropic and combined isotropic-kinematic hardening under proportional loading histories [9,11]

$$
\mathbf{n}_A = \mathbf{n}_s + \sqrt{\frac{3}{2}}\,\frac{I_1}{\sqrt{3J_2^o}}\,\frac{1}{3}\mathbf{I} \qquad (10)
$$

where $\mathbf{n}_s=(\mathbf{s}-\boldsymbol{\alpha})/\|\mathbf{s}-\boldsymbol{\alpha}\|$.

The matrix equivalent inelastic strain rate, $\dot{\bar{\epsilon}}^p$, is obtained by assuming that the inelastic work rate per unit volume of the matrix, $(1-\vartheta)R\dot{\bar{\epsilon}}^p$, and aggregate, $(\boldsymbol{\sigma}-\mathbf{A}):\mathbf{D}^p$, are equal (since voids do not contribute). This assumption leads to

$$\dot{\epsilon}^p = \frac{(\boldsymbol{\sigma}-\mathbf{A}):\mathbf{D}^p}{(1-\vartheta)R_d} = \frac{(\boldsymbol{\sigma}-\mathbf{A}):\mathbf{n}_p}{(1-\vartheta)R_d}\|\mathbf{D}^p\| \tag{11}$$

where R_d denotes the dynamic uniaxial yield strength [8] of the matrix and is obtained by inverting the kinetic equation, Eq. (8).

The evolution equation for damage (voids), Eq. $(5)_6$, implicitly specifies a void growth law, i.e.

$$\dot{\vartheta} = \sqrt{3}(1-\vartheta)\,\beta\,\|\mathbf{D}^p_d\| \tag{12}$$

Note that if β is specified, the void growth law is explicitly stated.

3 MODEL FRAMEWORKS FOR POROUS INELASTICITY

3.1 Associative Porous Inelasticity with F = $\hat{F}(I_1,J_2^o,R,\vartheta)$

The explicit expression for β in terms of F is obtained using the associative flow rule, $\mathbf{n}_p = \mathbf{n}_\sigma$. Then, by equating the volumetric and deviatoric components of $\mathbf{n}_p$ in Eq. (6) with $\mathbf{n}_\sigma$ in Eq. (7), we obtain

$$\beta = \frac{\sqrt{3}}{\sqrt{2J_2^o}}\,\frac{\partial F/\partial I_1}{\partial F/\partial J_2^o} \tag{13}$$

$$\dot{\vartheta} = \frac{3(1-\vartheta)}{\sqrt{2J_2^o}}\,\frac{\partial F/\partial I_1}{\partial F/\partial J_2^o}\,\|\mathbf{D}^p_d\| \tag{14}$$

For an elliptic functional form of F, Eq. (4), the explicit form for the void growth equation obtained from Eq. (14) is

$$\dot{\vartheta} = \sqrt{6}\,\frac{(1-\vartheta)h_2}{h_1}\,\frac{I_1}{\sqrt{3J_2^o}}\,\|\mathbf{D}^p_d\| \tag{15}$$

3.2 Non-associative Porous Inelasticity with F = $\hat{F}(I_1,J_2^o,R,\vartheta)$

This model framework combines a pressure-dependent flow potential of the elliptic type, for example, with an explicit void growth law such as Eq. (1). A number of explicit void growth models have been proposed in the literature [2-3]. For the Cocks and Ashby model [2], for example, the void growth factor is given by

$$\beta = \frac{\sqrt{2}}{3}\sinh\left(\frac{2(2-m)}{3(2+m)}\,\frac{I_1}{\sqrt{3J_2^o}}\right)\left[\frac{1}{(1-\vartheta)^{1/m+1}}-1\right] \tag{16}$$

For an elliptic F, β is given by

$$\beta = \sqrt{2}\,\frac{h_2}{h_1}\,\frac{I_1}{\sqrt{3J_2^o}} \tag{17}$$

A non-associative flow rule is obtained by selecting a pressure dependent function F which is different from the one used to derive β (i.e. $\mathbf{n}_p \neq \mathbf{n}_\sigma$).

3.3 Non-Associative Porous Inelasticity with F = $\hat{F}(J_2^o,(1-\vartheta)R)$

The specific equations of this very simple model framework with pressure-independent F are

$$\begin{aligned}
F &= \sqrt{3J_2^o} - (1-\vartheta)R \\
\mathbf{D}^p &= \sqrt{1+\beta^2}\,\|\mathbf{D}^p_d\|\,\mathbf{n}_p \\
\overset{\nabla}{\boldsymbol{\sigma}} &= \underset{\sim}{\mathbf{C}}(\vartheta):(\mathbf{D}-\mathbf{D}^p) - \sqrt{3}\,\beta\,\|\mathbf{D}^p_d\|\,\boldsymbol{\sigma} \\
\overset{\nabla}{\boldsymbol{\alpha}} &= C(b\mathbf{n}_s - \boldsymbol{\alpha})\,\|\mathbf{D}^p_d\| - \Omega^\alpha(\boldsymbol{\alpha})\,\boldsymbol{\alpha} \\
\dot{R} &= \mu^R(R^s - R)\,\|\mathbf{D}^p_d\| - \Omega^R(R) \\
\dot{\vartheta} &= \sqrt{3}(1-\vartheta)\,\beta\,\|\mathbf{D}^p_d\|
\end{aligned} \tag{18}$$

with the flow rule

$$\|\mathbf{D}^p_d\| = g\left(\left\langle\frac{F}{(1-\vartheta)V}\right\rangle\right) \tag{19}$$

4 VISCOPLASTIC MODEL

Applications of the three constitutive frameworks will be based on Bammann's viscoplastic model [10,12]. Recently, this model has been applied along with explicit void growth relations [2] to admit compressibility effects (partially damaged-coupled

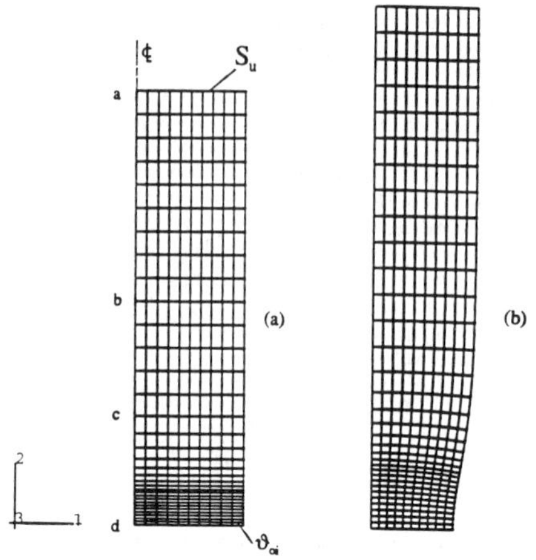

Fig. 1 Axisymmetric specimen: (a) Undeformed and (b) deformed finite element meshes.

TABLE I Void Growth Models for Porous Viscoplasticity

1 .**Explicit void growth models, Eq.(1)**

Cocks and Ashby [2]

$$\dot{\vartheta}=\left[\frac{1}{(1-\vartheta)^{1/m}}-(1-\vartheta)\right]\sinh\left(\frac{2(2-m)}{2+m}\frac{\sigma_h}{\bar{\sigma}}\right)\dot{\bar{\epsilon}}^p$$

2. **Implicit void growth models (Elliptic model, Eq.(4))**

Cocks [6]

$$h_1=1+\frac{2}{3}\vartheta\ ,\quad h_2=\frac{1}{2}\frac{1}{1+m}\frac{\vartheta}{1+\vartheta}\ ,\quad h_3=(1-\vartheta)^{1/(m+1)}$$

Sofronis and McMeeking [7]

$$h_1=(1+\vartheta)^{2/(m+1)},\quad h_2=\frac{1}{4}\left(\frac{m\vartheta^m(1-\vartheta)}{1-\vartheta^m}\right)^{2/(m+1)},\quad h_3=(1-\vartheta)^{1/(m+1)}$$

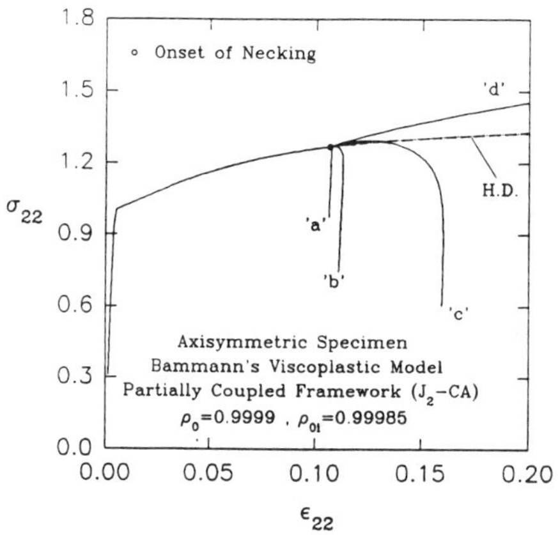

Fig. 2 Effect of necking on stress-strain response at some material particles along the axis of specimen.

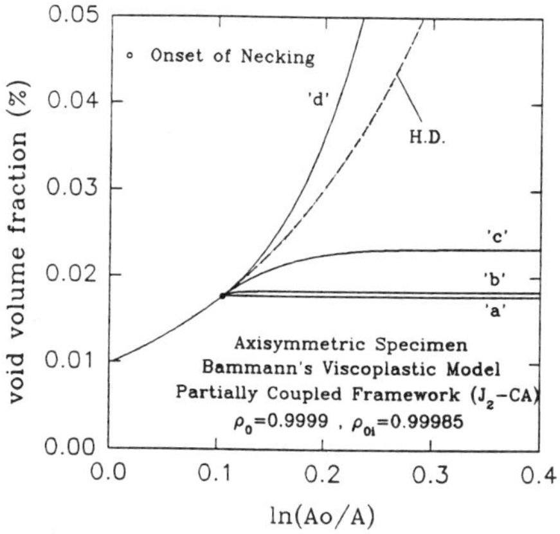

Fig. 3 Effect of necking on porosity evolution at some material particles along the axis of specimen.

framework) [12]. In this section, this model is extended to deal with porous (compressible) inelasticity in an associative and a non-associative frameworks. The viscoplastic flow rule is given by

$$\mathbf{D}^p=\sqrt{1+\beta^2}\,g\left(\left\langle\frac{F}{h_3 V}\right\rangle\right)\mathbf{n}_p \tag{20}$$

where

$$g=\sqrt{\frac{3}{2}}\,f\sinh\left(\frac{\sqrt{3h_1 J_2^u+h_2(I_1^u)^2}-h_3(R+Y)}{h_3 V}\right) \tag{21}$$

where $h_i=\hat{h}_i(\vartheta,m)$ (see Table 1) with $m=V/Y$; f, Y and V are temperature-dependent matrix material parameters [10]. The overstress invariants, I_1^u and J_2^u, are defined as [10,12]

$$I_1^u=\mathrm{tr}(\boldsymbol{\sigma}-\frac{2}{3}\mathbf{A})\ ,\quad J_2^u=\frac{1}{2}\mathrm{tr}(\mathbf{s}-\frac{2}{3}\boldsymbol{\alpha})^2 \tag{22}$$

In this model, the evolution equations for the hardening variables **A** and R are given by

$$\overset{\nabla}{\mathbf{A}}=\sqrt{\frac{3}{2}}\left((1-\vartheta)b\mathbf{n}_A-\frac{2}{3}\frac{k_d}{1-\vartheta}\|\mathbf{A}\|\mathbf{A}\right)\dot{\bar{\epsilon}}^p-\sqrt{\frac{2}{3}}\frac{k_s}{1-\vartheta}\|\mathbf{A}\|\mathbf{A} \tag{23}$$

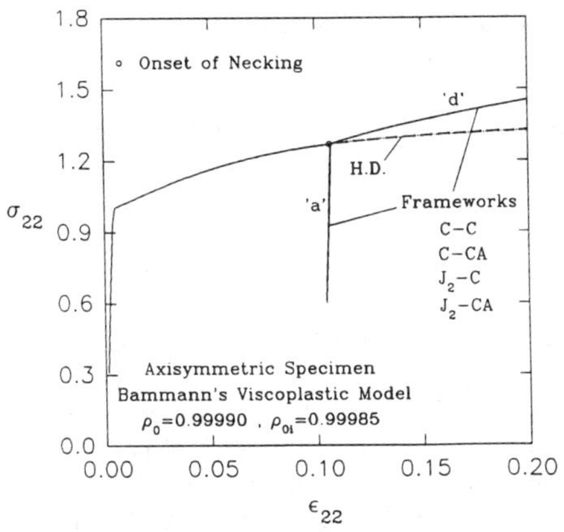

Fig. 4 Stress-strain response during neck development predicted by various constitutive frameworks.

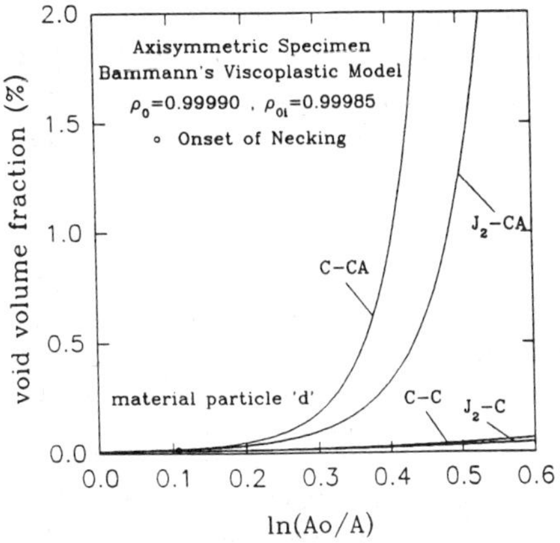

Fig. 5 Porosity evolution during neck development predicted by various constitutive frameworks.

$$\dot{R} = \sqrt{\frac{3}{2}}\left(\sqrt{\frac{2}{3}}B - \sqrt{\frac{2}{3}}K_d R^2\right)\dot{\bar{\epsilon}}^p - K_s R^2$$

where b, k_d, k_s, B, K_d, K_s are temperature-dependent matrix material parameters [10,12].

5 APPLICATIONS

Equations (20)-(23), in the context of the three model frameworks, have been implemented in the user material subroutine UMAT of the displacement-based finite element code ABAQUS [13], using a semi-implicit integration scheme [14]. Since different combinations of potentials and void growth factors are possible, a specific model computation will be denoted by a set of two letters in the form X-Y. The first letter, X, indicates the model for the pressure dependent flow potential while the second one, Y, denotes the model for the void growth factor.

In this work, we will use the associative C-C (C=Cocks [6]), non-associative C-CA (CA=Cocks and Ashby [2]) and partially damage-coupled J_2-C, J_2-CA (J_2=von Mises flow potential with the factor $(1-\vartheta)$ affecting R, Eq. $(18)_1$) frameworks to study the stress-strain response and porosity evolution of an axisymmetric tensile specimen during necking development. The calculations were performed at 21°C for 6061-T6 Al [12]. The elastic constants for this material are $E = 69\text{x}10^3$ MPa, $\nu = 0.33$. A uniform initial void volume fraction of $\vartheta_o = 1 - \rho_o = 10^{-4}$ is assumed. Necking is initiated using an initial perturbed value of ϑ, ϑ_{oi} (ρ_{oi}), at a specific location in the specimen. In this study, we use $\vartheta_{oi} = 1.5\text{x}10^{-4}$.

The finite element mesh of only 1/4 of the tensile specimen and the element where the initial imperfection ϑ_{oi} is introduced are shown in Fig. 1a. The mesh consists of 330 four-node quadrilateral elements type ABAQUS-CAX4. The initial length of the specimen is $2l_o$ ($l_o = 101.6\text{x}10^{-2}$ m) and the ratio of initial length to initial radius a is $l_o/a = 4$. A displacement-rate boundary condition of $2.54\text{x}10^{-2}$ m/s (strain rate in the order of $2\text{x}10^{-2}$ s^{-1}) is applied on surface S_u (see Fig. 1a). The deformed finite element mesh at an averaged axial strain of $\epsilon = 0.180$, where $\epsilon = \ln(1+u/l_o)$, is shown in Fig. 1b.

Figure 2 and 3 show the local stress-strain and porosity-$\ln(A_o/A)$ responses (A_o and A are the initial and current area of the specimen cross section, respectively) at material points located at positions 'a', 'b', 'c' and 'd' (see Fig. 1a) along the axis of the axisymmetric specimen. These results were obtained using the J_2-CA framework. Note that the local response at these material points starts to deviate from the initial uniaxial homogeneous deformation (H.D.) path at the onset of necking due to the induced triaxial stress state. Note also that when necking starts, elastic unloading initiates at the material points on the boundary S_u (e.g. point 'a', see Fig. 2), preventing further void growth (Fig. 3).

Figure 4 compares the prediction of the local stress-strain response at material points 'a' and 'd' and Fig. 5 gives the porosity evolution response at material point 'd', for the different model frameworks mentioned above. It is observed that the point of elastic unloading at material point 'a' (onset of necking) is predicted at the same strain level (approx. 0.1066) by all models. The stress-strain response at point 'd' after the onset of necking differs slightly among the models. However, the predicted porosity, which is of the order of 0.01 at the onset of necking for all models, changes *significantly* among models *after* necking initiates.

6 CONCLUSIONS

A general constitutive framework for porous viscoplasticity with isotropic damage was presented. This framework considered either explicit or implicit void growth models along with combined nonlinear isotropic-kinematic hardening rules in a hardening minus recovery format. Both associative and non-associative inelastic flow rules were considered. The non-associativity of the flow rule was due to compressibility (void growth) effects. Void nucleation effects were neglected.

Using the general framework, three particular model frameworks for porous viscoplasticity were identified. These were denoted as associative, non-associative and partially coupled frameworks. The void growth factor, β, was introduced here to characterize a specific void growth model. Bammann's partially coupled rate- and temperature-dependent state variable viscoplastic model, which was extended in this work to deal with associative and non-associative frameworks, was used to numerically study the effect of necking on the local material response of an axisymmetric tensile specimen. The numerical results showed no effect of the particular framework on the onset of necking, but strong effects on local post-necking behavior and resulting ductility.

7 ACKNOWLEDGEMENTS

The authors wish to acknowledge the support of the U.S. Army Research Office in this research, as well as support of the U.S. National Science Foundation. The authors are also grateful to Dr. D.J. Bammann of Sandia National Laboratories for fruitful discussions and for supplying viscoplastic model constants.

REFERENCES

[1] Rogers, H.C., "The Tensile Fracture of Ductile Metals," Transactions of the Metallurgical Society of AIME, Vol. 218, pp. 498-506, 1960.

[2] Cocks, A.C.F., and Ashby, M.F., "Intergranular Fracture During Power-Law Creep Under Multiaxial Stresses," Metal Science, Vol. 14, pp. 395-402, 1980.

[3] Lee, Y.S., "Modelling Ductile Damage Evolution in Metal Forming Processes," Ph.D. Thesis, Mechanical Engineering, Cornell University, Ithaca, N.Y., 1991.

[4] Gurson, A.L., "Continuum Theory of Ductile Rupture by Void Nucleation and Growth: Part I - Yield Criteria and Flow Rules for Porous Ductile Media," Journal of Engineering Materials and Technology, Vol. 99, pp. 2-15, 1977.

[5] Duva, J.M., and Hutchinson, J.W., "Constitutive Potentials for Dilutely Voided Nonlinear Materials," Mechanics of Materials, Vol. 3, pp. 41-54, 1984.

[6] Cocks, A.C.F., "Inelastic Deformation of Porous Materials," Journal of Mechanics and Physics of Solids, Vol. 37, pp. 693-715, 1989.

[7] Sofronis, P., and McMeeking, R.M., "Creep of Power-Law Material Containing Spherical Voids," ASME Journal of Applied Mechanics, Vol. 59, pp. S88-S95, 1992.

[8] Perzyna, P., "Fundamental Problems in Viscoplasticity," Advances in Applied Mechanics, G.G. Chernyi, et.al. (eds), Academic Press-New York, Vol. 9, pp. 243-377, 1966.

[9] McDowell, D.L., Marin, E., and Bertoncelli, C., "A Combined Kinematic-Isotropic Hardening Theory for Porous Inelasticity of Ductile Metals," International Journal of Damage Mechanics, Vol. 2, pp. 137-161, 1993.

[10] Bammann, D.J., "Modeling the Temperature and Strain Rate Dependent Large Deformation of Metals", Applied Mechanics Reviews, Vol. 43, No. 5, Part 2, E. Krempl and D.L. McDowell, eds, pp. S312-S319, 1990.

[11] Becker, R., and Needleman, A., "Effect of Yield Surface Curvature on Necking and Failure in Porous Plastic Solids", Journal of Applied Mechanics, Vol. 53, pp. 491-499, 1986.

[12] Bammann, D.J., Chiesa, M.L., Horstemeyer, M.F., and Weingarten, L.I., "Failure in Ductile Materials Using Finite Element Methods," Chapter 1 in Structural Crashworthiness and Failure, Ed. N. Jones and T. Wierzbicki, Elsevier Appl. Sci. Publ, New York, pp. 1-54, 1993.

[13] ABAQUS, Reference Manuals, Hibbit, Karlsson and Sorensen Inc., Pawtucket, Rhode Island, 1993.

[14] Moran, B., Ortiz, M., and Shih, C.F., "Formulation of Implicit Finite Element Methods for Multiplicative Finite Deformation Plasticity," International Journal for Numerical Methods in Engineering, Vol. 29, pp. 483-514, 1990.

Simulation of Materials Processing: Theory, Methods and Applications, Shen & Dawson (eds)
© 1995 Balkema, Rotterdam. ISBN 90 5410 553 4

Reflecting microstructural evolution in hardening models for polycrystalline metals

M.P. Miller & P.R. Dawson
Sibley School of Mechanical and Aerospace Engineering, Cornell University, Ithaca, N.Y., USA

D.J. Bammann
Center for Materials and Mechanics, Sandia National Laboratory, Livermore, Calif., USA

ABSTRACT: Within this paper, large strain inelastic deformation models for polycrystalline metals are examined. In particular, links between underlying stress state-dependent microstructural evolution and hardening formulations for both the micro- and macro-scale models are explored. A slip system hardening model which uses the instantaneous value of the average slip system shearing rate, $\dot{\Gamma}^*$, as a stress state delineator is proposed for use with a continuum slip polycrystalline formulation. On the macro-scale, two approaches are proposed for the evolution of the hardening variables. The first employs J_3^*, the third invariant of the overstress as a stress state delineator; the second uses $\dot{\Gamma}^*$. To determine the utility of the models, effective stress-stain data from compression and torsion experiments conducted on OFHC copper are correlated. The material response in plane strain compression is then predicted.

1 INTRODUCTION

Large inelastic strains common in deformation processing induce crystallographic texture, phase transformations, and dislocation substructures, as well as other changes in the underlying microstructure of a polycrystalline metal. The evolution of the microstructure is reflected in the macroscopically observed material behavior. When considering the stress-strain behavior of a material subjected to a particular loading, it is observed that the elevation of the flow stress during continued straining arises due to both geometric and material hardening mechanisms. Geometric (textural) hardening or softening occurs as the slip planes rotate relative to the loading axes. The effect of lattice rotation may be either to raise or lower the stress required for yielding. In tension and compression, texturing generally elevates the stress while in torsion the value is diminished. Material (strain) hardening, on the other hand, elevates the inherent resistance a particular material has to inelastic deformation. In the case of slip, this is essentially the critical resolved shear stress for slip on any of the active slip systems. Sources of material hardening include dislocation - dislocation interactions and the nucleation and evolution of dislocation substructures.

When modelling stress-strain behavior, accuracy is improved by linking the model more closely to the underlying physics. The formalism for continua with internal variables (cf. Rice, 1971) is employed by many current large strain models which are formulated on either a macro- or micro-scale. Pertinent variables for large strain inelasticity include the slip system hardening variables of continuum slip polycrystal models and the kinematic and isotropic hardening variables of macro-scale formulations. In the case of the polycrystal model, geometric hardening is an explicit part of the formulation and the evolution of the internal variables account for material hardening only. In macro-scale formulations, on the other hand, the evolution equations of the internal variables must account for both geometric and material hardening sources on all size scales.

Within this paper hardening models for both micro- and macro-scale formulations are presented. The utility of each model is ascertained by correlating compression and torsion experimental data with a consistent parameter set. Elastic deformation is not included in the models presented in the paper.

2 POLYCRYSTAL MODEL

Using polycrystal plasticity, macroscopic properties are determined by averaging the individual crystal responses over the ensemble of crystals that represent a material element (cf. Havner, 1992).

Polycrystal models commonly function by having the macroscopic motion imposed on the aggregate and partitioned among the crystals of the aggregate according to a specific rule. For FCC crystals the Taylor assumption of equal straining in all crystals often is invoked, making the process of partitioning the straining among the crystals trivial. With the deformation rate of each grain known through the partitioning, the stress in each crystal may be computed and then averaged over the aggregate to give the macroscopic stress. The stress computed in each crystal for this common deformation rate depends on its orientation and the nature of the slip on the active slip systems. The relationship between the resolved shear stress, τ^α, acting on a particular slip system (denoted by the superscript α) and the rate at which it shears is assumed here to follow a simple power law:

$$\tau^\alpha = \hat{\tau}\left(\frac{\dot{\gamma}^\alpha}{\dot{a}}\right)^m \tag{2.1}$$

where $\dot{\gamma}$ is the slip system shear rate and m is the rate sensitivity. The strength of the slip system is set by $\hat{\tau}$, which accounts for material hardening as the crystal deforms. The scaling between the slip system hardness and the macroscopic stress depends on the mode of loading for a particular texture. This scaling, known as the Taylor factor, exhibits a value of 3.06 in tension and 2.86 in torsion for a uniform distribution of crystal orientations (Bishop and Hill, 1951). Although equivalence depends on the characteristic slip system strength chosen (Mecking and Kocks, 1995) the average activity of the slip systems reflects this same trend. Greater normalized slip system shearing (where the normalizing variable is the effective deformation rate) is shown in tension or compression compared to torsion. We assume here that this difference in slip system activity will affect the hardening rates so that on the average the slip system strength $\hat{\tau}$ will increase more rapidly with higher normalized slip system shearing. This is accomplished by modifying the evolution equation for the strength to incorporate a dependence on the Taylor factor (Miller and Mika, 1995), as computed by average slip system shearing rate, $\dot{\Gamma}^*$, where

$$\dot{\Gamma}^* = \frac{1}{N}\sum_{C=1}^{N}\sum_{\alpha=1}^{n}\dot{\gamma}^\alpha/\bar{D}^p \tag{2.2}$$

where N is the number of crystals, n is the number of slip systems per crystal, and $\bar{D}^P$ is the effective plastic strain rate defined as

$$\bar{D}^p = \sqrt{\frac{2}{3}}\|\boldsymbol{D}^p\| \tag{2.3}$$

Here $\boldsymbol{D}^p$ is the plastic rate of deformation tensor. The evolution equation for the strength (Kocks, 1976) was altered to reproduce the difference in behavior between compression and torsion for copper. It was necessary only to modify the hardening coefficient, h, in the evolution equation

$$\dot{\hat{\tau}} = h\left(1 - \sqrt{\frac{\hat{\tau}}{\tau_s}}\right)\bar{D}^p \tag{2.4}$$

where

$$h = h_p\left(\frac{\dot{\Gamma}^*}{\dot{\gamma}_0}\right)^n \tag{2.5}$$

and, in general,

$$\tau_s = \tau_s(T, \bar{D}^p) \tag{2.6}$$

The comparisons between the experimental data and the polycrystal model computations are shown in Figure 1 where

$$\bar{\sigma} = \sqrt{\frac{3}{2}}\|\boldsymbol{\sigma}'\| \text{ and } \bar{\epsilon}_p = \int \bar{D}^p \mathrm{d}t \tag{2.7}$$

Here $\boldsymbol{\sigma}'$ is the deviatoric Cauchy stress tensor. Also shown in Figure 1 is the polycrystal model prediction for plane strain compression which compares qualitatively with experimental data (Ford, 1968). The polycrystal model automatically accounts for geometric hardening through the evolution of texture and its effect on the stress. In addition, the changes in the Taylor factor that accompany the evolving texture are explicitly included in the hardening law through the use of the current Taylor factor. The Taylor factor histories, as computed from the normalized slip system activities, are shown in Figure 2 for the compression, torsion and plane strain compression tests. The modified hardening law is capable of differentiating between the different loading states, although the predicted stress is saturating more quickly than the measured value for compression.

3 MACRO-SCALE MODEL

Macro-scale formulations make their constitutive assumptions on the scale of the test specimen, work piece, or structural member. By their very

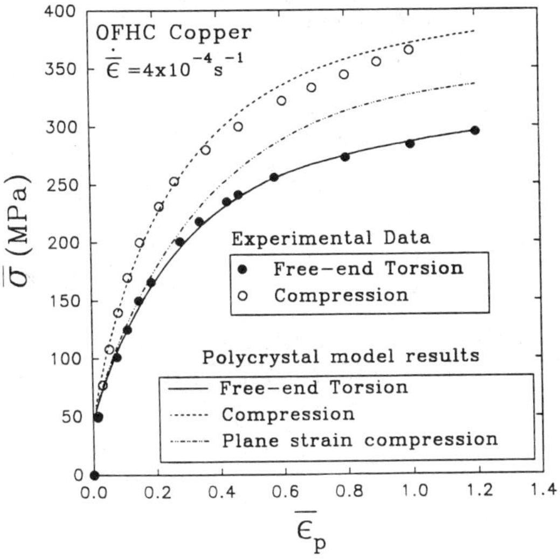

Figure 1. Effective stress-strain data for OFHC copper (Miller and McDowell, 1994) showing the polycrystal model correlations for compression and torsion. Also shown is the model prediction for plane strain compression.

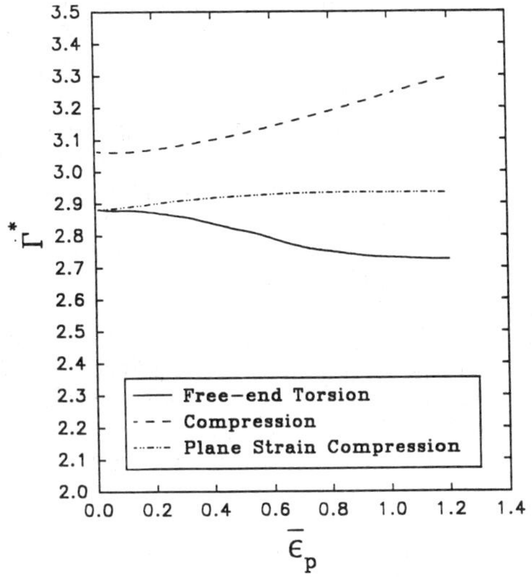

Figure 2. Average slip system shearing rate, $\dot{\Gamma}^*$, vs strain for compression, torsion, and plane strain compression as predicted by the polycrystal model.

nature, these models assume material homogeneity but can include anisotropic effects. Macroscale models are basically extensions of classical, infinitesimal plasticity into the finite strain regime. Both geometric and material hardening phenomena associated with features at all size scales must be accounted for within the evolution equations of isotropic and kinematic internal variables. While isotropic variables are used to reflect hardening sources such as elevation of dislocation density and strain-induced phase transformations, kinematic variables reflect directionally- dependent hardening and are used as a repository for deformation-induced anisotropy in general. The model presented here employs a deviatoric kinematic hardening variable, the second order tensor $\boldsymbol{\alpha}$, and a scalar isotropic variable, κ.

The rate-dependent flow rule is written as

$$\boldsymbol{D}^p = \frac{3}{2} f \sinh\left[\frac{\bar{\xi} - \kappa - Y}{V}\right] \frac{\boldsymbol{\xi}}{\bar{\xi}} \qquad (3.8)$$

where

$$\boldsymbol{\xi} = \boldsymbol{\sigma}' - \frac{2}{3}\boldsymbol{\alpha}$$

and

$$\bar{\xi} = \sqrt{\frac{3}{2}} \|\boldsymbol{\xi}\|$$

Evolution equations for $\boldsymbol{\alpha}$ and κ are written:

Table 1. Polycrystal model parameters used to correlate OFHC copper data shown in Figure 1.

Parameter	Units	Value
$\dot{a}$	s^{-1}	1.0
m	-	0.05
h_p	MPa	500.
$\dot{\gamma}_0$	-	2.86
τ_s	MPa	120.

$$\overset{\circ}{\boldsymbol{\alpha}} = h\boldsymbol{D}^p - r_d \bar{D}^p \boldsymbol{\alpha} \qquad (3.9)$$

and

$$\dot{\kappa} = H\bar{D}^p - R_d \bar{D}^p \kappa \qquad (3.10)$$

In general, parameters V, Y, f, h_α, and H, are temperature dependent and static thermal recovery terms are included in the evolution equations for $\boldsymbol{\alpha}$ and κ. Here temperature and static thermal recovery are omitted. Also omitted are dynamic recovery functions (Bammann, 1990); hence, r_d and R_d are temperature independent material parameters. The superscript $\circ$ in eqn(3.9) implies an objective rate for $\boldsymbol{\alpha}$ i.e.

$$\overset{\circ}{\boldsymbol{\alpha}} = \dot{\boldsymbol{\alpha}} - \boldsymbol{\omega} \cdot \boldsymbol{\alpha} + \boldsymbol{\alpha} \cdot \boldsymbol{\omega} \qquad (3.11)$$

The spin $\boldsymbol{\omega}$ is taken to be the spin of the stress-free configuration, i.e. $\omega = \boldsymbol{W} - \boldsymbol{W}^p$, where $\boldsymbol{W}$ and $\boldsymbol{W}^p$ are the continuum and plastic spin, respectively. From eqns (3.9) and (3.11) it may be observed that kinematic hardening will proceed differently in torsion compared to compression due to the spin terms in the evolution equation for $\boldsymbol{\alpha}$. Fitting data such as that in Figure 1 can be accomplished by manipulating the plastic spin expression as well as the amount of kinematic hardening used (cf. Bammann and Aifantis, 1987). However, since spin effects are implicitly linked to microstructural reorientation, this approach would assume that the hardening differences are only geometric in nature. In this paper, therefore, we assume $\boldsymbol{\omega} = \boldsymbol{W}$ and we use modest levels of kinematic hardening ($\bar{\alpha} < 0.25\bar{\sigma}$).

Two different approaches are taken here to model the effective stress level difference shown in Figure 1. First we introduce the third invariant of the overstress, J_3^*, into the formulation (Miller and McDowell, 1992, 1994)

$$J_3^* = \frac{1}{3}\mathrm{tr}\boldsymbol{\xi}^3 \tag{3.12}$$

We make the hardening parameters h_α and H both linear functions of J_3^* i.e.

$$h = h_0(1 + h_1\frac{J_3^{*2}}{J_2^{*3}}) \tag{3.13}$$

$$H = H_0(1 + H_1\frac{J_3^{*2}}{J_2^{*3}}) \tag{3.14}$$

where the second invariant of the overstress is defined as

$$J_2^* = \frac{1}{2}\mathrm{tr}\boldsymbol{\xi}^2 \tag{3.15}$$

For tension and compression $\frac{J_3^{*2}}{J_2^{*3}} = 4/27$ where for torsion $\frac{J_3^{*2}}{J_2^{*3}} = 0$. The correlation attained using this approach is shown in Figure 3 and the model parameters are given in Table 2. In contrast to the polycrystal prediction, the model prediction for plane-strain compression is indistiguishable from the torsion curve. This is because as in torsion, $J_3^* = 0$ during plane strain compression.

Finally, by way of pointing towards forms for future macro-scale hardening models, we introduce expressions for h and H which are functions of the micro-scale parameter introduced previously, $\dot{\Gamma}^*$. For torsion, compression and plane-strain compression, we fit polynomial expressions for $\dot{\Gamma}^*$ as

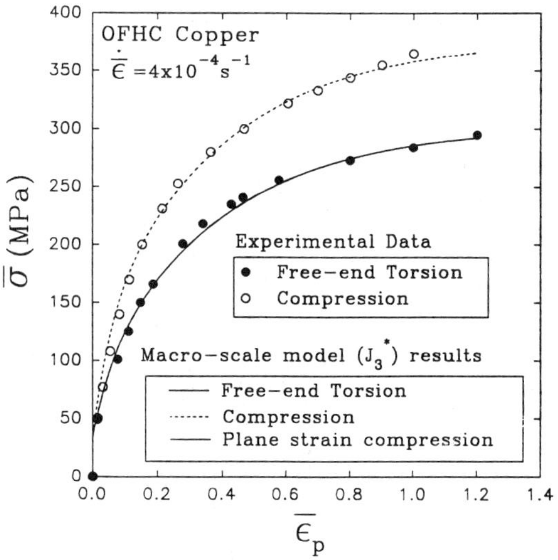

Figure 3. Copper data and macro-scale model predictions using J_3^*

Table 2. Macro-scale model parameters using J_3^* (eqns (3.13) and (3.14))

Parameter	Units	Value
f	s^{-1}	10^{-6}
Y	MPa	3.44
V	MPa	2.06
r_d	-	18.0
R_d	-	2.6
h_0	MPa	900.
h_1	-	5.4
H_0	MPa	570.
H_1	-	1.07

functions of strain as predicted by the polycrystal model and shown in Figure 2. Now we employ $\dot{\Gamma}^*$ in expressions similar in form to eqn (2.5) for the saturation values for $\boldsymbol{\alpha}$ and κ i.e.

$$h = h_\Gamma\left(\frac{\dot{\Gamma}^*}{\dot{\gamma}_0}\right)^g \tag{3.16}$$

and

$$H = H_\Gamma\left(\frac{\dot{\Gamma}^*}{\dot{\gamma}_0}\right)^G \tag{3.17}$$

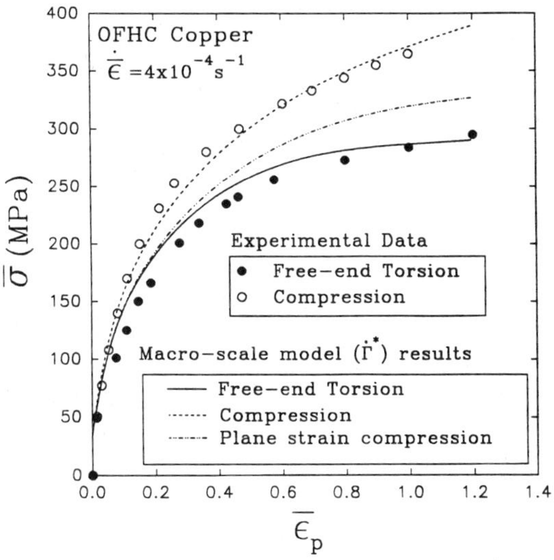

Figure 4. Copper data and macro-scale model predictions using $\dot{\Gamma}^*$

Table 3. Macro-scale model parameters using $\dot{\Gamma}^*$ (eqns (3.16) and (3.17)).

Parameter	Units	Value
f	s^{-1}	10^{-6}
Y	MPa	3.44
V	MPa	2.06
r_d	-	18.0
R_d	-	2.6
h_Γ	MPa	1250.
H_Γ	MPa	570.
g	-	2.0
G	-	2.0
$\dot{\gamma}_0$	-	2.86

The correlation of the torsion and compression data and the prediction of plane-strain compression using eqns (3.16) and (3.17) are shown in Figure 4. The parameter values are given in Table 3. While the correlation of the experimental data is not exact, the model does correctly distiguish plane strain compression from torsion. It is conceivable that a more precise correlation would be possible by altering the forms of the evolution equations eqns (3.9) and (3.10). The goal here, however, was to explore the utility of using a microscale parameter like $\dot{\Gamma}^*$ within existing forms for the evolution equations for $\boldsymbol{\alpha}$ and κ.

4 CONCLUSIONS

The stresses within a material subjected to a particular loading history depend on both the material and the loading geometry. To accurately model large strain material behavior under different loading situations, both material and geometric hardening need to include proper stress state-dependence. The stress state- dependence of textural hardening is implicit in micro-scale crystal plasticity formulations. To make the material hardening stress state-dependent, we made the slip system hardening formulation a function of a quantity related to the Taylor factor, $\dot{\Gamma}^*$. We used this formulation to successfully correlate compression and torsion equivalent stress-strain data as well as qualitatively predicting the material response in plane-strain compression.

In macro-scale models, geometric hardening is generally accounted for by varying the magnitude and objective rate of the kinematic hardening variable. We introduced stress state-dependent material hardening on the macro-scale in two ways. First, we introduced J_3^* into the evolution equations of the kinematic and isotropic hardening variable. We were able to correlate compression and torsion data. During plane strain compression, however, $J_3^* = 0$; hence, and the effective stress levels predicted during plane strain compression directly coincide with those of torsion. To further differentiate stress state a macro- scale equivalent of $\dot{\Gamma}^*$ is needed. We showed the effectiveness of employing such a parameter by using $\dot{\Gamma}^*$ values predicted by the polycrystal model in the macro-scale kinematic and isotropic hardening formulations.

REFERENCES

Bammann, D.J., 1990, "Modeling Temperature and Strain Rate Dependent Large Deformations of Metals," Applied Mechanics Reviews, vol. 43, no. 5, Part 2, s312.

Bammann, D.J. and Aifantis, E.C., 1987, "A Model for Finite- Deformation Plasticity," Acta Mechanica, Vol. 69, pp 97.

Bishop, J.F.W., and Hill, R., 1951, "A Theoretical Derivation of the Plastic Properties of a Polycrystalline Face-Centred Metal," Philosophical Magazine, ser. 7, vol. 42, pp 1298.

Dafalias, Y.F. and Cho H.-W., 1989, "Verification of the Plastic Spin Concept in Viscoplasticity," *Proceedings of Plasticity '89, 2nd International Symposium on Plasticity and its Current Applications*, Pergamon Press, Oxford-New York, pp. 287.

Ford, H., 1968, "The Working Properties of Metals and Their Relationship to Mechanical-Test Behaviour," Journal of the Institute of Metals, Vol. 96, pp. 65.

Havner, K.S., 1992, *Finite Plastic Deformation of Crystalline Solids*, Cambridge University Press, New York.

Kocks, U.F., "Laws for Work-Hardening and Low-Temperature Creep," J. of Eng. Mater. Tech., vol. 98, 76.

Mecking, H., and Kocks, U.F., 1995, "Taylor Factors for Very Anisotropic Materials," Submitted for publication in Scripta Metalurgica.

Miller, M.P. and McDowell, D.L., 1992, "Stress State Dependence of Finite Strain Inelasticity", *Microstructural Characterization in Constitutive Modeling of Metals and Granular Media*, ASME MD Vol. 32, G.P. Voyiadjis, Editor, 1992, pp 27.

Miller, M.P. and McDowell, D.L., 1994, "Stress State Dependent Deformation Behaviour of FCC Polycrystals," in *Numerical Predictions of Deformation Processes and the Behaviour of Real Materials*, Andersen, Bilde-Sorensen, Lorentzen, Pedersen, Sorensen, eds, Riso National Labs, Roskilde, Denmark, pp. 421.

Miller, M.P., and Mika, D.P., "Modelling Stress State-Dependent Deformation Behavior of Polycrystalline Metals," Proceedings of the Conference on Engineering Mechanics, Boulder CO., May 21, 1995

Rice, J.R., 1971, "Inelastic Constitutive Relations for Solids: An Internal Variable Theory and Its Application to Metal Plasticity, " Journal of the Mechanics and Physics of Solids, Vol. 19, pp. 433.

Simulation of Materials Processing: Theory, Methods and Applications, Shen & Dawson (eds)
© 1995 Balkema, Rotterdam. ISBN 90 5410 553 4

Implementation of an anisotropic friction law in a 3D finite element model of hot rolling

Pierre Montmitonnet & Arnaud Hacquin
École des Mines de Paris, Centre de Mise en Forme des Matériaux (CEMEF), Sophia-Antipolis, France

ABSTRACT: Friction has a place among the many parameters which influence lateral spread and stress states in hot rolling. Obviously, a frictional anisotropy potentially modifies the balance between forward and lateral flow. A friction law combining the characteristics of Coulomb and "viscoplastic" (=velocity dependent) friction is implemented in a viscoplastic finite element model, which itself uses an anisotropic viscoplastic behaviour law coupling Hill's anisotropy to Norton-Hoff viscoplasticity. The implementation is described, and examples show the potential importance of both frictional and material anisotropy on the mechanics of hot rolling.

1 INTRODUCTION AND BIBLIOGRAPHY

A lot of variables influence spread in hot rolling. The geometry of the system is the most important one (relative spread decreases when width to thickness ratio and width to contact arc length ratio increase). Material parameters also play an important part: strain rate sensitivity, work hardenability, yield stress dependence on temperature and thermal coupling. Their influence is more complex, as the sense of variation depends on geometry: for instance, increasing the strain rate sensitivity increases spread in slab and strip rolling, but decreases the dogbone defect (which is also a manifestation of spread) in edge rolling due to less localized strain. The same is true for friction, which increases spread when rolling of long products is considered (because contact length is larger than width), but decreases it in strip rolling (where the reverse is true).

Recently, attention has been drawn on the influence of material anisotropy on metal flow in bulk forming processes. Various models have been published using non quadratic yield criteria (Karabin et al. 1992), quadratic ones (Montmitonnet & Chenot 1995) or polycrystal plasticity models (Mathur & Dawson 1989; Beaudoin 1993 among others). In the flow competition from which spread originates, the "softest" direction is favored. This can result in a very large influence in certain cases (Montmitonnet et al. 1995). However, another source of anisotropy is friction. Although anisotropic friction has been studied in recent years, both theoretically (Felder 1988; Zmitrowicz 1989; He & Curnier 1993; Montmitonnet 1993) and experimentally (see (Felder 1986) for references), no application to large strain forming processes has ever been described. The purpose of the present paper is to implement a simple anisotropic friction law in a 3D FEM and perform a parametric study to show in which situations and to what extent frictional anisotropy may be an important parameter.

2 MODEL DESCRIPTION

2.1 *Anisotropic viscoplastic behaviour law*

The model is based on a flow formulation using Norton-Hoff viscoplastic behaviour. Rheological anisotropy has been introduced by coupling Norton Hoff strain rate dependent equations with Hill's quadratic yield criterion. The tensorial stress-strain rate relation is written (Bay & Chenot 1990; Montmitonnet & Chenot 1995):

$$s = 2K\left(\sqrt{3}\dot{\varepsilon}_{eq}\right)^{m-1} \mathrm{Dev}(C_o:\dot{e}) \qquad (1a)$$

s is the stress deviator tensor, $\dot{e}$ the deviatoric strain rate; "Dev" is the deviator operator, $\dot{\varepsilon}_{eq}$ the equivalent strain rate; K, the material consistancy and m, the strain rate sensitivity index, are strain and temperature dependent material parameters. (1a) can also be written:

$$s = 2K\left(\sqrt{3}\dot{\varepsilon}_{eq}\right)^{m-1} (C'_o:\dot{e}) \qquad (1b)$$

with

$$C'_{oijkl} = C_{oijkl} - \frac{1}{3} C_{oppkl}\, \delta_{ij} \quad \text{(sum on p)} \qquad (2)$$

$C_{o1111} = C_{oXX}$, $C_{o2222} = C_{oYY}$, $C_{o3333} = C_{oZZ}$
$C_{o2323} = C_{oYZ}$, $C_{o1313} = C_{oXZ}$, $C_{o1212} = C_{oXY}$
$C_{oijkl} = 0$ otherwise.

In the orthotropy axes (X,Y,Z), C_{oXX} and the like are easily derived from the well known Hill's criterion coefficients:

$$C_{oXX} = \sqrt{\frac{3}{\Sigma}}\,F,\quad C_{oYY} = \sqrt{\frac{3}{\Sigma}}\,G,\quad C_{oZZ} = \sqrt{\frac{3}{\Sigma}}\,H$$

$$C_{oYZ} = \frac{\sqrt{3\Sigma}}{L},\quad C_{oXZ} = \frac{\sqrt{3\Sigma}}{M},\quad C_{oXY} = \frac{\sqrt{3\Sigma}}{N} \tag{3}$$

with $\Sigma = FH+FG+GH$ (4)

The equivalent strain rate is given by:

$$\frac{3}{2}\,\dot{\varepsilon}_{eq}^2 = \dot{e}_{ij}:C_{oijkl}:\dot{e}_{kl} \tag{5}$$

The material consistency K is calculated using:

$$K = \left(\frac{1}{12\Sigma}\right)^{1/4} \left(\sqrt{3}\dot{\varepsilon}_{eq}\right)^{-m} \tag{6}$$

The anisotropic viscoplastic potential is finally written:

$$\Phi = \frac{K}{m+1}\left(2\dot{e}:C'_o:\dot{e}\right)^{(m+1)/2} = \frac{K}{m+1}\left(\sqrt{3}\ \dot{\varepsilon}_{eq}\right)^{(m+1)}$$

which shows the deep analogy with Norton-Hoff behaviour (Chenot & Bellet 1992).

In the orthotropy axes, the material is finally represented by 8 parameters: 6 C_{oIJ}'s, K and m. The complete definition of the material also requires the orientation of (X,Y,Z), i.e. 3 Euler angles, or the rotation matrix R between these systems. Starting from its initial value R_o, R may *for instance* be computed from:

$$^t R.\dot{R} = \Omega \tag{7}$$

(Ω is the antisymmetric part of the velocity gradient). C_o and C'_o are then replaced by:

$$C_{ijkl} = C_{omnpq} R_{mi}R_{nj}R_{pk}R_{ql} \tag{8a}$$
$$C'_{ijkl} = C'_{omnpq} R_{mi}R_{nj}R_{pk}R_{ql} \tag{8b}$$

2.2 *Anisotropic friction law*

Two contact conditions were initially available :
- sticking contact (v_g=0),
- "Norton" sliding-velocity-dependent friction law:

$$\tau = -\,\alpha_f K \|v_g\|^{p_f} \frac{v_g}{\|v_g\|} \tag{9}$$

where τ is the friction stress, v_g the sliding velocity, α_f and p_f two friction coefficients.

Two other, anisotropic friction laws have been added. They consider "orthotropic" friction, in that they are defined by two different friction coefficients in two orthogonal directions (a,b) of the tangent plane. Let Q be the transformation matrix between (t_L,t_T) and (a,b), where (t_L,t_T) are the unit tangent vectors in the longitudinal and transverse directions.

- the first friction law generalises Norton's law. It derives from the modified frictional potential Ω_f:

$$\Omega_f = \frac{\alpha_f K}{p_f+1}\left(A^2 + v_g.J_N.v_g\right)^{(p_f+1)/2} \tag{10}$$

where

$$J_N = Q.\begin{pmatrix} \left(\frac{2\alpha_a}{\alpha_a+\alpha_b}\right)^{2/(p+1)} & 0 \\ 0 & \left(\frac{2\alpha_b}{\alpha_a+\alpha_b}\right)^{2/(p+1)} \end{pmatrix} {}^tQ \tag{11a}$$

and $\alpha_f = \frac{1}{2}(\alpha_a+\alpha_b)$ (11b)

giving:

$$\tau = -\,\alpha_f K \left(A^2 + v_g.J_N.v_g\right)^{(p_f-1)/2} J_N.v_g \tag{12}$$

A is used to regularize the law at $v_g = 0$. This form implies that, if sliding occurs in direction a (resp. b),

$$\tau = -\,\alpha_a K \|V_g\|^{p_f}\, a \quad (\text{resp. } \tau = -\,\alpha_b K \|V_g\|^{p_f}\, b) \tag{13}$$

- the second law is an anisotropic friction law of the Coulomb type.

$$\tau = -\,\mu\, \sigma_n \left(A^2 + v_g.J_C.v_g\right)^{-1/2} J_C.v_g \tag{14}$$

where σ_n is the normal stress, and:

$$J_C = Q.\begin{pmatrix} \left(\frac{2\mu_a}{\mu_a+\mu_b}\right)^2 & 0 \\ 0 & \left(\frac{2\mu_b}{\mu_a+\mu_b}\right)^2 \end{pmatrix}.{}^tQ \tag{15}$$

$$\mu = \frac{1}{2}(\mu_a+\mu_b)$$

If sliding occurs in direction a (resp. b),

$$\tau = -\,\mu_a \sigma_n\, a \quad (\text{resp. } \tau = -\,\mu_b \sigma_n\, b) \tag{16}$$

This frictional behaviour is similar in nature to the one presented in (Montmitonnet 1993) ("associated" friction respecting the normality rule). In the present model, Coulomb's law may be limited at high friction by Norton's law, in a way similar to the friction model in (Bay & Wanheim 1976) based on a microplastic approach.

2.3 *Numerical implementation*

These behaviour equations are implemented in a 3D finite element context. As remeshing is not necessary in the cases studied, 8-node trilinear isoparametric bricks with 8 Gauss points are used. Incompressibility is enforced by a modified penalty technique (Chenot & Bellet 1992), with one central Gauss point. The resulting set of non linear equations is solved by Newton-Raphson's method. The stiffness matrix is easily deduced from the usual form, replacing $\dot{e}$ by $C'_0:\dot{e}$ (rheological term) and v_g by $J_N.v_g$ or $J_C.v_g$ (friction term). In eq.(14), σ_n is an estimate at the previous iteration. The model has two versions:

- incremental, with a one-step forward Euler coordinate updating scheme;
- steady state; the free surface updating scheme, based on the streamline method with correction of the contact surface, is described in (Hacquin et al 1995).

3 PRELIMINARY TESTS: UPSETTING

The incremental version has been used to model two kinds of upsettings with flat tools. The first one concerns a very tall cylinder (initial height / diameter ratio = 4); the second one is a thinner square (initial height / length ratio = 0.25). In both cases, an isotropic Norton Hoff behaviour has been selected, with K = 1. and m = 0.1 , with no thermal coupling, no work-hardening or softening. The friction law is of the anisotropic Coulomb-Norton type (Coulomb's law at low contact pressure, limited by Norton's friction law at high pressure).

3.1 *Cylinder upsetting*

A cylinder with diameter 2 and height 8 is reduced to height 2. By symmetry, only 1/8th is modelled. The upsetting speed is $V_z = -2$. The friction coefficients used are listed in table 1. x and y are two orthogonal directions in a section parallel to the tools. The ratio of μ_x and μ_y is 1 (isotropic friction), 2 (moderate anisotropy) and 10 (academic example).

Table 1. Friction coefficients used in cylinder upsetting. $p_f = 0.05$ in all cases

case nr	μ_x	μ_y	α_{fx}	α_{fy}
1	0.4	0.4	1.0	1.0
2	0.267	0.533	0.667	1.333
3	0.036	0.364	0.182	1.818

Figure 1 shows views from above of the three upset cylinders at maximal reduction. With isotropic friction, axisymmetry is exactly preserved; increasing the μ_y/μ_x ratio ovalizes the initial contact surface. However, the equatorial section remains roughly circular. Figure 2 gives the evolution of two perpendicular vertical sections for case 3, at three different stages of upsetting (41%, 62.5% and 75% reduction). Up to 60% reduction, ovalization is significant only in the vicinity of the contact surface; it extends to the equator only when the cylinder is flat enough. It can therefore be concluded that, contrary to rheological anisotropy (Montmitonnet & Chenot 1995), the geometrical effects of frictional anisotropy are restricted to areas close to the contact surface.

In this configuration, the influence of friction, therefore of frictional anisotropy, on stresses, strains and forces is negligible.

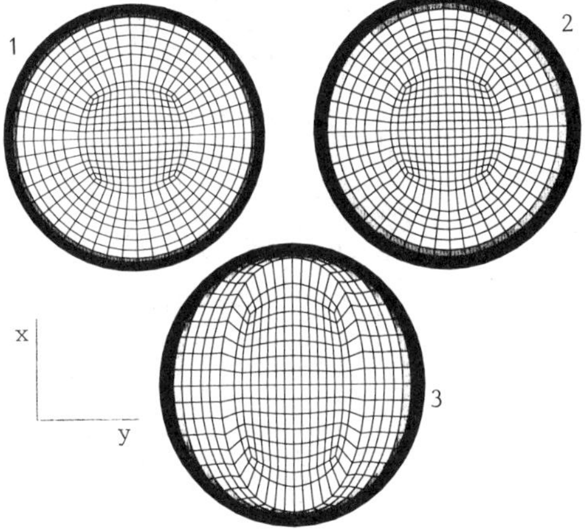

Figure 1 : comparison of cylinders upset (75%) with isotropic (case 1) or anisotropic friction (cases 2, 3)

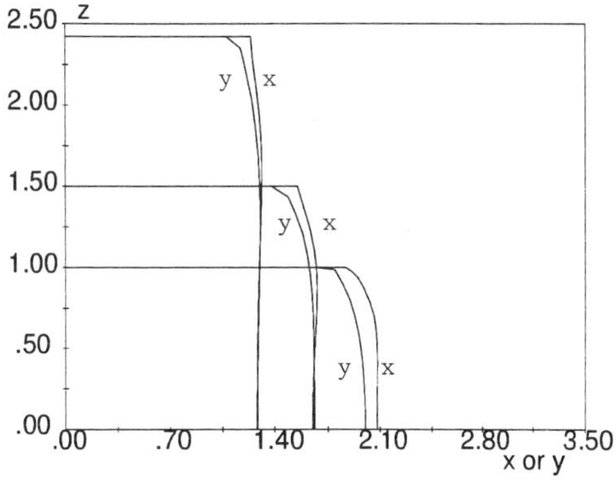

Figure 2 : cylinder upsetting under anisotropic friction, case 3 (see table 1). Comparison of sections in planes xOz (noted x) and yOz (noted y) at 41.5%, 62.5% and 75% reduction (initial half height 4 mm).

3.2 *plate upsetting*

A thin plate should therefore display much more conspicuous effects of frictional anisotropy. A square plate with side length 2 and height 0.5 is reduced to height 0.2 . By symmetry, only 1/8th is modelled. The upsetting speed is $V_z = -2$.

Table 2. Friction coefficients used in plate upsetting. $p_f = 0.05$ in all cases

case nr	μ_x	μ_y	α_{fx}	α_{fy}
1	0.2	0.2	1.0	1.0
2	0.133	0.267	0.667	1.333
3	0.036	0.364	0.182	1.818

Figure 3 shows views from above of the three upset plates at full reduction. Very large length differences result even from moderate anisotropic friction. As expected, large effects of frictional anisotropy are found with thin pieces, a configuration globally quite sensitive to friction.

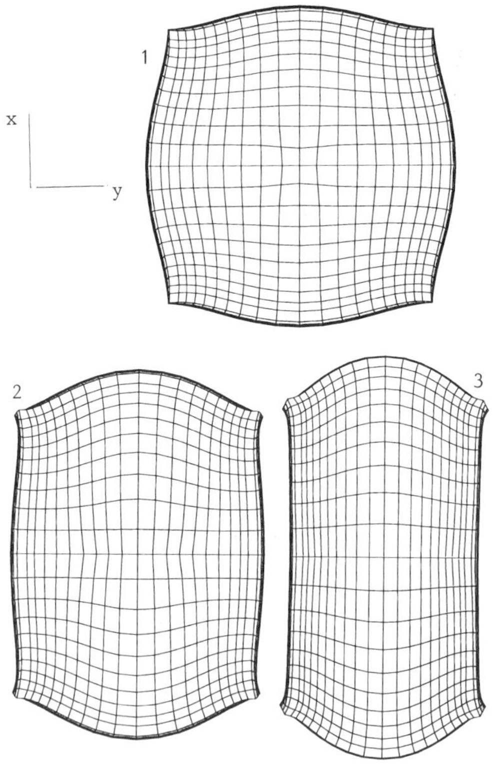

Figure 3 : comparison of plates upset (50%) with isotropic (case 1) or anisotropic friction (cases 2, 3)

Here, frictional anisotropy has a significant effect on normal stress and upsetting force: a low μ_x makes flow easy in the x direction, and the corresponding slope of the friction hill is reduced.

4 APPLICATIONS TO HOT ROLLING

From what precedes, it may be inferred that very different effects may be observed depending on the roll bite geometry. Two operations will therefore be examined, using the same material behaviour and type of friction law as in paragraph 3. In this paragraph, Ox is the rolling direction, y the transverse direction and z the vertical, through-thickness direction.

4.1 *Square bar rolling*

A 200 mm square bar is rolled (reduction 30%) with 800 mm diameter flat rolls. 1/4th is modelled by virtue of symmetries, using the steady state version. Bite length is ≈ 150mm, width 200mm, average thickness 170mm. Rolling speed is 2m/s. The friction in the longitudinal direction (μ_x) is kept equal to μ=0.4 (to avoid skidding), while the transverse friction coefficient (μ_y) varies (table 3).

Table 3. Friction coefficients used in bar rolling. $p_f = 0.10$ in all cases

case nr	μ_x	μ_y	α_{fx}	α_{fy}
1	0.4	0.4	1.0	1.0
2	0.4	0.2	1.333	0.667
3	0.4	0.8	0.667	1.333

Figure 4 gives the symmetrized configuration studied; the initial mesh is pictured, along with the converged mesh (5994 nodes, 4672 8-node bricks).

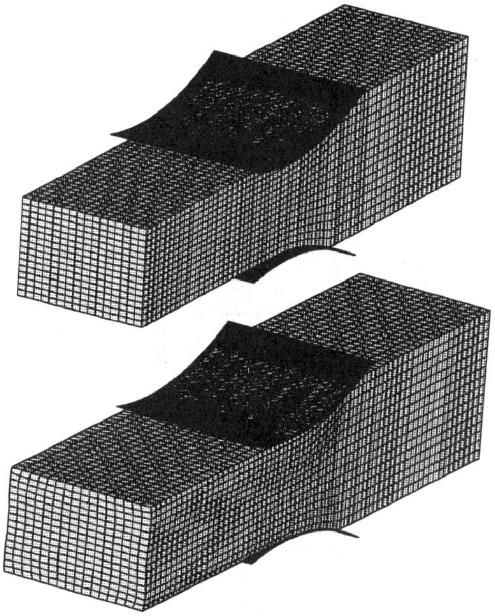

Figure 4 : initial and final mesh of piece and roll, steady state modelling of square bar rolling (case 1).

Figure 5 compares cross sections (in the yOz plane; x is the rolling direction). Again, it is found that increasing transverse friction hinders spread near the contact surface, but has little influence on spread in the horizontal plane of symmetry xOy. It should be noted that, in this type of rolling passes, increasing (isotropic) friction results in more spread.

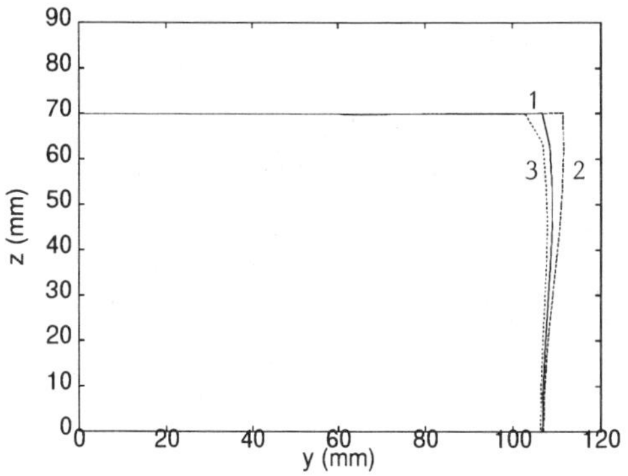

Figure 5 : rolling of a square bar, 1/4th of a cross section (initial half-widh=100mm). Comparison of isotropic (case 1) with anisotropic friction (case 2: transverse friction smaller than longitudinal friction; case 3: reverse).

Again, as expected when rolling this type of product, the influence of the transverse friction on normal stress, force and torque is negligible. Figure 6 thus gives the normal and (longitudinal) friction stresses on symmetry plane xOz for the 3 cases modelled; in this plane of course, $\tau_y = 0$.

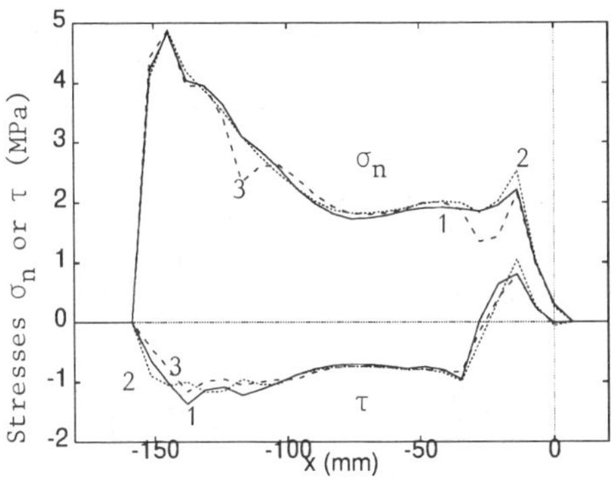

Figure 6 : contact stresses along the roll bite in plane xOz (friction hill) for cases 1, 2 and 3 of table 3.

4.2 *Hot strip rolling*

A 5x400 mm strip is rolled (reduction 20%) with 800 mm diameter flat rolls. 1/4th is modelled by virtue of symmetries, using the steady state version. Bite length is ≈ 19mm, width 400mm, average thickness 4.5 mm. Rolling speed is 2 m/s. The friction in the longitudinal direction is kept equal to μ_x=0.2 , while the transverse friction coefficient (μ_y) varies according to table 4.

Table 4. Friction coefficients used for strip rolling. p_f = 0.10 in all cases.

case nr	μ_x	μ_y	α_{fx}	α_{fy}
1	0.2	0.2	5.0	5.0
2	0.2	0.1	6.667	3.333
3	0.2	0.4	3.333	6.667

Figure 7 gives the configuration modelled (1/4th by symmetry); only the initial mesh is pictured (4615 nodes, 3360 8-node bricks), with the active part of the roll. Note the refinement in the vicinity of the roll bite, which is possible only in a eulerian model.

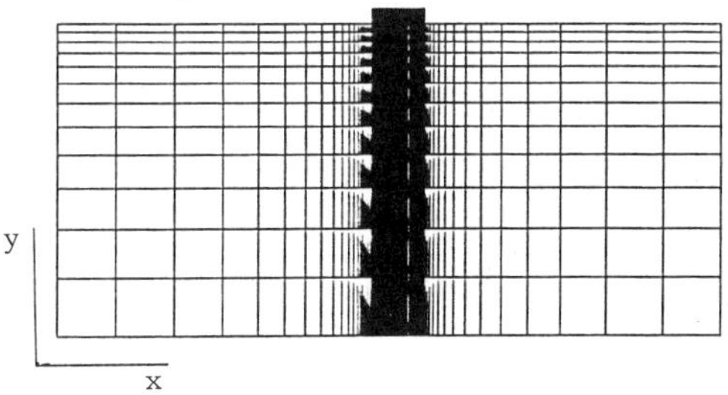

Figure 7 : initial mesh of piece and roll, steady state modelling of hot strip rolling.

Figure 8 compares cross sections (in the yOz plane). Differences are found, but it should be noted that spread is negligible in all three cases (less than 0.1%). Small discrepancies are observed on the friction hills on the symmetry plane xOz (figure 9). However, the transverse profile of rolling force f per unit width is significantly affected (figure 10): the larger μ_y is, the smaller f is in the centre, but the larger it is near the edges. This will result in changes in the roll deformation profile, thence in strip profile.

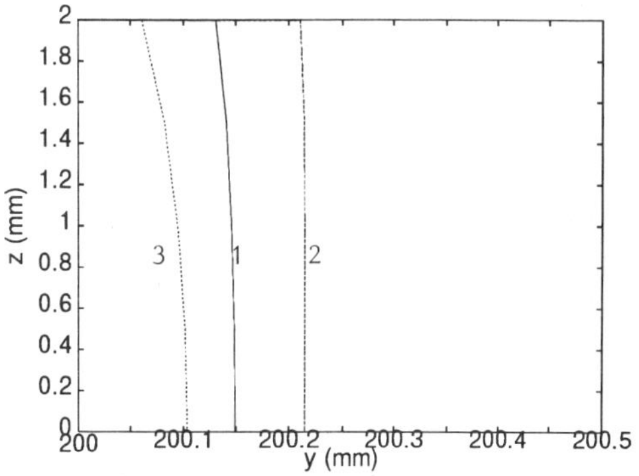

Figure 8 : hot strip rolling, 1/4th of a cross section (initial half-widh=200mm). Comparison of isotropic (case 1) with anisotropic friction (case 2: μ_y smaller than μ_x; case 3: μ_y larger than μ_x).

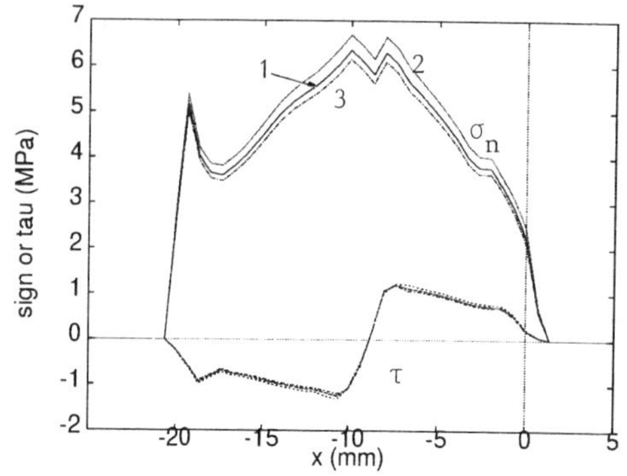

Figure 9 : contact stresses in strip rolling, along the roll bite in plane xOz (friction hill) for cases 1, 2 and 3 of table 4.

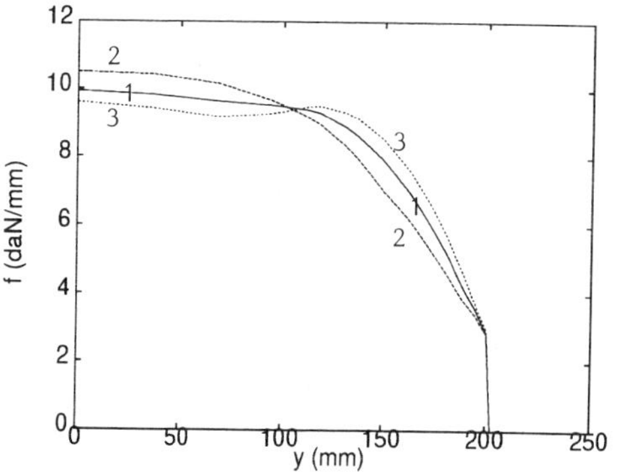

Figure 10 : distribution of the rolling force per unit width f in strip rolling.

The global rolling force and torque are not significantly affected. Differences are observed in the transverse velocity (under small transverse friction, the edge zone concerned by outward flow expands towards the centre), but also in the normal stress profiles (3D friction hill) reflected in integral form in figure 10, and in the longitudinal stress σ_{xx}.

5 CONCLUSION

A simple anisotropic friction law has been presented, its implementation in a 3D FEM with anisotropic viscoplastic material behaviour has been briefly described. It has been shown both in forging and rolling simulations that significant effects may be expected if (reasonable) frictional anisotropy is present. The flow competition in 3D forging simulations is all the more affected as the product is thinner. In 3D rolling modelling, the effect on spread has been found significant in situations where spread is anyway large (bar or thick slab rolling). In thin strip rolling, the dominant effect is probably on rolling force distribution; significant consequences on strip profile may be expected.

Experimental work on rolling equipment is now needed to investigate the existence and the amount of frictional anisotropy in real rolling situations.

REFERENCES

Bay, F. & J.L. Chenot 1990. Introduction of anisotropy in the numerical simulation of the forging process. Eng. Comput. 7: 235-240

Bay, N. & T. Wanheim 1976. Real area of contact and friction stress at high pressure sliding contact. Wear 38 : 201-209

Beaudoin, A.J. 1993. 3D Deformation process simulation with explicit use of polycrystal plasticity models. PhD Thesis, Cornell University, Ithaca, New York, U.S.A.

Chenot, J.L. & M. Bellet 1992. The viscoplastic approach for the finite element modelling of metal forming processes. In "Numerical modelling of material deformation processes", P. Hartley, I. Pillinger, C. Sturgess eds, Springer Verlag, London

Felder, E., 1986. Un modèle de frottement solide anisotrope. C.R. Acad. Sc. Paris T303, Série II, 8

Felder, E., 1988. Etude expérimentale de l'anisotropie de frottement de tôles d'acier doux. J. Méca. Th. Appl. 7, 4 : 479-504

Hacquin, A., P. Montmitonnet & J.Ph. Guillerault 1995. Coupling of roll and strip deformation in 3D simulation of hot rolling. Proc. NUMIFORM 95, P.R. Dawson et al., ds, Balkema, Rotterdam.

He, Q.C. & A. Curnier, 1993. Anisotropic dry friction between two orthotropic surfaces undergoing large displacement. Eur. J. Mech. A/ Solids 12, 5: 631-666

Karabin, M.E., R.E. Smelser & R. Becker 1992. Quasi 3D rolling simulations with anisotropy. Proc. Numiform 92, J.L. Chenot et al. eds, Balkema, Rotterdam.

Mathur, K.K. & P.R. Dawson, 1989. On modelling the development of crystallographic texture in bulk forming processes. Int. J. Plasticity 5: 67-94

Montmitonnet, P. 1993. Lois de frottement et déformation plastique. Matériaux et Techniques, nr 1,2,3: 8-21

Montmitonnet, P. & J.L. Chenot 1995. Introduction of anisotropy in viscoplastic 2D and 3D finite element simulations of hot forging. J. Mat. Proc. Techn. (in press)

Montmitonnet, P., P. Gratacos & R. Ducloux, 1995. Application of an anisotropic viscoplastic behaviour in 3D FE simulations of hot rolling. Submitted to J. Mat. Processing Technology

Zmitrowicz, A., 1989. Mathematical description of anisotropic friction. Int. J. Solids Structures 25: 837-862

Simulation of Materials Processing: Theory, Methods and Applications, Shen & Dawson (eds)
© 1995 Balkema, Rotterdam. ISBN 90 5410 553 4

Benchmarks for finite element modeling of cold forging processes with elasto-plastic microvoided materials

Philippe Picart, Ludovic Lazzarotto & Jérôme Oudin
Industrial and Human Automatic Control and Mechanical Engineering Laboratory, CNRS Research Unit, MECAMAT, University of Valenciennes and Hainaut-Cambresis, France

ABSTRACT : Two cold forging benchmarks which involve an elasto-plastic microvoided material are exposed in order to compare and validate the finite element modelisation and to show up the damage influence to develop and optimize cold forging processes. In that aim, two specific tests are proposed, the first one concerns the upsetting test of a cylinder which is clamped at its top and bottom ends, and second one a wheel bulk forging process. The numerical results obtained for a microvoid free material and microvoided materials with various damage parameter values are analysed and discussed. The evolution curves for local variables like the microvoid volume fraction, the effective plastic strain and the hydrostatic stress and for global variables like the workpiece geometry and the forging load are exposed in details.

1 INTRODUCTION

Many mechanical parts are obtained by cold forging processes. Due to simplicity, such processes are well adapted to mass production but needs to be improved to insure a good quality of final products. Most experiments are often very expensive and not adapted to the industrial requirements. In that purpose, finite element simulations are more and more frequently used to develop new processes or optimize them. In fact, most engineering materials such as steels and aluminium alloys are not perfected. Therefore it is essential to take into account the real constitution of material microstructure which is often microvoided and inclusions contents. An efficient prediction of the damage evolution during the forging sequence allows us to avoid defects initiation and control the final physical characteristics of the mechanical parts.

Up to now, a lot of microscopic observations have shown three main mechanisms to describe the microvoid evolution (Bennani 1993). First, new microvoids may appear by nucleation that mainly occurs either by decohesion at inclusion-matrix interfaces, or by inclusions or matrix fractures. Second, when the plastic strain goes on, the growth of existing microvoids is observed. Third, the coalescence of microvoids may appear and leads to the ductile rupture of the material. Thus, adapted constitutive equations have been proposed for each mechanism. The corresponding finite element formulations have been implemented in ASTRID code (Picart 1992).

In this paper, two benchmarks adapted to cold forging processes are proposed. The first one concerns the upsetting test of a cylinder which is clamped at its top and bottom ends in a tubular rigid die. For this original test, all required conditions for a significant development of the microvoid volume fraction are observed. The second one concerns a wheel bulk forging sequence that represents a good illustration of the industrial cold forging processes.

For each benchmark, several computations are achieved for a microvoid free material and several microvoided materials with various damage parameter values. So the influence of the damage evolution can be quantified and compared with a microvoid free material. A detail analysis of the evolution curves for local variables as the microvoid volume fraction, the effective plastic strain and the hydrostatic stress, and for global variables as the workpiece dimensions and the forging loads are proposed.

2 UPSETTING OF A CLAMPED CYLINDER

The upsetting test of a cylinder which is clamped at its top and bottom ends in a tubular rigid die represents an original test to evaluate and compare finite element modelisations adapted to cold forging processes including material damage evolution with large plastic strains. So the main mechanical consequences of the damage devolpment in mechanical parts can be show up .

Initially, the cylinder is 40 mm height and 15 mm diameter. Then, a 8.75 mm axial displacement is prescribed for the upper and lower punches which move towards each other. Thus the final height cylinder is reduced to 22.5 mm. In order to obtain good forging conditions, a fillet of 2 mm radius is imposed between the internal surface and the upper and lower parts of the tubular rigid dies (Figure 1).

The material involved is a 1070 steel which is annealed to eliminate the residual stresses which can appear during the achievement of the initial cylinder.

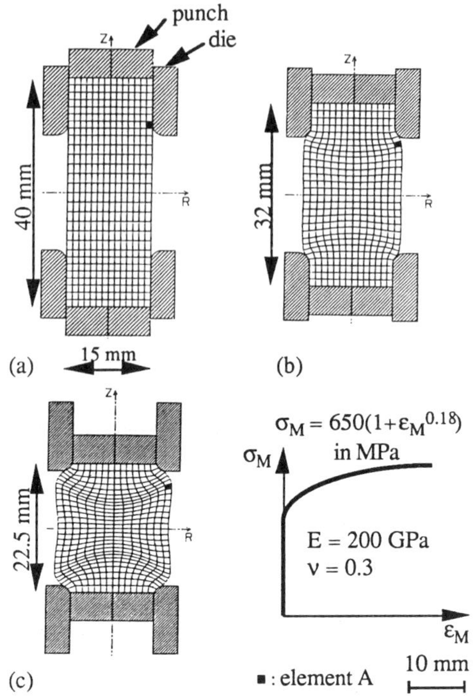

Figure 1. Upsetting test of a clamped cylinder. (a) initial mesh, (b) deformed mesh at 4 mm axial displacement of the punches, (c) final deformed mesh.

The elastic behaviour is given by a 200 GPa Young's modulus and a 0.3 Poisson's ratio. The plastic behaviour is described by the matrix yield stress evolution (1).

$$\sigma_M = 650\ (1+\varepsilon_M^{0.18}) \text{ in MPa} \qquad (1)$$

Due to symmetrical properties, the finite element analysis is limited to the upper right part of the workpiece and contained 98 axisymmetric Q4 elements with four integration points.

Sticking contact conditions at punch-cylinder interfaces and frictionless contact conditions at die-cylinder interfaces are used.

Four finite element computations are achieved. The first one (case 1) relies on a microvoid free material, the others concerns microvoided materials (cases 2-4). The finite element modelisation used to describe the microvoids evolution requires eight material parameters. Three of them are lied to the growth of the microvoid volume fraction; f_i, the initial microvoid volume fraction, q_1 and q_2, the Gurson's potential parameters. The nucleation phenomenon needs three material parameters; ε_N, the effective plastic strain at incipient nucleation, f_N, the potential nucleated microvoid volume fraction related to the

Table 1. Upsetting test of a clamped cylinder. Material parameters.

case	f_i	ε_N	f_N	S_N	f_{cr}	$\Delta\varepsilon$	q_1	q_2
1	microvoid free material.							
2	0.001	0.2	0.04	0.1	0.2	0.25	1.5	1.
3	0.01	0.2	0.04	0.1	0.2	0.25	1.5	1.
4	0.01	0.2	0.04	0.1	0.02	0.25	1.5	1.

inclusions volume fraction and S_N, the standard deviation of the inclusions distribution. Finally the coalescence modelisation uses two parameters; f_{cr}, the critical microvoid volume fraction at incipient coalescence and $\Delta\varepsilon$, the effective plastic strain increment leading to ductile rupture. The damage material parameter values used in finite element computations are summarized in Table 1.

The numerical results are presented through the analysis of local variables evolutions ; the effective plastic strain, the hydrostatic stress and the microvoid volume fraction, and global variables evolutions ; the geometry of the deformed cylinder and the axial loading.

2.1 *Local mechanical variables*

The final effective plastic strain distributions in the cylinder are exposed for all cases on figure 2. The maps obtained are very similar. The maximum values are always located at the diagonal part of the deformed cylinder. The effective plastic strain gradient between the diagonal part and the contact surface with the upper punch is more important than between the diagonal part and the equatorial diameter of the deformed cylinder. The most significant differences are obtained between case 1 and case 2, for which the final ranges are respectively : 0.004 to 0.941 and 0.003 to 0.948.

For the final hydrostatic stress maps, it appears that distributions are rather similar for microvoided materials, whereas some differences are noticed between the microvoid free material and microvoided materials (Figure 3). Nevertheless, the minimum value is always located at the upper and lower parts of the cylinder and a tension zone is observed in the vicinity of the maximum diameter of the deformed cylinder. The refined analysis of the tension zone shows that the maximum value is accurately located at the maximum diameter of the deformed cylinder for cases 2 and 3, and decreases when the initial microvoid volume fraction goes on. Indeed, the maximum value of the hydrostatic stress goes from 244 MPa in case 2, to 241 MPa in case 3. In the same way, the final maximum hydrostatic stress is located at the external surface of the cylinder but above the maximum diameter of the deformed cylinder. For case 4, the maximum value also decreases and goes from 262 MPa in case 1 to 246 MPa in case 4.

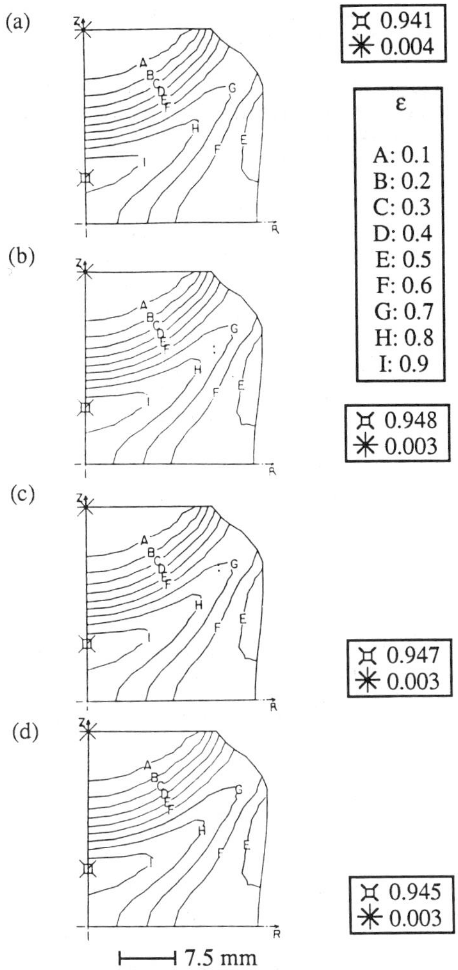

Figure 2. Upsetting test of a clamped cylinder. Final effective plastic strain distributions. (a) case 1, (b) case 2, (c) case 3, (d) case 4.

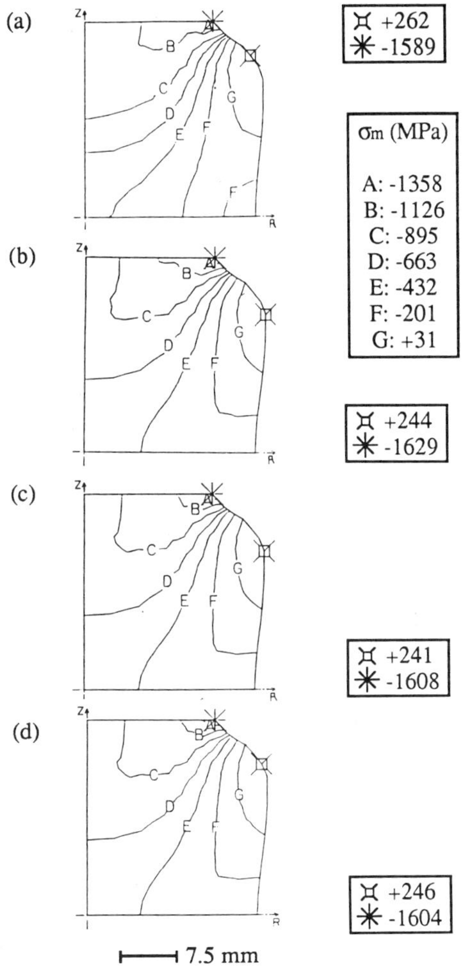

Figure 3. Upsetting test of a clamped cylinder. Final hydrostatic stress distributions. (a) case 1, (b) case 2, (c) case 3, (d) case 4.

As expected, a tension state and an important value of the effective plastic strain at the vicinity of the maximum diameter of the deformed cylinder permit the evolution of a damage zone in the same part of the cylinder (Figure 4).

The final maximum values of the microvoid volume fraction are equal to 0.01498, 0.02446 and 0.29734 respectively for cases 2, 3 and 4. Considering the large increase of the microvoid volume fraction, especially in case 4, it is very interesting to observe the evolution of the microvoid volume fraction during the forging sequence. The microvoid volume fraction evolution is observed at one integration point of the element A (Figure 1) which is located at the maximum diameter of the deformed cylinder. (Figure 5).

The microvoid volume fraction increases if a tension state is observed, it is important to specify that the tension state in element A appears at a 3.75 mm axial displacement of the punches. Moreover, the effective plastic strain in element A reaches the effective plastic strain at incipient nucleation, 0.2, (Table 1), at a 2.77 mm axial displacement of the punches. Consequently, as soon as the tension state in element A appears the microvoid volume fraction increases by both nucleation and growth phenomena (Figures 5 a, b and c). Then, as the critical value of the microvoid volume fraction at incipient coalescence, 0.2 (Table 1) is never reached in cases 2 and 3, the microvoid volume fraction always increases by both nucleation and growth until 0.01498 in case 2 and 0.02446 in case 3 (Figures 5 a and b). On the contrary, in case 4, the microvoid volume fraction reaches the critical microvoid volume

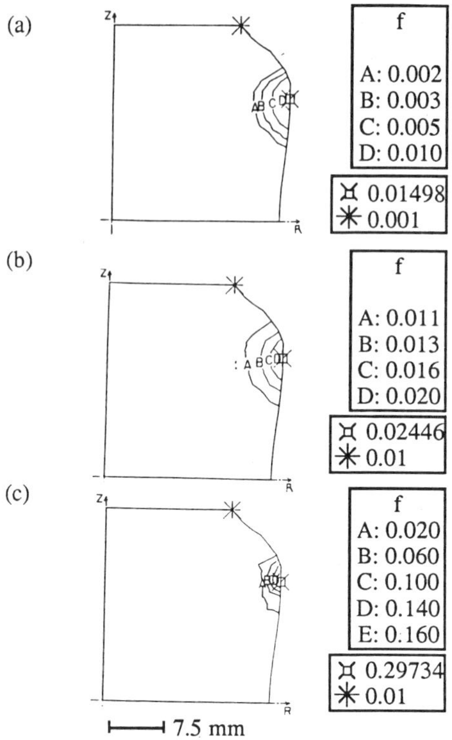

Figure 4. Upsetting test of a clamped cylinder Final microvoid volume fraction distributions. (a) case 2, (b) case 3, (c) case 4.

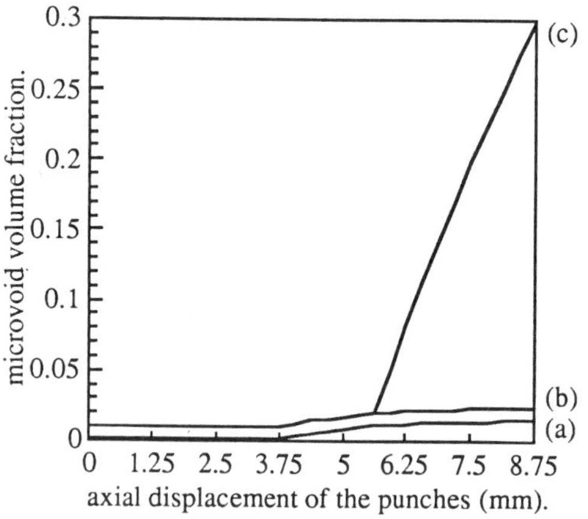

Figure 5. Upsetting test of a clamped cylinder. Microvoid volume fraction evolutions at element A : (a) case 2, (b) case 3, (c) case 4.

fraction at incipient coalescence, 0.02, at a 5.625 mm axial displacement of the punches, thus the microvoid volume fraction evolution strongly increases since the coalescence of microvoids is activated (Figure 5 c).

Table 2. Upsetting test of a clamped cylinder. Geometrical modifications of the cylinder.

case	equatorial radius (mm)	maximum radius (mm)	maximum radius axial position (mm)
1	9.995	10.479	7.940
2	9.962	10.521	7.968
3	9.968	10.510	7.958
4	9.964	10.517	7.953

Table 3. Upsetting test of a clamped cylinder. Axial loading modifications.

case	final axial loading (N)	final relative decrease (%)
1	356428	-
2	347794	2.4
3	343093	3.7
4	342432	3.9

When the axial displacement of the punches advances in the range 0 to 5.625 mm in cases 2 and 3, the microvoid evolution curves are very similar, moreover the gap between the two curves is roughly constant and goes from 0.009 to 0.0082 (Figure 5 a and b).

2.2 *Global mechanical variables*

The evolution of the damage zone in the vicinity of the maximum diameter of the deformed cylinder causes two main geometrical modifications compared with the microvoid free cylinder; first, the location of the final maximum diameter and second, the reduction of the final value of the equatorial diameter (table 2).

The evolution of the axial loading curves during the forging sequence is also affected by the damage development. A roughly constant gap between the curves is observed during the process. The analysis of the final values compared with the microvoid free material enables to estimate the damage influence (Table 3).

3 WHEEL BULK FORGING

The wheel bulk forging is a good illustration of cold forging process. The finite element computations achieved with microvoided material can bring many informations to optimize the design of the dies and define the initial geometry of the workpiece in order to avoid the appearance of macroscopic defects and control the physical and mechanical characteristics of the final products. The main interest of this investigation is to show up the influence and the evolution of the damage zone in the workpiece during the forgng process.

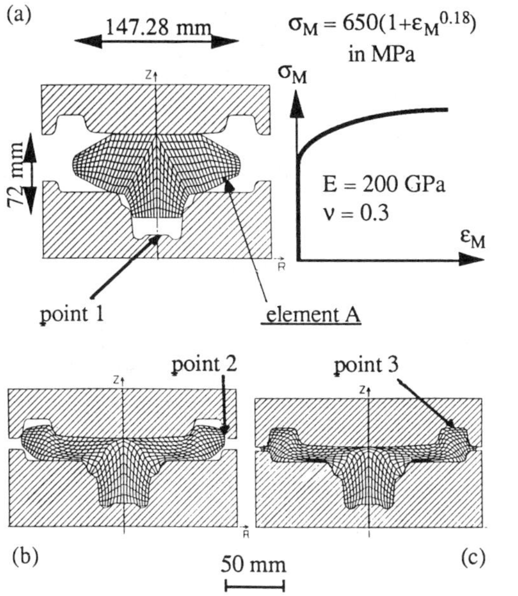

Figure 6. Wheel bulk forging. (a) initial mesh, (b) deformed mesh at about 31 mm axial displacement of the upper die, (c) final deformed mesh.

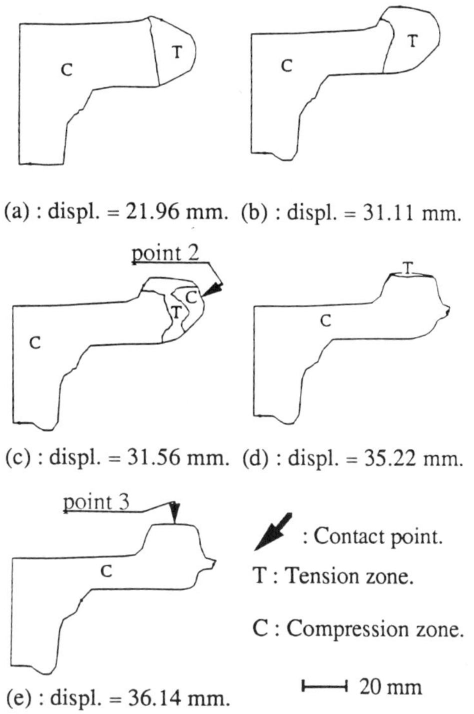

Figure 7. Wheel bulk forging. Evolution of the tension zone during the forging sequence.

Initially, the workpiece is 72 mm height and 147.28 mm diameter, the related finite element mesh is composed of 200 axisymmetric Q4 elements with four integration points. Sliding contact conditions are imposed at die/workpiece interfaces. The upper die goes down until 36.6 mm of axial displacement, a good filling of the dies' cavities is observed (Figure 6).

As for the previous benchmark, the material involved is a 1070 annealed steel. Two finite element computations are achieved ; first, with a microvoid free material and second with a microvoided material for which the material parameters values are : f_i=0.001, ε_N=0.2, f_N=0.04, S_N=0.1, f_{cr}=0.2, $\Delta\varepsilon$=0.25, q_1=1.5 and q_2=1.0.

The finite element simulations require 400 loading increments and five local remeshings with numerical interpolation of the mechanical variables.

During the forging sequence, two main phases are observed for the workpiece deformation; first, a radial expansion of the peripheral part of the wheel and the filling of the central cavity of the lower die (Figure 6 b) and second, the filling of the peripheral cavity of the upper die (Figure 6 c).

3.1 *Local mechanical variables*

As previously observed for the upsetting of a clamped cylinder, the evolution and the distribution of the effective plastic strain are rather close. The final ranges are ; 0.017 to 2.180 for the microvoid free material and 0.053 to 2.262 for the microvoided material.

The proposed analysis is focused on the comparison between the hydrostatic stress and the microvoid volume fraction evolutions during the forging process.

The radial expansion of the workpiece during the first phase leads to a tension state within the peripheral part of the wheel (Figure 7 a and b), therefore the microvoid volume fraction can increase. At the beginning of the second phase, the contact between the workpiece and the upper die at point 2 (Figure 6 b) reduces the tension zone (Figure 7 c). The microvoid volume fraction again increases and reaches its maximum value, 0.0207, at a 34.31 mm axial displacement of the upper die (Figure 8 a). Then the tension zone is limited at the upper peripheral part of the workpiece (Figure 7 d). Finally the contact with the upper die at point 3 (Figure 6 c) leads to a global compression state (Figure 7 e), so the damage development is stopped. The maximum value of the microvoid volume fraction is always equal to 0.0207 (Figure 8 b).

A refined analysis of the damage zone is now presented through the comparison between the evolution of the microvoid volume fraction (Figure 9 a) and the hydrostatic stress (Figure 9 b) at one integration point of the element A (Figure 6 a). This

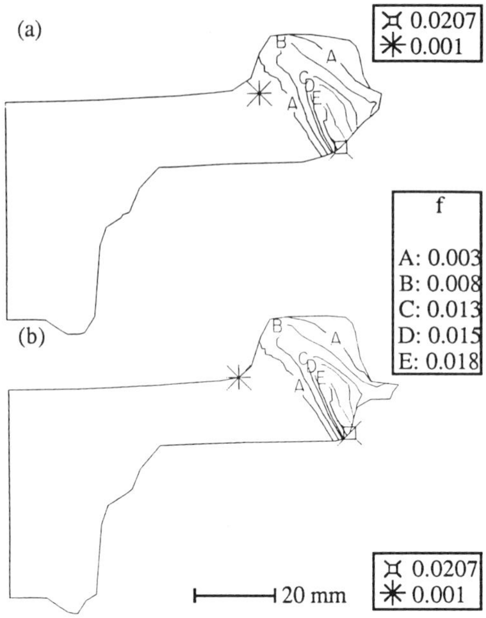

Figure 8. Wheel bulk forging. Microvoid volume fraction distributions. (a) at 34.31 mm axial displacement of the upper die, (b) final distribution.

analysis clearly shows the relation between these local mechanical variables. Three steps are observed. In a first step, a tension state produces the growth of existing microvoids. When the axial displacement of the upper die is about 25 mm, the effective plastic strain reaches 0.2, the value of the effective plastic strain at incipient nucleation, ε_N. The second step begins where nucleation and growth of microvoids are activated, the hydrostatic stress being positive. The third step starts at a 32 mm axial displacement of the upper die when the hydrostatic stress becomes negative, consequently the evolution of the microvoid volume fraction is stopped.

3.2 *Global mechanical variables*

The geometrical modifications can be analysed in four parts. First, the contact between the workpiece and the lower die at the point 1 (Figure 6 a) is observed at a 19.215 mm axial displacement of the upper die for the two materials. Second, the contact with the upper die at the point 2 (Figure 6 b) is observed at a 30.927 mm axial displacement of the upper die for the microvoid free material and 31.110 mm for the microvoided material. Third, the opposite process is noticed for the contact between the wheel and the upper die at the point 3 (Figure 6 c). Indeed, the contact at the point 3 is obtained when the axial displacement of the upper die reaches 35.68 mm for the microvoided material and 35.78 mm for the microvoid free material. Fourth, the final filling ratio

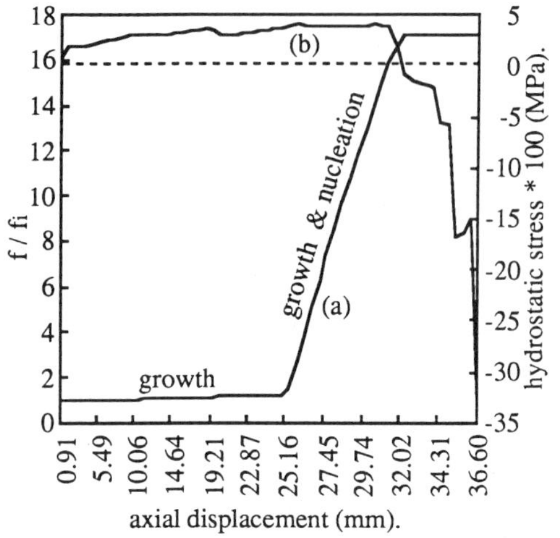

Figure 9. Wheel bulk forging. (a) microvoid volume fraction evolution in element A versus axial displacement of the upper die, (b) hydrostatic stress evolution in element A versus axial displacement of the upper die.

of the peripheral dies' cavities is better with the microvoided material.

The analysis of the axial loading evolutions during the forging process indicates that the required power for the microvoided material is less important than for the microvoid free material.

4 CONCLUSION

The upsetting test of a clamped cylinder and the wheel bulk forging sequence represents two efficient benchmarks well adapted to test and compare the damage finite element modelisation for cold forging processes.

The detail analysis of the finite element computations achieved by ASTRID code with microvoid free material and various microvoided materials permits to evaluate first, the influence of the damage material parameters and second the influence of damage on local variables as the effective plastic strain and the hydrostatic stress, but also on global variables as the workpiece geometry and the forging loading. These results show the interest of the finite elememt analysis including damage effects to develop and optimize the cold forging sequences without defects initiation.

REFERENCES

Bennani, B., Picart, P. and Oudin, J. 1993. Some basic finite element analysis of microvoid, nucleation, growth and coalescence, *Engineering computation* s 10: 409-421.

Picart, P., Bennani, B. and Oudin, J. 1992. Finite element simulation of void nucleation growth and coalescence in isotropic standard elasto-plasticity : application to cold forging, *Journal of Material Processing Technology* 32: 179-188.

Simulation of Materials Processing: Theory, Methods and Applications, Shen & Dawson (eds)
© 1995 Balkema, Rotterdam. ISBN 90 5410 553 4

Dislocation model for work hardening and recrystallization applied to the finite-element simulation of hot forming

M. Pietrzyk
Akademia Gorniczo-Hutnicza, Krakow, Poland

C. Roucoules
Monash University, Clayton, Vic., Australia

P.D. Hodgson
BHP Research, Melbourne Laboratories, Mulgrave, Vic., Australia

ABSTRACT: Application of the internal variable model to the simulation of hot forming is presented. The model is based on the numerical solution of differential equations describing the time dependence of three internal variables: the dislocation density, the recrystallized volume fraction and the grain size. A method of discretization of the problem and simulation of the metallurgical events of hardening, recovery, recrystallization and grain growth is presented. The model is implemented in the rigid-plastic finite-element program for rolling. The internal variable equations are solved along the flow lines during optimisation of the power functional. Advantages of the approach include accounting for the history of the process in the constitutive law and an ability to simulate complex microstructural phenomena including static, dynamic and metadynamic recrystallization and recovery. Validation of the model and typical examples of simulation are presented in the paper.

NOTATION

b - Burgers vector
D - austenite grain size
D_r - recrystallized grain size
d_s - dislocation cell size
G - volume distribution of dislocation density
l - dislocation mean free path
N - number of recrystallized grains per one old grain
n - number of intervals of dislocation densities
Q_m - activation energy for grain boundary mobility
Q_M - activation energy for recovery
Q_s - activation energy for self-diffusion
RV, DRV - static and dynamic recovery
RX, DRX - static and dynamic recrystallization
t - time
X - recrystallized volume fraction
x, y - coordinates
Z - Zener-Hollomon parameter
Δt - time step
$\Delta\varepsilon$ - strain increment
$\Delta\rho_0$ - length of the interval of dislocation densities
$\Delta\rho_i$ - level of the dislocation density in the ith interval
ε, $\dot{\varepsilon}$ - strain and strain rate, respectively
γ - fraction of subgrain boundary which is migrating
μ - shear modulus
ρ - dislocation density
ρ_{cr} - critical dislocation density for nucleation
σ - yield strength
σ_g - grain boundary energy
τ - energy per unit length of dislocation

1 INTRODUCTION

The description of material properties during the simulation of hot metal forming processes has been of interest for a long time. Metals subjected to plastic deformation at high temperatures and strains exceeding the critical strain for recrystallization may behave in various ways, depending on the history of deformation. Descriptions of the stress-strain relationship based on power law or hyperbolic-sine function often fail to describe the hot deformation behaviour. Therefore, in the recent years, modelling the deformation behaviour has been increasingly based on the physical occurrences in the material. The model for hot working and recrystallization based on the dislocation density was suggested by Sandstrom & Lagneborg 1975a & 1975b. More recently, similar approaches include Mecking & Kocks 1981, Estrin & Mecking 1984, Alden 1987 and Brown & Wlassich 1992. All of these publications give a description of the stress-strain relationship which considers the dependence of the process on the strain hardening and fractional softening. The objective of the present work is to implement the dislocation density model into a finite-element program which simulates metal flow in industrial forming processes. A first attempt, limited to the evaluation of the microstructural phenomena during hot rolling, has been presented by Pietrzyk 1993. This did not include the correlation between the microstructure of the metal and the flow stress. This relationship, as well as the validation of the approach, are the aims of the present work.

2 MATHEMATICAL MODEL

The model is based on an assumption that the evolution of stress during plastic deformation is determined by the evolution of dislocation populations. This leads to the concept that hardening is controlled by the competition of storage and annihilation of dislocations, which processes superimpose in an additive manner. The equations used for the evolution of the dislocation density are given in Table 1. Assuming that the mechanical strength of the obstacles to dislocations σ is related to the dislocation density as:

$$\sigma = \alpha\mu b\rho^{0.5} \tag{1}$$

allows us to derive the stress-strain relationship based on recovery processes. The model fails, however, when dynamic recrystallization is involved, or during the simulation of multistage processes when short interpass times result in partial recrystallization. Moreover, the evaluation of the constants in equation (3) - (8) presents difficulties. A suggestion is made at this stage that including more microstructural parameters into the model will allow it to: i) account for dynamic and static recrystallization, ii) validate the model and evaluate the material's constants using a larger data set.

The extension of the model includes the equations describing the kinetics of recrystallization and grain size. To simulate the complex microstructural phenomena, the dislocation density cannot be treated as an average value. Rather, the entire spectrum of the dislocation densities has to be considered. To allow this the distribution function $G(\rho,t)$ (Sandstrom & Lagneborg 1975a & 1975b), defined as a volume fraction which has a dislocation density between ρ and $\rho+d\rho$, is introduced. As a consequence, the equation which describes the evolution of the dislocation populations accounting for recrystallization is :

$$\frac{dG(\rho,t)}{dt} = \phi(\Delta\varepsilon) - g(\varepsilon) - \frac{v\gamma}{D} m\tau\rho G(\rho,t) \tag{2}$$

In equation (2), $\phi(\Delta\varepsilon)$ represents athermal storage (hardening), $g(\varepsilon)$ thermally activated softening (recovery) and $\tau = \mu b^2/2$. This equation is solved, for each interval of dislocation density, together with two equations describing the kinetics of recrystallization and grain growth (see Table 1). The fraction of migrating grain boundary, γ, in equation (6) is assumed to be determined by the nucleation rate at the beginning of the recrystallization and by grain impingement in the final stages. This leads to an assumption that γ is proportional to $X(1-X)$. All formulae used in the model are given in Tables 1 and 2. The differential equations describing evolution of dislocation populations, kinetics of recrystallization and grain size are solved in parallel with the simulation of the considered forming process.

3 NUMERICAL SIMULATION

Detailed description of the numerical solution is given by Pietrzyk 1994. Briefly, the approach is based on the division of the whole spectrum of the dislocation densities into a number of intervals measuring $\Delta\rho_0$ (Fig.1) Since all the intervals $\Delta\rho_0$ are equal, the volume fraction with the dislocation density between ρ_i and $\rho_i+\Delta\rho_0$, which is originally given as $G(\rho_i,t)\Delta\rho_0$, will be further referred to as simply G_i. As a consequence, the condition $\sum_{i=1}^{n} G_i = 1$ is always met during the simulation. Internal variables such as grain size and recrystallized volume fraction are given for each interval separately. Moreover, the average dislocation density in the interval is prescribed. As a consequence, the arrays $D(n)$, $\Delta\rho(n)$ and $G(n)$, where n is a number of the intervals, are introduced into the program. The process of deformation is divided into time steps Δt and five cycles of simulation are performed for each step. The process represented by the cycle, the variable which is involved, the direction of the simulation (towards increasing or decreasing dislocation densities), the condition for the cycle to be performed and the equation describing the process are given in Table 1. Hardening increases the level of the dislocation densities in the interval considered by $\Delta\rho_\varepsilon$ and, eventually, moves the distribution function towards right. Contrary, recovery decreases the level of the dislocation densities by $\Delta\rho_r$ and, eventually, moves the distribution function towards left. Both of these cycles may eventually shift the volume fraction G_i to the next (hardening) or previous (recovery) interval. New variables for hardening are calculated from:

$$D_{i+1} = \frac{D_{i+1}G_{i+1} + D_i G_i}{G_{i+1} + G_i}$$

$$\Delta\rho_{i+1} = \frac{\Delta\rho_{i+1}G_{i+1} + (\Delta\rho_i + \Delta\rho_\varepsilon - \Delta\rho_0)G_i}{G_{i+1} + G_i}$$

$$\Delta\rho_i = 0 \qquad G_{i+1} = G_{i+1} + G_i \qquad G_i = 0$$

Table 1. Cycles of the simulation performed at each time step of the solution

no	process	variables	direction	condition	Ref.	equation
1	hardening	$\Delta\rho, \rho$	$\xleftarrow{\rho}$	$\dot{\varepsilon} > 0$	Sandstrom & Lagneborg 1975a	$\frac{d\rho}{d\varepsilon} = \frac{1}{bl}$ (3)
2	recovery	$\Delta\rho, \rho$	$\xrightarrow{\rho}$	always	Sandstrom & Lagneborg 1975a	$\frac{d\rho}{dt} = -2\tau\rho^2 M_0 \exp\frac{-Q_M}{RT}$ (4)
					Estrin & Mecking 1984	$\frac{d\rho}{d\varepsilon} = -\rho k_{20} \exp\frac{Q_s}{RT}$ (5)
3	recrystallization	G, X, ρ	$\xleftrightarrow{\rho}$	$\rho > \rho_{cr}$	Sandstrom & Lagneborg 1975a	$\frac{dX}{dt} = \frac{\nu\gamma\tau}{D} G\rho m_0 \exp\frac{-Q_m}{RT}$ (6)
4	grain refinement	D	$\xleftrightarrow{\rho}$	$\rho > \rho_{cr}$	Sandstrom & Lagneborg 1975a	$\frac{dD}{dt} = -D\frac{dX}{dt}\ln N$ (7)
5	grain growth	D	$\xleftrightarrow{\rho}$	always	Sandstrom & Lagneborg 1975a	$\frac{dD}{dt} = \sigma_g \frac{m_0}{D} \exp\frac{-Q_m}{RT}$ (8)

Table 2. Equations in the model (Roberts & Ahlblom 1978, Sandstrom & Lagneborg 1975a)

cell size (free path)	mobile fraction of boundary	number of new grains per old	critical dislocation density
$l = d_s = \frac{K_d}{Z^q}$ (9)	$\gamma = (0.1 + X)^{q_1}(1 - X)\left(\frac{\rho}{\rho_{cr}}\right)^{q_2}$ (10)	$N = 4\gamma\left(\frac{D}{d_s}\right)^2$ (11)	$\rho_{cr} = \frac{8\sigma_g}{\tau l}$ (12)

By an analogy, the equations concerning $(i-1)^{st}$ interval are written for the recovery cycle (Pietrzyk 1994). The recrystallization cycle is performed on all intervals above the critical dislocation density for nucleation. The recrystallized volume fraction ΔX_i in the time interval Δt is calculated from equation (6) in Table 1. It is subtracted from G_i and appears as fresh material with the lowest dislocation density in the first interval. The following substitutions follow:

$$D_1 = \frac{D_1 G_1 + D_{ri}\Delta X_i}{G_1 + \Delta X_i}; \quad G_1 = G_1 + \Delta X_i; \quad G_i = G_i - \Delta X_i$$

Grain refinement and grain growth are performed for all intervals using equations in Table 1.

4 VALIDATION OF THE MODEL

The main difficulty in the application of the dislocation density model to the simulation of industrial forming processes is connected with the evaluation of the material parameters. It is assumed that these parameters can be determined by fitting the results of the finite element calculations with the measurements of stress-strain curves and kinetics of recrystallization in typical plastometric tests. Since all of these constants have a physical meaning and their dependence on temperature is known, the values can be easily extrapolated to other deformation conditions. The inverse technique described by Pietrzyk et al. 1994 is used here, with the objective function during optimization being defined as a sum of the squares of differences between the measured and calculated values of the considered parameter. The tests available for the validation included the measurements of the fractional softening (Hodgson et al. 1990) and the stress-strain curves (Laasraoui & Jonas 1991) for carbon-manganese steels. Since all of the considered tests have been performed under isothermal conditions at constant strain rate, the model can be validated without including it into the finite element program. Several material constants need to be determined. They are: dislocation cell size coefficients K_d and q, mobility of recovery coefficient M_0 or k_2, mobility of grain boundary coefficient m_0, number of recrystallized grains per old grain N, activations energies for recrystallization Q_m and recovery Q_M, coefficients describing the mobile fraction of the subgrain boundary q_1 and q_2, shear modulus μ, critical dislocation density for nucleation of recrystallization ρ_{cr}, grain boundary energy σ_g, coefficient α in equation (1) and Burgers vector b. Starting values of these constants were taken from the literature (Table 3). The temperature dependence of the shear modulus is given by Frost & Ashby 1982. Some of the material's parameters are known with very good accuracy, therefore, they were not considered as the unknowns. Nine constants (K_d, q, m_0, M_0, Q_m, Q_s, q_1, q_2 and α) were introduced in the optimisation and became the parameters in the inverse technique. The objective function, defined as the error between the measured and calculated softening curves and stress-strain curves, is very complex within the wider range of unknowns. Therefore, the search was performed in the vicinity of the values of the constants published in the literature.

Both versions of the recovery term, one dependent on ρ (eq.(5), Estrin & Mecking 1984) and the second on ρ^2 (eq(4), Sandstrom & Lagneborg 1975a) were considered. It was observed, however, that an approach based on equation (4) with Q_M = 400 kJ/mol and $M_0 = 5.7\ 10^7\ m^2(Ns)^{-1}$ (Roberts & Ahlblom 1978) underestimates the recovery for low temperatures. Therefore, only the recovery term described by equation (5) was used in further analysis.

5 ANALYSIS OF RESULTS

The inverse technique applied to determine the material constants using the stress-strain curves and softening curves yielded the values shown in Table 4. Minimum in the error function was not clearly defined and the searching procedure was slow. Optimisation was also tried using the stress-strain curves and the softening curves separately. The latter gave an apparent activation energy Q_s equal 24 kJ/mol for SRX and 14 kJ/mol for the stress-strain curves. Although the error was now below 5%, recrystallization was underestimated during deformation (Fig.2). Recovery became a dominant factor competing with hardening. Thus, this approach failed to model RX qualitatively during deformation.

The first approach which consisted of performing the optimisation on all experimental results led to an accuracy of 13%. Using the material constants in Table 4, the sensitivity of SRX to strain (Fig.3) agreed with the experimental data. Sensitivity of SRX to temperature, not presented here, shows that

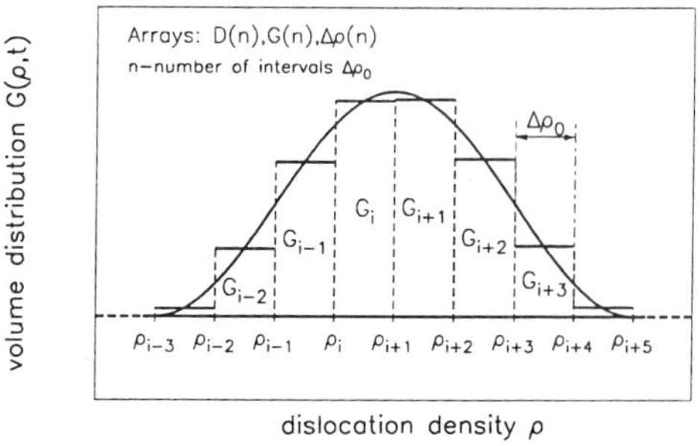

Fig.1. Schematic illustration of discretization of the volume distribution function.

at higher temperatures RX is predicted to be faster than measured. This is believed to be due to the high grain boundary mobility activation energy Q_m which is reported to be 360 kJ/mol (Roberts & Ahlblom 1978). The stress-strain curves are qualitatively in agreement with the experiments. Recovery was found to be an important component of hardening (Fig.4). However, the recrystallization during deformation was not very well fitted, especially at 2 s^{-1} (Fig.5).

6 FINITE ELEMENT CALCULATIONS

The dislocation density model discussed in previous sections has been included into the finite element program for flat rolling. The program, described by Pietrzyk & Lenard (1991), is based on the rigid-plastic flow formulation coupled with the steady state solution of the convection-diffusion equation. The solution of the microstructural model is performed along the flow lines for each iteration in the finite element procedure. The current local values of the strain increments, temperatures and strain rates are used in equations (2), (7), (8) and (9). The yield strength calculated from equation (1) for the mean dislocation density in each element is used in the finite element formulation. This yield strength is not based on the current values of the external variables. It accounts for the development of internal variables

Table 3. Material constants used as starting values

variable	b, m	k_{20}	Q_s, kJ/mol	m_0, $m^3(Ns)^{-1}$	Q_m, kJ/mol	σ_g, Jm^{-1}	q	α	ν
Ref.	Frost & Ashby 1982	Roucoules 1992		Roberts & Ahlblom 1978				Mecking & Kocks 1981	
value	$0.25\ 10^{-9}$	30	25	$3.5\ 10^4$	360	0.8	0.17	1	6

Table 4. Material constants determined from the optimization

variable	K_d, m/s	k_2	Q_s, kJ/mol	m_0, $m^3(Ns)^{-1}$	Q_m, kJ/mol	q_1	q_2	q	α
value	$0.26\ 10^{-3}$	34	20	$1.4\ 10^5$	340	0.5	2.	0.1	1.23

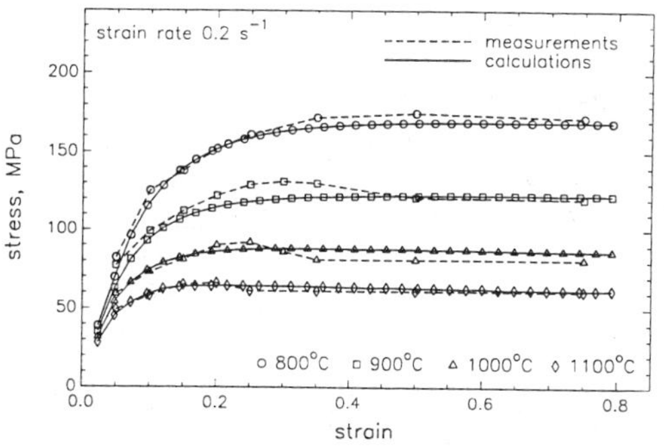

Fig.2. Stress-strain curves calculated for the constants obtained from the one-experiment optimization compared with the measurements, strain rate 0.2 s^{-1}.

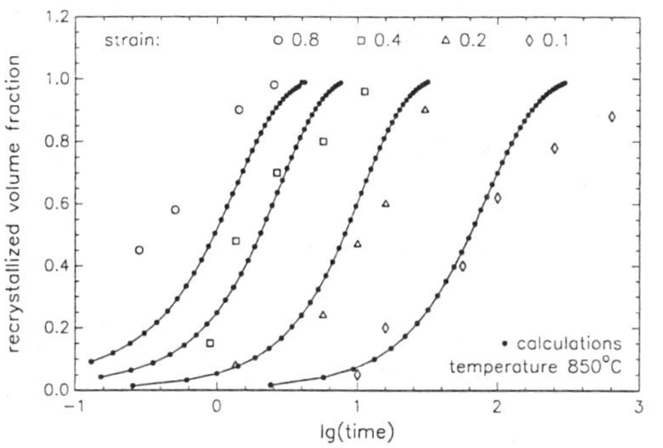

Fig.3. Comparison of calculations for constants in Table 4 with measurements of fractional softening.

and, in consequence, for the history of the process which includes deformation in previous passes.

Continuous hot strip rolling has been chosen to demonstrate the abilities of the model. Two subsequent passes in the finishing train are considered. The steel strip, 4 mm thick, is reduced by 20% and 25% with 0.5 s interpass time. Fig.6 shows the softening curves due to static recrystallization determined in three locations; centre, half distance between the centre and the surface and the surface zone. About 50% softening is predicted at the centre and 10% at the surface. As a consequence, the metal entering the next pass has an increased dislocation density. The calculated distributions of temperature, strain and the dislocation density in the roll gap in the second considered pass is shown in Fig.7. A nonuniform dislocation structure at entry, resulting from the nonuniform softening, affects the metal flow in a pass. Due to larger strains and lower temperatures, the largest dislocation density is observed at the surface zone.

The results were obtained with the internal variable model being used as a constitutive law. Since the rigid-plastic formulation applied here does not require the integration of the constitutive equations, the

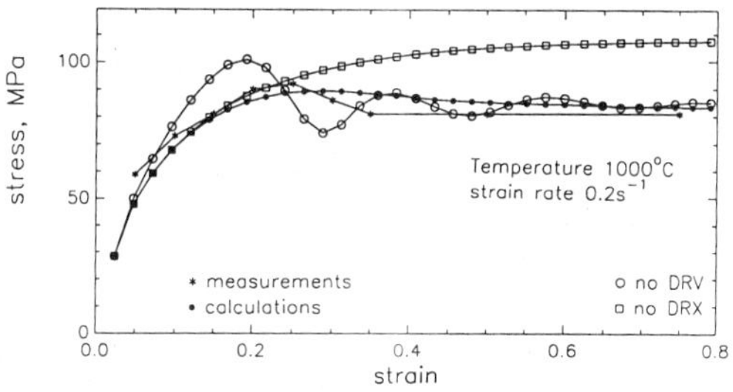

Fig.4. Stress strain curves measured and calculated with various terms in the model being active.

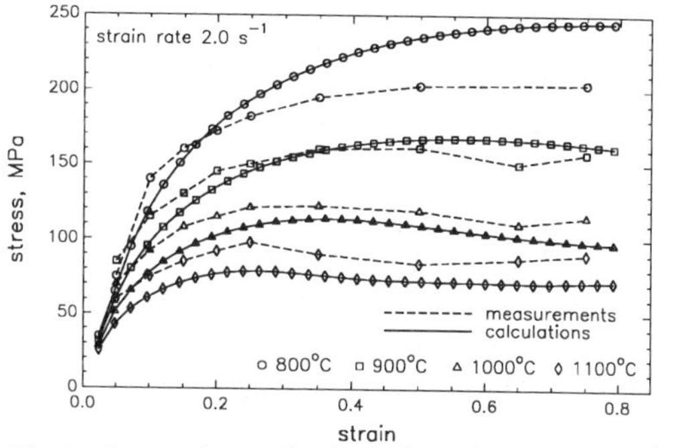

Fig.5. Comparison of calculations for constants in Table 4 with measurements of strain hardening.

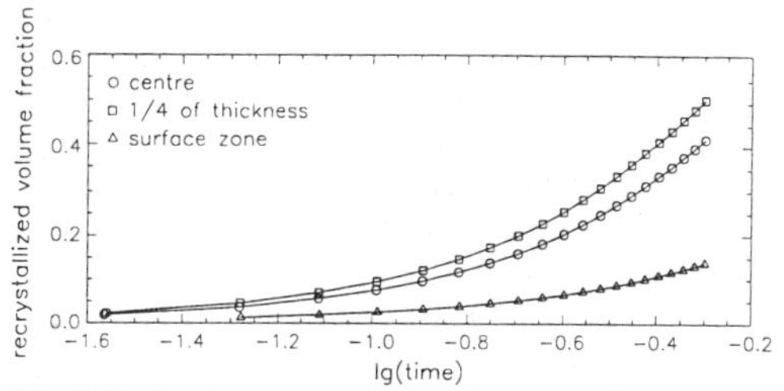

Fig.6. Softening curves during interpass time.

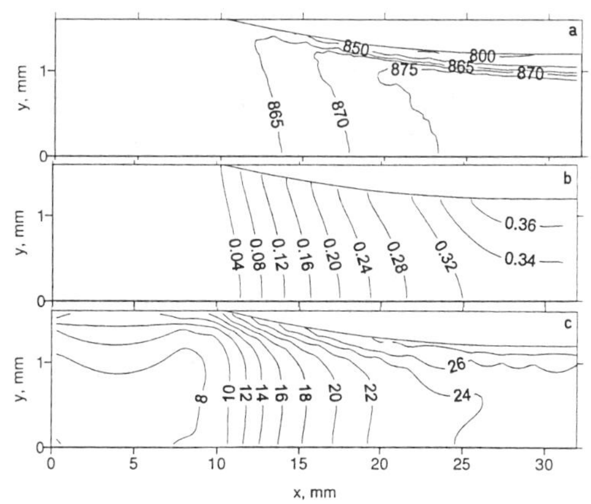

Fig.7. Distributions of temperature (a), effective strain (b) and dislocation density [x10^{14} m^{-2}] (c) in the roll gap during the second considered pass.

solution did not create difficulties. Some increase of the required memory and computing time was the main disadvantage, but it did not exceed 10%. Less than a hundred intervals of the dislocation density and this size of the arrays containing the variables G, D and $\Delta\rho$ are satisfactory. Indeed, the separate arrays are necessary for each flow line, but using only about 10-15 flow lines yields accurate solutions. The situation would become more complex for transient analyses of compression or forging, where the internal variable model would have to be solved in each element of the mesh resulting in an increase of both required memory and computing time.

The deformation history dependence of the yield stress is the main advantage of the presented approach; next is an ability of simulation of various microstructural phenomena. These include static, dynamic and metadynamic recrystallization and recovery, the grain growth and the grain refinement at varying temperatures and strain rates, computed by the finite element program.

7 CONCLUSIONS

Application of the internal variable approach to the simulation of the hot forming is described. The model includes a large number of the material constants which are determined from the experimental data obtained for the fractional softening and for strain hardening. The values of the constants compare well with the data published in the literature. Since all the constants in the model have a physical meaning with known dependence on temperature and strain rate, the model once validated under isothermal and constant strain rate conditions can be applied to the simulation of various complex deformation problems which involve strain rate and temperature variations. The developed model was implemented into the finite element program for rolling, enabling an analysis of the microstructural phenomena in the roll gap and during the interpass times.

The numerical solution does not present special difficulties. Some increase (about 10%) of the required memory and the computing time is observed. The internal variable model does not affect the convergence of the finite element procedure. More problems connected with the memory and computing time can be expected in the non-steady state analyses.

ACKNOWLEDGEMENTS

Financial assistance of the Australian Research Council and the Polish Committee for Scientific Research is gratefully acknowledged. Part of the work was performed during MP's sabbatical leave taken at Monash University in Australia.

REFERENCES

Alden, T.H. 1987. Theory of Mobile Dislocation Density: Application to the Deformation of 304 Stainless Steel. *Metall. Trans. A.* 18A: 51-62.

Brown, S. & J. Wlassich 1992. Application of Invariant Theory to Dynamic Recrystallization Constitutive Behaviour. *Metall. Trans. A.* 23A: 2091-2103.

Estrin, Y. & H. Mecking 1984. A Unified Phenomenological Description of Work Hardening and Creep Based on One-Parameter Models. *Acta Metall.* 32: 57-70.

Frost, H.J. & M.F. Ashby 1982. *Deformation Mechanism Maps.* Oxford: Pergamon Press.

Hodgson, P.D., A. Brownrigg & S.H. Algie 1990. The Static Recrystallization of Coarse and Fine Grained C-Mn Steels. In: T. Chandra (ed.), *Proc. Conf. Recrystallization'90*:541-546. Wollongong.

Laasraoui, A. & J.J. Jonas 1991. Prediction of Steel Flow Stress at High Temperatures and Strain Rates. *Metall. Trans. A.* 22A: 1545-1558.

Mecking, H. & U.F. Kocks 1981. Kinetics of Flow and Strain-Hardening. *Acta Metall.* 29: 1865-1875.

Pietrzyk, M. & J.G. Lenard 1991. *Thermal-Mechanical Modelling of the Flat Rolling Process*, Berlin: Springer-Verlag.

Pietrzyk, M. 1993. Numerical Simulation of Strain Hardening and Recrystallization in the Hot Forming Processes. In: R. Pichoir & P.J. Costa (eds), *Proc. EUROMAT'93, Physique IV.* 3: 1163-1170. Paris.

Pietrzyk, M. 1994. Numerical Aspects of the Simulation of Hot Metal Forming Using Internal Variable Method, *Metall. Foundry Eng.*, 20: 429-439.

Pietrzyk, M., J. Kusiak & M. Packo 1994. Application of the Finite Element Method to the Interpretation of the Axisymmetrical Compression Test. In: J. Bartecek (ed.), *Proc. Conf. Formability'94*: 339-346. Ostrava.

Robersts, W. & B. Ahlblom 1978. A Nucleation Criterion for Dynamic Recrystallization during Hot Working. *Acta Metall.* 26: 801-813.

Roucoules, C. 1992. *Dynamic and Metadynamic Recrystallization in HSLA Steels.* PhD thesis, McGill University, Montreal.

Sandstrom, R. & R. Lagneborg 1975a. A model for Hot Working Occurring by Recrystallization. *Acta Metall.* 23: 387-398.

Sandstrom, R. & R. Lagneborg 1975b. A Model for Static Recrystallization after Hot Deformation. *Acta Metall.*, 23, 1975, 481-488.

Simulation of Materials Processing: Theory, Methods and Applications, Shen & Dawson (eds)
© 1995 Balkema, Rotterdam. ISBN 90 5410 553 4

Stress-strain behaviour of commercial aluminium alloys under hot-working conditions

E. S. Puchi
School of Metallurgical Engineering and Materials Science, Central University of Venezuela, Los Chaguaramos, Caracas, Venezuela

A. J. McLaren & C. M. Sellars
Department of Engineering Materials, The University of Sheffield, UK

ABSTRACT: An analysis of the stress-strain and work-hardening behaviour of an Al-1%Mg alloy deformed under hot-working conditions has been conducted on the basis of three different exponential-saturation constitutive equations. It has been determined that the initial flow stress, the saturation or steady-state stress and the relaxation or transient strain can all be satisfactorily described in terms of the deformation temperature and strain rate over almost 5 orders of magnitude of the Zener-Hollomon parameter by means of the hyperbolic-sine relationship. It has been concluded that the constitutive equation proposed by Sah et al. (1969) describes much more closely the regimes of work-hardening observed than the other two equations examined, although all of them describe adequately well most of the stress-strain curve and strain-hardening behaviour for every deformation condition.

1. INTRODUCTION

Integrated microstructural and thermomechanical finite element models have been developed in recent years to predict the microstructural evolution and the associated thermomechanical response of commercial aluminium alloys during processing at elevated temperatures and high strain rates. One of the most important aspects of these models is having sound constitutive equations for the materials involved (Gelin et al. 1993). Most commercial aluminium alloys processed for the manufacture of sheet are produced by direct-chill casting and subsequent hot rolling operations which give rise to significant dynamic and static microstructural changes that may affect drastically the thermomechanical response of the material as well as the final mechanical strength, formability and geometrical features of the processed sheet. However, both mechanical and microstructural modelling and further optimisation of the hot working schedule of these materials requires a precise description of the functional relationships for the flow stress of the material in terms of the applied strain, rate of straining, temperature of deformation and current microstructure. The present investigation has been conducted in order to develop a rational procedure to analyse the stress-strain behaviour of commercial aluminium alloys deformed under hot working conditions and to compare three different constitutive equations, all with an exponential saturation form, in such a way that the flow stress of the material can be precisely described in terms of strain, strain rate and deformation temperature. Also, it is intended to provide simple and reliable relationships that could be implemented in any thermomechanical and microstructural model.

2. ANALYSIS AND DISCUSSION

The present study has been conducted employing a number of stress-strain curves obtained for Al-1% Mg alloy (AA1050 + 1% Mg) in the temperature range of 578-728 K, at strain rates varying between 0.25-25 s^{-1}. Details of the thermomechanical history of the material and testing techniques are given elsewhere (Shi et al. 1995a, 1995b). Figure 1 illustrates a representative number of effective stress-effective strain curves determined over the spectrum of deformation conditions mentioned above (Orsetti 1995). Such curves have already been corrected for the adiabatic temperature rise by means of the hyperbolic-sine relationship earlier proposed by Sellars and Tegart (1972), of the form:

$$Z = \dot{\varepsilon} \exp\left\{\frac{Q}{RT}\right\} = A[\sinh(\alpha\sigma)]^m \qquad (1)$$

In the above equation Z represents the Zener-Hollomon parameter, $\dot{\varepsilon}$ the effective strain rate, A a pre-exponential factor, α and m stress sensitivity parameters of the strain rate, σ the stress, Q the activation energy for deformation with a value of 156 $kJmol^{-1}$ for aluminium and aluminium alloys, R the Universal gas constant and T the absolute temperature.

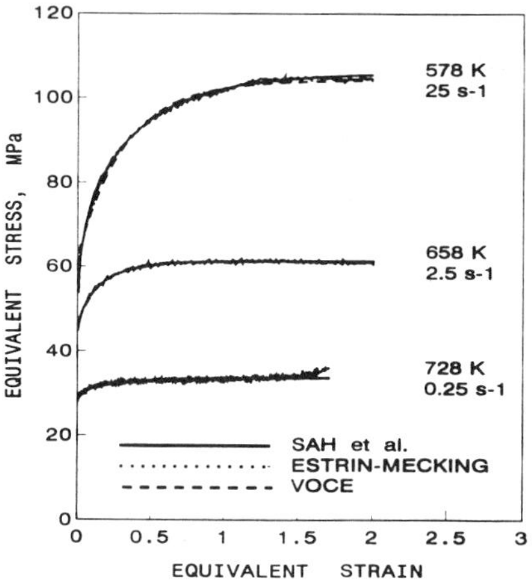

Fig. 1. Stress-strain curves for Al-1% Mg alloy corresponding to some of the experimental conditions. Least-square curves of the three models are also shown for comparison.

The temperature-corrected curves shown in figure 1 display the usual characteristics of those materials that soften by dynamic recovery during deformation at elevated temperatures. Initially, the flow stress rises sharply with the strain applied until it starts to saturate towards a final flow stress level significantly dependent upon temperature and strain rate. In order to describe this behaviour, three different constitutive equations with a saturation-exponential form have been employed: (a) The general exponential saturation equation originally proposed by Sah et al. (1969) to describe the flow stress behaviour of nickel during high temperature deformation:

$$\sigma = \sigma_0 + (\sigma_{ss} - \sigma_0)\left[1 - \exp\left(-\frac{\varepsilon}{\varepsilon_r}\right)\right]^n \qquad (2a)$$

In the present work it has been determined that the mean strain-hardening exponent n ≈ 0.5. (b) The classical exponential-saturation equation formulated by Voce (1948; 1955) in order to account for the achievement of a saturation stress level during deformation of metals even at low temperatures. This equation is represented by eqn. (2a) by setting n=1. (c) The constitutive equation put forward by Estrin and Mecking (1984) as part of a phenomenological treatment of plastic deformation of materials both under dynamic loading and creep conditions:

$$\sigma = \left[\sigma_{ss}^2 - \left(\sigma_{ss}^2 - \sigma_0^2\right)\exp\left(-\frac{\varepsilon}{\varepsilon_r}\right)\right]^{\frac{1}{2}} \qquad (2b)$$

In the above equations, σ represents the effective or equivalent tensile stress, ε the effective strain, σ_0 the flow stress at the initiation of plastic flow, σ_{ss} the steady-state or saturation stress and ε_r the relaxation or transient strain, that is to say, the strain required to achieve some particular value of the hardening fraction which varies depending upon the equation employed. The parameters σ_0, σ_{ss} and ε_r which appear in eqns. (2) can be readily determined by combining a Taylor series expansion of any of the above equations around an initial educated guess for each of the constants involved, together with the universally accepted least-squared method. In the particular case of the Sah et al. equation, the Taylor expansion around an initial solution σ_0', σ_{ss}' and ε_r' can be represented as:

$$\sigma = [\sigma]_0 + \left[\left(\frac{\partial\sigma}{\partial\sigma_0}\right)\right]_0\left(\sigma_0 - \sigma_0'\right) + \left[\left(\frac{\partial\sigma}{\partial\sigma_{ss}}\right)\right]_0\left(\sigma_{ss} - \sigma_{ss}'\right) + \left[\left(\frac{\partial\sigma}{\partial\varepsilon_r}\right)\right]_0\left(\varepsilon_r - \varepsilon_r'\right) + \left[\left(\frac{\partial\sigma}{\partial n}\right)\right]_0\left(n - n'\right) \qquad (3)$$

where, $[\sigma]_0$ and the partial derivatives in squared brackets are the constitutive equation and the derivatives all evaluated at the initial solution of the four parameters indicated as σ'_0, σ'_{ss}, ε'_r and n'. Considering that n=0.5, the above equation can be simplified and re-arranged as:

$$F = \sigma - [\sigma]_0 + \left[\left(\frac{\partial\sigma}{\partial\sigma_0}\right)\right]_0\sigma_0' + \left[\left(\frac{\partial\sigma}{\partial\sigma_{ss}}\right)\right]_0\sigma_{ss}' + \left[\left(\frac{\partial\sigma}{\partial\sigma_{ss}}\right)\right]_0\varepsilon_r'$$

$$= \left[\left(\frac{\partial\sigma}{\partial\sigma_0}\right)\right]_0\sigma_0 + \left[\left(\frac{\partial\sigma}{\partial\sigma_{ss}}\right)\right]_0\sigma_{ss} + \left[\left(\frac{\partial\sigma}{\partial\varepsilon_r}\right)\right]_0\varepsilon_r \qquad (4)$$

Once this linearisation technique has been applied, it is possible to determined an improved estimate of the parameters σ_0, σ_{ss} and ε_r by means of the least square method which minimises the sum of squares defined in this case as:

$$\Omega = \sum_{i=1}^{N} (F-F_i)^2 \qquad (5)$$

Differentiation of Ω with respect to σ_0, σ_{ss} and ε_r would yield a system of three linear equations and three unknowns:

$$\left(\frac{\partial F}{\partial \sigma_0}\right) = 0\,;\ \left(\frac{\partial F}{\partial \sigma_{ss}}\right) = 0\,;\ \left(\frac{\partial F}{\partial \varepsilon_r}\right) = 0 \qquad (6)$$

The solution of this system of equations the determination of an improved value for the initial estimates of the parameters involved. The iterative procedure thus implemented enables the calculation of the unique set of values of the parameters σ_0, σ_{ss} and ε_r which renders the sum of squares a minimum. This analysis was conducted for the three constitutive equations of interest to the present work, employing the experimental stress-strain data derived from the tests. For this purpose a computer program was written in BASIC to be run on a PC. The number of experimental points in every curve varied between about 150-350 depending upon the strain rate of the test and the convergence for the final solution was attained in few seconds. The analysis was also conducted excluding all the experimental points corresponding to an equivalent strain less than 0.05, where experimentally the strains in compression are difficult to defined accurately. The purpose of this particular study was to investigate whether the accuracy of the solution at higher strains could be further improved. In fact, it was observed that the accuracy did not improve but even became worse, particularly at the higher temperatures and lower strain rates when the relaxation strain is expected to be small and the saturation stress is attained rather quickly. Therefore, for the sake of improving the accuracy of the solution for the parameters involved in the constitutive equations it is important to take into consideration as many points as possible, particularly in the strain hardening range of the stress-strain curve. Figure 1 also illustrates the least square curves determined by this procedure for the three constitutive equations analysed. Once the parameters σ_0, σ_{ss} and ε_r have been calculated for any of the constitutive equations involved, the subsequent step is to determine a relationship between the stress parameters, deformation temperature and strain rate. As mentioned above in relation to the temperature correction of the experimental data, a suitable relationship is the hyperbolic-sine equation (1). Figure 2 illustrates such a relationship for both σ_0 and σ_{ss} employing the data derived by using the Sah et al. equation. The dotted lines that appear in the figure correspond to the experimental results determined by Orsetti Rossi (1995) in the same alloy, which are observed to agree reasonably well with the present results.

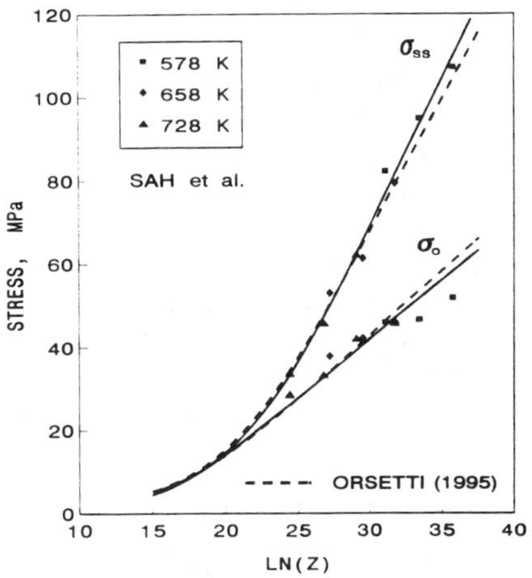

Fig. 2. Flow stress versus natural logarithm of the Zener-Hollomon parameter for Al-1% Mg alloy.

As it can be appreciated, the use of the hyperbolic-sine relationship provides a smooth transition from one stress regime to the other which makes it appropriate for use in the numerical algorithms and thermomechanical models for which it is intended. Two important remarks should be made in relation to this figure. In the first place, it must be mentioned that eqn. (1) should be strictly applied only to steady-state data for which it has been derived. Under steady-state conditions a dynamic equilibrium between the storage of dislocations associated to athermal work-hardening and the annihilation of dislocations associated with the strain rate and temperature dependent work softening (dynamic recovery in the present case) is established, which in principle would keep the dislocation density constant. However, as suggested by Luton and Jonas (1972), such a relationship could also be extended to deal with the initial flow stress since the dislocation density would remain approximately equal to that in the annealed state. On such a basis, figure 2 also contains data for σ_0 as a function of Z described by the same relationship with different values for the constants involved. The second aspect that should be mentioned refers to the fact that in order to avoid any failure of the computer codes where these constitutive equations might be implemented, both curves for σ_0 and σ_{ss} should approximate each other tangentially at low Z values maintaining the condition that $\sigma_{ss}>\sigma_0$. This situation could be fulfilled if:

$$\frac{d\sigma_{ss}}{d(\log(Z))} = \frac{d\sigma_0}{d(\log(Z))} \tag{7}$$

Both derivatives should be evaluated at a particular value of the stress, which in the present case was determined to be in the vicinity of 10 MPa. Therefore, if the initial description of σ_0 and σ_{ss} produces a cross over of the curves, the application of the above tangentiality condition would allow the determination of new values of the parameters m and A in eqn. (1) corresponding to the σ_0 equation. A final remark in relation to figure 2 is concerning the reasonably good alignment of the experimental points in relation to the predicted curve which justifies the use of a value of 156 kJmol^{-1} as the activation energy for deformation. Should the experimental points be aligned along different curves at the three testing temperatures, an activation energy other than the one employed in the present work would have to be used in the calculations. As it can be appreciated in the graph, the points corresponding to the saturation stresses at 578 K and 0.25 s^{-1} and 658 K and 2.5 s^{-1} strain rate are basically the only ones that look somewhat far from the mean curve. It is also important to mentioned that because the application of eqns. (2) requires temperature correction of the stress-strain data, the final values of the constants in the hyperbolic-sine relationship are obtained in an iterative way. The first estimation of such constants was calculated assuming that no adiabatic temperature rise has taken place during deformation. Once this estimation is available the corrected stress at any particular strain can be calculated by means of the hyperbolic-sine equation and the corresponding equations for α, m and the actual temperature of the specimen as a function of the strain applied. Subsequent iterations provide the final estimation of such constants. In the present case it was determined that after three iterations no significant changes occurred in the value of the parameters. As was previously stated, the relaxation or transient strain represents the strain required to achieved some hardening fraction, that is to say, some fraction of (σ_{ss}-σ_0). In the particular case of the Sah et al. equation, this hardening fraction is approximately 0.795, whereas for the Voce equation it is about 0.632. In the case of the Estrin-Mecking equation (as defined in the present context) such a fraction varies with the magnitude of the parameter (σ_{ss}-σ_0). As illustrated in figure 1 it is expected that, in general, the strain to achieve any particular hardening fraction shows a strong dependence on Z, increasing as Z increases. Figure 3 represents the relationship between the relaxation strain determined from the Sah et al. equation, and Z which could be adequately described by means of a hyperbolic-sine relationship similar to that employed in the description of the stress parameters.

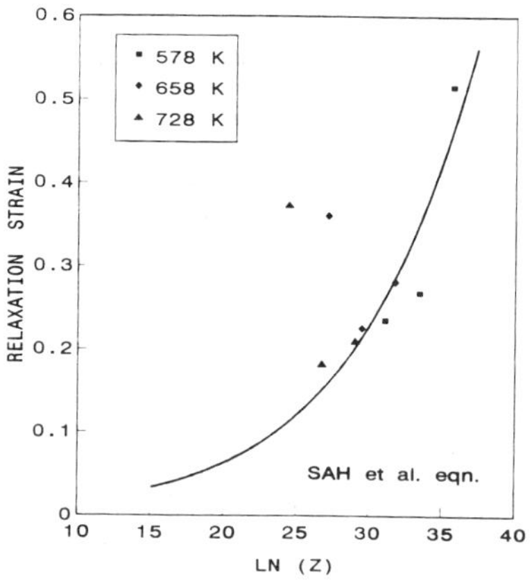

Figure 3. Relaxation strain determined from eqn. (2a) as a function of the Zener-Hollomon parameter.

Several comments should be made in relation to this figure. Firstly, there are basically two experimental conditions that do not conform to the relationship: 658 and 728 K at 0.25 s^{-1}. In both cases the relaxation strain appears to be greater than expected since at the lowest strain rate and higher temperatures the saturation stress is observed to be achieved after a small strain increment. A possible explanation to this fact is that under such conditions the stress-strain curves tend to flatten at a very early stage of the deformation process and therefore a slight variation in the hardening fraction could result in an significant error in the calculation of the strain to achieve it. However, for this same reason it is not expected that such uncertainties would have a major effect on modelling such stress-strain curves. For all the other conditions the experimental data is observed to scatter reasonably well about the line given by the hyperbolic-sine equation. Table I contains the all the data concerning the different constants involved in the determination of the stress parameters and the relaxation strain, assuming the validity of eqn. (2a), as a function of Z.

Table I. Different constants involved in the calculation of σ_0, σ_{ss}, and ε_r corresponding to eqn. (2a), by means of a hyperbolic-sine relationship as a function of temperature and strain rate.

	A, s^{-1}	α, MPa^{-1}	m
σ_0	4.35×10^7	10.16×10^{-2}	3.48
σ_{ss}	1.91×10^{10}	2.99×10^{-2}	4.67
ε_r	6.18×10^{17}	105.6×10^{-2}	7.76

A convenient way of evaluating how well each constitutive equation can describe the work-hardening behaviour of the investigated alloy is to represent the rate of strain-hardening as a function of the applied stress together with the theoretical predictions provided by each equation as shown in figure 4. Here the rate of work-hardening ($\Theta = d\sigma/d\varepsilon$) for the same stress-strain curves depicted in figure 1, is represented as a function of σ. As can be clearly observed, the differential plots present an initial stage where the rate of work-hardening diminishes markedly with stress, particularly dependent upon the deformation conditions. At 578K and $25s^{-1}$ for example it can be appreciated that Θ decreases from about 600 MPa to less than 80 MPa in a stress interval hardly greater than 20MPa. However, as can be observed in figure 1 the strain interval over which this change occurs is also quite small since most of the experimental points that form the stress-strain curve are between 80-107 MPa. Therefore, it can be concluded that in a differential form there appear to be several work-hardening regimes with a smooth transition between them. It also can be clearly observed that the Sah et al. equation (eqn. (2a)) follows quite closely this transition, whereas both the Estrin-Mecking and Voce equations fail to do so. In this particular sense therefore, the Sah et al. equation appears to be more suitable than the Estrin-Mecking equation and this in turn is observed to be better than the Voce equation. However, because of the fact that the first stage where the rate of change of the work-hardening rate with stress is more pronounced only spans a very small strain interval, such a failure has an insignificant effect in modelling the whole stress-strain curve for FEM computations of thermomechanical processing.

As observed in figure 1 the least-square curves corresponding to the three constitutive equations of interest to the present work describe quite satisfactorily both the stress-strain curves and the work-hardening behaviour. Finally, figure 5 illustrates an example of the computation of the stress-strain curves for 5 different experimental conditions, which are obtained by combining the constitutive equation (2a) with the hyperbolic sine-log Z relationships above referred to for σ_0, σ_{ss} and ε_r and the data presented in Table I. Two different experimental curves at 578K and $25s^{-1}$ have been included for comparative purposes. These are extreme examples of experimental scatter and show the present model reproduces quite satisfactorily the experimental stress-strain data, with statistical determination coefficients in the range of 0.97 for all the three equations.

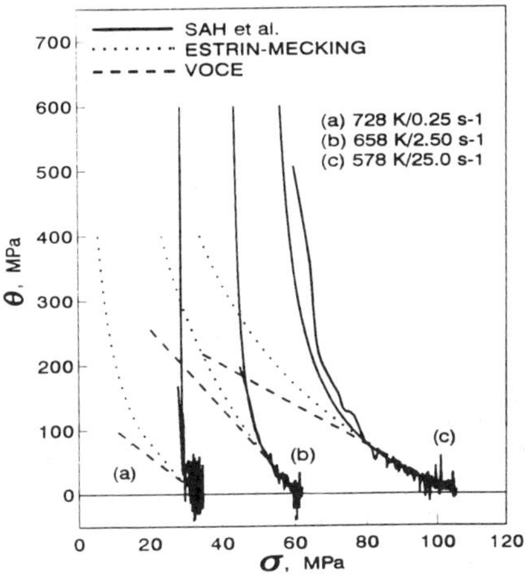

Fig. 4. Work-hardening rate versus applied stress for three deformation conditions.

CONCLUSIONS

The Al-1% Mg alloy investigated in the present work presents markedly different work-hardening stages. At first the rate of change of the strain-hardening rate with stress is very pronounced in contrast to the final stages where it appears to be

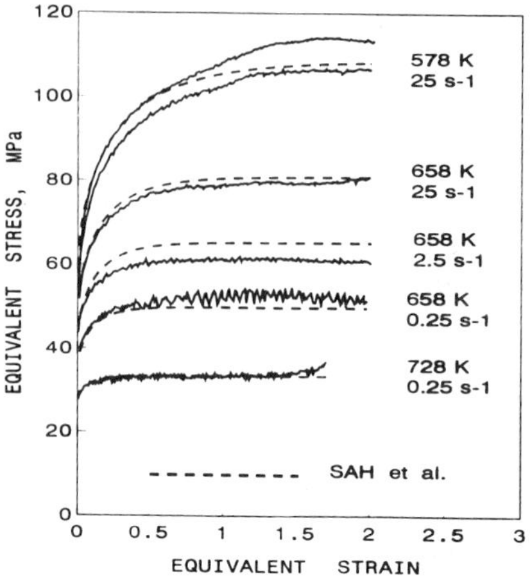

Fig. 5. Computed stress-strain curves under different deformation conditions.

more moderate. Among the three stress-strain relationships analysed in the present work, the constitutive equation proposed by Sah et al. appears to be the only one that follows quite well this work-hardening behaviour and therefore it is considered to be the most suitable equation for FEM computations and modelling thermomechanical processing. Nonetheless, because of the fact that the first stage of work-hardening spans a relatively small strain interval, both Voce and Estrin-Mecking equations also describe satisfactorily most of the stress-strain behaviour. It has also been determined that the hyperbolic-sine relationship is quite suitable for describing the strain rate and temperature dependence of the initial and saturation flow stresses and the relaxation strain over a wide range of Z values, which makes the whole approach here presented appropriate to be incorporated in any thermomechanical and microstructural model where it might be necessary. An optimised least squares curve fitting procedure has been developed to analyse the stress-strain behaviour of metals which undergo work-hardening and dynamic recovery when deformed under hot-working conditions.

ACKNOWLEDGEMENTS

The present investigation has been conducted with the financial support of the British Royal Society and the Venezuelan National Council for Scientific and Technological Research (CONICIT), Projects S1-2580 and RP-II-C-135. The authors also gratefully acknowledge the assistance of Mr. P. Orsetti Rossi and Dr. G. Baxter for the provision of the stress-strain curves analysed in the present work.

REFERENCES

Gelin J. C., Ghouati O. and Shahani R. 1993. Identification and modelling of constitutive equations for hot rolling of aluminium alloys from plane strain compression testing. In J. Beynon (ed.), 1st. Internat. Conf. on Modelling of Metal Rolling Processes: 239-255. The Institute of Materials, U. K.

Estrin Y. and Mecking H. 1984. A unified phenomenological description of work hardening and creep based on one-parameter models. *Acta Metall.* 32: 57-70.

Luton and Jonas J. J. 1972. Solute strengthening at high temperatures in zirconium-tin alloys. *Can. Met. Quart.* 11: 79-90

Orsetti Rossi P. 1995. Unpublished work, Department of Engineering Materials, The University of Sheffield, U. K.

Sah J. P., Richardson G. J. and Sellars C. M. 1969. Recrystallization during hot deformation of nickel. *J. Australian Inst. Metal,* 14: 292-297.

Shi H., McLaren A., Sellars C. M., Shahani R. and Bolingbroke R. 1995. Constitutive equations for high temperature flow stress of aluminium alloys. To be publihsed.

Shi H., McLaren A., Sellars C. M., Shahani R. and Bolingbroke R. 1995. Hot plane strain compression testing of aluminium alloys. To be publihsed.

Sellars C. M. and Tegart McG. 1972. La relation entre la résistance et la structure dans la déformation a chaud. *Mém. Sci. Rev. Met.*, 23: 731-746.

Voce, E. 1948. The relationship between stress and strain for homogeneous deformation. *J. Inst. Metals.* 74: 537-562.

Voce, E. 1955. A practical strain hardening function. *Metallurgia.* 51: 219-226.

Simulation of Materials Processing: Theory, Methods and Applications, Shen & Dawson (eds)
© 1995 Balkema, Rotterdam. ISBN 90 5410 553 4

Numerical modelling of metal formability using polycrystal plasticity

Y. Qiu & K.W. Neale
Université de Sherbrooke, Que., Canada

A. Makinde & S.R. MacEwen
Alcan International, Kingston, Ont., Canada

ABSTRACT: The formability of FCC polycrystalline metals is simulated numerically by a finite element analysis which incorporates crystal plasticity constitutive models. In particular, the influence of microstructure and textures are assessed. Critical strains for localized necking failures in thin sheets are determined for a variety of deformation paths. Various microstructural effects are considered, which include slip system hardening, material strain-rate sensitivity, initial texture, as well as the evolution of texture and microstructural hardening.

1 INTRODUCTION

Many factors are known to affect metal formability. These include strain hardening, material strain-rate sensitivity, microscopic structure, and anisotropy due to crystallographic texture. As most formability analyses to date have been based on phenomenological models of plastic behaviour, which do not explicitly include the basic physics of deformation, they are not entirely satisfactory for assessing formability. That is, although they are capable of adequately describing many features of a metal forming process, they generally cannot accurately predict whether or not failure will occur. Furthermore, such phenomenological models are inherently incapable of describing the effects of microstructure and its evolution with deformation on metal performance, nor can they link mechanical properties to evolving microstructures and textures. Since initial textures and their development during forming processes directly affect subsequent fabrication processes as well as the quality of products, their influence is very important for assessing metal formability. Metal forming modelling based on crystal plasticity theories, with the consideration of initial texture and its evolution, can thus be expected to be more accurate than modelling based on phenomenological constitutive theories.

In this work we simulate numerically the formability of FCC polycrystalline metals, and assess in particular the influence of microstructure. A finite element analysis has been developed which is based on crystal plasticity constitutive models. Various studies concerning metal formability have been conducted to predict the critical strains for localized instability failures. These include localized necking instabilities during sheet forming, shear band formation in plane strain deformation, as well as tensile instability in a round bar. Only the results of our sheet necking simulations are presented here.

The formability of sheet metals, as a general practice, can be assessed by examining the forming limit diagram (FLD) which depicts, in terms of the major and minor strains in the plane of a deformed sheet, the critical strain state for localized necking failure. The classical Marciniak-Kuczyński (M-K) theory (Marciniak and Kuczyński, 1967, 1973), with later extensions by Hutchinson and Neale (1978a, 1978b), have been successfully used in predicting FLD's for various materials. In this paper, we use a numerical procedure employing special purpose finite elements to calculate the FLD's of aluminum sheets. The M-K theory, which consists of simulating the evolution of a pre-existing imperfection in the sheet, is employed together with the polycrystal deformation theory. The Taylor model used previously by Harren *et al.* (1989) and which includes slip system hardening and elasticity of the grains, is employed locally at each material sampling point in the finite element analysis. In our

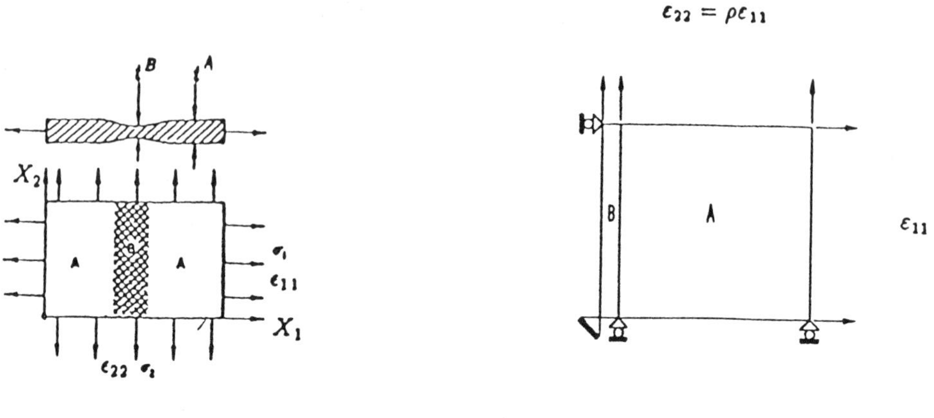

(a) (b)

Figure 1: (a) An imperfect sheet under biaxial stretching; (b) Finite element mesh and boundary condition used for FLD prediction.

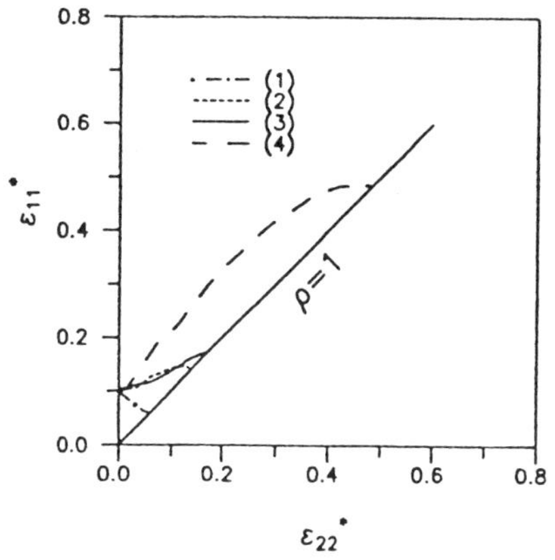

Figure 2: Predicted forming limit diagram for different initial textures: (1) ideal cube orientation; (2) AA1145 with x_1 as rolling direction and x_2 as transverse direction; (3) random orientation; (4) cube texture (10° spread). Initial imperfection $f = 0.99$.

work, main attention is focused on the influence of microstructure and microscopic properties on the FLD. Special interest is in the influence of crystallographic textures on forming limits. For this purpose, different initial textures are examined which include a random orientation distribution, the ideal cube orientation, the cube orientation distribution with a 10° Gaussian spread, and a texture distribution obtained experimentally for an annealed aluminum sheet alloy (AA1145). For each finite element sampling point, the analysis assumes an FCC polycrystal comprising different numbers of grains, varying from 100 to 328, to describe the various initial textures.

2 CONSTITUTIVE MODEL

The single crystal constitutive law employed is based on Peirce, Asaro and Needleman's (1983) formulation which accounts for (i) the polycrystalline nature of the metal, (ii) the microscopic deformation mechanism, assumed to be rate-dependent crystallographic slip, (iii) the evolution of crystallographic texture, (iv) finite deformation, and (v) anisotropic crystal elasticity. The constitutive law for a single crystal is of the following form:

$$\hat{\boldsymbol{\tau}} = \mathbf{L} : \mathbf{D} - \dot{\boldsymbol{\tau}}_o \tag{1}$$

where $\hat{\tau}$ is the Jaumann rate of Kirchhoff stress, **L** is the tensor of elastic moduli and **D** represents the strain-rate tensor. The term $\dot{\tau}_o$ accounts for the viscoplastic crystallographic slip rates $\dot{\gamma}^{(\alpha)}$ on the various slip systems α. The slip rate is taken to be governed by the power-law expression

$$\dot{\gamma}^{(\alpha)} = \dot{\gamma}_0 \; sgn \; \tau^{(\alpha)} |\frac{\tau^{(\alpha)}}{g^{(\alpha)}}|^{1/m} \tag{2}$$

where $\tau^{(\alpha)}$ is the resolved shear stress on slip system α and $g^{(\alpha)}$ is its hardness. In the present simulations, isotropic hardening is assumed and the slip system strength law is taken as:

$$g(\gamma) = 38.8(\gamma + 0.002)^{0.095} \text{ (MPa)} \tag{3}$$

based on experimental data for commercial purity aluminum (Becker *et al.*, 1993). Here γ is the total accumulated slip for all systems.

The polycrystal response is obtained by invoking the Taylor assumption for each sampling point in the finite element discretization. Thus, for each grain comprising the polycrystal at a sampling point, we assume that the deformation is identical to the macroscopic deformation at that point. By averaging the stress response and constitutive moduli over the number of grains, the macroscopic behaviour is determined.

3 METHOD OF ANALYSIS

A sheet under biaxial stretching is modelled using our finite element scheme. As in the M-K theory, the initial imperfection is introduced as an element B of reduced thickness, as illustrated in Figure 1. The imperfection parameter is denoted by $f = t_0^B/t_0^A \leq 1$. The orientation of the imperfection is chosen to be perpendicular to the x_1 axis. Only one quarter of the sheet is modelled due to the existing symmetry. Figure 1b) shows the two-element mesh employed in the present analysis.

A prescribed strain path, $\varepsilon_{22} = \rho\varepsilon_{11}$ where $0 \leq \rho \leq 1$, is imposed on the edge of the sheet. Identical material parameters and initial textures are assumed in both elements A and B. During stretching, the strain grows more rapidly in the weaker element B. The onset of a local instability is deemed to occur when the maximum strain increment in element A $d\varepsilon_{11}^A \to 0$ and the ratio of the strain increment in element B to that in element A becomes unbounded, i.e. $d\varepsilon_{11}^B/d\varepsilon_{11}^A \to \infty$. The corresponding strain level in element A represents the critical failure strain, denoted by ε^*.

The constitutive response is implemented by assuming that each material sampling point locally represents a polycrystalline aggregate consisting of N grains. That is, the Taylor model discussed in the previous section is employed locally at each sampling point in the finite element analysis. The same Euler angle distributions are input for each sampling point to describe the initial texture. The texture evolution is traced by computing the re-orientations of each grain during deformation. Crystal hardening is also updated at each step of the computation. The polycrystal constitutive description is implemented into the incremental virtual work equation and an updated Lagrangian finite element code is developed. An automatic time-stepping scheme (Van der Giessen and Neale, 1993) is incorporated in the finite element code to ensure numerical stability and to optimize computational time.

4 RESULTS AND DISCUSSION

For the simulations presented here, we have assumed a strain-rate sensitivity value $m = 0.016$ and a slip hardening exponent $n = 0.095$, as in Equation (3). The single crystal elastic constants are taken to be $C_{11} = 108.0$ GPa; $C_{12} = 62.0$ GPa; and $C_{44} = 28.3$ GPa, representing typical values for aluminum single crystals. Different grain numbers per sampling point are used for different initial textures: 100 grains for the initially random orientation, 100 grains for the 10° Gaussian distribution of orientations around the ideal cube position $\{100\} < 001 >$, and 328 grains for the annealed aluminum sheet alloy (AA1145).

Figure 2 shows the predicted forming limit diagrams for the different initial textures with an imperfection level $f = 0.99$. Only biaxial stretching cases ($\rho \geq 0$) are considered here. The influence of initial texture on the FLD, for the same material parameters and same initial imperfection, is very pronounced. It is seen that for the initial random texture, the forming limit strains remain approximately constant for small ρ values and increase as ρ increases. The best stretchability is found at the equibiaxial stretching state ($\rho = 1$). This is very close to the simulation results obtained by Zhou and Neale (1995) for a rigid-viscoplastic polycrystal with an initial random texture. For the tex-

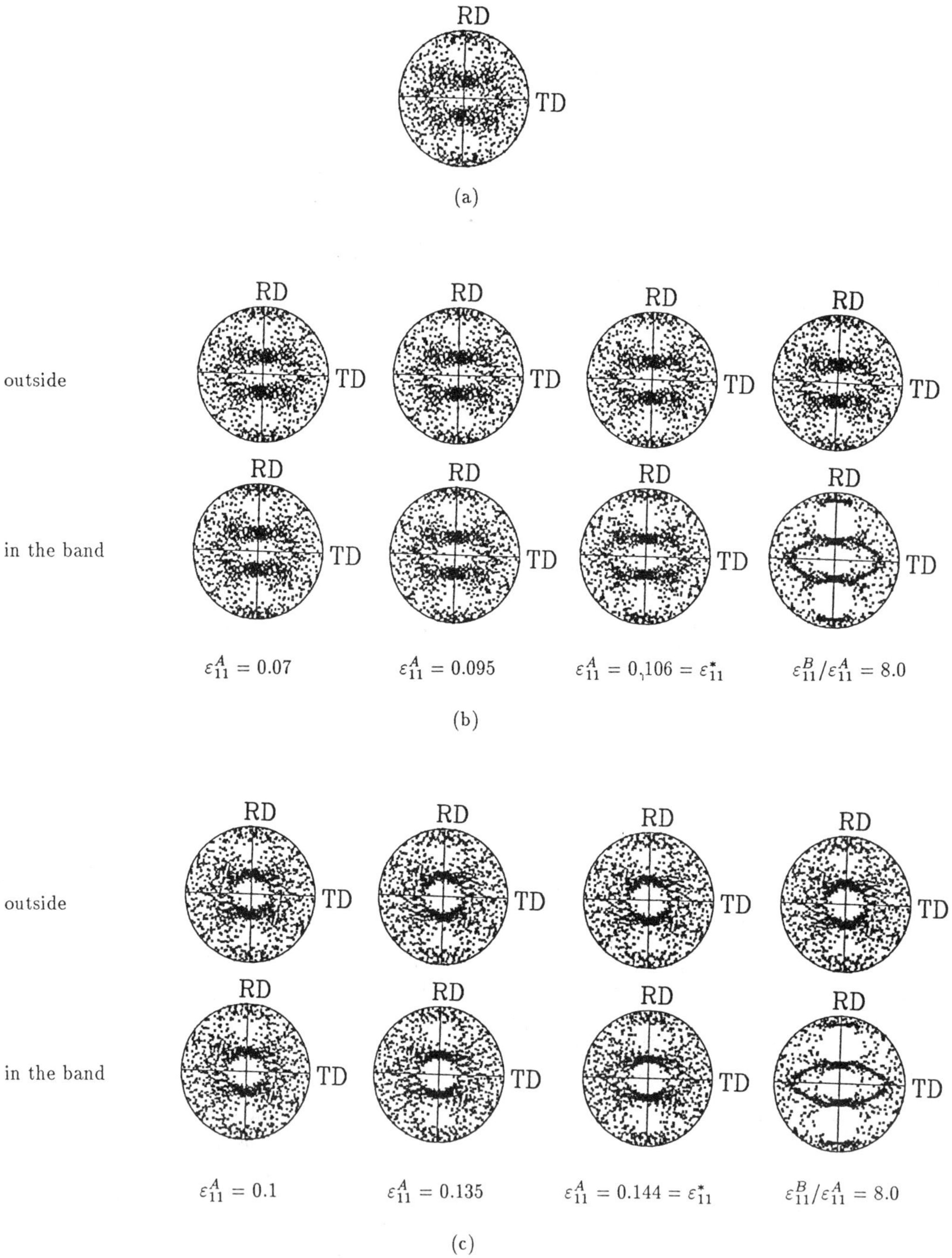

Figure 3: Predicted texture evolution for AA1145 aluminium alloy sheet. "RD" – rolling direction; "TD" – transverse direction. $f = 0.99$. 328 grains per sample point. (a) initial texture; (b) in-plane plane strain stretching ($\rho = 0$); (c) equibiaxial stretching ($\rho = 1$).

ture of the AA1145 sheet, we observe a ductility very similiar to that of the initial isotropic texture over most of the biaxial stretching range. However, its ductility becomes lower than that of the isotropic texture as ρ increases, and the greatest difference occurs at the equibiaxial stretching state. The lowest ductility found here is for the initial ideal cube orientation. In fact, the forming limit curve of the ideal cube orientation is identical to that given by Zhou and Neale (1995) for a polycrystal where each grain has the ideal cube orientation. Surprisingly, the forming limit curve of the 10° spread cube texture does not fall between the FLD of the ideal cube orientation and that of the isotropic texture. A cause of this may be the assumed initial angle of the imperfection groove. Experiments suggest that for equibiaxial stretching, the groove orientation is at 45° to the rolling direction for strongly cube-textured sheets (Wilson *et al.*, 1981). For in-plane plane strain tension ($\rho = 0$), there is no noticeable difference on the forming limit for the four initial textures considered here, as expected.

Figure 3 shows the evolution of the crystallographic textures of the AA1145 sheet at different deformation stages for in-plane plane strain and equibiaxial stretching. These are presented in the form of $< 111 >$ pole figures (both in the band and outside). As expected, the textures are much sharper in the bands than outside, especially when close to the critical strains, as the strains in the bands are then much greater than in the region A. For different strain paths, the same initial texture and same material parameters lead not only to different forming limits but also to different texture evolutions. Comparing the evolved textures to the initial ones, it is obvious that equibiaxial stretching produces greater texture changes than in in-plane plane strain stretching.

This example serves to illustrate the type of formability analysis that has been developed which incorporates crystal plasticity constitutive modelling. As mentioned previously, numerous other studies are currently underway in this topic.

ACKNOWLEDGEMENT

This work was supported by the Natural Sciences and Engineering Research Council of Canada (NSERC) and a grant from Alcan International Limited.

REFERENCES

Becker, R., Smelser, R.E. and Panchanadeeswaran S. (1993). Simulations of Earing in Aluminum Single Crystals and Polycrystals. *Model. Simul. Mater. Sci. Eng.* 1: 203–224.

Harren, S., Lowe, T.C., Asaro, R.J. and Needleman, A. (1989). Analysis of Large-Strain Shear in Rate-Dependent Face-Centered Cubic Polycrystals: Correlation of Micro- and Macromechanics. *Phil. Trans. R. Soc. Lond.* A328: 443–500.

Hutchinson, J.W. and Neale, K.W. (1978a). Sheet Necking - II. Time-Independent Behavior. In: D.P. Koistinen and N.-M. Wang (Ed.), *Mechanics of Sheet Metal Forming*, Plenum Publishing Corporation: 127–153.

Hutchinson, J.W. and Neale, K.W. (1978b). Sheet Necking - III. Strain-Rate Effects. In: D.P. Koistinen and N.-M. Wang (Ed.), *Mechanics of Sheet Metal Forming*, Plenum Publishing Corporation: 269–285.

Marciniak, Z. and Kuczyński, K. (1967). Limit Strains in the Processes of Stretch-Forming Sheet Metal. *Int. J. Mech. Sci.* 9: 609–621.

Marciniak, Z., Kuczyński, K. and Pokora, T. (1973). Influence of the Plastic Properties of a Material on the Forming Limit Diagram for Sheet Metal in Tension. *Int. J. Mech. Sci.* 15: 789–805.

Peirce, D., Asaro, R.J. and Needleman, A. (1983). Material Rate Dependence and Localized Deformation in Crystalline Solids. *Acta Metall.* 31: 1951–1976.

Van der Giessen, E. and Neale, K.W. (1993). Analysis of the Inverse Swift Effect Using a Rate-Sensitive Polycrystal Model. *Comp. Meth. App. Mech. Eng.* 103: 291–313.

Wilson, D.V., Roberts, W.T. and Rodrigues, P.M.B. (1981). Effects of Grain Anisotropy on Limit Strains in Biaxial Stretching: Part II. Sheets of Cubic Metals and Alloys with Well-Developed Preferred Orientation. *Metall. Trans.* 12A: 1603–1611.

Zhou, Y. and Neale, K.W. (1995). Predictions of Forming Limit Diagrams Using a Rate-Sensitive Crystal Plasticity Model. *Int. J. Mech. Sci.* (in press).

Simulation of Materials Processing: Theory, Methods and Applications, Shen & Dawson (eds)
© 1995 Balkema, Rotterdam. ISBN 90 5410 553 4

Analysis of finger growth during the flow of reacting liquid through porous fiber preforms

C.N.Satyadev & K.Jayaraman
Department of Chemical Engineering and Composite Materials and Structures Center, Michigan State University, East Lansing, Mich., USA

ABSTRACT: Viscous fingering during flow of a reacting liquid mixture through a porous fiber bed has been analyzed in this work. Spatial variations in monomer concentration are brought about not only by dispersion but also by the polymerization reaction within the porous medium. The combined effect of chemical reaction and dispersion on finger growth depends on the type of reaction as seen in the incremental rate of reaction with increasing extent of reaction, $S=dR/d\alpha$. When S is negative, the unstable region is diminished. If S is increasingly negative along the flow direction, and the reaction rate is of the order of the rate of dispersion, it is possible to eliminate fingers entirely. When S is positive, the unstable range of wavenumbers is increased with reaction.

1. BACKGROUND

Mold filling in liquid composite molding processes involves the flow of a polymerizing liquid mixture through an anisotropic porous medium comprised of fiber reinforcements. Microstructural defects in the final part may arise from a number of causes -- uneven distribution of the resin, voids embedded within the fiber bed and viscous fingering of fresh resin with a lower viscosity into aged resin of higher viscosity within the mold. The scale of the defects is important to determine whether they affect the mechanical performance of the molded part significantly.

During the mold filling stage of resin transfer molding (RTM) and structure-reaction injection molding (S-RIM) operations, there is a progressive increase of resin viscosity due to reaction from its monomeric state to the gel state. The gel conversion of the resin is an important factor; it is imperative that the mold be filled completely, before gelation occurs. It is well known that miscible displacement of a fluid by another fluid with lower viscosity leads to flow instabilities, causing the displacing fluid to finger through the displaced fluid (Homsy 1987). The work of Homsy and co-workers has established that dispersion within the porous medium is a stabilizing influence at higher wavenumbers or narrower finger widths. In the present application, spatial variations in monomer concentration are brought about not only by dispersion but also by the polymerization reaction within the porous medium. Hence it is important to understand the influence of these two factors on the finger widths most likely to grow during the mold filling process.

The fiber bed is, in general, anisotropic both for permeation of the resin through the preform and dispersion of the monomer in the reaction mixture. The permeability of the preform is determined by its physical structure and can vary according to the kind of fiber arrangement chosen, viz. random or aligned. The effective dispersion coefficient is dependent on the characteristics of both the porous medium and the resin. Due to the polymerization reaction, the molecular weight of the resin increases with time; this affects the dispersion of the monomeric resin (of low molecular weight) through the polymerized resin. The bulk dispersion depends on the injection velocity as well as the length scale of the fiber bed.

2. OBJECTIVES

The object of this work is to analyze viscous fingering during reactive filling of an end gated mold containing a pre-located porous fiber bed. The mechanism for growth or attenuation of fingers in flow of a reacting liquid mixture through a porous medium is explored here. This mechanism is directly related to the reaction kinetics of the resin and dispersion characteristics of the monomer through the polymerized resin. This work will also lead to criteria for attenuation of fingers in the presence of reactive and dispersive effects.

3. MODEL PROBLEM AND EQUATIONS

The model problem is posed in the context of multiple shot injection of two-component resins, mixed in-line, into a mold. A batch of fluid is injected into the mold and allowed to react for some time, during which it undergoes polymerization, with a rise in viscosity. At a later time (t=0), another shot of resin is injected. Due to the difference in ages of the two shots of fluid, a step jump $\Delta\alpha$ in the extent of reaction is observed at the interface of the two shots. This multiple shot approach was used by Losure (1994) to prepare molded composites with a variety of fingered microstructure; the mechanical properties of such composite specimens were found to be affected significantly by the details of the microstructure.

In order to delineate clearly the effect of chemical reaction on fingering, we pose the problem in an infinite porous medium, as in the work of Tan & Homsy (1986) which addressed dispersion alone. The defining equations are the equation of continuity, Darcy's law (for flow through porous media), an equation for the location of the moving flow front and a component mass balance written for the monomer concentration. The length and time scales chosen are D/U_0 and $D/U_0{}^2$; this choice eliminates the Peclet number from the equations. The component mass balance then is left with only one parameter, θ, which is the product of the Damkohler number and the Peclet number. The Damkohler number ($Da=t_f/t_R$) is the ratio of flow time scale to reaction time scale. The Peclet number (Pe) is the ratio of the dispersive time scale to the flow time scale. The product of these two represents a ratio of the dispersive time scale to the reaction time scale. Other groups arising in the scaled eigenvalue problem are $S=(dR/d\alpha)$ -- the incremental rate of reaction with increasing extent and $G=-d(\ln\lambda)/d\xi$, where -G represents the gradient of mobility (λ) along the flow direction.

Figure 1 depicts the flow domain and also the profiles of conversion, viscosity and S at the initial time, t=0. The differential balance equations, in a frame of reference moving with the constant injection velocity U_0, are given below.

$$\nabla \cdot \underset{\sim}{q} = 0$$

$$\underset{\sim}{q} = -(\bar{\lambda}\,\underline{\underline{K_D}} \cdot \nabla P) \qquad (1)$$

$$\frac{\partial\alpha}{\partial t} + \underset{\sim}{q}\cdot\nabla\alpha = \theta R(\alpha,T) + \nabla^2\alpha$$

A linear stability analysis leads to the following eigenvalue problem with two coupled equations.

$$[D^2 + \rho(\xi)\frac{\partial\bar{\alpha}}{\partial\xi}D - k'^2]\,q_\xi = k'^2\rho(\xi)\,(\Delta\alpha)\,\psi$$

$$[\sigma - \theta\frac{\partial\bar{R}}{\partial\bar{\alpha}} + k^2 - D^2]\,\psi = -\frac{1}{\Delta\alpha}\frac{\partial\bar{\alpha}}{\partial\xi}q_\xi \qquad (2)$$

The two coupled equations in (2) may be combined to obtain

$$(\sigma - \theta\frac{\partial\bar{R}}{\partial\bar{\alpha}} + k^2 - D^2)Lq_\xi = -\frac{\partial\bar{\alpha}}{\partial\xi}k'^2\,q_\xi$$

$$L \equiv [\frac{1}{\rho(\xi)}(D^2 + \rho\frac{\partial\bar{\alpha}}{\partial\xi}D - k'^2)] \qquad (3)$$

$$\rho(\xi) = -\frac{1}{\bar{\lambda}}\frac{d\bar{\lambda}}{d\bar{\alpha}}$$

In this eigenvalue problem, σ is the exponential growth rate of disturbances and k is the wavenumber. The boundary conditions associated with the eigenvalue problem are the decay conditions at $\pm\infty$, and the continuity conditions at the interface, on disturbances in velocity, coversion and pressure (cf. Tan and Homsy, 1986). It may be shown that the principle of exchange of stabilities holds for this problem and that the growth rates are real. When the growth rate is negative, fingers are damped.

The type of resin polymerization affects the stability of the mold filling process, through the parameter S. Some polymerization reactions involve consumption of inhibitor almost up to the gel point, the rate of which is independent of the monomer concentration (Gonzalez-Romero & Macosko 1985). For this zeroth order reaction, S=0, and the stability of the mold filling process is unaffected by the chemical reaction. In other polymerization reactions, the initial stage of the network forming process is of first order kinetics, which leads to a constant value of S. For

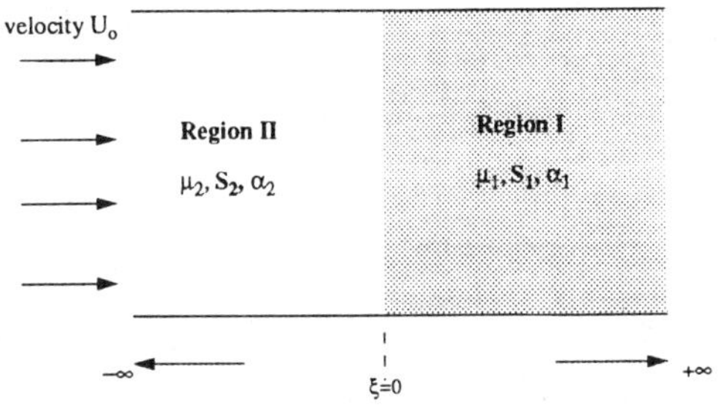

Fig. 1a Flow of reacting liquid through a porous medium (moving frame)

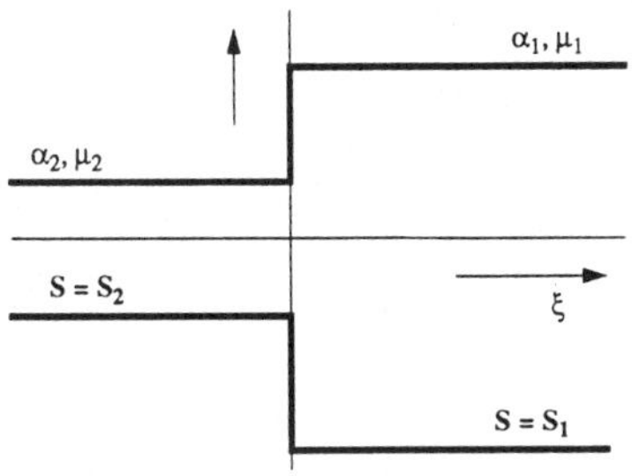

Fig. 1b Profiles of conversion, viscosity and S at t=0

many polymerization reactions S is negative and it becomes increasingly negative as the reaction progresses. For example, polyurethanes follow second order kinetics (Castro & Macosko 1982). Figure 2 shows the behavior of S for different types of chemical reactions, schematically.

4. INITIAL GROWTH RATE

The initial growth rate for an exponential viscosity-concentration relationship, with a step jump in concentration at time t=0, was analyzed by Tan & Homsy (1986). A similar approach has been used in this work to obtain the initial growth rates (at t=0) for the model problem with combined reaction and dispersion. The particular case in which viscosity varies exponentially with conversion, i.e. ρ=constant, is chosen for the following analysis. At t=0, the eigenvalue problem has a solution of the form given in eq. 4, for $\xi \neq 0$.

$$\begin{aligned} \xi<0:\quad & q_\xi = A_1 \exp(m_- \xi) + B_1 \exp(k' \xi) \\ \xi>0:\quad & q_\xi = A_2 \exp(-m_+ \xi) + B_2 \exp(-k' \xi) \end{aligned} \tag{4}$$

Upon applying the continuity conditions on disturbances at the interface, we obtain the following equation relating the growth rate to the wave number.

$$\frac{(m_+ + m_-)(k'^2 - m_+^2) - k'^2 \rho \Delta\alpha}{k' \rho \Delta\alpha} = -\frac{k'(m_- + m_+ M) + m_+(m_+ + m_- M)}{(k' + m_-)\ (1 + M)} \tag{5}$$

where

$$\begin{aligned} M &= \exp\ (\rho) \\ k'^2 &= b k^2 = b\ (k_x^2 + k_y^2) \\ m_+^2 &= \sigma + k^2 - \theta S_1 \\ m_-^2 &= \sigma + k^2 - \theta S_2 \end{aligned} \tag{6}$$

The parameters in the growth rate curves are θ, S_1 and S_2. The cutoff wavenumbers are defined as the wavenumbers at which the growth rate changes sign, implying a transition between stable and unstable regimes. Eq.5 may be solved for the cutoff wavenumbers k_c by setting σ=0. The roots of the algebraic equations in either case, are obtained by Newton-Raphson iterations. The initial guesses must be close to the roots, to obtain convergence of the numerical scheme. Suitable guesses for the lower cutoff wavenumber are obtained from the results of two different analyses. One was performed by Hickernell & Yortsos (1986) for non-dispersive displacement through porous media, and another was done for reactive filling without dispersion (Jayaraman et al 1992).

5. RESULTS AND DISCUSSION

The initial growth rate curves have been computed and plotted for several selected values of θ. The numerical results presented here were obtained with ρ=3 throughout. Two different types of reactions are examined in these results. For one type, the values of S_1 and S_2 are chosen to be negative, -3 and -1 respectively. The initial growth rate (σ) is plotted as a function of the wavenumber for four selected values of θ in Fig. 3. The curve for θ=0 corresponds to the flow process dominated by dispersion. It matches up well with the result of Tan & Homsy (1986). For this case, small fingers get smeared out due to dispersion, while large fingers remain undamped. This is in contrast to the case of θ≠0, where two different non-zero cutoff wavenumbers may be identified. As the rate of chemical reaction is increased relative to the rate of dispersion, by increasing θ, it is seen that only

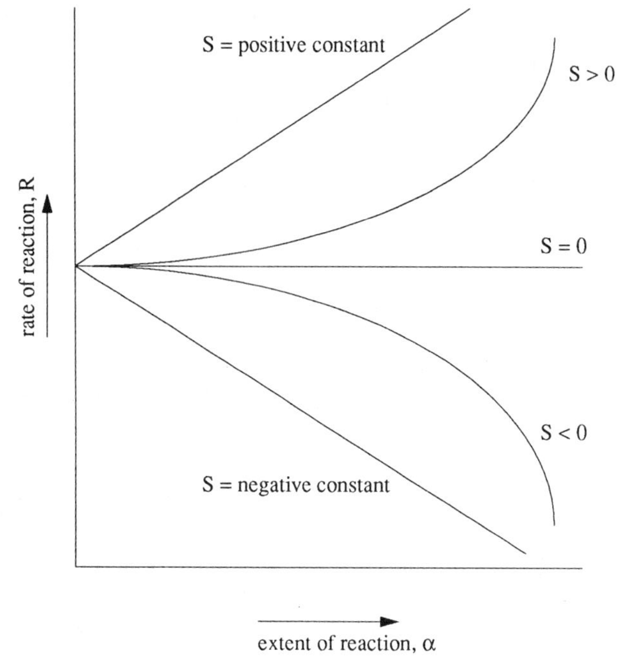

Fig. 2 Types of chemical reactions classified by S behavior

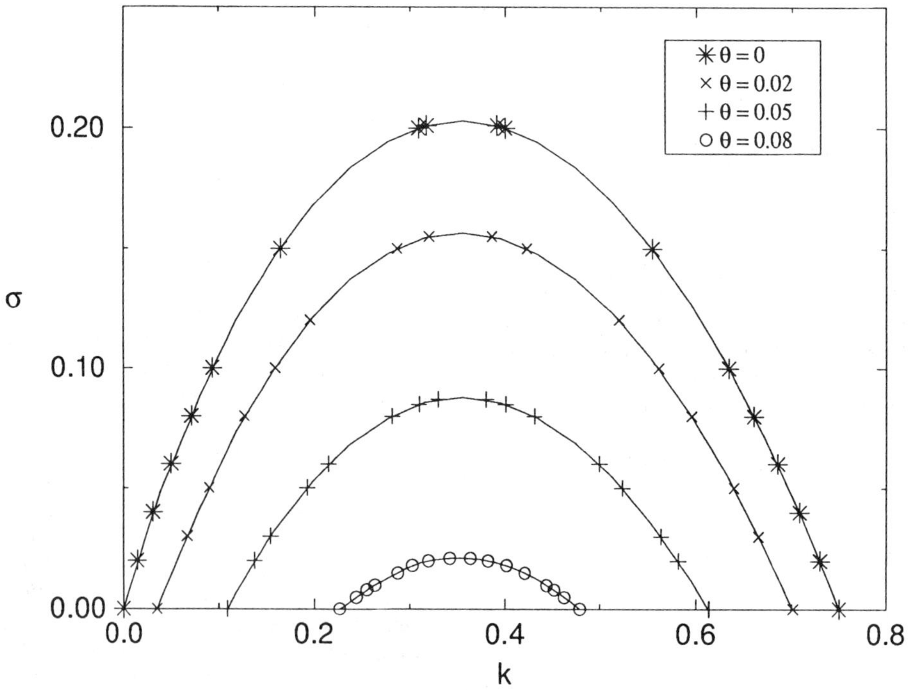

Fig.3 Growth constant versus wavenumber for ρ=3; S_1=-3; S_2=-1

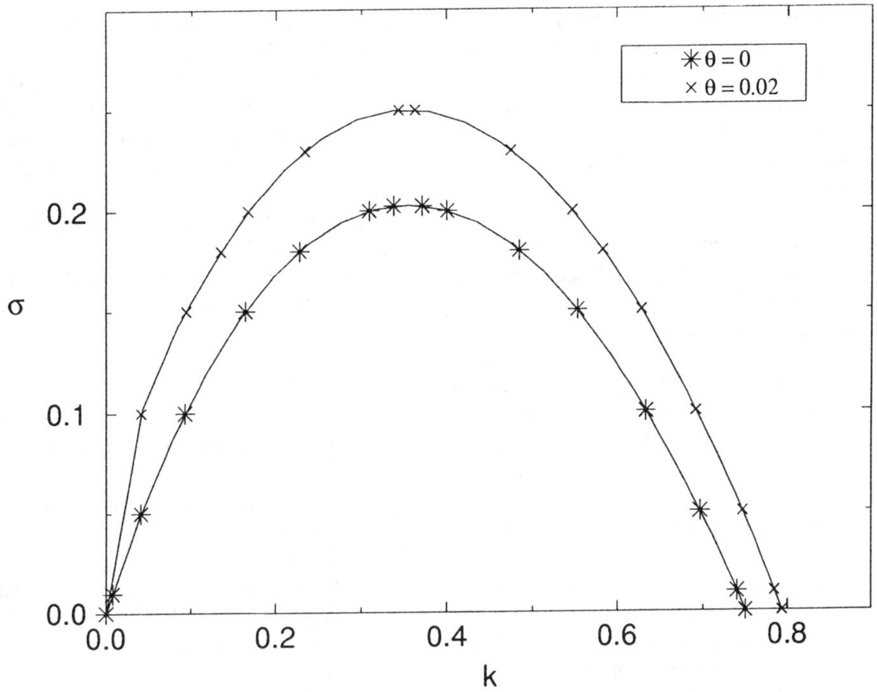

Fig.4 Growth constant versus wavenumber for ρ=3; S_1=+3; S_2=+1

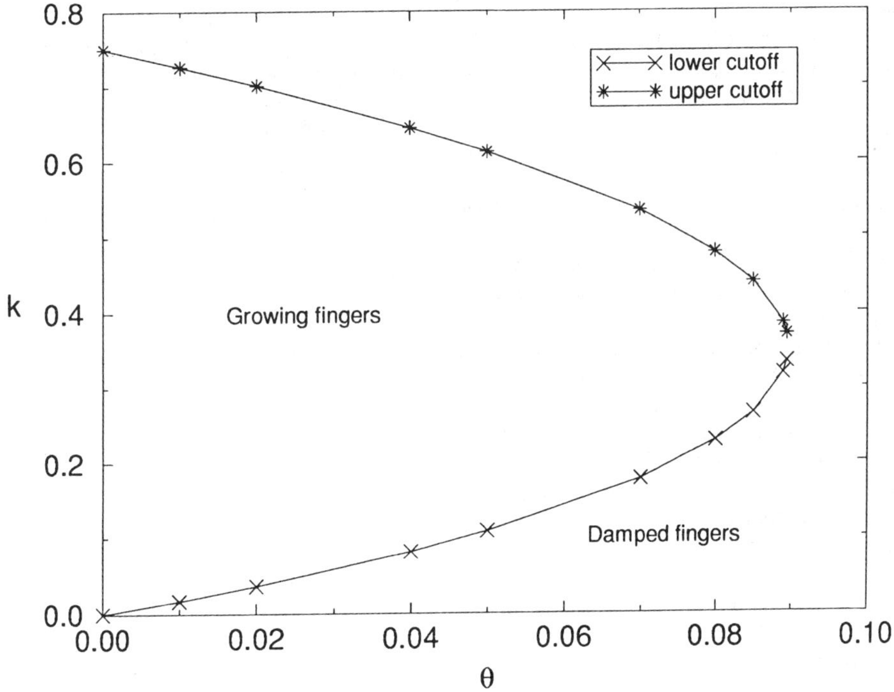

Fig.5 Marginal stability curves for ρ=3; S_1=-3; S_2=-1

a small range of intermediate finger sizes are unstable. The polymerization reaction combined with dispersion thus stabilizes a broader range of fingers. For the chosen values of S_1, S_2 and ρ (-3, -1 and 3 respectively) it is seen that, when the rate of reaction is about eleven times the rate of dispersion (θ>0.09), disturbances of all sizes decay.

For the other type of reaction with autocatalytic kinetics, the parameters S_1 and S_2 are both positive, chosen as +3 and +1 respectively. In this case, the unstable range of wavenumbers is larger with reaction than without reaction as seen in Fig. 4. Compared to the no-reaction case (θ=0), the upper cutoff wavenumber is greater for higher θ. Also, for any given wavenumber of the disturbance in the unstable range, the growth rate is higher for the θ>0 case. This means that positive S_1 and S_2 destabilize the flow process.

The marginal stability curve for the case with negative S_1 and S_2 is plotted in fig.5. For a given θ, fingers of wavenumbers within the envelope would grow, while the rest are damped out. The width of the unstable region decreases with a rise in θ. These results may be compared to results reported earlier for miscible displacement without reaction or dispersion. Hickernell and Yortsos (1986) have reported that without dispersion, any incipient fingers would invariably grow.

Computations are now in progress to obtain the solution of the eigenvalue problem at longer times (t>0) with frozen base state profiles.

6. CONCLUSIONS

The combined effects of chemical reaction and dispersion on the growth of flow instabilities in reactive flow through a fiber bed are strikingly different from the effect of dispersion alone on fingering. The result depends on the type of reaction as seen in the incremental rate of reaction with increasing extent of reaction, S. As the rate of chemical reaction is increased relative to the rate of dispersion, the process becomes increasingly more stable. If S is increasingly negative along the flow direction, and the reaction rate is of the order of the rate of dispersion, it is possible to eliminate fingers entirely. When S is positive, the unstable range of wavenumbers is increased with reaction. These effects are more pronounced when the spatial variation in S is steeper.

REFERENCES

Castro, J.M. & C.W.Macosko 1982. A.I.Ch.E. J. 28(2):250.

Gonzalez-Romero, V.M. & C.W.Macosko 1985. J. Rheol. 29(3):259.

Hickernell, F.J. & Y.C.Yortsos 1986. SIAM 74:93.

Homsy, G.M. 1987. Ann. Rev. Fluid Mech. 19:271.

Jayaraman, K., C.N.Satyadev & C.A.Petty 1992. presented at the Annual AIChE Meeting in Miami, Florida.

Losure, N.L. 1994. Ph.D. Dissertation, Michigan State University, East Lansing.

Tan, C.T. & G.M.Homsy 1986. Phys. Fluids 29(11):3549.

NOTATION

b	transverse to longitudinal permeability ratio
D	isotropic dispersion coefficient
k	wavenumber
$\underline{\underline{K}}_D$	Dimensionless permeability tensor
M	the initial mobility ratio
P	Pressure
$\underline{q}$	velocity in the moving reference frame
q_ξ	ξ-component of the velocity disturbance
R	rate of reaction
t	time
U_o	uniform injection velocity
α	conversion
$\Delta\alpha$	α_1-α_2
θ	(Da)(Pe)
ξ	length coordinate along the flow direction in the moving reference frame
λ	mobility
μ	1/mobility
ρ	d(ln μ)/dα
σ	exponential growth rate
ψ	conversion disturbance amplitude

Simulation of Materials Processing: Theory, Methods and Applications, Shen & Dawson (eds)
© 1995 Balkema, Rotterdam. ISBN 90 5410 553 4

The bulk processing of 2223 BSCCO powders

S.E.Schoenfeld, S.Ahzi & R.J.Asaro
Department of Applied Mechanics and Engineering Sciences, University of California, San Diego, La Jolla, Calif., USA

ABSTRACT: The anisotropic mechanical properties of densified BSCCO ($Bi - Sr - Ca - Cu - O$) powders are of paramount importance during thermo-mechanical processing of superconducting tapes and wires. The current work develops a micromechanically based material model for such densified powders. The deformation mechanisms of interest are crystallographic glide and porosity evolution, thus the model takes the form of a porous, elastic-viscoplastic polycrystal material theory. This has been achieved by coupling the modified Taylor polycrystal model of Schoenfeld et al. (1995a) in a generic way to yield surface type flow theories. It is shown here that the model can be used to describe the stress-strain behavior as well as texture and porosity evolution during the confined channel die compaction of 2223 BSCCO ($Bi_2Sr_2Ca_2Cu_3O_x$) powder compacts.

1. Introduction

Since the discovery of high temperature superconducting oxides, there has been great interest and research activity in the processing of superconducting films, tapes and wires. The fabrication of tapes or wires can be achieved via thermomechanical oxide powder-in-tube (OPIT) processing. Oxides of primary interest to the OPIT process are the Bismuth-based oxides, $Bi_2Sr_2Ca_1Cu_2O_x$ and $Bi_2Sr_2Ca_2Cu_3O_x$, which have superconducting transition temperatures of $\approx 85K$ and $110K$ respectively. This makes them attractive for use at temperatures up to $77K$.

The electronic properties in these materials are highly anisotropic. They preferentially superconduct (high critical current) on the basal planes of their orthorhombic crystalline structure. This implies that one of the most important conditions for obtaining high critical current in a polycrystal of these oxides is the alignment (texture) of the basal planes with the direction of the current (Uno et al. 1989, Jin et al. 1991). Other phenomena that play a role in reducing critical current in the polycrystalline oxide are: the presence of second phase particles, the presence of cracks, and low relative density (i.e. residual porosity). The shape and orientations of the grains as well as the grain boundaries also have an effect on the critical current. Optimum microstructures for high critical current were discussed by Asaro et al. (1992). The purpose of this paper is to show that the current model can be calibrated to predict deformation response and material microstructural states, at least in bulk form, in 2223 BSCCO powders.

2. Material Modeling

The macroscopic mechanical properties of the BSCCO powder are a product of the complex interaction of a number of deformation mechanisms. For the purposes of our model, the aggregate will be viewed as a porous, elastic-viscoplastic polycrystalline material. The lattice based deformation mechanisms will include both elastic stretching and rigid body rotations, as well as inelastic crystallographic glide. The inelastic crystallographic glide represents both dislocation motion and sliding following microcracking. The evolution of a void volume fraction contributes an inelastic, dilatational component to the macroscopic deformation. This is obtained by coupling the polycrystal stress state directly to a macroscopic flow theory for porous materials. The macroscopic velocity gradient, $\bar{\mathbf{L}}$, will admit the following additive decomposition;

$$\bar{\mathbf{L}} = \bar{\mathbf{L}}_{solid} + \bar{\mathbf{L}}_{void} = (\bar{\mathbf{L}}^p + \bar{\mathbf{L}}^*)_{solid} + \bar{\mathbf{L}}_{void} \;. \quad (1)$$

Where the velocity gradient of the solid, $\bar{\mathbf{L}}_{solid}$, contributes both an elastic part, $\bar{\mathbf{L}}^*_{solid}$, and a purely deviatoric inelastic part, $\bar{\mathbf{L}}^p_{solid}$. The velocity gradient due to void evolution, $\bar{\mathbf{L}}_{voids}$, will be a purely dilatational contribution to the inelastic flow.

The polycrystalline solid is represented by a large number of BSCCO single crystals with initially random or non-random distribution of the crystallographic orientations. Following the work of Asaro et al. (1992), these crystals deform by crystallographic gliding on the basal and lateral planes of the orthorhombic lattice. These gliding systems are (001)[100], (001)[010], (100)[010] and (010)[100]. The two lateral systems may represent sliding following microcracking on (100) and (010) planes rather than dislocation controlled slip. For simplicity, we neglect pressure sensitivity and friction due to sliding so that all the systems will be treated as slip systems obeying the Schmid rule. Note that only three of the four slip systems are independent; this implies that the three principal directions of the orthorhombic lattice are plastically inextensible. In other words, two degrees are missing for full kinematic freedom of the single crystals (Asaro et al. 1992, Ahzi et al. 1993). These two missing degrees of freedom can be represented by kinematic constraints on the plastic stretching tensor of the single crystal (Parks & Ahzi 1990, Schoenfeld et al. 1995a)

If we denote by $\mathbf{s}^\alpha$ and $\mathbf{m}^\alpha$ the unit vectors representing the slip direction and slip plane normal of the αth slip system in the current configuration, the plastic part of the velocity gradient for single crystal, $\mathbf{L}^p$, is defined as

$$\mathbf{L}^p = \sum_\alpha \dot{\gamma}^{(\alpha)} \mathbf{s}^\alpha \mathbf{m}^\alpha \quad (2)$$

where $\dot{\gamma}^{(\alpha)}$ is the rate of shearing on slip system α. The plastic inextensibility of the single crystal in the principal crystallographic directions can be represented by the following two independent constraint conditions (Parks & Ahzi 1990, Schoenfeld et al. 1995a):

$$\mathbf{L}^p \cdot \mathbf{C}^1 = 0 \text{ and } \mathbf{L}^p \cdot \mathbf{C}^2 = 0 \quad (3)$$

$$\mathbf{C}^1 = \mathbf{cc}; \;\; \mathbf{C}^2 = \mathbf{bb} - \mathbf{aa} \quad (4)$$

Where $\mathbf{a}$, $\mathbf{b}$ and $\mathbf{c}$ are the normalized vectors parallel to the crystallographic directions of the orthorhombic lattice.

If we denote by $\dot{\mathbf{n}}$ the rate of nominal stress, the single crystal elastic-viscoplastic behavior can be written in the following form;

$$\dot{\mathbf{n}} = \mathcal{K} : \mathbf{L} - \dot{\mathbf{S}} \quad (5)$$

where the fourth order tensor $\mathcal{K}$ represents the instantaneous modulus of the crystal and $\dot{\mathbf{S}}$ is a relaxation stress rate. Details of the derivation of the above constitutive relation are omitted here but could be found elsewhere (e.g. Asaro and Needlaman 1985, Nemat-Nasser & Obata 1986).

Now that the single crystal behavior is described, we suppose that the crystal is embedded in a polycrystalline aggregate that constitutes the solid part of the porous material. Further, we assume that this solid is subject to an average velocity gradient $\bar{\mathbf{L}}_{solid}$. The modified elastic-viscoplastic, Taylor type averaging scheme of Schoenfeld et al. (1995a) for crystals lacking five independent slip systems will be used to resolve the macroscopic stress state due to the solid. This model relates the velocity gradient of the single crystal to the average velocity gradient as,

$$\mathbf{L} = \mathcal{R} : \bar{\mathbf{L}}_{solid} \;. \quad (6)$$

The interaction tensor, $\mathcal{R}$, explicitly accounts for the constraint conditions of equation (3) which should be imposed on the single crystal. In terms of the constrained directions;

$$\mathcal{R} = \mathcal{P} : \langle \mathcal{P} \rangle^{-1} : (\mathcal{I} - \sum_i \mathcal{X}^i \mathcal{Q}^i) + \sum_i \mathcal{X}^i \mathcal{Q}^i \quad (7)$$

$$\mathcal{Q}^i = \frac{\mathbf{C}^i \mathbf{C}^i}{\mathbf{C}^i : \mathbf{C}^i} \quad (8)$$

$$\mathcal{P} = \mathcal{I} - \sum_i \mathcal{Q}^i \quad (9)$$

$$\mathcal{X}^i = \frac{\mathbf{C}^i : \mathbf{L}^*}{\mathbf{C}^i : \bar{\mathbf{L}}_{solid}} \quad (10)$$

where $\mathbf{L}^* = \mathbf{D}^e + \mathbf{W}^*$ and i summed from 1 to 2. $\mathbf{D}^e$ and $\mathbf{W}^*$ are the elastic stretching and spin tensors respectively. The above interaction law (6) prevents inelastic deformation in the constrained directions while maintaining a state of global compatibility. Details of the derivation of this interaction law are given in the work of Schoenfeld et al. (1995a).

Macroscopic equilibrium now gives the macroscopic constitutive law for the solid as;

$$\dot{\bar{\mathbf{n}}}_{solid} = \bar{\mathcal{K}} : \bar{\mathbf{L}}_{solid} - \dot{\bar{\mathbf{S}}} \tag{11}$$

where

$$\bar{\mathcal{K}} = < \mathcal{R} : \mathcal{K} > \ . \tag{12}$$

In the above equation < > denotes the volume average over the entire solid aggregate.

The densification of powder compacts involves various mechanisms such as plastic flow, creep and diffusion; all of which may contribute to the densification rate. Since we are presently interested in the room temperature processing of BSCCO, our discussion will be limited to densification by plastic flow; the dominant mechanism at low temperatures.

Fleck et al. (1992) considered powder compaction from an initially hand packed state where densification is controlled by limited contact between particles. The work generalizes a plastic potential for spherical particles under axisymmetric compression into, among other forms, a quadratic form for general loading purposes. The work extends from the limited contact regime to the fully plastic regime by interpolation of the plastic potential into one described earlier by Gurson (1977).

In the present work, we implement the quadratic form for the plastic potential (Φ) given by Fleck et al. (1992) throughout the low relative density regime, as well as their interpolation to the Gurson (1977) model at higher relative densities. If we denote by $\bar{\mathbf{D}}^p_{flow}$ the total plastic deformation rate (or macroscopic plastic deformation rate) that arises from the flow theory model, we assume normality of this plastic flow and write,

$$\bar{\mathbf{D}}^p_{flow} = \lambda \frac{\partial \Phi}{\partial \bar{\sigma}} \tag{13}$$

The plastic multiplier, λ, is computed by assuming that the plastic work of the porous materials is equal to that of the matrix, ie.

$$\bar{\mathbf{D}}^p_{flow} : \bar{\tau} = d\ \sigma_y\ \dot{\epsilon}^p_{solid} \tag{14}$$

where $\dot{\epsilon}^p_{solid}$ is the equivalent plastic strain rate of the matrix (solid). $\bar{\sigma}$ and $\bar{\tau}$ are the macroscopic Cauchy and Kirchoff stresses for the porous media. σ_y, the uniaxial yield stress of the fully dense aggregate, is calculated by inversion of the yield function $\Phi(\sigma_y) = 0.0$ such that we remain on the yield surface during loading of the aggregate. The evolution of σ_y is governed by the evolution of anisotropy and strain hardening in the polycrystalline aggregate that composes the solid. Using (13) and (14) we get

$$\lambda = \frac{d\ \sigma_y\ \dot{\epsilon}^p_{solid}}{(\partial\Phi/\partial\bar{\sigma}) : \bar{\tau}} \ . \tag{15}$$

We then use the volumetric part of $\bar{\mathbf{D}}^p_{flow}$ as the volumetric contribution of the voids to the macroscopic velocity gradient, ie;

$$\bar{\mathbf{L}}_{voids} = \frac{1}{3} tr\ (\bar{\mathbf{D}}^p_{flow})\mathbf{I} \tag{16}$$

and the rate of evolution for the relative density is now

$$\dot{d} = -d\ tr(\bar{\mathbf{L}}_{voids}) \ . \tag{17}$$

In the elastic-viscoplastic polycrystal model for the solid, the macroscopic stress state is used to calculate the resolved shear stress on each slip system. This resolved shear stress, in conjunction with the viscoplastic power law for slip, is used to resolve inelastic, deviatoric part of the velocity gradient in the single crystal. By averaging these deformations, we then calculate the inelastic deviatoric part of the macroscopic velocity gradient. Likewise, our implementation of the quadratic yield criteria uses the stress state in the solid to estimate the macroscopic, spherical part of the inelastic velocity gradient based on the macroscopic phenomenology of porous aggregates. More details concerning this material model and its numerical implementation can be found in Schoenfeld et al. (1995b).

3. Results and Conclusions

The above model has been calibrated to the confined channel die compaction experiment conducted by Schoenfeld et al. (1995b). The apparatus is shown schematically in Figure 1. Such an apparatus is designed to maintain a state of plane strain compression with confinement stress in the flow direction due to frictional contact between the plugs and the die walls, and between the plugs and the loading ram. The confinement stress is estimated using data from a separate experiment which measures the force required to push the copper plugs as a function of the applied load.

The observed flow properties are now assumed to be controlled by an initial resistance to sliding on each system. The evolution of this resistance

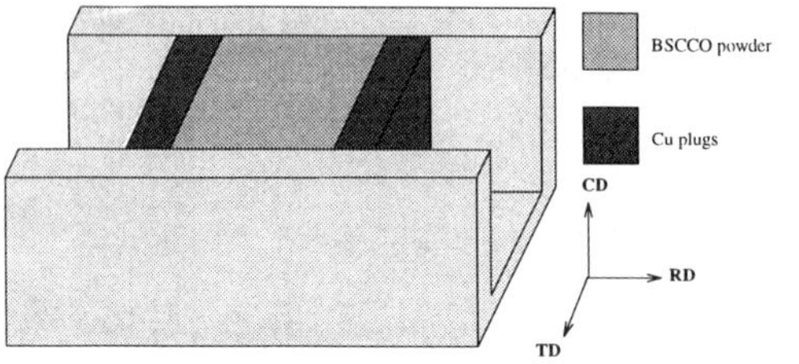

Figure 1: A schematic representation of the confined channel die apparatus.

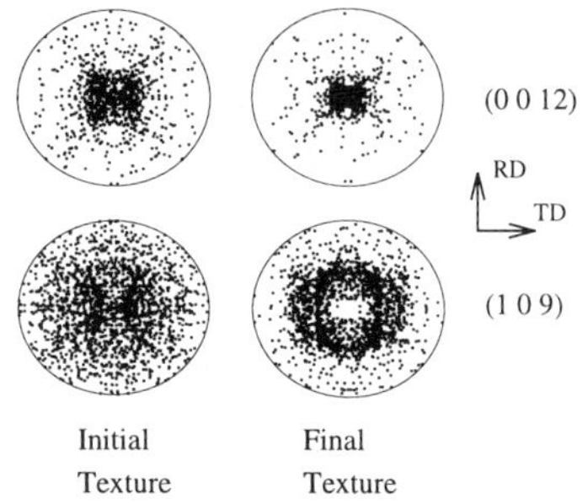

Figure 2: The estimated initial texture is shown by the (0 0 12) and (1 0 9) pole figures on the right while the final texture is shown on the left. RD and TD designate the rolling (channel) and transverse directions respectively.

is assumed to be linear, and any non-linear hardening in the solid is assumed to be due to increasing c-axis alignment with the compression direction. Since the only observed deviation from linear hardening at high densities shows an increase in hardening rate with increasing compressive strain, this is a reasonable assumption. For lack of experimental data concerning the true hardening properties on each system, the evolution of the resistance to shearing on each slip system, $\dot{g}^{(\alpha)}$, is set to be equivalent for all systems and proportional to the total accumulated strain on all the systems combined,

$$\dot{g}^{(\alpha)} = h \sum_{\beta} |\dot{\gamma}^{(\beta)}| \, . \tag{18}$$

For crystallographic slip on the basal planes of the unit cell, such hardening would be due to the increased density of dislocations and thus increased resistance to motion through a higher population of dislocations. Concerning the cracking followed by sliding on the lateral planes, such increased resistance would be due to increased contact area that is opened during the sliding that follows cracking on these planes. The hardening modulus, h, has been set here to $h = 162.0$ MPa which agrees with the initial slope (for compressive strains less than 0.15) of the experimental data for the solid.

Since there is always a degree of initial grain alignment due to pre-compaction, we have assumed that any compaction at very low density would most closely resemble that of axisymmetric compression. The initial grain alignment shown in Figure 2 gives the strain hardening required to simulate the experimental data. Once the desired strain hardening characteristics were achieved, the initial resistance to slip was scaled to give an initial macroscopic flow stress consistent with the experiment. For the experimental material data, this would correspond to $g_0^{(\alpha)} = 29.0$MPa. Stress and strain due to channel die compaction for the calibrated model is compared to the experimental observations in Figure 3

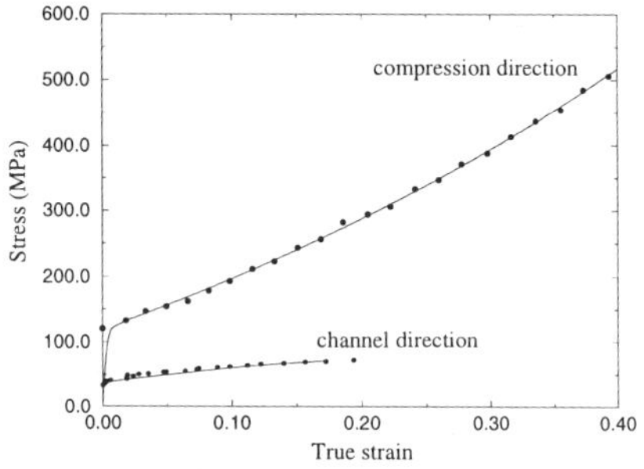

Figure 3: Experimental stress-strain response of compacted powder (circles) and stress strain response of model (solid line).

The compressive strain versus relative density has been calculated from the coupled porous-viscoplastic model using a hand-packed relative density of $d_0 = 0.50$ and the interpolation constants (Fleck et al., 1992) $d_1 = 0.86$ and $d_2 = 0.90$. Figure 4 shows these calculated results along with the experimental data.

In the preceding work, we have proposed a simple method of coupling macroscopic porosity models to the stress state from a detailed polycrystal model. Application of this model to the densification of 2223 BSCCO superconducting powder yields results in agreement with experimental observation.

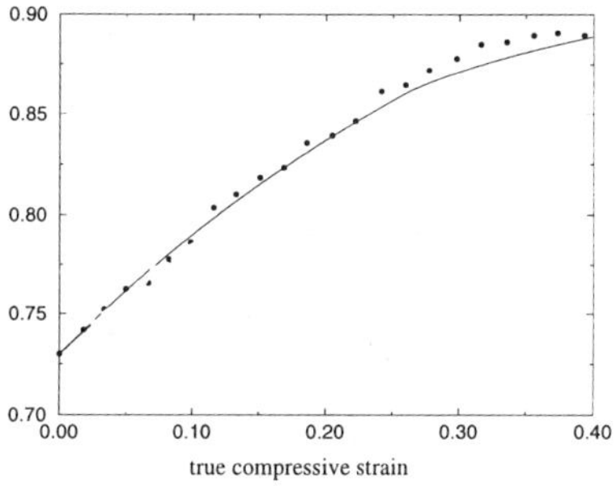

Figure 4: Experimental relative density vs. compressive strain response (circles) and the calculated relative density vs. compressive strain response (solid lines)

Acknowlegements

The Authors would like to thank J.O. Willis and W.R. Blumenthal for useful discussions. Work done at Los Alamos National Laboratory was performed under the auspices of the U.S. Department of Energy, Office of Energy Management. Support for this work was provided by the High Temperature Superconductivity Center at Los Alamos National Laboratory.

References

Ahzi, S., R.J. Asaro, and D. Parks 1993. Application of Crystal Plasticity Theory for Mechanically Processed BSCCO Superconductors. *Mechanics of Materials*15:201-222.

Asaro, R.J., S. Ahzi, W. Blumenthal, and A. DiGiovanni 1992. Mechanical Processing of High J_c BSCCO Superconductors. *Phil. Mag.*A66:517-538.

Asaro, R.J. & A. Needleman 1985. Texture Development and Strain Hardening in Rate Dependent Polycrystals. *Acta Metall.*33:923-953.

Fleck, N.A., L.T. Kuhn and R.M. McMeeking 1992. Yielding of Metal Powder Bonded by Isolated Contacts. *J. Mech. Phys. Solids*40:5:1139-1162.

Gurson, A. L. 1977. Continuum Theory of Ductile Rupture by Void Nucletion and Growth: Part 1 - Yield Criteria and Flow Rules for Porous Ductile Media. *Trans. ASME Ser. H, J. Engng Mater. Technol.*99:2-15.

Haldar, P. and L. Motowidlo 1992. Processing High Critical Current Density Bi-2223 wire and tape. *J. of Metals*44:54-58.

Jin, S., J.E. Graebner, T.H. Tiefel, R.B. Van Dover, and A.E. White 1991. High Critical currents in c-axis Textured $Bi-Sr-Ca-Cu-O$ superconductor Ribbons. *Physica C*177:189-196.

Nemat-Nasser, S. & M. Obata 1986. Rate-dependent, Finite Elasto-Plastic Deformation of Polycrystals. *Proc. Roy. Soc. Lon.* A407:343-475.

Sato, K.I, N. Shibuta, H. Mukai, T. Hikata, M. Ueyama and T. Kato 1991. Silver-shealded Bi(2223) wire and application. *Physica C*1990:50-52.

Sandhage, K.H., G.N. Riley and W.L. Carter 1991. Critical Issues in OPIT Processing of High-J_c BSCCO Superconductors. *J. of Metals*23:21-25.

Schoenfeld, S., S. Ahzi, and R.J. Asaro 1995a. Elastic-plastic Crystal Mechanics for Low Symmetry Crystals. *J. Mech. Phys. Solids, In Press.*

Schoenfeld, S., S. Ahzi, R.J. Asaro and W.R. Blumenthal 1995b. The Bulk Processing of 2223 BSCCO Powders; Part I: Densification and Mechanical Response *In Preparation.*

Parks, D.M. & S. Ahzi 1990. Polycrystalline Plastic Deformation and Texture Evolution for Crystals Lacking 5 Independent Slip Systems. *J. Mech. Phys. Solids*38:701-724.

Uno, N., N. Enomoto, H. Kikuchi, K. Matsumoto, M. Mimura and M. Nakajima 1989. The Transport Critical Current Property in High T_c Superconducting Wires. In I. Ishiguro and K. Kajmura (eds) *Advances in Superconductivity II (ISS'89)*

Simulation of Materials Processing: Theory, Methods and Applications, Shen & Dawson (eds)
© 1995 Balkema, Rotterdam. ISBN 90 5410 553 4

A planar isotropic yield criterion based on mechanical testing at multi-axial stress states

H. Vegter & P. Drent
Hoogovens Groep B.V., IJmuiden, Netherlands

J. Huétink
University of Twente, Enschede, Netherlands

ABSTRACT: In simulation models for sheet forming, the plastic behaviour of the material at multi-axial stress states is commonly described with a quadratic Hill yield criterion. The constants in this yield criterion are determined with uni-axial tensile tests. For the simulations of sheet forming, this description is not sufficiently accurate. Experimental information of tests at multi-axial stress states is necessary for a more advanced material model. A description is proposed here, which uses directly the test results at multi-axial stress states. A four point interpolation method is developed based on the pure shear point, uni-axial point, the plane strain point and the equi-biaxial point. With this description, the shape of the yield locus can be described easily as a function of strain. To obtain substantial improvement in the simulations, it was necessary to develop a more complex strain hardening law, which takes the plastic deformation state into account. The new material model is evaluated on a low carbon steel and on an aluminium 5000-alloy. Results of simulations with a finite difference model are compared with tests on axisymmetric products with a punch diameter of 75 mm. In the simulations, different material descriptions are compared with each-other.

1 DESCRIPTION OF THE MATERIAL MODEL

Most conventional descriptions of the material behaviour consist of two constitutive equations:

- The description of the (uni-axial) plastic flow stress as a function of strain, strain rate and temperature. Considering sheet forming, only strain hardening is included and usually, the effect of strain rate and temperature is neglected.
- The yield criterion for the translation of uni-axial to multi-axial stress states. More advanced models include planar anisotropy

In such a description the only strain history-effect is strain hardening, which is described as a function of the equivalent plastic strain:

$$F(\sigma_{ij},\theta) - \Phi_f(\overline{\varepsilon},\dot{\varepsilon},T) = 0 \qquad (1)$$

where:

F = Plastic yield function
Φ_f = Value of the plastic yield function for the onset of plastic deformation, i.e. the flow stress or an expression for the flow stress
σ_{ij} = stress tensor
θ = angle to the rolling direction
$\overline{\varepsilon}$ = equivalent plastic strain
$\dot{\varepsilon}$ = plastic strain rate
T = absolute temperature

From insights of the dislocation based theories for strain hardening and texture based yield loci, it is not likely that the plastic material behaviour is covered by these two separate equations [Dawson 1992, Aernoudt 1993]. Two arguments can be given:

- Strain hardening is dependent on the strain path during the plastic deformation. The dislocation substructure develops depending on the strain path history.
- The yield criterion depends on the crystallographic texture of the material. During plastic deformation, the texture changes due to the rotation of the slip planes in the lattice.

Therefore equation (1) must be extendend to:

$$F(\sigma_{ij},\varepsilon_{ij}^{his},\theta) - \Phi_f(\varepsilon_{ij}^{his},\theta,\dot{\varepsilon},T) = 0 \qquad (2)$$

where:

ε_{ij}^{his} = tensor describing strain history, which is a combination of the amount of deformation and the complete strain path

The strain tensor dependency (including strain-path history) in equation (2) is present in both the yield locus and the flow stress. In this work, the plastic behaviour of the material is derived from simple tests at straight strain paths. Not all the effects, which are included in (2), can be derived from these simple tests. For convenience, the following simplifications are proposed. Planar isotropy is considered and the effect of strain history is described by total strains, proportional equivalent plastic strain and (conventional) equivalent strain:

$$F(\sigma_{ij},\overline{\varepsilon_p}) - \Phi_f(\varepsilon_{ij},\overline{\varepsilon},\dot{\varepsilon},T) = 0 \qquad (3)$$

where:

$\overline{\varepsilon_p}$ = equivalent proportional strain, defined on the basis of proportional strain paths

ε_{ij} = total plastic strains

The stress strain data from the mechanical tests are transformed to lines of equal (conventional) equivalent strain in the stress space. These are considered to be the yield surface. A better but very laborious way to determine yield loci is the measurement of the stresses at the onset of plastic deformation on a pre-deformed material. In the following sections, only principal stresses and strains are considered, because the theory is used in axi-symmetric simulations. However, the theory still holds for a full set of stress components.

1.1 The yield criterion

A yield criterion is proposed on interpolation of the quantities derived directly from measured data. Four points of the yield criterion are considered to be the reference: Pure shear, uni-axial, plane strain and equi-biaxial. Between these points, a second order Bezier interpolation is used (Figure 1). For this interpolation, the tangent of the yield locus in these points must be known. The whole description is represented at the plane stress plane of the principal stress space:

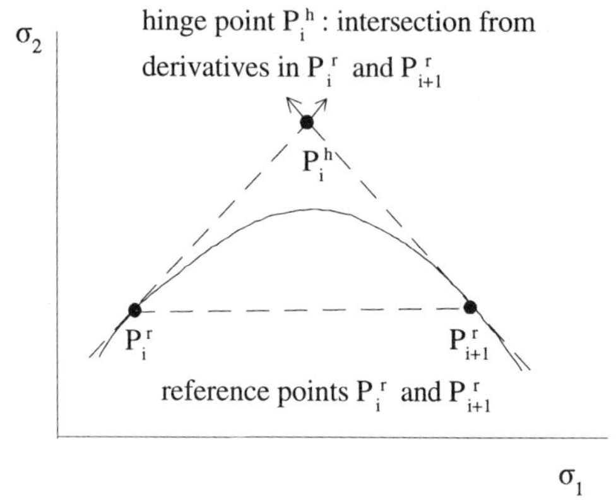

Figure 1: Second order Bezier interpolation (eq. 4)

$$\begin{pmatrix}\sigma_1\\ \sigma_2\end{pmatrix} = (1-\lambda)^2 \cdot \begin{pmatrix}\sigma_1\\ \sigma_2\end{pmatrix}_i^r + 2\cdot\lambda\cdot(1-\lambda)\cdot\begin{pmatrix}\sigma_1\\ \sigma_2\end{pmatrix}_i^h + \lambda^2\cdot\begin{pmatrix}\sigma_1\\ \sigma_2\end{pmatrix}_{i+1}^r \qquad (4)$$

where:
λ = interpolation parameter

The reference points are indicated with the subscript (r,i) in equation (4), and hinge points with the subscript (h,i). These hinge points are the intersection points of the slopes in two consecutive reference points, i and i+1. This description can easily be extended to more measured points. The yield locus is described as a function of deformation by the dependency of these reference points and R-value on deformation. An empirical formulation is chosen based on the following assumptions:

- The shape of the yield locus is dependent on the total strains but not on the way these strains are achieved. This assumption holds with the Taylor theory for the development of textures, in which the total strain is taken over by every individual crystal. The rotations of the slip planes depend on the total strains and do not necessarily depend on the way these total strains are achieved.
- The shape of the yield locus converges at large strains to an end shape. Implicitly, it is assumed that the texture of the material at large deformations at a straight strain path evolutes to a constant end texture.

It is obvious, that the shape of the yield locus must remain convex during all these operations. The following simple relation for the shape of the yield locus is assumed:

$$X = X_\infty + (X_0 - X_\infty)\cdot e^{-\omega\cdot\overline{\varepsilon}_p} \qquad (5)$$

where:

X = reference point i or tangent in point i; for the present description the stress points at pure shear, uni-axial state, plane strain state and equi-biaxial state and the R-value.

ω = steepness parameter

Subscripts ∞ and 0 mean the values at infinite and at zero strains, $\overline{\varepsilon}_p$

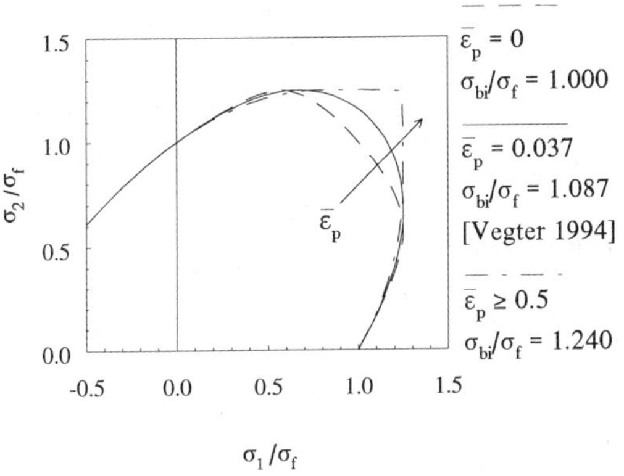

Figure 2: Development of yield loci of a low carbon steel with increasing (proportional) strain

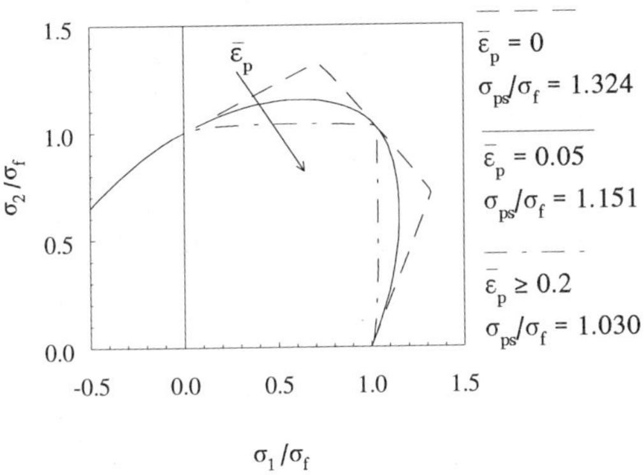

Figure 3: Development of yield loci of an Al-5000-alloy with increasing (proportional) strain

In figures 2 and 3, the development of the yield loci derived from the mechanical tests are given for a low carbon steel and an aluminium 5000 alloy. In these figures, three flow stresses are definied under particular plane stress conditions:

σ_{ps} = the axial plane strain flow stress,

σ_{bi} = the equi-biaxial flow stress.

σ_f = the uni-axial flow stress.

1.2 The work hardening description

The physical origin of work-hardening is caused by the multiplication of dislocations during plastic deformation. The existence of so called dislocation cells, which can be considered as a periodic distribution of dislocations, gives a realistic basis for the quantitative description of work hardening [Kuhlmann-Willsdorf, 1968]. Bergström [1969/70] derived a stress strain relationship, in which the effect of annihilation of dislocations by meeting of two dislocations of opposite sign was taken into account. Later on van Liempt [1994] added the effect of the change of shape of the dislocation cells, which is of importance for the description of strain hardening at large deformations such as rolling processes. A further modification is introduced here for taking the effect of dislocation cells with the strain tensor into account:

$$\begin{aligned}
\sigma_f &= \sigma_{f0} + \Delta\sigma_m\cdot y_d \\
\frac{dy_d}{d\overline{\varepsilon}} &= \Omega\cdot\left(f(\varepsilon_{ij}) - y_d\right) \\
y_{do} &= 1 - e^{-\Omega\cdot\varepsilon_0} \\
f(\varepsilon_{ij}) &= \frac{3\cdot e^{-\beta\cdot\varepsilon_i^{min}}}{e^{-\beta\cdot\varepsilon_1} + e^{-\beta\cdot\varepsilon_2} + e^{-\beta\cdot\varepsilon_3}}
\end{aligned} \qquad (6)$$

where:

σ_{f0} = yield stress

$\Delta\sigma_m$ = stress increase parameter for strain hardening

y_d = parameter $\propto \sqrt{\text{dislocation} - \text{density}}$

β = parameter describing strain hardening due to change of shape of the dislocation cells; if β = 1 the change shape of the cells is equal the macro-change of shape of the material by the total deformation.

Ω = annihilation parameter for the encounter of dislocations of opposite sign.

ε_0 = pre-deformation parameter resulting in a dislocation density parameter y_{do} at zero strain

ε_i^{min} = minimum of the strains $\varepsilon_1, \varepsilon_2, \varepsilon_3$

The previously described theory of work-hardening is based on the multiplication of immobile dislocations. The slip takes place by means of a relatively small amount of mobile dislocations, which also contribute to the flow stress. This can be considered as a dynamic part of the flow stress, which is incorporated in σ_{f0}-term of equation (6):

$$\sigma_{f0} = \sigma_0 + \sigma^* \qquad (7)$$

where:

σ_0 = yield stress of dislocation free material

σ^* = dynamic part of the flow stress.

Krabiell and Dahl [1982] derived the following relationship between the dynamic stress σ^* with strain rate and temperature, based on the thermal activated movement of mobile dislocations over the slip planes.

$$\sigma^* = \sigma^*_0 \cdot \left\{1 - \frac{\Delta G}{\Delta G_0}\right\}^{m'} \quad ,0 < \Delta G < \Delta G_0$$
$$\sigma^* = \sigma^*_0 \quad ,\Delta G = 0 \qquad (8)$$
$$\sigma^* = 0 \quad ,\Delta G = \Delta G_0$$
$$\Delta G = k \cdot T \cdot \ln\left(\frac{\dot{\varepsilon}}{\dot{\varepsilon}_0}\right)$$

where:

σ^*_0 = limit dynamic flow stress

ΔG = activation enthalpy

ΔG_0 = maximum activation enthalpy,

m' = power for the strain rate behaviour

k = Boltzmann constant = 8.617 . 10^{-5}eV/K

T = absolute temperature in (K)

$\dot{\varepsilon}_0$ = limit strain rate for thermal activated movement

2 DETERMINATION OF THE MATERIAL CONSTANTS

In order to obtain the various constants in equations (4-8), three kind of tests were carried out:

- Uni-axial tensile tests at the different strain-rates and temperatures for the determination of the constants: σ_0 , $\Delta\sigma_m$, Ω , σ^*_0 , m' and the R-value. The materials were tested in three sheet directions (0°, 45° and 90° to the rolling direction). In the present planar isotropic description, mean values are used.
- Plane strain tensile tests, for the determination of the plane strain point in the yield criterion (figure 2,3). In addition plane strain tests on pre-rolled material are carried out to obtain more insight in strain hardening at large deformations. (These last tests were indicated with plane strain 2x in figure 4).
- Hydraulic bulge test with a diameter of 90 mm for the determination of the equi-biaxial point of the yield criterion. The stress at the equi-biaxial point are calculated from the internal pressure and the curvature at the pole of the bulge.

Values of the constants are represented in table 1. In this table, σ_{sh} is definied as the positive stress component during plastic flow under pure shear conditions.

The four constants ΔG_0 , $\dot{\varepsilon}_0$ are taken from literature [Krabiell 1982, Vegter 1990, 1994]. The parameter ε_0 is fixed on a value of 0.005 for steel and 0.0 for aluminium, which corresponds with the amount of pre-deformation during skin passing.

In figures 4 and 5, stress strain curves for the different tests are represented for both the low-carbon-steel and the aluminium 5000 alloy. The data from these tests are described satisfactory with the present material model.

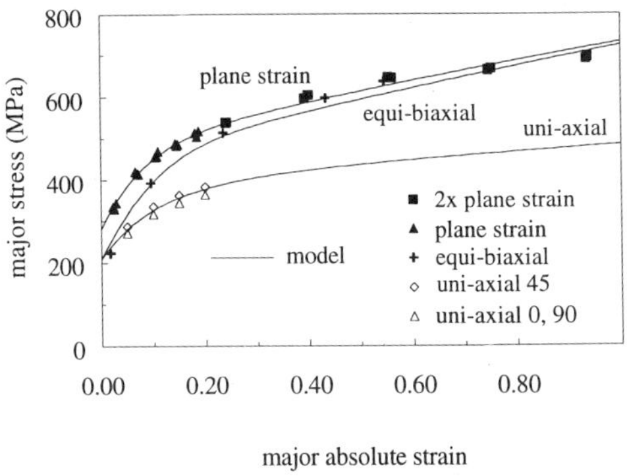

Figure 4: Comparison of test results with the present material description (constants in table 1) for low-carbon-steel

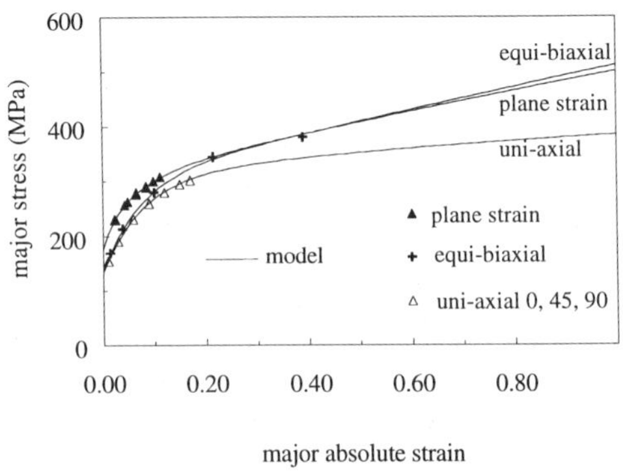

Figure 5: Comparison of test results with the present material description (constants in table 1) for Al-5000-alloy

Table 1: Values of the material constants in eqs. (4-8) for steel and aluminium

	Strain hardening constants (eq. 6)					Strain rate constants (eq. 8)			
material	σ_0 (MPa)	$\Delta\sigma_m$ (MPa)	β	Ω	ε_0 (fixed)	σ^*_0 (MPa)	ΔG_0 (eV) (lit.)	m' (fixed)	$\dot{\varepsilon}_0$ (s^{-1}) (lit.)
LC- steel	169.2	193.5	1	9.475	0.005	585.7	0.8	2.2	10^8
Al-5000	133.2	182.9	1	12.53	0.0	0	-	-	-

	Yield criterion (eq. 4,5) reference stress points and R-value							
material	R	σ_{sh}/σ_f	σ_{ps}/σ_f (∞)	σ_{ps}/σ_f (0)	ω_{ps}	σ_{bi}/σ_f (∞)	σ_{bi}/σ_f (0)	ω_{bi}
LC-steel	1.693	0.555	1.250	1.250	-	1.240	1.000	12.21
Al-5000	0.865	0.575	1.030	1.324	17.71	1.026	1.026	-

3 COMPARISON OF SHEET FORMING SIMULATIONS WITH EXPERIMENTS

For the present purpose, a relatively simple calculation method is applied, because this makes the introduction of different constitutive relations more easy. In this model, the following assumptions are made:
- One-dimensional discretization; only division into cells in the profile direction.
- Solution of the equation of the force equilibrium in the profile direction. No gradients in stress and strain over the thickness, which means that only membrane stresses are considered.
- Only plastic deformation is considered
- Coulomb friction is assumed.

For the verification of the sheet forming model, stretch forming tests were performed. The punch diameter was 75 mm with a radius of 22.5 mm. The punch velocity was 1 mm/s. The applied material was a low-carbon steel with properties according to table 1. Preserving oil was used as a lubricant. A coefficient of friction of 0.15 is supposed in the simulations.

To demonstrate the influence of the yield criterion on a stretching process, we only varied the equi-biaxial stress point as indicated in figure 2. Four types of simulations were carried out:

a $\sigma_{bi} = \sigma_{bi,0}$

b $\sigma_{bi} = \sigma_{bi,Hill}$ according to a former yield criterion with real powers [Hill 1979, Vegter 1994]

c $\sigma_{bi} = \sigma_{bi,\infty}$

d $\sigma_{bi} = \sigma_{bi,\infty} + (\sigma_{bi,0} - \sigma_{bi,\infty}) \cdot e^{-\omega_{bi} \cdot \varepsilon_p}$, variation with strain according to equation (5) and constants in table 1.

In figures 6 and 7 predictions of the thickness-distributions and force-displacement curves are compared with measurements. The predictions with the sheet forming model agree very well with the experiments for the last two models (case (c) and (d)). Some underestimation of stretching from the bottom is predicted in these models. The amount of stretching from the bottom is very large in the case a low value of the equi-biaxial stress is used (case (a) and (b)), resulting in a poor agreement with the experiments. In comparison with the former description with real powers (based on only the uniaxial point and the plane strain point) [Vegter 1990 1994], substantial improvement in the reliability of the simulations has been achieved.
The present model gives also satisfactory results for deep-drawing simulations. These are not further considered here, because the influence of the position of the equi-biaxial point of the yield criterion is demonstrated more clearly by means of stretch-forming simulations.

4 CONCLUSIONS AND RECOMMENDATIONS

The model based on a realistic description of the plastic behaviour at proportional deformations gives substantial improvement of the reliability of the predictions of axi-symmetric sheet forming processes in comparison with former work [Vegter 1991, 1994].

For the prediction of non proportional hardening processes, it is necessary to carry out multi strain path tests.

For further improvement of the material model for deep drawing purposes, the pure shear point (σ_2= -

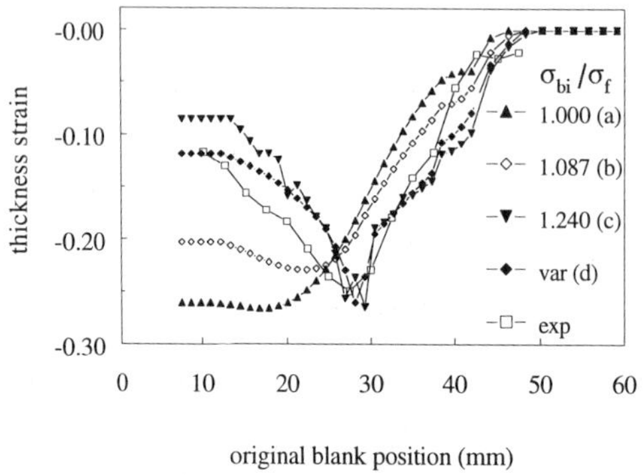

Figure 6: Comparison of calculated thickness strain distributions with measured ones during punch stretching

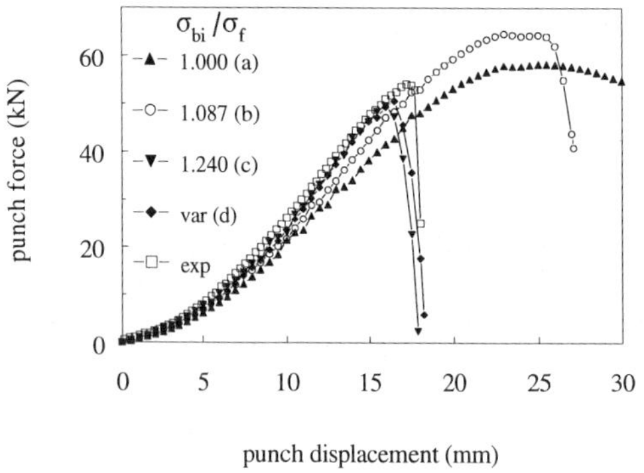

Figure 7: Comparison of calculated force displacement curves with experiments

σ_1) of the yield criterion has to be determined, because this point determines the draw in force from the flange into the die radius-region. A plate shear test provides possibly the information for this point [Huétink 1991, Vreede 1992].

In addition to the present work, texture based yield loci will provide more insight in the development of the plastic behaviour at multi-axial stress states [Aernoudt 1993, Dawson 1992].

ACKNOWLEDGEMENT

We want to thank mr. G. Monfort from C.R.M. Liège for the use of the hydraulic bulge tester

REFERENCES

Aernoudt E., Van Houtte P, Leffers T, 1993, "Deformation and textures of metals at large strains", In: Material science and technology, 6: 89-136.

Bergström Y, 1969/70, "A dislocation model for the stress strain behaviour of polycrystalline α-Fe with special emphasis on the variation of the densities of mobile and immobile dislocations", *Mat. Sci. Eng.*, 5: 179-192.

Dawson P.R, Beaudoin A.J, Mathur K.K, 1992, "Simulating deformation-induced texture in metal forming", In: Proceedings *Numiform '92*, edited by Chenot J.L, Wood R.D, Zienkiewicz O.C, Sophia Antipoles, France, 25-33.

Hill R, 1979, "Theoretical plasticity of textured aggregates", *Math. Proc. Camb. Ph. Soc.* 85: 179-191.

Huétink J, 1991, "Extension of anisotropic hardening elastic-plastic-theory to finite strains based on visco-elastic finite strain theory", *IUTAM-symposium 1991, University of Hannover.*

Krabiell A, Dahl W, 1982, "Zum Einfluss von Temperatur und Dehngeschwindigkeit auf die Streckgrenze von Baustahlen unterschiedlicher Festigkeit", *Arch. Eisenhüttenwesen.* 52: 429-436.

Kuhlmann-Willsdorf D, 1968, "Unified theory of stages II and II of workhardening in pure fcc metal crystals", In: "*Workhardening*", J.P Hirth and J. Weertman, eds, *Gordon and Breach*, N.Y. 8: 97-132.

Liempt P. van, 1994, "Workhardening and substructural geometry of metals", In: Proceedings *Metal Forming 94*, Hartley, P et. al, *University of Birmingham*, UK, 459-464.

Vegter H, 1991, "On the plastic behaviour of steel during sheet forming", Dissertation, *University of Twente*, The Netherlands.

Vegter H, 1994, "Influence of different material descriptions on simulations of axisymmetric sheet forming processes", In: Proceedings *Shemet '94*, Sing, U.P, et.al, *University of Ulster*, 165-175.

Vreede P.T, 1992, "A finite element method for simulations of 3D-dimensional sheet metal forming", Dissertation, *University of Twente*, The Netherlands.

Simulation of Materials Processing: Theory, Methods and Applications, Shen & Dawson (eds)
© 1995 Balkema, Rotterdam. ISBN 90 5410 553 4

A numerical and experimental investigation of the bendability of AA6000 extrusions formed in rotary draw bending

T.Welo & F.Paulsen
SINTEF, Materials Technology, Rich. Birkelands, Trondheim, Norway

S.R.Skjervold
Hydro Aluminium a.s., R&D Center-Karmøy, Håvik, Norway

SUMMARY: This paper presents three-dimensional, elasto-plastic finite element analyses of an industrial forming process—the so-called rotary draw bending method. The overall objective of the investigation is to determine the applicability of using numerical simulation for the prediction of geometrical tolerances and formability of thin-walled, rectangular hollow sections. Particular attention is placed on the effects of material properties, die radius and bend angle on (a) elastic springback and (b) the development of undesirable deformation modes like local post buckling and sagging of individual members of the profile. The computations have been validated by a number of tests conducted in an industrial bending machine. The results show that the simulated and experimental results compare quite well.

1 INTRODUCTION

The trend in the automotive industry is to reduce weight and increase the recycling value through the application of alternative materials such as aluminum alloys. This has called for new design concepts involving various applications of curved thin-walled extrusions (M. Isogai & S. Murakami 1993). These are commonly cold formed in stretch bending or rotary draw bending. Advances in supplying competitive solutions, however, presume basic insight into the fundamental variables that influence the bending process.

The major concern in industrial bending is to meet the geometrical tolerances specified. In this connection, basic mechanisms governing elastic springback and failure modes have to be considered. Elastic recovery is sensitive to the stress–strain curve of the material, indicating that extrusion parameters and time to bending must be consistent for alloys that age naturally at room temperature (Welo *et al.* 1994a). Springback effects must also be taken into account when designing the tools.

In order to minimize cross-sectional distortions owing to post buckling and sagging, the bending operation and profile geometry have to be correctly designed. Moreover, the strain hardening characteristic of the material will also to some extent enters into the discussion on local tolerances.

Numerical simulation is today's modern alternative to trial-and-error practice. Finite element analyses have been used successfully to reduce the lead time and optimize the product in a variety of metal forming operations. Application of FE analysis to industrial bending, however, represents a great computational challenge due to the combination of large-strain plasticity and elastic recovery, contact problems and geometrical complexity. The aim of this paper is therefore to demonstrate the applicability of using commercial FEA programs (MARC K6.1/MENTAT 2.1 1994) to real practical bending problems.

2 EXPERIMENTAL

2.1 Materials

Two aluminum alloys from the AA6000 series with different stress-strain characteristics were cast and homogenized using industrial production equipment. The billets were extruded under close control using full-scale extrusion parameters.

Uniaxial quasi-static tensile testing, accord-

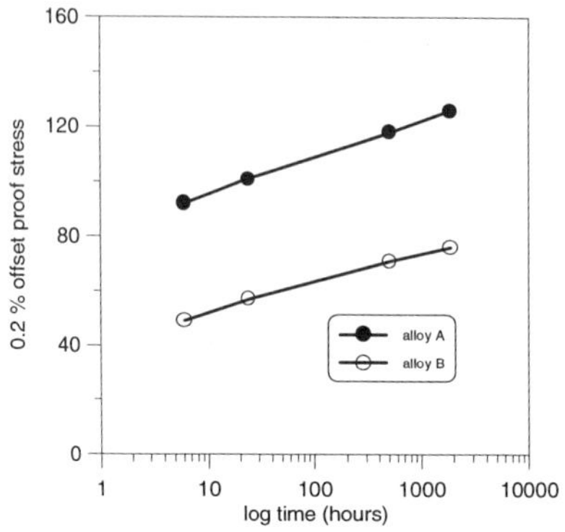

Figure 1. Room temperature stability of the two alloys.

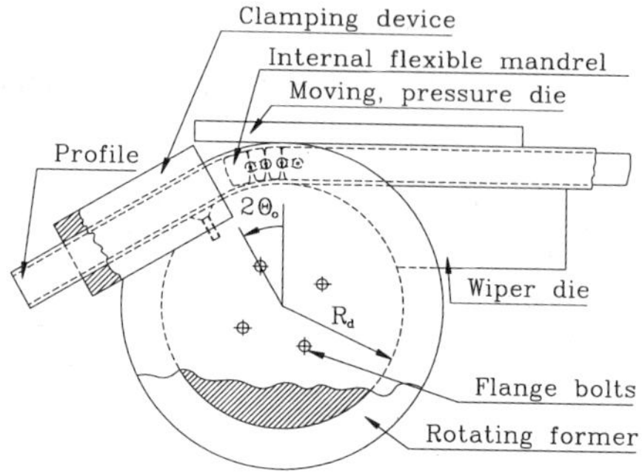

Figure 2. Schematic illustration of the HYDROMAX bending machine.

ing to the DIN-50145 standard, was conducted on specimens taken parallel to the extrusion direction. Samples were tested at various time after extrusion in order to characterize the room temperature stability of the alloys, see Figure 1. The data extracted from the force-elongation curve was fitted to the Ramberg-Osgood (1943) model, see Section 3, and used as input to the FE analyses. The material parameters are listed in Table 1. Here σ_u and ε_u are respectively the true stress and natural strain at the onset of necking.

Table 1. Material data.

Alloy†	E	a	σ_a	m	σ_u	ε_u
A	69,000	0.2	118.3	2.18	269.8	0.17
B	69,000	0.2	71.4	1.72	190.2	0.16
	MPa	–	MPa	–	MPa	–

†Data obtained approximately six weeks after extrusion.

2.2 Set-up and procedure

An experimental program was run six weeks after extrusion using the HYDROMAX bending machine at Hydro Aluminum Automotive Structures Tønder. The set-up being schematically illustrated in Figure 2. HYDROMAX is basically a rotary draw bending machine modified to accomplish production flexibilities.

Upon bending, the length of the profile is passed over the lubricated mandrel. The front-end portion of the profile is clamped to the former with the grip die. Then, the hole assembly is rotated together with the mandrel until the centroid of the rear linkage is located next to the former. From this stage, the mandrel is held at a fixed position until the required degree of bend is achieved.

The cross section of the extrusions is shown in Figure 3. The bending radii varied between 100 mm and 400 mm, while the bending angles were in the range of 10° to 60°.

Rotary draw bending is well-suited for tight-radius bending due to the mandrel and wiper dies, which give additional support upon bending. Since the bending moment is almost constant along the bend, the contact stresses are quite low provided that no clamping occurs between the profile and the mandrel. Friction between the profile and the stationary tool parts produces internal tensile actions that reduce springback and improve the buckling resistance.

The global bending angles of prime interest are $\Delta\theta_b$ and $\Delta\theta_o$, shown in Figure 4. The former is generally greater than the latter since elastic recovery occurs over the portion that is clamped between the pressure die and the wiper die. The geometry was characterized using a MESSWERK 3D coordinate machine. The radius of curvature was taken along a line marked on the mid-depth of the extrusion to avoid errors due to local distortions. The bending angle was characterized on the basis of points situated on the straight end portion and in regions *o–b*, meaning that the measured value will be something between $\Delta\theta_b$ and $\Delta\theta_o$.

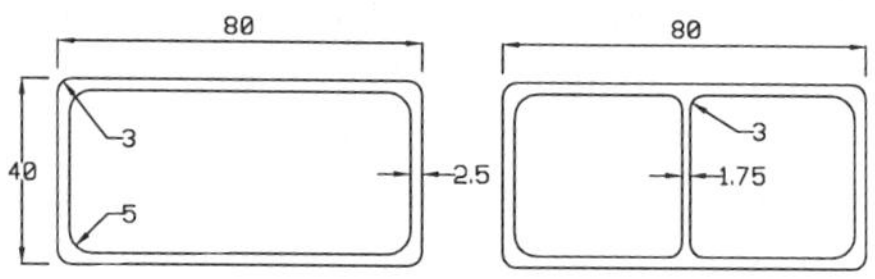

Figure 3. Geometry of cross sections.

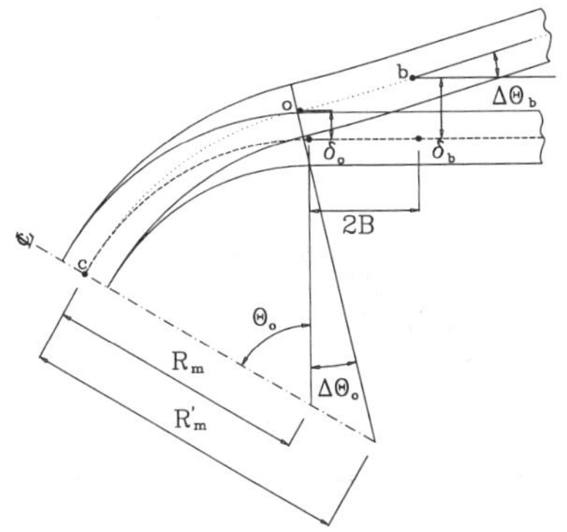

Figure 4. Main parameters describing deformation of the profile (half bend) before and after unloading.

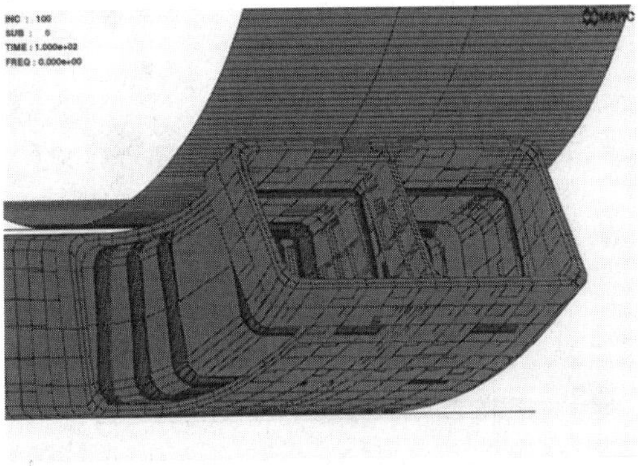

Figure 5. Finite element model.

3 MODELING

3.1 Constitutive equation

The Ramberg-Osgood (1943) model is commonly used to represent the stress–strain characteristic for heat treatable aluminum alloys. In the uniaxial case, the present relationship can be quoted as follows:

$$\varepsilon_n = \varepsilon_n^e + \varepsilon_n^p = \frac{\bar{\sigma}}{E} + \frac{a}{100}\left(\frac{\bar{\sigma}}{\sigma_a}\right)^{2m+1} \tag{3.1}$$

where σ_a and m are the a per cent offset proof stress and strain hardening parameter, respectively. By combining Eq. (3.1) with the well-known Prandtl-Reuss equations, the incremental constitutive law yields

$$d\varepsilon_{ij} = \frac{1+\nu}{E} ds_{ij} + \frac{1-2\nu}{E}\frac{d\sigma_{kk}}{3}\delta_{ij} + \frac{3(2m+1)}{400}\left(\frac{a}{\sigma_a}\right)\left(\frac{3J_2}{\sigma_a^2}\right)^m\left(\frac{dJ_2}{J_2}\right)s_{ij} \tag{3.2}$$

for the criteria,

$$J_2 = \frac{\bar{\sigma}}{3}, \qquad dJ_2 > 0$$

where E is Young's modulus, ν is Poisson's ratio, s_{ij} is the deviatoric stress, $\sigma_{kk}/3$ is the mean stress, J_2 is the second deviatoric stress invariant and $\bar{\sigma}$ is the equivalent von Mises stress. The parameters used in the analyses are listed in Table 1.

3.2 Finite element models

The rotary stretch bending process was analyzed by the implicit FE program MARC K6.1, using full integration linear brick elements to achieve an accurate representation of complex contact conditions. The assumed strain formulation was invoked to improve the bending behavior of the elements (F.J.H. Peters 1989). In this procedure, the interpolation functions are expressed with respect to the distorted element such that the prime bending modes are represented. Two elements were used across the thickness of individual cross-sectional members to get a realistic representation of local deformations (F. Haase 1993). One disadvantage of the assumed strain formulation, however, is the increased computational cost due to increased assembly of the stiffness matrix.

The simulations were divided into three series:

(i) rotary draw bending, no mandrel;

(ii) rotary draw bending with a rigid or stationary mandrel;

(iii) draw bending with a flexible or moving mandrel, applying tyings at the end.

In each case bending was analyzed by forcing the profile around the former die, applying non-linear constraint equations to represent the translation of each node at the extreme end of the profile. External traction was applied by imposing smaller deformation increments to the node tyings at the other extreme end; i.e., introducing combined bending and stretching. A conventional rotary draw bending method is represented by releasing the latter tying.

In Figure 5, a model of a double chamber profile with a flexible mandrel and a 100 mm

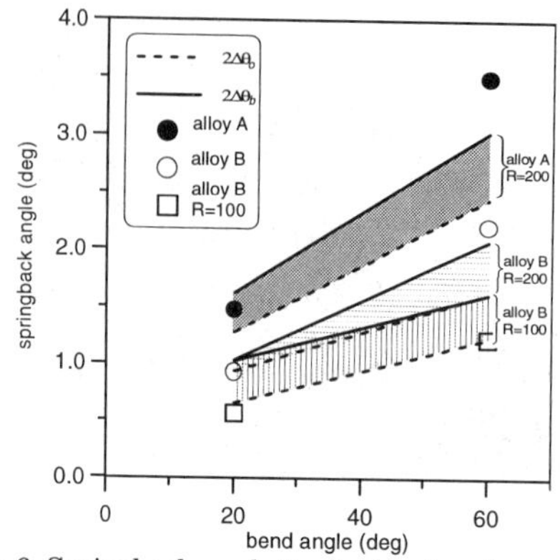

Figure 6. Springback angle for double chamber section of the two alloys; markers indicate mean test results.

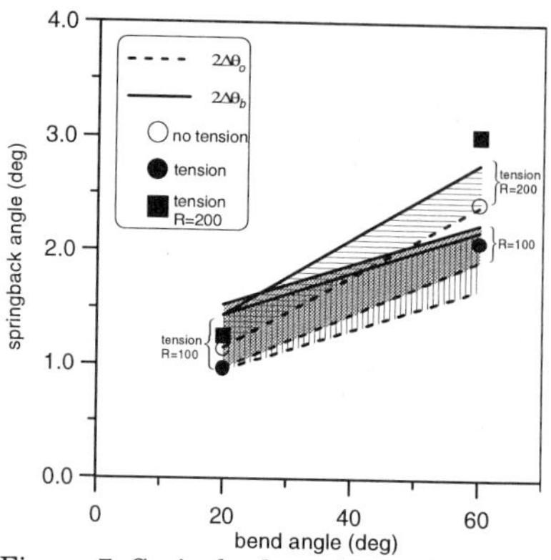

Figure 7. Springback angle for double chamber section (alloy A), including effect of tension.

die radius is illustrated at the final position. Taking advantage of symmetry, typically one half of the FE model consists of 2700 elements. The mandrel was made up by four rigid bodies. These were displaced according to a prescribed path made by the extremities of the mandrel when translating and rotating each body around the current position of each linkage. The assembly was imposed to the same speed as the profile until the centroid of the rear linkage was situated next to the former. At this point the mandrel became fixed, introducing a relative velocity between the profile and the tooling. When applying a finger mandrel, on the other hand, a rigid body was held stationary inside each chamber of the profile throughout bending.

Friction between the profile and the internal support was represented by a classical Coulomb model. No friction was applied to the other interfaces since the relative velocity being zero.

4 RESULTS AND COMPARISON

4.1 Elastic springback

The overall geometrical tolerances are of great interest in case the bent shape is to be fitted together with other parts. In practice, elastic springback is taken into account when designing the tooling and specifying an overbend. The main concern is therefore the repeatability in springback, rather than its magnitude.

Figure 6 shows elastic recovery as a function of bend angle for the two materials. Here the broken and full lines represent the predicted $2\Delta\theta_o$ and $2\Delta\theta_b$, respectively. The springback angle increases with increasing bend angle mainly as a result of the length of the bent portion of the profile. The same comments apply to the influence of the bending radius—although the bending moment increases with decreasing radius, provided the cross sectional distortions being small. It can also be observed that elastic recovery of alloy A is larger than that of alloy B. This is due to a higher current stress level at the instance of unloading for the former, indicating that the initial yield stress and the strain hardening characteristic are the two prime material parameters.

The elastic springback angle, Figure 7, is seen to decrease with increasing tension applied at the ends. This can be interpreted in terms of the cross section's interaction characteristic between flexural and axial actions: the bending moment required to produce a bend with a given radius decreases under a state of combined loading, as will the released curvature which is illustrated in Figure 8.

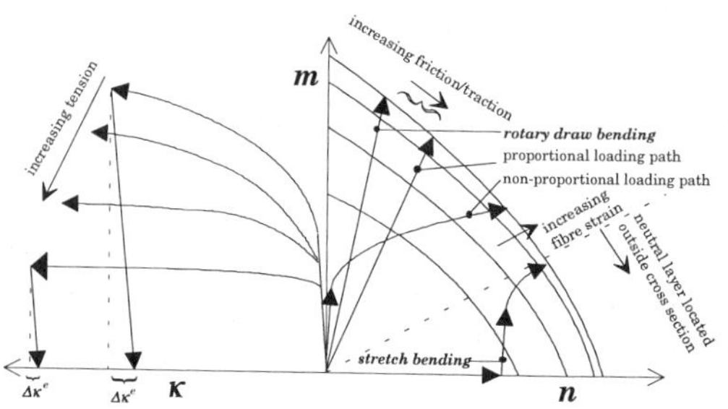

Figure 8. Springback curvature (κ^e) and loading path in the moment–axial space (m–n).

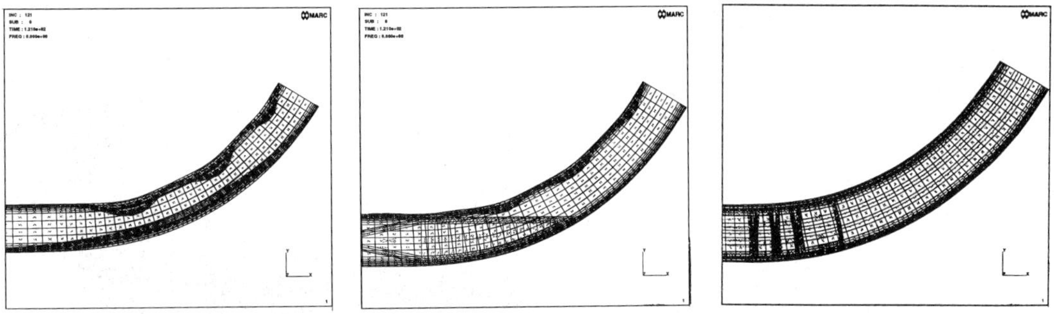

(a) (b) (c)

Figure 9. Development stages: (a) rotary draw bending; (b) rotary draw bending with rigid mandrel; draw bending with flexible mandrel.

In case the frictional conditions at the interfaces vary from time to time, the tolerances of the bent part will not be repeatable.

There is a tendency that the simulation slightly underestimates the springback for the largest bend angle (60°). However, the overall impression is that the numerical results compare reasonably well with the experimental ones.

If the external tension is sufficient high such that the neutral layer is located outside the cross section, elastic springback will be correspondingly low. Then, the current average slope in the stress-strain curve—rather than the stress level—represents the main material parameter with respect to overall geometrical tolerances. It is also worth noting that in case excessive cross-sectional distortions occur, elastic springback increases as compared to that of a faultless bend.

4.2 Local buckling and sagging

In bending of thin-walled extrusions, the most important failure mechanisms are post buckling and sagging of individual members. The former appears as one or several folds developing along the flange that is located at the internal face. The main cause for this mechanism is heavy compressive stresses acting along the curved flange; i.e., shell effects. Sagging appears as one single wave developing in the longitudinal direction of the external flange. This mechanism can be interpreted in the same way as post buckling, but the principal stress is reversed such that a state of bifurcation being omitted.

Close local dimensions are demanded in regions that are visible or are to be welded together with other parts, otherwise more flexible tolerances may be permitted. In production, internal supports or mandrels are commonly used to improve the appearance of the bend.

Figure 9 illustrates three different stages in the development of an industrial bending operation based on FE simulations. In case (a) excessive post buckling and sagging waves appear along the flanges of the extrusion. Applying a stationary mandrel, this gives perfect support to the flange at the external face (b), but folds are still observed to occur on the internal face. The draw bending method with a flexible mandrel and tyings applied on the ends (c), however, appears to provide an excellent bend.

Figure 10 shows the maximum post buckling and sagging depths for a single chamber section without mandrel. It is noticed that the bucking depth increases more steeply than the sagging depth at an early stage of the process. Once the first buckling mode is formed, the maximum buckling depth remain nearly stationary upon further bending. However, the sagging depth proceeds to increase during the entire process; in idealized pure bending, its maximum value is mainly limited to bending radius, but effects from the constraints at the extreme ends change this picture in rotary draw bending.

The number of wrinkles, i.e. buckling length, are correctly predicted in the FE analyzes—

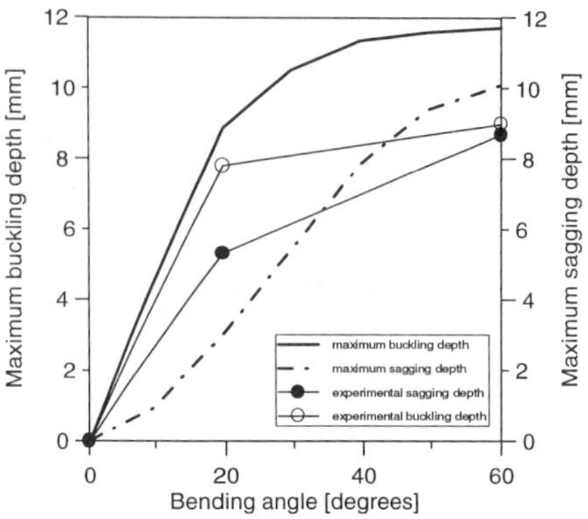

Figure 10. Maximum buckling and sagging depths vs. bending angle.

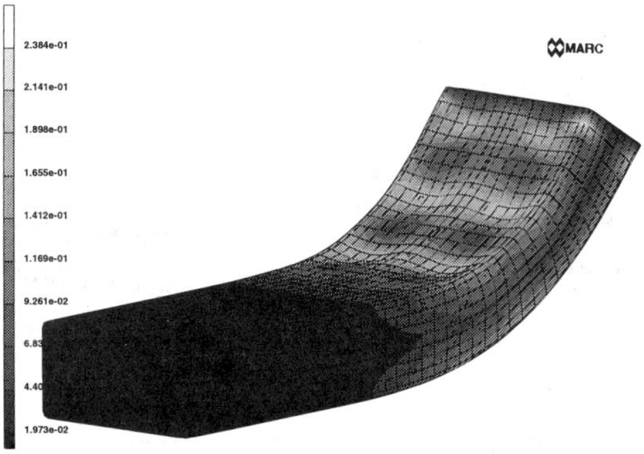

Figure 11. Post buckling modes predicted without internal mandrel.

even the magnitude of local deformations is estimated within reasonable accuracy. In the experiments, a flange was attached to the rotating former to provide additional side support to the webs upon bending. This device was neglected in the FE analyses, which may fully explain the overestimated cross-sectional distortions.

According to the deformation theory of plasticity (J. Lubliner 1990), the prime parameters with respect to the plate bucking stress are the instantaneous secant and tangent moduli of the stress-strain curve and the 'effective' width-to-thickness ratio of the member. This means that the initial yield stress and the current strain hardening coefficient of the material have to be considered. In profile bending, internal restraints, plate slenderness and bending radius also enter into the discussion on bendability.

The present results almost suppressed the difference in post buckling behavior of the alloy A and B extrusions. It appeared, however, that the latter provided slightly improved resistance to buckling owing to a steeper stress–strain gradient in the region of interest. This indicates that bending radius and slenderness of cross-sectional members are the prime influential parameters.

The sagging behavior is quite insensitive to the stress–strain characteristic, although anisotropy has been found to be a first-order parameter (T. Welo *et al.* 1994b). It was noticed that external tension increased the suck-in of the external flange. Additionally, there was found some interaction effect between post buckling and sagging modes in case excessive distortions occur. The flange width appeared to be the primary parameter determining the magnitude of the sagging depth.

In Figure 11, the post buckling modes for a double chamber section bent with a 125 mm die radius are illustrated. It is noticed that two almost identical waves are formed along the bend. Strain concentrations are indicated with light shadings in the buckled regions of the profile.

5 CONCLUSION

Based upon the present investigation, it is concluded that numerical simulation has been used successfully to quantify the influence of various parameters on geometrical tolerances in industrial bending.

REFERENCES

Haase, F. & Schilling, R. 1993. Proc. of MARC German user's meeting. Munich, Sept. 15–16.

Isogai, M. & Murakami, S. 1993. *J. Mater. Process. Technol.* 38: 635–654.

Lubliner, J. 1990, *Plasticity Theory*, New York: Macmillan Publ. Comp.

MARC User's Manuals 1994. MARC Analysis Research Corp. Palo Alto, U.S.A.

Peters, F.J.H. 1989. Proc. of FEM '89 Congress. Baden-Baden.

Ramberg, W. & Osgood W.R. 1943. NACA Tech. note no. 902.

Welo, T., Paulsen, F. & Brobak, T.J. 1994a. *J. Mater. Process. Technol.* 45: 173–180.

Welo, T., Paulsen, F. & Moen, K.E. 1994b. Proc. of Formability '94. Ostrava, Oct. 24–27.

Simulation of Materials Processing: Theory, Methods and Applications, Shen & Dawson (eds)
 ISBN 90 5410 553 4

Anisotropic finite element simulation of plane strain tests

J.Winters, A.Van Bael & P.Van Houtte
Department of Metallurgy and Materials Engineering, Katholieke Universiteit Leuven, Heverlee, Belgium

S.Munhoven & A.M.Habraken
Department MSM, University of Liège, Belgium

K.Mols
Onderzoekscentrum voor Aanwending van Staal (OCAS), Zelzate, Belgium

ABSTRACT:The behaviour of a number of steel sheet materials during a plane strain test was investigated. In an experimental setup, first the material properties have been determined, after which plane strain tests were performed. A clear influence of the plastic material properties was noticed. The material flow during these tests was simulated by means of an anisotropic elastic-plastic finite-element program. The anisotropic yield criterion that is used, is derived from polycrystal plasticity models. The input for such models consists of the material texture data and information on the crystal structure of the material. Evaluation of the results reveals the necessity for the use of such a yield criterion when simulating sheet metal forming.

1 INTRODUCTION

In industrial sheet forming operations, the strain mode at failure will be near plane strain in many, not to say most. Therefore, it is important for steel sheet manufacturers as well as for steel sheet users to be able to understand the material behaviour very precisely. During a research project conducted at KULeuven and at OCAS, a number of plane strain experiments on several sheet materials have been carried out.

The results of these experiments have been compared with numerical finite element simulations. In the past, the FE method has proved to be a very effective tool for predicting material flow during plastic forming operations. Although many effort has already been put in developing better ways for describing the material behaviour, most finite element codes still use rather simple yield criteria (e.g. von Mises [1], Hill[2], ...) which often prove to be valid for small areas of usage, but never show a general applicability. A need for better yield criteria, based on physical material properties, was felt[3].

Two different approaches have been proposed in the recent past. A first one is used by Dawson[4], who incorporates Taylor-Bishop-Hill models for crystal plasticity directly into the finite element code. This has shown to be a valuable approach, but rather computing time intensive. A second one was proposed by Van Houtte. In that method, a yield criterion in strain rate space is fitted to data that is calculated by means of the same TBH-models from texture data of the material under consideration. The method has been applied with success in many studies conducted in co-operation with the University of Birmingham [5].

This last approach is used in the present work, with the modification that now a yield criterion in stress space is derived from the TBH-models. This yield criterion is incorporated into the finite element code Lagamine[6], developed at the University of Liège.

Experiments and simulations show very good agreement.

2 EXPERIMENTS

About ten different steel sheet materials have been considered in the experimental scheme. For all of these, tensile tests have been performed in 3 different directions in the sheet plane (0°, 45° and 90° towards the rolling direction). The yield curves that were obtained have been fitted to a Swift-type hardening law. They have been used as an input to the finite element calculations (see further). The r-values calculated from these tensile tests, have been used to evaluate the quality of the calculated yield surfaces.

For the plane strain tests, a geometry was cho-

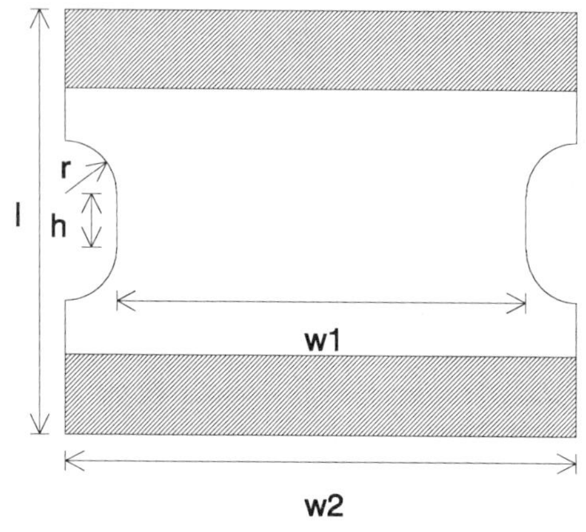

figure 1: Geometry of the plane strain specimens. The hatched zones are clamped. The dimensions are: l=160mm, r=30mm, w2=200mm, w1=160mm, h=20mm (type 1) or 10mm (type 2). The thickness of the specimen is equal to the sheet thickness.

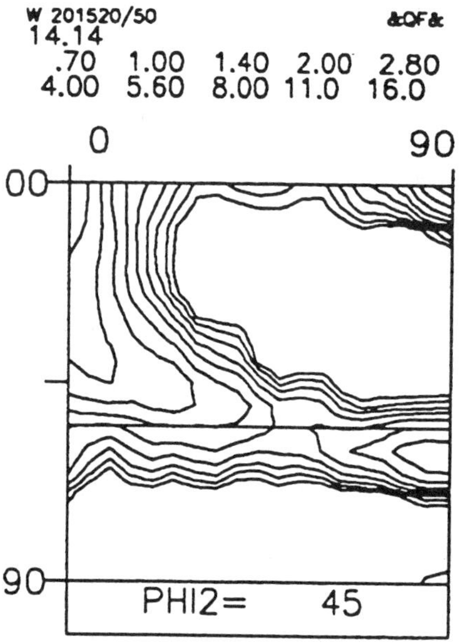

figure 2: An example of a Crystallographic Orientation Distribution Function. The figure shows a 2-dimensional section through the Euler-space (ϕ2=45°).

sen as in figure 1. After a few tests, the geometry was modified to obtain a better approximation of plane strain in the centre of the specimen. The tests were conducted up to a displacement of the clamping heads of 16mm. During the tests, the axial force was continuously measured as a function of the head displacement. On the specimens a square grid with a size of 2mm was etched. This was twice photographed after a head displacement of 9mm and 16mm, respectively. Measurement of the deformed grid was performed by an automatic measuring system [7]. The strains in the middle (narrow) section were calculated by dividing each of the squares into two triangles. From the position of their corners, the Green tensor was calculated and from this the in-plane true strains.

Due to the experimental situation, some slippage of the wide zones of the sheet specimens within the clamping heads has occurred. This had an influence on the measured deformations.

For all materials, the initial texture, i.e. the original orientation of the grains that are present in the sheet material, was measured by X-ray diffraction. The crystallographic orientation distribution function (CODF) [8] was calculated (see fig. 2). This served as an input for the calculation of the anisotropic yield loci that are used in the finite-element program.

3 FINITE ELEMENT SIMULATIONS

3.1 Material behaviour

It is assumed that the total deformation consists of an elastic and a plastic part. The elastic component is governed by Hooke's law. The plastic behaviour is derived from an anisotropic yield surface and a hardening law.

The anisotropic yield surface has the form of a series expansion of 6th order in stress space. By using the information of the CODF and of the slip systems for BCC materials in combination with TBH-models, it is possible to calculate average Taylor-factors of the polycrystal for a huge number of deformation modes. Using the geometrical method described by Mols [9], it is possible to derive from those the stress states that cause plastic deformation for those discrete deformation modes. By fitting the coefficients of the series expansion to these stress points with a least squares method in a similar way as proposed by Van Houtte [10], a (scaled) continuous yield surface in stress space is obtained.

Up to now, this surface is not updated during the simulation. This means that no texture evolution is taken into account.

Hardening is incorporated by making use of a Swift-type law ($\sigma = K(\varepsilon_0 + \varepsilon)^n$). This is not the 'raw' hardening law coming from the tensile tests, but it is recalculated using the mean Taylor-factor $\overline{M}$ to obtain a hardening law that gives the critical resolved shear stress as a function of the slip[11].

$$\tau = K(\gamma_0 + \gamma)^n$$

This critical resolved shear stress acts as a scaling factor for the yield surface.

The stress integration algorithm for the anisotropic yield criterion was newly developed. It is comparable with a trapezoidal integration rule. The algorithm makes use of a subincrementation technique [12].

The tangent matrix is calculated by a perturbation method. An analytical method, which would make the stress updating algorithm about ten times faster, is being developed.

3.2 Finite element formulation

The finite element program that is used, is a fully 3-dimensional finite-element program developed at the University of Liège ('Lagamine'). The element type that was used for the simulation of the plane strain tests is an 8-node hexahedral solid element type of the assumed strain method family with hourglass control and one point [13]. From this it is clear that with only one element layer in thickness direction, no variation of stresses or strains in thickness direction can be simulated. The influence of the difference in sheet thickness, which existed in the experimental trials, can not be modelled in this way.

3.3 Finite element discretisation and boundary conditions

The FE mesh that is used, is presented in figure 3. It is generated by means of the pre-processor I-Deas. Due to symmetry (the sheet material is supposed to be orthothropic), only one eighth of the sheet is modelled. This is valid since for all plane strain test simulations the tensile direction corresponds to the rolling direction. The boundary conditions were chosen such that first of all the symmetry was accomplished. The clamping heads were simulated by preventing the movement in width direction (Y) for the right most nodes and simultaneously imposing an axial movement (X) on the same nodes. The magnitude of the axial movement was automatically adapted to obtain fast convergence.

The results of the simulations have been evaluated in a very similar way as the experimental results.

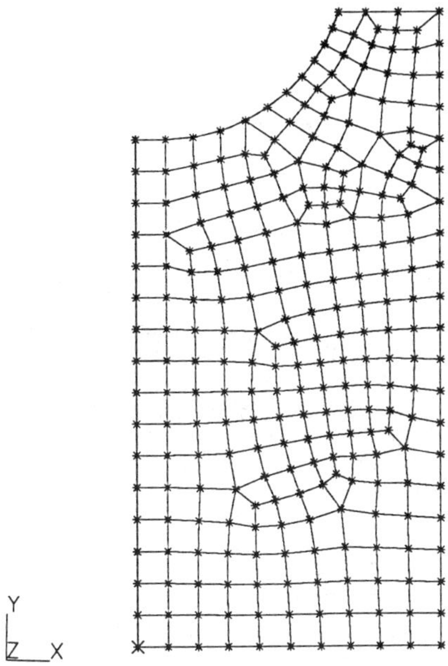

figure 3: FE mesh used for the simulation of the plane strain deformation test

table 1: parameters for Swift-type hardening law used in the FE-simulations:

material	K (MPa)	γ_0	n
E	192	0.021	0.195
von Mises	623	0.008	0.195
H	143	0.008	0.263

4 RESULTS AND DISCUSSION

Two of the materials that were used in the experimental scheme have been selected for simulation with the FE program. The respective anisotropic yield loci were calculated from the measured material texture, while the hardening laws were calculated from the tensile testing data.

Calculation times for an anisotropic simulation tend to be about four times longer than a comparable isotropic one.

The first material (material E in table 1) was a batch annealed low carbon steel with an experimental average r-value of about 1.2. The second material was a continuously annealed interstitial free steel (H) with an average r-value of about 2. In table 1, the parameters for the hardening laws are listed. Both for material E and H, geometry type 2 (see figure 1) was used. The hardening law used for the iso-

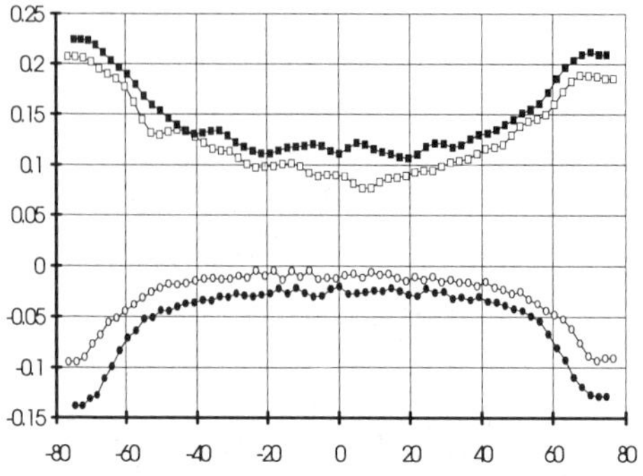

figure 4: raw experimental strain data for material E (open markers)and H (black markers) as a function of the position on the specimen; the upper two curves give the axial strain, the lower 2 the width strain

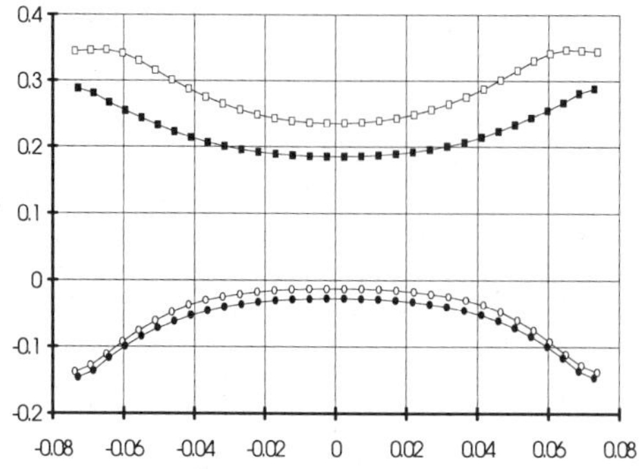

figure 5: as in figure 4, but strains are calculated from FE simulations

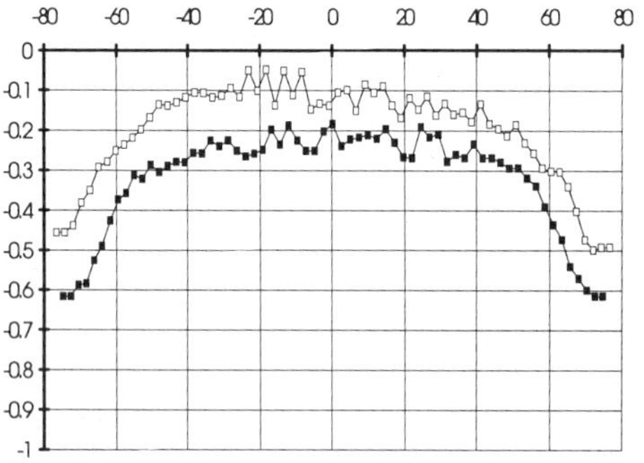

figure 6: ratio between experimental axial and width strains for material E (open markers) and H (black markers)

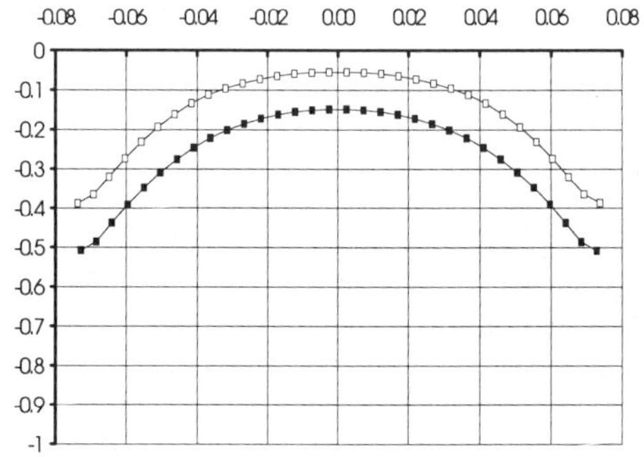

figure 7: as in figure 6, but strains are calculated from FE-simulations.

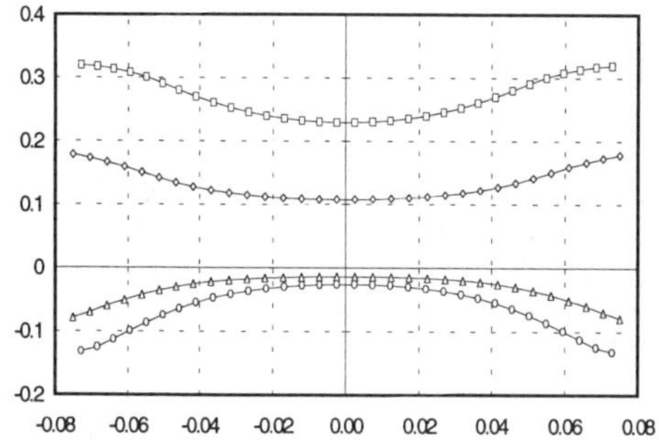

figure 8: simulated axial and width strains for an isotropic material (von Mises yield criterion) after 9 and 16mm displacement of the clamping heads; the hardening law is comparable with material E

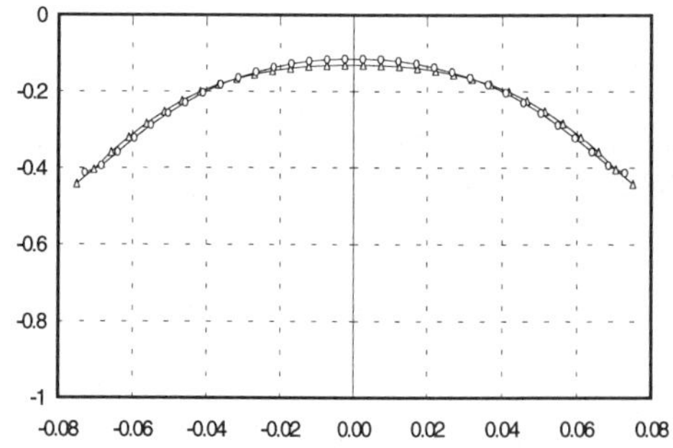

figure 9: ratio between axial and width strains for the isotropic (von Mises) material law; the deformation mode remains almost unchanged during the whole simulation. This holds too for the experiments

tropic simulation corresponded to material E.

For materials E and H the experimentally determined axial and transverse deformations after 9, resp. 16mm head displacement are shown by figure 4 as a function of the distance to the centre of the specimens. It is seen that the difference in material properties leads to a clear difference in material flow. The corresponding graphs for the FE simulations are shown in figure 5. Qualitatively, the same deformation profiles are obtained. The magnitude of the deformations measured in the experiments, is lower than in the simulations. The above mentioned slippage in the clamping heads, which is obviously not considered in our FE model, most probably causes this difference. This slippage depends probably also on the r-value, which might be an explanation for the simulated axial strain of material H being lower than that of E, while it is the inverse for the experimentally measured strains.

The agreement between experiments and simulations is even made more clear by drawing a graph of the ratio between width and thickness strain as a function of the position on the specimen. As previously reported by other authors (see e.g. Wagoner [14]), the deformation mode does almost not change during this kind of plane strain tests. Therefore, these graphs are less sensitive to the absolute magnitude of the deformations. In figures 6 and 7 the results of the experiments and the simulations, respectively, are plotted. Both the clear difference between the material flow in the experiments for E and H as the striking resemblance between the results of the simulations and the experiments can be noticed.

The influence of the anisotropic material behaviour can be seen at the edge of the specimens as well as in the centre. At the edge, the deformation is almost the same as in a tensile test. For the ratio $\frac{\varepsilon_2}{\varepsilon_1}$ this means a value of about 0.5 for material E and a value of about 0.66 for material H. Both in the experiments and in the simulations, these values are not exactly obtained.

In the centre of the specimen, the material with the highest r-value (H) does not at all come to a plane strain deformation state. Due to the high r-value, it is very difficult for the material to deform in the thickness direction. Material E with the lower r-value, though, gets much closer to a plane strain deformation state, since the elongation in tensile direction can be almost fully compensated by a reduction of the thickness.

The results of the isotropic simulation which are shown in figures 8 and 9, show a material behaviour very near to the one of material E, as could be expected. The differences that appear can be attributed to the difference in hardening behaviour of an isotropic and an anisotropic material.

5 CONCLUSIONS

An anisotropic yield criterion based on polycrystal plasticity was used in a FE program to simulate plane strain tests. From the comparison between experimental results and simulations, it was shown that a good agreement was obtained.

The material behaviour during sheet forming operations depends strongly on the plastic properties. If more than a rough approximation of the material flow is requested, an anisotropic yield criterion becomes a necessity. The advantage of the yield criterion that was used here is that it is based on physical material properties.

Refinements of the numerical algorithms for the integration of the stress during deformation and of the algorithms for the calculation of the tangent stiffness matrix, will certainly lead to acceptable calculation times.

6 ACKNOWLEDGEMENT

The experimental work presented here, was carried out in a project that was funded by the Flemish Government through the IWT.

7 REFERENCES

1) von Mises, R., Mechanik der festen Körper und plastisch deformablen Zustant , Göttinger Nachrichten Math.-Phys. Klasse, 1913

2) Hill, R., Theoretical plasticity of textured aggregates, Math. Proc. Cambridge Philos. Soc., vol. A193, 1979, pp. 179-191

3) Wilson, D.V., Representation of material behaviour in finite element methods of modelling sheet forming processes, Int. Mat. Rev., vol. 35, no. 6, 1990, pp. 329-348

4) Dawson, P.R., Beaudoin, A.J., Mathur, K.K., Simulating deformation-induced texture in metal forming, Proc. Numiform '92, Chenot et al., eds., Balkema, Rotterdam (The Netherlands), 1992

5) Van Houtte, P., Van Bael, A., Winters, J., Aernoudt, E., Hall, F.R., Wang, N., Pillinger, I., Hartley, P. and Sturgess, C.E.N.: Modelling of complex forming processes, in: Modelling of Plastic Deformation and Its Engineering Applications (Proc. 13th Risø International Symposium on Material Science), Andersen, S.I. et al., eds., Risø National Laboratory, Roskilde (Denmark), 1992, pp. 161-172.

6) Charlier, R., Approche unifiée de quelques problèmes non linéaires de mécanique des milieux continus par la méthode des éléments finis (grandes déformations des métaux et des sols, contact unilateral de solides, conduction thermique et écoulements en milieux poreux), (Collection publications de la Fac. des Sc. Appl., n°109), Fac. Sc. Appl., Univ. of Liège, Liège (Belgium), 1987

7) Vermeulen, M., A computer automated strain analyzer (CASA) for the investigation of sheet material deformation, BSME report 1412

8) Bunge, H.J., Texture analysis in material science, Butterworths, London, 1982

9) Mols, K., Van Praet, K., Van Houtte, P., A generalized yield locus calculation from texture data, in: Proc. Seventh Int. Conf. on Textures of Materials (ICOTOM 7), Brakman, C.M., Jongenburger, P. and Mittemeijer, E.J., eds., Netherlands Society for Material Science, Zwijndrecht (The Netherlands), 1984, pp. 651-656.

10) Van Houtte, P., Mols, K., Van Bael, A., Aernoudt, E., Application of Yield Loci Calculated from Texture Data, Textures and Microstructures, vol. 11, 1989, pp. 23-39.

11) Aernoudt, E., Van Houtte, P., Leffers, T., Deformation and Textures of Metals at Large Strains, in: Plastic Deformation and Fracture of Materials, Mughrabi, H., ed., (Vol. 6 of Materials Science and Technology: A Comprehensive Treatment, Cahn, R.W., Haasen, P., Kramer, E.J., eds.), VCH, Weinheim, Federal Republic of Germany, 1993, pp. 89-136.

12) Schreyer, H.L., Kulak, R.F., Kramer, J.M., Accurate numerical solutions for elastic-plastic models, J. Pressure Vessel Technology, vol. 101, 1979, pp. 226-234

13) Zhu, Y.Y., Contribution to the local approach of fracture in solid dynamics, PhD thesis, Fac. Sc. Appl., Univ. of Liège, Liège (Belgium), 1992

14) Wagoner R.H., Wang, N.-M., An experimental and analytical investigation of in-plane deformation of 2036-T4 aluminium sheet, Int. J. Mech. Sc., vol. 21, 1979, pp. 255-264

Simulation of Materials Processing: Theory, Methods and Applications, Shen & Dawson (eds)
© 1995 Balkema, Rotterdam. ISBN 90 5410 553 4

Property predictions for a short-fiber-reinforced, injection molded center-gated disk

N. Yan, B. N. Rao, J. D. Baldwin & M. C. Altan
School of Aerospace and Mechanical Engineering, University of Oklahoma, Norman, Okla., USA

ABSTRACT: The exact solution for the flow-induced orientation field in a steady, Newtonian, radially diverging flow between parallel plates is reviewed. The evolution of orientation structure is presented by a closed-form expression that depends on radial and through-the-thickness coordinates. In addition, the effect of inlet conditions on the final orientation structure is explicitly incorporated into formulation. The process-induced microstructure is then used to estimate the anisotropic stiffness properties of the molded part. Through-the-thickness and radial variations in stiffness predictions are presented by using a micromechanical formulation.

1. INTRODUCTION

The use of short-fiber-reinforced composite materials in a variety of industries has grown considerably in the last two decades. These material systems are most often fabricated by molding operations such as injection or compression molding. During molding processes, the fibers, mixed with a polymeric resin, are influenced by the non-Newtonian, non-isothermal flow kinematics within the mold cavity. The flow in molding operations is usually characterized by a low-Reynolds number (i.e., $Re \ll 1$) such that the fluid inertia can be neglected compared to viscous effects. Therefore, the filling simulation of a thin, injection molded part can be accurately performed with the Hele-Shaw flow formulation. Then, the fibers in the mold can be taken to move affinely with the fluid based on the local velocity vector. On the other hand, spatial changes in the velocity field over the length scale of a fiber (i.e., local velocity gradient tensor, u_{ij}) rotate and orient the fibers in certain directions, thus forming a so-called flow-induced orientation structure throughout the molded part. In a number of molded parts, the orientation field may be three-dimensional and its spatial variation within the part may be significant. Consequently, in order to predict the thermo-mechanical properties of a molded part, the flow-induced orientation structure needs to be completely and accurately characterized. Then, micromechanical models can be implemented to estimate the physical or mechanical properties.

In this study, we briefly present the closed-form solution of the orientation field in a center-gated disk (Altan and Rao 1995). The orientation structure is assumed to be formed by the steady flow of a radially diverging, Newtonian fluid containing hydrodynamically isolated fibers. Although such an analysis contains rather restrictive assumptions, the orientation field is obtained without using closure approximations that have been among the primary sources of errors in previous studies. Moreover, the predicted orientation field agreed very well (Altan and Rao 1995) with the available experimental data (Bay and Tucker 1992) obtained from an injection molded, center-gated disk.

2. FLOW AND ORIENTATION FIELD IN A CENTER-GATED DISK

In a center-gated disk, the fluid is introduced into the cavity from a narrow inlet gate at the center. The flow diverges between two parallel disks and decelerates as it moves away radially. The flow field is steady if the fluid is injected at a constant rate, and is characterized by the shear and extensional gradients through the gap width and on the flow plane, respectively. The flow geometry and the coordinate systems used for the velocity field and orientation descriptions are shown in Figure 1. The two-dimensional velocity and the three-dimensional orientation fields are described in cylindrical (i.e., x, z) and spherical (i.e., θ, ϕ) coordinate systems, respectively. In addition, half gap width h, and maximum velocity at the entry U_{max} are used as the length and velocity scales to nondimensionalize the flow parameters.

After assuming a fully-developed velocity profile at the entry of the disk (i.e., at $x = 1$), the nondimensional velocity profile can be expressed as

$$u\,(x,\,z) \;=\; \frac{1}{x}(1 - z^2) \qquad 0 \le z \le 1; \qquad 1 \le x \qquad \textbf{(1)}$$

Based on the Newtonian, steady velocity field given by Equation 1, the velocity gradient tensor u_{ij} on any x - z plane is obtained in rectangular coordinates as

$$u_{ij}(x,\,z) \;=\; \begin{bmatrix} -\dfrac{(1 - z^2)}{x^2} & 0 & -\dfrac{2z}{x} \\ 0 & \dfrac{(1 - z^2)}{x^2} & 0 \\ 0 & 0 & 0 \end{bmatrix} \qquad \textbf{(2)}$$

It should be emphasized that the velocity gradients are not spatially uniform, and as Equation 2 shows, each fiber experiences continuously changing shear and extensional gradients depending on its x and z location. For a slender fiber, a particle rotation tensor, E_{ij}, can be defined as

$$\frac{dE_{ij}}{dx} \;=\; \frac{1}{u}\, u_{ik}\, E_{kj} \qquad \textbf{(3)}$$

which can be solved for E_{ij} as

$$E_{ij}(x,\,z) \;=\; \begin{bmatrix} \dfrac{1}{x} & 0 & \dfrac{z(1 - x^2)}{x(1 - z^2)} \\ 0 & x & 0 \\ 0 & 0 & 1 \end{bmatrix} \qquad \textbf{(4)}$$

The exact solution of the particle rotation tensor E_{ij} would make it possible to obtain the three-dimensional orientation probability density function (OPDF) $\Psi(\bar{p})$ as

$$\Psi(\bar{p},\,x,\,z) \;=\; \frac{1}{4\pi}\,(\Lambda_{lm}\,\Lambda_{lj}\,p_j\,p_m)^{-3/2} \qquad \textbf{(5)}$$

where Λ_{ij} is the inverse of particle rotation tensor E_{ij} and $\bar{p}$ is the fiber orientation vector. Equations 4 and 5 characterize the orientation field completely throughout the center-gated disk from an Eulerian viewpoint. With the availability of a closed-form expression for the orientation distribution, exact values of the moments of the OPDF can be evaluated numerically. The second- and the fourth-order moments are commonly used (i.e., S_{ij} and S_{ijkl}, respectively) and are also known as orientation tensors. For the center-gated disk, the orientation tensors can be computed as

$$S_{ij} \;=\; \int_{\bar{p}} \frac{(E_{ik}\,p_k)(E_{jl}\,p_l)}{E_{mn}\,E_{mq}\,p_n\,p_q}\,\Psi(\bar{p},\,1,\,z)d\bar{p} \qquad \textbf{(6)}$$

$$S_{ijkl} \;=\; \int_{\bar{p}} \frac{(E_{ir}\,p_r)(E_{js}\,p_s)(E_{kt}\,p_t)(E_{lu}\,p_u)}{(E_{mn}E_{mq}\,p_n\,p_q)^2}\,\Psi(\bar{p},\,1,\,z)d\bar{p} \qquad \textbf{(7)}$$

where $\Psi(\bar{p},\,1,\,z)$ is the OPDF at the entry of a center-gated disk (i.e., at $x = 1$). From Equations 6 and 7, it is also possible to evaluate S_{ij} and S_{ijkl} components for any inlet orientation structure. This capability is useful in a center-gated disk where the fibers, instead of being isotropically oriented at the inlet, are mostly aligned in the flow direction as they pass through a vertical, narrow gate. However, in this study we prefer to use an isotropic orientation (i.e., S_{11}=S_{22}=S_{33}=1/3) at the inlet, a condition used most often in studies predicting the orientation field in a thin gap width cavity. Once the integrations are performed, the maximum orientation angle and the degree of fiber alignment with respect to maximum orientation angle can be simply obtained from the eigenvector and eigenvalues of the second-order orientation tensor, S_{ij}. In three-dimensional flow and orientation space, the components of S_{ij} can be used to construct an orientation ellipsoid, which is suitable for graphical presentation of orientation field in a complex flow.

3. MECHANICAL ANALYSIS OF THE MOLDED MICROSTRUCTURE

Following the analysis of Advani and Tucker (1987), we assume that each individual fiber in the molding is straight, with finite aspect ratio, and has no interaction with surrounding fibers. Under these assumptions, an isolated fiber and its adjacent matrix material can be treated as a transversely isotropic material system.

The Halpin-Tsai micromechanics model

(Halpin, et al. 1969, 1976) is used to express the mechanical behavior of the fiber-matrix system in terms of the properties of the constituent fiber and matrix materials. Specifically, the following five independent material parameters can be used to fully describe the behavior of the transversely isotropic material.

- Longitudinal tensile modulus, E_1

$$E_1 = \frac{1 + \xi_{11}\eta_{11}V_f}{1 - \eta_{11}V_f} E_m$$

$$\xi_{11} = 2\ (l/d) \qquad \eta_{11} = \frac{(E_f / E_m) - 1}{(E_f / E_m) + \xi_{11}} \tag{8}$$

- Transverse tensile modulus, E_2

$$E_2 = \frac{1 + \xi_{22}\eta_{22}V_f}{1 - \eta_{22}V_f} E_m$$

$$\xi_{22} = 2 \qquad \eta_{22} = \frac{(E_f / E_m) - 1}{(E_f / E_m) + \xi_{22}} \tag{9}$$

- Longitudinal shear modulus, G_{12}

$$G_{12} = \frac{1 + \xi_{12}\eta_{12}V_f}{1 - \eta_{12}V_f} G_m$$

$$\xi_{12} = 1 \qquad \eta_{12} = \frac{(G_f / G_m) - 1}{(G_f / G_m) + \xi_{12}} \tag{10}$$

- Transverse shear modulus, G_{23}

$$G_{23} = \frac{1 + \xi_{23}\eta_{23}V_f}{1 - \eta_{23}V_f} G_m$$

$$\xi_{23} = \frac{K_m / G_m}{(K_m / G_m) + 2} \qquad \eta_{23} = \frac{(G_f / G_m) - 1}{(G_f / G_m) + \xi_{23}} \tag{11}$$

-Longitudinal Poisson ratio, υ_{12}

$$\nu_{12} = \nu_f V_f + \nu_m V_m \tag{12}$$

The constituent shear and bulk moduli are computed from the familiar relations

$$G = \frac{E}{2\ (1 + \nu)} \tag{13}$$

$$K = \frac{E}{3\ (1 - 2\nu)} \tag{14}$$

In Equations 8-14, V_f is the fiber volume fraction, l is the fiber length and d is the fiber diameter. Subscripts f and m indicate the fiber and matrix material, respectively. The axis normal to the isotropic plane is coincident with the fiber axis, i.e., the 1-direction and the 2-3 directions define the plane of isotropy.

The constitutive relationship for a transversely isotropic material with the axis of material symmetry coincident with the 1-direction is (Jones 1975)

$$\begin{Bmatrix} \sigma_1 \\ \sigma_2 \\ \sigma_3 \\ \sigma_4 \\ \sigma_5 \\ \sigma_6 \end{Bmatrix} = \begin{bmatrix} c_{11} & c_{12} & c_{12} & 0 & 0 & 0 \\ c_{12} & c_{22} & c_{23} & 0 & 0 & 0 \\ c_{12} & c_{23} & c_{22} & 0 & 0 & 0 \\ 0 & 0 & 0 & \frac{1}{2}(c_{22}-c_{23}) & 0 & 0 \\ 0 & 0 & 0 & 0 & c_{66} & 0 \\ 0 & 0 & 0 & 0 & 0 & c_{66} \end{bmatrix} \begin{Bmatrix} \epsilon_1 \\ \epsilon_2 \\ \epsilon_3 \\ \epsilon_4 \\ \epsilon_5 \\ \epsilon_6 \end{Bmatrix} \tag{15}$$

Note that in Equation 15, the strains ϵ_1, ϵ_2, and ϵ_3 are the tensor strains ϵ_{11}, ϵ_{22}, and ϵ_{33}, respectively; ϵ_4, ϵ_5, and ϵ_6 are the engineering shear strains γ_{23}, γ_{13} and γ_{12}, respectively. By virtue of the isotropy in the 2-3 plane, $E_2 = E_3$, $\upsilon_{12} = \upsilon_{13}$, $\nu_{23} = \nu_{32}$ and $\nu_{31} = \nu_{21} = \nu_{12}\ (E_2 / E_1)$. In terms of the engineering quantities predicted by the Halpin-Tsai micromechanics equations, the elements of the transversely isotropic stiffness matrix c_{ij} are given by (Jones 1975)

$$\begin{aligned} c_{11} &= \frac{1 - \nu_{23}^2}{E_2^2 \Delta} \\ c_{12} &= \frac{\nu_{12} + \nu_{12}\nu_{23}}{E_1 E_2 \Delta} \\ c_{22} &= \frac{1 - \nu_{12}\nu_{21}}{E_1 E_2 \Delta} \\ c_{23} &= \frac{\nu_{23} + \nu_{12}\nu_{21}}{E_1 E_2 \Delta} \\ c_{66} &= G_{12} \end{aligned} \tag{16}$$

where

$$\Delta = \frac{1 - 2\nu_{12}\nu_{21} - 2\nu_{12}\nu_{21}\nu_{23} - \nu_{23}^{2}}{E_1 \, E_2^2} \qquad \textbf{(17)}$$

The third Poisson ratio, υ_{23}, is computed from Equation 13 using E_{22} and G_{23}.

The orientation averaging process requires that the "local" material properties, i.e., those of the single transversely isotropic fiber-matrix system, be weighted by the fiber orientation probability density function and integrated over the orientation space. Advani and Tucker (1987) proposed a set of orientation tensors that represent the dyadic products of fiber orientation vector $\bar{p}$ weighted by the fiber orientation probability density function $\Psi(\bar{p})$. The second- and fourth-order orientation tensors that arise from this computation are given by

$$\begin{aligned} S_{ij}(x, z) &= \int_{\bar{p}} p_i \, p_j \, \Psi(\bar{p}) \, d\bar{p} \\ S_{ijkl}(x, z) &= \int_{\bar{p}} p_i \, p_j \, p_k \, p_l \, \Psi(\bar{p}) \, d\bar{p} \end{aligned} \qquad \textbf{(18)}$$

Note that the connection between the orientation tensors and the molding conditions was given in Equations 6 and 7.

Because the stiffness tensor, given in matrix form by Equations 15-17, is related to the fixed 1-2-3 coordinate system, it can be transformed according to tensor transformation rules into different coordinate directions. To simplify the computation of the stiffness matrix in alternate coordinates, the stiffness tensor c_{ijkl} needs to be restated in terms of its invariant form. The invariant form of the transversely isotropic stiffness tensor is given by Spencer (1984) for an arbitrary orientation vector $\bar{p} = \{ p_1 \; p_2 \; p_3 \}$ as

$$\begin{aligned} c_{ijkl} &= \lambda\delta_{ij}\delta_{kl} + \mu_T(\delta_{ik}\delta_{jl}+\delta_{jk}\delta_{il}) \\ &+ \alpha(p_k \, p_l\delta_{ij} + p_i \, p_j\delta_{kl}) + \beta p_i \, p_j \, p_k \, p_l \\ &+ (\mu_L - \mu_T)(p_i \, p_k\delta_{jl} + p_i \, p_l\delta_{jk} + p_j \, p_k\delta_{il} + p_j \, p_l\delta_{ik}) \end{aligned} \qquad \textbf{(19)}$$

where δ is the Kronecker delta and repeated indices imply summation. In matrix form, the invariant transversely isotropic stiffness matrix is given by

$$\begin{bmatrix} \lambda+2\alpha+4\mu_L-2\mu_T+\beta & \lambda+\alpha & \lambda+\alpha & 0 & 0 & 0 \\ \lambda+\alpha & \lambda+2\mu_T & \lambda & 0 & 0 & 0 \\ \lambda+\alpha & \lambda & \lambda+2\mu_T & 0 & 0 & 0 \\ 0 & 0 & 0 & \mu_T & 0 & 0 \\ 0 & 0 & 0 & 0 & \mu_L & 0 \\ 0 & 0 & 0 & 0 & 0 & \mu_L \end{bmatrix} \qquad \textbf{(20)}$$

where λ, α, μ_L, μ_T, and β are the invariants related to the physical properties of the material. Comparison of the stiffness matrices (15) and (20) shows that

$$\begin{aligned} \lambda &= c_{23} \\ \alpha &= c_{12} - c_{23} \\ \mu_L &= c_{66} \\ \mu_T &= 1/2(c_{22} - c_{23}) \\ \beta &= c_{11} + c_{22} - 2c_{12} - 4c_{66} \end{aligned} \qquad \textbf{(21)}$$

Therefore, using the invariant representation of Equations 19-21, we may compute the required orientation-averaged, "global" stiffness matrix. The orientation average of the stiffness tensor is given by the integral

$$< C > = \int_{\bar{p}} c\,(\bar{p}) \, \Psi(\bar{p}) \, d\bar{p} \qquad \textbf{(22)}$$

Substituting for $c(\bar{p})$ from Equation 19 and noting the definition of the orientation tensors in Equation 18, the stiffness tensor for the composite in the *x-y-z* coordinate system is given by

$$\begin{aligned} < C >_{ijkl} &= \lambda\delta_{ij}\delta_{kl} + \mu_T(\delta_{ik}\delta_{jl}+\delta_{jk}\delta_{il}) \\ &+ \alpha(S_{kl}\delta_{ij}+S_{ij}\delta_{kl}) + \beta S_{ijkl} \\ &+ (\mu_L-\mu_T)(S_{ik}\delta_{jl}+S_{il}\delta_{jk}+S_{jk}\delta_{il}+S_{jl}\delta_{ik}) \end{aligned} \qquad \textbf{(23)}$$

It is interesting to note that after orientation averaging, the molded material takes on a monoclinic form with symmetry about the *y*-axis, which, in matrix form, is given by

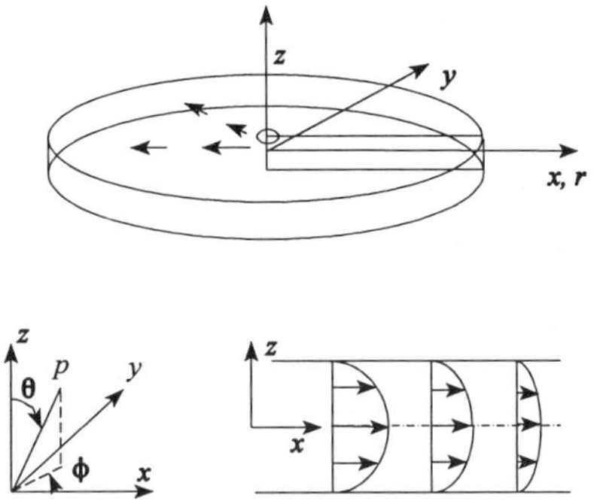

Figure 1: Center-gated disk geometry and coordinate system

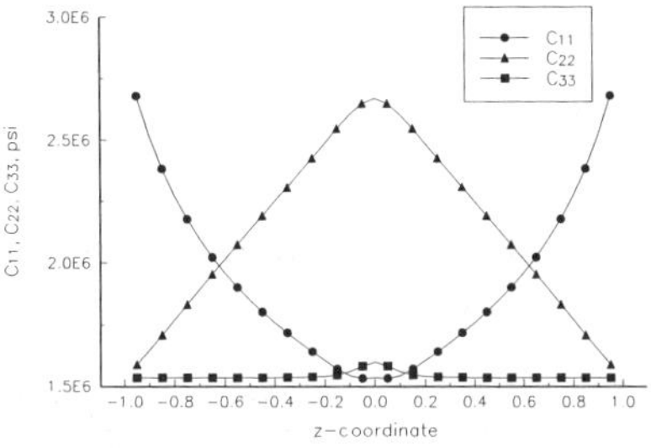

Figure 2: Variation of tensile stiffnesses, $x = 10$

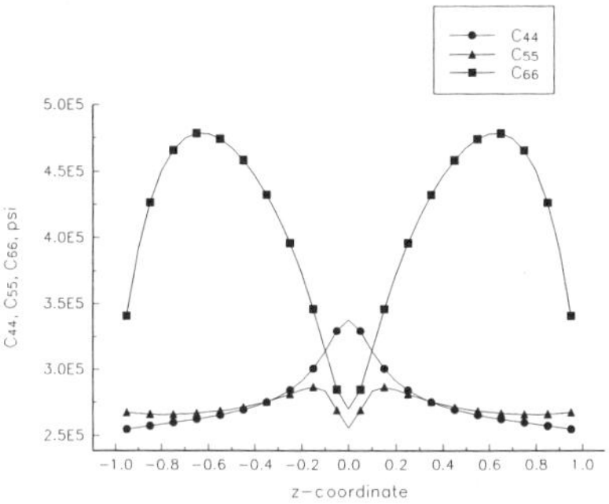

Figure 3: Variation of shear stiffnesses, $x = 10$

$$
\begin{Bmatrix} \sigma_x \\ \sigma_y \\ \sigma_z \\ \tau_{yz} \\ \tau_{xz} \\ \tau_{xy} \end{Bmatrix} = \begin{bmatrix} C_{11} & C_{12} & C_{13} & 0 & C_{15} & 0 \\ C_{12} & C_{22} & C_{23} & 0 & C_{25} & 0 \\ C_{13} & C_{23} & C_{33} & 0 & C_{35} & 0 \\ 0 & 0 & 0 & C_{44} & 0 & C_{46} \\ C_{15} & C_{25} & C_{35} & 0 & C_{55} & 0 \\ 0 & 0 & 0 & C_{46} & 0 & C_{66} \end{bmatrix} \begin{Bmatrix} \epsilon_x \\ \epsilon_y \\ \epsilon_z \\ \gamma_{yz} \\ \gamma_{xz} \\ \gamma_{xy} \end{Bmatrix} \quad \textbf{(24)}
$$

The stiffness terms are computed by direct evaluation of Equation 23.

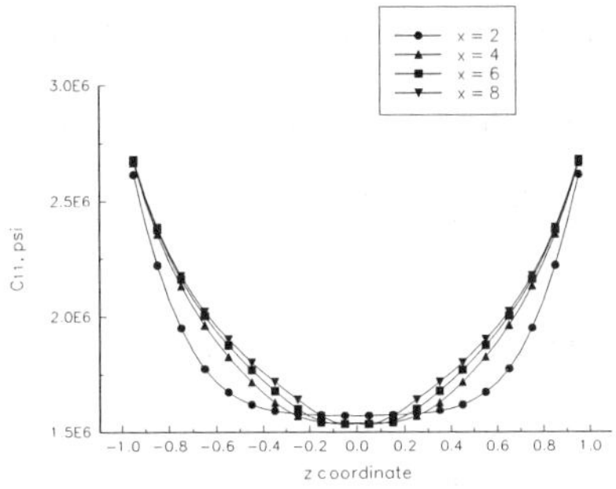

Figure 4: Variation of C_{11} along radial direction

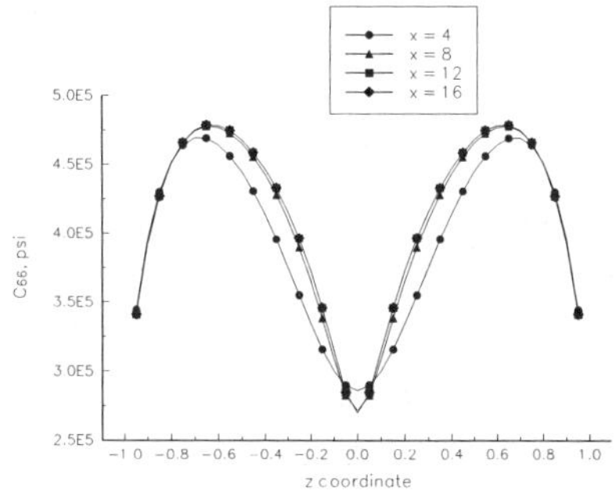

Figure 5: Variation of C_{66} along radial direction

4. EXAMPLE CALCULATION

To illustrate the mechanical property estimates based on fiber orientation, we consider a typical thermoplastic matrix material stiffened by glass fibers at a fiber volume fraction of 20 percent; the fibers are assumed to have an aspect ratio *l/d* of 100. The constituent material properties are as follows:

$E_f = 10.5 \times 10^6$ psi $\quad E_m = 0.5 \times 10^6$ psi

$\nu_f = 0.20 \quad \nu_m = 0.35$

For the non-dimensional center-gated disk geometry in Figure 1, the orientation tensors S_{ij} and S_{ijkl} were evaluated using Equations 6 and 7 at 39 points across the thickness (from $z = -0.95$ to $z = +0.95$) and at 20 stations along the radius. Equations 21 and 23 were then used to compute the elements of the local stiffness matrix, oriented with respect to the disk *x-y-z* axes.

Figure 2 shows the variation of the extensional stiffnesses (C_{11}, C_{22} and C_{33}) across the disk thickness at the fixed radius $x = 10$. It can be seen that C_{11} is largest at the edge of the disk, where the fibers are aligned in the radial direction, and smallest at the center, where the fibers are aligned in the tangential direction. These characteristics are consistent with observations from uniaxial fiber composites. C_{22}

exhibits a complementary variation, with low stiffness at the edges and high stiffness at the center. It is interesting to note that C_{33} is constant across the disk thickness, except for a small increase in stiffness near the center. We feel that this slight variation is due to the presence of some fiber alignment components in the *z*-direction at the center of the disk. The Figure also shows that the minimum values of these three stiffness components are coincident at about 1.5×10^6 psi.

Figure 3 shows the through-the-thickness variation of the stiffnesses coupling shear stresses and shear strains. The peaks in C_{66} correspond to regions in the disk where quasi-random fiber orientations are known to exist on the *x-y* plane. These regions are where the transition occurs from radially-aligned fibers to tangentially-aligned fibers. Although it appears that the stiffness at the wall is higher than at the center, recall that the outermost stiffness was computed at $z = \pm 0.95$. If the stiffness had been evaluated at the wall, the stiffnesses would have matched.

Figures 4 and 5 show the through-the-thickness variation in C_{11} and C_{66} as a function of position along the *x*-axis. Figure 4 shows that the stiffness decreases by 43 percent from the wall to the core. We also note that C_{11} tends toward an asymptotic minimum value at $z = 0$, $x \approx 8$ that is approximately 2 percent lower than at $x = 2$. The variation of C_{66}, Figure 5, also shows that the stiffness decreases by 44 percent from the wall to core. C_{66} reaches an asymptotic minimum value at $z = 0$, $x \approx 16$ that is approximately 6 percent lower than at $x = 4$.

5. CONCLUSIONS

Calculations have been given that show how an analytical prediction of the fiber orientation state in a center-gated injection-molded disk can lead to material stiffness estimates throughout the molding. The individual elements of the as-molded stiffness tensor were found to vary through the disk thickness predictably, according to the prevailing fiber orientation. The ability to establish the local material stiffness matrix is essential to a structural finite element analysis of the entire disk.

REFERENCES

Advani, S.G. & C.L. Tucker 1987. "The Use of Tensors to Describe and Predict Fiber Orientation in Short Fiber Composites." *J. Rheol.* 31(8), 751-784.

Altan, M.C. & B.N. Rao 1995. "A Closed-Form Solution for the Orientation Field in a Center-Gated Disk." Accepted for publication, *J. Rheol.*

Bay, R.S. & C.L. Tucker 1992. "Fiber Orientation in Simple Injection Moldings, Part 1: Theory and Numerical Methods." *Polym. Comp.* 13, 317-331.

Bay, R.S. & C.L. Tucker 1992. "Fiber Orientation in Simple Injection Moldings, Part 2: Experimental Results." *Polym. Comp.* 13, 332-341.

Halpin, J.C. 1969. "Stiffness and Expansion Estimates for Oriented Short Fiber Composites." *J. Comp. Mat.* 3, 732-734.

Halpin, J.C. & J.L. Kardos 1975. "The Halpin-Tsai Equations: A Review", *Polym. Eng. Sci.* 16(5), 344-352.

Jones, R.M. 1975. *Mechanics of Composite Materials*. New York: Hemisphere.

Spencer, A.J.M. 1984. "Constitutive Theory for Strongly Anisotropic Solids", in A.J.M. Spencer, (ed.) *Continuum Theory of the Mechanics of Fibre-Reinforced Composites*. New York: Springer-Verlag.

Simulation of Materials Processing: Theory, Methods and Applications, Shen & Dawson (eds)
© 1995 Balkema, Rotterdam. ISBN 90 5410 553 4

Calculation of polycrystal yield surfaces and forming limit diagrams of aluminum alloy sheets

Y.Zhou & J.J.Jonas
Department of Metallurgical Engineering, McGill University, Montreal, Que., Canada
M.Jain & S.R.MacEwen
Alcan International Limited, Kingston R&D Center, Ont., Canada

ABSTRACT: A simulation method is described for determining the shape and magnitude of the polycrystal yield surface. The method can also be employed to update the surface during the modelling of deformation processing. The numerical yield surfaces obtained in this way are *directly* applicable to the simulation of plastic deformation without requiring the use of analytic yield functions. This technique is employed in conjunction with the Marciniak-Kuczynski approach to predict forming limit diagrams for annealed Al-Mg sheets. The effects are examined of initial texture and texture evolution during deformation.

1 INTRODUCTION

Previous investigations (Hill, 1948; Neale & Chater, 1980; Barlat, 1987) have demonstrated that the formability of sheet metals is very sensitive to mechanical properties such as yield surface shape, strain or strain-rate hardening, and flow stress anisotropy. The forming limit is also affected by the microstructure as well as the texture that develops during prior processing. Efforts have been made in recent years to incorporate the plastic anisotropy that results mainly from crystallographic texture into the prediction of FLD's (Barlat, 1987; Barlat & Chung, 1993; Van Houtte & Tóth, 1993; Qiu *et al.*, 1994; Ratchev *et al.*, 1994; Tóth *et al.*, 1994; Zhou & Neale, 1994). These simulations are of two types: (a) in the first, the theory of polycrystal plasticity is embedded directly into a finite strain formulation as a constitutive theory; and (b) in the second, the anisotropy is assumed to be described by a yield function that is fitted to the polycrystal yield surface.

A method developed by Zhou *et al.* (1994) is used in this paper to determine numerical relations between the plastic strain rates and yield stresses of polycrystals. This method can be employed, not only to determine the shape and magnitude of the polycrystal yield surface, but also to update the surface during deformation. The numerical yield surfaces obtained in this way are *directly* applicable to the simulation of plastic deformation without requiring the use of analytic yield functions. This technique is employed in conjunction with the Marciniak-Kuczynski approach to predict the forming limit diagram of an annealed Al-Mg sheet. The prediction obtained in this way is in good agreement with the measured FLD.

2 THE CALCULATION METHOD AND RESULTS

Marciniak and Kuczynski (1967) first introduced the approach in which localized necking is initiated in a geometric imperfection initially present in the sheet as a narrow band. It is known that the M-K approach generally overpredicts forming limit strains. One cause of the overprediction is the approximate nature of the analytic description of the yield behaviour in sheet metals, since such yield functions, by their natures, do not provide good fits to the "true" polycrystal yield surfaces. Moreover, yield surface evolution during deformation is not generally considered in these calculations.

In order to overcome the above problems, we recently developed a method for determining numerical relations between the plastic strain rates and yield stresses for particular textured sheets (Zhou, *et al.*, 1994). This method is based on the

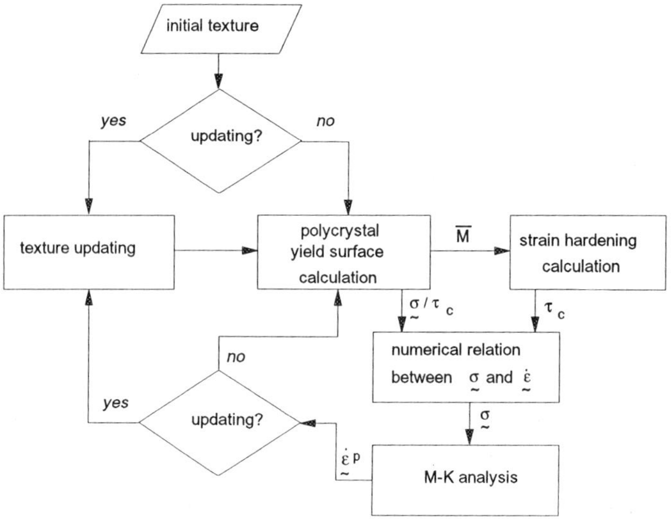

Fig. 1. Flow chart of a calculation using the present method.

Taylor theory of crystal plasticity and thus allows for the anisotropy that results from the presence of crystallographic texture. A flow chart of the method is illustrated in Fig. 1. There are three parts: texture updating, calculation of the normalized yield stress, and computation of the average strain hardening. This technique is described in detail in Zhou *et al.* (1994). The advantage of the method is that a polycrystal yield surface can be employed directly in the simulation of deformation processes, without the use of any analytic yield function. By taking texture evolution into account, the approach also permits the yield surface to be updated during the simulation with respect to the strain path that the sample is experiencing.

In the present work, the above calculation method was employed in conjunction with the Marciniak-Kuczynski approach to predict the forming limit diagram of an annealed Al-Mg sheet. The pole figures of the initial texture were measured by X-ray diffraction. The orientation distribution function (ODF) was then deduced using the series expansion method (Bunge 1982). The volume fractions of the components of the measured texture are listed in Table 1, from which it can be seen that the main orientations are the S, cube, cube-Goss (CG), Taylor, brass, Cu, cube-H (CH) and P_D. The FLD of sheet of 1 mm thickness was measured by conducting hemispherical punch tests on gridded sheet samples, as described by Jain and Allin (1994).

The full constraint rate-insensitive model was employed to calculate the Taylor factors and to simulate texture evolution. The following power-law hardening relation was assumed to describe the hardening behaviour of the sheet:

$$h = h_0\left(\frac{h_0\Gamma}{\tau_0 n}+1\right)^{n-1} \qquad (1)$$

where Γ is the average value of the total shear in each grain (see Zhou *et al.*, 1994, for

Table 1 Measured volume fractions of the texture components

Texture Component		Vol. Frac. (%)	Texture Component		Vol. Frac. (%)
S	{123}<634>	8.3	Brass	{011}<211>	4.5
Cube	{100}<100>	5.1	Cu	{112}<111>	3.8
CG	{021}<100>	5.1	CH	{001}<120>	2.3
Taylor	{4 4 11}<11 11 8>	4.8	P_D	{011}<8 11 11>	2.1

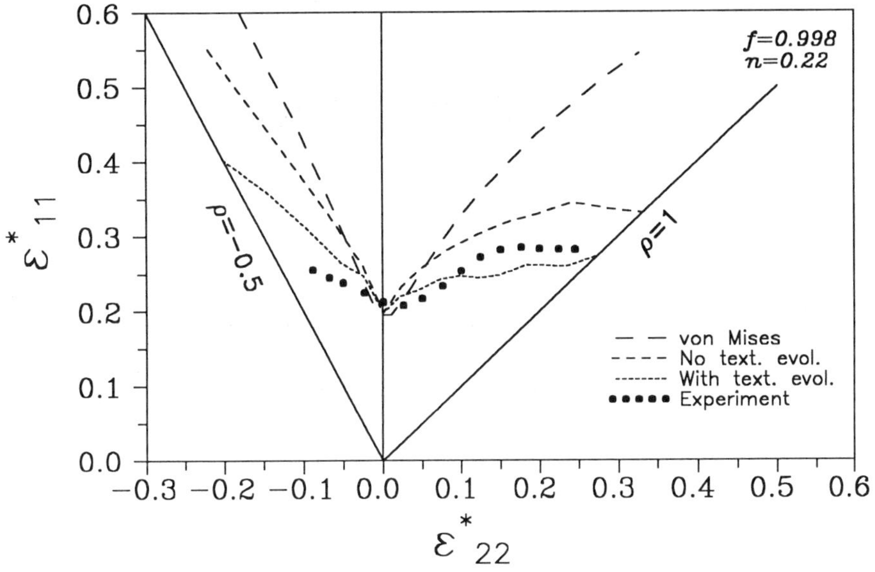

Fig. 2. Measured and predicted forming limit diagrams.

further information). The hardening parameters were specified as τ_0 = 20 MPa, h_0 = 2000 MPa and n = 0.22 In each calculation, the M-K imperfection was assumed to lie perpendicular to the major strain direction X_1. The magnitude of the M-K imperfection was arbitrarily set at $f = t_0^B / t_0$ = 0.998. Here t_0^B and t_0 are the thicknesses of the imperfection band and the sheet, respectively.

The measured and predicted forming limit diagrams are illustrated in Fig. 2. For comparison purposes, the FLD was also calculated under the condition that the texture does not evolve. It can be seen that the predicted FLD which takes texture evolution into account is in better agreement with the experimental FLD than when texture evolution is not considered. The forming limit strains predicted in the positive strain-ratio region are in good agreement with the measured ones, but are somewhat too large in the negative strain-ratio region. The closer the ρ-value is to -0.5, the larger is the discrepancy in the forming limit strain. The main source of error is probably the angle of inclination of the imperfection band, which can influence the forming limit strain, particularly in the negative ρ-value region. The influence of this parameter on the FLD is not considered here, but will be described in a future paper (Zhou *et al.*, 1995).

3 EFFECT OF TEXTURE ON THE FORMING LIMIT DIAGRAM

3.1 *Texture evolution*

It can be seen from Fig. 2 that texture evolution decreases the forming limit strain in both the positive and negative strain-ratio regions. The larger the absolute value of ρ, the greater is the drop in the respective forming limit strain. However, texture evolution does not affect the limit strain for the in-plane plane strain state.

From a macroscopic viewpoint, yield surface shape is one of the most important factors affecting forming limit diagrams. For the biaxial stretching region, Neale and Chater (1983) pointed out that a decrease in the sharpness of the yield locus promotes an increase in the limit strain. Yield surface shape in turn depends mainly on the texture. When the texture changes during deformation, the yield surface must change its shape as well. Predicted yield loci for the sheet at the forming limit strain are illustrated in Fig. 3 for the cases of ρ = -0.5 and 1. It is evident that the

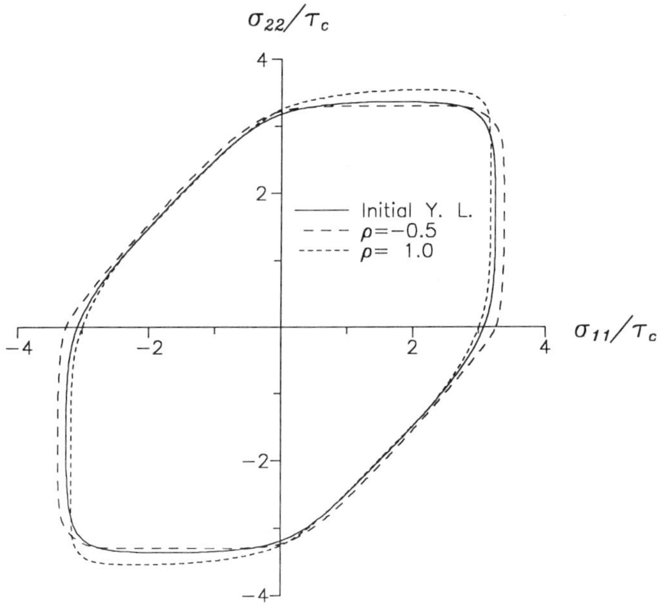

Fig. 3. Predicted yield loci at the forming limit strain for ρ = -0.5 and 1.

curvature of the yield locus in the neighbourhood of in-plane plane strain deformation is small and does not vary significantly with the texture. This is why neither the initial texture nor the evolution of the texture is able to affect the forming limit strain along this strain path. By contrast, the shape of the yield locus in the biaxial stretching region becomes sharper during deformation. Such sharpening reduces the formability of the Al-Mg sheet.

3.2 *Initial texture*

The FLD's shown in Fig. 4 reflect the influence of initial texture. One of the FLD's was calculated by taking a cube or recrystallized texture, instead of the measured one, as the initial texture of the sheet. This texture was measured by Savoie *et al.* (1994) in their experimental investigation of an annealed aluminum sheet. It can be seen that the forming limit strains for plane strain deformation are unaffected when the initial texture is changed from the measured one to the cube texture. It should be noted that the sheet containing the cube texture displays higher formability than the one containing the measured texture in both the positive and negative strain-ratio regions. The closer the ρ-value is to 1 or -0.5, the larger is the difference between the respective limit strains. By contrast, the measured FLD is close to the FLD predicted from the measured initial texture, particularly in the equibiaxial stretching region.

The predicted yield loci that correspond to these two textures are illustrated in Fig. 5. The more rounded nature of the yield locus pertaining to the cube texture is probably responsible for the higher formability predicted for this material. If the annealing process is controlled to increase the cube component of the initial texture, the formability of the Al-Mg sheet is therefore expected to be improved. By contrast, the limit strain for equibiaxial stretching of an FCC sheet that consists solely of grains of the *exact* cube orientation is only half that for plane strain deformation (Barlat, 1987; Zhou & Neale, 1994). This is because the yield locus of the cube orientation is identical to the well-known Tresca yield function. Thus it appears that when some scatter is present about the exact cube orientation, as is the case for the simulations described here, the more *rounded* nature of the yield surface, see Fig. 5, leads to a higher limit strain in the biaxial stretching region.

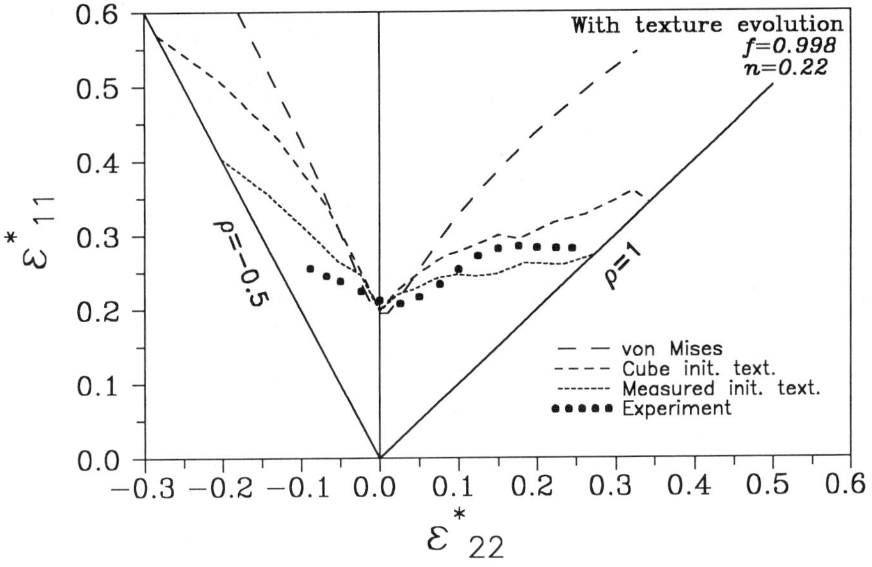

Fig. 4. Forming limit diagrams predicted using different initial textures.

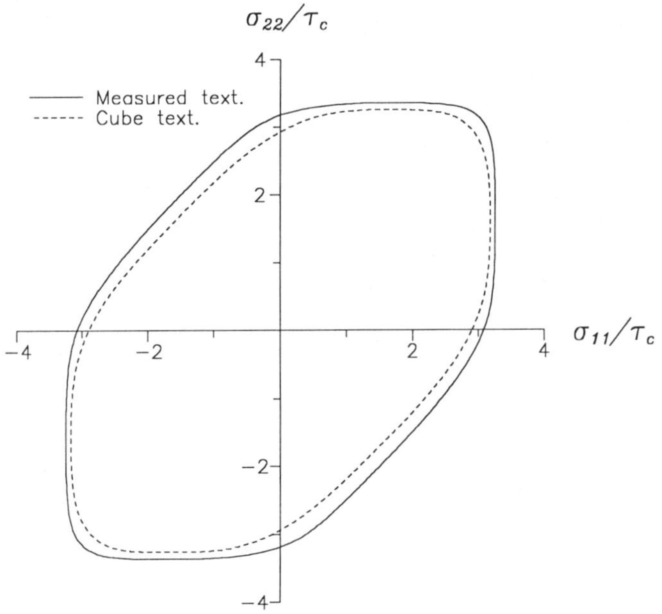

Fig. 5 Predicted yield loci for materials containing the cube and the measured initial textures.

4 CONCLUSIONS

In the simulations described here, the numerical relations between the plastic strain rates and yield stresses of textured aluminum sheets were determined and updated during deformation. This method was employed in conjunction with the Marciniak-Kuczynski approach to predict the forming limit diagram for an Al-Mg sheet. The predicted diagram is in good agreement with the experimental one.

The effects of texture on the FLD were also

examined. Neither the initial texture nor texture evolution affects the forming limit strain for in-plane plane strain stretching. However, texture evolution decreases the limit strain in both the positive and negative strain-ratio regions. It was also shown that increasing the cube component of the initial texture can improve the formability of Al-Mg sheet.

ACKNOWLEDGMENTS

This work was supported by the Natural Sciences and Engineering Research Council of Canada (NSERC) and Alcan International Limited (KRDC, Kingston, Canada).

REFERENCES

Barlat, F. 1987. Crystallographic texture, anisotropic yield surfaces and forming limits of sheet metals. *Mat. Sci. Eng.* 91:55-72.

Barlat, F. & Chung, K. 1993. Anisotropic potential for plastically deformed metals. *Modelling Simul. Mater. Sci. Eng.* 1:403-416.

Bunge, H.J. 1982. *Texture Analysis in Materials Science*. London:Butterworths.

Hill, R. 1948. A theory of the yielding and plastic flow of anisotropic metals. *Proceedings of the Royal Society of London*, A193:281-297.

Jain, M. & Allin, J. 1994. Analysis of experimental variability of forming limit diagrams. Presented at *39th Int. SAMPE Symp. & Exhibition*, Anaheim, CA, April 11-14.

Marciniak, Z. & Kuczynski, K. 1967. Limit strains in the processes of stretch-forming sheet metal. *Intl. J. of Mech. Sci.* 9:609-620.

Neale, K.W. & Chater, E. 1983. Limit strain predictions for strain-rate sensitive anisotropic sheets. *Int. J. Mech. Sci.* 22:563-574.

Qiu, Y., Neale, K.W., Makinde, A. & MacEwen, S.R. 1994. Numerical modelling of metal formability using polycrystal plasticity. In these proceedings.

Ratchev, P., Van Houtte, P., Verlinden, B., Neutjens, P. & Baartman, R. 1994. Calculation of forming limit diagrams of an Al-Mg alloy taking texture into account. In *Proc. 15th Risø Int. Symp. on Mat. Sci.*, S.I. Andersen *et al.* (eds), p.499-504. Roskilde:Risø National Laboratory.

Savoie, J., Zhou, Y., Jonas, J.J., MacEwen, S.R. & Perrin, R. (1994). Textures induced by tension and deep drawing in aluminum sheets. Submitted to *Acta Metall. & Mater.*.

Tóth, L.S., Hirsch, J. & Van Houtte, P. 1994. On the role of texture development in the forming limits of sheet metals. Submitted to *Int. J. Mech. Sci.*.

Van Houtte, P. & Tóth, L.S. 1993. Generalization of the Marciniak-Kuczynski defect model for predicting forming limit diagrams. In *Advances in engineering plasticity and its application*, W.B. Lee (eds), 1013-1020. Amsterdam: Elsevier Science.

Zhou, Y. & Neale, K.W. 1994. Predictions of forming limit diagrams using a rate-sensitive crystal plasticity model, *Int. J. Mech. Sci.*, in press.

Zhou, Y., Savoie, J. & Jonas, J.J. 1994. The application of polycrystalline yield surfaces to the prediction of forming limit diagrams. In *Proc. of the 15th Risø Int. Symp. on Mat. Sci.*, S.I. Andersen *et al.* (eds), p.627-633. Roskilde:Risø National Laboratory.

Zhou, Y., Jonas, J.J., Jain, M. & MacEwen, S.R. 1995. to be published.

Simulation of Materials Processing: Theory, Methods and Applications, Shen & Dawson (eds)
 ISBN 90 5410 553 4

Application of damage models in metal forming

Y.Y.Zhu & T.Zacharia
Metals and Ceramics Division, Oak Ridge National Laboratory, Tenn., USA

ABSTRACT: The development of damage models in the analysis of metal forming processes, to characterize the formability limits, is an important area of ongoing research. In this paper, two energy-based damage models for the simulation of crack initiation in metal forming processes are presented. The first one is an isotropic damage model with two damage variables. The second one is an anisotropic model with a damage characteristic tensor. The damage models are developed within the general framework of continuum thermodynamics for irreversible processes by identifying a proper set of internal variables together with their associated generalized forces. An approach is proposed to account for microcrack opening and closing. A viscoplastic regularization algorithm is used to take into account the strain rate effect and to improve numerical stability. Both models have been incorporated into the finite element code, LAGAMINE. The models were applied to simulations of upsetting of collar cylinders and non-isothermal hemispherical punch stretching. The results of the analyses were validated by comparing the finite element simulations with experimentally obtained data.

1 INTRODUCTION

When a material is formed by processes such as forging, rolling, drawing, etc., it experiences large plastic deformations. These deformations lead to material damage and degradation of mechanical properties. Numerical simulation of metal forming processes, to be realistic, needs to take into account damage phenomena. One major purpose of process design is to avoid, or at least minimize, damage growth. Detailed reviews of the application of damage models in metal forming processes can be found in Predeleanu (1987), Zhu (1992), and Gelin & Predeleanu (1992). Two major classes of damage models can be identified. The first one is a phenomenological approach inspired from Kachanov's model (1958) which is principally based on the average deterioration of material properties. The constitutive model is developed within the general framework of continuum thermodynamics. Only a few damage variables are involved, and the associated material parameters are easily identified by simple experiments. Applications of this kind of approach in metal forming processes have been realized by Predeleanu et al. (1986), Bontcheva et al. (1991), Zhu et al. (1992), and Neto et al. (1994). The second is a microscopic-macroscopic approach motivated from Guson (1977) which constructs homogeneous macroscopic material models using some considerations from non-homogeneous microscopic mechanisms and other additional simplified assumptions. This approach seems to be more realistic. However, to experimentally determine material parameters for these models is not an easy task since the micoscopic components are not well defined and are difficult to measure. Such an approach has been applied in metal forming to predict the occurrence of material defects (Gelin et al. 1987, 1989; Mathur et al. 1987; Groche et al. 1989; and Zavaliangos et al. 1991).

Recent experimental evidence indicate that structural failures are often associated with the development of anisotropic material properties even if the initial material properties are isotropic. In sheet metal forming in particular, the effect of anisotropy on the deformation characteristics may be appreciable and important because sheets are usually cold-rolled and have different properties in rolling and transverse directions. Therefore, complex phenomena have to be accurately taken into account, including anisotropic evolution of damage induced by the industrial process. The anisotropic damage models based on the extension of the Kachanov type have been developed by Cordebois & Sidoroff (1982), Lu & Chow (1990), Ju & Simo (1991), Zhu (1992), and Zhu et al. (1995). However, only a few models have been applied to metal forming processes (Zhu et al. 1995).

In this paper we present two models based on Kachanov's approach that have been recently developed. The first one is a fully coupled isotropic elastoplastic damage model with two damage variables (Zhu 1992). The second one is a fully coupled anisotropic elastoplastic damage model with a damage characteristic tensor (Zhu 1992; Zhu et al. 1995). Both models have been implemented in a finite element code, LAGAMINE, and this code has been used to model metal forming processes. The accuracy of the numerical predictions has been assessed by comparison with experimental data available in the literature.

2 ISOTROPIC DAMAGE MODEL

Ladevèze (1984) suggested a damage model with two scalar damage variables, d (deviatoric component) and δ (volumetric component), by which not only the elastic modulus but also Poisson's ratio can vary with the damage growth. In the present model, these two damage variables are used. The true stress tensor $\boldsymbol{\sigma}$ can be transformed into an effective stress $\bar{\boldsymbol{\sigma}}$, viz:

$$\bar{\boldsymbol{\sigma}}' = \frac{\boldsymbol{\sigma}'}{1-d}, \quad \bar{\sigma}_m = \frac{\sigma_m}{1-\delta} \tag{2.1a-b}$$

with $\boldsymbol{\sigma}'$ as the deviatoric stress tensor, and σ_m the hydrostatic stress.

For continuum damage models, various equivalence hypotheses have been proposed in order to transform the damage state into virgin state, such as strain equivalence (Ladevèze 1984), stress equivalence (Simo & Ju 1989), and elastic energy equivalence (Cordebois & Sidoroff 1982). From the viewpoint of energy conservation, the energy equivalence may be of more physical significance. In the present model, we propose an extension of this hypothesis, in the case of two damage variables d, δ model. We obtain the following relations between damaged material and virgin material:

$$\bar{\epsilon}' = \epsilon'(1-d), \; \bar{\epsilon}_m = \epsilon_m(1-\delta), \; \bar{\epsilon}^p_{eq} = \epsilon^p_{eq}(1-d) \tag{2.2a-c}$$

Here, ϵ' is the deviatoric strain tensor, ϵ_m is the average strain, and ϵ^p_{eq} is the equivalent plastic strain.

The yield function F_p used here is made of the energy-based Von Mises yield criterion with both isotropic and kinematic hardening, in the form:

$$F_p = \sqrt{\frac{(\boldsymbol{\sigma}'-\boldsymbol{\gamma}):(\boldsymbol{\sigma}'-\boldsymbol{\gamma}')}{4G(1-d)^2}} - R_0 - R(\alpha) \tag{2.3}$$

in which γ' is the deviatoric component of the shift stress tensor γ, R_0 denotes initial plastic hardening threshold, R is the plastic hardening threshold, and α is the accumulated plastic energy variable.

In order to describe the growth of damage, Ladevèze (1984) proposed an energy-based damage evolution criterion. Here, a modified energy-based damage evolution criterion is used (Zhu, 1992):

$$\begin{aligned} F_d &= -Y_d - \langle\tau\rangle Y_\delta - B_0 - B(\beta) \\ &= \frac{\boldsymbol{\sigma}:\boldsymbol{\sigma}}{2G(1-d)^3} + \frac{\langle\tau\rangle \sigma_m^2}{\chi(1-\delta)^3} - B_0 - B(\beta) \end{aligned} \tag{2.4}$$

With the definition,

$$\langle\tau\rangle = \begin{cases} \delta/d & (\sigma_m > 0) \\ 0 & (\sigma_m \le 0) \end{cases} \tag{2.5}$$

the difference of mechanical effects observed under tension and compression states can be described. Here, B_0 denotes initial damage strengthening, B is the damage strengthening threshold, β is the overall damage, and G and χ are the shear and bulk moduli. In eq.(2-4), Y_d and Y_δ are the thermodynamic forces associated with damage variables d and δ, respectively.

3 ANISOTROPIC DAMAGE MODEL

In this section, an energy-based anisotropic damage model proposed originally by Cordebois & Sidoroff (1982) is extended with some special considerations (Zhu 1992). The effective stress tensor written in the principal coordinate system of damage is:

$$[\bar{\sigma}_{11}\; \bar{\sigma}_{22}\; \bar{\sigma}_{33}\; \bar{\sigma}_{23}\; \bar{\sigma}_{31}\; \bar{\sigma}_{12}]^T = \boldsymbol{M}[\sigma_{11}\; \sigma_{22}\; \sigma_{33}\; \sigma_{23}\; \sigma_{31}\; \sigma_{12}]^T \tag{3.1}$$

where the damage effect tensor $\boldsymbol{M}(\boldsymbol{D})$ is a fourth-order symmetric tensor depending on the damage tensor $\boldsymbol{D}$ (Chow & Wang 1987):

$$\boldsymbol{M}(\boldsymbol{D}) = \mathrm{diag}\left[\frac{1}{1-D_1}\; \frac{1}{1-D_2}\; \frac{1}{1-D_3} \;\; \frac{1}{\sqrt{(1-D_2)(1-D_3)}}\; \frac{1}{\sqrt{(1-D_3)(1-D_1)}}\; \frac{1}{\sqrt{(1-D_1)(1-D_2)}}\right] \tag{3.2}$$

D_1, D_2, and D_3 are damage components in their principal axes.

According to the hypotheses of elastic energy equivalence (Cordebois & Sidoroff 1982), the effective elastic strain tensor is written as:

$$\bar{\epsilon}_e = \boldsymbol{M}^{-1}\epsilon_e \tag{3.3}$$

Following the thermodynamic rule of irreversible processes, the associated thermodynamic force can be written (Lu & Chow 1990):

$$\boldsymbol{Y} = -\boldsymbol{\sigma}^T\boldsymbol{M}\boldsymbol{C}_e^{-1}\frac{\partial \boldsymbol{M}}{\partial \boldsymbol{D}}\boldsymbol{\sigma} \tag{3.4}$$

where $\boldsymbol{C}_e$ is the elastic material stiffness tensor. For orthotropic materials, its inversion form is represented by a 6×6 matrix:

$$C_e^{-1} = \begin{bmatrix} \frac{1}{E_1} & -\frac{\nu_{21}}{E_2} & -\frac{\nu_{31}}{E_3} & 0 & 0 & 0 \\ -\frac{\nu_{12}}{E_1} & \frac{1}{E_2} & -\frac{\nu_{32}}{E_3} & 0 & 0 & 0 \\ -\frac{\nu_{13}}{E_1} & -\frac{\nu_{23}}{E_2} & \frac{1}{E_3} & 0 & 0 & 0 \\ 0 & 0 & 0 & \frac{1}{G_{23}} & 0 & 0 \\ 0 & 0 & 0 & 0 & \frac{1}{G_{31}} & 0 \\ 0 & 0 & 0 & 0 & 0 & \frac{1}{G_{12}} \end{bmatrix} \tag{3.5}$$

In the damage characterization of materials undergoing large plastic deformations, Hill's yield criterion is expressed in the following form:

$$F_p = \left\{\frac{1}{2}\sigma^T M H M \sigma\right\}^{1/2} - R_0 - R(\alpha) \tag{3.6}$$

The positive definite tensor $\boldsymbol{H}$ for orthotorpic materials is represented by a 6×6 matrix in the material principal coordinate system (Hill 1950).

In a similar way to arguments leading to plastic dissipative potential, one can assume that there exists a damage surface $F_d = 0$, which separates the damaged domain from the undamaged domain. An energy-based damage evolution criterion was proposed by Cordebois & Sidoroff (1982) as:

$$F_d = \left[\frac{1}{2} Y^T J Y\right]^{1/2} - B_0 - B(\beta) \tag{3.7}$$

in which $\boldsymbol{J}$ is the damage characteristic tensor. Recently, Zhu (1992) proposed the following generalized expression for $\boldsymbol{J}$:

$$J = \begin{bmatrix} 1 & \sqrt{J_2} & \sqrt{J_3} \\ \sqrt{J_2} & J_2 & \sqrt{J_2 J_3} \\ \sqrt{J_3} & \sqrt{J_2 J_3} & J_3 \end{bmatrix} \tag{3.8}$$

The terms J_2, J_3 can be determined by a concept of equating damage work (Zhu 1992) as:

$$J_2 = \frac{Y_1^2}{(D_{t2}/D_{t1})(Y_1^2 - Y_{10}^2) + Y_{20}^2} \tag{3.9a}$$

$$J_3 = \frac{Y_1^2}{(D_{t3}/D_{t1})(Y_1^2 - Y_{10}^2) + Y_{30}^2} \tag{3.9b}$$

where D_{ti} are the slopes of Y_i-D_i curves with component i, and Y_{i0}, Y_i are the initial and current equivalent damage energy release rates corresponding to component i, respectively.

It is generally accepted that only the tensile stress component contributes to microcrack growth in some materials. In this case the thermodynamic force eq.(3.4) can be expressed in the following form:

$$Y = -(\sigma^+)^T M C_e^{-1} \frac{\partial M}{\partial D} \sigma^+ \tag{3.10}$$

where σ^+ is the so-called tensile tensor which is obtained by a projection method of stress-spectral-decomposition (Zhu 1992).

4 VISCOUS REGULARIZATION

The present damage models are able to predict not only the crack initiation but also the propagation of damaged zones. However, further developments are needed, especially in the case of very localized zones, to handle the possible instabilities and bifurcation of the solution corresponding to local strain softening. Viscous regularization (Loret & Prevost 1990), which naturally introduces a length-scale into the dynamic initial boundary value problem, seems to be a powerful approach to solve localization problems associated with material softening. As an extension of the proposal of Loret & Prevost (1990) for softening elastoplasticity, we construct the viscous regularization of the present damage models as:

$$\sigma_{n+1} = \frac{\sigma_{n+1}^{(0)} + (\Delta t_n/\mu)\, \sigma_{n+1}^*}{1 + \Delta t_n/\mu} \tag{4.1}$$

$$q_{n+1} = \frac{q_n + (\Delta t_n/\mu)\, q_{n+1}^*}{1 + \Delta t_n/\mu}$$

in which $\boldsymbol{q}$ is the vector of state variables; $\sigma^{(0)}$ is the solution of the elastic predictor step; σ^*, q^* are the inviscid solutions of isotropic elastoplastic damage models, and μ is the viscosity coefficient. More details on this algorithm can be found in Zhu (1992).

5 NUMERICAL APPLICATIONS

5.1 *Upsetting of collar cylinders*

The upsetting of collar cylinders often results in lower fracture strains in comparison with the upsetting of a circular cylinder (Zhu 1992). For the present tests, two kinds of specimens were used: specimen with one flange and specimen with three flanges. All the tests were terminated when a surface crack could be detected with the naked eye. The isotropic damage law for aluminum was identified with a uniaxial tensile test (Zhu 1992). Fig.1 shows the experimental results of final deformation at fracture for both specimens respectively. The cracks appear to propagate inwards to a depth of 2 to 3 mm and cover the full height of the collar. The theoretical distributions of Von Mises stress and volumetric damage variable are shown in Fig.1. As expected, the maximum value of volumetric damage variable δ is located in the collar. In general, the

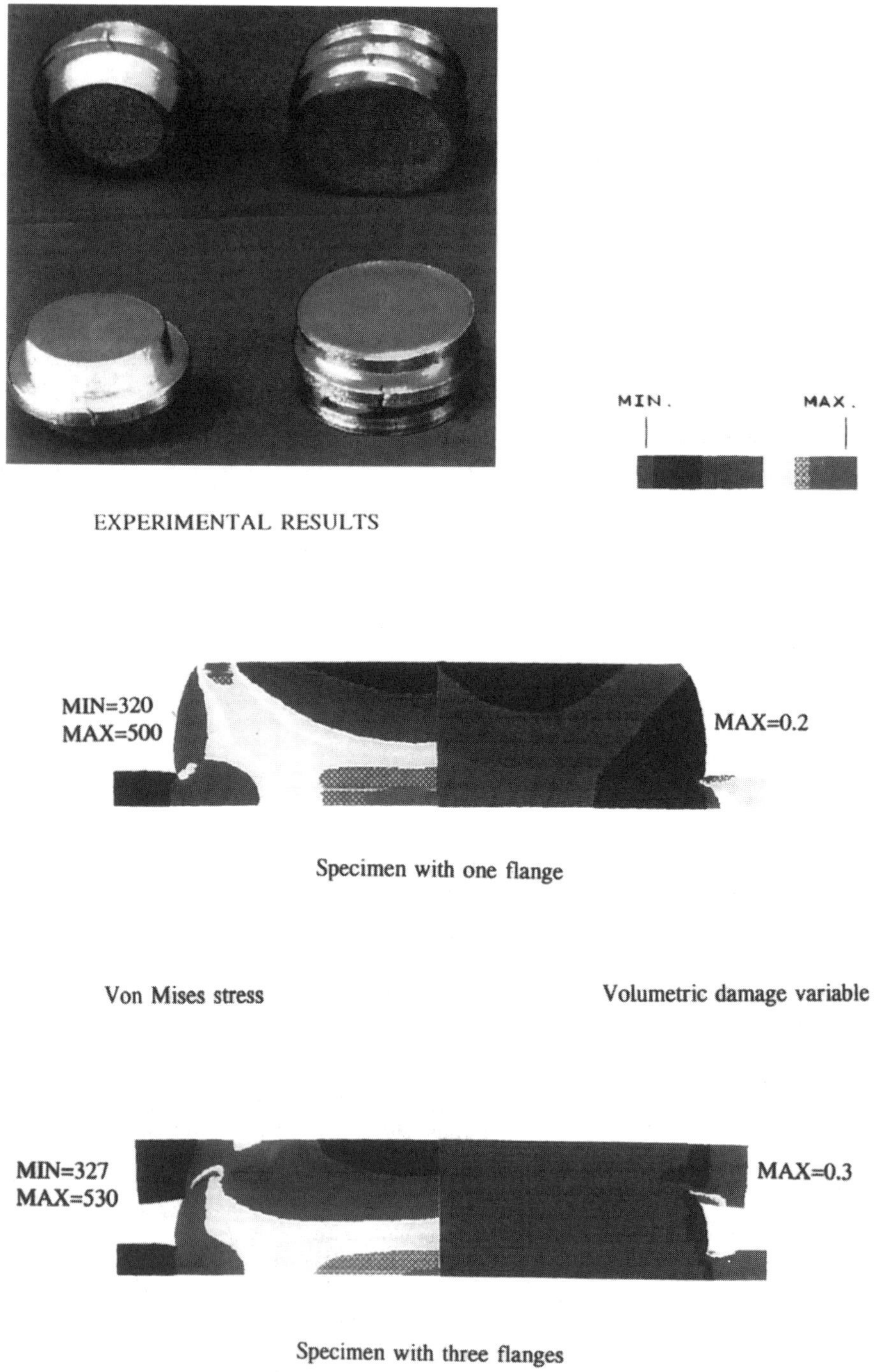

Fig 1. Upsetting of collar cylinders

hydrostatic stress σ_m becomes larger at the equatorial free surface. Cracks usually form in those regions where the hydrostatic state of stress is highly tensile, while the Von Mises stress may not be too high.

5.2 *Non-isothermal hemispherical punch stretching*

This example is aimed at applying the anisotropic damage model to solve coupled thermo-plasticity and damage problems related to sheet forming processes. To minimize complexity, temperature-independent material properties are used. However, the introduction of temperature-dependent material properties, leading to additional terms, does not present fundamental difficulties. Fig.2 shows the initial and deformed shapes of sheet. The punch and die are initially at room temperature and the sheet is initially heated to 348K. The heat transfer to tool and air and internal heat conduction in sheet are considered. Fig.2 also shows the distribution of damage components and the radial strain along the original position for the cases: (1) anisotropic with friction at a punch depth of 40 mm, (2) anisotropic with friction at a punch depth of 35 mm, and (3) isotropic without friction at a punch depth of 40 mm. In our simulation, we find that as the friction between the sheet and tool is decreased or the degree of anisotropy reduced, the peak damage and strain move toward the center of the sheet and the peak values are increased. In the case of isotropic without friction, the peak values are located near the center of the sheet, and the damage value is three times higher than that in the case of anisotropic with friction. In Fig.2 other previous experimental (Ghosh et al. 1975) and theoretical (Wang et al. 1978) results are also plotted. These results are relevant to a punch depth between 35 and 38 mm and show good agreement with the predicted results.

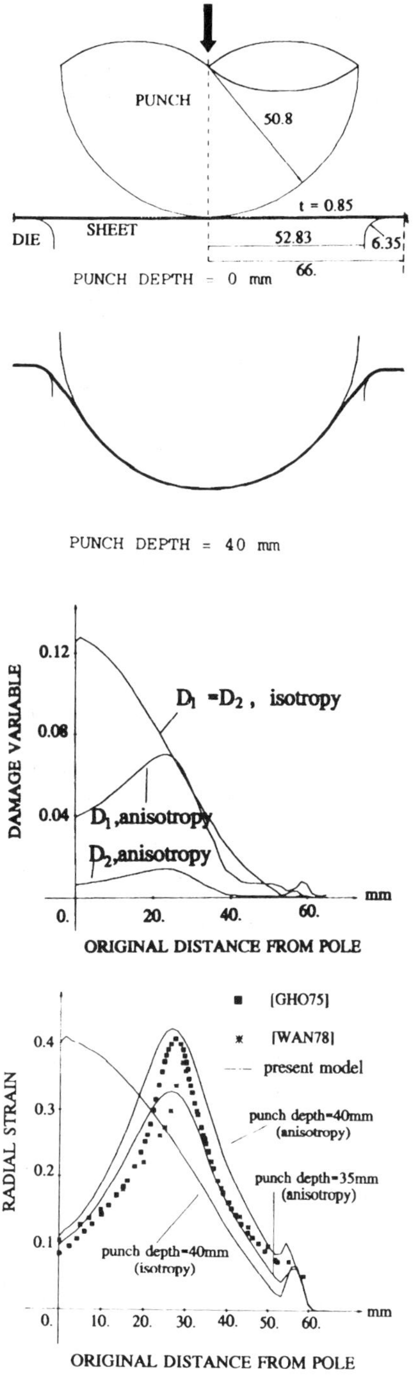

Fig. 2. Non-isothermal hemispherical punch stretching

6 CONCLUSIONS

Two energy-based elastoplastic damage models have been presented in this paper to characterize damage and crack growth. The constitutive models were developed within the general framework of continuum thermo-dynamics for irreversible processes. The models are capable of accommodating general nonlinear elasto-plastic response, the coupling of damage and plasticity, rate dependency, damage threshold, isotropic or anisotropic microcrack opening and closing. The numerical problem of mesh dependency is improved by means of viscous regularization algorithm.

The numerical simulations of metal forming processes and the results obtained concerning the distribution of damage variables clearly indicate the zones where defects can appear. Finite element programs based on the damage models should help predict damage evolution and defects occurrence in the simulation of metal forming processes where the quality of the final product is required.

Our further research will be focused on the extension of the present models to those with a non-local damage framework in order to completely avoid mesh dependency problems.

Acknowledgments - The authors would like to acknowledge Dr. G.M. Ludtka and Dr. J.H. Schneible for reviewing the paper. This research was supported in part by an appointment to the Oak Ridge National Laboratory Postdoctoral Research Associates Program administered jointly by the Oak Ridge National Laboratory and the Oak Ridge Institute for Science and Education. This research was also sponsored by the U.S. Department of Energy, Assistant Secretary for Energy Efficiency and Renewable Energy, Office of Industrial Technologies, Advanced Industrial Materials Program, under contract DE-AC05-84OR21400 with Martin Marietta Energy Systems, Inc.

REFERENCES

Bontcheva, N. & R.Lankov 1991. Numerical investigation of the damage process in metal forming. *Engng. Fract. Mech.* 40:387-393.

Chow, C.L. & J.Wang 1987. An anisotropic theory of continuum damage mechanics for ductile fracture. *Engng. Fract. Mech.* 27:547-558.

Cordebois, J.P. & F.Sidoroff 1982. Endommagement anistope en élasticité et plasticité. *J. Méc. Thé. App., Numéro spécial*. 45-60.

Gelin, J.C. 1987. Numerical analysis of strain rate and temperature effects on localization of plastic flow and ductile fracture - application to metal forming processes. *Computational Methods for Predicting Material Processing Defects*. Ed. by M.Predeleanu. Elsevier Science Publishers. 123-132.

Gelin, J.C. & M.Predeleanu 1989. Finite strain elasto-plasticity including damage-application to metal forming problems. *Processing of Numiform'89*. 151-157. Rotterdam: Balkema.

Gelin, J.C. & M.Predeleanu 1992. Recent advances in damage mechanics: Modelling and computational aspects, *Processing of Numiform'92*. 89-98. Rotterdam: Balkema.

Ghosh, A.K. & S.S.Hecker 1975. Failure in thin sheets stretched over rigid punches. *Metallurgical Transactions*. 6A:1065-1074.

Groche, P. & E.Doege 1989. Application of continuum damage mechanics to sheet-metal forming. *Processing of Numiform'89*. 33-67. Rotterdam: Balkema.

Gurson, A.L. 1977. Continuum theory of ductile rupture by void nucleation and growth - yield criteria and flow rules for porous ductile media. *J. Engng Mater. Technol.* 99:2-15.

Hill, R. 1950. *The mathematic theory of plasticity*. Oxford.

Kachanov, L.M. 1958. Time of the rupture process under creep conditions. *IZV. Akad. Nauk, Otd. Tech. Nsauk*. 8:26-31.

Ladevèze, P. 1984. Sur une théorie de l'endommagement anistropic. Rapport Interne no.34. LMT. Cachan.

Loret, B. & J.H.Prevost. 1990. Dynamic strain localization in elasto-(visco)plastic solids. Comp. Meth. Appl. Mech. Engng. 83:247-273.

Lu, T.J. & C.L.Chow 1990. On constitutive equations of inelastic solids with anisotropic damage. *J. Theory Appl. Fract. Mech.*. 14:187-218.

Mathur, K.K. & P.R.Dawson 1987. Damage evolution modeling in bulk forming processes. *Computational Methods for Predicting Material Processing Defects*. Ed. by M.Predeleanu, Elsevier Science Publishers. 251-262.

Neto, E.A. D.Peric & D.R.Owen 1994. A model for elastoplastic damage at finite strains: Algorithmic issues and applications. *Engng. Computations*. 11:257-281.

Predeleanu, M. J.P.Cordebois & L.Belkhiri 1986. Failure analysis of cold upsetting by computer and experimental simulation. *Proceedings of the NUMIFORM'86*. 277-282.

Predeleanu, M. 1987. Finite strain plasticity analysis of damage effects in metal forming process. *Computational Methods for Predicting Material Processing Defects*. Ed. by M.Predeleanu, Elsevier Science Publishers. 295-307.

Simo, J.C. & J.W.Ju 1989. On continuum damage elastoplaticity at finite strains. *Computational Mechanics*. 5:375-400.

Zavaliangos, A. L.Anand & B.F.Von Turkovich. 1991. Towards a capability for predicting the formatting of defects during bulk deformation processing. *Ann. CIRP*. 40:267-271.

Wang, N.M. & B.Budiansky 1978. Analysis of sheetmetal stamping by a finite element method. *J. Appl. Mech.* 45:73-82.

Zhu, Y.Y. S.Cescotto & A.M.Habraken 1992. A fully coupled elastoplastic damage modelling and fracture criteria in metal forming processes. *J. Mater. Proc. Technol.* 32:197-204.

Zhu, Y.Y. 1992. Contribution to the local approach of fracture in solid dynamic. Ph.D. dissertation, University of Liege, Belgium.

Zhu, Y.Y. & S.Cescotto 1995. A fully coupled elasto-visco-plastic damage theory for anisotropic materials. *Int. J. Solids Strut.* (accepted).

Simulation methods and large scale computation

Simulation of Materials Processing: Theory, Methods and Applications, Shen & Dawson (eds)
© 1995 Balkema, Rotterdam. ISBN 90 5410 553 4

Finite element analysis of forward/backward extrusion using ALE techniques

E.H.Atzema & J.Huétink
University of Twente, Department of Mechanical Engineering, Enschede, Netherlands

ABSTRACT: Arbitrary Lagrangian Eulerian finite element methods gain interest for the capability to control mesh geometry independently from material geometry. After explaining the underlying ideas of the method briefly, the problem of mesh management is touched upon. A possible way to control the distortion of the mesh is given and some results for an academic as well as an industrial problem are presented.

1. INTRODUCTION

Finite element techniques have found their way into simulations of forming processes as early as the 1970's. It was not until the early 80's however that solutions were found for the grid distortion that was a consequence of using the Lagrangian formulation to model the large deformations that occur in forming processes.

One solution to this problem is remeshing [Gelten, 1982], i.e. every number of steps or when the grid distortion has become too large a new mesh is constructed over the old one. The values of the old mesh are transferred to the new one using interpolation techniques. When extremely large deformations occur, remeshing must be carried out frequently which results in excessive numerical diffusion.

Another solution is to use an Arbitrary Lagrangian Eulerian (ALE) method. One implementation of this method is in the DIEKA code [Huétink, 1986]. In these methods the grid is no longer fixed to the material as in a Lagrangian formulation. Instead the grid has the ability to move independent of the material. Finite element codes of the ALE type thus possess the capability to control mesh geometry independently of material geometry. In forming processes with large deformations, such as extrusion, this is of invaluable advantage. The movement of grid points relative to the material can be viewed as a convection problem. In this way advantage can be taken of the considerable knowledge of convection problems gained in Computational Fluid Dynamics.

At the NUMIFORM '92 conference some applications of ALE methods have been shown, see e.g. [Tong, 1992]. Moreover from literature applications are known for rolling [Liu, 1991][Lugt, 1988] and Taylor bar impact [Ponthot, 1992]. In these papers emphasis is put on the ALE technique itself. However, to apply it succesfully to an industrial application, additional algorithms are needed to keep the grid regular. These algorithms provide the tools for managing the mesh and will henceforth be denoted as: mesh management.

2. ALE METHOD

There have been several implementations of the ALE technique. Some solve for the material displacement and the grid displacement at the same time, using the SUPG method [Brooks, 1982]. Others apply a Lagrangian step followed by an Eulerian step, albeit that the grid moves over the material instead of the other way around. The latter technique was developed by [Huétink, 1986] and will be briefly recapitulated here.

In this strategy an updated Lagrangian step to calculate material displacements is performed, which results in meshdistortion. Thereafter, using a mesh management strategy treated in the next section, grid displacements are obtained. The difference between material and grid displacement is convective displacement. To compute convection the gradient of a variable is needed. Lets e.g. take the equivalent plastic strain $\boldsymbol{\varepsilon}_p$, the new strain is

$$\begin{aligned}&\boldsymbol{\varepsilon}_p(\boldsymbol{x}^g+\Delta\boldsymbol{x}^g,t+\Delta t)= \\ &\boldsymbol{\varepsilon}_p(\boldsymbol{x}^g,t)+\Delta\boldsymbol{\varepsilon}_p^m+(\Delta\boldsymbol{x}^g-\Delta\boldsymbol{x}^m)\cdot\vec{\nabla}\boldsymbol{\varepsilon}_p\end{aligned} \tag{1}$$

Superscript m denotes the material increment and g

the value in a grid point. To obtain the gradient a continuous field is constructed. The integration point values for the dependent variables are extrapolated to the nodal points. From averaging nodal point values and subsequent interpolation by ψ^N a continuous field ε_p^* is obtained

$$\varepsilon_p^* = \psi^N \varepsilon_p^N \tag{2}$$

$$\vec{\nabla}\varepsilon_p^* = \vec{\nabla}\psi^N \varepsilon_p^N \tag{3}$$

Then the new strain is

$$\begin{aligned} &\varepsilon_p(\boldsymbol{x}^g + \Delta\boldsymbol{x}^g, t+\Delta t) = \\ &\psi^N(\boldsymbol{x}^g)\varepsilon_p^N(t) + \Delta\varepsilon_p^m + (\Delta\boldsymbol{x}^g - \Delta\boldsymbol{x}^m)\cdot\vec{\nabla}\psi^N\varepsilon_p^N \end{aligned} \tag{4}$$

As this procedure gives rise to oscillations local and global smoothing can be applied to obtain nodal point avarages. This, however, is over-diffusive, so a weighted sum of the integration point value and the smoothed value is taken

$$\begin{aligned} &\varepsilon_p(\boldsymbol{x}^g + \Delta\boldsymbol{x}^g, t + \Delta t) = \\ &(1-\alpha)\,\varepsilon_p(\boldsymbol{x}^g, t) + \alpha\,\psi^N(\boldsymbol{x}^g)\,\varepsilon_p^N(t) \\ &+ \Delta\varepsilon_p^m + (\Delta\boldsymbol{x}^g - \Delta\boldsymbol{x}^m)\cdot\vec{\nabla}\psi^N\varepsilon_p^N \end{aligned} \tag{5}$$

Now the new integration point values are obtained. The weight parameter α is bounded by

$$\frac{|\Delta\boldsymbol{x}^g - \Delta\boldsymbol{x}^m|}{l_e} \le \alpha \le 2\,\frac{|\Delta\boldsymbol{x}^g - \Delta\boldsymbol{u}^m|}{l_e} \tag{6}$$

Where l_e is the element length. To ensure accuracy convergence is checked after convection. For a more elaborate treatment see e.g. [Huétink, 1990] or for new developments [Huétink, 1992].

3. MESH MANAGEMENT

The key element in effective use of the ALE method in industrial forming processes is the determination of the grid displacements.

One strategy to handle this problem is treated here. Other strategies may be better suited for other problems. It is not the intention of the authors to provide a definitive answer to this issue, but to initiate a discussion on the topic.

The strategy treated here only takes into account the geometry. From at least one other strategy [Ponthot, 1992] it is known that also dependent variables and their gradients can be used to optimise the grid for accuracy. The latter objective is sometimes referred to as: r-adaptivity. As the simulations in [Ponthot, 1992] show large geometrical changes, but not as large as in e.g. backward extrusion, which we are aiming at, the current approach is restricted to geometrical manipulation.

The interior of the mesh depends on the surface which is first treated point by point. Locally the new surface is approximated by a smooth curve through the material position of the current node and its neighbours. With two auxiliary points an intersection can be made with the new surface determining the new position of the node, see figure 1. The difference between material position and the new position is the convective displacement. In this way a surface point can be kept on e.g. a tool corner.

Instead of intersecting, the line through the auxiliary points can be used to determine the direction of grid displacement, see figure 2, thus leading to a second way to determine the new position. This algorithm is very useful for keeping in place internal surfaces, as seen in figure 3.

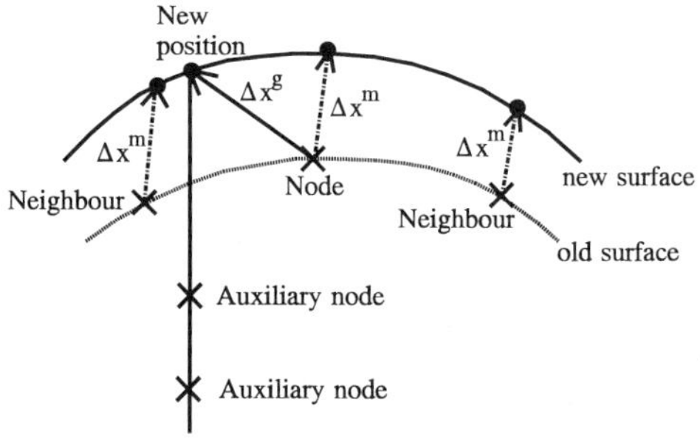

Fig. 1 Intersection

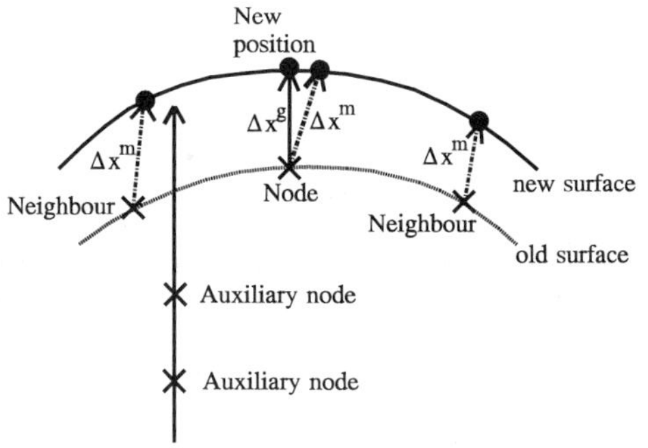

Fig. 2 Direction

Moreover specific points can be treated Eulerian, i.e. spatially fixed. After the surface is handled the interior is interpolated between the surfaces. Together with the two algorithms mentioned above this suffices for simulating the academic problem in section 5.

For the industrial problem in section 6 some

more mesh management algorithms were needed. In an additional algorithm the direction of the total displacement was in the mean direction of the displacements of the neighbours. As a last option it was possible to have the new location of the surface point as an orthogonal projection of an auxiliary point on the new surface.

It was noted in test calculations that the sweep to update the coordinates should always take place in stream direction. Failure to do so causes instabilities.

4. CONTACT MODEL

In simulations of forming processes contact and friction play an important role. For the academic problem featuring in the next section, friction is neglected and tool action is modelled by boundary conditions. For the industrial problem, in section 6, however, contact and friction have to be taken into account.

Before the model is treated in detail it is worth noting that any convective displacement of the grid orthogonal to the tool will produce changes in contact. However the mesh management can be devised in such a way that only convective displacements tangential to the tool are admitted.

If the distance between two bodies is given by a gap function g and the (normal) contact pressure by t_N, then the following holds

$$\begin{aligned} g &\geq 0 \\ t_N &\leq 0 \\ g \cdot t_N &= 0 \end{aligned} \tag{7}$$

When the deformations and rotations are small, a relatively simple formalism for contact can be used. If two points are in contact and $\boldsymbol{v}^i$ is the velocity of body i and $\boldsymbol{n}^i$ is the outward normal of the body at the same point, then

$$(\boldsymbol{v}^1 - \boldsymbol{v}^2)\cdot\boldsymbol{n}^1 \leq 0 \quad on\ \Gamma_c \tag{8}$$

This inequality states that when two points are in contact no relative velocity that will cause penetration can occur. In the contact area, Γ_c, when smooth surfaces are assumed, the outward normals of points 1 and 2 are opposite

$$\boldsymbol{n}^1 = -\boldsymbol{n}^2 \tag{9}$$

When the traction vector on the surface is $\boldsymbol{t}$, the normal traction is $\boldsymbol{t}_N$

$$\boldsymbol{t}_N = (\boldsymbol{n}\cdot\boldsymbol{t})\boldsymbol{n} = t_N\boldsymbol{n} \tag{10}$$

and the sliding traction $\boldsymbol{t}_T$

$$\boldsymbol{t}_T = (\mathbf{1} - \boldsymbol{nn})\cdot\boldsymbol{t} \tag{11}$$

For the normal traction as mentioned above only pressure is allowed

$$t_N^1 = t_N^2 = t_N \leq 0 \tag{12}$$

Looking at one body and relating the normal vector to it, taking the velocity $\boldsymbol{v}^{rel}$ of this body relative to the other body, the normal relative velocity is

$$v_N = \boldsymbol{v}^{rel}\cdot\boldsymbol{n} \tag{13}$$

Modelling contact with a penalty parameter β the contact pressure is a function of the gap

$$t_N = \beta\, g \tag{14}$$

The specific virtual work for contact then becomes

$$\delta w_c = t_N\, \delta v_N = \beta\, g\, \delta v_N \tag{15}$$

So in the weak form an additional term appears analogous to the surface traction

$$\delta W_c = \beta \int_{\Gamma_c} g\, \delta v_N\, d\Gamma \tag{16}$$

For an incremental formulation see [Huétink, 1990].

5. ACADEMIC PROBLEM

For this exercise the backward extrusion process was chosen by which e.g. toothpaste tubes are produced. A rather simple axi-symmetric geometry was chosen for simplicity of mesh generation. The mesh is divided in three parts, one under the punch, another that can freely flow out and a part under the outflow with the same height as the section under the punch.

The division can be viewed as internal surfaces A and B, see figure 3. Internal surface A is kept at the same height as the top of the billet. Internal

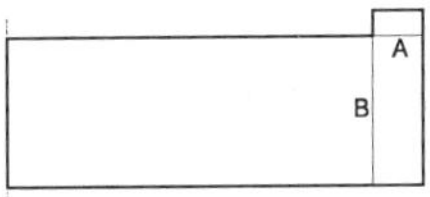

Fig. 3 Internal surfaces

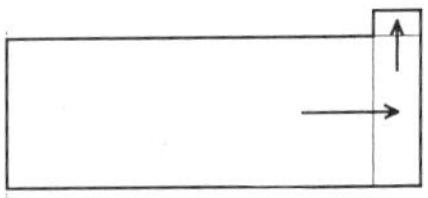

Fig. 4 Material flow

surface B is kept in line with the outflow. The elements under the punch maintain their width and thus lose material which will flow through the grid to the elements in the outflow section, see figure 4, where elements will grow in height.

In this simulation a relatively coarse mesh was used. The billet under the punch was modelled with 20 elements radially and 5 elements in thickness direction. In the outflow section 3 elements over the width were used and 20 over the height. The area under the outflow outside the punch therefore has 5 elements in thickness and 3 in width direction. This totals to 195 isoparametric

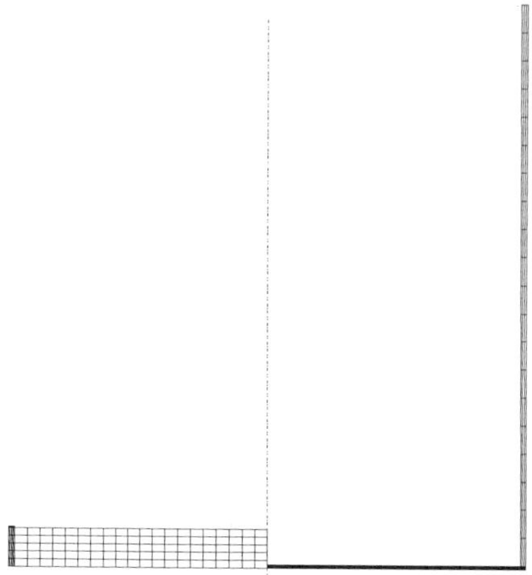

Fig. 5 *Left of centre: undeformed mesh*
Right of centre: deformed mesh

4-node axi-symmetric elements in which mean-dilatation was used. As a hardening model the Voce relation was used

$$\sigma_y = \sigma_0 + \delta\sigma\, e^{\left(\frac{\varepsilon^p{}_1 - \varepsilon^p}{\varepsilon^p{}_0}\right)} \qquad (17)$$

with σ=200 *MPa*; $\delta\sigma$=50 *MPa*; $\varepsilon^p{}_0$=0.1; $\varepsilon^p{}_1$=0.

Figure 5 will illustrate the deformed geometry vs. the undeformed geometry. Note that the deformed mesh is not representative of the deformations. The elements in the outflow section cannot be distinguished individually in the initial mesh whereas they are quite large in height in the deformed mesh.

Since the grid is decoupled from the material in an ALE method no direct clues to the deformation are given in the deformed mesh.

The deformed mesh and the initial mesh differ only 1% in volume, showing good accuracy of the algorithms presented in section 3.

Computer resources consisted of a DEC-5000/200 workstation (25 Mips/ 3.7 MFlops) on which the simulation took approximately 5 hours.

6. INDUSTRIAL PROBLEM

After the academic problem we tried a problem presented to us by industry. The model includes contact behaviour and frictional effects as well as a more complex shape. However, the material model is still elasto-plastic and no temperature effects were taken into account.

The benchmark comprised a combined forward/backward extrusion problem, for tooling see figure 6. A mounting pin is formed at the centre, below the billet, in a forward extrusion. At the same time the wall of the product is formed in a backward extrusion process. The mounting pin has a conical side and the outer edge of the bottom as well as the outer edge of the punch are rounded off. Moreover, the billet fits in the die easily, thereby leaving space to be filled later in the process. Since the billet does not fit within the rounded part of the die it gets stuck on the outer edge. All these effects represent highly non-linear contact-problems.

The die and the punch have a limited part contacting the outflowing wall-section of the product to calibrate the dimensions. Once again this represents a contact problem but now contact is lost rather then made. The material is in contact with a large part of the tools during the process and frictional effects are important. The geometry of the punch and die are illustrated in figure 6, the billet is 49.9 mm round and 7.4 thick.

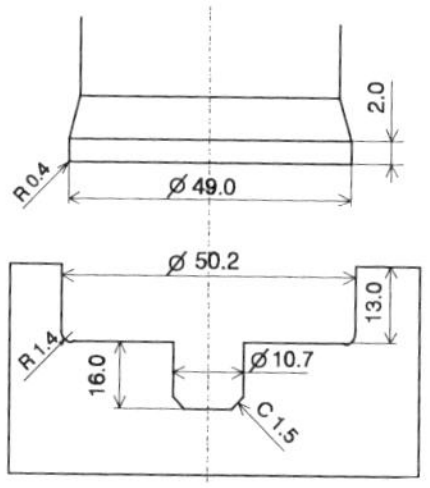

Fig. 6 *Tool geometry*

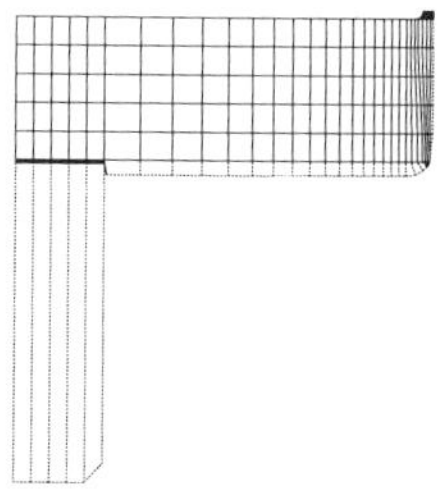

Fig. 7 *Initial mesh*

In the same way as in the academic problem where the outflow can be foreseen a large number of small elements is generated and internal surfaces are defined. The initial mesh is shown in figure 7; the bulk-elements are pictured by solid lines and the contact-elements are dotted.

Friction was modelled with the Coulomb model: μ=0.05. As the round-off of the die is such that the billet does not fit in initially, contact with the die occurs first at one point and is therefore critical. Firstly the corner of the billet is rounded to fit the die and then contact occurs at the bottom of the die. Since contact behaviour is highly non-linear this too is a critical phase in the calculation.

Plastic flow was modelled with the von Mises yield criterion and hardening with a Nadai formula

$$\sigma_y = C\,(\varepsilon^p)^n \qquad (18)$$

with the plastic parameters C=140 MPa; n=0.27 and elastic constants: E=70 GPa, ν=0.3

Four typical intermediate stages will be shown along with the final geometry. These stages can easily be recognised in the force-displacement diagram, figure 8.

In the first stage, A, contact is established at the bottom, but not yet on the outside edge of the billet and flow into the mounting pin is starting. As the diameter increases contact is made at the outer edge of the billet. When some material has flown into the mounting pin, stage B, the slope of the curve decreases, the force increases only slightly. At this stage already some flow in the wall section is seen, which can be attributed to friction. When the pin is filled completely, shortly after stage C, the force increases because material flowing out the wall section now has to flow a greater distance under the punch. Complete filling of the mounting pin has occurred at stage D. Between stage D and E (figure 16) the force slightly increases because of hardening and friction.

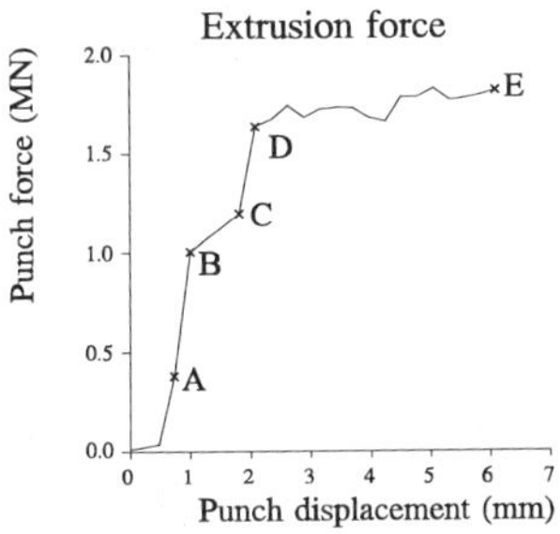

Fig. 8 Force displacement diagram

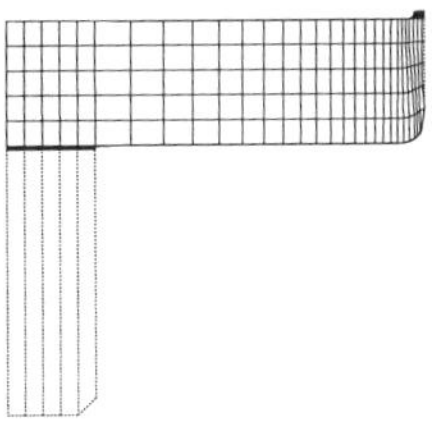

Fig. 9 Stage A

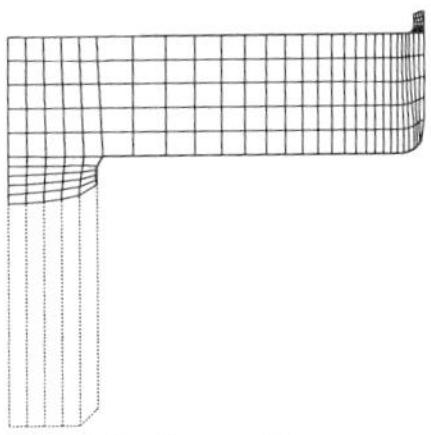

Fig. 10 Stage B

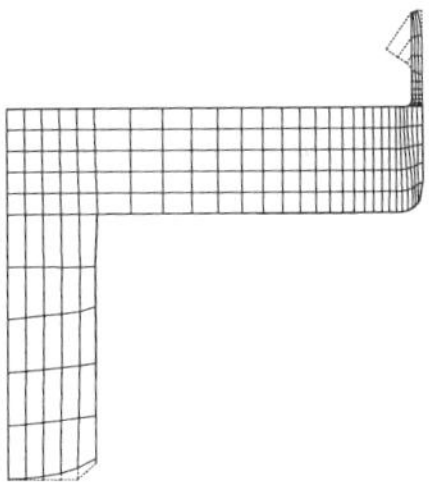

Fig. 11 Stage C

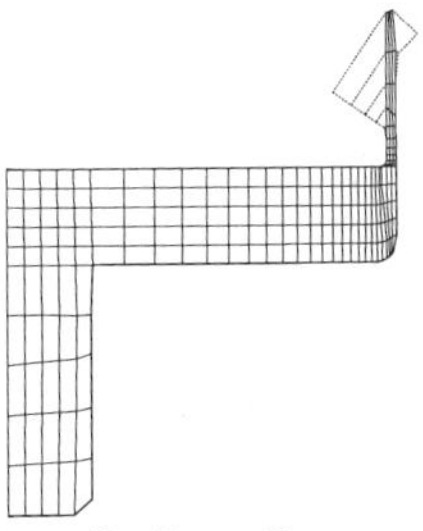

Fig. 12 Stage D

At this last part some noise can be seen on the plot. This is due to the contact elements in the outflow section, which model the geometry rather crudely. In some steps contact will be broken as an element moves up, resulting in a large change in

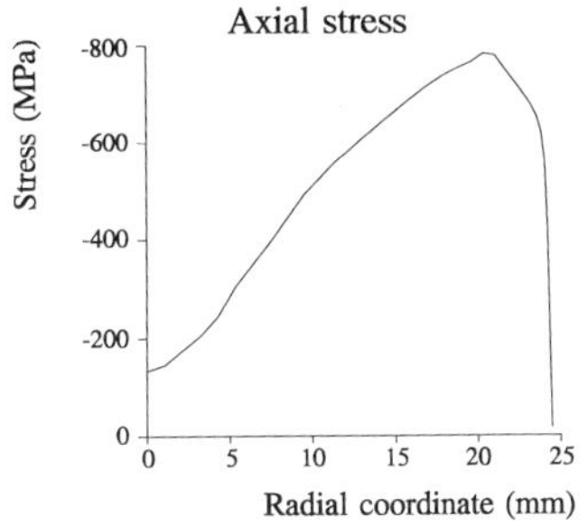

Fig. 13 Axial stress on top of billet

the force. Quite a severe amount of damping was applied in the contact behaviour so the release of contact smoothens out considerably.

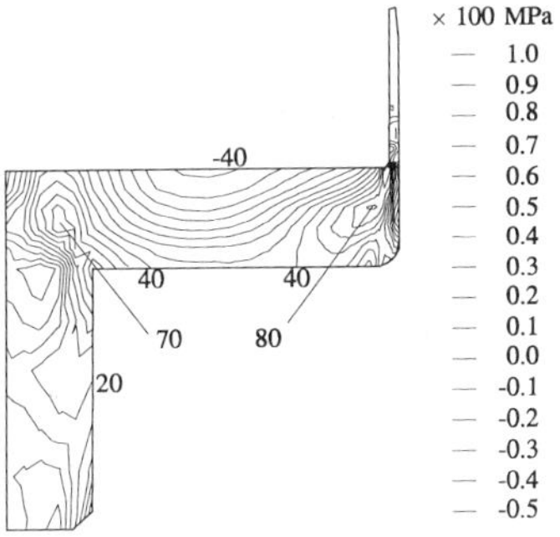

Fig. 14 Shear stress, stage D

In practice, it is observed that the topside of the billet draws in at the centre. In the simulation this does not show, however, the pressure that the punch exhibits on the billet is lowest at the centre. This can be seen in figure 13, which shows the axial stress on the top of the billet. From the shear stress, figure 14, clearly two shear bands can be established as well as wall friction. Particle tracking can be done by plotting lines that were of equal r-coordinate in the initial mesh, figure 15.

7 CONCLUSIONS

It is shown that backward and combined forward/backward extrusion can be simulated successfully with an ALE type finite element method. Contact and friction behaviour can be taken into account and the results can be explained satisfactorily.

A drawback of the method used is that all mesh

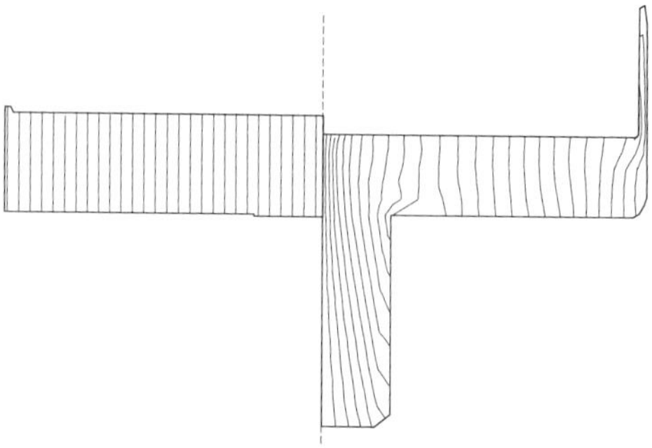

Fig. 15 Particle tracking, left: initial mesh
right: stage D

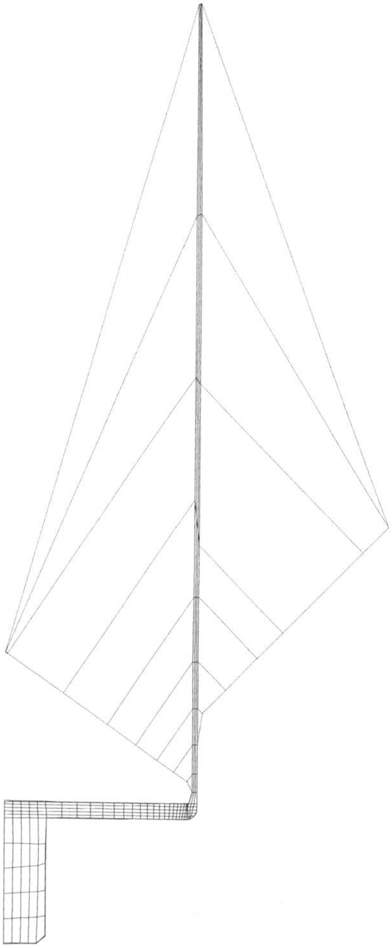

Fig. 16 Stage E

subdivisions must be present at the start of the calculation. This gives rise to large elements in the outflow and thus the geometry is only crudely approximated. The elements in the mounting pin become rather large too. By performing one or two remeshing steps this could be improved. This would introduce some extra numerical diffusion but not as much as when the simulation was carried out totally with an updated Lagrange code with remeshing. Thus ALE techniques and remeshing can be thought of as complementary. For a large part the ALE method represents a remeshing operation every step. An essential difference is that with an ALE method no excessive diffusion need occur and for very large deformations this method is therefore preferable.

8. REFERENCES

Brooks, A.N. & Hughes, T.J.R. 1982. *Streamline upwind/Petrov-Galerkin formulations for convection dominated flows with particular emphasis on the incompressible Navier-Stokes equations.* Comp. Meth. Appl. Mech. Eng. 32: 199-259

Gelten, C.J.M. & Konter, A.W.A. 1982. *Application of mesh rezoning in the updated Lagrangian method to metal forming processes*, in J.F. Pittman et.al. Numerical Methods in Industrial Forming Processes, Swansea UK: Pineridge Press

Huétink, J. 1986. *On the simulation of thermo-mechanical forming processes*, Dissertation University of Twente.

Huétink, J. 1990. *Progress in mixed Eulerian-Lagrangian finite element simulations of forming processes.* Int. J. Num. Meth. Eng. 30: 1441-1457

Huétink, J. & van der Helm, P.N. 1992. *On Euler-Lagrange finite element formulation in forming and fluid problems.* In J.-L. Chenot et. al. Numerical Methods in Industrial Forming Processes, 45-54, Rotterdam: Balkema

Liu, W.K., Chen, J.S., Belytschko, T. & Zhang, Y.F. 1991.*Adaptive ALE finite elements with particular reference to external work rate on frictional interface.* Comp. Meth. Appl. Mech. Eng. 93: 189-216

Lugt, J. van der 1988. *A finite element method for the simulation of thermo-mechanical contact problems in forming processes.* Dissertation University of Twente.

Ponthot, J.-Ph. 1992 *The use of the Eulerian-Lagrangian formulation including contact Applications to forming simulation via FEM.* In J.-L. Chenot et. al. Numerical Methods in Industrial Forming Processes, 293-300, Rotterdam: Balkema

Tong, L., Hora, P. & Reissner, J. 1992, *Application of the arbitrary Lagrangian-Eulerian method in the FE-simulation of 3D-bulk forming processes.* In J.-L. Chenot et. al. Numerical Methods in Industrial Forming Processes, 669-674, Rotterdam: Balkema

Simulation of Materials Processing: Theory, Methods and Applications, Shen & Dawson (eds)
© 1995 Balkema, Rotterdam. ISBN 90 5410 553 4

Rigid plastic modelling of hot metal forming in three dimensions using body fitted coordinates

A.W. Bush, I. Charity & G.S. Marshall
University of Teesside, Middlesbrough, Cleveland, UK

ABSTRACT: This paper describes the use of efficient Computational Fluid Dynamics (CFD) pressure correction based algorithms to predict material stress and deformation in the hot rolling process using collocated body fitted coordinates (BFC). A rigid plastic model based on the constitutive equations of Levy-Mises and von Mises yield condition is applied to the hot rolling of steel beams.

1 INTRODUCTION

This paper describes a computational study of the deformation of hot steel in rolling using the rigid-plastic constitutive relation. This has been adopted because in many metal forming processes the permanent plastic deformation dominates the recoverable elastic deformation.

The Levy-Mises constitutive equations relate the non-zero components of the deviatoric stress tensor to the strain rate tensor components. These are analagous to the Navier Stokes equations of fluid dynamics.

The solution of the coupled partial differential equations describing stress equilibrium, mass continuity and conservation of energy uses techniques more commonly found in computational fluid dynamics applications, the dominant feature being the imposition of mass conservation through the solution of a pressure correction algorithm. The grid is Eulerian and uses collocated body fitted coordinates, (Rhie & Chow 1983), (Coelho & Pereira 1992), to permit the modelling of complex geometries such as occur in section rolling. An Eulerian formulation is chosen because the processes of interest are steady state where a computational region fixed in space is most convenient.

The thermal and mechanical behaviour is highly coupled. The plastic deformation can generate large quantities of heat and the subsequent temperature rise affects the material flow behaviour through variations in the yield stress of the material.

The method has been applied to processes of interest to the steel industry, (Bradbury, Bush, Jones & Marshall 1992). The rolling of strip is often treated as a two dimensional process by neglecting edge effects. The strip rolling numerical predictions can be compared against approximate methods such as those of Sims (1954) and Orowan (1943), analytical solutions obtained using slip-line field theory, albeit for a limited number of geometries such as described in Crane & Alexander (1968) and Johnson & Mellor (1973), and of course engineering data, (Beveridge, Bush, Charity, Jones & Marshall 1993).

For truly three dimensional processes such as section rolling and extrusion through shaped dies no analytical methods exist and so direct comparison with engineering data is the only verification available.

2 GOVERNING EQUATIONS

The coupled partial differential equations describing mass conservation, force equilibrium and energy transfer are to be solved along with a constitutive equation describing the response of the material to applied stress. The mathematical description of material behaviour in the plastic state involves a relationship between the stress tensor, σ_{ij}, the strain tensor, ϵ_{ij} and/or the strain rate tensor, $\dot{\epsilon}_{ij}$. The strain in material consists of both a permanent plastic strain and a recoverable elastic strain. In hot rolling the permanent plastic deformation dominates the smaller recoverable elastic deformation. Thus the Levy-Mises rigid-plastic constitutive equations have been adopted. The von-Mises yield condition is used with the ma-

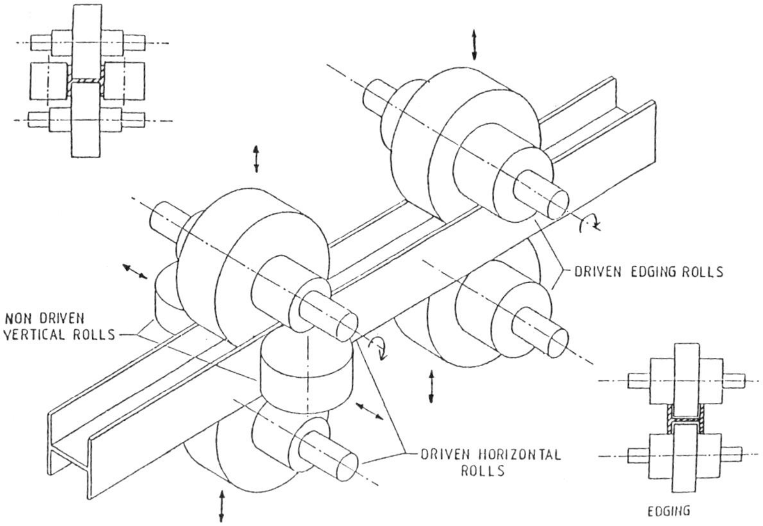

Figure 1. Overview of three dimensional beam rolling

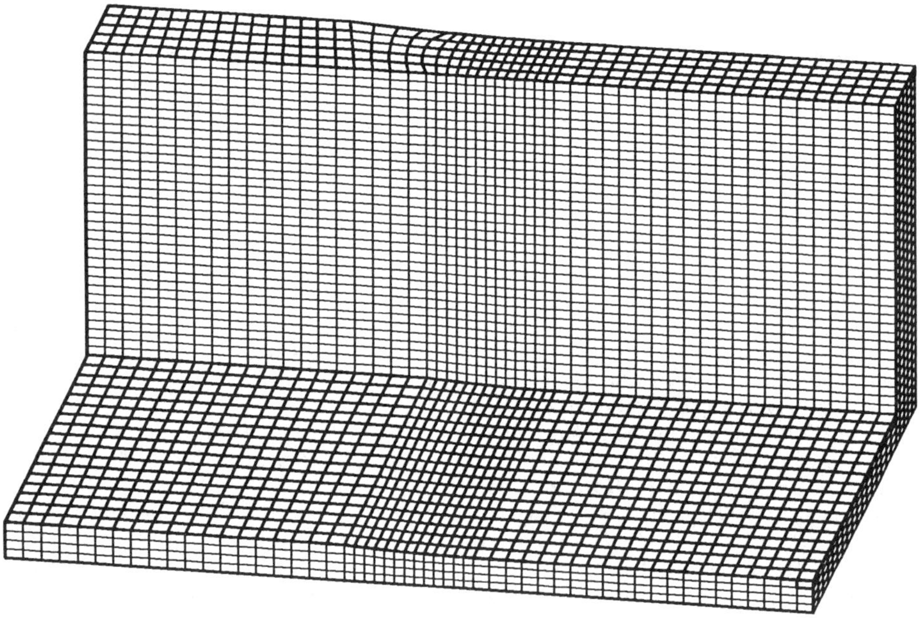

Figure 2. bfc grid for H beam, quarter symmetry assumed

terial behaving rigidly until a level of deviatoric stress is reached whereupon permanent deformation occurs.
The deviatoric stress tensor σ'_{ij} is linearly related to the strain rate tensor $\dot{\epsilon}_{ij}$,

$$2\dot{\epsilon}_{ij} = \dot{\lambda}\sigma'_{ij}(i,j = 1,2,3) \qquad (2.1)$$

where

$$\sigma'_{ij} = \sigma_{ij} - \sigma_{kk}/3$$

and

$$\dot{\epsilon}_{ij} = \frac{1}{2}\left(\frac{\partial v_i}{\partial x_j} + \frac{\partial v_j}{\partial x_i}\right)$$

with $-\sigma_{kk}/3$ defined to be the material pressure, as in the case of fluids. Here v_i is the velocity field and x_i are cartesian coordinates. The parameter $\dot{\lambda}$ plays the role of a reciprocal viscosity for a Newtonian fluid. Its value is velocity dependent and is determined by the yield condition.
The von-Mises yield condition is that plastic deformation occurs when the invariant $\sigma'_{ij}\sigma'_{ij}$ achieves the value $2k^2$ where k is the shear stress at which plastic yield occurs when the material is in a state of pure shear. In a general state of stress

$$\sigma'_{ij}\sigma'_{ij} = \frac{1}{3}[(\sigma_{11} - \sigma_{22})^2 + (\sigma_{22} - \sigma_{33})^2 + (\sigma_{33} - \sigma_{11})^2] + 2(\sigma_{12}^2 + \sigma_{23}^2 + \sigma_{31}^2) \qquad (2.2)$$

and therefore

$$k^2 = \frac{1}{6}\left[(\sigma_{11} - \sigma_{22})^2 + (\sigma_{22} - \sigma_{33})^2 + (\sigma_{33} - \sigma_{11})^2\right] + \sigma_{12}^2 + \sigma_{23}^2 + \sigma_{31}^2 \qquad (2.3)$$

Substituting equations (2.1) into (2.3) leads to the following expression for $\dot{\lambda}$,

$$\frac{1}{\dot{\lambda}} = \frac{k}{ESR} \qquad (2.4)$$

where ESR, the equivalent strain rate, is a function of velocity gradients.
In the approaches of Sims (1954), Ford & Alexander (1964), Johnson & Mellor (1973) among others the yield stress is assumed to remain constant throughout the arc of contact. This constant value is dependent on mean strain rates and bulk mean temperatures. The following expression for the yield stress, k, is typical of the type used,

$$k = A(\bar{\dot{\epsilon}})^n exp\left(\frac{B}{T}\right) \qquad (2.5)$$

where T is the mean bulk temperature and $\bar{\dot{\epsilon}}$ the mean strain rate of the material in the rollgap. In equation (2.5) A, B and n are constants for any given steel quality, for example Watts (1987) used the values A=6.54, B=3493.4 and n=0.168 which were obtained experimentally for type EN3 mild steel.
In hot rolling, particularly in beam rolling, the yield stress can vary significantly throughout the material because of surface chilling near the roll contact and heat generation due to plastic deformation. The models developed during this project permit the yield stress to vary according to the local strain rate and temperature.
For an incompressible material the conservation of mass is described by,

$$\frac{\partial u_i}{\partial x_i} = 0 \qquad (2.6)$$

The equilibrium of stresses and applied forces is given by,

$$\rho\frac{Du_i}{Dt} = \frac{\partial \sigma_{ij}}{\partial x_j} + F_i \qquad (2.7)$$

where ρ is the material density and F_i is the component of the applied forces.
When analysing problems of slow steady flow, such as occur in rolling, the inertia terms on the left hand side of equation (2.7) may be neglected in comparison with the stresses.
As thermal effects are important the energy equation, (2.8) below, must also be solved.

$$\rho c v_j\frac{\partial T}{\partial x_j} = \frac{\partial}{\partial x_j}\left(\kappa\frac{\partial T}{\partial x_j}\right) + \dot{q} \qquad (2.8)$$

where T, c, κ and $\dot{q}$ represent temperature, specific heat, thermal conductivity and heat source per unit volume respectively. These quantities vary throughout the material. The heat source in rolling is primarily due to the release of plastic deformation energy,

$$\dot{q} = \sigma'_{ij}\dot{\epsilon}_{ij}$$

The coupled system of equations (2.6)-(2.8) are solved for the three velocity components, pressure and temperature using the SIMPLE algorithm, (Patankar & Spalding 1972).

3 HOT ROLLING OF STEEL BEAMS

The finishing sequence in beam section rolling uses a series of reverse passes in a single stand finishing mill where reduction of flange and web thickness is achieved using two pairs of work rolls whose axes are perpendicular. A schematic of the rolling of a three dimensional H shaped beam is shown in figure 1.
The driven horizontal rolls provide the torque to the web. They are also in contact with the inside faces of the flanges. The vertical rolls are free

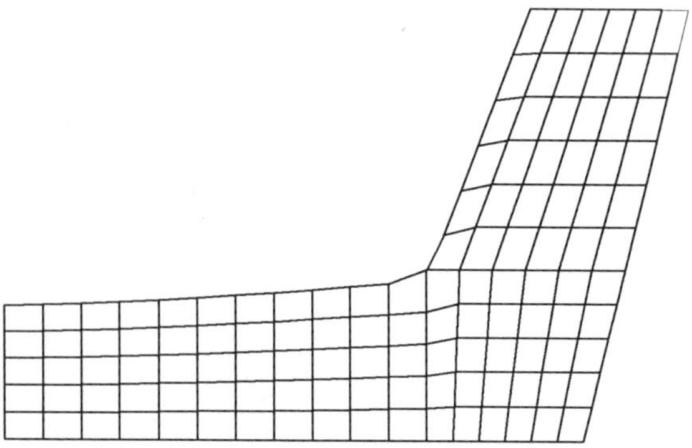

Figure 3. Cross section of H beam, quarter symmetry assumed

Table 1. Load and torque comparison for a beam

Pass	Web					Flange		
	red	Load (kN)		Torque (kNm)		red	Load (kN)	
	%	Eng	BFC	Eng	BFC	%	Eng	BFC
1	17.16	7379	8314	1412	2301	10.24	4358	6941
2	19.07	6137	6858	1296	1947	16.67	6986	9821
3	18.98	6898	6891	953	1371	16.81	4816	6555
4	17.61	5328	5991	1153	1501	15.74	6961	8214
5	18.34	6386	5714	895	999	18.83	5544	6939
6	15.79	3971	3401	1083	1433	16.78	7332	8442

to rotate about their axes so provide zero torque while exerting a compressive force on the material. Depending on the relative reductions material will flow from web to flange or vice versa.

The assumption of quarter symmetry leads to a significant reduction in the computational requirements. A typical bfc grid is shown in figure 2.

The mesh includes features such as roll crown, angled flanges and fillets at the web/flange junction. A cross section is given in figure 3 with the roll crown and junction exaggerated for illustrative purposes.

In beam rolling the surface temperature can vary from 1100^oC at the junction of the web and flange where the material flow and deformation is greatest to 700^oC on the web surface due to contact with the cooler work roll and the accumulation of scale. These large temperature variations create corresponding variations in yield stress which have a major influence on material flow, roll force and torque.

4 RESULTS AND DISCUSSION

To illustrate the initial results obtained using the code the six passes needed to roll a particular beam have been modelled. Table 1 compares the predicted and measured loads and torques, (no torque is applied to the vertical rolls).

The initial results are encouraging. The features which are considered important have been included in the model and the next stage is to verify the model against a range of beam geometries and steel types. In common with plate rolling the model provides a more accurate estimate of load than torque. The required torque is consistently overestimated. A possible cause of this effect is the boundary condition at the roll contact. We have used the no slip boundary condition on the grounds that if slip occurs then Coulomb friction would require the shear to equal μP where μ is a coefficient of friction and P is the normal stress. In hot rolling estimates of μ are in excess of 0.3 so that μP exceeds the maximum shear, k, which the material can support. Thus slip cannot occur and the sticking boundary condition is appropriate. There are regions at inlet and outlet where the normal stress is low and slip may occur. Including this effect may lead to better torque estimates. Other authors, for example Rowe, Sturgess, Hartley & Pillinger (1991), treat friction by means of an empirically determined friction factor, m, such that the surface shear takes a value $\pm mk$. It would be possible to investigate a range of rolling geometries and determine an appropriate choice of m, as a function of the rolling parameters, in order to achieve better torque predictions.

The deformation of material in the rollgap is an important aspect of the overall rolling process. The method developed is to be used within an overall simulation of rolling processes which will compare predictions obtained from a variety of analytical, approximate and numerical models. The intention is to obtain a better understanding of the important features involved and their effect on other aspects of the process, with a view to improving the on-line control system.

ACKNOWLEDGEMENTS

I. Charity gratefully acknowledges the financial support of British Steel.

A.W. Bush and G.S. Marshall gratefully acknowledge the support of the SERC through the provision of the cooperative research grant GR/G/20028.

REFERENCES

Beveridge, D., Bush, A., Charity, I., Jones, T. & Marshall, G. (1993), Predictions of loads and torques in hot rolling using body fitted coordinates, Presented at 1st Int. Conf. on Modelling of Metal Rolling Processes, 21-23 September 1993, Imperial College, London.

Bradbury, P., Bush, A., Jones, T. & Marshall, G. (1992), Computational fluid dynamics simulation of three dimensional section rolling, *in* 'Numerical Methods in Industrial Forming Processes', AA Balkema, pp. 707–712.

Coelho, P. & Pereira, J. (1992), 'Finite volume computation of the turbulent flow over a hill employing 2d or 3d non-orthogonal collocated grid systems', *Int.J.Numer.Methods Fluids* **14**, pp423–441.

Crane, F. & Alexander, J. (1968), 'Slipline fields and deformation in hot rolling of strip', *Jnl. Inst. Metals* **96**, 289–300.

Ford, H. & Alexander, J. (1964), 'Simplified hot rolling calculations', *Jnl. Inst. Metals* **92**, 397–404.

Johnson, W. & Mellor, P. (1973), *Engineering Plasticity*, Van Nostrand Reinhold, London.

Orowan, E. (1943), 'The calculation of roll pressure in hot and cold flat rolling', *Proc. Inst. Mech. Engrs.* **150**, 140–167.

Patankar, S. & Spalding, D. (1972), 'A calculation procedure for heat, mass and momentum transfer in three-dimensional parabolic flows', *Int.J.Heat Mass Transfer* **62**, 1787–1806.

Rhie, C. & Chow, W. (1983), 'Numerical study of the turbulent flow past an airfoil with trailing edge separation', *AIAA Jnl.* **21**, 1525–1532.

Rowe, G., Sturgess, C., Hartley, P. & Pillinger, I. (1991), *Finite Element Plasticity and Metal Forming Analysis*, Cambridge University Press.

Sims, R. (1954), 'The calculation of roll force and torque in hot rolling mills', *Proc. Inst. Mech. Engrs.* **168**, 191–200.

Watts, J. (1987), 'British steel technical report MW/810/1/87/A'.

Simulation of Materials Processing: Theory, Methods and Applications, Shen & Dawson (eds)
© 1995 Balkema, Rotterdam. ISBN 90 5410 553 4

A study on the friction occurring in the context of deep drawing

P. Chabrand & F. Dubois
Laboratoire de Mécanique et d'Acoustique, CNRS, Marseille, France

ABSTRACT: The aim of the present study was to investigate the effects of drawbeads on the restraining forces exerted by the blankholder, and especially to determine the effects of friction. Friction plays an important role in deep drawing processes and the friction coefficient depends on some parameters such as the normal pressure, the velocity and the viscosity of the lubricant. Before dealing with a dynamic friction coefficient in a further study, we analyse here the sensitivity of the blankholder restraining forces, assuming various friction coefficients between the blank and the flat and rounded parts of the tools. To deal with this problem, a 2-D finite elastoplastic model including contact with friction was developed. A specific algorithm was developed for evaluating the clearance between blankholder and die, in order to ensure that the normal force remains constant. The experimental and numerical results are compared and discussed.

1 INTRODUCTION

It is well known that in the draw forming of sheet metal parts, the restraining forces provided by the blankholder play an important role affecting the quality of the final product. The aim of the present investigation was two-fold. The first part deals with integrating an equivalent drawbead model into 3-D finite element simulations. As a matter of fact, it is not generally possible to take into account the exact drawbead shape in 3-D analysis because of the radii of the bead and the shoulders of the groove are too small, and to deal with the associated bending/unbending effects, a very large number of elements is required. To overcome this difficulty, a 2-D analysis of the drawbead has to be performed. The restraining forces calculated in this way will be used to provide an equivalent drawbead model which will be introduced into 3-D simulations.

In the second part of this study, it was proposed to develop a realistic, accurate analysis to provide a simple but efficient tool for designing drawbeads. Numerical simulations performed to investigate the effects of the bead geometry, material properties, friction coefficient and boundary conditions on the strain distribution and on the force/displacement relations. In particular, numerical analysis of the effects of parameters such as the radii of the groove shoulders, the radii of the upper (male) and lower (female) bead, rectangular or round bead shapes, and the clearance from the bead to the die edge can do away with the need for experimental studies, thus providing inexpensive information about the magnitude of the restraining forces, the true contact areas, the variation of constitutive parameters in the thickness of the sheet metal, and so on.

We shall focus in this paper on the effects of the friction occurring between the sheet metal and the blankholder on the restraining forces. To deal with this problem, a 2-D finite elastoplastic model including contact with Coulomb's friction was developed. To reduce the computational times, the tools are assumed to be undeformable and contact is assumed to occur between a deformable body and a rigid obstacle. Since the clamping force remains constant during the pulling phase, the clearance between the lower and upper parts of the blankholder is not known in our model. During the locking phase, the male bead is moved up until the prescribed clamping force is reached. Then, during the next pulling phase, the bending and unbending of the metal sheet lead to the occurence and developement of plastic areas which, until a steady state is reached, give rise to variations in the contact pressure. To be able to maintain a constant blankholder pressure, the clearance between the lower and upper parts of the blankholder has to be calculated. This constitutes a new variable which is introduced into the problem to be solved, and a specific algorithm was developed to deal with it.

The discussion consists of two comparisons between numerical and experimental studies. The first study focused on an experimental device in which the bead was replaced by a roller. The drawbead forces, as discussed by Nine (Nine 1978), have two components: a bending/unbending component and a deformation component due to the friction. In line with the pioneering studies by Nine (Nine 1978), a roller is used to separate these two components. The friction is then assumed to be eliminated, which makes it possible to assess the bending/unbending force. A fairly good agreement was found to exist between the experimental and numerical results obtained (Chabrand *et al* 1995) on Nine's apparatus. In the present study the test-piece was wound around the roller, just as it is around the bead, and was drawn in a second phase just as it is pulled outside the blankholder by the punch. The second example deals with a drawbead simulator with a constant clamping force. In both analyses the effects of various friction coefficients on the plane and round parts are discussed.

The paper is organized as follows. We will begin by describing the finite elastoplastic formulation. To deal with the frictional contact problem, several methods have been implemented and their performances were discussed in (Chabrand & Dubois 1995). Here we briefly describe Lemke's method (Chabrand *et al.* 1995), which is appropriate for solving the linear complementary problems arising from rewriting Coulomb's friction law. We will then describe the specific algorithm developed for calculating the clearance between the blankholder and the die to ensure that a constant normal force is maintened. The fourth and last part deals with the simulation of the two experimental devices. The experimental and numerical results will be compared and discussed.

2 EQUILIBRIUM EQUATIONS

In the displacement based finite element method, the discretized form of the equilibrium equations is used to calculate the estimated incremental displacement. A modified Newton-Raphson method is used to deal with the non-linear equations arising from the finite deformations, the constitutive equations and the frictional contact conditions. The loading is taken to consist of a sequence of steps, at each of which an iterative process is performed to solve the set of non-linear equations written on successive intermediate configurations. We denote Ω_n the reference configuration at the end of loading step n. Configuration Ω_{n+1} is computed iteratively from Ω_n (which is known from the previous loading step) by performing iterations i of the Newton-Raphson method with the corresponding intermediate configurations Ω^i_{n+1}.

Let u_n be the nodal displacement vector at the end of loading step n. We denote Δu^i_{n+1} the incremental displacement between configurations Ω_n and Ω^i_{n+1} and du^i the displacement between Ω^i_{n+1} and Ω^{i+1}_{n+1}. Hence :

$$\Delta u^{i+1}_{n+1} = \Delta u^i_{n+1} + du^i, \; u^{i+1}_{n+1} = u_n + \Delta u^{i+1}_{n+1} \quad (2.1)$$

The equilibrium equation of the configuration Ω^i_{n+1} is:

$$\begin{aligned}\left\{Res^i\right\} &= \left\{F^{int}\left(u_{n+1}\right)^i\right\} - \left\{F^{ext}_{n+1}\right\} - \left\{R^i\right\} \\ &= 0 \end{aligned} \quad (2.2)$$

where $\{Res^i\}$ denotes the equilibrium residual vector obtained by assembling the internal and external forces. $\left\{F^{int}\left(u_{n+1}\right)^i\right\}$ is the discrete load vector corresponding to the internal stresses, and $\left\{F^{ext}_{n+1}\right\}$ is the discrete load vector corresponding to the external forces excluding the contact forces. $\{R^i\}$ is the unknown vector of the contact forces. Using Newton's method for generalized equations (Klarbring & Bjorkman 1992), a linearized form of (2.2) is given by :

$$\begin{aligned}&[K_T]\{du\}^{i+1} = \\ &\quad -\left\{F^{int}\left(u_{n+1}\right)^i\right\} + \left\{F^{ext}_{n+1}\right\} + \{R^{i+1}\}\end{aligned} \quad (2.3)$$

where K_T is the tangent stiffness matrix consistent with constitutive equation integration and contact force calculations.

3 MATERIAL BEHAVIOUR

In the present study, the material is assumed to have an isotropic elastoplastic behaviour involving isotropic hardening. The model adopted is characterized by an intermediate configuration. It leads to the local multiplicative decomposition of the deformation gradient F into its elastic F^e and plastic F^p counterparts, as proposed by (Lee 1969):

$$F = F^e \; F^p \quad (3.4)$$

The constitutive laws of the model are determined by introducing the free energy function :

$$\Psi\left(b^e, \bar{\varepsilon}^p\right) = \Psi^e\left(b^e\right) + \Psi^p\left(\bar{\varepsilon}^p\right) \quad (3.5)$$

where the elastic potential Ψ^e is an isotropic function of the left Cauchy Green strain tensor :

$$b^e = F^e \; F^{e^T} \quad (3.6)$$

and where $\bar{\varepsilon}^p$ denotes the isotropic hardening variable and Ψ^p the hardening energy.

The behaviour between the intermediate and current configurations is given by a hyperelastic response. The Kirchhoff strain tensor is then given by the classical constitutive equation (Sidoroff 1982):

$$\tau = 2b^e \frac{\partial \Psi^e}{\partial b^e} \tag{3.7}$$

The intermediate configuration is up-dated by integrating plastic evolutionary laws. The plastic response of the material is described using the classical Von-Mises yield condition. The associative flow rule determined by the principle of maximum plastic dissipation is written as:

$$d^p = \lambda \sqrt{\frac{3}{2}} \frac{dev\tau}{|dev\tau|} \tag{3.8}$$

where d^p is the plastic strain rate, λ is the plastic multiplier and $dev\tau$ denotes the deviatoric part of the Kirchhoff stress. The yield function associated with (3.8) is defined as :

$$\sqrt{\frac{3}{2}}|dev\tau| - Y(\bar{\varepsilon}^p) = 0 \tag{3.9}$$

where $\bar{\varepsilon}^p = \int_0^t \sqrt{\frac{2}{3}\, d^p : d^p}\; dt$ is the equivalent plastic strain and $Y(\bar{\varepsilon}^p)$ the isotropic hardening law.

From the computational point of view, the above constitutive equations can be integrated using an elastic predictor/plastic corrector algorithm. We assume that at time t_n with the configuration Ω_n of the body, the state variables (F_n^e, F_n^p, τ_n) are known. Given Δu_{n+1} the displacement between the configurations Ω_n and Ω_{n+1}, we can perform an elastic prediction, assuming that the plastic variables remain frozen ($F_{n+1}^p = F_n^p$). The trial elastic part of the deformation gradient (3.4) as well as the trial elastic Kirchhoff stress (3.7) can be then computed. If the stress lies outside the elastic region, a plastic correction step is then carried out using a classical radial return algorithm (Simo & Taylor 1985).

4 FRICTIONAL CONTACT

The surface behaviour is described in terms of unilateral contact conditions involving Coulomb's friction. Let n be the outward normal unit vector to the rigid tool. We use the following decomposition into the normal and tangential components of the displacements (u) and the contact force vector (R) :

$$\begin{aligned} u_N &= u.n, \quad u_T = u - u_N n \\ R &= R_N n + R_T \end{aligned} \tag{4.10}$$

Let Δg_N be the gap function at the start of the loading step. The unilateral contact conditions can then be written in terms of the relative displacement (Δw) as follows :

$$\begin{cases} \Delta w_N = \Delta u_N + \Delta g_N \\ \Delta w_N \geq 0, \quad R_N \geq 0, \quad \Delta w_N . R_N = 0 \end{cases} \tag{4.11}$$

As a contact condition for the tangential direction, we take Coulomb's friction law, written below in an incremental form, in which Δg_T denotes the tangential incremental displacement of the rigid body and Δw_T the relative tangential displacement:

$$\begin{aligned} &\Delta w_T = \Delta u_T - \Delta g_T \\ &\|R_T\| \leq \mu R_N \begin{cases} \|R_T\| < \mu R_N \Rightarrow \Delta w_T = 0 \\ \|R_T\| = \mu R_N \Rightarrow \\ \Delta w_T = -\alpha R_T, \alpha \geq 0 \end{cases} \end{aligned} \tag{4.12}$$

As a means of dealing with the three-dimensional case, (Klarbring & Bjorkman 1988) have introduced a piecewise linear approximating Coulomb's friction law. This discretization procedure makes it possible to write the friction relations as complementarity conditions and then to set the problem as a linear complementarity one. In the present study, we shall restrict ourselves to a two-dimensional analysis. In this situation, Klarbring and Bjorkman's approach leads to introducing two new variables, λ and ϕ, which define the boundary of the Coulomb's cone. The Kuhn Tucker conditions for the frictional Coulomb problem can then be written in the following form:

$$\begin{cases} R_T \in C(R_N) \\ \phi_1(P_T, R_N) = -P_T + \mu R_N \\ \phi_2(P_T, R_N) = P_T + \mu R_N \\ \Delta w_T = -\lambda_1 \dfrac{\partial \phi_1}{\partial R_T} - \lambda_2 \dfrac{\partial \phi_2}{\partial R_T} \\ \lambda_m \geq 0, \; \phi_m \geq 0, \; \lambda_m \phi_m = 0 \; (m = 1, 2) \end{cases} \tag{4.13}$$

with $C(R_N) = \{P_T, \quad \phi_m(P_T, R_N) \geq 0, m = 1, 2\}$.

Reformulating the initial system (2.3) in terms of contact components (c) and free components (f), one obtains :

$$\begin{bmatrix} K_{T_{cc}} & K_{T_{cf}} \\ K_{T_{fc}} & K_{T_{ff}} \end{bmatrix} \begin{bmatrix} du_c \\ du_f \end{bmatrix} = \begin{bmatrix} F_c \\ F_f \end{bmatrix} + \begin{bmatrix} R_c \\ 0 \end{bmatrix} \tag{4.14}$$

with $F = F^{ext} - F^{int}$.

Thanks to a condensation procedure, the system (4.14) can be written as two connected ones :

$$K^* du_c = F^* + R_c \tag{4.15}$$

$$du_f = K_{ff}^{-1} F_f - K_{ff}^{-1} K_{fc} du_c \tag{4.16}$$

with :

$$K^* = K_{cc} - K_{cf} K_{ff}^{-1} K_{fc} \tag{4.17}$$

$$F^* = F_c - K_{cf} K_{ff}^{-1} F_f \tag{4.18}$$

The first system deals only with the normal and tangential components of the contact nodes which, upon introducing variables λ and ϕ, are all constrained by complementarity conditions. This linear complementarity problem can then be straighforwardly solved using a pivot algorithm such as Lemke's method (Chabrand *et al.* 1995). The second system deals with the nodes which are not involved in the contact. This is a non constrained problem in which the only unknown vector is the displacement one. Its solution obviously depends on the solution of the previous system, and can be obtained by classically solving a linear system.

5 BLANKHOLDER CONSTANT CLAMPING PRESSURE CONTROL

In the forming of sheet metal parts, the quality of the final products, in terms of the absence of wrinkles and necking, depends on the restraining forces exerted by the blankholder on the sheet.
We consider here the contact between a deformable body (the metal sheet) and a rigid obstacle (the tools). To ensure that the modelling is accurate as possible, the blankholder is taken to be a moving body, the (vertical) position of which is determined in order to exactly equilibrate the (vertical) loads acting on it by the jack pressure and the reactions due to the frictional contact with the sheet. Introducing this new degree of freedom ($\Delta\lambda$ assumed to be vertical (y)) leads, in the case of a node in contact with the blankholder, to the following parametric form of the normal and tangential relative displacements :

$$\Delta w_N = \Delta w_{N_0} - \Delta\lambda\ n_y \quad (5.19)$$

$$\Delta w_T = \Delta w_{T_0} - \Delta\lambda\ t_y \quad (5.20)$$

where Δw_{N_0} and Δw_{T_0} are the values determined at the beginning of the load step (4.11 and 4.12). Rewriting the system (4.15) in terms of components in contact with the blankholder (b, neq_b components) and others (o) :

$$\begin{bmatrix} K_{bb} & K_{bo} \\ K_{ob} & K_{oo} \end{bmatrix} \begin{bmatrix} \Delta w_{ob} \\ \Delta w_o \end{bmatrix} = \begin{bmatrix} F_b \\ F_o \end{bmatrix} + \begin{bmatrix} R_b \\ R_o \end{bmatrix} \quad (5.21)$$

We introduce into our system the new unknown ($\Delta\lambda$) and a new equation expressing the equilibrium of the blankholder : the sum of the vertical contact reactions (due to contact and friction) equals the jack pressure (F_{blh}). After some intermediate calculus (Dubois 1994), we obtain the following formal system :

$$\begin{bmatrix} K_{bb} & K_{bo} & K_{b\lambda} \\ K_{ob} & K_{oo} & K_{o\lambda} \\ K_{\lambda b} & K_{\lambda b} & K_{\lambda\lambda} \end{bmatrix} \begin{bmatrix} \Delta w_b \\ \Delta w_o \\ \Delta\lambda \end{bmatrix} = \begin{bmatrix} F_b \\ F_o \\ F_\lambda \end{bmatrix} + \begin{bmatrix} R_b \\ R_o \\ R_\lambda \end{bmatrix} \quad (5.22)$$

where :

$$K_{b_i\lambda} = \sum_{j=1}^{neq_b} K_{b_ib_j}\ k_j,\quad K_{o_i\lambda} = \sum_{j=1}^{neq_b} K_{o_ib_j}\ k_j$$

$$K_{\lambda b_j} = \sum_{i=1}^{neq_b} k_i\ K_{b_ib_j},\quad K_{\lambda o_j} = \sum_{i=1}^{neq_b} k_i\ K_{o_ib_j}$$

$$K_{\lambda\lambda} = \sum_{i=1}^{neq_b} k_i\ K_{b_i\lambda}$$

$$R_\lambda = \sum_{i=1}^{neq_b} k_i\ P_{b_i},\quad F_\lambda = \sum_{i=1}^{neq_b} k_i\ F_{b_i}$$

$$k_j = \begin{cases} n_{y_{ib}} & \text{if } j \text{ relates to the contact component of the node } ib \\ t_{y_{ib}} & \text{if } j \text{ relates to the friction component of the node } ib \end{cases}$$

In order to enforce $R_\lambda = F_{blh}$, we use Lemke's method both with the following obvious change of variable :

$$\Delta\tilde{\lambda} = \Delta\lambda + \alpha \text{ with } \alpha \gg 0 \quad (5.23)$$

$$\tilde{R}_\lambda = R_\lambda - F_{blh} \quad (5.24)$$

The couple($\Delta\tilde{\lambda}$, $\tilde{R}_\lambda$) must satisfy complementary conditions, due to the former change of variable $\Delta\tilde{\lambda} > 0$ and $\tilde{R}_\lambda = 0$.

6 APPLICATIONS

6.1 *Friction on bead test*

The principle of which is given in Figure 1, consists of : two flat dies which apply a constant clamping pressure, a central roller around which the test piece is wound, and two jaws which impose the winding and pulling.

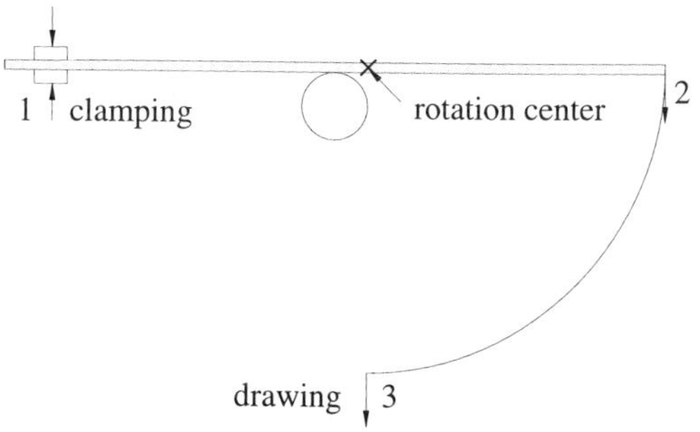

Figure 1. Friction on bead test : geometry

In the first phase, the test-piece is clamped on its left hand side by the flat dies. On its right hand side, the jaws prescribe on the test-piece a rotation around the central roller. When the required rolling-up angle has been reached in the second phase, the test-piece is drawn without any further angle variations. With this experimental device it is possible to change the restraining forces (by adjusting the clamping pressure and the friction between the flat dies), the roller radius, the rolling-up angle and the speed of the pulling. Based on experimental data, two friction coefficients were determined : $\mu = 0.17$ in the case of the flat dies (using flat dies test friction test) and $\mu = 0.14$ (using analytical arguments) in that of the roller.

Table 1. Comparison between experimental/numerical results : effects of the rolling-up angle

θ	45°	60°	75°	90°
Gap (%)	11.5	8	11.5	8

Table 1 gives the difference between the numerical and experimental results depending on the rolling-up angle. In all four cases, this difference was around 10%. Figure 2 gives close-up views of the deformed geometry around the roller after drawing with a 90 degree rolling-up angle and with various friction coefficients. The arrows indicate the contact forces and their length is proportional to the magnitude of the force. From these figures, it can be observed that higher friction coefficients increase the magnitude of the forces at the two extreme bending points and lead to a more uniform distribution of the contact between these points.

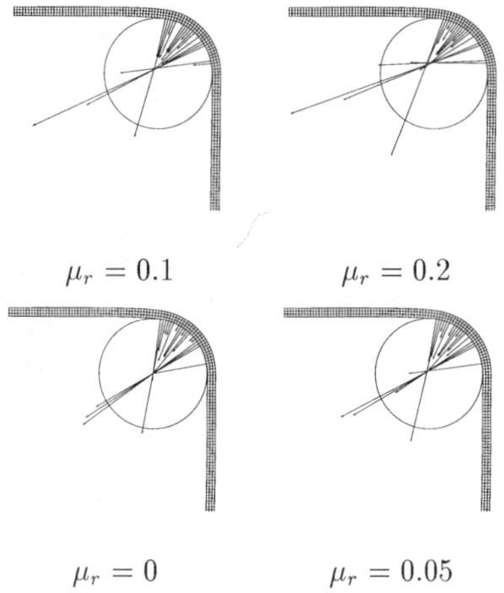

Figure 2. Friction on bead test : effects of the friction coefficient

Table 2. Comparison between experimental/numerical results : effects of the rolling-up angle

μ_r	0	0.05	0.1	0.14	0.2
Gap (%)	8	4	2.7	8	15

From table 2, it is possible to compare the numerical and experimental results on the restraining forces depending on the roller friction coefficient. It can be seen here that the best agreement was obtained with $\mu = 0.1$. The low contact pressure on the roller led here to a small friction coefficient. The friction therefore seems to fit a hydrodynamic or mixed lubrication regime rather than a limit one.

6.2 *Drawbead simulator test*

Here we will deal with simulating the drawbead shown in Figure 3. This figure gives the tool geometries and the material properties. The sheet metal was modelled with two layers of Q4/P0 elements. The analysis was carried out under the plane strain assumption. During the locking phase the lower (male) part of the blankholder moves up until the given clamping force is reached. The pulling phase which follows is characterized by a tangential displacement prescribed on the right hand side of the sheet, causing the sheet to bend and unbend as it is pulled throught the bead. The clamping forces used here were 500 daN and 1000 daN. Figure 4 illustrates the variations in clearance between the upper and lower parts of the blankholder. This distance becomes constant when a steady state is reached. The initial clearance variations are caused by the occurence and the development of plastic zones in the sheet metal. A steady state is reached when the parts of the sheet which were wound around the bead and the groove shoulders at the end of the locking phase are drawn outside the blankholder. At the end of the clamping (in the 1000 daN case), the

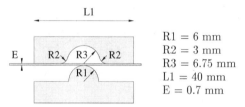

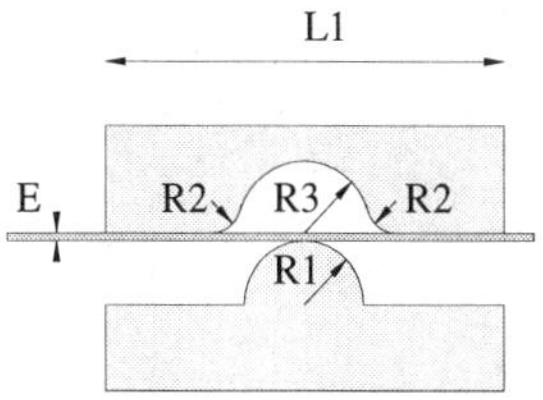

R1 = 6 mm
R2 = 3 mm
R3 = 6.75 mm
L1 = 40 mm
E = 0.7 mm

Young's modulus : 0.1909 10^6 Mpa
Poisson's coefficient : 0.3
Hardening law : $Y(\bar{\varepsilon}^p) = 673.6(0.010095 + \bar{\varepsilon}^p)^{0.17}$

Figure 3. Drawbead simulator : geometry and material data

clearance between the two tools is 0.74 mm and in the steady state, it remains greater (0.705 mm) than the sheet metal thickness (0.7 mm).

Under both clamping conditions (500daN and 1000daN), a constant friction $\mu = 0.2$ was used to describe the friction in the flat areas. The simulations were carried out with various friction coefficients for the bead and the groove shoulders, as illustrated in figure 5. The aim was to numerically determine the friction coefficient which would lead to the best agreement between the numerical and experimental results. With the 500daN clamping force, the experimentally measured restraining force was 600 daN, and with the 1000 daN force a restraining force of 800 daN was measured. In the clamping load range used here, a high fric-

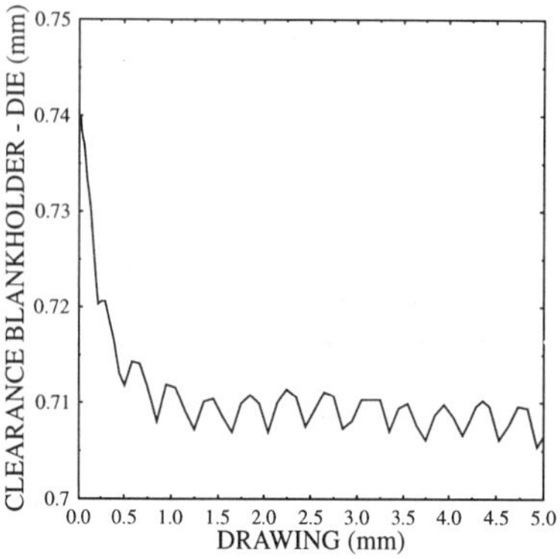

Figure 4. Drawbead simulator: clearance between the upper and lower parts of the blankholder

tion coefficient value was determined numerically to fit the experimental measurements. There are two reasons for adopting this procedure. First, neither of the two holding forces is sufficient to clamp the metal sheet between the flat parts of the tools, where there are only a few isolated contact points (see Figure 6), and hence the friction forces are almost negligible. Secondly, the high contact pressure on the round parts leads to a limit lubrication regime, contrary to what occured in the previous example, which yielded a higher friction coefficient.

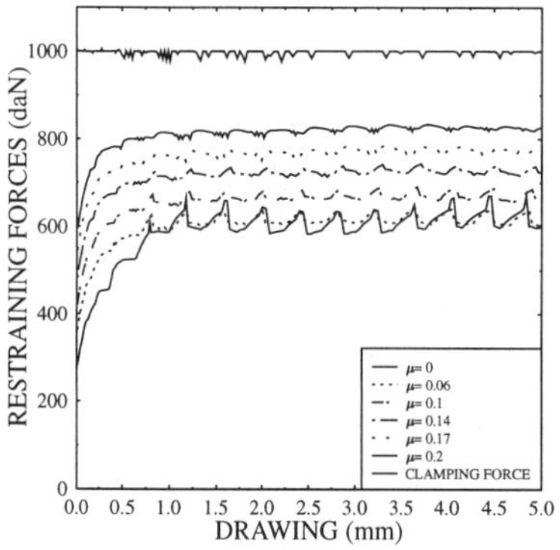

Figure 5. Drawbead simulator : Blankholder forces

CONCLUSIONS

A numerical tool for drawbeads analysis was developed. It was used here to study the friction on the bead. In a further study, we intend to use a variable friction coefficient depending on parameters such as the contact pressure, sliding speed,

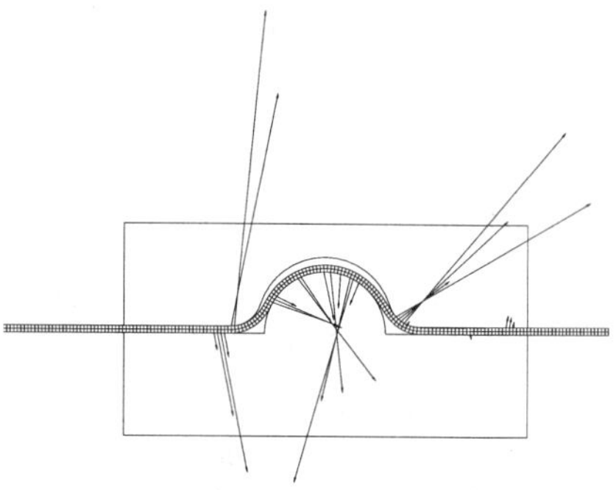

Figure 6. Drawbead simulator: deformed geometry and contact forces

lubricant viscosity, etc. On the other hand, we are interested in analysing the flattening and gearing of surface asperities as the first step in a study in which friction will be viewed as plastic deformation of surface asperities.

REFERENCES

Chabrand, P., F.Dubois and M.Raous 1995. Programmation mathématique pour le contact avec frottement et comparaison avec d'autres méthodes. *2e Colloque National en Calcul des Structures, 16-19 mai 1995,*Giens(var),France.

Chabrand, P., F.Dubois and J.C.Gelin 1995. Modelling drawbeads in sheet metal forming. *Submitted for publication.*

Dubois F. 1994. Contact, frottement, grandes déformations élastoplastiques. Application à l'emboutissage. *Ph. D. Thesis, Université Aix-Marseille II.*

Lee, E.H. 1969. Elastic-plastic deformations at finite strains. *J. Appl. Mech.*36:1-6.

Klarbring, A. & G.Bjorkman 1988. A mathematical programming approach to contact problems with friction and varying contact surface. *Comp. & Struct.*30:1185-1198.

Klarbring, A. & G.Bjorkman 1992. Solution of large displacement contact problems with friction using Newton's method for generalized equations. *Int. J. Num. Meth. Eng.*34:249-269.

Nine, H.D. 1978. Drawbead forces in sheet metal forming. In D.P. Koistinen and N.M. Wang(eds), *Mechanics of sheet metal forming*:179-211.Plenum Press.

Sidoroff, F. 1982. Les grandes déformations. *Rapport GRECO GDE*29.

Simo, J.C. & R.L.Taylor, 1985. Consistent tangent operator for rate-independant elastoplasticity. *Comput. Meth. Appl. Mech. Eng.*48:101-118.

Simulation of Materials Processing: Theory, Methods and Applications, Shen & Dawson (eds)
© 1995 Balkema, Rotterdam. ISBN 90 5410 553 4

3D metal forming applications of ALE techniques

Richard Couch, Rose McCallen, Ivan Otero & Richard Sharp
Lawrence Livermore National Laboratory, Calif., USA

Abstract: A three-dimensional arbitrary Lagrange-Eulerian (ALE) code is being developed for use as a general purpose tool for metal forming analyses. The focus of the effort is on the processes of forging, extrusion, casting and rolling. A prototype version of the software package, ALE3D, exists and is being applied to the enumerated processes. The status of the code is described. Several examples of application of the code to typical forming simulations are presented with discussions of the advantages and disadvantages of the ALE approach.

1 INTRODUCTION

Accurate simulation of large deformation metal forming provides a major challenge to developers of numerical modeling tools. This challenge presents itself both in the area of physical algorithm development and in the area of computational science. A general purpose simulation tool must be able to treat a wide array of physical phenomena, and must be sufficiently flexible in simulation strategy to adequately represent the various forming processes. For example, metal forming technologies provide scenarios in which the working material can be treated as a solid structure (Lagrangian-like) as well as those in which the material can be considered a fluid (Eulerian-like). Molten metal flow, as in a casting process, clearly falls in the latter category, but other solid phase, large-deformation processes such as extrusion can also be treated with the same computational techniques as those used for true fluids. Other processes such as sheet stamping and rolling also involve large deformations but are best treated in the Lagrangian format. In simulations of production processes it is not always possible to idealize the simulation to the point where only the dynamics of the work piece are important. Often the workpiece/die combination must be treated as a tightly coupled thermomechanical system where structural mechanics and heat transfer are important.

Based on these considerations we have chosen the arbitrary Lagrangian-Eulerian (ALE) format as the most powerful in terms of providing a general purpose tool. The potential advantage from ALE is that it provides the flexibility to treat a material in either a Lagrangian mode or one in which the material is allowed to move relative to the mesh (relaxation). Different materials within the computational mesh can thus be treated in the most appropriate manner. The relaxation option can also provide the further benefit of mesh refinement for improved accuracy or efficiency. More traditional approaches have treated large deformations using a Lagrangian representation combined with periodic remeshing, as required. The ALE approach assumes that continual remeshing using high order techniques provides improved accuracy and efficiency. The examples presented will illustrate the manner in which the technique has been applied to real forming simulations and indicate the generality of the approach. The forming methods illustrated here include forging, extrusion and rolling. The modeling presented is derived from a joint program initiated in collaboration with Alcoa and Sandia National Laboratory. We have chosen a 3D ALE code developed at LLNL called ALE3D (Sharp et al., 1994) as the vehicle to be used in these studies.

2 ALE3D CODE FEATURES

ALE3D is a finite element code which treats fluid and elastic-plastic response on an unstructured mesh. The grid may consist of arbitrarily connected hexahedra, shell and beam elements. There is no allowance for tetrahedra or wedge shaped elements. The mesh can be constructed from disjoint blocks of elements which interact at the boundaries via slide surfaces. This interaction can consist of pure sliding in which there are no tangential forces on interface nodes, or the nodes may be tied to inhibit sliding entirely, or a coulomb friction algorithm can be used. Voids may open or close between the surfaces as dictated by the dynamics of the problem, and there is an option to allow a block surface to fold back on itself (single-sided sliding).

The basic computational cycle consists of a Lagrangian step followed by an advection step. In the Lagrangian phase, nodal forces are accumulated and an updated nodal acceleration is computed.

Following DYNA3D (Hallquist, 1982), the stress gradients and strain rates are evaluated by a lowest order finite element method. A diagonal mass matrix is used. For second order accuracy a staggered space and time grid is also used. The stress tensor for the elastic plastic material strength model is integrated by an incremental strain method. The Jauman rate derivative is used for the stress tensor and the von Mises yield condition is applied.

At the end of the Lagrange phase of the cycle the velocities and nodal positions are updated. At this point several options are available. If the user wishes to run the code in a pure-Lagrangian mode, no further action is taken and the code proceeds to the next time step. If a pure-Eulerian calculation is desired, the nodes are placed back in their original positions. This nodal motion generates inter-element fluxes which must be used to update masses, velocities, energies, stresses and other constitutive properties. This re-evaluation process is referred to as advection, and can best be visualized as a remapping algorithm. In general, it is not sufficient to allow advection only within material boundaries. ALE3D has the ability to treat multi-material elements, thus allowing relaxation to take place across material boundaries.

The full potential of the ALE approach is realized when the code user has options available to tailor the evolution of the mesh in order to maximize either efficiency or accuracy. In the simplest implementation, the code is instructed to relax nodes only as required to eliminate large distortions in the mesh. A more powerful approach has the code relax nodes on the basis of some optimization scheme. The method currently implemented in ALE3D is a finite element-based equipotential method developed by R. Tipton (Tipton, 1992). This method accommodates weighting functions which can be used to optimize the mesh based on some defined criterion. The code currently allows weighting by pressure, by plastic strain, by material and along designated slip surfaces. The solution will result in a more highly resolved mesh in the volumes containing the elements with the higher weights. This provides a form of dynamic mesh refinement.

The heat conduction and implicit dynamic routines are still in a development phase. The calculations presented here were all performed using the explicit version of the code without heat conduction. In most instances the actual forming times are too long to be efficiently treated in an explicit formulation. One must either shorten the time scale of the forming event or artificially lower the sound speed of the materials. The latter approach has been chosen for these investigations. The sound speed is lowered by increasing the material density. If care is taken, the time step can be increased in some cases by orders of magnitude without inertial effects perturbing the solution.

3.1 EXAMPLE - FORGING

Figure 1 illustrates a forging configuration which serves as an appropriate vehicle for

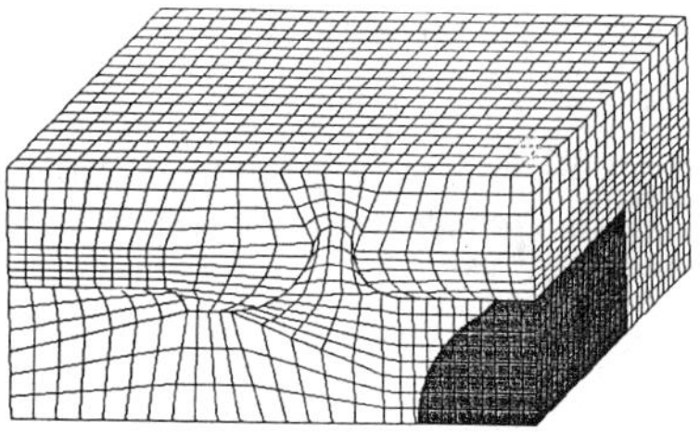

Fig. 1. Illustration of the initial mesh in the die piece, air/void and workpiece. The die remains Lagrangian. The region containing the workpiece will be allowed to relax throughout the calculation. There is a slide surface between the two regions. The workpiece is cylindrical, with the apparent flat top being a plotting artifact.

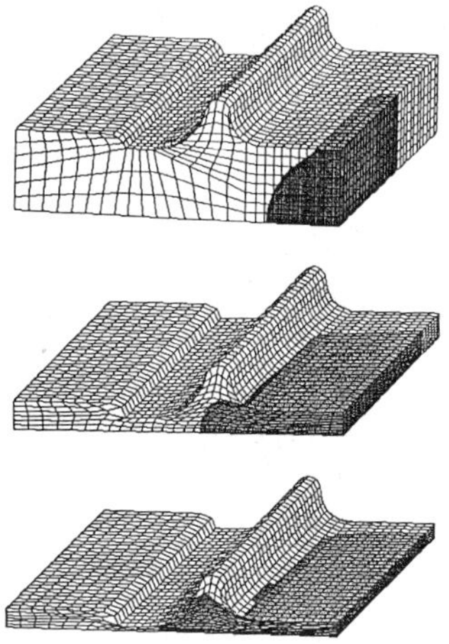

Fig. 2. An illustration of the evolution of the mesh and workpiece through time. The die is not shown and the air/void region is opaque so the workpiece can only be seen where it contacts the mesh surface.

considering ALE utility. It is a 3D representation of two die pieces closing on a cylindrical aluminum billet to form a beam section. This problem and the associated data are from a report by Smelser, Pishko and Klingensmith (1991). Taking advantage of three planes of symmetry, only one-eighth of the configuration need be modeled. The upper structure represents a die piece which is given a constant velocity at its top surface and allowed to compress the region below.

The die is treated Lagrangian with a slide surface between it and the lower region. The lower region represents the workpiece and the inter-die volume. This is meshed as a single region with the chosen zoning pattern selected after consideration of both the initial and final die configurations. The mesh is generated independently of the workpiece shape, with it, the shaded region, added by ALE3D at problem run time. This approach treats both the air and the workpiece as if they were fluids with an

Figure 3. The calculated part shape.

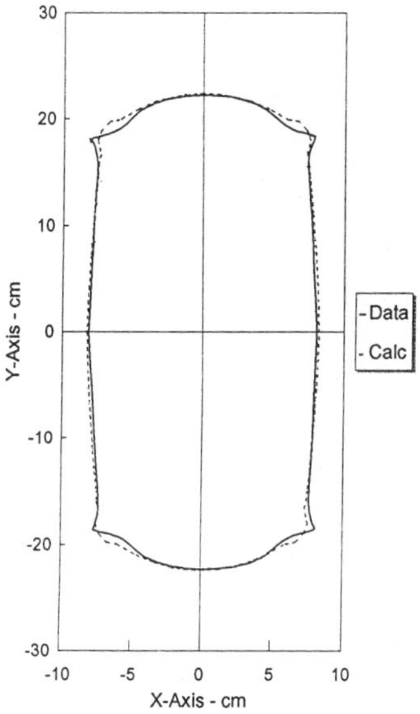

Figure 4. A comparison between calculated and measured part dimensions in the plan view.

interface maintained between them. The stresses and constitutive parameters of the workpiece are advected through the mesh as the material flows. The constitutive model for the aluminum is temperature, strain and strain rate dependent. A coulomb friction algorithm was operative. The results were very sensitive to the chosen friction parameters.

Figure 2 shows the evolution of the mesh and workpiece with time. The mesh shown above was used for illustration purposes. A more finely resolved mesh produced a part with the shape shown in Figure 3. Contours of the actual part exist at the mid plane reflection surface, at the location of the original edge of the cylindrical workpiece and in the plan view. The comparison between calculation and data in the plan view is shown in Figure 4. A friction coefficient of .3 was used with a limiting frictional stress of 15 Mp. These are reasonable values for these parameters, but they were chosen to provide the best fit to the data. The coulomb model applied here is too simplistic to expect any better agreement than that obtained. The intention here was merely to demonstrate that with reasonable material and friction descriptions the ALE approach could provide quality results in a complex 3D geometry.

3.2 EXAMPLE - EXTRUSION

The approach to extrusion modeling is similar to that employed in the forging simulation. The extrusion die and ram are treated as Lagrangian objects with slip surfaces providing the interaction with a relaxing mesh containing the work piece and the volume into which the extruded material may flow. This approach is illustrated in Figure 5. This is an idealized die geometry. Constraints were applied to prevent large deformation at the die orifice. An actual die configuration would have been much more massive to inhibit those distortions. The extruded configuration is shown in Figure 6. The ALE approach can treat the flow with little difficulty, but yet to be determined is the accuracy with which an actual extrusion process can be modeled. This process involves a highly coupled thermomechanical system. Predicting the exact die geometry with mechanical loading and thermal effects included is required in order to produce measurables like ram force and material flow. Another issue to be resolved is finding an approach that will allow efficient analysis of material flow throughout the entire charge run.

3.3 EXAMPLE - ROLLING

Figure 7 illustrates the geometry and mesh utilized in modeling a rolling pass with a 30% reduction. The rolls were given an applied angular velocity and allowed to relax to an equilibrium stress state under that rotation. The ingot was then pushed into the rolls with a velocity approximately equal to the linear rolling speed. Frictional forces then grabbed the ingot and pulled it through the rolls. The deformations were such that the mesh could remain Lagrangian throughout the calculation. This particular simulation depends critically on the contact algorithms which control the gap opening and closing between ingot and roller, and on the friction model assumed. In practice the product would be sensitive to the dynamics of the roll assembly. It serves to illustrate the requirement for a general purpose tool to posses a robust Lagrangian formulation and a significant structural analysis capability.

4 ISSUES RELATED TO THE ALE APPROACH

4.1 REMESHING

The hypothesis being investigated assumes that ALE can be used in its most general form. Objects that suffer little deformation can be treated

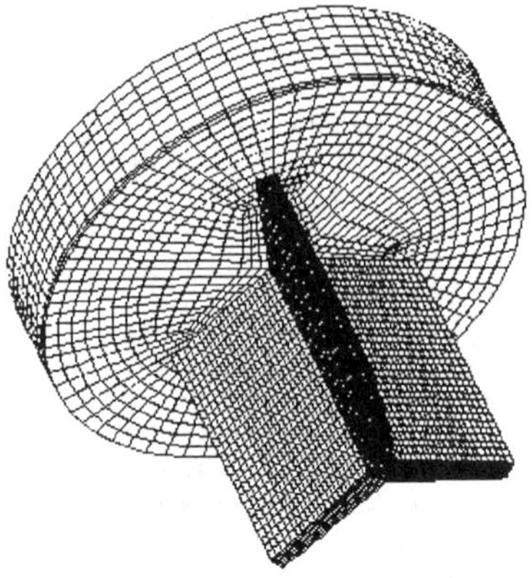

Figure 5. A Y-shaped part is to be extruded. Within the die are a cylindrical aluminum billet and a ram to drive the material. The mesh beneath the die provides volume in to which the extruded material can flow.

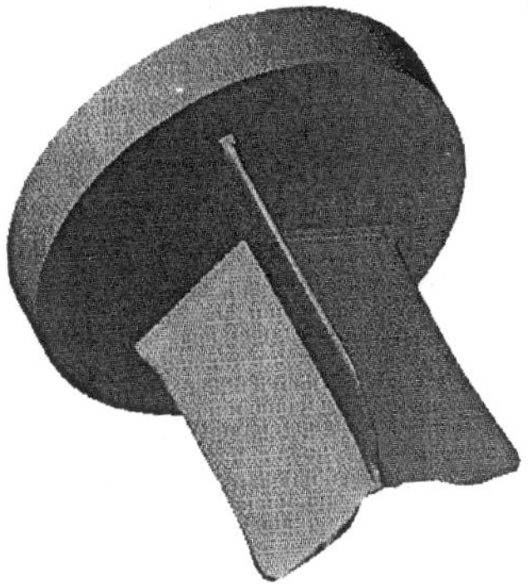

Figure 6. The shape of the extruded part.

Lagrangian, with contact discontinuities on their surface, as appropriate. Regions or materials that undergo large deformations will be treated with a continuous remeshing approach.

Assuming that the stresses, any constitutive parameters, such as plastic strain, and the material-void/air interface can be moved through the mesh with sufficient accuracy, the calculation can proceed to completion with minimally distorted elements and with no further intervention required. A more traditional approach would involve a Lagrangian mesh with an appropriate number of isolated remeshing operations. Some advantages of the ALE approach are that the dynamic remapping may be more accurate in preserving gradients in the quantities being remapped, and minimizing element distortion increases accuracy and minimizes computations. In an explicit mode the ALE calculation should run with a larger time step; implicit calculations should attain acceptable solutions with fewer iterations. ALE also brings a

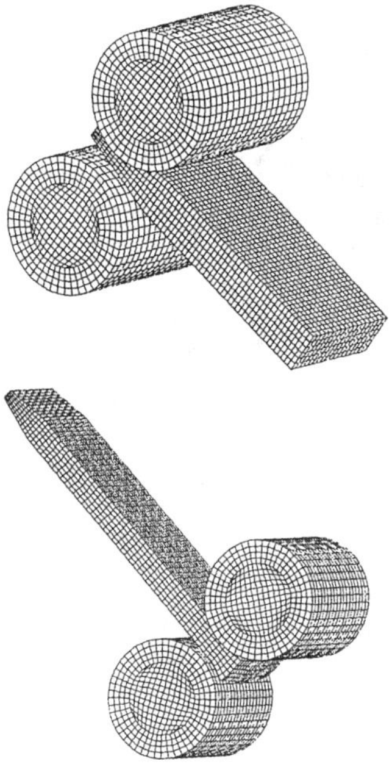

Figure 7. The initial and final rolling configurations. The calculation was Lagrangian with contact and friction forces pulling the ingot through the rolls.

certain robustness in that the imposed regularity of the mesh is less likely to allow a drift into unphysical configurations or to require user intervention. The disadvantages are that the ALE calculation may require a larger total number of elements, and increased memory and logic are required to treat the inter-element flows and the multi-material elements that are generated. There may also be situations where a predetermined mesh may not be adequate and a remapping to an altered mesh is required. Experience thus far indicates that for many practical metal forming scenarios the predefined mesh is adequate. Remeshing is most likely to be required if one is modeling a forming process with multiple steps having different geometries.

4.2 TIME INTEGRATION

Time scales for most metal forming problems are too long for efficient time integration by means of a standard explicit scheme time step limited by the Courant stability condition. An implicit approach or a technique which artificially forces an explicit code to run at a larger time step is required. Any conclusions reached about the utility of the ALE approach are independent of the time integration method. ALE codes are generally built on a combination of a Lagrangian step followed by a remap operation. The two need not run at the same time step. The size of the advection time step is determined by accuracy considerations which limit the amount of flux that can pass through an element interface in one integration period. The number of advection steps required is therefore a function of the strain history of the working material, and is independent of the actual time scale of the process being modeled. The plan is for ALE3D to have both implicit and explicit options available so that the user can choose the method most appropriate for the given application.

4.3 CONTACT AND SLIDE SURFACES

The interaction between the working material and whatever structure is involved should not be constrained by the geometric complexity of the structural surfaces. The precise dynamic path that the material will take cannot always be anticipated. The contact algorithm should be general enough that the specification of allowed contact surfaces can be consistent with the physically allowed degrees of freedom of the system being modeled. In the Lagrangian mode, the extreme, but not unreasonable, case, would allow any node on the surface of the working material to have the potential for coming in contact with any node on the surface of the containing structure. The contact logic can become quite expensive in this situation. Operating in the ALE mode, with a pre-positioned mesh, the association of fluid nodes with structural nodes is determined at problem generation time. This association or ordering of the nodes can change with time due to relative motion along the slide surface, but the reordering is straightforward because a node will always move from one element to an adjacent element. In this mode, the contact of one surface with another becomes a problem of material advection and the tracking of the material interface as it moves from element to element.

A more difficult problem arises if the situation calls for a void to open at an interface. This can occur in forging and casting scenarios. While closing of a material on a surface can be treated by traditional flow and interface tracking methods, the process is not automatically reversible when the material wants to separate from the wall. The technique can accommodate a switch to the Lagrangian mode either locally or globally, but this is not always an appropriate response. Contraction of a quiescent fluid on solidification, as arises in casting simulations, is one example of where that would be a reasonable approach. The more general response would be to allow a void to open by injecting a void-like material into the element to represent the gap that would exist between surfaces. The decision to open a void could be accurately made in the Lagrangian phase of the ALE time step. The node could then be placed back on the master surface creating a net volume flux into the element which would represent the injection of void. This is the approach that will be followed in ALE3D. The accuracy of such an approach needs to be demonstrated.

4.4 SURFACE INTERACTIONS

The principal surface interaction of interest is friction. There does not appear to be any fundamental limitation to the ALE method as applied to physically based friction models. The ALE format is complicates the tracking of history variables, for example, but this is no different in scope than the difficulties encountered in tracking material properties through a relaxing mesh. More complicated interactions such as crack formation are difficult to treat in any format. Certain failure models which generate strongly peaked damage functions are not readily made consistent with the continual remap approach.

4.5 HEAT CONDUCTION

For processes such as forging and extrusion where heat transport is either near steady state or is dominated by conduction within a material or at contact discontinuities, the various approaches are essentially identical. Operator splitting can be used in the heat transfer modeling. In this case the code need not know the difference between an ALE mesh and a Lagrangian mesh. Casting simulation, however, may be an instance where the ALE approach may provide interesting alternative approaches. Certainly, during the die filling process, the use of a pre-positioned mesh through which the fluid is transported simplifies the treatment of the complicated flow. A similar simplification could extend to heat conduction. Thermal contact algorithms may be simplified and the problem of transport across an air/void region could have a straightforward solution. More experience is required with thermomechanical simulations to understand the advantages and disadvantages of the ALE approach. Currently, there appears to be no fundamental flaw in the approach, and there could be significant simplifications which emerge.

4.6 MATERIAL DESCRIPTIONS

The ALE approach is quite capable of dealing with typical models where flow stress is a function of temperature, strain rate and history variables such as accumulated strain. Also, there is nothing that would preclude treatment of anisotropic materials. In general, material models can be more difficult to implement in an ALE code. Not all types of constitutive parameters are consistent with the continual remapping approach. It may be necessary, at times, to tailor the material model to the type of code structure in which it resides.

5 CONCLUSIONS

The ALE technique provides a robust method of dealing with both large and small deformation flows and fluid-structure interactions. As such, it has the potential to be quite useful in modeling metal forming processes. The ultimate utility of the approach will depend on the details of its implementation. The progress made thus far indicates that the emergence of a general purpose forming simulation tool from this approach is likely.

6 ACKNOWLEDGMENTS

The staff of Alcoa's Advanced Technology Center have been helpful in guiding this work. R. Becker has been particularly helpful in providing simulation data and interpretation. Work performed under the auspices of the U.S. Department of Energy by the Lawrence Livermore National Laboratory under contract No. W-7405-ENG-48.

7 REFERENCES

Hallquist, J. O., 1982, *Theoretical Manual for DYNA3D*, Lawrence Livermore National Laboratory, Livermore, CA, UCID-19401 draft.

Sharp, R., Anderson, S., Dube, E., Otero, I., Futral, S. 1994, *Users Manual for ALE3D*, Lawrence Livermore National Laboratory, Livermore, CA, unpublished.

Smelser, R.E., Pishko, R., Klingensmith, M.A., 1991, *Three Dimensional Forging Experiments - Case Studies*, Alcoa Technical Center, Alcoa Center, PA, unpublished.

Tipton, R.E., 1992, private communication, Lawrence Livermore National Laboratory, Livermore, CA.

Simulation of Materials Processing: Theory, Methods and Applications, Shen & Dawson (eds)
© 1995 Balkema, Rotterdam. ISBN 90 5410 553 4

Automatic remeshing in three-dimensional moving mesh finite element analysis of industrial forming

T. Coupez
École Supérieure des Mines de Paris, CEMEF, Sophia-Antipolis, France

ABSTRACT: The general mesh topology improvement method is presented and applied to the remeshing problem arising in industrial forming calculation. The improvement procedure presented in this paper, already used for "static" remeshing, is used here efficiently for dynamic remeshing. This new strategy is described in the context of forging simulation. The solver used tetrahedra and it is based on a stable mixed finite element interpolation providing a simple linear velocity field per tetrahedron. The flexibility of mesh of tetrahedra is then demonstrated by the very large deformations reached by our code. The remeshing is considered as the improvement of an existing mesh rather than the complete generation of a new one. The combination of boundary and internal mesh operations give the capability to create (and/or delete) boundary node and thus to perform automatically the mesh adaptation to the dynamic change of contact. It is shown how this approach succeed in automatic remeshing on industrial cases. The parallelization of this local improvement strategy in the context of partitioning method is performed by associating parallel remeshing and dynamic repartitioning.

1 INTRODUCTION

The mesh topology improvement method has been first developed for remeshing purpose in the the context of forging simulation (Coupez T. and Chenot J. L. 1992). It is a mesh generation method making unstructured mesh composed of tetrahedra but with also a particular ability to improve or to restore and now to adapt an existing mesh (Coupez T. 1994). It is being also presented at the same place, in the context of polymer forming calculation, addressing further issue on anisotropic mesh generation technique (Gobeau J.F. *et al* 1995).

Topological changes have been first introduced, in order to improve meshes (Talon J. Y. 1987), latter to enforce the boundary mesh recovery in Delaunay/Voronoi triangulation (Georges P.L. *et al* 1991). The efficient and very popular diagonal swapping is the simplest topological change in 2D. It is often used to improve a mesh, but it has been also used to make generalized Delaunay triangulation (Borgers C. 1990) also to optimize the shape of element in a different metric, which is the purpose of anisotropic mesh generation.

The mesh topology improvement method described here, may be seen as the generalization of such now classical method. It is a generalization for several reasons. The first one is that we propose a mesh topology operator, valid and efficient in 3D (there is no clear equivalent to the famous diagonal swapping in 3D). We add the possibility of creating node and deleting node. The size of the region of work is flexible. A deep generalization is a mesh topology (also said the connectivity of the mesh) independent from the geometry. That means that element edge or faces can cross each other. The figure 1 shows immediately what it is allowed within this approach. Obviously, the interest of such configuration is that the input of the improvement process can be a fully degenerate mesh and the output a restored and enriched mesh.

The starting point of improvement can be so far from a real mesh that an automatic mesh generator has been developed, based on a unique mesh topology operator (Coupez T. 1994).

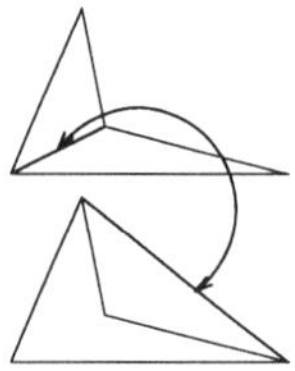

Figure 1. Allowed diagonal swapping of the mesh topology approach

Applications focus on remeshing in the context of large deformation calculations using the finite element method. Our 3D solver is mainly based on tetrahedra associated with a first order stable/stabilized mixed velocity/pressure formulation (Arnold D.N. *et al* 1984, L.P. Franca *et al* 1992). The strategy of remeshing presented requires less CPU effort consuming than the previously presented one (Coupez T. *et al* 1991, 1992). In this early work, the remeshing stages was first to create a new boundary mesh (set of facets) and last to create a mesh of tetrahedra fitting exactly a given boundary discretization. Therefore, the remeshing was a re-building process. The new remeshing scheme presented here, consists in improving the current mesh by operating almost simultaneously both on its boundary and its internal topology. The fully automatic remeshing is allowed without any user intervention. The remeshing stage is activated periodically ensuring that the mesh remains quite optimal. The computational effort of remeshing is greatly reduced with this strategy. Indeed, the cost of improvement is reduced when only few changes are required at each stage of remeshing.

2 THE GENERAL MESH TOPOLOGY IMPROVEMENT METHOD

2.1 *Mesh topology definition*

Let D be the number of nodes per element (3 for a triangle or 4 for tetrahedron). Let us consider $\mathcal{T}$ a set of sets of D integers. T a member of $\mathcal{T}$ is composed of D integers (the node numbers of element) . Its boundary ∂T is defined as a set of all sets of $(D-1)$ integer include in T (3 edges for triangles or 4 faces for tetrahedra) . Let us introduce the following sets associated with $\mathcal{T}$:

- the set of node indexes:
  $\mathcal{N} = \bigcup_{T \in \mathcal{T}} T$
- the set of faces (or edges if D=3) :
  $\mathcal{F} = \bigcup_{T \in \mathcal{T}} \partial T$
- the set of element having a set of nodes η in common:
  $\mathcal{T}(\eta) = \{T \in \mathcal{T}, \eta \subset T\}$

let us denote by $|A|$ the number of members of A when A is a finite set of integers.

With the above notations a mesh topology is defined by :

DEFINITION: $\mathcal{T}$is a mesh topology if $(|\mathcal{T}(F)| \leq 2, \forall F \in \mathcal{F})$
and $\partial\mathcal{T}$, the boundary of $\mathcal{T}$, is defined by

$$\partial\mathcal{T} = \{F \in \mathcal{F}, |\mathcal{T}(F)| = 1\}$$

2.2 Minimal volume theorem

The coordinate of nodes are generally referenced in an array of reals. The data of both the coordinate and the connectivity (mesh topology) is sufficient to define a mesh. The following describes precisely what it is needed in order to ensure a mesh.

The list of coordinate of node provides a mapping X from $\mathcal{N}$to $\mathbb{R}^d$ (d=2,3). $X(i)$ or X^i is the coordinate of node numbered i.

$T \in \mathcal{T}$, can be associated with $x(T)$, the triangle (tetrahedron) of vertices of coordinate X(T).

Let us denote by $|x(T)|$ the volume of domain $x(T)$.

THEOREM: If $x(\partial\mathcal{T}) = \partial\Omega$ and $\mathcal{T}$is a mesh topology then

$$\sum_{T \in \mathcal{T}} |x(T)| \geq |\Omega|$$

Moreover, if $(X, \mathcal{T})$ is a mesh of Ω

$$\sum_{T \in \mathcal{T}} |x(T)| = |\Omega|$$

In clear, it means :

- when the boundary is fixed (what we set by $x(\partial\mathcal{T}) = \partial\Omega$), a mesh topology and a list of coordinate give a set of tetrahedra such as, the sum of the volume of elements is always greater or equal than the volume of the domain to be meshed.
- When the boundary is fixed, a mesh topology associated with a list of coordinate give a mesh if the sum of the volume of elements is exactly the volume of the domain to be meshed. This volume can be exactly calculated by its boundary.

Consequently, he mesh topology must be constructed in order to minimize the volume of elements.

2.3 Automatic mesh generator

A very simple mesh topology generator can be defined when the data is a boundary mesh topology $\partial\mathcal{T}$. One particular node index is connected to every boundary faces, which can be summary by :

$$\mathcal{T}^*(n, \partial\mathcal{T}) = \bigcup_{F\in\partial\mathcal{T}} \{\{n\} \cup F\}$$

The choice of the node to be connected provides the set of topology to be tested. The computing effort is precisely controlled by the size of the region to reconnect as it appears in the following programs :

Program 1 :

```
repeat {
    for ( n ∈ N(T) )
        choose m ∈ list-of-test-number
        exchange :
        T ⟵ (T − T(n)) ∪ T*(m, ∂T(n))
    end for
}
```

In fact, exchanging conditions must be checked and are verified by introducing an envelope of tetrahedra in a slightly different algorithm (Coupez 92, 94)

REMARKS: A second algorithm, including the diagonal swapping in 2D, consists in considering a loop over the edge. Successive passes with these two kind of loop has been proven to be sufficient in every cases.

The "list-of-test-number" is composed of numbers of node connected to n : $\mathcal{N}(\mathcal{T}(\backslash))$. Internal nodes generation is performed by adding a new number in the list :

"list-of-test-number"=

$$\mathcal{N}(\mathcal{T}(n)) \cup \{min(i \in I\!N - \mathcal{N}(\mathcal{T}))\}$$

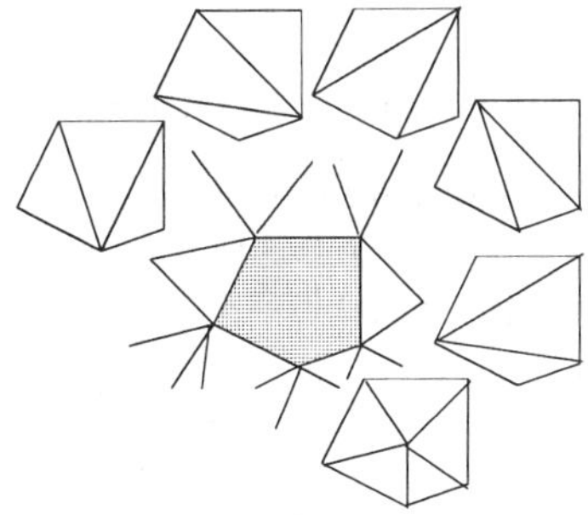

Figure 2. Possible mesh topologies generated by a list of test nodes

2.4 Mesh generator respecting a given boundary mesh

Every numbers in the "list-of-test-number" are associated with coordinate, except the possible new one : $\{min(i \in I\!N - \mathcal{N}(\mathcal{T}))\}$ which is then associated with the barycenter of the others.

A first subset is extracted :

$M \subset$ "$liste - of - test - number$"such that $min\{\sum_{T\in\mathcal{T}^*(i,\partial\mathcal{T}(n))} |x(T)|, i \in N(\mathcal{T}(n)) \cup \{|N(\mathcal{T})| + 1\}\}$ is reached $\forall m \in M$

It means that one selects the sub-topologies ensuring the minimal volume.

m is selected in M in order to optimize the quality of element with the following shape measure for tetrahedra:

$$c(X, T) = \frac{|x(T)|^2}{|x(\partial T)|^3}$$

Which can be calculated in Euclidean metric or not for obtaining anisotropic effects.

3 SURFACE MESH TOPOLOGY IMPROVEMENT

3.1 The local curvature conservation

Let us denote by $\alpha_X(n, T)$ the angle at vertex numbered n of triangle T for the mapping X. Then, the local curvature is given by an angle :

$\pi(X^n) = 360^o - \sum_{T\in\mathcal{T}(n)} \alpha_X(n, T)$

Each topological operation preserving the above local curvature preserves the geometry of the domain to be meshed. This conservation principle replaces the minimal volume criterion used previously in Euclidean metric. Rather than the general mesh topology operator defined in the previous section, we used a more specific set of operators for the surface.

3.2 Topological operation on curve surface

The swapping of diagonals on curve surface:

It is the simple topological change. Two adjacent triangles can be changed in two other adjacent triangles by exchanging the common edge. This efficient topological improvement process in 2D is generalized to 3D surface by using the local curvature criterion.

Boundary point deletion:

Only nodes connected to less than six other nodes (the degree of the node is six) can be deleted in our program. It seems to be sufficient. When a node is deleted, all the facets connected to it are also deleted. Therefore, it is necessary to replace them. The algorithm consists in trying all possible topologies. There are 1, 2, 5 ,14 possibilities for a node with 3, 4, 5 ,6 neighbouring nodes respectively.

Boundary point generation:
New boundary nodes may be created. they are created at the mid-point of edges. the two adjacent triangles associated with an edge where a mid-point is added are replaced by four triangles. This strategy is sufficient because the mesh topology is "smoothed" by the swapping of diagonals.

4 GENERALIZED MESH IMPROVEMENT METHOD

4.1 Degenerate element

Boundary changes are taken into account in the volume by inserting a layer of degenerate elements. These elements will be deleted and the mesh will be restored by the improvement process. For each topological change on the boundary the following is done:

- in case of swapping of diagonals the tetrahedron composed of the four nodes concerned is added or deleted if it already exists.
- in case of deletion, the vertex to be deleted becomes an internal node. Tetrahedra are added using the $\mathcal{T}^*(,)$ operator:
  $\mathcal{T}^*(node\ to\ be\ deleted, new\ part\ of\ boundary)$
- in case of boundary node addition, the $\mathcal{T}^*(,)$ operator is used again:
  $\mathcal{T}^*(created\ node, old\ part\ of\ the\ boundary)$

5 3D FORGING SIMULATION WITH FULLY AUTOMATIC REMESHING

5.1 The solver

The viscoplastic incompressible material is modeled by a power law (Norton Hoff) in the Finite element solver : FORGE3 .Large deformation are computed using an updated Lagrangian description and a moving mesh method. The total deformation is calculated step by step.

The results plotted on figure 5 is considered by us as a test case for remeshing in case of large deformation including flashes. It has been already used in a previous Numiform (Coupez, 1992) showing the capacity of the present strategy in capturing the flash.

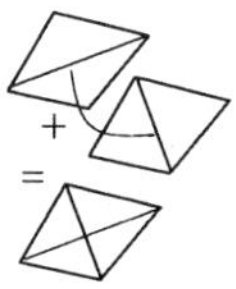

Figure 3. flat tetrahedron associated with the diagonal swapping

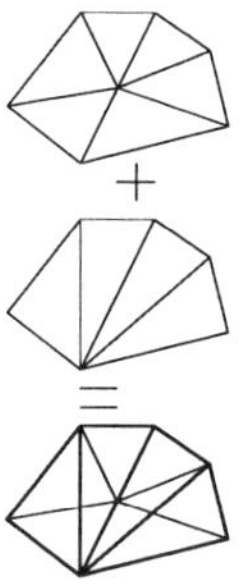

Figure 4. Status boundary/internal transformation of a node

It is clear on the examples plotted here that the topology of the mesh must be changed during calculation. The mesh of very simple initial part (here a cylinder) is deformed to the final automotive part complicated by the flash composed of extra material. The user has only to define the initial shape (here a cylinder) and the surface tools discretization. The periodical remeshing strategy allows us to obtain the final shape is obtained without human intervention. Moreover the mesh is adapted to the contact evolution.

The relative position of midpoint of boundary edges to the surface tools is tested for adaptation. If this midpoint is found inside the dies the point is created. An augmented Lagrangian method for solving the unilateral contact makes this strategy possible. Indeed, the node created inside the tool is enforced to move immediately on the contacting surface by adding the gap distance to impenetrability condition. A non trivial industrial applications is plotted on figure 7. This example has been provided by TRANSVALOR and runs with the commercial version of the code.

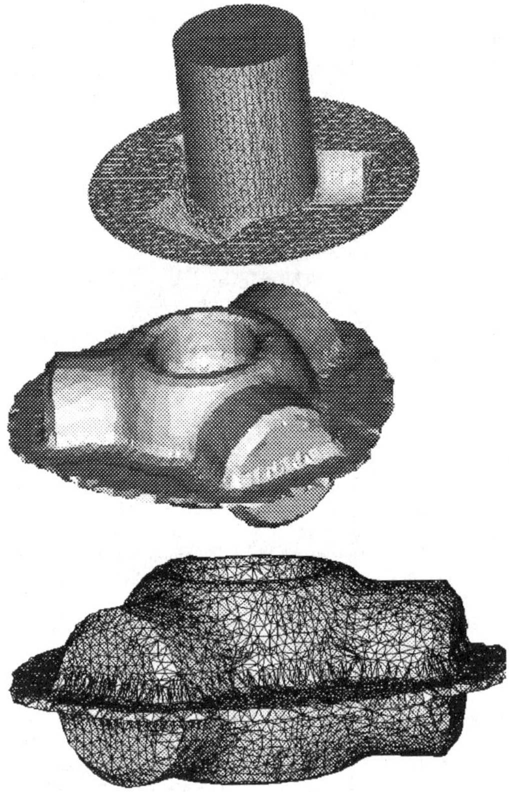

Figure 5. 3D Simulation of forging of a tripod with fully automatic remeshing. Initial shape (cylinder) with the lower surface tool and final deformed part with flash and mesh

6 PARALLEL REMESHING

6.1 Partitioning and iterative parallel solver

Our solver used an iterative solver mainly based on conjugate residual method with a simple diagonal preconditioning. A partitionning approach is considered for the parallelization of the matrix vector product . Consider a partition of the domain Ω:

$$\Omega = \bigcup_{p=1}^{nbP} \Omega^p$$

The global matrix A is split in a natural way and the matrix vector product can take place on each sub-domain :

$$A.x = \sum_{p=1}^{nbP} A^p.x^p$$

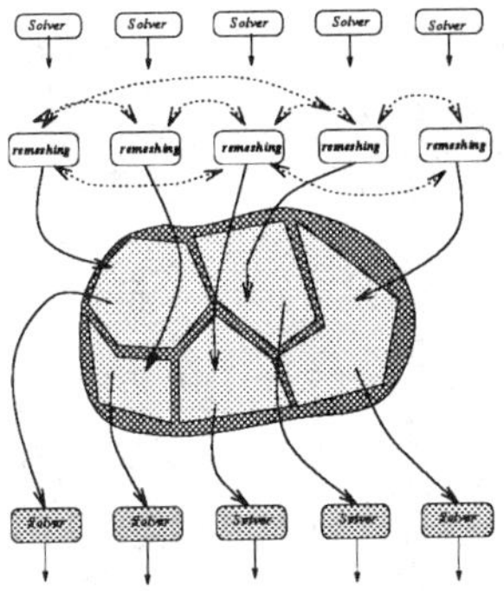

Figure 6. SPMD approach for parallel remeshing with partitioning

It requires to add the contribution of domains having a common boundary : the interface. On MIMD distributed memory architecture (SP2 machine or cluster of workstations for instance), interfaces play a crucial role on the efficiency and the sacalability. The parallel version of our code is based on a SPMD approach. It means that almost the complete serial code is loaded per processor. Each processor runs a complete version of the code, with a data set containing the triangulation of the surface of the tools and the mesh of a sub-domain associated with it. Since the remeshing procedure is associated with the code, the SPMD strategy involves automatically a first step of parallelization : the remesher is acting independently on each sub-domain (figure 6). The mesh topology at the boundary interface between two sub-domains must be shared. There is no particular difficulty in doing it with the mesh topology improvement method, even if two problems must be solved. The first one is the now classical problem of load balancing after remeshing or mesh adaptation : the number of elements per sub-domain (consequently the load of processor) can greatly change. The second one is that the interface are artificial and must be treated as the remaining of the mesh. The association of the parallel remesher with a step of repartitioning (dynamic repartitioning) provides a solution of both these two problems.

7 CONCLUSION

The 3D remeshing procedure presented here is based on a general mesh topology improvement method. It has been presented how to combine surface operations to the volume mesh. It is fully automatic and makes possible the numerical simulation of very large deformations of industrial part

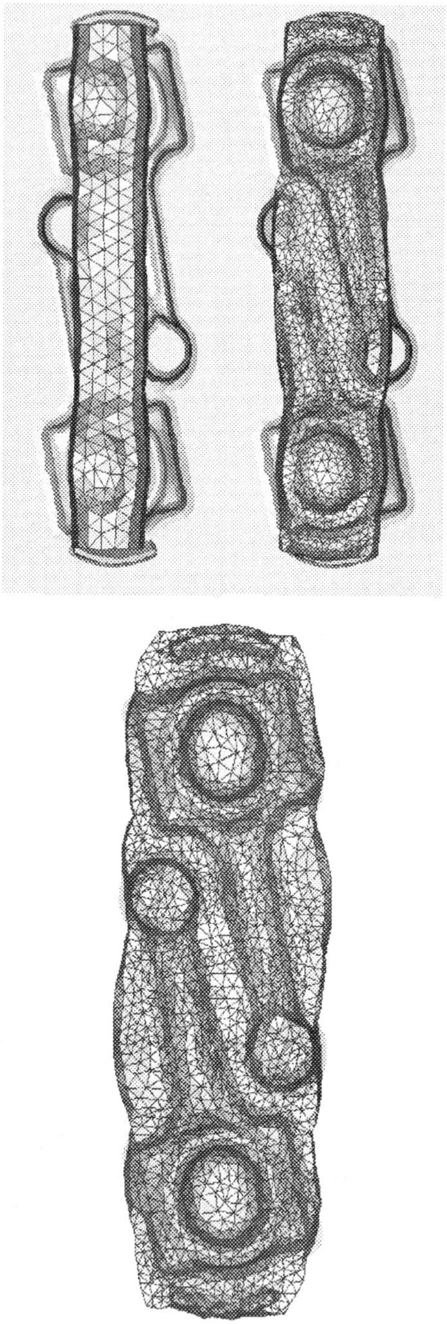

Figure 7. 3D Simulation of forging of a pair of connecting rods

without human intervention. The flash in forging, for instance, can be obtained. The computational effort required by the method depends upon the quality of the mesh to be remeshed. Therefore, a good strategy is to remesh as often as possible. The meshing method developed for mesh generation problems arising in forming calculation is very general. Other applications in CFD are under study. Indeed, the proposed method makes possible three-dimensional moving mesh methods. Moreover, because the method is local, its parallelisation is almost immediate. Nevertheless, it must be be related to the general difficulty of dynamic repartitioning.

REFERENCES

Borgers C. 1990 Generalized delaunay triangulations of non convex domains. *Comp. Math. Appl.*, 20(7):45–49.

Arnold D.N., Brezzi F., and Fortin M. 1984 A stable finite element for stokes equations. *Calcolo*, 21:337–344.

L.P. Franca, S.L. Frey, and T.J.R. Hughes. 1992 Stabilized finite element methods : Ii. the incompressible navier-stokes equations. *Comput. Meth. Appl. Mech. Engrg.*, 99:209–233.

Gobeau J.F., Coupez T., Vergnes B., and Agassant J.F. 1995. Computation of profile dies for thermoplastic polymers using anisotropic meshing. In P.R. Dawson, editor, *to appear in Numerical Methods in Industrial Forming Processes (NUMIFORM)*. A.A. Balkema 1995.

Georges P.L., Hecht F., and Saltel E. 1991. Automatic mesh generator with specified boundary. *Computer Methods in Applied Mechanics and Engineering*, 92:269–288.

Coupez T. 1994. A mesh improvement method for 3d automatic remeshing. In J. Hauser N.P. Weatherill, P.R. Eiseman and J.F. Thompson, editors, *Numerical Grid Generation in Computational Fluid Dynamics and Related Fields.* Pineridge Press.

Coupez T. and Chenot J. L. 1992. Large deformations and automatic remeshing. In E. Hinton D.J.R. Owen, E. Onate, editor, *Computational Plasticity (COMPLASIII)*. Pineridge Press 1992.

Coupez T. and Chenot J. L. 1992. Mesh topology for mesh generation problems. application to three-dimensional remeshing. In O.C. Zienkiewicz J. L. Chenot, R.D. Wood, editor, *Numerical Methods in Industrial Forming Processes (NUMIFORM)*. A.A. Balkema.

Coupez T., Soyris N., and Chenot J. L. 1991. 3d finite element modelling of the forging process with automatic remeshing. *Jour. of Materials Processing technology*, 27:119–133.

Talon J. Y. 1987. Algorithme d amelioration de maillage. Rapport technique, Artemis Imag, 25 1987.

Simulation of Materials Processing: Theory, Methods and Applications, Shen & Dawson (eds)
© 1995 Balkema, Rotterdam. ISBN 90 5410 553 4

An Eulerian formulation for the simulation of bulk massive forming processes

E. Doege & B. Laackman
Institute for Metal Forming and Metal Forming Machine Tools, University of Hannover, Germany

ABSTRACT: The simulation of bulk massive forming processes with the method of finite elements presently allows a realistic prediction of the material flow. However, the finite elements become so distorted that they lose their regular shape. Remeshing and rezoning techniques become necessary to avoid resulting numerical problems. Recently research is done in the development of such algorithms. Only a few attempts have been made in order to avoid remeshing generally. One possibility is to consider the material to be deformed as a fluid that flows through an Eulerian mesh. The elements keep their regular shape.

1 INTRODUCTION

The simulation of bulk massive forming processes with the method of finite elements presently allows a realistic prediction of the material flow. It becomes more and more an important contribution to avoid failure in forging. There are different techniques used in the formulation of finite elements for large plastic strain analysis, s. figure 1.

Usually for the simulation of such processes the Updated Lagrangian description is used. Thereby the mesh is incremented with the flow. However,

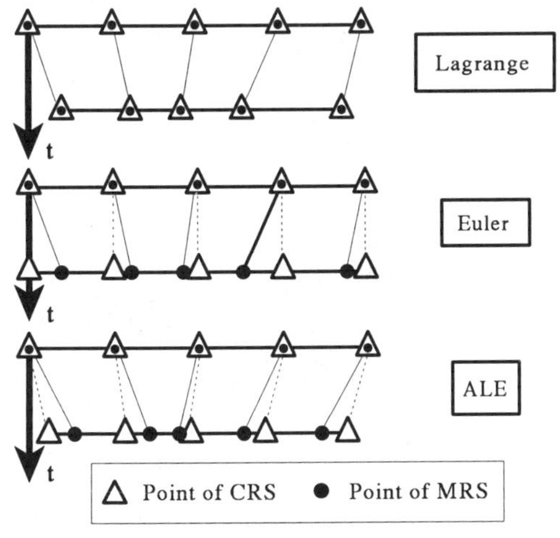

Figure 1: Different methods to describe the motion of a particle [1].

the elements become distorted during the calculation. This requires Remeshing and Rezoning procedures to keep the elements in a regular shape.

Another way used to describe forming processes is the Arbitrary Lagrangian-Eulerian (ALE) method. This method is frequently employed to simulate processes in gasses and fluids, and processes with fluid-structure interaction, e. g. [2, 3, 4, 5]. It was applied on forming processes for the first time by HUETINK [6] and SCHREURS [7]. The basis of the ALE is the use of a reference system which is not associated with the material (Lagrangian description) and has no fixed spatial position (Eulerian description).

A further approach is the method of pseudo-concentrations presented by THOMPSON et al. [8, 9, 10], s. figure 2.

Here a finite element mesh is defined which covers the entire region through which the material will flow. A pseudo-concentration M is then assigned to the nodes which are covered with either the true material ($M = 1$) or a ficticious one ($M = -1$). The free surface is then located at $M = 0$. The equation for M is obtained by setting its material derivative equal to zero, s. Eqn. 1.

$$\frac{\partial M}{\partial t} + u_i \frac{\partial M}{\partial x_i} = 0 \qquad (1)$$

The method presented here works with two reference systems, similar to the ALE. The first one is the Material Reference System (MRS) which is connected with the material to be deformed. The other one is the Computational Reference System (CRS) which is used for the calculation of the free surface, the material transport etc. . This definition is used by PONTHOT [11], too. It uses algo-

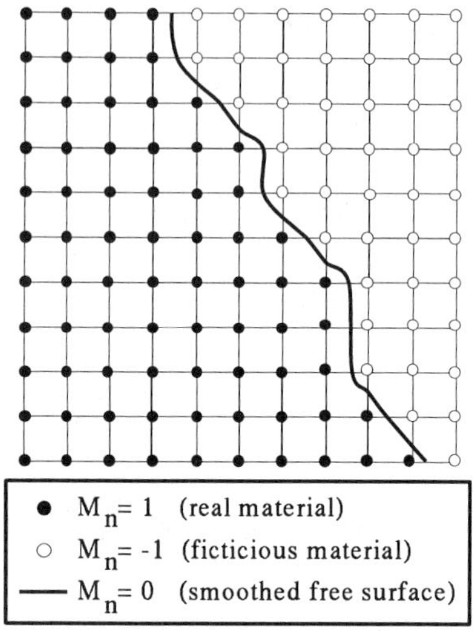

Figure 2: Pseudo-concentrations M_n; Free surface, smoothed.

rithms of the VOF–method (VOF = Volume Of Fluid) presented by NICHOLS et al. [12, 13].

The program developed for this purpose consists of two parts. The first part is an FEM-code with a rigid-plastic material behaviour (MRS). The other one is a transport module which calculates the flow through the CRS.

2 FINITE ELEMENT FORMULATION

Variational principle

The FEM-code is based on the variational principle by MARKOV [14]. It was extended by LUNG [15] with Lagrangian multipliers σ_h in order to take the condition of incompressibility into account, s. Eqn. 2. The Lagrangian multiplier σ_h corresponds to the hydrostatic pressure, s. Eqn. 3. First the formulation neglects friction at moving bounded walls for the following investigations.

$$\Pi_L = \int_V Y\dot{\varepsilon}_{eq}dV + \int_V \sigma_h\dot{\varepsilon}_{ii}dV - \int_{S^\sigma} \sigma_i^0 v_i dS \Rightarrow Extr. \qquad (2)$$

$$\sigma_h = \frac{1}{3}\sigma_{ii} \qquad (3)$$

Rigid-plastic material behaviour

The material behaviour is described using the Huber-Levy-v. Mises law, s. Eqn. 4.

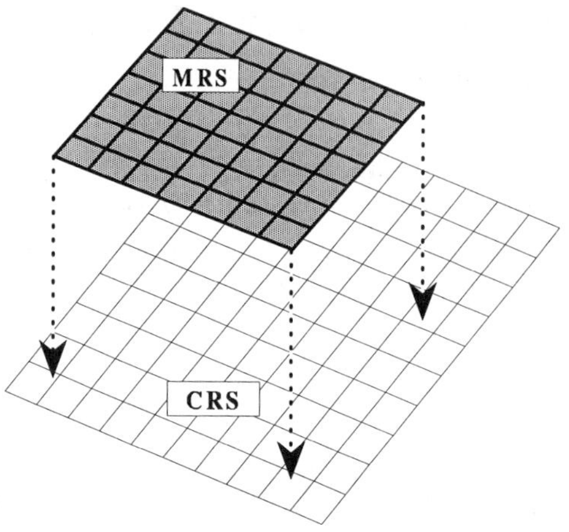

Figure 3: Mesh definitions: CRS and MRS

$$s_{ij} = \sigma_{ij} - \frac{1}{3}\sigma_{ii} = \frac{2Y}{3\dot{\varepsilon}_{eq}}\dot{\varepsilon}_{ij} \qquad (4)$$

Element types

The FEM-code uses two–dimensional (axisymmetrical and plain strain) three– and four–node elements, as they were developed by DUNG [16]. They use linear shape functions for the velocity, and are economic with regard to CPU time. The condition of incompressibility has to be fulfilled only in the center of mass [17].

3 MATERIAL TRANSPORT

The approach to calculate the transport of the rigid-plastic material through an Eulerian mesh is based on the VOF-method [12, 13]. There are some definitions to be declared in connection with the transport module.

Definitions

Meshes: There are two meshes. We call them CRS and MRS, s. figure 3. The CRS corresponds here to the Eulerian mesh, and the MRS to the Lagrangian mesh, as described above. The CRS is non-deformable and is used for the transport calculation. The MRS is the one used by the FEM-code. Consider the relation between them as a projection of the MRS on the CRS.

Concentration M: M is the concentration related to an element. An element is either completely ($M = 1$) or partially filled ($0 < M < 1$) with material, or is empty ($M = 0$). NICHOLS et

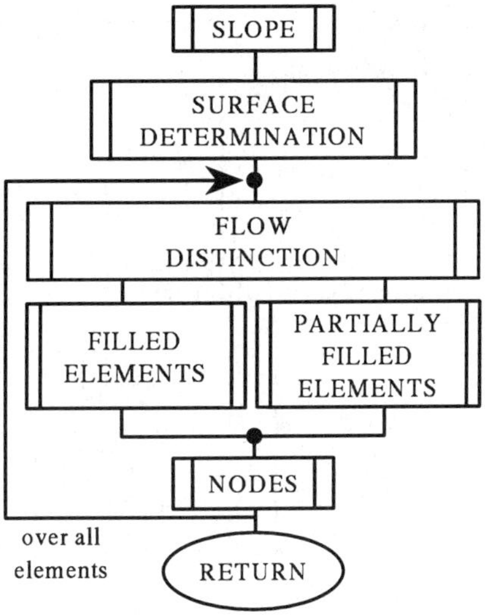

Figure 4: Material transport algorithm

al. [12] call it 'Fractional Volume of Fluid'.

Concentration M_n: M_n is the concentration related to nodes. A node is either covered with material ($M_n = 1$) or not ($M_n = 0$).

Boundary elements: They contain the free surface. They are characterised by

- $0 < M < 1$ or
- $M = 1$ and at least one neighbouring element with $M = 0$.

Algorithm

The algorithm to calculate the material transport through the CRS is shown in figure 4. The subscripts i and j correspond to figure 5.

The main topics are:

Slope: The free surface in the boundary elements is assumed to be made up of linear segments. The free surface can be represented either as a single–valued function $Z(x_i)$ or $X(y_j)$, depending on its orientation, s. Eqns. 5 and 6 [12].

$$\begin{aligned} Z(x_i) &= M_{i(j-1)}\delta y_{j-1} + M_{ij}\delta y_j \\ &\quad + M_{i(j+1)}\delta y_{j+1} \qquad (5) \\ X(y_j) &= M_{(i-1)j}\delta x_{i-1} + M_{ij}\delta x_i \\ &\quad + M_{(i+1)j}\delta x_{i+1} \qquad (6) \end{aligned}$$

The slope of the free surface then is the result of the differentiation. If the free surface is e. g. representable as $Z(x_i)$, one has to compute dZ/dx, s. Eqn. 7.

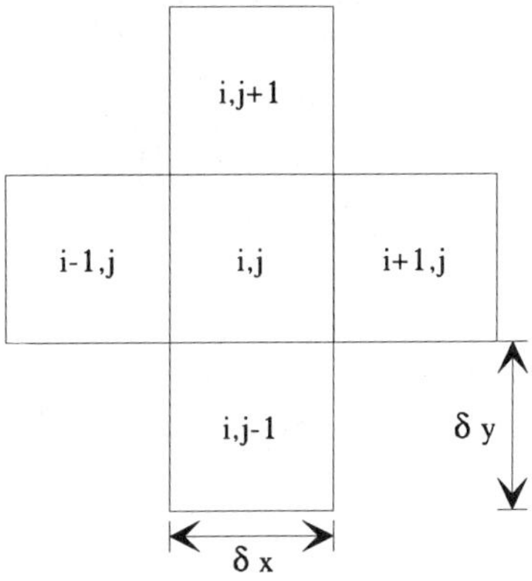

Figure 5: Free surface representation; Subscripts

$$\left(\frac{dZ}{dx}\right)_i = \frac{\frac{(Z_{i+1}-Z_i)\delta x_{i-1/2}}{\delta x_{i+1/2}} + \frac{(Z_i - Z_{i-1})\delta x_{i+1/2}}{\delta x_{i-1/2}}}{\delta x_{i-1/2} + \delta x_{i+1/2}} \qquad (7)$$

where, e. g.

$$\delta x_{i+1/2} = \frac{\delta x_{i+1} + \delta x_i}{2} \quad .$$

Surface determination: The distribution of M and M_n is used to determine the boundary elements. Then we use Eqn. 8 for the calculation of the intercept of the free surface with the y–axis (x–axis, resp.) of each element.

$$M_{ij}\delta x_i \delta y_j = \int_0^{\delta x} Z dx = [\frac{1}{2}x_i^2 a + b x_i]_0^{\delta x} \qquad (8)$$

Flow distinction: Here it is distinguished whether the material flows out of a filled or a partially filled element into the neighbouring elements.

Nodes: After finishing the transport calculation this function determines the new distribution of M_n.

4 COUPLING FEM–TRANSPORT

Figure 6 shows the coupling of the FEM-code with the transport algorithm. At the beginning of the analysis there are only completely full or empty elements in the CRS. The full elements are used for the determination of a velocity field at the nodal points in the MRS, considering the boundary conditions.

The velocity field is handed over to the transport module. There the new distribution of the material in the CRS is calculated. This results in partially filled elements containing the new loca-

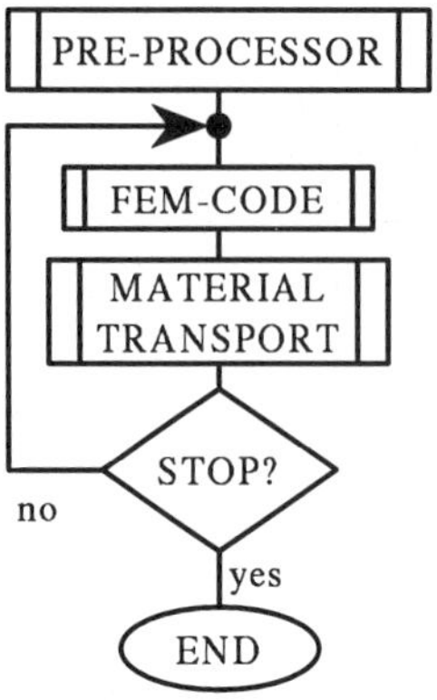

Figure 6: Coupling FEM–Transport

tion of the free surface. Afterwards it prepares a new mesh with adapted boundary conditions for the MRS (e. g. to take a moving die into consideration). This mesh preparation includes a slightly distortion of the boundary elements. Here a remeshing strategy for those elements is needed.

Having passed the transport algorithm for a second time the adapted boundary elements will take their original shape. Other (new) boundary elements will be adapted.

Remeshing

On the whole there are two steps to prepare the mesh for the next time step with the FEM–code (MRS). Figure 7 shows the procedure.

Firstly the points of intersection of the free surface with the element edges are assigned to certain nodes of the CRS. Secondly the corresponding nodes found this way will be displaced on the intersection points (i. e. the free surface), that is in horizontal and vertical direction. The result is that slightly deformed four–node elements limit the free surface.

5 CONCLUSIONS

The algorithm to calculate bulk massive forming processes presented here is an alternative to the time-consuming remeshing strategies. The coupling of the transport routines with an FEM-code works well provided that the history dependent variables are interpolated on the new mesh. The advantage of the used rigid-plastic material behaviour is that it neglects springback effects. These would lead to problems in the determination of the free surface. First investigations prove the model to be well applicable to calculate the material transport through an Eulerian mesh.

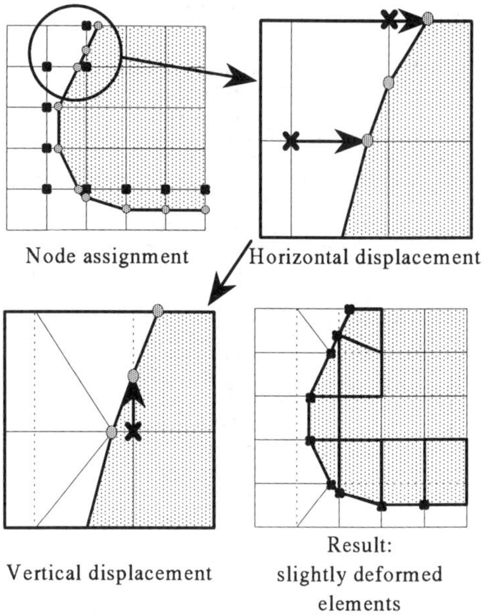

Figure 7: Preparing the mesh for the MRS

ACKNOWLEDGEMENTS

The authors gratefully acknowledge the support given towards the project Do 190/90–1 by the "Deutsche Forschungsgemeinschaft — DFG".

Additionally they would like to express their appreciation to Christian Jaeckel and Christoph Riewenherm for their excellent work in programming the routines of the FEM-code and the transport module.

REFERENCES

[1] HUERTA, A.; PIJAUDIER-CABOT, G.; BODÉ, L.: ALE Formulation in nonlocal strain softening solids. Proc. COMPLAS 3. Barcelona, Spain: 6th–10th April, 1992. 2249–2268.

[2] HIRT, C. W.; AMSDEN, A. A.; COOK, J. L.: An arbitrary Lagrangian-Eulerian computing method for all flow speeds. *J. comp. phys.* 14 (1974), 227–253.

[3] PRACHT, W. E.: Calculating three-dimensional fluid flows at all speeds with an Eulerian-Lagrangian computing mesh. *J. comp. phys.* 17 (1975), 132–159.

[4] STEIN, L.R.; GENTRY, R. A.; HIRT, C. W.: Computational simulation of transient blast loading on three-dimensional structures. *Comp. meth. appl. mech. eng.* 11 (1977), 57–74.

[5] HUGHES, T. J. R.; LIU, W. K.; ZIMMERMANN, T. K.: Lagrangian-Eulerian finite element formulation for incompressible

viscous flows. *Comp. meth. appl. mech. eng.* 29 (1981), 329–349.

[6] HUETINK, H.: On the simulation of thermomechanical forming processes. Dr.-Ing. thesis. TH Twente (The Netherlands): 1981.

[7] SCHREURS, P.: Numerical simulation of forming processes. Dr.-Ing. thesis. TH Eindhoven (The Netherlands): 1983.

[8] THOMPSON, E.: Transient analysis of metal forming operations using pseudo-concentrations. Proc. NUMIFORM '86. Gothenburg, Sweden: 25.–29.8.1986.

[9] THOMPSON, E.: Use of pseudo-concentrations to follow creeping viscous flows during transient analysis. *Int. J. Num. meth. fluids* 6 (1986), 749–761.

[10] THOMPSON, E.; SMELSER, R. E.: Transient analysis of forging operations by the pseudo-concentration method. *Int. J. Num. meth. eng.* 25 (1988), 177–189.

[11] PONTHOT, J.P.: The use of the Eulerian-Lagrangian FEM with adaptive mesh. Applications to metal forming simulation. Proc. COMPLAS 3. Barcelona, Spain: 6th–10th April, 1992. 2269–2280.

[12] NICHOLS, B. D.; HIRT, C. W.; HOTCHKISS, R. S.: SOLA-VOF: A solution algorithm for transient fluid flow with multiple free boundaries. Los Alamos Scientific Report LA–8355: August 1980.

[13] HIRT, C. W.; NICHOLS, B. D.: Volume of Fluid (VOF) method for the dynamics of free boundaries. *J. comp. physics* 39 (1981), 201–225.

[14] MARKOV, A. A.: Variational principles in the theory of plasticity. *Prikladnaja matematika i mechanika* 11 (1947) 3, 339–350.

[15] LUNG, M.: Ein Verfahren zur Berechnung des Geschwindigkeits- und Spannungsfeldes bei stationären starr-plastischen Formänderungen mit finiten Elementen. Dr.-Ing. thesis. TU Hannover (Germany): 1971.

[16] DUNG, N. L.: Ein Beitrag zur Berechnung instationärer starr-plastischer Formänderungen mit einer Finite-Element-Methode. Fortschr.-Ber. VDI-Z. 2, Nr 46: 1981.

[17] NAGTEGAAL, J. C.; PARKS, D. M.; RICE, J. R.: On numerically accurate finite element solutions in the fully plastic range. *Comp. meth. appl. mech. eng.* 4 (1974), 153–177.

Simulation of Materials Processing: Theory, Methods and Applications, Shen & Dawson (eds)
© 1995 Balkema, Rotterdam. ISBN 90 5410 553 4

Efficient error estimates for adaptive remeshing in 2D metal forming modelling

M. Dyduch, S. Cescotto & A. M. Habraken
Département MSM, Université de Liège, Belgium

ABSTRACT: An automatic, adaptive remeshing based on efficient Zienkiewicz-Zhu type error estimates with various smoothed fields and on a-posteriori finite element interpolation error estimate in 2D metal forming processes are presented. Some comparisons of the effectiveness of these errror estimates for cylinder upsetting and forging example are given. The local finite difference method for irregular grids is used to compute superconvergent nodal values of the smoothed fields.

1. INTRODUCTION

The updated Lagrangian description is generally adopted for non-steady metal forming processes, with dependency on material history eg. forging, rolling, drawing. An advantage of the Lagrangian description is that deformation history dependent variables (strain hardening) can easily be taken into account and the changing shape of the formed product can be followed. However one of the significant limitations of this approach, especially when very large strains are involved, is the progressive distortion, sometimes entanglement of the finite element mesh. In most cases it is necessary to redefine the mesh after some stages of deformation. Another reason to change the discretization and refine it, is when due to complicated die geometry, the initially (previously) generated mesh does not have locally enough degrees of freedom to describe the details of material flow, because of die boundary conditions. This usually happens when a part of the workpiece is entering in contact with the curved die border and therfore a much finer mesh is needed . To solve this problem the automatic, adaptive remeshing (h-type refinement) [1], [3], [4] is used. It consists in defining a new appropriate finite element mesh within the updated, deformed geometry and interpolating (mapping) the pertinent variables from one mesh to another in order to continue the simulation. Automation of the procedure is imperative, as it is frequently repeated.

This adaptive procedure can be usually split into the following parts :

1. error and mesh control techniques;
2. remeshing of the deformed structure-workpiece;
3. mapping of history dependent variables from the old to the new mesh;
4. restart of the simulation

Efficient, computationally cheap error and mesh control techniques are crucial for the analysis, as they predict when the simulation should be stopped and how the new mesh should be refined in order to settle the solution discretization errors within specified limits. This part of the procedure is guided by the Zienkiewicz-Zhu type (Z-Z) a posteriori error estimator [2], error indicators based on interpolation error and adapted for large deformation viscoplastic formulation [5] and finally by geometric measures of elements accuracy [4].

2. GOVERNING EQUATIONS FOR F.E. METAL FORMING MODELLING

2.1 Elastic-viscoplastic constitutive law for large strains in solids.

This type of law is very appropriate to simulate the behaviour of metals in forming processes, especially at high temperatures, as rate effects in inelastic deformation are included in this model, as well as isotropic hardening.

The constitutive equations can be written as :

$$\overset{\nabla}{\hat{\underline{\sigma}}} = 2G(\hat{\underline{D}} - \hat{\underline{D}}^{vp}), \quad \dot{\sigma}_m = \chi D_m$$
$$\hat{\underline{D}}^{vp} = \phi(F)\frac{\hat{\underline{\sigma}}}{\sqrt{J_2}} \tag{2.1}$$

with :

$$\phi(F) = B(F)^n, \quad F = \frac{\sqrt{J_2}}{k}$$
$$\dot{k} = \frac{2}{\sqrt{3}}\phi(F)h_k, \quad h_k = \frac{H_1}{(k)^m} \tag{2.2}$$

where :

$\hat{\underline{\sigma}}, \hat{\underline{D}}^{vp}$ are the deviatoric Cauchy stress tensor and the non elastic part of deviatoric strain rate tensor. In eq. (2.1) objective Jaumann stress derivative is adopted.
σ_m , D_m are the hydrostatic Cauchy stress and the hydrostatic strain rate
J_2 is the second invariant of the deviatoric stress tensor
B, n, H_1 m are inelastic parameters at the considered temperature and k_0 is the initial yield radius for the isotropic hardening variable k
χ, G are the bulk and shear moduli respectively

2.2 Contact friction law.

A penalty method is used for the contact formulation, allowing a small penetration of the workpiece into the die. The Coulomb friction law is adopted in which the friction stress is proportional to the contact pressure through the coefficient ϕ.

3. ERROR ESTIMATES FOR VISCOPLASTIC FEM FORMULATION

Among different error estimates available in finite element computations, two were chosen, that can be almost directly applied in nonlinear analysis (large strains, viscoplasticity). Computational efficiency (cost) is crucial here, as the errors are calculated at each step of the analysis.

3.1 Finite element interpolation error estimate

The computation of the interpolation error [10], [5] is based on a priori interpolation estimate, as well as on the superconvergent recovery of higher order derivatives of the finite element solution. The increment of error for an element in elastic domain is bounded by [10] :

$$\Delta_i E_e^2 \le C h_e^2 \int_{\Omega_e} (2G+2\lambda)(\Delta u_{1,11}^2 + \Delta u_{1,22}^2 + \Delta u_{2,11}^2 + \Delta u_{2,22}^2)\, d\Omega_e \tag{3.1}$$

where : i is the increment number, G is the shear modulus, λ is the Lame's constant, C is a scaling constant and h_e may be approximated by a constant times the area of the element. Second order derivatives of displacements must be computed in order to apply this estimator.
In viscoplastic analysis a modified form of interpolation error estimate can be given [1], [5] , the increment of error of an element can be approximated by :

$$\Delta_i E_e = \left(\int_{\Omega_e} ((\Delta\bar{\varepsilon}^p_{,x})^2 + (\Delta\bar{\varepsilon}^p_{,y})^2)\, d\Omega_e \right)^{1/2} \tag{3.2}$$

where $\Delta\bar{\varepsilon}^p_{,q=x,y}$ is the gradient of the equivalent inelastic strain increment. The error is cummulated in time, for n increments we obtain :

$$E_e = \sum_{i=1}^{n} \Delta_i E_e \tag{3.3}$$

In geometric representation, the error of a particular element may be viewed as the difference of the equivalent inelastic strain of the element and its adjacent neighbours.This error estimator is very suitable for local use as indicator (guiding the variable density mesh generator), but is less convenient for global error prediction. The main difficulty is how to obtain the gradients of equivalent plastic strain in a precise manner.

3.2 Zienkiewicz-Zhu type error estimates

A simple and efficient a-posteriori error estimator was introduced by Zienkiewicz and Zhu [2], first for linear analysis and later for metalforming modelling by the flow formulation [3] in the mixed form. The error in energy norm was given in the form :

$$\|e\|^2 = \int_\Omega (\hat{\underline{\sigma}}^h - \hat{\underline{\sigma}})(2\mu)^{-1}(\hat{\underline{\sigma}}^h - \hat{\underline{\sigma}})\, d\Omega + \int_\Omega (p^h - p)(\lambda + 2\mu/3)^{-1}(p^h - p)\, d\Omega \tag{3.4}$$

where :
$\hat{\underline{\sigma}}^h, p^h$ are f.e. solutions of the deviatoric stress and pressure respectively
$\hat{\underline{\sigma}}$, p are the corresponding exact values
μ, λ are Lamé viscosities.
In eq. (3.4) strain rates are expressed in terms of viscosity and stress (Newtonian fluid constitutive law). In general elastic-viscoplastic formulation of

the type given by equation (2.1) this is not possible and L_2 error norms should be used. Following L_2 norms for different variables : stress and strain rate are computed :

$$\|e\|_{\sigma}^{2}=\int_{\Omega}(\underline{\sigma}^{h}-\underline{\sigma})(\underline{\sigma}^{h}-\underline{\sigma})d\Omega \qquad (3.5a)$$

$$\|e\|_{\dot{\varepsilon}}^{2}=\int_{\Omega}(\underline{\dot{\varepsilon}}^{h}-\underline{\dot{\varepsilon}})(\underline{\dot{\varepsilon}}^{h}-\underline{\dot{\varepsilon}})d\Omega \qquad (3.5b)$$

and for energy norm with both fields :

$$\|e\|_{\sigma\dot{\varepsilon}}^{2}=\int_{\Omega}(\underline{\sigma}^{h}-\underline{\sigma})(\underline{\dot{\varepsilon}}^{h}-\underline{\dot{\varepsilon}})d\Omega \qquad (3.5c)$$

As the exact values of $\underline{\sigma},\underline{\dot{\varepsilon}}$ are unknown, they are replaced in eq. (3.4-3.5) by some recovered, higher order approximation of finite element solution : $\underline{\sigma}^{R},\underline{\dot{\varepsilon}}^{R}$. These recovered solutions can be obtained by various projections : local-global least square smoothing or by the recommended superconvergent patch recovery techniques [6]. This is of course a crucial step in error computation, as if a poor projection is applied, $\underline{\sigma}^{R},\underline{\dot{\varepsilon}}^{R}$ are not of higher order than f.e. solutions and the estimation is not reliable.
Our projection is based on the local finite difference method [7],[8] (section 4) and has already been used for Z-Z estimator with Norton-Hoff constitutive law by [9]. Its superconvergent recovery properties are demonstrated in [11] for analytical examples.
The global error bound used to stop the simulation and to activate the adaptive remeshing procedure is :

$$\eta=\frac{\|e\|}{\|u^{*}\|}<\bar{\eta} \qquad (3.6)$$

where for stresses (3.5a.), we have :

$$\|u^{*}\|_{\sigma}^{2}=\int_{\Omega}\underline{\sigma}^{R}\,\underline{\sigma}^{R}d\Omega \qquad (3.7)$$

and not as in [2], [3] :

$$\|u^{*}\|_{\sigma}^{2}=\int_{\Omega}\underline{\sigma}^{h}\underline{\sigma}^{h}\,d\Omega+\|e\|^{2} \qquad (3.8)$$

Respectively, similar formula are obtained when (3.5b) and (3.5c) are considered.
The permissible error per element is then computed :

$$\|e\|_{\sigma}^{pe}=\bar{\eta}\frac{\|u^{*}\|_{\sigma}}{\sqrt{M}} \qquad (3.9)$$

where M is the number of elements.

The element error indicators are obtained for each element according to :

$$\xi_{\sigma}=\frac{\|e\|_{\sigma}^{e}}{\|e\|_{\sigma}^{pe}} \qquad (3.10)$$

$\xi_{\sigma},\xi_{\dot{\varepsilon}},\xi_{\sigma\dot{\varepsilon}}$ serve as the guide for variable density mesh generation.

4. SUPERCONVERGENT DERIVATIVE RECOVERY BY THE LOCAL FINITE DIFFERENCE METHOD

Only the basic idea of this projection technique is described hereafter, for further details, references [7], [8] should be consulted. The local finite difference method applied to arbitrary irregular grids is used to compute superconvergent values of stress, strain rate and gradients of equivalent inelastic strain increments at element nodes from Gauss integration points.
Let us consider a patch of eight node isoparametric elements, with quadratic shape functions, as shown in fig.4.1. In order to find a continuous field of $\underline{\sigma},\underline{\dot{\varepsilon}}$ at nodes from integration points (four Gauss points in that case), we develop it into Taylor expansion around each node of the patch.

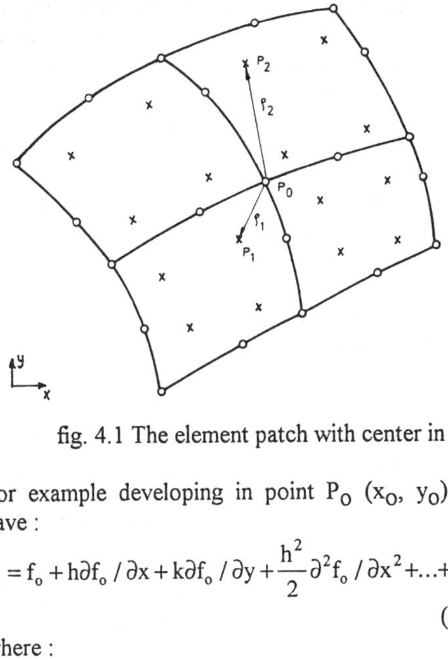

fig. 4.1 The element patch with center in P_0

For example developing in point P_0 (x_0, y_0), we have :

$$f_i=f_o+h\partial f_o/\partial x+k\partial f_o/\partial y+\frac{h^2}{2}\partial^2 f_o/\partial x^2+\ldots+0(\rho_i^{\,3}) \qquad (4.1)$$

where :

$$f_o=f(x_o,y_o)$$
$$f_i=f(x_i,y_i)$$
$$h=x_i-x_o$$
$$k=y_i-y_o$$
$$\rho_i=\sqrt{h^2+k^2}$$

The expansion (4.1) is done for every Gauss point "i" in the patch. The solution is obtained by the minimization of the weighted norm

$$B = \sum_{i=1}^{n}\left[(f_i - f_o - h_i\partial f_o/\partial x - k_i\partial f_o/\partial y - \dots)^2 w_i^2\right] \quad (4.2)$$

where n is the number of points in a patch :

$$\frac{\partial B}{\partial D_f} = 0 \quad (4.3)$$

From (4.3) we get an overdetermined system of equations with respect to unknown values :

$$D_f = \left[f_o, \partial f_o/\partial x, \partial f_o/\partial y, \partial^2 f_o/\partial x^2, \dots\right] \quad (4.4)$$

From numerical tests [11], we finally retained eight terms development in (4.1) and $w_i = 1/\rho_i^4$ (six terms and $w_i = 1/\rho_i^3$ in [7]). Usually the 16-20 nearest integration points were considered in a patch. The use of weighted values for each integration points and minimization efficiently prevents from singular or ill-conditioned system of equations, it also emphasizes the influence of nearest integration points to the considered node [8]. We would like to point out the similarity between this approach and the superconvergent patch recovery [6], the differences are the weighted norm and some kind of coordinate normalization. By this technique not only the function, but also its derivatives can be computed (eq.4.4), since they are necessary for the previously discussed interpolation error indicator.
Finally we obtain the continuous field of a considered variable in the element using its shape functions and the computed nodal values.

5. REMESHING OF THE DEFORMED STRUCTURE AND MAPPING

When the admissible global error value $\bar{\eta}$ of eq.(3.6) is reached, the remeshing module is activated, the updated shape of the workpiece (structure) is recovered and a variable density mesh generator provides a new adaptive mesh, according to the appropriate error indicators of section 3. In order to continue with the interrupted simulation, all history dependent variables are mapped from the old mesh to the new one. Here again the local finite difference is applied; this time we pass from Gauss points of the old mesh to the Gauss points of the new one.
The restart is then performed automatically.

6. NUMERICAL RESULTS

6.1 Cylinder upsetting

A steel cylinder upsetting problem is considered

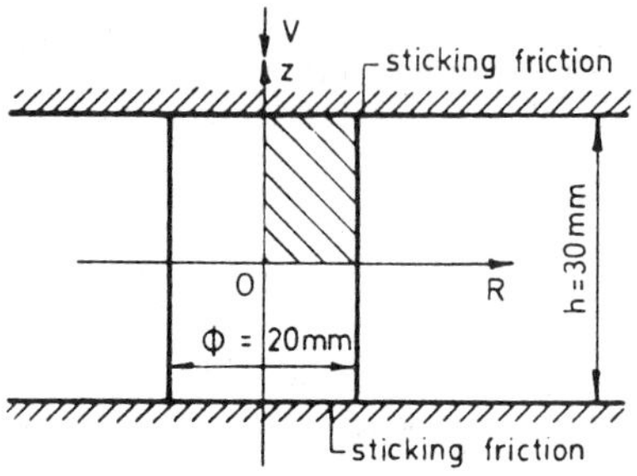

fig. 6.1 Cylinder upsetting

(fig.6.1.). This simple example has been already treated by many authors and is a kind of benchmark in forming analysis. In [3], it shows the performance of the Z-Z estimator, dealing with the distorted mesh at the folding corner.
Since it is a symmetrical workpiece only the top right quadrant of the cylinder needs to be analysed. The initial mesh is regular and composed of 110 eight node isoparametric elements (fig. 6.2). The total upsetting process is carried on up to 60 % height reduction. with constant velocity of 15mm/s. The constitutive law is that of section 2.1 with following parameters : E=1.2 $\times 10^5$Mpa, ν=0.3, n=9.259, B=0.034, $k=k_0=0.5\times10^5$MPa, no hardening is considered. In order to impose sticking contact, we set the friction coefficient in Coulomb law to ϕ =1.0. With these conditions the deformation is non-uniform. In figures 6.2-6.3 the initial and deformed mesh at 14% height reduction and the new adaptive mesh generated according to $\xi_{\dot\varepsilon}$ (eq.3.10) are shown respectively. The error indicators according to ξ_σ, $\xi_{\sigma\dot\varepsilon}$, $\xi_{\dot\varepsilon}$ and interpolation error indicator at 14% height reduction are presented in fig. (6.4-6.7). For Z-Z type estimator it can be noticed that $\xi_{\dot\varepsilon}$ gives the most sharp detection of the distortion and maximum error value in the upper folded corner. Error indicators in stresses or energy detect a large zone to refine and the maximum error is indicated far from the folded corner (fig.6.4). A very good agreement between Z-Z error indicator based on strain rate and interpolation error indicator can be noticed, although of course these variables do not contain the same information and only relative intensity can be compared. For global error prediction ($\bar{\eta}$=30%) only η_σ is retained for global error estimation due to its relatively rising monotonic character, $\eta_{\dot\varepsilon}$ and $\eta_{\sigma\dot\varepsilon}$ give rather meaningless results (large oscillations appear). This is also confirmed in other examples.

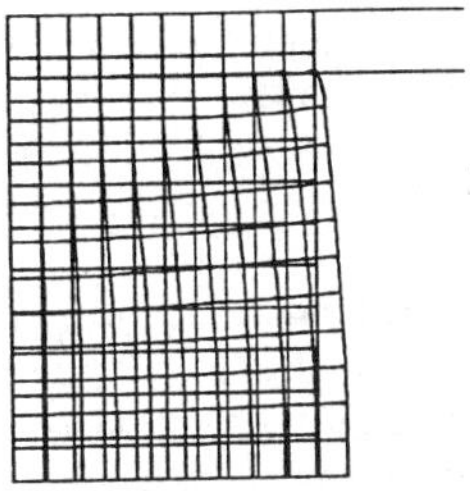

fig. 6.2 Initial and deformed mesh

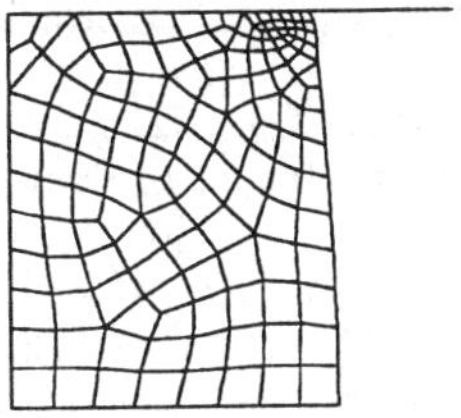

fig 6.3 New adaptive mesh according to $\xi_{\dot{\varepsilon}}$

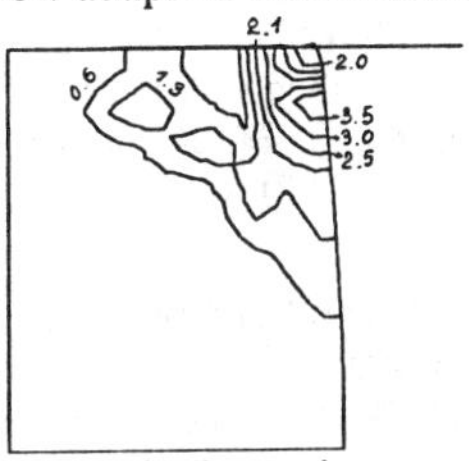

fig.6.4 ξ_{σ} error indicator in stress

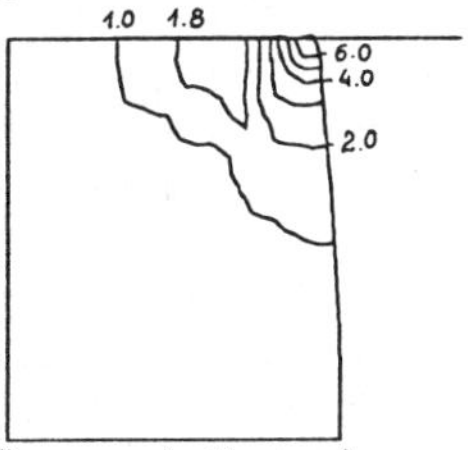

fig. 6.5 $\xi_{\sigma\dot{\varepsilon}}$ error indicator in energy

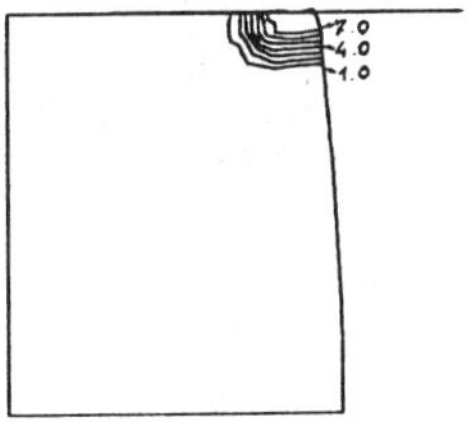

fig. 6.6 $\xi_{\dot{\varepsilon}}$ error indicator in strain rate

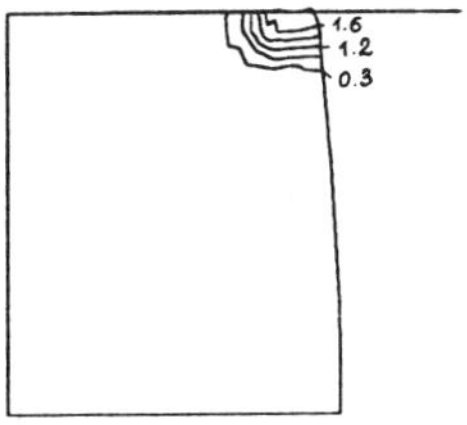

fig. 6.7 interpolation error indicator

6.2 Forging of a machine part.

Forging of a machine part with complicated die geometry is presented if figures 6.8-6.11. The structure is modelled in axisymmetric state with eight node quadratic quadrilaterals. The material data for steel at 1200°C are: $E=1.106\times10^5$Mpa, $\nu=0.35$, $n=4.878$, $B=7.934$, $k=k_0=0.2\times10^5$MPa, the coefficient in Coulomb law is $\phi=0.2$, constant velocity of 10mm/s is applied to the upper die. The limit global error $\bar{\eta}$ is fixed to 25%. In fig. 6.8-6.9 the initial and deformed meshes at an early stage of deformation are shown. Error indicators in strain rate and interpolation error indicator are given in fig.6.10-6.11 respectively. As for the preceeding example, good correlation between them can be observed. It should be noticed, that not only the currently deforming contact zone of the workpiece is indicated for refinement, but also the zone of important shear, where the elements of too big size are located. The corresponding error indicators in stress and energy give meaningless results for this simulation.

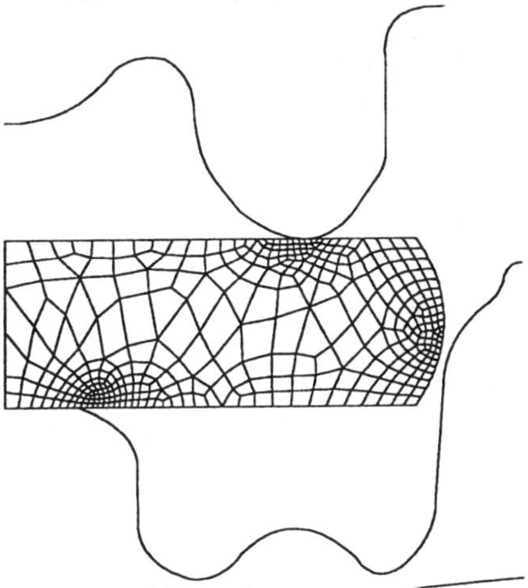

fig.6.8 Initial mesh

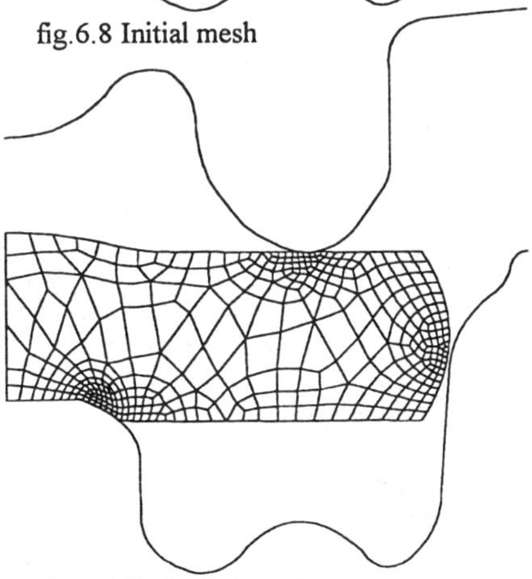

fig 6.9 Deformed mesh

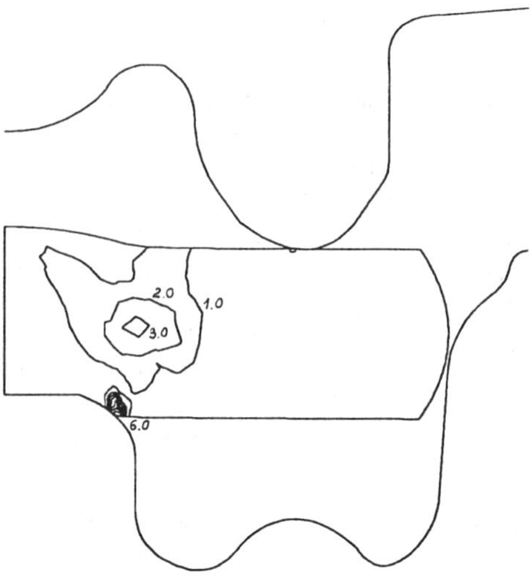

fig. 6.10 $\xi_{\dot{\varepsilon}}$ error indication in strain rate

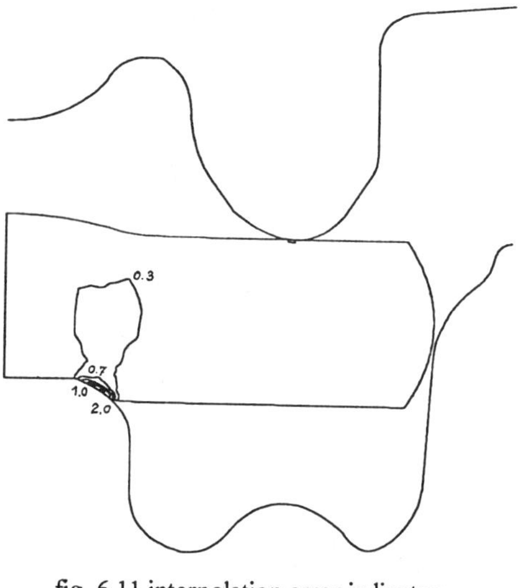

fig. 6.11 interpolation error indicator

7. CONCLUDING REMARKS

The local, finite difference method applied to Z-Z type estimators and interpolation type error estimator proved to be efficient in adaptive remeshing. For mesh refinement, error indicator in strain rate and interpolation error indicator giving similar results are used; for global error estimation η_σ in stresses is chosen. Significant differences in the error estimation have been found depending on the smoothed field (stress or strain rate). The error estimates considered here have been presented for the purpose of guiding the adaptive remeshing procedure. All the computations are independent of the adopted constitutive law.

REFERENCES

1. J.H. Cheng, Automatic adaptive remeshing for finite element simulation of forming processes, Int. J. Num. Methods Eng.,26,1-18, 1988
2. O.C. Zienkiewicz and J..Z. Zhu, A simple error estimator and adaptive procedure for practical engineering analysis , Int. J. Num. Methods Eng., 24, 337-357, 1987
3. O.C. Zienkiewicz, G.C. Huang and Y.C. Liu, Adaptive FEM Computations of Forming Processes-Application to Porous and Non-Porous Materials , Int. J. Num. Methods Eng., 30, 1527-1553, 1990
4. M. Dyduch, A.M. Habraken and S. Cescotto, Automatic adaptive remeshing for numerical simulations of metalforming, Comp. Methods Appl. Mech. Eng., 101, 283-298 ,1992
5. J. Bass and J.T. Oden, Adaptive Finite Element Methods for a Class of Evolution Problems in Viscoplasticity , Int. J. Engng. Sci, 25, 623-653 , 1987
6. O.C. Zienkiewicz and J.Z. Zhu, The Superconvergent Patch Recovery and A Posteriori Error Estimates. Part 1. The Recovery Technique, Part 2. Error Estimates and Adaptivity , Int. J. Num. Methods Eng., 33, 1331-1382 , 1992
7. T. Liszka and J. Orkisz, The Finite Difference Method at Arbitrary Irregular Grids and Its Application in Applied Mechanics , Comp. and Struct., 11, 83-95, 1980
8. J. Krok and J. Orkisz, Application of the Generalized FD Approach to Stress Evaluation in the FE Solution , Proceedings of Int. Conf. on Computational Mechanics, Tokyo, 1986.
9. L. Fourment and J.L. Chenot, Study of some error estimates for adaptive meshing in metal forming , Proceed. of the Complas III Conf., 1992
10. L. Demkowicz, Ph. Devloo and J.T. Oden, On an h type mesh refinement strategy based on a minimization of interpolation error, Comp. Meth. in Appl. Mech. Eng., 3, 67-89, 1985
11. M. Dyduch, Error estimation and adaptive remeshing in 2D metal forming problems, Ph. D. thesis, University of Liège, 1995

Simulation of Materials Processing: Theory, Methods and Applications, Shen & Dawson (eds)
© 1995 Balkema, Rotterdam. ISBN 90 5410 553 4

Shot peening simulated by two different finite element formulations

Jonas Edberg
MEFOS Metal Working Research Plant, Luleå, Sweden

Lars-Erik Lindgren
Luleå University of Technology, Sweden

Ken-ichiro Mori
Osaka University, Machikaneyama, Toyonaka, Japan

ABSTRACT: Two different finite element formulations are used for simulating the impact between a shot and a surface. The first is a viscoplastic finite element formulation where the residual deformation is transferred to a linear elastic formulation in order to calculate the residual stresses. The second is an elastoplastic finite element formulation for the nonlinear transient dynamic response of three-dimensional solids and structures. The results from the simulations are compared with experiments. The advantages, disadvantages, accuracy and limitations of the finite element methods compared are discussed. The results show that the two different finite element formulations calculate similar plastic deformation, but that the residual stresses predicted are larger using the viscoplastic finite element formulation.

1 INTRODUCTION

Shot peening is generally considered as a method to clean dirty and rusty surfaces before painting or other surface treatments. Shot peening also introduces residual stresses beneath the treated surface. It thus improves the fatigue resistance of the structure. It is therefore applied after welding or when compressive residual stresses in a surface region are needed. Finite element simulations are a very powerful tool for simulation of a large number of different metal forming processes. However, shot peening is a very complex process. Several thousands of small shots impact on the surface at a high velocity during shot peening. The size, shape, velocity and the properties of the shots may vary and the interaction between the shot and the workpiece has a great influence on the deformation behavior. Simulations must therefore be limited to a small number of shots that impact on a relatively small surface. Thereby giving a reasonable description of the residual stresses created. It will then be possible to understand the relations between the process parameters and the residual stresses created.

The first step in the simulation of shot peening is the simulation of one shot impacting on a surface. Such simulations have been carried out using an elastoplastic finite element formulation with an implicit time integrator (Al-Obaid 1990).

The elastoplastic finite element formulation can calculate not only plastic deformation in loading, but also residual stress distributions after unloading. However, small deformation increments are required in the simulation process to avoid accumulation of errors. This is even more pronounced in high speed deformation problems due to the high deformation rate. The time steps must therefore be small enough to accurately describe the velocity variations in time.

The explicit central difference method is used for the temporal integration in this paper. This method is only conditionally stable giving an upper limit on the time step size. For solid elements this limit is essentially the time required for an elastic stress wave to propagate across the shortest dimension of the smallest element in the mesh. In steel the acoustic wave speed is about 5200 m/sec. The time steps are by necessity very small. On the other hand, if this requirement is fulfilled, than accuracy is also achieved. The deformation rate is fully resolved.

In the rigid plastic or viscoplastic finite element method the stress components are directly determined in each deformation step without reference to the stress components in the previous steps. The inertia effect is introduced into the functional by using the accelerations derived from the change in velocity in the viscoplastic finite element formulation. Thus, the equilibrium equations obtained from the minimization of the functional are expressed as non-linear simultaneous equations of only the nodal velocities.

The above mentioned finite element formulations are used for simulating the impact between a shot and a surface in this paper. The first is a dynamic viscoplastic finite element formulation where the residual stresses are calculated by a linear elastic finite element code by using the stress distribution obtained from the viscoplastic analysis. These results are taken from Mori et.al(1994). The second is an explicit elastoplastic finite element formulation for the nonlinear transient dynamic response of three-dimensional solids and structures. These simulations are performed for this study. The advantages, disadvantages, accuracy and limitations of the finite element methods compared are discussed.

2 THEORETICAL ANALYSES

The analysis is performed using two different finite element formulations. Two slightly different finite element meshes are used in each code.

2.1 Viscoplastic finite element program

This method is formulated on the basis of the plasticity theory for a material with a slight compressibility (Osakada et al. 1982, Mori et al. 1982). The material is assumed to obey the yield criterion

$$\sigma_e^2 = \frac{3}{2}\sigma_{ij}'\sigma_{ij}' + g\sigma_m^2$$

where σ_{ij}' is the deviatoric stress, g is a small positive constant (0.01-0.0001), σ_m is the hydrostatic stress and σ_e is the effective stress. The effective stress is a function of the strain and the strain rate. The volume of the material is changed during plastic deformation as a function of the pressure sensitivity and thus the stress components $\{\sigma\}$ can be calculated directly from the strain rate components $\{\dot{\varepsilon}\}$ as

$$\{\sigma\} = [D^p]\{\dot{\varepsilon}\}$$

where $[D^p]$ is the constitutive matrix for the viscoplastic deformation. The effective strain rate $\dot{\varepsilon}_e$ for this material is expressed by

$$\dot{\varepsilon}_e^2 = \frac{2}{3}\dot{\varepsilon}_{ij}'\dot{\varepsilon}_{ij}' + \frac{1}{g}\dot{\varepsilon}_v^2$$

where $\dot{\varepsilon}_{ij}'$ is the deviatoric strain rate and $\dot{\varepsilon}_v$ is the volumetric strain rate. The condition of incompressibility is approximately satisfied by using a small positive value of g in the present method.

The residual stresses are calculated using a linear elastic finite element code from the stress distribution for the maximum contact force obtained from the viscoplastic analysis. The plastic deformation calculated by the viscoplastic finite element program after the maximum contact force is small. In static indention by a ball, the indentation force increases in loading and decreases only in unloading. This is also assumed for the shot. The stress distribution at the time of the maximum contact force is elastically released to calculate the residual stress distribution after the rebound of the shot. A more detailed description of the work using this method is presented elsewhere (Mori et al. 1994).

2.2 Elastoplastic finite element program

The finite element program DYNA3D (Hallquist 1989) obtained from Lawrence Livermore National Laboratory, USA, is used in the simulations. DYNA3D is an explicit code for the nonlinear transient dynamic response of three-dimensional solids and structures. It does not form and solve a large system of coupled equations typical of implicit codes and does not require iterations within each time step. The explicit central difference method is used to integrate the equations of motion in time. This method is only conditionally stable and stability is governed by a limit on the time step size. For solid elements this limit is essentially the time required for an elastic stress wave to propagate across the shortest dimension of the smallest element in the mesh. The acoustic wave speed is

$$c = \sqrt{\frac{E}{\rho}}$$

where E is the Young's modulus and ρ is the density. An 8-node solid element is used. All stresses/strains are underintegrated in DYNA3D. Therefore an hourglass viscosity is used to control zero energy modes. The contact algorithms used in DYNA3D are implemented in a way that permits unlimited sliding between the contact surfaces. They are based on a penalty formulation or a Lagrangian projection formulation and implemented in a completely symmetric way. The later means that the results do not depend on which side of the contact is called the master or the slave side. Surfaces may

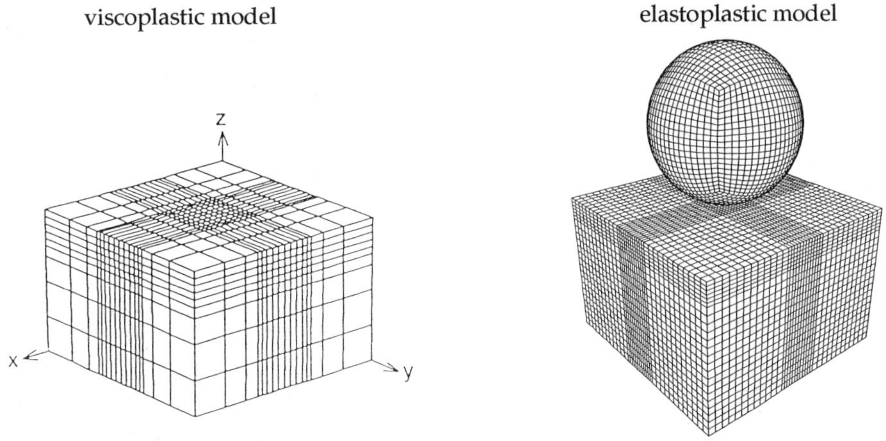

Figure 1. The two finite element meshes used in the simulations.

separate and come together in a completely arbitrary fashion.

2.3 Finite element models

Two slightly different finite element meshes are used in the codes. The first model refers to the model used in the viscoplastic finite element formulation called vp, and the second model refers to the explicit elastoplastic formulation called ep.

The shot impacts perpendicularly on the workpiece. Due to symmetry only a quarter of the shot and a quarter of the workpiece are analyzed. The three-dimensional models can be seen in Figure 1. The first and the second models have 810 and 17615 nodal points and 576 and 15452 elements respectively.

The shot has an initial velocity of 12.2 m/s at the beginning of the simulations. The workpiece in the models is 60.0 mm long, 60.0 mm wide and 40.0 mm thick before the shot impact on the surface. The displacements on the nodes at the bottom of the workpiece are constrained in all directions. The diameter of the shot is 40 mm and the gap between the shot and the workpiece is 1 mm when the simulation starts.

The simulation time is 400 micro seconds. This time is much longer than the time of contact between the shot and the workpiece (see the results of the simulations) but is necessary in the elastoplastic formulation in order to study the elastic stress wave propagation in the material after impact. The friction along the slideline between the shot and the workpiece is modelled as Coulomb friction. The friction coefficient is assumed to be 0.25 in all simulations.

2.4 Material model

The material of both the shot and the workpiece is steel. Young's modulus is assumed to be 210 GPa and Poisson's ratio is set to 0.3. The density of both the shot and the workpiece is set to 7790 kg/m^3.

In the viscoplastic model the flow stress in the material is described by

$$\sigma_y = A\varepsilon_e^{\alpha}\dot{\varepsilon}_e^{\beta}$$

where A is 1150.0, α is 0.11 and β is 0.01. The shot is assumed to be rigid.

In the elastoplastic model the yield condition according to von Mises and the associated flow rule is used. Variable isotropic hardening is assumed. The effective stress-effective plastic strain relation is given from linear interpolation between the values in Table 1. The strain rate dependence for plastic yielding is given by linear interpolation between the values in Table 2.

Table 1. Variable hardening model.

Effective plastic strain	Effective stress [MPa]
0.0	349.7930
0.0017	605.4187
0.0039	645.8964
0.0079	689.0803
0.0153	735.1515
0.0288	784.3029
0.0533	836.7406
0.0975	892.6842

Table 2. Strain rate scale factor.

Effective plastic strain rate [sec^{-1}]	Scale factor for effective stress [-]
0.0	0.9333
0.001	0.9333
0.0027	0.9425
0.0072	0.9519
0.0193	0.9613
0.0518	0.9708
0.1389	0.9805
0.3728	0.9902
1.0	1.0
2.6827	1.0099
7.1969	1.0199
19.307	1.03
51.7947	1.0403
138.9495	1.0506
372.7594	1.061
1000.0	1.0715

The data in Table 1 and Table 2 is optimized in such a way that this model describes the flow stress used in the viscoplastic model as good as possible. The yield stress is set to 349.793 MPa. The shot is assumed to be elastic or elastoplastic. Young's modulus is assumed to be 210 GPa and Poisson's ratio is set to 0.3 in the elastic case. The shot has the same material properties as the workpiece in the elastoplastic case.

3 EXPERIMENTS

All the simulations are compared with experiments performed at the Faculty of Engineering Science at Osaka University, Japan.

A steel sphere with a diameter of 40 mm is dropped from a height of 8 m down on a workpiece with a flat surface. The dimensions of the workpiece are different from those in the calculation. When the dimensions are sufficiently large, the dimensions do not influence plastic deformation. The workpiece is made of steel. The workpiece was placed on a concrete floor for steady support. The shot was removed after the rebound of the shot so the workpiece was only hit once. The velocity is calculated from the height of the free fall.

The shape of the obtained indent and the residual stresses on the surface of the workpiece are measured. The residual stresses are measured using X-ray diffraction method. The error in the measured residual stresses is about 10%. The spot diameter of the X-ray is about 0.15 mm.

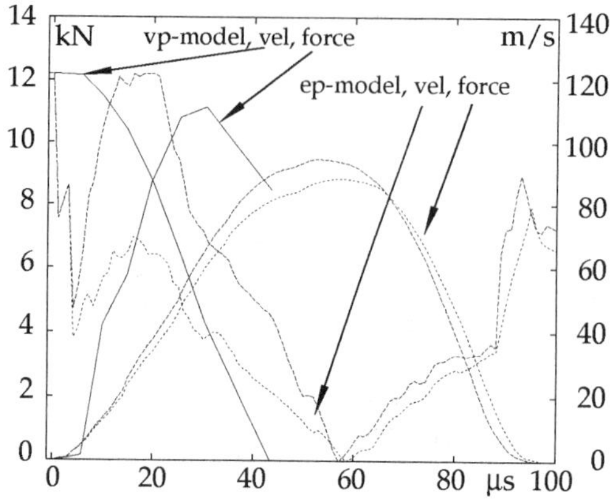

Figure 2. Velocity and contact force.

4 CALCULATED RESULTS

4.1 Computation time

The simulations using the viscoplastic formulation required 19 CPU minutes on a IBM RS 6000/3AT and the calculation of the residual stresses using the linear elastic formulation required 11 CPU seconds on a IBM RS 6000/3AT.

The simulations using the explicit elastoplastic formulation required 76 CPU minutes on a IBM RS6000/590. In this case the total simulation time was 400 microseconds using 8932 time steps.

4.2 Contact forces and velocity of shot

The changes in the velocity and in the total interfacial force for the two formulations are shown in Figure 2. In the viscoplastic model the time of contact is not calculated because it is not meaningful to continue simulation after the time of maximum interface force, but the time of maximum interface force is 30.5 microseconds after the first contact. In the elastoplastic model the time of contact between the shot and workpiece is 94 and 96 microseconds for elastic and elastoplastic shot respectively. The corresponding time of maximum interfacial force is 52 and 56 microseconds after the first contact.

The maximum interfacial force is 111.5 kN in the viscoplastic model. In the elastoplastic model the maximum interfacial force is 94.8 and 88.4 kN for elastic and elastoplastic shot, respectively.

The time of maximum interfacial force is

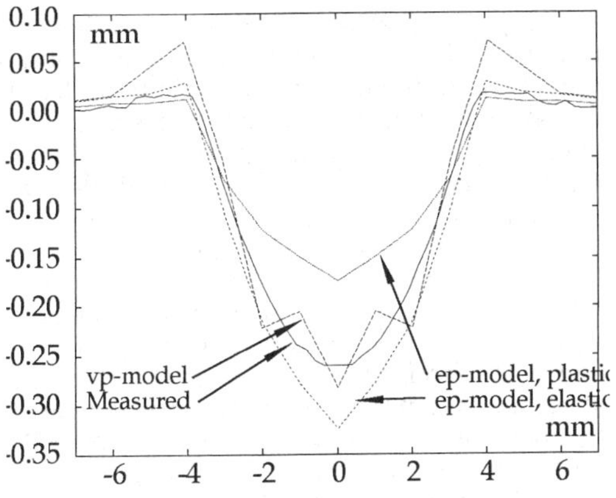

Figure 3. Shape of the deformed contact area.

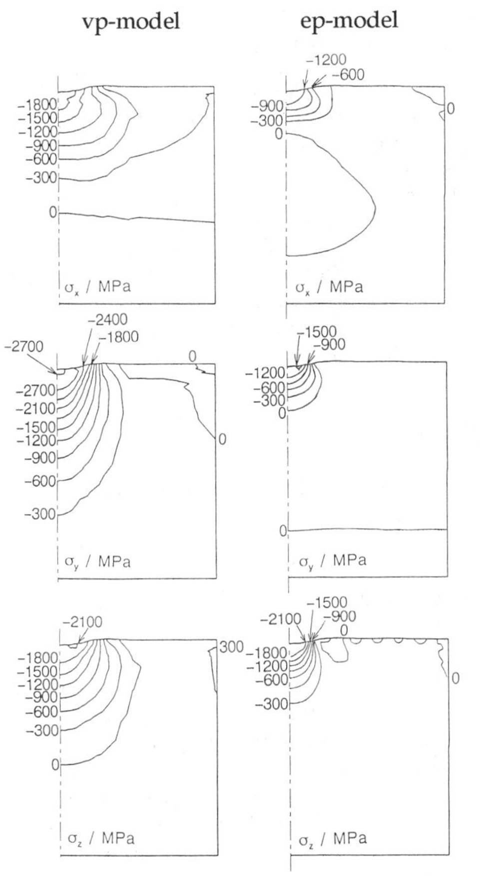

Figure 4. Stress at the maximum contact force.

reached before the velocity changes sign in the viscoplastic model, but in the elastoplastic model the maximum interfacial force is reached at about the same time as the velocity changes sign.

4.3 Shape of the deformed contact area

The deformed shape of the workpiece is shown in Figure 3 together with the experimental results. A round depression is caused by the impact of the shot at the centre of the workpiece. The calculated shapes agree well with the experiments. The elastoplastic model shows the largest difference with 0.325 and 0.173 mm residual deformation at the centre for the elastic and elastoplastic shot respectively. The viscoplastic model has residual deformation of 0.282 mm compared with the measured 0.26 mm.

The radius of the of the deformed zone is about 4 mm for all cases.

4.4 Stress at the maximum contact force

The distribution of the calculated stress components are compared at the time of the maximum contact force even though this time and the maximum contact force differ in the different models. The results are shown in Figure 4.

Both models calculate similar σ_x, σ_y and σ_z. The stresses are higher for the viscoplastic model. The regions of high stress levels are also larger in the viscoplastic model.

4.5 Residual stresses

The distributions of the calculated residual stress components are shown in Figure 5.

In this case both models also calculate similar σ_x, σ_y and σ_z. Again the stresses are higher for the viscoplastic model. The regions of high residual stresses are also larger in that model.

The viscoplastic model predicts larger negative σ_x-values on the surface than the elastoplastic model but the slopes of the curves are quite similar (not shown). The predicted σ_y-values on the surface are very similar for the two models.

5 COMPARISONS AND CONCLUSIONS

The most important difference between the two models is the size of the region in which there is

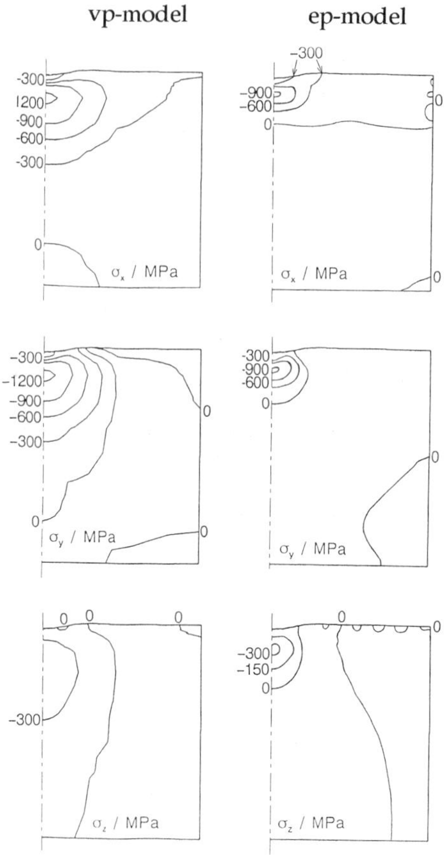

Figure 5. Residual stresses.

high residual stresses. This region is larger in the viscoplastic model. The size of the region in which there are high stress levels at maximum contact force is also larger in that model. This can probably be explained by differences in how the material is treated in the two models. In the viscoplastic model, the material is treated as a viscoplastic medium that has no yield strength. This means that there have to be plastic deformations in a larger region before the material can carry any loads. In the elastoplastic model there is a distinct point of yielding and the material can carry loads even if there is no plastic deformation. In this case the stresses are removed after unloading if there is no plastic deformation, but in the viscoplastic model all deformation is assumed to be plastic and therefore they contribute to the residual stress field. In the viscoplastic model it is hard to distinguish when the unloading starts, and therefore this model is a little more uncertain. The elastoplastic model actually shows that for some elements close to the area of contact, there is a very small plastic deformation after the time of maximum contact force. This of course has effects on the residual stress distribution. There is also a risk of overestimating the residual stress levels in the viscoplastic model due to the fact that there is no check if the residual effective stresses exceed the yield stress of the material.

There are also some important differences in the velocities and contact forces between the two models. The time of maximum contact force and the time when the shot velocity is zero are almost identical for the elastoplastic model. This is not the case in the viscoplastic model where the maximum contact force is reached about 10 microseconds before the time when the shot velocity is zero. The contact force is also higher and the contact time is shorter in the viscoplastic model.

For the two elastoplastic models there are some differences, especially in the final deformation, depending on the properties of the shot. If the shot and the workpiece are made of the same material, it is important to include the plastic deformation of the shot. This will be less important if the shot is harder than the workpiece.

The study performed shows that about the same accuracy can be obtained with the approach by Mori et.al(1994) as with a more general approach as in Dyna3d. The comparison is made for one working condition, so it is desirable to make the comparison for more working conditions.

REFERENCES

Al-Obaid, Y.F. 1990. Three-dimensional dynamic finite element analysis for shot peening mechanics. *Computers and Structures*. Vol. 36, pp. 681-689.

Hallquist, J O 1989. DYNA3D-Users manual. Report UCID-19592. Rev 5. Lawrence Livermore National Laboratory. USA.

Mori, K. K Osakada & T Oda 1982. Simulation of plane strain rolling by the rigid-plastic finite element method. *Journal of Mechanical Engineering Science*. Vol. 24, pp. 519-527.

Mori, K. & K Osakada 1994. Application of dynamic viscoplastic finite element method to shot-peening process. *Transactions of NAMRI/SME*. Vol. XXII, pp. 29-34.

Osakada, K. J Nakano & K Mori 1982. Finite element method for rigid-plastic analysis of metal forming - Formulation for finite deformation. *Journal of Mechanical Engineering Science*. Vol. 24, pp. 459-468.

Simulation of Materials Processing: Theory, Methods and Applications, Shen & Dawson (eds)
© 1995 Balkema, Rotterdam. ISBN 90 5410 553 4

Incremental mass conservation and adaptive remeshing for the thermo-mechanical coupling between workpiece and tool in non steady-state metal forming

L. Fourment, M.P. Miles & J.L. Chenot
CEMEF, URA CNRS, École des Mines de Paris, Sophia-Antipolis, France

ABSTRACT: For improving the accuracy of the finite element simulation of non steady-state forming processes we study two methods. Firs, we tackle the numerical mass variations which are generated by the time integration scheme and by the contact algorithm. An incremental mass conservation method is proposed. It is applied to an industrial backward extrusion problem. Then, we tackle the problem of mesh adaption for controlling the discretization error during non-steady forming processes. Zienkiewicz-Zhu error estimators are used, together with a Delaunay adaptive mesh generator. The procedure is applied to orthogonal cutting.

1 THERMO-MECHANICAL PROBLEM

1.1 *Problem statement*

The material is assumed to obey the isotropic Norton-Hoff constitutive equation:

$$s = \mu_{vp}\left(\dot{\varepsilon}\right)\dot{\varepsilon} = 2K\left(2\,\dot{\varepsilon}:\dot{\varepsilon}\right)^{\frac{m-1}{2}}\dot{\varepsilon} \qquad (1)$$

$$\text{where:}\quad K = K_0\left(\bar{\varepsilon}_0 + \bar{\varepsilon}\right)^n \exp\left(\frac{\beta}{T}\right) \qquad (2)$$

where s is the deviatoric stress tensor, $\dot{\varepsilon}$ is the strain rate tensor, $\bar{\varepsilon}$ is the cumulated strain and T is the temperature. The plastic deformation being incompressible, the mass conservation equation is written:

$$\text{div}(v) - 3\alpha_T\dot{T} = \text{tr}(\dot{\varepsilon}) - 3\alpha_T\frac{dT}{dt} = 0 \qquad (3)$$

where v is the velocity field and α_T is the thermal dilatation coefficient. At the interface between the workpiece and the tool, the contact condition is unilateral and a power law friction is considered:

$$\tau = \sigma n\,.\,t = -\,\alpha\,K\,\|\,\Delta v_t\,\|^{q-1}\,\Delta v_t \qquad (4)$$

where Δv_t is the tangential relative velocity at the interface.

The forming tools are assumed to remain elastic during the process:

$$\sigma = 2\mu\,\varepsilon + \lambda\,\text{tr}(\varepsilon)\,\mathbb{1} - \alpha_T\,(2\mu+3\lambda)\,(T-T_0)\,\mathbb{1} \qquad (5)$$

where ε is the strain tensor, λ and μ are the Lamé coefficients.

The thermal equation is written on any domain:

$$\rho\,c\,\frac{dT}{dt} = \text{div}(\,k\,\text{grad}(T)\,) + \sigma:\dot{\varepsilon} \qquad (6)$$

Boundary conditions are imposed on the free surface, where radiation and convection are dominant, and on the contact surface, where there is conduction as well as friction heat flux:

$$k\,\text{grad}(T)\,.\,n = h_c\,(T-T_a) + \frac{b}{b+b_a}\,\alpha\,K\,\|\Delta v_t\|^{q+1} \qquad (7)$$

where T_a and b_a are respectively the temperature and the effusivity of the adjacent part.

1.2 *Resolution method*

The mechanical problem is solved using a velocity based approach, as described by Bohatier et al. (1985). Meanwhile here, a mixed velocity-pressure formulation is considered. Then the equilibrium and the mass conservation equations are written:

$$\left\{\begin{array}{l} \displaystyle\int_\Omega s:\dot{\varepsilon}^*\,dw - \int_\Omega p\,\text{tr}(\dot{\varepsilon}^*)\,dw \\ \qquad\qquad \displaystyle - \int_{\partial\Omega c} \tau.v^*\,ds = 0 \\ \displaystyle\int_\Omega p^*\left(\text{tr}(\dot{\varepsilon}) - 3\alpha_T\dot{T}\right)dw = 0 \end{array}\right. \qquad (8)$$

The axisymmetrical problem is discretized using the standard finite element method. P2+P1 (quadratic+bubble velocity, linear pressure) triangular elements are used. Time integration is based up- one step explicit scheme:

$$X_{t+\Delta t} = X_t + V_t\,\Delta t$$

The unilateral contact co- nodes of the mesh. I-time t+Δt, it is orthogonaly projected on the tool surface. Accordingly, a node which belongs to the contact surface at time t is reprojected on the tool surface at time t+Δt, in order to remain in contact.

The classical Galerkin method makes it possible to solve the thermal problem and the resolution of the differential equation is based on a Crank-Nicholson scheme. In case of thermal shock, the heat capacity matrix is lumped.

The equilibrium equations are solved independently on the different domains, using a weak coupling algorithm: first the tools are assumed to be rigid and the deformation of the part is calculated, then the deformation of the tools is computed in order to balance the normal stresses which have been calculated on the part. On the other hand, the thermal equations are strongly coupled and a fixed point algorithm allows for iteratively computing the temperature on the different domains (Miles et al. (1994)).

2. INCREMENTAL MASS CONSERVATION

2.1 *The incremental mass conservation equation*

Several reasons can explain the numerical mass variations during simulation: the time discretization scheme (if $\alpha_T=0$, the velocity field v_t is incompressible but the transformation from t to t+Δt is not a constant volume transformation), the various projections due to the contact algorithm and to remeshing (the new domain does not exactly cover the initial domain). In order to tackle the first and main cause of volume losses, Coupez et al. (1990) have proposed the *incremental incompressibility method*. Then, reprojection errors were minimized by adapting the time step to contact, so reducing the penetration of the mesh into the tools. This way, the volume losses are less than 0.5%. Meanwhile, sometimes it is still too much. Therefore, we propose a generalization of this method for compressible materials and which allows for taking into account the volume losses due to projections.

ρ_t denotes the actual density of the material, at time t. Numerical densities, θ_t, are introduced to take into account the numerical variation of mass. It is defined according to:

$$\theta_t\, dw_t = \theta_0\, dw_0 \qquad (10)$$

where θ_0 is an initial numerical density (it is equal to 1 at the begining of the process). Then, the mass conservation equation is written:

$$dm_{t+\Delta t} = \rho_{t+\Delta t}\, dw_{t+\Delta t} = \rho_t\, dw_t \qquad (11)$$

·t: $$dw_{t+\Delta t} = J_t^{t+\Delta t}\, dw_t \qquad (12)$$

$$J_t^{t+\Delta t} = \det\left(\frac{\partial x_{t+\Delta t}}{\partial x_t}\right) \qquad (13)$$

according to (9): $J_t^{t+\Delta t} = 1 + \Delta t\, \mathrm{div}(v_t) + O(\Delta t^2)$ (14)

Taking into account the thermal dilatation of the material (3), (11-14) allows for writting:

$$\rho_{t+\Delta t} = \frac{\rho_t}{1 + 3\alpha_T \dot{T}\, \Delta t} \qquad (15)$$

On the other hand, equations (11-14) can also be written for the numerical densities:

$$\theta_{t+\Delta t} = \frac{\theta_t}{1 + \Delta t\, \mathrm{div}(v_t)} \qquad (16)$$

and according to (11-12): $\theta_t\, dw_t = \theta_0\, J_t^0\, dw_t$ (17)

Finally, we write that the numerical densities should be equal to the actual densities at time t+Δt:

$$\rho_{t+\Delta t} = \theta_{t+\Delta t} \qquad (18)$$

Then the *incremental mass conservation* equation is obtained:

$$\mathrm{div}(v_t) = \frac{1}{\Delta t}\left(\frac{\theta_0\, J_t^0}{\rho_t}\left(1 + 3\alpha_T \dot{T}\, \Delta t\right) - 1\right) \qquad (19)$$

This equation is substituted to (3) in (8b) and it is solved in a similar way. After remeshing, the actual densities are transfered onto the new mesh. Then they are noted ρ_0. The initial numerical density, θ_0, is calculated from the actual mass of the workpiece, M, and from the mass conservation equation:

$$M = \theta_0 \int_\Omega \rho_0\, dw_0 \qquad (20)$$

If we want to avoid the transfer of the actual densities onto the new mesh, then the actual volume before remeshing has to be computed:

$$V_{act} = \int_\Omega \frac{\theta_t}{\rho_t}\, dw_t \qquad (21)$$

Lastly, the initial numerical density is calculated from the volume of the mesh after remeshing, V_{rem}:

$$\theta_0 = \frac{V_{act}}{V_{rem}} \qquad (22)$$

2.2 Application to a backward extrusion problem

For a backward extrusion problem of a viscoplastic material in hot conditions, various algorithms are compared: a1/ (α_T=0) classical incompressibility algorithm, a2/ (α_T=0) incremental volume/mass conservation algorithm, a3/ ($\alpha_T=10^{-5}$ and $\alpha_T=5\ 10^{-5}$) incremental mass conservation algorithm. The workpiece and tools respectively are heated at 1050°C and 300°C. Initial and final configurations are given in figure 1. 30 remeshings have been required.

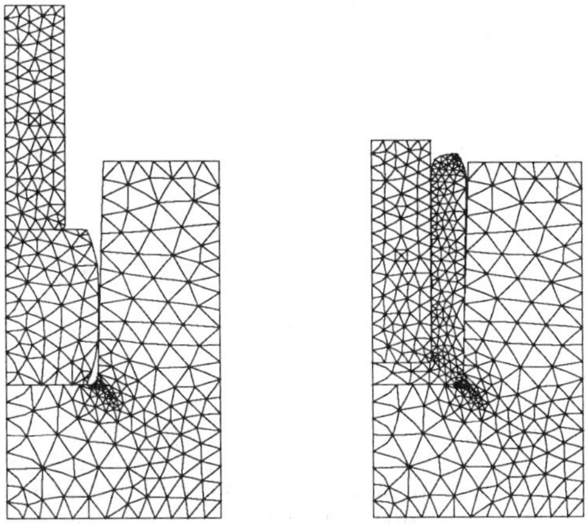

figure 1: meshes of the initial and final configurations, for the backward extrusion process

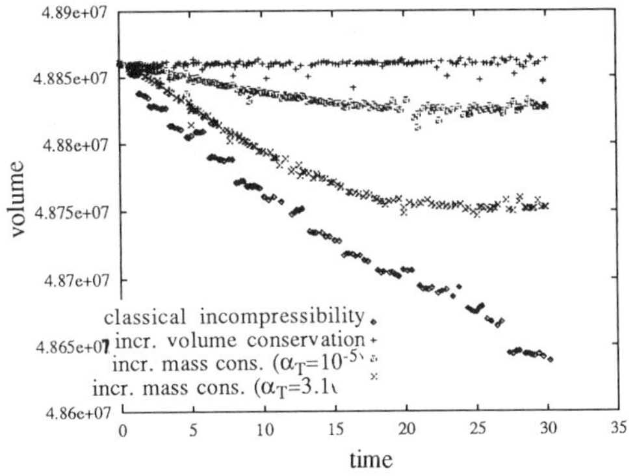

figure 2: volume variations during forging for the various algorithms (a1/, a2/, a3/) and for both values of α_T ($\alpha_T=10^{-5}$ and $\alpha_T=5\ 10^{-5}$)

Figure 2 shows the volume variations versus time. Using a1/, the final volume loss is equal to 0.455% ; a2/ makes it possible to reduce this value to 0.00572% - a perfect mass conservation is achieved. Taking into account the thermal dilatation of the material and using a3/, for both values of α_T, the final volume variations are -0.070% and -0.221% respectively. The ratio of these values is almost equal to the ratio of the α_T coefficients. In figure 2, oscillations can be noticed for a2/ and a3/. They correspond to the linearization of equation (14) and to the projection of nodes in contact. The volume loss of a1 is greater than the volume variation due to the thermal dilation of the material during forging which shows the relevance of considering the mass conservation equation in an incremental form.

3. DISCRETIZATION ERROR CONTROL

3.1 *Error estimators*

The discretization error is defined in terms of an energy norm for both the elasticity and plasticity problems. We use the Z^2 error estimators which have been proposed by Zienkiewicz et al. (1987) for elasticity problems (23) and by Zienkiewicz et al. (1988) for viscoplasticity problems (24):

$$\Theta_{el}=\left(\int_\Omega (\tilde{\sigma}_h-\sigma_h):(\tilde{\varepsilon}_h-\varepsilon_h)\ dw\right)^{1/2} \qquad (23)$$

$$\Theta_{vp}=\left(\int_\Omega \mu_{vp}(s_h)^{-1}(\tilde{s}_h-s_h):(\tilde{s}_h-s_h)\ dw\right)^{1/2} \qquad (24)$$

$\tilde{\sigma}_h$, $\tilde{\varepsilon}_h$, $\tilde{s}_h$ are continuous stress, strain and deviatoric stress tensors. Fourment et al. (1995) have numerically shown that the Orkisz smoothing method (which is described, for instance, in Fourment et al. (1994)) improves the accuracy of the Θ_{el} and Θ_{vp} estimations, and that the error due to estimation is less than 5% in elasticity and less than 10% in viscoplasticity.

3.2 *Adaptive remeshing*

When the estimated error Θ_{el} (for the tools) or Θ_{vp} (for the workpiece) is greater than a prescribed value Θ_p, a totally new mesh is generated (either for the tool or for the worpiece, or for both) and the calculations are repeated before moving to the next time step. An adaptive Delaunay mesh generator is used, as described by Fourment et al. (1994). The optimal element sizes h_n^e of the new mesh are functions of Θ_p, Θ_e (the contribution of element e to Θ), h_p^e (the element size of element e of the present mesh) and p (the convergence rate of the finite element method, which is taken here as 2):

$$h_n^e = h_p^e\, \Theta_p^{\frac{1}{p}}\, \Theta_e^{\frac{-1}{p+1}} \left(\sum_e \Theta_e^{\frac{2}{p+1}}\right)^{\frac{-1}{2p}} \qquad (25)$$

After remeshing, if the estimated error is still greater than Θ_p, then the procedure is repeated. It allows for solving pseudo-singular problems for which the convergence rate of the finite element method is not uniform.

3.3 *Application to orthogonal cutting*

Sekhon et al. (1993) have shown that non-steady orthogonal cutting can be simulated using a viscoplastic model. Standard material data for steel are used: K=125 10^3 kPa, m=0.1, α=0.5, p=0.1. Thermo-mechanical coupling between the workpiece and the cutting tool is considered. The process data

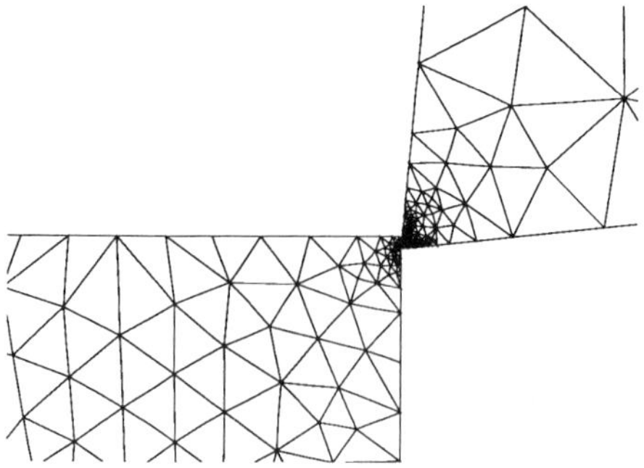

figure 3: beginning of the cutting simulation

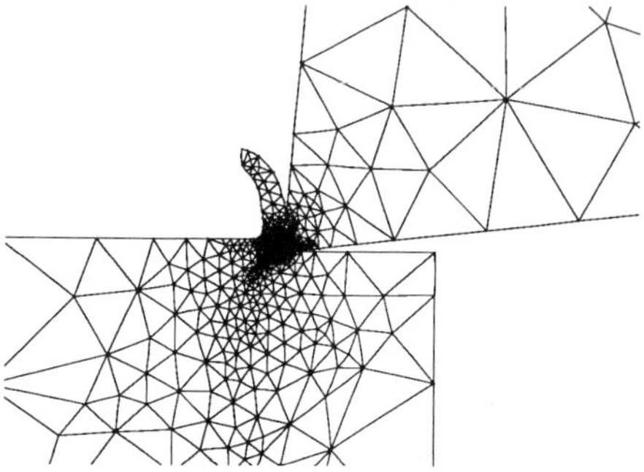

figure 4: formed chip during the quasi-steady state of the cutting simulation

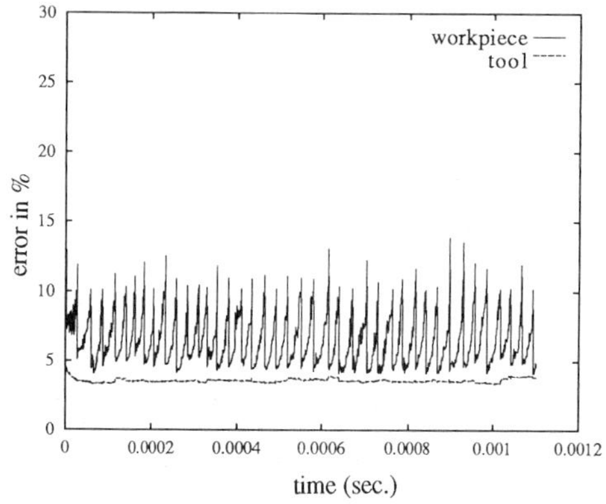

figure 5: estimated error versus time for the workpiece and the cutting tool

are the followings: cutting velocity=200m/mn, cutting angle=5°, cutting depth=0.4mm, radius of cutting edge=0.8mm (see figure 3). A total length of 5mm is machined before quasi-steady conditions are reached (figure 4). The error control procedure makes it possible to maintain the discretization error under

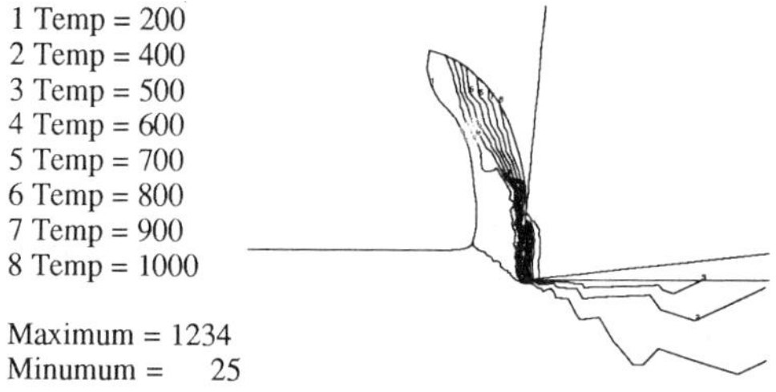

figure 6: isovalues of the temperature

10% for the workpiece and the cutting tool (figure 5). Isovalues of the equivalent strain rate shows a highly localized deformation zone. The tool temperature (figure 6).has a maximum value of 1220°C, which is in good agreement with experimental measurements.

ACKNOWLEDGEMENT

This work has been funded by the French Forging Profession and its Technical Center (CETIM).

REFERENCES

Bohatier C., Chenot J.L. 1985. Finite element formulation for non-steady state viscoplastic deformation. Int. J. Num. Eng. 21, 1697-1708.

Coupez T., Chenot J.L. 1990. Incremental volume control in F.E. calculation in non steady-state incompressible flow. Numerical Method in Engineering, Pande et al. (editors), Elsevier. 894-901.

Fourment L., Chenot J.L. 1995. Error estimators for viscoplastic materials: application to forming processes. Eng. Comp. to appear.

Fourment L., Chenot J.L. 1994. Adaptive remeshing and error control for forming processes. Revue Européenne des éléments finis. 3, n°2, 247-279.

Miles M.P., Fourment L., Chenot J.L. 1994. Calculation of tool temperature during periodic non steady-state metal forging. J. Material Processing Technology. 45, 643-648.

Sekhon G.S., Chenot J.L. 1993. Numerical simulation of continuous chip formation during non-steady orthogonal cutting. Eng. Comp. 10, 31-48.

Zienkiewicz O.C., Liu Y.C., Huang G.C. 1988. Error estimation and adaptivity in flow formulation for forming problem. Int. J. Num. Met. Eng. 25, 23-42.

Zienkiewicz O.C., Zhu J.Z. 1987. A simple error estimator and adaptive procedures for practical engineering analysis. Int. J. Num. Met. Eng. 24, 337-357.

Simulation of Materials Processing: Theory, Methods and Applications, Shen & Dawson (eds)
© 1995 Balkema, Rotterdam. ISBN 90 5410 553 4

Finite element analysis of thermomechanical coupling in finite deformation elasto-plasticity with applications to hot forming

J.C.Gelin & B.V.Nguegang
Laboratoire de Mécanique Appliquée R. Chaléat, URA CNRS, Université de Franche-Comté, Besançon, France

ABSTRACT: In the present paper, a strategy for solving transient thermoplastic problems is described. The heat transfer equation is constructed using a thermodynamic theory of deformation with the concept of internal variables. The general thermal and mechanical coupling phenomena are discussed by assuming material properties to be temperature dependent and accounting for the mechanical terms in non-stationary heat conduction equation. Time integration is carried out with the implicit scheme. Finally, a global product coupling algorithm, in which the unknown nodal displacements and temperature are solved with a two-step solution procedure, is suggested. The proposed fractional step method partitions the initial coupled problem into an isothermal elasto-plastic problem followed by a heat conduction problem at fixed configuration. Numerical applications involving full thermomechanical coupling conclude the paper.

1 INTRODUCTION

At elevated temperatures, metals have greater ductility and lower flow stresses. The workpiece initially heated at high temperature, cools during the forming processes but not uniformly, due to the heat conduction and to the heat radiation or convection with the dies and environment. On the other hand, a significant variation of the temperature field occurs mainly due to the conversion of the mechanical energy dissipation into heat, even in the cold working processes. Since the material properties are temperature-dependent, the analysis of the deformation during hot forming should be coupled with the heat transfer analysis.

In this paper, a finite element technique is presented using an updated Lagrangian approach to incrementally solve the quasistatic large displacement elastic-plastic problem, coupled with the thermal problem. The current configuration corresponds to the configuration at the begining of each increment and each iteration.

2 CONSTITUTIVE LAWS

Considering large elasto-plastic deformations, leads to the multiplicative split of the deformation gradient F into an elastic part F^e and a plastic part F^p (Lee 1969);

$$F = F^e F^p \tag{2.1}$$

The theory is restricted to elastic-plastic, isotropic material properties. To avoid unnecessary complexity, attention is focussed on temperature-independent heat capacity, conductivity and elastic coefficients. The introduction of temperature-dependent material properties, leading to additional non-linearities, does not present fundamental difficulties. The stress-strain relation chosen can be deduced from the expression of the free energy (Sidoroff 1982). Adopting a logarithmic measure e^e for the elastic strain (Sidoroff 1982, Gelin and Lochegnies 1990).

$$e^e = \frac{1}{2} \ln B^e \tag{2.2}$$

where $B = F^e F^{eT}$ is the elastic left Cauchy Green tensor, we obtain:

$$\sigma = C : e^e - 3\alpha\kappa (T - T_0) \mathbf{1} \tag{2.3}$$

where σ is the Cauchy stress tensor, α the coefficient of thermal expansion, κ the bulk modulus, T the absolute temperature, T_0 a given reference temperature, C the elastic tensor and **1** the second order identity tensor.

$$C = \frac{E}{1+\nu} \left(\mathbf{I} + \frac{\nu}{1-2\nu} \mathbf{1} \otimes \mathbf{1} \right) \qquad \kappa = \frac{E}{3(1-2\nu)} \tag{2.4}$$

where E is the Young modulus, ν the Poisson's coefficient and **I** the fourth order identity tensor. The constitutive assumption for heat conduction is given by Fourier's law. For isotropic materials, it is given by:

$$q = -k \operatorname{grad} T \qquad (2.5)$$

where q is the heat flux and k the conductivity coefficient. For the elastoplastic constitutive law, using standard notations we can formulate the von Mises flow criterion in terms of Cauchy stress tensor:

$$f = \|\operatorname{dev}\sigma\| - \sqrt{\frac{2}{3}}\sigma_0(\bar{\varepsilon}^p, T) \le 0 \qquad (2.6)$$

where isotropic strain hardening depending on the the equivalent plastic strain $\bar{\varepsilon}^p$ and thermal softening due to the temperature change T is assumed, $\|\operatorname{dev}\sigma\|$ is the norm of the Cauchy stress tensor. From the macroscopic normality rule, the plastic strain rate takes the form

$$D^p = \dot{\lambda}\frac{\partial f}{\partial \sigma} \qquad (2.7)$$

where $\dot{\lambda}$ is the plastic consistency parameter.

3 VARIATIONAL FORMULATION

The local form of the balance of momentum can be stated for the quasistatic case as follows:

$$\operatorname{div}\sigma + b = 0 \qquad (3.1)$$

where b is a given body force.
The thermal effects enter into the formulation via the balance of internal energy which in combination with Fourier's law leads to the thermal equation in differential form:

$$\rho c \dot{T} - \operatorname{div}(k \operatorname{grad} T) = \chi D_m - H + r \qquad (3.2)$$

with the mass density ρ, the heat supply per unit mass r, the specific heat capacity c and the Taylor-Quinney coefficient χ (Taylor and Quinney 1934, Farren and Taylor 1925) indicating how much of the plastic power D_m is converted into heat. The plastic dissipation can be shown to be a function of stress and plastic strain rate tensors as:

$$D_m = \sigma : D^p \qquad (3.3)$$

while the thermoelastic coupling term given by:

$$H = 3\alpha\kappa T \mathbf{1} : D^e \qquad (3.4)$$

depends on the elastic strain rate D^e.

Using the Galerkin formulation, the weak forms of the thermal and mechanical equations are as follows:

$$\begin{aligned} &\int_V \sigma : \nabla^s \eta \, dv - \int_V b\eta \, dv - \int_{\Gamma_t} \bar{\tau}.\eta \, ds = 0 \\ &\int_V \left[\rho c \dot{T} - \chi D_m + H\right] \omega \, dv + \\ &\int_V k \nabla\omega \nabla T \, dv - \int_V r\omega \, dv + \int_{\Gamma_q} \bar{q}\omega \, ds = 0 \end{aligned} \qquad (3.5)$$

here $\bar{\tau}$ is the traction vector specified on Γ_t of the boundary Γ and $\bar{q}$ the normal heat flux through the part Γ_q of Γ, η is an homogeneous displacement field associated to the displacement field u and ω an homogeneous temperature field associated to the temperature field T. System of equations (3.5) incorporates the following different thermomechanical coupling effects: the influence of plastic dissipation on the change of temperature, the influence of deformation on heat conduction and the influence of temperature on the stresses via the constitutive relation (2.3). The discretization of this fully coupled problem is made by the finite element method (Zienkiewicz 1977) and the solution of the resulting algebraic system of equations obtained by the product algorithm addressed in section 5.

4 TIME INTEGRATION

The integration of the plastic flow rule is done by the classical radial return mapping (Simo and Taylor 1986). The incremental plastic consistency parameter is determined by satisfying the yield condition (2.6) at the end of each increment. This leads to finding the root of a nonlinear equation which can be easily solved by Newton's method.
For the integration of the heat evolution equation, the Euler family method (Hughes 1983) is employed. It consists in the following equations

$$T_{n+1} = T_n + [(1-\theta)\dot{T}_n + \theta \dot{T}_{n+1}] \qquad (4.1)$$

where n denotes the time step number, Δt the time increment between t_n and t_{n+1}, $\theta \in [0, 1]$, $\theta = 0$ corresponds to the Euler forward integration scheme while $\theta = 1$ deals with the Euler backward scheme.

5 GLOBAL PRODUCT COUPLING ALGORITHM

The non linear system of field equations is discretized using the finite element method and the resulting system of equations is solved in two steps with Newton's method. The steps to be performed in the proposed product algorithm (Simo 1992) are as follows.

1- Initial conditions at t_n

Internal variables at quadrature points $\bar{\varepsilon}^p_n$.

initial conditions at nodal points $\{u_n\}$, $\{T_n\}$.

2- Mechanical step at fixed initial temperature:

let $\{u^i_{n+1}\}$ and $\{T_n\}$ be the solution at the i-iteration.

2.1 compute Cauchy stress deviator and incremental consistency parameter $\Delta\lambda^i_{n+1}$, and update the equivalent plastic deformation $\bar{\varepsilon}^{pi}_{n+1}$ by the return mapping algorithm at quadrature points;

2.2 compute the exact hessian of the mechanical part of (3.5) $[K_{mm}]$ and the corresponding mechanical residual $\{R_m\}$.

IF $\|R_m\| \geq$ TOL THEN: Continue iteration and update

$$u^{i+1}_{n+1} = u^i_{n+1} - \left[K^i_{mm}\right]^{-1}_{n+1} \{R^i_m\}_{n+1}$$

set i = i + 1 and go to 2.1.

ELSE update

$$u_{n+1} = u^i_{n+1}, \; \bar{\varepsilon}^p_{n+1} = \bar{\varepsilon}^{pi}_{n+1}$$

END IF

3- Thermal step at fixed updated configuration:

let $\{u_{n+1}\}$ and $\{T^i_{n+1}\}$ be the solution at the i-iteration

3.1 Compute Cauchy stress deviator and incremental consistency parameter $\Delta\lambda^i_{n+1}$, and update the equivalent plastic deformation $\bar{\varepsilon}^{pi}_{n+1}$ by the return mapping algorithm at quadrature points; then compute the internal dissipation and the thermoelastic coupling energy;

3.2 Compute the exact hessian of the thermal part of (3.5) $[K_{tt}]$ and the corresponding thermal residual $\{R_t\}$.

IF $\|R_t\| \geq$ TOL THEN: Continue iteration and update

$$T^{i+1}_{n+1} = T^i_{n+1} - \left[K^i_{tt}\right]^{-1}_{n+1} \{R^i_{tt}\}_{n+1}$$

set i = i + 1 and go to 3.1.

ELSE update

$$\bar{\varepsilon}^p_{n+1} = \bar{\varepsilon}^{pi}_{n+1}, \; T_{n+1} = T^i_{n+1}$$

END IF

The sequence of calculations to be performed in one time step are as follows. The first iterative loop solves the mechanical equations using the temperature known from the last converged step; then a loop over the thermal equations at fixed configuration is performed until convergence. Since the plastic dissipation is a function of the stress tensor the return mapping algorithm is included in the thermal code for its computation during the thermal iteration process.

6 NUMERICAL RESULTS

In the three numerical simulations that follows, the isotropic hardening law is given by the following general expression.

$$\sigma_0(\bar{\varepsilon}^p, T) = (k_0 + k_1\bar{\varepsilon}^p + k_2 A(\bar{\varepsilon}^p))\, B(T)$$
$$A(\bar{\varepsilon}^p) = 1 - \mathrm{Exp}(-\delta\bar{\varepsilon}^p) \quad B(T) = 1 - \omega_\Theta (T - T_0) \qquad (6.1)$$

k_0 is the initial yield stress, k_1 the linear hardening modulus, ω_Θ the thermal softening modulus, δ a hardening exponent and k_2 the saturation hardening modulus.

6.1 Plane strain tensile of a rectangular bar

The problem corresponds to the plane strain tensile test of a rectangular bar. The specimen considered in the numerical simulations has a width of 12.83 mm and a length of 53.33 mm, plane strain conditions are assumed. As thermal boundary conditions, the bar is assumed to be insulated along its lateral face, while the temperature is kept constant to the reference value T_0 on the upper and lower faces. Because of the symmetry, only the top rigth quadrant of the bar is analyzed, with the appropriate mechanical symmetry boundary conditions imposed. The material properties are listed in table 1. Figure 1 shows the geometry, the initial mesh and thermal boundary conditions for the plane strain tensile of a rectangular bar.

Table 1. Material properties for the tensile of a rectangular bar

E	206899.9 N/mm^2	ρ	7.8e-9 Ns^2/mm^4
ν	0.29	k	45 N/sK
k_0	450 N/mm^2	α	0.00001K^{-1}
k_1	129.24 N/mm^2	χ	0.9
k_2	715 N/mm^2	ω_Θ	0.002 K^{-1}
δ	16.93	T_0	293 K
c	0.46e9 mm^2/s^{2}K		

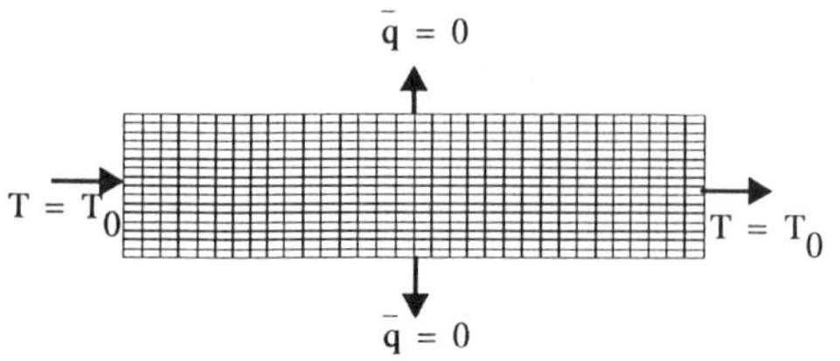

Figure 1. Initial mesh, geometry and thermal boundary conditions for the tensile test of a rectangular bar

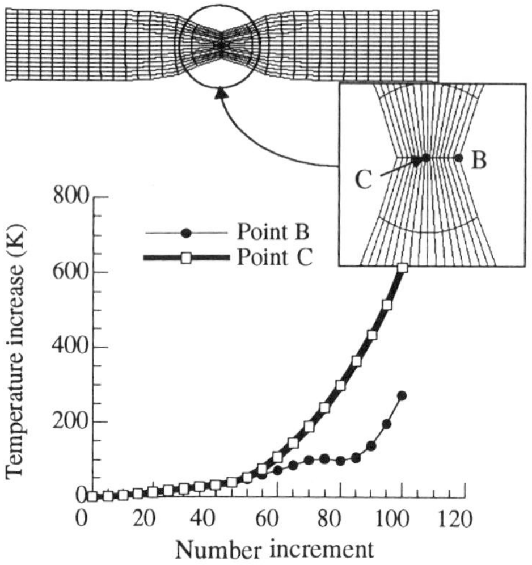

Figure 2. Evolution of the temperature increase at points B and C of the deformed mesh

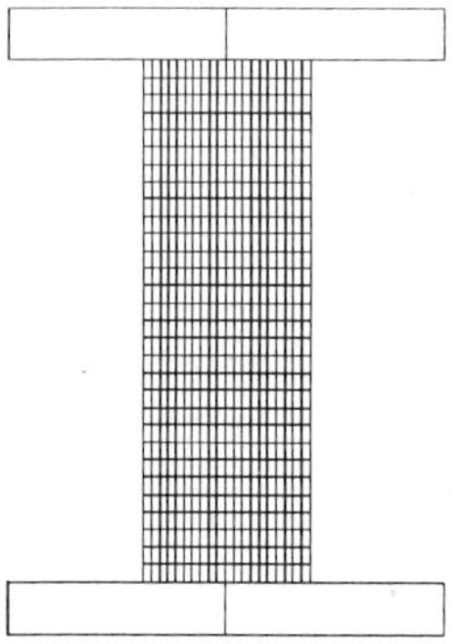

Figure 3. Initial mesh and geometry for the upsetting of a cylinder between two rigid plattens.

As time evolves, the structure is stretched with constant velocity v = 0.1 m/s applied in 100 equal increments until a final elongation of 10 mm is attained. For the thermal phase, the scalar integration coefficient taken is 0.5. In contrast to purely mechanical simulation (Argyris and Doltsinis 1979), no imperfection is needed to achieve shear banding. Shear bands appear systematically due to non uniform temperature distribution caused by heat conduction in the bar. To illustrate this physical phenomenon, we show in Figure 2 the evolution of the temperature distribution at the middle of the bar (points B and C). Our results are in agreement with results found in (Simo and Armero 1992) where the J_2-elastoplasticity model is used to simulate plane strain shear banding.

Table 2. Material properties used for the upsetting test

E	200000 N/mm^2	c	0.46e9 mm^2/s^{2}K
ν	0.3	ρ	7.8e-9 Ns^2/mm^4
k_0	300 N/mm^2	k	58 N/sK
k_1	500 N/mm^2	α	0.00001 K^{-1}
k_2	0 N/mm^2	χ	0.9
δ	16.93	ω_Θ	0.003 K^{-1}

6.2 The upsetting test

A billet with a diameter D = 10 mm and a heigth H = 30 mm is compressed between two flat parallel plattens moving towards each other with the same constant velocity. Sticking friction conditions are assumed between the specimen and the dies. Material parameters are listed in table 2.

Due to symmetry conditions only a quarter of the problem is meshed. As thermal boundary conditions, the system is assumed to be thermally insulated thus preventing heat exchange on the outer surface. Figure 3 shows the finite element discretization used. For the thermal problem, the simulation was carried out with the backward Euler time integration scheme which gives better results than other schemes. As in the previous example, the large plastic deformation cause a self heating of the specimen. The contour lines of the temperature increase and the equivalent plastic strain on the deformed geometry are shown in figure 4 and figure 5 for a compression of 66.67% attained in a total process time of t = 1s.

6.3 Axisymmetric extrusion of a billet

In this example of metal forming the extrusion of a billet through a conical die having a semi curve angle of 15° is investigated. The geometry of the problem is illustrated in figure 6.

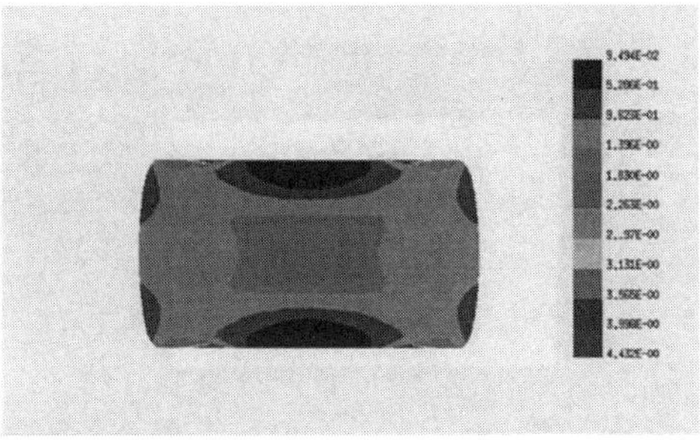

Figure 4. Contour lines of equivalent plastic strain on the deformed mesh

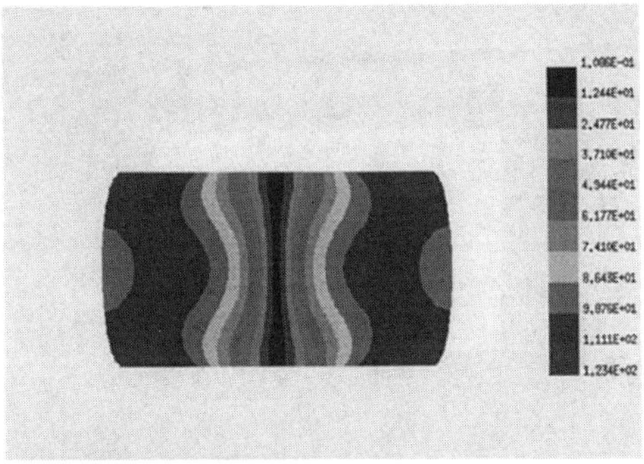

Figure 5. Contour lines of temperature increase on the deformed mesh. (°K)

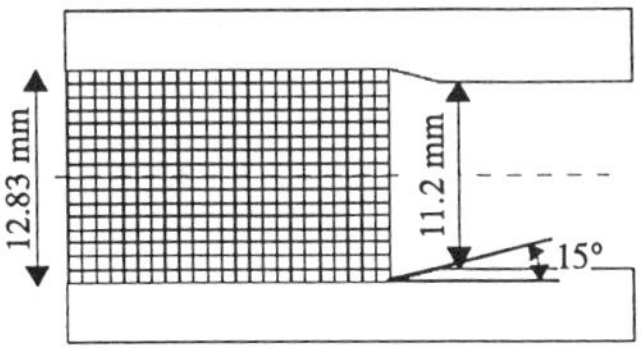

Figure 6. Undeformed finite element mesh with four node finite elements

The material parameters are identical with those used in test 6.1. The initial temperature is 293K. There is no heat transfer to the environment (natural boundary conditions). Only half of the problem is modelled because of symmetry . For simplicity no friction is assumed in the present study. For the thermal phase, the time integration coefficient considered is 0.5. The specimen is drawn through the die with a constant velocity of v = 1m/s and the analysis is performed in 120 equal time steps, each 0.00005 seconds long. The large deformations in the problem produce a plastic zone confined to the region of the billet possessing the largest curvature. In figure 7 we show the distribution of the temperature increase and the distribution of the equivalent plastic strain.

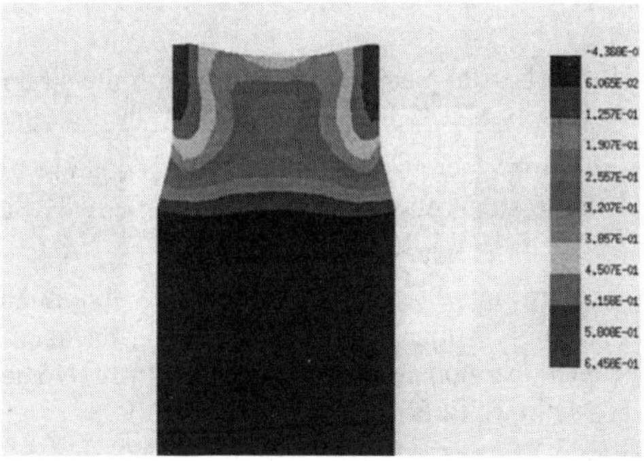

Figure 7. Temperature increase distribution (°K)

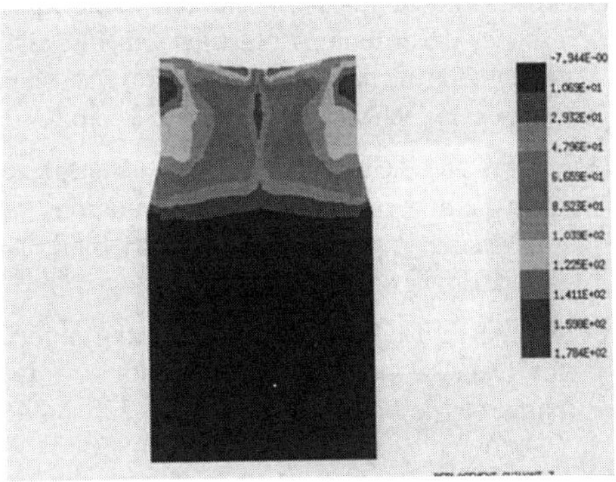

Figure 8. Equivalent plastic strain distribution

7 CONCLUSIONS

In this paper, we have proposed the formulation of a physically based model of thermoplasticity at finite strains. On the computational side which is the salient aspect of this work, an attractive feature of the proposed product formula algorithm lies in its simplicity which renders the scheme easily implementable with two independent mechanical and thermal finite element codes.

The only minor modifications required are the introduction of temperature-dependent material properties in the mechanical code, and the inclusion of a return mapping algorithm in the thermal code to compute the mechanical dissipation during the thermal iteration process.

The numerical examples presented demonstrate the efficiency of the algorithm in industrial situations.

8 REFERENCES

E.H Lee, Elastic-plastic deformation at finite strains, J. Appl.Mech, 36, 1-6, 1969.

F. Sidoroff, Incremental constitutive equation for large strain elasto-plasticity, Int. J. Engng. Sci, 20 - 1, 19-26, 1982.

F. Armero and J.C Simo, A new unconditionnally stable fractional step method for non-linear coupled thermomechanical problems, Int. J. Numer. Methods. Engrg, Vol 35, 737-766 (1992)

G.I Taylor and H. Quinney, The latent energy remaining in metal after cold working, Proc. Royal Soc Vol 143, pp. 307 - 326, 1934

J.C Gelin, O.Lochegnies, Simulations numériques en grandes déformations plastiques,In Physique et Mécanique de la Mise en Forme des Métaux, Presses du CNRS, Paris 1990, 553 - 562

J.H Argyris and J.St. Doltsinis. On the large strain inelastic analysis in natural formulation. Part 1 Quasistatic Problems, Comput. Methods Appl. Mech. Engrg. 20 (1979) 213 - 251

J.C Simo, R.L Taylor, A return mapping algorithm for plane planc stress elastoplasticity , Int. J. Numer. Methods. Engrg, Vol 22, 649-670. (1986)

O.C Zienkiewicz, The finite element method, Mc Graw Hill, 1977.

T.R.J Hughes, Analysis of transient algorithms with particular reference to stability behaviour, Computational Methods for transient analysis, Ed. T.Belytschko and T.J.R Hughes, Elsevier, pp. 67 - 153, 1983

W.S Farren and G.I Taylor, The heat developed during plastic extension of metals, Proc. Royal Soc Vol 107, pp.413 - 436, 1925

Simulation of Materials Processing: Theory, Methods and Applications, Shen & Dawson (eds)
© 1995 Balkema, Rotterdam. ISBN 90 5410 553 4

Fully automated analysis of 3-D metal forming processes

Nitin V. Hattangady, Lohitha Dewasurendra & Anil B. Chaudhary
UES, Inc., Dayton, Ohio, USA

ABSTRACT: Modeling of industrial forming processes like forging, extrusion, etc. for manufacturing components that need a full 3-D capability has not seen widespread use in the aerospace, automotive and related industries. An important factor contributing to this lack of use is the severe limitation imposed by the need for providing a new workpiece mesh, or, remeshing as it is commonly known. This paper presents a scheme developed and implemented within the code, ANTARES^{TM}, for automating the finite element modeling of metal forming processes that incorporates an automatic remeshing scheme to regenerate the workpiece mesh model during the analysis without user intervention. The modeling of a valve forging process is described.

1. Background

Until recently, process design of metal forming operations was considered more of an art rather than a science. Lack of mathematics that can predict the complex material flow can be identified as the primary reason for this. The process designer who is equipped with years of experience, typically solves issues in metal forming process design through costly trial and error. In many cases, the result is a process that is not the optimal, but the one that delivers the required product properties. Computer simulation of metal forming processes can effectively replace the costly shop-floor trials by fast, effective and accurate computer trials. Among the currently available techniques, the Finite Element Method can be identified as the most comprehensive and the one that is becoming very popular in industrial applications. Current high performance workstations equipped with effective analysis tools and modern visualization techniques can serve as a virtual forging plant that can not only simulate the actual forming process, but also give comprehensive details such as material flow within the workpiece, temperature distribution in the die and workpiece, required soak time, die/workpiece stresses, etc. These details can certainly assist the designer in altering the process design including the die geometries, forming sequence, heating sequence, etc. to significantly reduce the lead time.

Due to the severity of deformation that often takes place in bulk forming processes, the finite element mesh needs to be regenerated many times during the analysis process. Hence, automatic mesh generation plays a key role in the applicability of the finite element method for the analysis of the forming problems. Several software codes have incorporated the automatic mesh generation algorithms within their 2-D modeling capabilities which utilize the presence of rotational symmetry (axi-symmetric) and plane strain situations. However, the extension of the same capabilities to automate the modeling of 3-D problems has not been simple for reasons explained below.

2. Problem Definition

A typical approach adopted by most commercial Finite Element Analysis software codes involves use of eight node hexahedral elements (to represent the workpiece geometry) for the modeling of bulk forming processes like forging, extrusion, rolling, etc. In the modeling of these processes, the workpiece geometry, as represented by its mesh, *evolves* as it undergoes deformation. This entails regeneration of a new mesh to represent the workpiece geometry due to the presence of distorted elements. The frequency of remeshes depends, typically, on the severity of the deformation in the process and also the quality of the mesh provided to the analysis code. The workpiece must be remeshed under the constraint of maintaining the volume.

The standard technique for generating a new mesh has been to remesh manually using semi-automatic, or, mapped mesh generators that discretize six sided regions into hexahedral elements. However, the task of representing the deformed workpiece geometry into mappable regions is, in most cases, extremely complex, and has served as the major bottleneck in the widespread application of the 3-D process modeling technology. We have observed, in the modeling of 3-D forming processes, that a significant portion of the time required for process modeling (in some cases, upto 90%) is taken up by the manual remeshing task. Obviously, to automate the analysis, we would need to use an automatic mesh generator that discretizes an arbitrary domain automatically into eight noded hexahedral elements. This is a difficult task, to say the least and, currently, robust algorithms are not available for this purpose.

3. Approach

Clearly, from the discussion above, there are two different approaches to automating the modeling of forming operations. The first approach would involve developing an automatic mesh generator that discretizes any arbitrary domain into hexahedral elements. This approach would use the current analysis capabilities (using eight node hexahedral elements for modeling workpiece deformation). The second approach would be to modify the finite element formulation used within the analysis code to account for tetrahedral elements and use currently available, robust automatic mesh generation techniques for discretizing the deformed workpiece geometry into tetrahedral elements.

This paper presents our solution to automating the analysis of forming problems using the second approach described above. Note that the procedure described here forms the overall approach used within a commercial code, ANTARES, and hence, details of the algorithms, finite element formulation, etc. are not described.

4. Elements of the Automated System

The core of the automated system consists of three different modules: (i) the finite element solver, (ii) the geometry update procedure, and, (iii) the automatic mesh generator. Each of these modules addresses specific tasks and encompasses a wide area of research in itself. In addition, the overall procedure is managed by a control algorithm that monitors the process and invokes the different modules as needed.

Theoretically, the large deformation problems can be solved using different approaches including the penalty formulation with rigid-viscoplastic material behavior, elastic-viscoplastic formulation, etc. Almost all industrial hot-forming operations can be modeled using the rigid-viscoplastic material constitutive model which seems to be about four times faster than the elastic-viscoplastic model. Penalty formulation with rigid-viscoplastic material model is the simplest method and quite a few codes have implemented this approach in their capabilities for modeling large deformation problems. However, mathematical difficulties arise with certain classes of elements due to the incompressibility constraint inherent to the rigid-viscoplastic model. Consequently, only a limited selection of elements were available for the rigid-viscoplastic constitutive model based penalty formulation. On the other hand, the elastic-viscoplastic model involves rigorous mathematics and usually requires a considerably large computation time compared to the rigid viscoplastic analysis. In our automated system, the finite element formulation has been modified to handle tetrahedral elements and account for the incompressibility constraint. This allows us use of tetrahedral elements to represent the workpiece geometry and hence the use of currently available automatic mesh generation technology to discretize the domain into tetrahedral elements.

The geometry update procedure forms the link between the finite element analysis code and the mesh generator. This module builds the workpiece geometric representation based on the results from the analysis code and provides the necessary geometric information to the automatic mesh generator for further discretization of the workpiece domain. Note that, in 3-D analysis, when the analysis stops for a remesh, the workpiece surface mesh is likely to be distorted. This is due to the fact that most "action" that results in material movement is focussed around the workpiece (model) boundary.

Important issues that are addressed in updating the workpiece geometry involve the presence of die objects that the workpiece can/has come in contact. In other words, the interaction between different objects needs to be *consistently* handled. Hence, the portions of the workpiece boundary that have been identified as being in contact by the update procedure must also be identified as being in contact by the analysis code. Note that contact portions of the workpiece are to inherit the relevant portions of the die geometry.

There are a number of different algorithms available for automatic mesh generation. These include the advancing front (paving/plastering), Delaunay triangulation, domain decomposition (quadtree/octree based), etc. Each one of these is capable of discretizing a domain automatically into tetrahedral elements. The geometry information needed by the mesh generator, representing the boundary of the object to be meshed, is provided by the geometry update procedure which has built/updated the geometric representation of the workpiece based on the analysis results. The mesh data provided by the mesh generator is then optimized to reduce the bandwidth.

5. Example Problem

The Antares automated system for modeling 3-D forming problems has been tested on numerous industrial examples. One of these is a valve forging simulation where we have used the presence of symmetry to our advantage and modeled only one-eighth of the model. This simulation starts off with a quarter of a cylinder of radius 2.0 inches and takes it down to a center thickness of 0.07 inches. The problem was completed with 53 remeshes in about 18 hours on an HP 735 workstation. On an average, the workpiece has about 1500 nodes and 7000 elements with the number increasing to about 2400 nodes and 11000 elements in the final stage when there is a large amount of flash. The figures below show a starting, intermediate, final and the full mesh model for the deformed workpiece.

6. Conclusions

The implementation of an automated scheme for modeling of 3-D forming problems has eliminated the need for manual remeshing and provides a valuable tool for use in process design for metal forming. Reduction in overall simulation time of upto 80%-

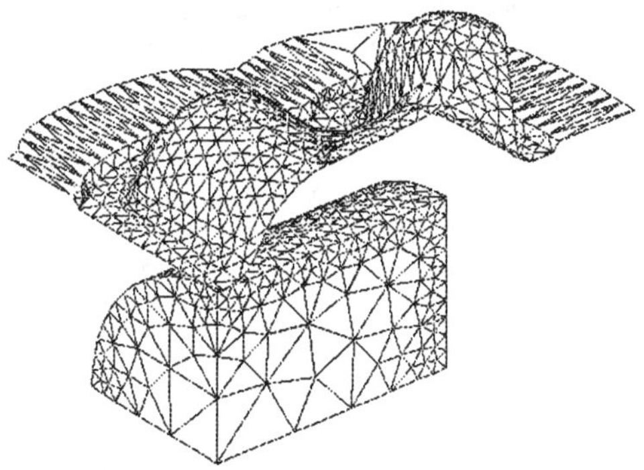

Figure 1. Initial Die-workpiece Mesh for Valve Forging Simulation

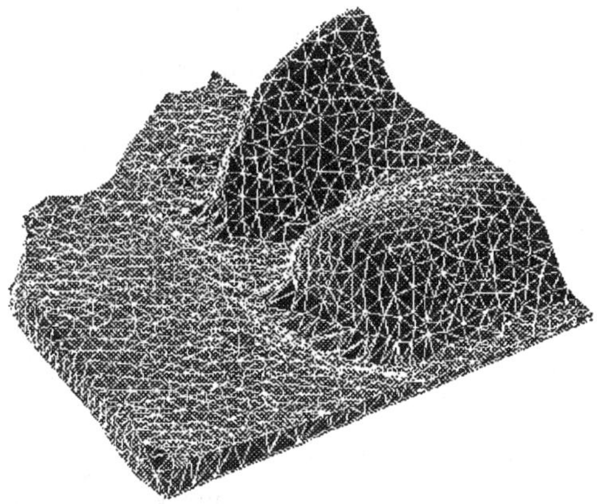

Figure 2. Final Workpiece Mesh for the Valve Forging Simulation (after 53 remeshes)

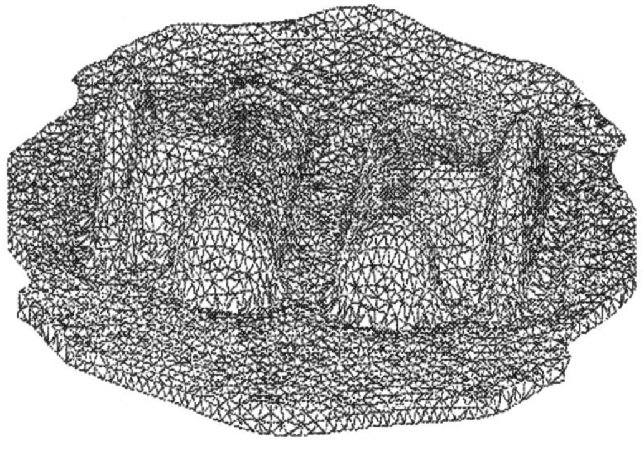

Figure 3. Mesh model for the Valve Forging (Full Mesh Model)

90% has been achieved in the modeling of some of these industrial problems. A typical approach in manual remeshing would be to use a coarse mesh so that the task of remeshing involves using data for a fewer points that define the workpiece boundary. However, with this automated problem, several problems were completed with meshes that were very fine thus giving a better quality of the solution.

7. References

1. ANTARES™ Primer Manual, UES, Inc., 4401 Dayton-Xenia Road, Dayton, OH 45432.

2. ANTARES™ Theory and Reference Manual, UES, Inc., 4401 Dayton-Xenia Road, Dayton, OH 45432.

3. Lee, C. H. and Kobayashi, S., "New Solutions to Rigid-Plastic Deformation Plastic Problems Using the Matrix Method", Trans. ASME, *J. Engg. for Industry*, 95, 1973.

4. Park, J. J. and Kobayashi, S. S., "Three Dimensional Finite Element Analysis of Block Forging", *Int. J. Mech. Sci.*, vol. 26, pp 165-176, 1984.

5. Babuska, I., "Error Bounds for Finite Element Method", *Numerical Math.*, vol. 16, pp 322-333, 1971.

6. Baehmann, P. L., Shephard, M. S., Ashley, R. A., Jay, A., "Automated Metal Forming Modeling Utilizing Adaptive Remeshing and Evolving Geometry", Computers and Structures, vol. 30(1/2), pp 319-325, 1988.

7. Cheng, J. H., "Automatic Adaptive Remeshing for Finite Element Simulation of Forming Processes", *Int. J. Numer. Meth. Engg.*, vol 26, pp 1-18, 1988.

8. Habraken, A. M., Cescotto, S., "An Automatic Remeshing Technique for Finite Element Simulation of Forming Processes", *Int. J. Numer. Meth. Engg.*, vol 30, pp 1503-1525, 1990.

9. Shephard, M.S. and Finnigan, P. M., "Toward Automatic Model Generation", *State-of-the-Art Surveys on Computational Mechanics*, A. K. Noor and J.T. Oden, editors, ASME, pp 335-366, 1989.

10. Baehmann, P. L., Collar, R. R., Hattangady, N. V., and, Shephard, M. S., "Geometry and Mesh Control for Automated Bulk Forming Simulations", *Proc. of Numerical Methods for Industrial Forming Processes*, pp 47-57, ASME Winter Annual Meeting, Anaheim, CA, Nov. 8-13, 1992.

11. George, P.L, "Automatic Mesh Generation - Application to Finite Element Methods", John Wiley & Sons, 1991.

12. Kobayashi, S., Oh, S., I., Altan, T., "Metal Forming and the Finite-Element Method", Oxford Series on Advanced Manufacturing, Oxford University Press, 1989.

Simulation of Materials Processing: Theory, Methods and Applications, Shen & Dawson (eds)
© 1995 Balkema, Rotterdam. ISBN 90 5410 553 4

Optimization of multi-stage metal forming processes using parallel computers

R. Kopp, R. Thomas & R. Bünten
Institute for Metal Forming (IBF), Technical University of Aachen, Germany

ABSTRACT: The simulation based optimization of metal forming processes requires a lot of iterative simulations. Especially when performing multi-stage simulations an enormous amount of calculating time is needed. The only solution to this problem is to use high-performance computers. Those achieve their power either on vectorisation or on parallelisation. In this paper a strategy is described to optimize metal forming processes with parallel computers. Due to the low need of communication between the parallel tasks, the concept is very efficient. Different optimization algorithms and optimization criteria can be implemented within the universal concept. Furthermore the number of used machines is easily scalable and robust against the loss of single processors. As an example the optimization of a two - stage forging process is presented. The simulation model used in this case is a finite - element - program that was developed at the IBF.

1) Introduction

In the past years the computing power of workstations has increased rapidly. Most university institutes which need large amounts of computing time for the simulation of processes own a number of those. Usually the workstations are connected by a local area network (LAN) and have access to the internet. Due to this connection, for example an ethernet cable it can be described as a loosely coupled workstation cluster. In those workstation clusters most services like authentification, fileserving and administration are done centrally. The IBF has a workstation cluster with eight IBM RS 6000 machines.

Due to the increase of accessible computing power the simulation of metal forming processes has become an important diagnosis tool for fault detection and for checking of new process designs. Many aspects of the forming process like grain sizes can be calculated. Finite element calculations of threedimensional processes with thermal coupling and automatic 3D - remeshing are possible in acceptable amounts of time. This leads to a more and more realistic simulation.

For future products the increasing economic advantages of using simulation tools make it necessary to use computers in any step of the development process. By that it will be possible to reduce the time for developing a new product, improving the properties of a product and saving energy and material.

2) Automatic optimization tools

Usually the optimization of metal forming processes requires a lot of iterative simulations. The initial process layout is simulated, the engineer looks at the results and makes changes until the simulation results satisfy the requirements. For creating an automatic optimization tool the simulation program has to be connected to an optimization module. The automatic optimization concept has the following specific needs:

- The simulation modell has to be robust against mistakes in the input data set. The program must not stop, it has to produce a specific error code for the optimization module.
- The quality of the process layout has to be measured by a target value after the simulation. With this a comparison between different process layouts is possible.
- The user has to define the criterium, when the algorithm should stop.
- The optimization module should include different algorithms to be flexible enough for different metal forming problems.
- The interfaces between simulation model and optimization algorithm should be flexible to adapt different simulation models and optimization modules.

The parameters which describe the process layout can be divided in variable parameters and fixed parameters. The fixed parameters for example describe material properties, which in the specific process layout can not be varied. Among the variable parameters like tool surface coordinates or reduction heights one has to pick out the so called design variables which have to be optimized. The set of design variables that induces the best target value is the optimization result. The number of simulations which are needed for the optimization depends on different points:

- The number of design variables.
- Wanted target value. How good has the solution to be?
- Shape of the target function. Has it local optima?
- Used optimization algorithm.

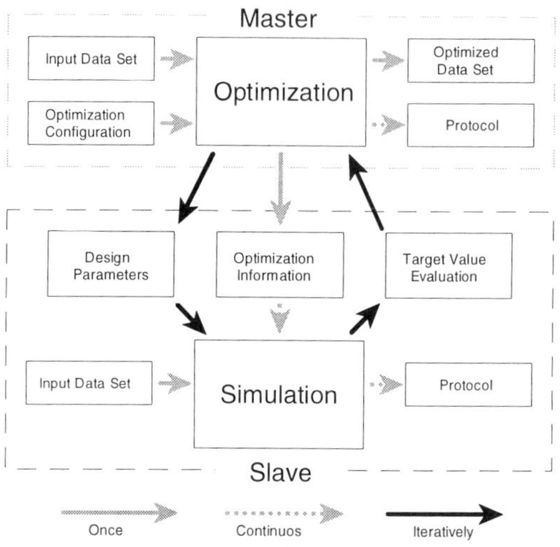

Figure 1: Interface and Data flow

The number of design variables depends on the number of stages of the process. A one step process is fixed in many terms due to the limitation of the resulting product. The tool shape can be varied little, for example radii. Most sizes can not be varied and a complex preform has to be produced for example by casting. In a multi stage process there are a lot of variable parameters. Except the tool shapes of the last stage the tool shape points and the intermediate reductions in height can easily be varied.

3) Parallel optimization concept

At the IBF several simulation tools for metal forming processes have been developed. The finite element program FINEL has been taken to connect with a self designed optimization module.

Figure 1 shows the interface between optimization and simulation. For an efficient use of the actual computing power at the IBF and with a view to further computer developments, a scalable parallel optimization concept was developed. Using the PVM (Parallel Virtual Machine) parallel programming libraries a parallel optimization program system was generated (PVM 1993). The system consists of two different programs, a master program and a slave program. The master program proceeds the optimization algorithm and feeds a variable number of slave programs with input data sets. In a configuration file the user can specify the number of needed slave tasks. The master starts those through PVM library calls (figure 2).

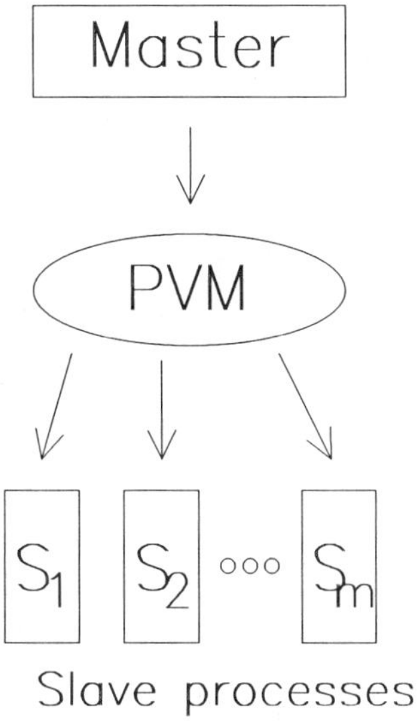

Figure 2: Process generation

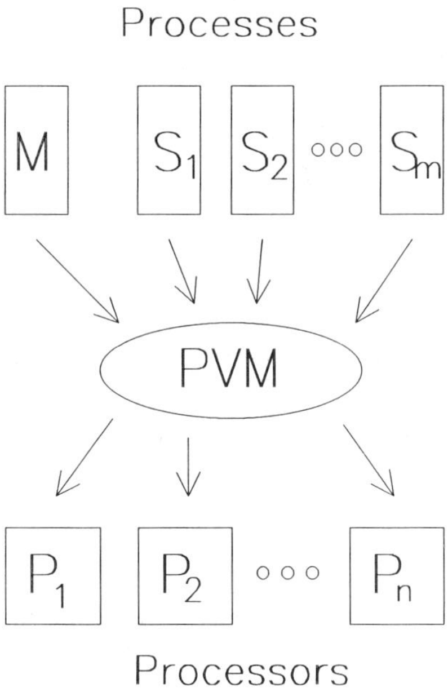

Figure 3: Process distribution

The distribution of the slave tasks over the available processors can be done in different ways. Either the user specifies a processor on which a program should run or PVM distributes them equally (figure 3). The load balancing is very simple and good. As soon as a slave task has finished its simulation, it calculates the target value and sends this by a PVM call to the master. The master task evaluates these results and generates a new set of design variables and sends this to the slave task (figure 4). The properties of this concept are little communication and good scalability.

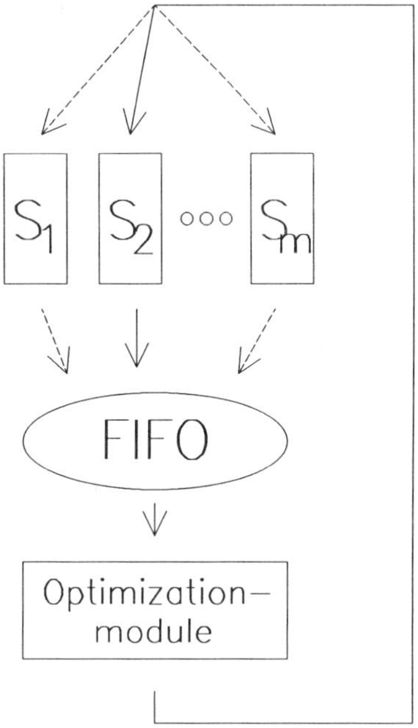

Figure 4: Load balancing

4) Optimization algorithms

The implemented algorithms follow three different basic strategies (Reklaitis 1983):

- Local search techniques
- Evolutionary algorithms
- Gradient strategies

Local search techniques are basically very simple, because they are 'greedy algorithms' and look for the optimum in every variation step. This has the advantage, that in a 'simple' target function the optimum is found in short time. The disadvantage is, that there is a high risc to get into a local optimum depending on the starting point. This effect can be avoided or toned down by creating an algorithm that has local search phases. One strategy could be, to start the local search from a number of randomly generated starting points. This will either lead to the knowledge that the target function likely has only local optimum or that there are many of those and we choose the best one. Such meta algorithms are good scalable and can be adapted to the specific problem.

In evolutionary strategies design variable sets are called individuals. Differently to local search here we have a number of actual individuals called population. Using the mechanisms recombination, mutation and selection new generations of individuals are generated, tested and eventually stored.

Gradient strategies basing on discrete tests use functions for interpolation or approximation of the target function. Then the new search directions and distances are calculated.

5) Example

An example for the optimization of a simple forging process in axisymmetric modelling is presented in figure 5. Due to the friction between tools and workpiece the workpiece bulges. Figure 5a shows the initial process layout and the resulting final shape. The optimization goal in this example is to reach a specific final shape. The desired final shape of the forged workpiece is cylindrical like the preform. In a one stage process there are many limitations. For example tool shapes can not be changed to much because the product which has to be manufactured requires the specific form.

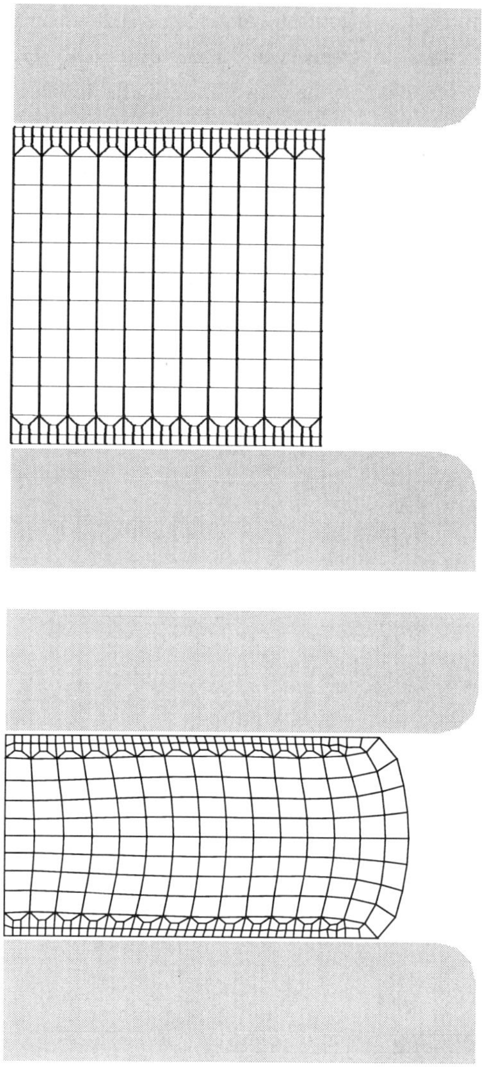

Figure 5a: Initial process

By dividing the process into two stages a variation of the tool shapes in the first stage and the first final height is possible. Figure 5b shows the stages of the optimized process layout. The second picture in figure 5b shows the configuration at the beginning of stage two.

In this example we varied 21 parameters. These are 10 points of the upper tool and 10 points of the lower tool in the first stage and the first reduction height. By that variation concept the tool shape can

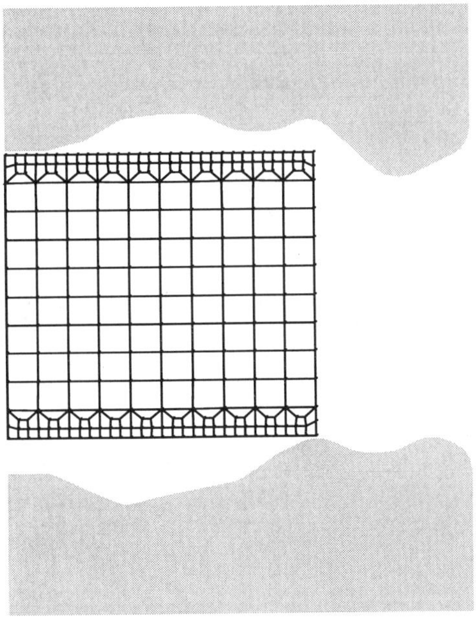

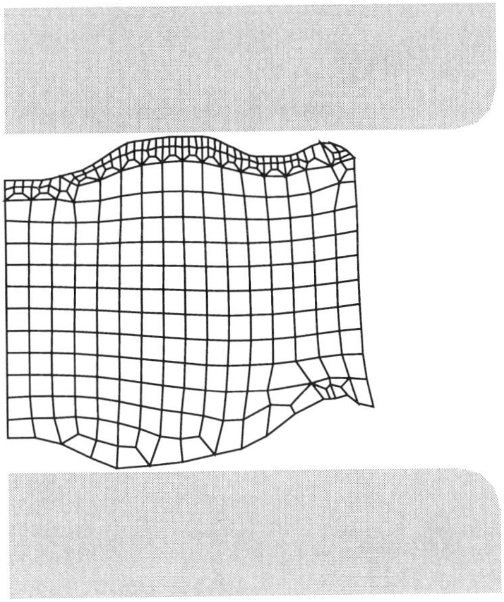

have sharp edges. This is in most cases not a senseful assumption for the process simulation and because of that in this example the edges are rounded automatically.

The quality of the optimization result depends on different aspects. The number of varied design variables corresponds to the needed number of

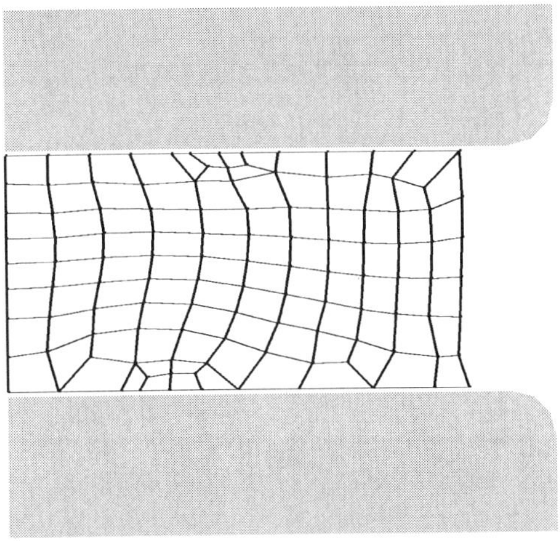

Figure 5b: Steps of the optimized process

simulations to reach an optimum. If this is a local or a global optimum can not be estimated due to the finite number of possible tests. A more complex algorithm can just reduce the possibility of ending in a bad local optimum. In this example the quality of the result is measured by the difference between the contour of the workpiece and the desired contour. Starting with a target value of 6.14 (figure 5a) the optimum in figure 5b with a target value of 0.37 was found in the 439th simulation using a local search algorithm. With the variation of only 6 points of the upper and lower first tool the target value was never better than 0.8.

6) References

(PVM 1993) Geist,A.; Beguelin,A.; Dongarra,J.; Jiang,W.; Manchek,R.; Sunderam,V.: PVM 3 User´s Guide and Reference Manual, Oak Ridge National Laboratory, USA 1993

(Reklaitis 1983) Reklaitis,G.V.; Ravindram,A.; Ragsdell,K.M.: Engineering Optimization, John Wiley and Sons, 1983

Simulation of Materials Processing: Theory, Methods and Applications, Shen & Dawson (eds)
© 1995 Balkema, Rotterdam. ISBN 90 5410 553 4

The simulation of texture evolution during bulk deformation processes using finite elements over orientation space

Ashish Kumar & Paul R. Dawson
Sibley School of Mechanical and Aerospace Engineering, Cornell University, Ithaca, N.Y., USA

ABSTRACT: In this work, polycrystal plasticity modeling based on a characterization of texture by an orientation distribution function (ODF) is applied to the simulation of steady-state bulk deformation processes. The texture of a macroscopic material element is characterized by an ODF, defined over a fundamental region of Rodrigues' parameter space. The evolution of the ODF is computed by applying finite element analysis to solve the ODF conservation equation over an explicit discretization of the fundamental region. The finite element ODF scheme is coupled with a spatial steady-state Eulerian finite element formulation to yield process simulations that involve finite element calculations at two length scales: at the macroscopic scale of the deformation process, over a spatial discretization; and at the microstructural level over the discretized fundamental region of orientation space. In the following, this multi-scale analysis is applied to simulate the cold rolling of aluminum on a data parallel architecture.

1 INTRODUCTION

Existing applications of polycrystal plasticity to the simulation of bulk deformation processes are derived primarily out of the discrete aggregate schemes (Mathur & Dawson 1989, Beaudoin *et al.* 1993), based on a characterization of the microstructure by a representative discrete aggregate of crystals. While such a characterization represents texture and derivative properties satisfactorily, it provides no basis for texture transformations such as differencing, projection or interpolation.

This shortcoming of discrete aggregate schemes is overcome naturally by polycrystal schemes based on a characterization of the microstructure by an ODF, expressing the distribution of crystal orientations over the space of orientation parameters. A new class of ODF polycrystal schemes was proposed in Kumar & Dawson 1994, based on a representation of the ODF by finite element piecewise polynomial functions over an explicit discretization of orientation space. The finite element ODF schemes were subsequently applied to simulate the deformation driven texturing of planar polycrystals, and more recently, to the texturing of three dimensional polycrystals (Kumar 1995).

In the following, we seek to apply the finite element ODF schemes to the simulation of bulk deformation processes, and in particular, to exploit the accompanying analytical structure to resolve spatial inhomogeneities of microstructural texture. Highly localized regions of inhomogeneity develop, for example, through the thickness of rolled sheet due to roll induced shearing at the surface. Studies, both experimental and computational, of the through thickness variation of texture in rolled sheet have been reported extensively in the literature.

2 POLYCRYSTAL MODELING

In the following, contrary to discrete aggregate polycrystal modeling, the microstructure associated with a macroscopic material element is characterized by an ODF defined such that the volume fraction of crystals contained within a part $\mathcal{R}^*$ of a fundamental region $\mathcal{R}$ of orientation space is,

$$v_f(\mathcal{R}^*) = \int_{\mathcal{R}^*} A(\boldsymbol{r}, t)\, dv \qquad (2.1)$$

Here, as in Kumar & Dawson 1994, under a particular parametrization of crystal orientations, fundamental region $\mathcal{R}$ is a subregion of the associated orientation space, irreducible under crystal symmetries.

Averages of orientation dependent quantities over the microstructure are developed from the ODF by orientation averaging as,

$$< f > = \int_{\mathcal{R}} f(\boldsymbol{r}) A(\boldsymbol{r}, t)\, dv \qquad (2.2)$$

This defines, for example, an aggregate stress from the stresses developed by individual crystals, or as in the following, an aggregate stiffness from the stiffnesses of individual crystals.

The evolution of an ODF defined in this manner is governed by the ODF conservation equation,

$$\frac{\partial A}{\partial t} + \nabla A \cdot \boldsymbol{v} + A \nabla \cdot \boldsymbol{v} = 0 \qquad (2.3)$$

where $\boldsymbol{v}$ is the reorientation velocity that drives the texturing. The conservation equation is accompanied by boundary conditions (Kumar 1995) that arise from symmetric relationships between orientations on the boundary $\bar{\mathcal{R}}$ of the fundamental region. In particular, $\bar{\mathcal{R}}$ can decomposed into subsurfaces $\bar{\mathcal{R}}^s$ and $\bar{\mathcal{R}}^m$, $\bar{\mathcal{R}} = \bar{\mathcal{R}}^s \cup \bar{\mathcal{R}}^m$ such that for every orientation $\boldsymbol{r} \in \bar{\mathcal{R}}^s$ there exists a symmetrically equivalent orientation $h(\boldsymbol{r}) \in \bar{\mathcal{R}}^m$. The consequent symmetries displayed by the ODF are then,

$$A(\boldsymbol{r}, t) = A(h(\boldsymbol{r}), t) \qquad (2.4)$$

for any $\boldsymbol{r} \in \bar{\mathcal{R}}^s$.

Two components must be supplied to determined the evolution of the ODF in (2.3). The first is a parametrization of crystal orientations, and the second, a particular set of crystal constitutive relations.

2.1 *Rodrigues' Parameter Space*

An attractive alternative to the conventional Euler angle parametrizations are the neo-Eulerian parametrizations of Frank 1988. These are constructed from the two invariants of a rotation, the axis of rotation $\boldsymbol{n}$ and the angle of rotation about the axis ϕ as,

$$\boldsymbol{r} = \boldsymbol{n}\, f(\phi) \qquad (2.5)$$

The parameters developed are symmetric relative to the sample axes, and are easily interpreted as they define the crystal orientation by a single rotation from the sample axes. More critically from the perspective of performing analyses over orientation space, the developed parameter spaces possess regular, singularity free metrics, and reduce in a regular fashion under crystal symmetries to yield geometrically simple fundamental regions.

Of particular merit among the neo-Eulerian parametrizations is the Rodrigues' parametrization, with $f(\phi) = \tan \phi/2$, marked by a certain rectilinearity. In particular, the bounding surfaces introduced under the symmetric reduction of Rodrigues' parameter space are always planar (Heinz & Neumann 1991). This can be seen in the fundamental region corresponding to the cubic symmetric group, shown in Figure 1. Moreover, the

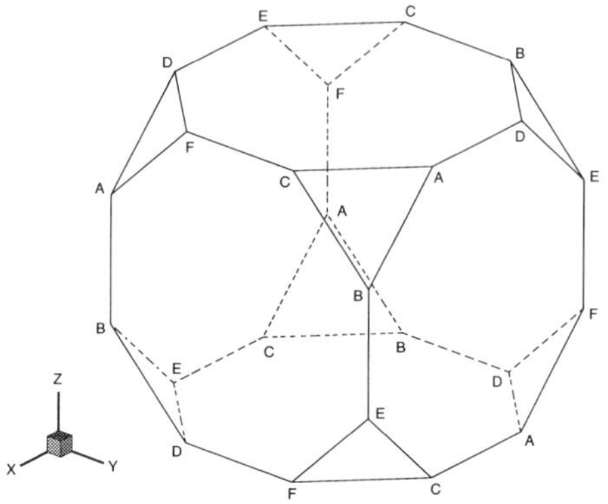

Figure 1. The fundamental region of Rodrigues' parameter space for the cubic symmetry group.

symmetries that relate orientations on the boundaries of such fundamental regions take on a particularly simple form. For the cubic fundamental region shown, orientations on opposing {100} planes are related following a rotation of $\pi/4$ about the corresponding <100> axis, and those on opposing {111} planes following rotations of $\pi/3$ about the corresponding <111> axis.

The metric tensor associated with Rodrigues' parameter space can be shown to be (Kumar 1995),

$$\boldsymbol{g} = \frac{1}{(1 + \boldsymbol{r} \cdot \boldsymbol{r})^2} \left((1 + \boldsymbol{r} \cdot \boldsymbol{r})\, \boldsymbol{I} - \boldsymbol{r} \otimes \boldsymbol{r} \right) \qquad (2.6)$$

and is free of singularities. Moreover, although the parameter space is non-Euclidean, the volumetric distortion experienced over the region of the parameter space spanned by the cubic fundamental region is mild.

2.2 *Constitutive Relations*

In the following we adopt the extended Taylor hypothesis and assume that crystals deform solely by crystallographic slip. In this situation, the crystal velocity gradient can be decomposed into a shearing along slip systems and a rigid lattice rotation,

$$\boldsymbol{L} = \boldsymbol{\Omega} + \sum_{\alpha} \dot{\gamma}^{\alpha} \boldsymbol{T}^{\alpha} \qquad (2.7)$$

where $\boldsymbol{\Omega}$ is the lattice spin, $\dot{\gamma}^{\alpha}$ the shear rate on slip system α and $\boldsymbol{T}^{\alpha}$ the Schmid tensor. A viscoplastic constitutive relation is used to relate the shear rate on a slip system to the resolved shear stress,

$$\dot{\gamma}^{\alpha} = a \frac{\tau^{\alpha}}{\tau} \left| \frac{\tau^{\alpha}}{\tau} \right|^{1/m - 1} \qquad (2.8)$$

where m is the rate sensitivity, τ is the slip system hardness, assumed identical for all crystals comprising the microstructure, and $\tau^\alpha = \boldsymbol{P}^\alpha \cdot \boldsymbol{\sigma}'$ is the resolved shear stress on slip system α, with $\boldsymbol{P}^\alpha = \operatorname{sym} \boldsymbol{T}^\alpha$.

An expression for the crystal stress follows (2.7) and (2.8),

$$\boldsymbol{\sigma}' = \mathcal{C}\,[\boldsymbol{D}'] \tag{2.9}$$

where stiffness $\mathcal{C}$ is obtained by the inversion,

$$\mathcal{C}^{-1} = \sum_\alpha \frac{a}{\tau} \left| \frac{\tau^\alpha}{\tau} \right|^{1/m-1} \boldsymbol{P}^\alpha \otimes \boldsymbol{P}^\alpha \tag{2.10}$$

The accompanying crystal rotation follows in the lattice spin, obtained from the skew part of (2.7),

$$\boldsymbol{\omega} = \operatorname{vect}(\boldsymbol{W} - \sum_\alpha \dot{\gamma}^\alpha \boldsymbol{Q}^\alpha) \tag{2.11}$$

where $\boldsymbol{Q}^\alpha = \operatorname{skew} \boldsymbol{T}^\alpha$, and the hardening rate from,

$$\dot{\tau} = \dot{\gamma}^* \, \Theta_0 \left(\frac{\tau_s - \tau}{\tau_s - \tau_0} \right) \tag{2.12}$$

where τ_s is the saturation hardness defined as,

$$\tau_s = \tau_{s0} \left| \frac{\dot{\gamma}^*}{\dot{\gamma}_s} \right|^{m'} \tag{2.13}$$

and $\dot{\gamma}^*$ is a representative rate of shearing for the microstructure, developed from the slip system shear rates and the ODF as

$$\dot{\gamma}^* = \int_{\mathcal{R}} A \sum_\alpha |\dot{\gamma}^\alpha| dv \tag{2.14}$$

3 FINITE ELEMENT FORMULATION

3.1 *Macroscopic Computations*

At the macroscopic level, a formulation for the deforming workpiece proceeds from statements of conservation of mass and momentum for the continuum. Under a particular characterization of the material, these together determine the macroscopic motion. For the polycrystal characterization employed here,

$$\boldsymbol{\sigma} = \mathcal{K}\,[\boldsymbol{D}'] + \frac{1}{\beta} \nabla \cdot \boldsymbol{u}\, \boldsymbol{I} \tag{3.15}$$

where $\mathcal{K}$ is the fourth order viscoplastic stiffness associated with the underlying polycrystalline microstructure. This is obtained by orientation averaging as,

$$\mathcal{K}(\boldsymbol{x}) = \int_{\mathcal{R}} A(\boldsymbol{x}, \boldsymbol{r})\, \mathcal{C}(\boldsymbol{x}, \boldsymbol{r})\, dv \tag{3.16}$$

where $A(\boldsymbol{x}, \boldsymbol{r})$ is the ODF at spatial coordinate $\boldsymbol{x}$, and $\mathcal{C}(\boldsymbol{x}, \boldsymbol{r})$ is the fourth order stiffness (2.10) of a crystal with orientation $\boldsymbol{r}$ at $\boldsymbol{x}$. Observe in (3.15) that a consistent penalty formulation is used, capable of representing an incompressible material in the limit as $\beta \to 0$.

A velocity-pressure finite element formulation is obtained from weak statements of the conservation of momentum and the modified conservation of mass,

$$\begin{aligned} &\int_\Omega \nabla \boldsymbol{\varphi} \cdot \boldsymbol{\sigma}\, dv - \int_\Gamma \boldsymbol{\varphi} \cdot \boldsymbol{t}\, dA = 0 \\ &\int_\Omega \psi\, (p + \frac{1}{\beta} \nabla \cdot \boldsymbol{u})\, dv = 0 \end{aligned} \tag{3.17}$$

Observe that the construction of (3.17) involves computations at both length scales as the macroscopic stiffness is developed from the crystal stiffnesses in (3.16) by an integration over the fundamental region. The developed discrete system is nonlinear and is solved iteratively for the macroscopic motion. Details of the implementation in a data parallel environment are available elsewhere (Beaudoin *et al.* 1993).

The solution of (3.17) for the macroscopic motion is coupled with a scheme for the evolution of microstructural state. In the case of the steady state deformations considered, this is performed in an effective manner by an integration of the ODF and the hardness along streamlines of the deformation (for descriptions of streamline integration see Dawson 1987; issues related to a data parallel implementation are discussed in Kumar 1995).

3.2 *Polycrystal Computations*

Along a streamline, the ODF is evolved using an updated Lagrangian scheme (Kumar 1995) to exploit characteristics of the ODF conservation equation for stability,

$$\int_{\mathcal{R}} \psi\, (\frac{1}{\Delta t}(A - \tilde{A}) + A \nabla \cdot \boldsymbol{v} - \nabla \cdot (\varepsilon \nabla A))\, dv = 0 \tag{3.18}$$

Here $\tilde{A}(\boldsymbol{r}, t) = A(\boldsymbol{r} - \boldsymbol{v}\Delta t, t - \Delta t)$ is the ODF projected back along characteristics over the time increment Δt. Artificial diffusion ε is introduced for stability against jump discontinuities of $\nabla \cdot \boldsymbol{v}$. The diffusivity used is of the form $\varepsilon = C h^\alpha R$, where h is the local element size and $R = |(A - \tilde{A}) \,/\, \Delta t + A \nabla \cdot \boldsymbol{v}|$ is the finite element residual.

In order to accommodate the symmetry boundary conditions associated with the ODF conservation equation (2.4), the fundamental region is discretized in a manner such that discretizations obtained on boundaries of the fundamental region mimic boundary symmetries (Figure 3). In this situation, approximations to satisfy the symmetry conditions are trivially obtained by explicitly identifying symmetrically equivalent boundary nodes.

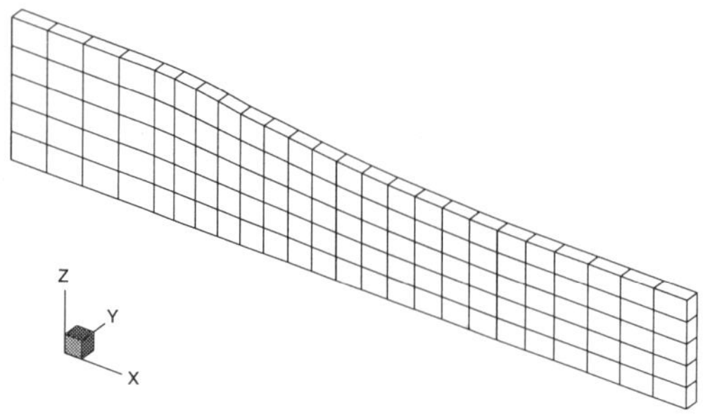

Figure 2. The finite element discretization of the workpiece.

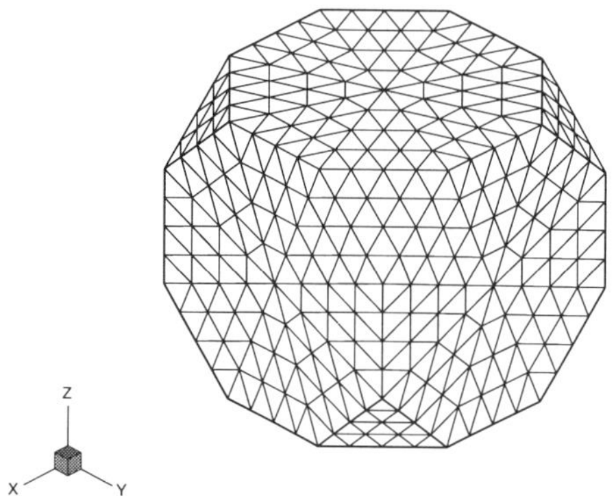

Figure 3. The finite element discretization of the cubic fundamental region of Rodrigues' parameter space.

4 RESULTS

The cold rolling of 1100 aluminum was simulated using the finite element polycrystal scheme. Aluminum sheet with initially uniform texture and hardness was rolled through a single pass with a nominal thickness reduction of 25%. The roll was taken to be circular and rigid, with an outer radius of 1.25 in and angular velocity 16.0 rad s^{-1}. The contact between the roll and the workpiece was modeled through the set of mixed boundary conditions,

$$\begin{aligned} \boldsymbol{u} \cdot \boldsymbol{e}_n &= 0 \\ \boldsymbol{t} \cdot \boldsymbol{e}_r &= \lambda (u_r - \boldsymbol{u} \cdot \boldsymbol{e}_r) \end{aligned} \qquad (4.19)$$

where $\boldsymbol{e}_n$ and $\boldsymbol{e}_r$ are the unit vectors in the normal and rolling directions, respectively. Frictional tractions $\boldsymbol{t} \cdot \boldsymbol{e}_r$ were assumed proportional to the relative tangential velocity between the roll and the workpiece $(u_r - \boldsymbol{u} \cdot \boldsymbol{e}_r)$, with friction coefficient λ assumed constant over the contact surface. A variation of the spline technique of Dawson 1987 was implemented to determine the workpiece surface upstream of the roll. The various rolling and material parameters used in the simulation are summarized in Table 1.

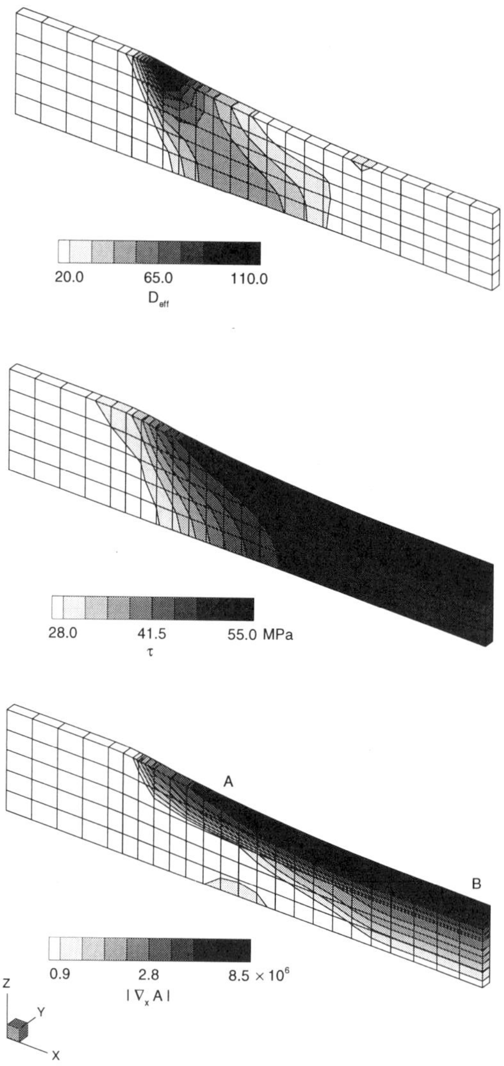

Figure 4. Contour plots of the effective deformation rate D_{eff}, the hardness distribution τ and the norm of the spatial gradient of the ODF $|\nabla_x A|$ computed over the workpiece.

The finite element discretization of the workpiece used in the simulation is shown in Figure 2. Due to symmetry about the centerline, only half the workpiece is discretized. Linear brick elements are used for the velocity, with a piecewise constant approximation for the pressure distribution. The ODF and the hardness over the workpiece were

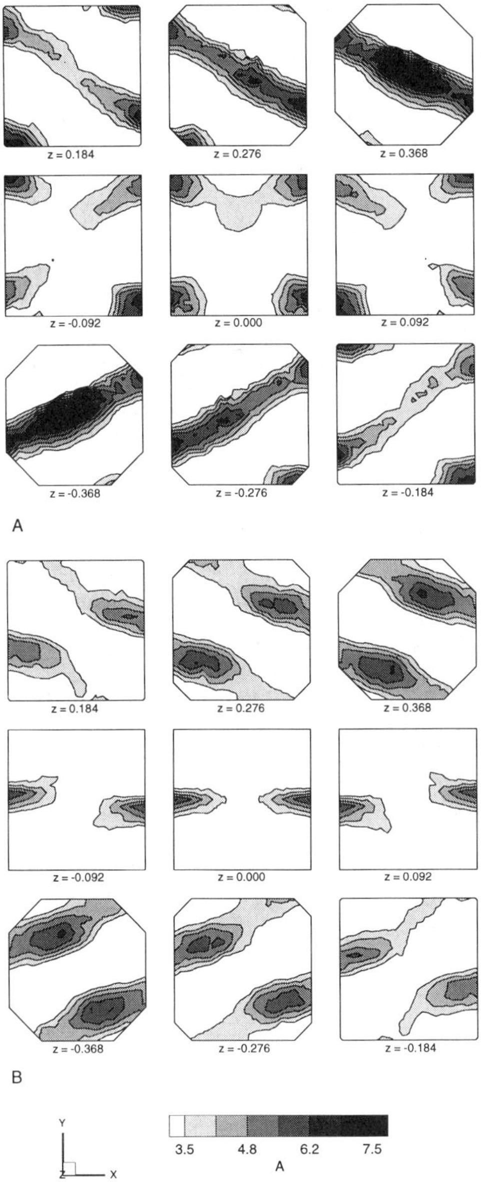

Figure 5. The ODF computed at a section under the roll displayed on sections through the fundamental region taken perpendicular to the normal direction (A at the surface, B at the centerline).

computed by streamline integration performed in steps of length less than 5% the characteristic element size. ODF evolution was carried out over the cubic fundamental region of Rodrigues' parameter space discretized into four noded tetrahedral elements (Figure 3).

The effective deformation rate, the hardness distribution and a measure of the spatial gradient of the ODF defined as,

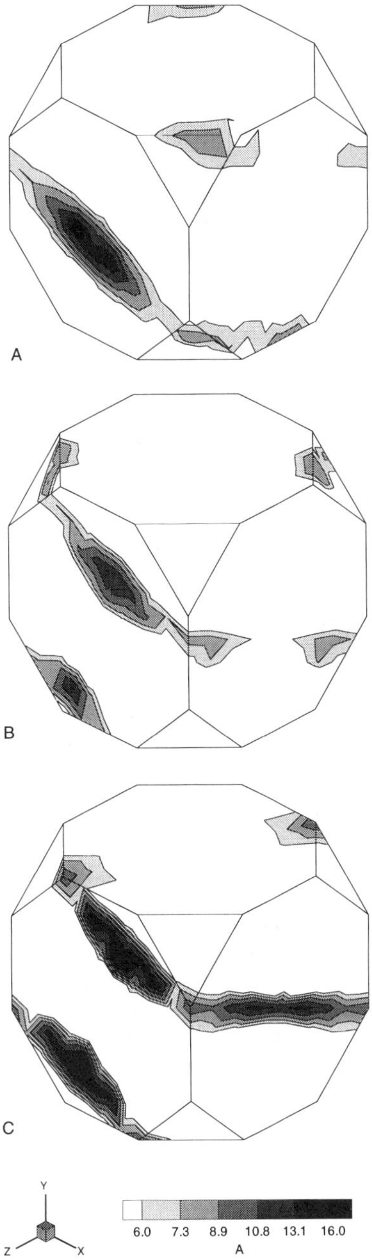

Figure 6. Surface contours of the ODF computed at the exit (A taken at the surface, B at the midplane, C at the centerline).

$$|\nabla_x A| = \int_{\mathcal{R}} |\nabla_x A(\boldsymbol{x}, \boldsymbol{r})| \, dv \qquad (4.20)$$

Table 1. Material and rolling parameters used in the simulation.

Parameter	Value
a	$1.0\,s^{-1}$
m	0.05
Θ_0	58.41 MPa
τ_0	27.17 MPa
τ_{s0}	61.80 MPa
$\dot{\gamma}_s$	5×10^{10}
m'	0.01
u_r	$0.508\,m\,s^{-1}$
λ	2.0×10^4 MPa s m^{-1}

computed over the workpiece are displayed in Figure 4. As expected, the microstructure hardens primarily within the deformation zone directly under the roll. Hardening is initiated earlier at the surface than at the centerline, and develops into a mild gradient through the workpiece thickness downstream of the roll.

Like the hardening observed, texture evolves primarily in the deformation zone directly below the roll, reflected in developing texture gradients through the workpiece thickness. In contrast to the hardening gradients, however, the texture gradients observed downstream of the roll are substantial. To examine the source of these gradients the ODF distribution computed is sampled at a cross-section beneath the roll (A in Figure 4). The centerline texture is comprised of a fiber that winds about the outside of the fundamental region, passing through the ideal Taylor, Goss and Brass orientations commonly observed in plane strain compression. In moving towards the surface, the deformation experienced varies from the plane strain compression developed at the centerline as shearing is induced by contact with the roll at the surface. This is reflected in the surface texture which is again composed of a fiber winding about the surface of the fundamental region, sections of which approach the {111} and <110> fibers typically seen in shear textures (Canova *et al.* 1983). This variation of texture across the workpiece cross-section is reinforced downstream of the roll (Figure 6 with ODFs extracted at B in Figure 4). Once again the texture is seen to vary through the thickness from a plane strain compression texture at the workpiece centerline, to a shear texture at the surface, displaying intermediate components at the midplane (B in Figure 6). Observe that the ODF gradient measure $|\nabla_x A|$ reflects the spatial texture gradients adequately.

ACKNOWLEDGEMENTS

This work was supported through Office of Naval Research contract NOOO 14-90-J-1810. Computing resources on the CM-5 were provided by the National Center for Supercomputing Applications, University of Illinois at Urbana-Champaign.

REFERENCES

G.R. Canova, U.F. Kocks and J.J. Jonas 1983. Theory of Torsion Texture Development. *Acta Metall.* 32:211-226.

P.R. Dawson 1987. On Modeling the Mechanical Property Changes During Flat Rolling of Aluminum. *Int. J. Solids Struc.* 23:947-968.

F.C. Frank 1988. Orientation Mapping. In J.S. Kallend and G. Gottstein (eds.), *Eighth International Conference on Textures of Materials* The Metallurgical Society.

K.K. Mathur and P.R. Dawson 1989. On Modeling the Development of Crystallographic Texture in Bulk Forming Processes. *Int. J. Plast.* 5:67-94.

A. Heinz and P. Neumann 1991. Representation of Orientation and Disorientation Data for Cubic, Hexagonal, Tetragonal and Orthorhombic Crystals. *Acta Cryst.* A47:780-789.

A. Beaudoin, K.K. Mathur, P.R. Dawson and G.C. Johnson 1993. Three-dimensional deformation process simulation with explicit use of polycrystalline plasticity models. *Int. J. Plast.* 9:833-860.

A. Kumar and P.R. Dawson 1994. The Simulation of Texture Evolution using Finite Elements over Orientation Space — I. Development. Submitted to *Comp. Methods Appl. Mech. Eng.*

A. Kumar 1995. Polycrystal Modeling with Finite Elements over Orientation Space. *PhD Thesis* Cornell University, Ithaca, NY.

Simulation of Materials Processing: Theory, Methods and Applications, Shen & Dawson (eds)
 ISBN 90 5410 553 4

Rigid bodies for metal forming analysis with NIKE3D

Bradley N. Maker
Lawrence Livermore National Laboratory, Calif., USA

ABSTRACT: Perhaps the most common approximation in engineering is that, relative to its neighbors, a system component is structurally rigid. This paper presents a development of the rigid assumption for use in nonlinear, implicit finite element codes. In this method, computational economy is gained by condensing the size of the associated linear system of equations, eliminating the processing of rigid elements, and reducing the overall nonlinearity of the problem.

1. INTRODUCTION

Finite element methods have traditionally been applied to problems where complexities in geometry or material behavior preclude closed form analytic solutions. The power of the finite element approximation has allowed engineers to include considerable detail in their structural response calculations, potentially eliminating the need for simplifying assumptions. However as engineers investigate processes with nonlinear material and geometric behavior, computational resources are quickly exhausted, and the rigid assumption once again becomes valuable. An excellent example is in sheet metal forming, where the assumption of rigid tooling is very reasonable.

Finite elements may be rendered rigid in several ways. The simplest method is to apply very large elastic moduli to an otherwise deformable element. This "brute force" method requires no code modifications, yet has several shortcomings. Rigid elements still need to be processed during stress, strain, and stiffness evaluations. Nodal degrees of freedom associated with rigid elements remain independently active. And large disparities in material properties lead to numerical errors.

This paper presents a rigid material implementation based on rigid body mechanics. Following the convention Benson and Hallquist [1986] used for explicit time integration, a group of finite elements may be defined to be a rigid body. The center of mass is computed, and six degrees of freedom are assigned to the body - three for translation and three for rotation. Nodal loads and boundary conditions are resolved onto the center of mass coordinates.

In this implicit implementation, motion of the rigid body is determined by solving a coupled set of equilibrium equations for the rigid body and any other deformable elements in the model. Finally, the current coordinates of rigid body nodes are updated using the center of mass coordinates and the rigid body assumption. The rotational motion is characterized using quaternion parameters, which also find application in computer graphics literature (Burger and Gillies [1989], Glassner [1990]).

This approach results in a method where stress, strain, and stiffness computations for rigid elements may be skipped. Numerical errors from disparate material properties are avoided. And most importantly, the total number of independent degrees of freedom in the model is reduced. This allows very large numbers of elements to be included in rigid bodies without adding significant computing cost to the model, and allows existing models to be executed more quickly by invoking the rigid assumption where appropriate.

2. FORMULATION

The following presents an integration of rigid body mechanics with the updated Lagrangian finite element formulation using implicit time integration. The procedure is applied to the inner-most loop of the nonlinear equilibrium solver, during stiffness and force computation, and after the linear equation solving procedure.

2.1. *Kinematics*

The position vector $\mathbf{x}$ of a finite element node point may be written as

$$\mathbf{x} = \mathbf{X} + \mathbf{u} \qquad (2.1)$$

where $\mathbf{X}$ is the initial coordinate of the point and $\mathbf{u}$ is its displacement vector. If the point is associated with a rigid body, then

$$\mathbf{x} = \overset{cm}{\mathbf{X}} + \mathbf{a} \qquad (2.2)$$

where $\overset{cm}{\mathbf{X}}$ is the current position of the center of mass, and $\mathbf{a}$ is the current vector from the center of mass to the point. The vector $\mathbf{a}$ may be written in terms of $\mathbf{a}_0$, its value in the undeformed or reference state, and a rotation matrix $\mathbf{\Lambda}$:

$$\mathbf{a} = \mathbf{\Lambda}\mathbf{a}_0 \qquad (2.3)$$

The incremental displacement relationships used in implicit finite element formulations are obtained by linearization of the above expressions :

$$\Delta\mathbf{u} = \Delta\overset{cm}{\mathbf{X}} + \Delta\mathbf{\Lambda}\mathbf{a}_0 \qquad (2.4)$$

Linearization of the rotation matrix $\mathbf{\Lambda}$, presented by Simo [1988], results in the convenient form:

$$\Delta\mathbf{\Lambda}\mathbf{a}_0 = \mathbf{R}\Delta\boldsymbol{\theta} = \begin{bmatrix} 0 & a_3 & -a_2 \\ -a_3 & 0 & a_1 \\ a_2 & -a_1 & 0 \end{bmatrix} \begin{Bmatrix} \Delta\theta_1 \\ \Delta\theta_2 \\ \Delta\theta_3 \end{Bmatrix} \qquad (2.5)$$

where $\mathbf{a} = \{a_1, a_2, a_3\}^T$, and $\{\Delta\theta_1, \Delta\theta_2, \Delta\theta_3\}$ are the components of the rigid body rotation increment expressed in global coordinates.

For a model containing both deformable and rigid components, the nodal degrees of freedom may be grouped, and the above expressions used to obtain a condensed set of unknowns:

$$\begin{Bmatrix} \Delta\mathbf{u}^D \\ \Delta\mathbf{u}^R \end{Bmatrix} = \begin{bmatrix} \mathbf{I} & \mathbf{0} & \mathbf{0} \\ \mathbf{0} & \mathbf{I} & \mathbf{R} \end{bmatrix} \begin{Bmatrix} \Delta\mathbf{u}^D \\ \Delta\overset{cm}{\mathbf{X}} \\ \Delta\boldsymbol{\theta} \end{Bmatrix} \Rightarrow \Delta\mathbf{u} = \mathbf{A}\Delta\hat{\mathbf{u}} \qquad (2.6)$$

where the $\hat{(\cdot)}$ superscript denotes a condensed degree of freedom vector. Substituting this expression into the discrete form of the principle of virtual work, we obtain expressions for the condensed finite element stiffness matrix and residual vector for the coupled deformable/rigid system:

$$\hat{\mathbf{K}}\Delta\hat{\mathbf{u}} = \hat{\mathbf{F}} \quad , \quad \hat{\mathbf{K}} = \mathbf{A}^T\mathbf{K}\mathbf{A} \quad , \quad \hat{\mathbf{F}} = \mathbf{A}^T\mathbf{F} \qquad (2.7)$$

This condensed system is passed to a standard linear equation solver, returning rigid body translation and rotation increments $\Delta\overset{cm}{\mathbf{X}}$ and $\Delta\boldsymbol{\theta}$. These are used to update the center of mass, and finally the nodal coordinates.

2.2. *Center of mass update*

The update phase proceeds by first updating the center of mass degrees of freedom, and then the positions of each node point in the rigid body. The center of mass update consists of a translational and rotational part. The translational degrees of freedom are trivially updated:

$$\overset{cm}{\mathbf{X}}_{(n+1)} = \overset{cm}{\mathbf{X}}_{(n)} + \Delta\overset{cm}{\mathbf{X}} \qquad (2.8)$$

We employ quaternion parameters to characterize the rotational configuration of the rigid body, allowing finite rotation increments in a single step without inducing deformation. These are updated as follows. First, the incremental quaternion parameters $(q_0, \mathbf{q})$ are computed from the rotation increments $\Delta\boldsymbol{\theta}$:

$$q_0 = \cos\alpha \quad , \quad \mathbf{q} = \beta\,\Delta\boldsymbol{\theta} \qquad (2.9)$$

where

$$\alpha = \tfrac{1}{2}\left(\Delta\theta_1^{\,2} + \Delta\theta_2^{\,2} + \Delta\theta_3^{\,2}\right)^{\frac{1}{2}}$$
$$\beta = \begin{cases} \frac{1}{2\alpha}\sin\alpha & , \quad \alpha \geq 10^{-5} \\ \frac{1}{2}\left(1 - \frac{\alpha^2}{6} + \frac{\alpha^4}{120} - \frac{\alpha^6}{5040}\right) & , \quad \alpha < 10^{-5} \end{cases} \qquad (2.10)$$

These are normalized according to

$$\gamma = \left(q_0^{\,2} + q_1^{\,2} + q_2^{\,2} + q_3^{\,2}\right)^{\frac{1}{2}} \quad , \quad q_i = {}^{q_i}\!/_{\gamma} \qquad (2.11)$$

The total quaternion parameters $(x_0, \mathbf{x})$ and $(y_0, \mathbf{y})$ characterize the total rotation of the rigid body from the undeformed or reference state to states *(n)* and *(n+1)*, respectively. The update is performed using quaternion algebra:

$$(y_0, \mathbf{y}) = (q_0, \mathbf{q}) \circ (x_0, \mathbf{x}) \qquad (2.12)$$

where

$$\begin{aligned} y_0 &= q_0 x_0 - \mathbf{q}\cdot\mathbf{x} \\ \mathbf{y} &= q_0\mathbf{x} + x_0\mathbf{q} + \mathbf{q}\times\mathbf{x} \end{aligned} \qquad (2.13)$$

The total quaternions are then normalized using the aforementioned procedure. From these, the current rotation matrix $\mathbf{\Lambda}$ may be computed:

$$\mathbf{\Lambda} = \begin{bmatrix} \left(y_0^{\,2} + y_1^{\,2} - \frac{1}{2}\right) & \left(y_1y_2 - y_3y_0\right) & \left(y_1y_3 + y_2y_0\right) \\ \left(y_2y_1 + y_3y_0\right) & \left(y_0^{\,2} + y_2^{\,2} - \frac{1}{2}\right) & \left(y_2y_3 - y_1y_0\right) \\ \left(y_3y_1 - y_2y_0\right) & \left(y_3y_2 + y_1y_0\right) & \left(y_0^{\,2} + y_3^{\,2} - \frac{1}{2}\right) \end{bmatrix} \qquad (2.14)$$

2.3. *Nodal update*

For consistency with the updated Lagrangian finite element formulation, the displacement and rotation vectors of each rigid body node must be determined.

With the center of mass degrees of freedom updated, the nodal displacements may be computed using (2.2), (2.3), (2.8), and (2.14) simply as

$$\Delta \mathbf{u} = \mathbf{x}_{(n+1)} - \mathbf{x}_{(n)} \tag{2.15}$$

The nodal rotation increments are equal to the center of mass rotation increments. These are computed similarly by differencing the total rotation vectors $\boldsymbol{\theta}_{(n+1)}$ and $\boldsymbol{\theta}_{(n)}$. The rotation vectors are extracted from the total quaternions using:

$$\boldsymbol{\theta} = \beta\, \mathbf{y} \tag{2.16}$$

where

$$\alpha = \left(y_1^{\,2} + y_2^{\,2} + y_3^{\,2}\right)^{\frac{1}{2}}$$
$$\beta = \begin{cases} \frac{2}{\alpha} \sin^{-1} \alpha \;, & \alpha \geq 10^{-6} \\ 0 \;, & \alpha < 10^{-6} \end{cases} \tag{2.17}$$

3. IMPLEMENTATION

The rigid body capability was implemented as a material attribute in the finite element code NIKE3D (Maker, Ferencz, and Hallquist [1991]). This proves convenient for users, since each part in a finite element model is typically assigned a unique material number as the model is constructed. Rigid bodies may be constructed from any of the element types available in NIKE3D - beams, shells, and/or solids.

The condensation procedure (2.7) is implemented at the element level. Processing of elements assigned rigid material types is skipped entirely. The element stiffness and internal force vector for all other elements are formed by the usual procedure. Next, deformable elements containing one or more rigid body nodes are passed to a condensation routine. This routine uses (2.5) and (2.6) to modify the element stiffness and internal force data prior to assembly.

Although the condensation procedure reduces the number of global unknowns in virtually every model, the element level matricies may in some cases be expanded. This occurs, for example, when a single node from a trilinear continuum element is associated with a rigid body. The three (translational) degrees of freedom of that node are then replaced by the six (translational and rotational) degrees of freedom of the rigid body, increasing the dimension of the element matricies by three. However, if several of the element nodes belong to the same rigid body, the net size of the element system is indeed reduced. In general, provision must be made for assembly of a variable sized element system. In the worst case of an eight node hexahedral element where each node is attached to a different rigid body, the element system expands from 24 to 48 degrees of freedom.

All of the boundary conditions in NIKE3D are available for use with rigid bodies. These include prescribed displacement, force, pressure, and body forces. In the case of prescribed displacements, the user must prescribe the motion of the center of mass of the rigid body. Nodal displacement boundary conditions are ignored. Slide surface or contact boundary conditions are also available for use with rigid bodies. Penalty parameters for these interfaces are chosen using reasonable values of elastic moduli which are entered as rigid body material properties.

4. APPLICATIONS

The following examples demonstrate the validity of the implementation, and some of the potential cost savings from the rigid assumption.

4.1 *Rotation of rigid block*

To test the accuracy of the rotational update, a simple block of solid elements was rotated a total of 360 degrees. Several simulations were performed, using from one to ten steps to complete the rotation. Figure 1 shows the model in several intermediate configurations. Deformation induced by errors in the update was measured by comparing initial and final geometry of the block. In all cases, normalized error was less than 10^{-6}.

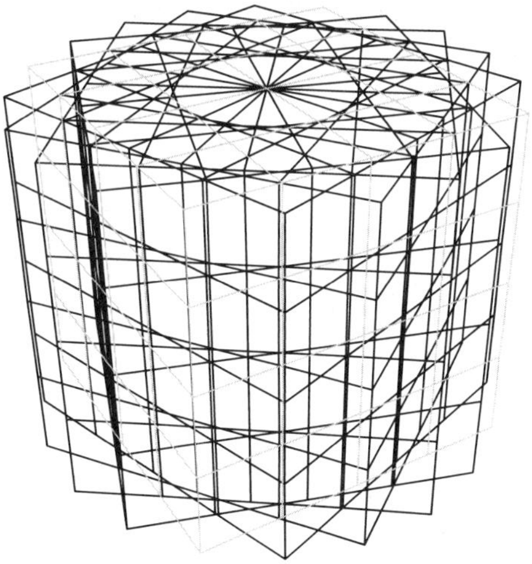

Figure 1 - In a test of the rigid body rotational update, a rigid block of elements is rotated 360 degrees using from one to ten steps. In all cases, normalized deformation of the body was less than 10^{-6}.

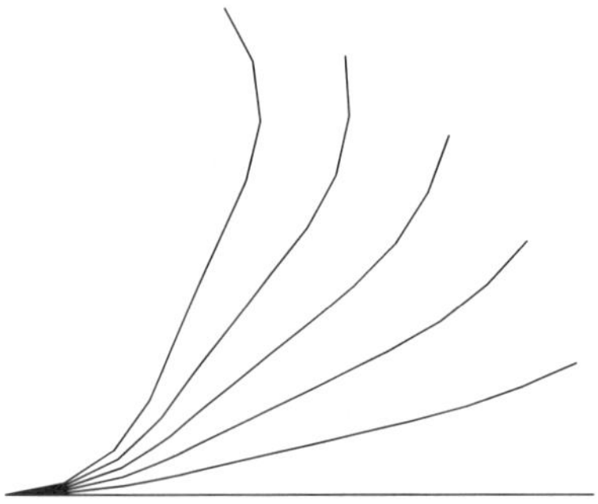

Figure 2 - The centerline of an elastic cantilevered beam loaded with an end moment is shown at several stages of deformation. The beam is discretized with ten elements, the central four elements being rigid. Load is applied in ten increments. The rigid behavior of the central portion of the beam through finite rigid body translation and rotation is demonstrated.

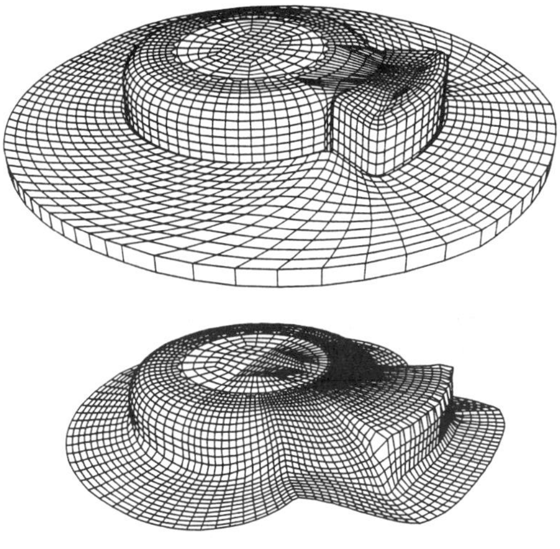

Figure 3 - Results from a NIKE3D hydroforming simulation show the tooling (top) and finished part (bottom). The tooling model is constructed using a single layer of eight node solid elements, making it an ideal candidate for the rigid material, which reduced runtime by over 90 percent.

4.2 *Elastic / rigid beam*

A simple cantilevered beam loaded with a tip moment is shown in Figure 2. The beam is discretized using ten elements. The four center elements are defined as rigid, while the remaining elements have elastic properties. Load is applied in ten increments. The beam centerline is shown for several stages during the loading. The rigidity of the center section of the beam through finite translation and rotation is apparent.

Table 1: Cost of hydroforming simulation using deformable or rigid tooling. The tooling (brick elements) consumed nearly half of the cost in the first model, and virtually none when the rigid material was used.

Model Characteristics:
sheet: 1650 shell elements
tooling: 1550 brick elements

		deformable	*rigid*
Linear System:			
equations		16000	10000
storage (Mword)		17.0	3.2
	(EBE)	6.6	
CPU Cost Breakdown			
initialization		0.5%	2.0%
slide surfaces		1%	53%
brick elements		47%	0%
shell elements		12%	21%
linear solver	(EBE)	40%	21%
Cray Y/MP cost		500 min.	37 min.

4.3 *Hydroformed sheet metal cover*

A hydroforming simulation first presented in [Maker, 1988] was repeated using the rigid material. Figure 3 shows the model used to form an aluminum sheet into a three dimensional pan. In the hydroform process, the male tooling emerges from an initially flat blank holder surface, while subsequent external pressure drives the sheet to conform to the tooling contour. The tooling and blank holder were modeled using one layer of solid eight node elements.

Table 1 shows the cost breakdown associated with two NIKE3D simulations: one with deformable tooling, the other with rigid tooling. The size of the linear system in the deformable tooling analysis necessitated the use of NIKE3D's EBE iterative linear equation solver [Ferencz, 1989], which dramatically reduces storage requirements by operating at the element level. Condensing the linear system using the rigid material assumption allowed the use of a traditional direct solver.

Applying the rigid material decreased the total simulation time by over 90%, from 500 down to 37 cpu minutes on our Cray Y/MP machine. A large portion of this savings was realized by skipping the processing of the solid elements which comprise the tooling, saving nearly 250 cpu minutes. The decreased size of the condensed linear system accounted for a savings of nearly 200 cpu minutes. In addition, the rigid tooling simplified enforcement of the contact constraint between the sheet and tooling, since only one side of this interface was deformable. This resulted in fewer iterations in the nonlinear equation solver while searching for equilibrium during each load step.

5. CONCLUSIONS AND FUTURE WORK

The rigid body capability has been implemented in NIKE3D for static analyses, and validated on a number of test problems. Significant cost savings have been obtained by reducing the number of degrees of freedom in the linear system, eliminating element processing for rigid elements, and by reducing the complexity of the contact problem. Additional work is required to extend the capability to dynamic problems, namely the solution of an additional (rotational) equilibrium equation, and the resulting (non-symmetric) set of linear equations.

REFERENCES

Benson, D.J. and J.O.Hallquist, "A simple rigid body algorithm for structural dynamics programs," *IJNME*, **22**, 1986

Burger, P. and D.Gillies, *Interactive Computer Graphics*, Addison-Wesley, 1989

Ferencz, R.M., "Element-by-element preconditioning techniques for large-scale, vectorized finite element analysis in nonlinear solid and structural mechanics", Ph.D. Thesis, Stanford University, 1989

Glassner, A.S., *Graphics Gems*, Academic Press, 1990

Maker, B.N., R.M.Ferencz and J.O.Hallquist, "NIKE3D: A nonlinear, implicit, three-dimensional finite element code for solid and structural mechanics - user's manual," University of California, Lawrence Livermore Lab Rept. UCRL-MA-105268, 1991

Maker, B.N., "Finite element modeling of a hydroformed sheet metal cover", University of California, Lawrence Livermore Lab Rept. UCID-21614, 1988

Simo, J.C., and Vu-Quoc, L., "On the dynamics in space of rods undergoing large motions - a geometrically exact approach", CMAME, 66, 1988, pp 125-161

Simulation of Materials Processing: Theory, Methods and Applications, Shen & Dawson (eds)
© 1995 Balkema, Rotterdam. ISBN 90 5410 553 4

Data transfer for 3-D h-adaptive thermal-elasto-plastic finite element analysis

J.M.J. McDill & A.S. Oddy
Mechanical and Aerospace Engineering, Carleton University, Ottawa, Ont., Canada

M.E. Klein
Fuel Engineering Branch, Atomic Energy of Canada Limited, Chalk River, Canada

ABSTRACT: Data transfer methods in h-adaptive mesh management are presented for 3-D, grading bricks used in thermal-elasto-plastic analysis. Data transfer by a simple method (S) which sacrifices solution accuracy for savings in computational time is contrasted with a more rigorous method (R) which maintains the accuracy of the solution during the transfer. In the simple method, during refinement, interpolation of nodal data and assignment of the nearest Gauss point data are made. During coarsening, superfluous nodal data are dropped and relevant Gauss point data are averaged. These make no attempt to minimize the difference between the data fields in the old and new meshes. The rigorous method computes the new data fields based on a least-square minimization of the error in the transferred nodal and Gauss point data. Both methods achieve data transfer with acceptable error, in simple problems. For problems with complex stress states, the rigorous method is more conservative. For coarsening in particular, the rigorous method is clearly superior. In general, a least-square minimization procedure is recommended to promote accurate data transfer.

1. Introduction

In many cases, a finite element (FEA) mesh may be too coarse to provide answers of sufficient accuracy in areas of concern. If a uniformly higher concentration of elements is used, the associated computational costs become large. Local refinement can increase the local accuracy of the solution while minimizing increases in computational costs.

Automatic finite element analysis (AFEA) addresses the concerns of accuracy and computational speed. Adaptive FEA and dynamic mesh-management algorithms modify the finite element mesh to increase the accuracy of the solution by refinement and/or improve the computational cost by coarsening. A mesh-management algorithm guides changes in the mesh to capture the effects of localized, transient, or moving phenomena.

McDill et al. [1] begin h-adaptive mesh management with a mesh which describes the geometry of the problem. An initial solution is used to identify elements that can be refined or coarsened. This can be likened to a family tree in which the element being divided is the parent, and the newly created elements, the children, are siblings. Alternatively, elements being combined are labelled the children and the new, larger element is identified as a parent. If the solution is rough; e.g., the gradient is discontinuous, then h-refinement is more efficient than p-refinement [2]. This occurs in the vicinity of point loads, material discontinuities and elasto-plastic boundaries.

Data transfer provides values for the history-dependent quantities in the new elements by projecting the nodal and Gauss point data from the old mesh. Conditions must be completely defined before the next solution step occurs [3]. Four different problems must be addressed: the transfer of nodal data and Gauss point data in refinement; and the transfer of nodal data and Gauss point data in coarsening.

The objective is to compare a simple method (S) of data transfer with a more rigorous method (R). S methods sacrifice solution accuracy for savings in computational time by using a means of data transfer where no attempt is made to minimize the difference between the data fields. An R method was developed to maintain the accuracy of the solution during the transfer of element data. The error is minimized at the expense of computational time.

2. Techniques for Data Transfer

Simply stated, a data field $T^*(x,y,z)$ exists for mesh 1, and is to be projected to mesh 2 to give the new field $T(x,y,z)$. Temperature and displacement are typical of nodal data transferred. The Gauss point data could be stresses, plastic strains, effective plastic and creep strains, solidification fractions or other history dependent data. There appear to be few references to data transfer, however interpolation [4-6] and a least squares approach [4][5] of data transfer are mentioned.

Heuristic methods can be used; e.g., in coarsening, the Gauss point data of the children can be averaged and copied to the appropriate Gauss points of the parent[7]. Similarly, in refinement, Gauss point data of a parent can be copied to the appropriate Gauss points in the children. Some algorithms remove nodes during coarsening. The associated data could be discarded, which affects the conservation of energy, momentum etc.

2.1 Interpolation

Interpolation is a fast method. Data are transferred from mesh 1 to 2 using the basis functions of mesh 1, by defining the coordinates of the new node within the original element. The value for the new node in mesh 2 is determined by

$$T_i = \sum_{j=1}^{n_1} P_j T_j^*$$

T_i is the value for the point in mesh 1 of the new node in mesh 2; P_j are the basis functions of the original element evaluated at the coordinates of the new node; T_j^* are the values for the element in mesh 1; n_1 is the number of nodes in mesh 1.

The transfer of Gauss point data is slightly more difficult. Data should first be projected to the nodes of the element before interpolating.

2.2 The Least Squares Method

The least squares approach (LSQ) minimizes the difference between the initial data field and the new field. It is useful where interpolation methods lead to a loss of energy. In a coarsening problem the nodal data from two or more elements are transferred to a new larger element. LSQ gathers the data to assemble equations which are solved to obtain the transferred data.

LSQ methods minimize the integral of the square of the difference between fields T^* and T over the domain of mesh 2.

$$\frac{\partial}{\partial T}\left(\frac{1}{2}\int\left(T-T^*\right)^2 d\Omega\right)=0$$

where T, T^* are $T=\sum_{i=1}^{n_2} N_i T_i$ and $T^*=\sum_{i=1}^{n_1} P_i T_i$

T_i are unknown nodal values in mesh 2; N_i are basis functions for mesh 2 evaluated at the Gauss points of mesh 2; T_i^* are the known nodal values for mesh 1; P_i are basis functions for mesh 1 evaluated at the Gauss points of mesh 2; n_1, n_2 are the number of nodes in mesh 1 and mesh 2 respectively; Ω is the domain of mesh 2.

The usual conversion to a computational space, inclusion of Gaussian quadrature and minimization with respect to the data field, T, gives:

$$\sum_{m=1}^{M} W_m \begin{bmatrix} N_1 \\ \bullet \\ \bullet \\ N_{n_2} \end{bmatrix} \begin{bmatrix} N_1 & \bullet & \bullet & N_{n_2} \end{bmatrix} \begin{bmatrix} T_1 \\ \bullet \\ \bullet \\ T_{n_2} \end{bmatrix} \det J_m =$$

$$\sum_{m=1}^{M} W_m \begin{bmatrix} N_1 \\ \bullet \\ \bullet \\ N_{n_2} \end{bmatrix} \begin{bmatrix} P_1 & \bullet & \bullet & P_{n_2} \end{bmatrix} \begin{bmatrix} T_1^* \\ \bullet \\ \bullet \\ T_{n_1}^* \end{bmatrix} \det J_m$$

M is the number of Gauss points in mesh 2; W_m are the Gauss weights. This expresses a linear set of equations,

$$[A][u]=[b]$$

$[A]$ is a $(n_2 \times n_2)$ matrix; $[u]$ is the vector of unknown values; $[b]$ is a vector of known values

3. Simple and Rigorous Methods

The h-adaptive mesh-manager in [1] uses a local process; i.e., only the area in which the mesh is being refined or coarsened is affected. The local zone includes those elements being refined and/or coarsened, and their immediate neighbours. The number of subelements in an element depends on the placement of midedge and midsurface nodes[8]. The introduction of new nodes by refinement, and the removal of existing nodes by coarsening, may change the number of subelements and Gauss points in neighbours. The Gauss point data in the neighbours must be transferred to the new subelement Gauss points.

3.1 Data Transfer for Refinement

Division of the 8-26 node parent into eight elements maintains shape regularity[1].

Nodal Data Transfer in Refinement

Interpolation and an LSQ solution which operates on a local zone were considered for the nodal data transfer. Constraints prevent nodal values on the boundary of the zone from changing. LSQ requires a minimum number of Gaussian integration points to prevent under-integration and a singular solution.

Interpolation and LSQ create the same field, but interpolation achieves this in far fewer steps [9]. The average of the nodal quantities is conserved in both cases. Interpolation is the method chosen for both S and R transfer of nodal data in refinement

Gauss Point Data in Refinement

Three candidate processes were considered for the transfer of Gauss point data during refinement. The first was duplication in which the Gauss point data of the parent are copied to the relevant Gauss points of the children. The others were a least squares projection and interpolation (LSP/IT), and a least squares projection and LSQ transfer (LSP/LSQ).

Duplication is a heuristic approach in which the parental Gauss point data are copied to the new Gauss points in the children. The method is computationally efficient and was selected as the S method for transferring Gauss point data during refinement.

For LSP/IT, the Gauss point data are projected to the eight corners of each subelement of the parent using a weighted least squares projection matrix[10]. The coordinates of the new Gauss point in the original element are calculated, the element basis functions evaluated and interpolation used. Each subelement is treated separately, with no averaging of the data at the nodes.

In LSP/LSQ, subelement data are projected to the corners of the subelement and the LSQ equations formulated. Since the equations are identical to those generated by interpolation[9], LSP/IT was chosen as the R method for Gauss point data in refinement.

3.2 Data Transfer for Coarsening

In h-coarsening, [1] creates a single new element from seven possible combinations of two, four, or eight 8-node elements. The number of subelements and the removal of redundant nodes determines the number of Gauss points in the coarsened parent.

Nodal Data Transfer in Coarsening

Two processes were considered for the nodal transfer data in coarsening; a heuristic drop approach and a least squares solution on a local domain.

In the drop method only data from nodes common to both meshes are preserved. Data from other nodes are discarded. This is simple and very fast but the average is not conserved especially when gradients are irregular. The drop method was selected as the S method for transferring nodal data during coarsening.

LSQ conserves data from the redundant nodes as it acts to minimize the difference between the old and new data fields. It works on an element-by-element basis. The neighbours of the elements being coarsened are part of the local domain in the LSQ formulation. LSQ was selected as the R method for nodal data transfer in coarsening.

Gauss Point Data in Coarsening

Three means of Gauss point data transfer for coarsening were considered: averaging the children's Gauss point data; a least squares projection followed by LSQ data transfer (LSP/LSQ); and a least squares projection followed by interpolation (LSP/IT).

Averaging a simple process in which the Gauss point data of the children are averaged and copied to the new parent was selected for the S approach.

LSP/LSQ is more expensive but no improvement over LSP/IT. In LSP/IT, the Gauss point data of the children are projected to the nodes. The parent Gauss points within each child are identified and the data are interpolated from the children's nodes on a child-by-child basis, preventing pollution of the projected data by averaging, with neighbouring data. LSP/IT was selected as the R method for coarsening.

In summary, for refinement both the S and R use interpolation for transfer of nodal data. Gauss point data are transferred by a duplication process for the S method while R uses LSP/IT. For coarsening, the S method discards nodal data at redundant nodes while the R method does a local LSQ.

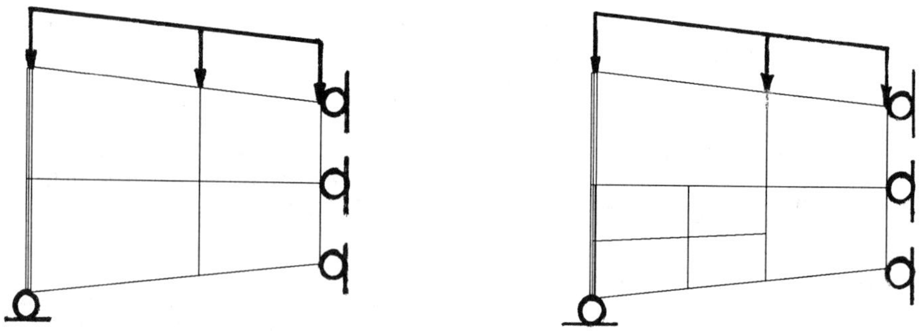

A) 8 element mesh B) 15 element mesh

Figure 1. Beam Bending Test;

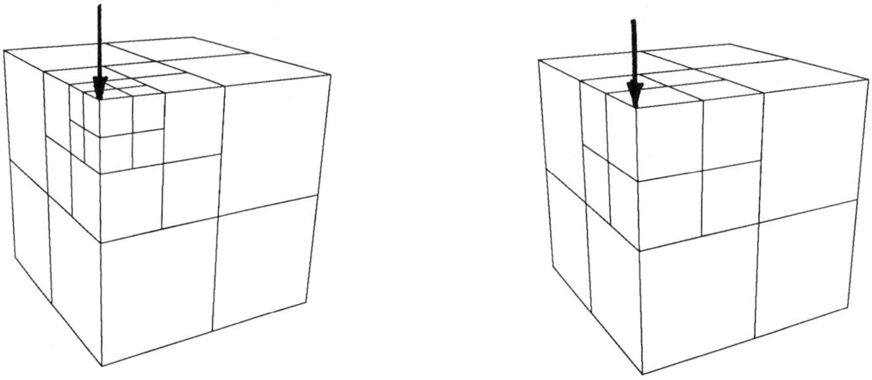

22 element mesh 15 element mesh

Figure 2. Point Load Test;

For Gauss point data the S method uses an averaging technique while R uses an LSP/IT.

4. Verification

The accuracy of data transfer was verified by testing problems with known analytical solutions. Error norms are computed based on the normalized difference between f the exact, and $\hat{f}$ the FEA solution. The G_1 error norm is ;

$$G_1 = \frac{\int \left(f - \hat{f}\right)^2 d\Omega}{\int f^2 d\Omega}$$

This is sensitive to the reference point; e.g., if the functions f and $\hat{f}$ are changed by addition of an arbitrarily large constant (i.e., ^{o}C to K, or rigid body motion), the value of G_1 will decrease. This difficulty is overcome by using an additional norm,

$$G_2 = \frac{\int \left(f - \hat{f}\right)^2 d\Omega}{\int \left(f - \bar{f}\right)^2 d\Omega} \quad \text{and} \quad \bar{f} = \frac{\int f d\Omega}{\int d\Omega}$$

Both G_1 and G_2 norms should be calculated. If both values are positive and small, a good fit is indicated.

Simple tests with uniform stress states showed that both the S and R methods caused no error whatsoever [11]. More complex tests with non uniform stress states are described here.

Beam Bending by Uniform Load

The bending of a narrow rectangular, simply supported beam, symmetric about the midspan, with a uniform load was selected. The dimensions are 40x20x1 and a distributed load of 10 is applied to one surface. Young's modulus is $30.\times10^6$ and Poisson's ratio is 0.3. The analytical solution is presented in [12].

FEA ——————— **Simple** ......□......□...... **Rigorous** ......▽......▽......

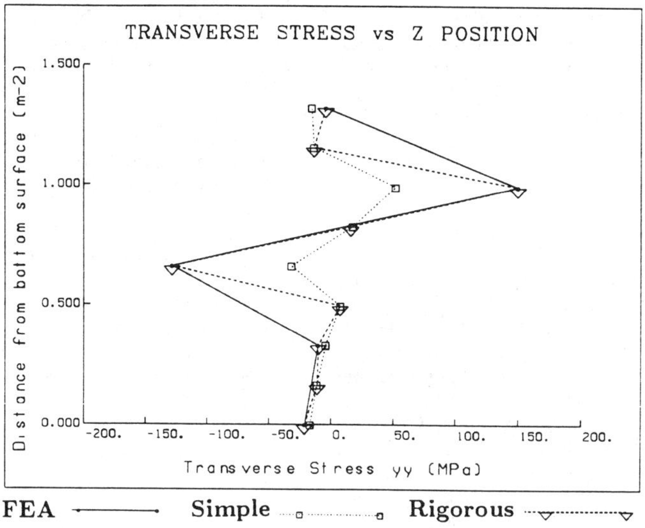

Figure 3. Weld Transverse Stresses; FEA, S and R Transfer in Refinement

The half beam was modelled using a mesh with eight elements (2x2x2) as shown in Figure 1. One element was refined.

Error norms are presented in Table 1. The G_1 displacement error norms for the data transfer methods are identical. The stress norms show that the S method has increased the error whereas the R method has not.

Table 1 Error Norms for Beam Bending: Refinement

Type	Displ G1(%)	Displ G2(%)	Stress G1 (%)	Stress G2 (%)
8-elem FEA	2.74	14.2	3.65	18.3
8-15 S	2.08	14.5	5.47	28.8
8-15 R	2.08	14.5	3.39	17.5

The beam test also studied the data transfer process during coarsening. The refinement procedures were reversed so that the 15-element mesh was coarsened to a single-element mesh. Similar differences for S and R transfer methods were seen[11].

Point Load on a Half Space

The next case selected was a point load applied at the centre of a half space. This has a $1/z^2$ singularity in σ_{zz} [12].

Several meshes were examined [11]. Two are shown in Figure 2. Symmetry allows a quarter space 10x10x10 to be used. No displacements were permitted on the planes of symmetry or the lower surface. A point load of 10 was applied at (0,0,0). Young's modulus is 2.0×10^{11} and Poisson's ratio is 0.3.

Initially a 15-element mesh, refinement creates a new 22-element mesh. The error norms for the S and R methods in refinement did not differ substantially but were better for the R method [11].

The data transfer code for coarsening also tested used this problem. The initial 22-element mesh was coarsened in two stages to obtain an 8-element mesh. Error norms are shown in Table 2 as are the FEA solutions. The results demonstrate that the R method is clearly superior, in some cases producing errors an order of magnitude less than the S method.

Table 2 Error Norms: Coarsening at Point Load

Type	Displ G1 (%)	Displ G2 (%)	Stress G1 (%)	Stress G2 (%)
22-elem FEA	0.33	1.21	28.7	29.9
22-15 S	9.15	34.15	5.47	5.95
22-15 R	1.32	4.93	1.09	1.18
15-elem FEA	0.39	1.46	29.3	31.8
15-8 S	161.5	639.6	8.30	9.91
15-8 R	2.42	9.58	3.77	4.50
8-elem FEA	0.51	2.02	30.2	36.0

Although this is a case of extreme coarsening, it does illustrate the potential risk of using the S method, particularly for the transfer of nodal data. Clearly the R method is the better data transfer process.

5. Thermal Elasto-Plastic Case

Refinement of the last step in a thermal-stress analysis of a plate weld was a final test. Weld conditions are described in [11]. Figure 3 shows the transverse stresses vs thickness at the weld centreline. The S method, with its simple assignment of Gauss point data tends to smooth the data. The R method with the nodal projection and reinterpolation better preserves the stress distribution

6. Contributions and Conclusions

Use of either an S or R method must consider the problem type, desired computational accuracy, and computational speed. In refinement, the S method will transfer data less accurately than the R method. The difference occurs only with Gauss point data since both methods use the same procedure for nodal data.

In coarsening, the S method was 5 to 40% faster than the R method in the transfer of data for these simple test cases. Clearly, the accuracy of the S method is greatly affected by the type of stress field in the problem. In the case of bending and stress singularities the S results may prove unacceptable. The point load is particularly informative showing the risk in using the S technique for transferring nodal data.

References

[1] McDill, J.M.J., A.S. Oddy and J.A. Goldak, An Adaptive Mesh-Management Algorithm for Three-Dimensional Automatic Finite Element Analysis, Trans. CSME, 15, 57-70 (1991).

[2] Szabo, B.A., Estimation and Control of Error Based on P-Convergence, Int. Conf. on Accuracy Estimates and Adaptive Refinements in Finite Element Computations (ARFEC), Lisbon, Portugal, June 1984.

[3] Dukowicz, J.K., Conservative Rezoning (Remapping) for General Quadrilateral Meshes, J. Comp. Phys., 54, 411-424 (1984).

[4] Cheng, J-H., Automatic Adaptive Remeshing for Finite Element Simulation of Forming Processes, Int. j. numer. methods eng., 26, 1-18 (1988).

[5] Cheng, J.-H. and N. Kikuchi, A Mesh Re-Zoning Technique for Finite Element Simulations of Metal Forming Processes, Int. j. numer. methods eng., 23, 219-228 (1986).

[6] Yang, Henry T.Y., M. Heinstein and J.-M. Shih, Adaptive 2D Finite Element Simulation of Metal Forming Processes, Int. j. numer. methods eng., 28 , 1409-1428 (1989).

[7] Jin, H., and N.E. Wiberg, Two-Dimensional Mesh-Generation, Adaptive Remeshing and Refinement, Int. j. numer. methods eng., 29, 1501-1526 (1990).

[8] McDill, J.M., J.A. Goldak, A.S. Oddy and M.J. Bibby, Isoparametric Quadrilaterals and Hexahedrons for Mesh-Grading Algorithms, Com. Appl. Num. Meth., 3, 155-163 (1987).

[9] Hinton, R. and J.S. Campbell, Local And Global Smoothing of Discontinuous Finite Element Functions Using A Least Squares Method, Int. j. numer. methods eng., 8, 461-480 (1974).

[10] Patel, B., Thermo-Elastic-Plastic Finite Element Analysis of Deformations and Residual Stresses Due to Welds , Ph. D. Thesis, Carleton University, Ottawa (1985).

[11] Klein, M.E., Development and Comparison of Data Transfer Methods for h-Adaptive Finite Element Analysis, M. Eng. Thesis, Carleton, 1992.

[12] Timoshenko, S.P., and J.N. Goodier, Theory of Elasticity, McGraw-Hill, New York, 49-52 and 399-403 (1970).

Simulation of Materials Processing: Theory, Methods and Applications, Shen & Dawson (eds)
© 1995 Balkema, Rotterdam. ISBN 90 5410 553 4

Effective application of different solvers to forming simulations

C.D. Mercer, J.D. Nagtegaal & N. Rebelo
Hibbitt, Karlsson & Sorensen, Inc., Pawtucket, R.I., USA

ABSTRACT: The goal of simulation software is to provide the designer with an accurate, reliable and efficient means to help optimize the product and its manufacturing process. No single solver has been found to be suitable to achieve this goal for different forming problems. The result is that there are different solvers for different applications, each with its own advantages and disadvantages. The objective of this paper is to illustrate the effective use of different solvers in the simulation of forming processes. In particular, a sheetforming process is most effectively simulated using a combination of an explicit solver for the forming simulation and a direct solver for the springback calculation.

1. INTRODUCTION

In the past two decades, significant progress has been made in the quest to provide designers with accurate, reliable and efficient software for the simulation of forming processes. Indeed, a large number of codes are now available to designers that range from special purpose codes (for example, for sheetforming processes only) to general purpose codes. Advances in the finite element software have been made primarily in two fields: first, in the development of accurate and reliable algorithms for materials and elements, and second, in the development of solver technology. The solver generally dictates the architecture of the code and the developments in this field have been closely linked to the progress that has been made in the hardware technology, such as code vectorization and parallelization.

There are two types of solvers that are used by most codes for the simulation of forming processes; those based on the direct solution of equations, and those based on explicit dynamics integration. The direct solvers were more popular in the earlier years as they could be used effectively on the available computers to solve 2-D problems using a rigid-plastic formulation. The progress in developing robust elastic-plastic algorithms with quadratic convergence (see Simo [1], [2]) and improvements in the contact algorithms (see Laursen and Simo [3]) coupled with the advances made in computer hardware, resulted in the codes solving 2-D problems very effectively. However, the extension to 3-D problems has not been straightforward. The major drawback involves the development of robust contact algorithms required to solve 3-D problems. The result has been that for direct solvers, the computer time required to solve any industrial forming problem, whether it is sheetforming or bulk forming, is too large to be practical. The codes based on an explicit dynamics solver do not have these limitations and have become highly efficient simulation tools, see Rebelo et al. [4], Matiasson et. al. [5] and the Numisheet conference proceedings [6]. The explicit dynamics codes have efficient contact algorithms and architectures that take maximum advantage of the vector and parallel capabilities in the hardware. These codes are now capable of performing simulations of complex forming problems in a reasonable time frame.

A third group of solvers is now receiving attention from code developers. These are iterative solvers which have been proven to be effective for large elastic problems, but have as yet not been widely used for highly nonlinear problems such as forming simulations. These solvers are not as robust as direct and explicit solvers and are only effective for a smaller range of problems.

In this paper, we first discuss how a direct solver and an explicit solver can be combined to provide an effective solution technique for the forming and springback simulation of the sheetforming process. Secondly, we illustrate the effectiveness of using an iterative solver for quasi-static forming simulations.

2. COMBINED SOLVER STRATEGY FOR SHEETFORMING PROBLEMS

A direct solver used in an implicit procedure with a Newton scheme to solve the quasi-static equilibrium equations is not effective when used for the simulation of sheetforming. The stress distribution can be accurately obtained, and when there is no contact involved, the springback can be obtained relatively easily. However, in 3-D problems the contact essentially controls the increment size and generally, very small increments are required to establish equilibrium implicitly. The result is that in terms of computer time it is impractical to simulate the forming process.

The explicit solver is more suitable for the simulation of sheetforming. It uses a procedure formulated with diagonal or lumped element mass matrices, and an explicit time integration operator. There is no solution required for a set of coupled equations as in the direct solver techniques, and the procedure is very efficient for large three-dimensional, nonlinear problems that are contact dominated, such as sheetforming. The analysis cost increases in direct proportion to the size of the mesh, whereas the direct solver cost increases with the square of the wavefront times the number of degrees of freedom.

The disadvantage is that the explicit method is conditionally stable and therefore requires very small time increments to ensure stability of the solution. For contact dominated problems this limitation is not too severe. However, when there is no contact and the solver is used for flexible structures that deform predominantly in their lowest frequency modes, as is typical for springback calculations, the analysis time can be significant, and often longer than the time required to simulate the forming process. Mass scaling techniques can be used to artificially speed up the forming simulation. By increasing the mass the stable time increment is increased and fewer increments are needed for the forming analysis. However, this does not help for the springback calculation since it will also increase the time needed to reach static equilibrium.

An effective strategy for the complete sheetforming simulation combines the advantages of both the direct and explicit solvers. It requires that the forming simulation be performed with an explicit solver, and that the subsequent springback analysis be performed with a direct solver. Provided there is no contact in the springback analysis, the direct solver will generally be a lot quicker than the explicit solver. This strategy has been successfully used by Karafillis and Boyce [7] in tooling and binder design.

The problem that one is faced with is how to take a body in contact and in a state of dynamic equilibrium, and transform it into a state of static equilibrium with no contact as efficiently as possible. In most cases nonlinear geometric effects need to be included in the computations, since (local) instabilities may occur during springback. If such an analysis is attempted in a single step, it will likely fail to find a solution. In order to overcome this problem, a relatively simple algorithm is used.

At the start of the springback analysis, an artificial stress state σ^* is applied (at the material points) that is in equilibrium with the stress state at the end of the explicit analysis σ^E. Because of this, the deformed sheet starts the springback analysis in static equilibrium. The artificial stress state is incrementally removed using a linear function over the analysis step so that at each increment static equilibrium is achieved using the standard Newton scheme. The incremental algorithm can be summarized as follows:

For each increment:

$$\sigma^*_{t+\Delta t} = \left(1 - \frac{t+\Delta t}{T_p}\right)\sigma^E$$

where $t+\Delta t$ is the current time and T_p is the time period of the analysis step.

Iterate using the Newton scheme:

$$\Delta u^{(i+1)} = \Delta u^{(i)} + \left(K^{(i)}\right)^{-1}\left(F^{(i)} - I^{(i)}\right)$$

where $K^{(i)}$ is the current tangent stiffness matrix, $F^{(i)}$ is the applied load vector, and $I^{(i)}$ is the internal force vector, assembled from the element internal force vectors $I_e^{(i)}$ calculated for each element as:

$$I_e^{(i)} = \int_{V_e} B^T\left(\sigma^{(i)} - \sigma^*_{t+\Delta t}\right)dV$$

where $\sigma^{(i)}$ are the current stresses. Note that in $K^{(i)}$, the initial stress stiffness contribution is based on the effective stress $\sigma^{(i)} - \sigma^*_{t+\Delta t}$. When the artificial stress state has been completely removed, complete springback will have been obtained with a self equilibrating residual stress state in the sheet.

This algorithm has been proven to be very effective in solving problems that have a large amount of

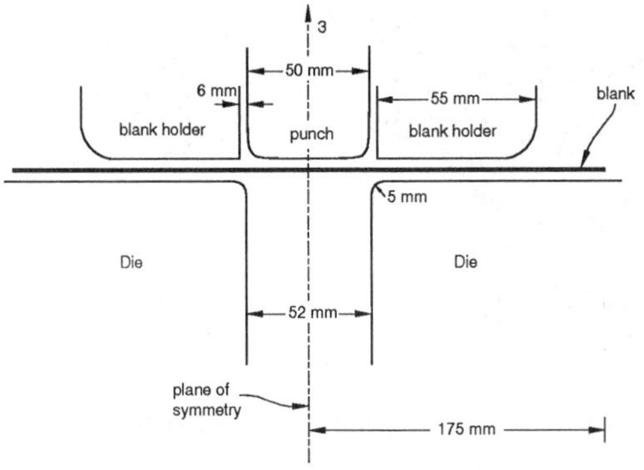

Figure 1: Cross-section showing geometry of the die, punch, blankholder and blank.

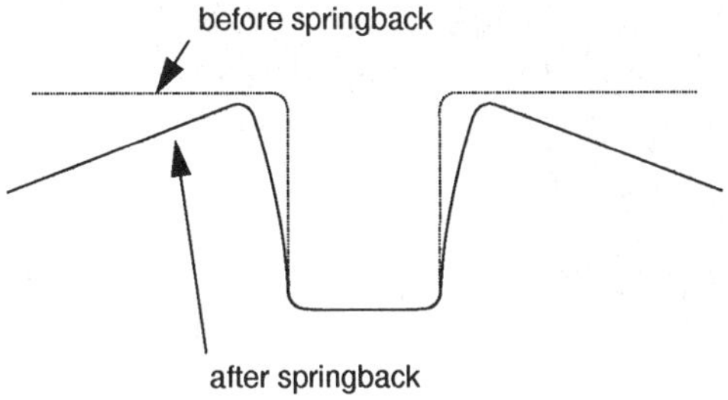

Figure 2: Blank before and after springback.

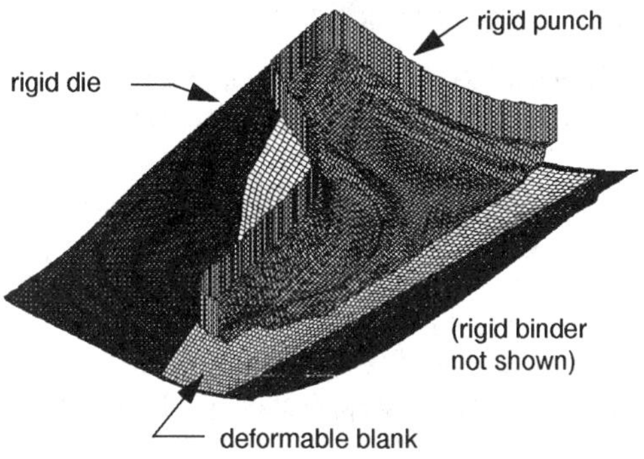

Figure 3: Sectioned view showing half of the model.

springback, as is illustrated in the following examples.

2.1 2-D drawbend problem

This problem is one of the benchmark tests for the Numisheet 93 conference [6]. The problem is essentially a plane strain problem and is modeled with a strip of shell elements. A cross-section of the die, punch, blankholder and blank geometry is shown in Figure 1.

The total blankholder force is 2.45 kN and the blank is made of mild steel. The forming simulation is performed with ABAQUS/Explicit [8]. The velocity of the punch is applied with a triangular amplitude function, starting and ending with zero velocity and with a peak velocity at the middle of the time period. This amplitude helps to minimize the dynamic effects, which is advantageous if the stress state is to be used for subsequent springback computations. A significant amount of springback occurs for this problem. The blank is very flexible and the fundamental vibration mode is low so that the springback calculation using the ABAQUS/Explicit solver takes approximately twenty times the analysis time required for the forming simulation. Using the strategy described above, the results from the forming simulation performed with ABAQUS/Explicit are imported into ABAQUS/Standard [9] where a static analysis is performed. The direct solver in ABAQUS/Standard completes the springback analysis in 14 increments and takes one sixth of the computer time needed by ABAQUS/Explicit for the forming analysis. The blank before and after springback is shown in Figure 2.

2.2 Car fender problem

The original problem for this example is also one of the Numisheet 93 benchmark tests [6]. It consists of a stamping for the front fender of a car. The model consists of 15900 shell elements, and approximately 24000 rigid elements to describe the die, punch and binder. The model is shown in Figure 3. The stamping simulation performed with ABAQUS/Explicit results in the deformed blank shown in Figure 4. Using the strategy described above, the springback analysis is performed with ABAQUS/Standard in 21 increments. The ratio of the computer time taken by the stamping simulation with the explicit solver, to the time taken for the springback analysis with the direct solver is 2.5:1. The amount of springback is depicted in Figure 5 with contours showing the out-of-plane displacements.

An often neglected aspect of metal forming is that there are usually additional operations after the forming process for which the designer requires analyses. For example, the residual stress state after springback may affect the performance of the part when in service. For these analyses, different solvers may be required.

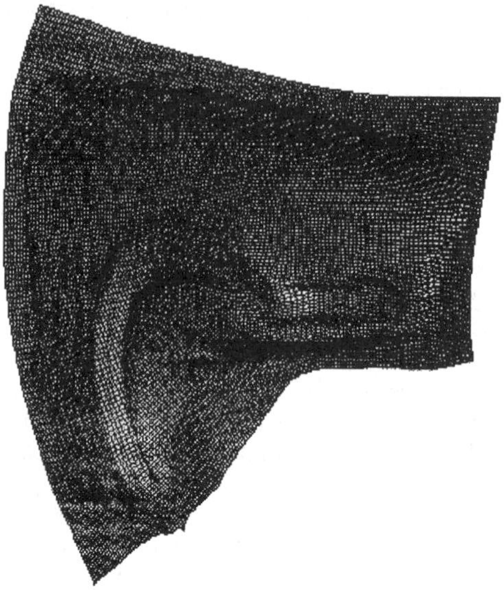

Figure 4: Deformed shape.

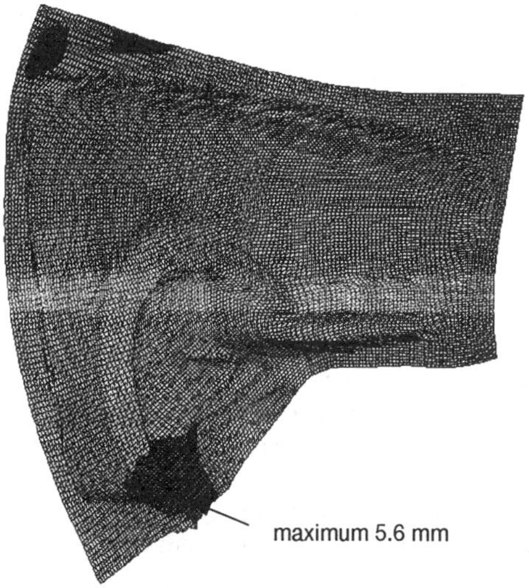

Figure 5: Contours of springback normal to the plane on which the blank is resting.

To illustrate this, we take the fender problem after springback has been obtained, and consider the following analyses:

1. Trimming to remove the flash.
2. Springback as a result of the trimming operation.

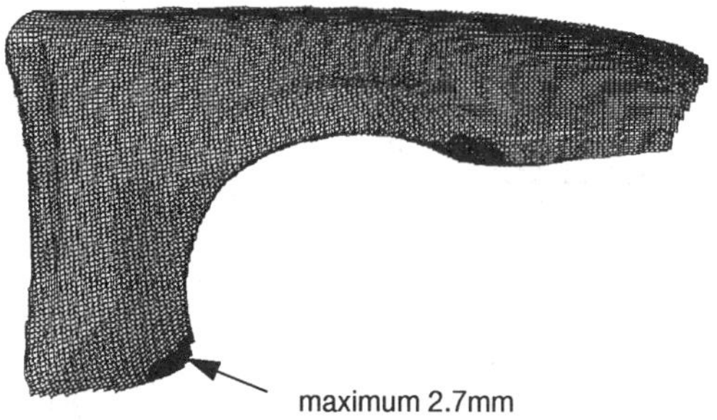

Figure 6: Contours of springback after trimming.

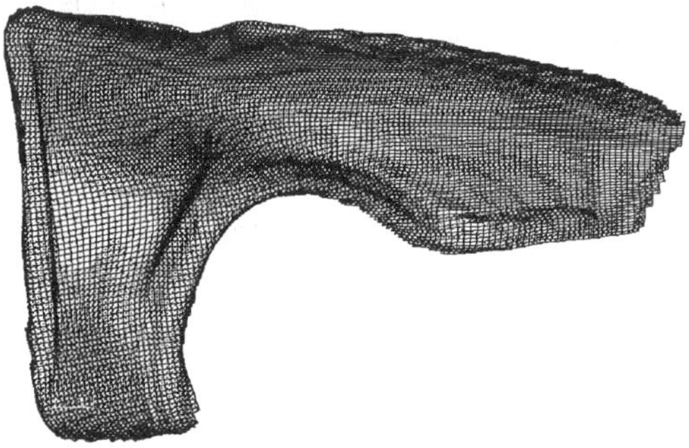

Figure 7: First mode of mounted fender.

3. Determination of the natural frequencies of the mounted fender.
4. Palming calculation to determine the structural load bearing characteristics of the mounted fender.

All of these analyses are performed with ABAQUS/Standard using either the direct solver or the eigensolver. The springback after the trimming is shown in Figure 6. The first natural mode of the mounted fender is shown in Figure 7. The palming analysis consists of the application of a localized distributed pressure, and the results are shown in Figure 8.

3. ITERATIVE SOLVER

Alternative solution strategies to using a direct solver for bulk forming simulations, are to use methods such as conjugate gradient methods, generalized minimal residual methods, etc. The main difficulty with these solvers is to obtain the desired efficiency while maintaining robustness of the solution process.

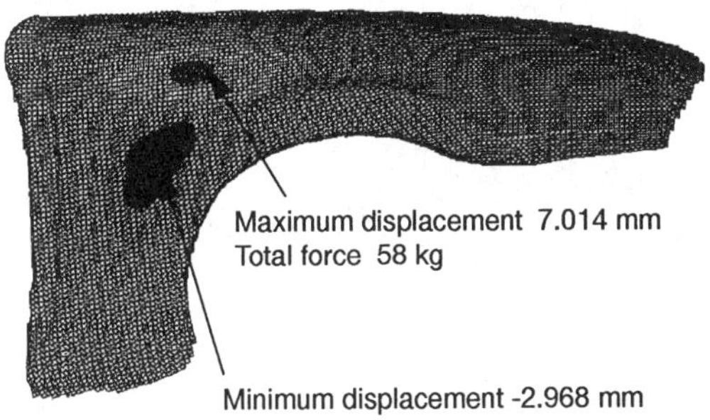

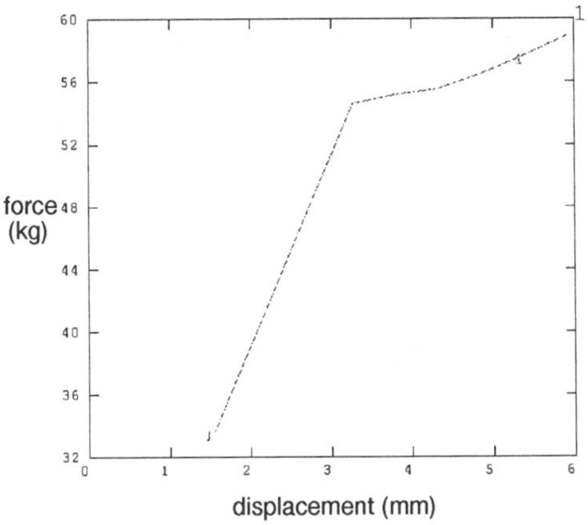

Figure 8: Palming analysis.

The conjugate gradient method is a well established method for solving sparse linear equations, and has been extended to solving nonlinear equations, see Shewchuk [10]. We have investigated the use of a diagonal preconditioned conjugate gradient solver for a quasi-static nonlinear analysis. This procedure has been implemented into the architecture of an explicit solver, so that a solution for a quasi-static problem can be obtained while maintaining the advantages of the explicit dynamics architecture. The nonlinear problem is solved incrementally, with static equilibrium obtained for each increment before advancing to the next increment. For each increment, the solution from the previous increment is extrapolated to form the initial search direction for the new increment. A line search is used to determine the minimum of the potential energy in the search direction. From this point, a new search direction is calculated that is conjugate to any previous search directions for the increment. The procedure continues to iterate until the static equilibrium equations have converged. A diagonal preconditioner is used to transform the equations into a more suitable form for solving with the conjugate gradient method.

Table 1: 2-D upsetting problem.

		Total number of variables		Relative CPU time	
Mesh size	RMS wavefront	Direct solver	CG solver	Direct solver	CG solver
12×12	51	391	345	0.7	1.0
50×50	175	5407	5209	1.5	1.0

Table 2: 3-D upsetting problem.

		Total number of variables		Relative CPU time	
Mesh size	RMS wavefront	Direct solver	CG solver	Direct solver	CG solver
10×10×10	597	4506	4011	1.2	1.0
20×20×20	1351	29358	27801	23.0	1.0

This solver is only effective for problems with a small eigenspectrum, which limits the application scope of the method (the eigenspectrum is the difference between the largest and smallest eigenvalues for the problem). For example, thin shell structures generally have a wide eigenspectrum and are not suitable for this solver, while bulk forming problems generally have a small eigenspectrum and are suitable.

To illustrate the effectiveness of the conjugate gradient solver (CG solver) for solving large 3-D bulk forming problems, we compare the computer times taken by the direct solver and the CG solver for a 25% upsetting of a billet modeled with different mesh densities in 2-D and 3-D.

The results for the 2-D model are given in Table 1 and the results for the 3-D model are given in Table 2.

The results show that there is not a big advantage to using the CG solver for 2-D problems and, in fact, the CG solver takes longer than the direct solver for the coarse mesh. The effectiveness of the CG solver is most pronounced for large 3-D problems with refined meshes.

4. CLOSING REMARKS

In this paper, we have illustrated the effective use of different solvers in the simulation of material forming processes. By combining the advantages of the explicit and direct solvers, an effective solution strategy is obtained for simulating the complete sheetforming process as well as subsequent analyses of in service loads. We also discussed the use of an iterative solver for simulating bulk forming processes. While this solver has good potential, it has not been proven yet as a robust solver for large industrial applications.

REFERENCES

[1] Simo, J. C., A Framework for Finite Strain Elastoplasticity based on Maximum Plastic Dissipation and Multiplicative Decomposition. Part I. Continuum Formulation, Comp. Meth. in Appl. Mech. and Eng., vol. 66, p. 199, 1988.

[2] Simo, J. C., A Framework for Finite Strain Elastoplasticity based on Maximum Plastic Dissipation and Multiplicative Decomposition. Part II. Computational Aspects, Comp. Meth. in Appl. Mech. and Eng., vol. 68, p. 1, 1988.

[3] Laursen, T. A. and J. C. Simo, A Continuum-Based Finite Element Formulation for the Implicit Solution of Multibody, Large Deformation Frictional Contact Problems, Int. J. for Num. Meth. Eng., vol. 36, p. 3451, 1993.

[4] Rebelo, N., J. C. Nagtegaal, L. M. Taylor, and R. Passmann, Comparison of Implicit and Explicit Finite Element Methods in the simulation of Metal Forming Processes, Proceedings of Numiform '92, Valbonne, France, p. 99, 1992.

[5] Matiasson, K., L. Bernspang, A. Samuelsson, T. Hamman, E. Schedin, and A. Melander, Evaluation of a Dynamic Explicit Approach Using Explicit Integration in 3-D Sheet Forming Simulation, Proceedings of Numiform '92, Valbonne, France, p. 55, 1992.

[6] Proceedings of 2nd International Conference, NUMISHEET '93, Isehara, Japan, ed. Makinouchi, A., E. Nakamachi, E. Onate, and R.H. Wagoner.

[7] Karafillis, A.P. and M.C. Boyce, Tooling and Binder Design for Sheet Metal Forming Processes Compensating Springback Error, submitted to Int. J. Machine Tools and Manuf.

[8] ABAQUS/Explicit User's manual, Hibbitt, Karlsson and Sorensen, Inc., 1994.

[9] ABAQUS/Standard User's manual, Hibbitt, Karlsson and Sorensen, Inc., 1994.

[10] Shewchuk, J. R., An Introduction to the Conjugate Gradient Method Without the Agonizing Pain, School of Computer Science, Carnegie Mellon University, 1994.

Simulation of Materials Processing: Theory, Methods and Applications, Shen & Dawson (eds)
© 1995 Balkema, Rotterdam. ISBN 90 5410 553 4

A parallel 3D ALE code for metal forming analyses

J. Robert Neely, Richard Couch, Scott Futral & Evi Dube
Lawrence Livermore National Laboratory, Calif., USA

ABSTRACT: A three-dimensional arbitrary Lagrange-Eulerian (ALE) code is being developed for use as a general purpose tool for metal forming analyses. The focus of the effort is on the processes of forging, extrusion, casting and rolling. The ALE approach was chosen as an efficient way to deal with the large deformations and complicated flows associated with these processes. A prototype version of the software package, ALE3D, exists and is being applied to the enumerated processes. The development of the code is being driven by the dual constraints of portability and extensibility. A general purpose simulation tool must be capable of running on a variety of platforms from single processor workstations to massively parallel platforms. It must also be configured to easily accommodate new physical models and parameters. The focus of this paper will be on computer science issues, with parallelization being the dominant issue. Long term goals will be described, as well as current status.

1 INTRODUCTION

This paper discusses a three dimensional arbitrary Lagrange-Eulerian (ALE) code being developed for use as a general purpose tool for metal forming analyses. The development of the code is being driven by the dual constraints of portability and extensibility. A general purpose simulation tool must be capable of running on a variety of platforms ranging from single processor workstations to massively parallel platforms. It must also be configured to easily accommodate new physical models and parameters. The focus of this paper is on computer science issues related to building such a code.

In section 2, we describe the concepts behind ALE, and the features of the ALE3D code. In section 3, we describe efforts to restructure ALE3D with the goal of improving parallelism, memory usage, scalability, extensibility, maintainability, and portability. In section 4, we will discuss the initial parallelization efforts applied to the existing sequential Fortran code. In section 5 we discuss current work in the area of domain decomposition of parallel problems. Section 6 presents some preliminary performance results gathered from parallel runs of up to 128 processors on a Meiko CS-2.

2 ARBITRARY LAGRANGE / EULERIAN (ALE)

Accurate simulation of large deformation metal forming provides a major challenge to developers of numerical modeling tools. This challenge presents itself both in the area of physical algorithms development and in the area of computational science.

ALE IN METAL FORMING ANALYSES A general purpose simulation tool must be able to treat a wide array of physical phenomena, and must be sufficiently flexible in simulation strategy to adequately represent the various forming processes. For example, metal forming technologies provide scenarios in which the working material can be treated as a solid structure (Lagrangian-like) as well as those in which the material can be considered a fluid (Eulerian-like). Molten metal flow, as in a casting process, clearly falls in the latter category, but other solid phase, large-deformation processes such as extrusion can be treated with the same computational techniques as those used for true fluids. Processes such as sheet stamping and rolling also involve large deformations but are best treated in the Lagrangian format.

Based on these considerations, we have chosen the arbitrary Lagrangian-Eulerian (ALE) format as the most powerful in terms of providing a general purpose tool. The potential advantage from ALE is that it provides the flexibility to treat a material in either a Lagrangian mode, or one in which the material is allowed to move relative to the mesh (relaxation). Different materials within the computational mesh can thus be treated in the most appropriate manner. The relaxation option can also provide the further benefit of mesh refinement for improved accuracy or efficiency. More traditional approaches have treated large deformations using a Lagrangian representation combined with periodic re-meshing, as required. The ALE approach assumes that continual re-meshing using high order techniques provides improved accuracy and efficiency.

ALE3D FEATURES ALE3D is a finite element code which treats fluid and elastic-plastic response on an unstructured mesh. The grid may consist of arbitrarily connected hexahedra, as well as shell and beam elements. There is no allowance for tetrahedra or wedge shaped elements. The mesh can be constructed from disjoint blocks of elements which interact at the boundaries via slide surfaces. This interaction can consists of pure sliding in which there are no tangential forces on interface nodes, or the nodes may be tied to inhibit sliding entirely, or a coulomb friction algorithm can be used. Voids may open or close between the surfaces as dictated by the dynamics of the problem, and there is an option to allow a block surface to fold back on itself (single-sided sliding).

The basic computational cycle consists of a Lagrangian step followed by an advection step. In the Lagrangian phase, nodal forces are accumulated and an updated nodal acceleration is computed. Following DYNA3D[1], the stress gradients and strain rates are evaluated by a lowest order finite element method, using a diagonal mass matrix. For more information on the simulation capabilities of ALE3D, please see the paper *3D Metal Forming Applications of ALE Techniques* in these proceedings[2].

3 CODE STRUCTURE AND FEATURES

The eventual goal of the ALE project is to design a code which is both efficiently portable in both sequential and parallel environments, as well as easily extensible in the context of adding new or different physical models or parameters. The existing ALE3D code has been used as a starting point, but in order to efficiently produce a code which has all of the features desired, it was determined that a complete restructuring of high level data structures and a partial rewrite of the existing code was necessary.

During the restructuring, the top levels of the call tree, and most of the integral data structures for the program are being written entirely in C, while low level numerically intensive physics packages will remain in Fortran. We will refer to this new version of the code as *ALE3D-C* throughout this paper when we wish to distinguish between it and the existing Fortran code.

Some of the reasons for making the decision to restructure the code include:

- *Parallelism* – Building the code to be more naturally decomposed into parallel components, while still being able to use the same code efficiently in both the sequential and parallel versions;
- *Extensibility* – Making addition of new features, models, and parameters easier;
- *Software Engineering* – Using capabilities of C to create a more easily understandable and maintainable interface to internal code modules.

Below, we discuss each of these issues in greater detail. Most of this work is in progress at the time this paper was written. Current status of work done to the existing Fortran code is discussed in section 4.

3.1 PARALLELISM

As the finite element is the fundamental computation element in the code, the natural choice for dictating the division of labor among parallel processors is to group elements in *domains* uniquely associated to each processor. A domain is defined as a subset of elements – to be computed in parallel with other domains.

By encapsulating the data that a domain needs to compute its portion of the problem into a single structure, the parallelism effort becomes easier, and much more efficient. However, this design heavily favors a coarse-grained approach to parallelism, with explicit message passing as the assumed method – which currently is the most portable, scalable and efficient method of parallelism on distributed memory architectures.

In addition, we are building into the new code the idea of *domain overloading.* This is the ability to have more than one domain being computed on each processor. There are several advantages to being able to do this:

First, it will be possible to run a problem which has been decomposed into D domains on any number of processors $d \leq D$, or effectively detach the number of physical processors required to run the problem from the decomposition. This is advantageous for restarting a problem on a different number of processors than the problem was decomposed into.

Secondly, domain overloading could help with cache effectiveness by limiting the size of loops as a function of the cache size. This is similar to the technique of *chomping* in the original code, where the goal was to reduce the size of transient work arrays, thus reducing the overall memory requirements.

Since the physics algorithms will be essentially unchanged from the original code, the prototype parallel version will be used as a model for parallelization (see section 4). However, no assumptions will be made about communication patterns – namely that they can be determined at the start of the simulation, and assumed to remain unchanged throughout the program run. Both slide surfaces and advection can cause new communication patterns to evolve throughout the problem run, either due to domains coming within proximity of each other on opposite sides of a slide surface, advection being added to elements in the problem, or a reversal of advective "flow" through the mesh.

3.2 EXTENSIBILITY

As the number of ALE3D users increased, it became apparent that these users were interested in defining their own grid variables and material models. In the existing version of ALE3D, making additions or changes to any of these aspects is cumbersome and prone to human error, as it involves changing the code in many different places.

Grid variables are values that describe some physical property of each element, and are stored as dynamically allocated arrays of domain length. In ALE3D-C, we store pointers to these grid variables in both a hash table and a standard linked list. This allows instant access to any variable from anywhere in the problem (via hashing on the name of the variable), as well as efficient access to all grid variables when operations on the entire set are to be performed (by looping through the list).

Material models are a representation of how certain materials behave under given conditions. Different users of ALE3D have different ways of thinking about material models. Being able to simply "plug in" whichever material model the user wants is currently very difficult in ALE3D. By restructuring the code, however, the decision on which material models to use is brought up to a high level in the code, instead of being strewn throughout the physics.

3.3 SOFTWARE ENGINEERING ISSUES

Although Fortran 77 is an excellent language for doing what it was designed to do (numerical computations), it lags behind in areas of software engineering. As a code becomes larger, good software engineering techniques are essential to keeping the code maintainable and expandable.

By using C at the top-level of the code, we can take advantage of many on the features it offers over Fortran 77, including: information hiding, modular structure, built-in dynamic memory allocation, user defined types, and better development tools. Although most of these features can be applied to a Fortran code through coding practices and external libraries, they require the programmers to follow certain rules which are not enforced by the language, and can be too easily subverted. In addition, C gives the programmer a much better chance of catching bugs at compile time, through judicious use of ANSI standards and prototyping.

Although we are writing the new code in C, we are currently planning on leaving much of the numerically intensive physics routines in Fortran.

Fortran 90 offers many of the features which we deemed desirable, but compilers are still rather rare, and it is still not known if Fortran 90 compilers can produce code as efficiently as the relatively mature C compilers which currently exist. In addition, future revisions of the program may involve object-oriented methods, which will be easier to employ from a C-based code.

4 PARALLELIZING ALE3D

A prototype parallelization of ALE3D has been accomplished incrementally using the existing Fortran code. The philosophy has been to work from the sources for the sequential version of the code, adding the parallelism as an integral part of the code, with special cases keyed to a logical flag only when necessary. This allows for validation testing against the sequential code at each step in the development process. The parallelization effort then reduces to:

- Identifying information that must be shared between domains;
- Identifying which domains must communicate information, and with whom;
- Identifying when the communication must occur in the algorithm;

Using this method of decomposition, each element in the problem will belong to one and only one domain. Nodes and faces however, will be on domain boundaries, and can belong to two or more domains.

As part of the parallelization effort, various domain decomposition algorithms are being explored and their relative performance analyzed (section 5). In addition, portability to various distributed memory platforms ranging from networks of workstations (NoW's) to massively parallel processors (MPP's) is required, which drove our decision on which parallel methods and libraries to use (section 4.4).

4.1 THE LAGRANGE STEP

The Lagrange step was relatively easy to parallelize. The original code, which was developed mainly for vector-class computers, contained mechanisms called *chomps*, which were used to keep loops over arrays at a length appropriate for the vector hardware.

The Lagrange step has only two points in the code where communication must occur at each timestep:

First, forces on a node are calculated by summing the partial forces from each of the node's corresponding elements. In the case where a node is on the boundary of a domain, communication of the partial forces must occur between the domains to which that node belongs.

Secondly, a global minimum reduction is performed to find the minimum time constraint across all the elements which defines ΔT for the next timestep. This type of communication operation is standard and optimized in many message passing libraries.

4.2 THE ADVECTION STEP

The advection step was more difficult to parallelize, mainly due to the fact that in order to calculate second order density distributions, each domain needs information from the first and second layer of neighboring boundary elements. Also, data dependencies across domains may be non-symmetric during advection for several reasons:

- Advection deals with material flowing through the mesh, and only second-order data from "upwind" elements is needed;
- The boundary between advecting and non-advecting elements may fall on a domain boundary;

In both these cases, data may be needed in domain D_1 from domain D_2, but not necessarily vise-versa. This means that the amount, and pattern, of data needing to be communicated at each timestep can change throughout the run of the program.

In our first cut at parallelization, we simplified this problem by ignoring flow direction when determining data dependencies, and simply sending data in a symmetric fashion. This approach will be optimized to send only data which will be used in ALE3D-C.

Information which must be communicated during various steps of the advection phase include: relaxed coordinates for boundary nodes, lists of potentially mixed elements for the domain boundary, partial sums of nodal mass fluxes and momentum fluxes, and partial sums of nodal masses.

4.3 SLIDE SURFACE CALCULATIONS

Parallelizing the slide surfaces is proving to be a challenge, as the assumption of fixed communication patterns is broken for elements along the slide surfaces. Two domains on opposite sides of a slide line may start out far apart from each other at the beginning of the simulation, yet at some point, the slide surface may bring these two domains into proximity which requires neighbor status, and thus communication.

As a first approach to the problem, the fixed communication is restored by requiring all elements on an entire slide surface to reside in a single domain. This is achieved through the domain decomposer (see section 5) prior to run-time. This method is seriously limiting to the effective parallelization of the code however, as the slide surface calculation can be a large fraction of the total computation done in a given problem. If a large fraction of the elements are slide surface elements, and the number of independent slide surfaces are small, this becomes a sequential bottleneck in the calculation. The choice of domain decompositions is also restricted by this requirement, which can inhibit load balancing for the Lagrange and advection portions of the calculation as well.

As a first refinement to the parallel algorithm, the domain decomposition restriction was removed by communicating information for a particular slide surface to a single processor (arbitrarily chosen as the one which initially contained the greatest number of slide surface nodes) that has been designated to calculate that slide surface.

A further improvement was to permit intersecting slide surfaces to be calculated by different processors. This requires sending data between the processors on which the slide surfaces reside during the slide surface calculation.

Additional improvements will be needed to provide a truly parallel slide surface calculation, and the optimal solution is still an open question.

4.4 MESSAGE PASSING LIBRARIES

Most recently, we have ported ALE3D to MPI [3] (*Message Passing Interface*), a standard defined by a committee of people from industry, universities, and national laboratories. Current trends show a wide and positive acceptance of MPI, although native implementations are not yet common.

Currently, our development work with MPI is being done on a Meiko CS-2 using the Argonne/Mississippi portable MPI implementation [4][5]. This implementation was designed to allow development work to go ahead with MPI while vendors create native implementations, and is available on most parallel environments as a library "on top of" an existing message passing system such MPSC, NX, or P4. We expect native ports of MPI to be ready on Meiko and Cray T3D platforms sometime in 1995.

The initial port to MPI was trivial, since the code mainly contained point-to-point communication (standard blocking sends and receives). Since then, we have been identifying places in the code which could benefit from both improved communication algorithms including non-blocking communication for faster communication and smaller buffering requirements, as well as some of the higher level functionalities of MPI, including *communication contexts*, which allow messages within a group to be guaranteed that they will not be received by other defined groups without requiring addition synchronization.

4.5 PARALLEL I/O

ALE3D uses a portable highlevel scientific database, Silo[6], which is based on PDB (Portable Database), for writing plot files and restart dumps. The files written are typically on the order of several megabytes in size. We have started making incremental changes for handling our I/O on parallel platforms.

Silo and PDB both expect the output files to be written sequentially, either by a single processor, or via strict serial access by multiple processors. Currently, we gather the data from the entire problem to a single processor which writes the restart and plot files. This creates a serious bottleneck in the code however, since $P-1$ processors are standing idle while a single processor writes to disk.

We are considering several optimizations to improve throughput of the output phase. One option, which will minimize idle time of processors not directly involved in writing output files, involves dedicating a single processor to performing the output. The output processor can then independently work at I/O without slowing down computation on the other nodes. This approach will work best when the problem is large and being executed by many processors, since $P-1$ processors will no longer sit idle waiting for one processor to finish.

In ALE3D-C, we are assuming that any single processor does not contain enough memory to store the entire problem, thus making the "gather" method impossible. Output will initially be performed in a "round robin" fashion, with each processor taking turns writing to the file with exclusive access. Alternatively, each processor may write it's own file, which will later be merged. Future versions of Silo may include parallel I/O capabilities, perhaps using the emerging MPI-IO efforts [7], which will hopefully give us access to true parallel I/O.

5 DOMAIN DECOMPOSITION

Initial approaches to domain decomposition included simple material decompositions, and radial partitions. In the former, each element of similar material type is placed in the same domain. In the latter, domains are created based on an element's distance from an origin point of the problem – essentially creating "shells" of domains around a central point. We have since started exploring published domain decomposition algorithms, specifically spectral methods for decomposing unstructured meshes.

Table 1: Parallel performance - 17k elements, RSB Decomp.

Num Procs	Time per ts (s) (Lag+Adv)	Speed-ups	Efficiency
1	27.4 (4.52+22.7)	1.0	100%
2	13.93 (2.86+11.06)	1.96	98%
4	6.10 (1.43+4.64)	4.50	112%
8	2.82 (0.71+2.09)	9.71	121%
16	1.56 (0.34+1.17)	17.6	109%
32	1.28 (0.18+0.91)	21.4	66%
64	.90 (0.12+0.73)	30.3	47%
128	.69 (0.1+0.50)	39.8	31%

Table 2: Parallel performance - 95k elements, RSB Decomp.

Num Procs	Time per ts (s) (Lag+Adv)	Speed-ups	Efficiency
1	309.7 (55.6 + 253.9)	1.0	100%
8	29.6 (4.89 + 24.7)	10.4	130%
16	12.1 (2.46 + 9.58)	25.5	159%
32	7.47 (1.33 + 6.08)	41.5	129%
64	4.98 (0.74 + 4.07)	62.2	97%

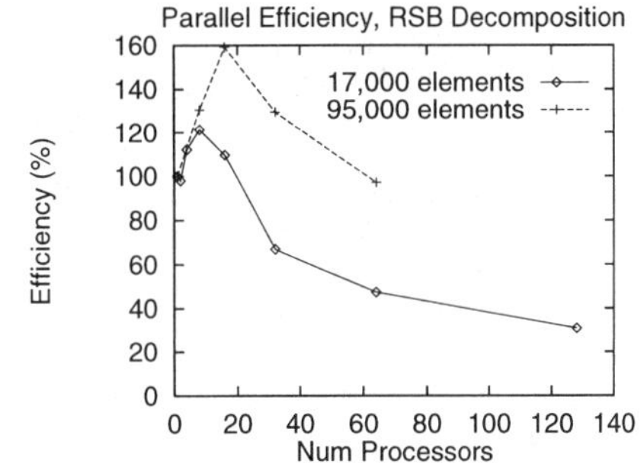

Figure 1: Efficiencies for 17k and 95k Element problems

One of the two major packages we have looked at is the RSB[8] algorithm, developed by Horst Simon at NASA. We are experimenting with two of his algorithms, including one which allows weighting of the nodes. The other package we have started experiments with is CHACO [9], developed by Bruce Hendrickson and Robert Leland at Sandia National Laboratories. On first impression, spectral methods seem to be fairly robust, although we need to gain more experience through analysis and empirical study.

Due to the current restrictions placed on slide surfaces (see section 4.3), it is sometimes advantageous to create a decomposition such that all elements on a slide surface reside in a single domain – thus saving the extra communication step of gathering the data at each timestep. This can be done during the decomposition using a *supernode* which essentially replaces all nodes on a slide surface with a single node. To help in load balancing, we implemented a *vote* option in which we retain the slide surface nodes along with the supernode during decomposition. Then, in the post-processing stage, we "vote" for the domain with the most slide surface nodes, setting the supernode, and thus the slide surface element to that domain.

6 PRELIMINARY PERFORMANCE RESULTS

Initial performance results were acquired on a Meiko CS-2, using between 1 and 128 processors, on problems ranging from 17,000 to 95,000 elements (small to moderate sized) with all elements advecting.

Tables 1 and 2 outline the parallel performance of ALE3D on the different sized problems. *Time per timestep* is a measure of the average wall-clock time spent in computing a single timestep in the problem (broken down into Lagrange and advection phases in parentheses). The last two columns outline the parallel speedups and efficiencies, using the single node run as a basis.

Efficiencies greater than 100% indicate super-linear speedups. In our case, we believe that our single processor run was artificially slowed down by ineffective cache usage, and/or excessive swapping of pages in virtual memory.

We do see however, that the efficiencies begin to drop off severely as the size of the problem per processor/domain approaches several thousand elements. At this point, communication costs can no longer be amortized by large amounts of computation.

As is consistent with our hypotheses of super-linear speedups outlined above, the "efficiency curve" of the larger sized problem reflects almost identical results as the smaller problem, scaled up by an almost constant factor. This is easily seen in figure 1.

7 CONCLUSIONS

We are concentrating effort into a major restructuring of the ALE3D code which will help us overcome many of the design issues that are inhibiting both the parallelization and extensibility of the existing ALE3D Fortran code. Written in C, ALE3D-C will attempt to initially reuse many of the existing Fortran modules. By designing a hierarchy of data structures based around the idea of a *domain* as a central unit of computation, we will make better use of available memory and cache. More robust analysis of communication patterns will also allow us to optimize communication in the advection and slide line routines. Also, by pulling some of the decision points in the code up to a higher level, and creating well defined modules with easy to understand interfaces, we aim to make the code more extensible for users who wish to add their own physical models and parameters.

We are currently able to run real problems on a parallel platform using a parallel version of ALE3D based on the original sequential version. Using some small to moderate sized industrial problems, we have started studying scalability of the parallel implementation, as well as the relationship of domain decomposers to the efficiency of a parallel run. Initial results taken on a Meiko CS-2 show good speedups for moderate numbers of processors.

References

[1] J. O. Hallquist. *Theoretical Manual for DYNA3D*. Livermore, CA, 1982. draft.

[2] Rich Couch, et al. 3d metal forming applications of the ale technique. In *Proceedings of the NUMIFORM '95 Conference*, June 1995.

[3] Message Passing Interface Forum. Mpi: A message-passing interface standard. Technical Report CS-94-230, University of Tennessee Computer Science Department, Knoxville, TN, May 5 1994.

[4] Nathan Doss, William Gropp, Ewing Lusk, and Anthony Skjellum. A model implementation of mpi. Technical Report MCS-P393-1193, MCS Division, Argonne Nat'l Laboratory, Argonne, IL 60439, 1993. in preparation.

[5] William Gropp and Ewing Lusk. An abstract device definition to support the implementation of a high-level point-to-point message-passing interface. Technical Report MCS-P342-1193, MCS Division, Argonne Nat'l Lab, Argonne, IL 60439, 1994.

[6] J.A Leibee and J.W. Long. *SILO User's Guide*. U.S. DOE, Lawrence Livermore Nat'l Laboratory, Oct 1994. Under contract W-7405-ENG-48.

[7] Peter Corbett, et al. Mpi-io: A parallel file i/o interface for mpi, version 0.2. Research Report 19841 (87784), IBM T.J. Watson Research Center and NAS Systems Division, Nov 1994. mpi-io@nas.nasa.gov.

[8] H. Simon. Partitioning of unstructured problems for parallel processing. *Computer Systems in Engineering*, 2(2/3):135–148, 1991.

[9] B. Hendricks and R. Leland. The chaco user's guide, version 1.0. Technical Report SAND93-2339, Sandia Nat'l Lab, Oct 1993.

Simulation of Materials Processing: Theory, Methods and Applications, Shen & Dawson (eds)
© 1995 Balkema, Rotterdam. ISBN 90 5410 553 4

Comparison of two Arbitrary Lagrangian-Eulerian formulations

A. Rodríguez-Ferran & A. Huerta
Departament de Matemàtica Aplicada III, ETS de Ingenieros de Caminos, Universitat Politècnica de Catalunya, Barcelona, Spain

ABSTRACT: In the context of fractional step methods, various algorithms for the time-integration of the governing equations in the Arbitrary Lagrangian-Eulerian formulation of nonlinear solid mechanics are presented. The algorithms have been implemented in CASTEM2000, an object-oriented code, and compared with the aid of two numerical tests.

1. INTRODUCTION

The key idea of the Arbitrary Lagrangian-Eulerian (ALE) formulation is allowing the mesh to have an arbitrary motion, independent of material motion. By doing so, it is possible to avoid the excessive mesh distorsion or the cumbersome description of material surfaces associated to purely Lagrangian or Eulerian formulations respectively. The ALE formulation, initially developed in fluid mechanics, has been extended to and successfully employed in nonlinear solid mechanics, [1-6].

The ALE governing equations contain convective terms induced by the relative mesh–material motion, which are similar to those encountered in the Eulerian formulation. These terms are numerically troublesome, and specific techniques are employed in fluid dynamics to handle them. In fact, the correct treatment of convection is a fundamental point in the ALE formulation.

An outline of paper is as follows. Section 2 briefly presents the ALE governing equations for quasistatic processes. The key aspect of the stress update (i.e., the time-integration of the constitutive equation) is treated in Section 3, where various algorithms are presented and discussed. Finally, Section 4 comments the implementation of the algorithms in an object-oriented code and compares them with the aid of two numerical tests.

2. ALE QUASISTATIC EQUATIONS

A detailed presentation of the ALE formulation of nonlinear solid mechanics may be found in [3]. Only a few basic notions in the context of quasistatic analysis will be presented here.

2.1 Kinetics

Three domains are involved in the ALE formulation: the usual material domain R_X and spatial domain R_x, and a *referential* domain R_χ. One-to-one transformations between these domains are provided by the equations of material and mesh motion

$$\boldsymbol{x} = \boldsymbol{x}(\boldsymbol{X}, t) \quad , \quad \boldsymbol{x} = \boldsymbol{x}(\boldsymbol{\chi}, t) \tag{1}$$

which yield the spatial position $\boldsymbol{x}$ at time t of a material particle $\boldsymbol{X}$ and a grid point $\boldsymbol{\chi}$ respectively. Material velocity $\boldsymbol{v}$ and mesh velocity $\hat{\boldsymbol{v}}$ are obtained by derivating (1) with respect to time holding $\boldsymbol{X}$ or $\boldsymbol{\chi}$ fixed respectively.

The relative motion between material and mesh is represented by the convective velocity

$$\boldsymbol{c} = \boldsymbol{v} - \hat{\boldsymbol{v}} \tag{2}$$

Acceleration $\boldsymbol{a}$ is expressed in the Lagrangian and the ALE formulation respectively as

$$\boldsymbol{a} = \left.\frac{\partial \boldsymbol{v}}{\partial t}\right|_{\boldsymbol{X}} = \left.\frac{\partial \boldsymbol{v}}{\partial t}\right|_{\boldsymbol{\chi}} + \boldsymbol{c}\frac{\partial \boldsymbol{v}}{\partial \boldsymbol{x}} \tag{3}$$

where $|_*$ means "holding $*$ fixed". As in the Eulerian formulation, two terms are needed to represent acceleration in the ALE formulation: the local acceleration $(\partial \boldsymbol{v}/\partial t)\,|_{\chi}$ and the convective acceleration $\boldsymbol{c}(\partial \boldsymbol{v}/\partial \boldsymbol{x})$.

2.2 Momentum balance

In the ALE formulation, the equation of momentum balance reads

$$\rho a_i = \rho \frac{\partial v_i}{\partial t}\bigg|_{\chi} + \rho c_j \frac{\partial v_i}{\partial x_j} = \frac{\partial \sigma_{ij}}{\partial x_j} + b_i \tag{4}$$

where ρ is the density, $\boldsymbol{\sigma}$ the Cauchy stress tensor and $\boldsymbol{b}$ the body force. As commented in the introduction, this governing equation has a convective term, in this case associated to the convective acceleration. A similar situation is found in the Eulerian, but not in the Lagrangian, formulation.

A common assumption in many problems in nonlinear solid mechanics is that inertia forces $\rho \boldsymbol{a}$ are negligible in comparison to the RHS of (4). The process is then called *quasistatic*, and the momentum balance reads

$$\frac{\partial \sigma_{ij}}{\partial x_j} + b_i = 0 \tag{5}$$

in either the Lagrangian, Eulerian or ALE formulation, [7]. Equation (5) shows that *there are no convective terms in the ALE momentum balance for quasistatic processes.*

2.3 Constitutive equations

In large strain solid mechanics, material behaviour is often described by a rate-form constitutive equation

$$\dot{\boldsymbol{\sigma}} = \boldsymbol{q}(\boldsymbol{v}, \boldsymbol{\sigma}) \tag{6}$$

relating the material rate of stress $\dot{\boldsymbol{\sigma}} = (\partial \boldsymbol{\sigma}/\partial t)\,|_{\boldsymbol{X}}$ to stress and a mesure of strain (symbolically put as velocity $\boldsymbol{v}$). The RHS of (6) also includes some terms that counteract the non-objectivity of $\dot{\boldsymbol{\sigma}}$, in such a way that an objective constitutive equation results, [8,9].

The material rate of stress may be related to the referential rate of stress by adding a convective term that accounts for the relative motion between material particle $\boldsymbol{X}$ and grid point χ (as done in (3) for velocity). Equation (6) is then rewritten as

$$\left(\frac{\partial \boldsymbol{\sigma}}{\partial t}\bigg|_{\boldsymbol{X}}\right) = \frac{\partial \boldsymbol{\sigma}}{\partial t}\bigg|_{\chi} + c_j \frac{\partial \boldsymbol{\sigma}}{\partial x_j} = \boldsymbol{q} \tag{7}$$

Equation (7) shows that *convective terms are present in the ALE constitutive equation.* This general statement is also valid for quasistatic processes, because the convective term in (7) is associated to the relative material–mesh motion, which is inherent to the ALE kinetics.

3. STRESS UPDATE

Equations (5) and (7) are commonly solved in an incremental/iterative fashion, the unknowns being the incremental material and mesh displacements, $\Delta \boldsymbol{u}$ and $\Delta \hat{\boldsymbol{u}}$, from one known equilibrium configuration Ω_n at time t_n to a new equilibrium configuration Ω_{n+1} at time t_{n+1}. In each iteration i, $\Delta \boldsymbol{u}^i$ are computed by linearising (5), and a certain rezoning strategy yields $\Delta \hat{\boldsymbol{u}}^i$. To check whether the corresponding Ω^i_{n+1} is the equilibrium configuration at time t_{n+1} (i.e., whether (5) holds), the constitutive equation (7) must be integrated in time.

Fractional step methods allow to treat material and convective terms in (7) separately, [7]. When updating stresses from t_n to t_{n+1}, two steps are performed: first, a Lagrangian step accounts for material effects, and stress is updated from ${}^n\boldsymbol{\sigma}$ (stress at time t_n) to ${}^L\boldsymbol{\sigma}$ (stress after the Lagrangian step); afterwards, a convection step deals with convective effects, updating stress from ${}^L\boldsymbol{\sigma}$ to ${}^{n+1}\boldsymbol{\sigma}$ (stress at time t_{n+1}).

3.1 The Lagrangian step

In the Lagrangian step, convection is neglected and only material effects are taken into account. This means that grid points χ are assumed to move attached to particles $\boldsymbol{X}$, that the constitutive equation is simply

$$\frac{\partial \boldsymbol{\sigma}}{\partial t}\bigg|_{\chi} = \boldsymbol{q} \tag{8}$$

and that the Gauss point-wise techniques of the Lagrangian formulation can be employed to time-integrate it. Two algorithms, discussed in detail in [9], have been employed.

Algorithm 1

The (elastic trial) stress increment is

$$^{n}\Delta\boldsymbol{\sigma} = \boldsymbol{C} : {^{n}\boldsymbol{E}} \tag{9}$$

where $\boldsymbol{C}$ is the modulus tensor and the $^{n}\boldsymbol{E}$ is the incremental Lagrange strain tensor referred to configuration Ω_n

$$^{n}\boldsymbol{E} = \frac{1}{2}\left({^{n}\boldsymbol{\Lambda}^T} \cdot {^{n}\boldsymbol{\Lambda}} - \boldsymbol{I}\right) \tag{10}$$

where $^{n}\boldsymbol{\Lambda}$ is the incremental deformation gradient relating Ω_n to Ω_{n+1}, see Figure 1. The (elastic trial) Lagrangian stress is then

$$\begin{aligned} {^{L}\boldsymbol{\sigma}} = & {^{n}J^{-1}}\, {^{n}\boldsymbol{\Lambda}} \cdot {^{n}\boldsymbol{\sigma}} \cdot {^{n}\boldsymbol{\Lambda}^T} + \\ & {^{n}J^{-1}}\, {^{n}\boldsymbol{\Lambda}} \cdot ({^{n}\Delta\boldsymbol{\sigma}}) \cdot {^{n}\boldsymbol{\Lambda}^T} \end{aligned} \tag{11}$$

where the Jacobian ^{n}J is defined as the determinant of $^{n}\boldsymbol{\Lambda}$, which is employed to push-forward both $^{n}\boldsymbol{\sigma}$ and $^{n}\Delta\boldsymbol{\sigma}$ into the new configuration Ω_{n+1}, [8,9]. After that, the plastic correction is performed via a return algorithm.

Algorithm 2

A more accurate algorithm is obtained by employing the Truesdell objective rate and the rate of deformation $\boldsymbol{d}$, evaluated in an intermediate configuration $\Omega_{n+\alpha}$, as a measure of strain, [8,9]. The resulting (elastic trial) Lagrangian stress is then

$$\begin{aligned} {^{L}\boldsymbol{\sigma}} = & {^{n}J^{-1}}\, {^{n}\boldsymbol{\Lambda}} \cdot {^{n}\boldsymbol{\sigma}} \cdot {^{n}\boldsymbol{\Lambda}^T} + \\ & {^{n+\alpha}J^{-1}}\, {^{n+\alpha}\boldsymbol{\Lambda}} \cdot (\Delta t\, \boldsymbol{C} : \boldsymbol{d})\,|_{n+\alpha} \cdot {^{n+\alpha}\boldsymbol{\Lambda}^T} \end{aligned} \tag{12}$$

where $^{n+\alpha}\boldsymbol{\Lambda}$ is the incremental deformation gradient relating the intermediate and the final configurations, see Figure 1, and $^{n+\alpha}J$ is its determinant. The approximation to $^{n+\alpha}\boldsymbol{d}$ needed in (12) is the symmetrized gradient of incremental material displacements with respect to the intermediate configuration.

Both algorithms 1 and 2 (with $\alpha = 0.5$, i.e., the mid-step configuration as the intermediate one) comply with the discrete version of the principle of objectivity.

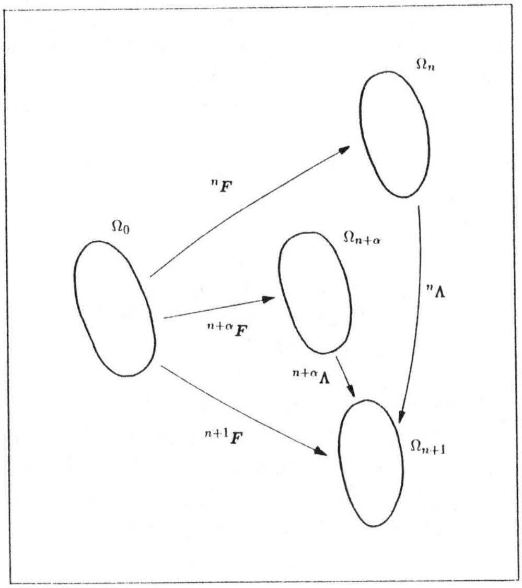

Figure 1: Deformation gradients in incremental analysis.

3.2 The convection step

In the convection step, the convective term in (7) is accounted for. The constitutive equation reads

$$\left.\frac{\partial\boldsymbol{\sigma}}{\partial t}\right|_{\boldsymbol{X}} + c_j \frac{\partial\boldsymbol{\sigma}}{\partial x_j} = \mathbf{0} \tag{13}$$

The stress gradient in (13) is numerically troublesome, because stress is often a discontinuous field and its gradient cannot be computed in a straightforward manner. In [5], the gradient of a continuous stress field, obtained through a smoothing procedure, is employed. A different approach has been preferred here, employing two algorithms where no gradient computations are needed.

Algorithm 1: interpolation procedure

The first algorithm is an adaptation of the interpolation procedures commonly employed in adaptive remeshing techniques, [7]. Equation (7) is recalled to rewrite (13) as

$$\left.\frac{\partial\boldsymbol{\sigma}}{\partial t}\right|_{\boldsymbol{X}} = \mathbf{0} \tag{14}$$

The splitting strategy of (8) and (14) admits the following geometrical interpretation, [2]: the stresses $^{L}\boldsymbol{\sigma}$ obtained in the Lagrangian step are referred to a fictitious Lagrangian mesh, attached to the material; in the convection step, this stress field –fixed in space because the particles do not move–, must be remapped to the true ALE mesh. For every Gauss point in the ALE mesh, two steps are needed to perform the remap: *i)* detect the element in the Lagrangian mesh containing the ALE Gauss point and *ii)* interpolate the stress field from the Lagrangian element to the ALE Gauss point.

This is a very general algorithm, since it does not require that the two meshes have the same topology. It is also a relatively expensive algorithm, especially the element search of step *i)*. However, if it is employed in the context of ALE rezoning (where the mesh topology does not change), the general algorithm can be adapted, making this search task easier. The basic idea is that information on the "neighbourhood" relations between elements can be produced right after mesh generation and used, with no change, through the whole analysis, [7].

Algorithm 2: Godunov-type technique
The second algorithm for the convection step is the extension of [10] to the Godunov-type technique developed in [3,6], and it will be presented just briefly here.

Let τ be a component of the stress tensor $\boldsymbol{\sigma}$. With the aid of the stress-velocity product, $\boldsymbol{Y} = \tau \boldsymbol{c}$, [1], equation (13) is put as

$$\left.\frac{\partial \tau}{\partial t}\right|_{\boldsymbol{\chi}} + \frac{\partial Y_j}{\partial x_j} = \tau \frac{\partial c_j}{\partial x_j} \qquad (15)$$

Equation (15) is a conservation equation, and the Godunov method is directly applicable if τ is a piecewise constant field. In the FEM context, this is the case if one-point integration quadratures are employed. Applying a weighted residual formulation to (15), an integral equation at the element level is obtained

$$\frac{\partial \tau}{\partial t} = -\frac{1}{2A}\sum_{s=1}^{N_s} f_s\,(\tau^c - \tau)\,[1 - \mathrm{sign}(f_s)] \qquad (16)$$

where τ is the stress component in the considered element, which has area A and N_s sides, τ^c is the stress component in the contiguous element across side s and f_s is the flux of convective velocity through side s.

Numerical time-integration of (16) over the time-increment Δt yields

$${}^{n+1}\tau = {}^{L}\tau - \frac{\Delta t}{2A}\sum_{s=1}^{N_s} f_s\,(\tau^c - \tau)\,[1 - \mathrm{sign}(f_s)] \qquad (17)$$

for each element.

If τ is not as piecewise constant field (i.e., for multiple-point quadratures), the extension of this algorithm presented in [10] can be used. The idea is to consider every element as formed by various subelements, each of them corresponding to the influence domain of one Gauss point. If 2×2 Gauss point quadrilaterals are employed, for instance, every element contains four subelements. In each subelement, τ is considered to be uniform, and represented by the Gauss point value. Thus, exactly the same method represented by (17) can be applied to the mesh of subelements to perform the convection step.

4. IMPLEMENTATION AND TESTING OF THE ALGORITHMS

CASTEM2000, an object-oriented code (OOC), [11,12], has provided the programming environment for this work. These codes store and handle information as objects far more complex than variables in conventional codes. Objects are created and manipulated by means of operators. The object–operator framework greatly eases the programmer's task, thus making OOCs especially suitable for the research and development stages. The basic idea is that objects directly represent the entities relevant to a FEM computation (mesh, stiffness matrix, nodal field...), so it is not necessary to translate them into more rudimentary variables.

4.1 Convection on a regular mesh

The two algorithms for the convection step are tested with the aid of a simple test, Figure 2. The mesh is formed of 10×10 unit eight-noded squares, with 2×2 Gauss points. The initial stress field is a cosinuidal hill of radius 3 centered in point P. To assess the 2-D behaviour of the algorithms, a uniform mesh velocity is prescribed in the direction of diagonal AC, Figure 2. Material velocity is null, so only convective effects are present (i.e., equation (13) must be solved). The time-step is set to $\Delta t = 0.25$.

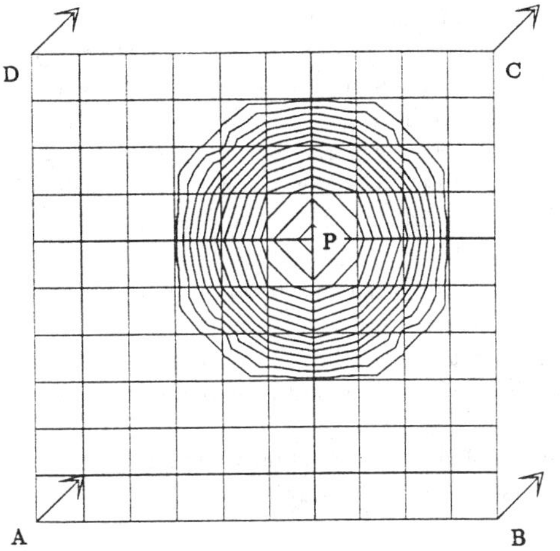

Figure 2: Mesh, initial stress field and mesh velocity for convection test.

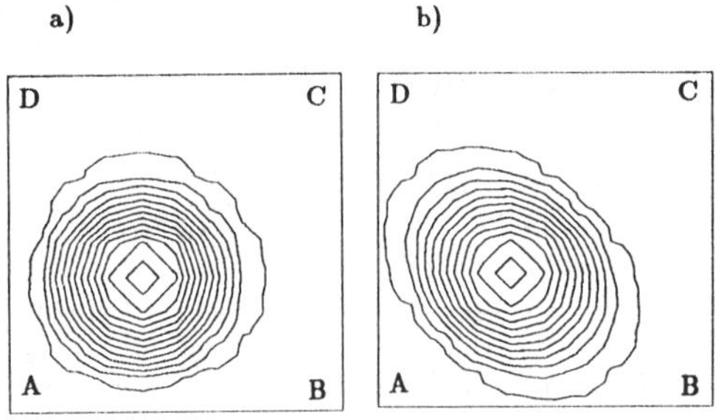

Figure 3: Stress field at $t = 2$. a) Algorithm 1. b) Algorithm 2.

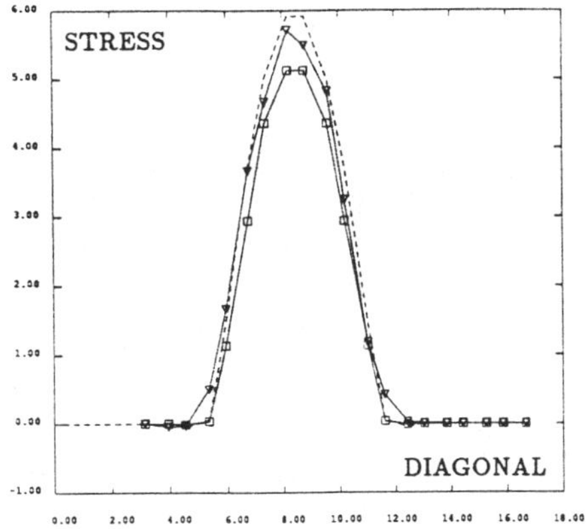

Figure 4. Stress profiles along diagonal AC: exact solution (dashed), algorithm 1 (triangles), algorithm 2 (squares).

Figure 3 shows the stress field at $t = 2$. For the interpolation procedure, the initial circular shape of the hill is well maintained, Figure 3a. For the Godunov-type update, on the contrary, the diameter normal to velocity direction is quite larger than in the input field. This is due to the lack of "corner convection" in algorithm 2: since convection is accounted for as fluxes through the element *sides*, (17), there is no stress transport between two elements that only share a corner node. This leads to the crosswind diffusion that alters the hill shape, and makes algorithm 2 more diffusive than algorithm 1. Figure 4 shows the stress profile along diagonal AC at $t = 2$ corresponding to the exact solution (identical to the initial profile, since there are no material effects), and to the two algorithms. Algorithm 2 shows a 13.5% reduction of the peak stress value, compared to only 3.6% reduction for algorithm 1.

4.2 Necking analysis

The various algorithms of Section 3 are further tested with a benchmark test in nonlinear solid mechanics: the necking of a circular bar, [13]. The bar, with a radius of 6.413mm and 53.334mm length, is subjected to uniaxial tension. The radius in the central part is slightly reduced (99% of given radius) to induce necking. Only a quarter of the specimen is modelled due to the symmetry.

The two algorithms for the Lagrangian step have been compared in a purely Lagrangian analysis, [9]. Algorithm 1 showed a much higher dependency on time-step than algorithm 2. For this reason, algorithm 2 has been employed here to perform the Lagrangian step.

Figure 5 shows the whole deformed piece after a 7mm pull for various cases. If a Lagrangian analysis is performed on a coarse mesh (50 elements), Figure 5a, the elements in the neck zone become very distorted, and a good definition of the deformed shape is not obtained. A possible solution is to use a fine mesh (320 elements) for the Lagrangian analysis, Figure 5b. An alternative approach, however, is to keep the coarse mesh and use the ALE formulation.

A simple ALE rezoning strategy has been employed: the upper part of the mesh, where element distorsion is reduced, Figure 5a, remains Lagrangian (i.e., convective velocity is null and there is no need of convection step). In the lower part, equal height of the elements is prescribed, thus avoiding the excessive distorsion

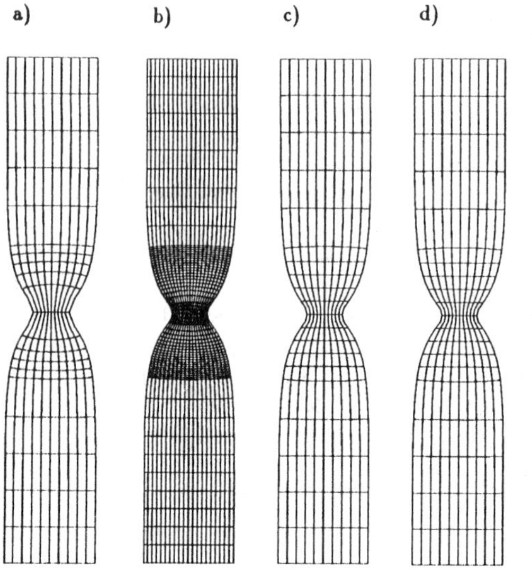

Figure 5: Deformed mesh after 7mm pull. Lagrangian analysis on a) coarse mesh and b) fine mesh. ALE analysis on coarse mesh with c) the interpolation procedure and d) the Godunov-type technique.

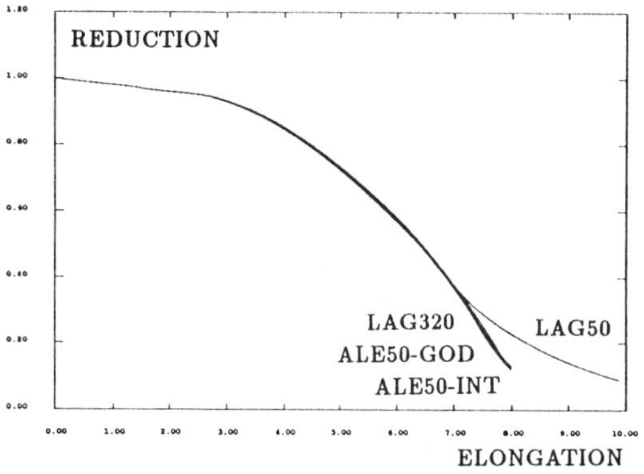

Figure 6: Radius reduction vs. elongation.

of Figure 5a. The two algorithms for the convection step yield very similar results, see Figure 5c (interpolation procedure) and Figure 5d (Godunov-type update).

A quantitative comparison of the four analyses is presented in Figure 6, which shows the radius reduction (ratio of current radius to initial radius) versus the piece elongation (in mm). Very similar results (in good agreement with the results in [13]) are obtained for the four analyses up to 7mm elongation approximately. Differences arise, however, if pulling proceeds. After 8mm pull, and taking the Lagrangian analysis with the fine mesh as the reference, the Lagrangian test with the coarse mesh overpredicts the neck radius by 85%. If the coarse mesh is combined with the ALE formulation, only a 5% overestimate and a 2% underestimate are obtained by performing the convection step with the interpolation procedure and the Godunov-type technique respectively. In this test, algorithm 2 does not show the crosswind diffusion of the previous test, because the convective velocity is approximately in the direction of the vertical coordinate lines, and not along the element diagonals.

REFERENCES

[1] Liu, W.K., T. Belytschko and H. Chang 1986. An Arbitrary Lagrangian-Eulerian Finite Element Method for Path-Dependent Materials. *Comp. Meths. Appl. Mech. Engrg.* 58:227-245.

[2] Benson, D.J. 1989. An efficient, accurate, simple ALE method for nonlinear finite element programs. *Comp. Meths. Appl. Mech. Engrg.* 72:305-350.

[3] Huerta, A. and F. Casadei 1991. Arbitrary Lagrangian-Eulerian Formulation for Large Boundary Motion in Nonlinear Continuum Mechanics. Int. Rep. MA006/1991, E.T.S. de Ingenieros de Caminos, Universitat Politècnica de Catalunya.

[4] Ghosh, S. and N. Kikuchi 1991. An Arbitrary Lagrangian-Eulerian Finite Element Method for Large Deformation Analysis of Elastic-Viscoplastic Solids. *Comp. Meths. Appl. Mech. Engrg.* 86:127-188.

[5] Huétink, J. and P.N. Van der Helm 1992. On Euler-Lagrange finite element formulation in forming and fluid problems. In J.L. Chenot et al. (eds), *Num. Meth. in Industrial Forming Processes*: 45-54. Rotterdam: Balkema.

[6] Huerta, A. and F. Casadei 1994. New ALE Applications in Non-Linear Fast-Transient Solid Dynamics. *Engineering Computations* 11:317-145.

[7] Rodríguez-Ferran, A. and A. Huerta 1995. ALE Quasistatic Analysis in an Object-Oriented Code. Fourth International Conference on Computational Plasticity (COMPLAS IV), Barcelona.

[8] Pinsky, P.M., M. Ortiz and K.S. Pister 1983. Numerical Integration of Rate Constitutive Equations in Finite Deformation Analysis. *Comp. Meths. Appl. Mech. Engrg.* 40:137-158.

[9] Rodríguez-Ferran, A. and A. Huerta 1994. A Comparison of Two Objective Stress Rates in Object-Oriented Codes. Monography No. 26, CIMNE Barcelona.

[10] Huerta, A., F. Casadei and J. Donea 1995. ALE Stress Update in Transient Plasticity Problems. the Fourth International Conference on Computational Plasticity (COMPLAS IV), Barcelona.

[11] *CASTEM2000 - Manuel d'utilisation* 1988. Rapport 88/176, Laboratoire d'Analyse Mécanique des Structures, Comissariat à l'Énergie Atomique.

[12] Bretones, M.A., A. Rodríguez-Ferran and A. Huerta 1995. La programación orientada al objeto aplicada al cálculo por el método de los elementos finitos. *Revista Internacional de Métodos Numéricos en Ingeniería* (to appear).

[13] Simo, J.C. 1988. A Framework for Finite Strain Elastoplasticity Based on Maximum Plastic Dissipation and the Multiplicative Decomposition. Part II: Computational Aspects. *Comp. Meths. Appl. Mech. Engrg.* 68:1-31.

Simulation of Materials Processing: Theory, Methods and Applications, Shen & Dawson (eds)
 ISBN 90 5410 553 4

Development of a penalty method contact algorithm and its application to sheet forming problem

Toru Shimizu & Toshio Sano
Mechanical Engineering Laboratory, Tsukuba, Japan

ABSTRACT: FEM analysis is becoming an important tool in metal forming industry. Many structure analysis codes are applied to simulate metal forming processes. In these codes, treatment of contact between tools and work piece is an important problem. We developed a penalty method contact algorithm which can treat contact and friction problem at the same time. This algorithm is incorporated into a 3-dimensional Rigid Plastic FEM code. In this code, the tool surface is expressed by a B-spline patch, which can describe smooth surface by few control vertices. Using this 3-dimensional code, square cup deep drawing process is analyzed. In this analysis, 8-node isoparametric solid elements are used. Reasonable results are obtained.

1 INTRODUCTION

Recently, FEM simulation of metal forming processes has become popular. In such simulations, contact treatment is an important problem. As one method to treat this contact problem, there is a penalty contact algorithm, that was applied to Rigid-Plastic FEM by Dalin[1]. This method solves the contact problem by minimizing the penalty function, which does not require complicate procedures. The method is extended by Shimizu [2],[3], such that it can treat the contact and friction problem simultaneously.

Now, this contact and friction algorithm is extended to the 3-dimensional case and applied to a Rigid-Plastic FEM code (RIPLE3D). In this code, the tool surface is expressed by a B-spline patch. The square cup deep drawing process is analyzed using this code. A sheet model which is expressed by 8-point isoparametric solid elements, having one element layer is thickness direction, is used. In most sheet forming analyses, the sheet is modeled using shell elements; there are few solution based on one-layer. solid elements[4],[5],[6]. Here, using these isoparametric solid elements, the whole drawing process is analyzed. Wrinkling of the sheet is observed in the case without blank holder.

2 FORMULATION OF 3-DIMENSIONAL CONTACT AND FRICTION ALGORITHM

2.1 *Formulation of flat rigid-body surface*

A 3-dimensional area is divided into two areas by a flat plane f=0, and the plane is defined by Eq.(1).

$$f = ax + by + cz + d = 0 \tag{1}$$

The area where f >0 is assumed a rigid-body area, and the area where f<0 is assumed an area where node can move freely. In this case, a penalty potential f1 is determined by Eq.(2).

$$f1 = (f + |f|)/2 = \frac{1}{2}(f + \sqrt{f^2}) \tag{2}$$

The penalty potential f1 is 0 at a free area, and f1 becomes greater than 0 at a rigid-body area. Using this penalty potential f1, another penalty potential fp is defined. The fp is determined by Eq.(3).

$$fp = p\, f1^2/2 \tag{3}$$

If the node N has the coordinates x, y, z, the penalty potential of the node becomes fp(x,y,z), and gradient of fp becomes force which works on the node. If the node is at the free area(f<0), force does not work on the node, and if the node is at the rigid-body area(f>0), force for the direction to the surface works on the node. The force which works on the node increases according to the penetration depth, and this potential works like an elastic boundary. By increasing the p value, this potential becomes to work like a rigid-body surface.

Φp is the functional of Rigid-Plastic FEM by penalty method. A deformation problem is solved by searching a stationary point of this functional. If the penalty method contact treatment is applied, the functional becomes like Eq.(4).

$$\Phi = \Phi p + \Sigma fp \qquad (4)$$

In rigid-plastic FEM, the node velocities [ux, uy, uz] are variable. Therefore, Eq.(5) should be used instead of Eq.(1) to define f.

$$f = a\,(x+ux\,\Delta t) + b\,(y+uy\,\Delta t) + c\,(z+uz\,\Delta t) + d \qquad (5)$$

2.2 *Formulation when friction works at surface*

Friction by the Coulomb-Amontons'low can be expressed using the penalty potential. A node is assumed to penetrate the surface to depth D, and slide in tangential direction over length L (see Fig.1). The contact force Fc which works on the node is determined by Eq.(6), the contact penalty potential Ec is determined by Eq.(7), and the penalty potential by friction is determined by Eq.(8). The total penalty potential is given by Eq.(9), where μ is friction coefficient.

$$Fc = p\,D = p\,f1 \qquad (6)$$

$$Ec = p\,D^2/2 = p\,f1^2/2 \qquad (7)$$

$$Ef = \mu\,Fc\,|L| = \mu\,Fc\,\sqrt{l^2 - f1^2} \qquad (8)$$

$$Ep = Ec + Ef = \frac{1}{2}\,p\,f1\,(f1 + 2\,\mu\sqrt{l^2 - f1^2}\,) \qquad (9)$$

Where $l^2 = (x-xo)^2 + (y-yo)^2 + (z-zo)^2$

Using the penalty potential of Eq.(9), the contact and friction problem are solved simultaneously. The contour of Eq.(9) where z is constant is shown in Fig.2. In actual analysis, however, this potential has the problem of convergence, because the gradient

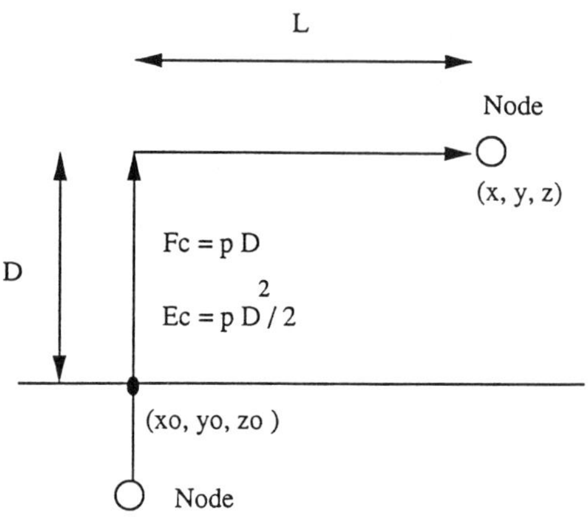

Fig.1 Contact and friction potential which works on node.

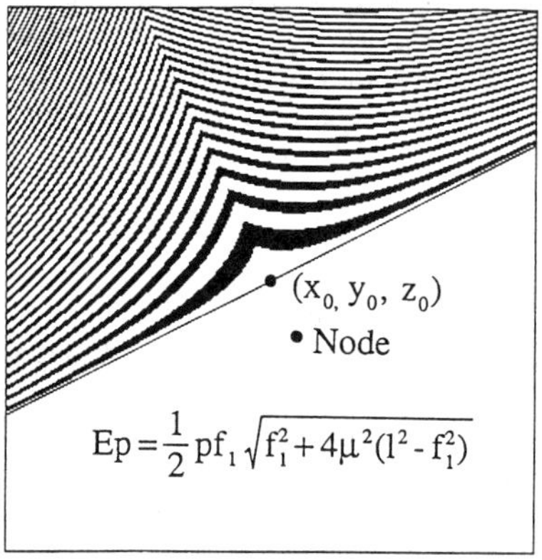

Fig.2 Contour of contact and friction potential.

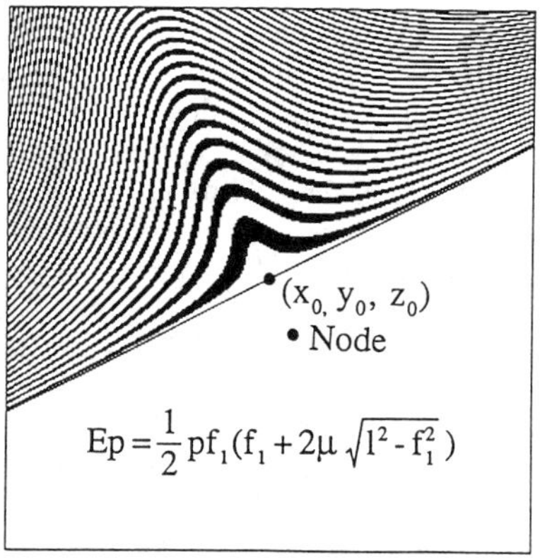

Fig.3 Contour of relaxed potential.

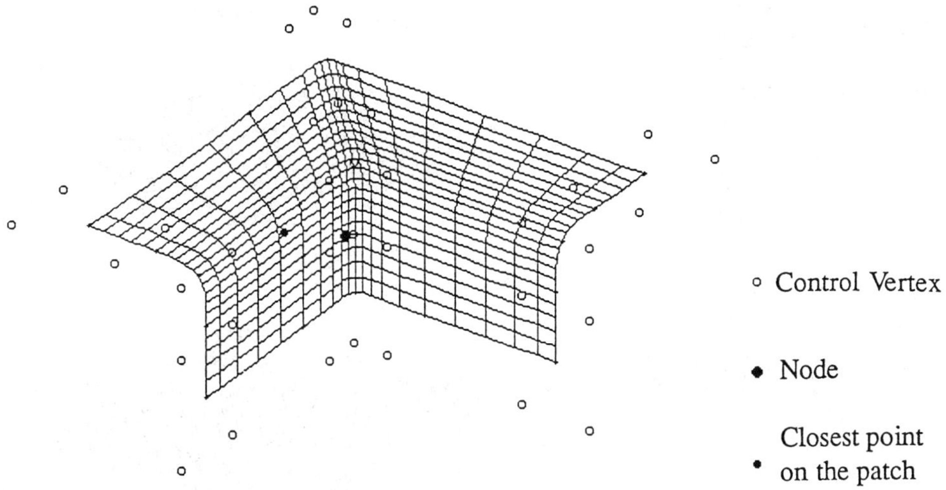

Fig.4 Example of uniform cubic B-spline patch.

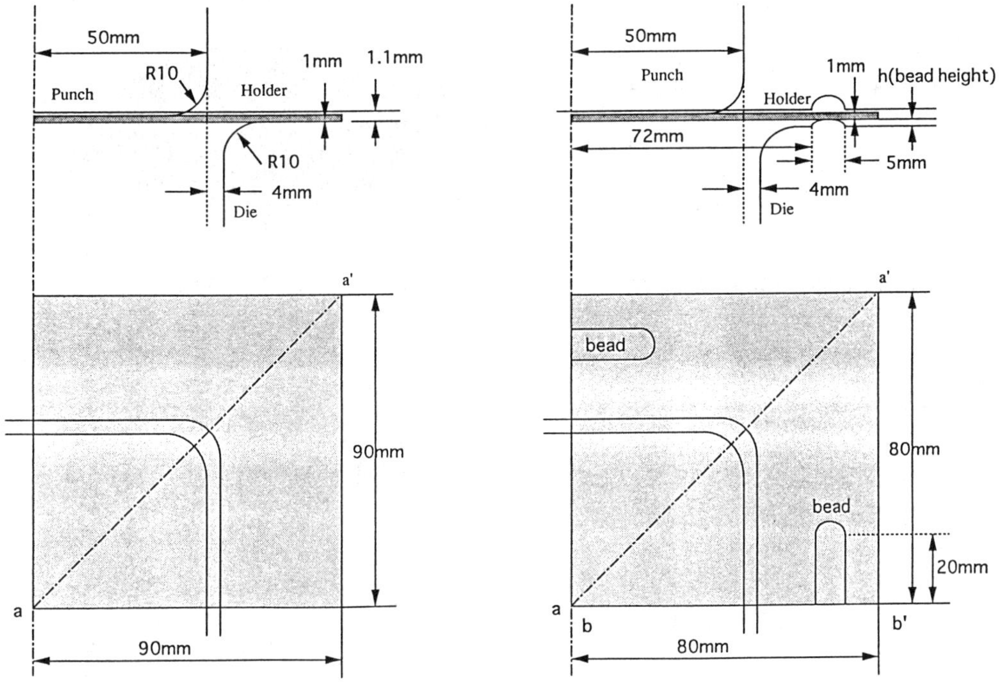

(a) Friction works between tools and sheet.

(b) Case there are draw-beads on tools.

Fig.5 Square cup deep drawing tools and sheet.

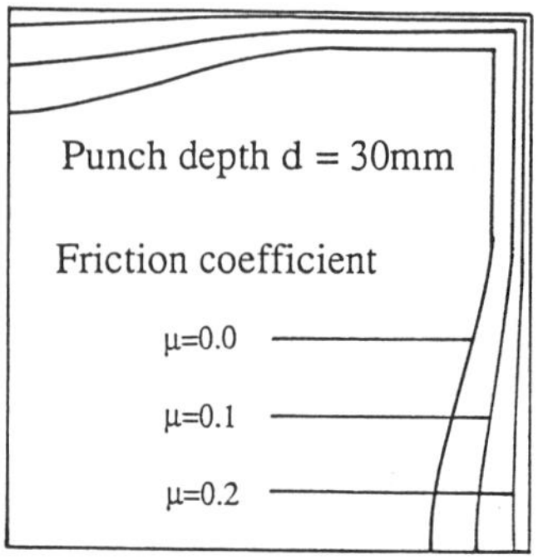

Fig.6 Flange shape where friction works.

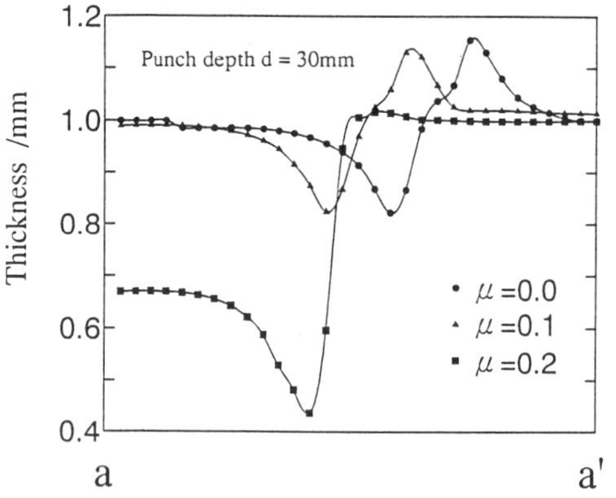

Fig.7 Thickness of the sheet where friction is changed.

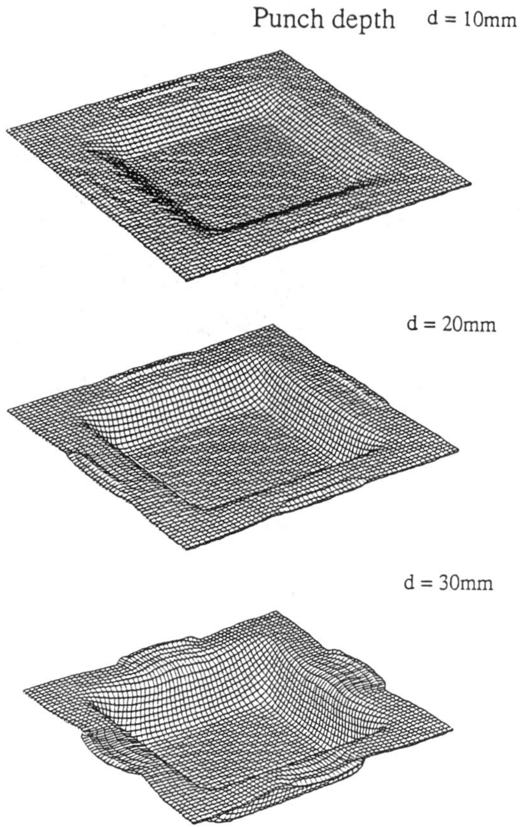

Fig.8 Forming process (bead height = 1mm).

cannot be calculated for certain values. To overcome this problem, the penalty potential Eq.(9) is relaxed like Eq.(10). Contour of the Eq.(10) is shown in Fig.3.

$$Ep = \frac{1}{2} p\, f1 \sqrt{f1^2 + 4\mu^2 (l^2 - f1^2)} \quad (10)$$

3 TOOL SURFACE EXPRESSED BY B-SPLINE PATCH

3.1 *B-spline patch*

The uniform cubic B-spline patch is used to express the tool surface. If coordinates of control vertices are defined by matrix [xij],[yij],[zij], the coordinates of the point on the B-spline patch are determined by Eq.(11). Fi(α) is defined by Eq.(12). Fj(β) is defined same way as Fi(α).

$$x_s = [F_i(\alpha)]\ [x_{ij}]\ [F_j(\beta)]$$
$$y_s = [F_i(\alpha)]\ [y_{ij}]\ [F_j(\beta)]$$
$$z_s = [F_i(\alpha)]\ [z_{ij}]\ [F_j(\beta)] \quad (11)$$

$$\begin{array}{lll} \text{if } i =< \alpha & F_i(\alpha) = 0 & \\ i-1 =< \alpha < i & F_i(\alpha) = & 0.5(t-1)^2 \\ i-2 =< \alpha < i-1 & F_i(\alpha) = & -t^2 + t + 0.5 \\ i-3 =< \alpha < i-2 & F_i(\alpha) = & 0.5\, t^2 \\ \alpha < i-3 & F_i(\alpha) = & 0 \end{array} \quad (12)$$

Where, $t = Int(\alpha) - \alpha$, and $Int(\alpha)$ is truncation of α.

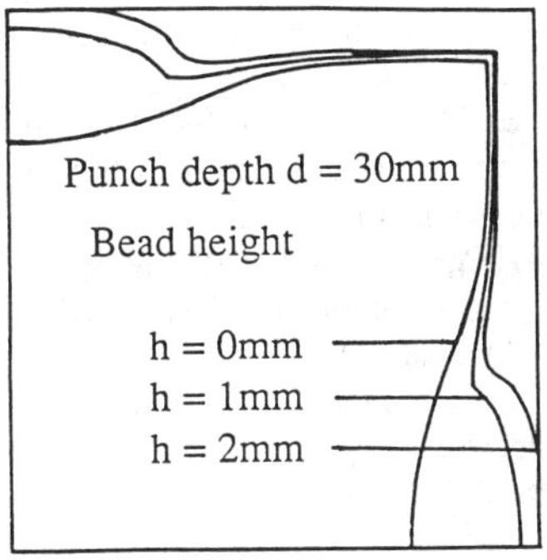

Fig.9 Flange shape where draw-bead height is changed.

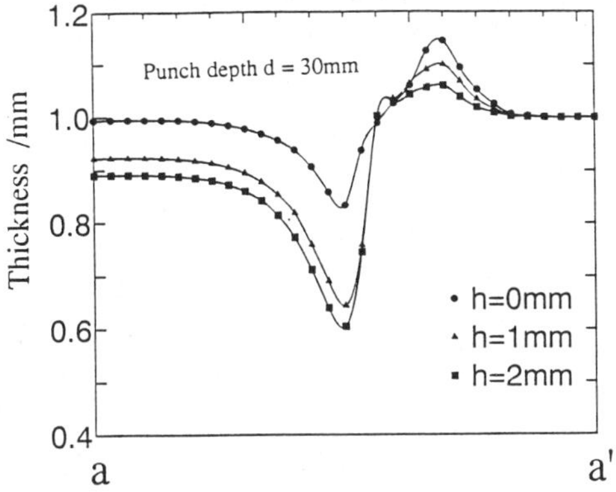

Fig.10 Sheet thickness along aa' where bead height is changed.

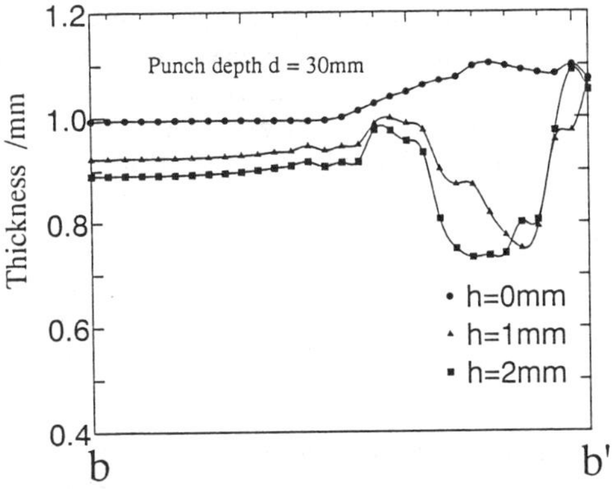

Fig.11 Sheet thickness along bb' where bead height is changed.

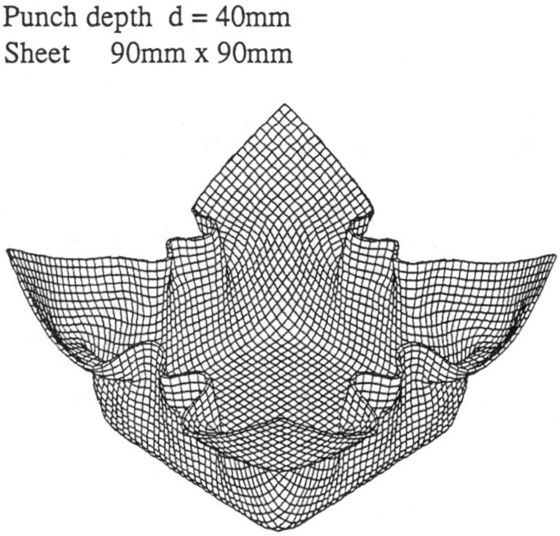

Fig.12 Sheet deformation where blank-holder is removed.

3.2 *Method to obtain closest point*

To apply the penalty method contact algorithm to the tool surface expressed by B-spline patch, an operation to search the closest point on the patch surface is necessary. The penalty potential is put so that its surface is accord with the tangential plane of the B-spline patch. The closest point is searched; then Eq.(13) adopts a minimum value. The Newton-Raphson method is used for this searching operation. If the spline patch has a complex shape, a direct searching method must be used for primary searching. After that, the Newton-Raphson method is used for searching the closest point.

$$fc = (x-xc)^2 + (y-yc)^2 + (z-zc)^2 \quad (13)$$

Where x, y, z are coordinates of the node, and xc, yc, zc are coordinates of the closest point.

4 SQUARE CUP DEEP DRAWING ANALYSIS

4.1 *Model*

The square cup deep drawing process is analyzed using a 3-dimensional code, including the penalty contact algorithm. An 8-node isoparametric solid element is used in this code as a standard element. Because of symmetry, only one quarter of the geometry is analyzed. The sheet is described by 30 x 30 elements; it has one layer in thickness direction. Drawing die, punch and blank-holder are expressed by the B-spline patch. Two cases are analyzed. First one is that friction works between sheet and tools.

Second one is that there are draw-beads on die and blank-holder and no friction on the tools. Sheet and tools are shown in Fig.5; (a) is the case where friction works; and (b) is the case where are draw-beads on die and blank-holder. The yield stress of the sheet is as in Eq.(14).

$$Y = 1.0 + \varepsilon^{0.2} \tag{14}$$

4.2 Resorts

The analysis is performed on a CRAY C-90 at Tsukuba Research Center ,AIST. It takes about 4 hours CPU time for each analysis. Flange shapes for some friction coefficients are shown in Fig.6. The thickness of the sheet along the line a-a' is shown in Fig.7. The second case concerned the deep drawing with draw-beads. The calculated forming proceeding is visualized in Fig.8 (draw-bead height: 1mm). Differences in the flange shape when the draw-bead height is changed is shown in Fig.9. Sheet thickness along aa' and bb' are shown in Fig.10 and Fig.11, respectively. In these analyses, the draw-bead works stronger than in the experimental case. The reason will be that the elements are not fine enough, so that the node cannot move smoothly over the draw-bead part. Also, a case without blank-holder is analyzed. Deformation of the sheet is shown in Fig.12. Wrinkling occurs at the flange part. It is known that this code can analyze unstable deformation.

5 CONCLUSION

A penalty method contact and friction algorithm is developed, and applied to the tool surface which is expressed by a B-spline patch. This contact treatment is suitable method for 3-dimensional forming analysis code. This code can analyze unstable deformation. This method and the code may become an efficient tool for metal forming analysis and forming process design.

REFERENCES

[1]. Dali, J.B. and Onate, E. (1989). "An automatic algorithm for contact problem: Application to sheet metal forming." Numerical methods in Industrial Forming Processes 89, Fort Collins, ...419-424.

[2]. Shimizu, T. and Sano, T. (1993). " Development of penalty method contact algorithm for rigid plastic FEM." Advanced Technology of Plasticity, Vol.3, Beijing, 1163-1168.

[3]. Shimizu, T. and Sano, T. (1995). " An application of a penalty method contact and friction algorithm to a 3-dimensional tool surface expressed by a B-spline patch." Journal of Materials Processing Technology, Netherlands, 48, 207-213.

[4]. Shimizu, T., Massoni, E., Soyris, N. and Chenot, J.L. (1991). "A modified 3-dimensional finite element model for deep drawing analysis." Winter meeting of the ASME, Atlanta.

[5]. Shimizu, T., Soyris, N., Massoni, E. and Chenot, J.L. (1992). " Development of a 3-D code for simulation of the deep drawing process." Computational Plasticity 92, Barcelona.

[6]. Massoni, E. and Chenot, J.L. (1992). " 3D finite element simulation of deep drawing process." Numerical Methods in Industrial Forming Processes 92, Sophia-Antipolis,503-507.

Simulation of Materials Processing: Theory, Methods and Applications, Shen & Dawson (eds)
© 1995 Balkema, Rotterdam. ISBN 90 5410 553 4

Quasi-static analysis of sheet metal forming processes on a parallel computer

S.C.Tang
Research Laboratory, Ford Motor Company, Dearborn, Mich., USA

Y.Hu
Automated Analysis Corporation, USA

ABSTRACT: One major problem in the application of the quasi-static analysis of sheet metal forming processes is long computing time for large finite element models. This paper presents a method to port Ford's quasi-static code to the supercomputer CRAY Y-MP C90 for a fast turnaround time by taking advantage of the vector and parallel operations. A numerical example shows that a typical automotive part with a model of 45000 degrees of freedom takes less than three hours for a draw depth of 141 mm on the supercomputer.

1 INTRODUCTION

Quasi-static formulation which is solved by implicit integration is the most reliable method to analyze sheet metal forming processes for automotive panels. However, the method has two major difficulties. The first is the convergence of the iteration for each increment of the tool travel (a load step) and the second is the long computing time for large finite element (FE) models which are occasionally needed to predict failure due to splitting in areas with complex geometry and wrinkling with a short wave length. The convergence difficulty has been addressed by Tang (1993a), but the long computing time for large models still remains an issue. A local 3-D model has been proposed by Tang (1993b), which can analyze areas with complex geometry and at the same time reduce the computing time to an acceptable range. In this paper, by taking advantage of concurrent computation, we propose improving the computing time by adapting the quasi-static code to a supercomputer of multiple processors with a shared memory.

2 QUASI-STATIC FORMULATION AND FE SOLUTION

Because the punch speed to form an automotive panel is relatively low (200 mm/sec), the inertia effect of a sheet mass can be neglected to simplify the solution procedure and avoid undesirable oscillation. This is called the quasi-static formulation. The equation of motion is then reduced to the equation of equilibrium:

$$\mathbf{P} = \mathbf{F} \tag{1}$$

where $\mathbf{P}$ is the internal force vector and $\mathbf{F}$ is the external force vector. It is a set of highly nonlinear simultaneous equations. Because of high nonlinearity and the flow theory of plasticity used in the analysis, an incremental solution procedure is used in this paper. The total punch travel distance is divided into a few hundred steps with an increment of about 1 mm (called a load step) for a typical automotive body panel.

Within a load step, the Newton-Raphson method of iteration is used to compute the displacement increment vector $\Delta\mathbf{U}$ in the following linear system:

$$\mathbf{K}_t \, \Delta\mathbf{U} = \Delta\mathbf{F} - \mathbf{R} \tag{2}$$

where $\mathbf{K}_t$ is the tangent stiffness matrix. The iteration reduces the residual force $\mathbf{R}$ so that the equation of equilibrium is satisfied for each load step. In sheet metal

Table 1. Computing time for linear simultaneous equation solution

Example	N	M	$NM^2/9.7E8$	CRAY/1	CRAY/8	IBM
#1	10404	306	1.0	2.4*(1.0)	0.51 (1.0)	31.4 (1.0)
#2	28125	375	4.1	8.7 (3.6)	1.72 (3.4)	129.9 (4.1)
#3	50000	500	12.8	24.3 (10.1)	4.37 (8.6)	390.6 (12.4)
#4	50625	675	23.7	40.1 (16.7)	6.19 (12.1)	728.9 (23.2)

Note: *Wall clock time in seconds
N = Number of linear simultaneous equations
M = Semi-band width
CRAY/1 means one processor on the CRAY C90
CRAY/8 means eight processors on the CRAY C90
IBM means IBM RS6000/560 workstation
Values in the parentheses normalized with respect to those in Example #1

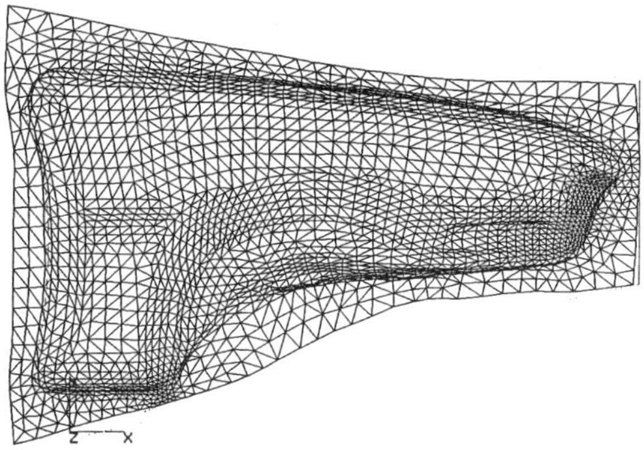

Fig. 1 Medium-size FE model

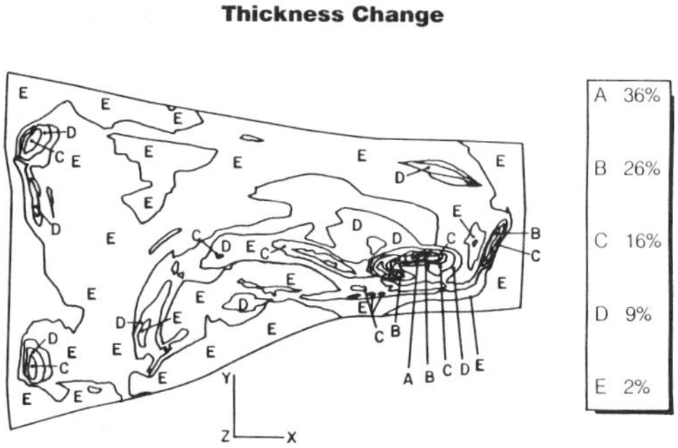

Fig. 2 Thickness change for the medium-size model

forming analysis, the external load increment is not explicitly specified in the contact area between the sheet and the tool surface. Equation (1) can be modified with contact conditions by Tang et al (1988).

3 SOLUTION ON A COMPUTER WITH PARALLEL PROCESSORS

In order to make the turnaround time for a computer analysis of a large FE model practical for design evaluation, we are proposing to port the analysis code to the CRAY Y-MP C90 (called C90 in this paper) supercomputer of 16 parallel processors with a shared memory. There are two major tasks in implementing the code to the C90. The first task is the element operation and the second is the solution of a set of simultaneous equations.

3.1 Element Operation

In the FE method, the element operation to establish an element tangent stiffness matrix, strain and stress and nodal force computations for each element can be performed independently. Therefore, it is relatively easy to port a finite element program to a parallel computer with a shared memory like the CRAY C90, to do the element operation concurrently. The most popular parallel FE analysis method invokes the substructure technique. In this method, the structure is divided into several substructures before the FE simulation. The simulation hence will be performed based on one substructure on a central process unit (CPU) assignment. The computational effort must be well balanced in this method. This is very difficult to achieve when CPU sharing on several

CPUs is not even. This is so called the static parallelization approach. In our application, a dynamic parallelization approach is employed to avoid any unbalanced computational effort distribution.

3.2 Solution of Simultaneous Equations

For quasi-static analysis of a sheet metal forming process which is a highly nonlinear problem, the Newton-Raphson iteration is used for each load step. In every iteration, a set of linear simultaneous equations from the global tangent stiffness matrix has to be solved. According to Bathe et al (1993), the number of numerical operations to solve a set of linear simultaneous equations is roughly proportional to NM^2, where N = number of degrees of freedom, M = semi-band width and roughly $M \propto N^{1/2}$ for an FE model. Therefore, the expense of solving the linear system in Eq. (2) increases as N^2. Using the vector and parallel processors in the C90 to bypass this bottle neck is the key to reduce the computing time. Since the global tangent stiffness matrix from the analysis is not necessarily positive definite for a thin shell in an unstable region, the Gaussian elimination instead of the Choleski decomposition is used to solve the linear system.

As described by Overman and Poole (1991), vector saxpy operations are more efficient than dot product operations on the Cray Y-MP. The reason behind this is that both addition and multiplication instructions are involved in vector saxpy operations and they can be carried out nearly simultaneously on vectors **X** and **Y**: a***X**+**Y**. On the other side, dot product operations also consist of both addition and multiplication instructions but the summation is slower because it is performed on a single scalar value: Sum = $X_1Y_1 + X_2Y_2 + \ldots + X_nY_n$. As a result, the gaxpy version Gaussian elimination is employed to replace the conventional outer product Gaussian elimination in this paper. In order to cooperate with the gaxpy version Gaussian elimination, we used varied bandwidth row wise storage for the global stiffness

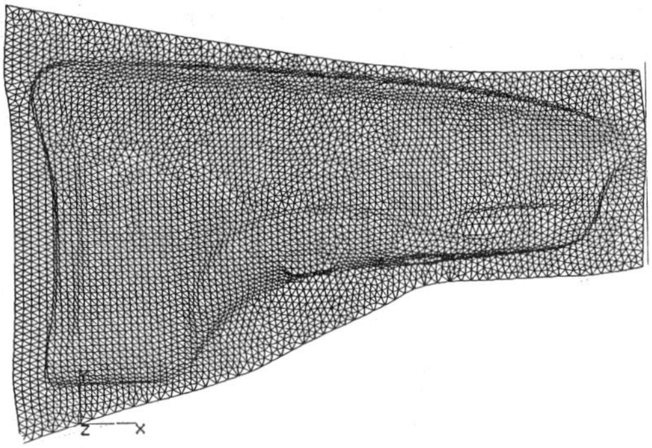

Fig. 3 Uniform FE model

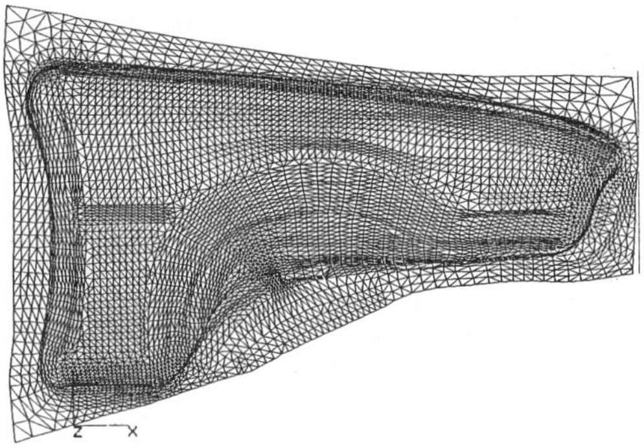

Fig. 4 Fine FE model

matrix while the traditional storage scheme of column wise sky line storage is abandoned.

Unrolling is another technique which exploits the computational efficiency of the C90. By using unrolling, multiple vector registers can be used effectively so that fetching and storing of the vectors can be speeded up. Since the calculation speed on the C90 is very fast, the memory access speed becomes more important and the ratio between memory access time and calculation time becomes relatively larger. Unrolling is one of the techniques used to reduce the memory access delay. In our application, an unrolling of value 9 is employed. Usually, trim is necessary to be carried out to shape the global tangent stiffness matrix so that the resulting matrix matches the unrolling requirement. Since an unrolling value of 9 matches our 9 degrees of freedom per node exactly for the thin shell element used in the code, we do not need to trim the global stiffness matrix further.

The parallel loop is placed on the second loop of the three loop LU factorization. Compiler directives CMIC$ DO PARALLEL and CMIC$ END PARALLEL are used to declare the parallel region. The C90 is a shared memory computer. The variable classification between the global variables and local variables is needed to ensure correct memory handling. The global variables then are stored in shared memory, while the local variables are stored in individual private CPU memory by using the compiler directives CMIC$ SHARED(var1,var2,...) PRIVATE(var1,var2,...) before the declared parallel region.

Using the above methods, excellent performance (3840 Mflops) is achieved in solving a system of 50000 equations on the C90 with 8 parallel processors. Table 1 shows the performance of the CRAY C90 computer and IBM RS6000/560 workstation for the equation solver. In the table, NM^2 for Example #4 is 23.7 times of that in Example #1. On the IBM workstation, the solution time is 23.2 times of that in Example #1. This proves that the number of numerical operations to solve a set of simultaneous equations is indeed proportional to NM^2. However, on the C90, the solution time for Example #4 is 16.7 times of that in Example #1 if a single process is used, while it is reduced to 12.1 times of that in Example #1 if 8 parallel processors are used. This factor (12.1) is only roughly one half of 23.7; therefore, the bottle neck for the solution time proportional to NM^2 can be bypassed.

4 NUMERICAL EXAMPLES

4.1 Medium-Size Model

As a first example, we analyzed a model of 3907 high performance thin shell elements for the fender in one of the NUMISHEET'93 (1993) benchmark tests on the C90. Using 16 parallel processors on the C90, the wall clock time, which is relevant for fast turnaround time, is 0.81 hour for a draw depth of 141 mm for a die

Table 2. Computing time using various numbers of processors

No. of processors	Wall clock time in hrs.
1	13.51
8	3.40
16	2.80

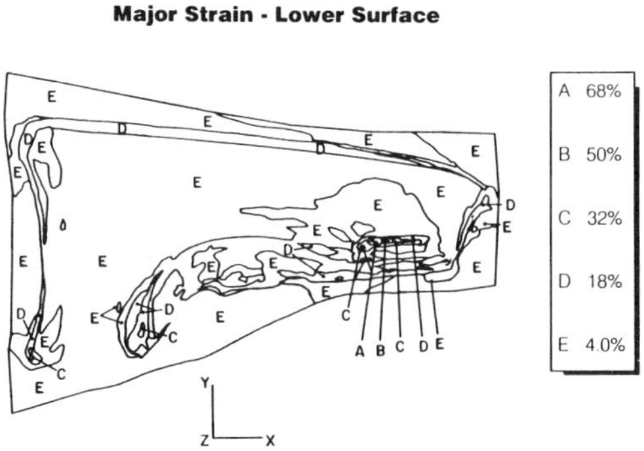

Fig. 5 Major principal strain for the fine model

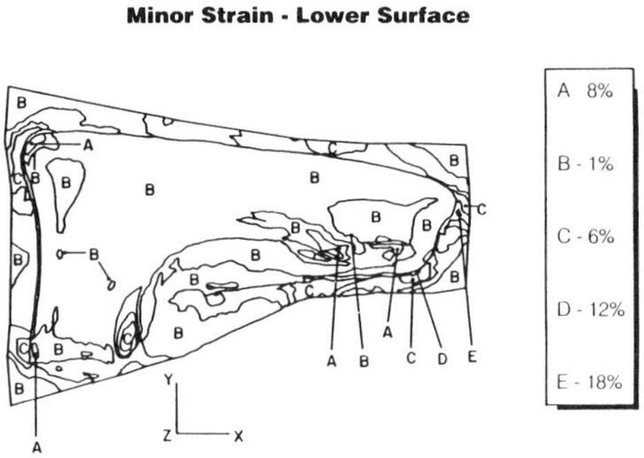

Fig. 6 Minor principal strain for the fine model

closure starting at the binder-wrap. It took 48 hours on the IBM RS6000/560 with the out-of-core storage and 11.5 hours on the RS6000/590 with the in-core storage. Figure 1 shows the model with 3907 elements. Figure 2 shows the thickness change for the sheet.

4.2 Model with a Uniform Mesh

For the purpose of measuring computing time only, we modeled the

Table 3. Wall clock time for the three models

Example	El #	Node #	N	M	Time in hr.
1	3907	2099	18081 (1.0)	486	0.81 (1.0)
2	10004	5117	46053 (2.5)	792	2.79 (3.4)
3	9828	4995	44955 (2.5)	801	2.80 (3.5)

Values in the parentheses normalized with respect to those in Example 1

fender with a uniform mesh of 10004 thin shell elements as shown in Fig. 3. Note that this model does not efficiently compute the forming strains and stresses. It took 2.79 wall clock hours to complete the same draw depth for a run on the C90 with 16 parallel processors.

4.3 Fine Model

For a better simulation, we modeled the geometrically complex areas in the same fender with more elements. A very detailed but efficient FE model was established. This model contains 9828 thin shell elements as shown in Fig. 4. Using the same number of processors on the C90, the wall clock time for a run of the same draw depth is 2.80 hours. Figures 5 and 6 show the major and minor principal strains at the bottom surface of the sheet. Figure 7 shows the thickness change for the sheet. The forming strains are slightly higher than those from the medium-size model. To show the efficiency of parallelization using the C90, we have also used one and 8 CPUs to analyze the same problem and the results are displayed in Table 2.

4.4 Discussion

For a comparison of the wall clock time while using 16 parallel processors on the C90, we summarized the results from these three models in Table 3. Note that the fine model took a little bit more time to complete the computer run because the semi-band is larger and it seems to take more iterations in the Newton-Raphson solution. Table 3 shows that the wall clock time for Example 3 is only 3.5 times as big as that in Example 1, while N^2 for Example 3 is 6.25 times of that in

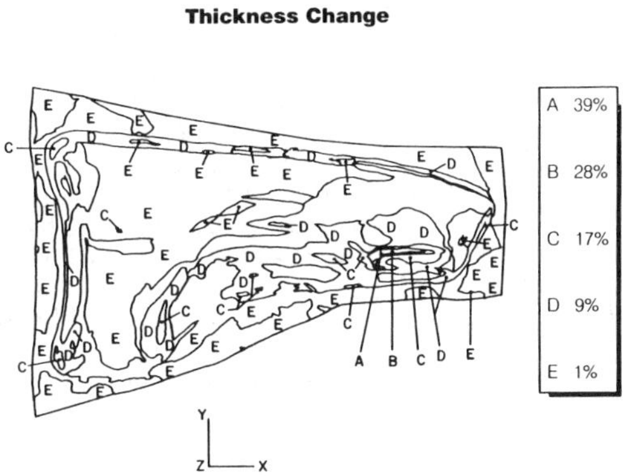

Fig. 7 Thickness change for the fine model

Example 1. Therefore, we can bypass the bottle neck of running time proportional to N^2 from Bathe et al (1993), where N = number of degrees of freedom in the FE model.

5 CONCLUSION

Taking advantage of the vector and parallel operations on the C90, we have demonstrated that the solution time for a set of linear simultaneous equations proportional to NM^2 can be much reduced and the overall wall clock time for analysis of a fine model in die closure is much smaller than that proportional to N^2. For a sufficiently detailed model as illustrated in Example 3, the wall clock time for a computer run is less than 3 hours which is a reasonable turnaround time for design evaluation. Note that the shell element used in these examples is a high performance element and we use 7 sampling points on the middle surface of the sheet and 7 layers through the thickness for numerical integrations to compute element forces and tangent stiffness. In

Ford, we are using the C90 to evaluate about 20% of panels which require detailed models in the quasi-static analysis.

REFERENCES

Bathe, K. J., et al 1993. Some recent advances for practical finite element analysis. Computers & Structures. 47: 511-521.

NUMISHEET'93 1993. Front fender stamping. Proc.: 521-621.

Overman, A. L. and Poole, E. L. 1991. Efficient multitasking of Choleski matrix factorization on CRAY supercomputers. NASA TM 4259.

Tang, S. C., et al 1988. Sheet metal forming modeling of automobiles body panels. Proc. 15th Biennial Congress of IDDRG: 185-193.

Tang, S. C. 1993a. Quasi-static analysis of sheet metal forming processes. Computer Applications in Shaping & Forming of Materials edited by M. Y. Demeri. TMS: 53-67.

Tang, S. C. 1993b. A local 3-D model for automotive sheet metal forming analysis. Advanced Technology of Plasticity edited by Wang and He. Beijing: 1647-1652.

Simulation of Materials Processing: Theory, Methods and Applications, Shen & Dawson (eds)
© 1995 Balkema, Rotterdam. ISBN 90 5410 553 4

Parallel performance of LS-DYNA3D on industrial applications

B.Wainscott
Livermore Software Technology Corporation, Calif., USA

ABSTRACT: Finite Element Analysis is computationally very demanding. The increasing size and complexity of models demands ever greater computing power. Several results are presented demonstrating that distributed memory parallel computers offer a viable alternative to traditional vector supercomputers.

1 Introduction

Finite Element Analysis codes are among the most computationally intense applications used in industry. As they continue to be more heavily utilized and the models used become ever more detailed, demands on both the software and hardware used are growing. LSTC is responding to this demand by developing a parallel version of the LS-DYNA3D program. Utilizing Massively Parallel Processor (MPP) machines, LS-DYNA3D is able to satisfy the needs of an ever more demanding community of users.

The decision to concentrate on MPP machines is a reasonable one. As the ability to produce high end components for traditional vector supercomputers increases, making it more and more difficult and expensive to increase the single CPU performance of such machines, the price and performance of desktop workstations causes them to be increasingly more attractive. Being able to harness several such machines to perform together on a single problem can result in very competitive performance at a fraction of the cost of a vector supercomputer.

While shared memory implementations of parallel computers have some advantages, the large scale potential for distributed memory MPP machines makes them a good platform for future development. A workstation network implementation of LS-DYNA3D has been tested, but the bandwidth and latency of Ethernet networks precludes impressive performance. Thus current work is focused on true MPP machines with their fast interprocessor communication. As increases in network technology begin to catch up with processor speeds, however, many applications that today only run on special MPP hardware will run well on networks of desktop workstations.

2 Description of problems

The uses for LS-DYNA3D are wide ranging. We have chosen three problems with which to demonstrate the capabilities and performance of the code on various MPP platforms.

Sedan Crash: This is the simulation of a sedan crashing head on into a rigid pole at 15.5 meters per second. (Figure 1.) The model has 26976 nodes, 28371 shell elements, 142 beam elements, and 124 materials. It has a single contact interface that comprises the whole of the vehicle forward of the apillar. There are 2 rigid walls in the problem: The pole, a cylindrical rigid surface, and the ground. In both cases the slave nodes consist of the portion of the car forward of the apillar.

VDI Crash: This problem was developed as the crash benchmark for the VDI (Society of German Engineers) conference at Würzburg 1994. It is a simulation of a vehicle impacting a rigid barrier head on at 8.3 meters per second (Figure 2.) The

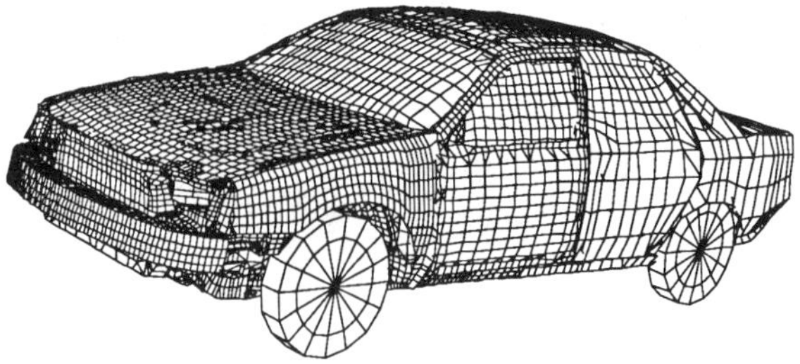

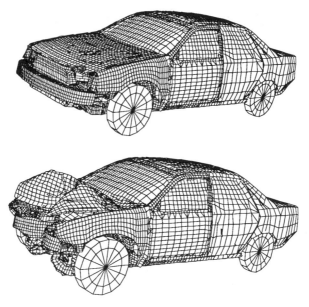

Figure 1: Sedan impacting rigid pole

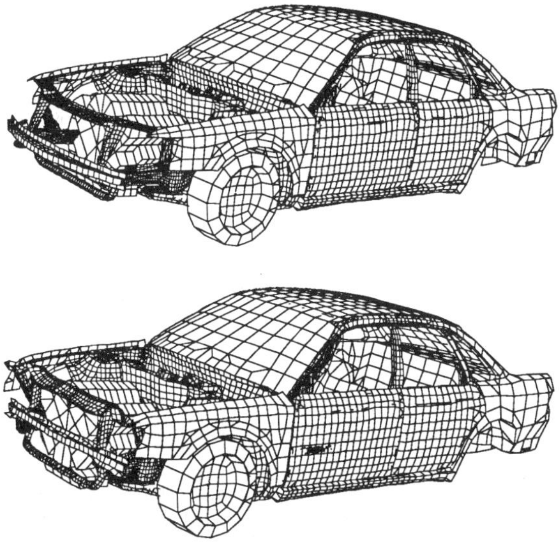

Figure 2: VDI Crash benchmark problem

model has 26949 nodes, 28007 shell elements, 216 beam elements, and 222 materials. There are 5 contact interfaces.

Metalforming: This problem was developed as the sheet metalforming benchmark for the VDI conference in Zürich 1991. The original problem was rather small by current standards, and so in this case the mesh of the blank material was refined, resulting in a total of 19,014 elements. There are 3 rigid tooling parts: a lower die, a binder, and an upper die.

Table 1: Timing comparison for parallel analysis of sedan impacting pole

Computer	Number of Processors						
	1	2	4	8	16	32	64
Cray C90 Vector version	9.03						
Cray T3D Incremental Speedup				25.98	14.99 1.73	9.37 1.60	5.36 1.75
IBM SP2 Incremental Speedup		23.8	13.65 1.74	8.17 1.67	5.57 1.47		
Intel Paragon Incremental Speedup				48.89	27.57 1.77	15.66 1.76	8.57 1.82

3 Results

Data was collected on three different MPP platforms: The Cray T3D, the IBM SP2, and the Intel Paragon. For comparison purposes, each of the problems was also run on a Cray C90 using the vectorized version of LS-DYNA3D.

Due to the length of time it takes to run these problems, and the difficulty of getting large blocks of dedicated time on large machines, the problems were not run to completion. Each problem was run for about a 5000 integration time steps on each machine.

One commonly used measure of the speed of LS-DYNA3D is "microseconds per zone cycle" (mzc) which is computed as follows. Let T be CPU time in microseconds, as measured over some number N of integration time steps. Let E be equal to the total number of elements in the problem (solid, shell, beam, and thick shell). Then mzc $= T/(N * E)$.

As an absolute measure of speed, mzc gives only a rough comparison between different models. The actual amount of computational work associated with a problem depends on many factors such as the material models, equations of state, number of shell integration points, number of contact interfaces, and so forth. But as a relative measure of the code, on a fixed problem, it is quite adequate. Of course, the number of microseconds per zone cycle changes during the course of a given problem, as conditions in the problem (e.g. contact or I/O) change. Thus care must be taken to compare numbers averaged over the same portion of a problem.

For the purposes of this paper, the speed of the program was measured as follows. Each problem was run for 5000 cycles, and the average mzc was computed. Typically, one of these problems run to completion takes about 50,000 to 70,000 cycles. But in terms of comparing overall performance, running the problems to completion is not necessary – a comparison of the mzc during the early part of a problem gives a good comparison of the overall performance.

For parallel applications, a traditional measure of performance is speedup. This is the ratio of how fast the program runs on P_n processors, as compared to running the same program on a single processor. For large problems, however, it is not always possible to run the program on a single processor. From a practical point of view, running a large problem on a single processor of a large MPP machine is not something a user would be interested in anyway. A more practical measurement of code efficiency can be described, from the users point of view, as "if I run on P_2 processors, how much faster will the problem run than if I run on P_1 processors?" If the speed of the program on P_i processors is S_i then the *incremental speedup* in going from P_1 to P_2 processors is given by S_2/S_1.

All problems were run with the same version of the MPP LS-DYNA3D program. Performance numbers for the IBM SP2 are regrettably not available for large numbers of processors, due to the scarcity of available CPU time. For the same

Table 2: Timing comparison for parallel analysis of VDI Crash benchmark

Computer	Number of Processors							
	1	2	4	8	16	32	64	128
Cray C90 Vector version	5.59							
Cray T3D Incremental Speedup				20.46	12.02 1.70	6.77 1.78	4.27 1.59	3.31 1.29
IBM SP2 Incremental Speedup		15.3	8.5 1.80	5.69 1.49	3.73 1.52			
Intel Paragon Incremental Speedup				39.94	22.56 1.77	12.27 1.84	7.21 1.70	

Table 3: Timing comparison for parallel analysis of VDI Metalforming benchmark

Computer	Number of Processors							
	1	4	8	16	32	64	128	256
Cray C90 Vector version	6.20							
Cray T3D Incremental Speedup				15.1	8.50 1.78	(4.67) (1.82)	(2.56) (1.82)	(1.58) (1.62)
IBM SP2 Incremental Speedup		19.87	9.56 2.08	5.37 1.78				
Intel Paragon Incremental Speedup				32.47	18.14 1.79	10.18 1.78		

reason, most of the data (on all platforms) was obtained during periods when other users may have been using other processors on the same system. This can have some effect on the timing, due to I/O and other system resource contentions.

Table 1 shows the results of the first test problem. As can be seen from the table, the MPP version of LS-DYNA3D is comparable to the C90 version in speed on about 32 processors on the T3D, 64 processors on the Paragon, and 8 processors on the SP2. Each doubling of the number of processors results in an overall speed increase of about 1.7 times.

Scaling in general on the SP2 does not extend well to higher numbers of processors, as is indicated by the incremental speedup beginning to drop off at only 16 processors here. This is primarily due to the communication hardware used being outmatched by the processor speed. The faster processor gives much shorter run times for a given number of processors, but as the number of processors increases the communication channels cannot keep up, and speedup drops off. Another symptom of this shows up in the I/O. As compared to the other test problems, this test problem has a significant amount of output. It is estimated that if output had been turned off for this problem, the 16 processor SP2 speed would be 5.01 mzc, a 10% improvement. For comparison, the T3D speed on 16 processors would be 14.88 mzc, a mere 0.7% difference from the measured rate.

Table 2 shows the results of the second test problem. Again, we see that LS-DYNA3D has a speed increase of about 1.7 times for each doubling of the number of processors, excepting the SP2. We see here a drop in scalability on the T3D when this problem reaches 64 and 128 processors. For larger problems, the range of scalability is extended, on any platform, up to higher numbers of processors. This is because the amount of communication does not increase, with problem size, as much as the overall amount of computation. Therefore the communication hardware can better keep up with the processors.

Table 3 shows the results of the final test problem. The numbers in parentheses are estimates, based on the currently available data combined with some data obtained a few months ago on the same problem. The superlinear speedup exhibited by the SP2 in going from 4 to 8 processors is attributable primarily to the way the problem was split up on 4 versus 8 processors, which may have resulted in poor load balance on 4 processors. Other users on the system may also have had some effect on the 4 processor timings. Due to the nature of the special contact algorithm used in this problem, which takes advantage of the rigid nature of the die tooling, good scaling is maintained up to very high numbers of processors. In fact, on the T3D using 256 processors, nearly 4 times the number of time step integration cycles per second are performed, as compared to the Cray C90.

4 Conclusion

Much work remains to be done to fully realize the potential of the MPP computer concept. As LS-DYNA3D is better tuned to the requirements of distributed memory machines, and better network hardware becomes available for engineering workstations, the MPP version of LS-DYNA3D will become very widely used, as an alternative to using traditional vector supercomputers. As the results displayed here demonstrate, complex problems can be solved very efficiently using the distributed memory/message passing paradigm on existing MPP hardware.

Simulation of Materials Processing: Theory, Methods and Applications, Shen & Dawson (eds)
© 1995 Balkema, Rotterdam. ISBN 90 5410 553 4

Design of an object oriented software for the computer aided simulation of complex forming processes

L.Walterthum & J.C.Gelin
Laboratoire de Mécanique Appliquée R. Chaléat, URA CNRS, Université de Franche-Comté, Besançon, France

ABSTRACT : In this paper, the advantages of object-oriented programming versus procedural programming in finite element computation codes for material forming computations are shown. Basic concepts on this programming mode are enuciated and are applied in the design of the EF++ program developped in the Applied Mechanics Laboratory at Besançon. This program is described and the description is followed by three forming process simulation examples.

INTRODUCTION

Nowadays, finite element codes have to process more and more complex simulations. The limits of FORTRAN or C programs will be soon reached: adding new functionalities to these already long programs requires an amount of work which increases exponentially with their sizes. What we often observe is a specialization or a splitting of these programs with a loss of their initial possibilities because the developer could not or did not have enough time to think about this problem. Thus we obtain a lot of executable programs which differ only by the solution algorithm they use or by a small difference in some constitutive law.
The object oriented conception method is better suited for large programming projects like finite element codes for metal forming computations. Thanks to the abstraction level reached with object oriented programming (OOP), the program architecture can be better organized and better prepared for future additions.

1 BASIC CONCEPTS OF OOP

OOP is a programming way in which the developper focusses mainly on concepts interacting one with another.
These concepts are called classes that are the fundamental bases of OOP. Representatives of these classes are called objects. Classes can represent diverse notions like matrices, integration points, elements or algorithms. The classes are similar to FORTRAN variables or C structures, but they possess much more possibilities for they not only contain data, they can also answer to messages by applying specific actions.
The data are called the attributes of the class. For a point, it could be for example its thre coordinates. The attributes can themselves be objects. The messages are called methods.
An essential characteristic of classes is the notion of encapsulation: the attributes are directly accessible only by this object or an other object of the same class. We so avoid all external interference and all global data organization which make the understanding of programs very difficult, as we see it too frequently in FORTRAN with the abusive use of "COMMONs". Also, the calling objects are not dependent on the implementation of the called methods.
That independence between objects makes the modularity and consequently the code readability better. This way of proceeding is near from the human way of thinking. As the objects contain their own attributes, it is not necessary to pass a lot of arguments in the messages (Zimmerman *et al.*, 1992)
Another important feature of OOP is inheritance. Inheritance allows grouping classes which have common methods and/or attributes. For example, in mechanics, plane stress elements, beam elements or shell elements have same attributes like material or geometry and same methods like rigidity matrix computing. For this reason, we can create a generic abstract class ELEMENT (called a parent class) which contains the common atrributes and methods and from which we can derive each particular element.
Lastly, the inheritance notion is often associated with the notion of polymorphism. Polymorphism allows having methods with different implementations to

share the same name. This possibility is very useful to keep a coherent interface for all classes derived from a same parent class. For example, stress computing differs from one material to another. Thanks to polymorphism, we can give the same name to the different methods. When this computing is called for a given material, the method attached to the right material is then executed by dynamic linking. This avoid a lot of tests.

2 ADVANTAGES OF OOP IN FINITE ELEMENT CODES

The advantages of OOP in finite element codes are illustrated by examples taken from the EF++ program written in C++ in the Applied Mechanics Laboratory in Besançon. EF++ was designed to perform multi-domains and multi-simulations analyses involving contact phenomena and adaptative remeshing. As example, EF++ is designed to automatically perform analyses involving thermomechanical computations in the workpiece and thermal and fatigue analyses in the dies and tools, or perform automatically subdomain partitions in order to take benefits from the efficiency of parallel computations, or to perform shape as well as material parameters identification and optimization (Gelin and Ghouati, 1995).

2.1 *Extendability*

Inheritance and polymorphism are essential for the extendability of the code. Generally, object-oriented programs are made of a kernel to which new classes (new features) can be appended. The additions are made thanks to derivation from base classes. That is why the implementation of these base classes is extremely crucial because only their virtual methods can be overrided by their derived classes. In the EF++ program, there are mainly 4 base classes :
- ELEMENT : from this class can be derived new finite elements, like plane strain, shell, beam, fluid, or thermal elements for example. Figure 1 shows a sample of the ELEMENT class hierarchy.
- MATERIAL : from this class can be derived new materials. It can be elastic, elasto-plastic or anisotropic visco-plastic materials for example. Figure 2 shows a sample of the MATERIAL class hierarchy.
- ALGORITHM : from this class can be derived new solution algorithms, like quasi-Newton, Euler or Newmark ones. Figures 3 shows a sample of the ALGORITHM class hierarchy.
- MATRIX : from this class can be derived new matrix types, like symmetrical, sparse or skyline

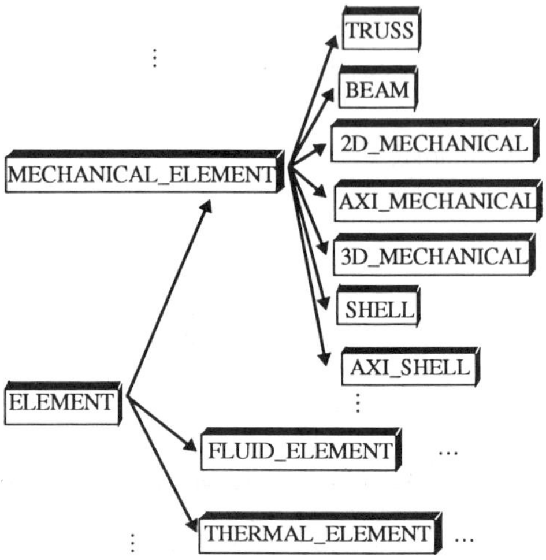

Figure 1. Hierarchy sample of the ELEMENT class

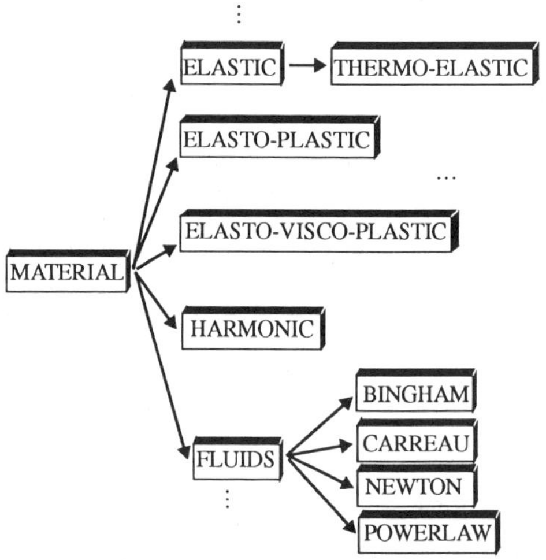

Figure 2. Hierarchy sample of the MATERIAL class

matrices for example. Figure 4 shows a sample of the MATRIX class hierarchy.

A C++ feature which greatly helps for extendability is associative arrays. Associative arrays allows the developper storing in it objects in it associated with a key-object and retrieving them with that key. The most useful associative arrays use a character string as the key. In finite element programs, lot of values are computed and stored at integration points. Some of these values are always the same, like the elementary volume or the flux, but lot of them differ according to the material or the element. Thus, it could rise a problem when one wants to store a

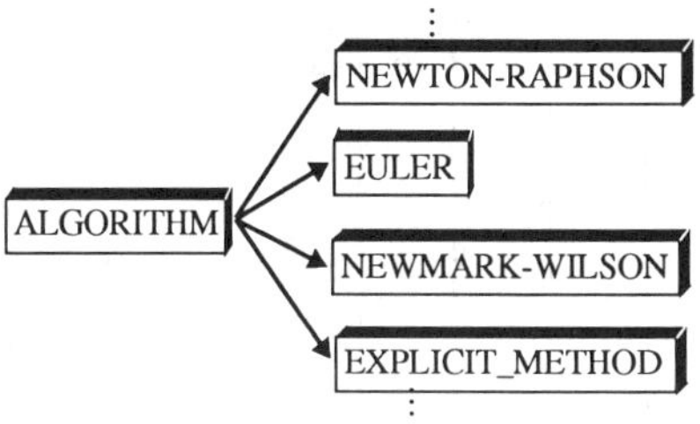

Figure 3. Hierarchy sample of the ALGORITHM class

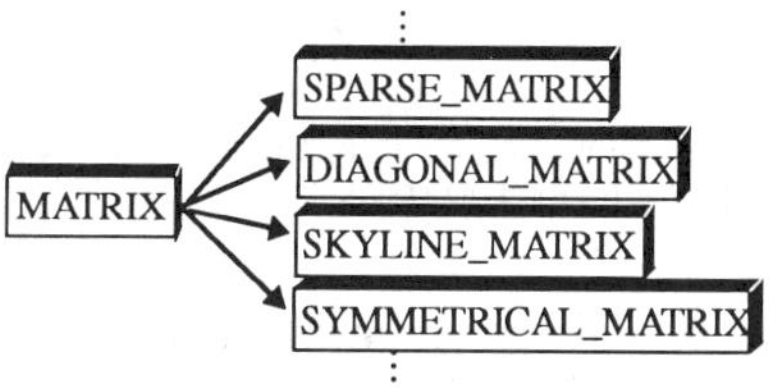

Figure 4. Hierarchy sample of the MATRIX class

Associative arrays to store data at an integration point:

```
Map<String, double> p_scalar;
Map<String, VECTOR> p_vector;
Map<String, MATRIX> p_matrix;
```

Storing a vector called "V1" at an integration point 'ip':

```
VECTOR V1 = ...;
ip->Store_Vector("V1", V1);
```

Retrieving the vector called "V1" at an integration point 'ip':

```
VECTOR& V1 = ip->Give_Vector("V1");
```

Figure 5. Using associative arrays to store and retrieve data at an integration point.

parameter which was not provided. In the EF++ program, thanks to associative arrays, the developper can store as many parameters (scalars, vectors, or matrices) he wants simply in giving them a key-name (see figure 5). Associative arrays are also used for nodal, material or algorithm parameters.

2.2 *Maintainability*

Maintainability is obtained by a good readability and an easy error detection.
In finite element programs, there are a lot of matrix computations. The C++ language allows the user to overload operators like addition, multiplication or

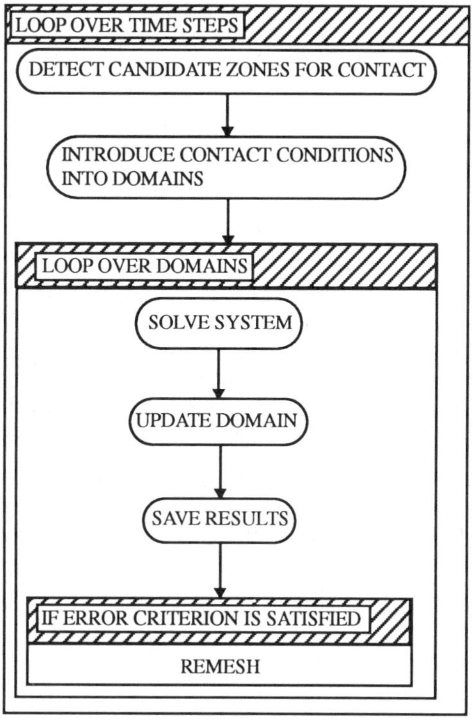

Figure 6. General solving algorithm

indexation. He can redefine their meaning for objects which do not use the same rules for these operations as scalars : for example matrices, vectors or complex numbers (Scholtz, 1992 and Zeglinsky *et al.*, 1994).
In C++, an operation on matrices can be written :

```
MATRICE A(10, 10);
MATRICE B(10, 10);
MATRICE C(10, 10);
MATRICE D(10, 10);
...
// Initialization of A, B, C, D
...
A = (B + C) * D;
```

In FORTRAN, the same operation would have been written:

```
      REAL*8 A(10, 10)
      REAL*8 B(10, 10)
      REAL*8 C(10, 10)
      REAL*8 D(10, 10)
      REAL*8 TEMPORARY(10, 10)
      ...
C Initialization of A, B, C, D
      ...
      ADDMAT(B, C, TEMPORARY, 10, 10)
      MULTMAT(TEMPORARY, D, 10, 10)
```

The C++ version is explicitly easier to read and to understand.
Also, compacity of code helps for readability. As

Table 1. Development time of some classes

Class	Development time
Harmonic material	2 hours
Newmark algorithm	0.5 day
Shell element	1.5 day

every object contains his own information through encapsulation, the user needs not to pass a lot of parameters when he sends a message to an object (see example above : the dimensions of the matrices are contained in the matrices themselves). Polymorphism helps not rewriting a function call for each type of object the function is applied to. For example, in the standard elementary rigidity matrix computing method listed below, the messages sent to the material in the integration loop are valid for each existing material.

```
p_rigidity.Cleanup();
int ipn = p_integr_point_number;
for(INTEGR_POINT* ip = p_integr_point;
     ipn--; ip++)
{
  MATRIX& D = p_material->
    Compute_Tangent_Operator_At(ip);
  MATRIX& B = ip->Give_Gradient();
  double dv = ip->Give_Elementary_Volume();
  p_rigidity += B.Transpose() * D * B * dv
}
```

Error detection is made easier for several reasons :
- better readability
- encapsulation : the different objects interact only through messages. One object cannot write spurious data into an other one. Thus it is generally not difficult to find an error because the error is confined into the class where it occurred.

2.3 *Modularity*

Encapsulation allows to have classes very independent one from another. The classes thus act like modules. Each class can be tested separately. Changes or improvements in one class method do not affect the other classes, even those which use that method. This is very useful for group programming. This modularity also helps the architecture designer in giving him more abstraction : he has only to think about what actions are to be performed and not how they must be performed. Figure 6 shows how the problem solving is done in the EF++ program.

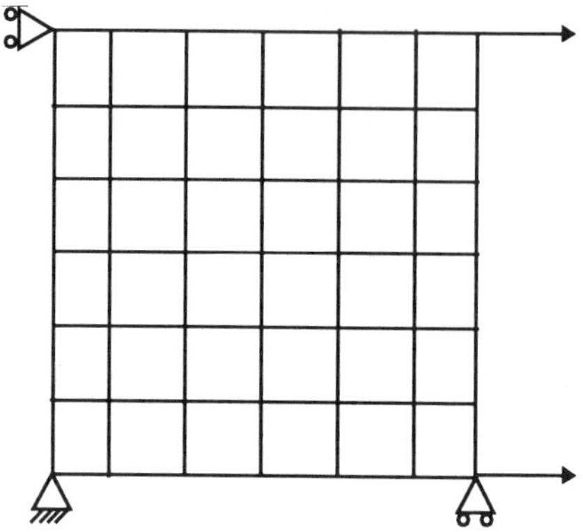

Figure 7. Simple tension test used to compare C++ and FORTRAN speed for finite element codes

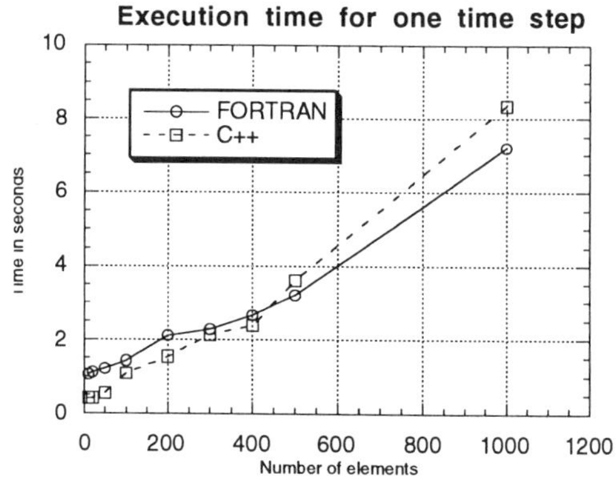

Figure 8. Computation time versus number of elements

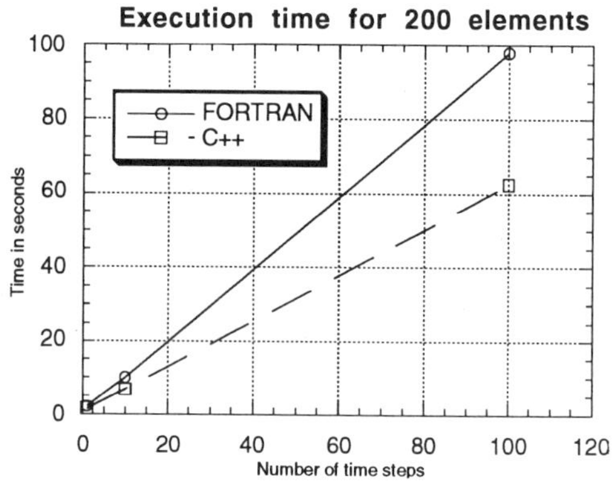

Figure 9. Computation time versus number of time steps

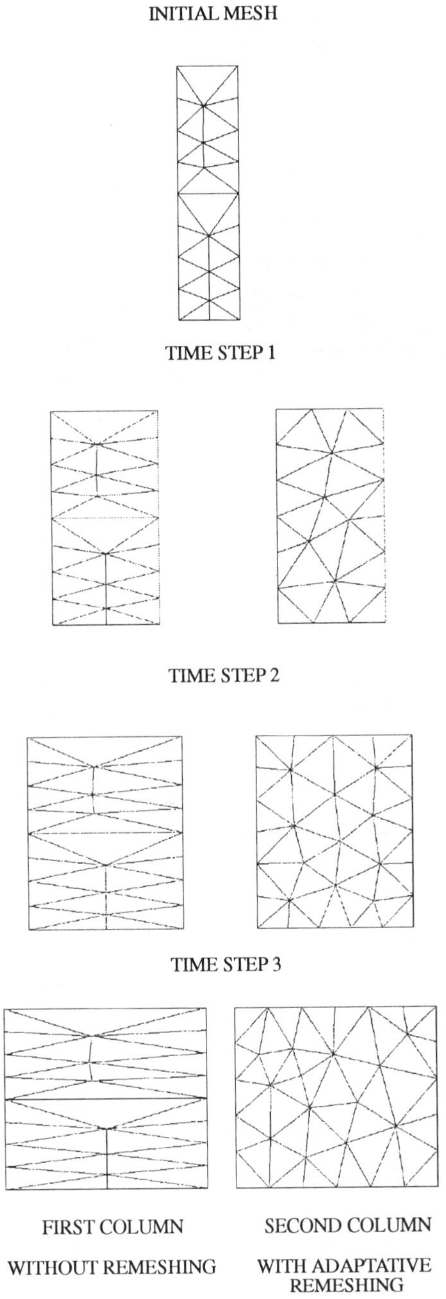

As the domain is compressed, distorsion of the elements appears.

With adaptative remeshing the elements remain homogeneous.

Figure 10. Compression with and without adaptative remeshing

2.4 *Reduced development time*

Because object oriented programs are easier to read, easier to debug, and easier to extend, development time (including design, programming, tests and debugging) are quite shorter, as it is shown in table 1.

2.5 *Computation time equivalent to FORTRAN*

A common idea is that what is gained in abstraction and in development time is lost in computation time. It is true that run-time type-checking and dynamical memory allocations are slower than static tests and static memory allocations, but this loss of time is not very significant compared to the time spent by the program in matrix and vector computations. Especially for the matrix and vector classes, low-level instructions, pointer arithmetics and high optimization were used. But the method interface remained convivial.

EF++ speed has been compared with the POLYFORM program (Daniel *et al.*, 1993). The two programs were tested in the same conditions: same problem, same values computed and stored in a file of the same format. The only difference is that the EF++ program does in addition preprocessing (mesh generation). The problem is a simple tension test (see figure 7) where the number of elements (see figure 8) and the number of time steps (see figure 9) vary.

3 EF++ application examples

The first example (figure 10) is a basic compression test in finite strains of a domain meshed with the Voronoi method. Two simulations are presented : the first one without remeshing and the second one with adaptative remeshing. We can observe that without remeshing a distorsion of the elements appear. This can raises numerical problems when the distorsion becomes too pronounced. But with adaptative remeshing, elements remain homogeneous.

The second example (figure 11) concerns a plane strain compression test of an aluminium alloy deformed at a temperature of 404 °C and with strain rate of 1 s^{-1}. The behaviour is described by an elasto-viscoplastic constitutive law with thermo-mechanical coupling. A three dimensionnal extension of this law is used in an hyperelastic based scheme for the evaluation of the stress tensor and internal variables (Ghouati 1994). Figure 11 represents the temperature contours for 50% height reduction. We can clearly see the temperature increase due to the mechanical dissipation, which follows the evolution of the plastic strain in the deformed zone.

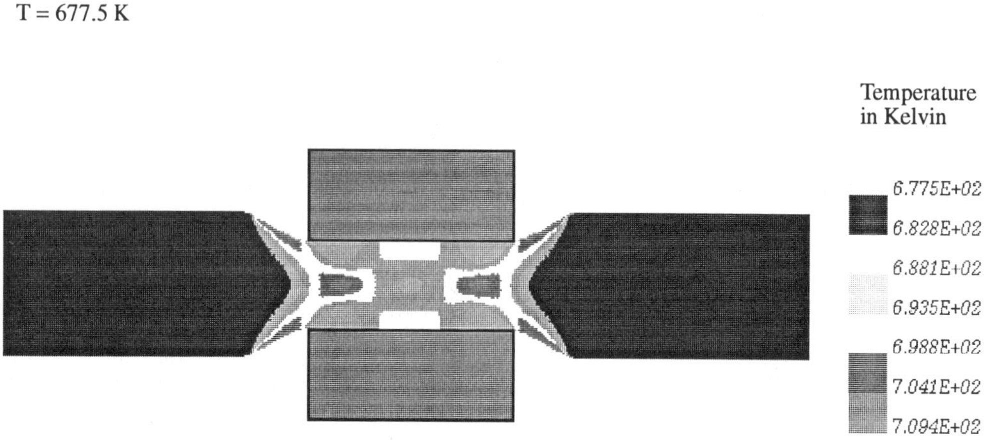

Figure 11. Plane strain compression of an aluminium alloy

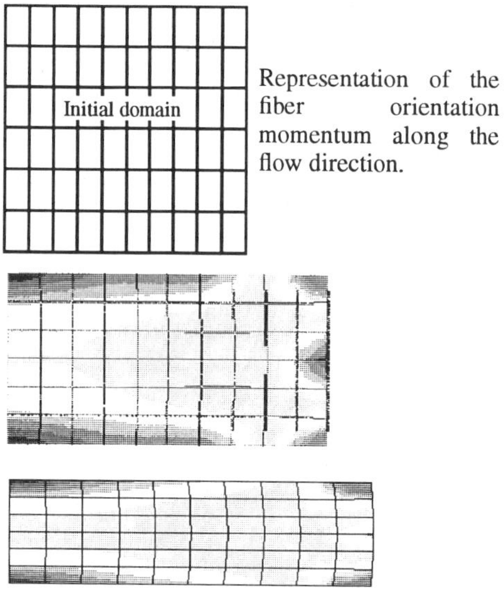

Figure 12. Flow simulation of a composite material

And the last one (figure 12) is the flow simulation of a composite material in liquid phase (Thévenin *et al.* 1994). The constitutive law realize a coupling between the orientation phenomena and the flow characteristics (pressure and velocity). Results shows the fiber orientation changes in the material.

CONCLUSION

This article has shown that for large developments like finite element codes, use of OOP and C++ is advantageous in comparison with classical languages like FORTRAN or C. Indeed, the higher level of abstraction reached and features like polymorphism and inheritance allows better and easier maintainance and extension of the code. This is very important as new constitutive laws, new elements or new algorithm appear or evolve continuously in the field of material processing simulations.

REFERENCES

Daniel J.L., Gelin J.C. and Paquier P. 1993. POLYFORM et DYNAFORM, logiciels 3D pour la simulation du forgeage et de l'emboutissage des matériaux. *Actes du Colloque National en Calcul des Structures*. Hermes, Paris, 1158-1166.

Gelin J.C. & Ghouati O. 1995. The inverse approach for the determination of constitutive equations in metal forming. To be published in *C.I.R.P. Annals* - Vol. 44/1.

Ghouati O. 1994. Identification et modélisation numérique directe et inverse du comportement viscoplastique des alliages d'aluminium. Thesis, University of Franche-Comté, France.

Scholtz S.P. 1992. Elements of an object-oriented FEM++ program in C++. *Computers&Structures* Vol. 43 n°3, 517-529.

Thévenin P., Perreux D., Varchon D. 1994. Flow simulation of chopped fibers reinforced thermoplastic composite in plate type geometry. *Advancing with Composites*. Milan 3-5 May 1994, 457-467.

Zeglinski G.W., Han R.P.S. and Aitchison P. 1994. Object oriented matrix classes for use in a finite element code using C++. *International Journal for Numerical Methods in Engineering*, vol. 37, 3921-3937.

Zimmermann T., Dubois-Pèlerin Y. and Bomme P. 1992. Object-oriented finite element programming : I. Governing principles. *Computer Methods in Applied Mechanics and Engineering* 98, 291-303.

Simulation of Materials Processing: Theory, Methods and Applications, Shen & Dawson (eds)
© 1995 Balkema, Rotterdam. ISBN 90 5410 553 4

Finite element simulation for determining tool motions in metal forming processes

Gang Yang, Ken-ichiro Mori & Kozo Osakada
Department of Mechanical Engineering, Faculty of Engineering Science, Osaka University, Japan

ABSTRACT: A method for determining multi-variable motions of tools in metal forming processes using the finite element method is proposed. In this method, the motions of the tools are determined by minimizing an objective function defined as a difference from the desired values such as the deforming shape, the stress and strain distributions in each deformation step of the finite element simulation. In minimizing the objective function, the relationships between the tool velocities and the nodal velocities are given by the equilibrium equations of the nodal forces. The present method is applied to the determination of the motions of the rolls for bending mild steel plates into circlar shapes. The objective function for this problem is expressed as a difference of the curvature distribution from the desired one. The factor of slipping between the plate and the rolls is also considered in the objective function.

1 INTRODUCTION

Although the finite element method is widely employed in the CAE (Computer Aided Engineering) systems for metal forming processes, forming conditions are optimized on the basis of the judgment of human experts by trial and error. It is desirable to develop a method for determining optimal working conditions automatically by using the finite element simulation. To meet this requirement, several studies for obtaining optimal working conditions with the finite element simulation have been carried out (Kitahara et al. 1977, Kobayashi 1984, Kusiak et al. 1989, Joun et al. 1993a, 1993b, Grandi et al. 1993, 1994 and Cheng et al. 1994). In these studies, initial working conditions such as the geometries of the dies and preforms were treated, but the working conditions during forming have not been optimized yet.

Osakada et al. (1993) and Yang et al. (1994) have proposed a method called controlled FEM simulation to determine an optimal motion of the tool in metal forming process. In this method, a control logic is installed into the finite element simulation. Non-steady-state deformation during forming is monitored in each deformation step of the finite element simulation and the motion of the tool is optimized on the basis of feedback control referring to the calculated results. Since the fuzzy reasoning was used as the control logic in this case, the applicability of this method is limited to the motion with a single variable.

In the present study, the controlled FEM simulation is extended to determine multi-variable motions of tools in the metal forming processes. The motions of the tools in each deformation step are optimized by minimizing an objective function defined as a difference from the desired values. In minimizing the objective function, the equilibrium equations of the nodal forces are used because the relationships between the tool velocities and the nodal velocities are found in the equations. As an example of application of the present method, the motions of the rolls in three-roll bending for forming a plate into a desired curvature distribution are determined.

2 METHOD FOR DETERMINING MOTIONS OF TOOLS

2.1 *Finite element simulation*

In the present method, the rigid-plastic finite element method based on the plasticity theory for a slightly compressible material is used to simulate the deformation behavior in metal forming processes

(Mori 1982 and Yang 1993). The nodal forces {p} for an element in a deforming material are obtained from the principle of virtual work as,

$$\{p\} = \left[\int_{V_e} [B]^T[D][B]dV\right]\{v\}\ , \tag{1}$$

where [B] is the matrix correlating the strain-rate with the nodal velocity, [D] is the matrix correlating the stress with the strain for plastic deformation and {v} is vector of the nodal velocity. The equilibrium equations at each nodal point are given as follows:

$$[K]\{v\} = \{F\}\ , \tag{2}$$

$$[K] = \int_V [B]^T[D][B]dV$$

where {F} is the vector of external force. The tool velocities are implicitly included in the equations as the boundary condition. For the given tool velocities, the nodal velocities are determined by solving the simultaneous equations (2). Since the nodal velocities and the tool velocities satisfy equations (2), the nodal velocities are considered dependent variables of the tool velocities:

$$\{v\} = \{v(\{u\})\}, \tag{3}$$

where {u} is the vector of the tool velocity.

2.2 *Controlled FEM simulation*

In the controlled FEM simulation, the motions of the tools are automatically optimized in each deformation step of the finite element simulation as shown in Fig.1. In the present study, the motions of the tools are determined by minimizing an objective function defined as a difference from the desired values such as the deforming shape, the stress and strain distributions. The objective function φ is expressed as a function of the nodal velocities and the tool velocities in the finite element simulation:

$$\varphi = \varphi(\{u\}, \{v(\{u\})\}) \rightarrow \min. \tag{4}$$

Since the nodal velocities are dependent variables of the tool velocities, the objective function is minimized by partially differentiating the function

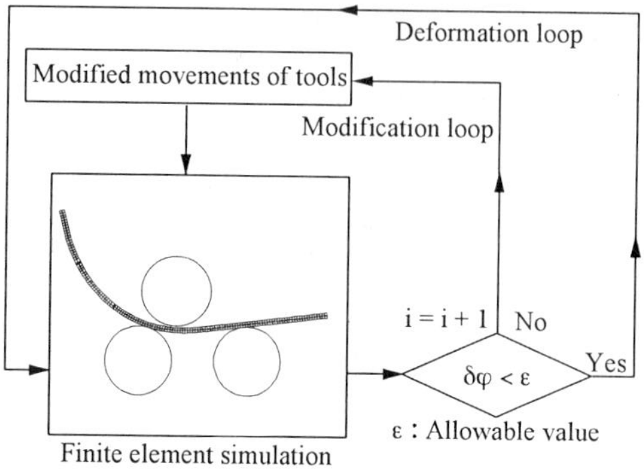

Fig. 1 Scheme of determination of motions of tools by controlled FEM simulation.

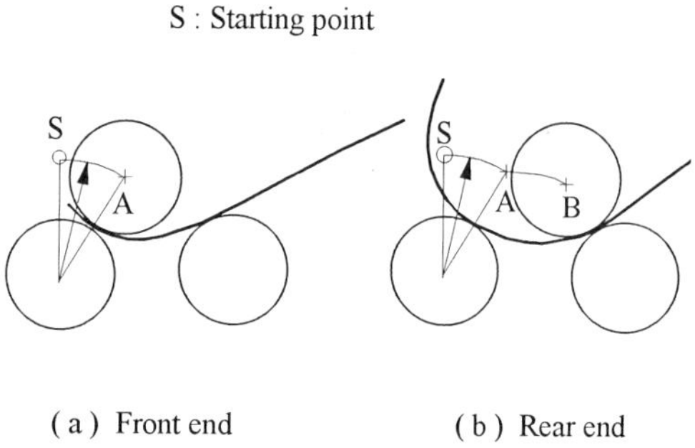

Fig. 2 Control of motions of rolls for preventing curvature decreases near front and rear ends of plate in three-roll bending process.

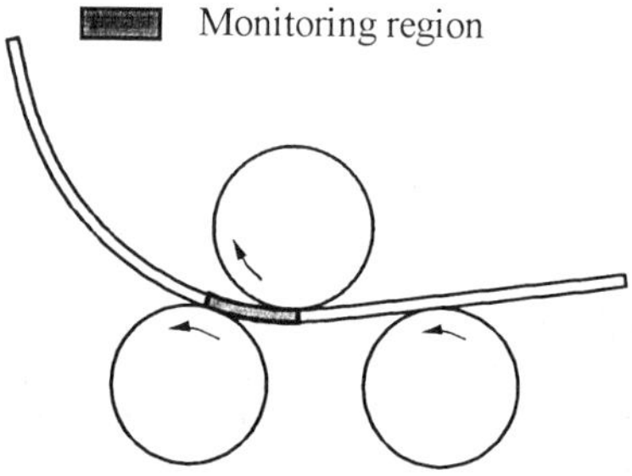

Fig. 3 Region for monitoring curvature in control of three-roll bending process.

with respect to the tool velocities and the derivatives are equated to zero

$$g_k = \frac{\partial\varphi}{\partial u_k} + \frac{\partial\varphi}{\partial\{v\}}\frac{\partial\{v\}}{\partial u_k} = 0,\ \ k = 1,...m \tag{5}$$

where m is the number of the tool velocities. Optimal tool velocities are determined by solving the non-linear simultaneous equations (5) iteratively by using the Newton-Raphson method. The simultaneous equations (5) are linearized as follows:

$$\left(g_k\right)_{i-1} + \left(\frac{\partial g_k}{\partial\{u\}}\right)_{i-1}\{du\}_i = 0, \tag{6}$$

$$\frac{\partial g_k}{\partial\{u\}} = \frac{\partial^2\varphi}{\partial u_k\partial\{u\}} + \frac{\partial^2\varphi}{\partial u_k\partial\{v\}}\frac{\partial\{v\}}{\partial\{u\}} + \frac{\partial^2\varphi}{\partial\{v\}^2}\frac{\partial\{v\}}{\partial\{u\}}\frac{\partial\{v\}}{\partial u_k}$$
$$+ \frac{\partial^2\varphi}{\partial\{v\}\partial\{u\}}\frac{\partial\{v\}}{\partial u_k} + \frac{\partial\varphi}{\partial\{v\}}\frac{\partial^2\{v\}}{\partial u_k\partial\{u\}}$$

where the subscript i denotes the iteration number. By solving equation (6), the corrective tool velocities $\{du\}$ are obtained, and the tool velocities are modified to approach the optimal ones.

2.3 Derivative of nodal velocity with respect to tool velocity

$\partial\varphi/\partial u_k$, $\partial\varphi/\partial\{v\}$, $\partial^2\varphi/\partial u_k\partial\{u\}$, $\partial^2\varphi/\partial u_k\partial\{v\}$, $\partial^2\varphi/\partial\{v\}\partial\{u\}$ and $\partial^2\varphi/\partial\{v\}^2$ in equation (6) are obtained directly from the derivatives of the objective function with respect to the tool velocities and the nodal velocities, respectively. $\partial\{v\}/\partial u_k$ is calculated by differentiating the equilibrium equations (2) of the nodal forces with respect to the tool velocity u_k in each deformation step of the simulation

$$\left(\frac{\partial[K]}{\partial\{v\}}\{v\} + [K]\right)\frac{\partial\{v\}}{\partial u_k} = 0. \tag{7}$$

For the nodal points in contact with the tool, the derivatives of the normal velocities are 1, and the others are 0

$$\frac{\partial v_n}{\partial u_n} = \begin{cases} 1 \text{ (node in contat with tool)} \\ 0 \text{ (node not contating with tool)} \end{cases}, \tag{8}$$

where v_n and u_n are, respectively, the normal components of the nodal velocities and the tool velocities.

$\partial\{v\}/\partial u_k$ is calculated by solving the simultaneous equations (7) under the boundary conditions given by equation (8). By solving equations (7) for all the tool velocities, $\partial\{v\}/\partial\{u\}$ is determined. $\partial^2\{v\}/\partial u_k\partial\{u\}$ is obtained by differentiating equations (7) with respect to all the tool velocities

$$\left(\frac{\partial[K]}{\partial\{v\}}\{v\} + [K]\right)\frac{\partial^2\{v\}}{\partial u_k\partial\{u\}} = -\left(\frac{\partial^2[K]}{\partial\{v\}^2}\frac{\partial\{v\}}{\partial\{u\}}\{v\}\right.$$
$$\left. + \frac{\partial^2[K]}{\partial\{v\}\partial\{u\}}\{v\} + \frac{\partial[K]}{\partial\{v\}}\frac{\partial\{v\}}{\partial\{u\}}\right)\frac{\partial\{v\}}{\partial u_k}. \tag{9}$$

The right hand side of equations (9) is calculated by the use of the solution of equations (7).

3 APPLICATION TO THREE–ROLL BENDING PROCESS

3.1 Three-roll bending of plate

The above explained controlled FEM simulation is applied to the determination of the motions of the rolls in plane-strain three-roll bending of plates. In this process, a plate is bent into a desired shape by locating the three rotating rolls to proper positions. To avoid the decreases in curvature near both ends of the bent plate, the relative positions of the three rolls are changed during bending as shown in Fig. 2.

To reduce the region of the curvature decrease near the front end of the plate, the top roll is moved from point S to point A on a circle with the identical center as the left bottom roll. In this motion, the gap between the top and left bottom rolls is equal to the thickness of the plate. The curvature around the front end is close to the desired one when the top roll reaches point A. The top roll, then, approaches point B near to the right bottom roll to bend the rear end of the plate. Because the curvature of the bent plate is influenced by the relative positions of the rolls and the spring back of the plate after bending, it is not easy to determine an optimal forming path from point A to point B.

In the conventional three-roll bending process, the two bottom rolls are driven and the top roll is idle. When the stroke of the top roll is too large, the plate slips over the surfaces of the rolls. To form a product with a large curvature, the top roll and one of the bottom rolls (left bottom roll in this paper) are driven and the other bottom roll (right bottom roll) is idle. It is better to drive the top roll because the frictional driving force for the top roll is larger than that for one of the bottom rolls.

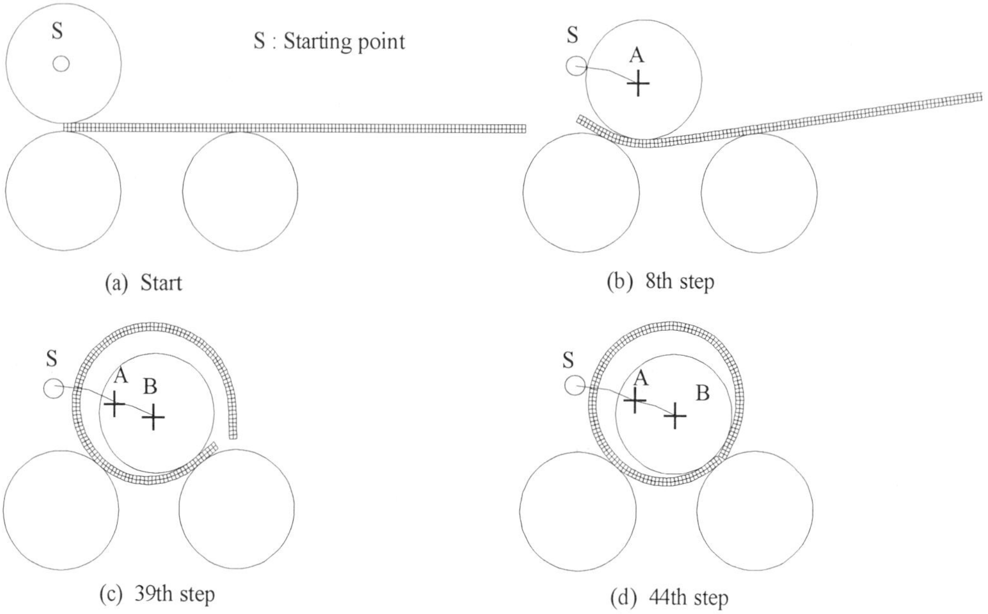

Fig. 4 Calculated distortions of grid pattern for bending plate into pipe in three-roll bending process.

3.2 Determination of motions of rolls

The forming path from point A to point B is determined by using the controlled FEM simulation. The positions of the two bottom rolls are fixed during the bending process, and only the top roll is moved. In this case, only the vertical velocity of the top roll is changed in the simulation, whereas the horizontal velocity is kept constant.

The motions of the rolls are optimized by monitoring the curvature distribution of the bent plate in the region shown in Fig. 3 to prevent the possible delay of the control in each deformation step of the simulation. The following objective function is defined to determine the vertical velocity of the top roll and the rotating velocities of the top roll and left bottom roll without slipping between the plate and the rolls

$$\varphi = \left[\left(\frac{1}{p_l}\sum_{k=1}^{p_l}\kappa_k\right) - \kappa_d\right]^2 + \alpha\left(\frac{u_t}{v_t}\right)^2 . \qquad (10)$$

In this equation, κ_k is the curvature of the nodal point on the surface in the monitoring region, κ_d is the desired curvature in the monitoring region, p_l is the number of the nodal points on the surface in the monitoring region, u_t is the rotating velocity of the top roll, v_t is the tangent velocity of the nodal point in contact with the top roll and α is a positive constant given with 0.75. The first term on the right hand of equation (10) represents the difference between the desired and calculated curvatures in the monitoring region. The second term represents a penalty to avoid non-feed of the plate, i. e., $v_t = 0$, when the plate is in a state of slipping and the second term becomes infinite.

Table 1 Working conditions and material properties used in controlled FEM simulation.

Thickness of plate / mm	6
Diameter of top roll / mm	160
Diameter of bottom roll / mm	160
Distance between bottom rolls / mm	130
Coefficient of friction	0.25
Flow stress / MPa	$\bar{\sigma} = 747\bar{\varepsilon}^{0.24}$
Young's modulus / GPa	210
Poisson's ratio	0.3
Diameter of circlar shape / mm	170

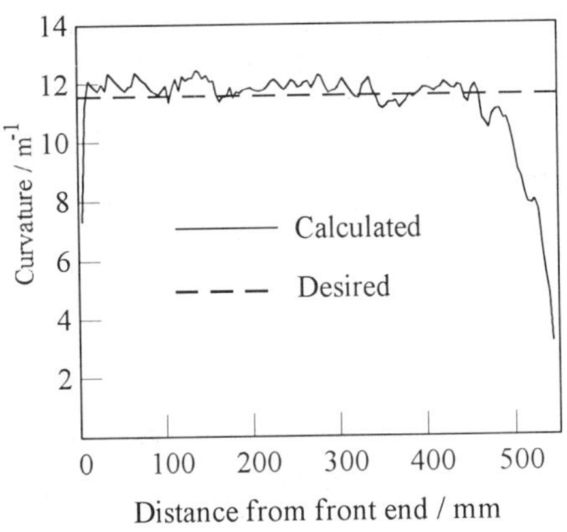

Fig. 5 Curvature distribution of bent plate obtained by controlled FEM simulation.

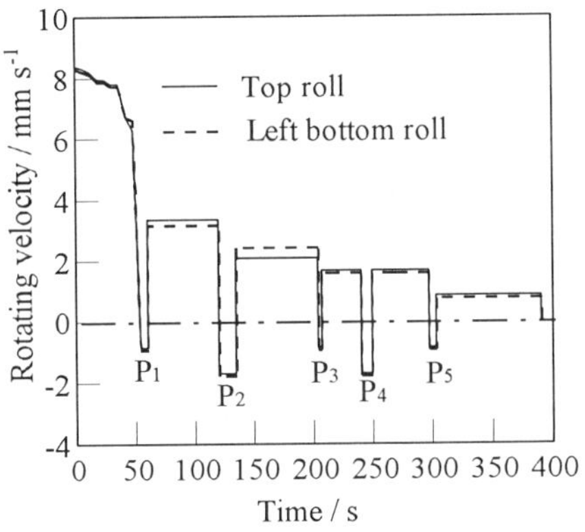

Fig. 6 Variations of rotating velocities of top and left bottom rolls with processing time obtained by controlled FEM simulation.

4 OPTIMAL MOTIONS OF ROLLS IN THREE-ROLL BENDING

4.1 Computational conditions

The motions of the rolls for bending a mild steel plate into a circular shape are determined by the controlled FEM simulation. The vertical velocity of the top roll and the rotating velocities of the top and left bottom rolls are optimized simultaneously in each deformation step of the finite element simulation for bending the plate into the desired shape without slipping. The working conditions and material properties used for determining the motions of the rolls are shown in Table 1. No lubricant is

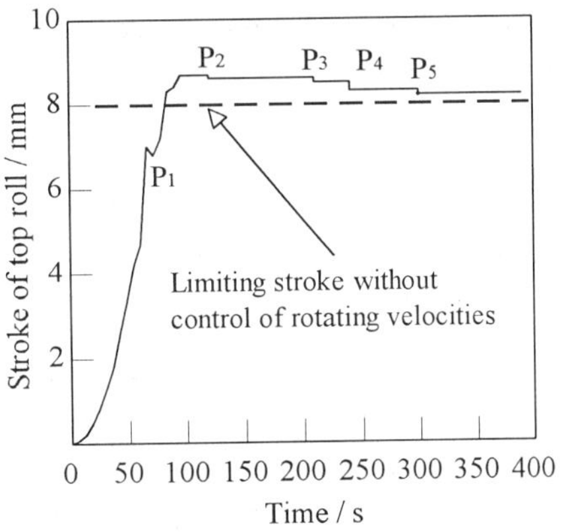

Fig. 7 Variation of stroke of top roll with processing time obtained by controlled FEM simulation.

applied at the interface between the plate and the rolls and the coefficient of friction is assumed to be 0.25.

4.2 Optimal motions of rolls

The calculated distortions of the grid pattern in the three-roll bending processes are shown in Fig. 4. The motion of the top roll from point A to point B is determined by monitoring the deforming shape of the plate in each deformation step. The plate is successfully bent into the desired shape by minimizing the objective function in the simulation.

The curvature distribution of the bent plate obtained by the controlled FEM simulation is illustrated in Fig. 5. The curvature of the bent plate is close to the desired one by controlling the motions of the rolls in the simulation. The decreases in the curvature are prevented around the front and rear ends of the bent plate.

Fig. 6 shows the rotating velocities of the top and the left bottom rolls obtained by the controlled FEM simulation. The rotating velocities of the top and left bottom rolls are almost the same. When slipping of the plate over the surface of the rolls is detected, the top roll is slightly lifted (as shown in Fig. 7) and the top and left bottom rolls are rotated in the reverse direction as denoted by P_1 -P_5 in Fig. 6.

The stroke of the top roll determined by the controlled FEM simulation for bending the plate into a pipe is shown in Fig. 7. The limiting stroke of the top roll is improved by controlling the vertical velocity of the top roll and the rotating velocities of the rolls in the simulation.

5 CONCLUSIONS

The controlled FEM simulation was extended to determine the multi-variable motions of the tools in metal forming. For the determination, the objective function was efficiently minimized by the use of the relationships between the tool velocities and the nodal velocities given by the equilibrium equations of the nodal forces in the simulation. Since the recent forming machines have high controllability for the motions due to the remarkable development of microcomputers, the present method is expected to enlarge the applicable range of the forming machine.

REFERENCES

Cheng, H., R. V. Grandhi & J. C. Malas 1994. Design of optimal process parameters for non-isothermal forging. *Int. J. Numer. Methods Eng.*, 37:155-177.

Grandhi, R. V., A. Kumar, A. Chaudhary & J. C. Malas 1993. State-space representation and optimal control of non-linear material deformation using the finite element method. *Int. J. Numer. Methods Eng.*, 36:1967-1986.

Grandhi, R. V. & R. Thiagarajan 1994. Model reduction and process control of thermomechanical behavior of non-linear material deformation. *Int. J. Numer. Methods Eng.*, 37: 3135-3152.

Joun, M. S. & S. M. Hwang 1993a. Optimal process design in steady-state metal forming by finite element method-I. Theoretical considerations. *Int. J. Mach.Tools Manufact.*, 33-1:51-61. Rotterdam: Balkema.

Joun, M. S. & S. M. Hwang 1993b. Optimal process design in steady-state metal forming by finite element method-II. Application to die profile design in extrusion. *Int. J. Mach. Tools Manufact.*, 33-1:63-70. Rotterdam:Balkema.

Kitahara, Y., K. Osakada, S. Fuji & R. Narutalki 1977. Analysis of deformation of plates in free forging using rigid-plastic finite element method. *J. Japan Sci. Tech. Plasticity*, 18:753-759.

Kobayashi, S. 1984. Approximate simulation for preform design in rolling. *Int. J. Mach. Tool Des. Res.*, 24:215-224.

Kusiak, J. & E. G. Thompson 1989. Optimization techniques for extrusion die shape design. *Proc. 4th NUMIFORM,* 569-574.

Mori, K., K. Osakada & T. Oda 1982. Simulation of plane-strain rolling by the rigid-plastic finite element method. *Int. J. Mech. Sci.*, 519-527.

Osakada, K., G. Yang & K. Mori 1993. Determination of optimum forming path in three-roll bending by combination of fuzzy reasoning and finite element simulation. *Ann. CIRP*, 42-1:291-294.

Yang, G., K. Mori & K. Osakada 1993. Finite element simulation of plane-strain three-roll bending of plates. *Proc. 4th ICTP*, Beijing, 3: 1596-11601.

Yang, G., K. Mori & K. Osakada 1994. Determination of forming path in three-roll bending using FEM simulation and fuzzy reasoning. *J. Mater. Process. Technol.*, 5:161-166.

Simulation of Materials Processing: Theory, Methods and Applications, Shen & Dawson (eds)
© 1995 Balkema, Rotterdam. ISBN 90 5410 553 4

A continuum yield criterion based on polycrystalline theory using a mapping technique

Jiun-Shya Yu & Antoinette M. Maniatty
Rensselaer Polytechnic Institute, Troy, N.Y., USA

ABSTRACT: An anisotropic, continuum yield criterion is developed based on microstructural behavior for polycrystalline metals. Since a single cubic crystal has cubic symmetry, an appropriate linear transformation representing this symmetry can be used to transform a general continuum isotropic yield criterion to an anisotropic yield criterion with cubic symmetry. The parameters in this transformation are determined using the theory of Bishop and Hill for single crystals. Then using information regarding the orientation distribution of the grains in a polycrystalline aggregate and using appropriate averaging techniques, a continuum yield criterion for an aggregate can be developed. This method has been found to be very simple and very efficient computationally. Good agreement has been found between the yield surfaces obtained with this method and those obtained with the less computationally efficient crystallographic approach for various textures. The yield criterion has also been found to predict the R-values fairly well.

1. INTRODUCTION

After metal forming and/or recrystallization processes, there appears in the polycrystalline metals the existence of preferred grain orientations, called 'texture', which has been recognized as one of the main sources of the anisotropy of polycrystalline metals. There are usually two different approaches to describe the yield surface of polycrystalline metals, the crystallographic approach and the continuum approach.

Since polycrystalline metals are composed of grains where each grain has the structure of a single crystal, the behavior of a polycrystalline metal is affected by that of each grain. In the crystallographic approach, the yield surface of a polycrystalline aggregate is determined from the yield properties of each grain in the aggregate which has the same yield surface but different orientation, in other words, texture data is used to construct the global yield surface. Bishop and Hill (1951a,b), using the principle of maximum work, outline the procedure for the crystallographic approach to calculate the yield stresses of polycrystalline aggregates where Taylor's assumption which models the interaction law between grains in an aggregate was utilized. The yield surfaces predicted from texture data are adequate in reflecting the anisotropy. However, since the establishment of whole yield surfaces will involve tremendous calculations, the crystallographic approach is very computationally intensive to apply to practical works, e.g., the finite element calculation of elasto-plastic problems.

Another approach to describe the plastic behavior of a polycrystalline aggregate is the continuum approach in which a phenomenological yield function is employed to depict the yield behavior. The yield criteria proposed by Tresca and Von Mises are the most popular ones to describe the isotropic yield surfaces. In addition to isotropic yield surfaces, the anisotropic yield surfaces have also been modeled by the continuum approach. Hill (1950) assumes a quadratic yield function and Barlat et al. (1991) gives a non-quadratic yield function to model the anisotropic yield surface, but both criteria can only describe the yield surfaces of materials which possess orthotropic symmetry. Karafillis and Boyce (1993) suggests a new generic isotropic yield function for polycrystalline metals, and an anisotropic yield criterion was developed as an extension of the isotropic criterion using the isotropic plasticity equivalent (IPE) concept. This anisotropic yield criterion can describe any kind of anisotropic yield surfaces. While the continuum expression of the yield surface is easier to use in practical applications, texture effect is not considered in this approach and experiments must be performed to obtain the parameters which appear in the yield criterion.

In this paper, with the aid of the IPE method proposed by Karafillis and Boyce (1993), a phenomenological yield function is developed for fcc polycrystalline metals with parameters determined from texture data. The fundamental feature of the IPE method is to find the linear transformation tensor which characterizes the anisotropic properties and decides the success of approximating the anisotropic yield surface. The linear transformation tensor is first constructed for a single fcc crystal of which the yield surface information developed by Bishop and Hill (1951a,b) is used. Then following Taylor's assumption, the linear transformation tensor for an fcc polycrystalline aggregate can be derived from a single fcc crystal and the grain orientations. The yield function of anisotropic polycrystalline metals can thus be determined.

2. ISOTROPIC PLASTICITY EQUIVALENT (IPE) METHOD

The basic idea of IPE is to establish the fourth order linear transformation tensor, L, which transforms the actual stress tensor, T, acting on an anisotropic material into a new stress tensor, S, called the IPE stress tensor. The new stress tensor, S, is then used in an isotropic yield criterion which represents the transformed yield surface of the original anisotropic material. In order to approximate the anisotropic yield surfaces well, both the accurate determination of the linear transformation tensor L and the appropriate choice of the isotropic yield function are important.

The isotropic yield criterion used here is the one proposed by Hosford (1972), which gives

$$|S_1 - S_2|^m + |S_2 - S_3|^m + |S_3 - S_1|^m = 2Y^m \tag{1}$$

where S_1, S_2 and S_3 are the principle stresses, Y is the uni-axial yielding strength and m is a constant which specifies different yield criteria. Here, the following value is taken as Y,

$$Y = 3.06\tau_c, \tag{2}$$

where τ_c is the yield strength of slip systems of fcc crystals. The reason to use this value is that the uniaxial yielding strength of an isotropic fcc polycrystal was found to be $3.06\tau_c$ as calculated by Bishop and Hill (1951a,b). In this paper, different values of m will be given to explore what is the best value for m. In addition to the isotropic yield criterion, the linear transformation tensor is also needed.

2.1 *IPE method for a single fcc crystal*

Since both the original stress tensor and IPE stress tensor are symmetric second order tensors, the linear transformation tensor of a single fcc crystal which possesses cubic symmetry can be expressed by a 6x6 symmetric matrix as

$$L = \begin{bmatrix} a & c & c & 0 & 0 & 0 \\ c & a & c & 0 & 0 & 0 \\ c & c & a & 0 & 0 & 0 \\ 0 & 0 & 0 & b & 0 & 0 \\ 0 & 0 & 0 & 0 & b & 0 \\ 0 & 0 & 0 & 0 & 0 & b \end{bmatrix}, \tag{3}$$

where a, b and c are unknown parameters. These parameters will be decided based on the yield surface of a single fcc crystal.

Bishop and Hill (1951a,b) propose that there are 56 vertices on the yield surface of a single fcc crystal and the stress state during yielding will normally lie in a vertex. The principle stresses of the IPE stress tensor S, which is obtained by transforming the yielding stress tensor of each vertex with linear transformation tensor L, are calculated. Vertices having the same principle stresses are grouped in the same category. There are five different categories with the corresponding principle stresses and the number of vertices as:

1. $\left\langle \frac{2}{3}(a-c)\tau,\ -\frac{1}{3}(a-c)\tau,\ -\frac{1}{3}(a-c)\tau \right\rangle$ 6 vertices
2. $\left\langle b\tau,\ -\frac{1}{2}b\tau,\ -\frac{1}{2}b\tau \right\rangle$ 8 vertices
3. $\langle 0,\ \pm b\tau \rangle$ 6 vertices
4. $\left\langle -\frac{1}{6}(a-c)\tau,\ \frac{1}{12}(a-c)\tau \pm \sqrt{\frac{1}{16}(a-c)^2 + \frac{1}{2}b^2}\,\tau \right\rangle$ 24 vertices
5. $\left\langle 0,\ \pm\frac{1}{2}\sqrt{(a-c)^2 + b^2}\,\tau \right\rangle$ 12 vertices

where $\tau = \sqrt{6}\tau_c$. It is noted that, instead of the individual values of a and c, only the value $a-c$ is important when the plastic behavior is concerned.

For each category, a residual function f_i can be defined by substituting the principle stresses into equation (1) and rearranging it as

$$f_i = |S_{1_i} - S_{2_i}|^m + |S_{2_i} - S_{3_i}|^m + |S_{3_i} - S_{1_i}|^m - 2Y^m, \tag{4}$$

then an objective function F can be defined as

$$F = 6f_1^2 + 8f_2^2 + 6f_3^2 + 24f_4^2 + 12f_5^2. \tag{5}$$

By minimizing the objective function F, we can obtain the optimum values of $a-c$ and b. Table 1 lists the corresponding values of $a-c$ and b for different exponents m.

Table 1: Optimal values of $a-c$ and b corresponding to different exponents m.

m	$a-c$	b
1.2	1.17945	0.76880
1.4	1.21090	0.78675
1.6	1.23209	0.79797
1.8	1.24630	0.80487
2.0	1.25566	0.80887

In a polycrystalline aggregate, each grain generally has a different orientation. The linear transformation tensors for different oriented grains can be express in indicial form as

$$L_{(s)ijkl} = R_{pi}R_{qj}R_{rk}R_{sl}L_{(c)pqrs}, \tag{6}$$

where R is the rotation tensor of each grain and $L_{(s)}$ and $L_{(c)}$ represent the linear transformation tensors expressed in global and crystal coordinates respectively.

If the orientations of every grain are known, i.e., the texture data are known, the linear transformation tensor of the polycrystalline aggregate can be obtained.

2.2 *IPE method for a polycrystalline aggregate*

The linear transformation tensor of a polycrystalline aggregate can be determined from that of each grain contained in the aggregate by averaging the effect of each grain. As derived by Karafillis and Boyce (1993), the relationship between the plastic rate of deformation tensors, D^p and d^p, can be expressed as

$$D^p = Ld^p, \tag{7}$$

where D^p is the plastic rate of deformation tensor applied to the anisotropic material, and d^p is the work conjugate to the IPE stress tensor. This expression ensures that the work dissipation rate of the anisotropic material is equal to that of the equivalent isotropic material.

If an aggregate is composed of n grains which may have different orientations, the linear transformation tensor of the aggregate can be derived by assuming that the work dissipation rates are the same for both anisotropic status and equivalent isotropic status. In the anisotropic status, the work dissipation rate of the aggregate, $\dot{W}$, can be calculated as the average value of work dissipation rate of each grain, which gives

$$\dot{W} = \frac{1}{n}\sum_{i=1}^{n} T^{(i)} \cdot D^{p(i)}, \tag{8}$$

where superscript i means ith grain. Using Taylor's assumption, equation (8) can be rewritten as

$$\dot{W} = \overline{T} \cdot D^p, \tag{9}$$

where $\overline{T}$ is the average stress tensor of the aggregate, i.e. $\overline{T} = \frac{1}{n}\sum_{i=1}^{n} T^{(i)}$, and $D^{p(i)} = D^p$.

The same work dissipation rate can be expressed from the isotropic plasticity equivalent status as

$$\dot{W} = \frac{1}{n}\sum_{i=1}^{n} S^{(i)} \cdot d^{p(i)}, \tag{10}$$

where $S^{(i)}$ is the IPE stress tensor and $d^{p(i)}$ is the work conjugate to $S^{(i)}$ for ith grain. Using the relation between d^p and D^p given by equation (7), equation (10) can be written as

$$\dot{W} = \frac{1}{n}\sum_{i=1}^{n} S^{(i)} \cdot L^{(i)-1} D^{p(i)} = \frac{1}{n}\sum_{i=1}^{n} L^{(i)-1} S^{(i)} \cdot D^p. \tag{11}$$

Since a continuum expression for the stress tensor is anticipated in the equivalent isotropic material, the IPE stress tensor for each grain is assumed to be the same as the stress tensor of the aggregate $\overline{S}$, which gives

$$S^{(1)} = S^{(2)} = \cdots = S^{(n)} = \overline{S}. \tag{12}$$

Substituting this expression into equation (11) and using the fact that both equation (9) and equation (11) represent the same dissipation rate, the following equation can be obtained

$$\overline{T} = \frac{1}{n}\left(\sum_{i=1}^{n} L^{(i)-1}\right)\overline{S}, \tag{13}$$

Equation (13) can be rewritten as

$$\overline{S} = L_{aggre}\overline{T} \quad \text{and} \quad L_{aggre} = \left[\frac{1}{n}\left(\sum_{i=1}^{n} L^{(i)-1}\right)\right]^{-1}, \tag{14}$$

where L_{aggre} is the linear transformation tensor of the aggregate. As equation (14b) indicates, the yield surface constructed in this way takes into account the influence of grain orientation on the anisotropic behavior.

3. R VALUE

The R-value is defined as the ratio of the transverse plastic strain increment to the thickness strain increment during a uni-axial tensile test. Usually, the tensile tests will be performed at different angles with respect to the rolling direction on the plane of a sheet and the R-value will be a function of the angle. This parameter can indicate the degree of anisotropy and is not easily measured from the yield surface.

Under the assumption of associated flow, the plastic strain increment $d\varepsilon_{ij}$ can be derived as

$$d\varepsilon_{ij} = d\lambda \frac{\partial \Phi}{\partial T_{ij}}, \tag{15}$$

where $d\lambda$ is a scalar that depends on the hardening behavior, T_{ij} is the corresponding stress component and Φ is the yield function.

Using the equivalent isotropic yield function, equation (1), we can calculate the IPE strain increment, $d^p\Delta t$ in terms of $d\lambda$. Then the real plastic rate of deformation tensor, D^p, can be determined from equation (7). Finally, the R-value can be expressed as

$$R = \frac{D_{22}}{D_{33}}, \tag{16}$$

where D_{22} and D_{33} are components of D^p. In this paper, the 1,2 and 3 axis represent the tensile direction, the transverse direction and the thickness direction, respectively.

4. NUMERICAL EXAMPLES

The yield surfaces of three commonly observed textures are considered first. These textures include the brass and copper textures which result from large deformation processes, and the cube texture which forms after recrystallization processes. Table 2 gives the Miller indices of the ideal orientation for these special textures. These textures are generated by assuming that the grain orientations are rotationally symmetric gaussian distributions about the ideal orientation of each texture, which can be expressed as

$$g(\omega)=g(0)\exp(-\frac{\omega^2}{2\omega_o^2}), \qquad (17)$$

where $g(0)$ is the density of the ideal orientation, ω_o represents the scatter width of the spread and $g(\omega)$ is the relative density of a given orientation rotated through an angle ω from the ideal orientation.

Following the procedure adopted by Lequeu et al. (1987a), the orientations of each grain are created according to equation (17) with $\omega_o=5^o$. Figure 1 show the {111} pole figures of these special textures and the random texture, and each figure is constructed by 400 grains

Table 2: Miller indices of ideal orientation for three special textures.

texture	Miller indices
brass	$\{110\}<1\bar{1}2>$
copper	$\{112\}<11\bar{1}>$
cube	$\{100\}<001>$

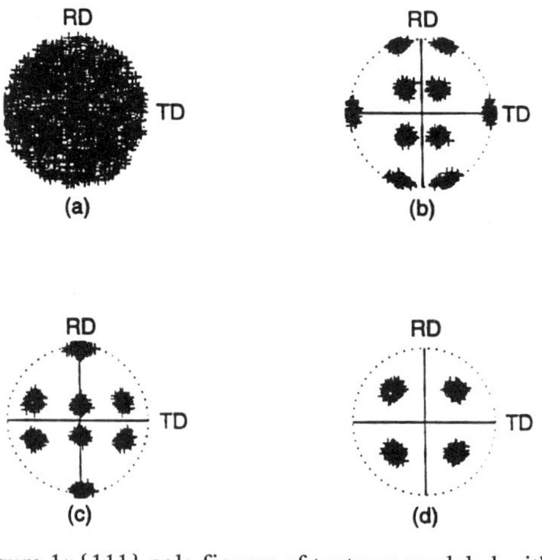

Figure 1: {111} pole figures of textures modeled with 400 grains by assuming rotationally symmetric gaussian distribution with $\omega_o=5^o$. (a) random (b) brass (c) copper (d) cube .

Barlat and Richmond (1987) use the Taylor/Bishop and Hill method to construct the yield surfaces of textures consisting of 50% special texture as shown in Figure 1(b)-1(d) and 50% random texture. We employ the IPE method to calculate the yield surfaces of the same textures considered by them. Figure 2 give the yield surfaces of these textures while four different values of component m are used, i.e. m=1.8, 1.6, 1.4, 1.2. Figure 3 show the projections of yield surfaces corresponding to different parameters s on the normalized $T_{11}-T_{22}$ plane for each texture while m=1.8 is used. The parameter s represents the ratio of shear stress to the yield stress for uni-axial tension in the rolling direction. The yield surfaces constructed by small m are more flat. This is reasonable since the yield function, equation (1), become linear when m approaches unity. The yield surfaces of brass texture depict that the yield stress in the transverse direction is larger than in the rolling direction. On the contrary, the yield surfaces of copper texture demonstrate a larger yield stress in the rolling direction. The cube texture shows a similar yield surface as a random texture. This is because the cube texture has the same orthotropic symmetric property with respect to both the RD and TD as displayed in Figure 1(d), which results in the yield surfaces to be symmetric about the line $T_{11}=T_{22}$.

Comparing Figures 2 and 3 with the results of Barlat and Richmond (1987) using Taylor/Bishop and Hill method which is a crystallographic approach to calculate the yield stress, the yield surfaces established by them are more flat and are similar to the yield surfaces constructed by IPE method with m=1.4. Since the yield surface of a single fcc crystal is composed of planes, it is not surprising that the re

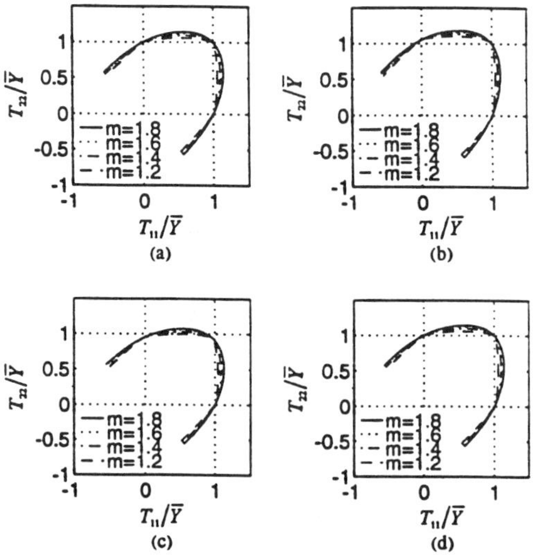

Figure 2: Yield surfaces corresponding to different values of m for textures composed of 50% random component and 50% following component, (a) random (b) brass (c) copper (d) cube .

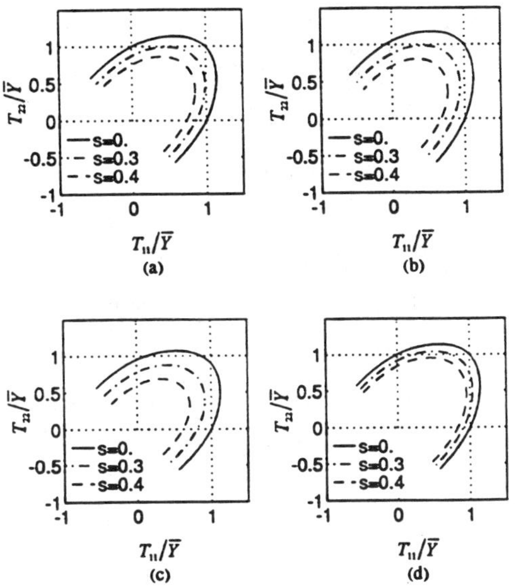

Figure 3: Yield surfaces corresponding to different shear stress ratio with m=1.8 for following textures, (a) random (b) brass (c) copper (d) cube . s is defined as $T_{12}/\bar{Y}$.

sults from the crystallographic approach will be more flat. On the other hand, in the IPE method, the yield surface of a single fcc crystal is modeled by equation (1), the modified surface may be curved or flat depending on what exponent value of m is used.

The degree of anisotropy is difficult to evaluate from the yield surface. The R values can be used to serve this purpose. To illustrate the effect of texture on R value, the R values are produced as a function of θ, which is the angle between the uni-axial tensile direction and the rolling direction. Figure 4 show the results for each special texture discussed above.where each texture has a strong symmetric gaussian distribution with $\omega_o = 5^0$. For the brass texture, the highest R value is found around $\theta = 50^o$ with $R \approx 1$ at $\theta = 90^o$ and $R \approx 0.5$ at $\theta = 0^o$. The highest R value of copper texture appears around $\theta = 40^o$ with $R \approx 1$ at $\theta = 0^o$ and $R \approx 0.5$ at $\theta = 90^o$. For the cube texture, the R-value distribution is symmetric about $\theta = 45^o$ with lowest value at $\theta = 45^o$ and highest value, R=1, at $\theta = 0^o, 90^o$. Comparing Figure 4 with the results of Lequeu et al. (1987b) who use a continuum function to model the plastic behavior of polycrystalline metals, the R-value distribution calculated by the IPE method with m=1.6-1.8 are quite similar to those obtained by Lequeu et al (1987b).

In the last example, we use the IPE method to evaluate the R values and the yielding strengths of a 2008-T4 sheet and compare the results with the experimental measurements performed by Lege et al. (1989). As measured by Lege et al., there are four major components which occupied 74% of the texture of the sheet sample, and the other 26% is a random component. Each component is expressed as a gaussian distribution density with the corresponding scatter width as given by Lege et al. (1989). We use 800 grains to model this texture. Figure 5(a) displays the predicted R values. The R values corresponding to m=1.8 give a similar trend comparing to the measured R values. The calculated yielding strengths which are normalized by the evaluated yield stress in the RD are shown in Figure 5(b) as a function of θ. The x's and o's represent the measured tensile and compressive yield stresses, which are normalized by the compressive yield stress measured in the RD. The yield stresses predicted by the IPE method are very close to the compressive yield stresses obtained from the experiment for $0^o \le \theta \le 60^o$, but the calculated results are lower than the test data after $\theta > 60^o$.

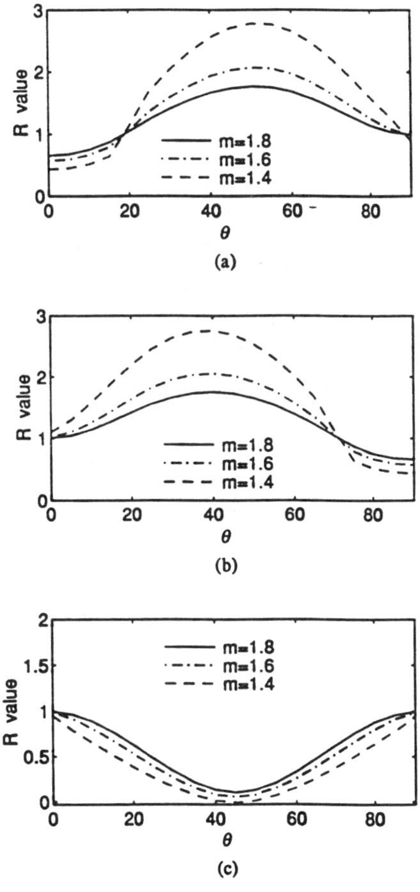

Figure 4: R values calculated for the following textures, (a) brass (b) copper (c) cube . m=1.8, 1.6, 1.4 are used for each case.

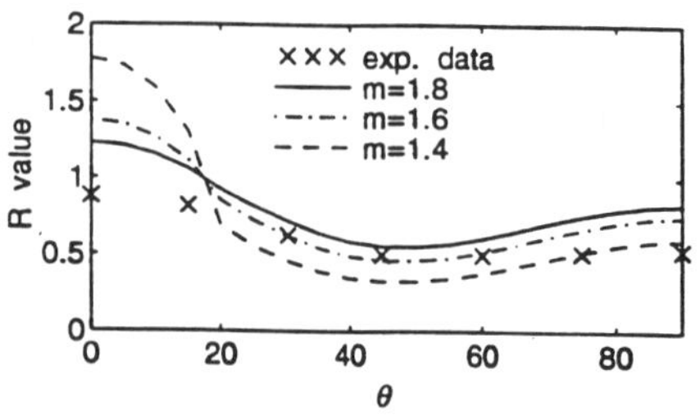

(a)

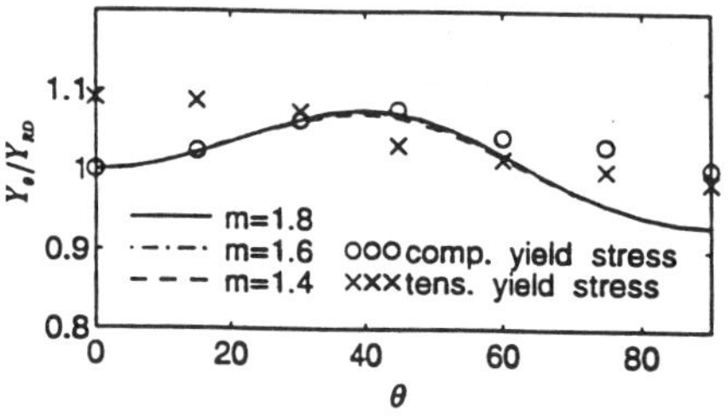

(b)

Figure 5: Results calculated by the IPE method compared with experimental data for 2008-T4 sheet. (a) R values (b) yield strengths .

5. CONCLUSION

A continuum yield criterion is developed to describe the yield behavior of polycrystalline metals while the influence of texture data is considered and used to determine the parameters appearing in the yield function. Constructing the yield surfaces in this way is very efficient computationally and can successfully approximate the yield surfaces calculated by the Taylor/Bishop and Hill method, a crystallographic approach. This yield criterion can also be used to calculate the R values which are in good agreement with those evaluated from traditional phenomenological yield functions.

ACKNOWLEDGEMENTS

This work was supported by the Henry Luce Foundation, Inc. and by the National Science Foundation through Grant DMI-9358123.

REFERENCE

Barlat, F., Lege, D. J. & Brem, J. C. 1991. A six component yield function for anisotropic materials., *Int. J. Plasticity* 7: 693-712.

Barlat, F. & Richmond O. 1987. Prediction of tricomponent plane stress yield surface and associated flow and failure behavior of strongly textured f.c.c. polycrystalline sheets. *Mater. Sci. Eng*. 95: 15-29.

Bishop, J. F. W. & Hill, R. 1951. A theory of the plastic distortion of a polycrystalline aggregate under combined stresses. *Phil. Mag*. 42: 414-427.

Bishop, J. F. W. & Hill, R. 1951. A theoretic derivation of the plastic properties of a polycrystalline face Centered metal. *Phil. Mag*. 42: 1298-1307.

Hill, R. 1950. *The Mathematical Theory of Plasticity*. Clarendon Press, Oxford 1950.

Hosford, W. F. 1972. A generalized isotropic yield criterion. *J. Appl. Mech.* 39: 607-609.

Karafillis, A. P. & Boyce, M. C. 1993. A general anisotropic yield criterion using bounds and a transformation weighting tensor. *J. Mech. Phys. Solids* 41: 1859-1886.

Lege, D. J., Barlat F. & Brem J. C. 1989. Characterization and modeling of the mechanical behavior and formability of a 2008-T4 sheet sample. *Int. J. Mech. Sci.* 31: 549-563.

Lequeu, PH., Gilormini, P., Montheillet, F., B. Bacroix, B. & Jonas, J. J. 1987. Yield surfaces for textured polycrystals-I. : crystallographic approach. *Acta Metall.* 35: 439-451.

Lequeu, PH., Gilormini, P., Montheillet, F., B. Bacroix, B. & Jonas, J. J. 1987. Yield surfaces for textured polycrystals-II. : analytic approach. *Acta Metall.* 35: 1159-1174.

Simulation of Materials Processing: Theory, Methods and Applications, Shen & Dawson (eds)
© 1995 Balkema, Rotterdam. ISBN 90 5410 553 4

Mixing analysis of highly viscous fluids in a numerically simulated wavy wall channel

Saeid Zerafati & David Bigio
Department of Mechanical Engineering, University of Maryland, College Park, Md., USA

ABSTRACT: Wavy wall channels (or screws) can be used as an efficient mixing geometry. This paper reports on the numerical considerations for modeling the wavy wall channel. We will also define two dimensionless parameters for this geometry, analyze the flow field solutions for several channels and comment on the effect of flow field changes on the mixing performance. In this study local and global mixing measures and efficiencies are developed for 3-dimensional flows. We will also determine the critical values of the dimensionless parameters to change the global mixing behavior of the system.

1. INTRODUCTION

The general domain in which this work resides is that of mixing in creeping flows. Mixing in this context refers to the stretch of an interfacial line, or area in a strain field.

Unlike many fluid mechanics problems in which the solution for the velocity field is considered to be the final answer, mixing analysis actually starts with the specification of the flow field. However, being able to model complex three dimensional flows is a pre-requisite to any mixing analysis.

In this research, mixing is investigated from both the local and global point of view. For the local analysis, mixing performance is evaluated for a single material particle as it moves through the flow field. The basis of analysis in this part are the evolution of the principal directions and values of the appropriate tensors. Different mixing efficiencies are defined and the mechanism of mixing is explained. This work, for the first time, examines the area evolution as a basis for the mixing performance in three dimensional flows.

In global analysis, mixing is simulated for an ensemble of $O(10^4)$ material points. Statistical analysis is performed on various measures to study the overall mixing behavior of the flow field. The approach is capable of finding the good and poor mixing regions of the flow field and relate the total mixing to the total energy input.

2. WAVY WALL CHANNEL

Spatial changes of the flow field directly affect the mixing performance of highly viscous fluids. Favorable changes can be generated by a combination of operating conditions and improvement of the shape of the mixing sections. Wavy wall screw is an interesting geometry that can be used as an efficient mixing device. The screw can be modeled as a wavy wall channel to facilitate numerical calculations.

Figure 1 shows the isometric view of a wavy channel. The vertical walls can be described by the following equation:

$$D(x) = W_0 + W \cos\left(\frac{2\pi x}{J}\right) \tag{1}$$

In this equation, W_0 is one half the average width of the cavity, W is the amplitude of the wall and J is the wave length between successive wall waves. In this analysis, $J = 2\pi$ and $W_0 = 2.0$. As can be seen in figure 1, polymer enters from one end of the channel with velocity $V_d = 0.3$ and exits with the same velocity from the other end. The top plate moves with velocity V_c causing the fluid to circulate as it flows through the channel.

Two dimensionless parameters were defined for this geometry:

$$\beta = \frac{W_0}{W} \tag{2}$$

and

$$f = \frac{2HJ}{27W_0^2}\frac{V_c}{V_d} \tag{3}$$

β represents the waviness of the channel and can be varied from zero for a straight wall channel to one for the maximum waviness of the channel. f represents the ratio of the dimensionless time of action of the boundary [2, 1] and the dimensionless time of redistribution in the cavity [4]. In this study, changes in f were generated by changing V_c/V_d.

A general purpose CFD package called FIDAP was used to solve the flow field. A boundary condition of $V_x = 0.3$ for the entrance and exit of the channel was chosen. For the top surface of the geometry, a no slip condition with $V_y = U$ was assumed. The required f parameters are obtained through varying U form 0.5 to 3.0 in several steps. No slip boundary condition with zero velocities were assumed for the rest of the channel surfaces. Each mesh has 7502 quadratic (i.e. 27 nodes) brick elements and 66125 nodes. For the analysis of this problem, the properties of a typical melted polymer was considered. This results in a Reynolds number of 10^{-3}.

3. FLOW FIELDS IN THE WAVY CHANNEL

In figure 2 a velocity vector plot for cross sections at $x = 2\pi$, $x = 3\pi$ and $x = 4.5\pi$ is shown. These vectors are drawn for the corner nodes of each element. For $x = 2\pi$, most of the velocity vectors are almost in the x plane while for $x = 3\pi$, they are almost perpendicular to this plane. This is due to the increase in the v_x with the decrease in the width of the channel. Circulation of the flow can be seen from the section at $x = 2\pi$. In general, the bottom part of the channel experiences very lower velocities. Therefore, particles tend to stay in this region for longer periods of time. For the section at $x = 4.5\pi$, a flow circulation zone as well as larger components of velocity in the x direction is observed.

To explain the behavior of the fluid in the wavy channel, the flow field for a geometry with $\beta = 0.5$ and $f = 1.16$ is studied in more detail. Figure 3a shows v_x contour for a section at z=1.5. As can be seen from this figure, v_x is lower for wider parts of the region due to the constant flow rate and the incompressibility assumption. Therefore, reducing the width should increase v_x. v_x is zero on the boundaries except for the entrance and exit where it is equal to 0.3. The center of the channel experiences higher velocities in the x direction and lower velocity gradients.

Figure 3b shows v_y for the same section as the previous figure. As can be noticed in this figure, for the top part of the geometry, v_y is mainly positive while for the bottom part it is mainly negative. v_y at $x = n\pi$, when n is an integer, is either totally positive or totally negative. For other x values it is mixed. v_y on the boundaries is zero except for the wavy part of the top boundary. The transition regions at the beginning and end of the wave can be seen clearly.

Figure 3c shows v_z contour for a section at z=1.5 and. v_z is symmetric w.r.t. the y axis and is zero on all the boundaries. For lower parts of the channel, maximum v_z occurs at $x = 2n\pi$ while for upper parts, it occurs at $x = (2n + 1)\pi$.

As can be seen in figure 3, the velocity field is spatially periodic in the x direction. For example, if the flow field for a geometry with 10 loops is modeled, the velocity field inside every loop will be the same as the velocity field for the middle loop of the geometry modeled in this study. Therefore, the middle loop of the channel can be cut numerically and be used as a building block for generating a channel with several loops.

Figure 4 shows v_x for a geometry with $\beta = 0.5$ and two different values of f. The cross sections are taken at $x = 2.5\pi$. For $f = 0.85$, the shape of the contours are close to the flow in a straight channel. Increasing f, will gradually distort the contours and enlarge the regions with $v_x \approx 0$. For $f = 2.33$, a small region of backward flow is observed in the top-right corner of the flow field. This causes a small circulation region. This region enlarges for $f = 3.49$, as can be seen in figure 4b. The reason for this circulation is that the fluid is being pushed against a converging wall at high velocity. This phenomena adds more complexity to the flow field and can possibly be a source of mixing improvement.

Generally, the bottom part of the channel experiences low velocities, so particles tend to spend more time in this region. Also, high values for v_x at the center of the channel causes low residence time for the particles in this region. These particles do not experience high velocity gradients therefore, poor mixing at the center of the channel can be expected.

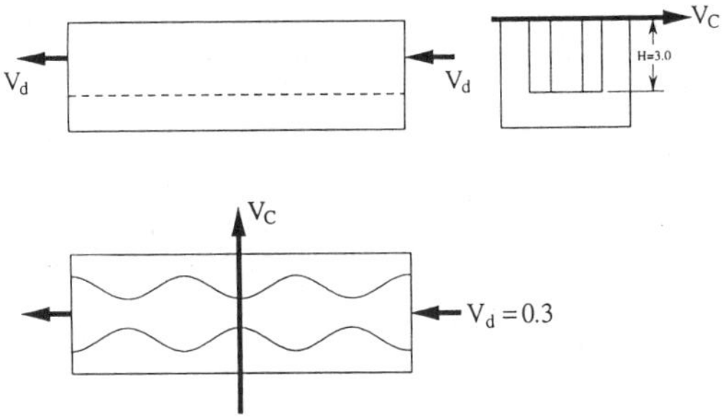

Figure 1: Isometric View of a 3-D Wavy Channel

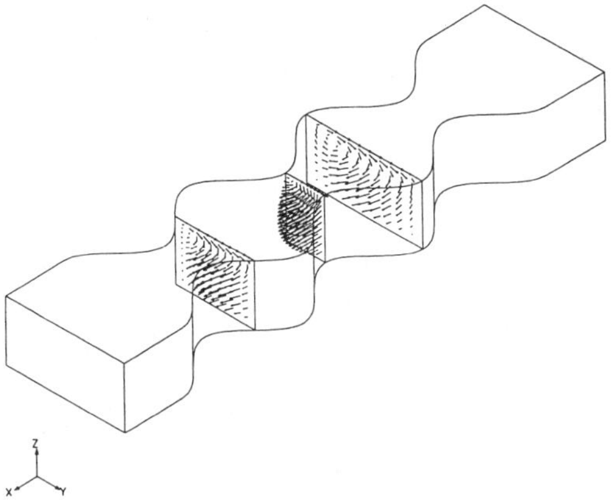

Figure 2: Velocity Vector Plot for Sections at $x = 2\pi$,$x = 3\pi$ and $x = 4.5\pi$

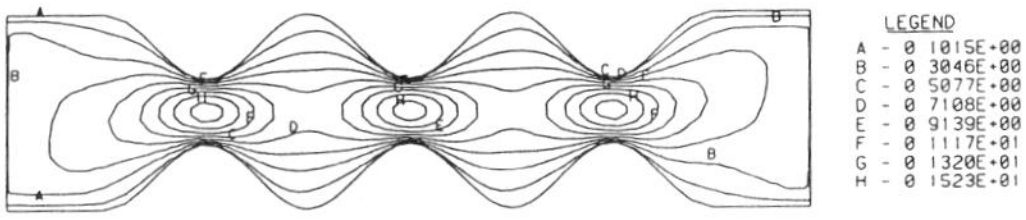

(a)

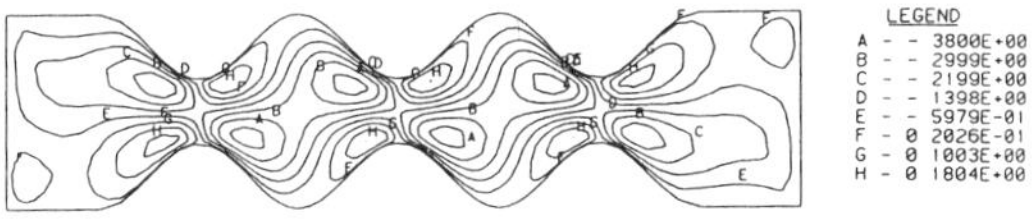

(b)

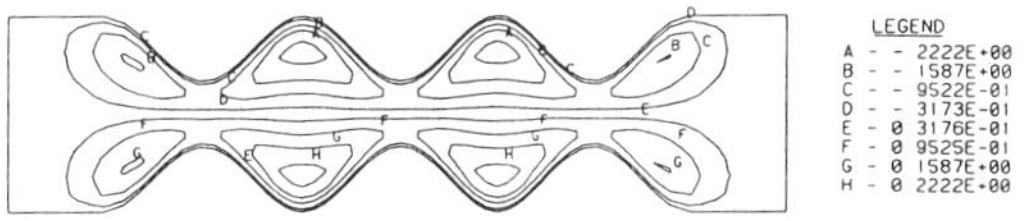

(c)

Figure 3: Velocity Components for a Geometry with $\beta = 0.5$. Sections are Taken at $z = 1.5$. (a)v_x, (b)v_y and (c)v_z

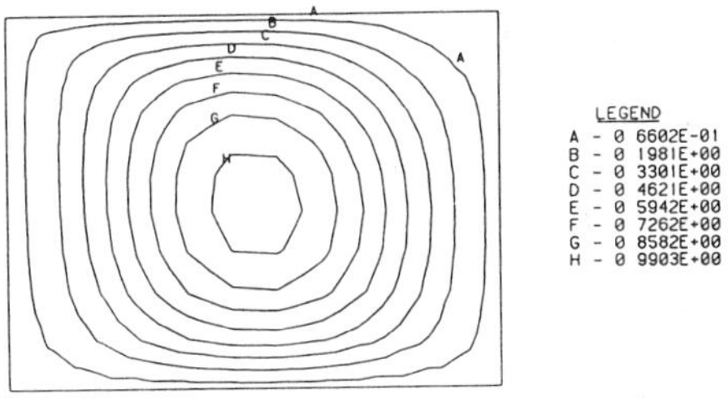

(a)

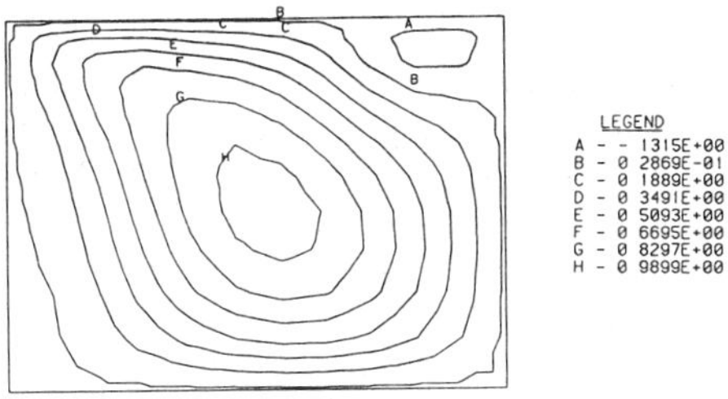

(b)

Figure 4: v_x Contours for a Geometry With $\beta = 0.5$, (a)$f = 0.85$ and (b)$f = 3.49$ at $x = 2.5\pi$

(a)

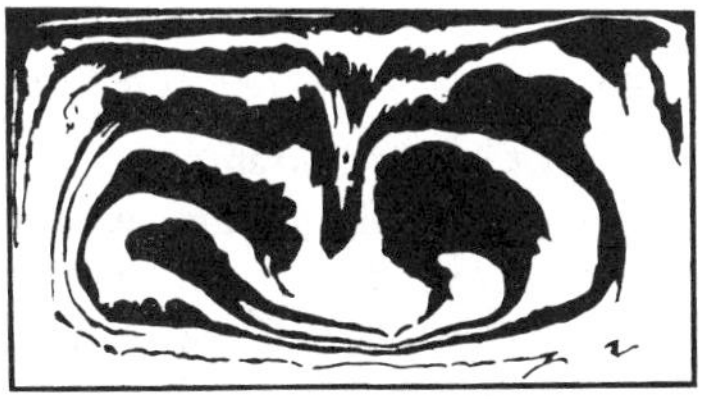

(b)

Figure 5: Comparison of Mixing in Two Different Sections

4. LAMINAR MIXING

Numerous measures and indices have been proposed to characterize a mixture's state of "mixedness". Many of these are indirect measures, such as the mixture's bulk electrical conductivity, the resistance of cured material to solvent or ultraviolet penetration, or some other gross property of the mixture. These measures are often directly applicable to the performance of a particular product, but while they may have direct technological application, they offer little insight into the mechanisms of mixing.

To clarify the relation between area growth and mixing, consider the samples from a wavy wall channel shown in figure 5. Obviously, the fluids in part (b) of the figure are mixed better than part (a). A simple observation reveals that the interface between black and white in part (b) is greater than part (a) and is the criteria for better mixing in this section. This explanation introduces the amount of interface between two fluids as a measure for mixing. Therefore, a flow field that increases the length of this interface *faster* is a more efficient mixer.

The samples in figure 5 are two dimensional snap shots of a three dimensional flow field. Naturally, the interface between black and white fluid shows as a line. In three dimensional flow fields, the interface between two fluids is a surface. In this case, *area growth* is the measure of mixing efficiency.

5. AREA GROWTH

The evolution of an area interface can be shown [3] to follow:

$$\mathbf{da} = \mathbf{dA}\,\mathbf{F}^{-1} \tag{4}$$

In this equation $\mathbf{dA}$ and $\mathbf{da}$ represent the area interface before and after deformation, respectively. $\mathbf{F}$ is the deformation tensor which relates the deformed and undeformed configurations. The corresponding equation for an element of arc is:

$$\mathbf{dx} = \mathbf{F}\,\mathbf{dX} \tag{5}$$

Comparing equations 4 and 5 shows that the evolution of line and area interfaces are totally different even for the same initial directions and deformations.

Obviously, stretch is a function of the initial position as well as the initial direction of the interface. To eliminate the dependence in the initial orientation, the area growth at each instant is computed as the average of 400 initial orientations of the interface uniformly spaced in a sphere around the initial location. In this way, the effect of the arbitrary initial direction is averaged out and average stretches can be compared only based on their initial position. Figure 6 shows the average area growth for the initial position of $z = 2.8$ in a geometry with $f = 1.16$ and $\beta = 0.5$.

6. EFFICIENCIES

Based on the physical picture described earlier, mixing is equivalent to the stretching of the material elements. If an area element $\mathbf{dA}$ is placed in a flow field, the specific rate of area generation ($\dot{\xi}/\xi$) can be calculated from:

$$\frac{\dot{\xi}}{\xi} = -d_{(\mathbf{m})} \tag{6}$$

in which ξ is the area growth and $d_{(\mathbf{m})}$ is the area stretching in the direction of $\mathbf{m}$. Since the specific rate of stretchings are bounded, we can use the Cauchy-Schwarz inequality to show that $\sqrt{\mathbf{D}:\mathbf{D}}$ is an upper bound for stretching values ($\mathbf{D}$ is the rate of deformation tensor). The overall mixing efficiency based on area growth is then defined as:

$$\eta_{e_\xi} = \frac{-\mathbf{m}^{\mathbf{T}}\mathbf{D}\,\mathbf{m}}{\sqrt{\mathbf{D}:\mathbf{D}}} \tag{7}$$

The area efficiency is defined as [6]:

$$\eta_{L_\xi} = \frac{\mathbf{m}^{\mathbf{T}}\mathbf{D}\,\mathbf{m}}{d_3} \tag{8}$$

In this equation d_3 is the smallest principal value of $\mathbf{D}$. The flow efficiency for an area interface can be defined as [6]:

$$\eta_{F_\xi} = \frac{-d_3}{\sqrt{\mathbf{D}:\mathbf{D}}} \tag{9}$$

It is important to notice that:

$$\eta_{e_\xi} = \eta_{L_\xi}\eta_{F_\xi} \tag{10}$$

Figures 7 and 8 show the efficiencies for a 5 loop wavy wall channel with $f = 1.16$ and $\beta = 0.5$. The interface starts at $z = 2.8$ with an initial orientation of (1,0,0). The sharp changes in η_{L_ξ} around $x = 15.0$ and $x = 28.5$ correspond to the top corners of the geometry. In this part, sudden rotation of the principal directions of $\mathbf{D}$ and the required time for the interface to respond to this change

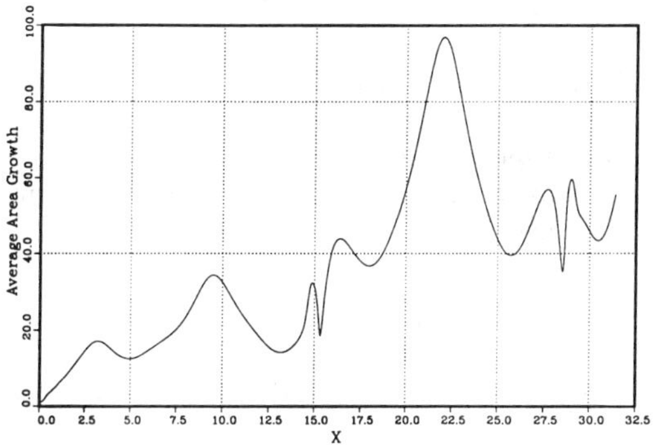

Figure 6: Average Area Growth for a Group of Interfaces Starting at $z = 2.8$ in a Geometry With $f = 1.16$, $\beta = 0.5$ and 5 loops

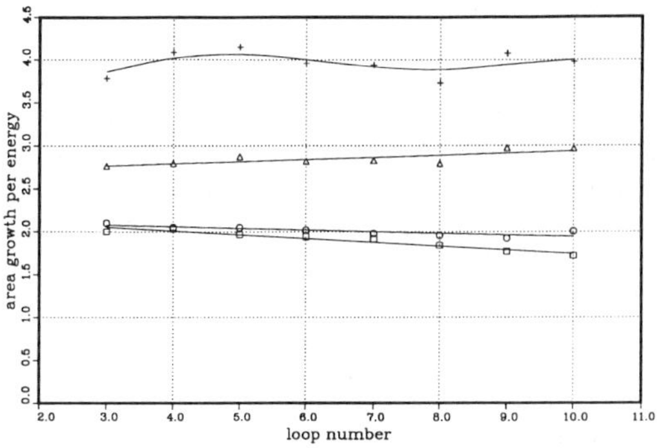

Figure 9: Average Area Growth per Energy as a Function of Loop Number. The Results Correspond to the Geometries With $f = 1.16$ and β Values of 0.0(Squares), 0.125(Circles), 0.25(Triangles) and 0.5(Plus Signs).

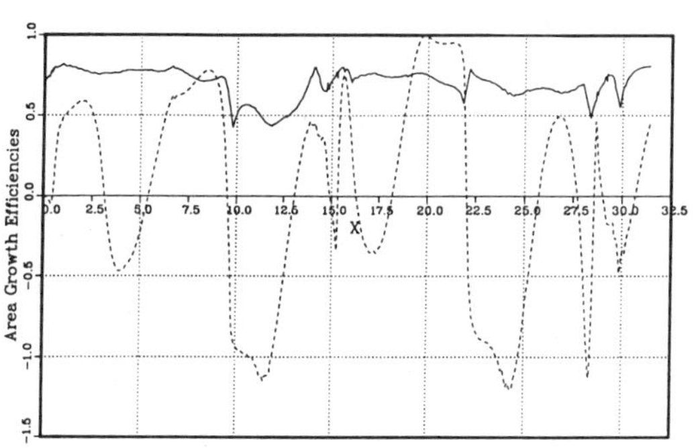

Figure 7: Mixing Efficiencies Based on Area Growth for a Geometry With $f = 1.16$, $\beta = 0.5$ and 5 Loops. The Interface Starts at $z = 2.8$ With an Initial Orientation of (1,0,0). Curves Correspond to η_{F_ξ} (Solid Line) and η_{L_ξ} (Dashed Line).

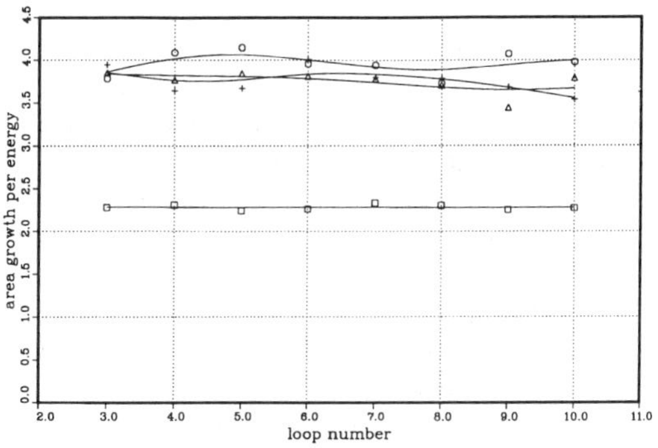

Figure 10: Average Area Growth per Energy as a Function of Loop Number. The Results Correspond to the Geometries With $\beta = 0.5$ and f Values of 0.58(Squares), 1.16(Circles), 2.33(Triangles) and 3.49(Plus Signs).

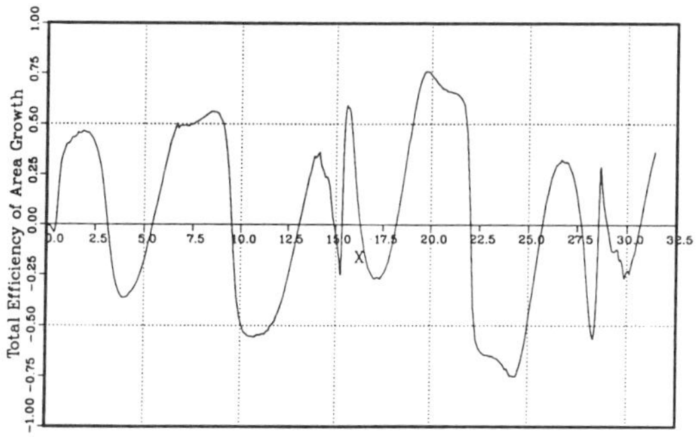

Figure 8: η_{e_ξ} For a Geometry With $f = 1.16$, $\beta = 0.5$ and 5 Loops. The Interface Starts at $z = 2.8$ With an Initial Orientation of (1,0,0).

causes the observed behavior of $\eta_{L\xi}$. For this interface, $\eta_{F\xi}$ values never get to the limiting values of 0.408 or 0.816. However, it nearly equals to 0.707 for many parts of the flow history, which is the efficiency for a two dimensional flow. Generally, $\eta_{F\xi}$ is low when the particle is near the bottom of the channel and also when the particle starts to move upwards toward the top part of the geometry.

Figure 8 shows $\eta_{e\xi}$ for the interface described in the previous figure. For $\eta_{e\xi} < 0.0$, the interface shrinks and for $\eta_{e\xi} = 0.0$, there is a local minimax in the area growth curve [5]. Therefore, $\eta_{e\xi}$ exactly predicts the behavior of the stretch.

7. MIXING PER ENERGY

In engineering applications, one important question is "how much energy is required to obtain a specific amount of mixing". A *mixing per energy* parameter can be created by relating the total area growth with the energy for mixing in incompressible viscous fluids yielding:

$$mpe = \frac{\xi_{av}}{\int_{t=0}^{T} \mathbf{D} : \mathbf{D} dt} \qquad (11)$$

In this case, at each initial position we used 400 interfacial directions and their average (ξ_{av}) was used in the above equation. To calculate area growth per energy for a section, the arithmetic mean of the 20000 values of mpe were calculated at that section.

Figure 9 shows area growth per energy for four geometries with $f = 1.16$ and β values of 0.0, 0.125, 0.25 and 0.5. For $\beta = 0.0$ (straight wall channel) and $\beta = 0.125$, the mixing efficiency is slightly decreasing. However, they both vary in a band between 1.8 and 2.1. As β increases to 0.25, mixing per energy level increases to almost 2.9. In this case, it slightly increases with the loop number. For $\beta = 0.5$, mixing per energy varies in a band between 3.8 and 4.1. The figure also shows that for $\beta > \beta_c$, increasing β will increase the level of mixing per energy and for $\beta < \beta_c$ the behavior is similar to a straight wall channel.

Figure 10 shows area growth per energy as a function of loop number for geometries with $\beta = 0.5$ and f values of 0.58, 1.16, 2.33 and 3.49. In this figure, only two levels of mixing per energy is observed. For $f = 0.58$, area growth per energy is constant at almost 2.3. f in this case is smaller than f_c. For the rest of the conditions area growth per energy varies between 3.5 and 4.1. In our study, we observed that for $f > f_c$, increasing f will increase the amount of mixing, but the mixing efficiency will not change. In other words, the improvement of mixing has been produced by putting more energy into the system. The mixing is better, but the efficiency does not change. This is somewhat different from the behavior of mixing per energy while changing β over its critical value. In that case, the more we increased β, the higher mixing per energy we observed. The reason is that the increase is generated by geometry changes, while the energy input has remained almost constant. This once more shows the power of geometry characteristics in generation of efficient mixing.

8. ACKNOWLEDGMENTS

The authors would like to acknowledge the Pittsburgh Supercomputing Center for the use of their CRAY supercomputer and software. This project was supported by GE, 3M and Welding Engineers companies.

9. REFERENCES

[1] H. Aref. Stirring by chaotic advection. *Journal of Fluid Mechanics*, 143:1–21, 1984.

[2] W. L. Chien, H. Rising, and J.M. Ottino. Laminar mixing and chaotic mixing in several cavity flows. *Journal of Fluid Mechanics*, 170:355–377, 1986.

[3] L.E. Malvern. *Introduction to the Mechanics of a Continuous Media.* Prentice-Hall Inc., 1969.

[4] C. J. Shearer. Mixing of highly viscous fluids: Flow geometries for streamlines subdivision and redistribution. *Chem. Eng. Sci.*, 28:1091–1098, 1973.

[5] S. Zerafati. *Local and Global Mixing Analysis of Complex Flow Fields: A Continuum Kinematics Approach to the Mixing in a Numerically Simulated Wavy Channel Flow.* PhD thesis, University of Maryland at College Park, 1994.

[6] S. Zerafati and D. Bigio. Area deformation as a basis for mixing prediction. *Society of Plastics Engineers ANTEC*, pages 310–317, 1994.

Design and inverse methods

Simulation of Materials Processing: Theory, Methods and Applications, Shen & Dawson (eds)
© 1995 Balkema, Rotterdam. ISBN 90 5410 553 4

Preform design in metal forming

S. Badrinarayanan & N. Zabaras
Sibley School of Mechanical and Aerospace Engineering, Cornell University, Ithaca, N.Y., USA

A. Constantinescu
Sibley School of Mechanical and Aerospace Engineering, Cornell University, Ithaca, N.Y., USA (On leave from: Laboratoire de Mécanique des Solides, CNRS URA, École Polytechnique, Palaiseau, France)

ABSTRACT: A major objective in metal forming is the determination of the initial shape of the workpiece (preform) and of the process parameters (e.g. the die shape) that lead to a final product with desired geometry and material properties. The solutions to these inverse problems are usually obtained by trial and error methods using the results of direct analysis for each set of preforms and process parameters. Sensitivity analysis facilitates a rigorous mathematical formulation and solution of preform and process design problems. In an earlier work (Badrinarayanan and Zabaras 1995), a sensitivity analysis was presented for determining the optimal shape of extrusion dies that leads to a desired material state in the final product. In this work, we concentrate on the formulation and the finite element solution of preform design problems. In particular, the objective is to design the initial shape of the workpiece that when it deforms under the action of a given die, results in a final product with a desired material state and geometry. Shape sensitivities are defined in a rigorous sense and the entire analysis is performed in a fully infinite dimensional setting. An example problem is solved in axisymmetric disk forging where the preform is designed such that, after forging with a flat die, a product with a minimum barreling effect is achieved.

1 INTRODUCTION

In a given forming process, the material state and geometry of the final product depend on several process parameters (loading conditions, geometry of the die surfaces, die lubrication conditions, geometry of the initial workpiece and other). Considering a fixed amount of deformation induced in a given forming process, one may want to control the process parameters in such a way that a final product with a desired material state and geometry can be achieved.

The design of forming processes can also be considered as the design of the initial workpiece and of the subsequent shapes at each of the forming stages known as *preforms*. A systematic study of these problems was done by Kobayashi and colleagues (see Kobayashi et. al. 1989 for a complete set of references). They introduced the so called 'backward tracing technique' and "traced backward the loading path in the actual forming process from a given final configuration."

The preform design and the die design problems can be formulated under a rigorous mathematical basis by posing them as optimization problems (Zabaras and Badrinarayanan 1993). The objective function for these optimization problems can be defined as an appropriate measure of the error between the desired final state and the numerically calculated state for a given set of design variables. In order to solve such optimization problems, one usually employs a sequential search method starting from a reference solution. Sensitivity analysis is a method widely used to evaluate the gradients of the objective functions. Sensitivities can be calculated either by employing finite differences, direct differentiation techniques or the adjoint variable method (Tortorelli and Michaleris 1994, Vidal and Haber 1993, Lee and Arora 1993). Since the problems under consideration are highly non-linear and history dependent, the direct differentiation method (DDM) is the most suitable one. In DDM, the governing equations are directly differentiated to obtain a set of field equations for the sensitivity fields.

We had earlier developed the DDM for die design problems (Badrinarayanan and Zabaras 1995). In these problems, the initial configuration of the body remains the same while the shape of the die surface changes. However, in the case of preform design, the initial configuration of the workpiece is the main unknown of the problem. To define the sensitivity of the deformation gradient, a reference configuration is introduced that is independent of the shape of the preform. A sufficiently smooth family of mappings are defined from the reference configuration to the preform. Optimization will be performed on this family of mappings.

The equilibrium equation is directly differentiated to obtain the corresponding equation for the sensitivity displacement field. A weak form of this equation is defined and solved using the FEM with exactly the same spatial and temporal discretizations as for the direct deformation analysis. In traditional literature for solving similar problems, the sensitivity problem is formulated on the discretized equations. In such formulations, the evaluation of the "force terms" and the application of the boundary conditions become extremely difficult. However, in the method

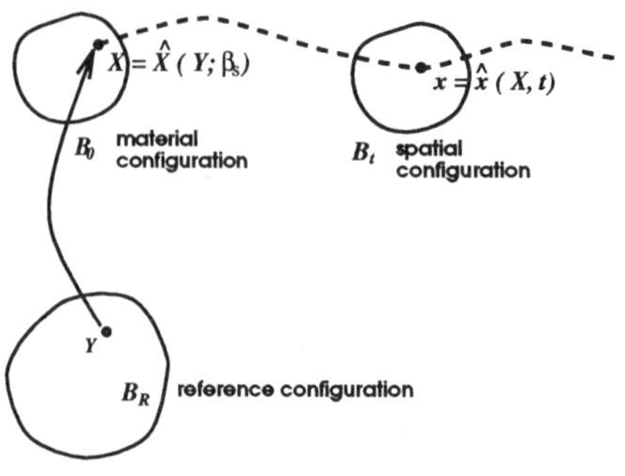

Figure 1: The reference, material and spatial configurations

proposed here, the boundary conditions and the "force terms" for the sensitivity problem, take a form similar to that of the direct analysis.

In the following, the definition of the sensitivity problem will be presented together with its weak form. In order to demonstrate the effectiveness of the present method, an example problem is solved to design the preform shape that when compressed with a flat die, results in a final product with minimum barreling effects.

2 THE CONFIGURATIONS

We consider the workpiece in the following configurations (see figure (1)):

- the *reference* configuration: $\boldsymbol{B}_R$.
  $\boldsymbol{Y}$ will denote an arbitrary point in this configuration. The points of this configuration do not have any material significance.
- the *material* configuration: $\boldsymbol{B}_0$, at time $t = 0$.
  This configuration represents the initial state of the material that undergoes the deformation process, and the material points of this configuration will be denoted by $\boldsymbol{X}$. The configuration $\boldsymbol{B}_0$ is completely described by a sufficiently smooth *reference map* of the form:

$$\boldsymbol{X} = \hat{\boldsymbol{X}}\,(\boldsymbol{Y}; \beta_s) \tag{1}$$

  where in general β_s are smooth functions that define the whole or parts of $\partial \boldsymbol{B}_0$. The above geometric map does not have a physical meaning and it is introduced only for the purpose of differentiating Lagrangian fields defined in the $\boldsymbol{B}_0$ configuration. All *Lagrangian* functions, $\boldsymbol{\Phi} = \hat{\boldsymbol{\Phi}}(\boldsymbol{X}, t)$, depend on the shape parameters β_s through the variable $\boldsymbol{X}$.
- the *spatial* configuration: $\boldsymbol{B}_t$ at time t.
  This represents the state of the body after an elapsed time t. The points in this configurations will be denoted by $\boldsymbol{x}$ and the motion of the body is described with respect to the material configuration by:

$$\boldsymbol{x} = \hat{\boldsymbol{x}}\,(\boldsymbol{X}, t) \tag{2}$$

  and with respect to the reference configuration by:

$$\boldsymbol{x} = \hat{\boldsymbol{x}}\left(\hat{\boldsymbol{X}}\,(\boldsymbol{Y}; \beta_s), t\right) = \tilde{\boldsymbol{x}}\,(\boldsymbol{Y}, t; \beta_s) \tag{3}$$

  All *Eulerian* functions, $\boldsymbol{\Phi} = \hat{\boldsymbol{\Phi}}(\boldsymbol{x}, t)$, depend on the shape parameters β_s through the variables $\boldsymbol{x}$ and $\boldsymbol{X}$.

3 THE SHAPE DERIVATIVES

The objective of this work is to optimize the design parameters β_s in order to obtain a desirable product after the deformation process. In the final product, the desirable qualities may be a required final shape, homogeneous material properties, or a desirable residual stress/plastic strain distribution. Solution to these preform problems can be generally achieved by minimizing with respect to β_s certain cost functionals based on the geometry, stress or state variable fields in the final product. Therefore, one needs to evaluate the derivatives of these variable fields in the direction of the design parameters.

We define the shape derivative of a variable field (Sokolowski and Zolesio 1992) as below:

> The *shape derivative*, $\overset{\circ}{\boldsymbol{\Phi}}$ of a scalar, vector or tensor valued function $\boldsymbol{\Phi}$ is the total Gateaux derivative of $\boldsymbol{\Phi}$ in the direction of $\Delta\beta_s$ computed at β_s:

$$\overset{\circ}{\boldsymbol{\Phi}}\,(\beta_s, \Delta\beta_s) = \frac{d}{d\lambda} \boldsymbol{\Phi}\,(\beta_s + \lambda \Delta\beta_s)\Big|_{\lambda=0}$$

The shape derivative can be understood as the difference between two fields representing the same physical quantity $\boldsymbol{\Phi}$, created by two different processes with slightly different parameters: β_s and $\beta_s + \Delta\beta_s$, i.e.:

$$\overset{\circ}{\boldsymbol{\Phi}}\,(\beta_s, \Delta\beta_s) = \boldsymbol{\Phi}(\beta_s + \Delta\beta_s) - \boldsymbol{\Phi}(\beta_s) + \mathcal{O}(||\Delta\beta_s||^2)$$

It should be noted that the comparison of the fields $\boldsymbol{\Phi}(\Delta\beta_s)$ and $\boldsymbol{\Phi}(\beta_s + \Delta\beta_s)$ is made at points of the reference configuration $\boldsymbol{B}_R$. The shape derivatives provide a measure of the change in a variable field due to a small change in the initial shape of the body.

4 THE DIRECT PROBLEM

The direct problem involves finding the time history of the deformation and the material state of a body deforming under the action of known external forces acting on it. The constitutive behavior follows the hyperelastic viscoplastic model developed by Brown et al. 1989. The computational aspects of the direct problem have been dealt with in Weber and Anand 1990 and Badrinarayanan and Zabaras 1993. Here, a brief review of the direct problem is presented to provide the necessary background for the sensitivity analysis.

The motion of the body was represented by a smooth mapping in equation (2). For this motion, the deformation gradient $\boldsymbol{F}$ defined over the material configuration is expressed as,

$$\boldsymbol{F} = \nabla\, \hat{\boldsymbol{x}}(\boldsymbol{X}, t) \tag{4}$$

This deformation gradient is decomposed as follows:

$$\boldsymbol{F} = \boldsymbol{F}^e\, \bar{\boldsymbol{F}}^p, \qquad \det \boldsymbol{F}^e > 0 \tag{5}$$

where, $\boldsymbol{F}^e$ is the elastic deformation gradient and $\bar{\boldsymbol{F}}^p$, the plastic deformation gradient, with $\det \bar{\boldsymbol{F}}^p = 1$. The elastic deformation gradient $\boldsymbol{F}^e$ has a unique polar decomposition given by:

$$\boldsymbol{F}^e = \boldsymbol{R}^e\, \boldsymbol{U}^e \tag{6}$$

The equilibrium equations can be expressed in the material configuration as:

$$\nabla \cdot \boldsymbol{P} + \boldsymbol{f} = \boldsymbol{0},\ \forall\ \boldsymbol{X} \in \boldsymbol{B}_0 \quad \text{and} \quad \forall\ t \in [0, t_f] \tag{7}$$

where $\boldsymbol{P}(\boldsymbol{X}, t)$ is the Piola Kirchoff stress measure (or the nominal stress) and $\boldsymbol{f}$ the body force in the material configuration:

$$\begin{aligned} \boldsymbol{P}(\boldsymbol{X}, t) &= \det \boldsymbol{F}\, \boldsymbol{T}\, \boldsymbol{F}^{-T} \\ \boldsymbol{f}(\boldsymbol{X}, t) &= \det \boldsymbol{F}\, \boldsymbol{b} \end{aligned} \tag{8}$$

Here, $\boldsymbol{T}$ is the Cauchy stress tensor field determined by the equilibrium equations and $\boldsymbol{b}$ the *known* body force field.

A linear hyperelastic law relating the rotation neutralized Kirchoff stress $\bar{\boldsymbol{T}}$ and the logarithmic strain is assumed as:

$$\bar{\boldsymbol{T}} = \boldsymbol{\mathcal{L}}^e\ [\bar{\boldsymbol{E}}^e] \tag{9}$$

where,

$$\bar{\boldsymbol{T}} = (\det \boldsymbol{U}^e)\ (\boldsymbol{R}^e)^T\ \boldsymbol{T}\ \boldsymbol{R}^e \tag{10}$$

and

$$\bar{\boldsymbol{E}}^e = \ln(\boldsymbol{U}^e) \tag{11}$$

with $\boldsymbol{\mathcal{L}}^e$ the isotropic elastic moduli.

The evolution of the plastic deformation gradient can be expressed as

$$\frac{d}{dt}(\bar{\boldsymbol{F}}^p)\,(\bar{\boldsymbol{F}}^p)^{-1} = \bar{\boldsymbol{D}}^p \tag{12}$$

where the rate of plastic deformation gradient $\bar{\boldsymbol{D}}^p$ is written as:

$$\bar{\boldsymbol{D}}^p = \dot{\bar{\epsilon}}^p\, \frac{3}{2}\frac{\bar{\boldsymbol{T}}'}{\tilde{\sigma}}$$

with the plastic strain rate $\dot{\bar{\epsilon}}^p$ defined as a scalar function:

$$\dot{\bar{\epsilon}}^p = f(\tilde{\sigma}, s) \tag{13}$$

and the equivalent stress, $\tilde{\sigma}$, given as:

$$\tilde{\sigma} = \sqrt{\frac{3}{2}\,\bar{\boldsymbol{T}}' \cdot \bar{\boldsymbol{T}}'} \tag{14}$$

Also, $\bar{\boldsymbol{T}}'$ is the deviatoric part of $\bar{\boldsymbol{T}}$ given by

$$\bar{\boldsymbol{T}}' = \bar{\boldsymbol{T}} - \frac{1}{3}\,\mathrm{tr}(\bar{\boldsymbol{T}})\ \boldsymbol{I} \tag{15}$$

Now, the evolution of the plastic deformation gradient can be re-written as,

$$\frac{d}{dt}(\bar{\boldsymbol{F}}^p)\,(\bar{\boldsymbol{F}}^p)^{-1} = \frac{3}{2}\,\frac{f(\tilde{\sigma}, s)}{\tilde{\sigma}}\ \bar{\boldsymbol{T}}' \tag{16}$$

Finally, the evolution of the scalar variable s (internal resistance to plastic deformation) is given as

$$\frac{d}{dt}(s) = g(\tilde{\sigma}, s) \tag{17}$$

In conclusion, for the above model, a deformed configuration of the body can be completely characterized by:

- the deformation gradient $\boldsymbol{F}$
- the triad of variables $\boldsymbol{V} = (\boldsymbol{T}, \bar{\boldsymbol{F}}^p, s)$

The functions $f(\tilde{\sigma}, s)$ and $g(\tilde{\sigma}, s)$ are experimentally determined for a particular material. Equations (2) and (4) – (17), together with the initial conditions on the scalar state variable s and the Cauchy stress $\boldsymbol{T}$, and the boundary conditions on the motion of the body, completely determine the direct problem. In a forming process, the boundary conditions are determined by the rigidity of the die, the die – workpiece interface friction conditions and the applied traction or displacement boundary conditions on the workpiece.

In an updated Lagrangian FEM formulation, one considers a sequence of time incremental problems from time t_n to $t_{n+1}, n = 0, 1, 2, \ldots$. The configuration of the body at time t_n is used as the reference configuration for the incremental process. Two sub-problems are defined: first the *kinematic incremental problem*, where given the triad $\boldsymbol{V}$ at the beginning and the end of the time step and the external forces acting on the body, one calculates the incremental deformation gradient; and second, the *constitutive problem* where one evaluates the triad $\boldsymbol{V}$ at the end of the step given the configuration of the body at the beginning and the end of the time step and $\boldsymbol{V}$ at the beginning of the step. The kinematic problem requires full linearization of the principle of virtual work and the calculation of the consistent material linearized moduli. The constitutive problem is handled with the radial return mapping. The kinematic incremental problem and the constitutive problem are coupled and must be solved iteratively within a time step.

5 SHAPE SENSITIVITY ANALYSIS

In the analysis performed below, we assume that the material state and deformation history for each parameter β_s (i.e. for each preform shape) are known from the solution of the corresponding direct problem. The dependence of the shape sensitivity fields on the material state history and deformation will not be shown explicitly.

Similar to the direct problem, the sensitivity problem is subdivided into two problems: the constitutive and the kinematic. In the *constitutive problem*, it is assumed that the sensitivity of the triad $\overset{\circ}{\boldsymbol{V}}_n$ is known at the beginning of a time step $t = t_n$. The objective is to determine the dependence of $\overset{\circ}{\boldsymbol{V}}_{n+1}$ on the sensitivity of the total deformation gradient $\overset{\circ}{\boldsymbol{F}}_{n+1}$. In the *kinematic problem*, using a principle of virtual work like equation for the sensitivity fields, one has to determine the sensitivity of the deformation gradient $\overset{\circ}{\boldsymbol{F}}_{n+1}$ at the end of the time step, knowing $\overset{\circ}{\boldsymbol{V}}_n$, the linear relationship between $\overset{\circ}{\boldsymbol{V}}_{n+1}$ and $\overset{\circ}{\boldsymbol{F}}_{n+1}$, and by applying appropriate boundary conditions for $\overset{\circ}{\boldsymbol{F}}_{n+1}$. Once $\overset{\circ}{\boldsymbol{F}}_{n+1}$ is calculated from the kinematic problem, the linear relationship derived in the constitutive problem can be used to obtain the sensitivity of the triad $\overset{\circ}{\boldsymbol{V}}_{n+1}$. The two problems are linearly coupled and together provide a *single linear* problem for the calculation of the sensitivities of both the deformation and the material state.

5.1 *Shape derivatives of constitutive equations*

The mathematical analysis for the derivation of the shape derivatives of the constitutive equations is similar to that developed for process derivatives as presented in Badrinarayanan and Zabaras 1995. Here, we provide only the final results in a concise form.

The shape derivative of the Cauchy Stress $\boldsymbol{T}$ is expressed as:

$$\overset{\circ}{\boldsymbol{T}} = (\det \boldsymbol{U}^e)^{-1} \boldsymbol{R}^e \, \overset{\circ}{\bar{\boldsymbol{T}}} \, (\boldsymbol{R}^e)^T - \operatorname{tr}\left(\overset{\circ}{\bar{\boldsymbol{E}}^e}\right) \boldsymbol{T} + \left(\overset{\circ}{\boldsymbol{R}^e} (\boldsymbol{R}^e)^T \boldsymbol{T} - \boldsymbol{T} \, \overset{\circ}{\boldsymbol{R}^e} (\boldsymbol{R}^e)^T\right) \tag{18}$$

where

$$\overset{\circ}{\boldsymbol{R}^e} (\boldsymbol{R}^e)^T = \overset{\circ}{\boldsymbol{F}^e} (\boldsymbol{F}^e)^{-1} - \boldsymbol{R}^e \operatorname{sym}\left((\boldsymbol{U}^e)^{-1} \operatorname{sym}\left((\boldsymbol{F}^e)^T \overset{\circ}{\boldsymbol{F}^e}\right)\right) (\boldsymbol{U}^e)^{-1} (\boldsymbol{R}^e)^T \tag{19}$$

$$\overset{\circ}{\boldsymbol{U}^e} = \operatorname{sym}\left((\boldsymbol{U}^e)^{-1} \operatorname{sym}\left((\boldsymbol{F}^e)^T \overset{\circ}{\boldsymbol{F}^e}\right)\right) \tag{20}$$

$$\overset{\circ}{\bar{\boldsymbol{E}}^e} = 4 (\boldsymbol{U}^e + \boldsymbol{I})^{-1} \overset{\circ}{\boldsymbol{U}^e} (\boldsymbol{U}^e + \boldsymbol{I})^{-1} \tag{21}$$

$$\overset{\circ}{\bar{\boldsymbol{T}}} = \mathcal{L}^e \left[\overset{\circ}{\bar{\boldsymbol{E}}^e}\right] \tag{22}$$

$$\overset{\circ}{\bar{\boldsymbol{T}}'} = \overset{\circ}{\bar{\boldsymbol{T}}} - \frac{1}{3} \operatorname{tr}\left(\overset{\circ}{\bar{\boldsymbol{T}}}\right) \boldsymbol{I} \tag{23}$$

and

$$\overset{\circ}{\bar{\sigma}} = \frac{3}{2} \frac{\bar{\boldsymbol{T}}' \cdot \overset{\circ}{\bar{\boldsymbol{T}}'}}{\bar{\sigma}} \tag{24}$$

The evolution equation for the direct problem can be directly differentiated to obtain the evolution equations for the sensitivity fields. Employing Euler backward integration, these evolution equations can be integrated to obtain:

$$\overset{\circ}{s}_{n+1} = \overset{\circ}{s}_n \, \exp(g_s \Delta t) + \frac{g_{\bar{\sigma}}}{g_s} \, \overset{\circ}{\bar{\sigma}}_{n+1} \left[\exp(g_s \Delta t) - 1\right] \tag{25}$$

and

$$\overset{\circ}{\boldsymbol{F}^p}_{n+1} = \exp\left(\Delta t \, \bar{\boldsymbol{D}}^p_{n+1}\right) \overset{\circ}{\boldsymbol{F}^p}_n + \left[\exp\left(\Delta t \, \bar{\boldsymbol{D}}^p_{n+1}\right) - \boldsymbol{I}\right] \left(\bar{\boldsymbol{D}}^p_{n+1}\right)^{-1} \overset{\circ}{\bar{\boldsymbol{D}}^p}_{n+1} \, \boldsymbol{F}^p_{n+1} \tag{26}$$

where

$$\exp\left(\Delta t \, \bar{\boldsymbol{D}}^p_{n+1}\right) = \bar{\boldsymbol{F}}^p_{n+1} \left(\bar{\boldsymbol{F}}^p_n\right)^{-1} \tag{27}$$

From the multiplicative decomposition we obtain,

$$\overset{\circ}{\boldsymbol{F}}_{n+1} = \overset{\circ}{\boldsymbol{F}^e}_{n+1} \, \bar{\boldsymbol{F}}^p_{n+1} + \boldsymbol{F}^e_{n+1} \, \overset{\circ}{\bar{\boldsymbol{F}}^p}_{n+1} \tag{28}$$

This equation can now be solved to express $\overset{\circ}{\boldsymbol{F}^e}_{n+1}$ as a linear function of $\overset{\circ}{\boldsymbol{F}}_{n+1}$. This in turn will imply that $\overset{\circ}{\boldsymbol{T}}_{n+1}$ can be written as a linear function of $\overset{\circ}{\boldsymbol{F}}_{n+1}$. For further discussion on many of the issues involved in the above derivation, refer to Badrinarayanan and Zabaras 1995.

5.2 *Shape derivative of the equilibrium equations*

We define the *reference* gradient of every material point $\boldsymbol{X}$ in the configuration $\boldsymbol{B}_0$ as below:

$$\boldsymbol{F}_R(\boldsymbol{Y}; \beta_s) = \nabla_Y \boldsymbol{X} = \frac{\partial \boldsymbol{X}}{\partial \boldsymbol{Y}} \tag{29}$$

We further define $\boldsymbol{L}_R$ as:

$$\boldsymbol{L}_R = \overset{\circ}{\boldsymbol{F}}_R \, \boldsymbol{F}_R^{-1} \tag{30}$$

We start from the local form of the equilibrium equation (7) on the material configuration $\boldsymbol{B}_0$, and determine the shape derivative of the equilibrium equation. In doing so, it has to be noted that all the spatial derivatives are evaluated with respect to the configuration $\boldsymbol{B}_0$. In the sequel, all spatial derivatives we mention are assumed to be evaluated with respect to $\boldsymbol{B}_0$.

It can be shown that

$$(\nabla \overset{\circ}{\cdot} \boldsymbol{P}) = \nabla \cdot \overset{\circ}{\boldsymbol{P}} + \nabla \cdot \left(\boldsymbol{P} \boldsymbol{L}_R^T\right) - \boldsymbol{P}\left[\nabla \cdot \boldsymbol{L}_R^T\right] \tag{31}$$

Thus, the shape derivative of the equilibrium equation becomes:

$$\nabla \cdot \left(\overset{\circ}{\boldsymbol{P}} + \boldsymbol{P} \boldsymbol{L}_R^T\right) - \boldsymbol{P}\left[\nabla \cdot \boldsymbol{L}_R^T\right] + \overset{\circ}{\boldsymbol{f}} = \boldsymbol{0}$$

$$\forall \ \boldsymbol{X} \in \boldsymbol{B}_0 \text{ and } \forall \ t \in [0, t_f] \tag{32}$$

where, $\overset{\circ}{\boldsymbol{P}}$ is obtained as

$$\overset{\circ}{\boldsymbol{P}} = \det \boldsymbol{F} \left[\operatorname{tr}\left(\overset{\circ}{\boldsymbol{F}} \boldsymbol{F}^{-1}\right) \boldsymbol{T} + \overset{\circ}{\boldsymbol{T}} - \boldsymbol{T} \left(\overset{\circ}{\boldsymbol{F}} \boldsymbol{F}^{-1}\right)^T\right] \boldsymbol{F}^{-T} \tag{33}$$

The weak form for the above equation (32) is given as:

$$\int_{\boldsymbol{B}_0} \overset{\circ}{\boldsymbol{P}} \cdot \frac{\partial \tilde{\boldsymbol{u}}}{\partial \boldsymbol{X}} \, dV_0 - \int_{\partial \boldsymbol{B}_{n+1}} \left\{\left(\overset{\circ}{\boldsymbol{F}} \boldsymbol{F}^{-1}\right) \cdot (\boldsymbol{I} - \boldsymbol{n} \otimes \boldsymbol{n})\right\} \hat{\boldsymbol{t}} \cdot \tilde{\boldsymbol{u}} \, dA_{n+1} =$$

$$\int_{\boldsymbol{B}_0} \left(\boldsymbol{P}\left[\nabla \cdot \boldsymbol{L}_R^T\right]\right) \cdot \tilde{\boldsymbol{u}} \, dV_0 + \int_{\boldsymbol{B}_0} \left(\boldsymbol{P} \boldsymbol{L}_R^T\right) \cdot \frac{\partial \tilde{\boldsymbol{u}}}{\partial \boldsymbol{X}} \, dV_0 + \int_{\partial \boldsymbol{B}_{n+1}} \left(\overset{\circ}{\hat{\boldsymbol{t}}} - \left(\boldsymbol{F} \boldsymbol{L}_R \boldsymbol{F}^{-1} \cdot (\boldsymbol{n} \otimes \boldsymbol{n})\right) \hat{\boldsymbol{t}}\right) \cdot \tilde{\boldsymbol{u}} \, dA_{n+1} \tag{34}$$

for admissible values of $\tilde{\boldsymbol{u}}$.

5.3 *The force terms*

Foe a finite element implementation of the weak form of the sensitivity equilibrium equations, one must evaluate the corresponding stiffness matrix and force vector. Here, we deal in some detail with the calculation of the force terms.

In order to evaluate the force terms, we need to calculate the quantities $\boldsymbol{L}_R$ and $\overset{\circ}{\hat{\boldsymbol{t}}}$. $\boldsymbol{L}_R$ depends on the variation in the design parameter β_s. In two dimensions, the reference configuration $\boldsymbol{B}_R$ is taken to be a unit square. For demonstration of ideas, let us consider the case (refer to figure (2))

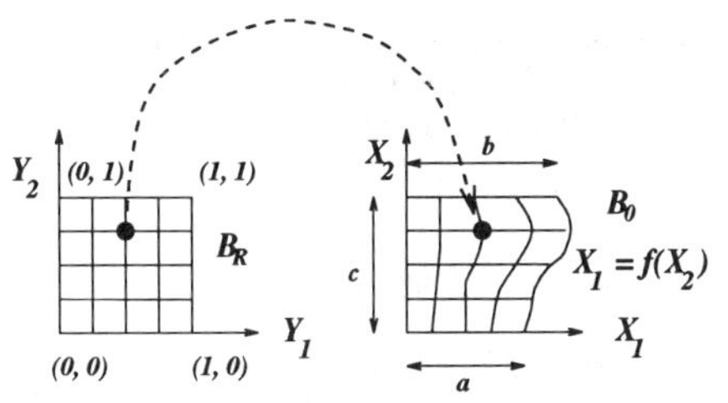

Figure 2: A mapping from the reference configuration to the material configuration

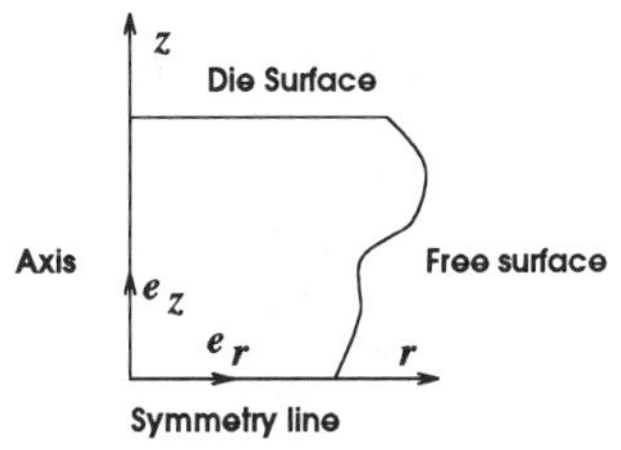

Figure 3: Preform shape for the upset forging

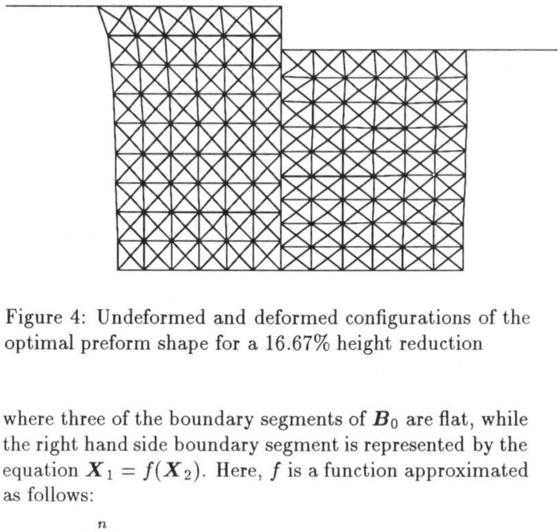

Figure 4: Undeformed and deformed configurations of the optimal preform shape for a 16.67% height reduction

where three of the boundary segments of $\boldsymbol{B}_0$ are flat, while the right hand side boundary segment is represented by the equation $\boldsymbol{X}_1 = f(\boldsymbol{X}_2)$. Here, f is a function approximated as follows:

$$f(\boldsymbol{X}_2) = \sum_{i=1}^{n} \beta_i \phi_i(\boldsymbol{X}_2) \tag{35}$$

with $f(0) = a$ and $f(c) = b$. Also, $(\phi_1, \phi_2, \ldots, \phi_n)$ are appropriately selected basis shape functions and $(\beta_1, \beta_2, \ldots, \beta_n)$ a set of n scalar parameters. Applying a linear transfinite mapping for this boundary representation (George 1991), one can introduce a mapping $\hat{\boldsymbol{X}}(\boldsymbol{Y};\beta_s)$ as below:

$$(\boldsymbol{X}_1, \boldsymbol{X}_2) = (\boldsymbol{Y}_1 f(c\boldsymbol{Y}_2), c\boldsymbol{Y}_2) \tag{36}$$

Therefore, for this case,

$$\boldsymbol{F}_R = \begin{bmatrix} f(c\boldsymbol{Y}_2) & c\boldsymbol{Y}_1 f'(c\boldsymbol{Y}_2) \\ 0 & c \end{bmatrix} \tag{37}$$

where f' denotes the derivative of f. If we perturb a parameter, say, β_k by $\Delta\beta_k$ while keeping all the other parameters constant, we can obtain $\overset{\circ}{\boldsymbol{F}}_R$ as

$$\overset{\circ}{\boldsymbol{F}}_R = \Delta\beta_k \begin{bmatrix} \phi_k(c\boldsymbol{Y}_2) & c\boldsymbol{Y}_1 \phi_k'(c\boldsymbol{Y}_2) \\ 0 & 0 \end{bmatrix} \tag{38}$$

From $\boldsymbol{F}_R$ and $\overset{\circ}{\boldsymbol{F}}_R$, one can calculate $\boldsymbol{L}_R$. Solving the weak shape sensitivity problem with this $\boldsymbol{L}_R$ induced by a change of $\Delta\beta_k$, one can evaluate at the direct solution corresponding to $(\beta_1, \beta_2, \ldots, \beta_n)$, the gradient of the objective function in the direction of the kth design variable (Badrinarayanan and Zabaras 1995).

Calculation of $\overset{\circ}{\hat{\boldsymbol{t}}}$ depends on the type of traction forces applied on the boundary. If the dependence of the $\hat{\boldsymbol{t}}$ on the geometry is well known, the shape derivative of $\hat{\boldsymbol{t}}$ can be easily evaluated. A method similar to that developed in Badrinarayanan and Zabaras 1995 can be applied when the traction condition is indirectly specified as in the case of sliding friction.

6 APPLICATION

We consider a preform design problem in open die forging of a cylindrical Aluminum workpiece, where we want to achieve a final product without barreling. The material constants are taken from Brown et. al. 1989. The die-workpiece interface is modelled with sticking friction. The cylinder is assumed to be axially symmetric and also symmetric about a plane parallel to the flat die surface. The geometry of the workpiece is shown in figure (3). The free surface is modelled using 3 pieces of cubic splines and 7 independent parameters β_i. Let the free surface of the deformed body in the final configuration be denoted by Γ_f. Then, the optimization problem can be written as

$$\min_{\beta_i} \int_{\Gamma_f} (\boldsymbol{x} \cdot \boldsymbol{e}_r - r_0)^2 \, dl \tag{39}$$

Subject to the constraint

$$\int_{\boldsymbol{B}_{0(\beta_i)}} dV = V_0 \tag{40}$$

where $\boldsymbol{e}_r$ is the unit radial vector, V_0, the volume of the cylinder and r_0 the radius of the cylinder that would have resulted while upsetting a cylindrical piece of volume V_0 under frictionless conditions. In order to obtain the optimal solution for this problem, we need to evaluate the gradients of $\boldsymbol{x} \in \Gamma_f$ with respect to the parameters β_i. We employ the sensitivity analysis developed in the previous sections to obtain the shape sensitivities and from them we compute the gradient of the objective function. The calculated sensitivities are within 1% compared to the finite difference solutions obtained using the direct results corresponding to two nearby preforms.

The optimization is performed using a modified BFGS algorithm. The solution is assumed to converge when $(||\Delta\beta_s||)/||\beta_s|| < 1\%$.

We start from the initial design of a flat free surface which results in barreling in the final product. A deformation corresponding to 16.67% reduction in height is im-

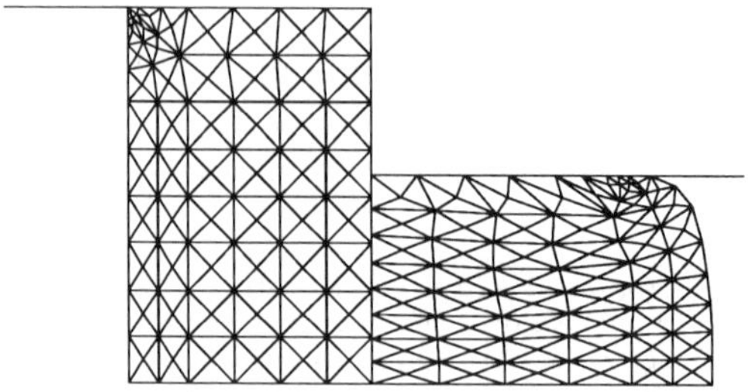

Figure 5: Undeformed and deformed configurations of the initial preform shape for a 44.44% height reduction

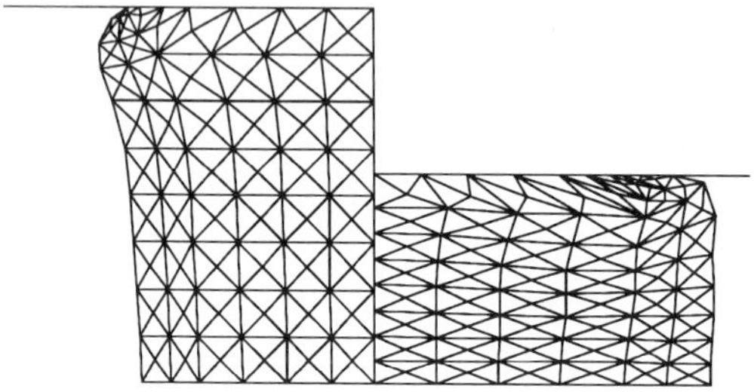

Figure 6: Undeformed and deformed configurations of the optimal preform shape for a 44.44% height reduction

posed and the preform shape that leads to a product with no barreling is obtained. The result is shown in figure (4). A similar problem was solved in Fourment and Chenot 1994 with a coarser mesh and for a simpler material model. The optimal preform shape reported there follows a similar trend to that of figure (4).

Let us now consider a substantial deformation (44.44% height reduction) which results in a fold over of the workpiece. The finite element mesh is reasonably fine in the region where fold over occurs. An optimal preform shape for this height reduction is obtained using the same optimization algorithm. The initial preform shape is shown in figure (5) and the optimal preform in figure (6). The preform shape obtained for this height reduction is considerably different from that of figure (4), especially in the regions where fold over occurs. With an even finer mesh near the fold over region and a better representation of the free surface, one, at the expense of a higher computational cost, could obtain in a fine detail the preform shape that leads to a perfectly flat free surface in the final product. Obviously, even in that case, the material properties will not be uniformly distributed in the final product.

7 CONCLUSIONS

A shape sensitivity analysis method has been developed for large deformation of hyperelastic viscoplastic solids that can be applied to preform design problems in metal forming. A principle of virtual work like equation is developed for the shape sensitivity fields. The calculated sensitivity fields are very accurate and satisfactory results were obtained while designing preform shapes for open die forging.

8 ACKNOWLEDGEMENTS

This work was funded by NSF grant DDM-9157189 to Cornell University. The computing was supported from the Cornell Theory Center.

REFERENCES

Badrinarayanan, S. and N. Zabaras 1993. A Two- Dimensional FEM Code for the Analysis of Large Deformations of Hyperelastic-Viscoplastic Solids, *Technical Report MM-93-05, Materials Processing Program, Sibley School of Mechanical and Aerospace Engineering, Cornell University.*

Badrinarayanan, S. and N. Zabaras 1995. A Sensitivity Analysis for the Optimal Design of Metal Forming Processes. Accepted in *Comp. Meth. Appl. Mech. Eng.*

Brown, S.B., K.H.Kim and L. Anand 1989. An Internal Variable Constitutive Model for Hot Working of Metals. *Int. J. Plasticity* 5:95-130.

Fourment, L. and J.L. Chenot 1994. The Inverse Problem of Design in Forging. *Inverse Problems in Engineering Mechanics, (edts. H.D. Bui et al.)* 21-28. Balkema, Rotterdam.

George, P.L. 1991. *Automatic Mesh Generation: Application to Finite Element Methods.* John Wiley and Sons, New York.

Kobayashi, S., S. Oh and T. Altan 1989. *Metal Forming and the Finite-Element Method.* Oxford University Press, New York.

Lee, T.H. and J.S. Arora 1993. Shape Design Sensitivity Analysis of Viscoplastic Structures. *Comp. Meth. Appl. Mech. Eng.* 108:237-259.

Fletcher, R. 1987. *Practical Methods of Optimization.* John Wiley and Sons, New York.

Sokolowski, J. and J.P. Zolesio 1992. *Introduction to Shape Optimization - Shape Sensitivity Analysis.* Springer-Verlag, New York.

Tortorelli, D.A. and P. Michaleris 1994. Design Sensitivity Analysis: Overview and Review. *Inv. Prob. in Eng.* 1:71-105.

Vidal, C.A. and R.B. Haber 1993. Design Sensitivity Analysis for Rate-Independent Elastoplasticity. *Comp. Meth. Appl. Mech. Eng.* 107:393-431.

Weber, G. and L. Anand 1990. Finite deformation constitutive equations and a time integration procedure for isotropic, hyperelastic-viscoplastic solids. *Comp. Meth. App. Mech. Eng.* 79:173-202.

Zabaras, N. and S. Badrinarayanan 1993. Inverse Problems and Techniques in Metal Forming Processes. *Inverse Problems in Engineering: Theory and Practice (edts. N.Zabaras et al.)* 65-76. ASME, New York, N.Y.

Simulation of Materials Processing: Theory, Methods and Applications, Shen & Dawson (eds)
© 1995 Balkema, Rotterdam. ISBN 90 5410 553 4

Open-loop control of a hot-forming process

Jordan M. Berg & James C. Malas III
USAF Wright Laboratories, WPAFB, Ohio, USA
Anil Chaudhary
UES, Inc., Dayton, Ohio, USA

ABSTRACT: A simulation study is performed to show that nonlinear finite-element process models can be integrated with material microstructural evolution equations ito develop an automated tool for the design of ram velocity profiles. The example presented is an isothermal forging of a complex disk shape from a γ phase titanium-aluminide alloy. The objective is to choose a ram velocity profile that maintains a desirable strain rate throughout the billet. For an arbitrarily complex given die and workpiece geometry, this objective is typically unattainable An acceptable compromise is obtained by controlling the strain rate only in regions of the billet where the material is undergoing microstructural transformation. The techniques developed in this paper can be generalized to a variety of design goals and forging parameters.

1 INTRODUCTION

Designing a complex forging based on repeated trials consumes valuable time and requires expensive equipment. The availability of reliable commercial nonlinear finite-element algorithms allows much of the iteration to be shifted to a computer, cutting costs. However, the process is still trial-and-error (now on a computer), and requires frequent input from a designer. This interaction can be further reduced. Once the designer has specified desired properties of the finished piece, and has indicated which process parameters are constrained and which can be freely varied, the search can be conducted automatically. The designer is freed from a tedious task, and productivity is increased. These gains are most dramatic when combined with automatic remeshing. With automatic remeshing now available for 3-D forgings, these techniques can be used on any part.

For the current effort, an isothermal forging is first expressed in an open-loop optimal control formulation. The design objective is written as a cost function to be minimized, subject to such constraints as are required. A wide variety of techniques exist to solve such problems. The choice of approach is strongly dependent on the specific form of the microstructural evolution equations. For the present case, the evolution equation for the fraction of material transformed is strongly dependent on effective strain, and only weakly dependent on strain rate. The equation describing the average transformed grain size has the opposite character.

This paper presents simulation results for this sample case. An axisymmetric cupped disk is to be formed from a TiAl alloy. The starting microstructure is specified. The process model is a nonlinear finite-element code. Die and preform shapes are fixed, and the forging is to be accomplished in a single step. The goal is to achieve the required shape, while transforming as much of the material as possible into a desired final microstructure. The variable control parameter to be optimized is the ram velocity.

2 DESIGN OBJECTIVES

The forging has two critical aspects. The first is shaping the billet. The die and preform geometries are specified. They are shown in Fig. 2.1. Figure 2.2 shows the situation when the ram has reached its full stroke of 2.8 in. Attaining the desired shape is a strict constraint, which is considered to have been exactly achieved when the ram reaches full stroke.

The second aspect, on which this study concentrates, is transforming the billet microstructure. The material considered here is Ti-48Al-2V. This material, in its γ phase, has high yield strength at elevated temperatures, making it extremely useful in aerospace applications. The γ alloy has a low ductility, however, and so is difficult to form and too brittle for many applications. The cast and hipped material has a lamellar structure, with alternating α_2 and γ layers. During deformation, the lamellar structure is transformed to one of equiaxed grains of pure γ and pure α_2 phase material (dynamic spheroidization). If the grains are small, then the material combines high temperature strength with improved ductility. The size of the spheroidized grains can be controlled using the deformation parameters. The design goal considered in this paper is achieving maximum spheroidization, and a minimal grain size.

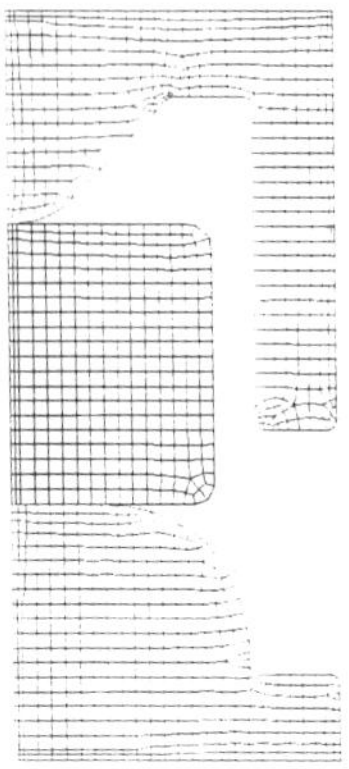

Figure 2.1. Undeformed forging geometry.

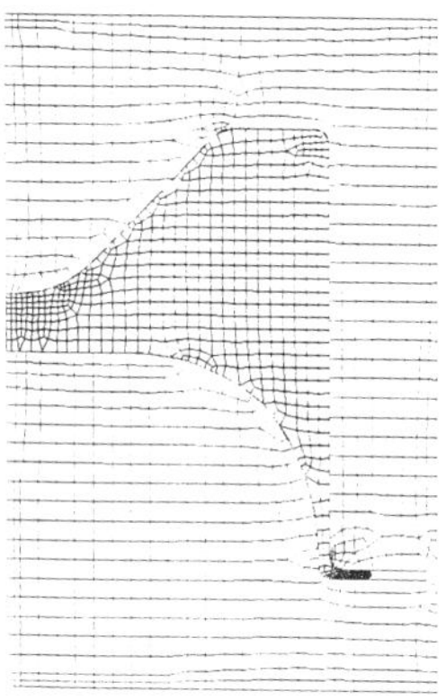

Figure 2.2. Forging geometry at full stroke.

Guillard (1994) has given desirable process parameters for isothermal, constant strain rate, compression. Guillard's work also indentifies a processing window in the strain-strain rate-temperature space to ensure material stability, and gives microstructural evolution equations that describe the average spheroidizeded grain size, and the fraction of material spheroidized, in terms of the strain, strain rate, and temperature.

The present work jointly utilizes the strain, strain rate, and temperature information output from the process simulation model in order to automatically arrive at the optimal ram velocity. Figure 2.3 shows the processing map at a strain of 0.7. The light gray region represents a likely processing window based on stability considerations, as discussed in Malas & Seetheraman (1992). The processing target can be further refined, using the techniques of optimal control theory, following the method of Malas, Irwin, and Grandhi (1993). The result, calculated by Frazier (1994) is the point indicated in Fig. 2.3. That is, the objective is to keep the material as close as possible to a strain rate of 0.3 1/min, at a temperature of 1100 C throughout the entire forging. The forging is assumed to be isothermal, so achieving the desired strain rate is the only objective.

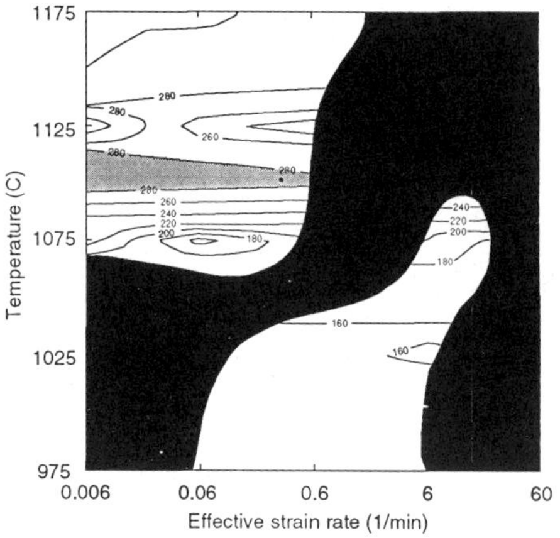

Fig 2.3. Deformation processing map at effective strain of 0.7. Unstable regions are blacked out, activation energy is background, processing window is shaded.

3 PROCESS MODEL

The macroscopic aspects of the forging are modeled using widely available commercial forming software, *Antares* (UES, Inc. 1993). This software uses a standard nonlinear implicit iterative finite element based computational strategy to perform forging analysis. The workpiece material is modeled as rigid-viscoplastic and a shear friction constitutive law is used for characterization of die-workpiece interface behavior.

4 NUMERICAL OPTIMIZATION

Given the control objective of obtaining a final forged microstructure that is completely spheroidized,and of the desired grain size, it is also necessary that the material stay within the pre-specified processing window. Since the progress of the forging is parameterized by the ram stroke, and the initial microstructure and the desired final microstructure is specified, it is viewed as a fixed end-point problem, where the cost function depends only on the final state, but the path is subject to inequality constraints. This approach requires that the microstructural evolution equations be integrated into the finite-element package. At the time this analysis was performed, that had not been done. Since then, it has been accomplished, and that software was used to evaluate the results of this study. However, a different method was used for the optimization.

As discussed in the previous section, the objective of the forging is to keep the strain rate at 0.3 1/min, at the temperature at 1100 C throughout the entire billet and at all times. In terms of the finite element process model, this means keeping every element in the billet at the optimal strain rate at every stroke increment. Except for an idealized frictonless upsetting, this will typically be impossible. Strain rates will vary across the billet, and most regions will be either above or below the desired value. The problem then becomes to achieve the minimum deviation from the desired process window, across the workpiece and throughout the stroke.

The best way to write the cost function for this problem is as the sum of the squares of the strain rate error for each element. At each step, then, the ram velocity that minimized this cost function would be determined. Another reasonable approach is to find the ram velocity at each step that drives the average strain rate across the billet to the desired value. These two formulations may seem similar, but they are quite different. The first is more physically meaningful for this problem. The second is far preferable numerically. The two will coincide when the strain rate distribution across the billet is uniform, otherwise the first method gives a lower value. It is important to note that if the strain rate variations throughout the billet is large, then neither technique will give good results. On the other hand, if the strain rate variations are small, then either method will work well. So if some means are found to reduce the strain rate variation, the approach with better numerical properties can be used, and it will give a good result.

The distribution of strain rates throughout the workpiece is strongly affected by the die/billet geometry. Once the geometry has been set, as is the case in this study, the designer has little further influence. However, the strain rate variation can be effectively reduced by only considering a portion of the billet. This is justified on the following grounds: The material is not transforming evenly. Some parts of the billet are nearly rigid, and are not transforming at all, while others may have been totally transformed already. Therefore it is possible to weight each point in the billet, based on the microstructural evolution equation describing its fraction transformed. That calculation is now discussed in detail.

Let $S(\varepsilon,\dot{\varepsilon},T)$ be the microstructural evolution equation describing the fraction of material transformed (spheroidized, in this case). The weighting of a piece of material will be proportional to the amount of material transforming during this stroke increment. That is,

$$W(\varepsilon,\dot{\varepsilon},T) = K\frac{dS}{dX}\Delta X \qquad (4.1a)$$

$$\approx K\frac{\partial S}{\partial \varepsilon}\frac{\Delta\varepsilon}{\Delta t}\frac{\Delta t}{\Delta X} \qquad (4.1b)$$

$$\approx K\frac{\partial S}{\partial \varepsilon}\dot{\varepsilon}V_{ram} \qquad (4.1c)$$

$$\approx K'\frac{\partial S}{\partial \varepsilon}\dot{\varepsilon} \qquad (4.1d)$$

where X is stroke, and t is time. This approximation assumes that the fraction spheroidized changes due to changes in strain only, and neglects effects due to changes in strain rate, and temperature. Temperature variations are not considered because of the isothermal assumption. Strain rate effects are considered to be small based on the character of the microstructural evolution equations. Also note that the experimental data underlying the expression for S are for constant strain rate compressions. As a result, the level of accuracy is difficult to determine for complex strain rate paths, without further extensive experimentation.

The microstructural evolution equation for S is as follows (Rack, 1994):

$$S(\varepsilon,\dot{\varepsilon},T) = 2061.38 + 7.017 \log \dot{\varepsilon} - 3.7908\ T + 56.84\ \varepsilon + 0.001776\ T^2 - 12.52\ \varepsilon^2 \qquad (4.2)$$

where T is the temperature specified in degrees C, log is base 10, and time is in seconds. It is understood that when the strain exceeds 2.27, the fraction transformed is one. So,

$$\frac{\partial S}{\partial \varepsilon}\dot{\varepsilon} = (56.84 - 25.04\ \varepsilon)\ \dot{\varepsilon} \qquad (4.3)$$

and,

$$W(\varepsilon,\dot{\varepsilon},T) = \begin{cases} (56.84 - 25.04\ \varepsilon)\ \dot{\varepsilon} & 0 \le \varepsilon \le 2.227 \\ 0 & 2.227 \le \varepsilon \end{cases} \qquad (4.4)$$

The microstructure model derived by Guillard is based on a curve fit to experimental data. That fit is strictly valid only in the region,

$$0.35 \le \varepsilon \le 2.03 \qquad (4.5a)$$

$$10^{-4}\ s^{-1} \le \dot{\varepsilon} \le 10^{-1}\ s^{-1} \qquad (4.5b)$$

$$1058\ C \le T \le 1142\ C \qquad (4.5c)$$

Note that this is particularly restrictive at the low end of the strain scale. Most forgings start with the billet material at close to zero strain. Furthermore, the weighting is at a maximum when the strain is zero, so these regions are simultaneously poorly modeled, and extremely important. This is an undesirable situation, but for now—pending the development of more suitable models—unavoidable.

The weighting function can be used in either the least squared error, or the mean strain rate, formulation. The remainder of this paper considers only the mean strain rate formulation. The cost function to be minimized is,

$$J = (\dot{\varepsilon}_{des} - \dot{\varepsilon}_{av})^2 \qquad (4.6)$$

where $\dot{\varepsilon}_{av}$ is a weighted average strain rate, given by

$$\dot{\varepsilon}_{av} = \frac{\Sigma\, W_i^n\, \dot{\varepsilon}_i}{\Sigma\, W_i^n} \qquad (4.7)$$

The computational advantage of this cost function springs from the fact that the weighted average strain rate can almost always be set to any desired value, by varying the ram velocity. Thus the problem of minimizing J reduces to the problem of finding the root of $\dot{\varepsilon}_{des} - \dot{\varepsilon}_{av} = 0$. The root finding problem is easy because Eqn. (4.7) is a monotonically increasing function of ram velocity, with a guaranteed zero crossing. The exponent, n, is a design parameter that allows the most rapidly spheroidizing regions of the material to be most heavily weighted. Note in particular that as n increases, the volume of material being actively controlled will decrease, and the strain rate in those regions will be very close to the target strain rate. When n is zero, the result is the unweighted average strain rate in the billet.

The root finding algorithm begins by bracketing the zero. Once bracketed, it switches to a general purpose routine called Brent's method (Press, et al. 1986). Brent's method switches between parabolic interpolation and bisection, depending on whether the parabolic fit is good. When the interval is within an acceptable tolerance the routine terminates. Figure 4.1 shows how Brent's method is integrated with *Antares*.

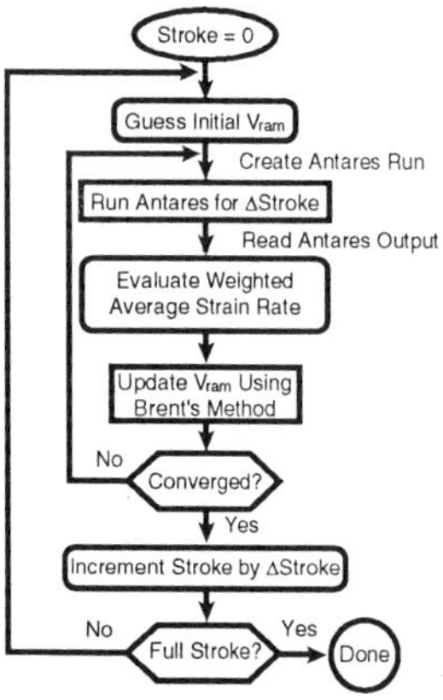

Figure 4.1. General scheme for optimizing ram velocity via Brent's method.

Note that Brent's method does not use derivative information. In principle, the derivatives can be obtained from the process simulation code. However, the total effort required to obtain such information in a succinct form is deemed to be extremely large, and therefore was not performed as part of this study.

5 EVALUATION MODEL

Once the cost function has been selected, and the optimizing ram velocity profile found, the result must be evaluated. At this point all simplifying assumptions should be dropped. The ideal test of the optimal ram velocity would be an actual forging, but this was not an available option. The main difficulty in evaluating the result of the optimization in simulation is that the microstructural evolution equations must be integrated into the finite-element simulation. As has been mentioned, these equations were derived for constant strain rate compression tests, so the limitations on their use in complex forging situations needs to be duly considered.

The microstructural evolution equations are Eqn (4.2), given previously, and (Rack, 1994):

$$g(\varepsilon,\dot{\varepsilon},T) = 248.22 + 142.97 \log \dot{\varepsilon} - 0.1284\, T - 59.82\, \varepsilon + 8.77 \log^2 \dot{\varepsilon} + 7\, \varepsilon^2 - 0.0833\, T \log \dot{\varepsilon} - 15.833\, \varepsilon \log \dot{\varepsilon} \qquad (5.1)$$

where g is the average transformed grain size, in microns, T is the temperature specified in degrees C, log is base 10, and time is in seconds. Again, the restrictions (4.5) apply to Eqn. (5.1).

As in the case of the weighting function, the low strain region is unavoidable, and critically important. There is, however, no alternative at the moment, short of actual experiments.

Equations (4.2) and (5.1) represent solutions to microstructural state equations for isothermal, constant strain rate compressions. What is required for the purposes of the evaluation model are equations governing behavior for nonconstant strain rate compressions, that is, expressions for $dS/dt(\varepsilon,\dot{\varepsilon},T)$ and $dg/dt(\varepsilon,\dot{\varepsilon},T)$.

$$\frac{dS}{dt}(\varepsilon,\dot{\varepsilon},T) = \frac{\partial S}{\partial \varepsilon}\frac{\partial \varepsilon}{\partial t} + \frac{\partial S}{\partial \dot{\varepsilon}}\frac{\partial \dot{\varepsilon}}{\partial t} + \frac{\partial S}{\partial T}\frac{\partial T}{\partial t} \qquad (5.2a)$$

$$\approx \frac{\partial S}{\partial \varepsilon}\dot{\varepsilon} \qquad (5.2b)$$

$$\approx (56.84 - 25.04\, \varepsilon)\, \dot{\varepsilon} \qquad (5.2c)$$

$$\frac{dg}{dt}(\varepsilon,\dot{\varepsilon},T) = \frac{\partial g}{\partial \varepsilon}\frac{\partial \varepsilon}{\partial t} + \frac{\partial g}{\partial \dot{\varepsilon}}\frac{\partial \dot{\varepsilon}}{\partial t} + \frac{\partial g}{\partial T}\frac{\partial T}{\partial t} \qquad (5.3a)$$

$$\approx \frac{\partial g}{\partial \varepsilon}\dot{\varepsilon} \qquad (5.3b)$$

$$\approx (-59.82 + 14\, \varepsilon - 15.833 \log \dot{\varepsilon})\, \dot{\varepsilon} \qquad (5.3c)$$

Neglecting the temperature effects is justified, if the forging is truly isothermal. Neglecting the strain rate term is justified if the strain rate varies slowly. It is preferable to include these effects, particularly for evaluation, but the spatial distributions of time derivatives of strain rate and temperature across the billet domain are not easily available from the process solver, due to the multiple remeshings performed throuhout the progression of the forging. Furthermore, the microstructural evolution equations

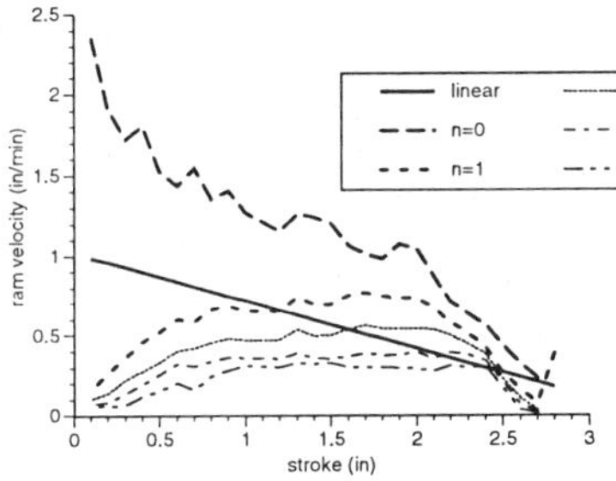

Fig. 6.1. Ram velocity profiles.

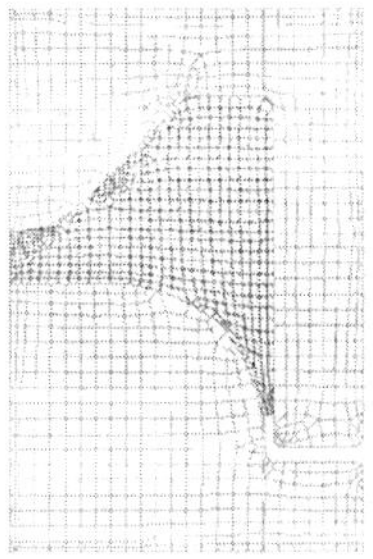

Fig. 6.2. Geometry at stroke of 2.6 in.

were derived for isothermal and constant strain rate conditions, so they are best used when these terms are small. With these restrictions understood, (5.2c) and (5.3c) are appended to the other state equations in the process solver. The final microstructure distribution can be displayed using a graphical post-processor.

Ideally the evaluation model would be improved through a better understanding of the microstructural evolution dynamics, and used directly for optimization. It would seem that the biggest payoffs for both analysis and synthesis of forgings would come from developing such codes.

6 RESULTS

The algorithm described in the preceding section was applied, for values of n equal to 0, 1, 2, 4, and 8. The resulting ram velocity profiles are shown in Fig. 6.1. Also shown in Fig. 6.1 is a linear velocity-stroke curve. This curve is obtained by modeling the forging as a simple frictionless upsetting, and using the ram velocity that gives the target strain rate.

The character of each solution can be illustrated by considering the strain rate distributions at a particular stroke. Figure 6.2 shows the geometry at a stroke of 2.6 in. Figure 6.3 shows the effective strain distribution at this stroke. The results for n = 8 are shown, but the effective strain is largely independent of the ram velocity profile. Figures 6.5 and 6.6 compare the strain rates in the billet at this stroke for the linear solution, and the case n = 4.

The results show the effectiveness of the algorithm. Because only a small portion of the material is transforming at the end of the forging—that in the rim of the "bowl"—the velocity profiles that do not explicitly account for the microstructural evolution do not adequately control the process. In particular, the unweighted average strain rate objective, and the linear profile both produce unacceptably high strain rates in the transforming region. Review of Fig. 2.3 shows that the material will be in the unstable region of the processing map. The result will be flow localization or cracking.

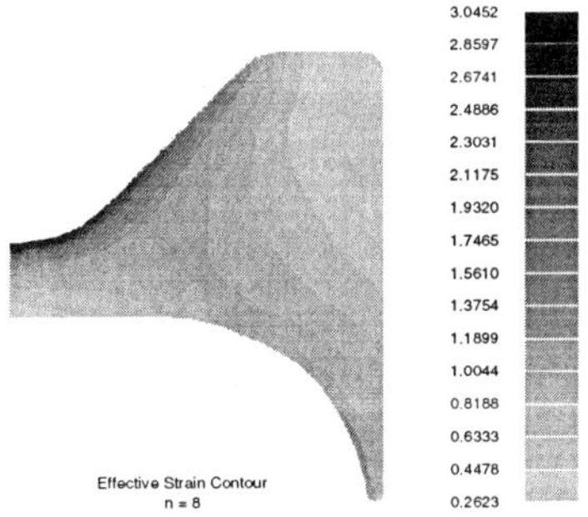

Fig. 6.3. Effective strain distribution at a stroke of 2.6 in, for n = 8 velocity-stroke curve.

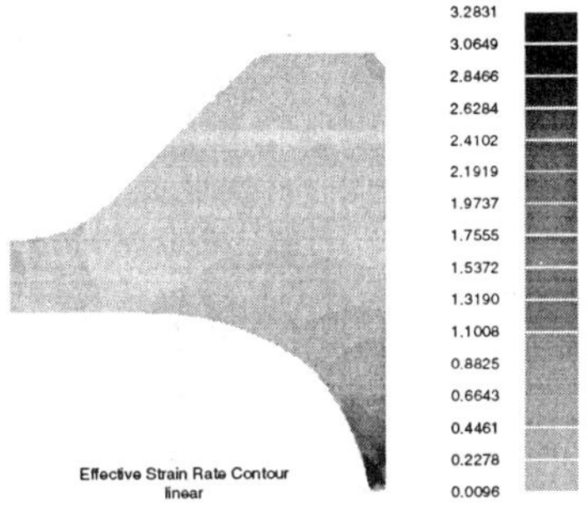

Fig. 6.4. Effective strain rate distribution at stroke of 2.6 in, for linear velocity-stroke profile.

On the other hand, the weighted strain rate objectives do significantly better. The cases of n = 4, and n = 8, both keep the transforming material within the process window. The rest of the billet will be out of the window, with strain rates orders of magnitude below the target. This corresponds to undesirable transformation mechanisms. But because very little of this material actually transforms, the quality of the end product is not significantly affected. Finally, Fig. 6.6 and 6.7 show the average transformed grain size, and fraction transformed, respectively, predicted for the case n = 4. The final microstructure is not very good. Since the ram velocity is, in a meaningful sense, optimal, further improvements will require redesign of the dies and/or preform.

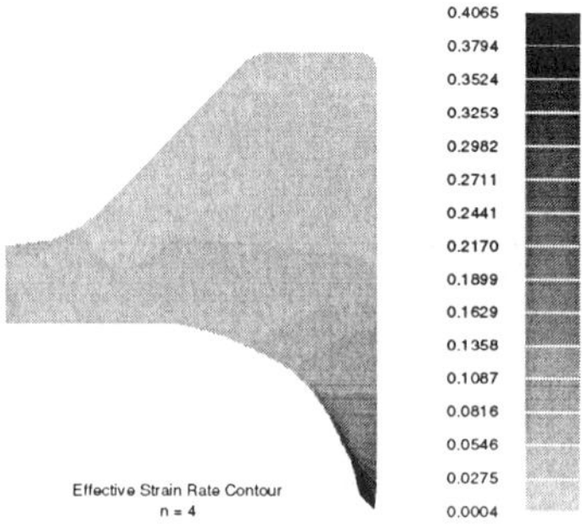

Fig. 6.5. Effective strain rate distribution at stroke of 2.6 in for case $n = 4$.

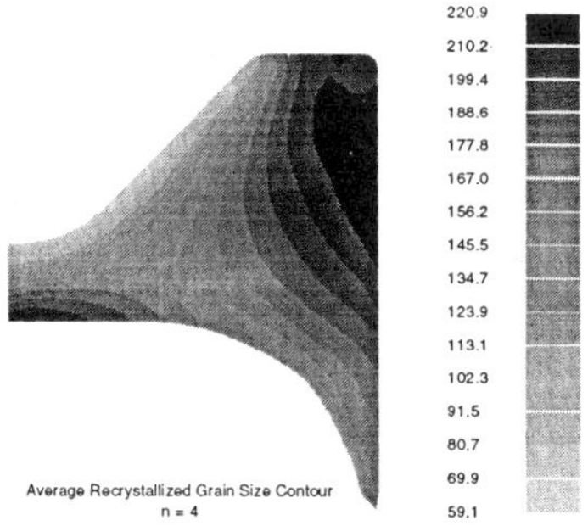

Fig. 6.6. Spheroidized grain size distribution at stroke of 2.6 in for case $n = 4$.

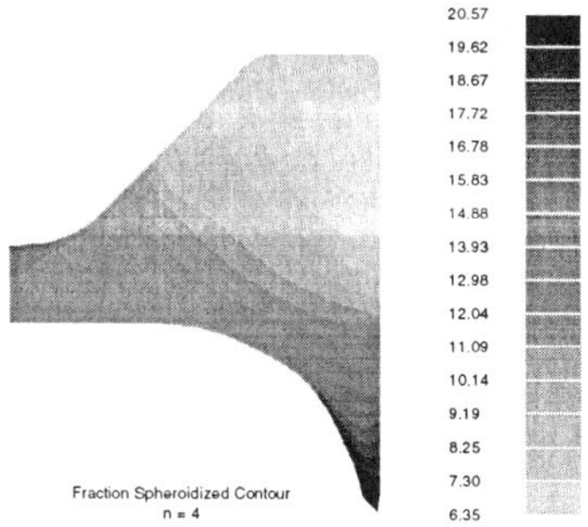

Fig. 6.7. Fraction recrystallized distribution at stroke of 2.6 in for case $n = 4$.

7 CONCLUSIONS

The integration of microstructural models and nonlinear finite element simulations adds a new tool to aid the forging process designer. This tool shows great promise in simulation studies. Once die and preform geometries are selected, the algorithm described generates ram velocity profiles that optimally produce a desired microstructure. The key innovation is a weighting function that concentrates on only those regions with a transforming microstructure. The degree to which these regions are emphasized is controlled by a weighting exponent. This exponent can be varied as a design parameter. The use of microstructural evolution equations incorporated into the finite element process model provides a very useful analysis tool for predicting the properties of the finished part.

REFERENCES

Frazier, W.G. 1994. Personal communication.

Guillard, S. 1994. *High Temperature Micro-Morphological Stability of the* ($\alpha_2 + \gamma$) *Lamellar Structure in Titanium Aluminides,* Ph.D. Thesis, Materials Science and Engineering Dept., Clemson University.

Malas, J.C. and V. Seetharaman 1992. Using Material Behavior Models to Develop Process Control Strategies, *JOM*, 44:8–13.

Malas, J.C. R.D. Irwin, and R.V. Grandhi 1993. An Innovative Strategy for Open Loop Control of Hot Deformation Process, *Journal of Materials Engineering and Performance*, 2:703–714.

Press, W.H., B.P. Flannery, S.A. Teukolsky, W.T. Vetterling 1986. *Numerical Recipes: The Art of Scientific Computing*: 251-254. Cambridge, Cambridge University Press.

Rack, H. 1994. Personal communication.

UES, Inc. 1993. *Antares User's Manual Version 3.0*, Dayton OH.

Simulation of Materials Processing: Theory, Methods and Applications, Shen & Dawson (eds)
© 1995 Balkema, Rotterdam. ISBN 90 5410 553 4

An inverse technique for the optimization of some forming processes

Ming-Fa Chen & Antoinette M. Maniatty
Rensselaer Polytechnic Institute, Troy, N.Y., USA

ABSTRACT: In this paper, we use an inverse problem solving technique to develop a numerical tool for the optimization of steady forming processes. In the solution procedure, the numerical algorithm employs an adjoint method for formulating the sensitivity coefficients and applies a gradient-based method for minimizing a least-square function; and the final step of the procedure examines the existence, stability, and convergence of the solution. This work presents some examples for the design of drawing and rolling processes. In particular, the problem focuses on optimizing the process parameters which produce the desired state variable distribution in the final product.

1. INTRODUCTION

A forming process can produce desired material characteristics in products by controlling the process parameters. These process parameters includes geometry, speed of both the tool and the workpiece, temperature, etc. In the design of forming processes, we attempt to search for a formulation that efficiently determines the optimum process parameters and produces the desired products. This design problem is usually formulated in a mathematical model and is referred to as an optimization problem.

Most studies in recent years formulate the problems described above as direct problems. The direct forming problem is mathematically based on a set of input information, including the domain, governing equations, constitutive relations, and the complete prescribed boundary conditions. After this information is provided, a solution for the deformation, stress, and internal state variable fields through the deformation zone is calculated.

A solution formulation developed for the optimization problem determines the optimum process parameters and satisfies the product specifications. The solution procedure starts with the product specifications as input, and then computes the process design parameters which will generate the desired product. This formulation for the design forming problem is effectively the inverse of the one in the simulation of direct problems. Hence, this class of problems is termed *inverse problems*.

In inverse problems, some studies have been devoted to the inverse problems in solid mechanics. Two-dimensional problems involving unknown boundary conditions are studied by Maniatty et al. (1989) for linear static elasticity, and by Zabaras and Schnur (1990) for elastoviscoplasticity. More recently, Karafilis and Boyce (1992) applied a linear inverse technique for designing die shapes for sheet forming using a springback calculation.

The solution procedure employs an iterative algorithm for solving the design forming problem. In the algorithm, the sensitivity analysis plays a key role to evaluate the solution direction such that the optimum solution can efficiently be determined. For the method of sensitivity analysis, both the direct variation method and the adjoint method were discussed by Panagiotis et al. (1994). Maniatty and Chen (1995) used the adjoint method to explicitly formulate the sensitivity coefficient for steady forming problems. This study uses this formulation to calculating the sensitivities with respect to geometry and speed.

2. PROBLEM DEFINITION

In the problems considered here, the given data consists of a specified state variable distribution in the exit region of the workpiece and incomplete boundary conditions; the output data is the optimum process parameters including geometry and speed. Solution of such problems involves two parts: solution of the direct problem and updating the process parameters for the optimal process. This work uses the Eulerian finite element method for the direct problem and a modified formulation based on inverse techniques for determining the update.

2.1 *Direct problem*

The direct problem is defined by the domain of interest, governing equations and complete boundary conditions given *a prior*. The formulation and solution procedure for the direct problem follows that given in Dawson (1987). Following the notation of Maniatty and Chen (1995), the resulting nonlinear systems of equations are of the following form:

$$\bar{k}(\bar{u}, \bar{s}) = \bar{f} \tag{1}$$

$$\bar{h}(\bar{u}, \bar{s}) = 0 \tag{2}$$

where equation (1) is the discretized equilibrium equation and equation (2) is the discretized evolution equation for the internal state variable. The vectors $\bar{u}$ and $\bar{s}$ are the vectors of nodal velocities and state variables.

A consistent penalty method (Engelman et al. 1982) is employed in equation (1) to enforce the incompressibility constraint. A staggered solution procedure is used for solving the nonlinear system.

2.2 *Optimization problem*

In this paper, the optimization problem is to determine the process parameters to generate a specified state variable distribution in the final product. Does the solution exist? In most cases, the answer is *NO* because the design variables are bounded in this problem. This study modifies the problem so that the optimum process parameters produce a state variable distribution that is as close as possible to a specified distribution.

The problem involves an incomplete domain and unknown boundary conditions on part of the boundary. We can define the problem in the same way as the direct problem, but include the unknown shape of part of the boundary $\partial\hat{B}$ which represents the tool contact zone. Let the shape of $\partial\hat{B}$ be defined by a vector of shape parameters b. The system equations of the optimization problem are now modified from equations (1) and (2):

$$\bar{k}(\bar{u}, \bar{s}, b) = \bar{f}(b) \tag{3}$$

$$\bar{h}(\bar{u}, \bar{s}, b) = 0. \tag{4}$$

In the finite element framework, the state variable can be discretized in the vector form with nodal values. The formulation for the optimization solution is then constructed by minimizing the least-square function with respect to the design variable, $\bar{b}$, as follows:

$$\mathrm{Min}\,(\bar{s}' - \hat{s}')^T(\bar{s}' - \hat{s}') \tag{5}$$

where $\bar{s}'$ is the vector of nodal state variables at the discrete locations in the exit region ∂B_e; and $\hat{s}'$ is the corresponding vector of desired state variables.

3. SOLUTION PROCEDURE

To solve equation (5) which is nonlinear and ill-posed, the solution procedure consists of three parts: sensitivity analysis, function minimization, and examination of the results.

3.1 *Nonlinear sensitivity analysis*

An sensitivity analysis measures how sensitive the design state field is with respect to design parameters and provides state gradient information that can be used in the minimization algorithm resulting from the optimization problem.

For the optimization problem described in section 2.2, the discrete sensitivities are expressed by the matrix:

$$C = \frac{1}{s_o}\frac{d\bar{s}'}{d\bar{b}} \tag{6}$$

where $\bar{b}_i = b_i/\hat{b}_i$, $\hat{b}_i$ is a reference value for the design variable b_i, and s_0 is the initial value of the state variable. The elements of this matrix define locally how the nodal state variables $\bar{s}'$ vary with the geometric parameters $\bar{b}$.

The adjoint method is used to determine the matrix C. For details of this method, see Maniatty and Chen. The resulting equation is

$$C = \bar{R}\left[\left(K_s - K_u H_u^{-1} H_s\right)^{-1} (F_b - K_b) + \left(H_u K_u^{-1} K_s - H_s\right)^{-1} H_b\right] \quad (7)$$

where

$$K_b = \frac{\partial \bar{k}}{\partial \bar{b}} \qquad K_u = \frac{\partial \bar{k}}{\partial \bar{u}} \qquad K_s = \frac{\partial \bar{k}}{\partial \bar{s}}$$

$$H_b = \frac{\partial \bar{h}}{\partial \bar{b}} \qquad H_u = \frac{\partial \bar{h}}{\partial \bar{u}} \qquad H_s = \frac{\partial \bar{h}}{\partial \bar{s}}$$

$$F_b = \frac{\partial \bar{f}}{\partial \bar{b}} \qquad \bar{R} = \frac{1}{s_0} R$$

In the above equations, R is a rectangular matrix which extracts only the components of the sensitivity matrix pertinent to the problem being considered.

3.2 *Function minimization*

The optimal solution is the minimum of the least-square function in equation (5) subject to imposed constraints. This study formulates the objective function as the following:

$$\Lambda_s = (\bar{s}' - \hat{s}')^T (\bar{s}' - \hat{s}') + \Gamma_s \quad (8)$$

where Γ_s is a constraint term. The construction of the constraint term affects the convergence of the minimization solution. This study uses the quadratic penalty form suggested by Fletcher (1975) (see the form in numerical examples), which is constructed by the Augmented Lagrange multiplier method.

The minimization of the function is performed by the quasi-Newton method with a line search scheme. The basic formula of this method is:

$$b^{(r+1)} = b^{(r)} + \alpha d^{(r)} \quad (9)$$

where $d^{(r)}$ is the search direction and depends on the sensitivity coefficients discussed in the previous section. The design variables of b are updated at each iteration until some convergence criteria is satisfied.

The gradient method is based on searching the solution in the steepest descent direction obtained from Taylor's expansion. The derivative of the function is calculated by the sensitivity matrix that is evaluated from the sensitivity analysis. In this framework, the multi-dimensional domain of the problem is converted into a one-dimensional variable. This process is a line search for the minimum of the function with respect to the scalar variable α. The line search is usually performed in an algorithm which logically searches for the line minimization and determines the value α. This study assumes that it is possible to bracket the minimum and uses the Brent's method (Michaleris et al. 1983) which logically keeps six function points for calculating a minimum in the bracketing interval by parabolic interpolation. The bracketing interval is reduced when a new minimum is evaluated. At the termination of the algorithm, the value α is the minimum solution where the size of the bracketing interval reaches a convergence criterion value.

3.3 *Examination of results*

Shape optimization problems are often ill-posed and therefore do not satisfy the usual conditions of existence, uniqueness, and stability. The nature of the constraints has a significant effect on the well-posedness or ill-posedness of the problem. This is true for the problems considered herein. For this reason, it is important to examine the results. An easy way to test for uniqueness and stability is to check if the solution converges to the same result with different initial guesses. While this is certainly not a proof and does not guarantee uniqueness and stability, it does provide additional confidence in the solution and may be satisfactory for engineering purposes.

For the class of problems considered, where a specified state variable distribution is sought in the solution, it is possible to have multiple solutions or no solution. Lack of uniqueness would not be considered a problem here because any solution that satisfies the constraints and is within the tolerance is assumed to be acceptable. It may also be possible that no forming geometry satisfying the geometric constraints gives a state variable distribution within the tolerance of the desired distribution. Then it is necessary for the designer to change the geometric constraints imposed.

4. NUMERICAL EXAMPLES

The following examples demonstrate the numerical algorithm for the design of drawing and rolling processes. In particular, these examples focus on optimizing the die shape for drawing and the three process parameters (roll radius, rolling speed, and reduction) for rolling. The two design problems use a hyperbolic sine law for the constitutive model, which is presented in Brown et al. (1989). The formed material used is 1100-Aluminum at 450 °C.

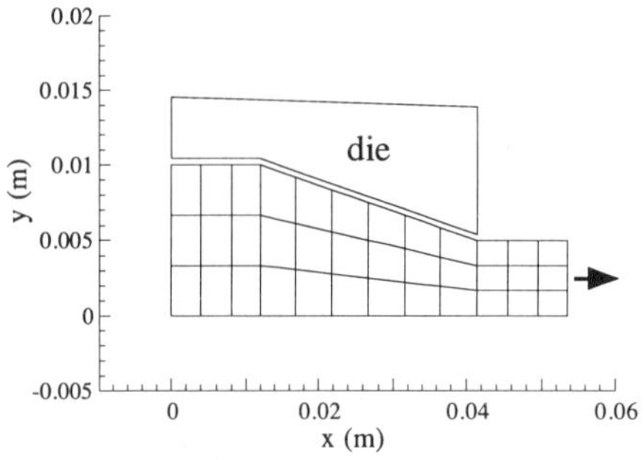

Figure 1. Geometry and dimension for drawing problem

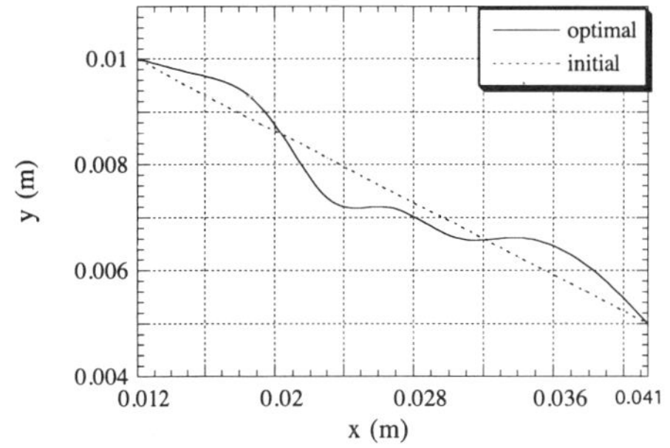

Figure 2. Optimal drawing die profile for the uniform state variable distribution (40 *MPa*)

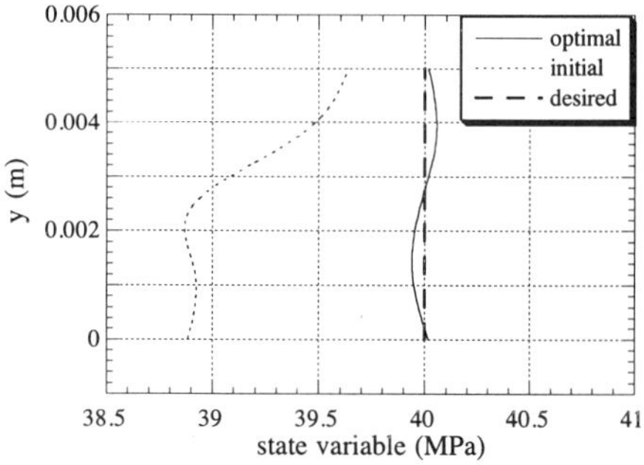

Figure 3. Results of state variable distribution for the optimization of drawing problem (40 *MPa*)

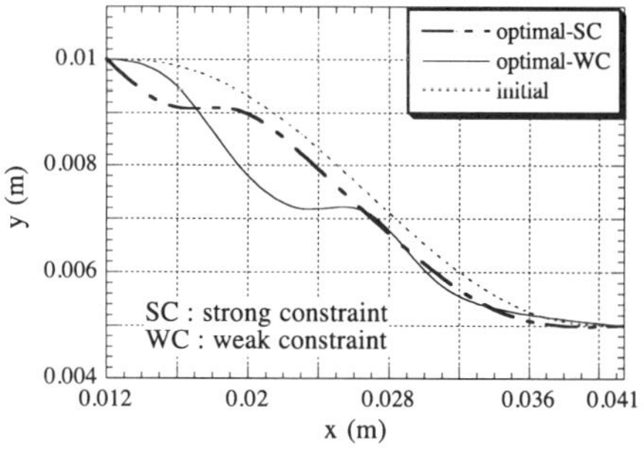

Figure 4. Optimal drawing die profile for the uniform state variable distribution (45 *MPa*)

The initial value of the state variable is 29.5 *MPa*. The friction coefficient β is assumed to be 0.5 $GPa\,(sm^{-1})$.

4.1 *Optimization of the die shape in drawing*

Figure 1 shows the geometry and dimension of the drawing problem assuming plane strain. The die is assumed to be initally straight. This example uses the y-coordinate parameters on the die contact zone as design variables. For this problem, a weak form of constraints is used, where the design variables are bounded by maximum and minimum values. The constraint functions are expressed in the forms:

$$\begin{aligned} G_i^d(\bar{b}) &= \bar{b}_i - \hat{\bar{b}}_u \le 0 & i &= 1, M_b \\ G_j^d(\bar{b}) &= \hat{\bar{b}}_l - \bar{b}_i \le 0 & j &= i + M_b \\ G_k^d(\bar{b}) &= \bar{b}_i - \bar{b}_{i-1} \le 0 & k &= i + 2M_b \end{aligned} \qquad (10)$$

where $\hat{\bar{b}}_u$ and $\hat{\bar{b}}_l$ are the upper and lower bounds being the values at the end points on boundary $\partial\hat{B}$, and M_b is the number of nodes on boundary $\partial\hat{B}$.

Now the objective function is written in the form:

$$\Lambda_d = (\bar{s}' - \hat{s}')^T (\bar{s}' - \hat{s}') + \frac{1}{2}\sum_{i=1}^{3M_b} \lambda_i \left(G_i^d(\bar{b}) + \frac{\upsilon_i}{\lambda_i} \right)^2 \qquad (11)$$

where λ_i and υ_i are the penalty parameter and the Lagrange multiplier, respectively.

For the drawing problem, an optimum die profile is shown in Figure 2. This die profile produces the most uniform distribution of the state variable at 40 *MPa*, shown in Figure 3. In examination of the result, this study simply investigated the uniqueness of the solution by giving different initial die profiles. It is shown that the solution depends on the following factors: (1) initial shape; (2) form of the constraint in the objective function; (3) penalty parameters for the constraint function $G_i(\bar{b})$; and (4) the bound values in equation (10). In another example, this work obtained two non-unique solutions for the optimum

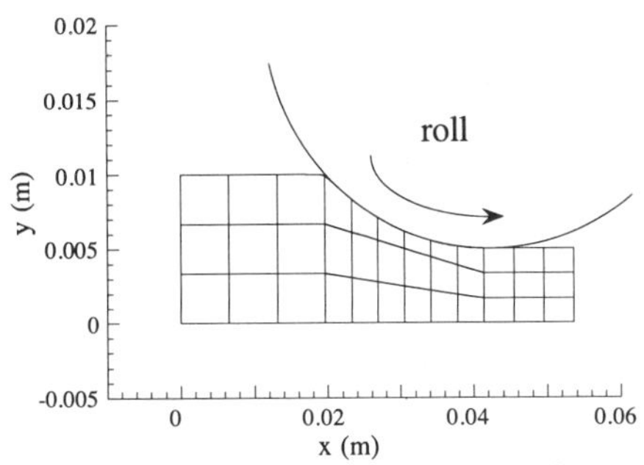

Figure 5. Geometry and dimension for rolling problem

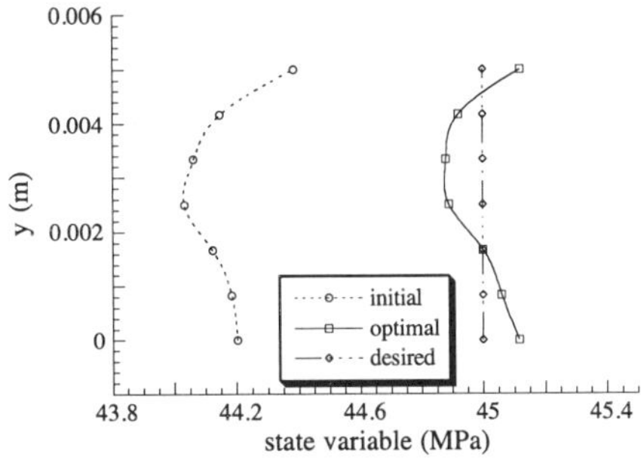

Figure 6. Results of uniform state variable distribution for rolling problem (45 *MPa*)

die profiles for the uniform distribution at 45 *MPa* for the state variable, shown in figure 4. The two solutions are solved with different forms of constraints which are a weak constraint and a strong constraint. The former is the same as in the previous example; and the latter is that the die shape is assumed to be fitted by a seventh order polynomial function.

4.2 *Optimization of the rolling process parameters*

Figure 5 shows the geometry and dimensions of the rolling problem. The design variables are roll radius, reduction, and rolling speed. The reference values for the design variables are the initial values. The objective function, Λ_r, is the same form as equation (12) in the drawing problem, but the constraint only restricts the design variable bound as the first two equations in equation (10). For this rolling problem, the optimum roll radius is 6.11 cm, reduction 39.02%, and rolling speed 7.70 cm/sec. The rolling process with these parameters produces the most uniform distribution at 45 *MPa* for the state variable, as shown in Figure 6. For the linear state variable distribution in Figure 7, the optimum roll radius is 2.24 cm, reduction 72.22 %, and rolling speed 19.94 cm/sec.

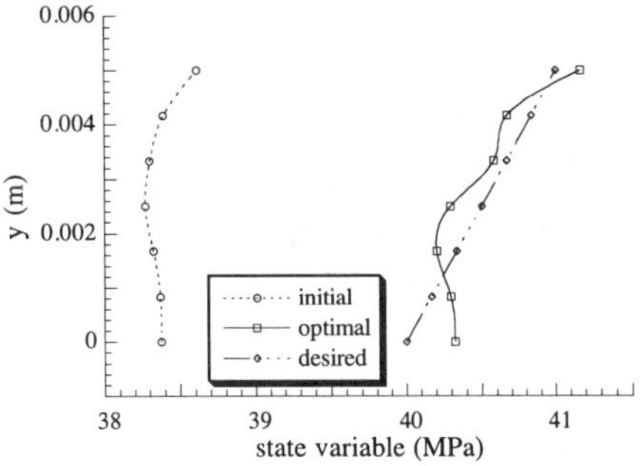

Figure 7. Results of linear state variable distribution for rolling problem

5. CONCLUSION

This study presents an inverse technique to solve for the optimum process parameters in drawing and rolling processes. The solution procedure includes sensitivity analysis, function minimization, and examination of the results. This work illustrates some design problems for drawing and rolling processes, where the goal is to obtain the desired state variable distribution in the formed material.

The examples presented focus on the simple forming problems which are studied for the development of the numerical tools for the design of forming processes. The results showed that the material characteristics of formed workpieces are sensitive to the process parameters, and demonstrated that the numerical algorithm based on the inverse technique can be used to determine the optimum process parameters.

The work will be extended for more complicated problems such as the shape optimization of extrusion and the consideration of thermal effects. In addition, the verification of results will be made by comparing with experiments, where the state variable can be represented by a physical measurement.

ACKNOWLEDGEMENTS

This work has supported by Henry Luce Foundation, Inc. and by the National Science Foundation through Grant DMI-9358123.

REFERENCES

Brown, S. B., Kim, K. H., and Anand, L. 1989. An internal variable constitutive model for hot working of metals. *Int. J. of Plasticity* 5: 95-130.

Dawson, P. R. 1987. On modeling of mechanical property changes during flat rolling of aluminum. *Int. J. of Solids Structures* 23: 947-968.

Engelman, M. S., Sani, R. L., Gresho, P. M., and Bercovier, M. 1982. Consistent versus reduced integration penalty methods for incompressible media using several old and new elements. *Int. J. for Numerical Methods in Fluids* 2: 25-42.

Fletcher, R. 1975. An ideal penalty function for constrained optimization. *J. Inst. Math. Optimization* 15: 319-342.

Karafillis, A. P., and Boyce, M. C. 1992. Tooling design in sheet metal forming using springback calculations. *Int. J. of Mechanical Sciences* 34: 113-131.

Kirkpatrick, S., Gelatt, C. D., and Vecchi, M. P. 1983. Optimization by simulated annealing. *Science* 220: 671-680.

Maniatty, A., Zabaras, N., and Stelson, K. 1989. Finite element analysis of some inverse elasticity problems. *J. of Engineering Mechanics* 115: 1303-1317..

Maniatty, A. M., and Chen, M. F. 1995. Design sensitivity analysis for steady metal forming processes. *Int. J. for Numerical Methods in Engineering*. submitted.

Michaleris, P., Tortorelli, D. A., and Vidal, C. A. 1994. Tangent operators and design sensitivity formulations for transient non-linear coupled problems with applications to elastoplasticity. *Int. J. for Numerical Methods in Engineering* 37: 2471-2499.

Zabaras, N. and Schnur, D. S. 1990. Solution of inverse elastic and elastoviscoplastic problems in two dimensions using the finite element method and spatial regularization method. In *NUMETA '90, Numerical Methods in Engineering: Theory and Applications*, Proceedings of the Third International Conference, University of Swansea, Swansea, Wales, Jan. 7-11.

Simulation of Materials Processing: Theory, Methods and Applications, Shen & Dawson (eds)
© 1995 Balkema, Rotterdam. ISBN 90 5410 553 4

A method for identifying forgibility parameters by combined experiments with finite element simulations

J.-H.Cheng & F.Wang
National Taiwan University, Department of Mechanical Engineering, Taipei, Taiwan

ABSTRACT: A computer-aided experimental procedure is proposed to determine the deformation limit of a material during a cold forging operation. Finite element simulations are used both in the material parameter identification and in the design of preform and forging processes.
Cracking in the domain or on the surface of a material is assumed to occur when the tensile strain energy of a deforming workpiece reaches a critial value, which is regarded as a material parameter to be determined from the suggested procedure. The criterion is implemented as a user's subroutine in the commerical finite element program, ABAQUS. With this extension, the program is then capable of predicting failure in a forging simulation. A closed-die forging operation on three different shapes of preform is selected to examine the validity of the development.

1. Introduction

Many finite element programs existed today are capable of simulating forging operations with complex geometries. However, very few, especially the commercial general purpose programs, are equiped to predict formability limit of a forging process, which if avaliable should be of great benefit to the designer in choosing the billet and the associated die.

A number of criteria have been proposed to characterize the deformation limits of metals [1-9]. Amongst these, Keeler and Backofen [1] pioneered the concept of forming limit diagram (FLD) in which the values of the major strain were plotted against the minor strain for some biaxial tensile stretching of sheet metals. The forming limit curve was obtained by connecting the points of strain combinations that produce instability or fracture.

Later, Kobayashi [2-3] and Lee and Kuhn [4] extended this idea to bulk forming and obtained similar results. Due to the ease of experimentation, upsetting of a cylindrical bar has been used extensively for this purpose. The cylinder bulges if friction exists at the die contact surfaces so that near the equator the circumferential strain (> 0) increases whereas the axial strain (< 0) decreases. The consequence is an increase in the secondary tensile stress in the circumferential direction, which can promote the fracture of the workpiece surface. Other test methods, such as rolling, bending, and plane-strain bending, have also been used to determine materials' workability [5].

Another method based on the void formation and void coalescence with a shear band was developed by McClintock for ductile fracture [6]. Thomason [7] noticed the correlation between the tensile plasitc instability of the workpiece before surface fracture and the miscroscopic observation of void growth, from which he proposed that plastic instability of the workpiece surface occurs if the following condition is met

$$\frac{d\sigma_\theta}{d\epsilon_\theta} = \sigma_\theta(2 + \frac{d\epsilon_z}{d\epsilon_\theta}) \qquad (1)$$

where σ_θ, ϵ_θ, and ϵ_z are circumferential stress, circumferential strain and axial strain, respectively. This form was adopted by Ettouney and Hardt [8] to test $A\ell$ 6061-$T6$ and obtained results similar to Kuhn's.

Recognizing that surface fracture is the consequence of some combination of tensile stress and plastic deformation, Cockcroft and Latham [9] suggested that fracture occurs when the tensile strain energy reaches a critical value, i.e.

$$\int_0^{\varepsilon_f} \bar{\sigma} \left(\frac{\sigma^*}{\bar{\sigma}} \right) d\bar{\varepsilon} = C_{cr} \qquad (2)$$

where ε_f is the fracture strain, $\bar{\varepsilon}$ is the equivalent strain, $\bar{\sigma}$ is the equivalent stress, $\frac{\sigma^*}{\bar{\sigma}}$ is the stress concentration factor due to the tensile stress σ^*, and C_{cr} is a constant determined experimentally for a given material, temperature, and strain rate.

In this paper, the Cockcroft-Latham criterion is adopted and implemented as a user's subroutine in the commercial general purpose finite element program, ABAQUS. The subroutine computes the integration in Eq.(2) and keeps tracking of those elements having experienced tensile stress histories. After reporting a failure message and the location of damage, the subroutine signals ABAQUS to terminate the analysis when the tensile strain energy reaches the critical state.

In order to determine the critical tensile strain energy for a particular material, a computer-aided experimental procedure is adopted. The procedure is based on the idea of inverse method [10] in which the difficult-to-measured quantities are identified from finite element simulations (regarded as theoretical models) of an experiment. With the material response and boundary constraints of the finite element model set as close to the test conditions as possible, the frictional behavior between the contact surfaces is adjusted and determined by an optimzation process. The converged finite element solutions provide the desired critical tensile strain energy.

Using the above procedure, the workability of several commonly-used aluminium alloys, such as 6061, 2014, and 7075, are evaluated. Verification of the program's efficacy and the applicability of the criterion is carried out by a closed-die forging operations of some purposely designed preforms. Prior to actual forging work, finite element solutions have reliably predicted possible fracture of some particular workpieces.

2. Finite Element Implementation of Fracture Criterion

The development of failure detection capability in ABAQUS, as illustrated in Figure 1, is made possible through the availability of UFCHK user's subroutine in the new release, Version 5.4. UFCHK is in the true sense an empty subroutine. However, when included in the input file, ABAQUS rewinds the output (binary) file and calls the subroutine at the end of each time increment. What the subroutine can do is completely up to the user. In this study, we first obtain the needed data, such as principal stresses and effective strain from the output file, for each finite element node location. If the largest principal stress is positive, then the following trapezoidal rule is used to compute the tensile strain energy

$$C_i = \int_0^{\bar{\varepsilon}_t} \sigma^* d\bar{\varepsilon} \approx \sum_0^{\Delta\bar{\varepsilon}_t} \frac{\sigma^*_{t-\Delta t} + \sigma^*_t}{2} \Delta\bar{\varepsilon} \qquad (3)$$

where $\bar{\varepsilon}_t$ is the effective strain of node i at time t, $\sigma^*_{t-\Delta t}$ and σ^*_t are, respectively, the largest tensile stresses at time t–Δt and t, and $\Delta\bar{\varepsilon}$ is the strain difference during the time increment Δt. The value of each node is accumulated until at a certain increment one of C_i exceeds the critical value C_c. The subroutine reports a message indicating the possibility of cracking and requests ABAQUS control to terminate the execution prematurely.

3. Computer-Aided Procedure for Evaluation of Forming Limit

The same routine shown in Figure 1 can also used to compute the critical tensile strain energy C_c for the material of interest. In this case, the program is run until the displacement reaches the same level as in the experiment right before cracking occurs.

The most common experiment for determining the forming limit is the upsetting of a cylindrical specimen [4]. Usually, etched grid lines are used to measure the strains induced on the surface of the specimen via a telescope, from which the tensile strain energy can be integrated (with the stress-strain relationship determined from a frictionless compression of a tensile test) along the strain path up to the point when the specimen fractures.

Another method is based on Bridgman's bulge correction factor [11] which allows the "true" stress be determined and thus the tensile strain energy be integrated for a typical material point on the fracture surface.

In this work, however, we propose a computer-aided experimental approach by applying finite element simulations to assist the determination of

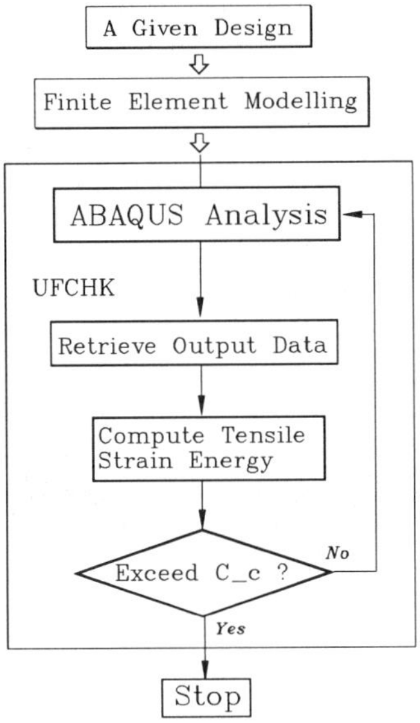

Figure 1: Implementation of cracking detection in ABAQUS program using UFCHK user's subroutine.

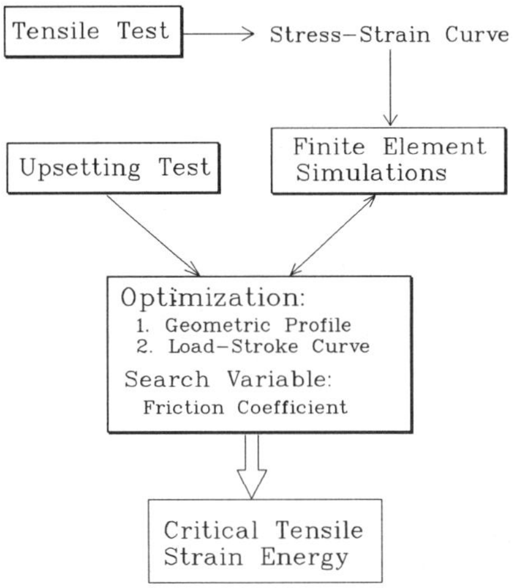

Figure 2: A computer-aided experimental procedure for evaluating material forming limits.

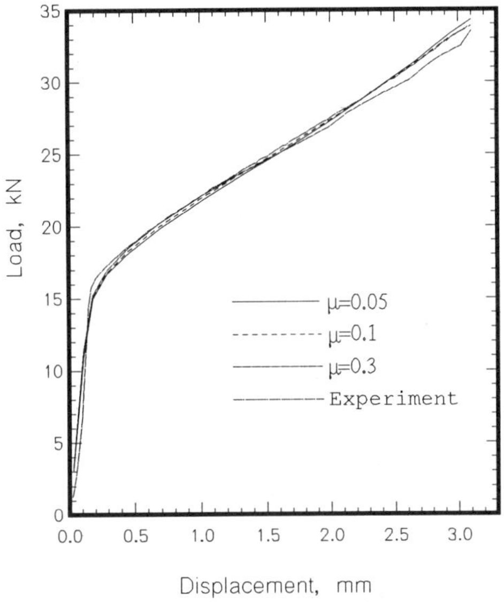

Figure 3: Load-displacement curves of an upsetting process, finite element solutions with different coefficients of friction (μ) are compared with experimental results.

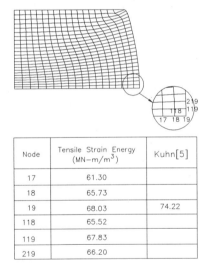

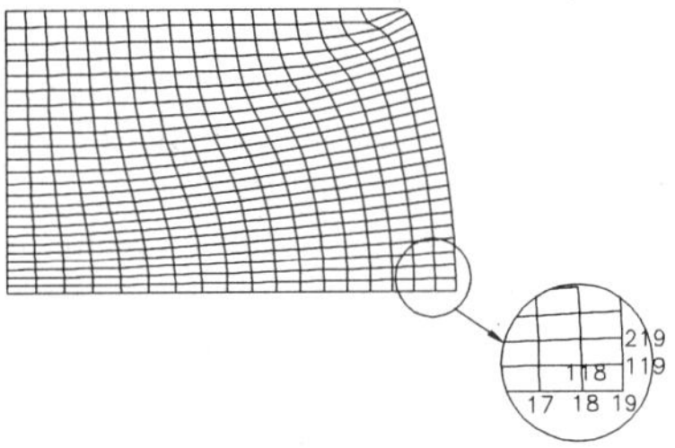

Node	Tensile Strain Energy ($MN-m/m^3$)	Kuhn[5]
17	61.30	
18	65.73	
19	68.03	74.22
118	65.52	
119	67.83	
219	66.20	

Figure 4: Finite element simulation of an upsetting process showing tensile strain energies near the equator.

the critical value, C_c. The process is illustrated in Figure 2. The stress-strain relation is obtained from a frictionless compression test or a uniaxial tensile test. In simulating an upsetting experiment, the coefficient of friction is an unknown, whose identification that yileds results matching the experiments is carried out as an optimization procedure.

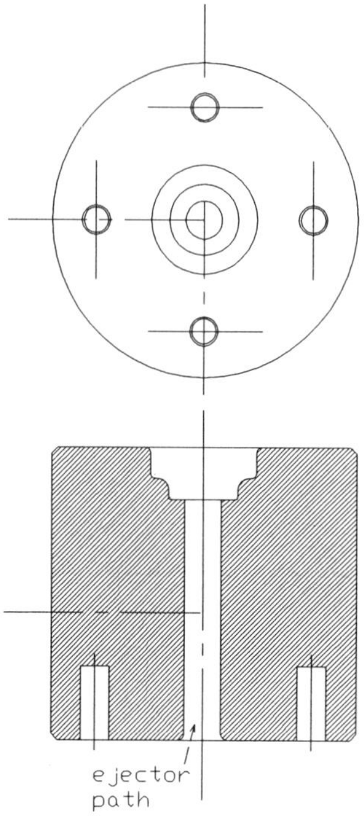

Figure 5: Lower die for an actual forging operation.

Basically two types of data from experiments are available for comparison, namely, the load versus displacement curve and the deformed shape. However, it is found that the former from finite element solution is sometimes not sensitive enough to the change of coefficient of friction to allow for a clear-cut indetification. This is shown in Figure 3. Therefore, the deformed profile, which is obtained from a least-square fit of several test results, is adopted as the optimal objective for the finite element solutions to converge to. The load-displacement relation is used only as an additional confirmation check.

The upsetting experiments were performed on an MTS material testing system. In an experiment, the loading frame exhibited a small amount of elastic deformation, which needs to be substracted from the measured displacement at the point of fracture. The integration of the tensile strain energy is carried out up to the corrected value.

Table 1 includes the diameters at the die-billet interface and the equator of the specimen at fracture. The measured data were obtained from a PJ-311 projector. The average coefficient of fric-

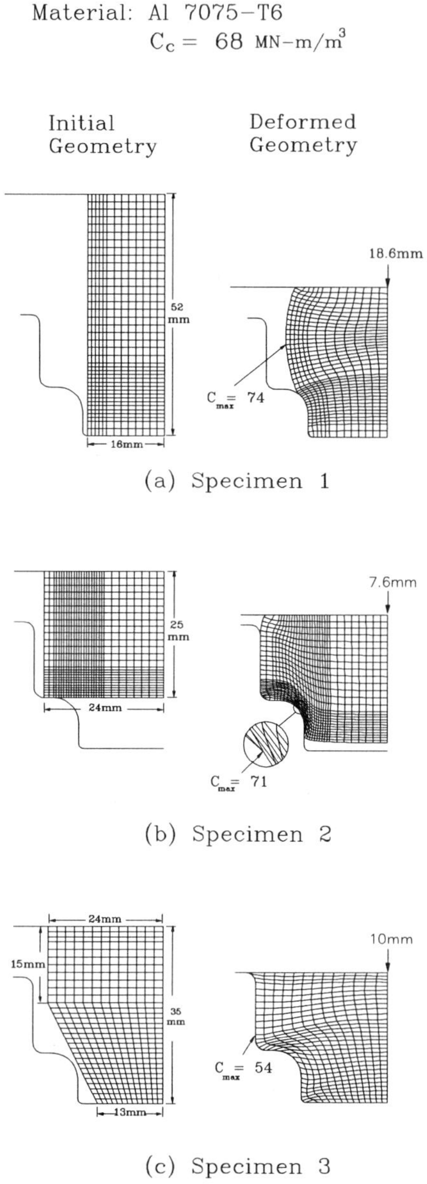

Figure 6: Finite element predictions of cold forging processes with three different preforms indicating only the last one can successfully produce the desired part.

tion were found to be close to 0.1, and are generally confirmed from the comparison shown in the table. The computed C_c for $A\ell 7075\text{-}T6$ with this coefficient of friction is about 68 MN-m/m^3 as shown in Figure 4. Only a quarter of the test specimen was modelled owing to the symmetry. Along with the listed tensile strain energies for

Figure 7: Forged parts, each corresponding to the specimen shown in Fig.6 (however, only the first one cracks).

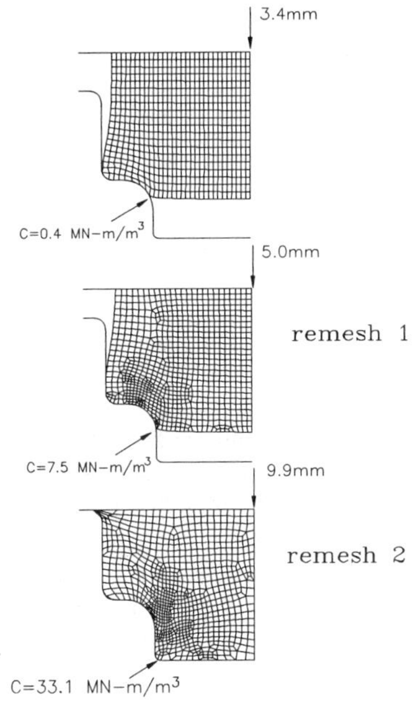

Figure 8: Finite element solutions of a forging process with automatic remeshing to overcome the element distortion problem.

the nodes around the fractured surface is the critical value computed using an approximated formula suggested by Kuhn, et al. [5].

Table 2 lists the critical tensile strain energies for 2024 and 7075 aluminium alloys as well as the maximum value for 6061 that is attainable from our testing system whose capacity is limited to 100 kN.

4. Verification of the Computer Implementation

In order to verify the proposed approach and illustrate the usefulness of the forgibility prediction, a die was designed as shown in Figure 5 to shape the part. The punch is simply a block with a flat surface. A 600 ton press was used to perform the forging operation. Both successful and failed designs were planned. Due to the high cost of the forging die, design changes were made by modifying the preforms instead of the dies. The material of the workpiece was $A\ell$ 7075-$T6$. Forging was executed at room temperature.

Before the acutal forging operation, finite element simulation of the process for each preform was performed to examine whether or not the workpiece would fracture. Figure 6 shows the analysis results of three different preforms. Due to symmetry, only half of the billet was modelled. The maximum tensile strain energies for each of the specimens included in the figure predict that the first two designs do not yield the desired product. The actual forged parts given in Figure 7, however, indicate that only the first one cracked. A careful analysis of finite element solutions of the second case revealed that the severe distortion of the element around the die corner has caused the

Table 1: Comparisons of experimental and finite element results from $A\ell$7075-$T6$ upsetting tests (length unit: mm)

	Measured	ABAQUS Results			
Friction Coefficient	?	0.05	0.1	0.15	0.2
Interface Diameter	6.99	7.247	6.972	6.811	6.649
Equator Diameter	7.838	7.724	7.835	7.892	7.914

Table 2: Critical tensile strain energies of three aluminium alloys (unit: MN-m/m^3)

Material	7075-T6	2024-T351	6061-T6
C_c	68.03	73.80	> 200

inaccurate prediction. The problem was re-run by incorporating an automatic remeshing procedure in the program [12]. The solutions with two remeshings are depicted in Figure 8 which clearly demonstrates the excellent agreement with the test result.

5. Conclusion

The incorporation of forgibility limit calculation in the finite element program can be of great value to the forging industries since the capability allows the designer to plan out the forging processes, including the intermediate annealing stages, prior to actual expensive cut-and-try cycles.

This paper described an implemention of the Cockcroft-Latham criterion in the general purpose finite element program, ABAQUS, to predict cracking during a forging simulation. A computer-aided experimental procedure, by setting up an optimization problem to select a suitable "average" coefficient of friction so that the geometric profile from the finite element solutions match the upsetting test result, was adopted to determine the critical tensile strain energy of a particular material.

Actual forging operations using a die on three different shapes of preform were performed to validate the finitc element predictions, which have generally confirmed the developments in this work.

Acknowledgement

This research has been sponsored by the Natioanl Science Council of the Republic of China under Grant No. NSC-82-0401-E-002-185.

References

1. Keeler, S. P. and Backofen, W. A., "Plastic Instability and Fracture in Sheets Stretched over Rigid Punches," *Trans. ASM* , Vol.56, 1963, p.25-48.

2. Kobayashi, S., "Deformation Characteristics and Ductile Fracture of 1040 Steel in Simple Upsetting of Solid Cylinders and Rings," *J. Eng. Ind., Trans. ASME*, Vol.92, p.391 (1970).

3. Kobayashi, S., "Workability of Aluminium Alloy 7075-T6 in Upsetting and Rolling," *J. Eng. Ind., Trans. ASME,*, Vol.98, pp.800-806 (1976).

4. Lee, P. W. and Kuhn, H. A., "Fracture in Cold Upset Forging-A Criterion and Model", *Met. Trans.*, Vol.4, p.969-974 (1973).

5. Kuhn, H. A., Lee, P. W., and Erturk, T., "A Fracture Criterion for Cold Forming," *J. Engr. Materials and Technology, Trans. of ASME*, Vol.95, p.213-218 (1973).

6. McClintock, F. A., "A Criterion for Ductile Fracture by the Growth of Holes," *J. Appl. Mech., Trans. ASME*, Vol.90, p.363 (1968).

7. Thomason, P. F., "The Use of Pure Aluminium as an Analogue for the History of Plastic Flow, in Studies of Ductile Frature Criteria in Steel Compression Specimens," *Int. J. Mech. Sci.*, Vol.10, p.501 (1968).

8. Ettouney, O. and Hardt, D. E., "A Method for In-Process Failure Prediction in Cold Upset Forging," *J. Engr. Indus.*, Vol.105, pp.161-167 (1983).

9. Cockcroft, M. G. and Latham, D. J., "Ductility and the Workability of Metals," *J. Inst. Metals*, Vol.96, pp.33-39 (1968).

10. Tanaka, M. and Bui, H. D. (Eds.), *Inverse Problems in Engineering Mechanics*, Springer-Verlag, (1993).

11. Bridgman, P.W., *Studies in Large Plastic Flow and Fracture*, McGraw-Hill, pp.9-37, 38-86, 181 (1952).

12. Cheng, J. H. and Liao, C. C., "Developing Automatic and Adaptive 2-D Remeshing Capabilities for Commercial Finite Element Programs," submitted to *Int. J. Num. Meths. Eng.*

Simulation of Materials Processing: Theory, Methods and Applications, Shen & Dawson (eds)
© 1995 Balkema, Rotterdam. ISBN 90 5410 553 4

Shape optimal design in forging

L. Fourment, T. Balan & J. L. Chenot
CEMEF, École des Mines de Paris, Sophia-Antipolis, France

ABSTRACT: Now that numerical simulation enables accurate study of the forging process, several authors have tackled the actual problem of forging, the problem of design. In this paper, we focus on the shape design for axisymmetrical problems. We describe a method of design of both the preforming tools and the initial part. We minimize the distance between the achieved and the required part. The optimization method is based on a BFGS algorithm. Shapes are discretized using spline functions. Gradients are calculated analytically. Numerical applications are discussed, for both one and two step forging operations, and for the shape optimization of both the initial part and the tools.

1 INTRODUCTION

In the field of forming by hot forging, tool design represents a large amount of the designing time. Moreover, it has a high responsibility in the quality of the final product. Actually, it is a multi-criteria study which is usually based on the industrial know-how. Recently, the numerical simulation of forging (Hans Raj et al. 1992, for instance) proved to be an efficient tool for reducing time and costs. Meanwhile, the thought process is not really modified. In the NUMIFORM'89 conference, Beckers et al. 1989 suggested to incorporate the forging software into a general purpose optimization algorithm. In the same conference, Kusiak et al. 1989 proposed to build a specific code for the shape optimization of the extrusion die. The gradients of the objective function were analytically calculated. The latter method, though more complex, appear to be more accurate and much more efficient. In Badrinarayanan et al. 1995 and in Fourment et al. 1994, the same kind of methodology was applied to non-steady processes. Meanwhile, at his level two approaches arise for the calculation of gradients. The previously quoted authors consider the optimization problem at a continuous level, differentiate the problem equations and then compute the gradients using the finite element method. Our approach is rather different. We consider the optimization problem at a discrete level and then differentiate the discrete problem equations.

2 THE FORGING PROBLEM

2.1 Axisymmetrical design problem

Usually, the forging of a part requires several operations. In this study, dedicated to axisymmetrical parts, we focus on a representative problem of the design process (figure 1). It consist of a two step forging sequence. Either the initial part has an arbitrary shape or the shape of the initial part is considered an unknown of the design problem. In the latter case, the shape is defined by a set of parameters p_1. The shapes of the preforming tools, for the first operation, are the main unknowns of the problem. They are defined by a set of parameters p_2. The shapes of the finishing tools, for the second operation, are supposed to be known. Actually, for a two steps forging process, these shapes are easily deduced from the shape of the intended final part.

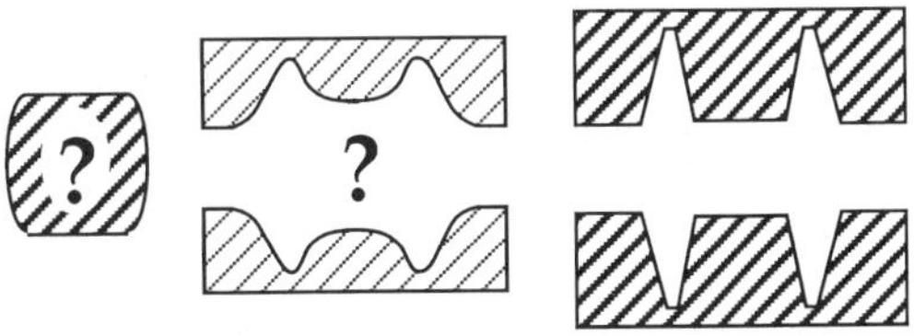

figure 1: representative problem for the design process: two step forging sequence ; the shape of the initial workpiece and of the initial toolings are unknown

2.2 Forging problem statement

For hot forging conditions, the elastic deformation can be neglected and the material is assumed to be homogeneous, isotropic, incompressible and to obey the Norton-Hoff law:

$$\mathrm{div}(v) = 0 \quad (1)$$

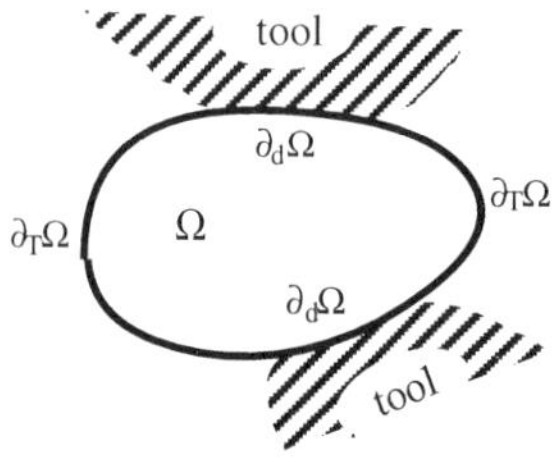

figure 2: workpiece and tool

$$s = 2K\left(2\,\dot{\varepsilon}:\dot{\varepsilon}\right)^{\frac{m-1}{2}}\dot{\varepsilon} \quad (2)$$

where v is the velocity, s is the deviatoric stress tensor, $\dot{\varepsilon}$ is the strain rate tensor, K is the consistency of the material and m is the viscoplasticity coefficient.

The tools which deform the workpiece are assumed to be rigid (figure 2). At the interface matter / tool, a power friction law is considered:

$$\tau = \sigma n\,.\,t = -\alpha K \,\|\Delta v_t\|^{q-1}\,\Delta v_t \quad (3)$$

$$\Delta v_t = (v - v_{tool})\,.\,t \quad (4)$$

where v_{tool} is the prescribed velocity of the tool, n and t are the normal and tangent of the surface of the workpiece, and α and q are the friction coefficients.

Inertia and gravity forces being neglected, at any time t, the equilibrium equations are written:

$$\mathrm{div}(\sigma) = 0 \qquad \text{on } \Omega \quad (5)$$

$$\sigma n = 0 \qquad \text{on } \partial_T\Omega \quad (6)$$

$$\begin{cases}(v - v_{tool})\,.\,n = 0 \\ \text{if } \sigma n\,.\,n < 0\end{cases} \qquad \text{on } \partial_d\Omega \quad (7)$$

$\partial_T\Omega$ and $\partial_d\Omega$ respectively represent the parts of $\partial\Omega$ upon which free-surface and unilateral contact conditions are imposed.

2.3 Finite element resolution

A weak velocity / pressure formulation of (1) and (5) is written: for any admissible velocity field v^* and any pressure field p^*,

$$\begin{cases}\int_\Omega s:\dot{\varepsilon}^*\,dw - \int_\Omega p\,\mathrm{tr}(\dot{\varepsilon}^*)\,dw \\ \qquad - \int_{\partial_d\Omega} \tau\,.\,v^*\,ds = 0 \\ \int_\Omega p^*\,\mathrm{tr}(\dot{\varepsilon})\,dw = 0\end{cases} \quad (8)$$

The finite element discretization is based on six noded quadratic triangular elements. Piecewise functions are used for the pressure interpolation.

$$x_h = \sum_k X_k N^k \quad ; \quad v_h = \sum_k V_k N^k \quad (9)$$

where N^k are the interpolation functions. A penalty method allows for eliminating the pressure variable. Then the discretized equilibrium equations are written:

$$\forall k, \qquad R_k(X,V) = 0 \quad (10)$$

$$\forall k \in \partial_d\Omega, \qquad (V_k - v_{tool}).n = 0 \quad (11)$$

Using the convenient B^k tensor, defined by equation (12), equation (10) is expanded as follows:

$$\dot{\varepsilon}_h = \sum_k V_k B^k \quad (12)$$

$$R_k(X,V) = \int_\Omega 2K\left(2\dot{\varepsilon}_h:\dot{\varepsilon}_h\right)^{\frac{m-1}{2}}\dot{\varepsilon}_h:B^k\,dw$$

$$+ \rho\sum_e\left(\int_{\Omega e} dw\right)^{-1}\int_{\Omega e}\mathrm{div}(v_h)\,dw\int_{\Omega e}\mathrm{tr}(B^k)\,dw$$

$$+ \int_{\partial_d\Omega}\alpha K\,\|\Delta v_{ht}\|^{q-1}\,\Delta v_{ht}.N^k\,ds \quad (13)$$

where ρ is a large numerical coefficient and e represents an element of the mesh. The non-linear set of equations (10,11)) is solved using a Newton-Raphson algorithm. The unilateral contact condition is imposed using the following simple algorithm: after calculation of the velocity field, for a given contact surface $\partial_d\Omega$, if the normal stress value of any node k belonging to $\partial_d\Omega$ is not compressive, then the node is released from $\partial_d\Omega$ and equations (10) are solved again with the new boundary conditions ; this is repeated until the unilateral contact condition (7) is satisfied for every node of $\partial_d\Omega$.

2.4 Time discretization and remeshing procedure

The time integration is based on a simple one-level explicit scheme. Therefore, the domain Ω_t, at time t, is updated into $\Omega_{t+\Delta t}$, at time t+Δt, according to:

$$\forall k \;,\quad X_k^{t+\Delta t} = X_k^t + V_k^t\,\Delta t \quad (14)$$

If a node penetrates the tool at time t+Δt, it is reprojected onto the tool surface. Accordingly, a node which belongs to $\partial_d\Omega$ at time t is reprojected onto $\partial_d\Omega$ at time t+Δt, in order to remain on the tool surface. Due to the large deformations of the metal during the process, the finite elements of the mesh get highly distorted. Therefore, new meshes have to be generated and an automatic mesh regenerator is used. Then the values of the state variables have to be transfered from the old mesh to the new mesh. The following simple algorithm proved to be efficient. For any node, n_{new}, of the new mesh, we search for the nearest node, n_{old}, of the old mesh and for a set of neighbouring nodes, n_{nei}, of n_{old}. For any transferable value φ, $\varphi(n_{new})$ is calculated by the weighted average of $\varphi(n_{old})$ and $\varphi(n_{nei})$.

3 THE OPTIMIZATION PROBLEM

3.1 Optimization parameters

The shapes of the initial workpiece and of the preforming tools are discretized using spline functions of the third order. The parameters $p=(p_1,p_2)$ of the optimization problem are the displacements of selected characteristic points of the spline in the normal direction (figure 3). Because the direction of the normal is not updated during optimization, the parameter definition depends upon the initial guess of the shape. Then, the spline curves are approximated by quadratic segments on the workpiece and by straight lines on the tools. It makes it possible to simplify the contact algorithm during the forging simulation.

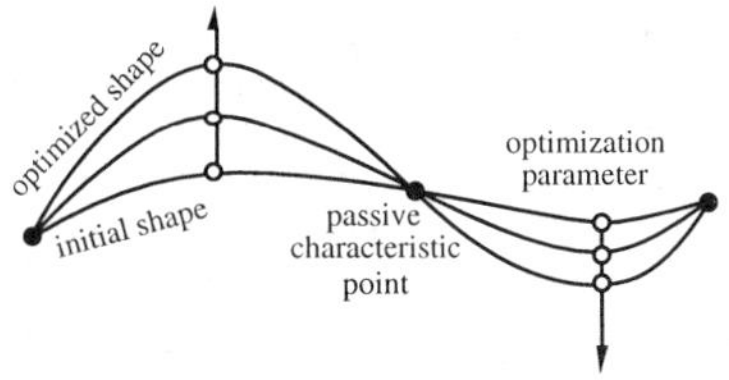

figure 3: definition of the optimization parameters for the cubic spline curve - active and passive points

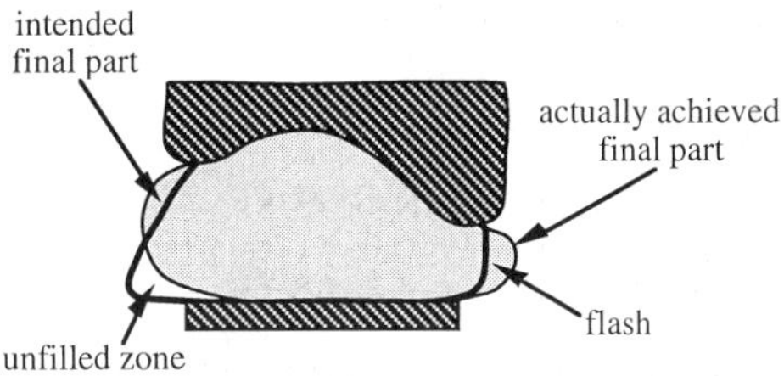

figure 4: intended shape, actually achieved shape, unfilled zone and flash

3.2 Objective functions

The main objective of forging is producing a part with the intended shape Ω_w. Then, our objective function is defined by the distance between this shape and the shape which has been actually achieved at the end of the second step of forging (figure 4):

$$\phi_d^1(p) = \int_{\partial\Omega} \| \pi(x) - x \|^2 \, ds \qquad (15)$$

where $\pi(x)$ is the projection of x onto the surface of the intended shape. Very often, there is a flash (figure 4) the shape of which has not to be optimized. Therefore, we shall not try to fit Ω_w exactly but to overlap it. So a modified objective function is introduced:

$$\phi_d^2(p) = \int_{\partial\Omega} \mathbb{1}_w(x) \, \| \pi(x) - x \|^2 \, ds \qquad (16)$$

where $\mathbb{1}_w$ is the characteristic function of Ω_w.

According to this, the optimization problem has become ill-posed and a regularization function should be introduced. The total energy of the process is chosen and then the lowest energy solution will be selected among all possible solutions:

$$\phi_e(p) = \int_0^T \left\{ \int_\Omega 2K\left(2\dot{\varepsilon}:\dot{\varepsilon}\right)^{\frac{m+1}{2}} dw + \int_{\partial_d\Omega} \alpha K \| \Delta v_t \|^{q+1} ds \right\} dt \qquad (17)$$

Therefore, two objective functions are considered:

$$\text{either:} \quad \phi_0^1(p) = \phi_d^1(p) \qquad (18)$$

$$\text{or:} \quad \phi_0^2(p) = \phi_d^2(p) + \omega\,\phi_e(p) \qquad (19)$$

where ω is a regularization parameter. ϕ_0^2 can also be described as a penalized objective function, for the minimization problem of ϕ_e under the inequality constraint $\phi_d^2=0$. In this case, $\frac{1}{\omega}$ represents a penalty coefficient.

After discretization, the objective functions are written:

$$\phi_d^1(p) = \sum_{k\in\partial\Omega} \| \pi(X_k) - X_k \|^2 L(X_k)^2 \qquad (20)$$

$$\phi_d^2(p) = \sum_{k\in\partial\Omega} \mathbb{1}_w(X_k) \| \pi(X_k) - X_k \|^2 L(X_k)^2 \qquad (21)$$

$$\phi_e(p) = \sum_{i=1}^{N} E^i(X^i,V^i)\,\Delta t^i \qquad (22)$$

$$\text{where:} \quad E^i(X^i,V^i) = \int_{\Omega^i} 2K\left(2\dot{\varepsilon}_h:\dot{\varepsilon}_h\right)^{\frac{m+1}{2}} dw^i + \int_{\partial_d\Omega^i} \alpha K \|\Delta v_{ht}\|^{q+1} ds^i \qquad (23)$$

where $L(X_k)$ is a segment length associate with the boundary node X_k and N is the total number of increments of simulation. From equations (20-22) we can notice that ϕ_0 is a function of X and V at all the time of the process:

$$j = 1, 2: \quad \phi_0^j(p) = \phi_0^j(\,X(p)\,,\,V(p)\,) \qquad (24)$$

3.3 Optimization algorithm

In short, the choice of an optimization algorithm depends on whether the gradient of the objective function can be computed or not. It is admitted that gradient methods are much more efficient, more particularly if a single evaluation of the objective function itself is quite time consuming. In this case, the BGFS algorithm (Dennis et al. 1983) seems to be high-performance. It is applied here to the minimization of ϕ_0. At each iteration, n, $\phi_0(p^{(n)})$ and $\frac{d\phi_0}{dp}(p^{(n)})$ are calculated, and $\frac{d^2\phi_0}{dp^2}(p^{(n)})$ is approximated according to the BFGS formula. A correction, $\Delta p^{(n)}$, is deduced and a line search procedure allows for finding λ such that $\phi_0(p^{(n)}+\lambda\Delta p^{(n)})$ is minimal. Finally, $p^{(n+1)} = p^{(n)}+\lambda\Delta p^{(n)}$.

4 THE SENSITIVITY ANALYSIS

4.1 Starting point

As mentioned in the introduction, our approach is based on the differentiation of the discretized problem equations. We have seen, in equation (24), that ϕ_0 is a function of $p=(p_1,p_2)$ which can be considered a function of X(p) and V(p). Therefore, the gradients of ϕ_0 can be broken down into derivatives with respect to X and derivatives with respect to V. For ϕ_0^2, for instance, and according to equations (19,21,22), $\frac{d\phi_0^2}{dp}$ is naturally broken down into:

$$\frac{d\phi_0^2}{dp}(p) = \frac{\partial\phi_d^2}{\partial X^N}(p)\frac{dX^N}{dp}(p) + \omega \sum_{i=1}^{N} \left(\frac{\partial E^i}{\partial X^i}(p)\frac{dX^i}{dp}(p) + \frac{\partial E^i}{\partial V^i}(p)\frac{dV^i}{dp}(p)\right)\Delta t^i \quad (25)$$

4.2 Incremental calculation of $\frac{dX}{dp}$ and $\frac{dV}{dp}$

Following the updated lagrangian approach which has been used in the forging problem, $\frac{dX^i}{dp}(p)$ and $\frac{dV^i}{dp}(p)$ are calculated in an incremental manner.

At increment i=0, which corresponds to time t=0:

$$\frac{dX^0}{dp_1}(p) = \frac{dX_{spl}}{dp_1}(p) \quad (26)$$

$$\frac{dX^0}{dp_2}(p) = 0 \quad (27)$$

where X_{spl} is the point of the spline curve which coincides with X. If X does not belong to the surface to be optimized or if the shape of the initial workpiece is not optimized, then the derivative is equal to 0 [1].

At increment i, which corresponds to time t, $\frac{dX^i}{dp}(p)$ is assumed to be known. $\frac{dV^i}{dp}(p)$ can be calculated after differentiation of the equilibrium equations of the discrete problem (10,11):

$$\forall k, \quad \frac{\partial R_k}{\partial X^i}\frac{dX^i}{dp} + \frac{\partial R_k}{\partial V^i}\frac{dV^i}{dp} = 0 \quad (28)$$

$$\forall k \in \partial_d\Omega_i, \quad \frac{dV_k^i}{dp} \cdot n = (V_k^i - v_{tool}) \cdot \frac{dn}{dp_2} \quad (29)$$

where $\frac{dn}{dp_2}$ is calculated after differentiation of the equations which define the normal of the spline curve. If the node X_k is not in contact with a part of the tool which is optimized, or if the tools are not optimized (for instance during the second forging operation), then the second term of equation (29) should be 0. Therefore, $\frac{dV^i}{dp}$ is a solution of a linear system (30). It can be noticed that the matrix of this system is similar to the matrix of the forging problem, which can be useful in terms of the computational cost of the gradients:

$$\frac{dV^i}{dp} = -\left(\frac{\partial R}{\partial V^i}\right)^{-1} \frac{\partial R}{\partial X^i}\frac{dX^i}{dp} \quad (30)$$

Finally, $\frac{dX^{i+1}}{dp}$ is deduced from the differentiation of equation (14):

$$\forall k, \quad \frac{dX_k^{i+1}}{dp} = \frac{dX_k^i}{dp} + \frac{dV_k^i}{dp}\Delta t \quad (31)$$

Meanwhile, if k belongs to the contact surface at time t+Δt (increment i+1), instead of equation (31) the following conditions must be written:

$$\frac{dX_k^{i+1}}{dp_2} \cdot n^{i+1} = \frac{dX_{spl}^{i+1}}{dp_2} \cdot n^{i+1} \quad (32)$$

and:

$$\frac{dX_k^{i+1}}{dp_1} \cdot n^{i+1} = 0 \quad (33)$$

but if the tool is not optimized: $\frac{dX_k^{i+1}}{dp} \cdot n^{i+1} = 0$ (34)

in any case:

$$\frac{dX_k^{i+1}}{dp} \cdot t^{i+1} = \left(\frac{dX_k^i}{dp} + \frac{dV_k^i}{dp}\Delta t\right) \cdot t^{i+1} \quad (35)$$

where n^{i+1} and t^{i+1} are respectively the normal and tangent of the tool surface at time t+Δt, X_{spl}^{i+1} is the

1 In this paper, we do not focus on the shape optimization of the initial wokpiece. We consider that it is only a secondary problem which can be used here for validating our method. Meanwhile, if large variations of the shape of the workpiece were to be considered, then the initial workpiece should be remeshed and the generated mesh would be a function of the parameter value. Therefore, the position of the internal nodes would be a function of p and equation (26) should be generalized for these nodes, which is not an easy problem.

coincidant point of the spline.

The calculation of $\frac{\partial R_k}{\partial X^i}$ has been described by Fourment et al. 1994 and will be further explained in a forthcoming paper. $\frac{dX_k^i}{dp}$ and $\frac{dV_k^i}{dp}$ are so calculated for each increment of the simulation. When the mesh becomes too distorted and when a new mesh is generated, then $\frac{dX_k^i}{dp}$ has to be transfered onto the new mesh, along with the forging state variables. The same algorithm, which has been described in section 2.4, is used.

4.3 Accuracy of the gradients

Fourment et al. 1994 have numerically studied the accuracy of the gradient calculation. The analytical values were compared to numerical values obtained by finite differences. Tests were carried out for a one step upsetting operation. The initial shape of the free-boundary of the workpiece was considered the unknown and was discretized with a spline function. The results showed that, at the end of the first increment, the relative error on $\frac{d\phi_e}{dp}$ is about 0.01%. At the end of forging, after 24 time increments, the error is about 0.7%. For $\frac{d\phi_d^2}{dp}$ it fluctuates between 0.5% and 10%, depending on the studied parameter. Therefore, the accuracy of the derivatives looked satisfactory.

5 NUMERICAL RESULTS

5.1 First academic problem of optimization of the initial shape of the part

A cylinder is upsetted between flat dies in a single operation. Sticking contact conditions are assumed (no friction). At the end of forging, the viscoplastic material bulges. We search for the initial shape of the free boundary which produces a cylinder with straight sides at the end of forging. The surface is controlled by 7 parameters of the p_1 type. The tools are not optimized. 14 iterations of the optimization algorithm are required to minimize the ϕ_0^1 objective function and to reach a relative accuracy of 10^{-6}. The tested shapes, before and after forging, are shown in figure 5.

5.2 Second academic problem of tool shape optimization

Again for a single step forging process, we tackle the problem of tool shape optimization. The initial workpiece is a straight cylinder which we do not try

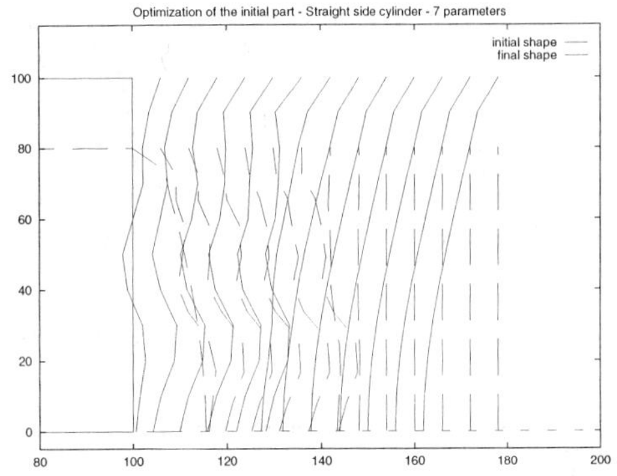

figure 5: free boundary shapes tested for the initial part design problem

to optimize. The tool shape is defined with a 4 points spline function. The academic problem is defined as follows. For an arbitrary value of the displacement of the spline points (parameter p_2) the workpiece is forged and the final shape of the part is stored. Then, it is considered as Ω_w, the intended final shape. Starting again with p_2=0, which corresponds a flat die, we look for the parameter values which make it possible to obtain again the Ω_w shape at the end of forging. In this case, there are friction, contact evolution and several remeshings procedures. 19 iterations of the optimization algorithm are required to minimize the ϕ_0^1 objective function with respect to p_2,and to reach a relative accuracy of 0.04%. Some of the tested tool shapes are shown in figure 6.

5.3 Preforming tool shape optimization for a two step forging operation

The forging of an axisymmetrical rib, from a cylindrical workpiece, is considered. Contact is assumed to be of a bilateral type. Simulation shows that the forging dies are not filled at the end of the process, if a preforming operation is not introduced. Therefore, a two step forging process is considered, and we search for the optimal shape of the preforming tool. The initial guess, a flat die, gives even poorer results. The shape of the tool is described by a 3 point spline function. Our objective function is ϕ_0^2 because a flash is allowed. Several remeshing steps are required. Amazingly, after 5 iterations of optimization, a perfect filling is achieved. Figure 7 shows the various shapes which have been tested.

6 CONCLUSION

Numerical results show the relevance of the proposed method. Firstly, gradients of the objective function

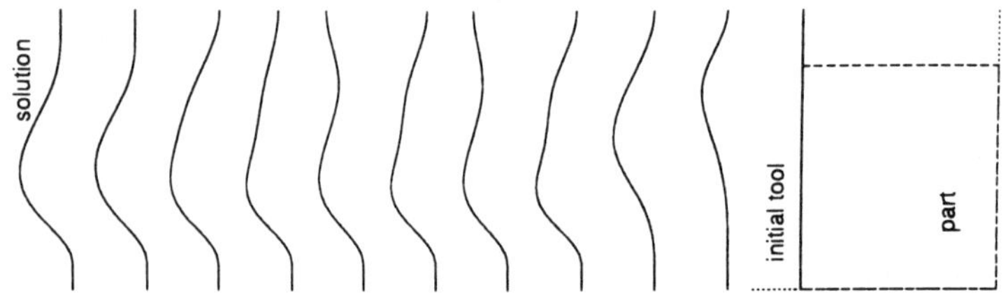

figure 6: tested tool shapes for the academic one step forging problem

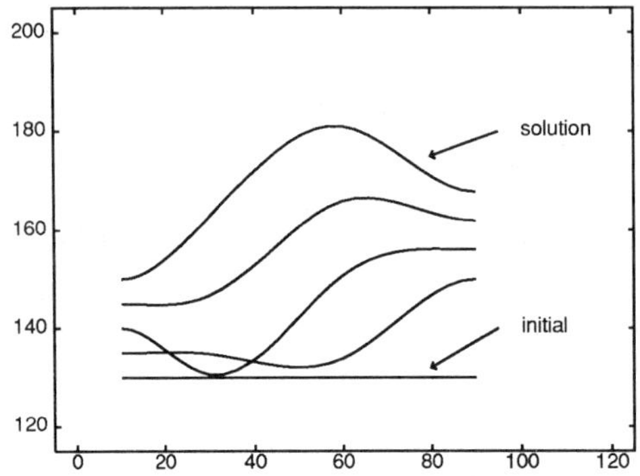

figure 7: preforming tool shapes tested for the two steps forging operation

are calculated with a satisfactory accuracy. Secondly, few iterations of the optimization method are sufficient to achieve a satisfactory solution, even when contact evolution and numerous remeshings are involved. It allows for tackling more complex industrial problems of preforming tool design.

ACKNOWLEDGEMENT

This work has been funded by the French Forging Profession and its Technical Center (CETIM).

REFERENCES

Badrinarayanan S., Zabaras N. 1995. A sensitivity analysis for the optimal design of forming processes, to appear.

Beckers M., Kopp R. 1989. A new approach to optimization of metal forming processes, *Numerical Method in Industrial Forming Processes*. 107-113. Rotterdam: Balkema.

Bohatier C., Chenot J.L. 1985. Finite element formulation for non-steady state viscoplastic deformation. *Int. J. Num. Eng.* 21:1697-1708.

Dennis J.E., Schnabel R.B. 1983. *Numerical methods for unconstrained optimization and non linear equations*. Englewood Cliffs: Prentice-Hall.

Fourment L., Balan T., Chenot J.L. 1995. Shape optimization for the forging process. *Fourth International Conference on Computational Plasticity*. Spain, Barcelonna 3-6 April.

Fourment L., Chenot J.L. 1994. The inverse problem of design in forging. *Inverse Problems in Engineering Mechanics*. 21-28. Rotterdam: Balkema.

Hans Raj K., Fourment L., Coupez T., Chenot J.L. 1992. Simulation of industrial forging of axisymmetrical parts. *Engineering Computations*. 9:575-586.

Kusiak J., Thompson E.G. 1989. Optimization techniques for extrusion die shape design. *Numerical Method in Industrial Forming Processes*. 569-574. Rotterdam: Balkema.

Simulation of Materials Processing: Theory, Methods and Applications, Shen & Dawson (eds)
© 1995 Balkema, Rotterdam. ISBN 90 5410 553 4

Constitutive parameter identification using a Computer Aided Rheology approach

A. Gavrus, E. Massoni & J. L. Chenot
Centre de Mise en Forme des Matériaux, URA CNRS, École des Mines de Paris, Sophia-Antipolis, France

ABSTRACT: An automatic parameter identification approach is developped for the analysis of thermoviscoplastic materials. First a finite element model for the tension or the torsion test is described. The inverse problem methodology is applied to our problem: giving experimental measurements on the sample, the model is able to determine the rheological parameters which allow to provide the best least square fit between experimental and computed data. The minimisation is performed with the Gauss-Newton method and the sensitivity terms are computed exactly by an analytical differentiation of the discrete equations. The method is illustrated with a tension and a torsion test example both from numerical and experimental data.

1 INTRODUCTION

The recent development of the material forming codes requires to introduce more and more sophisticated non-linear behaviour with thermomechanical coupling (Chenot 1991), microstructure evolution, and anisotropy (Amar 1993). The accurate determination of the material parameters describing the constitutive equations became a full scientific problem (Kopp 1992). In the classical approach, experimental tests (torsion, tension, compression) are analysed with a simple analytical model which allows to transform experimental results (forces, torques, etc...) into a stress-strain curve, and to determine the constitutive parameters with either a trial and error method, or an optimization procedure. However these analytical methods suffer from severe limitations: the basic hypothesis is a uniform strain (or a simple variation of the strain), and thermal coupling can be introduced only in a very approximate way. To improve the accuracy of the constitutive modelling, even when simplifying assumptions are no more valid (i.e. when large strain, large strain rate and important internal heating are present), a complex numerical simulation of the laboratory test must be used, which can cope with heterogeneous fields. The material parameters identification from rheological experiments can be formulated as solving an inverse problem (Tanaka 1992, Bui 1994, Ledesma 1991). The corresponding direct model is the computed relation of global forces, torques or sample shapes from physical parameters and the inverse formulation requires to determine the parametric vector which minimizes the deviation between computed and experimental data.

In this work we propose to apply the parameter identification principle to our finite element model (Moal 1993) able to simulate the torsion, tension or combined torsion-tension tests. An optimization module is coupled to the numerical code and a Computer Aided Rheology approach has been developed in order to identify simultaneously all the rheological parameters which describe a non-linear thermoviscoplastic constitutive law. The difference between measured and computed data is defined by a cost function formulated in the least square sense and the minimization algorithm uses a Gauss-Newton procedure. The main feature of this approach is that the sensitivity matrix, used in the iterative optimization module, is calculated in close form. Similar study is presented for viscoplastic problems by Mahnken (1994) which use only the mechanical sensitivity equation and by Gelin (1994) with a numerical evaluation of the cost function gradient. After the presentation of the numerical model, the material parameters identification principle is described. The numerical analysis and the results on the identified parameters of the thermoviscoplastic law with an anisothermal torsion and tension tests will be presented.

2 THE NUMERICAL SIMULATION MODEL

The rheological tests are simulated with a finite element method using a large strain thermoviscoplastic material behaviour (Moal 1993).

2.1 *Formulation of the thermomechanical model*

At high temperature the material is assumed to obey a general viscoplastic law (elasticity is neglected):

$$s = f(\bar{\varepsilon}, \dot{\bar{\varepsilon}}, T)\,\dot{\varepsilon} \qquad (2.1)$$

where s is the deviatoric stress tensor, $\dot{\varepsilon}$ the strain rate tensor, $\dot{\bar{\varepsilon}}$ the equivalent strain rate, $\bar{\varepsilon}$ the equivalent strain, and T the temperature.

In the case of a Norton-Hoff rheological law, with work hardening, equation (2.1) is rewritten:

$$s = 2K_1(\bar{\varepsilon}_0+\bar{\varepsilon})^n \exp(\beta/T)\,(\sqrt{3}\dot{\bar{\varepsilon}})^{m(T)-1}\,\dot{\varepsilon} \qquad (2.2)$$

where K_1 is the consistency of the material, n the strain hardening coefficient, m the strain rate sensitivity index, and β the thermal parameter.

The thermomechanical coupling appears in the rheological law through the temperature dependency of the consistency and sometimes of the rate sensitivity index m. The mechanical equilibrium is defined in a domain Ω with boundary ∂Ω divided in the subsets: $\partial\Omega_T$ where a stress vector T^d can be imposed and $\partial\Omega_V$ on which the velocity V^d is prescribed. Using the principle of virtual work (inertia forces and dilatation phenomena are neglected) the mechanical problem can be formulated approximately as finding the velocity field V which minimizes a convex functional Φ(V) (Moal 1993):

$$\begin{cases} \min\limits_{V\in \mathcal{V}} \Phi(V) \\ \mathcal{V} = \{V/\mathrm{div}(V) = 0,\ V = V^d \text{ on } \partial\Omega_V\} \end{cases} \qquad (2.3)$$

The incompressibility condition (div(V) = 0) is introduced via an augmented lagrangian formulation. The stress tensor deviator s is deduced from the Norton-Hoff constitutive law (2.2) and the hydrostatic pressure p is determined with the local equation of the mechanical equilibrium:

$$\begin{cases} \mathrm{grad}(p) - \mathrm{div}(s) = 0 \\ (s - p\mathbb{I})n = T^d \text{ on } \partial\Omega_T \end{cases} \qquad (2.4)$$

where n is the normal vector of $\partial\Omega_T$ and $\mathbb{I}$ the unity tensor.

The pressure equation (2.4) is formulated by (Moal 1993) as the minimization of a least square functional. Introducing the power dissipated through the plastic strain, the heat equation is written:

$$\rho c\frac{dT}{dt} = \mathrm{div}(k\ \mathrm{grad}T) + s : \dot{\varepsilon} \qquad (2.5)$$

where ρ is the material density, c the heat capacity, k the thermal conductivity and $\dot{w} = s : \dot{\varepsilon}$ is the rate of plastic work.

The temperature field is determined in order to satisfy the thermal equilibrium (2.5) with the following boundary conditions: the conduction with the tools or jaws on the contact surfaces, the convection and the radiation conditions on the free surfaces.

2.2 *Finite element discretization and resolution*

A spatial finite element discretization is introduced in the classical way (Moal 1993) and the unknown velocity V, the temperature field T and the unknown pressure p are expressed in terms of the nodal components. For isoparametric elements, the space coordinates X is expressed in terms of the nodal coordinate vector using the same interpolation functions.

The velocity vector which minimizes the convex functional Φ(V) must verify the non-linear equilibrium equations:

$$\frac{\partial\Phi}{\partial V}(V)=0 \qquad (2.6)$$

An iterative Newton-Raphson algorithm with a line search method is used. Finally the mechanical equilibrium (2.6) can be written:

$$\mathcal{R}(q, V, T, X, t) = 0 \qquad (2.7)$$

where q represents the set of parameters which define the viscoplastic law (2.2).

Using the Galerkin formulation we obtain the discretized form of the pressure equation (2.4) which can be reduced to the general form:

$$\mathcal{P}(q, V, T, p, X, t) = 0 \qquad (2.8)$$

Discretization of thermal equation (2.5) with the boundary conditions and using the Galerkin method, leads to an integral form of the heat transfer equation written formally:

$$\mathcal{T}(q, V, T, \dot{T}, X, t) = 0 \qquad (2.9)$$

If the domain at time t is Ω_t, the system of equations (2.7), (2.8), (2.9) is solved on this configuration with an incremental scheme. We obtain the following fundamental system which describes the direct model:

$$\begin{cases} \mathcal{R}(q, V^t, T^{t-\Delta t}, X^t, t) = 0 \\ \mathcal{P}(q, V^t, T^{t-\Delta t}, p^t, X^t, t) = 0 \\ \mathcal{T}(q, V^t, T^t, \dot{T}^t, X^t, t) = 0 \\ \dot{T}^t = \alpha\dfrac{T^{t-\Delta t}-T^{t-2\Delta t}}{\Delta t} + (1-\alpha)\dfrac{T^t-T^{t-\Delta t}}{\Delta t} \end{cases} \qquad (2.10)$$

where α is a numerical parameter characterising the thermal scheme.

The thermomechanical coupling is applied at each increment and the mechanical and thermal equations are solved separately. The temperature field is known at the beginning of each increment and is introduced in the thermomechanical equations. The velocity field is then determined by the solution of the first equation, the nodal pressure is computed with the second equation and finally we obtain the temperature field with a three level time stepping.

Giving the approximate nodal velocity vectors V^t, the new nodal positions at time $t+\Delta t$ are updated following the explicit Euler scheme:

$$X^{t+\Delta t} = X^t + \Delta t \, V^t \qquad (2.11)$$

Evolution of temperature, strain and stress are computed by this approach. Heat generation and strain localization can be simulated. Figure 1 shows that the distribution of temperature field corresponding to a tensile test for a notched sample is homogeneous only along a small area of the useful part.

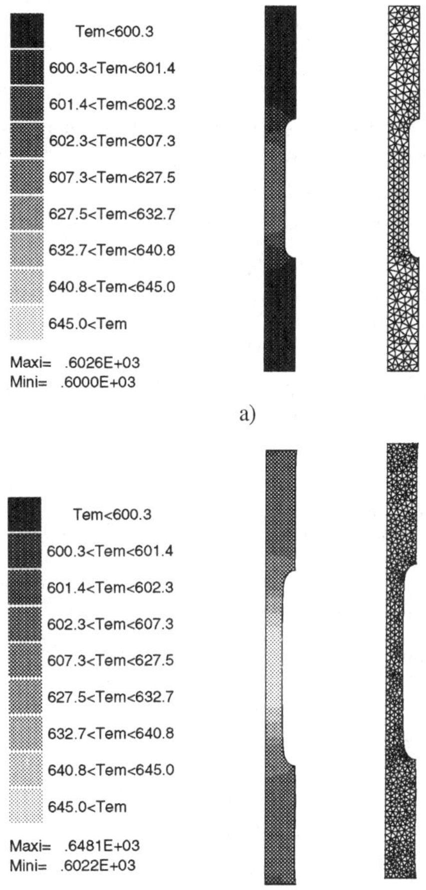

Figure 1. Temperature distribution and mesh for a tensile test of a notched specimen (speed = 10mm/s); a) initial mesh and iso-values at 0.1s; b) final mesh and iso-values at 1.1s.

An adaptative remeshing based on error estimate is absolutely necessary in large strain material forming simulation. Thus the tension simulation starts with an initial non-structural 6-nodes triangular mesh plotted in figure1-a). During the incremental computation the remeshing procedure leads to a new geometrical discretization with non-degenerate elements. An intermediate mesh with the corresponding temperature distribution is plotted in figure1-b).

The simulation code, presented above, permits also to compute the forces (for compression and tension tests) or torques (for torsion test) representing the quantities directly observable from the experimental device. The relation of these quantities can be expressed under the general form:

$$M^c(q, t) = \mathcal{F}(q, \upsilon, X, t) \qquad (2.12)$$

where υ is a compressed form for the state vector

$\upsilon = (V, T, p, \dot{T})$

3 THE IDENTIFICATION PROBLEM

A general method for systematic identification of physical parameters is presented by Beck (1977) and the solution is formulated as a determination of the parametric vector $q = (q_1, q_2, ..., q_r)$ (where r is the dimension of the parametric space) corresponding to the minimum of an objective function $Q(q, M^c, M^{ex})$ describing the difference between the experimental data $M^{ex} = (M^{ex}_1, M^{ex}_2, . . ., M^{ex}_b)$ (where b is the number of experimental data) and the computed data $M^c = (M^c_1, M^c_2, . . ., M^c_b)$ (for the same operating conditions).

In the torsion or tension test, the measured values are respectively the torques or the forces at different times, for several pilot speeds and several initial temperatures. In this case b is the number of experimental curves multiplied by the number of points per curve. For the Norton-Hoff material given by equation (2.2) the physical parameters are:
$q_1 = K_1$, $q_2 = m$, $q_3 = n$, $q_4 = \beta$.

3.1 Quantitative description of the inverse problem

The objective function is formulated with a weighting least square expression:

$$Q(q, M^c, M^{ex}) = \sum_{i=1}^{b} \beta_i \, [M^c_i(q) - M^{ex}_i]^2 \qquad (3.1)$$

where $\beta_i > 0$ represent the weighting factors. In this study β_i is chosen equal to $1/\sum_{j=1}^{b} (M^{ex}_j)^2$ for i = 1, b.
The nonlinearity of the objective function requires to use a robust algorithm to find the minimum. In this

paper a standard Gauss-Newton's algorithm has been used and it requires the following computation of the objective function derivatives:

$$\frac{\partial Q(q)}{\partial q_j} = 2(M^c - M^{ex})^T \frac{\partial M^c}{\partial q_j} \qquad j = 1, r \qquad (3.2)$$

where $\frac{\partial M_i^c}{\partial q_j} = S_{ij}$ (i = 1,b) is the term of the sensitivity matrix [S].

The principle of Gauss-Newton's algorithm is used in order to determine q that minimizes Q following an iterative procedure (Gavrus 1994). The main feature of the proposed method is the sensitivity matrix evaluation. This matrix can be estimated using numerical differentiation (Amar 1993, Cailletaud 1993, Courage 1990, Schnur 1992, Gelin 1994) but if the number of estimated parameters is large, it requires a considerable computing effort per iteration. Moreover, for large strain experiments we are interested in, remeshing is compulsory and may lead to a dramatic drop of accuracy when derivatives are obtained by a finite difference numerical approach. In this paper it is proposed the use of an analytical computation of the sensitivity matrix completely integrated in the finite element model.

3.2 *Sensitivity matrix evaluation*

In order to compute the sensitivity matrix, an analytical differentiation of relation (2.12) with respect to the parameter vector q is proposed:

$$[S] = \frac{\partial \mathcal{F}}{\partial q} + \sum_{i=1}^{k} \frac{\partial \mathcal{F}}{\partial \upsilon_i} \cdot \frac{\partial \upsilon_i}{\partial q} + \frac{\partial \mathcal{F}}{\partial X} \cdot \frac{\partial X}{\partial q} \qquad (3.3)$$

where k is the number of state variables.

The derivatives $\frac{\partial \mathcal{F}}{\partial q}$, $\frac{\partial \mathcal{F}}{\partial \upsilon_i}$ and $\frac{\partial \mathcal{F}}{\partial X}$ are obtained by analytical differentiation of $\mathcal{F}$. The main point is the determination of $\frac{\partial \upsilon_i}{\partial q}$ and $\frac{\partial X}{\partial q}$. In this study the state variables are : $\upsilon_1 = V$, $\upsilon_2 = T$, $\upsilon_3 = p$, $\upsilon_4 = \dot{T}$; the fundamental system (2.10) has to be differentiated with respect to constitutive parameters q. The following sensitivity system is obtained:

$$\begin{cases} \frac{\partial \mathcal{R}}{\partial q} + \frac{\partial \mathcal{R}}{\partial V}\frac{\partial V^t}{\partial q} + \frac{\partial \mathcal{R}}{\partial T}\frac{\partial T^{t-\Delta t}}{\partial q} = 0 \\ \frac{\partial \mathcal{P}}{\partial q} + \frac{\partial \mathcal{P}}{\partial V}\frac{\partial V^t}{\partial q} + \frac{\partial \mathcal{P}}{\partial p}\frac{\partial p^t}{\partial q} + \frac{\partial \mathcal{P}}{\partial T}\frac{\partial T^{t-\Delta t}}{\partial q} = 0 \\ \frac{\partial \mathcal{T}}{\partial q} + \frac{\partial \mathcal{T}}{\partial V}\frac{\partial V^t}{\partial q} + \frac{\partial \mathcal{T}}{\partial T}\frac{\partial T^t}{\partial q} + \frac{\partial \mathcal{T}}{\partial \dot{T}}\frac{\partial \dot{T}^t}{\partial q} = 0 \\ \frac{\partial \dot{T}^t}{\partial q} = \alpha \dfrac{\frac{\partial T^{t-\Delta t}}{\partial q} - \frac{\partial T^{t-2\Delta t}}{\partial q}}{\Delta t} + (1-\alpha)\dfrac{\frac{\partial T^t}{\partial q} - \frac{\partial T^{t-\Delta t}}{\partial q}}{\Delta t} \end{cases} \qquad (3.4)$$

with $\frac{\partial V^t}{\partial q} = 0$ on $\partial\Omega_V$

and $\frac{\partial p^t}{\partial q} = n^T \frac{\partial s^t}{\partial q} n$ on $\partial\Omega_T$.

It is important to notice that the derivatives of $\mathcal{R}$, $\mathcal{P}$ and $\mathcal{T}$ with respect to q, X and the state variables are computed analytically. If the explicit integration scheme is used the derivatives $\frac{\partial X^t}{\partial q}$, $\frac{\partial T^{t-2\Delta t}}{\partial q}$ and $\frac{\partial T^{t-\Delta t}}{\partial q}$ are known and the velocity derivative is obtained by the first sensitivity equation, the pressure derivative by the second and the temperature derivative with the last sensitivity equation. At the end of the increment $\frac{\partial X^t}{\partial q}$ is updated into $\frac{\partial X^{t+\Delta t}}{\partial q}$ taking the derivative of relation (2.11) with $\frac{\partial X^0}{\partial q} = 0$ (the initial configuration is known). Then the sensitivity matrix can be evaluated using the relation (3.3). Thus the objective function and his derivatives are simultaneously calculated as compared to the finite differences method which necessitates r+1 evaluations of $Q(q, M^c, M^{ex})$ and is often associated with numerical inaccuracy.

4 APPLICATION TO THE RHEOLOGICAL TEST

The inverse model described above permits to improve the rheological tests analysis in more complex situations: short or notched sample. In this section the constitutive parameter influence on the specimen shape evolution is neglected i.e. $\frac{\partial X}{\partial q} = 0$.

4.1 *Numerical study*

The first analysis for an identification problem are the convergence and the numerical stability. In order to achieve these studies, the identification principle is applied to an artificial experimental data numerically generated with the simulation model. To take into account the influence of the experimental measurement error, a randomly perturbed data is introduced:

$$M^c(q, t) = M^{c0}(q, t) + \varepsilon RAN(x) M^{c0}(q, t) \qquad (4.1)$$

a) The torsion test

The experimental principle of the torsion test consists in applying a rotating speed $\dot{N}$ to a specimen while recording the values of the torque C as a function of the number of revolution N. The results presented by Massoni (1994) show a very good convergence and stability for parameter identification in the case of torsion test. We shall focus now on the comparison between classical and FE (automatic) identification using an artificial experimental data. The simulation of the torsion test is made for a notched sample (L=51mm, R = 4.5mm, l = 20mm, r = 3mm) at different initial temperatures and rotation speeds (table 1) and for a material with given rheological parameters:

$K_1 = 2$ MPa, $m = 0.5$, $n = 0.5$ and $\beta = 6500$.

Table 1. The operating conditions defining the torsion test simulation.

test	a	b	c	d	e	f
rotation speed (rpm)	30	300	3000	30	300	3000
initial temperature	800	800	800	1000	1000	1000

The computed torques, perturbed with 5% random error, are used as "true experimental recordings". Introducing an initial guess of the parameters q according to : $K_1^0 = 1$ MPa, $m^0 = 0.3$, $n^0 = 0.3$ and $\beta^0 = 5000$, the identified rheological parameters obtained are plotted in table 2.

Table 2. Identification of parameters using a simulated data for an anisothermal torsion test of a notched sample.

	K_1	m	n	β	ε_r	it.
true values	2.	0.5	0.5	6500.	0.	-
estimated values	1.	0.3	0.3	5000.0	0.863	-
automatic identification	1.98	0.50	0.50	6545.5	0.015	7
analytical identification	3.64	0.47	0.29	5117.4	0.118	8

This Computer Aided Rheology method permits to find the rheological parameters with great precision: about 0.7% - 1%, compared with the analytical analysis where the level of identification error is about 6% - 80%. This difference between analytical and automatic identification is due to the fact that the analytical description can be applied only to the useful part of the specimen (r = 3mm and l = 20mm) and is not accurate enough for describing the plastic flow and the heat transfer in the whole specimen.

b) The tensile test

In the tensile test case a tension speed is applied to a specimen while recording the values of the forces F as a function of the elongation. The artificial experimental forces is obtained with a numerical simulation for a cylindrical specimen (L = 30mm, R = 10 mm) for different tension speeds, different initial temperatures (table 3) and for a given rheological parameters.

Table 3. The operating conditions defining the numerical simulation of the tensile test.

test	a	b	c	d	e
speed (mm/s)	1.	1.5	2.	1.5	2.
initial temperature	300	300	300	600	600

In table 4 identification results for two cases are plotted: with unperturbed data ($\varepsilon = 0\%$ corresponding to the convergence study) and with perturbed data ($\varepsilon = 5\%$ corresponding to the stability analysis).

Table 4. Identification of rheological parameters with the automatic procedure for the tension test.

parameters	K_1	m	n	b	ε_r
estimated values	1.5	0.2	0.1	3000.	13.32
identified (ε=0%)	1.0000	0.30001	0.2000	2000.0	$5x10^{-6}$
identified (ε=5%)	0.9729	0.30005	0.1814	2000.03	$1.5x10^{-2}$
true values	1.	0.3	0.2	2000.	0.

The nominal values of the rheological parameters is obtained with a great precision. The variation of parameters introduced by the disruption of the input data (forces) remains very small. Furthermore the convergence is obtained in few iterations (6-9) and without numerical oscillations in the iterative identification procedure (fig.2).

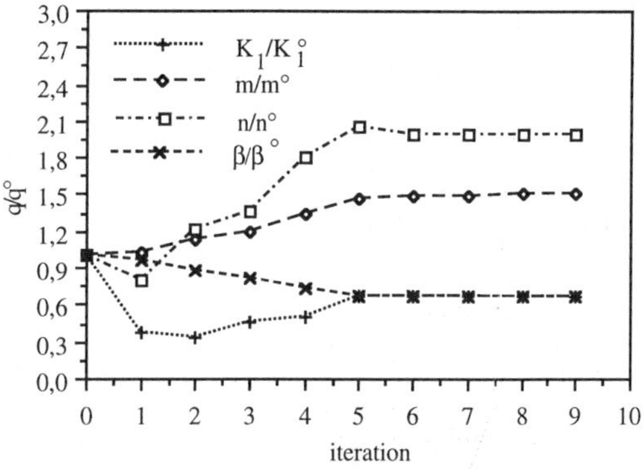

Figure 2. Parameter values variation until stagnation condition ($\Delta q/q° \leq 0.001$) in the identification procedure using unperturbated values of forces.

4.2 *Identification for a real experimental data*

A real experimental data is obtained from the torsion test of TA6V alloy for two rotating speeds (19.8 rpm and 198 rpm) and two initial temperatures (800°C and 900°C). For describing the softening phenomena characterising this material, the thermal dependency of strain rate sensitivity index: $m=m_0+m_1T$ and strain dependency of hardening coefficient: $n= n_0+ n_1\bar{\varepsilon}$ is introduced. The parametric vector q is defined, in this case by:

$q_1= K_1$, $q_2 = m_0$, $q_3 = m_1$, $q_4 = n_0$, $q_5 = n_1$, $q_6 = \beta$. The automatic identification is applied for this modified viscoplastic model and we can compute simultaneously all the rheological parameters (table 5).

Table 5. Identification of parameters using the experimental data for the torsion test of TA6V.

parameters	K_1	m_0	$m_1 10^{-3}$	n_0	n_1	β	ε_r [%]
estimated values	2.	0.1	0.	0.1	0.	5000	17.
identified values	0.2	0.28	-0.22	0.39	4.11	9189	4.

The identified parameters are introduced in the simulation code and a very good agreement between computed torques and experimental values can be observed (fig.3). The softening phenomena corresponding to the decreased values of torques can be simulated with the modified viscoplastic law.

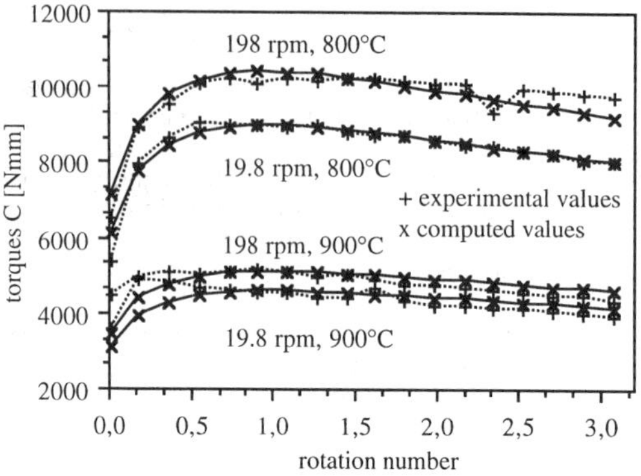

Figure 3. Comparison between experimental and computed torques for a torsion test of the TA6V.

5 CONCLUSIONS

A Computer Aided Rheology methology has been presented in this paper. A FE code, simulating the rheological tests, is coupled to an optimization module. In order to determine the constitutive parameters a cost function, formulated as an error between experimental measurement and corresponding computed data, is minimised. This numerical method permits to identify simultaneously and with a good accuracy all the rheological parameters for a given viscoplastic Norton-Hoff material. The main point of this identification strategy is that sensitivity coefficients are calculated in close form using analytical differentiation of governing equations with respect to the parameters.

The presented results demonstrates the stability of the identification algorithm which can be applied to a real experimental data.

In the future the influence of the parameters on the shape evolution will be introduced in the evaluation of the sensitivity matrix in order to extend the presented approach to the identification with the geometrical observations characterising the tension or compression tests.

REFERENCES

Amar, G. & J. Dufailly 1993. 'Identification and validation of viscoplastic and damage constitutive equations', *Eur. J. Mech., A/Solids*, 12 (2), 197-218.

Beck, J. V. & K. J. Arnold 1977. *Parameter Estimation in Engineering and Science*, John Wiley & Sons, New York.

Bui, H.D., M. Tanaka, M. Bonnet, H. Maigre, E. Luzzato, M. Reynier 1994. *Inverse Problems in Engineering Mechanics,* ISIP Symposium, Paris, Bui, Tanaka & al (Eds), Balkema.

Cailletaud, G. & P. Pilvin 1993. 'Identification, problèmes inverses: un concept modulaire', *Actes du Colloque National en Calcul des Structures*, Giens (France), pp. 770-787.

Courage, W. M. C., J. P. G. Schreurs & J. D. Janssen 1990. 'Estimation of Mechanical Parameter Values of Composites with the Use of Finite Element and System Identification Techniques', *Computers & Structures*, 34 (2), 231-237.

Chenot, J.L. 1991. 'Finite Element Modelling of Metal Forming: From Process Simulation To Physical Identification', *International Conference on Mathematical Modelling of Materials Processing,* Bristol, 23-25 sep.

Gavrus, A., E. Massoni, J.L. Chenot 1994. 'Computer aided rheology for non linear large strain thermo-viscoplastic behaviour formulated as an inverse problem', *International Symposium in Inverse Problem,* ISIP, Paris, 123-130.

Gelin, J.C.& O. Ghouati 1994. 'An Inverse Method for determing Viscoplastic Properties of aluminium alloys', *Journal of Materials Processing Technology,* 45(1), 435-440.

Kopp, R. & F. D. Philipp 1992. 'Physical parameter and boundary conditions for the numerical simulation of hot forming processes', *Steel Research*, 9, pp. 392-398.

Ledesma, A., A. Gens & E. E. Alonso 1991. 'Identification of parameters of nonlinear geotechnical models', *Computer Methods and Advances in Geomechanics*, 5, pp. 1005-1010.

Mahnken, R. & E. Stein 1994. 'Gradient-based methods for parameter identification of viscoplastic materials', *International Symposium in Inverse Problem,* ISIP, Paris, 137-144.

Massoni, E. 1994. *Identification automatique du comportement par analyse inverse*, Metal Forming and Plasticity Seminar, chapter 20, CEMEF, France, octobre.

Moal, A., E. Massoni & J. -L. Chenot 1993. 'A finite element model for the simulation of the torsion and torsion-tension tests', *Computer Methods in Applied Mechanics and Engineering*, 103, 417-434.

Schnur, D. S. & N. Zabaras 1992. 'An Inverse Method for Determining elastic material properties and a material interface', *International Journal for Numerical Methods in Engineering*, 33, 2039-2057.

Tanaka, M. & H.D.Bui 1992. *Inverse Problems in Engineering Mechanics,* IUTAM Symposium, Tokyo, Tanaka&Bui (Eds), Spinger Verlag.

Simulation of Materials Processing: Theory, Methods and Applications, Shen & Dawson (eds)
© 1995 Balkema, Rotterdam. ISBN 90 5410 553 4

Inverse identification methods for materials parameters estimation in metal forming

O.Ghouati & J.C.Gelin
Laboratoire de Mécanique Appliquée R. Chaléat, URA CNRS, Université de Franche-Comté, Besançon, France

ABSTRACT.
A sensitivity analysis is used along with an optimization process and a finite element method to establish an inverse identification algorithm for material parameters estimation in the non-linear range. The sensitivity analysis procedure is based on the direct differentiation of the scheme used in stress analysis wich is based on an hyperelastic formulation of the constitutive equations and a prediction-correction scheme with a multiplicative split for the deformation gradient. Application of this inverse identification algorithm to the determination of viscoplastic material parameters of an aluminium alloy is presented.

1. INTRODUCTION

The use of a material in metal forming requires a perfect knowledge of its behaviour. This consists on elaborating a set of constitutive equations representing physical processes occuring during the deformation. These constitutive equations involve various parameters specific to the material under study. For the so established model to be of a practical use, numerical values for these parameters have to be determined, using different measurements performed during one or several experimental tests. Very often though, measurements can not be used directly to this end but have to be converted first in some values actually involved in the constitutive equations. This conversion requires some assumptions and thus some approximations introducing errors in the identified parameters. Recently, new identification methods for material parameters estimation based on the theory of inverse problems were developed (Gelin 1995, Schnur 1992), to avoid the use of these approximations and to identify material parameters in a more accurate way. These inverse identification methods are based on optimization techniques that allow adjusting the material parameters so that the calculated response matches the measured one in a least squares sense, the calculated response being obtained by a finite element method (Gelin 1995). The optimization is performed using a Levenberg-Marquardt method, modified to account for physical constraints on the material parameters, introduced by the means of weighted penalty functions. In this optimization process, the gradient of the calculated response with respect to the material parameters have to be determined, requiring therefore a sensitivity analysis. Sensitivity analysis can be performed by three main methods: finite difference method, adjoint variables method and direct differentiation method. The first one is the most straightforward and the easiest to implement, but can also be computationally expensive and deficient in terms of accuracy and reliability. The other two methods are both based on analytic differentiation of the response as an implicit function of the model parameters. An important difference between the direct difference method and the adjoint variables method is that the direct differentiation method provides sensitivity information for the response fields at every point in the continuum, whereas the adjoint variables method only generates sensitivity information for one or more cost functions (Vidal 1991).

In this paper, we present an inverse identification method made up of an optimization method and a finite element method. A sensitivity analysis is performed using a direct differentiation method. In order to obtain highly accurate sensitivity predictions, special attention is paid to the details of the sensitivity formulation which must be fully consistent with the details of the finite element formulation and the algorithm used to calculate the stress tensor and internal variables. This algorithm is based on an hyperelastic scheme and uses a prediction-correction procedure to calculate the evolution of the internal variables (Ghouati 1994).The sensitivity formulation is obtained by direct differentiation of this scheme. A similar approach have been developed for sensitivity analysis using an implicit objective integration with a hypoelastic formulation (Leu 1994).

2. AN HYPERELASTIC BASED SCHEME FOR STRESS INTEGRATION IN THE ELASTO-VISCOPLASTIC RANGE

In this paper, an hyperelastic formulation of the constitutive equations developed by the authors (Ghouati 1994, Gelin 1994a) is used to determine the stress tensor and internal variables following a prediction-correction scheme and a multiplicative split of the deformation gradient. we summarize here the main lines of this formulation.

Under an assumption of plastic flow incompressibility, the plastic strain rate tensor is expressed as:

$$\mathbf{D}^p = \frac{3\dot{\varepsilon}^p}{2\sigma_f}\mathrm{dev}\mathbf{T} \tag{1}$$

where $\mathrm{dev}\mathbf{T}$ is the deviatoric part of the Cauchy stress tensor defined as $\mathrm{dev}\mathbf{T} = \mathbf{T} - \frac{\mathrm{tr}\mathbf{T}}{3}\mathbf{1}$ ($\mathbf{1}$ is the second order identity tensor) and $\dot{\varepsilon}^p = \sqrt{\frac{2}{3}}(\mathrm{dev}\mathbf{D}^p : \mathrm{dev}\mathbf{D}^p)^{1/2}$ is the equivalent plastic strain rate. In equation (1) σ_f stands for the flow stress of the material that depends on the equivalent plastic strain and internal variables.

It can be shown that in these conditions, the deviatoric part of the Cauchy stress tensor satisfies the following property (Ghouati 1994):

$$\frac{\mathrm{dev}\mathbf{T}}{|\mathrm{dev}\mathbf{T}|} = \frac{\mathrm{dev}\mathbf{T}^e}{|\mathrm{dev}\mathbf{T}^e|} \tag{2}$$

2.1. Elastic prediction step

For a configuration at time t_{n+1}, the total deformation gradient can be expressed in terms of the deformation gradient at time t_n and the incremental deformation gradient as:

$$\mathbf{F}_{n+1} = \mathbf{F}_u\mathbf{F}_n \tag{3}$$

where $\mathbf{F}_u = \mathbf{1} + \nabla_n u_{n+1}$.

We consider that at time t_{n+1}, for the elastic trial state, the plastic deformation gradient is fixed to be equal to the value at time t_n. This implies that we have (Ghouati 1994):

$$\mathbf{F}^{e(0)}_{n+1} = \mathbf{F}_{n+1}(\mathbf{F}^p_n)^{-1} \tag{4}$$

Considering relations (3) and (4) above, one obtains the elastic trial state for the left Cauchy-Green tensor $\mathbf{B}^e$:

$$\mathbf{B}^{e(0)}_{n+1} = \mathbf{F}_u\mathbf{B}^e_n\mathbf{F}^T_u \tag{5}$$

The choice of a logarithmic measure for the strains implies:

$$\mathbf{e}^{e(0)}_{n+1} = \frac{1}{2}\ln\mathbf{B}^{e(0)}_{n+1} \tag{6}$$

Finally, the Cauchy stress field is computed by the functional evaluation of the stress-strain relation as (Ghouati 1994):

$$\mathbf{T}^{e(0)}_{n+1} = \mathbf{C}^e : \mathbf{e}^{e(0)}_{n+1} \tag{7}$$

where $\mathbf{C}^e = K(\mathbf{1}\otimes\mathbf{1}) + 2G\left(\mathbf{I}_4 - \frac{1}{4}(\mathbf{1}\otimes\mathbf{1})\right)$, is the isotropic elasticity tensor.

2.2. Plastic correction step

The stress tensor is updated using the plastic flow by means of an Euler backward integration scheme and with the assumption that the isotropic elastic tensor remains fixed during the time increment (Gelin 1994a). The updating can be performed using the following nonlinear scalar equation:

$$\sqrt{\frac{3}{2}}|\mathrm{dev}\mathbf{T}_{n+1}| + 3G\dot{\varepsilon}^p\Delta t = \sqrt{\frac{3}{2}}|\mathrm{dev}\mathbf{T}^{e(0)}_{n+1}| \tag{8}$$

In this equation, the term $\sqrt{3/2}|\mathrm{dev}\mathbf{T}_{n+1}|$ must be replaced by the expression derived from the constitutive law.

The correction scheme described above can be performed in the elastoplastic case or the elastoviscoplastic case. In the elastoplastic case, this mechanical solution process can be defined as the radial return mapping.

3. SENSITIVITY FORMULATION

In this section, a sensitivity analysis of large deformation problems is proposed. The scheme developed is based on direct differentiation of the solution process for mechanics problem. A prime is used to denote the sensitivity of a variable.

The goal is to find the sensitivity of the Cauchy stress tensor $\mathbf{T}'_{n+1}$ at time t_{n+1} knowing the sensitivity $\mathbf{T}'_n$ at time t_n and the sensitivity of the incremental deformation gradient $\mathbf{F}'_u$, which can be expressed in terms of the sensitivity of the incremental displacement, $\mathbf{u}'$, as:

$$\mathbf{F}'_u = \nabla_n \mathbf{u}'_{n+1} \tag{9}$$

The scheme used for sensitivity analysis is similar to the mechanic solution process. First the sensitivity of the elastic prediction of the Cauchy stress tensor is evaluated, then a correction procedure is driven to compute the sensitivity of the corrected stress tensor.

For a configuration at time t_{n+1}, the sensitivity of the total deformation gradient can be expressed using equation (3) as follows:

$$\mathbf{F}'_{n+1} = \mathbf{F}'_u \mathbf{F}_n + \mathbf{F}_u \mathbf{F}'_n \tag{10}$$

The sensitivity of the elastic trial state for the left Cauchy-Green tensor is then evaluated by differentiation of equation (5):

$$\begin{aligned}&[\mathbf{B}^{e(0)}_{n+1}]' \\ &= \mathbf{F}'_u \mathbf{B}^e_n \mathbf{F}^T_u + \mathbf{F}_u [\mathbf{B}^e_n]' \mathbf{F}^T_u + \mathbf{F}_u \mathbf{B}^e_n [\mathbf{F}'_u]^T\end{aligned} \tag{11}$$

with $\mathbf{F}'_u$ given by equation (9) and $[\mathbf{B}^e_n]'$ determined form the differentiation of the expression of $\mathbf{B}_n{}^e$ as follows:

$$\begin{aligned}\mathbf{B}^e_n &= \mathbf{F}^e_n [\mathbf{F}^e_n]^T \\ [\mathbf{B}^e_n]' &= (\mathbf{F}^e_n)' [\mathbf{F}^e_n]^T + \mathbf{F}^e_n [(\mathbf{F}^e_n)']^T\end{aligned} \tag{12}$$

The corresponding sensitivity of strains are evaluated by differentiating the logarithmic measure introduced in equation (6), as follows:

$$[\mathbf{e}^{e(0)}_{n+1}]' = \frac{1}{2} [\mathbf{B}^{e(0)}_{n+1}]' [\mathbf{B}^{e(0)}_{n+1}]^{-1} \tag{13}$$

The sensitivity of the trial Cauchy stress tensor follows directly and expresses as:

$$[\mathbf{T}^{e(0)}_{n+1}]' = \mathbf{C}^e : [\mathbf{e}^{e(0)}_{n+1}]' \tag{14}$$

Once the sensitivity of the trial state is evaluated, an updating scheme similar to that used in the mechanic solution process, is performed to determine the sensitivity of the Cauchy stress tensor. This scheme concerns only the deviatoric part of the stress tensor as it is assumed that the volumetric part is unchanged:

$$\mathrm{tr}\mathbf{T}_{n+1} = \mathrm{tr}\mathbf{T}^{e(0)}_{n+1} \Rightarrow \mathrm{tr}\mathbf{T}'_{n+1} = \mathrm{tr}[\mathbf{T}^{e(0)}_{n+1}]' \tag{15}$$

In the mechanic solution process, the correction of the stress tensor is performed by solving the non-linear equation (8). For the sensitivity analysis, the corresponding equation is the following:

$$\sigma'_{n+1} + 3G [\dot{\varepsilon}^p_{n+1}]' \Delta t = (\sigma^{e(0)}_{n+1})' \tag{16}$$

with:

$$(\sigma^{e(0)}_{n+1})' = \frac{3}{2} \frac{\mathrm{dev}[\mathbf{T}^{e(0)}_{n+1}]' : \mathrm{dev}\mathbf{T}^{e(0)}_{n+1}}{\sigma^{e(0)}_{n+1}} \tag{17}$$

the difference is that this equation is non longer non-linear and can be solved in a straightforward way to assess σ'_f and $[\dot{\varepsilon}^p]'$, as following:

$$\begin{cases} \sigma'_{n+1} = \dfrac{\partial \sigma_{n+1}}{\partial \dot{\varepsilon}^p_{n+1}} [\dot{\varepsilon}^p_{n+1}]' \\ [\dot{\varepsilon}^p_{n+1}]' = \dfrac{(\sigma^{e(0)}_{n+1})'}{\dfrac{\partial \sigma_{n+1}}{\partial \dot{\varepsilon}^p_{n+1}} + 3G\Delta t} \end{cases} \tag{18}$$

Finally, the following equation is used to update the sensitivity of the Cauchy stress tensor:

$$\begin{aligned}[\mathbf{T}_{n+1}]' &= \eta'_{n+1} \mathrm{dev}\mathbf{T}^e_{n+1} \\ &+ \eta_{n+1} \mathrm{dev}[\mathbf{T}^e_{n+1}]' + \frac{1}{3} \mathrm{tr}\mathbf{T}'_{n+1} \mathbf{1}\end{aligned} \tag{19}$$

with

$$\eta_{n+1} = \frac{|\mathrm{dev}\mathbf{T}_{n+1}|}{|\mathrm{dev}\mathbf{T}^{e(0)}_{n+1}|} = \frac{\sigma_{n+1}}{\sigma^{e(0)}_{n+1}} \tag{20}$$

and

$$\eta'_{n+1} = \frac{\sigma'_{n+1} \sigma^{e(0)}_{n+1} - \sigma_{n+1} (\sigma^{e(0)}_{n+1})'}{(\sigma^{e(0)}_{n+1})^2} \tag{21}$$

4. APPLICATION TO MATERIAL PARAMETERS ESTIMATION

The solution of the parameter estimation problem involves adjusting the parameters in a finite element model until the calculated displacements, **u***, match the measured displacements, $\hat{\mathbf{u}}$, in the least squares sense. The agreement between the data and the model with a particular set of parameters is measured by a cost function. The calculated displacements, **u***, are obtained by a finite element method, using the algorithm above described to calculate the stress tensor and internal variables.

4.1. Non-linear least squares

The cost function S to minimize with respect to the parameters **m** is given by (Gelin 1995):

$$S(\mathbf{m}) = \frac{1}{2}\sum_{i=1}^{p} [\mathbf{r}_i(\mathbf{m})]^2 = \frac{1}{2}\mathbf{r}^T\mathbf{r} \qquad (22)$$

where p is the number of measurements. The residual vector, **r**, is defined by:

$$\mathbf{r} = \mathbf{u}^* - \hat{\mathbf{u}} \qquad (23)$$

Material parameters are subject to constraints, due to physical considerations. These constraints must be taken into account in the solution process, and can be written as:

$$\mathbf{C}_j(\mathbf{m}) \geq 0 \qquad j = 1, q \qquad (24)$$

where $\mathbf{C}_j$ are the constraint functions and q is the number of constraints.

The parameters are incorporated directly in the objective function, S*, as follows:

$$S^*(\mathbf{m}) = S(\mathbf{m}) + \sum_{j=1}^{q} \zeta_j(\mathbf{m}) \qquad (25)$$

where the weighted penalty functions, ζ_j, are the inverse barrier functions proposed by Carroll (1961):

$$\zeta_j(\mathbf{m}) = \frac{\omega_j}{\mathbf{C}_j(\mathbf{m})} \qquad (26)$$

with ω_j being the non-negative weights.

4.2. Parameter estimation method

The non-linear least squares problem is solved using a modified Levenberg-Marquardt method to account for the weighted penalty functions in the objective function, S* (Ravi 1990). Starting from an initial parameter guess, the modified Levenberg-Marquardt method determines a sequence of corrections to the parameters until convergence is achieved according to specified criteria. The parameter correction, $\mathbf{dm}^{(k)}$, at iteration k is calculated from the following system of equations (Gelin 1995, Schnur 1992):

$$\left[(\mathbf{J}^{(k)})^T\mathbf{J}^{(k)} + \lambda^{(k)}\mathbf{I} + \mathbf{H}^{(k)}\right]\mathbf{dm}^{(k)} = -(\mathbf{J}^{(k)})^T\mathbf{r}^{(k)} + \mathbf{g}^{(k)} \qquad (27)$$

where λ is the Levenberg-Marquardt parameter (a non-negative scalar), **J** is the jacobian matrix of S, and **g** and **H** contain the first and second derivatives, respectively, of the weighted penalty functions with respect to the parameters.

The Jacobian matrix is evaluated using the sensitivity analysis described in the previous section.

5. NUMERICAL EXAMPLES

As an application of the above described material parameters identification process, we present the determination of viscoplastic material parameters of an aluminium alloy in the hot working regime. The constitutive model used to represent such behaviour provides an expression of the flow stress as a function of the equivalent viscoplastic strain, the equivalent strain rate and the temperature, following the equation (Gelin 1994b):

$$\sigma_f(\mathbf{m}) = \sigma_0 + (\sigma_{ss} - \sigma_0)\left[1 - \exp\left(-\frac{K^2\varepsilon}{(\sigma_{ss} - \sigma_0)^2}\right)\right]^{1/2} \qquad (28)$$

where:

$$\sigma_0 = \frac{\sigma_Y}{1 + \left[\frac{\exp(Z^*)}{Z}\right]^m}, \quad \sigma_{ss} = \beta \,\mathrm{Argsh}\left[\frac{Z}{\exp(A)}\right]^{1/n}$$

and $Z = \dot{\varepsilon}^{vp} \exp\left[\frac{Q}{R\theta}\right]$.

The material parameters vector to identify is: $\mathbf{m} = [m_1, m_2, m_3, m_4, m_5, m_6] = [Z^*, m, \beta, A, n, K]$.

The inverse identification of the material parameters was performed using two different experimental tests, usually used in hot working modelling; plane strain compression test (PSC, figure 1) and axisymmetric compression test. This identification must take into account the friction coefficient present during these tests. The friction coefficient was obtained by inverse identification following the method described above.

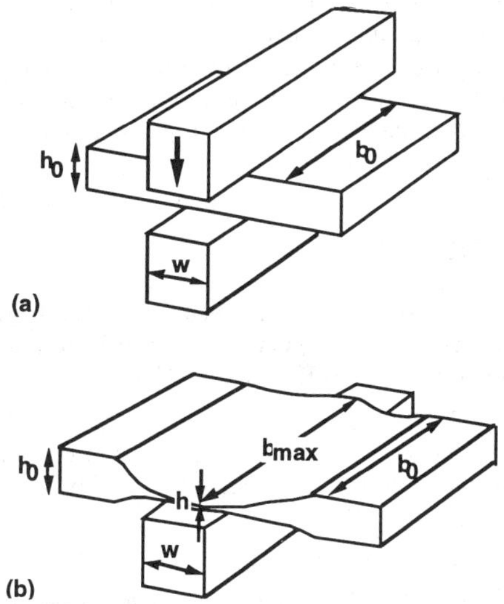

Fig.1• Schematic diagram of the PSC test (a)before (b)after deformation.

This identification takes profit from the fact that during PSC test, the lateral spread of the sample depends only on the friction coefficient (Ghouati 1994). The optimization is thus based on geometrical considerations, with the reduction of height Δh being the sollicitation and the lateral spread Δb_{max} being the response.

A similar approach can be used to identify the friction coefficient from axisymmetric compression test, where this time, the sollicitation is the reduction of height Δh and the response is the sample radius variation, ΔR.

Figure 2 represents the results at the end of the identification process for the lubricant used during the experiment (TEFLON). The value of the friction coefficient identified is μ=0.075. This value was used in the inverse identification of the material parameters.

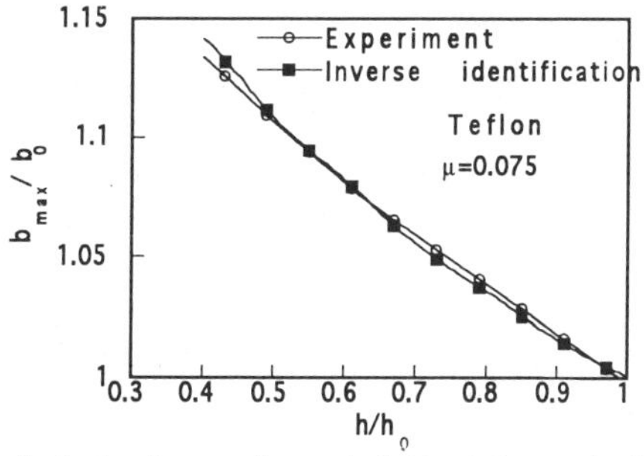

Fig.2• Lateral spread .vs. height at the end of the inverse identification of the friction coefficient.

Table 1 represents the constraints on the material parameters introduced in the solution process.

Table 1 : Parameter constraints

$Z^* > 0$ and $50 - Z^* > 0$
$m > 0$ and $1 - m > 0$
$\beta > 0$ and $100 - \beta > 0$
$A > 0$ and $50 - A > 0$
$n > 0$ and $1 - n > 0$
$K > 0$ and $500 - K > 0$

Figures 3 and 4 represent the experimental and calculated load-height reduction curves obtained at the end of the solution process, respectively for plane strain compression test and axisymmetric compression test of a 3104 aluminium alloy deformed at a temperature of 400°C and with a strain rate of 5 s^{-1}.

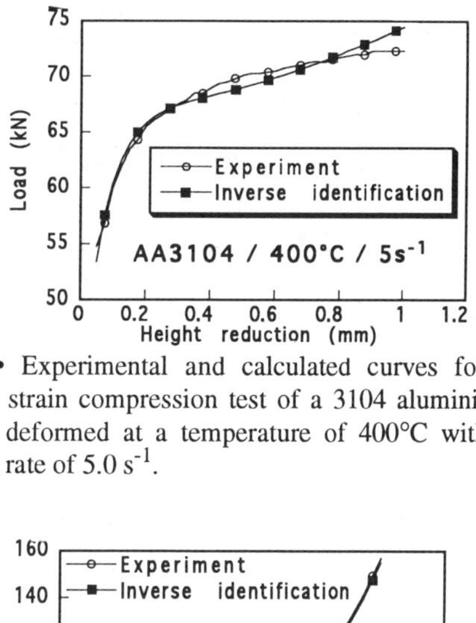

Fig.3• Experimental and calculated curves for a plane strain compression test of a 3104 aluminium alloy deformed at a temperature of 400°C with a strain rate of 5.0 s^{-1}.

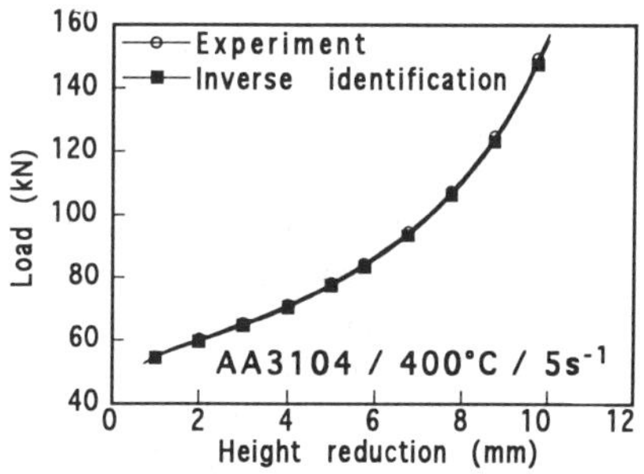

Fig.4• Experimental and calculated curves for axisymmetric compression test of a 3104 aluminium alloy deformed at a temperature of 400°C with a strain rate of 5.0 s^{-1}.

The results of the identification process for each test are reported in table 2 along with the initial guesses for the material parameters.

Table 2 : Initial and identified parameter values.

Parameter	Initial guess	PSC	Axi-sy mmetric compression
Z^*	40.0	30.7	30.65
m	0.1	0.21	0.22
β (MPa)	40.0	56.4	56.38
A	40.0	25.6	25.58
n	10.0	8.3	8.5
K (MPa)	300.0	264.8	264.78

Figure 5 respresents the evolution of the load sensitivity with respect to the different material parameters during the deformation process.

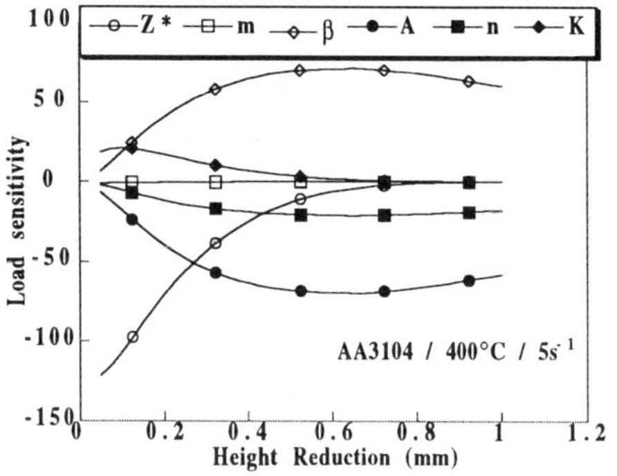

Fig.5• Load sensitivity .vs. height reduction at the end of the solution process.

6. CONCLUSION

The inverse identification method proposed in this paper allows an accurate material parameters estimation for metal forming processes regardless of the experimental test used. The procedure consists of an optimization method and a finite element method providing the response of the material to a given sollicitation. This method is shown to be very efficient when applied to material parameters estimation as it avoids any assumption concerning the deformation state. A sensitivity analysis is developed on the base of a direct differentiation method formulated in a way that is fully consistent with the algorithm of evaluation of the stress tensor and internal variables. This sensitivity analysis allows an accurate and costless determination of the gradient of the cost function with respect to material parameters.

REFERENCES

Carroll, C.W., 1961. The created response surface technique for optimizing nonlinear restrained systems, Operations Res., 9, 169-184.

Gelin, J.C., Ghouati, O. and Shahani, R., 1994a. Modelling the plane strain compression test to obtain constitutive equations of aluminium alloys, Int. J. Mech. Sci., 36, 773-796.

Gelin, J.C., Ghouati, O. and Shahani, R., 1994b. Apports de la simulation numérique pour l'identification du comportement rhéologique des alliages d'aluminium par essai de bipoinçonnement. J. Phys. III, France, 4, 685-706.

Gelin, J.C. & Ghouati, O., 1995. An inverse method for material parameters estimation in the inelastic range. To be published in Computational Mechanics.

Ghouati, O., 1994. Identification et modélisation numérique directe et inverse du comportement viscoplastique des alliages d'aluminium, Ph.D Thesis, Université de Franche-Comté.

Leu, L.J. & Mukherjee, S., 1994. Implicit objective integration for sensitivity analysis in non-linear solid mechanics, Int. Jour. for Num. Meth. in Eng., vol.37, 3843-3868.

Ravi, V. & Jennings, A.A., 1990. Penetration model parameter estimation from dynamic permeability measurements, Soil Sci. Soc. Am. J., **54**, 13-19.

Schnur, D.S. & Zabaras, N., 1992. An inverse method for determining elastic material properties and material interface, Int. Jour. for Num. Meth. in Eng., vol.33, 2039-2057.

Simo, J.C., 1992. Algorithms for static and dynamic multiplicative plasticity that preserve the classical return mapping schemes of the infinitesimal theory, Comput. Meth. Appl. Mech. Eng., 99, pp 61-112.

Vidal, C.A., Lee, H.S., Haber, R.B., 1991. The consistent tangent operator for design sensitivity analysis of history-dependent response, Comp. Syst. in Eng., 2, p. 509-523.

Simulation of Materials Processing: Theory, Methods and Applications, Shen & Dawson (eds)
© 1995 Balkema, Rotterdam. ISBN 90 5410 553 4

Inverse analysis with surface wave spectroscopy to determine material properties of thin film on substrate

Ikuo Ihara & Hideo Koguchi
Department of Mechanical Engineering, Nagaoka University of Technology, Niigata, Japan

ABSTRACT: An inversion technique is applied to the acoustic material signature of layered solids to determine material properties and thickness of layers. The inverse problem is reduced to the optimum problem of minimizing the difference between dispersion curves obtained theoretically and experimentally, where the dispersion curves are determined from the ultrasonic reflectivity of the layered solid. In this paper the basic numerical and experimental results for several layered materials are reviewed.

1 INTRODUCTION

Thin film coated materials are now widely used in the surface control engineering. In particular, hard coatings for softer surfaces are sufficiently effective to improve the strength of tool materials in the cutting and forming technologies. The strength of the coated materials is closely related to the elastic properties and the thickness of the coating film. The elastic properties and the thickness of the film deposited on a substrate are, however, difficult to measure. Robust techniques for determining the elastic properties and the thickness are not, as yet, well developed.

It is known that the velocity of surface acoustic wave(SAW) propagating over a thin film coated material depends on the materials properties of the layered structure(Brekhovskikh 1980). The frequency dependence of SAW velocity is very sensitive to the elastic properties, the density and the thickness of the film. Therefore, it is considered that SAW spectroscopy based on the velocity dispersion is a promising candidate for quantitative characterization of the thin film(Achenbach and Keshava 1967, Luukkala and Hattunen 1971, Nayfeh *et al.* 1981). Recently, many theoretical and experimental studies on the characterization of layered materials by SAW have already been made using an acoustic microscope (Kushibiki and Chubachi 1985, Weglin 1985). Inverse analyses with SAW spectroscopy for estimating thin film properties have also been studied(Karim *et al.* 1990, Kundu 1992, Kim *et al.* 1992). However, most of the studies have not been very successful in solving the inverse problem. This is mainly because the velocity dispersion data obtained by the use of a conventional acoustic microscope is insufficient for the accurate inverse analysis.

In this work, an efficient ultrasonic spectroscopic technique for accurately measuring the velocity dispersion and an inversion technique are combined to determine elastic properties, densities and thicknesses of thin films. In the present paper, to investigate the feasibility of the inversion of the present problem, experiments and analyses are demonstrated with metal and ceramic coated materials.

2 ACOUSTIC MATERIAL SIGNATURE OF LAYERED SOLID

The frequency dependence of SAW velocity in a layered half space, *i.e.*, the dispersion characteristics, have been studied extensively. Here the dispersion curve for the thin film coated material is determined from the ultrasonic reflectivity of the material. First, we present a short summary of the theoretical analysis of the ultrasonic reflection coefficient for the present problem(Chimenti *et al.* 1982, Brekhovskikh 1980).

Consider that an acoustic plane wave is injected from a liquid onto the surface of a thin elastic layer tightly bonded to an elastic half space as shown in Fig.1. To reduce the analysis to two dimensions, we assume that all particle motions in the medium are confined to the x-z plane. Also it is assumed that each of the media(layer, substrate and water) is isotropic.

For each media the potential functions Φ and Ψ are

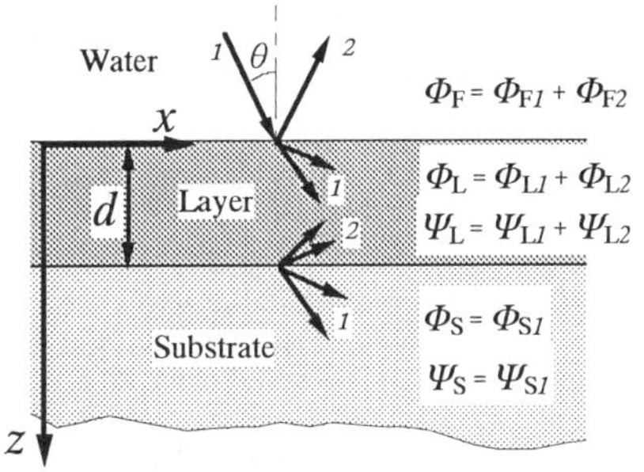

Fig. 1. Wave vectors of incident, reflected, and refracted plane waves at a layered system.

introduced, corresponding to longitudinal and transverse wave motions, respectively. The potentials Φ and Ψ are related to the particle displacements u and stresses σ by

$$\left.\begin{aligned} u_x &= \frac{\partial\Phi}{\partial x} - \frac{\partial\Psi}{\partial z} \\ u_z &= \frac{\partial\Phi}{\partial z} + \frac{\partial\Psi}{\partial x} \end{aligned}\right\} \tag{1}$$

$$\left.\begin{aligned} \sigma_{zz} &= (\lambda + 2\mu)\frac{\partial^2\Phi}{\partial z^2} + \lambda\frac{\partial^2\Phi}{\partial x^2} + 2\mu\frac{\partial^2\Psi}{\partial x\partial z} \\ \sigma_{xz} &= \mu\left(2\frac{\partial^2\Phi}{\partial x\partial z} + \frac{\partial^2\Psi}{\partial x^2} - \frac{\partial^2\Psi}{\partial z^2}\right) \end{aligned}\right\} \tag{2}$$

where λ and μ are the Lame constants which are related to the longitudinal and transverse wave velocities v_l and v_t through

$$\left.\begin{aligned} v_l^2 &= (\lambda + 2\mu)/\rho \\ v_t^2 &= \mu/\rho \end{aligned}\right\} \tag{3}$$

where ρ is the mass density, and the subscripts l and t refer to the longitudinal and transverse, respectively. Solutions of the field equations must satisfy the continuity conditions at the interfaces. The boundary conditions can be written as follows:

$$\left.\begin{aligned} &u_{Fz} = u_{Lz}, \sigma_{Fzz} = \sigma_{Lzz}, \sigma_{Fxz} = \sigma_{Lxz} = 0 && at\ z = 0 \\ &u_{Lx} = u_{Sx}, u_{Lz} = u_{Sz},\ \sigma_{Lzz} = \sigma_{Szz}, \sigma_{Lxz} = \sigma_{Sxz} && at\ z = d \end{aligned}\right\} \tag{4}$$

where the subscripts F, L and S refer to fluid, layer and substrate, respectively. Also the formal solutions for the incident and reflected wave potentials in the fluid, layer and substrate are given by

$$\Phi_F = (\phi_{F1}e^{i\alpha_{Fl}z} + \phi_{F2}e^{-i\alpha_{Fl}z})e^{i(\beta x-\omega t)} \tag{5}$$

$$\left.\begin{aligned} \Phi_L &= (\phi_{L1}e^{i\alpha_{Ll}(z-d)} + \phi_{L2}e^{-i\alpha_{Ll}(z-d)})e^{i(\beta x-\omega t)} \\ \Psi_L &= (\psi_{L1}e^{i\alpha_{Lt}(z-d)} + \psi_{L2}e^{-i\alpha_{Lt}(z-d)})e^{i(\beta x-\omega t)} \end{aligned}\right\} \tag{6}$$

$$\left.\begin{aligned} \Phi_S &= \phi_{S1}e^{i\alpha_{Sl}(z-d)}e^{i(\beta x-\omega t)} \\ \Psi_S &= \psi_{S1}e^{i\alpha_{St}(z-d)}e^{i(\beta x-\omega t)} \end{aligned}\right\} \tag{7}$$

where ϕ and ψ are unknown amplitudes, the subscripts *1* and *2* refer to incident and reflected waves, and d is the thickness of the layer, ω the angular frequency, α and β the wavenumbers for z and x components of the wavevector. The wave vector components are given by

$$\left.\begin{aligned} \alpha_{ij} &= (k_{ij}^2 - \beta^2)^{1/2} \\ k_{ij} &= \omega/v_{ij} \end{aligned}\right\} \tag{8}$$

where i=F,L,S and j=l,t; k is the wavenumber. The invariant x component of the wave vector is $\beta=k_{Fl}\sin\theta$ with the incidence angle θ and the wavenumber in the fluid k_{Fl}.

Here the reflection coefficient R is defined as the potential ratio(ϕ_{F2}/ϕ_{F1}) which is determined by solving the boundary condition equations which result from substitution of Eqs.(5), (6) and (7) with Eqs.(1) and (2) into Eq.(4).

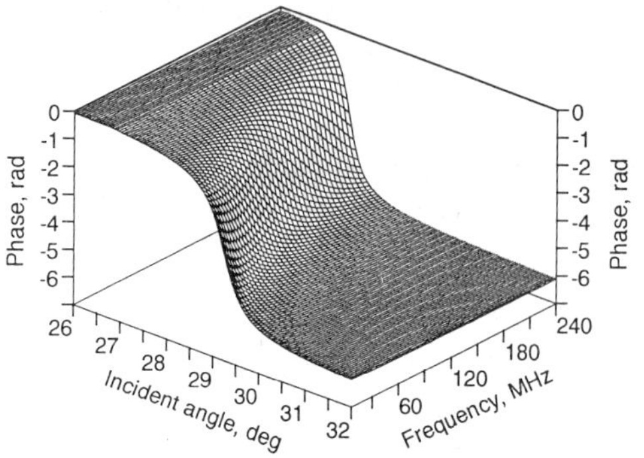

Fig.2. Calculated frequency dependence of phase of reflection coefficient as function of incidence angle, for TiN coated steel.

Fig. 2 shows the frequency dependence of the phase of the reflection coefficient calculated for TiN coated steel which is a typical cutting tool material, using bulk properties(E_L=411GPa,ν_L=0.236 and ρ_L=5300kg/m^3 for TiN; E_S=220GPa,ν_S=0.28 and ρ_S=8050kg/m^3 for steel, where E and ν are the Young's modulus and the Poisson's ratio, respectively). The phase curve at each frequency in Fig.2 is also indicated as a function of the incidence angle. In Fig.2 we can recognize a phase shift of nearly 2π rad at each frequency. Because the critical angle θ_c at which leaky surface acoustic wave(LSAW) is exited, are determined as the angle at which the slope of the phase curve is steepest, LSAW velocity v_{sw} at each frequency is then given by the Snell's law as

$$v_{sw}(f) = v_w/\sin\theta_c(f) \tag{9}$$

where v_w is the velocity of sound in the fluid, and f the frequency. Thus we can determine the dispersion characteristic of LSAW velocity. The dispersion in a multilayered structure can also be determined by using the modified above procedure.

3 INVERSE PROBLEM

Fig.3 illustrates the influence of the Young's modulus of the TiN on the LSAW dispersion curve for the TiN coated steel, where the normalized parameter fd(the product of the frequency f and the layer thickness d) is used as the abscissa. It is seen that the Young's modulus contributes much to the change of the dispersion curve. The dispersion curve is also sensitive to the density and the thickness of the TiN. These suggest the possibility of the prediction of layer properties utilizing the dispersion curve. Here the inverse problems are defined as the following two cases:

(1) For a given single layered system, estimating simultaneously the elastic constants and density of the layer from the dispersion curve, where the substrate properties and the layer thickness are known.
(2) For a given two-layered system, estimating the

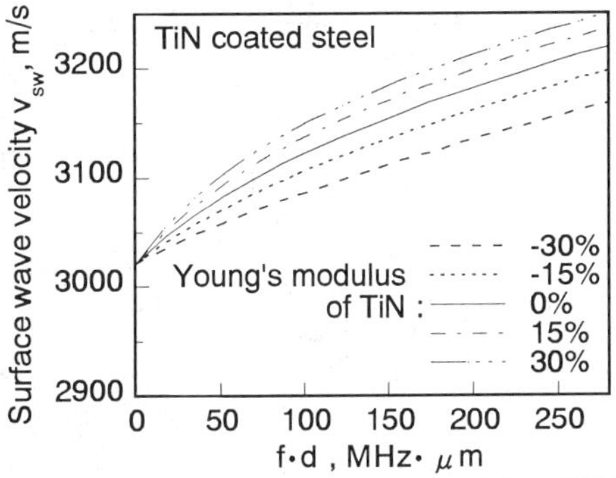

Fig. 3. Influence of change of Young's modulus of TiN on dispersion curve of TiN coated steel.

thickness of each layer from the dispersion curve, where material properties of the substrate and the layers are known.

The inverse problems are reduced to the optimum problem which determines a set of parameters that corresponds to the minimum of the mean error function r defined as

$$r = \frac{1}{N}\sum_{k=1}^{N} W(k)\left|v^*_{sw}(k) - v_{sw}(k)\right| \tag{10}$$

where $v^*_{sw}(k)$ and $v_{sw}(k)$ are discrete data of the calculated and the true(measured) dispersion curves, respectively, N is the number of data points, and $W(k)$ is weighting coefficient assigned to each data point. In the present analysis, $W(k)$ is taken as unity. The inversion procedure is summarized in Fig.4. The Complex method(Box 1965) which had been developed from the Simplex method of Spendley *et al.* (Spendley *et al.* 1962) is applied to the inverse analysis. This method, starting with 2n(n is the number of unknown parameters) feasible points which is given

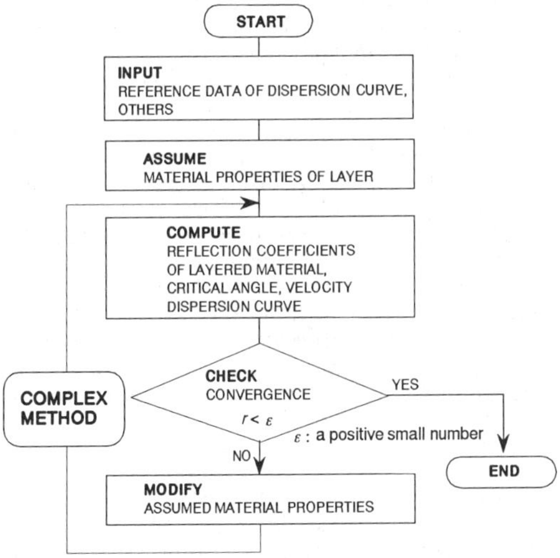

Fig. 4. Flow chart of the inverse analysis.

by using random numbers, determines the optimum within constrained regions by a typical descent method.

To investigate the stability and the accuracy of the solutions in the problems, the behaviors of the error function r in the solutions space to be explored are examined. Figures 5 and 6 show the contours of r in the vicinity of the true values for the case(1) and case(2), respectively. In Fig. 5(a) and Fig. 6, no local minimum point are observed except for the minimum which agrees with the true value. In such cases,

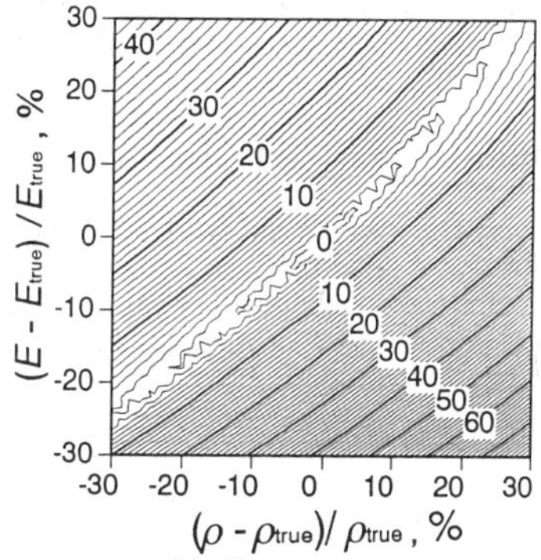

(a) $E - \rho$ space

(b) $E - \nu$ space

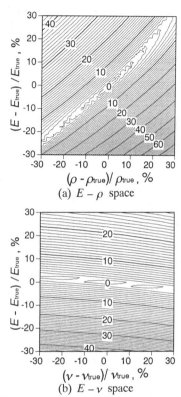

Fig.5. Countour plots of r for TiN/steel system, where N=20, fd range=20–300MHz·μm, and r=0 indicate true values.

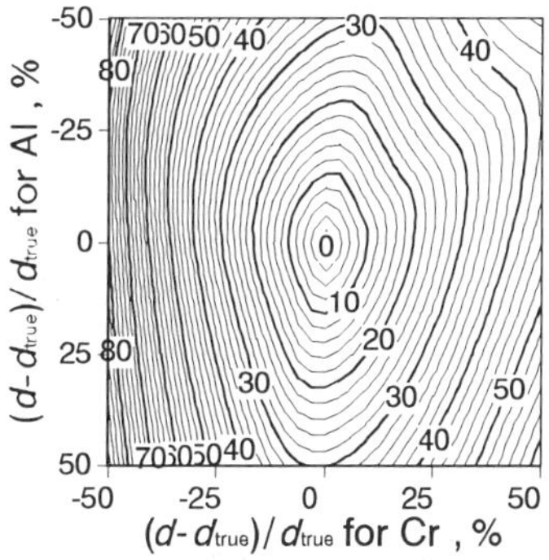

Fig.6. Countour plot of r for Al/Cr/glass system, where N=20, f range=0–200MHz, d_{Al}=2μm, d_{Cr}=2μm, and r=0 indicates true values.

therefore, it is considered that a stable solution may be obtained by the use of a suitable inversion technique. In the case of Fig.5(b), however, it must be difficult to find the stable solution for ν because of the presence of the long and narrow minimum area in the solution space. Using these contour plots, we can estimate the accuracy of the solutions. For the case(1), when r=1m/s, from Fig. 5(a) the estimation errors in Young's modulus and density are presumed to be about ±6% and ±7%, respectively. For the case(2), when r=1m/s, from Fig. 6 the estimation error in each thickness is presumed to be within ±2%. It should be noted here that the shapes of the contours depend on the range of fd or f in the dispersion curve used in the estimation.

4 NUMERICAL SIMULATION

A program for inverse analysis is tested, for the problem of the case(1), by processing simulated dispersion curve from theoretical calculation with a known set of parameters. The model is TiN coated steel with properties mentioned above. The conditions of the calculation are as follows: incidence angle 26– 31° with 0.01° step, fd 20–260MHz·μm, N 20 points, and constraints(upper and lower limits of the range to be searched) are within ±30% of a given true value for each parameter.

At first, the inverse analyses were carried out with ε of a sufficiently small value , *e.g.* ε=0.01m/s. It has been confirmed in this case that the estimated properties of E, ν and ρ agreed with their true values within errors of about 2%. Then, it took about 500 iterations before converging. Next, considering the measurement accuracy of the velocity, ε was set to 1m/s. The inverse analyses were independently repeated 25 times. Figure 7 shows all the obtained results which are expressed in the errors relative to true values. As predicted from Fig. 5, it can be seen from Fig.7 that E and ρ can be estimated with accuracy of about ±6% and ν can be hardly identified by the analysis. For the problem of the case(2), it was also checked according to the same procedure as above.

5 EXPERIMENTAL APPARATUS

The reflection coefficients were measured by an

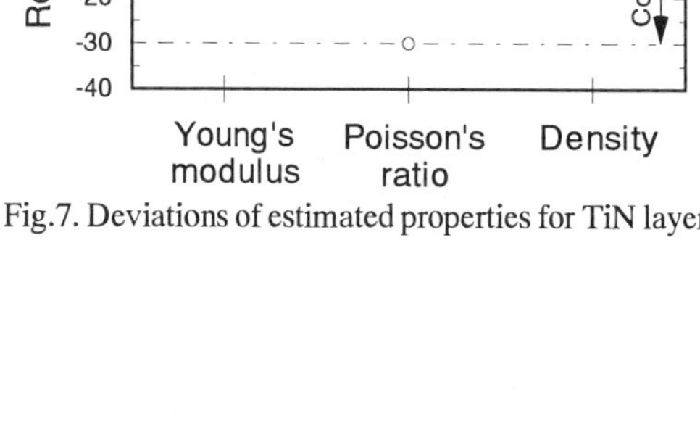

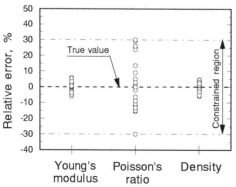

Fig.7. Deviations of estimated properties for TiN layer.

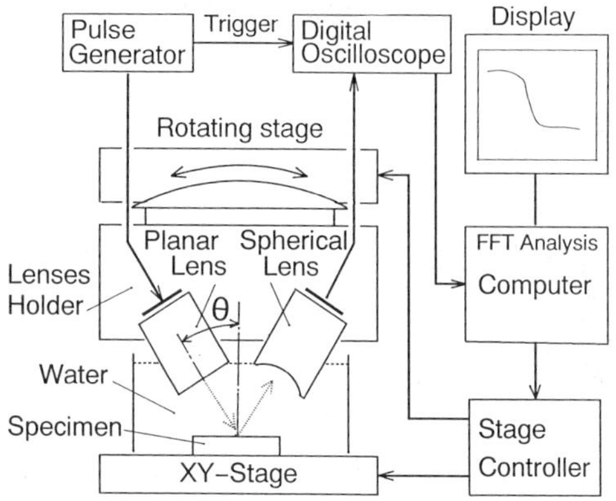

Fig.8. Block diagram of an ultrasonic spectrometer.

ultrasonic spectrometer(USM)(Ihara *et al.* 1994, Nakaso *et al.* 1994). Fig. 8. shows a block diagram of the USM. In the USM, broadband impulsive waves at incidence angle from 26 to 38° with 0.05° step are injected to the specimen surface through water. Applying Fourier transform analysis to the reflected waves, the reflection coefficient at the liquid-solid interface is measured as a function of frequency in the range from 20 to 130MHz. The dispersion curve of LSAW velocity is obtained from the phase curve at each frequency, as described above. The measurement accuracy of the velocity was established to be about 0.02%(Ihara *et al.* 1994). This means that the deviations of measured values at each frequency are within ± 1m/s.

6 EXPERIMENTAL RESULTS

6.1 Elastic properties and density of ceramic film

Two kinds of ceramic coated steels were prepared: TiN coated steel and TiC coated steel. TiN layer was formed on the tool steel(JIS type SKH51: E=213GPa,ν=0.288 and ρ=7830kg/m^3) by physical vapor deposition (PVD) with Ar+N_2 gases at 723K and 2x10^{-5}Torr. TiC layer was formed on the tool steel(JIS type SKD11: E=216 GPa,ν=0.288 and ρ=7800kg/m^3) by chemical vapor deposition(CVD) with H_2+CH_4 gases at 1293K and 560Torr. After the ultrasonic measurements were completed, the specimens were cross-sectioned and the layer thicknesses were measured by the optical microscope. Thicknesses of the TiN and TiC layers were about 3.3μm and 2.0μm, respectively.

The reflection coefficients for the specimens were measured by the USM. Figs. 9 shows the LSAW dispersion curves determined from the reflection coefficients for these specimens, respectively. A slight fluctuation is observed in the dispersion curve as shown in Fig. 9. This fluctuation was reproducible, and thought to be due to the mechanical distortion and the acoustic lens characteristic of the USM. In order to eliminate the errors resulting from these system characteristics, the dispersion curves were smoothed

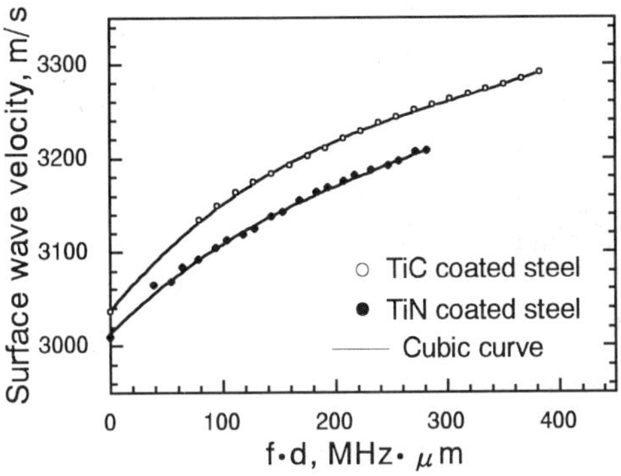

Fig. 9. Dispersion curves measured for TiN coated steel and TiC coated steel

Table 1. Estimated elastic constants and densities.

	Young'smodulus (GPa)	Poisson's ratio	Density, (kg/m^3)
TiN	359.2±8.86	0.235±0.0417	4828.5±116.9
TiC	540.3±54.9	0.225±0.0443	4706.5±579.4

by fitting a cubic curve to the measured points using the least squares method. We should notice that the smoothing is convenient but it has no theoretical background.

Using the smoothed dispersion curves, the elastic constants and densities of TiN and TiC layers were determined from the inverse technique discussed in Sec.3 and Sec.4. The inverse analyses were independently repeated 10 times. Table 1 shows the 95% confidence limits of the estimated values as $s \pm 1.96\sigma$, where s is the mean value and σ is the standard deviation. For the TiN coated steel, the inversion was successful: the mean error function r converged to a small value enough to obtain accurate estimation results(r<0.8m/s), before finishing the analysis. Thus, the Young's modulus and density were precisely estimated as expected from the results of numerical simulations shown in Sec.4. However, for the TiC coated steel, the deviations of the estimated values were larger than that for the TiN because the error function r converged insufficiently (1<r<1.7m/s). It should be noted that the TiC coated steel used here may violate the layered model assumed in the calculation: the TiC layer is not always homogeneous and isotropic.

6.2 Layer thickness dependence of dispersion curve

In order to investigate the effect of layer thickness on the LSAW dispersion curve, two kinds of coating specimens are employed: Au coated glasses with layer thicknesses of 0.08, 0.15, 0.30 and 0.72μm and TiN coated steels with layer thicknesses of 3.0, 5.3 and 9.0μm. Au layers were formed on the crown glasses (E=69.6GPa, ν=0.22 and ρ=2550kg/m^3) by DC magnetron spattering with Ar gas at 3x10^{-3} Torr. TiN layers were formed on the tool steels(JIS type SKD62: E=220GPa, ν=0.28 and ρ=8050kg/m^3) by PVD. Layer thicknesses were changed by adjusting the depositing time.

Figure 10 shows the dispersion curves measured for Au coated glasses and TiN coated steels. In each figure, line shows the theoretical result calculated with bulk properties of the materials. It is seen from Fig.10(a) that measured dispersion curves agree well with each other and they also correspond to the theoretical result. This suggests that the material properties of Au films used here are independent of the film thickness, and the properties of the films are almost the same as bulk properties of Au. The dispersion curves for the TiN coated steels shown in Fig. 10(b), on the other hand, have somewhat different shapes for different film thicknesses. This reveals that the material properties of the TiN films depend on the film thickness. It seems reasonable to suppose that the density of TiN film is independent of the film thickness because the thin layer has been grown under the same and stable conditions. Therefore, assuming that the value of density is fixed and is known(ρ_{TiN}=5300kg/m^3), the Young's moduli of the TiN films were estimated from the inversions of dispersion curves shown in Fig.10(b) according to the same procedure as described in Sec.6.1. The estimated Young's moduli are shown in Table 2 as average values obtained from five analyses. It is presumed from Fig.5(a) that errors in the estimated values are about ±10% because the

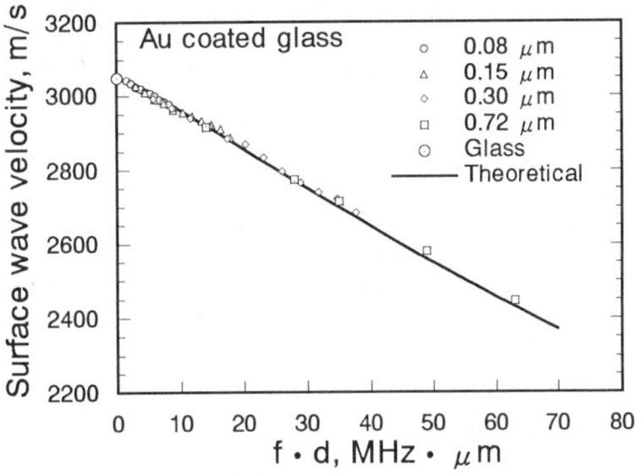

(a) Au coated glass

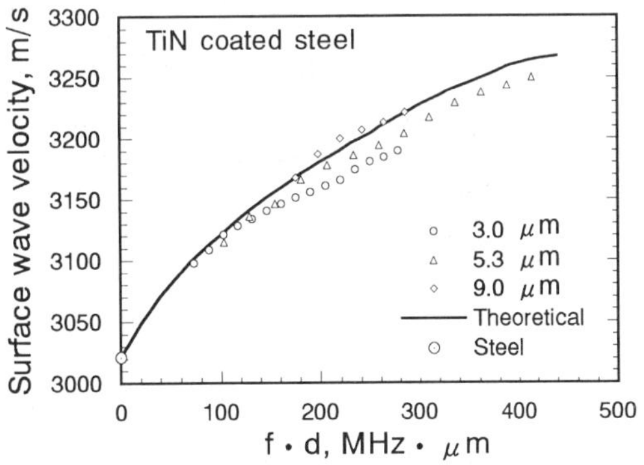

(b) TiN coated steel

Fig. 10. Dispersion curves measured for various TiN coated steels and Au coated glasses.

error function r after converging are 1.2–2.6m/s. Also, it should be noted that if the anisotropy in the layer medium is strong, an anisotropic model for the layered media should be incorporated into the inverse analysis. Nevertheless, the elastic constants estimated here may be a useful parameter for characterizing the average elastic property of thin layers.

Table 2. Estimated Young's moduli of TiN layers.

Coating thickness (μ m)	Young's modulus (GPa)
3.0	329.2
5.3	391.4
9.0	421.6

6.3 Layer thicknesses in a two-layered coating

In order to estimate the thickness of individual layers in a two-layered coating, the inverse analysis is carried out. The specimen is a glass with 3.5μm-thick coating. The coating is made of 1.3μm-thick aluminium over a 2.2μm-thick chromium layer. The layers were formed on the glass(Corning7059:E=67.5GPa, ν=0.28 and ρ=2760kg/m^3) by DC magnetron spattering with Ar gas at 7×10^{-3} Torr.

Figure 11 shows the dispersion curve measured for the two-layered specimen. We can see that the measured result agree well the theoretical result calculated with the bulk properties of the materials and the true values of thicknesses. Using the smoothed dispersion curve by the least squares method as mentioned in Sec.6.1, the inverse analyses were carried out on conditions of N=21, f=50–140MHz. The inverse analyses were independently repeated 10 times. Table 3 shows the 95% confidence limits and the relative errors($(d_{true}-d_{mean})/d_{true}$) of the estimated results. In this case, the error function r converged to the values of about 2.2m/s which was larger than those for ceramic coated steels in Sec.6.1; nevertheless, the deviations of the estimated values were relatively small as expected from Fig.6.

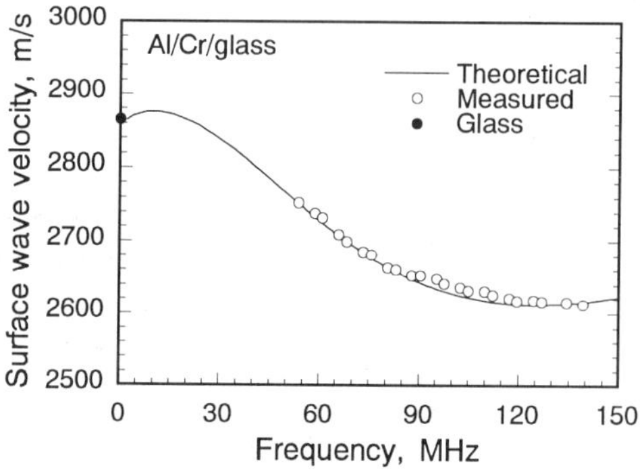

Fig. 11. Measured dispersion curve for Al/Cr/glass. Line shows the theoretical result.

Table 3. Estimated thicknesses of Al and Cr layers.

	Aluminium	Chromium
Thickness(μ m)	1.370±0.0268	2.112±0.0105
Relative error(%)	5.4	-4.0

7 CONCLUSION

It is reviewed how material properties and thicknesses of thin layer on a substrate can be obtained by inverting the LSAW dispersion curve. The complex inversion algorithm is incorporated for the purpose. It is also shown that a prior knowledge about the behaviors of the error function r in the solutions space is sufficiently effective to investigate the stability of solutions. This inverse technique will be an efficient way of quantitatively estimating materials properties and thicknesses of single and multiple layers.

ACKNOWLEDGMENT

We are grateful to Professor J. Kihara and Professor T. Aizawa of University of Tokyo and Dr. Y.Tsukahara and Mr. N.Nakaso of Toppan Printing Co.,Ltd for their helpful discussions. We also acknowledge Mr. K.Sasaki of Toshiba Tangaloy Co.,Ltd and Professor Y.Mutoh and Mr. K. Tokura of Nagaoka University of Technology for supplying the specimens. This research was partly supported by a Grant-in-Aid (No.05555175) for Scientific Research from the Ministry of Education, Science and Culture of Japan.

REFERENCES

Achenbach, J.D. and Keshava, S.P. 1967. Free Waves in a Plate Supported by a Semi-Infinite Continuum. *J. Appl. Mech.* June: 397-404.

Box, M.J. 1965. A New Method of Constrained Optimization and a Comparison with Other Methods. *Computer Journal.* 8-42:42-52.

Brekhovskikh, L.M. 1980. *Waves in Layered Media.* NewYork: Academic.

Chimenti, D.E. *et al.* 1982. Leaky Rayleigh Waves on a Layered Halfspace. *J. Appl. Phys.* 53-1: 170-176.

Ihara, I. *et al.* 1994. Microscopic Determination of Surface Wave Velocities in Heat Treated Steels by Ultrasonic Reflectivity Measurement. In Green, R.E. *et al.* (eds), *Nondestructive Characterization of Materials VI*: 87-94. NewYork: Plenum.

Ihara, I. *et al.* 1994. Approach for Estimating Material Properties of Thin Film on Substrate by Inverse Analysis with Surface Wave Spectroscopy. *Trans. Jpn. Soc. Mech. Eng.*,60-579: 2664-2671 (in Japanese).

Kushibiki, J. and Chibachi, N. 1985. Material Characterization by Line-Focus-Beam Acoustic Microscope. *IEEE Trans.* SU-32-2: 189-212.

Karim, M.R. *et al.* 1990. Inversion of Leaky Lamb Wave Data by Simplex Algorithm. *J. Acoust. Soc. Am.* 88-1: 482-491.

Kundu, T. 1992. Inversion of Acoustic Material Signature of Layered Solids. *J. Acoust. Soc. Am.* 91-2: 591-600.

Kim, J. O. *et al.* 1992. Elastic Constants of Single-Crystal Transition-Metal Nitride Films Measured by Line-Focus Acoustic Microscopy. *J. Appl. Phys.* 72-5: 1805-1811.

Luukkala, M. and Hattunen, M. 1971. Observation of Possible Leaky Plate Waves in a Plating Using Ultrasonic Reflectivity Measurements. *Appl. Phys. Lett.* 19-3: 56-57.

Nayfeh, A.K. *et al.* 1981. Ultrasonic Leaky Waves in the Presence of a Thin Layer. *J. Appl. Phys.* 52-8: 4985-4994.

Nakaso, N. *et al.* 1994. Measurement of Acoustic Reflection Coefficients by an Ultrasonic Microspectrometer. *IEEE Trans. Ultrason. Ferroele. Freq. Contr.*,41-4: 494-502.

Spendley, W. *et al.*1962. Sequential Application of Simplex Designs in Optimization and Evolutionary Operation. *Technometrics.* 4:441.

Weglin, D.D. 1985. Acoustic Micro-Metrology. *IEEE Trans.* SU-32-2: 225-234.

Simulation of Materials Processing: Theory, Methods and Applications, Shen & Dawson (eds)
© 1995 Balkema, Rotterdam. ISBN 90 5410 553 4

Tooling and binder design for 3-D sheet metal forming processes using springback calculations

Mary C. Boyce
Department of Mechanical Engineering, Massachusetts Institute of Technology, Cambridge, Mass., USA

Apostolos P. Karafillis
Department of Mechanical Engineering, Massachusetts Institute of Technology, Cambridge, Mass., USA
(Presently: General Electric Company, Research and Development Center, Schenectady, N.Y., USA)

ABSTRACT: Part shape error due to springback can be a manufacturing defect in sheet metal forming processes. This manufacturing problem can be corrected by using appropriate designs of the tooling and the binder shape together with a binder constraint. In the present work a methodology for tool and binder design based on inverse springback calculations is developed and demonstrated. The design procedure invokes finite element analysis of the manufacturing process. The proposed method was demonstrated for the case of forming of a three-dimensional channel geometry with an aluminum alloy sheet. The tooling needed to form the desired part was numerically designed using the proposed algorithm then constructed with the aid of CNC machining. The designed tooling was found to produce the desired part shape demonstrating both the accuracy of the numerical simulations and the success of the proposed tooling design algorithm.

INTRODUCTION

Sheet metal forming processes involve a combination of elastic-plastic bending and stretching deformation of the workpiece. These deformations can lead to large amounts of springback of the formed part upon unloading (e.g. removal of the punch and lifting of the blankholder). Springback has been examined as a manufacturing problem by a number of investigators, (for example Ayres [1], Liu [2], Stevenson [3], Wenner [4], Karafillis and Boyce [5]). Various methods for the reduction of the amount of springback have been proposed by Liu [2] and Ayres [1]. The purpose of these methods is to minimize the deviation of the produced part shape from the tooling shape which is identical to the desired part shape. Hardt and coworkers [6,7] and Karafillis and Boyce [5,8] developed tooling design methods to reduce the springback part shape error by tooling shape compensation. These methods were confined to two-dimensional shapes and/or sheet metal forming cases with negligible amounts of draw-in of the deformed blank. In the present method we present a method which can be used to design tools for three-dimensional geometries, compensating for part shape errors due to springback. Details on this method can also be found in Karafillis and Boyce [9].

THREE-DIMENSIONAL SPRINGBACK CALCULATIONS

The use of explicit dynamic finite element analysis to model sheet metal forming processes was suggested by Honecker and Matiasson [10]. In the case of large three-dimensional forming operations, the explicit dynamic analysis may render the simulation substantially faster as compared to the implicit static analysis, see also Matiasson et al. [11], and Rebelo et al. [12]. In the present work we use dynamic explicit analysis to simulate sheet metal forming during the punch loading motion, ABAQUS [13]. The internal stress distribution of the blank at the fully loaded stage was then transfered to a static implicit analysis in order to simulate the unloading of the blank due to the removal of the tool surfaces. Details on this formulation can be found in Karafillis and Boyce [9].

TOOLING AND BINDER DESIGN USING SPRINGBACK CALCULATIONS

The occurrence of large amounts of springback causes considerable shape deviation from the tooling shape, see also Karafillis and Boyce [5,8]. If shape fidelity is required from the produced part of a sheet metal forming process, the correct part shape can be obtained by appropriate modifications of the die and the binder shape.

In this section, we present a tooling design procedure, developed by Karafillis and Boyce [9], that compensates for part shape error due to springback. The considered forming processes are general three-dimensional cases of sheet metal forming processes with large amounts of draw-in in order to contain the developed stretching strains. The formed parts are considered in the fully unloaded configuration.

We now consider an initial finite element analysis of a stamping process, where the tooling shape is identical to the desired part shape. The shape of the blank in the finite element model is described by the displacement vector u_{l1}, whereas the material stress-strain response causes the development of the internal force vector I_{l1}. As the tooling is identical to the desired shape we will obtain $u_{l1} = u_{des}$, where u_{des} is the displacement vector of the desired shape. Springback occurs upon removal of all contact forces $F_{l1} = I_{l1}$, and the new displacement vector u_{ul1} of the blank is:

$$u_{ul1}(u_{l1}, I_{l1}) = u_{l1} + u_{sb1}(u_{l1}, I_{l1}) \quad (1)$$

where u_{sb1} is the displacement increment vector during springback. Also springback can be considered as the additional deformation described by u_{sb1} which occurs due to the application of $-F_{l1}$ on the blank in the fully loaded state and configuration resulting in a zero net load applied to the blank.

We now consider a process where the blank is deformed to a different shape u_{l2} which will springback upon unloading to a new shape u_{ul2}:

$$u_{ul2}(u_{l2}, I_{l2}) = u_{l2} + u_{sb2}(u_{l2}, I_{l2}) \quad (2)$$

We would like to have u_{ul2} to be identical to the desired shape:

$$u_{ul2} = u_{des} \quad (3)$$

where recall that

$$u_{l1} = u_{des} \quad (4)$$

giving the desired outcome of

$$u_{ul2} = u_{l1}. \quad (5)$$

In this case, u_{l2} would define the tooling shape that will produce the correct part shape, $u_{ul2} = u_{des}$. By rearranging Eqn. (2), we have:

$$u_{l2} = u_{ul2} - u_{sb2}(u_{l2}, I_{l2}) \quad (6)$$

By combining Eqns. (5) and (6) we obtain:

$$u_{l2} = u_{l1} - u_{sb2}(u_{l2}, I_{l2}) \quad (7)$$

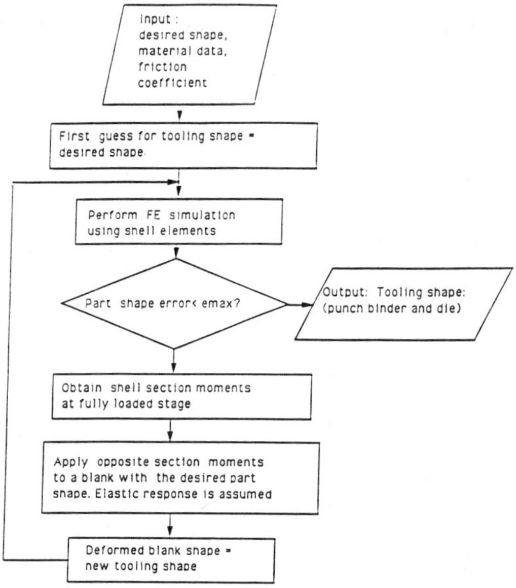

Fig. 1: Flowchart of the tooling design algorithm.

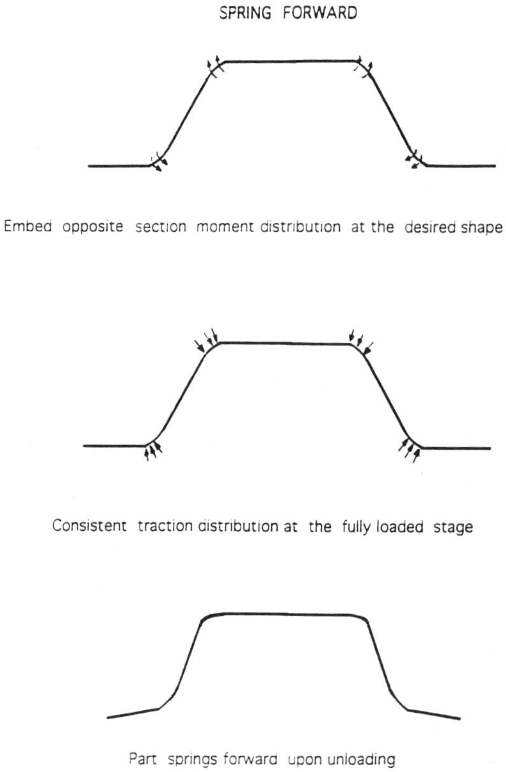

Fig. 2: Schematic graph of the "springforward" simulations.

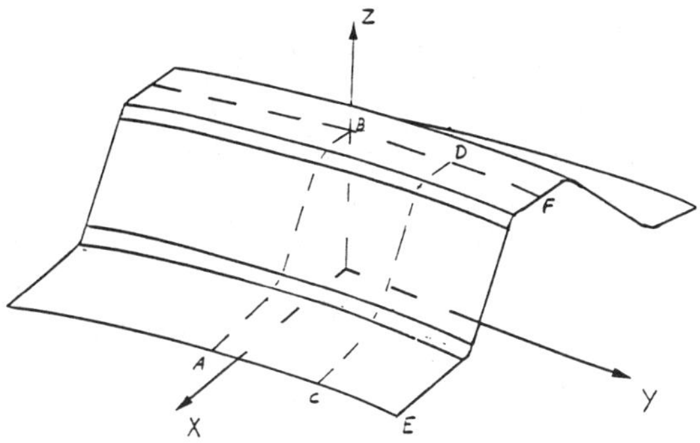

Fig. 3: The three-dimensional channel part.

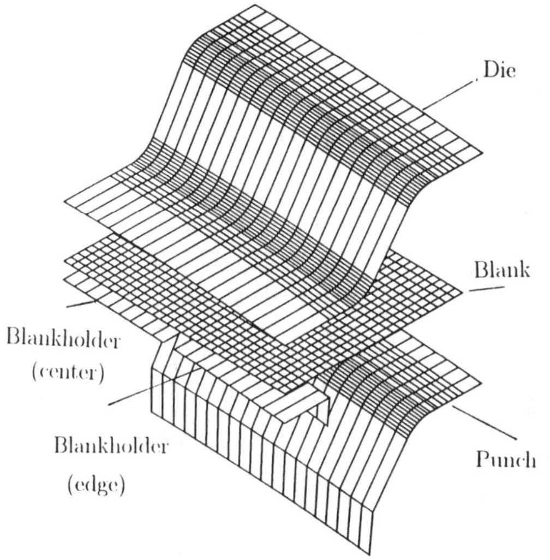

Fig. 4: Finite element model of the forming tools and the blank.

In order to obtain u_{l2}, we can perform an inverse simulation of springback where a blank defined by the displacement vector u_{l1} will be subject to the additional deformation $-u_{sb2}$. However, since we do not know $-u_{sb2}$, we can attempt to calculate u_{l2} in an iterative manner, until the part shape error is negligibly small. This procedure is effected by using the algorithm of Fig. 1. In order to perform the inverse springback simulation, we consider a blank at a shape defined by u_{l1}. We also consider a finite element analysis step where we constrain all the degrees of freedom of the nodes of the blank to remain constant, while simultaneously embedding into the material the internal forces $I_I^* = -I_{l1}$. The presence of the internal forces I_I^* results in the transmission of the external force vector $F_I^* = -I_{l1}$ to the nodes of the blank due to the constrained nodal degrees of freedom. The subsequent release of the nodes "springs forward" the blank to a new equilibrium position u_{l2}^* in a manner opposite from the one taking place during springback unloading, due to the opposite direction of applied loads, see also Fig. 2. However, all nodal loads are reduced to zero in the new equilibrium position u_{l2}, and therefore the development of high unbalanced bending moments is avoided.

The new deflected shape defined by u_{l2}^* is the first iteration on the designed tooling shape. If the part shape u_{ul2}^*, obtained after unloading of u_{l2}^*, is within an acceptable part shape error, then u_{l2}^* is the die shape which will produce the correct part shape. Otherwise we can repeat the inverse springback simulation, by loading the blank defined by u_{l1} with the internal force vector I_{l2} obtained by forming the blank with tools defined by u_{l2}^*. This procedure can be repeated until the part shape obtained shows negligible part shape error, see also the flow chart of Figure 1.

One problem encountered when using this method was the possible development of structural instabilities in the material, due to the application of $-I_{l1}$ to the blank. The internal force vector I_{l1} contains bending and membrane forces of the blank in a sheet metal forming process. Therefore if any tensile membrane forces are present, the application of $-I_{l1}$ causes the development of compressive membrane forces which may lead to fictitious structural level buckling instabilities. In order to ensure that buckling instabilities due to compressive membrane forces do not develop we decompose I_{l1} to the force vector obtained by the integration of the shell membrane section forces I_{l1m}, and the shell section bending moments I_{l1b}. As also observed and shown in an earlier section of this paper, the contribution of I_{l1m} is not important to springback phenomena. We therefore considered only the force vector $-I_{l1b}$ in our inverse springback calculations thereby excluding possible development of fictitious instabilities in the blank due to compressive membrane forces.

AN APPLICATION OF THE TOOLING DESIGN ALGORITHM

The tooling design algorithm developed in the previous section was used for the design of a part with a three-dimensional channel geometry. We will now demonstrate how our tooling design algorithm was used in order to design the forming tools that would produce the desired part.

We consider the forming process of the part depicted in Fig. 3. This part possesses a double-curvature three-dimensional geometry.

The FE model of this part and the tooling set-up is shown in Fig. 4, where only a quarter of the model is depicted. Due to the large size of the problem and to the complicated geometry of the tools, we modeled the loading portion of the forming operation by using dynamic explicit analysis only.

We note here that after preliminary dynamic explicit finite element simulations of the process we concluded that a divided binder was needed, see Fig. 4, in order to prevent excessive draw-in at the edge of the blank , (section E-F of Fig. 3). A vertical binder force of 1 KN was applied at the center segment of the binder whereas a vertical binder force of 10 KN was applied at the edge segment of the binder. A friction coefficient of 0.4 was assumed *a priori* for the interaction between the binder and the blank.

In the first cycle of our tooling design procedure, the tooling shape matched the desired part shape, see Fig. 5 (a), (b), (1st cycle). The obtained part shape, upon the occurrence of springback possessed a certain shape error, see also Figs. 6 (a), (b), (1st cycle). In order to reduce the part shape error, we modified the die shape following the algorithm of Fig. 1. Sections of the new die shape are depicted in Figs. 5 (a), (b), (2nd cycle). Note the changes in the binder area depicted in Fig. 5 (a). The tooling design compensates for part shape error due to springback by changing the tooling shape in both the punch area and the binder area. Sections of the part produced with the corrected tooling shape upon the occurrence of springback, as obtained by finite element analysis, are shown in Figs. 6 (a), (b) (2nd cycle). We see that the produced part is nearly identical to the desired part shape. Therefore, our tooling design algorithm has converged to the desired part shape in only one iteration.

An experimental verification of this tooling design was also performed by effecting the forming operation of a 2008-T4 aluminum blank with the designed forming tools. The tooling shape obtained from the finite element analysis, was first offset and interpolated and then transformed to G-code for CNC machining. The shape offset prior to the CNC machining was equal to the machining tool radius $\pm\frac{1}{2}$ thickness of the blank. Details

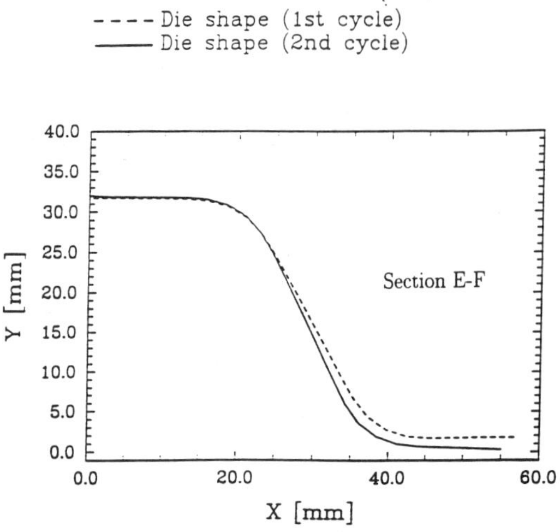

(a)

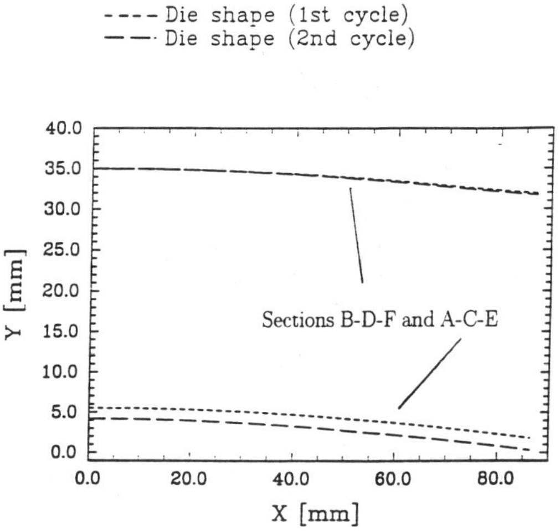

(b)

Fig. 5: Tooling shapes for the different cycles of the toolling design algorithm; (a) Section E-F, (b) Section A-C-E and B-D-F. Section names as in Fig. 3.

on the machining of the tools and the experimental procedures for the three-dimensional channel forming operation can be found in [14].

The produced part shape was measured in a Brown-Sharpe Coordinate Measuring Machine. The obtained section shapes are depicted in Figs. 6(a), (b), (2nd cycle, experiment). We see that the experimentally obtained parts are in very good agreement with the theoretical analysis, see Figs.

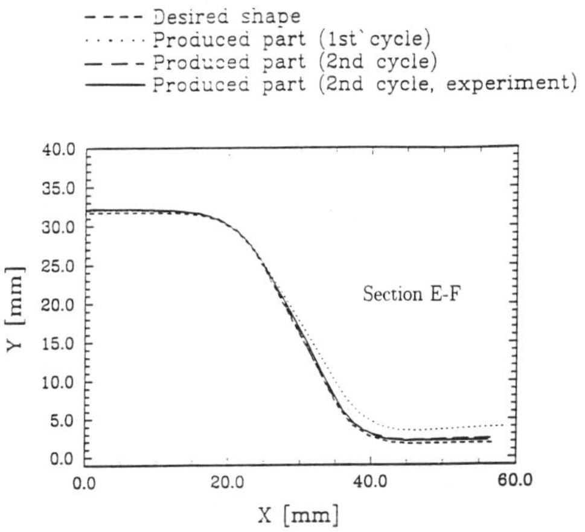

(a)

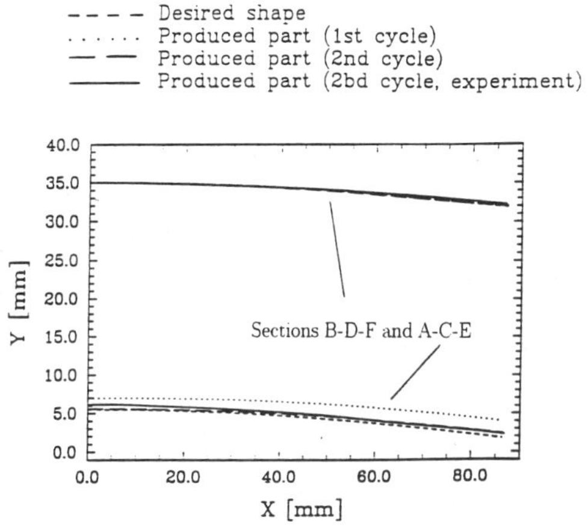

(b)

Fig. 6: Part shapes for the different cycles of the tooling design algorithm; (a) Section E-F, (b) Sections A-C-E and B-D-F. Section names as in Fig. 3.

6(a), (b), (2nd cycle, experiment). We see that the experimentally obtained parts are in very good agreement with the theoretical analysis, see Figs. 6(a), (b). The slightly higher amount of springback observed in our experiment is possibly due to the higher amount of draw-in obtained in the experiment (13.5 mm at point A of Fig. 3) compared with the draw-in obtained in the simulation (13.2 mm at point A of Fig. 3). However the experimentally obtained part shape is in good agreement with the desired part shape.

CONCLUSIONS

A tooling design algorithm for sheet metal forming which compensates for part shape error due to springback by modifying the tooling shape was presented. In this algorithm the internal force distribution on the part, obtained by finite element analysis, is used in order to perform an inverse simulation of springback of the fully loaded part. In other terms the fully loaded part, whose shape matches the desired part shape, is sprung forward thereby reversing the springback phenomenon. The sprung-forward shape is obtained to be the next tooling shape in an an iterative approach where the procedure is repeated until the desired part is obtained. We note that this tooling design algorithm requires shape changes of the tools in both the punch area and the binder area.

We used the tooling obtained by our design algorithm in order to produce an aluminum part with a three-dimensional geometry. The designed tooling was found to produce the desired part shapes with very small error. The design was also verified experimentally, where excellent agreement between simulation prediction and experimental results was obtained.

ACKNOWLEDGEMENTS

This research has been funded by the NSF-DDM-9202362 grant, Division of Manufacturing and Design. The 2008-T4 aluminum sheets were provided by ALCOA. Also, the help of Daniel Walczyk in machining the dies is greatly appreciated.

REFERENCES

1. Ayres, R. A., SHAPESET: A Process to Reduce Sidewall Curl Springback in High-Strength Steel Rails, *J. App. Metalworking*, **3**, no. 2, 127, (1984).

2.Liu, Y.C., The Effect of Restraining Force on Shape Deviations in Flanged Channels, *Trans. ASME, J. Eng. Mat. Tech.*, **110**, 389, (1988).

3. Stevenson, R., Springback in Simple Axisymmetric Stampings, *Metallurgical Transactions A*, **24A**, 925 (1993)

4. Wenner, A.L., On Work Hardening and Springback in Plane Strain Draw Forming, *J. Applied Metalworking*, **2**, 277 (1983).

5. Karafillis, A.P. and Boyce, M.C., Tooling Design in Sheet Metal Forming Using Springback Calculations, *Int. J. Mech. Sci.*, **34**, 113 (1992).

6. Hardt, D.E., Hale, M., Closed-Loop Control of a Roll-Straightening Process, *Annals of CIRP*, 33, (1984).
7. Webb, R.D., Hardt, D.E., A Transfer Function Description of Sheet Metal Forming for Process Control, *Trans. ASME, J. Eng. Ind.*, **113**, 44, (1991).
8. Karafillis, A.P. and Boyce, M.C., Tooling Design Accommodating Springback Errors, *Journal of Materials Processing Technology*, **32**, 499 (1992).
9. Karafillis, A.P., amd Boyce, M.C., Tooling and Binder Design for Sheet Metal Forming Processes Compensating Springback Error, *Int. J. of Machine-Tools and Manufacture*, in press.
10. Honecker, A. and Matiasson, K., Finite Element Procedures for 3D sheet forming simulation, *Proc. NUMIFORM '89*, 457 (1989).
11. Matiasson, K., Bernspang, L., Samuelsson, A., Hamman, T., Schedin, E., Melander, A., Evaluation of a Dynamic Explicit Approach Using Explicit Integration in 3-D Sheet Forming Simulation *Proceedings of NUMIFORM '92*, Valbonne, France, p. 55, (1992)
12. Rebello, N., Nagtegaal, J.C., Taylor, L.M., Passmann, R., Comparison of Implicit and Explicit Finite Element Methods in the Simulation of Metal Forming Processes *Proceedings of NUMIFORM '92*, Valbonne, France, p. 99, (1992)
13. ABAQUS 5.2, User's Manual, (1993)
14. Karafillis, A.P., Tooling Design for Three-Dimensional Sheet Metal Forming Using Finite Element Analysis, *Ph. D. Thesis, Dept. of Mechanical Engineering, Massachusetts Institute of Technology* (1994).

Simulation of Materials Processing: Theory, Methods and Applications, Shen & Dawson (eds)
© 1995 Balkema, Rotterdam. ISBN 90 5410 553 4

Design of die and its effect on wear in cup forming

L. Karlsson & M. Öqvist
Division of Machine Element, Luleå University of Technology, Sweden
L.-E. Lindgren & M. Näsström
Division of Computer Aided Design, Luleå University of Technology, Sweden

ABSTRACT: The effect of the geometric design of die on tool wear is studied. The Finite Element Method is used in the simulations. Two different geometries of the upper part of the die are compared. They can be described as a circular curve in the first case and as a tractrix curve in the second case. The latter is known to have some good properties with respect to springback and residual stresses of cup. A lower maximum punch force is obtained for the tractrix case.

1 INTRODUCTION

Reducing tool wear in sheet forming can be done by lubrication and appropriate geometric design of the tools. Various theories of elastic-plastic bending of sheet metals have been presented in the literature. Considerable effort has been undertaken to develop analytical models and understand the bending process. However, these models are not easily extended in order to study the interaction between the tool and the formed material. Experimental studies to reduce tool wear have been performed (Hsu 1993, Schedin 1993). Improving wear resistance of tool can be done by material selection. Punching tools are normally made of tool steel that is heat treated after machining. For high volume production it is common to use punches and dies made of cemented carbides to obtain extended time between exchange or regrinding of tools. Another option is to reduce wear by design of tool geometry and by lubrication. This can, of course, be combined with increasing wear resistance by material selection.

This study is a first step in a study aiming at reducing tool wear. The Finite Element Method is used to study the interaction between tool and formed material. A fictive case has been investigated. The main purpose is twofold, evaluate the computational tool and evaluate the model.

The conclusions are that Nike2d (Engelmann 1991a) in combination with Island (Engelmann 1991b) is very powerful. The industrial experience that the tractrix curve reduces tool wear is only weakly confirmed in the study as the contact forces are only reduced by about 10%. A better model for integrating the effect of parameters like force, relative velocity and time on wear is needed.

2. THE ANALYSIS OF PLASTIC BENDING

Several models for the deformation in plastic or elastoplastic bending have been used in sheet metal manufacturing processes. They do have the advantage of producing relations where the influence of specific parameters can be seen directly. However, they will never be as accurate as numerical models.

Hill (1950) presented a complete solution for pure bending in which deformation of a sheet metal is achieved by a couple applied along its length. In this analysis he predicted the movement of the neutral axis but no change in thickness for rigid-perfectly plastic materials under plane strain bending. Lubahn and Sachs (1950) analysed the cases of plane stress and plane strain bending in a similar manner to Hill's, but they predicted material thinning based on the improper assumption that the neutral surface will remain at fixed positions during small increments of bending curvature. Crafoord (1970) found that the thickness change of sheet during bending is negligible. That conlusion is reinforced by the finite element analysis of Oh and Kobayashi (1980). Dadras and Majlessi (1982) conducted an analysis on bending of rigid-strain-hardening materials. Weinmann et al. (1988) carried out an experimental study of Bauschinger effect on sheet metal under cyclic reverse pure bending in an attempt to establish a plastic stress-strain relationship. However, their model did not incorporate the Bauschinger effect as expected.

The residual stresses in deep drawing with circular and tractrix die profiles have been studied both experimentally and with the FE-code LS-Dyna2D by Danckert (1994). He achieved good agreement between the FE-simulations and the experimental results.

In this study we will simulate two different curve forms of the upper part of the die. The influence of a geometrical design of the tool is the emphasis in this study and will be followed by investigations of the effects of lubrication.

3 REDUCING TOOL WEAR

The industrial objective of research in forming processes is to reduce lead times and costs in the product development cycle and to increase production speed in the production of sheet metal parts. While these objectives apply primarily to the users of punching tools, the objective for the designers and manufactures of the tools is to enhance competitiveness among toolmakers by introducing a new and highly efficient design. The final goal is to simulate, produce and test a prototype tool in order to reduce tool wear in sheet forming. The models described in the previous chapter are not easily extended to study tool wear as it is related to the interaction between tool and formed material. Therefore the Finite Element Method has been chosen in this study.

Next to metal cutting machine tools, hydraulic and mechanical presses are the most common equipment in mechanical industries. Production speeds are generally high in press tools and the initial tool design, manufacturing and tool wear are extremely expensive and time consuming. One problem is that it is often impossible to finalise tool design before a number of parts have been produced and tested for functionality, accuracy and integration with other components in the product. Hence, for many years the challenge has been to cut down initial costs and shorten lead times in work preparation for such parts.

The first step in this work is to analyse the forming of an axisymmetric cup and investigate the influence of the geometry of the die on the contact forces between the sheet and the die. The Finite Element Method is used in the simulations. Two different geometries of the upper part of the die are compared. In the first case a normal circular radius is used and in the second case the tractrix curve is used. The latter is described in the next chapter. A study of how lubrication will reduce the tool wear in cup forming will be performed in future work. The dynamic behaviour focus on the interference between natural frequencies of punch and press. Working frequency of the machine will also be of interest in future studies.

Design modifications will be performed based on the results from computer simulations and experiments. A tool will be manufactured for testing in laboratory and in production.

4 TRACTRIX CURVE

Wang (1993a) state that "The tractrix bending die may be the optional one to reduce the sensitivity of bending angle variations to punch stroke variations, to improve the control of brake bending process, and part tolerance". A mathematical model of plane-strain bending of sheet and plate have also been studied by Wang (1993b). Drawing die with tractrix shape have been used by Kampus (1992). It is known by the designers of tools, that the curvature form in sheet metal forming techniques has an important influence on the tool wear.

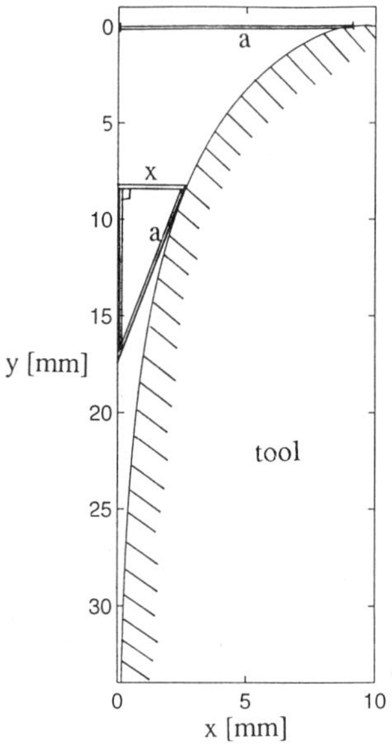

Figure 1. Used tractrix curve. Line with length "a" is always tangent to curve. x=0 corresponds to inner radius of tool which will approximately be the outer radius of the formed cup.

The derivation and definition of the tractrix curve follows. A tangent line to the tractrix curve has a fixed length ´a´ and its left end should always be at x=0, as shown in Figure 1. Pythagora's theorem and the use of uniform triangles gives the differential equation

$$\frac{dy}{dx} = \frac{\sqrt{a^2 - x^2}}{x}. \tag{1}$$

The solution of the equation is

$$y(x) = a\ln\left|\frac{a + \sqrt{a^2 - x^2}}{x}\right| - \sqrt{a^2 - x^2} \tag{2}$$

, where $0 \le x \le a$.
The curve is extrapolated as a straight line when the angle with the vertical axis reaches 3°.

5 COMPUTATIONAL MODEL

The Finite Element Method has been used very much for simulating drawing. Finite element codes have been based on either the flow formulation or the solid formulation. Explicit codes have gained ground in the latter case. The monograph by Oñate

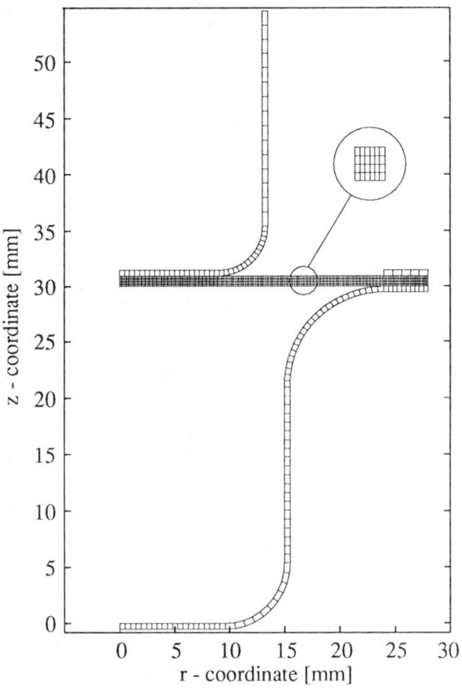

Figure 2. Finite element mesh for circular case.

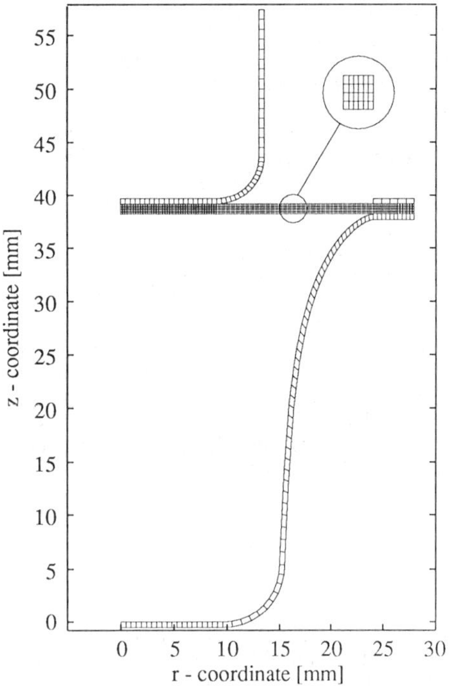

Figure 3. Finite element mesh for tractrix case.

(1993) gives an overview of the methods and applications. Danckert (1994) used the explicit finite element code LS-Dyna2d in simulations where the effect of the die geometry on the residual stresses of the formed cup was the main concern. He concludes that the tractrix curve is favourable but has the disadvantage of requiring an increased punch stroke.

Two computational models have been created by using the public domain codes from Lawrence Livermore National Laboratory. They are the mesh generator Maze (Hallquist 1983), the finite element code Nike2d (Engelmann 1991a) with the driver Island (Engelmann 1991b)] and the postprocessor Orion (Hallquist 1985a).

The two finite element models are shown in Figures 2 and 3. The first case has a die with a circular transition at the upper edge and the second case has a tractrix curve. They will be named circular and tractrix case below. They have 980 elements and 1341 nodes. The models are axisymmetric. The material is assumed to be elastic, except for the formed plate where an elastoplastic material model is used. The Young's modulus, E, is taken as 205GPa and Poisson's ratio as 0.3. Rate-independent plasticity with von Mises yield criteria and the associated flow rule is used. Isotropic, power law hardening is assumed. The yield stress is

$$\sigma_y = k\left(\varepsilon_0 + \bar{\varepsilon}^p\right)^n \tag{3}$$

and initial yield strain is

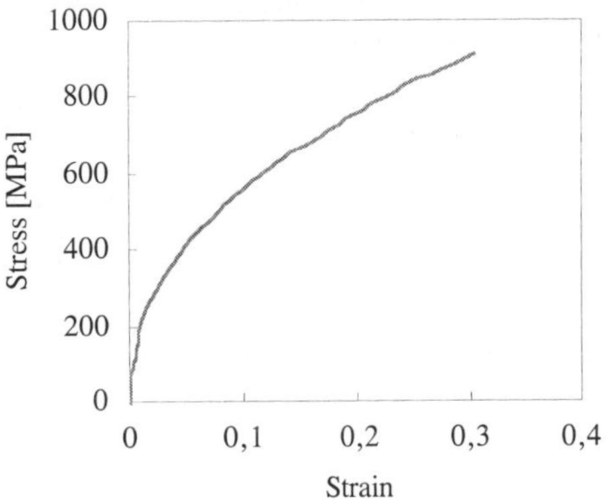

Figure 4. Stress-strain relation for power law hardening.

$$\varepsilon_0 = \left(\frac{E}{k}\right)^{\frac{1}{n-1}} \tag{4}$$

The value for k, the strength coefficient, is 1543 MPa and n, the hardening exponent, is 0.48. The effective plastic strain, $\bar{\varepsilon}^p$, is plotted versus stress in Figure 4.

The finite element formulation accounts for large deformations. The Green-Naghdi stress-rate (Johnson 1984) is used in combination with midpoint strain increment (Hughes 1980). The contact algorithm, (Hallquist 1985b), permits arbitrarily large sliding. The Coulumb friction law was applied with a friction coefficient of 0.15.

The driver Island proved to be an efficient tool together with Nike2d. The incremental and iterative procedure for the simulation applied both displacement and energy criteria for convergence checks. The time steps were cut down and the convergence criteria were tightened when encountering convergence problems during the solution procedure. Starting anew from previous converged steps is also possible. Several procedures for solving the equations are available in Nike2d. The BFGS method was chosen for this case.

Island compensated for one of the drawbacks using an implicit finite element formulation instead of an explicit code like Dyna2d (Hallquist 1988). Namely, the problem in attaining convergence in the solution process. Furthermore, it includes the capacity of optimising the punch speed for controlling the strain rate in the material as one can change load curves in relation to computed variables.

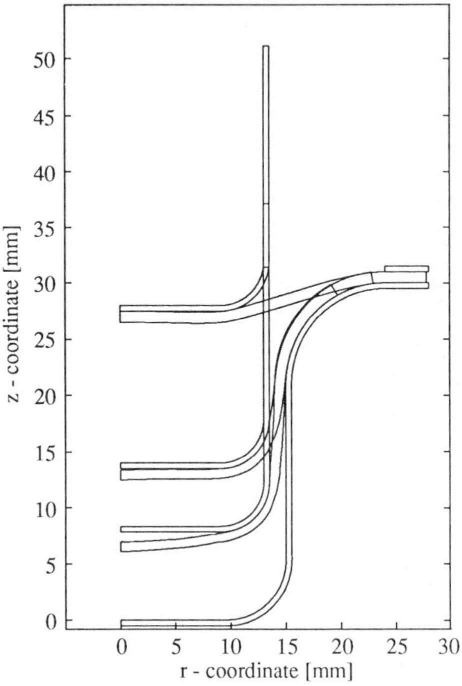

Figure 5. Deformations in circular case.

6. RESULTS

The deformation process for the two cases is shown in Figures 5 and 6. Note that the same punch travelling distance in the two cases do not correspond exactly to the same amount of drawing as the punch must travel longer in the tractrix case to achieve the same shape. The punch travel distance is 3.5, 17.5 and 23 mm for the different configurations of the formed material.

The effective plastic strain is about the same for the two cases, around 0.25, in the corner where the bending is concentrated.

The punch force, Figure 7, is somewhat higher in the circular case. Therefore the difference between the computed forces is less than what it looks like in the Figure. The force is probably not more than 10% higher in the circular case.

Noticeable is the problem of getting a flat bottom of the cup in the circular case. The deformation changes behaviour at a punch travel of 23 mm, see Figure 5 and it is not possible to recover a flat bottom later on in the process. This phenomena is also visible in the force diagram in Figure 7, especially for the radial force. This force decreases when the bottom of the cup "falls out".

The punch force, Figure 7, is somewhat higher in the circular case. Therefore the difference between the computed forces is less than what it looks like in the Figure. The force is probably not more than 10% higher in the circular case.

Noticeable is the problem of getting a flat bottom of the cup in the circular case. The deformation changes behaviour at a punch travel of 23 mm, see Figure 5 and it is not possible to recover a flat bottom later on in the process. This phenomena is also visible in the force diagram in Figure 7, especially for the radial force. This force decreases when the bottom of the cup "falls out".

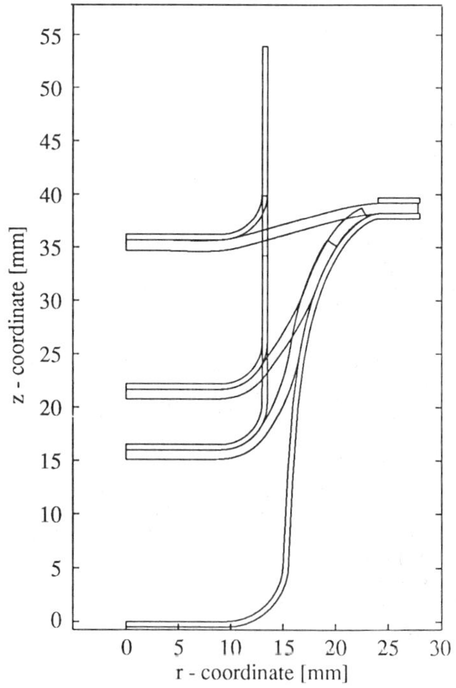

Figure 6. Deformations in tractrix case.

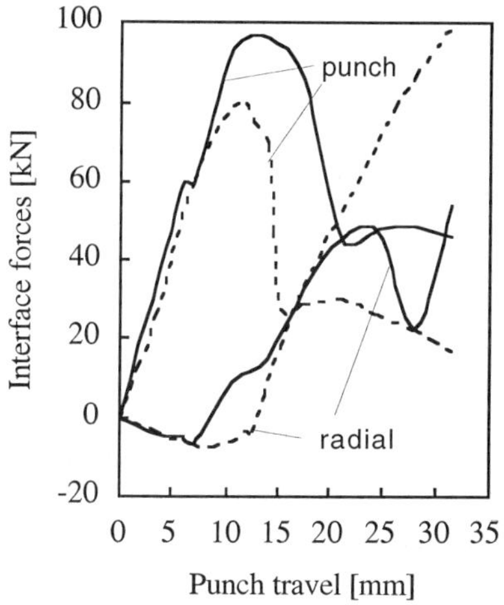

Figure 7. Total force on punch and radial force on die. Dashed line for tractrix case.

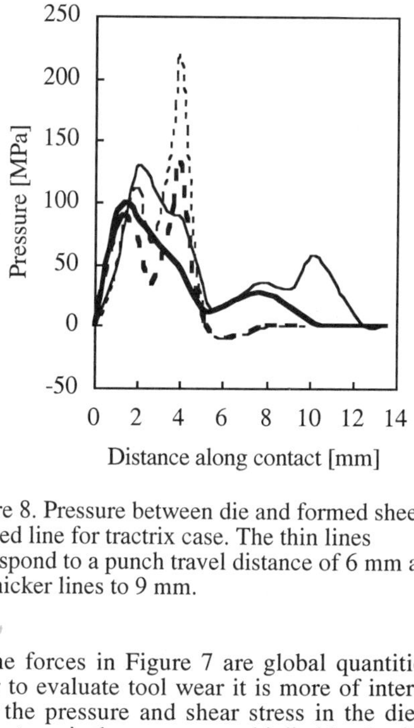

Figure 8. Pressure between die and formed sheet. Dashed line for tractrix case. The thin lines correspond to a punch travel distance of 6 mm and the thicker lines to 9 mm.

The forces in Figure 7 are global quantities. In order to evaluate tool wear it is more of interest to have the pressure and shear stress in the die. The shear stress is the pressure multiplied by the friction coefficient, due to the Coulomb friction model. The pressure is shown in Figure 8. The pressure is locally higher in the tractrix case.

7. CONCLUSIONS AND FUTURE WORK

The main conclusion is that the claim that the tractrix curve reduces tool wear is only weakly confirmed in the study as the contact forces are only reduced by 10%. A better model for integrating the effect of parameters like force, relative velocity and time on wear is needed.

A well documented experiment will be needed in order to verify the computational model and special attention will be given to the friction model. The Coulomb friction model will be extended along the same as in Edberg (1992 and 1994). Anisotropic material models with Bauschinger effect , available in Nike2d, will also be necessary for a real test case.

It is also found that Nike2d in combination with Island is very powerful. Thus the implicit finite element formulations gain ground versus the use of explicit codes that have found their place in simulation of material processing, Edberg (1993) and Häggblad (1992).

8. REFERENCES

Crafoord, R., PhD Thesis, Chalmers University of Technology, Göteborg, 1970.

Dadras, P. and Majlessi, S. A., *Trans. of the ASME*, Vol. 104, p.224, 1982.

Danckert, J., *Residual stresses in deep drawing*, Dept of Production, Aalborg University, 1994.

Edberg, J. and Lindgren, L.-E., Efficient three-dimensional model of rolling using an explicit finite-element formulation, *Comm. in Numerical Methods in Eng.*, Vol. 9, No. 7, pp. 613-627, 1993.

Edberg, J., L-E Lindgren and M Jarl, Wedge rolling test, *Journal of Materials Processing Technology*, Vol. 42, pp 227-238, 1994.

Edberg, J., Three-dimensional simulation of plate rolling using different friction models, *proc. of NUMIFORM'92*, September 14-18, 1992, Sophia Antipolis, France.

Engelmann, B., *NIKE2D - A Nonlinear, Implicit, Two-Dimensional Finite Element Code for Solid Mechanics*, Lawrence Livermore National Laboratory, UCRL-MA-105413, 1991a.

Engelmann, B. and Whirley, R. G., *ISLAND - Interactive Solution Language for an Adaptive NIKE Driver*, Lawrence Livermore National Laboratory, UCRL-MA-108721, 1991b.

Hallquist, J. O., *MAZE - An Input Generator for DYNA2D and NIKE2D*, Lawrence Livermore National Laboratory, UCID-19029, Rev 2, 1983.

Hallquist, J. O. and Levatin, J. L., *Orion: An interactive color post-processor for two-dimensional finite element codes*, Lawrence Livermore National Laboratory, UCID-19310, Rev 2, 1985a.

Hallquist, J. O., Goudreau, G. L. and Benson, D. J., Sliding interfaces with contact-impact in large-scale, Lagrangian computations, *Comp Methods Appl. Mech. Eng.*, Vol. 51, pp. 107-137, 1985b.

Hallquist, J. O., *DYNA2D - An explicit two-dimensional hydrodynamic finite element code with interactive rezoning and graphical display*, Lawrence Livermore National Laboratory, UCID-18756, Rev 3, 1988.

Hill, R. *Mathematical Theory of Plasticity*, Oxford Press, London, 1950.

Hsu, T. S., The "internal-variables" approach in friction modeling of computer simulation in sheet-metal forming processes, *Journal of Materials processing Technology*, Vol. 37, pp. 95-114, 1993.

Hughes, T. J. R. and Winget, J., Finite rotation effects in numerical integration of rate constitutive equations arising in large-deformation analysis, *Numer. Methods in Eng.*, Vol. 80, pp. 1862-1867, 1980.

Häggblad, H.-Å., Explicit versus implicit finite element simulation of metal powder compaction, *proc. of NUMIFORM'92*, September 14-18, 1992, Sophia Antipolis, France.

Johnson, G. C. and Bammann, D. J., A discussion of stress rates in finite deformation problems, *Int. J Solids Structures*, Vol. 20, No. 8, pp. 725-737, 1984.

Kampus, Z. and Kuzman, K., Experimental and numerical (FEM) analysis of deep drawing of relatively thick sheet metal, *Journal of Materials processing Technology*, Vol. 34, pp. 133-140, 1992.

Lubahn, J. D. and Sach, G., Trans. of the ASME, Vol. 72, p. 31, 1950.

Oh, S. I. and Kobayashi, S., Finite-element analysis of plane strain sheet bending, *Int. J. Mech. Sci.*, Vol. 22, pp. 583-594, 1980.

Oñate, E., *Numerical Simulation of Industrial Sheet Forming Processes*, Int. Center for Numerical Methods in Eng., Barcelona, Spain, Monografia CIMNE, No. 23, 1993.

Schedin, E., Control of friction in sheet metal forming can result in more stable production. *Materials & Design*, Vol. 14, No. 2, 1993.

Wang, C. T., Kinzel, G. and Taylan, T., Process Simulation and Springback Control in Plane Strain Sheet Bending, The Ohio State University, SAE Paper No 930280, 1993a.

Wang, C. T., Kinzel, G. and Taylan, T., Mathematical Modeling of Plane-Strain Bending of Sheet Plate, *Journal of Materials Processing Technology*, Vol. 39, pp. 279-304, 1993b.

Weinmann, K. J., Rosenberger, A. H. and Sanchez, L. R., *Annals of the CIRP*, Vol. 37, No. 1, p. 289, 1988.

Simulation of Materials Processing: Theory, Methods and Applications, Shen & Dawson (eds)
© 1995 Balkema, Rotterdam. ISBN 90 5410 553 4

A computer aided design – finite element – 3D measuring strategy for creep forming and/or pressing of glass skew surfaces

D. Lochegnies, P. Moreau, E. Francois & J. Oudin
Mechanical Engineering Laboratory, URA CNRS, University of Valenciennes, France

ABSTRACT : A new analysis strategy for creep forming and press bending by coupling computer aided design, finite element and three dimensional measuring is proposed. From the design product via Bezier curves, the manufacturer is now able to anticipate and adjust creep forming or press bending data base (specification of the size of the tools and the initial glass sheet geometry, forming time and temperature map at the furnace exit). The new results of our original strategy are fully illustrated for the forming of a reference automotive rear screen.

1 INTRODUCTION

The two basic deformation methods used in the flat glass industry are creep forming and press bending. In the first case, the sheet of glass is placed on a stainless steel ring (or skeleton) at the furnace exit ; under its own weight, the sheet bends until geometrically deformed, as instructed by the designer before toughening. In the second case, the sheet of glass is first placed on the skeleton for horizontal pressing, or held by tongs for the whole process for vertical pressing. Next, it goes out of shape under its own weight for a few seconds before being pressed between the skeleton and press covered with abestos and glass fiber. Finally, a second creep forming is carried out again before toughening.

Beyond esthetic aspects, the success of forming will be governed with respect to the deformed sheet contour line and especially to the designer slope evolution along the previous contour. These data are prescribed to the manufacturer via the designer's computer aided design models in Bezier curves ; the results of the forming are compared to three dimensional measurements. As a matter of fact, since the competition is becoming greater and greater, the glass manufacturer has to respond as quickly as possible to the feasibility of shaping up the sheet by forming parameter adjustments.

Previous numerical analysis were concerned with contact evolution at the glass/skeleton interface [1], with thermal evolution in glass, during forming and consequences on glass bulk properties [2-3] ; these give prominence to the importance of the viscosity and contact conditions at the time of forming. Other studies have integrated computer aided design (C.A.D.) into flat glass product manufacture [4-5]. C.A.D. is shown as an efficient tool, for the skeleton and mould design and manufacture, which produces an appreciable time saving and increased precision in manufacturing.

Coupling C.A.D. with finite element was first used [6] for mould geometries in the glass blowing industry (in this case, mould conception is achieved by taking into account thermal finite element analysis and essential parameters for a successful blowing).

Other means of coupling computer aided design with three dimensional measurements in the glass industry are proposed in [7] for the design of a perfume bottle. Using experimental model measurements with a stylus instrument measurement points are visualized with C.A.D. and modification is quick (time reduction about 30 to 70 %).

We propose an original strategy [8] using computer aided design, finite element models and 3D measurements ; it allows to determine the most suitable process (creep forming or press bending), the tools and initial glass sheet geometries, forming time and temperature map. For a reference automotive rear screen, creep forming and press bending processes are analysed. We constitute the optimum forming data base in three main steps :

- first the constitution of the initial data base (tools and glass sheet geometries, temperature map and forming times),
- next, the numerical validation and optimization of the manufacturer's choices. (With thermodependent viscoplastic finite element models, this step leads to optimum forming parameters, concerning skeleton bending radii and/or non-homogeneous heating adjustments [9]),
- and finally, the start of the sheet manufacture with optimum data base.

Throughout this strategy, a comparison between computations and measurements is made on the reference automotive rear screen via computer aided design.

Finally, we complete the analysis with a prediction of the glass product quality according to numerical and experimental analysis of creep forming and press bending.

2 STRATEGY DESCRIPTION

To succeed in creep forming and press bending, we propose an original procedure involving C.A.D. (Euclid software from Matra Datavision, running on a DEC personal workstation), finite element (Abaqus running on HP 700 PA-RISC) and 3D measuring machine (D.E.A. machine) to collect measurements in the specimen. Experiment data are collected on an IBM PC 486 computer via Tutor-P software developed by D.E.A. company. Two specific interfaces are developed between C.A.D. and finite element modules and between C.A.D. and 3D measurement modules for data communications on an ethernet network. These give, from IGES standard format for C.A.D. exploitation, data files which will be readable by the finite element software. Conversely, they collect finite element results (node position of the deformed mesh of the sheet) from Abaqus output files and the 3D measurements from the 3D machine with Tutor-P software.

The strategy proposed to the manufacturer is made up of three steps. The first step is to establish the initial forming data base after choosing the most suitable process (creep forming or press bending). With the C.A.D. module, using the designer's specifications in Bezier curves, the manufacturer obtains the initial geometry of the sheet by creep of the final sheet geometry and tools geometry in accordance with the slope evolution along the design contour line. With the C.A.D. mesh generation module, he then defines the mesh of the sheet with three- and four- node shell elements ; he finishes by the tool definition (skeleton and punch) with rigid surfaces from appropriate nodes of Bezier curves of design C.A.D. models. The initial data are finally obtained with the temperature in the sheet at the furnace exit, the bulk behaviour of the glass in forming temperature range and the forming time.

The mesh of the glass sheet and tools geometry are transferred to the finite element software through the C.A.D. - finite element interface. Three dimensional viscoplastic finite element models are developed for the numerical simulation of the sheet forming.

Results of the finite element viscoplastic computations, transferred through C.A.D. - finite element interface in C.A.D. module, are analysed and a comparison with design specifications is obtained.

The second step is the adjustment of the manufacturer's data base : the manufacturer proceeds to new numerical simulations and C.A.D. comparisons to reach the design objectives. Through numerical method, he is able to reduct the delays and costs linked to the feasibility research of a new product forming.

Finally, on successful completion, the optimum data base is used on the production line. Three dimensional measurements of the specimen are analysed with the 3D measurement - C.A.D. interface and final optimum adjustments are performed on the production line.

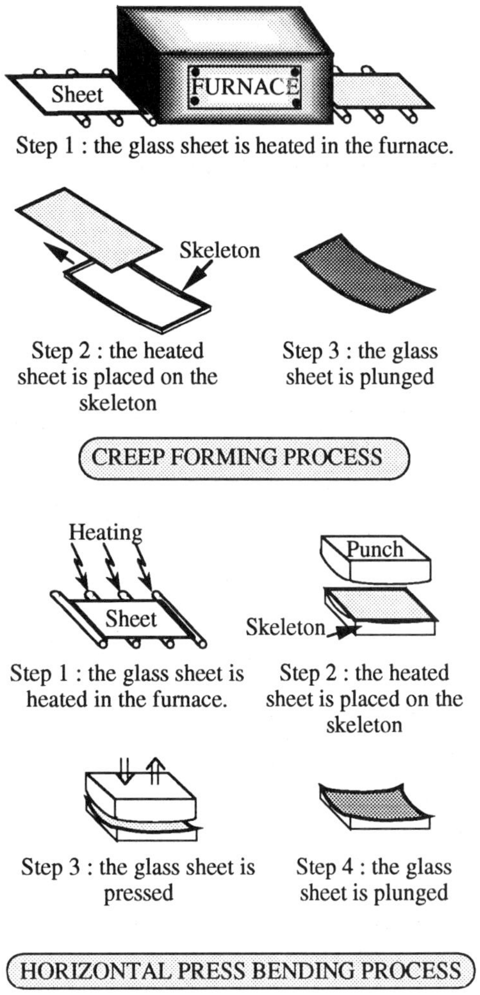
Sheet
FURNACE
Step 1 : the glass sheet is heated in the furnace.
Skeleton
Step 2 : the heated sheet is placed on the skeleton
Step 3 : the glass sheet is plunged
CREEP FORMING PROCESS
Heating
Sheet
Punch
Skeleton
Step 1 : the glass sheet is heated in the furnace.
Step 2 : the heated sheet is placed on the skeleton
Step 3 : the glass sheet is pressed
Step 4 : the glass sheet is plunged
HORIZONTAL PRESS BENDING PROCESS

Fig. 1 - Creep forming and press bending processes description

3 REAR SCREEN FORMING APPLICATION

From the creep forming process or hoizontal press bending process (Fig. 1), the manufacturer's aim is to obtain the final soda-lime silica glass rear screen (Fig. 2) composed of 14 Bezier curves. Using the creep of the final sheet geometry through C.A.D. and finite element modules, the initial dimensions of the glass sheet are 1220 mm x 495 mm (extreme dimensions) with 4 mm thickness.

Fig. 2 - Industrial profile of the rear screen automotive.

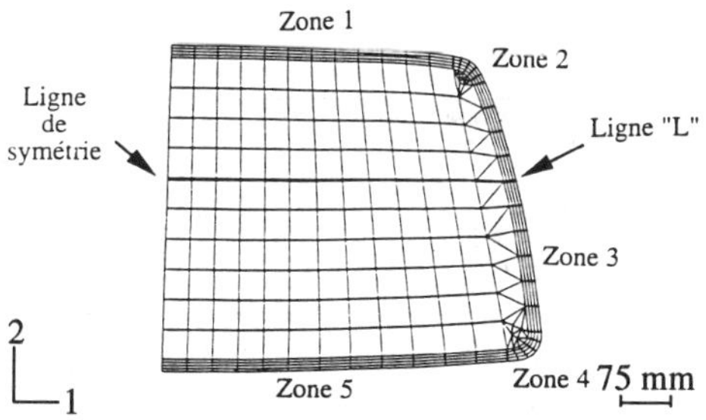

Fig. 3 - Mesh of the initial glass geometry

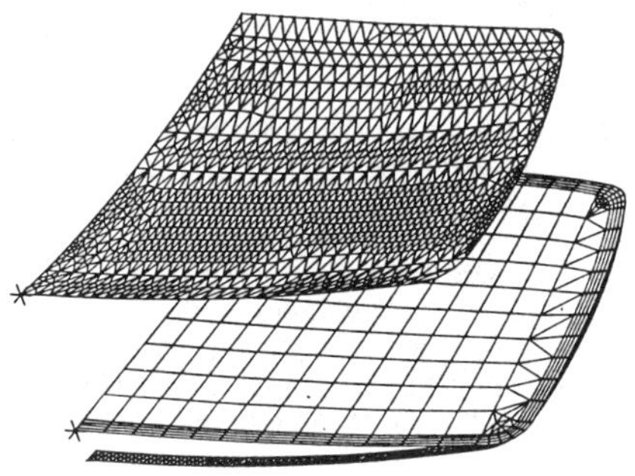

Fig. 4 - Skeleton and punch geometries

With the Euclid mesh generation module, the mesh of the glass sheet is made up of 296 quadrilateral four node shell elements (S4R5 for Abaqus) and 45 triangular three node shell elements (STRI35 for Abaqus) (Fig. 3). Because of symmetry, only one half of the sheet is modelled. Figure 3 shows the symmetry line, the longitudinal line "L" and the zones 1 to 5 which will be taken as references for the manufacturer in order to reset geometrically the manufacturer and the conception.

With slope evolution of the designer shape along the contour line, the AISI 316 stainless steel skeleton geometry (Fig. 4) is defined by means of 13 facets and 27 geometric points.

Punch geometry (Fig. 4) is determined from the final geometry of the sheet by a polynomial equation which is found from the design data (deflection at the center of the punch : 50.6 mm, slope evolution along the contour line). This equation is used in the RSURFU subroutine of Abaqus. The tools and initial glass sheet geometries are then transferred, with the help of C.A.D. - finite element interface to finite element module.

According to the manufacturer's knowledge, the sheet is heated as it moves in the furnace in order to obtain a quasi-homogeneous temperature in the 600°C 680°C range which is ideal for press bending and creep forming. Indeed, contact between the glass and the tools is considered as sliding, as experimentally observed in the 680°C glass temperature used. In this temperature range, the bulk glass behaviour is modelled by the Newtonian isochoric viscoplastic law given by

$$\sigma = 3\eta(\theta)\,\dot{\varepsilon}$$

where σ is the equivalent stress, $\dot{\varepsilon}$, the equivalent strain rate and η the soda lime silica glass viscosity. It is strongly dependent on temperature θ and is described by the Fulcher model [10]

$$\log(\eta(\theta)) = -9{,}3 + \frac{4630}{(\theta - 247)}$$

with η in N.mm^{-2} and θ in celsius.

Other soda lime silica characteristics are density : 2.5 10 $^{-6}$ kg.mm^{-3}, Young's modulus : 20000 N.mm^{-2}, considered constant in this temperature range, and Poisson's ratio : 0.22.

At this stage of the strategy, the first level is finished ; the second level using three dimensional viscoplastic finite element models continues in order, firstly, to determine the first initial temperature of the sheet (function of the forming time) and, secondly to optimize initial choices.

After several numerical simulations with Abaqus with a sheet temperature varying between 650°C and 680°C, the temperature of 675°C is retained for creep forming and press bending processes.

Indeed, this temperature allows to obtain forming times below 1 s (the manufacturer will reach high production rates) and to heat locally, if necessary, several areas of the soda lime silica glass sheet to a temperature below 680°C to reach the expected geometry [11]

For the creep forming process and with our strategy, the temperature 675°C is analysed for four different forming times (0.25s, 0.5s, 0.8s and 1s). The best results, in this case, are obtained with 675°C in the sheet and 0.8 s forming time. With both these values (temperature - forming time), the slope error is maximum in zone 5 (2.5° difference) and the error in the other zones varies from 0.5° to 2°.

For the press bending process, the best forming times are found with : first creep forming : 0.2 s, pressing : 0.4 s, second creep forming : 0.25 s. The slope error is more homogeneous according to the zones than the creep forming slope error. The maximum slope error is about 1.5° mainly because of

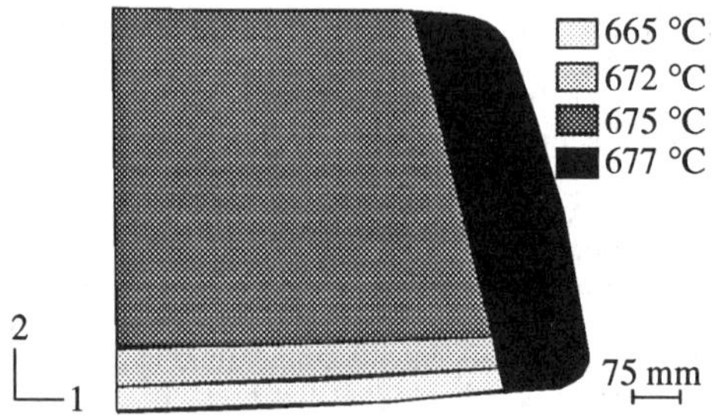

Fig. 5 - Non homogeneous temperature distribution in the rear screen for creep forming process

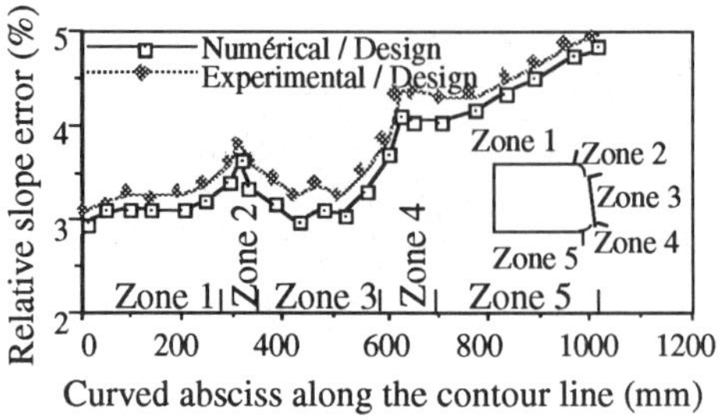

Fig. 7 - Slope relative error for creep forming process on experiment and computation with regard to design along the contour line

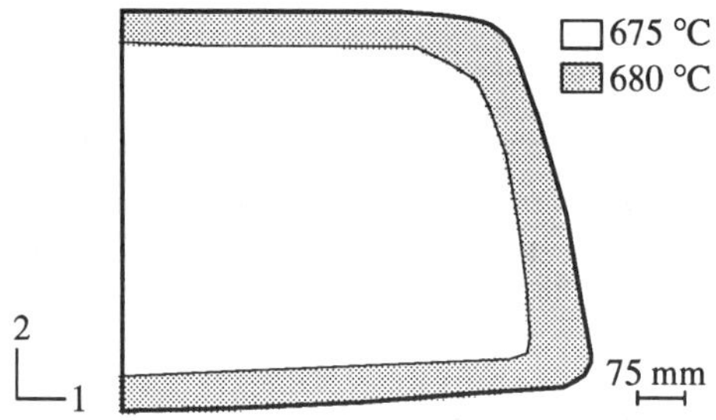

Fig. 6 - Non homogeneous temperature distribution in the rear screen for press bending process

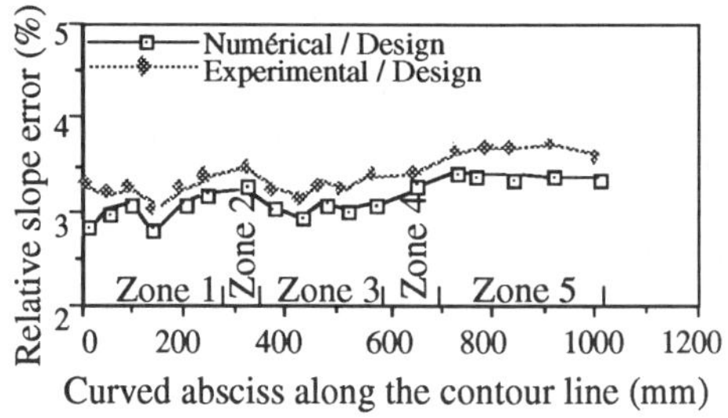

Fig. 8 - Slope relative error for press bending process on experiment and computation with regard to design along the contour line

the spring back effect after pressing.

In the two cases, the solution to reach thc designer's objectives is to create during the heating of the sheet, a non homogeneous temperature distribution at the furnace exit [12-13]. Based on the industrial possibilities (crease heaters in the furnace), the sides of the sheet of glass can be locally heated. This allows to increase slopes along the outline without modifying the total geometry of the sheet. Taking into account the manufacturer's knowledge in heating, the temperature maps proposed for the numerical analysis are described in figure 5 for the creep forming process and figure 6 for the press bending process.

These temperature maps produce significant modifications on the slopes of the final geometry of the sheet without change in deflection.

For creep forming and press bending, the slope error in the five zones no longer exceeds 0.5°, which corresponds to 5% relative error compared with the conception.

For this geometry, the manufacturer would rather use the press bending process because the production rate is higher than with the creep forming process and the surface quality is better.

According to our strategy, the last step of the manufacturer's work is the application, on the production line, of the previous forming data base : geometries of the initial sheet and tools, temperature maps and forming times. The first formed specimens are then measured with the 3D measuring machine : measuring points are selected, using the mesh node position, on the symmetry line, the "L" line and the whole contour line. With the 3D measurement - C.A.D. interface, the comparison between the specifications of the designer and the numerical results have been analysed. The two processes lead to relative errors in the slopes and deflection below 5%, which confirms numerical prediction (Fig. 7, Fig. 8)

4 CONCLUSIONS

An original strategy with computer aided design, finite element elastoviscoplastic models and three dimensional measurements have been proposed for the forming, from flat products, of a skew geometry by creep forming or press bending.

(i) C.A.D. allows to determine the initial geometries of the glass sheet and tools (skeleton and punch).

(ii) Coupling with finite element models allows to adjust the forming parameters (geometries of the sheet and tools, temperature maps, forming time (s)).

(iii) Finally, the optimum data base is validated by comparison with numerical prediction and experimental trials.

The optimization of the forming parameters is thus now possible in a quicker time than by an experimental method, reducing the adjustment of production means and the delay in knowing the feasibility of a new shape. This leads also to a better relationship between designer and manufacturer.

5 ACKNOWLEDGEMENTS

The present research work has been supported by the CNRS, the 'Conseil Régional Nord-Pas de Calais' and the 'Ministère de l'Enseignement Supérieur et de la Recherche'; the authors gratefully acknowledge the support of these institutions.

REFERENCES

[1] Costa, P. 1972. Untersuchung der thermischen und geometrischen Kontaktverhältnisse zwischen Glas und Metall bei der Formgebung heißen, zähflüssigen Glases. *Glastechn. Ber.* 6 : 277-287.

[2] Farag, I.H., Beliveau, M.J. and Curran, R.L. 1987. Heat transfer during glass forming. *Chem. Eng. Comm.* 52/21 : 21-32.

[3] Kent, R. and Rawson, H. 1971. An experimental and theoretical investigation of glass pressing. *Glass Tech.* 12/5 : 117-127.

[4] Wallis, J. 1986. Saint-Gobain and C.A.D./C.A.M. partners in productivity. *Glass* 63/5 : 157-167.

[5] Manchester, D. and Hawksworth, G. 1987. A C.A.D./C.A.M. solution of structural glass systems. *Glass Int.* 3 : 61-64.

[6] Hernandez-Luna, A.A. and Cardenas-Franco, L. 1988. Optimal design of glass molds using C.A.D./C.A.E. and response surface methodology techniques. *Comput. & Graphics* 12/3 : 391-399.

[7] Basset, F. 1990. Techniques verrières : repousser les limites. *Parf., Cosm., Ar.* 129-138.

[8] Moreau P. 1994. Nouvelle stratégie pour le formage de produits plats vers une géométrie complexe avec C.A.O, modèles éléments finis et métrologie. Application au fluage et au pressage de verre plat. *Thèse de Doctorat*

[9] Lochegnies, D., Moreau, P. and Oudin, J. 1994. Finite element strategy for glass sheet manufacture by creep forming. *Klei Glas Keramick* 15/12 : 311-315.

[10] Fulcher, G.S. 1925. Analysis of recent measurements of the viscosity of glasses. *J. Amer. Ceram. Soc.* 8/6 : 339-355.

[11] Moreau, P., Lochegnies, D. and Oudin, J. 1994. Une stratégie conception assistée par ordinateur - elements finis pour le fluage de surfaces à géométrie complexe. *StruCoMe 1994* 691-698

[12] Lochegnies, D., Vicart, T. and Oudin J. 1995. Finite element elastoviscoplastic models for optimization of horizontal and vertical press bending of sheets of glass. *will be published in Glass Technology.*

[13] Lochegnies, D., Francois, E. and Oudin, J. 1995. New computer aided design, finite elements modelling and 3D measuring of rear screen creep forming. *will be published in J. of Proc. Mech. Eng.*

Simulation of Materials Processing: Theory, Methods and Applications, Shen & Dawson (eds)
© 1995 Balkema, Rotterdam. ISBN 90 5410 553 4

Analysis and optimization of weakly coupled thermo-elasto-plastic systems with applications to weldment design

Panagiotis Michaleris
Edison Welding Institute, Columbus, Ohio, USA

Daniel A. Tortorelli
Department of Mechanical and Industrial Engineering and Department of Theoretical and Applied Mechanics, University of Illinois at Champaign-Urbana, Ill., USA

Creto A. Vidal
Departamento de Computacao, Universidade Federal do Ceara', Fortaleza, Brazil

ABSTRACT: A systematic approach for the design of weakly coupled thermo-elasto-plastic systems is presented. The Newton-Raphson iteration method is used in the solution process so that analytic design sensitivity formulations may be efficiently derived via the direct differentiation technique. Analysis and sensitivity analysis capabilities are combined with numerical optimization to form an optimum design algorithm. To demonstrate the algorithm, we optimally design a weldment with respect to manufacturing and service life aspects.

1 INTRODUCTION

Engineering design is an iterative process, in which a design is continuously modified until it satisfies the design goals. The increasing complexity of today's engineering designs renders the traditional "trial-and-error" design process, based on experience and intuition, insufficient and necessitates the utilization of more systematic methods such as design sensitivity analysis and numerical optimization.

Design sensitivity analysis and numerical optimization are successfully used to design minimum weight structures with linear behavior (cf. [1, 2, 3]). In the present work, this approach is used to optimize systems governed by the nonlinear equations of thermo-elasto-plastic systems. In the optimization problem, an initial design comprised of the shape, boundary conditions, loading and material properties, is parameterized by a set of design parameters. Generalized response functionals are then defined to quantify cost and constraint specifications. These response functionals depend on the design parameters both explicitly and implicitly through the system response, which is evaluated in the analysis, e.g. temperature, heat flux, displacement, plastic hardening, stress, strain and reaction forces. The optimization iterations commence when the analysis is performed and the computed response is used to evaluate cost and constraint functionals. An explicit sensitivity analysis is then performed to evaluate the derivatives of the cost and constraint functionals with respect to the design parameters. The numerical optimization program uses the values of the cost and constraint functionals and their sensitivities to modify the design parameter values to minimize the cost functional and satisfy the constraints. Convergence criteria are evaluated to determine if more iterations are required.

In this work, the analysis and sensitivity analysis for weakly coupled thermo-elasto-plastic systems are derived. The tangent operators of the Newton-Raphson solution processes are used to compute sensitivities accurately and efficiently via the direct differentiation method. The thermal analysis and sensitivity analysis are consistent with those presented in [4, 5]. The elasto-plastic analysis and sensitivity analysis follow the formulation presented in [6] for transient nonlinear coupled systems.

The analysis and sensitivity analysis are incorporated into a numerical optimization algorithm to design a weldment. The residual stresses which evolve during the welding process are predicted by solving a nonlinear transient thermo-elasto-plastic problem The sensitivities of the thermal and mechanical fields are efficiently computed via the direct differentiation method. Finally, the analysis and sensitivity analysis results are used to optimize a weldment design.

2 WELDEMENT DESIGN OPTIMIZATION

A butt weldment is optimized to improve its performance. For each optimization iteration, both manufacturing and service analyses are performed. In the manufacturing analysis, a thermal analysis and a thermal sensitivity analysis are initially performed to compute the thermal field and its sensitivity. Then, the mechanical analysis and mechanical sensitivity analysis commence to determine the thermal induced residual stress and its sensitivity. To simulate the product's service environment, the structure is subsequently loaded and sensitivities of the service stresses are computed. This information is used by the optimization algorithm to update the design. The optimization iterates until it converges to an optimum.

2.1 *Problem Statement*

The analysis models both the manufacturing process and the product's service. In the manufacturing process, two beads are simultaneously deposited on a 0.23% carbon mild steel rectangular plate—see figure 1. The welding direction is perpendicular to the plane of the figure and the shaded region represents the filler metal, also mild steel. After the plate has cooled, it is loaded with a tensile load $P = 7 \times 10^5$ N on the edges to simulate the product service load (cf. figure 2).

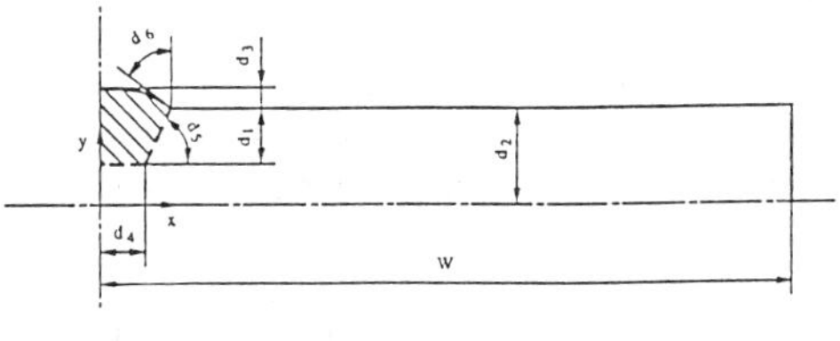

Figure 1. Quarter cross-section of a butt weldment.

Because of symmetry, only a quarter of the plate is modeled. The same finite element mesh is used for both thermal and mechanical analyses. It consists of 552 quadrilateral 8-noded elements—see figure 3. The mesh is parameterized using PATRAN [7] and a command language program (PCL [8])—see [5]. The shape of the welded plate is described by six shape parameters (d_1 through d_6)—see figure 1. The plate half width W and

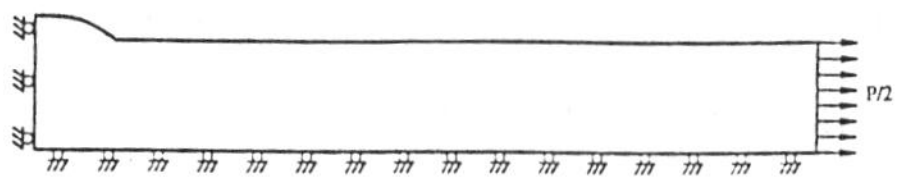

Figure 2. Mechanical analysis boundary conditions.

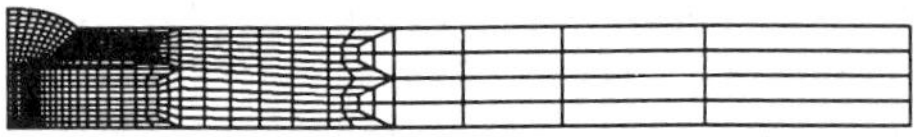

Figure 3. Finite element mesh.

length L are fixed at 140 and 100 mm, respectively.

The optimization problem is to determine the plate half thickness d_2, weldment shape and weldment parameters that minimize the weight of the welded loaded plate and simultaneously limit the von Mises stress level to 420 MPa. The cost functional is equal to the volume of the welded plate. The design parameter vector consists of the six shape parameters (see figure 1), as well as the welding energy input rate Q, and the torch speed v. The initial values of the design parameters, their upper and lower bounds, and their optimum values appear in table 1.

Table 1. Design parameter values of the weldment design optimization problem.

Design parameter	Initial value	Upper Bound	Lower Bound	Optimum value
d_1 (mm)	6	8	4	4.97
d_2 (mm)	15	20	10	10.00
d_3 (mm)	3	5	1.5	1.50
d_4 (mm)	5	8	3	3.29
d_5 (deg)	45	30	60	30.00
d_6 (deg)	50	70	0	69.64
Q (W)	32000	100000	1000	9531.73
v (mm/s)	6	10	2	4.96

2.2 *Thermal Analysis*

The thermal analysis is performed in the plane normal to the welding direction by assuming that the heat flow in the welding direction is negligible. This is a reasonable assumption for high welding speeds (cf. [9]). The internal heat generation r for the weldment is modeled by the double ellipsoid heat source model of reference [10],

$$r = \frac{6\sqrt{3}Qf}{abc\pi\sqrt{\pi}} e^{-\left\{\frac{3x^2}{a^2}+\frac{3(d_2-y)^2}{b^2}+\frac{3[z+v(\tau-t)]^2}{c^2}\right\}} \tag{2.1}$$

where a, b and c are the ellipsoid semiaxes, t is the time and τ is a lag factor defined by the position of the torch at time $t = 0$; f is an empirical coefficient equated to 0.6 and 1.6 ahead of and behind the torch, respectively (cf. [10]). The energy input rate Q, is evaluated from

$$Q = \eta V I \tag{2.2}$$

where η, V and I are the heat source efficiency, voltage and current, respectively.

The ellipsoid semiaxes a, b and c are defined as (cf. [10])

$$a = d_4 + d_2 \tan(d_5) \tag{2.3}$$

$$b = d_1 \tag{2.4}$$

$$c = \begin{cases} a, & \text{ahead of the torch} \\ 4a, & \text{behind the torch} \end{cases} \tag{2.5}$$

An enhanced version of the finite element package FIDAP [4, 5, 11] is used to compute the temperature field and its sensitivity. The time domain is discretized in 35 variable steps. The time step size is controlled by a local time truncation error algorithm (cf. [11]). The peak temperature contours obtained throughout the heat cycle of the initial design appear in figure 4. Their sensitivities with respect to the plate half thickness d_2 and energy input rate Q are shown in figures 5 and 6 respectively. As seen from these figures, the peak temperatures decrease by increasing the plate half thickness d_2, and increase by increasing the energy input rate Q.

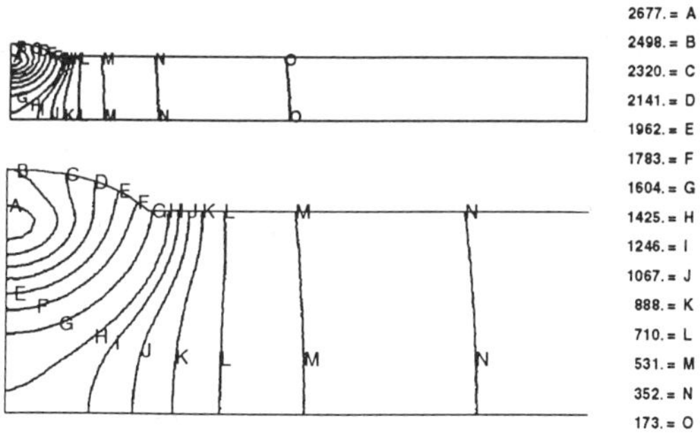

Figure 4. Peak temperatures of the initial design.

2.3 *Mechanical Analysis*

The plane stress assumption is used to model the welding process and subsequent service loading. Rate-independent, elasto-plastic material response with isotropic hardening is assumed. The

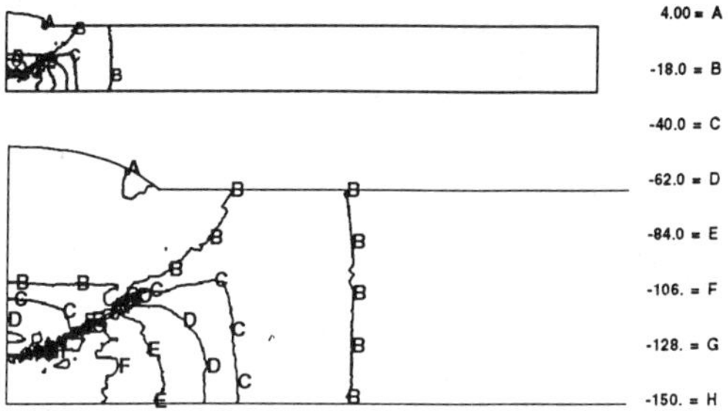

Figure 5. Peak temperature sensitivity of the initial design with respect to d_2.

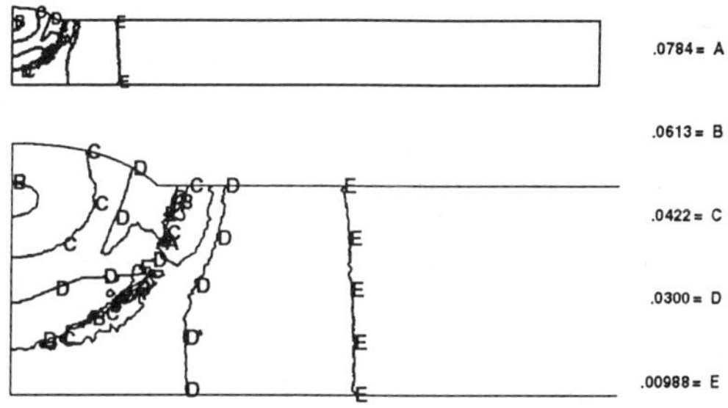

Figure 6. Peak temperature sensitivity of the initial design with respect to Q.

present model does not include solid state transformations. The small deformation isotropic hardening model is inadequate to evaluate the plastic hardening at high temperatures; therefore, it is assumed that no further hardening occurs for temperatures exceeding 600 °C. This assumption does not significantly affect the residual stress computations, since plastic strain energy is negligible at high temperatures [12].

The same finite element mesh is used throughout the analysis. For the simulation of the filler metal deposition, a small value for the Young's modulus and a large initial yield limit value are assigned to the filler metal before deposition. In this way, in the filler metal region the stress is essentially zero before deposition. Base metal properties are assigned to the filler metal elements as their temperature drops below the solidification temperature.

The contour plots of figures 7 and 8 illustrate the computed von Mises residual and service stress fields of the initial design, respectively. A peak residual stress of 443 MPa and service stress of 448 MPa are predicted. The sensitivities of the von Mises residual and service stress fields with respect to the plate half thickness d_2 and the energy

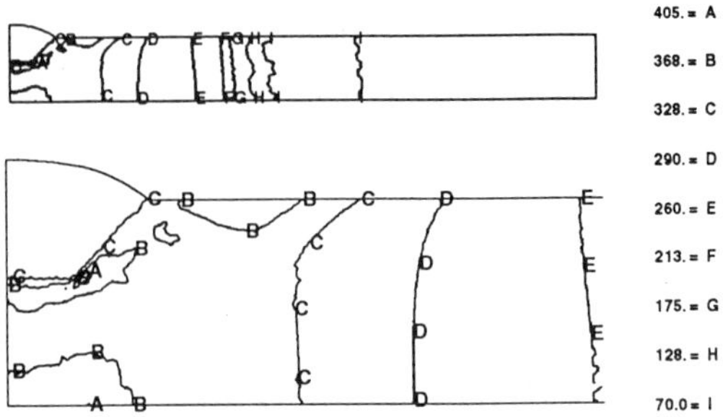

Figure 7. von Mises residual stress.

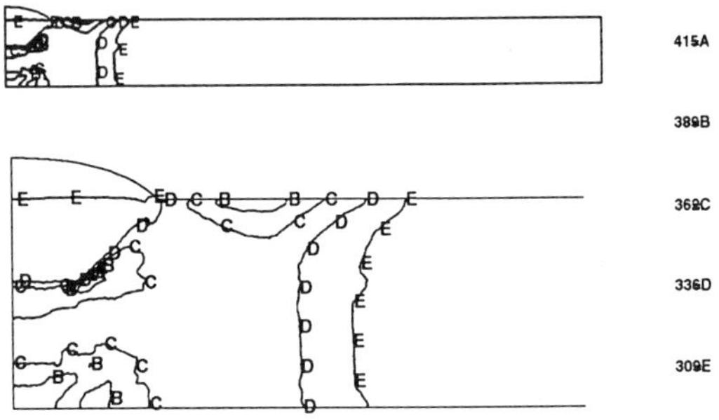

Figure 8. von Mises service stress.

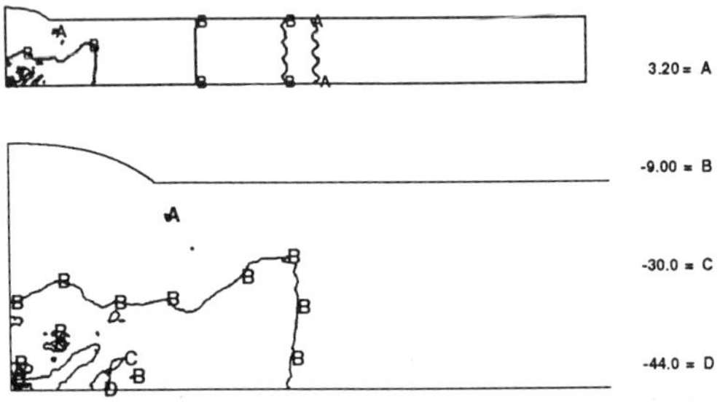

Figure 9. von Mises residual stress sensitivity with respect to d_2.

input rate Q are shown in figures 9 - 12. As seen in these figures, the von Mises residual and service stresses increase in the region of the peak von Mises stress, as the plate thickness increases. Elsewhere, in general, they decrease. The von Mises residual and service stresses will, in general, increase as the energy input rate Q increases.

2.4 *Design Optimization*

An optimization is performed to minimize the weld

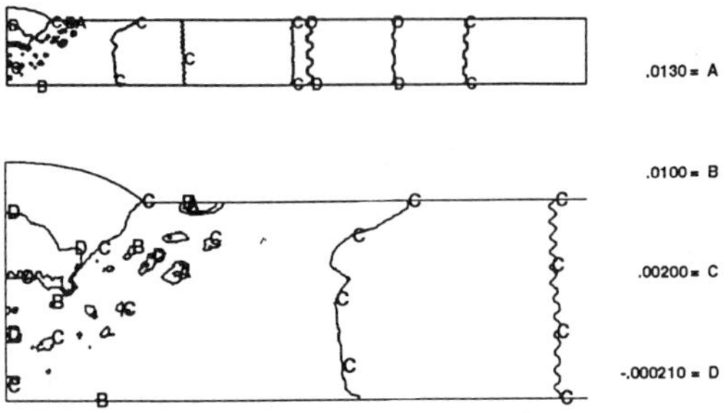

Figure 10. von Mises residual stress sensitivity with respect to Q.

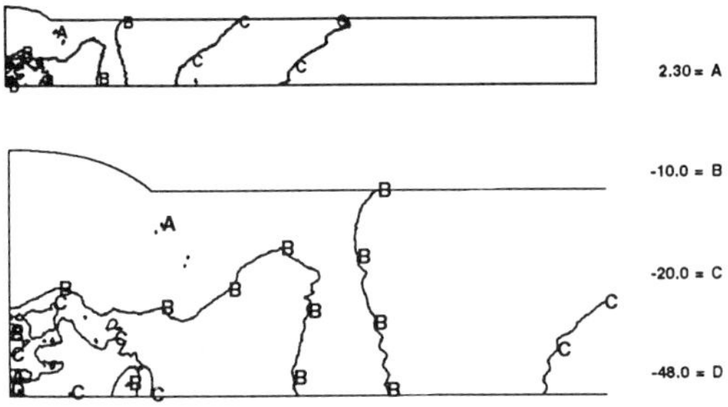

Figure 11. von Mises service stress sensitivity with respect to d_2.

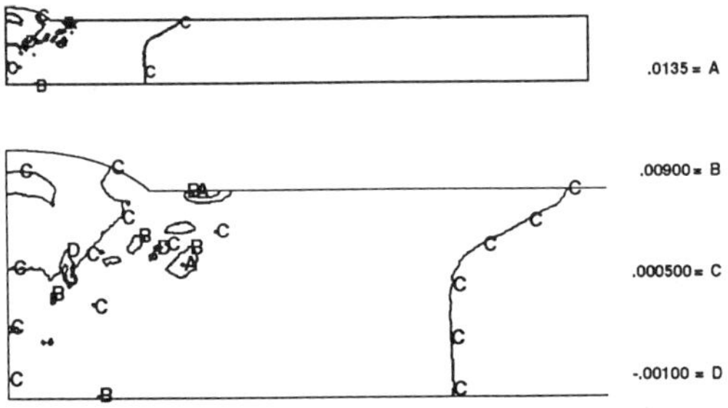

Figure 12. von Mises service stress sensitivity with respect to Q.

mass and constrain the peak temperature and von Mises stress levels which develop during the manufacturing process and product service loading.

Constraints are defined to ensure that the heat input is sufficient to perform the welding. These constraints require the peak nodal temperatures of the filler metal to exceed 1482 °C. To limit filler metal evaporation during the welding, additional constraints restrict the peak nodal temperatures of the filler metal to be less than 3000 °C.

The 420 MPa von Mises stress level limit, en-

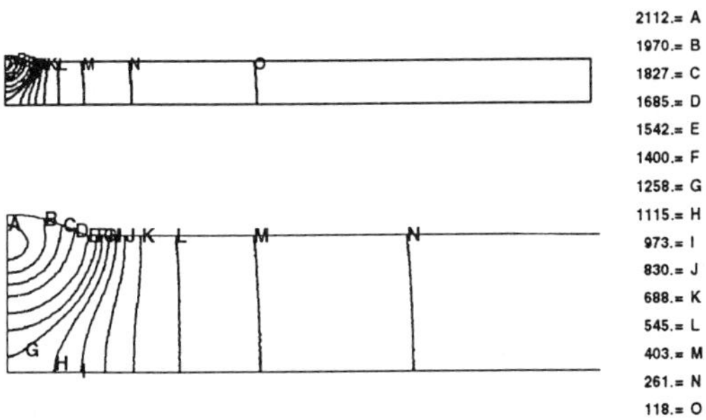

Figure 13. Peak temperatures of the final design.

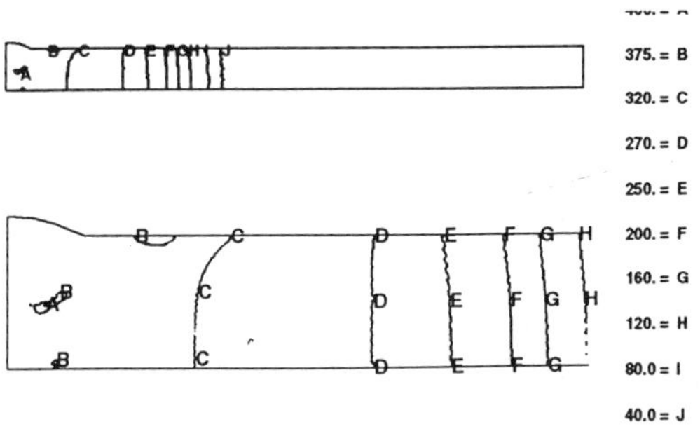

Figure 14. von Mises residual stress of the optimum design.

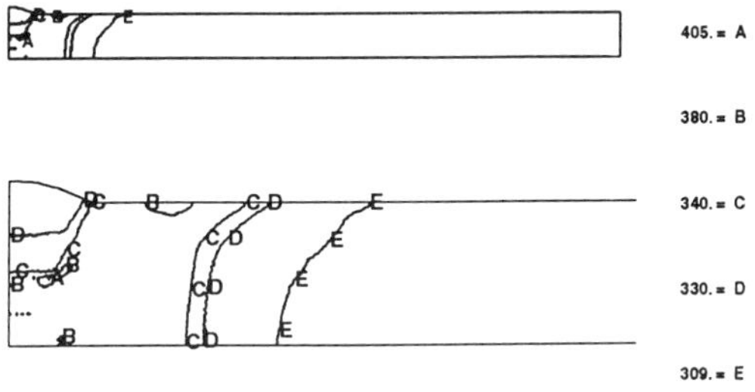

Figure 15. von Mises service stress of the optimum design.

forced over the entire structure, defines the remaining constraints for the optimization problem. The total number of stress constraints is reduced and the optimization efficiency is improved by using the mean von Mises stress of each element rather than the individual element Gauss point stress.

The numerical optimization is performed with the sequential linear programming algorithm of DOT [13]. Nine sensitivity analyses and fourteen analyses are required to obtain convergence. Each sensitivity analysis requires 10% of the analysis CPU time. The optimum values of the design parameters are denoted in table 1. Most of the design parameters attain their minimum or maximum allowable values (see table 1). The objective function, i.e. the welded plate area, is reduced from 2124 mm^2 to 1406 mm^2. The peak temperature contours throughout the heat cycle and the von Mises service stress contour of the optimum design are illustrated in figures 13 and 15, respectively. The peak von Mises residual stress is reduced to 411 MPa and the peak von Mises service stress is reduced to 413 MPa. Note that only those constraints related to the peak temperature field remain active at the optimum configuration.

3 CONCLUSION

A systematic design approach for systems governed by transient thermo-plastic problems is presented. The approach utilizes numerical optimization, the finite element method, and analytic sensitivity analysis. The analysis and sensitivity analysis capabilities are combined with numerical optimization to form the optimum design algorithm. The design of a weldment demonstrates the efficiency of the algorithm. The residual stresses which evolve during the heat cycle of the welding process are predicted by solving a nonlinear transient thermo-elasto-plastic problem. The product's service environment is simulated by subsequently loading and analyzing the structure. The weldment shape, energy input rate, and the torch speed are optimized to minimize the structure's weight and simultaneously limit the von Mises stress level.

ACKNOWLEDGMENT

The studies presented here were conducted under the support of the Edison Welding Institute for the project 8.5.2/13UIE1198 and the National Science Foundation through contract NSF DDM93-58/32NYI.

NOTE

This work will appear in the International Journal for Numerical Methods in Engineering

REFERENCES

[1] R. J. Yang, K. K. Choi, and E. J. Haug. Numerical Considerations in Structural Component Shape Optimization. *Transactions of*

the ASME, 107:334–339, 1985.

[2] A. D. Belegundu and S. D. Rajan. A Shape Optimization Approach Based on Natural Design Variables and Shape Functions. *Computer Methods in Applied Mechanics and Engineering*, 66:87–106, 1988.

[3] S. D. Rajan and A. D. Belegundu. Shape Optimal Design Using Fictitious Loads. *AIAA Journal*, 27(1):102–107, 1988.

[4] D. A. Tortorelli, M. M. Tiller, and J. A. Dantzig. Optimal Design of Advanced Parabolic Systems - Part I: Fixed Spatial Domain with Applications to Process Optimization. *Computer Methods in Applied Mechanics and Engineering*, 1993. To appear.

[5] D. A. Tortorelli, J. A. Tomasko, T. E. Morthland, and J. A. Dantzig. Optimal Design of Advanced Parabolic Systems - Part II: Variable Spatial Domain with Applications to Casting Optimization. *Computer Methods in Applied Mechanics and Engineering*, 1993. To appear.

[6] P. Michaleris, D. A. Tortorelli, and C. A. Vidal. Tangent Operators and Design Sensitivity Formulations for Transient Nonlinear Coupled Problems with Applications to Elasto-Plasticity. *International Journal for Numerical Methods in Engineering*, 37:2471–2499, 1994.

[7] P D A Engineering. *Patran Plus User's Manuals Vols I and II.* Software Products Division, Costa Mesa, CA, 1990.

[8] P D A Engineering. *Patran Command Language Guide.* Software Products Division, Costa Mesa, CA, 1990.

[9] B. A. B. Andersson. Thermal Stress in a Submerged-Arc Welded Joint Considering Phase Transformations. *Transactions of the ASME*, 100:356–362, 1978.

[10] J. Goldak, A. Chakravarti, and M. Bibby. A New Finite Element Model for Welding Heat Sources. *Metallurgical Transactions B*, 15B:299–305, 1984.

[11] M. S. Engelman. *FIDAP Theoretical Manual.* Fluid Dynamics International, Evanston, IL, 1987.

[12] A. P. Chakravati, L. M. Malik, and J. A. Goldak. Prediction of Distortion and Residual Stresses in Panel Welds. In *Computer modelling of fabrication processess and constitutive behaviour of metals*, pages 547–561, Ottawa, Ontario, 1986.

[13] VMA Engineering. *DOT User's Manual, Version 3.00.* Vanderplaats, Miura and Associates, Inc., Goleta, CA, 1992.

Simulation of Materials Processing: Theory, Methods and Applications, Shen & Dawson (eds)

ISBN 90 5410 553 4

FC-method: An inverse method to determine flow stress up to large strain range

Masahiro Michino & Mitsuyuki Tanaka
Nippon Metal Industry Co., Ltd, Sagamihara, Japan

Takashi Kitaoka
Shimadzu Corporation, Kyoto, Japan

ABSTRACT : An inverse analysis, named FC–method, has been applied to obtain flow stress of material up to large strain range. This procedure employs, as the object function of inverse analysis, the balance of external and internal powers which are given to a material during a deformation process and spent for the deformation respectively. The balance is explicit scalar functional of flow stress which is a function of some unknown material constants. Friction coefficient between the material and tools for deformation is also one of the unknown constants. Optimizing the balance is equivalent to solving the stationary problem of the functional and the solution gives the values of material constants and friction coefficient simultaneously. Ring compression of commercially pure aluminum was selected, as an example, to apply FC–method and the results demonstrated that the accuracy of this procedure is applicable enough.

1. INTRODUCTION

Reliability of a numerical analysis depends on accuracy of material properties used in the analysis. the most important property for analysis of a plastic forming process is the flow stress of material covering the whole strain range which appears in the process.

Uniaxial tensile test is the most common to measure flow stress of a material, since shaping the test pieces and the test process itself are easy. This method, however, cannot be applied to the range of large strain which appears during industrial forming, because this method is based on uniform elongation of test piece which is limited by necking or fracture.

In order to solve the problem of uniaxial tensile test, many kinds of compression tests are frequently adopted. Uniaxial compression test that uses cylindrical specimen and parallel dies is the most popular. This method is also based on uniform deformation of the specimen. The friction force between specimen and dies generates nonuniform deformation, when the deformation of specimen reaches the range of large strain. Many studies have been carried out for the purpose of reducing the effect of the friction. In the range of large strain, however, the effect of the friction still can not be negligible and accuracy of the flow stress is not enough even if the process is controlled carefully.

Osakada, Mori and Taniguchi (1981) proposed the constrained compression test, a compression test with a cylindrical specimen and grooved dies. Osakada, Shiraishi, Muraki and Tokuoka (1991) improved the constrained compression test using ring shaped test pieces. The constrained compression test is based on, instead of uniform deformation, the concept of average equivalent strain and average equivalent stress of the whole volume of the test piece. These averaged value are estimated from load–displacement curve and the result of the FEM analysis which executed beforehand with a certain flow stress. Therefore, the physical meaning of the flow stress obtained by this method is not clear and the accuracy is unknown when the flow stress is applied to actual forming processes in which the material has complicated distribution of strain.

Authors of this paper have been studying FC–method, an inverse method to determine flow stress of a material up to large strain range (Michino and Tanaka 1993, 1994). Because FC–method needs neither uniformity of deformation nor any particular concept like the constrain compression test, the flow stress in the range of large strain is possible to be obtained theoretically with good accuracy. In this paper, FC–method will be applied to ring compression of commercially pure aluminum and contrasted with the conventional uniaxial compression.

2. OUTLINE OF FC–METHOD

2.1 *Characteristics of FC–method*

Tomita, Hayashi and Tanaka (1992) determined flow stress of polymers through inverse analyses using FEM. Schnur and Zabaras (1992) solved an inverse problem to determine elastic material properties. Gelin and Ghouati (1994) expanded the inverse solution to the inelastic range. Since they used traction force or displacement at some points to construct the object function of inverse analysis, their object functions implicitly contain parameters to be determined. Thereby, they had to adopt finite difference approximation in order to minimize their object function of inverse analysis. In the finite difference approximation, parameters were perturbed one by one and an FEM analysis was executed corresponding to each perturbation at every iteration cycle.

On the other hand, FC–method employs the balance of external power given to a material and internal power spent for the deformation. This balance is the object function of inverse analysis. Thus object function of FC–method is expressed an explicit functional of flow stress. This is the point where FC–method is superior to predecessors', and all partial derivatives of that are consequently obtained simultaneously through one time execution of FEM analysis with suitable approximation. This approximation is only used for the purpose of updating the parameters to be determined through FC–method and flow stress of material is independent from this approximation. By FC–method, therefore, the cost of analysis to obtain flow stress with good accuracy is reduced.

In order to obtain flow stress in large strain range, the test piece of the objective material should be compressed. Friction between material and tools has large effect on deformation behavior in compression but the friction coefficient is not known beforehand. In FC–method, the friction coefficient can also be determined simultaneously with flow stress.

2.2 *Premises*

The following general premises are made to material, deformation and friction.

(1) Material is homogeneous and isotropic.
(2) Work hardening is free from strain path.
(3) Material yields accordingly to von Mises criterion.
(4) Deformation is isothermal.
(5) Friction force follows shear friction low.

Flow stress of a material is a function of equivalent plastic strain $\bar{\varepsilon}^p$, equivalent plastic strain rate $\dot{\bar{\varepsilon}}^p$ and temperature of the material T.

$$\bar{\sigma} = \bar{\sigma}(\bar{\varepsilon}^p, \dot{\bar{\varepsilon}}^p, T) \qquad (1)$$

When a suitable form of a function is selected so as to expresses the flow stress of the material, then eq.(1) is rewritten to eq.(1)'. This equation shows that the flow stress can also be regarded as a function of unknown constants C_i which depend on the material. To obtain the values of the unknown constants, which is the aim of FC–method, is equivalent to determine the flow stress of the material.

$$\bar{\sigma} = \bar{\sigma}(C_i) \qquad (i = 1 \sim j) \qquad (1)'$$

Shear friction coefficient m is also expressed as a function of C_{j+1}, the (j+1)th unknown constant, in order to satisfy the domain of shear friction coefficient, $0 \le m \le 1$.

$$m = \frac{1}{\pi}\tan^{-1}C_{j+1} + 0.5 \qquad (2)$$

2.3 *Object function of inverse analysis*

The external power w is the inner product of p, the vector of the loads necessary to deform, and v, the vector of the displacement velocities of the tools.

$$w = \sum pv \qquad (3)$$

And internal power e is the total of powers for the elastic and plastic deformation of the material and the frictions between the material and the tools, e^e, e^p and e^f respectively.

$$e^e = \int \sigma_{kl}\, \dot{\varepsilon}^e_{kl}\, dV \qquad (4)$$

$$e^p = \int \bar{\sigma}\, \dot{\bar{\varepsilon}}^p\, dV \qquad (5)$$

$$e^f = \frac{m}{\sqrt{3}} \int \bar{\sigma}\, |\Delta u|\, dS \qquad (6)$$

Where, σ_{kl} and $\dot{\varepsilon}^e_{kl}$ are components of stress and elastic strain rate respectively. Einstein's summation convention is adopted in expression of eq.(4). $|\Delta u|$ in eq.(6) is the norm of the relative sliding velocity between the material and the tool at the contact surface. Eq.(4) and (5) are integrated with respect to the whole volume of the material, and eq.(6) is integrated at every contact surface of the material to the tools.

Since external power w is invariably equal to internal power e at any time, t, of the deformation process, the balance of the external and the internal power leads to the following equation :

$$\varrho = \int (w - e)^2 dt = 0 \qquad (7)$$

In order that eq.(7) is true, unknown constants C_i in eq.(1)' and (2) must represent the real flow stress of the material and the real friction coefficient at the contact surface respectively. ϱ in eq.(7) is the object function of inverse analysis in FC–method.

Because eq.(7) is non–linear in the unknown constants C_i, minimization of ϱ by optimization of C_i is a kind of non–linear least squares problem.

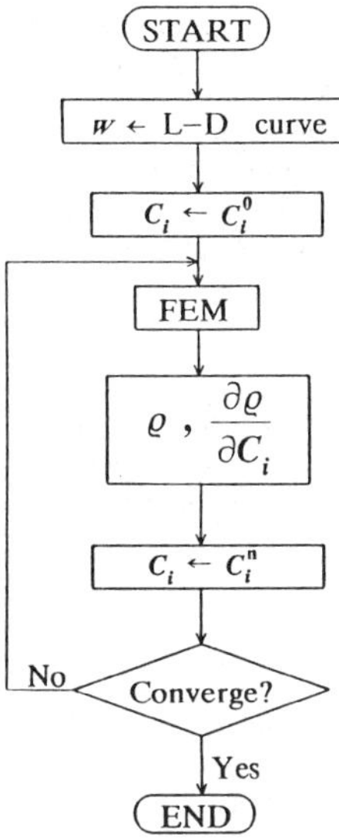

Fig. 1 Procedure of FC–method.

2.4 *Solution algorithm*

The non–linear least squares problem of FC–method is solved using the Davidon–Fletcher–Powell algorithm (Fletcher and Powell 1963). At the start of the (n)th iteration cycle, an FEM analysis is executed with constants $C_i^{(n-1)}$ which are obtained at the (n−1)th iteration cycle. Then $C_i^{(n)}$, the next improved values of unknown constants, are calculated using metric matrix $\boldsymbol{H}$ which is also obtained at the (n−1)th iteration. The matrix $\boldsymbol{H}$ is an approximation to the inverse of hessian matrix the components of which are the 2nd order partial derivatives. If it is the first iteration, n=1, C_i^0 are a set of initially guessed values and matrix $\boldsymbol{H}^0$ is a unit matrix.

(i)Using the result of the FEM analysis, $\varrho^{(n)}$ and g, the object function and the vector of partial derivatives respectively, are evaluated as functions of unknown constants C_i, where

$$g = \left\{ \frac{\partial \varrho^{(n)}}{\partial C_1}, \frac{\partial \varrho^{(n)}}{\partial C_2}, \cdots, \frac{\partial \varrho^{(n)}}{\partial C_{j+1}} \right\}^{T} \tag{8}$$

Each component of g is obtained using the next approximation.

$$\frac{\partial e}{\partial C_i} = \begin{cases} \displaystyle\int \frac{\partial \bar{\sigma}}{\partial C_i} \dot{\bar{\varepsilon}}^{p}\, \mathrm{d}V + \frac{m}{\sqrt{3}} \int \frac{\partial \bar{\sigma}}{\partial C_i} |\Delta u|\, \mathrm{d}S & (\, i = 1\text{~}j \,) \\ \displaystyle\frac{1}{\sqrt{3}} \frac{\partial m}{\partial C_i} \int \bar{\sigma}\, |\Delta u|\, \mathrm{d}S & (\, i = j+1 \,) \end{cases} \tag{9}$$

(ii) δ_i, the correction of unknown constants C_i, is calculated using eq.(10) and (11)

$$\boldsymbol{C}^{(n)} = \boldsymbol{C}^{(n-1)} + \lambda\, \boldsymbol{\delta} \tag{10}$$

$$\boldsymbol{\delta} = \boldsymbol{H}^{(n-1)}\, g(\boldsymbol{C}^{(n-1)}) \tag{11}$$

Where $\boldsymbol{C} = \{C_1, C_2, \cdots, C_{j+1}\}^T$ and $\boldsymbol{\delta} = \{\delta_1, \delta_2, \cdots, \delta_{j+1}\}^T$

Superscripts for the vector $\boldsymbol{C}$ and the matrix $\boldsymbol{H}$ are iteration numbers when those are obtained. Scalar value λ is determined through the line search with the interpolation using the 3rd order polynomial approximation, so as to minimize the value of the function $\varrho^{(n)}(\boldsymbol{C}^{(n-1)} + \lambda \boldsymbol{\delta})$.

(iii) The matrix $\boldsymbol{H}^{(n-1)}$ is updated to $\boldsymbol{H}^{(n)}$, using eq.(12), which is necessary to execute the next iteration.

$$\boldsymbol{H}^{(n)} = \boldsymbol{H}^{(n-1)} + \lambda \frac{\boldsymbol{\delta}\, \boldsymbol{\delta}^T}{g_S\, \boldsymbol{H}^{(n-1)} g_S^T} - \frac{\boldsymbol{H}^{(n-1)} (g_E - g_S)(g_E - g_S)^T \boldsymbol{H}^{(n-1)}}{(g_E - g_S)^T \boldsymbol{H}^{(n-1)} (g_E - g_S)} \tag{12}$$

where $g_S = g(\boldsymbol{C}^{(n-1)})$ and $g_E = g(\boldsymbol{C}^{(n)})$.

2.5 *Procedure to determine unknown constants*

Fig. 1 shows the procedure of FC–method. At the beginning of FC–method, a test piece of the material concerned is deformed and the external power by eq.(3) is calculated using the load–displacement curve measured during the deformation process. The deformation must contain the strain level to cover the range of strain concerned, but uniformity of the deformation is not necessary.

C_i^0, a set of initially guessed values of unknown constants, is substituted for eq.(1)' and (2), then the deformation of the test piece is analyzed by FEM using the initial flow stress and the initial friction coefficient. Using the results of the FEM analysis, the next improved assumption of the constants C_i are obtained through the algorithm in above section. Then the initial values of unknown constants are replaced with these values for the next iteration cycle. And the steps are iterated from the execution of FEM until a specified condition of convergence is satisfied.

3. EXPERIMENT

3.1 *Procedure*

In order to demonstrate effectiveness of FC–method and accuracy of the flow stress obtained through FC–method, ring compression of commercially pure

Table 1 Sizes of Test Pieces

	Inner Diameter (mm)	Outer Diameter (mm)	Height (mm)
Ring 1	24	48	16
Ring 2	24	48	24
Cylinder	-	40	40

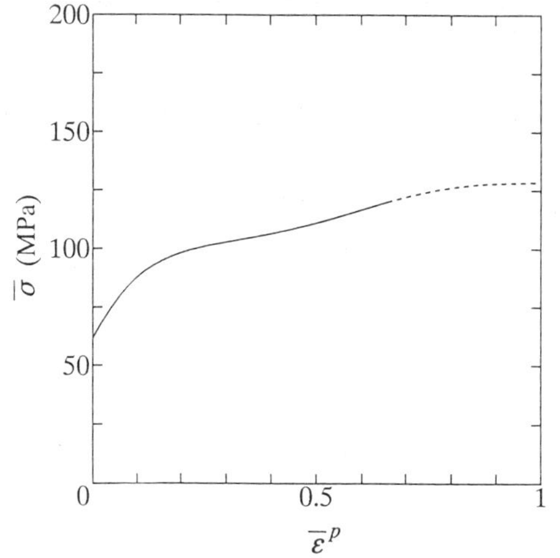

Fig. 2 Flow stress measured by the conventional compression with frequent lubrication.

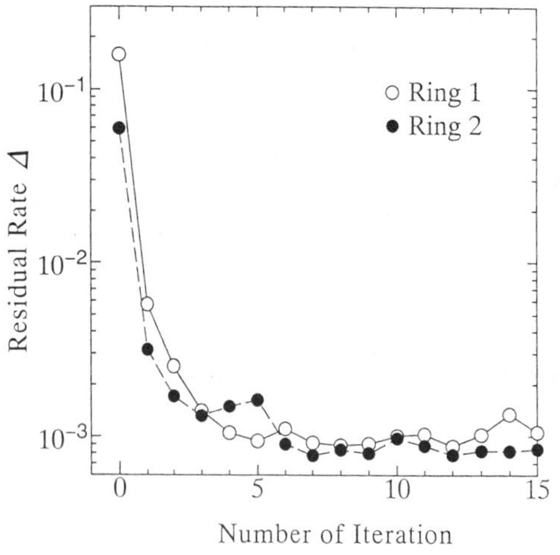

Fig. 3 Residual rate of the object function.

aluminum (ISO Al 99.7) was selected as an example. Test pieces were finished by machining after solution treatment at 610K. Two types of rings which had different aspect ratios as shown in Table 1 were compressed by parallel two plane dies with no lubrication. The compression of ring were continued up to 60% reduction in height.

The 5th order polynomial of equivalent plastic strain was adopted to the function form of flow stress in eq.(1)'. Flow stress and shear friction coefficient were determined for each ring. The residual rate Δ of the object function ϱ was defined and averaged value over the whole reduction range was used for test of convergence.

$$\Delta = \left(1 - \frac{e}{w}\right)^2 \tag{13}$$

Iterations were continued up to 15 times to check the stability of the residual rate.

On the other hand, conventional uniaxial compression of a cylindrical specimen, sizes of which are also shown in Table 1, was carried out to measure the flow stress of the material, $\overline{\sigma}_C$. In the conventional compression, contact surface of the specimen to dies and the whole surface of the dies were frequently lubricated at every 10% reduction in height.

3.2 *Conditions for FEM analyses in FC-method*

Because of symmetry of the problem, a quarter of the cross section of the ring was modeled using 4-nodes axi-symmetric element. The height of the cross section was divided into 10 and radius into 15 to construct FEM mesh. The die was modeled using a rigid segment of line. FEM analyses were executed up to 60% reduction in height with incremental reduction step of 1%. A general purpose code MARC was employed for FEM analyses.

Initial flow stress, $\overline{\sigma}_0$, was calculated from the load-displacement curve of the ring compression using the assumption of uniform deformation. This initial flow stress was calculated for each size of ring respectively. Initial shear friction coefficient was set to 0.5 for the both rings.

3.3 *Results and discussion*

Fig. 2 shows $\overline{\sigma}_C$, the flow stress measured by the conventional uniaxial compression. Bulging grew to appreciable level, in spite of frequent lubrication, when the reduction exceeds 50%. True strain of 0.7 was accordingly the limit of strain range where the flow stress $\overline{\sigma}_C$ was reliable.

Fig.3 shows Δ, residual rate of the object function ϱ, averaged over the reduction in height from 0 to 60% at each iteration of FC-method. The object function ϱ converged within 7 times of iteration and the residual rate was stable after that.

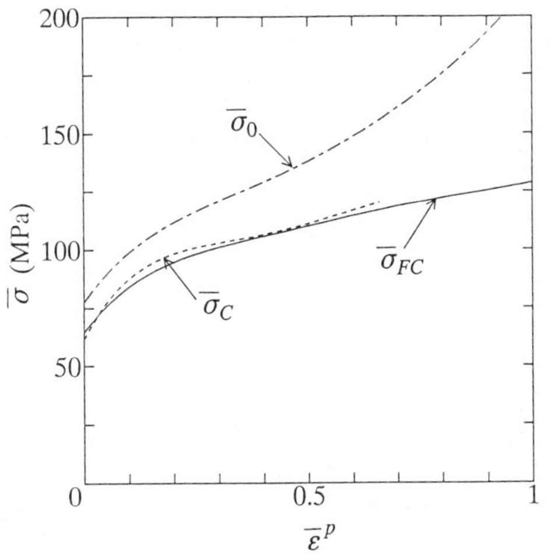

(a) Ring 1

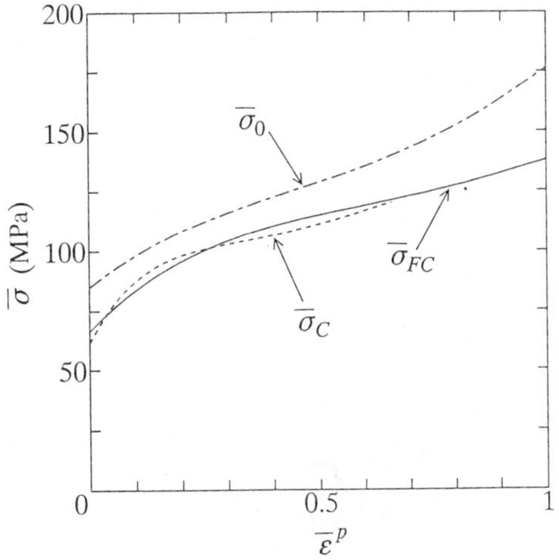

(b) Ring 2

Fig. 4 Comparison of $\overline{\sigma}_{FC}$ with $\overline{\sigma}_0$ and $\overline{\sigma}_C$.

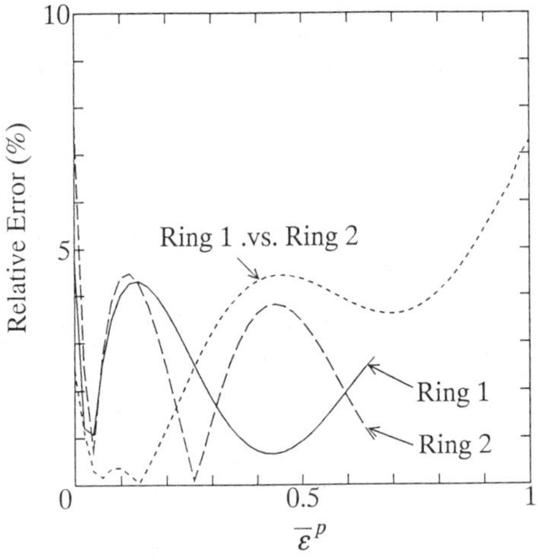

Fig. 5 Accuracy of $\overline{\sigma}_{FC}$ obtained for each ring.

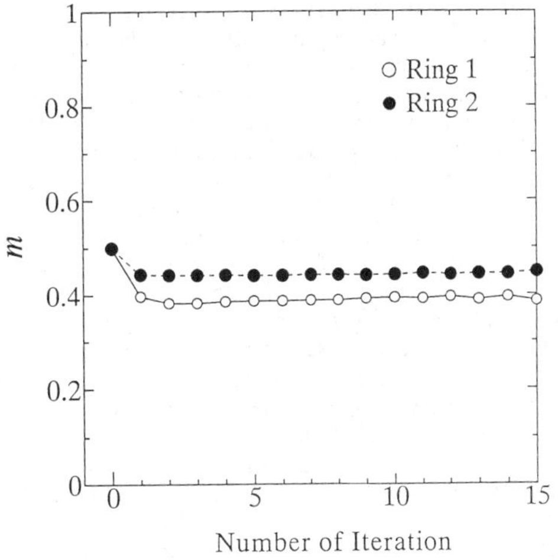

Fig. 6 m obtained at each iteration cycle.

Hereinafter in this paper, subscript FC to the flow stress $\overline{\sigma}$ and to shear friction coefficient m means that those were obtained through convergence of FC–method.

Fig. 4 shows comparisons of $\overline{\sigma}_{FC}$, flow stress obtained trough FC–method, with $\overline{\sigma}_0$, initial flow stress, and $\overline{\sigma}_C$, measured using the conventional compression, where Fig.4 (a) is the comparison for ring 1 and Fig.4 (b) is for ring 2 respectively. In each ring, $\overline{\sigma}_{FC}$ seemed to agree $\overline{\sigma}_C$.

Fig. 5 shows accuracy of $\overline{\sigma}_{FC}$ obtained for each ring. In the range of true strain less than 0.7, the accuracy was estimated using relative error of $\overline{\sigma}_{FC}$ to $\overline{\sigma}_C$ because 0.7 in true strain was the limit where $\overline{\sigma}_C$ was reliable in this study. The accuracy was also estimated using relative error of $\overline{\sigma}_{FC}$ between ring 1 and ring 2 in the whole range of strain from 0 to 1. The maximum error of flow stress was less than 5% within the strain range from 0 to 0,9. Therefore reliable range of strain in FC–method is limited by the maximum strain which is calculated from displacement with the assumption of uniform deformation.

Fig. 6 shows m, shear friction coefficient, obtained at each iteration cycle. The value of m converged within 3 times of iteration, which was earlier than convergence of the object function ϱ. The initial vale of m was not so different from the real value obtained through FC–method. This may be the reason of the faster convergence of m.

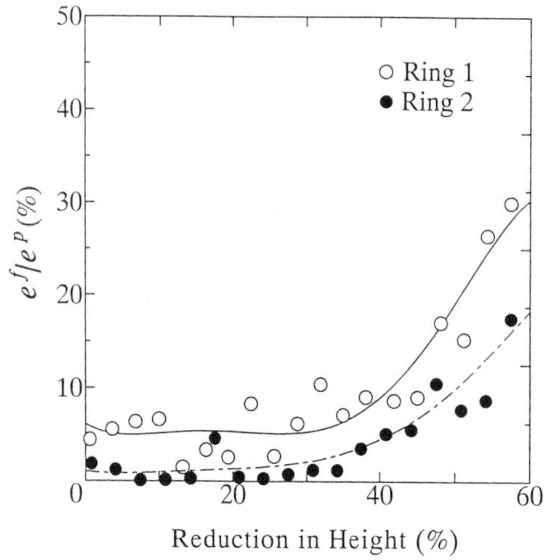

Fig. 7 Ratio of e^f to e^p.

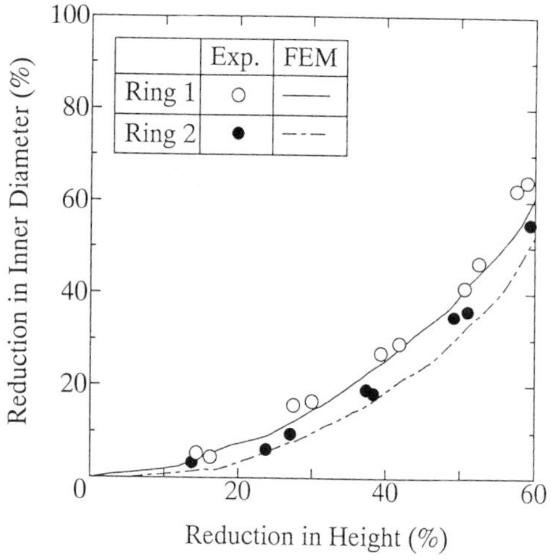

Fig, 8 Change of reduction in inner diameter by reduction in height.

The value of shear friction coefficient obtained in ring 1 and 2 was slightly different, less than 0.05, from each other. Fig. 7 shows ratio of e^f to e^p, ratio of the power spent for friction to the power for plastic deformation. The ratio in ring 2 was much smaller than that of ring 1 in the whole process of compression. The accuracy of m_{FC} depends on ratio of e^f/e^p, which is affected by aspect ratio of the specimen, because the total of the powers for the deformation and the friction was used to construct the object function in FC-method. The value of$m$ obtained for ring 1 was, thereby, more reliable than that for ring 2. It is suggested that the aspect ratio of ring 1 is suitable to FC-method.

Fig. 8 shows the changes of reduction in inner diameter by reduction in height, where circles indicate experiments and lines were calculated by FEM using $\overline{\sigma}_{FC}$ and m_{FC}. The calculated results show good agreement with those observed. Therefore, the accuracy of flow stress and shear friction coefficient obtained through FC-method, $\overline{\sigma}_{FC}$ and m_{FC}, is practically good enough.

Conclusions

An inverse method, named FC-method, has been applied to obtain flow stress of material up to range of large strain. In FC-method, a test piece is compressed to measure load-displacement curve in order to obtain external power of the deformation. And internal power is obtained as a function of material constants by FEM analysis. The balance of the powers is minimized by solving non-linear least squares problem. Then the flow stress of the material and shear friction coefficient between the material and the tools are given simultaneously. Ring compression of commercially pure aluminum was selected to demonstrate this method, which showed that this method to determine flow stress is effective.

References

Fletcher,R. and Powell,M.J.D. 1963. A Rapidly Convergent Descent Method for Minimization, *Comp. J.*, 5: 163-168

Gelin,J.C. and Ghouati,O. 1994. An Inverse Method for Material Parameters Estimation in the Inelastic Range, *Proc. WCCM* Ⅲ, 2:988-989.

Michino,M and Tanaka,M. 1993. Determination of Flow Stress by Convergence of Unknown Constants using Finite Element Method, *Proc. Int. Conf. Comp.-Assist. Mater. Des. Process Simul., 1993*: 201-206.

Michino,M and Tanaka,M. 1994. Determination of Flow Stress by Inverse Solution using Finite Element Method, *Proc. WCCM* Ⅲ, 2: 984-986

Osakada,K., Mori,K. and Taniguchi,N. 1981. A Method of Determining Flow Stress under Forming Conditions, *Ann. CIRP*, 30: 135-138.

Osakada,K, Shiraishi,M., Muraki,S. and Tokuoka,M. 1991. Measurement of Flow Stress by the Ring Compression Test, *JSME Int. J. Series I*, 30: 312-318.

Tomita,Y. Hayashi,K. and Tanaka,S. 1992. Identification of Constitutive Equation of Polymeric Bars with Instability Propagation, *Trans. Jpn. Soc. Mech. Eng. A*, 58:1859-1863.

Schnur,D.S. and Zabaras,N. (1992): An Inverse Method for Determining Elastic Material Properties and a Material Interface, *Int. J. Num. Meth. Eng.*, 33:2039-2057.

Simulation of Materials Processing: Theory, Methods and Applications, Shen & Dawson (eds)
© 1995 Balkema, Rotterdam. ISBN 90 5410 553 4

FEM – Micro genetic algorithms based approach to optimal design of multi-stage metal forming processes

Subir Roy
Altair Engineering, Inc., Troy, Mich., USA

Somnath Ghosh
Department of Engineering Mechanics, The Ohio State University, Columbus, Ohio, USA

Rajiv Shivpuri
Department of Industrial and Systems Engineering, The Ohio State University, Columbus, Ohio, USA

ABSTRACT: This paper describes a new method for design optimization of multi-stage metal forming processes. The selected forming processes are multi-pass cold wire drawing, multi-pass cold drawing of a tubular profile and cold forging of an automotive outer race preform. An adaptive Micro Genetic Algorithm (μGA) has been implemented for minimizing a wide variety of objective/cost functions relevant to the respective processes. Significant improvements in the simulated product quality and reduction in the number of passes has been observed as a result of optimization.

1 INTRODUCTION

Design by numerical simulation is a trial and error procedure, where the objective is to find a cost effective process to manufacture a defect free product by varying process parameters. It is extremely important to devise an efficient integrated analysis-design method that will minimize human intervention while maintaining a high level of accuracy in optimality considerations. In the recent past, biologically inspired Genetic Algorithms (GAs) optimization technique based on probabilistic transition rules, have been successfully implemented for a wide variety of problems in physical and social sciences, engineering, manufacturing and operations research, and computer sciences [1-3]. Micro Genetic Algorithms (μGA) have evolved [4,5] to reduce the large computation time required for Simple Genetic Algorithms (SGA) based optimization schemes. In this study, a design tool has been developed for multi-stage metal forming processes based on the coupling of Micro Genetic Algorithms for optimal design with finite element analysis of metal forming processes using general purpose packages. The main advantages of this technique are its modular nature and its capability to handle a large number of discrete and continuous variables.

2 SIMPLE GENETIC ALGORITHMS (SGA)

An optimal design problem can be generally represented as:

$$\text{minimize } f(\mathbf{d}) \qquad \mathbf{d} \in H$$
$$\text{subject to } g_i(\mathbf{d}) \leq 0 \quad i=1,2,\ldots,m \tag{1}$$

where $f(\mathbf{d})$ is the objective or cost function to be minimized, $\mathbf{d}$ represents the design variables to be optimized, H is the design space which determines the bounds for the design variables, $g_i(\mathbf{d})$ are the design constraints and m is the total number of such constraints. In the implementation of Simple Genetic Algorithms a constrained problem is first transformed to an unconstrained form, by associating a penalty term for all constraint equations. Eq. (1) is thus transformed to the form:

$$\text{minimize } f(\mathbf{d})+\gamma\sum_{i=1}^{m}\Phi[g_i(\mathbf{d})] \tag{2}$$

where $\Phi[g_i(\mathbf{d})]$ is a chosen penalty function and γ

is a penalty coefficient. A quadratic penalty function of the form:

$$\Phi[g_i(\mathbf{d})] = \{g_i(\mathbf{d})\}^2 \qquad (3)$$

is commonly used in literature [1, 6].

In the design optimization of multi-pass metal forming processes implementation of the Simple Genetic Algorithms involves the following steps:

i) Discretization of all chosen design variables:

The number of independent design variables depend on the process being considered. Typically in metal forming processes, the design parameters can be the preform dimensions, the total number of forming stages, the amount of deformation at each stage, the tool dimensions etc. Appropriate bounds should be imposed on each design variable depending on the process being considered. For example, in cold forging processes the length to diameter ratio of the billet should not exceed the limit so as to cause buckling. After identifying these bounds, each discretized design variable is coded into a string of "1's" and "0's" which is the binary equivalent of the actual value of the design variable. The length of a string, that is the number of bits that constitute a string, depends on the range and resolution with which each variable is being represented. The problems considered in this study involve multiple variables. A multi-parameter code can be constructed by partitioning each string into a number of compartments equal to the number of design variables as shown in Figure 1. Here the total length of the string depends on the bit size corresponding to each design variable and the maximum number of allowable forming stages.

ii) Evaluation of objective function values for each string in the initial population:

Typical objective functions that may be considered

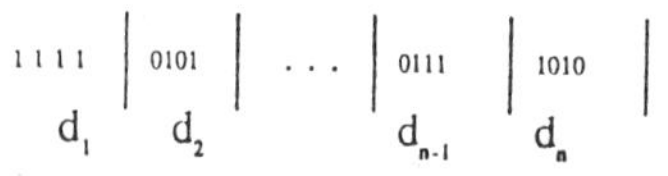

Figure 1 Construction of a multi-parameter binary coded string (d_i = i^{th} design variable)

for metal forming processes are,

(a) Difference between maximum and minimum effective plastic strains in the product, (b) Total energy consumption, (c) Maximum damage value, (d) Die filling etc. The values of the chosen objective function corresponding to the strings arbitrarily generated initially, can be obtained from the analysis module. Each string is next assigned a weight between 0 and 1 which is referred to as its fitness coefficient and is proportional to the magnitude of its objective function as discussed in [1]. A 0 fitness coefficient may be assigned to strings which are undesirable. For example, strings which correspond to design variables that cause necking of the workpiece in a multi-pass wire drawing process, may be assigned a zero value. For the multi-stage metal forming processes considered in this study, design constraints of the form described in Eq. (1) are not present. Hence penalty functions are not required for this study.

iii) Generation of new string populations with better fitness coefficients using GA operations:

There are three distinct operations involved in this process. They are (a) Reproduction, (b) Crossover and (c) Mutation.

(a) Reproduction is a selection procedure where the strings with higher fitness coefficients have a higher probability of being selected or even copied multiple times for mating in the next generation. The algorithmic implementation of this operation involves creation of a roulette wheel, with each string in the current population having a slot size proportional to its fitness. The wheel is then rotated to select a pool of strings for crossover and mutation. For a fixed population size, the number of such rotations is always constant.

(b) The crossover operation involves two steps. First, members resulting from the reproduction operation are randomly paired for mating. Second, a crossover site is selected randomly for each pair and all characters between that position and the end of the string for both mates are swapped for generating new strings.

(c) The mutation operation involves an occasional random alteration of one or more bits in the string. This operation prevents premature convergence. A schematic of the SGA mechanism is illustrated in Figure 2.

Since GA is a probabilistic search technique, a stopping criteria has to be fixed apriori. A common practice is to fix the number of trials in which no further improvement in the value of the objective function takes place.

The population size for SGAs being currently used in a wide variety of optimization problems range from 30 to 200. It is believed that a large population size helps in better schema processing and thus reduces the chances for premature convergence. This is a serious drawback for the present problem, considering the cost of objective function evaluations using a large deformation FEM analysis code. To overcome this problem, the concepts of Micro Genetic Algorithms (μGA) as proposed by Krishnakumar [4] have been utilized. A population size as small as 5 can be used successfully with μGA. However, the problem of relatively large number of objective function evaluations to arrive at near optimal solutions could still arise with this strategy. To overcome this, an adaptive μGA has been developed and implemented in this study.

3 IMPLEMENTATION OF THE ADAPTIVE MICRO GENETIC ALGORITHMS

The details of the μGA implementation used in the present study are discussed in Ref. [5]. The following steps are used:

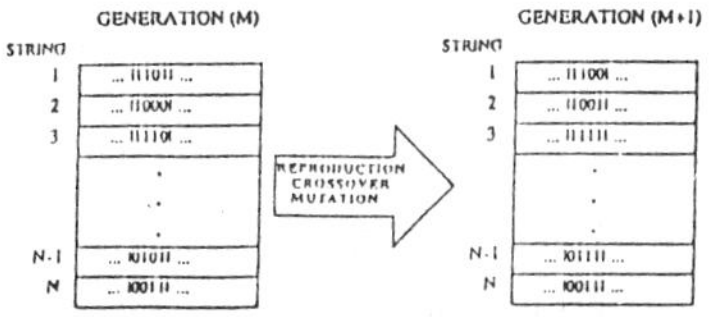

Figure 2 Schematic of the Simple Genetic Algorithm (SGA) mechanism

i) A small population of a very small number of strings, say 6, is randomly generated.

ii) The next generation is formed by choosing the best string from step 1 and generating 5 more strings randomly. The best or the fittest string is the one which has a minimum value for the objective function. The 6 chosen strings are grouped randomly into 3 pairs and crossed. This operation (referred to as μGA-1) is continued for a pre-determined number of times. This gives a good starting population for the next level of μGA.

iii) The best string from step 2 is crossed with 3 more randomly chosen strings using a deterministic tournament selection strategy. In this strategy, the pairing of mates for crossover is enforced between the strings without using a roulette wheel based probabilistic mating pool selection. This operation (referred to as μGA-2) is continued as long as better strings with lower values of objective functions are generated. The best string so far is always retained and passed on to the next mating pool to cross with 3 more randomly chosen strings. If better strings do not result, switchover is made to the next level (referred to as μGA-3) after a fixed number of trials. Switchover to μGA-3 is also made if strings with fitnesses close to the best string so far result.

iv) The population from step 3 is chosen and the genetic operations of reproduction, crossover and mutation are performed. Reproduction now is based on a roulette wheel selection strategy. However, the best string so far is always forced to enter the mating pool. This operation is continued until a nominal convergence measure is achieved. Nominal convergence is assumed if 3 or more identical strings result and if there are no other strings in the current population with fitnesses fairly close to the identical strings. When nominal convergence occurs switchover is made to μGA-2.

v) Steps 3 and 4 are continued until a global convergence measure is achieved. Global convergence is assumed if no significant improvement in the objective function takes place after a fixed number of trials.

The key feature of the μGA developed here lies in its adaptive nature. Switchover between μGA-2 and μGA-3 is constantly made depending on the distribution of the design variables in the current population. SGA operations like roulette wheel selection and mutation are still active (μGA-3) since they are effective in improving the solutions locally. On the other hand deterministic tournament selection strategy and pure crossover (μGA-2) which has been used in the μGA developed by Krishnakumar [4] is found to be more effective for global search. Thus, the constant switchover accelerates the convergence, especially when a large number of design variables are active at near optimal solutions.

Convergence to optimal solutions with very small tolerances may still be expensive due to relatively large number of objective function evaluations that occur as a result of slowing down of the convergence at near optimal solutions. However in the problems considered, the significance of substantially enhancing the existing design with limited number of trials is much more than reaching actual optima. Hence in such cases, a rich database of near optimal solutions is sufficient. The developed algorithm fulfills this objective satisfactorily and this is evident from the results of the numerical experiments discussed later.

4 DESIGN OPTIMIZATION OF SELECTED MULTI-STAGE METAL FORMING PROCESSES WITH MICRO GENETIC ALGORITHMS

Three different multi-stage metal forming processes have been considered to demonstrate the applicability of the developed technique. They are:

(i) Multi-pass cold wire drawing process where rods of different sizes are reduced using several dies to very small diameter wires. The objective function considered for minimization is the difference between the maximum and minimum effective plastic strains in order to obtain uniform

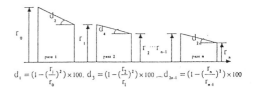

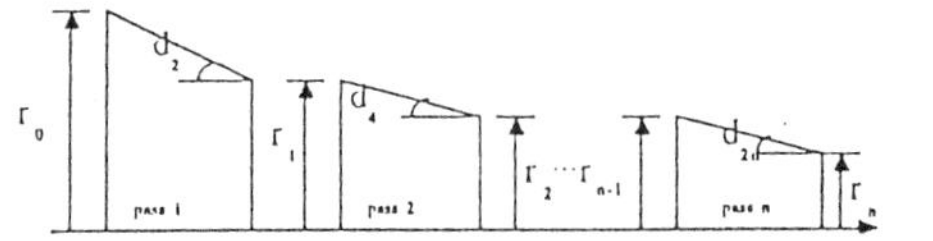

$$d_1 = \left(1 - \left(\frac{r_1}{r_0}\right)^2\right) \times 100,\ d_3 = \left(1 - \left(\frac{r_2}{r_1}\right)^2\right) \times 100 \ldots\ d_{2n-1} = \left(1 - \left(\frac{r_n}{r_{n-1}}\right)^2\right) \times 100$$

Figure 3 Description of chosen design variables in wire drawing (d_i = i^{th} design variable)

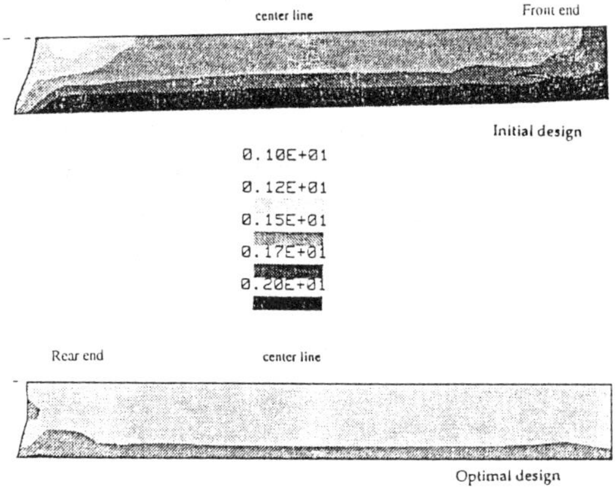

Figure 4 Improvement in effective plastic strain

Table 1 Evolution of optimal design variables

trial #	minima of strain difference	N	die semi-angle in passes (deg.)	reduction ratio (%)		energy (J)
				upto final pass	final	
initial	1.158	6	15.15.15.10.7.15	25.22.30.20.20	5.27	55.16
1-9	0.339	5	8. 9. 9. 5. 7	23. 23. 35. 31	6.66	41.24
10	0.337	5	8. 9. 10. 5. 5	23. 23. 35. 31	6.66	41.62
11	0.331	5	8. 5. 10. 5. 9	23. 23. 35. 35	0.917	42.82
12-14	0.326	5	8. 5. 10. 5. 9	23. 23. 35. 31	6.66	40.66
15-16	0.320	4	8. 5. 10. 5	23. 26. 35	32.98	43.50
17-19	0.265	4	5. 5. 10. 5	22. 26. 35	33.84	43.65
20 - 22	0.258	5	5. 5. 10. 5. 8	23. 26. 31. 35	2.88	41.17
23	0.243	5	5. 5. 6. 5. 8	23. 26. 31. 33	5.78	39.55
24-32	0.234	5	5. 5. 6. 5. 5	23. 26. 29. 33	8.43	38.86

distribution of mechanical properties. The objective function evaluation module is FEM package, NIKE2D [7]. The design variables are die angles, area reduction ratio at each pass and the total number of passes as shown in Figure 3. The improvement in effective plastic strain distribution

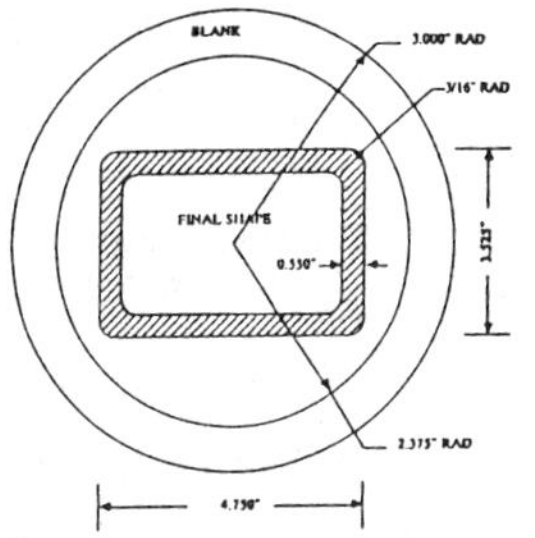

Figure 5 Billet and final shape for a cold drawn tubular profile

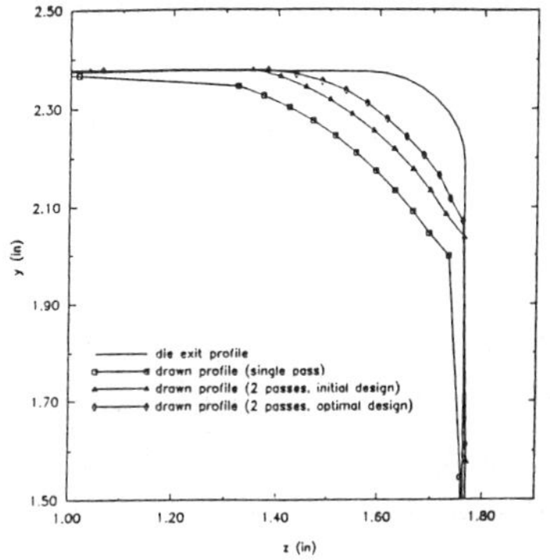

Figure 6 Improvement in corner filling

Table 2 Evolution of optimal design variables

trial #	minima of underfill (%)	preform outer radius (in)	preform inner radius (in)	die length (in)
initial	1.999	3.000	2.375	5.50
1	0.632	3.200	2.325	5.50
2	0.518	3.200	2.325	4.30
3-7	0.350	3.173	2.225	4.30

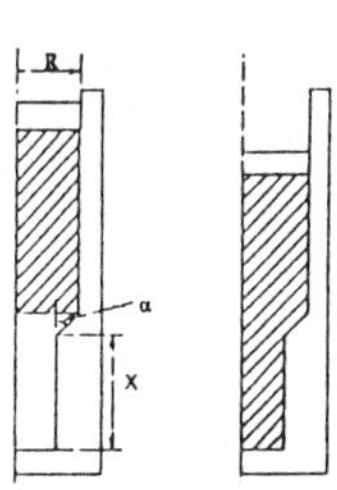

Figure 7 Design variable in extrusion stage

resulting from design optimization is shown in Figure 4. The evolution of the design variables is shown in Table 1.

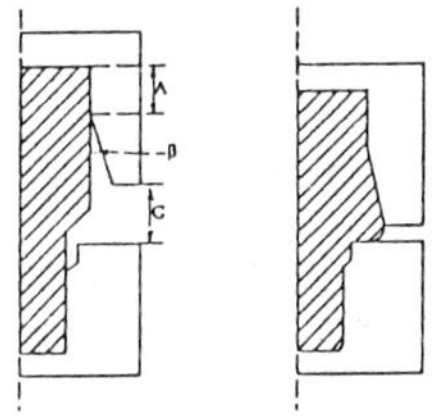

Figure 8 Design variable in coning stage

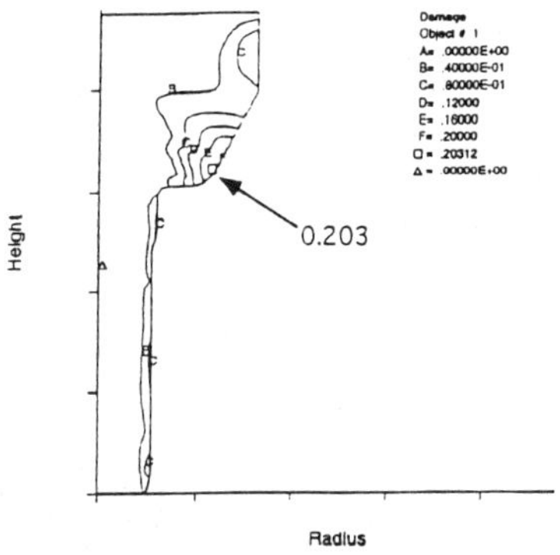

Figure 9 Damage values with initial design

(ii) Multi-pass profile drawing where a rectangular tubular profile with a small corner radius is cold drawn from a tubular preform with circular cross-section as shown in Figure 5. The preform dimensions shown in Figure 5 are the ones supplied by a major manufacturer and are referred to as the "initial design". The objective here is to maximize filling at the die corner. The objective function evaluation module is FEM package, DEFORM3D [8]. The design variables are the preform inner and outer radii and the die lengths. The total number of passes is fixed at 2. The improvement in corner filling with design optimization is shown in Figure 6. The evolution of the design variables at the first pass is shown in Table 2.

(iii) Cold forging of an outer race preform for an automotive constant velocity joint manufactured by a combination of one extrusion and a maximum of 2 coning operations. The objective here is to eliminate the probability of cracking based on the model for ductile fracture proposed by Cockroft and Latham [9]. The objective function evaluation

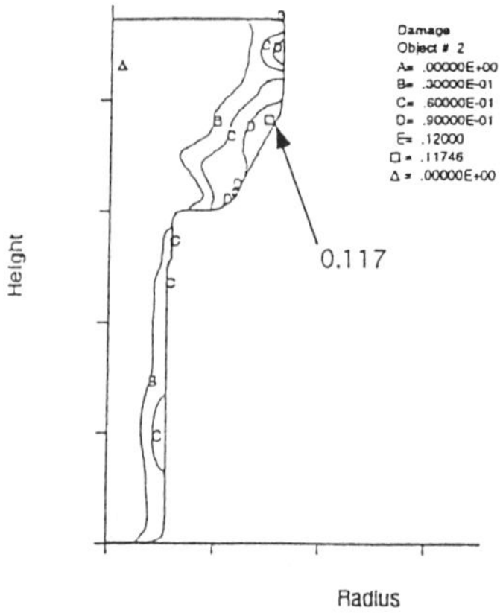

Figure 10 Damage values with optimal design

Table 3 Evolution of optimal design variables

Trial #	Minima of damage value	Process sequence	Design variables R,α,x,Λ,β,G
1	0.1740	3-step	34.925,53.000,110.900, 37.739,10.500,37.739
2-3	0.1310	3-step	37.306,57.667,109.260, 33.201,10.500,33.201
4-7	0.1251	3-step	37.306,53.000,105.980 32.790,10.500,32.790
8-12	0.1170	3-step	37.306,53.000,105.980 32.790,9.000,30.361

module is FEM package, DEFORM2D [8]. The design variables are the preform diameter, the extrusion angle and the length of the extrudate for the extrusion stage (shown in Figure 7), the total number of coning operations, the guide length, the cone angles and the gaps between the top and bottom dies for the coning stages (shown in Figure 8). The reduction in maximum "damage value" resulting from design optimization is shown in Figures 9 and 10. The evolution of design variables is shown in Table 3.

5 CONCLUSIONS

This study has demonstrated Micro Genetic Algorithms to be a powerful technique in the design optimization of multi-stage metal forming processes. It has been found to be effective in the optimization of a wide range of objective/cost functions and design variables. The algorithm is modular in nature and can be coupled to any analysis code without having to make costly modifications. It generates a rich knowledge base of design variables with corresponding cost function evaluations which may be of great value to the designer. Often in process sequence design, non-optimal solutions may be selected due to practical constraints such as lubrication or existence of prior tooling. The database generated by the proposed optimization approach helps in selecting "preferred non-optimal solutions". A significant advantage offered by μGA is its capability in optimizing the number of stages in a multi-stage process and thereby eliminating costly redundant operations.

ACKNOWLEDGMENTS

The authors of this paper would like to acknowledge the support and assistance provided by The Engineering Research Center (ERC) for Net Shape Manufacturing at The Ohio State University. Additional support in resources was provided by The Ohio Supercomputer Center.

REFERENCES

1. Goldberg, D.E., *Genetic Algorithms in Search, Optimization, and Machine Learning*, Addison-Wesley Publishing Company, Inc., 1989.

2. Davis, L., Editor, *Handbook of Genetic Algorithms,* Von Nostrand Reinhold, New York, 1991.

3. Mizugaki, Y., Hao, M., Sakamoto, M., and Makino H., "Optimal Tool Selection Based on Genetic Algorithm in a Geometric Cutting Simulation," Annals of CIRP, Vol. 43/1/1994, pp. 433-436.

4. Krishnakumar, K., "Microgenetic Algorithms for Stationary and Nonstationary Function Optimization," SPIE Proceedings, Vol. 1196, pp. 289-296, 1989.

5. Roy, S., *An Approach to Optimal Design of Multi-stage Metal Forming Processes by Micro Genetic Algorithms*, PhD Dissertation, September, 1994.

6. Adeli, H., and Cheng, N.T., "Integrated Genetic Algorithm for Optimization of Space Structures," J. of Aerospace Eng., ASCE, Vol. 6, No. 4, pp. 315-328, 1993.

7. Engelmann, B., and Hallquist, J.O., NIKE2D- A Nonlinear, Implicit, Two-Dimensional Finite Element Code for Solid Mechanics-User Manual, Lawrence Livermore National Laboratory, April 1991.

8. Oh, S.I., Wu, W.T., Tang, J.P., and Vedhanayagam, A., "Capabilities and Applications of FEM code DEFORM: The Perspectives of the Developer," J. Mater. Process. Technol., Vol. 27, pp. 25-42, 1991.

9. Cockroft, M.G., and Latham, D.J., "Ductility and Workability of Metals," J. Inst. Metals, Vol. 96, pp. 33-39, 1968.

Simulation of Materials Processing: Theory, Methods and Applications, Shen & Dawson (eds)
© 1995 Balkema, Rotterdam. ISBN 90 5410 553 4

Optimization and sensitivity analysis in polymer processing: Sheet extrusion die design

D. E. Smith & C. L. Tucker III
Department of Mechanical and Industrial Engineering, University of Illinois at Urbana-Champaign, Ill., USA

D. A. Tortorelli
Department of Mechanical and Industrial Engineering, Department of Theoretical and Applied Mechanics, University of Illinois at Urbana-Champaign, Ill., USA

ABSTRACT: A design methodology is presented which combines polymer process modeling, sensitivity analysis, and numerical optimization to design polymer sheeting dies. The methodology is applicable to processes of steady, isothermal creeping flow of purely viscous non-Newtonian fluids through thin cavities, in which the generalized Hele-Shaw approximation can be applied. Design variables are identified and cost and constraint functions are defined which measure the performance of the process. Analytical design sensitivities are derived which quantify the relationship between design variable variations and changes in the performance measures. The design sensitivities are numerically evaluated and combined with the sequential quadratic programming algorithm to efficiently search the design space for the optimal design parameter values. The optimization is used to minimize the pressure drop and reduce the exit velocity variations in polymer sheeting dies.

1 INTRODUCTION

Polymer extrusion is one of the most important manufacturing methods used today. Of particular interest here is the die design for producing polymer sheets with large width-to-height ratios.

Limited techniques are available for the extrusion die design. One-dimensional flow approximate geometry models have been used to design coat-hanger dies [1–3]. Expressions for die geometry parameters are derived to give a uniform exit velocity distribution at the die exit, however, pressure drop is not considered and no means is provided to study design alterations or material variations. Flow in extrusion dies has also been analyzed using numerical methods e.g., [4], however, such analyses do not provide a direct means for die design.

The objective of this research is to apply design sensitivity analysis and optimization to design polymer sheeting dies. In our approach, numerical simulations analyze isothermal flow in polymer sheeting dies. Once design variables and performance measures (i.e., cost and constraint functionals) have been defined, we derive and numerically evaluate analytical design sensitivities that quantify the relationship between the design parameters and the process performance measures. The design sensitivities are combined with gradient-based optimization algorithms to systematically search the design space for the optimal design parameter values.

2 MODELING AND SIMULATION

Before numerical optimization can be performed, an accurate model must be developed to predict the performance of a proposed design. In this study, processes with fluid flow through thin cavities of slowly varying cross section and arbitrary in-plane dimensions are considered. The generalized Hele-Shaw flow model of Hieber and Shen [5] simplifies the analysis for isothermal creeping flows of incompressible generalized Newtonian fluids [6].

2.1 *Conservation of Mass and Momentum*

Equations of mass and momentum conservation for the flow of polymer melt under the above assumptions reduce to [5]

$$\nabla \cdot S(P, h)\nabla P = 0 \quad \text{in} \quad D \tag{2.1}$$

where P is the pressure field over the domain of interest D, h is the local die cavity half-height, and ∇ represent the spatial gradient. The flow conductance S is a function of the pressure and is defined as [5]

$$S \equiv \int_0^h \frac{z^2\,dz}{\eta(\dot{\gamma}(z))} \qquad (2.2)$$

where for a power-law fluid, the viscosity η is a function of the strain rate $\dot{\gamma}$, given as $\eta(\dot{\gamma}) = m\dot{\gamma}^{n-1}$ [6]. Here m is the consistency index and n is the power law index.

The pressure is prescribed as $P = P^p$ on the boundary ∂D^p, where P^p is the prescribed die inlet and exit pressures. We impose $\frac{\partial P}{\partial n} = 0$ on impermeable boundaries and symmetry planes, where $\frac{\partial}{\partial n}$ denotes the normal derivative.

The governing differential equation and boundary conditions are converted to a variational statement, where the residual R is given by

$$R(P) = \int_D \nabla\overline{P} \cdot S\nabla P\,dA \qquad (2.3)$$

We note that P is a solution to equation 1 if $R(P) = 0$ and P satisfies the boundary conditions on ∂D^p. Here, $\overline{P}$ is an arbitrary weighting function with a zero value on ∂D^p.

In this research we consider isothermal flow of power-law fluids. Non-isothermal analysis and other generalized Newtonian fluid models may be included in this design methodology, however, they are not used in this study.

2.2 *Newton Raphson Iteration*

For shear thinning fluids [6], where $0 < n < 1$, the flow conductance S is a function of the pressure gradient [5] and equation 1 is nonlinear. Equation 1 is solved via the Newton Raphson iteration method. We chose this method since it converges quadratically and because it permits efficient computation of design sensitivities.

In the Newton Raphson method, the residual of equation 3 is expanded by a first-order Taylor series expansion about an initial assumed pressure field solution P^I

$$R(P^I + \delta P) \approx R(P^I) + \frac{\partial R}{\partial P}(P^I)\,\delta P \qquad (2.4)$$

Here, $\frac{\partial R}{\partial P}$ is called the tangent operator. Since we desire $R(P^I + \delta P) = 0$, the pressure correction δP is evaluated from

$$\delta P = -\left[\frac{\partial R}{\partial P}(P^I)\right]^{-1} R(P^I) \qquad (2.5)$$

The solution is updated, i.e. $P^{I+1} = P^I + \delta P$, and the iteration process is repeated until convergence is reached.

The Galerkin finite element method [7] is used to discretize the above equations. To be consistent with the flow approximation, the die cavity is described as a $2\frac{1}{2}$-D solid; a 2-D finite element mesh describes the in-plane cavity shape, and the cavity height is defined at each node. The domain D (cf. equations 3) is divided into element subdomains.

The pressure is interpolated over each element as $P = \mathbf{N}\cdot\mathbf{P}^N$, where $\mathbf{P}^N$ is a vector of element nodal pressures and $\mathbf{N}$ is the shape function matrix. The flow domain height h is also defined at the nodes and interpolated over each element via the same shape functions, i.e., $h = \mathbf{N}\cdot\mathbf{h}^N$, where $\mathbf{h}^N$ is a vector of element nodal heights.

Element pressure vectors $\mathbf{P}^N$ are assembled into a global pressure vector $\mathbf{P}$, the residual R (cf. equation 3) becomes the residual vector $\mathbf{R}$ and the tangent operator $\frac{\partial R}{\partial P}$ (cf. equation 4) becomes the tangent stiffness matrix $\frac{\partial \mathbf{R}}{\partial \mathbf{P}}$. The derivations to follow consider only this finite-dimensional representation.

3 OPTIMIZATION

We use numerical optimization to design polymer sheeting dies. Optimization algorithms systematically search the design space for a set of design variables $\boldsymbol{\phi}$ that minimize some cost function f. Here $\boldsymbol{\phi}$ is an N-dimensional vector where N is the number of design variables. Constraints are imposed based on process limitations or product specifications that render some $\boldsymbol{\phi}$'s infeasible.

We desire efficient optimization algorithms, because each design iteration requires one or more computer simulations of the polymer process. Higher-order algorithms are generally more efficient, i.e. they require fewer iterations; however, higher-order derivatives may be impractical to evaluate. A first-order method is used in this work since it requires far fewer function evaluations than zero-order methods such as genetic algorithms, simulated annealing and neural networks, and avoids costly higher-order sensitivities.

Most problems in engineering design are formulated as constrained minimization problems. It is convenient, however, to consider the unconstrained minimization algorithm of steepest decent [8], since it provides a starting point for developing more advanced methods and illustrates the importance of sensitivity analysis in optimization. In the steepest decent method the design is iteratively updated as

$$\boldsymbol{\phi}^{I+1} = \boldsymbol{\phi}^I - \alpha^I \nabla f(\boldsymbol{\phi}^I) \qquad (3.6)$$

where I is the iteration number, α^I is the step length and $-\nabla f(\boldsymbol{\phi}^I)$ represents the direction of largest decrease in the cost function f. For each iteration, the cost function f and its gradient ∇f must be evaluated. A one-dimensional search with multiple function evaluations is performed to determine the value of α_I to minimize $f(\boldsymbol{\phi}^I - \alpha^I \nabla f(\boldsymbol{\phi}^I))$. We emphasize that any movement in the design space is dictated by gradient information. The gradients ∇f are called *design sensitivities*. Each design sensitivity component is a measure of how much the cost function f varies when the design variable ϕ_i is changed.

4 SENSITIVITY ANALYSIS

Design sensitivities have many uses in engineering. Our present application is in design optimization, however, when sensitivities are used in a Taylor series expansion, the performance of a modified design can be predicted without additional analyses. Other sensitivity applications include reliability analyses and inverse problems. There are many simulation packages to perform analyses and numerical programming packages to perform optimizations; however, software to perform design sensitivity analyses is not commonly available.

For problems where the pressure $\mathbf{P}$ is the underlying response field, the cost and constraint functions may be expressed as

$$F(\boldsymbol{\phi}) = G(\mathbf{P}(\boldsymbol{\phi}), \boldsymbol{\phi}) \tag{4.7}$$

Here G is a performance measure of the process. Recall that the pressure field $\mathbf{P}$ is an implicit function of the design variables $\boldsymbol{\phi}$, hence G is both implicitly dependent on $\boldsymbol{\phi}$ through $\mathbf{P}$ and explicitly dependent on $\boldsymbol{\phi}$. Assuming sufficient smoothness, the gradient, or *design sensitivity*, of F is calculated from

$$\begin{aligned} \nabla F^T(\boldsymbol{\phi}) = \frac{DF}{D\boldsymbol{\phi}}(\boldsymbol{\phi}) &= \frac{\partial G}{\partial \mathbf{P}}(\mathbf{P}(\boldsymbol{\phi}), \boldsymbol{\phi}) \frac{D\mathbf{P}}{D\boldsymbol{\phi}}(\boldsymbol{\phi}) \\ &+ \frac{\partial G}{\partial \boldsymbol{\phi}}(\mathbf{P}(\boldsymbol{\phi}), \boldsymbol{\phi}) \end{aligned} \tag{4.8}$$

The difficulty of evaluating this expression arises from the presence of the derivative $\frac{D\mathbf{P}}{D\boldsymbol{\phi}}$, as this term is implicitly defined on the design through the system equation, cf. equation 1. The other terms, $\frac{\partial G}{\partial \mathbf{P}}$ and $\frac{\partial G}{\partial \boldsymbol{\phi}}$, are readily available, once the engineer defines the performance measure.

The finite difference method offers one alternative means for finding cost and constraint gradients. The sensitivity is approximated by a forward difference as

$$\frac{DF}{D\phi_i}(\boldsymbol{\phi}) \approx \frac{F(\boldsymbol{\phi} + \delta\boldsymbol{\phi}_i) - F(\boldsymbol{\phi})}{\delta\phi_i} \tag{4.9}$$

where $\delta\boldsymbol{\phi}_i$ is a zero vector with the exception of the i-th component which equals $\delta\phi_i$. Finite difference sensitivities are easily obtained; however, the method has severe disadvantages when used with the finite element method: 1) if $\delta\phi_i$ is too small then round-off errors result, 2) if $\delta\phi_i$ is too large then truncation errors occur, and 3) N additional nonlinear finite element analyses are required to compute the $\mathbf{P}(\boldsymbol{\phi} + \delta\boldsymbol{\phi}_i)$ that are required to evaluate $F(\boldsymbol{\phi} + \delta\boldsymbol{\phi}_i)$. As seen here, finite difference sensitivities are both inaccurate and inefficient.

In this research, we derive analytical sensitivity expressions for the cost and constraint functions. These analytical expressions are evaluated via the finite element method in the same manner used to compute the solution. The analytical sensitivity analysis, however, requires only a small fraction of the computational effort needed to complete the original analysis (cf. equation 5).

4.1 *Adjoint Design Sensitivity Analysis*

In the design problem considered here, the number of design variables greatly exceeds the number of cost and constraint functions. It is therefore computationally more efficient to use the adjoint sensitivity method [9]. To this end, we define an augmented functional as

$$\hat{F}(\boldsymbol{\phi}) = G(\mathbf{P}(\boldsymbol{\phi}), \boldsymbol{\phi}) - \boldsymbol{\lambda} \cdot \mathbf{R}(\mathbf{P}(\boldsymbol{\phi}), \boldsymbol{\phi}) \tag{4.10}$$

where $\boldsymbol{\lambda}$ is the adjoint variable vector to be determined in the sensitivity analysis. Here, $\hat{F} = F$ since all designs $\boldsymbol{\phi}$, must satisfy the governing equation, i.e. $\mathbf{R}(\mathbf{P}(\boldsymbol{\phi}), \boldsymbol{\phi}) = \mathbf{0}$. Differentiation of $\hat{F}$ with respect to the design variable ϕ_i, gives

$$\begin{aligned} \frac{D\hat{F}}{D\phi_i} &= \frac{\partial G}{\partial \mathbf{P}} \frac{D\mathbf{P}}{D\phi_i} + \frac{\partial G}{\partial \phi_i} \\ &- \boldsymbol{\lambda} \cdot \left(\frac{\partial \mathbf{R}}{\partial \mathbf{P}} \frac{D\mathbf{P}}{D\phi_i} + \frac{\partial \mathbf{R}}{\partial \phi_i} \right) \end{aligned} \tag{4.11}$$

where the arguments have been suppressed for conciseness. Since $\mathbf{R} = \mathbf{0}$ for all designs, its design derivative is also zero, i.e. $\frac{D\mathbf{R}}{D\phi_i} = \frac{\partial \mathbf{R}}{\partial \mathbf{P}} \frac{D\mathbf{P}}{D\phi_i} + \frac{\partial \mathbf{R}}{\partial \phi_i} = \mathbf{0}$. Therefore, comparing equation 8 and equation 11, we see that $\frac{DF}{D\boldsymbol{\phi}} = \frac{D\hat{F}}{D\boldsymbol{\phi}}$.

The augmented functional sensitivity is rewritten as

$$\begin{aligned} \frac{D\hat{F}}{D\phi_i} &= \left(\frac{\partial G}{\partial \phi_i} - \boldsymbol{\lambda} \cdot \frac{\partial \mathbf{R}}{\partial \phi_i} \right) \\ &+ \frac{D\mathbf{P}}{D\phi_i} \cdot \left[\left(\frac{\partial G}{\partial \mathbf{P}} \right)^T - \left(\frac{\partial \mathbf{R}}{\partial \mathbf{P}} \right)^T \boldsymbol{\lambda} \right] \end{aligned} \tag{4.12}$$

Here $\boldsymbol{\lambda}$ is arbitrary so we may select it to eliminate the coefficient of the $\frac{D\mathbf{P}}{D\phi_i}$ term. The resulting adjoint problem is

$$\boldsymbol{\lambda} = \left[\frac{\partial \mathbf{R}}{\partial \mathbf{P}} \right]^{-T} \left(\frac{\partial G}{\partial \mathbf{P}} \right)^T \tag{4.13}$$

Note that the adjoint analysis is linear whereas the original analysis is nonlinear. Once $\boldsymbol{\lambda}$ is evaluated, the unknown derivative $\frac{D\mathbf{P}}{D\phi_i}$ is eliminated from equation 12 and the sensitivity expression reduces to

$$\frac{DF}{D\phi_i} = \frac{D\hat{F}}{D\phi_i} = \frac{\partial G}{\partial \phi_i} - \boldsymbol{\lambda} \cdot \frac{\partial \mathbf{R}}{\partial \phi_i} \tag{4.14}$$

In finite element implementations, the transpose of the inverted tangent stiffness matrix is used to

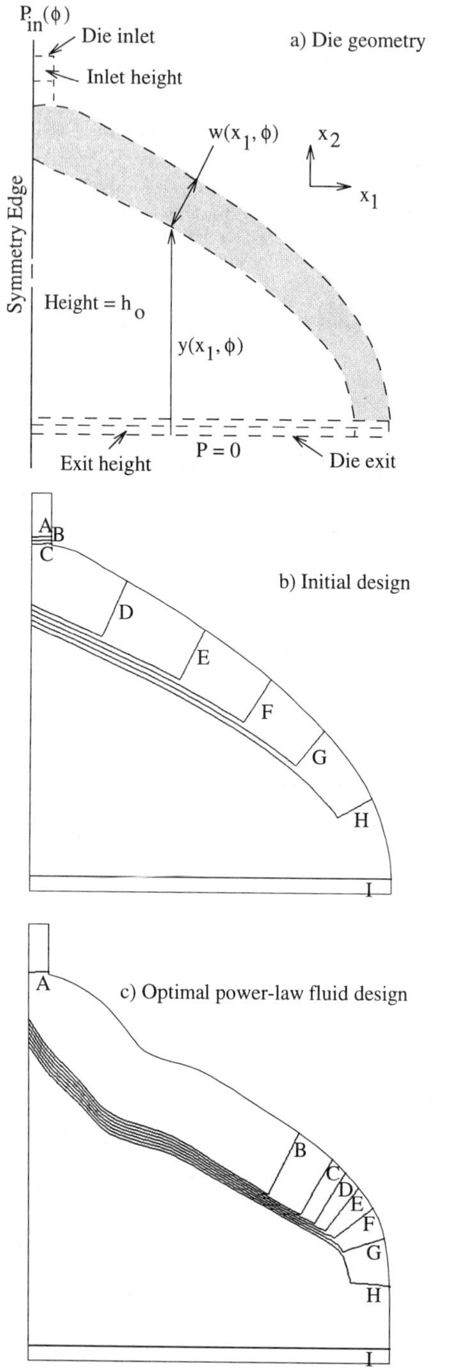

Figure 1. Geometry and half-height distributions for coat-hanger die design. Half-heights (mm): A=3.0, B=2.8, C=2.6, D=2.4, E=2.2, F=2.0, G=1.8, H=1.6,I=1.4

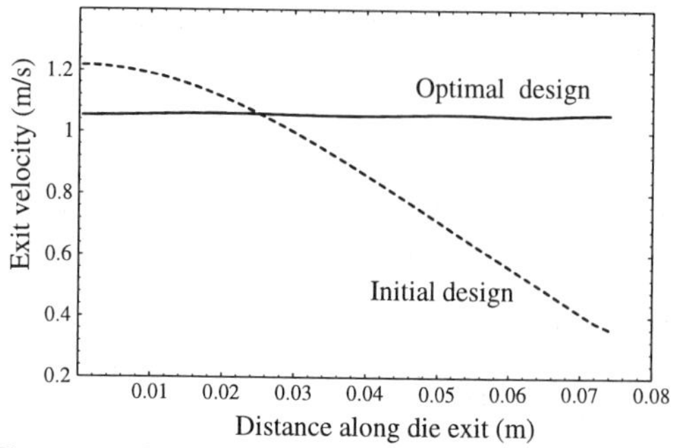

Figure 2. Coat-hanger die exit velocity distributions.

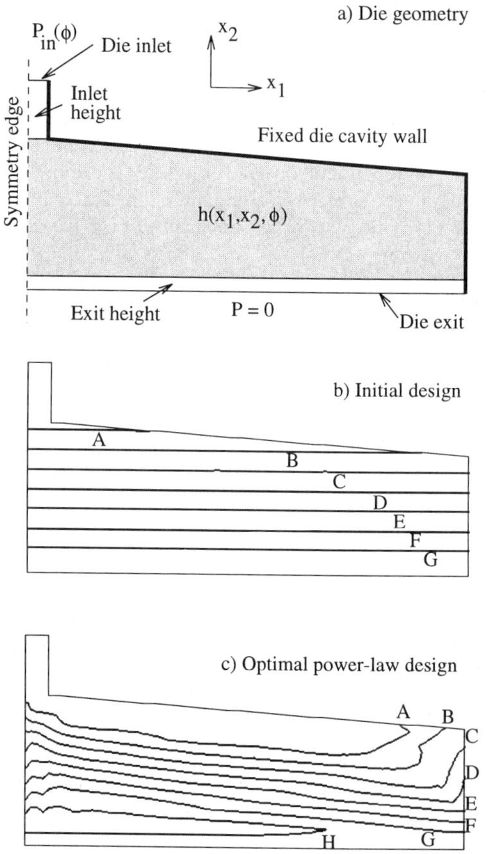

Figure 3. Geometry and half-height distributions for general-height die design. Half-heights (mm): A=3.85, B=3.43, C=3.01, D=2.59, E=2.17, F=1.75, G=1.33, H=.91

evaluate $\boldsymbol{\lambda}$. The efficiency of the method results because the inverted tangent stiffness matrix is available from the Newton-Raphson iteration used to evaluate $\mathbf{P}$ (cf. equation 5). Therefore, the adjoint method requires the solution of one adjoint problem for each cost or constraint function G (cf. equation 13) to evaluate $\boldsymbol{\lambda}$, whereupon equation 14 is evaluated for each design parameter ϕ_i. We emphasize that each adjoint sensitivity analysis is efficient since it only involves a load vector assembly for $\frac{\partial G}{\partial \mathbf{P}}$ followed by a back substitution into the existing decomposed tangent stiffness matrix.

We specialize the sensitivity analysis for sizing parameters (such as cavity height) and shape parameters (such as dimensions that define the in-plane shape of the flow cavity). Shape sensitivities are obtained via the domain parameterization method (see e.g., Tortorelli and Wang [10]). Variations in nodal positions with respect to shape design parameters are determined as described by Tortorelli, et al. [11].

5 EXAMPLE OPTIMIZATION PROBLEM

The design methodology discussed above is applied to design coat-hanger dies and dies with a general cavity height distribution (e.g., figure 1a and 3a, respectively). We consider the following nonlinear constrained optimization problem for minimizing the pressure drop across the die

$$\begin{array}{ll} \min\limits_{\boldsymbol{\phi}\in\Re^N} & f(\boldsymbol{\phi}) = P_{in} \\ \text{such that} & g_1(\boldsymbol{\phi}) = \int_{l_{exit}} \left(\frac{\overline{v}_2(\boldsymbol{\phi})}{\overline{v}_P} - 1 \right)^2 dl \leq \epsilon \\ & g_2(\boldsymbol{\phi}) = \frac{\|\nabla h(\boldsymbol{\phi})\|}{\|\nabla h\|_P} - 1 \leq 0 \end{array}$$

Design variables are given by $\boldsymbol{\phi} = \{P_{in}, h^I, b^J\}$, $I = 1, 2, ...M_h$, $J = 1, 2, ...M_b$, where P_{in} is the inlet pressure, h^I are the M_h die cavity height parameters, and b^J are the M_b in-plane die cavity shape parameters.

The cost function represents the pressure drop across the die, since $P_{out} = 0$ is fixed. The constraint g_1 is imposed to obtain a uniform exit velocity variation to within the tolerance ϵ where $\overline{v}_2$ is the x_2-component of the gapwise average velocity $\overline{\mathbf{v}} = -\frac{S}{h}\nabla P$ [5]. A zero value of g_1 indicates that the computed exit velocity $\overline{v}_2(\boldsymbol{\phi})$ equals the prescribed exit velocity $\overline{v}_P$ over the entire width of the exit l_{exit}, i.e. the exit velocity is everywhere uniform. Note also that $\overline{v}_P$ defines the total flow rate Q since the die exit area A_{exit} is fixed in the analysis, i.e., $Q = A_{exit}\int_{l_{exit}} \overline{v}_P\, dl$. Since $h(x_1, x_2)$ may be arbitrary, the constraint g_2, an explicit function on the design ϕ, is imposed to restrict the slope of h to within the prescribed value $\| \nabla h \|_P$.

In these examples (see e.g., figures 1a and 3a), inlet and outlet die half-heights are fixed at $4\,mm$ and $1\,mm$, respectively. The total die exit width is $150\ mm$ and the width-to-height ratio is 75. The desired flow rate is $.317E-3\frac{m^3}{s}$. Sheeting dies are designed for Newtonian fluids ($n = 1, m = .0009\,MPa\cdot s$) and power-law fluids ($n = .4, m = .00936\,MPa{\cdot}s^{0.4}$). The velocity variation tolerance is $\epsilon = 5E-4$.

5.1 *Example 1: Coat-Hanger Die Shape Design*

A coat-hanger sheet extrusion die is considered in the first example. The die cavity is parameterized in a manner that follows current industrial practice. Specifically, the Winter and Fritz [2] die is studied in which the in-plane die cavity shape is a function of design (i.e., $M_b \neq 0$) and the manifold half-height is a function shape (i.e., $M_h = 0$).

The design parameterization model of the Winter and Fritz [2] die is given in figure 1a, where 18 design parameters describe the flow channel in-plane shape y and the flow channel width w. In the shaded region of figure 1a, the flow-channel height h is related to the flow-channel width w through $h = \alpha w$ [2], where $\alpha = 0.2$. The constant-height region is assigned a height $h_0 = 1.5\ mm$. In the optimization, the flow channel height h is bounded above and below by $3.0\ mm$ and $0.0\ mm$, respectively. The finite element design model contains 745 nodes, 714 elements, 19 design variables, and one constraint g_1. The constraint g_2 is not used here since the spatial gradient of the height is controlled by the design parameterization.

The half-height distributions for the initial and optimal power-law fluid designs are shown in figures 1b and 1c, respectively. The initial design is taken from Winter and Fritz [2] for a Newtonian fluid.

Coat-hanger die results are summarized in table 1. The optimization procedure reduces the exit velocity variation from its initial values of 7.32 (Newtonian) and 7.72 (power-law) to $5E-4$. This corresponds to a significant decrease in the exit velocity variation as shown in figure 2 for the power-law fluid design. The pressure loss across the die is reduced by 47.5% for the Newtonian design and

Table 1. Coat-hanger results summary.

Material	Newtonian		Power-Law	
# ∇f calls	27		39	
# f calls	79		116	
Data	initial	optimal	initial	optimal
$P_{in}(MPa)$	116.2	62.4	10	10.16
g_1	7.32	$.5E-3$	7.72	$.5E-3$

increased slightly in the power-law design to satisfy the flow rate requirement (which is implicitly defined through $\overline{v}_P$).

The optimal power-law design was obtained with 39 gradient and 116 function evaluations (cf. table 1). The velocity variation decreases significantly in the first 10 optimization iterations, however, additional optimization iterations are required to fully satisfy the constraint g_1.

5.2 *Example 2: General-Height Die Design*

To obtain sheeting dies with a further reduced pressure drop, a general manifold height distribution is considered in the second example. The die cavity half-height $h(x_1, x_2)$ is interpolated using nodal height parameters h^I and the finite element shape functions. Therefore, a height design variable exists at every node in the shaded region of figure 3a. In discretized form, the constraint g_2 is enforced at the centroid of each element associated with the h^I and $\| \nabla h \|_P = 0.25$. The finite element design problem contains 630 nodes, 570 elements, 516 design variables, 1 velocity variation constraint, and 506 explicit height-gradient constraints. The initial half-height distribution varies linearly from a 4.0 mm inlet height to a 1.0 mm exit height. The die cavity in-plane shape is fixed (*i.e.*, $M_b = 0$).

General-height die optimization results are summarized in table 2. The optimization procedure reduces the exit velocity variation from its initial values of 6.84 (Newtonian) and 143.1 (power-law) to $4E-4$. The pressure loss across the die is reduced by 13.7% for the Newtonian design and increased slightly in the power-law design again to conform to the flow rate constraint. Optimal designs are obtained with 34 and 41 gradient evaluations for the Newtonian and power-law designs, respectively.

Manifold half-height distributions are shown for the initial and optimal power-law design in figures 3b and 3c, respectively. The optimal die design contains a deep channel along the back edge of the die cavity to facilitate the flow of the polymer toward the die's outer edge. Also, a restriction area in the die center is formed to reduce the flow directly in front of the inlet. The die cavity height just inside the die exit in the center region is, in fact, smaller than at the exit.

Table 2. General-height results summary.

Material	Newtonian		Power-Law	
# ∇f calls	34		41	
# f calls	101		125	
Data	initial	optimal	initial	optimal
$P_{in}(MPa)$	35	30.2	5	5.02
g_1	6.84	$.4E-3$	143.1	$.4E-3$

6 CONCLUSION

A sheeting die design method is developed using numerical modeling, sensitivity analysis and optimization. Sheet extrusion dies are designed for reduced exit velocity variation and minimal pressure drop. Examples of coat-hanger die designs and designs with a general manifold height distribution are given.

7 ACKNOWLEDGMENT

The authors wish to express their appreciation for the financial support provided by the National Science Foundation Grant DDM 92-15599.

REFERENCES

[1] Chung, C. I. and D. T. Lohkamp. Designing Coat-Hanger Dies by Power-Law Approximation. *Modern Plastics*, pages 52–55, March 1976.

[2] Winter, H. H. and H. G. Fritz. Design of Dies for the Extrusion of Sheets and Annular Parisons: The Distribution Problem. *Polym. Eng. Sci.*, 26(8):543–553, 1986.

[3] Liu, T. and L. Liu. A Unified Lubrication Approach for the Design of a Coat-Hanger Die. *Polym. Eng. Sci.*, 34(7):541–550, 1994.

[4] Vrahopoulou, E.P. A Model for Fluid Flow in Dies. *Chem. Eng. Sci.*, 46(2):629–636, 1991.

[5] Hieber, C. A. and S. F. Shen. A Finite-Element/Finite-Difference Simulation of the Injection-Molding Filling Process. *J. Non-Newtonian Fluid Mechanics*, 7:1–32, 1980.

[6] Bird, R. B., R. C. Armstrong, and O. Hassager. *Dynamics of Polymeric Liquids*, volume 1: Fluid Mechanics. John Wiley & Son, New York, NY, 2nd edition, 1987.

[7] Stasa, F. L. *Applied Finite Element Analysis for Engineers.* CBS College Publishing, New York, NY, 1985.

[8] Haftka, R. T. *Elements of Structural Optimization.* Kluwer Academic Publishers, Boston, MA, third edition, 1992.

[9] Haug, E. J., K. K. Choi, and V. Komkov. *Design Sensitivity Analysis of Structural Systems.* Academic Press, New York, NY, 1986.

[10] Tortorelli, D. A. and Z. Wang. A Systematic Approach to Shape Sensitivity Analysis. *Int. J. Solids Structures*, 30(9):1181–1212, 1993.

[11] Tortorelli, D. A., J. A. Tomasko, T. E. Morthland, and J. A. Dantzig. Optimial Design of Nonlinear Parabolic Systems. Part II: Variable Spatial Domain with Applications to Casting Optimization. *Computer Methods in Applied Mechanics and Engineering*, 113:157–172, 1994.

Simulation of Materials Processing: Theory, Methods and Applications, Shen & Dawson (eds)
© 1995 Balkema, Rotterdam. ISBN 90 5410 553 4

Optimal design of the die shape using nonlinear finite element analysis

Boris Štok & Aleš Mihelič
University of Ljubljana, Faculty of Mechanical Engineering, Slovenia

ABSTRACT: A finite element based approach to optimal process design in steady-state metal forming is considered. In particular, the optimization is applied to the problem of a wire drawing die design. Conical and streamline dies are investigated, their shape being parametrized by a polynomial function. Optimal design is effected by enforcing minimization of the forming energy consumption. The same algorithm is used also for determination of possible area reduction. In both investigated problems several technological and geometry constraints are taken into account. From the comparison between the numerical results and known solutions for rigid plastic material model, it is demonstrated that the solutions are trustful, and can be applied efficiently for more realistic material model not only in reducing the total energy consumption, but also in preventing the formation of defects and the die wear.

1. INTRODUCTION

The main preoccupation of any manufacturing research regarding a real industrial production is how to improve its actual quality and efficiency, both in technical and economical sense. Determination of the related process and design parameters, fulfilling these aspirations, is in principle equivalent to a solution of the correspondent optimization problem. Depending on the product requirements or customer's needs the optimization criteria concerning the same problem may differ, but whatever the process design criteria may be, it is only our thorough understanding of the physics governing the problem that can ensure an efficient optimal solution. When referring to metal forming processes it would not be possible to reliably design the material flow, or to predict and prevent the occurrence of eventual defects, without a sufficient knowledge of the influence that process variables, such as friction, material physical properties, and workpiece geometry, have on the process evolution.

In the past it was the empirical experience which predominantly governed the process design. But nowadays, with computer-aided techniques well established also in the metal forming industry, a designer should rely the refinement of his design much more on information obtained by the computer process simulations. These simulations, which are based mainly on the finite element computational models, are an efficient means for the identification of the process response under boundary conditions, imposed by the investigated tool geometry and process kinematics. Although these models have room for further improvement they are of great importance for a design engineer by enabling him a quantitative estimation of the spatial distribution and time evolution of the process variables.

Optimization in structural engineering is rather frequent in practice, mostly because of its relative simplicity due to the linear material behaviour. It is rather obvious that an optimal solution related to metal forming is directly related to the material flow. Unfortunately, a mathematical description of this process is far away of being simple. In fact the process is characterized by an extreme complexity due to nonlinearities arising from the inelastic behaviour of the formed material, large deformation and contact boundary conditions at the material-tool interface. Further, when the amount of heat, generated because of irreversibility of the plastic deformation and interfacial friction, is not negligible the material physical parameters exhibit an additional temperature dependence. Nevertheless, when considering an efficient integration of classical mathematical methods of optimization and the finite element concept, and attained reliability of the computational analysis of metal forming processes, it seems that despite all complexity application of optimization to the area of metal forming and to problems associated with the correspondent tool design is possible.

In this paper the optimization of extrusion processes is considered. Actually, the material flow in the cold wire drawing process is optimized by proper design of the die's geometry. Starting from an initial die design, an optimized design solution, based on nonlinear optimization procedures [1], is found. Two optimization objectives have been imposed separately, first, minimization of the energy consumption, and second, maximization of the possible wire reduction.

2. FORMULATION OF THE OPTIMAL DESIGN PROBLEM

Experience proves that geometry of a die used in cold extrusion and drawing processes is of extreme importance, its influence on the overall process performance being essential. Usually, conical dies are used, while their contour is chosen by experience, taking process parameters, such as area reduction, forming rate, material flow stress curve and interfacial friction conditions, somehow into account. Since it is evident that an optimized die design solution depends directly on the given process parameters, an optimal design problem could be set regarding this choice. Also, considering the fact that modern tool manufacturing enables production of relatively inexpensive dies of more complex shapes, it is not reasonable to limit our investigation to conical dies only.

In what follows we address two different optimization problems associated with optimal die design in wire drawing process. Regarding the imposed optimization objectives they can been stated as follows:

PROBLEM I:

MINIMIZE:	*total energy consumption*
Subject to:	*stress constraints*
	die geometry constraints

PROBLEM II:

MAXIMIZE:	*area reduction*
Subject to:	*stress constraints*
	die geometry constraints

Introducing s for the process state variables and d for the design variables, and $\Psi_0 = \Psi_0(s,d)$ for the objective function and $\Psi_i = \Psi_i(s,d)$ for the constraint functions, respectively, a mathematical formulation of the stated optimal problems can be written as

FIND d TO EXTREMIZE:

$$\Psi_0 = \Psi_0(s,d) \tag{1}$$

Subject to:

$$\Psi_i(s,d) \leq 0 \, , \quad i = 1,\dots,\alpha \tag{2}$$

$$\Psi_i(s,d) < 0 \, , \quad i = \alpha+1,\dots,\beta \tag{3}$$

The stress constraints (2) are associated with possible technological defects arising due to an inadequate stress field distribution and flow pattern. The die geometry constraints (3), however, impose the geometric consistency in the forming zone. Although the above equations define the framework within which an optimal solution is to be sought, this system is not sufficient. Actually, it has to be enlarged by taking governing equations of the process considered and the correspondent boundary conditions into account. Without entering into a detailed description of the thermomechanics of metal forming processes this set of equations can be written in the following form

$$R(s,d) = 0 \tag{4}$$

It is evident that this relation establishes an implicit dependence of the process state variables s on the design variables d

$$s = s(d) \tag{5}$$

The fulfilment of equation (4) in view of the optimal design problem (1-3) yields a solution of the considered problem. This solution can be solved, for instance, by a sequential quadratic programming (SQP) method which is considered to be one of the most reliable optimization methods [1]. In accordance with the method a quadratic subproblem is formulated and solved to get a search direction, which is followed afterwards by a line search to obtain an improved design solution.

3. DESIGN SENSITIVITY ANALYSIS

For an efficient optimal solution search strategy a sensitivity of the system response variables, i.e. state variables s, to changes in design variables d is required. Consequently, a sensitivity analysis of the objective and constraint functions is to be performed. When possible it is performed analytically by direct differentiating the response functions $\Psi(s,d)$. Analytical treatment is advantageous since it ensures highest accuracy and relatively low computational time. But in general, when the respective functions are complex numerical approaches have to be used.

The simplest numerical procedure is the one associated with the finite difference concept. Accordingly, the partial derivative $\frac{\partial \Psi}{\partial d}$ is approximated by

$$\frac{\Delta \Psi}{\Delta d} = \frac{\Psi(s^o, d^o + \Delta d) - \Psi(s^o, d^o)}{\Delta d} \tag{6}$$

where a perturbation from the current problem solution (s^0, d^0) is given by a step size vector Δd, the components of which are linearly independent.

4. NUMERICAL IMPLEMENTATION IN VIEW OF APPLICATION TO WIRE DRAWING

The analysis of the problem specified by equation (4) has to be treated numerically. It has been successfully proved several times also in our experience that the finite element method is reliable and appropriate to tackle high computational complexity encountered in process analyses [8,9,10].

With Lagrangian incremental elastic-plastic formulation assumed to model the material flow, the general purpose finite element program ABAQUS [11] has been employed to simulate the cold wire drawing process in the present study. The optimization has been performed upon the finite element analysis results and finite difference sensitivity analysis by the NLPQL computer code developed by K. Schittkowski [2].

A pre-processor that allows a finite element discretization of the problem domain and its automatic adaptive remeshing in accordance with the intermediate optimized design solutions has been designed.

Since wire drawing is mainly a steady-state process, some attention is needed when modelling this state numerically. In fact, plastic deformation of a control volume in a drawn wire is dependent on the stress history, which is in the considered case imposed implicitly by a time variation of the volume initial and boundary conditions. In order to identify properly the steady-state response, a careful tracing of the process time evolution is required from the very beginning. Drawing of a wire through a die is imposed in the numerical model (Fig. 1) by applying the displacement of the wire's front section incrementally.

For purposes of this investigation a die has been assumed rigid while friction at the die-workpiece interface has been modelled by assuming the Coulomb friction law with an upper bound τ_{max} on the frictional shear stress.

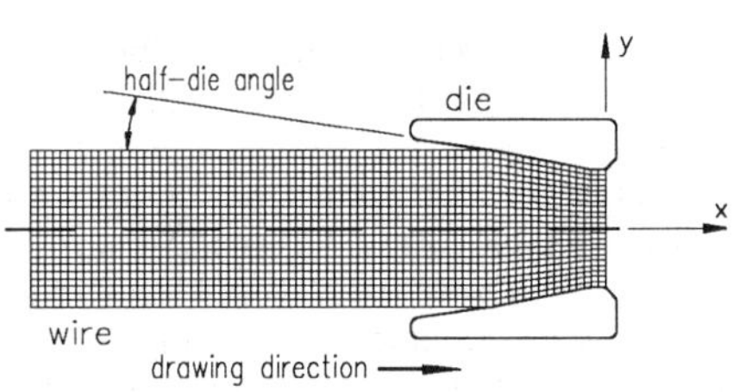

Figure 1: Finite element model of the wire drawing process

In order to perform an optimal die design a mathematical description of the die's contour is needed. In our case we have chosen a polynomial form

$$Y(x, a_0, a_1, \dots, a_n) = \frac{d_1}{2} - (x - x_1) \cdot tg\, a_0 - (x - x_1)(x - x_0)(a_1 + a_2 x + a_3 x^2 + \dots + a_n x^n) \tag{7}$$

where $Y(X)$ stays for the die radius at axial position x and d_1 for the final diameter of the drawn wire, while x_0 and x_1 are the inlet and outlet die coordinates in the axial direction. They can be readily determined from the area reduction data.

Evidently, the polynomial assumption (7) ensures smoothness of the die shape by function definition, while from the optimization point of view this selection results in relatively small number of design variables, which are in fact the polynomial coefficients a_k. Since description (7) includes also a linear variation of the die radius $Y(x)$ it can be used for analysing both, conical and streamline dies.

Finally, let us consider mathematical structures of the objective and constraint functions associated with the two specified optimal design problems.

In the Problem I the objective is to find such die angle of the conical die or such polynomial function for the streamline die, that the required forming energy be minimized. The drawing force being closely connected to the forming energy the optimization problem remains equivalent when referred to the drawing force minimization. Considering the numerical model the total drawing force is represented by the resultant of the nodal forces at the wire's front section

$$\Psi_0 = \sum_{i=1}^{N} F_i(a_k) \tag{8}$$

At this stage of investigations we have decided to consider only one stress constraint, actually the one associated with the prevention of the wire necking and subsequent collapse. The correspondent mathematical form takes the resultant of the Mises equivalent stresses in a cross section of the extruded wire as competent measure. Again, by considering the finite element discretization the discussed constraint can be written in the following form

$$\Psi_1 = \frac{\sum_j \sigma^j_{eqv}}{\sum_j C\sigma^j_{yield}} - 1 \le 0 \tag{9}$$

where summation is taken columnwise over the nodes of a reference cross section. The above constraint states that the actual stress state, measured by the sum of the equivalent stresses σ^j_{eqv}, should be less or equal to the sum of the correspondent yield stress state, measured by the sum of the current yield strength σ^j_{yield}, multiplied by a safety factor C.

Considering the die geometry constraints two inequalities bounding the die radius $Y(x)$ with respect to the feasible design can be set

$$\Psi_2 = \frac{2\cdot Y(x,a_k)}{D_0} - 1 \le 0 \tag{10}$$

In the Problem II the maximal area reduction is sought by considering fixed die design. In this

$$\Psi_3 = 1 - \frac{2\cdot Y(x,a_k)}{D_1} \le 0 \tag{11}$$

case the design variable is the initial wire diameter D_0.

For this case the objective function is formulated as

$$\Psi_0 = \left[1 - \left(\frac{D_1}{D_0}\right)^2\right]\cdot 100 \tag{12}$$

While the stress constraint is identical to the one in the Problem I, the geometry constraints reduce to only one of the following form

$$\Psi_2 = \frac{D_1}{D_0} - 1 < 0 \tag{13}$$

5. NUMERICAL EXAMPLES

For the present investigation the wire has been modelled by 75 x 10 four-node axisymmetric elements. Also, only for reasons of comparison with the published results [4,5], the wire material is assumed to behave elastic-ideally plastically, with the yield strength being $\sigma_{yield}=650\text{N/mm}^2$. For the same reason coefficient C=0.975 in the constraint equation (9) has been assumed.

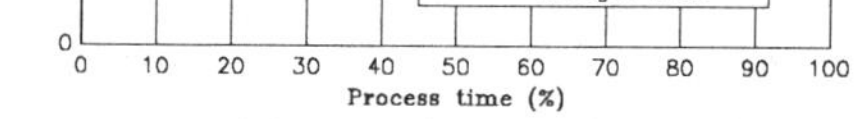

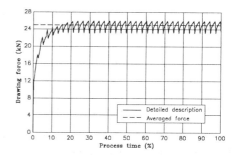

Figure 2: Typical oscillations observed by the FEM analysis

First, let us discuss some general findings. Using the default ABAQUS increment adjustment did not yield satisfactory results, since the wire often detached from the die at the die-land part. This happened even for the admissible selection of the area reduction and die half-angle. After some trials the increment size capable of reliable modelling was found to be approximately 20% of the average

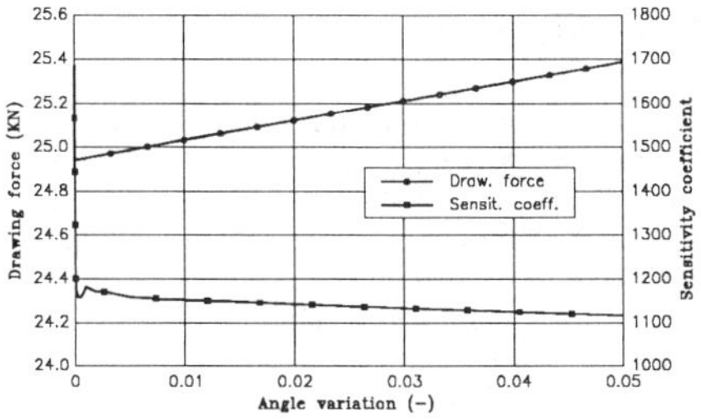

Figure 3: Function value and sensitivity results for small angle variation

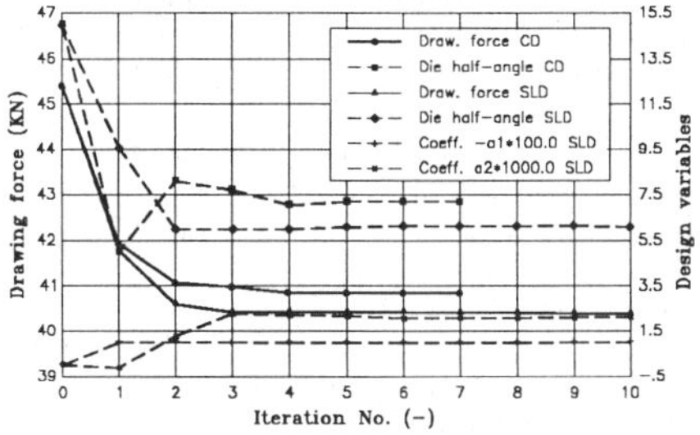

Figure 4: Iteration history and comparison for conical and streamline dies

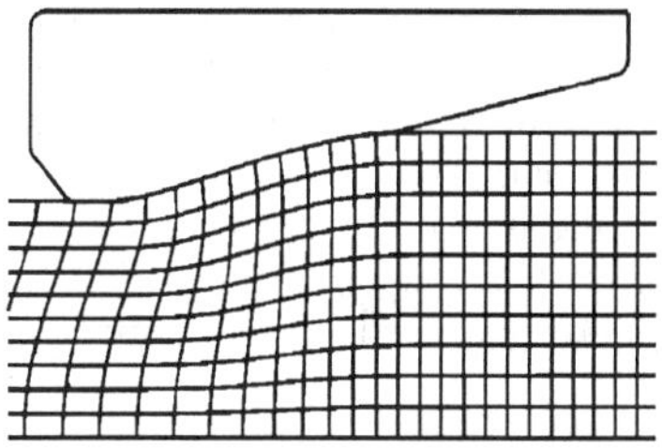

Figure 5: The optimal shape of the stream line die

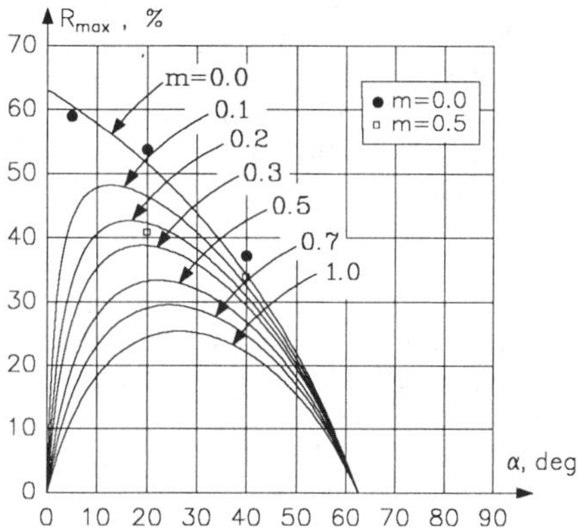

Figure 6: Comparison of the numerical and known Avitzur's solutions

finite element length.

The plotted curve in Fig. 2 exhibits "time/increment" dependence of the drawing force which is evidently a consequence of the discretization effect when modelling contact boundary conditions. The remedy to reduce these oscillations is the mesh and incrementation refinement.

With respect to the above discussed oscillating of the analysis results, the objective function has to be appropriately averaged in order to perform sensitivity and optimization at all. By doing this the only possibility to calculate the sensitivity analysis is by the finite difference approach. However, wherever possible derivatives are computed analytically.

To check the accuracy of the design sensitivity predicted by the finite difference scheme a series of analyses has been performed. The function value and sensitivity for the drawing force are shown in Fig.3. According to these investigations the step size vector is selected to be 0.1% of the design variable.

As an illustrative example for the Problem I, i.e. the minimization of the drawing force, the computer program based on the discussed theory is applied to the optimization of conical and streamline dies. The streamline die profile is described by three design variables.

In both cases the die half-angle is initialized to the value $\alpha=15.0^0$. The extruded diameter is 15mm, while the area reduction is 20%. The friction conditions on the die-wire interface are specified by friction coefficient $\mu=0.05$ and shear factor m=0.2. The results of the performed optimization are plotted for both, the conical and streamline die in Fig. 4. The performed optimization yields, considering the constraint functions as defined in section 4, optimal die half-angles of values $\alpha=7.20°$ and $\alpha=6.09°$ for the conical die and streamline die, respectively.

When comparing the optimized drawing forces which are actually 40.8kN for the conical die, and 40.4kN for the streamline die, it is rather surprising that the difference is so small, in fact only about 1%. However, when observing the graphical blow-up in Fig.5, these results become quite reasonable.

As an illustrative example for the Problem II, i.e. the maximization of the possible reduction, the computer code is applied only to the optimization of the conical die. The optimization was performed for several reductions and for two different friction conditions, namely, for m=0 and m=0.5. The results of the performed optimization are plotted in Fig. 6, and they agree quite good with known solutions [4,5], especially for the m=0.

6. CONCLUDING REMARKS

It is demonstrated in the present paper that the optimization techniques can be very efficient in controlling process parameters by allowing for an appropriate choice of die design. Consequently, significant cost savings and product improvements can be achieved. Hence, methods used in the structural shape optimization have been shown as useful tool also when applied to the design of wire-drawing dies.

The iterative optimization scheme requires evaluation of the design sensitivities, therefore an efficient method for calculating the design sensitivities is generally quite important. However, by a proper die shape representation the total number of the gradient analyses can be reduced. It has been demonstrated that by assuming a polynomial function with three design variables satisfactory results can be achieved.

It seems that the most time consuming part of needed computations is the actual finite element analysis. Namely, to consider the problem as a quasi-steady one, the analysis has to go on until the response quantities in a control volume become steady. Only then, the values needed for the optimization can be considered.

In the near future a series of the parametric studies should be made and compared to the existing solutions.

ACKNOWLEDGEMENT

The authors are very thankful to Prof. Martin P. Bendsøe and Prof. Niels Bay from the Technical University of Denmark, and to Prof. Karl Kuzman from the University of Ljubljana, Slovenia, for their interest in the work and the numerous ideas and hints they had. The research work was supported by the grant No. S24-0782-006/10440/93 of the Ministry of Science and Technology of the Republic Slovenia and by the grant within the TEMPUS JEP No. 1925-93/3 "Advanced Manufacturing Technology, Engineering Economy and CIM Oriented Techniques in Metal Forming". The latter enabled the second author's stay for 6 months at the Technical University of Denmark, where part of this work was done. All support is gratefully acknowledged.

REFERENCES

1. Kamat M.P., ed.: Structural Optimization: Status and Promise, AIAA, 1993.
2. Schittkowski K.: NLPQL - A Fortran Subroutine Solving Constrained Nonlinear Programming Problems: Annals of Operations Research, Vol. 5, 1985.
3. Kusiak J., Thompson E.G.: "Optimization techniques for extrusion die shape design": presented at NUMIFORM'89, pp. 569-574.
4. Avitzur B.: "Analysis of Wire Drawing and Extrusion Through Conical Dies of Small Cone Angle": Journal of Engineering for Industry, Feb. 1963, pp. 89-96.
5. Avitzur B.: "Analysis of Wire Drawing and Extrusion Through Conical Dies of Large Cone Angle": Journal of Engineering for Industry, Feb. 1964, pp. 305-316.
6. Thomsen S.G.: "Optimization of Dies for Axisymmetric Cold Forward Extrusion": Eng. Opt., Vol. 11, 1987, pp. 13-20.
7. Funke P., Cai Y., Heinemann G.: "Über die Genauigkeit von Finite-Elemente Berechnungen in der Umformtechnik, dargestellt am Beispiel des Stangenzugs": DRAHT, Vol. 42, 1991, pp. 241-246.
8. Štok B., Hudoklin A.: "How to Tackle the Compatibility Constraints in a Computational Solution of Frictional Contact Proble ": Finite Elements in Analysis an Design, Vol. 18, 1994, pp. 111-119.
9. Štok B., Mole N., Švajger M., Zajc M.: "Finite Element Modelling and Simulation of Electroslag Remelting Process", 3rd International Conference on Technology of Plasticity; 1990, Kyoto
10 Štok B., Mole N., Srna J.: "Finite Difference Thermo-Plastic Stress Analysis of Cyclically Thermally Loaded Problems" - Proceedings of Nonlinear Engineering Computations - NEC91; 1991, Swansea
11 ABAQUS - A General Purpose Finite Element Program for Nonlinear Static and Dynamic Analysis, Version 5.3: Hibbit, Karlsson and Sorensen, Inc., Providence, Rhode Island, 1993.

Simulation of Materials Processing: Theory, Methods and Applications, Shen & Dawson (eds)
© 1995 Balkema, Rotterdam. ISBN 90 5410 553 4

A parameterized physically based model for flatness control in six-high rolling mills

Olof Wiklund
MEFOS Metal Working Research Plant, Luleå, Sweden

W.Y.D.Yuen
Port Kembla Laboratories, BHP Research, BHP Steel, Australia

ABSTRACT

In order to take full advantage of a six-high rolling mill with several control options, it is necessary to use a model that can describe the influence and interaction of the bending and the axial shifting of the rolls using the current roll profiles. This was the purpose of developing a parameterized physically based model for flatness control in six-high rolling mills.

The bending, shearing and flattening of the rolls are calculated using the same theory as in a physically based segmented model. The main difference is that instead of pressure values at each segment, the unknowns are the coefficients in the polynomial expressions of the pressure distributions between the rolls.The above approach yields 8th order polynomials describing the deflection of the rolls and the flattening between the rolls. The flattening between the work rolls and the strip is described using influence functions, which works although the relation is non-linear, because it is independent of the unknown variables.

The roll interfaces are supposed to match (or to separate near the roll barrel ends when appropriate), but the polynomials cannot match perfectly in the general case because of the limited degrees of freedom included in the model formulation. Instead the integral of the square of the mismatch is minimized. This yields a system of 12 linear equations, which is solved very rapidly compared to a segmented model.

1 INTRODUCTION

Customer demands on flat rolled products have gradually become more stringent. The profile and flatness must be measured accurately and kept at the desired level by advanced control equipment. Six-high rolling mills are offering increased control compared to four-high mills. In a six-high rolling mill it is possible to control flatness and edge drop using axial roll shifting of the intermediate rolls and roll bending as well as to control the influence of wear using cyclic work roll shifting.

Physically based segmented models for simulation of the deformation of the roll stack in four-high and six-high rolling mills have been available for decades[1-3]. The models describe the bending, shearing, flattening, thermal expansion and the wear of the rolls and calculate the resulting transverse roll gap that will create the strip profile and flatness. These models help to understand the mechanisms and to optimize the roll profiles and the pass schedule. However, they are too slow for online control of a strip mill.

For mill automation there are simple analytical models based on relations that are purely empirical or highly simplified from physical principles. These online models need tuning and adaption to work properly and are not very good for setup or whenever the rolling conditions change beyond the regime they are derived for.

In order to take full advantage of a six-high mill with several control options, it is necessary to use a model that can describe the influence and interaction of the bending and the axial shifting of the rolls using the current roll profiles. This was the purpose of developing a parameterized physically based model for flatness control in six-high rolling mills. The model uses the same method as that for four-high rolling mills developed by BHP [4,5], with extension to six-high mills. It was very important to select carefully what to add or refine, and what to skip, in order to keep the model fast enough for online control. Segmented models have been used to assist in making those decisions.

The formulation of the model will indicate, that the present approach is different from most simulation models as well as control models. A similar but much simpler model for four-high rolling mills was presented several years ago [6].

2 MODEL FORMULATION

The flatness and relative thickness profile of the strip are supposed to be maintained or kept at a controlled level. Hence the pressure distribution between the strip and the work rolls is assumed to be known.

The roll profiles are known. The bending, shearing and flattening of the rolls are calculated using the same theory as in the segmented models. The main difference is that instead of pressure values at each segment, the unknowns are the coefficients in the polynomial expressions of the pressure distributions between the rolls. The model allows for 0th to 4th order terms. Symmetry about a point is assumed and odd terms are included in the model because of axial roll shifting in which case the operator side of the top rolls are similar to the drive side of the bottom rolls.

2.1 *Geometry, forces and pressure distributions*

The geometry, forces and pressure distributions in the six-high roll stand are shown in Figure 1. Because of symmetry it is sufficient to consider only the top half of the stand, namely the top backup roll (bur), the top intermediate roll (imr), the top work roll (wr) and the strip. The pressure distributions are defined in such a way that the constant term is the average over the contact length and includes higher order terms:

between strip and wr $p = p(p0,p2,x)$
between wr and imr $q = q(q0,q1,q2,q3,q4,x)$
between imr and bur $r = r(r0,r1,r2,r3,r4,x)$

Here x is denoting a coordinate along the roll axis. The origin is in the middle of the roll or strip (unless the roll is shifted axially).

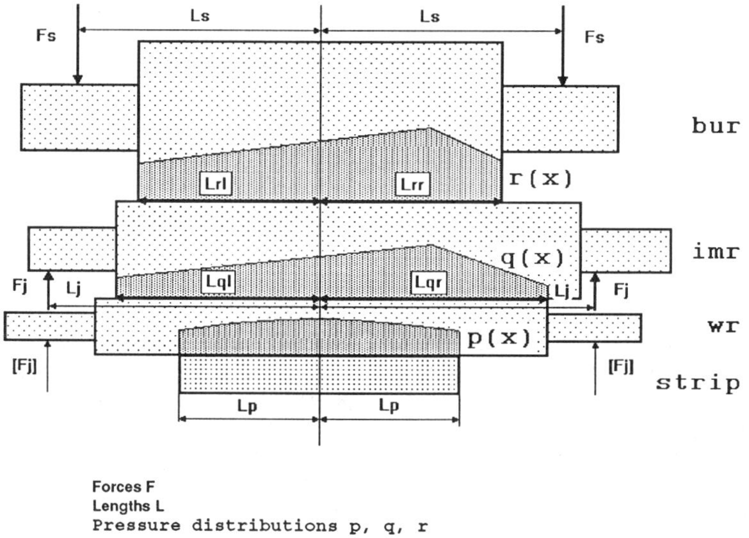

Figure 1: Roll stack geometry

Table 1: Sensitivities for work roll bending

wrb sensitivities	$\frac{\partial F_j}{\partial Fs}$	$\frac{\partial F_j}{\partial C_W}$	$\frac{\partial F_j}{\partial Taper}$	$\frac{\partial F_j}{\partial Shift}$
units	-	MN/m	MN	MN/m
Crown426 of MEFOS	0.0875	-1500	-30.0	-1.67
Shape of BHP	0.0470	-1200		-1.72
Jack6 without penalty	0.0397	-944	-38.3	-2.11
Jack6 with penalty 3232	0.0304	-632	-15.2	-2.75
Jack6 without taper/pen				-2.82

2.2 *Balance equations*

Equations are set up using vertical force and moment balance of each roll. The rolling force *Fs* and the bending force *Fj* can be expressed using their lever lengths *Ls* and *Lj*, the known coefficients *(p0, p2)* of *p* and the unknown coefficients of *q* and *r*.

2.3 *Matching of the roll and strip interfaces*

The interfaces of the rolls and the strip are supposed to match vertically (or to separate near the roll barrel ends when appropriate), but the polynomials cannot match perfectly in the general case because of the limited degrees of freedom included in the model formulation. Instead the integral of the square of the mismatch is minimized. The tilting of the outer rolls are unknown. Furthermore, the rolls are allowed to be shifted vertically in order to minimize the mismatches *dr, dq, dp*. The centre-lines of the rolls are described by deflection functions. The deflections are vertical and are denoted by functions *Y*.

The mismatch or interference at the contact between the rolls or the wr and strip must also consider the roll flattening and the total roll profiles (due to grinding, wear and thermal expansion) as well as the strip profile. The roll flattening between rolls contributes to coefficients of the unknowns, and must be solved analytically. The roll flattening between wr and strip as well as the roll profiles contribute only to the constant term and can be calculated numerically, which makes it possible to handle arbitrary profiles (edge drop phenomena, sharp wear edges, and tapered intermediate rolls).

The following mismatch functions between the rolls and strip are derived:

between bur and imr

$$f(x) = Y_b - Y_i + dr + Y_{(flat)bi} - C_b(x) - C_i(x)$$

between imr and wr

$$g(x) = Y_i - Y_w + dq + Y_{(flat)iw} - C_i(x) - C_w(x)$$

between wr and strip

$$h(x) = Y_w + dp + Y_{(flat)ws} - C_w(x) - C_{so}(x)$$

Symmetry about a point and assuming that the strip bends according to the work rolls eliminate the odd terms of *h(x)*. *Cs* is scaled from the input profile to *Cso* assuming that the relative profile is constant in order to obtain good flatness. *Cb, Ci* and *Cw* are the roll profiles (including grinding, axial shift, wear and thermal expansion). The following error functions are defined for each contact between the rolls and the strip. The integration is performed from the left end to the right end of each contact zone.

$$Er = \int_{-Lrl}^{Lrr} f(x)^2 dx \qquad \text{error bur-imr (m)}$$

$$Eq = \int_{-Lql}^{Lqr} g(x)^2 dx \qquad \text{error imr-wr (m)}$$

$$Ep = \int_{-Lp}^{Lp} h(x)^2 dx \qquad \text{error wr-strip (m)}$$

Equations are obtained by minimizing the error functions *Er, Eq, Ep* with respect to each of the unknown variables by setting each partial derivative = 0. This process yields a system of linear equations with 12 unknowns. Several equations are set up as linear combinations.

$$\frac{\partial Ep + \partial Eq + \partial Er}{\partial q2} = 0 \qquad \text{equation 11}$$

$$\frac{\partial Ep + \partial Eq + \partial Er}{\partial q4} = 0 \qquad \text{equation 12}$$

The linear combinations are merged using a

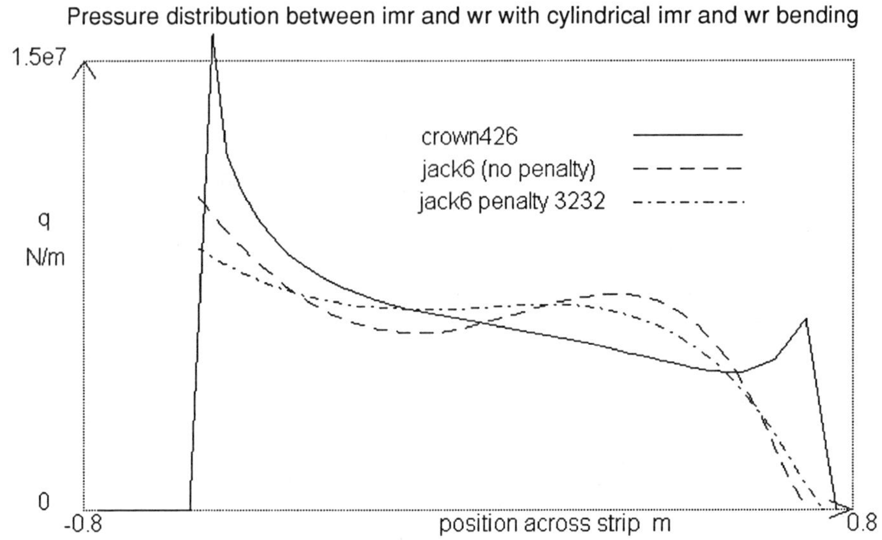

Figure 2: High pressure peak with cylindrical intermediate roll

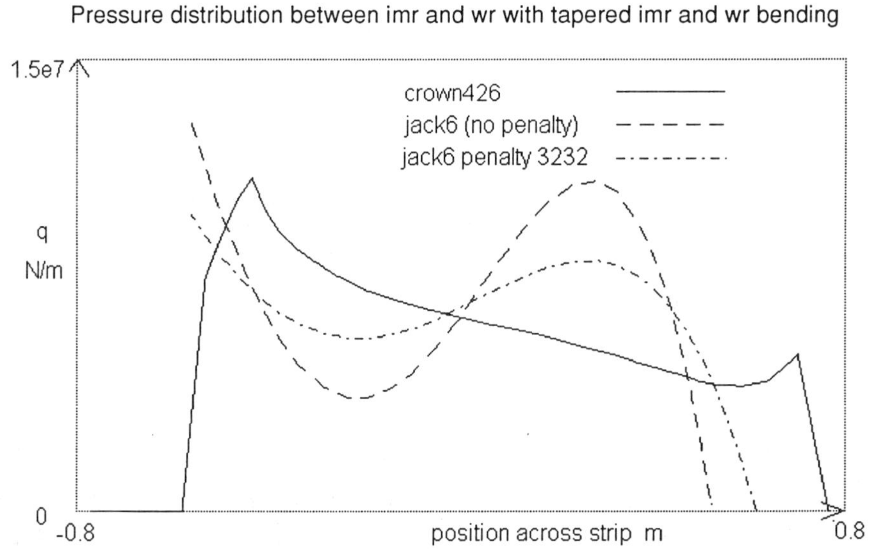

Figure 3: Reduced pressure peak with tapered intermediate roll

subroutine, that arranges the proper merging of the coefficients in the arrays representing the system of equations. Another subroutine creates a display of the mismatch error functions across the strip and the root mean square errors. During the development of the computer program, several linear combinations were tested for the 12 equations, since it was not obvious which error function(s) should be minimized with respect to each variable to obtain the best minimization of the three error functions $f(x)$, $g(x)$ and $h(x)$ at the same time.

2.4 *Roll separation*

In the case of bad mismatch between the rolls or the strip and wr, there can be a gap. The model cannot handle this in a direct way. Instead, the pressure comes out negative. This phenomenon can be handled near the edges by cutting off the portion outside the point where the pressure is 0 and iterating with integration inside that point. If there is a gap in the central parts of the contact zone, the program cannot handle it.

Because of the 4th order term, the pressure distributions q and r may vary steeply and create problems related to roll separation. One solution would be to eliminate those terms, but simulation with a segmented model shows significant 4th order terms. As a compromise the 4th order terms can be penalized. Increasing the penalties of the 4th order terms gradually changes the pressure distributions to 3rd order polynomials.The tilting of the rolls is also sensitive to problems of matching the rolls near the edges. The tilting terms can be penalized to make it possible to handle unsymmetrical taper or chamfer on the intermediate rolls.

3 TOOLS

The work was done on an IBM compatible 33 MHz 486-DX PC platform using DOS operating system. The model was implemented as a FORTRAN 77 program, which was readily compiled and debugged with Salford's and Lahey's compilers. Formulas and algorithms were derived using MAPLE, a system for symbolic mathematical computation. Simulation with the segmented models CROWN426 of MEFOS and SHAPE of BHP/IAS was used to validate the control model.

4 VALIDATION

In order to validate the parameterized model, sensitivities (partial derivatives) were compared to corresponding results from the segmented models CROWN426 of MEFOS and SHAPE of BHP/IAS. The sensitivities were calculated using a small perturbation of the rolling force (*Fs*), work roll crown (*Cw*), intermediate roll taper (*Taper*) and intermediate roll shift (*Shift*). Each sensitivity corresponds to the bending force (*Fj*) that optimizes the pressure distribution between the work roll and strip (and the flatness). The taper was defined as change of diameter / taper length (0.3 mm / 100 mm or no taper). Data from the 6-high reversing cold rolling mill of BHP New Zealand Steel were used to validate the parameterized model JACK6 for work roll bending. The result is presented in Table 1.

The difference between the two segmented models CROWN426 and SHAPE gives an indication of the accuracy that can be expected. Both models are validated with measured data from 4-high rolling mills. CROWN426 is also validated with measurements from 6-high rolling mills with intermediate roll shifting.

The pressure distribution between the intermediate roll and the work roll is illustrated in Figure 2 for cylindrical rolls and Figure 3 for tapered intermediate rolls. The results from the segmented model, CROWN426, show that the taper reduces the pressure peak. The 4th order polynomials cannot reproduce the complicated pressure distribution, that can be described by the segmented model. Figure 3 also illustrates excessive(?) cutoff at the right side.

When the pressure distributions differ like this, you cannot expect the calculated results, e. g. the sensitivities, to be the same.

5 DISCUSSION

The program using the parameterized physical model is much faster than segmented models for simulation, but it is not as accurate, because of the limited degrees of freedom. The absolute level of the predicted optimal bending force is expected to be offset, while the sensitivity is expected to be more reliable.

One problem of the model is that the polynomials of the pressure distributions are very steep near the ends of the contact zones between the rolls. For some ill-conditioned, but real roll profiles, the mismatch error functions will be too big if the high order terms are penalized. Sometimes cutting off the portion of the contact zone with negative pressure will cut off too much of the contact, yielding very awkward pressure distributions with a high peak near the cutoff point and negative pressure in the central parts of the contact zone.

Another problem is that the sensitivity to bending seems to be too small compared to simulation with segmented models. This problem may be created by the formulation with error functions rather than absolute matching at the contacts between the rolls.

Roll taper and roll shift should be optimized prior to bending either inside or outside the parameterized model, in order to create well-behaved pressure distributions and to improve the matching of the contact between rolls and strip.

The application to a certain rolling mill should start with extensive comparison of the results to measured data and simulation with segmented models in order to tune the absolute level (and maybe also tune the sensitivity). Once the model is tuned, it is a very fast and convenient off-line tool for optimizing the roll geometry or pass schedules. For on-line control, the input and output of the present program will be simplified before merging it into a flatness control system.

6 ACKNOWLEDGEMENTS

The parameterized model for flatness control in six-high rolling mills was developed and programmed at the Port Kembla Laboratories of BHP Research, Port Kembla, Australia.

We want to thank the National Swedish Board of Industrial and Technical Development for financial support of Mr Wiklund's stay at BHP.

7 REFERENCES

[1] N-G Jonsson et.al: "Crown and shape control in flat rolling using process models", Steel Times International, December 1992, p 548.

[2] O Wiklund et al: "Simulation of crown, profile and flatness of cold rolled strip by merging several physically based computer models", Proc. 4th International Steel Rolling Conference, Deauville, France, June 1-3, 1987, Vol, 2:E9

[3] O Wiklund et al: "Profile and flatness control methods for rolling of flat products simulated with MEFOS's physically based computer models", Proc. 33rd MWSP Conference, ISS-AIME, Vol. XXIX, St Louis, USA, 1991, p 363.

[4] W Y D Yuen, B Wechner: "A physical model for on-line roll bending set up in cold rolling", Proc. 5th International Rolling Conference, London, UK, 1990, p 234.

[5] W Y D Yuen, L M Miller: "An on-line model for profile and shape control in hot rolling", Proc 4th International Conference on Technology of Plasticity, Beijing, China, Sept 5-12, 1993, p 701.

[6] Y Misaka, T Yokoy: "Calculation of the roll deflection and its applications", Proc. ICSTIS, Suppl. Trans. ISIJ, Vol 11, 1971, p 680.

Simulation of Materials Processing: Theory, Methods and Applications, Shen & Dawson (eds)
© 1995 Balkema, Rotterdam. ISBN 90 5410 553 4

Iterative FEM die surface design to compensate for springback in sheet metal stampings

Longwu Wu
Cray Research Inc., Eagan, Minn., USA

Changqing Du & Li Zhang
Chrysler Corporation

ABSTRACT: Recent advances in FEM analysis of sheetmetal forming enable not only simulations of forming processes but also prediction of springback. Based on these capabilities of FEM analysis an iterative FEM die design algorithm was developed to systematically generate tooling geometry that produces parts with accurate shape after springback. The algorithm was tested by real automotive sheetmetal parts and proved to be able to generate appropriate die geometry provided an FEM package is available to give accurate enough results on forming and springback simulations.

1.INTRODUCTION

FEM analysis has been increasingly employed to simulate sheetmetal forming processes producing information on splitting and wrinkling that helps a great deal to die surface development engineers. Recently, researchers started to report successes in predicting springback in sheetmetal forming by FEM simulations [1,2]. The results of springback analysis offer some guidelines to compensate for springback in die surface development. However in most cases FEM analysts can only give one number such as over-bending or over-crowning rather than a complete description of correct tooling geometry. Compensation for springback in die surface development is still artisan's work with much guess involved.

Problems that die surface design engineers face in their everyday practice, as far as springback is concerned, are to determine the die geometry with given net shape of parts. These problems are just the inverse of what FEM analysts can do to date by simulations i.e. to determine net part shape with given die geometry. There are no direct solutions to these type of inverse problems. But, from a mathematical point of view, an appropriately designed iteration algorithm should be able to address the issue. Some authors reported preliminary work along this direction on a 2-D plane-strain part of simple geometry, a u-shaped channel [3]. Presented in this paper is an entire algorithm that is able to systematically generate complete geometric description of tooling with springback compensation taken into consideration.

Sheetmetal parts, when removed from dies after forming, are subject to springback due to resultant in-plane forces and moments throughout the sheet at the end of forming process. Those forces and moments, in a general term, can be denoted by a vector $\boldsymbol{f}$. FEM simulations can be conducted with $\alpha\boldsymbol{f}$, where $-1 < \alpha < 1$ is a scaler multiplier, applied to the formed part. When $\alpha = 1$, physical springback is simulated. However when $\alpha < 0$, the formed part will deform in a manner opposite to that of springback, referred to as "spring-forward." Apparently, die surface designs based on the shape after spring-forward will form a part with better geometric accuracy. With appropriately selected value of α, a die design can be obtained such that the embedded springback compensation will reduce geometry deviation of the part within tolerance.

Description of the iterative die surface design algorithm is given in section 2 of the paper. The algorithm was applied to two real automobile sheetmetal parts, a roof bow and a rail cap, and the results are presented in section 3 with some discussions. Section 4 contains brief conclusions.

2. DESCRIPTION OF THE ALGORITHM

The objective of the algorithm is to obtain a geometric description of tooling that will form parts of desired shape after springback. As in any iteration scheme, iteration zero starts with some initial guess which in this case is naturally the designed part shape. Therefore FEM mesh is first generated from the CAD file of the part design and simulation of forming is performed as FEM analysts usually do.

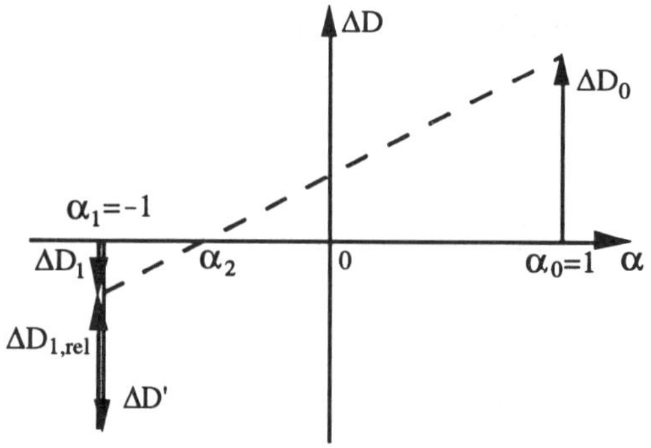

Figure 1. Deviation in Critical Dimension from Different Iterations

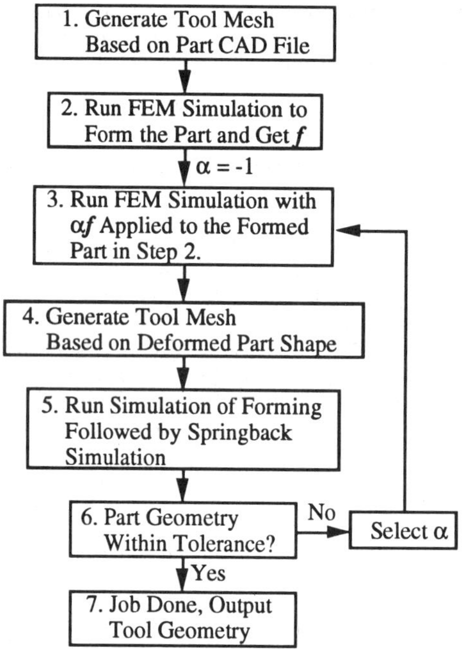

Figure 2. Flow Diagram of the Algorithm

After simulation of forming, two more simulations are conducted. The first one applies $\alpha_0 f$ to the formed part with $\alpha_0 = 1$ and the second one applies $\alpha_1 f$ with $\alpha_1 = -1$. Let D be a critical dimension that is the major concern in springback and ΔD be the deviation of that dimension from design value. The springback ($\alpha_0 = 1$) simulation results in deviation of ΔD_0 in the critical dimension. The full spring-forward ($\alpha_1 = -1$) simulation results in deviation $\Delta D'$ which is usually very close to the negative of ΔD_0. Those deviations and their relation to α are shown in Figure 1.

Based on the part shape resulted in the full spring-forward simulation, FEM mesh of tooling for the next iteration (iteration 1) is prepared. Simulation of forming is conducted and followed by a regular springback simulation resulting in $\Delta D_{1,rel}$ with respect to the die geometry used in this iteration. Considering that the die in this iteration is based on part geometry with deviation of $\Delta D'$ the net deviation of the part after springback is $\Delta D_1 = \Delta D_{1,rel} + \Delta D'$ as shown in Figure 1. If ΔD_1 is within tolerance, the die geometry data used in this iteration can be output and fed back to CAD/NC system for die cut.

In the case of ΔD_1 being still beyond tolerance more iterations may be carried out. Assuming a linear relation between ΔD and α, value α_2 can be determined as shown in Figure 1. Partial spring-forward simulation is performed with $\alpha_2 f$ applied to the part formed in iteration zero. The deformed part from the simulation is used to generate FEM mesh of tooling and steps described above are repeated until ΔD is within tolerance. Experiences of the authors show that three iterations are usually enough. The flow diagram of the algorithm is given in Figure 2.

3. APPLICATION EXAMPLES

The iterative FEM die surface design algorithm was applied to two real sheetmetal parts of automotive, a roof bow and a rail cap. The algorithm is independent of FEM codes used in the simulations of forming and springback/spring-forward. For thc convenience of the authors, an FEM package LS-DYNA3D/LS-NIKE3D from LSTC [4] was used in the work. LS-DYNA3D, an explicit FEM solid mechanics code for large deformations was used to do the simulations of forming. At the end of each run a file containing internal stresses of all elements was produced and multiplication of α was handled by a small utility program written by the authors. LS-NIKE3D, an implicit FEM code took the output from LS-DYNA3D and ran simulations of springback or spring-forward. A Cray C90 supercomputer was used for all the simulations for fast return time.

3.1 Roof Bow Application

A roof bow was selected as the first example to test the algorithm. This part is long in one dimension and has gentle curvature along the length. The geometry features cause severe springback in this part. There were two pieces in the tooling: upper die and lower die with identical mesh. Material of the part is high strength steel of gauge 1.3mm. Material properties are listed in Table 1 along with the material properties used for the next part, a rail cap. One quarter of the part formed by FEM simulation is shown in Figure 3.

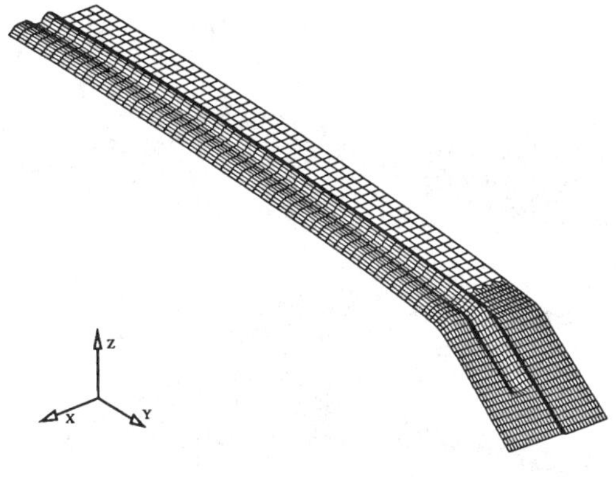

Figure 3. Roof Bow Formed by FEM Simulation

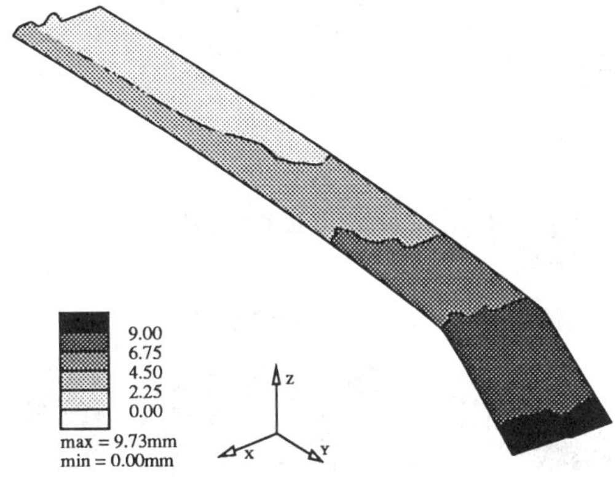

Figure 4. Springback in Z-direction of the Part Formed in Iteration 0

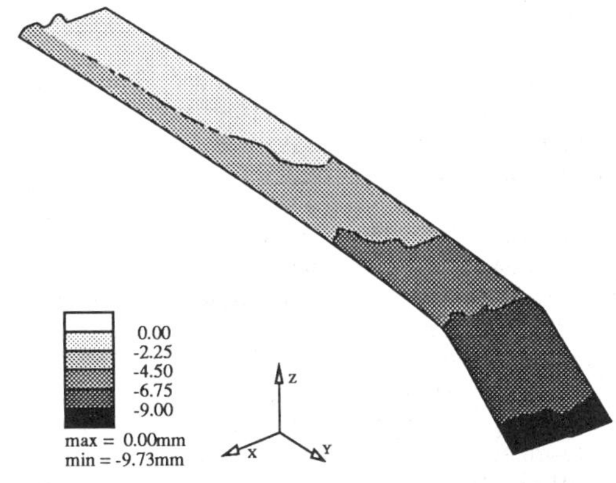

Figure 5. Spring-forward in Z-direction of the Part Formed in Iteration 0

Table 1. Material Properties Used in the Simulations

	Roof Bow	Rail Cap
Modulus(GPa)	207.0	207.0
Poisson Ratio	0.3	0.3
Y.S. (MPa)	310.0	344.0
T.S. (MPa)	379.0	427.0
n Value	0.17	0.17
r Value	1.2	1.2
Friction	0.125	0.125
Thickness (mm)	0.73	1.3

The center point of the part was selected as the reference point when simulations of springback and spring-forward were conducted. At this point, all three translational degrees of freedom and three rotational degrees of freedom were fixed in the simulations. Deviation in Z direction of the part formed in iteration 0, i.e. the springback obtained by letting α be 1, is shown in Figure 4. The end of the roof bow was 9.73mm off. The simulation of α being -1 resulted in a deformation distribution of similar pattern but in opposite direction as seen in Figure 5.

The Z-deviation of the part formed in iteration 1 is shown in Figure 6. Inaccuracy was reduced by about 63%. The part formed in iteration 2 had the maximum Z-deviation of only 1.38mm as shown in Figure 7. Often the scattering of springback of similar parts on a press floor may exceed this amount due to various reasons, therefore further iterations are not considered to be necessary. The predicted deviation distribution is very smooth over the entire part as seen in the figure.

In this example, the tooling mesh of iteration 1 and 2 was obtained by copying the mesh of parts formed in the previous iteration. This mesh generation method is simple and easy. However, in general, the mesh quality is poor especially in areas where large curvature occurs because the mesh is originated from part mesh. Nevertheless, for simple geometry like the roof bow, this method produced good enough results.

3.2 Rail Cap Application

The second part used to test the algorithm was a rail cap that had more complicated geometry and forming process. Figure 8 shows the exploded view of the initial mesh. The tooling was composed of several pieces including punch, die, pad, and flanging punch. Movement of the pad was first imposed by velocity loading curve and then controlled by upward pressure when the punch started pushing the pad. The flanging punch did not start moving until the forming of the major portion was completed. The formed part is shown is Figure 9.

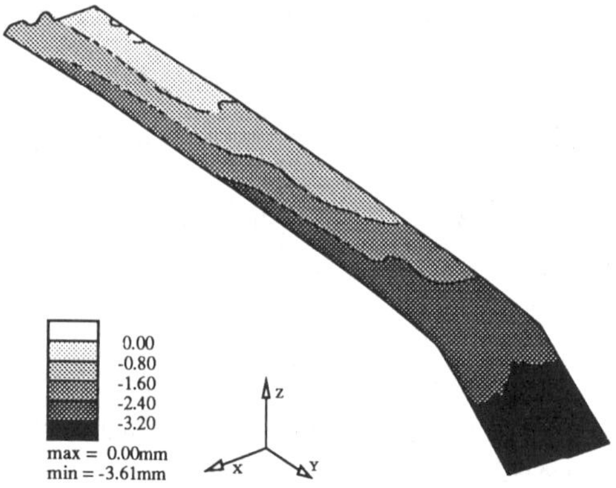

Figure 6. Z-deviation after Iteration 1

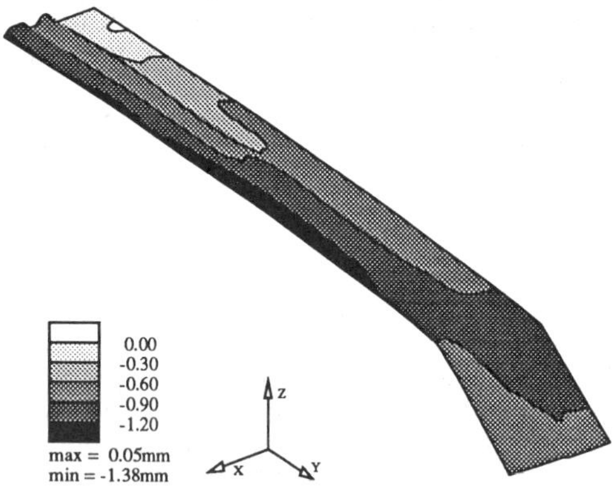

Figure 7. Z-deviation after Iteration 2

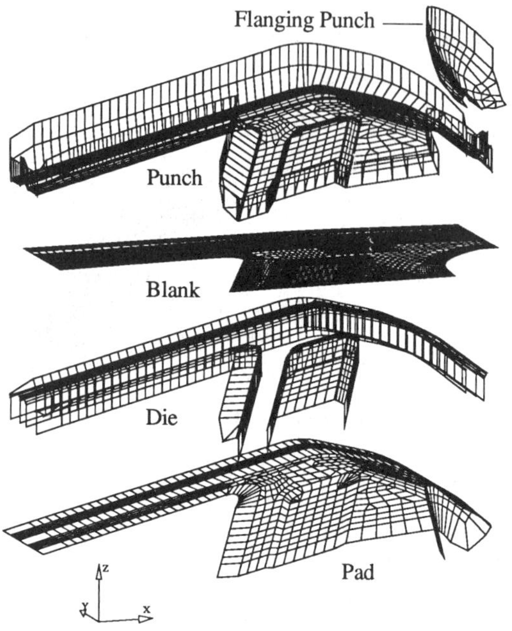

Figure 8. Initial Mesh of Rail Cap FEM Model

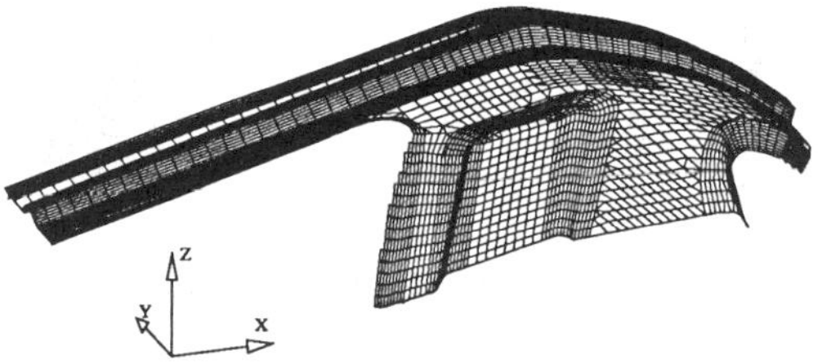

Figure 9. Rail Cap Formed by FEM Simulation

Shown in Figure 10 is the Z-deviation of the part obtained by springback simulation with appropriate boundary conditions imposed to three points labeled as A, B, and C in the figure. The maximum Z-deviation, 10.3mm occurred at the left end of the part therefore the deviation in that area was taken to determine the value of α in the following iterations. The Z-deviation at the left end was reduced to 2.88mm and 0.88mm in iteration 1 and iteration 2 respectively as shown in Figure 11 and Figure 12.

For this rail cap, like most sheet metal stampings, the tooling includes several pieces. On each piece, there are areas where the geometry coincides with the part geometry and contact between tooling and part occurs during forming. These areas are referred to as contact areas. Mesh generation method of the first example was attempted for iteration 1. The part mesh after spring-forward was broken into pieces according to the tooling configuration. Mesh in those pieces was used as mesh for the contact areas of the tooling. Then expansions and augmentations were made to complete the tooling mesh. The mesh quality was so bad that the springback simulation yielded meaningless results though the forming simulation seemed acceptable.

A special technique was developed to map the mesh of the contact areas in the tooling onto the part surface. Then the entire tooling mesh was generated based on those mesh patches. By this technique, not only the mesh quality is high but also the key geometric features of the original tooling mesh is reserved. The improved die surface design is represented by the tooling mesh used in the last iteration. With the key features in the tooling geometry reserved, it is much easier for the die surface development engineers to rework on it when the mesh file is turned back to them.

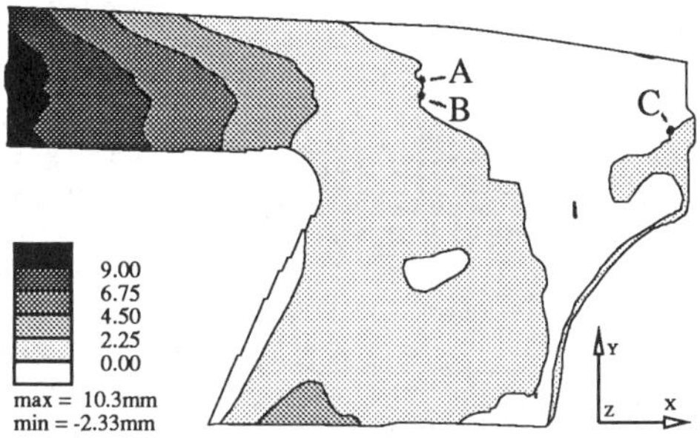

Figure 10. Springback in Z-direction of the Rail Cap Formed in Iteration 0

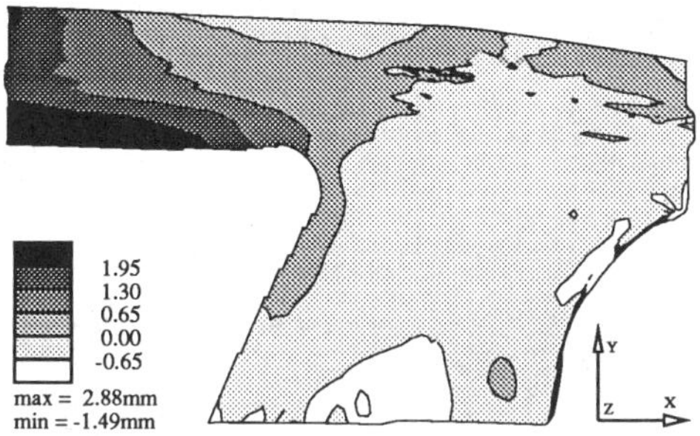

Figure 11. Z-deviation after Iteration 1

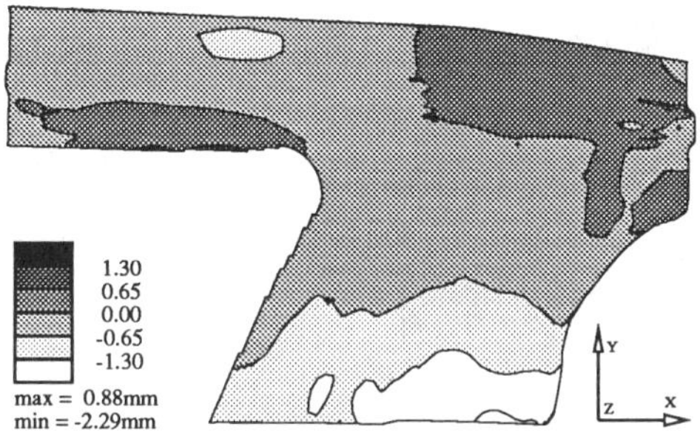

Figure 12. Z-deviation after Iteration 2

In the iterations, values of α are determined by the deviation of one critical dimension, therefore deviation of other dimensions may not be reduced as much for complicated geometry. This can be seen on Figure 12 that at the lower right region of the part the deviation was not changed significantly though it was within acceptable level. A possible remedy of this might be to make value of α vary over the part. This is left open for future research if needed.

Experimental validation of the algorithm on the rail cap is being planned.

4. CONCLUSIONS

An iterative FEM die surface design algorithm has been developed to produce a complete geometry description of stamping die surfaces with springback compensation included. Taking advantages of FEM analysis the algorithm simulates spring-forward, a process that is opposite to the physical springback, to generate new tooling surfaces. With the proper amount of spring-forward imposed on tooling geometry the formed parts will have dimensions within tolerance after springback. Applications of the algorithm to two real auto parts, a roof bow and a rail cap, demonstrates that the algorithm is able to handle parts of general three dimensional geometry and dimension deviation can be corrected to acceptable level.

To apply the algorithm, one needs to generate FEM tooling mesh based on the part geometry after spring-forward. A special technique was developed to map the original tooling mesh onto the part surface. The technique gives high quality tooling mesh and makes the work of die surface development engineers afterward much easier.

The algorithm works with any FEM codes that can simulate forming and springback. From the work conducted by the authors, it has been observed that the accuracy of springback simulation depends upon the forming simulation because the stresses in the panel at the end of the forming process are used to calculate springback. Therefore care should be exercised with punch velocity and tooling/blank mesh when explicit FEM code is employed for forming simulation.

The capability of the algorithm working on parts with severe bending and unbending is to be investigated.

REFERENCES

1. Wu, L., "Springback and Residual Stress Analysis for Sheet Metal Forming Processes by Dynamic Relaxation in Conjunction Finite Element Method," Proceedings of the 4th ICTP, Beijing, China, Sept. 5-9, 1993, pp1840-1845.

2. Du, C., Zhang, L. and Wang, N. M., "Springback Prediction in Sheet forming Simulation," SAE International Congress & Exposition, SP-1021, Feb. 28 - Mar. 3, Detroit, Michigan, 1994, pp113-123.

3. Boyce, M., "Finite Element Simulations in Mechanics of Materials and Deformation Processing Research," ABAQUS User's Conference, June 1 - June 3, Newport, Rhode Island, 1994.

4. Hallqquist, J. O., Stillman, D. W. and Lin, T. L., "LS-DYNA3D User's Manual," Apr. 1994.

Simulation of Materials Processing: Theory, Methods and Applications, Shen & Dawson (eds)
© 1995 Balkema, Rotterdam. ISBN 90 5410 553 4

Forging preform design using the inverse die contact tracking method

Guoqun Zhao, Ed Wright & Ramana V. Grandhi
Wright State University, Dayton, Ohio, USA

ABSTRACT: The inverse die contact tracking method utilizes both the forward and inverse finite element simulations to design the preform shapes in forging processes. The procedure starts with the forward simulation of a candidate preform into the final forging shape. A record of the boundary condition changes is produced by identifying when a particular segment of the die makes contact with the workpiece surfaces in forward simulation. This recorded time sequence is then optimized according to the material flow and the state of die fill to satisfy the requirement of material utilization and forging quality. The modified boundary conditions are finally used as the boundary condition control criterion for the inverse deformation simulation. Additionally, a procedure to determine process staging points using trial forward simulation is given. The preform design of an example plane-strain forging process is performed.

1 INTRODUCTION

With the introduction of the FEM based backward tracing method for preform design by Park (1983), several variations of this method have been studied for solving specific problems. Hwang (1986) studied the preform designs in disk forging process. Kim (1990) also applied the FEM backward tracing scheme for the design of preform shapes for an axisymmetric forging problem with H-shaped cross section. Recently, Kang (1990) established systematic approaches for the preform design in forging of an airfoil section blade as a two dimensional plane strain problem using back-tracing scheme. Zhao (1994) established a node detachment criterion for inverse deformation and preform design based on minimizing the forging shape complexity factor. Han (1993) applied optimization techniques by minimizing strain variance and other factors using the backward tracing method with velocities of a selected group of nodes taken as design variables.

The determination of billet-to-die contact boundary conditions for inverse simulation is a major concern. Strategy used for boundary condition control will influence the shape and the resulting performance of the preform design. In Kang's (1990) research, several candidate preform shapes were suggested and tried using FEM forward simulation. During the forward simulation, the changing boundary conditions were recorded as the time sequence in which the workpiece nodes were in contact with the dies. The recorded time sequence of boundary node contact was modified and then used as the boundary condition sequence for the inverse simulation. To retain continuity of the boundary node detachment sequence, the finite element mesh division used for the inverse simulation must be the same as that of the forward simulation. However, for complex forging processes, the mesh distortion can be severe enough to require remeshing. The rezoned mesh may require a different number of boundary nodes from the previous mesh and provide new boundary node locations. Therefore, the proper recording method should be established to take remeshing into account.

This paper presents an inverse die contact tracking method using both the forward and inverse simulation and expands on recent work by Kang (1990). The preform design of an example plane-strain forging process is developed. The number of stages is also determined by trial simulation.

2 PREFORM DESIGN METHOD

2.1 *Inverse die contact tracking method*

The inverse die contact tracking method is a node detachment control criterion for determining the boundary conditions for the inverse deformation simulation. The boundary condition time sequence for the inverse simulation can be derived from the forward simulation of a trial preform even if the trial preform does not satisfy the design objectives completely. This necessitates the development of a methodology for recording the boundary condition sequence which is independent of the

FEM meshing requirements and will account for momentary contact of boundary nodes. Once the time sequence is known, a method for modifying this sequence for the inverse simulation must be determined.

In the inverse die contact tracking method, the die surface is divided into a number of straight line or arc segments. During the forward simulation with a trial preform, the time at which each die segment comes into contact with the deforming body is calculated and recorded. In this way, it is not necessary to retain the original mesh division or the same number of boundary nodes. If a die segment was to repeatedly separate and make contact with the workpiece, the last moment of contact is considered for the record. To adequately reflect the change of boundary condition using the contact time of the die segments, the length of each die segment should be less than the minimum side length of the workpiece boundary elements. With this die segment spacing, there may be some segments which are between workpiece surface nodes and as such, will never make contact with a boundary node. The contact time for these middle die segments is determined through linear interpolation using the contact times of two nearest adjacent die segments as the end data points. This provides a continuous time sequence record for the segments which touch the workpiece between nodes.

The boundary condition sequence derived directly from the trial forward simulation must be modified according to die fill state and material flow characteristics if the trial preform does not satisfy the design objectives. The extent of the modification depends on how close the selected trial preform is to the desired shape. The following guidelines were developed and used for modifying the boundary conditions:

1. If there is an area of incomplete die fill, the contact time of the die segment at the underfilled corner of this area is then defined as the time that the upper die reaches the required final position. The contact time of the die segments at the both sides of this segment is gradually decreased to match the nearest adjacent segments where the contact time is defined.

2. For forging dies having more than one deep cavity, the contact time of the die segments in the earliest filled cavities should be reduced to allow all deep cavities be filled simultaneously at the end of forging process. If one or more of the deep cavities are not filled, the time sequence is modified as in step 1. This guideline can insure the reasonable material volume distribution for the finishing operation.

3. If there is large flash at the end of trial forward simulation, the contact time of die segments near flash area is adjusted to achieve the required flash size. The contact time of the outermost die segment within the flash gutter which touches the workpiece, is defined as the time that the upper die moves to the required final position.

Because of the modifications to the recorded boundary conditions, some of the workpiece boundary nodes could possibly penetrate into the dies during the inverse simulation. To prevent this from occurring, additional contact constraints are applied to these nodes. These additional contact constraints will remain in effect until these nodes do not tend to penetrate the dies. For a typical simulation time step, after the workpiece geometry is updated according to the converged velocity field, any nodes which have penetrated into dies are found by checking the coordinates in relation to the dies. These nodes are typically free surface nodes which have been detached at the previous steps. Contact constraints which permit motion only along the die surface are reapplied to these penetrating nodes. By returning back to the beginning state of previous step, the converged velocity field in the workpiece is again determined. The workpiece geometry and die position are then updated. After the current simulation step is finished, the additional constraints on the penetrating nodes are released at once. This checking process is performed continuously until the entire simulation is finished.

During inverse simulation, if a die segment is scheduled to separate from the workpiece in the current step according to the modified record, then the boundary nodes which lie in this die segment will be detached from the die. The last die segments which come into contact with the deforming workpiece during trial forward simulation become the first die segments to break away from the dies during inverse simulation. All the die segments except the initial contact segments will become free at the completion of the inverse simulation. The intermediate workpiece shape at the completion of inverse simulation is the preform shape.

2.2 *Staging*

Another important aspect of forging design is to determine the number of preform stages. When considering product cost, the required forging should be completed in as few operations as possible. Because of the extreme complexity in analyzing defects resulting from forging die design and a number of independent process variables involved, it is not now possible to develop a completely analytical procedure for defining staging points. A method for determining the forging stage number is presented based on the inverse preform design and trial forward simulation. The procedure is summarized as follows:

1. According to the required final forging and die shapes, the forward deformation simulation of a trial preform shape is done and the blocker dies are designed by using the inverse die contact tracking method.

2. By using the forward simulation, the material deformation in the blocking stage is evaluated starting from the initial billet (round or square). This is done for several cases by changing the ini-

tial billet size (ratio of height to diameter). If the desired blocker forging is successfully produced by using the selected initial billet, then no other preforming operations are necessary. Otherwise, another forging stage must be designed.

3. If an additional preform is required, the inverse die tracking method is performed using the blocker shape obtained in step 1 to determine the shapes of the buster dies.

4. The busting stage is checked by forward simulation using the initial billet shape. If the required busting forging is not realized, the above steps are repeated to design the next forging operation.

3 PREFORM DESIGN IN FORGING PROCESS

The example plane strain isothermal forging pattern is shown with the die segments numbered in Figure 1. The trial preform shape was designed using a knowledge-based blocker design system called BID (Vemuri, 1989). A strain rate hardening material having the constitutive relation $\bar{\sigma} = Y_0 \dot{\bar{\epsilon}}^{0.3}$ with $Y_0 = 10000$ *psi* ($68.93 MN/m^2$) was used in this example. A constant shear friction factor of 0.2 was

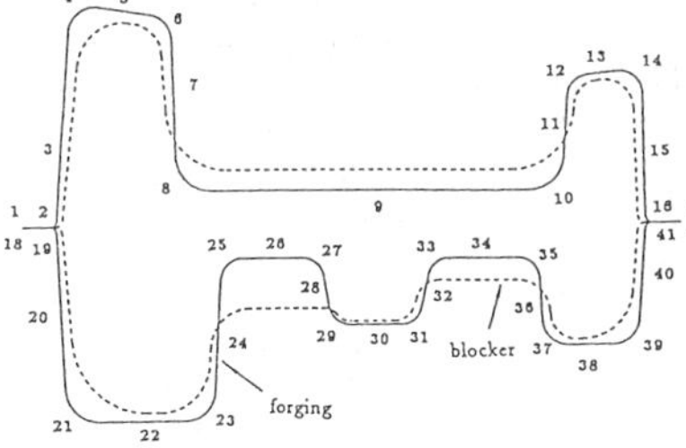

Fig 1. Final forging and blocker designed by Vemuri (1989).

assumed between the workpiece and the dies. The velocity of the upper die is −0.5 *in/min*, and the lower die is stationary. The design objective is to determine preform shapes and identify the number of stages which will produce a sound final forging, free of fold-over defects, underfill and with as little extrusion of flash as possible over the whole forging process.

3.1 *Blocker design*

Figure 2 shows the material flow patterns in forward simulation of final forging stage using the test preform shape shown in Figure 1. When the upper die moves to the required position as shown in Figure 2(b), the forging exhibits an excessive flash and incomplete die fill. Continued down motion of the upper die will completely fill the dies, but the forging will no longer meet the dimensional requirement for height. It is clear that a sound forging can

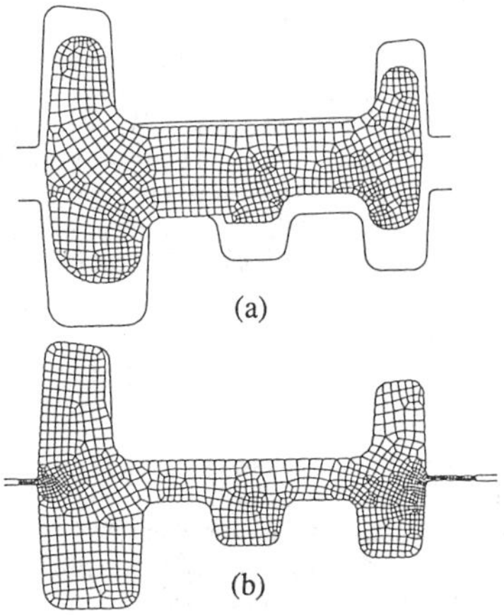

Fig 2. Forward simulation of final forging stage using preform shape of Fig. 1.

not be produced with the preform shape shown in Figure 1. However, from this trial forward simulation, useful material flow information is obtained. Both the upper die and lower die are divided into 648 die segments for recording the contact time at which the every die segment comes into contact with workpiece. Figure 3(a) is a plot of the recorded boundary condition changes of the upper die in forward simulation of final forging stage. For proper presentation and convenience of the mod-

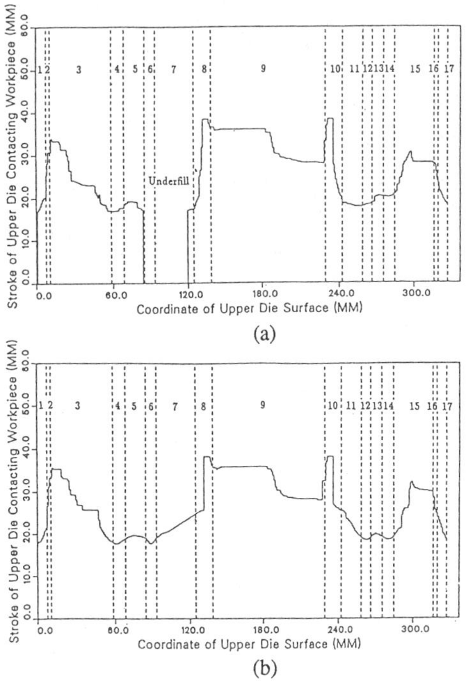

Fig 3. Boundary conditions of the upper die in forward deformation of final forging stage. (a): The recorded. (b): The modified.

ification of boundary conditions, the die boundaries are separated into regions and numbered corresponding to those shown in Figure 1. The record of contact time in relation to the upper die stroke is plotted along the die boundary length coordinates to illustrate the boundary condition change in forming process. The recorded boundary conditions are modified according to the guidelines 1 to 3 as shown in Figure 3(b) and then used to control the boundary node release during the inverse simulation in searching for the preform shape. Similarly, the recorded and modified boundary conditions changes for the lower dies can also be obtained.

Applying the modified boundary condition sequences of the upper and lower dies, the inverse simulation for design of blocker dies is easily preformed. Figure 4 displays the material flow patterns in inverse simulation of final forging stage,

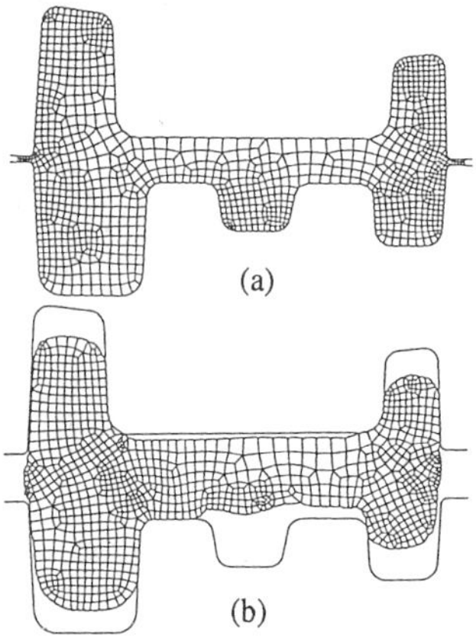

Fig 4. Metal flow patterns in inverse simulation of final forging stage.

starting from the final forging shape with a small flash volume and complete die fill. During the inverse simulation, two remeshing procedures were done due to mesh distortion. At the end of inverse

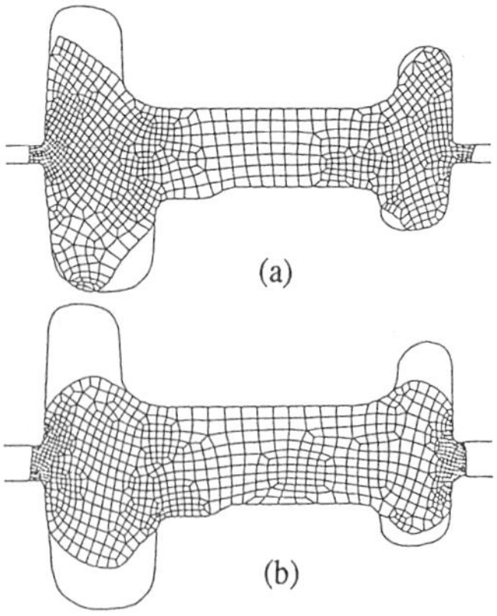

Fig 5. Trial simulation of blocking stage using two possible billet sections. (a): Square. (b): Round.

simulation, the intermediate forging shape shown in Figure 4(b) is obtained. Smoothing the boundary of the preform shape in Figure 4(b) results in the preform die shape shown in Figure 5.

3.2 *Buster design*

To determine if a single blocking stage satisfies preforming requirements, trial forward simulations of the blocking stage with two possible billet shapes and blocker dies designed by the inverse simulation are performed, as shown in Figure 5. From the simulation results, it can be seen that the formed blocker forging has a very large flash and the die cavity is difficult to fill completely, especially when using the round billet. It is also demonstrated that material trends to flow more toward the lateral direction resulting in a rapidly increasing lateral dimension for this plane strain problem. A busting operation which can distribute material volume reasonably in correspondence to die cavity volume will be needed to obtain a flashless blocker forging.

Figure 6(a) shows a suggested trial busting forging which can be used for the initial forward simulation. Figure 6 also illustrates the material flow patterns of the trial forward simulation. When the upper die stroke reaches the required value as shown in Figure 6(b), the blocker forging has little flash but still exhibits underfill in two upper side cavities. From this trial simulation, the change of boundary condition is obtained. The recorded boundary condition is modified slightly according to guidelines 1 and 2 and then used for the inverse simulation. Figure 7 shows the material flow patterns for the inverse simulation of blocking stage. When the inverse simulation is finished, a buster shape shown by Figure 7(b) is obtained. The buster dies shown in Figure 8 are obtained by smoothing the preform shape given by the inverse die tracking simulation.

Figure 8 shows the material flow patterns in

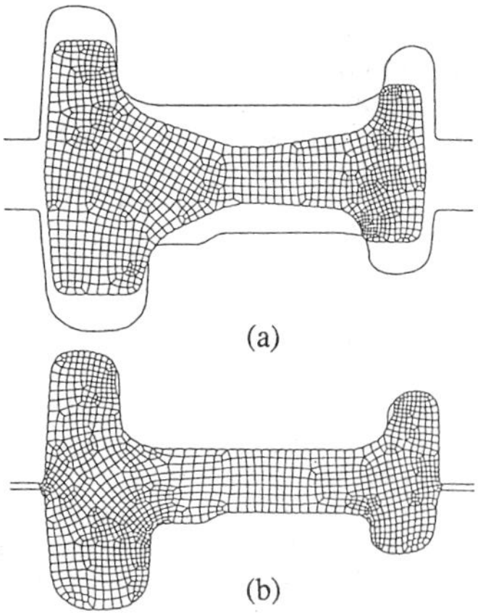

Fig 6. Trial simulation of blocking stage for recording the boundary conditions.

the test simulation of busting stage using the initial billet with square cross section. When the upper die moves to the required position, the buster forging still shows a little flash and the die cavity is still not completely filled. To produce a flashless buster forging, a fullering operation as shown in Figure 9 is required. After fullering, the workpiece is turned through 90 degrees counter clockwise and used as the billet for the busting operation. This completes the staging and preform design for this plane strain problem.

3.3 *Forward verification*

In order to verify the effectiveness of the preform design by using the inverse deformation, the forward simulation of whole deformation processes from the initial billet to final forging was completed. Figure 9(a) shows the initial square billet and the fuller dies. Figure 9(b) is the ending state of fullering operation. The fullered billet is turned through 90 degrees counter clockwise and used as the initial workpiece shape for busting stage. The material flow patterns in forward simulation of busting stage are shown in Figure 10 where Figure 10(a) is the initial state. Because of the function of fullering operation, the material can completely fill the die cavity without formation of flash. When the operation is finished, a flashless buster forging which completely fills the dies is obtained as shown in Figure 10(b).

Figure 11 shows the material flow patterns in

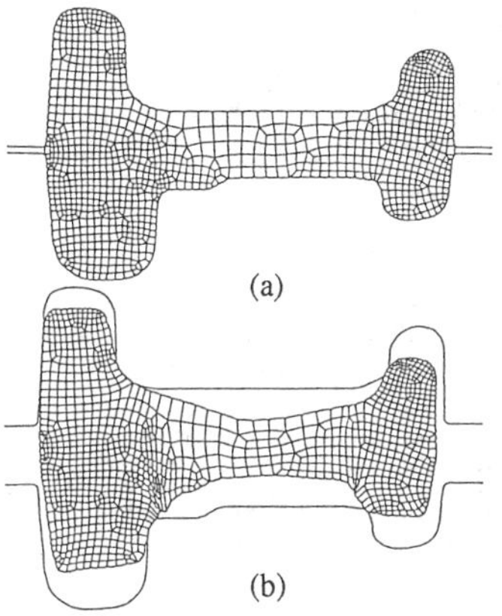

Fig 7. Metal flow patterns in inverse simulation of blocking stage.

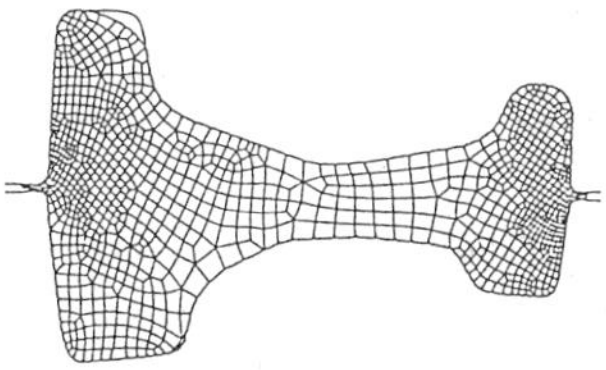

Fig 8. Trial simulation of busting stage using billet with square cross section.

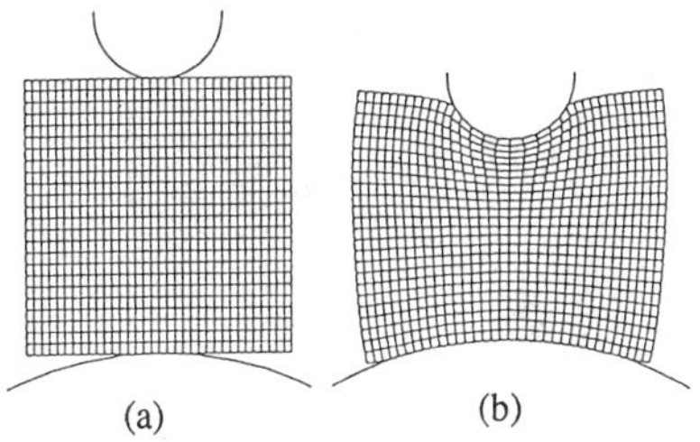

Fig 9. Forward simulation of fullering process from the initial square billet.

forward verification simulation of blocking stage. When the upper die moves to the required position, a nearly flashless blocker forging is formed. It is obvious that the both right and left hand die cavities are filled simultaneously. This indicates that a reasonable material volume distribution is achieved in the busting stage. Because so little flash is generated in the blocking operation, flash trimming operation can be eliminated. Figure 12 shows the forward verification of final forging process. Figure 12(a) shows the final forging dies and the initial billet for finish-forging. The material in central area has sufficient space to flow and thereby does not move to the side cavities. It is found that the central cavity and two side cavities are filled at same time. The right and left hand flash are nearly of the same volume. This resulted from the proper volume distribution achieved in the blocking operation which corresponds to the volume requirements of the die cavity. When the final forging process is finished, the workpiece has very little flash and completely fills the dies with no evidence of surface defects as shown in Figure 12(b). In comparison with the original blocker design, the width of the left-hand flash was reduced

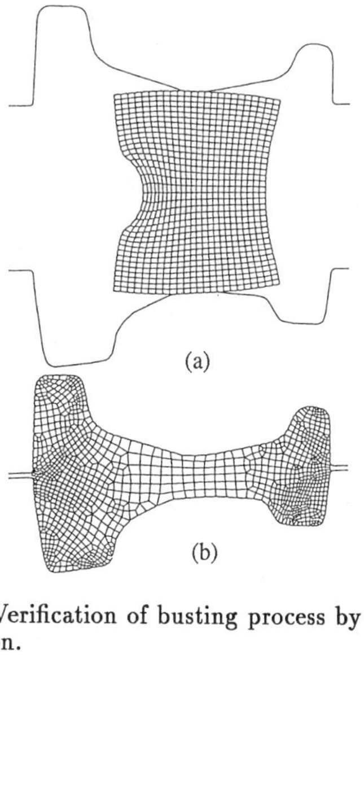

Fig 10. Verification of busting process by forward simulation.

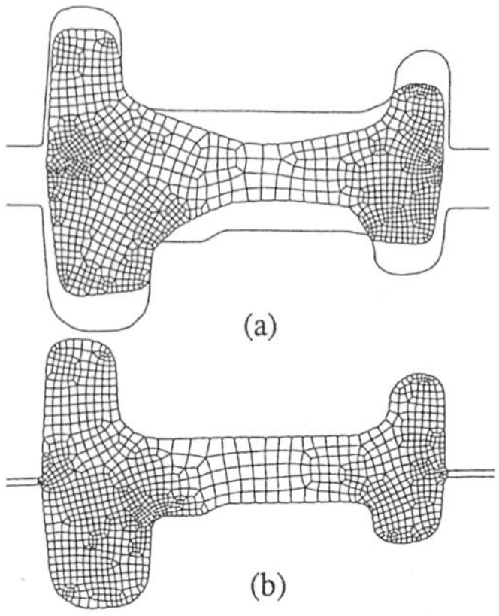

Fig 11. Verification of blocking process by forward simulation.

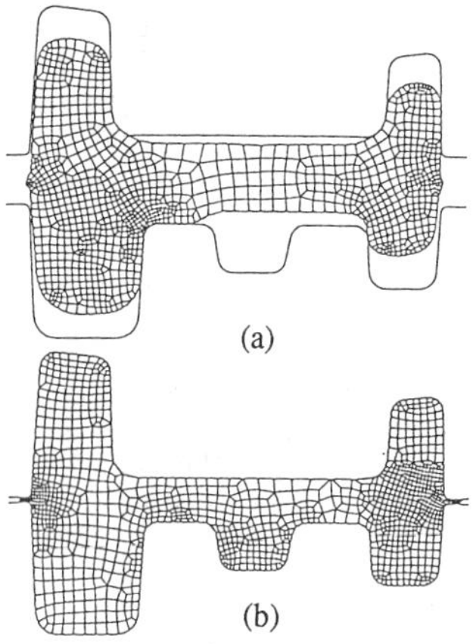

Fig 12. Verification of final forging process by forward simulation.

by one half and the right-hand flash was reduced by eighty percent. The final shape in Figure 12(b) is also matched very close with the initial shape for the inverse simulation in Figure 4(a).

Using this design methodology, it was determined that four stages (fullering, busting, blocking and finishing) are required to meet the design objectives. This number was primarily driven by the requirement of minimizing the flash volume while completely filling the dies.

4 SUMMARY REMARKS

This paper utilizes the inverse die contact tracking method to design the preform shapes of a typical plane strain problem. This method does not depend on the mesh division of workpiece and the number of boundary nodes. It can be used to design the preform shapes of forging processes including plane strain and axisymmetric deformation problems in which the deformation is severe and remeshing procedure is required. A method based on trial forward simulation to determine staging points is also presented. By using this preform design method, the fuller, buster, and blocker dies of a typical plane strain forging problem are designed. The number of stages is also determined by trial simulation. The effectiveness of preform design is verified via forward simulation. The analysis shows that a sound final forging can be produced using the present preform design methods.

Once a trial preform shape is selected, the inverse die contact tracking method improves the preform shape to more exact shapes. This trial preform shape can be obtained using the empirical guidelines or computer aided design methods. If the selected trial preform shape is far from what the design indicates, more extensive modification of the recorded boundary conditions would be required. If the design objectives can not be met by modification of boundary conditions, it is better to use the designed preform shape as the trial preform shape and repeat the entire tracking procedures. The manner in which the boundary condition time sequence is modified will influence the resulting performance of the preform shape given by the inverse simulation. Therefore, further research is needed to find a consistent mathematical approach for modifying the boundary condition sequence which will provide predictable results.

REFERENCES

Han,C.S., R. V. Grandhi & R. Srinivasan 1993. Optimum Design of Forging Die Shapes Using Nonlinear Finite Element Analysis. *AIAA Journal.* 31: 774-781.

Hwang,S.M. & S. Kobayashi 1986. Preform Design in Disk Forging. *International Journal of Machine Tool Design & Research.* 26: 231-243.

Kang,B.S., N. Kim & S. Kobayashi 1990. Computer Aided Preform Design in Forging of an Airfoil Section Blade. *International Journal of Machine Tools & Manufacture.* 30: 43-52.

Kim,N. & S. Kobayashi 1990. Preform Design in H-shaped Cross Section Axisymmetric Forging by Finite Element Method. *International Journal of Machine Tools & Manufacture.* 30: 243-268.

Park,J.J., N. Rebelo, & S. Kobayashi 1983. A New Approach to Preform Design in Metal Forming with Finite Element Method. *International Journal of Machine Tool Design & Research.* 23: 71-79.

Vemuri,K.R., S. I. Oh & T. Altan 1989. BID: A Knowledge-Based System to Automate Blocker Design. *International Journal of Machine Tools & Manufacture.* 29: 505-518.

Zhao,G., E. Wright & R. V. Grandhi 1994. Preform Design in Forging processes Using Nonlinear Finite Element Method. *Transactions of NAMRI/SME.* 22: 17-24.

Forming of metal sheet

Simulation of Materials Processing: Theory, Methods and Applications, Shen & Dawson (eds)
© 1995 Balkema, Rotterdam. ISBN 90 5410 553 4

OPTRIS: The complete simulation of the sheet metal forming

Frédéric Aberlenc, Jean-Luc Babeau & Pascal Jamet
Dynamic Software, Immeuble Ontario, Les Ulis, France

ABSTRACT: Developed by DYNAMIC SOFTWARE, OPTRIS is a software package dedicated to the modelization of sheet metal forming in an industrial environment.

Forming is not a single process and may be split into different stages : the gravity stage that leads to the correct initial shape of the sheet metal, the blankholder closing stage, the forming stage and the springback stage. Moreover, stamping is often a multi-step process and numerous specific technics are used, such as drawbeads, cutting during the process, ... Finally, aluminium is an investigation field for the industry.

In order to simulate the complete process in a realistic time for the industry and with the greatest easiness, some specific tools and algorithms have been developed in OPTRIS - the solver, based on the explicit finite element method combined with an automatic mesh refinement - and its pre- and post-processor FICTURE.

1 INTRODUCTION

For many mass production mechanical industries such as cars, domestic appliances or metal components manufacturing, the control of sheet metal forming process turned out to be a key factor of costs and development times reduction, in highly and increasingly competitive areas.

However, many complex physical phenomena may occur during this process, such as rupture due to excessive metal stretching, wrinckling due to local compression, and striction leading to locally insufficient thickness and poor mechanical resistance. To avoid these phenomena, the blankholder force and shape, the drawbead shape and position, as well as the shape of the initial sheet metal must be adjusted and optimized. The control of this process often requires years of experience from skilled craftsmen, and several months to determine each parameter and to finalize the forming operations for each new part. Such a trial and error method turns out to be prohibitively time-consuming for many industrials, particularly for car manufacturers who have to reduce the duration of their development cycles drastically.

In that sense, numerical simulation may be of great help to understand the interaction of complex physical phenomena, to reduce the number -and the cost- of experiments, and to support early design activities. However, a numerical approach of sheet metal forming has to meet severe industrial requirements : all of the physical parameters must be taken into account (contact, friction, lamination, strain hardening, rupture, striction, ...) within reasonable calculation times; graphic pre- and post-processing must be well-adapted to the specificities of sheet metal forming : a non-specialist user may be able to handle a three-dimensional object easily, interactively and quickly, to animate the whole process on the screen, as well as to follow a pre-defined methodology to define the process, to launch the computation and to analyse the results. This point is particularly important if the numerical tools are to be integrated in a team where people are used to manipulating real objects.

OPTRIS and FICTURE have been developed by DYNAMIC SOFTWARE in order to cope with such numerical and industrial requirements.

2 OPTRIS

• Overview:

OPTRIS is based on an explicit finite element method using an automatic mesh refinement. The sheet is modeled by 3- and 4-nodes plate elements, the die and the punch are modeled by rigid plate elements and the blankholder is modeled by non-rigid volumic elements to obtain very good contact areas and forces. Two plasticity laws (for steel and aluminium) are orthotropic and depend on anisotropic parameters. OPTRIS automatically detects penetration and the contact algorithm requires a few amount of CPU time. A COULOMB law is used to model friction. A global kinetic damping algorithm has been developed in OPTRIS to

simulate in a quick way the gravity and the springback stages. Drawbead equivalent elements have been introduced in OPTRIS.
Its high degree of vectorization and parallelization allows OPTRIS to take full advantage of the supercomputers and of the powerful workstations.

• Explicit time integration method:
The two hypothesis of such a method are:

a) use of a diagonal mass matrix M.
b) evaluation of velocities and accelerations using the central difference method:

X_n	:	coordinates array
$V_{n+1/2}$	:	velocities array
A_n	:	accelerations array
ΔT	:	time step

$$V_{n+1/2} = \frac{X_{n+1} - X_n}{\Delta T}$$

$$A_n = \frac{V_{n+1/2} - V_{n-1/2}}{\Delta T}$$

At cycle n :

$MA_n = F_n^{ext} - F_n^{int}$
F_n^{ext} : external forces array
F_n^{int} : internal forces array

Then,

$$M \frac{X_{n+1} - 2X_n + X_{n-1}}{\Delta T^2} = F_n^{ext} - F_n^{int}$$

Thus,

$$X_{n+1} = 2X_n - X_{n-1} + M^{-1} \Delta T^2 (F_n^{ext} - F_n^{int})$$

M is diagonal. This fact allows to compute immediately M^{-1} without any matrix inversion.

If F_n^{int} is a function of X_p (p less or equal to n), X_{n+1} becomes a function of X_p. (F_n^{ext} is user defined). Then X_{n+1} can be computed using results of previous steps without any matrix inversion (M is diagonal).

Let us describe the computation of F_n^{int} :

a) Evaluation of the strain increment $d\varepsilon$ between steps n-1 and n using X_{n-1} and X_n.

b) Evaluation of stresses using the stresses of previous step , $d\varepsilon$, and the plasticity law.

c) Evaluation of internal forces for each element using stresses.

d) Assembly of these internal forces in order to obtain F_n^{int}.

We will see later that the time step ΔT is necessary small. Therefore, the displacements of the nodes between two consecutive steps is small. This fact allows to compute the strain increment de between two consecutive steps using a small displacement theory without decrease of the precision.

• Stability of the explicit scheme
In order to simplify the demonstration, let us consider the linear case.

$F_n^{int} = K X_n$ where K is the stiffness matrix of the system.

We suppose that an error appears in the calculation of X (a truncation for instance) and we want to study the propagation of this error.

(X_n) is the solution with the error.
(Y_n) is the solution without the error.
$(E_n) = (X_n) - (Y_n)$ is solution of the homogeneous equation:

$$E_{n+1} + (\Delta T^2 M^{-1} K - 2) E_n + E_{n-1} = 0 \qquad (1)$$

In the basis of the eigenvectors of $M^{-1}K$, we obtain:

$$e_{n+1} + (\Delta T^2 \omega^2 - 2) e_n + e_{n-1} = 0 \qquad (2)$$

where ω^2 is an eigenvalue of $M^{-1}K$.

The characteristic equation of (2) is

$$x^2 + (\Delta T^2 \omega^2 - 2) x + 1 = 0 \qquad (3)$$

Consequently,

$$e_n = \lambda x_1^n + \mu x_2^n \qquad (4)$$

where x_1 and x_2 are the roots of (3) and l and m are complex.
If $\Delta T > 2/\omega$, x_1 and x_2 are two different reals and $x_1 * x_2 = 1$. Consequently, $|x_1|$ or $|x_2|$ is greater than 1 and (E_n) is unbounded.
If not, x_1 and x_2 are conjugated complex having an amplitude of 1., and (En) is bounded.

Consequently, (E_n) is bounded if and only if $\Delta T < 2/\omega_{max}$, where ω_{max}^2 is the greater eigenvalue of the system.

So the stability condition is : $\Delta T < 2/\omega_{max}$

Pratically, ΔT is very small. This fact allows to use a "small displacement" theory between two consecutive steps.

The error done on X is proportionnel to ΔT^2 and is consequently very small.

• Mesh refinement:

The respect of the geometry of the tools leads to use a very fine mesh for the blank in some critical areas as the fillets. However, the elements of the blank that are in contact with these areas are not known at the beginning, so that it is difficult to mesh correctly the blank: either you choose to use a fine mesh everywhere, which is CPU time consuming without any quality improvement, or you try to mesh the blank in an intelligent way which is a long job, and gives uncertain results. The solution was to introduce in OPTRIS an automatic adaptive mesh. During the computation, OPTRIS tests if the elements fullfill a geometrical criterium, defined by the user. The criterium used is based on the angle between the normals of two elements sharing a common edge.

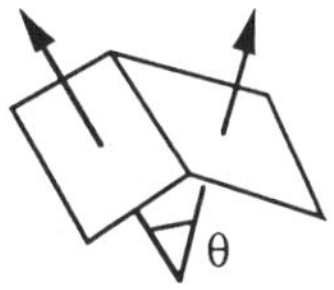

If the value of θ is greater than the user defined value θ_{max}, both the elements are split into four smaller elements as shown below.

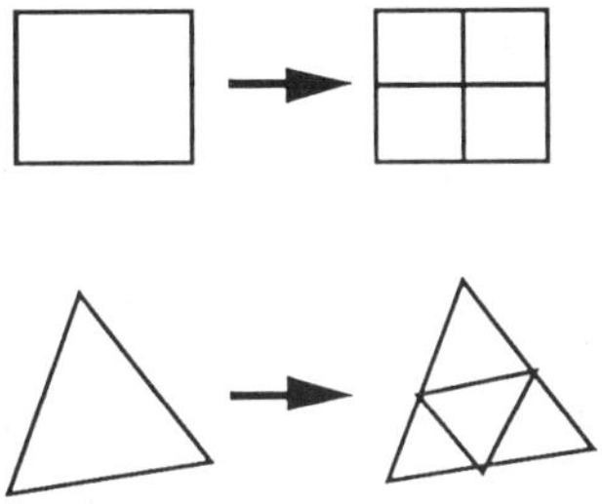

The criterium may be modified by the user at any moment during the computation and especially between the stages, allowing to use coarse meshes in the first stages (like the gravity stage) and fine meshes in the following stages (like the forming stages).

Some new nodes and elements have been inserted. The new elements, called sons, inherit the values of their father elements (stresses, strains, energies, ...). Some new nodes have been inserted in the middle of an edge of an element : these middle nodes must always interpolate the two nodes at the end of the edge.

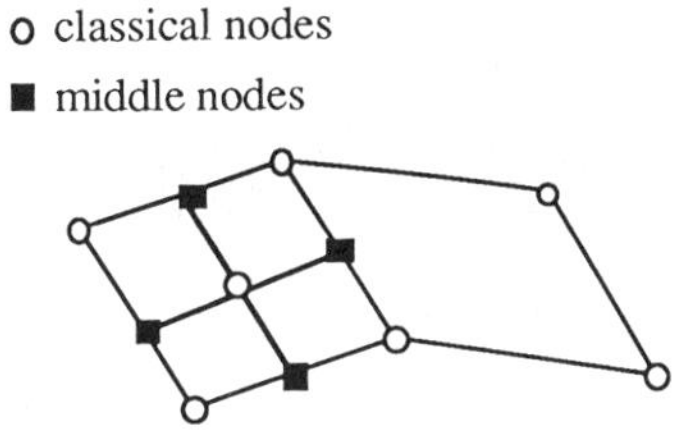

Once all the elements sharing this edge are split, the middle node becomes a classical node.

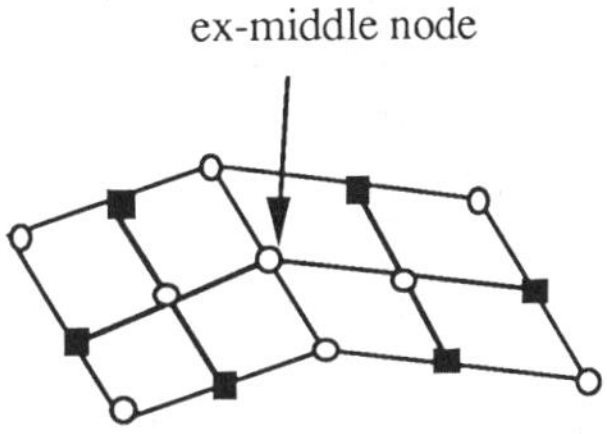

The benefits of this improvement are the following ones:
- Preparation time reduction: meshing the blank is not a long and hard job any more.
- CPU time reduction: the total number of elements is smaller.
- Quality improvement: in all the critical areas (and here only), the size of the elements is automatically small.

• The elements in OPTRIS:

- The blank: the blank is modelled by plate elements (3 nodes and 4 nodes elements) or by volumic elements.

The 4-nodes plate element used by OPTRIS is a C° bilinear plate element developed by T. Belytshko [2], based on the Mindlin plate theory. The sections normal to the midplane are rigid but can rotate.

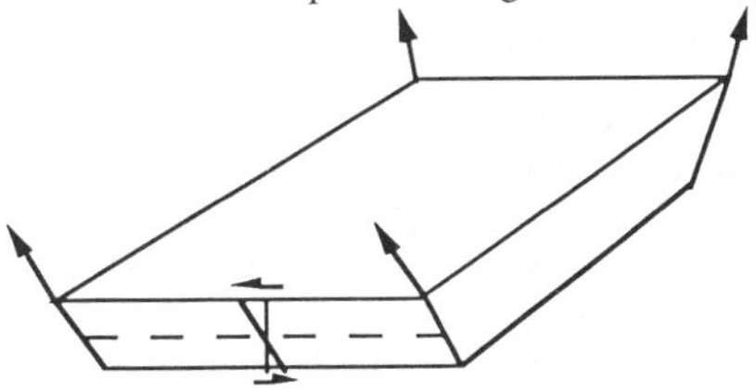

The displacements and the rotations of a point of the midplane is bilineary interpolated using displacements and rotations of the nodes:

$$u_m = \phi_I\,(r,s)\;u_I$$

$$\theta_m = \phi_I\,(r,s)\;\theta_I$$

where ϕ_I are the bilinear shape functions.

$\phi_I = 1/4\,(1+nr)(1+ms)$ where n and m equal 1 or -1 depending on I.

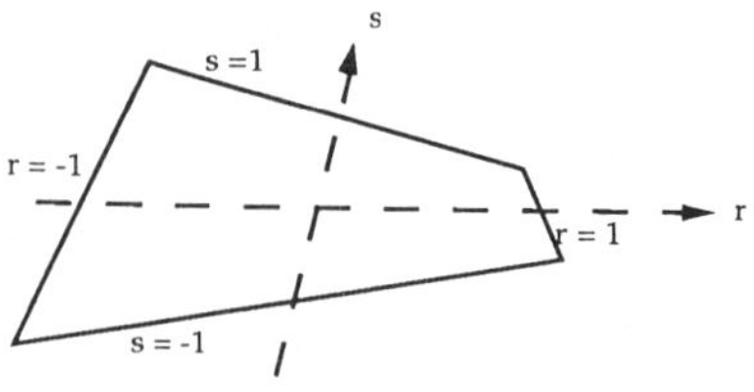

The displacement of any point M of the plate in a direction e in the midplane is:

$$u_M = u_m + z\,\theta_m\,e$$

where m is the nearest point of M in the midplane of the plate before any deformation of the plate.

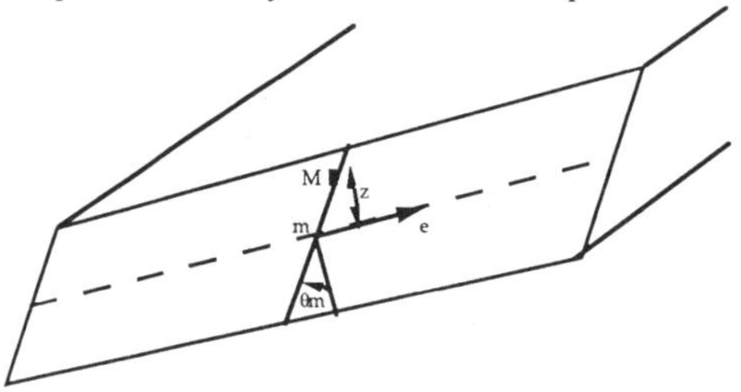

We work in a corotational system embedded in the plate. The strain tensor $(d\varepsilon_{ij})$ between two consecutive steps is computed using "small displacement" theory:

$$d\varepsilon_{ij} = 1/2\,(\partial_j\,u_i + \partial_i\,u_j)$$

The tensor $(d\varepsilon_{ij})$ of any point M of the plate can be written [2] under the following form:

$$(d\varepsilon_{ij}) = B_u\,(u_I) + B_\theta\,(\theta_I)$$

where B_u and B_θ depends on ϕ_I , $\partial_i\phi_I$ and z.

$(d\varepsilon_{ij})$ and the plasticity law allow to compute the stresses and the forces applied by the plate on its own nodes.
To obtain a very CPU efficient element, only one integration point is used on the surface of the plate. But the accuracy of the analysis depends greatly on the number of integration points through the thickness : OPTRIS uses five integration points to obtain correct stresses all through the plate thickness.

The volumic elements are used in OPTRIS when the Mindlin plate theory cannot be applied, for instance, when forming a very thick sheet metal with tools presenting very small radii. It appears for instance in the forming of a clutch part.

- The blankholder and the hydraulic jacks: the blankholder is a very important tool whose stiffness has to be taken into account to have a correct estimation of the contact areas and the contact forces. So it is modelled by volumic deformable elements and the force is, like in reality, applied by jacks located at particular points on the blankholder.

- The die and the punch: these tools are modelled by rigid plate elements. Only their shape is of interest.

• Zero energy modes
The use of a single integration point leads to "zero energy" modes of deformation for the 4 nodes plate elements and for the volumic elements. It means that the element can be deformed without modification of $(d\varepsilon_{ij})$. The next figure shows one of these modes for the 4 nodes plate element:

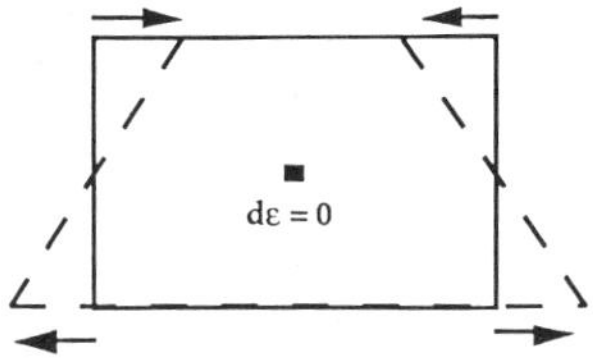

These modes can propagate in such way:

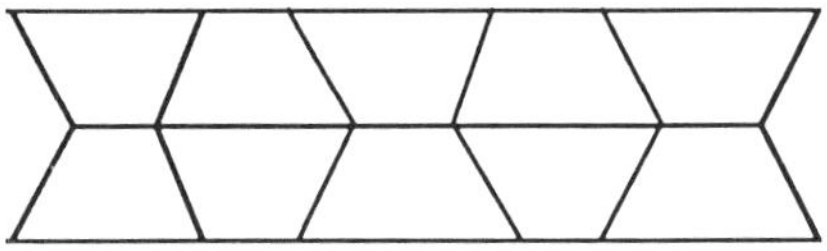

In order to stiffen the plate, B_u and B_θ are modified [2]. This modification adds a few amount of CPU time and does not change the precision.

•Plasticity Laws

Steel: In order to take into account the rolling of the sheet of metal, OPTRIS uses an orthotropic HILL criterium. The coefficients to be entered in OPTRIS are the Lankford coefficients and the Yield stress at 0°, 45° and 90°, which can be measured in three tensile tests.

Aluminium: the anisotropic plasticity of aluminium is well approximated with the BARLAT law. A software tool is supplied by DYNAMIC SOFTWARE to identify the coefficients of the law from the results of tensile tests.

• Contact algorithm

The time step ΔT is very small and consequently, the displacement of the nodes between two consecutive steps are small as compared to the thickness of the plate elements. The contact search is equivalent to checking the penetration of nodes inside the volume of all the elements. A normal force is applied between the element and the contacting node. Unlike the penalty method, the method used in OPTRIS to compute the contact force is such that the resulting normal force is exact: it leads to an exact estimation of the friction force, based on the normal force and the Coulomb friction coefficient.

• Springback

An algorithm of equilibrium search has been developed in OPTRIS in order to compute the springback of the blank after the stamping process. OPTRIS detects the maxima of the kinetic energy of the blank and stops all the nodes of the blank when a maximum is reached. This algorithm leads in a quick way to the solution corresponding to equilibrium.

• Drawbead model

The physical phenomenon which appears in a drawbead requires a very fine mesh if good results are expected. For this reason, an equivalent model of drawbead has been added. A restraining force is applied to the blank elements passing through the drawbead model. To define the restraining force of the equivalent drawbead, a small and quasi-automatic 2D simulation must be performed. A slice of 1 millimeter of the blank (meshed with volumic elements) is pulled through the drawbead : as results, we obtain the restraining force and the part of the force applied on the blankholder that is used to keep the drawbead closed.

3 APPLICATIONS

3-1 Stamping of a roof panel

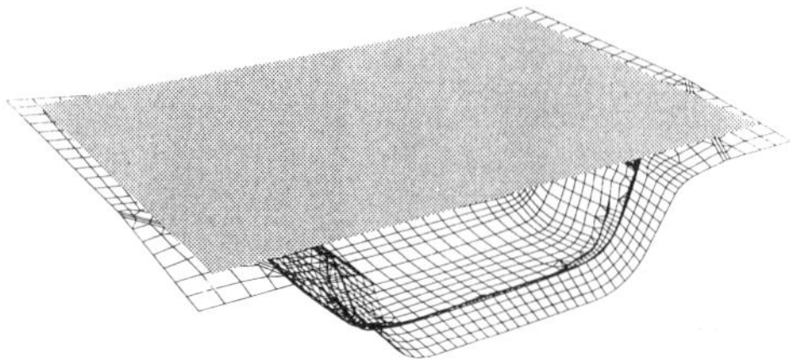

Initial shape

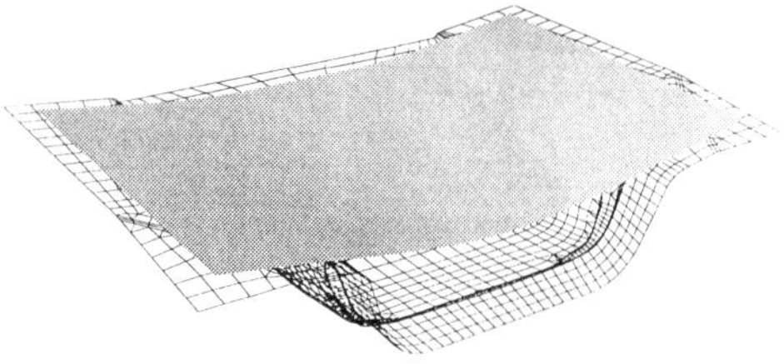

Shape after the gravity stage

Shape after the holding stage

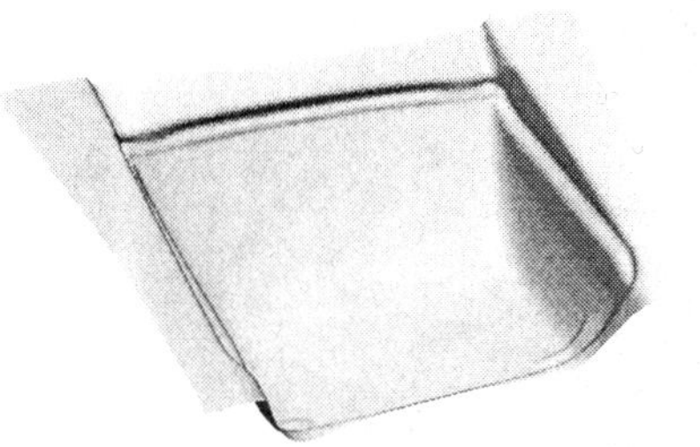

Shape at the end of the forming stage

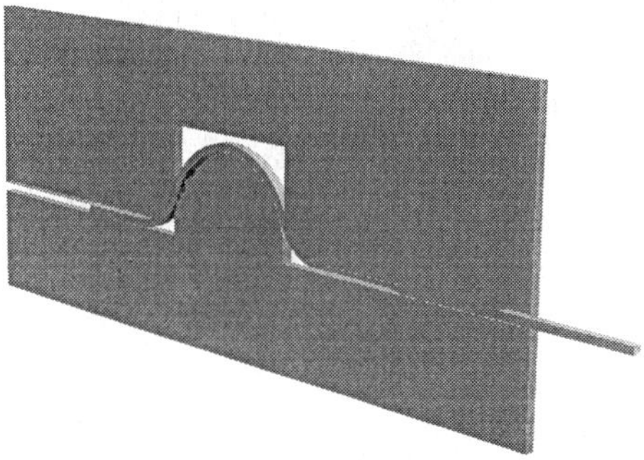

Pseudo 2D drawbead simulation

3-2 Forming of a wheel

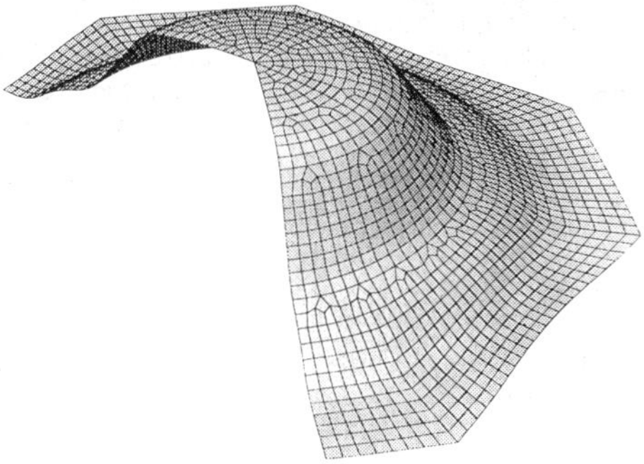

At the end of the first step

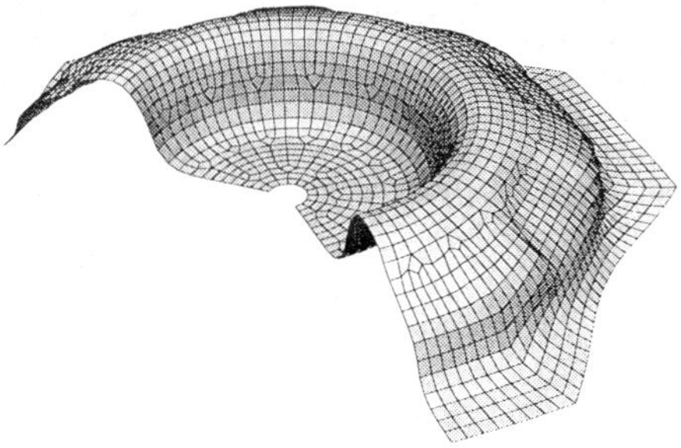

At the end of the second step

3-3 Forming of a front fender
(courtesy of PSA)

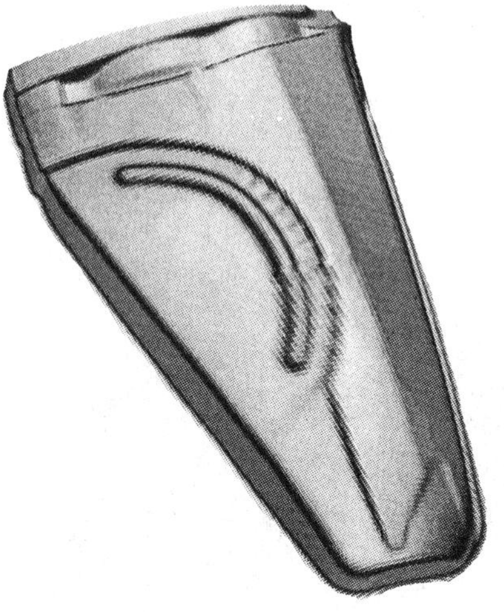

4 CONCLUSION

OPTRIS and its pre and post-processor FICTURE are already used in many companies and are considered as real industrial tools to simulate sheet metal forming. The new improvements, as the automatic mesh refinement described here, allow to solve realistic problems in an easier way (meshing the blank is not a difficult point anymore) and in a shorter delay due to the CPU time and preparation time reduction.

References

[1] D. Aubry, cours éléments finis, Ecole Centrale Paris.

[2] T. Belytschko, explicit algorithms for the non-linear dynamics of shells, Comp. Meth. Mech. 42 (1984) p 225-251.

[3] T. Belytschko, cours STRUCOME 89.

[4] TJR Hugues, reduced and selective integration techniques in finite element analysis of plates, nuclear engineering and design 46 (1978) p 679-706.

[5] T. Belytschko, A uniform strain hexaedron and quadrilateral with orthogonal hourglass control, IJNME 17 (1981) p 679-706.

[6] Newman, W.M. and R.F. Sproull, principles of interactive computer graphics, McGraw-Hill,New-York (1973).

[7] Shamos M.I., computational geometry, Springer Verlag (1979).

Simulation of Materials Processing: Theory, Methods and Applications, Shen & Dawson (eds)
© 1995 Balkema, Rotterdam. ISBN 90 5410 553 4

Modeling large deformation anisotropy in sheet metal forming

D.J. Bammann & M.L. Chiesa
Sandia National Laboratories, Livermore, Calif., USA

G.C. Johnson
University of California, Berkeley, Calif., USA

ABSTRACT

We consider a strain rate and temperature dependent, large deformation plasticity model in which two internal variables have been introduced to describe the state of the material. A scalar variable is introduced to describe the isotropic hardening or change in radius of the yield surface, while a tensor variable is introduced to describe any anisotropic hardening. The evolution equations for both variables are cast in a hardening minus recovery format, and a coupling between the variables is introduced through a parameter dependent upon the difference between previous and current loading directions. Analytic solutions can be determined for each variable for the case of isothermal, constant true strain rate, uniaxial stress. We utilize these analytic solutions to determine the parameters for the model. We then investigate whether these material properties for an annealed material can be used to model the anisotropic response of rolled sheet in the hydroforming process.

INTRODUCTION

The analysis of many manufacturing processes involves large material deformations. Generally, in these types of analyses, it is assumed that the initial short transient in the material response during yielding is unimportant in deformations at large strains, and a back extrapolated measure of the yield stress is employed. However, it has been well documented that the definition of the yield stress has a large influence upon the degree of softening which is predicted upon reverse loading [1]. A large offset definition of yield stress results in a prediction of a very small Bauschinger effect—the degree of softening which occurs during a load reversal, while a small offset definition results in the prediction of a larger Bauschinger effect. Hence, the definition of the yield stress directly affects the parameter determination for any constitutive model. In this work, we investigate the effect of the short transient associated with a Bauschinger effect upon the anisotropic deformation of an aluminum alloy in the sheet metal forming process. The model employed in the analysis is strain rate and temperature dependent, internal state variable model proposed by Bammann [2,3]. The parameters of the model are determined for annealed aluminum utilizing techniques which have been previously described [2,3,4,5]. These parameters are then utilized to predict the non-uniform final state of a rolled sheet. These final values of the state variables are then introduced as initial values in the analysis of a cup hydroformed from the sheet as described previously by Dawson et. al. [6].

CONSTITUTIVE MODEL

The aluminum behavior is represented by a rate dependent plasticity model. This model has a scalar hardening variable, κ and a tensor state variable, α. The scalar variable describes the hardening associated with longer transients and is isotropic in nature. The tensor variable acts as a back stress which evolves more quickly with straining. κ can be thought of as the radius of the yield surface and α as its center. Neglecting the effects of thermal expansion, one can write the model as summarized below. The plastic flow rule is chosen to have a strong nonlinear dependence upon the deviatoric stress σ',

$$\boldsymbol{D}_p = \sqrt{\frac{3}{2}} f(\theta) \sinh\left\{ \frac{\sqrt{\frac{3}{2}}|\xi'| - \kappa - Y(\theta)}{V(\theta)} \right\} \frac{\xi'}{|\xi'|} \tag{1}$$

where,

$$\xi' = \sigma' - \frac{2}{3}\alpha' \tag{2}$$

and $f(\theta)$ and $V(\theta)$ describe a rate dependence of the yield stress at constant temperature θ, and $\boldsymbol{D}_p$ is the plastic stretching or deformation rate tensor. The deviatoric stress σ', is defined by,

$$\sigma' = \sigma - \frac{1}{3}\mathrm{tr}(\sigma)1., \tag{3}$$

and the magnitude of a tensor is defined as,

$$|X| = \sqrt{X_{ij}X_{ij}}\,. \qquad (4)$$

A tensor variable α and a scalar variable κ have been introduced to describe the deformed state of the material. The evolution for the state variables α and κ are defined by,

$$\dot{\kappa} = H(\theta)\overline{D}_p - \{R_s(\theta) + R_d(\theta)\overline{D}_p\}\kappa^2 \qquad (6)$$

$$\overset{\circ}{\alpha} = h(\theta)D_p - \{r_s(\theta) + r_d(\theta)\overline{D}_p\}\overline{\alpha}\alpha, \qquad (7)$$

where,

$$\overline{\mathbf{X}} = \sqrt{\frac{2}{3}}|\mathbf{X}|. \qquad (8)$$

The basic character of this model, and the dependence of the response on the various constitutive parameters have been discussed in some detail in [3]. Each evolution equation is cast into a hardening minus recovery format, and each variable can recover dynamically or thermally. The hardening parameters h(θ) and H(θ) are functions of temperature and describe the initial linear post yield hardening which is observed in a true stress-true strain tensile test. The dynamic recovery mechanism (described by parameters R_d and r_d), results in a softening of the material response. It operates on the same time scale as the plastic flow and, therefore, introduces no additional rate dependence. At higher temperatures, thc recovery is dominated by diffusional climb of dislocations (described in the model by parameters r_s and R_s), which operates on a different time scale than the plastic flow, resulting in a strong rate dependence of the state variables. In this work, we modify the evolution of the variable, κ, by introducing a parameter which quantifies the effect of directional changes in the load path. We introduce a scalar parameter η, which is defined as the inner product of the tensor , α, and the direction of plastic flow. Hence, equation (6) is written,

$$\dot{\kappa} = (H_1(\theta) + H_2(\theta)\eta)\overline{D}_p - \{R_s(\theta) + R_d(\theta)\overline{D}_p\}\kappa^2 \qquad (9)$$

where,

$$\eta = \frac{\xi \cdot \alpha}{|\xi||\alpha|}. \qquad (10)$$

The inclusion of this parameter results in a more accurate prediction of the apparent softening which occurs after large changes in load paths. This is best illustrated by considering the experimental work on the large deformation of aluminum 1100 by Armstrong et al [7]. Figure 1 depicts the uniaxial compression of this material to a true axial strain of 6. This was achieved by loading the material until just prior to barreling, re machining, then reloading. In the same figure is the multi axial curve which is much softer than the uniaxial one. This curve is the result of loading a cube of material sequentially on the three orthogonal sides. Each loading resulted in a strain of 0.0075 and the sequence was continued until the same effective strain of 6.0 was achieved. Due to the continual large changes in loading path the material hardened at a much lower rate than in the case of uniaxial deformation. In the same figure, the model predictions are shown. The parameters were determined to fit this response, and with the addition of the coupling term the model is quite capable of capturing this type of response. A more accurate prediction of the early portion of the curve could possibly be achieved by assuming the tensor variable to be a much shorter transient, but this will be attempted when more reverse loading data is available to precisely determine the length of the transient.

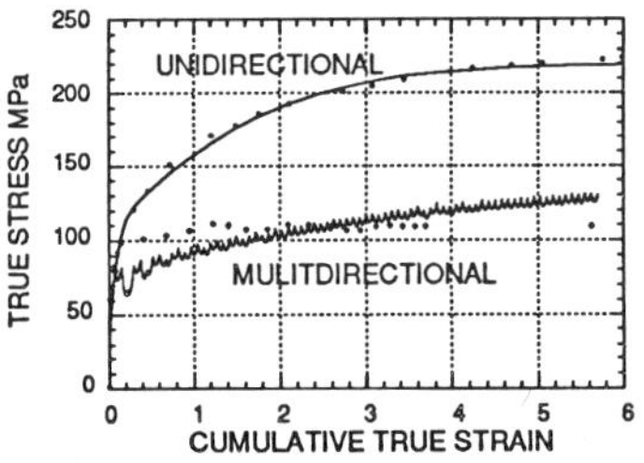

Figure 1 - Comparison of the model and data for aluminum 1100 for the case of uniaxial compression (labeled unidirectional) and sequential loading of a cube on the three orthogonal sides (multidirectional). Data from Armstrong et. al. [7]

IMPLEMENTATION

The model was implemented into the VEC-DYNA3D [8] finite element code. The implementation is based upon the formulation of a numerical flow rule at each time step derived from a numerical consistency condition. A detailed discussion of the implementation can be found in [9]. Several approximations could be used in the implementation into the explicit code since the time steps (and thus strain increments) are typically very small. Implementation into implicit codes would require more robust algorithms. For maximum efficiency, a radial return method as proposed by Krieg [10] was used to integrate the stress over a given time step. For this specific implementation, the value of η at the beginning of the time step was assumed constant (valid for very small strain increments) thus preserving the efficiency of the

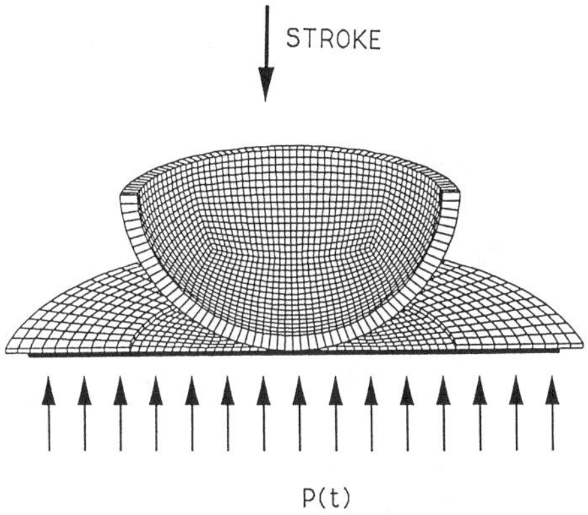

Figure 2. Initial setup of the hydroforming problem.

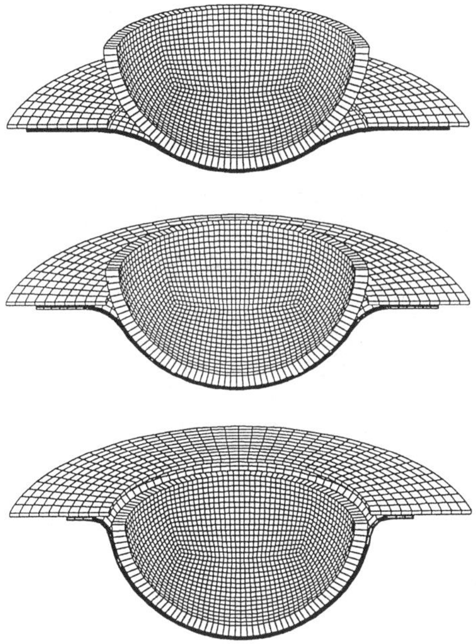

Figure 3. Deformation of the hemispherical cup.

standard radial return algorithm. Current work includes using an iterative cutting plane algorithm as discussed by Simo [11] that would be accurate for larger strain steps. The model, implemented with the radial return method, requires approximately 25 percent more computer time than a simple rate and temperature independent, linear hardening, plasticity model.

HYDROFORMING ANALYSIS

The model will now be utilized in the hydroforming of an aluminum cup. Utilizing the annealed parameters, the state of the sheet will be modeled assuming that the rolling process resulted in a state of plane strain compression. Both the same

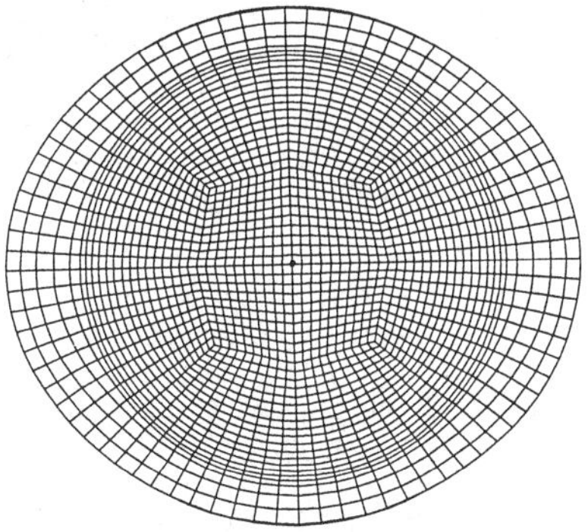

Figure 4. Prediction of overalling in the hydroformed part.

approach and assumption were used by Bammann and Dawson in the modeling of the rolling of a tapered copper plate [4]. Having estimated the initial values of the plane strain components of the tensor, α, these values are then input as initial conditions in the finite element code.

The VEC-DYNA3D program was used to model the hydroforming problem. Since an explicit code was used to model a quasi-static problem it was necessary to scale the material densities to increase the calculated time steps. Several densities were used to ensure that the dynamic effects were negligible. The initial setup for the problem is shown in Figure 2. The circular, rolled , aluminum plate was formed into a hemispherical cup as shown in Figure 3. The analysis predicted an ovalling of the part as shown in Figure 4. The actual tested part showed earring in addition to the ovalling. Current work includes further modification of the model to be able to predict earring effects.

REFERENCES

1. M. G. Stout, et.al., "Multiaxial Yield Behavior of 1100 Aluminum Following Various Magnitudes of Prestrain," *Int. J. Plasticity* , 1 (1985), pp. 163-174.

2. D. J. Bammann, "An Internal Variable Model of Viscoplasticity," *Int. J. Eng. Sci.*, 22 (1984), pp. 1041-1053 .

3. D. J. Bammann, "Modeling Temperature and Strain Rate Dependent Large Deformations of Metals," Appl. Mech. Rev., 1 (1990), pp. 312-318.

4. D. J. Bammann and P. R. Dawson, "Modeling the Initial State of a material and It's Effect on Further Deformation", *Material Parameter Estimation for Modern Constitutive Equations*, eds. L. A. Bertram, F. B. Brown, A. D. Freed, ASME NY,ASME-AMD 168 (1993), pp.13-20

5. D. J. Bammann, and A. R. Ortega, "The Influence of the Bauschinger Effect and Yield Definition on the Modeling of the Welding Processes, in Modeling of Casting," Welding, and Advanced Solidification Processes-VI, ed. T. S. Piwonka, V. Voller, L. Katgerman, TMS Publications, Warrendale, Penn., A. R., (1993), pp. 543-551.

6. A. J. Beaudoin, P. R. Dawson, K. K. Mathur, U. F. Kocks, D. A. Korzekwa, "An Application of Polycrystal Plasticity to Sheetforming", Computr Methods in Applied Mechanics and Engineering, 117 (1994), pp. 49-70

7 B. E. Armstrong, J. E. Hockett, O. D. Sherby, 1982 "Large Strain Multidirectional Deformation Of 1100 Aluminum at 300K", Journal Mech. Phys. Solids, 1/2(1982), pp. 37-58

8. J. O. Hallquist and D. W. Stillman, "VEC/DYNA3D User's Manual", LSTC Report No. 1018. (1991)

9. D. J. Bammann et al., "Failure in Ductile Materials Using Finite Element Methods", in Structural Crashworthiness and Failure, ed. by N. Jones and T. Wierzbicki, pub. by Elsevier Applied Science. (1993)

10. R.D. Krieg and D. B. Krieg, "Accuracies of Numerical solution Methods for the Plastic-Perfectly Plastic Model". *ASME J. Pressure Vessel Tech.*, 99 (1977,) pg. 510-515.

11. J. C. Simo and M. Ortiz, "A Unified Approach to Finite Deformation Elastoplastic Analysis Based on the Use of Hyperelastic Constitutive Equations", *Computer Methods in Applied Mechanics and Engineering*, 49 (1985), pg. 221-245.

Simulation of Materials Processing: Theory, Methods and Applications, Shen & Dawson (eds)
© 1995 Balkema, Rotterdam. ISBN 90 5410 553 4

The inverse approach including bending effects for the analysis and design of sheet metal forming parts

J.L.Batoz, Y.Q.Guo & F.Mercier
Division MNM/Laboratoire LG2mS, URA CNRS, Compiègne, France

ABSTRACT : The present state of development of the inverse approach to estimate the large strains of thin sheets obtained by deep drawing is briefly presented. In this paper we give more details on the procedure to take bending effects into account using a simple Discrete Kirchhoff Triangular Shell element. No additional dof are introduced compared to the membrane model. The results for steel and aluminium square cups (NUMISHEET 93 benchmarks) are presented. The drawing of a circular anisotropic blank with a cylindrical punch is studied showing that the bending stresses are important to predict the necking.

1 COMPUTER SIMULATIONS AT THE PRELIMINARY DESIGN STAGE

Since around 1985 research and development groups from the automotive industries, technological universities and software companies have recognized that computer simulations can be very helpful for the preliminary design of the initial blank (surface, contour, thickness, material properties) and of the tools and process parameters (blank holder characteristics, draw bead types and positions, lubrications,...). To be attractive at the preliminary design stage the computer simulations must be efficient in terms of computer time and used in a routine manner by technicians and practical engineers as early as possible in the whole design and production process [1]. The conditions just described are highly dependant on the quality of the interface between the finite element software and the CAD definition. Of course these computer simulations should also be as precise as possible to evaluate and predict the formability aspects of the final product despite the fact that questionable assumptions are made regarding the tool actions (with contact and friction), the constitutive equations and the important and complicated geometrical changes during drawing. Simplified procedures have been developed by a number of authors and have been named differently [2 to 13]. They are all based on the fact that the 3D (middle) surface of the useful part of the workpiece is known.

2 EVOLUTION OF OUR INVERSE APPROACH

A first version of our inverse approach (I.A.) was presented to the international community in 1989 [7]. The following aspects were considered :

- triangular CST membrane elements
- large elasto-plastic logarithmic strains
- total deformation theory of plasticity with transverse anisotropy
- implicit static algorithm for solving the nonlinear equilibrium equations (only two dof per node)
- vertical or normal nodal forces due to tool actions
- determination of the deformed (trimming) contour for an imposed initial flat blank.

Simplified external force vectors were established in [8, 9, 11, 12] to represent the influence of blank holder pressure with friction and the punch action with friction. Curved blankholder was considered in [12]. Plane anisotropy was taken into account in [8, 9, 11, 12].

The evaluation of our I.A. was performed since 1989. Some academic and classical benchmark problems were considered showing that the I.A. was not only very efficient in terms of computer time but also quite precise in a number of benchmark problems [8, 9, 13, 15]. In collaboration with RENAULT the numerical results were compared with experimental ones for a front wing part [10] and a cross member part [11, 12]. Overall satisfactory results were obtained.

In 1994 the improvements and developments of our I.A. were made on the following aspects :

- formulation of a thin shell triangular element in order to take the bending effects into account
- formulation of a four nodes warped membrane quadrilateral element
- evaluation of new algorithms and strategies to solve the nonlinear equilibrium equations (explicit static, dynamic relaxation,...).

In the next section we give more details on the first aspect since the approach is original and the applications are quite important.

3 A SIMPLE TRIANGULAR SHELL ELEMENT

For a large number of industrial applications the membrane effects are dominant but it is justified to consider bending effects when sharp tools are used. Also if wrinkling analysis and springback simulations are foreseen, then a knowledge of the variation of the stresses through the thickness is necessary.

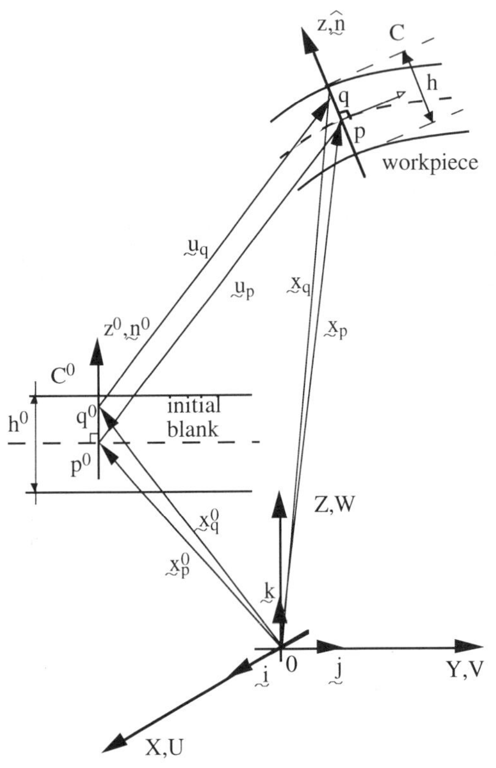

Figure 1. Geometry and kinematics

In the I.A. we know a priori the shape of the final drawn surface as well as the original initial surface (In this presentation we assumed that the original blank surface is flat, in the global plane XY). We partly know the displacement vector u_p of the material points on the middle surface since :

$$W = u_p \cdot k = \overline{W} \tag{1}$$

is known (figure 1). Since the normals of the middle surface of the blank and workpiece are known the rotations (of the normals) are also fully known. Therefore if bending effects are considered the virtual kinematically admissible displacement vector is given by :

$$u_q^* = U^* \, i + V^* \, j \tag{2}$$

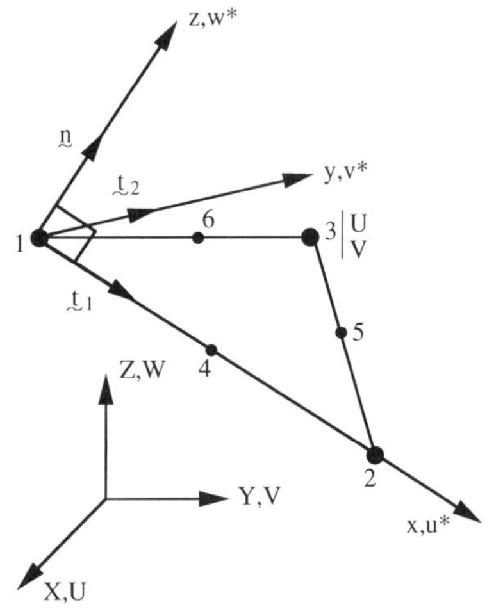

Figure 2. Global, local coordinates and dof.

This means that the global internal force vector (and residual force vector) will have only two dof per node for the membrane or the membrane plus bending model.

To define the internal force vector we consider the simple triangular shell element called DKT12 in [17] having three displacements at the corner nodes and one rotation around the edges at the mid-nodes (the DKT12 element is the simplest Discrete Kirchhoff Element formulated for thin shells. The linear stiffness matrix is the same as the Morley element [17]).

The six components internal force vector of an element of the discretized workpiece is given by :

$$\{f_{int}\} = [T]^T \left([B_m]^T \{N\} + [B_f]^T \{M\} \right) A \tag{3}$$

$$\underset{9 \times 6}{[T]} = \begin{bmatrix} [t] & & \\ & [t] & \\ & & [t] \end{bmatrix} \; ; \; [t] = \begin{bmatrix} t_{1X} & t_{1Y} \\ t_{2X} & t_{2Y} \\ n_X & n_Y \end{bmatrix}$$

$$\{N\} = \frac{h}{2} \int_{-1}^{1} \{\sigma\} \, d\zeta \; ; \; \{M\} = \frac{h^2}{4} \int_{-1}^{1} \zeta \{\sigma\} \, d\zeta$$

$$<\sigma> = < \sigma_x \quad \sigma_y \quad \sigma_{xy} > \; ; \; -1 \leq \zeta \leq 1$$

$t_{1X}, t_{1Y}, \ldots, n_Y$ are the direction cosines of the local coordinates axes x, y, z w.r.t. the X, Y global axes. A is the element area. h is the actual thickness. $\{\sigma\}$ contains the Cauchy stresses varying through the thickness and defined in the local coordinate system x, y, z. $\{N\}$ and $\{M\}$ are the membrane forces and bending moments.

$[B_m]$ (3 x 9) is the "[B]" matrix of the CST element whereas $[B_f]$ (3 x 9) relates the curvature components to the local virtual normal displacement components w^* of the DKT6 element. The terms of $[B_f]$ can be

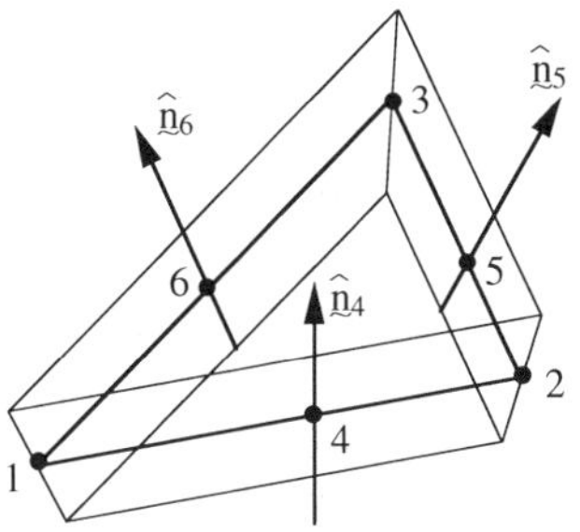

Figure 3. Thickness directions at mid-nodes

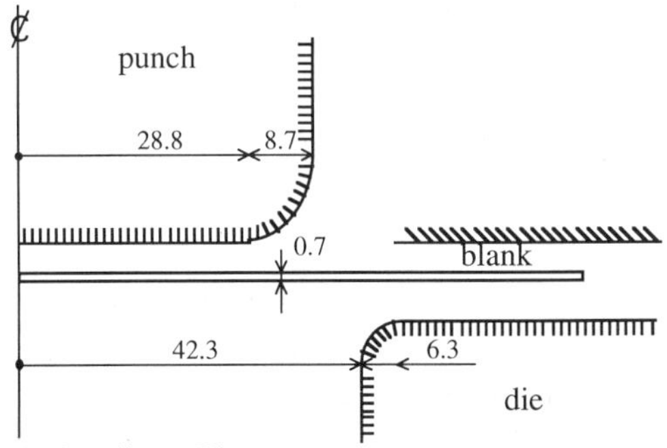

blank radius = 75 mm
$\bar{\sigma} = 729.09\ (0.01345 + \bar{\varepsilon})^{0.1657}$ MPa
$r_0 = 0.831$; $r_{45} = 0.949$; $r_{90} = 1.07$
Holder pressure : 6.5 MPa
friction (holder and die) : 0.13

Figure 4. Drawing with cylindrical punch (data)

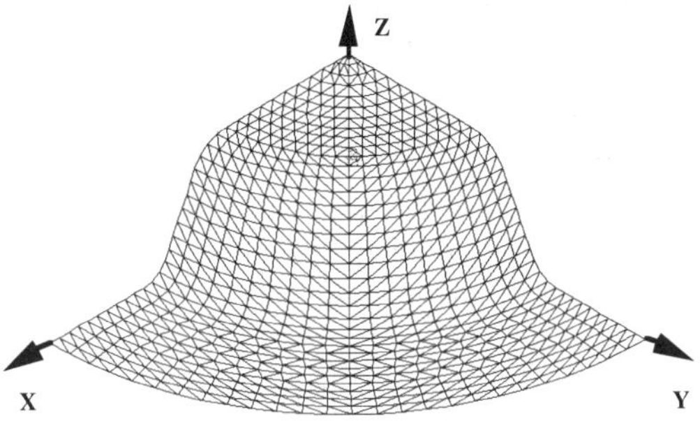

Figure 5. Finite element mesh

found in [14, 17]. Both $[B_m]$ and $[B_f]$ are constant per element. In Eq. 3, only h and $\{\sigma\}$ are changing during the iteration equilibrium process. Three or five Lobatto integration points are considered to evaluate the resultant forces at the centroid of each element.

The total deformation theory of plasticity is considered to relate the total stresses $\{\sigma\}$ to the total logarithmic strains $\{\varepsilon\}$. The constitutive equations can simply be written as :

$$\{\sigma\} = [H]\,\{\varepsilon\} \tag{4}$$

where [H] results from the assumptions of plane stress, in-plane and transverse anisotropy of the material (Hill anisotropic yield criterion is considered). [H] (3 x 3) depends upon the secant modulus of the uniaxial stress-strain curve and upon the transverse anisotropic coefficients r_0, r_{45}, r_{90} (see [13, 14, 21]).

The strain components at a point through the thickness are obtained from the strain tensor :

$$[\varepsilon] = \text{"ln"}\,[B]^{1/2} = [M]\left[\ln \lambda\right][M]^T \tag{5}$$

where [B] is the left Cauchy-Green tensor and $[\lambda]$, [M] its eigenvalues and eigenvectors. In the (eulerian) I.A. it is easier to compute the inverse deformation gradient $[F]^{-1}$ such that :

$$[B]^{-1} = [F]^{-T}[F]^{-1} \quad ; \quad \left\{dx_q^0\right\} = [F]^{-1}\,\{dx_q\} \tag{6}$$

For the present DKT12 shell element the inverse deformation gradient is defined as :

$$[F]^{-1} = \left[F_x^0\right]^{-1}[F_x]^{-1} \tag{7}$$

with :

$$\left[F_x^0\right]^{-1} = \left[\ \underset{\sim}{t}_1 - \underset{\sim}{u}_{p,x}\ \vdots\ \underset{\sim}{t}_2 - \underset{\sim}{u}_{p,y}\ \vdots\ \frac{1}{\lambda_3}\underset{\sim}{n}^0\right] \tag{8}$$

$$[F_x] = \left[\ \underset{\sim}{t}_1 + z\,\hat{\underset{\sim}{n}}_x\ \vdots\ \underset{\sim}{t}_2 + z\,\hat{\underset{\sim}{n}}_y\ \vdots\ \hat{n}\right] \tag{9}$$

$$z = \frac{1}{2}\zeta\,\lambda_3\,h^0 \quad ; \quad -1 \le \zeta \le 1$$

$\underset{\sim}{t}_1$, $\underset{\sim}{t}_2$: local orthonormal base vectors
h^0, $\underset{\sim}{n}^0$: initial thickness and unit normal vector of the blank ; λ_3 : thickness stretch (assumed to be constant through the thickness) ; $\hat{\underset{\sim}{n}}$: fiber or "pseudo normal" vector of the element (figure 3) such that :

$$\hat{\underset{\sim}{n}} = \sum_{k=4,5,6} N_k\,\hat{\underset{\sim}{n}}_k \tag{10}$$

$\hat{\underset{\sim}{n}}_k$ are the thickness (fiber) normal vectors at the mid-side nodes 4, 5, 6. These vectors are computed prior to calculations and results from the knowledge of the deformed surface. They can also be defined as the average of $\hat{\underset{\sim}{n}}^{(a)}$ and $\hat{\underset{\sim}{n}}^{(b)}$ (the unit normal vectors of the mid-surfaces of elements (a) and (b) sharing the mid-side node k).

N_k are semi- C^0 linear shape functions [17].

Since $\hat{\underset{\sim}{n}}$ is not constant per element, $\hat{\underset{\sim}{n}}_{,x}$ and $\hat{\underset{\sim}{n}}_{,y}$ represent the curvature effects. $[F_x^0]^{-1}$ represent the

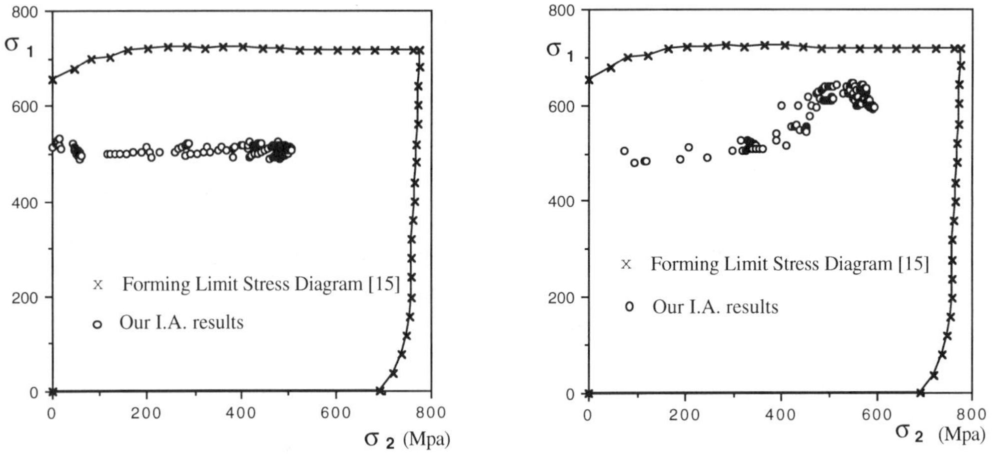

a) Membrane model

b) Membrane and bending model

Figure 6. Positive principal stresses for 22 mm punch stroke

membrane effects and $[F_x]^{-1}$ the bending contribution.

To compute the eigenvalues and eigenvectors of $[B]^{-1}$ we assume that $[B]^{-1}$ at $z \neq 0$ has a form similar to $[B]^{-1}$ at $z = 0$ i.e. we neglect the spurious transverse shear effects.

Full incompressibility is assumed and the thickness stretch is obtained from the principal stretches λ_1, λ_2 of the mid-surface :

$$h = h^0 \left(\lambda_1 \lambda_2\right)^{-1} \quad \text{at } z = 0 \qquad (11)$$

From Eqs 5 to 10, we can see that to evaluate the strain at any point through the thickness (at the centroid of the element) it is necessary to define $\hat{\underset{\sim}{n}}_4$, $\hat{\underset{\sim}{n}}_5$, $\hat{\underset{\sim}{n}}_6$ prior to calculation. Then from the estimation of λ_3 (Eq. 11) we can compute $[F_x]^{-1}$ (Eq. 9), $[F]^{-1}$ (Eq. 7), $[B]^{-1}$ (Eq. 6) and $[\varepsilon]$ (Eq. 5).

More details on the computation of bending strains and internal force vector are given in [14].

4 SOME NUMERICAL RESULTS

4.1 *Square cups deep drawing*

To evaluate the membrane model (CST elements) and the shell model (CST + DKT6 elements) several benchmark problems have been considered. In particular a detailed study of the deep drawing of square cups made of mild steel or aluminium was undertaken. The tool and blank dimensions, the material properties of the anisotropic sheets are defined in the NUMISHEET 93 proceedings [15].

The calculations were performed on a DEC 3000/500 (128 MB) workstation. The following meshes were considered for a quarter :

- for AL-15 and AC-15 : 856 elements, 468 nodes, 936 dof
- for AC-40 : 1036 elements, 563 nodes, 1126 dof

The trimming contours were obtained as well as the distribution of strains at any point assuming that the tool actions are replaced by nodal forces normal to the sheet. The results obtained by our implicit static I.A. code are reported in Tables 1, 2, 3 : membrane model without friction ; membrane model with friction (μ = 0.144) under the blankholder ; membrane + bending (M + B) model with $\mu = 0$; M + B model with μ = 0.144 under the blankholder. The draw-in values along the axes X and Y (DX and DY) and along the diagonal (DD) are reported as well as the min and max thickness and equivalent strain. The pourcentage of error is defined w.r.t the average experimental values taken from [15]. The values for h_{min} and h_{max} are compared to h^0 = 0.78 mm for steel and 0.81 mm for aluminium.

Some comments on the results of Tables 1, 2, 3 :

- if friction under the blankholder is included a small reduction of DX, DY, h_{max} and $\bar{\varepsilon}_{max}$ is observed (2 to 4 %)
- if bending effects are simulated for the present discretization the CPU time is increased by a factor of 2 to 3. The effect of bending is a reduction of DX, DY (3 to 5 %), a small reduction of h_{max} and $\bar{\varepsilon}_{max}$ but a small increase of h_{min} (around 2 %).

The largest variations of thickness (between the membrane model and the M + B model) are observed in the zones of highest curvatures. For the mild steel, 40 mm stroke, the thickness decrease reaches 4.2 % between both models and the difference in equivalent strain between the external and middle surface (bending effect) varies from - 9.4 % to 19 % (results obtained for the M + B model, without friction).

For the three cups we can see that our I.A. can predict quite well the draw-in values and strain

Table 1 : MILD STEEL 15 mm stroke ($h^0 = 0.78$ mm)						
	DX	DD	DY	h_{min}	h_{max}	$\bar{\varepsilon}_{max}$
experiment (average [15])	6.17	3.24	6.12			
numerical (average [15])	6.07 (- 1.6 %)	3.27 (+ 1 %)	6.10 (0.3 %)			
I.A. membrane $\mu = 0$; CPU = 1'36	5.51 (-10.7 %)	3.62 (11.9 %)	5.85 (-4.4 %)	0.72 (- 7.6 %)*	0.837 (7.3 %)	0.232
I.A. membrane $\mu = 0.144$; CPU = 1'35	5.48 (-11.25 %)	3.50 (8.05 %)	5.86 (-4.21 %)	0.714 (- 8.5 %)	0.834 (6.9 %)	0.233
I.A. membrane+bending $\mu = 0$; CPU = 3'42	5.32 (-13.7 %)	3.54 (9.4 %)	5.7 (-6.95 %)	0.714 (- 8.5 %)	0.835 (7 %)	0.224
I.A. membrane+bending $\mu = 0.144$; CPU = 3'42	5.16 (-16.3 %)	3.34 (3.05 %)	5.57 (-9.1 %)	0.707 (- 9.3 %)	0.832 (6.7 %)	0.225

Table 2 : MILD STEEL 40 mm stroke ($h^0 = 0.78$ m)						
	DX	DD	DY	h_{min}	h_{max}	$\bar{\varepsilon}_{max}$
experiment (average [15])	27.96	15.36	27.95			
numerical (average [15])	28.08 (0.4 %)	15.86 (3.2 %)	28.16 (0.7 %)			
I.A. membrane $\mu = 0$; CPU = 3'47	28.7 (2.6 %)	16.03 (4.4 %)	29.89 (6.9 %)	0.718 (- 7.9 %)	1.24 (59 %)	0.885
I.A. membrane $\mu = 0.144$; CPU = 3.05	28.37 (1.5 %)	15.53 (1.1 %)	29.66 (6.1 %)	0.697 (- 10.6 %)	1.20 (54 %)	0.816
I.A. membrane+bending $\mu = 0$; CPU = 10'29	27.27 (- 2.5 %)	15.27 (- 0.6 %)	28.5 (1.96 %)	0.704 (- 9.7 %	1.23 (58 %)	0.871
I.A. membrane+bending $\mu = 0.144$; CPU = 10'56	27.18 (- 2.8 %)	14.92 (- 2.9 %)	28.6 (2.32 %)	0.679 (- 12.9 %)	1.18 (52 %)	0.827

Table 3 : ALUMINIUM 15 mm stroke ($h^0 = 0.81$ mm)						
	DX	DD	DY	h_{min}	h_{max}	$\bar{\varepsilon}_{max}$
experiment (average [15])	5.35	3.02	5.47			
numerical (average [15])	5.63 (5.2 %)	3.03 (0.3 %)	5.64 (3.1 %)			
I.A. membrane $\mu = 0$; CPU = 1'33	5.74 (7.4 %)	3.37 (11.7 %)	5.71 (4.5 %)	0.678 (- 16 %)	0.889 (9.7 %)	0.208
I.A. membrane $\mu = 0.144$; CPU = 1'29	5.53 (3.41 %)	3.08 (2.17 %)	5.50 (0.5 %)	0.661 (- 18.4 %)	0.882 (8.9 %)	0.200
I.A. membrane+bending $\mu = 0$; CPU = 3'52	5.57 (4.1 %)	3.28 (8.75 %)	5.56 (1.65 %)	0.672 (- 17 %)	0.885 (9.25 %)	0.203
I.A. membrane+bending $\mu = 0.144$; CPU = 3'59	5.31 (- 0.7 %)	2.97 (- 1.5 %)	5.30 (- 3.1 %)	0.654 (- 19 %)	0.879 (8.5 %)	0.187

distribution for a small amount of CPU time.

4.2 *Drawing of a circular blank with a flat headed punch*

Forming limit stresses state diagram criterium have been proposed to predict necking by Arrieux, Brunet, Boivin et al [15, 16, 18]. A cylindrical cup drawing is described in figure 4. The material is a an anisotropic XD340 steel.

A punch displacement of 22 mm is considered and a quarter of the cup is discretized with 1250 elements, 672 nodes (Figure 5). For the M + B model our I.A. predicts a draw-in of 4.7 mm along X (the rolling direction) and of 5.72 mm along Y. (If membrane only is considered the values are DX = 5.25 mm ; DY = 6.45 mm).In figure 6 we report the values of the positive principal stresses σ_1 and σ_2 at the outer surface at the centroid of each element.

We can observe that some points (σ_1, σ_2) obtained with the M + B model are near the limit stress curves obtained experimentally [18]. These limit curves are associated with necking. The most critical state of stress is located near the limit of the flat head and punch radius. The experimental and numerical values (necking) reported in [15, 18] for the critical punch displacement is 22 mm.

We can observe that the present I.A. underestimate the maximum stresses compared to the incremental approaches [15, 18]. However figures 6a and b clearly show the importance of bending. The CPU time is 3'18" for the membrane model and 10'45" for the M + B model (on a DEC 3000/500).

5 CONCLUSIONS

In this paper we have briefly presented how we have modified the previous membrane model of the inverse approach to include the bending effects without increasing the number of nonlinear equations. This is done using a simple three node shell element. A number of benchmark problems have been considered to show the validity of the present formulation to estimate the draw-in displacements and the strains and stresses at any point through the thickness. The overall quality of the results is good for a very small amount of CPU time.

ACKNOWLEDGEMENTS

The help of Professor Brunet for the definition and interpretation of results for the cylindrical cup drawing and the help of Brigitte Duch for the preparation of the paper are duly acknowledged.

REFERENCES

[1] Roll, K., Gröber, M., 1994. Anforderungen der Praxis an Programme zur Umformsimulation. in *Metal Forming Process Simulation in Industry*, Int. Conf., Baden-Baden, Germany, 166-182.

[2] Gerdeen, J.C., Chen, P. 1989. Geometric mapping method of computer modelling of sheet metal forming", *NUMIFORM'89*, 437-444.

[3] Chung, K., Lee, D. 1984. Computer-aided analysis of sheet material forming processes. Vol. 1, *1st Int Conf. on Technology of Plasticity*, Tokyo, Japan, 660-665.

[4] Liu, S.D., Karima, M. 1992. A one step finite element approach for product design of sheet metal stampings. *NUMIFORM'92*. 497-502.

[5] Sklad, M.P., Yungblud, B.A., 1992. Analysis of multi-operation sheet forming processes, *NUMIFORM'92*, 543-547.

[6] Chung, K., Richmond, O., 1992. Sheet forming process design on ideal forming theory, *NUMIFORM'92*. 455-460.

[7] Batoz, J.L., Duroux, P., Guo, Y.Q., Detraux, J.M. 1989. An efficient algorithm to estimate the large strains in deep drawing", *NUMIFORM'89*, 383-388.

[8] Guo, Y.Q., Batoz, J.L., Detraux, J.M., Duroux, P. 1990. Finite element procedures for strain estimations of sheet metal forming parts, *IJNME*, 30, 1385-1401.

[9] Batoz, J.L., Guo, Y.Q., Detraux, J.M. 1990. An inverse finite element procedure to estimate the large plastic strain in sheet metal forming, *3rd ICTP*. Kyoto. 1403-1408.

[10] El Mouatassim, M., Detraux, J.M., Batoz, J.L., Guo, Y.Q. 1991. Application of an inverse FE - Procedure to sheet forming. *Int. Conference FE - Simulation of 3-D Sheet Metal Forming Processes in Automotive Industry,* Zürich, Switzerland. p. 647-654.

[11] Guo, Y.Q., Batoz, J.L., El Mouatassim, M., Detraux, J.M. 1992. On the estimation of thickness strains in thin car panels by the inverse approach. *NUMIFORM'92*. 1403-1408.

[12] Guo, Y.Q., Batoz, J.L., El Mouatassim, M., Detraux, J.M. 1992. Determination of the trimming contour under the curved blankholders with the inverse approach.*ATMMF'92*. Wuhan, China. p. 279-289.

[13] Batoz, J.L., Narainen, R., Duroux, P., Guo, Y.Q. 1993. Comparison of the implicit, explicit and inverse approaches for the estimation of the strain distribution in axisymmetrical thin sheets. *4th ICTP*. Beijing, Vol. 3. p. 1695-1700.

[14] Batoz, J.L., Guo, Y.Q., Mercier, F. 1995. Accounting for bending effects in sheet metal forming using the inverse approach. COMPLAS IV. Barcelona, Spain. (to appear).

[15] "NUMISHEET 93". 1993. *Proc. of 2nd Int. Conf. on the Numerical Simulation of 3-D Sheet Metal Forming Processes*, MAKINOUCHI et al. Editors, Isehara, Japan.

[16] Brunet, M., Arrieux, R., Nguyen Nhat, T. 1995. Necking prediction using forming limit stress surfaces in 3D sheet metal forming simulation. to appear in *NUMIFORM'95*.

[17] Batoz, J.L., Dhatt, G. 1992. *Modélisation des structures par éléments finis*, Volume 3, Coques, Hermès Editeur, Paris.
[18] Hage Chehade, Brunet, M., Boivin, M., Arrieux, R., Sept. 1991. Strictions en emboutissage. Prédiction par utilisation des courbes limites en contraintes. Matériaux et Techniques. p. 11-16.

Simulation of Materials Processing: Theory, Methods and Applications, Shen & Dawson (eds)
© 1995 Balkema, Rotterdam. ISBN 90 5410 553 4

Necking prediction using forming limit stress surfaces in 3D sheet metal forming simulation

M. Brunet
Laboratoire de Mécanique des Solides, INSA, Villeurbanne, France

R. Arrieux & T. Nguyen Nhat
Laboratoire de Mécanique Appliquée, ESIA, Université de Savoie, Annecy, France

ABSTRACT: For anisotropic sheet metal off-axis experiments have been carried out where the principal axes of strain do not coincide with the material's orthotropic axes. The off-axis tests were conducted with a flat-headed punch described previously by Marciniak. From different strain paths, it is shown that an intrinsic localized criterion is obtained with the calculated forming limit principal stresses, the angle between the orthotropic directions and the principal strain axes which gives a Forming Limit Stress Surface. Finite Element analyses with our simplified triangular shell element of the off-axis tests show that the proposed necking criterion is adequate for 3D. sheet metal forming simulation.

1 EXPERIMENTAL

In the experimental approach, the localized necking failures are measured and represented as a Forming Limit Diagram (F.L.D.). A F.L.D. is a representation of the critical combination of the two principal surface strains (major and minor) above which localized necking instability is observed. It is well known that these usual F.L.D. are strongly influenced by the strain path. Moreover many experiments carried out in our Laboratory on *anisotropic* sheets have shown that the Forming Limit Strain curves are not symmetric with respect to the first bisector in the orthotropic axes as it can be illustrated on the following Fig.(1) for titanium sheet /1/,/2/.

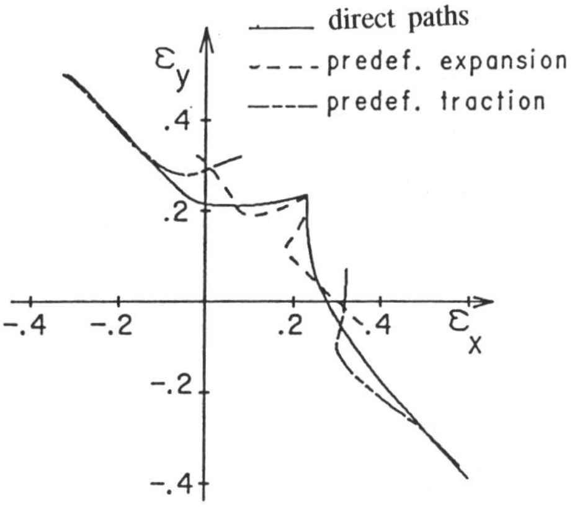

Fig.(1) F.L.D. in orthotropic axes

1.1 *Forming Limit Stress Curves*

We have already shown that the representation of the F.L.D. on a stress diagram is much more intrinsic than the usual Forming Limit Strain curves /1/,/2/,/3/. The Forming Limit Stress Curves are simply calculated from experimental measured strains $\Delta\varepsilon_x$, $\Delta\varepsilon_y$ and plastic incompressibility by a step by step integration of Hill's criterion and flow rule on each straight path in the orthotropic axes. The orthotropic flow theories proposed by Hill /4/ seem to be adequate for this purpose.

Noticing as usual :

$H = r_0/(1+r_0)$, $F = H/r_{90}$, $G = H/r_0$

the Hill's coefficients, where the r-values are the width-to-thickness strain increment ratio used to characterise sheet anisotropy. An isotropic material corresponds to the constant values $F=G=H=1/2$. The effective strain increment is given by /4/:

$$\Delta\varepsilon_{eq}^2 = [F(G\Delta\varepsilon_y-H\Delta\varepsilon_z)^2+G(H\Delta\varepsilon_z-F\Delta\varepsilon_x)^2+ H(F\Delta\varepsilon_x-G\Delta\varepsilon_y)^2]/(FG+GH+HF)^2 \quad (1)$$

and assuming hardening of the form:

$$\sigma_{eq} = K(c + \varepsilon_{eq})^n \quad \text{where} \quad \varepsilon_{eq} = \Sigma\, \Delta\varepsilon_{eq} \quad (2)$$

This implies the use of the simple model of isotropic hardening in which the initial yield surface is assumed to expand in stress space retaining its shape and orientation. The approach ignores strain-induced anisotropy and translation in the yield surface associated with the Bauschinger effect.

Provided that the loading is radially outward, and that the proportional increases in stress produce proportional plastic strains, isotropic hardening is a realistic assumption to make. In the case of orthotropy, the tensile tests provide the constants F,G,H that characterise the preservation of initial anisotropy. In this section it is required that the principal axes of stress coincide with the material's orthotropic axes and it is possible to derive an expression for the stresses of the form:

$$\sigma_x = \sigma_{eq} [(F+H)\Delta\varepsilon_x + H\Delta\varepsilon_y] / [\Delta\varepsilon_{eq}(FG+GH+HF)] \quad (3)$$

$$\sigma_y = \sigma_{eq} [H\Delta\varepsilon_x + (G+H)\Delta\varepsilon_y] / [\Delta\varepsilon_{eq}(FG+GH+HF)] \quad (4)$$

From these equations, the Forming Limit Stress Curves corresponding to the Forming Limit Strain Diagram shown on Fig.(1) have been obtained and presented on the following Fig.(2) :

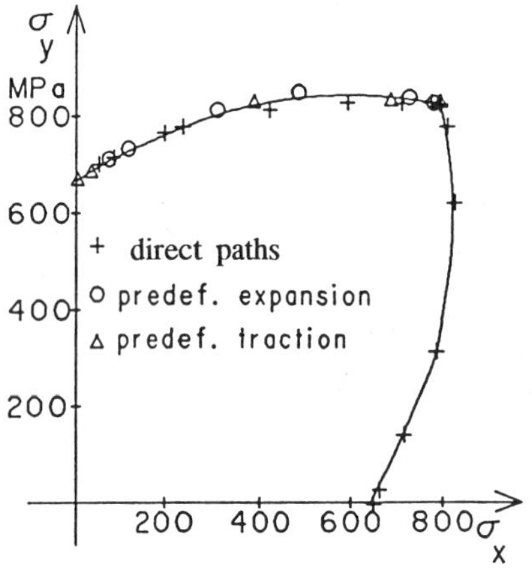

Fig.(2) F. L. Stress Curves in orthotropic axes

The intrinsic aspect of the curves obtained is not due to the strain hardening curve flattening on the large strain range because the reverse calculation of strains from the limit stress curves would give a very large dispersion which is not the case as it can be shown on Fig.(4) where the continuous line is the calculated limit strains from stresses on Fig.(3).

The results have been obtained from a mild steel sheet named SEDDK of thickness 0.67 mm and also used for the off-axis analyses with the following material properties:
Young' modulus: 198 GPa , Poisson' ratio: 0.3
True stress - true strain uniaxial curve (2):

K = 521.22 c = 0.00511 n = 0.224

Anisotropic coefficients:
H = 0.643 G = 0.357 F = 0.321 N = 1.221

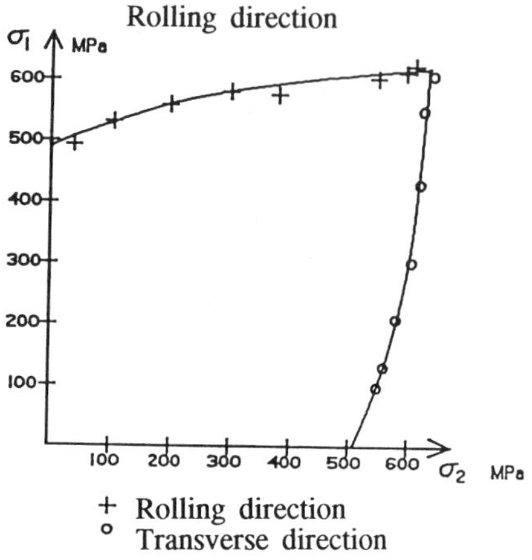

Fig.(3) F. L. Stress in orthotropic axes

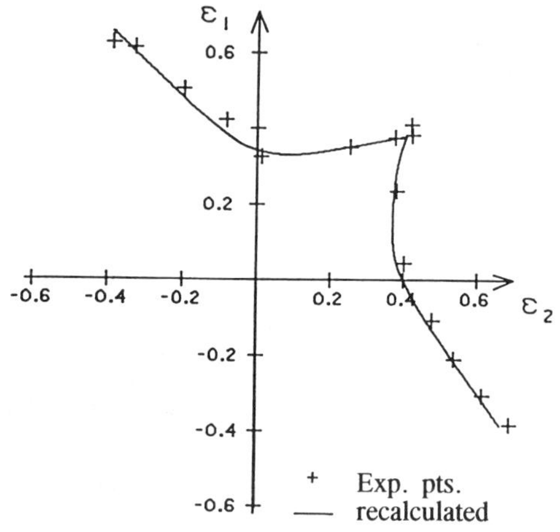

Fig.(4) F. L. Strain in orthotropic axes (direct path)

1.2 *Off-axis experiments*

The formability of sheet metals is defined as their ability to be deformed into complex shapes without cracking. In general formability depends on the formed shape, the forming conditions, the mechanical properties of the material and the principal axes of stress may not coincide with the material's orthotropic axes during the forming process.

In order to extend the Forming Limit Stress Curves concept for anisotropic sheets, experimental Marciniak's tests have been carried out as it can be shown on Fig.(5) where x,y are the principal strains directions of rectangular strips of various widths *l* and i,j are their material's orthotropic axes /5/.

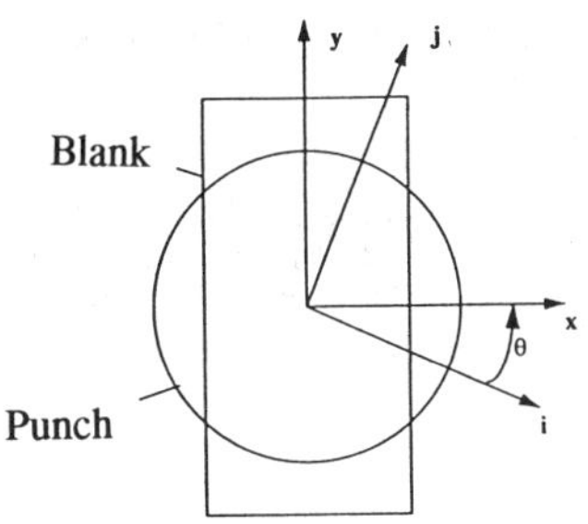

Fig.(5) Principal strain and orthotropic axes

In order to obtain in the experiments that the principal axes of strains may not coincide with the material's axes , the rectangular testpieces are cut (see Fig.(5)) such that $\theta = 0^\circ$, 30° , 45° , 60° and 90° with respect to the rolling direction. For each orthotropic direction , 8 testpieces with respectively 40 , 60 , 80 , 90 , 110 , 120 , 140 and 160 mm of width have been used to obtain different principal strain ratio. Experimental strains have been determined from changes to the diameter of circles etched onto the surface of the sheet in the vicinity of the flat center zone. The circular grid enables direct measurement of the principal strains with the onset of instability. Many refinements to the basic experimental Marciniak's method have been proposed to improve accuracy and to speed up the process of the data acquisition and analysis. Moreover, to avoid fracture of sheet metals on the die radius, a second blank is pasted on the testpiece with a circular hole at the center of about 25 to 40 mm as it is illustrated on the following Fig.(6) :

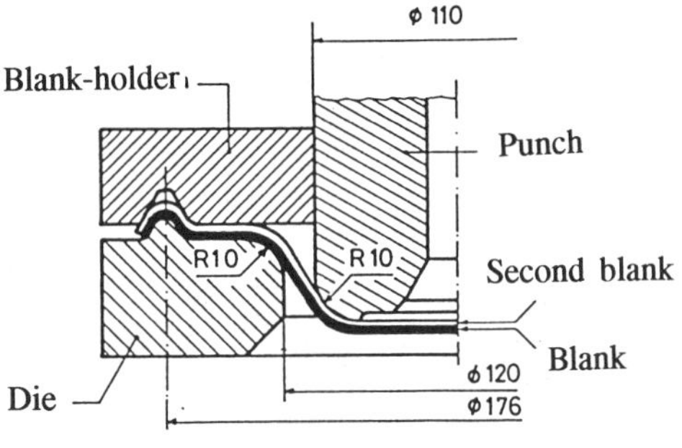

Fig.(6) Schematic illustration of the experiment

The direct linear strain path can be modified with predeformation of the circular blank (ϕ=160mm) up to 10 % in an equi-biaxial strain state. After that, the circular blank is cut in rectangular testpiece of various width with the axes inclined at 0°,30°,45°,60°,90° to the direction of rolling. The experiment is carried out again up to localized necking occurs.

When the longitudinal axes y of the testpiece is inclined at θ to the material j direction, the increments of principal strains transform according to:

$$d\varepsilon_{ii} = d\varepsilon_x \cos^2\theta + d\varepsilon_y \sin^2\theta \qquad (5)$$

$$d\varepsilon_{jj} = d\varepsilon_x \sin^2\theta + d\varepsilon_y \cos^2\theta \qquad (6)$$

$$d\varepsilon_{ij} = (d\varepsilon_x - d\varepsilon_y) \sin\theta \cos\theta \qquad (7)$$

The relationships between the strain increments and the stresses on each straight paths are :

$$\sigma_{ii} = \sigma_{eq}[(F+H)d\varepsilon_{ii} + Hd\varepsilon_{jj}]/[d\varepsilon_{eq}(FG+GH+HF)]$$

$$\sigma_{jj} = \sigma_{eq}[Hd\varepsilon_{ii} + (G+H)d\varepsilon_{jj}]/[d\varepsilon_{eq}(FG+GH+HF)]$$

$$\sigma_{ij} = d\varepsilon_{ij}\sigma_{eq} / N\, d\varepsilon_{eq} \qquad (8),(9),(10)$$

By carrying out these strain analyses for off axis solicitations we have found that the normal stresses in the principal strain axes are very near of the principal stresses σ_1 and σ_2 whatever is the angle θ between the orthotropic axes and the principal strain axes. The generalization of the forming limit stress curves gives the Forming Limit Stress Surfaces (σ_1, σ_2, θ) as it is illustrated in the following figures (7) and (8):

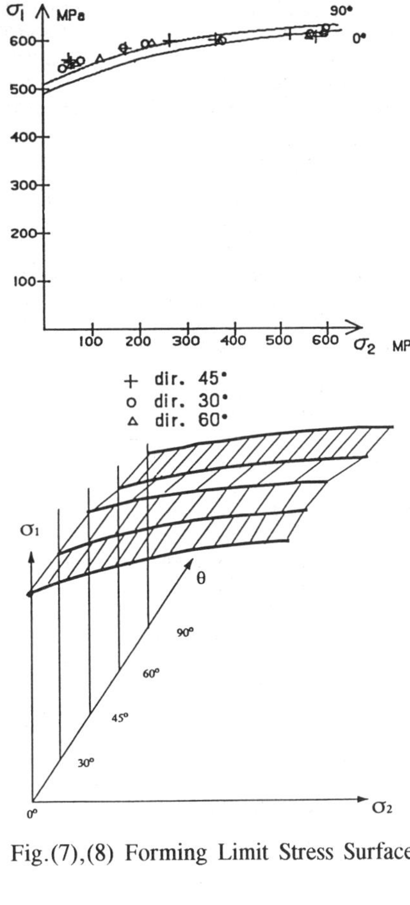

Fig.(7),(8) Forming Limit Stress Surfaces

2 NUMERICAL

Sheet metal forming is characterized by the need for extensive computer power and in this context triangular shell elements are very popular because they hold conceptual advantages over the more traditional and complex elements. The use of membrane elements only is not suitable for necking prediction because when tooling curvatures become large enough to cause significant bending strains, these elements no longer can predict deformation strains accurately. The finite elements for sheet modeling are often based on Reissner-Mindlin theory and designed for large displacement and large rotation analysis and can be used for large strains. This class of elements with linear C_o fields is based on the concept of decomposing the deformation into well-defined bending and shear modes. The decomposition is chosen so that the element behaves as closely as possible to a Kirchhoff C_1 element in the thin-structure limit.

All these elements made use of rotational degrees of freedom in order to take into account of the bending effects and bending elements using only out of plane translational degrees of freedom seems to receive more and more attention /6/,/8/. Some bending energy augmented membrane elements (B.E.A.M.) are sometimes proposed /7/ in order to maintain the advantages of membrane elements. The elements are hinged together at their common boundaries, the bending stiffness being represented by torsional springs of stiffness k resisting the rotation about the hinge side.

Our formulation which is presented in this section is also based on a collection of triangular flat elements with nodes at the vertices and three degrees of freedom per node but the strains are obtained in a consistent manner in the local basis of the element considered /6/,/9/. Each element may be considered as a six node macro-element but our procedure does not involve the inversion of a 3x3 matrix for each patch. Also, fictitious external nodes are needed to impose the boundary conditions at the edges /6/,/9/.

2.1 *A simplified triangular shell element*

We consider two initially coplanar elements which have a common boundary (ij) as it is shown in Fig.(9) where α_k corresponds to the out of plane rotation (ij) opposite node k of the corresponding element due to a relative displacement of the node. The small rotation along the common side is determined by assuming *a constant curvature* in Fig.(9) CH=R such that :

$$\theta_k = h_k/2R \qquad \text{and} \qquad \theta_m = h_m/2R \tag{11}$$

$$\theta_k = (\alpha_k+\alpha_m)h_k/(h_k+h_m) \tag{12}$$

$$\theta_m = (\alpha_k+\alpha_m)h_m/(h_k+h_m) \tag{13}$$

The strain normal to the boundary side at any depth z is given by:

$$\varepsilon_{nk} = -2z\theta_k/h_k = -2z(\alpha_k+\alpha_m)/(h_k+h_m) \tag{14}$$

and respectively for ε_{nm} . The factor 2 holds because for uniform stress and the area of the rectangle containing the triangle (ijk) is twice the area of the triangle.

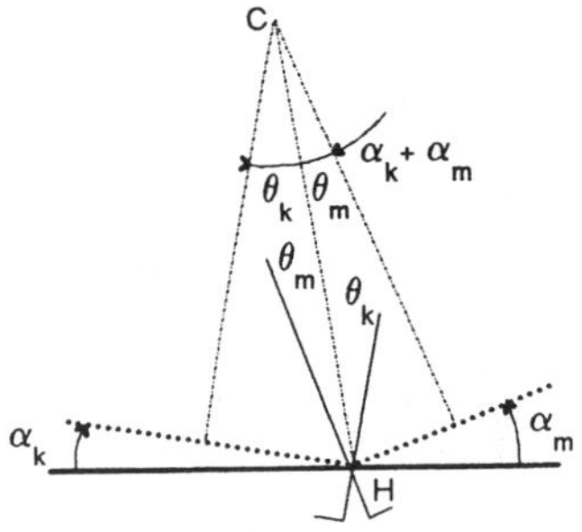

Fig.(9) Rotation angles

Consider triangle (1,2,3) which have its immediate neighbors (1,4,2),(2,5,3),(3,6,1) on Fig.(10) where (h_1,h_2,h_3), (q_1,h_4,q_2), (r_2,h_5,r_3) and (s_3,h_6,s_1) are respectively the heights of the triangles (1,2,3) ,(1,4,2), (2,5,3) and (3,6,1). Following eq.(14) the normal strains to the three sides of element (1,2,3) are given by the matrix relation :

$$\varepsilon_n = -z\,\mathbf{H}\,\alpha \tag{15}$$

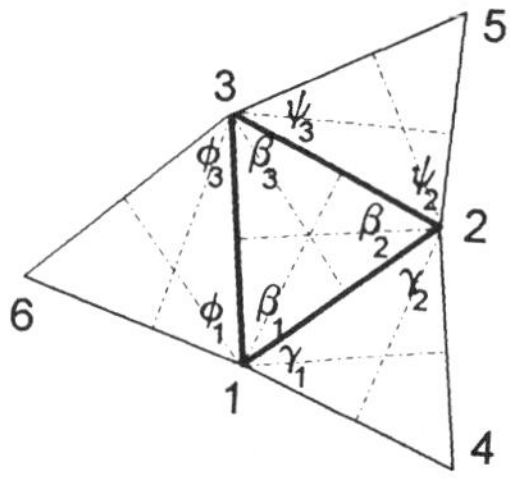

Fig.(10) Triangle (1,2,3) and neighbors

Referring to Fig. (9) and (10), the 6 mid side rotations α_i are related to the normal displacement w_i of the 6 apex nodes by a (6x6) **C** matrix /6/,/9/:

$$\alpha = \mathbf{C}\,w \tag{16}$$

We consider the triangular plate element in its arbitrary local co-ordinate (x,y,z) and we introduce in the usual way the components b_i and c_i of the outward pointing normals n_1,n_2,n_3 of sides (2,3), (3,1) and (1,2) respectively.

The strain in the local co-ordinates system are obtained by superposition such that :

$$\varepsilon = \begin{Bmatrix} \varepsilon_{xx} \\ \varepsilon_{yy} \\ \gamma_{xy} \end{Bmatrix} = \begin{bmatrix} b_1^2 & b_2^2 & b_3^2 \\ c_1^2 & c_2^2 & c_3^2 \\ 2b_1c_1 & 2b_2c_2 & 2b_3c_3 \end{bmatrix} \begin{Bmatrix} \varepsilon_{n1} \\ \varepsilon_{n2} \\ \varepsilon_{n3} \end{Bmatrix} = \mathbf{R}\ \varepsilon_n \quad (17)$$

The plane stress state in the (xy) plane is given by :

$$\sigma = \mathbf{D}\ \varepsilon \quad (18)$$

where **D** is the plane-stress matrix of isotropic or orthotropic elastic constants. For a virtual displacement δw inducing corresponding virtual strains $\delta\varepsilon$ at depth z from the middle surface :

$$\delta\varepsilon = -z\ \mathbf{RHC}\ \delta w = -z\ \mathbf{B}\ \delta w \quad (19)$$

The internal virtual work due to bending is given by the product of real stresses and virtual strains integrated throughout the volume of the element:

$$\delta W = \delta w^t \int_V z^2 \mathbf{B}^T \mathbf{D} \mathbf{B} dv\ \delta w = \delta w^t\ \mathbf{K}^b\ \delta w \quad (20)$$

where $\mathbf{K}^b$ is the element bending stiffness matrix. The equivalent nodal forces to the stresses are:

$$F_{int} = -\int_V z\ \mathbf{B}^T\ \sigma\ dv \quad (21)$$

which is to be used in non-linear analysis.

In the case of *static-implicit* large deformation analysis, the Newton-Raphson scheme used to solve the incremental equation has the following standard form:

$$\partial \mathbf{Q}(\mathbf{U}_{n-1})/\partial \mathbf{U}\ .\ \Delta\mathbf{U} = \mathbf{F} - \mathbf{Q}(\mathbf{U}_{n-1}) \quad (22)$$

where $\Delta\mathbf{U}$ is the perturbation of the displacement vector, **F** the external load vector and **Q** is the internal force vector, $\mathbf{U}_n$ is the current approximation for the displacement field and is given by:

$$\mathbf{U}_n = \mathbf{U}_{n-1} + \Delta\mathbf{U} \quad (23)$$

The iteration of equations (22) and (23) is carried out until the following equilibrium criterion is satisfied :

$$\|\mathbf{F} - \mathbf{Q}(\mathbf{U}_{n-1})\| \ /\ \|\mathbf{F}\| \le \varepsilon \quad (24)$$

where $\|.\|$ is a Euclidean vector norm and ε a very small constant, usually taken as around 10^{-4}.

At any stage in the incremental-iterative analysis the deformation angles between adjacent elements can be set for each side using the normal vectors of the elements which are obtained from the current updated nodal coordinates. A local Cartesian system (x,y,z) is constructed in the plane of each triangle. The shell bending element stiffness and the membrane element stiffness are generated and combined in this coordinate system. It is worth noting that there is no coupling in the local system because the bending and membrane have different degrees of freedom. The stiffness matrix and nodal force vector are then transformed from local coordinates to global coordinates before assembly. The strain rate is obtained in the local (x,y,z) orthonormal basis of each element which rotates rigidly as the element deforms.

In the following procedure, the stresses are computed and stored in the local system and we employ the rate elasto-plastic constitutive equations in the implementation which facilitate maintaining zero normal-stress incrementally. Due to the fact that the Hill function and the associate flow rule are not isotropic, the constitutive law based on this criterion is not objective. In order to assure the objectivity a co-rotating frame is used. The axes of orthotropy of the Hill criterion are updated by the rotation due to the spin rate:

$$\omega_{ij} = (\partial v_j/\partial x_i - \partial v_i/\partial x_j)/2 \quad (25)$$

Another possibility of rotating frame is the rotation of the polar decomposition **F** = **R U** which corresponds to the Green-Nagdhi objective stress rate while in the former case it is the co-rotational stress rate.

The first step in the elasto-plastic algorithm is to calculate the elastically induced trial stress. The trial stress is then tested to see if it is inside or outside the yield surface. If it falls within or on the yield surface the process is an elastic one. If the trial stress is outside, the radial return algorithm is employed to generate a stress state on the updated yield surface in the plane $\sigma_z = 0$. Usually fewer than 4 iterations are required for convergence /12/.

For the contact problems, there exists in general 3 approaches: the penalty method, the Lagrange multiplier method and the augmented Lagrange method. The penalty approach is employed here where the local stiffness and contact forces are calculated for solving the non-penetration condition between tools and sheet. For modelling the friction between tools and sheet, the Coulomb model is used which yields satisfactory results. As it can be shown on Fig.(11) where the Marciniak's tool pieces with a rectangular testpiece are modelled, the geometry is presented by facet triangular elements. For shell elements, there at least two potential contact surfaces, which are the two surfaces of the sheet.

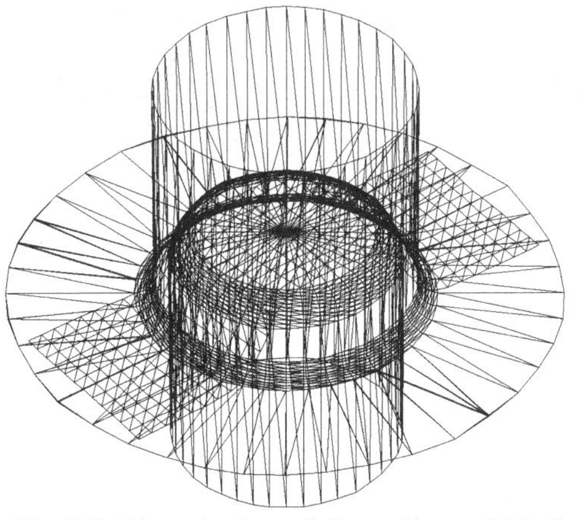

Fig.(11) Discretization of the tooling and blank

2.2 *F.E. Analysis*

Tools and blanks geometries used in F.E. analysis and experiments are shown in Fig. (5),(6) and (11). The rectangular sheet blanks were clamped firmly at the periphery by a lock bead. F.E. analysis was done for the complete formed part and the thickness of shell elements was twice in the zone where a second blank was pasted. The Forming Limit Stress Surface on Fig. (7),(8) has been introduced in the code as a set of data in order to detect necking numerically.

Fig.(12) shows the deformed mesh at necking and isovalues of major strains for a blank width of 60 mm and $\theta=45^{\circ}$. It is the same in Fig.(13) but for a blank width of 120 mm.

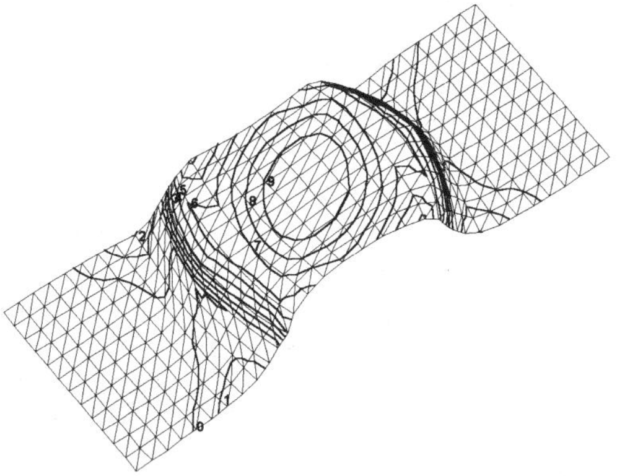

Fig.(12) Deformed mesh (l=60 mm $\theta=45^{\circ}$)
Major and minor strains at necking :
Measured: $\varepsilon_1=0.567$ $\varepsilon_2=-0.274$
Calculated: $\varepsilon_1=0.559$ $\varepsilon_2=-0.277$

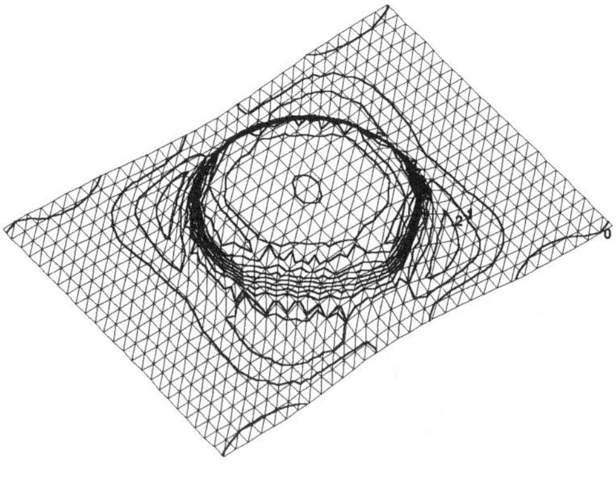

Fig.(13) Deformed mesh (l=120 mm $\theta=45^{\circ}$)
Major and minor strains at necking :
Measured: $\varepsilon_1=0.375$ $\varepsilon_2=0.216$
Calculated: $\varepsilon_1=0.371$ $\varepsilon_2=0.220$

The last example, but without comparison with experiments, is the deep drawing of a square cup based on the geometry of the *NUMISHEET'93* benchmark test which can be found in Ref. /6/. The material data are again the SEDDK mild steel, the thickness 0.67 mm is much lesser than the 0.78 mm of the benchmark mild steel. With a blank holder force of 19.6 kN and $\theta=0^{\circ}$, the numerical analysis shows a localized necking just above the punch nose corners for a punch travel of 26.1 mm.

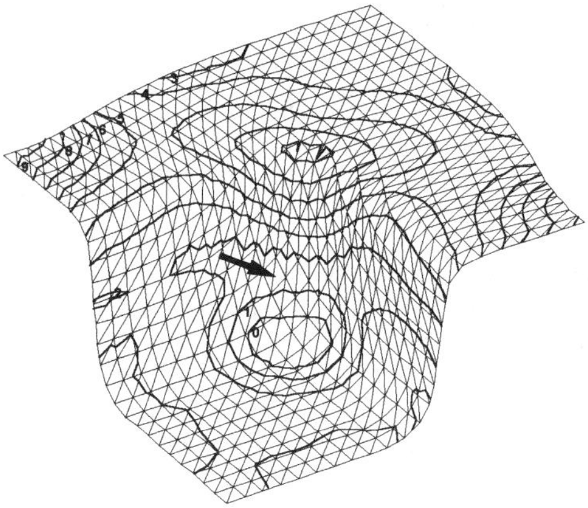

Fig.(14) Thickness strain at necking

REFERENCES

1 M. Brunet, I.H. Chehade. *V.D.I. Beritche no* 894 (1991), 185-197.
2 R. Arrieux, M.Boivin. *C.I.R.P. Annals*, Vol. 36/1 (1987), 195-198.
3 R. Arrieux, C.Bedrin, M. Boivin. *C.I.R.P. Annals*. Vol. 34/1 ,(1985), 205-208
4 R. Hill, *Proc. Roy. Soc.* 193 A, (1948) 281.
5 R. Arrieux, J.Nguyen Nhat, *18 th. IDDRG Congress*, Lisbonne , May (1994), 16-19
6 M. Brunet, F. Sabourin , *NUMISHEET'93, Ed. Makinouchi,A., Nakamachi,E., Onate,E., Wagoner,R.H.*, RIKEN, Tokyo, (1993), 229-238
7 D.Y.Yang, D.W.Jung, L.S.Song, D.J.Yoo and J.H.Lee, *NUMISHEET'93, Edt. Makinouchi,A., Nakamachi,E., Onate,E., Wagoner,R.H.*, RIKEN, Tokyo, (1993), 35-42
8 E.Onate, M.Cervera, *Engineering Computation* Vol.10, 6, (1993), 543-561
9 M.Brunet, F.Sabourin, *Metal Forming Process Simulation in Industry,* Baden-Baden, (1994), Germany, 28-30 Sept., 75-93
10 J.O.Hallquist, D.J.Benson, G.L.Goudreau. *F. E. Meth. Non Linear Prob.*, Ed. Bergan, Bathe, Wunderlich. Springer-Verlag (1986) pp. 465-479.
11 K.Schweizerhof, J.O.Hallquist. *V.D.I. Beritche* no 894,(1991),405-439
12 M.Brunet, *Engineering Analysis with B. E.* (1989), vol.6, 2, 78-83.

Simulation of Materials Processing: Theory, Methods and Applications, Shen & Dawson (eds)
© 1995 Balkema, Rotterdam. ISBN 90 5410 553 4

Optimization of sheet metal forming processes by instability analysis and control

J. Cao & M. C. Boyce
Department of Mechanical Engineering, Massachusetts Institute of Technology, Cambridge, Mass., USA

ABSTRACT: A wrinkling criterion defining the buckling wavelength/mode and initial buckling stress for a plate under lateral constraint is developed together with a proposed modeling method to capture wrinkling behavior accurately and robustly in the finite element analysis. Using this method, a conical cup forming process is simulated and compared to the experimental results for the cases of constant binder force. Excellent agreement is obtained. To optimize the process, a variable binder force history is designed in the simulation via wrinkling and tearing control feedback to the binder force. Experimental results have demonstrated that improved forming heights are obtained.

Introduction

Sheet metal forming involves many complicated issues, such as the elastic-plastic behavior of the material, the tribology of the interfaces, and the local and global instabilities of the material. For a given die shape, three of the major issues in forming a part are: First, the feasibility and formability – forming a part without wrinkling or tearing failure; Second, the efficiency – using the lowest cost and/or thinnest material; And third, the repeatability or robustness of the process which can be increased dramatically by using a closed-loop controller during the process (see for example, Lee and Hardt (1986), Fenn and Hardt (1993), Jalkh et.al (1993), Cao and Boyce (1994)). The first aspect, feasibility, has been studied recently and reported to be improved by changing the conventional forming procedure which generally sets the binder force at a constant level during the process, to a new process method where a variable binder force trajectory as a function of punch displacement is applied (for example, Hirose et. al (1990), Jalkh et al. (1993), Cao et al. (1994)). However, experimentally determining the proper trajectory is through trial and error approach and time-consuming.

Recent advances in both the finite element method and computer hardware have made the numerical simulation of full three-dimensional sheet forming a reality (NUMISHEET, IDDRG). However, little research has addressed the correct modeling of the initiation of wrinkling of the sheet under the constraint of a binder, although there have been several recent attempts (NUMISHEET (1994)). In this paper, a methodology for the numerical prediction of the initiation and propagation of wrinkling as a function of binder constraint using the finite element method will be reported. Our technique for accurately predicting wrinkling is then combined with a local tearing indicator in order to design an optimal variable binder force trajectory for a cup forming process. Both experimental and simulation results will be reported.

Problem Setup

In a sheet metal forming process, a sheet of material is plastically deformed into a desired shape. The manner in which the plastic deformation is imparted to the sheet is critical to the successful forming of a part and can be controlled by proper specification of the restraints imposed to the sheet during the process. The sheet can be restrained from entering the process zone by constraining it within a frictional, flat binder and/or through the use of drawbeads. Here, we will focus on the specific case of conical cup forming of a sheet of aluminum with a frictional binder.

A schematic of the cup forming process and a typical cup failure height vs. constant binder force curve is shown

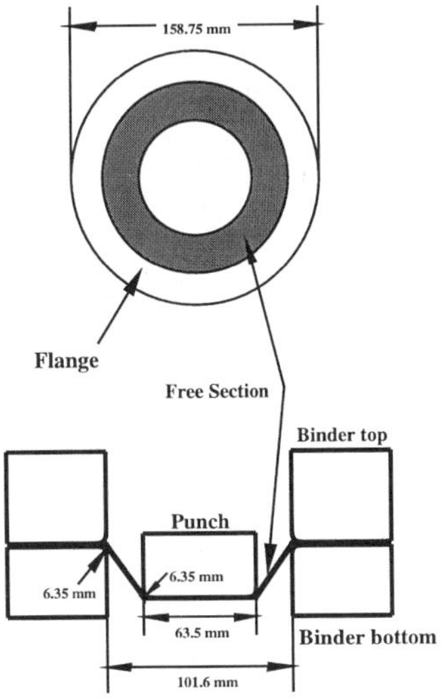

Fig. 1 Schematic of conical cup forming process

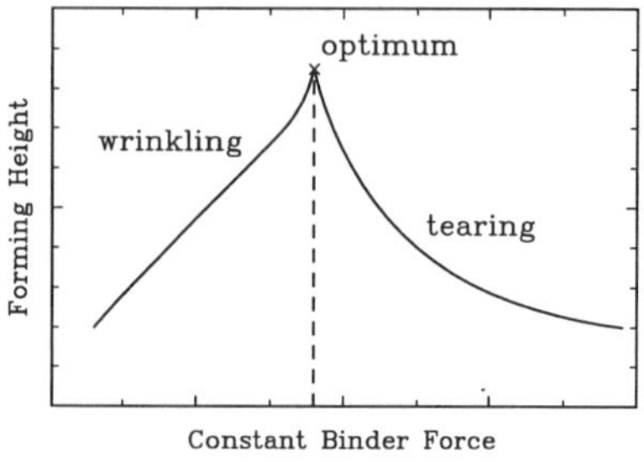

Fig. 2 A typical forming height diagram.

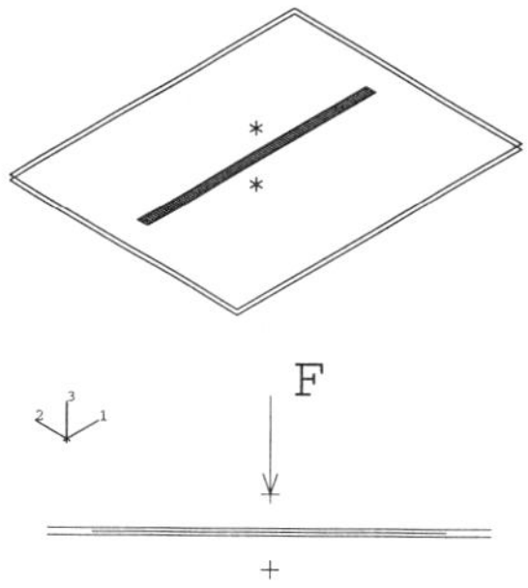

Fig. 3 Schematic of a plate under lateral constraint.

in Figs. 1 and 2. As shown, there exists an optimal constant binder force for a given friction condition which will result in the deepest cup before failure by wrinkling or tearing. If the binder force is too low, excessive material draw-in occurs and wrinkling cannot be suppressed due to the developed compressive hoop stress in the sheet under the binder. On the other hand, if the binder force is too high, not enough material draws-in and excessive stretching occurs near the punch producing tearing failure.

Buckling Criterion

Buckling of plates due to compressive stress in the plane has been studied by various researches. In 1974, a bifurcation theory for structures where the material is in the plastic range is detailed in Hutchinson (1974) where Hill's bifurcation criterion (1958) and Donnell-Mushtari-Vlasov (DMV) plate and shell theory were used. Hutchinson and Neale (1985) later extended this work to the conditions needed for the onset of wrinkling in doubly-curved sheet metal without lateral constraint. They found that the wrinkling is most likely a local instability problem depending on the local curvature and local stress states. This finding applies to plates and shells whose top and bottom surfaces are free of contact. The problem differs slightly when we consider the effect of a lateral constraint (constraint normal to the plane of the sheet) on buckling. As shown above, wrinkling constraint depends on, for example, the pressure applied normal to the sheet by the binder, as well as the local stress states and curvatures.

A wrinkling criterion for a plate under lateral constraint

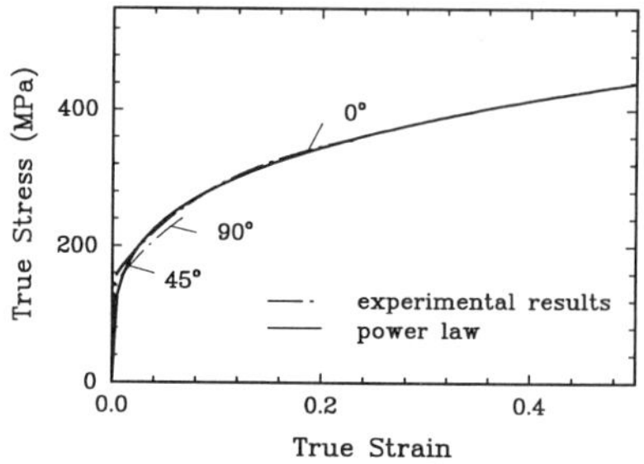

Fig. 4 Material behavior of aluminum Al2008-T4.

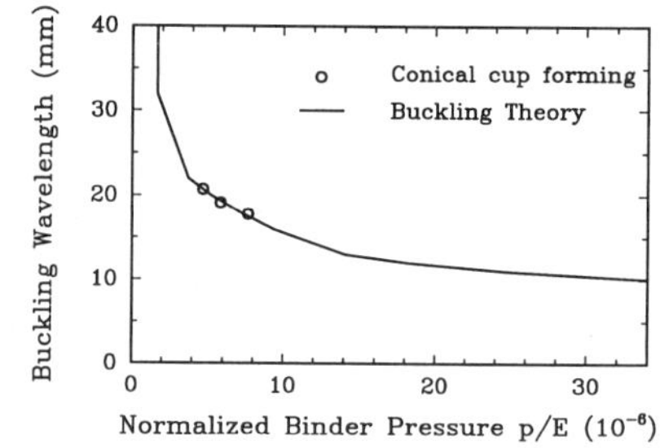

Fig. 5 Buckling Wavelength as a function of applied binder pressure.

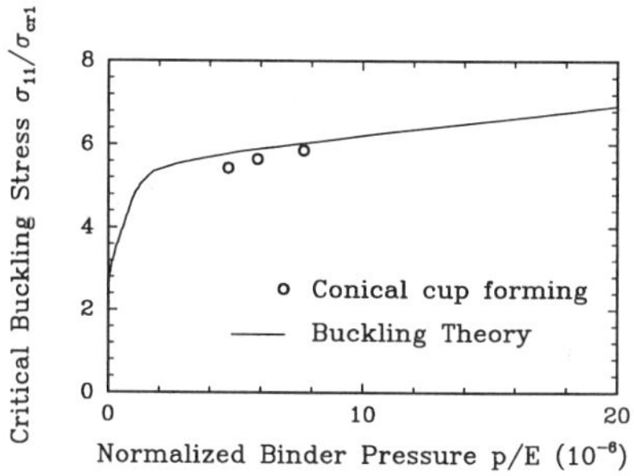

Fig. 6 Buckling stress as a function of applied binder pressure.

was developed by Cao and Boyce (1995) using a combination of energy conservation and finite element analysis where the buckling wavelength/mode and initial buckling stress are defined as a function of the applied binder pressure. Figure 3 shows the schematic of a plate under lateral constraint where the plate is modeled by finite strain four-node reduced integration shell elements (ABAQUS type S4RF). For the aluminum Al2008-T4, an elastic-plastic material with true stress and true strain relationship shown in Fig. 4, the calculated results are shown as solid lines in Figs. 5 and 6 where the buckling wavelength and initial buckling stress for an infinite plate are defined, respectively.

In order to simulate buckling behavior using finite element analysis, some form of imperfection is necessary to implement into the model in order to initiate the wrinkling numerically. Effects of various forms and distributions of defects on the initiation of buckling and post buckling behavior have recently been examined (Cao and Boyce, 1995).

a) undeformed mesh (scale=[1.,1.,100.])

b) $p_{binder}/E = 2.42E-6$

c) $p_{binder}/E = 4.82E-6$

d) $p_{binder}/E = 7.42E-6$

Fig. 7 Undeformed and deformed meshes of a plate under various binder pressures.

A model with randomly distributed imperfections in a form of thinned and mid-surface offset was proposed. Figure 7 shows the side view of an example undeformed mesh where the amounts of thinning and offset are 2% and 1% of the nominal plate thickness, respectively. Under various levels of binder pressure, the same model predicts various buckling behavior (Fig. 7). A summary of the simulated buckling wavelength and buckling stress as a function of applied binder pressure is shown in Fig. 8 together with those calculated from energy conservation. These results demonstrate the ability of the random imperfection finite element model to accurately capture buckling initiation and post-buckling behavior — import aspects needed to accurate numerical modeling of three-dimensional sheet forming processes.

Constant Binder Force

From the previous discussion, a model with randomly distributioned imperfections is capable of capturing both buckling and post-buckling behavior accurately. The same strategy is now used to model the buckling of conical cup forming problems. One quarter of the entire sheet is modeled using shell elements (S4RF) as shown in Fig. 8 where the darkened area is the location of the imperfections. Figure 9 shows the predicted wrinkling for three low binder force cases. By simulating wrinkling explicitly, the cup failure heights due to wrinkling can be easily determined in the simulations. The calculated initial buckling wavelength and stress are plotted in Figs. 5 and 6. The calculated wavelengths show excellent agreement with those we observed experimentally. Notice that the result obtained from the study of plate buckling is also applied to conical cup forming problem.

The tearing failure involves the localized stretching of the material near the punch. We have found (eg. Cao et. al) that by tracking the history of the state of straining of material points near the punch, the onset of tearing can be predicted. The material in this region will undergo a change

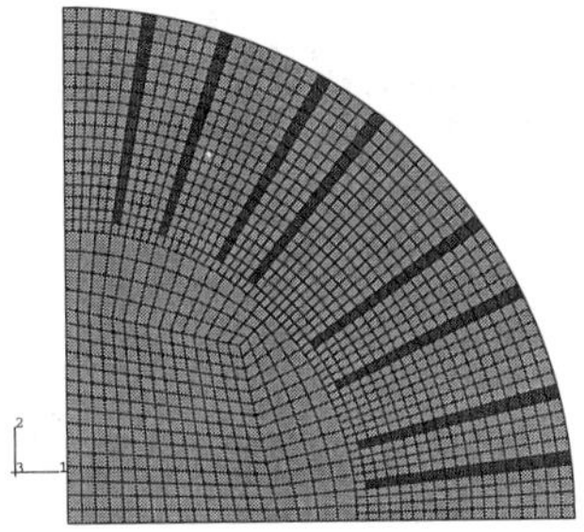

Fig. 8 Undeformed mesh of the sheet in the conical cup forming problems.

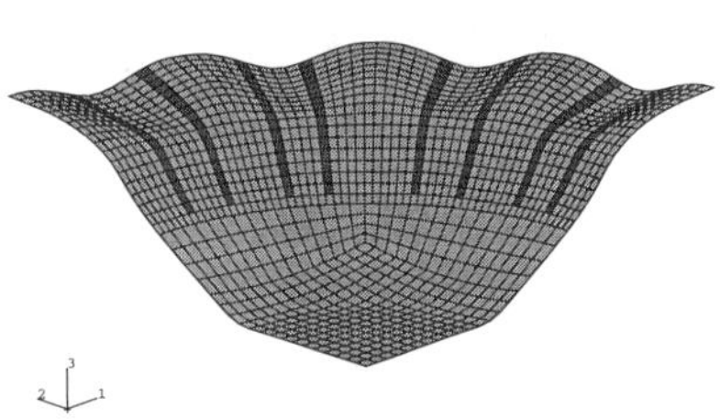

CBF = 2.2 kN PD=20.mm

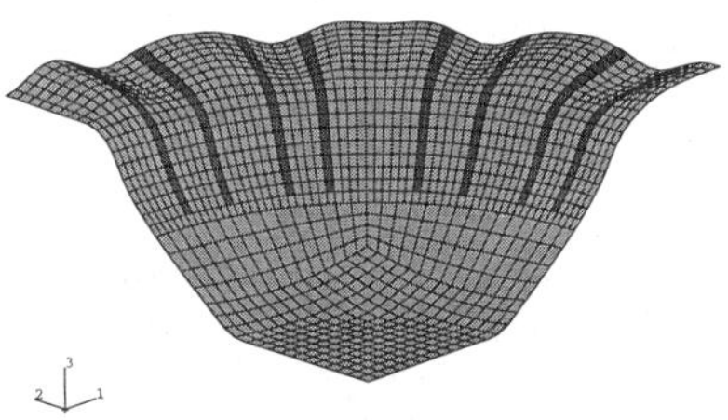

CBF = 2.8 kN PD=25.mm

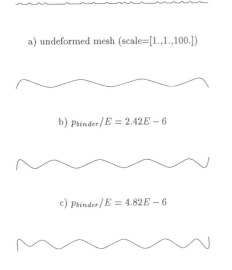

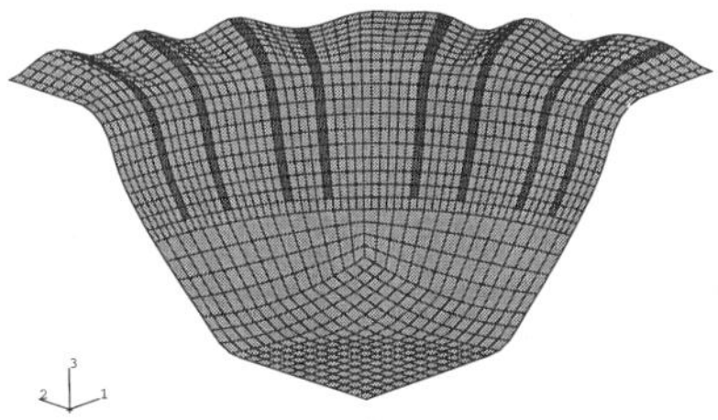

CBF = 4.0 kN PD=30.mm

Fig. 9 Deformed meshes of cups under various constant binder force cases.

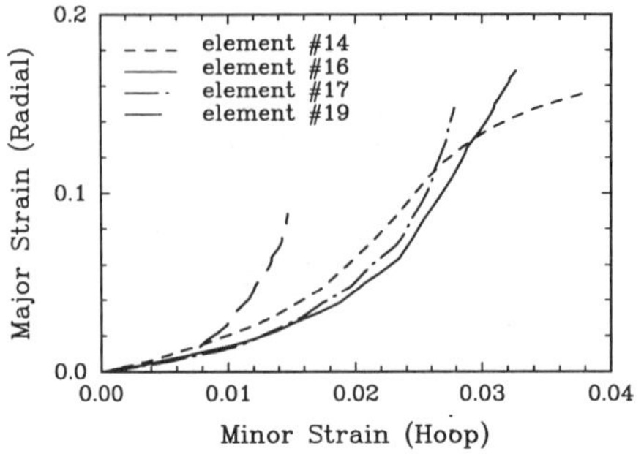

Fig. 10 Strain histories of points near the punch at a CBF=6.9N.

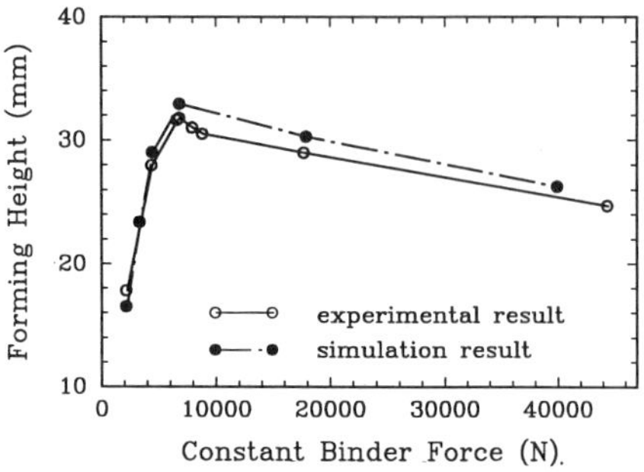

Fig. 12 Experimental and predicted cup forming height diagram.

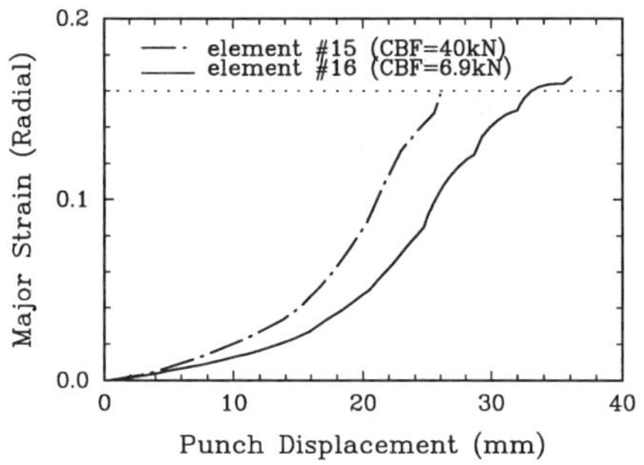

Fig. 11 Major strain histories of critical points at the CBF of 6.9kN and 40.kN.

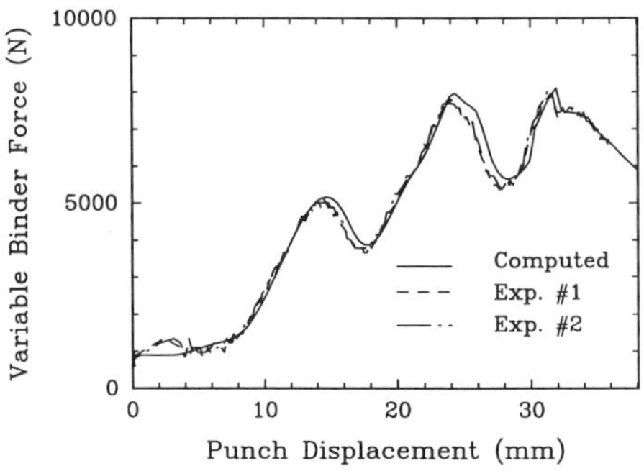

Fig. 13 Variable binder force history designed by the simulation.

in its state of straining from one of equi biaxial stretching to one of plane strain stretching which will then localize in one element. The strains can then be compared to forming limit diagrams. Figure 10 shows the state of strain history for various points near the punch for CBF=6.9N showing the transition to plane strain for element #17. Failure occurs at the height of $26mm$ for the $40kN$ case yet does not occur until $33mm$ for the $6.9kN$ case as shown by comparing major strain histories in Fig. 11. The experimental and predicted cup failure height vs. binder force is shown in Fig. 12. Excellent agreement is obtained.

Much recent work in sheet metal forming has focused on designing variable binder force histories. A variable binder force history can improve the formability of a given part shape with a given material by changing the material straining path and strain distribution within the part. The design of variable binder force histories has been accomplished through trial and error and is time consuming. Here we present a numerical approach using internal closed-loop control during the simulation to design appropriate variable binder force histories.

A first attempt at designing a binder force history was made in earlier work (Sim and Boyce, 1992) on the premise of avoiding tearing by avoiding local thinning. Unfortunately, buckling is a likely result of such an approach and our current focus is on using our capabilities to predict both tearing and buckling in designing the variable binder force history. During the simulation we monitor both the binder displacement (indicator of the onset of wrinkling) and the major strain in the punch radius region. Using a proportional-integral controller, if the behavior of either of these parameters deviates from the desired behavior, the error is fed back to the binder where the binder force is then changed in order to alter (control) and thus design a desirable, variable binder force history. A detailed description of the numerical model and control algorithm is available in Cao and Boyce (1994). The designed binder force history for the case of as lightly thicker stock of the same material used earlier is depicted in Fig. 13 where we see the gradual increase in binder force with punch displacement in order to suppress the occurrence of wrinkling within the binder. The periodical change in the binder force is associated with the change of wrinkling mode and selection of gain (Cao and Boyce 1995). The binder force then decreases giving the overall trajectory shown in Fig. 13. This trajectory was then used in experiments in a open-loop manner and found to give a forming height of $36.0mm$ where previous forming experiments on this stock of material failed at heights less than $31.mm$ for the constant binder force cases. Figure 14 shows the predicted deformed cup shape. Notice that the predicted number of total wrinkling waves in an entire cup is 32 compared to 34 observed in the experiments. This approach of using a simulation to intelligently design binder force histories shows great promise in improving forming of parts.

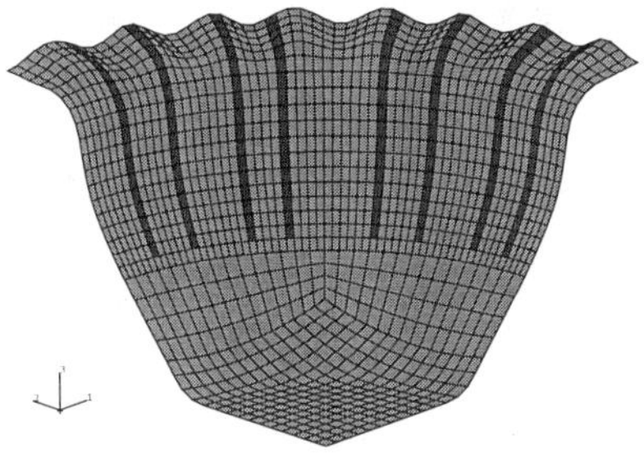

Fig. 14 Predicted cup shape for the variable binder force case at a punch displacement of 40.mm.

Conclusion

A wrinkling criterion was developed using a combination of energy conservation and finite element analysis. Meanwhile, a robust finite element model is proposed to capture the initiation of buckling and post-buckling behavior. Using this model combined with a tearing criterion based on the local strain history, the finite element model is able to accurately predict the experimental conical cup forming processes. Moreover, the model designed a successful variable binder force history using the buckling and tearing controllers, which results in a 16% increase in the forming height.

Acknowledgements

The research presented here was supported by the National Science Foundation through a PYI Award to MCB (No. DDM-9157899) and matching funds from ALCOA. We also gratefully acknowledge our long term collaboration with Prof. David Hardt's group in the Laboratory for Manufacturing and Productivity.

Reference

Cao, J., Bakkestuen,R.,Jalkh, P., Boyce, M and Hardt, D. (1994) "Improvement of Forming Height and Stability of Aluminum Parts Using Active Binder Control" *International Deep Drawing Conference 1994*, Lisbon, Portugal.

Cao, J. and Boyce, M. (1994) "Design and Control of Forming Parameters Using Finite Element Analysis," *Symposium on Computational Material Modeling, American Society of Mechanical Engineers*, Chicago, IL.

Cao, J. and Boyce, M. (1995) "Wrinkling Behavior of Rectangular Plates Under Lateral Constraint — Part I: Elastic Plate", submitted to *International Journal of Solids and Structures.*

Cao, J. and Boyce, M. (1995) "Wrinkling Behavior of Rectangular Plates Under Lateral Constraint — Part II: Elastic-Plastic Plate", submitted to *International Journal of Solids and Structures.*

Fenn, R. and Hardt, D.E., (1993) "Real-Time Control of Sheet Stability During Forming," ASME *Journal of Engineering for Industry*, Dec.

Hirose Y., Hishida, Furubayashi, T., Oshima, M., and S. Ujihara (1990) "Part I I: Application of BHF-Controlled Forming Techniques", *Proc. 4th Symposium of the Japanese Society for the Technology of Plasticity.*

IDDRG (1994) *International Deep Drawing Conference 1994*, Lisbon, Portugal.

Jalkh, P., Cao, J., Boyce, M.C.,and Hardt, D.E., (1993). "Effect of Constant a nd Variable Binder Force on Conical Cup forming of AL 2028 T4", SAE paper #930286, *Proc. SAE Congress, Sympos ium on Sheet Forming*, Detroit, March.

Lee, C., and Hardt, D.E.,(1986) "Closed-Loop Control of Sheet Metal Stability During Stamping", *1986 North American Manufacturing Research Conference*, May 28-30.

NUMISHEET (1993) *Numerical Simulation of Sheet Metal Forming 1993*, Tokyo, Japan.

Simulation of Materials Processing: Theory, Methods and Applications, Shen & Dawson (eds)
© 1995 Balkema, Rotterdam. ISBN 90 5410 553 4

Modelling drawbeads in 3D finite element simulations of the deep drawing process

B.D.Carleer, M.F.M.Louwes & J.Huétink
University of Twente, Department of Mechanical Engineering, Enschede, Netherlands

P.T.Vreede
Hoogovens Groep R&D, IJmuiden, Netherlands

ABSTRACT: In 3D simulations of the deep drawing process the drawbead geometries are seldom included. In this paper the development of an equivalent drawbead is described. In order to investigate the drawbead behaviour a 2D plane strain finite element model was used. For verification of this model experiments were performed. These analyses showed that not only the restraining force should be applied but also the strain changes. Both effects were implemented in an equivalent drawbead. The effect of using the equivalent drawbead is demonstrated with a few examples.

1 INTRODUCTION

Drawbeads are widely used in deep drawing processes to control the flow of the blank during the forming operation. In 3D finite element simulations the drawbead geometries are seldom included because of the small radii. These small radii require a very large number of elements and therefore large computer time. For this reason equivalent drawbeads are used.

Equivalent drawbeads are defined as fixed lines on the tool surface. At these lines a drawbead restraining force acts. This restraining force is assigned through a 2D simulation of the drawbead or through an experiment. This restraining force reaches its steady state value when a particle has been pulled through the entire bead [Cao]. However, in most of the finite element programs the drawbead force has been taken as the steady state value.

Different strategies were used for the equivalent drawbead. In Abaqus [Taylor] the drawbead restraining forces were applied as a set of frictional forces. A concentrated load was applied along the drawbead line and a drawbead friction coefficient was used. The normal force was chosen so that the tangential friction force matches the steady state value. In Pam-Stamp [Haug] the tangent resistance force is a sum of the resistance of bending of the sheet and the sliding resistance due to friction. The bending resistance depends on the geometry of the drawbead, its penetration depth and the sheet material. The sliding resistance is a function of the equivalent drawbead friction coefficient and the normal force. ITAS-3D [Kawka] uses an other strategy. Here the drawbead force introduces some additional stiffness. This drawbead stiffness matrix is assembled into the global stiffness matrix.

[Wouters] found that the effect of drawbeads is much more than adding a restraining force to the sheet; the whole material properties are modified in a significant manner. During the flow of the material through a drawbead the strain distribution changes and the material usually becomes thinner [Carleer]. Modelling drawbeads by only applying an additional restraining force does not incorporate these effects. For this reason an equivalent drawbead has been developed which incorporates the effects of sheet thinning and the change of the strain distribution.

The first part of this paper deals with the drawbead behaviour. In order to get more insight in the drawbead behaviour a 2D analysis has been performed. This analysis gave information on the drawbead restraining force and the strain changes. To verify the 2D analysis an extensive set of experiments was performed [Carleer]. With the results of this 2D analysis an equivalent drawbead has been developed. In the second part we focus on the equivalent drawbead. The influence of only applying the restraining force and also applying strain changes will be shown in some examples.

2 DRAWBEAD MODEL

In order to study the drawbead behaviour a 2D plane strain analysis has been performed. With this model it is possible to do a parameter study of different drawbead geometries and sheet materials in an accurate and time efficient way. For verification of this 2D model an experimental set-up was built.

Here we focus on one drawbead geometry and one sheet material. The dimensions of the drawbead are depicted in figure 1. The material properties of the sheet material were:

Ludwik value (C) = 551 N/mm^2

n-value (n) = 0.230

initial yield stress = 149 N/mm^2

blank thickness = 0.7 mm

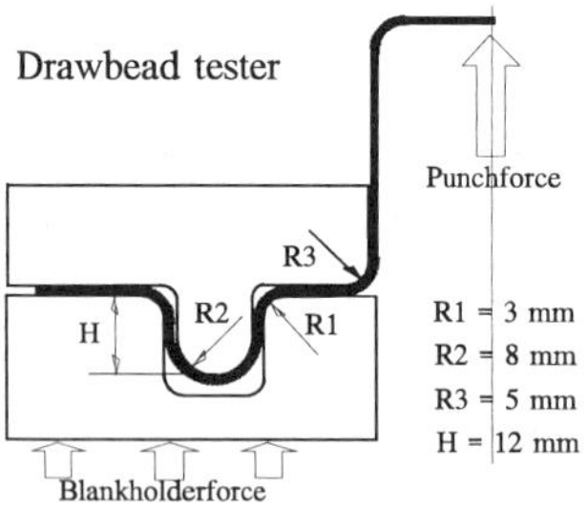

figure 1. Experimental setup of the drawbead tester.

2.1 *2D Plane Strain Analysis*

The analyses were carried out using the implicit finite element code DiekA which is developed at the University of Twente in co-operation with Hoogovens Corporate Research. This code has the possibility to use a mixed eulerian lagrangian formulation [Huétink]. This formulation has the opportunity to decouple the material displacement and the grid displacements. In this 2D drawbead model the mesh is fixed in flow direction, perpendicular to the flow the mesh is free to move. The advantage of this formulation is that the grid refinements remain at their place and the effects of sheet thinning can be described as well. Besides there is no need to model a large mesh in contrast with lagrangian formulation where the sheet can be pulled out of the drawbead.

The sheet was modelled with four node bi-linear plane strain elements. The number of elements in thickness direction was four and the total number of elements in the strip was 400. For contact description special contact elements were used [Huétink2]. The friction coefficient was 0.16. The mesh of the model is depicted in figure 2. The contact elements are also depicted.

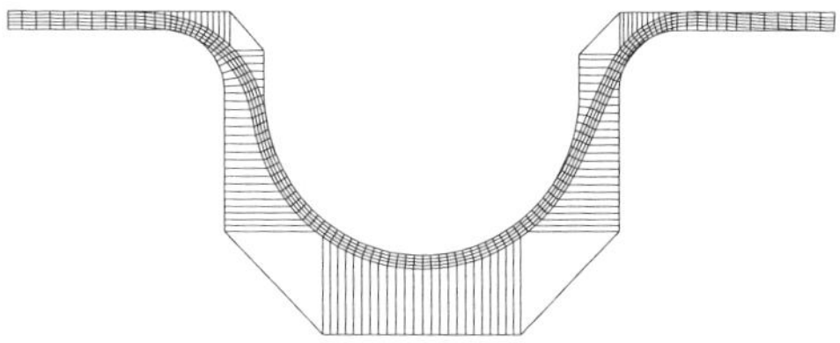

figure 2. Finite elements mesh 2D analysis.

In figure 3 the calculated force to pull the sheet through the drawbead and the tangential strain both as a function of the sheet displacement are printed. As can be seen the pulling force reaches its stationary value at 30 mm displacement. Due to the plane strain assumption and the almost incompressible material behaviour this is almost (except for the sign) the thickness strain. The stationary value of the tangential strain is reached after 35 mm displacement.

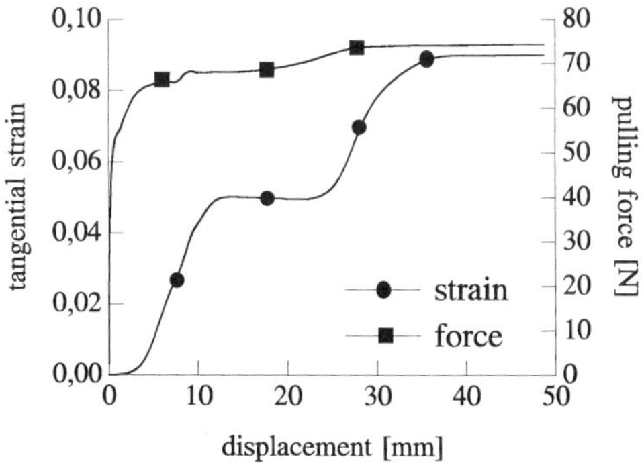

figure 3. Numerical results 2D analysis.

2.2 *Experimental Verification*

For verification of this 2D model an experimental set up had been built. At Hoogovens Corporate Research the experiments were performed at a fully equipped Erichsen press. The experimental set up of the drawbead tester is shown in figure 1. The surface roughness of the tools was about 0.25 μm. The applied lubricant was a deep drawing oil

with a viscosity of 0.04 Pas at room temperature. The blank was lubricated on both sides with help of a paint roller. The punch speed was kept constant at 3 mm/s.
After performing the experiment the thickness was measured at 10 locations of the strip (figure 4).

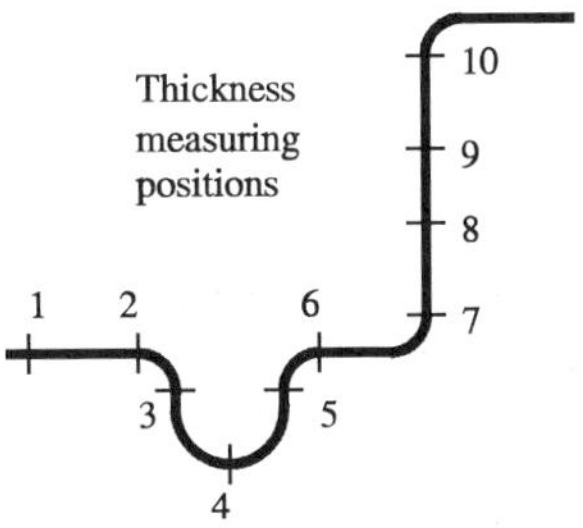

figure 4. Positions of the thickness measuring points.

In figure 5 the thickness strain is printed for the first six measuring points of the experiments, and the corresponding points in the FE simulation. At the entrance and the exit of the drawbead the strains show good agreement. In the drawbead small differences appear.

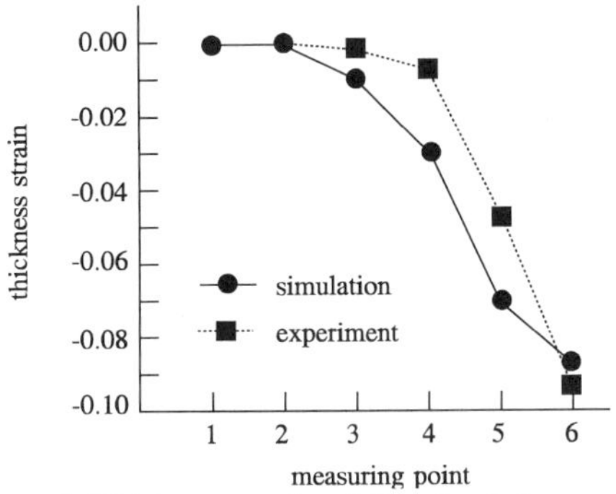

figure 5. Thickness strain of the simulation and the experiment.

To compare the stationary value of the experimental punch force with the FE calculation an additional force should be added to the calculated pulling force. This additional force is caused by the 90° bending between measuring point six and seven. The force needed for this 90° bend was calculated separately using the changed thickness and strain distribution. The measured punch force was 106 N/mm and the calculated punch force was 105 N/mm. These two punch forces agree very well. Therefore we can accept that the calculated drawbead restraining force is correct.

So, we have an accurate model to gain information of the drawbead strain and the drawbead force. This information will be used in the equivalent drawbead which is described in the next section.

3 EQUIVALENT DRAWBEAD

For economic reason the drawbead geometry has seldom been included, therefore an equivalent drawbead has been used. The drawbead has been defined as a line on the tool surface. If an element passes this drawbead line a restraining force and an additional strain has been added to the element. The restraining force is an additional force which is taken into account at the right hand side of the FE equations. The force is equally divided among the two nodes which element side cuts the drawbead line. The force is history dependent, the value is a function of the material which already passed the drawbead. First a simple model has been taken, the force linearly increases until the steady state value has been reached.

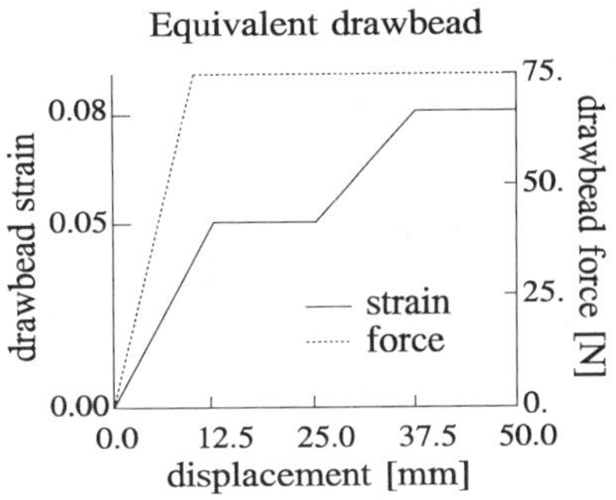

figure 6. Force and strain functions of the equivalent drawbead.

It is also possible to take into account the strain changes. When a node passes the drawbead line an amount of equivalent strain according to figure 6 is added to the nodal strain. The drawbead strain is also history dependant. For the strain there are two linear increasing parts and two constant parts. Both functions for force and strain are shown in figure 6, they are a result of the 2D analysis.

In the following sections some tests are described to verify and study the equivalent drawbead.

3.1 *Simple Strip*

The first test for the equivalent drawbead is pulling a strip through a die and blankholder as described in figure 7.

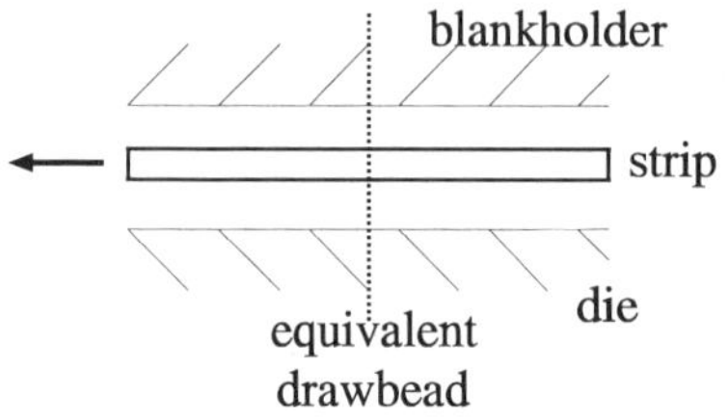

figure 7. Simple strip test.

The geometry, material and friction coefficient are exactly the same as used to verify the 2D drawbead model. In addition on the blankholder an equivalent drawbead has been defined. Due to the low blankholder force, the only resistance to sliding was caused by the drawbead. In this test we used the equivalent drawbead in two different ways. In the first simulation we only used the option of applying an additional restraining force to the elements which pass the drawbead line. In the second simulation we applied both the restraining force as well as the additional strain. The strip has been translated 50 mm. At the end of the translation the equivalent drawbead is at strip coordinate 50. The obtained thickness strains of both simulations are depicted in figure 8.

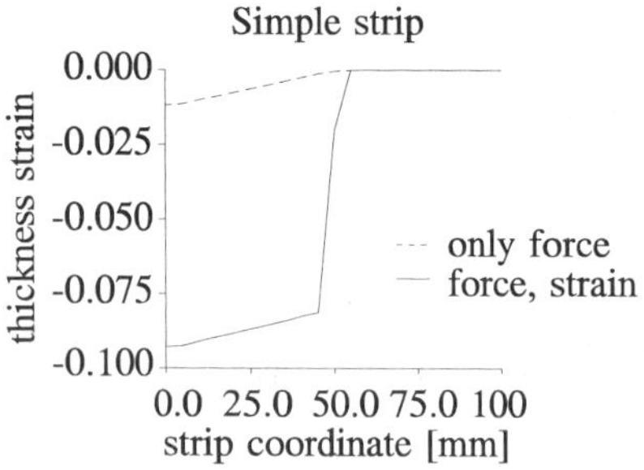

figure 8. First test for the equivalent drawbead, the thickness strain of the simple strip.

Comparing the two simulations there is no difference in the pulling forces. As we expect the strain distribution differs significantly. When neglecting the strain caused by passing the drawbead, the thickness strain is only 0.015. When including the additional strain caused by the drawbead, the thickness strain is about 0.09 and this agrees very well with the 2D analyses.

3.2 *Strip Drawing*

The next test is one step closer to deep drawing, namely strip drawing. The set up is identical to the drawbead experiment. For this simulation three node Mindlin elements with three integration points in thickness direction have been used. The equivalent drawbead has been defined as a line on the blankholder. The drawbead force and the drawbead strain are both history dependent as mentioned in figure 6 with the same steady state values as the simple strip test. Figure 9 shows the deformed finite element mesh with the original x-coordinate.

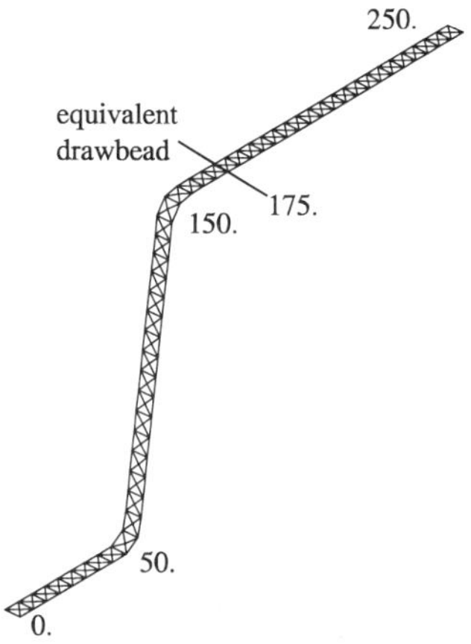

figure 9. The deformed mesh of the strip drawing test with initial x-coordinate.

In the first simulation both force and strain effects have been taken into account, in the second simulation only the force has been taken into account and the additional strain has been neglected. For comparison a simulation with no drawbead has been carried out. The punch force for both simulations with the drawbead showed no difference and agreed well with the experiments.

The thickness strain of the various simulations is shown in figure 10. At coordinate 50 there is a dip in the strain due to the bending around the punch radius. The die radius is situated at coordinate 150, and the equivalent drawbead is situated at coordinate 175. As can be seen in this figure the drawbead has considerable influence on the strain distribution in the strip. The only difference in both simulations with the drawbead is found in the material which has passed the drawbead line. The extra dip in thickness strain (A) in the simulation where we added force and strain is also seen in the experiments. For an accurate prediction the strain change due to passing the drawbead cannot be neglected in the equivalent drawbead.

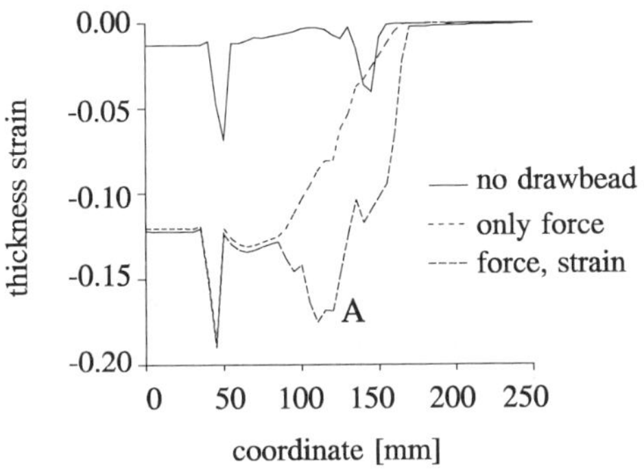

figure 10. Second test for the equivalent drawbead, the thickness strain of the strip drawing.

4 CONCLUSIONS

The 2D plane strain model works very satisfactory to study the drawbead behaviour. The agreement with the experiments is very good. So we have a reliable tool to predict the effects of a drawbead. These 2D analyses have been used to develop a 3D equivalent drawbead.

This 3D equivalent drawbead has been applied to a simple strip test and a strip drawing test. When only the drawbead force is applied in the equivalent drawbead, the strain changes due to passing the drawbead are not described well. The equivalent drawbead in which the force and strain effects have been included gives a good description of the real drawbead.

With this equivalent drawbead it is possible to include drawbeads in an accurate way. In the near future simulations of complicated deep drawing parts with drawbeads will be done.

ACKNOWLEDGEMENTS

The authors thank Pieter Drent for performing the experiments and Johan Beugels for carrying out some 2D analyses.

REFERENCES

Cao J. , M.C. Boyce
"Draw Bead Penetration as a Control Element of Material Flow"
SAE 930517, Sheet-Metal and Stamping Symposium, Detroit 1993

Carleer B.D., P.T. Vreede, M.F.M. Louwes, J. Huétink
"Modelling Drawbeads with Finite Elements and Verification"
Metal Forming 94, conference proceedings, Birmingham, 1994

Haug E., E. Di Pasquale, A.K. Pickett, D. Ulrich
"Industrial sheet metal forming simulation using explicit finite element methods"
FE-Simulation of 3D Sheet Metal Forming Process in Automotive Industry, conference proceeding , Zurich, 1991

Huétink J.
"On the Simulation of Thermo-mechanical Forming Processes"
Dissertation, University of Twente, 1986

Huétink J., P.T.Vreede, J.van der Lugt
"The simulation of contact problems in forming processes with a mixed euler-lagrangian FE method"
Numiform 89, conference proceedings, Colorado, 1989

Kawka M., N. Wang, A. Makinouchi
"Improving Draw Beads and Friction Models in Simulation of Industrial Sheet Metal Forming Processes"
Metal Forming Simulation in Industry, conference proceedings, Baden-Baden, 1994

Taylor L.M., J. Cao, A.P. Karafillis, M.C. Boyce
"Numerical Simulations of Sheet Metal Forming"
Numisheet 93, conference proceedings, Isehara, 1993

Wouters P., G. Montfort, J. Defourny
"Numerical simulation and experimental evaluation of the modifications of material properties in a drawbead"
IDDRG 94, conference proceedings, Lisbon, 1994

Simulation of Materials Processing: Theory, Methods and Applications, Shen & Dawson (eds)
© 1995 Balkema, Rotterdam. ISBN 90 5410 553 4

Using binder forming analysis to assist the design of sheet metal forming dies

Kuo-Kuang Chen
GM Research and Development Center, Warren, Mich., USA

Darryl J. Schmitz
GM Lansing Automotive Division, Mich., USA

ABSTRACT: A binder forming analysis was applied to two experimental dies: a decklid die and a fender die. The calculated binder wrap surface of the decklid die is basically cylindrical and agrees with the original design intent. No bifurcation is found by using several model variations. The calculated binder wrap surface for the fender die is almost cylindrical. Although this is not indicated by the orthogonal projection views of the punch opening line, the punch opening line does lie, more or less, on a cylindrical surface after it is rotated. This finding reiterates the importance of checking a punch opening line for closeness of a binder wrap to a cylindrical surface.

1 INTRODUCTION

A sheet metal forming die for autobody panels uses a curved binder surface to achieve an optimum forming condition for the panel (Stevenson, 1988). Therefore, sheet metal forming process for autobody panels consists of two forming processes: binder forming (or, binder wrap forming) and punch forming (or loosely, sheet metal forming). Each of the forming processes has its own distinct deformation characteristics. They are analyzed in general, by different numerical models and computer softwares.

Binder forming can be considered also as an intermediate step in sheet metal forming. The product of binder forming is a binder wrap which is to be further deformed by punch forming to make the part. From the analysis point of view, the binder wrap surface geometry describes the initial condition of punch forming. Examination of the binder wrap surface shape can also provide a designer with valuable information on how the die performs in a press (Mital, 1983).

Binder forming is a subject of substantial research and development in recent years (Chen, 1986 and Tang, 1985). It has been applied routinely to evaluate die design (Chen and Baran, 1993 and Balun, et al., 1993) and to be used in punch forming analysis (Chen and Campbell, 1994 and Tang and Chappuis, 1989). This paper presents findings obtained from recent applications of a binder wrap evaluation method (Chen, 1986). The bifurcation possibility of the binder wrap surface of an experimental decklid die was examined. A design guideline was obtained through the application to an experimental front fender die.

2 THE CALCULATION METHOD

The binder forming analysis method uses the geometry of the punch opening line, the blank thickness and its material property parameters as input. It uses an iteration procedure and finite element analysis in the calculation. It calculates the analysis domain (or, pseudo blank, the portion of the blank which goes in the die cavity to form the binder wrap) and the binder wrap surface geometry simultaneously. The latter is represented by a collection of (finite element node) points on the surface.

The calculation employs a particular displacement input procedure (called boundary vertical compliance method) to move the boundary of the analysis domain toward the punch opening line. The deformation of the pseudo blank (including in-plane displacement of boundary nodes and displacement of interior nodes) is calculated by the finite element method. An iteration procedure is used to improve the calculated results until a convergent solution is obtained.

Given a punch opening line, a planar initial analysis domain was obtained by flattening the punch opening line on the xy plane. For this purpose, a point (x,y,z) on the punch opening line is mapped to (x,y',0) while y' is obtained by maintaining the length of the punch opening line (Chen, 1988). Therefore, a one-to-one mapping exists between the given punch opening line and the boundary of the created initial analysis domain. The intention is to obtain an analysis domain which, when deformed, will lie inside the die cavity bounded by the punch opening line and have its boundary within a predetermined tolerance to the punch opening line.

A finite element mesh is drawn on the analysis domain. The pseudo blank is subjected to a displacement input described by the boundary vertical compliance method. A boundary node is displaced in the vertical (punch travel) direction by a displacement component such that it displaces the node to a horizontal plane passing through its corresponding point on the punch opening line. The horizontal displacements of the boundary nodes and the displacements of all the interior nodes are calculated by a finite element solver.

The finite element solver used is ABAQUS. The pseudo blank is modeled as a thin shell. For accuracy, quadrilateral (instead of triangular) elements are used in the model. After each finite element calculation, the positions of the boundary nodes are checked to determine the amount of correction needed for the next iteration. The maximum correction distance is obtained for all boundary nodes. If the maximum correction distance is less than a preassigned value, say 5 mm, then we claim that the iteration has converged. Otherwise, the analysis domain is updated and another iteration is carried out. In general, the iteration process converges rapidly.

After a convergent solution is obtained, the analysis domain geometry used last is the geometry of the pseudo blank and the deformed analysis domain defines the shape of the binder wrap. The binder wrap surface is specified by the coordinates of the displaced finite element nodes. A surface can be generated from these points to be used in punch forming analysis, such as in Chen and Campbell (1994).

3 BINDER WRAP OF A DECKLID DIE

Bifurcation in binder forming is an annoying phenomenon. Experience indicates that bifurcation in binder forming does occur in production environments which may cause shut down of the press line if a second mode of blank deformation occurs. Therefore, the possibility of bifurcation in binder forming should be avoided. Bifurcation in binder forming for a decklid die can happen (Chen and Baran, 1993). This paper revisits binder forming of a new experimental decklid die. Besides the calculation of the binder wrap surface geometry, we are particularly interested in the existence or nonexistence of bifurcation solutions in binder forming of the decklid die.

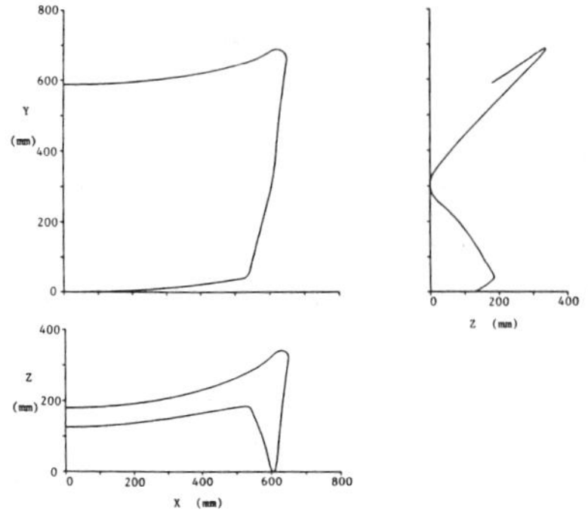

Figure 1. Orthogonal projection views of the punch opening line of the experimental decklid die.

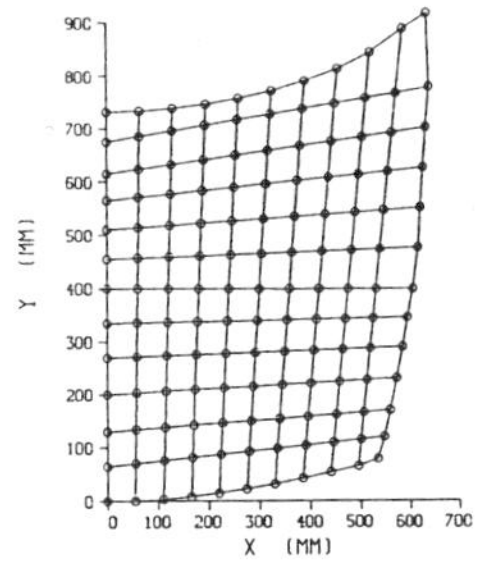

Figure 2. Initial analysis domain and finite element mesh used for the experimental decklid die.

Figure 1 shows the punch opening line of the experimental decklid die. Only half of the layout and punch opening line are shown in the figure, due to symmetry. An analysis domain was obtained by flattening the punch opening line in the xy plane. Then, a finite element mesh was drawn on the analysis domain (Figure 2). Again, due to symmetry, only half of the blank is shown. A steel blank with a thickness of 1.00 mm is used.

The blank is then deformed following a boundary vertical compliance displacement input procedure. The displacement input is applied in 200 equal linear increments. The resulting structural problem was solved. Convergence of the calculation method was then checked. It shows that a maximum boundary node correction is 75.9 mm. This large correction distance indicates that solution is not converged yet. Another iteration is performed.

After four iterations, a convergent solution is obtained. The maximum corrections of the boundary nodes are 75.9, 21.0, 10.9 and 3.23 mm respectively. The analysis domain and finite element mesh used in the last iteration is shown in Figure 3. The center line of the analysis domain has a length of 665 mm. The half analysis domain lies in a 640 by 842 mm rectan-

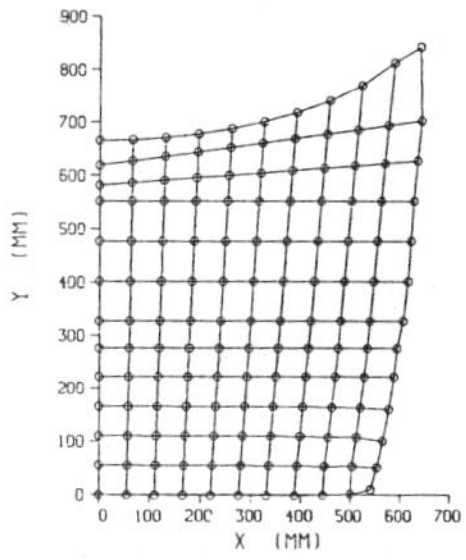

Figure 3. Converged analysis domain of the decklid die and finite element mesh used in analysis.

gle. The contour lines of the calculated binder wrap surface geometry, in its x-y projection view, is shown in Figure 4. Figure 5 shows a graphic display of the analysis domain and binder wrap.

The side view of the binder wrap surface is shown in Figure 6. The curves in this figure represent the deformed shape of the boundary lines between elements in Figure 3, which are oriented approximately in the y-direction The circles represents nodes on the boundary of the analysis domain. The curve passing through the circles is the punch opening line. With the

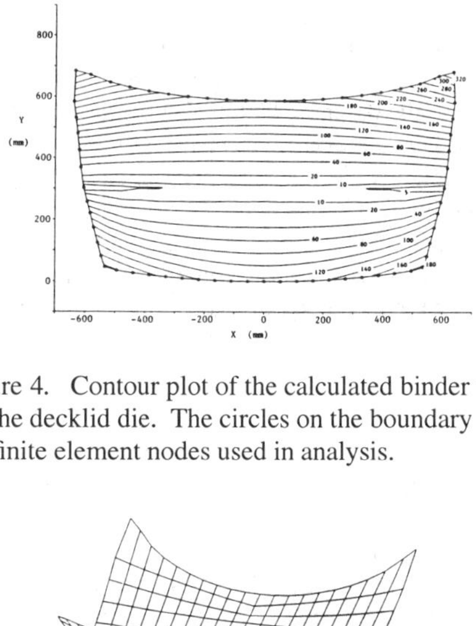

Figure 4. Contour plot of the calculated binder wrap for the decklid die. The circles on the boundary are the finite element nodes used in analysis.

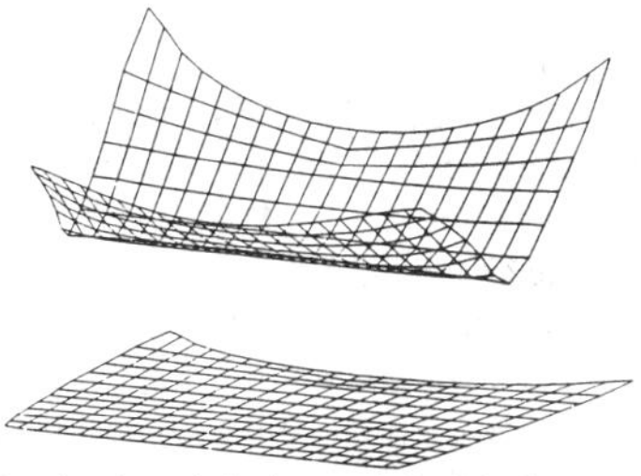

Figure 5. A pictorial view of the binder wrap for the decklid die. The analysis domain is shown in the lower figure.

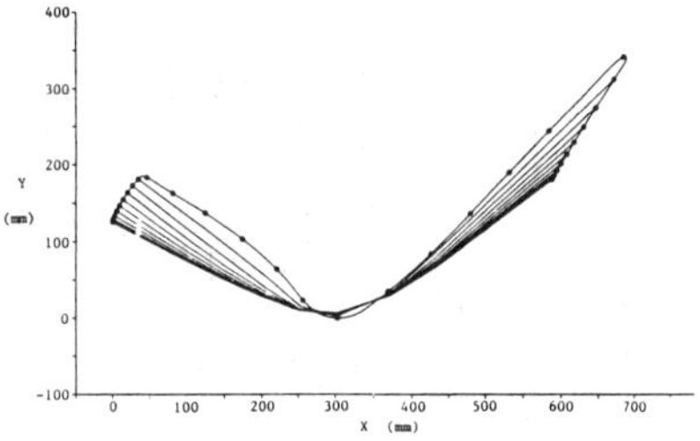

Figure 6. Side view of the binder wrap for the decklid die.

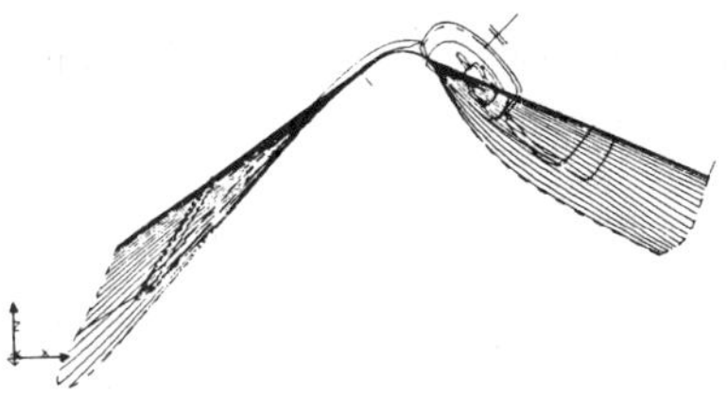

Figure 7. Side view of the designed binder wrap.

resolution of Figure 6, the deformed boundary nodes lie on the punch opening line. This indicates that the iteration solution has converged. Figure 7 shows the blank shape after the close of the binders intended by the designer. The binder wrap surface geometry is given by the central section of the curves which are enclosed by the punch opening line.

Comparing Figures 6 and 7, the calculated and intended binder wrap surface geometries are very close. The calculated geometry shows slightly less curvature along the center line of the blank. Thus, the length of the central cross-section may be slightly shorter than originally intended by the designer. On the whole, the designed binder surface fabricates a good unique binder wrap.

The above results were obtained using a particular finite element mesh. In order to ensure that the results do not depend on finite element used and there is no bifurcation possibilities, a second mesh is used. A different finite element mesh with 15 x 21 nodes (Figure 8) is drawn on the previously obtained converged analysis domain, for this purpose. Almost identical (by the resolution of the graphics) binder wrap surface geometry was obtained as shown by the contour lines and side views of Figures 9 and 10, respectively.

The blank thickness was then reduced to 0.8 and 0.6 mm respectively. Using the original mesh in Figure 3, the binder wraps were again obtained. The binder wrap surface made from the thinner blank shows slightly more bending and waviness than the

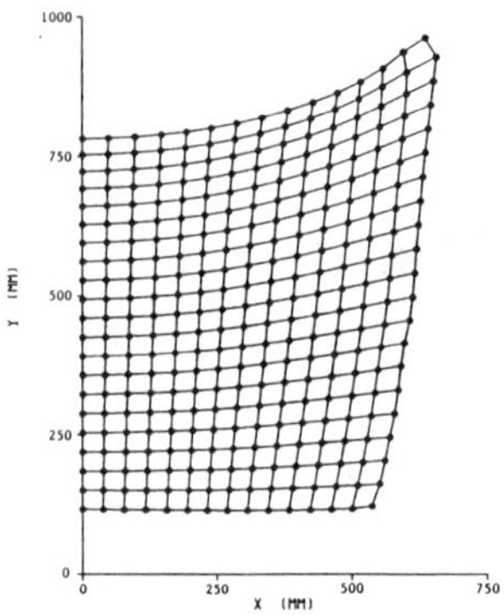

Figure 8. A second finite element mesh for the deck-lid die.

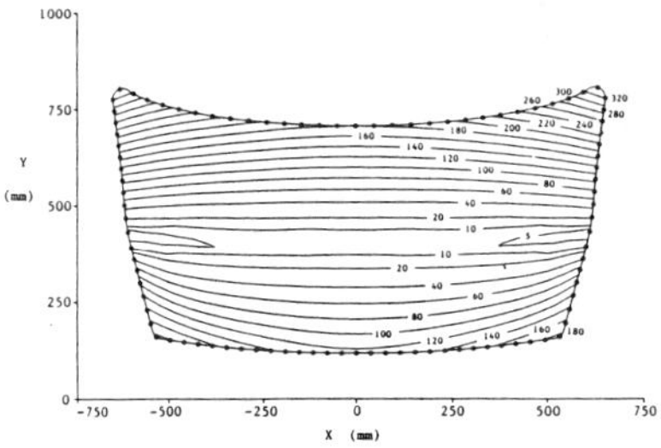

Figure 9. Contour plot of the binder wrap using the alternative finite element mesh.

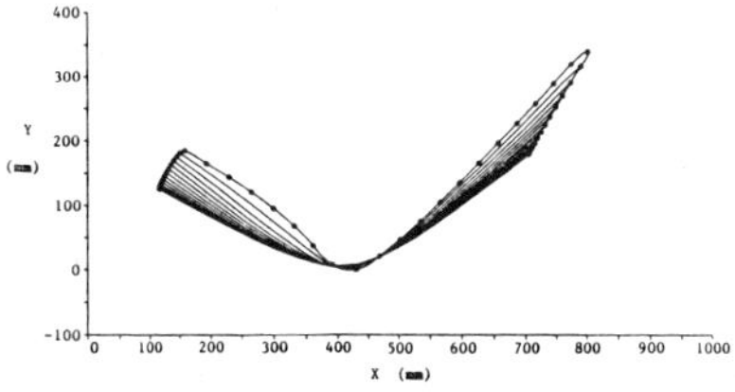

Figure 10. Side view of the binder wrap calculated using the alternative mesh.

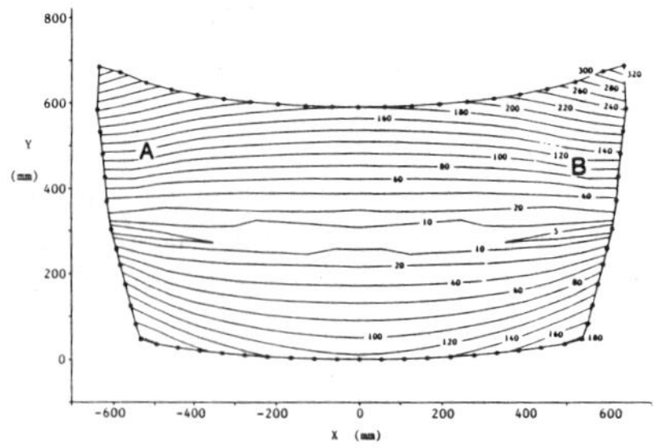

Figure 11. Calculated contour of binder wrap using a thinner blank of 0.6 mm thickness.

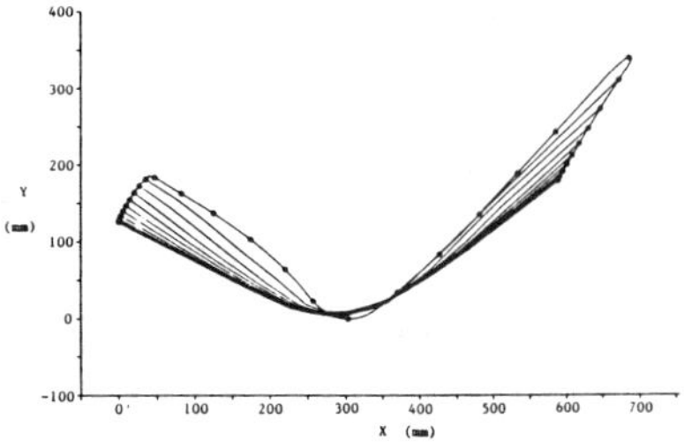

Figure 12. Side view of the binder wrap for the 0.6 mm thick blank.

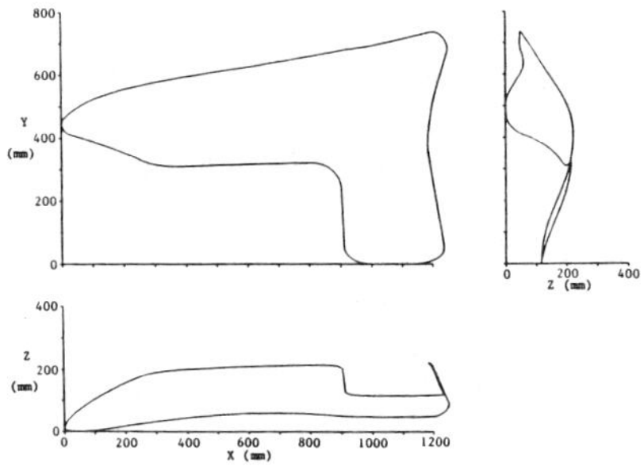

Figure 13. Orthogonal projection views of the punch opening line of the experimental fender die.

thicker blank cases. Figures 11 and 12 show the calculated results for the thin (0.6 mm) blank. The results for the 0.8 mm blank lies in between the 0.6 mm and 1.0 mm results. Comparing Figures 9 and 11, the bending and waviness of the thinner case can be detected in areas A and B near the upper corners in these figures. In general, for this particular die layout, a blank thickness variation from 0.6 to 1.0 mm does not affect the binder surface geometry significantly.

4 BINDER WRAP OF A FENDER DIE

A experimental fender die, whose binder wrap was judged initially to be difficult to make, was examined here. The orthogonal projection views of its punch opening line in the originally used cartesian coordinates as shown in Figure 13 is severely and doubly curved. The die may also be severely and doubly curved in these cartesian coordinates. The binder forming may be difficult. We used this die, initially, to demonstrate the applicability of the calculation model to this fender die to check out the model for a severely and doubly curved die. The binder wrap surface geometry was calculated. Some unexpected results were obtained by visualizing the binder wrap surface. A guideline for layout of sheet metal forming dies is

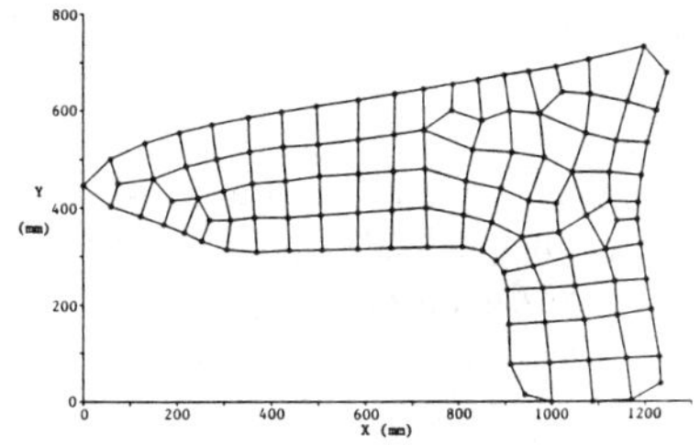

Figure 14. Initial analysis domain and finite element mesh used in analysis for the fender die.

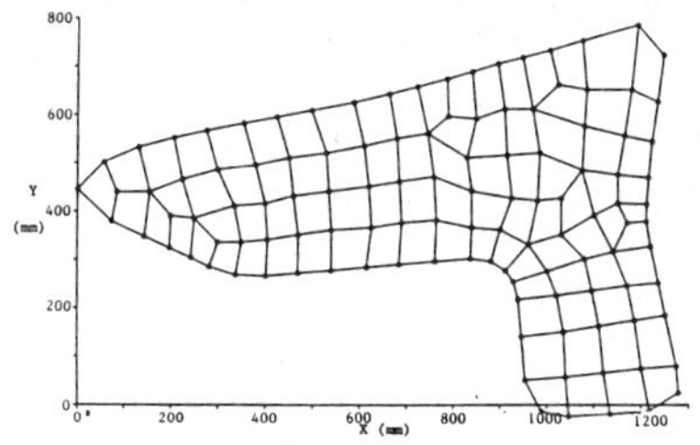

Figure 15. Converged analysis domain of the fender die and finite element mesh used in analysis.

obtained from this application.

The punch opening line of the fender die is shown in Figure 13. Since the punch opening line is doubly curved, the flattening of the punch opening line is non-trivial. For simplicity, the initial analysis domain used is assumed to coincide with the plan view of the punch opening line. The initial analysis domain and the finite element mesh used are shown in Figure 14. A 1.00 mm thick steel blank was used.

Following the routine calculation steps, the analysis domain was updated and deformation of the blank calculated in each iteration. The updated analysis domain was determined by the geometry of the boundary (finite element) nodes. The interior nodes were redistributed manually in each iteration. The maximum corrections for the boundary nodes are 82.6, 35.3, 4.4 and 0.4 mm for the first four iterations. A convergent solution is obtained in four iterations.

The analysis domain as well as the finite element mesh used in the last iteration is shown in Figure 15. We note that the dimension of the analysis domain in this figure is substantially bigger than that of Figure 14. This is the geometry of the blank which stays inside the die cavity after the binders are closed. The calculated contour lines of the binder wrap surface are shown in Figure 16. In this Figure, the majority of the contour lines have an inclination angle of approximately fifteen degrees measured from the x-axis. Therefore, the binder wrap surface has a basically

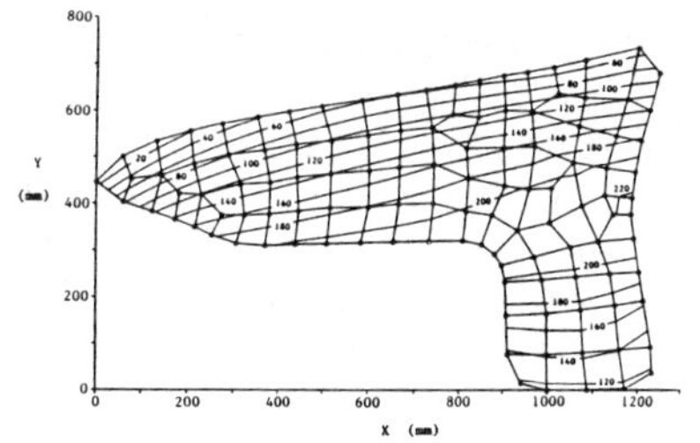

Figure 16. Contour plot of the calculated binder wrap for the fender die. These lines have basically an inclination angle of about 15 degrees.

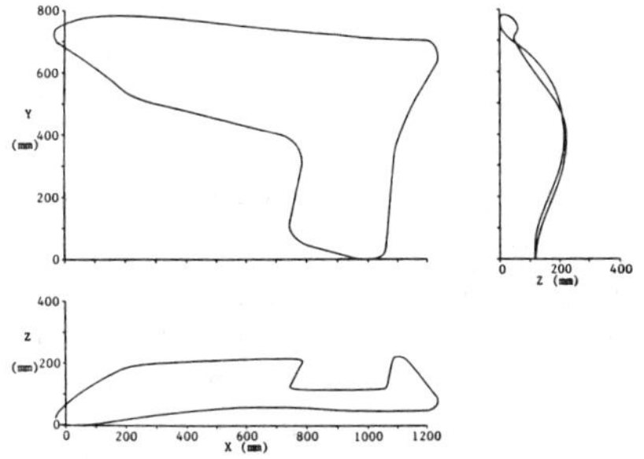

Figure 17. Orthogonal projection views of the punch opening line of the fender die plotted using a cartesian coordinate system which is rotated 15 degrees in the positive z-direction.

cylindrical shape. The cylindrical surface is so oriented such that its position follows that of the punch opening line. This can be visualized clearly if the punch opening line is rotated fifteen degrees in the xy plane (Figure 17). The xz side view of Figure 17 shows that the punch opening line lies more or less in a cylindrical surface.

The above shows that the orthogonal projection views of the punch opening line may not indicate the severity of binder wrap forming. The binder wrap surface of the fender die is almost cylindrical while the punch opening line (as viewed from a particular viewpoint) appeared to be severely and doubly curved. Therefore, in binder surface design, the punch opening line should be rotated and checked in a computer terminal. If any modification is needed, it should be done to make the punch opening line lie, as close as possible, in a cylindrical surface. This will ensure the binder wrap will be close to a cylindrical surface and the forming of a good part.

5 COMPUTNG REQUIREMENTS

The iteration procedure (Chen, et al., 1988) was followed in the calculations. Each iteration consists of an update (or, creation, in the first iteration) of the analysis domain and convergence check, and a finite element analysis. The convergence check and analysis domain updates run on an IBM mainframe computer instantaneously. The finite element solver used in this paper is ABAQUS which resides in a CRAY XMP mainframe computer. It takes 220 and 150 CPU seconds per iteration to solve the nonlinear structure deformation problem for the decklid and fender dies, respectively.

6 DISCUSSIONS

Binder forming for two experimental sheet metal forming dies were obtained routinely. The calculated binder wrap surface geometry can be used to evaluate die layout by predicting initial punch-blank contact point, etc., (Mital, 1983). Also, the binder wrap surface geometry provides the initial condition of the blank in punch forming analysis (Frey and Wenner, 1987 and Stoughton, 1985).

The binder wrap surface geometry of a decklid die and a fender die are calculated. Following a calculation procedure (Chen, 1988), the calculations are straightforward. For the decklid die, extended calculations were performed to ensure the accuracy of the calculation and to check for bifurcations. For both meshes used and for blank thickness between 0.6 and 1.0 mm, there were no indications for singularity in the solutions. Therefore, no bifurcation or wrinkling will occur in the binder wrap forming of this die.

For the fender die, the orthogonal projection views (displayed in the originally used coordinates) of the punch opening line do not suggest a cylindrical binder wrap until it was revealed by the calculated binder wrap surface. In die layout, punch opening line is obtained as a 'by-product' of the binder surface. However, since 1) a cylindrical binder wrap is a desired feature in die layout, 2) the binder wrap is determined by the punch opening line and 3) the punch opening line can be rotated and viewed easily in a computer screen, therefore, it is advantageous for a designer to view the punch opening line of a die layout in designing the binder surface. This suggests that in the process of die layout using computers, the punch opening line should be rotated on the screen such that it is designed as close to lie in a cylindrical surface as possible to ensure the forming of a good binder wrap and a quality part.

REFERENCES

Balun, T., P. Ling, M.M.K.Lou & S.C.Tang. 1993. Detection and elimination of wrinkles on an autobody panel by the binder set analysis. *SAE Paper #930515, SAE Transactions.*

Chen, K.K. 1986. An iteration method in binder wrap calculations. *NUMIFORM'86*: 321-326.

Chen, K.K. 1988. A binder wrap calculation procedure for sheet metal formability analysis. In B.V.Kiefer, et al (eds), *Materials in Manufacturing Processes*, ASME: 1-6.

Chen, K.K. & D.F.Baran. 1993. Bifurcation in binder wrap forming. *SAE Paper #930514, SAE Transactions.*

Chen, K.K. & M.P.Campbell. 1994. Using a geometric toolkit to link finite element calculations in sheet metal forming analysis. *SAE Paper #940748, SAE Transactions.*

Frey, W.H. & M.L.Wenner. 1987. Development and applications of a one-dimensional finite element code for sheet metal forming analysis. *Proc. of Interdisciplinary Issues in Material Processing and Manufacturing*. ASME. 1:307-319.

Mital, N.K. 1983. Prediction of binder wrap in sheet metal stampings using finite element method. *Proc. Int. Conf. Computers in Eng'g*, ASME: 71-77.

Stevenson, R. 1988. Development in the design and analysis of sheet metal stamping dies. *Product and Process Design*. ASME H00412:173-179.

Stoughton, T.B. 1985. Finite element modeling of 1008 AK steel stretched over a rectangular punch with bending effects. In N.M.Wang & S.C.Tang (eds), *Computer Modeling of Sheet Metal Forming Process*, TMS:143-159.

Tang, S.C. 1985. Verification and application of a binder wrap analysis. In N.M.Wang & S.C.Tang (eds), *Computer Modeling of Sheet Metal Forming Process*, TMS:193-208.

Tang, S.C. & L.B.Chappuis. 1989.Analysis of sheet metal forming processes by a general thin shell element. *NUMIFORM89*: 507-514.

Simulation of Materials Processing: Theory, Methods and Applications, Shen & Dawson (eds)
© 1995 Balkema, Rotterdam. ISBN 90 5410 553 4

Drawing of a curved wire

W. Daves & F. D. Fischer
Institute of Mechanics and Christian Doppler Laboratory for Micromechanics of Materials, University for Mining and Metallurgy, Leoben, Austria

ABSTRACT: With a fully tridimensional finite element model the drawing of a curved wire is simulated. In reality, in the most cases the wire is wound off from a coil and has, therefore, an initial curvature. This initial curvature influences the stress state and the distribution of the plastic deformation in the wire during and after drawing. Different radii of this coil are assumed, and the results of the simulation are compared. Furthermore it is checked how different types of plastic hardening influence the results. A damage parameter is incorporated for the detection of particular points where the danger of damage exists. For the reduction of inertia forces the die is moved not the wire. This feature allows the calculation with fictious higher densities and reduces strongly the computing times.

1 INTRODUCTION

The modelling of wire drawing by means of the finite element method has been dealt with by several authors, recently by Hoppe et al. (Hoppe 1994). Most of this work was done using an axisymmetric model. In reality the wire is not straight but curved before it is drawn through the die since it must be wound off from a coil. For the man at the drawing machine it is of essential interest if this initial curvature of the wire influences the mechanical properties and the curvature of the wire after drawing.

Within the work at hand a fully tridimensional model is presented for the drawing of a curved wire. These calculations are very time intensive and only possible using the new computer generation.

Following processes are modelled and compared:

- Axisymmetrical wire drawing,
- The drawing of initially curved wire.

The modelling is done by finite elements using an explicit integration scheme of ABAQUS/Explicit. Parameter studies are performed in dependence of the initial diameter of the wire coil. This parameter studies were only possible with a density of the material 100 times higher than the original one. It is checked that inertia effects have even no influence on the results.

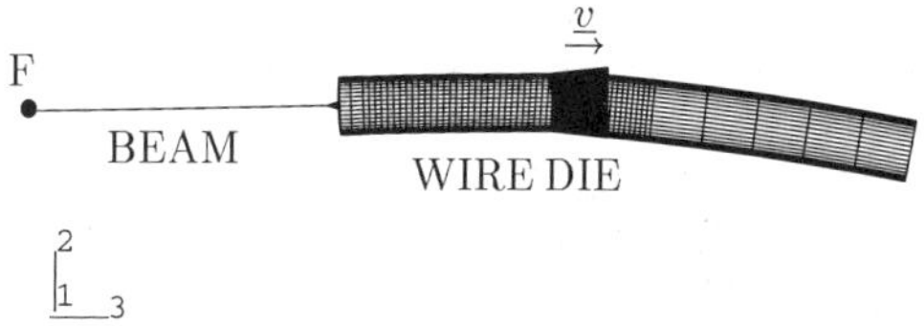

Figure 1. Finite Element Model

For a better understanding of the drawing process and more transparent representation of the results a damage indicator is programmed within the material subroutine of the ABAQUS/Explicit finite element code.

2 FINITE ELEMENT MODEL

A wire is modelled with an initial length of 100 mm. This wire is assumed to be wound off from a coil with various radii r taken as 0.5, 0.4 and 0.3 m. For all simulations the area reduction amounts 15 % with a die angle of $2\alpha = 12°$.

The finite element mesh is plotted in figure 1. 8287 tridimensional elements with 8 nodes and reduced integration are used for the mesh of the wire. The number of degrees of freedom is then

13899 for the whole model. The computing time with original density of the material would be 225 hours on an IBM 590 Risc machine.

On the front end of the mesh of the wire a circular beam is fixed which has the same diameter as the wire. Since this beam can be connected to the wire in one node only, additional truss elements are used to connect the beam with the wire. The top of the beam (point F) is fixed in direction 3, and is free in direction 1 and 2, see figure 1. The die is moved in 3 direction with a given velocity $\underline{v}$. At point F a reaction force is acting which corresponds to the drawing force. This feature reduces the kinetic energy in the system since only parts of the wire are accelerated.

Hu et al. (Hu 1994) define an error S for a quasistatic simulation using the difference between the strain distributions of the explicit dynamic and implicit static methods. They give a scaling relation for a specific value of the error S between the normalized material density

$$\rho_n = \frac{\rho}{k} \tag{2.1}$$

and a process velocity V

$$\rho_n V^{2-m} = C \tag{2.2}$$

where C is a constant. They assumed a strain rate sensitiv power law for the material

$$\sigma_x = k \varepsilon^n \dot{\varepsilon}^m \tag{2.3}$$

with σ beeing the true stress, ε the true strain and $\dot{\varepsilon}$ the strain rate. k,n and m are material parameters.

With relation 2 a relationship can be calculated between a critical process velocity V_{crit} and ρ_n. V_{crit} is defined as the process velocity at S = 1%.

If inertia effects can be minimized due to changes in the finite element model the error S is reduced, and V_{crit} increases. In the case wire drawing it is obvious that the wire has to be accelerated only due to the elongation during drawing when the rigid die is moved along the wire.

The beam with the same cross section as the wire at its front end allows for a more realistic simulation of the drawing of the wire out of the die. Fixing the cross section at the front end of the wire in the 1 - 2 plane produces too strong and unrealistic bending moments in the wire. Due to the beam, the reaction force works far away from the die without the use of additional tridimensional elements. Since the beam can move in 1 and 2 direction, the residual curvature of the drawn wire can be calculated. For the estimation of the influence of the cylindrical part of the die on the results two different dies are modelled. Figure 2 shows the meshes of the two dies which are used for the calculations.

The diameter of the wire befor drawing is given with 10 mm. For the most cases the tangent modulus E_T of the wire is 2000 MPa. The Young's

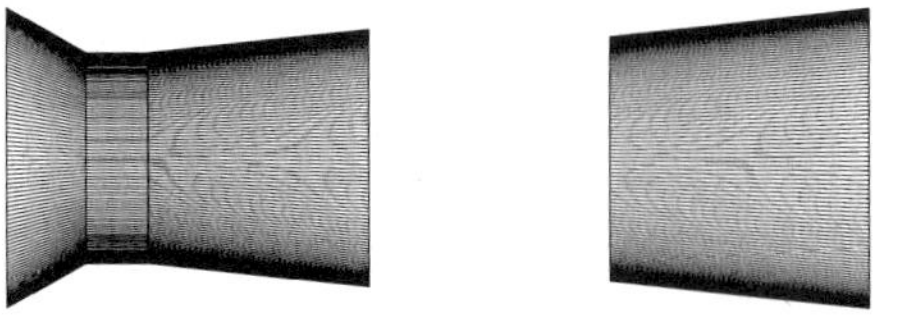

DIE 1 DIE 2

Figure 2. Finite Element Mesh of the Dies

modulus E = 210000 MPa and Poisson ratio = 0.3. The die is assumed to be rigid.

3 DAMAGE INDICATOR

During drawing of a wire ductile failure may occur. To indicate such an event in a computational study one has to use a material law formulation which describes the weakening of the material by voids. Frequently used formulations are based on the concept of Gurson (Gurson 1977) or on the concept of damage mechanics, e.g. published by Lemaitre et al. (Lemaitre 1990). A review on the literature dealing with the prognosis of ductile failure was given by Fischer et al. (Fischer 1995). They proposed a damage indicator D_i after Leckie (Gunawardena 1991).

The damage indicator D_i is defined by the relation

$$D_i = \int_0^{\varepsilon_{pv}} \frac{\exp(\frac{3}{2}\frac{\sigma_m}{\sigma_y})}{1.65\ \varepsilon_0}\, d\varepsilon_{pv}. \tag{3.4}$$

In this relation ε_{pv} denotes the equivalent plastic strain, σ_y the flow stress of the material, σ_m the average stress and ε_0 the critical strain. The determination of ε_0 from an uniaxial tensile test is described by Fischer et al. (Fischer 1995).

The evaluation of D_i is performed for each increment of the elastic - plastic analysis being programmed within a material subroutine for ABAQUS/Explicit. The damage indicator does not influence the results of the elastic - plastic analysis which means that a "perfect" structure is investigated. Therefore, D_i can only be used to indicate the start of a micro - crack.

4 RESULTS

4.1 *Dependence of the results on the material density*

Finite element programs with an explicit integration scheme allow for reduction of the computing

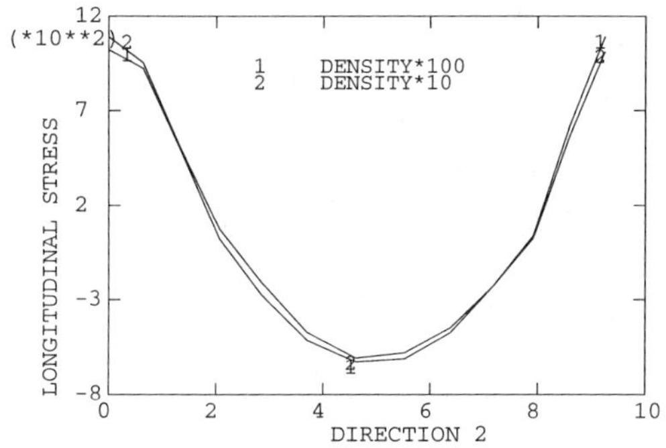

Figure 3. Dependence of the longitudinal stress in the drawn wire on the material density ρ

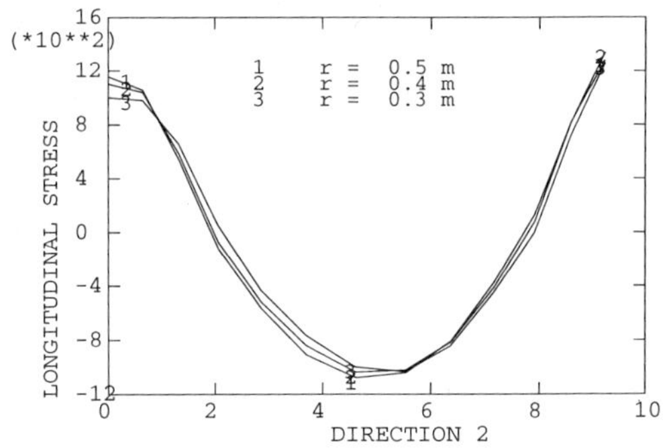

Figure 4. Dependence of the longitudinal stress on r

times by increase of both the fictious velocity of the process and the material density ρ.

The stable time increment Δt during the calculation is given by the relation

$$\Delta t \leq \min(L^{el}\sqrt{\frac{\rho}{\lambda + 2\mu}}), \tag{4.5}$$

where the minimum is taken over all elements in the mesh, L^{el} is a characteristic length associated with an element, and λ and μ are the effective Lamé constants for the material in the element (ABAQUS/Explicit User Manual 1993).

If ρ is increased one gets larger time increments and higher inertia forces. It must be checked if inertia forces influence the results. The comparison of calculations with 10 times and 100 times higher densities is shown in figure 3 demonstrating the longitudinal stress in direction 2 over the cross section of the wire in the 2-3 plane. First deviations of the results were found at ρ higher than 100 times of the original material density.

4.2 *Influence of the initial curvature*

The initial curvature of the wire is decribed by the radius r corresponding with the radius of the wire on the coil from which it is uncoiled. In the following diagrams the longitudinal stress and the equivalent plastic strain are plotted along the direction 2 over the cross section. With respect to direction 3 the diagram is positioned at a point of the wire where a steady state of the drawing process has been reached.

It was not possible to recieve the residual stresses after drawing for all calculated cases since the removal of the boundary conditions leads to dynamic oscillations of the wire. A force-free state would have been reached only after long computing times. Therefore, in the stationary state after passing the die the longitudinal stress and the equivalent plastic strain are plotted for r = 0.5, 0.4 and 0.3 meters in figures 4 and 5.

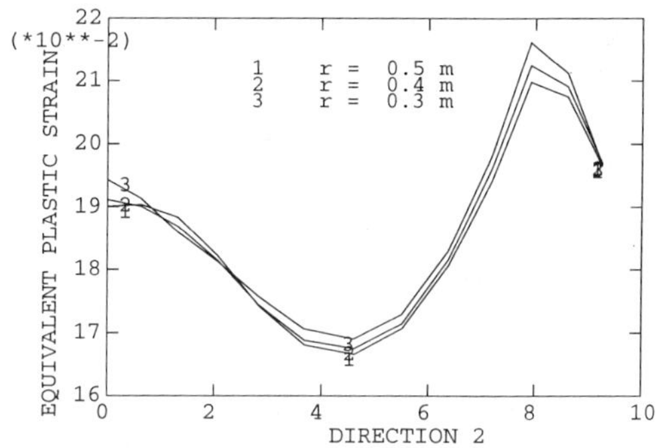

Figure 5. Dependence of the equivalent plastic strain on r

4.3 *Influence of kinematic and isotropic hardening*

In the center of the wire the longitudinal stress alters from a tensile state during deformation in the die to a pressure state outside the die after drawing. Therefore, an influence of the material description by an isotropic or kinematic theory can be expected. A bilinear stress - strain curve for the wire material is assumed with a tangent modulus E_T of 2000.

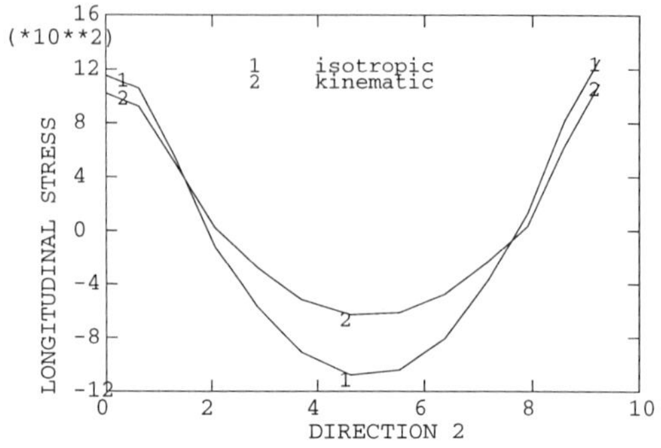

Figure 6. Dependence of the longitudinal stress on hardening

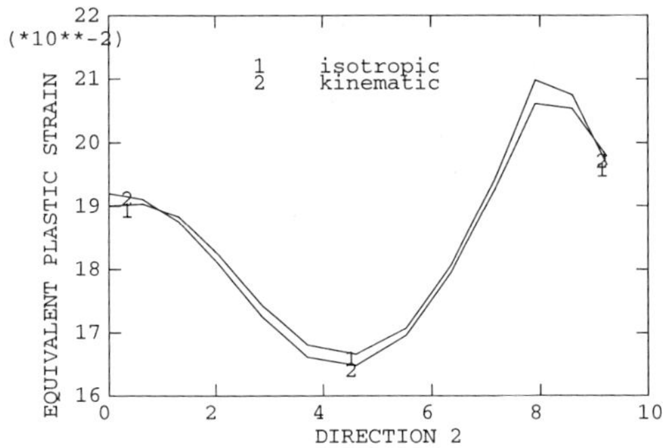

Figure 7. Dependence of the equivalent plastic strain on hardening

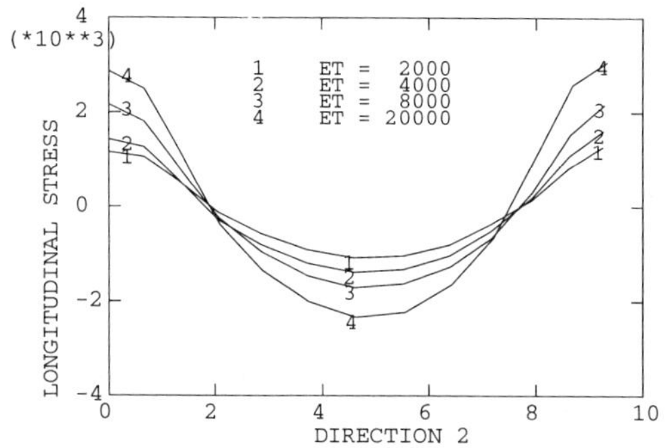

Figure 8. Dependence of the longitudinal stress on E_T

In the figure 6 and 7 the influence of kinematic and isotropic hardening on the longitudinal stress and the equivalent plastic strain is demonstrated. The diagrams are plotted at the same coordinates of the wire as above.

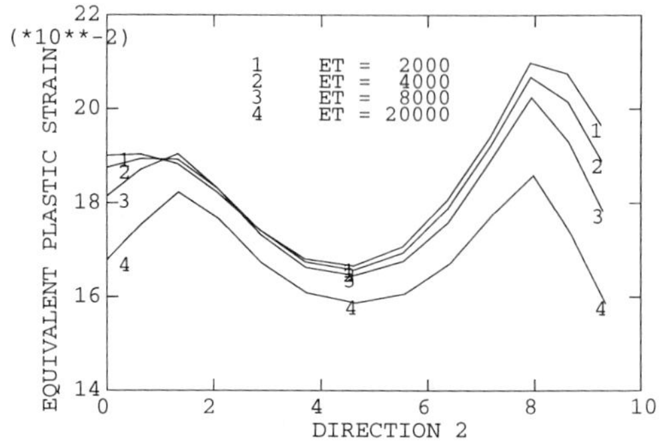

Figure 9. Dependence of the equivalent plastic strain on E_T

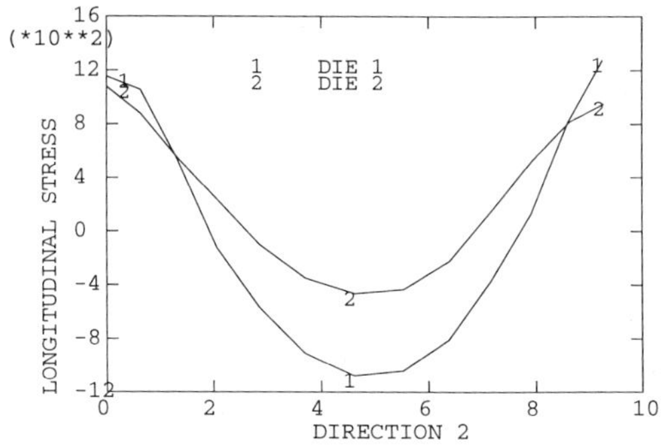

Figure 10. Dependence of the longitudinal stress on the die shape

4.4 *Influence of the tangent modulus of the wire material*

For r = 0.5 m the influence of the tangent modulus E_T of the wire material on the results is checked assuming isotropic hardening. In the figure 8 and 9 again the longitudinal stress and the equivalent plastic strain are plotted along the same coordinates as above.

4.5 *Results depending on the die shape*

The influence of the cylinder part of the die was analysed, too. The results for the dies 1 and 2 are plotted in figures 10 and 11, for the die shape see also figure 2.

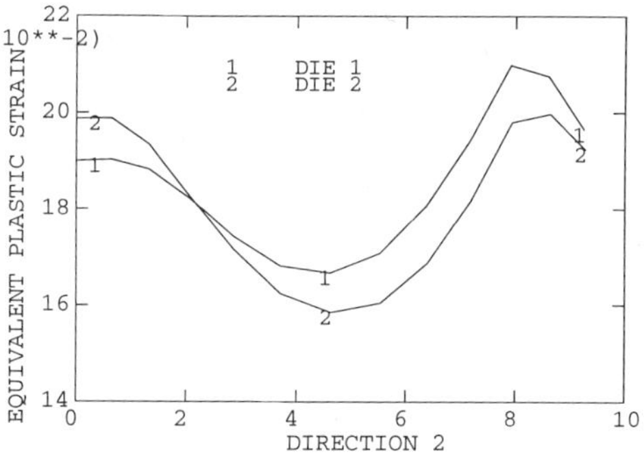

Figure 11. Dependence of the equivalent plastic strain on the die shape

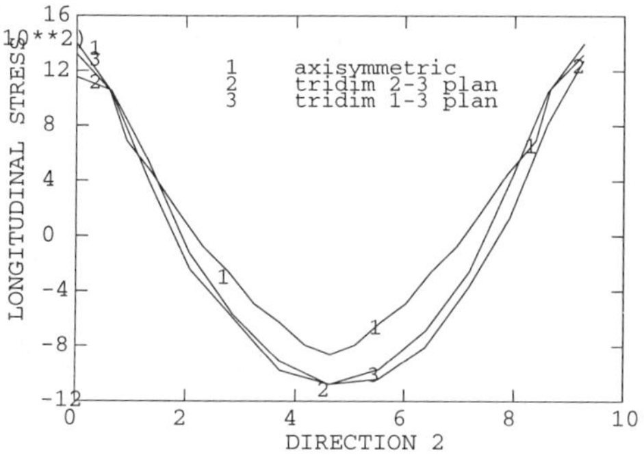

Figure 12. Comparison of axisymmetric and tridimensional calculations in the loaded state

4.6 *Comparison of an axisymmetric and a tridimensional calculation*

In figure 12 the longitudinal stress for an axisymetric and a tridimensional calculation are compared. For the tridimensional calculation the stress in the stationary state is plotted in the 2-3 and the 1-3 plane over the cross section of the wire.

Figure 13 shows the residual stress state in the wire after unloading.

4.7 *Influence of r on the damage indicator*

In figure 14 the damage indicator D_i is plotted over the cross section in the 2-3 plane. For the material parameter ε_0 the value 0.5 is chosen.

The results demonstrate nearly no influence of the initial curvature r on D_i. The most critical zone is situated in the center of the wire where a hydrostatic tensile stress is acting during the deformation process.

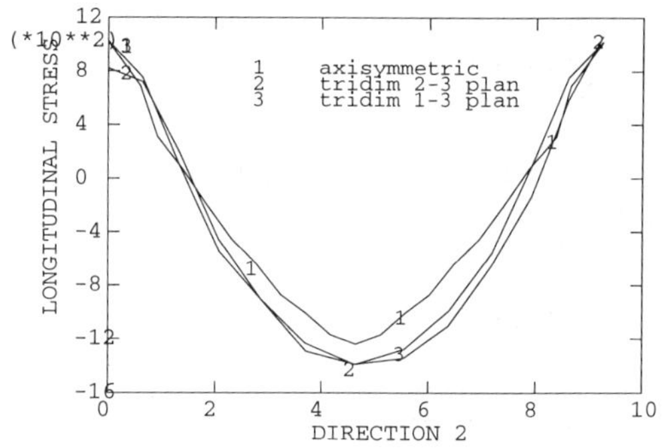

Figure 13. Comparison of axisymmetric and tridimensional calculations in the unloaded state

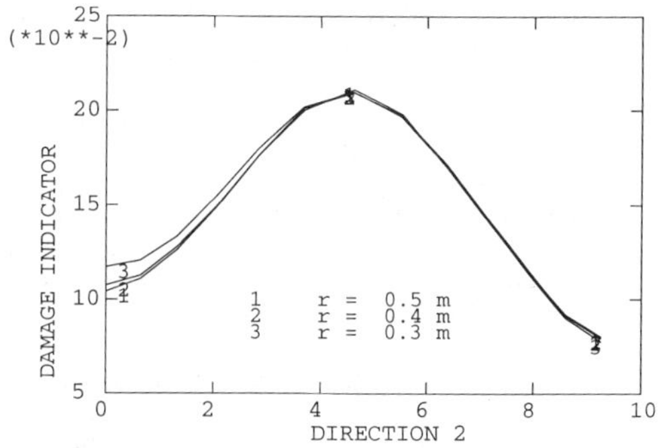

Figure 14. dependence of D_i on different initial radii r

5 CONCLUSION

The drawing of a curved wire is simulated, and parameter studies are performed. The calculations are very time consuming. many questions will only be answered with faster computers and more sophisticated finite element developments in the future.

Nevertheless, the study can demonstrate that an initial curvature of the wire has an influence on the results. For the chosen material and geometrical parameters this influence is not very significant. However the remarkable difference between the results for the two investigated die shapes may point out that geometrical variations of the die shape should be investigated in further studies.

A damage parameter was included in the finite element calculations. This parameter indicates a danger of ductile fracture in the center of the wire which is in accordance with the practical experience. There is almost no influence of the initial curvature on the values of the damage indicator.

It is further remarkable that an influence of the chosen type of hardening exists on the results.

For the reduction of inertia effects during deformation process simulations it is recommended to move the deformation tool and not the workpiece if this is possible.

REFERENCES

ABAQUS/Explicit User Manual 1993, Vers. 5.3, Hibbit, Karlsson & Sorensen, Inc., Pawtucket, Rhode Island, USA

Fischer, F.D., O.Kolednik, G.Shan and F.G. Rammerstorfer 1995. A note on Calibration of ductile failure damage indicator. To be published in *Int. J. Fracture.*

Gunawardena, S.R., S.Jansson and F.A.Leckie 1991. Transverse ductility of metal matrix composites. In G.K. Haritos, G. Newaz and Sh. Mall (eds.), *Failure Mechanisms in High Temperature Composite Materials.* AMD 122:23-30 New York: ASME.

Gurson A.L. 1977. Continuum theory of ductile rupture by void nucleation and growth. I. Yield criteria and flow rules for porous ductile media. *J. Eng. Mat. Tech.*: 99:1-15.

Hoppe, M. & R.Kopp 1994. Analyse des Eigenspannungszustandes in gezogenen Drähten und Stäben: In U. Kühn (ed), *Ziehen von Drähten, Stangen und Rohren*: 71-83, Oberursel: Deutsche Gesellschaft für Materialkunde.

Hu, X., R.H.Wagoner, G.S.Daehn and S.Ghosh 1994. Comparison of explicit and implicit finite element methods in the quasistatic simulation of uniaxial tension. *Comm. numer. Methods Eng.*: 10:993-1003

Lemaitre J. & J.L.Chaboche 1990. *Mechanics of solid materials.* Cambridge: Cambridge University Press.

Simulation of Materials Processing: Theory, Methods and Applications, Shen & Dawson (eds)
© 1995 Balkema, Rotterdam. ISBN 90 5410 553 4

Springback analysis in orthotropic sheet metal forming with an elasto plastic formulation using a kinematic hardening model

Frédéric Fenoglietto, Fabrice Morestin & Maurice Boivin
Laboratoire de Mécanique des Solides INSA de Lyon, Villeurbanne, France

Xiaolin Deng
Régie Renault Boulogne, Billancourt, France

ABSTRACT : The knowledge of the sheet metal's springback after forming is important to control the manufacturing processes. Nowadays, the importance of this issue increases because of the use of steel sheeting with high yield stress and also aluminium alloys. A mechanical theory has been developped and implemented in a software called *PLIAGE* in order to predict the final shape of the drawing. *PLIAGE* is a semi analytical software and does not use the F.E.M. formulation. So, the computation time is very short without being prejudicial to the quality of the results. (This software had given the best result for 2D spring-back determination at the NUMISHEET 93 [2] in Tokyo.) The calculations take into account the change of the Young's modulus versus plastic strain because of the importance of this parameter for springback computation. After the final stress state has been entirely computed, the residual radius of the identical history strain area is then calculated in order to determine spring-back. The complete shape, after spring-back, is rebuilt with the residual radius of each drawing area which has experienced the same strain path. The first versions of *PLIAGE* [8] did not take into account the orthotropic nature of the rolling sheets. This characteristic being very important, and causing significant differences on the springback, it is implemented in the new release of *PLIAGE*. The orthotropy complicates the model, by increasing the number of required parameters for the calculation. That's the reason why the developped theory leans on an orthotropic extension of the non linear kinematic hardening model proposed by Lemaître and Chaboche. We also present experimental results of 2D deep drawing and the comparison with theoretical data. *PLIAGE* has been developed with the help of RENAULT, the French car manufacturer.

1. INTRODUCTION

In order to predict the springback of the stamped sheet, *PLIAGE* must know the geometry of the deformation' story.
First, the forming process is simulated geometrically with an other software in order to find the areas of the drawing which have experienced the identical strain path. Fig.1. shows the distribution of these areas during the deep drawing of a Ω specimen [6].

Area 0 has no strain due to bending.
Area 1 is simply bent to R_1
Area 2 is bent to R_1 and then to R_2
Area 3 is only bent to R_3

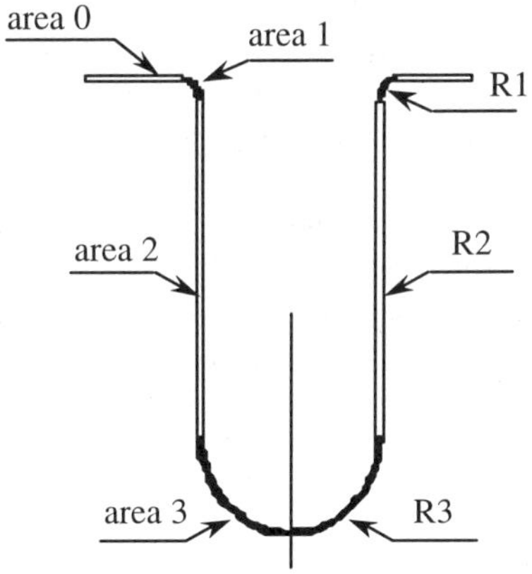

Fig. 1. Example of identical strain history areas

2. GENERAL ASSUMPTIONS AND NOTATIONS

The steel sheet may be considered as being composed of layers piled up in its thickness, see fig.2.

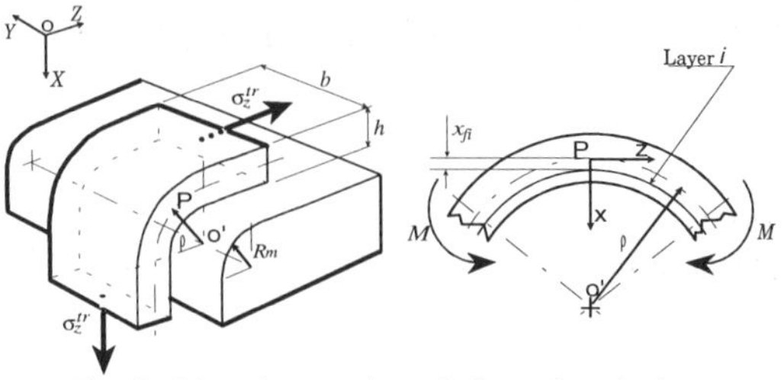

Fig. 2. Notations and steel sheet description

$[\sigma]$	stress tensor
$[\bar{\sigma}]$	reduced stress tensor
$[\varepsilon^p]$	plastic strain tensor
$[\varepsilon^e]$	elastic strain tensor
$[\bar{\varepsilon}]$	total strain tensor
$[\alpha]$	backstress
ρ	imposed radius
ρ_{res}	residual radius
l_0	initial length of one layer
l_t	length of the same layer after tensile
l_f	final length of layer after bending
h	thickness of the steel sheet
b	length of the steel sheet
x_{fi}	ordinate of layer i
C_{ii}, γ_{ii}	coefficients of the Lemaître and Chaboche law in i direction
F, G, H	Hill's coefficients
$E_{jj}(x_{fi})$	Young's modulus of layer i in j direction
υ_{ij}	Poisson ratio, i tensile direction, j deformation direction
ε_{zz}^{tr}	tensile strain
$\varepsilon_{eq/ii}^{p}$	equivalent plastic strain in i direction
M_f	bending moment
$d\lambda$	factor of proportionality
σ_{eii}	yield stress in i direction
σ_{zz}^{tr}	'z' tensile stress
σ_{yy}^{tr}	'y' tensile stress
$\bar{\sigma}_{eq/ii}$	reduced equivalent stress in i direction
f	loading surface's function
r_i	anisotropic ratio in i direction

The general hypothesis are as follows [3]:

H1 * The steel sheet is sufficiently wider than its thickness. So, this is a plane strain problem, in the direction of the sheet length : $\varepsilon_{yy} = 0$

H2 * As the steel sheet is considered to be thin, this is a plane stress problem in the direction of the sheet thickness. Therefore the normal stress on the steel sheet is zero : $\sigma_{xx} = 0$

H3 * All the calculations of the radius are made at the neutral layer of the steel sheet for $x_{fi} = 0$. The thickness and width of the sheet are uniform and remain constant during the bending process. The bending line is a circular arc.

H4 * All the stresses are applied in the orthotropic directions of the sheet, that is to say x,y and z directions of fig.2.

The calculations take into account the variation of the Young's modulus versus plastic strain [1], [5] because of the importance of this parameter for springback computation. As the sheet is orthotropic, the Young's modulus variation must be calculated in each direction then we note : j = y, z. The Young's modulus noted $E_{jj}(x_{fi})$ is evaluted for each layer with a plastic equivalent strain deduced from a kinematic work hardening law. The Young's modulus variation is approximated by a piecewise linear function. [9].

3. Calculation of the residual curvature for one identical strain history area

The initial stress state is known and the stress' increments are computed for each identical strain history area by solving the plasticity equations associated with the non linear kinematic hardening model based on an extension of the one proposed by Lemaître and Chaboche [4].

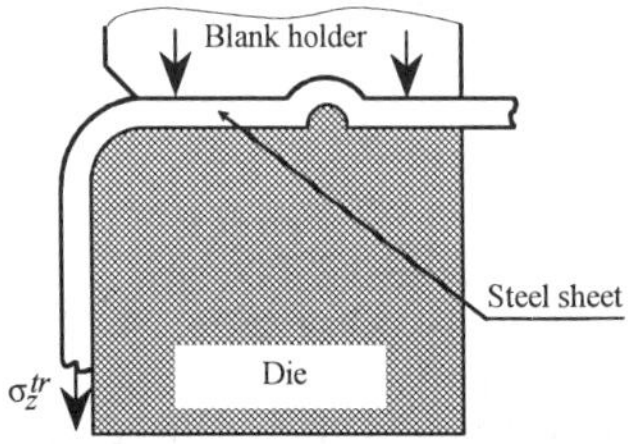

Fig.3. Tensile stress during deep drawing

3.1. Determination of the initial strain generated by tensile stress

During deep drawing, the applied blank holder pressure generates a tensile force. This stress must be taken into account for the springback determination [7].

First, the total initial strain is calculated. Before, the strain, noted ε_{zz}^{tr}, which is caused by the tensile force applied to the steel sheet must be estimated. This strain's calculation depends on the value of the 'plastic criterium'. Two cases have to be considered.

3.1.1. Elastic range

The value of ε_{zz}^{tr} must be find for $f(\sigma_{ij}) \le 0$. In the elastic range, the Hooke's relations give :

$$\begin{bmatrix} \varepsilon_{xx}^{tr} \\ \varepsilon_{yy}^{tr} \\ \varepsilon_{zz}^{tr} \end{bmatrix} = \begin{bmatrix} 1/E_{xx} & -\upsilon_{yx}/E_{yy} & -\upsilon_{zx}/E_{zz} \\ -\upsilon_{xy}/E_{xx} & 1/E_{yy} & -\upsilon_{zy}/E_{zz} \\ -\upsilon_{xz}/E_{xx} & -\upsilon_{yz}/E_{yy} & 1/E_{zz} \end{bmatrix} \cdot \begin{bmatrix} 0 \\ \sigma_{yy}^{tr} \\ \sigma_{zz}^{tr} \end{bmatrix} \quad (01)$$

and $\dfrac{\upsilon_{yx}}{E_{yy}} = \dfrac{\upsilon_{xy}}{E_{xx}}, \; \dfrac{\upsilon_{zx}}{E_{zz}} = \dfrac{\upsilon_{xz}}{E_{xx}}, \; \dfrac{\upsilon_{zy}}{E_{zz}} = \dfrac{\upsilon_{yz}}{E_{yy}}$

but, H1 is equivalent to $\varepsilon_{yy}^{tr} = 0$, thus

$$\frac{1}{E_{yy}} \cdot \sigma_{yy}^{tr} - \frac{\upsilon_{zy}}{E_{zz}} \cdot \sigma_{zz}^{tr} = 0 \text{ so } \sigma_{yy}^{tr} = \frac{\upsilon_{zy} \cdot E_{yy}}{E_{zz}} \cdot \sigma_{zz}^{tr} \quad (02)$$

From relations (01), we can write :

$$\varepsilon_{zz}^{tr} = -\frac{\upsilon_{yz}}{E_{yy}} \cdot \frac{\upsilon_{zy} \cdot E_{yy}}{E_{zz}} \cdot \sigma_{zz}^{tr} + \frac{1}{E_{zz}} \cdot \sigma_{zz}^{tr}$$

$$\text{so} \quad \varepsilon_{zz}^{tr} = \frac{1 - \upsilon_{yz} \cdot \upsilon_{zy}}{E_{zz}} \cdot \sigma_{zz}^{tr} \quad (03)$$

3.1.2. Plastic range

$f(\sigma_{ij}) > 0$. Each layer of the steel sheet is yield before bending action. The value of ε_{zz}^{tr} can only be calculated with an elasto-plastic constitutive law. To do that, the theory described in section 3.3. is used. The curvature increment $d\rho$ is set to zero and the value ε_{zz}^{tr} is increased until the force balance is verified. By this way, the value of the total strain noted $\tilde{\varepsilon}_{zz}^{tr}$ is calculated and then :

$$\varepsilon_{zz}^{tr} = \tilde{\varepsilon}_{zz}^{tr} \quad (04)$$

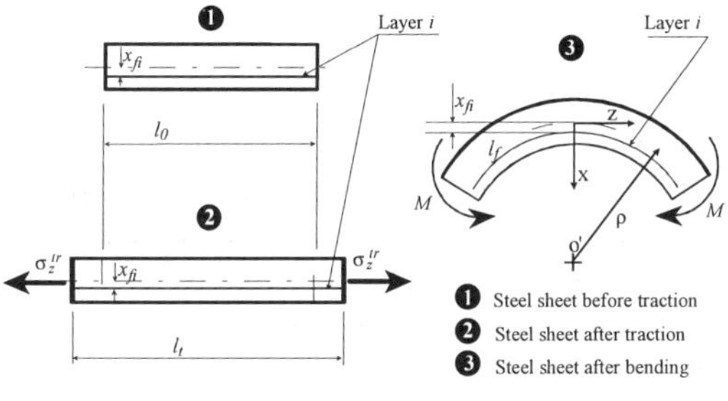

Fig.4. Notations

3.2. Geometrical determination of the total strain

ε_{zz}^{tr} is also the longitudinal strain of the mean layer of the steel sheet :

$$\varepsilon_{zz}^{tr} = \frac{l_t - l_0}{l_0} \quad (05)$$

The total natural strain $\bar{\varepsilon}_{zz}$ is given by :

$$\bar{\varepsilon}_{zz} = \ln(l_f / l_0) \quad (06)$$

The final length of the layer is geometrically determined, see fig.4:

$$l_f = \frac{\rho - x_{fi}}{\rho} \cdot l_t \quad (07)$$

then

$$l_f / l_0 = \frac{\rho - x_{fi}}{\rho} \cdot (1 + \varepsilon_{zz}^{tr}) \quad (08)$$

and so, the strain can be stated as :

$$\bar{\varepsilon}_{zz} = \ln\left[(\rho - x_{fi}) / \rho \cdot (1 + \varepsilon_{zz}^{tr})\right] \quad (09)$$

This expression is differentiated and gives :

$$d\bar{\varepsilon}_{zz} = \frac{\partial(\bar{\varepsilon}_{zz})}{\partial\rho} \cdot d\rho + \frac{\partial(\bar{\varepsilon}_{zz})}{\partial\varepsilon_{zz}^{tr}} \cdot d\varepsilon_{zz}^{tr} \quad (10)$$

$$d\bar{\varepsilon}_{zz} = \frac{x_{fi}}{\rho \cdot (\rho - x_{fi})} \cdot d\rho + \frac{1}{1 + \varepsilon_{zz}^{tr}} \cdot d\varepsilon_{zz}^{tr} \quad (11)$$

The strain increment along the z axis will be established from two incremental variables. The first noted $d\rho$ has a constant value and is given arbitrarily in order to be small against ρ . The second parameter, $d\varepsilon_{zz}^{tr}$, is calculated for each layer by the convergence of $\int \sigma_{zz} \cdot dz$ to the imposed tensile stress in an equilibrium loop. So, in order to verify this equilibrium, we must know the stress's distribution in the thickness. This is the aim of the following sections.

3.3. Determination of stress state in the thickness

As the model work with an orthotropic sheet, it must use the loading surface's function defined by HILL, which can be write with the kinematic formulation and H4 :

$$f(\bar{\sigma}_{ij}) = \frac{1}{2} \cdot F \cdot (\bar{\sigma}_{yy} - \bar{\sigma}_{zz})^2 + \frac{1}{2} \cdot G \cdot (\bar{\sigma}_{zz} - \bar{\sigma}_{xx})^2 + \frac{1}{2} \cdot H \cdot (\bar{\sigma}_{xx} - \bar{\sigma}_{yy})^2 - \frac{1}{2} = 0 \quad (12)$$

with $\bar{\sigma}_{ij} = \sigma_{ij} - \alpha_{ij}$

Lemaître and Chaboche proposed as incremental value of backstress the formulation :

$$d\alpha_{ij} = C \cdot d\varepsilon_{ij}^{p} - \gamma \cdot \alpha_{ij} \cdot d\varepsilon_{eq}^{p} \quad (13)$$

γ and C are positive parameters which characterise the material. They are deduced from tensile or cyclic tests. $d\varepsilon_{eq}^{p}$ is the equivalent plastic strain increment. However, this definition can be applied only in an isotropic situation, because it does not take care of the orthotropic directions.

The modification proposed is :

$$d\alpha_{ij} = C_{zz} \cdot d\varepsilon_{ij}^{p} - \gamma_{zz} \cdot \alpha_{ij} \cdot d\varepsilon_{eq/zz}^{p} \quad (14)$$

In this formulation appears the notion of direction. The choised direction is the tensile direction z because it's the best to lead the calculation.

The backstress increment, the plastic strain and the Hill strain are coupled with the Lemaître and Chaboche extended relation.

3.4. Calculation of the factor of proportionality $d\lambda$

The Drucker's rule can be written as follows :

$$d\varepsilon_{ij}^{p} = \frac{\partial f}{\partial \bar{\sigma}_{ij}} \cdot d\lambda \tag{15}$$

so, with (12)

$$d\varepsilon_{xx}^{p} = \left[G \cdot (\bar{\sigma}_{xx} - \bar{\sigma}_{zz}) + H \cdot (\bar{\sigma}_{xx} - \bar{\sigma}_{yy})\right] \cdot d\lambda \tag{16}$$

$$d\varepsilon_{yy}^{p} = \left[F \cdot (\bar{\sigma}_{yy} - \bar{\sigma}_{zz}) + H \cdot (\bar{\sigma}_{yy} - \bar{\sigma}_{xx})\right] \cdot d\lambda \tag{17}$$

$$d\varepsilon_{zz}^{p} = \left[F \cdot (\bar{\sigma}_{zz} - \bar{\sigma}_{yy}) + G \cdot (\bar{\sigma}_{zz} - \bar{\sigma}_{xx})\right] \cdot d\lambda \tag{18}$$

$$d\varepsilon_{xy}^{p} = d\varepsilon_{xz}^{p} = d\varepsilon_{yz}^{p} = 0 \tag{19}$$

The loading parameter $d\lambda$ can be obtain from the equation of the yield surface (12) and from the consistency condition. $df = 0$ (20)

thus

$$df = \sum_{ij} \frac{\partial f}{\partial \sigma_{ij}} \cdot d\sigma_{ij} + \sum_{ij} \frac{\partial f}{\partial \alpha_{ij}} \cdot d\alpha_{ij} = 0 \tag{21}$$

but

$$\frac{\partial f}{\partial \bar{\sigma}_{ij}} = \frac{\partial f}{\partial \sigma_{ij}} = -\frac{\partial f}{\partial \alpha_{ij}} = \frac{d\varepsilon_{ij}^{p}}{d\lambda} \tag{22}$$

then

$$df = -\sum_{ij} \frac{\partial f}{\partial \alpha_{ij}} \cdot d\sigma_{ij} + \sum_{ij} \frac{\partial f}{\partial \alpha_{ij}} \cdot d\alpha_{ij} = 0 \tag{23}$$

Substitution of equation (13) leads to :

$$df = -\sum_{ij} \frac{\partial f}{\partial \alpha_{ij}} \cdot d\sigma_{ij} + \sum_{ij} \frac{\partial f}{\partial \alpha_{ij}} \cdot \left(C_{zz} \cdot d\varepsilon_{ij}^{p} - \gamma_{zz} \cdot \alpha_{ij} \cdot d\varepsilon_{eq/zz}^{p}\right) = 0 \tag{24}$$

Relation (12) gives :

$$f(\bar{\sigma}_{eq/zz}) = 0{,}5 \cdot (F+G) \cdot \bar{\sigma}_{eq/zz}^{\ 2} - 0{,}5 = 0 \tag{25}$$

so $\bar{\sigma}_{eq/zz} = \dfrac{1}{\sqrt{F+G}}$ (26)

It can be noticed that the equivalent reduced stress in z direction is a constant. It can also be explained with consistency condition.

The plastic energy dissipated by a volume unit is

$$dW^{p} = \sum_{ij} \bar{\sigma}_{ij} \cdot d\varepsilon_{ij}^{p} = \bar{\sigma}_{eq/zz} \cdot d\varepsilon_{eq/zz}^{p} \tag{27}$$

however

$$\begin{aligned} dW^{p} &= \sum_{ij} \bar{\sigma}_{ij} \cdot d\varepsilon_{ij}^{p} \\ &= \begin{bmatrix} F \cdot (\bar{\sigma}_{yy} - \bar{\sigma}_{zz})^2 + G \cdot (\bar{\sigma}_{zz} - \bar{\sigma}_{xx})^2 \\ + H \cdot (\bar{\sigma}_{xx} - \bar{\sigma}_{yy})^2 \end{bmatrix} \cdot d\lambda \\ &= \left[2 \cdot f(\bar{\sigma}_{ij}) + 1\right] \cdot d\lambda = d\lambda \end{aligned} \tag{28}$$

with the help of (16) to (19) it can be written, with relation (25) at (27), as :

$$d\varepsilon_{eq/zz}^{p} = \sqrt{F+G} \cdot d\lambda \tag{29}$$

The equation (24) becomes with (15), (22) and (29):

$$-\sum_{ij} \frac{\partial f}{\partial \alpha_{ij}} \cdot d\sigma_{ij} + \sum_{ij} \frac{\partial f}{\partial \alpha_{ij}} \cdot \left[C_{zz} \cdot \left(-\frac{\partial f}{\partial \alpha_{ij}} \cdot d\lambda\right) - \gamma_{zz} \cdot \alpha_{ij} \cdot \sqrt{F+G} \cdot d\lambda\right] = 0$$

$$\sum_{ij} \frac{\partial f}{\partial \alpha_{ij}} \cdot d\sigma_{ij} = \sum_{ij} \frac{\partial f}{\partial \alpha_{ij}} \cdot \left[-C_{zz} \cdot \frac{\partial f}{\partial \alpha_{ij}} - \gamma_{zz} \cdot \alpha_{ij} \cdot \sqrt{F+G}\right] \cdot d\lambda$$

and finally, with simplification :

$$d\lambda = \frac{\dfrac{\partial f}{\partial \alpha_{yy}} d\sigma_{yy} + \dfrac{\partial f}{\partial \alpha_{zz}} d\sigma_{zz}}{\sum_{i} \dfrac{\partial f}{\partial \alpha_{ii}} \cdot \left[-C_{zz} \cdot \dfrac{\partial f}{\partial \alpha_{ii}} - \gamma_{zz} \cdot \alpha_{ii} \cdot \sqrt{F+G}\right]} \tag{30}$$

3.5. Calculation of the plastic strain increment

From equations (17), (18) and (30), plastic strain components can be written as :

$$d\varepsilon_{yy}^{p} = \begin{bmatrix} F \cdot (\bar{\sigma}_{yy} - \bar{\sigma}_{zz}) + \\ H \cdot (\bar{\sigma}_{yy} - \bar{\sigma}_{xx}) \end{bmatrix} \cdot \frac{\dfrac{\partial f}{\partial \alpha_{yy}} \cdot d\sigma_{yy} + \dfrac{\partial f}{\partial \alpha_{zz}} \cdot d\sigma_{zz}}{\sum_{i} \dfrac{\partial f}{\partial \alpha_{ii}} \cdot \left[-C_{zz} \cdot \dfrac{\partial f}{\partial \alpha_{ii}} - \gamma_{zz} \cdot \alpha_{ii} \cdot \sqrt{F+G}\right]} \tag{31}$$

$$d\varepsilon_{zz}^{p} = \begin{bmatrix} F \cdot (\bar{\sigma}_{zz} - \bar{\sigma}_{yy}) + \\ G \cdot (\bar{\sigma}_{zz} - \bar{\sigma}_{xx}) \end{bmatrix} \cdot \frac{\dfrac{\partial f}{\partial \alpha_{yy}} \cdot d\sigma_{yy} + \dfrac{\partial f}{\partial \alpha_{zz}} \cdot d\sigma_{zz}}{\sum_{i} \dfrac{\partial f}{\partial \alpha_{ii}} \cdot \left[-C_{zz} \cdot \dfrac{\partial f}{\partial \alpha_{ii}} - \gamma_{zz} \cdot \alpha_{ii} \cdot \sqrt{F+G}\right]} \tag{32}$$

3.6. Calculation of elastic strain increment

The elastic strain follow the Hooke's relations :

$$d\varepsilon_{yy}^{e} = \frac{1}{E_{yy}(x_{fi})} \cdot d\sigma_{yy} - \frac{\upsilon_{zy}}{E_{zz}(x_{fi})} \cdot d\sigma_{zz} \tag{33}$$

$$d\varepsilon_{zz}^{e} = -\frac{\upsilon_{yz}}{E_{yy}(x_{fi})} \cdot d\sigma_{yy} + \frac{1}{E_{zz}(x_{fi})} \cdot d\sigma_{zz} \tag{34}$$

3.7. Calculation of total strain

The equations of our issue are obtained by writting the equality of the total strain (35) with the one which is calculated in the equation (11) and given by H1 hypothesis :

$$d\varepsilon_{zz}^{p} + d\varepsilon_{zz}^{e} = d\bar{\varepsilon}_{zz} \text{ and } d\varepsilon_{yy}^{p} + d\varepsilon_{yy}^{e} = 0 \tag{35}$$

These two equations, with relations (31) to (34), can be written as :

$$d\bar{\varepsilon}_{yy} = \begin{bmatrix} F \cdot (\bar{\sigma}_{yy} - \bar{\sigma}_{zz}) + \\ H \cdot (\bar{\sigma}_{yy} - \bar{\sigma}_{xx}) \end{bmatrix} \cdot d\lambda + \frac{1}{E_{yy}(x_{fi})} \cdot d\sigma_{yy} - \frac{\upsilon_{zy}}{E_{zz}(x_{fi})} \cdot d\sigma_{zz} = 0 \tag{36}$$

$$d\varepsilon_{zz}=\begin{bmatrix}F\cdot(\bar{\sigma}_{zz}-\bar{\sigma}_{yy})+\\G\cdot(\bar{\sigma}_{zz}-\bar{\sigma}_{xx})\end{bmatrix}\cdot d\lambda-\frac{\upsilon_{yz}}{E_{yy}(x_{fi})}\cdot d\sigma_{yy}+\frac{1}{E_{zz}(x_{fi})}\cdot d\sigma_{zz}$$
$$=\frac{x_{fi}\cdot d\rho}{\rho\cdot(\rho-x_{fi})}+\frac{d\varepsilon_{zz}^{tr}}{(1+\varepsilon_{zz}^{tr})} \qquad (37)$$

From equations (14), (16) at (18) and (29) the expressions of the backstress increment can be extracted:

$$d\alpha_{xx}=d\lambda\cdot\left\{C_{zz}\cdot\begin{bmatrix}G\cdot(\bar{\sigma}_{xx}-\bar{\sigma}_{zz})+\\H\cdot(\bar{\sigma}_{xx}-\bar{\sigma}_{yy})\end{bmatrix}-\gamma_{zz}\cdot\alpha_{xx}\cdot\sqrt{F+G}\right\}$$
$$d\alpha_{yy}=d\lambda\cdot\left\{C_{zz}\cdot\begin{bmatrix}F\cdot(\bar{\sigma}_{yy}-\bar{\sigma}_{zz})+\\H\cdot(\bar{\sigma}_{yy}-\bar{\sigma}_{xx})\end{bmatrix}-\gamma_{zz}\cdot\alpha_{yy}\cdot\sqrt{F+G}\right\} \qquad (38)$$
$$d\alpha_{zz}=d\lambda\cdot\left\{C_{zz}\cdot\begin{bmatrix}F\cdot(\bar{\sigma}_{zz}-\bar{\sigma}_{yy})+\\G\cdot(\bar{\sigma}_{zz}-\bar{\sigma}_{xx})\end{bmatrix}-\gamma_{zz}\cdot\alpha_{zz}\cdot\sqrt{F+G}\right\}$$

A linear system is obtained with five parameters which are $d\sigma_{zz}$, $d\sigma_{yy}$, $d\alpha_{xx}$, $d\alpha_{yy}$, $d\alpha_{zz}$. It is solved for each layer of each identical history strain area. Then increment by increment, the stress state in sheet steel is entirely determined.

3.8. Residual curvature

After the final stress state was entirely computed, the residual curvature of the identical history strain area is calculated to predict springback. The Young's modulus is used for the calculation of the elastic curvature ρ_e which it is necessary to subtract from ρ (imposed curvature) in order to obtain the residual curvature ρ_{res}.

$$\frac{1}{\rho_{res}}=\frac{1}{\rho}+\frac{1}{\rho_e} \quad \text{and} \quad \rho_e=\frac{\int_{-h/2}^{+h/2}E_{zz}(x_{fi})\cdot b\cdot x_{fi}^2\cdot dx_{fi}}{M_f\cdot(1-\upsilon_{yz}\cdot\upsilon_{zy})} \qquad (39)$$

with $M_f=\int_{-h/2}^{+h/2}x_{fi}\cdot\sigma_{zz}\cdot b\cdot dx$ (40)

The complete shape, after springback, is rebuilt with the residual radius of each identical history strain area.

3.9. Calculation of residual stresses

The residual stresses for each layer are given by :

$$\sigma_{yyres}=\sigma_{yy}-\upsilon_{zy}\cdot\frac{E_{yy}}{E_{zz}}\cdot\left(\frac{12\cdot M_f\cdot x_{fi}}{b\cdot h^3}+\sigma_{zz}^{tr}\right) \qquad (41)$$

$$\sigma_{zzres}=\sigma_{zz}-\left(\frac{12\cdot M_f\cdot x_{fi}}{b\cdot h^3}+\sigma_{zz}^{tr}\right) \qquad (42)$$

4. THEORY / EXPERIMENT COMPARISON

The experimentation is a deep drawing with an Ω specimen, as on Fig. 1 [8]. The values of the differents radius are, R_1 = 2 mm, R_2 = infinite and R_3 = 25 mm. The formed sheet was a P280 steel (Orthotopic) and XES (Othotropic), the traction direction was the one of rolling. In order to demonstrate the importance of the utilisation of an orthotropic theorie, we use the two versions of *PLIAGE*. One noted *PLIAGE* doesn't take into account the orthotropic property of the considered sheet. The second noted *PLIAGE* (N), the last version, works with the new model and consider the orthotropy.

4.1. Material's characteristics

In table 1, we present the mechanical characteristic of XES and P280, determined with several tensile tests.

Table 1. Metal's characteristics

	C_{zz} MPa	γ_{zz}	$E_{1\,zz,yy}$ 0% MPa	$E_{2\,zz,yy}$ 5% MPa	r_0	r_{90}	σ_{ezz} MPa
XES	1348	7,4	205000	180000	1.6	1.6	163
P280	1821	9,6	200000	180000	1.5	1.8	287

4.2. Results

In tables 2 and 3 we expose the residual radius of each identical history strain area. Four drawings are made for XES and P280. The residual radius presented in the tables are the mean of the four experimental radius which are mesured with a scanner and image processing. Figures 5 and 6 show the differences in % between calculations and experiments.

4.2.1. XES

Table 2. XES results, the tensile stress is 139 MPa

	Area 1, Rres mm	Area 2, res mm	Area 3, Rres mm
Experiment	3.1	837.5	28.1
PLIAGE	3.4	1197.3	27.1
PLIAGE (N)	3.4	775.2	27.5

4.2.2. P280

Table 3. P280 results, the tensile stress is 160 MPa

	Area 1,Rres mm	Area 2,Rres mm	Area 3,Rres mm
Experiment	3.4	176.9	30.7
PLIAGE	3.5	222.9	29.9
PLIAGE (N)	3.5	189.5	30.6

4.3. Discussion

The mechanical theory implemented in *PLIAGE (N)* gives good results in every case. For area 1, the results are closely connected with the theory. There is no difference between the results given by the two

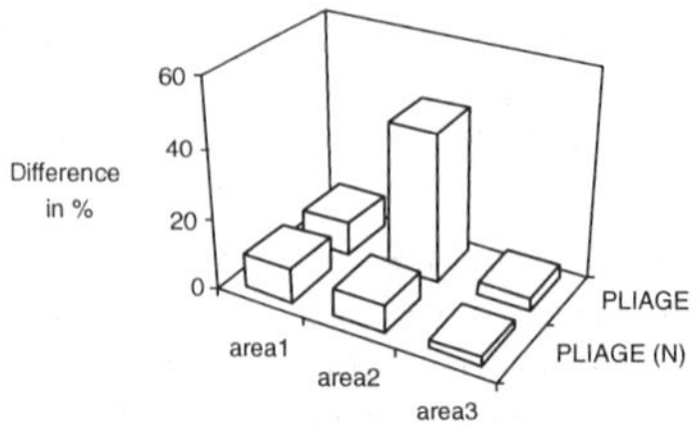

Fig. 5. Differences in % between calculations and experiments for XES.

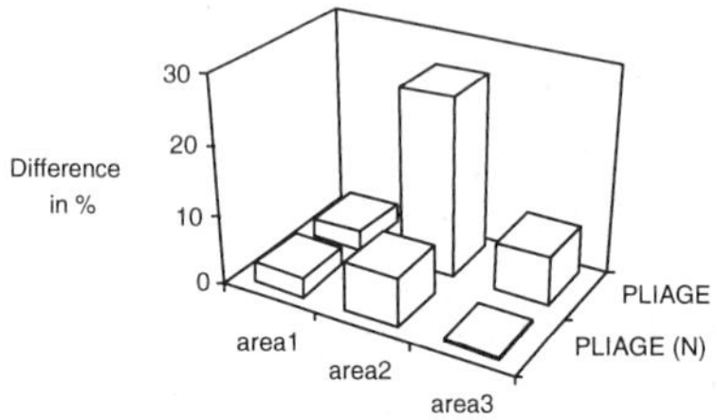

Fig. 6. Differences in % between calculations and experiments for P280.

versions of *PLIAGE* because the strain path is very basic, just one bending, then the orthotropic nature of the sheet is not really significant. The results concerning area 2 are very interesting. Here is demonstrated the necessity to consider the orthotropic nature of the sheet. The first version of *PLIAGE* gives less good results than the new one. For area 3, the results are good too, the punch radius R3 is tall, then there is no difference between the two versions of *PLIAGE* for the reason that we exposed with area 1. Nevertheless, previous works had shown that the first version of *PLIAGE* gives suitable results when the considered steel sheet is isotropic [2].

5. CONCLUSION

This theory applied for the orthotropic sheet gives good results seeing that the exact parameters of the stamping are given. *PLIAGE* was proved and validated by laboratory experiments in this case. *PLIAGE* can be used to simulate, in very short time, the variations of springback versus many parameters as thickness, tensile stress or material properties. *PLIAGE* predicts also the residual stresses, after springback, in the thickness of the steel sheet. The software can also calculate a tool shape, when there is only one bend, in order to obtain the correct geometry against springback. *PLIAGE* is interfaced with the CAD software EUCLID IS since one year and is used by the Metal Forming Process Department of RENAULT.

REFERENCES

[1] LEMS W.
1963 - "The change of Young's modulus after deformation at low temperature and its recovery" Ph. Dissertation, Delft

[2] MORESTIN F., BRUNET M.
1993 - "2D Simulations, B-Sim6" Proceedings of the second international conference NUMISHEET93. Tokyo (Japon), p 623-659.

[3] BOYER J.C., BOIVIN M.
1983 - "Calcul du retour élastique et des contraintes résiduelles lors du pliage des tôles sous tension". Rapport GRECO "Grandes déformations et endommagement", France.

[4] LEMAITRE J., CHABOCHE J.L.
1985 - "Mécanique des matériaux solides" Edition DUNOD, France

[5] MONTHEILLET F., MOUSSY F.
1986 - "Physique et mécanique de l'endommagement". Groupe de reflexion Endommagement GRECO Grandes Déformations et Endommagement. Les éditions de physique, France

[6] MORESTIN F.
1990 - "Préprocesseur informatique pour la détermination des zones de même histoire de déformation pour le logiciel de calcul du retour élastique PLIAGE v4.1" Rapport de DEA Mécanique - INSA de Lyon, France

[7] MARCINIAK Z. and DUNCAN J.
1992 - "Mechanics of sheet metal forming" Edward Arnold ISBN 0-340-56405-9 - London

[8] MORESTIN F.
1993 - "Contribution à l'étude du retour élastique lors de la mise en forme.des produits plats" Mémoire de Thèse de Mécanique. INSA de Lyon, 180 p.

[9] MORESTIN F., BOIVIN M.
1993 - "Variation of the Young's modulus with plastic strain, applying to elastoplastic software". 12th International Conference on Structural Mechanics in Reactor Technology. Stuttgart (Allemagne), 1993, Volume L. p. 141-146.

Simulation of Materials Processing: Theory, Methods and Applications, Shen & Dawson (eds)
© 1995 Balkema, Rotterdam. ISBN 90 5410 553 4

Deep drawing of square boxes: Analysis of the influence of geometrical parameters by numerical simulations and experimental tests

L. Fratini, S. Lo Casto & F. Micari
Dipartimento di Tecnologia e Produzione Meccanica, University of Palermo, Italy

ABSTRACT: In the paper the deep drawing process of square boxes is analysed by means of an explicit numerical model in order to highlight the influence of the geometrical parameters; in particular different punch and die radii have been taken into account, finding out conditions able to guarentee a successful execution of the process or, in turn, the insurgence of defects. The numerical results have been compared with the corresponding ones obtained by means of a set of experimental tests, showing a good fitting between the numerical prediction and the experimental verification.

1. INTRODUCTION

Deep drawing of non circular components is one of the most employed sheet forming operations, which is frequently applied in many industrial fields such as the mechanical, electrical, electronical and the aeronautical ones. Among the non-asymmetric deep drawing processes, the square box drawing is the most basic operation: this process is characterised by a complex deformation mechanics which differs appreciably from the one observed in the axysimmetrical drawing process, and which is almost impossible to be studied by means of the traditional analytical methods.

One of the most important problems in box drawing is the variation of metal flow rates along the straight walls and around the corners of the deforming flange; as a consequence of this, an uneven material distribution around the box walls is obtained. Only recently, by bonding or electrically etching small circles on the flat blank, details of metal flow have been revealed, as the circles deform into ellipses [1].

These observations have suggested that the process mechanics can be investigated by subdividing the part shape into basic contours. In this way the four box corners are analysed together, representing a cup shape complexively; on the other hand the four straight sides or walls present a similar behaviour [2], and consequently can be studied following the same considerations.

As a corner of the box is regarded, it represents one fourth of a cup: consequently the metal must undergo circumferential compressive stresses during the drawing process, to allow the material to flow over the die radius. As in cup drawing, defects such as wrinkling may occur due to the above stress state; to prevent these problems, the use of a blankholder is needed, as well as a good drawing lubricant to make easier the metal flow underneath the blankholder by reducing friction. At the same time, as the tensile load in the corners is too severe, the metal tears as in cup drawing and this defect can progress into the flat wall.

On the other hand, as the flat wall is regarded, no compression of metal is required to permit metal flow toward the die radius and in this way the process cannot properly be referred to as drawing. The process mechanics which occurs at the straight box walls is usually called bend and straighten: in fact after a first slip toward the die radius, the metal is bent and then it is straightened to create the flat wall of the box.

Only recently the application of the Finite Element Method has allowed a deeper knowledge of the stress states and the strain distributions occurring during the operation; by this way the deep drawing process of square boxes has been analysed by several researches in order to determine the formability conditions of the sheet as well as the optimal choice of the blank design [3-7]. Anyway, the influence of the die geometrical parameters on the process mechanics remained partially uninvestigated. On the other hand these parameters are very important in order to optimise the deep drawing technology and to avoid the insurgence of defects such as wrinkles or cracks which can affect the deep drawing processes.

For this reason in the paper the authors carry out a numerical analysis of the deep drawing process of square boxes with the aim to individuate the influence of the geometrical parameters on the metal flow. The numerical analysis is based on a finite element approach and in particular an explicit model, which is based on the analysis of the dynamic response of the workpiece under deformation, has been employed.

The finite element simulations have been carried out for several different punch and die radii values and it has been possible to find out both the geometrical conditions able to guarentee a successful execution of the process and the ones which cause the insurgence of defects such as tearing or wrinkling. The obtained results have been verified by means of a set of experimental tests in which several blanks have been drawn into square boxes.

2. THE NUMERICAL MODEL

The deep drawing process of square boxes has been analysed by means of a finite element explicit approach. This type of formulation has been chosen on the basis of the following considerations:

- a complex 3-D analysis is requested: in this case the explicit models offer clear advantages with respect to the implicit ones in terms of CPU times, since in order to calculate the vector of the nodal displacement at the end of each step of the deformation process only the solution of a set of linear and independent equations is required;
- one of the aim of the research is represented by the individuation of the geometrical conditions which cause the insurgence of shape defects in the flange or cracks in the wall of the drawn box: if the numerical simulation is carried out employing an implicit model it is not possible to receive clear indications about the occurrence of the above defects, since when plastic collapse occurs or instability problems arise the stiffness matrix becomes non-positive definite and the analysis stops. On the other hand the explicit codes do not require the inversion of the stiffness matrix and, consequently, the analysis goes on and supplies an easily observable prediction of defects;
- finally the analysed process is characterised by complex contact conditions: the contact algorithm must be able to update the boundary conditions taking into account the nodes which leave or come in contact with the tools. In this case the use of an implicit model requires a very small value of the time increment to be used in the analysis in order to avoid problems of uncertain contact which do not allow to the iterative procedure to converge; consequently a large number of steps is necessary to simulate the process and for each of them the construction and the inversion of the stiffness matrix is required with large CPU times and costs. On the other hand, the explicit models are based on a time integration scheme which makes the algorithm conditionally stable: consequently small time increments are always required, allowing a detailed analysis of the contact conditions.

In the following the main features of the employed explicit approach will be shown; more details can be found in the references [8-11].

The explicit approach is based on the equilibrium equations governing the dynamic response of a finite elements system:

$$[M]\{\ddot{U}\}+[C]\{\dot{U}\}+[K]\{U\}=\{R\} \qquad (1)$$

where [M], [C] and [K] are the mass, damping and stiffness matrices respectively; $\{R\}$ is the external load vector; $\{U\}$ $\{\dot{U}\}$ and $\{\ddot{U}\}$ are the displacement, velocity and acceleration vectors of the finite element assemblage.

The most important points which characterise an explicit approach can be summarised as follows:

- the central difference method is used as time integration scheme in order to solve equation (1);
- damping effects are generally neglected;
- a diagonal lumped mass matrix is assumed.

The use of the above assumptions allows to write the following governing finite element equation:

$$\{{}^{t+\Delta t}U\}=2\{{}^{t}U\}-\{{}^{t-\Delta t}U\}+[M]^{-1}\Delta t^{2}\left({}^{t}R-{}^{t}F_{int}\right) \qquad (2)$$

where the external loads vector and the internal forces have been pointed out (they are both known because they are referred to time t).

Equation (2) shows that in order to obtain the nodal displacements at the discrete time t+Δt the inversion of the stiffness matrix is no more required; the final equations are linear and, due to the assumption of a "lumped" mass matrix, they are independent and can consequently be solved one by one. A direct consequence of all these benefits can be found out in the reduced CPU time necessary for the simulations.

On the other hand, the most important shortcoming of the use of the central difference method as the time integration scheme is that this procedure is conditionally stable and requires the choice of a time increment Δt lower than the following critical value:

$$\Delta t_{cr} = \frac{L}{c} = \frac{L}{\sqrt{E/\rho}} \qquad (3)$$

being L the typical dimension of the element and c the sound speed in the workpiece material, which can be calculated on the basis of the material Young modulus E and the density ρ.

The stable time increments are typically very small compared to the natural time of forming processes and consequently a very large number of increments should be necessary to analyse the deformation path. In order to increase the simulation speed the die velocity has been artificially increased. Such a procedure requires a very accurate checking, since by this way the inertia effects augment their relevance.

3. THE EXPERIMENTAL TESTS

The material used in the tests was mild steel for deep drawing, received in the form of flat rolled drawing quality special killed sheets, 0.5mm by 2.0m by 1.0m, ASTM A622M type with 0.08% carbon (silicon 0.035%, manganese 0.48%, phosphorous 0.03%). The sheets were deep-drawn in the state as received. The uniformity of thickness (0.5mm) was checked: in particular the initial thickness distribution has shown a maximum discrepancy lower than 0.01mm. The square blanks, with a side of 80.0mm, were cut starting from strips the dimensions of which were 0.5mm by 80.0mm by 1.0m.

The blanks were deep-drawn to assume the shape of a square box by an inverted die shown in Fig. 1. The die is characterised by a square section (50.0mm edge) and presents a clearance of 0.55mm for side and a vertical corner radius equal to 8.0mm; the values of the punch radius Rp and of the die radius Rd are the variable geometrical parameters the influence of which on the process mechanics has been studied in the present research. In particular the values of the variable geometrical parameters Rp and Rd have been chosen according to the following experimental design: 3.0mm, 6.0mm and 9.0mm. One hundred tests were carried out for each couple of the above mentioned geometrical parameters Rp and Rd; it should be outlined that no cross tests have been performed, but in each set of tests the punch and the die had the same value of the radius. In a previous research on deep-drawing of axysimmetric

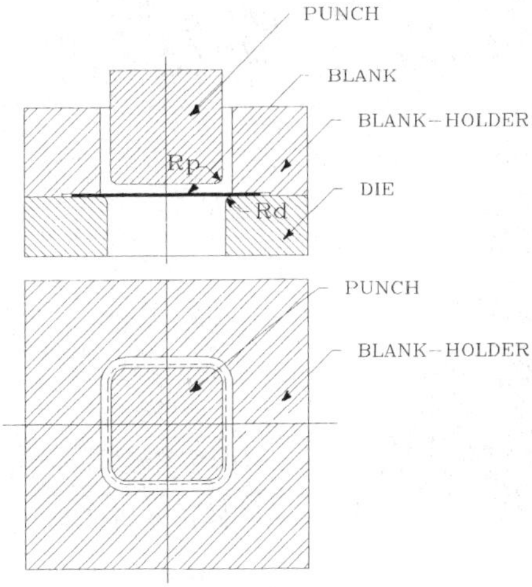

Fig. 1

Fig. 2 - The die mounted in the press by a die set

components [12] in fact, it has been found that the process mechanics is mainly governed by the value of the die radius, and that, fixed a particular value of Rd, the same results were obtained for all the Rp investigated.

The die was assembled in a die set with a spring activated blank-holder. Fig. 2 shows the die mounted in the press by a die set. The employed press was a single action screw press for tensile or compression tests with slow ram speed equal to 2.0mm/s. The square shaped box were deep-drawn up to punch displacement able to supply a box height equal to 16.0mm.

The boxes were checked for the presence of defects. All the deep-drawn components with Rp=Rd=9.0mm have shown the presence of wrinkles

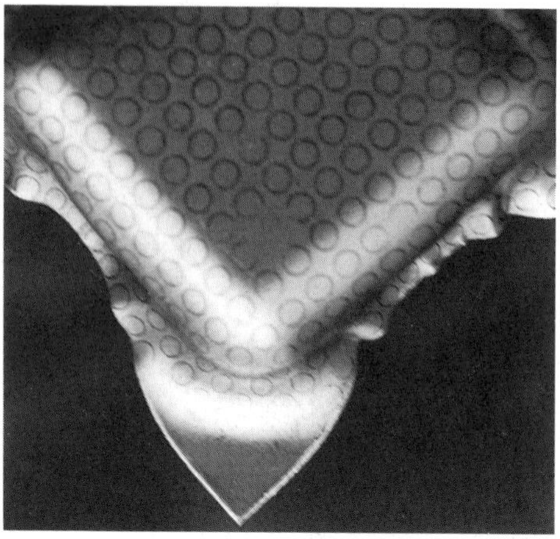

Fig. 3 - A deep drawn box with Rp=Rd=9.0mm

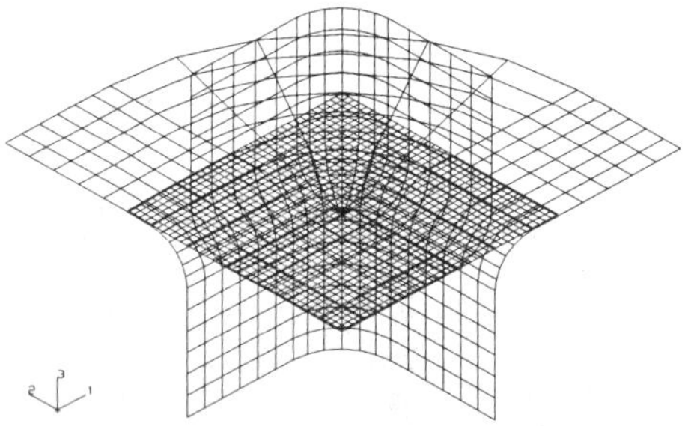

Fig. 4 The 3-D model. Rp=Rd=6.0mm

on the box flange, as can be observed in fig.3. On the other hand a significant thinning in the cup wall has been found for all the deep-drawn boxes with Rp=Rd=3.0mm; in particular for most of them this defect increased up to give rise to tearing.

Finally the choice of a punch and a die radius equal to 6.0mm allowed to obtain sound components, with neither wrinkles in the flange nor cracks in the box wall.

4. THE NUMERICAL RESULTS

The numerical simulations over the three different geometries have been carried out by means of full 3-D explicit analyses; due to simmetry considerations only a quarter of the system has been discretized by means of 1225 shell elements with 6 degrees of freedom per node as far as the blank is regarded, while 234, 144 and 324 four node rigid elements have been employed for the punch, the blankholder and the die respectively. Fig.4 shows the model for

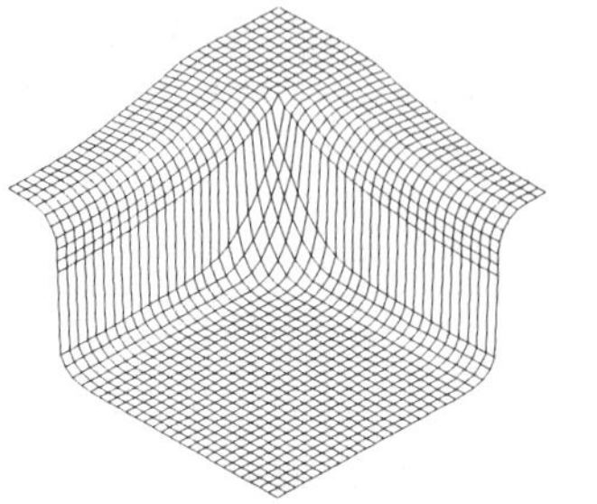

Fig. 5 The deformed mesh: Rp=Rd=3.0mm

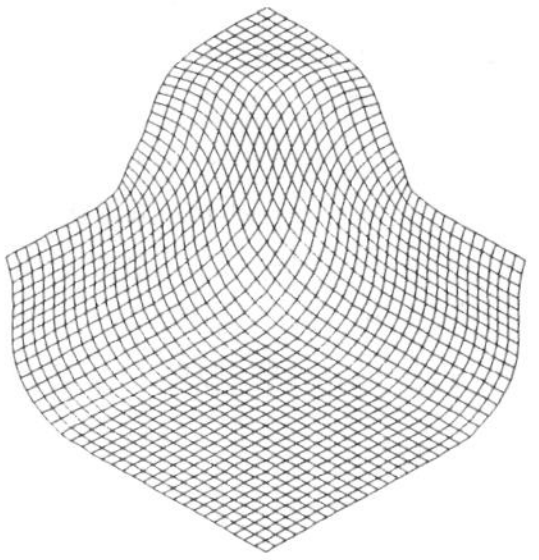

Fig. 6 The deformed mesh: Rp=Rd=6.0mm

the geometry with die and punch radii of 6.0mm: in order to simplify the figure only the blank, the punch and the die have been reported.

The non linear behaviour of the blank material has been accounted by means of the flow stress expression $\sigma_{eq} = 584\varepsilon_{eq}^{0.11}$ [N/mm^2]. Frictional effects have been considered by the Coulomb model with a friction coefficient equal to 0.25 at the punch-sheet, 0.125 at the die-sheet and finally 0.05 at the blankholder-sheet interface respectively.

The analysis of the process by an explicit model requires a proper choice of the time step increment: in particular the critical value of Δt was of the order of $5*10^{-7}$ s. Consequently the time step increment was assumed equal to $1*10^{-7}$ s. The punch velocity was fixed to 8.0m/s, three orders larger than the actual value. The total CPU time, necessary in order to complete the simulation, was about 6 hours and 30 minutes.

Figg.5-7 report the final deformed meshes obtained for the three geometrical conditions taken into account; the numerical results are in good agreement with the experimental ones: with Rp=Rd=3.0mm the elements in the box wall are largely stretched, while for Rp=Rd=9.0mm wrinkles occur in the box flange. Finally for Rp=Rd=6.0mm the numerical simulations do not show any occurrence of defects.

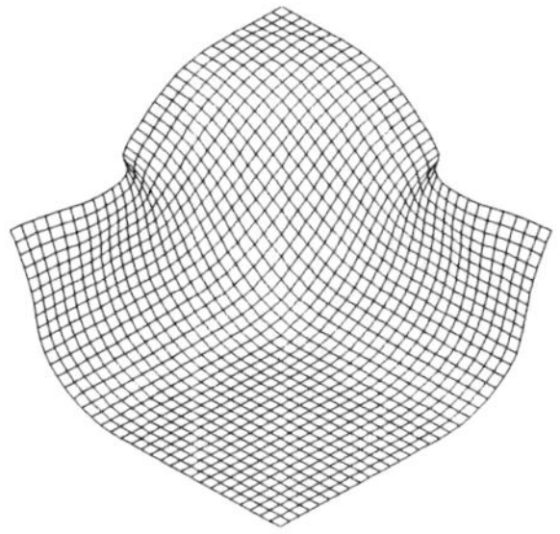

Fig. 7 - The deformed mesh: Rp=Rd=9.0mm

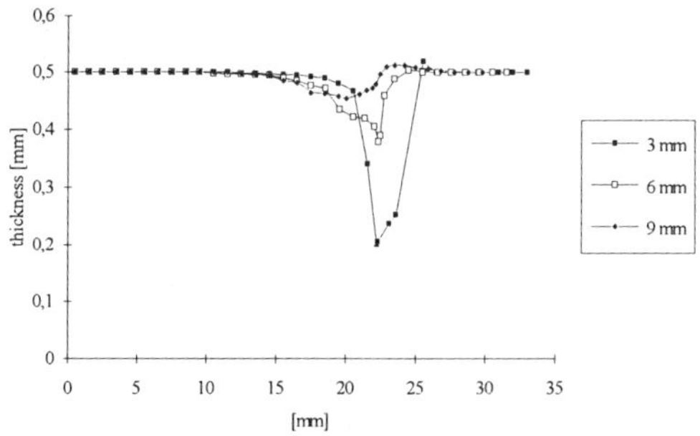

Fig. 8 - Thickness trend vs. the distance from the centre of the blank on the main diagonal

The above results are confirmed by the trend of the thicknesses measured on the main diagonal of the drawn box for the three analysed geometrical conditions: in the zone close to the corner of the box a large reduction of the thickness is observed for Rp=Rd=3.0mm, while it decreases at increasing the punch and the die radii.

5. ACKNOWLEDGEMENTS

This work has been made using MURST (Italian Ministry for University and Scientific Research) funds.

REFERENCES

[1] Doege, E., Boinski, F. 1994. Computer-aided design and calculation of deep drawn components. *Journal of Materials Processing Technology*. 46: 321-331.

[2] Eary, D. F., Reed, E. A. 1974. *Techniques of pressworking sheet metal*. Engewood Cliffs, New Jersey, Prentice-Hall.

[3] Kapinski, S. 1992. Influence of the punch velocity on deformation of the material in deep drawn flange. *Journal of Materials Processing Technology*. 34: 419-424.

[4] Doege, E., Sommer, N. 1987. Blank-holder pressure and blank-holder layout in deep drawing of thin sheet metal. *Advanced Technology of plasticity*. 11: 1305-1314.

[5] Sosnowski, W., Onate, E., Agelet de Saracibar, C. 1992. Comparative study on sheet metal forming processes by numerical modelling and experiment. *Journal of Materials Processing Technology*. 34: 109-116.

[6] Toh, C. H., Kobayashi, S. 1985. Deformation analysis and blank design in square cup drawing. *Int J. Mach. Tool Des. Res.* 25: 15-32.

[7] Kim, N., Kobayashi, S. 1986. Blank design in rectangular cup drawing by an approximate method. *Int J. Mach. Tool Des. Res.* 26: 125-135.

[8] Bathe, K. J. 1982. *Finite element procedures in engineering analysis*. Prentice-Hall

[9] Rebelo, N., Nagtegaal, J. C., Taylor, L. M., Passmann, R. 1992. Comparison of implicit and explicit finite element methods in the simulation of metal forming processes: *Proceedings of NUMIFORM '92*, pp. 99-108.

[10] Onate, E., Agelet de Saracibar, C. 1992. Alternatives for finite element analysis of sheet in metal forming problems: *Proceedings of NUMIFORM '92*, pp. 79-88.

[11] Alberti, N., Cannizzaro, L., Fratini, L., Micari, F. 1994. An explicit model for the analysis of bulk metal forming processes: *Trans. of NAMRI/SME*, pp. 11-16

[12] Alaimo, F., Lo Casto, S., Micari, F., Sunseri, M. 1994. Influence of the die and punch radii on deep drawing: Numerical analysis and experimental tests: *Proc. of CAPE 10*, pp. 177-187

Simulation of Materials Processing: Theory, Methods and Applications, Shen & Dawson (eds)
© 1995 Balkema, Rotterdam. ISBN 90 5410 553 4

A new analysis and verification method of the axisymmetrical deep drawing process

Bjarne G. Hansen, Niels Bay & Michael P. Malberg
Technical University of Denmark, Institute of Manufacturing Engineering, Lyngby, Denmark

ABTRACT: A new finite difference analysis of the axisymmetric deep drawing process is presented and experimentally verified. In the analysis the influence of strain hardening, axisymmetric anisotropy and the real contact area between the flange and the blankholder/die is taken into account. Furthermore a new method for validating the reliability of the analysis based on a comparison between calculated and experimentally measued thickness distribution of the cup wall is proposed. Finally a short discription of a sheet metal forming software program in which the analysis is incoorperated will be presented.

INTRODUCTION

The deep drawing process is a well known and very common sheet metal forming process in industry. Many simple as well as more complex shapes are deep drawn for boxes, cans etc.

A number of theoretical investigations of deep drawing have been developed over the years starting in the 1920-ies with the analyses of Sommer [1] and Siebel and Pomp [2]. A more comprehensive analysis was developed by Chung and Swift [3] in 1951. In this analysis the thickness of the flange was assumed to decrease from the rim towards the die corner and the blank-holder force was assumed to act only on the rim of the flange during drawing. The influence of material anisotropy was neglected. Later Woo [4] refined the analysis and implemented it on a finite difference form. The most important improvement compared to the previous analysis was that the blank-holder force was distributed over a certain area near the rim of the flange instead of at the rim itself. These improvements gave considerable better results when comparing with experiments.

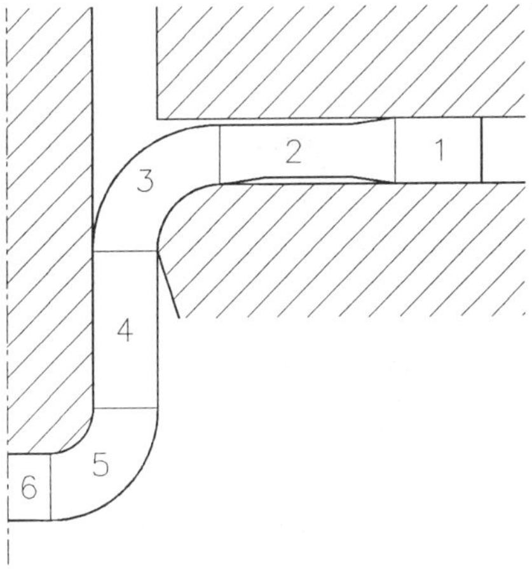

Fig. 1 The deep drawing process divided into six zones

BACKGROUND

The present analysis is based on the Finite Difference Method described by Woo [4] adopting the slab method of analysis. The influence of axisymmetrical anisotropy and the real contact area between flange and blankholder/die is taken into account.

The deep drawing process is divided into six different zones, Fig. 1.

In zone 1 the material is in contact with both the blankholder and the die and loaded by the blank-holder pressure. The sheet thickness in this zone is uniform.

In zone 2 the material is not in contact with the blankholder. The thickness of the material decreases gradually when moving towards center of the die.

In zone 3 the material is drawn over the die corner profile while the material in zone 4 is stretched in axial direction. In zone 5 the material is bended and stretched over the punch nose and in zone 6 the material is stretched over the flat punch face.

In the present analysis the most important zones, 1 - 3 are dealt with.

THE PRESENT ANALYSIS

General calculation method

The calculation starts by dividing the blank into a specified number of ring elements and assume an incremental drawing-in of the flange. For each ring element the force equilibrium is expressed starting with the element at the rim of the blank. After the calculation of the stress - strain state for the first element, the boundary conditions for the inner surface of this element are used as boundary condition for the surface of the next ring element. This procedure is continued until a stop criteria is reached. After this initial incremental drawning-in, the blank is once more applied with an incremental drawing-in and the calculation is again started from the rim going inwards. The whole procedure is repeated until the outer radius of the flange reaches the die profile radius. At this point the calculation is stopped and complete information about stresses, strains and forces are now known for all elements during the whole deformation process.

Physical relations

Expressions of the force equilibrium for the elements depend on the position of the elements in relation to the zones described previously. Before establishing the equilibrium equations for each zone some general equations are presented.

In the following incremental analysis the notation shown in Fig. 2 is used.

$$\sigma_{r,i}^{j}$$

Fig. 2 Notation used throughout the description of the analysis

where:

- j: stage number
- r: stress or strain type
  - r: radial direction
  - θ: circumferential direction
  - t: thickness direction
- i: element number

The incremental strains for the actual stage can be written as:

$$\Delta\varepsilon_{\theta,i}^{j} = \varepsilon_{\theta,i}^{j} - \varepsilon_{\theta,i}^{j-1},\quad \Delta\varepsilon_{t,i}^{j} = \varepsilon_{t,i}^{j} - \varepsilon_{t,i}^{j-1},\quad \Delta\varepsilon_{r,i}^{j} = -\Delta\varepsilon_{\theta,i}^{j} - \Delta\varepsilon_{t,i}^{j} \tag{1}$$

Inserting the incremental strains into the expression of the equivalent strain one obtains:

$$\Delta\varepsilon_{ækv,i}^{j} = \frac{1+R}{\sqrt{1+2R}}\sqrt{(\Delta\varepsilon_{r,i}^{j})^2 + \frac{2R}{1+R}\Delta\varepsilon_{r,i}^{j}\,\Delta\varepsilon_{\theta,i}^{j} + (\Delta\varepsilon_{\theta,i}^{j})^2} \tag{2}$$

The true strains are given by:

$$\Delta\varepsilon_{\theta,i}^{j} = \ln\left(\frac{r_i^j}{R_i}\right),\quad \Delta\varepsilon_{t,i}^{j} = \ln\left(\frac{t_i^j}{t_0}\right) \tag{3}$$

where R_i og t_0 are initial radius and initial thickness of the actual ring-element and r and t the local radius and thickness.

The flow-law for axisymmetrical anisotropic plate material is given by:

$$\frac{d\varepsilon_{ækv}}{(1+R)\sigma_{ækv}} = \frac{d\varepsilon_r}{(1+R)\sigma_r - R\sigma_\theta - \sigma_t} \tag{4}$$

$$\frac{d\varepsilon_{ækv}}{(1+R)\sigma_{ækv}} = \frac{d\varepsilon_\theta}{(1+R)\sigma_\theta - R\sigma_r - \sigma_t} \tag{5}$$

$$\frac{d\varepsilon_{ækv}}{(1+R)\sigma_{ækv}} = \frac{d\varepsilon_t}{2\sigma_t - \sigma_r - \sigma_\theta} \tag{6}$$

On incremental form this gives:

$$(\sigma_{\theta,i}^{j} - \sigma_{r,i}^{j}) = \frac{1+R}{1+2R}\frac{\sigma_{ækv,i}^{j}}{\Delta\varepsilon_{ækv,i}^{j}}(\Delta\varepsilon_{t,i}^{j} + 2\Delta\varepsilon_{\theta,i}^{j}) \tag{7}$$

$$(\sigma_{t,i}^{j} - \sigma_{r,i}^{j}) = \frac{1+R}{1+2R}\frac{\sigma_{ækv,i}^{j}}{\Delta\varepsilon_{ækv,i}^{j}}((1+R)\,\Delta\varepsilon_{t,i}^{j} + \Delta\varepsilon_{\theta,i}^{j}) \tag{8}$$

The stress - strain relation used in the calculation follows the Swift expression:

$$\sigma_{ækv} = C(B+\varepsilon_{ækv})^n \tag{9}$$

MODEL FOR ZONE 1

In this zone the following assumptions are made:

- The thickness of the flange is constant
- The blankholder force is known
- The radial stress $\sigma_r = 0$ at the rim

Volume constancy states:

$$(R_i^2 - R_{i+1}^2)\,t_o = (r_i^2 - r_{i+1}^2)^j\,t_1 \tag{10}$$

This is used to determine the current radius r_i:

$$r_{i+1}^j = \sqrt{(r_i^j)^2 - \frac{(R_i^2 - R_{i+1}^2)\,t_0}{t_1}} \tag{11}$$

The force equilibrium for the element in zone 1 can be written as, see Fig. 3:

$$r\,\sigma_r\,d\theta + 2\,\sigma_\theta\frac{d\theta}{2}dr = \frac{-2\mu\,\sigma_t\,r\,d\theta\,dr}{t_1} + (r+dr)(\sigma_r + d\,\sigma_r)d\theta \tag{12}$$

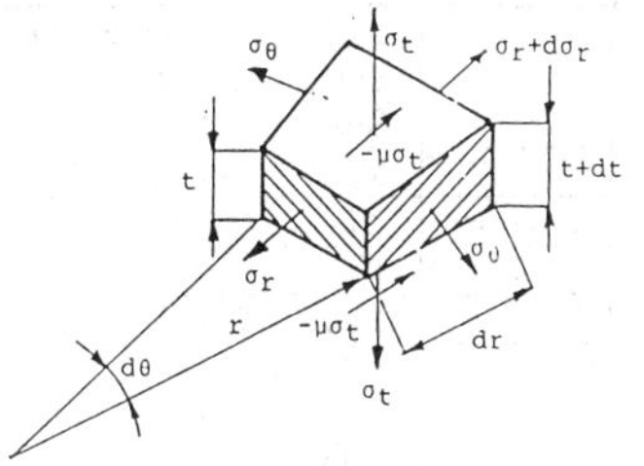

Fig. 3 Force equilibrium for the ring-element in zone 1

assuming $drd\sigma_r \approx 0$ Eq. (12) can be reduced to:

$$d\sigma_r = \frac{\sigma_\theta - \sigma_r}{r}dr + \frac{2\mu\,\sigma_t}{t_l}dr \tag{13}$$

which is written on incremental form as:

$$\sigma^j_{r,i+1} = \sigma^j_{r,i} + \int_{r_i}^{r_{i+1}}\left(\frac{\sigma_\theta - \sigma_r}{r}\right)^j dr + \frac{2\mu}{t_l}\int_{r_i}^{r_{i+1}}\sigma^j_t dr \tag{14}$$

Eq. (14) can be numerically integrated by using the trapezoidal rule:

$$\sigma^j_{r,i+1} = \frac{\sigma^j_{r,i} + \left[\frac{1}{2}\left[\left(\frac{\sigma_{\theta,i}-\sigma_{r,i}}{r}\right)^j_{i+1} + \left(\frac{\sigma_{\theta,i}-\sigma_{r,i}}{r}\right)^j_i\right] + \frac{\mu}{t_l}\left(\sigma^j_{t,i} + (\sigma_t-\sigma_r)^j_{i+1}\right)\right](r_{i+1}-r_i)^j}{1-\frac{\mu}{t_l}(r_{i+1}-r_i)^j} \tag{15}$$

In Eq. (15) the parameters on the right hand side are all known implying that $\sigma^j_{r,i+1}$ can be calculated. Inserting this calculated stress in Eqs. (7) and (8) $\sigma^j_{\theta,i+1}$ and $\sigma^j_{t,i+1}$ can be determined. Then the unit reactive force can be calculated:

$$\Delta F' = 2\pi\, r^j_{i+1}\sigma^j_{t,i+1}(r_i - r_{i+1})^j \tag{16}$$

The calculations continue in the next element and stops when one of the following conditions are reached:

- σ_t becomes less than 0
- r_i becomes less than the die bore radius.

By summing up the unit reactive force increments the total force $\Sigma F'$ can be obtained. This force is compared with the blankholder force and if the discrepancy between these two forces is satisfactory the calculations in zone 1 for this stage are finished. If not the thickness is changed and the calculations for this zone are performed again.

MODEL FOR ZONE 2

In this zone the following assumptions are made:

- The thickness is varying
- Plane stress conditions exist $\sigma_t = 0$.

The boundary conditions from the last element from zone 1 are used as input boundary condition for the first element in zone 2. The force equilibrium is almost identical with the one in zone 1, except that there is no friction in zone 2:

$$(\sigma_r t)^j_{i+1} = (\sigma_r t)^j_i + \int_{r_i}^{r_{i+1}}\left(\frac{(\sigma_{\theta,i}-\sigma_{r,i})^j\, t}{r}\right)dr \tag{17}$$

which by use of the trapezoidal rule gives:

$$(\sigma_r t)^j_{i+1} = (\sigma_r t)^j_i + \frac{1}{2}\left[\left(\frac{(\sigma_{\theta,i}-\sigma_{r,i})t}{r}\right)^j_{i+1} + \left(\frac{(\sigma_{\theta,i}-\sigma_{r,i})t}{r}\right)^j_i\right](r_{i+1}-r_i)^j \tag{18}$$

By volume constancy the correlation between the radius and the thickness in zone 2 becomes:

$$(R_i^2 - R_{i+1}^2)\,t_0 = (r_i^2 - r_{i+1}^2)\left(\frac{t_i + t_{i+1}}{2}\right) \tag{19}$$

From Eq. (19) r_{i+1} is determined, and from Eqs. (1) - (8) the strains: $\varepsilon^j_{t,i+1}, \varepsilon^j_{\theta,i+1}, \varepsilon^j_{r,i+1}, \varepsilon^j_{ækv,i+1}$, and the stresses: $\sigma^j_{ækv,i+1}, (\sigma_{\theta,i+1}-\sigma_{r,i+1})^j$ and $(\sigma_{t,i+1}-\sigma_{r,i+1})^j$ are determined. $\sigma^j_{r,i+1}$ is found inserting $\sigma_t = 0$. From Eq. (18) $(\sigma_r t)^j_{r,i+1}$ is calculated and this expression is compared with $\sigma_{r,i+1} \bullet t^j_{i+1}$. If the thickness t^j_{i+1} is assumed to be correct the next unit element can be calculated. If not the thickness is changed and the calculations for this element redone. The calculations in zone 2 are stopped when the die radius profile is reached.

MODEL FOR ZONE 3

In this zone the force equilibrium expressed on finite difference form is, see Fig. 4:

$$(\sigma_r t)^j_{i+1} = (\sigma_r t)^j_i - \int_{\theta_i}^{\theta_{i+1}}\left[\sigma^j_\theta\, t(\cos\theta + \mu\sin\theta) - \sigma^j_r\, t\cos\theta\right] * \frac{\rho'_1 d\theta}{r_c - \rho'_1\sin\theta} + \int_{\theta_i}^{\theta_{i+1}}\mu\,\sigma^j_r\, t d\theta \tag{20}$$

where $\rho'_1 = \rho_1 + \frac{t}{2}$ and

r_c = radius in the die + die profile radius

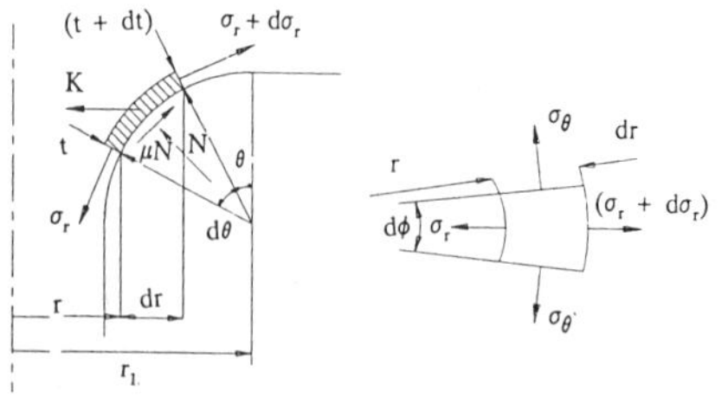

Fig. 4 Force equilibrium at the rounding of the die

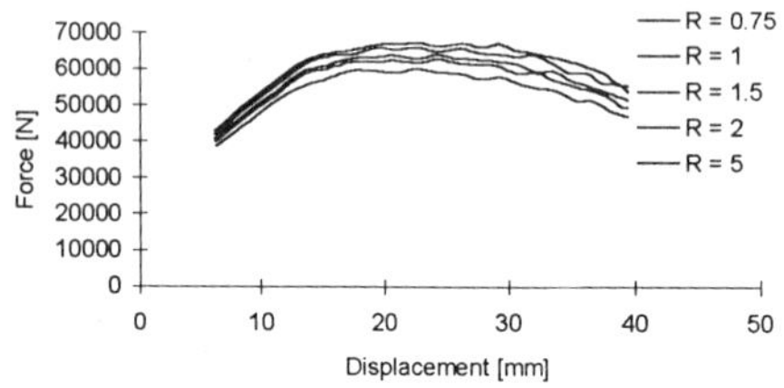

Fig. 5 Load - displacement curves with the normal anisotropi as parameter

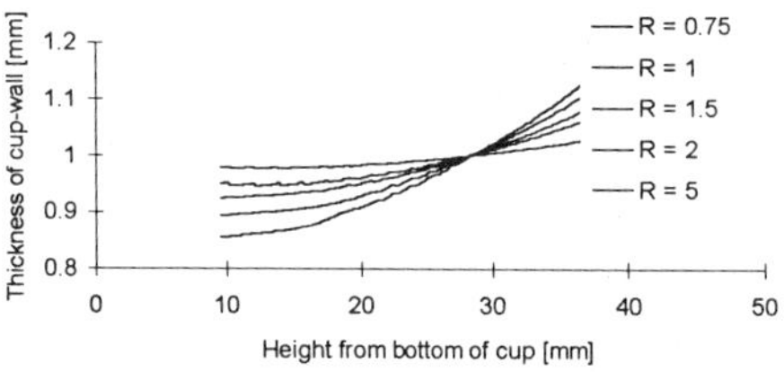

Fig. 6 Thickness distributon of the cup-wall with the normal anisotropi as parameter

From the volume constancy it is given that:

$$\left(R_i^2 - R_{i+1}^2\right) t_0 = 2\rho_1'^2\left[\frac{r_c}{\rho_1'}\left(\theta_{i+1}-\theta_i\right) + \left(\cos\theta_{i+1}+\cos\theta_i\right)\right]\frac{t_i + t_{i+1}}{2} \quad (21)$$

This is used to determine θ_{i+1}, setting the radius r_{i+1}, = $\left(r_c - \rho_1' \sin\theta_{i+1}\right)$. By use of Eq. (21) R_{i+1} can be calculated. From Eqs. (1) - (8) the strains: $\varepsilon^j_{t,i+1}, \varepsilon^j_{\theta,i+1}, \varepsilon^j_{r,i+1}, \varepsilon^j_{\text{ækv},i+1}$ and the stresses: $\sigma^j_{\text{ækv},i+1}, \left(\sigma_\theta - \sigma_r\right)^j, \left(\sigma_t - \sigma_r\right)$ can be found. As in zone 2, σ_t is assumed to be 0 from which $\sigma^j_{r,i+1}$ can be calculated. $\left(\sigma_r t\right)_{i+1}$ is found by Eq. (20) and compared with $\sigma^j_{r,i+1} \bullet t^j_{i+1}$. If the discrepancy is satisfactory small the applied thickness is adopted as the correct one, and calculations continue with the next element, otherwise the thickness is modified. The calculation in zone 3 stops when the current angle reaches $\pi/4$. The punch force can then be calculated as:

$$P = 2\pi r\left(\sigma_r t\right)\sin\theta \quad (22)$$

		Conv. D.D. Steel	Ferritic st. less steel	Ferritic st. less steel	Tita-nium
Number		st.14	1.4016	1.4520	-
Swift	C [N/mm2]	557	809	771	495
	n	0.25	0.18	0.25	0.15
	B	0.003	0.012	0.007	0.005
Anisotropy	0 °	1.87	0.55	1.08	6.43
	45 °	1.20	0.62	0.87	3.83
	90 °	1.58	1.28	1.72	1.77
	Normal	1.46	0.77	1.14	3.97
	Plane	0.53	0.30	0.53	0.27

Fig. 7 Materials data

LDR	Conv. D.D. Steel	Ferritic stainless steel	Ferritic stainless steel	Titan
	1.1403	1.4016	1.4520	-
Experiments	2.27	2.16	2.16	2.08
Theory	2.19	2.14	2.19	2.07

Fig. 8 Limit Drawing Ratio (LDR) for different materials

INFLUENCE OF ANISOTROPY

Very often the load - displacement curve from a deformation process is used to validate the reliability of a theoretical analysis. In many cases this method is satisfactory but in case of analysis of deep drawing of anisotropic material the validity of the analysis may not be sufficiently discovered. In Fig. 5 the load - displacement curves for deep drawing with varying anisotripy are shown calculated by use of the new analysis. It is noticed that the influence of the anisotropy rather limited.

A much more pronounced effect is seen when estimating the wall thickness distribution in the drawn cup as a function of the anisotropy, Fig. 6. In order to validate the present analysis this observation is applied in the following experimental verification.

EXPERIMENTAL VERIFICATION

To verify the theoretical analysis a series of experiments have been performed applying a conventional tool for deep drawing equipped with a data acquisition system connected to a PC computer. The data acquisition system is capable of measuring the punch travel, the blankholder force and the punch load. Besides these parameters geometrical parameters of the deformed cup have been measured comprising the thickness of the cup-wall, strains in the cup flange and springback of the cup-wall.

The theoretical analysis makes it possible to determine the strain and stress distribution, the punch load, the thickness distribution of the cupwall and the Limit Drawing Ratio (LDR).

Due to limited space only a few representative

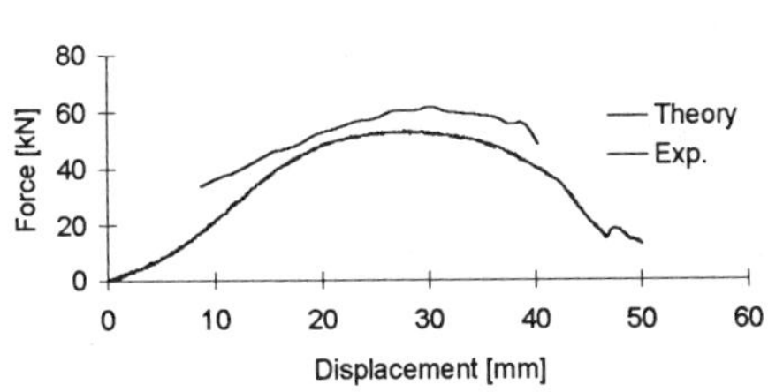

Fig. 9 Load - displacement curve for conv. deep drawing steel

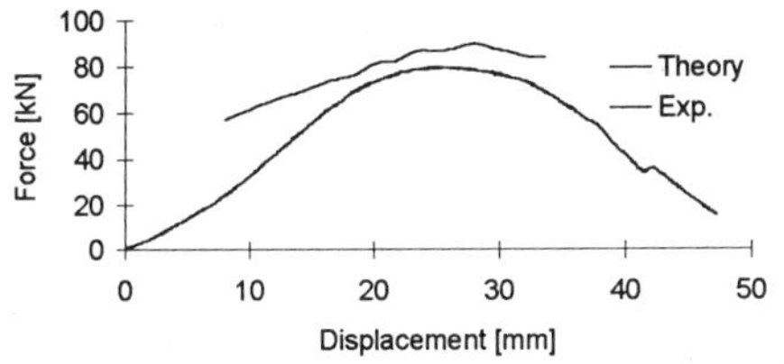

Fig. 10 Load - displacement curve for ferritic stainless steel, Werkst. No. 1.4016

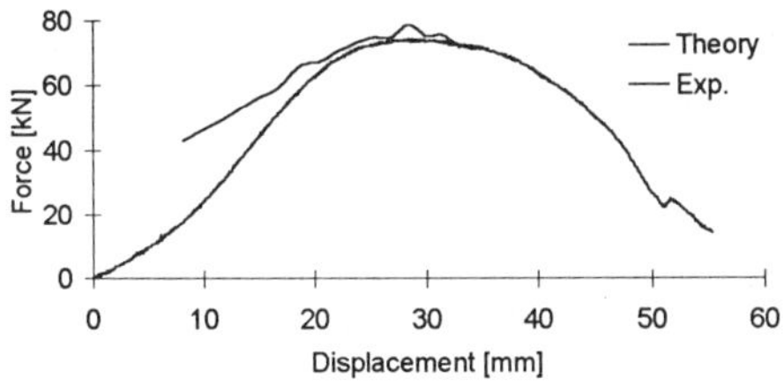

Fig. 11 Load - displacement curve for ferritic stainless steel, Werkst. No. 1.4520

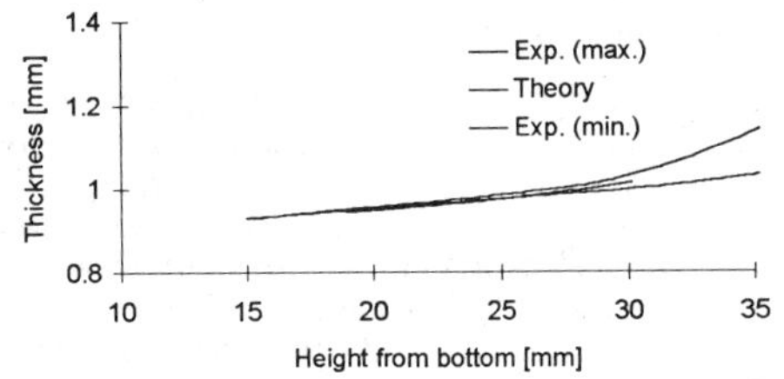

Fig. 12 Thickness distribution in the cup-wall for conv. deep drawing steel

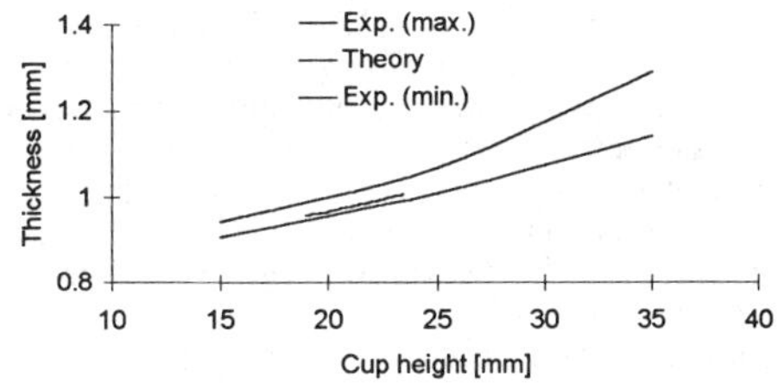

Fig. 13 Thickness distribution in the cup-wall for ferritic stainless steel, Werkst. No. 1.4016

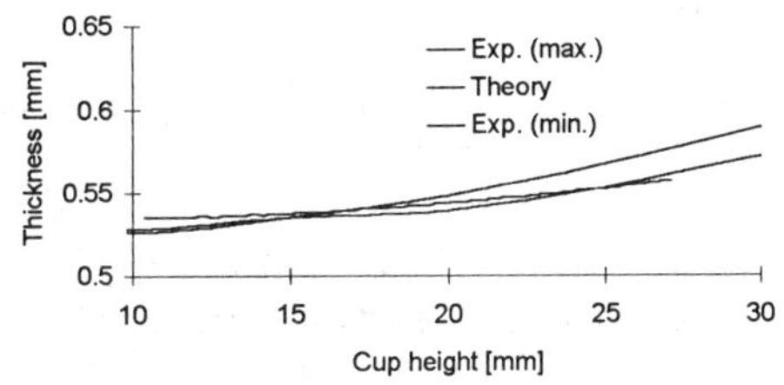

Fig. 14 Thickness distribution in the cup-wall for titanium

results will be presented in this paper. A more complete representation will appear in a later publication.

Materials for experiments

Experiments have been performed on different kind of materials including conventional deep drawing steel, austinitic stainless steel, ferritic stainless steel, aluminium, brass and titanium. To achieve the necessary materials data uniaxial tensile tests have been performed in three different directions to the rolling direction (0° , 45° and 90°) and with three subsequent tests for each direction. The mean values of the materials data involved in the analysis are shown in Fig. 7, from which is seen that the normal anisotropy of the materials shown is varying from 0.77 to 4.

Comparison between theory and experiments

Limit Drawing Ratio (LDR)

The comparison between experiments and theory as regards the Limit Drawing Ratio (LDR) for each material is shown in Fig. 8.

It is noticed that there is a rather good agreement between theory and experiments with a maximum deviation between measured and calculated values of 3.5%.

Load - displacement

The load displacement curves for three of the materials can been seen on Fig. 9 -Fig. 11. The agreement between theory and experiments is satisfactory.

Thickness in the cup-wall

Fig. 12 - Fig. 14 show comparison between calculated and measured wall thickness distribution in the cups drawn in three different materials. The experimentally determined distributions are shown by a maximum and minimum thickness curve caused by the plane anisotropy in the sheet. Very satisfatory agreement is noticed with the theoretical model supporting the validity of this.

CONCLUSIONS

A new finite difference analysis of axisymmetric deep drawing has been developed taking into account strain hardening, nornal anisotropy and real contact area between flange and die/blankholder. The theoretical analysis shows how variation in the anisotropy has large influence on the wall thickness distribution in the drawn cup, whereas the influence on the load - displacement curve is small. Comparison with experiments drawing steel, stainless steel and titanium shows good agreement as regards the estimated Limit Drawing Ratio, the load - displacement curves and the thickness distribution.

REFERENCES

[1] Sommer, H. M.: *Versuche über das Zieben von Hohlkorpern*. Forsch.-Arb. No. 286, (1926).

[2] Siebel, E. and Pomp, A.: *Über den Kraftverlauf beim Tiefziehen und bei der Tiefungsprüfung*. Mitt. Kaiser-Wilhelm Inst. Eisenforsch., Vol 11, p. 139, (1929).

[3] Chung, S. Y. and Swift, h. W.: *Cup-drawing from a flat blank*. Proceedings The Institution of Mechanical Engineers, Vol 165, p. 199, (1951).

[4] Woo, D. M.: *Analysis of the Cup-drawing Process*. Journal of Mechanical Engineering Science, Vol. 6, No. 2, p. 116, (1964).

Simulation of Materials Processing: Theory, Methods and Applications, Shen & Dawson (eds)
© 1995 Balkema, Rotterdam. ISBN 90 5410 553 4

FEM analysis of punch stretching and cup drawing tests for aluminum alloys using a planar anisotropic yield function

Y. Hayashida, Y. Maeda, K. Matsui, N. Hashimoto, S. Hattori & M. Yanagawa
Kobe Steel Ltd, Japan

K. Chung, F. Barlat, J.C. Brem, D.J. Lege & S.J. Murtha
Alcoa Technical Center, Pa., USA

ABSTRACT: FEM simulations of the punch stretching test and the cup drawing test were performed to evaluate the anisotropic yield function previously developed by Barlat et al. This yield function was implemented into the FEM code and the simulations were compared with the experiments. It was shown that this yield function provides a reasonable description of the plastic behavior of the aluminum alloys. The effect of the yield surface shape on the formability was also discussed.

1. INTRODUCTION

The use of aluminum alloy sheet for automotive bodies improves fuel efficiency and enhances the performance of the vehicles. However, further improvements in the formability of aluminum alloy sheet compared with current sheet steel are needed for more widespread use in this automotive application.

To understand the deformation of aluminum alloy sheet, a finite element method (FEM) has been developed recently using the ABAQUS/EXPLICIT code. An anisotropic yield function previously developed by Barlat et al. [1] was implemented into the code to represent the plastic yielding of aluminum alloys.

When sheets are formed, material elements usually experience stretching either in one or both principal directions. Such deformation states are typically obtained by deformation in punch stretching and cup drawing tests that are performed to investigate the forming limit and the limiting drawing ratio, respectively.

In this report, FEM simulations were performed for these two tests and the results were compared with experiments for Al-Mg alloy sheets to evaluate this yield function.

This program was performed as part of the cooperative research effort between Alcoa and Kobe Steel aimed at improving of the formability of aluminum alloys.

2. MATERIAL MODELING

The commercial ABAQUS/EXPLICIT code was applied for this FEM analysis.

Numerical models of the yield surface shape and the stress-strain curve, which represent material plasticity phenomena, are required as input for the FEM simulation.

The phenomenological six-component function developed by Barlat et al. [1] was adopted for the anisotropic yield surface description. The equation is:

$$\phi = |s_1 - s_2|^m + |s_2 - s_3|^m + |s_3 - s_1|^m = 2\bar{\sigma}^m \qquad (1)$$

$$L = \begin{bmatrix} (c_2+c_3)/3 & -c_3/3 & -c_2/3 & 0 & 0 & 0 \\ -c_3/3 & (c_3+c_1)/3 & -c_1/3 & 0 & 0 & 0 \\ -c_2/3 & -c_1/3 & (c_1+c_2)/3 & 0 & 0 & 0 \\ 0 & 0 & 0 & c_4 & 0 & 0 \\ 0 & 0 & 0 & 0 & c_5 & 0 \\ 0 & 0 & 0 & 0 & 0 & c_6 \end{bmatrix} \qquad (2)$$

s_1, s_2 and s_3 are the eigenvalues of the 3x3 matrix in which the components are given by the product $L{\cdot}\sigma$ where L is a 4th order tensor that is described as a 6x6 matrix, and σ is the stress tensor. L contains the coefficients c_i that describe anisotropy. When all the coefficients c_i are one, the yield function is isotropic. For example, if the exponent m in equation 1 is equal to two, an ellipsoidal yield surface corresponding the Mises criteria is defined. When the exponent m increases, the yield surface shape approaches a hexagonal one approaching the Tresca criteria. In the analysis, this yield function was implemented into the code through a user subroutine under the plane stress condition (c_4 and c_5 are not effective).

The Voce equation was adopted to describe the stress-strain curve:

$$\bar{\sigma} = A - B\exp(-C\bar{\varepsilon}) \qquad (3)$$

Three constants of the Voce equation were determined from the stress-strain curve at the relatively large strains attainable in biaxial bulge tests as described by Young et al. [2].

Table 1 Chemical composition

Material	Element /mass%					
	Si	Fe	Be	Ti	Mg	Al
A, B, C	<0.01	<0.01	<0.01	<0.01	4.70	Bulk

Table 2 Process variables for specimens

Material	Solution heat treatment	Cold rolling condition /%	Grain size /micron
A	530℃, 30sec	50	100
B	530℃,120sec	80	150
C	530℃,120sec	50	150

Table 3 Coefficients of Barlat's function

Material	c1	c2	c3	c4	c5	c6	m
A	1.04	1.04	0.96	1.00	1.00	0.98	5.28
B	1.14	1.11	0.88	1.00	1.00	1.01	6.92
C	1.06	1.02	0.98	1.00	1.00	0.90	5.85

Table 4 Constants of the Voce equation

Material	A /MPa	B /MPa	C
A	401.9	345.0	5.441
B	371.8	264.2	4.808
C	335.7	231.3	6.668

Table 5 Material properties

Density /g.cm^{-3}	2.64
Young's modulus /MPa	69000
Poisson's ratio	0.3

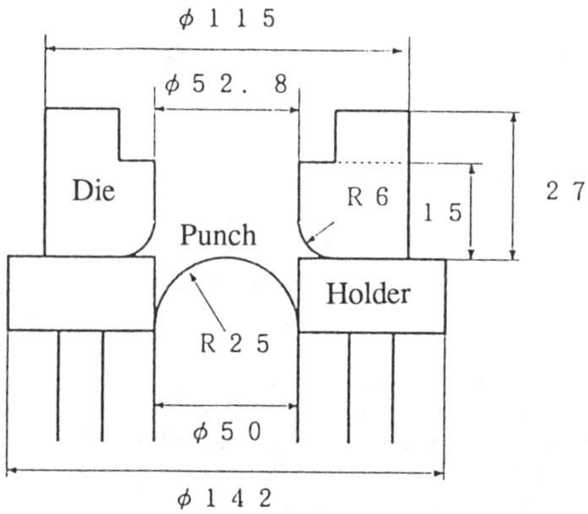

Figure 1 Punch stretching cup test

3. FEM VERIFICATIONS

3.1 *Material fabrication and measurements*

3.1.1 *Material fabrication*

Three Al-Mg alloy sheet samples [3] were used for verifications. Table 1 shows the chemical composition of these materials and Table 2 shows the process variables for the samples. Among the three materials, material B shows the strongest anisotropy in mechanical properties.

3.1.2 *Measurement of material constants*

From the isotropic work hardening assumption, the yield surfaces are defined by the set of stress states at yielding which cause the same amount of plastic work. The yield surfaces were measured for different levels of plastic work (stress and strain) and the coefficients were determined by the least squares method. The measurements were performed by a biaxial compression machine on cubic specimens made from laminated sheet samples, as described by Tozawa et al. [4] and Kanetake et al. [5]. Table 3 shows the seven coefficients of the yield function.

Regarding work hardening, Table 4 shows the constants of the Voce equation that were determined by the bulge test.

Table 5 shows the other material constants that are required for the simulation.

In order to effectively control dynamic effects for quasi-static simulation results, the mass densities $\rho_1=\rho_0 x10^9$ and $\rho_2=\rho_0 x10^7$ were used for stretching simulation and drawing simulation, respectively, where ρ_0 is the real mass density. Also, shell elements with nine integration points through thickness were used.

3.2 *Simulation of punch stretching cup test*

3.2.1 *Punch stretching cup test*

The stretching tests were performed to failure on specimens with different widths to measure the major and minor surface strains near the fractures. These strain combinations indicate the forming limits of different strain paths. The Forming Limit Diagram (FLD) for effective indicator of a sheet's stretchability, is constructed from these strain combinations.

Figure 1 shows the equipment for the stretching test. Circles with an initial diameter of 6.35mm were scribed on the specimens. Since circles exactly on the fractures could not be measured, the circle next to the fracture on the flange side (i.e. away from the center line) was measured. Polyethylene film of 95mm diameter was used as a lubricant between the punch and the blank. Blanks with thickness of 1mm, length of 120mm and width ranging from 40mm - 120mm were used. Table 6 shows the tool dimensions.

3.2.2 *Simulations*

Simulations were performed on the 1/4 symmetric area for materials A and B. The stretching test was simulated with two typical specimen sizes (a square:120mm x 120mm, a strip:120mm x 60mm). The strip and square specimens undergo plane strain deformation and biaxial deformation near the fracture, respectively. Regarding the Coulomb friction coefficients between the blank and the tools, 0.08,

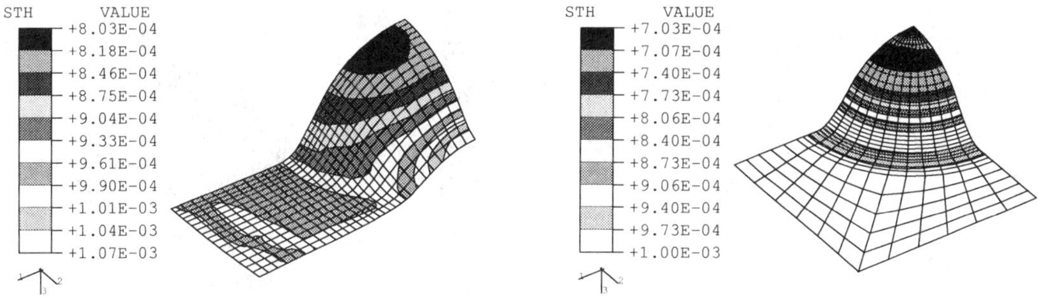

(a) Strip blank

(b) Square blank

Figure 2 Thickness distribution of blank (unit : m)

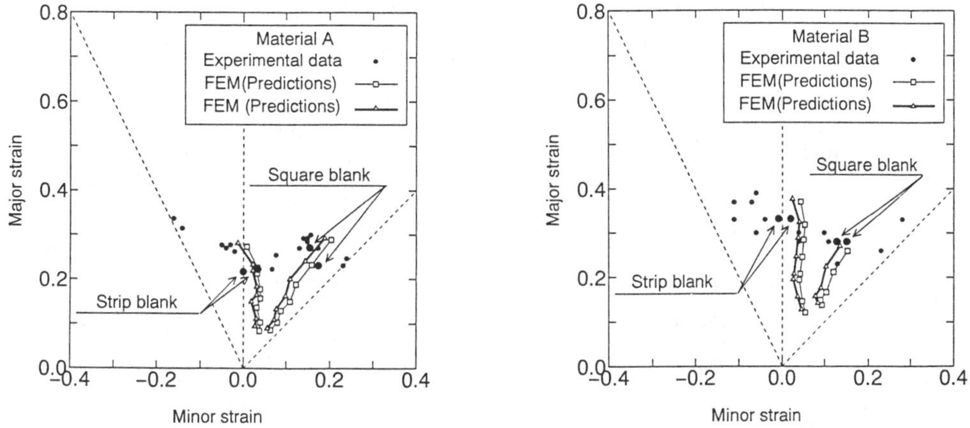

(a) Material A

(b) Material B

Figure 3 Comparison of FEM and experiments

Table 6 Tool dimensions

Punch diameter /mm	50.0
Punch radius /mm	25.0
Die diameter /mm	52.8
Die radius /mm	6.0
Blank holder force /Tonf	9.0
Lubricant (Punch only)	Polyethylene film
Velocity /mm·min^{-1}	13
Gauge diameter /mm	6.35
Blank shape /mm	120 x W x 1 W=40 to 120

0.15 and 0.25 were used for the punch, holder and die, respectively.

Figure 2 shows the thickness distribution of material A. The thinnest site of the strip blank is predicted to be 10.3mm from the specimen pole while that of the square blank is 12.7mm away from the pole.

3.2.3 *Comparison of experiments and simulation*

Figure 3 compares the computed and experimental major and minor strains of materials A and B. Each dot shows the experimental major and minor strains near fracture for all blank sizes. In particular, large dots show the experimental data with the strip and the square specimens. On the other hand, thin solid lines show the simulated major and minor strain paths (deformation path) of the thinnest element, and thick solid lines show the strain paths of the element that is next to the thinnest element (the element length is about 3mm).

The deformation paths of the element that were next to the thinnest element were used for the comparison of the experiment and the simulation. The figure shows reasonably good agreement. Adding to the above, the minor strains at the thinnest site seem to be higher than those near the fracture site.

Figure 4 shows the comparison of the major and minor strain distributions. In this figure, the comparison was performed for the square blank of material B in the transverse direction. The circular dots and the open squares show the experimental major and minor strains respectively. The experimental data were obtained from two samples. The solid line and the broken line show the simulated major and minor strains respectively. The comparison shows reasonably good agreement especially near the

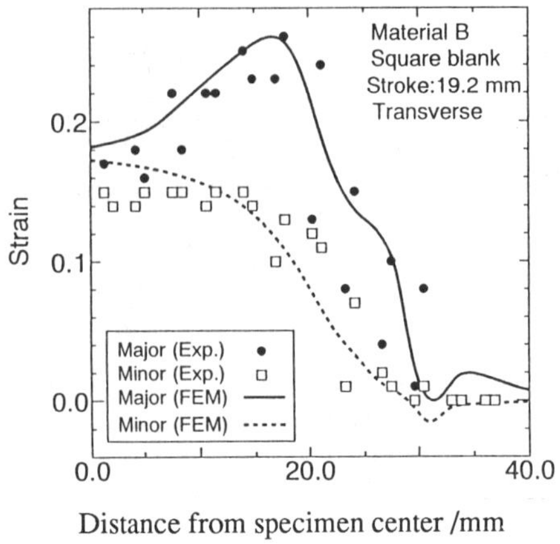

Figure 4 Comparison of strain distributions

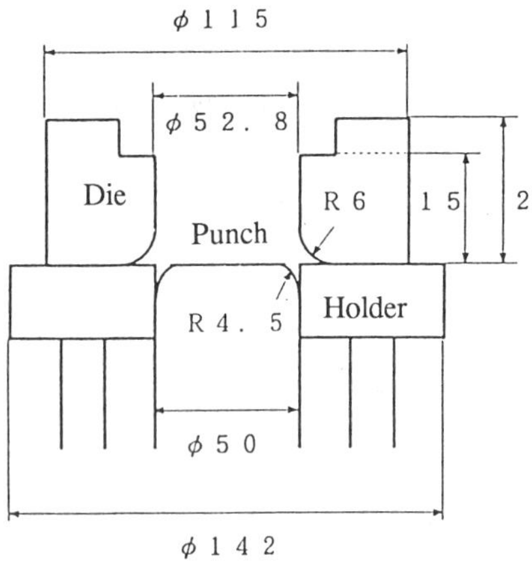

Figure 5 Cup drawing test

fracture where the major strain attains high values.

The comparison of the thickness distributions also shows good agreement.

3.3 *Simulation of cup drawing test*

3.3.1 *Cup drawing test*

The cup drawing test is used to evaluate the drawability of the material. Cup height variations, called earing, develop if the sheet samples have planar anisotropic properties.

Figure 5 shows the equipment for the cup drawing test. The blank diameter varied from 90mm to 105mm with a thickness of 1mm and punch diameter of 50mm. The blank holder force was constant (0.5 Tonf) throughout the process. Mineral oil was used for the lubricant.

Table 7 shows the tool dimensions and Table 8 shows the experimental results for the blank diameter of 102mm. In Table 8, O indicates a blank drawn without breaking and X indicates a failed cup.

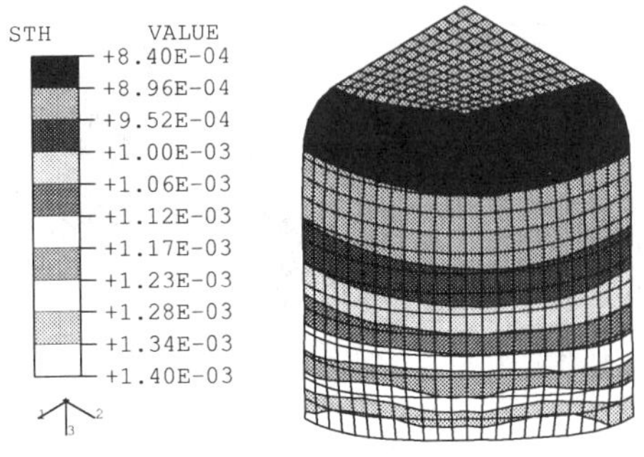

Figure 6 Thickness distribution of cup blank (unit : m, material C)

Table 7 Tool dimensions

Punch diameter /mm	50.0
Punch radius /mm	4.5
Die diameter /mm	52.8
Die radius /mm	6.0
Blank holder force /Tonf	0.5
Lubricant	Mineral oil
Velocity /mm·min^{-1}	850
Blank diameter /mm	90 to 105

Table 8 Results of drawing tests

Material	B	C
Drawing results	OOX	OOO

Blank diameter : 102mm
(Drawing ratio : 2.04)
O : successful drawn cup
X : failed drawn cup

3.3.2 *Simulations*

Simulations were performed on the 1/4 symmetric area for material B and C. Regarding the Coulomb friction coefficients between the blank and the tools, 0.15, 0.10 and 0.10 were used for the punch, holder and die, respectively.

Figure 6 shows the simulated thickness distribution for material C with blank diameter of 102mm. The completely drawn cup shows small earing of 2.4mm for the maximum cup height of 40.8mm. This small earing agrees with the nearly isotropic mechanical properties of this material.

In the case of material B, some elements that were located at the side wall near the punch corner were extremely stretched and their thickness reduction was up to 30%. These elements are known for their plane strain deformation. The simulated major strains reached 0.35 to 0.4 for these elements and were larger than the experimental major strains near the fracture on the strip blank of material B (0.33, see figure 3). This large major strain implies that this drawing ratio (initial blank diameter / punch diameter) is likely to cause failures in material B. This was also shown by the experimental results where one of the three blanks of material B could not be drawn.

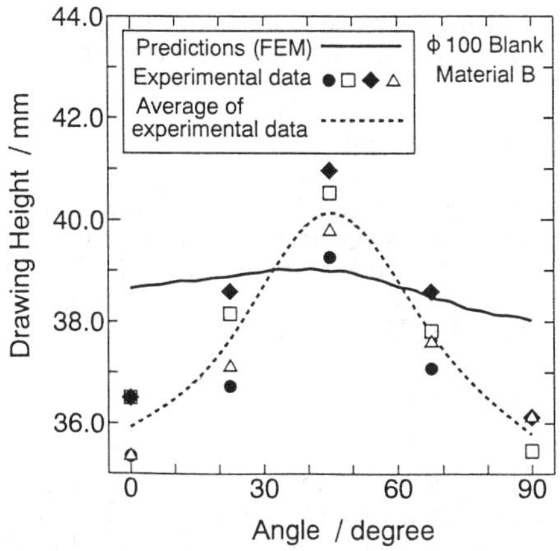

Figure 7 Comparison of earing height

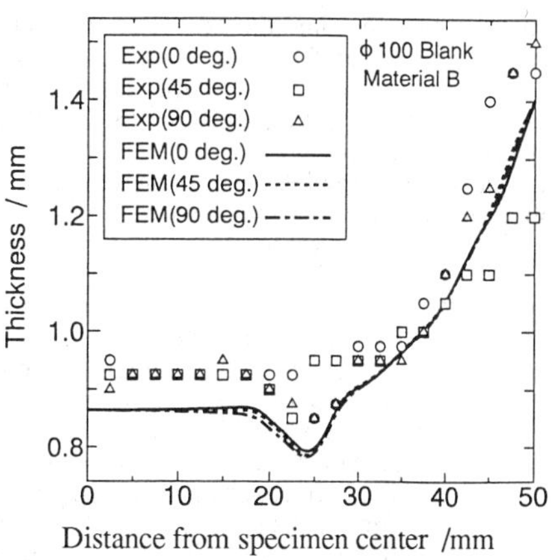

Figure 8 Comparison of thickness distribution

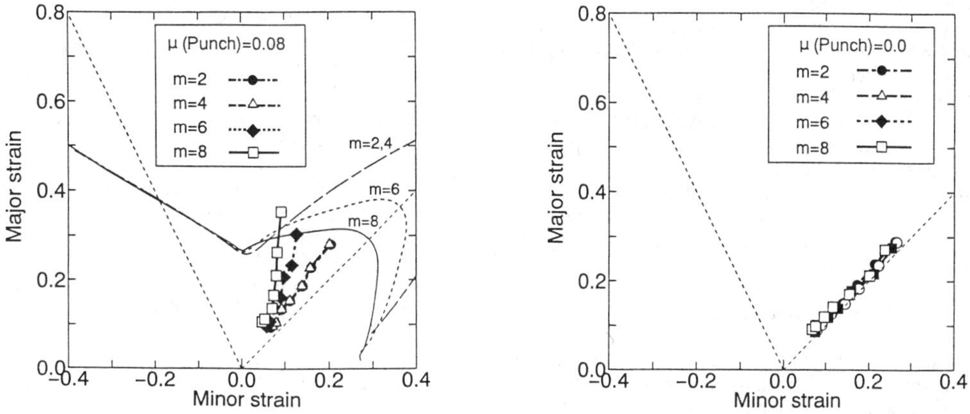

(a) non-zero friction case

(b) zero friction case

Figure 9 Effect of yield surface shape on deformation path

On the other hand, cup drawing of material B with a 100mm diameter blank was simulated. In this case, the thickness reduction of the blank remained under 25% and the major strains of the elements at the side wall near the punch corner were predicted to be from 0.25 to 0.3. Therefore, it is implied that drawing is possible for this blank diameter for material B. Experiments also showed that the blanks with 100mm diameter were completely drawn.

3.3.3 *Comparison of experiments and simulation*

Figure 7 shows the simulation/experiment comparison for the earing height of material B, the material which shows the strongest anisotropy and the largest experimental earing. The comparison is performed for the drawn cup with an initial blank diameter of 100mm. The experimental data were obtained from four samples. The simulated cup shows little earing while the experimentally drawn cup shows maximum earing of 5.1mm for maximum cup height of 41.0mm.

Figure 8 shows the comparison of the thickness distribution in the cup. The comparison was performed in the directions of 0, 45 and 90 degrees to the rolling direction. The trend of the distribution shows agreement. However, the directionality of the thickness distribution that is related to the earing is not observed in the simulation results.

4. EFFECTS OF YIELD SURFACE SHAPE ON FORMABILITY

A parametric study of the yield function can increase the understanding concerned with the effect of yield surface shape on strain distribution.

For the parametric study, isotropic square blanks were used, whose yield functions were defined with different exponents m (2, 4, 6 and 8). However, the same c_i (all of them equal to one) were assumed in each case. The punch stretching simulations were performed using these yield function constants while the other material constants were obtained from material A.

Figure 9 shows the simulated deformation paths at the thinnest site when the yield surface shape is changed. To clarify the effect of friction, further simulations without friction between the punch and the blank were performed.

Without friction, no yield surface shape effect can be observed on the deformation paths, which nearly coincide with the line where the major strains and minor strains are equal.

However, in the case of non-zero friction (μ=0.08), deformation paths shift towards the plane strain area as the exponent m increases. The thinnest element on the blank is at a distance of 10 to 20mm from the center in the case of non-zero friction, while it is located at the specimen center in the case of zero friction. Therefore, two reasons possibly explain why the deformation paths shift towards the plane strain area. The first is that the friction force stretches the blank between the contact points and the flange, and the thinnest site moves towards the flange where the minor strains are lower than those at the center of the blank. The second is that the directions of strain increments move towards plane strain deformation on the yield surface.

In the non-zero friction case, the predicted FLD curves calculated using the MK model [6] are also plotted. They decrease in the biaxial stretching range as the exponent m increases.

These parametric survey imply that higher m exponents results in lower the formability because they lower forming limits and direct strain increments towards plane strain.

5. SUMMARY/CONCLUSIONS

In the punch stretching simulations with two different blank widths, the deformation paths near the thinnest sites showed reasonably good agreement with the experimental results. The cup drawing simulations also verified that the critical blank size depends on the material. These verification results imply that this anisotropic yield function provides a reasonable description of the plastic behavior of the aluminum alloys.

However, the earing of the cup drawn for material B could not be predicted. These discrepancies imply that further improvement of the yield function is needed to predict the earing behavior.

The punch stretching simulations were performed to investigate the effect of the yield surface shape on the formability. These simulations with non-zero friction showed that with higher values of exponent m the deformation paths at the thinnest sites move towards the plane strain area. There is a certain amount of friction in the real punch stretching process and the plane strain deformation path is usually inferior in formability. Therefore, the material that has a higher exponent m shows lower formability under biaxial stretching.

For example, crystal plasticity suggests that the aluminum alloy sheets have a yield function exponent m that is higher than that of steel sheets [7] [8]. Thus, these results imply that aluminum alloy sheets will experience more plane strain deformation and show inferior formability under biaxial stretching conditions.

ACKNOWLEDGMENT

The authors thank Dr. E. Usui and Dr. K. Iwai (Kobe) and Dr. W.G. Truckner (Alcoa) for their support of this work. Also, thanks to Dr. R. Becker (Alcoa) for his implementation of the yield function into the ABAQUS/EXPLICIT code.

REFERENCES

[1] Barlat, F., Lege, D.J. and Brem, J.C. 1991. A six-component yield function for anisotropic materials. *Int.J.Plasticity*. 7:693-712.

[2] Young, R.F., Bird, J.E. and Duncan, J.L. 1981. An automated hydraulic bulge tester. *J.Applied Metalworking*. 2:11-18.

[3] Hattori, S., Maeda, Y., Yanagawa, M., Hayashida, Y., Matsui, K., Chung, K., Barlat, F., Lege, D.J., Brem, J.C., Murtha, S.J. 1994. *Alcoa/Kobe J4 report*.

[4] Tozawa, N. and Nakamura, M. 1992. A biaxial compression testing method for thin sheets. *Plasticity and Processing*. 13:538-541.

[5] Kanetake, K., Tozawa, Y., Kato, T. and Aiba, S. 1981. Effect of texture on deformation behavior of aluminum sheets. *J.Japan Institute of Light Metals*. 31:307-312.

[6] Marciniak, Z. and Kuczinski, K. 1967. Limit strains in the processes of stretch-forming sheet metal. *Int.J.Mech.Sci*. 9:609-620.

[7] Hosford, W.F. 1985. Comments on anisotropic yield criteria. *Int.J.Mech.Sci*. 27:pp.423-427.

[8] Lian, J. and Chen, J. 1991. Isotropic polycrystal yield surfaces of B.C.C. and F.C.C metals: crystallographic and continuum mechanics approaches. *Acta Metall. Mater*. 39:2285-2294.

Simulation of Materials Processing: Theory, Methods and Applications, Shen & Dawson (eds)
© 1995 Balkema, Rotterdam. ISBN 90 5410 553 4

Numerical simulation of rectangular cup drawing processes with drawbeads

Hoon Huh
Korea Advanced Institute of Science and Technology, Taejon, Korea

Soo Sik Han
Pohang Iron and Steel Co., Kyungbuk, Korea

ABSTRACT: Rectangular cup drawing processes are numerically simulated by a modified membrane finite element method. The simulation examines the effect of drawbeads on the deep drawing process. The numerical result demonstrates that drawbeads control the amount of drawin from the flange region into the die cavity and the strain distribution of the deep drawn cup.

1 INTRODUCTION

Finite element simulation of sheet metal forming processes has gained general popularity with a burst of development in finite element techniques (Makinouchi 1993). The finite element method has been fortified by various numerical techniques such as the geometric description for tools and dies, the treatment of contact and friction between tools and dies in order to simulate three-dimensional sheet metal forming processes. Yet the last obstacle for the simulation seems to be the treatment of drawbeads, which demands large memory size and many iterations for contact in a solution procedure. Such simulation with drawbeads is hardly carried out by some finite element codes in which drawbeads are replaced by corresponding boundary conditions. Simulation of deep drawing processes with drawbeads has been carried out for an axisymmetric or plane strain case with continuum elements(Cao 1992, Heege 1992), but not yet for three-dimensional cases with membrane elements. In this case, the convergence and the computational efficiency is the most challenging issue to solve.

In modelling of the real process, a membrane element is regarded as more preferable rather than a shell element because of the computing efficiency and the contact treatment. Nevertheless, a membrane element has a disadvantage of disregarding the bending effect during the deformation, which leads to inaccuracy in computing the forming load and the deformed shape and sometimes to inability in simulating a class of bending dominant problems. To overcome such deficit in using a membrane element, some modifications may be required in the finite element formulation procedure to take the bending effect into account. One of them could be to make a membrane element fortified with the advantage of a shell element by considering the bending effect. This idea is adopted in the finite element formulation procedure. The strain energy term in a variational formulation is decomposed into the term due to the mean stretching and the term due to the bending. The two kinds of strain energy are calculated separately and added together within a membrane element. The remedy correctly enhances the flexural rigidity not only within an element but also among neighboring elements, which is almost zero especially in the ordinary four-node bilinear quadrilateral membrane element.

In this paper, rectangular cup drawing processes with drwabeads are successfully simulated by a modified membrane finite element method. For the validity of the simulation result, experiments were carried out and their result is compared with the simulation one. The simulation examines the effect of drawbeads on the deep drawing process. The numerical result demonstrates that drawbeads control the amount of drawin from the flange region into the die cavity and the strain distribution of the deep drawn cup. The drawbead makes the thickness distribution more uniform to prevent localized necking. It also makes the amount of overall thickness strain larger than that without drawbeads. The drawbead causes increase in the drawing load since it interferes the drawin of sheet metal.

2 MODIFIED MEMBRANE FINITE ELEMENT FORMULATION

The deformation of sheet metal can be described with the following boundary value problem. At a time t_o, the shape of the surface and the distribution of the effective strain are supposed to be given or to have already been determined. The necessary and sufficient condition for the stress field in the sheet metal to be in equilibrium at time $t_o + \tau$ is given from the principle of virtual work

$${}^{\tau}\delta W = \int_{{}^0V} {}^{\tau}\sigma^{ij}\, \delta\varepsilon_{ij}\, dV \tag{1}$$

where σ^{ij} is the second Piola-Kirchhoff stress tensor and ε_{ij} is the Lagrangian strain tensor. The second Piola-Kirchhoff stress tensor is turned into the Cauchy stress for an incompressible material in the convected coordinate system. The sheet metal undergoes bending and unbending around the die profile while it deforms and the bending effect should be taken into account for accurate analysis. When the sheet metal is in the bending condition, the strain tensor can be considered to consist of the two terms(Huh 1994, Han 1995)

$$\delta\varepsilon_{ij} = \delta\varepsilon^m_{ij} + \delta\varepsilon^b_{ij} \tag{2}$$

where ε^m_{ij} is the membrane strain, which has uniform distribution throughout the thickness due to the stretching deformation and ε^b_{ij} is the bending strain, which has linear distribution throughout the thickness due to the bending deformation. By using eqn. (2), Equation (1) can be rewritten as

$$\begin{aligned} {}^{\tau}\delta W &= \int_{{}^0A}\int_{-\frac{t}{2}}^{\frac{t}{2}} {}^{\tau}\sigma_{ij}\, \delta\varepsilon^m_{ij}\, dz\, dA \\ &+ \int_{{}^0A}\int_{-\frac{t}{2}}^{\frac{t}{2}} {}^{\tau}\sigma_{ij}\, \delta\varepsilon^b_{ij}\, dz\, dA \end{aligned} \tag{3}$$

The strain ε^b_{ij} can be expressed in terms of the curvature from the geometrical compatibility as

$$\varepsilon^b_{ij} = z\,\kappa_{ij} \tag{4}$$

where z is the distance from the neutral surface and the curvature κ_{ij} can be expressed as a second derivative of the vertical displacement w at the neutral surface. Substituting eqn. (4) into eqn. (3), the bending energy term is obtained in terms of the bending moments and the curvature. Equation (3) can be rewritten as

$$\begin{aligned} {}^{\tau}\delta W &= \int_{{}^0A} {}^{\tau}\sigma^m_{ij}\, \delta\varepsilon^m_{ij}\, {}^0t\, dA \\ &+ \int_{{}^0A} {}^{\tau}M_{ij}\, \delta\kappa_{ij}\, dA \end{aligned} \tag{5}$$

where σ^m_{ij} is the mean stress throughout the thickness, which corresponds to the membrane strain. In order to obtain the curvature κ_{ij} for the finite element approximation, we have to use at least a quadratic element. The high order element, however, has a problem in convergence especially with contact treatment and has less computational efficiency than a linear element. To overcome these problems, we propose a new method which has the advantage of a linear element while the bending effect is still considered. The curvature is approximated as the difference of the kink angle of the two neighboring elements. Although, the mesh system is not generally constructed to coincide with the direction of the principal bending moment, this assumption has the physical reliability as an upper bound. By using this method, the bending energy term can be approximated as

$$\begin{aligned} &\int_{{}^0A} {}^{\tau}M_{ij}\, \delta\kappa_{ij}\, dA \cong \\ &\int_{{}^0A} {}^{\tau}M_1\, \delta\kappa_1\, dA + \int_{{}^0A} {}^{\tau}M_2\, \delta\kappa_2\, dA \end{aligned} \tag{6}$$

where M_1 and M_2 are the bending moments corresponding to κ_1 and κ_2 , defined in the normal direction of each edge boundary of a four-node element, respectively.

3 FINITE ELEMENT APPROXIMATION IN THE CONVECTED COORDINATE SYSTEM

3.1 *Membrane energy*

The natural Lagrangian strain in the natural convected coordinate system is given by

$$\varepsilon_{ij} = \frac{1}{2}\left[\frac{\partial u^\alpha}{\partial \xi^i}\frac{\partial x^\alpha}{\partial \xi^j} + \frac{\partial x^\alpha}{\partial \xi^i}\frac{\partial u^\alpha}{\partial \xi^j} + \frac{\partial u^\alpha}{\partial \xi^i}\frac{\partial u^\alpha}{\partial \xi^j} \right] \tag{7}$$

where ξ^i is the natural convected coordinates. Using the relationship between the stress and the strain-rate which is obtained from the Hill's quadratic yield criterion and the associated flow rule, the membrane energy term in eqn. (5) is expressed as(Yang 1986)

$$\begin{aligned} \int_{{}^0A} {}^{\tau}\sigma^m_{ij}\, \delta\varepsilon^m_{ij}\, {}^0t\, dA &= \int_{{}^0A} {}^{\tau}\bar{\sigma}\, \delta(\Delta\bar{\varepsilon})\, {}^0t\, dA \\ &= \delta\bar{U}^T Q^m(\bar{U}) \end{aligned} \tag{8}$$

The membrane energy in eqn. (8) is linearly proportional to the thickness variation.

3.2 *Bending energy*

When two elements adjoin to each other, the curvature can be expressed by the kink angle of two neighboring elements approximately as

$$\begin{aligned} \kappa_i &= \frac{1}{L_i}\left(w^{(1)}_{,t} - w^{(2)}_{,t}\right) \\ &= \frac{1}{L_i}\left(\mathbf{n}^{(1)}_n \cdot \frac{\partial \mathbf{H}}{\partial n^{(1)}_t} - \mathbf{n}^{(2)}_n \cdot \frac{\partial \mathbf{H}}{\partial n^{(2)}_t} \right) \mathbf{U} \end{aligned} \tag{9}$$

where $\mathbf{n}_n$ is a unit vector normal to the element

surface and $\mathbf{n}_t$ is a unit vector tangential to the element surface and normal to the folded boundary of two neighboring elements. By using the eqn. (9), the bending energy term in eqn. (6) can be expressed as

$$\int_{^0A} [{}^{\tau}M_1\delta(\Delta\kappa_1) + {}^{\tau}M_2\delta(\Delta\kappa_2)]\, dA = \delta\bar{U}^T Q^b(\bar{U}) \quad (10)$$

where

$$Q^b(\bar{U}) = \sum_{n=1}^{N} \int_{^0A_v^{(n)}} C\, B_4^T B_4\, U\, dA \quad (11)$$

where N is a total number of boundary lines between elements in the whole domain, A_v is the area inside the virtual element, and C is the stiffness modulus of bending. In a sheet bending problem, it can be assumed that one component of the bending moment induces a corresponding component of strain without affecting other components of strain. For the reason, this kind of bending problems frequently adopts the Frobenius norm as a yield criterion for bending and calculates the relation between the moment and the curvature(Huh 1991). In the case, C is expressed as

$$C = \frac{t^3}{18}\frac{\bar{\sigma}}{\Delta\bar{\varepsilon}}\frac{(1+r)(2+r)}{(1+2r)} \quad (12)$$

where t is the thickness of a sheet metal. It is noted that the bending energy in eqn. (12), is triply proportional to the thickness variation.

The variational expression of eqn. (5) is approximated and minimized so that the following algebraic equation can be obtained:

$$Q^m(\bar{U}) + Q^b(\bar{U}) = \mathbf{P}(\bar{U}) + \mathbf{F} \quad (13)$$

where **P** is a residual vector and **F** is a load vector. This nonlinear algebraic equation can be solved for U iteratively by the Newton-Raphson method.

4 NUMERICAL RESULTS AND DISCUSSION

The amount of drawin from the flange region into the die cavity in deep drawing processes are generally controlled by the blank holding force and appropriate drawbeads. Both of them actually control the rate of metal flow into the die cavity during the forming process. The blank holding force, however, does not control the flow of metal adequately since the restraining force comes from the friction and the thickness variation. In the real process, the rate of metal flow is controlled by various types of drawbeads depending on the amount of drawin into the die cavity. There are many studies on the role and behavior of drawbeads (Wang 1991), whilst there are few studies on the effect of drawbeads in real deep drawing processes (Cao 1992). To investigate the effect of drawbeads, numerical simulation has been carried out for rectangular cup drawing processes with drawbeads by a modified membrane finite element code named CASHE.

The material used in simulation and experiments is cold rolled steel sheet for auto-body stamping whose stress-strain characteristic is expressed as $\bar{\sigma}=58.78(0.000903+\bar{\varepsilon})^{0.274}$ (kgf/mm^2) and Lankford value r is 1.679. The process variables used in the simulation are as follows:

thickness of sheet: t = 0.69mm
size of sheet blank: 170mm×130mm
punch dimension: 120mm×40mm
punch radius: 9mm
length of drawbead: 100mm
radius of drawbead: 5mm
height of drawbead: 3mm
blank holding force: 2500kgf
Coulomb coefficient of friction: μ = 0.112

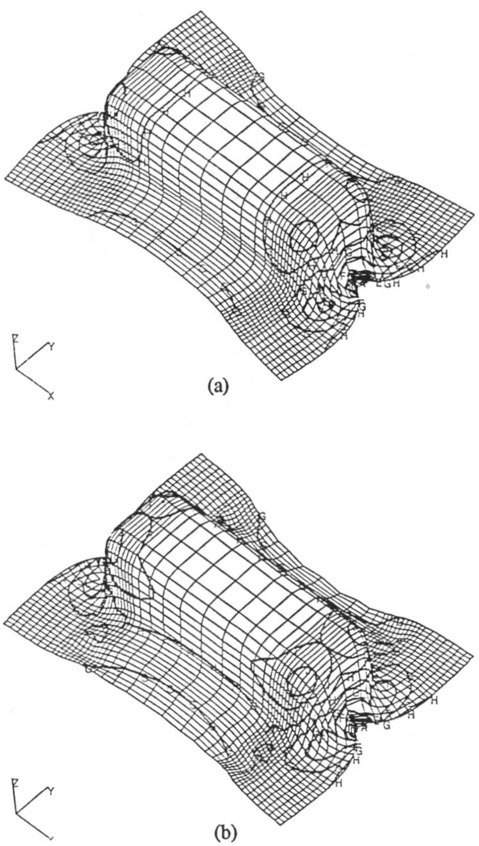

Fig. 1. Deformed configuration and the thickness strain distribution of a rectangular cup: (a)without drawbeads; (b)with drawbeads. (strain level: A = 0.325 to J = - 0.125)

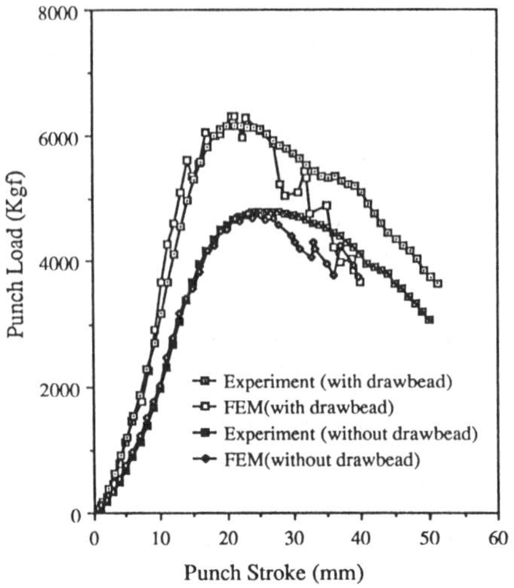

Fig. 2. Drawing load in rectangular cup drawing with and without drawbeads.

A quarter of the sheet blank is discretized into 484 elements and 529 nodal points for finite element simulation. The flange region and the end region of drawbeads is discretized with relatively fine meshes as shown in Fig. 1. The computation is executed with HP710/50 workstation and the CPU time was 18300s with drawbeads and 16030s without drawbeads.

The deformed shape calculated is shown with finite element meshes in Fig. 1. The first one without drawbeads shows that its flange part is drawn into the die cavity more than that with drawbeads. It is well identified in Fig. 5. The thickness strain distribution indicates that the rectangular cup with drawbeads has more uniform but larger distribution than that without drawbeads. Both figures show that the maximum thickness strain is located in the longitudinal flange region and the minimum thickness strain is at the corner of a rectangular cup.

The variation of the drawing load with respect to the punch stroke is calculated and shown in Fig. 2. The drawing load increases by 30% when there are drawbeads in comparison with the load when there is no drawbead. The drawing load calculated is in good agreement with the load obtained from experiments. This result proves the validity of the simulation method with a modified membrane element.

When there is no drawbead in the flange part, sheet metal is merely drawn with bending and forms a rectangular cup during the process. As a result, the thickness strain does not vary much along the transverse direction as shown in Fig. 3a. On the other hand, when there are drawbeads, sheet metal is restrained from being drawn into the die cavity. As a result, the thickness strain in a rectangular cup increases and varies much along the transverse

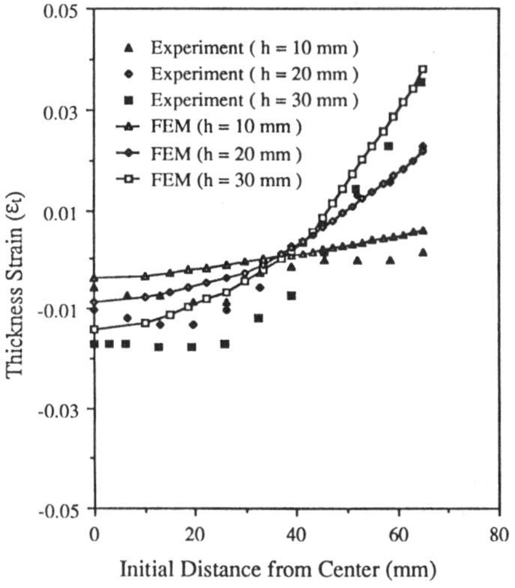

(a)

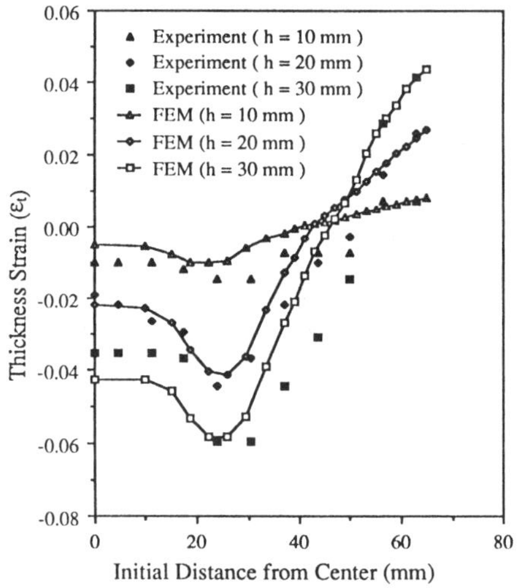

(b)

Fig. 3. Distribution of the thickness strain along the transverse direction in rectangular cup drawing: (a)without drawbeads; (b)with drawbeads.

direction as shown in Fig. 3b. The amount of the thickness strain increases gradually with the deep drawing process because of the restraining force of drawbeads. The thickness strain along the longitudinal directon shows little variation but becomes larger with drawbeads as shown in Fig. 4. Figures show that the thickness strain in the longitudinal flange becomes distinctively large as the deep drawing process progresses.

The deformed edge contour in Fig. 5 demonstrates the amount of drawin with respect to the punch

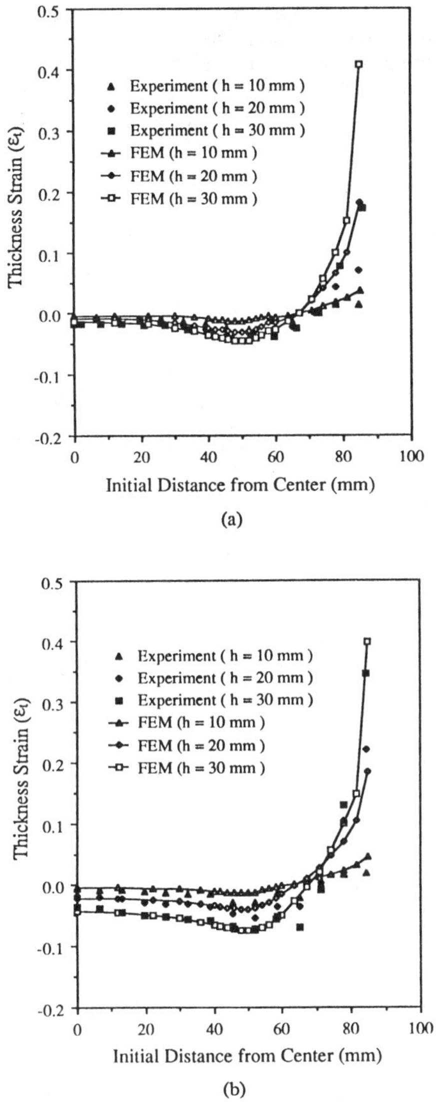

Fig. 4. Distribution of the thickness strain along the longitudinal direction in rectangular cup drawing: (a)without drawbeads; (b)with drawbeads.

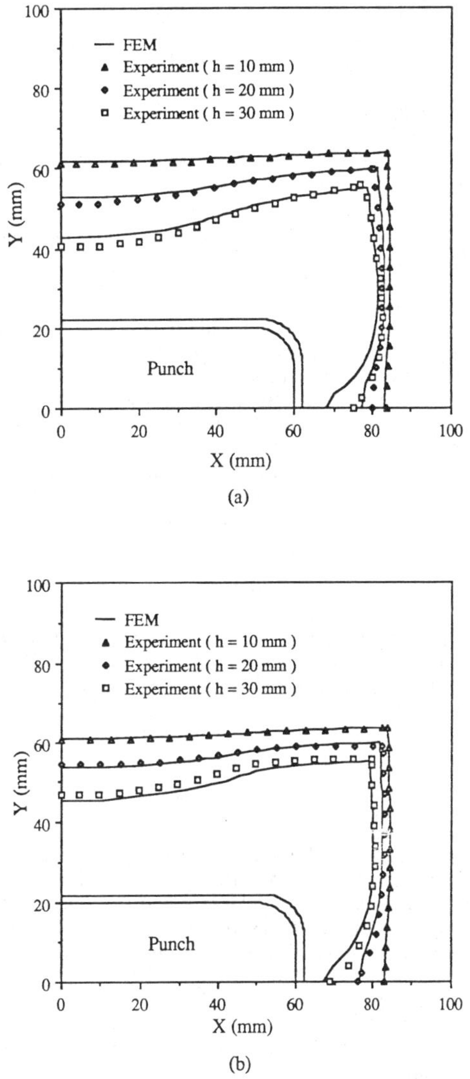

Fig. 5. Deformed edge contour in rectangular cup drawing: (a)without drawbeads; (b)with drawbeads.

stroke. Figures well explain the role of drawbeads which restrain sheet metal in the flange part from being drawn into the die cavity. The numerical result is in good agreement with the experimental result throughout the all edge contour except that near the longitudinal flange especially when there is no draebeads. It is because the thickness strain becomes very large near the longitudinal flange with deformation as shown in Fig. 4 and the blank holding force is concentrated in that region. As a result, the sheet metal in the transverse flange is drawn more and sheet metal in the longitudinal flange is drawn much less than calculated results which did not consider this phenomenon.

The present analysis, considering the bending effect, provides accurate prediction of the punch load and the thickness strain as well as valuable information about the deep drawing process including the drawbead effect. In addition to the above advantages, the present algorithm enhances the convergence in the solution procedure and prevents the numerical buckling or distortion of finite elements, both of which are very important in the analysis of a real complicated sheet metal forming process.

5 CONCLUSION

The effect of drawbeads in the deep drawing process is investigated with simulation of a rectangular cup drawing process by a modified membrane finite element method. The finite element method applied takes the bending effect into account and shows good convergence in such simulation. The method is proved to accurately simulate the deep drawing process with drawbeads by the experimental result. The simulation result demonstrates the role of drawbeads in controlling the amount of drawin and the strain distribution. The drawbead makes thickness distribution more uniform to prevent localized necking and tearing when it is properly located. It also makes the amount of overall thickness strain larger than that without drawbeads, which is very important in auto-body stamping.

ACKNOWLEDGEMENTS

This study was partially funded by the Korean Science and Engineering Foundation. The authors wish to gratefully acknowledge their support during this work.

REFERENCES

Cao, H.-L. & Teodosiu, C. 1992. Numerical simulation of drawbeads for axisymmetric deep drawing process. Proc. NUMIFORM '92: 439-448.

Han, S. S. & Huh, H. 1995. Modified-membrane finite-element simulation of square cup drawing processes considering influence of geometric parameters. J. Mater. Process. Technol. Vol. 48: 81-87.

Heege, A. & Alart, P. 1992. On an implicit contact-friction algorithm dedicated to 3D sheet forming simulation. Proc. NUMIFORM '92: 479-484.

Huh, H., Han, S. S. & Yang, D. Y. 1994. Modified membrane finite element formulation considering bending effects in sheet metal forming analysis. Int. J. Mech. Sci. Vol. 36, No. 7: 659-671.

Huh, H. & Yang, W. H. 1991. A general algorithm for limit solutions of plane stress problems. Int. J. Solids Structures Vol. 28, No. 6: 727-738.

Makinouchi, A., Nakamachi, E., Onate, E. & Wagoner, R. H. 1993. Benchmark problems: verification of simulation with experiment. Proc. 2nd Int. Conf. of NUMISHEET '93: 373-666.

Wang, N.-M. & Shah, V. C. 1991. Drawbead design and performance. J. Mater. Shaping Technol. Vol. 9:21-26.

Yang, D. Y. & Kim, Y. J. 1986. A rigid-plastic finite element formulation for the analysis of general deformation of planar anisotropic sheet metals and its applications. Int. J. Mech. Sci. Vol. 28: 825-840.

Simulation of Materials Processing: Theory, Methods and Applications, Shen & Dawson (eds)
© 1995 Balkema, Rotterdam. ISBN 90 5410 553 4

Accurate simulation of springback in 3D sheet metal forming processes

D.Joannic & J.C.Gelin
Laboratoire de Mécanique Appliquée R. Chaléat, URA CNRS, Université de Franche-Comté, Besançon, France

ABSTRACT: Numerical simulations of deep drawing processes, request in addition to an accurate behaviour description, an accurate prediction of the final geometry. The objective of this paper is to describe numerical procedures to evaluate stamping or deep drawing processes and in particular springback of final parts. Such a tool able to compute by a finite element method the springback of a shape would provide to the designer an alternative to experimental design tooling. Numerical examples are presented that illustrate the possibilities of the approach developed.

1 INTRODUCTION

In sheet metal forming industries as automotive industries, the control of sheet metal forming processes all along the operation is a crucial point. The development of numerical models becomes an important task to reduce time and final costs of the products. On the other hand, in order to help the designer, numerical procedures are used as a tool to investigate tool design or process sequences which are of main importance for the final shape. Nowadays, methods only based on practical knowledge are not sufficient.

Several authors have made research on prediction of the amount of springback. Some studies deal with theoretical, either by finite element calculation or simplified methods, and/or experimental methods, as influence of blank holder force, to compute or eliminate effects of springback (Gardiner, 1958; Tang, 1987).

The objective of this paper is to describe numerical procedures to accurately evaluate behaviour and the springback amount during unloading stage.

The present paper dealing with deep drawing processes considers this problem as a process of large elastoplastic strains and finite rotations (Daniel, Gelin, Paquier, 1993). Two different approaches are presented: a quasi-static implicit one which is described by an updated Lagrangian formulation with a consistent tangent matrix, and a transient explicit one where the diagonal mass matrix is computed. The plastic properties of the sheet metal are either isotropic, described by a von Mises yield condition, or anisotropic described in using a rotated frame subjected to follow the anisotropy axes. Isotropic and kinematic hardening are taken into account with an associated flow rule. Treatment of the contact between the sheet and the tools used a master-slave scheme where penetration have to be detected first. Then conditions are adjusted after reaching the equilibrium. The finite element code used to simulate the sheet metal forming is based on three or four nodes isoparametric shell elements, the sheet being treated as a thin shell. These elements are well suited for geometrical non-linearities. Based on isoparametric continuum kinematics, shear locking is avoided by using a mixed interpolation of the transverse shear strain components (Boisse, Daniel, Gelin, 1994).

2 FINITE SHELL ELEMENTS

The shell finite elements used are included in the so-called degenerated approach (Ahmad, 1971). In order to insure the continuity of the geometrical description, pseudonormal vectors X^i are defined at each node, see figure 1.

Position at point M is defined as the following form:

$$x(M) = \sum_{i=1}^{n} N_i(\xi,\eta)\bar{x}_i + \zeta \sum_{i=1}^{n} N_i(\xi,\eta)\frac{h^i}{2}X^i \quad (2.1)$$

where h^i is the shell thickness at node i.

An orthogonal frame $(V^i_{10}, V^i_{20}, V^i_{30})$ is defined at each node with $V^i_{30} = X^i_{30}$. The nodal rotation that transforms X_0^i to X^i is given by:

$$\theta^i = \theta_1^i V_{10}^i + \theta_{20}^i V_{20}^i \qquad (2.2)$$

that gives the displacement interpolation:

$$u(M) = \sum_{i=1}^{n} N_i(\xi,\eta)\bar{u}_i + \zeta \sum_{i=1}^{n} N_i(\xi,\eta)\frac{h^i}{2}(-\theta_1^i V_{20}^i + \theta_2^i V_{10}^i) \qquad (2.3)$$

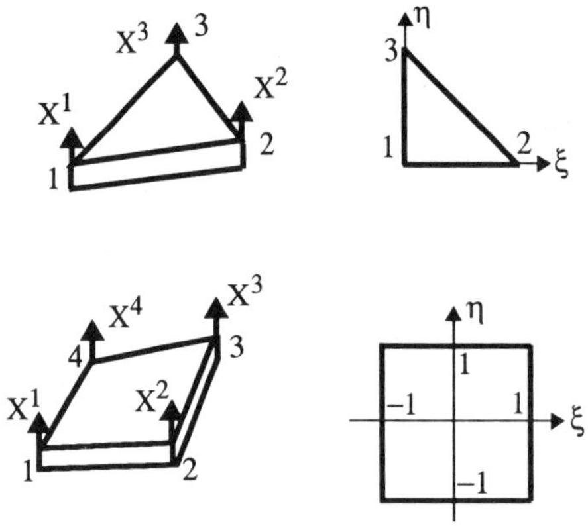

Figure 1: Three or four node elements

The material covariant frame (g_1, g_2, g_3) in the reference element is given by:

$$g_1 = \frac{\partial x}{\partial \xi} \qquad g_2 = \frac{\partial x}{\partial \eta} \qquad g_3 = \frac{\partial x}{\partial \zeta} \qquad (2.4)$$

The interpolation of the in-plane strain components are classically derived from nodal displacements (Boisse, Daniel, Gelin, 1994):

$$E_{\alpha\beta} = \frac{1}{2}\left(\frac{\partial u}{\partial \xi_\alpha} g_{\beta o} + \frac{\partial u}{\partial \xi_\beta} g_{\alpha o} + \frac{\partial u}{\partial \xi_\alpha}\frac{\partial u}{\partial \xi_\beta}\right) \qquad (2.5)$$

$\alpha,\beta =$ 1 or 2.

In order to avoid shear locking, an intermediate covariant frame related to the sides of the elements is introduced at each node i, see figure 2:

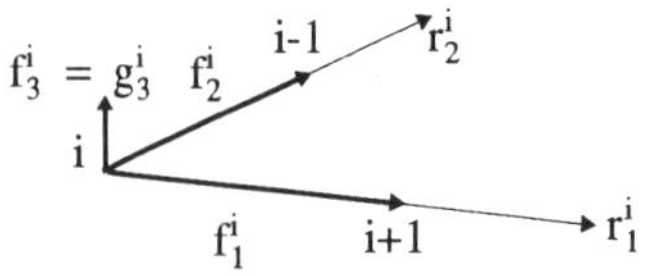

Figure 2: Nodal covariant vectors related to the element sides

$$f_1^i = \frac{\partial x}{\partial r_1^i} \qquad f_2^i = \frac{\partial x}{\partial r_2^i} \qquad f_3^i = g_3^i \qquad (2.6)$$

Using the contravariant frame (f^1, f^2, f^3) with:

$$f^k f_l = \delta_l^k \qquad (2.7)$$

the Green Lagrange tensor can be written:

$$E = E^*_{kl} f^{k0} \otimes f^{l0} \qquad (2.8)$$

The transverse shear strain components along the sides are then given by:

$$E^*_{\alpha 3} = \frac{1}{2}(f_\alpha g_3 - f_{\alpha 0} g_{30}) \qquad (2.9)$$

Along each sides, these components are assumed to be constant and equal to the mean value at the mid-point of the side:

$$2E^{*i}_{\alpha 3} = \left[\frac{\partial u}{\partial \xi}\right]_m (f^i_{\alpha 0})_m + \left[\frac{\partial u}{\partial r^i_\alpha}\right]_m (g_{30})_m + \left[\frac{\partial u}{\partial \xi}\right]_m \left[\frac{\partial u}{\partial r^i_\alpha}\right]_m \qquad (2.10)$$

The different mean quantities $(.)_m$ are calculated from the definition of $(f^1_{10}, f^2_{20}, f^3_{30})$ and from the geometrical and kinematic interpolations (2.1) and (2.3). The elements built on this formulation are efficient and do not present any shear locking (Boisse, Daniel, Gelin, 1994).

3 MECHANICAL MODELLING

3.1 Equilibrium formulation

Described by an updated Lagrangian formulation, the equilibrium equations are satisfied at the end of each step, i.e. the configuration obtained at the end of the previous loading step is considered as the initial configuration for the load step. The principle of virtual work can be written on this initial configuration using S (the second Piola-Kirchhoff stress tensor), E (the Green Lagrange strain tensor), f the surface load vector:

$$G(u,\eta) = \int_{\Omega_n} S:E(\eta)\,dV_n - \int_{(\Gamma_f)_n} \bar{f}\cdot\eta\,dS_n = 0 \qquad (3.1)$$

$G(u,\eta)$ being a non linear equation in term of displacement, a Newton method is used to solve equation (3.1):

$$D_u [G(u,\eta)]^i \Delta u^i = -G(u,\eta)^i \qquad (3.2)$$

where $\Delta u^i = u^{i+1} - u^i$ is the solution obtained at the iteration with:

$$(D_u[G(u,\eta)]^i)\Delta u^i =$$

$$\left[\frac{\partial S}{\partial E}\right]^i : E(\Delta u^i) : E(\eta)\, dV_n + \int_{\Omega_n} S : E(\Delta u^i)\, dV_n \qquad (3.3)$$

In order to respect the convergence properties of the Newton method, a tangent consistent operator is computed $C^i = [\partial S / \partial E]^i$ (Boubakar, 1994).

In the case where the acceleration are take into account, the virtual work principle can be written in an Eulerian manner as:

$$\int_{\Omega_n} \sigma : \nabla^s \eta\, dV_n + \int_{\Omega_n} \rho \gamma \eta\, dV_n - \int_{(\Gamma_f)_n} \bar{f} \cdot \eta\, dS_n = 0 \qquad (3.4)$$

After discretization, the integral form (9) is transformed as:

$$M_n \gamma_n + C_n v_n + F^{int}(u_n) - F_n^{ext} = 0 \qquad (3.5)$$

where γ_n is the nodal acceleration vector, v_n the nodal velocity vector and u_n the nodal displacement vector, with:

$M_n = \int_{\Omega_n} \rho N^T N\, dV_n$ the mass matrix,

$F^{int}(u_n) = \int_{\Omega_n} B^T \sigma\, dV_n$ the internal nodal force,

$F_n^{ext} = \int_{(\Gamma_f)_n} N^T \bar{f}\; dS_n$ the external nodal force

$C = \alpha M$ the damping matrix is taken to be proportional to the mass matrix (Boulmane, 1994).

Nodal accelerations vector, respectively velocities and displacements are obtained by:

$$\begin{aligned} \gamma_n &= M_n^{-1}(F_n^{ext} - K(u_n) - C_n v_n) \\ v_{n+1/2} &= v_{n-1/2} + (t_{n+1} - t_n)\gamma_n \\ u_{n+1} &= u_n + (t_{n+1} - t_n) v_{n+1/2} \end{aligned} \qquad (3.6)$$

In the particular case where M is a diagonal matrix, the solution process run more easy and fast.

3.2 *Elastic-Plastic constitutive equations*

To describe the behaviour of sheet metal, we use a rotated formulation defined as:

$$\dot{\sigma}_a = C^e : \{\dot{d}_a - \dot{d}_a^p\} \qquad (3.7)$$

where:

$$\begin{aligned} \sigma_a &= a^T \sigma a \\ \dot{d}_a = a^T D a \qquad & \dot{d}_a^p = a^T D^p a \end{aligned} \qquad (3.8)$$

The strain rate tensor D, the plastic part D^p and the Cauchy stress tensor are defined in the Lagrangian configuration by the anisotropic rotated frame defined by the orthogonal tensor a (Boubakar, 1994).

The anisotropic initial plastic behaviour is defined by the Hill criterion and the associated flow rule:

$$\begin{aligned} f(\sigma_a, X, s) &= \frac{1}{2}\sigma_x : H : \sigma_x - \frac{1}{2}\sigma_0^2(s) \\ \dot{S} = \dot{\lambda} H : \sigma_x \qquad & \dot{d}_a^p = \dot{\lambda} H : \sigma_x \\ \dot{s} &= \dot{\lambda}\sqrt{\sigma_x : H : \sigma_x} \end{aligned} \qquad (3.9)$$

where: $\sigma_x = \sigma_a - X$.

The isotropic strain hardening law uses the scalar variable s and the tensor S. To take into account the kinematic hardening law we introduce the associated variable X. Tensor H describes anisotropy is well defined in the anisotropic frame and furthermore constant. The axes of orthotropy of the Hill criterion are updated by the rotation of the polar decomposition F=RU.

The incremental laws of anisotropic elastoplasticity are obtained by integrating previous equations (3.7),(3.8) and (3.9) during the increment $[t_n, t_{n+1}]$. By denoting $(.)_n$ values at beginning increment and $(.)_{n+1}$ values reached at the end of increment, we can write:

$$\begin{aligned} (\sigma_a)_{n+1} &= (\sigma_a)_n + \Delta\sigma_a \\ s_{n+1} = s_n + \Delta s \qquad & S_{n+1} = S_n + \Delta S \end{aligned} \qquad (3.10)$$

In order to compute the stress state at the end of the loading increment a numerical predictor-corrector scheme is used. In a first stage an elastic stress state is computed according to:

$$\begin{aligned} (\sigma_a)^*_{n+1} &= (\sigma_a)_n + C^e : \Delta d_a \\ s_{n+1} = s_n \qquad & S_{n+1} = S_n \end{aligned} \qquad (3.11)$$

followed by a plastic correction if the following con-

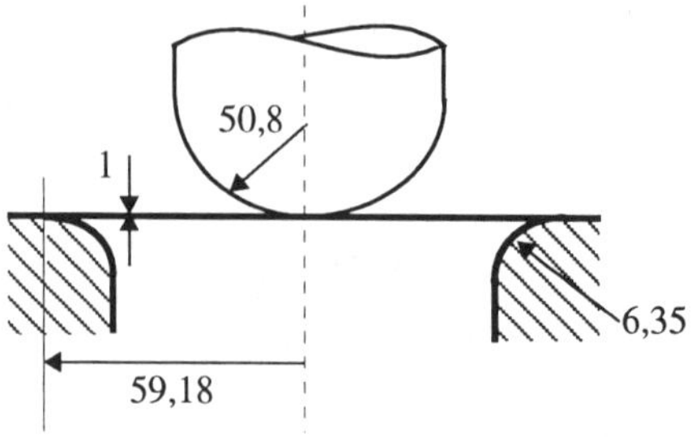

$E = 69000 MPa$

$\gamma = 0{,}3$

$\bar{\sigma} = 589\,(10^{-4} + \bar{\varepsilon})^{0{,}216}\ MPa$

Figure 3: Geometry and material properties.

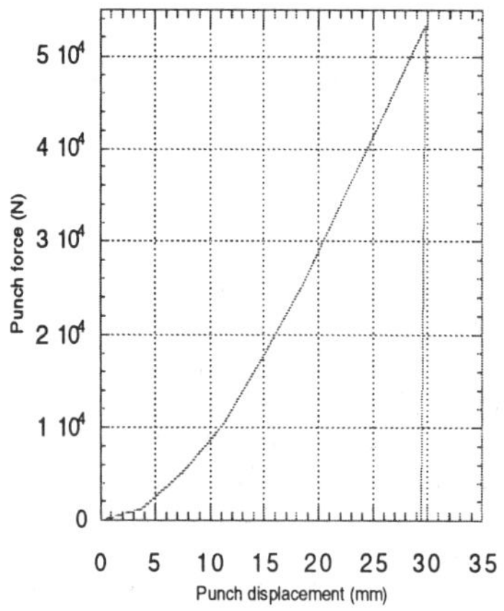

Figure 4: Load versus punch displacement with springback computation

dition is not satisfied:

$$f\left((\sigma_a)_{n+1}, X_{n+1}, s_{n+1}\right) = 0 \qquad (3.12)$$

In the case of a kinematic linear hardening ($\dot{X} = b d_a^p$), the integration algorithm is defined as (Boubakar, 1994):

$$(\sigma_a)_{n+1} = [I + b\Delta\lambda H]\,(\sigma_x)_{n+1} + X_n$$

$$s_{n+1} = s_n + \Delta\lambda\sqrt{(\sigma_x)^T_{n+1} H\,(\sigma_x)_{n+1}} \qquad (3.13)$$

$$S_{n+1} = S_n + \Delta\lambda H\,(\sigma_x)_{n+1}$$

where:

$$(\sigma_x)_{n+1} = (I + \Delta\lambda\,(C^e + bI)\,H)^{-1} \left[(\sigma_a)^*_{n+1} - X_n\right] \qquad (3.14)$$

Then the plastic behaviour described by the yield condition:

$$f\left((\sigma_a)_{n+1}, X_{n+1}, s_{n+1}\right) = 0$$

which is a non linear equation in $\Delta\lambda$ is solved by the Newton method.

3.3 Springback prediction

In sheet metal forming, springback prediction is important to evaluate the conformity of the desired shape (Karafillis, Boyce, 1992). The amount of springback is the difference between the final shape and the shape after the punch travel. Indeed, after the removal of tools, the deformed part shape undergoes another significant geometric changes. In other terms, springback is an additional deformation of the material which happens during the unloading. This change in the blank shape depends as well as material properties than as technological parameters: restraining force, friction between the sheet and the tools. The final part results in the deviation of the final products including dimensional errors as curvature and sidewall curls and internal defects as residual stresses and strains.

The unloading process is primarily elastic; however, the sheet could be bent and unbent around die and punch corners and secondary yielding can occur. Then an elastoplastic recovery would take into account a better strain path.

To predict geometry of formed parts, the springback modelling proposed with the quasi-static approach is obtained in a natural manner. At its beginning, tools are removed and all contact nodal forces are recovered. Then, all these contact node forces are decreased up to vanish by a proportional unloading using the same incremental method as for the drawn process. So, the sheet is set free from contact nodes. Sometimes it is also necessary to add new conditions to avoid rigid body motion. Another springback modelling is possible: at the end of the load stage, we give an opposite motion to the tools and computation continues until no contact with the sheet metal.

4 NUMERICAL RESULTS

4.1 Benchmark test

The first one is the well known benchmark test pro-

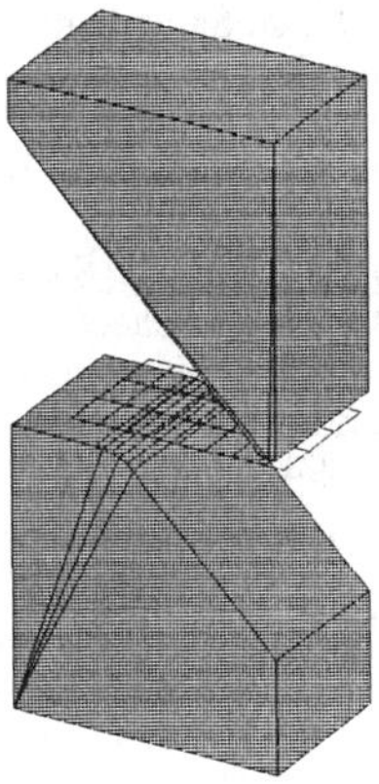

Figure 5: Mesh of the sheet metal, die and punch at the initial stage

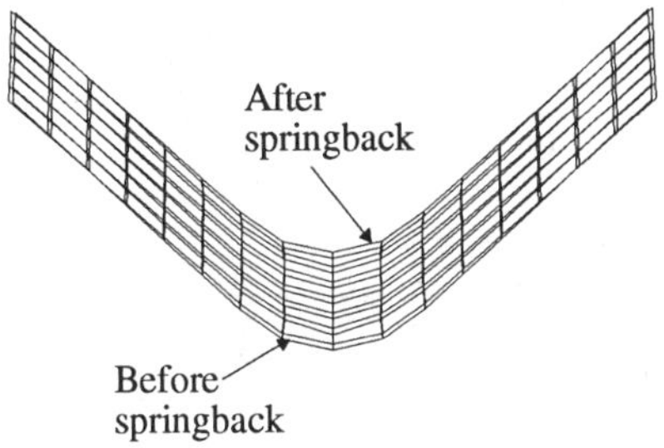

Figure 6: Deformed shape at 4.5mm and after springback

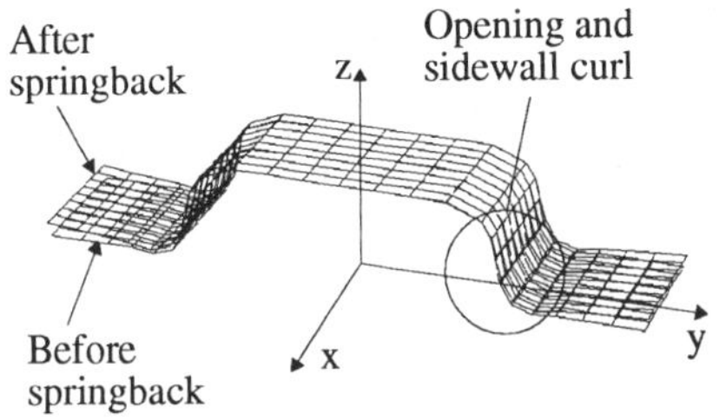

Figure 7: U-profile before and after unloading

posed for Numiform'89 conference (Lee, Wagoner, Nakamachi, 1990). This is a punch stretching with an hemispherical punch where the sheet metal is rigidly clamped around its external boundary. The geometry and the material properties are given in figure 3. Figure 4 shows the punch load displacement curve including springback computation. Results of springback effects obtained to a punch travel of 30 mm and then the punch have been removed are in a very good agreement with literature (Onate, Agelet de Saracibar, 1992). The final part shape have gone up under the punch of about 4 mm and the geometrical changes due to springback are insignificant because of the hoop stifness of the workpiece.

4.2 V-bending

This other example have been proposed by Takizawa and al. (1991). The problem to solve is a V-bending one. Geometry of tools and initial mesh are given in figure 5. Dimensions of the sample are 16x10 mm f and the thickness is 1 mm. Punch radius is 0.6 mm and die radii are 2 and 6 mm. Young modulus is 200GPa, Poisson 's ratio 0.3 and the work hardening properties are $\bar{\sigma} = 500\,(\bar{\varepsilon}^p + 00102)^{0,2}$ that represents the Swift's law with the yield stress equal 200 MPa.

Figure 6 shows the deformed mesh at the end of the loading stage and after springback. Punch displacement is 4.5mm and springback angle obtained is about 1.1° which is the springback angle computed by Takizawa. The final shape is more open and have gone up under the punch.

4.3 Draw-bending

The last example is taken from Ladreydt (1992) and concerns the forming of an U-profile corresponding to a punch stroke equal 20 mm. This problem has been analysed to show the deformed mesh at the end of the punch travel and at the final stage after springback.

The material of the sheet is defined by : E = 160 GPa, $\nu = 0.3$, $\bar{\sigma} = 616{,}3\,(0{,}00285 + \bar{\varepsilon}^p)^{0,1747}$, the following geometrical parameters are: length of the sheet 95 mm, thickness of the sheet 1,5 mm, punch length 41 mm and punch radius 6 mm with 1,8 mm clearance between the punch and the die.

Numerical results obtained are in agreement with the experimental ones. Indeed, figure 7 shows curvature and sidewall curl of the flange between the die and the punch.

The springback produces important change on the geometry of the shape and in particular a large opening of the U-profile which is a tendency usually observed on such U-profile. This example proves that the springback procedure proposed can evaluate the amount of springback in deep drawing processes in a realistic manner.

5 CONCLUSION

A finite element code is described to simulate stamping or deep drawing operations all along the complete process. This 3-D simulation code based on a thin shell element uses both a quasi-static fully implicit procedure and a transient explicit one.

To design appropriate tools in sheet metal forming, it is necessary to predict the amount of springback of parts. In the present paper, a simple method using contact forces in the quasi-static procedure achieve to evaluate accurately the final geometry of the shape. Some results presented on characteristic examples are satisfying. Other examples have to be done in order to validate the code. In case of success, this finite element code would reduce development of deep drawing experiments and tooling costs.

6 REFERENCES

Ahmad S., Irons B.M., Zienkiewiecz O.C., 1971, Analysis of thick and thin shell structures by curved finite elements, *Int. J. for Num. Meth. in Eng.*, 2, 419-451.

Boisse Ph., Daniel J. L., Gelin J. C., 1994, *A C° three-node shell element for non linear structural analysis*, *Int. J. for Num. Meth. in Eng.*, Vol. 37.

Boubakar M. L., 1994, Contribution à la simulation numérique de l'emboutissage des tôles. Prise en compte du comportement élastoplastique anisotrope, Ph. D. Thesis, University of Besançon.

Boulmane L., 1994, Application des techniques implicites-explicites de la dynamique transitoire à la simulation numérique en mise en forme, Ph. D. Thesis, University of Besançon.

Daniel J. L. - Gelin J. C. - Paquier P., 1993, POLYFORM- Logiciel 3D pour la simulation du forgeage et de l'emboutissage des matériaux, In National Congress on Structural Analyses, Hermès, Paris, pp. 1158-1166.

Gardiner F.J., 1958, The springback of metals, Trans. ASME 79.

Ladreydt Th., 1992, Modélisation bidimensionnelle et simulation numerique des processus de mise en forme des tôles dans l'industrie automobile, Ph. D. Thesis, I.N.P. Grenoble.

Lee J.K., Wagoner R.H., Nakamachi E., 1990, A benchmark tests for sheet metal forming analysis, Report n° ECR/NSM-5-90-22,NSF ERC, The Ohio State University.

Karafillis A.P. and Boyce M.C., 1992, *T*ooling design in sheet metal forming using springback calculation, *Int. J. Mech. Science*, vol. 34 n°2, pp 113-131.

Santos A., Makinouchi A., Takizawa H., Mori N., 1992,Simulation of springback behavior of 3D sheet bending processes by elasto-plastic FEM, Proceedings of the Third International Conference on Computational Plasticity, Fundamentals and Applications, Parts II, Barcelona, Spain, pp 1333-1343.

Onate O. and Agelet de Saracibar C., 1992, Alternatives for finite element analysis of sheet metal forming problems, *In Numerical Methods in Industrial Forming Processes*, Balkema, pp 79-87.

Takizawa H., Makinouchi A., Santos A., Mori N., 1991, Simulation of 3D sheet bending processes, Proceedings of F.E. Simulation of 3-D sheet metal forming processes in automotive industry, Zürich. VDI Report n° 894, pp 167-184.

TANG S.C., 1987, Analysis of springback in sheet forming operation, *Advanced Technology of Plasticity*, 1, pp 93-197.

Simulation of Materials Processing: Theory, Methods and Applications, Shen & Dawson (eds)
© 1995 Balkema, Rotterdam. ISBN 90 5410 553 4

Some advances in FEM simulation of sheet metal forming processes using shell elements

M. Kawka
The Institute of Physical and Chemical Research – RIKEN, Wako, Saitama, Japan (On leave from: Warsaw University of Technology, Poland)

A. Makinouchi
The Institute of Physical and Chemical Research – RIKEN, Wako, Saitama, Japan

ABSTRACT : This paper has been devoted to the development of the ITAS-3D FEM code. This code is widely used in Japanese industry to simulate sheet metal forming processes. Two important features of the efficient simulation are discussed : frictional contact problem and prediction of a ductile failure. In the ITAS-3D code contact phenomena are described in an efficient way by means of a boundary condition - "spring" element. In order to improve accuracy of a ductile fracture prediction, a precise thickness updating algorithm for shell elements is implemented. Both problems - friction and thickness updating are verified by experiment - hemispherical punch stretching - Nakazima test. A reasonably good agreement has been observed between numerical and experimental results.

1 INTRODUCTION

A competition between producers in the car industry has demanded to enhance product quality, reduce production costs and make design and planing processes shorter. One of the way to achieve these aims is to integrate FEM codes used in simulation of sheet metal processes into the CAD/CAM environment. Many of a car manufactures already implemented the FEM simulation into the decision making process of a company. Despite of successes, there are still unsolved or needed improvement problems. Efficient simulation of a contact behavior and prediction of a sheet metal ductile failure are some of them.

A precise description of the contact phenomena is necessary for successful simulation of stamping process. For example the tangential, friction forces on a sheet - tool interface can severely affect formability of stamped parts. On the other hand, many of implicit FEM codes suffer problems with a convergence due to instability of contact/uncontact and stick/slip behavior. In our paper we present a robust approach to describe contact phenomena by means of a boundary condition element. A constitutive equation of this element is equivalent to the friction law defined for a tool-sheet interface.

Failure by necking and rupture is one of the most important, limiting phenomenon in sheet metal forming. Diffuse necking is a form of the thickness strain localization, so the precise thickness updating algorithm is crucial for prediction of a ductile failure. In our paper we present an improved algorithm used to update thickness of shell elements.

The other problem is to find the most appropriate finite element for simulation of a ductile failure. We analyze behavior of different shell elements with various integration schemes including the element with modified and improved stabilized integration scheme. Finite elements and numerical procedures are validated by comparison of experimental and numerical results of a hemispherical punch stretching test. A good agreement can be observed.

Results presented in our paper are obtained using the semi-explicit FEM code ITAS-3D. This code is continuously developed by Sheet Metal Forming Simulation Research Group in Japan and is widely used in Japanese automotive industry to simulate sheet metal forming processes.

2 FORMULATION

The updated Lagrangian rate formulation is the base of the incremental elastic-plastic finite element code ITAS-3D. The principle of virtual velocity describes a rate form of the equilibrium equation and boundary conditions. The form of the principle is very similar to the form proposed by (McMeeking & Rice 1975) :

$$\int_V \left\{ \left(\overset{J}{\sigma}_{ij} - 2\sigma_{ik}D_{ik} \right)\delta D_{ij} + \sigma_{jk}L_{ik}\delta L_{ij} \right\} dV = \int_{S_t} \dot{f}_i \delta v_i dS$$

While deriving this equation we took into account that

for sheet metal forming processes $\det(\partial x/\partial x_o) \cong 1$. As a consequence we have $\tau^J = \sigma^J$ where τ^J is the Jaumann derivative of Kirchhoff stress and σ^J is the Jaumann derivative of Cauchy stress. Here, δv is the virtual velocity field satisfying the boundary condition $\delta v=0$ on S_u. V and S denote respectively, the region occupied by the body and its boundary at time t. S_t is a part of the boundary S, on which the velocity is prescribed. L is the gradient of the velocity field ($L=\partial v/\partial x$), D and W are respectively the symmetric and anti symmetric parts of L.

A small strain linear elasticity and the large deformation rate-independent work-hardening plasticity was assumed. The constitutive equation can be written as in the form :

$$\sigma^J_{ij} = C^{ep}_{ijkl} D_{kl} = C^{ep}_{ijkl} L_{kl}$$

where C^{ep}_{ijkl} is the tangent constitutive matrix. Introducing the above constitutive equation to the variational principle the final form of the principle of virtual velocity is obtained :

$$\int_V D_{ijkl} L_{kl} \delta L_{ij} dV = \int_{S_t} \dot{f}_i \delta v_i dS \qquad (1)$$

where :

$$D_{ijkl} = C^{ep}_{ijkl} + \Sigma_{ijkl}$$

$$\Sigma_{ijkl} = \frac{1}{2}\cdot\left(\sigma_{jl}\delta_{ik} - \sigma_{ik}\delta_{jl} - \sigma_{il}\delta_{jk} - \sigma_{jk}\delta_{il}\right)$$

3 FINITE ELEMENT EQUATIONS

The above stated problem is solved in a standard way: Eqa.(1) is integrated from time *t* to $t+\Delta t$ where Δt is a small time increment. The displacement increment, the Jaumann stress increment, and the increment of the displacement gradient are written as :

$$\Delta u = v\cdot\Delta t, \quad \Delta\sigma^J = \sigma^J\cdot\Delta t, \quad \Delta L = L\cdot\Delta t, \quad etc.$$

and all rate quantities are simply replaced by incremental quantities, assuming that rates are kept constant within an incremental step. When performing a standard finite element discretization procedure, the variational principle can be replaced by a system of algebraic equations :

$$K\cdot\Delta u = \Delta F + \Delta C \qquad (2)$$

where K is the elastic-plastic stiffness matrix. The terms ΔF and ΔC come from the right hand side of Eqa.(1), where the derivative $\dot{f}$ has been replaced by the expression $\dot{f} = \dot{f}\cdot e + f\cdot\dot{e}$. The term $\Delta F = \Delta f\cdot e$ denotes the increment of the external force vector and the term $\Delta C = f\cdot\Delta e$ expresses the rotation of the total force vector during the incremental step.

In the ITAS3D code a semi-explicit approach to the solution of Eqa.(2) is applied. A stiffness matrix K is described at time *t* and is regarded as constant within the time increment Δt. The so-called "r-min" method (Yamada *et al.* 1968) is employed to impose a limitation on a size of the time step

Some number of iteration are needed to solve Eqa.(2) due to the fact that the ΔC term can be calculated only if the final configuration is known (a change of the local coordinate base vector Δe has to be found).

4 ELEMENTS

It seems to be very natural to use elements based on the shell theory to simulate sheet metal forming processes. Elements based on the Mindlin-Reissner theory which demands only C^0 continuity dominate the shell analysis. The most popular elements among them is the family of degenerated shell elements.

The idea of a degenerated shell element comes from the work of (Ahmad *et al.* 1970). This type of element is based on the Mindlin-Reissner theory : deformation field is defined by means of the two independent variables - the displacement and the rotation of the mid surface nodes. The 4 node quadrilateral element with bilinear shape functions is preferred for its effectiveness and its simplicity of formulation.

The kinematics of the degenerated shell element are defined in terms of the shape functions by invoking the isoparametric hypothesis. Thus geometry and velocity field can be written as :

$$x_i(\xi,\eta,\zeta) = N^\alpha(\xi,\eta)\cdot x_i^\alpha + N^\alpha(\xi,\eta)\cdot\frac{1}{2}\cdot\zeta\cdot h^\alpha\cdot p^\alpha$$

$$v_i(\xi,\eta,\zeta) = N^\alpha(\xi,\eta)\cdot v_i^\alpha + N^\alpha(\xi,\eta)\cdot\frac{1}{2}\cdot\zeta\cdot h^\alpha\cdot\theta_i^\alpha \qquad (3)$$

or

$$v_i(\xi,\eta,\zeta) = \overline{N}^\alpha_{ri}(\xi,\eta,\zeta)\cdot\overline{v}^\alpha_r$$

where v^α and θ^α are respectively the velocity and the fiber rotation velocity for node "α". Symbols $\overline{N}^\alpha$ and $\overline{v}^\alpha$ mean the generalized shape function and the generalized fiber velocity $\overline{v}^\alpha = [v^\alpha, \theta^\alpha]$. Index "r" takes the value of 1 to 5 (i.e. $\overline{v}^\alpha_r = [v_1, v_2, v_3, \theta_1, \theta_2]$). The components of the unit nodal vector are written as p_i^α. Using Eqa.(3), expression defining the elemental stiffness matrix K can be obtained :

$$K^{\alpha\beta}_{rs}(\xi,\eta,\zeta) = \int_V \overline{N}^\alpha_{ri,j}(\xi,\eta,\zeta)\cdot D_{ijkl}\cdot\overline{N}^\beta_{sk,l}(\xi,\eta,\zeta)dV$$

where the partial differentiation operator $\partial/\partial x$ is designated be a comma ",j".

Fully integrated stiffness matrix (8 - integration points for a 4 - node shell element) performs very well for moderately thick shell situation. However, when working with thin shell structures, some parts of the stiffness matrix relate to membrane or shear deformation artificially dominate the total stiffness matrix (membrane or shear locking). Consequently too stiff solution is obtained. The most popular approach to overcome this problems is to use a special integration technique. In the ITAS-3D code the most reliable results are obtained for elements with integration scheme based on assumed strain field for shear components - ASF and elements with stabilization matrix - SM.

Assumed strain field - ASF : to avoid the problem of shear locking and at the same time to retain the rank of the stiffness matrix the concept of a substitute shear strain field has been proposed by (Dvorkin & Bathe 1984). For the four node degenerated shell element, transverse shear components of the velocity gradient are calculated using a special shape functions. The stiffness matrix for the ASF approach is fully integrated so there are no problems with zero energy modes of deformation. The only disadvantage is its low efficiency; the element is therefore expensive in computation as a full integration scheme has to be used : $2 \times 2 \times 2$ integration points .

Stabilization matrix - SM : to suppress spurious deformation modes which affect solution when reduced integration is used, the concept of stabilization matrix has been introduced (Liu *et al.* 1985). This concept comes from the assumption that the velocity gradient tensor L_{ij} can be approximated by Taylor's series around point $(0,0,\zeta)$ (in a natural coordinate system). If the above assumption is used to calculate the stiffness matrix K the following expression can be obtained $K = K^o + K^s$, where K^o is the standard stiffness matrix calculated at the central point of element $(0,0,\zeta)$ and K^s is the stabilization stiffness matrix. In our version of stabilization matrix, an assumed strain field is used to calculate shear components (an idea similar to the ASF strain field) and number of integration points can vary from 1 to 4 through thickness.

5 FRICTION

The friction phenomena can affect sheet metal forming processes crucially. Formability of stamped parts, surface defects and springback are one of the most important problems related to the friction. Over the last few years many papers have been published in a field of theoretical and computational aspects of the friction phenomena. But the problem of effective implementation of friction to FEM codes is still far from being solved.

Frictional contact model implemented to the ITAS-3D code is based on a modified framework of theory of plasticity (Michalowski & Mroz 1978, Wriggers *et al.* 1989). Following a standard way for formulation of elastic-plastic models, the friction constitutive relation can be written as (Kawka & Makinouchi 1984) :

$$f_i^{\nabla} = C_{ij} \cdot \tilde{v}_j \tag{4}$$

where : $\tilde{v}$ is the relative velocity between two bodies, f^{∇} is the co-rotational rate of the contact force vector : $f_i^{\nabla} = \dot{f}_i - \tilde{W}_{ij} \cdot f_j$ ($\tilde{W}$ - is the spin tensor) and C is the constitutive tensor. The co-rotational rate of force vector f^{∇} has one normal and two tangential components to the contact surface - subscripts '1' and '2' mean tangential directions to the contact surface, '3' means normal direction. The constitutive tensor C has two "material" constants (isotropic behavior of friction in the contact plane is assumed) : tangential stiffness - E_t, and normal stiffness - E_n.

To implement the frictional contact conditions to the ITAS-3D code, we improved a concept of the "spring" contact element - (Vreede 1992).

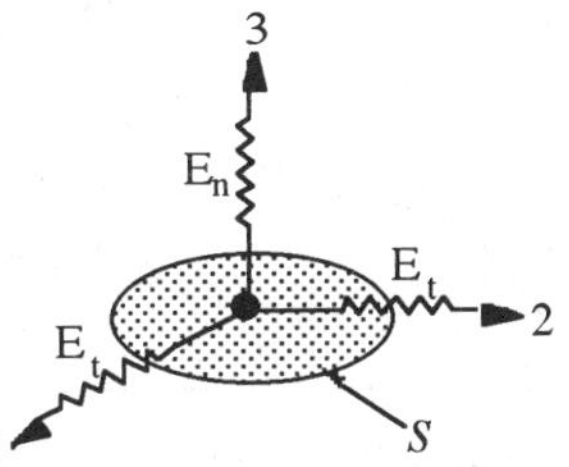

Fig. 1 The "spring" contact element
(*S* - an area of surface surrounding node)

For all nodes in contact with tool, the "spring" elements are used - Fig.1, while Eqa.(4) serves as a constitutive equation of this element (normal direction to the tool surface is coincident with the direction '3' of the contact spring element). Combining Eqa.(2) and Eqa.(4) and integrating over an area of surface *S* for all elements we get :

$$(K + K^{fr}) \cdot \Delta u = K^{fr} \cdot \Delta u^t + \Delta C \tag{5}$$

where K^{fr} is the friction stiffness matrix, Δu^t is the tool's displacement vector.(given at the beginning of the incremental step). The friction matrix K^{fr} (size 3x3) exists only for degrees of freedom corresponding to nodes being in contact with tool.
An implementation of friction in the form of Eqa.(5) seems to be very suitable for static explicit FEM codes because solution can be obtained without any iteration (all the elements of K^{fr} matrix are known at the beginning of incremental step). Some complications arise due to the fact that the friction constitutive matrix K^{fr} is unsymmetric. To deal with this problem Eqa.(5) is rearranged to separate unsymmetric terms related to the change of the normal contact force. Then the final form of FEM

equation is obtained :

$$(K+K^{fr^*})\cdot \Delta u = K^{fr^*}\cdot \Delta u^t + \Delta N + \Delta C \qquad (6)$$

where K^{fr^*} is the symmetric part of the friction matrix K^{fr} and the vector ΔN expresses the change of friction forces due to the change of normal load (unsymmetric terms). Some number of iteration is needed to solve Eqa.(6) as the vector ΔN can be calculated only in an implicit way. Additionally, a predictor-corrector radial return scheme is applied to integrate the friction constitutive equation - so changes in the friction state : sticking to sliding and opposite are treated in a very efficient way. An additional advantage of using Eqa.(6), is that normal stiffness parameter - E_n is eliminated from the friction constitutive equation and does not need to be identified.

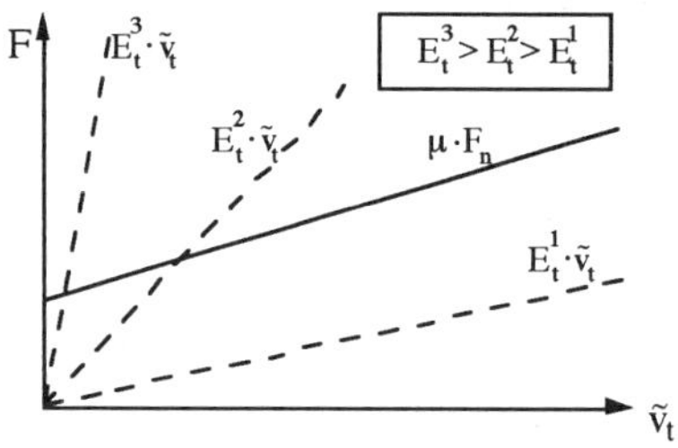

Fig.2. Identification procedure - E_t
(F_n - normal force, μ - friction coefficient,
$\tilde{v}_t$ - tangential component of relative velocity $\tilde{v}$)

Any calculation involving the above friction model needs to identify a value of the tangential stiffness parameter - E_t. This parameter can affect strongly behavior of the friction phenomena. If too small value of E_t is chosen (E_t^1 - Fig.2) then the "spring" contact elements remains in a "sticking" state - no slip occurs. Too big value of E_t - (E_t^3 - Fig.2) affects stability of calculation - many changes of friction state (stick to slip and opposite) may occur and as a consequence, time of calculation can be significantly increased.

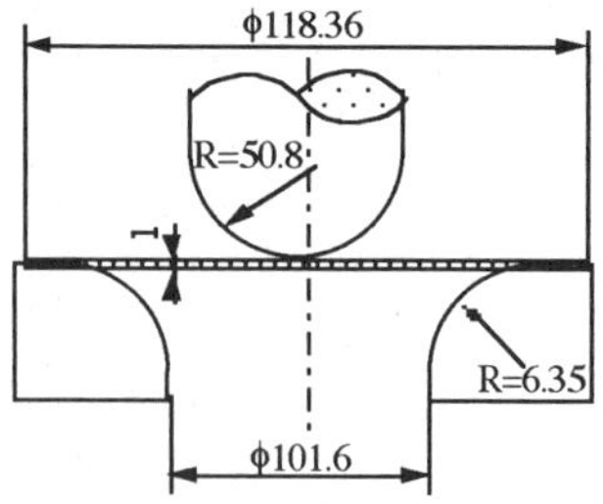

Fig.3 The OSU benchmark test - geometry

In practice, we use numerical procedure to identify E_t parameter. As example of this procedure, results of the OSU hemispherical punch stretching test are presented in this paper - Fig.3 (Lee *et al.* 1990).

A coarse mesh with 300 elements was used in FEM simulation (friction coefficient μ=0.3). Table 1 and Fig.4 presents results of calculation using various values of the tangential stiffness E_t. For E_t values higher then E_t^3 no major changes in results can be observed. For big value of E_t (E_t^4) time of calculation is much longer than for other values, so finally we chose E_t^3 as a value of the tangential stiffness parameter E_t in FEM simulation.

Tab.1 The OSU benchmark test
Identification of the E_t parameter

	Tang. stiffness E_t	Max. force [kN]	Solution time T/T_1
1	1	72.88	1
2	10	59.92	1.59
3	50	59.28	1.61
4	100	59.22	2.28

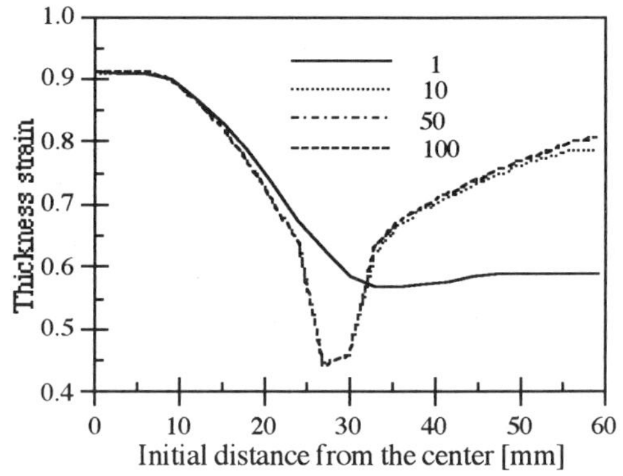

Fig.4. The OSU benchmark test
Identification of the E_t parameter.

6 STRAIN LOCALIZATION

Sheet metal thickness variations is of great importance for stamping technology. The most common defect in that field results from the localization of the thickness strain which leads to a breakage or to a ductile failure. In the ITAS-3D code a special algorithm is used to deal with description of shell element's geometry and to precisely update thickness of shell elements.

To assure conservation of the element's volume during deformation process, thickness of element is stored and updated at each integration point. At first strain increments $\Delta\varepsilon_{11}$ and $\Delta\varepsilon_{22}$ are evaluated (a plane stress assumption is imposed on constitutive relation). Then thickness strain increment $\Delta\varepsilon_{33}$ is calculated, taking into account plastic incompressibility condition. That strain increment is used to update thickness h_q at the integration point "q". As a consequence the following equation is always valid for the two incremental steps n and $n+1$:

$$\sum_q \det(J_q)^n = \sum_q \det(J_q)^{n+1}$$

J_q stands for Jacobian at the integration point q. Since element's thickness is stored at integration points, before formulation of the stiffness matrix additional procedure is used to extrapolate thickness from integration points h_q to nodes h^α :

$$h^\alpha \xleftarrow[N^\alpha]{} h_q$$

Extrapolation is done using the same shape function N^α which are used in definition of geometry of the degenerated shell element - Eqa.(3).

To verify applicability of the ITAS-3D code to the prediction of the strain localization phenomena, simulation and experiments of a hemispherical punch stretching test were performed. That test is often used to evaluate the basic formability of automotive sheet metals.

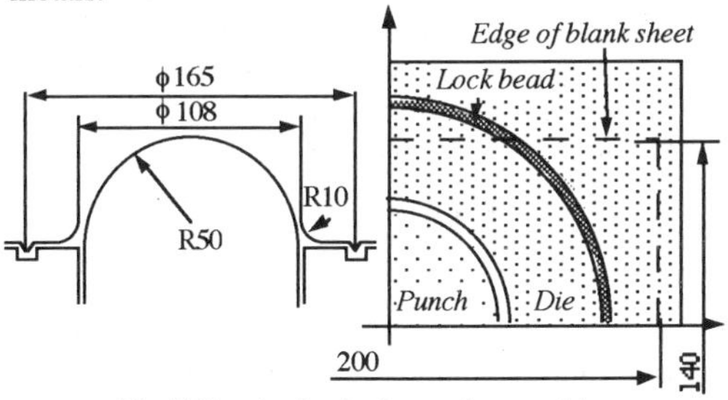

Fig.5 Hemispherical punch stretching
Tools and blank sheet geometry

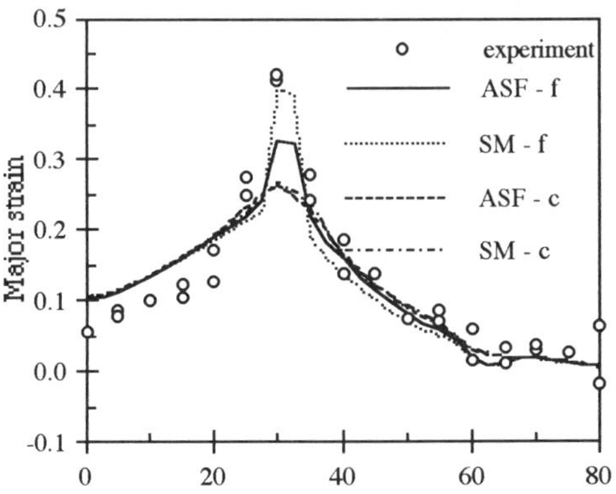

Fig.6 Major strain distribution in the hemispherical punch stretching test (LDH=35.7mm).
ASF and SM elements; f - fine mesh, c - crude mesh

In that test formability of sheet metals is measured by the LDH *(Limit Dome Height)* parameter - defined as a punch stroke at which breakage of a specimen occurs. Tools and specimen geometry used in experiment and simulation is given in Fig.5. Experiment was done using SPCE steel - thickness 0.8 mm. The corresponding material parameters used in FEM simulation were : yield stress - 169 MPa and stress-strain curve $\sigma_p = 568 \cdot (0.0078 + \varepsilon_p)^{0.249}$MPa. Friction coefficient μ=0.14 was measured experimentally using a flat-die sliding test. Distribution of principal strain in experiment was found by measurement of scribed circles (initial diameter 5 mm) before and after deformation. Two types of mesh were used in simulation - crude mesh (elements 5x5 mm initially), and fine mesh (elements 2.5x2.5 mm initially). Results are given in Fig.6.
Good correlation of experiment and simulation results can be observed. Both types of elements ASF and SM gave adequate accuracy in prediction of localized deformation field. Accuracy is increased when a fine mesh is used. In case of a fine mesh, SM element gave very good results, better than ASF element. That situation can be explained by reduced number of integration points in SM element what accelerates strain localization phenomena in a small area of few elements.

Additional experiments and simulations were conducted to verify friction model implemented to the ITAS-3D code. Hemispherical punch stretching tests were done for two completely different friction condition : using degreased blank sheet surface (no lubricant) and very good lubricant (Teflon film). Punch stroke was 30 mm. In a case of Teflon film, friction coefficient μ=0.04 was measured in a flat-die sliding test.

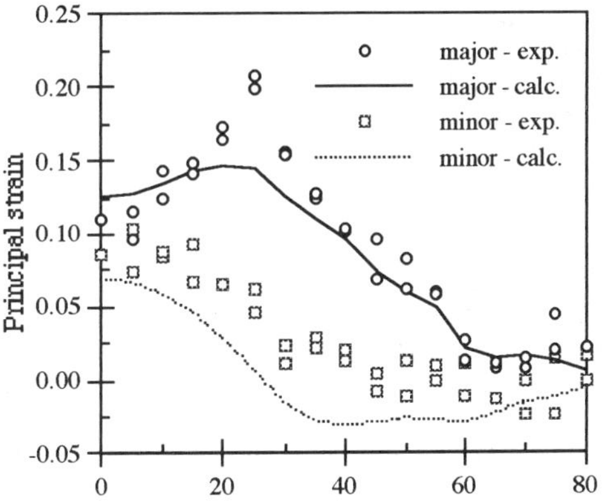

Fig.7 Principal strain distribution
in the hemispherical punch stretching test (μ=0.04)

For degreased sheet metal surface friction coefficient was found in range μ=0.2-0.35 (due to problems with stick-slip behavior of degreased metal sheet). Results of experiment and calculation (using coarse mesh of SM elements) are given in Fig.7,8.

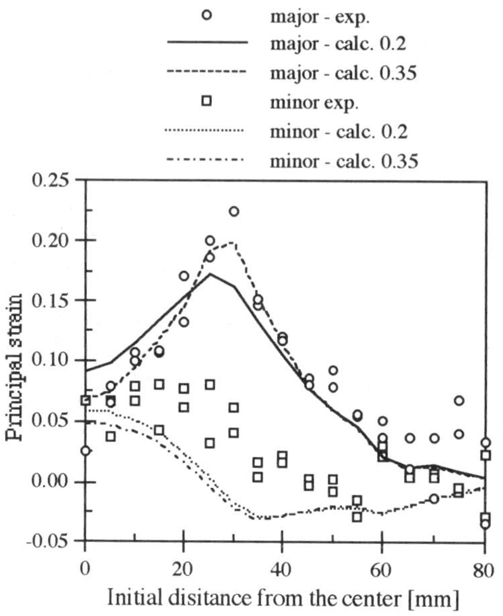

Fig.8 Principle strain distribution
in the hemispherical punch stretching test
(μ=0.2 and μ=0.35)

Reasonable correlation between results of experiment and simulation can be observed. Some discrepancy in distribution of minor strain can be explained by difficulties in precise measurement of scribed circles on specimen's surface and difficulties in imposing boundary conditions corresponding to a lock bead. Simplified constitutive material model (isotropic properties were assumed for calculations) can be also a source of discrepancy.

CONCLUSIONS

Hemispherical punch stretching test was numerically analyzed by the ITAS-3D FEM code. Results of calculation agree quite well with the experimental results. That proves usefulness of numerical algorithms implemented to the ITAS-3D code and dealing with friction and thickness localization phenomena.

ACKNOWLEDGMENT

Authors would like to thank Mr. T.Yoshida from Steel Research Laboratories of Nippon Steel Corporation for providing experimental results of the LDH test.

REFERENCES

Ahmad S., Irons B.M. & Zienkiewicz O.C. 1970. Analysis of thick and thin shell structures by curved finite elements, *Int. J. Num. Meth. Eng.* 2: 418-451.

Dvorkin E.M. & Bathe K.J. 1984. A continuum mechanics based four-node shell element for general non-linear analysis, *Eng. Comp.*, 1: 77-88.

Kawka M. & Makinouchi A. 1994. FEM in simulation of Sheet Metal Forming Processes. In M.Pietrzyk *et al.* (eds), *Huber's Yield Criterion in Plasticity* : 241-266. Krakow: Akademia Gorniczo-Hutnicza.

Liu W.K., Ong J.S.J. & Uras R.A. 1985. Finite element stabilization matrices - a unification approach, *Comp. Methcds Appl. Mech. Eng.* 53: 13-46.

McMeeking R.M. & Rice J.R. 1975. Finite element formulation for problems of large elastic-plastic deformation, *Int. J. Solids Structures* 11: 601-616

Yamada Y., Yoshimura N. & Sakurai T. 1968., Plastic stress-strain matrix and its application for the solution of elastic-plastic problems by the finite element method, *Int. J. Mech. Sci.* 10: 343-354.

Michalowski R. & Mroz Z. 1978. Associated and non-associated sliding rules in contact friction problems, *Arch. Mech.* 30: 259-276.

Wriggers P., Vu Van T. & Stein E. 1989. Finite element formulation of large deformation impact-contact problems with friction, *Comput. Struct.* 32: 319-331.

Vreede P. 1992. *A finite element method for simulation of 3-dimensional sheet metal forming,* P.T.Vreede, Enshede, 1992.

Simulation of Materials Processing: Theory, Methods and Applications, Shen & Dawson (eds)
© 1995 Balkema, Rotterdam. ISBN 90 5410 553 4

Finite element analysis and experimental verification for drawbead drawing processes

Changman Kim
Daewon Junior College, Jecheon, Korea

Youngseok Im
Daewoo Electronics Co., Ltd, Seoul, Korea

Naksoo Kim
Hong Ik University, Seoul, Korea

Daegyo Seo
Sung Kyun Kwan University, Suwon, Korea

ABSTRACT: Theoretical and experimental drawing characteristics for the single circular and the single square drawbeads are discussed. During the blank holding process, the strain distributions on the upper and the lower skins of specimens, and the die reactional forces are calculated by the F.E.M. The simulational results are compared with the experimental ones. The effects of the drawing length and the drawing angle on the drawbead restraining forces and the strain distributions in the drawn specimens are investigated and the calculations are compared with the experiments. It is concluded that the theoretical simulations and results would be useful for the prediction of real cases.

1 INTRODUCTION

The drawbeads installation is essential not only for fabricating automobile panel but also for the many kinds of general purpose press works in industries. The sheet formabilities could be improved by the drawbeads installation because the locations, shapes, and size of drawbeads could control metal flow, and hence minimize the defects during the forming processes. In fact, a small variation of drawbead shapes would largely affect the success or failure of sheet metal forming. The installation of drawbeads is widely applied in automobile industry because the selection of shapes and sizes of drawbeads and their installations are relatively easy compared to other methods, such as controlling the blank holder. Therefore, many studies on the drawbeads have been performed in the developed countries of automobile industry. For example, these nations have been establishing data bases for the various drawbead shapes to be used in the plants. A study on drawbeads had been initiated with single circular drawbead by Nine in 1978 and 1982. Wang derived the drawbead restraining forces by means of mathematical model in 1982 and Nine had performed experiments with inserting urethane in drawbead cavity. In 1983, Levy had improved Wang and Nine's study on drawbead. In 1984, Yellup and Painter predicted the restraining forces by the computer modelling of shallow drawbead system. From 1986 to 1987, Maker and Samanta studied the relations on the drawbead restraining force-displacement by the elastic-plastic shell element model. In 1988, Weinmann analyzed the circular drawbead by assuming it as a roller geometrical concepts. Also, Stoughton had considered the bending effects of sheet and studied the drawbead restraining forces and blank holding forces in connection with the frictional coefficients. Since 1988, Ujihara, Sakamoto and Frubayashi have established expressions on drawbead restraining forces and strains approximately based on the belt theory. Recently, automobile companies have been performing the studies on the combined drawbeads. It could be easily guessed that it would need more time and financial expenses for the design and production of drawbeads with the trial and error method based on the experiences of skilled engineers only. In the case of sheet metal forming process, during the first step(binder wrap), specimen is formed to drawbead shape by moving upper die, and appropriate blank holding forces are applied. During the second step, specimen is drawn through drawbead part by deep drawing process. In this study, we considered two steps, i.e. the drawbead forming (the first step) and the drawing at flange part (the second step). We used the single circular and the single square drawbeads which are used most generally in practical industries. The restraining forces and deformed phenomena of specimens predicted by F.E.M are compared with the experimental results for each step.

2 MODELLING OF DRAWBEAD

The F.E.M is formulated into two ways, i.e. elastic-plastic and rigid-plastic according to the constitutive equation of material. It is impossible to calculate the residual stresses by the rigid-plastic F.E.M only, however it could be efficient and hence it is used generally in the large deformation problems. In this study, we have made basic assumptions for the drawbead analysis by using isoparametric four-node solid elements which are easy to be handled and efficient in the calculations: 1) In drawbead forming, we assumed plane strain deformation by disregarding width strain because it is negligible. 2) Specimen is of isotropic and rigid-plastic material which has no differences for the planar and normal anisotropy, and follows the general flow rule. 3) Tool is all rigid body and coulomb frictional law is satisfied at the tool-specimen interfaces.

2.1 Governing equations

In this study, inertial effects, body forces and geometry changes of tools due to deformation are neglected. Therefore the velocity and stress fields are satisfied as the following conditions.

(i) Equilibrium equations:

$$\frac{\partial \sigma_{ij}}{\partial x_j} = 0 \qquad \text{in V} \quad i,j=1,2,3, \tag{1}$$

or $\sigma_{ij,j} = 0$,

where σ_{ij} are stress tensors, x_j are Cartesian rectangular coordinates of a current point of the sheet at time t.

(ii) Boundary conditions:

$$\begin{aligned} \upsilon_i &= \overline{\upsilon_i} && on\ S_u \\ t_i &= \sigma_{ij}\, n_j && on\ S_F \end{aligned} \qquad (S_u+S_F=S_{Total}), \tag{2}$$

where t_i is unit surface forces(traction), n_j is outward unit normal vector.

(iii) Constitutive equations

$$\dot{\varepsilon}_{ij} = \frac{3}{2}\frac{\dot{\bar{\varepsilon}}}{\bar{\sigma}}\sigma'_{ij} , \tag{3}$$

where $\dot{\bar{\varepsilon}} = (\frac{2}{3}\dot{\varepsilon}_{ij}\dot{\varepsilon}_{ij})^{1/2}$ is an effective strain rate and $\bar{\sigma} = (\frac{3}{2}\sigma_{ij}\sigma_{ij})^{1/2}$ is an effective stress.

(iv) Compatibility equation:

$$\dot{\varepsilon}_{ij} = \frac{1}{2}(\upsilon_{i,j} + \upsilon_{j,i}) , \tag{4}$$

where $\dot{\varepsilon}_{ij}$ are strain rate tensors.

(v) Incompressibility condition:

$$\dot{\varepsilon}_{kk} = 0 . \tag{5}$$

(vi) Flow stress

$$\bar{\sigma} = C(\varepsilon_0 + \bar{\varepsilon})^n . \tag{6}$$

Equation (6) is work hardening law of Swift, where C, ε_0, n are material parameters.

2.2 Weak form of the boundary value problem

The weak form of the boundary value problem including boundary conditions, the equilibrium equations, and the incompressibility conditions can be written as

$$\int_V \sigma_{ij}\delta\dot{\varepsilon}_{ij}dV - \int_{S_F}\overline{t_i}\delta\upsilon_i dS + \int_V k\,\dot{\varepsilon}_{kk}\delta\dot{\varepsilon}_{kk}dV = 0 , \tag{7}$$

where k is a penalty constant(a very large positive constant).

2.3 Finite element discretization and assembly

Fully discretized form of nonlinear system of equations may be written as

$$\begin{aligned} \Phi(\hat{\upsilon}) &= \sum_{m=1}^{M}\Phi(\hat{\upsilon})^{(m)} \\ &= \sum_{m=1}^{M}\Big[\int_{V^{(m)}}\frac{2}{3}\frac{\bar{\sigma}}{\dot{\bar{\varepsilon}}}B^TDBdV\,\hat{\upsilon}^{(m)} \\ &\quad + \int_{V^{(m)}}kB^Tcc^TBdV\,\hat{\upsilon}^{(m)} - \int_{S_F^{(m)}}t^TNds\Big] = 0 , \end{aligned} \tag{8}$$

where B is a matrix composed of the derivatives of the shape functions, c is a constant vector, D is a coefficient matrix, and $\sum_{m=1}^{M}$ represents the assembly over all volume and surface of each element. Also, the global form of equation (8) can be written as follows:

$$\Phi(\hat{\upsilon}) = F(\hat{\upsilon})\hat{\upsilon} + G\hat{\upsilon} - h = 0 , \tag{9}$$

where $\hat{\upsilon}$ is a nodal velocity vector and the other notations are summarized in equations (10) ~ (14).

$$F(\hat{\upsilon}) = \int_V \frac{2}{3}\frac{\bar{\sigma}}{\dot{\bar{\varepsilon}}}B^TDBdV , \tag{10}$$

$$G = \int_V kB^Tcc^TBdV , \tag{11}$$

$$h(\hat{\upsilon}) = \int_{S_F} t^TNdS , \tag{12}$$

$$t = -\mu p\Big(\frac{2}{\pi}\tan^{-1}\big(\frac{|\upsilon_r|}{a}\big)\Big)\frac{\upsilon_r}{|\upsilon|} , \tag{13}$$

$$\upsilon_r = N(\hat{\upsilon} - \upsilon_{die}) , \tag{14}$$

In the above equation, υ_r is the relative velocity, υ_{die} is the velocity of tool, a is a small constant of order 10^{-4}~10^{-5}, N is a matrix of shape functions, and μ is a friction coefficient.

2.4 Iterative methods in nonlinear system

(i) Direct iterative method

This method can be used to find an initial guess for the Newton-Raphson iterative method. From equation (9),

$$\{F(\hat{v}) + G\}\hat{v} = h(\hat{v}) \tag{15}$$

Equation (15) can be written in terms of a stiffness matrix K and a load vector L,

$$K_{dir.}\,\hat{v} = L_{dir.}\ , \tag{16}$$

where $K_{dir.} = F + G$ and $L_{dir.} = h$. Since $K_{dir.}$ is a nonlinear function of $\hat{v}$, we use iterative method to find a reseaonable solution as following:

$$K_{dir.}^{\ i}\ \hat{v}^{i+1} = L_{dir.}^{\ i}\ , \tag{17}$$

where superscript i denotes the number of iterations.

Direct iterative method usually gives a reliable initial guess for the Newton-Raphson iterative method, but can not be used to find the fully converged solution because of its slow convergence near the exact solution. The iterations will be terminated if the solution of linearized equation (17) satisifies the following condition.

$$\frac{\| \hat{v}^{i+1} - \hat{v}^{i} \|}{\| \hat{v}^{i} \|} \leq \epsilon_{norm} \tag{18}$$

(ii) Newton-Raphson iterative method

The Newton-Raphson method can be applied to solve equation (9) by means of Taylor's expansion as in the following.

$$\Phi(\hat{v}_{exact}) = \Phi(\hat{v}_{guess} + \Delta\hat{v}) \tag{19}$$

$$\cong \Phi(\hat{v}_{guess}) + \left.\frac{\partial \Phi}{\partial \hat{v}}\right|_{\hat{v}_{guess}} \Delta\hat{v} = 0$$

Linearized equation system (19) can be used in iterative method,

$$\Phi(\hat{v}^{i}) + \left.\frac{\partial \Phi}{\partial \hat{v}}\right|^{i} \Delta\hat{v} = 0\ , \text{ and} \tag{20}$$

$$\hat{v}^{i+1} = \hat{v}^{i} + \alpha\Delta\hat{v}\ , \tag{21}$$

Where α is a deceleration coefficient and i is the number of iterations. At the first iteration, $\alpha \doteqdot 0.1$ is used. Equation (21) can be written in a form containing the stiffness matrix and the load vector as follows:

$$K^{i}_{N-R}\,\Delta\hat{v} = L^{i}_{N-R}\ , \tag{22}$$

where $K^{i}_{N-R} = \left.\dfrac{\partial \Phi}{\partial \hat{v}}\right|^{i}$

and $L^{i}_{N-R} = -\Phi(\hat{v}^{i}) = -F(\hat{v})\hat{v} - G\hat{v} + h(\hat{v})$.

The numerical solution found by the Newton-Raphson method can be regarded as accurate enough if the following conditions are satisfied.

$$\frac{\| \Delta \hat{v}^{i+1} \|}{\| \Delta \hat{v}^{i} \|} \leq \epsilon_{norm}, \quad \| \Phi(\hat{v}^{i}) \| \leq \Phi_{norm} \tag{23}$$

Usually, the following conditions are used:

tolerance of functional norm(Φ_{norm})= 10^{-3},

tolerance of fractional norm(ϵ_{norm})= 10^{-5}.

2.5 Updating the configuration

Once the boundary value problem including the system of equation is solved, it is necessary to update the current configuration. The updating method is similar to Euler's method used in the initial value problem. During the updating procedure, all the necessary changes in boundary conditions, such as changes in the contact with tool surface, should be considered.

$$x^{n+1} \xleftarrow{\text{b.c. handling during } \Delta t} x^{n} + \Delta t\, v^{n} \tag{24}$$

Where x^{n} denotes the current configuration at time t and x^{n+1} denotes the updated configuration at time ($t+\Delta t$).

Since plastic deformation at a given time t depends on the history of deformation, the total effective strain should be defined. In this study, the total effective strain is updated as follows:

$$\bar{\varepsilon} = \int_{t_0}^{t_0+\Delta t} \dot{\bar{\varepsilon}}\, dt = \bar{\varepsilon}_0 + \Delta\bar{\varepsilon} = \bar{\varepsilon}_0 + \dot{\bar{\varepsilon}}\Delta t\ , \tag{25}$$

where $\bar{\varepsilon}_0$ is represent total effective strain up to time t_0.

3 EXPERIMENT

A material used in experiments is Zn oil coated sheet(SPC3J-3J/3J) of 0.65mm in thickness which is widely used to the press forming of automobile. The mechanical properties and dimensions of specimens are shown in Table 1.

3.1 Drawbeads forming test

Figure 1 shows the shapes and dimensions of single circular and single square drawbead. The upper male die moves to the lower female die step by step with 1.0mm, 3.5mm and 4.9mm in drawbead depth as the blank holding forces applied. The longitudinal logarithmic strain at upper and lower skins of specimen are measured by using tool microscope connected with P.C. for the formed states at each step. The reactional forces of upper die during blank holding process are measured by load cell and the results are compared with numerical results.

Table 1. Mechanical properties, dimension of specimen and input parameters

EXPERIMENT	Specimen dim.	Experimental conditions	
	WxLxT	Friction coefficient	Test velocity (mm/min.)
Drawbead forming	50x220x0.65	0.05	3 ~ 5
Drawbead drawing	50x220x0.65	0.1	50
flow stress	$\overline{\sigma} = C(\varepsilon_0 + \overline{\varepsilon})^n$		
	$C = 50.81\ Kg_f/mm^2$, $\varepsilon_0 = 0.0001$, $n = 0.25$		

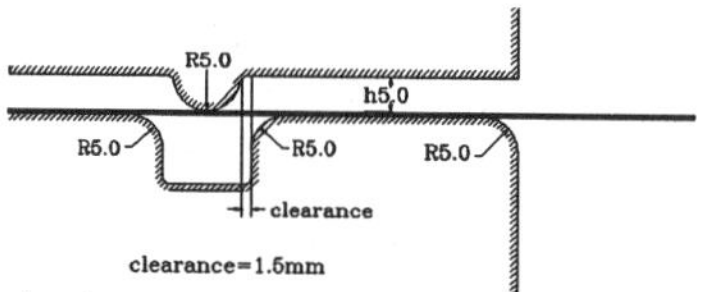

a) single circular drawbead

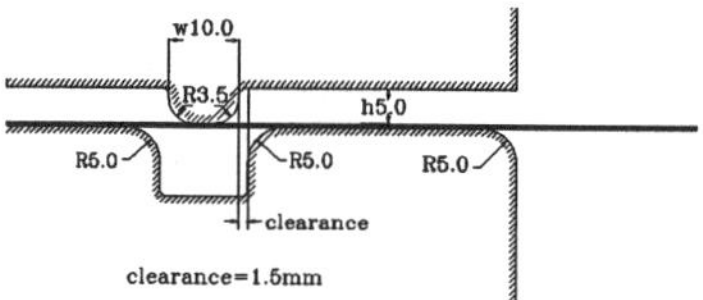

b) square draw bead

Fig. 1. Drawbeads geometry and dimension.

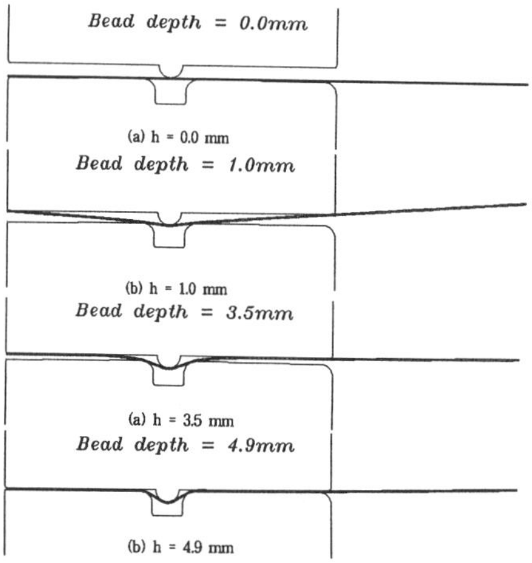

Fig. 2. Deformed specimen by predicted F.E.M.

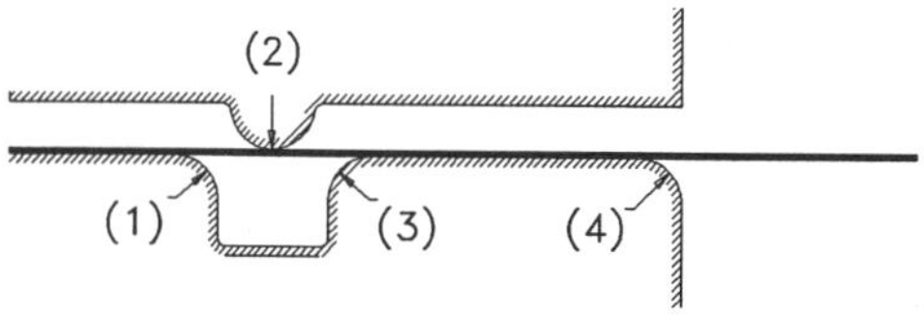

Fig. 3. Selected positions for the strain measurement.

3.2 Drawbeads drawing test

In the procedure of drawing test, we inserted drawbead into the apparatus and pressed with oil jack. We grasped specimen tip with a part of grips onto the drawing apparatus and drew the specimen out. The drawing velocity of specimen is 50mm/min. and drawing angle is varied to 0 ° ,30 ° and 60 ° respectively. The strain distributions of specimens drawn by each drawing angles are measured with tool microscope. Also, the drawbead restraining forces for the drawing length are measured with load gauge of drawbead apparatus.

4 NUMERICAL AND EXPERIMENTAL RESULTS

Figure 2 shows blank holding processes predicted by the finite element analysis with one layer element at the 0, 1.0, 3.5 and 4.9mm of drawbead depth. Figure 3 shows the detailed locations in the parts at which the strain distributions are measured. In the cross-section shown in figure 3, the points (1) and (3) are the left and right corner parts of female drawbead, the point (2) is the radius part of male drawbead, and the point (4) is the corner part of tool die. Figure 4 shows the variation of the reactional die forces for the increasing drawbead depth from 0 to 4.9mm. Figure 5 and 6 show the comparison of theoretical strain distributions with the experimental results in the case of blank holding with 4.9mm of drawbead depth. In figure 5, parts(1) and (3) are extended and part (2) is compressed during the blank holding process for the upper skin of specimen. In the other hand, in figure 6, part (2) is extended, and parts (1) and (3) are compressed for the lower skin of specimen. Figure 7 shows the drawn phenomena of specimen with the 0 ° ,30 ° and 60 ° of drawing angles. Figure 8 shows the restraining forces-drawing length curve in the condition of 0 ° drawing angle and 4.9mm drawbead depth. Figure 9 shows the upper skin strain distribution of specimen drawn with 0 ° drawing angle and 60mm drawing length. In this case, two layer of elements are used for the theoretical analysis, and the results have good agreements with the experimental results. Figure 10 and 11 show the strain distribution on the upper skin of specimen for the drawing length 20mm and drawing angle 30 ° and 60 ° ,respectively. In the figures, it should be noted that the strains at the point (4) are larger with drawing angle 60 ° than those with drawing angle 30 ° , in spite that the strains at the other locations do not show much difference with varied drawing angle. Figure 12 shows the predicted strain distributions on the upper skin for the three kinds of drawing angles with the drawing length 20mm and drawbead depth 4.9mm. In the figure, the strain distributions on the upper skin near the profiled part (points (1), (2), (3)) show coincidences with varied drawing

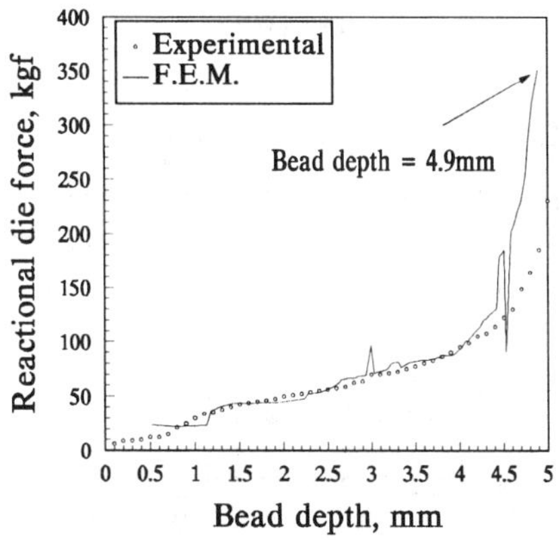

Fig. 4. Reactional die force for the drawbead depth.

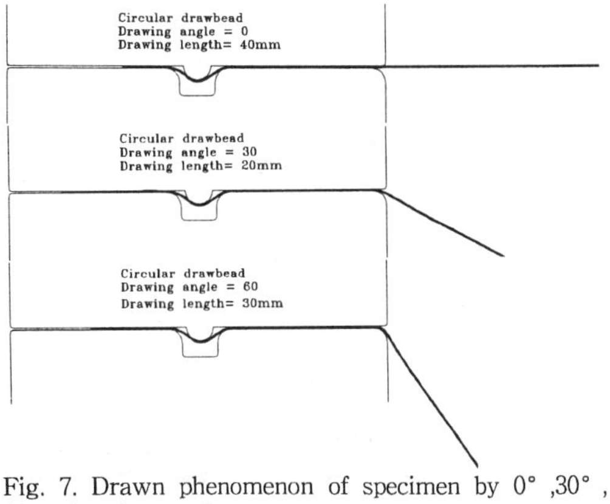

Fig. 7. Drawn phenomenon of specimen by 0°, 30°, 60° drawing angle.

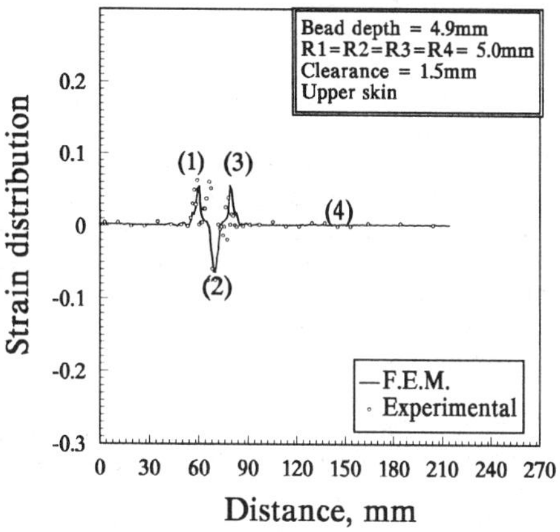

Fig. 5. Strain distribution of upper skin for the 4.9mm, drawbead depth.

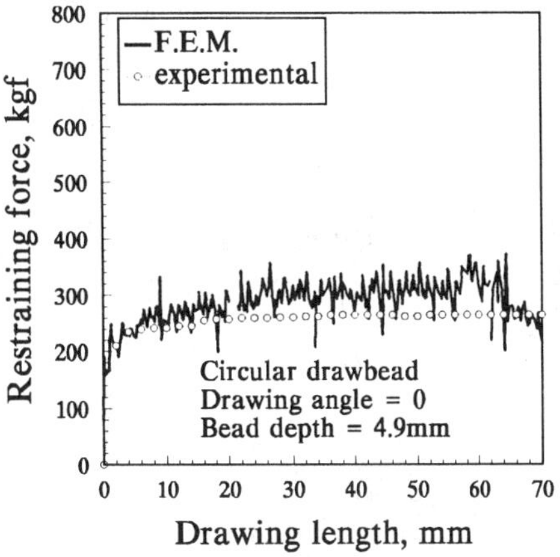

Fig. 8. Circular drawbead restraining forces for the drawing length.

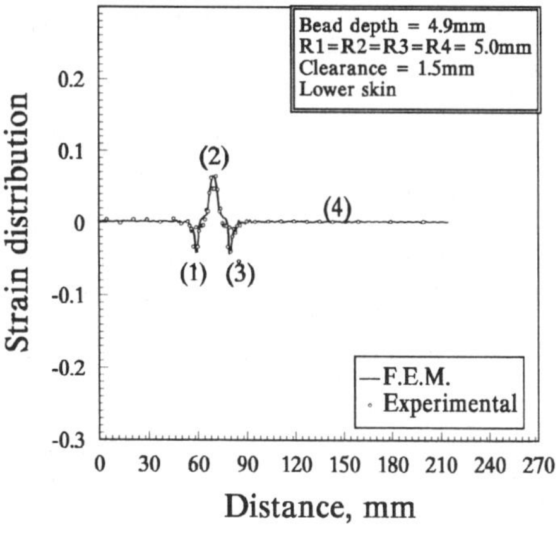

Fig. 6. Strain distribution of lower skin for the 4.9mm, drawbead depth.

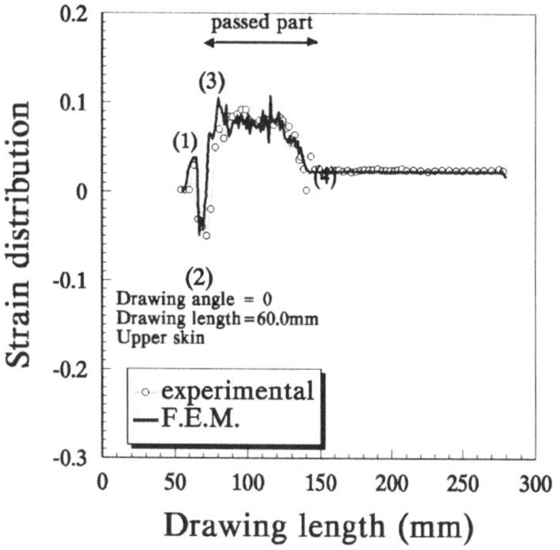

Fig. 9. Upper skin strain distribution for the 60.0mm drawn specimen and 0° drawing angle.

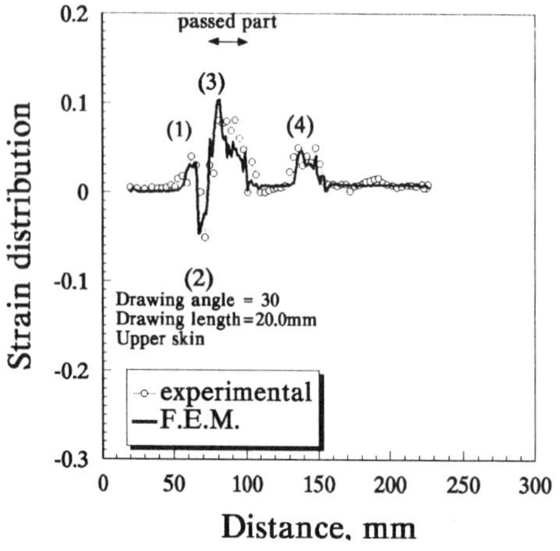

Fig. 10. Upper skin strain distribution for the 20.0mm drawn specimen and 30° drawing angle.

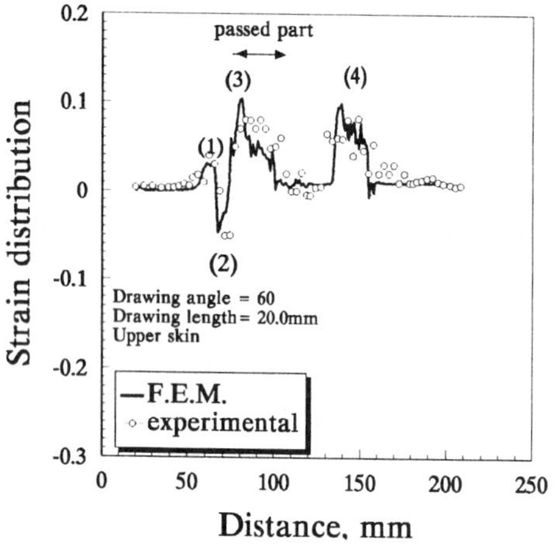

Fig. 11. Upper skin strain distribution for the 20.0mm drawn specimen and 60° drawing angle.

angles. But at the tool die corner, the strain seems to be proportional to drawing angle. Hence the largest strain at point (4) is obtained with 60° of drawing angle.

5 CONCLUSIONS

The predicted reactional die forces to the variation of drawbead depth are in good agreement with the experimental results. However, in the last step, the theoretical results for the single circular drawbead showed larger values than the measured ones. In the flange parts except the drawbead parts, the strains are nearly zero. For the drawbead parts, the strain distributions of theory and experiments are in

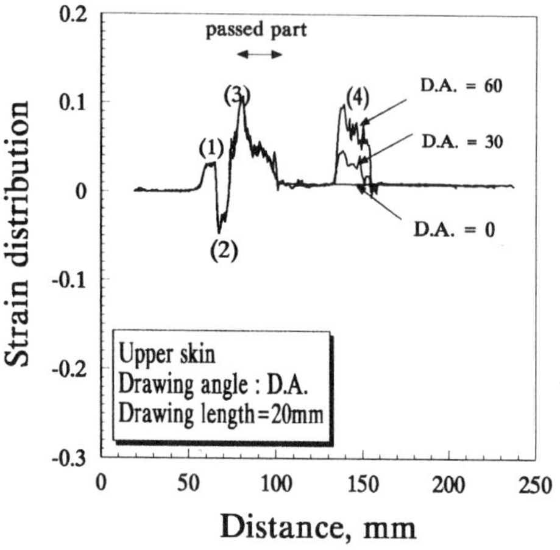

Fig. 12. Theoretical upper skin strain distribution of 20mm drawn specimen by 0°, 30°, 60° drawing angle.

good agreement. Therefore, the usefulness of 2-D rigid-plastic finite element analysis has been verified for the drawbead forming and drawing processes. Concludingly, we can expect that it is possible to predict the strains for the drawn specimens to the arbitrary drawing length.

Acknowledgements

This paper has been supported by Korea Foundation of Science & Technology and Kia Motors Corp. The authors would like to appreciate for their supports.

REFERENCES

H.D. Nine, 1978. Drawbead Forces in Sheet Metal Forming. Mechanics of Sheet Metal Forming : 179~211.

H.D. Nine, 1982. The Applicability of Coulomb's Friction Law to Drawbeads in Sheet Metal Forming. J. of Applied Metal Working, Vol.2, no.3: 200~210.

N.M.Wang, 1982. A Mathematical Model of Drawbead Forces in Sheet Metal Forming. J.of Applied Metal Working., Vol.2, no.3: 193~199.

H.D. Nine, 1982. New Drawbead Concepts for Sheet Metal Forming, J.of Applied Metal Working. Vol.2, no.3: 185~192.

B.S. Levy, 1983. Development of a Predictive Model for Draw Bead Restraining Force Utilizing Work of Nine and Wang. J. of Applied Metal Working Vol.3, no.1: 38~44.

J.M. Yellup, 1984. Modelling of Sheet Metal Flow Through a Drawbead. 13th IDDRG: 166~177.

J.M. Yellup, 1985. The Prediction of Strip Shape and Restraining Force for Shallow Drawbead Systems. J. of Applied Metal Working, Vol.4, no.1: 30~38.

L.R. Sanchez, K.J.Weinmann, 1988. A General Computer Model for Plane Strain Sheet Flow and its Application to Flow between Circular Drawbeads. 15th IDDRG: 217~226.
N.Triantafyllidis, B. Maker, and Shyam S.K.Samanta, 1986. An Analysis of Drawbeads in Sheet Metal Forming : Part I -Problem Formulation. J. of Eng. Materials and Technology, Vol.108: 321~327.

Simulation of Materials Processing: Theory, Methods and Applications, Shen & Dawson (eds)
© 1995 Balkema, Rotterdam. ISBN 90 5410 553 4

Numerical simulations and benchmarks of 3-D sheet metal forming using LAGAMINE program

K.P.Li, S.Cescotto & A.M.Habraken
MSM Department, University of Liège, Belgium

H.Bruneel
OCAS, Zelzate, Belgium

ABSTRACT: The finite element program LAGAMINE for analyzing 3-D Sheet Forming processes has been developed at MSM department, University of LIEGE for many years now and has been tested against various benchmarks and experiments. In this paper, numerical simulations of Sheet Metal Forming will be performed by 8-node 3-D mixed elements and by 4-node quadrangular 3-D shallow shell elements. Various material laws (including isotropic elasto-plastic *Von Mises*' law, elastic anisotropic plastic *Hill*'s law, elasto-visco-plastic law, etc.) are available. Die, punch and blank holder are assumed rigid, with the geometry discretized by triangular facet elements. The Coulomb's friction law is used to model the friction between sheet and tools. A nonsymmetric Newton-Raphson scheme is used to solve the equation system. Two examples, namely the deep drawing of a square cup and a rectangular box both made of mild steel, are simulated with these two different types of elements. For the first example, the numerical results are compared with experimental ones. The influences of different meshes and different types of elements are discussed.

1. INTRODUCTION

With the development of large, powerful computers and the good comprehension of element's behavior undergoing large inelastic deformations, the finite element method plays a very important role in real 3-D engineering designs not only because it can reduce the cost of design but also because it can help engineers to better control the quality of the products. This requires that the results given by a finite element program be accurate, robust and the analysis be economical (for examples, the CPU, the central memory, the time for preparing the input data, etc.). In the context of finite element method, to meet these requirements, it is essential to choose and develop appropriate elements, material constitutive laws and solution of the inequality constraints induced by the complex contact problem. All these options depend on the sound understanding of the element behavior in case of large elastoplastic deformations. The LAGAMINE program is developed for this purpose. It was originally created to simulate rolling processes in the 80's and now can be used in 3-D large elastoplastic deformation analysis with or without contact problem for different kinds of materials (metals, soils, rocks, etc...) by static or dynamic procedure. In this paper, we just limit ourselves to the simulation of sheet forming processes by static implicit procedure. For this purpose, two different kinds of elements are developed, one is a solid element, the other is a shell element and some special attention is paid to the contact searching algorithm in order to reduce the computational costs.

2. THE CHOICE OF ELEMENTS FOR SHEET FORMING PROCESSES

2.1. General Concepts

The sheet forming processes can be modeled using both shell elements or brick solid elements. There are some disadvantages for using conventional brick (8-node) elements in large strain analysis for sheet forming processes: in their standard isoparametric form, these elements present "shear locking" and "volumetric locking" for bending dominated and (quasi-) incompressible problems respectively. Thus it would be advantageous to model the sheet using shell elements based upon geometric considerations. But it should be recognized that the use of solid elements has several advantages. 1.) Large deformation formulation can be easily developed for this kind of elements. 2.) The implementation of new material constitutive laws can be easily realized. 3.) Classical contact algorithms can be used directly. Considering this, we have developed a mixed 8-node brick solid element in which there is no "shear" and "volumetric" locking problems and adapted a 4-node shallow shell element developed by Ph. Jetteur [2]. From some numerical comparisons, we can see that both kinds of elements can give very satisfying results for sheet forming processes.

2.2. JET3D - A Mixed 8-Node Brick Element

2.2.1. *Variational Principle*

The updated *Lagrangian* rate formulation is used as a base of the incremental elasto-plastic finite element undergoing large deformations. The starting point is a rate form of the virtual work equation very similar to that proposed by McMeeking and Rice [6]:

$$\int_v \{(\dot{\sigma}^J_{ij} - 2\sigma_{ik}D_{kj})\delta D_{ij} + \sigma_{jk}L_{ik}\delta L_{ij}\}dv = \int_{s_t} \dot{f}_i \delta v_i dS$$

here, $\dot{\sigma}^J_{ij} = \dot{\sigma}_{ij} - W_{ik}\sigma_{kj} + W_{kj}\sigma_{ik}$ is the *Jaumann* derivative of the *Cauchy* stress tensor σ_{ij}, L_{ij} is the velocity gradient, D_{ij} and W_{ij} are the symmetric and skew-symmetric parts of L_{ij}, respectively.

2.2.2. *Mixed 3-D Element with Hourglass Control*

For the displacement approach, the deformation gradients of a standard 8-node brick element can be written briefly as,

$$\frac{\partial u_i}{\partial x_j} = \left(b_j^T + \gamma_k^T h_{k,j}\right) U_i$$

where h_1=*st*; h_2=*rt*; h_3=*rs*; h_4= *rst* with (r,s,t) the isoparametric coordinates; U_i is the nodal displacement vector; x_i are the *cartesian* coordinates; b_i (i=1,2,3) corresponding to the standard interpolation function evaluated at the center of the element represent the mean deformation gradients and γ_i (i=1,2,3,4) represent the hourglass mode deformation gradients (see [1] for more details) with $\int_V h_{i,j}\, dv = 0$.

In order to avoid incompressibility locking and shear locking, the strain and stress fields are chosen as:

$$\dot{\underline{\varepsilon}} = \dot{\bar{\underline{\varepsilon}}} + \underline{\underline{h}}_{,\alpha}\dot{\underline{\varepsilon}}^* ; \dot{\underline{\sigma}} = \dot{\bar{\underline{\sigma}}} + \underline{\underline{h}}_{,\alpha}\dot{\underline{\sigma}}^* \text{ and } \nabla\dot{\underline{u}} = \left(\underline{\underline{\bar{B}}} + \underline{\underline{B}}^*\right)\dot{\underline{U}}$$

where: $\dot{\underline{\varepsilon}}$ and $\dot{\underline{\sigma}}$ are the strain rate and stress rate vectors respectively. $\underline{\underline{\bar{B}}}$ is the conventional strain-displacement matrix evaluated at the center of the element; $\underline{\underline{h}}_{,\alpha}$ and $\underline{\underline{B}}^*$ are the hourglass generalized interpolation functions, which are chosen in such a way that the element presents neither incompressibility locking nor shear locking [1]. $\dot{\bar{\underline{\varepsilon}}}$ and $\dot{\bar{\underline{\sigma}}}$ are the constant strain and stress rates evaluated at the center of the element while $\dot{\underline{\varepsilon}}^*$ and $\dot{\underline{\sigma}}^*$ are called the hourglass generalized strain and stress rates respectively which can be determined by using the *Hu-Washizu* principle [1].

This gives the following relations:

$$\dot{\bar{\underline{\varepsilon}}} - \underline{\underline{\bar{B}}}\dot{\underline{U}} = 0 \quad \text{and} \quad \underline{\underline{H}}\dot{\underline{\varepsilon}}^* - \underline{\underline{H}}\underline{\underline{\Gamma}}\dot{\underline{U}} = 0$$

$$\underline{\underline{C}}\dot{\bar{\underline{\varepsilon}}} - \dot{\bar{\underline{\sigma}}} = 0 \quad \text{and} \quad 2G\underline{\underline{H}}\dot{\underline{\varepsilon}}^* - \underline{\underline{H}}\dot{\underline{\sigma}}^* = 0$$

with: $v\underline{\underline{\bar{B}}}^T\dot{\bar{\underline{\sigma}}} + \underline{\underline{\Gamma}}^T\underline{\underline{H}}\dot{\underline{\sigma}}^* = \dot{\underline{F}}$

where $\underline{F}$ is the nodal force vector and $\underline{\underline{H}} = \int_v \underline{\underline{h}}^T_{,\alpha}\underline{\underline{h}}_{,\alpha}\, dv$ is the anti-hourglass matrix. $\underline{\underline{C}}$ is the elastic stiffness matrix. $2G = \frac{E}{1+\nu}$ and

$$\underline{\underline{\Gamma}}^T = \begin{vmatrix} \underline{\gamma}_1 & \underline{0} & \underline{0} & \cdots & \underline{\gamma}_4 & \underline{0} & \underline{0} \\ \underline{0} & \underline{\gamma}_1 & \underline{0} & \cdots & \underline{0} & \underline{\gamma}_4 & \underline{0} \\ \underline{0} & \underline{0} & \underline{\gamma}_1 & \cdots & \underline{0} & \underline{0} & \underline{\gamma}_4 \end{vmatrix}$$

With the aid of the updated *Lagrangian* rate formulation, this element has successfully passed large elastic strain patch-tests.

2.2.3. *Material Laws for This Element*

For this element, since in the strain and stress formulation there exists both constant and variable terms over the element, some special considerations are required. For instance, in elastoplastic analysis, the mean plasticity criterion is taken over the whole element. So that it takes into account the influence of constant stresses and the anti-hourglass stresses simultaneously,

$$\sigma^2_{eq} = \frac{1}{V}\int_v \left(\underline{\sigma}^T \underline{\underline{A}}\underline{\sigma}\right) dv$$

where $\underline{\underline{A}}$ is the material property matrix depending upon the isotropic or anisotropic case. For isotropic case, it can be written simply as,

$$A = \frac{1}{2}\begin{vmatrix} 2 & -1 & -1 & 0 & 0 & 0 \\ -1 & 2 & -1 & 0 & 0 & 0 \\ -1 & -1 & 2 & 0 & 0 & 0 \\ 0 & 0 & 0 & 6 & 0 & 0 \\ 0 & 0 & 0 & 0 & 6 & 0 \\ 0 & 0 & 0 & 0 & 0 & 6 \end{vmatrix}$$

In the anisotropic case, for objectivity, a local axes system embedded in the element is used with an incremental updated formulation. The material is supposed to be initially orthotropic and remains so during deformation. Then the initial anisotropy tensor, corresponding to *Hill*'s quadratic yield condition, can be determined by,

$$\bar{\sigma}^2 = F(\sigma_{11} - \sigma_{22})^2 + G(\sigma_{22} - \sigma_{33})^2 + H(\sigma_{33} - \sigma_{11})^2$$
$$+ L\sigma^2_{12} + M\sigma^2_{23} + N\sigma^2_{13}$$

therefore,

$$A = \begin{vmatrix} F+H & -2F & -2H & 0 & 0 & 0 \\ -2F & F+G & -2G & 0 & 0 & 0 \\ -2H & -2G & G+H & 0 & 0 & 0 \\ 0 & 0 & 0 & M & 0 & 0 \\ 0 & 0 & 0 & 0 & N & 0 \\ 0 & 0 & 0 & 0 & 0 & L \end{vmatrix}$$

So the yield function can be written in an unified form as :

$$f = \sigma^2_{eq} - \sigma^2_y = 0$$

where σ_y is the current yield stress and its evolution with deformation is characterized by an isotropic piecewise work-hardening law.

If f < 0, the stress state is purely elastic and the final stresses can be directly calculated by Hooke's law.

When f = 0, the stress state is the elastoplastic and an implicit algorithm, namely the *radial return* method (elastic predictor, plastic corrector) is activated.

2.3. COQJ4 - A 4-Node 3-D Shallow Shell Element

2.3.1. *General Concepts about This Element*

This is a corrotational 4-node quadrangular 3-D shallow shell element with six degrees of freedom per node [2] for large displacement and large rotation analysis.

In this element, the membrane behavior is the classical isoparametric one. It can be obtained directly from the 4-node isoparametric functions,

$$\underline{\varepsilon}_l = \underline{\underline{B}}\underline{U}_l$$

where $\underline{\varepsilon}_l$ and $\underline{U}_l$ are the membrane strain and nodal displacement vectors in the local axes respectively; $\underline{\underline{B}}$ is the classical strain-displacement matrix.

For the flexural behavior, a discrete *Kirchhoff* theory is adopted. The bending deformations are obtained as,

$$\underline{\chi} = \underline{\underline{B}}_{\chi}\underline{q}_{wl}$$

where $\underline{\chi}, \underline{q}_{wl}$ are the flexural deformations and displacements in the local axes respectively; $\underline{\underline{B}}_{\chi}$ is the corresponding interpolation matrix.

For the initial curvature, the *Von Karman-Marguerre* approximation is used, which gives the coupling between membrane and flexural behaviors:

$$\varepsilon_{ij}^{M} = \frac{1}{2}\left(u_{i,j} + u_{j,i} + w_{,i}w_{,j}\right)$$

It can be shown that such an element converges towards the deep shell solution [2].

To avoid membrane locking, a special process is used, that is, the mean strains of the edges and diagonals are computed by taking into account the initial deflection. Then $\varepsilon_{11}^{M}, \varepsilon_{22}^{M}$ and $\varepsilon_{12}^{M} + \varepsilon_{21}^{M}$ are calculated as functions of the strains of edges and diagonals.

Another special point of this element is that the sixth degree of freedom, that is the "*drilling degree*", is interpolated as a real degree by using the hierarchical interpolation form, which relates the nodal rotations θ_{zi} to the hierarchical membrane displacements to get the non-conventional displacement interpolation [3], described hereafter:

$$u^{h} = \sum_{I=5}^{8} N_I(r,s)\frac{l_{jk}}{8}\left(\theta_{zk} - \theta_{zj}\right)n_{jk}$$

where l_{jk} is the length in the side of nodes J-K and n_{jk} is its normal; N_I is the Serendipity shape function defined by [3]:

$$\begin{cases} N_I = \frac{1}{2}\left(1 - r^2\right)\left(1 + s_I s\right) & \text{for } i = 5,7 \\ N_I = \frac{1}{2}\left(1 + r_I r\right)\left(1 - s^2\right) & \text{for } i = 6,8 \end{cases}$$

The thickness is updated during the deformation process either by plane stress condition or by the incompressible volumetric condition.

2.3.2. *Material Laws for This Element*

For the material constitutive laws, the classical *Von Mises*' yield function and also *Hill*'s criterion under plane stress condition is implemented for isotropic material or cross-anisotropic material respectively. Numerical integration through the thickness is required to give the resultant stresses and the material stiffness matrix. For economical purpose, an anisotropic *Ilyushin*'s type law is also adopted [4] in this element, which uses the resultant stresses in the yield criterion directly. So it does not require numerical integration through the thickness, gives satisfying results, uses much less central memory for recording the variable data and less CPU too. It can be written as below,

$$f = Q_{nn} \pm \frac{1}{\sqrt{3}}Q_{nm} + Q_{mm} - \sigma_y^2$$

with:

$$Q_{nn} = \frac{1}{n_0^2}\left(\alpha_{11}N_x^2 + \alpha_{22}N_y^2 + \alpha_{33}N_{xy}^2 - 2\alpha_{12}N_xN_y\right)$$

$$Q_{nm} = \frac{1}{n_0 m_0}\left(\alpha_{11}N_xM_x + \alpha_{22}N_yM_y + \alpha_{33}N_{xy}M_{xy} - \alpha_{12}\left(N_xM_y + N_yM_x\right)\right)$$

$$Q_{mm} = \frac{1}{m_0^2}\left(\alpha_{11}M_x^2 + \alpha_{22}M_y^2 + \alpha_{33}M_{xy}^2 - 2\alpha_{12}M_xM_y\right)$$

and $n_0 = h\sigma_{y0}$; $m_0 = \frac{h^2}{4}\sigma_{y0}$ and h is the actual thickness.

3. CONTACT PROBLEM

3.1. General Considerations

In previous sections, we have shown the elements and their material laws chosen for the sheet forming processes. Additionally, metal forming analysis is complicated by the requirement to impose nonlinear boundary conditions on the deformable workpiece. These boundary conditions are imposed generally by a non-penetration condition between workpiece and tools.

In LAGAMINE, this non-penetration condition is realized by the penalty method, which can be implemented quite easily in the code without modifying the whole structure of the program, by means of contact interface elements [5]. The geometry of tools is represented by 3-D triangular facet elements which can be easily adapted to any arbitrary 3-D geometric form of tools. Both rigid-deformable as well as deformable contact can be modeled. In the rigid-deformable case, tools are considered to be rigid.

3.2. Frictional Contact Treatments

For modeling the large slide between tools and sheets during the forming processes, a modified *Coulomb*'s model is used with an elastoplastic formulation (fig. 3.1), which yields very satisfying results.

Special attention has been paid to the contact treatments for shell elements since for such elements, there are at least two potential contact surfaces, which are the two surfaces of the sheet.

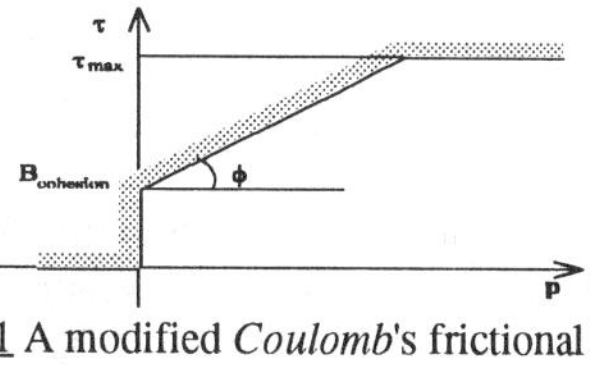

Fig. 3.1 A modified *Coulomb*'s frictional model

3.3. Contact Searching/Detecting Algorithms

In LAGAMINE, in 3-D contact searching/detecting algorithms, a special neighborhood relation has been established at the beginning of the simulation. Thus during the forming processes, for each contact interface element, the contact searching begins at the segment in which previous contact has taken place: if the possible contact relation is confirmed and then combined with the false-contact criterion the contact treatment based on the penalty method is realized; if the possible contact condition is not confirmed, the searching will start at the closest neighborhood; if there is no possible neighbor for this segment (triangular facet element) then the searching will continue for the whole segments which define the tools. In this way, all contact interface elements can be tested.

But this searching algorithm has a drawback: for elements which have no physically impossibility of contact with the tools, the searching will still go throughout all segments defining the tools; therefore useless contact searching will be made.

In order to solve this problem, a sub-zone contact searching algorithm is added: that is, for each tool, possible contact zones are firstly established either by user or by program during the simulation. Then for each contact interface element, contact searching begins firstly to see if it belongs to the possible contact zones. If it does, further contact detecting and treatment is needed. Otherwise, no contact searching will be done.

4. NUMERICAL EXAMPLES

4.1. A Square Cup Deep Drawing

4.1.1. *General Presentation*

In order to test the elements mentioned before, we have chosen the simulation of a square cup deep drawing as a numerical example to show the capability and efficiency of the elements.

The structure (Fig. 4.1) and all others conditions are the same as those used in NUMISHEET'93 [7]. They are shown below.

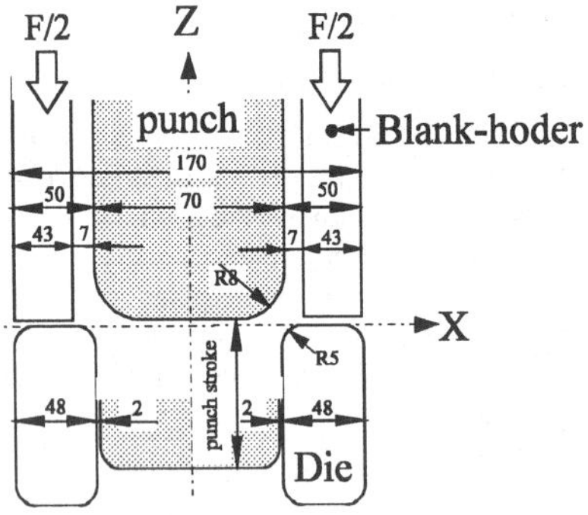

Fig. 4.1 Square Cup deep drawing

Blank holding force is 19.6 KN. Initial blank size is 150 × 150 (mm). The material used is mild steel with the following parameters: E=206 Gpa, ν=0.3, σ_y=157.1 N/mm^2 , the anisotropic parameter r=1.77, the friction coefficient μ=0.144, initial thickness h_0=0.78mm.

For this simulation, by symmetry, only a quarter of the structure is discretized. Tools are presented by 265 triangular facet elements (see Fig 4.2).

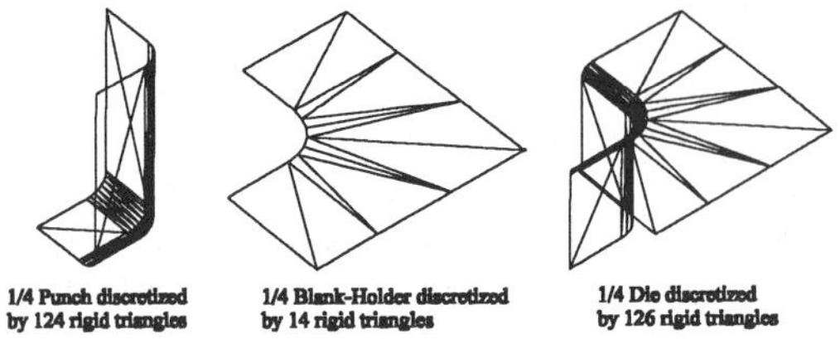

Fig. 4.2 The discretizations of the tools

4.1.2. *Simulations by JET3D Elements*

Since sheet forming problems can be considered as membrane dominated problems and no springback analysis is required for this simulation, we just use one element over the thickness (one layer modeling), but in order to show the mesh sensibility and the efficiency of the elements we have tested two meshes. The first mesh has 256 (16×16) JET3D elements (total 578 nodes); The second one has 900 (30×30) JET3D elements (total 2094 nodes). The results obtained by these two meshes will be shown at section 4.1.4.

4.1.3. *Simulations by COQJ4 Elements*

For this kind of element, we have just used the first mesh to compare the numerical results with JET3D element and 5 integration points through the thickness of the sheet. The mesh has 256 (16×16) COQJ4 elements (total 289 nodes). The results obtained by this mesh will be shown at next section.

4.1.4. *Numerical Results For Square Cup Deep Drawing*

The computer hardware is: DEC/ALPHA 3000/400. The CPU and number of steps for the entire simulation are listed in the following table.

Table 4.1: CPU and steps for the entire simulation

Element	mesh	CPU (min.)		Steps
		Contact search	Total	
JET3D	mesh_1	3	45	172
	mesh_2	7.5	258.5	171
COQJ4	mesh_1	6	75.5	217

Table 4.2: Draw-in for a square cup deep drawing

Element	mesh	Stroke 15mm (mm)		Stroke 40mm (mm)	
		DX	DD	DX	DD
JET3D	mesh1	6.45	3.39	29.70	16.24
	mesh2	6.61	3.39	30.40	16.83
COQJ4	mesh1	6.67	3.81	31.47	22.20
Exp.[7]		6.17	3.24	27.96	15.26

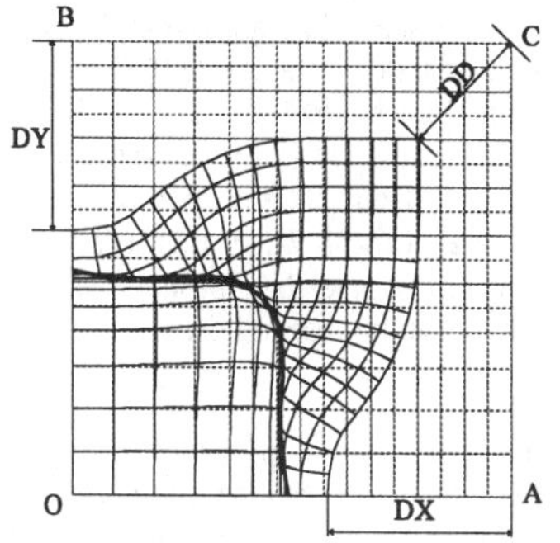

Fig. 4.3 The initial and final configurations and the definition of draw-in.

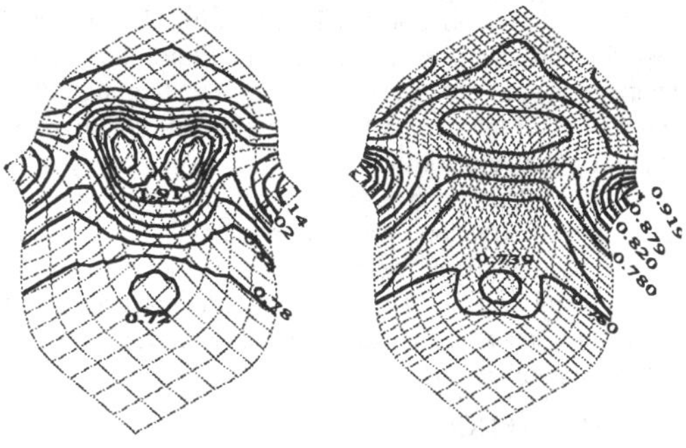

COQJ4 -Mesh1 JET3D -Mesh2

Fig. 4.4 The thickness contour in deformed sheet.

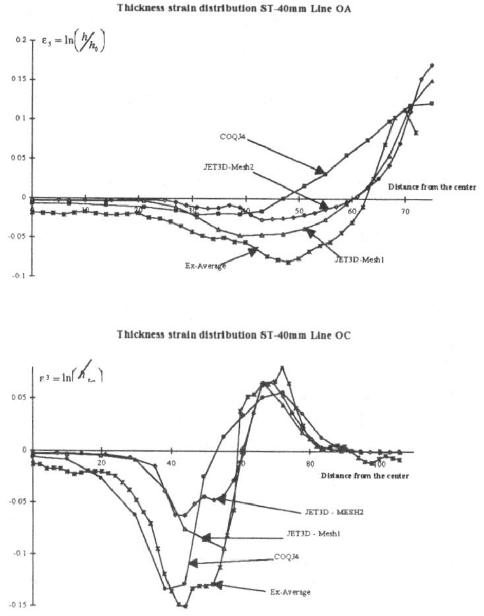

Fig. 4.5 Thickness strain distribution at punch stroke=40mm

From these tables and figures, it is seen that for the same mesh and the same simulation, JET3D elements are more cost-efficient than COQJ4 elements. For the draw-in, at line OA and line OB, all the meshes and types of element have given almost the same values. But at line OC, there are large differences between the results obtained by JET3D elements and those by COQJ4 elements. After comparison with the average experimental results, it is seen that JET3D elements give much better draw-in than COQJ4 elements. For the thickness strain distribution, it should be pointed out that for JET3D elements, the thickness and the thickness strain are calculated *a posterior* during the simulation, the sheet thickness for JET3D has no meaning, only the node coordinates are updated. So that for the calculation of the thickness and the thickness strain, they must be computed after the simulation. On the contrary, for the shell element, the thickness is updated during the simulation, so the thickness strain can be obtained quite directly. Nevertheless the distributions of the thickness strain given by JET3D elements are nearly the same as those given by COQJ4 elements.

4.2. OCAS benchmark

For the second example which is proposed by OCAS company, the structure and all other conditions are defined as: blank holding force is 64 kN (16 kN for ¼ region); initial blank size is 400×500 mm with thickness 0.8mm for the material ST15; so that the material properties are given as: E=206 Gpa; ν=0.3; the friction coefficient is 0.15; with the initial yield stresses: σ_x=167, σ_y=174; σ_z=208, τ=96 Kpa in rolling, transversal and thickness direction respectively. The geometry for this simulation is shown in Fig. 4.6.

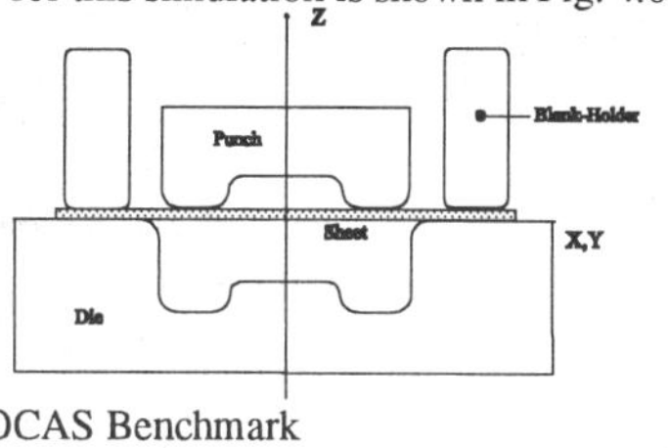

Fig. 4.6 OCAS Benchmark

Only a quarter of the whole structure is discretized. Tools are represented by 586 triangle facets (see Fig. 4.7).

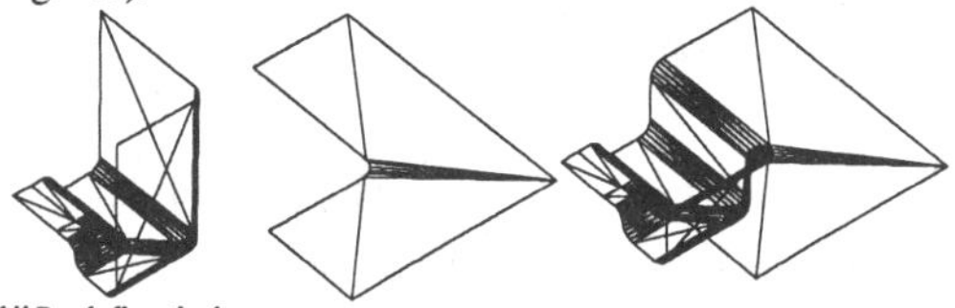

1/4 Punch discretized by 249 rigid triangles 1/4 Blank-holder discretized by 9 rigid triangles 1/4 Die discretized by 328 rigid triangles

Fig. 4.7 The discretizations of the tools

For the static implicit procedure, only the JET3D elements are used with two different meshes: one has 1122 (33×34) JET3D elements, the other uses 1330

(35×38) JET3D elements. For the first mesh, the classical contact searching algorithm is used; but for the second mesh the sub-zone contact searching is activated. The CPU and number of steps for this simulation are listed in table 4.3. It should be remarked that in experiment the sheet was broken before the punch stroke has reached the desired 75 mm and in numerical simulation none of these two simulations can reach this value since the loading step is too small and no convergence has been found which is often the case associated with static implicit procedure as already mentioned by other researchers. The remedy is: using more refined mesh or introducing the remeshing technique or using dynamic explicit procedure instead of static implicit procedure (which exists already in LAGAMINE and has been used in the sheet forming processes, but we will not discuss it in this article).

Table 4.3 CPU and steps for OCAS benchmark

Mesh-1: 1122 JET3D elements:

Punch Stroke	CPU (min.)		Iter.	Steps
	Contact search	Total		
50	4694	9527	3691	1470
69.4	11486	23622	9269	4086

Mesh-2: 1330 JET3D elements:

Punch Stroke	CPU (min.)		Iter.	Steps
	Contact search	Total		
50	411	3743	2281	851
66.1	2431	9360	4655	1875

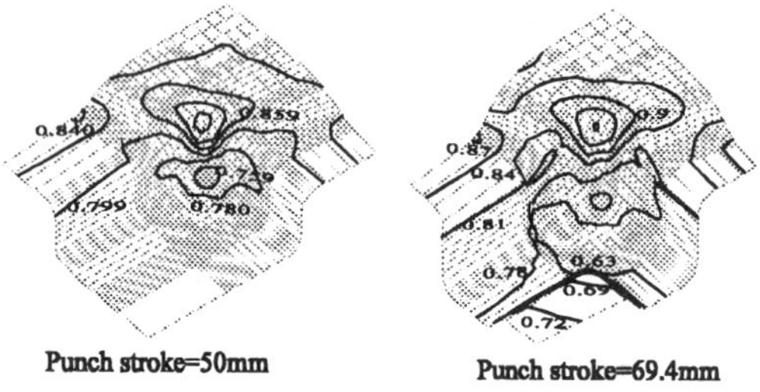

Fig.4.8 The thickness contour in mesh-1

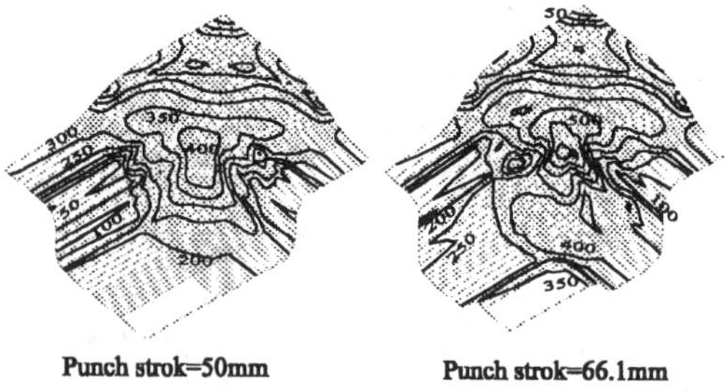

Fig.4.9 Von Mises stress contour in mesh-2

The thickness distribution contour obtained by the first mesh is shown in fig. 4.8 and the *Von Mises* equivalent stress distribution contour obtained by the second mesh is drawn in Fig 4.9.

In this example, the emphasis is hold on the contact searching algorithm. From Table 4.3, we can see that if sub-zone contact searching algorithm is not used, the contact searching CPU is about a half of the total CPU of the simulation (in the case of mesh 1); But if the sub-zone contact searching algorithm is used, then the contact searching CPU is from a tenth to a fourth of the total CPU (in the case of mesh 2), so that it is very important for us to develop or use the optimal contact searching algorithms in the future researches.

5. CONCLUSION

In this paper, a square cup deep drawing is simulated by different elements. It can be concluded that JET3D elements can give results as good and even better than those given by COQJ4 elements and is more economical than the last one. Both elements allow to handle large draw-in and give reliable results at reasonable times and cost for the complex tools shapes. From the second example, we can see that the contact searching algorithm plays a very important role in the simulations by static implicit procedure and it will be even more important in dynamic explicit procedure since in this procedure no resolution of equation system is required, then the ratio of CPU used in contact searching will be more important, so that using or developing the optimal contact searching algorithms plays and will play a key role in real industrial applications.

6. REFERENCES

1. K.P. Li, S. Cescotto and Ph. Jetteur: "An element with hourglass control for the large deformation analysis in three-dimension", Proc. of 3rd int. Conf. on Computational Plasticity, Fundamentals and Applications, 6-10 April 92, Barcelona, Spain, 1992
2. Ph. Jetteur: "Non-Linear shell element based on Maguerre theory", IREM Internal report 85/5, Ecole Polytechnique fédérale de Lausanne, Lausanne, Dec. 1985
3. Ibrahimbegovic A. and F. Frey: "A Study of Continuum Mechanics Rotations. Part II: Discrete Formulation of Membrane Problems", LSC Internal report, 91/3, Ecole Polytechnique Fédérale de Lausanne, Lausanne, 1991
4. J.C. Simo and J.M. Kennedy: "Finite Strain Elastoplasticity in stress Resultant Geometrically exact shell", in Proceeding COMPLAS II, (eds. D.R.J. Owen, E. Hinton and E. Onate), 651-670, Barcelona, Spain, 1989
5. R. CHARLIER and S. CESCOTTO: "Modélisation du phénomène de contact unilatéral avec frottement dans un contexte de grande déformation", J. of Theoretical and Applied Mechanics, Special Issue, Suppl. N° 1 to Vol. 7, 1988
6. R.M. McMeeking and J.R. RICE, "Finite-element formulation for problems of large elastic-plastic deformation", Int. J. Solids Struct., vol.11, pp.601-616, 1975
7. NUMISHEET'93, Proc. of 2nd int. Conf. on Numerical Simulation of 3-D Sheet Metal Forming Processes - verification of simulation with experiment-, Isehara, Japon, 1993

Simulation of Materials Processing: Theory, Methods and Applications, Shen & Dawson (eds)
© 1995 Balkema, Rotterdam. ISBN 90 5410 553 4

Finite-element analysis of earing using non-quadratic yield surfaces

Roger W. Logan
University of California, Lawrence Livermore National Laboratory, Calif., USA

ABSTRACT: During deep draw cupping, the phenomenon known as earing may occur as the cup wall is formed, resulting in a periodic variation of cup wall height around the perimeter of the finished cup. This is generally due to planar anisotropy of flow in rolled sheet product. It is generally observed that the anisotropy parameter R will vary in the plane of the sheet when ears are observed in cupping, with a parameter ΔR describing the variation of R in the plane of the sheet. For many common textures in face-centered and body-centered materials, the ears form relative to the sheet rolling direction at 0° and 90° around the perimeter if $\Delta R>0$, and at -45° and $+45^o$ if $\Delta R<0$. There is extensive experimental evidence that ear height shows a linear correlation with $\Delta R/R$, but attempts to duplicate this using the finite-element method are highly dependent on both the methodology and yield surface used. It was shown previously that using a coarse mesh and the quadratic Hill yield surface tends to greatly under predict earing. In this study, we have used two different finite-element codes developed at LLNL to examine the predicted earing using both quadratic Hill and alternative non-quadratic yield surfaces. These results are compared to experimental data and conclusions drawn about the most desirable closed-form yield surfaces to duplicate the observed earing phenomena.

1. INTRODUCTION AND MOTIVATION

The forming of thin sheets is an important manufacturing process which is used in the production of a wide range of products from beverage cans to auto body parts to lightweight airframe and military aerospace components. There are two primary goals for the engineering analysis of a sheet metal forming process. First, analysis aims to reduce the trial and error in tooling and process design, and thereby reduce material waste and lead times to produce a new part. Second, analysis aims to influence the design of the desired part for ease of manufacture. Both of these goals ultimately lead toward the objective of faster production of better parts at minimum cost.

One of the causes of material waste in stamping is the often unanticipated formation of ears on the periphery of the flank from which the part is formed. The formation of ears in simplest form results from the stamping of a circular blank into a cylindrical cup. The essence of this process is illustrated by and studied numerically with the tooling and blank finite-element meshes in Fig. 1.

This tooling is used with codes such as the explicit finite-element code DYNA3D to follow the formation of cylindrical cups, with ear formation due to planar anisotropy as shown in Fig. 2. The final cup in Fig. 2 clearly illustrates the potential material waste due to ear formation on the cups. The magnitude of the observed earing is often similar to the level as shown in Fig. 2. However, it is usually below this level for two reasons. The first is that the example shown was intentionally given contrived material properties of $R=4$, $Q_{ab}=1$, and $P=4$, where R, Q_{ab}, and P are the width-to-thickness strain ratios in tension tests on the sheet at 0, 45, and 90 degrees to the rolling direction. This level of planar anisotropy is higher than commonly observed in most body-centered cubic (bcc) or face-centered cubic (fcc) metals, but may be sometimes achieved in other crystal structures. A second reason for the large ears shown in Fig. 2 is the use of the 1948 Hill criterion (Hill, 1948) for anisotropic flow. It has been observed (Yang and Kim, 1986) that use of the 1948 Hill criterion can lead to overprediction of earing. Thus, an important motivation for this work was to examine, with fully 3-D simulations, the predicted vs. observed earing using both the 1948 Hill quadratic yield criterion and an alternative (Hosford, 1979) criterion in an effort to allow more accurate predictions for this simple case of cylindrical cups, as well as the stamping of complex parts where the proper blank shape to counteract earing is not at all obvious.

In this study we have chosen to begin with the Lawrence Livermore National Laboratory (LLNL) collaborator version of DYNA3D, our explicit, transient dynamics Lagrangian finite-element code (Whirley and Hallquist, 1991).

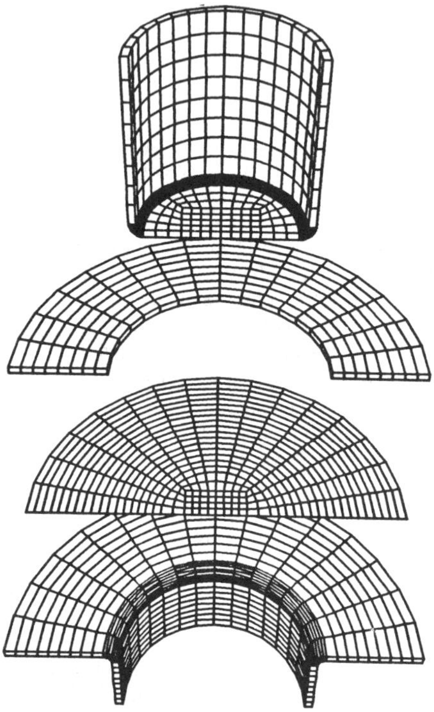

Fig. 1. Tooling (blank and 'rigid' punch, die, and blankholder) used to model cup drawing.

However, this traditional implicit method produces a system of nonlinear algebraic equations which are then solved at each load step using a linearization and iteration technique. The implicit formulation yields a large linear matrix equation which must be solved at each iteration of each load step. Since many sheet forming problems have large bandwidths arising from large workpiece/tooling contact areas, these matrices may tax even the storage capacity of supercomputers. This situation is being overcome with recent improvements in solution methods and storage capability, yet there remains a rivalry between the use of explicit and implicit codes for sheet stamping. We find it is often efficient to implement a methodology in our explicit DYNA family first, bearing in mind that the formulation must be appropriate for subsequent implementation into the implicit NIKE family as well. Such is the case in this work, which examines the usefulness of a new yield

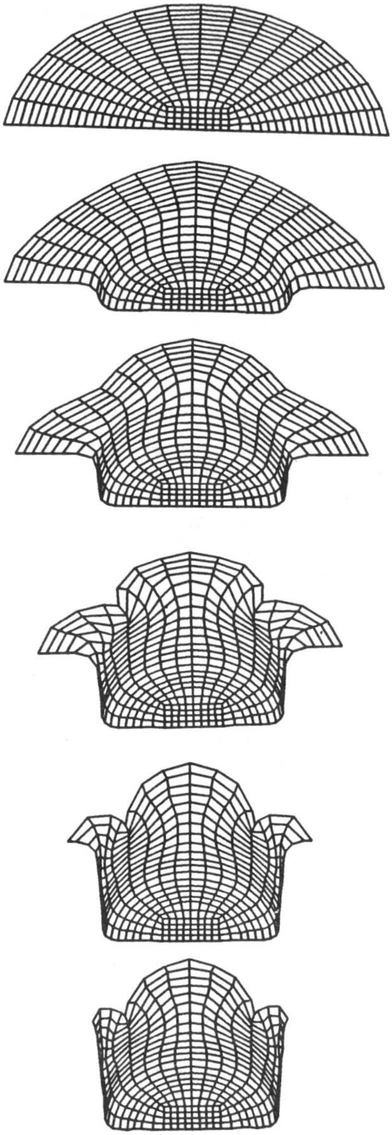

Fig. 2. DYNA3D simulation of earing development during deep drawing, R, Q_{ab}, P = 4,1,4, and a =2.

surface implemented into an internal W-DYNA version of our DYNA code family.

It was shown in a previous work (Whirley, Engelmann, and Logan, 1992) that the cupping problems shown below can be solved effectively by applying loads slowly to minimize dynamic effects, so nearly quasistatic solutions may be obtained.

This work uses implementations of anisotropic yield surfaces in our DYNA family of codes to show that alternate, non-quadratic yield criteria can provide more accurate predictions of the earing observed in stamping processes.

2. YIELD SURFACE IMPLEMENTATION

This section describes the form and implementation of the yield surfaces implemented into the LLNL public DYNA3D (1948 Hill as Model 33) and the internal W-DYNA (1979 Hosford as Model 33x). An overview of the explicit finite element approach used in DYNA3D is given elsewhere (Whirley, Engelmann, and Logan, 1992). Material behavior enters heavily into the update of the internal force vector for an element e :

$$f_{n+1}^{int,e} = \int_{\Omega_e} B^T \sigma_{n+1} d\Omega_e, \tag{1}$$

using B , is the "strain-velocity operator," and the updated Cauchy stress σ_{n+1} . Next, the global force vector f_{n+1}^{int} is found by assembling contributions from all elements. This completes the update of all quantities from time $t=t_n$ to time $t=t_{n+1}$. The stress tensor term in eqn. (1) depends in principle on the material constitutive equation chosen. One of the most straightforward and commonly used is the 1948 Hill equation for anisotropic plastic flow:

$$\bar{\sigma}^2 = \frac{F(\sigma_b - \sigma_c)^2 + G(\sigma_c - \sigma_a)^2 + H(\sigma_a - \sigma_b)^2 + D}{R+1} \tag{2}$$

Eqn. (2) relates the effective stress to the three normal components of Cauchy stress, with the term D containing the shear stress terms:

$$D = 2L\sigma_{bc}^2 + 2M\sigma_{ca}^2 + 2N\sigma_{ab}^2 \tag{3}$$

The values for the constants in Eqns. (2) and (3) can be expressed in terms of the strain ratios R, Q, and P as described above, with the following additional relations needed:

$$F=R/P \tag{4}$$
$$G=1 \tag{5}$$
$$H=R \tag{6}$$

$$L=(Q_{bc}+\tfrac{1}{2})(R+1) \tag{7}$$
$$M=(Q_{ca}+\tfrac{1}{2})(R+Z) \tag{8}$$
$$N=(Q_{ab}+\tfrac{1}{2})(1+Z) \tag{9}$$

In addition to being comparatively straightforward to implement, the quadratic 1948 Hill criterion permits the relatively simple calculation of the ratio, $X(\theta)/X$, of the yield stress in a direction at an angle in the plane of the sheet to the rolling direction, as well as the calculated R-value in that direction:

$$\frac{X(\theta)}{X} = \left[\frac{R+1}{2Ns^2c^2 + R(c^4 - s^4) + c^4 + Zs^4}\right]^{1/2} \tag{10}$$

$$R(\theta) = \frac{R + ((2Q_{ab}+1)(1+Z) - Z - 1 - 4R)s^2c^2}{Zs^2 + c^2} \tag{11}$$

$$\text{where:} \quad \begin{aligned} c &= \cos(\theta) \\ s &= \sin(\theta) \end{aligned} \tag{12}$$

Eqns. (10)-(11) are relatively easy to comprehend, but show several trends that are not usually borne out by experimental data. For example, at 45 degrees to the rolling direction, 1948 Hill predicts (if $R=P$) :

$$\frac{X(45^\circ)}{X} = \left[\frac{R+1}{Q_{ab}+1}\right]^{1/2} \tag{13}$$

Eqn. (13) shows a high dependence of yield stress on orientation in the plane of the sheet. This dependence is a likely factor in overprediction of earing as in Fig. 2, as the strong material in the 45 degree direction tends to pull in to form the wall of the punch (forming a deep trough), while compressing the 0 degree and 90 degree walls (forming high ears). One yield criterion which in principle should show better agreement with experiment in the 1979 Hosford equation, extending 1948 Hill to a non-quadratic form with values of the exponent a in the range of a =8 for fcc (Hosford, 1979), and a =6 for bcc (Logan and Hosford, 1980) metals:

$$\bar{\sigma}^a = \frac{F(\sigma_2 - \sigma_3)^a + G(\sigma_3 - \sigma_1)^a + H(\sigma_1 - \sigma_2)^a}{R+1} \tag{14}$$

This equation predicts much milder dependencies of strength ratios in stress states and directions other than tension in the rolling direction. For example, in biaxial tension, where both , $\sigma_a = \sigma_b = B$, both 1948 Hill and 1979 Hosford predict the relatively simple value:

$$\frac{B}{X} = \left(\frac{R+1}{Z+1}\right)^{1/a} \tag{15}$$

Clearly, values of $a > 2$ will give a much milder dependence often observed experimentally. Further, if we assume the R-value orientation dependence as in eqn. (11), the orientation dependence of uniaxial flow stress is much milder. This ratio is important in the flange of the cup during draw-in, and thus indicates promise for greater accuracy in earing calculations.

However, difficulties arise in the implementation of the 1979 Hosford criterion for cases in other than principal stress/strain space. This stems from the lack of shear terms in the criterion. Eqn. (13) must remain in principal stress space to be used without spurious results and non-convexity problems. To do so, we must make several approximations in updating the Cauchy stress tensor (expressed as a vector of six) when using eqn. (13). First and foremost is a rotation to the principal stress coordinate space. This does not normally coincide with either the material (rolling and transverse direction) coordinate system, nor with the axes of principal strain. Our first assumption is that the axes of principal stress and strain coincide, although for planar isotropy we know that they normally will not. However, this assumption, which leads us to ignore cross-terms in the constitutive matrix, is believed to lead only to small errors for the degree of anisotropy observed in most sheet metals. This is the assumption we will use below to update the stresses. The first step in doing so is a calculation of contact stresses and updated elastoplastic stresses as follows:

$$\sigma_i^c = \sigma_i^n + C_{ij}^e d\varepsilon_j^e \qquad (16)$$

$$\sigma_i^N = \sigma_i^c + C_{ij}^{ep}(d\varepsilon_j - d\varepsilon_j^e) \qquad (17)$$

Here, $d\varepsilon_j^e$ are the elastic portions of the strain increment, and C_{ij}^e is the elastic constitutive matrix. To obtain the updated stresses, we proceed further by applying the remainder of the strain increment $(d\varepsilon_j - d\varepsilon_j^e)$ using the elastoplastic matrix C_{ij}^{ep}:

$$C_{ij}^{ep} = C_{ij}^e - \frac{C_{ij}^e q_j (C_{ij}^e q_j)^T}{p_i q_i + q_i^T C_{ij}^e q_j} \qquad (18)$$

The yield surface F directly affects the calculation of the matrix C_{ij}^{ep}, since

$$q_i = \frac{dF}{d\sigma_i} \qquad (19)$$

$$p_i = -\frac{dF}{d\varepsilon_i} \qquad (20)$$

In the following section we will demonstrate the effect of the chosen yield surface (Eqn. (2) or (13)) on the extent of earing in cupping.

3. EARING DEVELOPMENT AND FINITE-ELEMENT PREDICTIONS

To demonstrate the correlation of DYNA earing simulations with experiments, we compare the extensive data obtained by (Wilson and Butler, 1962) with that obtained using simulations with either a =2 or a =8 in eqn. (21). Numerous runs with $R = P$ were made as outlined in Table 1 below:

TABLE 1. Data Used in DYNA Simulations of Earing.

a	*R*	*Q*	*P*	*h*	*Δh*	*ΔR/R*	*Δh/h*
2	4.0	1.0	4.0	7.32	2.98	1.20	.407
2	2.0	1.0	2.0	7.20	1.92	0.67	.267
2	1.5	1.0	1.5	7.22	1.20	0.40	.166
2	0.5	1.0	0.5	7.08	-1.50	-0.67	-.212
2	1.0	2.0	1.0	7.18	-1.72	-0.67	-.240
2	0.5	2.0	0.5	7.26	-2.40	-1.20	-.330
2x	4.0	1.0	4.0	7.62	2.34	1.20	.312
8	4.0	1.0	4.0	7.84	1.54	1.20	.196
8	2.0	1.0	2.0	7.64	0.73	0.67	.096
8	1.0	2.0	1.0	7.78	-0.79	0.67	-.102
8	0.5	2.0	0.5	7.66	-1.27	-1.20	-.166
2m	2.0	1.0	2.0	7.48	0.36	0.67	.048

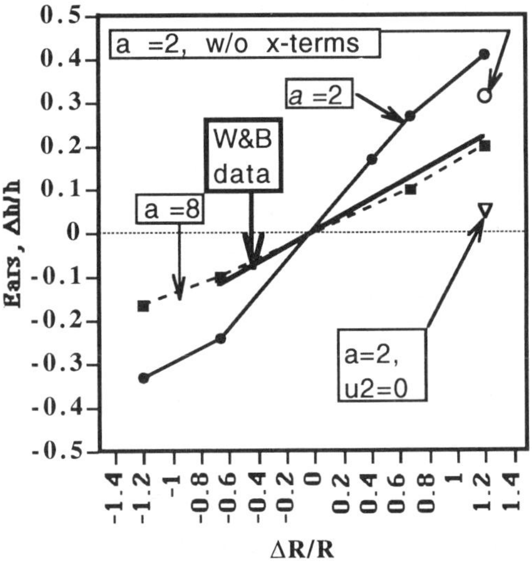

Fig. 3. Plot of earing (Δh/h vs. ΔR/R) including Wilson and Butler data compared to DYNA results with a =2 and a =8.

These results are plotted as Δh/h vs. $\Delta R/R$ in Fig. 3, along with the original data from Wilson and Butler. It is strikingly clear that the use of a =2 (identically 1948 Hill) overpredicts earing by about a factor of two. In contrast, the use of a =8 matches almost exactly the observed earing data; in fact a =8 seems to under predict earing slightly. However, recall that we were forced to ignore cross-terms in the matrix C_{ij}^{ep} above when using a =8. We do not need to do this when a =2, but we can choose to do so to examine the effect.

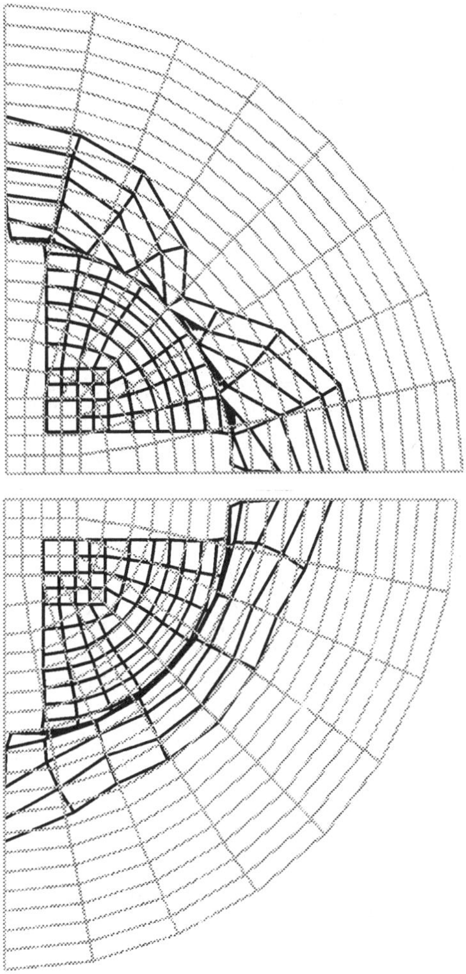

Fig. 4. Quarter-symmetry view of ear development in full 3-D (upper), and with hoop motion restricted (lower). Grid of original blank shown as background.

There is indeed some difference, as shown by the data labeled 'a =2x' in Table 1 and in Fig. 3. Thus, if the proper cross-terms existed for a =8, we might expect to see slightly more calculated earing, and perhaps slightly over predict the data instead of slightly under predicting it. Either way, the value of a =8 is much more accurate than a =2.
Several variations of runs were made in the process of compiling the DYNA comparisons in Fig. 3. These included variations of the SVE (Selective Velocity Enhancement) factor from 20x to 50x, with very little difference noticed in earing. The runs presented here were run at SVE=20, and otherwise the parameters used previously (Whirley, Engelmann, and Logan, 1992). Meshing consisted of 24 nodes radially along the blank, and 9 nodes around the quarter-symmetry hoop direction from 0 to 90 degrees. Meshing of 12x9 or even 12x5 nodes gave reasonable ear development early on, but 5 nodes in the hoop direction showed difficulties in completing the cupping operation. A much more important fact is that in the DYNA3D codes there is obviously a degree of freedom in the hoop direction in the flange. Early simulations by the author (Logan, 1984) considerably under predicted earing, even though a =2 was used in the MARMOT code of that work. However, a key constraint of the MARMOT sheet forming code was the assumption of no displacement in the hoop direction. When sliding boundary planes were employed in DYNA to preclude motion in the hoop direction at θ=0, 22.5, 45, 67.5, and 90 degrees, earing is cut drastically (see the data for '2m' in Table 1 and 'u2=0' in Fig. 3). This explains why the early MARMOT code under predicted earing even with 1948 Hill, and shows the importance of hoop displacement and shear in the earing process. This is borne out graphically in Fig. 4, which compares ear development at R, Q, P = 4,1,4 with a =2. In the upper view, hoop motion is allowed, and lines of constant θ become distorted as the ears form. In the lower view, hoop motion is precluded as described above, and both earing and overall θ-direction motion are greatly reduced.

4. CONCLUSIONS

Using full 3D finite-element simulations of the cupping process, it is demonstrated that use of the 1948 Hill criterion (a =2) will greatly overpredict earing compared to typical experimental data. This requires the use of a suitable mesh and large motions in the hoop direction. Use of the 1979 Hosford criterion with a =8 gives very good correlation with the data of Wilson and Butler. It is possible that other yield criteria such as the generalized Hill (Hill, 1979) or the tricomponent criterion (Barlat and Lian, 1989) would show similar improvements in agreement in earing calculations such as these.

ACKNOWLEDGMENTS

The author wishes to thank Dr. W.F. Hosford at U. Michigan for assistance in the a>2 implementation, and to T.P. Slavik of LLNL for help with the never ending tasks of tensor coordinate transformations. This work was performed under the auspices of the U.S. Department of Energy by the Lawrence Livermore National Laboratory under contract W-7405-Eng-48.

REFERENCES

Barlat, F. and Lian, J. (1989), "Plastic Behavior and Stretchability of Sheet Metals, Part I: A Yield Function for Orthotropic Sheets Under Plane Stress Conditions", Int. J. Plasticity 5, pp. 51-66.

Hill, R. (1948), *The Mathematical Theory of Plasticity,* Clarendon Press, Oxford.

Hill, R. (1979), "Theoretical Plasticity of Textured Aggregates", Math. Proc. Camb. Phil. Soc. , 85, p.179.

Hosford, W.F. (1979), "On Yield Loci of Anisotropic Cubic Metals", Proc. 7th N. Amer. Metal Working Research Conf., p. 191, SME, Dearborn.

Logan, R.W. and Hosford, W.F. (1980), "Upper-Bound Anisotropic Yield Locus Calculations Assuming <111> -Pencil Glide", Int. J. Mech. Sci. 22, p.419.

Logan, R.W. (1984), "Sheet Formability: Simulation and Experiment", PhD Thesis, Univ. Michigan.

Whirley, R.G., and Hallquist, J.O. (1991), "DYNA3D: A Nonlinear, Explicit, Three-Dimensional Finite Element Code for Solid and Structural Mechanics- User Manual," University of California, Lawrence Livermore National Laboratory, Report UCRL-MA-107254.

Whirley, R.G., Engelmann, B.E., and Logan , R.W. (1992), " Some Aspects of Sheet Forming Simulation Using Explicit Finite Element Techniques", CED-Vol. 5, AMD Vol. 156, "Numerical Methods for Simulation of Industrial Metal Forming Processes", ed. M.J. Saran, A.B. Pifko, N. Kikuchi, and K.K. Tamma, ASME.

Wilson, D.V. and Butler, R.D. (1962), "The Role of Cup-Drawing Tests in Measuring Drawability", J. Inst. Metals 90, p. 473.

Yang, D.Y. and Kim, Y.J. (1986), "A Rigid-Plastic Finite-Element Formulation for the Analysis of General Deformation of Planar Anisotropic Sheet Metals and its Applications", Int. J. Mech. Sci. 28, No. 12, pp. 825-840.

Simulation of Materials Processing: Theory, Methods and Applications, Shen & Dawson (eds)
© 1995 Balkema, Rotterdam. ISBN 90 5410 553 4

An industrial finite element code for one-step simulation of sheet metal forming

M. El Mouatassim & B. Thomas
Renault Direction de l'Ingénierie Véhicule, Boulogne Billancourt, France

J.-P. Jameux
Renault Direction de la Recherche, Boulogne Billancourt, France

E. Di Pasquale
Simtech, Paris, France

ABSTRACT: This paper describes recent advances in one-step sheet metal forming simulation, jointly carried out by RENAULT and SIMTECH researchers. The aim of the research was to develop an accurate, user-friendly system that enables the user to check quickly the feasibility of a stamped part and the effectiveness of a tool design solution. Important factors controlling the sheet metal forming operation, such as material plasticity, friction and restraining system, can be taken into account easily and accurately. In the automotive industry, it is expected that part designers and die engineers, without prior knowledge of numerical methods, will profit of this system, with significant reduction of costs and delays of product development.

INTRODUCTION

Present-day automotive market, with its fierce competition and demanding customers, prompts the industry to shorten feasibility study and tool development for stamped body parts, in addition to reducing weight and to using material of increasingly high mechanical properties.

Sheet metal forming, extensively used in the automotive industry, is a very difficult process to control, due to the great number of parameters affecting the process, including the geometry of the part, of the die addendum and of the blank holder, the type of material and the applied restraining forces.

In the traditional methodology of stamping development, these parameters are adjusted during the tool try-out phase, through numerous, costly and time-consuming experiments. This places stamping in the critical path for vehicle planning schedules.

In order to achieve shorter lead times while improving product quality and reducing costs, RENAULT is shifting from sequential work to simultaneous engineering. New forms of work organisation are being introduced to enable the "product designer" and the "process designer" to work in conjunction with each other far upstream in the development of a stamped part. The extensive use of numerical simulation at any stage of product development is meant to foster this joint work.

This paper describes the development and the validation of SIMEX, a finite element code for analysis of stamped parts, based on the so-called inverse or one-step approach. Such an approach enables the user to obtain quickly and easily an assessment of the formability of a part and of the effect of proposed tool design solution.

PRODUCT-PROCESS VALIDATION

The objective of RENAULT program for numerical simulation of sheet metal forming is the development and operational use of a software package for product-process validation. Product-process validation is achieved by the simultaneous design of the automotive part and of the stamping process needed to produce it, starting from the component design. At this stage, the inverse analysis of the stamping process of the formed part, and constant dialogue between the "product designer" and the "process designer" are essential. They enable the part design to be adapted so that to improve the part formability while retaining its functional characteristics. At this stage, the goal of the activity is to anticipate and reduce the number of part design modifications.

Once a robust design of the part has been achieved, the actual die engineering work (process design) may start. Such work consists essentially of the design of additional surfaces, such as run-offs and blank holder. The process designer has at his or her disposal a number of dedicated software tools, for the optimisation of stamping direction, the design do die addendum, the check of the uniformity of the drawing depth and, finally, the protection of the visible areas of the outer panels. Throughout all the process design, inverse simulation can be used to validate design choices, and to compare the effect of different proposed solutions.

After all the additional surfaces have been designed, and before the construction of test tools, if deemed necessary, the numerical simulation of the whole process is required. This phase is actually a numerical die try-out, aiming at assessing the performance of a given die design. In order to achieve the precision needed for such task, incremental codes are used.

RENAULT engineers have measured the impact of the extensive application of numerical simulation techniques in the development of a new vehicle. Experience has shown a reduction by approximately 50% the cost and lead time associated to tool design and try-out.

OUTLINE OF INVERSE APPROACH

The starting point of inverse simulation is the Finite Element model of the stamped part (Guo et al., 1990, El Mouatassim et al., 1991). Inverse simulation algorithms enable the user to find the position of the nodes of the blank in its original, horizontal plane, ensuring the equilibrium of the stamped blank under tool forces and restraining actions. At each iteration, stresses and strains of the blank are computed from the displacement field.

The equilibrium search is carried out without any intermediate position, whence the denomination *one-step*.

Inverse simulation seeks to impose the equilibrium of the final configuration of the stamped part. As a consequence, strain paths are assumed to be linear (radial strain path) and the effect of the history of deformation of the material is neglected. The advantage of such an approach is that a stamped part can be analysed without prior definition of run-offs and blank holder surfaces.

INVERSE vs. INCREMENTAL APPROACH

As mentioned above, incremental codes constitute an alternative to soft tooling in the last phase of product development, before the manufacturing of hard tools and the series production. One-step codes, on the other hand, are suitable for use in the product design phase as well as in the first stages of die design.

A comparison between inverse and incremental approach is outlined in the following table:

	incremental	inverse
model (mesh)	- tools - flat blank	stamped part
ease of use	low	high
EWS CPU time 20000 elem.	ca. 40h	ca. 2h
application: part design process design	 no late	 yes early
information	all	no springback no visible wrinkles no strain path no force evolution

THE SIMEM2 AND SIMEX CODES

SIMEM2™ is an integrated system for sheet metal forming analysis, developed by RENAULT. SIMEM2™ includes a dedicated user interfaces (the Project Organiser), an automatic mesh generating system, and SIMEX™, a finite element solver jointly developed by RENAULT and SIMTECH. This system is integrated with the dedicated graphic pre- and post-processor SIMVISION™ (DYNAMIC SOFTWARE), or, in alternative, can be interfaced to commercial graphic pre- and post-processors.

SIMEM2™ Project Organiser is a standard MOTIF application that guides the user in the process of simulation of a stamped part. As SIMEM2™ has been conceived for use by personnel that is not trained in computer simulation technology, the Project Organiser is designed to follow the user in each step of the analysis, preventing the user from encountering most problems that currently face the inexperienced user.

SIMEM2™ Automatic Mesh Generator enables the user to build a Finite Element model of a stamped part in a few minutes, starting from CAD data. The main advantage of SIMEM2™ Automatic Mesh Generator is that it needs no, or very little, preliminary preparation of the CAD surfaces. It has been developed to handle industrial CAD data, including overlapping surfaces, holes, micro surfaces etc....

SIMEX™ is a dedicated finite element program for sheet metal forming simulation using the inverse approach, jointly developed by RENAULT and SIMTECH. SIMEX™ can take into account most of the significant parameters of stamping process, such as metal plasticity, friction and restraining forces. Its computation speed, robustness and ease of use make it particularly suitable for use in industrial environment by users with little or no knowledge of numerical simulation.

SIMVISION™ is a graphic pre- and post-processor, designed to make full use of 3D graphic capabilities of recent Engineering Work Stations.

THE RENAULT TWINGO DASHPOT CUP

Choice of the component

The RENAULT TWINGO dashpot cup was selected because of the simplicity of its CAD definition and the difficulty of its drawing. Two passes were in fact needed to obtain a part in compliance with the specifications. The thickness variations on the stamped part were sufficiently large (about 20%) for the comparison with computed results to be significant.

Experimental and numerical results regarding the two stages of the forming operation of this part have been presented in a former paper

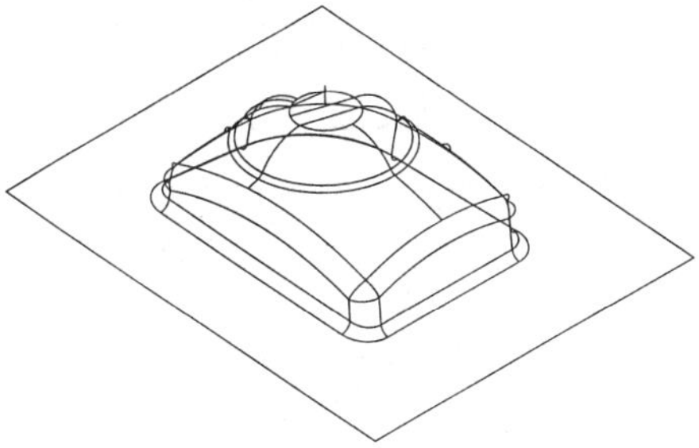

Figure 1 : CAD model of TWINGO dashpot cup

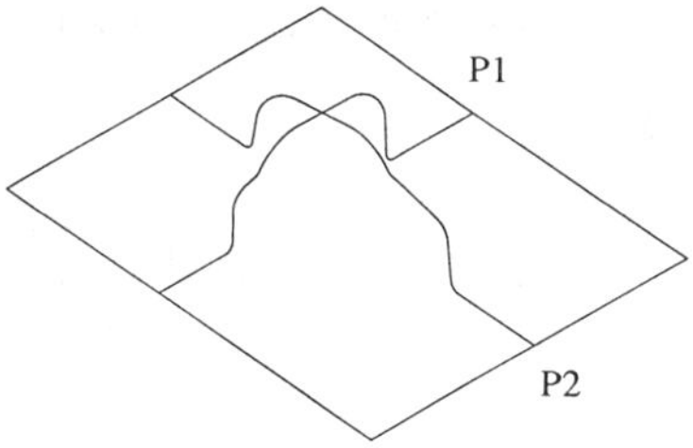

Figure 2 : measurement profiles

(El Mouatassim et al., 1994). In this paper, only the first pass (preform) is analysed (Figure 1). Analysis of multi-step operations using the inverse approach is under development.

Experimental measurements

To perform the stamping tests in suitable experimental conditions, centring pins were installed on the tools to ensure the reproducibility of the blank position on the blank holder surface.
During the preform operation, the die descends and applies, first, a counter-force to balance the 400 kN blank holder load. Then, the die continues its descending movement, deforming the blank until it reaches the end of the die travel point.
The tests are performed without blank holder adjusting shims. Accordingly, the total force developed by the blank holder is applied to the blank.
The thickness measurements were performed at RENAULT using an ultrasonic device. Residual thickness was measured on the profiles P1 and P2 (Figure 2).

Material model

The material used for the manufacturing of the component is a high-strength steel, modelled using a Krupkowsky-Swift hardening law:

$$\sigma = k(\varepsilon_0 + \varepsilon)^n$$

Normal anisotropy of the material is considered, through an average Lankford coefficient R.
For the material considered, the material characteristics provided to the code are:

Young modulus	210000 Mpa
Yield stress	277
Swift coefficient K	624
Swift coefficient ε_0	0.0099
Swift coefficient n	0.176
Lankford coefficient R	1.0

Input Coulomb friction coefficient between the blank and the tools is equal to 0.15.

Numerical model

For the purpose of SIMEX analysis, the model of the stamped part (Figure 13) has been split into two regions:
Under the punch 7217 elements
Under the blank holder 3932 elements
The portion of material under the punch is subjected to friction conditions. The code identifies the points at which the material rubs against the punch and applies the friction force (normal force multiplied by friction coefficient). As for the portion of material under the blank holder, it is subjected to the appropriate restraining load. Drawbeads and other concentrated restraining device are treated in a similar fashion.
The computation took 20 minutes on a SGI R4400 workstation.

SIMEX VALIDATION

SIMEX validation on the RENAULT TWINGO dashpot cup aimed at assessing two features of the code:

a) the ability to reproduce the experimentally measured results;
b) the ability to reproduce the effect of the variation of the process variables.

Numerical vs. experimental results

Figures 3 and 4 show, respectively, the profiles of thickness on the measurement line P1 and P2. From the analysis of the results, the following remarks can be made:

a) SIMEX follows quite accurately the pattern of the thickness profile, picking up the zones where the material is thinned (traction) or thickenned (compression) and those where concentration of strain takes place. The identification of the areas of strain concentration is possible because the code takes into account material plasticity and friction. The relative error between numerical and experimental results does not exceed 10%, except in some points where the error can be as much as 20%.

b) The most important discrepancy between numerical and experimental results can be found in the first portion of the two lines. Here the experimental results show two peaks of

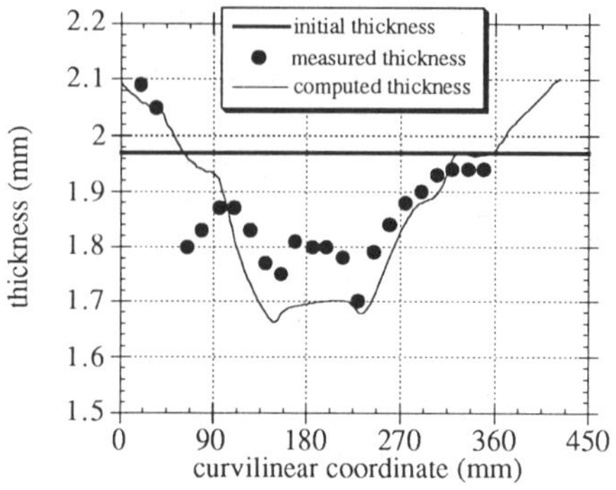

Figure 3 : Profile P1, measured vs. SIMEX results

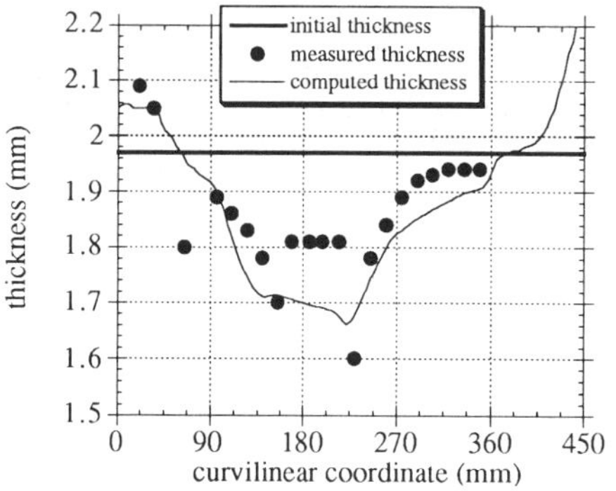

Figure 4 : Profile P2, measured vs. SIMEX results

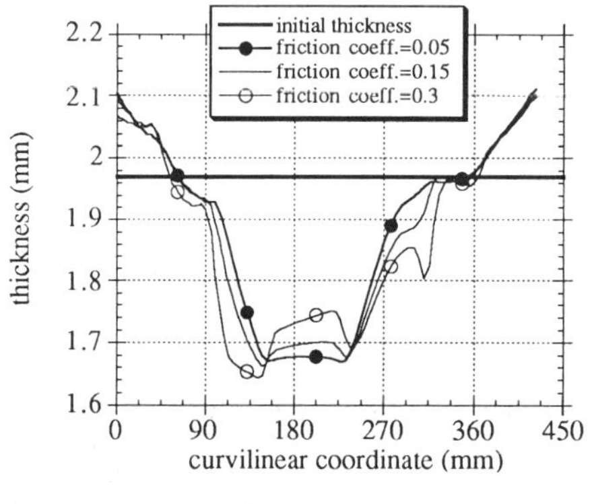

Figure 5 : Profile P1, effect of friction

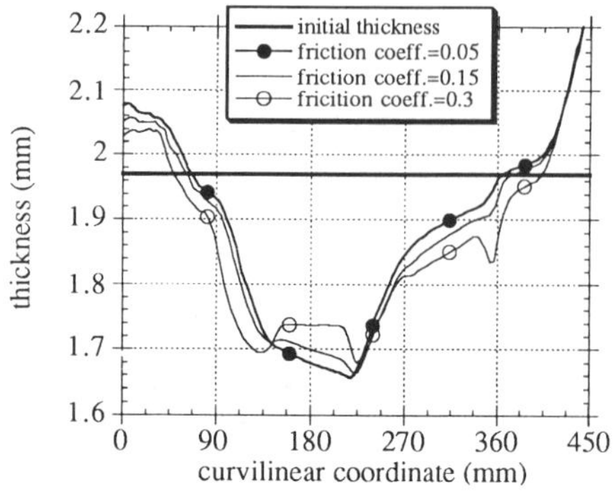

Figure 6 : Profile P2, effect of friction

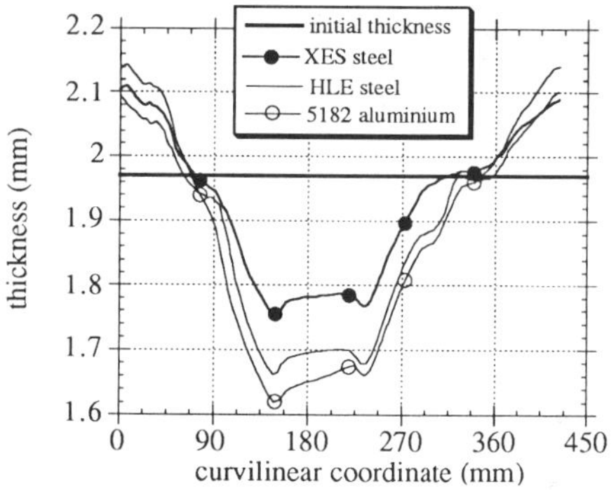

Figure 7 : Profile P1, effect of materials

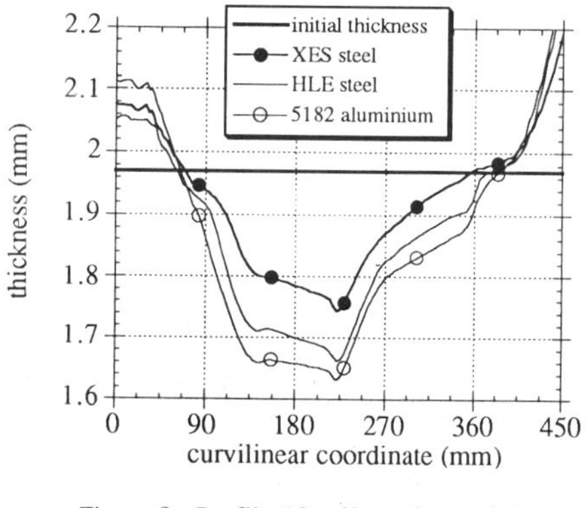

Figure 8 : Profile P2, effect of materials

deformation, while the code detects only one. This discrepancy is probably due to the approximation of the inverse formulation, which neglects the history of deformation.

Effect of friction

Three friction coefficients have been considered between the blank and the punch:

low friction	0.05
service friction	0.15
high friction	0.30

The analysis of the thickness profile along the measurement lines (Figures 5 and 6) shows that thinning, which, for low values of friction, is spread along the top portion of the cup, concentrates along the border of the top dome as friction increases. This is in agreement with

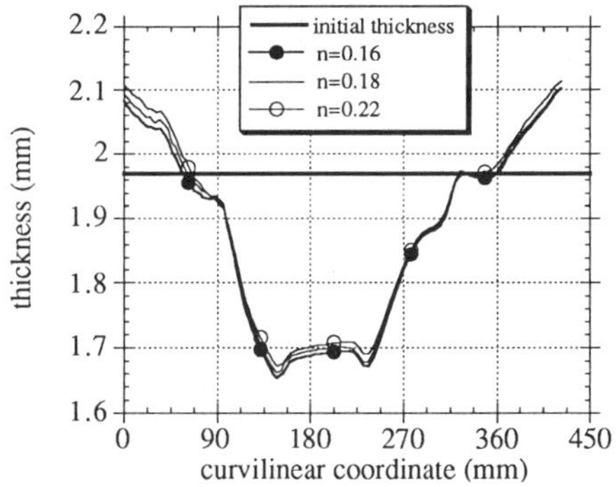

Figure 9 : Profile P1, effect of hardening coeff. n

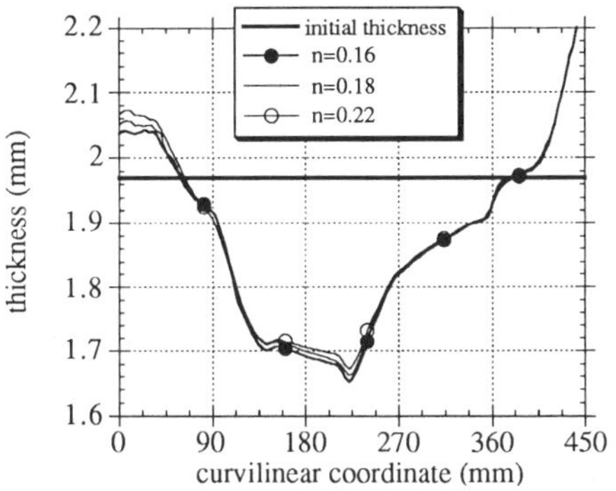

Figure 10 : Profile P2, effect of hardening coeff. n

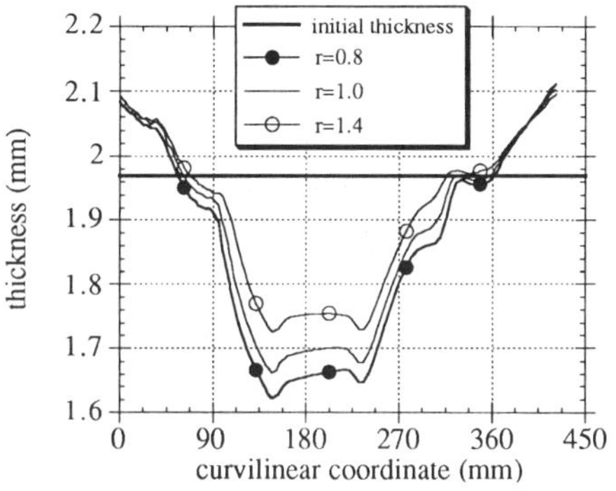

Figure 11 : Profile P1, effect of Lankford coeff. r

experience and with published results on similar although simpler geometries (Lee et al. 1990).

Effect of material characteristics

First, the results, relative to different industrial material types have been analysed. The following materials have been selected:

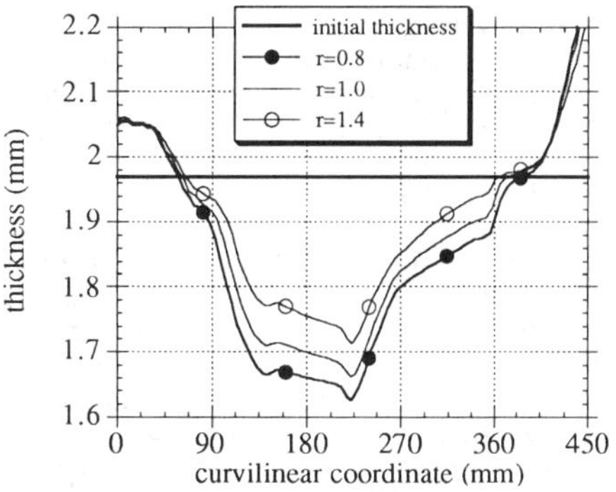

Figure 12 : Profile P2, effect of Lankford coeff. r

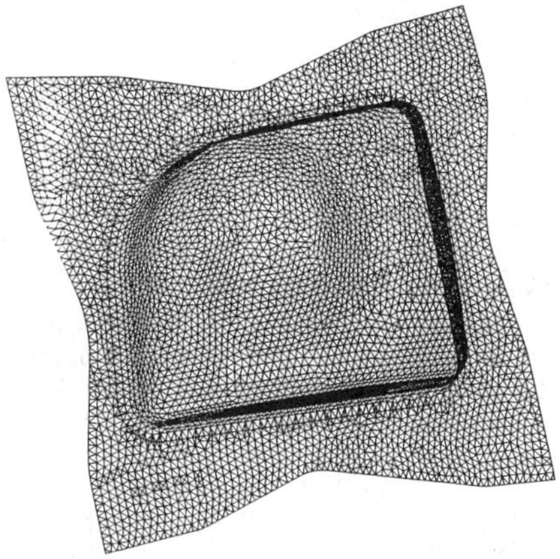

Figure 13 : mesh of the TWINGO dashpot cup

	σ_y	K	n	r
deep drawing steel	200	546	0.20	1.70
high strength steel	277	624	0.18	1.00
Aluminium-5182	137	540	0.29	0.72

Figures 7 and 8 show the thickness profiles along the measurement lines. From deep drawing steel to aluminium, the deformation tends to increase and to concentrate. This is in accordance with prior experience on such components (deep drawn).

It is also interesting to investigate the decoupled effect of the rheological characteristics of the material. In this connection, some simulations have been carried out, varying separately the hardening coefficient n and the Lankford coefficient r.

Tree values of n and r have been considered:

	n	r
low value	0.16	0.8
average value	0.18	1.0
high value	0.22	1.4

The results are shown in figures 9 through 12.It can be pointed out that an increase of both n and r leads to a more homogeneous distribution of thickness, i.e. to a safer, better quality component.

CONCLUSIONS

The joint work by RENAULT and SIMTECH on the development of SIMEX^{TM} has produced a code that models the effect of the most important factors affecting sheet metal forming.

Good correspondence between numerical and experimental results has been observed, within the approximations proper to the inverse approach.

SIMEM2 and SIMEX are effective aids for the product designer and for the first stage of the work of the die designer.

REFERENCES

Guo, Y.Q., Batoz, J.L., Detraux, J.M., Duroux, P., Finite element procedures for strain estimation of sheet metal forming parts, Int. J. Num. Meth. Eng., 30 (8) 1385-1401, December 1990.

El Mouatassim, M., Detraux, J.M., Batoz, J.L., Guo, Y.Q., Application of inverse FE-procedure to sheet forming, VDI Berichte 894, 1991.

El Mouatassim, M., Jameux, I.P., Thomas, B., Mehrez, F., Milcent. G., The simulation of multi-operation deep-drawing process at RENAULT with PAM-STAMP, J. Mat. Proc. Tech., 45 (1994) 317-322.

Lee, J.K., Wagoneer, R.H., Nakamachi, E., A benchmarck test for sheet forming analysis, Center for Net Shape Manufacturing Report No. ERC/NSM-S-90-22, Columbus, Ohio, July 1990.

Simulation of Materials Processing: Theory, Methods and Applications, Shen & Dawson (eds)
© 1995 Balkema, Rotterdam. ISBN 90 5410 553 4

Application of an anisotropic yield locus based on texture to a deep drawing simulation

S. Munhoven & A. M. Habraken
Department MSM, University of Liège, Belgium

J. Winters, R. Schouwenaars & P. Van Houtte
Department MTM, Katholieke Universiteit Leuven, Heverlee, Belgium

ABSTRACT: When it comes to representing the mechanical behaviour of polycrystalline engineering materials during forming processes with sufficient accuracy, simple analytical yield locus expressions often do not prove to be very successful. In this paper the implementation of a more appropriate anisotropic yield locus based on crystallographic texture in a finite element code is presented. Some modifications added to the latter in order to simulate correctly general three-dimensional anisotropy in elasto-plasticity are described. Application to a cup drawing operation and comparison with experimental results show that the capability to account for texture evolution through deformation history needs to be added to the model.

1 INTRODUCTION

The plastic properties of many engineering materials reveal a more or less pronounced anisotropy. Experimental evidence for this can be found for example in the non-unit r-values of steel sheets in tensile tests and their dependency on the angle between the rolling direction and the longitudinal axis of the test sample. Another well known phenomenon caused by plastic anisotropy are the ondulations of the rim of a cup drawn from a circular piece of sheet material with the help of axisymmetric tools. An example of this so-called earing in cup drawing is simulated in the present paper. In some other forming processes, the external shape of the product may remain unaffected, while only the flow field and the resulting stress distribution will reveal the presence of plastic anisotropy (Van Houtte 1992).

In any case, the material behaviour needs to be described precisely. As the peculiar shape of the yield loci of many polycrystalline engineering materials can not be represented accurately by means of a quadratic expression, such as the widely used Hill yield criterion (Hill 1979), efforts have been made in recent years to base their description more directly on the internal structure, mainly on their non-random crystallographic texture. One approach proposed to achieve this aim consists in incorporating a material description at crystallographic scale into a macroscopic finite element (FE) continuum description, implying massive data parallel computations (Dawson *et al.* 1992, Beaudoin *et al.* 1994). Alternatively, one could choose to define a macroscopic yield surface description consistent with the texture of the polycrystal, which might in turn easily be incorporated in a FE model (Van Houtte *et al.* 1989, Van Bael *et al.* 1991).

In the present work, an approach of the second type is used to implement a yield locus description coming from the KULeuven in a FE code developed at the University of Liège and to simulate a cup drawing operation.

2 ANISOTROPIC FINITE ELEMENT ANALYSIS

Usage of anisotropic materials in metal forming processes was so far limited to two-dimensional simulations or special element types in the FE code LAGAMINE (Li *et al.* 1995). In the context of the present work some features are added for the general usage of anisotropic yield loci based on texture.

2.1 *Incorporation of texture based yield loci*

The complete flowchart of a model using a yield locus based on crystallographic texture is shown at figure 1. Experimental X-ray texture pole figure measurements serve as starting point. A transition module uses the corresponding crystallographic orientation distribution function (CODF) (Bunge 1982) and crystallographic plasticity theory (Taylor-Bishop-Hill

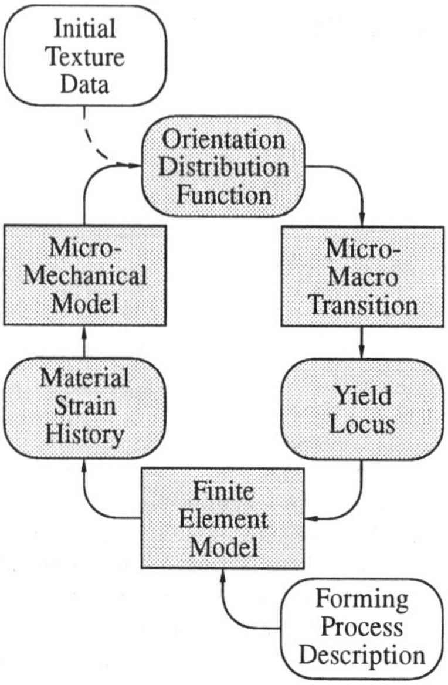

Figure 1. Complete flowchart for the usage of a yield locus based on crystallographic texture.

model) to compute the so-called average Taylor factors for a large number of plastic deformation modes. A sufficiently continuous yield locus is then constructed upon these discrete macroscopic data points, either by least squares fitting of a series expansion (Van Houtte *et al.* 1989) or by an appropriate spline interpolation scheme (Van Houtte 1994).

As polycrystal texture is, in general, subjected to evolution through deformation history, the so shaped yield locus will, however, lose its validity beyond certain moderate strains and require updating. A micro-mechanical model, ajusting the CODF according to the FE results, followed by a new micro-macro transition are used to achieve this aim. The corresponding module is under construction. Although texture evolution in sheet forming with moderate strains is often of secondary importance (Beaudoin *et al.* 1994), results found in the present work show that, for the particular material used (a cold rolled interstitial free steel), it seems required.

In fact only the shape of the yield locus results from the transition procedure described above, while its size is determined by the critical resolved shear stress τ_c (Van Bael *et al.* 1991). The latter may for example be assumed to evolve according to a Swift-type hardening law as a function of the slip (Winters *et al.* 1995)

$$\tau_c = K(\gamma_0 + \gamma)^n \tag{2.1}$$

which may be fitted to tensile test results.

2.2 *Local axis and local velocity gradients*

Let $\boldsymbol{x} = \boldsymbol{x}(\boldsymbol{X}, t)$ be the mapping of the initial configuration Γ of the solid to the current one $\gamma(t)$. The Jacobian matrix

$$\boldsymbol{F} = \frac{\partial \boldsymbol{x}}{\partial \boldsymbol{X}} \tag{2.2}$$

is called the deformation gradient.

Furthermore, let the material behaviour be described by a set of constitutive equations of the form

$$\begin{aligned} \dot{\boldsymbol{\sigma}} &= \boldsymbol{f}_\sigma(\boldsymbol{\sigma}, q, \boldsymbol{L}), \\ \dot{q} &= f_q(\boldsymbol{\sigma}, q), \end{aligned} \tag{2.3}$$

where q are state variables (scalars, vectors, tensors, ...) and

$$\boldsymbol{L} = \frac{\partial \dot{\boldsymbol{x}}}{\partial \boldsymbol{x}} = \dot{\boldsymbol{F}}\boldsymbol{F}^{-1} \tag{2.4}$$

are velocity gradients with respect to the current coordinates in the deformed configuration γ. The symmetric and skew-symmetric parts of $\boldsymbol{L}$, $\boldsymbol{D} = \frac{1}{2}(\boldsymbol{L} + \boldsymbol{L}^T)$ and $\boldsymbol{W} = \frac{1}{2}(\boldsymbol{L} - \boldsymbol{L}^T)$, are the rate of deformation and spin tensor.

In general, a non-linear FE analysis involves a step-by-step procedure, and a strain path will have to be choosen between consecutive configurations, $\gamma(t_A)$ and $\gamma(t_B)$ ($t_B = t_A + \Delta t$), in order to integrate the constitutive equations (2.3). In this context, a constant velocity gradient path seems to be a good choice (Cescotto 1992), leading to

$$\boldsymbol{L} = \frac{1}{\Delta t} \ln(\boldsymbol{F}_{AB}), \qquad (t_A < t < t_B), \tag{2.5}$$

with $\boldsymbol{F}_{AB} = \boldsymbol{F}_B \boldsymbol{F}_A^{-1}$ the incremental deformation gradient tensor.

The yield locus based on texture (say the functions $\boldsymbol{f}_\sigma$ and f_q), however, is linked to a certain material reference system. Since the latter is subject to distorsions during deformation, for computational purposes an alternative suitable so-called local reference system has to be found. For the sake of simplicity it is supposed to coincide initially with the global cartesian frame used in the FE code and to be later, at least locally, only time dependent (rotation gradients neglected). Global and local axis are thus linked by a proper-orthogonal tensor $\boldsymbol{R}$ ($\boldsymbol{R}\boldsymbol{R}^T = \boldsymbol{R}^T\boldsymbol{R} = \boldsymbol{I}$). In a similar way, the velocity gradient corresponding to a local description is then defined as follows :

$$\boldsymbol{F}' = \frac{\partial \boldsymbol{x}'}{\partial \boldsymbol{X}'}, \qquad \boldsymbol{L}' = \frac{\partial \dot{\boldsymbol{x}}'}{\partial \boldsymbol{x}'} = \dot{\boldsymbol{F}}'\boldsymbol{F}'^{-1}, \tag{2.6}$$

a $'$ identifying variables relative to local axis. Among other possible choices, a local reference frame may now be defined by requiring the local velocity gradient $\boldsymbol{L}'$ to be not only constant, but also symmetric (spin free, $\boldsymbol{W}' = 0$). For a given initial rotation $\boldsymbol{R}_A$ and global incremental deformation gradient $\boldsymbol{F}_{AB}$, the corresponding final rotation $\boldsymbol{R}_B$ and the tensor-exponential of $\boldsymbol{L}'$ can be shown to form the right polar decomposition of $\boldsymbol{F}_{AB}\boldsymbol{R}_A$.

The following computational procedure uses this definition of a local reference system :

1. Rotate the initial stress tensor $\boldsymbol{\sigma}_A$ from global to local axis :

$$\boldsymbol{\sigma}'_A = \boldsymbol{R}_A^T \boldsymbol{\sigma}_A \boldsymbol{R}_A \tag{2.7}$$

2. Compute the constant symmetric local velocity gradient $\boldsymbol{L}'$ and the final rotation $\boldsymbol{R}_B$:

$$\begin{aligned} \boldsymbol{L}' &= \frac{1}{2\Delta t} \ln\left((\boldsymbol{F}_{AB}\boldsymbol{R}_A)^T (\boldsymbol{F}_{AB}\boldsymbol{R}_A)\right) \\ \boldsymbol{R}_B &= (\boldsymbol{F}_{AB}\boldsymbol{R}_A) \exp\left(-\boldsymbol{L}'\Delta t\right) \end{aligned} \tag{2.8}$$

3. Integrate the local constitutive equations :

$$\begin{Bmatrix} \boldsymbol{\sigma}'_B \\ q'_B \end{Bmatrix} = \begin{Bmatrix} \boldsymbol{\sigma}'_A \\ q'_A \end{Bmatrix} + \int_{t_A}^{t_B} \begin{Bmatrix} \boldsymbol{f}'_\sigma(\boldsymbol{\sigma}', q', \boldsymbol{L}') \\ f'_q(\boldsymbol{\sigma}', q') \end{Bmatrix} dt \tag{2.9}$$

4. Rotate the final stress tensor $\boldsymbol{\sigma}'_B$ from local back to global axis :

$$\boldsymbol{\sigma}_B = \boldsymbol{R}_B \boldsymbol{\sigma}'_B \boldsymbol{R}_B^T \tag{2.10}$$

In moderate tension and shear tests with superimposed large rigid-body rotations, this procedure performs very well. However, being related to the well known Jaumann-Zaremba stress rate ($\boldsymbol{W}' = 0$ involves $\boldsymbol{W} = \dot{\boldsymbol{R}}\boldsymbol{R}^T$), although explicitly no objective stress rate is used, under very large shear deformations meaningless oscillating results are found (c.f. Szabó & Balla 1989). In the present context, this is however not too serious a problem, since the yield locus is supposed to require updating before such bad behaviour occurs, making it possible to reset the rotations.

Since the incremental deformation gradient $\boldsymbol{F}_{AB}$ for a given step is in general computable at element level and in the absence of other special requirements, implementation of the above described formulation in any existing FE code causes no major problems. Additional storage is required only for the rotation $\boldsymbol{R}$, as opposed to another approach taken into consideration, using a local reference frame based on the right polar decomposition $\boldsymbol{F} = \boldsymbol{R}\boldsymbol{U}$ of the total deformation gradient. Indeed, in the latter, both the rotation $\boldsymbol{R}$ (different from the previous one) and the right stretch $\boldsymbol{U}$ refer to some absolute configuration and thus need to be stored. On the other hand, it would be free from the above mentioned flaw.

2.3 *Computation of the incidence constitutive tensor*

The configuration $\gamma(t_B)$ at the end of a step being a priori unknown, the nodal coordinates $\{x_B\}$ describing it have to be determined by solving the non-linear discretized equilibrium equations, which may be expressed in terms of nodal out-of-balance forces $\{F_B\}$ (Cescotto 1992). In general this is achieved in an iterative fashion, using the linearized relations

$$\{dF_B\} = [K_T]\{dx_B\} \tag{2.11}$$

where $[K_T]$ is the tangent stiffness matrix, containing a contribution, among others, of the material behaviour in form of the so-called incidence tensor $\boldsymbol{C}_T$ (c.f. Cescotto 1992 for details). The latter may be seen as an expression of the influence of variations $d\boldsymbol{L}$ of the global velocity gradient $\boldsymbol{L}$ on the final stress state $\boldsymbol{\sigma}_B$:

$$d\boldsymbol{\sigma}_B = \boldsymbol{C}_T d\boldsymbol{L}\Delta t. \tag{2.12}$$

Depending on the constitutive equations as well as on the numerical integration scheme, its analytical form is in general unreachable and it is mostly computed through successive numerical perturbations of the components of $\boldsymbol{L}$, requiring each a new integration of the constitutive equations.

In the present formulation, however, only the local velocity gradient $\boldsymbol{L}'$ appears explicitly; equivalent perturbations of the strain path may nevertheless be achieved through pre-multiplication of the global incremental deformation gradient tensor $\boldsymbol{F}_{AB}$ by $(\boldsymbol{I} + d\boldsymbol{L}\Delta t)$. Steps 2 through 4 needing now to be repeated, the computational overhead is somewhat higher. The latter might be slightly reduced by the usage of approximate formulae replacing (2.8) in the first of these steps. On the other hand, an analytical local incidence tensor together with a consistent local-to-global rotation scheme could provide an alternative formulation. Both possibilities are under investigation.

3 APPLICATION TO CUP DRAWING

The computational procedures described in the previous section are used to simulate a cup drawing operation from a circular piece of steel sheet using axisymmetric tools described in figure 2.

3.1 *Numerical simulation*

The presence of orthotropic sample symmetry allows to limit the discretization to one quarter of the blank. A single layer of 8-node hexahedral solid elements

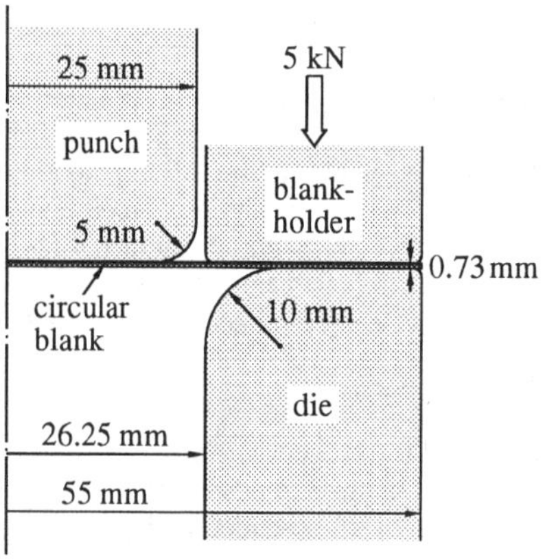

Figure 2. Radial section of the cup-drawing process.

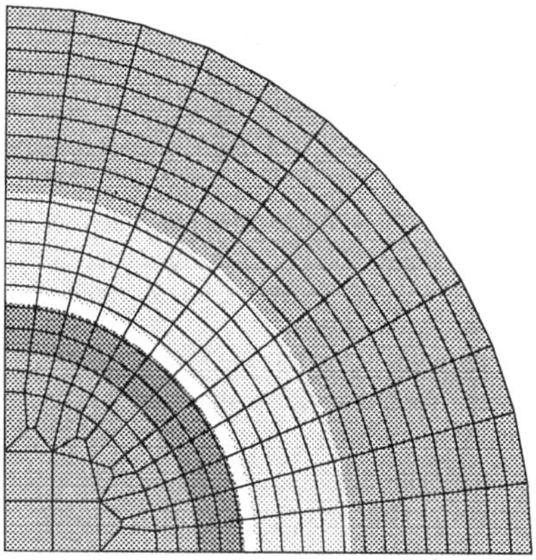

Figure 3. Discretization of the blank (236 elements, 528 nodal points) with and indication of the tool positions.

of the assumed strain method family with one-point quadrature and hourglass control is used (figure 3). Axisymmetric three-dimensional tool segments being not yet available, the die, punch (figure 4) and blank-holder (twenty-four segments) are represented with the help of triangular facets.

The material behaviour is derived from initial texture measurements using a sixth order series expansion in the stress space as yield locus description (Winters *et al.* 1995). A Swift-type law (equation (2.1)) incorporates hardening, while the elastic part of the response is supposed to follow Hooke's law. Respect of the contact conditions between blank and tools is enforced using a penalty method and friction is assumed to be governed by Coulomb's law (Li *et al.* 1995). The values used for the different parameters in the FE simulation are summarized in table 1.

Until a punch stroke of 50 mm, a force of 5 kN is applied to the blank-holder, just as in the underlying

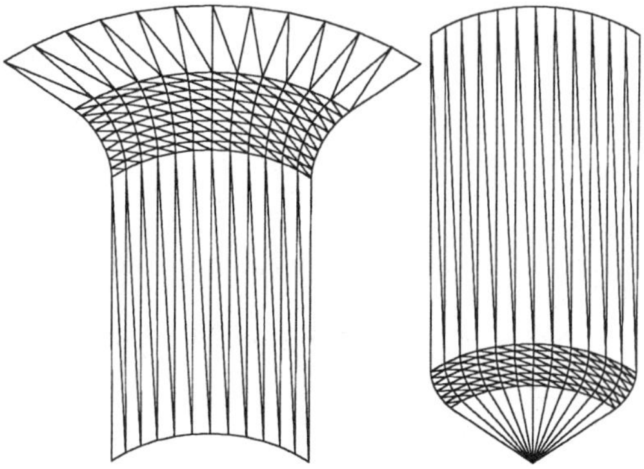

Figure 4. Discretization of the die (240 segments) and the punch (180 segments).

Table 1. Summary of material parameters used in the FE simulation.

Young's modulus	E	=	206 GPa
Poisson's coefficient	ν	=	0,3
Hardening factor	K	=	144 MPa
Hardening deformation	γ_0	=	0.008994
Hardening exponent	n	=	0.247
Contact penalization	K_p	=	500 MPa/mm
Friction coefficient	ϕ	=	0.15

experiment; afterwards, mainly for numerical reasons, it is supposed to be fixed.

4 RESULTS AND DISCUSSION

Figure 5 shows an orthogonal projection of the deformed mesh on a bisecting plane at different strokes. As can be seen, the blank-holder prevents wrinkling quite efficiently. The progressive earing of the drawn cup can be observed, its lower part remaining perfectly axisymmetric.

Equivalent plastic strain in the final configuration is plotted on the deformed mesh at figure 6. The lack of symmetry appearing in its distribution leads to four ondulations in the rim of the drawn cup which differ slightly in position and amplitude (figure 7). This appears more clearly on figure 8 representing the percentage earing, defined in a similar fashion as by Beaudoin *et al.* (1994) as :

$$e = \frac{h - h_{\min}}{h_{\min}} \qquad (4.13)$$

Although representing qualitatively the same phenomenon, the overplotted experimental results are obviously different. Indeed the same sheet material,

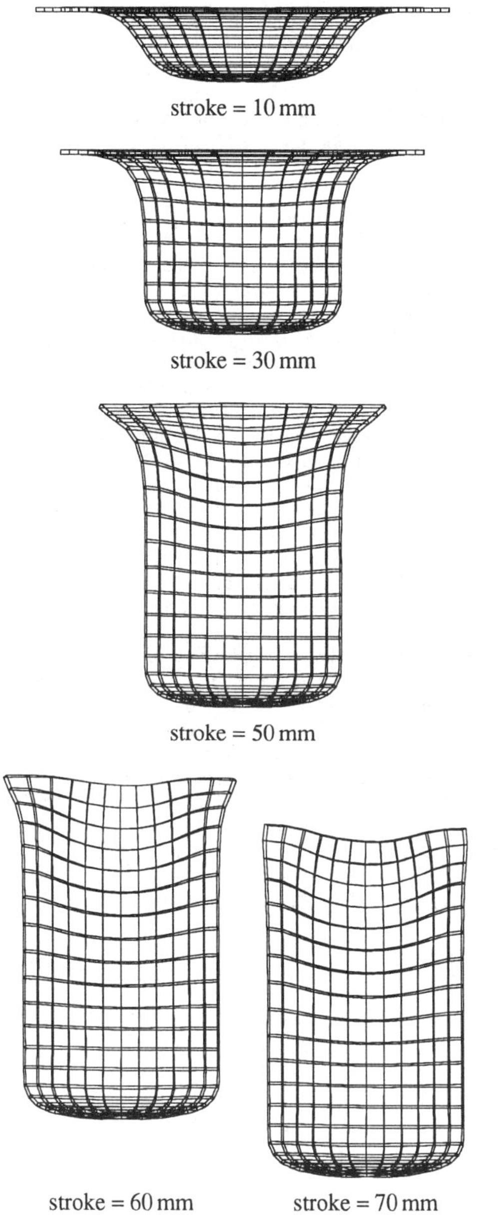

Figure 5. Orthogonal projection of the deformed mesh on a bisecting plane at different strokes.

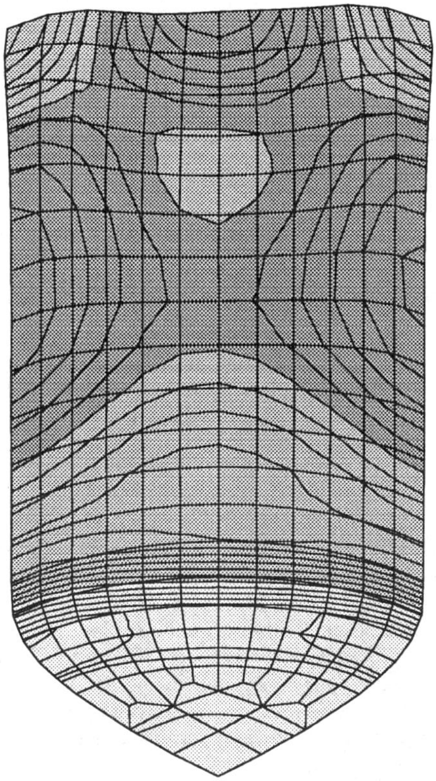

Figure 6. Lines of equal equivalent plastic strains in the drawn cup.

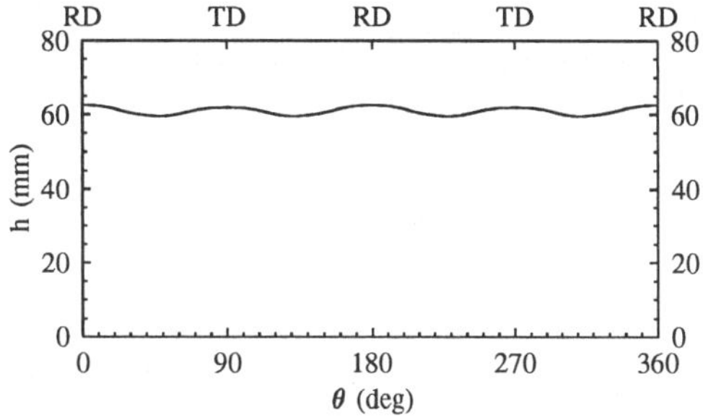

Figure 7. Earing of the drawn cup. Results have been mirrored to represent an entire cup.

formed from an interstitial free steel, is used both for the experiment and for the texture measurments underlying the yield locus, but no updating of the latter is performed during the simulation. Numerical results correspond therefore to the initial crystallographic texture of the material subsequent to its cold rolling and which is known to lead to four-eared cups. On the other hand, in reality, deep drawing is shown to induce significant texture changes to this particular type of material, with the progressive appearance of a new preferred crystal orientation (Daniel *et al.* 1993). In turn, the resulting drawn products evolve to cups with six ears instead of four, a slight tendency of which can already be observed in the experimental data shown.

While the usage of larger samples will be enough to provide clearer experimental evidence for this phe-

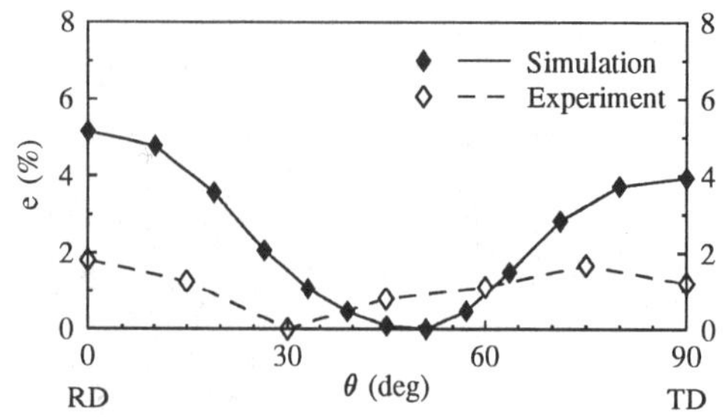

Figure 8. Simulated and experimental percentage earing *e*.

nomenon, the introduction of texture updating in the numerical model, closing the loop described at figure 1, is essential for its simulation.

5 CONCLUSIONS

Modifications have been added to the FE code LAGAMINE in order to allow general three-dimensional anisotropy of constitutive laws. An anisotropic yield locus based on (initial) crystallographic texture has successfully been incorporated. The need for additional extensions aiming to account for deformation induced texture evolution was proven by a cup drawing simulation. Further experimental and numerical investigations are underway.

ACKNOWLEDGMENT

The first author is pleased to acknowledge the support of his work provided by the F.N.R.S. (Belgian National Scientific Research Fund).

REFERENCES

Beaudoin, A.J., P.R. Dawson, K.K. Mathur, U.F. Kocks & D.A. Korzekwa 1994. Application of polycrystal plasticity to sheet forming. *Comput. Methods Appl. Mech. Engrg.* 117:49-70.

Bunge, H.J. 1982. *Texture analysis in material science.* London: Butterworth.

Cescotto, S. 1992. Finite deformation of solids. In P. Hartley, I. Pillinger & C. Sturgess (eds), *Numerical modelling of material deformation processes:* 20-67. London: Springer-Verlag.

Daniel, D., J. Savoie & J.J. Jonas 1993. Textures induced by tension and deep drawing in low carbon and extra low carbon steel sheets. *Acta metall. mater.* 41:1907-1920.

Dawson, P.R., A.J. Beaudoin & K.K. Mathur 1992. Simulating deformation-induced texture in metal forming Application of Polycrystal Plasticity to Sheet Forming. In , J.L. Chenot, R.D. Woods & O.C. Zienkiewicz (eds), *Numerical methods in industrial forming processes. NUMIFORM 92:* 25-33. Rotterdam: Balkema.

Hill, R. 1979. Theoretical plasticity of textured aggregates. *Math. Proc. Cambridge Philos. Soc.* A193:179-191.

Li, K.P., S. Cescotto & A.M. Habraken 1995. Numerical simulations and benchmarks of 3-D sheet metal forming using LAGAMINE program. In *Numerical methods in industrial forming processes. NUMIFORM 95.* Rotterdam: Balkema.

Szabó, L. & M. Balla 1989. Comparison of some stress rates. *Int. J. Solids Structures* 25:279-297.

Van Bael, A., P. Van Houtte, E. Aernoudt, F.R. Hall, I. Pillinger, P. Hartley & C.E.N. Sturgess 1991. Anisotropic finite-element analysis of plastic metalforming processes. *Textures and Microstructures* 14-18:1007-1012.

Van Houtte, P. 1992. Anisotropic plasticity. In P. Hartley, I. Pillinger & C. Sturgess (eds), *Numerical modelling of material deformation processes:* 84-111. London: Springer-Verlag.

Van Houtte, P. 1994. Application of plastic potentials to strain rate sensitive and insensitive anisotropic materials. *Int. J. Plasticity* 10:719-748.

Van Houtte, P., K. Mols, A. Van Bael & E. Aernoudt 1989. Application of yield loci calculated from texture data. *Textures and Microstructures* 11:23-39.

Winters, J., A. Van Bael, S. Munhoven, A.M. Habraken, K. Mols & P. Van Houtte 1995. Anisotropic finite element simulation of plane strain tests. In *Numerical methods in industrial forming processes. NUMIFORM 95.* Rotterdam: Balkema.

Simulation of Materials Processing: Theory, Methods and Applications, Shen & Dawson (eds)
© 1995 Balkema, Rotterdam. ISBN 90 5410 553 4

The application of novel numerical methods to multiple state forming operations in the manufacture of thin walled steel cans

I.E. Nichols
British Steel Tinplate, Port Talbot, UK

G. Louwerse
Hoogovens Groep, B.V., IJmuiden, Netherlands

A.J.L. Crook
Rockfield Software Ltd, Swansea, UK

ABSTRACT: Within the steel can making industry, commercial and environmental pressures have resulted in a drive towards the manufacture of lighter and thinner cans, for the beer and beverage market, without sacrificing performance standards. A joint R&D project by a group of steel manufacturers comprising Rasselstein A.G., British Steel Tinplate and Hoogovens Groep B.V. was established. Rockfield Software provided the finite element expertise to complement the theoretical and experimental programmes of the steel manufacturers.

Critical aspects in the production of lightweight steel cans include forming operations on both the top end and the base of the can. In order to reduce the overall cost of the can, the initially open top end is reduced in diameter in a multiple stage die forming operation, known as necking, and a smaller end is then used to seal the can. The base of the can is formed into a concave dome in a reverse redraw operation. The shape of the dome is designed to withstand internal pressure in the filled can, and to support the can under axial loads.

Novel finite element techniques have been required to enable the simulation of these processes, including:
- Advanced process control to enable simulation of multi-stage processes, e.g. changes of tooling.
- Accurate contact algorithms and friction models.
- Models including large finite strains.
- Dynamic 3D algorithms for assessment of wrinkling during forming operations and assessment of denting on can performance.

The paper focuses on the application of this technology to the can forming processes and to the assessment of can performance. The results of the finite element simulations are compared to and validated with experimental data.

1 INTRODUCTION

Since the introduction of the steel DWI (Drawn and Wall Ironed) beverage can manufacturing process in the late 1960's, there has been a drive to reduce the unit cost of production. In addition to this there are environmental issues concerning the entire packaging industry, and as a consequence there are legislative and public pressures to reduce the overall packaging content of consumables. The DWI beverage can faces competition from other packaging materials such as glass and plastic, and in particular the steel DWI can faces competition from aluminium.

In this context three European tinplate manufacturers, Rasselstein A.G., British Steel Tinplate and Hoogovens Groep B.V., decided to collaborate in a joint Research and Development project. The Ultimate Can project was conceived, a programme of theoretical and practical development to maximise the potential of the steel DWI can. As part of this work a programme to develop Finite Element Analysis techniques to predict aspects of can forming and performance behaviour was initiated with Rockfield Software.

The production of a DWI beverage can comprises a number of stages:

1. Blank and Draw: a circular blank is cut from a strip of tinplate and drawn into a shallow cup.
2. Redraw and Wall Ironing: the shallow cup is redrawn to the final can diameter and the walls are ironed by three or four ironing rings in a single operation, to thicknesses approximately 35% of the original.
3. Doming: at the end of the wall ironing operation a concave dome is formed in the base of the can by a reverse redraw operation.
4. Trimming and Washing: The cans are trimmed to the correct height, and the forming lubricants are washed from the cans.
5. Decoration and Lacquering: an external decorative and protective coating is applied to the can, and internal spray coatings of lacquer are applied to the inside of the can.
6. Neck Reduction: the top end of the can is reduced in diameter by a series of spin and or die forming operations to facilitate the use of a reduced area (and cost) end closure.
7. Filling: the can is filled with a carbonated beverage.

8. Closing: the end closure is sealed on to the end of the can.

The metal costs comprise approximately 60% of the final cost of a steel DWI can, compared to 10% for a car. The greatest savings can therefore be gained by reducing the metal content of the can, i.e. lightweighting. Critical aspects in the production of a lightweight steel can include the neck reduction operation and the dome design.

The focus of the FEA work is on that of die necking and dome performance. In the past design processes have been largely empirical, and evaluation of new designs is costly in terms of tooling and lead times.

2 IMPLEMENTATION

The programme was to develop three major tools:

- A 2D axisymmetric project to evaluate dome performance.
- A 2D axisymmetric project to evaluate multi-stage die necking operations.
- A 3D project to assess the probability of wrinkling during die necking.

Customised pre-processing facilities have been developed. A parametric model of the doming tooling, and of the 3-D necking tooling, enables quick and efficient modifications to be made to existing tooling designs. The effect of the modifications can then be assessed.

In order to develop these models the implementation of new features within ELFEN* was required. These included:

- Mixed Enhanced Strain Elements for 2D and 3D Structural Analysis.
- Contact (general deformable surface / deformable surface).
- Process Control for Multi-stage analyses.
- Explicit Dynamic Solution Algorithms for 3D wrinkling analyses.

2.1 *Mixed Enhanced Strain Elements*

The assumed strain mixed elements developed mainly by Simo and co-workers [1,2] have been implemented in ELFEN. These linear enhanced strain elements provide three major advantages:

- "Spurious locking" problems when materials are nearly incompressible are avoided.
- Improved accuracy in coarse and distorted meshes.
- To provide effective linear elements for use with contact algorithms (quadratic elements are notoriously poor for contact problems).

The elements are essential for the can forming and necking analyses, where the can is subjected to finite deformations in the presence of contact constraints.

* Elfen is the finite element analysis software developed by Rockfield Software and used in these analyses.

2.2 *Contact*

All stages of the manufacturing process require procedures for representing the response of the tool / sheet interface. The majority of the tools may be treated as rigid, however in certain cases, e.g. the ironing dies, the elastic deformation of the tools is significant. A general deformable surface / deformable surface contact algorithm is therefore required.

The algorithm implemented is an extension of the 3-D deformable / rigid penalty method proposed in [3], with the friction forces following a Coulomb friction law. The formulation is based on a return mapping algorithm that enables consistent linearisation of the field equations, leading to an implicit scheme with a non-symmetric tangent matrix. The return mapping algorithm is also utilised for the explicit scheme.

2.3 *Multi-stage Implementation*

Multi-stage analysis is required for the can forming applications. The geometries of the necking / blending dies and knockouts change at each stage, and the dome reversal loading is applied after the dome is formed and the tooling has been removed.

Multi-stage implementation required the facilities to:

- Add and remove portions of the finite element mesh.
- Add and remove load cases and load functions.
- Add and remove contact surfaces.

2.4 *Solution Algorithms*

The 2-D dome performance and 2-D die necking simulations are performed using an implicit solution algorithm.

The 3-D die necking simulation, required to predict wrinkling, is performed using a non-linear dynamic solution algorithm. An explicit dynamic solution algorithm is utilised.

Efficient damping algorithms have been developed to remove noise due to high frequency oscillations, and reduce the inertia effects of low frequency modes of deformation. Damping can be used in the 3-D necking problem even though a dynamic solution algorithm is utilised. This is because the inertia effects are negligible in practice due to the small can mass, and relatively low speed.

3 DOME PERFORMANCE

The concave dome that forms the base of the can is designed to fulfil three primary requirements:

- To provide a stable base, which allows one can to stack on top of another.
- To maintain its shape and integrity when the can is pressurised at filling.

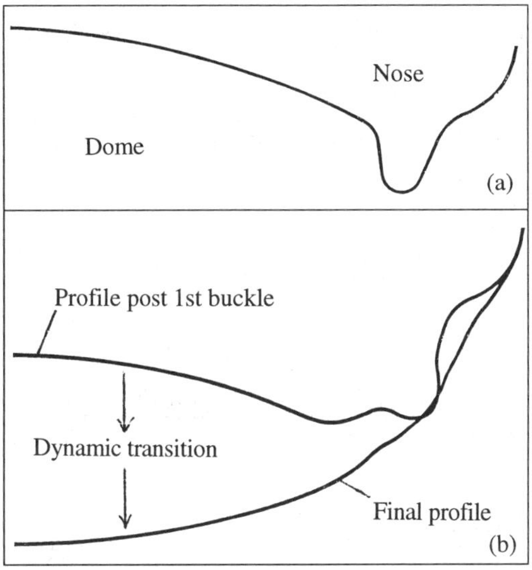

Figure 1 a,b.

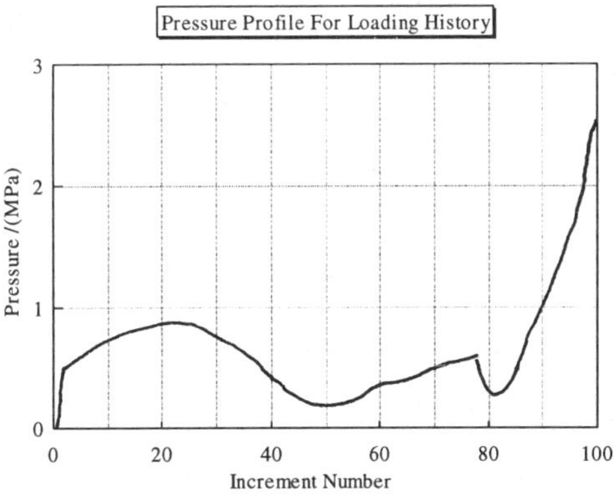

Figure 2.

Table 1.

Starting Gauge /(mm)	Experimental Dome Reversal Pressure /(MPa)	FEA Prediction Dome Reversal Pressure /(MPa)
0.23	0.725	0.726
0.24	0.755	0.745
0.26	0.847	0.851

- To support the can when subjected to axial loads during manufacture, transportation and filling.

The "S" shaped outer section of the base profile is designed to stack on the sealed end closure of another can. The concave dome and nose sections of the base (fig.1a) are designed to withstand the internal pressure present during and after filling.

Can manufacturers have reduced the gauge of tinplate strip used in canmaking, and the steel in the dome has become thinner as a result. The metal thickness in the base is essentially the same as the starting gauge, exhibiting approximately 3-5% thinning at the centre of the dome. In recent years existing commercial dome designs have been nearing their limit with regard to dome reversal performance. In addition to this a reduction in the diameter of the end closure has resulted in a need to design domes with new stacking parameters.

3.1 *Dome Reversal*

The initial approach toward predicting the dome reversal properties of the can, was to utilise a dome profile obtained from a manufactured can. Estimates of thickness (starting gauge 0.23mm) and yield strength variation throughout the dome were based on published data [4]. The initial yield strength and hardening curve was that of the steel used in the manufactured can. The material model is elasto-plastic with non-linear hardening. A normal face loading was applied to the internal element faces, and an arc-length, variable load incrementation utilised. The problem is axisymmetric about the y axis. The first peak buckling pressure (fig.2) is the dome reversal pressure. The form of the dome, post first buckle (fig.1b), is different to that observed in practice. This is because the reservoir of air in the can during the test causes a dynamic transition to a fully reversed state (fig.1b). The predicted value of 0.726MPa compared very favourably to a mean experimental value of 0.725MPa.

Further simulations for starting gauges of 0.24mm and 0.26mm also gave encouraging results (table 1).

Although these simulations proved to be accurate, an estimate of localised work hardening and thickness distribution in the dome was required for each case. The ability to model dome formation prior to assessing the dome reversal characteristics was considered necessary. The work hardening and thickness distribution would be calculated automatically during the forming stages, and effects of changes in tool design could be assessed.

3.2 *Dome Formation and Reversal*

The dome is formed in a reverse redraw operation at the end of the wall ironing process, with the can still on the ironing punch. There are three tool pieces to be modelled in the dome forming operation (fig.3a), and consequently three contact surfaces. The initial configuration is chosen to be that at the end of the ironing operation, immediately prior to doming. The lower portion of the ironed can is modelled on the punch, just in contact with the doming tooling. The material is of constant thickness over the dome area and is not work hardened. The effects of forming the can over the punch nose during the redraw and ironing operations are disregarded.

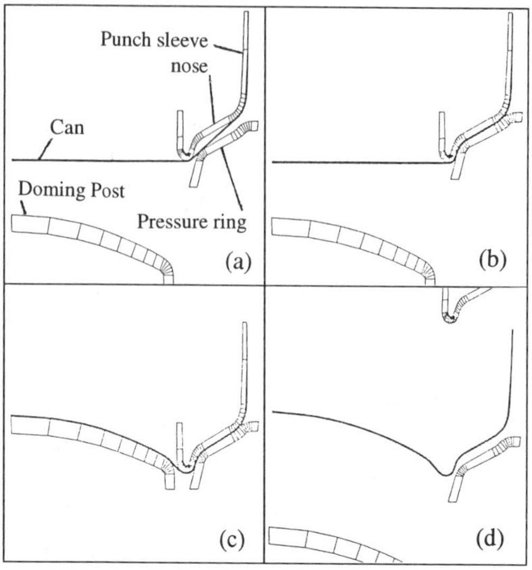

Figure 3 a,b,c,d.

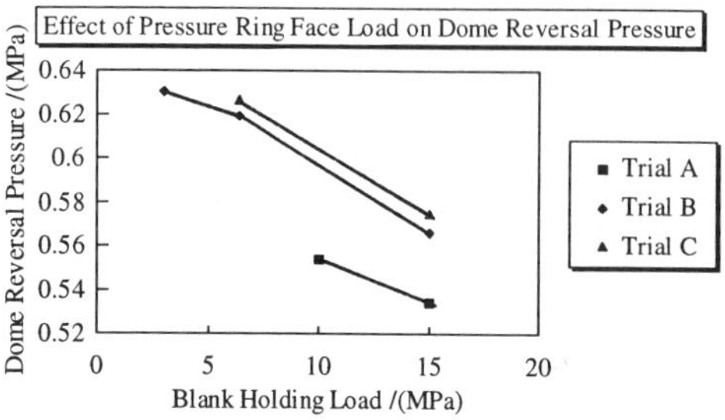

Figure 4.

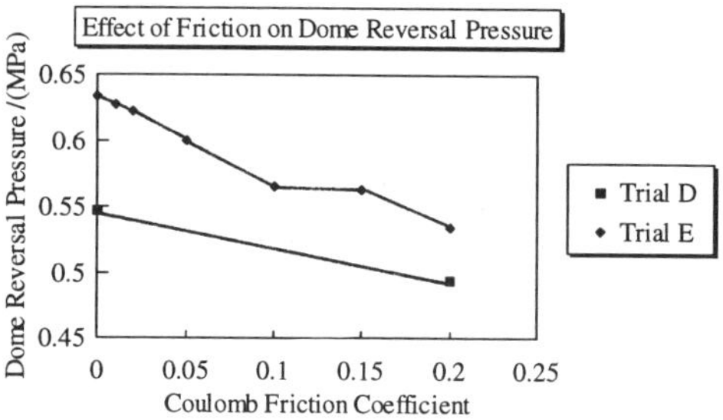

Figure 5.

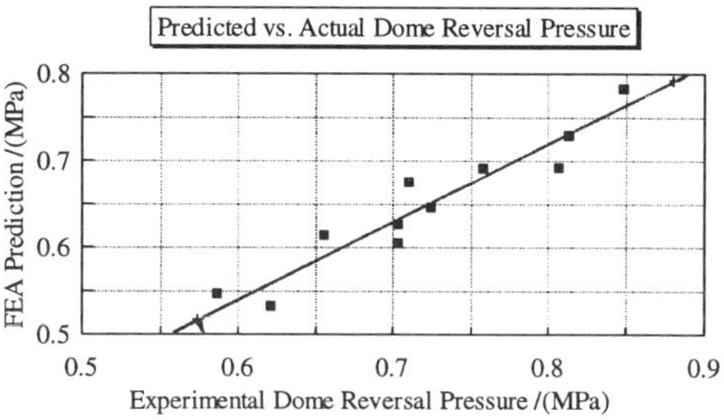

Figure 6.

Both surfaces of the can and the surface of the pressure ring are deformable. The surfaces of the punch sleeve / nose and doming post are rigid. The problem is axisymmetric about the y axis. The analysis is divided into four load cases:

1) A vertical face loading is applied to the pressure ring, forming the "S" shaped outer section, and providing the blank holding load for the reverse redraw (fig.3b).
2) A prescribed vertical displacement and rigid surface movement is applied to the doming post, reverse redrawing the base into the dome profile (fig.3c).
3) The tooling (including the punch sleeve / nose) is withdrawn from the can allowing springback to occur (fig.3d).
4) A normal face load is applied to the internal surface of the can, simulating the dome reversal test.

FEA results have been consistently lower than experimental results by approximately 5% - 15%. The initial use of high blank holding loads (pressure ring face load = 15 MPa) and friction coefficients (μ = 0.2) resulted in lower dome reversal pressures. Varying these parameters revealed an effect on dome reversal pressure (fig.4, fig.5).

In practice the pressure ring face load should be selected to give the correct blank hoiding load. Experimental measurements of doming forces give results of the order 20kN. For most pressure ring geometries, face loads are of the order 6MPa - 10MPa. Friction coefficients for contact with carbide tooling during ironing have been shown to lie in the range 0.01 - 0.25 [5]. The lowest coefficients apply to hydrodynamic lubrication at the ironing die, and coefficients for contact at the punch are above 0.1. A friction coefficient of 0.1 has been chosen for the majority of analyses.

A good correlation between predicted values for the dome reversal pressure, and actual experimental results has been established (fig.6). Analysis times are of the order 3 hours (HP 715).

4 DIE NECKING

The "as trimmed" DWI can has what is known as a 211 diameter (i.e. the outside diameter of an end closure sealed onto the can would be $2^{11}/_{16}$"). The cost and mass of the end closure used to seal the can is significantly reduced by the use of a smaller diameter 206, or latterly 202, end. The top end of the can is reduced in diameter (necked) to accept the smaller ends. This is achieved by a number of die and or spin forming operations. The successive diameter reductions result in a necked can, with either a "206" or a

"202" neck profile.

The continual lightweighting drive has resulted in thinner can walls. And combined with the introduction of the 202 end, the necking process has become much more critical. The work reported here concentrates on the axisymmetric die necking process.

The die necking operation consists of several stages in which the end diameter is successively reduced. The resulting neck profile can be either smooth (preferred) or stepped. A small amount of wax is applied to the outside of the can before necking, to reduce friction. Different tooling is used at each stage, and a multi-stage analysis is required, with several load cases at each stage.

The tooling for each stage consists of a die and a knockout (fig.7a). The can is pushed into the die by means of a pusher supporting the base, and is pressurised internally to prevent the can walls collapsing under the axial necking load. The knockout moves slightly faster than the can, so the net effect is to withdraw the knockout. The gap between the die and the knockout is critical, the radial clearance is of the order 10μm + wall thickness.

The can wall is not of uniform thickness. The top 10mm - 20mm of the can is nearly 2x the thickness of the majority of the wall, with a 10mm linear transition between the two. This thickness profile enables the top of the can to be necked and sealed with the end closure, whilst reducing the metal content in the rest of the wall.

4.1 *Multistage 2-D Simulation*

A 2-D multistage simulation of the necking process was implemented to enable necking forces and final neck profiles to be evaluated. A four stage die pre-necking operation is modelled.

The top 40mm of the can is modelled. The top wall thickness is 0.115mm, the mid wall thickness 0.070mm. The top wall and transition lengths are 15mm and 10mm respectively. The movements of the can and tooling, and the internal air pressure are defined in pseudo-time for each stage.

A simplified set of material properties is used Young modulus = 205×10^3MPa, Poisson ratio = 0.3, Yield Stress = 750MPa, linearly hardening to 770MPa at unit strain. Coulomb friction coefficients are 0.1 on the knockout and 0.18 on the die.

Results are obtained for diametrical springback, can height, and axial necking load (table 2). The shape of the neck at each stage is also predicted (fig.7b, fig.7c). Each stage takes from 1.5 to 3.5 hours (HP 715) to compute.

The results do not appear to be in good agreement. However, they are of the correct order of magnitude and do show trends. Different techniques between experimental and FEA methods for measuring the springback may account for this difference. The FEA values are calculated from measurements taken at the very top of the can, which flares outward, whilst experimental values are obtained using a plug gauge. Higher values for the FEA result would therefore be expected.

Table 2.

Stage	Springback /(mm)		Height /(mm)	Axial Load /(kN)	
	FEA	Exp.	FEA	FEA	Exp.
1	0.289	0.135	40.03	1.524	2.8
2	0.335	0.209	40.01	2.027	2.74
3	0.336	0.195	39.94	2.312	2.8
4	0.248	0.035	39.88	1.395	2.6

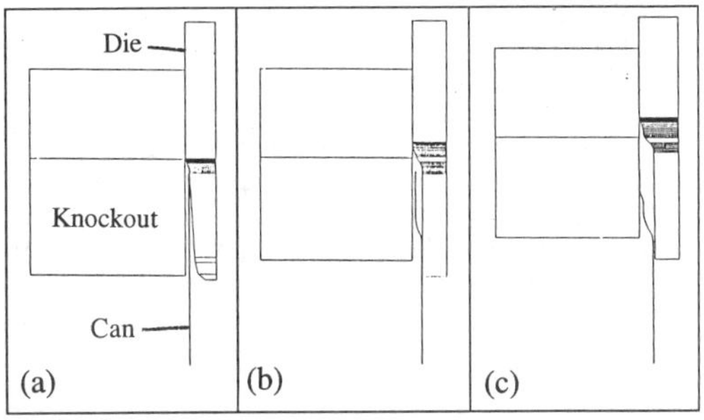

Figure 7 a,b,c.

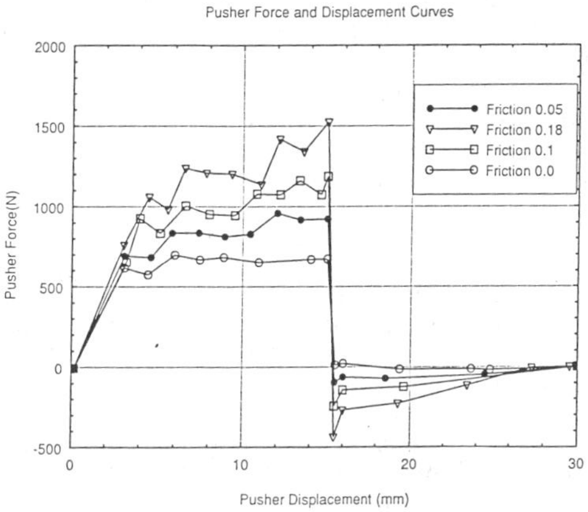

Figure 8.

Experimentally the axial load is measured by a transducer in the pusher. This load reading will include a reaction to the air pressure in the can equal to the product of the pressure and the area of the open end. Considering stages 1 and 4, a pressure of 0.4MPa in the can would account for the difference in results.

The effect of Coulomb friction coefficient at the can - die interface was also investigated. The friction coefficient at the knockout was maintained at 0.1, whilst values of 0.0, 0.05, 0.1 and 0.18 were used at the die. Only the first stage was studied.

Both springback and can height showed only slight variation. Values ranged from 0.285mm - 0.289mm for springback, and 40.03mm - 40.05mm for final can height. Results for the peak pusher force, or axial load, range from 700N to 1.5kN as friction is varied (fig.8).

4.2 *3-D Simulation of Wrinkling*

Experimental work on die necking thin walled cans indicated that wrinkling of the neck was of increasing concern as wall thicknesses were reduced. A 3-D simulation of the necking process was implemented to enable assessment of the probability of wrinkling.

This analysis, unlike the 2-D analyses, which use an implicit analysis, utilises an explicit dynamic solution algorithm. The can is pushed into the necking die and the deformation computed. The buckling or wrinkling is then produced as a direct consequence of the simulation process.

In practice two broad types of wrinkles are observed, small "faceted" evenly spaced wrinkles around the whole circumference, or small vertical wrinkles evenly spaced at close centres. Consequently only ${}^{1}/_{8}$ or ${}^{1}/_{4}$ of the circumference requires modelling. A 40° x 10mm segment of the top wall of the can has been modelled and contains 6000 continuum elements. A 40° segment of the knockout and die is also modelled.

A more complex material model has been utilised, with an accurate yield stress and non-linear hardening curve. Coulomb friction coefficients are 0.0.

The internal pressure is applied as a step function at time 0. The can is given an initial velocity, and then a prescribed velocity at the lower end. The movement of the knockout is specified as a rigid surface movement.

Simulations of 1st stage die necking operations have been conducted with a can wall thickness of 0.120mm. A linear 2μm variation in thickness, from one end of the segment to the other, has been included. This variation simulates the top wall thickness variation observed in real cans, which takes the form of two non-concentric circles. The total thickness variation around a can top wall is of the order 10μm. The 2μm variation here corresponds to 9μm over the whole circumference.

The simulations do successfully produce wrinkling of the correct form (fig. 9), but not at realistic die - knockout clearances. In order to produce wrinkles a clearance of the order 130μm + wall thickness is required, an order of magnitude too large. At the time of writing little success has been achieved with realistic gaps. However, the results do show that the die geometry does affect the form and severity of wrinkling.

The severity of wrinkling is increased by the use of a smaller entry radius and a larger bore radius. This is in broad agreement with experimental observations. Likewise, an increase in the severity of wrinkling is observed when the neck angle is increased. Analysis times are of the order 20 hours (HP 715), for 80000 time steps.

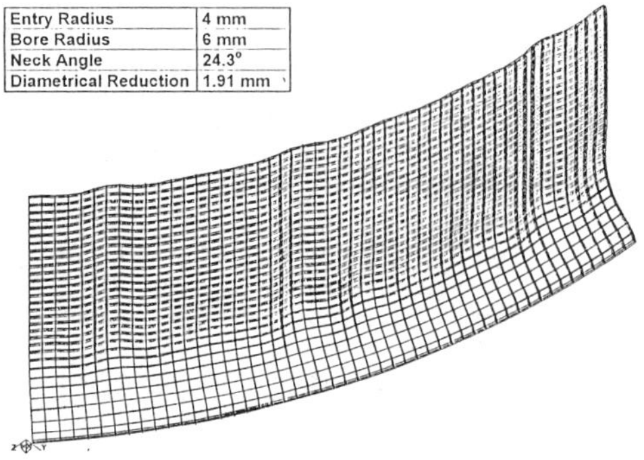

Figure 9.

5 SUMMARY AND CONCLUSIONS

The software development undertaken as part of this programme has demonstrated the successful application of finite element techniques to aspects of can forming and performance evaluation.

Accurate design tools that enable the prediction of dome reversal pressure, final neck form and necking forces are available. These reduce lead times and costs incurred with changes in tooling design, and with changes in can and material specifications. In addition a tool is available to assess implications for wrinkling, and the effect of tooling design, in the die necking process.

REFERENCES

[1]J.C.Simo and M.S.Rifai,
A class of Mixed Assumed Strain Methods and the Method of Incompatible Modes,
IJNME, Vol.29, pp1595-1638, 1990.

[2]J.C.Simo and F.Armero,
Geometrically non-linear enhanced strain mixed methods and the methods of incompatible modes,
IJNME, Vol.32, 1991.

[3]D.Peric, and D.R.J. Owen,
Computational model for 3-D contact problems with friction based on the penalty method,
IJNME, Vol.35, pp1289-1309, 1992

[4]C.G.Carson III, T.E.Johnson Jr., E.W.Gilbreath Jr.,
Material and design factors affecting the manufacture of lightweight drawn-and-ironed tinplate cans

[5]P.Deneuville and R.Lecot,
The study of friction in ironing process by physical and numerical modelling,
J. Mater. Process. Technol., Vol.45, pp625-630, 1994

Simulation of Materials Processing: Theory, Methods and Applications, Shen & Dawson (eds)
 ISBN 90 5410 553 4

A joint project for the numerical simulation of 3D sheet metal forming processes with quasi-static or dynamic approaches

B.Tathi

Renault Joint Research Project, Boulogne-Billancourt, France

ABSTRACT: The paper presents methods and results obtained in a joint research project coordinated by RENAULT automotive company for the numerical simulation of 3D sheet metal forming processes. The joint project involves several groups of french scientists contributing together to construct a software (SIMEM3) able to perform numerical simulation and integrated into the CAD Renault environment (pre and postprocessing, data bases, ...).The present paper only deals with the mechanical problems as large elastic-plastic deformations, contact and dry friction, and blank holder behaviour. Numerical results obtained on automotive sheet metal parts illustrate the possibilities of the developed software.

I INTRODUCTION

The joint project is divided into 5 subgroups working at different places and dealing with different subjects as listed below:

-contact and friction at LMGC Montpellier,
-contact and drawbead modelling at LMA Marseille,
-simplified shell elements at UBO Brest,
-shell elements and solution procedures at LMA Besançon,
-Renault Research Group which coordinates the project with the help of Benallal software company.

This paper describes the development and presents the validation of SIMEM3.

II 3D SHELL ELEMENTS FOR SHEET METAL FORMING

Our goals here are to underline recent developments and possibilities offered by two new shell elements recently proposed for the simulation of sheet forming processes, both in using a classical implicit quasi-static formulation, or in using an explicit transient dynamic formulation. The first one considers an approximate shell description where the membrane and bending effects are uncoupled during the solution process. The bending contribution being calculated from some discrete geometrical approximation and added in the rigidity and internal force vector to overcome convergence problems generally encountered with pure membrane elements. The second one is a C^0 triangular element and is based on an exact shell description and on an assumed strain interpolation of shear strain components. Such an element is well suited for nonlinear simulations due to the low interpolation order and mainly due to the fact that the element can be used with single set of integration points in the mid plane. An other major point is that the triangular element can be used to discretize arbitrary shaped surface geometries.

II.1 Semi-finite element

The computational time in sheet metal forming simulation is an important parameter. With the classical shell finite element models, the high number of degrees of freedom is often an inconvenience. For this reason, a new concept of shell finite element for the simulation of deep drawing is implemented into SIMEM3, which results into a significant reduced computational time.

In this model, an approximated bending behaviour is added to a membrane behaviour. The bending behaviour is calculated from Love-Kirchhoff assumptions and from a finite difference technique between adjacent elements and therefore we named this element "Semi Finite Element" (SFE) [1]. It allows to keep C^0 continuity without introducing more variables than the three classical displacements. An approximate curvature is defined between two adjacent facets in using the average of both contiguous normal vectors. During time increments, a rotation is taking place between two adjacent facets and makes the definition of the curvature tensor variation possible. It should be noted that in the same way, another model taking into account the coupling effects between both membrane and bending behaviours also gives good results [2].

II.2 C^0 shell elements with mixed interpolation of shear strains

The elements considered are based on Ahmad kinematics [3] associated to an assumed strain method in order to avoid transverse shear locking [4][5].

The position vector of a point of the shell is defined from the associated point on the mid-surface and its position on the pseudonormal vector. Pseudonormal vectors are defined at each node in order to enforce the continuity of the description along the element edges. It results that the position vector is interpolated with a consistent shell description using the classical linear (three nodes) and bilinear (four nodes) shape functions.

The displacement vector of a point in the thickness of the shell is the difference between its initial position and its current one. If the incremental rotations are small enough in a loading step, the displacement vector is interpolated as the sum of the mid-surface displacements and the displacements due to the rotation of the pseudo-normal vector. The same interpolation functions are used for the displacement and the position. When performing a non-linear calculation, the coordinates at the end of the loading step are obtained by adding the element displacements to the original position, so the use of the same interpolation functions for both displacements and geometry is a consistent solution approach. The shell element has five degrees of freedom per node, three displacement components in the global coordinate system and two rotation components in a frame constructed from the mid-surface.

Because achievement of an efficient element needs a modification of the interpolation of transverse shear components, the "transverse" notion leads to work in a coordinate system involving the plane tangent to the shell. The proposed formulation uses the material covariant frame associated to the reference elements definition.

Considering the state at the beginning of the load step as the initial configuration, the Green Lagrange strain tensor E has five significant components (E_{33} associated to $S^{33}=0$, plane stress condition, does not appear in displacement calculation). It is yet a posteriori calculated using the behaviour law in order to update the thickness of the shell after each loading step. The in-plane strain components are classically derived from nodal displacements.

A direct interpolation of the transverse shear strain tensorial components is made from the values of those components at the nodes of the element. The nodal values of the out plane shear strain components are obtained from the out of plane shear strains in the direction of the side of the element that are assumed to be constant [4][5]. When making this assumption, only three conditions are imposed per element in order to get the Kirchhoff solution ($E_{\alpha 3}= 0$). While the nullity of the shear components given by derivation of the displacements which are linear in ξ imposes six conditions per element. As a consequence the element built on this so called assumed strain method avoid shear locking and can be used with one point quadrature, as well in the implicit non-linear quasi-static case [6] as in the explicit transient one [7].

III SOLUTION PROCEDURE

One considers first the case when contact is not occuring.

III.1 Quasi-static assumption: implicit scheme

Using standard space discretization the equilibrium equation may be written

$$Res(u) = F_{int}(u) - F_{ext}(u) = 0$$

where u is the vector of nodal displacements, $F_{int}(u)$ represents the internal forces, F_{ext} is the nodal contribution of the external forces. The expression of Res(u) is derived from the virtual work principle: for a set of test functions,

$$\forall \eta \quad G(u,\eta) = 0$$

$$G(u,\eta) = \eta^T(\Sigma_e \int B_e{}^T S_e \, dv - F_{ext}),$$

where the integral is to be understood on each element, and the sum is performed on all elements. The matrix B_e is the strain interpolation matrix on each element defined by,

$$D_u[E]\eta = B_e(u_{en})\eta_{en} \, .$$

The equilibrium equation is linearized using,

$$D_u G(u,\eta) \, \Delta u = \eta_n [K_\sigma + K] \, \Delta u_n \, ,$$

where K_σ is the initial stress stiffness matrix and K is the tangential stiffness matrix. These matrices are computed from the geometrical and kinematic interpolations in taking into account modification of transverse shear interpolation. The equilibrium solution u(I+1) at some load step I+1, is sought as the limit of $u_{j+1}=u_j+\Delta u_j$ using a Newton method on each load step. The Newton iteration procedure may be summarized as

$$\Delta u_{j+1} = \Delta u_j - (D_u Res(u_j))^{-1} \, Res(u_j)$$

III.2 Transient dynamic: implicit and explicit schemes

Performing time and space discretization (for instance as above), the dynamical equation may be linearized, between times I, I+1, and may take the following general form: the increment of the node displacement vector $\Delta u(I+1)$ at time I+1 is found as the limit of a sequence Δu_j satisfying,

$$\Delta u_{j+1} = \Delta u_j + W_j \, (-M(\Delta u_j - \Delta u(I))/h^2 + F_{int\,j} + F_{ext})$$

where $h=\Delta t$ is the time step, M is the mass matrix, $-M(\Delta u_j-\Delta u(I))/h^2$ represents the acceleration terms, F_{int} represents the internal forces, F_{ext} represents the external forces. The matrix W_j is a regular matrix called here after the corrector matrix. For instance, in the quasi-static assumption, i.e. M=0, a possible choice is $W_j = (K_j)^{-1}$, where K is the derivative of the residual Res = $F_{int\,j}$ with respect to X. Then, the above formula generates the Newton Raphson method iterates as above. Another example is the dynamic case using an implicit Euler time integration scheme, and a first order expansion of Res with respect to X. The corrector matrix takes the form $W_j=(K_j+M/h^2)^{-1}$. When K_j is ill-conditionned, the term M/h^2 improves the conditionning of the corrector matrix. If the time step is small enough, one may use a quasi-Newton method with the diagonal matrix h^2M^{-1} as corrector matrix. The computation of the tangent matrix K_j is thus omitted and one avoids to solve a large linear system. One notices that if a single such quasi-Newton iteration is performed, the scheme is an usual explicit time integration scheme, as in [11]. There is thus a number of possible choices, from fully implicit schemes to fully explicit schemes, to deal with the dynamical equation. An explicit scheme for the dynamical equation

$$-Ma - Cv + F_{int} - F_{ext} = 0,$$

where a is the acceleration term and $C=\alpha M$ is some dumping matrix may be derived using a finite central difference time interpolation, [9]

$$u_{n+1} = u_n + \Delta t_{n+1} v_{n+1/2} ,$$

$$v_{n+1/2} = v_{n-1/2} + 1/2 (\Delta t_{n+1}+\Delta t_n) a_n ,$$

$$a_n = M^{-1} (- Cv_n + F_{int_n} - F_{ext}) .$$

IV FRICTIONAL CONTACT MODEL

Usual unilateral conditions are adopted: the gap is positive, the normal stress is negative and vanishes when there is no contact (Signorini's condition). Coulomb's law is adopted. It relates the normal pressure, the tangential friction stress, and the relative velocity. The discrete frictional contact forces are supposed to be concentrated on nodes candidates to contact and to follow as well Signorini's condition and Coulomb's law. Then Signorini's condition writes at each node

$$u_N \geq 0, r_N \geq 0, u_N \ r_N = 0,$$

where u_N is the gap and r_N is the normal component of the reaction force. Coulomb's law may be written

$$||r_T|| \leq \mu r_N \quad \Delta x_T \neq 0 \Rightarrow r_T = \mu r_N \quad \Delta u_T/||\Delta u_T|| ,$$

where r_T is the tangential friction force and $\Delta u_T/\Delta t$ is the sliding velocity, (Δt is the time step).

When frictional contact occurs, the representative R of frictional contact forces has to be written at the left hand side of the dynamical equation together with F_{int} and F_{ext},

$$\Delta u_{j+1}=\Delta u_j+W_j(-M(\Delta u_j-\Delta u(I))/h^2+F_{int_j}+F_{ext}+R_{j+1})$$

Notice that the approximation R_{j+1} has been adopted rather than R(I), the representative of frictional contact forces induced (but not deduced) from $\Delta u(I)$, and rather than R_j induced (not deduced) from Δu_j. When using the term R(I) or R_j, the numerical scheme is explicit with respect to frictional contact forces, while using R_{j+1} the numerical scheme is fully implicit with respect to frictional contact forces.

Several methods to deal with *unilateral conditions* are available: a) the unilateral condition may be treated using a penalisation method which is equivalent to introduce a normal stiffness. b)Signorini's condition being written as a complementary condition, variational inequation technics may be used, such as the gradient method with projection. A refinement of this method is Rosen's method [12]. c) The so-called classical status method is also implemented: the null gap at some nodes candidates to contact is prescribed imposing the corresponding boundary condition, while the gap is free at other nodes. When the equilibrium is reached, the null gap nodes are relaxed if the normal reaction force r_N is found to be attractive, while the free nodes are set to null gap status if they are found to interpenetrate. The solution is reached after some iterations when the status is found unchanged [13].

The previous methods a), b), c) are used when the quasi-static approximation may be adopted. d) When the dynamical equation is used, either because dynamical effects occur in the deep drawing process, or as a regularization technique in a ill-conditionned quasi-static problem the following unilateral condition, said "velocity Signorini's condition" may be used [9],[17]:

if a node is expected to come into contact:

$$\Delta u_N \geq 0, r_N \geq 0, \Delta u_N r_N = 0 ,$$

where $\Delta u_N/\Delta t$ is the normal relative velocity. It may be proven [14] that this relation is equivalent to Signorini's condition. When the masses are concentrated on nodes, the relation derived from the dynamical equation using an explicit integration time scheme, provides straightforwardly the value of r_N such that the above relation is satisfied.

Several methods to deal with *Coulomb's friction,* similar to those used for unilateral conditions are also available: a') A regularized or smooth graph of Coulomb's graph may be used. In this case the reaction force r_T may be written as a (strongly non linear) function of the sliding velocity, or deriving from a potential. In this case the friction force is treated in the same way that internal forces. b') Coulomb's law satisfy the maximum dissipation principle:

$$||r_T|| \leq \mu r_N , \forall s_T \text{ such that } ||r_T|| \leq \mu r_N \quad (s_T-r_T)\Delta u_T \geq 0.$$

Variational inequation technics may be used, as well as the Rosen's method [15]. c') A status method is possible to operate but it has not been used in SIMEM3. d') When the dynamical equation is used, as mentionned above in d), the computation of the friction force r_T is also straightforward.

The so called "successive equilibrium method",[16], allows to deal simultaneously with Signorini's condition and Coulomb's law. It is based on the fact that the local reaction force r^α on some candidate node α, may be computed easily when provisional values are known on other candidate nodes $\beta \neq \alpha$. The free relative velocity at the node α, i.e. the relative velocity assuming no contact force is first estimated. The reaction force is thus obtained using a few iterations of a generalised Newton's method on a 6x6 non differentiable system, with 6 unknowns which are the components of the pair of the 3-vectors, Δx^α, r^α, and with 6 equations equivalent to Signorini's condition and Coulomb's law. The values at each candidate node may thus be actuated, using the value just computed, and the process is applied successively on each candidate node until a convergence criterium is satisfied. This relaxation method is a kind of non linear Gauss-Seidel method. The loop computing the reaction forces and the representative R_{j+1} is nested within the loop dealing with the dynamical equation. When the corrector matrix W is chosen as $W=h^2M^{-1}$ the computation of the reaction force is straightforward.

V TREATMENT OF ELASTIC-PLASTIC BEHAVIOUR

A correct description of the material behaviour is an important aspect for sheet metal forming simulations. In this study, the anisotropy of the plastic behaviour is concerned. Using the new class of three and four node shell elements, a stress calculation algorithm in the case of Hill's orthotropic plasticity including orthotropy directions updating is used [10] and described below.

The macroscopic behaviour model considered is based on the use of a rotational objective derivative defined from the initial microstructure rotation governing the anisotropic direction update. The expression of the constitutive laws in a frame rotated by this rotation permits to built a numerical scheme including an elastic prediction and a plastic correction for the stress calculation. The elastic prediction insures the incremental objectivity and a second order accuracy. The correction stage is based on the Euler-backward method and leads to solve a non linear equation using a local Newton method. Consistency and stability of the proposed scheme are insured. In the proposed approach, shell finite elements are used to model the blank. The proposed stress calculation algorithm takes into account of the mixed interpolation used in the shell element formulation in order to avoid shear locking and of the zero normal stress assumption.

VI BLANKHOLDER CONTROL AND DRAWBEADS EFFORTS

During deep drawing of sheet metal, the blankholder plays a crucial role conditioning wrinkling or sheet necking by imposing shearing forces under the blankholder and controlling the gap between the blankholder and the sheet metal. In order to perform the blankholder modelling, it is considered as a moving rigid body whose position is calculated in order to exactly equilibrate the loads acting on with the constraint associated to the blankholder load [9]. The main difficulty in such an approach is to find the blankholder position and an approach is proposed in [9].

In the forming of sheet metal parts, the quality of the final products, in terms of the absence of wrinkles and necking, depends on the restraining forces exerted by the blankholder and the drawbeads.

The material is assumed to have an isotropic elastoplastic behaviour involving isotropic hardening. The model adopted is characterized by an intermediate configuration [18]. The behaviour between the intermediate and current configurations is given by a hyperelastic response. The intermediate configuration is up-dated by integrating plastic evolutionary laws, using an elastic prediction - plastic correction method. In order to insure that the behaviour is accurately described with low degree of interpolation elements (Q4/P0), a special procedure is used to deal with the incompressibility of the plasticity, based on a three-field formulation [19].

The surface behaviour is described in terms of unilateral contact conditions involving Coulomb friction. Various numerical methods are implemented, including both mathematical programming methods, such as linear complementarity method (Lemke's method) and minimization with projection (Gauss-Seidel, Rosen), and regularization (penalization, augmented Lagrangian) techniques. All these methods have been tested and compared on several examples [18], [19]. No general conclusions can obviously be drawn about the efficiency of the methods, but Lemke's algorithm turned out to be a suitable method for dealing with those two-dimensional frictional contact problems where the number of contact nodes was not too large. With Lemke's algorithm both the displacements and the contact forces can be straightforwardly computed without performing any additive iterations (direct method). Furthermore, there are no parameters to be tuned and there is no need to use a penalty parameter as in the augmented Lagrangian and penalty methods, which is an advantage in the case of an industrial code.

An algorithm is proposed with which the vertical force applied by the blankholder on the sheet can be maintained constant [19].

In order to study the working of the blankholder during the metal forming process, with specific emphasis on the effects of the friction and those of the drawbeads, drawbead simulators have been designed by Renault. We have performed a simulation of Renault's apparatus with the view to comparing experimental and numerical data.

The tool geometry and properties of the material are given in figure 1. The sheet was modelled with two layers of Q4/P0 elements.

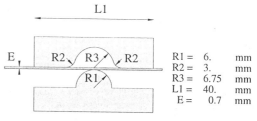

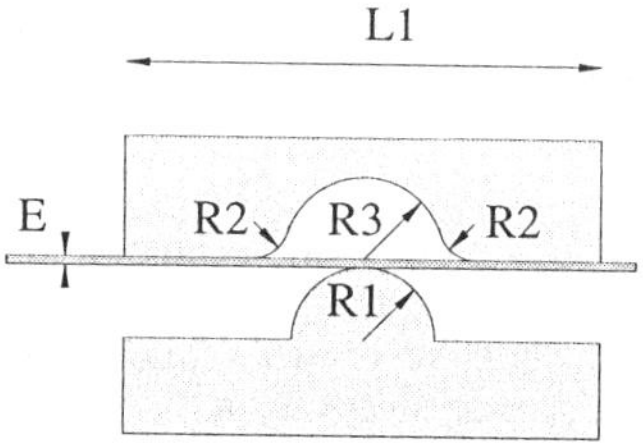

R1 =	6.	mm
R2 =	3.	mm
R3 =	6.75	mm
L1 =	40.	mm
E =	0.7	mm

Young's Modulus : $0.1909\ 10^6$ MPa

Poisson Ratio : 0.3

Hardening law : $\sigma_0 = 673.6\ (0.010095 + \varepsilon_p)^{0.176}$

Friction coefficient on flat part: 0.17

Friction coefficient on rounded part : 0.0 to 0.17

Figure 1 : geometry and material data.

During the locking phase the blankholder moves up until the prescribed force is reached. At the end of the locking, a tangential movement is prescribed on the right hand side of the sheet. During this second phase, the blankholder vertical force is kept constant

(1000daN). The resistance of the strip to be moved through the drawbead is caused by friction and bending and unbending. With SIMEM3, a fine analysis of this phenomenon can be carried out by studying the changes in the contact state and in the material's behaviour particularly in the thickness of the sheet. A fairly good agreement was found to exist between the numerical results and experimental measurements.
Figure 2 gives the restraining forces as a function of the drawing. The results shown in figure 2 illustrate, at a given friction coefficient between the flat parts of the tools and the sheet, the influence of various friction coefficients between the rounded parts and the sheet on the restraining forces.

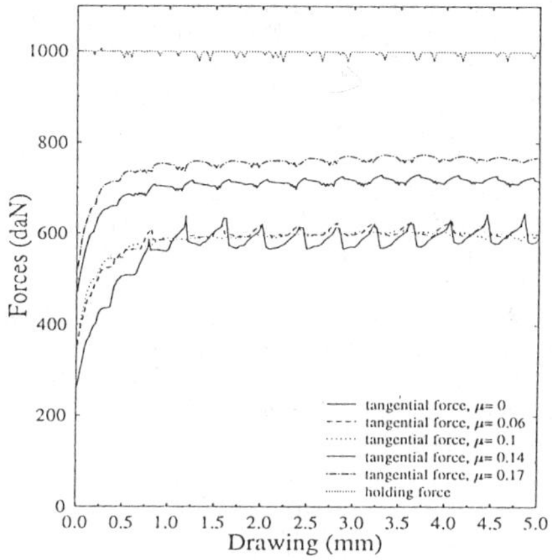

Figure 2 : Restraining forces versus drawing.

These 2D results can be used in 3D simulations to account the effects of the drawbeads.

VII APPLICATION

In order to check the reliability of SIMEM3 software, the dashpot cup of RENAULT Twingo was selected for the simplicity of its CAD definition and the difficulty of its drawing. The cup is stamped with two passes (preforming and finish forming). Experimental validation has been described in [20]. In this paper, only the preform is analysed. Figure 3 represents a schematic view of the tool surfaces and profiles P1, P2 along which the thickness was measured.

Geometry and material data :

- -Dimemsion of the blank : 380*420 (mm)
- -Thickness : 1.98mm
- -Friction coefficient : 0.15
- -Blankholder force : 400kN
- -Young's modulus =207 GPa
- -Poisson's ratio =0.3
- -Hardening law : $\bar{\sigma} = 624\,(0.008 + \bar{\varepsilon}\,)^{0.176}$ MPa

This example is used to evaluate two frictional contact methods (Status method and successive equilibrium method)

The numerical results calculated with two methods are

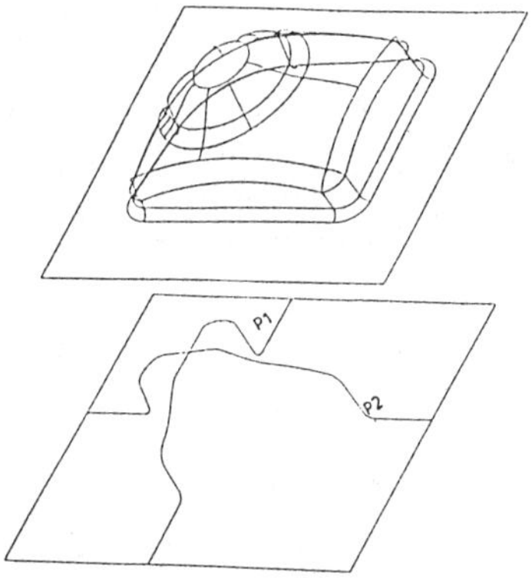

Figure 3 : CAD Definition of tools and Profiles P1,P2

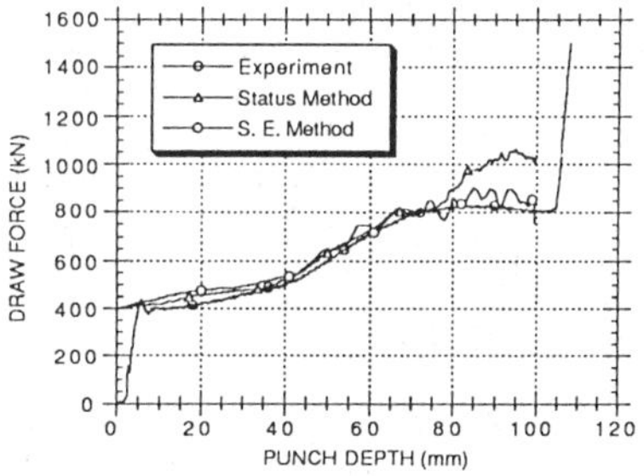

Figure 4 : Stamping force

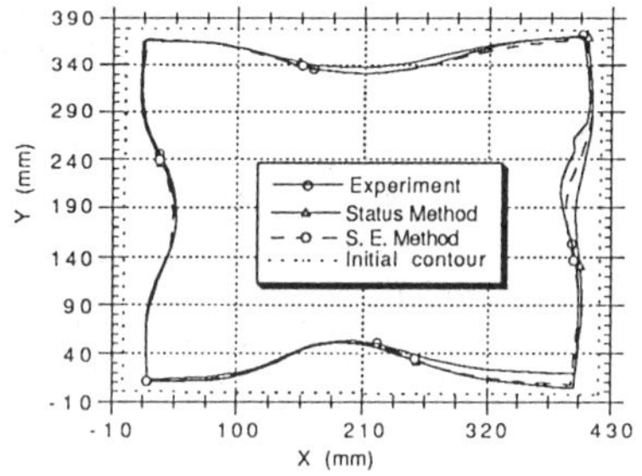

Figure 5 : Draw-in

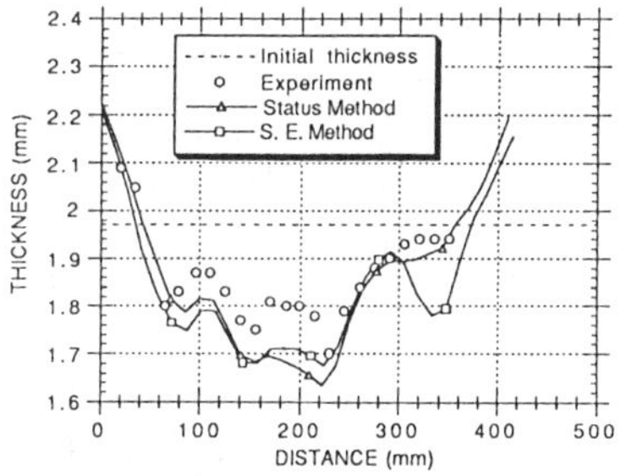

Figure 6 : Thickness along P1

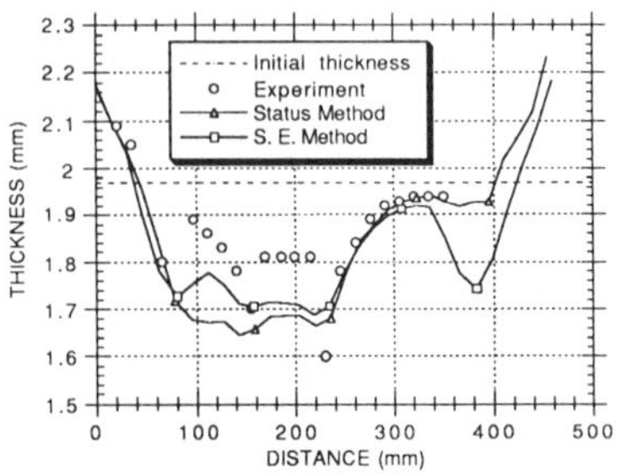

Figure 7 : Thickness along P2

almost similar in draw-in (figure 4) and in stampingforce except at the end of stamping (figure 5) where the results of the second method (Successive equilibrium method) follow nearly the experimental curve. Figures 6 and 7 show a comparison between the thickness measured at 0mm from the end of travel point along the profiles P1, P2 and those calculated with the SIMEM3 software. The numerical results are in a good agreement with the experimental results (Figures 4, 5, 6 and 7).

VIII CONCLUSION

The SIMEM3 software is now used for the simulation of the industrial sheet metal parts in order to evaluate the different methods and the degree of reliability of the proposed simulation strategies. The example presented in application shows that both quasi-static approaches and transient dynamic ones can give accurate results in comparison with experiments.

REFERENCES

[1] RIO G., TATHI B., HORKAY F. *"Introducing bendin rigidity in a finite element membrane sheet metal forming model"*, Proceeding : Large Plastic Deformations, A.A.Balkena (Eds), Rotterdam 1993.

[2] RIO G., TATHI B., LAURENT H. , *"A new efficient finite element model of shell with only three degrees of freedom per node. Applications to industrial deep drawing test"*, Proceeding : IDDRG'94, "Recent Developments in Sheet Metal Forming", Lisbon (Portugal), mai 1994.

[3] AHMAD S., IRONS B.M., and ZIENCKIEWICZ O.C. *"Analysis of thick and thin shell structures by curved finite elements"*, Int. J. for Num. Meth. in Eng., Vol. 27, 271-283, 1971.

[4] BOISSE P., DANIEL J.L. , GELIN J.C. *"A simple isoparametric three node shell finite element,"* Computers and Structures, 44-6, 1992, 1263-1273.

[5] BOISSE P., DANIEL J.L. and GELIN J.C. *"A C^0 three node shell element for nonlinear structural analysis"*, Int. J. for Num. Meth. in Eng., Vol. 37, 1994.

[6] P. BOISSE, J.L. DANIEL and J.C. GELIN *"Numerical simulations of 3D sheet metal forming employing new Ahmad shell finite elements"*, J. Mater. Proc. Technology, 34, 1992, 117-124.

[7] GELIN J.C., BOULMANE L., POISSE P. *"Quasi- static implicit and transient explicit analyses of sheet metal forming in using a C^0 three node shell element"* to be published in J. Mater. Proc. Technology, 1995.

[8] HUGHES T.R.J. *"The Finite Element Method - Linear static and Dynamic Finite Element Analysis"*, Prentice Hall, 1986.

[9] BOULMANE L. *"Application des techniques inplicites-explicites de la dynamique transitoire à la simulation numérique en mise en forme"*, PhD Thesis, University of Besançon, 1994.

[10] BOUBAKAR M.L.*"Contribution à la simulation numérique de l'emboutissage des tôles. Prise en compte du comportement élastoplastique anisotrope"*, PhD Thesis, University of Besançon, 1994.

[11] JEAN M.*"Numerical methods for three dimensional dynamical problems"*, Proceedings of the conference Contact Mechanics 93, Computational Mechanics Publications, Southampton Boston, ed. M.H. Aliabadi, C.A. Brebbia, 1993

[12] MEHREZ F. *"Modélisation du contact frottement dans la simulation numérique de l'emboutissage"*, Thèse de l'Université Paris 6.

[13] JEAN M., HORKAY F., MEHREZ F. *"Contact and friction problems in numerical simulation of deep drawing"*, in Proc. of the EUROMECH 273 conference, La Grande Motte 1990.

[14] MOREAU J.J. *"Unilateral contact and dry friction in finite freedom dynamics"*, Non Smooth Mechanics and Applications, CISM Courses and Lectures, Springer Verlag, Sept. 1987.

[15] JEAN M., JOURDAN F., TATHI B. *"Numerical dynamics for the simulation of deep drawing"*, Proceeding IDDRG'94 "Recent Developments in Sheet Metal Forming", Lisbon (Portugal), mai 1994.

[16] JEAN M., MOREAU J.J. *"Unilaterality and dry friction in the dynamics of rigid body collections"*, Proceeding of Contact Mechanics International Symposium, Lausanne, Swizerland Oct.1992.

[17] JEAN M. *"Frictional contact in collections of rigid or deformable bodies: numerical simulation of geomaterial motions"*, to be published in Mechanics of Geomaterial Interfaces, Elsevier Science Publishers B.V. ed M. Selvadurai.

[18] CHABRAND P., DUBOIS F. 1994.*"Numerical methods for dealing with frictional contact problems in finite elastoplastic deformation"* Second Contact Mechanics International Symposium. London : Plenum.

[19] DUBOIS F. 1994.*"Contact, Frottement, Grandes déformations élastoplastiques. Application à l'emboutissage" Ph. D. Thesis, Université Aix-Marseile II.*

[20] EL MOUATASSIM M., JAMEUX J.P., THOMAS B., MEHREZ F. *"The simulation of multi-operation deep drawing process at RENAULT with PAM STAMP"* Proceeding of the 5th International Conference on Metal Forming, Sept 1994.

Simulation of Materials Processing: Theory, Methods and Applications, Shen & Dawson (eds)
© 1995 Balkema, Rotterdam. ISBN 90 5410 553 4

3D-thermomechanical simulation of laser forming*

F. Vollertsen & S. Holzer
Chair of Manufacturing Technology, University of Erlangen, Germany

ABSTRACT: Sheet metal forming with laser-induced internal forces yields complex shapes without rigid tools. For this innovative process three different mechanisms are explained. The finite-element-simulation helps as well to calculate as to illustrate these mechanisms. The first one is based on a high temperature difference between the irradiated and the non-irradiated sheet surface, which leads to a strain difference and is called temperature-gradient-mechanism. It is used to bend the free leg towards the laser beam along a straight line. The second one allows both, to bend away from and towards the laser beam. It does not derive from a gradient, but from the creation of a partly plastical wrinkle and is therefore called buckling mechanism. It is essential that both effects are distinguished carefully by the choice of the laser parameters in order to gain the desired result. The third mechanism enables both, creation of a three-dimensional shape from a plain sheet and the treatment of very stiff structures, e.g. profiles. This straightening method easy to use works by shortening the irradiated zone and is called upsetting mechanism. For all different effects numerical models are presented and applications are shown.

1 STATE OF ART

Bending with thermal stresses has been known for some time and is often used in forming thick plates in ship building [1,2] because of the lack of external forces. Several empirical formulas are given [3], but mainly the complexity of the process inhibits a complete analytical solution for the use in process control. The laser offers a much better repeatability than manual flame bending and is also suitable for thin sheets [4] and small foil parts [5], because it is possible to concentrate the laser energy on a small area. The laser beam offers the possibility of generating different effects because of the good control of the power. The reason for all laser forming processes is the thermal expansion [6,7] of the heated area. Yet three different types of forming mechanisms are known:

Bending plain sheets along a straight line towards the laser beam can be done by the temperature-gradient-mechanism (TGM). Many applications of this process are known [8-12] because it is easy to reproduce this effect, but only few systematic experiments [13] were done. Analytical models, which offer a good solution for the bending angle, were shown in [14-16]. Numerical methods are necessary as shown in [17-19] in order to consider thermal dependencies of the material parameters and the by nature instationary character of the process.

The second process is called buckling-mechanism (BM)[20]. It allows both to bend towards and away from the laser beam, depending mainly on the precurvature of the sheet. Applications of this effect are given in [21-23]. For the understanding of this mechanism the physical and numerical model shown in [24] is also very helpful. An analytical model is presented in [25].

The third mechanism is similar to flame bending. It works by upsetting the whole thickness in the heated zone and is called upsetting mechanism (UM). This one is suitable for treating profiles as well as correcting [26] and shaping them [27-29]. Furthermore plain sheets [30,31] can be treated to gain a three-dimensional curvature.

In the following paragraph some examples for the usage of laser shaping are given. All above-mentioned mechanisms are used.

For all these mechanisms the finite-element simulation can be used as a calculation method. In this article the special problems of the simulation of these different mechanisms are explained.

* Report from the Chair of Manufacturing Technology, Prof. Dr.-Ing. Dr. h.c. Manfred Geiger

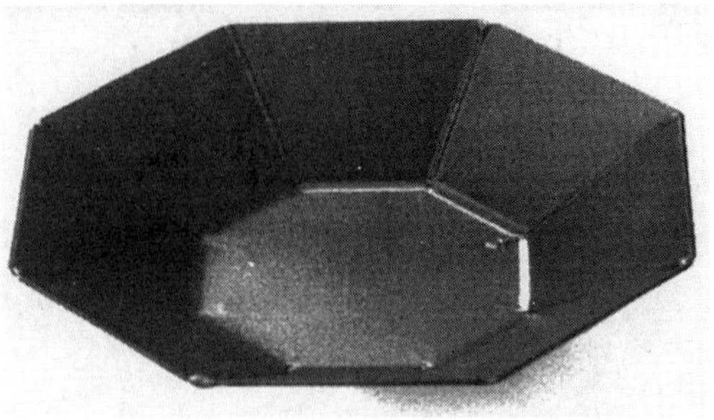

Fig. 1: Eight-sided saucer

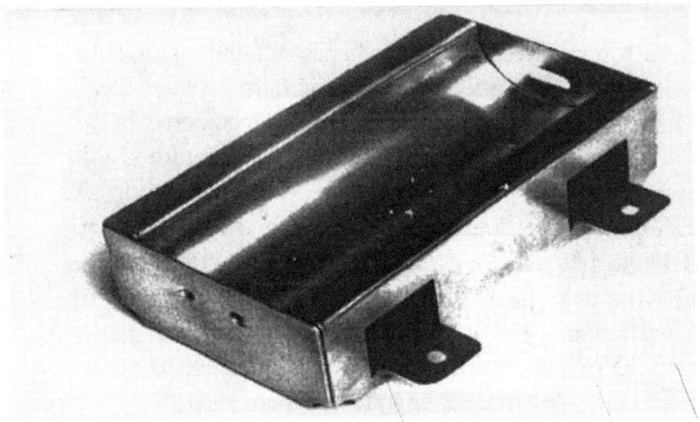

Fig. 2: Lamp housing

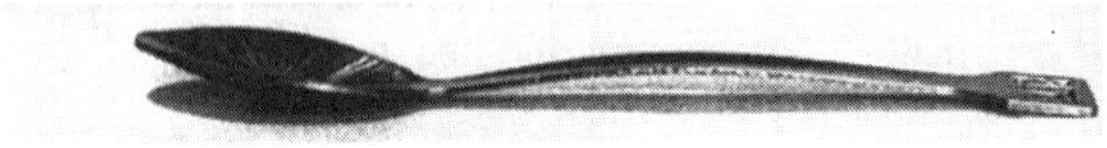

Fig. 3: Spoon

2 EXAMPLE SPECIMENS

The following examples are used in order to show the flexible potential of the laser forming process. It has to be mentioned that even if some of these parts could be produced with conventional methods – which are cheaper in mass production – this process has a great advantage, because no tool is needed. So the following parts are also examples for rapid prototyping.

For the example in fig. 1 a 3-axes laser machine with 750 W beam power was used. No fixed but a magnetic clamping is necessary and complete processing (cutting, bending and welding) in one machine is possible. The bending operation takes six minutes.

Fig. 2 shows the possibility to produce closed shapes with very different bend angles and radii. For production in a 3-axes laser machine it was necessary to turn round this part several times manually. This leads to a fabrication time of 2.5 h. If an effective cooling and a handling system were used, this time should be less than half an hour [23].

Fig. 3 demonstrates the possibilities of cutting and three-dimensional forming of a plain sheet in one clamping. Here the production time is eight minutes.

3 SIMULATION METHODS

Process simulation with the FEM method yields good results when some special rules are obeyed. Because of the high temperature and stress gradients only fully-integrated 3D-elements show proper results. These elements do not use approximations like shell elements, for instance. In order to limit the number of elements in the simulation a sophisticated mesh-refinement has to be used. So only in the heated area four layers of integrated 3D-elements are used. In the cold areas there is only one layer of these elements with a geometrically similar shape. The transition between these areas is suitable done by h-refinement. Of course this cannot be done by commercial pre-processors, but it delivers a stable convergence for most parameters because of a well-conditioned stiffness matrix. In ABAQUS it is easy to define the necessary restrictions for the not connected nodes for entire NODE SETs by a multi-point-constraint (MPC). The heat flow is described in an USER SUBROUTINE because this enables the whole irradiation to be defined in one single STEP even if the heat source affects varying elements because of the movement. By this definition it is also very easy to vary laser beam power, beam geometry and path feed rate. For custom parameters the thermal field affects the stress field, but there is normally no interaction in the other direction because of the very small strains. So it is allowed to separate thermal and stress analysis, which reduces the computation time substantially.

4 DESCRIPTION OF THE MECHANISMS

It must be pointed out that the mechanisms are very different and need different machining parameters even if they are caused by the same tool – the laser.

4.1 Temperature-Gradient-Mechanism

Using the TGM demands a high temperature difference between the irradiated and the non-irradiated surface, typically about half of the melting temperature. A fast heating with high heat fluxes through the surface is especially required for thin sheets in order to gain a thermal gradient. This is due to the good heat conductivity of metals. Because of the demand for such high energy sources this effect was not known before high power lasers were available. The material properties have a strong influence on the process [32,33], because the thermal expansion of the heated zone is hindered by the cold surrounding material, thus leading to compressive stresses. In most metals the yield limits decrease with increasing temperature and so the heated area begins to flow plastically. After cooling the irradiated surface is shorter than the non-irradiated surface, because this plastic strain and this strain difference causes the bending angle. Analytical models must use some simplifications in order to gain a complete formulation. These models calculate an approximation of the strains in a two-dimensional cross section. For a quasi-stationary process in the middle of a long bending edge they represent a good solution for the bending angle.

A numerical model is more suitable in order to gain the influence of the thermal dependency of the material properties or the instationary behaviour of the process. For the FEM-simulation the high temperature and stress gradients are a severe difficulty. It is essential to have at least eight integration points in the thickness direction. The surrounding cold material cannot be omitted, because it has to raise the compressive stresses on the heated zone. Even for a relatively small model with only 3640 elements for one half of a 80 x 80 x 2 mm³ wide specimen the stress calculation needs more than 40 hours CPU-time on a CRAY YMP-EL, because we have to calculate more than 100 time steps in this example. But on the other hand, we found several interesting explanations for some experimental results with this model.

First there is the relation between process speed and bend angle (fig. 4). It is shown that an increased path feed rate and a raised laser beam power leads to a higher bending angle even if the totally coupled energy is constant. The simulation could reveal the reason: For low path feed rates the bottom is also heated by the thermal conductivity (fig. 5). So there is also a compressive stress which results in a relatively small strain difference between irradiated and non-irradiated surface. When we use high speed the non-irradiated surface remains at the initial temperature and the resulting strain difference, which causes the bending angle, is higher than for low speeds. In this picture there are two further interesting points. First there is a restrain in the heated zone only for high speed, caused by a very high temperature in this area and therefore a very low yield stress. So during cooling also small stresses generated by thermal contraction lead to a

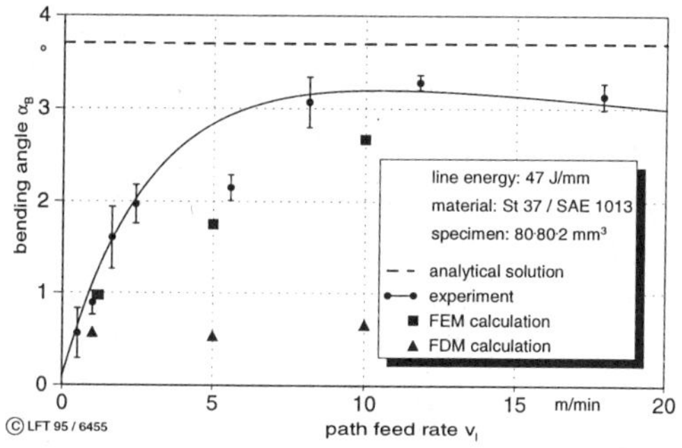

Fig. 4: Influence of Process Speed

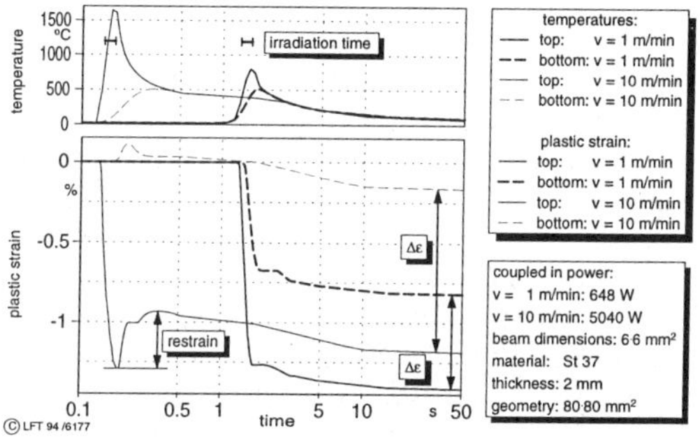

Fig. 5: Temperature and Strain

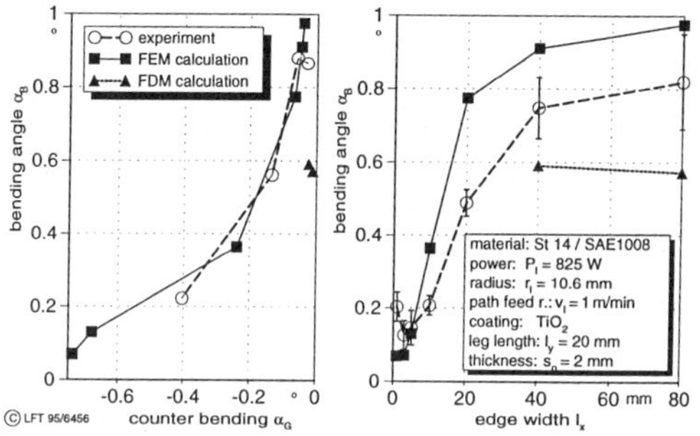

Fig. 6: Influence of Length and Counter Bending

plastic strain. The second effect is that the compression of the heated zone for v_l= 10 m/min is not much higher than for v_l= 1m/min, although the temperature is much higher. We have to observe the so-called counter bending. Because of the thermal expansion of the top the sheet first bends a little bit away from the laser beam. The counter bending for high path feed rates is higher than for low speed, because nearly the whole edge is heated at the same time, resulting in a high bending moment. The counter bending reduces

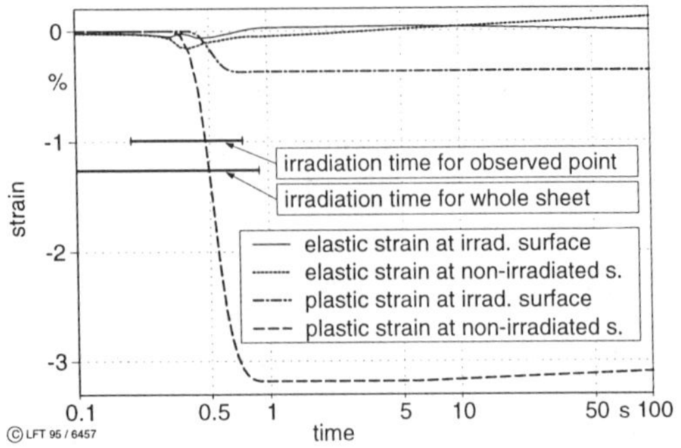

Fig. 7: Strain at a Convex Curvature

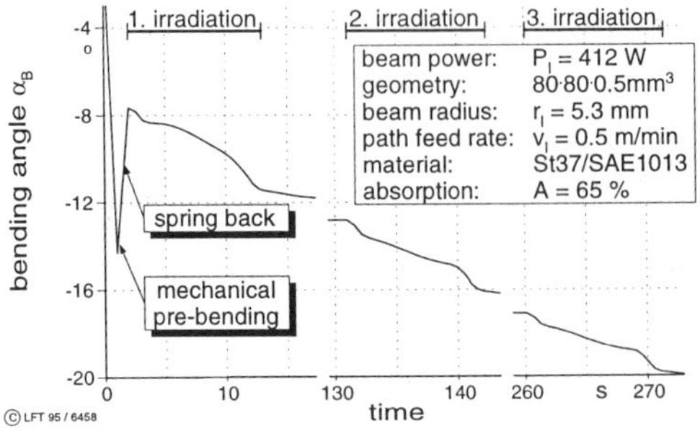

Fig. 8: Multiple Convex Bending

the compressive stresses in the irradiated surface, causing relatively small plastic strain even for high temperatures in this case. In spite of this reduction of the plastic strain in the surface we get a higher bending angle, as the plastic strain in the lower layer is very small compared to that of the sheet treated with the lower path feed rate.

The influence of the counter bending is visible for the dependency on the width of the edge (fig. 6): The bending angle diminishes with a declining width. This is also an effect of an increasing counter bending, which increases with declining resistance of the edge.

4.2 The Buckling Mechanism

The buckling mechanism depends on the occurrence of a partly plastic buckle of the irradiated area. The path feed rate and the intensity are chosen lower than for the TGM in order to gain this buckle. An area relatively wide compared to the sheet thickness is heated up slowly and because of the expansion of this area a buckle occurs. It is possible to give an empirical formula for the limit between TGM and BM, but the parameters should be chosen carefully in order to have a single effect which can be calculated. In the experiment the direction into which this buckle rises is determined by an initial curvature and the user of this process should only use pre-curved specimens. This can be done mechanically or preferably by the TGM with a single irradiation. In an application of the BM it is very useful that the bending direction does not depend on the irradiation side but mainly on the pre-curvature. This fact also allows to bend closed shapes. So it is also interesting to simulate this process, but there are some additional obstacles. First, normally thin specimens are used. This results in very flat 3d elements, which are unfavourable for numerical stability. In ABAQUS there are no advantages of using shell elements, because they use reduced integration, which causes mistakes where dealing with non-linear gradients. The second problem is the geometric approximation of the buckle. It is necessary to have a high number of elements in order to allow the locally high curvature in the buckle. In [24] a physical and numerical model for the development of the buckle are presented. This numerical model provides only a good solution if an adequate number of elements is used. If the number of elements is too small, no buckle occurs and the solution is completely different from reality. This is a very severe difficulty, because not only the accuracy is affected but the general solution by the discretisation. So much more elements are needed for the simulation of the BM than for the TGM. In fig. 7 the strains for the numerical model for BM are shown. It is worth to note that in this case the plastic strain at the non-irradiated side is significantly higher than on the irradiated side while the elastic strains are very small. The strain difference leads to a bending angle away from the laser beam. Thus the treatment can be repeated to a desired extent as shown in fig. 8. This is very important because also bending angles over 180 degrees can be obtained by this way: The pre-curvature is produced by mechanical operation. The relaxation leads to a high spring back. The subsequent multiple convex bending operation is spring-back free.

4.3 Upsetting Mechanism

In order to shape three-dimensional parts another mechanism must be used. The main problem is that these structures are very stiff and therefore very high external forces would be required in order to gain a deformation. By the aid of laser shaping it is possible to shorten the heated area only by internal compressive stresses. Therefore parameters with a relatively

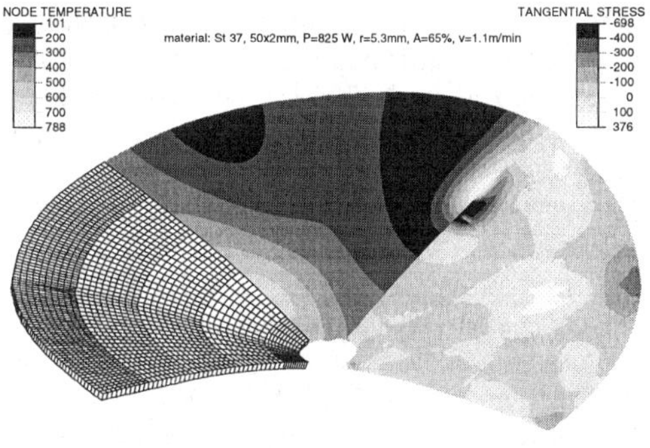

Fig. 9: Circular Plate during multiple Irradiation

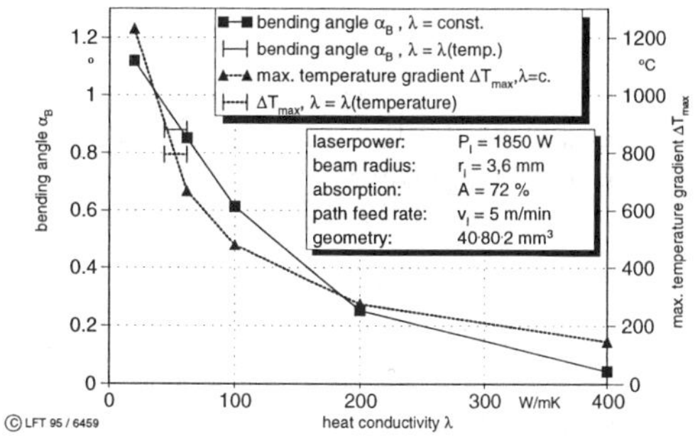

Fig. 10: Influence of the Heat Conductivity

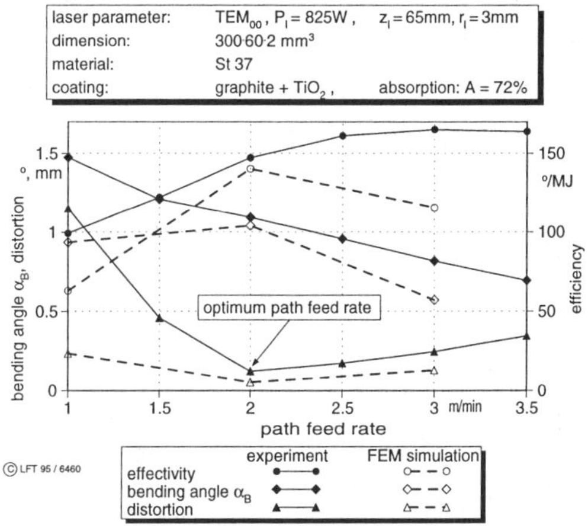

Fig. 11: Process Optimisation by FEM

low path feed rate and a small beam diameter are used, because the whole thickness has to be heated up but must not buckle. There is a difficulty to find a simplification for this process because a local heat treatment causes global stresses and a global change of the shape. So two-dimensional or quasi static models cannot be used. Like shown at the other examples, there is no problem in calculation of the heat distribution in the specimen, but there is a difficulty in the stress analysis. Unlike that for the other mechanisms it is not allowed to use a local refinement just around the heat source, because there are very high stresses also at far distances from the heat source and there is a global curvature. To gain this curvature the discretisation must not be too stiff and a high number of elements is necessary. In fig. 9 the mesh, temperature and Mises stress distribution for a circular plate during laser forming is shown.

5 APPLICATIONS OF THE SIMULATION

Laser forming requires knowledge of suitable parameters. Because of the complexity of the process it is very difficult to give a complete instruction for the choice of the parameters. Even if only one part has to be produced, it is better to simulate first the process than to find out the parameters by trial and error.

In research the FEM is a useful tool to find out connections between the parameters, e.g. the influence of the thermal conductivity. For bending St 37 (SAE 1013) with the TGM it is shown in fig. 10, that compared to the real temperature depending conductivity, the bending angle decreases with an increasing conductivity because of the smaller temperature gradient. Such a variation is impossible in experiment without changing other parameters, too. So the FEM enables investigation of one separated parameter.

Another application of the FEM on the TGM is the optimisation of this process. Three different aims are possible: to gain the maximum bending angle in one single irradiation, to gain maximum effectivity of the laser beam power or to gain an exact shape. The result of this investigation is shown in fig. 11. In experiment it is convenient that the bending angle has a maximum for the lowest speed for a constant beam power, because then the highest energy input is applied. We also see that the efficiency increases with increasing speed. The interesting point is the curvature of the sheet. For low speeds we get a distortion similar to welded parts because of stresses in direction of the laser path. For high speeds the sign of curvature is contrary to that of low speeds. This is very interesting, because this effect is not known from welded parts. Figure 11 reveals that there is a set of parameters which result in an optimal shape with a minimum of distortion. Such an optimisation is not possible by two-dimensional models but only by the FEM.

6 CONCLUSIONS

- Laser forming may be simulated using FEM. Problems in these simulations are the high computing times due to the large number of steps necessary.
- A wrong discretisation may lead to completely wrong results concerning the mechanism predicted by FEM compared to the real mechanism. So one has to care for a suitable choice of the elements. Fully-integrated brick elements were found to be the best ones for the work done here as they are defined with the lowest number of assumptions of the process.
- The FEM may be used to point out some effects or changes in material parameters which cannot be observed in experiments. This might be used for process optimisation.

7 ACKNOWLEDGEMENT

The authors are grateful to the Deutsche Forschungs Gesellschaft who supports this scientific research within the framework of the project Ge 530/13.

8 REFERENCES

[1] Iwasaki, Y & Hira be, T. & Taura, Y. & Tookura, A. & Shioda, H. 1983. Study on the Forming of Hull Plate by Line Heating Method. In: *Hull Plate Forming Process:* 17-34. University of Michigan.

[2] Suhara, T. Studies on Thermo Plastic Working: Bending of Beam of Rectangular Cross Section. 1983. In: *Hull Plate Forming Process:* 1-16. University of Michigan.

[3] Masubichi, K. 1992. Studies at M.I.T Related to Applications of Laser Technologies to Metal Fabrication. *Proc. of LAMP'92*. Nagaoka: 939-946.

[4] Geiger, M. & Vollertsen, F. & Deinzer, G. 1993. Flexible Straightening of Car Body Shells by Laser Forming. *SAE PAPER 930279*. Warrendale USA.

[5] Rebhan, T. & Becker, W. 1994. Präzisionsbiegen u. -richten von Mikrobauteilen. *EPP 9 (1994):* 22-23.

[6] Namba, Y. 1986. Laser Forming in Space. In C. P. Wang (ed.), *Proc. of Lasers 1985/86*: 403-407.

[7] Namba, Y. 1987. Laser Forming of Metals and Alloys. *Proc. of LAMP'87:* 601-606. Osaka

[8] Frackiewicz, H. & Trampczynski, W. 1992. Shaping of Tubes by Laser Beam. *Proc. of 25th ISATA:* 373-380.

[9] Frackiewicz, H. 1993. Methods and Problems of Metal Sheet Forming by Laser. *European Machining:* 67-68.

[10] Vaccari, J. 1993. The Promise of Laser Forming. *American Machinist June (1993):* 36-38.

[11] Deinzer, D. & Vollertsen, F. 1994. Welding and Straightening using a laser beam. *Laser und Optoelektronik 26 (1994) 3:* 48-52.

[12] Tönshoff, H. K. & Berendt, A. H. & Rosenthal, A. R. 1994. Laser Based Cutting, Tensioning and Straightening of Saw Blades - A Flexible Production Line. In M. Geiger & F. Vollertsen (eds.), *Laser Assisted Net Shape Engineering:* 337-344. Bamberg: Meisenbach.

[13] Scully, K. 1987. Laser Line Heating. *Journal of Ship Production 4 (1987) Vol. 3:* 237-246.

[14] Geiger, M. & Vollertsen, F. 1993. The Mechanisms of Laser Forming. *Annals of the CIRP Vol. 42:* 301-304.

[15] Vollertsen, F. 1994. An Analytical Model for Laser Bending. *Lasers in Engineering 1994 Vol. 2:* 261-276.

[16] Vollertsen, F. 1994. Model for the Temperature Gradient Mechanism of Laser Bending. In M. Geiger & F. Vollertsen (eds.), *Laser Assisted Net Shape Engineering:* 371-378. Bamberg: Meisenbach.

[17] Vollertsen, F. & Geiger, M. & Li, W. M. 1993. FDM- and FEM-Simulation of Laser Forming: A Comparative Study. *Advanced Technology of Plasticity 1993 Vol. III, Proc. of:* 1793-1798.

[18] Alberti, I. & Franiti, L. & Micari, F. 1994 Numerical Simulation of the Laser Bending Process by a Coupled Thermal Mechanical Analysis. In M. Geiger & F. Vollertsen (eds.), *Laser Assisted Net Shape Engineering:* 327-336. Bamberg: Meisenbach.

[19] Geiger, M. & Holzer, S. & Vollertsen, F. 1994. Laserstrahlbiegen - Simulation eines 3-dimensionalen, thermomechanischen Prozesses. In: *Metal Forming Process Simulation, Vol. I:* 335-352. Mönchengladbach: UBL

[20] Vollertsen, F. & Komel, I. & Kals, R. 1994. Laser Bending of Steel Foils for Micro Parts by the Buckling Mechanisms - A Model. *Modelling and Simulation on Materials Science and Engineering (in print) 5 (1994) 18:.*

[21] Arnet, H. & Vollertsen, F. 1994. Extending Laser Bending for Generation of Convex Shapes. *Submitted to: Journal of Engineering Manufacture 1994.*

[22] Vollertsen, F. & Geiger, M. 1994. System Analysis for Laser Forming. *Submitted to: NAMRC XXIII 1995.*

[23] Geiger, M. & Vollertsen, F. 1994. Rapid Prototyping für Aluminium-Bleche. In: *VDI-Berichte Nr. 1080:* 293-298

[24] Holzer, S. & Arnet, H. & Geiger, M. 1994. Physical and Numerical Modelling of the Buckling Mechanism. In M. Geiger & F. Vollertsen (eds.), *Laser Assisted Net Shape Engineering:* 379-386. Bamberg: Meisenbach.

[25] Vollertsen, F. 1994. Mechanisms and Models for Laser Forming. In M. Geiger & F. Vollertsen (eds.), *Laser Assisted Net Shape Engineering:* 345-360. Bamberg: Meisenbach.

[26] Frackiewicz, H. 1993. Laser Metal Forming Technology. *Proc. of FABTECH INTERNATIONAL'93:* 733-747. Dearborn: SME.

[27] Polish Academy of Science Warsaw. Laser Bending

Shapes Future. *Professional Engineering:* 24
[28] Geiger, M. & Kraus, J. & Vollertsen, F. 1994. Laserstrahlumformen räumlicher Bauteile. *Bänder Bleche Rohre 11 (1994):* 26-37.
[29] Geiger, M. & Kraus, J. & Vollertsen, F. 1994. Werkstoffeigenschaften von Profilen nach dem Laserstrahlbiegen. *Bänder Bleche Rohre 12 (1994):* 18-21.
[30] Geiger, M. & Vollertsen, F. & Arnet, H. 1994. Laserumformung von Blechen. In: *Tagungsband zur Sächsischen Fachtagung Umformtechnik 1994:* 24./1-13
[31] Vollertsen, F. & Holzer, S. 1994. Laserstrahlumformen. *VDI-Z 136 (1994) Nr. 1/2:* 35-38
[32] Vollertsen, F. & Geiger, M. 1993. Laser Beam Bending of Ferrous and Nonferrous Alloys. *Blech Rohre Profile 40 (1993) Nr. 9:* 666-670
[33] Sprenger, A. & Vollertsen, F. & Steen, W.M. & Watkins. 1994. Influence of Strain Hardening on Laser Bending. In: M. Geiger & F. Vollertsen (eds.), *Laser Assisted Net Shape Engineering:* 361-370. Bamberg: Meisenbach.
[34] Holzer, S. & Vollertsen, F. 1994. Laserstrahlbiegen - Ein berührungsloses Blechumformverfahren. *wt-Produktion und Management 84 (1994):* 8-10

Simulation of Materials Processing: Theory, Methods and Applications, Shen & Dawson (eds)
© 1995 Balkema, Rotterdam. ISBN 90 5410 553 4

Impact analysis of binder rings during stretch draw operation

X. Amber Wang, Sreedhar Vangavolu & S.C.Tang
Ford Research Laboratory, Ford Motor Company, USA

ABSTRACT: In a typical automotive stretch draw operation, the upper binder is driven downwards by a press outer ram and impacts the lower binder with a certain velocity. The binder impact usually generates a significant amount of pressure. This impact pressure is transmitted back to the press, usually called press peak load, and can be so high as to damage both dies and presses. The present work studies the impact process including calculation of press load and binder stresses by finite element method. The dies/binders are modeled by 3-D solid elements. The ABAQUS code with implicit algorithm is used for the finite element analysis. The effects on press load and binder stresses by various process variables, such as cylinder characteristics, draw bead stiffness and binder geometry are investigated.

INTRODUCTION

Press and die breakage is a major equipment failure mode in automotive stamping operations. The resulting production down time is responsible for millions of dollars loss. The problem has become magnified with utilization of stretch draw dies in the stamping operation. The increasing reports of press and die breakage have drawn attention from both production and research organizations. In the past few years, there were a few reports and memos concerning the press breakage issue [1,2,3], but no thorough investigation has been reported. Furthermore, binder impact has been identified as a major contributor to the problem, but there is no rigorous analysis of the binder impact process available in the literature.

Stretch draw dies can be categorized into two types a) three piece and b) four piece. A four piece die is chosen for our studies because it is the type more prone to an overloading problem from binder impact. A four piece die consists of an upper binder, a lower binder, an upper die and a lower punch as shown in Fig.1. The lower binder is supported by nitrogen cylinders which act as air springs. A typical stretch draw operation is a three step process. First, the upper binder ring is driven downwards by a press outer ram and closes on the lower binder ring to set draw beads and hold the blank between the binders. Secondly, the upper and lower rings move together and stretch the blank over the lower punch. In the final step, the upper die moves down and closes the die cavity over the lower die. The focus of the present work is on the first step, i.e., the binder impact process, because press load signatures showed that previously unknown and dangerously high loading occurred during the binder impact process.

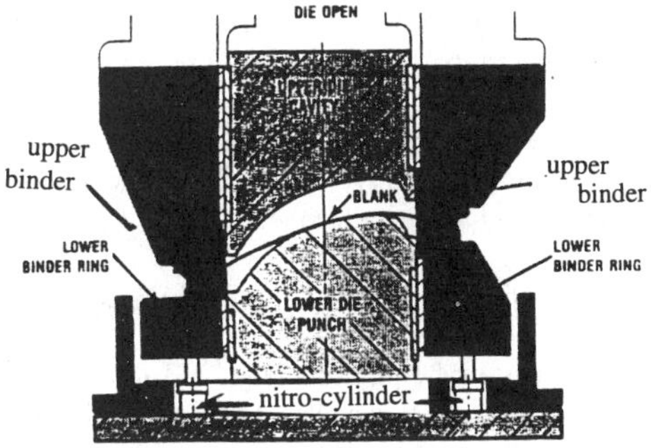

Fig. 1 A four piece draw die

The immediate consequence from binder impact is the tremendous pressure at the binder interface. The pressure induces high stress in the binders which causes binder failure. The pressure is also transmitted up to the press and can cause damage. Therefore, press load and binder stresses, as measurements in predicting press and die failure respectively, are important elements in the analysis. Key variables influencing the press load and binder stresses are: a) upper binder velocity; b) binder weight, c) the number of nitrogen cylinders and their tonnage; d) draw bead characteristics, and e) binder geometry. The Present study will not only analyze and establish a mathematical model to simulate the impact process, but also quantitatively determine the press loads and calculate the binder stresses. The effects of the aforementioned variables on press load and binder stresses will be investigated.

Dynamic impact analysis is performed on three different geometrical models. One of them is an automotive panel die constructed from IGES line data, while the other two consists of simpler geometry for case studies. The binders are modeled by 3-D solid elements using ARIES. The ABAQUS STANDARD code is used in carrying out the finite element analysis. An implicit integration scheme is employed in the dynamic algorithm. Equilibrium is assured at the end of each time step which makes the end results reliable.

The finite element model is described in the next section. Finally, numerical results are presented and discussed.

MODEL DESCRIPTION

The system being modeled is a 3-D nonlinear dynamic system. It consists of upper and lower binders, nitrogen cylinders and draw beads.

The top surface of the upper binder is directly connected to the press outer ram. The history of its movement, including displacement, velocity and acceleration, are therefore specified according to the press kinematics. The nitrogen cylinders are mounted on the fixed die shoes. The lower binder is supported by the nitrogen cylinders and pushed down against nitrogen cylinders by the upper binder.

The nitrogen cylinders are self-contained and pre-charged to a certain pressure. They are simulated by nonlinear springs. The stiffnesses are calculated from static test data. From our preliminary dynamic test, there is no significant difference from static test results.

Nonlinear springs between the upper and lower binders at bead locations are introduced in order to simulate the draw bead setting effect. The characteristic of the draw bead in our model is called bead stiffness. It is defined as the bead setting force divided by the bead height. The bead setting force is the minimum average force required to form the beads which has units of N/mm. It is obtained by dividing the total force to set the beads by the total length of the beads. The magnitude of the bead setting force varies from 88 N/mm to 210 N/mm depending on bead design parameters such as bead shape and height.

It should be noted that draw bead effect can not be neglected, although the size of a draw bead itself is very small and seemed to be neglected from geometrical point of view. Different beads result in different press load. Although the draw bead setting occurs before the binder impact, it affects the state of the whole system at the instant of the impact which in turn determines the impact pressure.

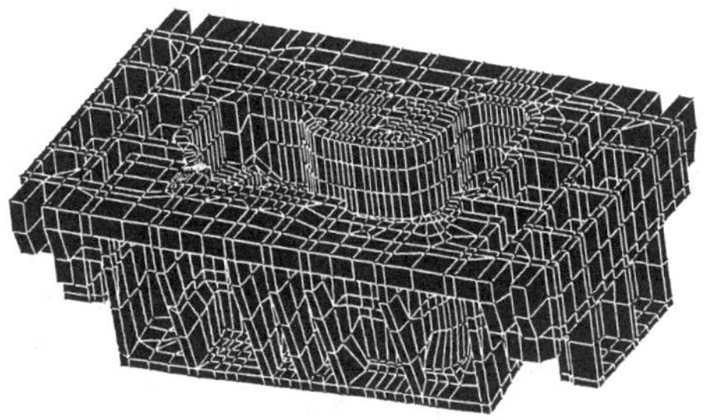

Fig. 2a Upper binder

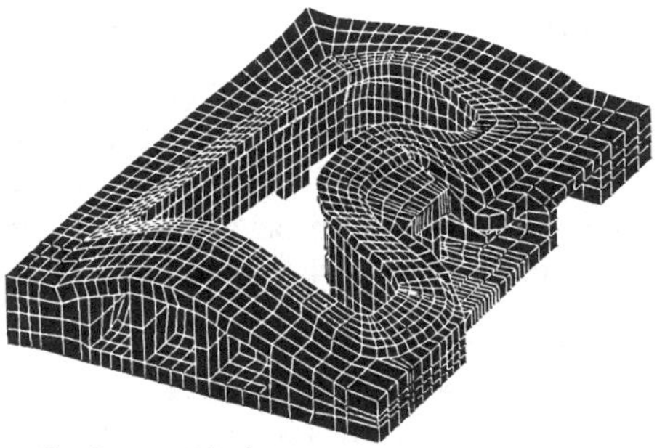

Fig. 2b Lower binder

Binders are usually made of cast iron. The following material constants are used in the present work. Young's modulus, E = 1.03E+5 N/mm^2, Poisson ratio = 0.211, mass density = 0.72E-05 Kg/mm^3. Considering the computational cost, the materials are assumed only to be subjected to elastic deformations, which is adequate for our analysis.

The finite element meshes analyzed are modeled primarily with linear eight-noded brick elements. Interface elements are used in the contact/interface area between upper and lower binders. Nonlinear spring elements are used to simulate the nitrogen cylinders and draw beads. Fig. 2a & 2b illustrates the meshes of a fender die. Fig. 2a shows both upper and lower binder and Fig. 2b shows the lower binder. This model consists of 10,500 nodes, 6300 brick elements, 50 six-noded wedge elements, 350 interface elements, 220 spring elements between upper and lower binders, and 24 spring elements to model nitrogen cylinders supporting the lower binder. A 6 mm gap between the upper and the lower binder is set which is the draw bead height. Dynamic impact analysis is performed in three different meshes which are modeled as described above.

RESULTS

There are many variables involved in the binder impact process. The variables of interest are press load and binder stresses. Press load is defined as the total force transmitted back to the press, which is the summation of the reaction forces on the top binder surface. The influence on press load and binder stresses from factors, such as, press velocity at impact, binder weight, the number of nitrogen cylinders and draw bead stiffness, are investigated.

In the following, the maximum value of the press load is referred as impact peak load. The nitrogen cylinder pressure is usually measured in tons and is referred to as tonnage. The velocity of the upper binder top surface is specified as shown in Fig. 3.

Fig. 4 shows a typical press load history curve plotted against time starting from the instant of the binder impact. The impact peak load appears 8.3 milli-seconds after initial impact. It can be seen that the press load dampens out as time progresses. The oscillations reflect the stress wave propagation.

The effect of press speed on the impact peak load has been investigated. For example, on the fender binder model, the press initial velocity is reduced by 10%, 20% and 30% respectively. Fig. 5 shows the variation of the peak load versus the press velocity. It is readily seen that reducing press velocity can very effectively reduce the impact peak load.

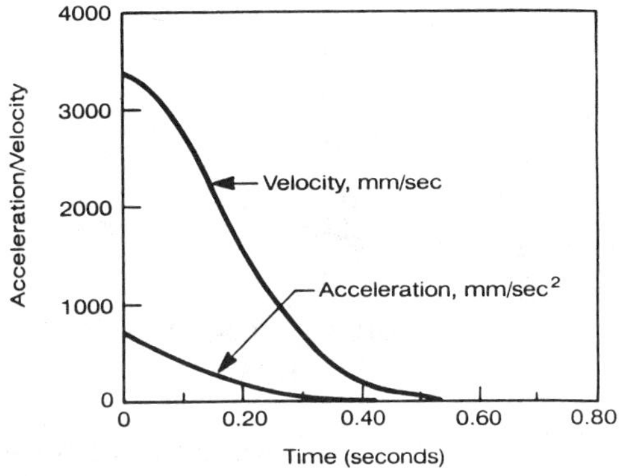

Fig. 3 Acceleration & Velocity Input Curve

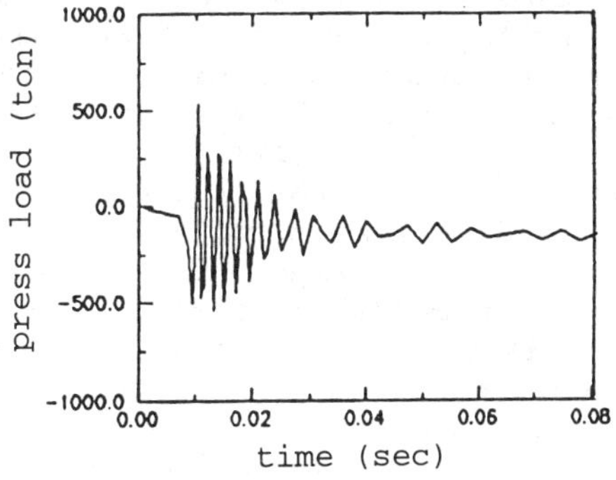

Fig. 4 Press load history (ton vs. second)

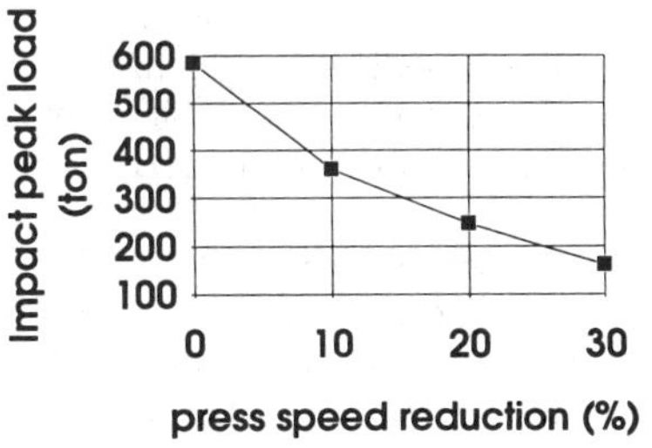

Fig. 5 Effect of press speed on impact peak load

In order to show the effect of die weight on the impact peak load, the material density of the fender binder is reduced to 75% and 50% of the original density of the cast iron. As shown in Fig.6, the press impact load decreases significantly as the density decreases. Lighter binders are desirable in order to reduce the impact peak load.

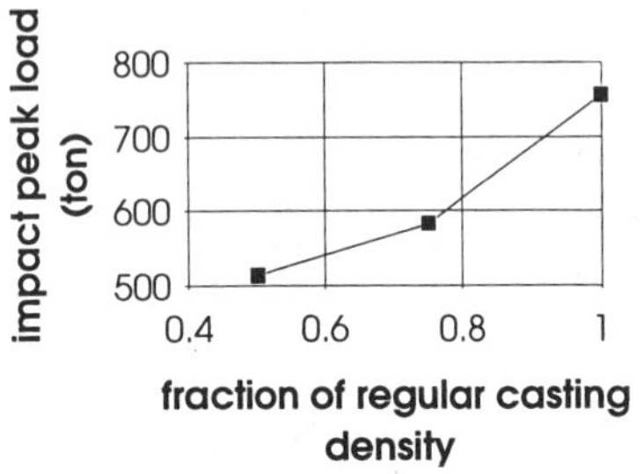

Fig. 6 Effect of binder weight on the impact peak load

When the upper binder makes contact with the lower binder quasi-statically, the press load comes mainly from the supporting cylinder pressure. Therefore, the total cylinder press is usually taken as static press load. However, in automotive stamping operations, the upper binder impacts the lower binder dynamically. The press load then is much higher than the static press load, as shown in Fig.7. In previous practice, this static press load was compared with press load capacity to determine whether there is press overloading. That is the main reason that the overloading has been overlooked in the past.

Fig. 8 illustrates the correlation between the cylinder tonnage with the impact peak load for a given draw bead stiffness and number of cylinders. The fender binder model is used for the illustration where the given bead stiffness is 123 N/mm and total number of cylinders is 24. In Fig. 8a, impact peak load is plotted against cylinder tonnage. It is readily seen that the impact peak load does not monotonically increase with the cylinder tonnage before cylinder tonnage reaches 4.75 ton. That is attributed to the fact that, only a portion of the beads are closed when cylinder pressure is less than 4.75 ton. The draw bead closure condition at the impact is shown in Fig. 8b. Only after the cylinder pressure reaches 4.75 ton, the draw beads are closed at 100% and the impact peak load increases with cylinder tonnage. From a process design point of view, a sufficient cylinder pressure is required so as to ensure 100% draw bead closure to achieve quality stamped parts. On the other hand, a minimum cylinder tonnage is desirable to minimize the impact peak load because, as indicated in Fig. 8a, the lower the cylinder tonnage, the smaller the impact peak load. Therefore, for given draw beads, by appropriately choosing cylinder tonnage, impact peak load can be minimized under the constraints of 100% draw bead closure. In this particular case, from Fig. 8a & 8b, the optimum cylinder pressure is approximately 4.75 ton.

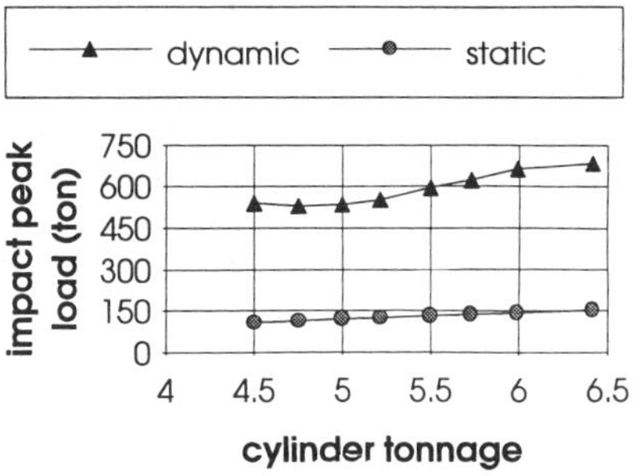

Fig. 7 comparison between the dynamic and the static press load

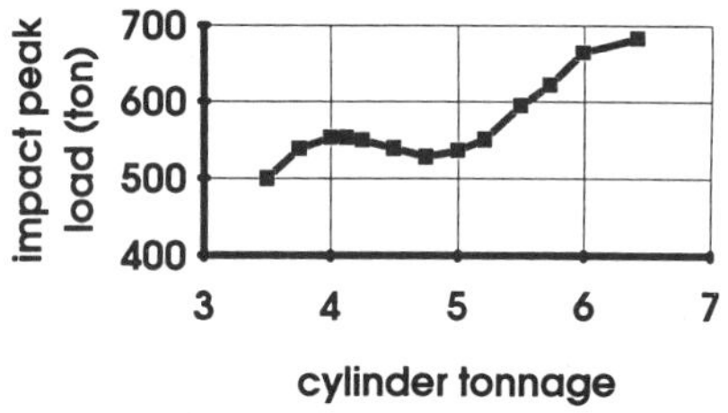

Fig. 8a Effect of cylinder tonnage on the impact peak load

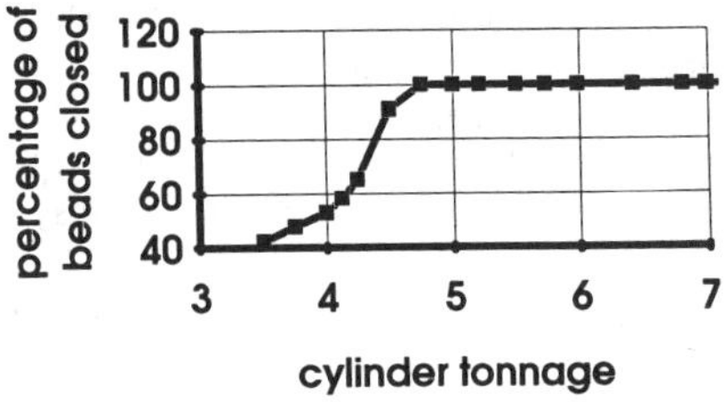

Fig. 8b Draw bead closure condition

The effect of draw bead stiffness on impact peak load is also investigated. Again, the binder fender model is analyzed. The impact peak load versus bead stiffness is plotted in Fig. 9. As is evident from the figure, the stiffer draw bead results in reduced impact peak loads. Additionally, the effect of the bead stiffness on the impact peak load is influenced another factors, For example, the cylinder tonnage. That is illustrated in Fig. 10. In that particular case, when smaller cylinders (cylinder pressure = 5.25 ton) are used, the bead stiffness effect on impact peak load is very little. It should be pointed out that increasing bead stiffness is not always encouraged. The reason being that is, if the beads are too stiff, the beads may not be closed in which case the quality of the stamped part becomes questionable.

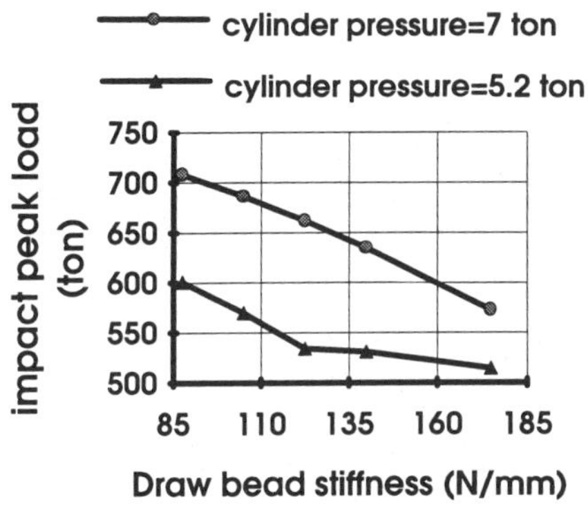

Fig. 9 Impact peak load vs. bead stiffness

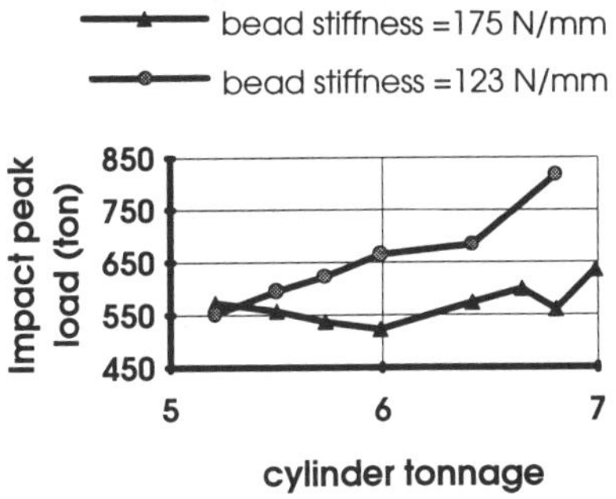

Fig. 10 cylinder tonnage influence on the relationship between the bead stiffness and the impact peak load.

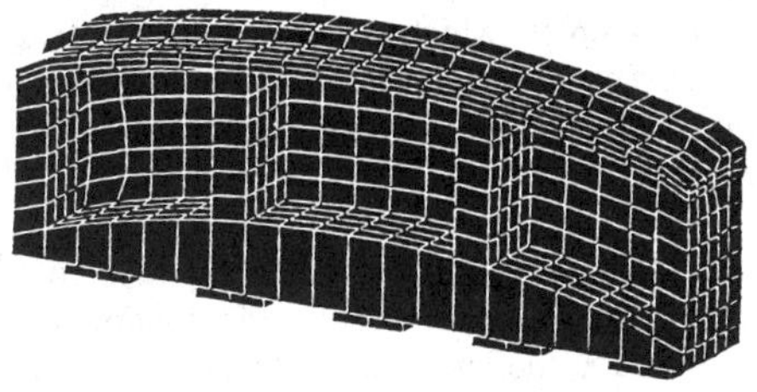

Fig. 11a Lower binder ring with a curved surface

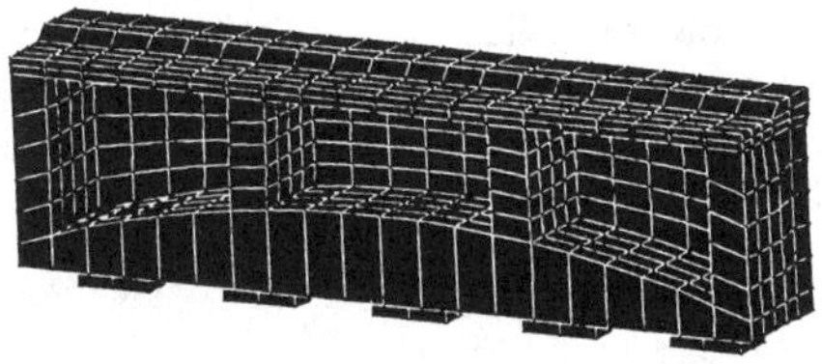

Fig. 11b Lower binder ring with a flat surface

Another factor influencing the impact load is the curvature of the binder surface at the interface/contact area. From production practice, it is observed that the impact of a roof binder is generally more severe than a fender binder. To investigate that, binder impact analyses are performed on two carefully constructed binders as shown in Fig. 11a & 11b. The two have the same weight and dimensions except that one of them has a flat binder surface and the other has a curved one. The curved binder surface consists of a portion of a cylinder of 635 mm in diameter. The impact peak load from the flat binder is 14% higher than the curved one.

From binder stress studies of magnitude and distribution, it is noticed that: a) stress concentrations occur at the supporting area of the nitrogen cylinders, b) due to the dynamic nature of the problem, stress waves propagate vertically along the press moving direction as does the maximum stress location and c) binder stress may exceed material yield stress. The above phenomena are common for all models investigated. For the fender binder, the maximum stress is higher than the yield stress, which indicate potential material failure. The stress concentration is much higher in two of the corner cylinders than the others for the fender model.

The present FEA results not only calculate the total load exerted back to the press, it also provides information on the distribution of the press load at the interface of the upper binder and the press as shown in Fig. 11. It can be concluded that at four corners, where the upper binder is attached to the press through connecting rods, different loads are expected. They can be calculated based on reaction force distribution at the press/binder interface. There was a misconception that the uneven loading at four corners is solely due to die misalignment. In the past, press design has been based on equal loading at four corners. The present results can be useful in press design.

REMARKS

The present finite element analysis of the binder impact will enable us to quantify the press load due to dynamic impact and identify the effect of draw beads and cylinder characteristics on press load in the dynamic system. Therefore, it can be used to predict press & die failures and obtain optimal press and die designs.

REFERENCES

1. 'Effects of Nitrogen Systems on Press/Die Interactions and Loads', *The Budd Company Technical Report*, January, 1992.
2. 'Taking the Shock Out of Stretch Draw', *General Motor Report*, May 1993.
3. 'Fluid Power Combines with Nitrogen Cylinders to Provide Dynamic Solutions', *Metal Forming*, Sept., 1993

Simulation of Materials Processing: Theory, Methods and Applications, Shen & Dawson (eds)
© 1995 Balkema, Rotterdam. ISBN 90 5410 553 4

Finite element simulation of sheet metal forming by using non-parametric tool description with automatically refined patches

D.Y.Yang & J.W.Yoon
Department of Precision Engineering & Mechatronics, Kaist, Korea

D.J.Yoo
R-D Department of Production Engineering, KIA Motors, Korea

ABSTRACT : An improved nonparametric tool description based on successive refined nonparametric patches is proposed and the related criterion for refinement is also discussed. In the proposed scheme, any required order of tool surface conformity can be achieved by employing successive refinements according to the suggested criterion. By using the suggested adaptive tool refinement technique based on the nonparametric patch tool description, the locally refined nonparametric tool surface with economic memory size and sufficient accuracy as well as with favorable characteristics for contact treatment can be obtained directly from the parametric patch related with commercial CAD system. Computation is carried out for a chosen complex sheet forming example of an actual autobody panel in order to verify the validity and the efficiency of the developed tool surface description.

1. INTRODUCTION

One of the most important factors to be considered for the analysis of sheet metal forming processes is the tool surface description for arbitrarily-shaped sheet metal parts. There have been suggested various schemes for tool surface description ; i.e. the finite element mesh approach[1], the piecewise linear approach[2], parametric patch approach[3], and non-parametric patch approach[4].

In the finite element mesh approach, the tool surfaces are divided into finite elements and the tool is treated either as an elastic body or a rigid body. The method has often been used as the most popular technique for describing a complex tool surface. However, the raw surface CAD data must be divided into finite elements by time-consuming graphic operations. Moreover, a great number of elements are required in order to describe the original tool surface with sufficient accuracy.

In the piecewise linear approach often called by point data approach or triangular jewel patch approach, an arbitrary tool surface is represented by multiple triangular planes generated by equi-spaced or nonequi-spaced rectangular array of grid points on the x-y plane. The approach does not need to store information relative to element number or connectivity, since regular distribution of points creates the necessary relation among entities. Therfore, the method is simple to implement. Some part which has vertical walls, however, cannot be described by the method.

The parametric patch approach is the most convenient method in describing arbitrarily-curved surface of real industrial parts. Furthermore, the tool surface data generated in a commercial CAD system can be directly transmitted into the input file for the FE analysis. However, this approach needs a more complex contact search algorithm as compared with other approaches and it needs the large computation time if the number of parametric patches used becomes larger.

The non-parametric patch approach introduces cubic blending functions as in the case of the parametric patch to allow C^1 continuity. In the approach, an array of equi-spaced or nonequi-spaced rectangles is required to construct the whole tool surface. This method is efficient in contact treatment so it is robust and fast in computation. As in the case of piecewise linear approach some parts which have vertical walls cannot be described by this method. From the viewpoint of dimensional accuracy, the nonparametric patch approach based on the equi-spaced point description requires a great number of points to describe the original tool suface accurately, which requires a large size of hardware memory capacity.

Therefore, in order to maintain the advantage of nonparametric patch approach and to overcome some drawbacks, an improved nonparametric tool description with locally refined patches is proposed. In the method, any required order of tool surface conformity can be obtained by successive refinements from the parametric patches. The locally refined nonparametric tool surface of economic memory size and sufficient accuracy with robust contact search algorithm can be obtained directly from the parametric patch related with a commercial CAD system.

The objective of the present study is to introduce a general method for tool surface description by locally refined nonparametric patch approach which can be employed so as to treat almost all kinds of complex tool data encountered in the industrial parts. In addition, efficient contact search algorithms are suggested to make the locally refined nonparametric patch approach realistic for FE analysis. Computation is carried out for a chosen sheet forming example to verify the validity and effectiveness of the proposed method.

2. THEORY

2.1 Parametric patch approach

A smooth surface of arbitrary complex shape which cannot be described with a single analytical expression can be represented in a piecewise continuous fashion using parametric patches. Tool surfaces of real industrial parts such as automobile body panels are commonly described in the CAD system by assembling parametric patches as shown in Fig. 1. Each surface patch is a function of two parameters, u and v. Thus the Cartesian coordinates of a point at any parametric coordinates will be determined as follows [5] :

$$r(u,v) = x(u,v)\mathbf{i} + y(u,v)\mathbf{j} + z(u,v)\mathbf{k} \qquad (1)$$

where x(u,v), y(u,v) and z(u,v) depend on the type of surface patch used. The parametric patch approach has the following merits and limitations when it is applied to FE analysis.

The advantages are ;

i) direct and efficient data transfer between CAD and FEA

ii) C^1 or C^2 continuity of the tool surface can be obtained.

The limitations are ;

i) Complex contact algorithm and enormous computation time for the treatment of tool-sheet contact

ii) Existence of untrimmed surfaces in the raw CAD data or IGES file.

iii) Existence of several kinds of surface entities. It is important to define a standard in order to assure easy data compatibility.

Therefore, in spite of many advantages of parametric patch approach, the aforementioned limitations induce some modification of the original parametric CAD data for simpler and more efficient contact treatment in FE analysis. There are many variations of the method concerning the data transfer from the parametric patch approach to the tool surface descriptions for FE analysis as shown in Fig. 2 ; finite element mesh approach, piecewise linear approach, nonparametric patch approach, etc. The most important point in any procedure is that the conversion should be performed efficiently without loss of surface accuracy. Among these approaches, the nonparametric patch approach is one of the most convenient and efficient method to describe the arbitrarily-shaped tool surface.

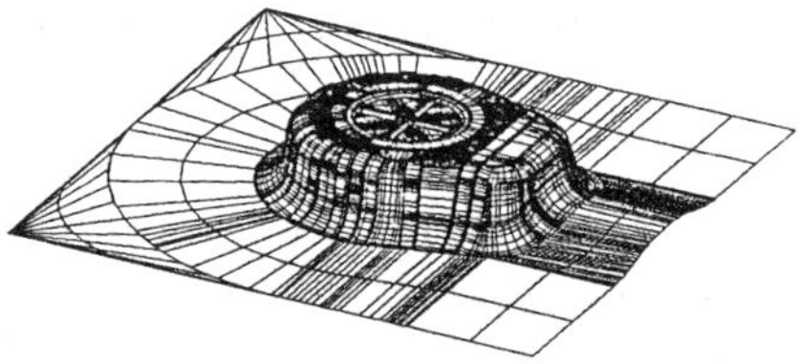

Fig. 1 Typical example of complex curved tool surface composed of parametric patches

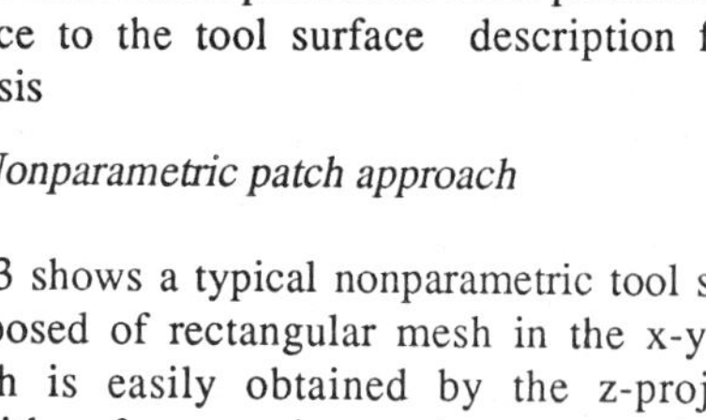

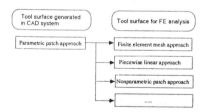

Fig. 2 Conversion procedures from parametric tool surface to the tool surface description for FE analysis

2.2 Nonparametric patch approach

Fig. 3 shows a typical nonparametric tool surface composed of rectangular mesh in the x-y plane which is easily obtained by the z-projection algorithm from equi-spaced or nonequi-spaced rectangular array of grid points on the x-y plane to the original parametric patches. Not as in the parametric case, each patch is a function of x and y.

$$z = f(x,y) \qquad (2)$$

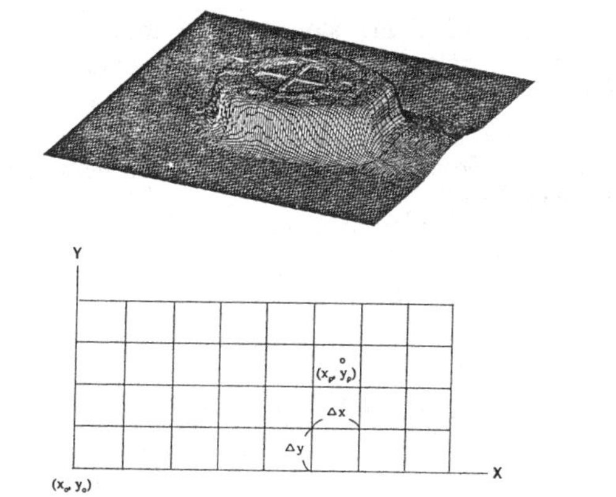

Fig. 3 Typical example of complex curved tool surface composed of non-parametric patches

The nonparametric patch approach has the following advantages and limitations when it is applied to FE analysis.

The advantages are ;

i) high speed of contact search / reduced computation time,

ii) easy data generation for complex geometries from the parametric patches, and

iii) C^1 continuous smoothness of the tool surface

The limitations are ;

i) array of huge size of dimension needed to describe tool surface / a large size of hardware memory capacity

ii) difficulty in describing vertical surfaces

Therefore, in order to overcome some drawbacks of nonparametric patch approach, an improved nonparametric tool description with locally refined patches is proposed.

2.3 Adaptively refined nonparametric patch

2.3-1 Tool surface description

Suppose that we have a rectangular mesh defined with (I+1)x(J+1) grid points. Then the Cartesian coordinates of a point at any non-parametric patch can be written as :

$$Z^{ij} = \sum_{m=0}^{3}\sum_{n=0}^{3} a_{mn}\left(\frac{x-x_i}{h_i}\right)^m \left(\frac{y-y_j}{k_j}\right)^n , \qquad (3)$$

$$i \in [0:I] \text{ and } j \in [0:J]$$

for a mesh $x_i \leq x \leq x_{i+1}$, $y_j \leq y \leq y_{j+1}$.

In eq. (3), $h_i = x_{i+1} - x_i$ and $k_j = y_{j+1} - y_j$.

The coefficients a_{mn} in eq. (3) are related to the patch corner points and derivatives as follows

$$A = CH_iQK_jC^T \qquad (4)$$

where

$$A = [a_{mn}] \ , \ C = \begin{bmatrix} 1 & 0 & 0 & 0 \\ 0 & 0 & 1 & 0 \\ -3 & 3 & -2 & -1 \\ 2 & -2 & 1 & 1 \end{bmatrix}$$

$$H_i = \begin{bmatrix} 1 & 0 & 0 & 0 \\ 0 & 1 & 0 & 0 \\ 0 & 0 & h_i & 0 \\ 0 & 0 & 0 & h_i \end{bmatrix}, \quad K_j = \begin{bmatrix} 1 & 0 & 0 & 0 \\ 0 & 1 & 0 & 0 \\ 0 & 0 & k_j & 0 \\ 0 & 0 & 0 & k_j \end{bmatrix}$$

$$Q = \begin{bmatrix} Z_{i,j} & Z_{i,\, j+1} & Z_{y,\, ij} & Z_{y,\, i,\, j+1} \\ Z_{i+1,\, j} & Z_{i+1,\, j+1} & Z_{y,\, i+1,\, j} & Z_{y,\, i+1,\, j+1} \\ Z_{x,\, ij} & Z_{x,\, i,\, j+1} & Z_{xy,\, ij} & Z_{xy,\, i,\, j+1} \\ Z_{x,\, i+1,\, j} & Z_{x,\, i+1,\, j+1} & Z_{xy,\, i+1,\, j} & Z_{xy,\, i+1,\, j+1} \end{bmatrix}$$

The derivative $Z_x^{(m)}$, $Z_y^{(m)}$ at four points of a master patch can be obtained by globally interpolating all of the grid points of master patches as the cubic Ferguson surface in both x and y directions. Then, $Z_x^{(s)}$, $Z_y^{(s)}$ at the slave patches are obtained by linear interpolation from the derivatives of the master patch. In the present work, the cross-boundary derivatives Z_{xy} at four corner points of each patch are assumed to be absent.

2.3-2 Refinement Criterion

Refinement is carried out by subdividing each rectangular master patch in four slave patches with half side lengths. The procedure for refinement can be repeated recursively, so that many slave patches can be obtained from a single master patch (see Fig. 4).

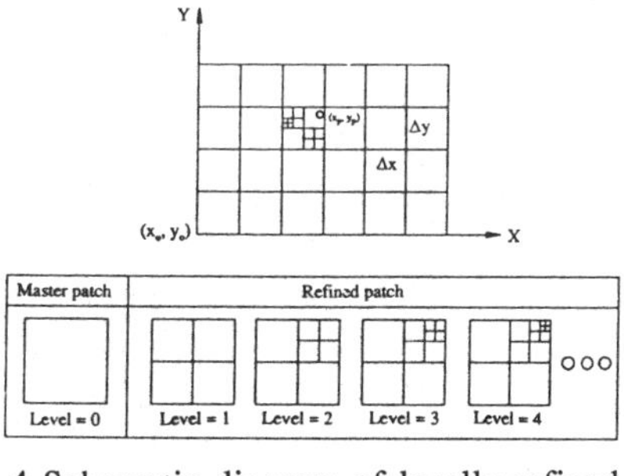

Fig. 4 Schematic diagram of locally refined non-parametric tool description

For the adaptive refinement, two criteria are devised as follows :

$$Z_x \geq slope_{crit} \qquad \text{or} \quad Z_y \geq slope_{crit} \tag{5}$$
$$\rho \leq \rho_{crit}$$

where Z_x and Z_y are the derivatives in x and y directions and ρ is the radius of curvature at a cartesian point of a master patch. Proper $slope_{crit}$ and ρ_{crit} should be input according to the required accuracy of the problem.

2.3-3 Contact search algorithm in adaptively refined nonparametric patch approach

The contact search of the locally refined nonparametric patch approach is simpler and faster in computation than that of the parametric patch approach. Fig. 4 shows the schematic diagram for contact search.

i) Patch selection

The master patch to which a node should be projected in the Z-direction can be found as follows :

$$m = \text{int}\left(\frac{x-x_o}{\Delta x}\right) + 1 \ , \ n = \text{int}\left(\frac{y-y_o}{\Delta y}\right) + 1 \tag{6}$$

ii) Projection

(case 1) The level of a master patch = 0 :

The projection points are obtained from eq. (3)

(case 2) The level of a master patch ≥ 1 :

The slave patch to which a node should be projected can be found from the candidate slave patches corresponding to the master patch which is now concerned by the simple direct searching process. Then, the projection points are obtained from eq. (3).

Therefore, in case of LEVEL ≥ 1, the number of slave patch, the coordinate and derivatives at the corner points of each slave patch corresponding to the master patch should be stored properly before the FE analysis.

2.3-4 Normal scheme for adaptively refined nonparametric tool description

In each contact iteration, the nodal position should be projected so as to conform to the tool surface along a specified path. Various projection schemes in this stage can be used. Among them, the Z-projection method is efficient in computation time since Z-coordinate can be found easily at any given coordinate point of x and y as shown in eq. (3) in the case of the (locally refined) nonparametric patch approach. However, this method has a problem of convergence, if the slope of the tool surface becomes very large. Therefore, in order to maintain the numerical efficiency of Z-projection, an improved Z-projection scheme based on the mixed normal is proposed as the projection scheme as shown in Fig. 5.

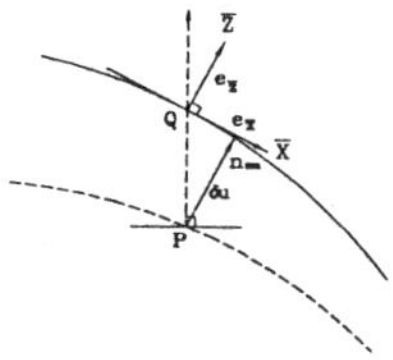

Fig.5 Z-projection combind with mixed normal

3. ANALYSIS OF STAMPING OF A TRUNK FLOOR

A trunk floor is one of the sheet parts that make up the body of a car. This part is one of difficult sheet parts to be formed, since it has steep slopes and sharp edge corners. In the present analysis, the tool surfaces are described with parametric patches and locally refined non-parametric patches to investigate the relative accuracy and numerical efficiency of both tool description methods. For the parametric patch approach, the raw CAD surface data described with a great number of parametric patches are constructed in the commercial CAD system and then directly delivered into the input file for the forming analysis.

For the locally-refined nonparametric patch approach, the locally refined nonparametric tool surface can be obtained from the parametric tool surface according to the suggested algorithm. Fig. 6 shows the parametric and locally-refined nonparametric tool surface. From the viewpoint of dimensional accuracy, the nonparametric patch approach based on the equi-spaced point description requires at maximum 16 times the dimensional array in order to obtain the same accuracy of the locally refined nonparametric patch since the 4 times successive refinement by the suggested criterion have been carried out in this example.

The material and process variables used in the analysis are as follows :

sheet material : cold rolled steel

stress-strain characteristics:

$$\bar{\sigma} = 519.2\,(0.0019 + \bar{\varepsilon})^{0.27} \text{ MPa}$$

Lankford value : r = 1.72

sheet thickness : 0.8 mm

Coulomb coefficient of friction : 0.12

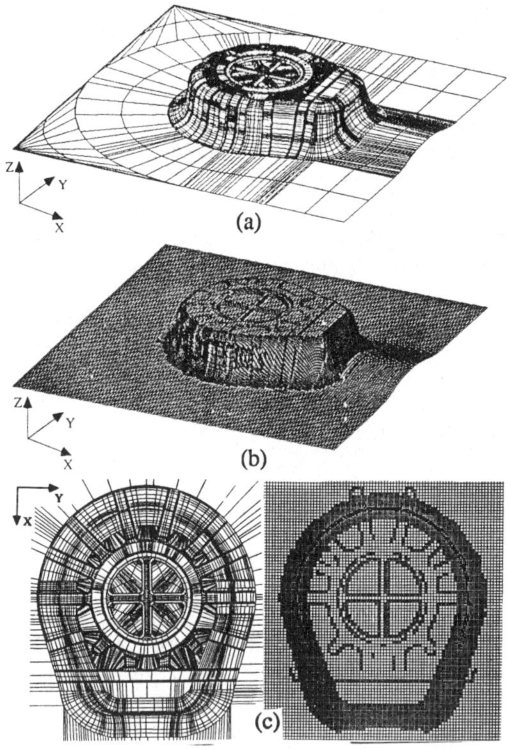

Fig. 6 Schematic view of the tool surfaces for stamping analysis of a trunk floor : (a) parametric patches, (b) locally refined nonparametric patches, (c) Comparision of parametric patches and the present theory on the X-Y plane

For the analysis, 1681 nodes and 1600 4-node rectangular elements were employed. In the case of the parametric patch approach, the sheet surface normal scheme was used to treat the tool-sheet contact. In the case of the locally-refined nonparametric patch approach, the Z-projection scheme combined with the mixed normal was employed. The restraint force due to the draw bead was applied to the nodes located at the outer region of the blank according to the type of the draw bead at each blank side. In this work, the binderwrap was calculated using the geometric non-linear elasticity FE code developed by the authors[6].

Fig. 7-(a) shows the calculated deformed configuration at the final punch stroke (222mm) in the case of parametric patch approach, and Fig. 7-(b) shows the calculated deformed configuation at the same stroke in the case of locally-refined nonparametric patch approach. From the comparison of the simulated results, the deformed configuration and the amount of draw-in in both approaches were almost identical. Fig. 8 shows the contour of the thickness strain distribution at the final punch stroke. Fig. 9 shows the thickness strain distribution at one section denoted by sections A-A in Fig. 9 at the final stroke. The overall tendency of strain distribution is similar to each other, but some minor differences were also observed. All these deviations may be considered as inherent results originating from the differences in tool surface description and related normal scheme. The comparative numerical details between parametric patch approach and non-parametric patch approach for the trunk-floor are summarized in TABLE 1. From the comparison of numerical results, the locally refined nonparametric patch approach is found to be far more efficient than the parametric patch approach if the number of nodes and parametric patches used becomes larger.

Therefore, the locally-refined nonparametric patch approach with the suggested refinement criterion and contact search algorithm provide the good choice for the general tool surface description of complex industrial sheet parts from the viewpoint of computational efficiency, tool surface conformity, hardware capacity and solution accuracy.

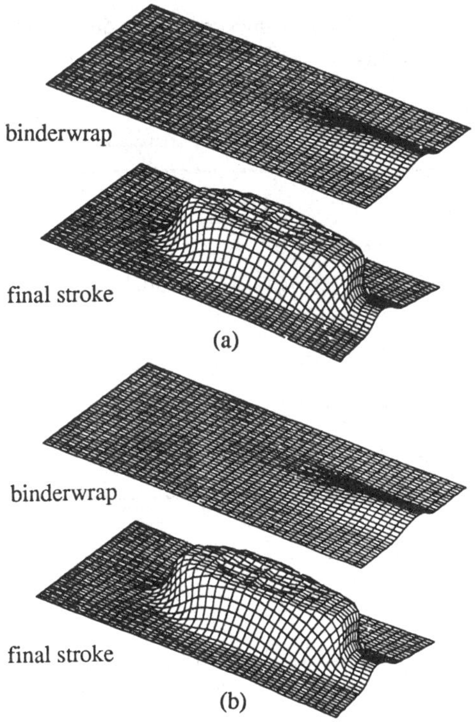

Fig. 7 Computed results for deep drawing of a trunk floor : (a) parametric patch approach, (b) locally refined nonparametric patch approach

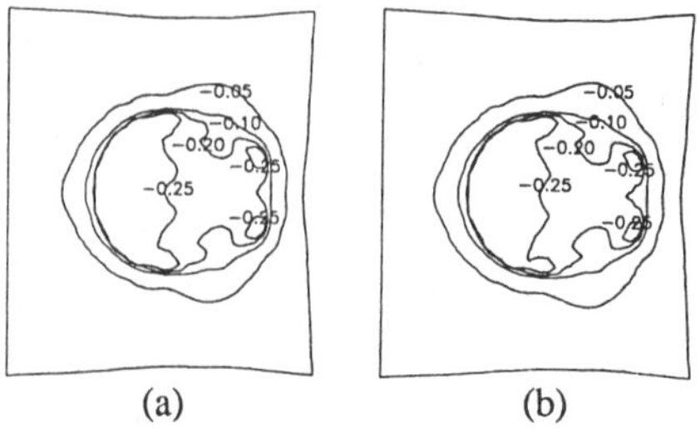

Fig. 8 Thickness strain contour for a trunk floor at the final step : (a) parametric patch approach, (b) locally refined nonparametric patch approach

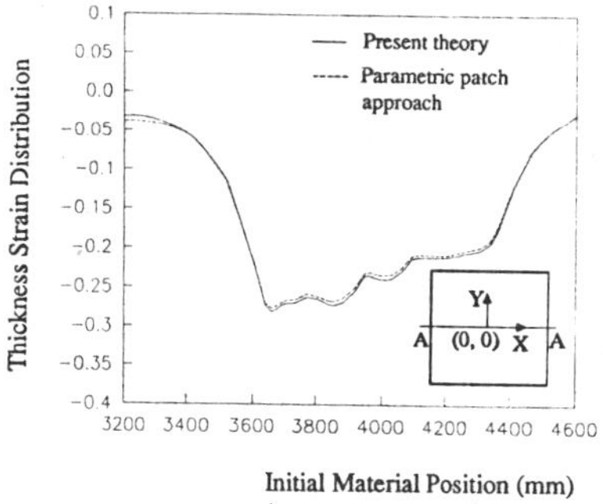

Fig. 9 Comparision of thickness strain distribution for deep drawing of a trunk floor on the section A-A

TABLE 1.

Comparision of numerical results for stamping of a trunk floor panel

	Parametric tool	Locally refined non-parametric tool
Projection scheme	sheet surface normal	Z-projection combined with mixed normal
Total no. of d. o. f.	5043 (1681 nodes)	5043 (1681 nodes)
Average iteration no. per step	8	9
Total no. of step	222	222
CPU (sec) per iteration (RISC 6000)	37	16
Total CPU (minites)	1,110.0	495.0
Max.punch height (mm)	222.0	222.0

4. CONCLUSION

The locally refined nonparametric patch approach in tool surface description and the related contact search algorithm have been applied to stamping processes of actual auto-body panel with arbitrarily complicated tool surface geometry. The possibility of the general use of the proposed tool surface description and the robustness of the related contact search algorithm have been confirmed through the chosen industrial example.

From the numerical results, it has been shown that i) the locally refined nonparametric patch approach is found to be more efficient than the parametric patch approach in computation time and contact search algorithm, when the number of parametric patches used becomes larger.

ii) the locally nonparametric patch approach has been shown to be superior to nonparametric patch approach from the viewpoint of dimensional accuracy and surface conformity.

The proposed method can be thus effectively applied to the tool surface description for FE analysis of complex sheet metal parts.

ACKNOWLEGEMENTS

The authors appreciate the courtesy of K.H. Kim for having provided parametric tool surface data.

REFERENCES

[1]Schweizerhof, K. and Hallquist, J.O. 1991. Explicit integration schemes and contact formulations for thin sheet forming. *In Proc. VDI* (edited by Altan, T. et.al.). Zurich, Switzerland : 405-439.

[2] Makinouchi, A., Nakamachi, E., and Nakagawa, T. 1991. Development of CAE system for auto-body panel forming die design by using 2-D and 3-D FEM. *Annals of the CIRP.* 40 : 307-310.

[3] Yoo, D.J., Song, I.S., Yang, D.Y., and Lee, J.H. 1994. Rigid - plastic finite element analysis of sheet metal forming processes using continuous contact treatment and membrane elements incorporating bending effect. *Int. J. Mech. Sci*. Vol.36. No.6 : 513 - 546 .

[4] Yang, D.Y.,Yoo, D.J., Song, I.S., and Lee, J.H. 1994. Investigation into tool surface description for finite element analysis of three - dimensional sheet metal forming processes. *J. Mater. Process. Technol.* 45 : 267-273.

[5] Faux, I.D. and Pratt, M.J. 1979. Computational geometry for design and manufacture. *Ellis Horwood.*

[6] Song, I.S. , Yoo, D.J. , Yoon, J.W., Yang, D.Y., Huh, H. , and Lee, J.H. Submitted. Finite element analysis and design of binder wraps for automobile sheet metal parts using surface boundary condition. *J. Mater. Engng. Perf.*

Metals – General topics

Simulation of Materials Processing: Theory, Methods and Applications, Shen & Dawson (eds)
© 1995 Balkema, Rotterdam. ISBN 90 5410 553 4

Numerical simulation of sheet cutting

A. Abdali, K. Benkrid & P. Bussy
LMCAO, INSSET, Université de Picardie Jules Verne, Pôle Regional de Modélisation, Saint-Quentin, France

ABSTRACT: Numerical simulation of sheet metal cutting is a non linear problem with plasticity, damage, contact and friction. To resolve this problem, we use a new stable elastoplastic damage model for large transformations. The algorithm used is the large time increment method where the loading path is considered as only one increment. Many numerical simulation are presented.

1.INTRODUCTION

Metal cutting or punching are forming processes that have seldom been modelled up to now. The optimization of these processes has always been left to the engineer. Nevertheless, numerical simulation of metal forming by the finite elements method could be very useful to the engineer. The modelling of cutting and punching problems involves a large number of non linearities which have, up to now, led to several difficult problems. Firstly, there are geometrical non linearities with large displacements and large deformations, in addition to conditions of unilateral contact. Then there are non linearities in material behaviour such as plasticity, damage or friction between dies and sheet. It then remains to solve the problem involving unsteady non-linear inequations, by a choice of formulation and an appropriate algorithm. The work presented here treats two of these difficulties, the first being material non linearity, the second being choice of algorithm. Plasticity and viscoplasticity have been very well modelized in the past, whereas damage has not. Existing damage models do not allow satisfactory calculations, because they lead to numerical instabilities. We propose a new stable elastoplastic damage model for large transformations. This is the extension of the elastic damage model developed by Benkrid and Bussy, stable to generalized standard materials in its formulation for small displacements. In this model we introduce a new scalar damage variable X related to the usual damage D by X=D/1- D. The choice of a damage convex written with the stress tensor ensures stability. We chose the non linear geometric model proposed by P. Ladevèze. It introduces a rotation R defining the corotational referential. We have to write our damage model into the corotational configuration of the geometrical model. The implementation of this model of in our finite element computational code gives stable calculations . The second part of this work presents the algorithm used. This is the version due to Bussy of the large time increment method proposed by P. Ladevèze. Unlike classical algorithms (e.g Newton, Newton-Raphson, etc), it does not cut up the loading path into small increments; instead the whole loading path is considered as only one increment. This strategy makes it easy to integrate the limit points and gives good prediction of the post-critical behaviour. This algorithm has been used by Bussy to simulate sheet metal forming. Illustrations of sheet metal cutting and punching processes are given.

2.MODELIZATION AND PRESENTATION OF THE NON LINEAR PROBLEM WITH CONTACT AND FRICTION.

2.1 Modelization of problem

Figure 1 presents the structure Ω_0 in its referential configuration at time t=0 and its actual configuration Ω at time t. The structure undergoes large displacements, large deformations and a large rotation. $\partial_1\Omega_0$ is the part of boundary when the displacements u_d are given. On $\partial_2\Omega_0$ the loading F_d is given. $\partial_3\Omega_0$ presents the set of points which come into contact with a tool. f_d is a body force density.

2.2 Presentation of problem

The geometrical non linear problem with contact and friction can be written in the following form:

Find S=[(F, v), (P, q)] function of t and M_0 verifying over the whole time increment [0,T]:
(F,v) such that:

$$u_{|\partial_1\Omega_0} = u_d \;,\; v = u_{|\partial_3\Omega_0} \;,\; F = I + \frac{du}{dM_0} \qquad (1)$$

u regular , F is the linear tangent mapping.

(P,q) such that: $\forall$ t, $\forall$ u*$\in$ U = {u, $u|_{\partial_1\Omega_0} = 0$}
u* "kinematically admissible to zero"

$$\int_{\Omega_0} Tr[\, P^T \, F^*]\, dV + \int_{\partial_3\Omega_0} q\, v^*\, dS = \int_{\partial_2\Omega_0} F_d\, u^*\, dS + \int_{\Omega_0} f_d\, u^*\, dV \quad (2)$$

(F,v) and (P,q) verifying the constitutive relations:

$$\dot{F}_t = \mathcal{A}\,(P_\tau\,,\, \tau \le t) \quad (3)$$

$$\mathcal{R}\,(q\,,\, \dot{v}) \quad (4)$$

Where $\mathcal{A}$ is a characteristic function of the material and $\mathcal{R}$ is the contact and friction relation. v and q are the displacement and the loading vectors on $\partial_3\Omega_0$.
$P = J\,\sigma\,F^{-T}$ is the Piola Kirchhoff tensor. J = Det F and σ is the Cauchy stress tensor.
This problem formulation reveals, on the one hand admissibility relations (1) and (2) which are linear, on other hand the constitutive relations (3) and (4) which regroup all the non linearities of problem.

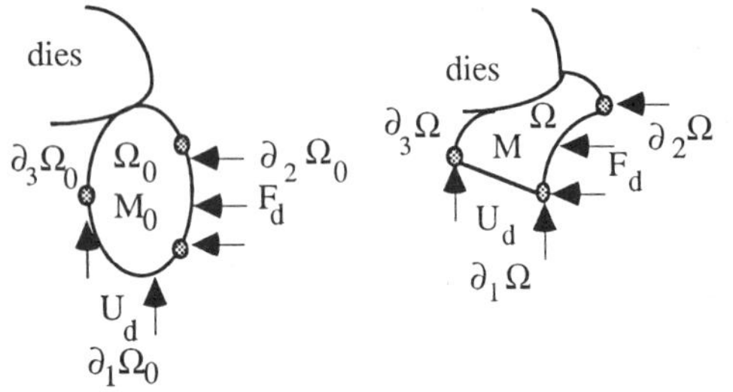

Figure 1

3. COROTATIONAL MODEL

Our problem is defined for large transformations. To write the objective constitutive law, we chose the non linear geometric model proposed by P. Ladeveze. This model introduces a rotation R defining the corotational referential. This rotation locally transforms the configuration Ω into a rotated configuration $\underline{\Omega}$. R is associated to the rotation rate and defined by the differential equation and the initial conditions:

$$\dot{R}^T R = \Omega = \frac{1}{2}(L - L^T)$$
$$R\,|_{t=0} = I$$

with $L = \dot{F}\,F^{-1}$ and $F = I + \frac{du}{dM_0}$
Ω is the rotation rate.

With this rotation we obtain the strain rate and the stress tensor in the rotated configuration:

$$\begin{cases} \underline{D} = RDR^T \\ \underline{\tau} = R\tau R^T \end{cases}$$

with $D = \frac{1}{2}(L + L^T)$
Remark:
The stress rate in the rotated configuration $\underline{\Omega}$ is the rotated Jaumann rate of the Kirchhoff stress.

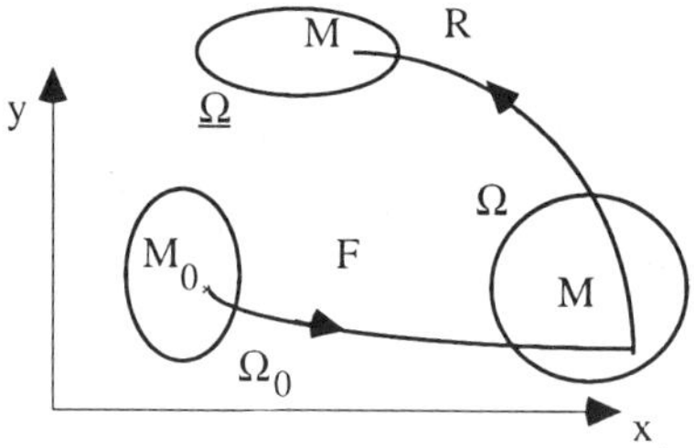

Figure 2

4.ELASTOPLASTIC DAMAGE MODEL IN LARGE DISPLACEMENTS.

In this presentation, we propose a new stable elastoplastic damage model for large transformations. This is the extension of the elastic damage model developed by Benkrid and Bussy in small displacements.
We assume linear isotropic elasticity coupled with damage, and assume an isothermal process and isotropic damage. The model is defined by the specific free energy in the following form:

$$\rho\psi(\underline{\Sigma},X,p) = \rho\psi_1(\underline{\Sigma},X) + \rho\psi_2(X) + \rho\psi_3(p)$$

$\rho\psi_1(\underline{\Sigma},X)$ is the elastic damage free energy where ρ is the mass density and $\underline{\Sigma}$ is the strain tensor.

$$\rho\psi_1(\underline{\Sigma},X) = \frac{1}{2}\,Tr\,(\underline{\Sigma}\,\frac{C_e}{1+X}\,\underline{\Sigma})$$

X is the new scalar damage variable. The relation between this new variable and the usual damage variable D, which is bounded by 0 and 1, is:

$$1 - D = \frac{1}{1+X} \quad \text{from which } X = \frac{D}{1-D}$$

δD must verify the condition $D + \delta D < 1$; then δX is not boundered, which avoids all numerical problems. $\rho\psi_2(X)$ and $\rho\psi_3(p)$ are convex and positive functions.

The associated thermodynamic forces are:

$$\underline{\tau}=\rho\frac{\partial\psi}{\partial\underline{\Sigma}}\ ,\quad Y=\rho\frac{\partial\psi}{\partial X}\ ,\quad R=\rho\frac{\partial\psi}{\partial p}$$

$$\underline{\tau}=\frac{C_e}{1+X}\underline{\Sigma}\ ,\ Y=-\frac{1}{2}\mathrm{Tr}\left(\underline{\Sigma}\frac{C_e}{(1+X)^2}\underline{\Sigma}\right)+Z$$

with $Z=\rho\frac{\partial\psi_2}{\partial X}$

$R=\rho\frac{\partial\psi_3}{\partial p}$ is the isotropic hardening variable and p is the plastic strain. The dissipation can be written:

$$\mathrm{Tr}\,[\underline{\tau}\,\underline{D}]-\rho\dot{\psi}\geq 0$$

$\underline{D}=\underline{D}e+\underline{D}_p$ where $\underline{D}e$ is the elastic strain rate and $\underline{D}_p$ is the plastic strain rate.

We suppose that the strain rate $\underline{D}e$ is split into a reversible part with X constant $\underline{D}_r$ and an irreversible part $\underline{D}_i$:

$$\underline{D}e=\underline{D}r+\underline{D}_i$$

The Clausius-Duhem inequality is written in the corotational configuration by:

$$\mathrm{Tr}\,(\underline{\tau}\,\frac{\underline{D}_i}{2})+\mathrm{Tr}\,(\underline{\tau}\,\underline{D}_p)-Z\dot{X}-R\dot{p}\geq 0\ .$$

Which is verified if we can write:

$$\mathrm{Tr}\,(\underline{\tau}\,\frac{\underline{D}_i}{2})-Z\dot{X}\geq 0\ \text{ and }\ \mathrm{Tr}\,(\underline{\tau}\,\underline{D}_p)-R\dot{p}\geq 0$$

For that, we introduce the potential $\varnothing^*(\underline{\tau},R,Z)$:

$$\varnothing^*(\underline{\tau},R,Z)=\varnothing^*p\,(\underline{\tau},R)+\varnothing^*_x\,(\underline{\tau},Z)$$

where $\varnothing^*_p\,(\underline{\tau},R)$ is the indicator function of the plastic convex defined by :

$$f_p\,(\underline{\tau},R)=\mathrm{Tr}\,(\frac{3}{2}\underline{S}\ \underline{S})^{\frac{1}{2}}-R-R_0\leq 0$$

$\varnothing^*_x\,(\underline{\tau},Z)$ is the indicator function of the damage convex:

$$f_x\,(\underline{\tau},Z)=\frac{1}{2}\mathrm{Tr}\,(\underline{\tau}\,C_e^{-1}\,\underline{\tau})-2Z-Z_0\leq 0$$

R_0 and Z_0 are respectivly the plastic and the damage limits. $\underline{S}$ is the deviator tensor of $\underline{\tau}$. The complementary laws are:

$$\underline{D}_p=\lambda_2\frac{\partial f_p}{\partial\underline{\tau}}\ \text{ and }\ \dot{p}=-\lambda_2\frac{\partial f_p}{\partial R}=\lambda_2$$

$$\frac{\underline{D}_i}{2}=\lambda_1\frac{\partial f_x}{\partial\underline{\tau}}\ \text{ and }\ \dot{X}=-\lambda_1\frac{\partial f_x}{\partial Z}=2\lambda_1$$

5.NUMERICAL TREATMENT

The large time increment method is introduced by P. Ladeveze. For the numerical resolution we use the version due to Bussy. We have an iterative algorithm with four fields s=(F,v,p , q).

Each iteration is split into two steps. Firstly a non linear but local step which enable us to calculate a solution $\hat{S}\in\Gamma$ and secondly a global but linear step which gives a solution $S_{n+1}\in A_d$. The solution of problem is : $S\in A_d\cap\Gamma$.

Γ is the set of solutions verifying the constitutive law. A_d is the set of solutions verifying the admissibility. The iterative pattern of the large time increment method gives for each iteration a new solution estimation for all loading history. The global iterative process stops when the error is small. Therefore the two steps on our algorithm are the following:

Local step: $S_n=[(F_n,v_n),(P_n,q_n)]\in A_d$

find $\hat{S}=[(\hat{F},\hat{v}),(\hat{P},\hat{q})]$, such that:

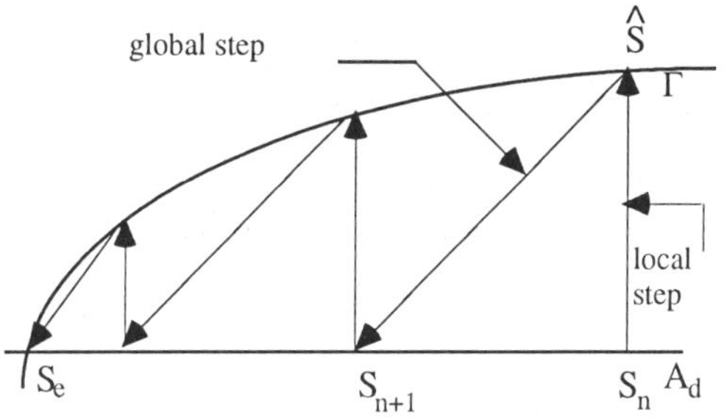

figure 3 : Algorithm

$$\hat{F}=\dot{F}_n \qquad \text{in } \Omega$$

$$\hat{q}-\dot{q}_n=-k(\hat{v}-\dot{v}_n) \qquad \text{on } \partial_3\Omega$$

$[(\hat{F},\hat{v}),(\hat{P},\hat{q})]\in\Gamma$, verifying (3) and (4)

Global step:

find $S_{n+1}=[(F_{n+1},v_{n+1}),(P_{n+1},q_{n+1})]\in A_d$ and verifying:

$$\hat{P}-\dot{P}_{n+1}=C_e(\hat{F}-\dot{F}_{n+1}) \quad \text{in } \Omega \qquad (5)$$

$$\hat{q}-\dot{q}_{n+1}=k(\hat{v}-\dot{v}_{n+1}) \quad \text{on } \partial_3\Omega \qquad (6)$$

with the admissibility relation

$$\forall t\ \forall u^* \in U, \int_{\Omega_0} Tr[(\dot{P}_{n+1}-\dot{P}_n)^T F^*]\ dV + \int_{\partial_3\Omega_0} (\dot{q}_{n+1}-\dot{q}_n)\ v^*\ dS=0$$

C_e is the linear elastic tensor and k is a positive parameter which has the dimension of stiffness. The equation (5) and (6) are search directions, which are algorithm parameters. The variables are in the following form:

$$\dot{u}_{n+1}(t, M_0)-\dot{u}_n(t, M_0) = \sum_{i=1}^{p} g_i(t)\ w_i(M_0)$$

$$\dot{F}_{n+1}(t, M_0) - \dot{F}_n(t, M_0) = \sum_{i=1}^{p} g_i(t)\alpha_i(M_0)$$

$$\dot{v}_{n+1}(t, M_0) - \dot{v}_n(t, M_0) = \sum_{i=1}^{p} g_i(t)\gamma_i(M_0)$$

with $\alpha_i = \frac{dw_i}{dM0}$ and $\gamma_i = w_i\Big|_{\partial_3\Omega_0}$

$$\dot{P}_{n+1}(t, M_0) - \dot{P}_n(t, M_0) = \sum_{i=1}^{p} h_i(t)\beta_i(M_0)$$

$$\dot{q}_{n+1}(t, M) - \dot{q}_n(t, M) = \sum_{i=1}^{p} h_i(t)\delta_i(M_0)$$

$g_i(t)$ and $h_i(t)$ are the time functions defined on [0,T], $w_i\ \alpha_i\ \beta_i$ and δ_i depend only on the space variable M_0. A finite element discretization is used to describe this space function. In practice p∈ [2,5].

On the global step the resolution of linear equation system is independant of time.

6. NUMERICAL EXAMPLES

The new elastoplastic damage model presented in large transformations is implemented in our code. We use the kinematic model proposed by P.Ladeveze and the large time increment method in its version due to Bussy to resolve the problems with geometrical non linearities, contact and friction.

The simulation presented in planar calculation uses the isoparametric triangle element with six nodes and three Gauss points. The first of these examples is an illustration of our algorithm and our constitutive law. This is a simulation of a test piece traction in plane stress. The second example is a first test of the simulation of sheet cutting.

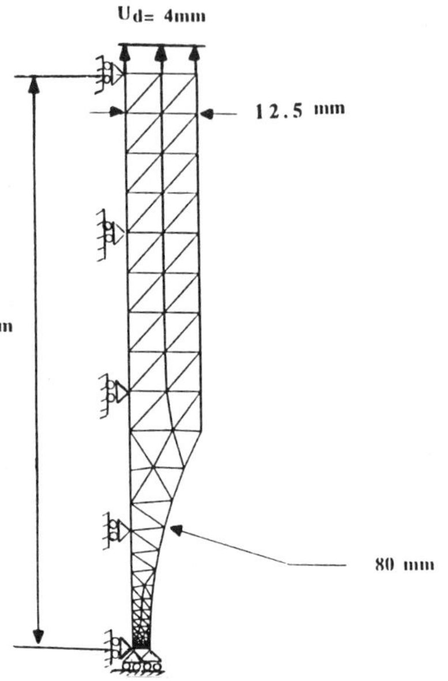

figure 4 : nodes 269, elements106

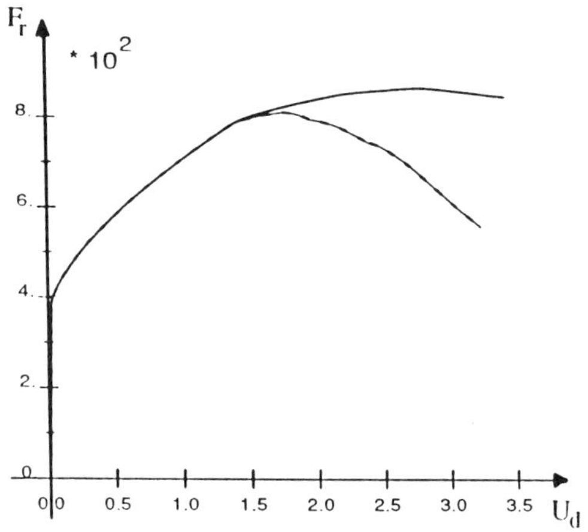

figure 5: Force behaviour F_r versus U_d

6.1 A test piece traction

Figure 4 describes the mesh and boundary conditions. The material is defined by its elastic caracteristics: Young's modulus E= 210.000 MPa, Poisson coefficient ν = 0.29. The isotropic hardening curve is defined by $R(p) = R_0 + k\ p^n$
with k=416 MPA , n=0.406, R0=125 MPA.

The function Z(X) is given by : Z = ln(1+X)/A. Where A is a localization factor A=0.0025

In figure 5, the top curve presents the force-displacement behaviour of the test piece in the

Figure 6

Figure 7

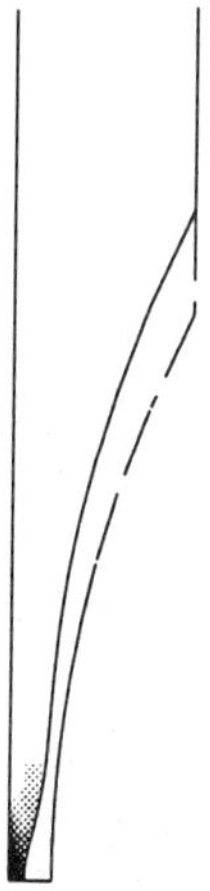

Figure 8

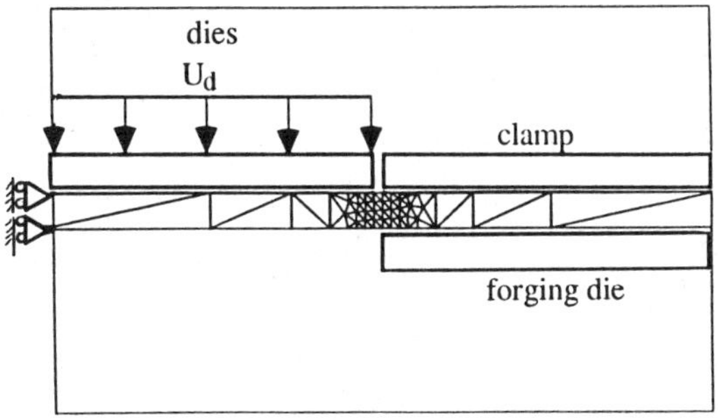

Figure 9 : nodes 86 , 211elements

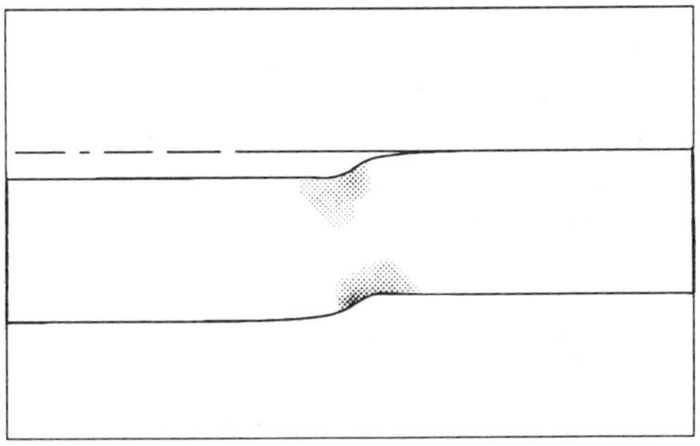

Figure 10 : $U_d = 0.1$mm

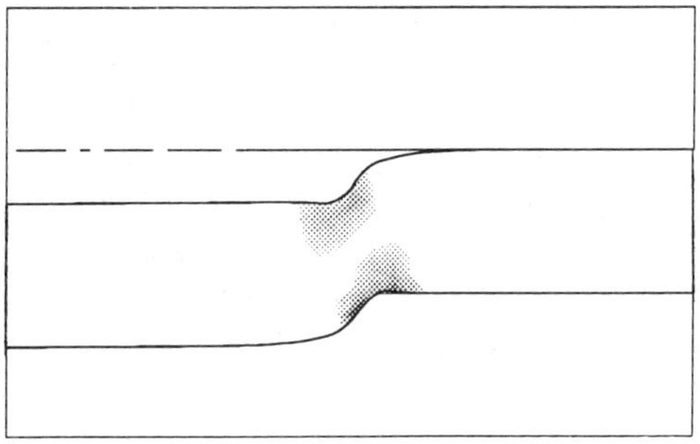

Figure 11 : $U_d = 0.2$mm

elastoplastic case. The bottom curve shows the behaviour when we take into account damage in our calculation. In this case the reaction force decreases rapidly.

Figures 6 and 7 show respectively for an applied displacement $U_d = 3.4$ mm the stress distribution on the test piece in the elastoplastic and elastoplastic damage cases. These figures show the deformation obtained at scale1. We see clearly the influence of damage on the localization of the deformation as well as the non negligible role of geometrical non linearities in the representation of the striction.

Figure 8 show the damage zone.

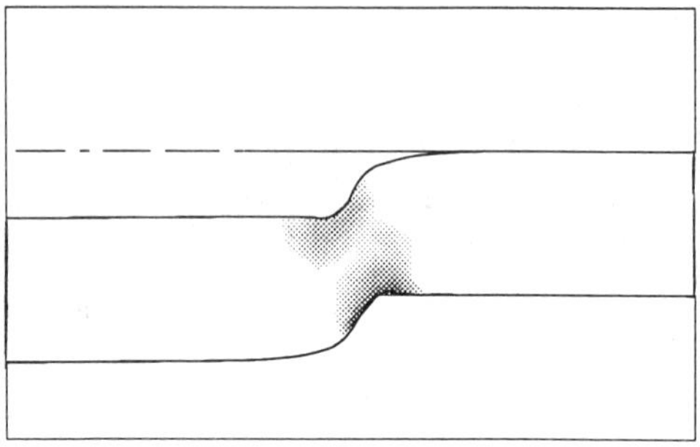

Figure12 : $U_d = 0.3$ mm

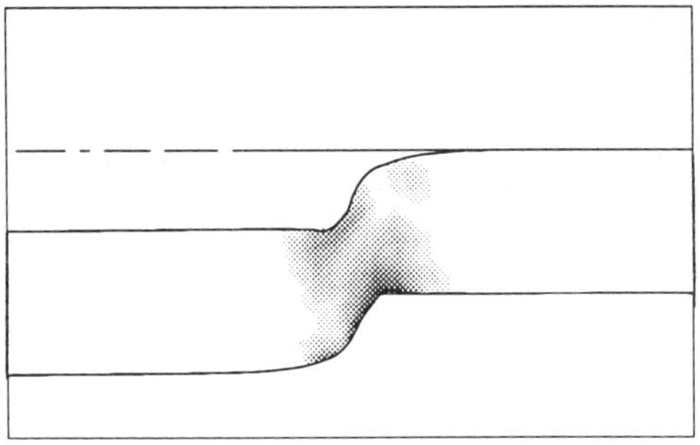

Figure 13 : $U_d = 0.45$mm

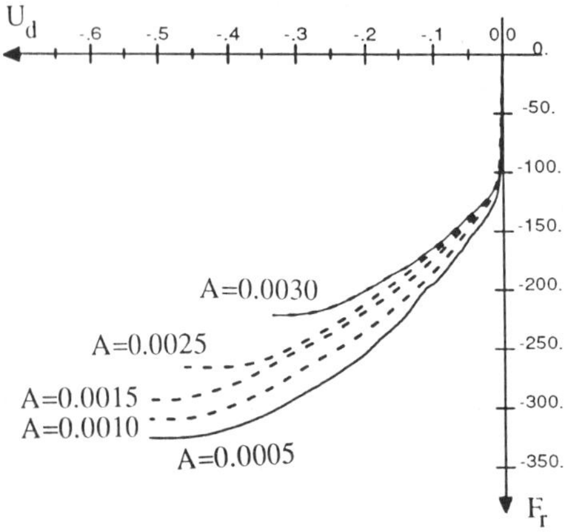

Figure 14 : Force behaviour F_r versus U_d

6.2 Example of the sheet cutting

The geometrical structure, the mesh used and the boundary conditions are described in figure 9.The material is defined by its elastic caracteristic:
Young's modulus E= 206.000 MPa.
Poisson coefficient $\nu = 0.3$. The isotropic hardening curve is defined by $R(p) = R_0 + k\, p^n$

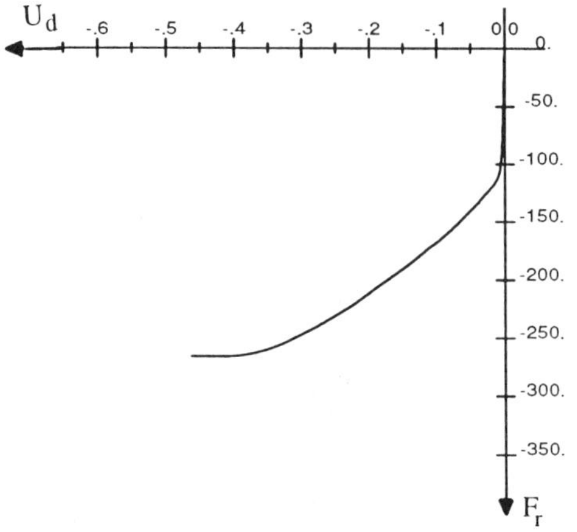

Figure 15 : Force behaviour F_r versus U_d

With k=557.66 MPA , n=0.24, R_0=188.41 MPA. The function Z(X) is given by : Z = ln(1+X)/A. The calculation is realised in plane deformation. For this example the maximal applied displacement is equal to 0.45 mm. The force-displacement behaviour is given in figure 15. The isodamage is presented for the applied displacements 0.1mm, 0.2mm, 0.3mm and 0.45mm (figures 10, 11, 12, 13). Figure 14 shows different responses obtain when we vary the localization coefficients of the damage law A.

REFERENCES

Benkrid.K, Bussy. P , Abdali. A: On the stable damage model for metal forming processes, Birmingham 94, J. Mater. Process. Technol. pp 521-526, 1994

Bussy. P : Sur un modèle d'endommagement stable, Rapport Interne du LMCAO n°2, INSSET, Saint-Quentin, 1992.

Boisse. P, Bussy. P, Ladevèze. P : A new approach in non linear mechanics : the large time increment method, Int. J. Num. Meth Eng., vol 29, pp 647-663-1990.

Bussy P.: Sur un algorithme à grand incrément de temps pour la gestion du contact en grandes déformations, Rapport Interne n° 93, L.M.T. Cachan, 1989.

Bussy P., Rougée P., Vauchez P.:The LArge Time Increment Method for numerical simulation of metal forming processes, Proceedings NUMETA, Swansea U.K, 1990.

Bussy P., Liu B., Vauchez P.: New algorithm for numerical simulation of metal forming process", Proc. NUMIFORM 92, pp 433-438 Sophia-Antipolis, Valbonne, France, 1992.

Bussy P., Coffignal G., Ladevèze P. : Reliability and optimisation of non linear finite element analyses, NUMETA, pp 659-667, Swansea, 1985.

Kachanov L. M.: Time of the rupture process under creep conditions, T.V.Z. Akad. Nauk, S.S.R. Otd. Techn. Nauk., 8, pp 26-31, 1958.
Lemaitre J : A course on Damage Mechanics, Springer-Verlag, 1992.
Elmenni M., Dubois M., Gelin J.C.: Traitement des problèmes numériques liés à la modélisation par éléments finis du processus de découpage des métaux, STRUCOME 93, Paris, pp 567-578.

Simulation of Materials Processing: Theory, Methods and Applications, Shen & Dawson (eds)
© 1995 Balkema, Rotterdam. ISBN 90 5410 553 4

Joint project of the German cold forming industry to improve an existing FE-code

G. H. Arfmann & M. Twickler
CPM Gesellschaft für Computeranwendung, Prozeß- und Materialtechnik mbH, Herzogenrath, Germany

The main scientific research in the field of two-dimensional Finite-Element-Computation was carried out in the nineteen-eighties. The german government funded a project to do further developments in this field of application. The result is a special programme for forging processes, that can be used on PC, has an integrated pre- and post-processing, an integrated remeshing and that is easy to handle, even by users with no specific knowledge in FEM.

The main scientific research work in the field of two-dimensional Finite Element computation was carried out successfully in the nineteen-eighties. In particular, the problem of automatic remeshing was solved. Over the same period, the computing performances of personal computers, introduced in the first third of the eighties, increased to a point where they were suitable for numerical simulation methods. As a result, Finite Element metal-forming simulations for PC applications were developed despite the especially high computing performances involved.

One such system is CAPS-Finel. This system grew out of research work carried out at the Aachen University of Technology (IBF/RWTH). Since 1989 it has been further developed for practical use in the PC environment by CPM GmbH.

Apart from these developments in computer performance and scientific FEM techniques, European industry has been characterised by a trend towards the use of new numerical methods even in small and medium-sized enterprises. Constant cost pressure, rising wage and wage-related costs, competition from low-wage economies and the need to maintain a technological lead all meant that leading industrial companies and industrial associations increasingly turned to the use of new, modern computer-based methods. One such activity was an initiative by the German Fasteners Association aimed at developing or sophisticating CAD and simulation software so that it could be applied industrially by the members of the association. In this context, the suitability of FE systems for simulation tasks was examined and a project funded by the Federal Ministry for Science and Technology Transfer was called into being. One task of this project was to continue the development of CAPS-Finel, with the aim of producing a practically-relevant simulation tool for cold forming operations. The present paper summarises the objectives of this project and outlines some interesting results (Fig. 1).

Contents of the project
"***Metal Forming, Field: Cold Forming***",
funded by the german government

- consideration of local friction
- spring loaded dies,
- technology-controlled remeshing
- automatic incrementation
- several dies
- automatic computer-runs for multi-stage processes
- precise analysis of machine characteristics

Fig. 1: Contents of project

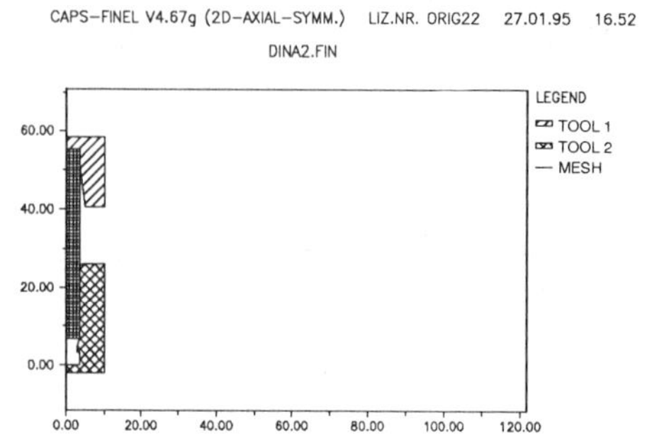

Fig. 2: Initial geometry

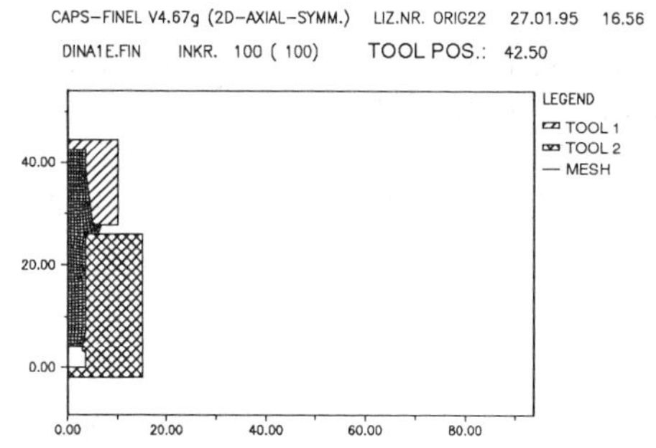

Fig. 3: Final geometry (without fixing)

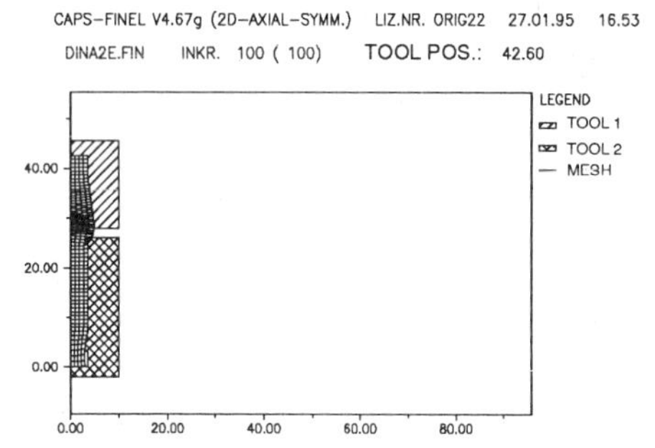

Fig. 4: Final geometry (with fixing)

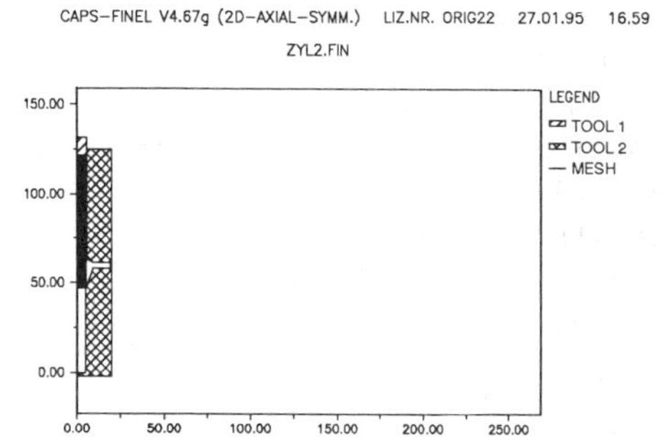

Fig. 5: Initial geometry

The main requirement for using FE software in practical industrial applications is a simple, easily-understood user interface for the line operator. As well as providing this user interface, it is necessary to simplify the sometimes relatively complex approaches used in such systems and make their nomenclature accessible for the line engineer. The FE systems also need additional features enabling them to model real manufacturing conditions without time-intensive scientific work during system use. Modifications of the friction laws are one aspect of these necessary adaptations and additions. A local friction description of the die is advisable, so that wear states and similar factors can be taken into account. It is also necessary to allow for differing machine characteristics, so that the user can base operation of the system on knowledge of his own machine type and so that the system can run automatically using sinus-oidal or free curves for the machines. As a matter of course, it must be possible to specify any pre hardening effects from earlier processes or drawing operations, and to run multistage processes in a single job. Finally, it must be pos-sible to simulate even relatively complex die systems consisting of several dies, some of them partly resilient. These features were added to CAPS-Finel in the course of the project. The following sections outline examples of a number of applications. One key aim was to model local friction.

Friction modules make it possible to model practical effects like local wear or intentional roughening or fixing of the formed part in the die. Varying frictional behaviour may also cause differences in local heating on die surfaces, leading, for example, to incorrect results on punches in reverse extrusion. This example shows a simulation of the usual fixing of the material. The example is taken from production and is a prestage in screw manufacturing, the pre-upsetting operation. For large height/ diameter ratios which reach critical limits, industrial practice is to choose the diameter in the upper die a few hundredths smaller than the diameter of the wire. During the forming process, this causes a slight moulding of the head zone before the actual forming process begins. This fixes or holds the material in the upper die. For the sake of simplicity, this fixing effect is modelled as local adhesion in the model. Fig. 2 shows the initial situation.

The next figure (fig. 3) shows the result of the forming process without fixing. There is no reduction, and the material emerges above the die. It is impossible to manufacture the part.

Fig. 4 shows the forming process with fixing. There is normal reduction and good forming in the head zone.

Pre-hardening of a wire also has a considerable influence on material behaviour in a forming process. A frequent problem for industry is that wires from different suppliers cannot be used in the same forming process. Assuming that order options like material strength, residual yield and minimum reduction were all identical, the explanation is often sought in surface treatment and coating. In reality, however, the nature of any predrawing process plays a decisive role. Depending on the drawing die, there may be widely differing strain distributions in the wire. The next example shows the influence of pre-hardening on the forming process, once again at the pre-upsetting stage (Fig. 5).

In the first case, the wire is relatively homogeneous. The result is a non-functional process. The material is not reduced and emerges in the head zone (Fig. 6).

In the second case, a wire with typical pre-hardening from a drawing operation is being used (Fig. 7).

The result is good reduction and proper filling in the head zone (Fig. 8).

Another project task was to allow for precise machine characteristics. An exact knowledge of movement sequences and speeds in forming machines is important in hot forming operations, where the material flow is heavily dependent on the current machine speed. The yield curves of steel materials are generally very speed-dependent in hot forming processes. However, the speed influence also plays an important part in cold forming, e.g. in analysing damage probability, crack initiation etc. Required machine power may also differ greatly depen-

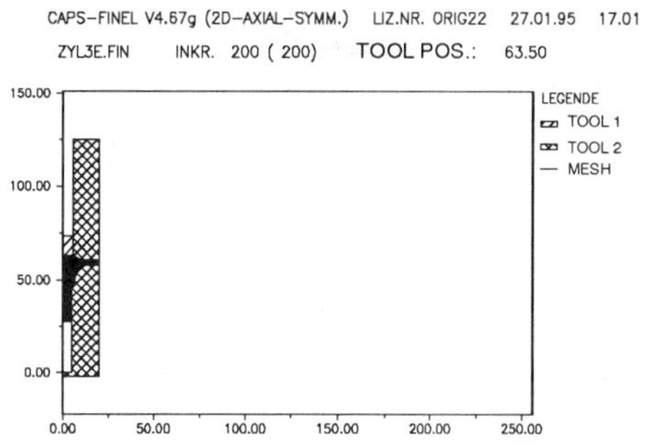

Fig. 6: Final geometry (without pre-hardening)

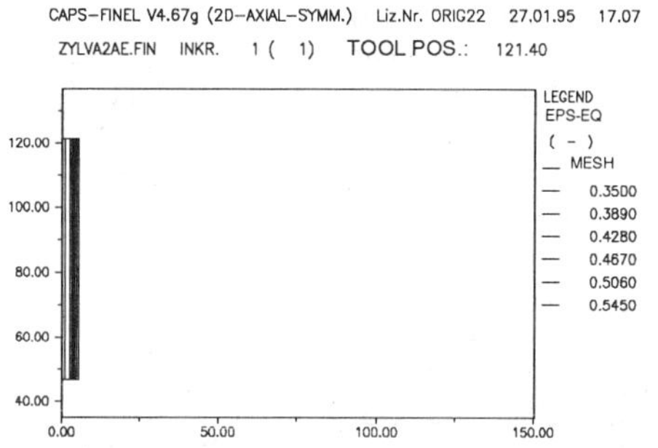

Fig. 7: Distribution of pre-hardening

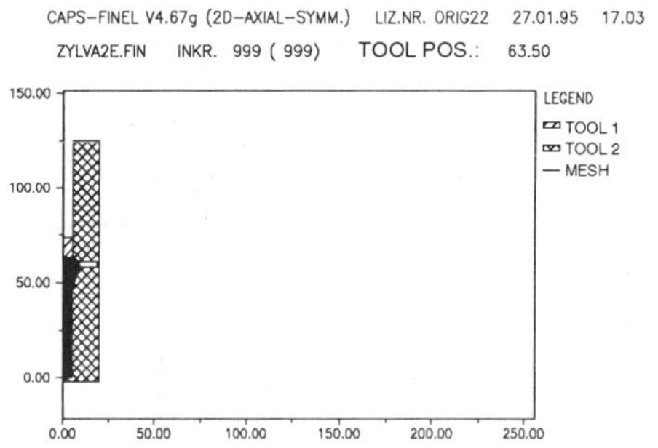

Fig. 8: Final geometry (with pre-hardening)

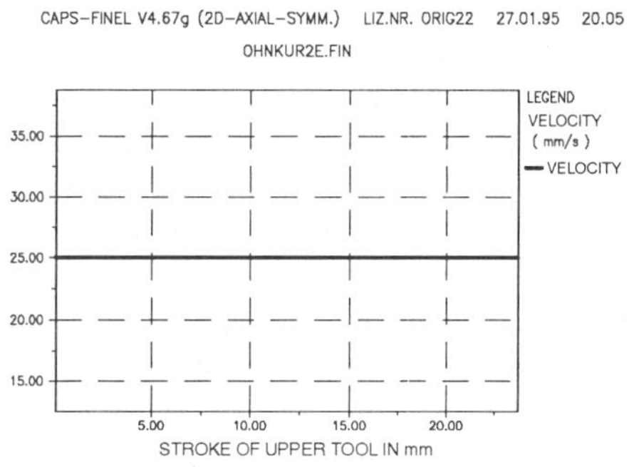

Fig. 9: Velocity vs. stroke (v=const.)

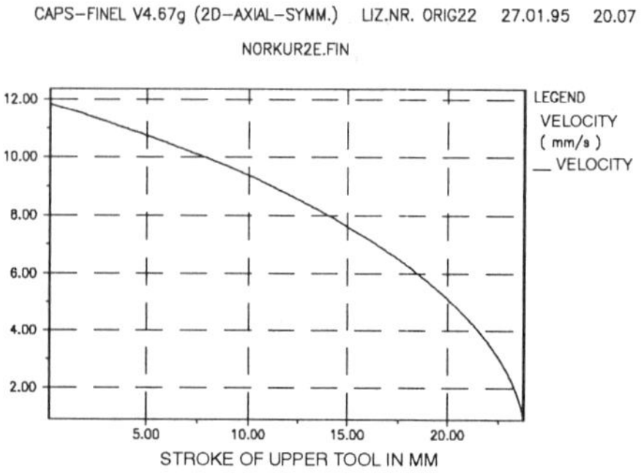

Fig. 10: Velocity vs. stroke (crank press)

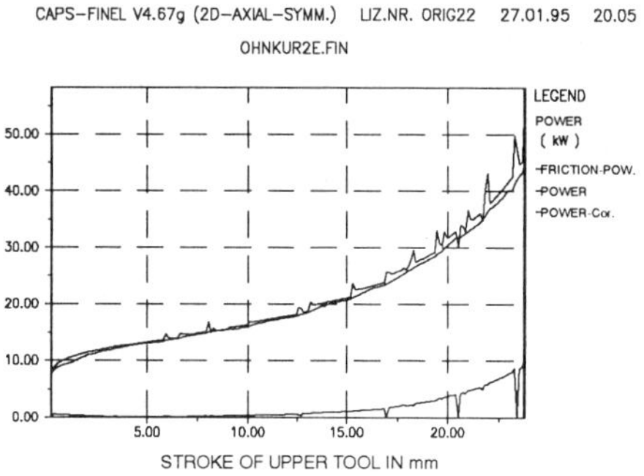

Fig. 11: Power vs. stroke (v=const.)

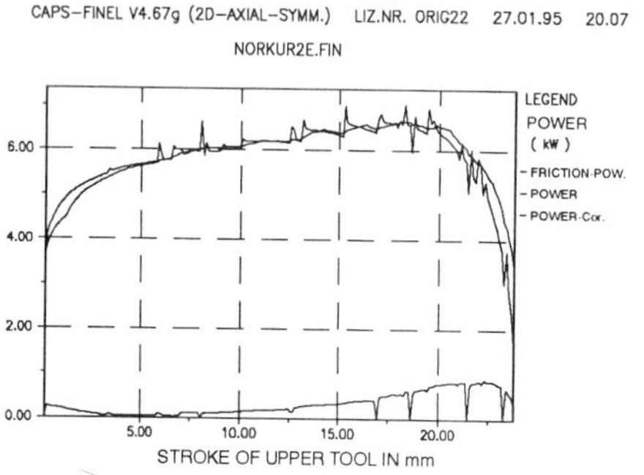

Fig. 12: Power vs. stroke (crank press)

ding on the motion characteristic.

As part of the project, the ability to specify such machine characteristics was realised in CAPS-Finel for typical cold-forming machines. For example, crank presses were described by specifying the crank radius and number of revolutions per minute and the slide path. Machines with special characteristics differing from these sinusoidal curves can be modelled using suitable path-time tables and stored in the reference database. Figures 9 and 10 clearly indicate the influence of these characteristics.

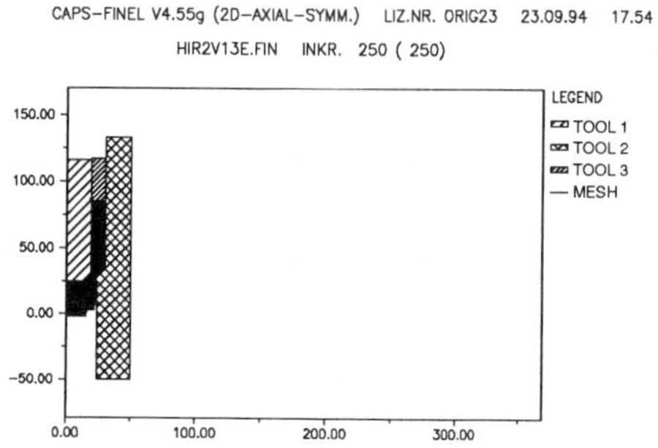

Fig. 13: Initial geometry (2 upper tools)

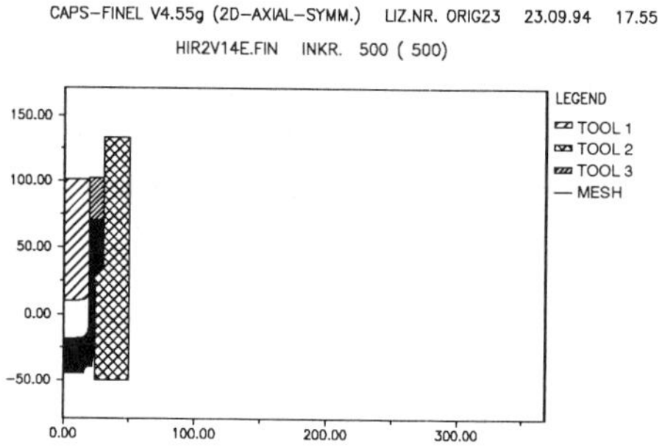

Fig. 14: Final geometry (2 upper tools)

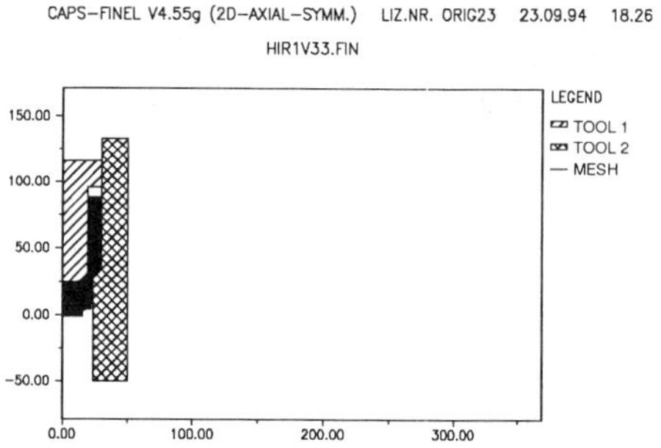

Fig. 15: Initial geometry (1 upper tool)

Fig. 9 is based on the characteristic of a press with v = constant,

Fig. 10 on the characteristic of a crank press. The different characteristics are expressed in the wide differences in required overall power input (Fig. 11 and 12) for a specific forming process.

Resilient dies or dies with relative motion due to air pressure are frequently used in real cold-forming. In conventional FEM, modelling relationships of this kind would entail time-intensive descriptions of the complex phenomena involved, possibly using different elements. By

agreement with the industrial companies in the project, die descriptions were realised as follows in CAPS-Finel: up to two upper and lower dies with relative motion or different resiliences are defined; the resiliences may also be prestressed. Successful studies were carried out with these modelling functions during the project itself. Figures 13 and 14 show an example of a forward reducing operation with a resilient inner die.
The task was to design the spring constants correctly. The main problem in this forming operation is to maintain the bottom geometry internally ahead of the punch. If a non-resilient die is used (Fig. 15),
material can flow ahead of the punch (Fig. 16) or tear away at the punch (Fig. 17).

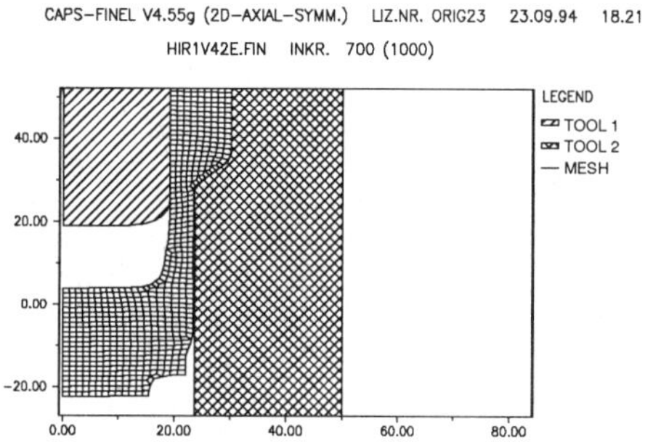

Fig. 16: Intermediate geometry (1 upper tool)

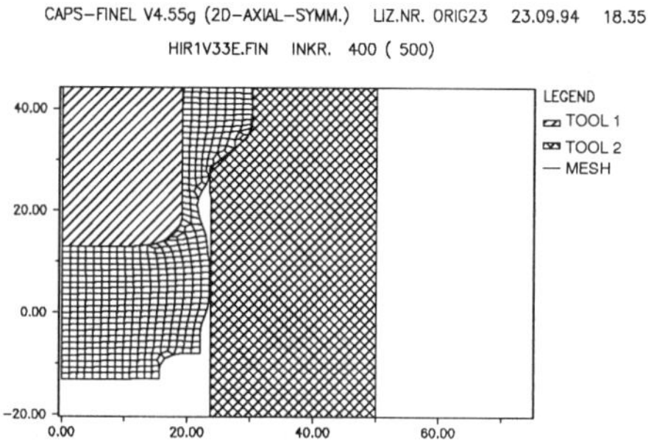

Fig. 17: Intermediate geometry (1 upper tool)

A suitable, resilient punch is essential for an optimum process.
In general, it may be stated that five years of experience with CAPS-Finel on small computers in small and medium-sized companies have demonstrated the usefulness of FEM as a simple tool for the design engineer, that FEM need not be labour-intensive, that results are realistic and that such systems can be used for process-near operations even on low-cost computers.

References

1) G. H. Arfmann, M. Twickler
Prozeßsimulation
Umformtechnik 28(1994)3,S. 165 ff.

2) G.H. Arfmann
Abschluß der Arbeiten zum BMFT-Projekt
"Kaltmassivumformung - Simulation"
Umformtechnik 28 (1994)1, S. 49

3) H. Voigt
Kaltformgerechte Gestaltung von Verbindungselementen mittels FEM Simulation
Deutscher Schraubentag
19./20.04.1994, Darmstadt

4) CAPS-Finel Handbuch
CPM Gesellschaft für Computeranwendung, Prozeß- und Materialtechnik mbH,
Herzogenrath 1994

5) G.H. Arfmann
Kosten minimieren - Simulationssysteme fürs Umformen verkürzen Entwicklungszeiten
Technischer Ausschuß DSV 1992

6) G.H. Arfmann, M. Twickler
Simulation von Umformprozessen in der Industrie
Metal Forming Process Simulation in Industrie
International Conference and Workshop
28.-30. September 1994, Baden-Baden, Germany

7) G. H. Arfmann
Praktische Umsetzung der FEM-Simulation für die Kaltmassivumformung
VDI, 22./23.02.1995, Düsseldorf

8) Neue und verbesserte Verfahren der Kaltmassivumformung
Verbundprojekt im Rahmen des Programms Fertigungstechnik 1988 - 1992
Kernforschungszentrum Karlsruhe, 1994

Simulation of Materials Processing: Theory, Methods and Applications, Shen & Dawson (eds)
© 1995 Balkema, Rotterdam. ISBN 90 5410 553 4

Mixed finite element model for forging processes with adaptive mesh refinement

José César de Sá, Luísa Costa Sousa & Maria Luísa Madureira
Department of Mechanical Engineering, University of Porto, Portugal

ABSTRACT: Forging processes are simulated by a mixed finite element model, in which the material is considered to have a rigid viscoplastic behaviour and the independent variables are velocities, pressure and deviatoric stresses. Interface elements both for 2-D and 3-D are used to deal with friction and contact. A recovery process of nodal stresses is used to improve the solution and to establish error estimators based on energy norms and point-wise stress errors which trigger an adaptive mesh refinement.

1. INTRODUCTION

The aim of this work was to develop a general model for both cold and hot bulk forging that could be easily used by the industry. As very often these processes can only be successfully executed within a narrow range of forming conditions, the possibility of accurately be able to simulate them can be crucial to their improvement and to reduce their costs as, presently, forming industries rely heavily on prototyping.

The code developed uses a rigid viscoplastic formulation, based on Perzyna´s model.

Despite the disadvantage of increasing the solution variables mixed methods are an effective way to solve constraint problems. That disadvantage can be eliminated if convenient iterative methods are used with the benefit of having continuous fields for stresses. In this work a mixed finite element model is used where the independent variables can be velocities, deviatoric stresses and pressures .

The finite element mesh, during the simulation of the forming process, can be very distorted or badly defined in certain zones, and therefore a mesh refinement or regeneration may be needed in order to have reliable results or even to be able to perform the simulation completely. An adaptive mesh refinement algorithm was developed, triggered by an error estimator based in the energy norm evaluated from the differences between the recovered values and those given by numerical solution.

Experimental examples were carried out to assess the code developed.

2. CONSTITUTIVE EQUATIONS

Assuming a rigid, isotropic, strain hardening viscoplastic incompressible deformation, with a flow rule based on the Von Mises criterion and the Perzyna model (Rebelo and Kobayashi(1980)) the constitutive equations are :

$$s_{ij} = \frac{2}{3}\frac{\bar{\sigma}}{\dot{\bar{\varepsilon}}}\dot{\varepsilon}_{ij} \quad (2.1) \qquad p = -\frac{\sigma_{ii}}{3} \quad (2.2)$$

$$\bar{\sigma} = Y_0(\bar{\varepsilon})\left[1 + \left(\frac{\dot{\bar{\varepsilon}}}{\gamma}\right)^n\right] \quad (2.3) \qquad \dot{\varepsilon}_{ii} = 0 \quad (2.4)$$

being σ_{ij} the stress tensor, p the pressure, s_{ij} the deviatoric stress tensor, $\dot{\varepsilon}_{ij}$ the infinitesimal strain rate, $\bar{\sigma}$ the effective stress, $\bar{\varepsilon}$ the effective strain, $\dot{\bar{\varepsilon}}$ the effective strain rate, $Y_{To}(\bar{\varepsilon})$ the static yield stress and γ and n material constants.

3 MIXED FORMULATION

Mixed methods are an effective way to solve constraint problems, but usually bring to the solution system of equations a significant increased number of unknowns and solution difficulties due to the diagonal zeros in the matrices of those systems. However, if appropriate iterative methods are used these disadvantages can be eliminated.

Governing equations and natural boundary conditions for the viscoplastic deformation problem

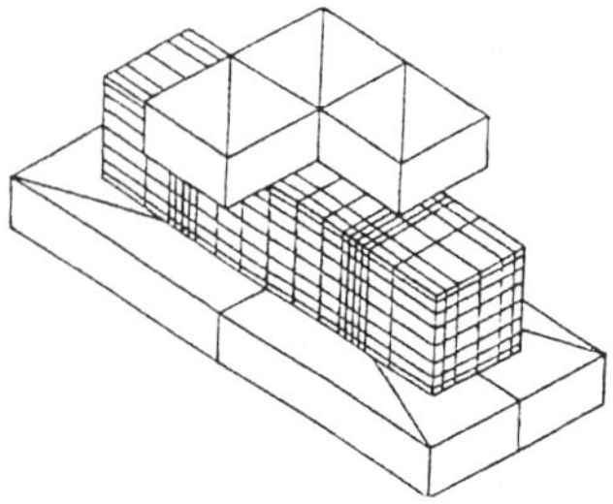

Figure 1 - 3-D forging : initial mesh and dies.

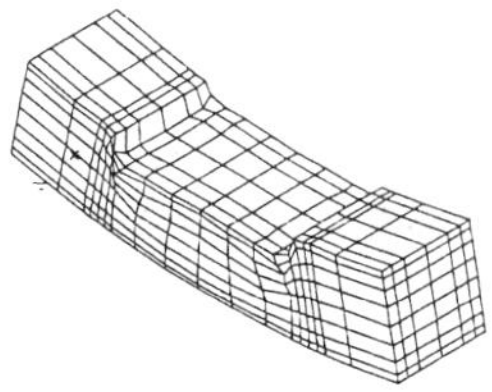

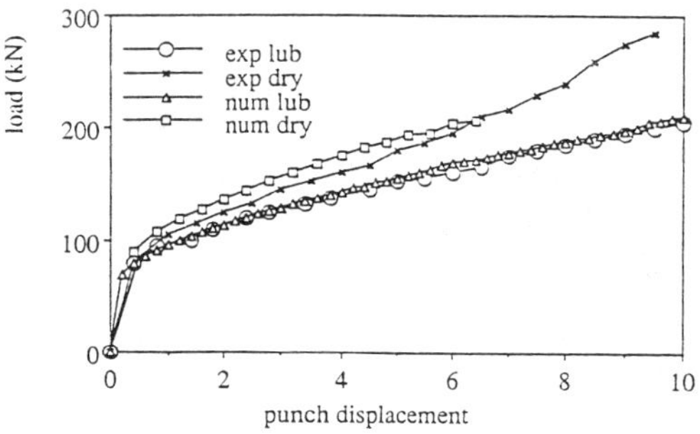

Figure 2 - 3-D : experimental and numerical results.

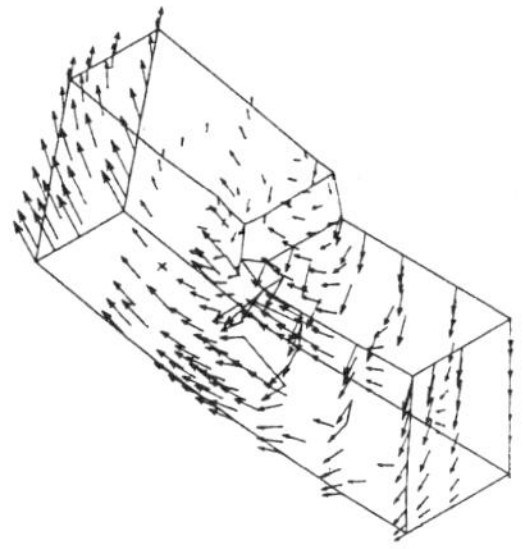

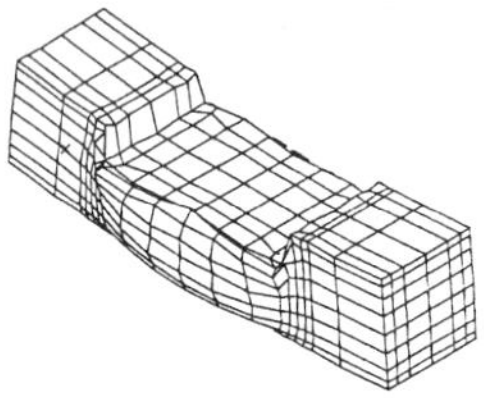

Figure 3 - 3-D forging : deformed shape, 34% height reduction, dry-friction.

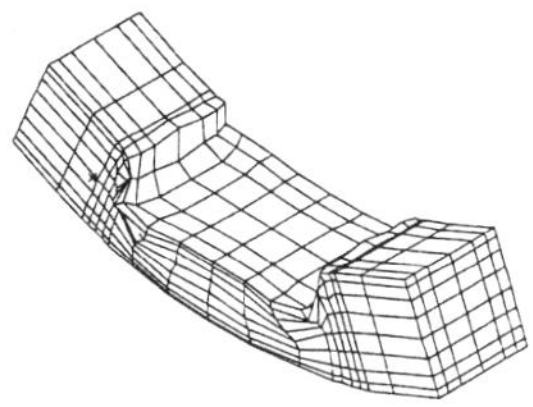

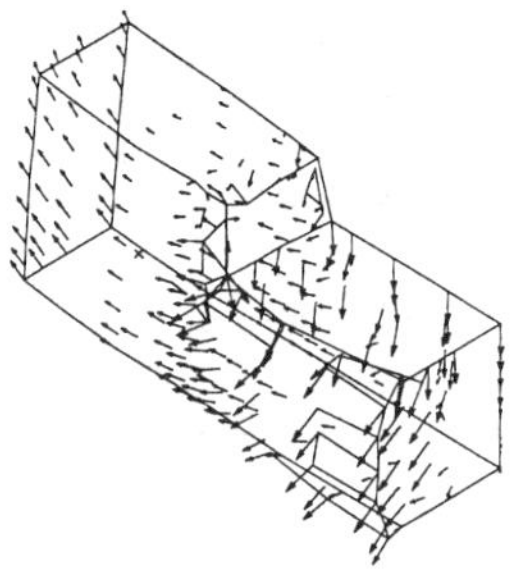

Figure 4 - 3-D forging : velocity field, 34% height reduction, dry-friction.

Figure 5 - 3-D forging : deformed shape, 25% height reduction, lubricated.

Figure 6 - 3-D forging : velocity field, 25% height reduction, lubricated.

Figure 7 - 3-D forging : deformed shape, 50% height reduction, lubricated.

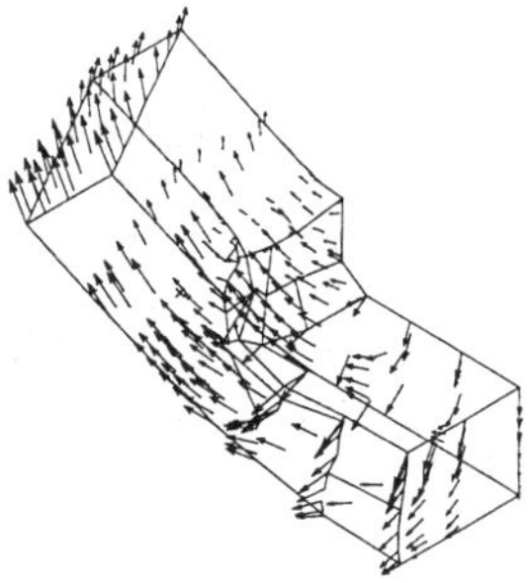

Figure 8 - 3-D forging : velocity field, 50% height reduction, lubricated.

can be evaluated from simplified variational principles derived from the principle of virtual work assuming a local linearization. Toyoshima (1985) introduced a functional whose variational principle gives a mixed formulation, having as independent variables the velocities, the pressure and a continuous field of stresses, $\hat{s}$, from which the deviatoric stresses can be evaluated:

$$\phi_\Gamma(\dot{u}_i, \hat{s}_{ij}, p) = \int_\Omega ((1+\beta)\tfrac{1}{2}\dot{u}_{i,j} D_{ijkl} \dot{u}_{j,i} + \beta\tfrac{1}{2} s_{ij} C_{ijkl} \hat{s}_{ij} - \beta\tfrac{1}{2}(\dot{u}_{i,j} + \dot{u}_{j,i})\hat{s}_{ij})\, d\Omega - \int_\Omega p\, \dot{\varepsilon}_{ii} d\Omega - \int_\Omega b_i \dot{u}_i\, d\Omega - \int_{\Gamma_o} t_i \dot{u}_i\, d\Gamma \quad (3.1)$$

being b_i and t_i body forces and surface tractions.

Stationary conditions of ϕ_Γ give for $\beta = 0$ the Euler equations of a two field solution, with the unknowns being $\dot{u}$ and p and for $\beta=-1$ the Euler equations of a three field solution, with the unknowns being $\dot{u}$,p and s, the deviatoric stresses. For different values of β the deviatoric stresses are evaluated from:

$$s_{ij} = -\beta \hat{s}_{ij} + (1+\beta) D_{ijkl} \tfrac{1}{2}(\dot{u}_{k,l} + \dot{u}_{l,k}) \quad (3.2)$$

If, for the solution of the variational problem associated with the functional ϕ_Γ, a standard finite element approximation is used, the following system of equations is obtained:

$$\begin{bmatrix} \beta Q_1 & -\beta Q_2 & 0 \\ -\beta Q_2^T & Q_3 & Q \\ 0 & Q^T & 0 \end{bmatrix} \begin{bmatrix} s \\ \dot{u} \\ p \end{bmatrix} = \begin{bmatrix} 0 \\ f \\ 0 \end{bmatrix} \quad (3.3)$$

in which

$$Q_1 = \int_\Omega N_s^T N^{-1} N_s \, d\Omega \quad (3.4)$$

$$Q_2 = \int_\Omega m^T N_s^T B \, d\Omega \quad (3.5)$$

$$Q_3 = (1+\beta)\int_\Omega B^T D\, B\, d\Omega = (1+\beta)\, K_d \quad (3.6)$$

$$Q = \int_\Omega B^T m\, N_p d\Omega \quad (3.7)$$

$$m^T = [1\,1\,1\,0\,0\,0] \quad (3.8)$$

$$f = \int_\Omega N^T b \, d\Omega + \int_{\Gamma_\sigma} N^T t \, d\Omega \quad (3.9)$$

The system of equations (3.3) can be re-written as:

$$\begin{bmatrix} \beta Q_1 & 0 & 0 \\ -\beta Q_2^T & K_d & Q \\ 0 & Q^T & -\alpha I \end{bmatrix} \begin{bmatrix} s \\ \dot{u} \\ p \end{bmatrix} = \begin{bmatrix} \beta Q_2 \dot{u} \\ f - \beta K_d \dot{u} \\ -\alpha p \end{bmatrix} \quad (3.10)$$

where α is a very small value, and can be iteratively solved by the following scheme:

$$\begin{bmatrix} \beta Q_1 & 0 & 0 \\ -\beta Q_2^T & K_d & Q \\ 0 & Q^T & -\alpha I \end{bmatrix} \begin{bmatrix} s_i \\ \dot{u}_i \\ p_i \end{bmatrix} = \begin{bmatrix} \beta\, Q_2 \dot{u}_{i-1} \\ f - \beta K_d \dot{u}_{i-1} \\ -\alpha\, p_{i-1} \end{bmatrix} \quad (3.11)$$

For each iteration we may obtain the variables by the expressions:

$$s_i = Q_1^{-1} Q_2 \dot{u}_{i-1} \quad (3.12)$$

$$u_i = \left[K_d + \frac{1}{\alpha} Q\, Q^T\right]^{-1} \left[f - Q\, p_{i-1} - \beta K_d u_{i-1} + \beta Q_2^T s_i\right] \quad (3.13)$$

$$p_i = p_{i-1} + \frac{1}{\alpha} Q\, u_i \quad (3.14)$$

4. FRICTION AND CONTACT

A very important variable during all the forming process is the friction between the workpiece and the tools, which influences the forces and the energy required for deformation.

One common and simple model for friction uses a friction factor m_f, obtained experimentally, being the mean interfacial or frictional shear stress then expressed as :

$$\tau_f = m_f \frac{Y_0}{\sqrt{3}} \quad (4.1)$$

The finite element model uses friction elements, both for 2-D and 3-D problems, with zero thickness in which the tangential stress is equated by:

$$\tau = \frac{\tau_f}{\|\Delta u_r\|} \Delta u_r \quad (4.2)$$

where Δu_r is the relative tangential velocity between the tools and the workpiece.

These elements are also used to prevent penetration of the nodes of the workpiece mesh into the elements used to discretise the dies. This is done by a penalty formulation that imposes the condition of equal nodal velocities of the tools and the workpiece at the interface.

5. ASSESSMENT OF THE CODE

The code developed has been assessed by carrying out different experimental tests, some of what have already been reported (Costa Sousa and César de Sá 1992, 1993)). Here, an experimental simulation of an open-die 3-D forging with a flat die of a prismatic block of an aluminium alloy 1050A H14, carried out by Rodrigues (1993), is presented.

The example provides interesting contact and friction effects. Two cases were tested with different friction conditions : a) with lubrication, $m_f = 0.05$, and b) with dry friction, $m_f = 0.5$. The friction factors were obtained experimentally. The finite element mesh, for half of the workpiece, is represented on Figure 1. In both cases good

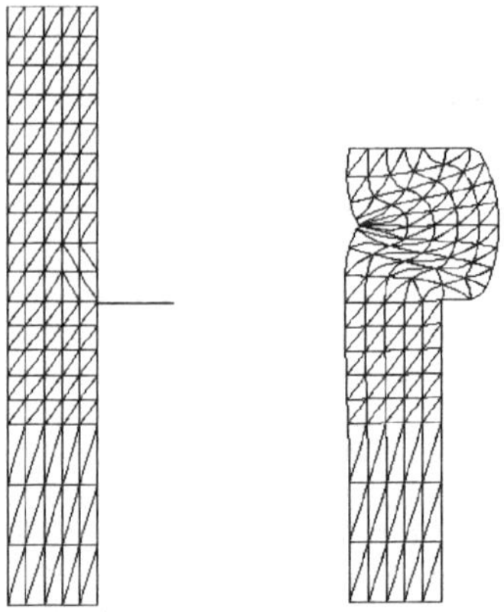

Figure 9 - Cylinder compression: initial mesh and final shape

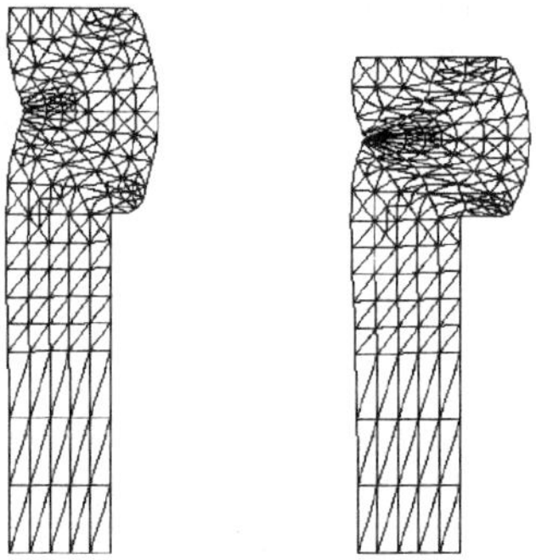

Figure 10 - Cylinder compression: element subdivision at 38% height reduction and final shape.

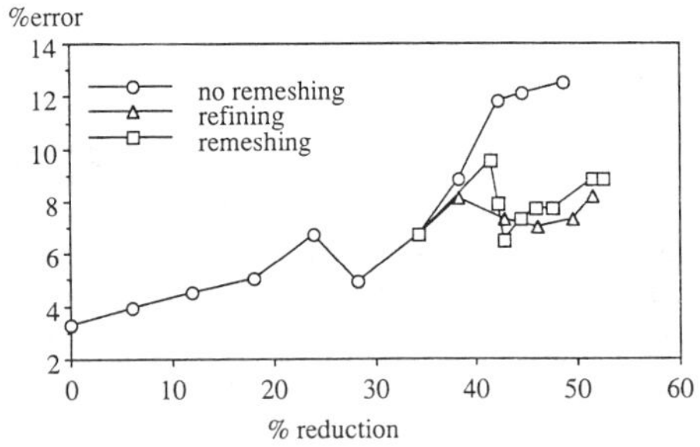

Figure 11 - Cylinder compression: error evolution

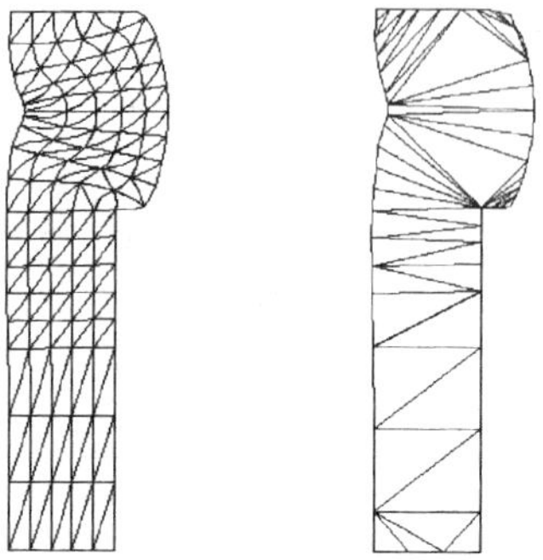

Figure 12 - Cylinder compression: boundary triangulation

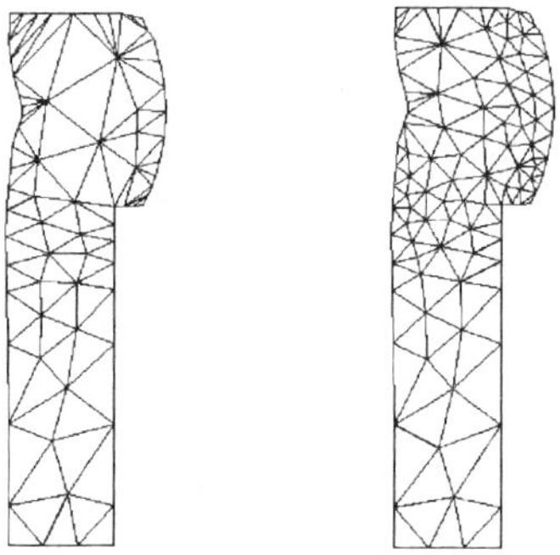

Figure 13 - Cylinder compression: intermediate triangulations.

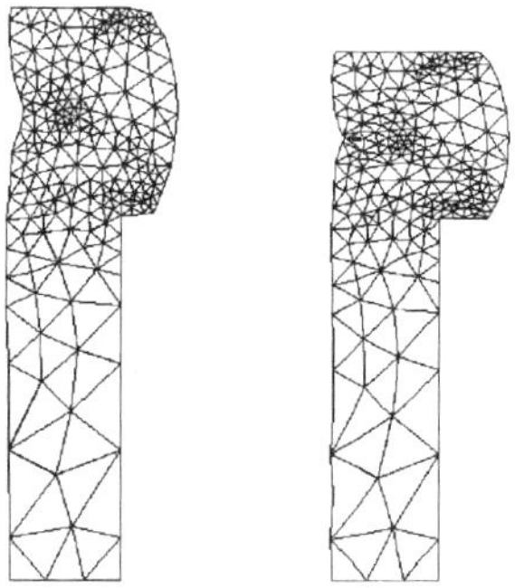

Figure 14 - Cylinder compression: adaptive mesh and final shape.

agreement was obtained between the numerical and experimental solutions, as is shown in Figure 2.

In Figure 3 it can be seen the initialisation of folded regions in the dry friction example. This effect is not present in the lubricated example which is shown in Figure 5 and Figure 7. The non deformed regions rotate and lose contact with the lower die, as was observed experimentally. In Figures 4, 6 and 8 this effect is stressed by representing the velocity field for one quarter of the workpiece.

6. ADAPTIVE MESH REFINEMENT

In the simulation of forming problems the finite element mesh can become very distorted or badly defined in certain zones due to stress and strain concentrations and extended deformations. In order to have reliable results and very often in order to proceed with the simulation an adaptive mesh refinement is needed. This requires a definition of an error estimator and a criterion to accept the new mesh.

The errors estimators used in this work are based on Zienkiewicz-Zhu error estimators (Zienkiewicz et all. 1987, 1988, 1989), due to the simplicity of implementation and reliability (Ainsworth et all. 1989).

Here an error energy norm is obtained by

$$||e|| = \left(\int_\Omega (s^* - \bar{s})\ D^{-1}(s^* - \bar{s})\ d\Omega\right)^{\frac{1}{2}} \qquad (6.1)$$

being s^* a smooth continuous field of deviatoric stresses and $\bar{s}$ a. discontinuous solution obtained from the f.e.m. solution; the "exact" total energy dissipation norm is approximated by

$$||u|| = \left(\int_\Omega \bar{s}\ D\ \dot{\varepsilon}\ d\Omega + ||e||^2\right)^{\frac{1}{2}} \qquad (6.2)$$

In the code developed, if a $\dot{u}$, p solution is used the smooth continuous field for the deviatoric stress is evaluated with a local patch recovery technique (Zienkiewicz and Zhu 1992, Wiberg and Abdulwahab 1993), based on the evaluation of the derivatives of $\dot{u}$ on their best sampling points and considering a continuous polynomial expansion on a patch, around the node where recovery is desired.

By defining an admissible energy error norm globally, a percentage η of the exact energy dissipation norm, and element-wise, imposing an equal distribution of the error by all M elements of the mesh (Oñate and Castro 1991),

$$\frac{||e||}{||u||} \leq \eta \qquad (6.3) \qquad\qquad ||e||_e^2 \leq \frac{\eta\ ||u||^2}{M} \qquad (6.4)$$

the optimal size for an element in the new mesh is obtained by:

$$h_{new} = h_{old}\left(\frac{\eta\ ||u||}{M^{\frac{1}{2}}\ ||e||_{e,old}}\right)^{\frac{1}{2}} \qquad (6.5)$$

This equation is used in the remeshing and refining techniques developed in this work, for 2-D problems and six-noded triangular elements, to define the desirable element size and control the node spacing.

Two remeshing techniques were developed : the first one simply subdivides the elements of the old mesh in more elements according to equation (6.5) ; the second one constructs a complete new mesh. In this case the algorithm begins by generating nodes on the boundary and proceeds by creating a boundary triangulation. Then new nodes are placed in the interior of the triangles trying to create a Delaunay triangulation, ideal to guarantee well shaped triangles. This is done using an interior node placement algorithm proposed by Frey (1987). The previous triangulation is then changed, by removing the original triangle and its neighbours, and new triangulations are performed until the element size requirements are met. In the end the mesh is smoothed by relocating some nodes, without changing the connectivities, by the Laplacian smoothing technique (Cavendish 1974, Sezer and Zeid 1991).

An example of the compression of an aluminium alloy cylinder, previously reported (Costa Sousa and César de Sá 1992), was used to assess the improvement of the adaptive mesh refinement in the numerical solution.

In Figure 9 the initial mesh of half of the cylinder is represented and the final shape attained with this mesh at 49% of height reduction of the cylinder free zone. In Figure 10 is shown the refinement of the mesh, operated at 38% height reduction, by element subdivision and the final shape obtained at 52% height reduction. With this refinement technique the error can be considerable diminished as it is shown in Figure 11.

The next figures illustrate the different stages of the remeshing technique used. In Figure 12 the original mesh, at 38% height reduction, is shown. At this stage the remeshing technique begins by a boundary triangulation. In Figure 13 two intermediate stages are shown, in which Delaunay triangulations were performed but with the optimal element sizes not attained. In Figure 14 it can be seen the optimal mesh after a node relocation smoothing and the final compression stage performed with this mesh, a 53% height reduction. The error evolution for this case is also presented in Figure 11. The initial higher value for the error may be caused by the transfer of the variable values from the old to the new mesh but it rapidly decreases to a similar value of the previous technique and the simulation can proceed a little further in this case.

7. CONCLUDING REMARKS

Mixed finite element methods can be very effective in the simulation of forging problems if adequate iterative methods are used to avoid the disadvantage of increasing the solution variables.

Adaptive mesh refinements techniques can be important tools in reducing numerical errors, associated with mesh distortion and coarseness, provide that reliable error estimators and effective mesh redefinition algorithms are used.

ACKNOWLEDGEMENTS

The support provided by Junta Nacional de Investigação Científica, Portugal, through the project PBIC/C/TPR/1288/1992 is gratefully acknowledged.

We also would like to thank the Instituto Superior Técnico, Secção de Tecnologia Mecânica, Lisbon, Portugal, for the experimental facilities provided.

REFERENCES

Ainsworth, M., Zhu. J. Z., Craig, A. W. & Zienkiewicz, O.C. 1989, Analysis of the Zienkiewicz-Zhu a-posteriori error estimator in the finite element method, Int. J. Num. Methods Eng., 28, 2161-2174.

Cavendish,J.C. 1974, Automatic triangulation of arbitrary planar domains for the finite element method, Int. J. Num. Methods Eng., 8, 679-696.

Costa Sousa, L. & César de Sá, J. 1992, Finite element modelling of thermo-mechanical forming processes using a mixed method, In: J. L. Chenot et al., editors, *Numerical Methods in Industrial Forming Processes - Numiform'92*, 229-236, A. A. Balkema

Costa Sousa, L. & César de Sá, J.M. 1993, Finite Element Simulation of Forming Processes, In F. Navarrina & M.Casteleiro, editors, *Métodos Numéricos en Ingeniería*, 1490-1499, SEMNI, Sociedad Española de Métodos Numéricos en Inginiería.

Frey, W.H.. 1987, Selective refinement: a new strategy for automatic node placement in graded triangular meshes, Int. J. Numer. Methods Eng., 24, 2183-2200.

Oñate, E. & CASTRO, J. 1991, Adaptive mesh refinement techniques for structural problems, In: E. Oñate et all, editors, *The Finite Element Method in the 90's. A book dedicated to O. C. Zienkiewicz*, CIMNE-Barcelona, Springer-Verlag.

Rebelo, N. & Kobayashi, S. 1980, A coupled analysis of viscoplastic deformation and heat transfer - I - Theoretical considerations, Int. J. Mech. Sci., 22, 699-705.

Rodrigues, J.M.C. 1993, Theoretical and experimental analysis of cold forming, Ph.D. Thesis , Instituto Superior Técnico, Lisbon, Portugal.

Sezer, L. and Zeid, I 1991, Automatic quadrilateral/triangular free-form mesh generation for planar regions, Int. J. Num. Methods Eng., 32, 1441-1483.

Toyoshima, S. 1987, Iterative mixed methods and their application to the analysis of metal forming processes, Ph. D. Thesis, University of Wales, Swansea.

Wiberg, N. E. & Abdulwahab, F. 1993 Patch recovery based on superconvergent derivatives and equilibrium, Int. J. Numer. Methods, Eng., 36, 2703-2724.

Zienkiewicz,O.C. & Zhu, J.Z. 1987, A simple error estimator and adaptive procedure for practical engineering analysis, Int. J. Numer. Methods Eng., 24, 337-357.

Zienkiewicz, O.C., Liu, Y. C. & Huang, G. C. 1988, Error estimation and adaptivity in flow formulation for forming problems, Int. J. Numer. Methods Eng., 25, 23-42.

Zienkiewicz, O.C., Liu, Y. C. & Huang, G. C. 1989, Error estimation and convergence rates for various incompressible elements, Int. J. Numer. Methods Eng., 28, 2191-2202.

Zienkiewicz, O.C. & ZHU, J.Z. 1992, The superconvergent patch recovery and a posteriori error estimates. Part1: The recovery technique, Int. J. Numer. Methods Eng., 33, 1331-1364.

Zienkiewicz, O.C. & ZHU, J.Z. 1992, The superconvergent patch recovery and a posteriori error estimates. Part 2: Error estimates and adaptivity, Int. J. Numer. Methods Eng., 33, 1365-1382.

Simulation of Materials Processing: Theory, Methods and Applications, Shen & Dawson (eds)
© 1995 Balkema, Rotterdam. ISBN 90 5410 553 4

Modelling and simulation of grinding process by finite element method

C.F.Cheng & M.M.F.Yuen
Department of Mechanical Engineering, The Hong Kong University of Science and Technology, Clear Water Bay, Kowloon, Hong Kong

ABSTRACT: This paper describes the modelling and simulation of surface grinding process using 2-D plane strain finite element analysis and material plastic deformation criteria. The finite element analysis result is verified against experimental data.

1. INTRODUCTION

Grinding is used in finish machining of precision components requiring good surface and fine tolerance. Different models have been applied to the grinding process: topography model, chip thickness model, grinding force model, grinding energy Model, temperature model, and surface roughness model as reviewed by Tonshoff [1]. However, they are mostly empirical models established from the grinding tests.

Many researchers also tried to derive mathematical formulation from basic physical principles, finite element method had been applied in the cutting process analysis. Riddle [2] discussed the effect of the failure criterion in the numerical modelling of orthogonal metal cutting, and used the effective plastic strain as the material failure criterion. While Shih and Yang [3] presented the application of a finite element analysis in the simulation of plane strain orthogonal metal cutting process with continuous chip formation. In this work, the separation criterion was based on the distance between the tool tip and the nodal point connecting the two elements ahead of the cutting tool. Komvopoulos [4] emphasised the effect of factors, such as plastic flow of the workpiece material, friction at the tool-workpiece interface, and wear of the tool, on the cutting process. Wayne [5] developed an expert system for design, analysis, and production of cutting tools by using three-dimensional finite element analysis. The separation formulation used was derived from a fracture mechanics-based debonding finite element located between the chip and workpiece. The debonding element provided a mechanism for gradual release of the chip from the workpiece material. However, all the orthogonal metal cutting models were focused on single-point cutting, and are not applicable to the multiple-point cutting of the grinding process.

In this paper, we developed the process models and simulated the grinding process by the finite element method. The following parameters were reviewed: cutting force, stress and strain distribution. The cutting force was verified by experimental data and the reasonable agreement has been obtained.

2. PROBLEM FORMULATION

2.1 Modelling Assumption

In this paper, we focused on the surface grinding without cross-feed. The following assumptions were used in the model:

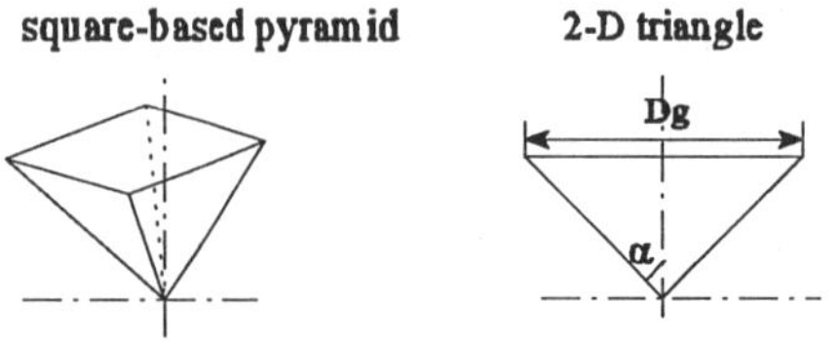

(Two factors of grit model: Grit Size Dg and Rank Angle α)

Fig 1. Wheel Grit Shape

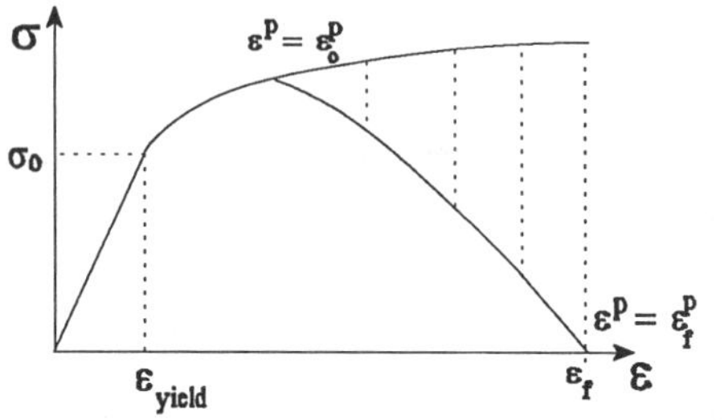

σ_0 — yield stress ε_{yield} — strain at yield point
ε^p — equivalent plastic strain
ε_f — strain at plastic failure point
$\varepsilon^p{}_0$ — equivalent plastic strain at start of failure
$\varepsilon^p{}_f$ — equivalent plastic strain at failure

Fig 2. Stress - Strain curve using the ductile failure model.

- The grinding machine structure is a rigid body, there is no deflection between the wheel centre and the machine bed;
- The grits on the grinding wheel surface are modelled as uniformly distributed multiple cutters;
- Each grit in the cutting zone are subjected to a vertical and a horizontal force;
- The workpiece material has the same yield stress in tension and compression.
- The workpiece can be modelled by a 2-D plane strain model.

2.2 Grinding Wheel Topography Modelling

The properties and performance of the grinding wheel depend on the type of abrasive grain material, the grit size, the bond material, the properties of abrasive and bond, and the porosity [6]. Two factors are considered in the grinding wheel topography modelling: the spatial distribution of abrasive grits over the wheel surface and their morphology.

In this study, we assumed that the abrasive grits are uniform square-based pyramid, and for 2-D problems, the grits take the shape of triangles as shown in Fig 1. A grit size *Dg* of 0.33 mm and a rank angle α of 60°was selected base on the properties of abrasives provided by Salmon [7]. The wheel is modelled by connecting the triangular grits, and is represented by a 2-D mesh with rigid body movement only.

2.3 Finite Element Modelling

Since the finite element mesh of the workpiece undergoes large plastic deformation and rigid body motion, the equilibrium equation of finite element formulation can be expressed as:

$$K\Delta U = F$$

where, K is the summation of standard stiffness matrix and the initial strain stiffness matrix; ΔU is the vector of incremental displacement corrections; F is the vector of applied loads.

To investigate the cutting performance of the grinding process, we have interest in the Mises equivalent stress and equivalent plastic strain.

The Mises equivalent stress is defined as

$$q = \sqrt{\tfrac{3}{2} S_{ij} S_{ij}}$$

where, S_{ij} is the deviator stress, defined as

$$S_{ij} = \sigma_{ij} + p\delta_{ij}$$

where, σ_{ij} is the stress and the p is the pressure stress:

$$p = -\tfrac{1}{3}\sigma_{ii}$$

and δ_{ij} is the Kronecker delta.

The equivalent plastic strain is the total accumulation of plastic strain used in isotropic hardening plasticity theory.

$$\varepsilon^p = \int \sqrt{\tfrac{2}{3} d\varepsilon_{ij} d\varepsilon_{ij}}$$

where, ε_{ij} is the plastic strain.

The horizontal (Ux) and Vertical (Uy) direction displacement at the base of the workpiece are constrained, as the workpiece is mounted on the machine bed. The cutting forces, including a horizontal force *Fh* and a vertical force *Fv*, are obtained by the nodal reaction forces, expressed by:

$$F_h = \sum_{n=1}^{N} RF1_n \qquad F_v = \sum_{n=1}^{N} RF2_n$$

where, $RF1_n$ and $RF2_n$ are the reaction forces of node n in the horizontal and vertical direction respectively; and N is the total number of nodes that located at the bottom of the mesh.

2.4 Material Separation Criteria

It is very important to formulate the material separation criterion in the simulation so that elements in the cutting zone may be removed. The separation criteria that we adopted is the equivalent plastic strain failure mode. The model is based on the material property and unlike the distance tolerance criteria [3], [4] it is difficult to formulate the relation between the distance tolerance and the material properties.

The equivalent plastic deformation criteria can be described by the following rules (Fig 2):

If $\varepsilon = \varepsilon_{yield}$,
then the material point starts to yield.

If $\varepsilon^P = \varepsilon_0^p$,
then the material point starts to be fail.

If $\varepsilon^p \geq \varepsilon_f^p$ or $\varepsilon \geq \varepsilon_f$,
then the material totally fails.

If all the nodal points in the element fail,
then the element is removed from the mesh.

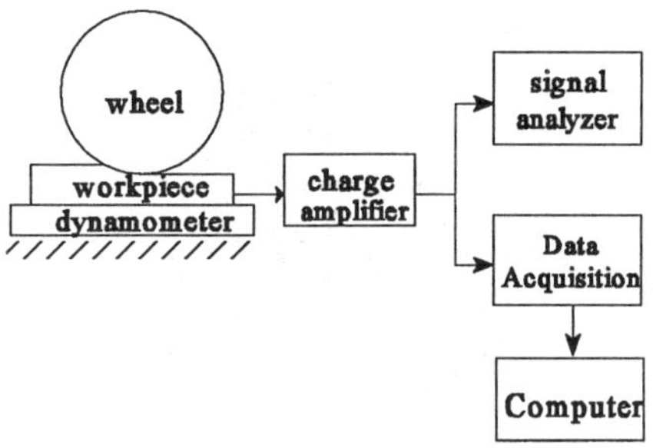

Fig 3. Experimental Setup

Table 1. Cutting Condition of Grinding Process

Grinding Machine	818ASD Surface Grinding Machine
Grinding Type	Surface Plunge Grinding
Wheel Type	RA46J8V 205x13x31.85
Wheel Diameter D	205 mm
Wheel Width B	13 mm
Wheel Speed V	2850 rpm 30m/s
Workpiece Material	S50C
Workpiece Thickness b	10 mm
Table traverse rate	0.24m/s
Depth of Cut d	20 um
Coolant Type	HYSOL X
Coolant Flow Rate	5.85 L/min

Table 2. Material Specification of workpiece S50C

Modulus of Elasticity E	193.5 GPa
Modulus of Rigidity G	79.3 GPa
Poisson's ratio r	0.293
Yield strength σ_{yield}	387 MPa
Density ρ	7.81 Mg/m^3

3. SIMULATION RESULT AND EXPERIMENTAL VERIFICATION

ABAQUS/Explicit was used for simulating the grinding process. The result of the finite element model and simulation was compared against experimental data. The layout of the experiment is figured in Fig 3. The cutting forces, horizontal force *Fh* and vertical force *Fv*, were measured using a 3-component dynamometer. The detail grinding condition is tabulated as Table 1. And the Table 2

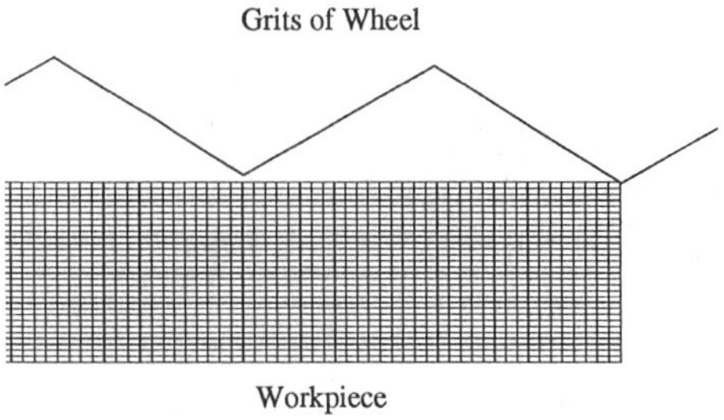

Fig. 4: Initial finite element mesh

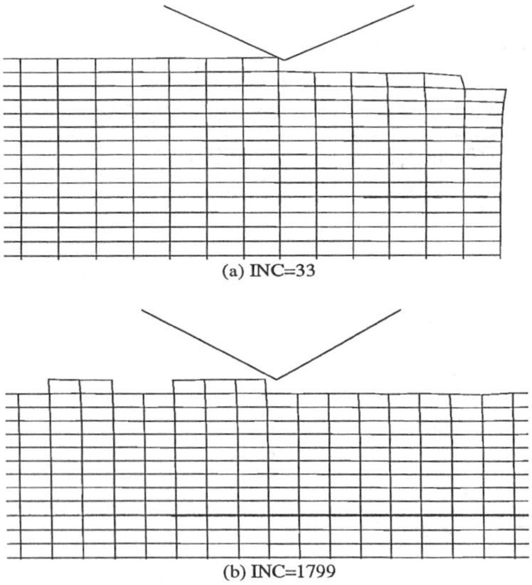

Fig 5. Deformation plot at: (a) INC=33; (b) INC=1799

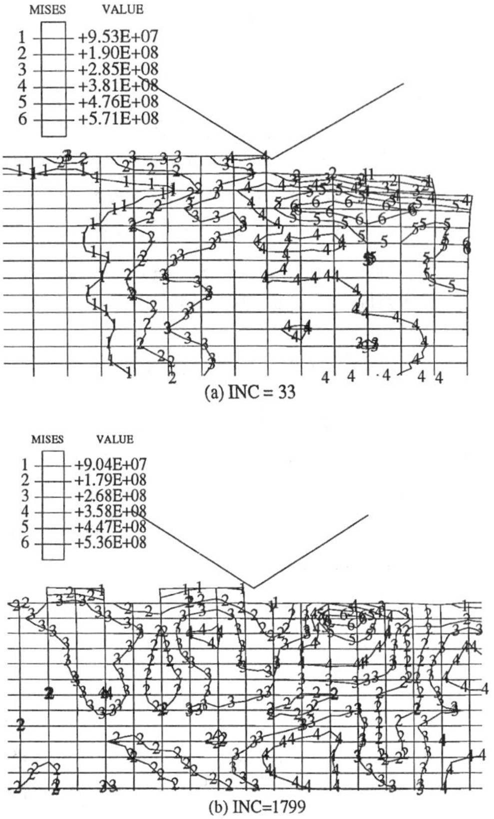

Fig 6. Von Mises stress contour at: (a) INC=33; (b) INC=1799

describe the characteristics of the workpiece material (S50C) specification at room temperature.

The initial finite element mesh and tool geometry was shown in Fig 4. The element type chosen for the workpiece is a 2-D plane strain element CPE4R. The workpiece model covered a section of 3mmx0.1mm, which enabled the simulation of the cutting action in the contact zone with a reasonable CPU time. The element size is 10um*5um with a depth of cut at 20um, so that the top 4 layers' elements are expected to be removed. While in the simulation process, the critical value of the equivalent plastic strain was chosen to be 0.2.

The deflection, Mises stress, and equivalent plastic strain states are given in Fig 5-7 at different increments: (a) increment number INC=33, just one grit in contact with the workpiece; (b) increment number INC=1799, 6 grits in contact with the workpiece. The removed elements are not displayed in the plots.

The measured cutting forces had an average value of 5.2N per mm width of cut in horizontal direction, and had an average value of 10.41N per mm width of cut in vertical direction. The finite element simulation result and the experimentally measured cutting forces were shown in Fig 8, and there is a general agreement in terms of magnitude.

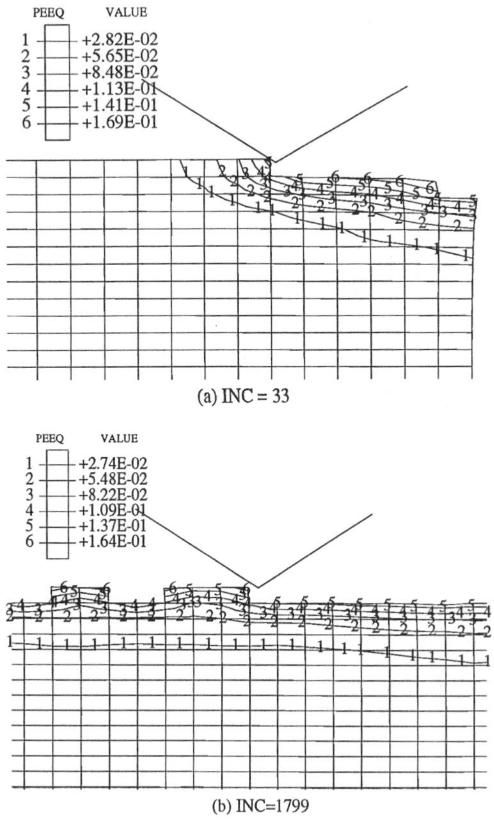

Fig 7. Contour of equivalent plastic strain at (a) INC=33; (b) INC=1799

4. DISCUSSION

Finite element method had been applied in the modelling and simulation of the surface grinding process. The following observations have been made:

(1) The von Mises stress distribution demonstrates that relatively higher stress arises in the region under the contacted grits, resulting in residual stress in the upper part of the workpiece. The stress fluctuation insider the workpiece can be related to the separation of the removed elements.

(2) From the plot of equivalent plastic strain, the plastic flow at the top of the ground workpiece is higher in the region adjacent to the removed material.

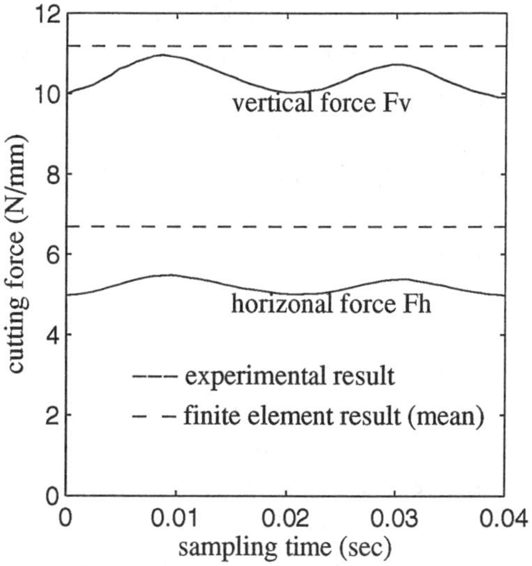

Fig 8. Cutting Forces: finite element model result vs experimental result.

This demonstrates that the separation criteria can be used as a good basis for modelling the grinding process.

(3) The model is a preliminary study on the feasibility of modelling the grinding process using the plastic strain failure criteria. The grinding forces are found be within a reasonable range. However , there are other factors such as wheel elasticity which need to be included to make the model realistic.

5. REFERENCE

[1] H.K. Tonshoff, J. Peters, I. Inasaki and T. Paul, "Modelling and Simulation of Grinding Process", Annals of the CIRP, Vol. 41, No. 2, 1992, pp677-688.

[2] Robert A. Riddle, "The Effect of The Failure Criterion in The Numerical Modelling of Orthogonal Metal Cutting", Research Report of Lawrence Livermore National Laboratory, April 1987.

[3] Albert J. M. Shih, S. Chandrasekar and Henry T. Y. Yang, "Finite Element Simulation of Metal Cutting Process With Strain-Rate and Temperature Effects", Fundamental issues in machining, ASME PED, Vol. 43, 1990, pp11-24.

[4] K. Komvopoulos and S. A. Erpenbeck, Finite Element Modelling of Orthogonal metal Cutting, Transactions of the ASME, Journal of Engineering for Industry, Vol. 113, August 1991, pp253-267.

[5] Steven F. Wayne, David A. O'Neil and Charles Zimmerman, "An Expert System for Design, Analysis and Production of Cutting Tools", Advances in Hard Material Production, 1992, pp1-15.

[6] S. Malkin, Grinding Technology: Theory and Applications of Machining with Abrasives. Ellis Horwood Limited, 1989.

[7] Stuart C. Salmon, Modern Grinding Process Technology, McGraw-Hill Inc., 1992.

[8] Hibbitt, Karlsson & Sorensen, Inc., ABAQUS Explicit User's Manual, Version 5.3, 1993.

Simulation of Materials Processing: Theory, Methods and Applications, Shen & Dawson (eds)
© 1995 Balkema, Rotterdam. ISBN 90 5410 553 4

Numerical simulation of hot powder forging of connecting rod

Ashoka G.K. Jinka, Lionel Fourment & Michel Bellet
Centre de Mise en Forme des Matériaux (CEMEF), UA CNRS, École Nationale Supérieure des Mines de Paris (ENSMP), Sophia-Antipolis, France

ABSTRACT : This paper presents a 3D finite element model of the thermomechanical phenomena occuring during the hot powder forging of connecting rod. The paper briefly reports a new development of three field variable formulation for dealing with mechanical calculations. The three field variables considered are the velocity, volumetric strain rate and Lagrangian multiplier/pressure. From the deformation point of view the original compressible material transforms into incompressible domain as the forging process advances. To cater for this phenomenon the third field variable switches from Lagrangian mutipler to pressure in the analysis. The material model considered is of modified von Mises type with a built-in pressure sensitivity and a modified viscoplastic type of friction law has been used. A transient thermal model has been developed and which is solved using implicit time integration scheme. The coupling between the heat transfer analysis and the mechanical resolution has been achieved in a staggard manner. $P_2P_1P_1$ 10-node tetrahedral finite elements have been used for numerical analysis. Utilizing the model developed here, we have simulated the hot forging of connecting rod.

1 INTRODUCTION

The emerging interest in automotive industry for structural components resulting from forming of porous materials either directly from powders which are obtained by automization process or from sintered preforms is becoming popular route of manufacturing industrial artifacts. Owing to the fact that the powder metallurgy (P/M) forming processes enable savings in energy and material in addition to quality improvement particularly in the automotive industry the weight reduction will directly result in favourable operative condition of the engine in terms of reduced exhaust gas emission and vibration. This translates into less fuel consumption in the operative life of automobiles. In this context connecting rod being one of the main moving parts of the engine, any weight reduction will lead to lighter oscillating masses which result in favourable influence on running of the engine. However, because the connecting rod is a bulk produced component the economic realisation of the P/M forming route is in highly competition with conventional processes such as die casting or forging. In order to help in the process of meeting the economic needs and further to understand the material behaviour under P/M forming process conditions the present study envisages to model the hot forging of a connecting rod by developing a non-isothermal finite element model.

The numerical simulation of the hot forging process aims at the prediction of relative density distribution and temperature field. Heither to numerical simulation of porous preforms were largly confine to two-dimensional parts (Jinka 1992) and three-dimensional study of non-dense metal forming is still a rarity. However, a so called velocity based approach (Chenot et al. 1990) was adopted to study the three-dimensional isothermal forming of porous connecting rods. They utilized a pressure modified visco-plastic von Mises yielding with a modified viscoplastic friction law. Although the obtained results demonstrated the success of the finite element approach for analysis of 3-D porous parts, the model predictions were plagued with problems of contact & kinematic tool movement. A point of further difficulty arises when the porous domain gets densified it essentially behaves like an incompressible solid and during this stage of deformation their model transformed into a penalty method not satisfying Babuska-Brezzi (BB) condition.

After an initial survey of knowledge base and understanding of the deformation behaviour of non-dense material, several issues need to be addressed in the modelling of porous metal forming, following points are worth noting in this direction:

1. The computational model should be capable of predicting accurate density distributions in the domain.
2. The initial non-dense material transforms into dense incompressible solid as the forging process advances. This element of change over in the behaviour of material should be considered while developing suitable models for numerical analysis. It is futher noted that the numerical modeling of incompressible solids prone to problems.
3. The plastic yielding of the porous material has been known to be a function of both deviatoric stresses and pressure. A suitable material model to be used should

take care of this material behaviour.
4. The frictional phenomena play an important role in the success of the forged components. To consider this a suitable friction model is required.

Taking into account these various aspects, the present study reports the development of a numerical model for the study of densification process during hot powder forging. We have developed a new three variable finite element model based on a perturbed Lagrangian formulation augmented with a weak form of dilatational deformation. Utilizing a modified von Mises type visco-plastic yield criterion, the model has velocity, Lagrangian multiplier/pressure and volumetric strain rate as three fundamental field variables. For discretization purpose, we have used 3D tetrahedral elements with velocity variables interpolated quadratically whereas Lagrangian multiplier/pressure and volumetric strain rate variables are interpolated linearly. A non-linear visco-plastic friction model has been utilized to chacterise the friction between the powder domain and the die wall. A powder based non-isothermal transient heat model has been developed with thermal conductivity characterized as a function of temperature and relative density. Also in the thermal model, heat generation due to plastic work has been incorporated. The solution of non-linear set of heat equations has been achieved by a three level (two step) time integration scheme. A thermo-mechanical coupling procedure is required to couple the mechanical and thermal calculations and in the present study this has been achieved. Although the developed model is quite general in nature but presently, the application is directed to simulation of hot forging of sintered non-dense aluminium connecting rod preforms. The obtained results are discussed in terms of density profile plots and temperature distribution plots.

2 MECHANICAL ANALYSIS

2.1 *Isotropic compressible material model*

Traditionally, the dense metallic material is considered to be homogenous and isotropic. The plastic yielding characteristics can be expressed in a criterion such as,

$$f(\sigma,\sigma_0) = \bar{\sigma} - \sigma_0 = 0 \qquad (1)$$

where σ_o is the yield stress of fully dense metal and $\bar{\sigma}$ the equivalent von Mises stress having the following definition:

$$\bar{\sigma} = \sqrt{\frac{3}{2}\mathbf{S}:\mathbf{S}} \qquad (2)$$

where the deviatoric stress tensor **S** associated with the Cauchy stress tensor σ can be given as:

$$\mathbf{S} = \sigma + p_m\mathbf{I} \qquad (3)$$

where p_m represents the mean stress or pressure as:

$$p_m = -\frac{1}{3}\,\mathrm{Trace}(\sigma) \qquad (4)$$

The essential characteristics of such a yield criterion is that the plastic yielding is independent from pressure and dependent only on the shear effects. For a non-dense material this hypothesis is not valid since the porous material has been known to be sensitive to both shear and volumetric deformation (Jinka 1992). The adaptation of this observation necessitates an incorporation of pressure sensitivity in the yielding hypothesis. Many investigators have proposed differing material models, predominate among them were Green (1972), Shima and Oyane (1976). Based on these observation, Abouaf (1985) proposed an adaptation of equations (1) & (2) for the deformation analysis of porous material, basically providing a modified expression which can be expressed as,

$$\bar{\sigma}^2 = 3c\,J_2 + f\;J_1^2 \qquad (5)$$

where material parameters c and f are to be determined via experiments as a function of relative density. Also in equation (5) the first invariant of stress tensor σ represented by J_1 and whereas J_2 represents the second invariant of the deviatoric stress tensor **S** with definitions, such as,

$$J_1 = \mathrm{Trace}(\sigma)$$

$$J_2 = \frac{1}{2}\mathbf{S}:\mathbf{S}$$

$$\mathbf{S} = \sigma - \frac{1}{3}J_1\mathbf{I} \qquad (6)$$

The yield function of the form given by equations (1) & (5) brings in the explicit influence of the hydrostatic pressue in addition to the deviatoric stress component and essentially considered as an extension of von Mises`s concept of yielding of fully dense materials to non-dense materials such as powder material. Schematically such a yield function is ellipsoidal in shape with relative density as single variable of the yield function.

The aforementioned concept of yielding can be expressed in the form of Norton-Hoff equation as applied to compressible porous material deformation (Abouaf 1985) by the viscoplastic potential φ, as:

$$\varphi(\dot{\varepsilon}) = \frac{K}{m+1}(\sqrt{3}\,\rho\dot{\bar{\varepsilon}})^{m+1} \qquad (7)$$

where ρ is the relative density, K is the consistency, m represents the strain rate sensitivity index of the dense material and the equivalent strain rate $\dot{\bar{\varepsilon}}$ is expressed as:

$$\dot{\bar{\varepsilon}}^2 = \frac{1}{\rho^2}\left[\frac{2}{3}\left(\frac{1}{c}\dot{\varepsilon}:\dot{\varepsilon} + \left(\frac{1}{6f} - \frac{1}{3c}\right)\mathrm{Trace}(\dot{\varepsilon})^2\right)\right] \qquad (8)$$

In the above definition of equivalent strain rate we note here that the Trace($\dot{\varepsilon}$) is equal to volumetric strain rate ($\dot{\theta}$), so that the viscoplastic potentional (i.e., equation (7)) is a function of velocity and volumetric strain rate only.

Using the viscoplastic potential φ, the stress tensor can be obtained as :

$$\sigma = \frac{\partial\varphi}{\partial\dot{\varepsilon}} = \frac{\partial\varphi}{\partial\dot{\bar{\varepsilon}}}\frac{\partial\dot{\bar{\varepsilon}}}{\partial\dot{\varepsilon}} \qquad (9)$$

or,

$$\sigma = 3K\left(\sqrt{3}\rho\dot{\bar{\varepsilon}}\right)^{m-1}\left[\frac{2}{3c}\dot{\mathbf{e}} + \frac{1}{9f}\dot{\theta}\,\mathbf{I}\right] \qquad (10)$$

where $\dot{\mathbf{e}}$ represents the deviatoric strain rate tensor.

In the present study the strength (as represented by consistency K) of the material caters both for the strain hardening and thermal softening effects of powder deformation behaviour by considering that

$$K = K_0(\bar{\varepsilon}_0 + \bar{\varepsilon})^n exp\left\{\frac{\beta}{T}\right\} \qquad (11)$$

where the equivalent strain is defined by

$$\bar{\varepsilon} = \int \dot{\bar{\varepsilon}}\, dt \qquad (12)$$

and n represents the work hardening coefficient, T the temperature and β is a constant and defined as the ratio of activation energy for volume diffusion Q_V and the gas constant R,

$$\beta = \frac{Q_v}{R} \qquad (13)$$

2.2 *Modified visco-plastic friction model*

In the deformation of powder two types of friction occur : the friction between the deforming powder and the die and the inter-grain friction. The friction between the deforming powder body and the die wall is taken into account by considering a non-linear visco-plastic law which can be expressed as a friction potential as:

$$\varphi_{fr} = \frac{\alpha K}{p_f+1}\left\|\frac{\mathbf{v}_g}{c^{1/2}}\right\|^{p_f+1} \qquad (14)$$

where $\mathbf{v}_g$ denotes the relative velocity, p_f is the frictional sensitivity parameter and α is the friction factor.

2.3 *A new three field varible finite element model*

From the deformation point of view, the powder domain is compressible at the begining of the process, however as the forming process progresses the material gets densified and tends to become incompressible. Due to this it is necessary to device a method to take care of this fact while formulating a numerical procedure. In other words one needs to have a computational model which is valid right from compressible to incompressible material behaviour. It is also well known that the finite element modelling of incompressible material has been a problem if one is not careful about the choice of elements and algorithms used in the analysis. In present work we have utilized the *Hu-Washizu* principle for casting the finite element model satisfying BB condition.

Considering these issues we have utilized an augmented Lagrangian formulation of the following type:

$$L(\mathbf{v},\dot{\theta},p) = \tilde{\Phi}(\mathbf{v},\dot{\theta}) + \int_\Omega p\left(\dot{\theta} - \text{Trace}(\dot{\varepsilon}(\mathbf{v}))\right) dV \qquad (15)$$

where we have considered the material deformation as an independent additive behaviour into total deformation part and dilatation part, proposed by equation (8). Also the Lagrangian term on R.H.S. of equation (15) provides a constraint which imposes the weak form of volumetric strain rate relation in the discrete field. In equation (15), $\tilde{\Phi}$ is given by,

$$\tilde{\Phi}(\mathbf{v},\dot{\theta}) = \int_\Omega \varphi\, dV + \int_{\partial\Omega_c} \varphi_{fr}\, dS - \int_{\partial\Omega_T} \mathbf{T}^d.\mathbf{v}\, dS$$

By virtue of equations (9) and (15), the above equation can be written as:

$$\tilde{\Phi}(\mathbf{v},\dot{\theta}) = \int_\Omega \frac{K}{m+1}(\sqrt{3})^{m+1}\left[\frac{2}{3c}\dot{\varepsilon}:\dot{\varepsilon} + \left(\frac{1}{9f} - \frac{2}{9c}\right)\dot{\theta}^2\right]^{\frac{m+1}{2}} dV + \int_{\partial\Omega_f} \frac{\alpha K}{p_f+1}\left\|\frac{\mathbf{v}_g}{c^{1/2}}\right\|^{p_f+1} dS - \int_{\partial\Omega_T} \mathbf{T}^d.\mathbf{v}\, dS$$

It is of interest to note that a Lagrangian (second term on the R.H.S. of equation (15)) of this type has been proposed by Simo et al. (1985) and Coupez (1993). The model due to Simo et al.(1993) has been developed in two dimensions with discontinous interpolation for volumetric strain rate and pressure. In fact the Simo et al. model recommended a priori kinematic splitting of total strain tensor into volumetric part and deviatoric part for use in the incompressible material deformation. Further this development has resulted in discontinous approximations for auxiliary variables leading to so called *`B-bar procedure`*. The functional as represented by equation (15) can be viewed as a form of the *Hu-Washizu* principle. However, in the present formulation it has to be noted that the Lagrangian multiplier p in compressible deformations transforms into pressure (i.e., mean stress) in the case of incompressible deformations, further leading to a different procedure compared to that of Simo et al. (1985). The question of distinction between the Lagrangian multiplier and the pressure is evident from the definition of σ (equation (10)), the pressure can be expressed as:

$$p_m = 3K\,(\sqrt{3}\rho\dot{\bar{\varepsilon}})^{m-1}\left(\frac{2}{3c}\right)\text{Trace}(\dot{\varepsilon}) + p \qquad (16)$$

where the expression for Lagrangian multiplier (p) can be obtained by the resolution of equilibrium equation associated with p, as:

$$p = -\,3K\,(\sqrt{3}\rho\dot{\bar{\varepsilon}})^{m-1}\left(\frac{1}{9f} - \frac{2}{9c}\right)\dot{\theta} \qquad (17)$$

The expression for principal of virtual work (PVW) can be derived by first derivative of the above Lagrangian framework and the equilibrium equation is a non-linear equation in velocity, volumetric strain rate and Lagrangian multiplier/pressure which needs to be solved, after discretization into finite elements (in the present investigations we have used 10 noded tetrahedral 3D elements) by considering quadratic interpolations for velocity and whereas for other two variables are interpolated linearly and such an element is denoted by $P_2P_1P_1$ quadratic terahedral 3D element, as shown in Figure 1. The discret linear equations have been solved by adopting the full Newton-Raphson procedures. A detailed discription of the development of the three field finite element model is given in a separate publication (Jinka et al. 1995).

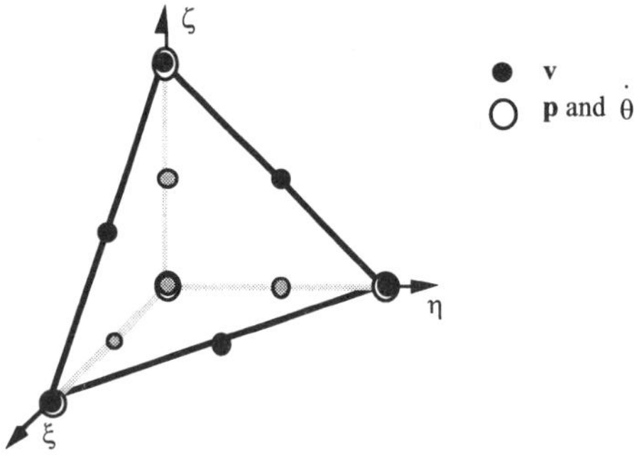

Figure 1: A $P_2P_1P_1$ 10-node tetrahedral element

3. THERMAL ANALYSIS

3.1 *Heat transfer equation*

The heat transfer is less rapid in a porous body than in a dense body. In the present study we have utilized empirical relations to modify the isotropic thermal conductivity for porous material, (Li et al.1987) as:

$$\frac{\kappa}{\kappa_s} = \rho^k \qquad (18)$$

where κ_s and κ represent the thermal conductivity of dense and porous metal (which is having a relative density, ρ), respectively and the exponent k is a material constant which can be determined from experiments.

The Fourier classical conduction law is considered on domain Ω, then we have :

$$\rho_p c_t \frac{dT}{dt} = \mathrm{div}\,(\kappa\,\mathbf{grad}\,T) + \delta w/\delta t \qquad (19)$$

where κ is the thermal conductivity, c_t the specific heat, ρ_p the specific density of the porous material and $\delta w/\delta t$ an internal power source. The parameters κ, c_t, and ρ_p may depend on T and ρ . The source of heat generation is the conservation of plastic work which can be expressed as:

$$\frac{\delta w}{\delta t} = \sigma:\dot{\varepsilon}$$

where σ represents the stress tensor and $\dot{\varepsilon}$ the total strain rate tensor.

Different boundary conditions can be handled on the external faces of the domain Ω :

- prescribed heat flux : $\Phi_{imp} = -\kappa\,\mathbf{grad}\,T\,.\,\mathbf{n}$ (20)

where **n** is the outward normal unit vector

- convection : $-\kappa\,\mathbf{grad}\,T\,.\,\mathbf{n} = \alpha\,(T-T_{ext})$ (21)

where T_{ext} is the external temperature

- radiation : $-\kappa\,\mathbf{grad}\,T\,.\,\mathbf{n} = \varepsilon\sigma\,(T^4-T_{ext}^4)$ (22)

where ε is the material emissivity, σ the Stefan constant. It is to be noticed that (22) can be written like (21) after defining, by linearization,

$$\alpha_r = \varepsilon\sigma\,(T^2 + T_{ext}^2)\,(T + T_{ext})$$

- prescribed temperature : $T = T_{imp}$ (23)
- conduction exchange with tooling (die and punch) domain Ω_j: the conductive flux is then prescribed as

$$-\kappa\,\mathbf{grad}\,T\,.\,\mathbf{n} = h_j\,(T - T_j) \qquad (24)$$

where h_j is the thermal resistance of the interface between Ω and Ω_j.

By adopting a three level two step time integration scheme, the space discretization of equation (19) is done by the finite element Galerkin method, leading to the classical following set of non-linear differential equations in which **T** is the vector of nodal temperatures as:

$$\mathbf{C}\frac{d\mathbf{T}}{dt} + \mathbf{K}\,\mathbf{T} + \mathbf{Q} = 0 \qquad (25)$$

4. THERMO-MECHANICAL COUPLING

At each time increment, the thermo-mechanical coupling eminates from two sources: firstly due to dependence of the mechanical behaviour on temperature in addition to relative density, and secondly through heat source term due to plastic work dissipation during thermal calculations. Also in addition to dependence of thermal parameters (conductivity and heat capacity) on the temperature and relative density. By virtue of these aspects the coupling is achieved in a staggered manner. For any given time increment, with known values of initial temperature, velocity, relative density, rheological parameters and geometry, the mechanical problem is first solved for nodal values of velocity, volumetric strain rate and Lagrangian multiplier/pressure. Next making use of these obtained nodal velocities (i.e., in the calculation of dissipated plastic work), the thermal resolution is carried out to obtain nodal temperature. Subsequently, updating of configuration, such as geometry and density, is performed as follows:

$$\Omega_{n+1} = \Omega_n + \mathbf{v}_n\,\Delta t_n \text{ and } \rho_{n+1} = \rho_n + \dot{\theta}_n\,\Delta \mathbf{t}_n \qquad (26)$$

5 APPLICATION

5.1 *Numerical simulation of connecting rod*

Using the above described model we have carried out the numerical simulations and the details are given below:

Currently, the various material parameters used were taken from the literature (see Grazzini 1991), such as:

$$f = 0.36\,\frac{(1-\rho)}{(\rho-0.64)}\,,\; c = 1 + 12.14\,f,$$

$\rho_{initial}$=0.8(Uniform),

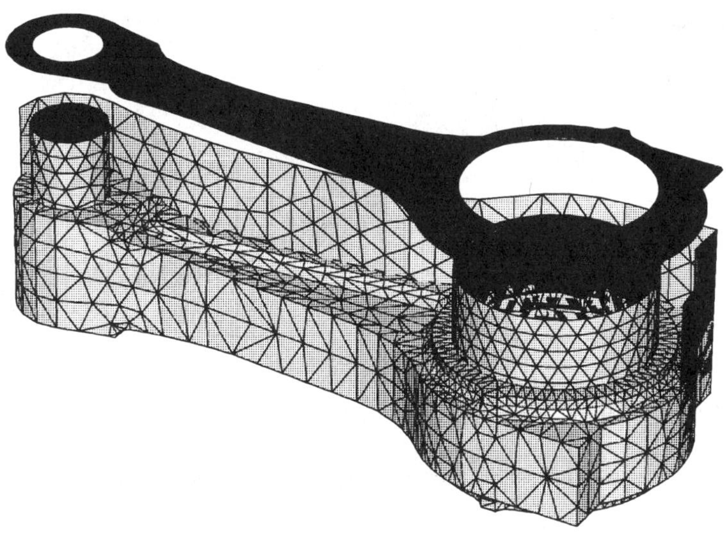

Figure 2 : Finite element mesh of connecting rod with tooling

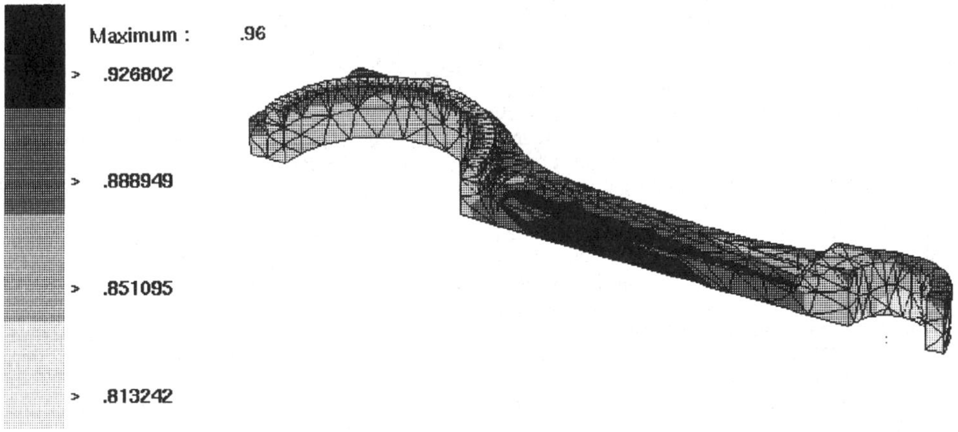

Figure 3 : Relative density distribution in the connecting rod after 0.0312 sec.

$m = 0.25$, $K = 54.5$ MPa s^m , $\alpha = 0.1$, $p = 0.25$,
$\beta = 4.2E{+}03$ K,
Preform : $\kappa = 2.5E{+}05$ Kg mm/s^3 K,
$c_t = 1.037E{+}09$ mm^2/s^2 K, $\rho_s = 2.7E{-}06$ Kg/mm^3
Tool : $h_{ij} = 6.0E{+}03$ Kg/s^3 K,
Free surface : $h_c = 6.0$ Kg/s^3 K, $\varepsilon_r = 0.7$
Initial temperature of the preform = 600 C
Initial temperature of the die (T_{tool}) = 200 C
Initial height of the preform = 16.06 mm.

A automatic tetrahedral mesh generator MAT (Coupez 1991) has been used for the discretization of con-rod and globally the computational domain consists of 3972 number of nodes with 2158 number of 10-node tetrahedral elements. At every iteration of the numerical problem, total number of equations solved were equal to 15,759. The convergence of each time increment was typically reached within 8 iterations.

The obtained result for the connecting rod with the tooling configuration as shown in Figure 2 is illustrated in figures 3 & 4. These results display the relative density and temperature distribution plots in the con-rod at time equal to 0.0312 sec during the forging of the rod.

From a study of Figure 3 one could realise that the relative density distribution as predicted by the model is quite realistic in the non-thermal regime of forming process. The density variation in the rod is quite appearent. It can be noted from Figure 3 that, the relative density varies across the domain. Since the lateral flow of the powder preform is prevented from the die, essentially the variation of relative density is

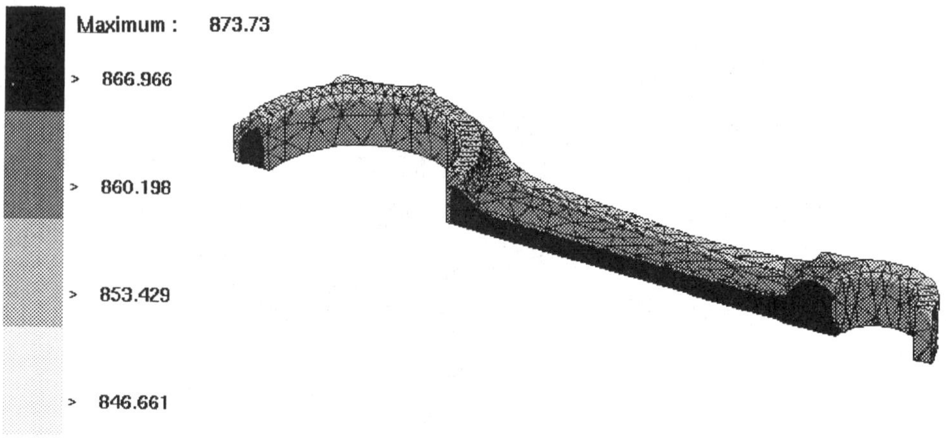

Figure 4: Temperature distribution (deg. K) in the connecting rod after 0.0312 sec.

due to friction on two fronts : a sliding type friction between the punch & the top side of preform and also sliding friction along the die walls & the lateral sides of preform. Due to the limited lateral flow the relative density distribution could have been homogeneous in areas away from the die and punch contact, however the complex shape of the con rod itself would cause the density gradients due to hindrance of the material flow. This can be clearly seen in the relative density distribution plot (Figure 3). These density gradients induce non-uniform plastic work which in turn result in temperature gradients. This occurs due to thermo-mechanical coupling which essentially couples the mechanical and thermal calculations. Also the induced thermal gradients in turn affects the mechanical deformation, due to reasons of coupling. However it should be noted that these gradients also depends strongly on the thermal characteristics of the preform itself at any given temperature. The temperature distribution as shown in Figure 3 provides further evidence of non-homgeneous material deformation during forging.

6 CONCLUSION

A new 3D finite element model has been presented for the numerical analysis of the hot forging stage in the compaction forming processes. A powder based transient thermal analysis has been outlined. The presented method invoked a framework for the thermomechanical coupling. The developed numerical model has been successfully applied to the hot forging of connecting rod.and a first attempt to model the deformation in a complete connecting rod has been achieved.

ACKNOWLEDGEMENTS

The authors of this paper acknowledge the financial support provided by the EEC through a Brite/Euram project :BRE2 CT 92-0220 (Project no. 5460). Also we take this opportunity to thank all our partners and in particular RENAULT and PEAK for fruitful collaboration. We are deeply indebted to Dr. T. Coupez and Ms. C. Gay of Cemef, and Dr. R. Ducloux of Transvalor for many discussions and help at various stages of this study.

REFERENCES

Abouaf, M. 1985. Modélisation de la compaction de poudres métalliques frittées. PhD thesis. Institut National Polytechnique de Grenoble (in French).

Chenot, J. L., F. Bay & L. Fourment 1990. Finite element simulation of metal powder forming. *International Journal of Numerical Methods in Engineering*. 30:1674-1674.

Coupez, T. 1993. Personal communication.

Coupez, T. 1991. Grandes transformations et remaillage automatique. PhD thesis. Ecole Nationale Superieure des Mines de Paris.(in French).

Grazzini, H. 1991. Etude expérimentale du comportement rhéologique de milieux granulaires à constituants plastiquement déformables: Comparaison de poudres de plasticine et d' alliages d' aluminium. PhD thesis. Ecole Nationale Superieure des Mines de Paris. (in French).

Green, R. J. 1972. A plasticity theory for porous solids. *International journal of mechanical science*. 14:215-224.

Jinka, A. G. K. 1992. Finite element simulation of powder compaction forming processes. PhD thesis. University College of Swansea (University of Wales).

Jinka, A. G. K., M. Bellet & F. Fourment 1995. A new three-dimensional finite element model for the simulation of hot forging of powder metallurgy connecting rods. *Computer methods in applied mechanics and engineering*. (communicated).

Li, W. B., M. F. Ashby & K. E. Easterling 1987. On densification and shape change during hot powder pressing. *Acta Metallurgy*. 35: 2831-2842.

Shima, S. & M. Oyane 1976. Plasticity theory for

porous materials. *International journal of mechanical science*. 18:285.

Simo, J. C., R. L. Taylor & K.S Pister 1985. Variational and projection methods for the volume constraint in finite deformation elasto-plasticity. *Computer methods in applied mechanics and engineering*, 51:177-208.

Simulation of Materials Processing: Theory, Methods and Applications, Shen & Dawson (eds)
© 1995 Balkema, Rotterdam. ISBN 90 5410 553 4

Rigid-viscoplastic finite element analysis of hot square die extrusion of complicated profiles with flow guides and lands by arbitrary Lagrangian-Eulerian formulation

Y.S. Kang & D.Y. Yang
Department of Precision Engineering & Mechatronics, Kaist, Korea

ABSTRACT: Based on arbitrary Lagrangian-Eulerian (ALE) description, a rigid-plastic finite element formulation for deformation analysis and the ALE scheme for temperature analysis is presented for rigid-viscoplastic materials. The formulation is applied to the analysis of square die extrusion of noncircular sections such as square and H sections, which are not amendable to exact analysis with updated Lagrangian approach. The computaional results for the extrusion of a square section are compared with those obtained by the updated Lagrangian finite element approach. In square die extrusion, die lands and flow guides play a important role on metallic flow. The effects of flow guides and die lands are numerically investigated for a complicated section of a H section. Finite element computations are carried out to assess the effects of important process and die design parameters. It is thus shown that the present finite element formulation is effective in analyzing hot square die extrusion of complicated sections such as including a H-section.

1. INTRODUCTION

The process of extrusion through square die extrusion is widely used to produce long profile sections. In recent years, the demand for complicated shapes has been increased and therefore the control of the process and the prediction of mechanical properties have become more important. Square die extrusion is normally carried out at the elevated temperature. Thus, the involved mechanics is basically non-steady and analysis should be made for the non-steady state throughout the entire process, which is not amenable to analysis due to the complicated metal flow in the non-steady hot extrusion process using a square die. Furthermore, the three-dimensional analysis is required, since the usual practical profiles are mostly asymmetric in the cross-sectional shape. Some systematic analyses for three-dimensional extrusion through continuous dies(Yang 1989, Lee 1990) have been made, but the number of research works for hot square die extrusion is quite limited treating only the steady state of the process.

In the finite element analysis of metal forming processes, the updated Lagrangian approach has been widely used to simulate the non-steady state problems. The method has been effectively used. However some difficulties have arisen from abrupt flow change as in extrusion through square dies. In the present work, hot square die extrusion of complicated profiles is analyzed by the rigid-plastic finite element method. The problem is treated as a nonsteady state and the arbitrary Lagrangian-Eulerian description is used due to the abruptly turning flow at the die aperture. The elastic-plastic arbitrary Lagrangian-Eulerian method has been generally used for nonlinear solid mechanics(Liu 1986, Ghosh 1991). In analyzing square die extrusion of profiles, the rigid-viscoplastic finite element method is advantageous from the viewpoint of computational efficiency. In the present work, a finite element formulation for deformation analysis and an ALE scheme for temperature analysis are presented for rigid-viscoplastic materials. The developed finite element program is applied to the analysis of square die extrusion of non-circular sections. The computational results are compared with those from the updated Lagrangian finite element analysis. For the sake of computational economy and conveniency, the modular remeshing technique(Yoon 1990) is used for the design of the mesh structure in finite element computation.

In square die extrusion of complicated profiles, land parts and flow guides play important roles in controlling the metal flow properly. The effects of lands and flow guides are investigated through the simulation of the process. The thermal state affects greatly the product quality in hot extrusion requiring the analysis of temperature distribution. In the present work, the temperature distribution is also analyzed in the framework of rigid-viscoplastic finite element computation. As computational examples, profile shapes such as a square and H sections are chosen.

2. METHOD OF ANALYSIS

2.1 *Derivation of variational equation*

The method used in the present work is based on the rigid-viscoplastic formulation with ALE description. The material is assumed to be isotropic and incompressible. The inertia effect is neglected.

The general momentum equations are written as follows:

$$(\tau_{ij})_{,j} + b_i = \rho \dot{v}_i \qquad (1)$$

where bi is body force and ρ is density.

The superposed dot and asterisk denote the time derivative with respect to material and fixed referential coordinates, respectively. The relationship between these two derivatives for velocity are represented as:

$$\dot{v}_i = \overset{*}{v}_i + c_j(v_i)_{,j} \quad \text{where } cj = Vj - Wj \qquad (2)$$

Where V is material velocity and W is mesh velocity. The constitutive equation during forming is represented by:

$$\tau'_{ij} = \frac{2}{3}\frac{\bar{\sigma}}{\dot{\bar{\varepsilon}}}\dot{\varepsilon}_{ij} \qquad (3)$$

The virtual work principle results in the following equation by use of the equations (1), (2) and (3):

$$\Pi = \int \bar{\sigma}\delta\dot{\bar{\varepsilon}}\, dV + K^* \int \varepsilon_v \delta\varepsilon_v dV + \int \rho(v_j - w_j)\, v_{i,j}\delta v_i dV - \int T_i \delta v_i dS = 0 \qquad (4)$$

where δvi is an arbitrary variation of the velocity field compatible with the boundary condition.

2.2 *Derivation of finite element equations*

The velocity and the strain can be expressed with respect to the nodal velocity as following:

$$v_i = N_{i\alpha}V_\alpha, \quad \delta v_i = N_{i\alpha}\delta V_\alpha$$

$$\dot{\varepsilon}_i = A_{ij}N_{j\alpha}V_\alpha = B_{i\alpha}V_\alpha$$

$$\dot{\bar{\varepsilon}} = \sqrt{\frac{2}{3}k_{\alpha\beta}V_\alpha V_\beta}, \quad k = \mathbf{B}^T\mathbf{B} \qquad (5)$$

$$\varepsilon_v = C_i B_{i\alpha}V_\alpha = Q_\alpha V_\alpha, \quad C_i = [1,1,1,0]$$

$$\delta\dot{\bar{\varepsilon}} = \frac{2}{3}\frac{k_{\alpha\beta}V_\beta}{\dot{\bar{\varepsilon}}}\delta V_\alpha$$

where $N_{i\alpha}$ is interpolation function and V_α is nodal velocity component. By sustituting equation (5) into equation (4), the final equation is obtained as follows:

$$\Pi = \sum_{m=1}^{M}\left[\int_{V^{(m)}} \frac{2}{3}\frac{\bar{\sigma}}{\dot{\bar{\varepsilon}}} k_{\alpha\beta}V_\beta dV + K^* \int_{V^{(m)}} Q_r V_r Q_\alpha dV + \int_{V^{(m)}} \rho(N_{j\beta}V_\beta - W_j)V_{i,j}N_{i\alpha}dV - \int_{S_t^{(m)}} T_i N_{i\alpha}dS\right] = 0 \qquad (6)$$

where M is the number of elements. Each term of equation (6) is defined as Π1, Π2, Π3 and Π4. Since equation (6) is a nonlinear system of equations, the Newton-Raphson method is applied and it is solved iteratively as in the following:

$$\sum_{m=1}^{M}\left[\frac{\partial\Pi_1^{(m)}}{\partial V} + \frac{\partial\Pi_2^{(m)}}{\partial V} + \frac{\partial\Pi_3^{(m)}}{\partial V} - \frac{\partial\Pi_4^{(m)}}{\partial V}\right]_{(n-1)} \Delta V_{(n)} = -\sum_{m=1}^{M}[\, \Pi_1^{(m)} + \Pi_2^{(m)} + \Pi_3^{(m)} - \Pi_4^{(m)}\,]_{(n-1)} \qquad (7)$$

Where, (n) and (n-1) denote current and previous iterations, respectively.

2.3 *Method for temperature analysis*

In the ALE description, the temperature distribution

Table 1. Thermal properties and heat transfer coefficients for Al-1100

Coeff	k N/(sec·°C)	ρc N/(mm²·°C)	h N/(sec·mm·°C)	h_{lub} N/(sec·mm·°C)	e
DIE	19.012	3.77	0.015	35.02	0.1
W.P. (Al 1100)	242.148	2.427	0.003		0.15

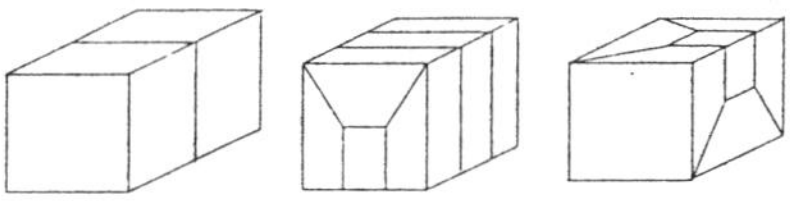

(a)

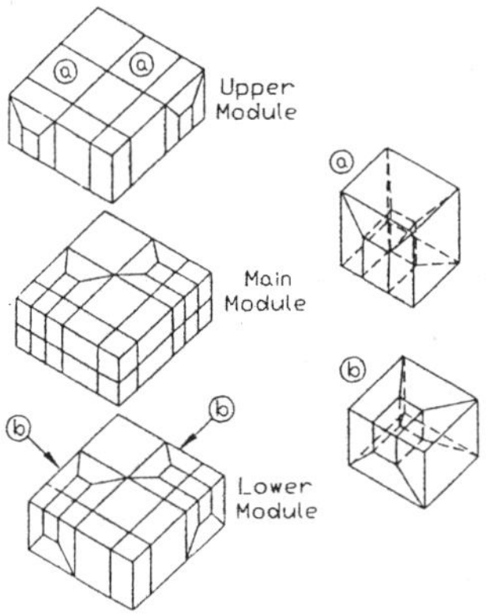

(b)

Figure 1 Selected basic modules and orifice adaptive module

is generally obtained by an energy balance equation as:

$$\rho c_p \frac{\partial T}{\partial t} \Big|_\chi + \rho c_p (V_i - W_i)\, T_i = k^* \bar{\sigma} \dot{\bar{\varepsilon}} + kT_{,ii} \qquad (8)$$

The first term of the left hand side of equation (8) can not be implimented exactly. Considering a computational time, temperature distribution is computed as following(Tong 1992):

i) construction of pseudo material domain by using the converged velocity field.

ii) calculation of heat flux and temperature for pseudo material domain by using Lagrangian heat equation(Yang 1992).

iii) updating a material variables(temperature and heat flux) to grid points as follow:

$$T^g_{(n)} = T^g_{(n-1)} + \Delta^g T$$

where,

$$\Delta^g T = \Delta^m T + \Delta t \, c_k \frac{\partial T}{\partial x_k} \qquad (9)$$

where superposed g and m represent the grid domain and the material domain, respectively.

3. NUMERICAL ANALYSIS

The extrusion process is basically a non-steady state and especially for hot extrusion the characteristics of non-steady state are dominant. In the present work, the analyses of both deformation and temperature are carried out by the finite element method as a non-steady state problem. Due to the large computation time in three-dimensional analysis, computation is carried out by decoupling the temperature analysis from the analysis of the deformation. The initial temperature of the workpiece, the die and the atmospheric temperature are taken to be 450 °C , 420 °C and 17 °C, respectively. The friction factor is taken to be 0.3. The punch velocity is assumed constant as 10mm/sec. The workpiece material and the die material used in simulation is AISI Al-1100 and H13, respectively. The flow stress is given by the following expression:

$$\bar{\sigma} = C(T)(\dot{\bar{\varepsilon}})^{m(T)} \qquad (10)$$

where C and m are interpolated with respect to temperature(Altan 1983).

The thermophysical properties and heat transfer coefficients are listed in Table 1. The design of mesh is very important in the finite element analysis. Mesh generation using the modular concept is used in the present work because the method must take into account the computational efficiency, geometry and physical characteristic of plastic flow. The method can be effectively applied to the present problem due to its geometric and physical characteristics. The mesh system is composed of unit zones, called as modules. The basic unit modules selected for the present analysis are shown in Figure 1(a). Figure 1(b) shows the combined modules (orifice-adaptive modules). The combined modules are designed for square die extrusion considering the flow at the orifice. Constant shear friction and Coulomb friction is applied for the frictional boundary condition.

4. NUMERICAL RESULTS AND DISCUSSIONS

4.1 Extrusion of a square section

As a first example, computations are carried out for

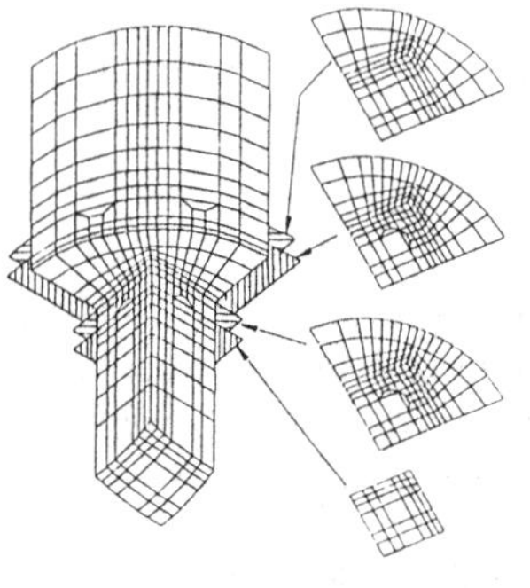

Figure 2 Mesh configuration of workpiece

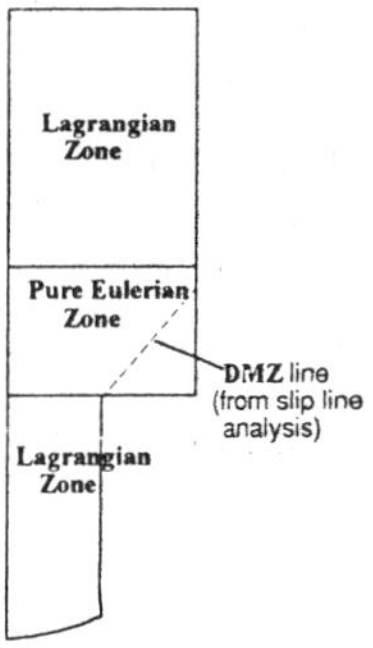

Figure 3 Division of extrusion billet into three zone

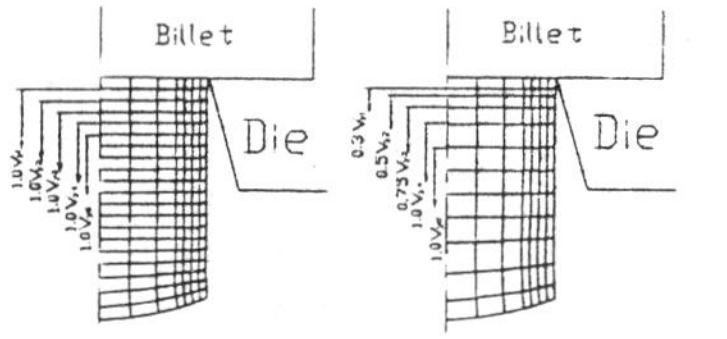

(b) after mesh control (a) before mesh control

Figure 4 Mesh control at the extruded part

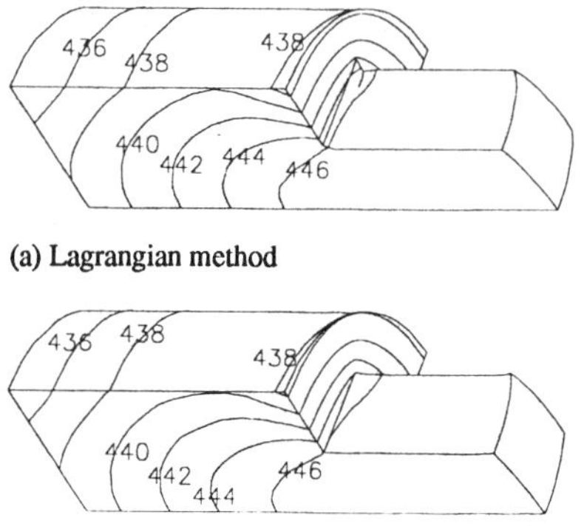

(a) Lagrangian method

(b) ALE method

Figure 5 Comparision of temperature distribution for the two methods

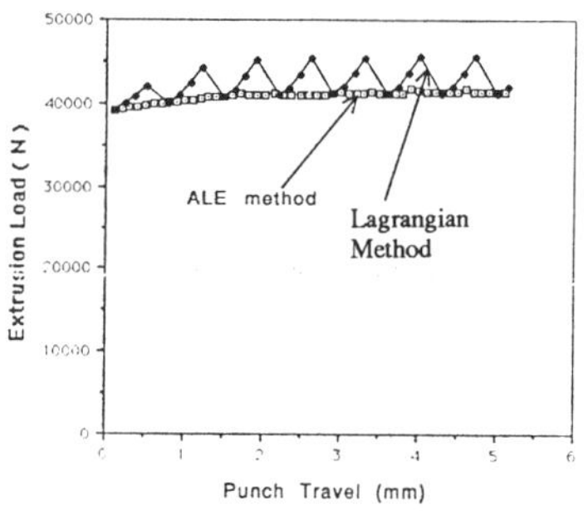

Figure 6 Load-punch travel curve

the extrusion of a square section without flow guide and without land. Considering symmetry and mesh quality, one fourth of the billet and the dies has been simulated. Figure 2 shows the mesh of the workpiece at the punch travel of 5.10mm. The mesh has been constructed by combining modules in Figure 1. About forty percent of the nodal points for the workpiece are located near the orifice in order to describe the metal flow properly.

Computations are carried out for two methods of analysis with the same condition and the same mesh system; Lagrangian FEM with automatic remeshing technique and the aribitrary Lagangian-Eulerian FEM based on the present formulation. Figure 3 shows the division of the extrusion billet according to the kinematic description. Considering the computational time, mesh velocity is controlled at the extruded part as in Figure 4.

Figure 5 shows the distribution of temperaure for the above two methods at the punch travel of 5.10mm. The ALE FEM result agree nearly with the Lagranginan FEM result. The effective strain distribution also shows same tendency when comparing the methods. The ALE method required more computational time almost by a factor three. Figure 6 shows the load-punch travel curves for the two method. From a viewpoint of load and volume loss, the ALE method provides the improved result, because the method possible keeps good mesh quality and is free from the interference with the die.

4.2 *Extrusion of H-section*

In this example, the flow guide and the die land is considered for the die construction. The flow guide and the die land play important roles for controlling

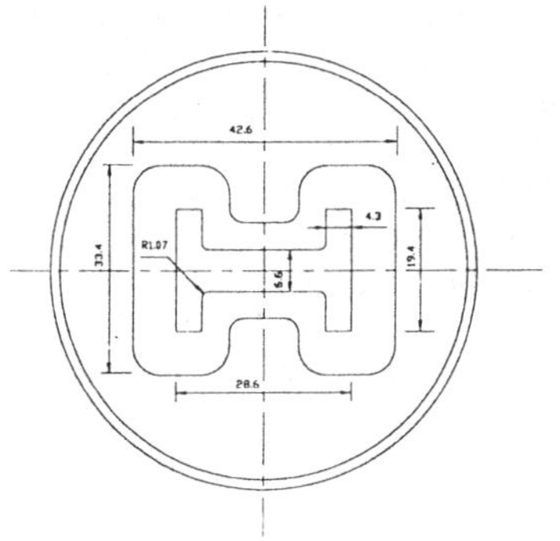

Figure 7 Die dimensions and orifice profile

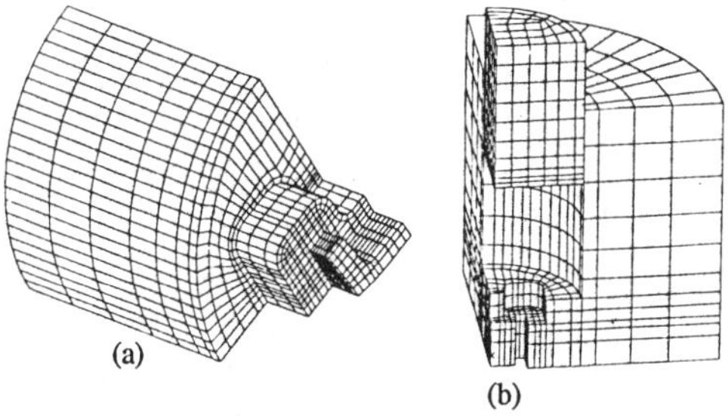

Figure 8 Mesh system for the extrusion of H-section

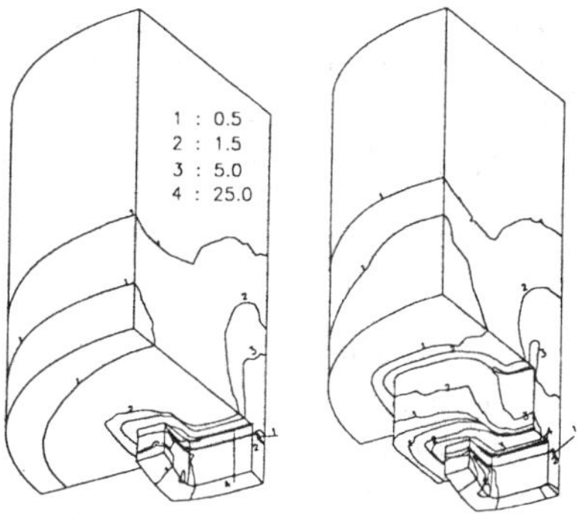

(a) (b)

Figure 9 Comparision of the effective strain rate between the two methods

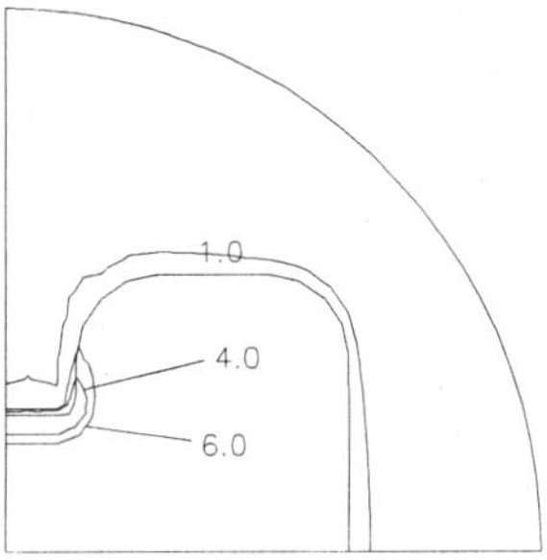

Figure 10 Distribution of effective strain rate on the just above the flow guide

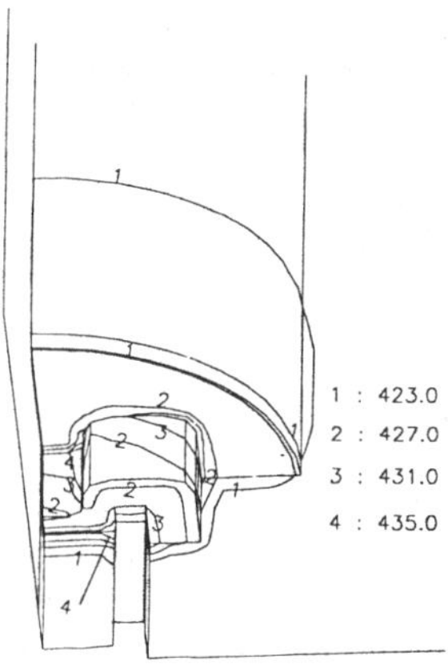

Figure 11 Temperature distribution around the flow guide

the flow in die design. Figure 7 shows the dimensions of the flow guide and the orifice profile. The die land is chosen to have a constant length for the sake of computational convenience. In the example, computations are carried out to assess the effect of a flow guide. Figure 8(a) shows the intermediate mesh for the workpiece. Figure 8(b) shows the mesh of the die and the punch. Considering symmetry and mesh quality, one fourth of the billet and the dies has been simulated. The method of mesh control for the ALE description is the same as in extrusion of a square section. From the results, the case with a flow guide exhibits somewhat more uniformed velocity field. Figure 9(a) shows the distribution of the effective strain rate for extrusion of a H-section without a flow guide and figure 9(b) corresponds to the case without a flow guide. When the flow guide is employed in the die system, the deformation region is separated into two zones and highly deformed region, i.e region of high deformation gradient, is decreased. Figure 10 shows the distribution of effective strain rate on the plane just above the flow guide. Figure 11 shows the temperature distribution of the die and the punch around the flow guide and the die orifice. When considering the temperature distribution in figure 11 and the distribution of effective strain rate in figure 10, temperature is higher where the effective strain rate is high in the workpiece. It is due to the heat generation by friction and deformation. In the extrusion with a flow guide, the computed extrusion load is higher by about 10% than the other case without a flow guide due to the increase of total redundant work.

5. CONCLUSION

From the comparison between the ALE method and the Lagrangian method, the ALE method has been shown to be more recommendable in spite of large computaton time due to its increased exactness of solution. As a tool of mesh generation, modular remeshing technique has been proved to be effectively used in the analysis of square die section since the physical chracteristies and computational efficiency are properly considered. In the result of extrusion of a H-section, the case with a flow guide has slightly improved uniform velocity distribution. It is thus expected that extrusion with a flow guide in combination with a relevant variation of die land would improve the product quality. Through the present work, it has been shown that the ALE method of analysis for hot square die extrusion of complicated noncircular sections would help to design the process with reliability.

REFERENCES

Altan, T., Oh, S.I. and Gegel, H.L. 1983. Metal Forming: Fundamentals and Applications. Americal Soc. for Metals. Chap.4

Ghosh, S. and Kikuchi, N. 1991. An Arbitrary Lagrangian-Eulerian Finite Element Method for Large Defromation Analysis of Elastic-Viscoplastic Solids. Comp. Meths. Appl. Mech. Engng. 86 : 127-188

Lee, C.M., Yang, D.Y. and Kim, M.U. 1990. Numerical Analysis of Three-Dimensional Extrusion of Arbitarily Shaped Sections by the Method of Weighted Residuals. Int. J. Mech. Sci. 32-1 : 65-82

Liu, W.K., Belytschko, T. and Chang, H. 1986. An Arbitrary Lagrangian-Eulerian Finite Element Method for Path-Dependent Materials. Comput. Meths. Appl. Mech. Engrg. 58 : 227-245

Tong, L., Hora, P. and Reissner, J. 1992 Application of the Arbitrary Lagrangian-Eulerian Method in the FE-Simulation of 3-D Bulk Forming Processes. Proc. 4th NUMIFORM. Valbonne. France. : 669-674

Yang, D.Y., Kang, Y.S. and Cho, J.R. 1992. Finite Element Analysis of Three-Dimensional Hot Extrusion of Sections through Continuous Dies considering Heat Transfer, Proc. 4th NUMIFORM. Valbonne. France. : 687-692

Yang, D.Y., Lee, C.M. and Yoon, J.H. 1989. Finite Element Analysis of Steady -State Three-Dimensional Extrusion of Sections Through Curved Dies. Int. J. Mech. Sci. 31-2 : 145-146

Yoon, J.H. and Yang, D.Y. 1990. A Three-Dimensional Rigid-Plastic Finite Element Analysis of Bevel Gear Forging by Using a Remeshing Technique, Int. J. Mech. Sci. 32-4 : 277-291

Simulation of Materials Processing: Theory, Methods and Applications, Shen & Dawson (eds)
© 1995 Balkema, Rotterdam. ISBN 90 5410 553 4

Flow of solid metal during extrusion: Three-dimensional simulations by finite element method

Manabu Kiuchi, Jun Yanagimoto & Victor Mendoza
Institute of Industrial Science, The University of Tokyo, Japan

ABSTRACT: Results of a study on the characteristics of the flow material during three-dimensional extrusion are presented. The numerical results are obtained by using a newly developed 3-D rigid-plastic FEM code. The extrusion processes of polygonal, angle and channel cross-section products through flat-faced dies are chosen as examples to demonstrate its simulation capability. Values of average extrusion pressure and metal flow characteristics are obtained as a function of the geometrical complexity of the cross-section of the product and the eccentricity of the die. Presented results are discussed mostly by reasoning and comparing to general observation in practice. Future work will include the simulation of more complex cross-section products including asymmetric shapes and comparison with other results.

1. INTRODUCTION

Extrusion is a manufacturing process used in many industries. Despite its broad application, the design of the tools and parameters of the process has been made by intuitive and empirical techniques which are laborious and expensive. Most of the dies designed by this kind of empirical approach for axisymmetric and non-axisymmetric extrusion are flat-faced with various die bearing lengths in order to keep extrudates straight. In general, this kind of die may suffer serious inconvenients, some examples are: high extrusion pressure, excessive redundant work, non-uniform metal flow and then non-uniform properties of the product. However, flat-faced dies are relatively easy to design, manufacture, handle, use and in most cases this practice provides the advantage of attaining high productivity.

If the flow is not controlled properly in a flat-faced die, then, die modification has to be carried out. Therefore, reliable techniques for metal flow analysis, theoretical die design approaches are of great value in reducing much of the tool tryout work.

Various upper bound solutions have been proposed for the three-dimensional problems in the last two decades. For example, Nagpal and Altan (1975) analyzed extrusion of products with non-axisymmetric cross-sections by using the concept of stream function. Their analysis was limited to simple cross-sections such ellipse, and it is difficult to find stream functions for extrusion for more complex cross-sections. Yang and Lee (1978) found kinematically admissible velocity fields by using the concept of conformal mapping. Gunasereka and Hoshino (1981, 1985) suggested a new method for extrusion of regular polygonal sections from round billets using the concept of stream lines and proportionate deformation in plastic flow. Kiuchi et al. (1981) proposed generalized formulas of the kinematically admissible velocity fields that can be used in any kind of extrusion analyses.

For complicated cross-sections, however, more detailed reliable informations are still required. The rigid-plastic FEM which has proved to be superior to the classical methods due to their flexibility, and ability to obtain detailed solutions is recommended. Despite this, a limited number of works have been reported. Yang (1989, 1992, 1994) reported the application of the rigid-plastic FEM using recurrent boundary conditions to the nonsteady-state extrusion of a helical gear-shape section through a curved die.

In the present paper, a three-dimensional rigid-plastic finite element code, called COPRESS (Kiuchi et al. 1994) is applied to the analysis of various extrusion processes: regular polygonal shapes and rectangular-angle-channel shapes. Also, for the second group the aspect ratio is varied in a very wide range. Another important variable studied in this work is the position of the die opening with respect to the billet axis.

Results on the average extrusion pressure, the characteristics of metal flow in terms of velocity distribution, stream lines distribution and grid distortion patterns are presented. Also, from a series of simulations, the optimal position of the die opening is predicted.

2. APPLICATION TO EXTRUSION OF POLYGONAL BARS, RECTANGULAR BARS, ANGLES AND CHANNELS

Extrusion processes of four kinds of products were simulated: polygonal, rectangular, angle and channel sections. They were assumed to be carried out from round billets through flat-faced dies (Fig. 1).

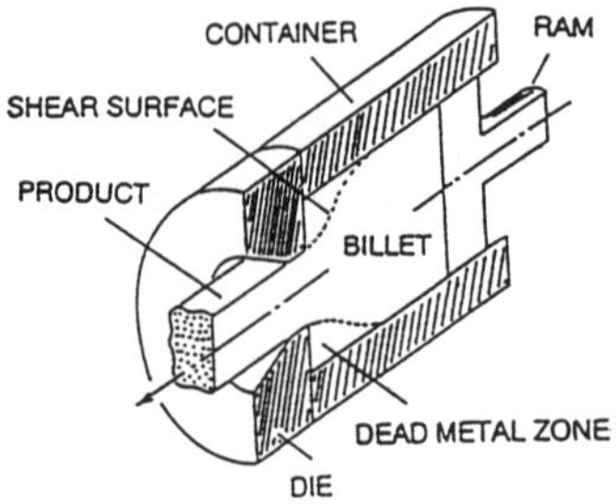

Fig. 1 "Round-3 D cross section" extrusion process through a flat-faced die.

The process conditions under which the computations were carried out are summarized in Table 1. The stress-strain relationship of the workpiece is expressed by the following rate dependent equation

$$\bar{\sigma} = 12\, \dot{\bar{\varepsilon}}^{0.13} \qquad [\mathrm{Kg/mm^2}]. \qquad (1)$$

Table 1. Extrusion process conditions for computation.

(i)	Container length:	DZC=100 mm
(ii)	Container radius:	R_O= 50 mm
(iii)	Bearing length:	Z_B= 30 mm
(iv)	Punch velocity:	V_{Z_C}= 10 mm/s
(v)	Friction coefficient:	μ = 0.1

For rectangular, angle and channel cross-sections, the parameters for computational models are listed in Table 2, 3 and 4, respectively. The geometrical complexity of the product, *i. e.* cross sectional of the die opening, is defined by the aspect ratio, A.R., expressed by the ratio "DYE/DXE" for rectangular and angle sections and by the ratio "2DYE/DXE" for channel sections.

Table 2. Computational models of rectangular sections from round billets

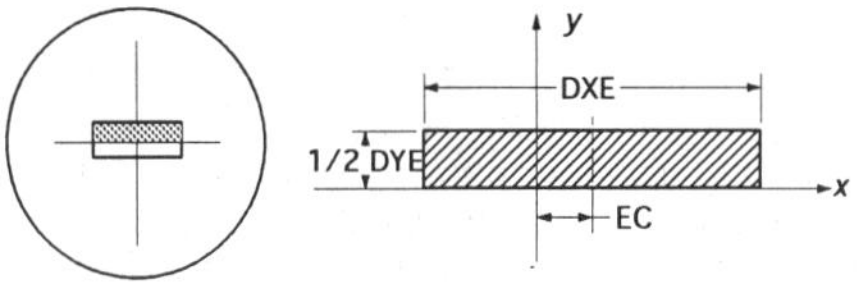

DXE (mm)	DYE (mm)	DYE/DXE
20.00	10.00	1/2
40.00	5.00	1/8
48.99	4.08	1/12
56.57	3.54	1/16

Table 3. Computational models of angle sections from round billets

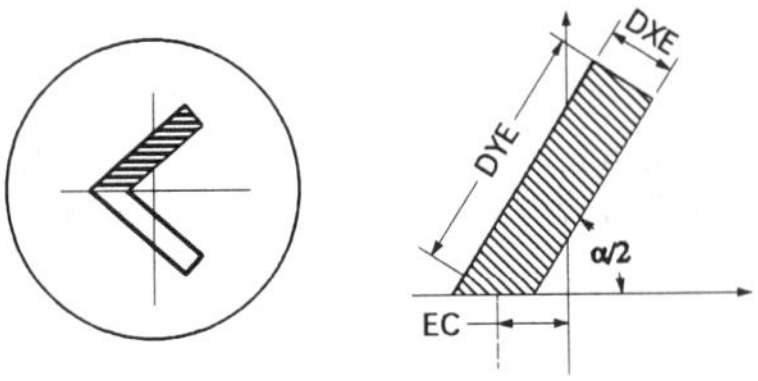

DXE (mm)	DYE (mm)	DYE/DXE	α (°)
10.00	20.00	2	90
5.00	40.00	8	90
4.08	48.99	12	90
10.00	20.00	2	120
5.00	40.00	8	120
4.08	48.99	12	120

Table 4. Computational models of channel sections from round billets

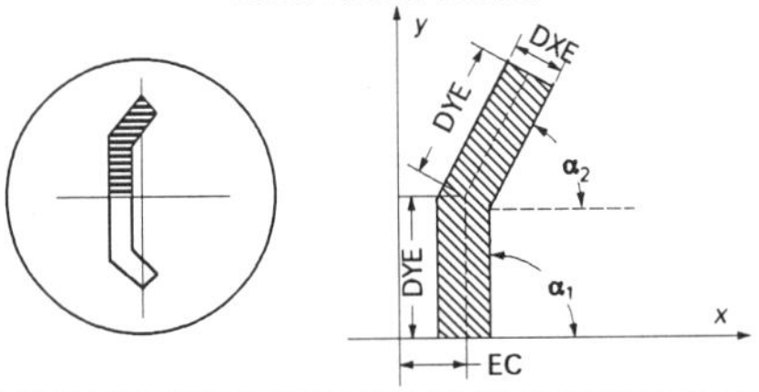

DXE (mm)	DYE (mm)	2 DYE/DXE	α_1 (°)	α_2 (°)
10.00	10.00	2	90	60
5.00	20.00	8	90	60
4.08	24.50	12	90	60
10.00	10.00	2	75	45
5.00	20.00	8	75	45
4.08	24.50	12	75	45

For rectangular sections, the eccentricity of the die "EC" is varied from 0 to $\frac{3}{5}R_O$ with variations of $\frac{1}{5}R_O$. In some cases, because of the geometry of the tooling, it is not possible to simulate eccentricities larger than $\frac{2}{5}R_O$. Because of the geometrical characteristics, for angle and channel sections, it is interesting to determine the optimum eccentricity of the die in order to get the minimum extrusion pressure. Therefore, the eccentricity of the die is varied from $-\frac{3}{5}R_O$ to $\frac{2}{5}R_O$.

Also, in order to compare the effect of the shape of the cross-section and the aspect ratio on the metal flow, the extrusion ratio (cross-sectional area of billet/cross-sectional area of extruded product) is kept constant for each of the products above described.

The contact algorithm implemented in COPRESS requires a discretization of the tool surface into a mesh of spatial triangular elements. A typical mesh sytem used for tooling, container, bearing and ram, is shown in Fig. 2. In the particular case shown in the figure, the model consists of 178 triangular elements interconnected at 128 nodal points.

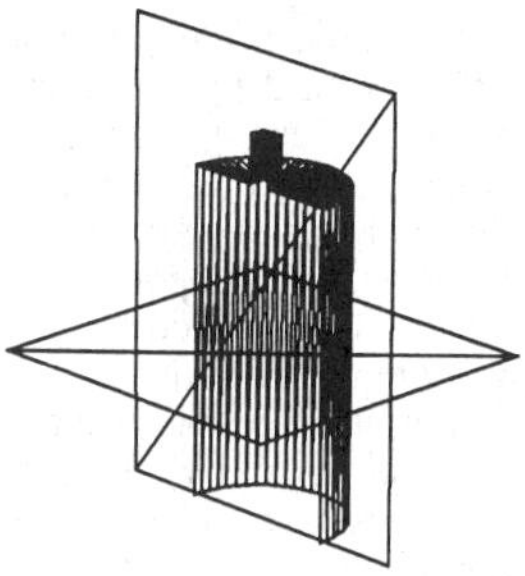

Fig. 2 Tool discretization for three-dimensional extrusion simulation

The most typical polygonal cross-sections used in the industry are: square, pentagonal, hexagonal and octagonal. Fig. 3 shows a typical finite element layout used for FEM calculations for square and hexagonal sections. Also, the same information for rectangular, angle and channel cross-sections is presented in Fig. 4. The finite element model includes a total of 980 eight-node hexahedral elements for the container portion. In bearing and extruded product portion, the number of elements varies in according with the geometrical complexity of the product's cross-section. In each figure are shown the complete models, but in actual calculations only one half of the geometry is considered due to the symmetry.

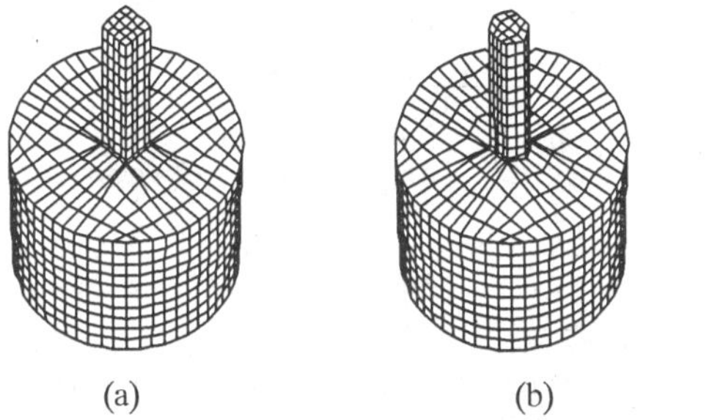

Fig. 3 Finite element mesh of working material for computation. Regular polygonal cross-sections: (a) square; (b) hexagonal.

3. RESULTS AND DISCUSSIONS

- Results on average extrusion pressure

The energy computed can be converted to the average extrusion pressure (P_{Av}) by using the relation given by Eq. (2)

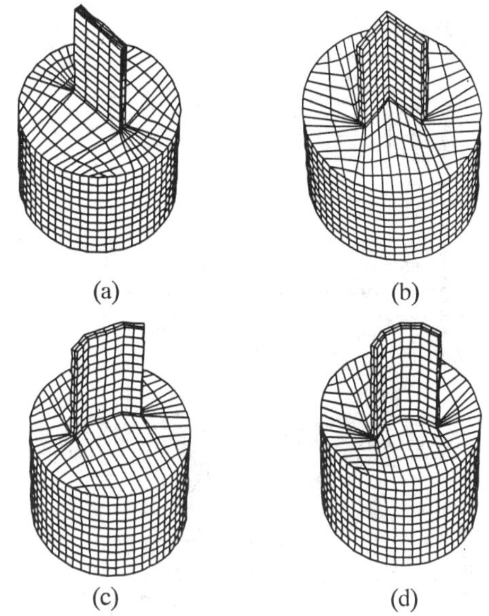

Fig. 4 Finite element mesh of working material for computation: (a) rectangular; (b) angle; (c) channel; (d) channel.

$$P_{Av} = \frac{\phi}{\pi R_O^2 V_{Z_C}} \qquad (2)$$

where ϕ is the functional corresponding to the FEM formulation (Kiuchi et al. 1994) and expresses the total extrusion power.

- Extrusion of regular polygonal cross-sections

The round to regular polygonal section is the most typical industrial extrusion process. The average extrusion pressure for 39.3 extrusion ratio is shown in Fig. 5. For this kind of extrusion the plot shows that as the number of sides increases, the deformation approximates to round extrusion, i.e. axisymmetric extrusion. Therefore, the deformation is more uniform and the internal energy component decreases.

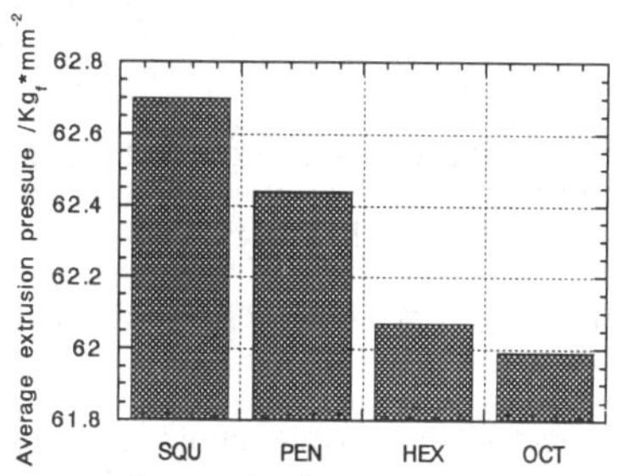

Fig. 5 Average extrusion pressure in extrusion of regular polygonal cross-sections.

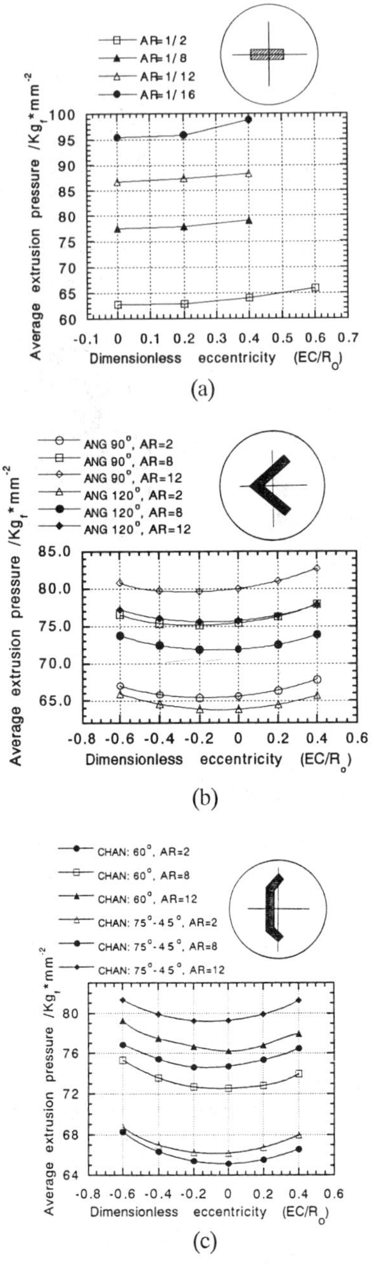

Fig. 6 Variation of average extrusion pressure with the aspect ratio and eccentricity of the die for: (a) rectangular; (b) angle; (c) channel.

- *Extrusion of rectangular, angle and channel cross-sections*

For rectangular cross-sections the plots of the average extrusion pressure variation with the eccentricity of the die, shown in Fig. 6 (a), indicate an important increment when the aspect ratio DYE/DXE decreases. However, the eccentricity of the die shows only a slight influence on the extrusion pressure.

Angle and channel sections are more complicated and less symmetric. Extrusion pressures required for them are shown in Fig. 6 (b) and (c), respectively. Also, it is well known that the pressure required to extrude a given section depends on material variables and geometric variables. Among other variables, in die design, the position of the die opening with respect to the billet axis also affects the pressure required. Therefore, it is important to determine the eccentricity of the die at which the pressure required is minimum. Both Figs. show the variation of the average extrusion pressure with the eccentricity of the die. From these plots or by using an interpolation technique, it is possible to determine the optimum eccentricity for a given geometry, Tables 5 and 6.

Table 5. Optimum eccentricity values for angle sections

DYE/DXE	α (°)	Opt. Eccentricity (EC/R_O)
2	90	-0.169
8	90	-0.300
12	90	-0.270
2	120	-0.098
8	120	-0.115
12	120	-0.152

Table 6. Optimum eccentricity values for channel sections

2 DYE/DXE	α_1 (°)	α_2 (°)	Opt. Ecce. (EC/R_O)
2	90	60	-0.011
8	90	60	-0.051
12	90	60	0.051
2	75	45	-3.817
8	75	45	-1.988
12	75	45	-5.762

As would be expected, comparison among the predicted extrusion pressure values for each of the cross-sectional shapes shows higher values for 90° angle sections and for products with high aspect ratio values. This trend can be explained by the functional corresponding to the FEM formulation (Kiuchi et al. 1994), as a consequence of the increase of internal power of deformation due to the more drastic change of the flow direction inside the plastically deforming zone.

For angle and channel sections, the plots of extrusion pressure variation with the eccentricity of the die indicate that a minimum value of extrusion pressure is reached at some die eccentricity. This pressure is reached at some die eccentricity. This value may be defined as the optimal die eccentricity. The optimality arises where the flow direction of the deforming material suffers the less abrupt changes.

- Results on velocity and streamlines distribution

The general trend of the velocity distribution is similar for each of the analyzed cross-sectional shapes. A typical example is given in Fig. 7 where the component velocity V_Z is plotted in all the volume.

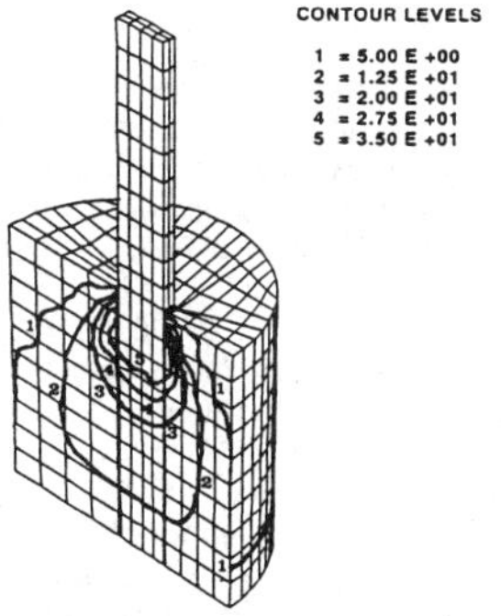

Fig. 7 Distribution of the component velocity V_Z.

Other important features of the velocity distribution can be better appreciated in three-dimensional plots, as shown in Fig. 8. In these plots, the nodal point velocities corresponding to the die entrance plane are plotted as function of the nodal point coordinates, then it is possible to see a three-dimensional surface representing the velocity distribution for this specific plane.

At every position the axial velocity V_z decreases or increases due to the frictional retardation at the die surface or to the contribution of the preferential tendencies of the flow material, that is, at locations far away from the wall of the die where the frictional effect has no longer an important influence. The velocity distributions tends to be more uniform for regular polygonal sections. Also, non uniformity in the velocity distribution increases with the aspect ratio. Higher aspect ratio causes higher local deformations, then the components V_x and V_y decrease which in turn leads to the lowering of the axial velocity V_z on account of material incompressibility. Also, such a nonuniform velocity distribution is expected to produce greater grid distortion.

For small aspect ratio rectangular sections, an interesting feature is observed in the velocity distributions, for example, with reference to Fig. 8 (b), it is clearly seen that the velocity distribution shows a double peak, that is, smaller values are at central positions. However, when the "DYE/DXE" ratio increases, this tendency lessens. This effect can be explained by observing the stream lines patterns shown in Fig. 9, where detailed differences of the

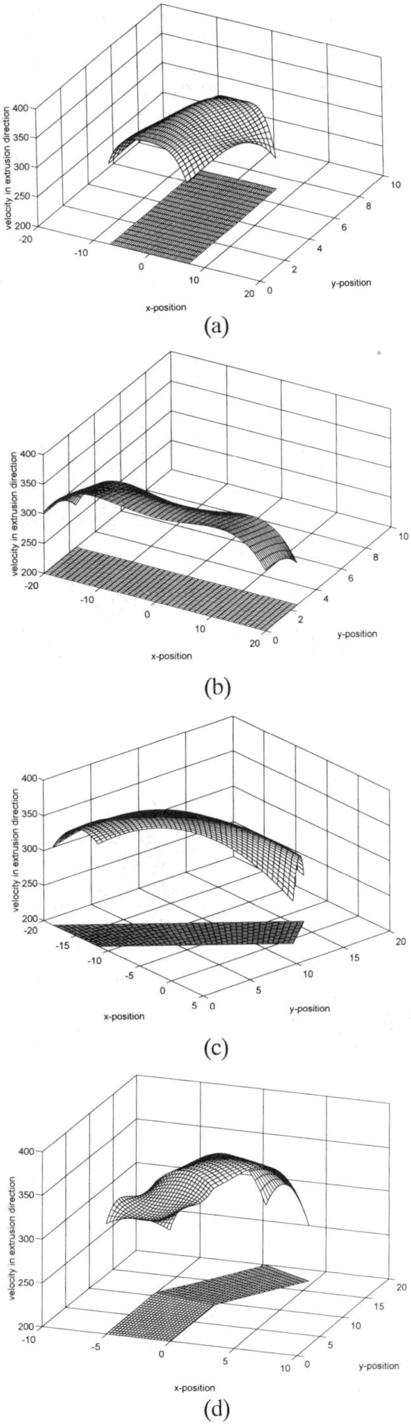

Fig. 8 Three-dimensional plot of V_z at the entrance of the die: (a) square; (b) rectangular; (c) angle; (d) channel.

metal flow are more clearly indicated. Fig 9 (a) shows, for example, the stream lines distribution pattern for rectangular sections with "DYE/DXE"=1/2 and 1/8, two views are shown: top and isometric. In top views, it is possible to appreciate the preferential tendencies of the metal flow. As might be expected, for "DYE/DXE"=1/2 case, there is some degree of stream lines concentration at the center of the cross-section, therefore, higher velocities. On the contrary, for "DYE/DXE"=1/8 case, Fig. 9 (b), the highest concentrations of stream lines are located towards the lateral positions of the die, which corresponds to the double peak of the velocity distribution.

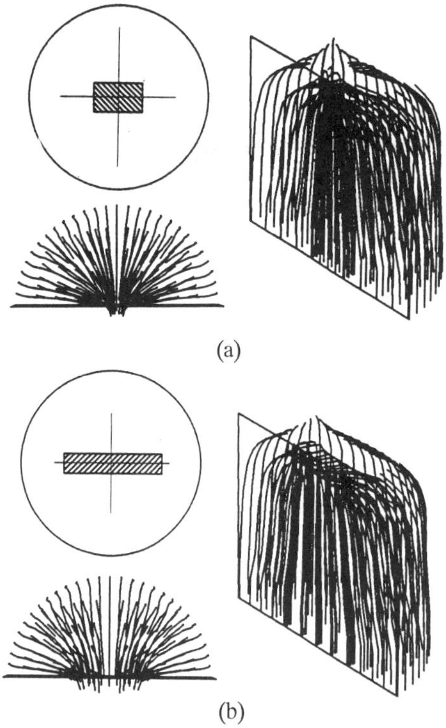

(a)

(b)

Fig. 9 Stream-lines distribution: (a) rectangular cross-section DYE/DXE=1/2, EC=0 mm; (b) rectangular cross-section DYE/DXE=1/8, EC=0 mm.

4. CONCLUDING REMARKS

This paper presents the application of a three-dimensional rigid-plastic finite element code called COPRESS System. Employing the system as an effective tool, a comprehensive investigation into the deformation characteristics was carried out. Extrusion of polygonal, angular and channel sections was simulated and their flow characteristics were summarized and discussed in terms of velocity distributions, stream lines distributions and grid distortion patterns. Although the die geometry is still simple, the investigation illustrates quantitatively the effect of geometrical complexity on the deformation mechanics. It is concluded that COPRESS System is indeed an effective tool which provides useful information on the detailed deformation characteristics for various process variables. In order to ascertain the accuracy of the solutions presented in this work, however, an extensive research work is still needed in order to evaluate the effect of the other important variables on the metal flow during three-dimensional extrusion.

REFERENCES

Nagpal, V., and Altan, T., "Analysis of the Three- dimensional Metal Flow in Extrusion of Shapes with the Use of Dual Stream Functions," Proceedings of the 3rd. North American Metalworking Research Conference, Pittsburgh, Pa., 1975, pp. 26-40.

Yang, D. Y. and Lee, C. H. "Analysis of Three-dimensional Extrusion of Sections through Curved Dies by Conformal Transformation," International Journal of Mechanical Science, 20, 1978, pp. 541-552.

Hoshino, S. and Gunasekera J. S. "An Upper Bound Solution for the Extrusion of Square Section from Round Bar through Converging Dies," Proceedings of the 21th MTDR Conference, 1980, pp. 97-105.

M. Kiuchi et. al: Proceedings of the 20th MTDR Conference, (1981), 523.

Kiuchi, M., Kishi, H. and Ishikawa, M. "Study on Non-symmetric Extrusion and Drawing," Proceedings of the 12th North American Manufacturing Research Conference, 1981, pp. 523-532.

Gunasekera, J. S. and Hoshino, S. "Analysis of Extrusion of Polygonal Sections Through Streamlined Dies," Trans. ASMEJournal of Engineering for Industry, 107, (1985), pp. 229-233.

Yang D. Y., Lee, C. M. and Yoon, J. H. "Finite Element Analysis of Steady-state Three-dimensional Extrusion of Sections through Curved Dies," International Journal of Mechanical Science, 31, 1989, pp. 145-156.

Yang, D. Y., Kang, Y. S. and Cho, J. R., "Finite Element Analysis of Three-dimensional Hot Extrusion of Sections through Continuous Dies Considering Heat Transfer," Numerical Methods in Industrial Forming Processes, Chenot, Wood and Zienkiewics (eds), 1992, pp. 687-692.

Yang, D. Y.,"Investigation into Non-Steady Three-dimensional Extrusion of a Trocoidal Helical Gear By the Rigid-Plastic Finite Element Method," Annals of the CIRP, 43, (1994), pp. 229-233.

Kiuchi, M., Yanagimoto, J. and Mendoza, V., "Flow of Solid Metal during Extrusion: I Three-dimensional Simulations by Finite Element Method," Proceedings of the 1994 JSTP Spring Conference, 1994, pp. 627-630.

Kiuchi, M., Yanagimoto, J. and Mendoza, V., "Flow of Solid Metal during Extrusion: II Three-dimensional Simulations by Finite Element Method," Proceedings of the 1994 JSTP Autumn Conference, 1994, pp. 683-686.

Simulation of Materials Processing: Theory, Methods and Applications, Shen & Dawson (eds)
© 1995 Balkema, Rotterdam. ISBN 90 5410 553 4

Modelling the compaction of metallic powders with ductile particles*

Sriram Krishnaswami & Juan R. L. Trasorras
Concurrent Technologies Corporation, Johnstown, Pa., USA

ABSTRACT: A study of compaction of metallic powders is presented. The powder is assumed to be rate independent isotropic elastic-plastic. Resonant frequency measurements are used to characterize the elasticity. The yield function is characterized by triaxial tests while the particle hardening is determined from simple compression tests of fully dense specimens. This model is implemented in a finite element scheme to simulate the two and three dimensional compaction of powders. The integration scheme for the constitutive law is a stable, fully implicit backward-Euler operator, with the kinematics being decoupled from the constitutive part. The finite element predictions of density and compaction loads from two dimensional simulations show good agreement when compared to experimental results for Ancorsteel 1000 powder. The history of tooling motions is shown to influence the final density distribution in the compaction of a multilevel part. Finally, fully three dimensional compaction simulation results are presented.

1 INTRODUCTION

Compaction of powders is widely used to produce components for the automotive and appliance industries. The process involves two stages: 1) compaction of powder in a die, and 2) sintering. In this work, we concentrate on finite element modeling of the first stage. The compaction process depends critically on the plastic deformation characteristics of the powder, friction between the sliding particles and between the particles and the die, the tooling motions imposed and the elastic recovery of the compact after release of loads. Dimensional control, density distribution and compact cracking are critical issues in the manufacturing process. The complex nature of the compaction process requires the use of numerical methods for simulation. This allows the design of tooling motions needed to produce crack-free compacts with the desired density distribution and to increase tool life.

2 CONSTITUTIVE MODEL

The finite element simulation of powder compaction has received considerable attention in the literature (*e. g.*, [13, 10, 3, 2, 15, 8, 14]). In this work, we focus on the simulation of the cold compaction of ductile powders with irregular particle morphology, such as iron, stainless steel or copper powders. We use an isotropic-elastic, isotropic hardening rate independent plastic model for the powder particles. The powders yield according to a yield function. Plastic strain increments are assumed to be given by normality, with the yield function being the plastic potential. The instantaneous state of hardening of the particle is given by the uniaxial flow strength of the particle $\bar{\sigma}_y$, which increases with strain according to:

$$\dot{\bar{\sigma}}_y = h\dot{\bar{\epsilon}}^p \ . \tag{1}$$

From the microscopic particle behavior, we con-

[a]This work was conducted by the National Center for Excellence in Metalworking Technology, operated by Concurrent Technologies Corporation, under contract to the U.S. Navy as part of the U.S. Navy Manufacturing Science & Technology Program.

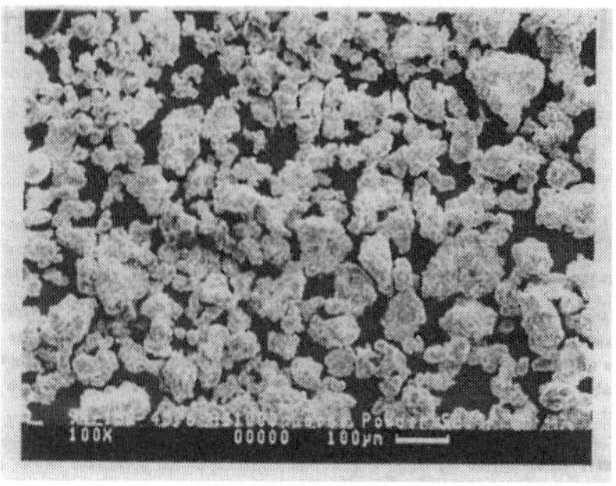

Figure 1: Scanning Electron micrograph of Ancorsteel 1000 powder.

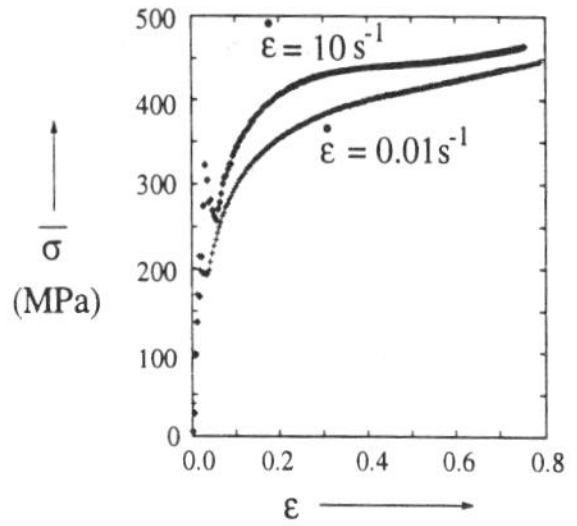

Figure 2: Uniaxial constant strain rate compression flow curves for fully dense Ancorsteel 1000 at 93°C. The specimens were produced by hot isostatic pressing of compacts.

struct a macroscopic theory. The transition is assumed to yield macroscopic continuum quantities of stress $\mathbf{T}$, rate of deformation $\mathbf{D}$ [4] and a relative density ρ. The relative density is defined as the ratio of the density of the powder compact in the unstressed state to the density of the particle in the unstressed state. All macroscopic plasticity is assumed to be solely due to microscopic particle plasticity. No contribution is expected from the particle sliding or rearrangement. This is a good approximation if the particle morphology is irregular. The powder morphology for Ancorsteel 1000 is shown in figure 1 at a magnification of 100X. These water atomized particles are seen to be irregularly shaped. Neglecting plastic work contributions from sliding and assuming that $\bar{\sigma}_y$ is not a function of position, the proper definition of the macroscopic quantities [4] gives

$$\mathbf{T} \cdot \mathbf{D}^p = \rho \bar{\sigma}_y \dot{\bar{\epsilon}}^p \ . \tag{2}$$

Macroscopic yielding is represented by a scalar yield function, ϕ, parameterized by ρ, the macroscopic deviatoric stress modulus S, pressure p, and the particle deviatoric stress modulus $s = \sqrt{\frac{2}{3}}\bar{\sigma}_y$. ϕ is relevant only when the powder particle is at yield. The powder is assumed to yield in a rate independent fashion. This assumption is valid for the Ancorsteel 1000 powder as is shown in figure 2. We have chosen the macroscopic yield function ϕ to be of the form:

$$\Phi = S^2 + b(\rho)p^2 - c(\rho)s^2 = 0 \tag{3}$$

The functions $b(\rho)$ and $c(\rho)$ need to be experimentally determined. Using triaxial compaction, we obtain the following expressions for Ancorsteel 1000 Iron powder:

$$b(\rho) = \frac{0.13\rho^{-5.60}}{(\rho + 0.62)^{1.35}} \qquad c(\rho) = \rho^6 \ . \tag{4}$$

The isotropic elastic modulus of powder compacts decreases with decreasing density. A solid premixed particulate lubricant is usually added to the powder to reduce friction with the tooling, and to allow compaction to proceed at low enough loads to prevent cracking. The amount of reduction in the elastic modulus strongly depends on the amount of lubricant, with the reduction in modulus increasing with increase in the amount of lubricant. The effect of compact density on compact elastic modulus obtained from resonant frequency measurements of prismatic bars is shown in figure 3.

Assuming that the plastic rate of strain vector for the particle is given by normality , it is possible to prove macroscopic normality [4]. Thus,

$$\mathbf{D}^p = \Lambda \mathbf{grad}(\phi). \tag{5}$$

During the deformation, the conservation of mass is expressed as:

$$\dot{\rho} = -\rho \mathrm{tr}\mathbf{D}^p \ . \tag{6}$$

3 FINITE ELEMENT IMPLEMENTATION

An isotropic hypoelastic formulation is used to define the evolution of the stress state as a function of the rate kinematics. Invariance of stresses under superposed rigid body rotations requires the use of *rotation neutralized* quantities [7, 11]. The constitutive model outlined in section 2 has been implemented into the following codes: an implicit two dimensional code NIKE2D [5], an explicit three dimensional code DYNA3D [6] and an implicit three dimensional code WOLF [12].

The objective rate of change of stress **T**, is given by

$$\overset{\triangledown}{\mathbf{T}} = \hat{\mathcal{C}}(\rho)[\mathbf{D} - \mathbf{D}^p] \qquad (7)$$

where the rate of stretching tensor, **D** is the symmetric part of the velocity gradient tensor, **L**. **D** is additively decomposed into elastic and plastic parts. Letting $\hat{\mathcal{C}}$ represent the isotropic elastic moduli of the aggregate, the objective rate of stress, $\overset{\triangledown}{\mathbf{T}}$, is defined by:

$$\overset{\triangledown}{\mathbf{T}} = \dot{\mathbf{T}} - \mathbf{WT} + \mathbf{TW} \qquad (8)$$

where **W** is obtained by the solution of the appropriate initial value problem [7].

The constitutive equations stated above are sufficient to determine the value of the parameter Λ under any prescribed deformation. The constitutive integration scheme used in this work is similar to [1] and is explained in [9].

4 RESULTS

We now illustrate the simulation of typical two and three dimensional pressing of powder compacts. In all the models, friction is modeled as Coulombian with a friction factor of 0.2.

4.1 *Compaction of a Bushing*

An axisymmetric bushing of Ancorsteel 1000 powder was compacted on a 150 ton mechanical press. The geometry of the part and tooling is shown in figure 4. The simulation was carried out using the powder properties of Ancorsteel 1000 powder using constant elastic properties. Two cases with different elastic moduli were studied, one corresponding to the elastic modulus of fully dense powder and the other corresponding to 0.75 weight percent Zinc Stearate lubricant. As compaction starts due to top punch motion, the density increases near the top of the powder compact (figure 5a), due to the friction between the compact, core-rod, die and the top punch. With continued top punch motion, the rest of the bushing densifies with the top being always more dense than the bottom. At the end of top punch motion, the bottom of the bushing starts to densify preferentially with bottom punch stroke (figure 5b). The compact expands as it exits the die cavity thereby decreasing the compact density (figure 5c). The amount of expansion decreases with increasing elastic bulk modulus. Thus the resulting density after compaction increases with increasing elastic bulk modulus, as shown by the final density profile in figure 6. Figure 6 also shows that the FEM predictions are in good agreement with experimental results of density obtained by sectioning and weighing compacted samples.

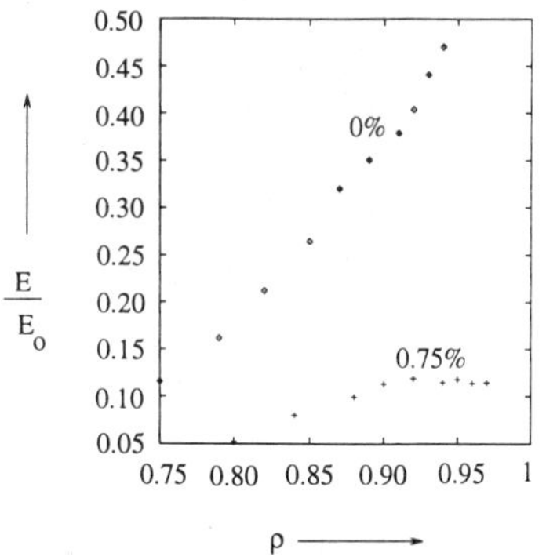

Figure 3: Effect of compact density and Zinc Stearate lubricant on the normalized elastic modulus. $E_o = 200GPa$ is the elastic modulus of fully dense and sintered Ancorsteel 1000 powder. The percentages indicate the weight percent of lubricant.

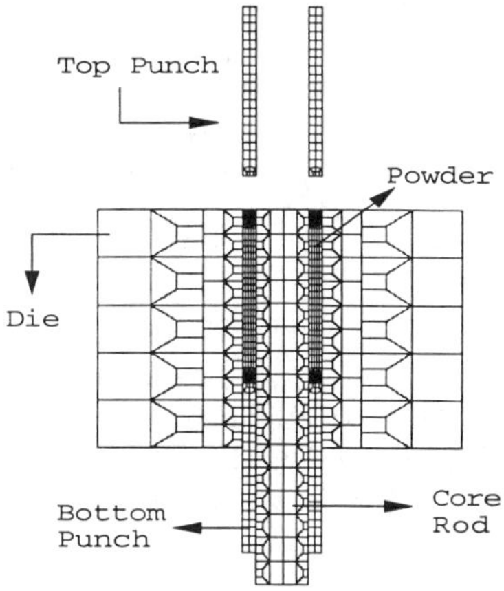

Figure 4: Finite Element mesh of the powder and tooling for bushing compaction with a stationary die. Punches are shown in the fill position. Bushing OD = 0.75in, ID = 0.50in, final height = 1in.

4.2 *Compaction of a Cylinder*

The compaction of a cylinder of Ancorsteel 1000 powder lubricated with 0.75 weight percent Zinc Stearate is shown next. The compact has a diam-

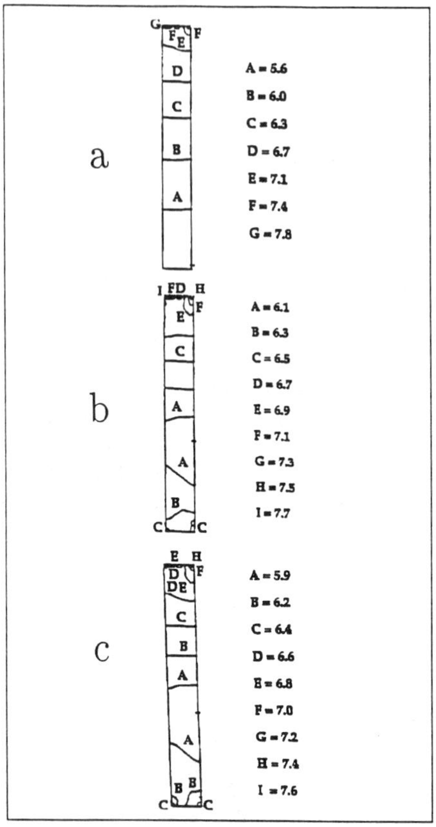

Figure 5: Evolution of density with stroke. a) Distribution at end of top punch motion. b) Distribution just as the compact exits the die cavity. c) Distribution after the core rod is stripped.

eter of 1.0in and a height to diameter ratio of 0.47. The compaction proceeds from an initial constant relative density of 0.42 to a final average relative density of 0.90. The load stroke predictions from this simulation are compared against experimental results from a specimen compacted with top punch motion on a Tinus-Olsen machine. This comparison is shown in figure 7 and shows the accuracy of both the constitutive properties for the powder and the finite element implementation.

4.3 *Compaction of a Multi Level part*

We now show the compaction of an axisymmetric multilevel part. The geometry is shown in figure 8a and is meant to be representative of cluster

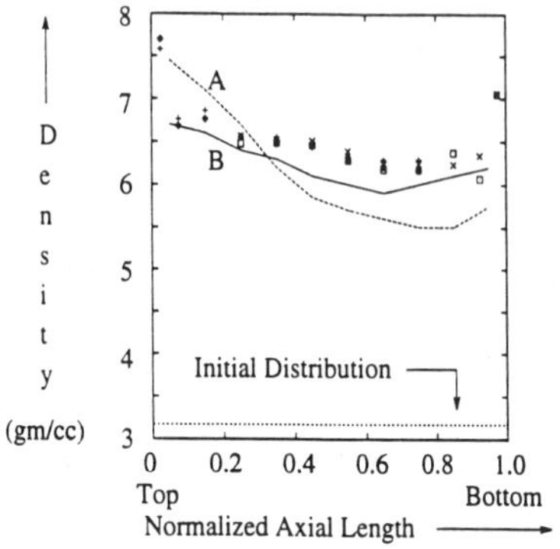

Figure 6: FEM predictions of averaged axial density distribution for the bushing. The solid curve marked as A denotes predictions using fully dense sintered powder elastic properties. The dashed line marked as B denotes predictions using elastic properties of 75 percent dense unsintered powder properties. The points represent experimental values from four different specimens.

gears commonly used in appliances.

The part is pressed with top and bottom outer punch motions, and floating die. The bottom inner punch punch is held stationary. Ancorsteel 1000 powder lubricated with 0.75 weight percent Zinc stearate is used. The part is pressed from an initial relative density of 0.41 to a final average relative density of 0.82. The gear section has an outer diameter of 1.5in, an inner diameter of 0.395in and is 0.12in long. The pinion has an outer diamter of 0.0395in and is 0.2in long. The purpose of this simulation is to illustrate the effect of two different punch displacement histories on the final density distribution.The two different punch motion histories are shown in figure 8b. Both cases have the *same final punch positions.* The final density profiles for both cases are shown in figure 8c.

In the first case, the lower outer punch drops at a relatively low speed. This results in an early increase in density of the gear section. It also leads to a region of minimum density (approximately 0.79) at the plane between the pinion and the gear. This is a location where cracks are likely to appear. In the second case, the lower outer punch drops more rapidly. This results in an early increase in density in the pinion. The plane between the pinion and the gear becomes a region of high density (approximately 0.83).

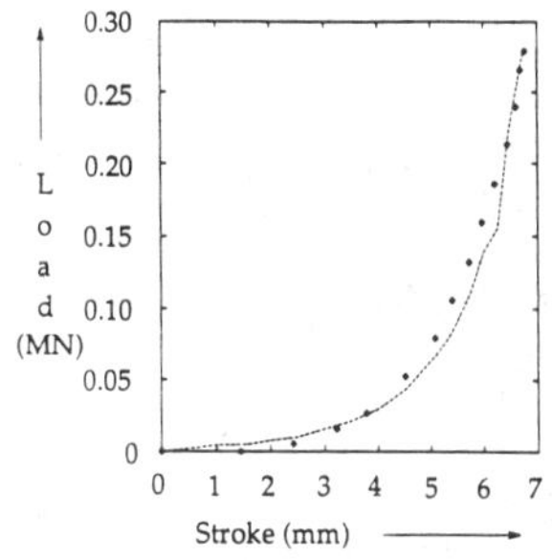

Figure 7: FEM load-stroke results for compaction of a cylinder. The points represent experimental values from compaction in a Tinus-Olsen mechanical testing machine.

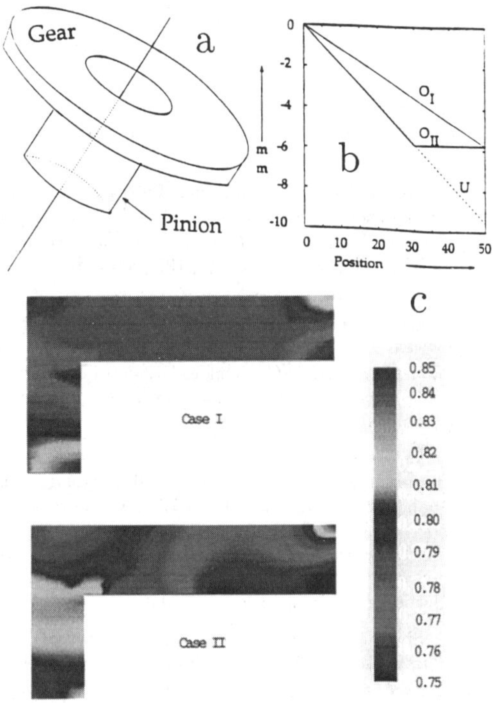

Figure 8: a) Multilevel geometry. b) Punch motion history for two different cases. O_I and O_{II} denote the lower outer punch motions. The upper punch motion is the same for both cases and is denoted by U. c) Effect of punch motion history on density profiles.

4.4 *Compaction of a three dimensional part*

The three dimensional compaction simulation of a link made of Ancorsteel 1000 powder is presented next. The powder properties account for 0.75 weight percent of Zinc Stearate as lubricant. The geometry of the powder and tooling is shown in figure 9. Compaction is achieved by top punch motion alone. The lower dies and the core rods remain stationary. The density distribution at an intermediate stage is shown in figure 10. Compaction with stationary bottom dies results in a substantially higher density results in the arm region as compared to the hub region.

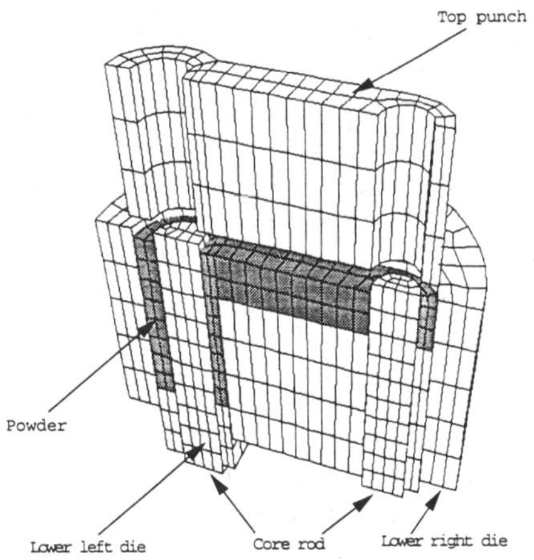

Figure 9: Three dimensional finite element mesh of powder and tooling for modeling the pressing of a link. The model was run using the explicit implementation in DYNA3D.

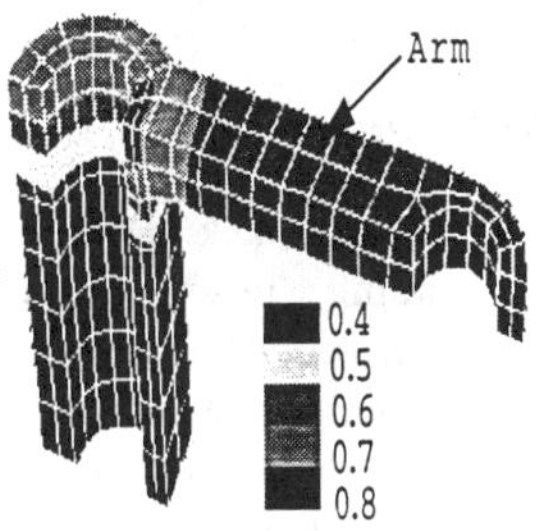

Figure 10: Density distribution in the powder compact for the geometry shown in figure 9.

5 CONCLUSIONS

We have presented a constitutive model for the simulation of the cold compaction of metallic powders with ductile irregular particles. The model was calibrated for Ancorsteel 1000 iron powder. The constitutive model was integrated and implemented in a finite element scheme using a hypoelastic isotropic plastic scheme.

The finite element predictions of loads and density distributions were found to be in good agreement with experiments conducted with Ancorsteel 1000 powder. The history of punch motions was found to have a significant effect on the density distribution both during and after compaction. A three dimensional simulation of a complex geometry was also carried out to illustrate the capabilities of the finite element code.

The experimental results suggest that the effect of the density on the elastic properties plays a critical role on the final density distribution through the amount of elastic relaxation during ejection. This effect can be linked to the common problem of the appearance of cracks during the ejection process. This mechanism is currently being investigated.

6 ACKNOWLEDGEMENTS

The authors wish to acknowledge Mr. Tom Prucher for providing the experimental data for the variation of elastic modulus with density, Mr. Vince Godby for conducting the experimental determination of the density variation in the bushing and Mr. Steve Armstrong for setting up the finite element models.

REFERENCES

[1] N. Aravas. On the numerical integration of a class of pressure-dependent plasticity models. *International Journal for Numerical Methods in Engineering*, 24:1395–1416, 1987.

[2] S.B. Brown and G.G. Weber. A constitutive model for the compaction of metal powders. In *Modern Developments in Powder Metallurgy*, volume 18–21, pages 465–476. MPIF, Princeton, New Jersey 08540, 1988.

[3] J. Crawford and P. Lindskog. Constitutive equations and their role in the modeling of the cold pressing process. *Scandinavian Journal of Metallurgy*, 12:277–281, 1983.

[4] A.L. Gurson. Continuum theory of ductile rupture by void nucleation and growth: Part (i)—yield criteria and flow rules for porous ductile media. *ASME Journal of Engineering Materials and Technology*, page 2, 1977.

[5] J.O. Hallquist. NIKE2D–a vectorized, implicit, finite deformation, finite-element code for analyzing the static and dynamic response of 2-D solids. Technical Report UCRL-19677, Lawrence Livermore National Laboratory, Livermore, California 94550, 1983.

[6] J.O. Hallquist. Theoretical manual for DYNA3D. Technical Report UCID–19401, Lawrence Livermore National Laboratory, Livermore, California 94550, 1983.

[7] T. J. R. Hughes. Numerical implementation of constitutive models: Rate independent deviatoric plasticity. In S. Nemat-Nasser, R. Asaro, and G. Hegemier, editors, *Theoretical Foundations for Large-Scale Computations of Nonlinear Material Behavior*, pages 29–57. Martinus Nijhoff Publishers, Boston, 1983.

[8] A. Jinka. R. Lewis, and D. Gethin. Finite element simulation of powder compaction via the flow formulation. In L. Peaese and R. Sansoucy, editors, *Advances in Powder Metallurgy*, volume 1, pages 123–144. American Powder Metallurgy Institute, Princeton, New Jersey, 1991.

[9] Sriram Krishnaswami and Juan R. L. Trasorras. Finite element simulation of powder compaction. *in preparation*, 1995.

[10] Y. Morimoto, T. Hayashi, and T. Takei. Mechanical behavior of powders in a mold with variable cross sections. *International Journal of Powder Metallurgy and Powder Technology*, 18(1):129–145, 1982.

[11] J.C. Nagtegaal and F.F. Veldpaus. On the implementation of finite strain plasticity equations in a numerical model. In J.F.T Pittman, R.D. Wood, J.M. Alexander, and O.C. Zienkiewicz, editors, *Numerical Methods in Industrial Forming Processes (Proc.)*, pages 351–371. John Wiley and Sons Ltd., New York, 1984.

[12] Rajesh Parameswaran, Shankar Rachakonda, and Juan. R. L. Trasorras. WOLF1.1 - A non-linear implicit two and three dimensional finite element code for the solution of thermomechanical boundary value problems - User Manual, Concurrent Technologies Corporation, Johnstown, PA 15904, 1994.

[13] S. Shima. *A Study of Forming of Metal Powders and Porous Metals*. PhD thesis, Kyoto University, 1975.

[14] J. R. L. Trasorras, S. Armstrong, and T. J. McCabe. Modeling the compaction of steel powder parts. In *Advances in Powder Metallurgy and Particulate Materials - 1994*, volume 7, pages 33–50. American Powder Metallurgy Institute, Princeton, New Jersey, 1994.

[15] J. R. L. Trasorras, T.M. Krauss, and B.L. Ferguson. Modeling of powder compaction using the finite element method. In T.G. Gasbarre and W.F. Jandeska, editors, *Advances in Powder Metallurgy*, volume 1, pages 85–104. American Powder Metallurgy Institute, Princeton, New Jersey, 1989.

Simulation of Materials Processing: Theory, Methods and Applications, Shen & Dawson (eds)
© 1995 Balkema, Rotterdam. ISBN 90 5410 553 4

The bending of Incoloy-800 tubes and the resultant internal stresses

B.W. Leitch
Materials and Mechanics Branch, AECL, Whiteshell Laboratories, Pinawa, Man., Canada
C.N.Tomé
Reactor Materials Research Branch, AECL, Whiteshell Laboratories, Pinawa, Man., Canada

ABSTRACT

A nickel alloy, INCOLOY-800, is used for the heat exchanger tubes of steam generators in some CANDU nuclear power generation stations. The mechanical process to produce the 180° bend from initially straight tubes, induces high residual stress in the bend and there is a concern that these stresses, in conjunction with the normal operating stress, could contribute to fatigue cracking and localized corrosive thinning of the tubes. The two mechanical processes of compression bending and draw/rotary bending that are used to form the U-bend are simulated using an explicit, finite element program. As the bend is formed, internal stresses and thinning of the tube wall thickness takes place. A good agreement is found between the calculated residual stresses and thinning values and experimentally determined values using neutron and X-ray diffraction.

INTRODUCTION

A recent non-destructive examination (NDT) of the thin-walled tubes in the steam generator system of a power generation station indicated a possibility that stress corrosion was very active within the generator. The NDT examination of the U-bend region of the heat exchanger tubes indicated that the wall thickness of the bend was significantly thinner than the nominal wall thickness of the straight tubes. It was reasoned that the process of manufacturing the bend created residual stresses that in conjunction with the normal operating stress would accelerate the stress corrosion action in the U-bend area and produce a subsequent thinning of the U-bend wall thickness.

A number of experimental tests (Holden, et al. 1988; Holt, et al. 1984; Winegar 1980) have been carried out to determine the residual stresses and internal strains in the U-bend. This paper describes the analytical simulation of the manufacturing process and the resultant stress and wall thinning that it produces.

MECHANICAL BENDING PROCESS

There is a variety of manufacturing processes to produce the U-bend section of the steam generator tubes (Murphy & Winegar 1991). The U-bends are created from straight pipes of 14.6 mm mean diameter and a nominal wall thickness of 1.27 mm. The U-bend radius is 73 mm. The two principal methods of producing a U-bend are compression bending, Figure 1(a) and the draw/rotary bending, Figure 1(b). The compression bending creates the U-bend by rotating the wiper shoe about a fixed form. In the draw/rotary process, the straight pipe is rigidly clamped to the bend former. The bend former and straight tube are then rotated together, forming the U-bend against the fixed pressure die. There is a third technique known as the modified compression bending process, Figure 1(c), which is identical to the draw/rotary method but the clamping die holds the workpiece further down on the straight section of the U-bend former. It is assumed that the modified compression and the draw/rotary bending processes are very similar in their operation, that the draw/rotary analysis would produce acceptable results for both techniques.

ANALYTICAL MODELLING

The analytical simulation of these processes were modelled using the explicit, finite element program LS-DYNA3D (Hallquist, 1994). Figure 2 shows an exploded view of the different parts of the finite

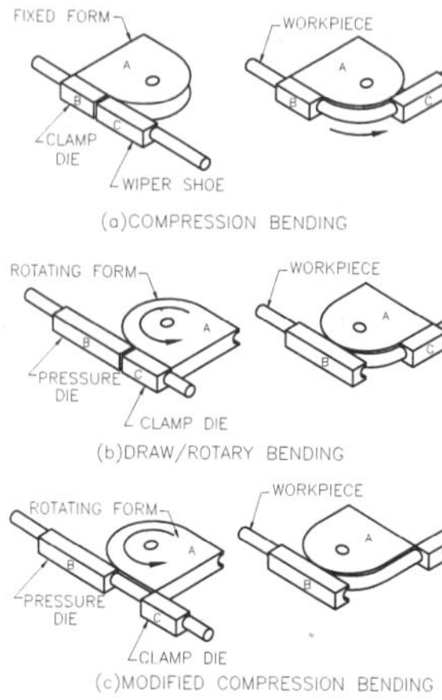

Fig 1. Mechanical Bending Processes

element model. All of the parts were constructed using 4-noded quadrilateral shell elements. Rigid material properties were assigned to the pressure die and U-bend former. Elastic-plastic material properties of a typical steam generator tube material (INCOLOY-800) were applied to the initially straight tube of the finite element model. Table 1 contains the relevant material properties. Membrane thinning was applied to the workpiece elements. Frictionless contact was assumed between the workpiece and the rigid former and wiper/pressure die.

Table 1. Material Properties

Young's Modulus, E	180 GPa
Poisson's Ratio	0.32
Flow Stress	448 MPa
Hardening Modulus, E_T	0.001
Density	8800 kg/m^3

By suitable manipulation of the boundary conditions in the model, compression (Figure 3(a)) and draw/rotary (Figure 3(b)) bending processes can be simulated.

RESULTS AND DISCUSSION

Shell thickness and axial stresses were measured at the center of the 180° bend, after each bending simulation had been completed and a full 180° arc had been circumscribed. Figure 4 contains the tube thickness measurements for the compression and draw/rotary bending analyses. The wall thickness values have been normalized with respect to the

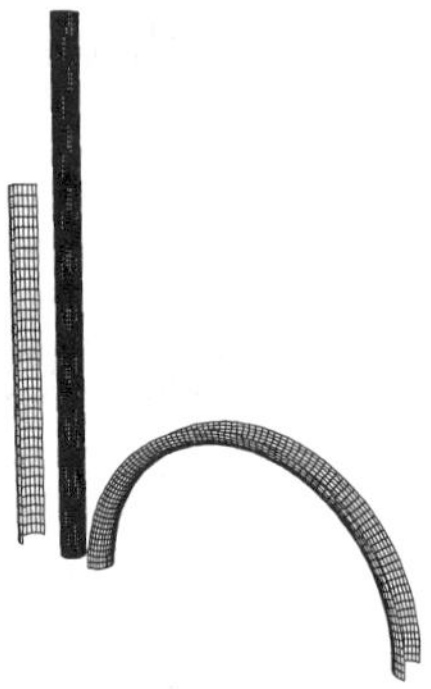

Fig. 2. Finite Element Model (Exploded View)

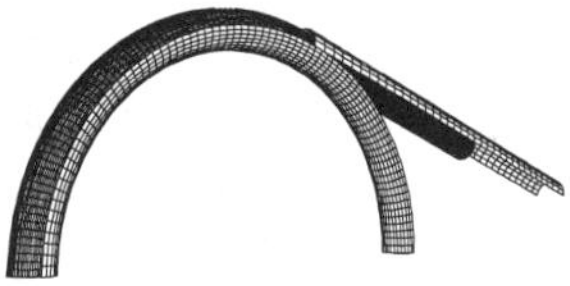

Fig. 3(a). Compression Bending

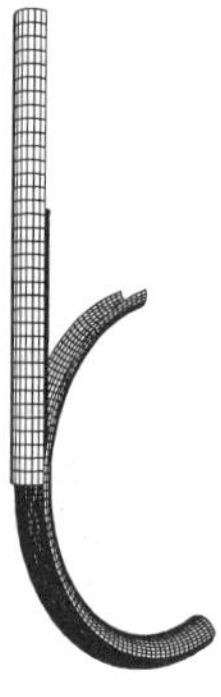

Fig. 3(b). Draw/Rotary Bending

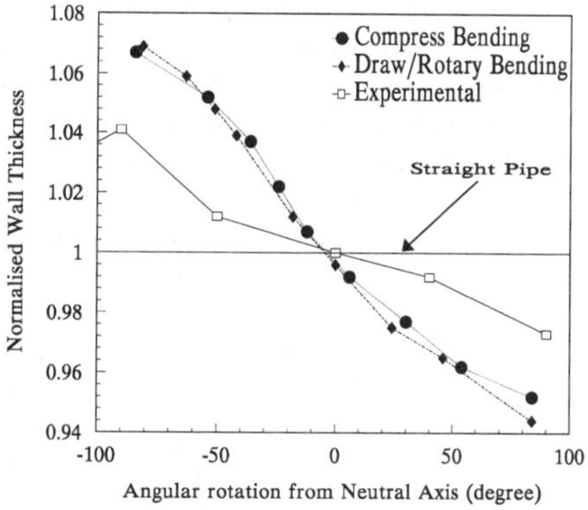

Fig. 4. Normalized Thickness Measurements at the 90° Position of the U-bend

nominal wall thickness of the straight pipe. The thickness values were measured at the 90° position of the 180° U-bend and are shown in Figure 4. The ±90° position are the extrados and intrados, respectively, of the U-bend where the curved neutral axis of the bend is denoted as the 0° position.

The normalized wall thickness values of the compression and draw/rotary bending simulations indicate that there is no difference between the two bending processes. The nominal wall thickness of the straight tube is 1.27 mm, and as shown in Figure 4, the intrados and extrados thickness values are approximately centered about the normalized value. This nominal value occurs at the neutral axis position of the U-bend. At the inside of the bend (i.e., -ve rotation from the neutral axis) there is an increase in the shell thickness, whilst at the top, outside of the U-bend there is a thinning of the tube thickness. The normalized, experimental thickness values (Holden, 1993), show the same wall-thinning trend, from inside to outside and has a maximum wall thickness of 1.27 mm. The experimental measurements were performed on tubes that had a nominal, straight tube wall thickness of 1.2 mm. Taking into account the different nominal wall thicknesses of the experimental and analytical tubes, the analytical modelling is a reasonable simulation of the actual bending process. The bending process does create a thinning of the tube wall section.

The bending process will create stresses in the U-bend which could be a potential accelerator for stress corrosion. Axial stress values were extracted from each bending model at the 90° position of the U-bend. These results are given in Figure 5. As expected, there is a region about the neutral axis position where the axial stress reverses from tensile to compressive values. Experimental neutron diffraction stress measurements (Holden, et al. 1988) of the U-bend are also included in Figure 5. Measurements of lattice spacings were made for [111] and [002] crystallographic planes oriented along the axial direction of the tube. The discrepancy between [111] and [002] normal stress is due to the fact that the deformation is not uniform, but depends on the individual grain orientation. The experimental stress measurements were determined from a U-bend made from 16 mm outside diameter, a 1.2 mm wall thickness and a band radius of 121 mm. The finite element simulations were performed on a tube of 12.7 mm outside diameter, a 1.27 mm wall thickness and a U-bend radius of 146 mm. The simulations did not take into account the elastic spring-back at the completion of the bending process, when the dies are separated. An approximate elastic spring-back stress was calculated and the finite element stress values were adjusted to include the elastic recovery. Taking into account these slight geometric differences, the experimental and analytical stress results show a reasonable agreement.

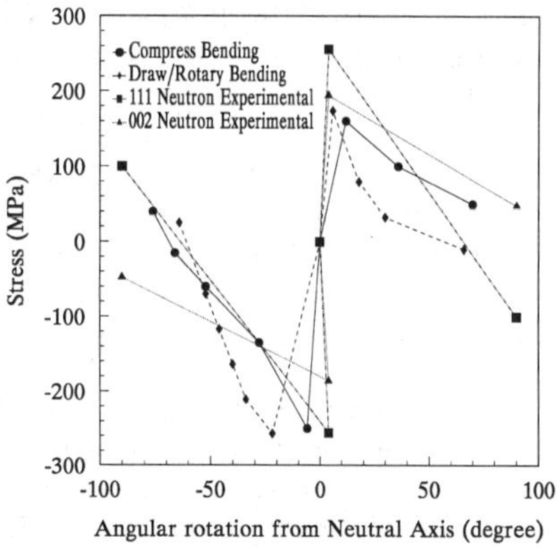

Fig. 5. Axial Stresses at the 90° Position of the U-bend

CONCLUSIONS

Numerical simulations of the two main processes for forming steam generator U-bends have shown that they produce the same type of bend, with similar distributions of the wall thickness and internal stress values.

Allowing for the slightly different geometrical experimental testpieces and the ideal conditions of the numerical simulation, the analytical results show an acceptable agreement with actual measurements.

Additional work is being planned to incorporate a polycrystalline material model into the FE simulation, in order to account for anisotropy effects.

ACKNOWLEDGEMENTS

The authors would like to thank Rudy R. Verderber and Khanh D. Bui of Livermore Software Technology Corporation and Dennis K. Lam of Cray Research Inc., for their valuable knowledge and assistance.

REFERENCES

Hallquist, J.O. 1994. *LS-DYNA3D User's Manual.* Livermore Software Technology Corporation, Livermore, California, U.S.A.

Holden, T.M. 1993. Chalk River Laboratories, Atomic Energy of Canada Limited. Unpublished results.

Holden, T.M., R.A. Holt, G. Dolling, B.M. Powell and J.E. Winegar 1988. *Characterization of Residual Stresses in Bent Incoloy-800 Tubing by Neutron Diffraction.* Metallurgical Transactions A, Vol. 19A, pp. 2207-2214.

Holt, R.A., G. Dolling, B.M. Powell, T.M. Holden and J.E. Winegar 1984. *Proceedings of the 5th Riso International Symposium on Metallurgy and Materials.* Riso National Laboratory, Denmark, pp. 295-300.

Murphy, E.V. and J.E. Winegar 1991. *Residual Stresses in Incoloy-800 Steam Generator Tubes.* Atomic Energy of Canada Report, AECL CANDU Operations, Toronto, Canada.

Rothman, M.F. (ed) 1988. *High-temperature Property Data: Ferrous Alloys.* ASM International, Metals Park, OH, U.S.A.

Winegar, J.E. 1980. *X-Ray Measurements of Residual Stress in Metals at Chalk River Nuclear Laboratories.* Atomic Energy of Canada Report, AECL-6961.

Simulation of Materials Processing: Theory, Methods and Applications, Shen & Dawson (eds)
© 1995 Balkema, Rotterdam. ISBN 90 5410 553 4

A finite element modelling for the inertia welding process

Elisabeth Massoni & Alain Moal
Cemef, École Nationale Supérieure des Mines de Paris, URA CNRS, Sophia-Antipolis, France

ABSTRACT: A finite element simulation of inertia welding of two similar parts is presented. Thermomechanical coupling accounts for the temperature dependence of material behaviour, but also -and primarily- of friction, which is the governing factor of the process. Due to the fast deceleration at the end of the process, inertial forces have been included in the equilibrium equation. Severe mesh distortions are experienced, mainly along the weld interface which necessitates remeshing; an efficient algorithm has been used. Numerical and experimental results are compared.

1 INTRODUCTION

Inertia welding is a very high-performance method for joining different parts (Wang 1975) particularly for plane motor shafts. The parts are forced to rub against each other and the frictional heat generated softens the material near the part interface. This complex process involves the interaction of thermal, mechanical and metallurgical phenomena. Most of the models simulate solely the transient temperature distribution during inertia welding (Wang 1970) or during conventional friction welding process using analytical solution (Francis 1985), finite difference method (Cheng 1962) & (Cheng 1963) or finite element model (Atsuta 1985). Very few papers analyse the coupling between thermal and mechanical effects during conventional friction welding (Sluzalec 1990).

The present paper outlines a more complete description by taking into account the thermomechanical effects in the numerical simulation of the inertia welding process of two similar pieces (Moal 1992).

The material is assumed to be isotropic, incompressible and to obey the viscoplastic Norton-Hoff constitutive law in which consistency and strain rate sensitivity are temperature dependent. Two models for the friction law, based on experimental measurements, are proposed. The mechanical problem is solved considering the virtual power principle including inertia terms. An extended 2D formulation is used including the computation of the three components of the velocity field with an axisymmetric domain.

The temperature field is computed separately for each time step. The mechanical and thermal resolutions are linked through the dissipated deformation energy, the friction heat flux on the contact surface and the temperature dependence of the constitutive and friction laws. The phase changes are taken into account. The heat transfer equation with the boundary conditions (conduction, convection and radiation) is discretized by the finite element method. An efficient remeshing algorithm is implemented. The mechanical and thermal resolutions have already been checked by comparison with high speed torsion experiments (Moal 1993). In this paper the simulation of inertia welding is compared with actual industrial welds.

2 THE THERMOMECANICAL MODEL

2.1 *Mechanical equations*

Dynamic equilibrium is written through the principle of virtual power:

$$\forall\, v^* \in V^*, \quad \int_\Omega (\rho\gamma\, v^* - \sigma : \dot{\varepsilon}^*)d\Omega =$$

$$\int_{\partial\Omega_f} \tau_f\, \Delta v^*\, dS \; + \; \int_{\partial\Omega} T\, v^*\, dS \qquad (1)$$

σ is the Cauchy stress tensor, $\dot{\varepsilon}^*$ the virtual strain rate tensor, ρ is the volumic mass and γ the acceleration vector, $\partial\Omega_f$ is the subdomain of the boundary where contact of the two pieces occurs, τ_f is the friction shear stress. The material is assumed to be homogeneous, isotropic and incompressible and the stress deviatoric part follows the viscoplastic Norton Hoff constitutive equation:

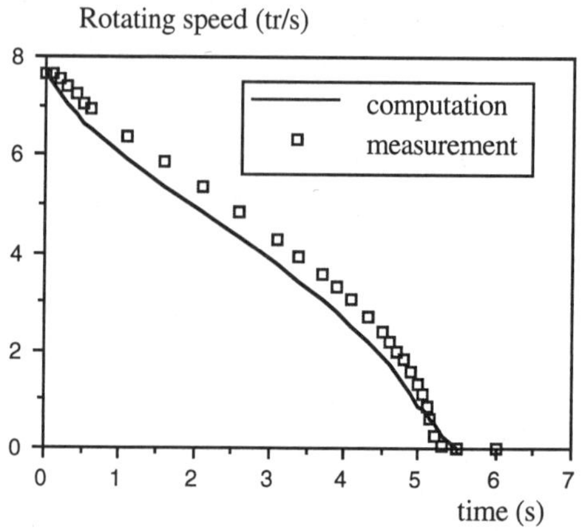

Figure 1. Evolution of the rotating speed during the inertia friction welding process.

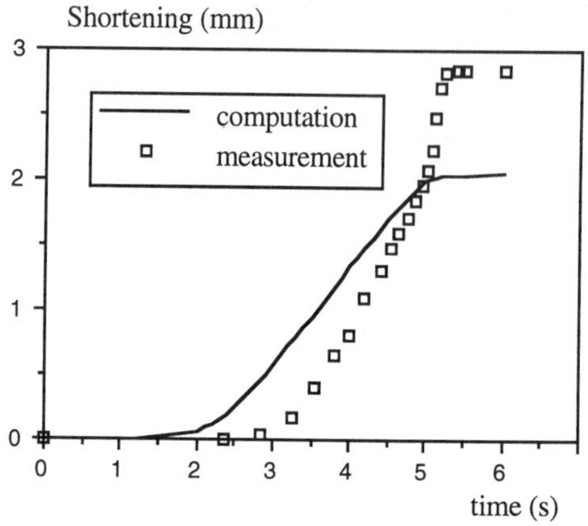

Figure 2. Evolution of the piece shortening during the inertia friction welding process.

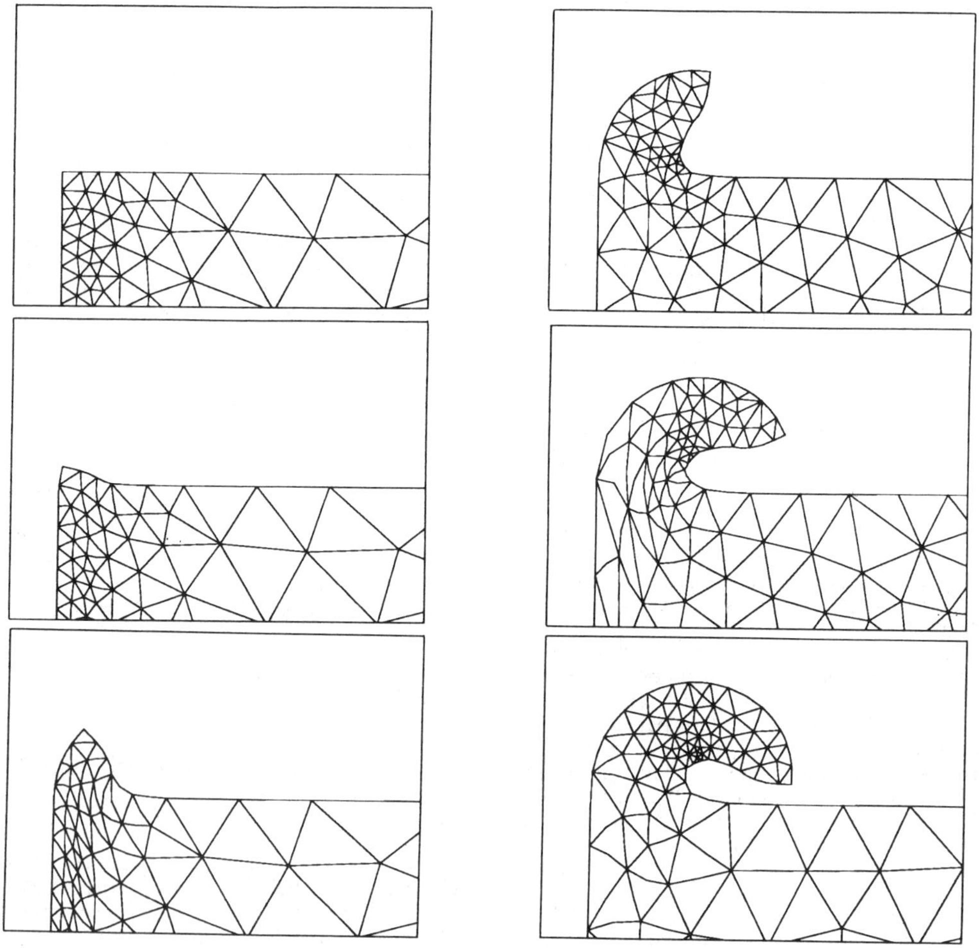

Figure 3. Successive deformation of finite element meshes during the whole process.

$$\sigma = 2\,K\left(\sqrt{3}\,\dot{\bar{\varepsilon}}\right)^{m-1}\dot{\varepsilon} - p\,\mathbb{I} \qquad (2)$$

where p is the pressure, $\mathbb{I}$ is the identity tensor, K is the material consistency, m the strain rate sensitivity index and $\dot{\bar{\varepsilon}}$ the equivalent plastic strain rate. K and m are thermodependent and experimentally determined (Soucail 1992).

Inertia welding is a process in which friction is the most important parameter. This is the reason why tribological experiments have been performed to determine an accurate friction model. Two models have been tested: the first (depending on the temperature) (Moal 1992) is described by equation (3) and the second (depending on the rotating velocity) by equation (4):

$$\tau_f = -\alpha\, p\ \Delta V_g \left|\Delta V_g\right|^{q-1} \qquad (3)$$

ΔV_g = the sliding velocity between the two pieces
q = the friction rate sensitivity index.
p = the prescribed pressure p_i when $T<T_c$ (T_c critical temperature)
p = the consistency parameter K(T) when $T \geq T_c$

$$\tau_f = -\alpha\, p_i\, g\,(\Delta V_g)\frac{\Delta V_g}{\left|\Delta V_g\right|} \qquad (4)$$

$\Delta V_g > [\Delta V_g]_c \Rightarrow\ g\,(\Delta V_g) = 1$
$\Delta V_g \leq [\Delta V_g]_c \Rightarrow\ g\,(\Delta V_g)$ = polynomial function
$[\Delta V_g]_c$ = critical value of the relative rotating velocity
p_i = the prescribed pressure.

2.2 *Heat transfer equation*

The heat transfer equation is represented by the following lagrangian equation:

$$\rho C(T)\ \frac{dT}{dt} = \sigma{:}\dot{\varepsilon} + \text{div}\,(k\ \text{grad}\ T) \qquad (5)$$

where $\rho C(T)$ represents the heat capacity, k the thermal conductivity and $\sigma{:}\dot{\varepsilon}$ the rate of internal heat generation during plastic deformation. The boundary conditions are:

on the contact surfaces the friction flux is prescribed:
$-\,k\ \text{grad}T \cdot n_c = -\,\Phi$

on the free surface , convection and radiation give:
$-\,k\ \text{grad}T \cdot n_f = -\,(h_{conv} + h_{rad})\,(T - T_a)$

T_a is the outside temperature, h_{conv} and h_{rad} are respectively the convection and radiation coefficient, n_c and n_f are respectively the exit normal vector to the contact surface and free surface.

3 NUMERICAL RESOLUTION

We solve separately mechanical and thermal equations. But thermal and mechanical problems are strongly coupled via the the thermodependence of rheological parameters and the power dissipated by plastic deformation and friction.

3.1 *Mechanical equation*

a/ Time discretization

The time discretization follows an updated lagrangian approach. We use a semi-implicit integration scheme to compute the acceleration at time t γ_t knowing the velocity field at time t and $t-\Delta t$, V_t and $V_{t-\Delta t}$:

for $\theta \in [0,1[$

$$V_t = V_{t-\Delta t} + \Delta t\left[\theta\,\gamma_{t-\Delta t} + (1-\theta)\gamma_t\right] \qquad (6)$$

$$\gamma_t = \frac{1}{1-\theta}\left(\frac{V_t - V_{t-\Delta t}}{\Delta t}\right) - \frac{\theta}{1-\theta}\gamma_{t-\Delta t} \qquad (7)$$

Then the configuration can be updated at time $t+\Delta t$ using the explicit integration scheme:

$$X_{t+\Delta t} = X_t + V_t\,\Delta t + \frac{\Delta t^2}{2}\left[(1-\theta)\,\gamma_t + \theta\,\gamma_{t-\Delta t}\right] \qquad (8)$$

This scheme is unconditionally stable, but the choice of a short time increment makes the accumulated error on the whole time interval acceptable.

b/ Spatial discretization

The spatial discretization is made with axisymmetric 2D elements. More details on this finite element discretization can be found in (Moal 1993). The three components of the velocity field in cylindrical coordinates have been introduced as $V = (rV_r, V_\theta, V_z)$ (three degrees of freedom per node of a 2D mesh). The time and spatial discretization of equation (1) leads to a non linear system solved by an iterative full Newton Raphson Method.

3.2 *Thermal part*

The space discretization of equation (5) is done by the finite element Galerkin method, leading to the below non-linear system in terms of the nodal unknown namely the temperature:

$$C(\mathrm{T})\frac{\mathrm{dT}}{\mathrm{dt}} + K(\mathrm{T})\,\mathrm{T} + F(\mathrm{T}) = 0 \qquad (9)$$

where C is the heat capacity matrix, K the conductivity matrix, F the thermal load vector. In order to avoid instabilities due to thermal chock experienced with the Galerkin formulation, a special condensation of matrix C was used as described in (Chenot 1992). The resolution of the heat equation uses the following time integration scheme:

$$\mathrm{T} = \mathrm{a}\,\mathrm{T}_{t-\Delta t} + \left(\frac{3}{2} - 2\mathrm{a} - \mathrm{b}\right)\mathrm{T}_t + \left(\mathrm{a} - \frac{1}{2} + \mathrm{b}\right)\mathrm{T}_{t+\Delta t}$$

$$\frac{\mathrm{dT}}{\mathrm{dt}} = \mathrm{b}\frac{\mathrm{T}_{t+\Delta t} - \mathrm{T}_t}{\Delta t} + (1-\mathrm{b})\frac{\mathrm{T}_t - \mathrm{T}_{t-\Delta t}}{\Delta t} \qquad (10)$$

for a Dupont scheme: $\mathrm{a} = \frac{1}{4}$ and $\mathrm{b} = 1$,

for an implicit scheme: $\mathrm{a} = 0$ and $\mathrm{b} = \frac{3}{2}$.

4 ADAPTIVE REMESHING

During inertia welding of even simple pieces, such large geometrical distortions are found locally that remeshing is necessary. There are different geometrical criteria to start the remeshing algorithm (Coupez 1992) & (Fourment 1992):

- elements are too degenerated,
- the curvature of an element on the boundary is too large,
- the finite element estimated errors overtake the tolerated maximum values.

One of them has been specifically developped for inertia welding and added to the geometrical curvature which is one of the remeshing criteria: a pseudo-curvature derived from the thermal gradient (Moal 1992) is defined; when introduced in the remeshing algorithm this curvature generates small elements along the weld interface where thermal gradient is high. However this criterion is not directly linked to finite element errors and only operates on the node along the boundary.

For finite element errors a mechanical error estimator is used (Zienkiewicz 1988) and for the thermal problem an error estimator based on the evaluation of the discontinuity of the temperature gradient (Huang 1989) is implemented .

5 NUMERICAL RESULTS

The material used is a Nickel base alloy. The initial rotating velocity is 460 rpm and the inertia moment of the flywheel is 171 kg.m^2. The tubular pieces are set in contact under a pressure of 360 Mpa. The length of each piece is 21 mm, the mid-wall diameter is 90 mm and the thickness is 10 mm. Since the two pieces are similar in geometry and nature, the problem was considered symmetric, only one tube is modelled with half the rotating

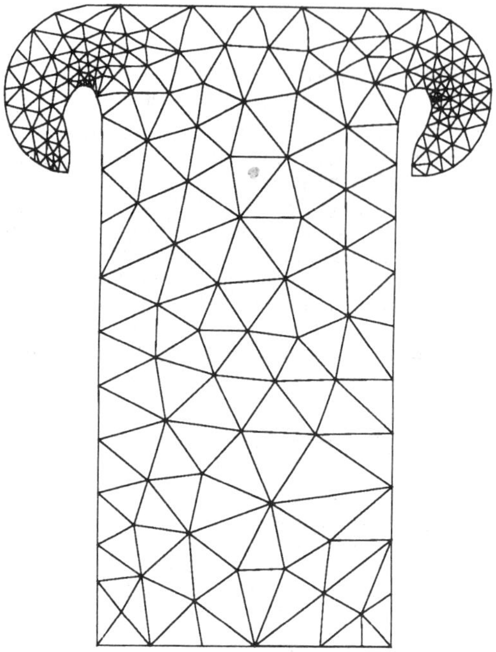

Figure 4. The whole mesh at the end of the process

speed. The friction is represented by equation (3) and remeshing is only based on geometrical criterion.

Concerning the rotating velocity, figure 1 shows that the numerical model is able to determine the time when the velocity of the rotating part comes to zero. At this time it was noticed (numerically and experimentally) a temperature increase (up to 1250°C) due to friction shear stress increase. In this range of temperature, the alloy is in the melting interval and the rheological, thermal and friction parameters are not accurately determined. This can explain that, on figure 2, the begining of the shortening is well predicted but the final is under estimate. Furthermore friction law depends at high temperature on the material consistency used.

Figure 3 represents several successive remeshings. Only half the cross section of a wall of the tube is represented here. The mesh density depends on the sum of the geometrical curvature and the thermal gradient " pseudo-curvature". Therefore, fine elements are found in the curved, folding areas, which allows a precise description of this important feature of the process. The whole file mesh is plotted on figure 4 and it can be seen that the internal fold is larger than the external one, which is also found in experiments.

In Figure 5 and 6 equivalent strain rate and temperature maps are presented for an another numerical simulation. The material is an Astroloy for exactly the same operating and geometrical condition than the previous results except the inertia moment of

■	Tem< 600.
	600.<Tem< 800.
	800.<Tem<1000.
	1000.<Tem<1100.
	1100.<Tem<1150.
	1150.<Tem<1175.
	1175.<Tem<1200.
	1200.<Tem<1225.
	1225.<Tem

Maxi= .1261E+04
Mini= .2337E+02

■	EbP< .300
	.300<EbP<1.000
	1.000<EbP<1.800
	1.800<EbP<2.700
	2.700<EbP<3.600
	3.600<EbP<4.500
	4.500<EbP<5.400
	5.400<EbP<6.300
	6.300<EbP

Maxi= .7464E+01
Mini= .1528E-04

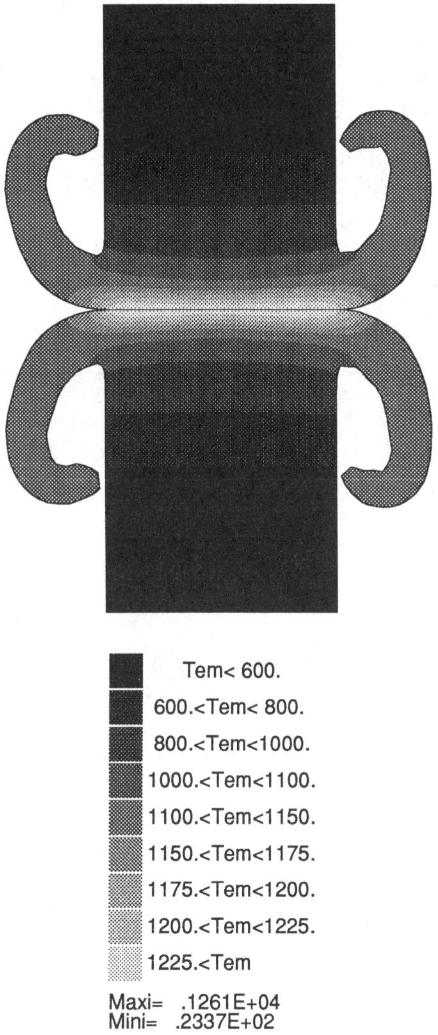

Figure 5. Map of temperature isovalues at t= 4.565 s

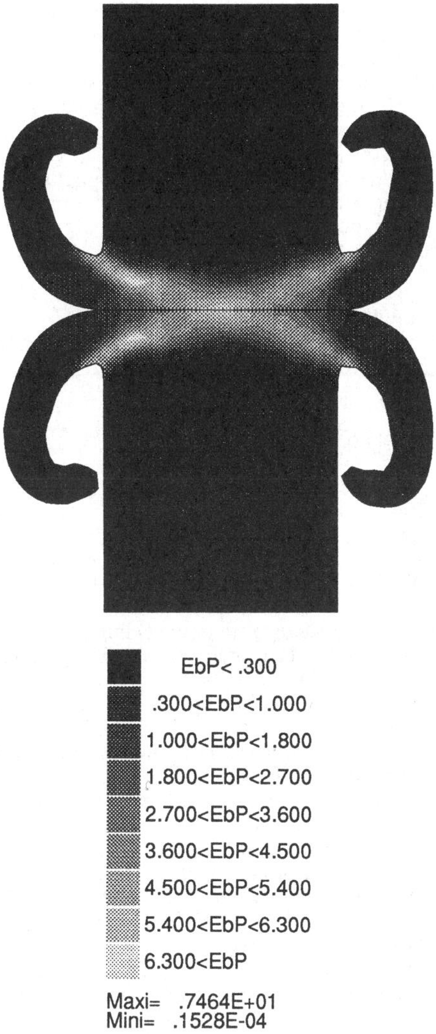

Figure 6. Map of $\dot{\bar{\varepsilon}}$ isovalues at t= 4.565 s

the flywheel equal to 102 Kg.m^2. Concerning the numerical model the equation (4) has been chosen for friction modelisation (based on the rotating velocity) and the remeshing is based on mechanical and thermal error estimator. At the end of the process (t = 4,565s) the maximum temperature (1261°C) is reached in the center of the weld interface. The temperature gradient is very large there and it correlates with the strain rate.

6 CONCLUSION

The reliability of the proposed model to simulate inertia welding process has been shown. It provides a reasonable representation of the temperature, strain fields, deceleration and the upset from the beginning to the end of the process. The coupled thermomechanical approach well describes the material behaviour. Nevertheless, more elaborated rheological study, taking into account some microstructural effects, could perform the metallurgical effects description, especially at the end of the process.

Two differents friction laws, based on experimental observations has been tested. The more accurate introduces the rotating velocity effect.

The recent developments in numericals methods like remeshing algorithm and lumping heat capacity matrix have been implemented in this code. The remeshing algorithm, not only based on geometric criterion, but also on an thermal and mechanical error indicator has permitted to force a refinement near the weld interface and increase the precision of the thermal description.

The future developments are the simulation of welding of two different materials (different nature) with different geometries. This will necessitate to perform contact algorithm.

For studying cooling of the welded pieces, a more sophisticated constitutive law must be introduced with elastic terms, in order to predict residual stresses.

Acknowledgements

We wish to acknowledge financial and technical supports provided by SNECMA.

REFERENCES

Wang, K.K. 1975. Friction welding, WRC Bulletin 204.

Wang, K.K. & P. Nagappan 1970. Transient temperature distribution in inertia welding Steels. Welding Research Supplement. 419-426.

Francis, A. & R.E. Craine 1985. On a model for frictioning stage in friction welding of thin tubes. Int. J. Heat Mass Transfer. Vol 28. n°9: 1747-1755.

Cheng, C.J. 1962. Transient temperature distribution during friction welding of two dissimilar materials in tubular form. Weld. Journal, Res. Suppl. 41. (12), 542-550.

Cheng, C.J. 1963. Transient temperature distribution during friction welding of two dissimilar materials in tubular form. Weld. Journal, Res. Suppl. 41. (5): 233-240.

Atsuta, T., S. Yamashita & T. Araki, 1985. Heat transfer simulation by FEM during friction welding. Kawasaki Heavy Industries. IIW Doc. III: 807-85.

Sluzalec, A. 1990. Thermal effects in friction welding. Int. J. Mech. Sci. vol 32, n°6: 467-478.

Moal, A., E. Massoni & J.-L. Chenot 1992. A finite element modelling for the inertia welding process. Proc. of Int. Conf. on Computational Plasticity. Barcelona. Pineridge Press. Owen, D.R.J. et al. (eds) vol 1: 289-300.

Moal, A., E. Massoni & J.-L. Chenot 1993. A finite element model for the simulation of the torsion and torsion-tension tests. Comp. Meth. in Applied Mech. and Eng. 103:417-434.

Soucail, M., A. Moal, L. Naze, E. Massoni, C. Levaillant & Y. Bienvenu, 1992. Microstructural study and numerical simulation of inertia friction welding of Astroloy. 7th International Symposium on Superalloys, Seven Springs (USA).

Chenot, J.-L., Y. Tronel & N. Soyris, 1992. Finite element calculation of thermo-coupled large deformation in hot forging. Int. Conf. on Comp. Meth. in Heat Transfer. L.C. Wrobel et al. (eds), Comp. Mech. Publications & Elsevier. 493-511.

Coupez, T. & J.-L. Chenot 1992. Large deformation and automatic remeshing. Proc. of Int. Conf. on Computational Plasticity. Barcelona. Pineridge Press. Owen, D.R.J. et al. (eds) vol 1: 10771087.

Fourment, L. & J.-L. Chenot 1992. Study of some error estimators for adaptive remeshing in metal forming. Proc. of Int. Conf. on Computational Plasticity. Barcelona. Pineridge Press. Owen, D.R.J. et al. (eds) vol 1: 199-211.

Zienkiewicz, O.C., Y.C. Liu & G.C. Huang. 1988. Error estimation and adaptivity in flow formulation for forming problems. Int. Journ. Num. Meth. Eng. 25: 23-42.

Huang, H.C.& R.W. Lewis 1989. Adaptive analysis for heat flow problems using error estimation techniques. Sixth International Conference for Numerical Methods in Heat Problems. Pineridge Press; Swansea, U.K. 1029-1044.

Moal, A. & E. Massoni 1994. Finite element simulation of the inertia welding of two similar parts. Accepted for publication in Eng. Comp. D.R.J. Owen & E. Hinton (eds). Pineridge Press.

Simulation of Materials Processing: Theory, Methods and Applications, Shen & Dawson (eds)
© 1995 Balkema, Rotterdam. ISBN 90 5410 553 4

Simulation of complex aluminium extrusion using an arbitrary Eulerian Lagrangian formulation

H.G. Mooi & J. Huétink
University of Twente, Department of Mechanical Engineering, Enschede, Netherlands

ABSTRACT: The purpose of this study is to gain insight in design and control parameters of aluminium extrusion by FE simulations. This is done with a thermal FE code based on an ALE algorithm. A short overview is given of the material models that are used. Furthermore, a stationary state solver for temperature and plastic strains is developed, which seemed to be very useful because transient temperature calculations are too much time consuming. At last, examples are given of the extrusion of a tube.

1. INTRODUCTION

During aluminium extrusion, the extrusion process itself, the life cycle and the behaviour of the dies are variable and difficult to control. The efficiency of extrusion is strongly influenced by the necessity of repeated corrections due to wear-out and plastic or creep deformation of the die. Up to now, the corrections were performed purely based on empirical knowledge. The aim of this study is to try to simulate (complex) aluminium extrusion by means of FEM in order to predict different process variables. For these simulation both the aluminium flow (in terms of e.g. pressure, temperature and residual stresses) as well as the deformation of the die (elastic, plastic and creep) should be examined. Of course, the flow through the die and the deformation influence each other.

The calculations are performed with a arbitrary Lagrangian Eulerian code (Dieka), which was developed by Huétink c.s. during the last decade.

2. ARBITRARY LAGRANGIAN EULERIAN METHOD (ALE)

As the name indicates, in the Arbitrary Lagrangian-Eulerian (ALE) method the user can make his own choice whether a grid point is completely Lagrangian or Eulerian or arbitrary. This means that the material displacements are disconnected form the grid point displacements. So, also convective changes have to be calculated and both the material rate of change of a magnitude f. $\dot{f}^m$ as well as the rate of change in a grid point $\dot{f}^g$ have to be examined. These rates read, respectively:

$$\dot{f}^m = \left(\frac{df}{dt}\right)^m = \frac{\partial f}{\partial t} + \dot{\underset{\sim}{x}}^m \cdot \vec{\underset{\sim}{\nabla}} f \tag{1}$$

$$\dot{f}^g = \left(\frac{df}{dt}\right)^g = \frac{\partial f}{\partial t} + \dot{\underset{\sim}{x}}^g \cdot \vec{\underset{\sim}{\nabla}} f \tag{2}$$

where $\dot{\underset{\sim}{x}}^m$ and $\dot{\underset{\sim}{x}}^g$ represent the material and grid velocity, respectively. The magnitude f may represent any material associated quantity such as strains, stresses and temperatures. The terms in eqs. (1) and (2) should be evaluated in the middle of time increment. Subtracting eq.(2) from eq.(1), and integrating the result in time from t_0 to $t_0+\Delta t$ a time incremental procedure is obtained:

$$f(\underset{\sim}{x}^g + \Delta\underset{\sim}{x}^g, t+\Delta t) = f(\underset{\sim}{x}^g, t) + \Delta f^m + (\Delta\underset{\sim}{x}^g - \Delta\underset{\sim}{x}^m)\cdot\vec{\underset{\sim}{\nabla}} f \tag{3}$$

This can be seen as a Taylor expansion of:

$$f(\underset{\sim}{x}^g + \Delta\underset{\sim}{x}^g, t+\Delta t) = f(\underset{\sim}{x}^g + \Delta\underset{\sim}{x}^g - \Delta\underset{\sim}{x}^m, t) + \Delta f^m \tag{4}$$

Mark that if the material displacement is equal to the grid displacement ($\Delta\underset{\sim}{x}^g = \Delta\underset{\sim}{x}^m$) the updated method is accomplished; if the grid displacement equals zero ($\Delta\underset{\sim}{x}^g = 0$) the Eulerian method is accomplished. The last term in eq.(3), the convection, complicates the calculation considerably, because the gradient of f is needed. The difficulty to calculate the gradient is, firstly, that the stresses and strains generally are discontinuous over the element boundaries. Therefore, the gradient at element level is not reliable. Furthermore, if linear elements are used the stress and strain field calculated by means of this displacement field is constant on element level. In this case the gradient completely vanishes. A solution for these problems, developed by Huétink (1986, 1992), is the construction of a continuous field for f by averaging the nodal point values of f of adjacent elements. The gradient of f can be calculated out of this continuous field. This is some kind of smoothing and can be shown to be numerical diffusion (Hinton, 1974). More details of the smoothing procedure together with more details on e.g. crosswind diffusion diminishing can be found in Huétink (1992).

In the FE code used, convection is not directly implemented in the stiffness matrix since that leads to an asymmetric system. Instead, they are accounted for in the corrector stage of a predictor-corrector algorithm. So, first in an updated calculation (predictor step) the material displacement increments are calculated. After this the convection of the dependent quantities is explicitly calculated at element level using the calculated velocities.

3. MATERIAL MODELS

In order to model aluminium extrusion many different material models are to be used. Firstly, the flow is dominated by the plastic deformation rate and the temperature; hardening plays a minor role. Furthermore, for the final shape of the extruded product the residual stresses in the outflow are important. Therefore, also an elastic part has to be accounted for. On the other hand, the elastic effects in the bulk forming can be neglected [Zienkiewicz (1991), p. 524]; also to save calculation time. At last, the deformation of the die is governed by elastic and creep deformation. So, many material models should be available in the FE code that is used.

Since extrusion is dominated by strain rate hardening, these models are described below. For rate dependent models the following constitutive law is used:

$$\underline{\sigma} = \eta\, \underline{d} + p\underline{I} \qquad (5)$$

where $\underline{\sigma}$ the Cauchy stress tensor, $\underline{d}$ the deformation rate tensor, p the total pressure and $\underline{I}$ the second order unity tensor. η denotes some kind of viscosity which is a function of material parameters, process parameters and the time; depending on the kind of material behaviour to be described. The pressure p in eq.(5) is calculated by means of a penalty. In literature the penalty is often chosen very high in order to fulfil the incompressibility constraint. In this case numerical problems can be encountered since the condition of the resulting stiffness matrix gets very bad. In our opinion a kind of (elastic) bulk modulus C_b can serve as the penalty very well. In this way numerical problems are not only avoided, but also some elastic compression is accounted for which is neglected in case of plastic incompressibility. The bulk modulus can be chosen higher than the pure elastic bulk modulus, also to prevent the fact that pure compression takes too much calculation steps. The linearized pressure at the end of an increment reads:

$$p = p_0 + \Delta p = p_0 + C_b\, tr\Delta\underline{\epsilon} \qquad (6)$$

where $\underline{\varepsilon}$ is the strain tensor. For Von Mises material behaviour the following equation for the equivalent stress σ_{eq} can be derived:

$$\sigma_{eq} = \sqrt{\frac{3}{2}\underline{s}:\underline{s}} = \eta\sqrt{\frac{3}{2}\underline{d}:\underline{d}} = \frac{3}{2}\eta\dot{\epsilon}_{eq} \qquad (7)$$

where $\underline{s}$ is the stress deviator and $\underline{d}$ is the deformation rate tensor.

3.1 Rate dependent material behaviour

The material behaviour of hot aluminium can be well described by a curvefit of the form:

$$\sigma_{eq} = \sigma_{ref}(\epsilon_{eq})\ \dot{\epsilon}_{eq}^{\,n} \qquad (8)$$

where σ_{ref} and n are material parameters depending on the temperature; σ_{ref} may be history dependent. This model is widely accepted for modelling the behaviour of hot aluminium and is in slightly different forms known under different names as:
- Ostwald de Waele law (Zienkiewicz, 1991),
- Lévy-Mises equations (Kobayashi, 1989) or
- Norton-Hoff law (Chénot, 1992).

Comparing eq.(7) and eq.(8) a formula for the viscosity for deformation rate dependent material can be derived:

$$\eta(\dot{\epsilon}_{eq}, \epsilon_{eq}) = \frac{2}{3}\sigma_{ref}(\epsilon_{eq})\,\dot{\epsilon}_{eq}^{\,n-1} \qquad (9)$$

As special cases of this model rigid plasticity

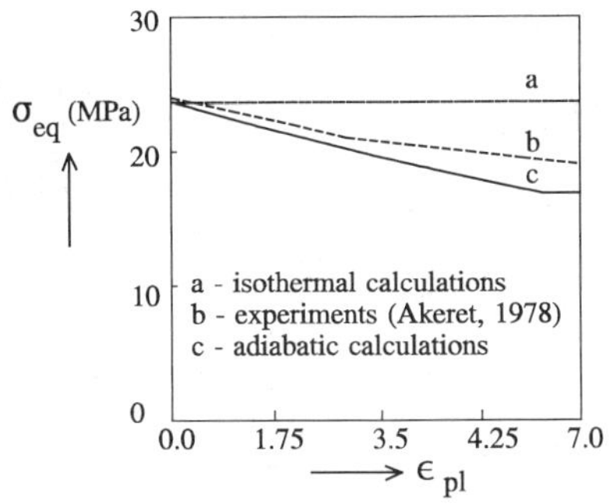

Fig. 1 Material testing: experiment vs. calculations

(n=0) and Newtonian viscosity (n=1) can be recognized.

For aluminium rather elaborate flow stress measurements are documented in Akeret (1978). The results in this Atlas, though, suggest a softening effect upon high deformation (equivalent strain up to 700%). In our opinion softening is merely due to the fact that the experiments are not isothermal (as suggested). So, the decrease of the flow stresses is almost only caused by heating of the specimen by dissipation. Simple calculations with a constant flow stress (no softening nor hardening) confirmed this assumption. For these calculations one element was deformed while it was isolated (adiabatic). Of course, this does not resemble the experiments completely, because no special isolation was applied at the experiments. Nevertheless, the experiments were performed so quickly that all the heat can never be transported away. At the calculations the flow stress decreased about 25% upon 700% plastic strain, whereas the flow stress at the experiments of Akeret decreased about only 15% (see Fig. 1).
This confirmed the assumption that the flow stress at the experiments probably only decreases because of the temperature increase, that was calculated to be 50 °C. Therefore, the viscosity in eq.(9) is not taken to be dependent on the accumulated strain ε_{eq} in this study (curve a in Fig. 1). It should be stressed that, if necessary, the ε_{eq} can be taken into account in the FE code that is used.

3.2 Creep

Creep can be described as a first estimate by a powerlaw also, the Baily-Norton law
(see Kraus, 1980). The primary and secondary creep are both approximated in this case by:

$$\epsilon_{cr} = A(\epsilon_{cr})\, \sigma^m\, t^n \tag{10}$$

where A, m and n are also a function of temperature; A may be history dependent. By taking the derivative with respect to time (assuming that $\sigma = \sigma_0$ = constant during a step) an equation for the strain rate is obtained:

$$\dot{\epsilon}_{cr} = A\, \sigma^m\, n\, t^{n-1} \tag{11}$$

Comparing eq.(11) and eq.(7) the viscosity η in the constitutive equation (eq.(5)) then changes to:

$$\eta = \frac{2}{3}(n A)^{-1/m}\, \dot{\epsilon}_{cr}^{(1-m)/m}\, t^{(1-n)/m} \tag{12}$$

This is called the time hardening formulation.

Also, a strain hardening formulation can be derived by eliminating the time from eq.(10) and eq.(11). In this case the viscosity η in the constitutive equation gets:

$$\eta = \frac{2}{3} A^{-1/m}\, n^{-n/m}\, \dot{\epsilon}_{cr}^{(n-m)/m}\, \epsilon_{cr}^{(1-n)/m} \tag{13}$$

These models can be built in very easily but have some serious disadvantages. Firstly, because no limit stress is applied under which no creep occurs, the creep deformation will overestimate the real creep. Also, forces can hardly be prescribed with this model which is of course not acceptable for creep. Finally, this model needs extremely many iterations.

A solution for these disadvantages is to make use of elasto-viscoplastic models combined with a first order approximation for the creep strain as were developed by Van der Lugt (1988). These models do not only use a constant stress, but also a first order term of the stresses. Unfortunately, the creep law of Van der Lugt needs many unknown material parameters, such as dislocation density.

For creep laws like eq. (10) the first order models can be worked out also. The starting point is a Taylor expansion of the creep strain rate:

$$\dot{\epsilon}_{cr} = \dot{\epsilon}_{cr,0} + \frac{\partial \dot{\epsilon}_{cr}}{\partial \sigma}\dot{\sigma} + \frac{\partial \dot{\epsilon}_{cr}}{\partial \epsilon_{cr}}\dot{\epsilon}_{cr} \tag{14}$$

Gathering the terms with the creep strain rate yields:

$$\dot{\epsilon}_{cr} = \left(1 - \frac{\partial \dot{\epsilon}_{cr}}{\partial \epsilon_{cr}}\right)^{-1} \left\{ \dot{\epsilon}_{cr,0} + \frac{\partial \dot{\epsilon}_{cr}}{\partial \sigma}\dot{\sigma} \right\} \tag{15}$$

which can be used in an incremental procedure to update ε_{cr}. The derivatives in this equation can be derived from the creeplaw eq.(10) and its time derivative eq.(11). An elasto-viscoplastic model based on the general creeplaw eq. (10) are under development.

For creep modelling in general experimental creep data is not much available, especially not for the materials used here. Current investigations are focused on obtaining material creep data in order to be able to carry out FE calculations.

4. STATIONARY CALCULATIONS

For extrusion a grid that is almost fixed can be adopted very well. Only for the contact regions displacement of the mesh has to be allowed. So, the calculated data (such as equivalent plastic strain, temperature, stresses, etc.) have to be convected through the mesh. Assuming that extrusion is a stationary process, in principal all convected quantities can be calculated in one step. Opposed to using the multi-step transient calculations to reach the stationary state, this can save a considerable amount of computer time. In this study the 1-step stationary solver for temperature is considered, out of which a first try to solve the equivalent stationary plastic strain distribution in one step is developed.

4.1 Stationary temperature calculations

The thermal equilibrium reads:

$$\rho C_p \dot{T} - \lambda \nabla^2 T - q = 0 \qquad (16)$$

where ρ is the mass density, C_p the specific heat, λ the coefficient of heat conduction and q the heat production per unit of volume. For stationary calculations the partial derivative with respect to time vanishes, leading to the following equation:

$$\rho C_p \frac{\partial T}{\partial t} = -\rho C_p \underset{\sim}{v} . \vec{\nabla} T - \lambda \nabla^2 T - q \equiv 0 \quad (17)$$

As can be seen, for temperature a convection-diffusion problem has to be solved. About this problem very much is written in literature the last 2 decades (see Hughes (1982) and Zienkiewicz (1991)). In order to solve the stationary equation eq.(17) with a transient code the term with partial derivative with respect to time must be put to zero and the convection terms should be added in the stiffness matrix (see also Sloot, 1994). The thermal equation are solved fully coupled with the mechanical system in Dieka. So, the stationary 1-step calculation can only be performed after the mechanical part of the calculation is stationary; only then the source term q is stationary. Path dependency during this temperature calculation is neglected this way.

As described in chapter 2, in our ALE code convection is not solved directly in the system of equations but calculated in the corrector stage of a predictor-corrector algorithm. Therefore, up to now no upwind techniques (such as SUPG) were developed, since they were not necessary. Nevertheless, a stationary solution of eq.(17) can be computed by means of a normal Galerkin method as long as the element Péclet number
($Pe = ½(v\ L_{el})/(\lambda/\rho C_p)$; L_{el} is the local element length) is smaller than 1. By this method much calculation time can be saved, since for stability very small time increments are needed. As an example the stationary temperature distribution near the outflow of a real extrusion process is calculated (the outflow of the mesh in Fig. 4; see also section 5 for more details). The 1-step method was applied after 200 transient steps that were necessary to yield a stationary source. The results are shown in Fig. 2.

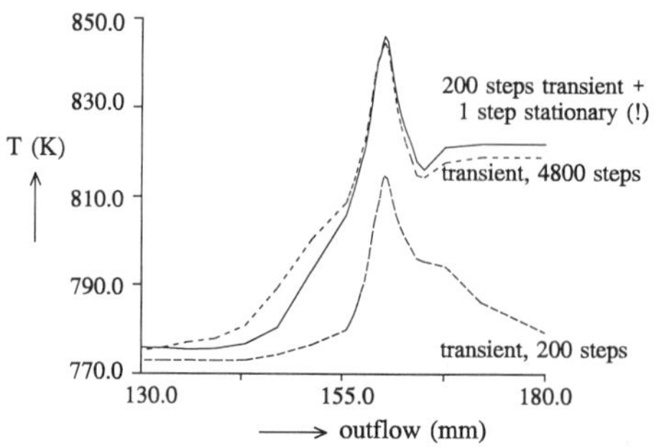

Fig. 2 Stationary temperature distribution near the outflow (transient vs. 1-step calculation)

It can be seen that the stationary solver gives a very good estimate for the steady state temperature in the outflow, whereas many calculation steps are saved (in the order of many thousands). It is stressed that for the transient calculation the stationary state is not reached yet! The wiggles on the 1-step solution and the fact that the 1-step method probably overestimates the level are caused by the fact that the requirement $Pé < 1.0$ is not fulfilled in the whole mesh.

4.2 Stationary strain calculations

Whereas stationary temperature calculations are well known, for plastic strain stationary calculations are not often applied. It is questionable in what extent these calculations are allowable, since the plastic strain and the stress fields influence each other (are path dependent) and, therefore, in what extent the deformation path may be omitted. Nevertheless, because the flow pattern at extrusion does not change very much upon hardening, this method of the plastic strain may be permitted. Then, the stationary state which is calculated this way is a good approximation of the transiently calculated state.

The starting point of stationary plastic strain calculations is the material derivative of the plastic strain (compare eq.(1)) out of which the following stationary equation can be derived:

$$\frac{\partial \epsilon_{eq}}{\partial t} = -\underset{\sim}{v}.\underset{\sim}{\vec{\nabla}}\epsilon_{eq} + \dot{\epsilon}_{eq} \equiv 0 \qquad (18)$$

where $\underset{\sim}{v}$ is the velocity and ε_{eq} the plastic strain. The $\dot{\epsilon}_{eq}$ indicates the source term of the plastic strain. This equation is completely analogous to the stationary temperature equation (eq.(17)). So, the plastic strain can be solved stationarily as a separate degree of freedom. This method can be seen as a streamline integration of the plastic strain.

As a first try, the thermal set of equations is used. In order to do so the thermal solver has to be changed in the following way:

- the ρC_p-term in the convective term in eq.(17) is set to 1.0;
- dissipation and thermal expansion are omitted (the mechanical system is not coupled with the plastic strain equation);
- the conductivity term can be used as a sort of numerical diffusion.

Very preliminary results are shown in Fig. 3. The stationary plastic strain distribution in a row of elements is shown. The six elements in the middle of the row are sheared every step, while the material is convected through the mesh (see also Fig. 3, where the incremental displacement per step is depicted by the dotted lines). The exact level of the stationary plastic strain is $1/\sqrt{3}$.
From Fig. 3 can be concluded that the method works very good for this simple test problem. Future research aims at more complex meshes.

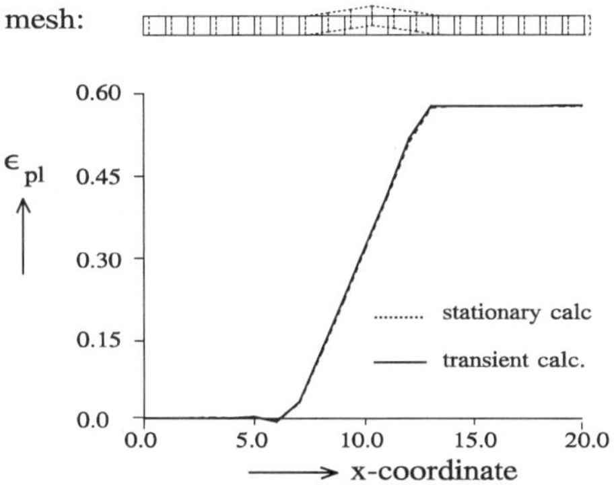

Fig. 3 stationary strain calculation

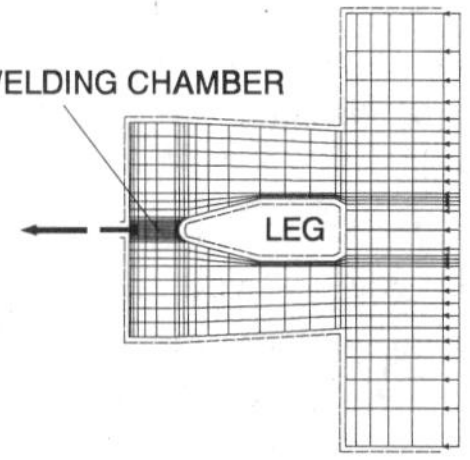

Fig. 4 Mesh for 2D flow calculation

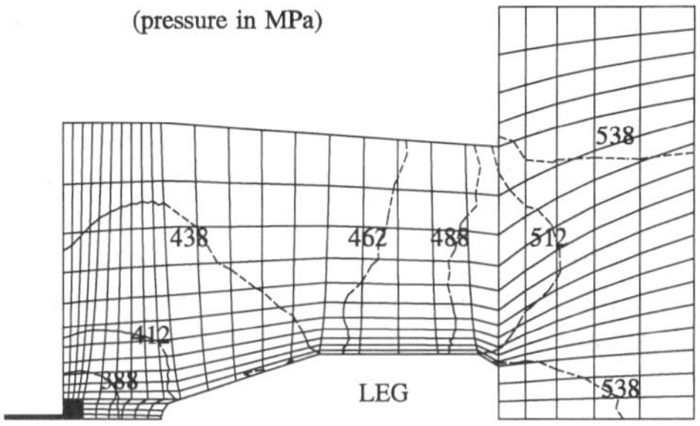

Fig. 5 Pressure distribution

5. RESULTS

In this section some results are given of several calculations that were performed on different aspects of extrusion of a circular tube. Up to now, the calculations of the flow and the calculation of the deformation of the die were performed separately. Tubes are extruded by means of a hollow die: the inner surface is formed by a core which is suspended by three or more legs in the die. So, the aluminium flow is split by legs, and after the legs the flow has to weld together again (in so-called welding chambers, see Fig. 4).

The friction forces on the legs form an important part of the total loading of the die. So, a finite element mesh to simulate the flow around the leg is considered (see Fig. 4). It is assumed that it is sufficient to calculate the flow two-dimensionally because the flow perpendicular to the extrusion direction between two legs is negligible. So, plane strain elements can be used. Only one half of the model has to be considered because of symmetry. The dotted lines represent the area with so-called contact elements (based on a penalty formulation), modelling the contact and Coulomb friction. Displacements were prescribed in the inlet. A strain

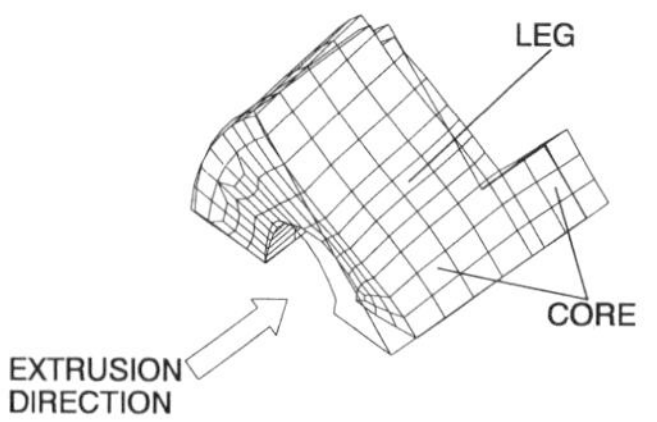

Fig. 6 3D mesh of a part of the extrusion die

rate dependent material model was used. Some results are shown in Fig. 5.
The straight pressure profile indicates a flow that is mainly governed by friction forces at the walls (Poisseuille distribution). The largest pressure drop was found, as expected, near the outlet. In a separate calculation a three-dimensional model of the die (one eighth, because of symmetry, see Fig. 6) was considered.
The loads put on the three-dimensional model, the friction on the leg as well as the pressure on the core, were calculated by means of the two-dimensional calculation. The (elastic) deflection of the core is calculated to be a few tenths of a millimetre (see Fig. 6 ; the plotted deflections are exaggerated!). Because this deflection directly influences the formed profile it is very important for the design of the die. Also, the calculation demonstrated that no plastic deformation of the die occurs.

Creep calculations with the viscous model predicted creep displacements in the same order (within one and a half hour of creeping time). This gave qualitative insight in these phenomena.

At the moment also the extrusion of other real profiles are simulated.

6. CONCLUSIONS

The ALE formulation is very suitable for the simulation of the aluminium extrusion. Interesting control and design parameters (exact temperatures, pressures and deformation of the die) can be achieved. In order to avoid huge three-dimensional models it seems to be a good solution to split the calculation by considering the flow through the die and the deformation of the die separately. Of course, this assumption has to be verified in the future. Stationary state solvers for temperature as well as plastic strain already give promising results, but certainly need further development. Probably, SUPG methods should be adopted in order to bypass the Pé-number requirement.

7. ACKNOWLEDGMENTS

BOAL B.V., De Lier in The Netherlands, is gratefully acknowledged for the cooperation and financial support in the Eureka 'Die Design Control' project. The authors want to thank Mr. Sloot for his contribution to the stationary state solver as well as Ms. Van de Moesdijk for carrying out the extrusion calculations in section 5.

8. REFERENCES

Akeret, R., Jung, H., and Scharf, G., 1978, '*Atlas of hot working properties of nonferrous metals*', Vol.1, Deutsche Gesellschaft für Metallkunde (DGM).

Chenot, J.-L., 1992, '*FE Modelling of Metal Forming: recent Achievements and Future Challenges*', in: Proc. of 3rd Int. Conf. on Comp. Plasticity, pp 1019-1036, Owen e.a. (eds.), Pineridge Press, UK.

Brooks, A.N. and Hughes, T.J.R., 1982, 'SUPG formulations for Convection dominated flows with particular emphasis on the incompressible NS equations', *Comp. Meth. in Appl. Mech. and Eng.*, vol.32, pp 199-259.

Hinton, E. and Gampbell, J.S., 'Local and global smoothing of discontinuous finite element functions using least square method', *Int.J.Num.Meth. Eng.*, Vol.8 (1974), pp. 461-480.

Huétink, J., 1986, '*On the simulation of thermomechanical forming processes*', Dissertation, University of Twente, The Netherlands.

Huétink, J. and Helm, P.N. van der, 'On Euler-Lagrange finite element formulation in forming and fluid problems', In: Chenot J.-L. e.a., '*Num.Meth.in Industr. Forming Proc.*', (1992), Balkema Rotterdam, pp. 45-54.

Kobayashi, S., Oh, S. and Altan, T., 1989, '*Metal Forming and the Finite Element Method*', Oxford University Press, New York.

Kraus, H., 1980, '*Creep Analysis*', John Wiley & Sons, US.

Lugt, J. van der, 1988, '*A finite Element Method for the simulation of thermo-mechanical contact problems in forming Processes*', dissertation, University of Twente, The Netherlands.

Sloot, E.M., 1994, '*Convergence Acceleration for ALE Methods*', Post-Graduate report, University of Twente, The Netherlands.

Zienkiewicz, O.C. and Taylor, R.L., 1991, '*The Finite Element Method*', fourth edition, McGraw-Hill Book Company (UK).

Simulation of Materials Processing: Theory, Methods and Applications, Shen & Dawson (eds)
© 1995 Balkema, Rotterdam. ISBN 90 5410 553 4

Material flow into arbitrarily shaped cavities in closed-die forging of long components: A plane strain UBET-simulation compared with FEM and model material experiments

Ulf Ståhlberg & Jin Hou
Materials Forming, Department of Materials Processing, Royal Institute of Technology, Stockholm, Sweden

ABSTRACT: Accurate die- and preform design in closed-die forging of components requires reliable mathematical simulation of the material flow into the tool cavities. In the present work, cross sections of long components are treated for plane strain conditions. The cavities are of arbitrary shape and the filling of them has been studied by means of UBET (Upper Bound Elemental Technique). Three types of quadrilateral elements have been used. A detailed description regarding the formulation of the model is presented. The computer program which has been developed includes automatic determination of the contact/boundary conditions and automatic generation of the UBET-elements. The structure of the program and some simulation results are presented. A constitutive equation characterizing the strain rate hardening of the work material, is included in the program in a simplified way. In order to verify the theory, model material experiments have been carried out. Velocity vectors obtained by a special photographic technique are compared with those predicted by the UBET-model. A commercial FEM-software, FORM-2D, has also been used for comparison. Satisfactory agreement has been found between the UBET-model, the FEM-software and the experiments.

INTRODUCTION

Considering closed-die forging analysis, the Upper Bound Elemental Technique (UBET) and the Finite Element Method (FEM) are the most commonly used.

The original idea of the UBET-method was proposed by Kudo[1] in 1960s. Lately McDermott and Bramley[2] have adapted the concept and made some extensions of the method considering the shape of the elements and the plastic deformation within them. Because of its advantages regarding short calculation times and easy handling of complicated geometrical boundary conditions, UBET has lately been widely used for analysing closed-die forging processes. The works by Bramley et al.[2-5], Keife[6], Kiuchi et al.[7-10] and Christensen[11] represent some examples. The first 3-D analysis based on UBET was presented in 1989 by Lugora and Bramley[12,13]. During 1992-1993 Lin and Wang[16-18] have presented a new UBET, the upper-bound flow function technique and 1994 Qin[19] published a general triangular element.

Assuming that no traction is applied on the free surface (1)-(3) represent central UBET-equations [14]

$$\dot{W}=\sum_{i=1}^{n_e}\dot{W}_{pi}+\sum_{i=1}^{n_s}\dot{W}_{si} \tag{1}$$

$$\dot{W}_{pi}=\int_{V_i}\sigma_{ei}\dot{\varepsilon}_{ei}dV_i \tag{2}$$

$$\dot{W}_{si}=\int_{S_i}\tau v_i^* dS_i \tag{3}$$

FORMULATION OF THE UBET-MODEL

The material is supposed to obey the von Mises yield criterion. Plane strain conditions are assumed. The problem is treated for a constant shear stress between the tools and the workpiece

In order to facilitate the programming, only 3 types of general UBET-elements are proposed, Fig.1(a). An example showing how the workpiece is divided into elements is given in Fig.1(b).

Element 1 is used for workpiece regions where the material is in contact with both the upper and lower dies. Element 2 and 3 are used for simulating the filling of the die cavity when both the left and right sides of the dies are in contact with the workpiece, Fig.1(b). Element 2 is utilized for regions characterized by one contact surface and element 3 for regions where the upper and lower surfaces of the workpiece are free.

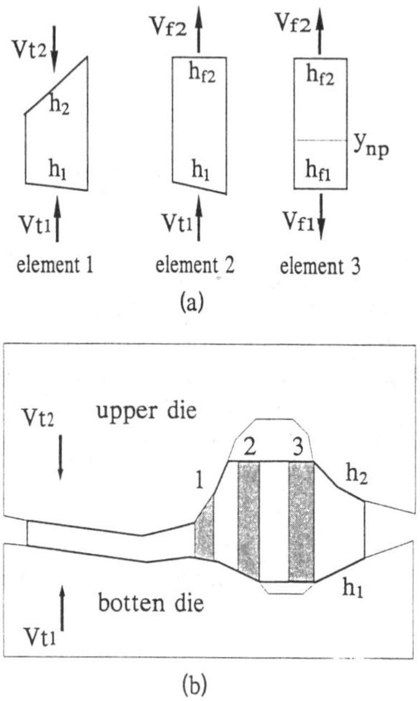

Fig.1 UBET- elements

To begin with the velocity fields of the elements are determined.

Assuming that the material flow in x- direction is uniformly distributed along the height of the *element 1*, v_x is described by equation 4.

$$v_{x1}=(v_{t1}-v_{t2})\frac{x-x_{np}}{h_2-h_1} \tag{4}$$

The strain rate component $\dot{\varepsilon}_{x1}$ is obtained from equation 4 and is given in equation 5.

$$\dot{\varepsilon}_{x1}=\frac{\partial v_{x1}}{\partial x}=(v_{t1}-v_{t2})\frac{1}{h_2-h_1}[1-\frac{h_2'-h_1'}{h_2-h_1}(x-x_{np})] \tag{5}$$

Considering the conditions of constancy of volume and boundary conditions at $y=h_1$ and $y=h_2$, the velocity v_{y1} is obtained.

$$v_{y1}=v_{t1}\frac{h_2-y}{h_2-h_1}+v_{t2}\frac{y-h_1}{h_2-h_1}+(v_{t1}-v_{t2})\frac{x-x_{np}}{h_2-h_1}(\frac{y-h_1}{h_2-h_1}h_2'+\frac{h_2-y}{h_2-h_1}h_1') \tag{6}$$

In *element 2*, the velocity component at the free surface in y- direction, v_{f1} or v_{f2} , is determined by the velocities at left and right borders of the element. In the same way as for element 1 and using the boundary condition $v_{x2}=v_{xL}$ when $x=x_L$, the velocity field for element 2 can be obtained.

When the bottom die is in contact with the workpiece, the velocity field is given by equations 7 and 8,

$$v_{x2}=(v_{t1}-v_{f2})\frac{x-x_L}{h_{f2}-h_1}+v_{xL}\frac{h_{2L}-h_{1L}}{h_{f2}-h_1} \tag{7}$$

$$v_{y2}=v_{t1}\frac{h_{f2}-y}{h_{f2}-h_1}+v_{f2}\frac{y-h_1}{h_{f2}-h_1}+[(v_{t1}-v_{f2})\frac{x-x_L}{h_{f2}-h_1}+v_{xL}\frac{h_{2L}-h_{1L}}{h_{f2}-h_1}]\frac{h_{f2}-y}{h_{f2}-h_1}h_1' \tag{8}$$

where v_{xL} is a known velocity determined from the velocity field of the neighbouring element on the left side.

When the upper die is in contact with the workpiece, the velocity field is given by equations 9 and 10.

$$v_{x2}=(v_{f1}-v_{t2})\frac{x-x_L}{h_2-h_{f1}}+v_{xL}\frac{h_{2L}-h_{1L}}{h_2-h_{f1}} \tag{9}$$

$$v_{y2}=v_{f1}\frac{h_2-y}{h_2-h_{f1}}+v_{t2}\frac{y-h_{f1}}{h_2-h_{f1}}+[(v_{f1}-v_{t2})\frac{x-x_L}{h_2-h_{f1}}+v_{xL}\frac{h_{2L}-h_{1L}}{h_2-h_{f1}}]\frac{y-h_{f1}}{h_2-h_{f1}}h_2' \tag{10}$$

The free surface velocity in y- direction, v_{f1} and v_{f2} are calculated by equation 11,

$$v_{f2}=\frac{v_{xL}(h_{2L}-h_{1L})-v_{xR}(h_{2R}-h_{1R})}{x_R-x_L}+v_{t1}$$

$$v_{f1}=v_{t2}-\frac{v_{xL}(h_{2L}-h_{1L})-v_{xR}(h_{2R}-h_{1R})}{x_R-x_L} \tag{11}$$

where v_{xR} is an unknown variable which will be determined by minimization.

Assuming that the workpiece is homogeneously deformed in x-direction, the velocity component v_x in *element 3* is given by equation 12,

$$v_{x3}=v_{xL}+(v_{xR}-v_{xL})\frac{x-x_L}{x_R-x_L} \tag{12}$$

where v_{xR} is an unknown variable. In the same way as above, the velocity component in y-direction can be obtained from the condition of constancy of volume and is written as equation 13,

$$v_{y3}=-(v_{xR}-v_{xL})\frac{y-y_{np}}{x_R-x_L} \tag{13}$$

where y_{np} is another unknown variable.

In order to determine the best velocity fields the total power dissipation is minimized with respect to the unknown variables.

The power dissipation due to *plastic deformation* in each element is defined by equation 2. For a certain temperature, the effective stress is a function of the effective strain rate and is determined from the constitutive equation of the workpiece material. The effective strain rate for plane strain conditions is defined by the general equation 14.

$$\dot{\varepsilon}_e=\sqrt{\frac{2}{3}(\dot{\varepsilon}_x^2+\dot{\varepsilon}_y^2+2\dot{\varepsilon}_{xy}^2)} \tag{14}$$

$$\dot{\varepsilon}_x=\frac{\partial v_{xi}}{\partial x}$$

$$\dot{\varepsilon}_y=\frac{\partial v_{yi}}{\partial y} \tag{15}$$

$$\dot{\varepsilon}_{xy}=\frac{1}{2}(\frac{\partial v_{xi}}{\partial y}+\frac{\partial v_{yi}}{\partial x})$$

The strain rate components and the powers for all the elements are calculated. The integral in equation 2 for element 1 and element 2 are solved numerically.

The power dissipation due to *shear between workpiece and dies* is calculated according to the general equation 3,

$$\dot{W}_{fi}=m\int_{x_L}^{x_R}k|v_{xi}|(1+h_j'^2)dx \tag{16}$$

where τ=mk. j=1 for contact with the bottom die and j=2 for the contact with the upper die. Because the yield stress is a function of strain rate, the yield shear stress k in equation 16 varies along the contact surface. In order to avoid calculating the integral in equation 16 numerically, the yield shear stresses k is substituted by a mean value k_e which is calculated by equation 17.

$$k_e=\frac{1}{x_R-x_L}\frac{1}{\sqrt{3}}\int_{x_L}^{x_R}\sigma_e(\dot{\varepsilon}_e|_{y=h_j})dx \tag{17}$$

After completing the integral and substituting k by k_e, the power dissipation in equation 16 can be written in a different form.

The power dissipation due to *shear between UBET-elements* is calculated according to the general equation 3, where τ=k.

$$\dot{W}_{di}=\int_{h_1}^{h_2}k|v_{yi}-v_{yi+1}|_{x=x_i}dy \tag{18}$$

The yield shear stress k in equation 18 is also substituted by a mean yield shear stress along the boundary between two elements. k_e is calculated according to equation 19.

$$k_e=\frac{1}{h_2-h_1}\frac{1}{\sqrt{3}}\int_{h_1}^{h_2}\sigma_e(\dot{\varepsilon}_e|_{x=xi})dy \tag{19}$$

By substituting h_1 by h_{f1} and h_2 by h_{f2} , the above equations 18 and 19 can be used for calculating the power due to shear between the different kinds of elements.

The *total power* is calculated by summing up all the powers dissipated.

$$\dot{W}=\sum\dot{W}_{pi}+\sum\dot{W}_{fi}+\sum\dot{W}_{di} \tag{20}$$

The following constitutive equation of the workpiece

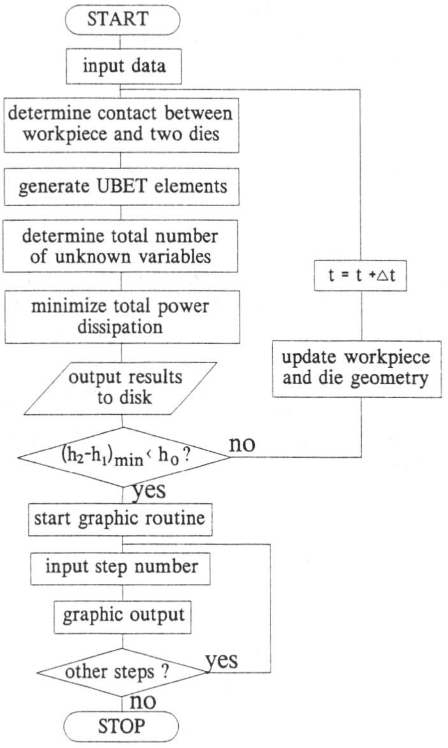

Fig.2 Flow chart of the computer program

material is used

$$\sigma_e = \sigma_0 \dot{\varepsilon}_e^{a_0} \qquad (21)$$

COMPUTER PROGRAM

The program is written in FORTRAN. It works on IBM or IBM compatible PC under DOS. The flow chart of the program is given in Fig.2. A quasi-Newton algorithm in the NAG mathematic library is chosen to minimize the total power dissipation with respect to all the unknown variables. In order to simulate a whole forging process, the calculations are carried out in a stepwise manner.
The geometry of the workpiece is updated according to equation 22.

$$\begin{aligned} x_{kj}^{t+\Delta t} &= x_{kj}^{t} + v_{xi}^{t}\Delta t \\ y_{kj}^{t+\Delta t} &= y_{kj}^{t} + v_{yi}^{t}\Delta t \end{aligned} \qquad (22)$$

SIMULATION RESULTS

Simulations were carried out for the forging of a crank-shaft cross-section. σ_0=8 MPa and a_0=0.2 were used as input parameters of the material. The velocity of the bottom die was v_{t1}=0, and the velocity of upper die v_{t2}=-1 mm/sec. A friction factor m=0.3 was chosen. The simulation was divided into 30 steps of calculation, starting with the die-workpiece contact and finishing at a distance of 15 mm between the upper and bottom dies. Graphic output results after some of the steps are shown in Fig.3.

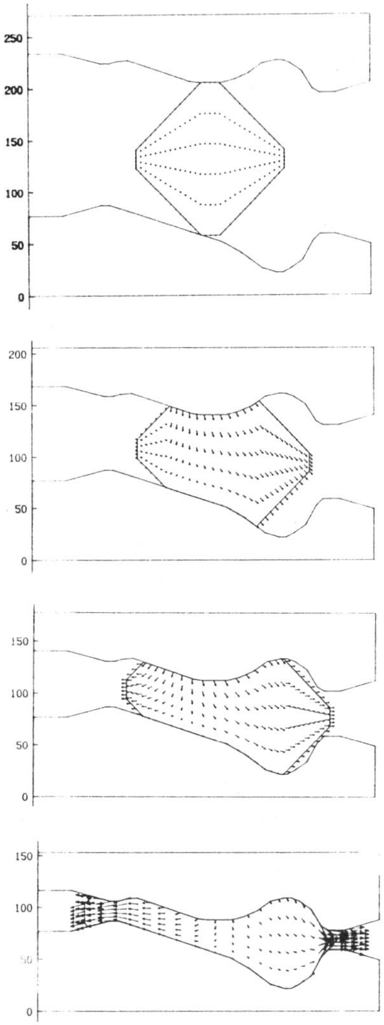

Fig.3 Velocity vectors predicted by the UBET-model during four subsequent forging steps

Experiments were carried out using plasticine as the model material. In order to secure plane-strain conditions, the workpiece and dies were placed between a well lubricated steel and a glass plate. The surface of the workpiece cross-section had been furnished with a painted dot grid network. A camera

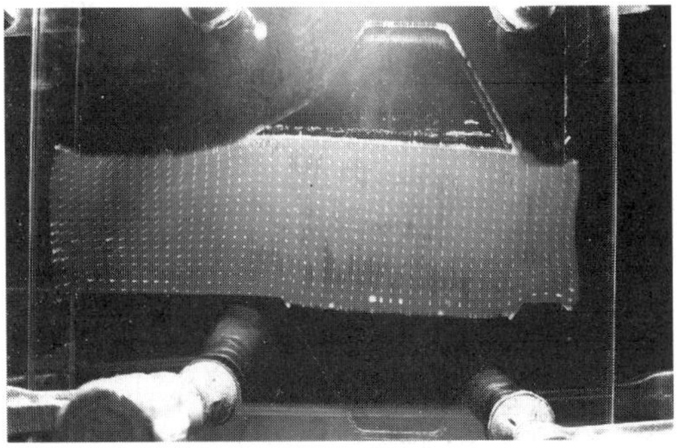

Fig.4 Material flow due to experiment

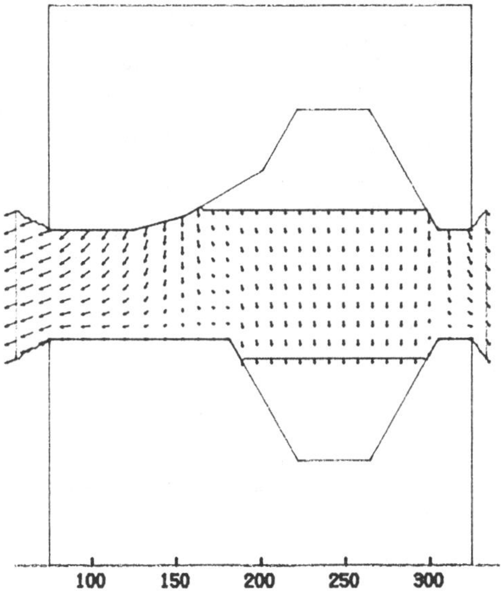

Fig.5 Velocity vectors predicted by the UBET-model

was fixed above the glass plate to photograph the tracks of the dots during deformation. The forging was started and then the film was exposed. After a short time, when a certain track length had been obtained, the press was stopped. The exposure continued for an extra instant of time in order to get brighter dots for identification of the final positions of the tracks. This procedure can be used for approximate determination of the velocity vectors. Fig.4 shows the result.

The UBET-model and the finite element software FORM-2D[15] were then utilized for calculating the material flow for the same conditions as those of the experiment. The results are shown by Fig.5 and Fig.6.

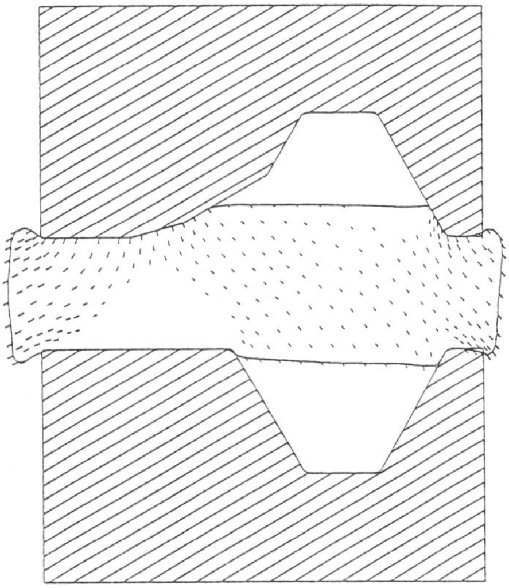

Fig.6 Velocity vectors predicted by FEM [15]

CONCLUSIONS

A plane strain model, meant for closed die forging of long components, has been formulated to simulate the material flow into arbitrarily shaped cavities. The model is based upon UBET and three kinds of quadrilateral elements are used. A computer program has been developed based upon the UBET-formulation. The constitutive equation of the material is included in the program in a simplified way. Results from the UBET-model have been compared with finite element simulations and model material experiments. The agreement between the results are found to be satisfactory.

ACKNOWLEDGEMENTS

The authors are indebted to NUTEK, NIF, OVAKO Arvika AB and IMATRA Kilsta AB for financial support.

REFERENCES

[1] H. Kudo, Some analytical and experimental studies of axisymmetric cold forging and extrusion-I, *J. Mech. Sci,*, 2(1960) 102-127

[2] R.P. McDermott and A.N. Bramley, Forming analysis - a new approach, *Proceedings of 2nd NAMRC*, Madison, Wisconsin, May 20-22, 1974, 35-47

[3] A.S. Gramphorn and A.N. Bramley, Computer aided forging design with UBET, *proceedings of 18th Int. Machine Tool Design & Research Conf.*, London, Sept. 14-16, 1977, 714-727

[4] F.H. Osman, A.N. Bramley and M.I. Ghobrial, Forging and preform design using UBET, *Advanced Technology of Plasticity*, vol.1(1984) 563-568

[5] M.I. Ghobrial, F.H. Osman and A.N. Bramley, Forging flash design with UBET, *Proceedings of 13th NAMRC*, Berkley, California, 1985, 151-158

[6] H. Keife, A new technique for determination of preforms in closed die forging of axi-symmetric products, *proceedings of 25th Int. Machine Tool Design & Research Conf.*, Birmingham, 1985, 473-477

[7] M. Kiuchi and Y. Murata, Simulation of contact pressure distribution on tool surface by UBET, *proceedings of s Int. Machine Tool Design & Research Conf.*, Swansea, Sept. 8-12, 1980, 13-20

[8] M. Kiuchi, Complex simulation system of forging based on UBET, Annals of the CIRP, 35(1986) 147-150

[9] M. Kiuchi and A. Karato, Simulation of flow behaviours of billet to fill side-flush cavity of die - application UBET to non-axisymmetric forging I, *J. Japanese Soci. Techn. Plasticity*, 26(1985) 307-315 (in Japanese)

[10] M. Kiuchi and A. Karato, Simulation of flow behaviours of billet to fill side-flush cavity of die - application UBET to non-axisymmetric forging II, *J. Japanese Soci. Techn. Plasticity*, 26(1985) 503-511 (in Japanese)

[11] P. Christensen, Computer aided design of forging dies using the upper bound elemental technique, Licentiate thesis, Technical University of Denmark, 1987

[12] C.F. Lugora and A. N. Bramley, Three dimensional analysis of closed-die forging process - part 1: formulation, Proc. Instn. Mech. Engrs. Part B, 203(1989) 33-37

[13] C.F. Lugora and A. N. Bramley, Three dimensional analysis of closed-die forging process - part 2: application, Proc. Instn. Mech. engrs. Part B, 203(1989) 38-42

[14] B. Avitzur, *Metal Forming: Processes and Analysis*, McGraw-Hill, New York, 1979

[15] G. Goon, N.V. Biba, O.B. Sadykhov, S.A. Stiebounov and A.I. Lishny, Automated finite element software system FORM-2D for analysis and design of forging processes, Open-die and closed-die forging technology, No.9-10, 1992, 4-7 (in Russian)

[16] Y.T. Lin and J.P. Wang, An investigation into the constant friction factor with the upper-bound flow function elemental technique, J. Materials Proc. Techn. 36(1992)57-68

[17] Y.T. Lin and J.P. Wang, A new upper-bound elemental technique approach to axisymmetric metal forming processes, Int. J. Mach. Tools Manufact, 33(1993)135-151

[18] Y.T. Lin and J.P. Wang. A thermal analysis of cold upsetting with the upper-bound flow function elemental technique, J. Materials Proc. Techn. 37(1993) 283-294

[19] J. Qin, An upper bound approach to plane strain problems using a general triangular element, J. Materials Proc. Techn. 40(1994)263-270

NOTATIONS

a_0	exponent for strain rate hardening curve
h_0	minimum distance between two dies when calculation stops
h_1,h_2	y-coordinates of bottom and upper contact surfaces
h_{f1},h_{f2}	y-coordinates of bottom and upper free surfaces
h_{1L},h_{2L}	y-coordinates of bottom and upper contact surfaces at $x=x_L$
h_j'	first derivative of, $h_j'=dh_j/dx$, $j=1$ for h_1 and $j=2$ for h_2
k	yield shear stress
k_e	mean yield shear stress
m	friction factor, $0\leq m\leq 1$
n_e	total number of elements
n_s	total number of velocity discontinuity surfaces
S_i	area of velocity discontinuity surface (including the contact surfaces between tools and the workpiece)
t	time
Δt	time increment
T_i	prescribed traction
V	total volume of the deformation zone
V_i	volume of element i
v_{f1},v_{f2}	velocities of bottom and upper free surfaces in y-direction
v_i	velocity component
v_{t1},v_{t2}	velocity of bottom and upper dies
v_{xi},v_{yi}	velocity components for element number i in x and y directions
v^*	velocity discontinuity
v_j^*	velocity discontinuity along S_i
$\dot{W}$	total power dissipation
$\dot{W}_{pi}$	power dissipation due to plastic deformation within element i
$\dot{W}_{si}$	power dissipation due to shear along boundary i
$\dot{W}_d$	power dissipated due to shear between elements
$\dot{W}_f$	power dissipated due to friction between the workpiece and dies
$\dot{W}_t$	power supplied by prescribed body traction
x,y	Cartesian coordinates
x_L,x_R	x- coordinate of the left and right edges of an element
x_{np}	position of neutral plane in x- direction
y_{np}	position of neutral plane in y- direction
$\dot{\varepsilon}_e$	effective strain rate
$\dot{\varepsilon}_{ei}$	effective strain rate for element i
$\dot{\varepsilon}_x,\dot{\varepsilon}_y$	strain rate components
$\dot{\varepsilon}_{xy}$	shear strain rate
σ_0	yield stress for static condition when $\dot{\varepsilon}_e=0$
σ_e	effective stress
τ	shear stress at velocity discontinuity surfaces, $\tau=k$. $\tau=mk$ for sliding contact between workpiece and dies.

Simulation of Materials Processing: Theory, Methods and Applications, Shen & Dawson (eds)
© 1995 Balkema, Rotterdam. ISBN 90 5410 553 4

Prediction of shape distortions of hard metal parts by numerical simulation of pressing and sintering

D.-Z. Sun & H. Riedel
Fraunhofer-Institut für Werkstoffmechanik, Freiburg, Germany

ABSTRACT: To calculate shape distortions of powder compacts during sintering, both die pressing and sintering processes were simulated for two hard metal parts by using the finite element method. The modified Drucker-Prager model with cap hardening and a micromechanical model based on the diffusion along grain boundaries were applied to simulate die pressing and sintering, respectively. The inhomogeneous density distributions caused by the pressing were computed and the shape changes of the parts during the subsequent sintering were predicted. A comparison between the numerical and experimental results for the first analyzed part shows that the undesired shape distortions of the powder compact can be very well calculated by the simulation method. The second example demonstrates that the numerical simulation can be used to optimize pressing schedule and punch forms.

1 INTRODUCTION

Die pressing of the powder and subsequent sintering of the powder compact are two essential steps in powder technology. Due to the variation of cross-section of a compact and the friction effect die pressing usually results in an inhomogeneous density distribution which causes shape distortions of the part during sintering. Later corrections of the shape of ceramic parts by hard-machining are very expensive. A favorable way to approach the desired shape is optimizing the forms of pressing tools and the pressing schedule. We expect that the efficiency of the optimizing process can be significantly increased by employing numerical simulation, especially the finite element method. Most of the numerical or analytical work about the forming processes of powder compacts concentrated only on one step, either on die pressing or on sintering. Brekelmans et al. (1991) and Brown et al. (1994) have analyzed the yield behavior of powder compacts under die pressing. Abouaf et al. (1988), Riedel (1990) and Du and Cocks (1992) have modeled the sintering (or hipping) step. Riedel and Sun (1992) have simulated the whole forming process of a hard metal cutting tool. In several cases the numerical predictions could be at least partly compared with experimental results and the achieved agreement was usually satisfactory. This encouraged an enhancement of the activities to simulate the pressing and sintering processes. The present paper describes two material models applied to simulate die pressing and sintering and shows the capability of the numerical method to predict the shape distortion observed after sintering by means of two examples.

2 MATERIAL MODELS

2.1 *Drucker-Prager cap model for die pressing*

The yield behavior of the powder compacts was described by the Drucker-Prager model with cap hardening which is available in the finite element program ABAQUS as a standard material model. The cap yield surface is given by

$$\Phi = \left[(p-p_a)^2 + \left(\frac{Rt}{1+\alpha-\alpha/\cos\beta}\right)^2\right]^{1/2} - \tag{1}$$

$$- R(d+p_a\tan\beta)=0 \; .$$

Here p is the hydrostatic pressure, t is a

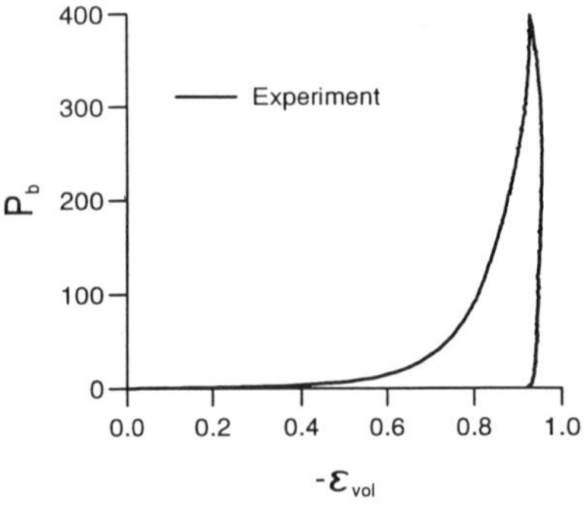

Fig. 1. Measured cap hardening of a hard metal powder.

combination of the second and third stress invariants (the third being neglected in our calculations; then t is the second invariant). R is a material parameter controlling the shape of the cap and β is the inclination angle of the failure line. The parameter α defines a smooth transition surface between the shear failure line and the cap. The material cohesion is given by the parameter d. Finally, p_a is defined as

$$p_a = \frac{p_b - Rd}{1 + R\tan\beta} , \qquad (2)$$

where p_b is the flow pressure for a given volumetric plastic strain under isostatic loading conditions.

The compaction behavior of two hard metal powders used for the study was partly determined by performing uniaxial die pressing tests. As long as no triaxial test results are available, the parameters in equation (1) are estimated to be $\alpha = 0$, $d = 0$, $\beta = 65°$ and $R = 0.5$. While the first two of these parameters usually play no role in die pressing simulations, the latter two determine the flow behavior under stress states near uniaxial compaction. They were estimated based on literature data on the radial pressure in a die. In uniaxial compaction, p_b is related to the axial stress by $p_b = 0.7435\ |\sigma_a|$ when R and β are chosen as specified above. The measured hardening behavior of one powder is shown in Fig. 1. With the parameters and the hardening curves, the model is completely specified.

2.2 *Constitutive equations for sintering*

The mechanical behavior during sintering is described by a linear viscous law. For isotropic material the general form is

$$\underline{\dot{\varepsilon}}^{vp} = \frac{\underline{\sigma}'}{2G} + \underline{1}\frac{tr(\underline{\sigma}) - 3\sigma_s}{9K} , \qquad (3)$$

where $\underline{\dot{\varepsilon}}^{vp}$ is the strain rate tensor, $\underline{\sigma}'$ is the deviatoric part of the stress tensor, $tr(\underline{\sigma})$ is the trace of the stress tensor, $\underline{1}$ is the unit tensor, G and K are the shear and bulk viscosity, respectively, and σ_s is the sintering stress; G, K and σ_s depend on the current relative density ρ, on the temperature and on the transport mechanism that supports densification.

Although liquid phase sintering contributes substantially to the densification of hard metals, we use a simple model for solid state sintering (Riedel, 1990, Riedel and Sun, 1992). If grain boundary diffusion is the dominant transport mechanism, this model gives

$$G = \frac{\sqrt{3}kT(1-\omega)^3 d^3}{48\,\Omega\,\delta D_b} \qquad (4)$$

$$K = \frac{4G}{3}\left(\frac{1-\rho_1}{1-\rho}\right)^{(m+1)/(m+5/3)} \qquad (5)$$

$$\sigma_s = \sigma_{s1}\left(\frac{1-\rho_1}{1-\rho}\right)^{1/(3m+5)} . \qquad (6)$$

Here Ω is the atomic volume, δD_b is the grain boundary diffusion coefficient, k is Boltzmann's constant, T is the absolute temperature, d is the grain facet diameter, ρ is the relative density, ρ_1 is a reference density (which is chosen as $\rho_1 = 0.6$), m is a parameter that characterizes the shape of the pore size distribution function (it is chosen as $m = 0.5$), σ_{s1} is the sintering stress at the reference density, and ω is the voided area fraction of grain boundary. It is related to the density by

$$\omega = 3^{3/4}\left(\frac{1-\rho}{\pi f(\psi)}\right)^{1/2} , \qquad (7)$$

where ψ is the dihedral angle at the pore tip (which is chosen as $\psi = 90°$) and $f(\psi)$ is the area of the pore normalized by the area of a circle with equal radius of curvature (for $\psi = 90°$, $f(\psi) = 1$). The density can be considered as an internal variable of the constitutive model. Its evolution equation is deduced through mass conservation

$$\dot{\rho} = -\rho \; tr(\underline{\dot{\epsilon}}^{vp}) \; . \tag{8}$$

This constitutive model was implemented in the finite element program ABAQUS as a user-supplied material model. An updated Lagrangian formulation was applied to account for geometric nonlinearities due to the large volumetric strains. To increase the efficiency and stability of the calculations, a tangent modulus method to determine the stiffness matrix and to integrate the variables was developed following Sester (1995). Now we outline the procedure of the formation of a tangent stiffness matrix. Following Peirce et al. (1984), we choose an implicit scheme:

$$\underline{\dot{\epsilon}}^{vp} = (1-\theta)\; {}^{t}\underline{\dot{\epsilon}}^{vp} + \theta \; {}^{t+\Delta t}\underline{\dot{\epsilon}}^{vp} \tag{9}$$

where θ is a parameter which can range from 0 to 1, with $\theta=0$ corresponding to a simple Euler time integration scheme. The sintering strain rate at $t+\Delta t$, ${}^{t+\Delta t}\dot{\epsilon}^{vp}$, can be approximated by using a Taylor series expansion:

$${}^{t+\Delta t}\underline{\dot{\epsilon}}^{vp} = {}^{t}\underline{\dot{\epsilon}}^{vp} + \frac{\partial \underline{\dot{\epsilon}}^{vp}}{\partial \underline{\sigma}} \Delta \underline{\sigma} + \frac{\partial \underline{\dot{\epsilon}}^{vp}}{\partial \rho} \Delta \rho \tag{10}$$

The relation between the stress increment $\Delta\underline{\sigma}$ and the total strain increment $\Delta\underline{\epsilon}$ is given by

$$\Delta \underline{\sigma} = \underline{C}_{e} \; (\Delta \underline{\epsilon} - \Delta \underline{\epsilon}^{vp}) \tag{11}$$

where C_e is the tensor of elastic moduli. Using equations 8 to 11, $\Delta\underline{\sigma}$ can be solved as:

$$\Delta \underline{\sigma} = \underline{C}\left\{\Delta \underline{\epsilon} - \Delta t \; {}^{t}\underline{\dot{\epsilon}}^{vp} + tr(\Delta \, {}^{t}\underline{\dot{\epsilon}}^{vp}) \; / \; A\right\} \tag{12}$$

with

$$\underline{C} = \left\{ \underline{C}_{e}^{-1} + \theta \Delta t \left[\frac{\partial \underline{\dot{\epsilon}}^{vp}}{\partial \underline{\sigma}} - \frac{\partial \underline{\dot{\epsilon}}^{vp}}{\partial \rho} \cdot \frac{\partial tr(\underline{\dot{\epsilon}}^{vp})}{\partial \underline{\sigma}} \; / \; A \right] \right\}^{-1}$$

$$A = \frac{1}{\theta \Delta t \rho} + \frac{\partial tr(\underline{\dot{\epsilon}}^{vp})}{\partial \rho} \; .$$

Obviously, the matrix $\underline{C}$ is the tangent moduli matrix required to yield the stiffness matrix for finite element calculations. Equation 12 was used not only to form the stiffness matrix but also to integrate the stresses and the internal variables. The consistency of the integration procedure with the determination of the stiffness matrix guaranteed good convergence behaviour.

3 APPLICATION TO CUTTING TOOL 1

3.1 *Simulation of die pressing*

The modeled cutting tool is made of a hard metal powder. The fill density, the green density and the theoretical density after sintering are 3.1, 7.5 and 13 g/cm^3, respectively. The form of the considered part looks like a house. It is oriented upside down in the powder press. Half of the cross-section of the house shaped part is shown in Fig. 2. The geometry of the cross-section does not change in the long direction, except that the "front walls" of the house are inclined by an angle of 7° to the vertical direction. Figure 2 a shows the original and deformed two-dimensional meshes for the pressing simulation.

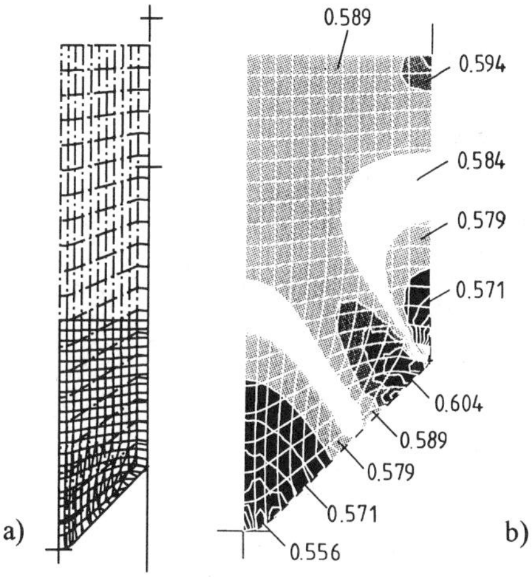

Fig. 2. a) Finite-element mesh before and after die pressing, b) Density distribution in the green body.

Plane-strain conditions were assumed, since little powder flow in the long direction of the house is expected. The left side of the mesh was defined as the symmetry plane. The die wall and the surface of the lower punch were modeled with rigid surfaces. The measured friction coefficient, $\mu=0.18$, was applied to simulate the contact processes between the powder and the die or the punch. In Fig. 2a it can be recognized that with forcing the upper punch down the die was also moved towards the bottom. However, the displacement of the die is only half of that of the upper punch. Fig. 2b shows the calculated distri-

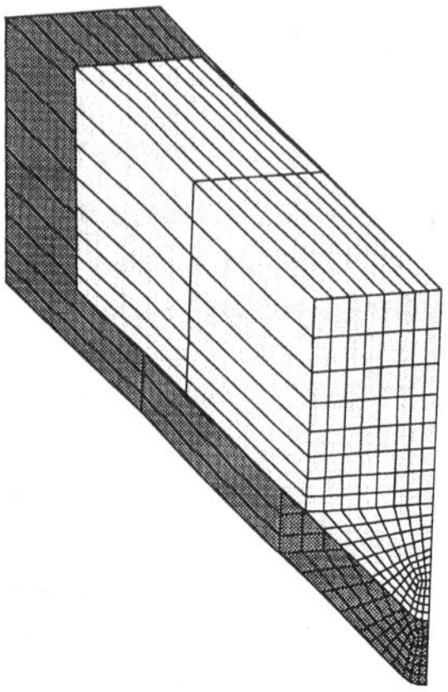

Fig. 3. One quarter of the house-shaped part before and after sintering.

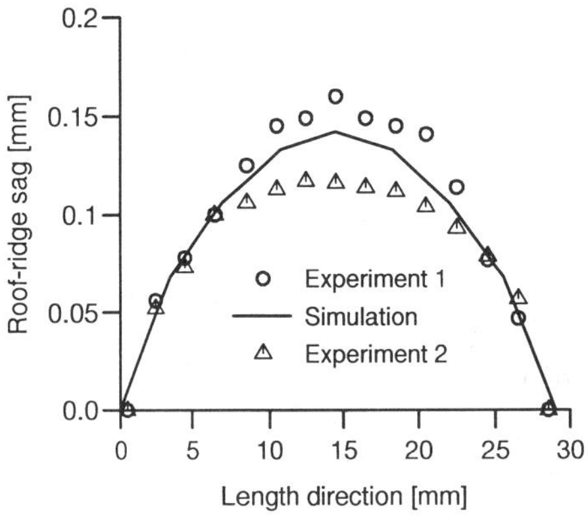

Fig. 4. Sag of the roof ridge along the length direction.

bution of the volumetric strain after pressing which can be converted into the relative density. It is evident that the lowest density is found in the region close to the roof ridge of the "house" and the highest at the transition from the lower punch to the die wall. In spite of the down movement of the die, a density gradient at the die wall is caused by the friction force and the inclined surface of the lower punch.

To determine the effects of the movement of the die on the density distribution, a simulation with fixed die was performed. It is found that the density distribution is slightly more homogenous, when the die is moved.

3.2 *Simulation of sintering*

Sintering was simulated as an isothermal process. The effect of this simplification on the results for the shape distortion is expected to be small. In this case, the parameters $\Omega\delta D_b/(kTd^3)$ and σ_{s1} only determine the time scale and the stress level; they can be given arbitrary values, if the time to sintering is not required from the simulation. By using a post-processing routine, the volumetric strain was converted into relative density by $\rho = \rho_s \exp(-\varepsilon_{vol})$, where ρ_s is the relative fill density, and transferred to the mesh for the sintering simulation. The case with moving die is considered.

Figure 3 shows a quarter of the house-shaped part with the 3D finite-element mesh, before and after sintering. As a consequence of the green-density inhomogeneity, the part undergoes shape distortions during sintering. However, they are relatively small and hardly visible in Fig. 3. Therefore, the deviations from the desired shape are shown more clearly in Fig. 4 and Fig. 5. The predictions are compared with measurements on parts sintered by Krupp Widia. The shape of the parts was determined with a profilometer, which projects the magnified contour of the specimen on a screen together with a coordinate system with a micrometer scale.

Due to the low green density in the "roof" of the "house", the house as a whole bends in its long direction during sintering, so that the bottom gets convex and the "roof-ridge" sags by some 150μm. Figure 4 shows the sag of the roof ridge. Apparently the numerical simulation predicts the measured curvature well.

Another consequence of the distortion in the long direction is a change of the inclination of the "front wall" of the house . The desired value is 7°, but after sintering a value of 7.95° is predicted, which coincides with the measured values nearly perfectly. The green density exhibits a gradient along the roof from about 7.01 g/cm^3 at the roof-ridge to about 7.8 g/cm^3 near the lower edge of the roof (the "gutter"). This leads to a reduction of the roof inclination angle, which is intended to be 45°. The predicted angle after sintering is 44.1°, in nearly perfect agreement with the measurement. The deviation from the intended shape is shown clearly in Fig. 5. Here the steeply decreasing initial part corresponds to the rounded roof tip, while the dotted line represents the intended 45° slope of the roof. Even at this magnification, the agreement of the predicted and measured contours is good.

4 APPLICATION TO CUTTING TOOL 2

4.1 *Simulation of die pressing*

The form of cutting tool 2, which is made of a different hard metal powder, is shown in Fig. 6. In the numerical simulations the part was pressed with two different pressing schedules: a) one-sided pressing only from the top, b) two-sided pressing from the top and the bottom with the same pressing volume for each direction. The position of the die modeled with rigid surfaces was fixed. The one-sided pressing was first simulated with a three-dimensional mesh by using the implicit finite element program ABAQUS/Standard. Due to the convergence difficulties the computation could not be performed to the required punch displacement of 3.4 mm. To solve the problem we carried out three 2D analyses (plane strain) on planes normal to the long direction of the part instead of a full 3D analysis. We located one plane in the cylindrical section and two planes in the section with varying width. The 2D results were then interpolated to generate a 3D green density distribution. This is a valid approximation, if little powder flow occurs in the long direction of the part. The converged 3D calculation validated this practical method.

An additional problem encountered in the simulation of the two-sided pressing is that the lower punch presses the powder into a widening cross-section. As a consequence the powder is first pushed upwards as a solid column and looses contact with the inclined faces of the die. As the process goes on, the powder in this region is unloaded elastically and the stress state moves from the cap to the failure line of the Drucker-Prager model. This leads to an unstable collapse of the freely standing powder column. Due to this instability problem we were not able to make the program ABAQUS/Standard converge to the required punch displacements despite 2D simulation. This numerical problem was avoided by using ABAQUS/Explicit which does not require the iteration process because of the concept of performing a large number of small time increments in an efficient manner.

Figure 6 shows the interpolated green density distributions. The mesh is that used in the sintering simulation. After one-sided pressing, the powder is more highly densified on the right-hand side of the cross-section due to the slope in the die shape. For the same reason the average density at the thin end is higher than in the cylindrical part of the specimen (Fig. 6a). Interestingly, the material in the right upper corner assumes a lower density than its vicinity. This is caused by large shear deformations when the

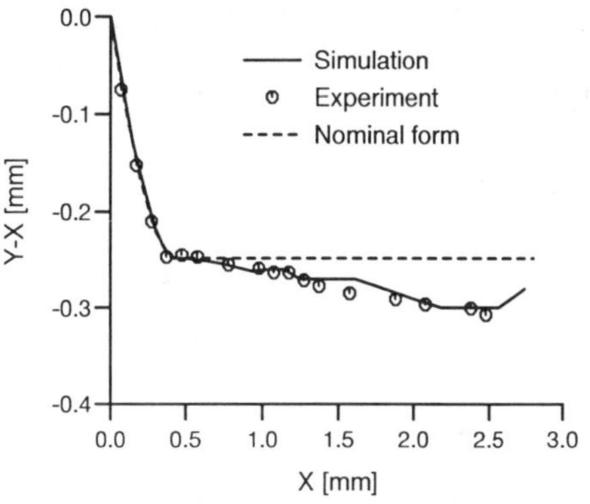

Fig. 5. Deviation of the roof inclination from the intended 45° line.

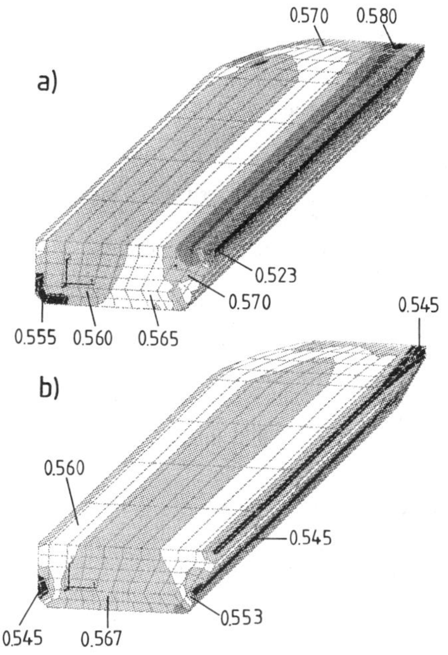

Fig. 6. Calculated green density distribution for a) one-sided, b) two-sided pressing.

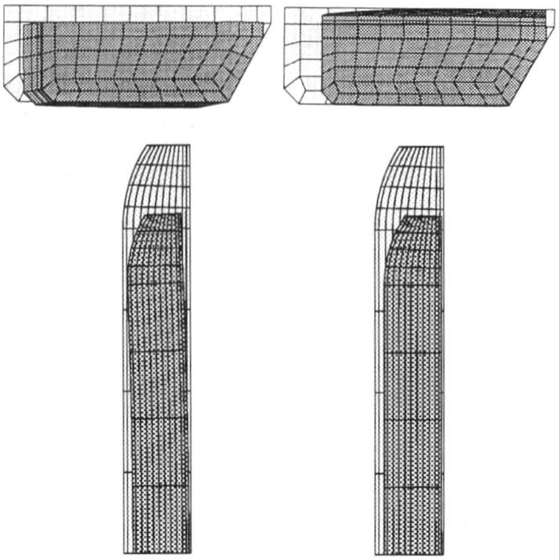

a) b)

Fig. 7. Shrinkage and shape distortions during sintering for a) one-sided pressing b) two-sided pressing.

material approaches the transition to the inclined part of the die. Figure 6b shows that the great density difference between the right-hand side and the left-hand side can be avoided by two-sided pressing. The density gradient in the vertical direction observed after the two-sided pressing can certainly be reduced by adjusting the ratio of the displacement of the upper punch to that of the lower punch.

4.2 *Simulation of sintering*

Sintering was simulated using the 3D mesh shown in Fig. 6 with 488 twenty-node elements. Three points on the cross-section displayed in Fig. 6 are constrained to suppress rigid-body motions and rotations, so that this plane is the reference plane. Figure 7 shows two aspects of the part before and after sintering. As expected, the density inhomogeneity caused by one-sided pressing results in a distinct sintering distortion in the plane of the part, which is not observed in the case of two-sided pressing. The part compacted by two-sided pressing shows a more pronounced out-of-plane distortion, which could be reduced, if the volume displaced by the lower punch was smaller than by the upper punch.

CONCLUSIONS

Die pressing and sintering were modeled by using the finite element method. Shape distortions observed after sintering as a consequence of inhomogeneous green density distributions caused by die pressing were predicted with a high accuracy. This numerical method can be used to optimize pressing schedule and the shape of pressing tools.

Acknowledgements. The authors would like to thank the German Ministry for Science and Technology and the companies Krupp Widia and Cérametal for financial support.

REFERENCES

Abouaf, M., J.L. Chenot, G. Raisson and P. Baudin 1988. Finite-element simulation of hot isostatic pressing of metal powders. *Int. J. Numer. Methods Eng.* 25: 191-212.

Brekelmans,W.A.M., J.D. Jansson and A.A.F. van de Ven 1991. An Eularian approach for the die compaction process. *Int. J. Numer. Methods Eng.* 31: 509-524.

Brown, S. and G. Abou-Chedid 1994. Yield Behaviour of Metal Powder Assemblages. *J. Mech. Phys. Solids,* 42: 383-399

Du, Z.-Z. and A.C.F. Cocks 1992. Consitutive Models for the Sintering of Ceramic Components - I. Material Models. *Acta Metall. Mater.,* 40: 1969-1979.

Peirce, D., C. F. Shih and A. Needleman 1984. A tangent modulus method for rate dependent solids, *Comp. Struc.* 18: 875-887.

Riedel, H. 1990. A constitutive model for the finite-element simulation of sintering distortions and stresses, in: Ceramic *Powder Science III,* G.L. Messing et al., Eds. American Ceramic Society, Westerville: 619-630.

Riedel, H. and D.-Z. Sun 1992. Simulation of die pressing and sintering of powder metals, hard metals and ceramics, *Numerical Methods in Industrial Forming Processes.* Chenot. Wood & Zienkiewicz (eds) Balkema, Rotterdam: 883-886.

Sester, M. 1995. Anwendung fortschrittlicher Werkstoffmodelle im Rahmen der Kontinuumsmechanik, IWM-report, to be published.

Simulation of Materials Processing: Theory, Methods and Applications, Shen & Dawson (eds)
© 1995 Balkema, Rotterdam. ISBN 90 5410 553 4

An integrated modeling system for heat treating and machining processes

T.C.Tszeng, W.-T.Wu & J.P.Tang
Scientific Forming Technologies Corporation, Columbus, Ohio, USA

ABSTRACT: Heat treating and machining operations are the integral stages of a typical process for manufacturing structural components. The changes of properties imparted in the heat treating cycles may actually dominate the final properties of the components. A poor heat treatment processing would lead to unsatisfactory microstructure, properties, residual stresses, geometric accuracy, or even cracking in the manufactured components. Distortion during the subsequent machining is also of concern, particularly in the manufacturing of high performance, high precision, high added-value structural components. It is understood that the aforementioned problems are usually the direct and accumulated results of complex interactions between thermal, mechanical, and metallurgical phenomena in the manufacturing processes. An easy remedy to the problem is not always obtainable without a thorough understanding of the underlining phenomena. This paper describes the development of an integrated process modeling system capable of simulating metallurgical and thermal-mechanical evolution during the heat treating and subsequent machining processes. The developed modeling capabilities are demonstrated by two examples in processing of steel and superalloy components.

1. INTRODUCTION

A typical structural component made of superalloy and other material systems would go through cycles of forming, heat treatment, and machining operations in the manufacturing stage. In the heat treating operation, the components are usually given a solution treatment followed by rapid quench, and, for age-hardenable materials, a subsequent aging treatment to improve the physical and mechanical properties. In order to achieve the desirable combination of microstructure, properties, residual stresses and dimensional accuracy in the final product through cycles of heat treating and machining, the process engineers have to adjust the heating rates, holding temperatures, holding times, cooling rates, machining strategy, etc. from their own experience. Since many of the design requirements are interacting with each other, it is not possible just to single out one problem and cure it without affecting other characteristics of the components. In other words, a comprehensive understanding and full control of the process is the better solution to most heat treating problems.

For property improvements, heat treating of superalloys and many other material systems can be grouped into three essential operations: solution treatment, quench, and age hardening. To help clarifying the problems in the heat treating processes, each of these operations are briefly explained as follows:

<u>Solution Treatment</u>

The purpose of solution treating is to dissolve the second phase particles to prepare the solid solution for developing various microstructures in the following quenching stage, and aging treatment for age-hardenable alloys. Several metallurgical processes would occur in this stage, including second phase dissolution, metadynamic or static recrystallization, and grain growth. The grain size is one of the most important factors in determining the properties of heat treated components.

A similar process, annealing is used to reduce hardness and increase ductility to facilitate forming or machining. It is important in this process to control the degree of recrystallization and grain size for the designated resultant characteristics and properties.

<u>Quench</u>

Quench is the operation to let the solution treated component cool from the high temperature and thereby control the microstructure. For age-hardenable alloys, the subsequent quenching is to maintain the state of supersaturation at room temperature by suppression precipitation. Due to the severe temperature gradient and other structural changes during quenching, the stress state can be damaging to components. The impact of quench design on cracking susceptibility during heat treating of a superalloy component was illustrated by the interesting example of Wallis et al [1989]. Also, the stress may promote over-aging during the subsequent aging treatment.

For some material systems (like ferrous alloys), the metallurgical changes during quench would result in different thermal-mechanical responses, and therefore lead to different deformation and stress state in the components. One of the most notorious example is the diffusionless martensitic transformation in steels, which usually gives rise to excessive stress level and/or deformation during the quench stage.

Age Hardening

For age-hardenable alloys, aging is the operation performed at intermediate temperatures to develop the maximum strength by precipitation of a dispersed second phase throughout the matrix. The important considerations include: precipitate volume fraction, and mean and distribution of precipitate size. In many superalloy applications, a two-step or multistep aging treatment is used to control the size distribution of γ' or γ'' precipitates. In this case, the first aging treatment at a higher temperature is to precipitate or control the morphology of grain boundary carbide for improved ductility. The second aging treatment at a lower temperature leads to a finer precipitate for better strength

In summary, there are complex thermal-mechanical and metallurgical processes occurring in the components during the heat treating processes. Since the changes in properties imparted in the heat treating cycles may, in many cases, dominate the final properties of the part, a poorly designed heat treatment procedure would lead to unacceptable mechanical properties in terms of hardness, tensile properties, fracture toughness, fatigue resistance, etc.

The overall objective of this work is to provide such a computer modeling system for analyzing the metallurgical and thermal-mechanical responses during heat treating and machining processes of components made of superalloy and other high performance material systems. The Simulation Engine, as the core of the modeling system, will have the computational capabilities in handling:

- Evolution of various metallurgical processes
- Heat transfer analysis
- Stress and distortion analysis

In the present work, we performed a preliminary study of several aspects of the modeling system. Specifically, we focused on the following areas which have the immediate impact to the technology:

- Heat treating of steel components
- Distortion in superalloy components after heat treating and subsequent machining

The Simulation Engine is consisted of two major modules: thermal-mechanical and metallurgical. The thermal-mechanical module is essentially a FEM computational model with thermal-elastoplastic and constituent-based transformation plasticity, whose theoretical foundation and other applications will be reported in a separate paper [Tszeng et al, 1995a]. The metallurgical module is a computational model for calculating various metallurgical phenomena, like phase transformation, precipitation, etc.

In the following, we will show two examples of using the modeling system for a steel system and a superalloy, respectively. The first example demonstrates the phase transformation as well as the mechanical responses, the latter example is mainly related to the distortion during machining processes. While not shown in this paper, the validation of the thermal-mechanical and metallurgical calculations can be found in Tszeng [1994] and Tszeng et al. [1995a,b].

2. HEAT TREATING OF STEEL COMPONENTS

In this section, we considered the heat treatment of a SAE 1042 (CK45) carbon steel cylinder of 30-mm-diameter × 100-mm-length. The cylinder is austenized at 830°C for 30 min. and followed by oil quench to achieve a combination of good strength and ductility. When the specimen is continuously cooled from the austenitizing temperature, the austenite can decompose into ferrite, pearlite and bainite through diffusion transformation. If the cooling rate is high enough, there may be austenite transformed into martensite when the temperature is dropped below the martensite transformation temperature. The properties of the component are highly dependent upon the content of each phase. While determining the progress of phase transformation during the quench process worthies the effort by itself, the formation of a new phase, particularly the martensite, has a definite impact to the deformation and stress state in the component. The progress of phase transformation is calculated by using the rate form of the generalized Avrami equation combined with a Isothermal Transformation Diagram. The details of the metallurgical module with a broader range of applications will be discussed in a separate paper [Tszeng, et al, 1995b].

We use the IT Diagram corresponding to an austenitizing temperature of 1050°C for 1 min. [Vander, 1991]. The martensite transformation is 350°C. The different austenitizing temperatures (1050 vs. 830°C) and times (1 vs. 30 min.) can have impact on the carbide dissolution and austenite grain size during austenitization, thereby on the subsequent transformation kinetics. For the present study, we use the IT Diagram of Vander [1991] just for the reason of convenience and model development. In a real industrial application, one should use the appropriate IT Diagram corresponding to the same austenitization condition as that of the actual process. In this computational model, we can utilize the thermophysical and mechanical properties of each constituent phase. For the material of consideration, SAE 1042, the properties of each phases are given by Schroder [1984]. The volume change for the considered carbon content is 0.0125 for bainite and 0.0147 for martensite.

When austenitized steel is quenched in a liquid medium such as water, oil, or an aqueous polymer, at least three types of heat transfer mechanism occur during cooling: vapor blanket, boiling and convection. Each of the heat transfer phenomena represents a very difficult problem to be modeled [e.g., Hetsroni, 1990]. In fact, there are very few attempts reported in the open literature to utilize these sub-models for heat treatment analysis. Instead, most of the heat transfer studies pertaining to heat treatment processes have been focused on the "cooling curves" in different ways. In these practices, the cooling histories at one or more points inside the test sample in the quenching process are measured. The relationship between the surface heat transfer coefficient (HTC) and the surface temperature are then determined. The inverse heat conduction analysis is one of the methods to determine the surface HTC [e.g., Beck and Osman, 1992].

In general, the HTC shows a peak value at a temperature in the boiling stage. In this study, we used a rather simple representation of Majorek et al. [1992] for the HTC at the surface point, which is characterized by several parameters as shown in Figure 1. T_b is the boiling point of the quenchant, and T_L is approximately the temperature at which the vapor blanket breaks down and boiling starts. The parameters of $T_{h,max}$, T_L, T_b, H_{max}, H_L and H_b are determined by lab tests for the specific quenchant. Also, these parameters can be location dependent. In this study, we used the data of Majorek et al. [1992], i.e., $T_{h,max}$ = 500°C, T_L=700°C T_b= 250°C, H_{max} = 5 KW/m^{2}K, $H_L = H_b$ = 0.5 KW/m^{2}K. The calculated temperature at two points, A and B, on the midplane of the cylinder is shown in Figure 2 (A on the axis, and B on the surface). The maximum principal stress at these two points are shown in Figure 3, in which the starting of martensite transformation is indicated. The competition between thermal shrinkage and volume expansion of mainly martensitic transformation is obvious in this figure. The distortion is shown in Figure 4 which indicates the major distortion is the length shortening.

Martempering is a heat treatment process which controls the quenching rate such that a spatially uniform martensitic transformation is occurred (nothing to do with tempering). The benefits include less distortion, less residual stress and lower cracking susceptibility. A common practice for this process is to maintain the quenchant at a temperature in the range of martensite temperature. When the temperature in the component becomes uniform, the component is slow cooled to allow the martensitic transformation continue slowly and homogeneously. In this study, we used an oil temperature of 200°C for the first 100 second of quench, and 20°C in the remaining of the quench. When the oil temperature is raised to a higher temperature, the parameters for determining the HTC as of Figure 1 will be changed as well. Because of the scope of this study, we keep using the same parameters as the previous calculation of straight quench in oil of 20°C. The calculated phase contents at the end of quenching are shown in Figure 5. There are traces of ferrite and bainite in the inner portion of the specimen (see Figure 5), although martensite is definitely the most abundant phase. The temperature history at points A and B is again shown in Figure 2. Because of the less severe quenching, the temperature drops at a lower rate in martempering. Figure 3 also shows the results of stress history of both cases. While the stress level is comparable

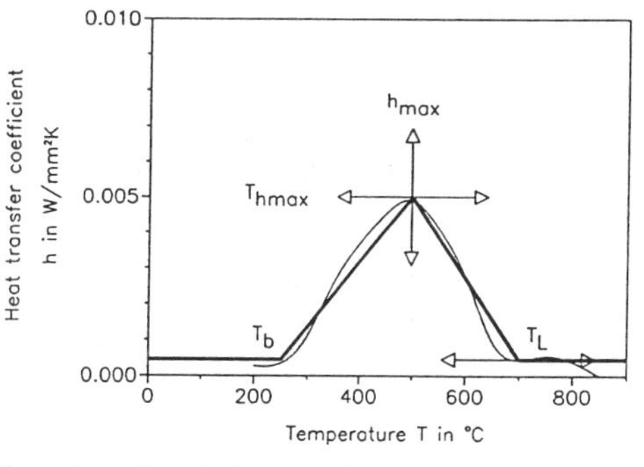

Figure 1. Description of the surface heat transfer coefficient during quenching in liquid quenchants [Majorek, et al, 1992].

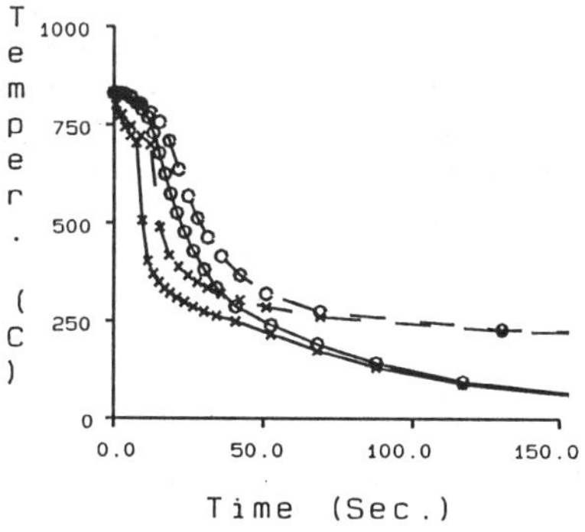

Figure 2. The temperatures at two points on the midplane of the SAE 1042 steel cylinder initially at 830°C. Circle = center, Cross = surface. Solid lines : straight oil quench, broken lines: martempering.

for both cases of straight oil quench and martempering, the residual stress is much lower in the case of martempering. The resulted distortion of Figure 4 indicated that martempering leads to a less shortening in the length. This is exactly one of the benefits we expect from martempering.

3. DISTORTION DURING MACHINING OF A HEAT TREATED COMPONENT

The example we reported is a nickel-based superalloy disk with 1-in.-ID and 6.6-in.-OD. The geometry are shown in Figure 6. The thermophysical and mechanical properties are given in Table 1. The component is solution treated at a temperature of 2100°F. After the solution treatment, the component is quenched to the room temperature and subjected to a machining operation which removes the portion shown in Figure 6. One of the design objectives is to minimize the distortion as a result of the material removal during the machining operations.

The thermal and mechanical responses during the quenching operation is readily calculated by using the thermal-elastoplastic material model in the DEFORM system, whose theoretical foundation and applications will be reported in a separate paper [Tszeng et al, 1995a]. Although the material model has been modified to include the influences of metallurgical processes in a special test module (see Section 3), there is no need to take these influences into consideration for the concerned superalloy. The determination of grain size, precipitation etc. will be discussed elaswhere [Tszeng, et al., 1995b]. We also assumed that the component at the end of solution treatment at 2100°F is completely recrystallized and free of stress and plastic strain. Further, we did not include the creep strain in the calculations, which is the topic under development by the authors.

The component is quenched in a relatively effective cooling medium with a constant and uniform heat transfer coefficient, $h = 1.157 \times 10^{-4}$ Btu/sec/in^{2}F, and a constant environment temperature of 70°F. The time step is adjusted continuously in the way that the maximum temperature change in one step is about 15°F. The calculated residual stress distribution is shown in Figure 7 for the tangential components.

The distortion during the machining operation is directly related to the level and distribution of the residual stress in the component right before the machining. In order to identify the major factors which contribute to the distortion during material removal, we employed a quasi-incremental approach for calculating the distortion. In this approach, the material is assumed to be removed in several sequential passes. In each pass, a designated portion of material is

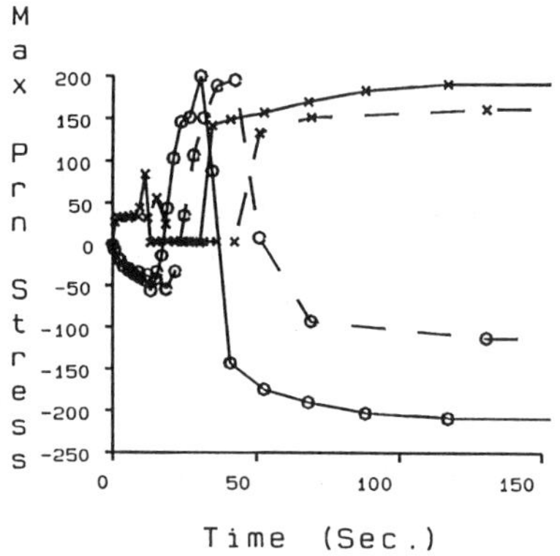

Figure 3. The max. principal stress at two points on the midplane of the of the SAE 1042 steel cylinder, Circle = center, Cross = surface. Solid lines : straight oil quench, broken lines: martempering.

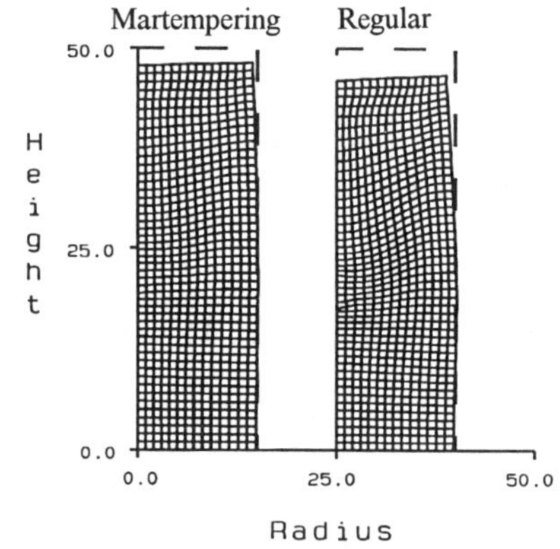

Figure 4. Distortions (magnified by 50 times) at the end of quenching of the SAE 1042 steel cylinder.

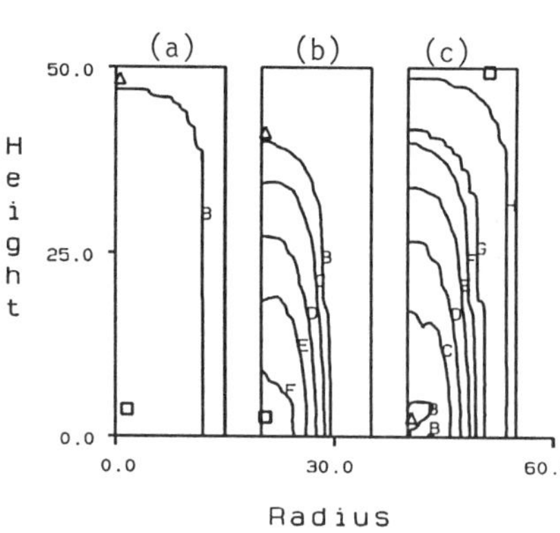

Figure 5. The phase content of (a) Ferrite, B = 0.01; (b) Bainite, B = 0.01, F = 0.05; and (c) Martensite, B = 0.88, H = 0.94 in the steel cylinder by martempering.

removed simultaneously. Figure 8 shows the calculated distortion at the end of each pass. Obviously, the biggest displacement is at the ID of the disk, which remains at a very low magnitude until the eighth pass. This pass corresponds to the stage when the top shoulder of the outer rim is removed (Figure 8h). The stress state in the shoulder region is mainly compressive in the hoop direction, as indicated in Figure 7 at the beginning of machining and right before this portion is removed (not shown). When this portion is removed, there is a net bending moment acting on the newly machined "hinge", resulting in the observed movement at the ID.

With the preceding observation regarding the relation between the stress state in the shoulder region and the distortion at the inside rim at the end of machining, it may be desirable to reduce the compressiveness of hoop stress in that region. One approach is to apply a less conductive coating on the top face of the component. The postulation is that the top face will cool at a lower rate such that the stress state would become more tensile at the end of quenching. To test this approach, the heat transfer coefficient at the top surface (between point B and C in Figure 6) is assumed to possess a much lower value of 1.0 $\times$ 10^{-5} Btu/sec/in^{2}F while the other surfaces remain the same h = 1.157 $\times$ 10^{-4} Btu/sec/in^{2}F. The results at the end of quenching indeed show a less compressive stress state at the shoulder region than that of Figure 8. Figure 9 shows the calculated distortion after machining. With the treatment on the top surfaces, the distortion is reduced by some extent.

Two-Step Quenching

For a component subjected to irregular material removal, a uniform distribution of residual stress does not lead to zero distortion. However, a lower level of residual stress certainly would result in less distortion during and after machining. To examine this possibility, a series of numerical experiments were conducted with the following quench design: the environment temperature is kept at an elevated temperature for the first 100 second, and changed

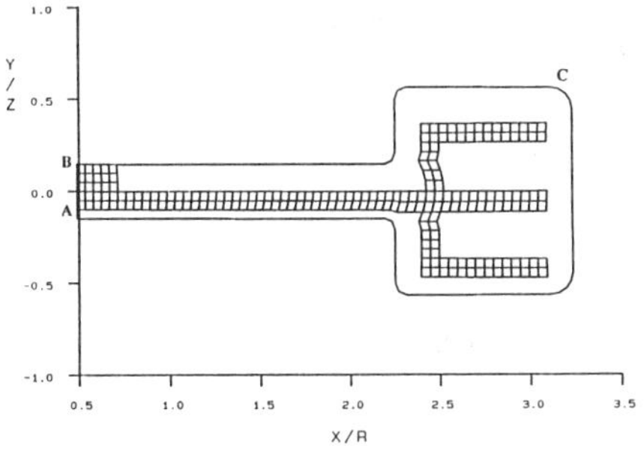

Figure 6. The geometry of a nickel-based superalloy disk initially at 2100°F. After quenched to the room temperature, the unmeshed portion will be removed by machining.

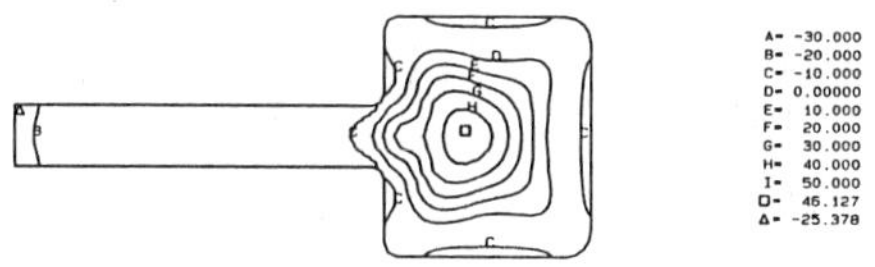

Figure 7. The residual hoop stress in the disk after the disk is quenched to the room temperature.

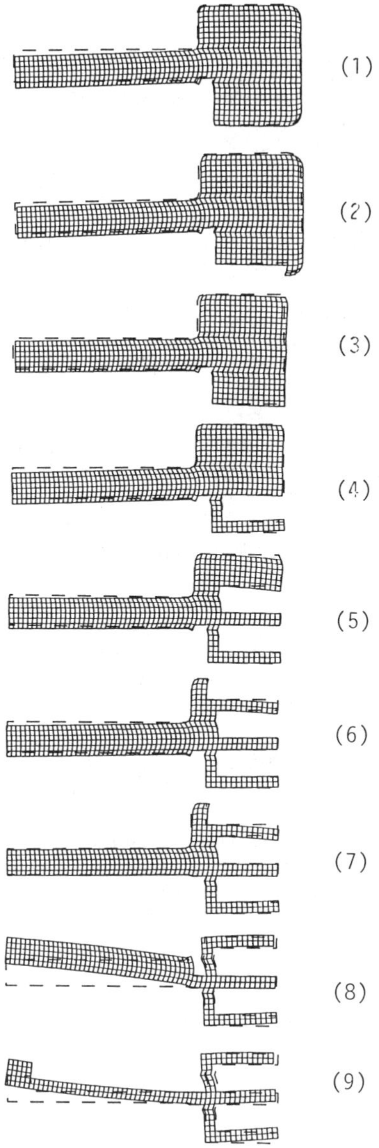

Figure 8. Distortions (magnified by 50 times) of the disk during a nine-pass removal operation. The largest distortion occurred at the eighth pass.

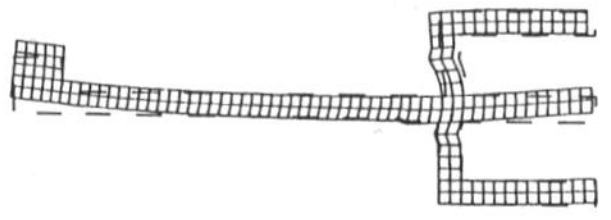

Figure 9. Distortion of the disk after machining when the top surface is applied with a insulation coating.

TABLE 1. Material Properties. E = Young's modulus, ν = Poisson's ratio, Y = yield stress, H = strain hardening coefficient, Cp = specific heat (Btu/lb.F), k = thermal conductivity (Btu/in.sec.F), α = thermal expansion coefficient.

TEMP.(F)	E(Ksi)	ν	Y(Ksi)	H (Ksi)
68	29.96	0.312	170.0	2996.0
1300	22.50	0.342	130.4	1700.0
1500	20.15	0.356	126.3	1940.0
1700	19.95	0.358	88.5	658.0
1900	13.50	0.430	38.6	282.0
2000	6.20	0.450	8.9	110.0
2050	4.25	0.450	6.8	18.8
2100	3.80	0.450	5.6	29.8

TEMP.(F)	Cp	$k(10^{-4})$	$\alpha\,(10^{-6})$
68	0.09	1.214	5.968
300	0.096	1.533	6.424
500	0.104	1.753	6.817
700	0.114	1.956	7.142
900	0.126	2.184	7.366
1100	0.139	2.436	7.569
1300	0.154	2.701	7.826
1500	0.170	2.944	8.191
1700	0.188	3.142	8.694
1800	0.198	3.220	8.997
1900	0.208	3.281	9.333
2000	0.218	3.329	9.697
2100	0.229	3.370	10.08
2200	0.240	3.409	10.48

to the room temperature in the rest of the quenching. According to the previous calculations, we found that the plastic deformation always occurs within the first 100 second. We postulate that a milder quenching in that period of time might lead to less plastic deformation and therefore lower level of residual stress. We conducted the numerical experiments by using different environment temperatures ranging from 900°F to 1800°F in the first 100 second of quenching. The axial displacement at point A (50 times) in Figure 6 is shown below as a function of the environment temperature:

T_{ENV}(F)	900	1200	1500	1800
Δy_A	0.051	0.041	0.0065	0.025

For all of the considered two-step quenching, the distortion is less severe than those resulted from the previous quench designs. The axial displacement reaches the minimum at an environment temperature of 1500°F. In fact, the overall distortion is very small with this quench.

4. CONCLUDING REMARKS

This paper reports some preliminary results of the modeling system for heat treatment processes. Two examples are used to show the various technical isses in modelig the quenching and subsequent machining of structural components. While this modeling system is still in development, the approach seems to be promising. Two remarks are noted pertaining to the reported results in this paper.

1. It is not the purpose of this paper to report the establishment of a generalized rule for minimizing the distortion during the quenching and machining of a component. However, the present examples indicated that the quench design has a great influence on the distortion in the components.

2. Creeping is significant in many heat treating processes of superalloy components. First, the inclusion of creep strain may be crucial in many cases for the calculated results to be accurate. This is important in our study of residual stress and dimensional changes due to stress relaxation by stress relieving or machining. Second, creeping is the major mechanism during stress relieving which is a common operation for many superalloy components.

Acknowledgments

This project is partially supported by a U.S. Air Force SBIR award (No. F33615-94-5204) with Dr. L. Semiatin as the Project Engineer. We would like to acknowledge the contribution of Dr. S. Srivatsa at GE Aircraft Engines.

REFERENCES

Beck, J. V. and Osman, A. M., 1992, "Analysis of Quenching and Heat Treating Processes Using Inverse Heat Transfer Method," *Quenching and Distortion Control* (ed. G. E. Totten), ASM Internation, pp. 147-153.

Hetsroni, G. (Ed.), 1990, *Heat Transfer 1990*, Hemisphere Publishing Co., p. 129.

Majorek, A., et al, 1992, "The Influence of Heat Transfer on the Development of Stresses, Residual Stresses and Distortions in Martensitically Hardened SAE 1045 and SAE 4140," *Quenching and Distortion Control* (ed. G. E. Totten), ASM Internation, pp. 171-179.

Schroder, R, 1984, "Some Influences on the Development of Thermal and Residual stresses in Quenched Steel Cylinders with Different Dimensions', *Proceedings of International Symposium on the Calculation of Internal Stresses in Heat Treatment of Metallic Materials*, Vol. 1, 1984, Linkoping, Sweden (eds. E. Attebo and T. Ericsson), pp. 1.-22.

Tszeng, T. C., U.S. Air Force SBIR Monthly Reports (Contract No. F33615-94-5204).

Tszeng, T. C., Wu, W.-T. and Tang, J. P., 1995a, "A Critical Comparison of Elastoplastic and Rigid-Plastic Approaches to Metal Forming Applications," in preparation.

Tszeng, T. C., Wu, W.-T. and Tang, J. P., 1995b, "Computational Models for Metallurgical Evolution in Superalloys during Heat Treatment Processes," in preparation.

Vander, G. F. (ed.), 1991, "Atlas of Time-Temperature Diagrams for Irons and Steels," page. 118, ASM Internation.

Wallis, R. A., et al., 1989, "Modeling the Heat Treatment of Superalloy Forgings," *Journal of Metals*, February, pp. 35-37.

Simulation of Materials Processing: Theory, Methods and Applications, Shen & Dawson (eds)
© 1995 Balkema, Rotterdam. ISBN 90 5410 553 4

Finite element analysis of tubular energy absorber in automobiles

P.T.Vreede, H.B.van Veldhuizen & S.P.Kragtwijk
Hoogovens Groep, IJmuiden, Netherlands

B.D.Carleer
University of Twente, Enschede, Netherlands

ABSTRACT: The use of a crash unit has been studied to decrease damage on automobiles when a minor collision takes place (with velocities lower than 15 km/h). A crash unit is considered which consists of an inverting tube. Aluminium and steel are used and the results of finite element calculations are compared with experiments. The calculated results are qualitative as well as quantitative in good accordance with experimental data.

1 INTRODUCTION

An important vehicle requirement is the ability to protect occupants against injury when a collision takes place. Therefore many parts outside the passenger compartment are designed to absorb energy easily during a crash. To avoid considerable damage of a car when the collision is relatively small, the use of a crash unit was studied.

Hoogovens Groep BV is originally a Dutch company with factories in the Netherlands, Belgium and Germany. Hoogovens Groep supplies steel as well as aluminium. To give advise on material selection and optimal use of materials supplied by Hoogovens a study to the inversion process was carried out.

The crash unit considered in this article consists of a tube and inverting die and a punch according to the principle scheme of figure 1.

The crash unit is mounted between the car body and the bumper and the tube inverts after a certain force level is exceeded. This inversion produces very large strains which form the energy absorbing mechanism. The amount of kinetic energy which can be dissipated depends on the material characteristics and the invertube geometry.

Numerical models are very efficient tools in analyzing the process by varying several material and geometrical parameters. To validate the numerical model invertube experiments were carried out.

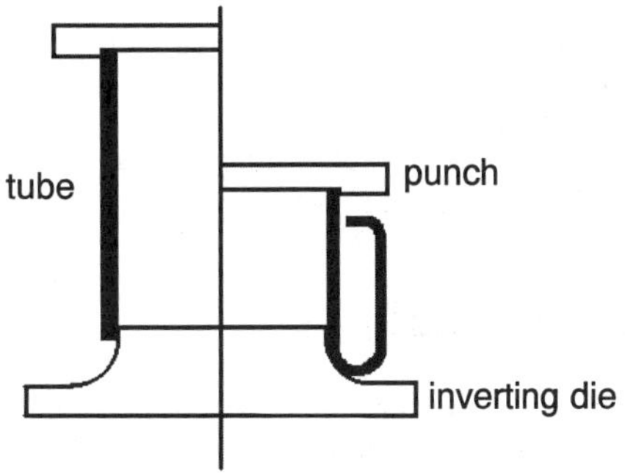

Fig. 1 Undeformed and deformed specimen.

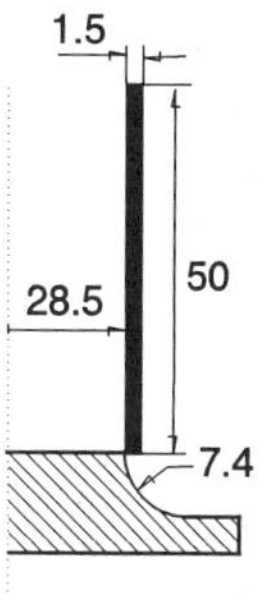

Fig. 2 Dimensions of specimen and tool in mm.

2 GEOMETRY

Based on considerations found in literature, we applied one invertube geometry in this article with dimensions given in figure 2.

3 NUMERICAL MODEL

The simulations were carried out with help of the finite element code Dieka which is developed at the University of Twente in cooperation with Hoogovens Corporate Research. This code has been applied successfully for simulations of hot and cold rolling, wire drawing, extrusion and deep drawing. A special feature of the program is the ability to apply the mixed Eulerian Lagrangian method. This means that the locations of the nodal points can be updated independently from the material displacements. However, we applied the updated Lagrangian formulation in the calculations of the present article.

A stiffness matrix is composed (implicit formulation). Because the experiments are performed at low speeds, the inertia forces are much smaller than the deforming forces. Therefore mass effects are not included in the model.

3.1 Material Modelling

In this study we compared the performance of a steel tube with an aluminium one. To obtain the relevant information, tensile tests were carried out on specimens cut out of the tube.
A Voce-like relation was used to describe the stress strain relation: $\sigma = \sigma_0 + \Delta\sigma\,(1 - \exp(-\epsilon/\epsilon_0))$.
After transforming the measured data to true stresses and strains the following parameters were found after curve fitting, see table 1. The comparison of the fit with the measured stress-strain curves is represented in figure 3.
The measurements of the tensile test can only be obtained in the areas until the plastic instability. In case of a multi-axial stress state, a criterion is needed to determine at which combination of stresses the material switches from elastic to plastic behaviour. A well-known planar isotropic yield criterion is the Hill criterion. In terms of the three principal stresses it can be written as:

$$\phi = R(\sigma_1-\sigma_2)^2 + (\sigma_2-\sigma_3)^2 + (\sigma_1-\sigma_3)^2 - (R+1)\sigma_{yield}^2 = 0$$

The *R*-value is defined to be the ratio between tangential and thickness strain when a tensile test is carried out. The tube is formed by extrusion and the texture gives rise to planar isotropic behaviour with the tube surface as plane of isotropy. The data given in table 1 are obtained from tensile tests with specimens taken out of the plane of isotropy.
From experiments carried out in our laboratorium we observed that aluminium exhibits a more Tresca-like yield criterion with increasing strains. We did not model such a behaviour and since the plane strain yield stress may be over-estimated with a Hill criterion, the process force may be over-estimated too.

Table 1

Properties	Aluminium	Steel
σ_0	191	293
$\Delta\sigma$	66.5	605
ϵ_0	0.068	0.405
R-value	0.6	1.4

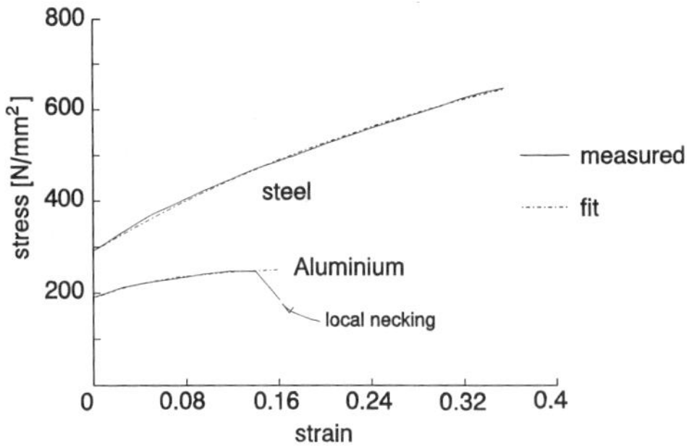

Fig. 3 Stress-strain curves, measured and fitted with Voce-relation.

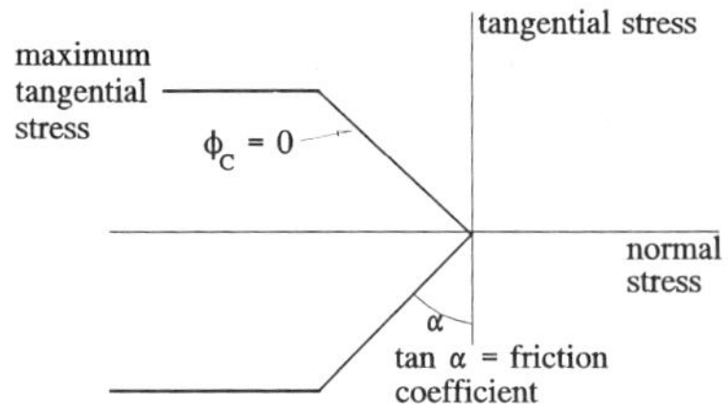

Fig. 4 The contact stress yield surface.

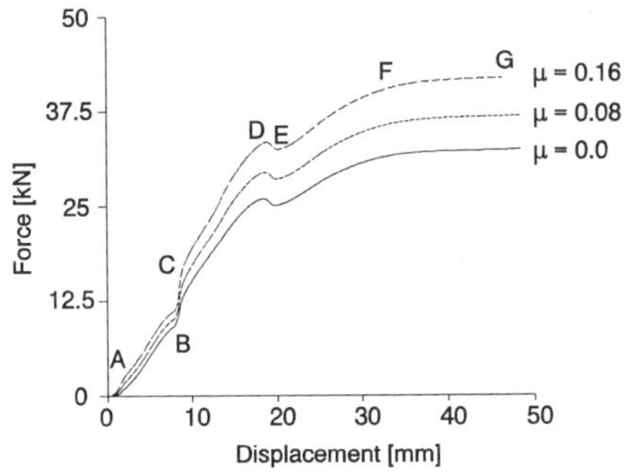

Fig. 5 Force-displacement curves of inverted aluminium tubes as a function of the Coulomb friction coefficient.

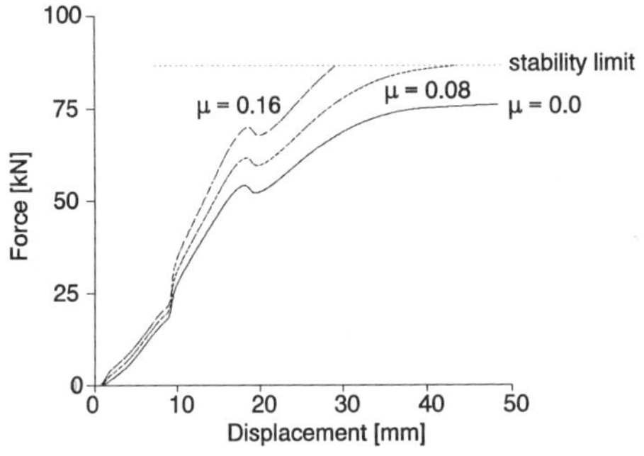

Fig. 6 Force-displacement curves of inverted steel tubes as a function of the Coulomb friction coefficient.

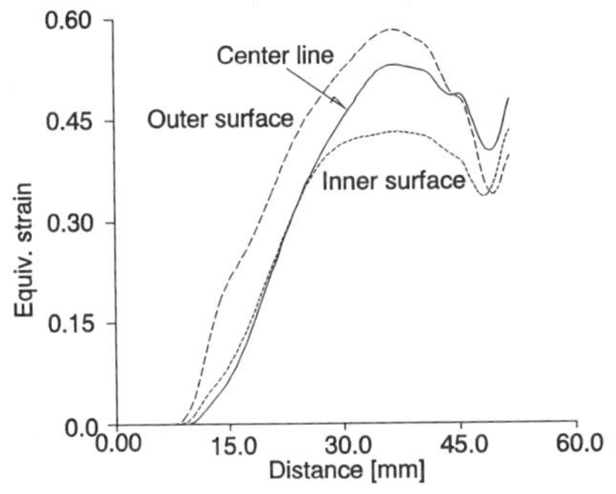

Fig. 7 Strain distribution along line of tube.

3.2 Contact Description

To describe contact between two bodies, interface elements were used. These elements are connected between the tools and the tube. If the elements are open, the contact elements have no effect. The penalty method is applied: when penetration occurs a normal force in the interface elements is generated. When the overlap increases the normal force increases proportionally.
If friction is included a tangential displacement will give a tangential stress. After a critical level of the tangential stress has been reached the bodies slide relative to each other. The stick-slip constitutive behaviour is introduced in a way which is analogous to the description of elasto-plasticity. In two dimensions, tangential and normal, this is:

$$\phi_C = \tau - \mu \sigma_n$$

where μ is the constant friction coefficient. The surface $\phi_C = 0$, figure 4, bounds the region in the contact stress space in which stick occurs. If sliding occurs the contact stress must satisfy the condition $\phi_C = 0$.

4 NUMERICAL RESULTS

The tube was modelled with 6 layers of 100 four-node isoparametric plane strain elements. In figure 5 and 6 the force displacements diagrams of aluminium and steel are presented. In comparison with the experiments the advantage of a model is the possibility to vary parameters such as friction. To examine the influence of friction we varied the Coulomb friction coefficient from 0.00 via 0.08 to 0.16.
The finite element meshes of the successive stages of the process are printed in figure 12. The capitals 'A' up to 'G' correspond with the position in the force displacement diagram of figure 5.
In figure 6 we observe a maximum process force of 87 kN. Both calculations terminate when this force is reached because the geometry of the not-deformed tube collapses.
In figure 12 the contact elements are printed too.

It is evident that the stress strain relation largely affects the process force diagram. Since the stress strain curves of a tensile test are extrapolated (see figure 7), prudence is required when examining the results. From not-reported invertube calculations we experienced a large sensitivity of the process force when the material data are varied.

Most of the energy absorption is caused by the deformation of the tube, friction, however, plays a role also.

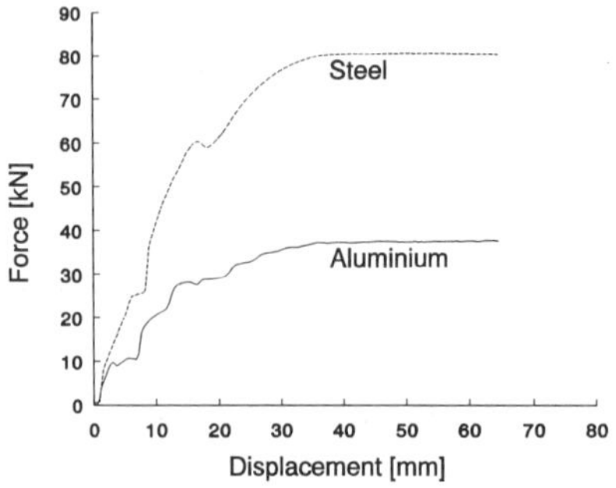

Fig. 8 Process force diagrams of aluminium and steel.

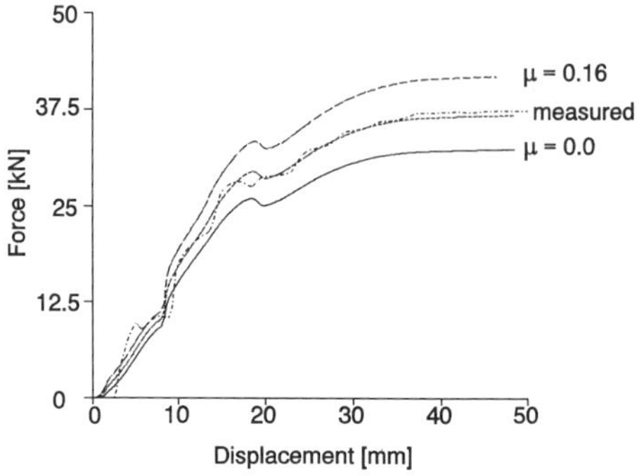

Fig. 10 Process force diagrams of aluminium. Measured and predicted.

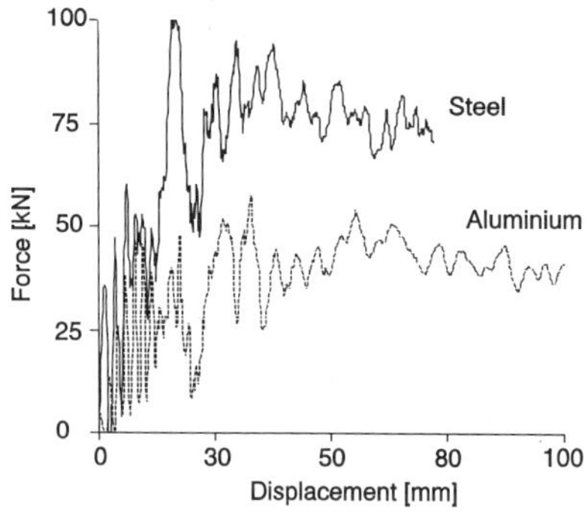

Fig. 9 Force displacement diagram of impact test. Speed was 30 km/h.

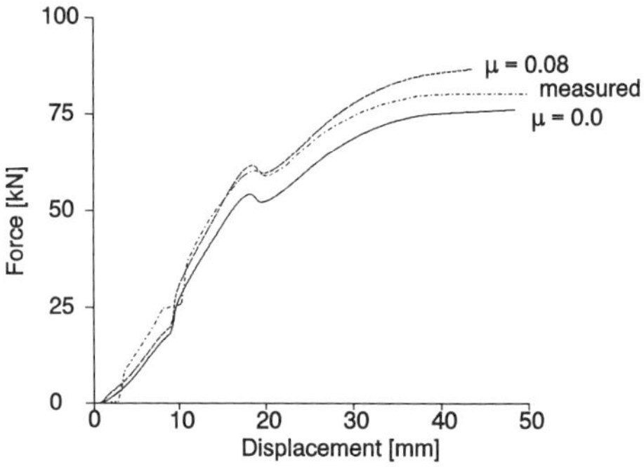

Fig. 11 Process force diagrams of steel. Measured and predicted.

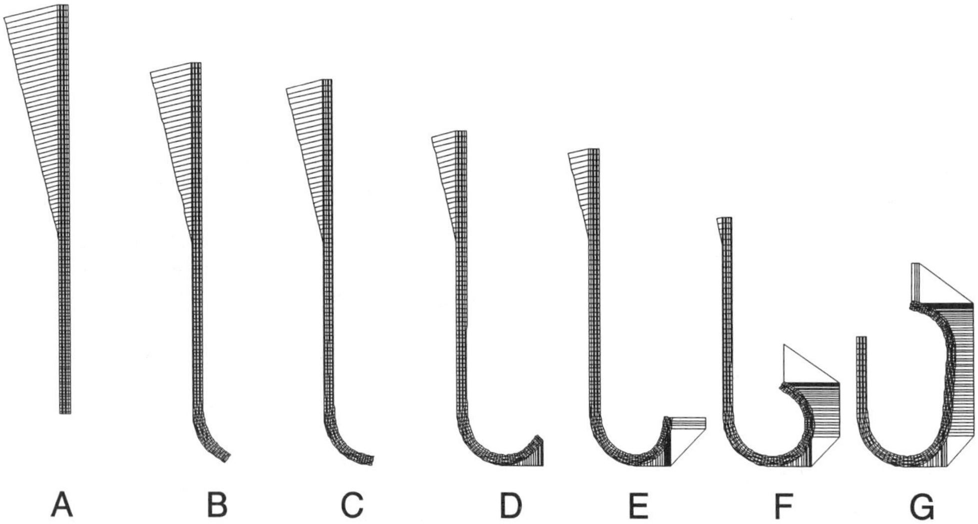

Fig. 12 Deformed finite element meshes of the successive stages of the inverting process.

5 EXPERIMENTAL VERIFICATION

5.1 Setup

The specimens were tested in a MTS fatigue test machine. Because of the limited capacity of the test machine, the process speed was kept constant on 0.05 m/s, much lower than during crash conditions. To control the friction the area of contact between tube and tool was sprayed with PTFE, also known as teflon.
In figure 8 the measured process forces are given.
In addition high speed tests were carried out with a drop hammer rig. The force displacement diagram is given in figure 9, the large fluctuations are caused by the mass-spring behaviour of hammer and crash unit.

5.2 Results evaluation

In figure 10 and 11 we compiled figure 5, 6 and 8 to obtain a clearer comparison between the calculations and measurements.
It is obvious that the calculations with a low Coulomb friction coefficient (0.04 - 0.08) give best correspondence with the experiments. This friction coefficient is very plausible bearing in mind that lubrication with teflon is applied.
Also near positions BC and DE in figure 5 and 12 the process force is predicted accurately. The force displacement diagram of the drop hammer test is, except for the fluctuations, similar to the low speed test. The energy absorption is the integral of the force and the fluctuations therefore hardly influence the energy absorption.
Experiments turned out that the process considered in this study is instable when the process force becomes too large. Hence, a low amount friction is required to stabilize the present inverting process.

6 CONCLUSIONS

The invertube is a simple and efficient means of providing uniform energy absorption. A finite element model was used to estimate the process force and the predicted results were very well in line with the measurements. The contribution of the friction to energy absorption is relatively small, however, we observe instabilities (numerical and experimental) when the process force exceeds a certain limit due to too much friction. In the next future this study will be extended to other geometries to obtain a construction which is more stable with respect to friction.

7 LITERATURE

Beltran, Goicolea, *Large strain plastic collapse: A comparison of explicit and implicit solutions*, Complas 2 1989 Barcelona, pg 1125-1136.

J. Ph. Ponthot and M. Hogge, *On Relative Merits of Implicit/Explicit Algorithms for transient Problems in Metal Forming Simulation.* Metal Forming Process Simulation in Industry, Baden Baden, Germany, 28-30 September 1994.

Huétink, J, Vreede, P.T. and Lugt, J. van der, *The simulation of contact problems in forming processes using a mixed-Eulerian-Lagrangian finite element method*, Numiform '89, eds. Thompson ea.

Al-Hassani, S.T.S., Johnson, W., Lowe, W.T., *Characteristics of inversion tubes under axial loading*, Journal Mechanical Engineering Science Vol 14, No 6 1972.

Kitazawa, K, Seino, J, *Shapes and sizes of inverted tubes*, Journal of the JSTP, vol 34, no 390. (1993-7) (In Japanese).

Searle, J.A., Brabin, E.J. *The Invertube*, Mira bulletin No 2, 1970 March/April.

Simulation of Materials Processing: Theory, Methods and Applications, Shen & Dawson (eds)
© 1995 Balkema, Rotterdam. ISBN 90 5410 553 4

Two- and three-dimensional finite element modelling of the industrial tube forging process

X.C.Wang, A.M.Habraken & S.Cescotto
Department MSM, University of Liège, Belgium

ABSTRACT: The complete process for the fabrication of billets of an airplane is simulated using an automatic numerical procedure. This procedure is composed of an introduction of necessary information for describing the tube and the hammers, a 2D F.E. calculation (including a mesh generation) and two stages of 3D F.E. calculation (including a mesh regeneration, a physical information (stresses, strains, internal variables, etc.) transfer and the replacements of the hammers' system). Bodner's elasto-visco-plastic model is used in the simulation. It is calibrated for our specific forged metal. In order to predict the processing defects, a fully coupled two-variable damage theory is used to extend the classical Bodner's model to include damage. Satisfactory numerical results have been obtained for the industrial forging process modelling.

1. INTRODUCTION

Typical phases for the fabrication of the billets of an airplane in industries is illustrated in Fig. 1.

This fabrication phase is principally an operation of rotary forging which is composed with three different stages: preliminary forging; pre-crushing and final forging. According to the form of the forged end of the tube, one such stage may be distinguished as the round-round one (as it is the preliminary forging) or round (square)-square one (the other two stages of forging). The rotary forging process are realized with the help of a forging machine with a hammers' system as is shown in Fig. 2. For round (square)-square forging the hammers and the tube rotate with the same velocity while for round-round forging there is a relative rotation between them. During the fabrication process, this hammers' system is removed and replaced one after another.

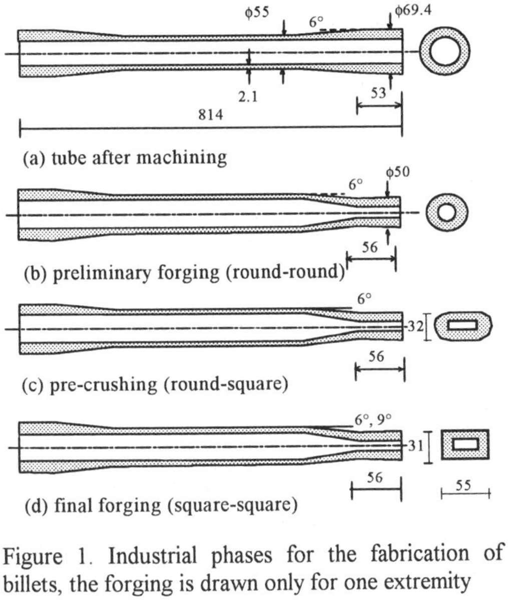

Figure 1. Industrial phases for the fabrication of billets, the forging is drawn only for one extremity

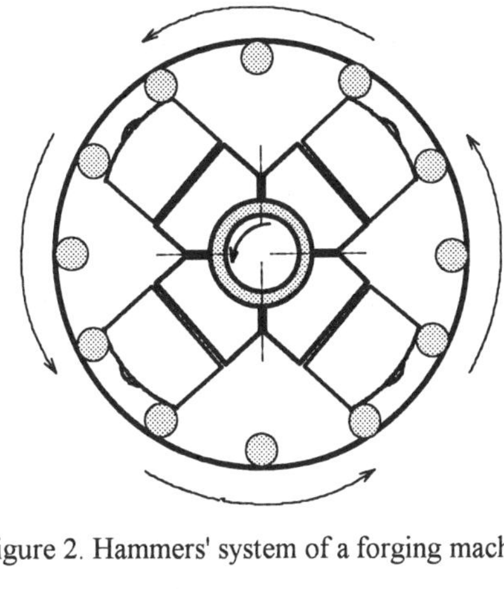

Figure 2. Hammers' system of a forging machine

To optimize the industrial fabrication process and to control the product quality, several results are often required: 1. exact shape of forged pieces; 2. distribution of stresses and plastic deformation; 3. distribution of the temperature and 4. prediction of possible initiation of fracture or small scale geometrical defects during the forging process. For getting these useful results, the finite element method is a suitable numerical tool.

Analysis of the rotary forging process is of great practical importance. An automatic simulation procedure has been established for a complete 2D/3D finite element simulation and some industrial forging process has been modelled. In this paper, we will report some obtained results.

2. AUTOMATIC SIMULATION PROCEDURE

A rotary forging, either round-round one or round (square)-square one, is in fact fully three dimensional. But with a suitable choice of a strategy representing numerically the hammers' movement, the round-round forging may be simulated approximately in an axisymmetric state. Several such strategies have been proposed (Wang *et al* 1994a). In this axisymmetric state and in some simplified conditions, an analytical solution has also been obtained for round-round rotary tube forging (Wang *et al* 1994b). Both the analytical and numerical results have shown their good agreements with experimental results and industrial records. For the round (square)-square forging, we can not do this axisymmetric simplification.

For simulating the compelete industrial forging process shown in Fig. 1, which is composed with the successive round-round, round-square and square-square forging, a 2D/3D finite element simulation procedure is set up. We simulate the preliminary forging in an axisymmetric state and the following ones in three dimensional state. At each stage, we have two sub-operations, in the first one, the hammers move closer the axis of the tube and forge it (charging), while in the second one, the hammers separate from it (discharging). Between two forging stages, a replacement of the hammers' system is necessary. It may be accompanied with, in the transition between 2D and 3D, a mesh regeneration and a physical information (stresses, strains and internal variables, etc.) transfer.

This fully automatic simulation procedure is summarized in Fig. 3. It starts by introducing the necessary information for describing the forged tube and the hammers, the simulation may realized automatically and an industrial process can be simulated completely from the beginning to the end without any interruption unless necessary.

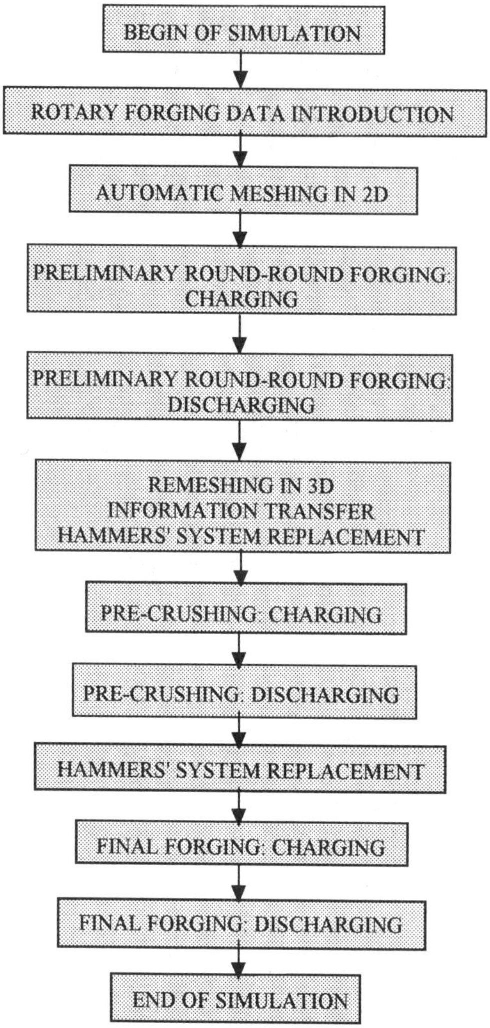

Figure 3. Automatic simulating procedure

3. BODNER'S CONSTITUVE LAW

The Bodner's elasto-visco-plastic model (Bodner & Partom 1975) is used in our simulations to describe the metal constitutive behavior. This model represents one of first unified constitutive model formulated explicitly without a yield criterion. Its general formulation is based on the decomposition of the total strain rate into elastic and inelastic components which are both non-zero for all loading and unloading conditions:

$\dot{\varepsilon}_{ij} = \dot{\varepsilon}^{e}_{ij} + \dot{\varepsilon}^{p}_{ij}$

where $\dot{\varepsilon}_{ij}$ is the symmetric part of the velocity gradient tensor: $\dot{\varepsilon}_{ij} = (L_{ij} + L_{ji})/2$. The generalized Prandtl-Reuss flow law is used together the volume conservation to establish a tensorial relation between the flow stress and inelastic strain rate:

$\dot{\varepsilon}^{p}_{ij} = \lambda S_{ij}$

where S_{ij} is the deviatoric stress component and λ is calculated by:

$$\lambda = \frac{D_0}{\sqrt{J_2}} \exp\left(-\frac{1}{2}\left(\frac{Z^2}{3J_2}\right)^n\right)$$

in which, D_0 corresponds to an assumed limit value of plastic strain rate in shear for large J_2; n is directly related to rate sensitivity, with lower n indicating higher sensitivity; Z is a total scalar hardening variable, it is the sum of isotropic and directional components:

$Z = K + D$

with the directional component D is given by:

$D = \beta_{ij} : u_{ij}$

where β_{ij} is the component of a symmetric second order tensor and:

$$u_{ij} = \frac{\sigma_{ij}}{\sqrt{\sigma_{ij} : \sigma_{ij}}}$$

Isotropic and directional hardening are both supposed to occur under the action of two simultaneously competing mechanism: a hardening process proceeding with deformation and a softening or recovery process evolving with time:

$$\dot{K} = m_1(K_1 - K)\dot{W}^p - A_1 K_1 \left(\frac{K - K_2}{K_1}\right)^{r_1}$$

$$\overset{\nabla}{\beta}_{ij} = m_2(D_1 u_{ij} - \beta_{ij})\dot{W}^p - A_2 K_1 \left(\frac{\sqrt{\beta_{ij}:\beta_{ij}}}{K_1}\right)^{r_2} \frac{\beta_{ij}}{\sqrt{\beta_{ij}:\beta_{ij}}}$$

where $\dot{W}^p$ is the plastic work rate. In this equation, some material parameters are included. Their physical meanings are given in Table 1. There are totally 14 material parameters in Bodner's model. But for most metal forming problems, the situation is normally simpler, the material parameters may be reduced to 7, they are: E, ν, D_0, K_0, K_1, m_1 and n, of which K_0 and n are temperature-dependent.

The Jaumann objective rate $\overset{\nabla}{\beta}_{ij}$ is used for large strain in above equation. It is defined as:

$$\overset{\nabla}{\beta}_{ij} = \dot{\beta}_{ij} + (\beta_{ik}\Omega_{kj} - \Omega_{ik}\beta_{kj})$$

with Ω_{ij} is the component of the antisymmetric part of the velocity gradient tensor.

Table 1. Material parameters of Bodner's model

PARAMETERS	UNITS	PHYSICAL MEANINGS
E	Mpa	Young's Modulus
ν	-	Poisson's ratio
D_0	s^{-1}	assumed limit plastic-shear strain rate
D_1	Mpa	directional hardening coefficient
K_0	Mpa	initial isotropic hardness
K_1	Mpa	maximum isotropic hardness
K_2	Mpa	minimum isotropic hardness
A_1	s^{-1}	recovery coefficient of isotropic hardness
A_2	s^{-1}	recovery coefficient of directional hardness
m_1	Mpa^{-1}	hardening exponent of isotropic hardness
m_2	Mpa^{-1}	hardening exponent of directional hardness
r_1	-	recovery exponent of isotropic hardness
r_2	-	recovery exponent of directional hardness
n	-	strain rate sensitivity coefficient

4. CALIBRATION OF BODNER'S MODEL

The material constants of Bodner's model are easy to be determined by common experiments. A general procedure was proposed by Chan *et al* (1988) for calibrating Bodner's elasto-visco-plastic model. This procedure is used in our research to determine the material parameters of Bodner's model for AISI 304 steel which is used in industries for rotary forging.

4.1. Experiments

A series of constant strain-rate uniaxial compression tests were defined for the calibration of Bodner's model. The specimen is a cylinder with its diameter ϕ_0 = its height H_0 = 15mm. The temperatures of the tests are respectively: room temperature, 100°C, 200°C, 300°C, 400°C and 500°C. For each temperature, 5 different strain rates are utilized, they are: $\dot{\varepsilon}$ = 0.1, 1.0, 5.0, 10.0 and 20.0 s^{-1}. In order to correct the experimental data for removing the machine rigidity effect and removing the friction effect in evaluating specimen stress and strain, two other kinds of tests are realized:

- The machine rigidity test or so-called "empty test" (platen against platen without specimen) is realized for room temperature with $\dot{\varepsilon} = 0.1\ s^{-1}$ until $F_{max} = 50$ T.

- The ring compression tests for evaluating the friction coefficient are realized for each temperature with the external diameter ϕ_{ext} =18mm, the internal diameter ϕ_{int}=9mm and the height H_0=9mm. But only one strain-rate ($10s^{-1}$), is used. This means that we suppose the friction coefficient varies only according to temperature but remains unchanged with different strain-rates.

All these tests are realized in the experimental center "Hall de Metallurgie " of CRM (Centre de Recherches Metallurgiques, Belgium) with the servohydraulic testing machine MTS. They are controlled with the help of *Hot Deformation Simulator* which adjusts and guarantees the constant strain-rate during the tests.

4.2. Experimental results

The experimental results are recorded during the compression process. The complete results are given under a form of numerical card (Magnée 1994). The original data (force F, displacement d, etc.) allow us to develop and to exploit one or another theoretical material model.

A maximum deflection of 0.4 mm at the maximum force of 50 T is recorded in the machine rigidity test. This deflection can not be neglected. With the recorded force and the corresponding displacement of the rigidity test, we can get a curve between them. To remove the machine rigidity influence from the recorded experimental data of systematic uniaxial compression tests, this curve is represented, with the help of the least-square method, in a polynomial of 3rd order:

$$F = a_0 + a_1\delta + a_2\delta^2 + a_3\delta^3$$

and we suppose that the machine deflection is the same for all tests in spite of their different temperature and different strain-rates.

For removing the friction influence over the stress evaluation, the classical Orowan' relation is used from the evaluated friction coefficient μ.

The real stress and strain can be calculated from the recorded testing force F, its corresponding machine deflection δ and recorded upper platen displacement d as follows:

$$\sigma = \frac{12F(H_0 - d + \delta)^2}{\pi R_0^2 H_0\left(3(H_0 - d + \delta) + 2R_0\mu\sqrt{H_0/(H_0 - d + \delta)}\right)}$$

$$\varepsilon = \ln\left(\frac{H_0}{H_0 - d + \delta}\right)$$

Table 2. Material parameters of Bodner's model for AISI 304 steel at room temperature

$E = 2.05\times105$ Mpa	$\nu = 0.3$
$D_0 = 10^8\ s^{-1}$	$K_0 = 665.04$ Mpa
$K_1 = 1850$ Mpa	$K_2 = 665.04$ Mpa
$m_1 = 0.01$	$n = 2.737$

The stresses and strains obtained in this way are a series of discrete points. In order to realizing mathematical computation, a two-piece (one for the small strain range and another for the large strain range) polynomial approximation of 3 or 4th order is used for each σ - ε curve.

4.3. Material parameters

Various series of material constants are obtained for different temperatures (Wang 1994). But the heating effect is not important in our forging cases, a coupling thermal-mechanical analysis is not necessary. We give so only the parameters in room temperature in Table 2. Here neither the directional hardening effect nor thermal recovery effect is not taken into account and the corresponding parameters are zero.

5. BODNER'S MODEL WITH DAMAGE

An one-scalar-variable damage model has been used to extended Bodner's law to include damage (see, e.g. Bodner,1985; Bodner & Robin 1986). In this model, the damage does not influence the elastic answer but reduces the hardening. Another damage theory used to develop Bodner's model with damage was presented recently by Habraken (Habraken *et al* 1995). This damage theory has been used by Zhu (Zhu et al 1992) to develop an elastoplastic damage model and gave a good prediction of the crack initiation for all fracture modes. The basic principles of this theory are:

- Two scalar parameters d and δ (deviatoric and volumetric components) are used. According to Ladevèze (1984), these two variables are necessary to modify the elastic modulus and Poisson's ratio with the damage growth.
- The energy equivalence (Cordebois & Sidoroff 1979), which may be of more physical significance, is used.

Under these hypotheses, we can obtain the following relations between damage material state and virgin material state:

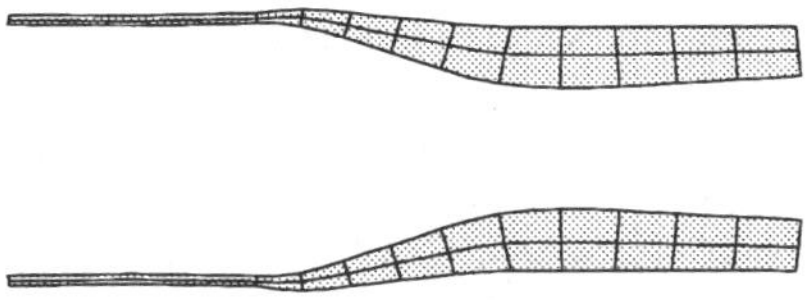

Figure 4. Forged extremity after the preliminary round-round forging

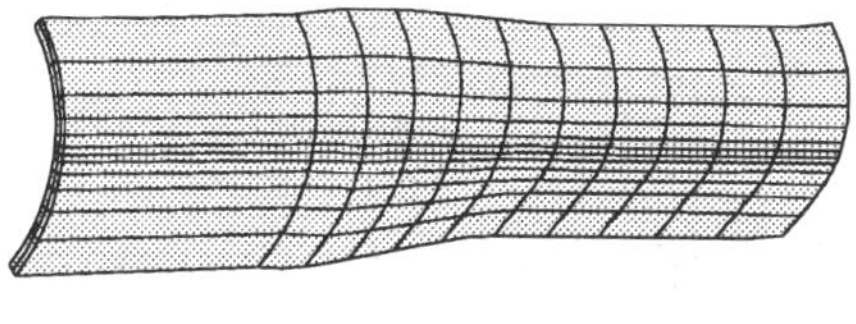

Figure 5. Forged extremity after the pre-crushing

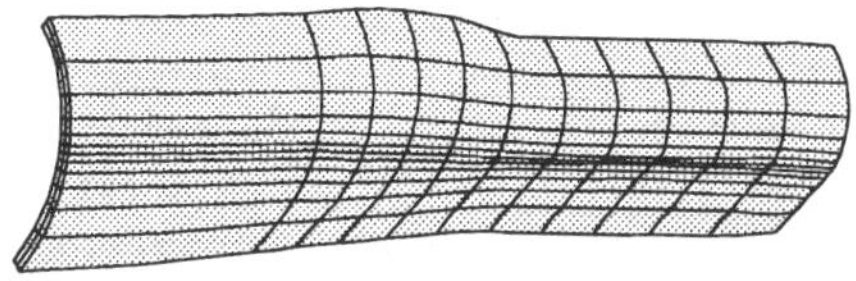

Figure 6. Forged extremity after the final forging

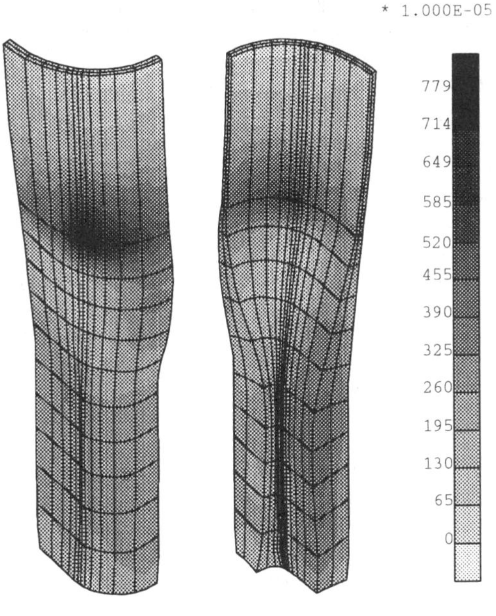

Figure 7. Damage distribution after the final forging

$$\bar{\hat{\sigma}}_{ij} = \frac{\hat{\sigma}_{ij}}{(1-d)}, \quad \bar{\sigma}_m = \frac{\sigma_m}{(1-\delta)}$$

$$\bar{\hat{\varepsilon}}^e_{ij} = (1-d)\hat{\varepsilon}^e_{ij}, \quad \bar{\varepsilon}^e_m = (1-\delta)\varepsilon^e_m, \quad \bar{\dot{\varepsilon}}^p_{ij} = (1-d)\dot{\varepsilon}^p_{ij}$$

In these equations, ^ denotes the deviatoric components; ¯ denotes the effective components; subscript $_m$ denotes the hydrostatic components; superscripts ^{e} and ^{p} denote respectively the elastic and plastic components.

Contrary to the damage theory used by Zhu (Zhu *et al* 1992), as Bodner's model assumes no yield surface and for keeping this characteristic for the damage evolution, no damage evolution surface has been used. Following Lemaître's approach (Lemaître & Chaboche 1985), the damage variables evaluate as follows (Habraken *et al* 1995):

$$\dot{d} = \frac{1}{2(1-d)}\left(\frac{<\chi(\sigma)-\sigma_D>}{A(1-d)}\right)^r \dot{p}^{r/2}, \quad \dot{\delta} = <\tau>\dot{d}$$

with:

$$<\tau> = \frac{\delta}{d} \text{ if } \sigma_m > 0 \text{ and } <\tau> = 0 \text{ if } \sigma_m \le 0$$

$$<\chi(\sigma)-\sigma_D> = \chi(\sigma)-\sigma_D \text{ if } (\chi(\sigma)-\sigma_D) > 0$$

$$= 0 \qquad \text{if } (\chi(\sigma)-\sigma_D) < 0$$

$$\chi(\sigma) = (1-\beta)\sqrt{3J_2} + 3\beta\sigma_m$$

$$J_2 = \frac{1}{2}(\hat{\sigma}_{ij}\hat{\sigma}_{ij}), \quad \dot{p} = \frac{3}{2}\dot{\varepsilon}^p_{ij}\dot{\varepsilon}^p_{ij}$$

where σ_D is a threshold stress value for the damage increase; r, s, A material constants.

Using above damage theory for Bodner's model, the flow law can be written as:

$$\bar{\dot{\varepsilon}}^p_{ij} = \lambda\,\bar{\hat{\sigma}}_{ij}, \quad \lambda = \frac{D_o}{\sqrt{\bar{J}_2}} \exp\left(-\frac{1}{2}\left(\frac{Z^2}{3\bar{J}_2}\right)^n\right)$$

Or, using the true stress and strain tensors:

$$\dot{\varepsilon}^p_{ij} = \lambda\hat{\sigma}_{ij}, \quad \lambda = \frac{D_o}{\sqrt{J_2}(1-d)} \exp\left(-\frac{1}{2}\left(\frac{Z^2(1-d)^2}{3J_2}\right)^n\right)$$

6. NUMERICAL RESULTS

The numerical modelling of the fabrication process shown in Fig. 1 is realized using the automatic simulation procedure described in Section 2 (Fig. 3). The forged metal is steel AISI 304, with its Bodner's parameters shown in Table 2. For the sake of simplicity, only the right extremity forging is simulated. About a half length of the tube is discretized. Figs. 4, 5and 6 give the tube shapes of the forged extremity after each stage.

For prediction of the material defect evolution in the forging process, this problem is also simulated using Bodner's law with damage, presented in Section 5. Since we have not yet the corresponding damage parameters of this damage model for the specific forged metal (the experiments are in the course of completion), we use another series of parameters calibrated on the law proposed by Benallal *et al* (1991) for their notched steel bar. The corresponding parameters for Bodner's damage model are given in Table 3.

Fig. 7 shows the predicted damage distribution after the final forging. Here the dangerous damage area is well indicated.

Table 3. Parameters for Bodner's damage model

E = 2.03×105 Mpa	$\nu = 0.3$
$D_0 = 10^8$ s^{-1}	$K_0 = 1213.54$ Mpa
$K_1 = 2026.38$ Mpa	$K_2 = 1213.54$ Mpa
$m_1 = 0.0065$	n =2.373
$\sigma_D = 1017.30$ Mpa	r = 0.0
s = 1.0	A = 1.0

7. CONCLUDING REMARKS

An automatic simulating procedure is used for the modelling of the principal fabrication process of the billets. The mastership of complex rotary forging process through mathematical modelling of mechanical rheological and thermal (not considered in this paper) phenomena occurring during forging is the goal of our research. Some special difficulties exist. They are, for example:

1. 3D simulation with strong non linearities, including the unilateral contact and (thermo)-elasto-visco-plastic material with damage;
2. taking account of high strain-rate and inertia (not considered in this paper) effects;
3. evaluation of the initiation of fracture or small scale geometrical defects during the forging process;
4. development of a robust strategy for thermo-mechanical coupling analysis (not included in this paper).
5. if the heating can not be neglected, development of robust strategy for thermo-mechanical coupling analysis.

The obtained finite element simulating results are useful for us to optimizing the industrial rotary tube forging process and to control the product quality, including the exact shape of the forged tube; optimal contribution of the stresses and plastic deformations, which are sometimes the causes of the geometrical defects of products. The use of Bodner damage model makes it possible to study the coupling strain-rate/damage effect and to predict the material defects during the forging process.

All computations reported in this paper are realized with the non linear finite element code LAGAMINE developed in MSM Department of University of Liège, Belgium.

ACKNOWLEDGEMENTS: The authors thank "La Région Wallonne (Belgium)" for its financial support and TAC (Technical Airborne Components, Belgium) for its technical support.

REFERENCES

Benallal, A., R. Billardon & J. Lemaître 1991. Continuum Damage Mechanics and Local Approach to Fracture: Numerical Procedures. *Comp. Meth. Appl. Mech. Engng.* 92:141-155

Bodner, S.R. & Y. Partom 1975. Constitutive Equations for Elastic-viscoplastic Strain-Hardening Material. *J. Appl. Mech.* 42:385-389

Bodner, S.R. 1985. Evolution Equations Anisotropic Hardening and Damage of Elastic-viscoplastic Materials. *Plastic Today.* A. Saczak & G. Biandsi (eds).:471-482. Elsevier

Bodner, S.R. & M.B. Rubin 1986. A Unified Elastic-viscoplastic Theory with Large Deformations. *Deformation of Solids.* J. Gittus *et al* (eds).:129-140. Elsevier

Chan, K.S., S.R. Bodner & U.S. Lindholm 1988. Phenomenological Modelling of Hardening and Thermal Recovery in Metals. *J. Eng. Mater. & Tech.* 110:1-8

Cordebois, J.P. & Sidoroff 1979. Damage Induced Elastic Anisotropy, *F. Euromech 115.* Villared de Lans. France

Habraken, A.M., Y.Y. Zhu, R. Charlier & X.C. Wang 1995. A Damage Model for Elasto-visco-plastic Materials at Large Strains, to appear in Proc. of Complas 4. Barcelona. Spain

Ladevèze, P. 1984. Sur une Théorie de l'Endommagement anisotrope, Internal Report no. 34. LMT. Cachan. France

Lemaître, J. & J.L. Chaboche 1985. *Mécanique des Matériaux Solides.* Dunod. Paris

Magnée, A. 1994. Etude et Analyse des Propriétés Métallurgiques d'Ecoulement Plastique de l'Acier Inoxydable Cr-Ni, AISI 304. Technical Report. CRM, Belgium

Wang, X.C., Y.Y. Zhu & S. Cescotto 1994. Strain-rate-sensitivity in Rotary Tube Forging Problems: Finite Element Modelling with Bodner's Constitutive model. *Proc. of 3rd Belgian Nat.*

Cong. of Theor. & Appl. Mech.:314-317. Liège. Belgium
Wang , X.C., Y.Y. Zhu and S. Cescotto 1994. Analytical Solutions and Finite Element Modelling of Rotary Tube Forging. Proc. of *Metal Forming Process Simulation in Industry.*:249-269. Baden-Baden. Germany
Wang, X.C. 1995. Experimental Calibration of Bodner's Elastoviscoplastic Model for Steel AISI 304 Steel, Internal Report. MSM Department. University of Liège. Belgium
Zhu, Y.Y., S. Cescotto & A.M. Habraken 1992. A Fully Coupled Elastoplastic Damage Modelling and Fracture Criteria in Metal Forming Processes. *J. Mater. Proces. Techn.*: 32

Simulation of Materials Processing: Theory, Methods and Applications, Shen & Dawson (eds)
© 1995 Balkema, Rotterdam. ISBN 90 5410 553 4

Numerical analysis on an extrusion process of a bimetallic rod

Masahiko Yoshino & Takahiro Shirakashi
Tokyo Institute of Technology, Japan

ABSTRACT: A frictional element is proposed to simulate the frictional mechanism on a disconnected interface by the Rigid Plastic FEM. The element is applied to unsteady state simulations on extrusion processes of disconnected bimetal rods, which showed not only reasonable frictional properties but also good agreement with experimental data. The element is also applied to steady state simulations on extrusion processes. These simulations reveal effects of extrusion conditions on deformation processes, and also clarify the possibility of continuous extrusion of a bimetallic rod.

1 INTRODUCTION

Bimetallic materials, such as a super conductive wire, a stainless steel clad plate, fiber reinforced metal etc., are widely used for high quality machinery elements, and will be one of the most important materials for industry. However their production processes are very difficult. In their processes, fracture or separation between both materials are often seen and they will cause the loss of their reliability. These troubles will come from difference of the mother materials, because the difference of their mechanical property causes unstable friction and unstable deformation, and these instabilities can develop ununiformed deformation.

In order to study the failure mechanism of a bimetallic material, many investigators have tried to simulate forming processes of disconnected bimetallic rods [1]-[4]. Although the FEM approach will be useful for this purpose, a usual FEM often encounters a calculation deadlock on their simulation. In a usual FEM simulation the frictional condition, such as slide or stick, must be determined before calculation. However, the condition may not be possible to be predetermined for an interface of a disconnected bimetallic rod. The FEM program must search an adequate frictional condition by the trial and error procedure so that stress state satisfies the frictional mechanism on the interface. The calculation is often stopped by iteration of the frictional condition between slide and stick.

In the present paper, we propose a new method to simulate the frictional mechanism without the unnecessary iteration using the Rigid Plastic FEM. We also apply the method to extrusion processes of bimetallic rods to discuss effects of forming condition. We will also discuss both unsteady and steady state of extrusion processes.

2 SIMULATING SYSTEM

2.1 *Frictional element*

The Rigid Plastic FEM for a compressive porous material proposed by Osakada et al. is adopted to the simulation [5]. It can simulate deformation of an incompressible material approximately when the density of the porous material is almost equal to 100%. In the following simulation the density of 98% is applied.

In order to simulate the frictional process on the interface without the iteration of the frictional condition we propose a new frictional element. Figure 1 shows the frictional element for an axisymmetric analysis schematically. The frictional element is a thin isoparametric element, and is inserted between two ordinary elements. The element has characteristics like a smooth thin film, which is hardly compressed to its thickness direction (η) but easily deformed to its shear direction (ξ–η). It is so thin that its shear deformation can be regarded as slide between two materials. On its shear deformation, the element consumes the same energy as the frictional work on the interface. Consequently,

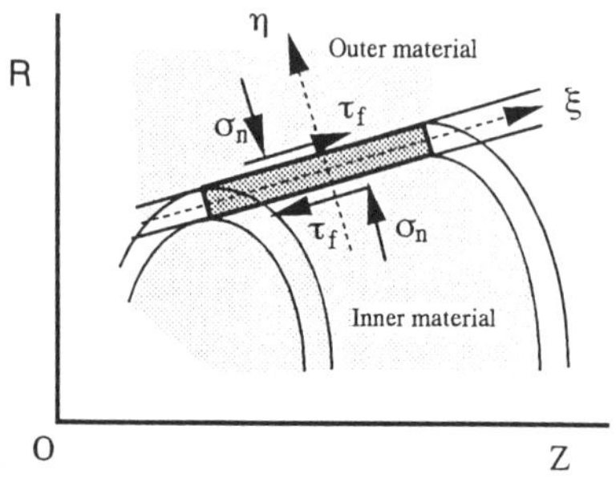

Figure 1 Details of composition of a fictional element.

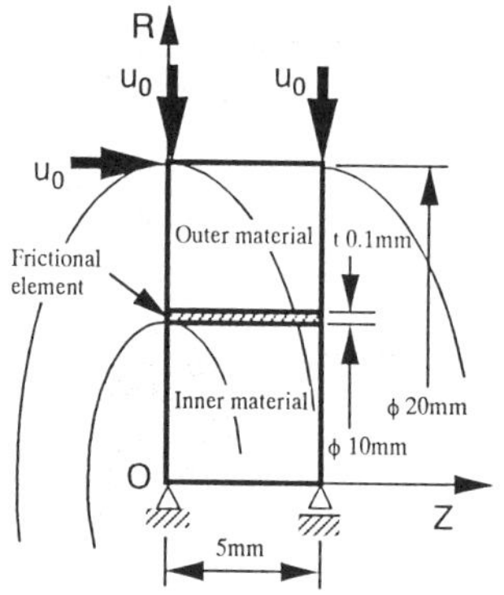

Figure 2 Simple element model of a bimetallic rod.

the energy dissipation rate of the frictional element is formulated as follows:

$$W_f = \rho \int_{V_f} \sigma_0 \dot{\varepsilon}_f \, dV + \int_{S_f} |\tau_f \dot{\gamma}| \, t \, dS \qquad (1)$$

where ρ is the relative density of the element, and σ_0 is its flow-stress, which is equivalent to one of the mother materials. τ_f is frictional stress on the interface, which will be explained in section 2.2. γ is shear strain-rate of the element, and t is its thickness. ε_f is pseud-strain-rate of the frictional element, and expressed by following equations:

$$\dot{\varepsilon}_f = \rho^{k-1} \left(\frac{4}{9} \dot{\varepsilon}_\eta^2 + f^2 \dot{\varepsilon}_v^2 \right)^{1/2} \qquad (2)$$

$$f = \frac{1}{2.49 \, (1 - \rho)^{0.514}} \qquad (3)$$

$$k = 2.5 \qquad (4)$$

where ε_η and ε_v are the thickness strain-rate and the volume one respectively, and k is a constant.

By converging the dissipation rate of the total deformation energy to minimum, the Rigid Plastic FEM procedure can realize an adequate frictional mechanism on the frictional element even if the frictional condition is not determined preliminarily. This is because the second term of the equation (1) works as a penalty function that determines frictional condition. If τ_f is large, the term restricts the slide of the elements very small, because it requires larger frictional work than the energy dissipation by shear deformation of the mother materials. This small sliding-state represents stick-state of the interface [6]. Besides if τ_f is small enough, the equation determines an adequate shear rate γ, i.e., the slide occurs between the mother materials, when shear stress in the material is balanced with τ_f because of the upper bound theorem.

2.2 *Frictional mechanism*

As a simple model is convenient to discuss the property of the element, a three-element model shown in Figure 2 is examined. This model of a bimetal rod has a frictional element between two conventional elements of the same flow stress, i.e., Y_1=Y_2=100 MPa. The outer element is pushed to its axial direction at the speed of u_0=1.0 mm, when it is compressed to its radial direction at the same speed simultaneously.

For this simulation the Coulomb's friction low is not adequate because it often overestimates frictional stress τ_f when thrust stress σ_n is very high. Hence the following frictional function whose τ_f never exceeds shear yield stress of a mother material is applied to the simulation [7]:

$$\tau_f = \frac{\sigma_y}{\sqrt{3}} \left\{ 1 - \exp\left(\frac{-\sqrt{3}\, \sigma_n \mu}{\sigma_y} \right) \right\} \qquad (5)$$

where σ_y is yield stress of a mother material, and μ is a frictional coefficient. When σ_n is very low, τ_f increases almost linearly as same as the Coulomb's friction low, whereas when σ_n becomes high, τ_f approaches shear yield stress.

Figure 3 shows the calculated variation of shear-stress τ against μ on the interface of the tested model. Shear-rate γ and frictional stress τ_f of the frictional element are also shown in the figure. When μ is small, τ_f restricts τ to be equal to τ_f, and shear-rate γ is very large, which means the slide occurs between the outer and inner materials. Besides, when μ becomes larger than 0.07, τ_f grows larger than τ though τ stays around 6.0 MPa. In this range of μ, shear-rate γ is almost zero, and this can be regarded as the stick state of the mother materials.

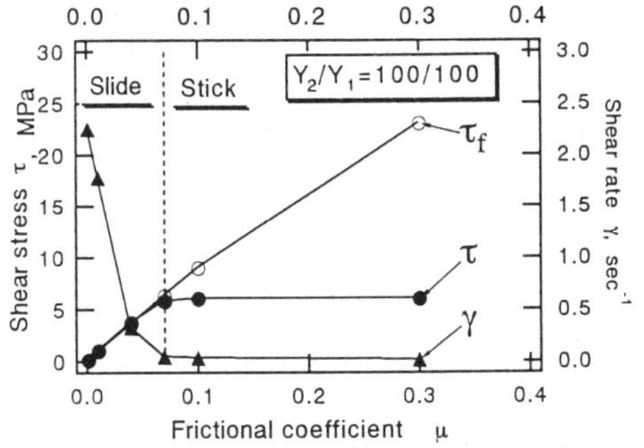

Figure 3 Effect of frictional coefficient on shear stress and sliding rate on interface of a simple bimetallic rod.

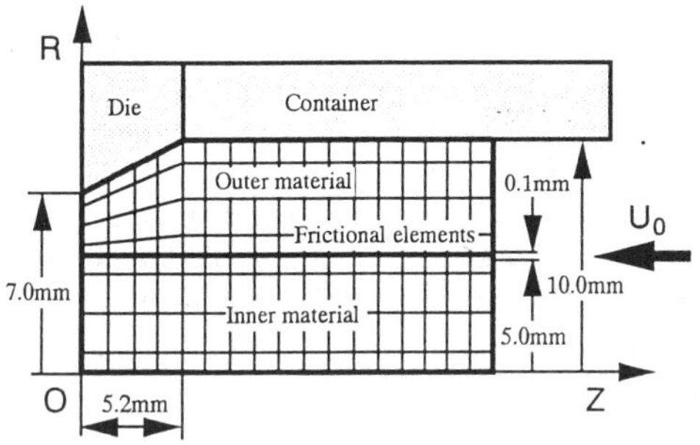

Figure 4 Finite element mesh used for unsteady state simulation of extrusion process of a bimetallic rod.

As shown in the prescribed simulation the frictional element can realize the frictional mechanism adequately, and frictional condition, i.e., stick state or slide state, is determined naturally without any preliminary assumption on the frictional condition. Accordingly the frictional element is available to a simulation on a extrusion process of a bimetallic rod.

3 SIMULATION ON EXTRUSION PROCESS

3.1 *Unsteady state on extrusion process*

Unsteady extrusion processes of some bimetallic rods are simulated using the frictional element on the updated Lagrangian system. Figure 4 shows construction of the FEM mesh used for the simulation. The materials are pure copper and pure aluminum, whose flow stress variations are calculated from the equivalent strain ε_{eq} by the following equations:

Pure Al: $Y_1 = 69 + 98\ \varepsilon_{eq}$ MPa (6)

Pure Cu: $Y_2 = 270 + 127\ \varepsilon_{eq}$ MPa (7)

which were determined experimentally from uniaxial compressive tests. Simulations are made on two combinations of these materials; one is a combination of pure copper for the outer material and aluminum for the inner material which is indicated by Cu/Al, and the other is the converse combination which is indicated by Al/Cu. The sizes of the rod are shown in the figure, and the half die angle is about 30 degrees. Frictional coefficient μ between the outer and inner materials is 0.4, and friction on the die is neglected.

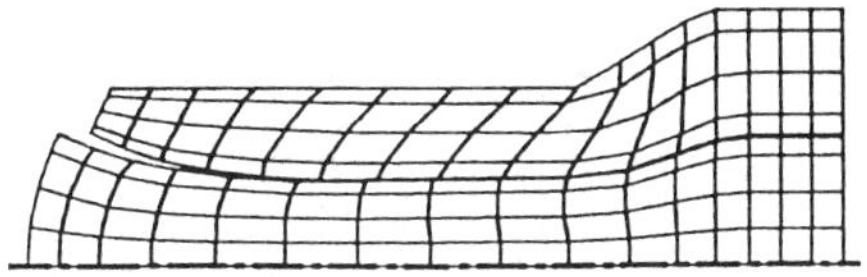

Figure 5 Calculated distortion of the mesh of a bimetallic rod extruded for 10 mm

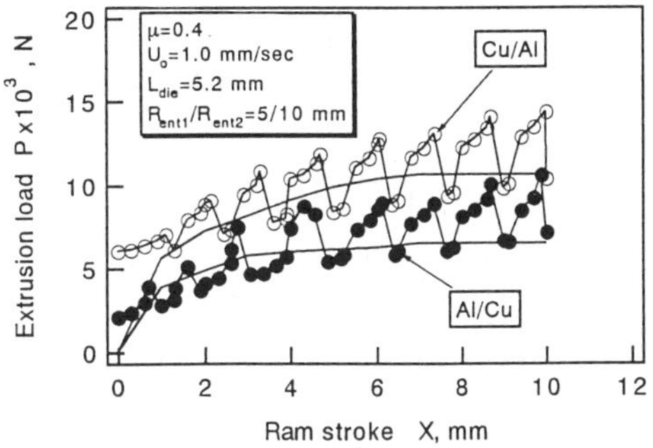

Figure 6 Variation of extrusion load of bimetallic rods constructed with copper and aluminum.

Figure 5 shows calculated distortion of the FEM mesh of the Cu/Al combination when both the outer and inner materials are extruded at the same speed for 10 mm on ram stroke. The figure shows that the outer material is deformed as well as the inner one though the outer material has much harder flow-stress than the inner one. Of cause the soft inner aluminum is extruded more than the outer copper, slide between the materials also occurs. Another important point is that no invasion of nodal points on the interface to the other material is seen in the figure. This is attributed to stiffness of the frictional element against compression. The frictional element requires no other control for the simulation to prohibit the invasion of nodal points.

Figure 6 shows variation of extrusion loads. The solid curves in the figure indicate experimental loads, and the symbols do calculated loads. Although

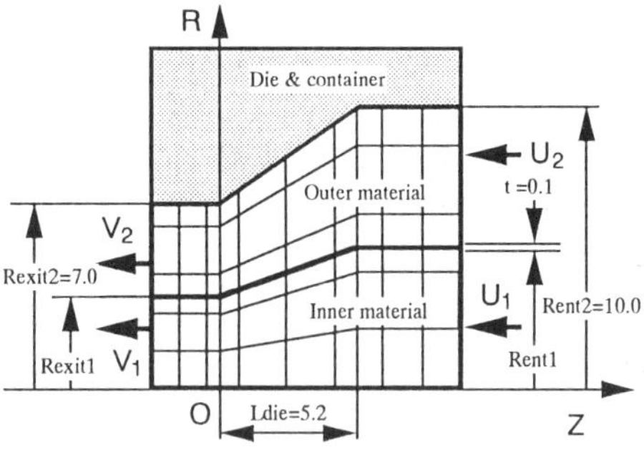

Figure 7 Finite element mesh used for steady state simulation of extrusion process of a bimetallic rod.

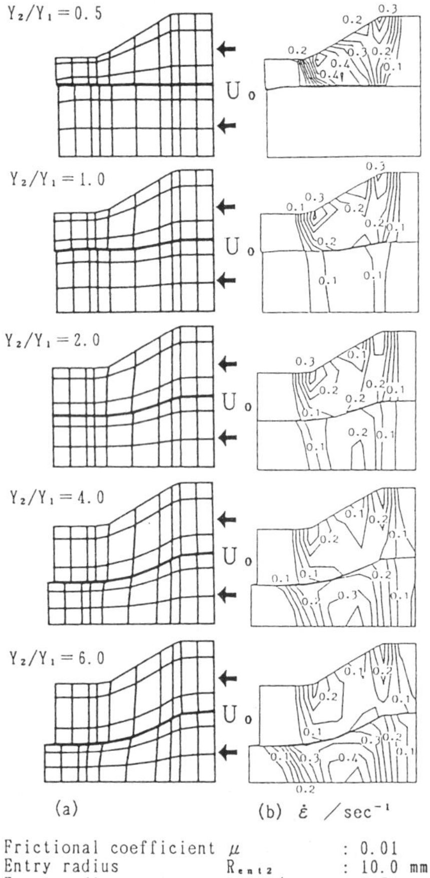

Frictional coefficient	μ	: 0.01
Entry radius	R_{ent2}	: 10.0 mm
Entry radius ratio	R_{ent1}/R_{ent2}	: 0.5
Exit radius	R_{ext2}	: 7.0 mm
Die length	L_{die}	: 5.2 mm

Figure 8 Distortions of the mesh and its strain-rate distributions at steady state on a extrusion process.

the calculated loads vary periodically, their averages almost correspond with the experimental data. We can conclude this simulating method shows good agreement with experimental data quantitatively.

3.2 *Steady state on extrusion process with restricted inner material*

The FEM simulation is extended to steady state simulation on extrusion processes. Figure 7 shows construction of the FEM mesh used for the simulation. Each inner and outer material is pushed to its axial direction on their rear ends at speeds of U_1 and U_2 respectively. In order to achieve steady state easily, the iterative convergent method is employed for the simulation [8]. In the method nodal velocities of the mesh are calculated by the FEM analysis, and then the mesh distortion on a 0.3 sec time increment is determined. Then all nodal points are taken back to the original Z-location when the FEM mesh is reconstructed in accordance with the new interface location because the interface are moved by the distortion. Frictional stress on the interface is determined using equation (4) from the thrust stress σ_n for the next step calculation. By repeating these procedures the simulation procedure derives an adequate velocity field and the interface shape for steady state extrusion.

First, deformation of the bimetallic rod is examined in the case that both inner and outer materials are pushed at the same speed U_0 (=1.0 mm/sec). Figure 8 shows some of (a) deformation patterns and (b) strain rate distributions of the bimetallic rod effected by difference of the flow-stress ratio Y_2/Y_1. The frictional coefficient μ is 0.01, and the sizes of the rod are as shown in the figure. Both the inner and outer materials are assumed as rigid plastic materials. According to the figure, when the flow-stress ratio Y_2/Y_1 is small such as 0.5, no strain-rate distribution is seen in the inner material. The inner material is not deformed but only the outer one is deformed, and the difference of the deformation causes a large slide between the two materials. The inner material works like a mandrel bar in a pipe extrusion process. When the flow-stress ratio becomes larger, such as the ratio is larger than 2.0, deformation of the inner material becomes larger and the inner material is extruded more than the outer one.

Figure 9 shows variation of the exit radius ratio R_{ext1}/R_{ext2} against the flow-stress ratio Y_2/Y_1 obtained from the simulation, where solid symbols indicate the frictional coefficient μ is 0.01 and open symbols indicate μ is 1.0. The dotted line shows

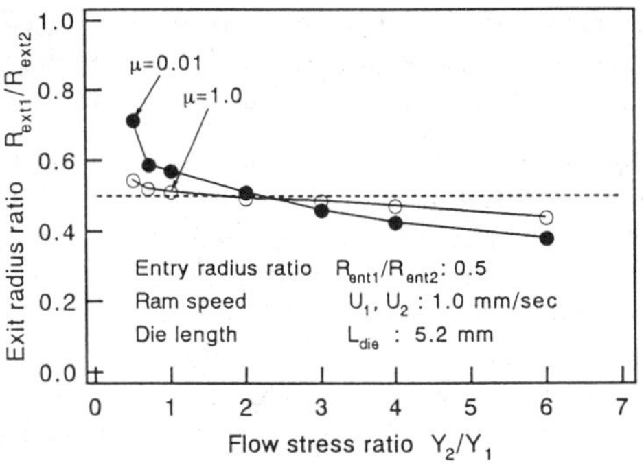

Figure 9 Variation of the exit radius ratio against the flow-stress ratio of the extruded bimetallic rods.

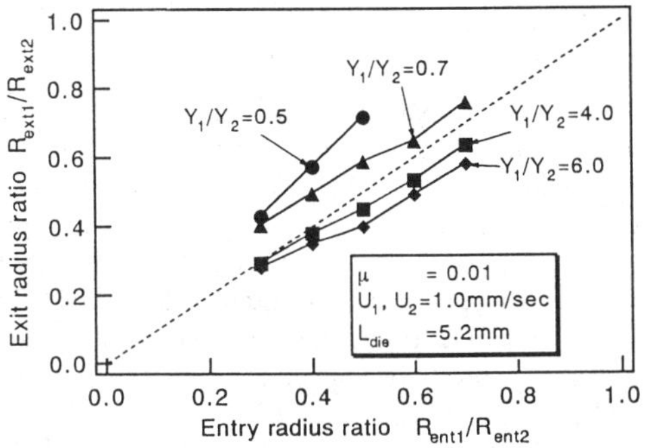

Figure 10 Effects of the flow-stress ratio on relationship between the exit radius ratio and the entry radius ratio of the bimetallic rods.

R_{ext1}/R_{ext2} =0.5, which means the radius ratio does not change through the extrusion. Both symbols show that R_{ext1}/R_{ext2} decreases with increase of Y_2/Y_1. The outer material is extruded more than the inner one when Y_2/Y_1 is smaller, but the inner is extruded more than the outer when Y_2/Y_1 becomes larger. The curve of the solid symbols does not cross to the dotted line at Y_2/Y_1=1.0. This means that deformation of the two materials are not equivalent even if their flow-stresses are equivalent, and the slide occurs on the interface. The curve crosses to the dotted line almost at Y_2/Y_1=2.0, and the outer material must have twice harder flow-stress than inner material to deform uniformly. The variation of R_{ext1}/R_{ext2} is smaller under the large friction (μ=1.0) than that under small friction (μ=0.01). This is because friction on the interface restricts the slide between the two materials, and this restriction derives equivalent deformation of the inner and outer material.

Figure 10 shows effects of the flow-stress ratio

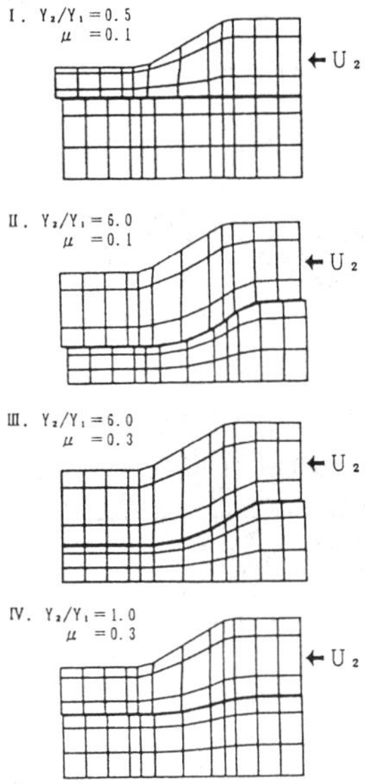

Figure 11 Four typical deformation patterns on steady state of extruding bimetallic rods.

Y_2/Y_1 on the relationship between the exit radius ratio R_{ext1}/R_{ext2} and the entry radius ratio R_{ent1}/R_{ent2}. The dotted line in the figure indicates that the radius ratio does not change by the extrusion, and that no slide occurs on the interface. According to the figure, when Y_2/Y_1 is small, symbols lie above the dotted line, where the outer material is extended more than the inner one. On the contrary when Y_2/Y_1 is great, symbols lie below the dotted line, where the inner material is more extended. These characteristics are confirmed qualitatively from experimental results.

3.3 *Steady state on extrusion process with free inner material*

Secondly, deformation of the bimetallic rod is examined in the case that only the outer material is pushed at the speed of U_2=1.0 mm/sec. Figure 11 shows four typical patterns of the FEM mesh deformation obtained from the simulation. The type I appears when the outer material is softer than the inner one (Y_2/Y_1=0.5) and the frictional coefficient μ is small. In this case the inner material is not deformed at all and is extruded at the same speed of the entry speed of the outer as if a mandrel is inserted in a pipe. The type II appears when the outer material is much harder than the inner one (Y_2/Y_1=6.0) and the frictional coefficient μ is small (μ=0.1). In this

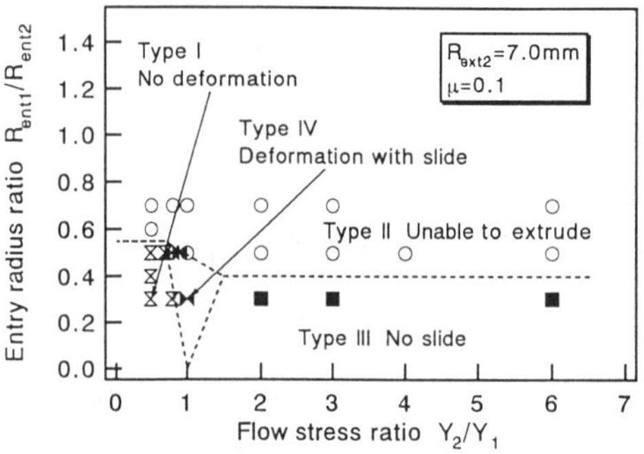

(a) frictional coefficient μ=0.1

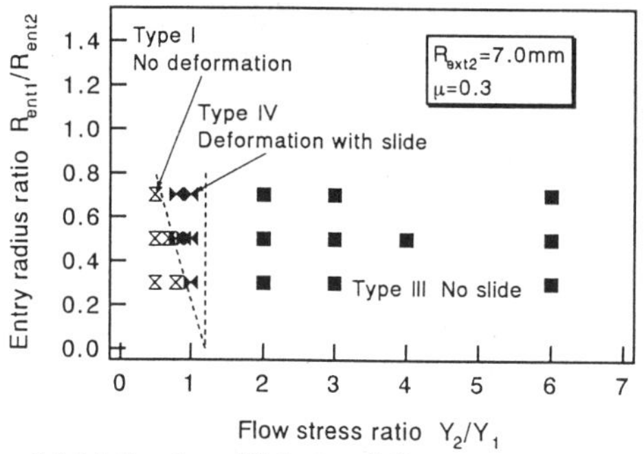

(b) frictional coefficient μ=0.3

Figure 12 Effect of entry radius ratio and flow-stress ratio on the extrusion pattern of bimetallic rods.

case only the outer material is extruded but the inner one is not extruded. The materials slip on the interface but the inner material remains undeformed. The type III appears when the outer material is much harder than the inner one (Y_2/Y_1=6.0) and the frictional coefficient μ is large (μ=1.0). In this case the inner material is driven into the die because of high friction and is deformed as well as the outer material. The exit speed of the inner material is equal to the exit speed of the outer material, and no slide occurs on the exit side of the bimetallic rod. This is the only possible state that both inner and outer materials are extruded continuously. The type IV appears when flow-stress of both materials are almost equivalent (Y_2/Y_1=1.0) and the frictional coefficient μ is large (μ=1.0). This is the medium state between the type II and type III because the inner material is extruded by the friction but its exit speed is not equal to that of the outer material.

Figure 12 (a) and (b) show effects of the flow-stress ratio Y_2/Y_1 and the entry radius ratio R_{ent1}/R_{ent2} on the extrusion pattern. In the figure (a) when μ=0.1, type III is seen in the range of Y_2/Y_1>2.0 and R_{ent1}/R_{ent2}<0.4, but type II arises in the range of R_{ent1}/R_{ent2}>0.4 where extrusion is impossible. Type I appears when Y_2/Y_1<1.0 and R_{ent1}/R_{ent2}<0.5, and type IV appears between the ranges of type I and type III. In the figure (b) when μ =0.3, the extrusion pattern hardly depends on the entry radius ratio but only on the flow-stress ratio. Transitions of the extrusion pattern are seen almost at Y_2/Y_1=0.8 and Y_2/Y_1=1.5, where the pattern changes from type I to type IV and from type VI to type III respectively. According to these figures, a bimetallic rod should have greater flow-stress ratio than about 1.5 to be extruded steadily because only type III does not arise difference of the exit speeds of both materials.

4 CONCLUSION

1) The frictional element is proposed to simulate the frictional mechanism on a disconnected interface by the Rigid Plastic FEM simulation without unnecessary numerical iteration of frictional condition.
2) The frictional element is applied to the simulation on extrusion processes of disconnected bimetallic rods. The simulation on unsteady state of extrusion processes show good agreement with experimental results.
3) The frictional element is also applied to the simulation on steady state of extrusion processes. The simulation revealed four extrusion patterns of a bimetallic rod, and clarified the possibility of stable extrusion of a bimetallic rod.

REFERENCES

1. A.K. Tayal & R.Natarajan 1981 *Int. Mach. Tool Des. Res.*, 21-3/4: 227.
2. L.A. Pacheco & J.M. Alexander 1982 *Num. Meth. Ind. Forming Process*: 205.
3. B. Avizer, R. Wu, S. Talbert & Y. T. Chaw 1982 *Trans. ASME J. Eng. Ind.*, 104: 293.
4. C. Liu, P. Hartley, C.E.N. Sturgess & G.W. Rowe 1985 *Int. J. Mesh. Sci.*, 27-7/8: 531.
5. K. Osakada, J. Nakano & K. Mori 1982 *Int. J. Mech. Sci.*, 24-459.
6. J.T. Oden & E.B. Pires 1983 *Trans. ASME J. Appl. Mesh.*, 50: 63.
7. Shirakashi, T. & Usui, E. 1973 *J. Japan Soc. Prec. Engg.*, 39-9
8. Shirakashi, T. & Usui, E. 1974 *Proc. Int. Conf. Prod. Eng. Tokyo.*

Rolling of metals

Simulation of Materials Processing: Theory, Methods and Applications, Shen & Dawson (eds)
© 1995 Balkema, Rotterdam. ISBN 90 5410 553 4

Finite element simulation and experimental analysis of rolling of bimetallic rods

M.Glowacki
University of Mining and Metallurgy, Krakow, Poland

H.Dyja
Technical University of Czestochowa, Poland

L.N.Lesik
Technical University of Donieck, Ukraine

ABSTRACT: Theoretical and experimental investigations of rolling of steel rods covered with copper are presented. The mathematical model of the process is based on the finite element solution for the generalized plane strain approach. Metal flow during rolling of bimetallic rods in oval-vertical oval-oval series is simulated. The behaviour and flow of the copper sleeve at various stages of the rolling process are analyzed. Experimental part of the project concerns rolling of bimetallic rods, 8 mm in diameter, in several roughing and final passes. Only the results of investigation of the first four roughing passes are presented in this paper. The stock material was a round steel, 34 mm in diameter, covered with 2.0 mm copper layer. The explosive technique was used to bond the layers. Rolling was performed in laboratory mill with the roll diameter of 300 mm and in the industrial type, light section mill in Donieck. The samples were heated before the rolling to the uniform temperature of 920°C. A piece of the sample was cut of after each pass. Analysis of the cross section of the samples allowed the investigation of the deformation of the copper layer and the uniformity of its thickness on the circumference of the rod. The results of the experiments were compared with the finite element calculations and confirmed good predictive ability of the model. Selected results of the temperature, strain and strain rate distribution on the cross section of the rods are presented.

1. INTRODUCTION

New technologies in telecommunication, power industry, military technology, electronics engineering and other branches of industry need new materials. Bimetallic steel-copper wires have been produced combining high strength of steel and good conductivity and corrosion resistance of copper. The cold drawing technology supplies high quality products but needs suitable quality of the billet. There are a few methods of steel-copper charge production for wire works. The most efficient is rolling of explosively bonded steel rod with copper tube. This method was applied for production of the stock for the rolling process which has been investigated by the authors of present paper. The main experiment was performed in the laboratory mill with the roll diameter of 300 mm and in the industrial, light section mill, in Donieck, Ukraine. The subject of present work was to investigate the rolling process of bimetallic rods 8 mm in diameter. The stock material for the process was 38 mm in diameter. The rolling schedule consist of several oval-vertical oval-oval roughing and finishing passes. Only the first four roughing passes were investigated using theoretical and experimental methods.

2. MODEL OF THE ROLLING PROCESS

The oval-vertical oval-oval rolling is fully three dimensional process. The process consists of several passes including ingot turnings. The computer simulation of such a process is very time consuming and requires a lot of computer memory (Glowacki 1986, Glowacki & Pietrzyk 1989). Mathematical model was developed using generalized plane strain approach. Detailed description of this model is given in (Glowacki 1990). Experimental substantiation of the model has already been performed for sheet rolling (Glowacki & Madej 1991) and elongation rolling (Madej & Glowacki 1992) showing good predictive ability of the model in application to the rolling of monometallic rods.

The main assumption in the model is the decomposition of the process into several subsequent steps connected with subsequent material locations in the roll gap. It is assumed that the strains in the rolling direction are distributed uniformly across the sample. This assumption results in constant components of strain and strain rate tensors in the rolling direction and in zero components of the shear strain and shear strain rate related to that direction. Only the cross section of the strip is analyzed using non-steady state approach keeping in mind that the whole process is three- dimensional.

2.1 *Mathematical formulation of the model*

The solution is performed for the rigid-plastic body which obeys Levy-Mises flow rule:

$$\sigma = \frac{2}{3}\frac{\sigma_i}{\dot{\varepsilon}_i}\dot{\varepsilon} \qquad (1)$$

In the equation **(1)** σ is the deviatoric part of stress tensor, doted ε denotes strain rate tensor and doted ε_i is effective strain rate. Effective stress σ_i is calculated from the strain hardening curve which takes into account effective strain , effective strain rate and temperature. The friction stress at the contact surface is given by:

$$|\tau| = m\frac{\sigma_p}{\sqrt{3}} \qquad (2)$$

where m is the friction factor and σ_p is the yield stress. The basic part of the solution is the optimization of the work functional given by:

$$W = \int_V \sigma_i \dot{\varepsilon}_i \, dV + \lambda \int_V \nabla \mathbf{v} \, dV + \int_S \tau \, |\mathbf{v}| \, dS \qquad (3)$$

where λ is the penalty coefficient, $|\mathbf{v}|$ is the relative velocity at the tool-material interface, S is the boundary surface and V is the control volume.
According to Malinowski and Lenard (1993) constant penalty coefficient λ was replaced by various value:

$$\lambda = \frac{E\Delta t}{6(1-2\nu)} \qquad (4)$$

where E is the Young's modulus, ν is the Poison's modulus and Δt is time interval. The Poison's modulus is very close to 0.5. For steel it was assumed to be 0.499 and for copper, which is much more soft, to be 0.4999.

The finite element approach is applied to the simulation of the heat transfer within the deformation zone. The approach is based on the non-steady state solution of the general diffusion equation (Zienkiewicz 1981). The equation for heating and cooling problems is:

$$\nabla^T(k\nabla T) + Q = \rho c_p \frac{\partial T}{\partial t} \qquad (5)$$

where T is the temperature, Q is the heat generation rate due to plastic deformation, k represents heat conduction coefficient, c_p designates the specific heat and ρ is the material density. Solution of the equation **(5)** must satisfy the boundary conditions:

$$k\frac{\partial T}{\partial n} = \alpha\,(T_0 - T) \qquad (6)$$

where n is the coordinate normal to the boundary surface, α is the heat transfer coefficient and T_0 is the ambient temperature.

2.2 *Discretization*

Discretization of thermal and mechanical problems is accomplished in usual finite element manner. The solution is performed using finite element mesh with 4-node linear elements in the axial cross-section. Temperature and velocity at any point inside an element are described by the product of nodal values and shape functions.

$$\bar{\mathbf{v}} = \mathbf{N}\,\mathbf{v} \qquad \mathrm{T} = \mathbf{N}\,\mathbf{T} \qquad (7)$$

where $\bar{\mathbf{v}}$ is the velocity vector and T is the temperature at any point of an element, **N** is the shape functions matrix, **v** is the nodal velocities vector and **T** the vector of nodal temperatures.

3. COMPUTER PROGRAM

Computer program in FORTRAN 77 was written for the developed model. The program allows the calculation of the distribution of temperature, strains and strain rates during rolling. Program can be run on workstations or on personal computers. Its source version is about 11,000 lines in length. Compilation time on IBM PC/486 with Microsoft FORTRAN compiler is about 5 min. Execution time depends on the number of nodes in the mesh, accuracy of the iterative procedure and on the number of steps. It varies from around 20 minutes to few hours per pass.

4. EXPERIMENT

The stock material was round 40H steel rod with the diameter of 34 mm covered with M1-E copper layer. The chemical composition of the steel and the copper are given in Tabl. 1 and Tabl. 2 respectively. The thickness of copper layer was 2 mm and the whole bimetallic rod was 38 mm in diameter. The samples were heated before rolling to the temperature 920°C. Rolling schedule consisted of six roughing and several finishing passes. The product was 8 mm in diameter. In the current investigation only the first four roughing passes are considered. After each pass a part of the billet was cut out and sample cross-section was selected. The view of all cross-sections is shown in Fig. 1. In this figure the deformation of the copper layer can be seen.

Table 1. Chemical composition of 40H grade steel

C	Si	Mn	S	P	Cr, Ni
0.41	0.27	0.65	0.04	0.04	0.3

Table 2. Chemical composition of M1-E grade copper.

Cu	Al	Fe, Pb, Si, Zn	Sb, Bi, As, Ni, Sn
100	0.1	0.005	0.002

5. RESULTS AND DISCUSSION

In order to compare numerical and experimental results, the simulation of the rolling of bimetallic rod was performed. The cross section of the billet was divided into 126 elements. Steel and copper layers were divided separately. The simulation included billet turnings. It was assumed that recrystallization in interpass times is completed. It means that the deeformation in i-th pass has no influence on pass no i+1. The temperature was simulated during both deformation and interpass time.

The comparison of the computational results with the experimental data was performed measuring the thickness of the copper layer in different places on the circumference of the rod. The measurement was done for several directions described by angle with respect to the x-axis. The corelation betwen the measured and computed copper layer thichness for passes no 1 to 4 are presented in the Figs. 2-5.

The mathematical model allows the calculation of the distribution of strain, strain rate and temperature. Figs. 6-9 show the strain distribution in the cross-section of the rod for passes 1 to 4. The figures show the changes of the cross-section of the billet in subsequent passes. The turnings of the billet after each pass are presented in the figures as well. Figs. 10-13 present temperature maps after passes and Figs. 14-15 strain rate distribution at the end pass 1 and pass 4.

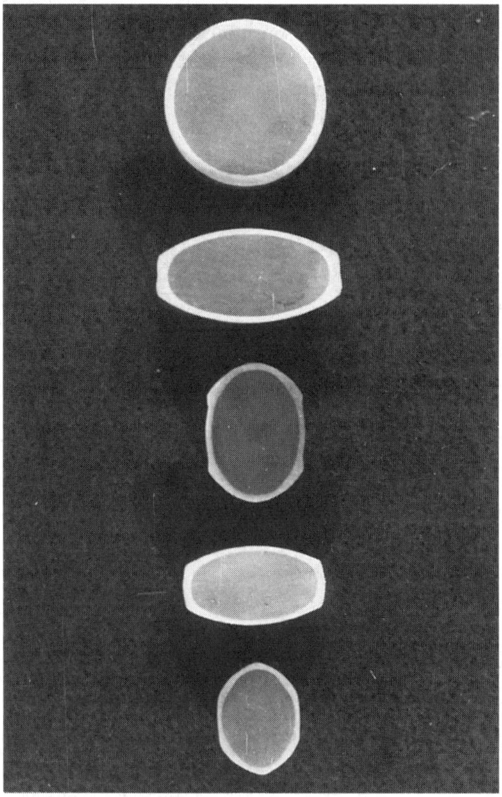

Figure 1. View of the copper sleeve thickness in subsequent passes by rolling of bimetallic rod.

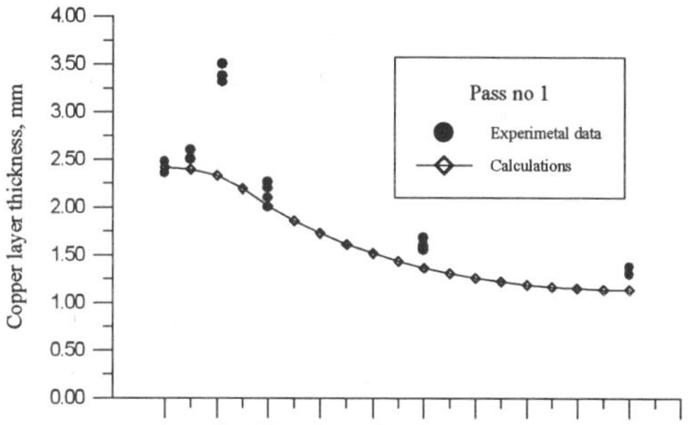

Figure 2. Comparison between measured and calculated copper layer thickness on the circumference of the oval after pass no 1.

In almost whole length of circumference the agreement betwen measured and computed values of thickness is good. Only in one direction for each pass the difference betwen experimental and theoretical data is significant. In this position the copper sleeve is extremely irregular and creates small hill which decreases in subsequent passes. This phenomenon is not predicted by the presented mathematical model.

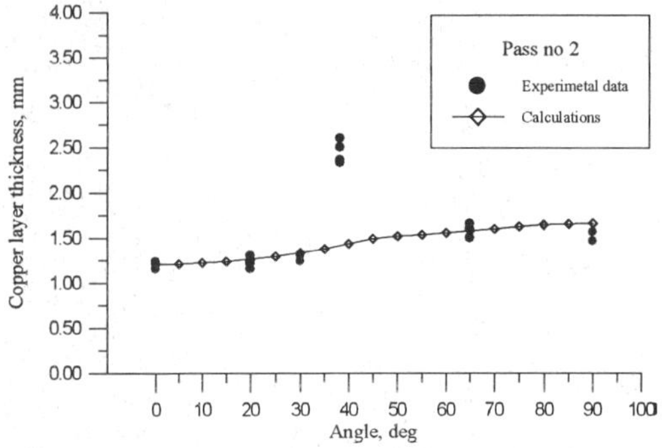

Figure 3. Comparison between measured and calculated copper layer thickness on the circumference of the oval after pass no 2.

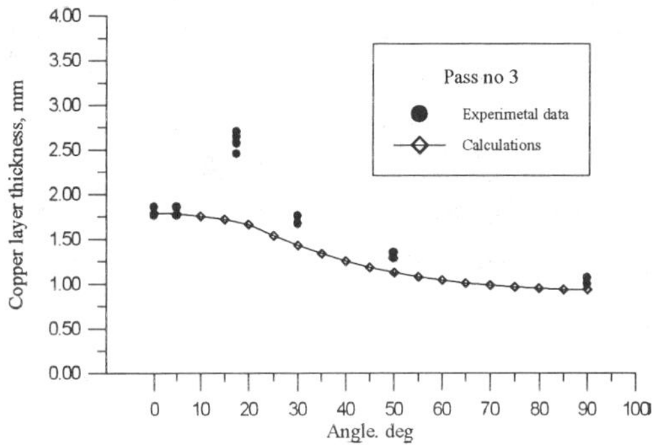

Figure 4. Comparison between measured and calculated copper layer thickness on the circumference of the oval after pass no 3.

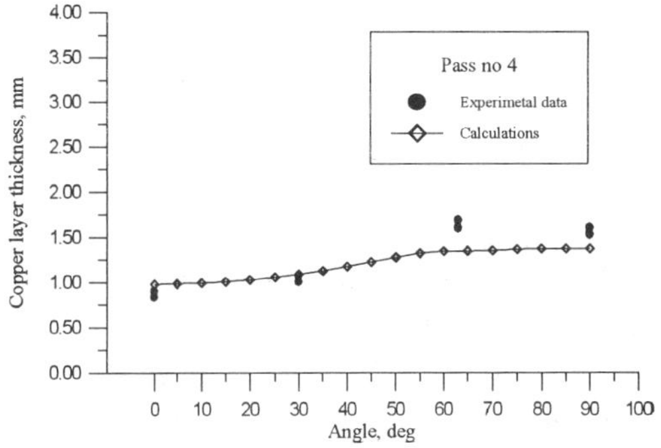

Figure 5. Comparison between measured and calculated copper layer thickness on the circumference of the oval after pass no 4.

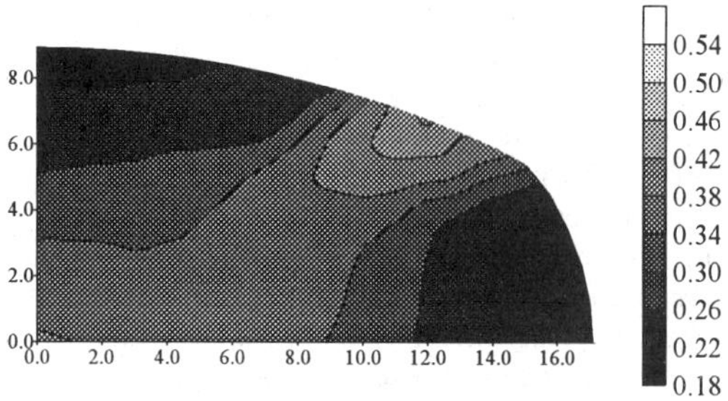

Figure 6. Strain distribution in the cross-section of the billet after pass no 1.

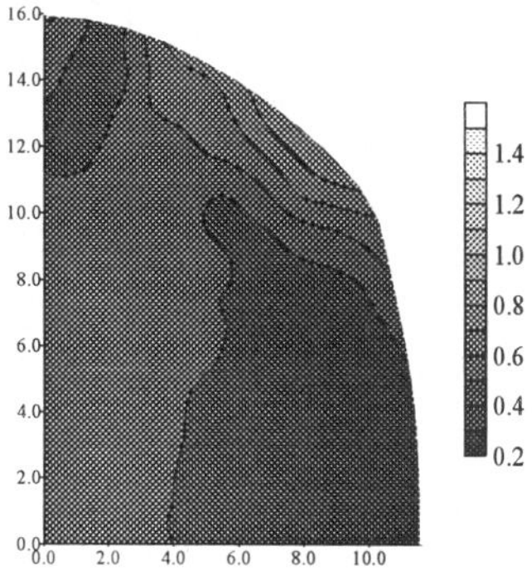

Figure 7. Strain distribution in the cross-section of the billet after pass no 2.

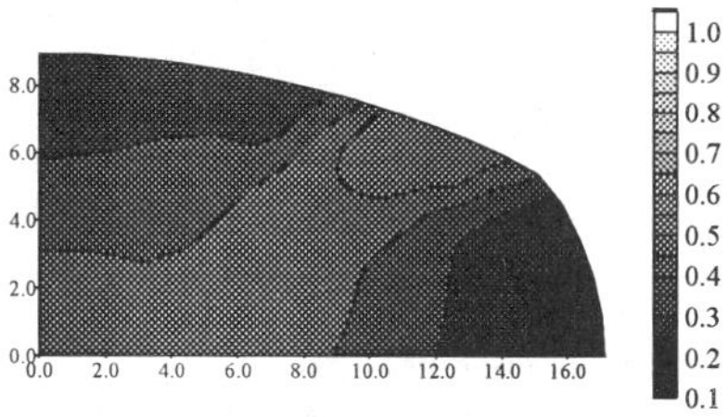

Figure 8. Strain distribution in the cross-section of the billet after pass no 3.

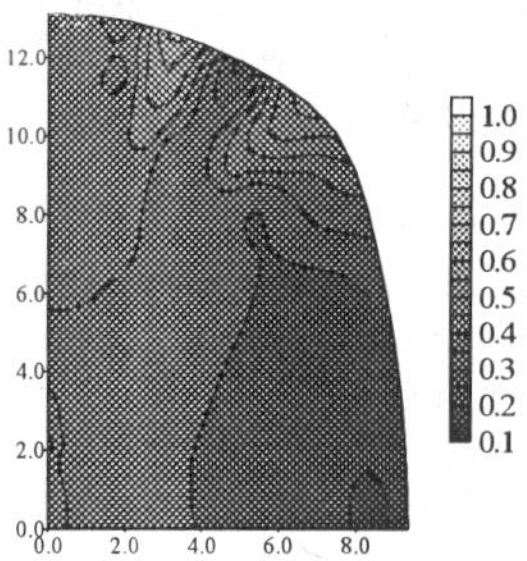

Figure 9. Strain distribution in cross-section of the billet after pass no 4.

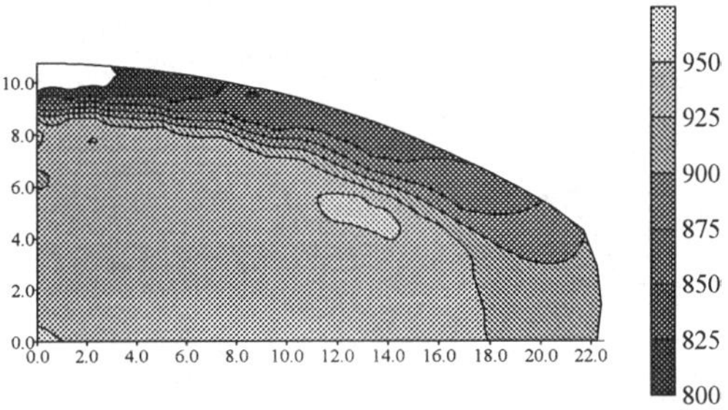

Figure 10. Temperature distribution in the cross-section of the billet after pass no 1.

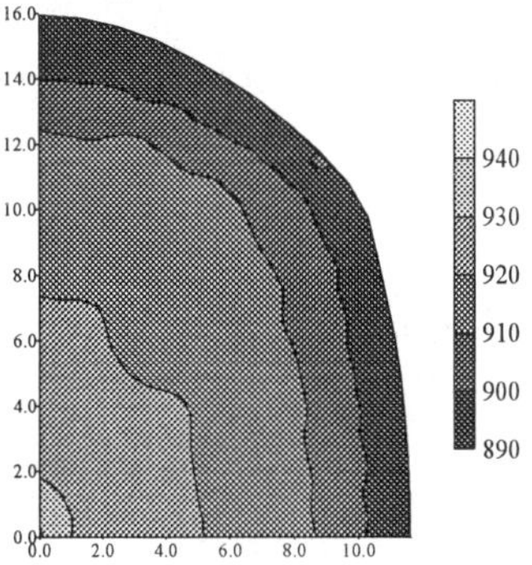

Figure 11. Temperature distribution in the cross-section of the billet after pass no 2.

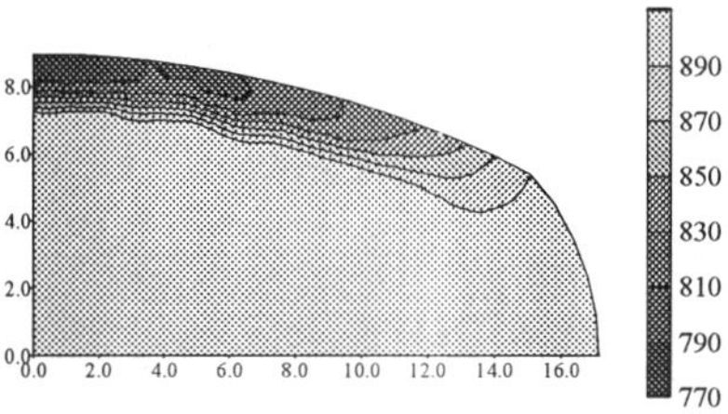

Figure 12. Temperature distribution in the cross-section of the billet after pass no 3.

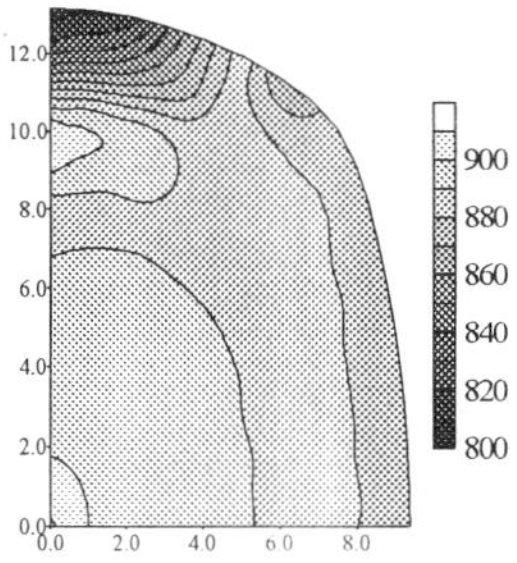

Figure 13. Temperature distribution in the cross-section of the billet after pass no 4.

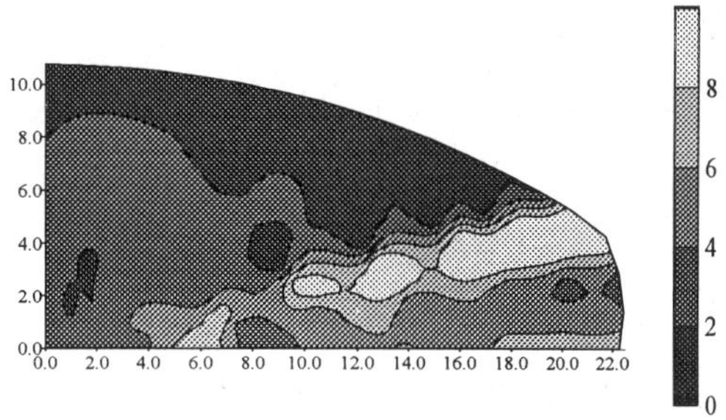

Figure 14. Strain rate distribution in the cross-section of the billet at the end of pass no 1.

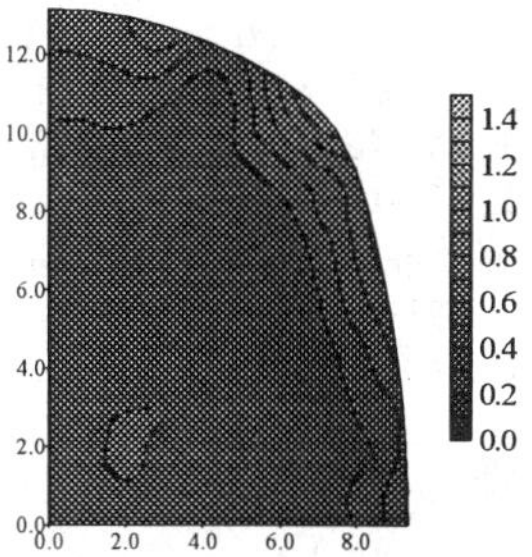

Figure 15. Strain rate distribution in the cross-section of the billet at the end of pass no 4.

6. CONCLUSIONS

The comparison of the calculated and measured thicknesses of copper sleeve confirms good predictive ability of the theoretical model. Generalized plane strain approach, on which the model is based, can be useful for the simulation of metal flow and distribution of strain, strain rate and temperature for steel and copper layer. The phenomenon of copper sleeve hill which appears on the circumference of the rod must be analysed and appropriate improvement of the theoretical model will be done in the future.

The optimization of the work functional must be made with great accuracy of controling of volume losts. Because of big difference betwen yield stresses for both metals the influence of deformation energy of the copper layer on the work functional is less significant than the steel deformation energy. It leads to copper sleeve decay and only well convergent optimization methods can avoid this problem. The strain hardening curves for bimetal materials must be determine with good accuracy from the appropriate tests. In the case of errors in the strength tests, the metal flow simulation can be incorrect.

ACKNOWLEDGEMENTS

The financial assistance of the Polish Committee of Scientific Research (grant no 7 0669 91 01) is gratefully acknowledged.

REFERENCES

Glowacki M. 1986, Possibilities of Application of the Finite Element Methods to the Analysis of Three-Dimensional Metal Forming Processes. *Metal. Odlew.*, 12:159-175.

Glowacki M. & Pietrzyk M. 1989, Experimental Substantiation of Rigid Plastic Finite Element Modelling Three Dimensional Forming Processes. *J. Mech. Work. Techn.*, 19:295-303.

Glowacki M. 1990, Thermal-Mechanical Model of Rolling of Shapes in 4-Roll Grooves. *Metal. Odlew.*, 16:541-562.

Glcwacki M. & Madej W. 1991, Experimental Substantiation of Thermal-Mechanical Model of Sheet Rolling Process. *Hutnik*, 58:337-340. (in Polish)

Madej W. & Glowacki M. 1992, Experimental Validation of Thermal-Mechanical Model of Elongation Rolling Process. *Hutnik*, 59:283-286. (in Polish)

Malinowski Z. & Lenard J.G. 1993, Experimental Substantiation of an Elastoplastic Finite Element Scheme for Flat Rolling, Comput. *Meths Appl. Mech. Engrg.* 104:1-17.

Zienkiewicz O.C. 1981, Finite Element Methods in Thermal Problems. In R.W. Lewis, K. Morgan and O.C. Zienkiewicz (eds), *Numerical Methods in Heat Transfer*:1-25, John Wiley & Sons Ltd.

Simulation of Materials Processing: Theory, Methods and Applications, Shen & Dawson (eds)
© 1995 Balkema, Rotterdam. ISBN 90 5410 553 4

Coupling of roll and strip deformation in three-dimensional simulation of hot rolling

Arnaud Hacquin & Pierre Montmitonnet
École des Mines de Paris, Centre de Mise en Forme des Matériaux (CEMEF), (URA CNRS), Sophia-Antipolis, France

Jean-Philippe Guillerault
Pechiney Centre de Recherches de Voreppe, France

ABSTRACT: A three-dimensional strip rolling model is presented. The rolling stand elastic deformation as well as thermal modeling of the Work Roll are coupled iteratively to a Eulerian Finite Element computation of the strip flow and temperatures. We also describe a hot rolling simulation of a thin strip of aluminum. It emphasizes the need to strongly couple thermal and mechanical phenomena in both the strip and rolls when aiming at thickness profile or flatness defects predictions.

1 INTRODUCTION

The need for a three-dimensional rolling model arises when dealing with spread, metallurgical heterogeneity or thickness deviations along the strip width. The latter concern, on which we focus in this paper, is of great significance in the thin strip rolling industry, as it is related to flatness defects.

Thickness profile defects result not only from the rolling stand elastic deformation under rolling stress, but also from the Work Roll thermal crown due to a temperature heterogeneity along the roll axis. Since the distributions along the strip width of stress and heat input to the roll themselves depend on thickness deviations, -all the more as the strip is thin and hard-, modeling such defects may require a close-coupling between the stand and strip models.

Pawelski et al (1985) combined a semi-analytical rolling stand elastic deformation model with rolling stress predictions by a succession of slab methods across the strip width. More recently, Yamada et al (1992) and Yanagimoto & Kiuchi (1992) introduced the Finite Element Method in the strip to improve bite length and stress predictions. The roll thermal crown was an input parameter to their models.

Though many 3-dimensional thermal models of the Work Roll are reported in the literature, only 2-dimensional models have been coupled to the strip flow so far (Pietrzyk 1990).

We have developed a steady-state model which couples a Eulerian FEM for the plastic flow in roll gap with a thermal-elastic rolling stand model based on an accurate, fast-computing combination of analytical and 2-dimensional FE Methods.

2 OUTLINE OF THE MODEL

Thermal-mechanical coupling between the strip and rolling stand is based on a fixed-point scheme (fig.1)

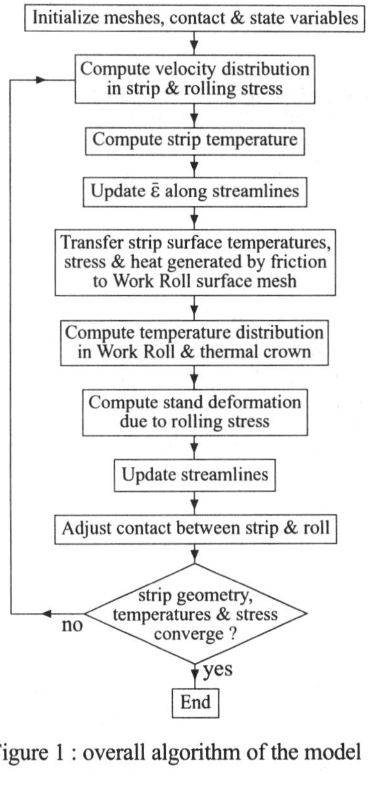

Figure 1 : overall algorithm of the model

2.1 *Strip deformation*

The velocity field in the strip is computed by Forge3, a general purpose 3-dimensional Finite Element Model of forming processes. Our rolling version relies on linear hexahedral elements and a penalty method to enforce incompressibility. Any viscoplastic rheologies associated to a Mises or Hill yield criterion may be simulated. In the case of hot rolling of aluminum, yield stress σ varies with temperature T, strain $\bar{\varepsilon}$ and strain rate $\dot{\bar{\varepsilon}}$ according to a Sellars-Teggart law with Voce hardening :

$$\sigma = \sigma_0 + (\sigma_{ss} - \sigma_0)\sqrt{1 - \exp(-\lambda\bar{\varepsilon})} \quad (1)$$

where $\sigma_0 = \sqrt{3}\, Ke^{\beta/T}\,(\sqrt{3}\,\dot{\bar{\varepsilon}})^m$

and $\sigma_{ss} = A\,\mathrm{argsh}\left(\frac{\dot{\bar{\varepsilon}}\, e^{\beta s/T}}{Z_0}\right)^{m_s}$

λ, K, β, m, A, βs, Z_0, m_s are material parameters.

Friction at the strip-roll interface can be modeled by a Norton law involving the sliding velocity v_s :

$$\tau = -\tau_N.\frac{v_s}{|v_s|},\ \tau_N = \alpha K' e^{\beta/T} v_s^p \quad (2)$$

or a Coulomb law with a Norton upper bound

$$\tau = -\tau_{NC}.\frac{v_s}{|v_s|},\ \tau_{NC} = \mathrm{Min}(\tau_N, \mu\sigma_n) \quad (3)$$

T, $\bar{\varepsilon}$ and normal stress σ_n appearing in equations (1-2) are those computed at the previous iteration.

Finally, the Principle of Virtual Work rate writes :

$$\int_\Omega \sigma : \dot{\varepsilon}^* = \int_{\partial\Omega} S.v^* \quad (4)$$

which has the discrete form

$$\int_\Omega \sigma(v_j.N_j) : B_i = \int_{\partial\Omega} S.N_i \quad (5)$$

where N_i and B_i denote Galerkin shape functions at degree of freedom i and their symmetrized gradients.

The system of eq. (5) relative to free degrees of freedom is solved by a Newton-Raphson algorithm. Computation after convergence of the left-hand side R_i of (5) enables to assess the distribution of rolling stress S. Indeed, if we assume that $S = S_j.N_j$, nodal stresses S_j are obtained by solving

$$\int_\Omega N_i.N_j.S_j = A_{ij}.S_j = R_i \quad (6)$$

Heat generated by friction $P_f = -\tau.v_s$, which serves as boundary condition for the heat transfer models in both the strip and work roll, is another output of Forge3. It is primarily computed at strip surface integration points and is then processed into nodal values P_{fj} by a least-squares method that results in

$$A_{ij}.P_{fj} = R'_i \quad (7)$$

Both S and P_f are passed to the work roll mesh by simply projecting roll nodes onto the strip surface and interpolating nodal values.

2.2 *Free boundaries and state variables of strip*

The geometries of the free surfaces are obtained iteratively using the well-known streamline method. We also make inner rows of nodes converge towards streamlines. Therefore, $\bar{\varepsilon}$ may be updated at each iteration by simply integrating $\dot{\bar{\varepsilon}}$ from one integration point to the other.

We have preferred a FE Method to update temperatures because of transverse diffusion. Modeling thin-strip rolling results in high element Peclet numbers, $\frac{vl_x}{a}$, which may make the standard Galerkin FEM oscillate. l_x denotes the mesh length in the rolling direction and a the diffusivity. So we relie instead on the SUPG formulation (Streamline-Upwind Petrov-Galerkin) of Brooks&Hughes (1982) which is based on modified test functions $\tilde{N}_i$:

$$\tilde{N}_i = N_i + \alpha\frac{a}{v^2}\, v.\nabla_x N_i \quad (8)$$

α is a dimensionless function of the Peclet numbers in the x, y and z directions.

That method is particularly efficient in our cases since the velocity v is everywhere parallel to the mesh directions.

Finally, the equations to be solved write :

$$\rho c v.\nabla T = k\Delta T + \sigma:\dot{\varepsilon} \quad (9)$$

with boundary conditions

$$k\nabla T.n = h_b(T_w - T) + \frac{b_p}{b_w + b_p} P_f \quad (10)$$

on the bite surface and

$$k\nabla T.n = h_o(T_o - T) + \varepsilon\sigma(T_o^4 - T^4) \quad (11)$$

elsewhere.

b denotes the effusivity $\sqrt{\rho c k}$ of the strip (p) or work roll (w) and T_o the outside temperature.

2.3 *Temperature distribution in the work roll*

The roll is approximated by a true cylinder of length L and radius R. The equation of steady-state heat transfer in the roll rotating at ω can be written as

$$\rho c r \omega \frac{\partial T}{\partial \theta} = k\Delta T \quad (12)$$

with boundary conditions

$$k\left.\frac{\partial T}{\partial r}\right|_{r=R} = h_b(T_p - T) + \frac{b_w}{b_w + b_p} P_f \quad (13)$$

: bite surface

$$k\nabla T.n = h_a(T_o - T) \quad (14)$$

: convection with atmosphere

$$k\left.\frac{\partial T}{\partial y}\right|_{\substack{y=L/2 \\ r<R_n}} = h_n(T_n - T) \quad (15)$$

: heat removal by roll neck of radius R_n

$$k \left.\frac{\partial T}{\partial r}\right|_{r=R} = h_s(T_s - T) \quad (16)$$

: heat removal from M water-sprayed areas represented by their extensions in θ and y

$$(\theta,y) \in \bigcup_{i=1}^{M} [\theta_{i1},\theta_{i2}]x[y_1,y_2] \quad (17)$$

The above system of equations is linearized at each iteration. A global 3-dimensional FEM was not considered here because of computing time and storage limitations. Instead we make the following simplifications :

a) $\frac{\partial T}{\partial \theta} = 0$ if $R-r > \delta = 4\sqrt{\frac{a}{\omega}}$ i.e. T does not depend on θ beyond the thermal skin of thickness δ

b) $\frac{\partial T}{\partial y} = 0$ if $R-r < \delta$ i.e. we neglect axial heat fluxes within the skin layer since δ is never larger than a few millimeters.

We can now decouple the problem into one axisymmetrical core model and plane models in each of the N cross sections n of the roll skin (fig.2). These models are linked by the unknown distribution of interface temperatures T_{in} at depth δ, which are computed as follows :

a) The core model is summarized, once discretized, by a relation between T_{in} and heat fluxes Φ_n entering sections n of the roll :

$$T_i = C.\Phi + B \quad (18)$$

Matrix C and vector B are computed by running N+1 times a 2D FE model.

b) skin models in sections n give relations :

$$\Phi_n = d_n T_{in} + e_n \quad (19)$$

At most 2N different runs of a 2D SUPG FE model are required to compute coefficients d_n and e_n.

Combination of (18) and (19) results in a system from which we obtain T_i. The remainder of the temperature field is then computed.

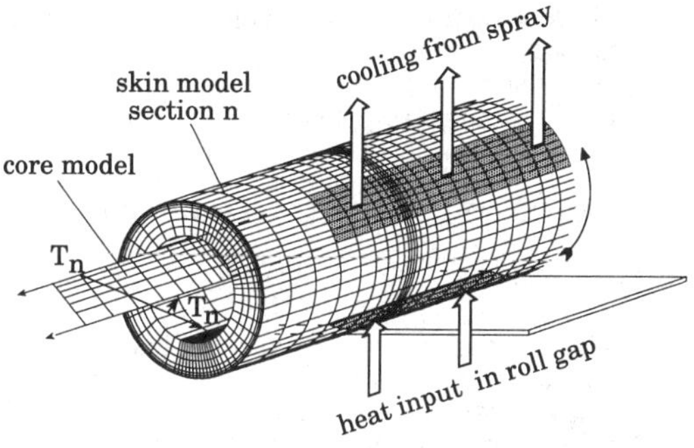

Figure 2 : thermal model in Work Roll

The 2D SUPG model we have developed in the skin layer is in quite a good agreement with analytical results from Patula (1981) who used Kelvin expansions to deal with a uniform heat input in the roll gap. No more than 100 nodes along the circumference and 15 throughout the skin thickness are required to achieve a good accuracy.

2.4 *Work roll deformation under rolling stress*

We have developed a semi-analytical slab method to compute 2,4 or 6-high stands elastic deformations. It was validated experimentally on a 4-high test mill (Hacquin et al 1994) and through comprehensive Finite Element comparisons (Hacquin et al 1995).

In this model, the main component of strip indentation into the work roll is given by the Boussinesq theory of elasticity :

$$\Delta z_m = \int_{\mathcal{B}} \frac{1-\nu^2}{\pi E} \frac{S_z(v,w)}{\sqrt{(x-v)^2+(y-w)^2}} dvdw \quad (20)$$

which requires an accurate characterization of the distribution of stress S_z over the bite surface $\mathcal{B}$. Here, FEM-predicted S_z is discretized onto a square grid of an estimation of $\mathcal{B}$:

$$S_z = S_{zjl} N_{jl} \quad (21)$$

where N_{jl} are Q1 shape functions of the roll surface mesh. Δz_{mik} are then related to S_{zjl} using equation (20), which involves 4th-order influence coefficients that are computed only once and stored in memory for reuse at each iteration. All other components of deformation only depend on sums of S_{zjl} along the rolling direction (from Saint Venant's principle).

Finally, the change in boundary conditions for the FEM in the strip due to the whole roll thermal-elastic deformation is taken into account at each iteration using a relaxation technique to prevent oscillations.

2.5 *Contact adjustment*

Contact between strip and Work Roll may be lost owing to either roll deformation or integration of streamlines. The first step in restoring contact consists in determining the new bite geometry along each longitudinal row. Bite entry is defined as the *horizontal* projection onto the roll of the strip surface node previously assumed at entry. As for bite exit, the sign of normal stress σ_{ni}, obtained by rotating S, is tested at previous exit node i. If tensile, bite is reduced up to where σ_n vanishes (in fact, a relaxation factor of .2 is used). Otherwise, it can be extended downstream geometrically by iteratively processing nodes : node i is projected onto the roll, assumed in contact and the downstream remainder of the row is translated accordingly to ensure mesh continuity. If this results in a penetration of node i+1, i+1 is

Table 1. Input parameters.

Strip geometry		Friction law		Rheology (high strength aluminum)	
Entry thickness	4 mm	μ	0.2	K	0.6 MPa
Exit thickness (y=0)	3 mm	p	0.15	m	0.1
Entry width	1400 mm	α	0.3 $(mm.s^{-1})^{-p}$	β	2836 K
Front/back traction	0	K'	0.823 MPa	λ	7.42
				A	58.735 MPa
Rolling stand geometry		Work Roll thermal parameters		m_s	0.256
Barrel length	2300 mm	ρ	7.8 10^3 $kg.m^{-3}$	β_s	22 577 K
WR diameter	600 mm	c	.46 $kJ.kg^{-1}.K^{-1}$	Z_0	3.62 10^{12} s^{-1}
WR radial crown	-100 µm	k	15 $W.m^{-1}.K^{-1}$	Strip thermal parameters	
WR rotating speed	5 $m.s^{-1}$	b	13 $kJ/(m^2.K\sqrt{s})$	Entry temperature	300 °C
BUR diameter	1200 mm	θ_1	45 ° (following bite exit)	ρ	2.62 10^3 $kg.m^{-3}$
dist. between screws	1600 mm	θ_2	115 °	c	1.07 $kJ.kg^{-1}.K^{-1}$
Young's modulus E	210 GPa	h_s	10 $kW.m^{-2}.K^{-1}$	k	240 $W.m^{-1}.K^{-1}$
Poisson's ratio ν	0.3	T_s	30 °C	b	26 $kJ/(m^2.K\sqrt{s})$
		y_2	750 mm	h_1	20 $kW.m^{-2}.K^{-1}$
		Thermal expansion	15.10^{-6} K^{-1}		

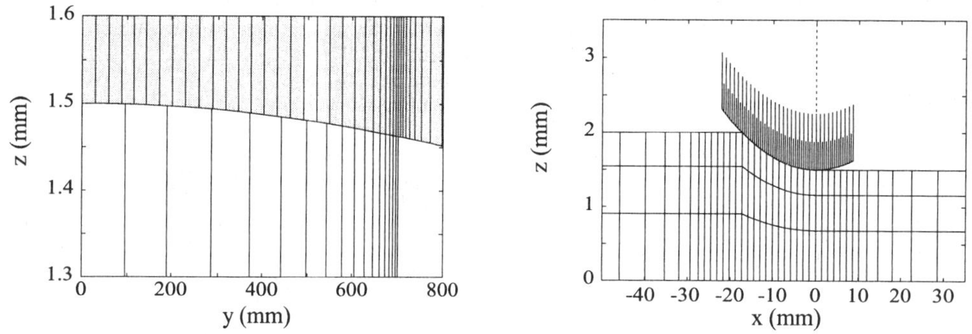

Figure 3 : undeformed meshes of strip and roll.

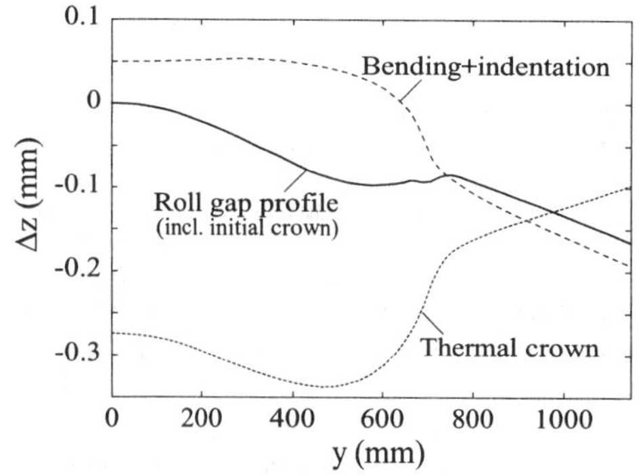

Figure 4 : components of Work Roll deformation

projected and the process continues until a free node is found. The second step consists in remeshing the rolling direction to fit nodes to the new bite extremities but still minimizing distortions of the pre-an post-deformation areas obtained by streamline integration.

As for inner streamlines, a smoothing algorithm was developed to prevent kinks.

3 EXAMPLE OF HOT ROLLING SIMULATION.

3.1 *Numerical data*

The selected configuration is typical of the last stand of a hot rolling tandem mill of aluminum. Table 1 lists the parameters of the model.

We found that the model required a downstream length equivalent to 1.75 width to stabilize the geometry of the exit cross section. An optimal mesh

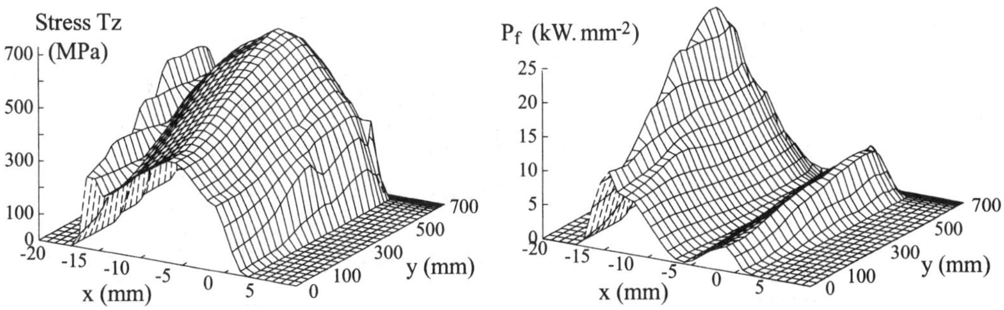

Figure 5 : boundary conditions for rolling stand thermal-elastic model. a) stress b) heat generated by friction

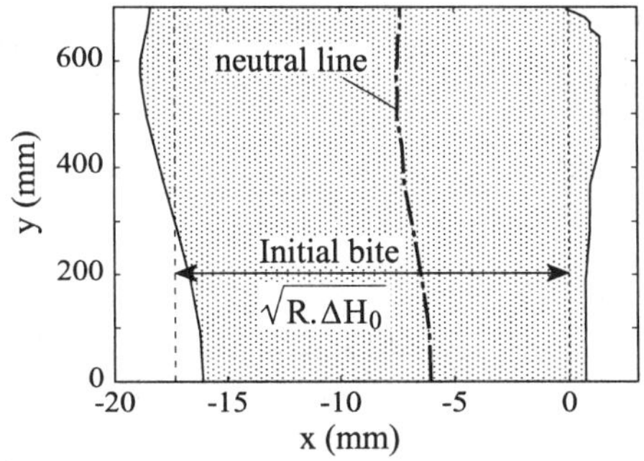

Figure 6 : bite geometry

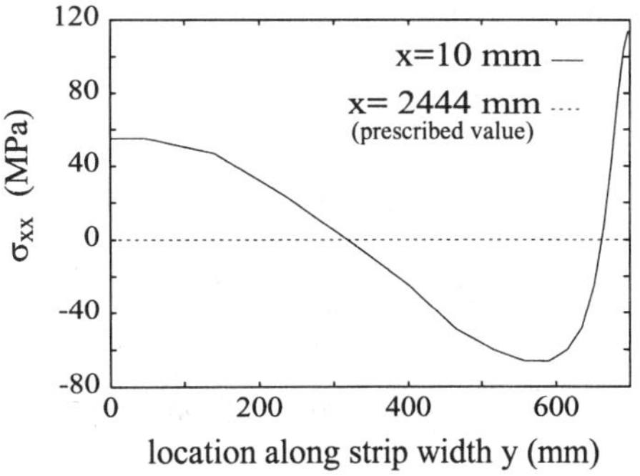

Figure 7 : longitudinal stress profile after bite exit.

of the system was then determined by trial and error (fig.3). 5832 nodes were used in the strip.

3.2 *Results and discussion*

Convergence of the simulation is achieved within 35 iterations and a computing time of about 2 hours on IBM Risc6000/580 machines. Loss of mass due to possible inaccuracies of the penalty or streamline methods amounts to no more than 0.01% of the mass flow rate.

As shown by figure 4, the superposition of Work Roll bending, indentation and thermal crown bring about a significant deviation of reduction in thickness -from 25% at mid-width to 30% at edges-, which is typical of an ill-controlled process (no roll bending).The Forge3-computed boundary conditions for the rolling stand model are plotted in figure 5. Heat generation by friction and contact through conductance h_b amounted respectively to 64 and 36% of the overall heat input to the roll.

The sensitivity to the thickness deviation of rolling stress and friction-generated heat demonstrate how closely coupled thermal and mechanical mechanisms in the strip and rolling stand are. We believe that sensitivity is enhanced by velocity rearrangements up- and downstream of the bite area. Indeed, in 2-dimensional thin strip rolling models (Ford et al 1951), the rolling force per unit width roughly varies as the bite length. Therefore, the maximum deviation one can expect from y-decoupled such models is of the order of 20%, as estimated by figure 6, whereas it nearly doubles -from 535 t.m^{-1} at y=0 to 1053 t.m^{-1} at y=580 mm- in our 3D model. Figure 7 suggests that the actual distributions of front and back tractions at the very bite entry and exit are critical to the rolling force distribution.

Figures 8 to 10 illustrate the above-mentioned velocity rearrangements. The most striking feature is undoubtedly a negative spread of 1 mm which we believe would in reality shift to flatness defects. The plane of symmetry z=0 imposes flatness in our simulation.

Thickness defects give rise to energy-dissipative gradients of longitudinal velocity at bite exit (fig.9). The homogenization of the velocity profile is achieved gradually mainly through that of the thickness profile (fig.8) ; the spread reverses sign after exit from bite so that the process globally results in no significant width alteration.

Approximate incompressibility formula $v_e = \frac{h_n . v_n}{h_e}$

provides an easy interpretation of the metal flow upstream of bite, if we denote by $v_e(y)$ the longitudinal velocity at bite entry and by v_n the velocity of the neutral point prescribed by the roll

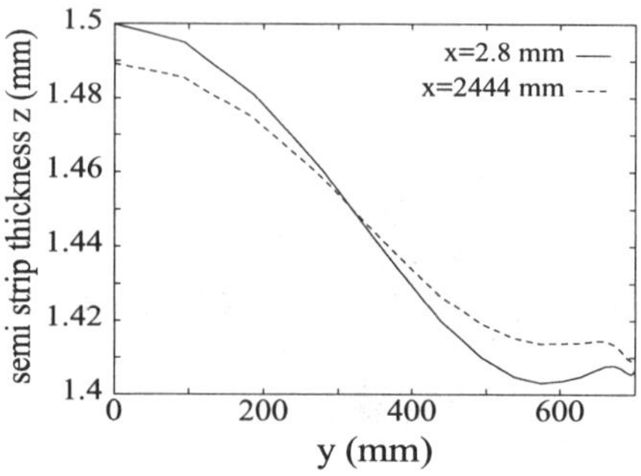

Figure 8 : evolution of strip transverse profile

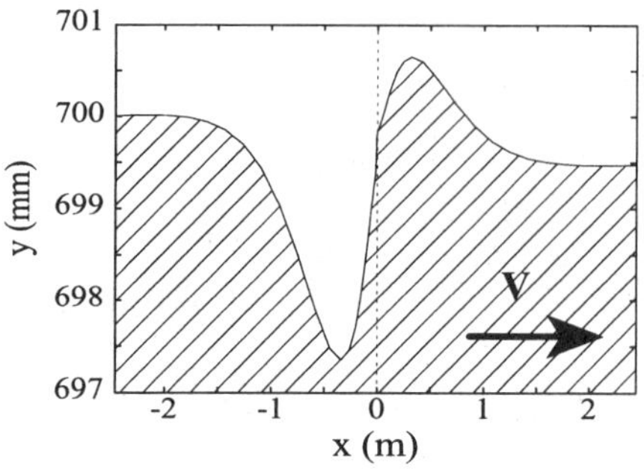

Figure 10 : spread profile (distorted by scales)

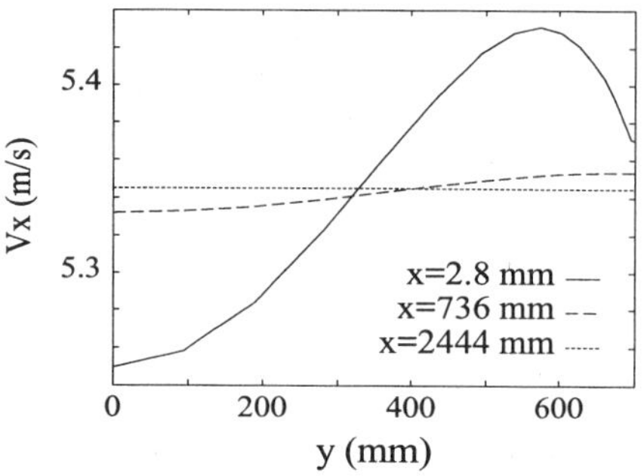

Figure 9 : Longitudinal velocities in strip

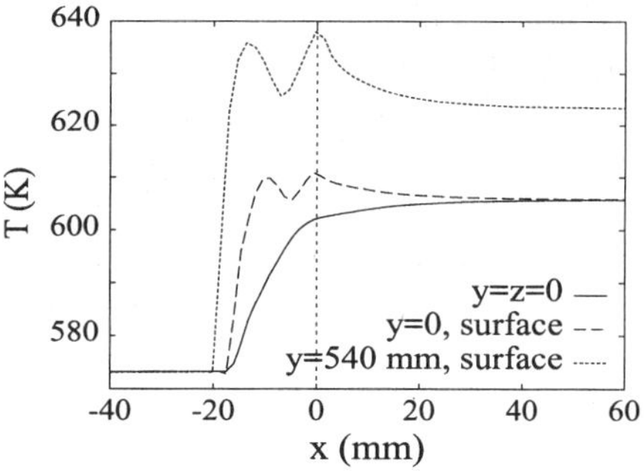

Figure 11 : evolution of strip temperatures

rotation. The decrease in h_n near the edge requires at entry that metal be slower near the edge than at mid-width. Since velocity is uniform far upstream, metal slows down near the edge whereas it accelerates at mid-width. The process is therefore similar to that downstream and results indeed in a maximum compressive stress σ_{xx} and an increase in thickness (120 μm higher than at mid-width) around y=500 mm. In fact, in spite of the neutral point moving backward near the edge, metal upstream of the bite area bears the brunt of the heterogeneity in thickness reduction.

The increase in heat generated by friction and deformation near the edge also gives rise to temperature deviations along the strip width (fig.11), significant enough to have metallurgical consequences. Figure 11 shows how successfully the SUPG technique erases oscillations in spite of element Peclet numbers as high as 100 in the roll gap.

4 CONCLUSION

The coupled model sheds new light on the velocity rearrangements outside roll gap when reduction in thickness is not uniform across the width. That complex velocity field alters dramatically the distributions of rolling force, which is why

1) predicting the thickness profile of thin strips requires a close-coupling between the rolling stand deformation and the strip flow

2) a three-dimensional model to predict the distribution of rolling force should be preferred to decoupled 2D models in different cross sections of the work roll.

The distribution along the strip width of heat generated by friction at the strip-roll interface is also very sensitive to thickness deviations, which may result in unexpected work roll thermal crown profiles.

Moreover, since that model enables an accurate prediction of longitudinal stress, it could, if combined with a buckling criterion, provide new opportunities of predicting flatness defects.

Acknowledgement : The authors are grateful to Pechiney Company for supporting this project and allowing publication of results.

REFERENCES

Brooks, A.N. & T.J.R. Hughes 1982. *Computer Meth. in Appl. Mech.and Eng.* 32:199-259

Ford, H., F. Ellis & D.R. Bland 1951. *J.Iron& Steel Inst.* May 1951:57-72

Hacquin, A., P. Montmitonnet & J.P. Guillerault 1994. Metal Forming 94: 199-206. Elsevier.

Hacquin, A., P. Montmitonnet & J.P. Guillerault 1995. submitted to *Int. J. Mech.Sci.*
Patula, E.J. 1981. *J.Heat Transfer* 103:36-41
Pawelski, O., W. Rasp & H.Teutsch 1985.*Steel Research* 56:327-331
Pietrzik, M. & J.G. Lenard 1990. *J.Mat.Proc. Techn.*22:177-190
Yamada, K., S. Ogawa & M. Ataka 1992. NUMIFORM 92, 755-761. Balkema.
Yanagimoto, J. & M. Kiuchi 1992. NUMIFORM 92, 763-768. Balkema.

Simulation of Materials Processing: Theory, Methods and Applications, Shen & Dawson (eds)
© 1995 Balkema, Rotterdam. ISBN 90 5410 553 4

Analysis of deformation behavior of slabs in edge-rolling by three-dimensional rigid-plastic finite element method

Shin-ya Hayashi & Taneharu Nishino
Nippon Steel Corporation, Osaka, Japan

Ken-ichiro Mori
Osaka University, Japan

Kazuo Watanabe
Nippon Steel Corporation, Chiba, Japan

ABSTRACT: Three-dimensional plastic deformation of slabs in edge-rolling for producing preforms of H-beams is simulated by the rigid-plastic finite element method. In the edge-rolling, the cross-section of the rolled slab has a dog-bone shape: the thickness near both edges of the width is partially large due to the concentration of plastic deformation. The effects of the work-hardening characteristic and the strain-rate sensitivity in the flow stress, the roll geometry, the coefficient of friction and the temperature distribution on the dog-bone shape of the rolled cross-section are examined. The calculated dog-bone shapes are in good agreement with the experimental ones for steel and plasticine. The dog-bone shape becomes remarkable as the work-hardening and the strain-rate sensitivity exponents decrease and the coefficient of friction increases. The geometry of the roll groove also has a great influence on the dog-bone shape.

1 INTRODUCTION

Steel slabs supplied for hot strip rolling are mainly produced by continuous casting. From the point of view of productivity, the cross-sectional dimension of the continuous casting slabs is fixed to be constant, and thus the width of the slab is changed by edge-rolling processes in subsequent hot rolling. In edge-rolling, the width of the slab is reduced by a pair of vertical rolls. The thickness near both edges in the width of the rolled slab becomes large because of the concentration of plastic deformation near the edges. The cross-sectional shape of the rolled slab is called a dog-bone shape. Since the dog-bone shape is deformed by subsequent horizontal rolling to be rectangular, the efficiency of rolling deteriorates due to the increase of the slab width. On the other hand, the dog-bone phenomenon is positively utilized for producing H-beams by rolling, i.e. in edge-rolling for H-beams the width of the slab is reduced by a pair of horizontal rolls and the flanges of an H-beam are formed by increasing the thickness near the edges of the slab. In particular, edge-rolling processes from slabs are effective in producing H-beams with a large cross-section, and thus the range of the H-beam size is extended by the use of edge-rolling. Since the shape of the rolled cross-section varies with the reduction in width and the groove geometry, it is important to estimate the deformation behavior during edge-rolling of slabs in the roll pass design of H-beams.

The finite element method has been applied to analysis of complex three-dimensional deformation in shape rolling processes (Bertrand et al.(1986), Mori et al.(1989), Park et al.(1990), Hayashi et al. (1992) and Iguchi et al. (1993)). The finite element method is suitable for simulating deformation behavior in edge-rolling of slabs because workpieces in edge-rolling have large free surfaces in comparison with those in shape rolling. Mori et al. (1982, 1984, 1987) have calculated three-dimensional deformation in edge-rolling of slabs by the rigid-plastic finite element method using simplified three-dimensional elements. Although the simplified elements have the advantage of a short computing time, the accuracy of the calculated results is not so high as that for the fully three-dimensional simulation.

In the present study, edge-rolling processes of slabs for producing H-beam preforms are simulated by the three-dimensional rigid-plastic finite element method. The effects of the work-hardening exponent, the strain-rate sensitivity exponent, the roll geometry, the coefficient of friction and the temperature distribution on the dog-bone shape of the rolled cross-section are examined.

2 METHOD OF SIMULATION

2.1 *Rigid-plastic finite element method*

The rigid-plastic finite element method is used to simulate three-dimensional deformation of slabs in edge-rolling. The present method is formulated on the basis of the plasticity theory for a material with slight compressibility (Osakada(1982)), and the following functional Φ for the material is minimized to obtain the exact solution.

$$\Phi=\int_V \left\{\int_0^{\dot{\bar{\varepsilon}}} \bar{\sigma}\, d\dot{\bar{\varepsilon}}\right\} dV+\int_{Sf} \tau_f\, \Delta v\, dS-\int_{St} \overline{T_i}\, v_i\, dS \quad \cdots(1)$$

where $\bar{\sigma}$ is the equivalent stress, $\dot{\bar{\varepsilon}}$ is the equivalent strain-rate, τ_f is the frictional shear stress, Δv is the relative velocity between the workpiece and roll, $\overline{T_i}$ is the front or back tension and v_i is the velocity. The equivalent stress is expressed by :

$$\bar{\sigma} = f \cdot \bar{\varepsilon}^{n} \cdot \dot{\bar{\varepsilon}}^{m} \qquad \cdots(2)$$

where f is the function depending on material properties and $\bar{\varepsilon}$ is the equivalent strain. In this method, not only the work-hardening characteristic but also the strain-rate sensitivity in the flow stress are taken into consideration.

2.2 *Treatment of steady-state deformation*

Steady-state deformation in edge-rolling of slabs is simulated by the steady-state scheme in the rigid-plastic finite element method. In this scheme, the deforming shape and flow stress distribution obtained by tracing stream lines are iteratively renewed until attaining to a steady state. The computing time for the steady-state scheme is much shorter than that for the non-steady-state scheme, because transient deformation before attaining to a steady state is not calculated.

To calculate multi-pass edge-rolling processes, the shape of the rolled cross-section is transferred to the next pass. The shape of the cross-section after rolling is equivalent to that before rolling in the next pass.

3 SIMULATION OF EDGE-ROLLING

3.1 *Computational conditions*

Three-dimensional deformation in multi-pass edge-rolling of slabs with grooved rolls shown in Figure 1 is simulated by the rigid-plastic finite element method. The computational conditions used for the simulation are given in Table 1. The work-hardening exponent, the strain-rate sensitivity exponent, the groove geometry, the coefficient of friction and the temperature distribution are changed to examine the effects on the dog-bone shape of the rolled cross-section. The underlined values in Table 1 are chosen as the standard ones in the comparison. It is assumed that the work-hardening undergone in previous passes is eliminated by recovery and recrystallization between passes, i.e. there is no accumulation of the work-hardening for the previous passes. In actual edge-rolling processes, the temperature is sufficiently high to cause recovery and recrystallization.

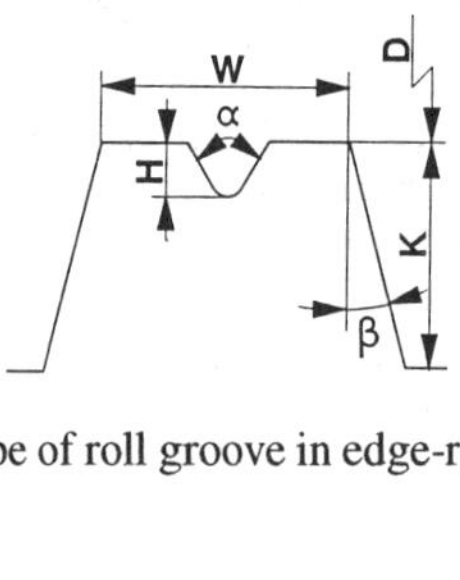

Figure 1. Shape of roll groove in edge-rolling of slab.

Table 1. Rolling conditions used for simulation.

Dimensions of slab	Width /mm	1500
	Thickness /mm	300
Dimensions of grooved roll	Roll diameter *D*/mm	888
	Length of groove bottom *W*/mm	<u>300</u>,450,1000
	Groove depth *K*/mm	350
	Wedge height *H*/mm	90
	Wedge angle α / °	60
	Side wall angle β / °	25
Reduction in width per pass /mm		60
Number of passes		10
Width after final pass /mm		1020
Flow Stress <u>(Hot steel)</u> $\bar{\sigma} = f(T,C) \cdot \bar{\varepsilon}^{n} \cdot \dot{\bar{\varepsilon}}^{m}$ (N/mm²) (Misaka(1967)) where $n=\underline{0.21} \pm 0.10$, $m=\underline{0.13} \pm 0.10$ $f(T,C)=9.81 \cdot \exp\{0.126-1.75C+0.594C^2+(2851+2968C-1120C^2)/T\}$ T:absolute temperature(K), C:carbon content(=0.15%) (Lead) $\bar{\sigma}=30.5 \cdot \bar{\varepsilon}^{0.30} \cdot \dot{\bar{\varepsilon}}^{0.07}$ (N/mm²) (Nakajima(1972)) (Plasticine) $\bar{\sigma}=0.259 \cdot \dot{\bar{\varepsilon}}^{0.19}$ (N/mm²) (Nakajima(1972))		
Coefficient of friction μ		0.25, <u>0.35</u>, 0.45
Temperature distribution of steel /℃		<u>1200 uniform</u>, nonuniform
Rotating speed of roll /rpm		10

(The underlined values are standard)

3.2 *Comparison with experimental results*

The three-dimensional element mesh after the convergence in the steady-state scheme for the first pass of the hot steel slab is shown in Figure 2. The slab is divided into isoparametric hexahedral elements with eight nodes to simulate complex three-dimensional

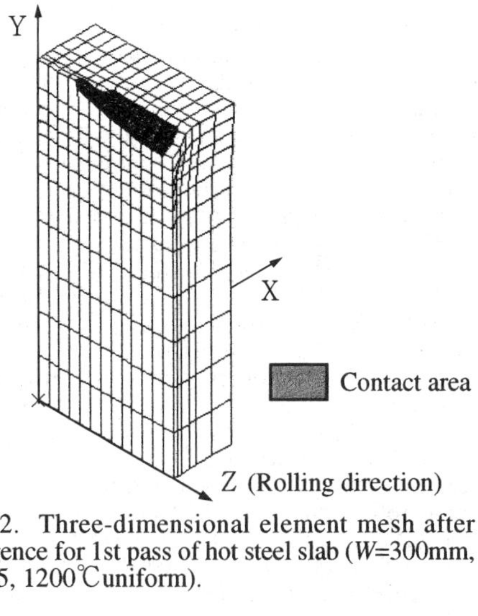

Figure 2. Three-dimensional element mesh after convergence for 1st pass of hot steel slab (*W*=300mm, μ =0.35, 1200℃uniform).

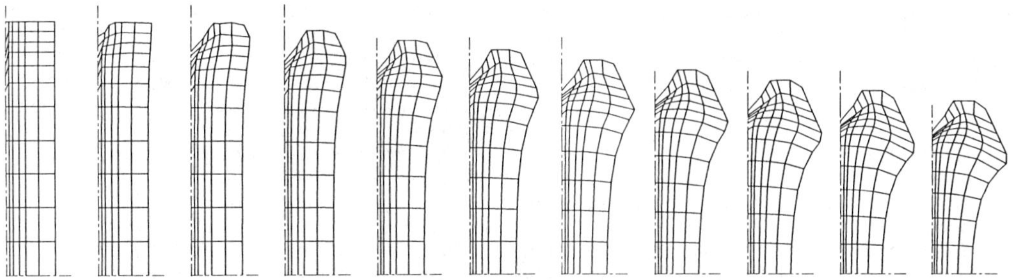

(a) Initial (b) 1st (c) 2nd (d) 3rd (e) 4th (f) 5th (g) 6th (h) 7th (i) 8th (j) 9th (k) 10th

Figure 3. Meshes calculated in each pass of edge-rolling of hot steel slab (W=300mm, μ =0.35, 1200℃ uniform).

deformation in the roll gap. Only a quarter of the workpiece is calculated because of the symmetry of deformation. The area in contact with the roll exhibits a complex profile.

The calculated mesh of the rolled cross-section in each pass for the hot steel slab is illustrated in Figure 3. In the first and second passes, only the center of the slab edge is indented with the wedge of the roll groove and the width of the slab is not reduced. Due to the increase in thickness near the edge of the width, the rolled cross-section exhibits a dog-bone shape. The dog-bone shape becomes remarkable as the pass number increases. The dog-bone phenomenon in edge-rolling is suitable for forming the slab into an H-shape.

The calculated cross-sectional shape after 6 passes for the hot steel slab is compared with the experimental one in Figure 4. The thickness near the edge of the width is increased by edge-rolling, whereas the thickness near the center is slightly decreased. The calculated cross-sectional shape is in good agreement with the experimental one.

A comparison between the calculated and experimental material flows in the cross-section after 4 passes for the plasticine slab is shown in Figure 5. Although

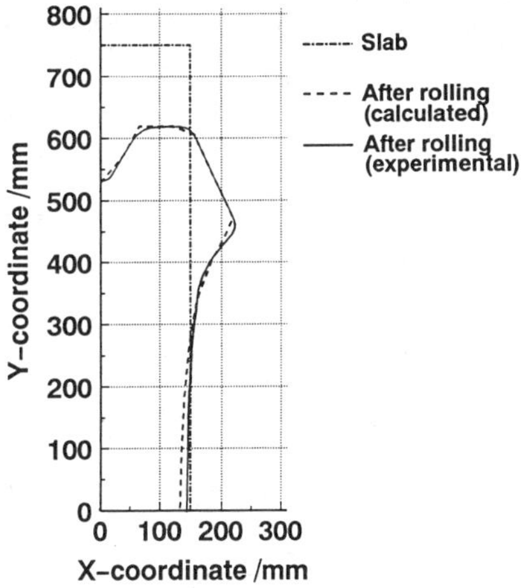

Figure 4. Comparison between cross-sectional shapes obtained by calculation and experiment after 6 passes for hot steel slab (W=300mm, μ =0.35, 1200℃ uniform).

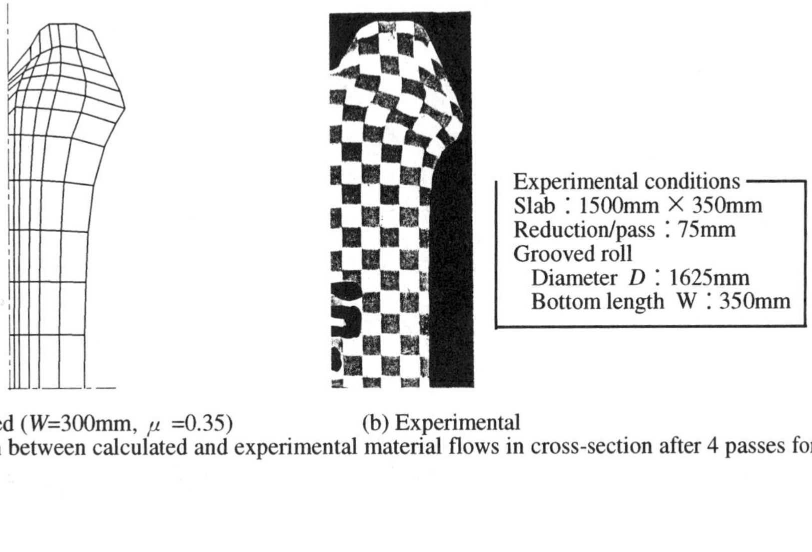

(a) Calculated (W=300mm, μ =0.35) (b) Experimental

Figure 5. Comparison between calculated and experimental material flows in cross-section after 4 passes for plasticine slab.

rolling conditions for the calculation and experiment are a little different, both results are almost similar. Large shear deformation is caused in the region near the maximum thickness.

3.3 *Effect of flow stress*

The effect of the work-hardening exponent n on the dog-bone shape of the rolled cross-section in edge-rolling is given in Figure 6. The growth of the flange is dependent on the work-hardening exponent. The dog-bone shape becomes remarkable as the work-hardening exponent decreases. The concentration of plastic deformation is relaxed as the work-hardening exponent increases. The equivalent strain distributions after 10 passes for n=0.11 and 0.31 are shown in Figure 7. The equivalent strain is induced by only the final pass because of recrystallization between passes. The maximum strain for n=0.31 is smaller than that for n=0.11.

The effect of the strain-rate sensitivity exponent m on the dog-bone shape of the rolled cross-section after 10 passes is shown in Figure 8. This tendency is similar to that for the work-hardening exponent.

The work-hardening and the strain-rate sensitivity exponents differ depending on the materials as shown in Table 1. The cross-sectional shapes after 10 passes calculated for the hot steel, lead and plasticine slabs are compared in Figure 9. The fill into the roll groove for the plasticine slab is larger than those for the hot steel and lead slabs. This is due to the small value of the work-hardening exponent in the plasticine slab. Since the cross-sectional shapes for the hot steel and lead slabs exhibit good agreement, the lead is suitable for a model material for hot steel rolling.

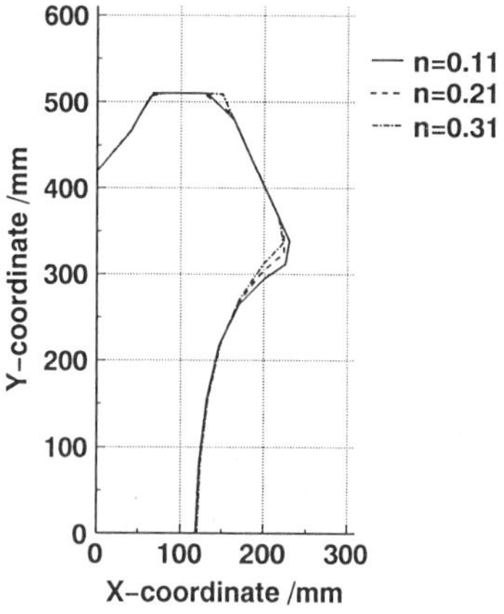

Figure 6. Effect of work-hardening exponent on dog-bone shape after 10 passes (m=0.13, W=300mm, μ =0.35, 1200℃ uniform).

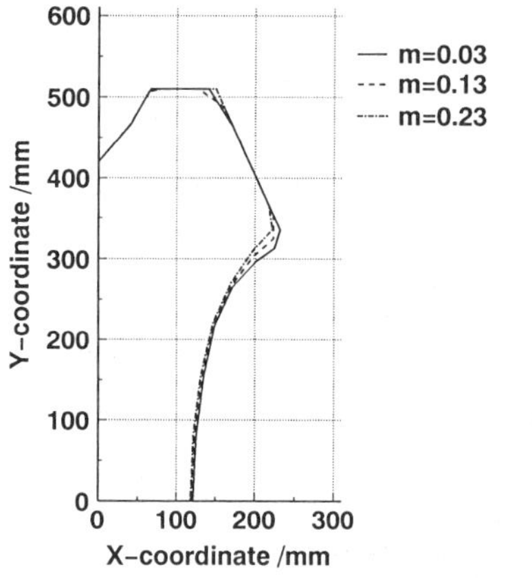

Figure 8. Effect of strain-rate sensitivity exponent on dog-bone shape after 10 passes (n=0.21, W=300mm, μ =0.35, 1200℃ uniform).

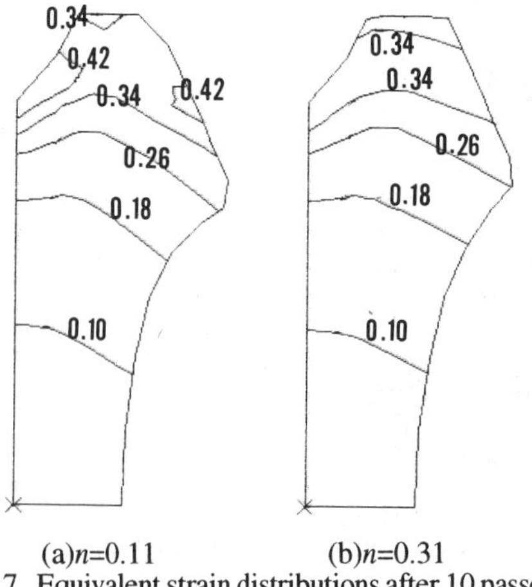

(a)n=0.11 (b)n=0.31

Figure 7. Equivalent strain distributions after 10 passes (m=0.13, W=300mm, μ =0.35, 1200℃ uniform).

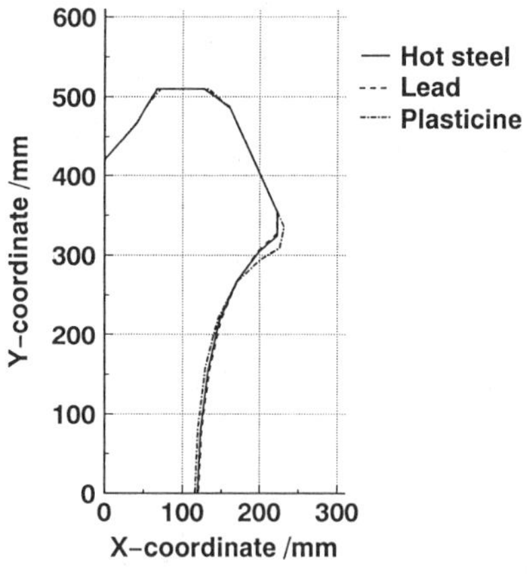

Figure 9. Cross-sectional shapes after 10 passes for hot steel, lead and plasticine (W=300mm, μ =0.35).

3.4 *Effect of geometry of roll groove*

Since the dog-bone shape is greatly influenced by the geometry of the roll groove, the length of the groove bottom is changed in the simulation. Figure 10 shows the effect of the length W of the groove bottom on the dog-bone shape. In the case of W=1000mm, the slab is not in contact with the side walls of the roll grooves, i.e. there is no obstruction to the growth of the flange. As the length of the groove bottom increases, the flange of the cross-section grows. In addition, the thickness near the center of the rolled slab increases relatively.

The variations of the cross-sectional area of a flange with the pass number for different lengths of the groove bottom are given in Figure 11. The cross-sectional area of the flange is defined in reference to the original thickness 300mm of the slab. The increasing rate of the area becomes small with the increase of passes. This tendency becomes more evident as the length of the groove bottom decreases. The cross-sectional area of the flange also becomes smaller. It is found that the H-shape of the cross-section can be controlled by appropriately changing the groove geometry and the reduction in width.

3.5 *Effects of coefficient of friction and temperature distribution*

The effect of the coefficient of friction μ on the dog-bone shape is shown in Figure 12. The dog-bone shape becomes slightly larger as the coefficient of friction increases.

The effect of the temperature distribution of the cross-section on the dog-bone shape is illustrated in Figure 13. The result for the uniform distribution of

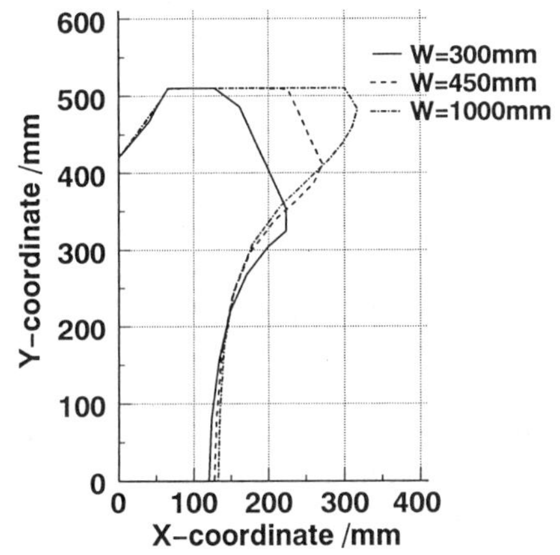

Figure 10. Effect of length of groove bottom on dog-bone shape after 10 passes (hot steel, μ =0.35, 1200℃ uniform).

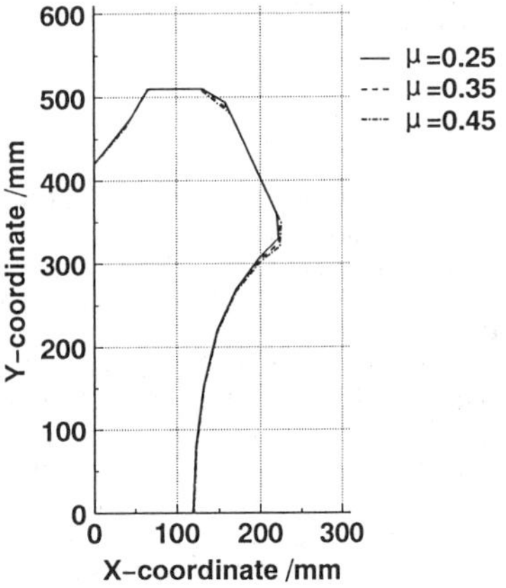

Figure 12. Effect of coefficient of friction on dog-bone shape after 10 passes (W=300mm, hot steel, 1200℃ uniform).

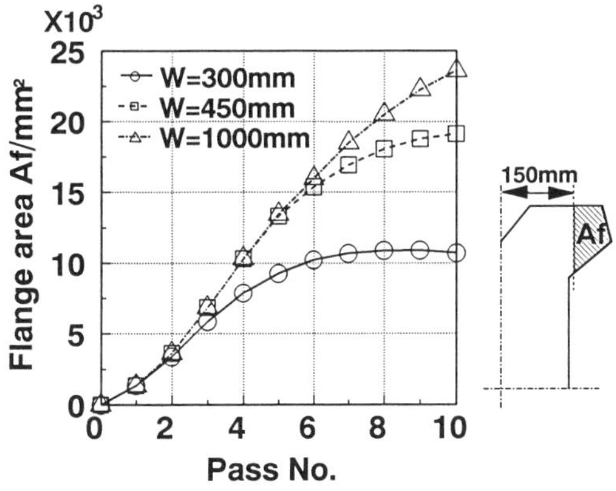

Figure 11. Variations of cross-sectional area of flange with pass number for different lengths of groove bottom.

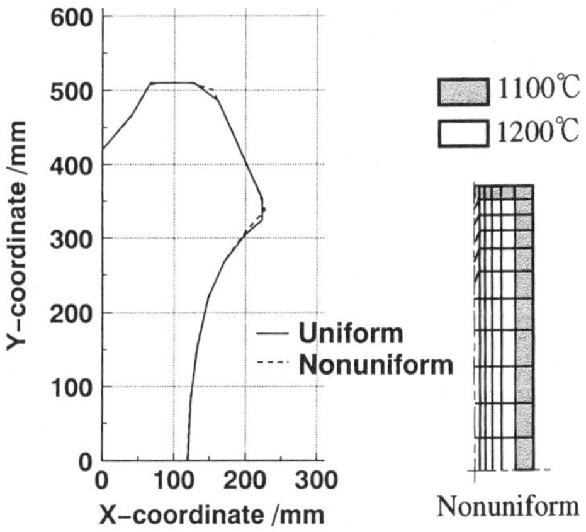

Figure 13. Effect of temperature distribution on dog-bone shape after 10 passes (W=300mm, hot steel, μ =0.35).

temperature is compared with that for nonuniform distribution due to the cooling in Figure 13. The difference of the dog-bone shapes between the two temperature distributions is relatively small.

4 CONCLUSIONS

The deformation behavior in edge-rolling of slabs for producing preforms of H-beams was investigated by the use of the three-dimensional rigid-plastic finite element method. As a result, the following findings were obtained.
(1) The dog-bone shapes are accurately calculated by the present simulator.
(2) The dog-bone shape becomes remarkable as the work-hardening and the strain-rate sensitivity exponents decrease and the coefficient of friction increases.
(3) The H-shape of the cross-section can be controlled by appropriately changing the groove geometry and the reduction in width.

REFERENCES

Bertrand, C., C. David, J.L. Chenot & P. Buessler 1986. Stresses calculation in finite element analysis of three-dimensional hot shape rolling. *NUMIFORM'86* (ed. K. Mattiasson et al.): 207-212. Balkema.

Hayashi, S., S. Ida, T. Fujimoto, K. Yamada, S. Hamauzu, M. Toda & T. Miki 1992. Analyses of shape rolling and cold forging of bars and wire rods by rigid-plastic finite element method. *Nippon Steel Technical Report*. 53: 67-74.

Iguchi, T., H. Hayashi & I. Yarita 1993. Stress analysis of H-beam universal rolling by rigid-plastic FEM. *1st Int. Conf. on Modelling of Metal Rolling Processes*: 707-719.

Misaka, Y. & T. Yoshimoto 1967. Formularization of mean resistance to deformation of plain carbon steels at elevated temperature. *J. Japan Soc. Tech. Plasticity*. *8*: 414-422. In Japanese.

Mori, K. & K. Osakada 1982. Simulation of three dimensional rolling by the rigid-plastic finite element method. *Numerical Methods in Industrial Forming Processes* (ed. J.F.T. Pittman et al.): 747-756. Pineridge Press.

Mori, K. & K. Osakada 1984. Simulation of three-dimensional deformation in rolling by the finite-element method. *Int. J. Mech. Sci.* 26: 515-524.

Mori, K., K. Osakada, H. Nikaido, T. Naoi & Y. Aburatani 1987. Finite element simulation of non-steady state deformation in rolling of slab and plate. *4th Int. Steel Rolling Conf.* *2*: F.6.1-F.6.8.

Mori, K. & K. Osakada 1989. Finite element simulation of three-dimensional deformation in shape rolling. *NUMIFORM'89* (ed. E.G. Thompson et al.): 337-342. Balkema.

Nakajima, K. & K. Watanabe 1972. Study of characteristics on caliber rolling in model rolling. *J. Japan Soc. Tech. Plasticity*. 13: 751-760. In Japanese.

Osakada, K., J. Nakano & K. Mori 1982. Finite element method for rigid-plastic analysis of metal forming - formulation for finite deformation. *Int. J. Mech. Sci.* 24: 459-468.

Park, J.J. & S.I. Oh 1990. Application of three dimensional finite element analysis to shape rolling processes. *Trans. ASME, J. Eng. Ind.* 112: 36-46.

Simulation of Materials Processing: Theory, Methods and Applications, Shen & Dawson (eds)
© 1995 Balkema, Rotterdam. ISBN 90 5410 553 4

A slab-FEM simulation of the material flow in cold rolling of continuously cast thin strips

Jin Hou, Ulf Ståhlberg & Hans Keife
Materials Forming, Department of Materials Processing, Royal Institute of Technology, Stockholm, Sweden

ABSTRACT: Continuous casting of thin strips is a high potential process in the future steel production. One common problem today is that the thickness variations of the cast strips are too high and difficult to control by casting operations. It was found in measurements that one kind of thickness unevenness is characterized by longitudinal ridges and valleys on the upper surface of the strip. During cold rolling of such a strip, the local material flow in width direction, from a ridge to its neighbouring valleys, is severe. This has been confirmed by experiments. In the present work a theoretical model based on slab-FEM (Slab-Finite Element Method) is proposed for analyzing the local 3D material flow. The principle of the model and its formulation is presented. A tailor-made computer program for simulating the material flow from the ridges to the valleys has been developed. Theoretical results are in satisfactory agreement with rolling experiments. It was concluded that this kind of imperfection, under certain conditions, is possible to eliminate by cold rolling.

1 INTRODUCTION

Direct casting of thin steel strips is a high-lighted research project. It has high potential benefits such as energy saving, reduced production cost, low investment cost and increased production yield.

A number of strip casting methods have been developed or are under development today. One of them is by Jacobsson, Hollinger and Fredriksson[1] at Metal Casting, Department of Materials Processing, Royal Institute of Technology (KTH). A schematic picture of the KTH-caster is shown in Fig.1.

Thickness variation of the cast strip is regarded as a common problems related to most casting methods[2-4] and it is difficult to control during strip casting operations. One route to solve this problem is to introduce rolling as the consequent step.

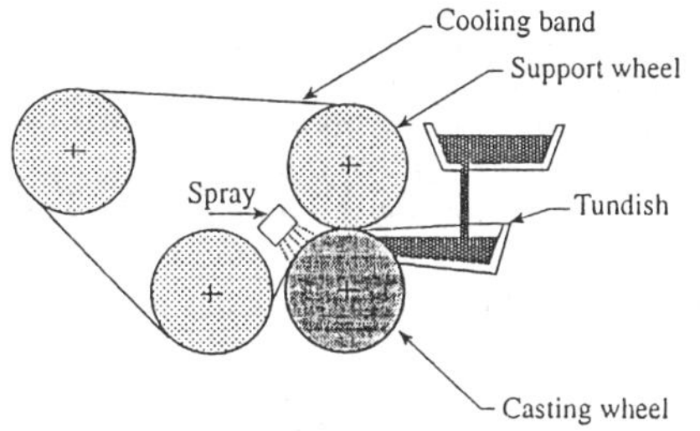

Principle drawing of the caster at KTH after reconstruction.

Fig.1 The strip caster developed at KTH[1]

The main aim of present work is to study the possibility to roll thin strips in as-cast conditions and to improve the thickness tolerance. The thicknesses of strips were below 3 mm and obtained from the KTH-caster (Fig.1). Based on observations and measurements[5] of the cast strips, it has been found that a common thickness unevenness appears as longitudinal ridges and valleys on the upper surface. Rolling of such kind of strip involves complicated 3D material flow. In order to carry out a theoretical analysis, the ridges and valleys were simplified to regular shapes and a slab-FEM model was proposed. A tailor-made computer program has been developed based on the model. The results of the numerical simulations are in satisfactory agreement with experiments. It was found that the longitudinal ridges and valleys can be eliminated by rolling.

2 THE SLAB-FEM MODEL

In earlier works, two kinds of slab-FEM formulations have been developed. One is proposed by Kiuchi and Yanagimoto[6] and the other one by Kim and Altan[7]. In Kim's model, the 2D FEM is utilized under the "generalized" plane-strain conditions. Coupling with slab method makes the simulation of 3D deformation in shape rolling more rapid

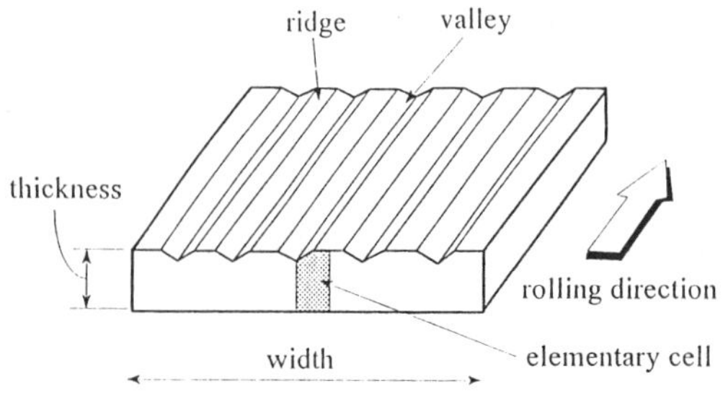

Fig.2 Ridges and valleys on the upper surface of a strip

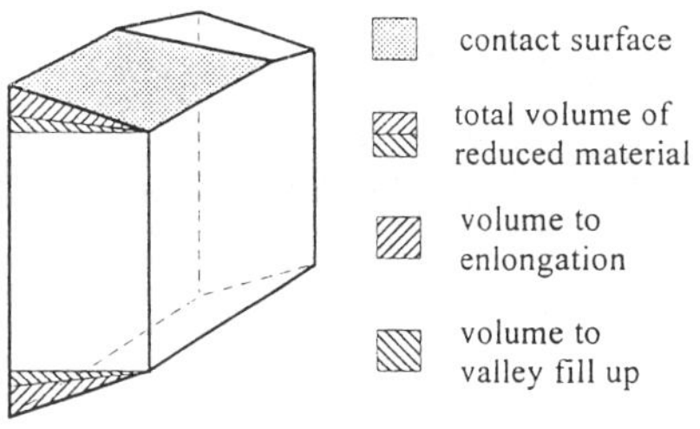

Fig.3 The division of reduced material in rolling and width directions

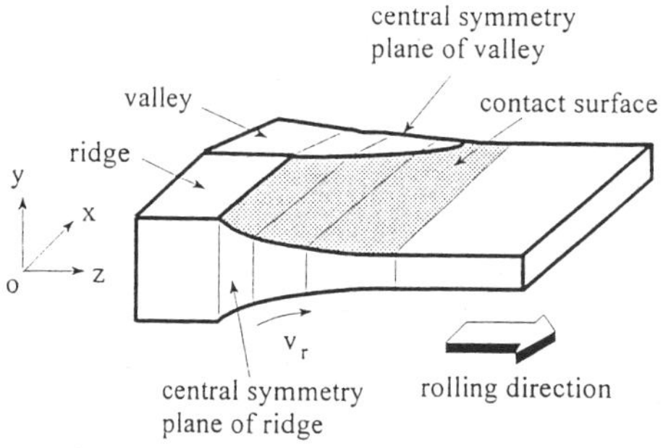

Fig.4 An elementary cell for analysis

compared to 3D FEM. The computer program TASKS has been developed based on their slab-FEM model. TASKS has been successfully used for different shape rolling and extrusion processes[8-11].

When a strip with uneven surface is rolled, material may flow from the ridges to the valleys during rolling. A valley can be closed within the deformation zone. For such a case, it is difficult to apply the same slab-FEM and same convergence criterion as that used in TASKS. In this work, a different type of slab-FEM has been proposed.

2.1 Basic assumptions

The following basic assumptions are introduced in order to set up the model and to couple the slab method with FEM.

a) The ridges and valleys on the upper surface of the strip are simplified to regular shapes, Fig.2. Considering one cross-section, all the valleys and ridges have the same geometry.

b) For any slab element, the volume of material from thickness reduction can always be divided into two portions. One contributes to the elongation (rolling direction) of the slab and the other one contributes to the fill up of the valley (width direction). The deformations related to the two portions are independent, as shown in Fig.3. This is based on the same kind of idea that was proposed by Ståhlberg[12].

c) The material flow in the width direction is restricted to take place between the longitudinal vertical symmetry planes of the ridges and valleys. No material flows across the symmetry plane of any ridge or valley. Thus the model only needs to treat one elementary cell which includes half a ridge and half a valley, Fig.4. The width of the strip will not change during the deformation.

d) The elongation is uniform over any vertical cross-section. The cross-section will remain flat and its orientation will be unchanged during rolling, Fig.4. Based on this assumption, "generalized" plane-strain conditions are assumed for the x-y plane (rolling direction) and the y-z plane (width direction).

e) The shear stress between rolls and strip is constant, $\tau = mk$.

f) Flattening of the rolls is taken into account and the arc of contact is assumed to keep a circular shape during rolling.

2.2 2D FEM

In rolling of strips with ordinary reduction in thickness, the plastic deformation is usually high and consequently the elastic strains negligible. It is acceptable to idealize the strip to be a rigid-plastic material. In this work, the rigid-plastic FEM[13] (flow formulation) is utilized. The basic principle and equations of FEM are roughly described below.

Among the kinetically admissible velocities u_i, the actual solution minimizes the following functional.

$$\Pi = \int_V \sigma_e \dot{\varepsilon}_e dV - \int_{S_F} F_i u_i dS \tag{1}$$

The "kinetically admissible velocities" means that u_i satisfy the velocity boundary conditions and incompressibility. The boundary conditions are:

$$v_i = v_i^* \quad on \quad S_u$$
$$F_i = F_i^* \quad on \quad S_F \tag{2}$$

Equations (1) and (2) define a boundary value problem. The actual solution should minimize the functional in equation (1), namely, the first-order variation of the functional vanishes.

$$\delta\Pi = \int_V \sigma_e \delta\dot{\varepsilon}_e dV - \int_{S_F} F_i \delta u_i dS = 0 \tag{3}$$

A convenient way to remove the incompressibility constraint is to use its penalized form as

$$\delta\Pi = \int_V \sigma_e \delta\dot{\varepsilon}_e dV + K_p \int_V \dot{\varepsilon}_v \delta\dot{\varepsilon}_v dV - \int_{S_F} F_i \delta u_i dS = 0 \tag{4}$$

The mean stress is calculated by

$$\sigma_m = \frac{1}{2} K_p \dot{\varepsilon}_v \tag{5}$$

where, K_p can be interpreted as a constant similar to the bulk modulus.

Four-node linear quadrilateral elements[13] are utilized for descretization. The discrete form of equation (4) is obtained for a cross-section of an elementary cell with area A, and given in the equation (6).

$$\int_A \frac{2}{3}\frac{\bar{\sigma}}{\dot{\varepsilon}_e}[B]^T[D][B]\{v\}dA + K_p \int_A [B]^T\{c\}\{c\}^T[B]\{v\}dA - \int_{Lf} \{N\}^T\{f\}dL = 0 \tag{6}$$

2.3 Slab method

When a flat strip is rolled, equation (7) can be written by considering force balance on a slab element in the horizontal (z) direction and making use of the yield criterion,

$$d\sigma_z^0 = \frac{1}{h}(\sigma_e dh \mp 2\tau dz) \tag{7}$$

where (-) sign stands for the backward slip zone and (+) sign for the forward slip zone. The neutral plane is determined numerically by calculating σ_z^0 from both inlet and outlet.

In the present work, the upper surface of the strip is not flat and the contact area changes along the deformation zone (Fig.4). A slab element is taken from the deformation zone and shown in Fig.5.

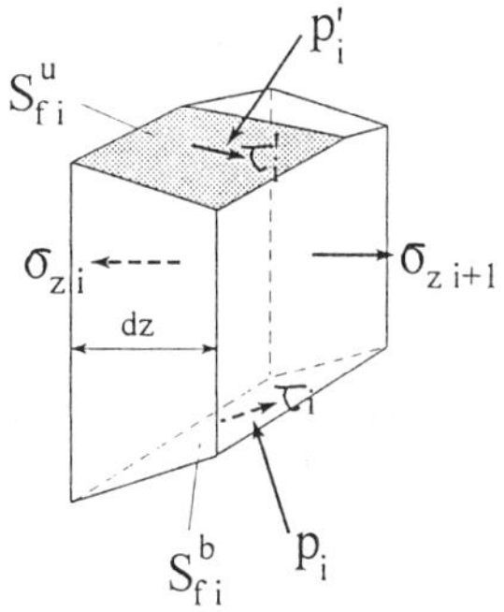

Fig.5 A slab element from deformed ridge and valley

It clear from the figure that ordinary slab method is difficult to apply for such a case. Another difficulty is that the upper contact area of the element depends on the FEM calculation. Iterations are needed between slab method and FEM to determine stresses in the rolling direction. In order to solve the problem, it is assumed that z_{np} is the same as that calculated by σ_z^0, and the following approximations have been introduced.

For $z < z_{np}$, the following equations are used.

$$d\sigma_{zi} = \frac{S_{fi}^u + S_{fi}^b}{2S_{fi}^b} d\sigma_{zi}^0$$
$$\sigma_{zi+1} = \sigma_{zi} + d\sigma_{zi} \tag{8}$$

Usually, the neutral plane is located inside an element. σ_z at the position of neutral plane, $z=z_{np}$, is calculated by extrapolation using the stress on the incoming side.

For $z > z_{np}$, σ_z is approximated by the equation (9).

$$\sigma_{zi} = \sigma_{zi}^0 - \frac{z_L - z_i}{z_L - z_{np}}(\sigma_{znp}^0 - \sigma_{znp}) \tag{9}$$

As having been shown in Fig.3, a volume distribution coefficient C_z is introduced for each slab-element. When $C_z=1$, all the materials reduced in thickness go to elongation, which means a plane-strain deformation in the y-z plane. When $C_z=0$, all the reduced material fill up the valley and the current slab will not elongate, this means a plane-strain deformation in the x-y plane.

For a cross-section i, the strain rate in rolling direction can be calculated by equation (10), where v_{ti} is approximated by the vertical component of the roll velocity and given by (11).

$$\dot{\varepsilon}_{zi} = \frac{v_{ti}(b_{ui}+b_{bi})C_{zi}}{A_i} \quad (10)$$

$$v_{ti} = v_r \frac{z_L - z_i}{R} \quad (11)$$

For the 2D FEM calculation, the vertical movement of the roll at the cross-section i is modified according to C_{zi}

$$v_{ti}^F = v_{ti}(1 - C_{zi}) \quad (12)$$

Using the strain rate given in (10), the stress in the rolling direction can be calculated as follows,

$$\sigma_{zi}^{FEM} = \frac{2}{3} \frac{\int_{A_i} \sigma_{ei} dA}{\int_{A_i} \dot{\varepsilon}_{ei} dA} \dot{\varepsilon}_{zi} + \frac{\int_{A_i} \sigma_{mi} dA}{A_i} \quad (13)$$

where the effective stress and effective strain rate are calculated from 2D FEM. The second term is the mean value of the hydrostatic pressure over the whole area of cross-section i. The convergence criterion for each cross-section is

$$|\sigma_{zi}^{FEM} - \sigma_{zi}| < \varepsilon_0 \quad (14)$$

A detailed formulation of the slab-FEM model is described in [14].

3 THE COMPUTER PROGRAM

The non-linear equation (6) can be solved either by a direct-iterative method or by the Newton-Raphson method. A detailed description of the numerical techniques can be found in [13].

For the element number i, the coefficient C_{zi} is adjusted in order to satisfy the convergence requirement after each iteration.

$$C_{zi}^{j+1} = C_{zi}^{j} - \beta_s \frac{dS^j}{dS^j - dS^{j-1}} (C_z^j - C_z^{j-1}) \quad (15)$$

where the deceleration coefficient β_s is between 0.3 to 0.8, dS^j is calculated by equation (16).

$$dS^j = \sigma_{zi}^j - \sigma_{zi}^{FEMj} \quad (16)$$

The computer program is written in Fortran. The 2D FEM part is developed based on the sample program provided by Kabayashi et al. in [13]. The main flow chart of the program is shown in Fig.6.

A preliminary version of the post processor has been developed for PC computers. The graphic presentation of mesh distortion, vectors of material flow and contact pressures are incorporated.

The data related to the experiments[5] were input

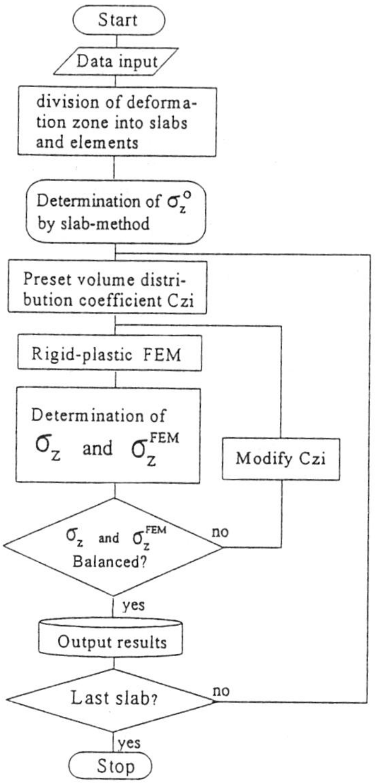

Fig.6 Flow chart of the computer program

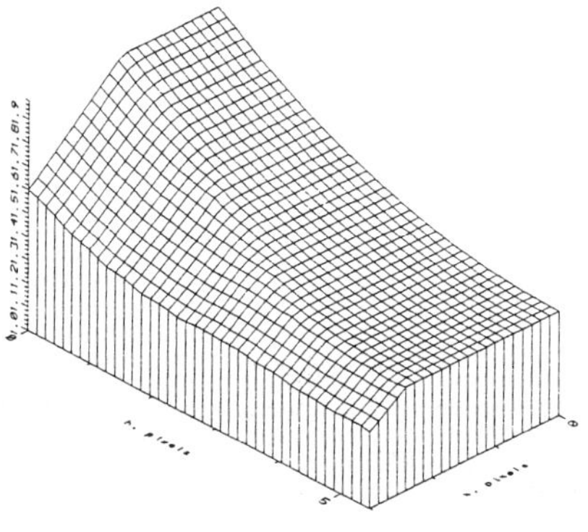

Fig.7 Deformation of a ridge and valley during rolling

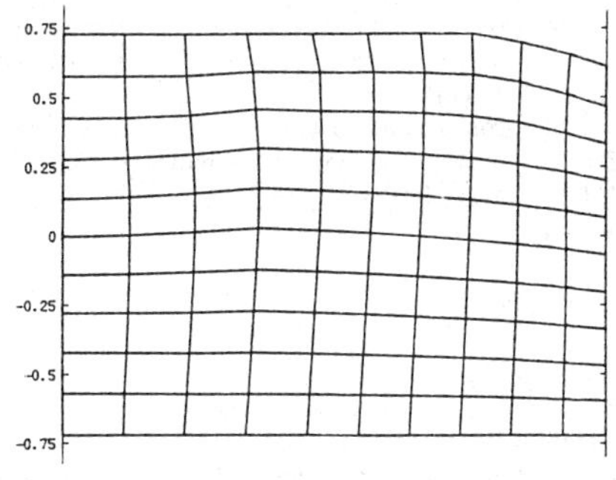

Fig.8 Distorted FE mesh in a cross-section

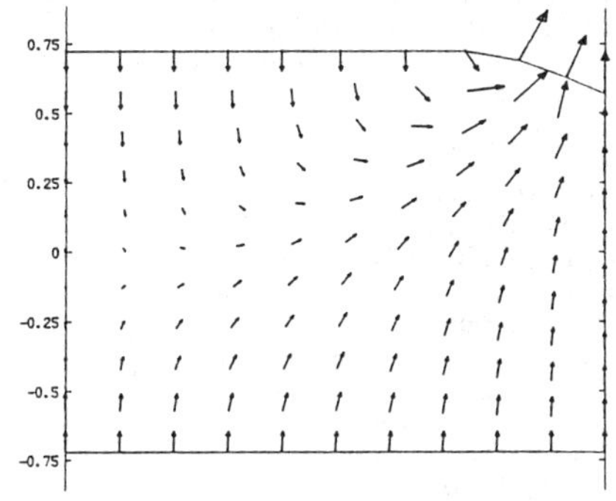

Fig.9 Calculated local material flow from a ridge to its neighbouring valley

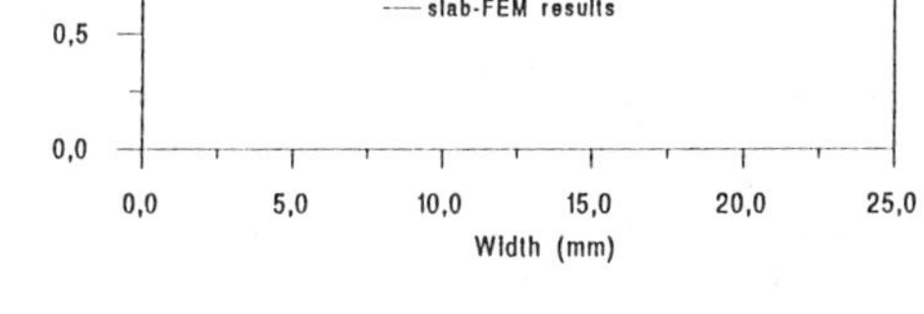

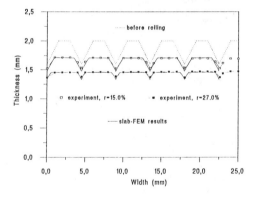

Fig.10 Comparison of slab-FEM results and measured thickness profiles of aluminum sheet after rolling

to the computer and simulated by the program described above. The time of simulation is dependent on the initial guess of C_{zi} and on the fill up of valleys. For typical examples under the following conditions

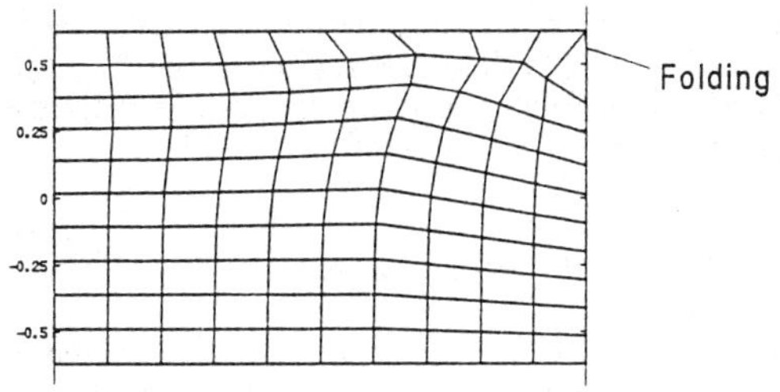

Fig.11 FE mesh after rolling showing the folding of side surfaces of a valley

- 9 slab elements (10 cross-sections) in the deformation zone in the rolling direction,
- 100 FEM elements on each cross-section,
- without good initial guess of C_{zi},
- valleys filled up at the last slab or not filled up,

the calculation time is about 3.5 hours on a PC (IBM486, 25MHz). Calculation time is much shorter on the modern workstations. One of the above examples is calculated on an IBM workstation (RISC/6000-3BT), the calculation time is only about 4 minutes.

4 RESULTS

4.1 Graphic output

By the post processor (PC version) of the program, some results can be shown graphically. A surface view of the deformation of a ridge and valley during rolling is shown in Fig.7.

The mesh distortion and material flow in each cross-section can also be plotted. One example of a distorted mesh after certain degree of deformation is shown in Fig.8. The calculated local material flow from a ridge to its neighbouring valley in the width direction is shown in Fig.9.

4.2 Comparison between theory and experiments

In order to be able to compare the theoretical results with experiments, the calculated results of one ridge and one valley after rolling are plotted repeatedly. One comparison is shown in Fig.10. It is clear that the slab-FEM simulations and the experiments are in satisfactory agreement.

The agreement between theory and experiment is good when the reduction per pass is not high (<20%). As the reduction per pass increases, the agreement becomes poor.

The folding of side surfaces of valleys has been observed in experiment[5]. In slab-FEM simulation, this kind of result has also been obtained.The mesh distortion after rolling, Fig.11, can be used to determine the folding. This problem occurs when the angle of side surface α is larger than 45 degrees.

5 CONCLUSION

A simplified slab-FEM model has been proposed to simulate the 3D deformation of strips during flat rolling. A computer program has been developed based on the formulation. The theoretical results are in satisfactory agreement with experiments.

6 ACKNOWLEDGEMENT

The authors are indebted to NUTEK (Swedish Board for Industrial and Technical Development) for the financial support of the project.

REFERENCES

[1] N. Jacobsson, B. Hollinger and H. Fredriksson, On the development of a thin stripcaster, Scand. J. Metallurgy, 22(1993), 75-82

[2] K. Shibuya and M. Ozawa, Strip casting techniques for steel, ISIJ International, 31(1991), No.7, 661-668

[3] J. Herbertson, R.I. Olivares and W. Blejde, Requirements of refining processes when casting thin strips, Steel Times, Nov. 1992, 520-523

[4] Å. Sjöström, Bandgjutning, Valsverksdagarna, Nov. 15-16, Sigtuna, Sweden

[5] J. Hou and U. Ståhlberg, Cold rolling of continuously cast thin strips - Part 1: experimental study, ISSN 1104-7119, TRITA-MB-NUT-0001

[6] M. Kiuchi and J. Yanagimoto, Computer aided simulation of shape rolling processes, Proceedings of 16th NAMRC, 1987, 34-40

[7] N. Kim, S.M. Lee, W. Shin and R. Shivpuri, Simulation of square-to-oval single pass rolling using a computationally effective finite and slab element method, J. Eng. for Ind., 114(1992) 329-335

[8] N. Kim, S. Kobayashi and T. Altan, Three-dimensional analysis and computer simulation of shape rolling by the finite and slab element method, Int. J. Mach. Tools Manufact., 31(1991) 553-563

[9] S.M. Lee, W. Shin and R. Shivpuri, Investigation of two square-to-round multipass rolling sequences by the slab-finite element method, Int. J. Mach. Tools Manufact., 32(1992) 315-327

[10] K. Sawamiphkdi and G.D. Lahoti, Application of slab-finite element method for improvement of rolled bar surface quality, Annals of the CIRP, Vol.43/1 (1994), 219-222

[11] H.W. Shin, D.W. Kim and N Kim, A simplified three-dimensional finite-element analysis of the non-axisymmetric extrusion processes, J. Materials Proc. Tech., 38(1993) 567-587

[12] U. Ståhlberg, Influence of spread and stress on the closure of a central longitudinal hole in the hot rolling of steel, J. Mech. Working Tech., 13(1986) 65-81

[13] S. Kobayashi, S.I. Oh and T. Altan, *Metal forming and the finite-element method*, Oxford University Press, 1989, New York

[14] J. Hou, Cold rolling of continuously cast thin strips - Part 2: S-FEM modelling and simulation, ISSN 1104-7119, TRITA-MB-NUT-0002

NOTATIONS

A	area of a cross-section
b_b	width of the ith cross-section at contact surface with bottom roll
b_u	width of the ith cross-section at contact surface with upper roll
$[B]$	strain rate matrix
C_z	volume distribution coefficient
$\{c\}$	volumetric strain rate vector
$[D]$	effective strain rate coefficient matrix
F_i	components of external stresses or tractions
F_i^*	prescribed traction
$\{f\}$	nodal force vector
h	thickness at a cross-section
k	yield shear stress
K_p	penalty constant
L_f	contact width of a cross-section
m	friction factor
$\{N\}$	shape function
S_F	surface where traction is prescribed
S_u	surface where velocity is prescribed
u_i	velocity components
V	volume of deformation zone
v_i^*	velocity discontinuity
v_r	tangential velocity of roll
v_t	components of the roll velocity in the vertical direction
$\{v\}$	nodal velocity vector
z_L	roll contact length
z_{np}	position of neutral plane
α	contact angle (Fig.5)
β_s	deceleration coefficients
ε_0	a small positive number to control convergence
$\dot{\varepsilon}_e$	effective strain rate
$\dot{\varepsilon}_v$	volumetric strain rate
$\dot{\varepsilon}_z$	strain rate component in the rolling direction
σ_e	effective stress or yield stress
σ_m	mean stress
σ_z, σ_z^0	stress component in the rolling direction
τ	shear stress
Π	functional

subscript

i	number of cross-section
n	total number of slab element
$n+1$	total number of cross-section

superscript

b	bottom
F	related to FEM calculation
j	number of iteration
u	upper

Simulation of Materials Processing: Theory, Methods and Applications, Shen & Dawson (eds)
© 1995 Balkema, Rotterdam. ISBN 90 5410 553 4

Thermo-plastic finite-element modelling of the rolling of a hot titanium ring

Z.M. Hu, I. Pillinger & P. Hartley
Solid Mechanics & Process Modelling Group, School of Manufacturing and Mechanical Engineering, The University of Birmingham, UK

S. McKenzie & P.J. Spence
Rolls Royce plc, Derby, UK

ABSTRACT: This paper describes finite-element analysis of the rolling of a hot titanium-alloy ring using an elastic-plastic updated Lagrangian approach. A hybrid finite-element mesh model has been developed to deal with the special deformation characteristics of the process, and boundary-surface and friction-layer techniques have been used to model the interaction between the various tools and the ring. As the process is carried out at high temperature, a thermo-mechanical approach has been adopted. The results show that the techniques are capable of successfully modelling the ring rolling process.

1. INTRODUCTION

Ring rolling is a versatile metal forming process for manufacturing seamless annular forgings that are accurately dimensioned and have circumferential grain flow. It usually requires less input material than alternative forging methods and is applicable to production in any quantity, so it is widely used in industry, particularly in the aerospace and automotive sectors. However, compared with other rolling processes, relatively little attention has been paid to the modelling of this process.

The ring rolling mill consists essentially of a freely mounted mandrel and a continuously rotating driving roll (the main roll). Guide rolls are usually employed to maintain the circular shape of the ring and axial rolls are also used to obtain the required height or to reduce geometric defects such as 'fish tailing' of the cross section. A typical schematic picture of a radial-axial rolling mill with guide rolls is shown in Fig. 1.

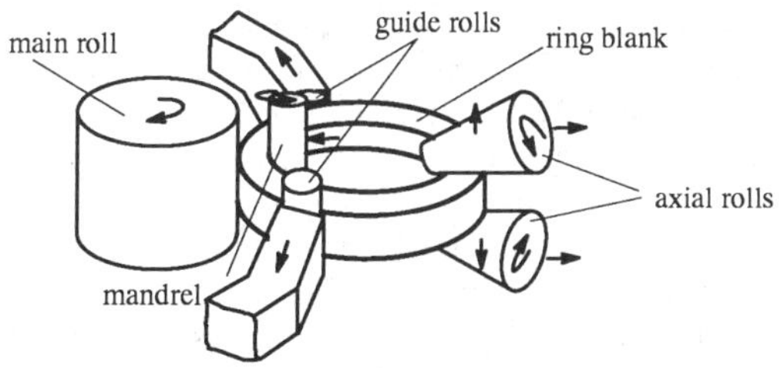

Figure 1. Schematics of ring rolling

Ring rolling differs from other more conventional rolling processes in the following ways:

1. The two main deforming rolls are of different sizes (the main roll is much larger than the mandrel), and only the main roll is driven while the mandrel is idle.
2. Non-steady-state deformation occurs throughout the process.
3. The mandrel moves towards the main roll relatively slowly, resulting in a large number of working rotations or 'passes' through the roll gap.

A series of experimental studies of ring rolling have been carried out by Johnson *et al.* as summarised by Mamalis *et al.* (1976). Recently, a new rolling strategy was investigated by Kluge *et al.* (1994) to obtain homogeneous strain distribution when rolling rings on radial-axial ring rolling mill. The slip-line field (Hawkyard *et al.* 1973), the upper-bound method (Yang *et al.* 1980 & Ryoo *et al.* 1983), the UBET (Hahn *et al.* 1994) and Hill's general method of analysis (Lugora *et al.* 1987) have been used to study the force or material flow during the ring rolling process.

Recently, there have been a number of examples in which the FE method is used to study ring-rolling processes. As the main deforming region is restricted to the vicinity of the roll gap, it is possible to reduce computational time by considering only this part of the ring. Yang *et al.* (1988) described a rigid-plastic analysis for plane-strain ring rolling. Ring rolling between plain cylindrical rolls was studied by Xu *et* al. (1991) who described an analysis of the 3-D flow in the roll gap. Yang *et al.* (1991) used the same technique to study profile ring rolling of a T-shaped section, but the final shape of the ring was difficult to predict owing to the assumption that the process was a steady-state operation. Such part-ring mesh techniques cannot model the effect of those rolls (guide rolls and axial rolls) outside the main deforming area. Hu and Liu (1992) used the Arbitrary Lagrangian Eulerian (ALE) formulation to model the ring rolling process using the whole ring. The guide rolls were also included by means of two concentrated forces acting on the side of the ring.

Kim *et al.* (1990) reduced the computational time by using two mesh systems for a full 3-D analysis of the ring-rolling process. One very finely divided mesh was fixed to the rotating and deforming workpiece and stored information about the workpiece, including geometry and deformation history. Another stationary mesh system, which was finely divided in the main deforming zone and coarse elsewhere, was used for the actual FE calculation. The geometry of the FE mesh was changed periodically to match that of the the fine mesh system. Since the FE mesh contained far fewer elements than the fine mesh, the computation time was reduced considerably. This method has been refined by the present authors (Hu *et al.* 1994) in their analysis of steel ring rolling to reduce the possibility of any error introduced when interpolating data between the two meshes and is used in the three-dimensional analysis of titanium ring rolling presented in this paper.

2. FINITE-ELEMENT FORMULATION AND THE HYBRID MESH MODEL

The rolling of the titanium ring was modelled using the FE program *epfep3*[1]. This uses an updated-Lagrangian approach with a large displacement/finite-strain formulation.

The material is assumed to be isotropic elastic-plastic and the constitutive expression is derived from the Prandtl-Reuss flow rule and von Mises's yield criterion. The thermal properties of the material can also be included to conduct a thermo-plastic simulation. Full details are given by Rowe *et al.* (1991).

The hybrid mesh model (Hu *et al.* 1994) was used for the analysis, which reduced computational time to about an eighth of that required for a uniformly fine mesh while still providing sufficient accuracy in

[1]©The University of Birmingham 1987

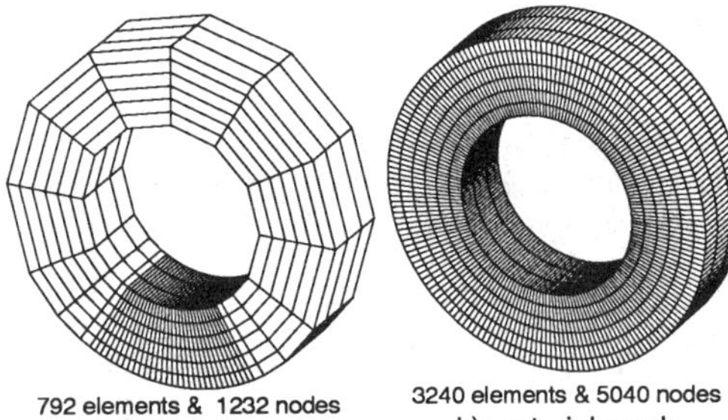

Figure 2. Hybrid FE mesh model

the main deforming region. As shown in Fig. 2, the ring was divided into a very fine mesh – the material mesh. This mesh was, however, not used for the FE computation, but instead was used to store all the information about the geometric shape, stress and strain, temperature etc. The material mesh was related to the material points of the ring and rotated and deformed together with the material. Another mesh – the computational mesh – was fine in the main deforming zone and relatively coarse elsewhere. This was used for the actual FE calculation. It was not spatially fixed as was that used by Kim *et al.* (1990), but underwent incremental deformation together with the material mesh; the fine mesh was re-formed by selecting nodes from the material mesh at the start of each increment of deformation. These were always chosen to position the fine mesh in the main deforming area. Hence both meshes had the same structure in this region.

As all the nodes of the computational mesh also belonged to the material mesh, the information about deformation history could be transferred between the nodes of the two meshes without any need for interpolation and without, therefore, incurring any interpolation error. At the end of each increment of the FE analysis, information was transferred from the computational mesh back to the material mesh in order to update the deformation history. Since both meshes had the same structure in the main deforming area, interpolation of transferred data was only needed outside this region. Moreover, the interpolation was carried using the *changes* in values rather than the values themselves. Because the changes in stress, strain etc. were small outside the main deforming zone, this minimised the interpolation error. This is especially important in ring rolling due to the large number of increments that are needed.

During the rolling analysis, the mesh underwent large deformation, particularly in the outer layer of

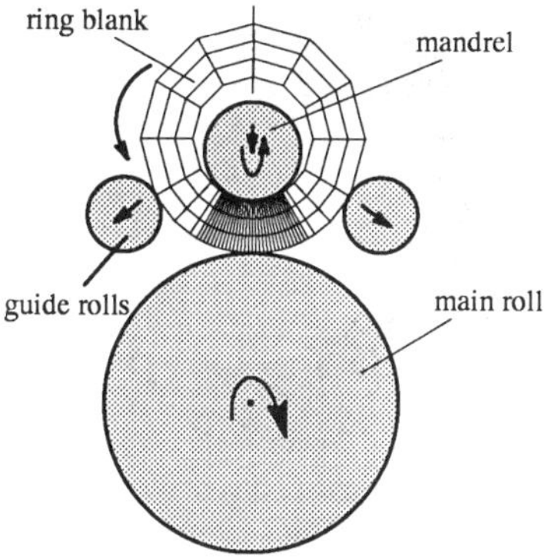

Figure 3. FE mesh and tool surfaces

Table 1
Material and working parameters

TOOLS		
main roll	radius=420 mm	circumferential speed=1000 mm/s
mandrel	radius = 85 mm	translation speed =1.86 mm/s
guide rolls	radius = 85 mm	
RING BLANK		
material: titanium alloy	rolling temperature: 930 °C (1203.16 K)	
elastic modulus E = 55 GPa	Poisson's ratio ν = 0.33	
outer diameter = 372 mm	inner diameter = 200 mm	
height = 178 mm		
flow stress	$\sigma = 120\ \dot{\epsilon}^{0.2} e^{4600\left(\frac{1}{T}-\frac{1}{1273}\right)}$ MPa	
THERMAL PARAMETERS		
material thermal conductivity	$k = 14\ J\,s^{-1}m^{-1}K^{-1}$	
interface thermal conductivity	$k_b = 140\ J\,s^{-1}m^{-2}K^{-1}$	
material thermal capacity	$c_v = 3.52 \times 10^6\ J\,m^{-3}K^{-1}$	
material emissivity	e= 0.6	
material convection coefficient	$h = 17.2\ J\,s^{-1}m^{-2}K^{-1}$	

the ring, and the elements gradually became distorted. The material mesh was therefore regenerated at the start of each revolution of the ring.

3. TOOL MODELLING

As shown in Fig.1, the mandrel is in contact with the inner surface of the ring and the main roll is in contact with the outer surface. The main roll rotates around its stationary axis, while the mandrel is moved, at a specific feed rate, radially against the inner surface of the rotating ring. The rotation of the ring, in turn, drives the idle mandrel. As the mandrel approaches closer to the main roll, the ring is caused to expand to its required size.

Guide rolls are also used to control the roundness of the ring. They are situated around the ring periphery and in contact with the outer surface of the ring. The

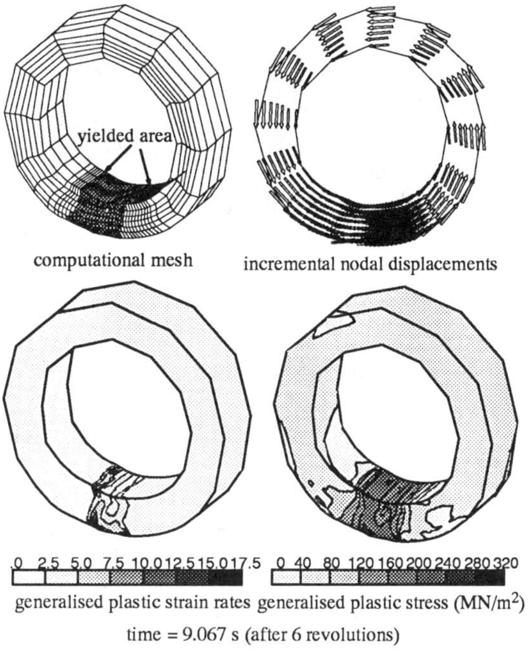

Figure 4. Deformation state of the rolled ring

guide rolls are usually mounted in hydraulically controlled arms and move outwards as the ring expands, ensuring that the ring does not lose its roundness and remains in the correct position relative to the longitudinal axis of the mill. In addition, another set of axial rolls is often used in order to control the height of the ring. Axial rolls were not included in the analysis presented here.

Since the ring contacts the various tools in isolated and very localised regions, the boundary friction and contact conditions need to be modelled very carefully. In the analysis presented in this paper, the 'boundary surface' and 'friction-layer' techniques (Pillinger *et al.* 1988 & Rowe *et al.* 1991) were used to model the interaction between the tools and the ring. These assume that the tool surfaces are rigid and can be represented by a small number of simple primitive surfaces such as planes, cones, spheres etc. Each primitive surface has associated with it an initial position and orientation, an incremental displacement and/or rotation, a friction factor, a temperature and some heat transfer coefficients. In the present case, the main roll, the mandrel and the guide rolls were represented by cylindrical boundary surfaces, as shown in Fig. 3.

In the ring-rolling process, the main roll rotates constantly around its stationary axis. The ring comes into contact with the roll when entering the gap and

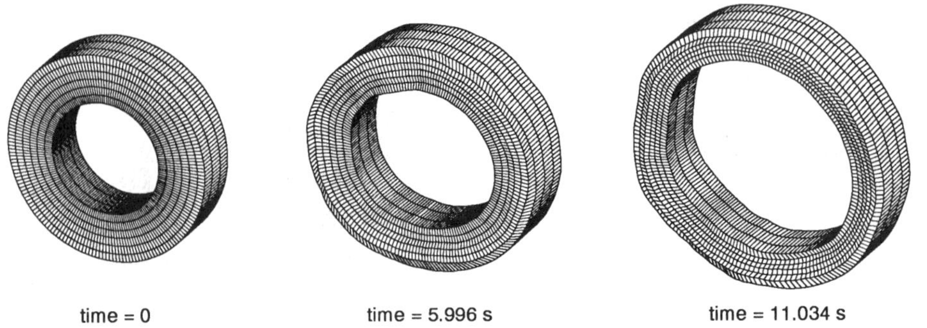

Figure 5. Deformed material mesh

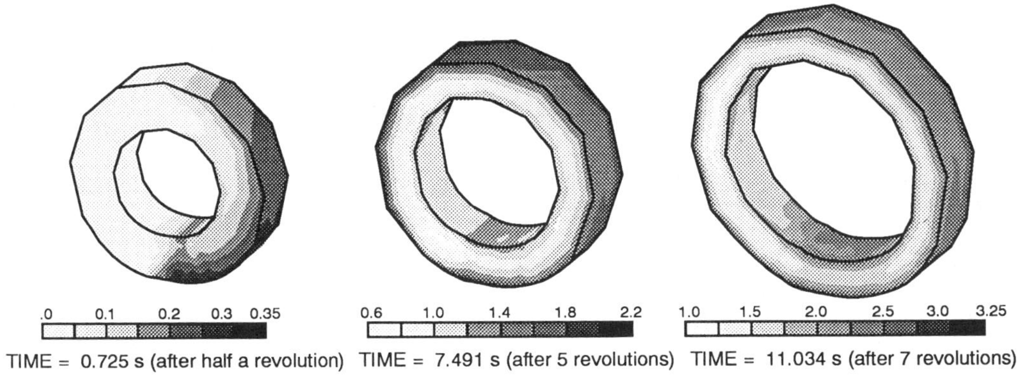

Figure 6. Plastic strain in the rolled ring

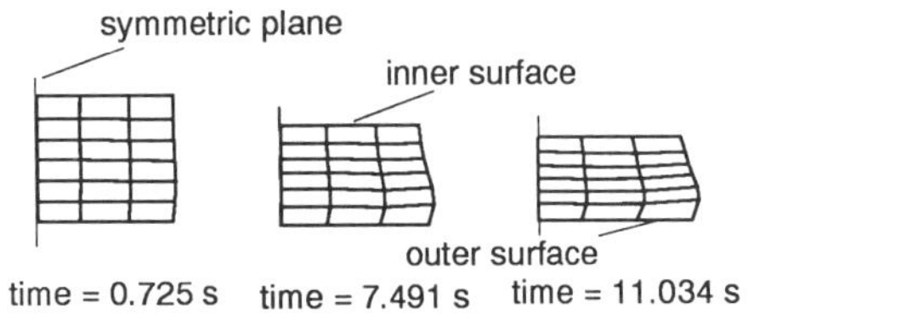

Figure 7. Geometrical changes on the cross section

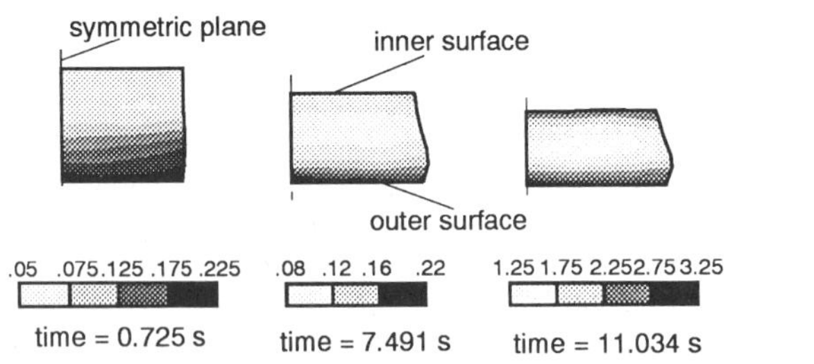

Figure 8. Generalised plastic strain on the cross section

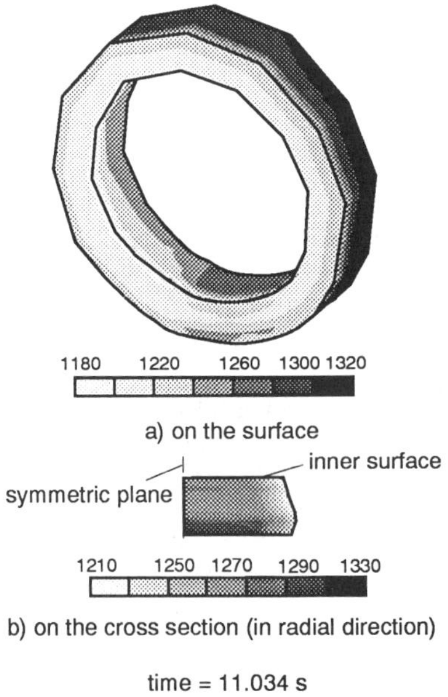

Figure 9. Temperature distributions

separates from the roll after passing through the roll gap. In the FE analysis, the main roll was represented by a frictional cylindrical surface which rotated by a specified angle during each increment. Since no lubricant is used in the process, a high friction factor *m* of 0.75 was specified for the surface.

In the actual process, the mandrel is idle. It moves towards the main roll, pushing the ring against the main roll. Its rotation with the ring is caused only by the friction between the ring and itself. Hence, the mandrel cannot be modelled the same way as the main roll. In this case, as there is no relative slip between the workpiece and the mandrel, it was sufficient for the mandrel to be represented by a frictionless boundary surface. It rotated around its axis which, during the analysis, moved towards the main roll. Since the surface was frictionless, there was no tangential, which was equivalent to the restraint imposed by a freely-rotating mandrel.

In the FE analysis, the guide rolls were also represented by frictionless cylindrical boundary surfaces. As in the case of the mandrel there is no relative slip between the ring and guide rolls in the actual process, and so a frictionless surface provided the contact constraint on the direction of nodal movement. The function of the guide rolls is to maintain the roundness of the ring. In the process, their movement is controlled, as the ring expands, by specifying the variation of pressure in an hydraulic ram acting on the arms on which the rolls are situated. This is difficult to simulate in the FE analysis and instead the movement of guide rolls was modelled by changing their positions to ensure that the ring did not lose its roundness and remained in a central position relative to the longitudinal axis of the mill. Specifically, the current diameter and the position of the centre of the ring were predicted from the reduction in the ring thickness, increase in the ring height and material volume constancy. The current centre and diameter were then used to define a circle on the circumference of which the guide rolls were to be positioned at the start of each increment of the FE analysis.

4. THERMO-PLASTIC ANALYSIS OF THE PROCESS

The analysis was carried out for a hot titanium-alloy ring with a starting temperature of 930 °C. The material and working parameters (Yurttas 1991, Sweeney 1993) are shown in Table 1. Because of symmetry, only half of the workpiece was modelled. The hybrid mesh model used consisted of 792 and 3240 elements in the computational and material meshes respectively. Approximately 24 hours of CPU time were required to model one revolution of the ring, corresponding to 900 increments in the FE analysis, on an HP710 workstation. It is noted that the computational cost is still relatively high even though the hybrid mesh model is used.

As the ring is rolled at high temperature, a thermo-plastic three-dimensional analysis was carried out. The thermal calculation included the heat generated inside the ring due to plastic work, the heat generated due to the frictional force on the interface, the heat conduction inside the workpiece, the heat conduction between the workpiece and the tools, and the heat radiation and convection on the surfaces.

Fig. 4 shows the deformed computational mesh, projected vectors of the incremental nodal displacements, and the contours of generalised plastic strain rates and generalised stress for the rolled ring after 6 revolutions. It clearly demonstrates that the ring was driven and guided by the rolls to move in a reasonable way in the FE simulation and that the deformation took place mainly in the gap between the main roll and the mandrel. Only the elements currently in the gap between the main roll and the mandrel were deforming plastically. The plastic strain rate was concentrated there, especially near the inner and outer surfaces, and so were the maximum values of the generalised stress.

As the deformation proceeded, the thickness of the ring was gradually reduced and its diameter expanded. The deformed shapes of the ring during the process can be see from the material mesh, as shown in Fig. 5.

As the plastic strain rate was always concentrated in the gap between the main roll and the mandrel, especially near the outer surface, the plastic strain in the material was accumulated in such a way that the maximum strain was eventually located around the outer periphery of the ring. Fig. 6 illustrates the development of this strain pattern. The pattern in the roll gap remained approximately the same throughout the process, although the level increased and this strain distribution was carried with the ring as it rotated. After the first revolution, the plastic strain contours formed a series of concentric circles, with the maximum strain occurring at the outer surface of the ring.

Figs. 7 - 8 show the changes to the geometrical shape, plastic strain and temperature on the cross section of the ring located between the main roll and the mandrel. It can be seen that a 'fish tail' was gradually formed. This kind of inhomogeneous deformation arises due to the plastic deformation taking place mostly in the outer layers of the material, with much of the centre remaining elastic. This leads to a lack of fill around the centre of either the axial or radial faces of the ring. With the introduction of axial rolls, such flow defects arising in the radial roll gap can be expected to be reduced. The plastic strain contours on the cross section, shown in Fig. 8, were very uniformly distributed along the axial direction, while in the radial direction, the values were higher near the outer surface, leading to the series of concentric circles already shown in Fig. 6.

The temperature distribution, shown in Fig. 9, is a result of two opposing effects. Heat is generated in the ring due to plastic work of deformation and, to a large extent, frictional work at the interfaces with the rolls. Meanwhile heat is lost from the ring due to the conduction into the rolls and radiation and convection at the exposed surfaces. Since, in ring rolling, only a small part of the workpiece is in contact with the rolls, the predominant effect here is that of convection and radiation. So heat is generated throughout the deforming region, but lost only at the surface. As a result, the centre of the ring, especially the region in the roll gap, becomes hotter while temperature at the surface decreases slightly.

5. CONCLUSION

Using a new hybrid mesh model and specialised boundary surface and friction-layer techniques, a full three-dimensional thermo-mechanical FE analysis of the rolling of a hot titanium ring has been successfully carried out.

As the whole ring is taken into consideration, the process can be analysed from the start to the end and all the tools involved can be included in the analysis, while the use of a finite-element mesh which is finely divided only in the main deforming region means the computational time is greatly reduced compared with an analysis using a uniformly fine mesh.

6. ACKNOWLEDGEMENTS

The authors are grateful to Rolls-Royce plc and Platestahl (Germany) for financial support. They are also grateful to the CVCP committee and the School of Manufacturing and Mechanical Engineering of the University of Birmingham for the research scholarship (ORS) and the additional support respectively.

REFERENCES

Hahn, Y.H. & Yang, D.Y. 1994. UBET analysis of the closed-pass ring rolling of rings having arbitrarily shaped profiles. *J. Mater. Process. Technol.* 40:451-463

Hawkyard, J.B., Johnson, W., Kirkland, J. & Appleton, E. 1973. Analysis for roll force and torque in ring rolling, with some supporting experiments. *Int. J. Mech. Sci.* 15:873-893

Hu, Y.K. & Liu, W.K. 1992. ALE finite element formulation for ring rolling analysis. *Int. J. Num. Methods Eng.* 33:1217-1236

Hu, Z.M., Pillinger, I., Hartley, P., McKenzie, S. & Spence, P.J. 1994. Three-dimensional finite-element modelling of ring rolling. *J. of Mater. Process. Technol.* 45:143-148

Kim, N., Machida, S. & Kobayashi, S. 1990. Ring rolling process simulation by the three dimensional finite element method. *Int. J. Mach. Tools & Manufact.* 30:569-577

Kluge, A., Lee,Y.-H., Wiegels, H. & Kopp, R. 1994. Control of strain and temperature distribution in the ring rolling process. *J. Mater. Process. Technol.* 45:137-141.

Lugora, C.F. & Bramley, A.N. 1987. Analysis of spread in ring rolling. *Int. J. Mech. Sci.* 29:149-157

Mamalis, A.G., Johnson, W. & Hawkyard, J.B. 1976. Ring-rolling: literature review and some recent experimental results. *Metallurgia and Metal Forming.* 132-140.

Pillinger, I., Hartley, P. & Sturgess, C.E.N. 1988. Modelling of frictional tool surfaces in finite-element metal forming analyses. In Chenot, J.L. & Oñate, E. (eds), *Modelling of Metal Forming Processes.*:85-92. Kluwer Academic Publishers.

Rowe, G.W., Sturgess, C.E.N., Hartley, P. & Pillinger, I. 1991. *Finite-element Plasticity and Metal Forming Analysis.* Cambridge University Press.

Ryoo, J.S. Yang, D.Y. & Johnson, W. 1983. Ring rolling; the inclusion of pressure roll speed for estimating torque by using a velocity superposition method. *Proc. 24th MDTR Conf.* Manchester:19-24

Sweeney, K.E., Kluge, A., Wiegels, H. & Kopp, R. 1993. Simulation of the temperature field along the cross-section of a ring during the ring rolling process. *Steel Research.* 64:604-610

Xu, S.G., Lian, J.C. & Hawkyard, J.B. 1991. Simulation of ring rolling using a rigid-plastic finite element model. *Int. J. Mech. Sci.* 33:393-401

Yang, D.Y., Ryoo, J.S., Choi, J.C. & Johnson, W. 1980. Analysis of roll torque in profile ring rolling of L-sections. *Proc. 21st Int. MTDR Conf.* London:69-74

Yang, D.Y. & Kim, K.H. 1988. Rigid-plastic finite element analysis of plane strain ring rolling. *Int. J. Mech. Sci.* 30:571-580

Yang, D.Y., Kim, K.H. & Hawkyard, J.B. 1991, Simulation of T-section profile ring rolling by the 3-D rigid-plastic finite element method. *Int. J. Mech. Sci.* 33:541-550

Yurttas, Z. 1991. Private communication, Platestahl, Stahlwerk Plate, Ludenscheid-Platehof, Germany.

Simulation of Materials Processing: Theory, Methods and Applications, Shen & Dawson (eds)
© 1995 Balkema, Rotterdam. ISBN 90 5410 553 4

A coupled thermo-mechanical force model for sheet rolling

R. E. Johnson
The University of North Carolina at Charlotte, N.C., USA

M. E. Karabin
Alcoa Technical Center, Pa., USA

ABSTRACT: An improved force and torque model accounting for thermal effects is the subject of this report. The goal was to develop a more accurate model for reversing mills. To accomplish this a systematic asymptotic formulation was used to deal with the combined thermo/mechanical problem. A key feature to account for was the effect of through thickness temperature variations on roll force, torque and slip. This manifests itself primarily through its effect on the flow stress. The cool thermal boundary layer in the sheet near the roll surface causes a substantial increase in the flow stress that affects the force and torque. Several modeling assumptions were used to cast the expression for roll force in a form that may be used easily. Comparisions are made with experiments conducted on a laboratory mill.

1 INTRODUCTION

This report develops a simple thermo/mechanical force model for preset and control applications in hot rolling. The goal is to capture the effects of temperature on force, torque and slip, and to evaluate through thickness variations in the deformation. The focus is on 2D plane strain and several modeling assumptions are used to simplify the coupled problem and cast the theory into a form analogous to classical rolling theories (Johnson, 1992; Johnson & Smelser, 1992). Both flow stress dependent friction (fsf) and relative slip friction (rsf) are considered.

In qualitative terms the situation in hot rolling is as follows. A hot slab enters the mill making contact with the relatively cold rolls (Figure 1). Since the processing speed is fairly high, the temperature transition from the hot core of the sheet to the cold rolls occurs in a narrow layer near the roll/sheet interface termed the thermal boundary layer. The core of the sheet remains close to the incoming sheet temperature and is nearly constant through the thickness, although it increases slightly in the processing direction due to deformation heating. The rolls are at a uniform temperature, except for a thin layer near the roll/sheet interface where they heat up slightly due to contact with the hot slab. For simplicity the model developed here assumes that the rolls are at a constant temperature. The effect of the thermal boundary layer in the rolls is neglected and only the thermal boundary layer in the sheet is considered. To account for the fact that the temperature of the roll surface contacting the sheet is slightly higher than the measured roll temperature, an estimate of the interface temperature is used in practice.

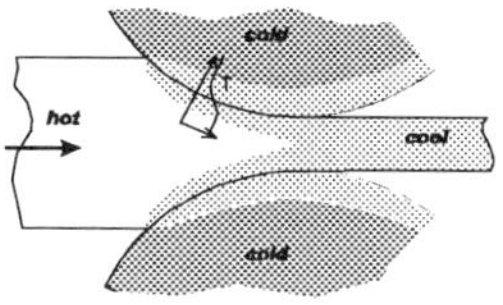

Figure 1: Sketch of thermal boundary layer structure.

In response to the temperature distribution through the sheet thickness the flow stress is fairly constant in the core of the sheet and increases rapidly as the cool roll surface is approached. The spatial variation of the flow stress is one of the predominant effects of temperature. The expected variations of temperature and flow stress through the thickness of the slab are sketched in Figure 2.

In the following sections a force model is developed that accounts for temperature induced variations in the flow stress. The model also includes the effect of rate dependence. First the asymptotic formulation of the deformation problem is reviewed highlighting where temperature effects enter the problem. Following this is an analysis of the temperature field that includes a model of the deformation induced heating and a simplified thermal boundary layer analysis. To evaluate the force an approximate expression for the flow stress that accounts for temperature and rate dependence is developed. At the end of the development several modeling assumptions are made to cast the final force expression in a form analogous to the classical isothermal force theory. This makes it easy to apply in practice.

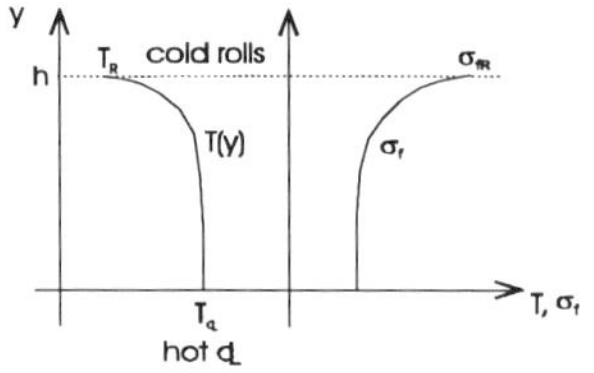

Figure 2: Typical through-thickness temperature and flow stress variation.

2 DEFORMATION MODELING

This section summarizes the basic asymptotic formulation of the deformation problem and highlights places where temperature effects will manifest themselves. The starting point are the dimensionless governing equations for 2D plane strain, incompressible sheet rolling

$$-\frac{\partial p}{\partial x}+\frac{\delta}{\beta}\frac{\partial s_{xx}}{\partial x}+\frac{\partial s_{xy}}{\partial y}=0 \tag{1}$$

$$\delta^2\frac{\partial s_{xy}}{\partial x}-\frac{\partial p}{\partial y}+\frac{\delta}{\beta}\frac{\partial s_{yy}}{\partial y}=0 \tag{2}$$

$$\delta\frac{\partial u}{\partial x}+\frac{\partial v}{\partial y} \tag{3}$$

The physics of the deformation problem is governed by two parameters. The thinness parameter $\delta = \hat{h}_0/\hat{\ell} << 1$, which is the ratio of the half-gauge at the entrance to the roll bite length, and the shear parameter $\beta = \hat{\tau}_o/\hat{s}_o$, which is the ratio of the characteristic shear stress to the characteristic longitudinal deviatoric stress. Carets denote dimensional quantities. The equation set is completed with the yield condition or flow rule and constitutive equations

$$\lambda s_{xx}=\frac{\partial u}{\partial x},\ \ \delta\lambda s_{yy}=\frac{\partial v}{\partial y} \tag{4}$$

$$\delta\beta\lambda s_{xy}=\frac{1}{2}\left(\frac{\partial u}{\partial y}+\delta\frac{\partial v}{\partial x}\right) \tag{5}$$

$$s_{xx}^2+\beta^2 s_{xy}^2=\sigma_f \tag{6}$$

The case of moderate shear, which is defined by $\delta << \beta << \delta^{-1}$, such that $\delta/\beta << 1$ and $\delta\beta << 1$,is the focus of attention here. The second of these restrictions implies from (5) that $u \approx u^{(0)}(x)$. The equilibrium equations suggest that all of the field variables should be expanded as

$$f=f^{(0)}+\frac{\delta}{\beta}f^{(1)}+... \tag{7}$$

The leading order description is a balance between shear stress and pressure given by

$$-\frac{\partial p^{(0)}}{\partial x}+\frac{\partial s_{xy}^{(0)}}{\partial y}=0,\ \ \frac{\partial p^{(0)}}{\partial y}=0 \tag{8}$$

Consequently, $p^{(0)}$ is a function of x and this gives $s_{xy}^{(0)}=\frac{dp^{(0)}}{dx}y$.

The friction model prescribes the stress applied tangent to the roll surface. Symbolically taking this prescribed tangential stress to be S_x the requirement is $s_{xy}^{(0)}+\frac{\delta}{\beta}s_{xy}^{(1)}+2\frac{\delta}{\beta}h's_{xx}^{(0)}\approx S_x$ on $y=h$. After substituting the stress expansions, the leading order friction condition becomes $s_{xy}^{(0)}=S_x$ on $y=h$ and therefore

$$\frac{dp^{(0)}}{dx}=\frac{S_x}{h(x)} \tag{9}$$

The two friction models being used in this study are $S_x=\sigma_f(x,h)$ for flow stress dependent friction (fsf) and $S_x = U-u(x,h)$ for relative-slip friction (rsf). In these two cases the characteristic shear stress $\hat{\tau}_o$ is $m\hat{\sigma}_{f0}$ or $\kappa\hat{u}_0$, respectively (where m and κ are friction coefficients and $\hat{\sigma}_{f0}$ and $\hat{u}_0$ are the entrance values of flow stress and sheet speed). In the case when $S_x=\sigma_f$ the cold roll surface would cause a higher value of the flow stress compared to the isothermal case and this will yield a correspondingly higher pressure and separating force contribution. Temperature effects have less influence on $p^{(0)}$ for the rsf case, since the friction in this case does not depend on the surface value of the flow stress.

The shear and longitudinal deviatoric stresses from (6), (8), and (9) are

$$s_{xy}^{(0)}=S_x\frac{y}{h},\ \ s_{xx}^{(0)}=\sqrt{\sigma_f^2-\beta^2S_x^2\left(\frac{y}{h}\right)^2} \tag{10}$$

Due to the intimate dependence of the longitudinal deviatoric stress on flow stress, temperature effects arise in any circumstances where the role of the longitudinal deviatoric stress is important.

The constitutive equation (5) leads to the conclusion that the velocity field is a function of x at leading order

$$u=u^{(0)}(x)+\frac{\delta}{\beta}u^{(1)}(x,y)+... \tag{11}$$

The leading order sheet velocity is determined from global mass conservation for steady rolling and given by $u^{(0)}=1/h(x)$.

The correction to the velocity field follows from the constitutive equations (5) and (4). Eliminating the plastic multiplier from these and retaining the largest terms gives

$$\frac{\partial u^{(1)}}{\partial y}=\beta^2\frac{s_{xy}^{(0)}}{s_{xx}^{(0)}}\frac{du^{(0)}}{dx} \tag{12}$$

This can be integrated explicitly in some cases. For example, using expressions for the stresses given in (10), assuming $\sigma_f=\bar{\sigma}_f$ =constant, noting that

S_x depends only on x, and applying global mass conservation $\int_0^h u^{(1)}dy = 0$ yields

$$u^{(1)} = \frac{2h(x)\,\bar{\sigma}_f}{S_x}\frac{du^{(0)}}{dx}\left\{g\left(\tilde{\beta}\right) - \sqrt{1-\tilde{\beta}^2\left(\frac{y}{h}\right)^2}\right\} \quad (13)$$

where $g(z) = \frac{1}{2}\left[\sqrt{1-z^2} + z^{-1}\arcsin z\right]$ and $\tilde{\beta} = \beta S_x/\bar{\sigma}_f$.

At second order, the equilibrium equations are

$$-\frac{\partial p^{(1)}}{\partial x} + \frac{\partial s_{xx}^{(0)}}{\partial x} + \frac{\partial s_{xy}^{(1)}}{\partial y} = 0 \quad (14)$$

$$-\frac{\partial p^{(1)}}{\partial y} + \frac{\partial s_{yy}^{(0)}}{\partial y} = 0 \quad (15)$$

Equation (15) suggests the definition $N(x) = p^{(1)} - s_{yy}^{(0)}$. This is the first correction to the vertical stress acting on the rolls. In addition, from incompressibility we have $s_{xx}^{(0)} = -s_{yy}^{(0)}$. Using these, and the friction boundary condition at second order yields the equation governing the force correction term

$$\frac{dN}{dx} = \frac{2}{h}\frac{d}{dx}\left(h\bar{s}_{xx}^{(0)}\right) - \psi\frac{u^{(1)}(x,h)}{h} \quad (16)$$

where

$$\bar{s}_{xx}^{(0)} = \frac{1}{h}\int_0^h \sqrt{\sigma_f^2 - \beta^2 S_x^2\left(\frac{y}{h}\right)^2}\,dy \quad (17)$$

and $\psi = 0$ for the fsf model and $\psi = 1$ for rsf model. The shear parameter and dimensionless friction function for the two friction models are $\beta = m$, $S_x = \sigma_f$ and $\beta = \kappa\hat{u}_0/\hat{\sigma}_{f0}$, $S_x = U - u$ respectively.

One of the key places where thermal effects manifest themselves is apparent in (16) and (17). In contrast with the isothermal theory, it is seen that the through-thickness variation in the flow stress influences N and consequently the roll force. The flow stress is generally substantially larger at the roll surface than in the interior due to the cooling effect of the cold rolls. Furthermore, for the fsf model, the value of flow stress at the roll surface controls the friction and this has an additional direct influence on N. When evaluating (17) note that the shear stress must be monitored to identify when and if the shear stress equals the yield value. This places a limit on the maximum acceptable value of shear stress at the roll surface. For points in the bite where the shear stress might exceed this maximum value, the shear stress boundary condition is replaced by requiring the shear stress to equal the maximum admissible value.

Integrating (17) to determine the average longitudinal deviatoric stress is difficult because the flow stress depends on the effective strain rate and temperature T. To deal with this complexity a Taylor series expansion about the state of homogeneous deformation and the entrance temperature $T = 1 \equiv T_0$ is used

$$\sigma_f = \sigma_{fo} + (D - D_o)\left.\frac{\partial\sigma_f}{\partial D}\right|_{D_o,T_0} + (T - T_0)\left.\frac{\partial\sigma_f}{\partial T}\right|_{D_o,T_0} + \ldots \quad (18)$$

where $\sigma_{fo} = \sigma_f(D_o, T_0)$ and $D_o = du^{(0)}/dx$. Using the constitutive relations and the expansion for the velocity field, the effective strain rate for an incompressible material made dimensionless using $\hat{u}_0/\hat{\ell}$ can be approximated as

$$D \approx \left[1 + \frac{1}{2}\beta^2\left(\frac{S_x}{\sigma_{fo}}\frac{y}{h}\right)^2\right]\left|\frac{du^{(0)}}{dx}\right| \quad (19)$$

Assuming $D - D_o$ and $T - T_0$ are small, equation (17) is approximated as

$$\bar{s}_{xx}^{(0)} = \frac{\sigma_{fo}}{h}\int_0^h \sqrt{1 - \tilde{\beta}^2\Lambda\left(\frac{y}{h}\right)^2 + \chi(T - T_0)}\,dy \quad (20)$$

where

$$\Lambda = 1 - \frac{|D_o|}{\sigma_{fo}}\left.\frac{\partial\sigma_f}{\partial D}\right|_{D_o,T_0} \quad (21)$$

$$\chi = \frac{2}{\sigma_{fo}}\left.\frac{\partial\sigma_f}{\partial T}\right|_{D_o,T_0} \quad (22)$$

$$\tilde{\beta} = \beta\frac{S_x}{\sigma_{fo}} = \begin{cases} m & ,\ \text{fsf} \\ \beta\,(U - u(x,h))/\sigma_{fo} & ,\ \text{rsf}\end{cases} \quad (23)$$

Expression (20) is used in (16) to compute N which contributes to the roll force.

3 ROLL FORCE

Since the vertical stress is given by $\tau_{yy} = -p + \frac{\delta}{\beta}s_{yy} = -p - \frac{\delta}{\beta}s_{xx}$, the roll force is given in dimensional terms by

$$\begin{aligned}\hat{F} &= \delta^{-1}\hat{\tau}_o\hat{W}\hat{\ell}\left(\int_0^1 p^{(0)}dx + \frac{\delta}{\beta}\int_0^1 N dx\right) \\ &\equiv \delta^{-1}\hat{\tau}_o\hat{W}\hat{\ell}\left(F_0 + \frac{\delta}{\beta}F_1\right) \quad (24)\end{aligned}$$

where $p^{(0)}$and N are determined from eqns (9) and (16) where $\hat{W}$ is the width of the sheet and $\hat{\ell}$ is the length of the roll bite. The temperature variation in the roll bite manifests itself in the roll force primarily by altering the flow stress. As noted earlier, the flow stress has a significant impact on the longitudinal deviatoric stress and therefore N, and can also alter the friction at the roll-sheet interface.

4 THERMAL MODELING

The dimensionless form of the energy equation is

$$Pe\left(\delta u^{(0)}\frac{\partial T}{\partial x}+v^{(0)}\frac{\partial T}{\partial y}\right)=\delta Br\Phi+\delta^2\frac{\partial^2 T}{\partial x^2}+\frac{\partial^2 T}{\partial y^2} \tag{25}$$

where $Pe = \rho C_P \hat{u}_0 \hat{h}_0/k$, $Br = 2\hat{s}_o \hat{u}_0 \hat{h}_0/k\hat{T}_0$, and the heat generation term is $\Phi = \lambda\sigma_f^2$. Boundary conditions include prescribed interface and entrance temperatures $T = T_I$ on $y = h$ and $T = 1$ at $x = 0$, respectively. In addition to these there is the symmetry condition on $y = 0$. To circumvent the difficulty associated with the coupling between the deformation and temperature fields, the heat generation term is approximated as.

$$\Phi \cong \sigma_{fo}F\left(\frac{y}{h}\right)\left|\frac{du^{(0)}}{dx}\right| \tag{26}$$

where a model of the through thickness variation is $F\left(\frac{y}{h}\right)=1+\frac{1}{2}\tilde{\beta}^2\left(\frac{y}{h}\right)^2$.

Since the Pe is generally large, conduction can be neglected in (25) and the approximate solution is

$$T \equiv T_{outer} = T_0 + \frac{Br}{Pe}\sigma_{fo}F\left(\frac{y}{h}\right)\ln\left(\frac{1}{h(x)}\right) \tag{27}$$

Note that (27) satisfies the condition at the entrance to the roll bite, but violates the roll surface boundary condition $T = T_I$. Consequently, thermal boundary layers exist near the surface of the rolls that allow the temperature to make the adjustment from the core value (outer solution) to the relatively cool interface temperature at the roll surface, i.e., $T_{BL} = T_{outer}(x,h) + \tilde{T}$. Conduction is important in these boundary layers.

To obtain a simple expression for the boundary layer temperature that can be used easily in (20), the von Karman integral technique is employed. Integrating the energy equation through the thermal boundary layer and using mass conservation gives the integral conservation law for the temperature perturbation $\tilde{T}$ in the boundary layer

$$\delta Pe\frac{d}{dx}\int_{h-b(x)}^{h}u^{(0)}\tilde{T}dy=\left.\frac{d\tilde{T}}{dy}\right|_{y=h} \tag{28}$$

where $b(x)$ is the edge of the thermal boundary layer.

The simplest approximation to the temperature is a linear distribution that equals the interface temperature at the roll surface and the outer temperature at the edge of the boundary layer. This gives for the perturbation temperature valid in the thermal boundary layer $h-b \le y \le h$

$$\tilde{T} = (T_I - T_{outer}(x,h))\left[1-\frac{h-y}{b(x)}\right] \tag{29}$$

Substituting the linear approximation (29) into (28) and integrating yields the boundary layer thickness

$$b(x)=\frac{2}{\sqrt{\delta Pe u^{(0)}\Delta T}}\left\{\int_0^x u^{(0)}\Delta T dx\right\}^{1/2} \tag{30}$$

where $\Delta T \equiv T_{outer}(x,h) - T_I$. It is interesting to note the when there is no heat generation and the interface temperature is constant this result is identical to the exact solution for the boundary layer temperature field.

5 FIRST ORDER FORCE CONTRIBUTION

5.1 *Flow stress dependent friction (fsf)*

Thermal effects have a significant effect on $p^{(0)}$ when flow-stress dependent friction is used. In this case, using the expansion (18) for the flow stress yields,

$$\frac{dp^{(0)}}{dx}=\pm\frac{\sigma_f(x,h)}{h(x)}=\frac{\sigma_{fo}+\Delta\sigma_{fI}}{h(x)} \tag{31}$$

where $\Delta\sigma_{fI}=\frac{1}{2}m^2|D_o|\frac{\partial\sigma_f}{\partial D}+(T_I-T_0)\frac{\partial\sigma_f}{\partial D}$ is the increase in the flow stress at the roll/sheet interface due to the added shear and lower surface temperatures. Since $\sigma_{fo}+\Delta\sigma_{fI}$ is generally a function of x, (31) may be difficult to integrate. However, if the average value is used then the leading order contribution to the net roll force in the absence of any front or back tension is

$$F_0=\int_0^1 p^{(0)}dx=\frac{(\bar{\sigma}_{fo}+\Delta\bar{\sigma}_{fI})}{\sqrt{\Delta h h_1}}\{(1-x_N)\left[2\tan^{-1}\left(\sqrt{\frac{\Delta h}{h_1}}(1-x_N)\right)-\tan^{-1}\left(\sqrt{\frac{\Delta h}{h_1}}\right)\right]+\frac{1}{2}\sqrt{\frac{h_1}{\Delta h}}\ln\left(\frac{h_1}{h_N^2}\right)\} \tag{32}$$

where $h_1 = 1-\Delta h$, x_N is the neutral point where the sheet speed equals the roll speed and $h_N = h(x_N)$.

5.2 *Relative-slip friction (rsf)*

In the case of relative-slip friction when $S_x \cong U - u^{(0)} = U - 1/h(x)$ thermal effects have no effect on $p^{(0)}$. With circular rolls and no front or back tension the leading order force contribution becomes

$$F_0=\int_0^1 p^{(0)}dx=-\frac{1}{4\Delta h h_1}\ln h_1-\frac{x_N}{2h_1}+\left(U-\frac{1}{2h_1}\right)\left\{\frac{x_N-1}{\sqrt{\Delta h h_1}}\tan^{-1}\left(\sqrt{\frac{\Delta h}{h_1}}\right)+\frac{1}{2}\sqrt{\frac{h_1}{\Delta h}}\ln\left(1+\frac{\Delta h}{h_1}\right)\right\} \tag{33}$$

6 SECOND ORDER FORCE CONTRIBUTION

Since one goal is to develop simple closed form force expressions, (16) must be cast in a form that will allow N to be easily evaluated. Two key simplifying features are made to accomplish this task. The flow stress is approximated by its average value, and the linear approximation for temperature developed in Section 4 is used. Defining $\xi = y/h$, $\bar{s}_{xx}^{(0)}$ is written as

$$\begin{aligned} \bar{s}_{xx}^{(0)} &\equiv \bar{\sigma}_{fo} G(x) \qquad (34) \\ &= \bar{\sigma}_{fo} \int_0^h \sqrt{1 - \tilde{\beta}^2 \Lambda \xi^2 + \chi (T - T_0)} d\xi \end{aligned}$$

where the temperature outside the boundary layer ($y \leq h - b$) and inside the boundary layer ($h - b \leq y \leq h$) is

$$T = \begin{cases} T_{outer}(x, y) \\ T_{outer}|_h + (T_I - T_{outer}|_h)\left[1 - \frac{h-y}{b(x)}\right] \end{cases} \qquad (35)$$

where T_{outer} is given by (27).

The temperature field requires breaking the integral G defined in (35) into two pieces, a core or outer contribution and a boundary layer contribution. After integration this yields

$$\begin{aligned} G &= \left[1 + \bar{\sigma}_{fo} \chi \frac{Br}{Pe} \ln h^{-1}\right]^{1/2} \{g(m_{eff}) \qquad (36) \\ &- \frac{b}{h}\sqrt{1 - m_{eff}^2}\Big\} + \frac{2b}{3c_1 h}\left[(c_0 + c_1)^{3/2} - c_0^{3/2}\right] \end{aligned}$$

where

$$m_{eff}^2 = \tilde{\beta}^2 \left\{ \frac{\Lambda - \frac{1}{2}\bar{\sigma}_{fo}\chi \frac{Br}{Pe} \ln h^{-1}}{1 + \bar{\sigma}_{fo}\chi \frac{Br}{Pe} \ln h^{-1}} \right\} \qquad (37)$$

$$g(m) = \frac{1}{2}\left[\sqrt{1 - m^2} + m^{-1}\sin^{-1}(m)\right] \qquad (38)$$

$$c_0 = 1 - \tilde{\beta}^2 \Lambda + \chi (T_{outer}|_h - T_0)$$

$$c_1 = \chi (T_I - T_{outer}|_h)$$

Note that in the classical isothermal fsf theory $G = g(m_{eff})$ and $m_{eff} = m$.

The equation for N is now

$$\frac{dN}{dx} = \frac{2\bar{\sigma}_{fo}}{h}\frac{d}{dx}(hG) - \psi \frac{u^{(1)}(x, h)}{h}$$

However, considerable complexity is still present in this equation making it difficult to integrate. The simplest result for N emerges if a constant value for G is used by replacing the arguments appearing within G by their average values. This approximate function is denoted as $\bar{G}$. Furthermore, noting that the velocity correction (13) for relative slip friction can be written as $u^{(1)} = 2h\bar{\sigma}_{fo} dg/dx$, leads to the following results for N; for $x \leq x_N$

$$N = 2\bar{\sigma}_{fo}\left[\bar{G}(1 + \ln h) - \psi (g - g_0)\right] \qquad (39)$$

and for $x \geq x_N$

$$N = 2\bar{\sigma}_{fo}\left[\bar{G}(1 + \ln(h/h_1)) - \psi (g - g_1)\right] \qquad (40)$$

where $g = g\left(\tilde{\beta}\right)$ and g_0 and g_1 are the values at the entrance and exit, respectively. Using the additional approximation $g\left(\tilde{\beta}\right) \cong 1 - \frac{1}{6}\tilde{\beta}^2$, gives the force contribution F_1 after integrating through the bite as

$$\begin{aligned} F_1 &= \int_0^1 N dx = 2\bar{\sigma}_{fo}\bar{G}\left[(1 - x_N)\ln h_1^{-1} - \bar{\epsilon}_o + 1\right] \\ &+ \frac{\psi \beta^2}{3\bar{\sigma}_{fo}}\Big\{x_N(1 - 2U) + (1 - x_N) u_1 (u_1 - 2U) - \frac{1}{2h_1} \\ &+ \frac{2U - 1/(2h_1)}{\sqrt{\Delta h h_1}} \tan^{-1}\left(\sqrt{\frac{\Delta h}{h_1}}\right)\Big\} \qquad (41) \end{aligned}$$

where the average strain in the bite is

$$\bar{\epsilon}_o = 2\left\{1 - \sqrt{h_1/\Delta h}\tan^{-1}\left(\sqrt{\Delta h/h_1}\right)\right\} \qquad (42)$$

7 NEUTRAL POINT

The force depends on the location of the neutral point x_N. If the incoming sheet speed and roll speed are known then the location of the neutral point is found from $u(x_N) = 1/h(x_N) = U = \hat{U}/\hat{u}_0$. Alternatively, if the entrance and exit tensions are prescribed then the neutral point is found from continuity of the stress

$$\left[p^{(0)} + \frac{\delta}{\beta}N\right]_{x_N^+} = \left[p^{(0)} + \frac{\delta}{\beta}N\right]_{x_N^-} \qquad (43)$$

In either case, this determines the neutral point in terms of the reduction, flow stress, entrance and exit tensions, thinness parameter, and friction parameter.

8 THROUGH-THICKNESS DEFORMATION

A useful tool for exploring the validity of the theory is the deformation of pins inserted vertically through the sheet. According to the theory, the deformation of an initially vertical material line segment after passing through the roll bite is

$$\begin{aligned} \Delta x &= \frac{\delta}{\beta}\int_0^{t_{exit}}\left[u^{(1)}(x, y) - u^{(1)}(x, 0)\right] d\tau \\ &\cong \frac{\delta}{\beta}\int_0^1 \frac{\left[u^{(1)}(x, y) - u^{(1)}(x, 0)\right]}{u^{(0)}(x)} dx \qquad (44) \end{aligned}$$

Using the velocity field given in (13)

$$\Delta x = 2\bar{\sigma}_{fo}\frac{\delta}{\beta}\int_0^1 \left\{1 - \sqrt{1 - \tilde{\beta}^2\left(\frac{y}{h}\right)^2}\right\}\frac{h^2(x)}{S_x}\frac{du^{(0)}}{dx}dx \qquad (45)$$

Table 1: Operating conditions for tests

CASE	I	II	III	IV
Entry Gage (in.)	0.945	0.944	1.774	1.890
Exit Gage (in.)	0.654	0.797	1.192	1.679
Width (in.)	4.725	4.727	5.663	5.658
Entry Temp. (°F)	836	835	850	863

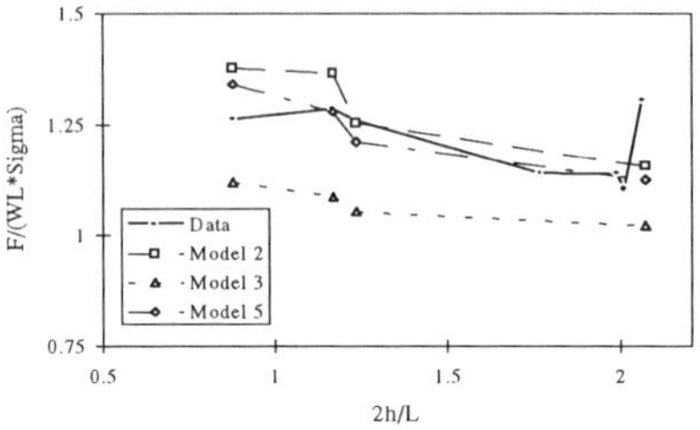

Figure 3: Laboratory mill force results.

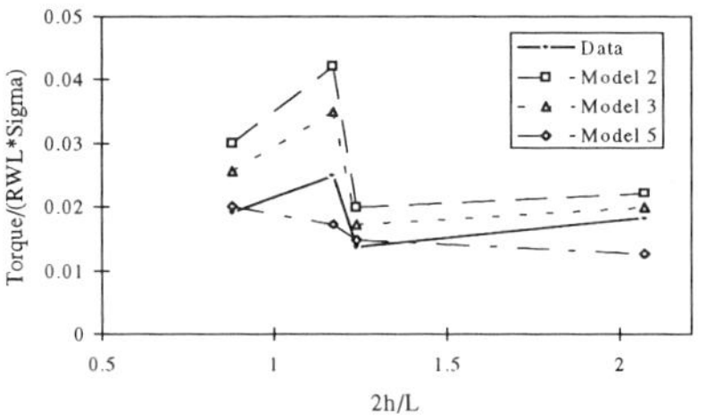

Figure 4: Laboratory mill torque results.

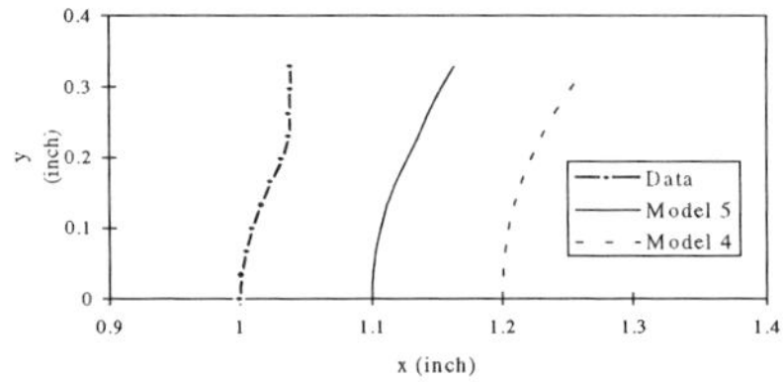

Figure 5: Measured and predicted pin distortions.

9 RESULTS

Experiments were conducted on a laboratory mill. Operating conditions are given in Table 1. The roll diameter is 7.885 in., exit sheet speed is 11.9 in./s, and roll temperature is 250 °F. Two different strip sizes and two nominal reductions were used; 5 in.× 1 in. and 6 in..× 2 in. and heavy (30%) and light (15%) reductions. Output consists of separating force, torque, slip and pin deformation. In the figures, Model 2 refers to the model using the fsf conditions including through thickness thermal effects. Models 3, 4 and 5 are for the rsf case. Model 3 uses only the leading order velocity in (23). Models 4 and 5 include the first order velocity correction to compute a more accurate surface shear stress. This requires calculating the surface velocity from (13) using Newton's method. In addition, Model 5 is enhanced using a bi-constant flow stress; one value outside the thermal boundary layer and a larger value within the boundary layer.

The force $\hat{F}/(\hat{W}\hat{\ell}\hat{\sigma}_{fo})$ is plotted versus $2\hat{h}_0/\hat{\ell}$ in Figure 3. Models 2 and 5 exhibit the best agreement, while Model 3 consistently underpredicts the force. This is apparently due to an inaccurate friction prediction since it uses only the leading order velocity terms. Dimensionless torque results, $\hat{T}/(\hat{R}\hat{W}\hat{\ell}\hat{\sigma}_{fo})$, are shown in Figure 4. The results are somewhat inconclusive. Model 5 appears to be best, but only marginally. This is an indication that improved friction models are needed.

Typical pin deformations are shown in Figure 5. Nonuniform deformation is clearly evident. The models predict the deformation in the sheet core, but in the thermal boundary layers the agreement is not as good. Model 4 uses a uniform flow stress to calculate $u^{(1)}$ and Model 5 uses a bi-constant flow stress as noted above. The predicted boundary layer thickness is about 30% of the sheet thickness. The experiments suggest that the boundary layer is thicker and more resistant to shear than that predicted by the theory.

10 SUMMARY

This report develops a force model for hot rolling that accounts for the effects of temperature and rate-dependence on the roll separating force. These effects make their presence felt via their influence on the flow stress. The model is developed for flow stress dependent friction and relative-slip friction. To include thermal effects, an approximate temperature field in the sheet is used. The temperature field consists of a thermal boundary layer near the roll sheet interface and a nearly uniform temperature in the core. The core temperature rises as the material moves through the roll bite due to heat generation associated with the deformation. A model of the heat generation term is developed. One objective is to cast the roll force results in a simple form that would allow them to be used easily. This is accomplished by making some modeling assumptions that simplify the equation, but capture the primary effects of temperature.

REFERENCES

Johnson, R.E. 1992. Asymptotic Methods for Manufacturing Processes, short course notes used at ATC Nov.1992.

Johnson, R.E. & R.E. Smelser 1992. An asymptotic formulation of shear effects in two dimensional rolling. J. Materials Proc. Tech. 23: 311-318.

Simulation of Materials Processing: Theory, Methods and Applications, Shen & Dawson (eds)
© 1995 Balkema, Rotterdam. ISBN 90 5410 553 4

Finite element simulation of four roll bar rolling

S. Kihara
Sumitomo Heavy Industries, Ltd, Research & Development Center, Japan

S. Ochi
Sumitomo Heavy Industries, Ltd, Industrial Machinery Division, Japan

K. Hiraki
Niihama National College of Technology, Department of Mechanical Engineering, Japan

K. Mori
Osaka University, Department of Mechanical Engineering, Japan

ABSTRACT : A three dimensional rigid–plastic finite element method program is developed to analyze the steady–state characteristics of bar rolling by 2, 3 and 4–roll mills. The finite element method is formulated on the basis of plasticity theory for a material with slight compressibility. By applying the simulator to the steady–state analysis, the characteristics of the 2 and 4–roll rolling such as the lateral spread ratio at the roll gap, rolling load, rolling torque, equivalent strain and equivalent strain rate are numerically and experimentally investigated as far the single pass and the oval–to–round multipass.

1. INTRODUCTION

To meet the recent demands for the precise roundness of the cross–section and the dimensional accuracy for each lot of rolled bar and rod products, the high precision rolling technology is required. From the view point of the fact that the performance of the 4– roll rolling is more desirable than that of the 2 or 3– roll rolling, the basic characteristics of the 4– roll rolling such as lateral spread ratio, the working load and torque has been experimentally evaluated by the author(1991, 1993). Furthermore the authors (1992) have shown that the highly elaborated rolling process is realized by 4–roll rolling and wide range of large reduction is achieved by employing oval–to–round multipass.

The details of characteristics of the bar rolled by the 4–roll mill still remain unclear, although the several numerical investigations such as 2–roll and 3–roll rolling analyses have been reported by Mori(1990a) and Yanagimoto(1990b). A computer program to simulate the three dimensional steady–state bar deformation in the 2,3 and 4 roll rolling is developed by the authors(1994a, 1994b). The single and multipass rolling simulations are carried out by using the developed computer program. Accuracy of the computer simulation results is verified by comparing with the experimental results. This paper describes the characteristics of 4– roll rolling.

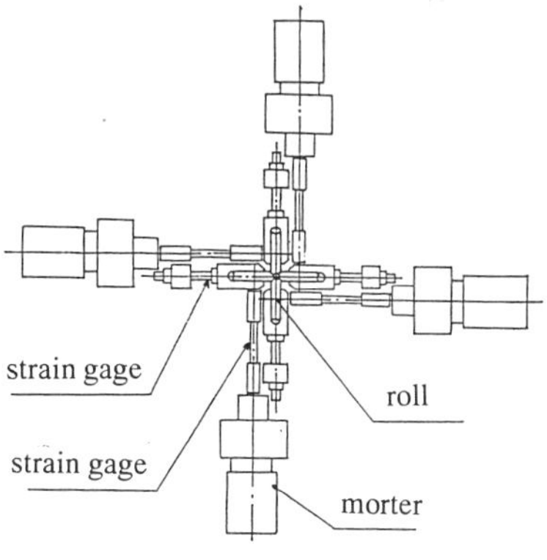

Fig. 1 Illustration of the test mill.

2. METHOD OF SIMULATION AND EXPERIMENTAL PROCEDURE

2.1 Method of simulation

The three dimensional rigid–plastic finite element method program was developed to analyze the steady–state characteristics of pipe and bar rolled by the 2–roll, 3–roll and 4–roll mills. The FEM is formulated on the basis of the plasticity theory for a material with the slight compressibility proposed by Mori(1990a). The correct velocity fields are given by minimizing the functional Φ:

$$\Phi = \int_v \sigma \dot{\bar{\varepsilon}} \, dv + \int_s \tau \Delta v \, ds - \int_{St} T v \, dS \quad (1)$$

where σ is the equivalent stress. $\dot{\bar{\varepsilon}}$ is the equivalent strain rate. τ is the shear stress acting on the friction surface on which the relative velocity between roll and workpiece is Δv. T is traction subjected to the surface St where the velocity is v.

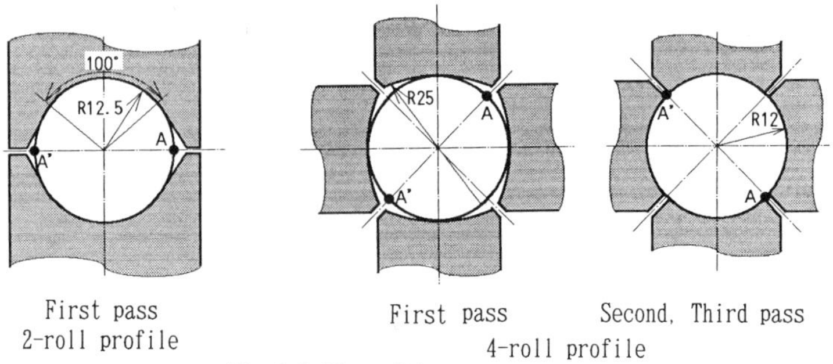

Fig.2 Roll profile.

2.2 Experimental procedure and computational conditions

Figure 1 illustrates the test mill. The 2 and 4–roll rolling test was conducted on the mill of 195mm roll diameter at a revolution speed of 7.49rpm. The rolling process is carried out under the dry friction condition at the room temperature. Figure 2 shows the roll profile. One is the 2–roll profile of which curvature is 12.5mm for a single rolling pass. The other is the 4–roll profile of the oval roll of 25mm curvature for the first pass and of the round roll of 12mm curvature for the second and the third pass. The workpiece is 25mm in diameter and made of 6% Pb–Sb alloy. Material flow stress is experimentally determined by compression test of a billet.

Owing to the symmetry of the process, only one–quarter of the workpiece is considered for the 2–roll rolling process simulation and one–eighth of the workpiece is considered for the 4–roll rolling simulation. The workpiece is divided into 10 layer along the longitudinal direction, with 64 elements in 2–roll modeling and with 101 elements in 4–roll modeling in each layer. The roll is treated as a rigid surface.

3. RESULTS AND DISCUSSIONS

3.1 Characteristics of single pass rolling conducted by 2 and 4–roll mill

The lateral spread ratio and the reduction ratio in area are shown in Fig.3. Symbol I, ○, and solid lines are the experimental and calculated results respectively, where μ is the coefficient of friction. The experimental lateral spread ratio is the average values at both sides of the point ("A,A' " in Fig.2) on the diagonal line of the cross section of the rod. As a well–known fact, the lateral spread ratio in the case of 2–roll rolling increases by the increase in the reduction ratio of area. On the other hand, when adopting a 4–roll rolling process as the rolling method, spreading is restrained and shrinking tends to occur. The calculated spread ratios, where the coefficients of friction are 0.2, 0.15 and 0.12, are compared with the experimental values. It is seen that the spread ratio is influenced by the friction on the interface between the workpiece and the roll, and the

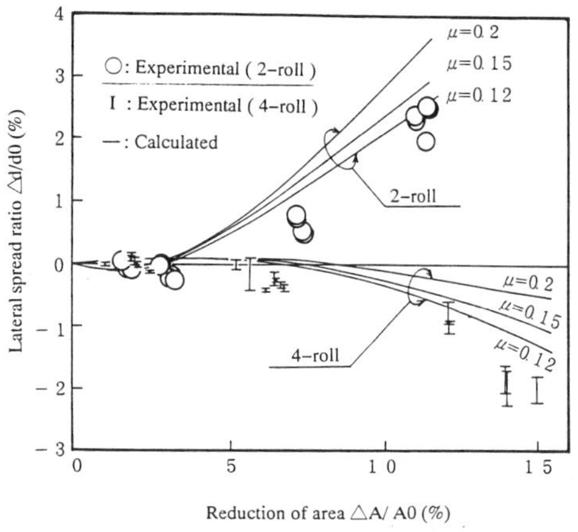

Fig.3 Lateral spread ratio vs. reduction of area.

experimental ratio is coincide with the calculated values of the coefficient of friction 0.12. The calculated lateral spread ratio of the workpiece rolled by 2–roll mill at the 11% reduction of area is 2.5%, which is in good agreement with experimental values. At the 15% reduction of the area by 4–roll mill, the calculated shrinking ratio in the condition of μ= 0.12 is −1.3%, whereas the experimental value is −1.8% ~ −2.2%. Although the calculated value is smaller than the experimental value about 30%, it gives values for the shrinking ratio showing in general a good agreement. These mean that the elongation ratio obtained by the 4–roll mill is larger than that obtained by the 2–roll mill and the 4–roll mill can give the workpiece more precise roundness of the cross section.

The spread ratio increases as the coefficient of friction becomes large. In the case that the reduction area of workpiece rolled by 2–roll rolling is 11%, the spread ratio at the coefficient of friction μ = 0.12 is 2.5%, the ratio at μ=0.15 is 2.75% and the value at μ=0.2 is 3.3%. In the case of 4–roll rolling at the reduction of area 15%, the shrinking ratio at μ = 0.12 is −1.3% and the value at μ = 0.2 is −0.5%. The lateral spread ratio rolled by 4–roll mill is largely affected by friction compared to that of 2–roll mill.

Figure 4,5 show the rolling load vs. the reduction of area and the total rolling torque vs. the reduction of area respectively. The total torque is calculated from the sum of the each roll force, that is, the sum of two roll forces in 2–roll rolling and the sum of four roll forces in 4–roll rolling. Symbols ○ and □ are the experimental results of 2–roll and 4–roll rolling respectively. Solid lines are the calculated results at the coefficient of friction 0.12. As is seen in the figure, both the calculated rolling load and the torque are good agreement with the experimental values. The predicted rolling force and torque at the coefficient of friction 0.15 and 0.2 are equal to that of 0.12. They are independent of friction condition on the interface between the roll and the workpiece.

Figure 6 ~9 show the deformed shape of the segmental overviews and the cross–sectional pro–

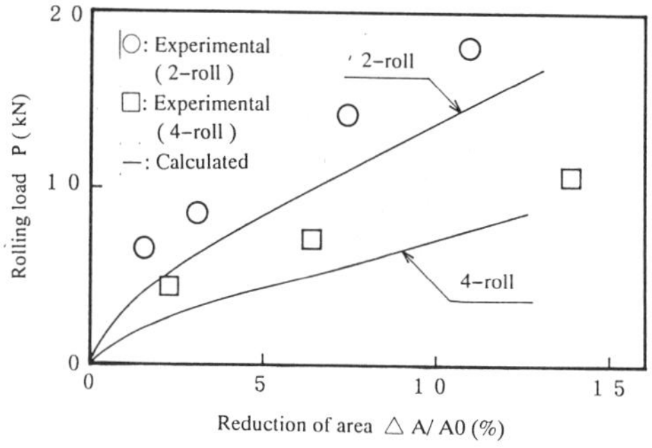

Fig. 4 Calculated and experimental rolling loads at the 2 and 4-roll rolling.

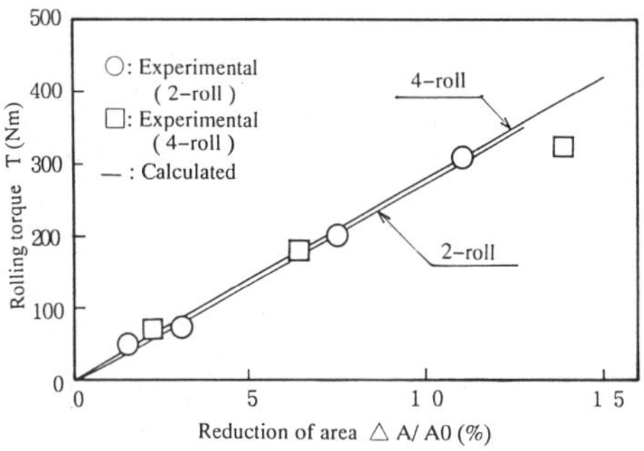

Fig. 5 Calculated and experimantal rolling torque at the 2 and 4-roll rolling.

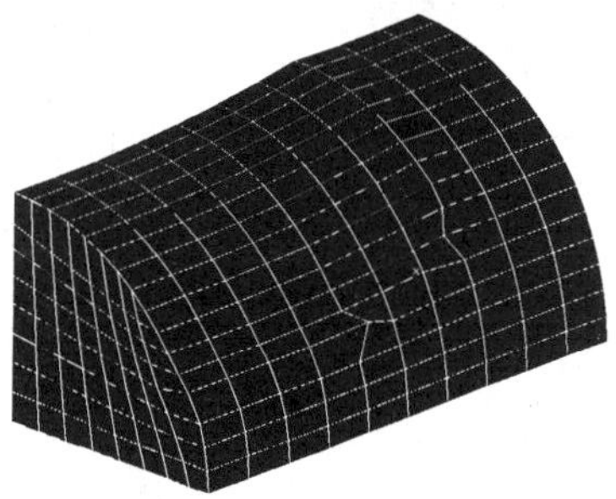

Fig. 6 Three dimensional deformed profile by the 2-roll rolling.
(reduction of area : 11.59%)

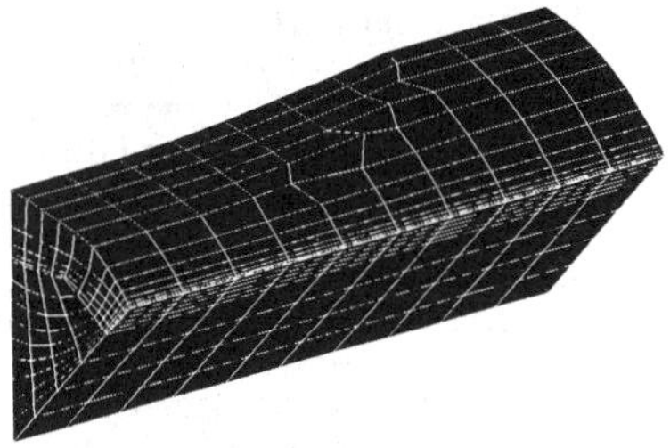

Fig. 7 Three dimensional deformed profile by the 4-roll rolling.
(reduction of area : 10.99%)

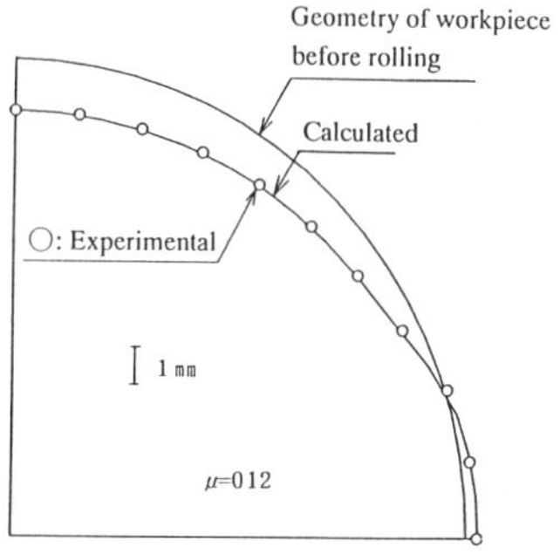

Fig. 8 Cross sectional deformed profile by the 4-roll rolling.
(reduction of area : 11.59%)

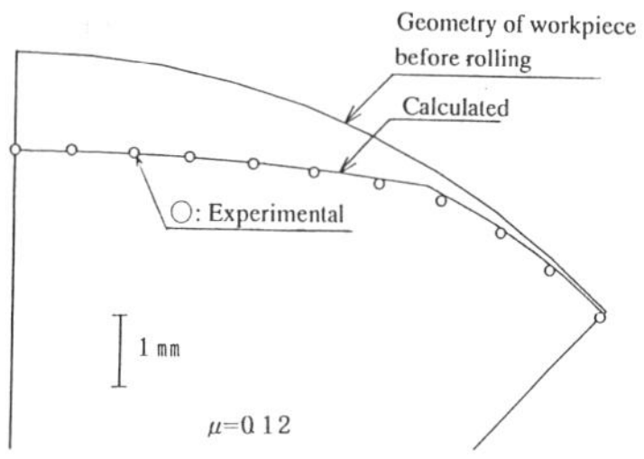

Fig. 9 Cross sectional deformed profile by the 4-roll rolling.
(reduction of area : 10.99%)

files of the workpiece when the reduction ratio in the area is 11.59% at the 2–roll rolling and 10.99% at the 4–roll rolling. It is also seen that, in the case of the 2–roll rolling, the lateral spread is occurred at the roll gap. On the other hand, when adopting 4–roll rolling, negative spread, that is, shrinking tends to occur. The calculated results are in good coincide with the experimental values.

3.2 Characteristics of 4–roll multipass rolling

The characteristics of the oval–to–round pass performed by the 4–roll mill is investigated. Roll pass at the first, the second and the third pass were designed by considering the lateral shrinking ratio to minimize the deviation from the roundness of the rod. Figure 10 shows the relationship between the calculated deviation from the roundness and the distance of roll grooves at the first/second pass, when the target diameter of the product is 22mm. The employed distance between roll grooves at the first pass is 22.1mm ~ 22.35mm and the one at the second pass is 22mm. The simulation was stopped at the second pass when the diameter at the roll gap is smaller than 22mm. Consequently, when the distance of the roll grooves at the first pass is set to be 22.2mm and that of the second/third pass is 22mm, the accuracy 322 μm of the deviation from the roundness is realized at the second pass and that of 160 μm is realized at the third pass. In this manner, the theoretical optimized value at the second and the third pass is obtained by considering the shrinking ratio.

The curves of the deviation from the roundness of rod versus the target diameter of the product are plotted in Fig.11. The solid line is the calculated results at the second/third pass and the symbol □ are the experimental results at the second pass. In the case when the target diameter of the products is 23mm and 22mm, the accuracy 1% and 1.45% of the deviation from the roundness at the second pass are realized respectively. In the third pass, when the exit diameter is set to be 23mm and 22mm, the accuracy of the roundness is 0.43% and 0.72%. The 22% reduction in area is achieved, when the diameter is set to be 22mm. From the mentioned above, it is numerically verified that the 4–roll rolling achieves large reduction in area by adopting the oval–to–round pass design. The calculated deviation from the roundness at the second pass are well coincide with the experimental results which are shown in the symbol □. This leads to the conclusion that the numerical results exactly simulate the behavior of the 4–roll multipass rolling. The figure also shows that if the allowable deviation from the roundness is 200 μm, the size free range of the workpiece is 2.6mm of which diameter is from 24mm to 21.4mm at the third pass.

Figure 12 shows the calculated deformed workpiece in cross–section at the third pass and the cross–sectional geometry of the roll at the second/third pass relative to the workpiece when the exit diameter of the workpiece is set to be 22mm.

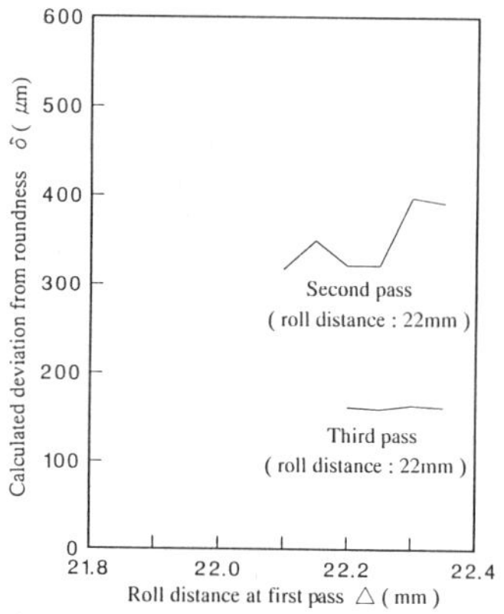

Fig.10 Calculated deviation from the roundness when the target diameter of the products is 22mm.

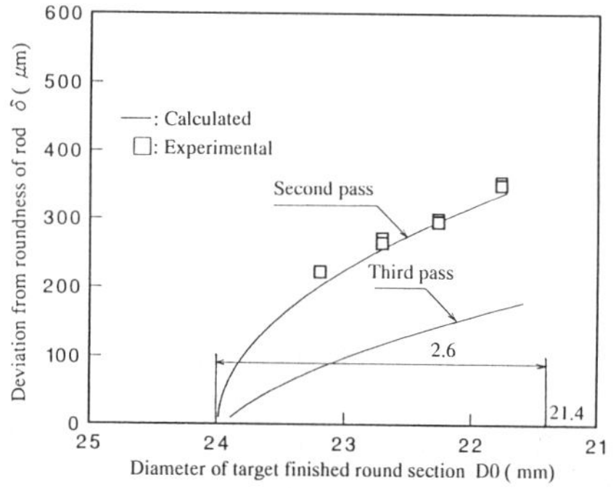

Fig.11 Deviation from the roundness of the rod vs. the target diameter of the products.

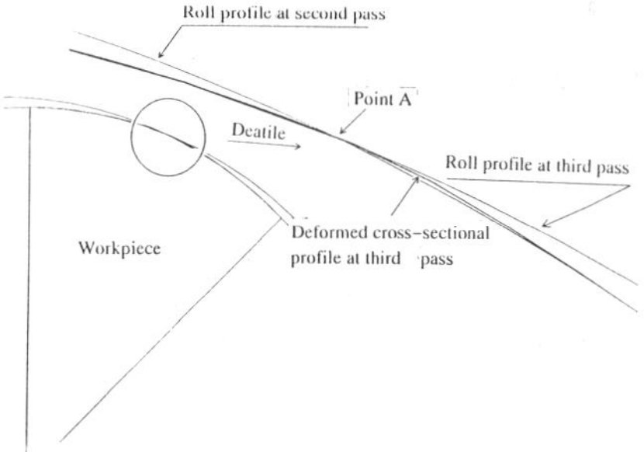

Fig.12 Calculated cross–sectional deformed workpiece at the third pass and the geometry of the roll.

Point "A", which is geometrically determined intersection of the roll at the second and the third pass, means the theoretical deviation from the roundness of the workpiece. The geometrically determined diameter is 22.14mm and its deviation from the roundness from the roundness is 140 μm. The shape calculated by the finite element method bulges at the roll gap of the third pass. This leads to the results that the calculated deviation from the roundness, which is 160 μm, becomes larger than that obtained by roll geometry. Figure13,14 show contour of the

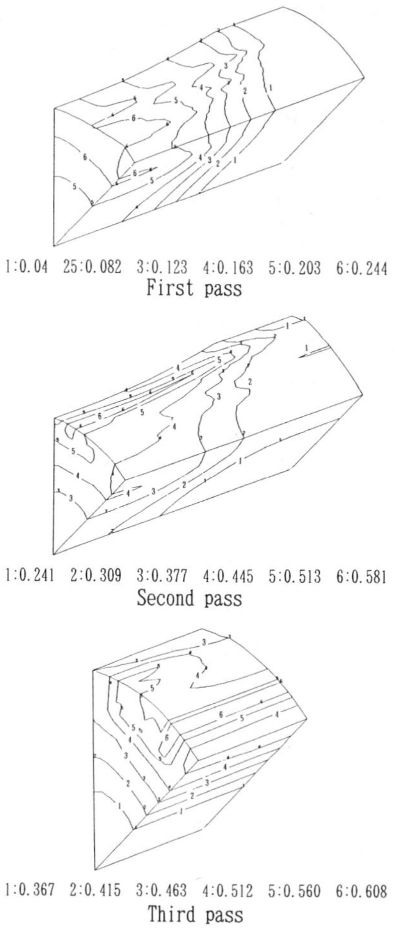

Fig. 13 Equivalent strain.
(target diameter of the products : 22mm)

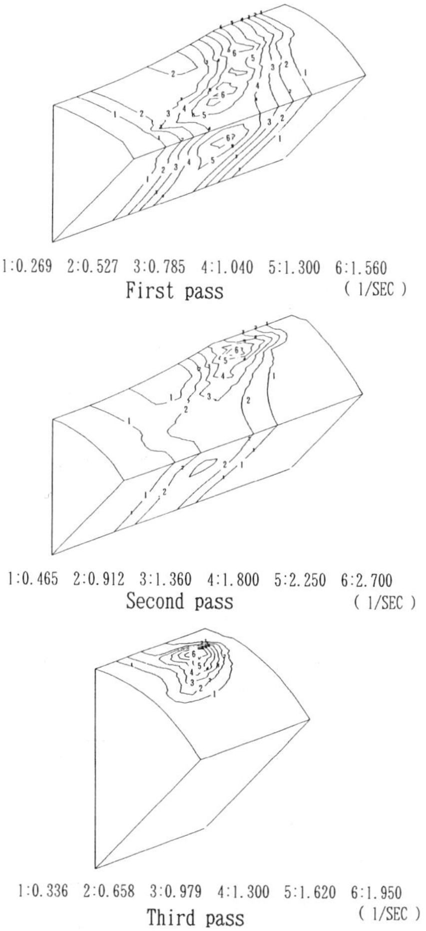

Fig. 14 Equivalent strain rate.
(target diameter of the products : 22mm)

equivalent strain and the strain rate at each passes, when the exit diameter of the workpiece is set to be 22mm. The equivalent strain is higher in the region near the rolled surface of the rod than in the center of the cross-section.

3.3 Influence of tension in the 4-roll rolling

Figure 15 shows the lateral spread ratio at the roll gap as a function of tension. Area reductions of 7.6, 10.13 and 13.66%are plotted in the figure, where the reduction is the value at which the tension equal to 0 MPa. Solid lines and broken lines are the results in the case when the front tension and the back tension are loaded respectively. It is seen that the spread ratio is not so affected by the front tension, whereas the ratio is greatly affected by the back tension. In the reduction 13.66%, the lateral shrinking ratio decreases at the rate of 0.36% when the back compres-

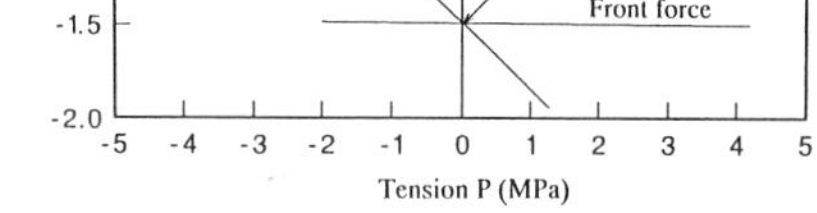

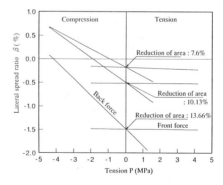

Fig. 15 Lateral spread ratio at the roll gap as a function of tension.

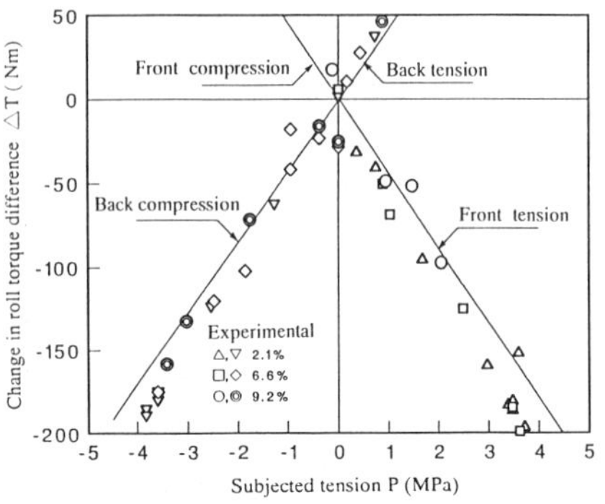

Fig. 16 Change in the roll torque difference with the subjected tension.

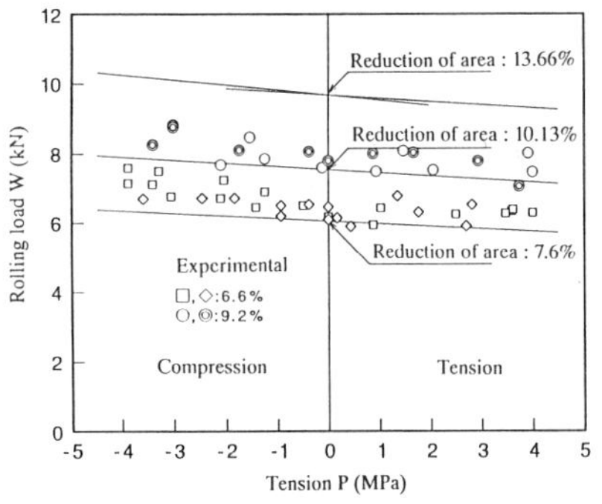

Fig. 17 The rolling load aginst the tension.

sion decreases 1 MPa. Results obtained from the simulation coincide well with the experimental values obtained by Hiraki(1993). These concluded that it is necessary to control the tension sensitively in tandem rolling.

Figure 16 shows the change in the roll torque difference △T with the subjected tension, where △T is the difference between the roll torque (T0) at the free tension and the torque (T) when the tension is subjected. Three kinds of the reduction in area 7.6, 10.13 and 13.66% are plotted in the figure. The the roll torque differences are independent of the tension and these are good agreement with the experimental results. The rolling load at the reduction of area 7.6, 10.13 and 13.66% are plotted against the tension in Fig.17. The loads decrease when the back and the front tension increase . These are also good agreement with the experimental results.

4. CONCLUSIONS

The simulator for the steady-state scheme based on the three dimensional rigid-plastic finite element method has been developed and used for the analysis of 4-roll rolling. The characteristics of the 4-roll rolling are numerically and experimentally evaluated and the following are the conclusions.

(1) The lateral spread of the workpiece rolled by 2 roll mill occur. On the other hand, when adopting a 4-roll rolling process as the rolling method, negative spreading, that is, shrinking tends to occur.
(2) The 4-roll mill gives the workpiece higher elongation than 2 -roll mill.
(3) The lateral spread ratio is greatly affected by the condition of friction on the interface between the roll and the workpiece.
(4) 4-roll rolling achieves high accuracy and large reduction in area by adopting the oval-to-round multipass design.
(5) The lateral spread ratio at the roll gap is affected by the back tension, whereas that is not so affected by the front tension.

REFERENCES

Hiraki,K., Inami,Y., Suzuki,S., Ochi,S. and Kamoto,S., 1992. The Proceedings of the 1992 Japanese Spring Conference for the Tech. of Plasticity: 335-338.

Hiraki,K., Inami,Y., Ochi,S. and Yoshida,K., 1993. Japan Soc. Tech. Plasticity 44 : 127-130.

Inami,Y., Hiraki,K., Suzuki,S., Ochi,S. and Kamoto,S., 1991. J. Japan Soc. Tech. Plasticity 42 : 437-440.

Kihara,S., Ochi,S., Hiraki,K. and Mori,K., 1994a. The Proceedings of the 1994 Japanese Spring Conference for the Tech. of Plasticity: 323-326.

Kihara,S., Ochi,S., Hiraki,K. and Mori,K., 1994b. J. Japan Soc. Tech. Plasticity 45: 415-418.

Mori,K. and Osakada,K., 1990a. TRANS. OF THE JSME, 56-525: 268-273.

Yanagimoto,J. and Kiuchi,M., 1990b. J. Japan Soc. Tech. Plasticity 41 : 45-48.

Simulation of Materials Processing: Theory, Methods and Applications, Shen & Dawson (eds)
© 1995 Balkema, Rotterdam. ISBN 90 5410 553 4

Anisotropic 3-D modelling of hot rolling and plane strain compression of Al alloys

P.Noat, P.Montmitonnet & Y.Chastel
CEMEF, Sophia-Antipolis, France

R.Shahani
Pechiney Centre de Recherches de Voreppe, France

ABSTRACT : The effect of flow stress anisotropy is rarely considered in metal forming problems. Here, anisotropic constitutive equations have been implemented in a 3-D finite element code using Hill's (1948) quadratic criterion. Hill's parameters have been determined by tensile tests at room temperature for a recrystallised 1XXX and 5XXX alloys. A Sellars and Teggart constitutive visco plastic law has been derived from plane strain compression (PSC) tests for both alloys. The anisotropy of the constitutive behavior has been found to influence the force and the lateral spread during PSC, as well as the edge profile at the exit of a rolling pass.

1. INTRODUCTION

Knowing precisely the constitutive equation and the anisotropy at high deformation is of great importance to achieve a FE simulation of metal forming processes. Anisotropy is a major phenomenon during sheet metal forming as it is responsible for the formation of ears during deep drawing. During hot rolling, it may influence the width and the edge profile of the sheet [1]. Two kinds of approaches have been proposed so far to take into account the anisotropy of the material behavior. The first approach consists in implementing an anisotropic criterion in FE codes initially based on the isotropic Von Mises criterion. This has been done in a viscoplastic model [1] [2] using the 1948 Hill's quadratic criterion [3] or a non-quadratic Barlat criterion [4]. Crystallographic texture is the main cause of anisotropic rheology for aluminium at high temperature. Spatial variations of the flow stress is generally in the range of 10-15% whereas stress measurements are conducted with an error of 3-5%. This makes a precise determination of the parameters of a Hill's or a Barlat's criterion difficult. It is all the more true at high temperature as stress values are very sensitive to temperature variations. An alternative method is to derive these parameters from the measurement of Lankford parameter [5]. Another difficulty in this approach is to check that the criterion which has been chosen is accurate to represent the constitutive anisotropy of the material.
The second approach is to include polycrystalline models into a FE code. It allows for the determination of stress-strain rate relationship and the prediction of textural evolution with the deformation from an initial crystallographic texture. This approach has been applied to rolling of aluminium alloys by Dawson *et al.* ([6], [7]). Parallely, recent studies on single crystals point out the possible activation at high temperature of crystallographic slip systems different from the classical (111) <110> slip systems of FCC crystals [8]. Work is still in progress to validate the relationship between texture and anisotropy of constitutive behavior using polycrystalline models. Morever this approach leads to large computational time for 3-D problems.
In this study, we describe the introduction of a quadratic Hill's criterion in a viscoplastic F E code. A constitutive Sellars and Teggart law has been used to take into account the variations of flow stress as a function of temperature and strain-rate. This law has been derived from PSC (plane strain compression) tests at different temperatures and strain rates for two families of aluminium alloys. For these alloys, the Hill's parameters were determined by measuring the stress and the Lankford coefficient during tensile tests at room temperature. Rolling and PSC have been simulated using these experimental data. The predictions of the force and the spreading during the test have been compared for isotropic and anisotropic materials.

2. FORMULATION

2. 1 Principles

The code is based on a flow formulation. The flow field is calculated by solving the principle of virtual power for any virtual kinematically admissible velocity field v*.

$$\int_{\Omega(t)} \sigma:\dot{\varepsilon}^* \, dV - \int_{d\Omega(t)} T.v^* dS = 0 \qquad (1)$$

σ being the stress tensor, $\dot{\varepsilon}^*$ the strain rate tensor calculated from v*, T the surface stress and Ω the material domain with boundary $d\Omega$.

It could be written as :

$$\int_{\Omega(t)} \sigma(v^i N^i):N^i_{,j} \, dV = \int_{d\Omega(t)} T.N^j dS \qquad (1')$$

where N^i are Galerkin shape functions and a comma denotes derivation. Incompressibility is enhanced via a penalty method.

The formulation can be chosen to be Eulerian or Updated Lagrangian. For the Eulerian approach, the displacement of the free surface is calculated for the surface to remain normal to the flow field. The Lagrangian approach leads to a time discretized form. At time t, after the end of the increment Δt, the position vector X is updated :

$$X(t+\Delta t) = X(t) + V(t)\,\Delta t$$

The temperature is calculated by solving the equation:

$$\rho\, C \frac{dT}{dt} = k\, \Delta T + \dot{W}$$

$\dot{W} = \sigma:\dot{\varepsilon}$ being the plastic dissipated power.

Boundary conditions include the heat dissipated by friction on the die-material interface, the conduction between the material and the die, and the convection on the free surface. A Coulomb law is used to take into account the friction between the die and the material :

$$\tau = -\mu\, \sigma_n \frac{\Delta Vt}{|\Delta Vt|}$$

σ_n being the normal stress at the surface of the tool, μ the Coulomb coefficient, ΔV_t the sliding velocity.

2.2 Rheology

A Sellars and Teggart [9] constitutive law with a Voce hardening law has been introduced in the code, this law depends on 8 parameters :

$$\bar{\sigma} = \sigma_0 + (\sigma_{ss} - \sigma_0)\,(1 - \exp(-C\varepsilon))^{1/2} \qquad (2)$$

with $\sigma_0 = K\sqrt{3}\ (\sqrt{3}\dot{\varepsilon})^{m_0}$, $\sigma_{ss} = 1/A \ \mathrm{argsh}\left(\left(\frac{Z}{Z_0}\right)^{ms}\right)$

$$Z = \dot{\varepsilon} \exp\left(\frac{\beta}{T}\right) \quad \text{and} \quad C = C_0 - \frac{C_1}{T}$$

ε the generalized deformation, σ_0 being the stress at $\varepsilon = 0$, σ_{ss} the steady state flow stress, $\bar{\sigma}$ the plastic stress, and T the temperature.

By analogy with the Von Mises criterion, the Hill's criterion can be written as :

$$f(\sigma) = F(\sigma_{zz} - \sigma_{yy})^2 + G(\sigma_{zz} - \sigma_{xx})^2 + H(\sigma_{yy} - \sigma_{xx})^2 + 2L\sigma_{yz}^2 + 2M\sigma_{xz}^2 + 2N\sigma_{xy}^2 = 2\sigma_t^2$$

F, G, H, L, M, N are non-dimensional parameters. The stress σ_t can be considered as the equivalent tensile stress (σ_{eq}) if we impose the normative relationship :

$$\Sigma = FG + GH + FH = 3$$

We apply the principle of normality to the material :

$$\dot{\varepsilon} = \lambda \frac{df}{d\sigma}$$

and we obtain a tensorial relationship between the stress and the strain rate deviator :

$$s = \frac{1}{6\lambda} \mathrm{dev}\,(C:\dot{e})$$

with $C_{ijkl} = C_{0ij}\,\delta_{ik}\delta_{jl}$

$C_{0xx} = F$, $C_{0yy} = G$, $C_{0zz} = H$

$C_{0yz} = 3/L$, $C_{0xz} = 3/M$, $C_{0xy} = 3/N$

The value of λ is given by the egality of the plastic power : $s:\dot{\varepsilon} = \sigma_{eq}\,\dot{\varepsilon}_{eq} = 4\,\lambda\,\sigma_{eq}2$

By analogy with the isotropic case, we choose $\dot{\varepsilon}_{eq}$ as :

$$\dot{\varepsilon}_{eq}^2 = 2/3\ \dot{\varepsilon}:C:\dot{\varepsilon}$$
$$= 2/3\left(F\dot{\varepsilon}_{xx}^2 + G\dot{\varepsilon}_{yy}^2 + H\dot{\varepsilon}_{zz}^2 + 6/L\dot{\varepsilon}_{yz}^2 + 6/M\dot{\varepsilon}_{xz}^2 + 6/N\dot{\varepsilon}_{xy}^2\right)$$

Finally, we obtain the relationship :

$$s = \frac{2\,\sigma_{eq}}{3\,\dot{\varepsilon}_{eq}} \mathrm{dev}\,(C:\dot{e}) \qquad (3)$$

We can easily verify that a unit value for F, G, H and a value of 3 for L, M, N leads to the isotropic Von Mises criterion.

A more detailed presentation is given in ref. [3] [10].

The strain rate and thermal dependency of the stress is taken into account by identifying σ_{eq} to $\bar{\sigma}$:

$$\sigma_{eq} = \bar{\sigma} = \sigma_0 + (\sigma_{ss} - \sigma_0)\,(1 - \exp(-C\varepsilon))^{1/2}$$

with the definition of eq. 2

This equality is used in eq. 3 and introduced in eq. 1 to solve the principle of virtual power using a Newton-Raphson scheme.

3. FLOW STRESS MEASUREMENTS OF Al ALLOYS

At high temperature the flow stress depends on temperature and strain rate. As a first approach to investigate the directional dependency of the flow stress, we make the hypothesis that the constitutive anisotropy can be related only to the crystallographic texture. Providing that the active crystallographic slip systems will remain the classical (111)<110> slip system for FCC metal even at high temperature (this point is discussed in ref.[8]), we can assume that this anisotropy will not depend on temperature. Thus, we can determine the Hill's coefficients by conducting mechanical tests at room temperature.
The coefficients of a Sellars and Teggart constitutive law described in eq.2 have been fitted from the results of plane strain compression tests at different temperatures and strain rates for a 1XXX and a 5XXX aluminium alloy. Temperature and strain rate were measured on-line during the test (strain rate was obtained from the speed of the punch). These data were used to calculate the stress using the constitutive equation determined from the experimental data. The form of the law allows for a good fitting of experimental curves even at very high -or very low- strain rates as it can be observed on fig1.

Tensile stresses and Lankford coefficients were measured at seven angles relative to the rolling direction on two recrystallised sheets, a 1XXX and a 5XXX aluminium alloy. The tensile tests were conducted at room temperature. The recrystallised 1XXX alloy presented a strong Cube texture. The 5XXX recrystallised sheet showed a nearly random texture with no dominant component and the sheet was considered as isotropic.
A least square method has been used to extract the Hill's parameters from the measured stresses and Lankford coefficients by means of the relations :

$$\sigma_H^2(\alpha)=\frac{2\sigma_t^2}{(H+G)+(F-G)\sin^4(\alpha)+(2N-3H-G)\sin^2(\alpha)\cos^2(\alpha)}$$

$$r(\alpha)=\frac{H-(F+G+4H-2N)\sin^2(\alpha)\cos^2(\alpha)}{F\sin^2(\alpha)+G\cos^2(\alpha)}$$

We consider the functional :

$$\Phi=\Sigma_{i=1,7}(1-\eta)\left(\frac{\sigma_H(\alpha i)-\sigma_{exp}(\alpha i)}{\sigma_t}\right)^2+\eta(r_H(\alpha i)-r_{exp}(\alpha i))^2$$

σ_t being the equivalent stress at room temperature. It could be calculated as the average of stress measurements: $\sigma_t=\Sigma_{i=1,7}\sigma_{exp}(\alpha)^2$,
σ_{exp} and r_{exp} being the experimental stress and the Lankford coefficient, σ_H and r_H the values calculated from the Hill's parameters.

The relevant parameters F, G, H and N minimize the functional Φ; they can be obtained by solving the following linear system :

$$\frac{\partial\Phi}{\partial F}=\frac{\partial\Phi}{\partial G}=\frac{\partial\Phi}{\partial H}=\frac{\partial\Phi}{\partial N}=0$$

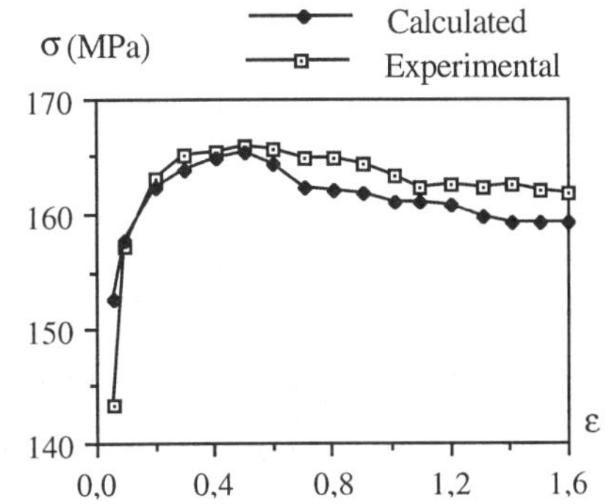

Fig 1 : Experimental and calculated stresses using the constitutive equation during PSC test at 300°C and 0.1 s^{-1} for a 5XXX aluminium alloy.

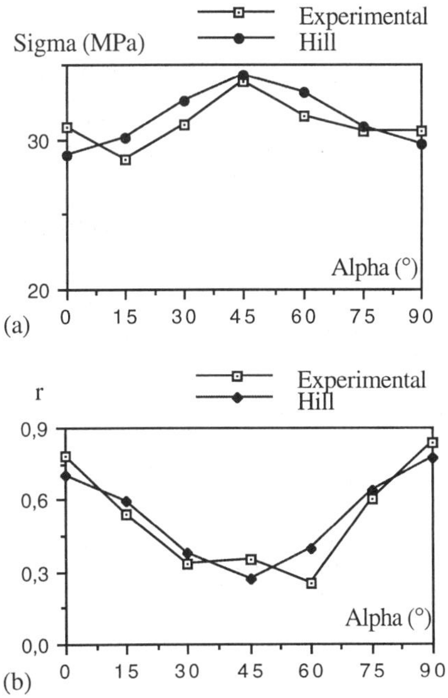

Fig 2 : Yield stresses (a) and Lankford coefficients (b) measured and calculated from the as-determined Hill's parameters at seven angles relative to the rolling direction for a 1XXX recrystallised alloy.

Table 1 : Hill's parameters for a 1XXX recrystallized alloy.

F	G	H	N
1.23	1.34	0.85	1.97

It can be seen in Fig. 2 that the Hill's criterion gives a good description of the anisotropy for this alloy. The values obtained for a recrystallized 1XXX alloy are given in table 1 (Ox is the rolling direction, Oy the transverse direction and Oz the normal direction).

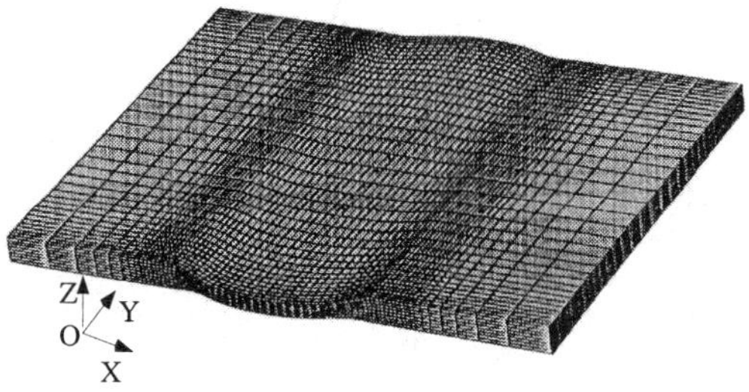

Fig 3 : A deformed mesh after PSC.

4 APPLICATIONS

4.1 FEM Modelling of PSC

The constitutive equations and the Hill's parameters determined as previously indicated were used to simulate the PSC test. A deformed mesh is shown in fig.3; it contains 26400 nodes; but due to the symmetry of the process, the calculation is done on one eighth of the sample. A real test is simulated with a constant strain rate $\dot{\varepsilon}$.
The characteristics of the test were the following :
$\dot{\varepsilon} = 0.1 s^{-1}$, initial thickness $h_0 = 3.1$ mm, final thickness $h_f = 1.6$ mm , initial temperature T = 300°C

During the PSC test, the force is very sensitive to the friction coefficient. To determine the constitutive equation, friction during the test was minimised using a PTFE film as a lubrificant and the Coulomb coefficient was assumed to be equal to zero. To verify this hypothesis, FEM simulations were conducted using several Coulomb coefficients ranging from 0.0 to 0.2 and the calculated force curves were compared with the experimental curve. The results are presented in fig 4. We observed that the curve obtained for a Coulomb coefficient equal to zero matches well the experimental data. With that coefficient, the difference between the experimental and calculated curves is of the same order of magnitude as the one obtained when fitting the rheological law with the same coefficient. However, we cannot conclude from these results that the Coulomb's coefficient is strictly equal to zero during PSC test lubrificating with a PTFE film because the constitutive law implemented into the FEM code was determined making this same hypothesis.
The spreading of the sample after PSC is another parameter which proves to be very sensitive to friction. Gelin *et al.* [11] conducted a 3-D FEM simulation of PSC. They related the spreading to the Coulomb coefficient which gives a value of μ slightly higher (0.05) for aluminium under the same experimental conditions. FEM calculations showed that spreading after PSC depends only on friction and is nearly insensitive to the type of alloy. But the simulations were conducted with an isotropic code and mechanical anisotropy was not considered [11].

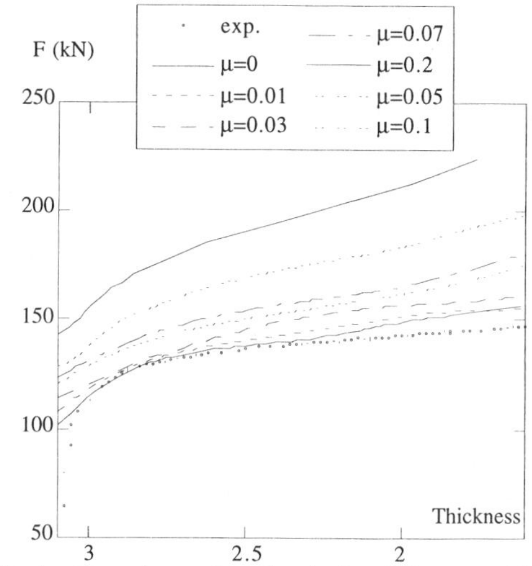

Fig 4 : Experimental and calculated forces with different Coulomb's friction coefficients (μ) for PSC of a 5XXX alloy.

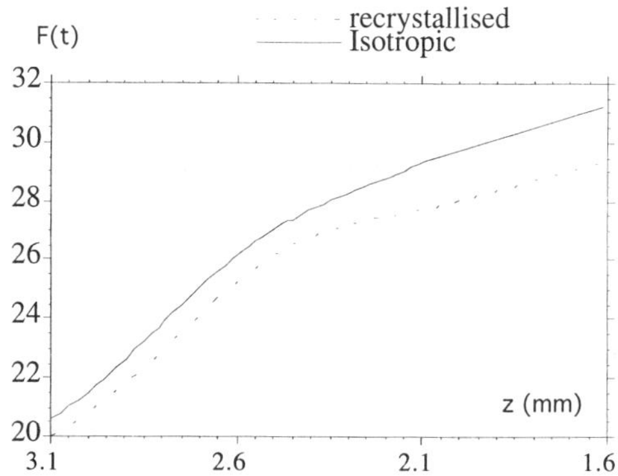

Fig 5 : Computed force for a 1XXX alloy

We carried out the same simulation using different friction coefficients for isotropic 1XXX and 5XXX alloys and a recrystallised 1XXX alloy with the Hill's parameters determined as explained

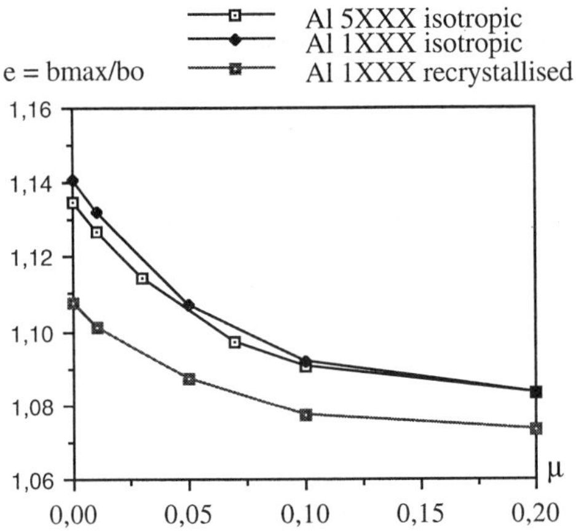

Fig. 6 : Spreading of the sample after PSC for different alloys and friction coefficients μ (bmax = maximum final width and b_0 = initial width). Simulations at T = 300°C, $\bar{\varepsilon} = 0.1 s^{-1}$.

previously. In fig.5, the calculated force with a μ coefficient egal to zero is displayed for a non-textured (isotropic) and a recrystallised 1XXX alloys. The curve for the isotropic alloy is higher by nearly 10% . The force can be related to the hardness along the normal axis of the die-metal interface (0z). Using Hill's criterion we obtain :

$$\sigma_{0zz} = \sqrt{\frac{2}{F+G}}\,\sigma_{eq}$$

$\sigma_{0zz} = \sigma_{eq}$ for the isotropic sheet

$\sigma_{0zz} = 0.88\,\sigma_{eq}$ for the recrystallised sheet.

This is the order of magnitude of the difference observed on the simulated force curves (fig. 5).

Fig. 6 shows that the spreading is sensitive to the mechanical anisotropy. For two isotropic alloys, the slight difference observed at low friction was related to the strain rate sensitivity of the flow stress. Spreading was significantly higher for the isotropic alloy than for the recrystallised alloy. This observation was confirmed by preliminary experimental trials, although the observed lateral spread was lower than that calculated in Figure 6. This sensitivity decreases with increasing values of the μ coefficient. During the test there is a competition for the material to flow along the Ox and the Oy direction to accommodate the deformation imposed by the punch. Morever the spreading -i.e. the deformation along the Oy direction- will produce a shearing in the Oxy plane. Then, the importance of spreading during the test can be related to the difference of hardness in these two directions, σ_{0xx} and σ_{0yy}, and to the value of the plastic shearing stress σ_{0xy} (which is determined by the value of the N parameter). Higher values of σ_{0yy} and σ_{0xy} will lead to a lower spread of the sample after the test. The maximum value of the shearing stress was found to be larger than the value of the deviatoric stress along the Oy direction (12.9 MPa vs 8.87 MPa for the isotropic 1XXX alloy) which means that the material undergoes a shearing strain higher than the tensile Oy strain. On that example it is difficult to separate both influences. With the two sets of parameters we used, we obtain for the stresses on the Oxy plane :

$$\sigma_{0xx}=\sqrt{\frac{2}{H+G}}\,\sigma_{eq}\ ;\ \sigma_{0yy}=\sqrt{\frac{2}{H+F}}\,\sigma_{eq}\ ;\ \sigma_{0xy}=\sqrt{\frac{1}{N}}\,\sigma_{eq}$$

Table 2 : Plastic stresses in the 0xy plane for a 1XXX alloy

	σ_{0xx}	σ_{0yy}	σ_{0xy}
Isotropic	σ_{eq}	σ_{eq}	$0.577\ \sigma_{eq}$
Recrystallised	$0.957\ \sigma_{eq}$	$0.978\ \sigma_{eq}$	$0.712\ \sigma_{eq}$

For the recrystallised alloy both values of σ_{yy} and σ_{xy} will tend to minimise the spreading ($\sigma_{0yy} > \sigma_{0xx}$ and σ_{0xy} is higher for the recrystallised sheet) which matches the trend of the calculated results.

4.2 FEM Modelling of the rolling process

The same analysis has been conducted for rolling for the two 1XXX alloys. The geometry of the rolling pass was the following :

Initial thickness	30 mm
Final thickness	20 mm
Width of the sheet	1000 mm
Radius of the roll	300 mm
Speed of the roll	3.33 m s^{-1}

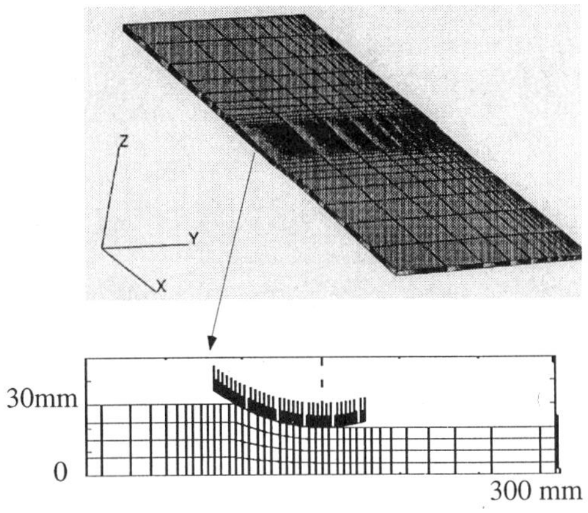

Fig 7 : Initial mesh for an Eulerian FEM simulation of rolling.

Contrary to PSC, rolling is a steady state process, ie. a permanent regime is expected to be reached a certain

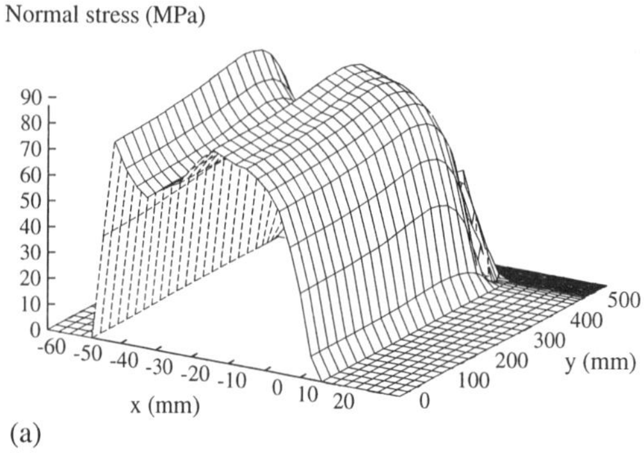

(a)

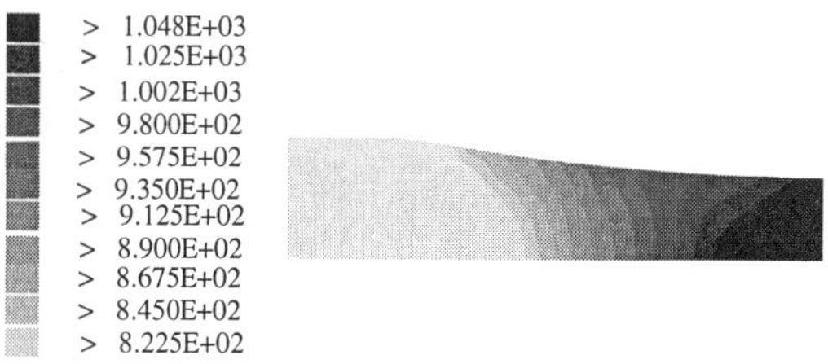

(b)

Fig 8 : Normal stress (a) and speed along the 0x direction (b) during rolling of a 1XXX aluminium alloy

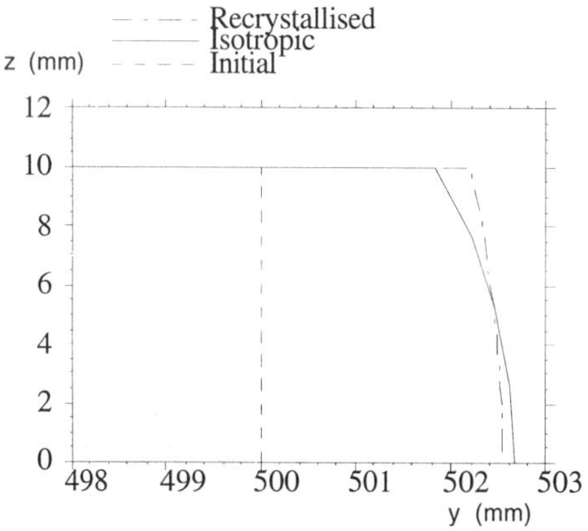

Fig.9: Computed edge profile after rolling for a 1XXX aluminium alloy

time after engaging the sheet. This allowed us to use the Eulerian version of our FEM code starting with the geometry displayed in fig 7. This mesh has 13400 nodes but, due to the symmetry of the process only a quarter is considered.

We perform the simulation for a 1XXX family alloy, non textured (isotropic) and recrystallized. The roll was considered as rigid (elastic and thermic bending was not taken into account) and thermal exchange was modelled by convection at the die-material interface. The calculations show that the flow field and the friction hill (fig 8) are nearly insensitive to the anisotropy of the constitutive behavior. The spreading and the force were found to be less sensitive than for PSC. Still, the computed force is slighty higher for the isotropic alloy (390 t/m vs 378 t/m). The average final spread is more or less the same for both sets of Hill's parameters but the edge profiles are different. The bulge is larger for the 1XXX isotropic alloy; this could be due to the value of the Hill's coefficients relative to shearing (L, M, N) which maximise the shearing stress for the recrystallised 1XXX alloy.

CONCLUSIONS

A strategy to determine anisotropic constitutive laws and implement them in a FEM code has been proposed. The use of Hill's criterion allowed a good representation of the anisotropy of the constitutive behavior for a recrystallised 1XXX alloy. The sensitivity to anisotropy during plane strain compression and rolling were investigated by means of FEM simulations. The spreading and the force after the PSC test have been found significantly influenced by anisotropy. For the rolling case under study, the effect is less conspicuous, but still visible. This effect could be related to the fact that the reduction was smaller during the computed rolling pass than during the PSC test. However, the material undergoes several rolling passes and the effects will accumulate.

REFERENCES

[1] M. E. Karabin R. E Smelser & R. Becker 1992. NUMIFORM92, 725-730. Rotterdam : Balkema

[2] P. Montmitonnet & J. L Chenot 1995. *J. Mat Proc. Tech.* (in press)

[3] R. Hill 1948., *Proc. Met. Soc.* A 193: 281-297

[4] F. Barlat, D. Ledge & J. C. Brem 1991. *Int J. Plast.* 7: 693-712

[5] American Society for Testing of Material, "Standart Test Method for Plastic Ratio r for Sheet Metal" Designation E-517-92 A

[6] K. K. Mathur, P. R. Dawson & U. F. Kocks 1990. *Mech. of Mat.* 10: 183-202

[7] Y. Chastel 1993., Ph.D., University of Cornell (USA)

[8] C. Maurice 1993., thesis -in french- Ecole des Mines de St Etienne (France)

[9] F. Bay & J. L. Chenot 1990. *Eng. Comp.* 7: 235-240

[10] C. M. Sellars & W. J. McG Teggart 1966. *Mem. Sci. Rev.* 63: 731

[11] J. C. Gelin, O. Ghouatti & R. Shahani 1994. *Int J. Mech. Sc.* 9: 773-796

Simulation of Materials Processing: Theory, Methods and Applications, Shen & Dawson (eds)
 ISBN 90 5410 553 4

Through-thickness variations in multi-stand rolling

S. Rogers & D.J. Browne
Alcan International Ltd, Banbury Laboratories, UK

ABSTRACT: During hot rolling, there can be considerable variation in strain rate, temperature and strain histories through the thickness of the material passing through the roll bite. For the aluminium alloys considered here, this can have a strong influence on both the local microstructural and physical properties. Using the finite element model, Hickory, developed at Cornell, simulations of single stand rolling have been performed and comparisons with experiments made. The effects of different friction and heat transfer conditions at the roll-sheet interface have been investigated.

By considering the cooling in the region between stands using a finite difference model, it has been possible, by linking the two models, to simulate multi-stand rolling. This leads to asymmetries in the temperature profile, although the strain and strain rate distributions remain reasonably symmetrical. Predictions of the thermo-mechanical history through the thickness of the material will be shown.

Introduction

In order to understand the complex behaviour of metals as they are rolled down to final gauge, it is necessary to have good predictive methods for calculating the temperature, strain-rate and strain distributions. These quantities are difficult to measure and hence a mathematical model of the process is needed. The model needs to be able to calculate the plastic flow occurring in the material coupled with the associated temperature changes. Langrangian methods have been used extensively for this sort of problem in the past [1,2]. However, this method is computationally expensive with remeshing and many time steps usually required throughout the calculation. A more attractive method for solving essentially steady state problems is the Eulerian method which solves the problem on a fixed domain. Examples of this method are given in references 3,4 and 5.

A suitable code for this purpose is Hickory, a finite-element program developed at Cornell University.

As commercial rolling operations involve several passes, the model needs to be able to transfer the results of one calculation to the next calculation. Also thermal and mechanical effects can occur between two passes (due to interstand cooling) and this also needs to be modelled.

Mathematical Formulation

The formulation is Eulerian and the equations for the laws of conservation of mass and momentum can be written as ;

$$\frac{\partial \rho}{\partial t} + \mathrm{div}(\rho \mathbf{u}) = 0 \qquad (1)$$

$$\mathrm{div}(\sigma^T) + \mathbf{b} = 0 \qquad (2)$$

Here ρ is the density, $\mathbf{u}$ is the velocity, σ is the Cauchy stress tensor, $\mathbf{b}$ is a body force (inertia terms being neglected) and t is time . For metal deformations the flow can be considered incompressible and hence equation (1) becomes;

$$\mathrm{div}(\mathbf{u}) = 0 \qquad (3)$$

The kinematic equation is ;

$$\mathbf{d} = \mathrm{grad}(\mathbf{u}) \qquad (4)$$

where $\mathbf{d}$ is the deformation rate tensor.

The constitutive equation is of the form;

$$\sigma' = 2\mu\mathbf{d}' \qquad (5)$$

where

$$\sigma' = \sigma + p\mathbf{I} \text{ and } \mathbf{d}' = \mathbf{d} \qquad (6)$$

which are the deviatoric stress and strain rate tensor respectively . Also μ is the shear viscosity and p ($= -\mathrm{Tr}(\sigma)/3$) is the hydrostatic pressure.

Defining the equivalent stress (with a similar equation for the equivalent strain rate) as;

$$\sigma' = \mathrm{sqrt}(3/2\ \mathrm{Tr}(\sigma^T.\sigma')) \qquad (7)$$

the relationship between σ' and d' then becomes;

$$\sigma' = 3\mu d' \qquad (8)$$

Equation (7) represents the relationship between flow stress and equivalent strain rate . For non-isotropic materials the viscosity, μ, is replaced by a viscosity tensor. Assuming that the relationship between flow stress and equivalent strain rate (and other variables) is given as;

$$\sigma' = f(d',T,\kappa) \qquad (9)$$

where T is the temperature and κ is a state variable, then the viscosity, μ, is given by;

$$\mu = f(d',T,\kappa)/(3d') \qquad (10)$$

Equations (2), (4), (5), (6) combine to give;

$$\mathrm{div}(2\mu\ \mathrm{grad}(\mathbf{u})) + \mathbf{b} = \mathrm{grad}(p) \qquad (11)$$

Equations (3), (10) and (11) form the basic equations for the mechanical part of the formulation .

The equation for heat transfer within an Eulerian frame of reference is;

$$\rho c \frac{dT}{dt} = \mathrm{div}(k\mathrm{grad}(T)) + Q \qquad (12)$$

where c is the specific heat and k is the thermal conductivity of the material . Q is the amount of work generated by plastic work and is given by;

$$Q = \alpha \mathrm{Tr}(2\mu\mathbf{d}^T\mathbf{d}') \qquad (13)$$

Here, α is the amount of plastic work converted to heat and is taken to be 0.95 .

The finite element formulation of the above equations is the standard variational approach (based on the principle of virtual work) together with a consistent penalty function approach to the pressure coupling problem. (see [3] for details) . Elements commonly used in 2-d models are 6 node triangles or 8 node quadrilaterals (with corresponding elements in 3-d).

The numerical technique to solve the above set of non-linear equations (together with appropriate boundary conditions) is that of a modified Newton-Rhapson scheme. This combines the iterative scheme of the conjugate gradient method with the matrix approach of a Newton-Rhapson scheme. Care is sometimes needed with the solution technique to ensure convergence. Different options are available within the models.

Constitutive Equations

Equation (9) describes the constitutive model used. It specifies the flow stress as a function of strain rate, temperature and a state variable, κ. The choice of state variable is not always obvious and needs a good knowledge of the material being modeled. Variables such as hardness, grain size and porosity have been used previously.

Generally an additional equation is required to determine the evolution of the state variable and this is usually of the following form;

$$\frac{D\kappa}{Dt} = g(d',T,\kappa) \qquad (14)$$

These forms contain various empirical constants which can be fitted to experimental stress strain data. The forms that are applied to rolling problems are given later. Equation (14) can be solved either by a standard Galerkin technique or by integrating (14) along a streamline to determine the state variable evolution. This second method has been used in the results shown later.

Streamlines ($\mathbf{x}=\mathbf{x}(t)$) are determined by integrating the velocity;

$$\frac{\partial \mathbf{x}}{\partial t} = \mathbf{u} \qquad (15)$$

Equations (14) and (15) are solved using predictor-corrector methods.

Applications to Rolling

To extend the model to rolling applications, suitable boundary conditions must be applied and the appropriate constitutive model used.

Referring to figure 1 which shows a typical mesh (for a symmetric case) used for a reduction of 48.3%, the boundary conditions are as follows .

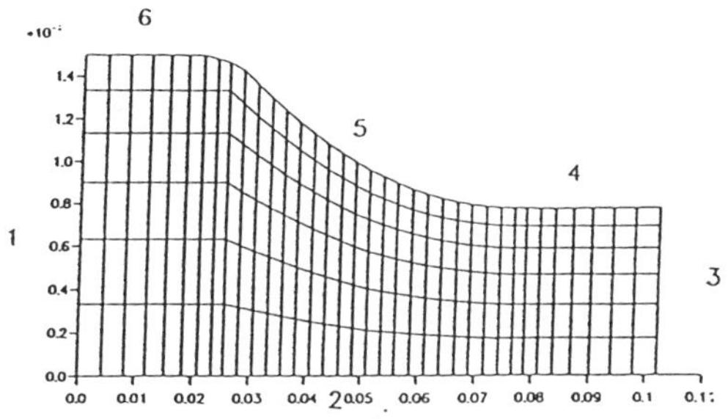

Figure 1. A Typical Finite Element Mesh

1) On boundary 1, the inlet boundary, the incoming temperature and state variable distributions are prescribed.

2) On boundary 2, which defines the symmetry about the centre line of the slab, the vertical velocity is zero and there are no tangential stresses. This boundary is also adiabatic.

3) On boundary 3, the adiabatic outlet boundary, there are usually no stresses, although the effect of exit tensions can be added.

4) Boundaries 4 and 6 are assumed to be adiabatic free surfaces, although heat transfer to air by means of a heat transfer coefficient is possible.

5) On boundary 5, the interface with the roll, the boundary conditions are more problematic. The friction with the roll, which drives the process, is still not completely understood. The friction model assumes the following;

$$\tau = \alpha\sigma'(u_r - u_s) \qquad (16)$$

where τ is the shear stress at the interface, α is a friction factor (between 0 and ∞), u_r is the roll surface speed and u_s is the speed of the slab tangential to the rolls. An alternative friction model is;

$$\tau = \alpha\sigma'(\mathrm{sgn}(u_r - u_s) \qquad (16a)$$

The thermal boundary condition is of the following form;

$$k\frac{\partial T}{\partial n} = h(T_r - T_s) \qquad (17)$$

where n is the normal to the interface, h is a heat transfer coefficient, T_r is the roll surface temperature and T_s is the slab surface temperature. Typically h is in the region of 100kW/m^2/°C.

The roll surface temperature is not considered to be fixed and is calculated using a simple one dimensional model [7]. This assumes that the inlet roll temperature is known, the circumferential conduction is ignored, and that only a thin surface layer of the roll is affected.

The constitutive model used is the PUCHI model and takes the following form;

$$\sigma = \sigma_0 + (\sigma_{ss}-\sigma_0).(1-\exp(-C^*\epsilon))^m$$

where $d'\exp(Q_0/(RT)) = A_0\exp(B_0\sigma_0)$

and $d'\exp(Q_{ss}/(RT)) = A_{ss}\exp(B_{ss}\sigma_{ss})$ (18)

Here σ_0 is the flow stress at zero strain, σ_{ss} is the steady state flow stress, ϵ is the accumulated plastic strain, and R is the gas constant. $C,m,Q_0,A_0,B_0,Q_{ss},A_{ss}$ and B_{ss} are fitted parameters. Data from plane strain compression tests was used to obtain a fit to equation (18) for the alloy 3004 [1].

The plastic strain, ϵ, is used at the state variable, κ, here and hence the evolution equation is;

$$\frac{D\kappa}{Dt} = d' \qquad (19)$$

Interstand Temperature Model

For hot tandem rolling, where consecutive reductions over up to four stands are

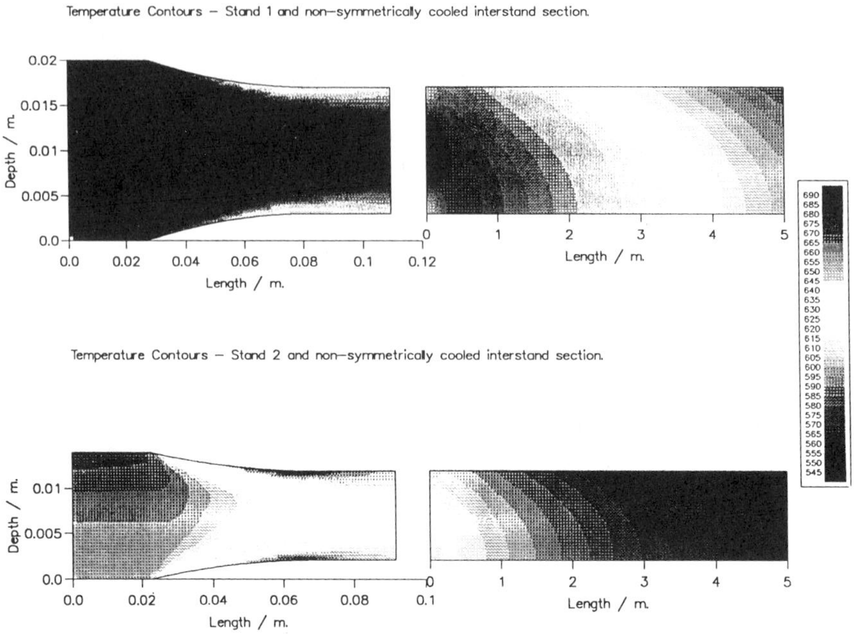

Figure 2. Simulation of Multi-Stand Rolling

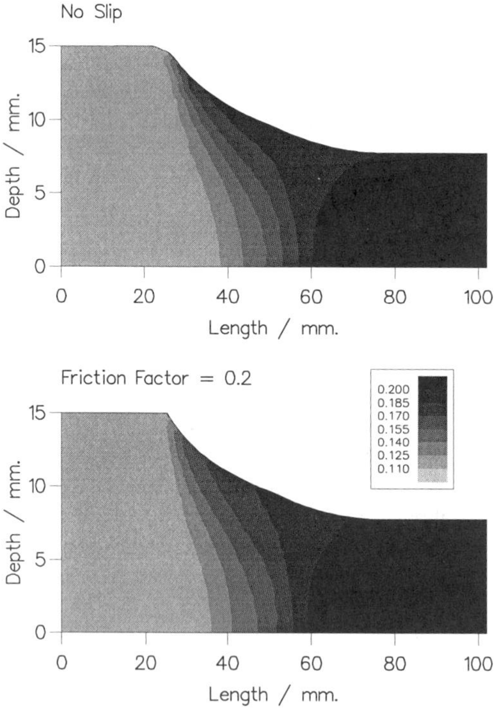

Figure 3. Velocity Dependence (m/s) on Friction.

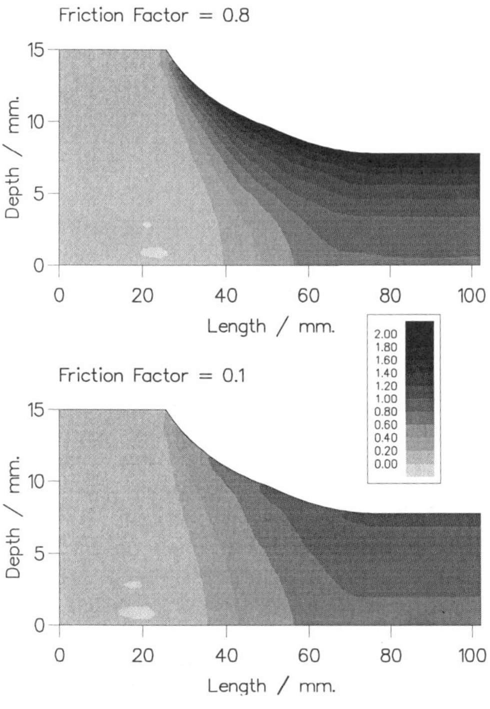

Figure 4. The effect of Friction on Predicted Effective Strain.

considered, significant cooling of the sheet can take place between rolls set by as much as five metres apart.

The finite difference method was employed to compute the interstand temperature field. Such an approach would essentially permit a fine mesh and short runtimes. The differential equation describing heat transfer by conduction, and a constant flow of mass, is shown below:

$$\rho c\frac{\partial T}{\partial t}+\frac{\partial}{\partial x}(u\rho cT)=\frac{\partial}{\partial x}(k\frac{\partial T}{\partial x})+\frac{\partial}{\partial y}(k\frac{\partial T}{\partial y}) \qquad (20)$$

where ρ is the density, c is the specific heat capacity, u is the velocity of the rolled strip, k is the thermal conductivity, and T is the temperature.

Numerical evaluation of equation (20) was performed using an Alternating Direction Implicit (A.D.I.) finite difference solver [2,3,4]. Such a technique reduces the solution of the set of simultaneous equations to the inversion of a tri-diagonal banded matrix. This can be performed directly without the need for iterative relaxation techniques. The stability of the solution method is insensitive to the timestep, Δt, and can therefore be applied to either transient or steady state problems.

Equations (21) and (22) show the two finite difference equations requiring solution for each timestep.

$$T^{n+1}_{i,j}(a_{ay}+a_{by}+\rho^{n}_{i,j}c^{n}_{i,j}\frac{\Delta x\Delta y}{\Delta t})-a_{ay}T^{n+1}_{i,j+1}-a_{by}T^{n+1}_{i,j-1}= a_{ax}(T^{n}_{i+1,j}-T^{n}_{i,j})+a_{bx}(T^{n}_{i-1,j}-T^{n}_{i,j})+T^{n}_{i,j}\rho^{n}_{i,j}c^{n}_{i,j}\frac{\Delta x\Delta y}{\Delta t} \qquad (21)$$

$$T^{n+2}_{i,j}(a_{ax}+a_{bx}+\rho^{n}_{i,j}c^{n}_{i,j}\frac{\Delta x\Delta y}{\Delta t})-a_{ax}T^{n+2}_{i+1,j}-a_{bx}T^{n+2}_{i-1,j}= a_{ay}(T^{n+1}_{i,j+1}-T^{n+1}_{i,j})+a_{by}(T^{n+1}_{i,j-1}-T^{n+1}_{i,j})+T^{n+1}_{i,j}\rho^{n+1}_{i,j}c^{n+1}_{i,j}\frac{\Delta x\Delta y}{\Delta t} \qquad (22)$$

where Δt is the time increment for each half of the timestep.
The subscripts i and j refer to the nodal coordinates in the cartesian system and the superscripts, n, $n+1$, and $n+2$, denote the time. From the form of the equations, each with three unknowns, the banded nature of the solution matrix can be appreciated when all nodes are considered. The coefficients, a, calculate the degree of heat transfer by convection, radiation, conduction and mass transfer at a point in space.

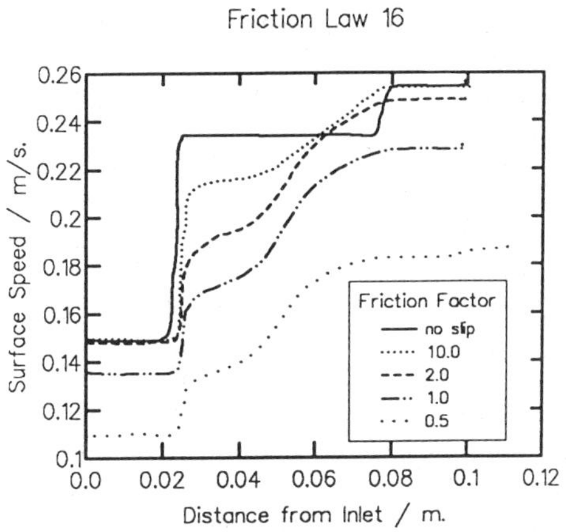

Figure 5. Surface Speed - Friction Law 16.

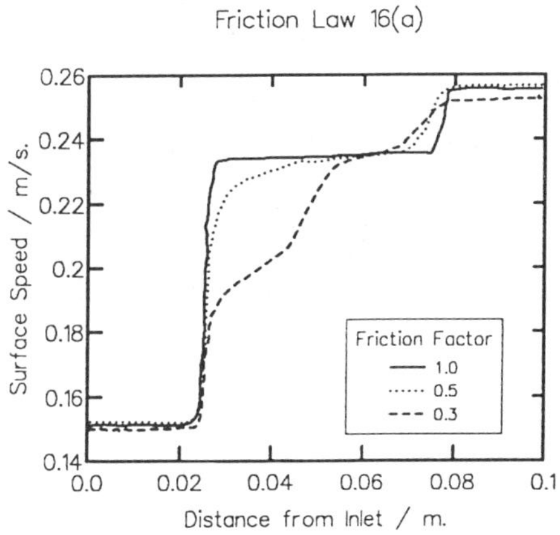

Figure 6. Surface Speed - Friction Law 16(a).

Results

Figure two shows a contour map of temperature in the roll bite for the first two stands of a tandem mill, as well as the interstand region between the two stands and the region after the second stand. Note that there is a large change of horizontal scale between the plots to the left and to the right of the figure. The length of the roll bite is about 80mm whereas the interstand region is several metres in length. The interstand cooling is generally asymmetric due to coolant remaining on the surfaces.

It can be seen in stand two that the temperature asymmetry remains in the material with temperature differences up to 30 °C between the two surface

Figure three shows contours of the horizontal component of velocity for two different friction conditions. The first is for the case of sticking friction and the second is using law 16(a) with a friction factor of 0.2. The effect of surface shear in the first case is quite noticeable, whereas the second case shows a more homogeneous reduction.

Figure four shows the predicted equivalent strain distribution for friction factors of 0.8 and 0.1 using law 16(a). The higher value creates a large amount of shear at the surface, and a subsequent high gradient of strain through the specimen thickness at the exit plane.

Figures five and six show the variation of surface speed with distance from the inlet for the two friction laws (16) and (16a) with different friction factors. The increase in speed is less for the low friction factors and there is no neutral point for the low friction coefficients for law 16(a).

Summary

A model has been presented to predict the through thickness variaration of temperature and equivalent strain in multi-stand tandem mills. The model shows that the temperature distribution is not symmetric about the pass line, and that the strain distribution can have large through-thickness gradients when the friction is high. The model has been compared, together with a model that estimates friction and heat transfer coefficients in the roll bite, to some experimental results which are the subject of a companion paper in this conference. Future work includes developing and linking micro-structural models to the model described in this paper.

References

1. H.L. Yiu et al. 1991. *The use of plane-strain compression testing to simulate the evolution of hot-rolled microstructures in aluminium alloys. Hot deformation of Aluminium Alloys*. Edited by T.G.Langdon et al. The Minerals, Metals and Materials Society.

2. Nagtegaal, J.C. 1982. On the implementation of inelastic constitutive equations with special reference to large deformation problems. *Comput. Meth. Appl. Mech. Engng.* Vol. 33,p469

3.Dawson P. 1987. On modelling of mechanical property changes during flat rolling of aluminium. *Int.J.Solids Structures* Vol 23, No 7 pp 947-968

4.Yu S. and E.G. Thompson. 1989. A direct Eulerian finite element method for steady state elastic plastic flow. *Numiform 89.* Edited by E.G. Thompson et al. Balkema, Rotterdam.

5. Ponter A.R.S. et al. 1993. Using the Eulerian finite element method to model hot metal rolling. *Modelling of metal rolling processes*. Institute of Materials (1993). Bourne Press Ltd.

6. Anand L. 1982. Constitutive equations for the rate-dependant deformation of metals at elevated temperatures.
ASME J.Engng Mater. Technol. 104,12.

7. Croft, D.R., Lilley, D.G.(1977) *Heat Transfer using Finite Difference Equations.* Applied Science Publishers, London, 1977.

8. Gerald, C.F., Wheatley, P.O. (1989) *Applied Numerical Analysis*. Addison-Wesley Publishing Company, Reading..

9. Suhas, V. (1980) *Numerical Heat Transfer and Fluid Flow.* Patanker, McGraw-Hill, 1980.

Simulation of Materials Processing: Theory, Methods and Applications, Shen & Dawson (eds)
© 1995 Balkema, Rotterdam. ISBN 90 5410 553 4

Simulating deformation-induced texture evolution during cold rolling

G. B. Sarma & P. R. Dawson
Sibley School of Mechanical and Aerospace Engineering, Cornell University, Ithaca, N.Y., USA

ABSTRACT: Polycrystal models have enjoyed considerable success in predicting deformation-induced texture evolution. In such models, the link between quantities at the continuum and crystal levels can have a significant bearing on the final textures. This aspect is examined by computing texture evolution during cold rolling of silicon steel using two different linking assumptions. The two models differ in the assignment of the macroscopic deformation to the crystals of the corresponding aggregate. Comparison of the texture predictions using a uniform and a non-uniform distribution of the deformation shows that the latter assumption leads to better match with experimental data.

1 INTRODUCTION

In recent years polycrystal models have been used with considerable success to predict the texture evolution which accompanies plastic deformation of a metal. An important issue in using these models is the link between the continuum (macroscopic) and crystal (microscopic) length scales. In most applications the deformation is discretized at the continuum level. However, since the constitutive response is modeled at the crystal level, some assumptions which relate the deformations across the two length scales become necessary.

In this article we consider two such assumptions and compare the predictions of texture development using these assumptions with experimental data. A discrete collection of weighted crystal orientations is used to represent the texture at a continuum point. The two modeling assumptions differ in the manner in which they assign deformation rate values to the individual crystals. The first one, based on the well known Taylor (1938) model, leads to a uniform distribution of the macroscopic deformation rate among the underlying crystals. The second model accounts for the interaction of a crystal with its neighbors to compute a different deformation rate for each crystal in the aggregate.

These models have been employed to study the evolution of texture during cold rolling of silicon steel. The textures were computed by integrating the evolution equations for the crystal orientations along streamlines of the flow. The streamlines at specific levels through the thickness were determined based on finite element simulations of the multi-pass operation. In the following sections we describe the polycrystal model used for computing the texture evolution and present the two assumptions used for the macro-micro linking. Following that we briefly discuss the rolling simulations and the streamline calculations. We show results from the calculations using the two models, and compare these predictions with experimental measurements.

2 POLYCRYSTAL MODEL

To characterize the textures, an aggregate of crystals is assumed to exist at each material point at the continuum (macroscopic) scale. Each crystal is described by its orientation and the hardness of its slip systems. Deformation experienced by the material point leads to changes in the orientations of the crystals, resulting in texture evolution.

Assuming the deformation of each crystal to be accommodated by rate-dependent slip, a power law is used to relate the rate of shearing on the α slip system to the resolved shear stress (Asaro & Needleman 1985, Mathur & Dawson 1989):

$$\dot{\gamma}^{(\alpha)} = \dot{\gamma}_0 \left| \frac{\tau^{(\alpha)}}{\tau_0} \right|^{\frac{1}{m}-1} \frac{\tau^{(\alpha)}}{\tau_0}. \tag{2.1}$$

The above relation is used to develop a constitutive law at the crystal level through the symmetric

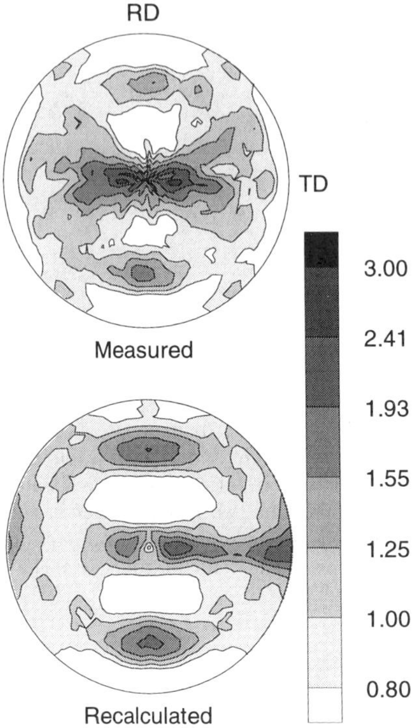

Figure 1. Initial <110> pole figures at the surface. Contour levels indicate multiples of random texture intensity.

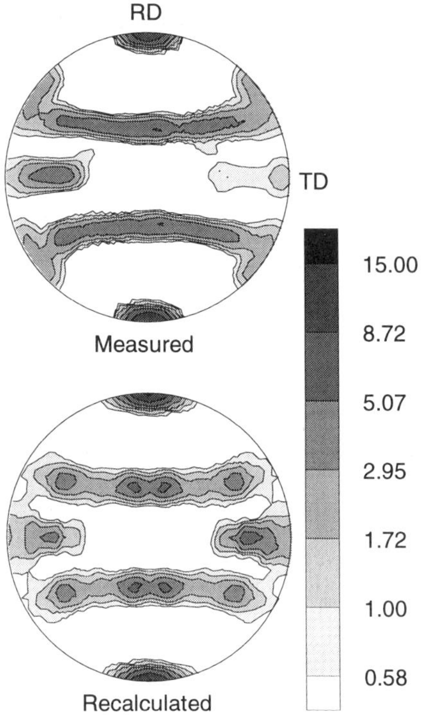

Figure 2. Initial <110> pole figures at the center plane. Contour levels indicate multiples of random texture intensity.

part of the Schmid tensor $\boldsymbol{P}^{(\alpha)}$:

$$\boldsymbol{D}^c = \left[\sum_\alpha \frac{\dot{\gamma}_0}{\tau_0} \left| \frac{\tau^{(\alpha)}}{\tau_0} \right|^{\frac{1}{m}-1} \boldsymbol{P}^{(\alpha)} \otimes \boldsymbol{P}^{(\alpha)} \right] \boldsymbol{\sigma}'^c, \quad (2.2)$$

which is solved to obtain the crystal deviatoric stress $\boldsymbol{\sigma}'^c$ for a given deformation rate $\boldsymbol{D}^c$.

Knowledge of the crystal stress permits calculation of the resolved shear stress $\tau^{(\alpha)}$, and hence the shearing rate $\dot{\gamma}^{(\alpha)}$ using equation (2.1), for each slip system. The crystal reorientation rate $\dot{\boldsymbol{R}}^*$ is computed from the crystal spin $\boldsymbol{W}^c$ (Mathur & Dawson 1989):

$$\dot{\boldsymbol{R}}^* = (\boldsymbol{W}^c - \sum_\alpha \dot{\gamma}^{(\alpha)} \boldsymbol{Q}^{(\alpha)}) \boldsymbol{R}^*, \quad (2.3)$$

where $\boldsymbol{Q}^{(\alpha)}$ is the skew-symmetric part of the Schmid tensor.

An important step leading to the calculations outlined above is relating the deformations at the continuum and crystal levels. The deformation applied to the material point is known in terms of its velocity gradient. Computation of the velocity gradient experienced by each crystal requires some constitutive assumptions. We discuss two such assumptions which have been used for this purpose.

2.1 *Taylor model*

The Taylor (1938) model assumes the deformation of each crystal to be the same as the deformation of the continuum material point. In terms of the crystal kinematics, this translates to the deformation rate $\boldsymbol{D}^c$ and the spin $\boldsymbol{W}^c$ experienced by each crystal being identical to the corresponding macroscopic values (Asaro & Needleman 1985, Mathur & Dawson 1989):

$$\boldsymbol{D}^c = \boldsymbol{D}, \quad (2.4)$$

$$\boldsymbol{W}^c = \boldsymbol{W}. \quad (2.5)$$

2.2 *Partition model*

The assumption of a uniform distribution of the deformation among the crystals often is oversimplified. In reality various factors, such as the crystal orientation, shape, size, and interaction with neighboring crystals, play a role in determining the deformation of a crystal. The detailed study of a model polycrystal (Sarma & Dawson 1995) showed

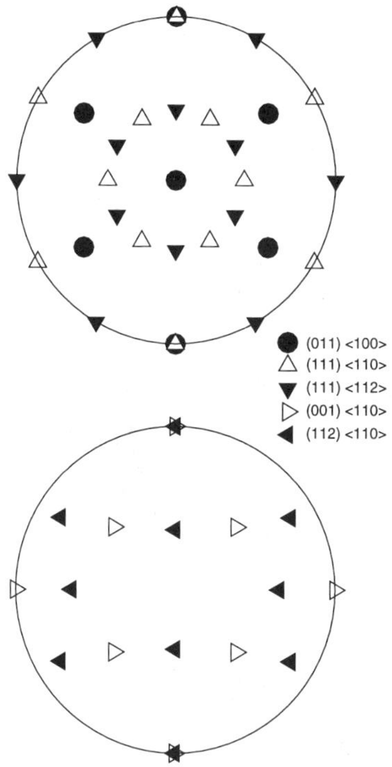

Figure 3. Ideal texture components in the <110> projection.

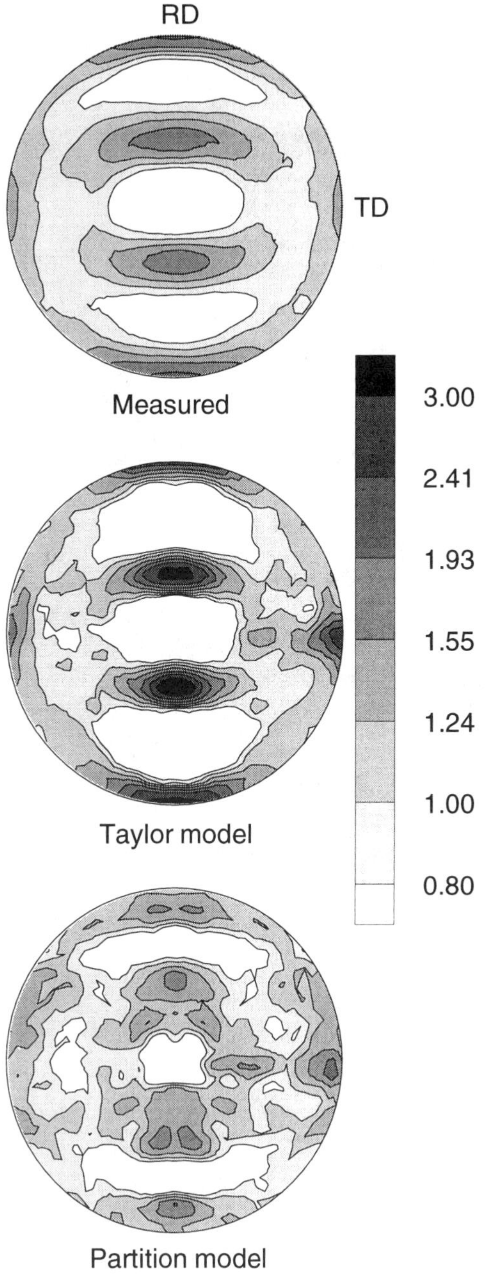

Figure 4. <110> pole figures at the surface after 40% reduction. Contour levels indicate multiples of random texture intensity.

that interaction among crystals plays a dominant role in determining the spread of the macroscopic deformation among the underlying crystals. Based on this study a new model has been developed to partition the deformation of the continuum point among the individual crystals.

The procedure consists of generating a set of deformation rate values which fall on a Gaussian distribution about the applied macroscopic value. Assignment of these values to the individual crystals is based on the following construct which accounts for interaction of each crystal with a set of neighbors. The crystals are assumed to be arranged in the form of a three-dimensional grid. A neighborhood of each crystal is defined as the six crystals which share a common face with the given crystal. Based on the crystal deviatoric stress a compliance tensor is computed for each crystal using equation (2.2). The average compliance of its neighborhood is computed as

$$\langle \mathcal{S} \rangle = \frac{1}{n} \sum_{i=1}^{n} \mathcal{S}^i, \tag{2.6}$$

where n is the number of crystals comprising the neighborhood. A relative compliance of the given crystal is then computed as

$$\mathcal{Q}^c = \mathcal{S}^c \langle \mathcal{S} \rangle^{-1}. \tag{2.7}$$

This relative compliance may be used to calculate a deformation rate for that crystal. However, the average value over all crystals of such a deforma-

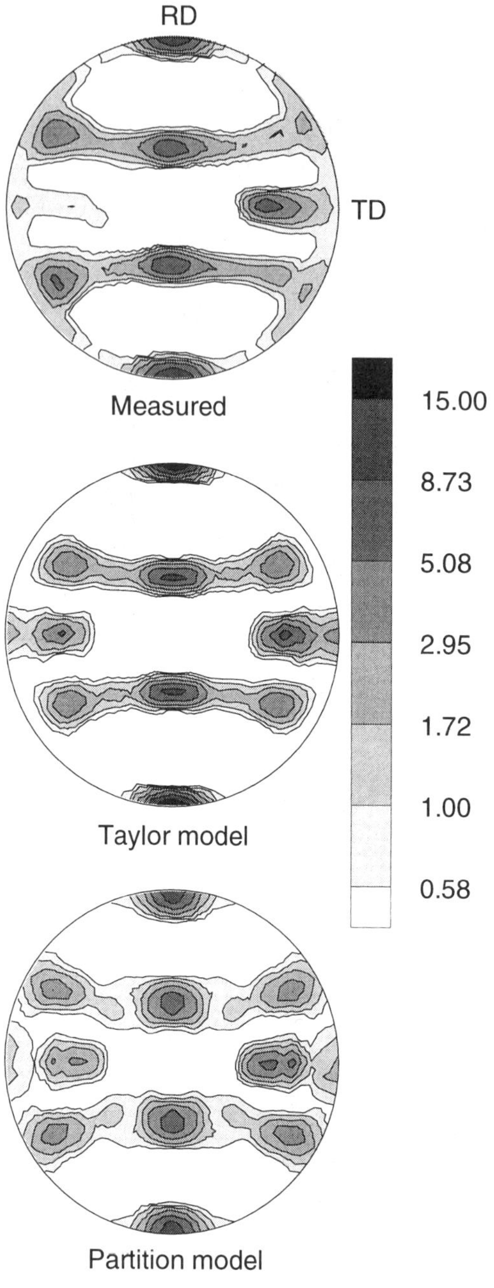

Figure 5. <110> pole figures at the center plane after 40% reduction. Contour levels indicate multiples of random texture intensity.

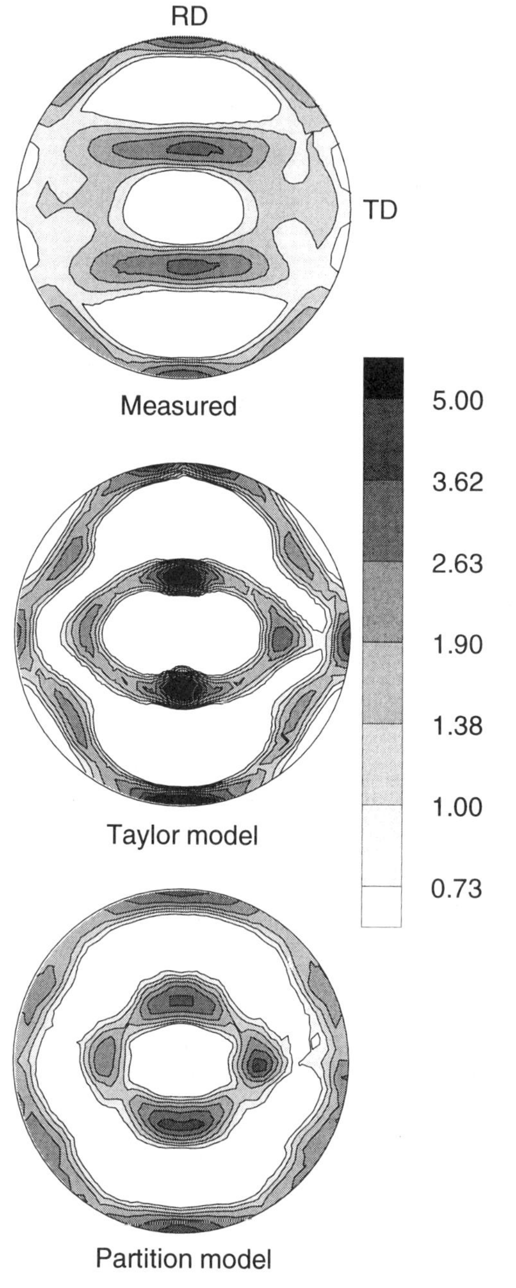

Figure 6. <110> pole figures at the surface after 80% reduction. Contour levels indicate multiples of random texture intensity.

tion rate will in general not equal the macroscopic value. To satisfy this condition the relative compliance is normalized by its average value over all crystals:

$$\langle \mathcal{Q} \rangle = \frac{1}{N} \sum_{c=1}^{N} \mathcal{Q}^c, \tag{2.8}$$

$$\mathcal{P}^c = \mathcal{Q}^c \langle \mathcal{Q} \rangle^{-1}. \tag{2.9}$$

The deformation rate of the crystal is then computed as

$$\mathbf{D}^c = \mathcal{P}^c \boldsymbol{D}. \tag{2.10}$$

This calculation results in a different deformation rate for each crystal in the aggregate. Some of the $\mathbf{D}^c$ values can have extremely large deviations from the macroscopic value; in addition, the spread of the deformation rates does not always follow a normal distribution. The role of this calculation is to provide the relative magnitudes of the deformation rates for the different crystals. For each component of the deformation rate the values obtained using the above calculation and the values from the normal distribution are sorted. A one-to-one correspondence between the two sorted sets of values is used to assign deformation rate components from the normally distributed set to each crystal. The crystal with the highest value from the calculation using equation (2.10) is assigned the highest value from the normal distribution, and so on. While the deformation rate varies from crystal to crystal, the spin is still given by equation (2.5).

3 APPLICATION TO COLD ROLLING

Cold rolling of Fe-3%Si was simulated using the finite element method, assuming steady state conditions and using an Eulerian formulation. Multiple passes were used to reduce the original strip thickness of 2 mm by 80%, with each pass contributing 7–10% reduction. Sliding friction conditions were assumed at the interface between the roll and the workpiece. Details of the process are given in Chastel (1992).

A complete description of the procedure used to compute the velocity field may be found in Dawson (1987). Once a converged velocity field is obtained, streamlines are extracted at the surface and the center plane of the workpiece. At these levels the streamlines are traced starting from the exit all the way to the entrance to the control region. Along each step of the streamline path the velocity gradient is computed and saved, along with the coordinates and velocity at that point. This provides the deformation history of a material point along that streamline. The evolution equation (2.3) for the crystal orientation is integrated along the streamlines to determine how the texture evolves along the surface and the center plane. By repeating this exercise for every pass the texture after different amounts of reduction is obtained.

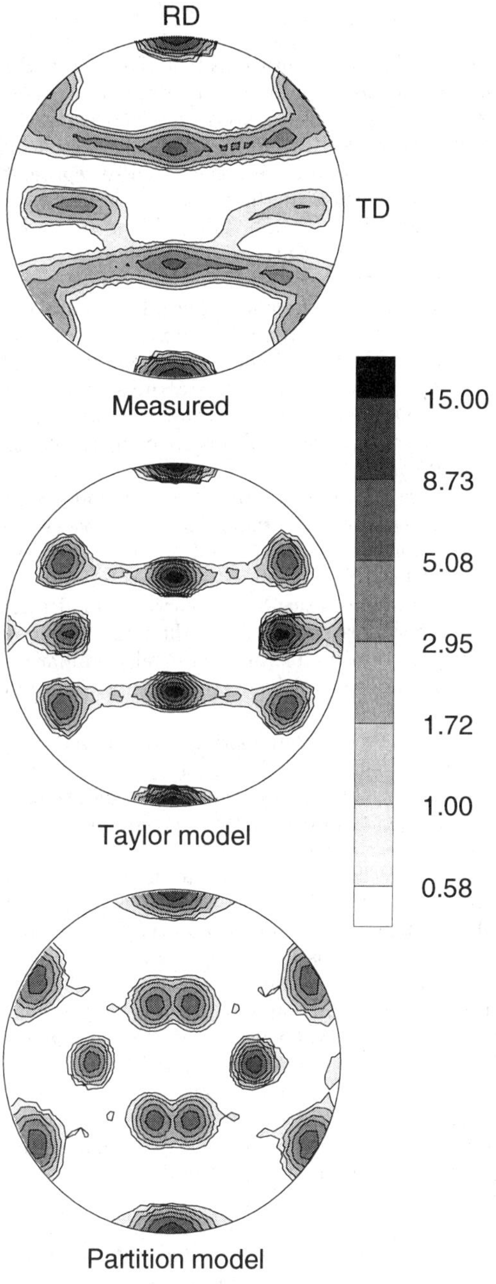

Figure 7. <110> pole figures at the center plane after 80% reduction. Contour levels indicate multiples of random texture intensity.

3.1 *Texture evolution*

The initial set of orientations required to start the texture calculations were generated from experimental measurements (Salsgiver 1989) using the PopLA software (Kallend *et al.* 1991). Two sets of weighted orientations were generated to represent the initial textures at the surface and the center plane respectively (Kocks 1994). Figures 1 and 2 show the initial measured and recalculated <110> pole figures at the surface and the center plane respectively. The weighted crystals capture the essential features of the initial textures. The surface shows presence of shear texture components while the center plane shows features of a plane strain compression texture (Dawson *et al.* 1993). Figure 3 shows some ideal texture components for comparison.

Texture predictions using the two models at the surface after 40% reduction in thickness are shown in figure 4 as <110> pole figures. The Taylor model matches the measured texture but with higher peak intensities. In the partition model the peaks show greater spread but the intensity levels match the measured values very well. Similar results are observed at the center plane, as shown in figure 5.

Textures after 80% reduction at the surface and the center plane are shown in figures 6 and 7 respectively. As before the Taylor model predicts sharper textures whereas the partition model matches the measured intensities. In addition, the location of certain peaks is better matched by the partition model. The texture at the surface after 80% reduction shows presence of the (111)<110> component which is shifted away from the circumference in the Taylor model prediction. In the partition model prediction as well as the experimental data, this component is close to the circumference. A similar trend is observed in the center plane texture, where the (112)<110> component is shifted closer to the circumference in the partition model as compared to the Taylor model. The non-uniform distribution of the deformation rate among the individual crystals based on interactions with neighbors enables the partition model to predict more diffuse textures as well as certain components in the proper locations.

The discretization of the initial textures to obtain weighted crystals may introduce some errors, which persist and appear in the computed texture evolution. The recalculated initial texture at the surface shows presence of peaks near the circumference at 0° and 180°, which are not present in the measured texture. These peaks remain in the computed textures after 40% and 80% reduction (figures 4 and 6), and are more prominent in the Taylor model prediction. The initial texture at the center plane has two peaks near the (112)<110> component close to the center which appear in the partition model prediction after 80% reduction (figure 7). These results show that textures computed using polycrystal models are subject to errors introduced by using the discrete crystals approach, which must be considered when comparing the textures with experimental measurements.

ACKNOWLEDGEMENTS

Support for this work was provided by the Office of Naval Research under contract NOOO 14-90-J-1810, and by the member companies of the Deformation Process Simulation Consortium at Cornell University. Helpful discussions with Donald Boyce are gratefully acknowledged.

REFERENCES

Asaro, R.J. & A.Needleman 1985. Texture development and strain hardening in rate dependent polycrystals. *Acta Metall.*33,6:923-953.

Chastel, Y.B. 1992. *Ph.D. Dissertation.* Cornell University.

Dawson, P.R. 1987. On modeling mechanical property changes during flat rolling of aluminum. *Int. J. Solids Struct.*23,7:947-968.

Dawson, P.R., Y.B. Chastel, G.B. Sarma & A.J. Beaudoin 1993. Simulating texture evolution in Fe-3%Si with polycrystal plasticity theory. *Proc. First Inter. Conf. on Modeling Rolling Processes.* London:The Institute of Metals.

Kallend, J.S., U.F.Kocks, A.D.Rollett & H.-R.Wenk 1991. Operational texture analysis. *Mater. Sci. Engrg.*A132:1-11.

Kocks, U.F. 1994. *Private communication.*

Mathur, K.K. & P.R.Dawson 1989. On modeling the development of crystallographic texture in bulk forming processes. *Int. J. Plast.*5:67-94.

Salsgiver, J. 1989. *Private communication.*

Sarma, G.B. & P.R.Dawson 1995. Effects of interactions among crystals on the inhomogeneous deformations of polycrystals. *In preparation.*

Taylor, G.I. 1938. Plastic strain in metals. *J. Inst. Metals.*62:307-324.

Simulation of Materials Processing: Theory, Methods and Applications, Shen & Dawson (eds)
© 1995 Balkema, Rotterdam. ISBN 90 5410 553 4

Three-dimensional analysis of long-product rolling using rigid-plastic finite element method

K. Seki, K. Yamada, S. Ida & S. Hamauzu
Nippon Steel Corporation, Futtsu, Japan

ABSTRACT: A three-dimensional finite element code is developed to investigate long-product rolling. The code is based on the rigid-plastic FEM with the foundation on the theory of plasticity for slightly compressible solid. Introducing a correction method of material streamline profile to the code and devising a versatile pre-processing code to obtain initial material streamlines profile adequate for steady-state analysis of the rolling, the code enables us to analyze the deformation of various types of rolling processes precisely and efficiently. Computations for rolling of hollow bloom, angle and widening rolling of H-beam are done, and numerical results such as cross-sectional shape of material and stress distributions are given as well as effects of rolling conditions of these observations.

1 INTRODUCTION

Rolling of long-products, such as section, bar, and wire rod involves complicated and diverse material and roll caliber geometry, assemblies of the rolls. The material during rolling exhibits complex three-dimensional deformation. This causes various problems that are difficult to solve experimentally in the course of clarification of rolling characteristics.

Recently, the rigid-plastic finite element method (FEM) is gaining attention as an effective numerical method for analyzing rolling process. A lot of investigations have been done using the rigid-plastic FEM for various types of three-dimensional rolling problems, such as tube rolling (Mori 1985), bar or shape rolling (Bertrand 1986, 1989, Mori 1989, Lee 1989, Yanagimoto 1993). One of the authors has previously developed a three-dimensional rigid-plastic finite element code to investigate mandrel rolling of seamless tube, and studied the characteristics of the tube deformation during rolling and dependency of them on various rolling conditions (Yamada 1989).

This paper centers on the analysis of long-product rolling using the rigid-plastic FEM. Considering variety of rolling process of long-product, versatility of analysis code is indispensable for practical use. The authors have expanded the correction method of material streamline profile introduced in the code for mandrel rolling to handle various shapes of material and rolls, and have devised a versatile pre-processing code to obtain initial streamline profile of material and boundary conditions adequate for the analysis. Finally an accurate and versatile system for steady-state analysis of rolling of long-products has been established.

2 METHOD OF ANALYSIS

The analysis of material deformation uses the rigid-plastic finite element method based on the theory of plasticity for slightly compressible solid, which obeys the yield criterion (Osakada 1982):

$$\bar{\sigma} = \sqrt{\frac{3}{2}\sigma'_{ij}\sigma'_{ij} + g\sigma_m^2} \quad , \qquad (1)$$

where $\sigma'_{ij} = \sigma_{ij} - \delta_{ij}\,\sigma_m$, σ_m is the hydrostatic stress and g is a small positive constant (in this paper, g = 0.01). Introducing slight change of volume during deformation, the stress can be calculated directly from the strain rate:

$$\sigma_{ij} = \frac{\bar{\sigma}}{\dot{\bar{\varepsilon}}}\left\{\frac{2}{3}\,\dot{\varepsilon}_{ij} + \delta_{ij}\left(\frac{1}{g} - \frac{2}{9}\right)\dot{\varepsilon}_v\right\} \quad , \qquad (2)$$

where

$$\dot{\bar{\varepsilon}} = \sqrt{\frac{2}{3}\,\dot{\varepsilon}'_{ij}\dot{\varepsilon}'_{ij} + \frac{1}{g}\,\dot{\varepsilon}_v^2} \quad , \qquad (3)$$

$$\dot{\varepsilon}'_{ij} = \dot{\varepsilon}_{ij} - \frac{1}{3}\,\delta_{ij}\,\dot{\varepsilon}_v \quad , \qquad (4)$$

ε_v is volumetric strain, the entity with " $\dot{}$ " denotes strain rate, and the entity with " $\bar{}$ " denotes the equivalent stress or strain.

The functional Φ to be minimized is given by

$$\Phi = \int_V \bar{\sigma}\,\dot{\bar{\varepsilon}}\,dv - \int_S T\cdot u\;ds \quad , \qquad (5)$$

where T is the traction vector prescribed on the boundary surface S and u is the velocity vector. If the surface S is on the roll-material interface, T is the frictional stress and u is the relative velocity. On the roll-material interface, mixed frictional condition is employed. Linear isoparametric solid elements with 8 nodes are used to formulate this problem.

In the beginning of the deformation analysis, a profile of the material surface should be assumed to incorporate the boundary conditions. In the steady-state analysis, this profile should be consistent with the material streamline profile obtained by integrating nodal velocity vectors. Since these two profiles do not coincide with each other in general, one of the authors has previously developed a correction method of the streamline profile for mandrel rolling (Yamada 1989) to correct this discrepancy. Versatility for various types of material shape and rolling mills and ease of computation are required to simulate rolling of long-product. Considering these requirements, the authors have expanded this correction method and devised a versatile pre-processing code.

Fig. 1 shows the overall flow of analysis and the correction method of material streamline profiles and boundary conditions. In the present method, material streamline profiles and boundary conditions are modified by two stages of correction.

In the first stage, iterative calculation is performed to determine whether each of material surface nodes is contact with the roll or not. Profiles of material streamline and boundary conditions of material surface nodes are corrected according to following procedure.

1. Material streamline profiles are obtained by integrating velocity vectors under the assumed boundary conditions.

2. If free surface nodes assumed in the last correction step penetrates into the roll, constraint of the node are corrected.

3. If contact nodes assumed in the last correction step are apart from the roll, constraint of first contact nodes along the streamlines is changed to free.

4. Position of contact nods are changed to the roll surface.

In the second stage, position of first contact nodes along the streamlines is exactly determined by adjusting the position of the nodes along the rolling direction. If a streamline of material surface break through a roll surface, position of first contact node on the streamline is stepped backward. If a material streamline with contact nodes is apart from the roll, position of first contact node is stepped forward.

In both stages, internal nodes and free surface are modified to make streamline profiles smooth.

In pre-processing code for analysis, initial material streamline profiles and boundary conditions are calculated automatically from several numerical values that indicate cross-sectional shape of material before rolling, roll caliber profile and assembly of rolls. In some analytical problem, material has two surfaces these deformations are related with each other. In these problems, consideration of 'shearing' deformation into the calculation of initial material streamline profiles is important to improve stability of convergency and accuracy of the analysis.

Fig. 2 shows schematic diagram of procedure to determine initial material surface profile adequate for the analysis. Cross-sectional shape of material is extruded step-by-step along rolling direction. In each step, penetrated parts of material surface are moved to the roll surface. When the one side of the two related surfaces of material penetrates to roll, the material is sheared off. Considering the shearing, we

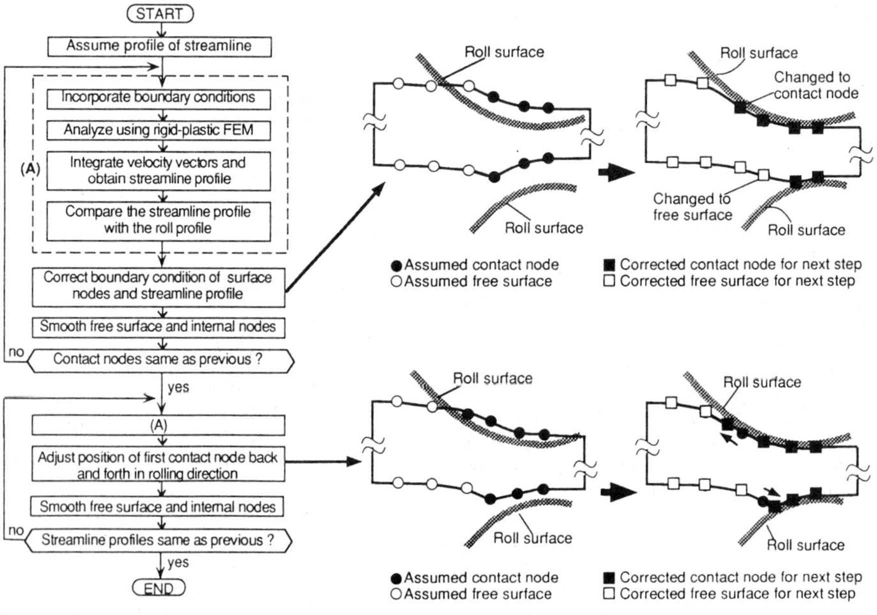

Fig. 1 Overall flow of analysis and correction method of material streamline profile

modify the profile of 'opposite' side surface of material according to the movement of the penetrating side surface.

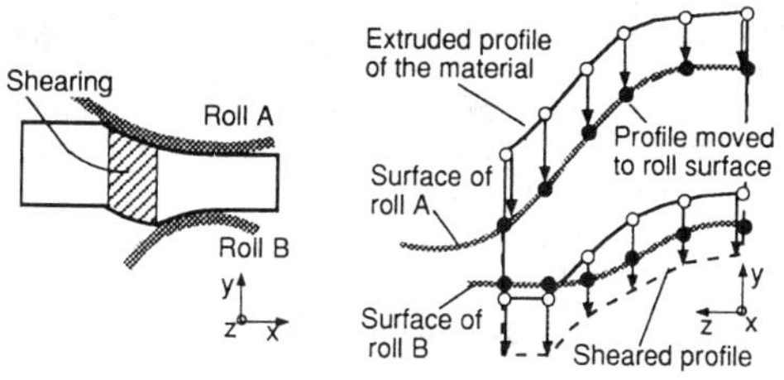

Fig. 2 Schematic diagram of procedure to determine initial material surface profile.

3 ANALYSIS OF LONG-PRODUCTS ROLLING

3.1 Analysis of rolling of hollow bloom

Rolling of a bloom with liquid core in a continuos casting process is effective for reducing center-line segregation and center porosity. On the other hand the rolling sometime causes internal crack of the bloom. These phenomena are concerned with deformation of the bloom especially with motion of liquid core induced by solid shell deformation and stress generating near the inner surface of solid shell. Although a simulation of rolling of material with liquid core has been reported (Pawelski 1992), the phenomena at large reduction condition under which joining of upper and lower solid shell occurred has not been studied. Assuming that motion of liquid core during rolling is not effective to the shell deformation at such high reduction rate, we have analyzed deformation of hollow bloom without liquid core.

Fig. 3 shows the analytical model of rolling of hollow bloom. This problem is regarded as the deformation with two symmetric planes. Besides upper roll, an imaginary flat tool is settled on a horizontal symmetric plane to simulate joining of upper and lower surface of hole. When nodes on upper surface of the hole are contact with the tool, velocity vectors of the nodes are constrained to the toll surface with no friction.

In order to verify accuracy of the analysis code, computed results have been compared with experimental results using plasticine. Table 1 shows experimental conditions as well as computational conditions.

Fig. 4 shows evolution of cross-sectional shape of the bloom during rolling. At first, the shape of the hole turns to oval from circle. Then, upper and lower surface of the hole starts joining from the end of width. Finally, upper and lower surface fully joins at the condition. Fig. 5 shows change of height of the hole at width center during rolling. Rate of decrease of the hole height becomes higher as initial diameter of the hole increase. The computational results show fairly good agreement with experimental results.

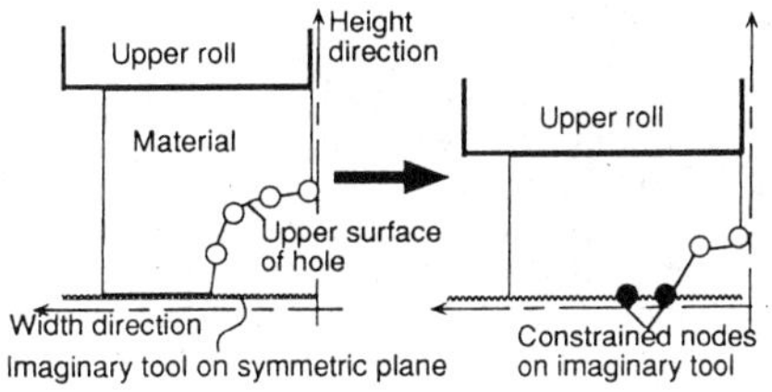

Fig. 3 Analysis model of rolling of hollow bloom.

Table 1 Computational conditions of rolling of hollow bloom.(comparison with experiments)

Initial dimension of bloom / mm :	110 X110
Initial diameter of hole / mm :	15, 25, 35
Roll diameter / mm :	350
Caliber geometry :	Flat
Reduction in height / % :	40
Periheral speed of rolls / m · min^{-1} :	1.0
Coefficient of friction :	0.3(lubrication of talc)
Yield stress / MPa :	$\sigma = 0.182\ \varepsilon^{0.11}$

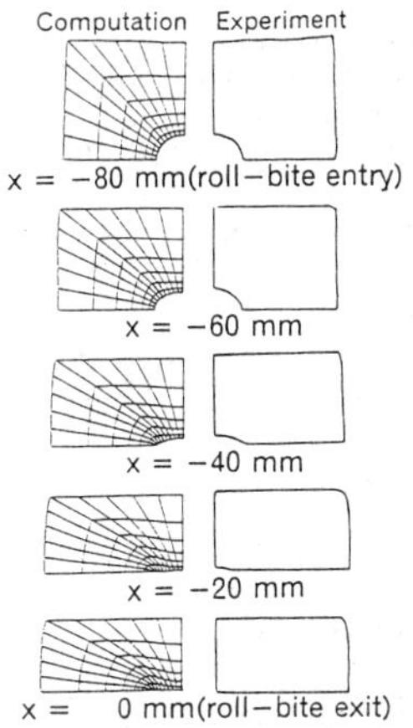

Fig. 4 Evolution of cross-sectional shape of bloom. (initial diameter of the hole = 25 mm)

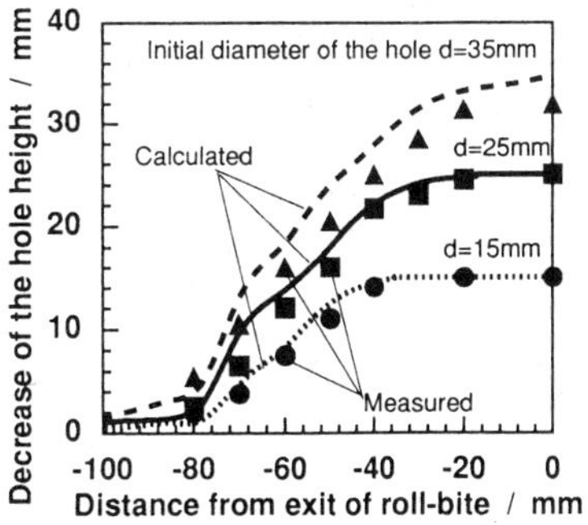

Fig. 5 Change of height of the hole during rolling.

To investigate effect of roll caliber geometry on deformation of the hole, computation has been done according to the computational conditions shown in Table 2 and roll caliber geometry shown in Fig. 6.

Fig. 7 shows evolution of area of the hole along the rolling direction. The figure shows that roll caliber seems not so effective on closing deformation of the hole within the calculated condition.

Fig. 8 shows evolution of normalized longitudinal stress at width center of surface of the hole along rolling direction. The stress is normalized by yield stress at the portion. The stress is maximum at middle of roll-bite and tends to compressive after joining of upper and lower surface. The longitudinal stress barely varies with the roll caliber geometry.

Fig. 9 shows evolution of normalized transversal stress at width center of the hole surface along rolling direction. The stress is maximum at entrance of roll-bite and tends to compressive after the joining. The compressive stress generated after the joining increases, as using constrictive or flat type caliber rolls.

Table 2 Computational conditions of rolling of hollow bloom

Initial dimension of bloom / mm :	220 X220
Initial diameter of hole / mm :	30, 50, 70
Roll diameter / mm :	700
Reduction in height / % :	40
Periheral speed of rolls at width center / m · min^{-1} :	2.0
Coefficient of friction :	0.3
Yield stress / MPa :	$\sigma = 72\,\varepsilon^{0.21}\,\dot{\varepsilon}^{0.13}$

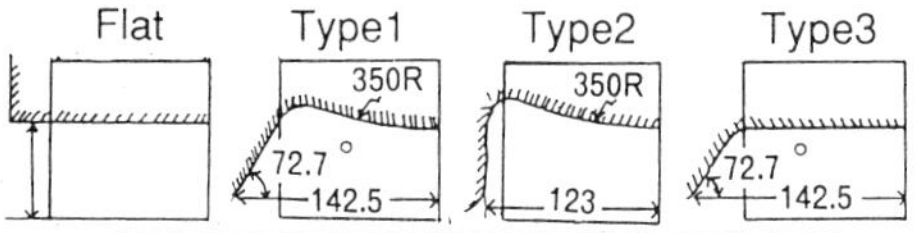

Type	Flat	1	2	3
Profile of width center	Flat	Arc	Arc	Flat
Constriction	None	Weak	Firm	Weak

Fig. 6 Geometry of roll caliber.

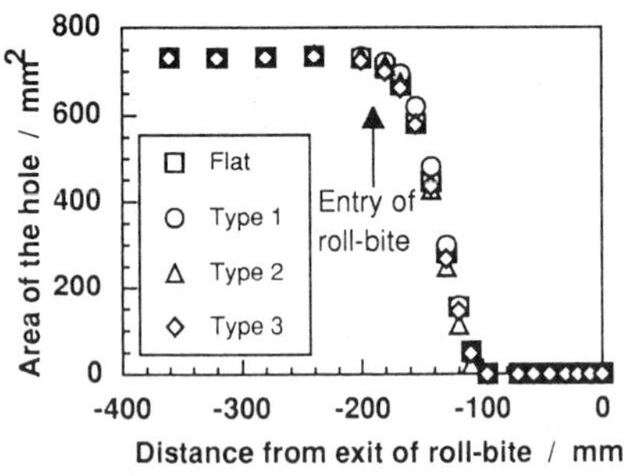

Fig. 7 Evolution of area of hole during rolling (initial diameter of the hole = 30 mm).

Fig. 10 shows variation of maximum of theses normalized stresses with initial diameter of the hole and roll caliber geometry. Maximum of longitudinal stress is as large as yield stress, and barely varies with initial diameter of the hole and the roll caliber geometry. Maximum of transversal stress increases, as initial diameter of the hole increases. When arc type caliber is adopted, the maximum of transversal stress is larger than the other case.

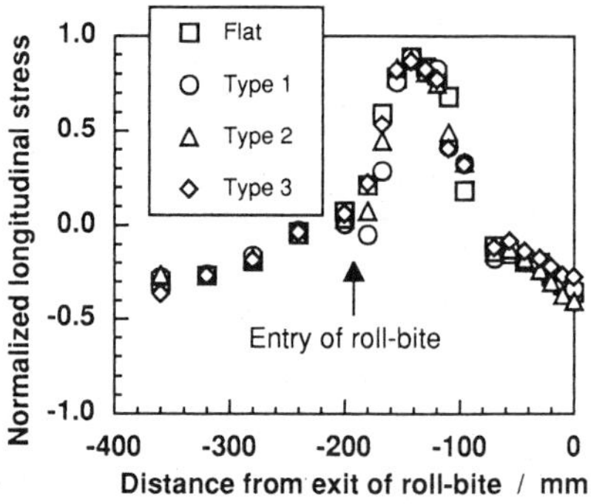

Fig. 8 Evolution of longitudinal stress along rolling direction. (initial diameter of the hole = 30 mm)

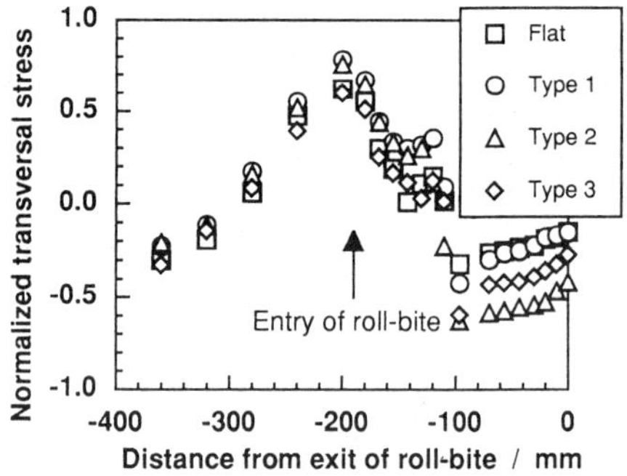

Fig. 9 Evolution of transversal stress along rolling direction.(initial diameter of the hole = 30 mm)

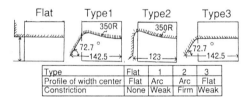

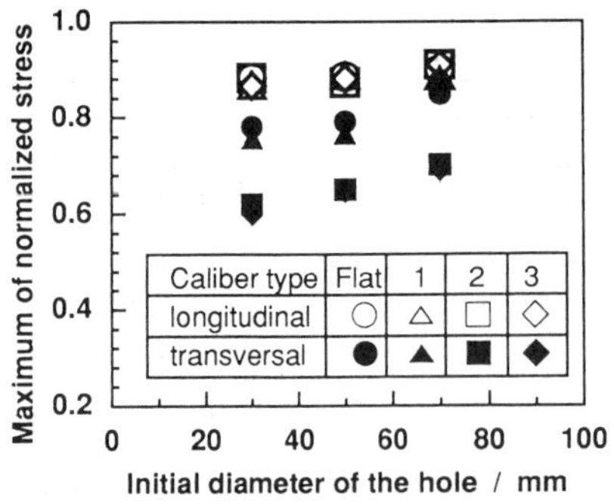

Fig. 10 Variation of maximum of stress with initial diameter of the hole and roll caliber geometry.

3.2 Analysis of rolling of angle

In rolling of section with asymmetry in height direction, the pass-line height effects on rolling characteristics. To investigate this effect, computation of angle rolling has been done according to computational condition shown in Fig. 11 and Table 3. In this analysis, pass-line height has been defined to zero when the material starts contact with upper and lower roll at the same time and has been defined to positive value when the material starts contact with upper roll before with lower roll.

Fig. 12 shows variation of cross-sectional shape of material after rolling with pass-line height. Height and width of rolled material increase, as the pass-line height decreases.

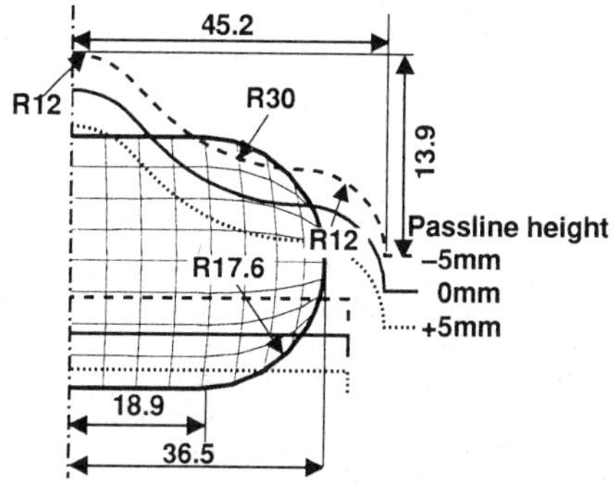

Fig. 11 Cross-sectional shape of material before rolling and the profile of rolls.

Table 3 Computational conditions of angle rolling.

Roll diameter at width center / mm :	120 (upper)
	130 (lower)
Periheral speed of rolls at width center	
/ m · min^{-1}:	4.0
Passline height / mm :	-5, 0, 5
Coefficient of friction :	0.3
Yield stress / MPa :	$\sigma = 82\,\varepsilon^{0.21}\,\dot{\varepsilon}^{0.13}$

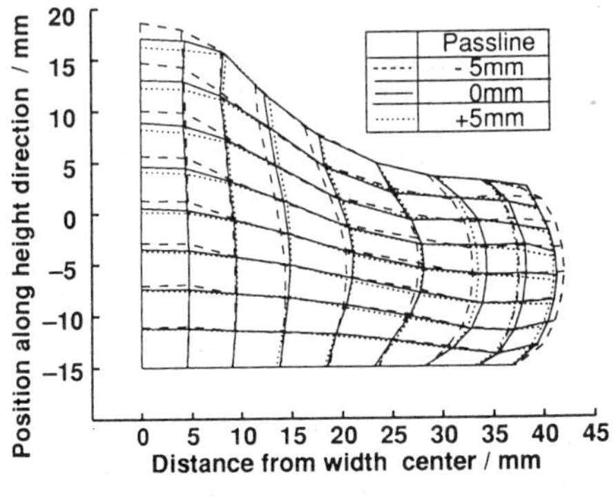

Fig. 12 Variation of cross-sectional shape of angle with pass-line height.

3.3 Analysis of widening rolling of H-beam

Recently new rolling process of H-beam has been developed. In this process, web height of an H-beam is freely adjustable by use of a skew roll mill. Fig. 13 shows schematic diagram of the skew roll mill. The mill has four rolls those axes are slanted from ordinary direction with cross angle α in horizontal plane, and with inclination angle β in vertical plane.

To investigate effect of cross angle on deformation of the H-beam, computation has been done according to computational conditions shown in table 4.

Fig. 14 shows variation of cross-sectional shape of H-beam after rolling with the cross angle. As the cross angle increases, slant of flange increases and the width of flange increases.

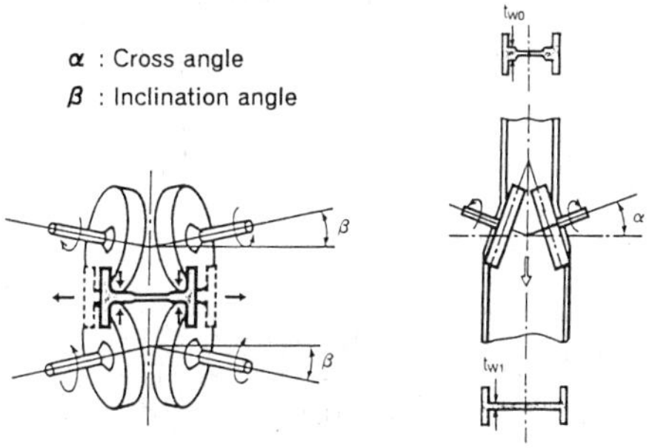

Fig. 13 Schematic diagram of the skew roll mill.

Table 4 Computational conditions of widening rolling of H-beam

Web height X Flange width X Web thickness /	
Flange thickness / mm	570X200X9.8 /14.7
Roll diameter / mm :	900
Cross angle / degree :	0, 5, 10
Inclination angle / degree :	5
Web height expasion / mm :	30
Coeffivient of friction :	0.3
Yield stress / MPa :	$\sigma = 150\,\varepsilon^{0.21}\,\dot{\varepsilon}^{0.13}$

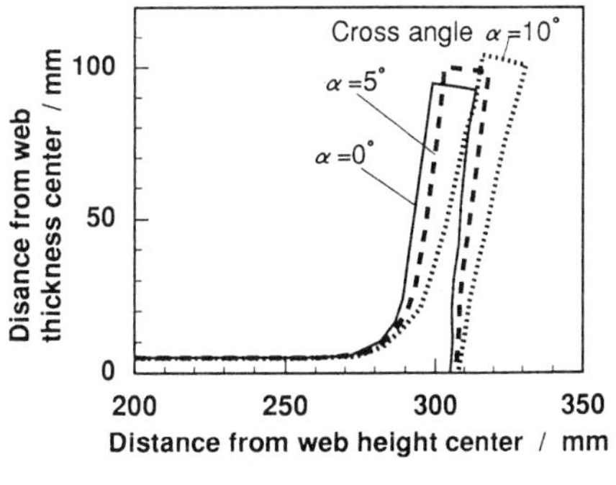

Fig. 14 Variation of cross-sectional shape of H-beam with cross angle.

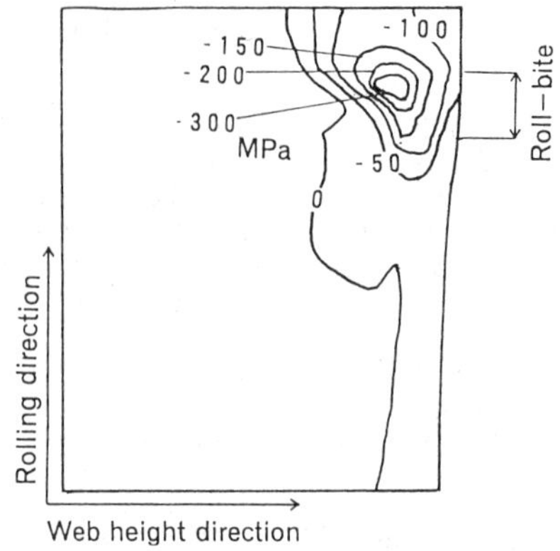

(a) Cross angle $\alpha = 0$ degree.

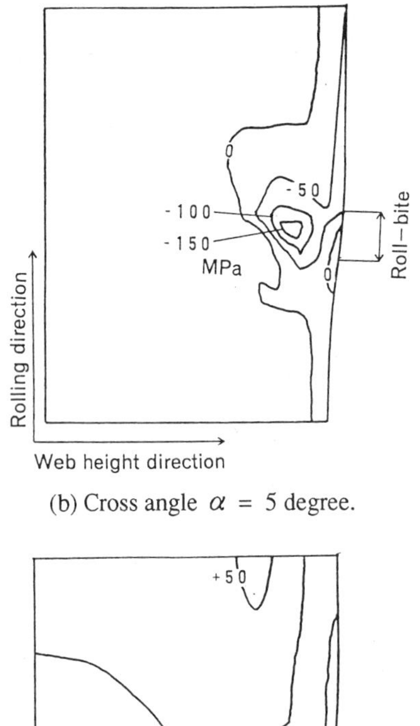

(b) Cross angle α = 5 degree.

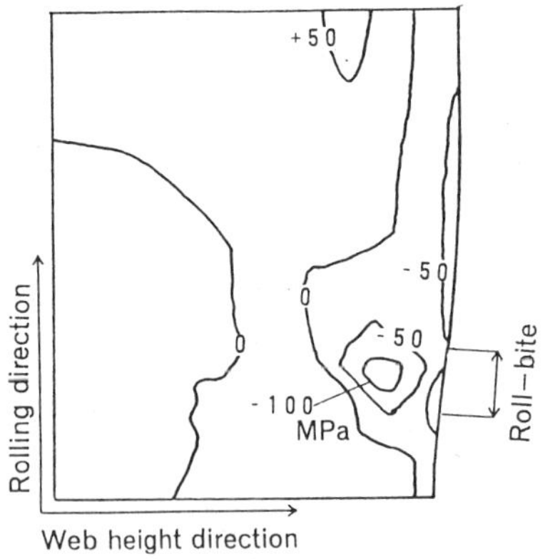

(c) Cross angle α = 10 degree

Fig. 15 Variation of longitudinal stress at center of web thickness with cross angle.

Fig. 15 shows variation of longitudinal stress distribution on center plane of web thickness with the cross angle. The distribution of longitudinal stress varies as the cross angle changes. Especially, the compressive stress near end of web height at the exit of roll-bite decreases as cross angle increase. This result is agree with experimental result that web waviness decreases as the cross angle increases.

4 CONCLUSION

Using rigid-plastic finite element method, a versatile system for analysis of long-product has been developed.

The system enables us to simulate almost all process of long-product rolling. We believe that the system will be applied to development of new process as well as to optimization of rolling conditions of current process.

REFERENCES

Bertrand, C., David, C., Chenot, J.L. and Buessler, P. 1986. Stress calculation in finite element analysis of three-dimensional hot shape rolling. *NUMIFORM 86*,207-212. Balkema.

Bertrand, C., Montmitonnet, P., Chenot, J.L., Fau, F. and Buessler, P. 1989. A three-dimensional analysis of hot rolling with a steady state thermomechanical approach. *NUMIFORM 89*, 303-308. Balkema.

Lee, Y.S., Dawson, P.R. and Dewhurst, T.B. 1989. Bulge predictions in steady state bar rolling processes. *NUMIFORM 89*, 323-330. Balkema.

Mori, K., Nakadoi, K., Mihara, Y., Hirakawa, T., Osakada, K. and Fukuda, M 1985. Finite element simulation of three-dimensional deformation in rolling of seamless pipe. *Proc. 3rd Int. Conf. Steel Rolling*, Tokyo: 375-382.

Mori, K. and Osakada, K. 1989. Finite element simulation of three-dimensional deformation in shape rolling. *NUMIFORM 89*, 337-342. Balkema.

Osakada, K., Nakano, J. and Mori, K. 1982. Finite Element Method for Rigid-Plastic Analysis of Metal Forming (Formulation for Finite Deformation). *Int. J. Mech. Sci.*, Vol. 24, 459-468.

Pawelski, O., Rasp, W. and Cremer, B. 1992. Numerical simulation of rolling of continuously cast slab with liquid core. *NUMIFORM 92*, 743-747. Balkema.

Yamada, K., Ogawa, S., Hamauzu, S. and Kikuma, T. 1989. Three-dimensional analysis of mandrel rolling using rigid-plastic finite element method. *NUMIFORM 89*,375-380. Balkema.

Yanagimoto, J., Kiuchi, M. and Inoue, Y. 1993. Characterization of wire and rod rolling with front and back tensions by three-dimensional rigid-plastic finite element method. *4th ICTP*, 764-769.

Simulation of Materials Processing: Theory, Methods and Applications, Shen & Dawson (eds)
© 1995 Balkema, Rotterdam. ISBN 90 5410 553 4

Characterization of angle rolling with two-roll mills and three-roll mills

Jun Yanagimoto & Manabu Kiuchi
Institute of Industrial Science, University of Tokyo, Japan

Kazuyoshi Shibata
Aichi Steel Works, Ltd, Japan

ABSTRACT: Three-dimensional general purpose FEM simulator for rolling, named CORMILL system, was applied to new angle shape rolling process with three-roll mill. Three-dimensional deformation of workpiece rolled by three-roll mill was compared with that rolled by conventional two-roll mill. Effects of three-roll mill on decrease in roll wear and workpiece wear were investigated and discussed.

1 INTRODUCTION

In order to design rolling mills, roll profiles, pass schedules and operation conditions, three-dimensional deformation of workpieces under rolling should be known. In order to develop new shape rolling processes, usually, test rolling mills and profiled rolls should be manufactured for trial experiments. Quite a long time and high costs are necessary for these trial experiments. If trial experiments are replaced by numerical experiments, time and costs necessary to develop new rolling processes will be greatly reduced.

Three-dimensional FEM is now becoming a practical tool to predict deformation of workpieces under various rolling processes. The authors have developed a general purpose FEM simulator for rolling processes. This simulator, which is named 'CORMILL System', has been applied to strip rolling (Yanagimoto 1991, Yanagimoto 1992b), bar/wire rolling (Yanagimoto 1992a, Yanagimoto 1993) and shape rolling (Yanagimoto 1994). The results for multi-pass shape rolling (Yanagimoto 1994) demonstrated that accurate predictions for three-dimensional deformation and rolling forces are possible by using CORMILL System.

In this paper, results of application of CORMILL System to angle shape rolling with three-roll mills will be presented. A series of simulations for three-roll angle shape rolling are done in order to clarify deformation characteristics of workpieces under this rolling comparing to those under conventional shape rolling with two-roll mill.

2 MATHEMATICAL MODEL

Fig.1 shows block diagram of CORMILL System for multi-pass rolling processes. Three-dimensional steady-state geometry of workpiece at a roll gap in each pass is analyzed by repeating three-dimensional rigid-plastic finite element analysis under assumed boundary conditions and contact analysis alternatively. Nodal coordinates and equivalent strain distribuion at exit cross-section are transferred to those at entrance cross-section of next pass. If cross-sectional mesh distortion becomes too large, cross-sectional nodal coordinates at the exit cross-section are remeshed in order to avoid descretization error which may occur in next pass (Yanagimoto 1994).

In the following, numerical results with respect to a single pass of angle rolling with three-roll mill and two-roll mill will be shown and compared with each other. $x-$, $y-$, $z-$ axes coincide with lateral, vertical and longitudinal directions, respectively.

2.1 *Rigid-plastic finite element analysis*

Three-dimensional velocity distributions under assumed contact areas are analyzed by Lagrange multiplier rigid-plastic FEM. Variational principle is expressed by equations (1) and (2).

$$\Phi = \int_V \sigma'_{ij}\dot{\varepsilon}_{ij}dV - \int_{S_F} T_i\dot{u}_i dS + \int_V \lambda\dot{\varepsilon}_{kk}dV \quad (1)$$

$$\delta\Phi = 0 \text{ for any } \delta\dot{u}_i, \delta\lambda \quad (2)$$

The boundary conditions at entrance cross-section and exit cross-section are expressed by following equations. These boundary conditions are obtained by assuming that no bending and twisting occur outside the region for analysis.

$$\dot{u}_x = 0, \dot{u}_y = 0, \dot{u}_z = (\text{uniform,unknown}) \quad (3)$$

By solving equations (1)-(3) for assumed contact areas between workpiece and both rolls, the velocity of each nodal point can be calculated.

2.2 *Contact analysis*

As a first step of contact analysis, nodal coordinates on each stream line are calculated by integrating nodal velocities obtained by three-dimensional rigid-plastic finite element analysis under assumed contact areas. Contact areas between both rolls and workpiece change in every contact analysis before three-dimensional deformation of workpiece at a roll gap reaches steady-state. Changes in contact areas cause geometrical interference.

Geometrical interferences of nodal coordinates on roll surfaces are modified in every contact analysis as is shown in Fig.1. Nodal coordinates inside the deforming workpiece are corrected by linear interpolation of geometrical interferences on upper and lower surfaces of workpiece. These modifications of geometrical interferences in every contact analysis are helpful to get steady-state solution for asymmetric shape rolling processes (Yanagimoto 1994). As these modifications of geometrical interferences will become close to zero if deformation of workpiece reaches steady-state, they do not affect the accuracy of steady-state solution.

2.3 *Velocity of idle roll*

Idle rolls are often used in shape rolling mills. **Fig.2** shows schematic illustration of roll arrangement of three-roll angle rolling mill. In practical three-roll angle rolling mills, a pair of idle upper rolls may be desirable as simpler driving mechanism is required for these mills. Torque of these idle rolls should be close to zero when deformation of workpiece reaches steady-state. This condition can be satisfied by changing upper roll velocities after every contact analysis (see Fig.1). Upper roll velocity is optimized until upper roll torque becomes less than 1% of lower roll torque.

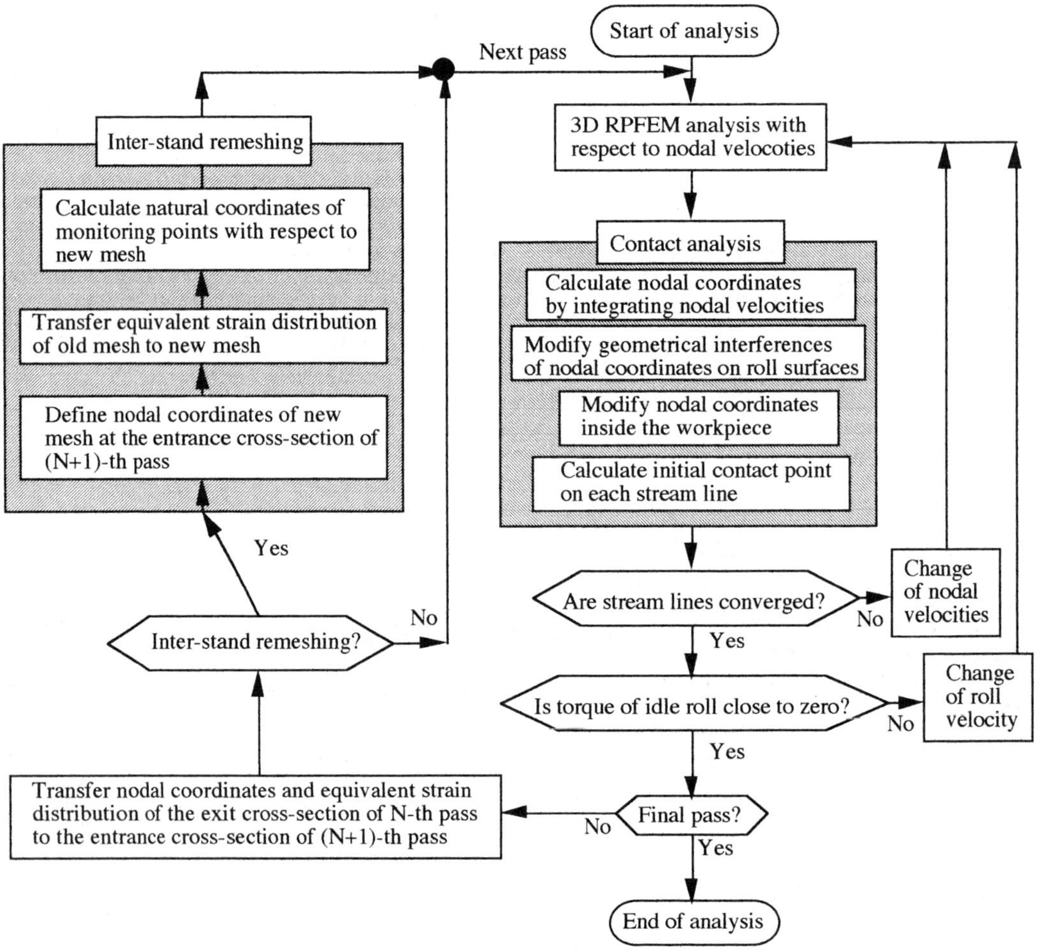

Fig.1 Block diagram of CORMILL System for multi-pass rolling processes.

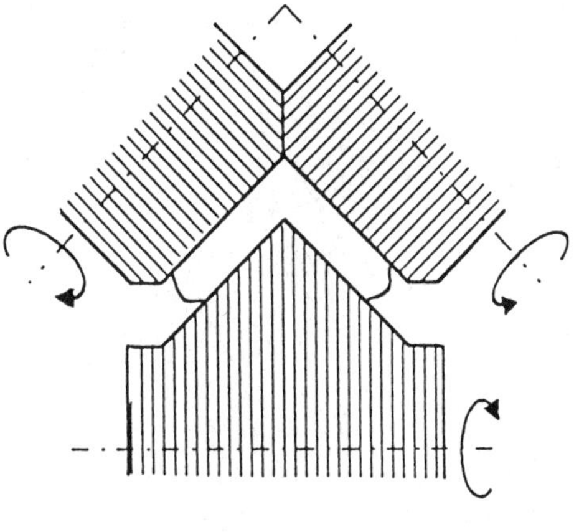

Fig.2 Schematic illustration of roll arrangement of three-roll angle rolling mill.

Table 1 Employed rolling conditions.

Angular velocity of upper roll /RPM	Idle (Three-roll mill) 50 (Two-roll mill)
Angular velocity of lower roll /RPM	50
Flow stress / **kgf mm** $^{-2}$	$\bar{\sigma} = 18.7\bar{\varepsilon}^{0.21}\dot{\bar{\varepsilon}}^{0.13}$
Friction coefficient	$\mu = 0.3$
FE mesh system $N_X \times N_Y \times N_Z$	$15 \times 4 \times 12 (L = 100)$ $11 \times 4 \times 12 (L = 50)$

Table 2 Geometries of rolls and workpiece.

θ_1	90 deg.	θ_2	90 deg.
h	40 mm (L = 100) 21 mm (L = 50)	R_3	4 (2) (5) (6) mm
		R_4	10 mm
R_1	2 (10) mm	t_2	10.5, 12, 14 mm
R_2	10 mm	L	50, 100 mm
t_1	10 mm		* (--) : For numerical analysis of overfill or underfill.
W	2 (1) (4) mm		
D_1	600 mm		
D_2	600 mm		

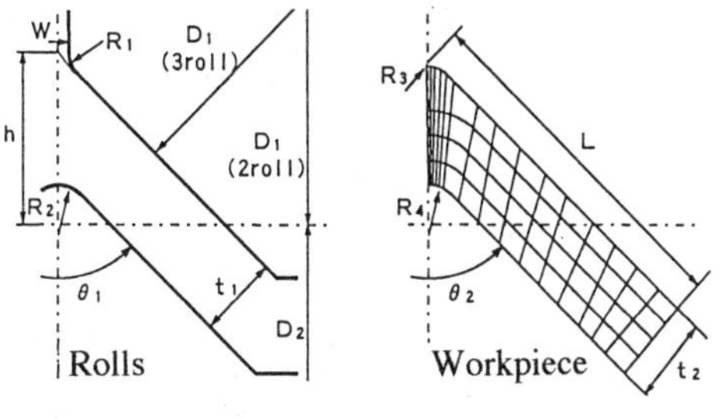

Fig.3 Geometries of rolls and workpiece.

3 NUMERICAL RESULTS

Usually, two-roll mill is used in angle shape rolling. Wears of rolls and rolled workpiece at edge portion are one of the problems in conventional angle rolling with two-roll mills especially for hot angle rolling of stainless steels. The excess in relative slip due to the difference in roll diameter and roll velocity in width direction of workpiece may be the biggest reason for these wears. If three-roll mill is introduced, relative slip between upper rolls and workpiece decreases at edge portion and then wears of rolls and rolled workpiece at this portion decrease. But overfill of workpiece into gap between upper rolls should be prevented as it is directly related to the profile of corner of angle section.

As there are quite a few study for three-roll angle rolling, effects of three-roll mill on decrease in roll wear and workpiece wear and that on increase in overfill into gap between upper rolls are now left unclarified. Therefore, CORMILL system is applied to a series of simulations for three-roll angle shape rolling, and numerical results are compared with those for two-roll angle shape rolling.

Table 1 shows employed rolling conditions for numerical case studies. **Table 2** and **Fig.3** show geometries of rolls and workpiece. These conditions, which correspond to those in finishing pass, are employed to clarify the effects of type of rolling mill, leg length of workpiece and reduction on three-dimensional deformation of workpiece such as relative slip and overfill. Roll profiles at the exit of roll gap used for rolling with three-roll mill and two-roll mill are same with each other. Because of symmetry in width direction, only a half of the workpiece is taken as the region for analysis.

It is widely known that wears of workpiece and rolls in hot forming are affected by relative slip between workpiece and rolls and rolling pressure. In the following, slip energy per unit width $W_{s/b}$ is introduced to evaluate wears of workpiece and rolls. This parameter reflects both relative slip and rolling pressure. From numerical values calculated by CORMILL Systsm, $W_{s/b}$ can be computed by equation (4):

$$W_{s/b} = \int_t \int_z \tilde{u}_s \tau_f dz dt \tag{4}$$

where τ_f is frictional stress, $\tilde{u}_s$ is relative velocity and t is contact time between workpiece and rolls. In order to evaluate the effects of relative slip and rolling pressure separately, slip length L_s and rolling force per unit width $P_{/b}$ are introduced. They are expressed by equations (5) and (6):

$$L_s = \int_t \tilde{u}_s dt \tag{5}$$

$$P_{/b} = \int_z p dz \tag{6}$$

where p is rolling pressure.

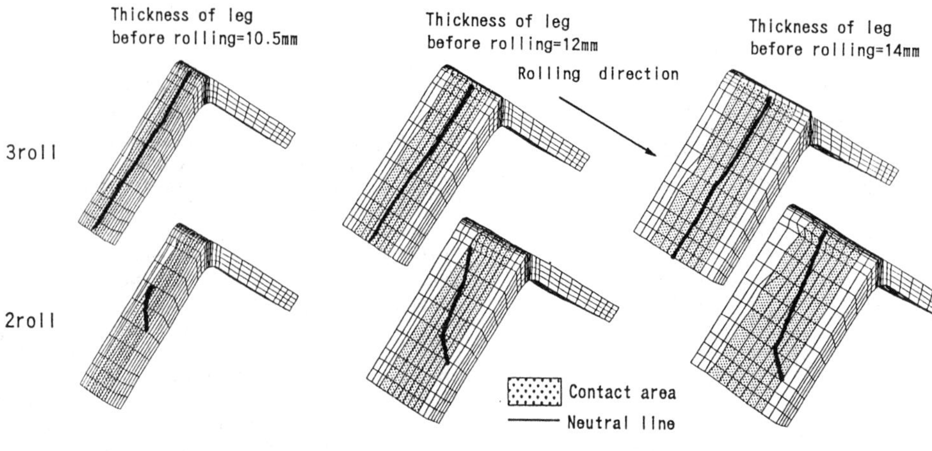

Fig.4 Three-dimensional geometry of rolled workpiece and neutral line.

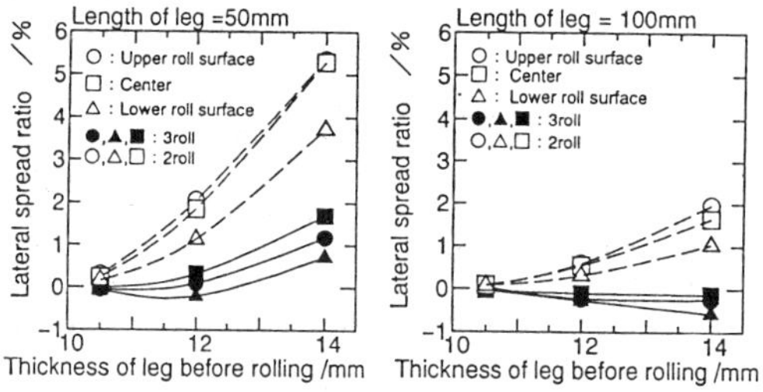

Fig.5 Lateral spread ratio.

3.1 *Three-dimensional geometry of workpiece, neutral line and lateral spread*

Fig.4 shows three-dimensional geometries and neutral lines rolled by three-roll mill and two-roll mill. Position of neutral line in rolling direction is almost uniform in width direction of workpiece when three-roll mill is used, but it distributes in width direction of workpiece when two-roll mill is used. This is due to the difference in roll diameter and roll velocity in width direction of workpiece in two-roll mill. When three-roll mill is used, workpiece overfills into gap between upper rolls at higher reductions. Typical overfill is obtained in the rolling with three-roll mill when thickness of leg before rolling is 14mm.

Fig.5 shows relationships between thickness of workpiece before rolling, which corresponds to thickness reduction, and lateral spread ratio. Width of angle before rolling used to normalize lateral spread is that at the center of thickness. Lateral spread of workpiece rolled by three-roll mill is smaller than that rolled by two-roll mill. This is due to the fact that upper roll of three-roll mill does not have a velocity component in width direction of workpiece (see Fig.9).

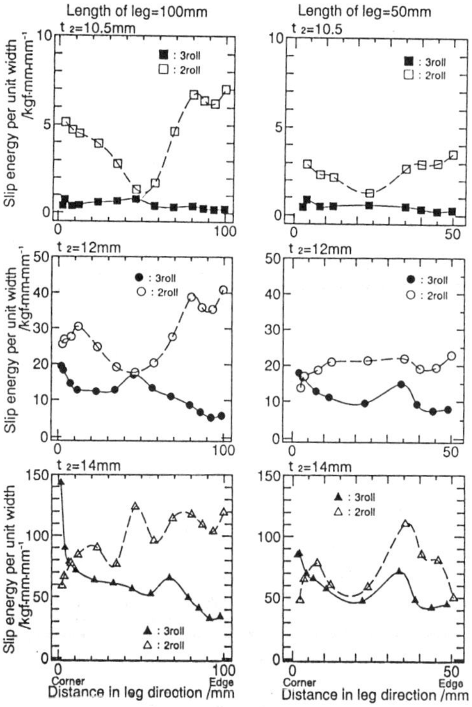

Fig.6 Slip energy per unit width of workpiece on upper roll surface.

3.2 *Evaluation of roll wear and workpiece wear*

1) *Slip energy per unit width* **Fig.6** shows change in slip energy per unit width at upper roll surface. Generally, slip energy per unit width of workpiece rolled by three-roll mill is less than that rolled by

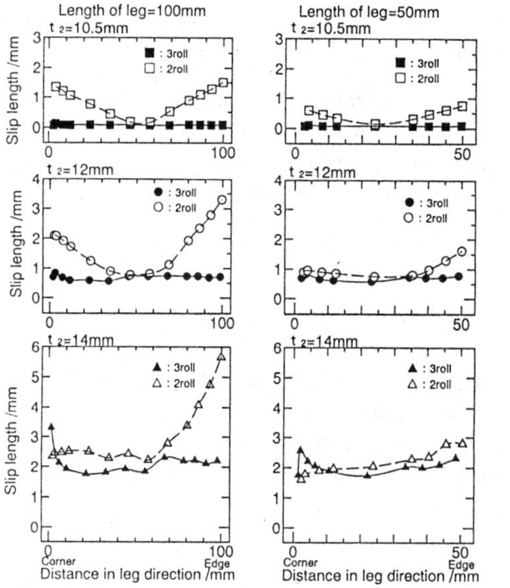

Fig.7 Slip length of workpiece on upper roll surface.

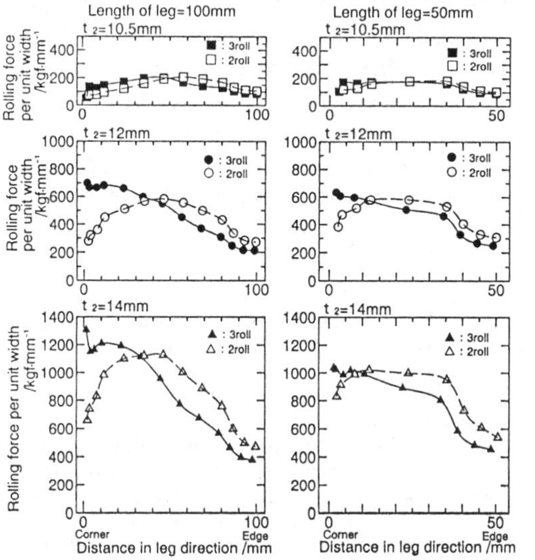

Fig.8 Rolling force per unit width acting to upper roll.

two-roll mill. It increases according to the increase in legth of leg (width) before rolling. Especially at edge portion, slip energy per unit width drastically decreases when three-roll mill is used. In the prectical angle rolling with two-roll mill, wears of workpiece and rolls often occur at edge portion of workpiece on upper roll surface. Then decrease in wears can be expected at this portion if three-roll mill is introduced.

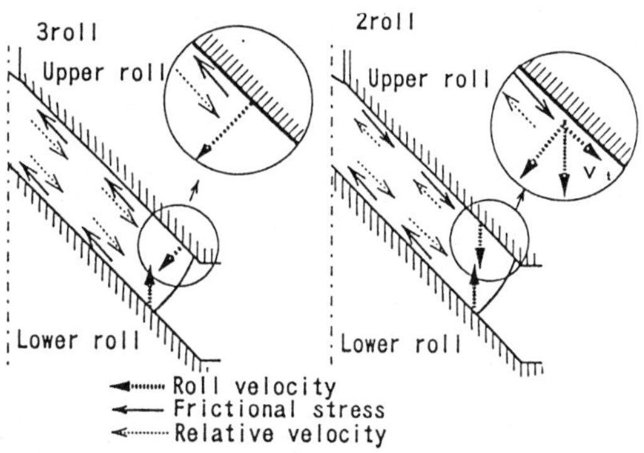

Fig.9 Comparisons of cross-sectional roll velocity, frictional stress and relative velocity between three-roll rolling and two-roll rolling.

2) *Slip length* **Fig.7** shows comparison of slip lengths between three-roll mill and two-roll mill. Slip length of workpiece rolled by three-roll mill is more uniform in leg (width) direction compared with that rolled by two-roll mill. Slip length of workpiece rolled by three-roll mill is less than that rolled by two-roll mill. These difference arise from difference in the position of neutral line shown in Fig.4. At higher reduction and at the center of workpiece, slip length of workpiece rolled by three-roll mill becomes longer than that rolled by two-roll mill. This is due to the overfill of workpiece rolled by three-roll mill.

3) *Rolling force per unit width* **Fig.8** shows rolling force per unit width acting to upper roll. It shows maximum value at the center of workpiece when three-roll mill is used. On the contrary, it shows maximum value at the middle position of leg when two-roll mill is used. **Fig.9** shows comparisons of directions of cross-sectional roll velocity, frictional stress and relative velocity between three-roll mill and two-roll mill. Upper roll velocity of three-roll mill does not have a cross-sectional velocity component in tangential direction of roll profile, but upper roll velocity of two-roll mill has a cross-sectional velocity component in tangential direction of roll profile. As frictional stresses acting to upper rolls are directed toward invers directions with each other, rolling pressure of workpiece rolled by three-roll mill and two-roll mill distribute differently in leg direction of workpiece.

3.3 *Evaluation of overfill in three-roll angle rolling*

In angle shape rolling with three-roll mills, it is impossible to eliminate gap between both upper rolls. Therefore overfill of workpiece into this gap may occur if rolling mill and roll profile are not correctly designed. In this section, the effects of gap between upper rolls W, upper corner radius of workpiece before rolling R_3 and corner radius of upper roll R_1 (see Fig.3) on overfill or underfill will be shown. In

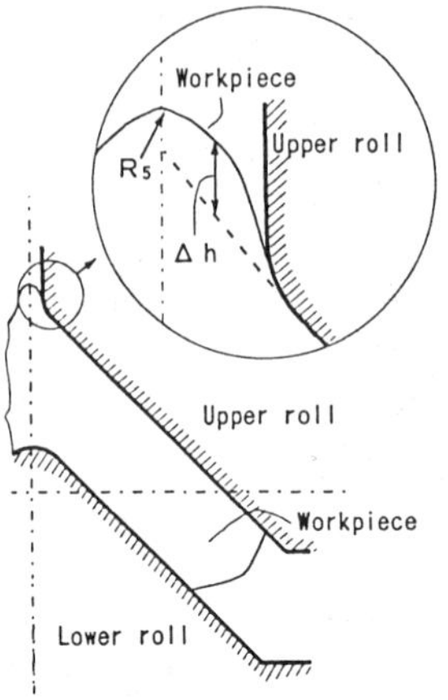

Fig.10 Overfill into gap between upper rolls.

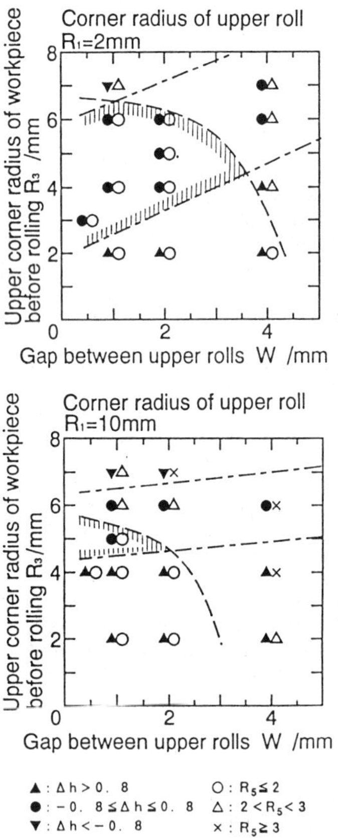

Fig.11 Range of rolling conditions without overfill and underfill.

order to evaluate overfill (or underfill) of workpiece in gap portion, ideal profile of workpiece, which is shown by broken line in **Fig.10**, is defined first and then the maximum difference of workpiece profile from ideal profile Δh is introduced. In addition, corner radius of rolled workpiece R_5 is used to evaluate shape of workpiece at gap portion. Regarding JIS (Japan Industrial Standards) for angle section bars, following criteria is introduced to judge the quality of rolled product.

$$\Delta h \leq 0.08 t_1 = 0.8 \text{ [mm] and } R_5 \leq 2 \text{ [mm]} \quad (7)$$

Fig.11 shows the effects of W, R_3 and R_1 on the quality of rolled product. The hatched areas show the range of W and R_3 in which rolled product satisfies the above mentioned criteria. If corner radius of upper roll R_1 is 10mm, it is possible to manufacture product which satisfies above mentioned criteria only within small variations of W and R_3. On the contrary, if corner radius of upper roll R_1 is 2mm, it is possible to manufacture product with good quality within larger variations of W and R_3. In this case, if gap between upper rolls W is less than 2mm, upper corner radius of workpiece before rolling R_3 can change from 4mm to 6mm. As upper corner radius of workpiece before rolling R_3 may vary in every workpiece in practical shape rolling operation, the above mentioned combination of W and R_1 should be chosen for three-roll rolling mill and roll profile.

4 CONCLUSION

CORMILL System has been applied to a series of simulations of angle shape rolling with three-roll mill and two-roll mill. Through the investigation, it becomes clear that wears of workpiece and rolls at edge portion can be decreased if three-roll mill is introduced. Three-roll angle shape rolling may be also effective to control the width (length of leg) of rolled product as lateral spread of workpiece is small compared with that rolled by two-roll mill.

One of the disadvantages for three-roll angle rolling is overfill or underfill of workpiece which may occur in gap between upper rolls. If gap between upper rolls and corner radius of upper roll are optimized, products with good quality, which means that products whose overfill and underfill lie under industrial tolerance, can be obtained even if three-roll mill is used.

REFERENCES

Yanagimoto, J. & Kiuchi, M. 1991. Three-dimensional coupled simulation of strip and shape

rolling processes, *Transactions of NAMRI/SME* **19**:15-22.
Yanagimoto, J. & Kiuchi, M. 1992a. Characterization of wire and bar rolling by three-dimensional rigid-plastic finite element method, *Transactions of NAMRI/SME* **20**:3-9.
Yanagimoto, J. & Kiuchi, M. 1992b. Three-dimensional simulation system for coupled elastic/rigid-plastic deformations of rolls and workpieces in strip rolling processes, *Proceedings of NUMIFORM'92:* 763-768.
Yanagimoto, J., Kiuchi, M. & Inoue, Y. 1993. Characterization of wire and rod rolling with front and back tensions by three-dimensional rigid-plastic finite element method, *Proceedings of 4th ICTP*:764-769.
Yanagimoto J. & Kiuchi, M. 1994. Three-dimensional Rigid-plastic FE Simulation System for Shape Rolling with Inter-stand Remeshing, *Proceedings of International Conference for Metal Forming Process Simulation in Industry*: 219-237.

Simulation of Materials Processing: Theory, Methods and Applications, Shen & Dawson (eds)

 ISBN 90 5410 553 4

A comparison of finite element models with experimental measurements during hot rolling of aluminium

H.L.Yiu, D.J.Browne & S.Rogers
Alcan International Limited, Banbury Laboratories, UK

ABSTRACT: Two finite element models were used to predict the deformation and temperature distributions in aluminium slabs during hot rolling. Similar results were obtained from either model when appropriate surface boundary conditions for heat transfer and friction were used. Predictions agree well with experimentally measured temperature profiles, speeds and strain accumulated during rolling, although at sub-surface regions, higher strain values were predicted by the models.

1. INTRODUCTION

During rolling, the strain, strain rate and temperature of metal passing through the roll-bite vary from the surface to the centre of the rolled slab. Knowledge of the distribution of these rolling parameters within the slab is important from the point of view of process and microstructural control. It is well known that these parameters determine the microstructure and properties of the hot-rolled material.

Over the past decade, several numerical techniques have been developed for process modelling [1] [2]. When applied to the hot rolling process, the finite element method can include different material constitutive models relating stress, strain, stain rate and temperature. Realistic boundary conditions i.e. friction and thermal effects between the roll and work material must be taken into account. The purpose of the present work was to examine the accuracy of finite element modelling in the prediction of metal flow and temperature evolution during hot rolling of aluminium.

The finite element models used in this work were developed at Cornell University, USA [3] and Leicester University, UK [4], to predict thermomechanical solutions. In order to evaluate these models, experimental rolling trials were performed using commercial purity Al and an Al-Mg-Mn alloy. The strain histories of elements through the thickness of the slab were analysed using a visioplasticity technique, whereas temperature histories were determined from embedded thermocouples.

2. DESCRIPTION OF THE MODELS

Both the Cornell and the Leicester computer models are 2-dimensional Eulerian models developed to predict local temperature, strain and strain rate as a function of position in the roll bite. Analysis is based on a steady state flow and therefore does not consider the initial threading operation or the eventual departure of the end of the slab from the roll gap. This process is suited to the Eulerian approach whereby the metal flows through the stationary FE mesh. The models directly couple the mechanical and thermal aspects of the problem, allowing the heat produced during deformation of the metal to change the temperature of the metal. It is assumed that 95% of plastic work is converted to heat which resides in the metal. The material is modelled as visioplastic in both models i.e. velocities are dependent variables to be calculated. Plane strain conditions are assumed in the calculation. Basically, the models give thermomechanical solutions that satisfy the conservation of mass, balance of momentum, and steady-state convective-diffusion heat flow subjected to the mechanical and thermal boundary conditions. The detailed mathematical formulation can be found in references 3,4 and 5, and the paper by Rogers and Browne published at this Conference.

Linear triangular elements are used in the Leicester Model, whereas rectangular elements are employed by the Cornell Model. Apart from this the two models differ mainly in the treatment of the friction and heat transfer boundary conditions at the work/roll interface. At the interface with the roll, the heat flux to the roll is given by:

$$K \frac{\partial T}{\partial n} = h\ (T - Tr)$$

where Tr is the temperature of the roll interface determined by using the phantom roll method [6] in the Leicester Model and a finite difference method in the Cornell Model. K is the thermal conductivity, n is the normal to the interface and h is a heat transfer coefficient. In both models, there is usually assumed symmetry about the centre line of the slab represented by an adiabatic boundary. Free surfaces and the exit plane from the roll bite are treated as insulated surfaces. At the interface, the boundary conditions are also complicated by the presence of friction which drives the process. The Leicester Model is based upon the proposition that the velocity of the surface of rolled stock (Vs), along the arc of contact with the roll, is a weighted average of the roll velocity (Vr) and the initial slab velocity (Vo). Thus the frictional behaviour is described by a velocity - compensated model :

$$Vs = \frac{hoVo}{hCos\theta}(1 - f) + fVr$$

where f represents the friction factor, θ is the angle that the node makes with the exit plane, ho is the initial slab thickness and h the slab thickness at the node in question. When $f = o$, the slab surface speed is that required to maintain volume conservation; but the friction factor should be above some threshold value to prevent slipping in the model. When $f = 1$, the slab surface speed is equal to the roll surface speed and this represents the case of sticking friction.

In the Cornell Model, Orowan's law of friction is employed, which states that the shear stress due to friction cannot exceed the yield stress of the material in shear. Thus, $o < f < 0.5$, the value of 0.5 representing the sticking condition, since the flow stress in shear is one-half of the flow stress in tension or compression.

To characterise the material behaviour mathematically, a modified Voce-type equation was used in this work:

$$\sigma = \sigma_o + (\sigma_{ss} - \sigma_o)\ [1 - exp\ (- C\varepsilon)]^m$$

The material constants in this equation were determined by fitting the stress (σ) and strain (ε) data obtained from plane-strain compression testing of each alloy over a range of temperatures and strain rates.

3. EXPERIMENTAL DETERMINATION OF THERMOMECHANICAL HISTORIES.

Commercial purity aluminium and an Al-Mg-Mn alloy were given a homogenisation treatment, cooled to the test temperature and hot rolled using the Banbury Laboratory Mill. Initial slab temperatures of 450°C for the pure Al and 485°C for the alloy and reductions of 40 and 48% were studied. The rolls, of diameter 368mm, were at a nominal initial temperature of 20°C, and were rotated at a constant speed of 10 rev/min.

Temperature evolution was measured on slabs of dimensions 30mm x 75mm x 200mm. These were instrumented with thermocouples placed 1mm below the surface, at the quarter thickness and along the central plane. The thermocouple wires were Ni-Cr/Ni-Al, encapsulated in one millimetre diameter steel sheaths. A pull-wire linear encoder, which gave 1000 signals per recording, was used to measure the horizontal velocity at the centre of the specimen.

In order to obtain the flow pattern and strain evolution, some rolling slabs were fitted with gridded inserts (8mm wide), so that the gridded face lay along the central plane of the slab. A square grid geometry was chosen, with grid lines scribed at 1.5mm intervals. Theses specimens were partially rolled to expose the deformed grid pattern within the roll bite. In these tests, the rolling mill was stopped during the reduction and raised to release the specimen. Figure 1 shows the pattern of deformation resulting from a 40% reduction during rolling of pure aluminium. Assuming no lateral flow, the equivalent strain generated during rolling can be calculated at any position within the roll-bite from the distortion of the grid relative to the original grid [7]. In this work, the positions of the nodes in the grid pattern (e.g. Figure 1) were digitised with respect to a given set of reference axes. The total accumulated equivalent strain was then calculated by summing the infinitesimal increments of equivalent strain ($d\varepsilon$) along the strain path taken by the element using the following equation:

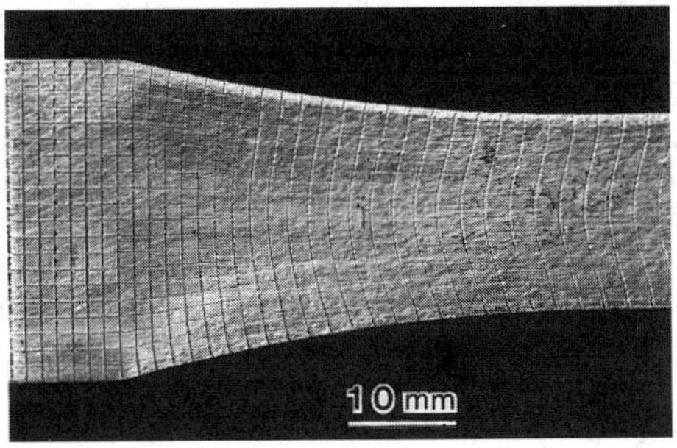

Figure 1. Deformed network after rolling aluminium slab with a single pass of 40% reduction.

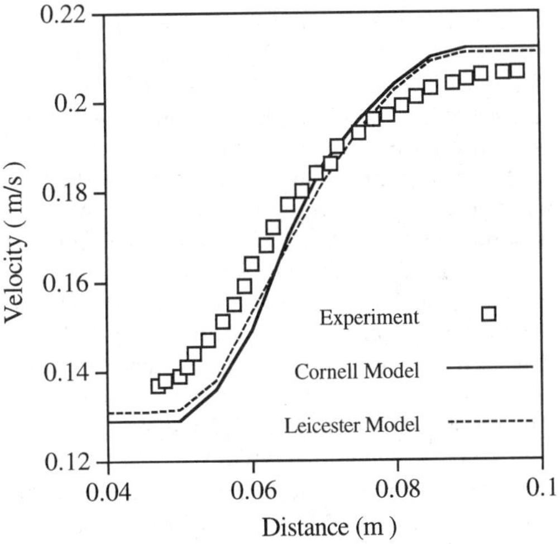

Figure 2. Computed and experimental velocities at the slab centre.

$$\overline{\varepsilon}_{total} = \sum_{1}^{n} d\,\overline{\varepsilon} = \sum_{1}^{n} \sqrt{\frac{2}{3}}\,(d\varepsilon_{ij}\;d\varepsilon_{ij})^{\frac{1}{2}}$$

4. NUMERICAL & EXPERIMENTAL RESULTS FOR PURE ALUMINIUM

As a result of frictional forces drawing the material into the roll, it is observed in Figure 1 that the grid lines transverse to the section are bent into a concave shape, the concavity being in the rolling direction. In the simulations discussed in this section a friction factor of 0.3 and 0.5 was used in the Cornell Model and Leicester Model, respectively.

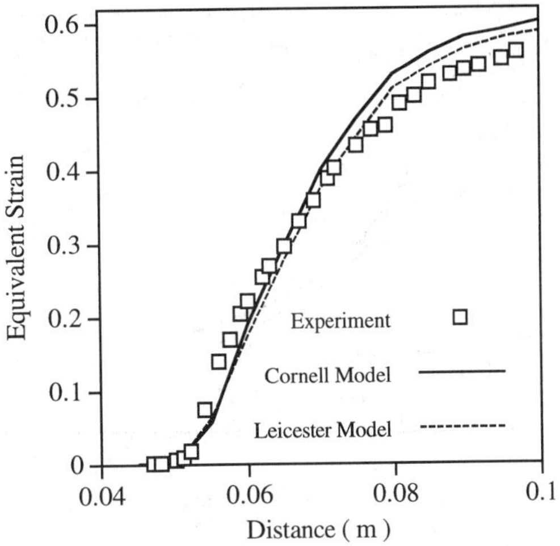

Figure 3. Computed and experimental equivalent strains at the slab centre.

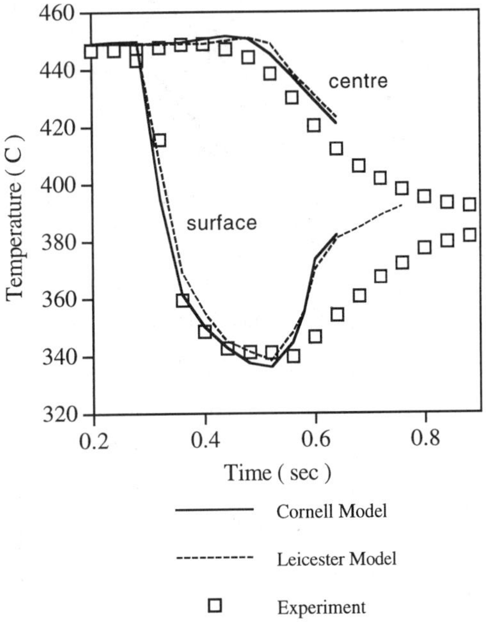

Figure 4. Measured and computed temperature histories for pure aluminium.

Figure 2 shows the model and experimental velocities at the slab centre. A good agreement is obtained between the models and with the experimental measurements. The velocity gradient representing the acceleration in the bite is fairly well matched, as are the entry and exit velocities.

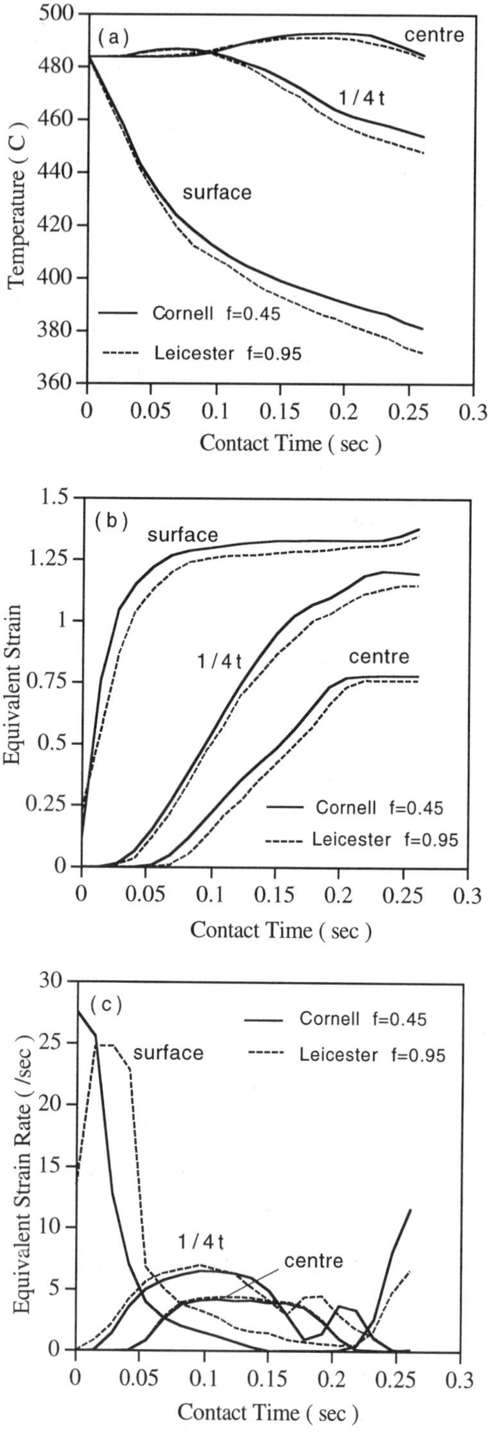

Figure 5. Predicted variation of (a) temperature (b) equivalent strain and (c) strain rate with contact time for Al-Mg-Mn alloy rolled by 48% reduction.

It will be noted that the friction conditions would have little effect on central velocity profiles. This was confirmed by using friction factors of 0.3 and 0.5 (sticking friction) in the Cornell model and 0.2 and 0.8 in the Leicester Model. The difference between model and experiment in the exit velocity is explained by the fact that in practice the material never reached the exit velocity which during steady state flow is dictated by mass conservation.

Figure 3 shows the evolution of strain at the centre line of the slab as rolling proceeds, comparing modelled strain values with those as determined from the experimental method. It can be seen that, for both models, comparison is in reasonable agreement with measurements. Equivalent strain reaches a plateau in Figure 3 corresponding to the final accumulated equivalent strain at the centre as the metal leaves the roll bite. As in the case of velocity comparison, the difference between experimental and computed strains becomes larger (with maximum $\Delta \varepsilon \approx 5\%$) when approaching the roll exit.

From an initial study of the effect of the heat transfer coefficient, the best agreement between measured and calculated temperature-time profiles was obtained for a heat transfer coefficient value of 23 $kW/m^2/K$ for both FE models with roll surface temperature calculations included in the prediction. This heat transfer coefficient lies within the values of 15 to 30 $kW/m^2/K$ reported by Smelser and Thompson [8] for rolling of aluminium.

Figure 4 shows the predicted and measured temperature profiles at the centre and surface of the slab (heat transfer coefficient = 23 $kW/m^2/K$). As soon as deformation has commenced, the surface temperature drops rapidly to a minimum due to the chilling effect of the rolls, the minimum temperature here indicating the time at which the slab leaves the bite. The predicted gradient of surface temperature (0.4mm below surface) with time is in very good agreement with experiment for the region of material in contact with the roll. In both models, the free surface line of the slab in the exit region is treated as an insulated surface instead of being cooled by air. Therefore, the models would be expected to predict a slightly greater recovery of temperature at the surface and immediate subsurface regions once the workpiece leaves the rolls, as shown in Figure 4. In specimens used to record the deformation characteristics of the reduction in the form of the distorted grids (e.g. Figure 1), the discrepancies in the surface temperature after rolling are even greater. This was due to the extra contact time of the partly rolled specimen with the rolls resulting in exaggerated cooling towards the end of the reduction.

At the centre, a small temperature rise of 2-3°C is calculated by the models and this is in agreement with the measurement. Under adiabatic conditions, the plastic work will give rise to a temperature increase of 4.5°C.

5. NUMERICAL & EXPERIMENTAL RESULTS FOR Al-Mg-Mn ALLOY

Figures 5a, 5b and 5c show a comparison made between local temperature, strain and strain rate predictions of the Cornell Model and the Leicester Model for a 48% reduction of the Al-Mg-Mn alloy. It is clear that both models are predicting similar profiles of these parameters as a function of position in the roll bite. In Figure 5a, again the temperature rise due to the release of deformation energy is predicted in the centre of the slab. However, the chilling influence of the roll penetrates the slab thickness and swamps the deformation heating effect. This dictates that the bulk temperature of the slab will decrease after passing through the roll bite. This, indeed, was the case observed during experimental rolling (Figure 4).

Both models predict the rapid straining of the slab surface material relative to the centre material upon initial contact (Figure 5b). The lagging behind of the slab centre is manifest in the forward bending of the vertical grid lines near the slab surface as shown in Figure 1 for the Al slab.

Flow in the roll bite is also characterised by maximum values of strain rates in the regions near the entry and exit points (Figure 5c). Similar results have been reported in modelling hot rolling of low carbon steel [9]. The surface strain attains a plateau at about the roll bite centre (Figure 5b). This and the corresponding low strain rates (Figure 5c) suggest that little deformation is taking place in these areas. Thus, these areas appear to be a dead metal zone within which the neutral zone lies. On the contrary, material at the slab centre achieves much higher strain rates in the roll bite centre (Figure 5c).

It is clear from the predictions of Figure 5b that there is a positive strain gradient throughout the rolling process. The strain distribution at the exit plane as calculated by the FE models are plotted in Figure 6 to compare with the strain distribution as determined from the visioplasticity method. Here the predicted and measured strains agree well in the interior of the slab from the centre to the quarter-

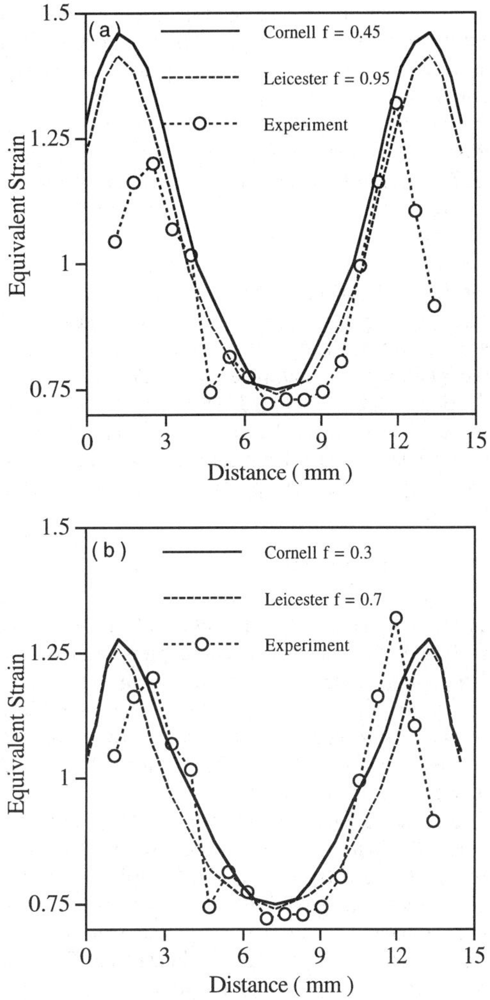

Figure 6. Computed and measured equivalent strain as a function of through-thickness position at the exit plane for 48% reduction of Al-Mg-Mn alloy; (a) high friction factors and (b) medium friction factors used in the models.

thickness position regardless of the value of the friction factor used.

However, both models predict higher strains than the measured values at sub-surface layers. Discrepancies, though reduced to a lesser degree, exist when the friction factor is decreased from 0.45 to 0.3 in the case of the Cornell model and from 0.95 to 0.70 in the case of the Leicester Model (Figure 6). Further reduction of the friction factor will simply produce discrepancies in strains at the quarter-thickness position.

In experimental measurement one source of error could be the out-of-plane flow observed from surface to the quarter-thickness positions. Errors could also come from digitising the gridded specimen which was reduced by a high reduction of 48%. Nevertheless, the models predict the right trends in strain profile through the slab thickness.

It is interesting to note that towards the exit side the slab surface is moving faster than the rolls and friction acts to restrain the material from leaving the bite. As a result, the sub-surface material is acted upon by two opposite shear forces at the roll exit, a forward shear from the forward motion of the slab and a backward shear due to surface friction. The combined effects of these shear forces is to produce the maximum strain at a sub-surface layer and a local maxima in strain rate at the exit region, as predicted by the models (Figure 5c).

6. CONCLUSIONS

The requirement of a finite element model of hot rolling is to make an accurate prediction of the temperatures, strains and velocities at every point in the region of deformation. The two FEM models studied in this work seem to give reasonable results in describing the deformation and temperature distributions within the roll bite. Some discrepancies, however, have been observed in the strain distribution at the surface region. This may be partly attributed to the simple boundary conditions employed. It is at the roll contact that the boundary conditions are least well known, particularly in terms of the interfacial friction. Most friction measurements to date have been obtained from external force determinations [10] and they relate little to processing. It would appear that greater use could be made of the resulting strain and velocity distributions in determining the friction effects from the rolled metal.

REFERENCES

1. Pitman, J.F.T. et al. Eds. 1984. Numerical Analyses of Forming Processes, New York: J. Wiley & Son.
2. Zienkiewicz, O.C. 1977. The Finite Element Method. New York: McGraw-Hill. 3rd ed.
3. Dawson, P.R. 1987. Int. J. Solids Structures.Vol 23, No 7, pp 747-968.
4. Beynon, J.H. et al. 1987. "Computational Methods for Prediction Materials Processing Defects" Elsevier: Ed. Predeleanu, Cachan. 19-28.
5. LeMat, N. et al. 1994. J. Mater Process Technol. 45: 631-636.
6. Sellars, C.M. & Kawai, R. 1993. Proc of Int Conf on "Modelling of Metal Rolling Processes". London: 648.
7. Jain. et al. 1988. J Mater Shaping Technol. 5: 243-248.
8. Smelser, R.E. & Thompson, E.G. 1987. Proc. symp. on Advances in Inelastic Analysis. Boston: ASME Winter Annual Meeting. AMD-88, PED-28, pp 273-282.
9. Silvonen, A., Malinen, M., Korhonen, A.S. 1987. Scand J of Metallurgy. 16: 103-108.
10. Petty, D.M. 1994. J Mater Process Technol 45: 7-12

Simulation of Materials Processing: Theory, Methods and Applications, Shen & Dawson (eds)
© 1995 Balkema, Rotterdam. ISBN 90 5410 553 4

Numerical modelling of cold rolling of metal plate adopting a general friction model

W. Zhang, M.J.P.A. Peeters & N. Bay
Institute of Manufacturing Engineering, Technical University of Denmark, Lyngby, Denmark

ABSTRACT: In the present paper a numerical analysis of the plate rolling process is carried out introducing Wanheim and Bay's (1974, 1978) general friction model. The analysis is based on the slab method and an operating program is worked out using the Finite Difference (FD) method. Experimentally measured load and torque are compared with the calculated varying the strip material, the lubricant and the reduction. For further verification of the general friction model, a method has been adopted comparing the theoretically calculated and experimentally measured backward and forward slip.

1 INTRODUCTION

Plate rolling is a widely applied metal forming process for production of metal plates, sheets and strips. Rolling with large diameter rolls results in heavy influence from friction on the pressure distribution in the roll gap, load and torque and it is thus importance to apply a realistic friction model in the theoretical analysis in order to obtain realistic results. Analyses by Alexander (1972) prove this.

In the present paper a general friction model developed by Wanheim and Bay (1974, 1978) is adopted for modelling of the plate rolling process. A method is applied for verifying the friction model by comparison of the theoretically calculated and the experimentally measured backward and forward slip. Schey (1983) has earlier pointed out, that the average coefficient of friction can be determined by measuring the forward slip, but as far as the authors know, literature does not report any application of this method for verifying or rejecting a friction model for plate rolling.

Experiments were carried out and rolling parameters were measured in order to verify the theoretical analysis. By comparison of the calculated and measured rolling load and torque, the numerical modelling of rolling is validated. By comparison of the calculated and measured backward and forward slip the general friction model applied is tested.

2 GENERAL FRICTION MODEL BY WANHEIM AND BAY

The friction laws adopted for metal forming analyses are normally either Coulomb's law:

$$\tau = \mu q \tag{1}$$

or the law of constant friction:

$$\tau = fk \tag{2}$$

Wanheim and Bay (1974, 1987) have shown that neither of these two laws are generally valid, and developed a general friction model which can be applied at both low and high normal pressures. This general friction model is expressed by:

$$\tau = \alpha f k \tag{3}$$

which combines the two friction laws, since α, the real contact area ratio of the rough specimen surface, being plastically deformed towards the smooth tool surface, at low normal pressures is proportional to the normal pressure thereby leading to Coulomb's law; at high normal pressures α goes towards 1 and Eq. (3) leads to the law of constant friction.

Transition between the two friction laws is continuous as expressed in the theory. In order to simplify the model, the two laws (1) and (2) are combined, reducing the transition zone to a point. The transition point between Eq.(1) and Eq.(2) is determined adopting the general friction model. According to this the relationship between the friction coefficient and the friction factor is given by (Christensen *et al*, 1986):

$$\mu = \frac{f}{1+\frac{\pi}{2}+\arccos f+\sqrt{1-f^2}} \tag{4}$$

The relationship between the flow stresses in tension and in shear is according to von Mises yield criterion:

$$\sigma_0 = \sqrt{3}\,k \tag{5}$$

By combining Eq. (1) with Eq. (2), the critical normal pressure expressing the transition point can be obtained:

$$\frac{q*}{\sigma_0} = \frac{1}{\sqrt{3}}\left(1+\frac{\pi}{2}+\arccos f+\sqrt{1-f^2}\right) \tag{6}$$

Fig.1 shows the simplified general friction model. The friction stress τ normalized by division with the flow stress in pure shear k is plotted as a function of the normal pressure q normalized by division with the flow stress σ_0 and with the friction factor f as a parameter.

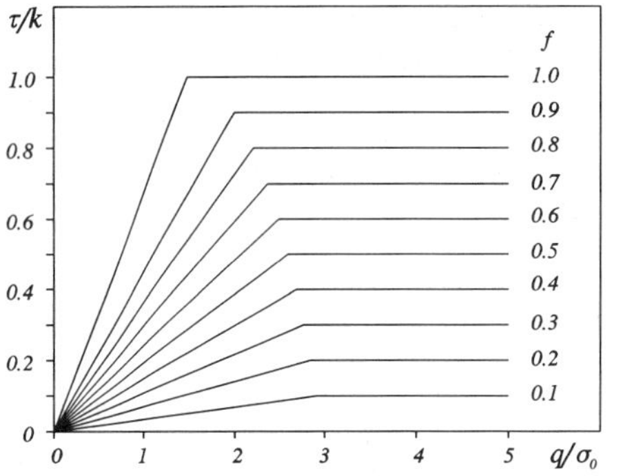

Fig. 1: Friction stress as a function of normal pressure with friction factor as a parameter.

The transition point between the two friction laws is listed in Fig. 2 as a function of the friction factor f.

3 THEORETICAL MODELLING

3.1 *General assumptions*

A. Plane strain deformation.
B. Circular contact arc.
C. Along a vertical plane the normal stress and the metal flow are uniform.
D. Workpiece material follows von Mises yield criterion.

3.2 *Stress-Strain Curve of Material*

Work hardening of the metal is taken into account by applying the stress-strain curves following Swift's equation:

$$\sigma_0 = C(B+\varepsilon)^n \tag{7}$$

The local equivalent strain is expressed by:

$$\varepsilon = \frac{2}{\sqrt{3}}\ln\frac{h_0}{h_x} \tag{8}$$

and the equivalent flow stress in plane strain rolling is:

$$s_0 = \frac{2}{\sqrt{3}}\sigma_0 = \frac{2}{\sqrt{3}}C\left(B+\frac{2}{\sqrt{3}}\ln\frac{h_0}{h_x}\right) \tag{9}$$

f	q^*/σ_0
0.1	2.91
0.2	2.84
0.3	2.77
0.4	2.68
0.5	2.59
0.6	2.48
0.7	2.36
0.8	2.20
0.9	2.00
1.0	1.48

Fig. 2: Critical values of normal pressure.

3.3 *Roll Flattening*

Flattening of the rolls is taken into account according to the formula of Hitchcock (1935). The deformed roll diameter D' is given by:

$$D' = D\left[1+\frac{16(1-\nu^2)P}{\pi E(h_0-h_1)}\right] \tag{10}$$

where D is the undeformed roll diameter, P is the load, ν and E are the Poisson's ratio and Young's modulus for the roll material and h_0 and h_1 are the initial and final strip thickness.

3.4 *Geometry of the deformation zone*

As seen in Fig. 3, the length of the deformation zone can be calculated by:

$$l = \sqrt{\frac{1}{2}D'(h_0-h_1)-\frac{1}{4}(h_0-h_1)^2} \tag{11}$$

If the increment is dx, then the surface taper angle of the deformation zone at point x is:

$$\theta = \arctan\left[\frac{\sqrt{\frac{D'^2}{4}-(l-x-dx)^2}-\sqrt{\frac{D'^2}{4}-(l-x)^2}}{dx}\right] \tag{12}$$

According to the direction of the friction in the

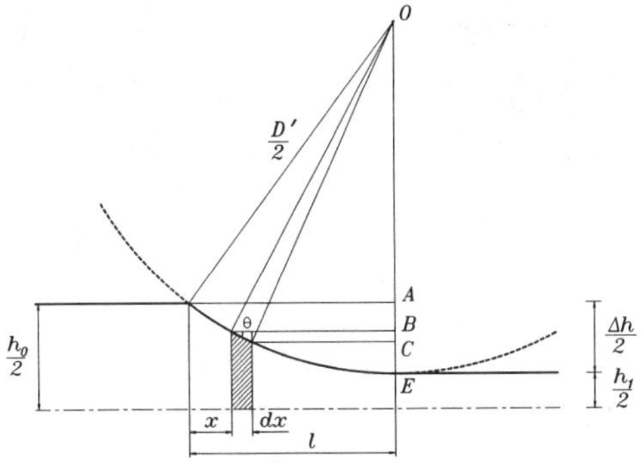

Fig. 3: Geometry of the deformation zone.

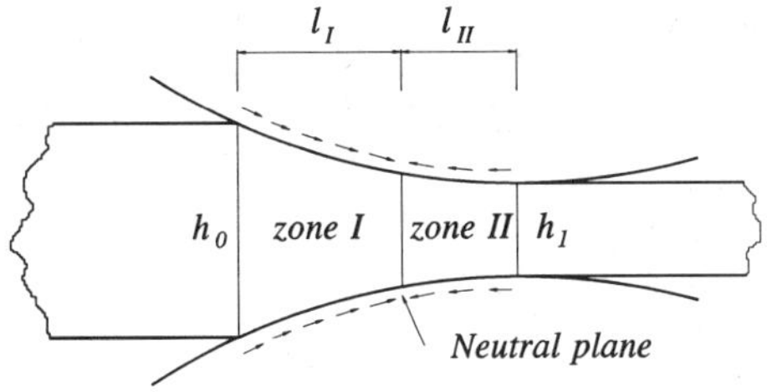

Fig. 4: Division of deformation zone.

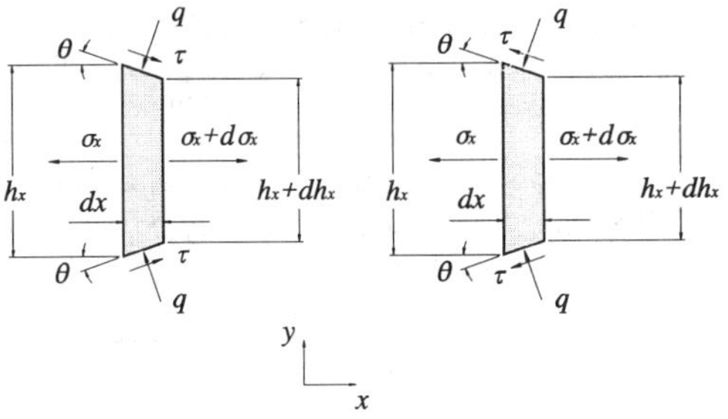

(a) zone I *(b) zone II*

Fig. 5: Slabs of stress state in zone I and zone II.

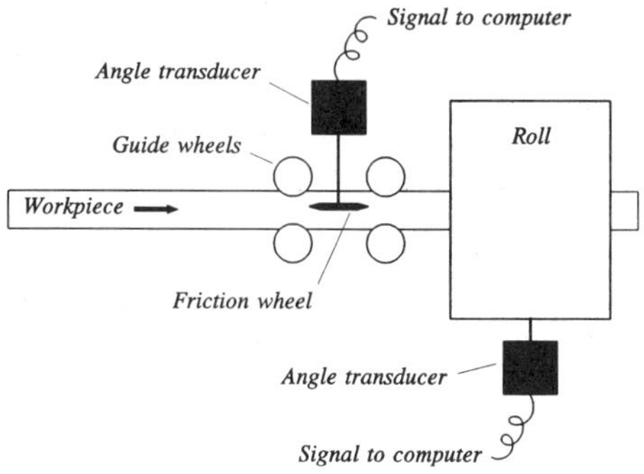

Fig. 6: Backward and forward slip measurement system.

too/workpiece interface, the deformation zone can be divided into two sub-zones: zone I, the backward slip zone, and zone II, the forward slip zone. The division of these two zones is the neutral point, see Fig. 4.

3.5 *Development of mathematical model*

As seen in Fig. 5, force equilibrium in the horizontal and the vertical directions give the following two equations:

$$(\sigma_x+d\sigma_x)(h_x+dh_x)-\sigma_x h_x-2q\tan\theta dx\pm 2\tau dx = 0 \quad (13)$$

$$\sigma_y dx+q dx\pm\tau\tan\theta dx = 0 \quad (14)$$

where the upper sign refers to zone I and lower sign to zone II. Von Mises yield criterion gives:

$$\sigma_x-\sigma_y = s_0 \quad (15)$$

and the geometrical conditions require:

$$dh_x = -2\tan\theta dx \quad (16)$$

Combining Eqs. (14)-(17), results in:

$$-\frac{dq}{dx}h_x \mp \frac{d\tau}{dx}h_x\tan\theta \pm 2\tau(1+\tan^2\theta) + \frac{ds_0}{dx}h_x - 2s_0\tan\theta = 0 \quad (17)$$

From now on the calculation has to be carried out in two different cases: low normal pressure and high normal pressure.

Low normal pressure, $q/\sigma_0 < q^*/\sigma_0$

At low normal pressure the friction stress is calculated according to Eq. (1), and the following equation is derived from Eq. (17):

$$\frac{dq}{dx} = \pm 2\mu\left(\frac{1+\tan^2\theta}{1\pm\mu\tan\theta}\right)\frac{q}{h_x} + \frac{1}{1\pm\mu\tan\theta}\left(\frac{ds_0}{dx}-2\tan\theta\frac{s_0}{h_x}\right) \quad (18)$$

High normal pressure, $q/\sigma_0 \geq q^*/\sigma_0$

At high normal pressure the friction stress is calculated according Eq. (2), and the following equation is derived from Eq. (17):

$$\frac{dq}{dx} = \left(1\mp\frac{1}{2}f\tan\theta\right)\frac{ds_0}{dx}+\left[\pm f(1+\tan^2\theta)-2\tan\theta\right]\frac{s_0}{h_x} \quad (19)$$

Numerical methods have to be used to solve Eqs. (18) and (19). The vertical pressure distribution σ_y is

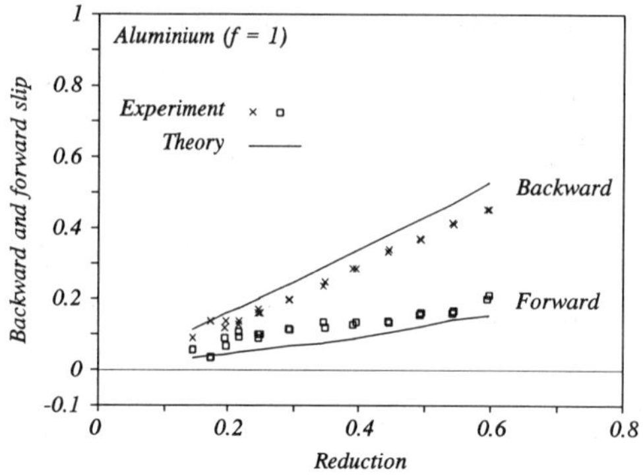

Fig. 7: Comparison of calculated and experimental backward and forward slip. Rolling of aluminum with dry friction (f = 1).

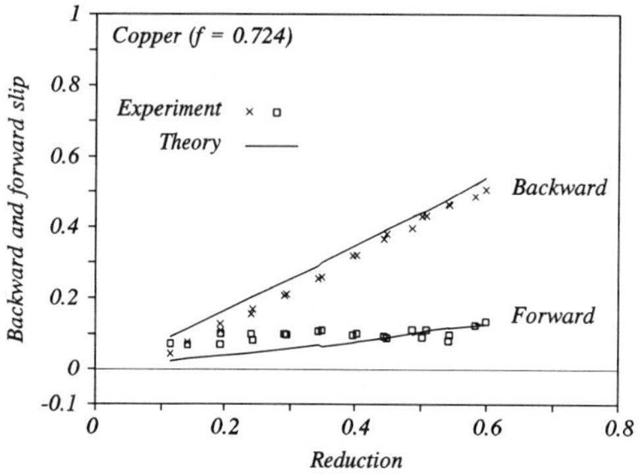

Fig. 8: Comparison of calculated and experimental backward and forward slip. Rolling of copper with dry friction (f = 0.724).

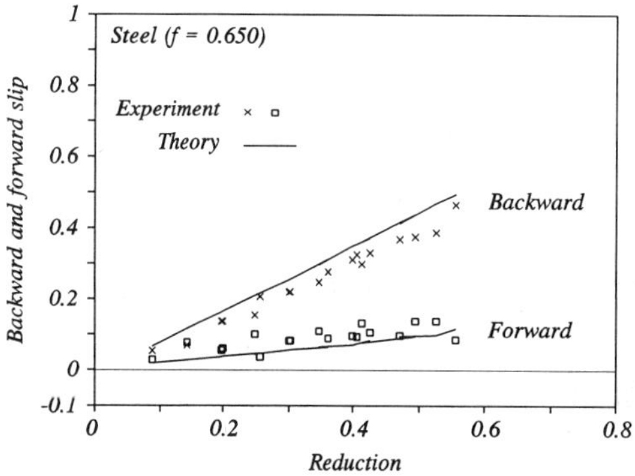

Fig. 9: Comparison of calculated and experimental backward and forward slip. Rolling of steel with dry friction (f = 0.650).

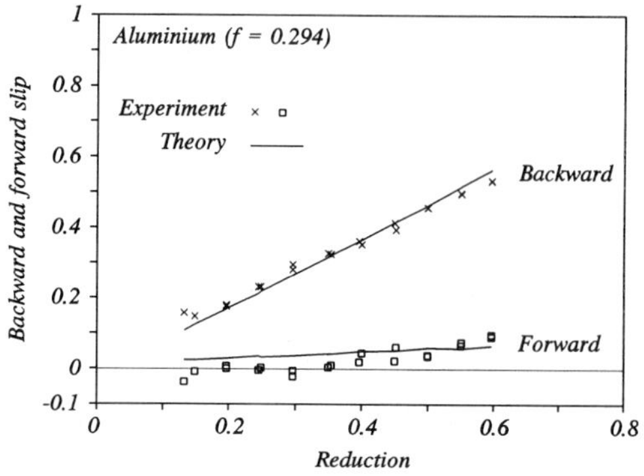

Fig. 10: Comparison of calculated and experimental backward and forward slip. Rolling of aluminum with grease as lubricant (f = 0.294).

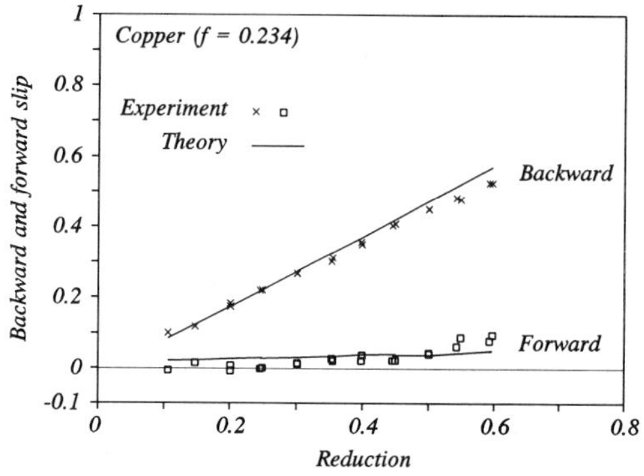

Fig. 11: Comparison of calculated and experimental backward and forward slip. Rolling of copper with grease as lubricant (f = 0.234).

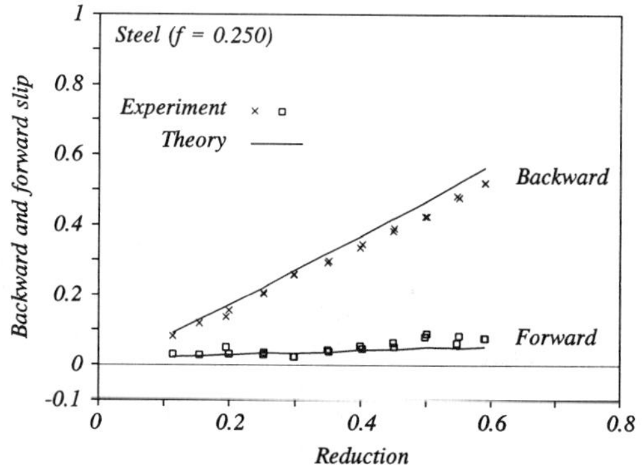

Fig. 12: Comparison of calculated and experimental backward and forward slip. Rolling of steel with grease as lubricant (f = 0.250).

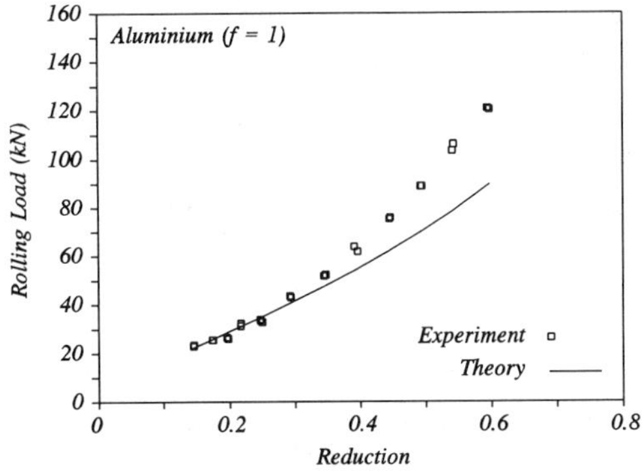

Fig. 13: Comparison of calculated and experimental load. Rolling of aluminum with dry friction (f = 1).

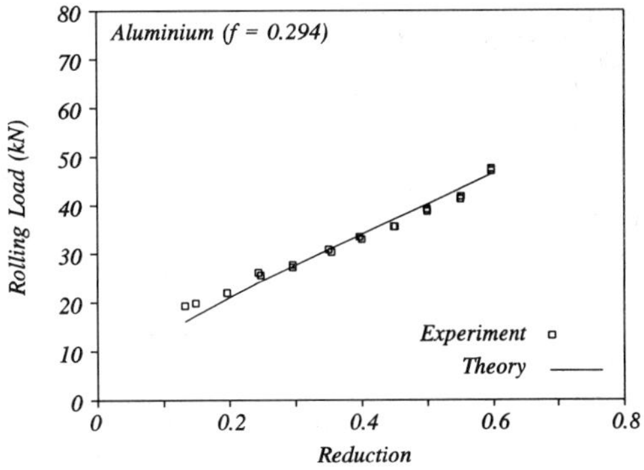

Fig. 16: Comparison of calculated and experimental load. Rolling of aluminum with grease as lubricant (f = 0.294).

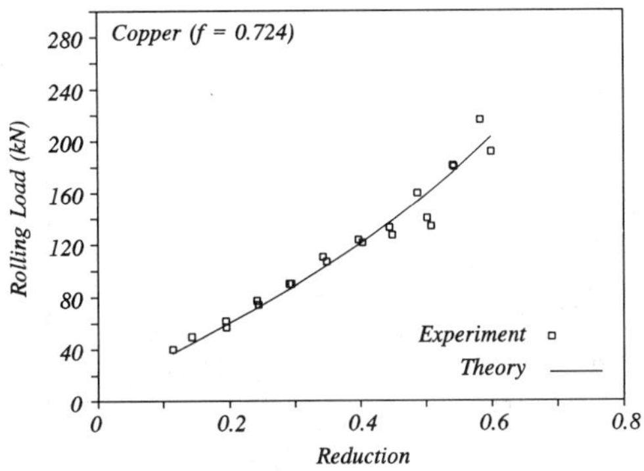

Fig. 14: Comparison of calculated and experimental load. Rolling of copper with dry friction (f = 0.724).

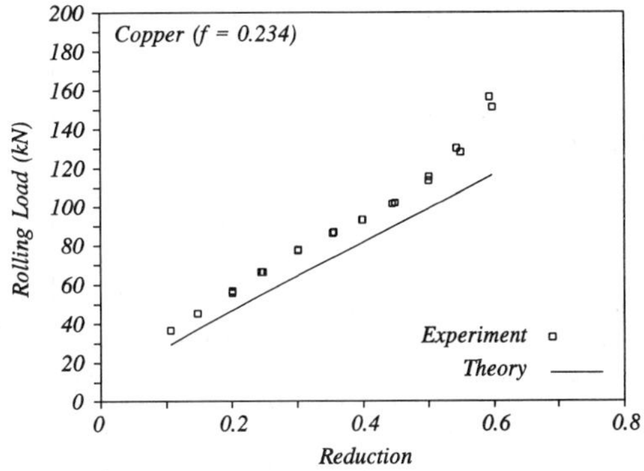

Fig. 17: Comparison of calculated and experimental load. Rolling of copper with grease as lubricant (f = 0.234).

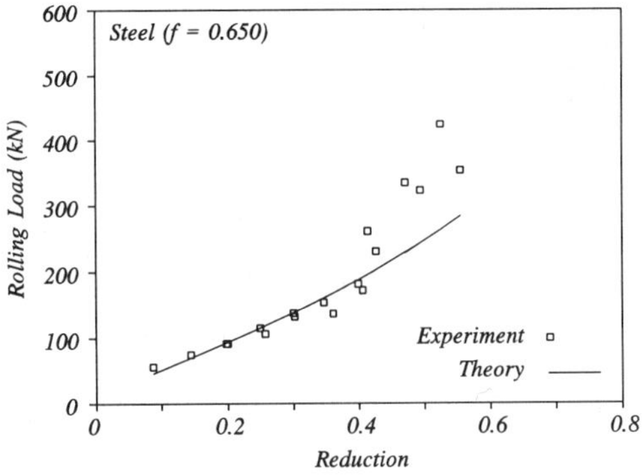

Fig. 15: Comparison of calculated and experimental load. Rolling of steel with dry friction (f = 0.650).

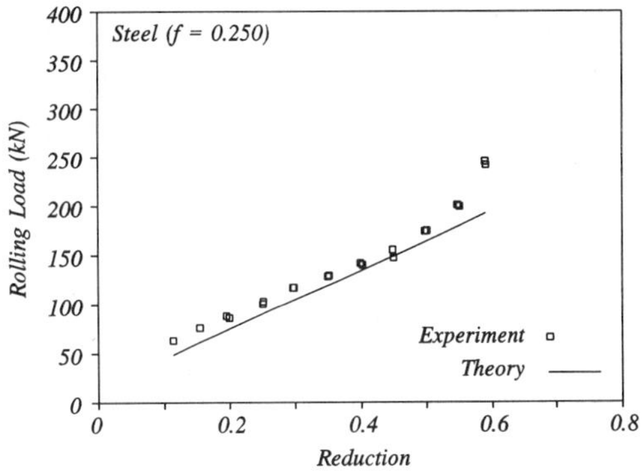

Fig. 18: Comparison of calculated and experimental load. Rolling of steel with grease as lubricant (f = 0.250).

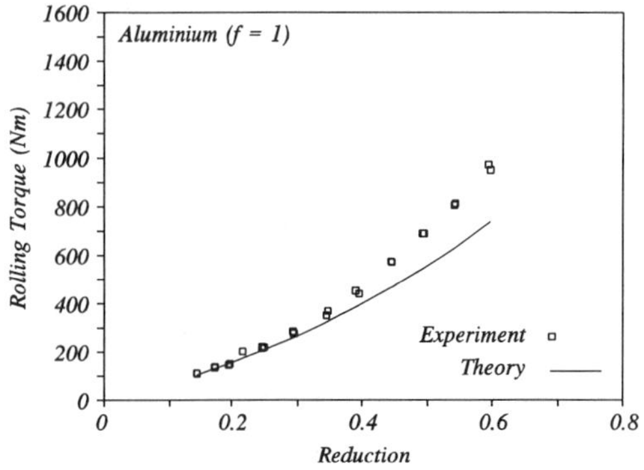

Fig. 19: Comparison of calculated and experimental total torque. Rolling of aluminum with dry friction (f = 1).

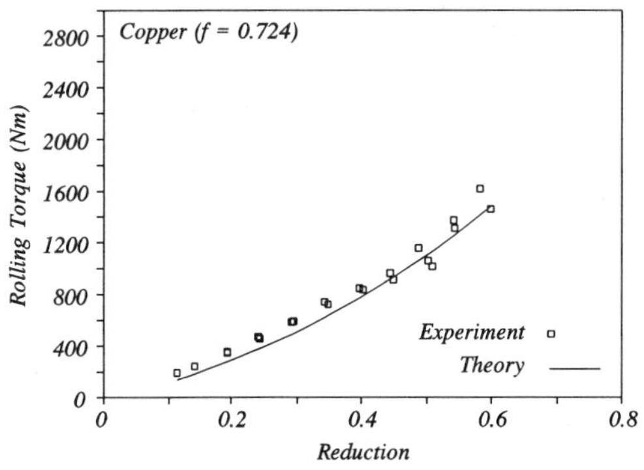

Fig. 20: Comparison of calculated and experimental total torque. Rolling of copper with dry friction (f = 0.724).

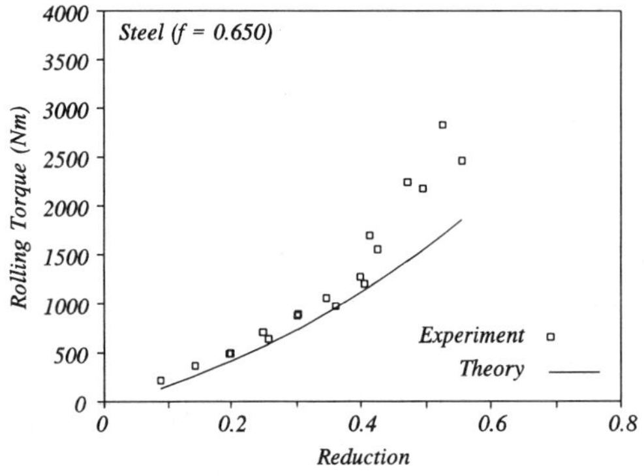

Fig. 21: Comparison of calculated and experimental total torque. Rolling of steel with dry friction (f = 0.650).

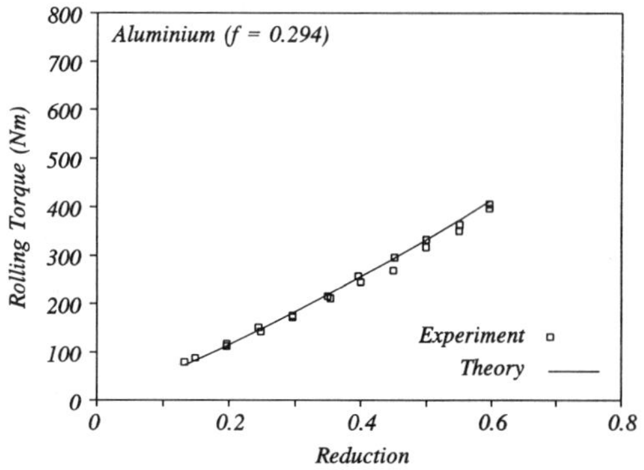

Fig. 22: Comparison of calculated and experimental total torque. Rolling of aluminum with grease as lubricant (f = 0.294).

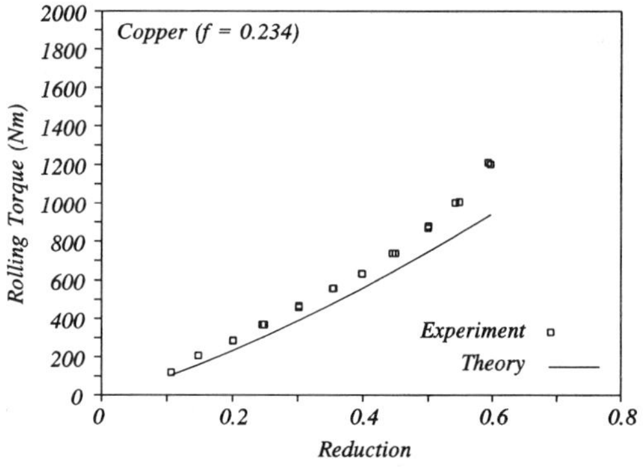

Fig. 23: Comparison of calculated and experimental total torque. Rolling of copper with grease as lubricant (f = 0.234).

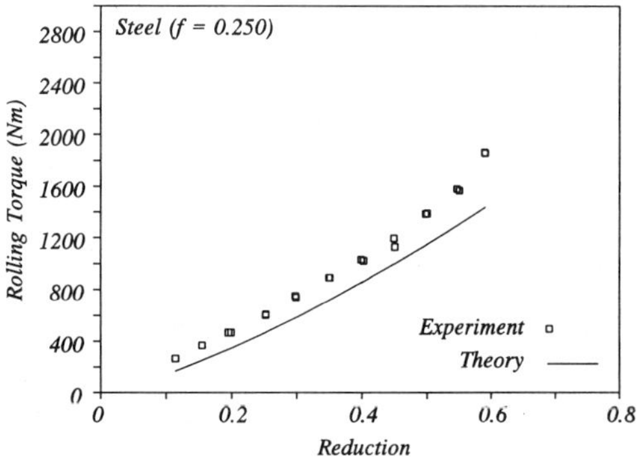

Fig. 24: Comparison of calculated and experimental total torque. Rolling of steel with grease as lubricant (f = 0.250).

then calculated by Eq. (14).

Calculation of the normal pressure distribution is carried out in two opposite directions, through zone I from the entrance side and through zone II from the exit side of the roll gap. The neutral point is determined as the point where the normal pressure is identical in the two calculations. Thus the lengths of zone I (l_I) and zone II (l_{II}), see Fig. 4, can be determined.

3.6 *Boundary conditions*

If the back tension is given by σ_b, then Eqs. (1), (14) and (15) give the boundary condition at the entrance of the deformation zone:

$$q_{(x=0)} = \frac{s_0 - \sigma_b}{1 + \mu \tan\theta} \tag{20}$$

If the front tension is given by σ_f, the boundary condition at the exit of the deformation zone will be:

$$q_{(x=l)} = \frac{s_0 - \sigma_f}{1 - \mu \tan\theta} \tag{21}$$

3.7 *Calculation of rolling parameters*

3.7.1 Calculation of rolling load and total torque

After calculation of the distribution of pressure, the rolling load and total torque can be obtained by integration along the deformation zone (Johnson *et al*, 1973):

$$F = b\int_0^l \sigma_y dx \tag{22}$$

$$T = \frac{1}{2} bDD' \int_0^{\phi} \sigma_y \theta \, d\theta \tag{23}$$

where ϕ is the roll gap angle for the whole deformation zone expressed by:

$$\phi = \arcsin\left(\frac{2l}{D'}\right) \tag{24}$$

3.7.2 Calculation of backward and forward slip

When the position of the neutral plane is obtained, the dimension of the workpiece at the neutral point will be known, and the backward and forward slip can be calculated knowing the roll speed, V_R (Wusatowski, 1969):

$$S_b = \frac{V_R \cos\phi - V_0}{V_R \cos\phi} \tag{25}$$

$$S_f = \frac{V_1 - V_R}{V_R} \tag{26}$$

where the entrance and exit speeds of the workpiece are calculated according to the volume constancy:

$$V_0 = \frac{b_n h_n}{b_0 h_0} V_R \tag{27}$$

$$V_1 = \frac{b_n h_n}{b_1 h_1} V_R \tag{28}$$

4 EXPERIMENTAL VERIFICATION

4.1 *Experiments*

Experiments were carried out for validation of the theoretical model. Three materials with the following stress strain curves were used:

aluminum, annealed: $\sigma_{oAl} = 142\varepsilon^{0.091}$
copper, annealed: $\sigma_{oCu} = 399\varepsilon^{0.215}$
steel, annealed: $\sigma_{oFe} = 641\varepsilon^{0.233}$

applying two different friction conditions: 1) dry friction, workpiece and rolls cleaned with acetone, and 2) lubrication with grease. The friction factor for each metal and lubrication condition was estimated by the ring test method (Hansen *et al*, 1988).

The rolling load and torque were measured by strain gauge transducers.

In order to validate the general friction model, backward and forward slip were determined by online measuring the peripheral speed of the rolls and the entrance speed of the workpiece. Two photo electric angle transducers were used to measure the speed of rotation of the roll and the specially designed friction wheel, contacting the workpiece prior to the roll gap, as shown in Fig. 6.

Knowing the peripheral speed of the rolls and the entrance speed of the workpiece, the backward and forward slip can be calculated. A PC was connected with the angle transducers for online collecting data and then calculating the backward and forward slip afterwards according to Eqs. (25) and (26).

4.2 *Comparison between calculation and experiments*

Figs. 7, 8 and 9 show the comparison between calculated and experimental backward and forward slip with dry friction for aluminum, copper and steel respectively. Good accordance is found for copper and steel. The larger discrepancy in case of aluminum is probably due to the influence of heavy sticking between the roll and workpiece surfaces. The agreement is improved with lubrication, as shown in Figs. 10, 11 and 12. The good agreement between calculated and experimental backward and forward slip supports the validity of the general friction model.

Figs. 13, 14 and 15 show the comparison of calculated and experimental rolling load with dry friction for aluminum, copper and steel. Good agreement is found for copper and steel. Again the influence of sticking in case of aluminum seems to cause some discrepancy. The agreement is improved with lubrication except for copper, as seen in Figs. 16, 17 and 18. Rather similar results are noticed for the torque, as shown in Figs. 19 to 24.

5 DISCUSSIONS AND CONCLUSIONS

A numerical analysis of plate rolling has been developed introducing a general friction model applicable at low as well as high normal pressures. The analysis has been experimentally verified by comparison between theoretically and experimentally determined load and torque, rolling aluminium, copper and steel under dry friction as well as lubricated conditions.

A new method of verifying the friction model has been applied by comparing calculated and measured backward and forward slip. Good agreement was found.

REFERENCES

Alexander, J. M. 1972. On the theory of rolling. *Proc. Roy. Soc. London.* A326:535.

Bay, N. 1987. Friction stress and normal stress in bulk metal forming processes. *Journal of Mechanical Working Technology.* 14:203-223.

Christensen, P., H. Everfelt and N. Bay 1986. Pressure distribution in plate rolling. *Annals of the CIRP.* 35:141-146.

Frederiksen, M. S. 1987. *Friktionsundersøgelser ved Pladevalsning.* M.Sc-thesis. Technical University of Denmark.

Hansen, P. H., N. Bay and P. Christensen 1988. Analysis of the ring compression test using a general friction model and the upper bound elemental technique. *Proceed. 16th NAMRC,* Urbana, Illinois, 41-47.

Hitchcock, J. 1935. Roll neck bearings. *ASME Res. Publ.*

Johnson, W. and P. B. Mellor 1973. *Engineering Plasticity.* Van Nostrand Reinhold Company. London.

Schey, J. A. 1983. *Tribology in Metalworking. Friction, Lubrication* and Wear. American Society for Metals. 260-262.

Wanheim, T., N. Bay and A. S. Petersen 1974. A theoretically determined model for friction in metal working processes. *Wear.* 28:251-258.

Wanheim, T. and N. Bay 1978. A model for friction in metal forming processes. *Annals of the CIRP.* 27:189-194.

Wusatowski, Z. 1969. *Fundamentals of Rolling.* Pergamon Press. Oxford. 146-147.

Zhang, W. 1994. *Bond Formation in Cold Welding of Metals.* Ph.D.-thesis. Technical University of Denmark.

Polymer – General topics

Simulation of Materials Processing: Theory, Methods and Applications, Shen & Dawson (eds)
© 1995 Balkema, Rotterdam. ISBN 90 5410 553 4

Finite element simulation of meltlines

R. Akkerman, G. Rekers & J. Huétink
University of Twente, Department of Mechanical Engineering, Enschede, Netherlands

ABSTRACT: A study of meltlines is presented, based on finite element calculations using a Euler-Lagrange method to describe the moving free surface. For the instationary thermal and mechanical contact behaviour between polymer and mould, use is made of a penalty formulation implemented in a contact element. For the description of contact between two polymer flows (meltlines) the model has been enhanced using reptation theory describing the healing between the two fronts. Thus, a quantitative prediction can be made concerning the development of notches as well as the strength of the meltline during the filling and cooling stages of injection moulding. Finally, the results of the simulations are compared to experiments.

1 INTRODUCTION

In numerical simulations of injection moulding of polymers often the lubrication or Hele Shaw approximation is being used. This implies that effects of fountain flow, meltlines, flow around corners and sudden changes of thickness of the cavity are neglected or described poorly. However, these phenomena have a significant influence on the distribution of molecular orientation, on the internal stresses and hence the warpage of the finished product.

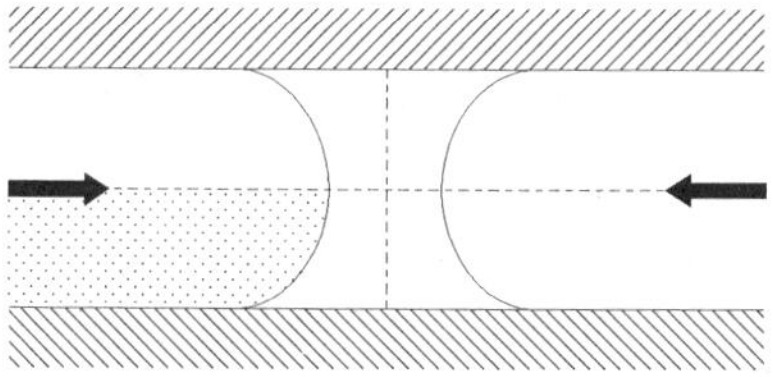

Figure 1 Two colliding flow fronts (schematically).

Meltlines are present in many injection moulded products. They may be caused by multiple gates or by splitting of the flow, e.g. by inserts in the mould. Two colliding melt fronts, as depicted in Figure 1, will weld to a certain degree. A meltline influences the optical and mechanical quality of the product. It may be visible because of differences in brilliance, colour, distribution of fillers or because of a v-notch. In general, this is undesired.

The mechanical quality of the product is influenced by imperfect bonding over the meltline and may be worsened by stress concentrations due to a v-notch. The notch may be generated by a number of different mechanisms. Because of trapped air or increasing viscosity near the mould, the injection pressure may be insufficient to close the gap between the melt fronts. Or, the gap may have been closed, but torn open due to thermal stresses. Numerical simulation may help to reveal the origin of a v-notch.

However, experiments show notches of order 1-10 μm, in samples of order 1 mm thickness, a ratio of 1:100-1000. An accurate numerical description of an evolving notch will thus require a very fine grid and is not feasible with our current solution strategies. Zooming in on the closing notch would be a feasible alternative. The "interior boundary conditions" must be approximated in that case.

Disregarding the notch for the time being, we concentrate on the interior of the meltline. Looking on a microscopic scale (due to the fountain flow phenomenon) the macromolecules will be oriented parallel to the meltline before the melt fronts collide. By a diffusion mechanism the molecules will cross the interface after collision has taken place. Increasing the temperature or decreasing the pressure will accelerate this process of healing, giving a higher mobility of the molecular chains.

Experimental and theoretical work can be found in literature. Implementing a model for healing in a contact element of a finite element package enables the simulation of the forming of a meltline. As a result the strength of the meltline may be predicted.

2 HEALING THEORY

Using reptation theory [de Gennes, 1971] the average interpenetration distance χ of the molecular chains over the interface can be modelled as

$$\frac{\chi}{\chi_\infty} = \left[\frac{t}{\tau_{rep}}\right]^{1/4}, \quad (t < \tau_{rep}), \tag{1}$$

where the reptation time τ_{rep} is the time needed to reach a homogeneous situation [Wool, 1981], [Bastien, 1991]. The tensile strength of the meltline, σ, is assumed to be proportional to the interpenetration depth, so

$$\sigma = \sigma_\infty \left[\frac{t}{\tau_{rep}}\right]^{1/4}, \quad (t < \tau_{rep}). \tag{2}$$

Here σ_∞ is the strength of the homogeneous material.

The parameters are influenced by the molecular weight M of the polymer. Theoretically the reptation time τ_{rep} can be shown to be proportional to M^3, and the tensile strength σ_∞ of the homogeneous material varies linearly with $M^{1/2}$.

Apart from this, the reptation time depends on the temperature T. We use the shift factor given by the WLF equation,

$$\tau_{rep}(T) = a(T)\tau_{rep,0}, \quad \text{with } a(T) = \exp\left[\frac{-c_1(T-T_{ref})}{c_2+T-T_{ref}}\right]. \tag{3}$$

Pressure dependency has been neglected at this stage. Assuming thermorheologic simplicity leads to a temperature-dependent evolution equation for the local strength of the interface

$$\dot{\sigma}(T) = \frac{1/4\,\sigma_\infty^4}{a(T)\,\tau_{rep,0}\,\sigma^3}, \quad (\sigma < \sigma_\infty), \tag{4}$$

which can be integrated numerically, with temperature varying in time.

3 FINITE ELEMENTS

The polymer was modelled with a viscous/elastic material description. The material model for deviatoric deformations is simple: a (shear thinning) viscous fluid when the temperature is above T_g,

$$\boldsymbol{\sigma}^d = 2\eta \boldsymbol{D}^d$$

$$\text{with } \eta = \eta_0 \exp\left[\frac{-c_1(T-T_{ref})}{c_2+T-T_{ref}}\right]\left(1+(s\dot{\gamma})^2\right)^{\frac{n-1}{2}}, \tag{5}$$

with shear rate $\dot{\gamma}$, where again a WLF-shift describes temperature dependence. Below T_g a linear elastic model is used, with a constant shear modulus G. In this stage the effects of flow induced molecular orientation on the healing behaviour have been neglected.

The isotropic part of the constitutive model is described using a Tait equation, a fit published by Zoller [1982], conveniently reproduced as

$$\begin{aligned} v(p,T) &= v(0,T)\left[1 - c\ln\left[1+\frac{p}{B(T)}\right]\right], \\ v(0,T) &= v_{g,0} + A(T-T_{g,0}), \\ B(T) &= B_0\exp(-B_1 T), \end{aligned} \tag{6}$$

where v is the specific volume and p the hydrostatic pressure and with different constants (A, B_i) above and below T_g respectively.

These material laws have been implemented in a thermomechanically coupled Euler-Lagrange finite element code, developed at the University of Twente (see e.g. [Huétink et al, 1990], [Akkerman, 1993]).

Contact was modelled using a penalty formulation: during contact the force normal to the surface is described elastically. In other words a penalty is set on overlapping surfaces; a repulsive force is generated when the *gap* between the surfaces is negative. A parallel viscous damper smoothens the open/closure behaviour. In the tangential plane the Coulomb criterion for dry friction can be used, in case of contact. The latter may be interpreted as an elastoplastic model with non-associated flow [Vreede, 1992], see Figure 2.

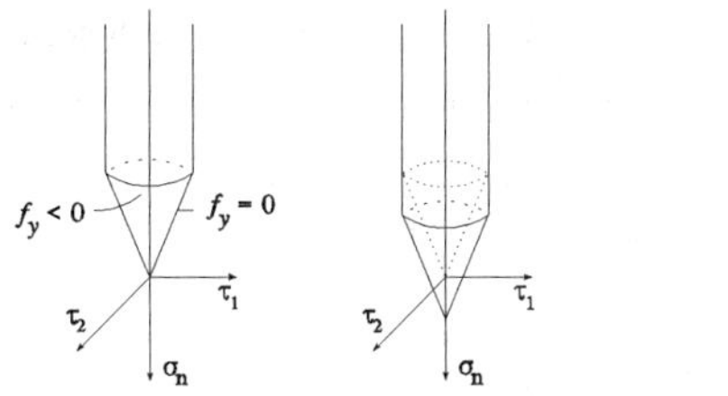

Figure 2 Slip surface in contact stress space. Right: shift of the surface due to developed strength.

The yield condition for the conical and the cylindrical part in Figure 2 can be specified as a limiting value of

$$f_y = \tau_1^2 + \tau_2^2 - \mu^2 \sigma_n^2, \qquad -\mu\sigma_n \leq \tau_{max};$$
$$f_y = \tau_1^2 + \tau_2^2 - \tau_{max}^2, \qquad -\mu\sigma_n > \tau_{max}. \tag{7}$$

When no slip occurs ($f_y<0$), the tangential stress is proportional to the relative displacement of the two bodies in contact (linear elastic model). In case of slip ($f_y=0$), the flow rule is most easily explained in a two-dimensional example:

$$\tau = \mu\sigma_n, \qquad -\mu\sigma_n \leq \tau_{max};$$
$$\tau = \tau_{max}, \qquad -\mu\sigma_n > \tau_{max}. \tag{8}$$

Here τ_{max} represents the radius of the cylindrical part in Figure 2. Healing is incorporated in this model by simply shifting the yield surface in the positive σ_n direction, thus admitting a tensile normal stress over the contact surface.

Thermal equilibrium for the contact element is solved by

$$\boldsymbol{q}\cdot\overleftarrow{\nabla} - \boldsymbol{\sigma} : \boldsymbol{D} + \boldsymbol{\sigma}_e : \boldsymbol{D}_e = 0, \tag{9}$$
$$\text{with } \boldsymbol{q} = -\alpha\,\Delta T\,\boldsymbol{e}_n,$$

incorporating a heat transfer coefficient of the interface, α, and heat generation due to friction. The subscript 'e' denotes the elastic part of the stress and the deformation rate respectively. In the normal direction this heat balance is equivalent to

$$-\alpha\Delta T - \sigma_{xy}\Delta v_{\tan} + \sigma_{xy}^e \Delta v_{\tan}^e = 0, \tag{10}$$

expressed in the differences in temperature and sliding velocity on both sides of the contact element.

Finally, the healing model is implemented by integrating the strength to

$$\sigma(t+\Delta t) = \sigma(t) + \tfrac{1}{4}\sigma_\infty^4 \cdot \int_t^{t+\Delta t} \frac{1}{\tau_{rep}(T)\,\sigma^3}\,dt \tag{11}$$

during contact, when $T>T_g$. The accumulation of strength stops when the homogeneous tensile strength σ_∞ has been reached.

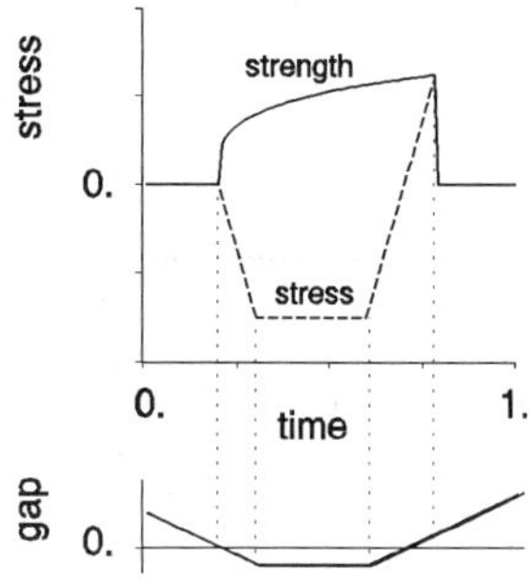

Figure 3 Contact with healing.

The behaviour of the element is illustrated in Figure 3. At a temperature above T_g a contact element is closed. When the gap is negative, the normal stress varies linearly with the gap. Strength starts building up from the moment of closure.

When the element is opened, a tensile stress can be sustained until the strength is exceeded. Strength and stress drop to zero. This discontinuous behaviour can be smoothened with the parallel viscous damper.

4 A MODEL EXPERIMENT

To check the validity of the healing model injection moulding experiments have been carried out at DSM Research. The mould in Figure 4 was filled with different materials under a number of conditions. The double gating caused a meltline in the centre of the specimen. Notches and mechanical properties were investigated by Eijpe [1990] and Carleer [1992]. One of the experimental conditions was used for a series of simulations. The pressure in the mould was recorded, and stylized as in Figure 5.

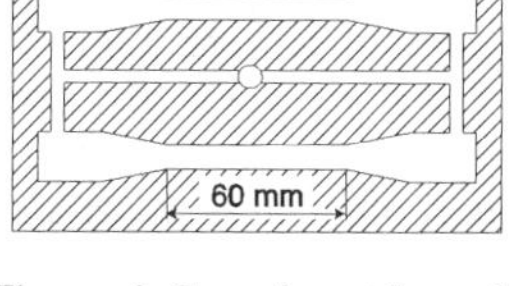

Figure 4 Experimental mould.

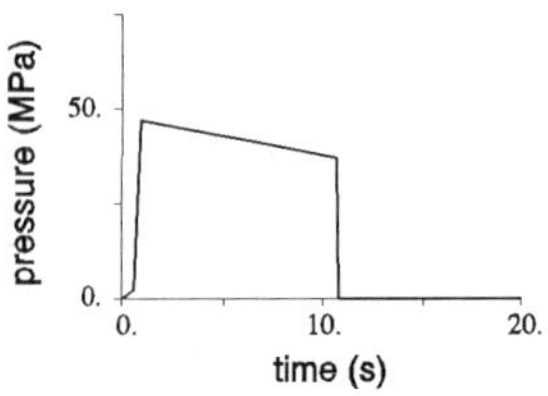

Figure 5 Course of pressure in the mould.

Table 1 Process data for healing experiment.

process parameters	
injection time	0.5 s
flow rate	23.2 10^{-6} m^3/s
injection temperature	503 K
mould temperature	333 K
holding time	10 s
cooling time	25 s
dimensions	
cross section specimen	3x10 mm

Further parameters are given in Table 1.
The material used was polystyrene (Styron 634).

5 SIMULATION RESULTS

The evolution of the meltline has been simulated by considering a cross section over the thickness of the product. A plane strain simulation was performed, eliminating effects across the width of the channel. With the given (symmetric) process conditions it is sufficient to model a quarter of the geometry (see Figure 1, dotted region).

The material data of Styron 634 are given in Table 2. For the Tait equation the data of Styron 678E were used, because the data of Styron 634 were not available. However, for a flow simulation the pvT-data are not strictly necessary so this is of minor importance. The pvT-behaviour becomes essential during the packing stage when extra material is added under high pressure to compensate for thermal shrinkage. This phase mainly determines the residual thermally induced stresses in an injection moulded product.

Previous simulations of the temperature dependent fountain flow phenomenon (Akkerman et al, 1993) showed a very limited temperature decrease in the polymer near the flow front. The thermal problem is dominated strongly by heat convection, even for this moderate flow rate.

Table 2 Material properties for healing simulation.

Polystyrene (Styron 634), deviatoric		
η_0	3.1 kPa s	
s	75 1/s	
n	0.25	
c_1	14.57	
c_2	94 K	
T_{ref}	423 K	
G	1300 MPa	
PS (Styron 678E), isotropic (Flaman, 1990)		
c	.0984	
$T_{g,0}$	373 K	
$v_{g,0}$	0.9758 10^{-3} m^3/kg	
	glass	rubber
A	.23 10^{-6} m^3/kgK	.58 10^{-6} m^3/kgK
B_0	259 MPa	167 MPa
B_1	3.0 10^{-3} 1/K	3.6 10^{-3} 1/K
PS, reptation (Wales, 1976)		
σ_∞	50 MPa	
$\tau_{rep,0}$	0.8 s	
c_1	20.4	
c_2	374 K	
T_{ref}	407 K	
thermal (materials and interface)		
	polymer	steel
λ	0.139 W/mK	36,35 W/mK
c_p	1.73 kJ/kgK	0,46 J/kgK
ρ	(pvT-data)	7830 kg/m^3
α	3 kW/m^2K	

Thus, for this study we simply started with a simulation of the closing of the gap between two melt fronts. Initially, the mould was at a constant temperature of 333 K and the polymer at 503 K. In between a layer of contact elements was situated, simulating the heat transfer resistance between the two materials. During the time-stepping procedure heat was exchanged between both materials.

The frame of reference is fixed to the mould wall. The left hand side of the computational domain is kept at the same location, the lower side is stretched until the gap has closed. Only the mould behind the contact point (between mould and polymer) is taken in the computational domain. Thus, a symmetric situation is created

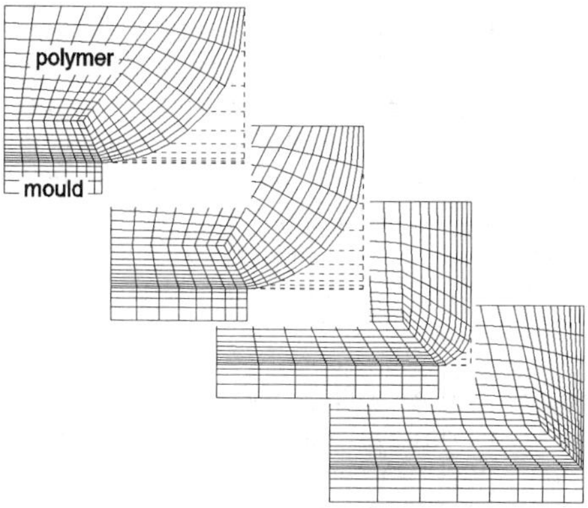

Figure 6 Closure of the gap in a number of timesteps. The dotted lines represent contact elements.

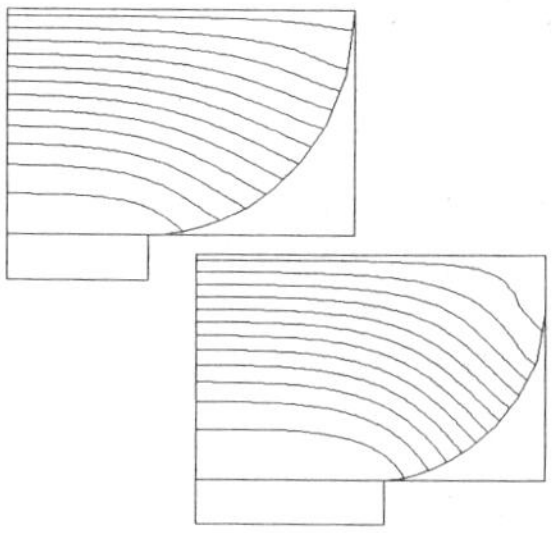

Figure 7 Streamlines during closure of the gap.

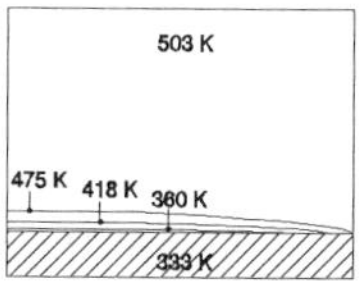

Figure 8 Temperature distribution at the moment of closure.

when the gap has closed, preserving the same element connectivity throughout the simulation.

The flow of the material can be illustrated by a streamline plot.

Near the meltline material is pushed towards the wall. As a result the temperature at the meltline after closing the gap is still almost equal to the entrance temperature, as depicted in Figure 8.

To simulate the development of the meltline strength, the presented mesh is still far too course. As the temperature near the meltline is almost independent of location, a semi 1D-model can be made to analyse the meltline strength with enough refinement to describe the details.

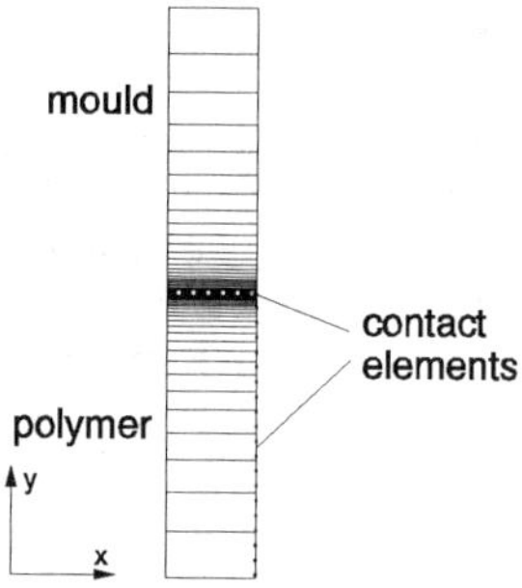

Figure 9 Element distribution for 1D simulation.

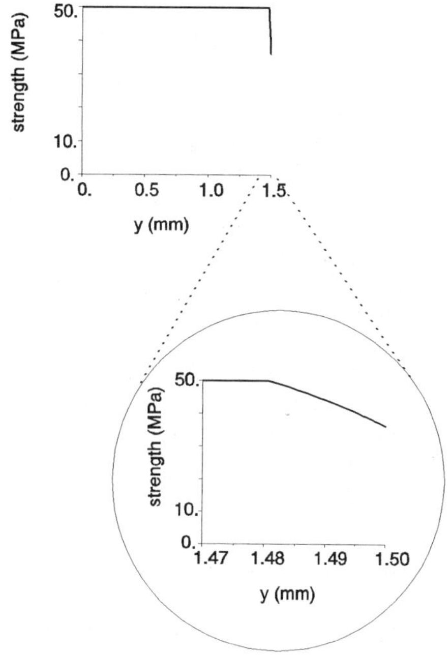

Figure 10 Final strength distribution (1D-simulation, 80 elements, gradually refined with $h_1 = 1000 \cdot h_{80}$).

We solved the thermal problem, again starting from a cold mould (333 K) and hot polymer (503 K). In between again a contact element was placed, with a high heat transfer coefficient of 3 kW/m^2K (as in Douven, 1991). The healing elements were connected to the right hand side, following the temperature history of the polymer.

In the core the maximum strength was reached in 0.013 s. Due to cooling a very thin layer of partial healing was found on the surface. This minimum strength converged with mesh refinement to a value of 36.3 MPa.

6 DISCUSSION

Quasistatic bending and tensile tests on the injection moulded specimens always resulted in unstable crack growth on the meltline. The process conditions under consideration lead to a tensile strength of 34 MPa. We might interpret the simulated tensile strength as a continuum stress-criterion for crack initiation. The simulated and experimental values are quite in the same range. Of course, the residual stresses will still have to be taken into account in this comparison.

Current research is directed towards accurate prediction of the internal stresses caused by injection moulding. Sofar, results indicate a tensile stress at the surface of a few MPa. Experimental and simulation results differ by a factor 2 at this stage, however.

The results have a great sensitivity to the value of the heat transfer coefficient between polymer and mould. For "infinite" values the strength reached at the surface drops to 6.4 MPa. For an accurate prediction reliable data have to be available. Note that the heat transfer is influenced by the surface conditions, such as roughness and contamination.

In a subsequent stage the orientation resulting from the flow simulation can be taken into account in the description of the healing process. The thermomechanically coupled problem can be solved, but research is continued to refine the results.

ACKNOWLEDGEMENTS

The support of DSM Research is gratefully acknowledged. Especially thanks to Markus Bulters and Arie Schepens for initiating the meltline research.

REFERENCES

Akkerman, R., G. Rekers, B.D. Carleer et al 1993. Thermomechanical contact elements for healing in polymers. In M.H. Aliabadi & C.A. Brebbia (eds) *Contact Mechanics, computational techniques*, first international conference on contact mechanics 93: 103-110. Southampton: Computational Mechanics Publications.

Akkerman, R. 1993. *Euler-Lagrange simulations of nonisothermal viscoelastic flows* (thesis). Enschede: University of Twente.

Bastien, L.J. & J.W. Gillespie jr 1991. A non-isothermal healing model for strength and toughness of fusion bonded joints of amorphous thermoplastics. *Polymer Engineering and Science*. 31:1720-1730.

Carleer, B.D. 1992. *Healing in finite elements, simulation of meltlines in injection moulded products* (graduate report, in Dutch). Enschede: University of Twente.

Douven, L.F.A. 1991. *Towards the computation of properties of injection moulded products: flow and thermally induced stresses in amorphous thermoplastics* (thesis Technical University Eindhoven). Enschede: Febodruk.

Eijpe, M.P.I.M. 1990. *Visibility of meltlines* (graduate report, in Dutch). Enschede: University of Twente.

Ferry, J.D. 1970. *Viscoelastic properties of polymers* (2nd edition). New York: Wiley.

Flaman, A.A.M. 1990. *Build-up and relaxation of molecular orientation in injection moulding* (thesis). Eindhoven: Technical University Eindhoven.

Gennes, P.G. de 1971. Reptation of a polymer chain in the presence of fixed obstacles. *Journal of Chemical Physics*. 55:572-579.

Huétink, J., P.T. Vreede, & J. van der Lugt 1990. Progress in Mixed Euler-Lagrange FE simulation of forming processes. *International Journal of Numerical Methods in Engineering* 30:1441-1457.

Vreede, P.T., M.F.M. Louwes, J. Huétink & N.A.J. Langerak 1992. Contact behaviour modelled with interface elements and tool description. In J.-L. Chenot, R.D. Wood & O.C. Zienkiewicz (eds), Numiform 92 *Proceedings of the 4th international conference on numerical methods in industrial forming processes*:565-570. Rotterdam: Balkema.

Wales, J.L.S. 1976. *The application of flow birefringence to rheological studies of polymer melts* (thesis). Delft: Delft University Press.

Wool, R.P. & K.M. O'Connor, 1981. A theory of crack healing in polymers. *Journal of Applied Physics*. 52:5953-5963.

Zoller, P., 1982. A study of the pressure-volume-temperature relationships of four related amorphous polymers: polycarbonate, polyarylate, phenoxy and polysulfone. *Journal of Polymer Science*. 20:1453-1464.

Simulation of Materials Processing: Theory, Methods and Applications, Shen & Dawson (eds)
© 1995 Balkema, Rotterdam. ISBN 90 5410 553 4

The manufacture of fiber preforms for liquid composite molding (LCM)

Richard B. Dessenberger & Charles L. Tucker III
Department of Mechanical and Industrial Engineering, University of Illinois, Urbana, Ill., USA

ABSTRACT: In the manufacture of fiber-reinforced composites parts by liquid composite molding processes, the fabrication of the fiber preform is critical to both part quality and process efficiency. In this study, a mathematical model for analyzing fiber preforming is discussed. Such models may be used to optimize the process in order to reduce defects and obtain the desired mechanical properties in the preform.

1 INTRODUCTION

Liquid composite molding (LCM) is the name given to a class of processes for producing fiber-reinforced polymeric parts in final shape. A preform of dry reinforcing fibers is placed inside a closed mold, and then a liquid resin is injected into the mold to fill the spaces between the fibers. The resin then cures, forming a stiff, solid composite that can be removed from the mold. Commercial LCM processes include Resin Transfer Molding (RTM) and Structural Reaction Injection Molding (S-RIM).

Mathematical models for mold filling and curing have been developed to help engineers design and control LCM processes (Aoyagi et al., 1992, Lin et al., 1993; and Tucker, Dessenberger, and Bakharev, 1994). Advani et. al (1994) give an overview of the isothermal simulations and Dessenberger and Tucker (1994) review non-isothermal modeling.

The manufacture of the fiber preform is termed preforming. Preforming techniques can be classified into two classes: sheet forming, and short fiber deposition (Lindsay, 1993). Only sheet forming is considered in this investigation. Here a planar sheet of fibers is deformed by stretching and pressing to conform to the surface of the mold. Random fiber mats, woven fiber mats, and combinations of the two are used in preforms manufactured by sheet forming operations.

Preforming effects production efficiency and part quality but has received far less attention than mold filling. The deformation imposed during preforming introduces porosity and permeability variations which affect mold filling, determine mechanical properties in the final part through re-orientation of the fibers, and often result in defects (tears and wrinkles) in the preform. Mathematical models for the manufacture of fiber preforms will allow engineers to optimize the process to develop defect free fiber preforms, obtain the desired porosity and mechanical properties in the final part, and investigate the relationship between deformation and permeability.

Most of the investigations on fiber preforming focus on the sheet forming of biaxially woven fabrics. Van West et al. (1990) and Van der Weeen (1991) studied the draping of biaxially woven fabrics on arbitrarily curved surfaces. Rudd et al. (1993) used draping algorithms to simulate the deformation occurring during the preforming of biaxially woven mats. Results of the model include fiber orientation, preform thickness, and in-plane permeability. These investigations suggest that the preforming of biaxially woven fabrics is fairly well understood.

Unfortunately, the same can not be said about the preforming of non-woven fiber mats. Fong et al. (1994) conducted preforming experiments on thermoformable random fiber mats to identify relationships between deformation modes and defects (tears and wrinkles). The results indicate that such relationships are very complex and the development of a good preforming operation requires an in-depth understanding of the forming process.

This study focuses on the manufacture of non-woven fiber preforms by the sheet forming process. The ultimate goal of this research is to develop the theory required to simulate the production of a three-dimensional fiber preform from a planar sheet of fibers. In this paper, we describe the theory behind such a simulation.

2 FORMING THEORY

There are two approaches that one may take in modeling sheet forming processes: mechanical deformation analysis and ideal forming theory. A mechanical deformation analysis provides a complete history of the deformation and stresses in the sheet during the forming process, (Tucker 1995). However, the calculations are intense and require a great deal of material property data and constitutive modeling.

In this study, we take the second approach to modeling sheet forming. This approach may be called a "kinematic analysis" or a type of ideal forming theory. Ideal forming theory considers only the initial and final states of the sheet and does not provide any information regarding the deformation path or the stresses required to perform the operation. Hence, our calculation is a type of inverse method that determines the best possible way to create the desired three-dimensional shape, rather than a direct simulation of the mechanics of forming.

2.1 Mathematical Modeling

Assume that the thickness of the sheet of fibers is much smaller than its planar dimensions so that we may regard it as a two dimensional sheet. The two-dimensional sheet of fibers is described by a set of material coordinates X^i which correspond to the laboratory coordinates x_i at some reference time (See Figure 1). Initially, the sheet is chosen to lie in the $X^3=0$ plane, such that each point in the sheet is described by a set of coordinates (X^1, X^2). The final shape of the preform is determined by the tool surface, which is described by a set of surface coordinates (u^1, u^2) which are given functions of the spatial coordinates

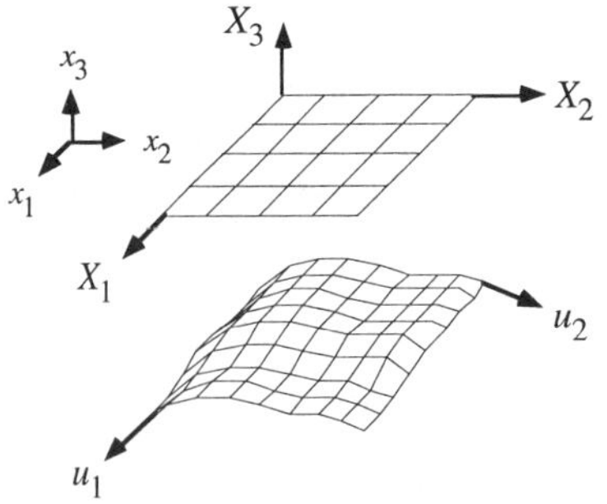

Figure 1. Initial and final shape of the fiber sheet.

$$u^\alpha = u^\alpha (x_i) \tag{2.1}$$

Note, that Latin indices i,j,k take on values of 1,2,and 3 while Greek indices α, β, γ take on the values of 1 and 2 only.

In the ideal forming theory, we seek to conform (map) each material point in the initial flat sheet to a point on the surface of the tool. Thus, one must calculate the mapping

$$X^\alpha = X^\alpha (u^\beta) \tag{2.2}$$

Once this mapping is established, the deformation of the sheet can be described by the Cauchy strain tensor C_{ij}. The Cauchy strain tensor is referenced to the three-dimensional laboratory coordinates, but we are only interested in the strain in the plane of the sheet. The in-plane strain is described by the surface Cauchy strain tensor

$$\tilde{C}_{\alpha\beta} = \frac{\partial X^\gamma}{\partial u^\alpha} \frac{\partial X^\gamma}{\partial u^\beta} \tag{2.3}$$

$\tilde{\underset{\sim}{C}}_{\alpha\beta}$ is referenced to the surface coordinates. Since $\tilde{C}_{\alpha\beta}$ is symmetric it possesses eigenvalues and eigenvectors. Let $\underset{\sim}{\mathbf{e}}_{(1)}$ and $\mathbf{e}_{(2)}$ be the two eigenvectors of $\tilde{C}_{\alpha\beta}$. Note that these vectors are tangent to the tool surface, and these vectors can be used to establish a local Cartesian coordinate system. In this local system the surface Cauchy strain tensor is diagonal

$$\tilde{C}_{\alpha\beta} = \begin{bmatrix} \dfrac{1}{\lambda_{(1)}^2} & 0 \\ 0 & \dfrac{1}{\lambda_{(2)}^2} \end{bmatrix} \tag{2.4}$$

where $\lambda_{(1)}$ and $\lambda_{(2)}$ are the principal stretch ratios. The principal stretch ratios are the ratio of initial material length to final material length along the direction of the corresponding eigenvector. Hence, one the mapping described in Eqn. (2.2) is established the local strains can be calculated and expressed in terms of the two principal stretch ratios.

2.2 Optimal Mapping

Assume that the mat is isotropic (or at least transversely isotropic in the plane of the fibers) so that the local state of strain is completely described by the stretch ratios defined in Eqn. (2.4). This is a good assumption for the random fiber mats used in the majority of RTM preforms. If we consider preform defects to be wrinkling and tearing of the fiber mat, one can determine the states of strain at which these are likely to occur by evaluating the formability of the mat.

The formability of a sheet is characterized with a forming limit diagram, similar to those used in sheet metal forming (See Figure 2). The forming limit diagram divides the strain space of $\lambda_{(1)}$ and $\lambda_{(2)}$ into various regions where defects are most likely or least likely to occur. Figure 2 shows that wrinkling is likely to occur if one of the strains is negative, and tearing is likely when one of the strains is too large. Experiments are needed to evaluate the points U, P, and B which mark the boundaries of the acceptable region of principal strains. In particular, point U can be found from a uniaxial elongation test, P from pure shear, and B from biaxial stretching.

Once the forming limit diagram is established, one can define a "good" forming process as one that keeps the local strain at every point in the preform away from the boundaries UP and UPB. To achieve the optimal mapping, one seeks to minimize the integral of a formability function over the entire tool surface S,

$$\Pi = \int_S W\,(\lambda_{(1)}, \lambda_{(2)})\, dS = \text{minimum} \qquad (2.5)$$

W is a formability function which is a measure of the local deformation in the mat relative to the allowable deformation. The formability function must be experimentally determined using data from the forming limit diagram. W is a scalar function of the two principal stretch ratios. It has a minimum at the point of no deformation ($\lambda_{(1)} = \lambda_{(2)} = 1$) and increases monotonically with distance from this point. Furthermore, W should increase rapidly in the vicinity of the boundaries of the acceptable forming region. Appropriate forms for this function are to be investigated as part of this project. Note that Eqn. (2.5) penalizes wrinkles and tears but does not rule them out altogether.

By defining the optimal mapping in this manner, we are guaranteed that an optimal mapping will

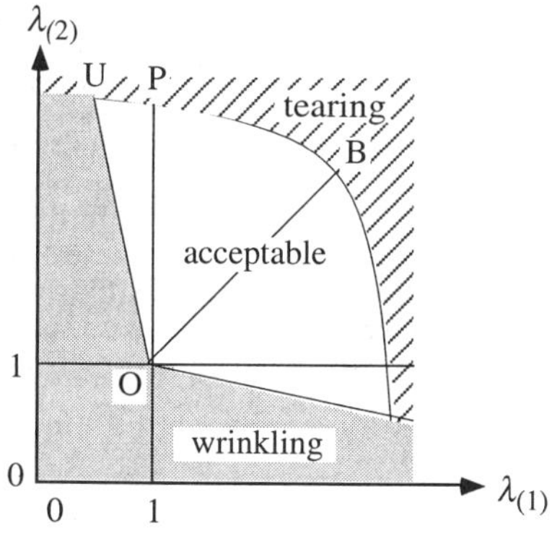

Figure 2. Forming limit diagram showing regions of acceptable deformation.

occur even if defects are present in the preform. Plotting the formability function over the tool surface provides information regarding the location and type of defects present in the preform. Large values of W indicate the presence of defects and the precise location at which they occur, while the local values of $\lambda_{(1)}$ and $\lambda_{(2)}$ show whether the problem is wrinkling or tearing. This type of data is very useful in the design of fiber preforms. Plotting the material coordinates ($X^1(x_i)$, $X^2(x_i)$) from the mappings given in Eqns. (2.1) and (2.2) reveals the initial shape of the fiber sheet required to manufacture the ideal preform.

3 AXISYMMETRIC FORMING

In this section, the ideas of the ideal forming theory are illustrated by considering the preforming of an axisymmetric sheet onto an axisymmetric tool surface. The axisymmetric tool surface is described in cylindrical coordinates (r, θ, z) as

$$r = r(s) \quad \text{and} \quad z = z(s) \qquad (3.1)$$

Here s is the distance from the symmetry axis along the tool surface. The surface coordinates describing the tool surface are s and θ. Using cylindrical coordinates for the material coordinates (R, Θ), the preform is mapped onto the tool surface by specifying

$$R = R(s) \quad \text{and} \quad \Theta = \theta \qquad (3.2)$$

The principal stretch ratios for this special case of axisymmetric mapping are found from Eqn. (2.4) and are given by

$$\lambda_{(1)} = \frac{ds}{dR} \quad \text{and} \quad \lambda_{(2)} = \frac{r(s)}{R(s)} \qquad (3.3)$$

Physically, $\lambda_{(1)}$ and $\lambda_{(2)}$ are the stretch ratios in the s and θ directions, respectively.

In the simulation, the tool surface is discretized using one-dimensional elements originating at the axis of symmetry and proceeding along the arc length s. At each node r, z, and s are known and the material coordinate R is unknown. The inner boundary at $s = 0$ is always mapped to $R = 0$ in the axisymmetric mapping, but one is free to apply a boundary condition at the outer edge where $s = s_{max}$. Here, one can apply a fixed value of R to specify the radius of the initial planar sheet, a fixed value of $\lambda_{(1)}$ indicating that tension is applied at the outer radius, or the node can be free to move about as required by the optimization. The choice of boundary condition and the actual conditions realized in real preforming processes is to be investigated as this study progresses.

The simulation starts with an initial guess for the material coordinate R at each node. Using the initial values of R and element shape functions, the principal stretch ratios are calculated (Eqn. (3.3)) at the midpoint of each element and the formability function W is integrated over the element. For this example we use a simple formability function

$$W = (\lambda_{(1)} - 1)^2 + (\lambda_{(2)} - 1)^2 \quad (3.4)$$

Note that this function does not penalize large deformation and hence, it does not discourage wrinkling or tearing. However, it is smooth and has a minimum at the point (1,1) so it will illustrate the concepts of the ideal forming theory. The selection of formability functions is one of the key topics being investigated in this study. An iterative procedure is used to minimize Π, which is just the integral of W over the elements (Eqn. (2.5)). At each node, R is updated using a golden section search technique to minimize Π in the adjacent elements, and the routine cycles through the nodes repeatedly until a global minimum is obtained.

As an application, consider the preforming of a spherical cap (a segment of a sphere). Figure 3 shows the tool surface and the relationship between the spatial coordinates of the tool surface and the initial material coordinates R. In this example, a "free" boundary condition is specified at the outer node ($s = s_{max}$) which produces the condition $\lambda_{(1)} = 0$ along the outer edge of the preform. The results for the spherical cap are shown in Figure 4. As one might expect, the material is stretched biaxially ($\lambda_{(1)}$ and $\lambda_{(2)} > 0$) near the pole of the cap ($s = 0$). Near the edge of the preform ($s = 0.5$), the material contracts in the hoop direction ($\lambda_{(2)} < 0$). Figure 4 shows that the formability function W is large where the deformation is great, and small elsewhere. The results indicate that this particular preform would probably have trouble with wrinkling around the outer edges and tearing (or at least thinning of the preform) near the pole.

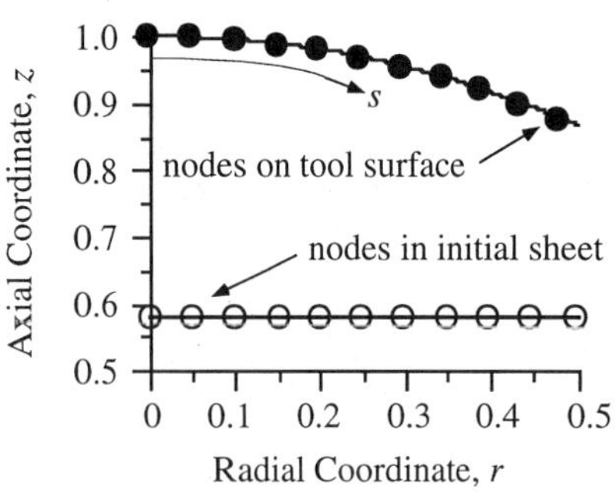

Figure 3. Nodes representing the spherical cap, and the calculated positions in the initial sheet.

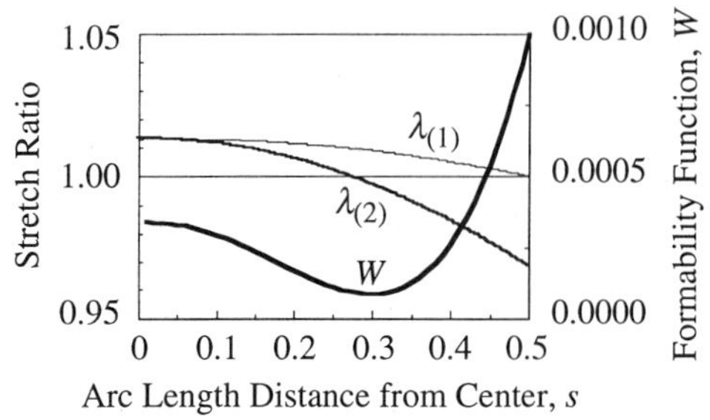

Figure 4. Local values of the stretch ratios and the formability function for the spherical cap.

4 SUMMARY

In this paper, an ideal forming theory for simulating the manufacture of non-woven fiber preforms was described and an application to axisymmetric preforming was presented. The direct results of this analysis include the initial un-deformed shape of the planar mat, the local value of the formability function over the mold surface and the principal strains in the preform. These results can be used to locate and determine the type of defects found in the preform. This information can help in the design of fiber preforms and improve mold filling simulations by providing more accurate permeability data.

Extension of this theory to arbitrary tool surfaces and realistic material behavior is required for this technique to be useful as an optimization tool. Experimental investigations to develop forming limit diagrams for random fiber mats are underway and should provide more insight into the selection of more appropriate formability functions. Further work is scheduled on the application of the theory to arbitrary shaped tool surfaces. Ultimately, this theory would also link material deformation to local preform permeability. At that point, preforming simulations could be integrated with simulations for mold filling and resin curing to provide a complete design and optimization tool for LCM processes.

ACKNOWLEDGMENTS

Support for this research is provided by the National Science Foundation, grant no. DMI 94-15051. Richard Dessenberger is supported by an Integrated Manufacturing Predoctoral Fellowship from the Department of Energy.

REFERENCES

Advani, S. G., M. V. Bruschke & R. S. Parnas 1994. Resin Transfer Molding Flow Phenomena in

Polymeric Composites. in *Flow and Rheology in Polymer Composites Manufacturing* S. G. Advani ed., Amsterdam; Elsevier: 465-515

Aoyagi, H., Uenoyama, M., & Güçeri, S. I. 1992. Analysis and Simulation of Structural Reaction Injection Molding (SRIM). *International Polymer Processing* 7:71-83.

Dessenberger, R. B., & C. L. Tucker III 1994. Thermal Dispersion in Resin Transfer Molding," in *Advances in Computer-Aided Engineering (CAE) of Polymer Processing.* K. Himasekhar, V. Prasad, T.A. Osswald, and G. Batch eds. 283:21-40. New York: ASME HTD.

Fong, L., J. XU, & L. J. Lee 1994. Preforming Analysis of Thermoformable Glass Fiber Mats–Deformation Modes and Reinforcement Characterization. *Polymer Composites* 15:134-146.

Lin, R. L., L. J. Lee & M. J. Liou 1993. Mold Filling and Curing Analysis in Liquid Composite Molding. *Polymer Composite.* 14:71-81.

Lindsay, K. L. 1993. Automation of preform fabrication makes SRIM viable for volume parts. *Modern Plastics* 70:48-51.

Tucker, C. L. 1995. to appear. Sheet Forming of Composites in *Advanced Composites Manufacturing* T. G. Gutowski ed. Butterworths.

Tucker, C. L., R. B. Dessenberger & A. S. Bakharev 1994. *TIMS Users's Manual, Version 3.0.* Department of Mechanical and Industrial Engineering, University of Illinois.

Van Der Weeen, F. 1991. Algorithms for draping fabrics on doubly-curved surfaces. *International Journal for Numerical Methods in Engineering.* 31:1415-1426.

Van West, B. P., R. B. Pipes, & M. Keefe 1990. A Simulation of the Draping of Bidirectional Fabrics over Arbitrary Surfaces. *Journal of the Textile Institute.* 81:448-460.

Simulation of Materials Processing: Theory, Methods and Applications, Shen & Dawson (eds)
© 1995 Balkema, Rotterdam. ISBN 90 5410 553 4

Fibre orientation calculation based on a two-dimensional analysis of mould filling in injection moulding of short fibre reinforced polymers

E. Devilers, M. Vincent & J. F. Agassant
Centre de Mise en Forme des Matériaux, URA CNRS, École des Mines de Paris, Sophia-Antipolis, France

ABSTRACT : The orientation of short fibres during the filling of a disk cavity is studied. The flow kinematics is obtained using a finite element method with a moving mesh technique. The orientation is calculated with a decoupled method. Results show the effect of the flow near the material front and at the sprue-disk junction.

1 INTRODUCTION

Short fibre reinforced thermoplastic polymers present interesting mechanical properties, and are processed with conventional techniques such as injection moulding. Nevertheless anisotropic mechanical properties are often obtained because fibres get oriented by the flow. Moreover, as the coefficients of expansion of the fibers, usually glass, and of the matrix are very different, part warpage is commonly observed after cooling. This is why fibre orientation phenomena have received considerable attention. The very different effects of shear flows, which preferentially lead to orient fibres in the flow direction, and of extensional flows, which lead to a stable orientation perpendicular or parallel to the flow according to the sign of the elongational rate are now well known. The aim of this work is to calculate and predict fibre orientation in injection moulding.

Jeffery (1922) proposed a system of equations for the evolution of orientation of a single particle immersed in a Newtonian fluid, knowing the general flow kinematics which is supposed to be only locally perturbed by the particle. Folgar and Tucker (1984) proposed to add to Jeffery's equations a diffusion term which represents fiber interactions in highly concentrated industrial fiber suspensions (typically 30 % by weight, 15 % by volume), but the experimental determination of the model parameter is difficult. Experimental studies have shown that rheological properties and flow kinematics are a function of fibre orientation (Laun 1984, Kamal & Mutel 1989, Ausias et al. 1992). This behaviour has been modeled by anisotropic constitutive equations, mostly for dilute or semi-concentrated Newtonian suspensions (Dinh & Armstrong 1984, Lipscomb et al. 1988). Flow kinematics and fibre orientation must be determined with a coupled procedure which has mainly been applied to steady isothermal flows (Papanastasiou & Alexandrou 1987, Ausias et al. 1994).

Injection moulding is an unsteady non-isothermal process. Geometries encountered in injection moulding are often complex. This is why orientation equations have been associated with existing mould filling simulations, assuming an equivalent shear thinning behaviour law for the composite. The results are in qualitative agreement with experiments. Nevertheless, apart from the assumptions concerning the constitutive laws, these models are based on the lubrication approximations and give only a poor description of the kinematics especially where large thickness variations occur, such as in gates. Furthermore, near the flow front there is an important velocity component in the thickness direction, which is neglected with the lubrication approximations. This effect, known as fountain flow, is responsible for the orientation especially at the surface of the moulded part. In this work, we will use a decoupled approach based on a precise two-dimensional flow kinematics calculation. The application to a center gated disk permits to show the influence of the flow at the entrance and near the material front.

2 FLOW KINEMATICS CALCULATION

The present method is based on the work of Magnin et al (1992,1994). Neglecting inertia and assuming incompressibility leads to the following equilibrium and mass conservation equations for viscous fluids :

$$\nabla.(2\,\eta\,\dot{\varepsilon}) - \nabla p = 0 \qquad (1)$$

$$\nabla.v = 0 \qquad (2)$$

where p and v are the pressure and the velocity, and

where the rate of strain tensor $\dot{\varepsilon}$ is defined as

$\dot{\varepsilon}_{ij} = (v_{j,i}+v_{i,j})/2$. The viscosity η is constant for a Newtonian behaviour, or depends on the second invariant of the rate of strain tensor for a shear-thinning behaviour. Here the flow is supposed isothermal. The boundary conditions are : (i) Poiseuille velocity profile at the entrance; (ii) sticking contact along the mould walls; (iii) no stress at the free surface.

A Galerkin finite element method is used and a penalty treatment of the incompressibility condition eliminates pressure as an unknown. The velocity is approximated using quadratic six-nodes triangular elements. Only the filled domain is meshed and the free surface nodes determine the front location. The position of the front nodes X_k is updated between time t and t+Δt using the following relation :

$$X_k(t+\Delta t) = X_k(t) + v_k(t)\Delta t \qquad (3)$$

In order to minimize mesh distorsion, a mesh velocity is determined at each time step : nodes far from the flow front hardly move, as in an Eulerian method, whereas free surface nodes move at a velocity, which is close, but not exactly equal, to the material velocity, as in a Lagrangian method. After several time steps, elements are highly distorted despite the regularization procedure, and remeshing is performed (Coupez & Chenot 1992). The evolution of fibre positions during mould filling is obtained by solving the following differential equation with a second order implicit scheme :

$$\frac{dX}{dt} = v(X(t),t) \qquad (4)$$

3 FIBRE ORIENTATION CALCULATION

The orientation of a population of fibers is represented by second and fourth order orientation tensors a_2 and a_4 :

$$a_{ij} = \int_p p_i p_j \psi(p,t)dp \qquad (5)$$

$$a_{ijkl} = \int_p p_i p_j p_k p_l \psi(p,t)dp \qquad (6)$$

where p is a unit vector aligned with fibers, and ψ is the probability distribution function. The equation of change in fibre orientation is given by Jeffery (1922) :

$$\frac{da_2}{dt} = a_2\Omega - \Omega a_2 + \lambda(\dot{\varepsilon} a_2 + a_2\dot{\varepsilon} - 2\dot{\varepsilon}:a_4) \qquad (7)$$

where Ω is the vorticity tensor defined as $\Omega_{ij}=(v_{j,i}-v_{i,j})/2$, λ is a function of the fibre aspect ratio r, defined as $(r^2-1)/(r^2+1)$ and d/dt is a material derivative. The fourth order orientation tensor appearing in Equation 7 is approximated as a function of the second order one. Among the different forms available (Advani & Tucker 1990), the results presented in this paper are based on the simple, yet quite accurate, quadratic approximation :

$$a_{ijkl} = a_{ij}\, a_{kl} \qquad (8)$$

At time t, the velocity gradient tensor $\nabla u(X(t),t)$ and $a_2(X(t),t)$ are known. The particle's position $X(t+\Delta t)$ and $\nabla u(X(t+\Delta t),t+\Delta t)$ at the next time step are known. $a_2(X(t+\Delta t),t+\Delta t)$ is obtained with an iterative procedure using :

$$a_2^{k+1}(X(t+\Delta t),t+\Delta t) = a_2(X(t),t) + \left[\frac{da_2}{dt}(X(t),t)+\left(\frac{da_2}{dt}\right)^k(X(t+\Delta t),t+\Delta t)\right]\frac{\Delta t}{2} \qquad (9)$$

where

$$\left(\frac{da_2}{dt}\right)^k(X(t+\Delta t),t+\Delta t) = \frac{da_2}{dt}(a_2^k(X(t+\Delta t),t+\Delta t),\nabla u(X(t+\Delta t),t+\Delta t)) \qquad (10)$$

The procedure is initialized explicitly. The convergence is tested by comparing each component of the orientation tensor between two successive iterations. Two or three iterations are usually sufficient to obtain the convergence.

4 RESULTS

The results were obtained for a polymer exhibiting Newtonian behaviour. The fiber aspect ratio is r=48, so that λ=0.999. A disk shaped cavity is studied, 4 mm thick with a radius of 70 mm .

4.1 *Fibre orientation near the flow front*

In this case, the sprue has not been taken into account, and the geometry and axis are represented on Fig. 1a.

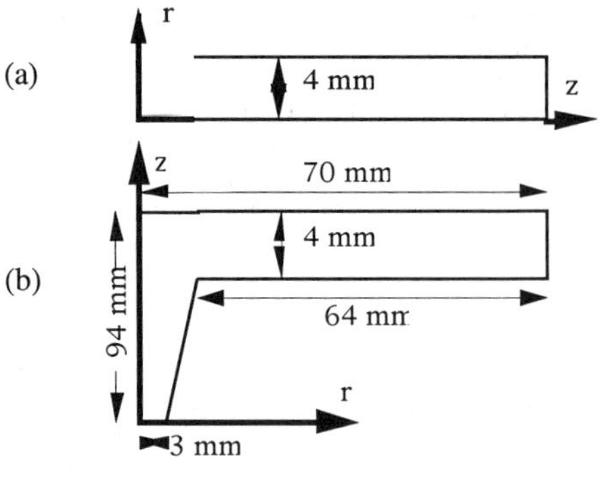

Figure 1 : (a) Disk cavity ; (b) sprue + disk cavity

The initial fibre orientation is isotropic in space, that is $a_{rr}=a_{zz}=1/3$. Figure 2 shows the evolution of the dimensionless position in the thickness during mould filling for a population of fibres entering the mould at z=2.2 mm. It moves first at constant z with velocity higher than the average flow front velocity. Then it enters the fountain region at a radius of about 35 mm, and z increases, which means that fibres move toward the surface. When they are close to the surface, z becomes constant again, and the fibers move slowly.

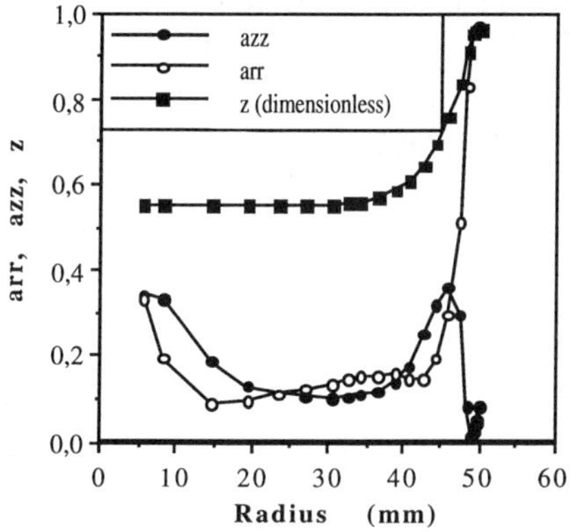

Figure 2 : Evolution of the fibre orientation in the disk cavity.

Of course thermal effects which are not taken into account here, cause solidification of the material, and hence fix the final orientation.

Figure 2 shows that first arr and a_{zz} decrease to about 0.1 : fibres are oriented almost in the plane of the disk and perpendicular to the flow direction. This is the influence of the elongational flow. Then in the fountain region, arr increases to 0.95. In the mean time, a_{zz} reaches a maximum of about 0.35 for a radius of about 46 mm. As a_{rr} has a similar value, this corresponds roughly to a random in space orientation. After, a_{zz} decreases to low values (between 0 and 0.1), so that fibres are well oriented in the flow direction, and parallel to the plane of the disk.

4.2 *Fibre orientation in the sprue and in the disk*

The geometry and axes are shown on Figure 1b. The initial orientation is random in space. Figure 3 shows the evolution of z, a_{rr} and a_{zz} during mould filling for fibers initially located near the sprue axis (r=0.05 mm) and at z=25 mm.

In the sprue (Fig. 3a), a_{rr} and a_{zz} decrease to values very close to 0, which means that fibers get oriented perpendicular to the sprue axis and to the r direction. This is due to the elongational flow. This is in agreement with experiments (Vincent & Agassant 1985). At the junction with the disk, fibres move towards the opposite mould wall, and in the mould (Fig. 3b) they move 0.22 mm from the wall.

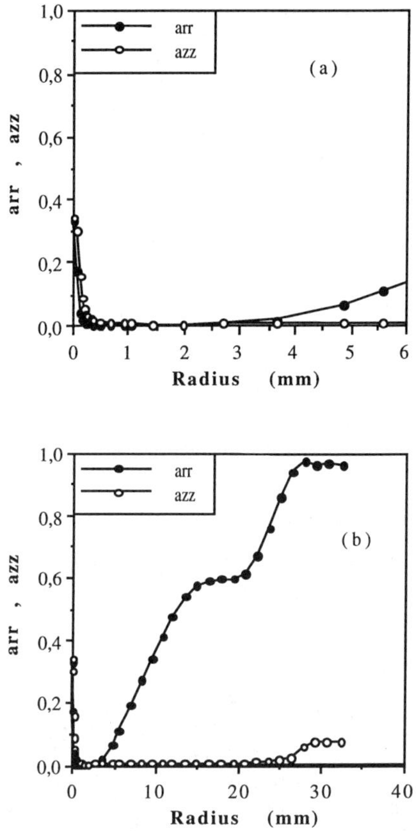

Figure 3 : Evolution of the orientation for fibers initially located near the sprue axis. (a) zoom on the sprue region ; (b) sprue + disk

They are submitted to shear flow, and they tend to orient in the flow direction.

The case of fibres initially located at r=1,5 mm and z=21 mm is shown in Figure 4. In the sprue, fibres get aligned in the flow (z) direction because of shear flow. This is again in agreement with experiments (Vincent & Agassant 1985). At the junction with the disk, they orient in the disk plane (a_{zz} tends to 0.05). In the disk they move 0.5 mm away from the midplane, they are submitted to an elongational flow, and they are oriented nearly perpendicular to the flow direction (a_{rr} is around 0.25).

5 CONCLUSIONS

The fountain flow near the material front clearly influences the orientation near the surface. The influence of the junction between the sprue and the disk has also been shown. A precise description of the phenomena will be possible when thermal effects are taken into account. This simulation will make it possible to test the physical models for fibre orientation and the constitutive laws.

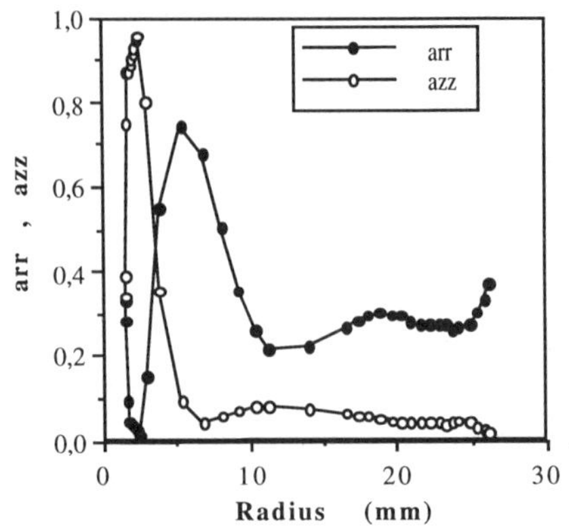

Figure 4 : Evolution of the orientation for fibres initially located at an intermediate position between the sprue wall and its axes.

ACKNOWLEDGEMENTS

We thank Aérospatiale and the Commission of the European Community for technical and financial help.

REFERENCES

Advani, S.G. & C.L.Tucker 1990. Closure approximations for three-dimensional structure tensors. *J. Rheol.* 34:367-386

Ausias, G., J.F.Agassant & M.Vincent 1994. Flow and fiber orientation calculations in reinforced thermoplastic extruded tubes. *Intern. Polym. Proc.* 9:51-59

Ausias, G., J.F.Agassant, M.Vincent, P.G.Lafleur, P.A.Lavoie & P.J.Carreau 1992. Rheology of short glass fiber reinforced polypropylene. *J. Rheol.*36:535-542

Coupez, T. and J.L.Chenot 1992. Large deformations and automatic remeshing. In D.R.J.Owen, E.Onate & E.Hinton (eds), *Proceedings of the 3rd Int. Conf. on Computational Plasticity* :1077-1088. Swansea:Pineridge Press

Dinh, S.M. & R.C.Armstrong 1984. A rheological equation of state for semiconcentrated fiber suspension. *J. Rheol.*28:207-227

Folgar, F. & C.L.Tucker 1984. Orientation behavior of fibers in concentrated suspensions. *J. Reinf. Plast. Composites* 3:98-119

Jeffery, G.B. 1922. The motion of ellipsoidal particles immersed in a viscous fluid. *Proc. Roy. Soc. London*, Ser A, 102:161-179

Kamal, M.R. & A.T.Mutel, 1989. The prediction of flow and orientation behavior of short fiber reinforced melts in simple flow systems. *Polym. Compos.* 10:337-343

Laun, H.M. 1984 orientation Effects and rheology of short glass fiber reinforced thermoplastics. *Coll. Polym. Sci.* 262:257-269

Lipscomb, G.G., M.M.Denn, D.U.Hur & D.V.Boger 1988. The flow of fiber suspensions in complex geometries. *J. Non-Newt. Fluid Mech.*26:297-325

Magnin, B. 1994. Modélisation du remplissage des moules d'injection pour polymères thermoplastiques par une méthode eulérienne-lagrangienne arbitraire. PhD Thesis, Ecole des Mines de Paris, France

Magnin, B., T.Coupez, M.Vincent & J.F.Agassant 1992. Numerical modeling of injection mold filling with an accurate description of flow front. In J.L.Chenot, R.D.Wood & O.C.Zienkiewick (eds.), *Numerical Methods in Industrial Forming Processes*:365-370. Rotterdam:Balkema

Papanastasiou, T.C. & A.N.Alexandrou 1987. Isothermal extrusion of non-dilute fiber suspensions. *J. Non-Newt. Fluid Mech.*25:313-328

Vincent, M. & J.F.Agassant 1985. Experimental and theoretical study of short fiber orientation in diverging flows. *Rheol. Acta* 24:603-610

Simulation of Materials Processing: Theory, Methods and Applications, Shen & Dawson (eds)
 ISBN 90 5410 553 4

A numerical formulation for the filler particle distribution during infiltration of fiber preforms

M. Erdal & S. Güçeri
University of Illinois at Chicago, Ill., USA

ABSTRACT: An accumulation-detachment filtration model is incorporated into a two-dimensional flow solution technique, using Darcy's law, to model the impregnation/resin transfer molding (I/RTM) of particle filled preceramic polymers into fiber preforms. Boundary Fitted Coordinate System (BFCS) in conjunction with the Finite Difference method (FDM) is used during solution. The results of the filtration model are used to update the flow field continuously while the mold filling proceeds. The method is capable of accounting for inherent nonlinearities present due to the coupled nature of flow and filtration parameters.

INTRODUCTION

With the development of low-viscosity preceramic polymers such as Blackglas^{TM}, the impregnation/resin transfer molding (I/RTM) technique stands to be one of the most promising techniques for manufacturing high temperature ceramic composite structures with reliable, cost-effective properties. In a typical RTM process, 2-D or 3-D fibrous structures (fiber preform assemblies) are placed in a mold (tool) cavity in the shape of the part to be produced. The mold/tool is closed and the polymeric matrix resin is injected to impregnate the preform. Cure of the resin takes place after the mold cavity is filled, followed by pyrolysis. However, the material loss during pyrolysis leads to a porous structure that needs to be densified for satisfactory structural performance. This can be accomplished through several reimpregnation and pyrolysis cycles at the expense of cost-effectiveness. One way of reducing the number of cycles and obtaining near-net shapes during the process is the inclusion of inert and/or active filler particles in the polymer during impregnation.

The presence of inert particles, such as SiC leads to the reduction of void fraction since they do not undergo any loss of mass during pyrolysis. A well-dispersed slurry containing agglomerate-free, sub-micron sized particles was found to result in high packing densities during impregnation into 3-D woven preforms [1]. The active particles, such as Ti have an additional advantage that they experience a volumetric expansion after reaction with the products of pyrolysis [2].

The inclusion of filler particles in the polymeric resin introduces additional variants into the process. A major transport issue is the prevention of particle filtering during impregnation and the achievement of a uniform particle distribution when the mold is completely filled. The fiber architecture, particle size and volume fraction, the rheological characteristics of the filled polymer are coupled leading to a nonlinear problem. The permeability tensor becomes a function of suspension characteristics as well as preform properties. Furthermore, the inevitable filtering of particles will lead to a nonuniform permeability distribution that will change both spatially and temporally.

In order to achieve reliable, high performance end-products, the filler particle distribution throughout the part must be uniform and filtration minimized. This requires the development of flow and filtration models to understand the underlying transport mechanism of impregnation with filler particles. A two-dimensional particle filtration model is presented to predict the filler particle distribution in the flow during impregnation, coupled with the flow model for resin transfer molding. The governing equations are solved using boundary fitted coordinate system (BFCS) together with the finite difference method (FDM). For geometric flexibility, physical domain is mapped onto a computational domain where the governing equations are discretized and solved. The permeability of the flow within the scope of this work, remains constant and uniform throughout the field.

FORMULATION

Main Flow Model

The basis of the main flow formulation is taken from Coulter (3,4) who presented the first comprehensive RTM study based on 2-D, isothermal, Newtonian Darcy flow. Since the impregnation process is similar to flow through porous media, the flow of preceramic polymer through the fiber preform can be modeled macroscopically by Darcy's law. For anisotropic media , Darcy's law is

$$\underset{\sim}{V} = -\frac{1}{\mu}(\underset{\approx}{K} \cdot \underset{\sim}{\nabla} P) \quad ....(1)$$

where $\underset{\sim}{V}$ is the velocity vector, μ the viscosity, P the pressure and $\underset{\approx}{K}$ the symmetric permeability tensor. Within the scope of this work, only isotropic cases are considered.

The flow can be modeled as isothermal and non-reacting therefore one other equation is sufficient for solution. This equation is the continuity equation given by

$$\underset{\sim}{\nabla} \cdot \underset{\sim}{V} = 0 \qquad (2)$$

Substituting equation (1) into equation (2) results in the governing equation

$$\underset{\sim}{\nabla} \cdot \left[\frac{1}{\mu}(\underset{\approx}{K} \cdot \underset{\sim}{\nabla} P)\right] = 0 \qquad (3)$$

which is a modified form of Laplace equation. This formulation is easily applicable to three- as well as two-dimensional flow configuration.

The pressure at the flowfront is assumed to be constant leading to a Dirichlet type boundary condition

$$P_{flowfront} = 0 \qquad (4)$$

The impermeable boundaries on the side walls across which there is no flow result in a Neumann type condition

$$(\underset{\approx}{K} \cdot \underset{\sim}{\nabla} P) \cdot \underset{\sim}{n} = 0 \qquad (5)$$

where $\underset{\sim}{n}$ is the normal vector on the walls.

At the inlet gate, either a prescribed pressure or a prescribed flowrate condition exist, leading to a Dirichlet or Neumann type condition

$$P_{inlet} = f(y) \qquad (6.a)$$

$$(\underset{\approx}{K} \cdot \underset{\sim}{\nabla} P) \cdot \underset{\sim}{n}_{\,inlet} = f(y) \qquad (6.b)$$

where f is some known function and y the coordinate direction along the width of the inlet gate. The description of the RTM process and the corresponding boundary conditions are presented in Figure 1. The solution is performed by the previously developed numerical code, TGMOLD [5]. The details of this formulation are well presented by Friedrichs [6,7]

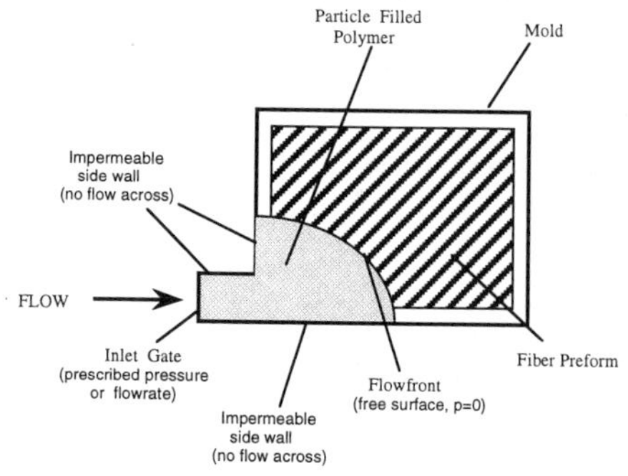

Figure 1. Description of the Process and the Boundary Conditions

Filtration Model

The filtration of filler particles through the fiber preform during impregnation can be assumed to have a mechanism similar to deep-bed filtration which involves "the frontal advancement of a working layer in which effective filtration takes place " [8]. For this purpose, the one-dimensional accumulation-detachment filtration model of Adin and Rebhun [8] is adapted and extended to simulate the two-dimensional filtration of filler particles through the preform. The material balance for particle concentration is given by

$$\frac{\partial \sigma}{\partial t} + \underset{\sim}{V} \cdot \underset{\sim}{\nabla} C = 0 \qquad (7)$$

where σ is the particle volume deposited to the preform per unit preform volume (specific deposit) and C the particle concentration (volume by volume) in the polymer. The effects of diffusional gradients and the concentration change with time are neglected in the present formulation [8,9].

The accumulation rate is modeled through the filtration kinetic equation

$$\frac{\partial \sigma}{\partial t} = V \alpha\, C\, (F - \sigma) - \beta\, \sigma |J| \qquad (8)$$

where V and J are the scalar velocity and pressure gradient in the direction of the flow, F the theoretical filter capacity, α the accumulation coefficient and β the detachment coefficient. The pressure gradient in the direction of the flow can be expressed as

$$J = (\underset{\sim}{V} \cdot \underset{\sim}{\nabla} P) / \, [\underset{\sim}{V} \cdot \underset{\sim}{V}]^{1/2} \qquad (9)$$

For isotropic preform , equation (9) can be written as

$$J = -\frac{\mu V}{K} \qquad (10)$$

Placing equation (10) into equation (8), the final form of the filtration kinetic equation is obtained

$$\frac{\partial \sigma}{\partial t} = V\left[\alpha C (F - \sigma) - \beta \frac{\sigma \mu}{K}\right] \qquad (11)$$

The theoretical filter capacity as well as the two filtration coefficients, must be experimentally determined. Without experimental data, the filter capacity is set to a constant value greater than fiber volume fraction. It is worthwhile to note that the particle size will become the dominating factor in the determination of theoretical filter capacity. The boundary and initial conditions for solution are

$$C_{inlet} = f(y) \qquad (12.a)$$

$$\sigma_{initial} = 0 \qquad (12.b)$$

respectively, where once again, f is a known function and y is the coordinate direction along the width of the gate. For a continuously deforming flow domain with a moving flowfront, equation (12.b) implicitly imposes the no-deposit condition at the flowfront. Solution of equations (7) and (11) for the initial condition results in

$$C = C_{inlet} \exp(-\alpha F s) \qquad (13)$$

where s is the direction of flow path. Equation (13) gives the concentration for the flowfront that has traveled a distance of s and is a specific case of the deep-bed filtration model of Ives [9], used for initiation of a filtration run in a non-evolving, rigid domain. After each advancement of the flowfront, the free passage through which the polymer can flow is modified by incorporating the effect of deposited filler particles as

$$v_{f_{new}} = v_{f_{old}} + \sigma \qquad (14)$$

where v_f is the total solid volume fraction in the domain (initially, fiber volume fraction).

NUMERICAL IMPLEMENTATION

Boundary fitted coordinate system (BFCS) in conjunction with the finite difference method (FDM) is employed for the solution of the governing equations [3,4,6,7]. The irregularly shaped domain in the physical x,y space is mapped onto a regularly shaped domain in the computational ξ, η space, providing geometric flexibility (Figure 2). The mesh grids are generated repeatedly and deform continuously to conform to the new flow domain.

Mesh Generation

To avoid interpolation to determine the location of the flowfront, a moving mesh approach is selected in which grids are generated at each flowfront advancement. Elliptic grid generation PDE for two-

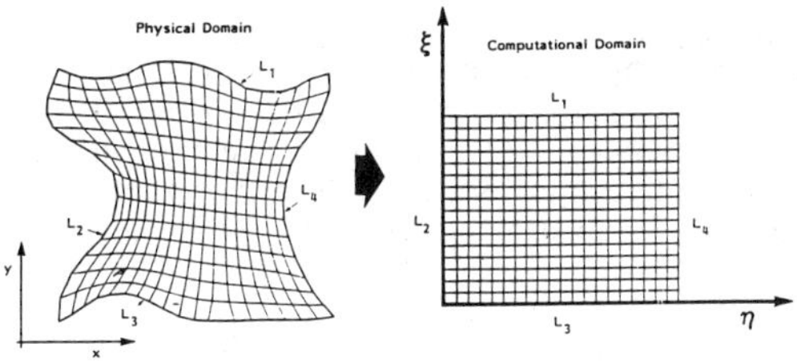

Figure 2. Physical and Computational Domains

dimensional problem are given as [10]

$$\nabla^2 \xi(x,y) = P(\xi, \eta) \qquad (15.a)$$

$$\nabla^2 \eta(x,y) = Q(\xi, \eta) \qquad (15.b)$$

where P and Q are the grid control functions that enable grid line concentration. Equations (15.a) and (15.b) are solved by interchanging the dependent and independent variables so that the physical coordinates x,y, corresponding to the computational node ξ, η of the generated mesh can be determined. The nodal positions along the boundaries of the physical domain are prescribed leading to Dirichlet type conditions.

Discretization

In order to use BFCS for irregularly shaped domains, the governing equations and Neumann type boundary conditions need to be transformed to the regularly shaped computational domain. The geometrically conservative form for the first derivative of a function f in two-dimensional computational space is given by [6]

$$f_x = \frac{1}{J}[(f y_\eta)_\xi - (f y_\xi)_\eta] \qquad (16.a)$$

$$f_y = \frac{1}{J}[-(f x_\eta)_\xi + (f x_\xi)_\eta] \qquad (16.b)$$

where J is Jacobian of the transformation. Higher order derivatives are similarly derived by replacing f by f_x or f_y in equations (16.a) and (16.b). The nodal spacing in the computational space does not affect the physical space solution, therefore for convenience, the node spacings in computational domain are taken as unity. The main flow governing equation (equation 3) is discretized using central-differencing whereas the particle conservation equation (equation 7) is discretized by upwind-differencing scheme in the flow direction. The accumulation rates are directly determined on the grid nodes from equation 11.

Solution

The discretized governing equation for resin transfer

molding is solved by successive over-relaxation method (SOR). Once the pressure values on the grid nodes are determined, the velocities are calculated from equation (1) using a geometrically conservative transformation. The new flowfront location is determined by a first order explicit scheme

$$\underset{\sim}{r}^{\,new} = \underset{\sim}{r}^{\,old} + \underset{\sim}{V}_{flowfront}\,\Delta t + O[(\Delta t)^2] \qquad (17)$$

where $\underset{\sim}{r}$ is the position vector of the flowfront in the physical domain.

The next step is the determination of particle concentration values from the already calculated flow field on the old mesh. The concentration values on the upcoming, new flowfront must be determined therefore, the flowfront line of the new mesh is added as the last grid line, onto the old mesh where equation (7) is solved for concentration. The concentration values on the new mesh nodes are then determined through interpolation. The specific deposit values are updated using the first order scheme

$$\sigma^{new} = \sigma^{old} + \frac{\partial \sigma}{\partial t}\,\Delta t + O[(\Delta t)^2] \qquad (18)$$

The updated values are also transformed onto the new mesh through interpolation. It should be noted that the flowfront deposit value is zero.

RESULTS AND DISCUSSION

The model is first tested on a simple rectangular mold for one-dimensional flow. The inlet gate extends across one side of the domain and the incoming flowrate is constant. The flow has a filler particle concentration of 0.3 (volume by volume) at the gate and the fiber volume fraction and the permeability of the domain are constant. The results of the one-dimensional run for various filtration coefficients at the instant of fill are presented in Figure 3. The particles present in the resin at the instant of fill and those deposited to the preform are added up to yield the final concentration in the mold, i.e. $C+\sigma$. Although the majority of the domain behaves consistently with one-dimensional filtration, there appears to be slight accumulation near the sidewalls close to the gate. The velocities calculated in the main flow solution show certain discrepancies near the walls due to numerical differentiation of pressure in equation 1. These errors are small by comparison however they become significant in the formulation of concentration, resulting in the accumulation of the particles near the walls at the early stages of advancement. Consistent with the model, a higher filtration rate is observed for the high accumulation coefficient run, leading to lower concentration values at the flowfront. For the low detachment coefficient run, the concentration decay along the flow path is more gradual with less filtration.

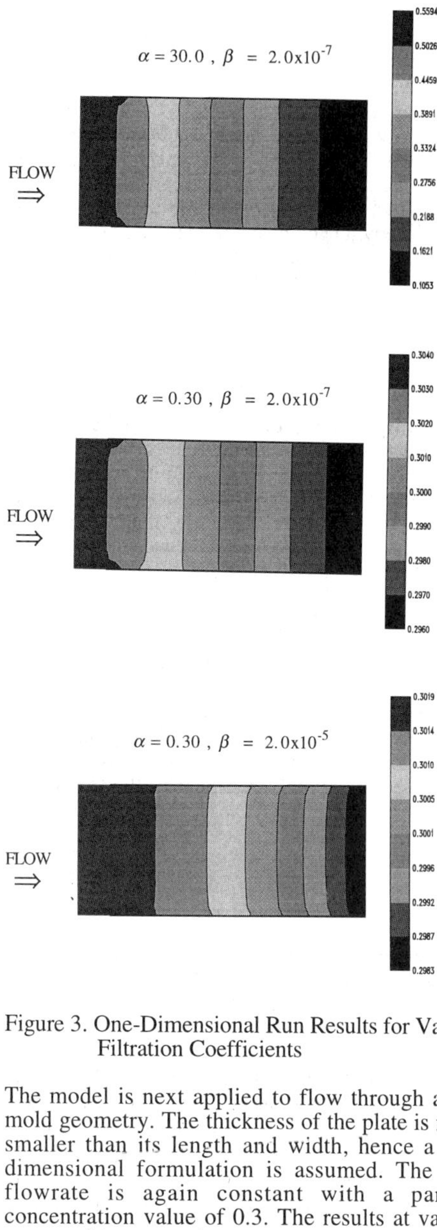

Figure 3. One-Dimensional Run Results for Various Filtration Coefficients

The model is next applied to flow through a 2-D mold geometry. The thickness of the plate is much smaller than its length and width, hence a two-dimensional formulation is assumed. The inlet flowrate is again constant with a particle concentration value of 0.3. The results at various stages of impregnation are presented in terms of total concentration (concentration in the fluid plus the deposit on the preform) in Figure 4. As the flowfront is advanced, the filtration rate slows down due to lower concentration values, resulting in a more gradual decrease in the overall concentration. This is clearly seen by comparing the first two results. The sharp contours extending inward from the endwall on the right-hand side of the mold do not realistically reflect a physical

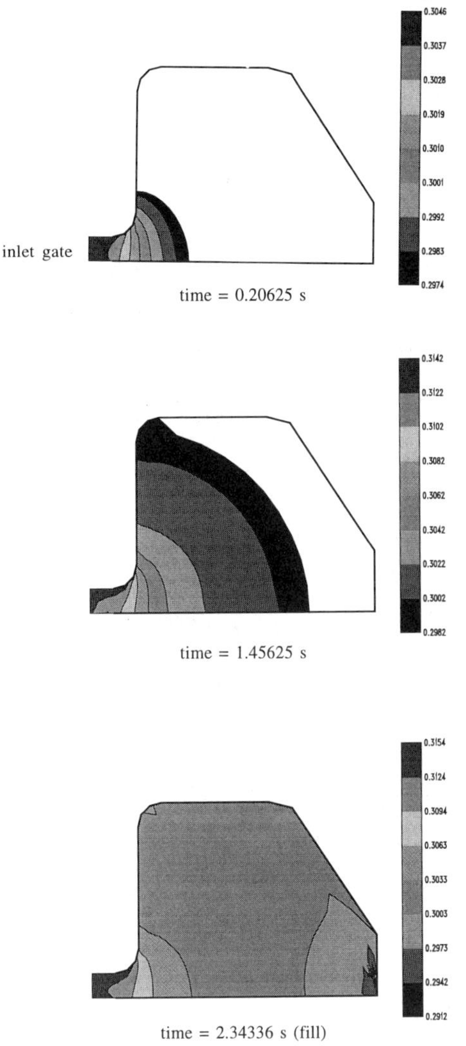

Figure 4. Flowfront Progression during Flow Through Two-Dimensional Mold

distribution. A smoother transition to a lower concentration contour would be expected. However the existence of backward velocities impose a limitation on the filtration model. In addition, numerical differentiation errors in the determination of velocities and a fast deforming rather coarse mesh grid structure tend to cause instabilities. A physically generated mesh that follows the actual paths could help reduce these discrepancies.

The overall results for this preliminary formulation are within satisfactory limits. All possible nonlinear relations between with the flow and filtration parameters can be handled in a single equation (equation 11) without complicating the discretized governing equation during numerical solution. However, there are certain aspects that need to be improved and a better understanding of transient, multi-dimensional filtration must be presented. Experimental data and numerical formulation must be coupled to determine the empirical coefficients in the filtration model. The contribution of particle size and existence of agglomerates and bridging effects must be accounted for through experimental analysis. Finally, the reimpregnation stages deal inevitably with anisotropic media therefore a relation between the preform permeability and filtration parameters must be constructed on a reliable basis and further expanded to the anisotropic case.

REFERENCES

1. Sim,S.M. and Kerans,R.J.,"Slurry Infiltration of 3-D Woven Composites", Cer. Eng. Science, Vol.13, No.9, pp. 632-641, 1992.
2. Greil,P. and Seibold,M., "Active Filler Controlled Pyrolysis (AFCOP) - A Novel Fabrication Route to Ceramic Composite Matreials", Adv. Comp. Mat. Cer. Trans., Vol.19, pp.43-49, 1991.
3. Coulter,J.P. and Güçeri,S., "Resin Impregnation During the Manufacturing of Composite Materials Subject to Prescribed Injection Rate", J. Rein. Plas. Comp., Vol.7, pp.200-219, 1988.
4. Coulter,J.P., Resin Impregnation During the Manufacturing of Composite Materials, Ph.D.thesis, Dept. of Mech. Eng., Un. of Delaware, 1988.
5. TGMOLD - A Computational Code for the Analysis of Injection Molding, RIM, RTM, and SRIM, S.Güçeri, Mech.Eng.,Univ. of Illinois, Chicago, IL, 1995.
6. Friedrichs,B., *Modeling of 3-D Flowfields in Injection Molding and Resin Transfer Molding Processes*, Ph.D.thesis, Dept. of Mech. Eng., Un. of Delaware, 1993.
7. Friedrichs,B. and Güçeri,S., "A Hybrid Numerical Technique to Model 3-D Flow Fields in Resin Transfer Molding", submitted to Pol. Proces. Sci., 1993.
8. Adin, A. and Rebhun, M., "A Model to Predict Concentration and Head-Loss Profiles in Filtration", Amer. Water Works Assoc. J., Vol.69, pp.444-453, 1977.
9. Ives,K.J., "Mathematical Models of Deep Bed Filtration", in K.J.Ives, editor, *The Scientific Basis of Filtration*, pp.203-224, Proceed. NATO Adv. Stu. Inst. Ser., Noordhoff, Leyden, 1975.
10. Güçeri,S., "Finite Difference Solution of Field Problems", in C.L.Tucker,III, editor, Fundamentals of Computer Modeling for Polymer Processing, Ch.5, pp.141-236, Hanser, New York, 1989.

Simulation of Materials Processing: Theory, Methods and Applications, Shen & Dawson (eds)
© 1995 Balkema, Rotterdam. ISBN 90 5410 553 4

Finite-element simulation of the stretch blow molding process

K. Hartwig & W. Michaeli
Institut für Kunststoffverarbeitung (IKV), RWTH-Aachen, Germany

ABSTRACT: In the stretch blow molding process, hollow articles with excellent qualities due to biaxial deformation are produced. The final wall thickness distribution mainly depends on the preform- and mold-design and on the temperature distribution in the preform prior to the molding process. A theoretical model is presented that allows investigation of the combined effect of thermal preform conditioning and the molding phase on the wall thickness distribution. Both the thermal conditioning and the stretch blow molding are simulated using the finite-element method and are coupled by describing a temperature-dependent material behavior.

1 INTRODUCTION

The stretch blow molding process is used to produce hollow articles for applications mainly in the packaging sector. Outstanding mechanical, optical and barrier properties are achieved through the biaxial orientation resulting from forming through the combination of pressure and the action of the rod. The processed materials are predominantly polyethylene terephthalate (PET) polyvinylchloride (PVC) and polypropylene (PP). In volume terms, PET clearly dominates because of its use as a packaging material for carbonated drinks. When processing the amorphous PVC, the improvement in properties during stretching results from an orientation of the polymer chains. The semi-crystalline PET is briefly heated during processing so that the degree of crystallization is very low and, as a result, the improvement in properties is due to a heavy orientation of the polymer chains, as it is with PVC. In the case of the semi-crystalline PP, on the other hand, the crystalline areas become oriented so that during the manufacture of PP preforms, a certain crystallization must be allowed. Stretching of the PP preforms takes place subsequently at temperatures just below the crystalline melting point [1].

A distinction is made between two types of processes with regard to the thermal history of the material. With the one-step process the preform is cooled to demolding temperature in the injection mold, thermally treated in a conditioning station and then directly molded. In the two-step process, on the other hand, the injection molding process is completely separate from the conditioning and stretching process. After the injection molding process, the preforms are cooled to room temperature and are not heated up again and stretched until some time later.

One very important quality characteristic of stretch blow parts is the final wall thickness distribution. This is dependent on the geometry of the molding and the preform, on the temperature distribution immediately before molding and on the rod movement and pressure profile during molding. The required preform geometry for a given bottle is currently designed on the basis of experience. Costly and very time-consuming trials are then carried out to establish the necessary process control for this preform geometry that leads to the desired wall thickness distribution. In the long term, this procedure is not competitive with the onset of decreasing development times and the increasing variety of products.

For this reason, it is necessary to develop a theoretical model which describes the quality characteristics resulting from the process as a function of the process parameters. By applying the finite-element method [FEM], the stretch blow molding process can be simulated so as to calculate the wall thickness distribution resulting from the process. Relevant process parameters in the one-step process are the temperatures in the injection mold and in the conditioning station, the conditioning and equilibrium times and the characteristics of the rod movement and pressure control. In the two-step process, the number and individual temperature of the infra-red heating elements, the movement of the preform, the cooling through the fans and the equilibrium time are the predominant parameters for the heating-phase.

Approaches for simulating the stretch blow molding process are described by HAESSLY and RYAN [2] and later by HAESSLY, RYAN and GARCIA REJON, DERDOURI and KHAYAT [3].

An axially symmetrical preform is inflated, first by varying its wall thickness distribution and later with a predefined temperature distribution. The axial stretching process, which superposes the inflation phase in the stretch blow molding process, is not taken into account in these approaches. CHUNG [4] and SCHMIDT et al. [5] simulated the stretch blow molding process for an axially symmetrical PET bottle. CHUNG calculated the resultant wall thickness distribution for three different velocity profiles of the rod movement. For his simulation calculations he used an elasto-plastic materials law to take into account the material behavior dependent on the deformation rate. SCHMIDT et al. describe the deformation process in the mold for a Newtonian liquid. Both approaches are based on a uniform temperature of the preform. All the approaches, with the exception of the one put forward by CHUNG, describe the deformation process with the membrane or shell theory. The reheating stage of the two-step process has been simulated by DIRADDO, GARCIA REJON and NGUYEN [6], assuming radiative heat transfer between two grey-surfaces. Reflection at the preform surface is neglected and absorption of layers in the preform-wall is not accounted for.

The aim of the work described below is to develop a model to simulate the stretch blow molding process. It takes account of the temperature distribution in the preform, both in the axial and in the radial directions. Therefore the thermal conditioning is simulated for both processes, the one- and the two-step process. For the reheating simulation of the two-step process the radiative heat-transfer is simulated for real bodies, accounting for the wavelength dependence of emission and absorption. Because the wall thicknesses of the preforms are quite large, the model is not based on a shell or membrane state. For the FE simulation, the preform is therefore discreted in both the axial and the radial directions.

2 SIMULATION

On the basis of algorithms for describing heat exchange processes, the temperature distribution in the preform immediately before the molding process can be calculated. The molding process can be simulated in a similar way to the FE simulation of the blow molding and thermoforming process. The decisive difference between the stretch blow molding process and the blow molding or thermoforming process is that the thickness of the article to be deformed cannot be neglected when compared with its external dimensions. This means that the molding simulation cannot be based on a shell or membrane state, and for this reason, a three-dimensional temperature field in the preform must be considered. Whereas in the thermoforming and blow molding process, we can work from a two-dimensional temperature field over the semi-finished product or parison, the stretch blow molding simulation has to account for a temperature profile over the wall of the preform.

For the simulation, the stretch blow molding process is divided into two part processes, namely

- the heating-up process and
- the stretching process.

In the simulation, the link between these two steps is created through a description of the temperature-dependent material behavior. Fig. 1 shows the procedure and the boundary conditions for the stretch blow molding simulation.

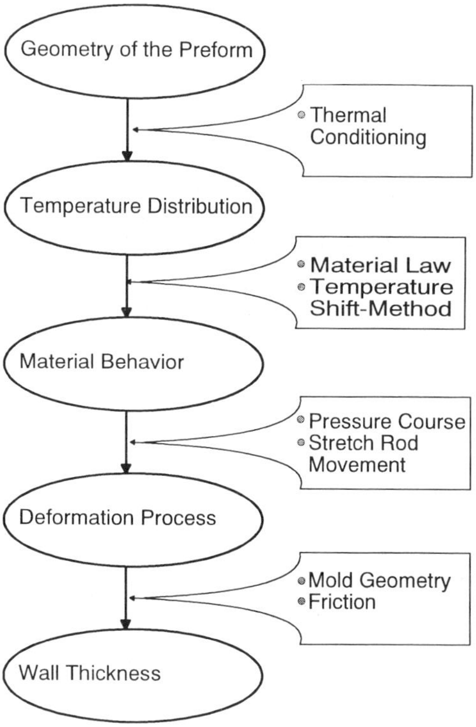

Fig. 1 Simulation of the stretch blow molding process

The three-dimensional temperature distribution in the preform is calculated on the basis of the preform geometry. The calculated temperature distribution must be converted into a description of the resultant material behavior by applying a suitable material law and a temperature shift-method. This results in a three-dimensional distribution of the material behavior; the coefficients of the material law are assigned to each preform area according to its temperature. After this, the deformation process is simulated for the characteristics of the rod movement and the pressure profile. With the description of the mold geometry and the contact properties between the hot preform and the cold mold, we get the wall thickness distribution of the molding.

2.1 *Simulation of the heating-up*

Because of the differences in the process, the

thermal simulations are different for the one- and two-step process. The thermal simulations presented below were performed using the commercial finite-element programm ABAQUS/Standard [7].

2.1.1 Two-Step Process

The temperature distribution in the preform prior to molding is calculated for the two step process, accounting for the radiation, conduction and convection. The modeling is illustrated in Fig. 2.

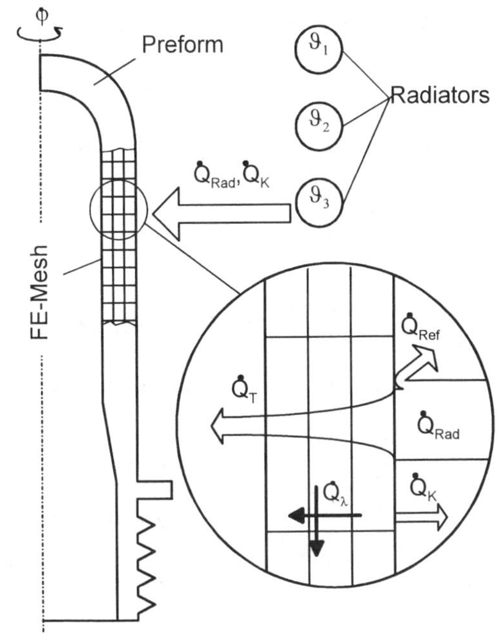

Fig. 2 Modeling of the heating-up

The modeling takes into account the reflection, absorption and transmission of the radiative heat, the convective cooling at the surface due to the fans and the conductive heat-transfer in the preform-wall.

The radiative heat is described with a sine-function in order to account for the rotation ψ of the preform. The reflection is described with the Fresnel-law. The absorbed heat-flux for each element, $\dot{q}''_E$, is modeled as a function of the wave-length, λ, and the distance which the wave passes through the preform, d. The absorbed heat-flux can be described by the following formula:

$$\dot{q}''(d)_E = \sum_{n_R}\sum_{\lambda} \Phi_{RE}\cdot \dot{q}''(\lambda)_R \cdot e^{-k(\lambda)d} \quad (1)$$

The angle factor, Φ_{RE}, describes the geometry dependent ratio between the emitted radiation, $\dot{q}''_R$, and the incoming radiation at the surface of the absorbing body. Due to the absorption in the outward layers, the absorbed heat flux decreases with increasing, d. Both the absorption-coefficient of the preform material, $k(\lambda)$, and the emitted heat-flux of the radiators, $\dot{q}''_R(\lambda)$, are modeled as a function of the wave-length. The heat-flux for each element is summed-up for all the element/radiator combinations, n_R, and over the whole interval of the wave-length, λ. An algorithm has been programmed to calculate the absorbed heat-fluxes for variable geometries of the preform and the radiators, and for variable spectra of emission and absorption.

The summed heat-flux for each element is written to a file, which can be processed with the convective boundary conditions and the thermal material behavior by ABAQUS/Standard. The convective heat transfer coefficient is modeled as a constant over the entire height of the preform and length of each heating element, but can be modeled height- and time dependent. At the inside wall of the preform adiabatic conditions are assumed and no heat loss through the clamping of the thread is calculated. Furthermore the equilibration time is accounted for.

In practical experiments with an infrared camera, the surface temperature of a preform has been measured under production conditions and was compared to numerical results. Fig. 3 shows the comparison of the measured and calculated temperature profiles over the preform heights.

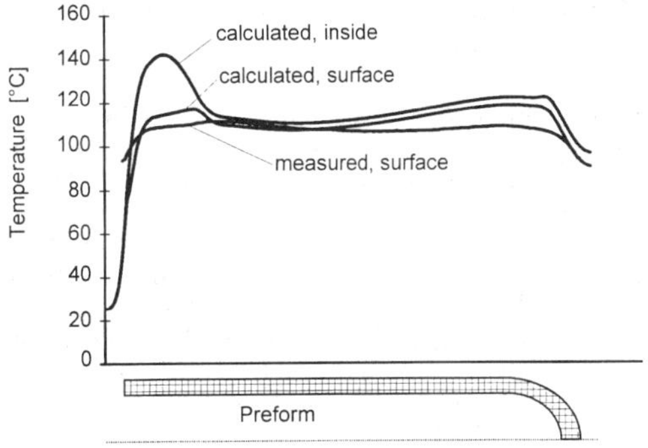

Fig. 3 Measured and calculated (inner and surface) temperature profile

Two calculated curves are shown, one being the outside-wall temperature and one being that of the inside-wall. The calculated and measured surface temperatures show good agreement. The calculated temperatures in the bottom and top area of the preform are slightly higher than the measured ones, while the curves correspond well in the middle of the preform.

2.1.2 One-Step Process

The modeling is presented here taking an axially symmetrical beaker of polypropylene as an example.

It is initially assumed that the preform is transported into the conditioning station with a constant temperature distribution of 165 °C. In the conditioning station, the preform is thermally treated by three separate heating circuits with oil tempe-

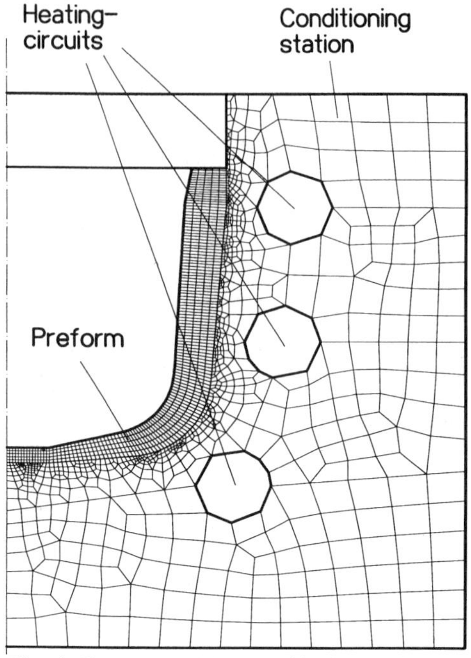

Fig. 4 Geometry of the conditioning station and preform

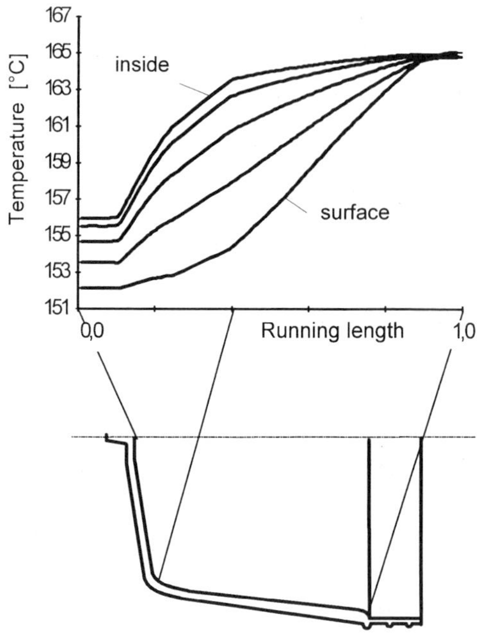

Fig. 5 Temperature distribution in the preform

ratures between 142 °C and 147 °C (Fig. 4). In order to ensure a good heat transfer between the conditioning element and the preform, the preform is pressed, through internal pressure, against the wall of the mold. In the simulation, contact is therefore assumed to be ideal; the temperature of the inner mold wall is identical to that of the outer wall of the preform. The heat transfer coefficients ,α_i, between the inner pipe wall of the conditioning circuits and the conditioning medium are determined by the oil volume flow,$\dot{V}$, and the material data, $Pr=f(\eta,\lambda,c_p)$, of the oil.

Adiabatic states are assumed to exist inside the preform. The material data of the polypropylene (ρ,λ,c_p) are first assumed to be constant and averaged for the temperature range between the initial temperature and the temperature of the conditioning oil.

With the large number of pieces produced and thus the extremely short cycle times during production, we can assume approximately stationary thermal conditions for the conditioning station. The stationary temperature distribution of the conditioning station was therefore first calculated with initial temperatures of 165 °C for the preform and 150 °C for the conditioning station. The equilibrium temperature of the conditioning station was then used as the initial temperature distribution for the heating-up simulation. Because of the poor thermal conductivity of polypropylene, a maximum radial temperature difference of approx. 6 - 8 °C is found in the preform after 20 s. The maximum axial temperature difference is also approx. 8 °C. The axial temperature distribution, which results after 20 s for various layers in the preform wall, is shown in Fig. 5. The high temperatures of the preform in the neck area result from the high temperatures of the conditioning station in this area. The preform retains approximately its initial temperature here.

2.2 *Description of the material behavior*

For simulating the molding process, every preform element is, on the basis of the thermal calculations, assigned a material behavior corresponding to the calculated element temperature.

The material behavior in the molding simulation is described by the hyperelastic material law according to Mooney/Rivlin. It is assumed that the entire deformation energy is converted into dimensional change energy; during the molding, no flow processes take place and, as a result, no stress relaxations occur in the material. This condition is confirmed by the observations of MEISSNER [8], HAESSLY [9] and WOLLSCHLÄGER [10].

The material law is calibrated with measured stress/strain data. WEINAND [11] measured these values for polypropylene at temperatures of between 150 °C and 165 °C at uniaxial strain rates of 400 s^{-1} and 500 s^{-1}. A temperature shift is applied to the material parameters and consequently the mechanical properties of all the preform areas are characterized according to their temperature.

$$\sigma\,(\vartheta_1)\big|_{\dot{\varepsilon}_1} = \sigma\,(\vartheta_2)\big|_{\dot{\varepsilon}_1} \cdot a_T\big|_{\vartheta_1}^{\vartheta_2} \qquad (2)$$

The temperature shift factor, a_T, is determined by the WLF function. The coefficients of the WLF function are calculated by non-linear regression for the temperature range under analysis.

As an example Fig. 6 shows a comparison between the measured stress/strain characteristics of polypropylene and shifted values.

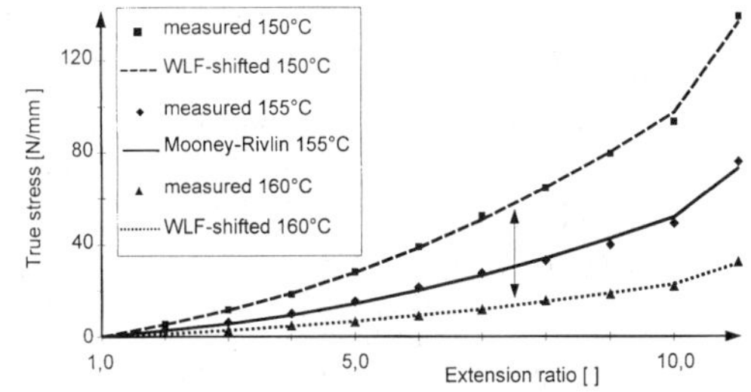

Fig. 6 Comparison between measured and calculated stress strain behavior (Mooney-Rivlin /WLF)

In Fig. 6 the Mooney/Rivlin coefficients were calculated from the measurements at 155 °C and, using these, the stress data were determined for given strains. The conformity between the measured stress data and those calculated using the Mooney/Rivlin formula is, as the figure shows, very good. In addition, the temperature shift coefficients were calculated for the shift in the curve from 155 to 160 °C and to 150 °C and these were then applied to the Mooney/Rivlin law. Here, too, conformity between the temperature-shifted Mooney/Rivlin approximation and the real material behavior is very good.

2.3 *Simulation of the molding process*

The molding process was simulated with the commercial finite-element program ABAQUS/Explicit [7], which applies the dynamic equilibrium condition to the structure under analysis. This features an explicit time integration, so that no stabilization iterations at each increment are needed. No tangent stiffness matrix (Jacobian) needs to be calculated, rendering this method faster than the implicit scheme used in ABAQUS/Standard.

The time integration rule is the central difference rule and the computational efficiency of the algorithm is obtained by using diagonal element mass matrices [12].

In the simulation, the forming process through the combination of pressure and the action of the rod is described. Any desired profile can be given both for the rod speed and for the pressure. The rod movement was defined in such a way that its speed first rose linearly up to 1.2 m/s and then decreased again linearly. The pressure profile during the molding was modeled in three steps. Up to the complete axial stretching through the rod, an initial pressure was applied which increased linearly to approx. $3 \cdot 10^5$ Pa. After this, the molding pressure was switched on, increasing linearly to approx. $15 \cdot 10^5$ Pa; it remained at this level until molding was complete. The characteristics of the rod velocity and pressure profile are shown in Fig. 7.

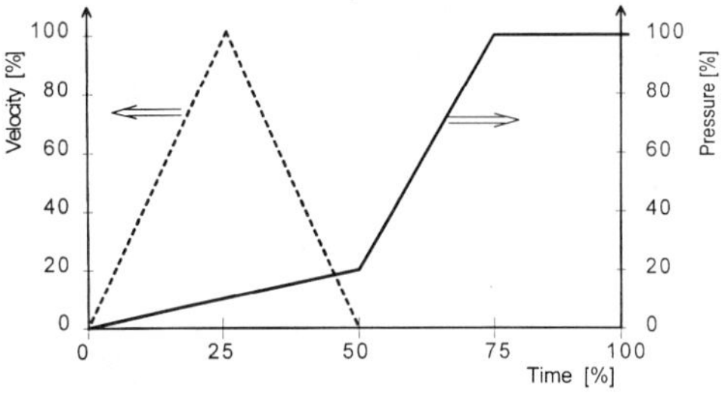

Fig. 7 Characteristics of the rod velocity and the pressure course

Complete contact is assumed between the rod and the preform and between the mold and the preform. The permissibility of this assumption has been shown by HARMS [13] for the extrusion blow molding process. With the much higher pressures prevailing in the stretch blow molding process and in view of the good surface quality of the components, full-area contact can also be assumed in the stretch blow molding process.

3 RESULTS

The calculated molding process is shown for a PP beaker, which is produced in the one-step process (Fig. 8). The rod first stretches the preform in the axial direction. Because of the higher temperatures in the neck area, the material is also softer in this region and is pulled out of this area. The initial pressure prevents constriction at this point and also prevents any sticking of the melt to the mandrel, but it is too low to produce any radial forming. Only when the axial stretching is more or less complete, is the molding pressure applied and the preform pressed completely against the mold wall. The influence of the temperature-dependent material behavior is shown clearly by the relatively uniform deformation. The considerable differences in the wall thickness distribution of the preform are compensated by the combined rod and pressure molding and by the defined distribution of the material behavior.

Due to the clamping at the thread of the preform in the mold and due to the contact between the rod and the base area, the preform cools quickly in these areas. In the simulation, this cooling is taken into account through the definition of stiffer material properties locally, resulting in higher local wall thicknesses.

4 CONCLUSIONS AND OUTLOOK

The stretch blow molding process can be simulated using the finite-element method (FEM) and the described assumptions and boundary conditions. The deformation profile in Fig. 8 shows that the material

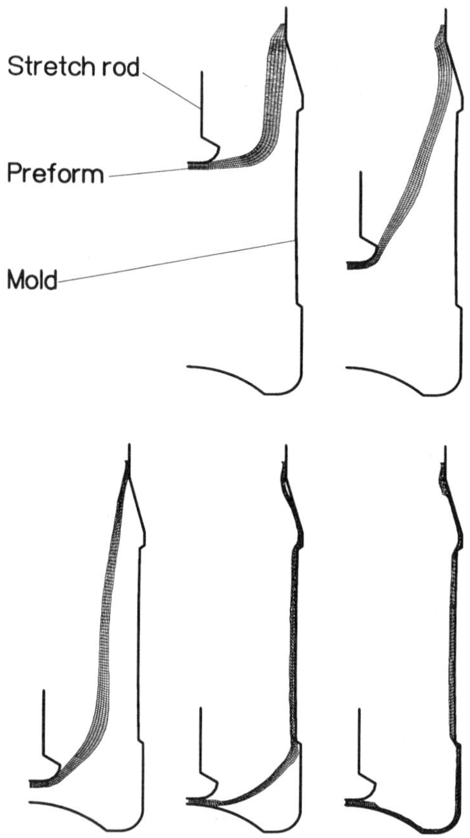

Fig. 8 Simulated molding sequences

properties in the preform can, in principle, be correctly described for various temperatures and thus for various areas of the preform.

For the thermal calculations of the one-step process, a further project takes into account the temperature distribution in the preform resulting from the injection molding process. By using suitable software (e.g. IKV-CADMOULD), both the heat of dissipation during molding and the cooling during the holding pressure phase and cooling period can be taken into account. Another important task to be solved in future will be the measurement of the material data. These data must be ascertained at process-relevant deformation rates and temperatures. A new rheometer of the bubble-inflation type, is currently being constructed at the IKV for this purpose.

REFERENCES

1. Esser, K.: Untersuchung zur Prozeßführung beim Streckblasen von PP, Ph.D. Thesis at the RWTH-Aachen, 1985
2. Haessly, W. P. and M.E. Ryan: Finite Element Modeling of the Injection Blow Molding Process, *Society of Plastics Engineers, Proceedings of the Annual Technical Conference (ANTEC)*, New York, USA, 1-4 May 1989, pp. 934-937
3. Garcia-Rejon, A. and A. Derdouri and R. Khayat and M. E. Ryan and W. P. Haessly: Effect of Initial Modulus Distribution on the Final Thickness of a Blown Preform, *Society of Plasics Engineers, Proceedings of the Annual Technical Conference (ANTEC)*, New York, USA, 5-9 May 1991, pp. 836-839
4. Chung, K.: Finite Element Simulation of PET Stretch/Blow-Molding Process, *Journal of Material Shaping Technology*, 7 (1989) 4, pp. 229-239
5. Schmidt, F. M. and J. F.Agassant and M. Bellet and G. Denis: Numerical Simulation of Polyester Stretch-Blow Molding Process, *Numerical Methods in Industrial Forming Processes*, Balkema, Rotterdam, 1992
6. DiRaddo, R. W. and A. Garcia Rejon and K. T. Nguyen: Finite Element Modeling of the Heat Transfer Dynamics for the Reheating Stage for Injection Blow Molding, *Society of Plasics Engineers, Proceedings of the Annual Technical Conference (ANTEC)*, New Orleans, USA, 9-13 May 1993, pp. 1890-1895
7. ABAQUS/Standard and -Explicit Theory Manual, Version 5.2, Hibbit, Karlsson & Sorensen Inc., Rhode Island, USA, 1992
8. Meissner, J.: Dehnungsverhalten von Polyaethylenschmelzen, *Rheologica Acta* 10 (1971), pp. 230-242
9. Haessly, W. P. and M. E. Ryan: Experimental Study and Finite Element Analysis of the Injection Blow Molding Process, *Polymer Engineering and Science*, 33 (1993) 19, pp. 1279-1287
10. Wollschläger, D. Spritzblasen, VDI-Verlag, Düsseldorf, 1976
11. Menges, G. and D. Weinand: Werkstoffkennwerte teilkristalliner Materialien fuer das Thermoformverfahren, DFG-Research Report Me-272/155, IKV, RWTH-Aachen,1987
12. Rebelo, N. and J.C. Nagtegaal and L.M. Taylor: Comparison of implicit and explicit finite element methods in the simulation of metal forming processes, *Numerical Methods in Industrial Forming Processes*, Balkema, Rotterdam, 1992
13. Harms, R.: Modellbildung zur Beschreibung der Formgebung beim Extrusionsblasformen - Eine praktische und numerische Analyse des Formbildungsprozesses, Ph. D. Thesis at the RWTH-Aachen, 1993

Simulation of Materials Processing: Theory, Methods and Applications, Shen & Dawson (eds)
© 1995 Balkema, Rotterdam. ISBN 90 5410 553 4

Computer-aided design of coathanger dies through 3-D finite element analysis

Chin-Ching Huang
National Kaohsiung Institute of Technology, Taiwan

Yeh Wang
Tunghai University, Taichung, Taiwan

ABSTRACT: A rational die design procedure is presented in this study for determining the optimum flow channel geometry of a coathanger die which consumes the minimum pressure drop, and produces a uniform transverse flow rate distribution. The results indicate that for a particular polymer melt and a particular aspect ratio of slit there may exist an optimum combintion of the manifold radius and the manifold angle from which the flow channel geometry of a coathanger die can be uniquely determined.

1 INTRODUCTION

Dies for flat film extrusion are one of the most important classes used in the industrial operartion. These dies usually consist of two basic components, namely the flow transition zone and the die land plus some die adjusting elements. The transition zone can be seen in most profile extrusion dies; and in flat dies this zone transforms a circular entrance into a slit exit. Various flow channel geometries of transition zone have been proposed in the literature for producing uniform films and sheets. The coathanger preform dies are the most popular among the current designs.

Proper design of the flow channel geometry for extrusion dies has long been pursued in industry. There are usually two primary objectives of die design. One is to yield the highest possible throughput at the lowest possible pressure drop, and another is to produce the extrudate within the preset dimensional tolerance. The former objective requires the variation of the cross-sectional area in a flow channel be smooth, and there be an appropriate die length which consumes the minimum pressure drop. The latter objective requires the contour of the cross section of the flow channel take a dumbbell-like shape in order to yield uniform distribution of the transverse flow rate for flat dies (Wang 1991).

Numerous studies based on the classical lubrication approximation analysis (Michaeli, 1984) have been published to define the geometry of the manifold. Though analytical solutions could be obtained, they can neither provide detailed information on the entrance geometry, nor predict the proper combination of the transition zone and the die land that would yield suitable die characteristics. In recent years there have been an increasing use of the finite element method in fluid dynamics problems because of its ease in dealing with the complex geometrical boundaries. The use of the finite element method certainly improves design accuracy and reduces the time spent in the trial-and-error process. A rational procedure is proposed here for optimizing a coathanger die. It is hoped that this study will provide some design aids on the basis of finite element simulation, and eventually lead to an efficient procedure for coathanger die designs.

2 FLOW OF POWER-LAW FLUIDS IN A COATHANGER DIE

We first analyze the coathanger

die with a manifold angle α of 17°. The three-dimensional parallel projection of the wire frame structure of a quarter of the coathanger die is shown in Fig. 1. Due to symmetry

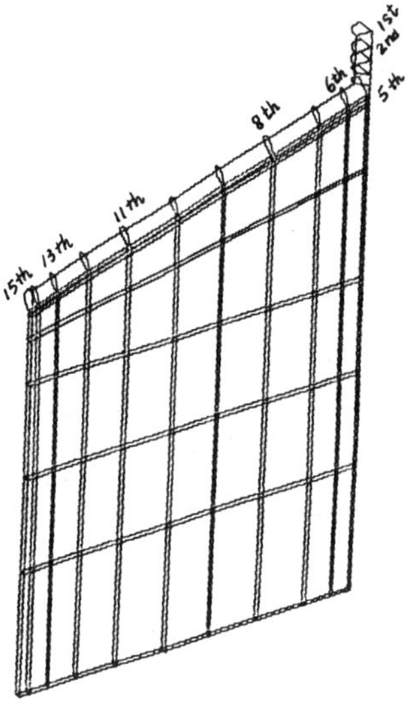

Fig.1 Parallel projection of a coathanger die with Rm^*=0.67 and α=17°

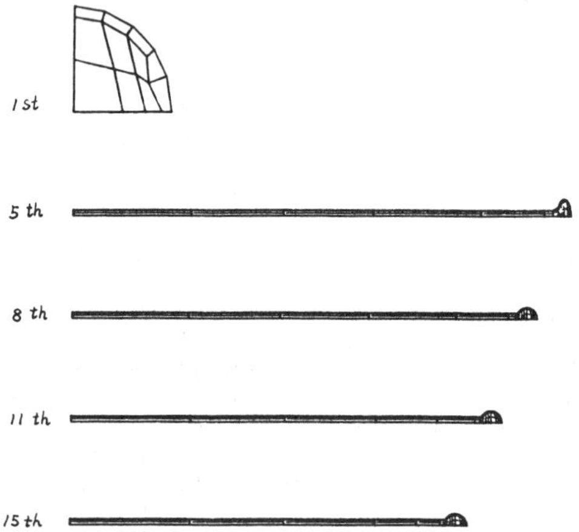

Fig.2 Side views of the cross sections in the transition zone.

of the die only one quarter of the flow domain is considered. There are 15 stations along the transverse direction with the first station beginning at the top which is the circular entrance of the die. There are 52 brick elements between

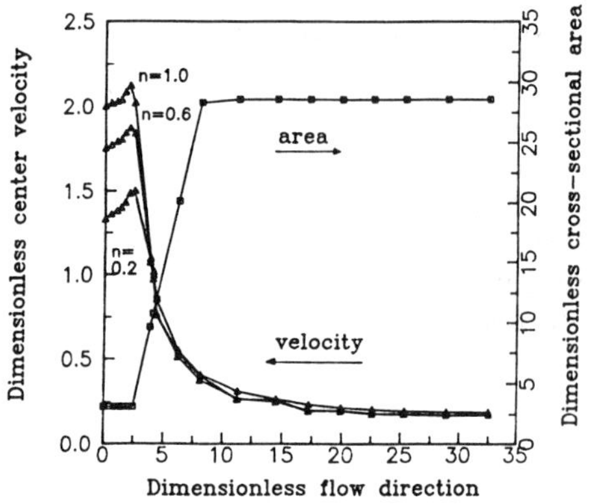

Fig. 3 Center velocity and cross-sectional area variation along the flow path.

each pair of these stations. Fig. 2 shows the side views of the stations in the transition zone. The transition zone begins at z^*=2.5, which is the entrance of the manifold, and ends at z^*=9.02, which is the beginning of the slit zone. The contour of these stations clearly take a dumbbell-like shape. The die manifold is a cylindrical channel with circular cross section of constant area, and the dimensionless radius R_m^* is 0.67. The dimensionless slit opening is 0.67, and the aspect ratio of slit is 64. The dimensionless die length is chosen as 32.5 for convenience. All dimensions are normalized with the entrance radius R_e which is 4.8 mm.

The dimensionless center velocity profiles for the power-law fluids with power-law exponents n=1, 0.6 and 0.2 are plotted in Fig. 3. Note that the dimensionless velocity is defined as $\underline{v}/v_{av}$, and v_{av} is the average velocity at the die entrance. Thus the dimensionless volumetric flow rate Q^* is π in all calculations. The variation of the cross-sectional area is also included. As expected, the magnitude of the center velocity decreases as n becomes smaller. There are three distinct regions. In the entrace region, starting from z^*=0 to 2.5, the cross section is circular, and the center velocity profiles are flat for all fluids. Next in the transition zone, starting from z^*=2.5 to 9.02, the cross-sectional area increases while the center velocity decreases. Finally in the die land region, starting from z^*=9.02 to 32.5, the cross section is a slit, and the

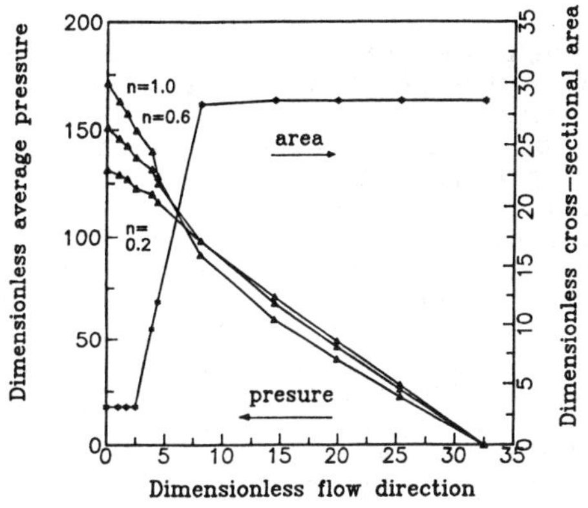

Fig. 4 Average pressure and cross-sectional area variation along the flow path.

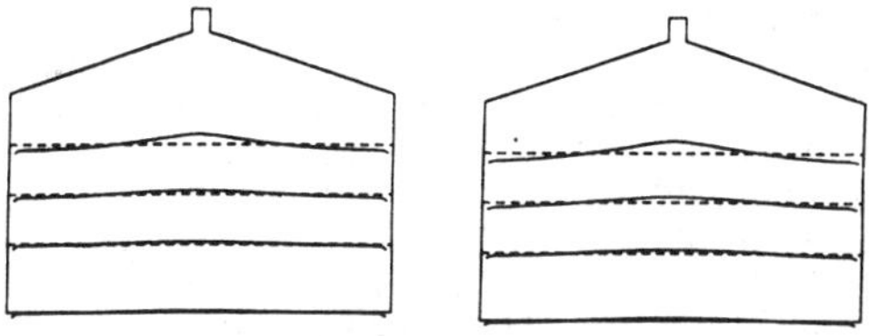

a. Newtonian fluid b. power-law fluid with n=0.6

Fig. 5 Transvers flow rate distribution inthe coathanger die

center velocity becomes flat again. The dimensionless average pressure for power-law fluids with n=1, 0.6, and 0.2 is plotted against the flow direction in Fig. 4. The dimensionless average pressure is calculated by

$$P^*_{av} = \int_{S^*} p^* dS^*/S^* \qquad (1)$$

Here S^* is the dimensionless cross-sectional area at each station, and

$$p^* = (p/K)(R_e/v_{av})^n \qquad (2)$$

where K is the consistency index in the power-law model. The average pressure decreases monotonically from the entrance to the exit for all cases. It can be seen that as the power-law index decreases, the pressure drop across the die decreases correspondingly.

We then calculate the transverse flow rate distribution at each station in the coathanger die. The dimensionless transverse flow rate q^* is calculated from

$$q^* = \int_0^{H^*} 2v^*_z dy^* \qquad (3)$$

where H^* is the dimensionless half-thickness; and y^* is the dimensionless thickness coordinate. The results for n=1 (Newtonian fluid) and 0.6 are shown in Fig. 5a and 5b respectively. The solid lines represent the distribution of the transverse flow rate, and the dashed lines represent the dimensionless average transverse flow rate of $Q^*/2W^*$, where W^* is the half die width. As can be seen from this two plots, the transverse flow rate distribution is very uneven in the transition zone. However, as the melt flows down the die land, the flow of polymer melt becomes fully developed, and the distribution becomes flat again for both fluids.

3 EFFECT OF MANIFOLD ANGLE α

In order to make quantitative comparisons, we also calculated the arithmetic mean deviation index, MDI, defined in the following equation

$$MDI = \frac{\sum^{l} |q_i/(Q/2b) - 1|}{l} \qquad (4)$$

Here q_i represents the transverse flow rate at the ith node; and l is the total number of nodes used for claculation. Usually twenty-five nodes are used at each station. When there is no difference between q and Q/2W, the MDI vanishes, i.e., the distribution of the transverse flow rate is perfectly uniform; and the distribution becomes worse as the magnitude of MDI increases. Note that in real dies the MDI can never vanish because of the no-slip boundary condition such that there is no flow at the die wall. The calculated MDI's along the flow path are presented in Fig. 6 for $\alpha=0^o$, 17°, and 30°. It can be seen that in all dies the MDI's decrease as the melt flows towards the exit. However, at the exit of the transition zone for $\alpha=0^o$ ($z^*=2$), the MDI is the worst, because the transition zone is the shortest; and as expected, the MDI at the exit of the transition zone for $\alpha=30^o$ is the lowest, because the transition zone is the longest. Another interesting observation is that the MDI for all cases level off to the same value. We may define this point as the end of die land; accordingly the overall die length required to reach uniform transverse flow rate distribution could be determined. The average pressure profiles for

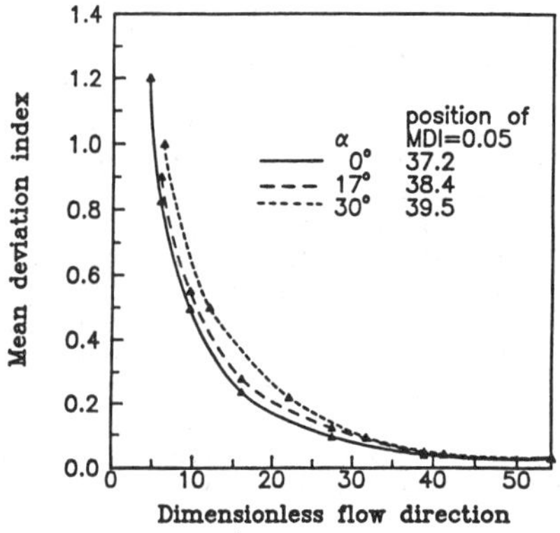

Fig.6 Mean deviation index along the flow path for a power law fluid with n=0.6, α=0°,17°,30°

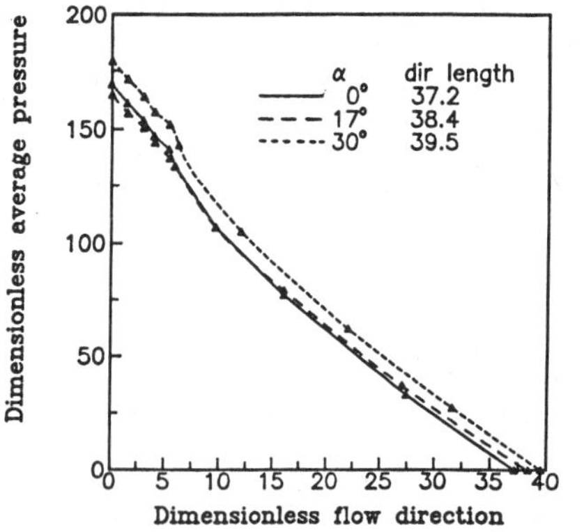

Fig. 7 Average pressure profile along the flow path for a power-law fluid with n=0.6, α=0°,17°,30°

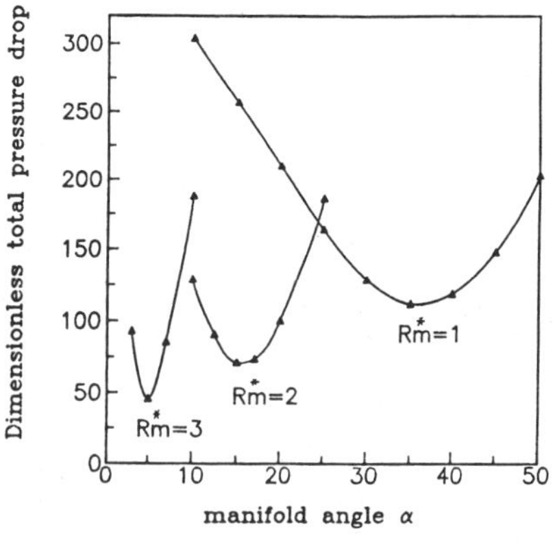

Fig. 8 Dimensionless overall pressure drop vs. manifold angle α of various manifold radius of coathanger dies foe a power-law fluid with n=0.6

all cases are shown in Figure 7. Though the die length is the shortest for $\alpha=0^{o}$, the lowest overall pressure drop occurs at $\alpha=17^{o}$ as shown in this figure. It is because the pressure drop through the transition zone is higher for large manifold angle; on the other hand, the die land becomes longer for small manifold angle. According to the two die design objectives, namely the uniform distribution and the minimum pressure drop across the die, we think the optimum manifold angle is close to $\alpha=17^{o}$ from the results of the computer trials.

4 OPTIMUM DIE LENGTH

Now we are in a position to discuss how to determine the optimum die length. Following the previous discussions, one should be able to find an optimum manifold angle corresponding to a certain manifold radius. Next we want to ask if it is possible to determine the proper transition length, which satisfies the aforementioned design objectives, for a particular material and a particular aspect ratio.

The overall pressure drops across the coathanger dies at various manifold radii are plotted in Figure 8 against the manifold angle. The minimum pressure drop for $R_m^*=1$ occurs at $\alpha=35^{o}$; and for $R_m^*=2$ and 3 the minimum pressure drop occurs at $\alpha=15^{o}$ and 5° respectively. It can be seen that the pressure drop at $R_m^*=3$ and $\alpha=5^{o}$ is the lowest among all cases. From the results of the computer trials, we think the combination of the dimensionless manifold radius of 3 and the manifold angle of 5° constitutes the optimum design of the coathanger dies which satisfy the design objectives. Though the trial-and-error procedure is tedious, it will not take too much time for an experienced user to achieve the optimum die design for a particular polymer melt.

5 CONCLUDING REMARKS

From the previous analysis, the overall pressure drop may be written in a functional form:

$$p^* = p^*(n, R_m^*, \alpha, H^*, W^*) \qquad (5)$$

Here we fix the material properties n, and the slit aspect ratio H/W, and the remaining unknowns are R_m and α. However, it is a formidable task to find an explicit form of this function for a die of irregular geometric boundary. Therefore, a trial-and-error procedure is proposed in this study in order to determine the optimum combination of R_m and α which minimize the implicit function in Eqn (5). The steps involving in the minimization process are summarized as follows:

1. Determine the material properties, and specify the slit aspect ratio.
2. Guess a manifold radius R_m, and calculate the pressure drop for various manifold angle α. Determine α which yields the minimum pressure drop corresponding to the guessed value of R_m^*.
3. Try other manifold radii, and determine the optimum combination of R_m and α which consumes the lowest pressure drop among all choices.

A rational fishtail die design procedure is presented in this study. The method is particularly suited for computer implementation and may encompass the study of different die geometries. It also has the advantage of easily accommodating eventual future developments in the analysis, because each design step is dealt with independently. As the computer becomes widely used for die design, the numerical cacualtion will be of increasing importance. The present effort represents one step towards this end.

ACKNOWLEDGEMENT

Partial support from the National Science Council of Taiwan under grant NSC-82-0405-E-029-013 is gratefully acknowledged.

REFERENCES

Wang, Yeh and H.-Y. Tsay 1994. Computer-Aided Geometric Design of the Preform Dies for Flat Film Extrusion through 3-D Finite Element Analysis, *Polym. Eng. Sci.*, 31:1090-1100 .

Michaeli, W. 1984. *Extrusion Dies*, Munich:Hanser Publishers.

Simulation of Materials Processing: Theory, Methods and Applications, Shen & Dawson (eds)
© 1995 Balkema, Rotterdam. ISBN 90 5410 553 4

Closed-loop optimization and integrated numerical analysis of the blow moulding process

Denis Laroche, Robert W. DiRaddo & Linda Pecora
Industrial Materials Institute, Boucherville, Que., Canada

ABSTRACT: Several applications of the blow moulding process require a high level of final part performance that necessitates stringent process control limits. Process simulation is a valuable tool for obtaining this final part performance, by predicting the final part quality for given moulding conditions. One of the limitations of such an approach is the need for simulating the entire process sequence (parison formation, clamping, inflation and solidification) in order to account for the contribution of the individual process phases. In addition to the difficulties in performing the integrated process simulation, the ability to predict the optimal process conditions is a formidable problem. A methodology that performs an inverse modelling prediction is proposed in this work. The methodology, by employing a closed loop optimization procedure, predicts the necessary moulding conditions inputs for a desired final part output. The approach includes an integrated clamping and inflation simulation.

1 INTRODUCTION

The blow molding process has seen a rapid growth due to the development of new markets in the automotive, sports & leisure, transportation and packaging industries. This success is due to the development of novel processing techniques, such as multilayer processing and 3 XY blow moulding, as well as the use of new materials. These new higher-end applications have increased performance requirements, such as resistance to typical stress applications under service conditions. The most critical parameter for optimization of the process is the final part thickness distribution since it will, to a large extent, determine the structural characteristics of the final part. The part must also be designed to ensure minimum resin usage and minimum cooling times.

The part thickness distribution is dependent on the four stages of the process, namely parison formation, parison clamping, parison inflation and part solidification. An adequate optimization, for obtaining a high quality part, should ultimately involve an optimization of the entire process. Parison formation involves the extrusion of the polymer melt through an annular die. The extrusion is a highly complex phenomenon that it is affected by swell, sag and parison programming. The parison is then clamped and inflated to take the shape of the enclosing mould. Part solidification includes the density change and crystallinity formation within the part.

In order to integrate the entire process, the four individual stages must be considered consecutively. A closed loop analysis of the four process phases, ending with a stress analysis of the final part under service conditions, is proposed in the long term. Once the stress analysis is performed, the results are compared to a specified result, and the difference is fed back in the loop to allow for correction of the initial parison formation. However, for the short term, only the relatively simple case of parison clamping and inflation is considered here. The required parison dimensions for a given final part thickness are predicted.

Such an automatic closed loop optimization approach allows for minimum process set-up times and tooling (die/mould) changes (DiRaddo 1993). The approach can be applied to any situation where one needs to predict the input process conditions for a specified output result.

2 INTEGRATED NUMERICAL ANALYSIS

Once the parison is extruded to the desired length, two mould halves close onto the parison, clamping it into a deformed shape. The deformed or clamped parison is then inflated to take the shape of the surrounding mould.

The integrated analysis performed in this work, combines the clamping and inflation simulations, that predicts the final part thickness distribution for a given parison thickness distribution. Intermediate parison shapes during the clamping and the inflation, part blowability and minimum pressure requirements can also be obtained from the simulation. The incorporation of the clamping simulation, facilitates the subsequent use of inflation simulation for its applicability to complex parts, including parts with handles and parts with multiple parting lines.

The deformation of the parison during the clamping and inflation stages is generally modelled with a hyperelastic solid behaviour, due primarily to the short deformation times involved (DeLorenzi 1991, Bellet 1992, Laroche 1995).

The parison is assumed to be thin enough to be modelled with membrane elements. This implies that the clamping stage, where only a small pressure is applied, becomes difficult to converge due to the small out-of-plane membrane stiffness. The choice of an appropriate initial stress and minimum internal pressure during this phase improves the convergence of the algorithm.

3 CLOSED-LOOP OPTIMIZATION

A closed-loop optimization approach is analogous to classical process control systems that employ a feedback configuration. The approach involves modifying a manipulated process input, A_i, in order to move a given process output, B_i, closer to its desirable target value, B_t, with the output error, E_i, being defined as follow:

$$E_i = B_i - B_t \tag{1}$$

The process input is modified iteratively until the error function is below a pre-specified limit (close to zero). The value of the process input, at this iteration, is the desired solution.

The feedback control system calculates a new process input, A_{i+1}, by employing a user defined function of the output error. For proportional closed loop optimization, the updated process input is defined as

$$A_{i+1} = A_i - \alpha\, E_i \tag{2}$$

where α is the user defined proportional gain.

A proportional-integral closed loop optimization system (PI) takes advantage of the history of previous iterations to improve the solution convergence by minimizing the solution error offsets from zero, present with the proportional closed loop optimization. The updated process input is therefore

$$A_{i+1} = A_i - \alpha\, E_i - 0.5\beta\, (\, E_i + E_{i-1}\,) \tag{3}$$

where β is the user defined integral term.

A proportional-integral-derivative closed-loop optimization system (PID) also takes advantage of the history of previous iterations to improve the solution convergence by dampening oscillations obtained with PI systems. The updated process input is therefore

$$A_{i+1} = A_i - \alpha E_i - 0.5\beta(E_i+E_{i-1}) - \gamma(E_i-E_{i-1}) \tag{4}$$

where γ is the user defined derivative term.

The three user defined parameters, (α, β and γ) are set so as to assure convergence, optimize the solution convergence time and minimize the error offset. The reader is referred to References (Stephanopoulos 1984, DiRaddo 1993) for further details on the approach.

The preliminary closed-loop optimization approach described here considers only one process target. However, when several targets are considered simultaneously, it becomes necessary to employ a weighted approach on the individual targets. This is also planned for in subsequent work.

The performance of a closed loop control optimization system is also dependent on the ability to estimate the transfer function that relates the process input, A and the process output B. In this specific case of a blow moulding clamping/inflation simulation the process output is defined at each integration volume of the whole inflated part. Local values for the tuning parameters, α, β and γ, may also be defined to assist in the solution.

In addition, there are cases when there are not as many maniputable process variables as there are output process variables. In such cases, the updated process inputs are obtained by an averaging over the parison volume. These controllable process parameters, $C_{i+1}{}^{k}$, are obtained using the following relation:

$$C_{i+1}{}^{k} = \Sigma\, (\, A_{i+1}{}^{j}\, w^{jk}\,) \tag{5}$$

where A_{i+1}^j are the local updated input at position j and w^{jk} are their respective weights.

For the clamping/inflation simulation looked at here, where the process input is the initial parison thickness distribution and the process output is the final part thickness distribution, a proportional closed loop optimization is proposed as a first step. Since the clamping/inflation phenomenon is highly non-linear, the estimated transter function is determined by a local blow-up ratio distribution throughout the part. Consequently, in order to minimize convergence time, the proportional gain varies throughout the geometry, and is defined as a function of the individual local blow-up ratios.

Therefore, Equation (2) is modified to yield

$$A_{i+1}^j = A_i^j - \alpha_u \, \alpha_m^j \, E_i^j \qquad (6)$$

where, α_u, is the user defined proportional gain and α_m^j , the gain defined by the inflation model, is obtained from Equation (7), for each integration point, j.

$$\alpha_m^j = A_i^j / B_i^j \qquad (7)$$

where A_i^j and B_i^j are the initial membrane thickness and the inflated membrane thickness at integration point j.

For the clamping/inflation stage of the process, the number of maniputable process inputs is dependent on the ability to locally control the parison geometry. If the parison formation can adequately be manipulated to have a specified local thickness distribution, Equation 5 can be then simplified to yield

$$C_{i+1}^k = A_{i+1}^j \qquad (8)$$
$$k=j$$

An example of this application is a programmed parison, where the thickness can only be varied in the axial direction. In such a case, C_{i+1}^k corresponds to the parison thickness profile.

4 NUMERICAL EXAMPLES

In this work, two different cases are evaluated, by employing the clamping/inflation simulation, in conjunction with the closed loop optimization approach. The target is a pre-defined final part thickess distribution and the updated process input being considered is the parison thickness distribution. At this point in the work, a proportional control system is employed, with a

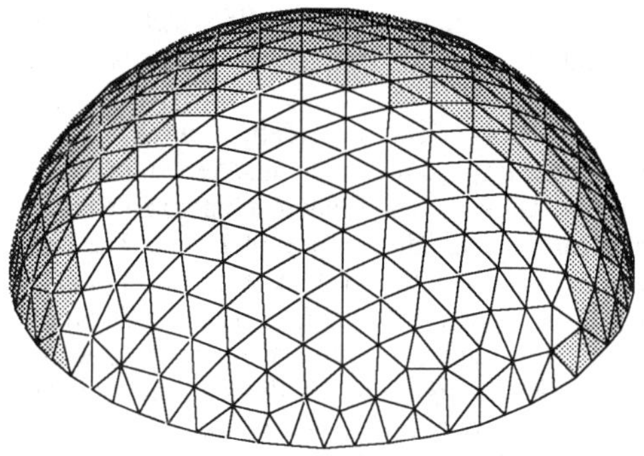

Figure 1: Final part after inflation into an hemispheric mould.

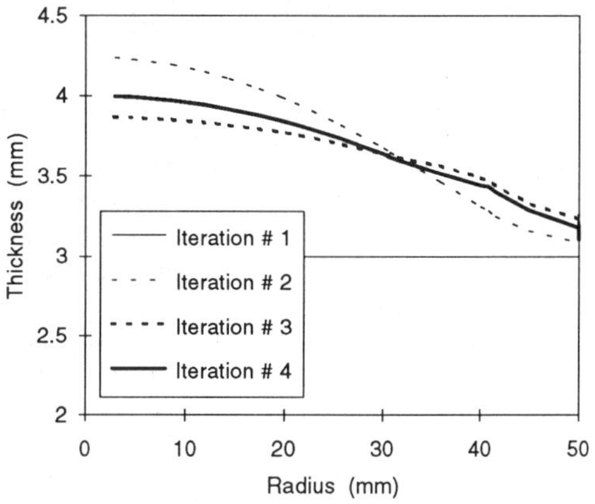

Figure 2: Membrane initial thickness distribution: Case 1.

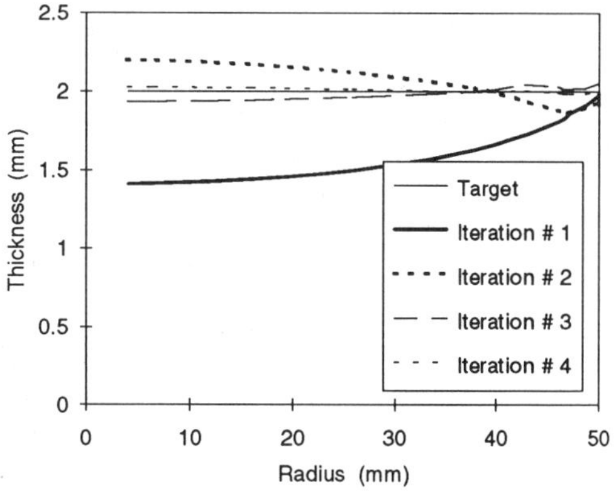

Figure 3: Membrane thickness distributions for subsequent iterations: Case 1.

proportional gain defined at each element integration point, as described by Equations 6 and 7.

The first case considered is a circular membrane inflated into an hemispheric mould. The solution target is an inflated membrane having a uniform thickness distribution of 2 mm. The initial membrane has a uniform thickness distribution of 3 mm. The inflated membrane is shown in Figure 1.

In this analysis, the membrane thickness distribution is updated locally and the thickness correction is optained by using Equation 8. The closed-loop optimization system was employed iteratively with a gain α_m^j defined by the local blow-up ratios as given in Equation 7 and a user defined proportional gain α_u set to 1. The results for the initial thickness distribution of the membrane and the part are shown in Figures 2 and 3, respectively.

The results demonstrate the rapid convergence of the closed-loop optimization system. In addition, the individual iteration solutions tend to oscillate to the final solution. Therefore, the inclusion of a derivative term or a decreased proportional term would be useful so as to minimize the oscillations.

The second case considered is a cylindrical parison successively clamped by two mould halves and inflated to a final container. This case will be validated experimentally in the near future, at our facilities, and the results will be presented at the podium presentation. A mould half is shown in Figure 4 while the completely inflated part is shown in Figure 5.

The parison diameter is assumed constant (value of 56mm) while the thickness can be variable along the axial direction. The initial parison thickness distribution is set to a uniform value of 11 mm as an initial guess. This is a typical value obtained on a PLACO 3XY coextruder blow moulding machine available at IMI.

For this case, the parison thickness profile along the axial direction is updated using Equation 5 where the local correction terms A_{i+1}^j are summed with a weight w^{jk} that is linearly dependant on the axial position from the controllable points C_{i+1}^k. In this case, due to the geometric nature of the problem, the non-linearity of the inflation is much more important than in the case of an inflated membrane. The introduction of a non-uniform parison thickness profile can also cause difficulties in the estimation of the transfer function.

As a first approximation, the closed-loop optimization system was employed iteratively with

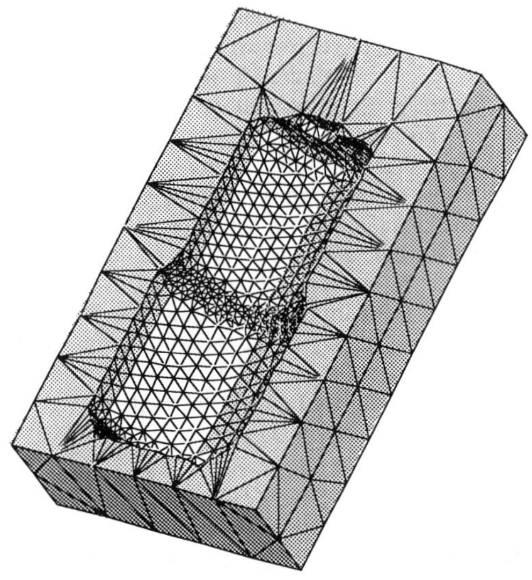

Figure 4: Half mould employed for clamping and inflation simulation.

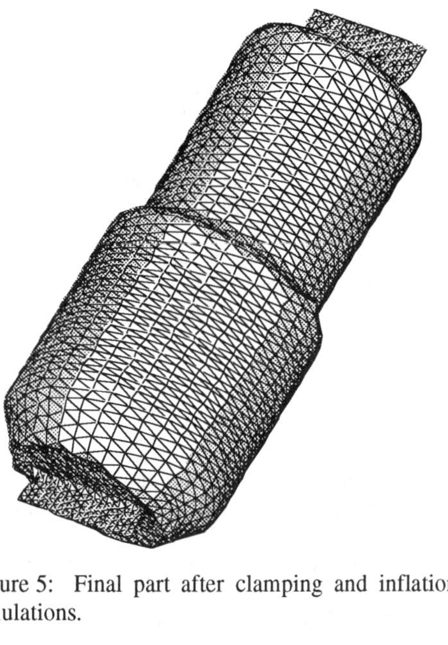

Figure 5: Final part after clamping and inflation simulations.

a gain α_m^j, given by Equation 7, and a user defined proportional gain α_u set to 1. A lower level constraint of a parison thickness of 8 mm is pre-defined for the axial extremities. Preliminary results for the parison thickness profile and the inflated part thickness are shown in Figures 6 and 7, respectively.

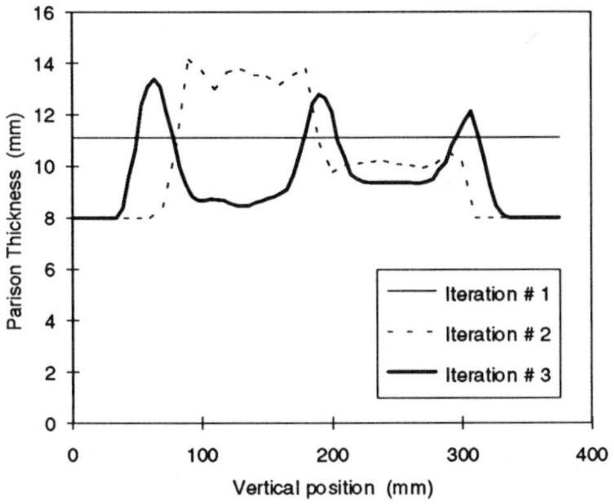

Figure 6: Parison thickness distribution: Case 2

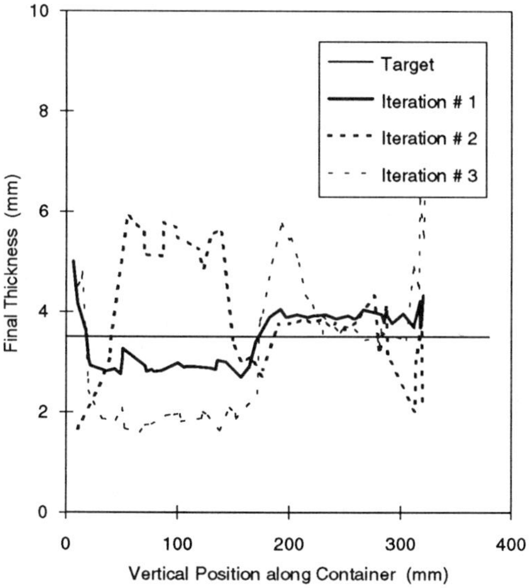

Figure 7: Final part thickness distribution: Case 2.

The results indicate that the algorithm is diverging. The clamping and inflation simulations are highly sensitive to non-uniformities in the parison thickness distribution. Small parison thickness variations are amplified in the final part thickness distribution. A smaller value for the user defined proportional gain α_u and a derivative term, will be used in future trials, to assure a better convergence.

5 CONCLUSION

A methodology that performs an inverse modelling prediction was proposed in this work, with the aim of optimizing the process parameters in blow moulding simulations. The technique, in the short term, employs a proportional closed loop optimization. However, modified PI and PID closed loop optimization approaches are proposed in the future. Future work also entails tuning of the user defined optimization parameters, as well as the inclusion of the remaining process sequences.

Two cases were evaluated in this study, in an effort to demonstrate the validity of the approach for an actual process case of blow moulding. The results were promising for the case of an inflated membrane into a simple geometry. However preliminary tests with a clamped and inflated parison have shown poor convergence and will require further investigations in the selection of appropriate proportional gain values.

REFERENCES

Bellet,M., Corsini,C., Assaker,D., Mercier,P. and Wouters,P., SPE ANTEC Tech. Papers, 38, 114 (1992).

DeLorenzi,H.G. and Nied,H.F., Modeling of Polymer Processing, (Ed. A.I. Isayev) Carl Hanser Verlag, Munich (1991).

DiRaddo,R.W. and Garcia-Rejon,A., Computers Chem. Engng., 17, 8 (1993).

Laroche,D., Pecora,L. and DiRaddo,R.W., Submitted to ANTEC '95.

Stephanopoulos,G., Chemical Process Control, Prentice-Hall, Englewood Cliffs, NJ (1984).

Simulation of Materials Processing: Theory, Methods and Applications, Shen & Dawson (eds)
© 1995 Balkema, Rotterdam. ISBN 90 5410 553 4

Finite element simulation of viscoelastic flow through a converging channel

G. Rekers, R. Akkerman & J. Huétink
University of Twente, Department of Mechanical Engineering, Enschede, Netherlands

ABSTRACT: A number of well-known nonlinear viscoelastic constitutive models of the differential type are fitted within the theory of irreversible thermodynamics. Two of these models, the Leonov model and the Giesekus model, are used to describe the flow behaviour of an LDPE and an HDPE in a converging slit channel. These constitutive models have been implemented in a transient finite element code using an Arbitrary Lagrange-Euler method, being particularly suitable for the description of history dependent processes involving free surface movements. Some occurring numerical problems are shortly discussed, such as an objectivity preserving numerical integration scheme. Finally, the calculated flow induced stresses are compared to the stress fields measured using flow induced birefringence.

1 INTRODUCTION

Various differential constitutive equations are formulated to describe the nonlinear viscoelastic behaviour of an initially isotropic polymer melt. All models have the same linear visco-elastic limit, but exhibit differences in the nonlinear viscoelastic region. This nonlinear behaviour plays an important role in polymer processing. For instance, the flow induced residual stresses in injection moulded products, caused by nonlinear viscoelastic flow during the injection, have a large influence on for example the mechanical, thermal and optical pro-perties of the finished product (Baaijens et al. 1991). The nonlinear behaviour is also important in many other processes, as e.g. film blowing (Tas 1994) and extrusion (Goublomme et al. 1992).

2 CONSTITUTIVE EQUATIONS

To describe the instationary behaviour of a highly viscous fluid (low Reynolds numbers) use is made of the mechanical equilibrium, in absence of body forces stating

$$\boldsymbol{\sigma}\cdot\overleftarrow{\nabla} = \boldsymbol{0}. \tag{1}$$

For the description of polymer melts and solutions often the generalized Jeffreys model is used, in which a number of Maxwell modes is connected in parallel with a Newtonian viscosity. By using a set of Maxwell elements a set of relaxation times can be described. Usually, the viscous damper is only temperature dependent and does not depend on the deformation. In the remaining part of this section firstly the kinematics and the basic model assumptions of a single Maxwell mode are discussed. Secondly, a number of well-known constitutive models describing incompressible nonlinear viscoelastic behaviour is fitted into this formulation.

To incorporate the nonlinear viscoelastic behaviour in the continuum mechanics approach, for each Maxwell mode, besides the reference and current configuration, the relaxed configuration (Leonov 1976) or natural reference state (Besseling 1968) is introduced to make it possible to determine the reversible part of the deformation of each Maxwell mode. Using this assumption the deformation gradient $\boldsymbol{F}$ can multiplicatively be decomposed into a reversible (elastic) part $\boldsymbol{F}_e$ and an irreversible (plastic) part $\boldsymbol{F}_p$: $\boldsymbol{F}=\boldsymbol{F}_e\cdot\boldsymbol{F}_p$. Thereupon an objective elastic strain measure can be defined in the shape of the elastic Finger tensor $\boldsymbol{B}_e=\boldsymbol{F}_e\cdot\boldsymbol{F}_e^T$. The evolution of the elastic Finger strain is found to be

$$\dot{\boldsymbol{B}}_e = \boldsymbol{L}_e \cdot \boldsymbol{B}_e + \boldsymbol{B}_e \cdot \boldsymbol{L}_e^T = (\boldsymbol{L}-\boldsymbol{D}_p) \cdot \boldsymbol{B}_e + \boldsymbol{B}_e \cdot (\boldsymbol{L}^T - \boldsymbol{D}_p), \tag{2}$$

in which $\boldsymbol{L}$ is the gradient of velocity tensor, $\boldsymbol{L}_e$ is the elastic part of the gradient of velocity tensor and $\boldsymbol{D}_p$ is the plastic part of the rate of deformation tensor.

In order to model compressible nonlinear viscoelastic material behaviour we want the isotropic deformations to be elastic and only the deviatoric deformations to be viscoelastic. When $tr\boldsymbol{D}_p=0$ the evolution equation for $\boldsymbol{B}_e$ also describes the law of mass conservation, but when $tr\boldsymbol{D}_p \neq 0$ the density variations are not described by the elastic Finger strain $\boldsymbol{B}_e$ anymore. Therefore, the strain split of the elastic Finger strain proposed by Huétink (1986) is introduced

$$\boldsymbol{B}_e^r = \boldsymbol{B}_e - J_e^{2/3} \boldsymbol{I}, \tag{3}$$

in which $J_e = det(\boldsymbol{F}_e)$. In case of purely isotropic deformations $\boldsymbol{B}_e^r = \boldsymbol{0}$. At this point the basic assumption of the theory of Leonov (1976) is introduced, which states that in polymeric liquids there always exists an 'elastic limit', a quasi-equilibrium situation is achieved on very rapid (instantaneous) deformations. Therefore, we suppose the thermodynamic state variables to be $\boldsymbol{B}_e^r$, $J=det(\boldsymbol{F})$ and T. Furthermore, a purely isotropic part is splitted from the free energy:

$$\psi(\boldsymbol{B}_e, J, T) \equiv \psi_d(\boldsymbol{B}_e^r, T) + \psi_v(J, T). \tag{4}$$

In case of rubber elasticity, the free energy function describing non-isotropic deformations is

$$\psi_d(\boldsymbol{B}_e^r, T) = \frac{RT}{M_e} I_{\boldsymbol{B}_e^r} = \frac{\mu(\rho, T)}{2\rho}(I_{\boldsymbol{B}_e} - 3J_e^{2/3}), \tag{5}$$

in which

$$I_{\boldsymbol{B}_e} = tr(\boldsymbol{B}_e)$$

The shear modulus μ is a function of the temperature T and the density ρ:

$$\mu(\rho, T) = \frac{\rho T}{\rho(T_{ref}) T_{ref}} \mu_{ref} = c_T \mu_{ref}. \tag{6}$$

Using the quasi-equilibrium approach the constitutive equation can be worked out to

$$\boldsymbol{\sigma}(\boldsymbol{B}_e^r, J, T) = c_T \mu_{ref} \boldsymbol{B}_e^r - p(J, T)\boldsymbol{I}, \tag{7}$$

in which $p(J,T)$ can be specified without knowing the free energy function

$$\dot{p} = -C_b \, tr(\boldsymbol{D}) + \alpha C_b \dot{T}, \tag{8}$$

in which α is the coefficient of thermal expansion and C_b is the bulk modulus. These parameters can be taken constant, but they can also be assumed to be a function of the temperature and the determinant of the deformation tensor J. The isotropic deformation is thus modelled as being time independent.

To complete the description the plastic rate of deformation has to be specified. By using different definitions for the plastic deformation rate a number of well-known constitutive equations can be found (Leonov 1992):

Upper Convected Maxwell (UCM) model

$$\boldsymbol{D}_p = -\frac{1}{2\theta}(\boldsymbol{B}_e^{-1} - \boldsymbol{I}). \tag{9}$$

Leonov model

$$\boldsymbol{D}_p = \frac{1}{4\theta}(\boldsymbol{B}_e - \boldsymbol{B}_e^{-1})^d. \tag{10}$$

Giesekus model

$$\boldsymbol{D}_p = \frac{1}{2\theta}\Big(\alpha(\boldsymbol{B}_e - \boldsymbol{I}) - (1-\alpha)(\boldsymbol{B}_e^{-1} - \boldsymbol{I})\Big). \tag{11}$$

In this equation is θ the relaxation time and the parameter α in the Giesekus model is a numerical fitting parameter introduced to get a better description of the nonlinear material behaviour. Note that it is also possible to incorporate the upper convected version of the Phan-Thien Tanner (PTT) model in our formulation. The PTT model including the Gordon-Showalter convected time derivative cannot be included, because it is totally non-equilibrium.

In this paper only isothermal polymer flow is considered, but non-isothermal flow can simply be incorporated in the description above, so the free energy is specified for the constitutive models used.

To obtain a good description of the rheological behaviour of the LDPE and HDPE melts, it is necessary to introduce a multi-mode approach. The nonlinear relaxation properties can be described by a set of mutually non-intersecting relaxation modes,

$$\boldsymbol{\sigma} = -p\boldsymbol{1} + \sum_{i=1}^{n} \boldsymbol{\sigma}_{M,i} , \tag{12}$$

in which $\sigma_{M,i}$ represents the thermodynamic stress of each Maxwell mode. When also additional viscous friction is incorporated, the generalized Jeffreys model is obtained.

3 FINITE ELEMENT FORMULATION

The constitutive models described are implemented in a transient Galerkin finite element code (updated Lagrange formulation), in which the weak form of the balance of momentum, equation (1), is approximated. By extending the updated Lagrange formulation with an Euler step, the resulting Arbitrary Lagrange Euler (ALE) formulation (Huétink 1992) can be applied to fluid flow problems as well as solid mechanics. Thus, the algorithm is based on a split Lagrangian and Eulerian step.

Every time step the evolution of the elastic Finger strain (equation (2)) has to integrated. If any integration scheme is directly applied to this equation, incremental objectivity is not necessarily preserved. To obtain incremental objectivity the recipe of Baaijens (1993) is applied. Firstly, an invariant tensor $\bar{\boldsymbol{B}}_e$ corresponding with the objective tensor $\boldsymbol{B}_e$ is defined

$$\bar{\boldsymbol{B}}_e = \boldsymbol{F}^{-1} \cdot \boldsymbol{B}_e \cdot \boldsymbol{F}^{-T} . \tag{13}$$

Secondly, the evolution equation (2) is rewritten by using the upper convected time derivative and by recognizing that for the constitutive models used $\boldsymbol{D}_p$ is a linear function of $\boldsymbol{B}_e$ and $\boldsymbol{B}_e^{-1}$:

$$\overset{\triangledown}{\boldsymbol{B}}_e = -2\boldsymbol{D}_p \cdot \boldsymbol{B}_e . \tag{14}$$

Using the invariant tensor $\bar{\boldsymbol{B}}_e$, the evolution equation can be elaborated to

$$\dot{\bar{\boldsymbol{B}}}_e = \boldsymbol{F}^{-1} \cdot (-2\boldsymbol{D}_p \cdot \boldsymbol{B}_e) \cdot \boldsymbol{F}^{-T} . \tag{15}$$

Integration of this form using a generalized midpoint rule yields

$$\frac{\bar{\boldsymbol{B}}_{e_{n+1}} - \bar{\boldsymbol{B}}_{e_n}}{\Delta t} = \boldsymbol{F}_{n+\alpha}^{-1} \cdot (-2\boldsymbol{D}_p \cdot \boldsymbol{B}_e)_{n+\alpha} \cdot \boldsymbol{F}_{n+\alpha}^{-T} . \tag{16}$$

Noting that $\dot{\boldsymbol{F}} = \boldsymbol{L} \cdot \boldsymbol{F}$ and assuming that $\boldsymbol{F}_n = \boldsymbol{1}$ (updated Lagrange formulation), it is proposed to calculate the deformation tensor as follows, assuming that the gradient of velocity is constant during a time step:

$$\boldsymbol{F}_{n+\alpha} = e^{\alpha \Delta t \boldsymbol{L}} = \sum_{k=0}^{\infty} \frac{(\alpha \Delta t \boldsymbol{L})^k}{k!} . \tag{17}$$

When $\Delta t \boldsymbol{L}$ is small, the series is converging and the deformation gradient can be calculated up to machine accuracy. Moreover, for the choice of $\alpha = 0.5$ in equation (16) the algorithm is second-order accurate, but it has to be solved iteratively. It is also possible to combine equation (16) with a fourth-order accurate Runge Kutta method.

The convection of the elastic Finger strain $\boldsymbol{B}_e$ during each time step, the Eulerian part of the algorithm, is considered independent of the integration scheme. For the simulations presented in this paper an upwind scheme is used (Akkerman 1993), but it is also possible to incorporate a more accurate convection scheme, such as the Law-Wendroff scheme or a TVD scheme using flux limiters. However, this part of the algorithm is still being investigated.

At the end of each time step the mechanical equilibrium has to be obtained. Therefore, a Newton-Raphson method is applied to solve the weak form of equation (1) (Akkerman 1993). In a finite element context it can also be said that this iterative procedure must find a zero for the vector of nodal unbalance forces.

4 FLOW IN A CONVERGING SLIT CHANNEL

For all models the relaxation time spectrum is fitted on dynamical mechanical experiments (DMS) in the linear viscoelastic region. This spectrum is independent of the model used, because in this region they predict identical behaviour. For both LDPE (Tas 1994) and HDPE

Table 1: Discrete relaxation time spectra for HDPE at 160°C and LDPE at 190°C.

HDPE		LDPE	
θ_i(s)	μ_i(Pa)	θ_i(s)	μ_i(Pa)
1.07e-3	2.18e5	7.70e-5	2.72e5
1.13e-2	1.64e4	7.05e-4	1.05e5
7.42e-2	2.59e3	5.13e-3	6.02e4
4.71e-1	2.56e2	3.59e-2	3.16e4
2.79e0	1.96e2	2.42e-1	1.37e4
9.13e0	2.87e0	1.58e0	4.52e3
		1.01e1	1.01e3
		7.20e1	1.46e2

(DSM Research, The Netherlands) this spectrum is given in Table 1.
The non-linearity parameter in the Giesekus model was obtained by fitting the results of stress build-up and relaxation experiments (LDPE) and steady shear data (HDPE). For HDPE $\alpha=0.6$ and for LDPE $\alpha=0.35$.

At DSM Research birefringence measurements were used to obtain the stress fields in a converging slit channel with the dimensions as given in Figure 1 (Tas, 1994).

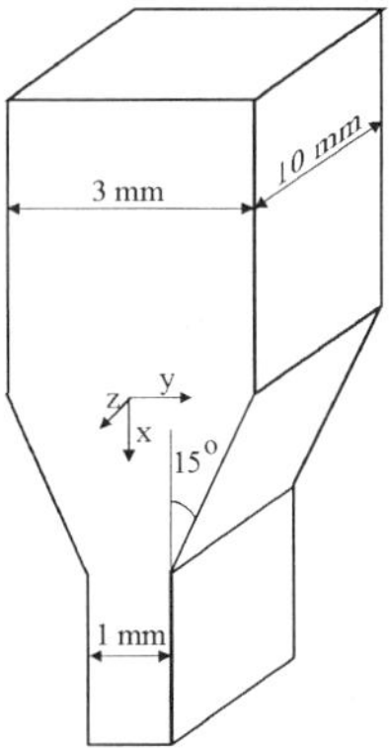

Figure 1: Geometry of the converging channel.

The experiments were carried out at 160°C for HDPE and 190°C for LDPE (De Bie et al. 1994). A more extensive description of the experimental work can be found in the graduate reports of Kikstra (1994) and De Bie (1994). In this paper the results of these measurements are used to evaluate the value of the results of the numerical simulations done with our finite element code.

In the simulations, a 2-D flow field (plane strain) is assumed with no flow in the z-direction. Because the problem is symmetric with respect to the x-axis only one half of the converging channel is modelled. The problem is meshed using bilinear interpolation 4-node isoparametric elements with the 8 nodal displacements as degrees of freedom. To prevent volume-locking the dilatation is integrated using one integration point. The finite element mesh used is represented in Figure 2, but it is noted that the in- and outflow are only partially depicted in this figure.

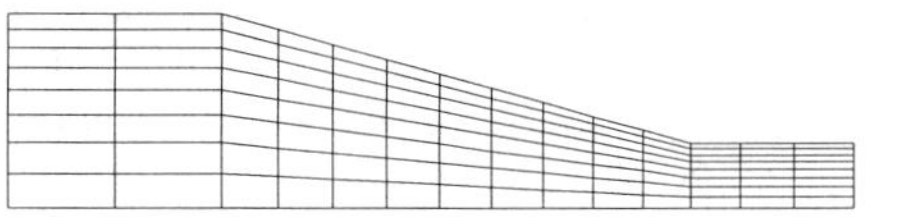

Figure 2: Part of the finite element mesh.

The boundary conditions that are used in the simulation are: stick at the wall, a prescribed velocity profile at the inflow. The inflow velocity profile has to change during the transient calculations: starting from an elasticity-free state the velocity profile develops from a parabolic profile to a flatter profile depending on the material properties and the flow rate (Rekers et al. 1993). Furthermore, incompressibility is assumed by taking a high value for the bulk modulus, i.e. C_b=1.e10 Pa (penalty formulation). Also, the flow is assumed to be isothermal.

The experiments for HDPE were carried out at a flow rate of 21.4mm^3/s. The calculated shear stresses σ_{xy} and first normal stress differences N_1 are compared to the measurements at a cross section in the contraction, x=2.20mm (Figure 3), and at a cross section in the outflow, x=7.70mm (Figure 4).

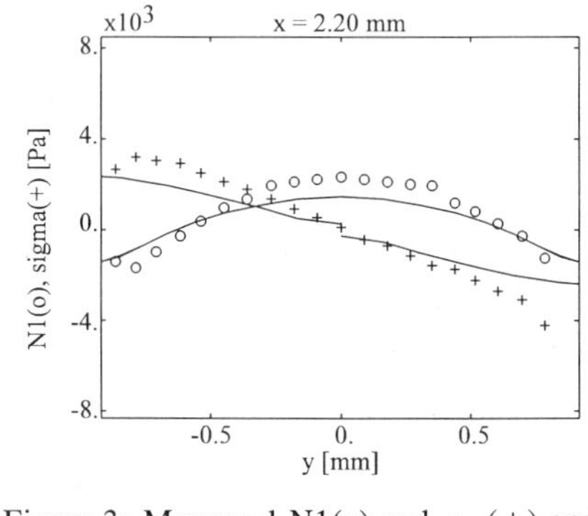

Figure 3: Measured N1(o) and σ_{xy}(+) and calculations with Leonov(-) and Giesekus(--) for HDPE.

At the centre line, y=0mm, of the converging slit the flow type is pure planar elongation. Hence, the first normal stress difference at the centre line is due to planar elongation only. Both measured and calculated N_1 are represented in Figure 5.

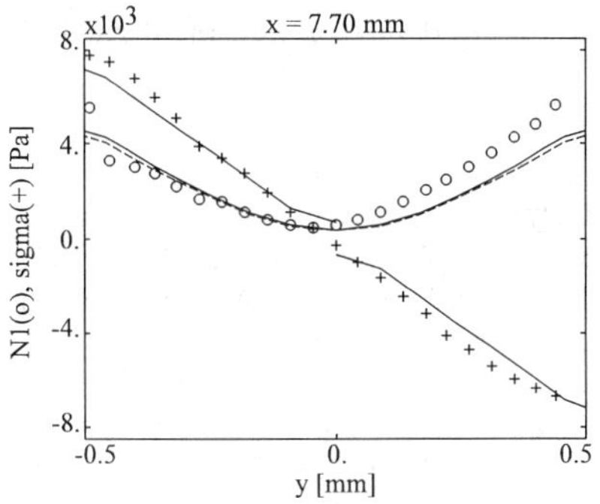

Figure 4: Measured N1(o) and σ_{xy}(+) and calculations with Leonov(-) and Giesekus(--) for HDPE.

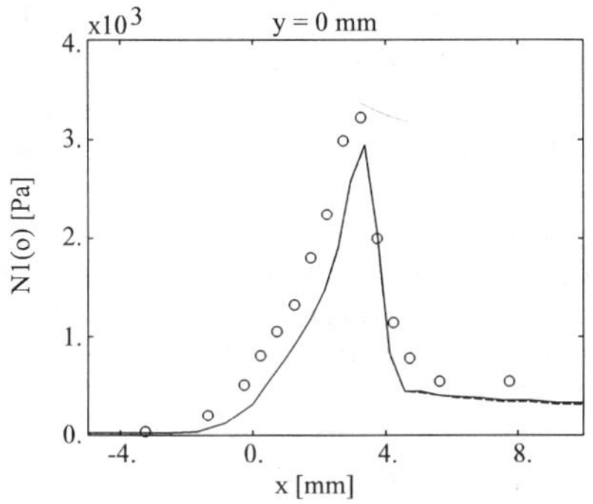

Figure 5: Measured N1(o) and calculations with Leonov(-) and Giesekus(--) for HDPE.

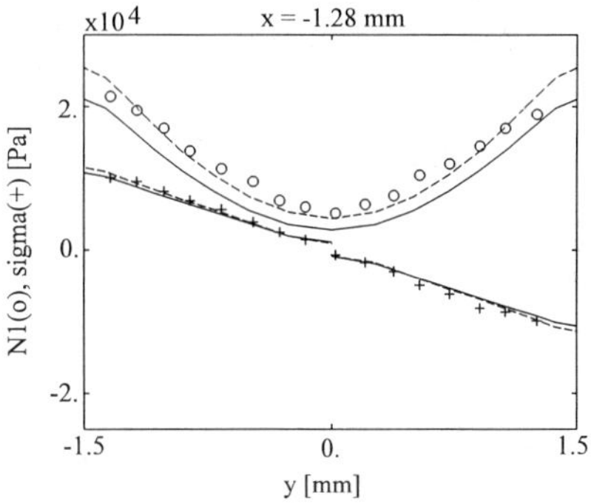

Figure 6: Measured N1(o) and σ_{xy}(+) and calculations with Leonov(-) and Giesekus(--) for LDPE.

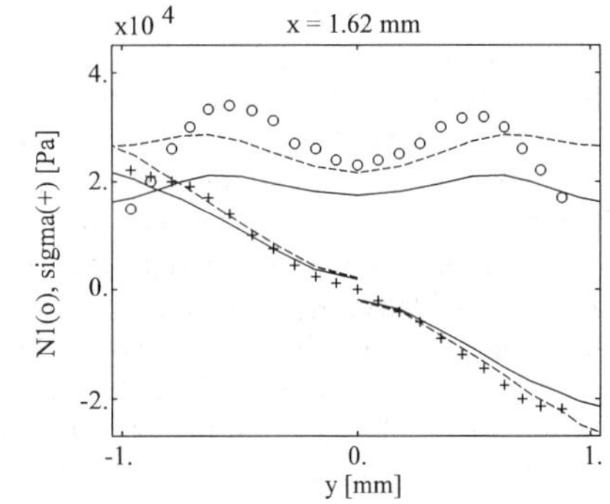

Figure 7: Measured N1(o) and σ_{xy}(+) and calculations with Leonov(-) and Giesekus(--) for LDPE.

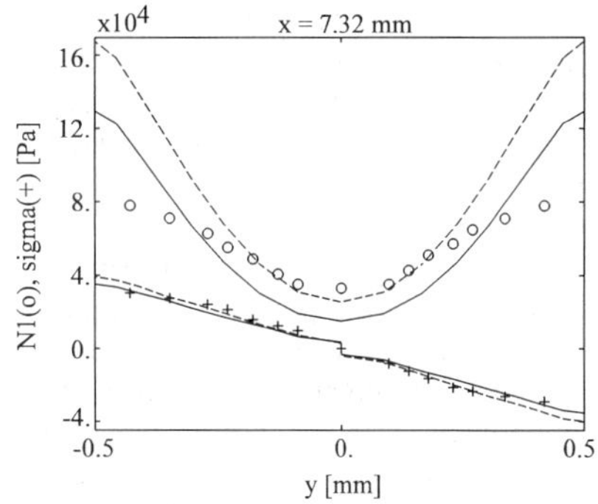

Figure 8: Measured N1(o) and σ_{xy}(+) and calculations with Leonov(-) and Giesekus(--) for LDPE.

The experiments for LDPE were carried out at a flow rate of 17.2mm³/s. The calculated shear stresses σ_{xy} and first normal stress differences N_1 are compared to the measurements at a cross section in the inflow, x=-1.28mm (Figure 6), at a cross section in the contraction, x=1.62mm (Figure 7), at a cross section in the outflow, x=7.32mm (Figure 8). The measured and calculated first normal stress differences at the centre line are depicted in Figure 9.

DISCUSSION AND CONCLUSIONS

For HDPE, it can be seen that both models predict nearly the same shear stresses and first normal stress differences on all points of the mesh. The predictions for the shape of the shear stresses and first normal stress differences is qualitatively very good. Quantitatively, the calculations give better results in the outflow region than in the contraction.

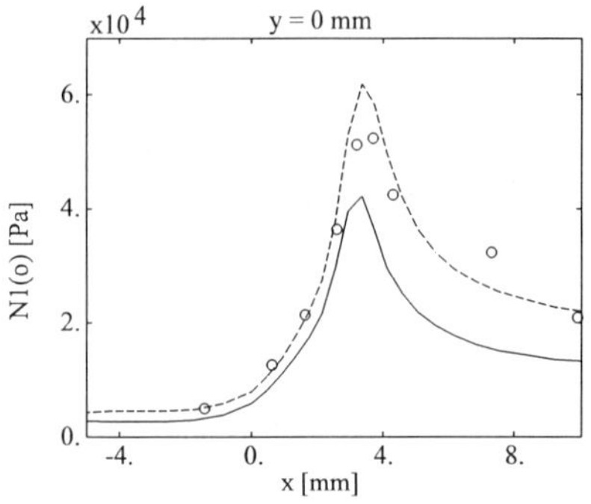

Figure 9: Measured N1(o) and calculations with Leonov(-) and Giesekus(--) for LDPE.

For LDPE, it is clear that the shear stresses and the first normal stress differences are better predicted by the Giesekus model than the Leonov model. In the inflow, the predicted first normal stress differences using the Giesekus model are quantitatively in good agreement with the measurements, but in the contraction and in the outflow only qualitative agreement is obtained. A possible explanation for this difference is that in this region wall slip is present. The first normal stress differences build up along the symmetry line are quite well predicted, but at the inflow the first normal stress differences are not equal to zero, as it should be for a fully developed shear flow.

Comparing the results of HDPE and LDPE, it is observed that the LDPE exhibits a higher level of elasticity which may explain that the predictions for LDPE are less accurate than for HDPE. Comparing the shape of the predicted first normal stress difference profiles at a cross section in the contraction, for HDPE a parabolic shape is found and for LDPE a maximum is found between the wall and the centre.

Concluding this paper it can be said that the elasticity of polymeric liquids is very important for predicting its flow behaviour. For the materials and flow rates investigated the High Weissenberg Number Problem was not present, but for other materials or higher flow rates it may be present. It is observed that some of the predictions are quite good, but there is still a lot of work to be done to get quantitatively good predeictions.

ACKNOWLEDGEMENTS

The support of the Dutch research programme "IOP Polymeren" and of DSM Research is gratefully acknowledged. Especially thanks to Marnix van Gurp, Arie Schepens, Han Slot and Gerard Vlogman of DSM for our valuable discussions.

REFERENCES

Akkerman, R. 1993. *Euler-Lagrange simulations of nonisothermal viscoelastic flows* (thesis). Enschede.

Baaijens, F.P.T. & L.F.A.Douven 1991. Calculation of flow-induced residual stress in injection moulded products, *Applied Scientific Research* 48:141-157. Kluwer Academic Publishers.

Baaijens, F.P.T. 1993. An U-ALE formulation of 3-D unsteady viscoelastic flow, *International Journal for Numerical Methods in Engineering* 36:1115-1143. John Wiley & Sons.

Bie, F.O. de, P.P.Tas, W.P.Kikstra & M. van Gurp 1994. Viscoelastic flow in a converging slit channel: rheo-optical measurements and numerical simulation. *Proceedings of the Fourth European Rheology Conference* 471-473. Sevilla.

Bie, F.O. 1994. *Visco-elastische stromingen in een convergerend kanaal: reo-optische experimenten en numerieke simulatie* (graduate report in Dutch). Eindhoven.

Besseling, J.F. 1968. A thermodynamic approach to rheology, *IUTAM symposium of irreversible aspects of continuum mechanics 1966*:16-53. Springer Verlag.

Goublomme, A., B.Draily & M.J.Crochet 1992. Numerical predicton of extrudate swell of a high density polyethylene, *Journal of non-Newtonian fluid mechanics* 44:171-195. Amsterdam: Elsevier Science Publishers.

Huétink, J. 1986. *On the simulation of thermomechanical forming processes* (thesis). Enschede.

Huétink, J. & P.N. van der Helm 1992. On Euler-Lagrange finite element formulation in forming and fluid problems, *Numerical Methods in Industrial Forming Processes* 4:45-54. Rotterdam: Balkema.

Kikstra, W.P. 1994. *Rheo-optical measurements and numerical simulations of a HDPE flow through a converging slit channel* (graduate report). Enschede.

Leonov, A.I. 1976. Nonequilibrium thermodynamics and rheology of viscoelastic polymer media, *Rheologica Acta* 15:85-98. Steinkopff Verlag.

Leonov, A.I. 1992. Analysis of simple constitutive equations for viscoelastic liquids, *Journal of non-Newtonian fluid mechanics* 42:323-350. Amsterdam: Elsevier Science Publishers.

Rekers G., R.Akkerman, J.Huétink 1993. Finite element simulations of local effects in nonisothermal viscoelastic and viscous flows. *Finite Elements in Fluids* 8: 1093-1102.

Tas, P.P. 1994. *Film blowing: from polymer to product* (thesis). Eindhoven.

Simulation of Materials Processing: Theory, Methods and Applications, Shen & Dawson (eds)
© 1995 Balkema, Rotterdam. ISBN 90 5410 553 4

Finite element simulation of the extrusion blow-molding process

Andres Rodriguez-Villa, Jean-François Agassant & Michel Bellet
École des Mines de Paris, CEMEF, Centre de Mise en Forme des Matériaux, UA CNRS, Sophia-Antipolis, France

ABSTRACT: This paper presents a 3D finite element membrane model of the extrusion blow-molding process. General aspects of the code are discussed before a detailed derivation of Maxwell's constitutive equation in curvilinear coordinates is presented. The last paragraph deals with a remeshing procedure recently added to the code and its application to local refinement on grounds of a geometrical criterion.

1. INTRODUCTION

The extrusion blow-molding process, described schematically in figure 1, has been developed for many years in the plastics industry [1] to manufacture hollow polymer components such as containers, bottles or automotive fuel tanks.

The primary step is the extrusion of a cylindrical parison which is then submitted to a combination of the following loads:

a. stretching of the bottom end of the parison.
b. closure of the pinch bars on the parison.
c. pre-blowing by internal pressure.
d. closure of the two half-molds.
e. blowing by internal pressure.

It is worth noting that the thickness profile of the final part as well as the final stress distribution depends not only on material rheology and the mold geometry, but also on the sequence of application of these different loads. There are other optimization factors the production engineer can modify in order to achieve a part with prescribed thickness and controlled stress distributions : the initial thickness of the parison (non-uniform thickness can be obtained by parison programming), the rate of inflation (in order to control the rate of deformation), etc. It is obvious that the help of a numerical simulation of the process is highly profitable to its optimization, especially if it is used at an early stage in the conception of the forming operation when a great number of design and process variables can still be changed. The present study is relevant to the final stage of the extrusion blow molding process, starting with an extruded parison of known shape and thickness.

Many numerical simulation models have been developed by various authors. Most of them have used membrane finite element formulations, such as De Lorenzi and Nied [2] who developed 2D and 3D approaches. Because of the relatively high strain rates, a "rubberlike" behavior for the molten polymer

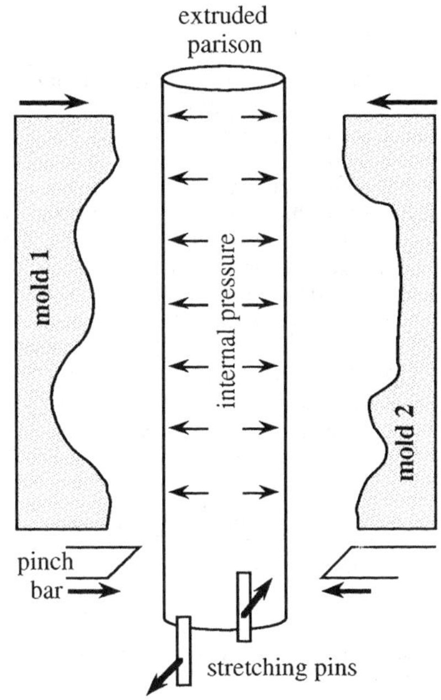

Figure 1: schematic description of the extrusion blow-molding process.

is assumed (non-linear Ogden or Mooney-Rivlin hyper-elasticity). The contact with the mold surface is considered sticking. The same kind of approach has been adopted by Wriggers and Taylor [3], Küppers and Michaeli [4], Diraddo et al. [6]. Others have used liquid-like viscoelastic models of the integral type: Kouba and Vlachopoulos [7], Warby and Whiteman [8], Shrivastava and Tang [9]. As shown by Rachik et al. [5] who compared different material behaviors, rheological models play a neglegible role in thickness prediction at least in the simple mold geometries that were tested. Hence, Bellet et al. [10] have developed a finite element membrane simulation software (Tform3) including either linear elasticity or elasto-viscoplasticity which gives satisfying thickness predictions on industrial components. However, the material model is expected to be very influent on the inflation kinetics and on the stress distribution during and at the end of the blowing stage: this last point is particularly important regarding the study of the cooling of the component and the evaluation of the residual stresses before use. This is the reason why, like Schmidt [11], a viscoelastic model of the differential type - Maxwell's model as a first approach - has been implemented in the Tform3 software: this particular point will be detailed in section 3 of the present paper, which also focuses on the development of a remeshing procedure of the finite element mesh of the parison, in order to be able to deal with formings of arbitrary shape for industrial application.

2. MECHANICAL FORMULATION

2.1 Membrane equilibrium

Considering that usual plastics parisons are thin with respect to their curvature radii, bending effects are assumed to be small compared to expansion effects and membrane equations can be used to describe the deformation and the equilibrium of the deformed parison. The deformed parison is then assumed to be a geometric surface and material points are identified by two convected material coordinates (θ^1 and θ^2) which remain constant throughout the whole process. For a cylindrical parison, which is extruded along the z direction, θ^1 is the initial angle θ_0 and θ^2 the initial z_0 coordinate. For each material point, the position vector $\mathbf{x}$ and the associated local thickness h depend on θ^i. A covariant tangential vector basis ($\mathbf{g}_1$,$\mathbf{g}_2$,$\mathbf{g}_3$) is then defined: $\mathbf{g}_1$ and $\mathbf{g}_2$ are tangent to θ^1 and θ^2 lines, and $\mathbf{g}_3$ is the local unit normal vector

$$\mathbf{g}_1=\frac{\partial \mathbf{x}}{\partial \theta^1} \;;\; \mathbf{g}_2=\frac{\partial \mathbf{x}}{\partial \theta^2} \;;\; \mathbf{g}_3=\frac{\mathbf{g}_1\wedge\mathbf{g}_2}{\|\mathbf{g}_1\wedge\mathbf{g}_2\|} \tag{1}$$

The virtual work principle is used to write the equilibrium of the deformed parison: for any arbitrary velocity field $\mathbf{v}^*$, we have:

$$\int_{\Omega} \sigma^{ij}\dot{\varepsilon}^*_{ij}\, h\, dS - \int_{\Omega_p} P\mathbf{g}_3.\mathbf{v}^* dS - \int_{\Omega_c} \mathbf{t}.\mathbf{v}^* dS + \int_{\Omega} \rho\, \gamma.\mathbf{v}^* h\, dS = 0 \tag{2}$$

where σ^{ij} are the contravariant components of the Cauchy stress tensor (plane stress: i,j=1,2), $\dot{\varepsilon}^*_{ij}$ are the covariant components of the virtual strain rate tensor associated with $\mathbf{v}^*$, $\mathbf{t}$ is the friction stress vector applied to the sheet regions Ω_c contacting the mold, P is the inflation pressure applied to the domain Ω_p of the deformed surface Ω, and γ is the acceleration vector.

2.2 Time discretization

The model is incremental: at each time increment [t,t+Δt], starting with a balanced deformed configuration Ω at time t, the problem consists in determining the unknown equilibrated state Ω' at time t+Δt. We will denote "prime" the variables at t+Δt.

First of all, an implicit time integration scheme on the time increment [t,t+Δt] is used for acceleration. Hence the unknown velocity field $\mathbf{v}'$ at t+Δt is expressed by:

$$\mathbf{v}' = \mathbf{v} + \Delta t\, \gamma' \tag{3}$$

Therefore the equilibrium of the material configuration at time t+Δt consists of equation (2), including the constitutive equations, the boundary conditions and the expression of γ' deduced from equation (3):

$$\forall \mathbf{v}^*: \int_{\Omega'} \sigma'^{ij}\dot{\varepsilon}^*_{ij}\, h'\, dS' - \int_{\Omega'_p} P'\mathbf{g}'_3.\mathbf{v}^* dS' - \int_{\Omega'_c} \mathbf{t}'.\mathbf{v}^* dS' + \int_{\Omega'} \rho \frac{\mathbf{v}'-\mathbf{v}}{\Delta t}.\mathbf{v}^* h'\, dS' = 0 \tag{4}$$

In this equation, the new configuration Ω' is deduced from the previous one by the following updating scheme of positions, which is consistent with (3):

$$\mathbf{x}' = \mathbf{x} + \Delta t\, \mathbf{v} + \frac{\Delta t^2}{2}\gamma' = \mathbf{x} + \frac{\Delta t}{2}(\mathbf{v} + \mathbf{v}') \tag{5}$$

Consequently, equation (4) may be solved in terms of the new velocity field $\mathbf{v}'$, provided that we can express $\mathbf{t}'$, h' and σ', as functions of $\mathbf{v}'$. These relations are exposed hereunder and the solution procedure in 2.3.

2.2.1 *Contact*

The contact condition applied to a point of Ω_c in contact with one of the moving molds can be either sticking (the point will have a prescribed velocity equal to the velocity of the mold) or sliding: in this case, the tangential friction stress is supposed to be given by Coulomb's friction law, in which the normal stress is the inflation pressure. Therefore, at t+Δt, we will have:

$$\mathbf{t}' = -\mu\, P'\, \mathbf{v}'/\|\mathbf{v}'\| \tag{6}$$

where μ is the friction coefficient.

2.2.2 *Incompressibility. Thickness updating*

The new local thickness h' is deduced from the volume conservation equation. To do this, the material is considered to be incompressible. Taking into account that the elementary surface dS is given by $dS = \sqrt{\det(\mathbf{g})}\, d\theta^1 d\theta^2$, where $\mathbf{g}$ is the covariant metric tensor ($g_{ij} = \mathbf{g}_i \cdot \mathbf{g}_j$), this yields :

$$h' = h \sqrt{\det(\mathbf{g}) / \det(\mathbf{g}')} \tag{7}$$

2.2.3 *Material behavior*

At this stage, the material behavior model is considered only to be of the general simple form:

$$\sigma = f(\dot{\varepsilon}) \tag{8}$$

At time t+Δt, this yields:

$$\sigma' = f(\dot{\varepsilon}') = f\left(1/2(\nabla \mathbf{v}' + (\nabla \mathbf{v}')^T)\right) = g(\mathbf{v}') \tag{9}$$

The specific expression of this relation for Maxwell's viscoelastic model will be discussed in section 3.

Finally, injecting equations (5) to (9) into the equilibrium equation (4), leads to a non-linear equation with **v'** as unknown:

$$\mathcal{R}(\mathbf{v}') = 0 \tag{10}$$

This equation is discretized in space and solved using a finite element method as explained in the next paragraph.

2.3 Spatial discretization by the finite element method

The initial parison surface is meshed with two-dimensional finite elements (linear triangles or linear quadrangles) the nodes of which are convected. The details of the finite element implementation are given in [12-13]. The discretization of equilibrium equation (4) or (10) at every time step leads to the resolution of a non-linear system, the unknown of which is **V'** the vector of nodal velocity components:

$$R(\mathbf{V}') = 0 \tag{11}$$

This system of equations is solved iteratively by the Newton-Raphson method: the tangent stiffness matrix used is the symmetric part of the exact tangent stiffness matrix $\partial \mathbf{R}/\partial \mathbf{V}'$. The exact computation of this matrix requires the derivation of the discretized expressions of equations (5-9) with respect to **V'**. In particular, it needs the determination of the exact tangent modulus $\partial\sigma'/\partial\dot{\varepsilon}'$ when solving the incremental behavior law at each integration point of each finite element.

The simulation code can handle several rigid molds. Their surfaces are discretized in triangular elements in order to evaluate the parison points' relative positions easily at each time increment. On each mold surface, different regions with different contact conditions can be specified.

2.4. Industrial application

The model has been efficiently used these recent years to predict thickness distributions in industrial applications using a linear elastic constitutive equation [10].

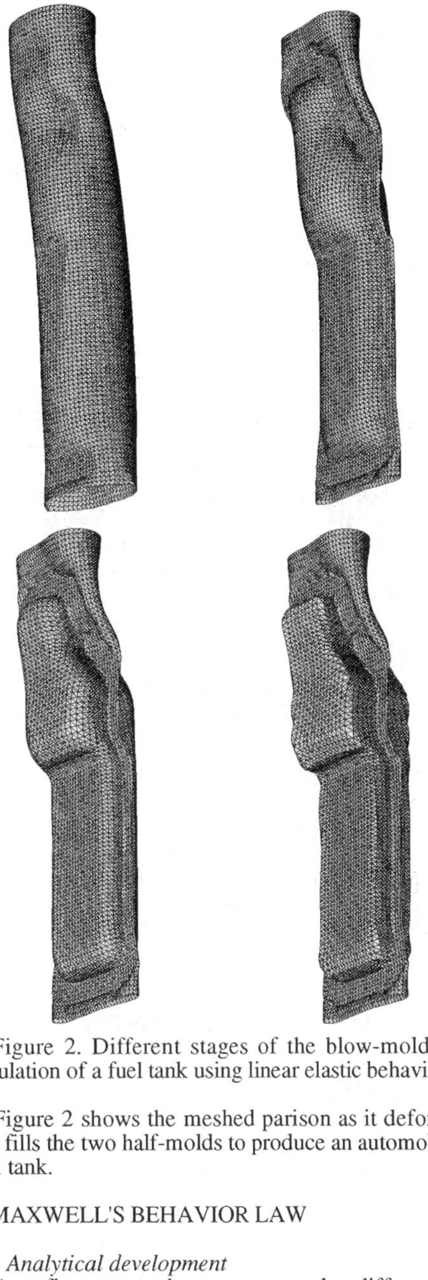

Figure 2. Different stages of the blow-molding simulation of a fuel tank using linear elastic behavior

Figure 2 shows the meshed parison as it deforms and fills the two half-molds to produce an automobile fuel tank.

3. MAXWELL'S BEHAVIOR LAW

3.1. Analytical development

As a first approach to more complex differential viscoelastic behavior laws, Maxwell's model has been chosen and implemented in Tform3. In three-dimensional tensorial form, the constitutive equations are:

$$\sigma = -q\,\mathbf{I} + \tau$$

$$\tau + \lambda \frac{\delta\tau}{\delta t} = 2\eta\dot{\varepsilon} \tag{12}$$

where:

- q is a scalar field with dimensions consistent with a pressure field
- $\mathbf{I}, \tau$, and $\dot{\varepsilon}$ are the identity, the extra-stress and the deformation rate tensors
- λ and η are the relaxation time and the viscosity
- $\frac{\delta}{\delta t}$ denotes upper-convected (Oldroyd) derivation

Written in terms of contravariant components of the tensors involved, these equations become:

$$\sigma^{ij} = -q\, g^{ij} + \tau^{ij}$$

$$\tau^{ij} + \lambda \left(\frac{\delta \tau}{\delta t}\right)^{ij} = 2\eta\dot{\varepsilon}^{ij} \tag{13}$$

in which i,j = 1, 2, 3 and g^{ij} are the metric tensor's contravariant components. In the tangent refererential (1) in which the equilibrium equation has been written, these components have particular values in the third (normal) direction: $g^{13}=g^{31}=g^{23}=g^{32}=0$ and $g^{33}=1$. Furthermore, the membrane approximation assumes $\sigma^{i3}=0$. All this reduces Maxwell's set of equations to

$$\sigma^{ij} = -\tau^{33} g^{ij} + \tau^{ij} \tag{14.1}$$

$$\tau^{ij} + \lambda\left[\frac{\delta\tau}{\delta t}\right]^{ij} = 2\eta\dot{\varepsilon}^{ij} \qquad i,j=1,2 \tag{14.2}$$

$$\tau^{33} + \lambda\left[\frac{\delta\tau}{\delta t}\right]^{33} = 2\eta\dot{\varepsilon}^{33} \tag{14.3}$$

In previous work, Bellet [12] has shown that the upper-convected or Oldroyd derivation of a tensor in a cartesian referential is mathematically strictly equivalent to its lagrangian derivation in a material coordinate system such as the one used in our model.

This is true as long as the referential's vector along which the derivation takes place is convected. In equation (14.2) this condition is satisfied, vectors $\mathbf{g^1}$ and $\mathbf{g^2}$ do follow the matter's deformation as it takes places. But this is *not* true for equation (14.3), since vector $\mathbf{g^3}$ is a *unit* vector and thus, does not account for the deformation occuring in this direction. Upper-convected (Oldroyd) differentiation must be applied in this equation to achieve objectiveness.

Introducing this data into set of equations (14):

$$\sigma^{ij} = -\tau^{33} g^{ij} + \tau^{ij} \tag{15.1}$$

$$\tau^{ij} + \lambda\frac{d\tau^{ij}}{dt} = 2\eta\dot{\varepsilon}^{ij} \tag{15.2}$$

$$\tau^{33} + \lambda\left[\frac{d\tau^{33}}{dt} - 2\,\dot{\varepsilon}^{33}\tau^{33}\right] = 2\eta\dot{\varepsilon}^{33} \tag{15.3}$$

Mass conservation enables us to link $\dot{\varepsilon}^{33}$ to the plane components of the deformation rate tensor, thus closing the system of equations:

$$\mathrm{trace}(\dot{\varepsilon}) = \dot{\varepsilon}_{33} + g^{mn}\dot{\varepsilon}_{mn} = 0 \tag{16}$$

where $\dot{\varepsilon}_{mn}$ are the covariant components of the deformation rate tensor and m,n=1,2. For computational reasons it is better to use covariant components of $\dot{\varepsilon}$, which are linked to the contravariant components through:

$$\dot{\varepsilon}^{ij} = g^{im}g^{jn}\dot{\varepsilon}_{mn} \tag{17}$$

Introducing (16) and (17) into (15), and eliminating τ^{ij} in this set of equations in order to comply to (8), we obtain:

$$\sigma^{ij} + \lambda\left[\frac{d\sigma^{ij}}{dt} + \tau^{33}\frac{dg^{ij}}{dt}\right] = 2\left[\eta E^{ijmn} + \lambda F^{ijmn}\tau^{33}\right]\dot{\varepsilon}_{mn}$$

$$\tau^{33} + \lambda\frac{d\tau^{33}}{dt} = -2\left[\eta + \lambda\tau^{33}\right]g^{mn}\dot{\varepsilon}_{mn} \tag{18}$$

where $F^{ijmn}=g^{ij}\,g^{mn}$ and $E^{ijmn}=F^{ijmn} + g^{im}\,g^{jn}$

3.2. Resolution of the incremental behavior

We can now establish the particular expression of relation (9) for Maxwell's constitutive equation. This is done applying an implict scheme to (18):

$$\sigma'^{ij} + \xi\left[\sigma'^{ij} - \sigma^{ij} + \tau'^{33}(g'^{ij} - g^{ij})\right] = 2\left[\eta E^{ijmn} + \lambda F^{ijmn}\,\tau'^{33}\right]\dot{\varepsilon}'_{mn}$$

$$\tau'^{33} + \xi\left[\tau'^{33} - \tau^{33}\right] = -2\left[\eta + \lambda\tau'^{33}\right] g^{mn}\,\dot{\varepsilon}'_{mn} \tag{19}$$

where $\xi=\lambda/\Delta t$, or, in a handier formualtion:

$$\sigma'^{ij} = \frac{\xi}{1+\xi}\left[\sigma^{ij} - \tau'^{33}(g'^{ij} - g^{ij})\right] + \frac{2}{1+\xi}\left[\eta E^{ijmn} + \lambda F^{ijmn}\,\tau'^{33}\right]\dot{\varepsilon}'_{mn}$$

$$\tau'^{33} = \frac{\xi\tau^{33} - 2\eta g^{mn}\,\dot{\varepsilon}'_{mn}}{1+\xi+2\lambda\, g^{mn}\,\dot{\varepsilon}'_{mn}} \tag{20}$$

As presented hereabove, resolution of the discretized virtual work principle by a Newton-Raphson method requires calculating the derivative of σ' with respect to $\dot{\varepsilon}'$. This tangent modulus $\partial\sigma'/\partial\dot{\varepsilon}'$ results directly from the simple but fastidious derivation of (20).

3.3. Validation

The bench test used to evaluate the precision of the viscoelastic model explicited hereabove is that of an infinitely long cylindrical membrane expanding under a linear pressure load P(t)=Kt. The evolution of the cylinder's adimensional radius has been compared to the solution found by the extension of a previous analytical model [14]. Figure 3 shows an excellent

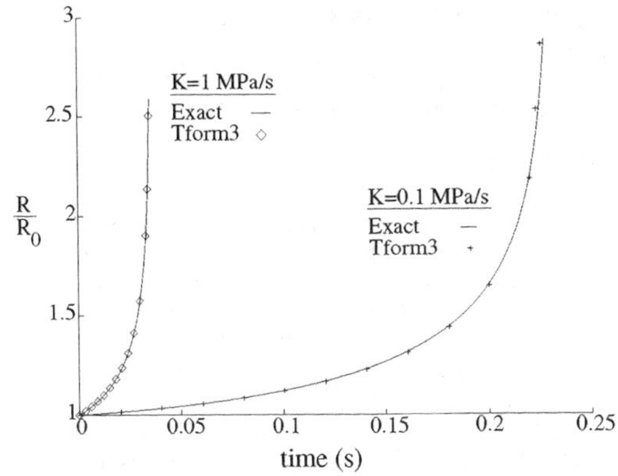

Figure 3. Validation of Mawell's behavior in Tform3 by comparison with the exact solution to the free inflation of a cylindrical membrane (initial radius R_0=50 mm, initial thickness e_0=4 mm).

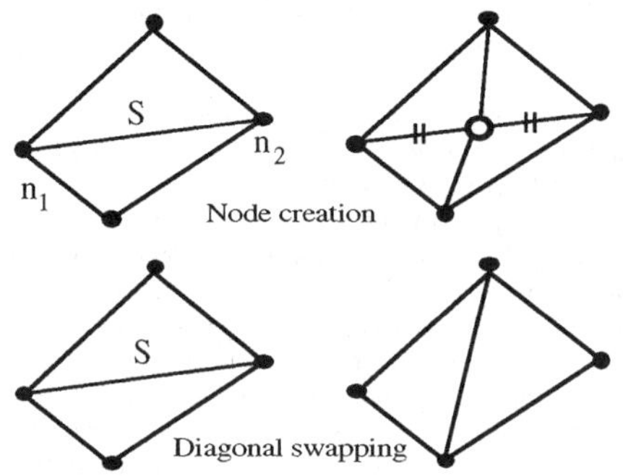

Figure 4. Basic topological operations

agreement in the results found by both models for two different load rates.

4. REMESHING PROCEDURE

Local large deformations and complex molds are usually encountered in the extrusion blow-molding process. In the context of a finite element simulation, these features make it preferable to begin the calculation with a great number of elements (fig.2). However, the initial small size of these elements might not be useful until the final steps of the computation. Therefore a remeshing procedure that refines the finite element mesh where and when necessary as the simulation is in progress is a profitable tool to have in terms of computational time and precision.

At Cemef, a 3D remesher has been developed and effectively used these recent years [15] in industrial forging simulations by the finite element code Forge3. The tool can be parted into two functional blocs: a linear-triangle surfacic mesher for closed 3D surfaces used to discretize the boundary surface of the 3D domain, and a volumic mesher which uses the meshed surface as input. In the context of our blow-molding membrane calculations, we have extended the surfacic mesher to 3D surfaces with edges, such those at the top and bottom parts of a parison (fig.1) and given it the ability to localize the refinement of the mesh given a certain criterion.

4.1. Basic principles of the meshing procedure

Tform3's surfacic mesher uses two basic topological operations: node creation at the mid-point of a segment (element side) and diagonal swapping (fig. 4). The segment and its two adjacent elements can be seen as a *sub-topology* and the remesher adopts a strategy of topological improvement which is detailed hereunder.

Mesh segments and their corresponding sub-topologies are examined one by one. Node creation is tested if, among other conditions, the size h of the segment is greater than a user-defined value h_{min}. The quality of the resulting sub-topology is then evaluated. Diagonal swapping is also attempted and the quality of the result compared to that of the first operation.The best sub-topology is stored and the next sub-topolgy can be processed. Details on how the measurement of the quality of a given sub-topology is performed can be found in [15]. Let us nevertheless note its two components: element aspect and geometrical precision with reference to the surface of the initial mesh.

4.2. Triggering of the remeshing procedure and refinement criterion

We have used a geometrical criterion, curvature C, both to determine when to remesh and where to refine our finite element discretization. Given the way the remeshing procedure examines the mesh, a natural choice is to define the curvature of a segment. We have tested the following definition:

$$C(S) = \alpha(\mathbf{N}(n_1), \mathbf{N}(n_2)) \tag{21}$$

where S is a segment linking nodes n_1 and n_2 (fig.4), **N** denotes the mean unit normal at a node and α the angle in between two vectors. Remeshing is called for whenever the maximal curvature observed exceeds a user-defined value C_{crit} and refinement is performed where curvature is higher than a user-defined fraction β of C_{crit}.

4.3. Localized refinement

Local refinement can be achieved to some extent by slightly modifying the topological improvement procedure: we specify a *local*, instead of a global, minimum size $h_{min}(S)$ over which segment S may be split by the topological improvement strategy. This local size is linked by a quadratic law to C(S) and the global extreme values of segment size H and curvature C:

$$h_{min}(S)=H_{MIN}+(H_{MAX}-H_{MIN})\left[1-\frac{C(S)-C_{MIN}}{C_{MAX}-C_{MIN}}\right]^2 \tag{22}$$

Obviously, not every node creation will be an

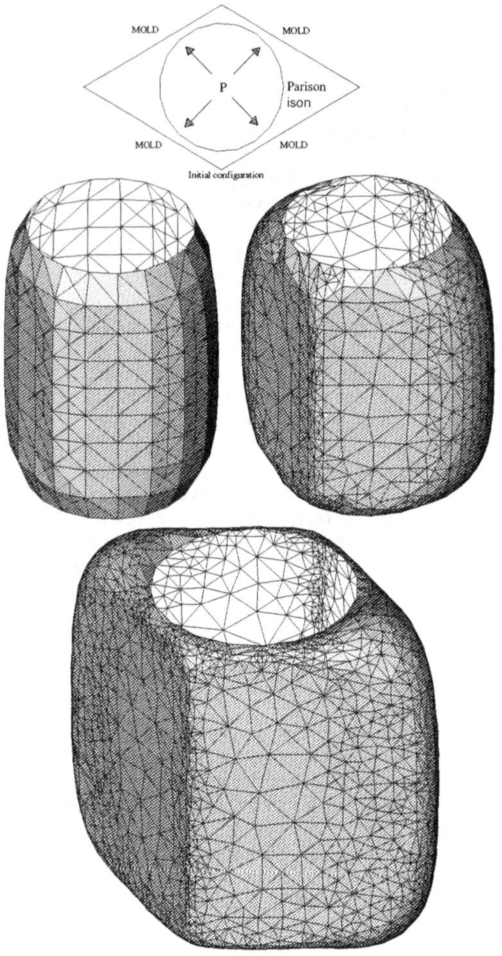

Figure 5. Meshes at different stages of the inflation and close-up of the highly refined regions

improvement to the quality of the topology. Therefore it is necessary to force this operation regardless of the quality of the result. Nevertheless, a combination of forced creation and topological improvement runs allows the obtention of an acceptable mesh, such as those shown in the parison inflation test at figure 5.

CONCLUSION

After a brief recall of the finite element membrane formulation, the implementation of a differential viscoelastic law and of a remeshing procedure have been examined in order to increase the predictive power of finite element code Tform3.

Future developments will aim at effectively implementing a more general differential viscoelastic law, at optimizing our control on the remeshing procedure in order to achieve better meshes and at adding a node elimination operation to the procedure to increase its efficiency.

ACKNOWLEDGEMENT

The developments in this report have been supported by Solvay and Elf-Atochem.

REFERENCES

1 Rosato. Blow-molding Handbook, Carl Hanser, 1989.
2 DeLorenzi H.G. and Nied H.F. Finite element simulation of thermoforming and blow molding. Progress in polymer processing, A.I.Isayev ed., Hanser Verlag, 1989.
3 Wriggers P. and Taylor R.L. A fully non-linear axisymmetrical membrane element for rubber-like materials, Eng. Comp., vol.7, pp.303-310, 1990.
4 Küppers M. and Michaeli W. Simulation of the stretching process of complex geometries with respect to thermoforming process, Institut fur Kunststoffverarbeitung, RWTH Aachen.
5 Rachik M., Roelandt J.M. and Batoz J.-L. Simulation numérique du soufflage et du thermoformage des plastiques, to be published in Journal Européen des Polymères.
6 Diraddo R.W. and Garcia-Rejon A. Experimental and theoretical investigation of the inflation of variable thickness parisons Polym. Eng. Sci., Vol. 34, pp.1080-1089, 1994.
7 Kouba K. and Vlachopoulos J. Modeling of thermoforming and blow-molding. Theoretical and Applied Rheology, vol.1, Elsevier, pp.374-376, 1992.
8 Warby M.K. and Whiteman J.R. Finite element model of viscoelastic membrane deformation, Comp. Meth. Appl. Mech. Engng, vol.68, pp.33-54, 1988.
9 Shrivastava S. and Tang J. Large deformation analysis of non-linear viscoelastic membranes with reference to thermoforming, J. Strain Anal., vol.28, pp.21-51, $.
10 Bellet M., Corsini C., Assaker D., Mercier P. and Wouters P. A numerical model of the extrusion blow-molding process, J. Reinforced Plastics and Composites, vol.12, pp.498-505, 1993.
11 Schmidt F.M., Agassant J.F. and Bellet M. Viscoelastic simulation of P.E.T stretch/blow molding, to be published in J. of Non-Newt. Fl. Mech., March 1995
12 Bellet M. Modélisation numérique du formage superplastique de tôles (Numerical modelling of the superplastic forming of metal sheets), PhD dissertation (in french), Ecole des Mines de Paris, 1988.
13 Bellet M., Massoni E. and Chenot J.L. Numerical simulation of thin sheet forming processes by the finite element method, Eng. Comp., vol.7, p21, 1990.
14 Rodriguez-Villa A. Simulation numérique de l'extrusion-soufflage des polymères (Numerical simulation of polymer extrusion blow-molding), DEA Degree report (in french), Ecole des Mines de Paris, 1994.
15 Coupez T., Grandes transformations et remaillage automatique (Large deformations and automatic remeshing), PhD dissertation (in french), Ecole des Mines de Paris, 1991.

Simulation of Materials Processing: Theory, Methods and Applications, Shen & Dawson (eds)
© 1995 Balkema, Rotterdam. ISBN 90 5410 553 4

A design tool for blow molding and thermoforming processes

N. Santhanam & K. Himasekhar
AC Technology, Ithaca, N.Y., USA

K. K. Wang
Sibley School of Mechanical and Aerospace Engineering, Cornell University, Ithaca, N.Y., USA

ABSTRACT: An important concern in blow molding and thermoforming processes is the prediction and control of thickness distribution on the molded part. In this paper, we present a finite-element based CAE design tool that can be used to evaluate processing parameters and provide essential information, i.e., thickness distribution, to determine part performance in blow molding and thermoforming.

1 INTRODUCTION

The processes of blow molding and thermoforming are used to produce a wide variety of plastic components ranging from large, thin-walled structural parts to small, commodity items such as plastic bottles [1,2]. In these processes, it is important to predict and control the thickness distributions on the molded part in order to satisfy design thickness requirements and optimize material usage. Computer Aided Engineering (CAE) techniques that have played a key role in part and process design in other polymer processing operations [5] are now being extended to the above processes [3,4]. A CAE design tool based on a nonlinear finite-element simulation for three-dimensional geometries is presented in this paper. In this tool, the polymer is taken to behave like a hyperelastic membrane and the mold surfaces are assumed to be rigid. Portions of the polymer that contact the mold surfaces are assumed to be subsequently fixed to those surfaces. The solution is based on a total Lagrangian formulation with a full-Newton iteration for equilibrium calculations. Contact between the polymer nodes and the mold surfaces is handled automatically without user intervention. Important design information that can be obtained using this tool includes the thickness distribution on the molded part, cooling time, and details of evolution of part shape and thickness distribution during processing. A case study is presented to illustrate the efficacy of the CAE tool.

1.1 *Process description*

In blow molding, a parison or preform of hot plastic is inflated into a mold cavity by the application of pressure. Prior to the application of pressure, there might be movement of the mold halves or stretch devices which can contact the polymer. Thermoforming is a secondary forming process in which a sheet of thermoplastic is heated and inflated into a cavity of desired shape through the application of pressure and vacuum. As in blow molding, the inflation stage might be preceded by movement of mold parts and stretch devices. Mechanics of both processes are quite similar and involve large deformations of a hot polymer membrane; contact between the mold and the polymer is also possible throughout the entire process. The thickness variation of the final part depends on how the inflation proceeds. Typically, regions that contact the mold surfaces first are subsequently prevented from moving and stretching, and regions that undergo substantial deformation before they contact the mold walls exhibit significant thinning.

1.2 *Assumptions*

In the CAE tool, the polymer is taken to be a membrane; this assumption is appropriate for thin-walled shell structures fabricated by these processes. The thickness of the structure is taken to be much smaller than its other dimensions and its bending resistance is neglected.

The plastic material is taken to behave in a "rubbery" manner, i.e., it exhibits nonlinear, elastic, incompressible behavior. There is significant experimental evidence that supports the use of this model to the blow molding and thermoforming processes [2]. Furthermore, use of a nonlinear viscoelastic material model for polymer behavior would entail considerable effort both from the point of view of experimental characterization and computation.

The mold surfaces are taken to be rigid and polymer nodes in the finite element mesh that contact the mold surfaces (elements) are taken to be subsequently fixed to the latter. This assumption is relaxed for processes that involve stretching devices off which the polymer can lift; in these cases, the polymer node is freed from the mold surface if the normal force exerted by the mold surface on the polymer node "pulls" the latter toward the former.

2 FINITE ELEMENT FORMULATION

Using the principle of virtual work and equating the work done by external forces (due to pressure) to the strain energy stored in the membrane, we can derive the equation for static equilibrium of the membrane as follows:

$$\int p\, j\, n_i v_i\, dA = \int H\, J\, S_{ij}\, F_{ki}\, V_{kj}\, dA \qquad (1)$$

where p is the applied pressure, j is the ratio of the deformed to the reference surface area, n_i is the i-th component of the normal to the membrane, v_i is the i-th component of the virtual displacement, dA is the area in the reference configuration, H is the undeformed thickness of the membrane, J is the ratio of the undeformed to the reference surface area, S_{ij} is the ij-th component of the second Piola-Kirchoff stress tensor, F_{ki} and V_{kj} are components of the real and virtual deformation gradients, respectively (summation is carried out over repeated indices above).

Applying a finite-element discretization, we obtain:

$$\sum_m \int p\, j\, \{N\}^T \{n\}\, dA \;=\; \left[\sum_m \int H\, J \left\{\frac{\partial N}{\partial X}\right\}^T \{S\} \left\{\frac{\partial N}{\partial X}\right\} dA \right] \{x\} \qquad (2)$$

where the summation is over m elements in the finite element mesh, {N} is a matrix of shape functions, {n} is the element normal, $\left\{\frac{\partial N}{\partial X}\right\}$ is a matrix of derivatives of shape functions with respect to coordinate axes tangential to the undeformed surface, {S} is a matrix of the second Piola-Kirchoff stresses, and {x} is the vector of nodal coordinates [2]. Since the term within the square brackets on the right-hand side of the above equation and both j and {n} on the left-hand side are dependent on the nodal coordinates ({x}), the above equation is nonlinear. It is linearized and solved incrementally using a full-Newton iterative scheme [2,6].

2.1 *Special concerns*

Since the material behavior can be such that the membrane can continue inflating with steady or even decreasing applied pressure, special numerical techniques have been adopted to handle this possibility [2,7].

Also, due to the possibility of contact

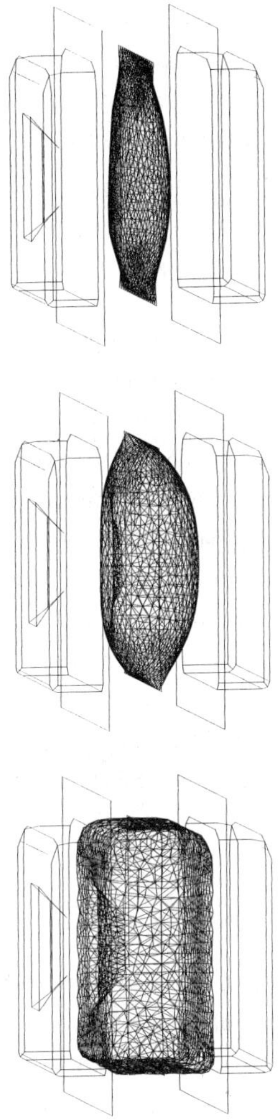

Fig. 1. Inflation of the parison (mold movement not shown).

between the polymer and rigid tool surfaces, equilibrium iterations have to be continued if new contacts are detected after convergence is achieved during any load increment. The solution terminates when all polymer nodes contact mold surfaces or a pre-set displacement limit is attained.

3 CASE STUDY

The following case study illustrates the accuracy of the design tool for the prediction of thickness distribution in a blow molded bottle. The geometry is approximately that of a rectangular parallelepiped with dimensions: 400mm x 350mm x 180mm. There is a sharp projection on one of the mold halves which creates a baffle on the molded bottle. The material used was a commercial grade high-density polyethylene. The geometry of the parison after the mold halves close was determined experimentally and used as the initial configuration of the parison. The initial thickness of the parison was 8.8mm at the processing temperature of 190°C. A hyperelastic Ogden material model [8] was fitted to true-stress vs. strain data generated using a Meissner-type uniaxial rheometer.

3.1 *Results*

The inflation sequence is shown in Figure 1; the movement of the mold halves are not shown for clarity. It can be seen from this figure that the initially partially-concave shape of the clamped parison fills out subsequently; contact occurs with the shallow side walls early. The sheet can stretch near the sharp edge of the projection and further thinning can occur even after contact is made with the edge. Figure 2 shows a comparison between the predicted and measured thickness along a diagonal slice across the bottle. It should be noted that the results from the simulation have been generated considering the cooling that takes place from the processing to room temperature. As can be seen from the figure, comparison between the predicted and measured thickness distribution is excellent. The thickness along this slice is highest at the feature-free shallow side wall where contact occurs early and the thickness is 'frozen' in quickly. On the other side-wall with the baffle, the thickness is slightly lower due to the additional area on to which the polymer stretches out. The corners of the bottle, which are contacted by the polymer last, correspond to areas that undergo the maximum stretching and are consequently thinner. Figure 3 shows the thickness measurement and prediction along the

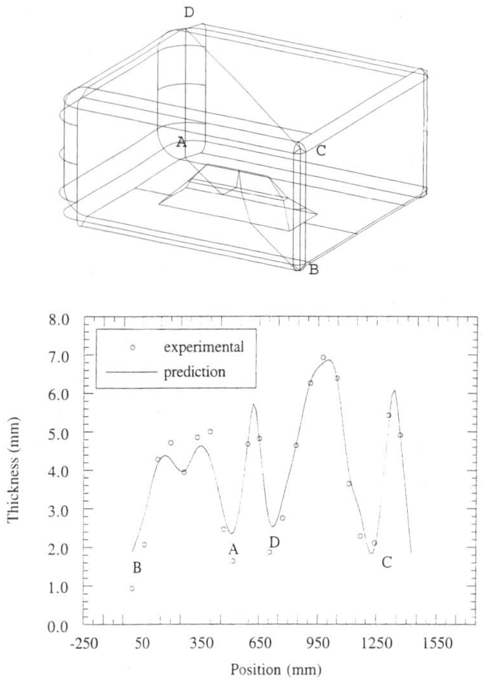

Fig. 2. Comparison between predicted and measured thickness along diagonal slice.

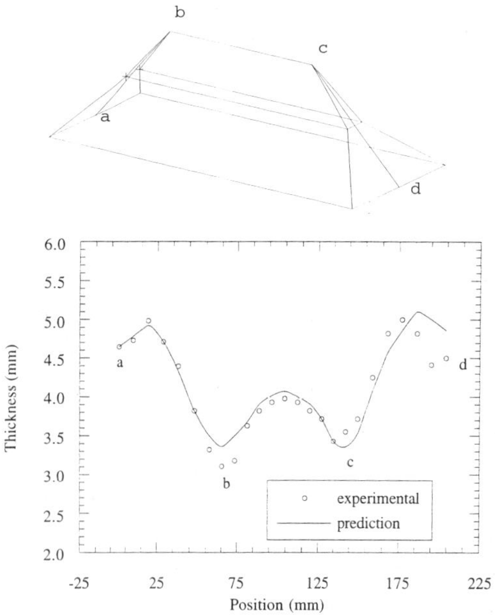

Fig. 3. Comparison between predicted and measured thickness along a cut across the baffle.

baffle on the side wall. Once again, the predicted thickness distribution compares very well with the experiments. Maximum thinning along this slice occurs at the two end points of the sharp edge where there is a maximum possibility of material stretch.

4 CONCLUSIONS

A CAE tool for blow molding and thermoforming processes has been developed. It can assist the part and process designers in: predicting thickness variations on the molded part, estimating optimum processing conditions, and improving part quality while minimizing part weight. A major benefit in employing this tool is that the iterative design process can be performed on the computer rather than by constructing prototype molds and performing extensive experimentation. The accuracy of the predictive tool, as demonstrated for one example, is quite acceptable for the above design purposes.

ACKNOWLEDGMENTS

Very special thanks are due to Mr. Toshiyuki Tajiri of Mitsubishi Kasei Corporation who provided all the experimental data, performed the simulation runs using the design tool, C-MOLD Blow Molding and Thermoforming, and granted us permission to reproduce the results in this paper. The authors would like to thank AC Technology for granting permission and for allocating resources for the publication of this work. The authors would also like to thank Dr. deLorenzi of GE CR&D, one of the key developers of PITA/STAT, a simulation tool developed by GE CR&D and GE Plastics, which formed the basis for C-MOLD Blow Molding and Thermoforming.

REFERENCES

1. Throne, J.L., "Thermoforming", Hanser Publishers, 1986.

2. deLorenzi, H.G., Nied, H.F., "Finite Element Simulation of Thermoforming and Blow Molding", *Progress in Polymer Processing*, Isayev, A.I., ed., p. 117, Hanser Publishers, 1991.
3. Kouba, K., Vlachopoulos, J., "Modeling of 3D Thermoforming", *SPE ANTEC*, p. 114, 1992.
4. Bellet, M., et al, "A Numerical Model of the Extrusion Blow-Molding Process", *Journal of Reinforced Plastics and Composites*, p. 498, 1993.
5. Chiang, H.H., et al, "Integrated Simulation of Fluid Flow and Heat Transfer in Injection Molding for the Prediction of Shrinkage and Warpage", *Journal of Engineering Materials and Technology*, Vol. 115, p. 37, 1993.
6. Zienkiewicz, O.C., Taylor, R.L., "The Finite Element Method", Vol. 2, McGraw-Hill Book Company, 1989.
7. Haessley, W.P., Ryan, M.E., "Finite Element Modeling of the Injection Blow Molding Process", *SPE ANTEC*, p. 934, 1989.
8. Twizell, E.H., Ogden, R.W., "Non-linear Optimization of the Material Constants in Ogden's Stress-Deformation Function for Incompressible Isotropic Elastic Materials", *Journal of the Australian Mathematical Society Series B*, p. 424, 1983.

Simulation of Materials Processing: Theory, Methods and Applications, Shen & Dawson (eds)
© 1995 Balkema, Rotterdam. ISBN 90 5410 553 4

A calculation of extrudate swell by FEM using marker particles

Yoji Shimazaki
Tokai University, Kanagawa, Japan

Ryoji Hanada & Sachio Nakamura
The Yokohama Rubber Co., Ltd, Kanagawa, Japan

ABSTRACT

Die swells of fluid flowing through Monsanto Processability Tester(MPT) are calculated. A finite element method using six-node triangular isoparametric element is used to analyze a transient flow motion in which marker particles are introduced to represent the motion. Marker position are determined by the use of triangular area coordinate system. Three simple examples of extruding process for Newtonian fluid flows are described.

1 INTRODUCTION

A finite element method using six-node triangular isoparametric element is applied to analyze a transient Newtonian fluid flow motion in which marker particles are introduced to represent the motion. For determining the marker positions in an element, area coordinates of the triangular element is used. Shiojima et al. [1,2] have introduced the area coordinates of a linear triangular element, which is utilized for finding the new marker position, into a four-node quadrilateral isoparametric element. The two kinds of interpolation systems, however, give complicated solution algorithm and may give unexpected errors on calculated results. Also, this method requires finer mesh arrangement to obtain an accurate result. This implies that a large amount of extra storage may be required for the analysis. With the six-node element, the determination of the new marker position in an element becomes simpler compared with those using the linear triangular element. It is because that we can make use of three mid-node points in each side of the triangle to the corresponding three area coordinates. In other words, if one sign of area coordinates for the marker is minus, it moves the direction of one of the three mid-node point. The only two elements share the mid-node point, we can easily determine the new element for the marker. Also the velocities at the six nodes obtained by the calculation can be directly used for the interpolation of the marker velocity [3]. The markers can be arranged when the mesh is generated.

In this paper we describe a simple extrusion process of the Monsanto Processability Tester(MPT). Marker particles are used to show the extruded free surfaces. The flow patterns are visualized by arranging markers inside the transient flow.

2 GOVERNING EQUATIONS

The time dependent creeping flows of incompressible Newtonian fluids in rectangular Cartesian coordinates are

equilibrium ;

$$\rho u_{i,t} = \sigma_{ij,j} + b_i \qquad (1)$$

continuity (incompressible fluids) ;

$$u_{i,i} = \varepsilon_{ii} = 0 \qquad (2)$$

constitutive relationships ;

$$\varepsilon_{ij} = (u_{i,j} + u_{j,i})/2 \qquad (3)$$

$$\sigma_{ij} = -p\delta_{ij} + \sigma'_{ij} \quad (4)$$

$$\sigma'_{ij} = 2\mu\varepsilon_{ij} \quad (5)$$

boundary conditions ;

$$u_i = \bar{u}_i \quad (6)$$

$$\nu_j\sigma_{ij} = \bar{T}_i \quad (7)$$

where ρ is the density, μ is the viscosity coefficient, u_i is the velocity component in the x_i-direction, b_i is the body force per unit volume, p is the pressure, σ_{ij} is the total stress, σ'_{ij} is the deviatoric stress, ν_j is the component of unit outward normal vector on the boundary.

3 FINITE ELEMENT METHOD

3.1 *Velocity calculation*

We interpolate the velocity and the pressure as

$$u_i = N_k u_{ki} \quad , \quad p = M_m p_m \quad (8a, 8b)$$

where, u_{ki}, p_m are the velocity and the pressure at nodal points, respectively. N_k and M_m are the shape functions. Galerkin's method applied to Equation(1), with the boundary conditions of Equation(7), gives

$$\rho\int_V N u_{i,t} dV + \int_V N_{,j}\sigma_{ij} dV = \int_S \nu_j\sigma_{ij} ds + \int_V N b_i dv \quad (9)$$

We now approximate Equation(2) using the small number $k(10^{-12})$. The penalty function method[4] gives

$$\frac{1}{k}\int_V M u_{i,i} dv + \int_V M p dv = 0 \quad (10)$$

Substitution of p in Equation(10) into Equation(9) gives, in matrix form,

$$[C]\{u_{,t}\} + [K]\{u\} = \{F\} \quad (11)$$

where $[C]$ is the matrix obtained from first term and $[K]$ is from second term of Equation(9), respectively. $\{F\}$ is from the right hand side of Equation(9).
The θ-method[5] for the time derivative in Equation(11) gives

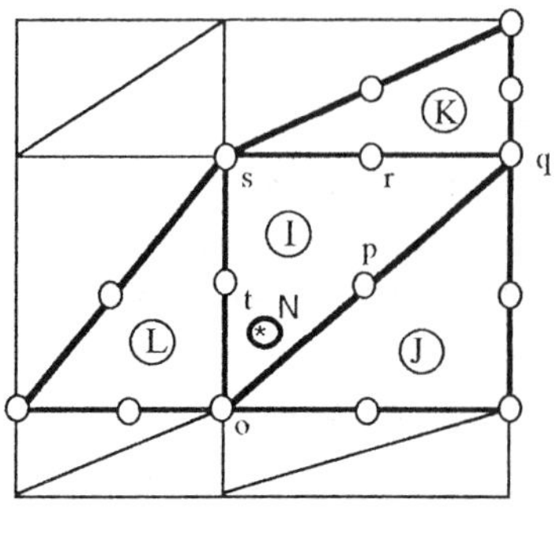

Figure 1. Marker N and Element I, J, K, L

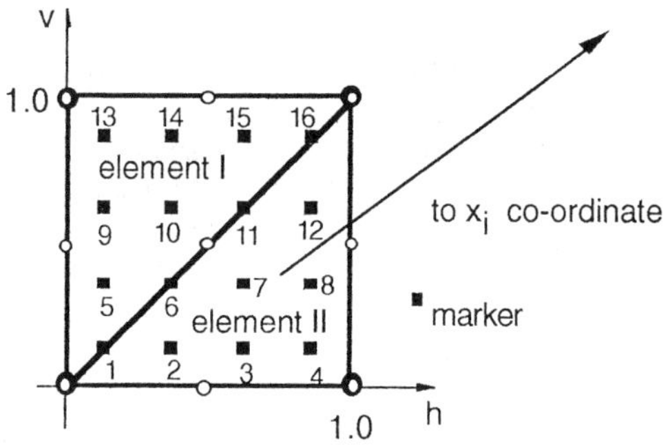

Figure 2. Markers in element I and II

$$\left[\frac{1}{\Delta t}[C] + \theta[K]\right]\{u\}_{t+\Delta t} = \{F\}_{avg} + \left[\frac{1}{\Delta t}[C] + (1-\theta)[K]\right]\{u\}_t \quad (12)$$

$$\{F\}_{avg} = \left\{(1-\theta)\{F\}_t + \theta\{F\}_{t+\Delta t}\right\} \quad (13)$$

3.2 *Marker method*

We now introduce marker particles in finite elements, which represent the fluid flow motion. Figure 1 shows a position of the marker at time t in a six-node triangular isoparametric element. The marker position at time $t+\Delta t$ becomes

$$(x_i)_{t+\Delta t} = (x_i)_t + \int_t^{t+\Delta t} (u_i)_t dt \quad (14)$$

$$(u_i)_t = [N_k]\{u_{ki}\}_t \quad (15)$$

where, $(x_i)_t$ is the marker position and $(u_i)_t$ is the velocity at time t, and u_{ki} is the velocity at the nodal point. With the shape functions N_k expressed by the area cordinate (ξ, η, ζ) and the nodal point position x_{ki} by the global coordinate, the marker position $(x_i)_{t+\Delta t}$ becomes

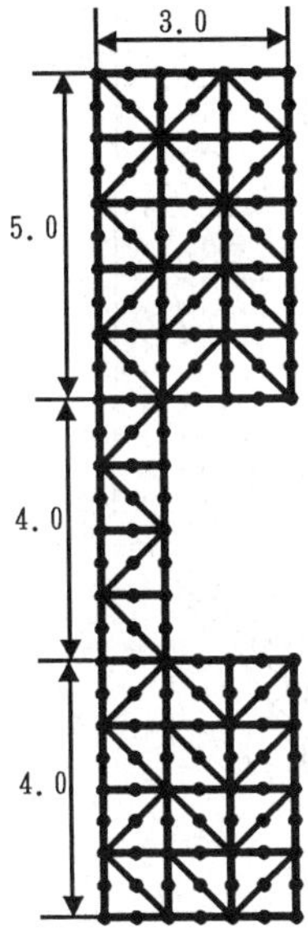

Figure3 Finite element mesh (analysis 1)

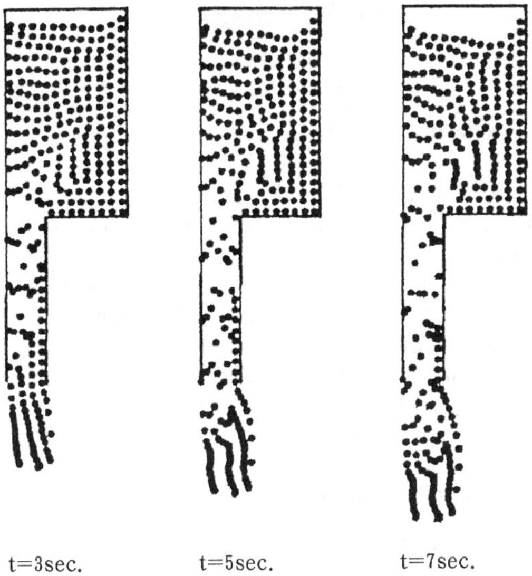

Figure4 Calculated Marker distribution

$$(x_i)_{t+\Delta t} = [N_k]\{x_{ki}\} . \quad (16)$$

Because $(x_i)_{t+\Delta t}$ is known from Equation(14) and x_{ki} is coordinate values of the six nodal points in an element, we can calculate the marker position in the triangular coordinate (ξ, η, ζ) by Newton-Raphson method. According to the signs of calculated

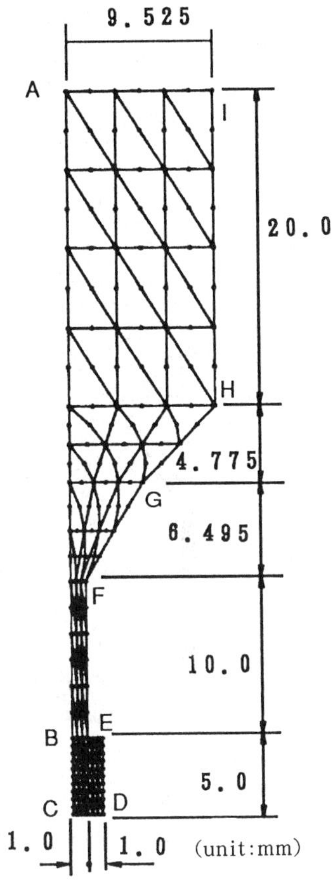

Figure5 Finite element mesh (analysis 2)

ξ, η and ζ, the new element of the marker can de determined.

3.3 *Automatic marker generation*

The markers are arranged automatically. For each marker, we have to know it's coordinate x_i and the element it belongs to. These informations are obtained when the finite element mesh is generated. Figure 2 shows a unit square element to be mapped into x_i co-ordinates. The element is divided into two triangles. In the figure, 16 markers are arranged in two elements.

4 ANALYSES

4.1 *Analysis 1*

Figure 3. shows a simple axisymmetric extrusion problem. The problem consists of

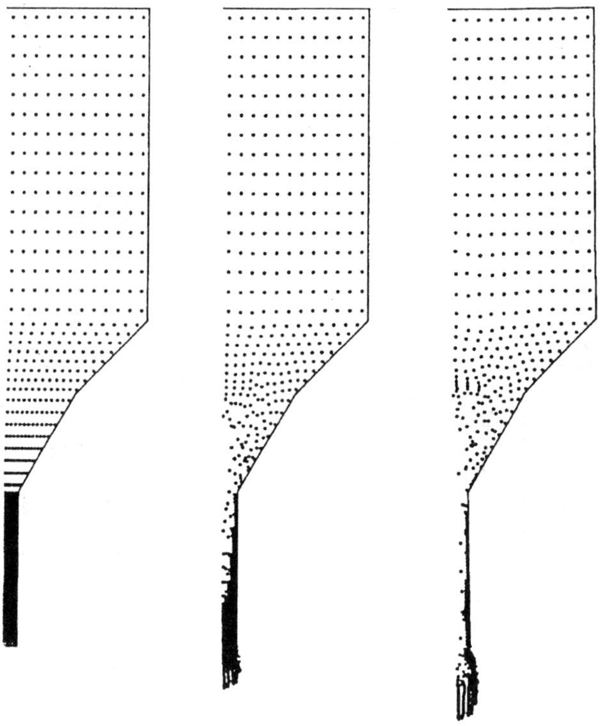

Figure6 Calculated Marker distribution in reservoir

62 elements and 161 nodal points. The number of markers arranged in the region is 304.

The material properties used for Newtonian fluid are $\rho=1.0$, $\mu=1.0$. We choose $\theta=0.75$ and $\Delta t=0.5$. The Δt is divided into from 20 to 50 intervals to obtain better accuracy and to avoid numerical instability. To proceed with the transient analysis, we assume the free surface is physically a line consisting of the forefront markers and also the place to specify the boundary conditions is located on the boundary line between the element having a forefront markers and the element having not. It should be noted that the traction force along the side is zero. Then we do not specify any boundary values and the assumption does not affect the visualization of the free surface as a marker front.

Figure 4 shows the result obtained by this method. The swell of fluid extruded through this sample region and also the flow pattern of the flow can be investigated with this method.

4.2 *Analysis 2*

Figure 5 shows a geometry and finite element mesh for MPT equipment. For illustrative purpose, we do not include any body forces of the fluid. The number of elements and nodal points in the region are 129 and 297, respectively. The material properties used in this example are $\rho=1.0$, $\nu=1.0$ and we choose $\Delta t=0.005$. The number of markers arranged in the region is 744. The result obtained is shown in Figure 6.

4.3 *Analysis 3*

We now solve a problem of practical importance. The same geometry as in the analysis 2 is used in this sample. In this analysis we include body force of the fluid in the direction of extrusin. The problem consists of 480 elements and 1093 nodal points. The number of markers arranged in the region is 1920. The material properties used in this analysis are

$\rho = 1.1\times10^{-9}$ kg/m^3

$\mu = 9740$ Pa・s

body force $= 1.1\times10^{4}$ N/m^3

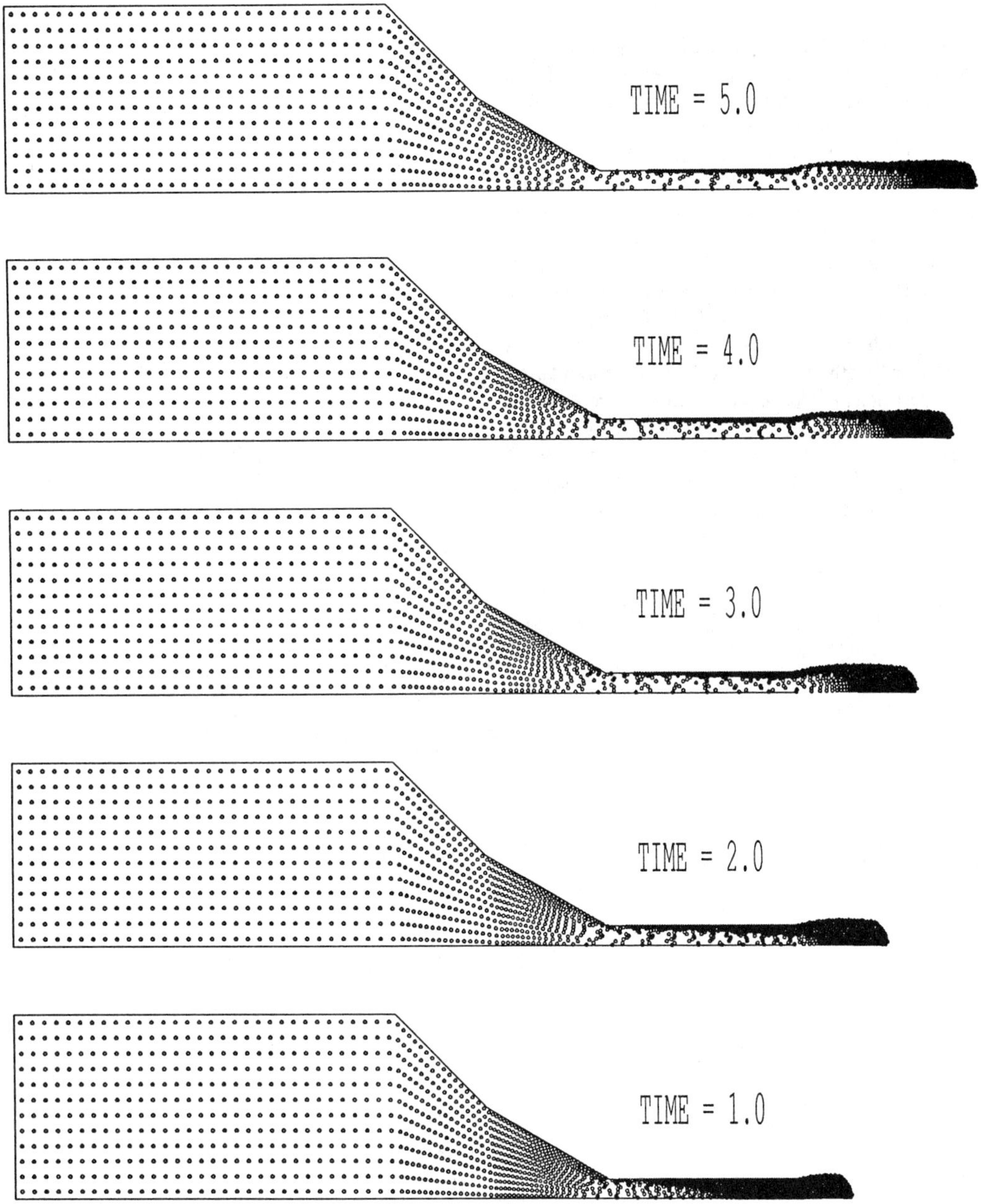

Figure7 Calculated Marker distribution in reservoir for analysis 3

and we choose $\Delta t=0.001$.

The results obtained are shown in Figure 7. Calculated die swell of the fluid through MPT is about 35%.

5 CONCLUSION

Marker particles were used to represent the transient incompressible creeping flows of Newtonian fluid. The six-node triangular (second order) element approximation was used to move the marker from element to element. A simple example of extruding process and the swell of the fluid flowing through Monsanto Processability Tester were determined.

REFERENCES

(1) Shiojima, T., Shimazaki, Y. and Daiguji: Finite element analysis of creeping flows using marker particles, Int. J. Num. Meth. Fluids, vol. 11, pp. 397-404, 1990.

(2) Shiojima, T., Shimazaki, Y. and Daiguji: Analyses of transient viscous fluid creeping flows with free surfaces, Proc. JSME, No. 86-1356A, pp. 2365-2369, 1987.

(3) Shimazaki, Y.: A visualization of flows by FEM using six-node triangular element, To appear in Int. J. Num. Meth. Fluids.

(4) Kikuchi, F. and Navarro, M. P.: An iteration method for the mixed formulation of parameter dependent problems related to the Stokes equations, Comp. Mech., pp. 141-151, 1986.

(5) Zienkiewicz, O. C.: The finite element method, Third ed., McGraw-HILL (1977).

Simulation of Materials Processing: Theory, Methods and Applications, Shen & Dawson (eds)
© 1995 Balkema, Rotterdam. ISBN 90 5410 553 4

Numerical analysis of viscous flows with free boundaries

J. Socolowsky
FB Technik (FG Mathematik), Fachhochschule Brandenburg, Brandenburg/Havel, Germany

ABSTRACT: Free boundary value problems for flows of heavy, viscous, incompressible, capillary liquids describing some coating processes are numerically investigated. The problems are assumed to be stationary and two-dimensional. A numerical method for such flow problems is analysed permitting the *simultaneous* computation of all field variables as well as of all free interfaces. Similar strategies are known in the literature as NEWTON's method (cf. e.g. [So-88],[KiScr-84],[SaBrDe]). The efficiency of the method is given by two coating flow problems which had been solved numerically. Dynamic contact lines are included as well as multiple-layer flows.
Furthermore, the paper gives results on the convergence of the presented procedure. The convergence results depend on corresponding statements for the existence and uniqueness of weak solutions to the associated continuous problems in unbounded flow domains. Such solutions exist in appropriate function spaces as weighted SOBOLEV or HÖLDER spaces. Some first error estimates for the numerical solutions are given, too.

1 THE MATHEMATICAL MODEL

The present paper deals with free *boundary value problems* (= BVPs)for isothermal and nonisothermal multiple-layer flows (cf.Fig.1). A numerical method which allows the simultaneous determination of all field variables and of all free boundaries by one iteration cycle is analysed. Thus the present paper represents a continuation and extension of earlier studies of the author (cf.[So-88],[So-89]).

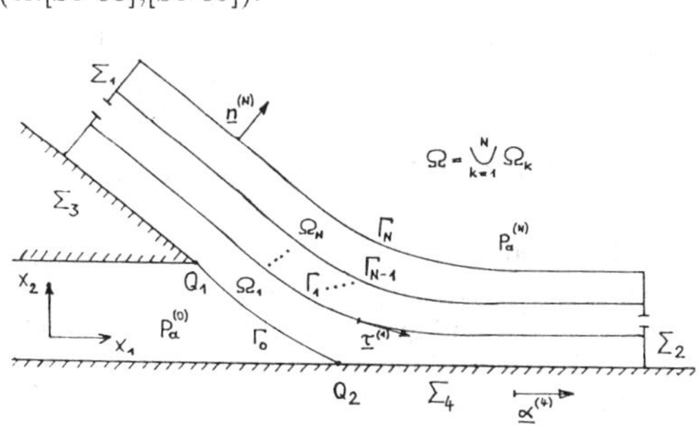

Fig.1 Flow domain of a multiple-layer slide coating process

Let us consider a typical flow domain of a coating flow occuring in some kinds of industrial forming processes. Fig.1 shows such a domain. The meaning of all symbols from Fig.1 and in the following BVP is given in the list of symbols (cf. Section 5). For all $x \in \Omega$ the governing *equations* (= eqs.) of the nonisothermal free BVP are

the eq. of motion

$$\kappa(v \cdot \nabla)v + \nabla p - \nabla \cdot S(v,\theta) - \kappa Fr^{-1} e_g = 0, \quad (1.1)$$

the eq. of continuity

$$\operatorname{div} v = \nabla \cdot v = 0 \quad (1.2)$$

and the eq. of energy

$$\mu\kappa(v \cdot \nabla\theta) - \frac{\lambda}{Fo}\nabla^2\theta - \mathcal{A}\, S_{ij}(v,\theta)\,\frac{\partial v_i}{\partial x_j} = 0. \quad (1.3)$$

The *boundary conditions* (= BCs) for the velocity v describe the no-slip at the walls Σ_j ($j = 3, 4$), the given inflow and outflow profiles at Σ_j ($j = 1, 2$) and the continuity of v crossing the unknown free interface Γ_m

$$v\,|_{\Sigma_j} = \alpha^{(j)}(x) \qquad (j = 1, \dots, 4), \quad (1.4)$$

$$[v(x)]\,\|_{\Gamma_m} = 0 \qquad (m = 1, \dots, N-1). \quad (1.5)$$

Furthermore, we have the well-known kinematic and dynamic BCs. The last one gives a balance of the stress

$$v \cdot n\,|_{\Gamma_m} = 0, \quad (1.6)$$

$$[T(v,\theta)\,n]\,\|_{\Gamma_m} - We_m \frac{dt}{ds} = 0 \; (m = 0, \dots, N) \quad (1.7)$$

Analogously, we are given prescribed temperature profiles at the inflow and outflow regions as well as some kind of heat transfer conditions at the walls and at the free boundaries

$$\theta \mid_{\Sigma_j} = \beta^{(j)}(x) \qquad (j = 1,2), \qquad (1.8)$$

$$a_j \frac{\partial\theta}{\partial n} + (\theta - \theta_w^{(j)}) \mid_{\Sigma_j} = 0 \qquad (j = 3,4) \qquad (1.9)$$

$$[\theta] \parallel_{\Gamma_m} = \left[\lambda \frac{\partial\theta}{\partial n}\right] \parallel_{\Gamma_m} = 0 \, (m = 1, \ldots, N-1), (1.10)$$

$$\frac{\partial\theta}{\partial n} + d_m(\theta - \theta_a^{(m)}) \mid_{\Gamma_m} = 0 \quad (m = 0, N). \qquad (1.11)$$

The symbol $[v(\bar{x})] \parallel_{\Gamma_m}$ for $\bar{x} \in \Gamma_m$ is defined by the formula

$$[v(\bar{x})] \parallel_{\Gamma_m} = \lim_{y \to \bar{x}, y \in \Omega_m} v(y) - \lim_{z \to \bar{x}, z \in \Omega_{m-1}} v(z).$$

In Eq.(1.3) as well as in the following study the convention of sums for double subscripts is valid (cf. the letters i and j). The temperature θ is made dimensionless using the relation

$$\theta = (\tilde{\theta} - \tilde{\theta}_0)/(\tilde{\theta}_w^{(4)} - \tilde{\theta}_0), \quad (\tilde{\theta}_w^{(4)} \neq \tilde{\theta}_0). \qquad (1.12)$$

Let us note that in this paper all symbols without a tilde are dimensionless. The constants $a_j, d_m, \theta_a^{(m)}, \theta_w^{(j)}$ and the functions $\alpha^{(j)}, \beta^{(j)}$ in the BCs (1.4)-(1.11) are heat transfer coefficients and they are given. From the definition of the ratios κ, μ and λ of densities, specific heat capacities and thermal conductivities, resp., it follows that for all $x \in \Omega$ one has

$$0 < \kappa(x) \leq 1, \; 0 < \mu(x) \leq 1, \; 0 < \lambda(x) \leq 1. \quad (1.13)$$

The BVP (1.1)-(1.11) is stationary and two-dimensional. It represents a generalization of a model for an single-layer, nonisothermal slot coating process given in [So-89]. Such problems are important in coating processes as well as in modeling mixing processes in internal mixers. Some one-dimensional or other simplified models of mixing processes were studied in [MePo]. The following rheological constitutive law

$$T(v,\theta) = -pI + S(v,\theta) =$$
$$= -pI + f(x,v,\theta)\, D(v), \qquad (1.14)$$

where f denotes a generalized viscosity, characterizes many practically important fluids. The function f can also be piecewise constant, i.e. given by a table (cf. e.g. [So-88],[So-89]). For mixing processes we have (using the notation $f_m := f \mid_{\Omega_m}$)

$$f_m(x,v,\theta) = \frac{\kappa}{Re_m^*} \exp(\gamma_m\theta) \mid 4J_2 \mid^{(q_m-1)/2}, \quad (1.15)$$

where $\gamma_m, q_m > 0$ and Re_m^* are constant. By J_2 we mean the second invariant of the tensor $D(v)$. In case (1.14),(1.15) Eq.(1.3) takes the form

$$\mu\kappa(v \cdot \nabla\theta) - \frac{\lambda}{Fo}\nabla^2\theta -$$
$$-2\frac{Br^*}{Fo} \exp(\gamma\theta) \mid 4J_2 \mid^{(q+1)/2} = 0. \qquad (1.16)$$

In Eq.(1.16) the generalized BRINKMANN number Br^* depends on the number m of the fluid layer, too.

Isothermal free BVPs which are similar to BVP (1.1)-(1.11) have been numerically studied by many authors (cf. e.g. [ChrScr-89], [ChrScr-92], [CuSch], [KiScr-83], [KiScr-84], [PrScoTav] and the references given therein). In these studies the well-known NAVIER-STOKES eqs. were considered for single-layer flows. In [So-88], [So-89] results for isothermal multiple-layer flows are presented. Nonisothermal viscous flows with free boundaries were numerically studied in [LuTa], [RiIl], [So-89], [So-90], [SuPhTa] and in many other papers.In [LuTa],[SuPhTa] non-NEWTONIAN fluids were incorporated for the extrusion process. In [SuPhTa] the *Finite Element Method* (= FEM) was coupled with the *Boundary Element Method* (= BEM). Further studies including non-NEWTONIAN fluids and/or solidification processes one can find in [AlAnPa], [AlTo], [CuSch], [KaHrVl]. Analytical results on the existence and uniqueness of solutions to nonisothermal free BVPs are presented by the author (cf. [So-89], [So-94]) and others.

Many authors apply the following strategy for solving free BVPs for viscous flows. In a first step the position of the free boundary is fixed and the flow fields, i.e. v, θ, p, are computed neglecting either the kinematic BC (1.6) or the normal component of the dynamic BC (1.7). In the second step the calculated flow fields and the neglected BC are used to generate a new shape of the free boundary. Then this procedure is continued until convergence is reached. The method that we have applied (cf. also [ChrScr-89], [ChrScr-92], [KiScr-83], [KiScr-84]) is an alternative since we need only one iteration cycle and hence the computational time is smaller. In [SaBrDe] a similar FEM - strategy is applied in order to solve a special free BVP in crystal growth (i.e. a solidification process).

The model (1.1)-(1.11) was partly developed by the author (cf.[So-89]). For the case of a single-layer nonisothermal slot coating process first numerical results were presented in [So-90].

2 THE CONCEPT OF A WEAK SOLUTION

In order to apply the GALERKIN-method in form of the FEM we have to transform the eq. of motion (1.1), the energy eq.(1.3) and the continuity eq.(1.2) into variational eqs. In the same way as in [So-88], [So-90] we multiply Eq.(1.1) by a differentiable function $\Phi(x)$ and subsequently we integrate this eq. over the flow domain Ω. The integration by parts with respect to the highest derivatives of v, p is separately realized in each fluid layer Ω_m. Using the natural BC (1.7) and after integration by parts over Γ_m we obtain for functions Φ_1 vanishing on Σ

$$\int_\Omega [\kappa\Phi_1(v\cdot\nabla)v + T(v,\theta)\nabla\Phi_1 - \kappa\Phi_1 Fr^{-1}e_g]\,dx +$$
$$+\int_{\Gamma_N}\Phi_1 p_a^{(N)}n^{(N)}\,ds - \int_{\Gamma_0}\Phi_1 p_a^{(0)}n^{(0)}\,ds +$$
$$+\sum_{m=0}^{N}\int_{\Gamma_m} We_m t^{(m)}\frac{d\Phi_1}{ds}\,ds = 0. \tag{2.1}$$

The integral representation of Eq.(1.2) yields as in [So-88], [So-90]

$$\int_\Omega \xi(x)\,\mathrm{div}\,v\,dx = 0. \tag{2.2}$$

Eq.(1.3) is multiplied by the same function Φ as Eq.(1.1) and then it is integrated over Ω. Integration by parts with respect to the highest derivatives of θ is applied. Since $\nabla\theta\cdot n = \partial\theta/\partial n$ and using the natural BCs (1.9)-(1.11) we obtain finally

$$\int_\Omega \{\Phi_T(x)[\mu\kappa(\mathbf{v}\cdot\nabla\theta) - \mathcal{A}S_{ij}(v,\theta)\,\partial v_i/\partial x_j] +$$
$$+\lambda Fo^{-1}(\nabla\theta\cdot\nabla\Phi_T)\}\,dx +$$
$$+\sum_{j=3}^{4} a_j^{-1}Fo^{-1}\lambda_1\int_{\Sigma_j}\Phi_T(\theta-\theta_w^{(j)})ds -$$
$$-d_0^{-1}Fo^{-1}\lambda_1\int_{\Gamma_0}\Phi_T(\theta-\theta_a^{(0)})\,ds +$$
$$+d_N^{-1}Fo^{-1}\lambda_N\int_{\Gamma_N}\Phi_T(\theta-\theta_a^{(N)})\,ds = 0. \tag{2.3}$$

In the last eq. test functions Φ_T are used which vanish on the boundary part $\Sigma_1\cup\Sigma_2$.

3 DISCRETIZATION BY FEM AND CONVERGENCE RESULTS

In the weak formulation (2.1)-(2.3) of the free BVP (1.1)-(1.11) the orders of derivatives of v, p and θ are 1, 0 and 1, resp.. Thus we are able to use the simplest finite elements - namely piecewise linear shape functions for v, θ and piecewise constant shape functions for p.

For the triangulation of the flow domain and for the approximation of free boundaries by means of so-called "spines" (or "support-lines") we refer to [ChrScr-92], [KiScr-83], [KiScr-84], [SaBrDe], [So-88] and others. Let us only emphasize that each triangle (or element) is subdivided into 4 congruent triangles (or quarter-elements) as shown in Fig.2.

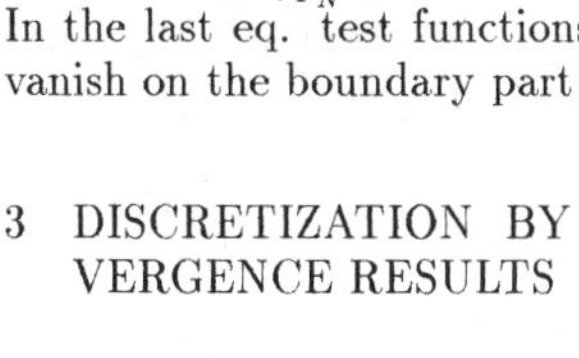

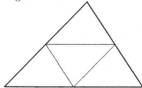

Fig.2 Element with nodes and quarter-elements

Thus each element contains 6 nodes. Let $\Phi_k(x_1,x_2)$ be a piecewise linear function which equals 1 at the node k and vanishing at all other nodes. On each quarter-element Φ_k is linear. The function $\xi_j(x_1,x_2)$ is identically equal to 1 on the element numbered by j and vanishes on all remaining elements. Then the velocity v, the pressure p and the temperature θ are approximated by

$$\hat{v}(x) = \sum_{k=1}^{NG} v^k\Phi_k(x), \qquad \hat{p}(x) = \sum_{j=1}^{NE} p_j\xi_j(x),$$
$$\hat{\theta}(x) = \sum_{k=1}^{NG}\theta_k\Phi_k(x), \tag{3.1}$$

where v^k, θ_k and p_j denote the approximated values of v, θ and p at the node k and on the element j, respectively. As shown in [GiRa] the approximation (16) for v, p fulfills the well-known BABUSKA – BREZZI condition (cf.[Ba], [GiRa]) and hence it is stable in some sense. The temperature θ does not have influence this consideration. In the following we restrict the study to the case of rheological constitutive law (1.14). Using each shape function Φ_k as test function Φ_1 (or Φ_T) in Eq.(2.1) (or (2.3)) and each shape function ξ_j as test function ξ in Eq.(2.2) we obtain a system of discretized scalar eqs. for the $(3*NG+NE)$ scalar unknowns $v_1^k, v_2^k, \theta_k, p_j$.

Under the stiffness matrix of element KE we understand that submatrix of the JACOBIAN of the whole system of discretized eqs. concerning all derivatives of discretized eqs. associated with the element KE with respect to all degrees of freedom of element KE.

If we consider the case of N liquid layers without interrupted "spines" (cf.[So-88],[So-90]) then the order NCN of stiffness matrix becomes

$$NCN = 3\cdot 6 + 4\cdot(N+1) + 1. \tag{3.2}$$

Obviously, for $N = 1$ we get $NCN = 27$. It is important to state that the presented procedure permits the analytical calculation of all contributions in the stiffness matrix. Here the partial derivatives of all residuals with respect to the geometric unknowns are included, too [So-89]. If the starting solution to the full NEWTON iteration scheme is sufficiently good then the convergence of the numerical solution is acceptable. The order of convergence is quadratic. The problem of error estimates, i.e. the convergence of the numerical solutions to the theoretical one as the discretization parameter h tends to 0, is still open even for the case of a single-layer NEWTONIAN flow in a bounded domain (cf. [PrScoTav]). The existence and uniqueness of the exact solution in appropriate SOBOLEV or HÖLDER spaces could be proved for many cases (cf. e.g. [So-89], [So-90] and the references therein). But - as far as we know - error estimates for free BVPs in Hydrodynamics (of viscous liquids) are not yet available.

PRITCHARD et al. (cf. [PrScoTav]) gave the following working hypothesis

$$\| v - v_h \|_{H^1(\Omega)} + h \| p - p_h \|_{L^2(\Omega)} \leq$$
$$\leq Ch^2 (\| v \|_{H^3(\Omega)} + \| p \|_{H^2(\Omega)}),$$
$$\| \xi - \xi_h \|_{L^r(\Omega)} \leq Ch^3 \| \xi \|_{W_r^3(\Omega)}, \; (1 \leq r \leq \infty). \quad (3.3)$$

In (3.3) the symbol ξ denotes representation function of the (one) free surface. The author is working in this field, too. Unfortunately, the corresponding study is not yet finished.

4 NUMERICAL EXAMPLES

A computer code was developed by our group some years ago. This code was written in FORTRAN and it works on personal computers, on work stations as well as on large computers. The numerical algorithm described above was implemented. A first (non-complete) description was given in [So-89]. The actual version of the program permits the computation of multiple-layer, two-dimensional, stationary and nonisothermal flows with free interfaces or surfaces. The fluids may satisfy a rheological constitutive law of the form (1.14) with distributed material properties. For the case (1.15) a special subroutine including all analytical gradients of the functions f_m was programmed.

Consider firstly a curtain coating process for two NEWTONIAN liquids of the same thickness. The ratio of density is $\rho_2/\rho_1 = 9/10$. For the dimensionless numbers we get $Re_1 = 1.25 \times Re_2$, $We_0 = We_2 = 2 \times We_1$. These numbers are small. The difference $p_a^{(2)} - p_a^{(0)}$ of the ambient atmospheric pressures is 6.728. In Fig.3 a sketch of the velocity behaviour is shown at various "support-lines" numbered by 1, 5,8,9 and 14. One can see that the thickness of the lower fluid layer reduces while the thickness of the upper layer is increasing.

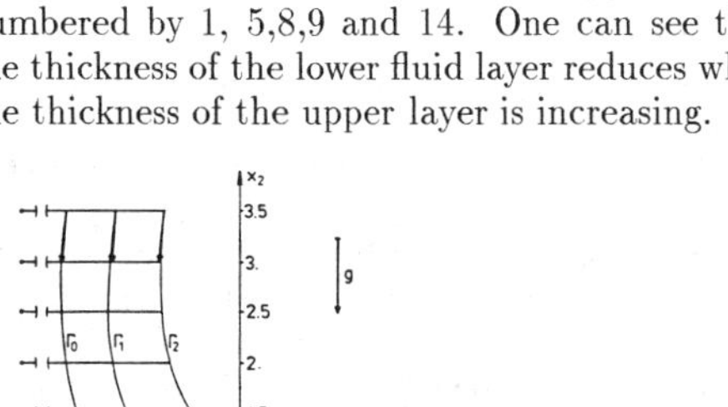

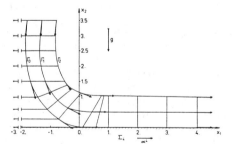

Fig.3 Two-layer curtain coating process with support-lines

Fig.4 shows the absolute values of the velocity at the main nodes lying on the free inter- or surfaces and on the moving wall. The numbers are the numbers of the corresponding "spines". Note that at the 9th "spine" we find a "bend".

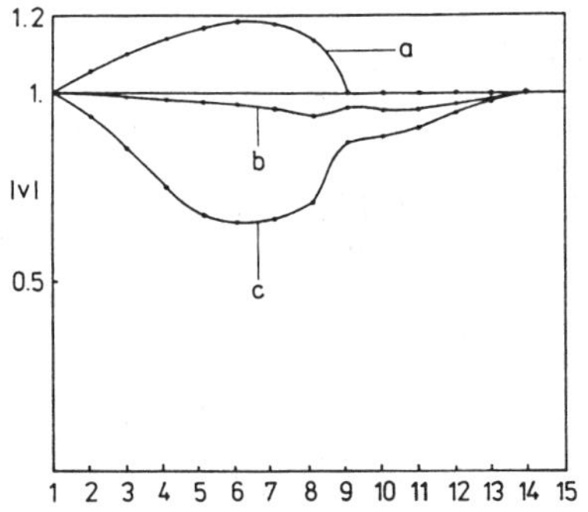

Fig.4 Absolute values of the velocity at the main nodes (a – on the lower boundary $\Gamma_0 \cup \Sigma_4$, b – on the free interface Γ_1, c – on the free surface Γ_2)

The pressure varies at most on the free surface Γ_2 and it has a singularity at the element near the dynamic contact point.

Now consider an example of a nonisothermal slot coating process for an NEWTONIAN fluid. It is based on model (1.1)-(1.11) for $N = 1$ fluid layer. The walls are of constant temperature. In this case the corresponding boundary integrals in Eq.(2.3) vanish and the arising BCs of the first kind with respect to θ are handled as well as BC (1.4) in a well-known manner (cf. [So-88]).

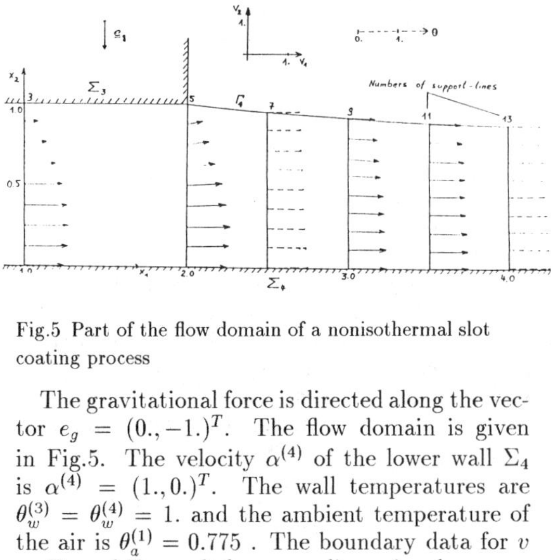

Fig.5 Part of the flow domain of a nonisothermal slot coating process

The gravitational force is directed along the vector $e_g = (0., -1.)^T$. The flow domain is given in Fig.5. The velocity $\alpha^{(4)}$ of the lower wall Σ_4 is $\alpha^{(4)} = (1., 0.)^T$. The wall temperatures are $\theta_w^{(3)} = \theta_w^{(4)} = 1.$ and the ambient temperature of the air is $\theta_a^{(1)} = 0.775$. The boundary data for v on Σ_1 and Σ_2 result from one-dimensional approximate solutions for a flux of 0.8 representing simultaneously the final layer thickness as $x_1 \to +\infty$. The temperature is linearly interpolated on Σ_1, Σ_2 using BC (1.11) (for $m = 1$) at the upper endpoint of Σ_2. Further we have $d_1 = 10.0$. The parameters We_1, Br_1 are small and Fo is about 500.

Modified NEWTON's methods were applied for solving the arising algebraic system of eqs. The numerical convergence was good. The principal shape of the free boundary Γ_1 and the main velocity behaviour are shown in Fig.5. Furthermore, one can see the nearly linear temperature profile along two "spines".

Table 1: Velocities, pressures and temperatures on the free boundary at various "spines" (No.)

No.	v_1	v_2	$\|v\|$	$p - p_a^{(1)}$	θ
1	0.0	0.000	0.0	2408.80	1.0
	0.0	0.000	0.0		1.0
2	0.0	0.000	0.0	1743.40	1.0
	0.0	0.000	0.0		1.0
3	0.0	0.000	0.0	1015.80	1.0
	0.0	0.000	0.0		1.0
4	0.0	0.000	0.0	298.34	1.0
	0.0	0.000	0.0		1.0
5	0.0	0.000	0.0	-208.19	1.0
	0.139	0.209	0.251		0.917
6	0.356	0.143	0.383	- 760.22	0.878
	0.447	0.057	0.451		0.867
7	0.526	-0.017	0.527	-313.19	0.853
	0.577	-0.051	0.580		0.848
8	0.621	-0.088	0.627	-129.34	0.839
	0.654	-0.094	0.661		0.836
9	0.686	-0.110	0.695	-67.10	0.828
	0.712	-0.102	0.719		0.826
10	0.739	-0.105	0.747	-48.01	0.820
	0.762	-0.098	0.768		0.817
11	0.786	-0.089	0.791	-40.39	0.812
	0.806	-0.075	0.809		0.809
12	0.828	-0.070	0.831	-34.57	0.805
	0.845	-0.056	0.847		0.802
13	0.865	-0.048	0.866	-44.88	0.799
	0.908	-0.032	0.909		0.799
14	0.932	-0.023	0.932	-6.32	0.798
	0.955	-0.012	0.956		0.799
15	0.966	-0.010	0.966	5.98	0.799
	0.979	-0.006	0.979		0.799
16	0.984	-0.006	0.984	8.14	0.799
	0.992	-0.001	0.992		0.799
17	1.0	0.000	1.000		0.800

Table 1 gives a survey of the velocities, pressures and temperatures at all nodes (and on all elements, resp.) on the upper boundary $\Sigma_3 \bigcup \Gamma_1$ of the flow domain. The singularity of p in the neighbourhood of the separation point is characteristic. Since the numbers Re and Fo are small in this example the convective term in Eq.(1.1) and/or the conductive term in Eq.(1.3) do not have significant influence to the numerical results. Only small changes in the shape of Γ_1 and in the flow fields were obtained.In future computational work it is necessary to realize computations of multiple-layer flows with more than 3 layers in order to justify the effectiveness of the presented procedure.

5 LIST OF SYMBOLS

a) Geometric notations (cf. Fig.1)

Q_1, Q_2 - contact points
N - number of liquid layers
Ω_m - liquid layer ($m = 1, \ldots, N$)
Ω - flow domain ($\Omega = \bigcup_{m=1}^{N} \Omega_m$)
Σ_j - inflow and outflow region ($j = 1, 2$), rigid walls ($j = 3, 4$)
Γ_m - m - th free surface or free interface ($m = 0, \ldots, N$)

b) Tensors

I - identity tensor
$D(v)$ - rate of strain tensor with components $D_{ij}(v) = 0.5(\partial v_i/\partial x_j + \partial v_j/\partial x_i)$
$S(v,\theta)$ - deviatoric stress tensor
$T(v,\theta)$ - stress tensor

c) Vectors and vector fields

x - coordinates ($= \tilde{x}/\tilde{L}$, $\tilde{L}$ - char. length)
$v(x)$ - velocity ($= \tilde{v}/|\tilde{w}|$, $\tilde{w}$ - char. velocity)
e_g - directional (unit) vector of the gravitational force
$\alpha(x)$ - boundary data for the velocity
$t^{(m)}, n^{(m)}$ - tangential and normal (unit) vectors to the m-th interface (oriented as in Fig.1)

d) Scalars and scalar fields

$p(x)$ - pressure field ($= \tilde{p}/(\tilde{\rho}_{\max}|\tilde{w}|^2)$)
$\tilde{\theta}_0, \tilde{\theta}_w^{(4)}$ - reference and wall (on Σ_4) temp.
$\theta(x)$ - temperature field (cf. Eq.(1.12))
$p_a^{(0)}, p_a^{(N)}$ - atmospheric pressures outside Γ_0, Γ_N
$\theta_a^{(0)}, \theta_a^{(N)}$ - ambient temp. outside Γ_0, Γ_N
$\theta_w^{(3)}, \theta_w^{(4)}$ - wall temp. of the boundaries Σ_3, Σ_4
$\beta^{(j)}(x)$ - boundary data for the temp. on Σ_j ($j = 1, 2$)
$\tilde{\rho}_m, \tilde{\nu}_m, \tilde{\lambda}_m$ - density, kinematic viscosity and thermal conductivity of the m-th liquid layer
$\tilde{c}_{p,m}$ - specific heat capacity of the m-th liquid layer
κ, μ, λ - ratios of densities, specific heat capacities and thermal conduct. defined by $\kappa(x) = \tilde{\rho}_m/\tilde{\rho}_{\max}$ as $x \in \Omega_m$, where $\tilde{\rho}_{\max} = \max_{k=1,\ldots,N}(\tilde{\rho}_k)$ and analogously for μ and λ
$\tilde{\sigma}_m$ - surface or interface tension for the m-th interface
γ_m, q_m - power law indices of the m-th liquid layer (cf.Eq.(1.15))
J_2 - second invariant of the tensor D ($= 0.5 D_{ij} D_{ij}$)
$\tilde{g}, s$ - gravity acceleration, arc length
NG, NE - numbers of nodes and elements, resp.

e) Dimensionless groups

Re_m - REYNOLDS number of the m-th liquid layer ($= |\tilde{w}|\tilde{L}\tilde{\nu}_m^{-1}$)
Re^* - generalized REYNOLDS number
Fr - FROUDE number ($= |\tilde{w}|^2 \tilde{L}^{-1} \tilde{g}^{-1}$)

Fo - FOURIER number
$(= |\tilde{w}|\tilde{L}\tilde{\rho}_{\max}\tilde{c}_{p,\max}^{-1}\tilde{\lambda}_{\max}^{-1})$

$\mathcal{A}$ - $(= |\tilde{w}|^2\tilde{c}_{p,\max}^{-1}(\tilde{\theta}_w^{(4)} - \tilde{\theta}_0)^{-1})$

Br_m - BRINKMANN number
$(= \tilde{\rho}_m\tilde{\nu}_m|\tilde{w}|^2\tilde{\lambda}_{\max}^{-1}\tilde{\rho}_{\max}^{-1}(\tilde{\theta}_w^{(4)} - \tilde{\theta}_0)^{-1})$

Br^* - generalized BRINKMANN number

We_m - WEBER number of the m-th interface
$(= \tilde{\sigma}_m/(|\tilde{w}|^2\tilde{L}\tilde{\rho}_{\max})^{-1})$

References

[AlAnPa] Alexandrou, A.N., Anturkar, N.R. and T.C. Papanastasiou (1989). An inverse finite element method with an application to extrusion with solidification. *Int. J.num. meth. fluids* 9: 541-555.

[AlTo] Alexandrou, A.N. & G. Torikoglu (1992). Thermo-viscous effects in extrusion of planar jets with solidification. *Intern. Polymer Processing* VII: 32-37.

[Ba] Babuska, I. (1971). Error bounds for finite element method. *Numer. Math.* 16: 322-332.

[ChrScr-89] Christodoulou, K.N. & L.E. Scriven (1989). The fluid mechanics of slide coating. *J. Fluid Mech.* 208: 321-354.

[ChrScr-92] Christodoulou, K.N. & L.E. Scriven (1992). Discretization of free surface flows and other moving boundary problems.*J. Comp. Physics* 99: 39.

[Ch] Chung, T.J. (1982). *Finite Elemente in der Strömungsmechanik*. Leipzig: Fachbuchverlag.

[CuSch] Cuvelier, C. & R.M.S.M. Schulkes (1990). *Some numerical methods for the computation of capillary free boundaries governed by the NAVIER-STOKES equations*: 1-82, Report 90-11, TU Delft.

[GiRa] Girault, V. & P.-A. Raviart (1986). *Finite element methods for NAVIER-STOKES equations, theory and algorithms.* Berlin-Heidelberg-New-York: Springer.

[KaHrVl] Karagiannis, A., Hrymak, A.N. and J. Vlachopoulos (1989). Three-dimensional non-isothermal extrusion flows. *Rheologica Acta* 28: 121-133.

[KiScr-83] Kistler, S.F. & L.E. Scriven (1983). Coating flows. In *Computational Analysis of Polymer Processing*: 243, J.R.A. Pearsen and S.M. Richardson, Eds., London and New York: Applied Science Publishers.

[KiScr-84] Kistler, S.F. & L.E. Scriven (1984). Coating flow theory by finite element and asymptotic analysis of the Navier-Stokes system. *Int. J. num. meth. fluids* 4: 207-229.

[LuTa] Luo, X.-L.& R.I. Tanner (1987). A pseudo-time integral method for non-isothermal viscoelastic flows and its application to extrusion simulation. *Rheologica Acta* 26: 499-507.

[MePo] Meissner, K. & B. Poltersdorf (1992). Model development for an internal mixer. *Intern. Polymer Processing* VII: 3-14.

[PrScoTav] Pritchard, W.G., Scott, L.R. and S.J. Tavener (1992). Viscous free-surface flow over a perturbed inclined plane. *Phil. Trans. Roy. Soc. London A* 340: 1-45.

[RiIl] Rivkind, V.Ja. & A.V. Il'in (1988). On an approximate solution method to a non-isothermal flow of a liquid layer. *Numerical methods (Izdat. Univ. St. Petersburg)* 15: 98-104 [in Russian].

[SaBrDe] Sackinger, P.A., Brown, R.A. and J.J. Derby (1989). A finite element method for analysis of fluid flow, heat transfer and free interfaces in CZOCHRALSKI crystal growth. *Int. J. num. meth. fluids* 9: 453-492.

[So-88] Socolowsky,J.(1988). Eine verallgemeinerte Leitlinienmethode zur Berechnung mehrschichtiger Strömungen nichtlinear-viskoser Fluide.*J. Appl. Math. Phys.(ZAMP)* 39: 221-232.

[So-89] Socolowsky, J. (1989). *Mathematische Untersuchungen freier Randwertaufgaben der Hydrodynamik viskoser Flüssigkeiten.* Habilitation thesis: 1-218, TH Merseburg.

[So-90] Socolowsky, J. (1990). A method of support lines for the computation of multi-layer flow-problems with free boundaries by FEM. In *Proceedings of the International FEM-Congress Baden - Baden*: 305-316, Ed. IKOSS GmbH.

[So-94] Socolowsky, J. (1994). Existence and uniqueness of the solution to a free boundary value problem with thermocapillary convection in an unbounded domain. *Acta Applicandae Mathematicae* 37: 181-194.

[SuPhTa] Sugeng, F., Phan-Thien, N. and R.I. Tanner (1987). A study of non-isothermal non- NEWTONIAN extrudate swell by a mixed boundary element and finite element method. *J. Rheology* 31: 37-58.

Simulation of Materials Processing: Theory, Methods and Applications, Shen & Dawson (eds)
© 1995 Balkema, Rotterdam. ISBN 90 5410 553 4

Modeling of coextrusion flows

A.Torres, A.Rincon, A.N.Hrymak & J.Vlachopoulos
CAPPA-D, Department of Chemical Engineering, McMaster University, Hamilton, Ont., Canada

ABSTRACT: The coextrusion of polymer melts is used to manufacture sheet and film products. Research in the area has examined two major problem classes, layer uniformity across the breadth of the die and layer stability to transient disturbances. A 3-D finite element formulation is employed to analyze the flow of bicomponent coextrusion flows by solving the weak form of the momentum, continuity and energy conservation equations, simultaneously with the interfacial kinematic boundary condition. The simulation of the transient problem is complicated by the fact that there are distinct types of instability observed in practice.

1. INTRODUCTION

The coextrusion process is an important process for products with tailored properties. Different polymer melts are brought together in a die to form a layered structure. Each layer may have a role due to its physical properties or cost in the final product. The different polymer phases are separated by internal interfaces and generally the polymer layers remain distinct continuous phases in the final product.

In the study of coextrusion fundamentals, two basic problem classes have been studied which relate to the position of the internal interfaces between adjacent polymer melts. The first class of problems relates to the issue of maintaining uniform layer structure and thickness within the die. The second class of problems deals with the stability of the interface over time and whether disturbances in the flow will lead to the growth or dissipation of irregularities in the interface.

In the layer uniformity problem, the less viscous fluid will tend to displace the more viscous fluid near the walls of the die. Experimental investigations have consistently shown that the viscosity difference between adjacent polymer melts is the major factor in the displacement phenomenon. The difference in layer thickness near the die wall has serious consequences on product yield. Given that the less viscous polymer melt will displace the more viscous melts along the die runner wall and along the feed channel walls of the coat hanger die, there is considerable opportunity for serious nonuniformity of the desired layer structure. The material near the edges of the die must be cut from the sheet or film and either discarded or recycled (Michaeli, 1992).

Researchers have often relied on computer simulations to understand the fundamentals of coextrusion flows. For the problem of two-layer flow through dies, Mitsoulis (1986), Mavridis et al. (1987), Nordberg and Winter (1988, 1990), Luo and Mitsoulis (1990) have developed two-dimensional finite element codes. Karagiannis et al. (1990) used a three-dimensional finite element model to examine the steady-state flow.

A number of complications arise when a full three-dimensional simulation of the coextrusion process is attempted. The models are nonlinear due to coupling between the energy and momentum conservation equations, as well as shear-thinning and thermal effects on apparent viscosity. The presence of internal interfaces means that the flow domain must be calculated with the other state variables.

One of the major difficulties in a model for coextrusion flows is the specification of the correct boundary condition at the contact line (the intersection of the interface separating two fluids with a solid (die) wall). It is well known (Dussan,

1976; Davis, 1983) that introduction of the common no-slip boundary condition at the wall produces a breakdown of the numerical solution. The presence of the fluid/fluid/wall contact line region creates a mathematical singularity if the no-slip boundary condition is applied.

Experimental and mathematical approaches have been followed to tackle the problem of interfacial instability. Several papers have studied the interfacial instability by examining the viscosity difference between the two fluids (Yih, 1967), critical interfacial shear stress (Schrenk et al., 1978; Mavridis and Shroff, 1994), mismatch in the normal stress between the layers (Chen, 1991; Larson 1992) and flow disturbances (Wilson and Khomami, 1992,1993). However, none of these studies explain completely all the observed phenomena. A number of authors have analyzed the problem of interfacial instability by using a linear stability analysis to determine which flowrate and viscosity ratio values characterize stable or unstable flows (Yih, 1967; Su and Khomami, 1992a, 1992b). The major limitation of the linear stability methodology is that only simple geometries can be modelled. Complex geometries, such as those in industrial coextrusion dies, are not amenable to standard linear stability analysis techniques.

2. GOVERNING EQUATIONS

For analysis of layer uniformity, the flow of molten polymers inside the die is modeled assuming steady state, creeping, incompressible flow with no body forces. The governing equations are the momentum and energy equations (without inertia terms) and the continuity equation, which can be written in dimensionless tensorial form as:

$$\nabla \cdot \boldsymbol{\tau}_k - \nabla P_k = \mathbf{0} \quad , \quad k=I,II \tag{1}$$

$$Pe\left(\boldsymbol{v}_k \cdot \nabla T_k\right) = \nabla^2 T_k + Br\, \boldsymbol{\tau}_k : \nabla \boldsymbol{v}_k \tag{2}$$

$$\nabla \cdot \boldsymbol{v}_k = \mathbf{0} \tag{3}$$

where τ is the extra stress tensor, P the pressure, v the velocity vector. A Generalized Newtonian constitutive equation is used to model the stress-strain relation for polymers:

$$\tau_{ij} = \mu_k \left(\frac{\partial v_i}{\partial x_j} + \frac{\partial v_j}{\partial x_i} \right) \tag{4}$$

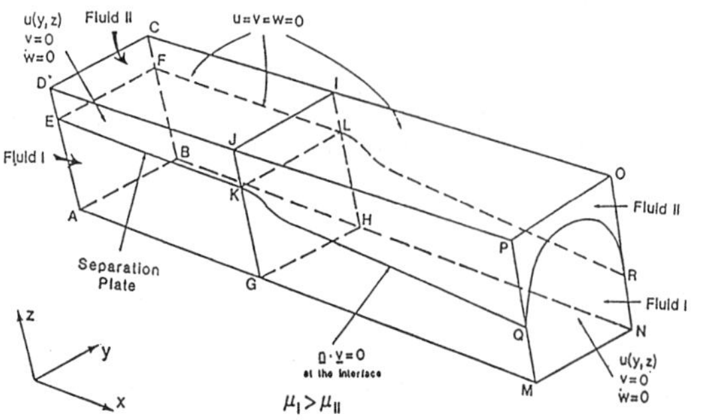

Figure 1 Schematic of coextrusion flow and boundary conditions in a typical die.

Indices k = I and II refer to the lower and upper layer, respectively (Figure 1). The viscosity dependence on shear rate and temperature is modelled either as Newtonian or fit to a Carreau-like model, of the form:

$$\mu(\dot{\gamma}) = (\mu_0 - \mu_\infty)\left(e^{-\beta T}\right)\left[1 + (\lambda \dot{\gamma})^2\right]^{\frac{(n-1)}{2}} \tag{5}$$

$$\dot{\gamma} = \sqrt{\frac{1}{2} II_{\dot{\gamma}}} \qquad II_{\dot{\gamma}} = \dot{\gamma}_{ij}\dot{\gamma}_{ji}$$
$$\dot{\gamma}_{ij} = \frac{\partial u_j}{\partial x_i} + \frac{\partial u_i}{\partial x_j} \tag{6}$$

The dimensionless groups used are:

$$\boldsymbol{x} = \frac{\boldsymbol{x}'}{L} \quad \boldsymbol{u} = \frac{\boldsymbol{u}'}{U} \quad P = \frac{P' L}{\mu_{01} U} \quad \boldsymbol{\tau} = \frac{\boldsymbol{\tau}' L}{\mu_{01} U}$$
$$\mu = \frac{\mu'}{\mu_{01}} \quad Pe = \frac{\rho C_p U L}{k} \quad Br = \frac{\mu U^2}{k(T_w - T_\infty)} \tag{7}$$

where L is a characteristic length (usually the die height), U is the total average velocity (total flowrate over total cross-sectional area for both fluids), μ_{01} is the zero-shear viscosity of fluid 1, C_p is the specific heat of fluid 1, k is the thermal conductivity of fluid 1, ρ is the density of fluid 1, T_w and T_∞ are the wall and ambient temperatures, respectively, and a superscript denotes a dimensional variable. All coordinates (x,y,z) in the figures are dimensionless values.

The problem becomes complete with the specification of the boundary conditions. A typical die is shown in Figure 1. At the die inlet (area ABCD), fully developed flow is assumed. At the die walls (areas ADPM, BCON, ABNM and DCOP), fixed temperature and the no-slip condition is applied, except for the contact line (KQ and LR), where boundary conditions for the contact line must be specified. At the interface, interfacial tension is neglected since viscous forces are extremely large. Thus the dynamic and kinematic boundary conditions at the interface are:

- Equilibrium of forces:

$$n \cdot \sigma_I = n \cdot \sigma_{II} \; ; \; t_1 \cdot \sigma_I = t_1 \cdot \sigma_{II} \; ; \; t_2 \cdot \sigma_I = t_2 \cdot \sigma_{II} \qquad (8)$$
$$\sigma_k = -P_k \cdot I + \tau_k \quad , \quad k=I,II$$

- Continuity of tangential velocities and vanishing velocity normal to the interface:

$$t_1 \cdot v_I = t_1 \cdot v_{II} \; ; \; t_2 \cdot v_I = t_2 \cdot v_{II} \qquad (9)$$
$$n \cdot v = 0$$

where σ is the total stress tensor, n is the outward normal vector to the interface and t_1 and t_2 are the tangential vectors at the interface. At the die exit (area MNOP), zero traction and zero cross flows (v = w = 0) are imposed.

3. FINITE ELEMENT FORMULATION

The flow domain is discretized into 27-node triquadratic brick elements or 8-node trilinear brick elements. Within each element, geometry, velocities and pressure are approximated by:

$$\phi^{(e)} = \sum_i N_i(\xi,\eta,\zeta)\phi_i \quad ; \quad \phi_i = x_i,\ v_i,\ T_i \qquad (10)$$
$$P^{(e)} = \sum_i N_{Pi}(\xi,\eta,\zeta)P_i$$

where T_i are the nodal temperatures, v_i is the nodal vector of velocities, x_i is the nodal coordinate vector, P_i are the nodal pressures and N_i and N_{Pi} are quadratic and linear interpolation functions, respectively (Karagiannis, 1989).

The Galerkin finite element is applied to reduce the set of differential equations (1,2,3) and the interface boundary conditions (8,9) to a set of algebraic equations representing the element residuals and $X = [\, V^T, P^T, h^T, T^T \,]$ is the vector of unknowns:

$$R(X) = 0 \qquad (11)$$

The pressure discontinuity at the interface is handled by using the double node technique (Mavridis et al., 1987) and the spine technique is used for the free-surface update (Karagiannis et al., 1990). Symmetry conditions were used (x-z is a symmetry plane) to reduce the size of the solved problem. The overall system of equations is solved using a Newton method with analytical gradients.

4. INTERFACE SHAPE AT THE CONTACT LINE

Introduction of the common no-slip boundary condition leads to a multivalued velocity at the contact point and to an infinite shear stress, which is unrealistic. The behaviour of the two-phase fluid near a solid wall (the contact point region) is not understood completely. A number of approximations have been used over the years to alleviate the stress singularity.

One attempt to circumvent the problem was to extrapolate the free surface from the flow domain to the wall (Dheur and Crochet, 1987; Karagiannis et al., 1990) . The no-slip model and several forms of localized slip models, were used by Torres et al., (1993) to capture more complicated interface shapes and to consider wall effects in polymer coextrusion. It was found that a simple linear slip model is sufficient to capture wall effects. By assuming that the contact line displaces only in the direction of encapsulation, the linear slip boundary condition reduces to (Silliman and Scriven, 1980):

$$u_{sl} = \beta\, \tau_w \qquad (12)$$

where u_{sl} is the slip velocity, β is the dimensionless slip coefficient (typical values are between 0.001-0.1, which correspond to dimensional values of 10^{-8} to 10^{-6} m/(Pa.s); Torres et al., 1993), and τ_w is the wall shear stress. The slip boundary condition is applied only at the nodes lying on the contact line. In all other regions of the surface J, no-slip at the wall is assumed. For the slip model, values of the shear stress at the wall are used in the calculations. The slip boundary conditions are introduced in the surface integral in the z-momentum equations only for the nodes at the contact line.

4.1 *Numerical Results*

A comparison between "stick" (no movement of the contact line) and slip models was done with a very low slip coefficient (B=0.0001), thus recovering, in part, the no-slip feature. Results for a test problem with the die geometry of Figure 1 (Newtonian fluids, μ_1/μ_2=2.5, Q_1/Q_2=13.2) are shown in Figure 2.

The predicted interface position at the die exit is insensitive to the contact line condition beyond 0.025 dimensionless units away from the die wall. The interface shape for the slip model shows the characteristic bending of the stick model at small values of the slip coefficient. This suggests that a higher slip coefficient is needed to eliminate the bending near the contact line. A very thin film of entraining liquid I (called "primary" film by Dussan, 1976) can be seen near the die walls for the stick model. Since the encapsulation by the less viscous fluid is energetically preferred, an interface 'bending effect' is seen. The thickness of this film is in qualitative agreement with the predictions of Durbin (1988). The predicted degree of encapsulation is the same for the extrapolation method and the stick model. The solution was unaffected by further grid refinement near the wall.

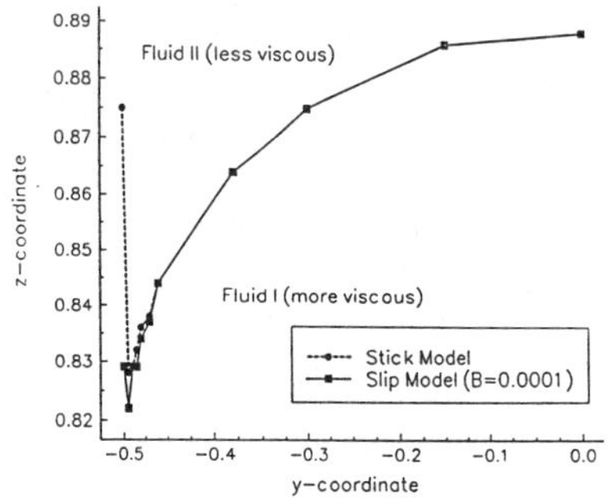

Figure 2 Comparison of interface deformation at the contact line assuming no contact line movement as compared to slip at the contact line.

5. TEMPERATURE EFFECTS

5.1 *Inlet and wall temperature effects*

The effects of variations in inlet and wall temperatures are shown for the cases of negligible Pe number (Pe=0) as compared to Pe=2000 (typical of extrusion flows). The case of Pe=0 is intended as a limiting case to establish the influence of the viscous dissipation term in the final interface shape. The dimensionless coefficient for the viscosity dependence in temperature is b=5 and the reference temperature is 250°C. The dimensionless slip coefficient is β=0.1. Variations of 30°C on the two inlet temperatures and wall temperatures were introduced and each variation studied in turn.

The temperature affects the viscosity, thus there is a local viscosity ratio that depends on temperature. Due to the temperature profile inside the die, there is also a viscosity profile. So the term viscosity ratio loses its meaning as an *overall* flow parameter, since it varies locally. Figure 3 shows the final interface shape for the case of Br=0.11 and Pe=2000. In all these cases, a flat temperature profile is assumed at the inlet and fixed temperature at the die walls.

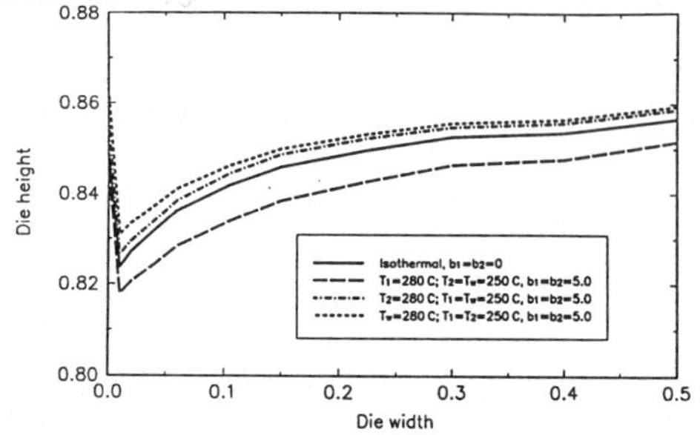

Figure 3 Effect of inlet and wall temperatures on final extrudate shape.

Inlet and wall temperatures can be used to adjust the final interface shape. Wall temperatures are easier to control than inlet temperatures. Inlet melt flows typically have a temperature profile across the die inlet, sometimes with substantial temperature differences across the die cross-section, which are difficult to predict.

5.2 *Effect of the sensitivity of viscosity to temperature.*

Figure 4 shows the final interface shape for the same flow conditions as in the previous section. The inlet temperature of fluid 2 (280°C) is greater than the inlet temperature of fluid 1 and the die wall temperature, which are both 250°C. The dimensionless b (viscosity thermal coefficient) for both fluids is varied from 0.0 to 10.0. A coefficient value of b=0.0 uncouples the dependence between the energy and momentum equations. Assuming a reference temperature of T_0=250°C and T_∞=0°C, a b coefficient of 10 will have a dimensional value of 0.04°K^{-1}. Typical polymer values for b lie between 0.01 and 0.04 (Vlcek and Vlachopoulos, 1993). As in the previous section, the biggest deviation in the final interface shape from the isothermal case occurs near the die wall. The inlet temperature profiles are assumed uniform in these simulations. There is little difference in the final interface shape between uniform and fully developed inlet thermal conditions in the computations.

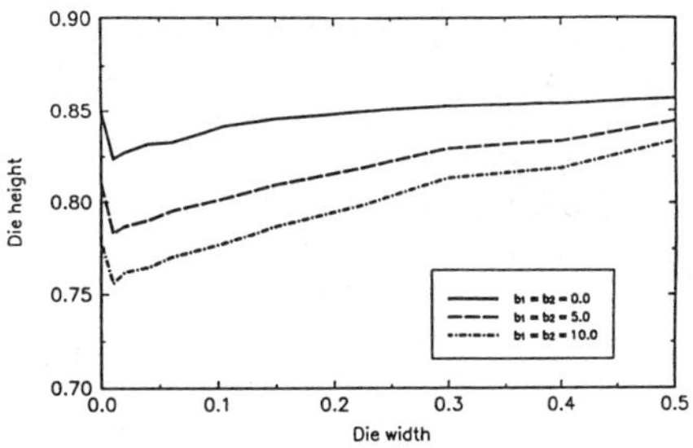

Figure 4 Extrudate final profile as a function of temperature effect on viscosity.

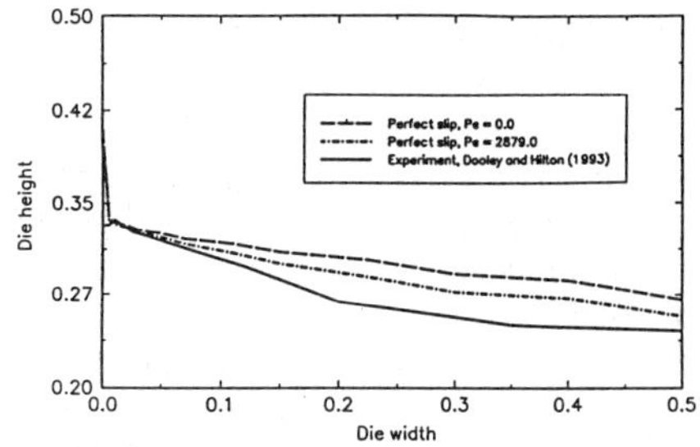

Figure 5 Final interface shape for polycarbonate coextrusion - simulation and experiment.

6. EXPERIMENTAL RESULTS

Dooley and Hilton (1993) undertook experiments with polycarbonate resins. The coextruded two-layer structure had a 20% cap layer and 80% substrate layer. The characteristic length of the die (L/D) used in the work reported here was approximately 32:1. The coextrusion was performed using the same material in each layer to minimize viscosity differences between layers. One of the layers was pigmented black so the interface could be seen. All the experimental runs were performed at the approximately same wall shear rate of 30-40 sec^{-1}. The entrance angle of the feeding channel is 30°. Details of the coextrusion process and die configuration can be found in Dooley and Hilton (1993). The polycarbonate coextrusion process shows very little layer rearrangement. The interface deformation is negligible at a distance of a few L/D and there is no appearance of a primary film along the die wall.

Torres et al. (1993) used an isothermal model to simulate the experimental work. Numerical simulations, with the inclusion of thermal effects, were also carried out to model this flow problem. A Newtonian model for viscosity was employed because the power-law index of the polycarbonate resin was nearly unity (n=0.92). The dimensionless thermal parameters that govern this problem are Pe=2879, Br=0.11 and b=9.635 (b'= 0.041 1/°K, assuming T_0-T_∞=235°C). The computed interface at L/D=3 is compared to the experimental result. After L/D=2, the calculated interface attained a stable position and is not deformed significantly. The experimental interface position also deforms within a few L/D and becomes completely stable at approximately L/D=10. Figure 5 shows the results of calculated interface positions and experimental measurements.

7. TRANSIENT EFFECTS

The governing equations are modified as follows:

$$\rho\frac{DV}{Dt}=\nabla\cdot\tau-\nabla P \tag{13}$$

The material derivative of the velocity in equation (13) is:

$$\rho\frac{DV}{Dt}=\rho\frac{\partial V}{\partial t}+\rho V\cdot\nabla V \tag{14}$$

Nodes move depending on the movement of the interface along a fixed line in space (spine). This corresponds to a mixed Eulerian-Lagrangian formulation. However, the time derivative is an Eulerian time derivative i.e., the nodal velocity field is defined for nodes fixed in space. Therefore, the time derivative (∂/∂t) is transformed to a time derivative which follows the moving nodes along the spines (δ/δt).

The boundary conditions assume fully developed flow at the die inlet for each one of the two channels. A sinusoidal perturbation in time (t) of the fully developed velocity profile entering feeding channel I is imposed. At the exit plane zero traction and zero cross flows (v=0) are imposed at each time step. However, the interface moves at that plane in order to satisfy continuity.

A Galerkin method is used to reduce the governing differential equations and the kinematic boundary condition to a first order system of differential equations written as:

$$\begin{gathered}[C]\{\dot{U}\}+[K]\{U\}=\{F\}\\ \{U(t_0)\}=\{U_0\}\end{gathered} \tag{15}$$

The Euler method is used for the transient simulation together with a Newton-Raphson procedure to solve the non-linear problem of (15).

8. CONCLUSIONS

A fully three dimensional finite element formulation has been used to model the coextrusion process for two polymer layers. The calculated interface profile depends strongly on the contact line boundary conditions.

A change in wall or inlet temperatures can change the interface position depending on the thermal dependence of the fluid viscosity. The most important parameter is the viscosity sensitivity coefficient b, which can drastically affect the final interface shape. The effect of this coefficient is more pronounced near the die wall, where the highest viscous heat generation occurs.

The model was tested against an experiment in which a polycarbonate/polycarbonate resin was coextruded, assuming perfect slip at the contact line. Computer simulations for these types of problems (with Generalized Newtonian constitutive relationship) show that the interface does not change appreciably after an L/D of approximately 2 for all cases tested and agreement between computed and experimentally measured interface profiles is very good.

ACKNOWLEDGEMENTS

The authors acknowledge the financial support of the Ontario Centre for Materials Research and of Investigación y Desarrollo, C.A. (INDESCA)Venezuela. A. Torres and A. Rincon are on leave from INDESCA.

REFERENCES

Chen, K.P. 1991. Elastic Instability on the Interface in Couette Flow of Viscoelastic Liquids. *J. Non-Newt. Fluid Mech.* 40:261.

Davis, S.H.1983. Contact-Line Problems in Fluid Mechanics. *J. Appl Mech*, 50:977-982.

Dheur, J., & M.J. Crochet 1987. Newtonian Stratified Flow Through an Abrupt Expansion. *Rheol Acta*, 26:401-413.

Dooley, J. & B.T. Hilton 1993. Coextrusion Layer Rearrangement in Different Geometric Channels. SPE ANTEC, New Orleans, LA, 3354-3364.

Durbin, P.A. 1988. Considerations on the Moving Contact-Line Singularity, with Applications to Frictional Drag on a Slender Drop. *J Fluid Mech.* 197:157-169.

Dussan, E.B. 1976. The Moving Contact Line: The Slip Boundary Condition. *J Fluid Mech*, 77:665-684.

Karagiannis, A. 1989. *Modelling of Single Component and Bicomponent Extrusion Flows*. Ph. D. Thesis. Dept. of Chemical Engineering. McMaster University.

Karagiannis, A., et al. 1990. Three-Dimensional Studies on Bicomponent Extrusion. *Rheol Acta*, 29:71-87.

Larson, R.G. 1992. Instabilities in Viscoelastic Flows. *Rheol. Acta* 31:213-263.

Luo, X.L. & E. Mitsoulis 1990. A Finite Element Study of Viscoelastic Effects in Double-Layer Coextrusion of Polymer Melts. *Adv. Polym. Tech.*, 10:47.

Mavridis, H. et al. 1987. Finite-Element Simulation of Stratified Multiphase Flows. *AIChE Journal*, 33:410-422.

Mavridis, H. & R.N. Shroff 1994. Multilayer Extrusion: Experiments and Computer Simulation. *Polymer Eng. and Sci.*, 34:559.

Michaeli, W. 1992., *Extrusion dies for plastics and rubber*. New York:Hanser.

Mitsoulis, E. 1986. Extrudate Swell in Double-Layer Flows. *J Rheol*, 30:S23.

Nordberg, M.E. & H.H. Winter 1988. A Simple Model for Non-Isothermal Coextrusion. *Polym. Eng. Sci.*, 30:408.

Nordberg, M.E. & H.H. Winter 1990. Fully Developed Multilayer Polymer Flow in Slits and Annuli. *Polym. Eng. Sci.*, 28:444.

Schrenk, W.J. et al. 1978. Interfacial Flow Instability in Multilayer Coextrusion. *Polymer Eng. and Sci.*, 18:620.

Silliman, W.J. & L.E. Scriven 1980. Separating flow near a static contact line: slip at the wall and shape of a free surface. *J Comp. Phys*, 34:287.

Su, Y.Y. & B. Khomami 1992a. Interfacial Instability of Multilayer Viscoelastic Fluids in Slit and Converging Channel Die Geometries. *J. Rheo.* 36:357.

Su, Y.Y. & B. Khomami 1992b. Stability of Multilayer Power Law and Second Order Fluids in Poiseuille Flow. *Chem. Eng. Commun.* 109:209.

Torres, A., et al. 1993. Boundary conditions for contact lines in coextrusion flows. *Rheol Acta*. 32: 513. Errata 1994. *Rheol Acta*. 33:241.

Vlcek, J. & J. Vlachopoulos 1993. *VISCOFIT: User manual*. Polydynamics Inc., Canada and Compuplast International, Czech Republic.

Wilson, G.M. & B. Khomami 1992. An experimental investigation of interfacial instabilities in multilayer flow of viscoelastic liquids: Part I. *J. Non-Newt. Fluid Mechanics*. 45:355-384.

Wilson, G.M. & B. Khomami 1993. An experimental investigation of interfacial instabilities in multilayer flow of viscoelastic liquids: Part II. *J. Rheology*. 37:315-339. Part III. *J. Rheology*. 37:341-354.

Yih, C.S. 1967. Instability Due to Viscosity Stratification. *J. Fluid Mech.* 27:337.

Simulation of Materials Processing: Theory, Methods and Applications, Shen & Dawson (eds)
© 1995 Balkema, Rotterdam. ISBN 90 5410 553 4

Boundary Integral Equations for unclosed surfaces: An application in profile extrusion simulations

Thanh Tran-Cong
Faculty of Engineering and Surveying, University of Southern Queensland, Toowoomba, Qld, Australia
Nhan Phan-Thien
Department of Mechanical and Mechatronic Engineering, The University of Sydney, N.S.W., Australia

ABSTRACT: A Boundary Integral Equation (BIE) formulation for elasticity problems with mixed boundary conditions, proposed by Parton and Perlin (1984), is implemented in this paper using quadractic boundary elements. The formulation is specialised to Stokes flow problems by setting the Poisson ratio to 0.5 in the relevant kernels, and is used to simulate profile extrusion processes. The free surface of the extrudate is found by tracing the path line of a particle exiting from the die lip. An example of the extrusion of a Newtonian fluid out of a square die is presented. The formulation is extended to problems involving non Newtonian fluids.

1 INTRODUCTION

A direct application of the reciprocal theorem, with the singular Kelvin solution playing the role of the second solution, leads to the so-called Direct Boundary Element Methods (DBEM), where the unknowns are the boundary displacement and traction. Its applications cover a wide range of engineering problems in many fields, which has been detailed in many excellent texts (for example, Barnerjee and Butterfield 1981, Brebbia *et al* 1984, Beer and Watson 1992, to name a few). Despite the success of the DBEM, a drawback of this method is that the resulting integral equations are not always of the second kind, which lead to a numerically ill-posed discrete system of linear algebraic equations. This problem prevents the use of iterative solvers which are neccessary for large scale simulation. A different approach can be taken by formulating the integral representation in terms of a fictitious surface density which lead to the so-called Indirect Boundary Element Method (IBEM) (for example, Banerjee and Butterfield 1981.) However this formulation still does not result in integral equations of the second kind. Formualtions that lead to integral of the second kind for Stokes problems are extensively discussed in Kim and Karrila (1991) and for elasticity problems in Phan-Thien and Kim (1994). In this paper we are concerned with non-linear problems with mixed boundary conditions and discuss a formulation that leads to integral equations of the second kind.

2 INTEGRAL EQUATIONS FOR UNCLOSED SURFACES

Parton and Perlin (1984), with reference to earlier Russian literature, suggest that the displacement field be sought as a sum of a single layer and a double layer on unclosed surfaces S_u and S_t:

$$u_i(\mathbf{x}) = \int_{S_u} K_{ij}(\mathbf{x},\mathbf{y})\varphi_j(\mathbf{y})dS(\mathbf{y}) + \int_{S_t} G_{ij}(\mathbf{x},\mathbf{y})\varphi_j(\mathbf{y})dS(\mathbf{y}), \quad \mathbf{x} \in V, (1)$$

where $G_{ij}(\mathbf{x},\mathbf{y})$ is twice the Kelvin fundamental solution; the traction associated with the Kelvin state is $K^*_{ij}(\mathbf{x},\mathbf{y})$ and $K_{ij}(\mathbf{x},\mathbf{y}) = K^*_{ji}(\mathbf{y},\mathbf{x})$ (see, for example, Phan-Thien and Kim 1994); $\varphi_j(\mathbf{y})$ is the unknown surface potential; S_u is an unclosed surface where the displacement is prescribed, and S_t is an unclosed surface where the traction is given. These surfaces need not be singly-connected, but they are non-overlaped and form a covering of the bounding surface of the domain.

The traction vector corresponding to this representation is

$$t_i(\mathbf{x}) = T_i\left[\int_{S_u} \mathbf{K}(\mathbf{x},\mathbf{y})\cdot\boldsymbol{\varphi}(\mathbf{y})dS(\mathbf{y})\right] + \int_{S_t} K_{ji}(\mathbf{y},\mathbf{x})\varphi_j(\mathbf{y})dS(\mathbf{y}), \quad \mathbf{x}\in V, (2)$$

where T denotes the traction operator, i.e., $T_i[\mathbf{u}]$ returns the traction from the displacement $\mathbf{u}$. Allowing $\mathbf{x}\to S_u$, where the prescribed displacement is $u_i^{(0)}(\mathbf{x})$, and noting the jump in the double layer (Phan-Thien and Kim 1994), one has

$$u_i^{(0)}(\mathbf{x}) = \pm\varphi_i(\mathbf{x}) + \int_{S_u} K_{ij}(\mathbf{x},\mathbf{y})\varphi_j(\mathbf{y})dS(\mathbf{y}) + \int_{S_t} G_{ij}(\mathbf{x},\mathbf{y})\varphi_j(\mathbf{y})dS(\mathbf{y}), \quad \mathbf{x}\in S_u. (3)$$

On S_t, where $t_i^{(0)}(\mathbf{x})$ is the prescribed traction, the first integral on the right-hand side of (2) is regular, since the integrand is always finite and the order of integration and differentiation can be interchanged; the only jump comes from the traction of the single layer, and one obtains

$$t_i^{(0)}(\mathbf{x}) = \mp\varphi_i(\mathbf{x}) + T_i\left[\int_{S_u} \mathbf{K}(\mathbf{x},\mathbf{y})\cdot\boldsymbol{\varphi}(\mathbf{y})dS(\mathbf{y})\right] + \int_{S_t} K_{ji}(\mathbf{y},\mathbf{x})\varphi_j(\mathbf{y})dS(\mathbf{y}), \ \mathbf{x}\in S_t. \ (4)$$

In the above equations the upper sign in front of $\varphi_i(\mathbf{x})$ corresponds to the exterior problem and the lower sign to the interior one. The boundary integral equations thus obtained, (3)–(4), are of the second kind, and therefore are well-posed. The traction of the double layer kernel is given explicitly in Phan-Thien and Kim (1994).

3 INTEGRAL REPRESENTATION FOR NON-HOMOGENEOUS PROBLEMS

In this section we outline an extension of the above formulation to deal with non-linear problems. Parton and Perlin (1984) also define an elastic volume (Newtonian) potential:

$$U(\mathbf{x}) = \int_V G_{ij}(\mathbf{x},\mathbf{y})\phi_j(\mathbf{y})dV(\mathbf{y}), \qquad (5)$$

which satisfies Navicr's equations

$$\mu\left(\frac{1}{1-2\nu}\nabla\nabla\cdot\mathbf{u} + \nabla^2\mathbf{u}\right) = -2\phi(\mathbf{x}). \qquad (6)$$

For non-linear problems such as those involving the flow of non-Newtonian fluids, we can take the pseudo body force approach. In this approach we begin by writing the total stress tensor as the sum of a linear and a non-linear part

$$\boldsymbol{\sigma} = \boldsymbol{\sigma}^H + \boldsymbol{\varepsilon} = \lambda\nabla\cdot\mathbf{u}\mathbf{1} + \mu\left(\nabla\mathbf{u} + \nabla\mathbf{u}^T\right) + \boldsymbol{\varepsilon}, \qquad (7)$$

where μ is a conveniently chosen Lame's constant. It follows that the right hand side of Navier's equations for the newly defined Hookean material with the constitutive equations given by $\boldsymbol{\sigma}^H$ is

$$\rho\mathbf{b} + \nabla\cdot\boldsymbol{\varepsilon},$$

where ρ is the material mass density and $\mathbf{b}$ is the body force per unit mass. Thus by defining the volume potential density as

$$\phi = \frac{1}{2}\left(\rho\mathbf{b} + \nabla\cdot\boldsymbol{\varepsilon}\right), \qquad (8)$$

we can obtain an integral representation for a non-homogeneous problem by adding to the representation (3)–(4) the volume potential (5):

$$u_i^{(0)}(\mathbf{x}) = \pm\varphi_i(\mathbf{x}) + \int_{S_u} K_{ij}(\mathbf{x},\mathbf{y})\varphi_j(\mathbf{y})dS(\mathbf{y}) + \int_{S_t} G_{ij}(\mathbf{x},\mathbf{y})\varphi_j(\mathbf{y})dS(\mathbf{y}) + \int_V G_{ij}(\mathbf{x},\mathbf{y})\phi_j(\mathbf{y})dV(\mathbf{y}), \ \mathbf{x}\in S_u (9)$$

and

$$t_i^{(0)}(\mathbf{x}) = \mp\varphi_i(\mathbf{x}) + T_i\left[\int_{S_u} \mathbf{K}(\mathbf{x},\mathbf{y})\cdot\boldsymbol{\varphi}(\mathbf{y})dS(\mathbf{y})\right] + \int_{S_t} K_{ji}(\mathbf{y},\mathbf{x})\varphi_j(\mathbf{y})dS(\mathbf{y}) + \int_V G_{ij}(\mathbf{x},\mathbf{y})\phi_j(\mathbf{y})dV(\mathbf{y}), \ \mathbf{x}\in S_t. (10)$$

4 NUMERICAL APPROXIMATION BY BOUNDARY ELEMENTS

In the present study the integral representation for linear homogeneous problems is solved numerically by a boundary element approximation of the geometry and the surface density. Quadratic isoparametric quadrilateral boundary elements (8-node) are used. A family of continuous serendipity elements and edge-discontinuous elements is implemented in the program (see figures 1–4) where the circles denote the functional nodes associated with the numbered geometric nodes.

The shape functions for continuous elements can be found in most finite and boundary element texts

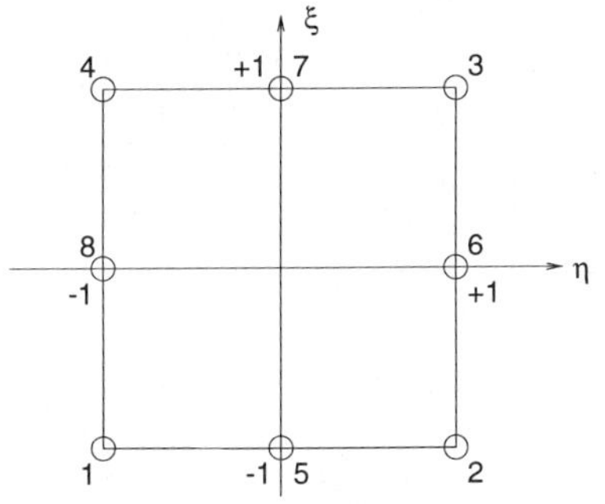

Figure 1: 8-node quadrilateral continuous element.

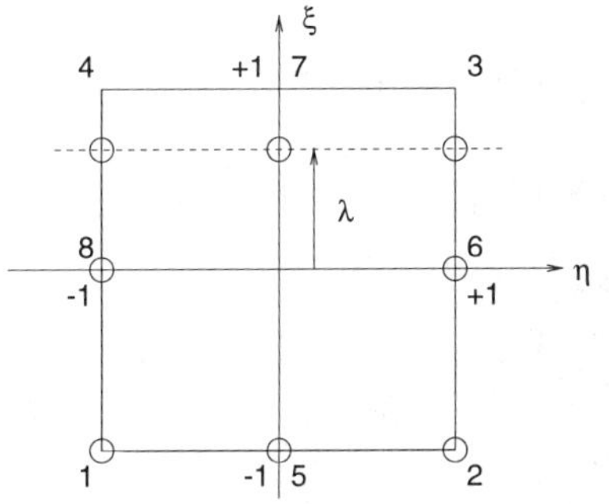

Figure 2: 8-node quadrilateral element of type I with one discontinuous edge.

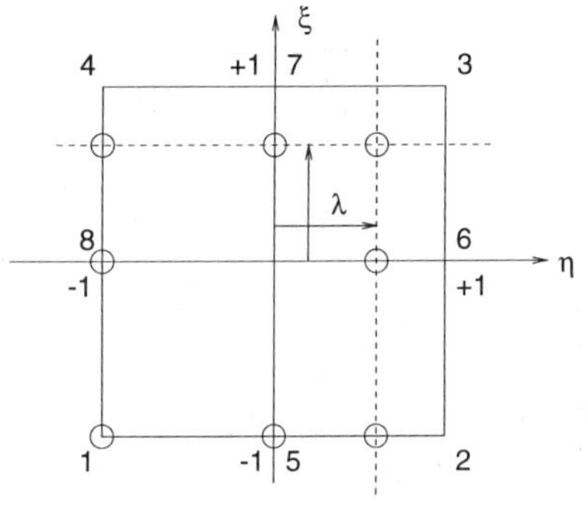

Figure 3: 8-node quadrilateral element of type II with two discontinuous edges.

and are included here for convenient reference. High order discontinuous elements have been used in other applications (Brebbia *et al* 1984, Patterson and Sheikh 1984, Mi and Aliabadi 1992, 1994, for example). The shape functions of discontinuous elements used in this study can be easily derived from the standard serendipity element shape functions and given in the following sections, where N^α is the shape function associated with node α, whose homogeneous coordinates are $\eta_\alpha = s$ and $\xi_\alpha = t$.

4.1 *Shape functions for 8-node serendipity quadrilateral elements.* See figure 1.

$$N^\alpha(\eta,\xi) = \frac{1}{4}(1+s\eta)(1+t\xi)(s\eta+t\xi-1),$$

for $\alpha = 1,2,3,4$.

$$N^\alpha(\eta,\xi) = \frac{1}{2}(1-\eta^2)(1+t\xi), \quad (\alpha = 5,7).$$

$$N^\alpha(\eta,\xi) = \frac{1}{2}(1+s\eta)(1-\xi^2), \quad (\alpha = 6,8).$$

4.2 *Shape functions for 8-node one-edge discontinuous quadrilateral elements: type I.* See figure 2.

$$N^\alpha(\eta,\xi) = \frac{(1+s\eta)(\lambda-\xi)(s\eta-\xi-1)}{2(\lambda+1)},$$

for $\alpha = 1,2$.

$$N^\alpha(\eta,\xi) = \frac{(1+s\eta)(1+\xi)(\lambda s\eta+\xi-\lambda)}{2\lambda(\lambda+1)},$$

for $\alpha = 3,4$.

$$N^5(\eta,\xi) = \frac{(1-\eta^2)(\lambda-\xi)}{(\lambda+1)},$$

$$N^7(\eta,\xi) = \frac{(1-\eta^2)(1+\xi)}{(\lambda+1)},$$

$$N^\alpha(\eta,\xi) = \frac{(1+s\eta)(1+\xi)(\lambda-\xi)}{2\lambda},$$

for $\alpha = 6,8$.

4.3 *Shape functions for 8-node two-edge discontinuous quadrilateral elements: type II.* See figure 3.

$$N^1(\eta,\xi) = \frac{(\lambda-\eta)(\lambda-\xi)(-\eta-\xi-1)}{(\lambda+1)^2},$$

$$N^2(\eta,\xi) = \frac{(1+\eta)(\lambda-\xi)(\eta-\lambda\xi-\lambda)}{\lambda(\lambda+1)^2},$$

$$N^3(\eta,\xi) = \frac{(1+\eta)(1+\xi)(\eta+\xi-\lambda)}{\lambda(\lambda+1)^2},$$

$$N^4(\eta,\xi) = \frac{(\lambda-\eta)(1+\xi)(-\lambda\eta+\xi-\lambda)}{\lambda(\lambda+1)^2},$$

$$N^5(\eta,\xi) = \frac{(1+\eta)(\lambda-\xi)(\lambda-\xi)}{\lambda(\lambda+1)},$$

$$N^6(\eta,\xi) = \frac{(1+\eta)(1+\xi)(\lambda-\xi)}{\lambda(\lambda+1)},$$

$$N^7(\eta,\xi) = \frac{(1+\eta)(\lambda-\eta)(1+\xi)}{\lambda(\lambda+1)},$$

$$N^8(\eta,\xi) = \frac{(\lambda-\eta)(\lambda-\xi)(1+\xi)}{\lambda(\lambda+1)}.$$

4.4 *Shape functions for 8-node discontinuous quadrilateral elements: type IV.* See figure 4.

$$N^\alpha(\eta,\xi) = \frac{1}{4\lambda^3}(\lambda+s\eta)(\lambda+t\xi)(s\eta+t\xi-\lambda).$$

for $\alpha = 1,2,3,4$,

$$N^\alpha(\eta,\xi) = \frac{1}{2\lambda^3}(\lambda^2-\eta^2)(\lambda+t\xi), \quad (\alpha = 5,7).$$

$$N^\alpha(\eta,\xi) = \frac{1}{2\lambda^3}(\lambda+s\eta)(\lambda^2-\xi^2), \quad (\alpha = 6,8).$$

In all cases the surface geometry is approximated by elements of the serendipity type.

5 AN ITERATIVE SOLVER

The discretization of the integral equations (3)–(4) results in an algebraic system with fully populated matrix of coefficients

$$\mathbf{A}\,\mathbf{x} = \mathbf{b} \tag{11}$$

In order to employ the conjugate gradient method, Doblare (1987) suggests that the system can be modified as follows

$$\mathbf{A}^T\mathbf{A}\,\mathbf{x} = \mathbf{A}^T\mathbf{b} \tag{12}$$

The coefficient matrix of the modified system is symmetric and we use the conjugate gradient algorithm reported by Jennings and Malik (1978). In practice the matrix multiplication $\mathbf{A}^T\mathbf{A}$ need not be performed. In fact the system matrix is assembled and stored on disk row by row and it turns out that only vector multiplications are required as we step through the algorithm.

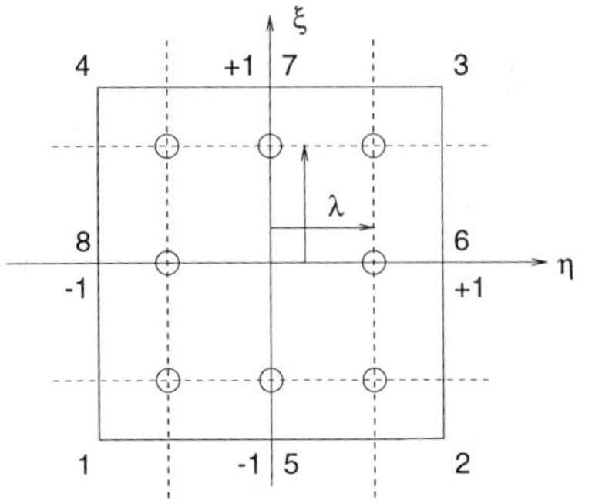

Figure 4: 8-node quadrilateral element of type IV with four discontinuous edges.

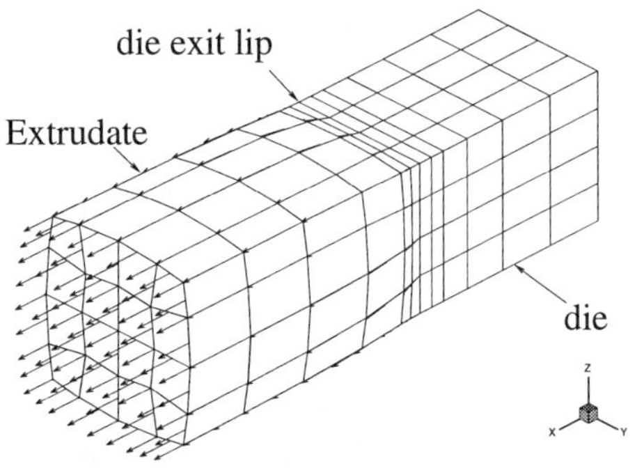

Figure 5: Boundary velocity field.

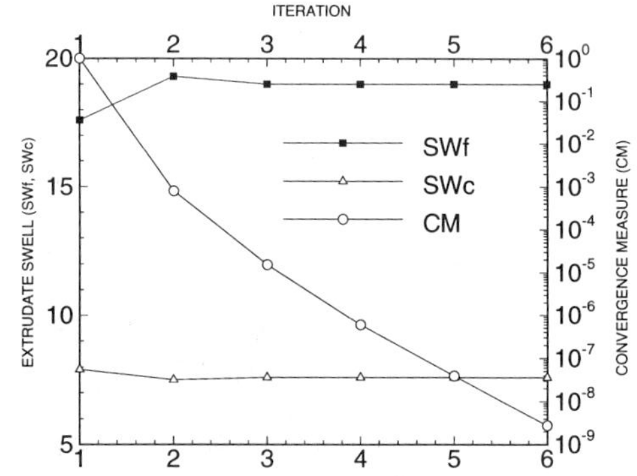

Figure 6: Extrudate Swell and convergence characteristics. SW_f is the percentage swelling corresponding to the middle of the flat face of the die and SW_c corresponding to the corner

6 AN EXAMPLE

The numerical discretization of equations (3)–(4) is implemented and the program is used to sim-

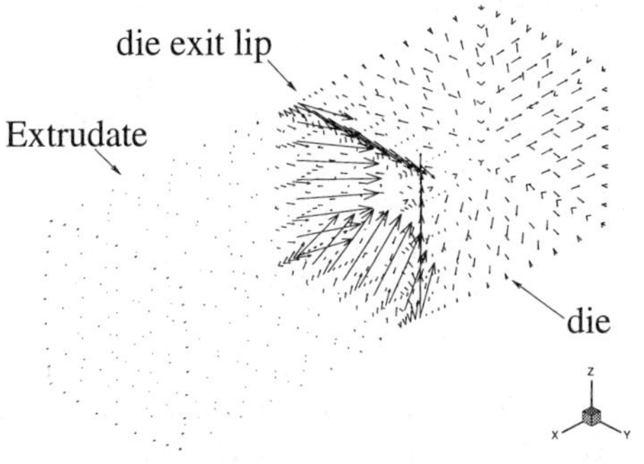

Figure 7: Surface potential.

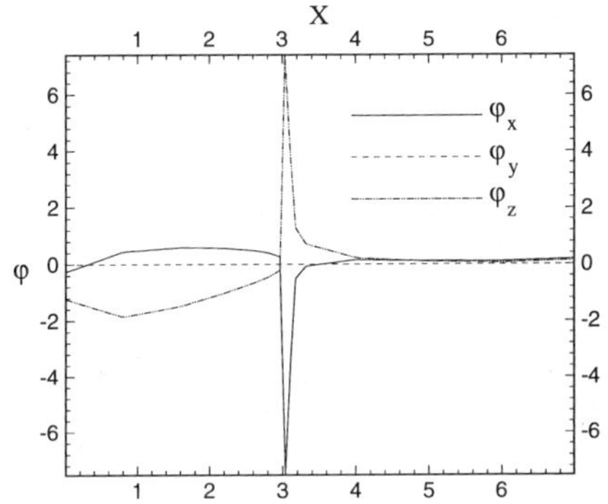

Figure 8: Surface potential along the intersection between XZ-plane and the $-Z$-surface of the domain (see the preceeding figure.)

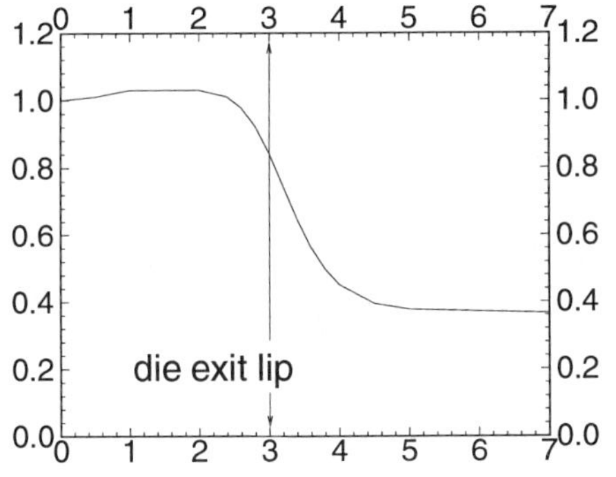

Figure 9: The centerline velocity as a function of the distance from the die inlet.

ulate the extrusion of a Newtonian fluid out of a square die. Although the present problem has symmetry, a full mesh is used to test the stability of the iterative solver. The resulting system has 2622 unknowns. The convergence tolerance is set to 10^{-15} and the modified conjugate gradient solver performs well, giving a converged solution after 2502–2786 iterations (the range corresponds to different stages of free surface iteration). The results are almost identical to those given by a direct Gaussian elimination solver.

The free surface is obtained iteratively by a pathline method (Bush and Phan-Thien 1985, Tran-Cong and Phan-Thien 1988). The results obtained are depicted in figures 5-9. Figure 5 shows a three dimensional view of the final mesh with the velocity field superimposed. In figure 6, it can be seen that the free surface iteration converges fairly rapidly and the final extrudate swells by 19% at the middle of the flat face and 7.6% at the corner. These results compared reasonably with those obtained by DBEM of Tran-Cong and Phan-Thien (1988) who obtain the corresponding swells of 18.0% and 3.4%. The discrepancy in the results obtained by two different integral formulations could be due to the fact that the kinematics are more sensitive to a fast variation in the potential, which is singular at the die exit, but less sensitive to the traction field, which is also singular at the die exit. Figures 7-8 show the rapid variation in the potential near the die exit. Figure 7 is a vector plot of the potential, illustrating its orientation along the $X-$axis. Figure 8 is the plot of the three components of the potential along the line of intersection of the $XZ-$plane and the surface with negative Z unit normal. Note that $\varphi_y = 0$ along this line. It is clear that a strong singularity in both φ_x and φ_z exists near the exit, suggesting the need for a mesh refinement. This is being carried out at present. The results for the same problem obtained by Karagiannis *et al* (1988) using finite element method are 18.4% and 2.9% respectively, with a finer mesh than used here. Finally figure 9 shows the variation of the velocity along the centerline of the die from the die inlet to the extrudate. A slight acceleration of the centerline velocity between the inlet and the exit lip is predicted in this study, in agreement with previous DBEM study by Bush and Phan-Thien (1985), but in disagreement with the FEM results of Karagiannis *et al* (1988), where no acceleration is predicted.

CONCLUDING REMARK

It is demonstrated that an alternate integral representation for elasticity or Stokes problems, based on potentials on unclosed surfaces, can be effectively discretized and solved using iterative method. An extension of the formulation to deal with non-homogeneous and/or non-linear problem is also presented. This will be the basis for future work on non-linear problems.

ACKNOWLEDGEMENT

This work is supported by Australian Research Council Grants.

REFERENCES

Banerjee, P.K. and Butterfield, R. 1981. *Boundary Element Methods in Engineering Sciences* London:McGraw-Hill.

Beer, G. and Watson, J.O. 1992. to Finite and Boundary Element Methods for Engineers Chichester:Wiley.

Brebbia, C.A., Telles, J.C.F. and Wrobel, L.C. 1984. *Boundary Element Techniques: Theory and Applications in Engineering* Berlin:Spinger-Verlag.

Bush, M.B. and Phan-Thien, N. 1985. Three dimensional viscous flow with a free surface: flow out of a long square die. *J. Non-Newt. Fluid Mech.* 18:211-218.

Doblare, M. 1987. Computational aspects of the Boundary Element Method. In C.A. Brebbia (ed), *Topics in Boundary Element Research* Vol.3:51-131. Berlin: Spinger-Verlag.

Jennings, A. and Malik, G.M. 1978. The solution of sparse linear equations by the conjugate gradient method. *Intl. J. Num. Meth. Eng.* 12:141-158.

Karagiannis, A., Hrymak, A.N. and Vlachopoulos, J. 1988. Three-dimensional extrudate swell of creeping Newtonian jets. *AIChE Journal* 34:2088-2094.

Kim, S. and Karrila, S.J. 1991. *Microhydrodynamics: Principles and Selected Applications* Boston:Butterworth-Heinemann.

Mi, Y. and Aliabadi, M.H. 1992. Dual boundary element method for three-dimensional fracture mechanics analysis. *Engng. Anal. Boundary Elements* 10:167-171.

Mi, Y. and Aliabadi, M.H. 1994. Three-dimensional crack growth simulation using BEM. *Computers & Structures* 52:871-877.

Parton, V.Z. and Perlin, P.I. 1984. *Mathematical Methods of the Theory of Elasticity.* Moscow:MIR Publishers.

Patterson, C. and Sheikh, M.A. 1984. Interelement continuity in Boundary Element Method. In C.A. Brebbia (ed) *Topics in Boundary Element Research* Vol.1:123-141. Berlin: Spinger-Verlag.

Phan-Thien, N. and Kim, S. 1994 *Microstructures in Elastic Media: Principles and Computational Methods.* Oxford:Oxford University Press.

Tran-Cong, T. and Phan-Thien, N. 1988. Three dimensional study of extrusion processes by Boundary Element Method. Part 1: an implementation of high order elements and some Newtonian results. *Rheologica Acta* 27:21-30.

Simulation of Materials Processing: Theory, Methods and Applications, Shen & Dawson (eds)
© 1995 Balkema, Rotterdam. ISBN 90 5410 553 4

Numerical simulation of the thermoforming of multi-layer polymer sheets

Marie-Hélène Vantal, Bernard Monasse & Michel Bellet
École des Mines de Paris, CEMEF, Centre de Mise en Forme des Matériaux, UA CNRS, Sophia-Antipolis, France

ABSTRACT: This paper presents a 3D finite element model of the thermoforming process. The polymer material is assumed to obey the viscoplastic law proposed by G'Sell and Jonas, the coefficient of which have been identified for a polystyrene. The implementation of this constitutive model, the heat transfer coupling and the proposed multi-layer approach are detailed in the paper. Axisymmetric validation tests and an application to the thermoforming of an industrial component are also reported.

1 INTRODUCTION

The thermoforming process consists in heating a polymer sheet and shaping it inside a mold. The deformation results from a pressure cycle eventually coupled with the use of a moving tool. The forming of thin products made of various polymer materials is carried out by this method: rubbery amorphous, solid semi-crystalline or multi-layer composites. The heating temperature depends upon the selected polymer. At low temperature, the forming is limited by the too high rigidity of the sheet. On the contrary, at high temperature, the sheet deforms by gravity and the forming is very difficult to control. Hence, the thermoforming is operated above the glass transition temperature (T_g), in the rubber-like behavior domain for amorphous polymers, and close to the melting temperature for semi-crystalline polymers.

The main problem of the process is the thinning in the corners of the parts, which leads to a decrease of the mechanical properties of the shaped components. The optimization of the final thickness profile is generally achieved by trial-and-error, changing the design of the component, the polymer material, and the process parameters such as the heating temperature distribution, the mold temperature, the pressure curve, or using a punch for deep parts. The numerical modelling should then result in a more efficient optimization of the process.

Various numerical simulation models have been proposed (Warby & Whiteman 1988, De Lorenzi & Nied 1991, Kouba et al. 1992, Shrivastava & Tang 1993). Regarding the application to three-dimensional formings, they are generally based on the membrane mechanical approximation, associated with the finite element method. Those computations use either hyperelastic (De Lorenzi & Nied 1991, Shrivastava & Tang 1993) or viscoelastic (Warby & Whiteman 1988, Kouba et al. 1992) constitutive equations, without any heat transfer coupling and a sticking contact hypothesis at polymer-tools interface.

It should be pointed out that this latter assumption has no experimental evidence. Most of the time, it is used just because the computation is isothermal and cannot account for the decrease of the polymer temperature and its "freezing" after tool contact. Another thermomechanical coupling is the high self heating source term due to the high strain rates. As the behavior of polymers is known to be highly temperature-dependent, it seems essential to couple heat transfer and mechanical models.

In the present paper, we describe a 3D finite element model with membrane approximation to predict the deformation. Concerning the contact conditions with punch and mold, either sliding contact with Coulomb's friction law or sticking contact can be considered. The paper is focused on the thermal evolution effects during the forming, and on the approach to multilayer sheet forming.

2 MATERIAL BEHAVIOR IDENTIFICATION

The one-dimensional constitutive equation initially proposed by G'Sell and Jonas (1979) has been selected in the context of polymer thermoforming:

$$\bar{\sigma} = K_p(T)\left[1 - \exp(-w\bar{\varepsilon})\right] \exp\left(h\bar{\varepsilon}^2\right) \dot{\bar{\varepsilon}}^m \qquad (1)$$

where T is the temperature, $\dot{\bar{\varepsilon}}$ the von Mises equivalent strain-rate, $\bar{\varepsilon}$ the von Mises equivalent strain. K_p, w, h and m are material parameters.

This law accounts for the material behavior of a great number of polymers, either amorphous or semi-crystalline, in a large temperature interval. Duffo et al. (1994) have identified parameters for polypropylene by means of tensile tests on sheets, yielding:

$$K_p = 5.31\ 10^{-3} \exp\left[2.85\ 10^{-3} / T\right] \text{ MPa.s}^m$$
$$w = 20,\ h = 0.4,\ m = 0.087$$

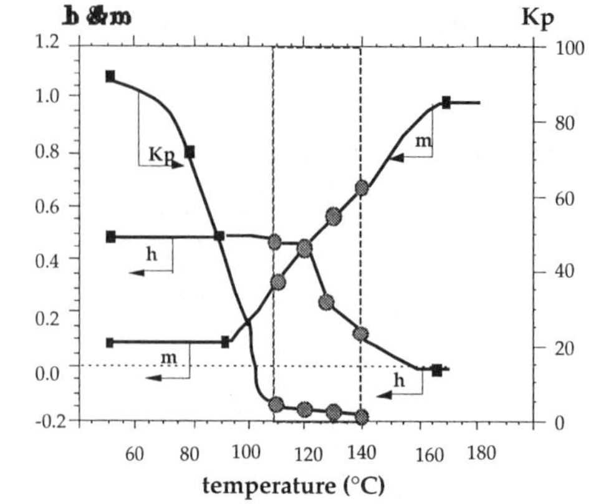

Fig.1 - Coefficients m, h and K_p(MPa.s^m) of equ. (1) for polystyrene vs temperature. (Grey dots identified by tensile tests, black squares issued from literature).

We have studied the material behavior of a polystyrene compound, by uniaxial tensile testing at constant axial strain rate. Tests have been carried out at four constant temperatures: 110, 120, 130 and 140°C (which cover the forming temperature domain of this polymer) and five different strain-rates between 10^{-4} and 10^{-2} s^{-1}. The associated determination of the coefficients of equation (1) shows the transition from a solid-like behavior to a liquid-like one: see figure 1. The consistency K_p of the material decreases suddenly above T_g and keeps low values ($\approx$2) in the experimental domain. The so-called viscoelastic coefficient w is found low ($\approx$3.7) and independent on temperature. Finally, it has appeared that the basic equation (1) has to be modified, in order to include a temperature dependency of the strain-rate sensitivity coefficient m and the hardening coefficient h.

3 MECHANICAL FORMULATION

3.1 Membrane equilibrium

According to membrane mechanical assumption, the deformed sheet is considered as a geometric surface, neglecting flexion and transverse shear. A material point is identified by two curvilinear coordinates: θ^1 and θ^2 which are, in the case of an initial sheet in xy plane, the initial x and y coordinate respectively. At any point of the deformed sheet (vector **x**) the local tangent basis is defined by:

$$\mathbf{g}_1 = \frac{\partial \mathbf{x}}{\partial \theta^1} \qquad \mathbf{g}_2 = \frac{\partial \mathbf{x}}{\partial \theta^2} \qquad \mathbf{g}_3 = \frac{1}{\|\mathbf{g}_1 \wedge \mathbf{g}_2\|}\,\mathbf{g}_1 \wedge \mathbf{g}_2 \quad (2)$$

The equilibrium of the deformed sheet is expressed by the principle of virtual work (without inertia):

$$\forall \mathbf{v}^* \int_{\Omega} \sigma^{ij} v^*_{i|j}\, e\, dS - \int_{\Omega_p} P\mathbf{g}_3.\mathbf{v}^* dS - \int_{\Omega_c} \mathbf{t}.\mathbf{v}^* dS = 0 \quad (3)$$

where σ^{ij} are the covariant components of the Cauchy stress tensor (plane stress: i,j=1,2), $._{|j}$ denotes covariant derivation with respect to θ^j , e is the sheet thickness, **t** are the friction stresses on regions Ω_c contacting the mold, P is the inflation pressure applied to the domain Ω_p of the deformed surface Ω.

3.2 Time discretization. Incremental procedure

Starting with a balanced configuration Ω at time t, the problem consists in determining the unknown equilibrated configuration Ω' at t+Δt. Variables at t+Δt are denoted "prime". By application of eq.(3) at t+Δt, we have, for any velocity field $\mathbf{v}^*$:

$$\int_{\Omega'} \sigma'^{ij} \mathbf{v}^*_{i|j}\, e' dS - \int_{\Omega'_p} P'\mathbf{g}'_3.\mathbf{v}^* dS - \int_{\Omega'_c} \mathbf{t}'.\mathbf{v}^* dS = 0 \quad (4)$$

This equation is solved for the incremental displacement field **u** between Ω and Ω', provided that **t**', e' and σ' can be calculated from **u**. Those relations are exposed hereunder and the resolution in 3.3.

• *Contact and friction:* as regard the contact condition applied to Ω_c, it may be either sticking (no relative velocity with respect to the mold) or sliding: in this case, the tangential stress is supposed to be given by the Coulomb's friction law (coefficient μ), in which the normal stress is the inflation pressure:

$$\mathbf{t}' = -\,\mu\, P'\,(1 / \|\mathbf{u}\|)\,\mathbf{u} \quad (5)$$

• *Thickness updating:* the new local thickness e' is deduced from material incompressibility. Denoting **g** the metric tensor ($g_{ij} = \mathbf{g}_i.\mathbf{g}_j$), we have:

$$e' = e\,\sqrt{\det(\mathbf{g})/\det(\mathbf{g}')} \quad (6)$$

• *Resolution of incremental constitutive equations:* the one-dimensional constitutive equation (1) can be written as a classical viscoplastic power law:

$$\bar{\sigma} = k\,\dot{\bar{\varepsilon}}^{m} \quad (7)$$

Hence, the flow rule derives from a viscoplastic potential Q. Under isotropy assumption, it yields:

$$\dot{\varepsilon} = \frac{\partial Q}{\partial \sigma} = \frac{\partial Q}{\partial \bar{\sigma}}\frac{\partial \bar{\sigma}}{\partial \sigma} = \dot{\bar{\varepsilon}}\,\frac{\partial \bar{\sigma}}{\partial \sigma} \quad (8)$$

In convective curvilinear coordinates, we have:

$$\dot{\bar{\varepsilon}}^2 = 2/3\; \dot{\varepsilon}^{ij}\dot{\varepsilon}_{ij} \qquad \bar{\sigma}^2 = 3/2\; s^{ij}s_{ij} = \sigma^T \mathbf{A} \sigma \quad (9)$$

$$A_{ijkl} = 3/2\; g_{ik}g_{jl} - 1/2\; g_{ij}g_{kl} \quad (10)$$

The constitutive equation can then be written:

$$\dot{\varepsilon} = (\dot{\bar{\varepsilon}}/\bar{\sigma})\,\mathbf{A}\,\sigma = (1/k)\,\dot{\bar{\varepsilon}}^{(1-m)}\,\mathbf{A}\,\sigma \quad (11)$$

A semi-implicit time integration scheme is used over the increment. The incremental strain tensor $\Delta\varepsilon$, the covariant components of which depend on the displacement **u** according to (12) is written as (13):

$$\Delta\varepsilon_{ij} = 1/2\,(u_{i|j} + u_{j|i} + u_{m|i}u^{m}_{|j}) \quad (12)$$

$$\Delta\varepsilon = \Delta t\left[(1-\eta)\,\dot{\varepsilon} + \eta\,\dot{\varepsilon}'\right] \qquad (13)$$

Equations (11-13) clearly permit to deduce the new local stress tensor σ', knowing the displacement field $\mathbf{u}$. Practically, a fully implicit scheme (η=1) is used.

3.3 Resolution: finite element method (f.e. m.)

Finally, injection of (5-6) and (11-13) in the equilibrium equation (4) leads to a non-linear equation for the displacement field $\mathbf{u}$. Its spatial discretization by f.e.m. (linear triangles or quadrangles) is detailed by Bellet (1988, 1990). At every time step a non-linear system for the vector $\mathbf{U}$ of the nodal displacements is solved by the Newton-Raphson method, with a consistent tangent matrix.

4 HEAT TRANSFER RESOLUTION

Considering the thinness of polymer sheets, the short processing times, and the low diffusivity of polymers, it can be assumed that heat transfer is essentially one dimensional across the thickness of the sheet (Vantal 1995). Consequently, s being the coordinate in the thickness direction ($s=\theta^3$), the 1D heat transfer equation can be expressed:

$$\rho c \frac{dT}{dt} = \frac{\partial}{\partial s}\left(\lambda \frac{\partial T}{\partial s}\right) + \sigma : \dot{\varepsilon} \qquad (14)$$

where ρ is the specific mass, c the heat capacity and λ the heat conductivity. The following boundary conditions are accounted for. At sheet/air interface:

$$-\lambda \frac{\partial T}{\partial s}\,\mathrm{sgn}(\mathbf{n}) = h_{conv}\,(T - T_{air}) \qquad (15)$$

where h_{conv} is the coefficient for heat exchange by convection, T_{air} the air temperature and sgn($\mathbf{n}$) is ± 1 depending on the orientation of the outward normal unit vector $\mathbf{n}$. At sheet/tool interface, due to the much higher diffusivity of metals, we will assume that the surface temperature of the polymer sheet is prescribed to the interface temperature given by:

$$T_{inter} = (b_{tool} T_{tool} + b_{sheet} \overline{T}_{sheet}) \,/\, (b_{tool} + b_{sheet}) \qquad (16)$$

where b is the thermal effusivity $\sqrt{\lambda \rho c}$ and $\overline{T}_{sheet}$ is the average temperature of the sheet in the thickness. The initial temperature profile is assumed to be known at the beginning of the process.

Equation (14) is discretized in space and time, at each integration point of membrane finite elements, using a Galerkin 1D f.e.m. (figure 2) and a semi-implicit time integration scheme (Vantal 1995).

The coupling between the mechanical and the thermal resolution is carried out at each time increment as explained in figure 3.

5 MULTI-LAYER FORMULATION

The multi-layer approach proposed here is based on the following assumptions:

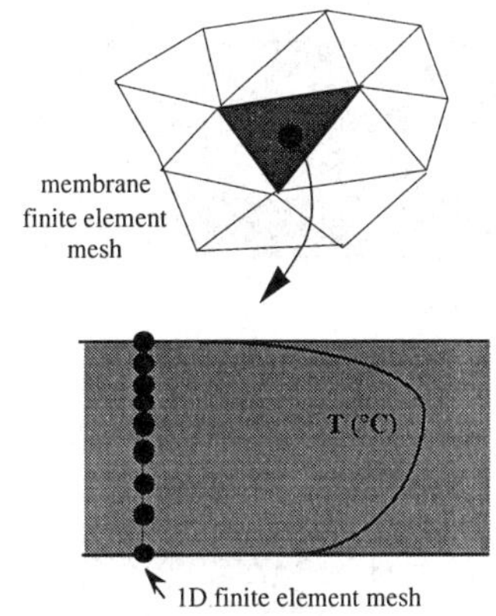

Fig.2 - 1D approach for thermal coupling

<u>A/ Thermal resolution</u>
- for all membrane elements
  - for all integration points of the element
    - perform a 1D f.e. computation of the new temperature profile T' across the thickness. The source term is given by the mechanical resolution of previous time step.
    - for all integration points ipth of 1D mesh update the temperature-dependent coefficients: $K'_p(T')$, m'(T') and h'(T').

<u>B/ Mechanical resolution</u>

B1) computation of the residual vector $\mathbf{R}(\mathbf{U})$
- for all membrane elements
  - for all integration points of the element
    - for all integration points ipth of 1D mesh solve the constitutive equations for σ'(ipth) and $\partial\sigma'/\partial\Delta\varepsilon$(ipth), using the updated values of material coefficients: K'_p(ipth), m'(ipth) and h'(ipth)
    - compute thickness-averaged values $\langle\sigma'\rangle$ and $\langle\partial\sigma'/\partial\Delta\varepsilon\rangle$ and sum in residual vector.

B2) iterative Newton-Raphson procedure to solve **R(U)=0**

Fig.3 - Thermo-mechanical coupling algorithm.

- each layer is submitted to the same deformation as the mean surface of the sheet. This is consistent with the membrane approach in which transverse shear and flexion are neglected. Hence, the thickness ratio of the different layers remains constant during the process.
- accordingly, the mechanical and thermal contact are assumed perfect (no sliding, no thermal contact resistance) at interfaces between layers.

The algorithm is identical to the one for heat transfer coupling, except that all thermal and mechanical parameters used at integration points in thickness (ipth) will now depend upon the material in which they are located (see fig. 4). Each layer material has an identified temperature dependent behavior law.

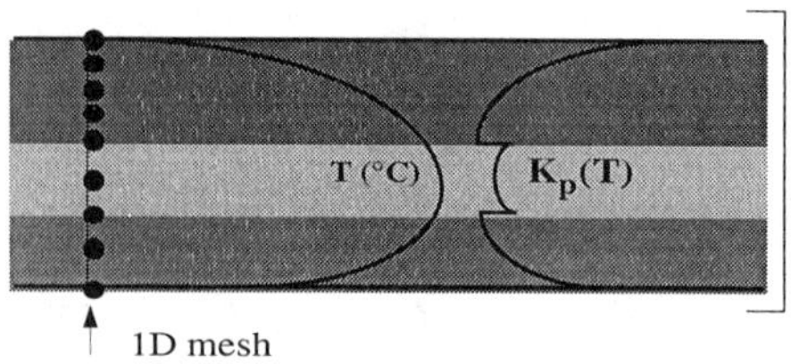

Fig.4 - Multi-layer formulation (schematic).

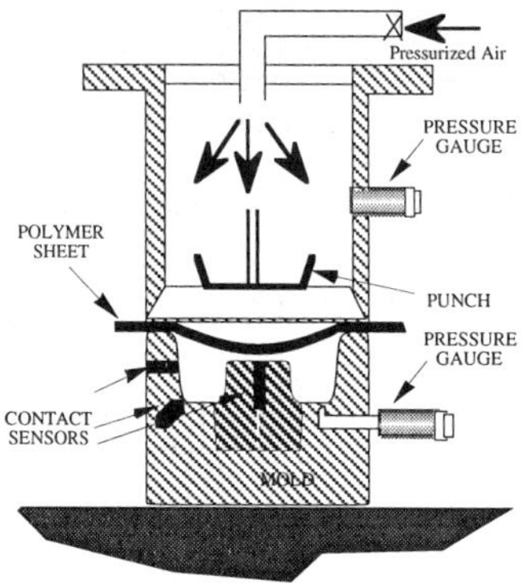

Fig. 5 - Experimental thermoforming system

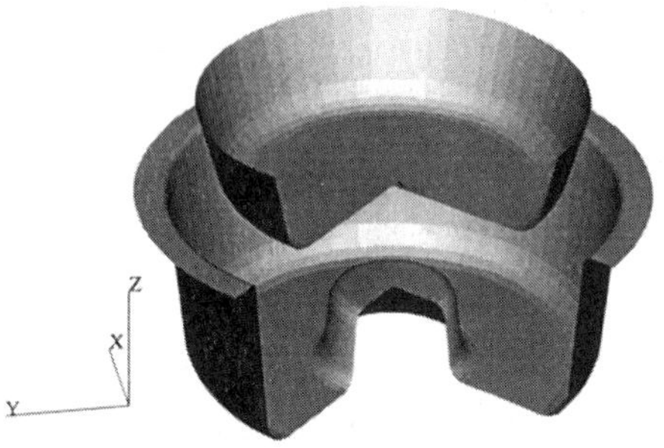

Fig. 6 - Mold and punch view.

Such a formulation is expected to be more precise than the reduction of the multi-layer to a single "equivalent" material, especially when steep temperature gradients appear in the sheet thickness when one side of the multi-layer contacts the tooling.

6 TEST OF THE MODEL

6.1 Experimental: axisymmetric thermoforming test

Formings were carried out on a small machine equiped with contact sensors and pressure gauges.

The mold geometry is axisymmetrical (diameter ≈ 140 mm, depth 60 mm), including a central insert in the bottom (see fig. 6) which makes it very sensitive both to cooling and friction effects. The forming parameters are the followings: aluminium mold, 20°C, air temperature = 20°C, linear pressure vs time: 0.6 $MPa.s^{-1}$, initial sheet temperature 130°C. The one-layer polystyrene sheet is initially 1.1 mm thick. Its orientation due to extrusion is relatively high, and it involves an evolution of the thickness during infrared heating as it is clamped. Consequently, the measured final thickness profile depends upon the measurement direction (extrusion or transverse direction, see fig.7).

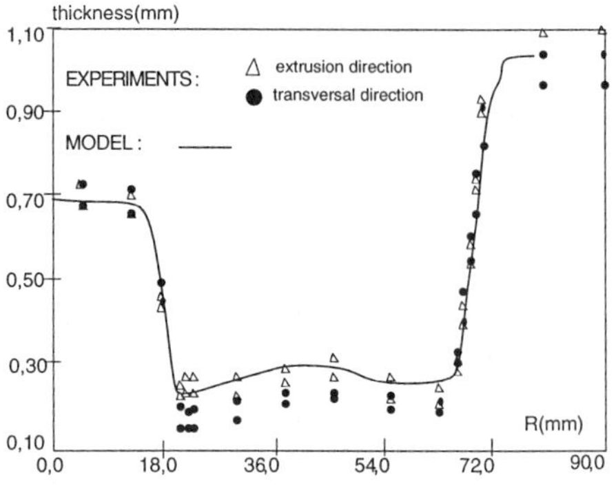

Fig. 7 - Comparison between measured and computed thickness distribution.

Table 1

contact time (s) at sensor	1	2	3
experiment	0.095 s	0.400 s	0.510 s
computation	0.102 s	0.311 s	0.390 s

6.2 Numerical simulation of the test

Due to axisymmetry, only a narrow sector has been meshed with 160 triangles. We have used the rheological data of fig.1 and the following values of thermal parameters: $\rho c = 1.92\ 10^6 J.m^{-3}.K^{-1}$, $\lambda = .2\ W.m^{-1}.K^{-1}$, $h = 33\ W.m^{-2}.K^{-1}$. A Coulomb friction has been accounted for (μ=0.4). The predicted final thickness profile, using the coupled thermomechanical model, is in good agreement with the measurements (figure 7).

Regarding kinematics, it should be noticed that the measured differential pressure was found very low. The use of these values as a prescribed boundary condition (P(t) in equation (3)) has lead to much slower forming rates than the actual ones. However, if the measured upper pressure is used instead of the differential pressure, then computed forming times are in agreement with experimental ones, as shown in table 1. This unexpected result needs more investigation.

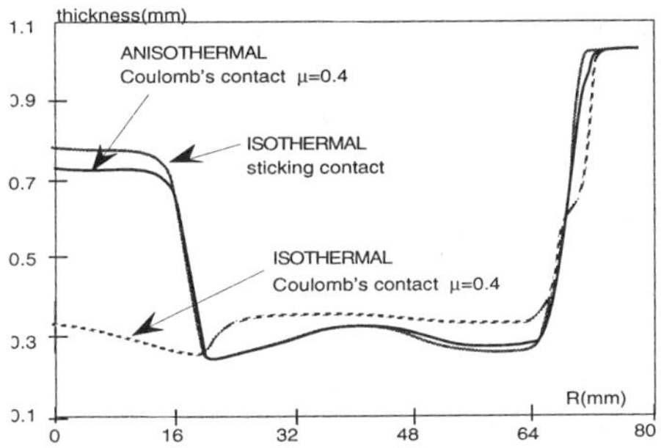

Fig.8 - Computed thickness for three models.

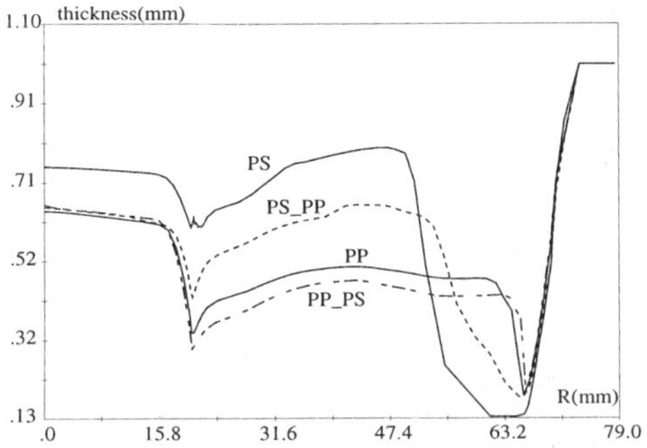

Fig.9 - Computed thickness profiles.

6.3 Influence of thermal coupling

On figure 8, three different computed thickness profiles have been plotted, using three different computational options.

As already said, the assumption of sticking contact has no clear experimental evidence. However, as shown on figure 8, it is the only means to get reasonable results if an isothermal model is used. For example the combination (isothermal ; high friction) yields completely erroneous thicknesses (far too low on the insert, despite the high friction coefficient)!

On another hand, the use of the present non-isothermal formulation permits to decouple clearly the frictional and the thermal effects: the quick cooling of the polymer when contacting the mold increases the consistency and the hardening coefficient of the material for the contact zones, as the strain-rate sensitivity coefficient decreases. The evolution of these coefficients localize the deformation in the warmer zones: the strain-rate values are almost zero in the "frozen" zones. The non-isothermal model is much more representative of the local phenomena in thermoforming and is able to account independently for interface tribology and heat transfer.

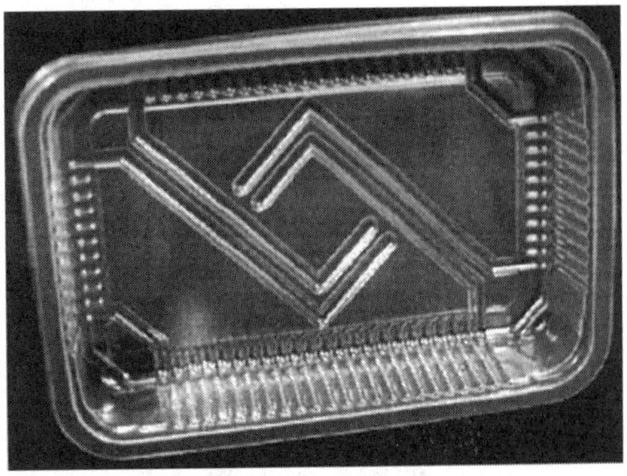

Fig.10 - Selected polypropylene test part.

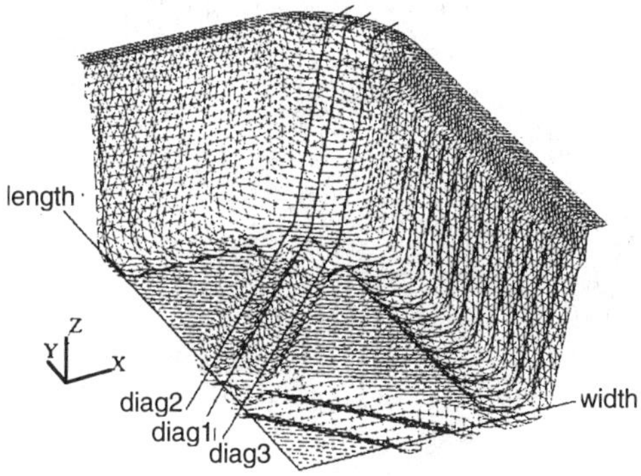

Fig.11 - Final deformed finite element mesh and selected directions for thickness comparison.

Table 2 - Predicted forming times.

PP only	PP_PS PP on top	PS_PP PS on top	PS only
0.59 s	0.50 s	0.39 s	0.45 s

6.4 Multi-layer application

We have simulated the forming of a bi-layer: 80% polystyrene (PS), 20% polypropylene (PP), total thickness 1mm. The rheological parameters for both polymers are those given in section 2. The sheet is formed successively with a punch (z=-30mm at 0.15 s) and pressure (linear increase of 0.8 MPa between 0.15 s and 0.65 s). The initial temperatures are 20°C for tools and 150°C for the sheet.

It is shown in table 2 and figure 9 that the deformation of the bilayer is not a simple addition or average of the deformation of each layer component with the same thickness. In addition, the results clearly depend on which material is on top.

This is due to the fact that the cooling effect is different for PS or PP: the consistency of PS

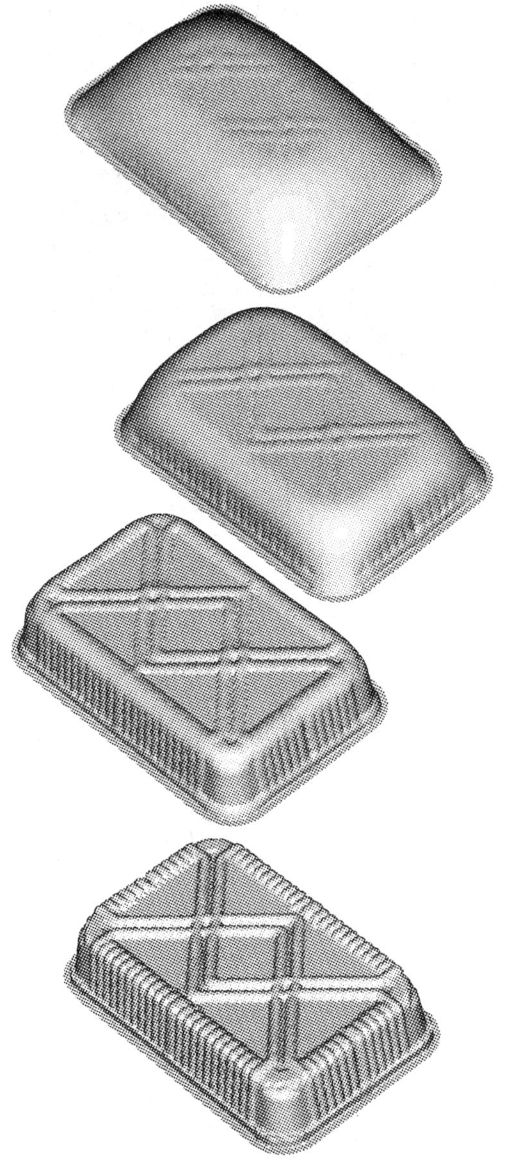

Fig. 12 - Computed deformed sheet

decreases suddenly above T_g whereas it follows an Arrhenius law for PP. Also m and h are temperature dependent for PS whereas constant for PP.

7 APPLICATION: INDUSTRIAL FORMING

In order to test the robustness and the results of the code , the forming of a one-layer shallow component for food packaging has been studied (fig. 10).

The material is polypropylene, initially 0.475 mm thick. The industrial forming conditions have not been

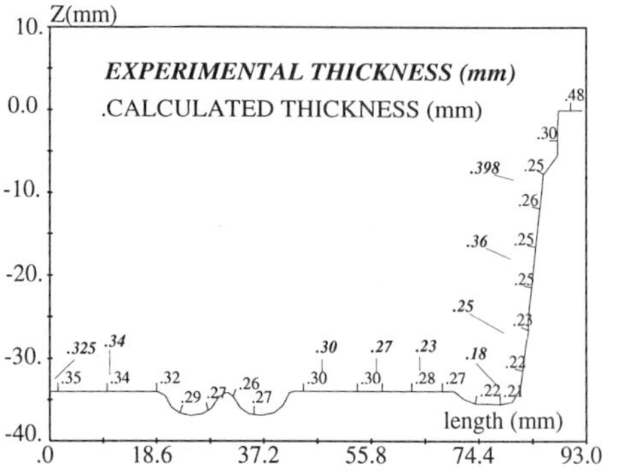

Fig. 13 - Example of comparison between measured and computed thickness ("length" on figure 12).

accurately measured, but the following figures can be considered realistic: aluminum mold (5°C), linear pressure reaching 0.7MPa at 0.5s, air temperature 20°C, initial sheet temperature 150°C.

Only a quarter of the symmetrized actual forming has been computed, using 7591 and 6463 triangles for the sheet and mold meshes respectively. A Coulomb's friction law (μ=0.4) is assumed at the mold surface, except near the edge where sticking contact is prescribed. As shown on fig. 12, the central bulging of the sheet is very fast, reaching the bottom at 0.04s. The forming is then slowed down until the end.

Figure 11 shows the deformed finite element mesh at the end of the process. Iso-values of temperature (not shown here) indicate that during forming the free regions are the warmer (close to the initial temperature), and that in contacting regions, the longer the contact duration, the cooler is the sheet.

The final thickness of actual parts has been measured along the five directions mentioned on figure 11. An example of comparison with the computed values is shown in figure 13. Experimental points are issued from measurements on two different parts, yielding four values per point. The experimental dispersion has been found high (up to 18% of the average value). The average of relative errors between computed and average experimental thickness is good: 10% at the bottom, 18% at wall.

8 CONCLUSION

Thermomechanical coupling and a multi-layer approach have been implemented in a finite element membrane model. They are shown to be very efficient tools to improve the predictive character of finite element simulations of thermoforming.

9 ACKNOWLEDGEMENT

The experimental part of this work has been done at Elf-Atochem (Cerdato) which has also supported this study.

10 REFERENCES

Bellet M. 1988. Modélisation numérique du formage superplastique de tôles. Doctorate thesis (in french), Ecole des Mines de Paris.

Bellet M., Massoni E. and Chenot J.L. 1990. Numerical simulation of thin sheet forming processes by the finite element method. Eng. Comp., 7, p. 21.

De Lorenzi H.G. & Nied H.F. 1991. Finite element simulation of thermoforming and blow molding. Modelling of Polymer Processing, A.I.Isayev (ed.), Hanser Verlag, chap.5, pp. 118-171.

Duffo P., Monasse B., Haudin J.M., G'Sell C. & Dahoun A. 1994. Rheology of polypropylene in the solid state. J. Mater. Sci., 29.

G'Sell C. & Jonas J.J. 1979. Determination of the plastic behavior of solid polymers at constant true strain rate. J. Mat. Sci., 14, pp. 583-591.

Kouba K., Bartos O. & Vlachopoulos J. 1992. Computer simulation of thermoforming in complex shapes. Polym. Eng. Sci., 32, pp. 699-704.

Shrivastava S. & Tang J. 1993. Large deformation finite element analysis of viscoelastic membranes with reference to thermoforming. J. Strain Anal., 28, pp. 21-51.

Vantal M-H. 1995. Etude numérique et expérimentale du thermoformage des polymères". Doctorate thesis (in french), Ecole des Mines de Paris.

Warby M.K. & Whiteman J.R. 1988. Finite element model of viscoelastic membrane deformation. Comput. Struct., 68, pp. 33-54.

Simulation of Materials Processing: Theory, Methods and Applications, Shen & Dawson (eds)
© 1995 Balkema, Rotterdam. ISBN 90 5410 553 4

Process simulation of polymer blow molding by 3-D rigid visco-plastic FEM

K.F.Zhang
Institute of Physical and Chemical Research (RIKEN), Wako, Saitama, Japan (On leave from: Harbin Institute of Technology, People's Republic of China)

A.Makinouchi, S.Wang & T.Nakagawa
Institute of Physical and Chemical Research (RIKEN), Wako, Saitama, Japan

ABSTRACT: A rigid visco-plastic FE code has been developed to simulate the industrial 3 dimensional blow molding processes. This code can simulate successive operations : the pre-blowing of parison, the closing of tools to trap the parison inside molds and the blowing of air in the parison. As examples forming process of a square container and a gasoline tank used for an automobile are simulated, and simulation results are compared with experimental results.

1 INTRODUCTION

In the blow molding process of polymer containers, control of the thickness distribution of products is always important in the tool and process design stage. Recently, due to the technical advance, complex shaped and big sized polymer products such as automotive gasoline tanks are formed by the blow molding. The FE simulation is expected to be a powerful means, predicting thickness distribution of formed products without manufacturing actual tryout tools, and thus reducing tool design time and cost. A rigid visco-plastic FEM code is developed to meet this demand. The non Newtonian fluid flow assumption is used to represent the behavior of polymer during deformation.

The procedure of blow molding process generally can be divided into two or three steps. In the first step, air pressure is applied to expend the extruded pipe-shaped parison to a suitable size (pre-blowing), then in the second step, the tools are closed to trap the parison inside molds (tool closing), and in the final step the parison is enforced to contact to the tool surface by the blow of air (blowing). The simulation should be perform all of these processes (Vlachopoulos 1992), (Debbaut 1994), (Tanaka 1994).

2 NUMERICAL MODELING

2.1 Rigid visco-plastic formulation

Deformation of polymers at elevated temperature is modeled by incompressible non-Newtonian fluid and is assumed to be deformed under a constant temperature, while inertia forces are neglected. The relations between equivalent stress and equivalent strain rate for the non-Newtonian material can be written as follows:

$$\bar{\sigma} = k\dot{\bar{\varepsilon}}^{m} \qquad (1)$$

Where $\bar{\sigma}$ is equivalent stress, $\dot{\bar{\varepsilon}}$ is equivalent strain rate, k is material constant, m is a strain rate hardening index. The constitutive equation is :

$$\sigma'_{ij} = \frac{2\bar{\sigma}}{3\dot{\bar{\varepsilon}}}\dot{\varepsilon}_{ij} \qquad (2)$$

Where $\dot{\varepsilon}_{ij}$ and σ'_{ij} are respectively strain rate tensor and stress deviator tensor. If we rewrite Eqa.(2) in matrix form and involve the plane stress assumption then it becomes

$$\sigma = aD\dot{\varepsilon}$$

In which

$$\sigma = \left(\sigma_{11}\ \sigma_{22}\ \sigma_{12}\ \sigma_{23}\ \sigma_{13}\right)^{T}$$

$$\dot{\varepsilon} = \left(\dot{\varepsilon}_{11}\ \dot{\varepsilon}_{22}\ 2\dot{\varepsilon}_{12}\ 2\dot{\varepsilon}_{23}\ 2\dot{\varepsilon}_{13}\right)^{T}$$

$$D = \begin{bmatrix} 2 & 1 & 0 & 0 & 0 \\ 1 & 2 & 0 & 0 & 0 \\ 0 & 0 & 1/2 & 0 & 0 \\ 0 & 0 & 0 & 1/2 & 0 \\ 0 & 0 & 0 & 0 & 1/2 \end{bmatrix}, \qquad a = \frac{2\bar{\sigma}}{3\dot{\bar{\varepsilon}}}$$

The following FEM equation is derived from rigid visco-plastic variational principle (Zhang 1993, 1994) :

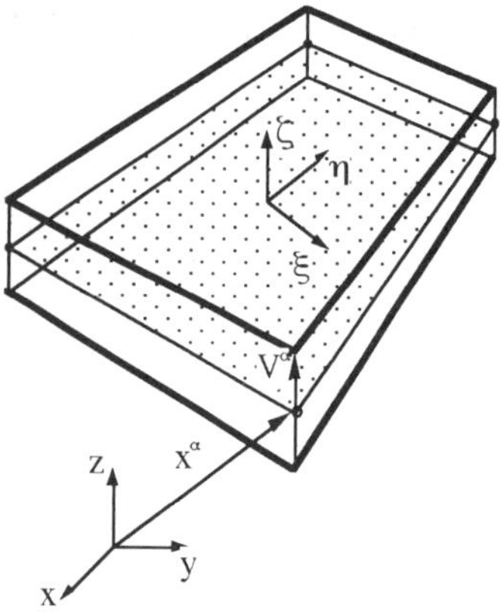

Fig.1 4 node degenerated shell element

$$\frac{2}{3}k\left(\int_V b_{(k-1)} K dV + \int_V c_{(k-1)} H_{(k-1)} H^T_{(k-1)} dV\right)\Delta U_k = $$
$$= -\int_V a_{(k-1)} H_{(k-1)} dV + \int_{S_t} N^T P dS \quad (3)$$

In which

$$b_{(k-1)} = \dot{\bar{\varepsilon}}^{(m-1)}_{(k-1)} \qquad c_{(k-1)} = \frac{2(m-1)}{3}\dot{\bar{\varepsilon}}^{(m-3)}_{(k-1)}$$

$$K = B^T D B \qquad H_{(k-1)} = K U_{(k-1)}$$

Where B is strain rate-displacement rate matrix, U and ΔU are displacement rate vector and it's incremental vector, N is shape function and P is surface traction vector.

$$U_k = U_{(k-1)} + \alpha \cdot \Delta U_k$$

The coefficient α is in range of 0.1~0.5 in this program, subscripts k and k-1 correspond to iteration number.

Using a standard rigid visco-plastic FEM approach for solid elements, additional condition (conservation of volume) has to be imposed by means of Lagrangian multiplier method or penalty method. However, in case of a degenerated shell element, it is not necessary to introduce such a condition to Eqa.(3).

2.2 Four node degenerated shell element

4 node degenerated shell element (Ahmad *et al.* 1970) with 8 integration points is implemented to our code - Fig.1. A plane stress assumption is used

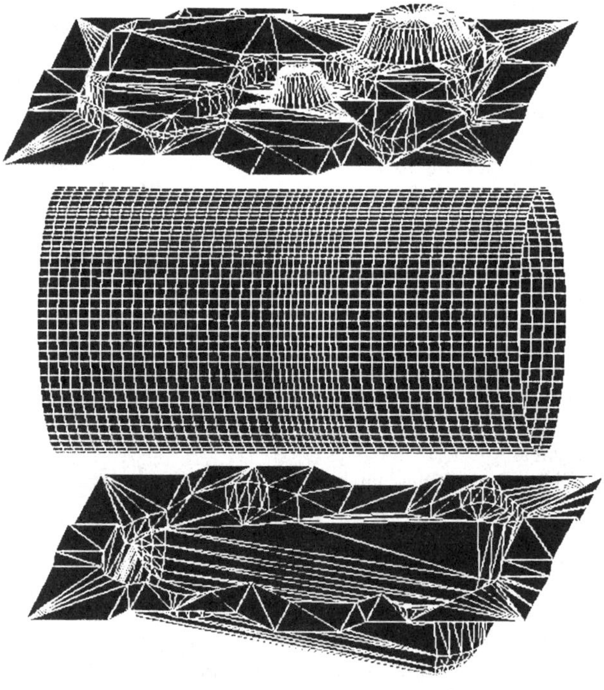

Fig.2 Tools used in blow molding process

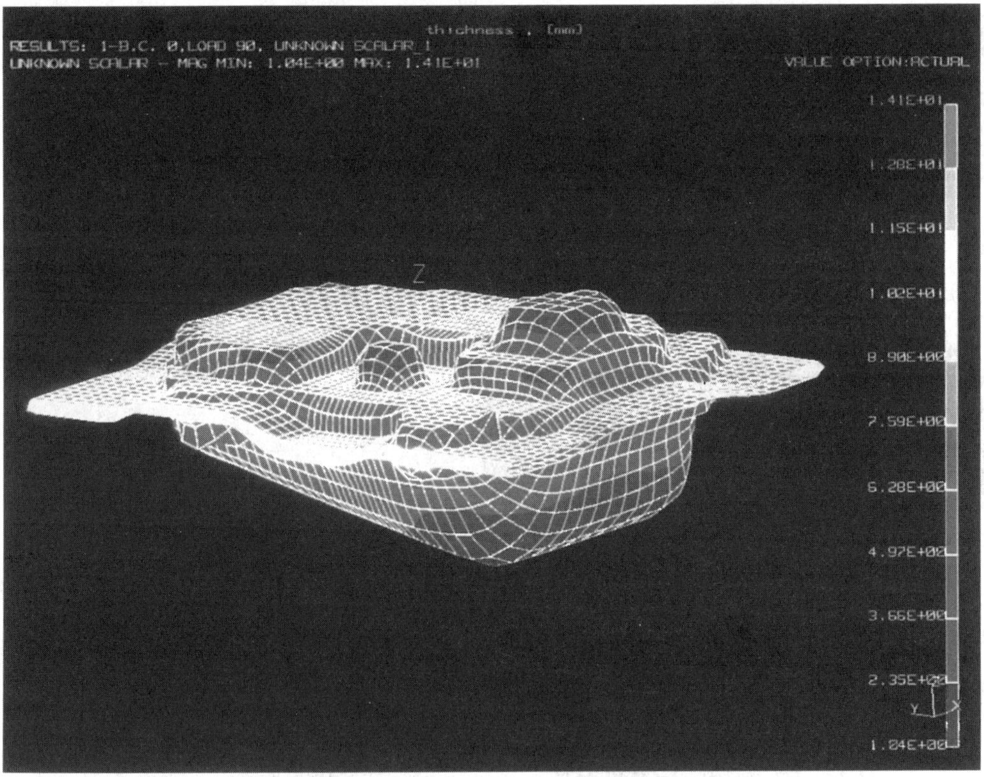

Fig.3 Thickness distribution

to modify a constitutive matrix.

Coordinate and velocity of arbitrary point of shell element can be described by following equations :

$$x(\xi,\eta,\zeta) = N^{\alpha}(\xi,\eta)\cdot x_i^{\alpha} + N^{\alpha}(\xi,\eta)\cdot\zeta\cdot V_i^{\alpha}$$

$$v(\xi,\eta,\zeta) = N^{\alpha}(\xi,\eta)\cdot v_i^{\alpha} + N^{\alpha}(\xi,\eta)\cdot\frac{1}{2}\cdot\zeta\cdot h^{\alpha}\cdot\theta_i^{\alpha}$$

where v^{α} and θ^{α} are respectively the velocity and the fiber rotation velocity for node "α". N is a shape function, V is a nodal fiber vector. Symbols XYZ and $\xi\eta\zeta$ mean respectively global and natural coordinate systems.

2.3 Contact search algorithm

Contact between parison and tool is formulated as a node-surface contact. Complex tool surface geometry is described in 3D space by means of triangular elements. An efficient contact search algorithm is implemented using this tool description.

For situation when node gets into contact with tool time increment is chosen in such a way that at the end of step node is precisely on a tool surface.

Another very important feature of this code is that when node gets into a contact with tool surface, sticking contact is assumed. That assumption follows experimental observation (Imamura 1993).

3 NUMERICAL EXAMPLE

During the design of automobile's gasoline tank various problems such as assembling flexibility, resistance to oil vibration etc. have to be considered. As a consequence, shape of gasoline tank is usually very complex, which make it difficult to predict thickness distribution. To check applicability of our code to process simulation, analysis of gasoline tank was done. As shown in Fig.2, the mold consists of two parts, upper and lower and geometrical shape of tools is approximated by an assembly of triangles. Parison is described by 3312 quadrilateral degenerated shell elements. Simulation was done on HP9000/735 workstation. Calculated thickness distribution is shown in Fig.3. Some stages of deformation process are shown in Fig.4. From this example, one can find that our code gives reasonable prediction of thickness distribution, even for such complex deformation process.

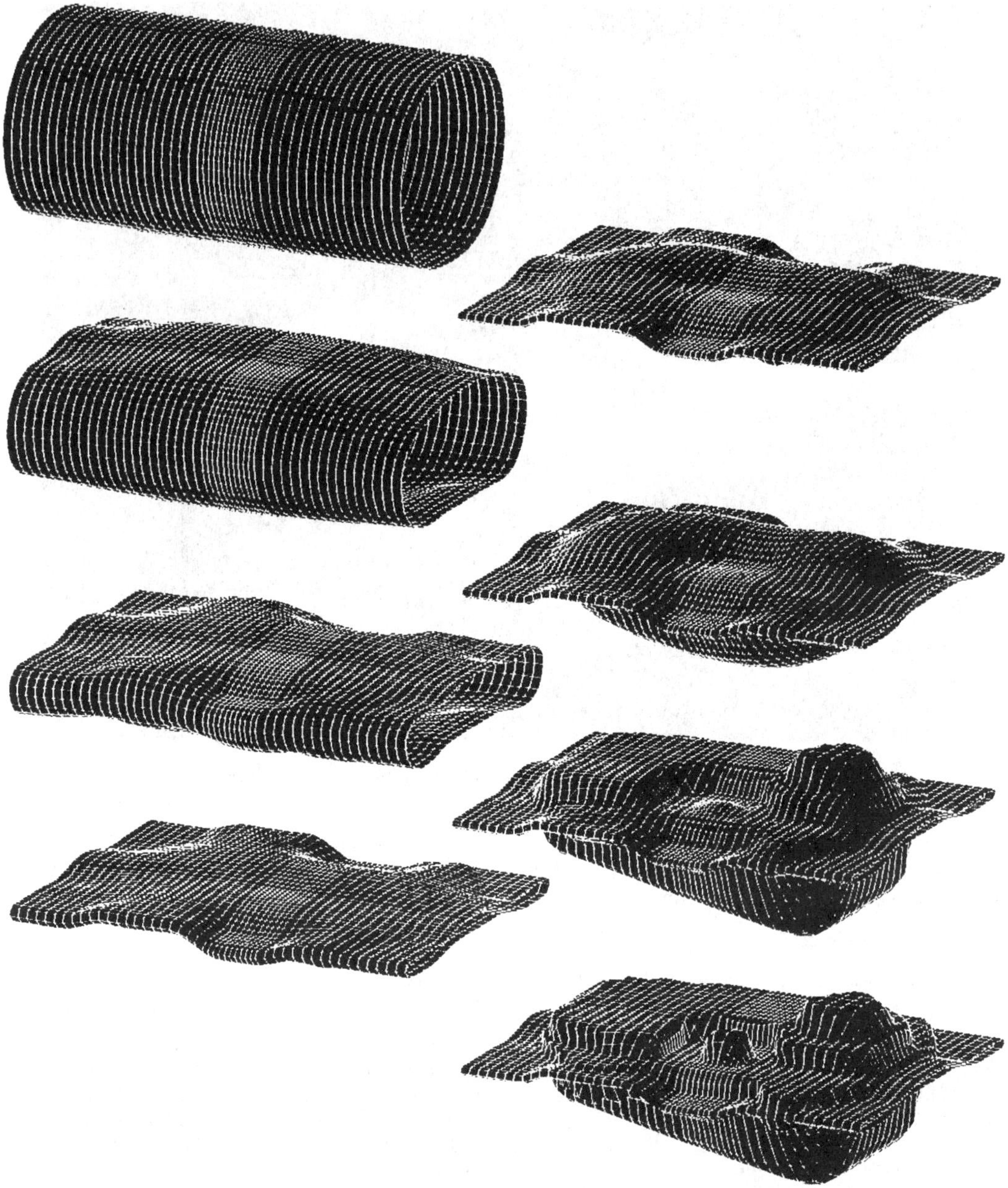

Fig.4 Deformed shapes in compression and blowing stages

4 VERIFICATION OF SIMULATION RESULTS

4.1 Experiment

To validate accuracy of the code, simulation results were compared with experiment for blow molding of a box shaped part. Experimental process consisted of four steps. In the first step, polymer at elevated temperature was extruded to a pipe shaped parison. Thickness along the longitudinal direction was not homogeneous. Fig.5 shows the measured thickness distribution .

Then, the one end of the parison was compressed and the second stage named pre-blowing took place. The parison was blown with a pressure of 2 kg/cm^2, to expand to a suitable size, as is shown in Fig.6.

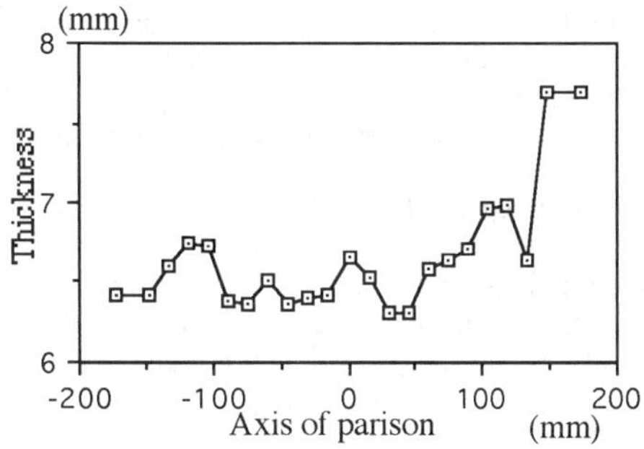

Fig.5 Thickness distribution along the longitudinal direction of parison

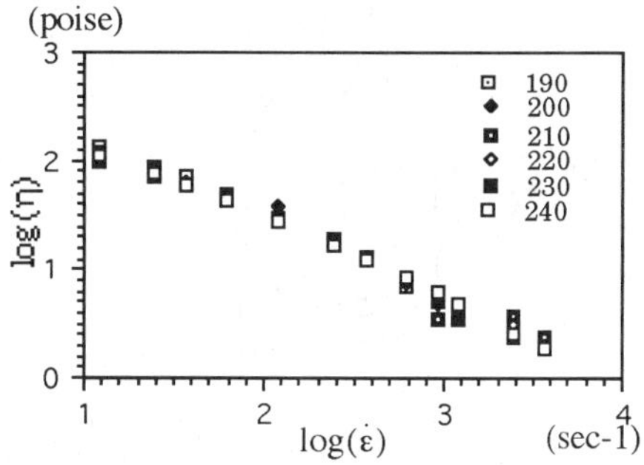

Fig.7 Relation between $\dot{\varepsilon}$ and η

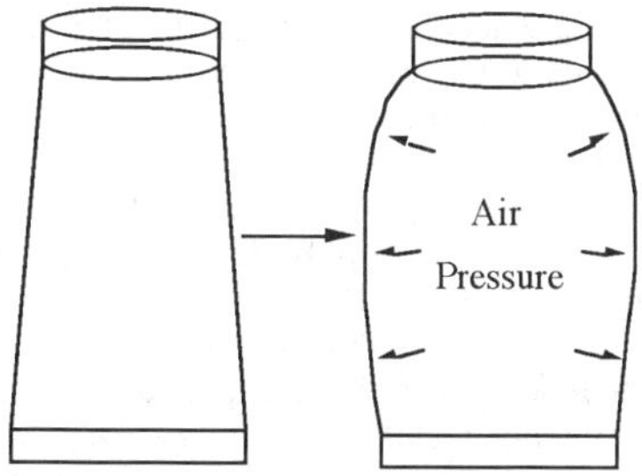

Fig.6 Pre-blowing

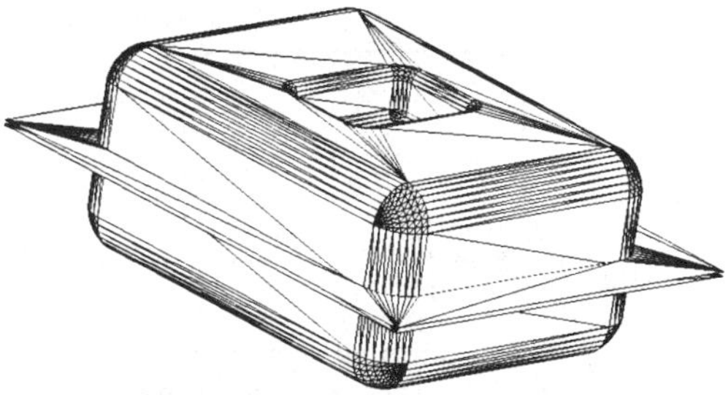

Fig.8 Mold

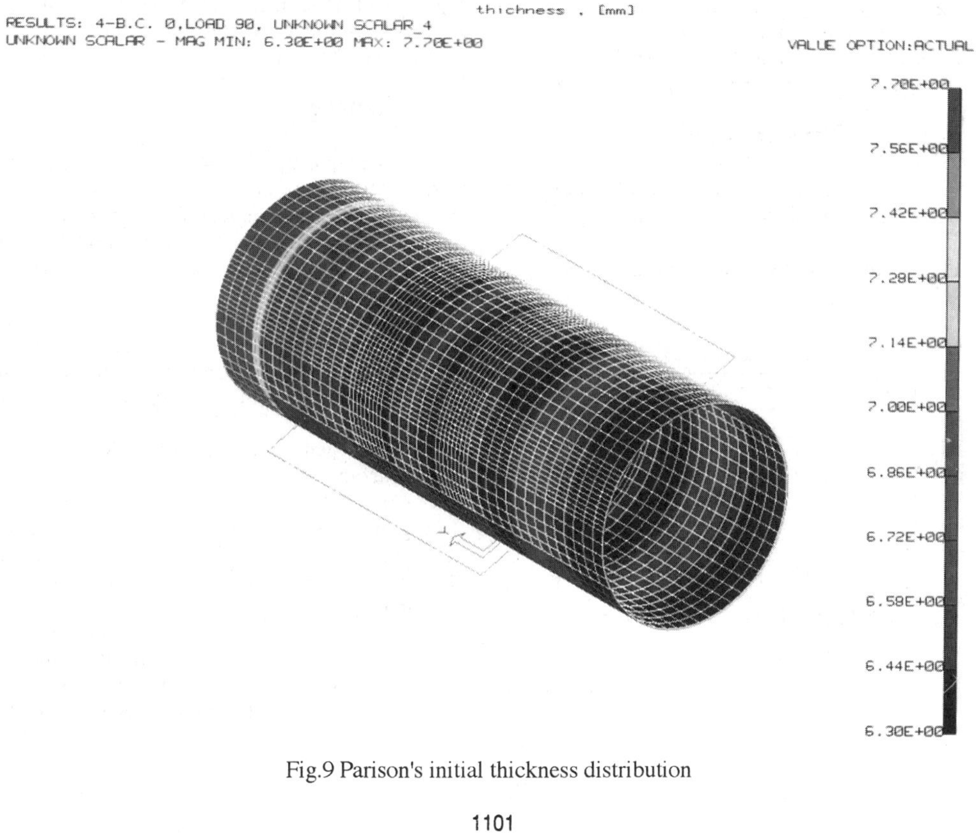

Fig.9 Parison's initial thickness distribution

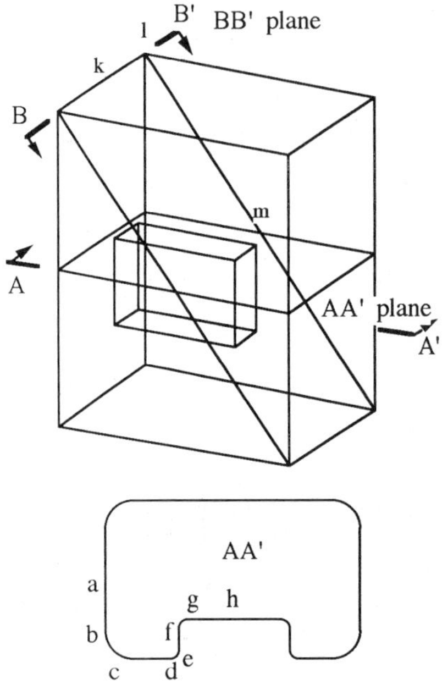

Fig.10 The cutting plane AA' and BB' and the position of points measured

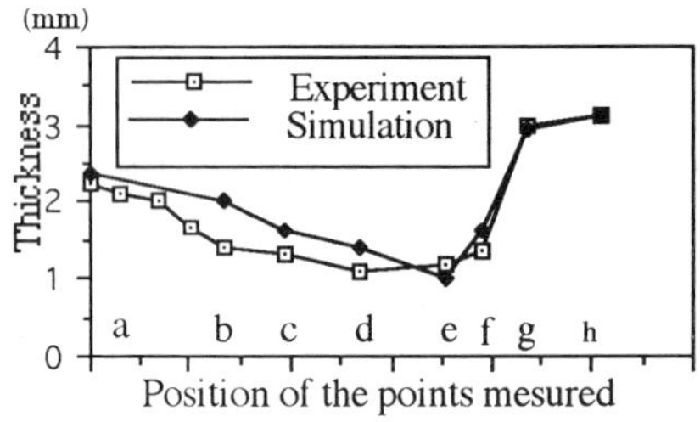

Fig.11 Thickness distribution on AA' plane

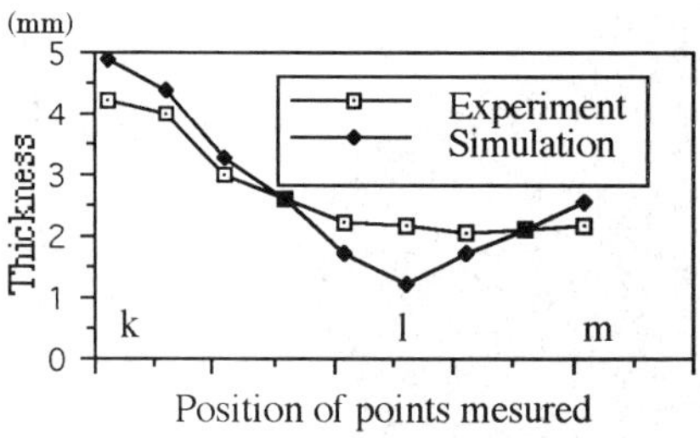

Fig.12 Thickness distribution on BB' plane along k-l-m line

In the third step, the pre-blown parison was completely compressed by edge between two molds. In the last blowing stage a pressure of 5 kg/cm^2 was applied inside the trapped parison.

4.2 Material parameters

Standard tests were carried out to find material parameters k and m. Relation between material's viscosity η and shear rate $\dot{\varepsilon}$ is presented in Fig.7.

4.3 Simulation process

Mold described by triangular elements is presented in Fig.8. Parison consisted of 3519 shell elements, whose initial thickness distribution was defined as shown in Fig.9 according to experimental thickness data. After molding, simulation process was divided into three stages: pre-blowing, mold closing and final blowing.
In the pre-blowing process, to simplify simulation, parison was blown with both ends fixed in the longitudinal direction

4.4. Comparison between simulation and experimental results

Comparison of experimental and simulation results of thickness distribution on planes AA' and BB' (Fig.10) can be seen in Fig.11 and Fig.12. Rather good correlation of experiment and numerical results can be observed.

5 CONCLUSION

A rigid visco-plastic FE code has been developed to simulate industrial three dimensional blow molding processes. Main features of the code are:

- Deformation of polymers at elevated temperature is modeled by the incompressible non-Newtonian fluid.
- The 4 node degenerated shell element with 8 integration points is used, treating the bending stiffens of the parison properly, which some times plays important role in a history of parison - tool contact.
- Contact between parison and tool is formulated as node-surface contact. Complex tool surface geometry is described in 3D space by means of triangular elements. Efficient contact search algorithm is introduced. Parison-mold contact behavior is assumed to be sticking condition which follows experimental observations.
- The code can simulate successive operations : the tool's closing to trap the parison inside molds and blow molding. Results of two processes are presented : blow molding of a square container

and blow molding of a gasoline tank. Good agreement between measured and simulated results can be observed.

ACKNOWLEDGMENT

Authors would like to thank Dr.Yamabe, Dr.Takahashi and Mr.Terada from Nissan Motor Co. for providing experimental results.
Authors' special thanks to Dr.M.Kawka for his helpful suggestions on this work.

REFERENCES

Ahmad S., Irons B.M., Zienkiewicz O.C. 1970. Analysis of thick and thin shell structures by curved finite elements., *Int. J. Num. Meth. Eng.*, 2, 418-451

Debbaut B.*et al.* 1994. Blow molding: a 3-D fluid approach. *In : Proc. of The Third World Congress on Computational Mechanics IACM WCCM3* :1580-1581.

Imamura S.*et al.* 1993. Direct observation of blow molding process from inside the parison. *Journal of J.JSPP.* 6:797-802.

Tanaka F., R.Nakano R. & Tanaka T. 1994. Computer simulation for blow molding. In *Proc. of The Third World Congress on Computational Mechanics IACM WCCM3* : 1578-1579.

Vlachopoulos J. 1992. Some experiences in creating and using software for polymer processing and design. In J.L.Chenot, R.D.Wood, and O.C.Zinkiewicz (eds) *NUMIFORM'92* : 117-122.. Rotterdam : Balkema.

Zhang K.F., Makinouchi A., Nakagawa T. 1994. Visco-plastic FEM simulation of blow molding process. In *Proc. JSTP Spring Conf.* : 113-116.

Zhang K.F., Makinouchi A., Nakagawa T. 1993. Visco-plastic FEM simulation of blow molding process. *Journal of J.JSPP.* 156-162.

Polymer injection molding and casting of metals

Simulation of Materials Processing: Theory, Methods and Applications, Shen & Dawson (eds)
 ISBN 90 5410 553 4

Modelling the casting process and predicting residual stresses

C. Bailey, P. Chow, G. Taylor, M. Cross & K. Pericleous
Centre for Numerical Modelling and Process Analysis, University of Greenwich, London, UK

ABSTRACT: In this paper we describe a 3D integrated model which simulates the shape casting process where effects such as thermal convection, solidification, deformation are calculated in a coupled manner. The paper concentrates on the modelling issues involved in predicting how a cast component will deform within a surrounding mould and how the effect of an evolving air gap at the cast mould interface effects the local solidification times.

1. INTRODUCTION

During the casting process a cold mould is filled with hot molten material, which cools and solidifies (Campbell 1991). As the cooling progresses thermal gradients evolve within the casting . In the liquid regions of the cast this will result in thermal currents, hence fluid flow that will redistribute heat around the component. Before the onset of solidification the casting is in full thermal contact with the mould, but in the solidified state the cast may move away from the mould and develop an air gap. The formation of this air gap will effect the heat loss at the cast/mould interface and hence incluence the way in which the casting solidifies and develops. To simulate the casting process the computational modeller must address the following types of analysis :

- Fluid flow for mould filling and residual thermal convection.
- Solidification and evolution of latent heat.
- Deformation, hence stress analysis to predict geometrical soundness of cast shape.
- Porosity formation and microstructure.

There is a high degree of coupling between the above phenomena. For example the convective currents are dependent on the temperature distribution which is influenced by the geometric deformation as heat losses are restricted across the cast/mould interface due to the formation of a gap. Casting is obviously a truly multi-physics phenomena which requires robust and efficient computational tools and algorithms (Cross 1992, 1993)

Deformations within solidifying castings enclosed within a deformable or non-deformable mould is a complex modelling scenario which involves non-elastic deformations especially at high temperatures and changing boundary conditions due to the contact effects with the mould (Thomas 1993). The type of constitutive laws governing the deformation of metals at high temperature has been the investigation of several authors where the general opinion is that the material deformation is governed by an elasto-viscoplastic constitutive law (Perzyk 1985). When the liquid material solidifies it is not stress free but develops residual stresses immediately after solidification which are equal to the liquid melt pressure. To include this effect Zabaras (1991,1990) used a hypoelastic-viscoplastic constitutive law which enables the melt pressure to be incorporated within the formulation as an initial condition. Predicting the effect of the melt pressure on the solidifying shell of a casting is important in predicting when the air gap forms at the cast/mould interface and whether this pressure will break through the thin shell and result in remelting of the casting at this point (Richmond 1990). Bellet (1993) used a hypoelastic-viscoplastic formulation to predict the deformation in an casting which included contact analysis between the casting and the mould.

Coupling thermal convection, solidification and deformation analysis has been attempted (Bailey 1992) where non-elastic strains where ignored. This paper presents extensions on this work for a three dimensional plate casting which includes viscoplastic strain effects.

2 GOVERNING EQUATIONS

The general equations governing the casting process are momentum, enthalpy and stress.

2.1 Fluid flow and heat transfer

The equations governing the conservation of fluid flow and heat transfer over a three-dimensional domain are (Chow 1993),

Momentum

$$\frac{\partial(\rho U)}{\partial t} + \nabla.(\rho \underline{V} U) = \nabla.(\gamma \nabla U) - \frac{\partial p}{\partial x} + S_U$$

$$\frac{\partial(\rho V)}{\partial t} + \nabla.(\rho \underline{V} V) = \nabla.(\gamma \nabla V) - \frac{\partial p}{\partial y} + S_V$$

$$\frac{\partial(\rho W)}{\partial t} + \nabla.(\rho \underline{V} V) = \nabla.(\gamma \nabla W) - \frac{\partial p}{\partial z} + S_W$$

Continuity

$$\frac{\partial \rho}{\partial t} + \nabla.(\rho \underline{V}) = 0$$

Energy

$$\frac{\partial(\rho h)}{\partial t} + \nabla.(\rho \underline{V} h) = \nabla.(k \nabla(h/c)) + S_h$$

Where U, V and W are the velocity components, with $\underline{V}$ the resultant velocity perpendicular to the cell surface. p is the pressure, h is the enthalpy, ρ, γ, c and k are the material density, viscosity, specific heat and thermal conductivity respectively. The source terms, S_U, S_V, S_V and S_h contain the sources necessary for the modelling of the solidification process. For momentum these sources are; buoyancy, to characterised the natural convection of the molten material as it cools and Darcy, to terminate the velocity components as the liquid material solidifies. For the energy equation the source is the latent heat release, which depicts the energy releases as the material changes phase (Voller 1991).

2.2 Stress

The Equilibrium equations for stress neglecting body forces are:

$$\sigma_{ij,j} = 0.0 \qquad i,j = x,y,z$$

where σ_{ij} are the cartesian components of the Cauchy stress tensor. The stresses that occur during the casting process are due to the elastic strains which using a hypoelastic formulation are given by

$$\dot{\underline{\sigma}} = [D]\dot{\underline{\varepsilon}}^{el}$$

where $\dot{\underline{\sigma}}$ and $\dot{\underline{\varepsilon}}^{el}$ are the stress and elastic strain rates respectively and D is the temperature dependent elastic material properties matrix. The elastic strain rate is given by

$$\dot{\underline{\varepsilon}}^{el} = \dot{\underline{\varepsilon}} - \dot{\underline{\varepsilon}}^{th} - \dot{\underline{\varepsilon}}^{vp}$$

where $\dot{\underline{\varepsilon}}$, $\dot{\underline{\varepsilon}}^{vp}$ and $\dot{\underline{\varepsilon}}^{th}$ are the total, visco-plastic and thermal strain rates respectively. The viscoplastic strain rate is assumed to be defined by the Perzyna model (1966):

$$\dot{\underline{\varepsilon}}^{vp} = \frac{d\underline{\varepsilon}^{vp}}{dt} = \lambda \left\langle \frac{\sigma^{eq}}{\sigma^{y}} - 1 \right\rangle^{\frac{1}{N}} \frac{3}{2\sigma^{eq}} \underline{s}$$

where σ^{eq}, σ^{y}, λ and s are the von-mises stress, yield stress, fluidity and deviatoric stress respectively.
The total strain rate is related to the displacement rate via :

$$\dot{\underline{\varepsilon}} = [L]\dot{\underline{d}}$$

where L is the differential matrix and $\dot{\underline{d}}$ are the displacement rates.

3 DISCRETISATION

All the above equations are discretised on an unstructured mesh using Finite Volume discretisation procedures (Chow 1993, Onate 1994, Bailey 1995). For the fluid flow and energy equations the control volumes are cell centred while for the stress analysis the control volumes are set up using a node centred format. It should be noted that the CFD and stress analysis use the same mesh. The dependent variables resulting from this discretisation are the fluid velocity components, enthalpy, and the solid displacement rates.

The stress and displacements at time step n+1 are approximated using Eulers method.

$$\underline{\sigma}^{n+1} = \underline{\dot{\sigma}}^{n}\Delta t + \underline{\sigma}^{n}$$

$$\underline{d}^{n+1} = \underline{\dot{d}}^{n}\Delta t + \underline{d}^{n}$$

Integrating the equilibrium equations and using the above approximation plus the constitutive law and strain-displacement rate relationships results in a set of discretised equations. The resulting discretised equations can be written in the general format:

$$[A]\Phi = b$$

Where Φ's are the dependent variables (i.e velocity, enthalpy and displacement rates) and b are the source terms. These system of equations are solved using a suitable iterative solver in correction format. The complete solution procedure for the above coupled equations is;

1) Given initial/current temperatures and cast geometry, start new time step.
2) Solve the fluid flow and enthalpy equations (Chow 1993, Voller 1991)
3) Calculate temperature, $T = h / c$, and any other temperature dependent variables, such as liquid fraction and material properties.
4) Return to step 2 and repeat the procedure until a convergence is obtained.
5) Calculate temperature change ΔT during this time step.
6) Calculate the displacement rates associated with this change in temperature. (If Change in contact state occurs then recalculate displacement rates until contact state remains the same.)
7) Update displacements and stresses.
8) If simulation not finished return to step 1 to carry out another time step.

It should be noted that when steps (5) and (6) are completed the mesh (x,y,z) is updated to $(x+d_x, y+d_y, z+d_z)$ and in step (2) the equations are rediscretised over this new mesh. The initial stress which equals the melt pressure is given when an element becomes reasonably solid such that σ^n = -p where p is the local melt pressure (Zabaras (1991), Richmond (1990)).

4 BOUNDARY CONDITIONS

It is important when predicting how a cast shape develops to be able to model the thermal changes correctly. To do this account must be taken of the air gap formation at the mould/cast interface. For this analysis coincident nodes have been used to model the interface. These type of nodes enable the cast to move freely away from the mould. The usual procedure for thermal modelling at the interface is to assume a convective heat transfer expression;

$$\frac{\partial T}{\partial n} = h_{eff}(T_{metal} - T_{mould})$$

where the heat transfer co-efficient is given by

$$h_{eff} = \frac{k_{gap}}{\Delta_{gap}}$$

where k_{gap} is the thermal conductivity of the gap medium (assumed air) and Δ_{gap} is the gap distance. During this analysis the mould is assumed to be non-deformable and hence the contact analysis although still time consuming becomes much more simplistic. A node is labelled as being in contact when it penetrates the mould and the stress solution is recalculated with this constraint in place. If the stress condition at a contacting node becomes tensile then the constraints at these nodes are released.

5 RESULTS

Figure one shows the mesh used to model a casting which is assumed to be instantly filled at a temperature of 700°C. The youngs modulus E(t) and yield stress σ_y(T) were assumed to vary with temperature where

E(T) = -143E+06 * (T - 600)
σ_y(T) = -0.87E+06 * (T -600)

The liquidus and solidus temperatures are 612 and 532 respectively and the coefficient of thermal expansion and poissons ratio are taken as constant and equal to 0.3 and 0.0001. Figure two shows the thermal convection after 30 seconds where it can be seen that extensive flow is occuring at the feeder which is distributing hotter material throughout the plate. Figure three shows the effect of airgap

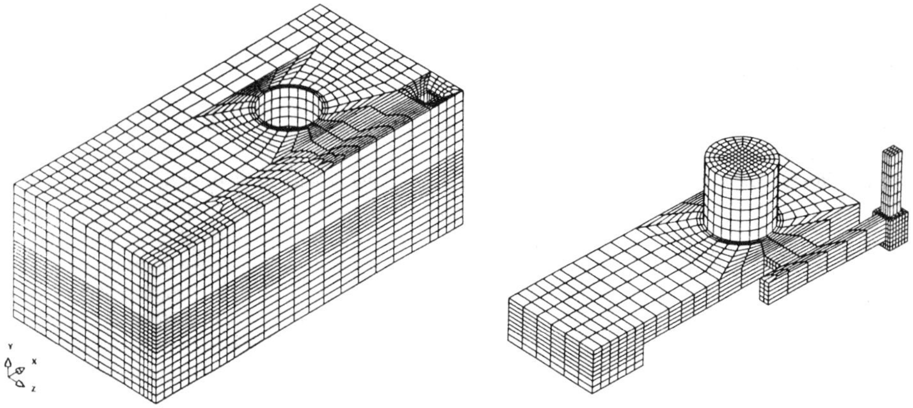

MOULD　　　　CASTING

Figure One　Mould and casting meshes.

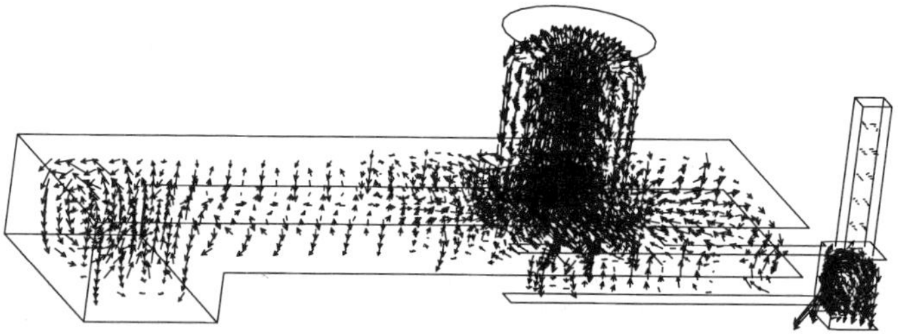

Figure Two　Thermal convection after 30 seconds.

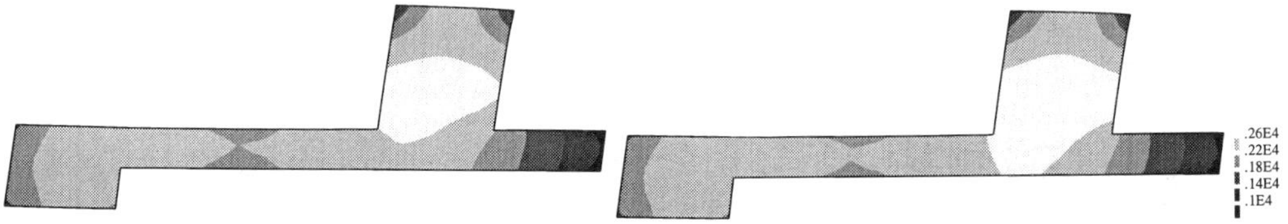

No deformation　　　　Deformation

Figure Three　Solidification times with and without deformation.

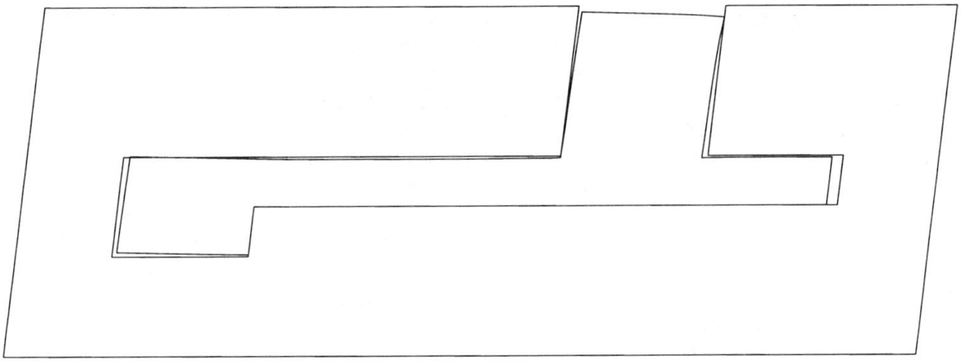

Figure Four Airgap formation along centre section of casting.

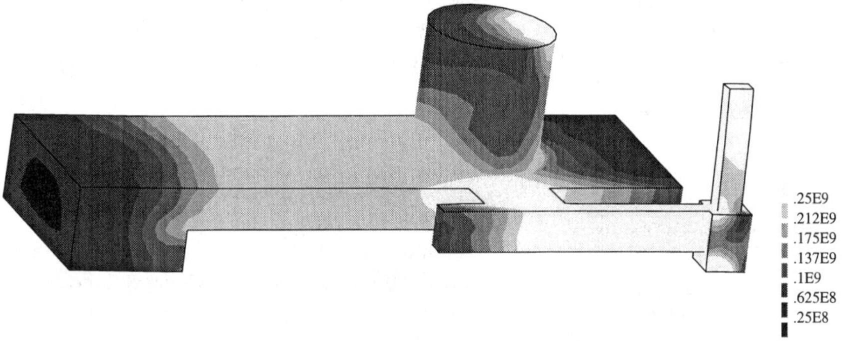

Figure Five Effective (Von Mises) stress contours.

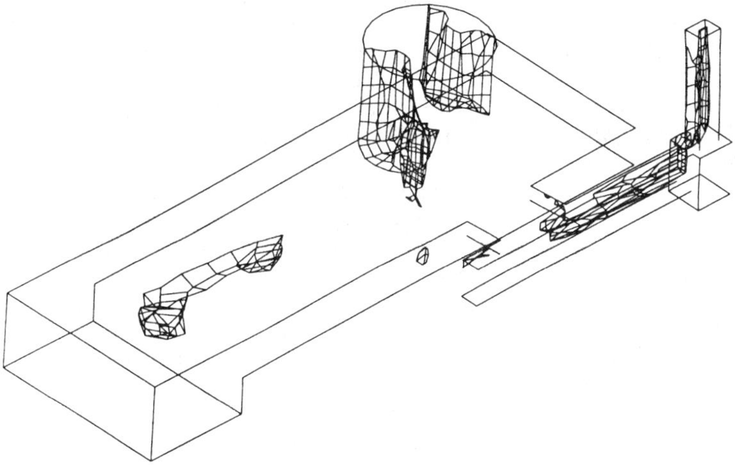

Figure Six Porosity prediction using NIYAMA.

formation on local solidification times where it can be seen that the casting solidifies at a faster rate when a deformation analysis is neglected. Figure four and five show the size of the air gap and the magnitude of the von-mises stress throughout the casting where the largest stress magnitudes are located between the feeder and the larger end section of the plate. This is due to the contact effects between the cast and the mould in this region. Figure six shows the likelihood of porosity using the NIYAMA criteria. Future work will include transformation strains and compare against experimental data.

AKNOWLEDGEMENTS

The authors would like to acknowledge the financial support of the EPSRC , AEA technology, Rolls Royce and BAe. Also Casting Technology International (Cti) and the National Physics Laboratory (NPL) for discussions on castings.

REFERENCES

Campbell, J. 1991, *Castings*, Pub Butterworth-Heinemann Limited, Oxford.

Chow P. 1993, *Control volume unstructured mesh procedure for convection-diffusion solidification processes*, PhD thesis, University of Greenwich.

Voller V.R. & Swaminathan C.R, 1991, General source-based method for solidification phase change, *Numerical Heat Transfer*, B, 19, 175-189

Thomas B.G. 1993. Stress modelling of casting processes : An overview, *Modeling of Casting, Welding and Advanced Solidification Processes - VI*, Ed T S Piwonka, V Voller and L Katgerman, TMS, 519-534

Bailey C, Fryer Y, Cross M & P Chow, 1991. Predicting the deformation of castings in moulds using a control volume approach on unstructured meshes, *Mathematical Modeling of Material Processing*, Ed M Cross, J F Pittman, Pub OUP, 259-272.

Bellet M, Menai M, Bay F, P Schmidt & Svensson I. 1993. Finite element modelling of the cooling phase in casting processes, *Modeling of Casting, Welding and Advanced Solidification Processes VI*, Ed T S Piwonka, V Voller and L Katgerman, Pub TMS, 639-645.

Zabaras N, Ruan Y & O Richmond, 1991. On the calculation of deformations and stresses during axially symmetric solidification, *Journal of Applied Mechanics (ASME)*, Vol 58, 865-871.

Zabaras N & O Richmond. 1990. Analysis and finite element approximation of deformations and thermal stresses in solidifying bodies, *Computer modelling and simulation of manufacturing processes* , B Singh and Y T Im, Eds., Winter Meeting (ASME), Dallas Texas, 89-105.

Perzyna P. 1966 Fundamental problems in viscoplasticity, *Adv in Applied Mechanics*, 9, 243-377

Richmond O, L Hector & J Fidy, 1990. Growth instability during nonuniform directional solidification of pure metals, *J Applied Mechanics (ASME)*, 57, 529-536.

Perzyk M.A. 1985. Validity of constitutive equations used for calculation of stresses in cooling castings, *Materials Science and Technology*, Vol 1, 84-92, TMS, 313-328, (1988).

Niyama E, T Uchida, M Morikawa & S Saito. 1982. Method of shrinkage prediction and its application to steel casting practice, AFS Cast Metals Research Journal, p 52-63.

Cross M. 1993. Development of novel computational techniques for the next generation of software tools for casting simulation, *Modeling of Casting, Welding and Advanced Solidification Processes VI*, Eds T S Piwonka, V Voller, L Katgerman, Pub The Minerals, Metals and Materials Society, p 115-126.

Cross M, C Bailey, P Chow & K Pericleous, 1992 Towards an integrated control volume unstructured mesh code for the simulation of all the macroscopic processes involved in shape casting, in *Numerical Methods in Industrial Processes NUMIFORM 92*, Eds J-L Chenot, R D Wood and O C Zienkiewicz, Balkema p 787-792.

Onate E, M Cerrera & O.C Zienkiewicz, 1994. A Finite volume format for structural mechanics, *Int. Jnl Num Meth in Engng*,37, 181-201.

Bailey C & M. Cross. 1995 A Finite Volume Procedure to Solve Elastic Solid Mechanics Problems in Three-Dimensions on an Unstructured Mesh, *Int. Jnl Num Meth in Engng*, 10.

Simulation of Materials Processing: Theory, Methods and Applications, Shen & Dawson (eds)
© 1995 Balkema, Rotterdam. ISBN 90 5410 553 4

An adaptation of the boundary element method for modeling gas injection molding

F.S.Costa & W.Thompson
Centre for Computer Integrated Manufacture, Swinburne University of Technology, Hawthorn, Vic., Australia

C.Friedl
Moldflow Pty. Ltd, Kilsyth, Vic., Australia

ABSTRACT: A flow simulation for Gas Injection Molding of polymer was formulated using the boundary element technique. Non-linear polymer melt flow was handled by local domain integrals. The two dimensional formulation presented in this paper was used to study the cross section of the flow field and gas bubble location.

In adapting the boundary element method to model Gas Injection Molding it was necessary to include a technique for creating sub-regions to split the domain. It was found that using sub-regions allowed improved convergence and decreased the program storage requirements. Results are presented for the gas bubble location in practical parts.

1 INTRODUCTION

Gas Injection Molding (GIM) is a relatively new manufacturing process that produces plastic parts with a hollow core. It is derived from the conventional injection molding process. Initially the mold is partly filled with polymer. The hollow core is formed when an inert gas, usually Nitrogen, is injected under pressure into the polymer to complete the filling and packing processes. Once the plastic material has solidified, the pressurized gas is released just before opening the mold and ejecting the part. Computer Aided Engineering software simulation tools currently used for conventional injection molding analysis do not provide a complete picture of the GIM process.

To simulate where the gas core goes during the mold filling process, a two dimensional non-linear boundary element method was used. To apply this method to Gas Injection Molding simulation, it was necessary to develop a capability for automatically sub-structuring the solution domain. An iterative process for accurately applying the boundary conditions at free surfaces was also implemented.

2 GAS INJECTION MOLDING PROCESS

The Gas Injection Molding process provides benefits in material saving and improved processing capabilities, but its use is currently being limited by a lack of molding experience and thorough design analysis tools. GIM is a more complex manufacturing process than conventional injection molding of plastic because it involves a greater number of manufacturing stages and process variables. The flow is also made more complex by the combination of low and high viscosity fluids.

Detailed descriptions of the process are provided by Eyerer et al. (1993), Shah (1991) and Rusch (1989).

2.1 Reasons for using GIM

The most obvious reason for using Gas Injection Molding is that considerable material saving may be possible. Thick sectioned parts can be made hollow. Minimizing the amount of raw material used has both economic and environmental advantages. Gas Injection Molding has been used in thick section parts such as handles, armrests and towel rails.

GIM also provides advantages in thin walled parts, overcoming a number of the problems which can arise in conventional injection molding. The injection pressure needed to complete the filling of a part is much lower in GIM because the gas has such a low viscosity. A lower injection pressure will minimize the residual stresses in the part, which are the cause of warping. Lower injection pressure may also mean that a smaller machine is sufficient for part manufacture.

Other advantages of the GIM process given by Costa (1995) are:

-Lighter parts with equivalent structural strength.

-Large, bulky parts can be filled and packed easily.
-Lower clamp tonnage required for certain parts.
-Reduction of Cooling times for thick parts.
-Removal of sink marks (Liou et al., 1990).
-Thick and thin sections can be present in the same part.

Disadvantages of the GIM process are: the increased equipment cost; that the gas vent hole remains on the part surface; and higher clamp tonnage for some parts. Also, GIM causes increased complexity in part and mold design. This problem is compounded by a lack of experience in GIM process design. Gas Injection Molding introduces the additional design and process parameters of location and size of gas channels; number and placement of gas injection points; gas pressure; amount of polymer to inject; delay time between polymer fill and gas fill; and the holding time before releasing gas pressure.

3 GOVERNING EQUATIONS

The gas was considered to be an inviscid fluid. Therefore, the gas pressure was assumed constant throughout the gas bubble at any instant. The domain to be solved was the flow equations in the polymer melt with gas pressure as a boundary condition.

The governing equations are the momentum and mass conservation equations. These were simplified by a dimensional analysis (Costa, 1995) which showed that for GIM the equations could be reduced by removing the convection, inertia and body force terms:

$$\eta \frac{\partial^2 v_i}{\partial x_j \partial x_j} - \frac{\partial p}{\partial x_i} = 0 \qquad (1)$$

$$\frac{\partial v_j}{\partial x_j} = 0 \qquad (2)$$

These simplifications are normally also made in conventional injection molding analyses (Kennedy, 1993 and Hieber and Shen, 1980). However, conventional injection molding analyses also make the further assumption that Hele-Shaw flow exists throughout the part. The Hele-Shaw assumptions for laminar flow between parallel plates dictate that no polymer melt can flow towards a cavity wall. This also means that the pressure is constant through the thickness of the part. For thin shell injection molded parts these assumptions are valid for all flow behind the flow front. These assumptions have allowed conventional injection molding simulations to be simplified to a shell analysis with a flow resistance integral taken through the part thickness.

At the flow front however, it is known that some polymer flows from the centre of the cavity and is deposited near the cavity walls, in what Rose (1961) first called a fountain flow effect. Similarly, the region at the tip of the gas core, where gas is displacing polymer melt, has a component of melt velocity perpendicular to the cavity walls. To understand the path and location taken by the gas bubble, it was necessary to study this region fully. For this reason, the Hele-Shaw assumptions were not used. A further motive for not making this assumption was that many GIM parts contain thick sections and so can not be accurately model as thin shells even in conventional injection molding. Equations (1) and (2) were solved fully, without further simplification, in order to predict the location of the gas core accurately. Due to the complex geometry of most real polymer parts, an analytic solution was not possible. A two dimensional boundary element simulation was used to model the part cross-section and predict the location of the gas core.

The non-Newtonian behavior of the polymer melt was modeled as a Generalized Newtonian fluid,

$$\tau_{ij} = \eta(\dot{\gamma}, T)\dot{\gamma}_{ij} \qquad (3)$$

where τ is the tensor of viscous stresses due to deformation, $\dot{\gamma}$ is the deformation rate tensor, and η is the non-Newtonian viscosity which is itself a function of deformation rate and Temperature (T). The temperature field was solved explicitly at each time step using a finite element method on a triangular grid. This triangular grid was already present for the discretisation of the domain integral for the boundary element scheme. This finite element temperature solution was a standard convection-diffusion upwinded formulation, which will not be discussed further here. The viscosity models used were the power law and the Moldflow second order models. The power law model, extended to include temperature dependence, is given by:

$$\eta(\dot{\gamma}, T) = m(\dot{\gamma})^n e^{cT} \qquad (4)$$

Where m, n and c are material constants.

4 BOUNDARY ELEMENT FORMULATION

The boundary element formulation for slow, incompressible, non-Newtonian flow was similar to that of Bush and Tanner (1990). This involved a base Newtonian linear solution and a non-linear domain integral perturbation to account for the non-Newtonian material behavior. The choice of the base viscosity η^N is arbitrary, however errors in the numerical approximation were minimized by choosing the viscosity to be representative of the viscosity values near the cavity wall where shear rates were highest. Such a choice was suggested by Bush and Phan-Thien (1984) and Bush and Tanner (1990).

The velocity at an internal or boundary point (ξ) is available as:

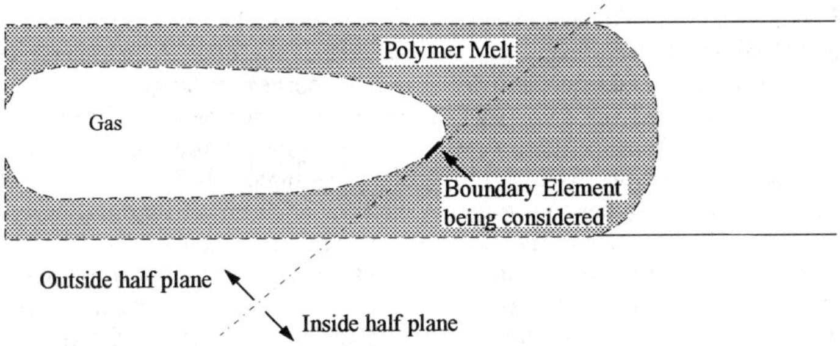

Figure 1. Concave Region with half planes shown

$$c_{ik}(\xi)v_k(\xi) = \int_\Gamma \left[g_{ik}(X-\xi)t_k(X) - h_{ik}(X-\xi)v_k(X)\right]d\Gamma(X) + \int_\Omega \left[\frac{\partial g_{ik}}{\partial x_j}(X-\xi)\varepsilon_{kj}(X)\right]d\Omega(X) \quad (5)$$

where: Ω is the domain, Γ is its boundary, $c_{ik}(\xi) = C(\xi)\delta_{ik}$, $C(\xi)=1$ if ξ is in Ω. $C(\xi)$ is between 0 and 1 on Γ, depending on the local curvature. v_i is a velocity component and t_i is the boundary traction component. The non-linear part of the stress field was defined as:

$$\varepsilon_{ik} = \left[\eta(\gamma, T) - \eta^N\right]\dot{\gamma}_{ik} \quad (6)$$

The fundamental solutions g_{ik} and h_{ik} are given by Dargush and Banerjee (1991).

4.1 Non-Linear Convergence

The boundary was discretised into linear constant value boundary elements while the domain integral was divided into linear triangular domain cells. An internal velocity point was located at the vertices of each triangular cell.

An iterative cycle was used to converge the domain stress and boundary velocity solutions. A system of equations was formed by assembling Equation (5) at each boundary element. The starting estimate of ε was zero, so the boundary system was solved without any domain integrals. From this boundary solution, the velocity at all internal points was obtained again using Equation (5). This velocity field was used to find the deformation rates and non-Newtonian viscosity values for each domain cell. The domain integrals were then computed and the boundary solution recalculated with the estimate of domain integral from the previous iteration. This cycle was repeated until the changes in boundary solution, velocity field, stress fields and domain integrals were below a user specified tolerance.

It was usually found to be necessary to allow relaxation factors on the iterative changes. This was to ensure the non-linear iterations were stable and converged to a solution. Commonly, the changes in velocity field, non-linear stress and domain integral were relaxed.

4.2 Free Surface

On a free surface the applied pressure was the known boundary condition. The velocity solution was required to move the free surface at the next time step. However, the pressure was first used to determine the tractions which would be the known boundary conditions in the numerical solution. The traction on a free surface is given by:

$$t_i = \tau_{ij}n_j - pn_i \quad (7)$$

where p is the pressure and n_i is a component of the outward pointing normal. The deformation rates and viscous stresses were not known before obtaining a solution. The estimate of the traction boundary condition was updated using the current estimate of viscous stress during each iteration cycle. The viscous stress at the free surface was assumed to be the value calculated for the adjacent domain cell.

4.3 Sub-Regions

The boundary element method considers the influence of each boundary element on all other boundary elements in the same domain. If non-linear effects are modeled through domain integrals, then each domain cell also contributes to the solution at a boundary element. Convergence was found to be more difficult in domains which were curved. This was because a domain cell could be close to a boundary element and influencing its solution even when they were not directly connected. These difficulties were not observed for Newtonian fluid models, where the domain integral was zero. The entry of gas into a cavity caused exactly this type of curved domain geometry. For example in Figure 1, a domain cell above the gas bubble was not directly connected to a boundary element below the bubble, even though the influence term was based on the distance between their centroids. Although in a complete domain, this does not cause any net error, it did cause an ill-conditioning in the system.

These numerical difficulties were overcome by dividing the solution domain into sub-regions. This created new internal boundaries which required additional boundary elements. In assembling the boundary system of equations, only influence terms from the local sub-region were included.

Internal boundary elements had no known boundary conditions, but appeared in two sub-regions. Therefore, it was possible to arbitrarily assign one sub-region to use Equation (5) cast for the boundary traction values and the other to use the equation cast for boundary velocity values. The two sub-regions were linked by the shared boundary values on the common internal boundary elements.

During the simulation process, the domain changed shape as the gas core advanced. At each time step, the domain was remeshed. It was therefore necessary to develop an algorithm to automatically divide a concave domain into non-concave sub-regions.

Each boundary element was considered in turn to divide the two dimensional domain into two half-planes, one on the outside and the other on the inside. The outside was defined by the outward pointing normal for that boundary element. Figure 1 shows these half planes.

By examining the proportion of the boundary elements in the same sub-region which were located on the outside half-plane, a measure of a region's concave nature was obtained (Ψ).

$$\Psi = \frac{1}{N}\sum_{n=1}^{N}\frac{1}{2}\left[1 + \mathrm{sgn}\left(n.\left(X(n) - \xi\right)\right)\right] \tag{8}$$

where N is the number of boundary elements in the region, X(n) is the centroid of a boundary element n, and ξ is the centroid of the boundary element used to defined the inside and outside half planes. The sign function is:

$$\begin{aligned} \mathrm{sgn}(z) &= +1, \quad z \geq 0 \\ &= -1, \quad z < 0 \end{aligned} \tag{9}$$

A user specified threshold was used on ψ. If any element in a sub-region exceeded this threshold, then the region was deemed to be sufficiently concave to require dividing into sub-regions. The check was repeated on the new sub-regions and further divisions made as appropriate.

Once it was determined that an internal boundary was needed, the algorithm for automatically dividing concave domains also needed to determine the optimum location for the division. Ideally a concave domain would be divided into two non-concave sub-regions. If the concave geometry was caused by an internal corner or the gas bubble tip, then the internal boundary should be started from this point. Elements with a highest values of ψ were considered. If the sharp tip angle was formed by this element and either element adjacent to it, then this point was used to commence the internal boundary. Otherwise, some other location with a sharp tip angle and high ψ was sought. The division was then made along a vector pointed to by the tip, as shown in Figure 2. Without this minimum tip angle, many unnecessary internal boundaries were formed which did not effectively eliminate the concave regions.

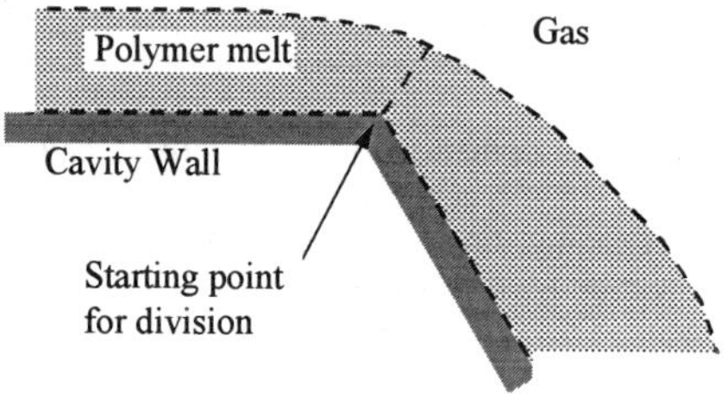

Figure 2. Direction of sub-region division.

Although the primary reason for dividing the domain into sub-regions was to improve convergence, the machine memory required to run the numerical analyses was also reduced by the use of sub-regions. The non-linear boundary element method stored the influence terms for each domain cell on each boundary element in the same region. If the domain was split, then fewer domain cells per element needed to be stored. If these influence factors had not been stored, but recalculated during each iteration, the computation time would still have been improved by using sub-regions.

5 NUMERICAL EXAMPLES

5.1 Convergence for a moderately concave domain

In situations with moderate non-Newtonian behavior and a moderately concave domain, an improvement in the speed of convergence was obtained if sub-regions were used. Figure 3 shows the cross-section

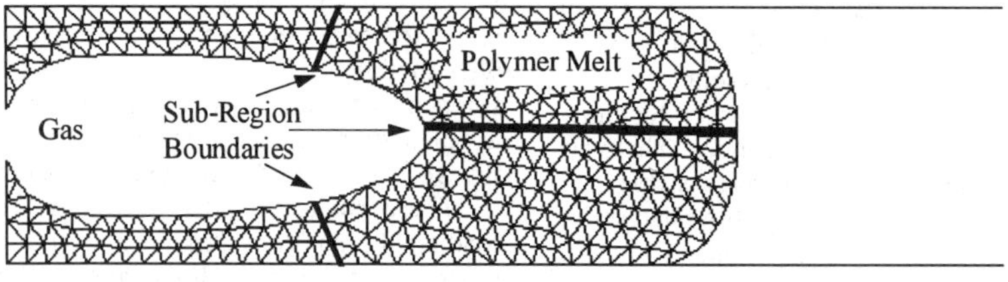

Figure 3. Moderate non linear polymer flow with concave domain.

of a centrally gated gas core entering a straight slit. It represents the position at a particular time step during filling. A power law viscosity model was used with n=0.5 and no temperature dependence.

In Figure 3, the domain is shown divided into four sub-regions. The same problem was also solved with both two sub-regions and one domain region respectively. The two sub-region model had only the long internal boundary seen in Figure 3. The single domain region model had no internal boundaries.

Table 1 shows the number of iterations required for convergence by each model for various relaxation factors (r.f.). The change in non-linear stress relaxation factor was always 1.0. The tolerance set on all relative changes was 0.1E-5.

As can be seen from the table, all three models successfully converged to a solution. In general however, the single region model was slowest to converge because of the instability created by the concave domain. The lowest relaxation factor of 0.1 ensured that even the single domain model was stable. However, in this case, the multi-region models were slightly slower to converge, possibly because there were a greater number of boundary elements in the calculation.

For relaxation factors of 0.5, none of the models converged within 2500 iterations, suggesting that the iterations were unstable. However, the multi-region models were closer to converging than the single domain model.

Table 1. Convergence rates for moderate non-linear problem

Vel r.f.	Dom r.f.	Single Region	Two Sub-regions	Four Sub-regions
		Number of Iterations to converge		
0.1	0.1	222	252	290
0.2	0.2	129	117	139
0.3	0.3	93	83	89
0.4	0.4	113	59	64
0.5	0.5	*did not conv*	*did not conv*	*did not conv*

5.2 Convergence for a highly concave domain

A second slit problem was considered. The four sub-region model is again shown in Figure 4. The domain is highly concave because the gas core has penetrated further. Again single and two sub-region models were also considered for this geometry. A highly non-Newtonian fluid was modeled with a power law index of n=0.3 and no temperature dependence.

The highly non-linear viscosity and highly concave domain resulted in greater difficulty for convergence. Therefore, the changes in the non-linear stress field were also relaxed. Table 2 shows the number of iterations required to converge all relative changes to below 0.1E-5. Notice that the single region model did not converge in 1,500 iterations and that the best performance over a range of relaxation factors was for the four sub-region model. Relaxation factors 0.8 and above were too high for any of the models to be stable. The other analyses that did not converge for the four sub-region model were probably caused by a pattern of changes oscillating across the sub-regions.

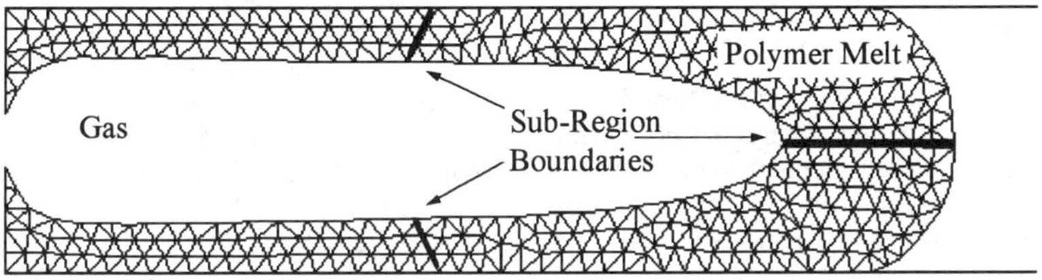

Figure 4. Highly non-linear polymer with highly concave domain

Table 2. Convergence rates for highly non-linear problem

Velocity r.f.	Non-linear Stress r.f.	Domain Integral r.f.	Single domain	Two sub-regions	Four sub-regions
			Number of Iterations to Converge		
0.1	0.1	0.1	*did not conv*	*did not conv*	*did not conv*
0.2	0.2	0.2	*did not conv*	*did not conv*	916
0.3	0.3	0.3	*did not conv*	*did not conv*	433
0.4	0.4	0.4	*did not conv*	330	*did not conv*
0.5	0.5	0.5	*did not conv*	363	250
0.6	0.6	0.6	*did not conv*	1413	163
0.7	0.7	0.7	*did not conv*	1130	156
0.8	0.8	0.8	*did not conv*	*did not conv*	*did not conv*

Figure 5. Sub-regions used to simulate a towel rail part.

5.3 Gas Entering A Typical Part

A typical part manufactured using Gas Injection Molding is the towel rail shown in Figure 5. Shown at an intermediate stage of filling, the highly concave domain created by the penetration of the gas bubble was automatically divided into sub-regions using the algorithm described above. The parameters used were that ψ exceed 0.5 and the tip angle be at least 0.65 radians. A power law viscosity model with shear and temperature dependence was used. (m=100, n=0.5, c=-0.03) The total part length is 42cm and the gas injection pressure was 10MPa. The polymer melt was injected at 250°C. Because this part is cylindrical and bulky, the two dimensional analysis was not too far removed from what happens in the actual part.

These towel rail results demonstrate the importance of fully modeling the polymer flow to predict the location of the gas core through the cross section. It can be seen that the gas takes the shortest path through corners. If too little polymer is left deposited on inside corners, some part designs may not achieve the design requirements for structural strength. The aim of polymer processing simulations has always been to allow the part and mold designer to test designs early and minimize the cost of mold trials and modifications.

6 CONCLUSIONS

The simulation method for Gas Injection Molding of plastics has been presented using a non-linear perturbation to the Boundary Element Method. It has been shown that the numerical procedure's convergence properties and efficiency can be improved by dividing concave domains into sub-regions.

The usefulness of a full simulation of Gas Injection Molding which provides accurate information concerning the location of the hollow gas core has been demonstrated. Further development to allow a full three dimensional simulation of the entire mold is underway.

ACKNOWLEDGMENTS

The authors wish to thank their colleagues at Moldflow for their technical support and encouragement.

REFERENCES

Bush M.B. and Tanner R.I. 1990. Boundary Element Analysis of Slow Non-Newtonian Flow. In Banerjee P.K. and Morino L. (eds), *Boundary Element Methods in Non-Linear Fluid Dynamics; Developments in Boundary Element Methods No. 6* :Ch 8. England: Elsevier.

Bush M.B. and Phan-Thien N. 1984. Drag Force on a Sphere in Creeping Motion through a Carreau Model Fluid. *J. non-Newtonian Fluid Mechanics.* 16:303-313.

Costa, F.S. 1995. Numerical Simulation of Gas Injection Molding of Polymer. *Ph.D. Thesis.* Swinburne University of Technology, Australia.

Dargush G.F. and Banerjee P.K. 1991. A Boundary Element Method for Steady Incompressible Thermoviscous Flow. *Int. J. Numerical Methods in Engineering.* 31:1605-1626.

Eyerer, P., Maertins, R., and Buerkle, E. 1993. Injection moulding with internal gas pressure. *Kunststoffe German Plastics.* 83:7-13.

Hieber, C.A., and Shen, S.F. 1980. A finite element/finite difference simulation of the injection-molding filling process. *J. non-Newtonian Fluid Mechanics.* 7:1-32.

Kennedy P. 1993. *Flow Analysis Reference Manual.* Moldflow, Melbourne, Australia.

Liou M.J, Ramaghanda D.M, Ishi K. and Hornberger I. 1990. Sink Marks Study in Injection Molded Parts. *SPE Annual Technical Conference proceedings.* 36:288-291.

Rose, W. 1961. Fluid-fluid interfaces in steady motion. *Nature.* 191:242-243.

Rusch, K.C. 1989. Gas-assisted injection molding-The new age of plastic molding technology. *SPE Annual Technical Conference proceedings.* 35:1014-1018.

Shah, S. 1991. Gas injection molding: Current practices. *SPE Annual Technical Conference proceedings.* 37:1494-1506.

Simulation of Materials Processing: Theory, Methods and Applications, Shen & Dawson (eds)
© 1995 Balkema, Rotterdam. ISBN 90 5410 553 4

Optimal casting design

J.A. Dantzig, S.A. Ebrahimi & D.A. Tortorelli
Department of Mechanical and Industrial Engineering, University of Illinois, Urbana, Ill., USA

ABSTRACT: We consider the optimal rigging design for metal castings in this work. We combine numerical process simulation with numerical optimization to systematically improve casting riser shapes, in order to minimize riser volume while still ensuring product quality. The importance of using analytical design sensitivities to make the optimization efficient is demonstrated using an example problem in superalloy investment casting.

1 INTRODUCTION

Casting, one of the oldest forming processes still in use today, owes its widespread use to its ability to make complex shaped parts cheaply and reproducibly, and to the process's adaptability to mass production. For these reasons, castings are widely used in the transportation industries for engine and vehicle components.

Typically, a cast product begins with an engineer designing the part based on its expected service requirements. The product design is then passed on to the manufacturing engineer for construction of casting patterns and production. The casting pattern differs from the product in several important respects: The pattern is larger than the final product to allow for thermal contraction, cores may be added to achieve holes in the part, and a set of gates and risers are added to ensure proper feeding of the casting during solidification.

Finding an acceptable casting pattern and rigging design can be an expensive and time consuming process, particularly if a trial-and-error approach is used, wherein a series of prototypes are created and tested. To this end, simulation has become an important tool for the evaluation of trial designs. Much effort has been expended to develop computer codes which faithfully simulate the casting process. [1, 2] These codes are capable of handling the nonlinear material property models, complex geometries and sophisticated boundary condition models which characterize the casting process. These capabilities come at a price, however. The geometric model may take man-months to create, and the analysis time can be an entire day, or longer.

Once the analysis is completed, the engineer must sift through the data to determine if the design is acceptable, and if not, he must attempt to improve it. Many codes embody *criterion functions* which quantitatively measure the quality of the casting, based on simulation results. Unfortunately, once problems are identified, remedies are not always easily obtained. It is difficult to envision the effects of changes in either the product or process designs have on other areas of a complex part. The combination of this fact with the long analysis time makes casting design using simulations a very challenging task.

Here, we systematize the casting design process, by linking simulations with numerical optimization. In the sections which follow, we first describe the design environment and discuss the link between analysis codes and numerical optimization software. We then discuss means for minimizing the number of designs evaluated (owing to the lengthy simulation times) using explicit design sensitivity analysis. We conclude with an example application in superalloy investment casting.

2 DESIGN ENVIRONMENT

2.1 *Terminology*

We formalize the design process by first introducing *design variables*. These variables may include any of the parameters which define the process, including both processing parameters (*e.g.*, melt superheat, mold preheat ...) and product design parameters (*e.g.*, dimensions). These design variables are collected into a *design vector*, $\mathbf{b}$.

We also assume the existence of a quality metric, G, defined such that G decreases as product quality increases. An example of a function with this characteristic is

$$G = (T_i - \bar{T}_i)^2 \tag{1}$$

where T_i represents the response computed using the current design at point i, and $\bar{T}_i$ represents the desired solution at the same point. We may also define a series of equality or inequality constraints, F, in a similar manner. Example constraints include

$$0.8\text{mm} \leq b_6 \leq 1.0\text{mm} \tag{2}$$

where b_6 represents a dimension, and

$$|\nabla T|_{t_f} \leq 2\text{K/mm} \tag{3}$$

which represents a constraint on the temperature gradient magnitude at solidification time t_f at a particular location.

We then define a formal optimization problem: Minimize the cost function, G, subject to the enumerated constraints, by varying the design variables, $\mathbf{b}$. The optimization is carried out by a numerical algorithm which systematically traverses design space in search of the optimum.

Many algorithms exist for performing the optimization. See Vanderplaats [3] for an excellent introduction and discussion. It is important to remember, however, that each function evaluation, *i.e.*, simulation, represents a great expense in this application. We must therefore reject the genetic algorithms and other approaches which typically require large numbers of function evaluations even for a few design variables, and use instead the more traditional Quasi-Newton methods. The latter methods are most applicable because existing design rules often produce designs not too far from acceptable. Additionally, the design space is often restricted by equipment and the product requirements.

Quasi-Newton methods require gradients in design space for the cost function, G, and constraints, F, *i.e.*, the *design sensitivities*. If these gradients are not available from the simulation, then the optimizer attempts to construct them through a series of perturbations of the design parameters. This "finite difference" approach is inefficient, because it requires an additional simulation for each design parameter, and it is potentially inaccurate due to roundoff or truncation errors deriving from the perturbation size. Indeed, in the example problem given later, the optimal rigging design of a casting riser with 13 design variables requires over 150 function evaluations to find the optimal design when no analytical sensitivities are used.

Fortunately, there is a better way. In the next section, we describe methods for explicitly computing the design sensitivities. These sensitivities are exact, and they are computed efficiently, using information in the simulation. The same example produces the design in just 24 function evaluations when this method is followed.

2.2 *Design Sensitivity Analysis*

We represent the governing equation as the generalized residual for a parabolic system [4]

$$\mathbf{R}(\mathbf{u}^t(\mathbf{b}), \mathbf{u}^{t-\Delta t}(\mathbf{b}), t, \mathbf{b}) = \mathbf{0} \tag{4}$$

where $\mathbf{R}$ is called the residual, $\mathbf{u}^t$ represents the solution at the current time, t, and $\mathbf{u}^{t-\Delta t}$ represents the solution at the previous time. We have included $\mathbf{b}$ in the argument list to emphasize the dependence of the solution on the design. The system of equations is solved using Newton-Raphson iteration [5]

$$\frac{\partial \mathbf{R}(\mathbf{u}_i^t(\mathbf{b}), \mathbf{u}^{t-\Delta t}(\mathbf{b}), t, \mathbf{b})}{\partial \mathbf{u}}(\mathbf{u}_{i+1}^t - \mathbf{u}_i^t) = -\mathbf{R}(\mathbf{u}_i^t(\mathbf{b}), \mathbf{u}^{t-\Delta t}(\mathbf{b}), t, \mathbf{b}) \tag{5}$$

where $\mathbf{u}_i^t$ represents the ith estimate for the solution at the current time. The quantity $\frac{\partial \mathbf{R}}{\partial \mathbf{u}}$ is often called the tangent matrix. Equation (5) is solved iteratively until convergence is attained. A variety of algorithms are available for solving the linearized algebraic system in Equation (5), including LU decomposition and numerous conjugate gradient methods. The advantage of the Newton-Raphson approach to solving the primal problem is its quadratic convergence rate. The method has a further advantage in that the tangent matrix, which has been described as "a treasure-trove" [6], may be used to great advantage in computing the design sensitivities.

The optimization algorithm will require gradients of both G and F, however, their evaluation is similar and we therefore consider only G. Let us assume that

$$G = G(\mathbf{u}^f(\mathbf{b}), \mathbf{b}) \tag{6}$$

where $\mathbf{u}^f$ represents the solution at a particular time in the analysis. More general forms are easily derived from this one. Differentiating G with respect to $\mathbf{b}$ yields

$$\frac{dG}{d\mathbf{b}} = \frac{\partial G}{\partial \mathbf{b}} + \frac{\partial G}{\partial \mathbf{u}^f}\frac{d\mathbf{u}^f}{d\mathbf{b}} \tag{7}$$

The definition of G permits evaluation of the partial derivatives of G in Equation (7), however the *response sensitivity*, $\frac{d\mathbf{u}^f}{d\mathbf{b}}$, requires further analysis. Tortorelli, *at al.* [7] present both adjoint and direct differentiation approaches for the response sensitivity evaluation. They show that the direct method is best suited to the transient problems we consider here, and we therefore present this method in abbreviated form.

Differentiation of Equation (4) with respect to the jth design variables, b_j, yields

$$\frac{d\mathbf{R}}{db_j} = \mathbf{0} = \frac{\partial \mathbf{R}}{\partial b_j} + \frac{\partial \mathbf{R}}{\partial \mathbf{u}^t}\frac{d\mathbf{u}^t}{db_j} + \frac{\partial \mathbf{R}}{\partial \mathbf{u}^{t-\Delta t}}\frac{d\mathbf{u}^{t-\Delta t}}{db_j} \tag{8}$$

where we have used the fact that Equation (4) must be satisfied for all designs. Rearrangement of Equation (8) yields

$$\frac{\partial \mathbf{R}}{\partial \mathbf{u^t}} \frac{d\mathbf{u}^t}{db_j} = -\frac{\partial \mathbf{R}}{\partial \mathbf{b_j}} - \frac{\partial \mathbf{R}}{\partial \mathbf{u^{t-\Delta t}}} \frac{d\mathbf{u}^{t-\Delta t}}{db_j} \tag{9}$$

We note the similarity of Equations (5) and (9), where both equations have a vector of unknowns multiplied by the tangent matrix, and a known right-hand side.

We may use this similarity to our advantage in computing the sensitivities. At each time step, we first solve the primal problem (Equation (4)) by LU decomposition. Once convergence is attained, we then form the right-hand side vector for each design variable, as given in Equation (9), and obtain the associated response sensitivity by an inexpensive back-substitution in the already decomposed matrix. Once the response sensitivities are known the design sensitivities, $\frac{dG}{d\mathbf{b}}$, are readily computed from Equation (7). Note that the sensitivity analysis is also transient, reflecting the fact that changing design variables at one time affects the solution at all future times.

2.2.1 *Example: Nonlinear transient conduction*

As an example of the analysis just presented, we consider nonlinear heat conduction, with $\mathbf{u} = \mathbf{T}$, the nodal temperatures. The governing partial differential equation for a Fourier material is

$$\rho c_p \frac{\partial T}{\partial t} = \nabla \cdot (k \nabla T) \tag{10}$$

where ρ is the density, c_p is the specific heat and k is the thermal conductivity. Reducing Equation (10) to Galerkin weighted residual form in the usual way [5] yields

$$\mathbf{R} = ([\mathbf{C}] + \Delta t[\mathbf{K}])\{\mathbf{T}\}^t - [\mathbf{C}]\{\mathbf{T}\}^{t-\Delta t} \tag{11}$$

where $[\mathbf{C}]$ and $[\mathbf{K}]$ are the capacity and conductance matrices, respectively, and $\{\mathbf{T}\}^t$ and $\{\mathbf{T}\}^{t-\Delta t}$ are the solution vectors at the current and previous times, respectively. Equation (9) becomes, in this form

$$\frac{\partial \mathbf{R}}{\partial \mathbf{T^t}} \frac{d\mathbf{T}^t}{db_j} = -\frac{\partial [\mathbf{C}]}{\partial b_j}(\{\mathbf{T}\}^t - \{\mathbf{T}\}^{t-\Delta t}) - \tag{12}$$

$$\Delta t \frac{\partial [\mathbf{K}]}{\partial b_j}\{\mathbf{T}\}^t + [\mathbf{C}]\frac{d\{\mathbf{T}\}^{t-\Delta t}}{db_j} \tag{13}$$

Computing the derivatives of the matrices on the right-hand side of Equation (13) requires knowledge of the context of the design variable b_j. We discuss this point further in the next section, but in the spirit of presenting an example, let us suppose that the thermal conductivity is a design variable. (This often occurs in identification problems, where the simulation is intended to actually determine k.) We have then that

$$\frac{\partial [\mathbf{C}]}{\partial b_j} = \mathbf{0}$$

$$\frac{\partial [\mathbf{K}]}{\partial b_j} = \int_V \mathbf{B}^T \frac{\partial k}{\partial b_j} \mathbf{B} dV$$

When b_j represents the shape, a more complex, but readily computed form appears. [8]

3 IMPLEMENTATION

The flow of the optimization process is illustrated below, showing cases where the analysis code must be treated as a "black box" in Figure 1, and where the source is available to be modified for explicit sensitivity analysis in Figure 2. In both cases, an interface is used to define design variables, compute the objective and constraints, and communicate with the optimizer. The interface sets up and runs the analysis for a given design, then computes the objective and constraints (and their sensitivities if available) and passes them to the optimizer. The optimizer in turn evaluates these data within the selected algorithm and returns a new design for evaluation. This cycle continues until an optimum is reached.

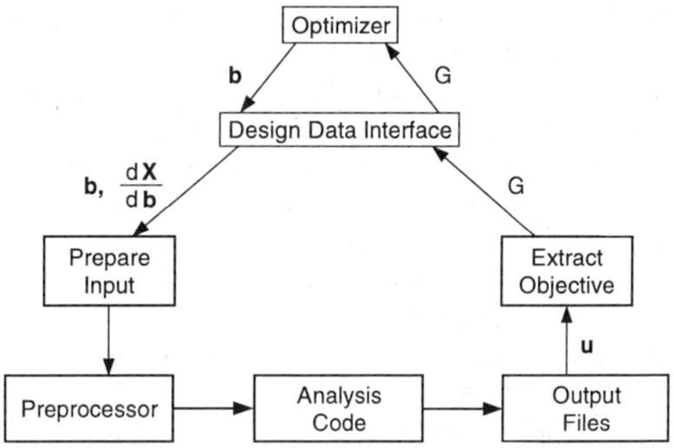

Figure 1: Schematic of optimization process with and without analytical sensitivities.

Most of the work involved in linking the simulation with the optimizer is routine: reading and writing data files, passing arguments between subroutines, *etc.* As we saw in the previous example, it is necessary to pass information about the meaning of each design variable, as well as its value, between the design environment and the analysis code. There are many possible data structures suitable for this task, and the implementation is straightforward, if somewhat tedious. There is an important conceptual issue to consider, however, when the design variables do not appear directly in the simulation. For example, this occurs in shape optimization.

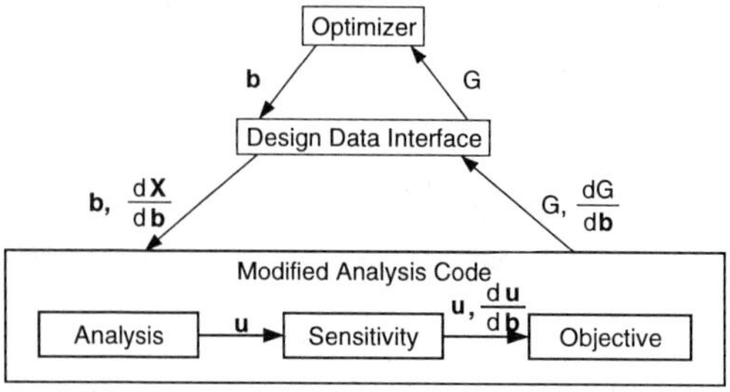

Figure 2: Schematic of optimization process with analytical sensitivities.

The natural design variables are used in shape optimization to parameterize the design, *e.g.,* the length of an edge, the angle between two surfaces, *etc.* While these parameters are readily available to the designer, there is usually no provision for them in the analysis code. A typical finite element analysis program understands geometry only through nodal coordinates and element connectivity. Higher level abstractions of these data, such as dimensions, are not needed in the analysis, and are hence unavailable.

To handle this problem, we determine the relationship between dimensional changes, $d\mathbf{b}$, and nodal coordinate movements, $d\mathbf{X}$, represented as $\frac{d\mathbf{X}}{d\mathbf{b}}$. This quantity has been called the "design velocity" [9], however this terminology can be confusing, especially in the context of fluid dynamics, so we will refer to it simply as $\frac{d\mathbf{X}}{d\mathbf{b}}$. A number of methods may be used to compute $\frac{d\mathbf{X}}{d\mathbf{b}}$. If the mesh has been defined parametrically, it may be possible to differentiate the mesh mapping function and thereby determine $\frac{d\mathbf{X}}{d\mathbf{b}}$ [10]. A more general method, applicable to unstructured grids, uses a series of virtual displacements at design keypoints to "deform" the mesh. [11] In either case, as the optimization proceeds, $\frac{d\mathbf{X}}{d\mathbf{b}}$ may be used to update the geometry based on the new design:

$$\mathbf{X}' = \mathbf{X}_0 + \frac{d\mathbf{X}}{d\mathbf{b}}(\mathbf{b}' - \mathbf{b}_0) \tag{14}$$

We use this approach in the example problems presented in the next section.

4 EXAMPLE: SUPERALLOY INVESTMENT CASTING

We consider the solidification of an axisymmetric superalloy investment casting, illustrated in Figure 3. The part is a ring with a protrusion, and a gate and riser are included to ensure directional solidification from the part into the riser. The ring is made as an investment casting, where superheated metal is poured in vacuum into a preheated shell mold. Cooling is effected by radiation from the shell exterior to a cold furnace, modeled as a nonlinear convective boundary condition, *viz.*

$$q_{rad} = h(T, T_{furn})(T - T_{furn}) \tag{15}$$
$$h = \sigma\epsilon(T^2 + T_{furn}^2)(T + T_{furn}) \tag{16}$$

Radiation view factors were not considered in the analysis.

The original gate and riser design were prepared by our collaborators at investment casting companies, and analysis of the solidification of the original design confirmed the feasibility of the design, as illustrated in Figure 4. It also showed that the riser was overdesigned, i.e., it was larger than required to effect directional solidification into the riser.

We then posed the optimization problem: minimize the riser volume, subject to the constraint that the casting freezes directionally from the casting into the riser. The design variables were illustrated in Figure 3. We present results for the optimization obtained using FIDAP [12] and direct differentiation sensitivities. Nineteen constraints of the form

$$T_{96}(1750) - T_{75}(1750) \leq 0 \tag{17}$$

were constructed to require monotonically increasing temperatures at a particular time (1750 seconds) along the centerline into the riser. Here, $T_{96}(1750)$ refers to the temperature at node 96 at time 1750 seconds. The $\frac{d\mathbf{X}}{d\mathbf{b}}$ field was obtained using the natural design variable approach described earlier. Equation (14) was used to update the mesh through the optimization. The original design was so far from optimal, however, that the original mesh became too distorted before the optimum was obtained. We therefore stopped at the intermediate shape shown, remeshed and built a new $\frac{d\mathbf{X}}{d\mathbf{b}}$ field.

The optimization was performed using ADS [13], employing the method of feasible directions. Four line searches were performed, each consisting of one analysis to determine the analytical sensitivities, followed by four or five analyses to establish the minimum along the selected direction. The optimal design was thus obtained after a total of 24 designs. The sensitivity analyses each took approximately one hour on a Sun SPARCstation 5, while the analyses without sensitivities each ran approximately 30 minutes, yielding a total of a little over 14 hours. The same casting was analyzed using ProCAST [14], using finite difference sensitivity analysis, since we did not have access to the source code. In this case a slightly different constraint was imposed, namely that the freezing time of the indicated nodes was required to advance in sequence along the centerline into the riser. Once again, mesh distortion required us to stop part way

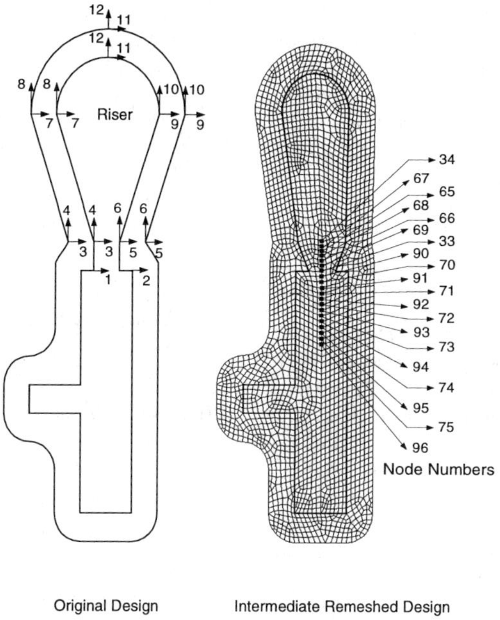

Figure 3: Original and intermediate designs of axisymmetric example casting, including the design variables shown on the orignal design, and the nodal locations where directional solidification is enforced. The casting centerline is approximately two part thicknesses to the right in the figure.

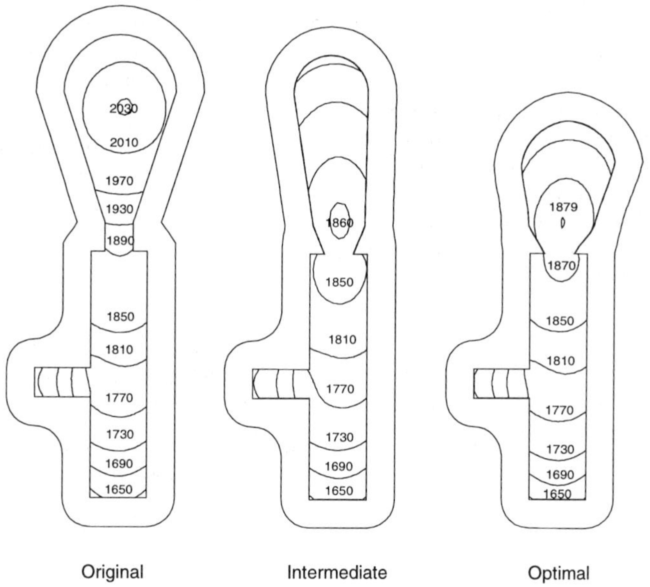

Figure 4: Solidification isochrones for the original, intermediate and optimal designs.

through the optimization to remesh. The analysis and optimization proceeded similarly to the optimization using FIDAP, producing very similar results. In this case, however, since explicit sensitivity data were not available, the optimal design was reached in 152 designs. The finite difference sensitivity analyses each took approximately four hours on the SPARCstation 5, and the individual function evaluations took about 20 minutes each, for a total of a little under two days.

5 DISCUSSION

The results shown in the example problem demonstrate several important aspects of the use of optimization with solidification simulation. First, they show that feasible casting rigging specifications may often be severely overdesigned. Both optimizations reduced the riser weight by approximately 30%. Studying the optimal designs shows that the optimized design cants the riser toward the centerline in both designs. This achieves a weight reduction due to the smaller volume associated with the smaller radius, which is an obvious design improvement, at least in hindsight!

The results also demonstrate that there is a difference between what is possible and what is practical. If we consider what will happen with more complex castings, we may anticipate that the number of design iterations will not change significantly (for the same number of design variables), and the total time will therefore be proportional to the analysis time for each design. For complex castings where the analysis takes one day to complete, the optimization would require a few weeks when explicit sensitivities are computed, and nearly six months without them. We recognize that both of these may be considered too long to be practical at present. However, we also note that most of the design improvement is found in the first few line searches, and the total time to reach a useful answer may be somewhat shorter.

6 ACKNOWLEDGMENT

This work was performed within the Investment Casting Cooperative Arrangement (ICCA), funded by ARPA under contract HC T1S073326.

REFERENCES

[1] M. Rappaz and K. W. Mahin. *Modeling of Casting, Welding and Advanced Solidification Processes - V.* TMS-AIME, Warrendale, PA, 1991.

[2] T. S. Piwonka, V. Voller, and L. Katgerman. *Modeling of Casting, Welding and Advanced Solidification Processes - VI.* TMS-AIME, Warrendale, PA, 1993.

[3] G.N. Vanderplaats. *Numerical Optimization Techniques for Engineering Design: with Applications.* McGrawHill, New York, 1984.

[4] Daniel A. Tortorelli, Michael M. Tiller, and Jonathan A. Dantzig. Optimal design of nonlinear parabolic systems - part I: Fixed spatial domain with applications to process optimization. *Computer Methods in Applied Mechanics and Engineering*, 113:141–155, 1994.

[5] R.D. Cook. *Concepts and Applications of Finite Element Analysis.* Wiley, New York, 1981.

[6] L. E. Scriven, 1993.

[7] D.A. Tortorelli, R. B. Haber, and S. C.-Y Lu. Design sensitivity analysis for nonlinear transient thermal systems. *Computer Methods in Applied Mechanics and Engineering*, 75:61–78, 1990.

[8] D. A. Tortorelli and Z. Wang. A systematic approach to shape sensitivitiy analysis. *Int. J. Solids Structures*, 30(9):1181–1212, 1993.

[9] K. K. Choi and K.-H. Chang. A study of design velocity field computation for shape optimal design. *Finite Elements in Analysis and Design*, 15:317–341, 1994.

[10] D.A. Tortorelli, J.A. Tomasko, T.E. Morthland, and J.A. Dantzig. Optimal design of nonlinear parabolic systems - part II: Variable spatial domain with applications to casting optimization. *Comp. Meth. Appl. Mech.*, 113:157–172, 1994.

[11] T. E. Morthland, P. E. Byrne, D. A. Tortorelli, and J. A. Dantzig. Optimal riser design for metal castings. *Met. Mater. Trans. B*, page to appear, 1995.

[12] Fluid dynamics International, Evanston, IL. *FIDAP Theoretical Manual*, 1987.

[13] G.N. Vanderplaats. *ADS - A FORTRAN Program for Automated Design Sythesis.* Naval Postgraduate School, Monterey, CA, 1984.

[14] Universal Energy Systems, Annapolis, MD. *ProCAST*, 1994.

Simulation of Materials Processing: Theory, Methods and Applications, Shen & Dawson (eds)
© 1995 Balkema, Rotterdam. ISBN 90 5410 553 4

A numerical model for cavity filling during gas-assisted injection molding

D.M.Gao, K.T.Nguyen & G.Salloum
Industrial Materials Institute, National Research Council Canada, Boucherville, Que., Canada

ABSTRACT: This paper presents a numerical model to simulate the mold filling as well as the gas penetration during the Gas-Assisted Injection Molding process. A Finite Element model combining with control volume approach is developed to calculate the pressure field, to track the flow fronts and to determine the gas penetration during the filling phase. A complex part is simulated to demonstrate the capabilities of the model.

1 INTRODUCTION

The Gas-Assisted Injection Molding(GAIM) process, as an innovative technology, is being increasingly used in the plastics industry. This technique consists of a partial injection of polymer melt followed by an injection of pressurized gas (usually nitrogen) into the core of the melt via the nozzle, the runner, or directly into the cavity. The gas would generally hollow out a network of specially designed thicker-sectioned channels leading to savings of polymeric material. As a result, the cooling time can be reduced. Since the gas transmits a fairly uniform pressure distribution as it penetrates the part, the pressure required to fill the whole cavity is much lower than in the conventional injection molding process. Consequently the clamping force is reduced due to lower pressure level. The GAIM process has other advantages such as reducing the sink marks, more uniform shrinkage and lower residual stresses.

This paper presents a numerical model based on a hybrid FEM/Control volume approach to simulate cavity filling as well as the gas penetration during the GAIM process. The simulation software allows the prediction of the location of polymer fronts, the gas penetration and the thickness fraction of skin polymer over the entire part during the filling process.

The capabilities of the numerical model are demonstrated through the analysis of the filling of a complex part. The influence of different molding parameters such as the percentages of polymer fill and different dimensions of the gas channels is in estigated for an experimental mold.

2 THEORETICAL MODELS

Gas-Assisted Injection Molding is a very complicated process because the process involves dynamic interaction between two completely different substances flowing into a generally complex cavity. In consequence, the knowledge about the conventional injection molding is no longer enough to handle this new process, particularly in the design of the gas channel system and optimizing the operation conditions.

On the basis of the above consideration, numerical modeling becomes evidently a attractive tool to help complete the lack of the knowledge in this field.

2.1 *Governing equations*

In this study, the polymer melt is considered as a generalized Newtonian fluid. Since most parts produced by GAIM globally have a shell like geometry, the lubrication approximation could be used for modeling the global flow behavior in GAIM. This leads to the well-known Hele-Shaw equations given by [1]:

$$\nabla \bullet S \nabla P = 0 \tag{1}$$

$$\rho C_p(\frac{\partial T}{\partial t}+u\frac{\partial T}{\partial x}+v\frac{\partial T}{\partial y}) = \frac{\partial}{\partial z}k\frac{\partial T}{\partial z} +\Phi \qquad (2)$$

Where P is the pressure, T is the temperature, ρ is the density, C_p is the specific heat, k is the thermal conductivity, Φ is the viscous dissipation, x and y are the coordinates of the middle plane, z is the gapwise direction and S is the fluidity defined by:

$$S=\int_0^h \frac{z^2}{\eta}dz \qquad (3)$$

In the energy balance, the heat conduction in the flow direction is neglected since the thickness of the cavity is much smaller than the other two dimensions. Furthermore, the convection in the gapwise direction is also not considered.

As far as gas channels are concerned, special cares should be taken in order to capture the gas penetration. The Hele-Shaw equation is solved only in regions completely filled by the polymer melt. More discussion is given in the following sections of the paper.

2.2 *Boundary Conditions*

2.2.1 *Hele-Shaw Model*

The Hele-Shaw equation is solved to obtain the pressure field subject to appropriate boundary conditions discussed below.

For a GAIM process, two different situations should be considered :

I) *Polymer filling phase*

In this phase, the polymer melt fills the mold up to a desired percentage. The Hele-Shaw equation is solved over the polymer filled regions. Three types of boundaries are considered: i) Moving flow fronts; ii) Polymer injection gates; iii) Lateral walls

i) At the moving front, it is assumed that the pressure is constant. For the sake of simplicity, this constant is set to zero;

ii) At the gate, the flow rate or the pressure is specified as a function of time depending on the chosen molding conditions; and

iii) On the lateral wall, it is required that the normal component of the velocity vanish. It should be noted that due to the lubrication approximation, the full no-slip condition is lost at the lateral wall. This is justified by the fact that the no-slip effect extends only a thin layer of order h into the cavity.

II) *Gas penetration phase*

After one switches the polymer injection to gas injection, the pressurized gas is penetrating into the thick gas channels without any extra polymer melt injecting into the cavity. The boundary conditions at the following boundaries need to be specified : 1) Moving flow fronts; 2) Interface between gas and polymer; 3) Lateral walls.

For the moving fronts and the mold walls, the same conditions are kept as in the pervious phase.

For the interface between gas and polymer, it is considered that the gas region is of single phase pure substance without any mixture with the polymer. Consequently, it is reasonable to assume that the gas transmits the pressure uniformly everywhere in the gas region including the interface with polymer. This is justified by the very low speed of gas during GAIM.

The Hele-Shaw equation is solved in the domain between the polymer flow fronts and the interface gas-polymer. A uniform pressure is specified at the interface gas-polymer. This pressure is generally determined by the operating conditions or by the specification of gas injection machines.

2.2.2 *Thermal system*

Since the conduction in the x and y directions is neglected in the energy equation, only the temperatures on the walls of the mold need to specified.

At the gate the temperature is assumed to be uniform and equal to the melt temperature. At the flow front, the fountain flow carries the material from the core and deposits it on the cavity wall. A scheme has been developed [2] to account for the effect of the fountain flow on the temperature field using a "flat front turnover rule". In this work, the temperature at the flow front is taken as the temperature in the core of the cavity.

3. NUMERICAL IMPLEMENTATION

3.1 *The pressure and energy equations*

The pressure equation is solved using the finite element method. A 3 node element was chosen to approximate the pressure. Details concerning the numerical implementation are given in [3,4]

The energy equation is discretized using the finite difference method. The temporal derivative of temperature is approximated by backward finite difference. The thickness of the part is divided into several layers to evaluate the conduction term in the gapwise direction. The viscous dissipation term and the convection term are calculated using a upwinding technique to ensure stability of the solution[1,4].

3.2 *Algorithm for flow front advancement*

The major challenge in modeling the filling phase of a polymer injection process is the successful tracking of the flow fronts. In case of a GAIM process, this aspect becomes extremely critical and complicated because of the existence of two different moving interfaces: one for polymer fronts, the other for gas-polymer interface. Tracking accurately both of moving interfaces is the key component of the filling analysis in a GAIM process.

In this work a control volume approach is employed to track the flow front advancement as well as the interface gas-polymer[5]. This approach was originally developed for finite difference method [6]. Since the last decade, this approach, integrated with the finite element concept, has been used increasingly to solve free surface and moving boundary problems [7]. The control volume approach uses a fixed mesh system with flow fronts passing through it. The principal advantage of this approach is its ability to handle complex flow fronts of arbitrary shapes (distortion, separation, etc.). Multiple fronts problem can also be modeled by this approach.

There are three regions to differentiate for modeling the filling phase of a GAIM process(see figure 1).

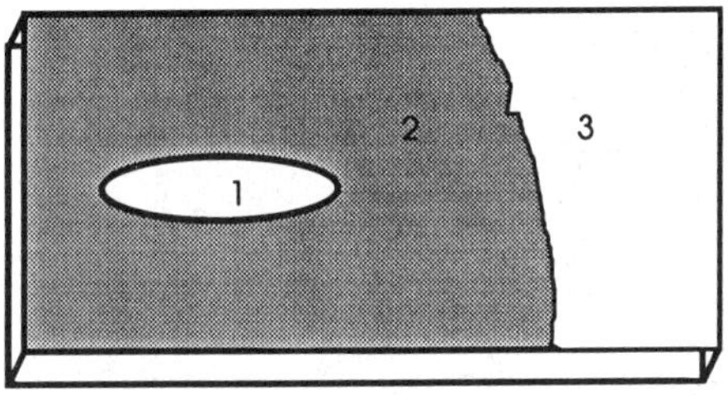

Figure 1 A simple scheme of different regions for a GAIM process

1- Gas penetrated region
2- Polymer filled region
3- Empty or unfilled region

Region 2 is completely occupied by the polymer melt and region 3 is empty. Region 1 is initially filled by polymer, then the gas penetrates into the polymer melt and creates a gas core which is enveloped by a polymer skin layer. This skin layer arises from two mechanisms : 1) freezing of the polymer due to its contact with the cold mold; 2) adherence between the polymer flow and the solidified layer.

A thickness fraction of the skin polymer Fs is associated for each control volume in order to represent these regions in the filling phase of a GAIM process. Fs is defined as the ratio of the thickness of skin polymer and the total thickness.

The value of Fs is between 0 and 1. An Fs value of 1 represents an element completely filled by polymer and 0 an empty element. $0 < Fs < 1$ represents an element where the gas penetrates through and a certain fraction of polymer is created as skin. With the above concept, for every time step, a calculation of pressure equation (1) is performed in order to obtain the velocity distribution in the polymer filled domain. The flow rate at the polymer fronts and at the gas-polymer interface is calculated from the velocity field, and then the value of Fs of each element can be evaluated by applying the principle of the conservation of mass. Finally, a newly filled domain can be easily defined as well as the gas penetrated region using the updated Fs field. The temperature equation (2) is solved next to obtain the temperature distribution in the gapwise direction, the frozen layer thickness and the new effective cavity thickness. This effective cavity thickness is then used in the pressure equation for the next time step. An iterative procedure is performed up to the end of the filling phase.

4 CASE STUDY

In order to demonstrate the capabilities of the numerical model, an experimental mold is simulated under different molding conditions. The geometry of the mold and gate locations are shown in figure 2. This mold has 6 polymer injection gates which can be opened simultaneously or in any sequence. In this work, only gates 5 and 6 are used for polymer injection and the gas injection location is indicated in figure 2. The gas channels can be identified based on the thickness distribution shown in figure 3 (black region).

Simulations were performed for three cases to investigate the effect of different parameters on gas penetration : I) polymer pre-fill of 85% and gas channel thickness 3/8"; II) polymer pre-fill of 95% and gas channel thickness 3/8"; and III) polymer pre-fill of 85% and gas channel thickness is increased to 3/4". The flow rate of polymer injection is 198 cm^3/sec and the gas pressure is 650 psi. Due to the limit of the size of the article, only the thickness fraction field at the end of fill is used for discussion.

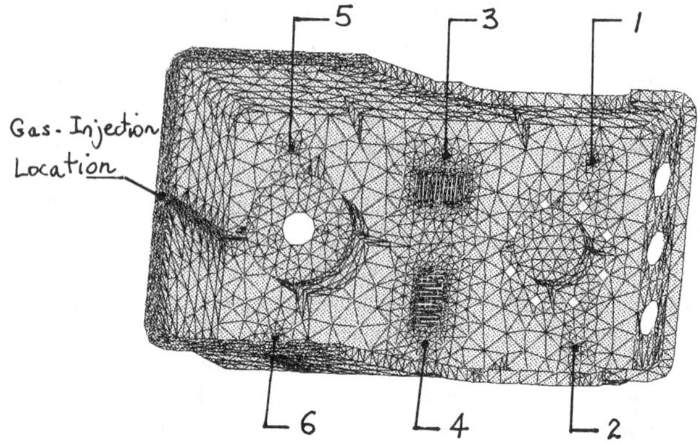

Figure 2. Mold geometry and gate locations

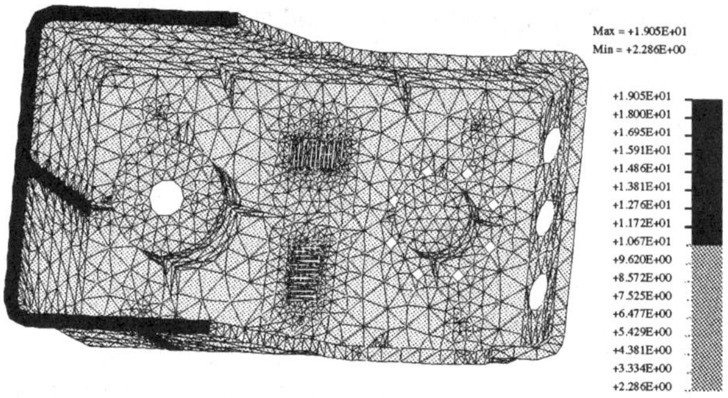

Figure 3. Thickness distribution in mm

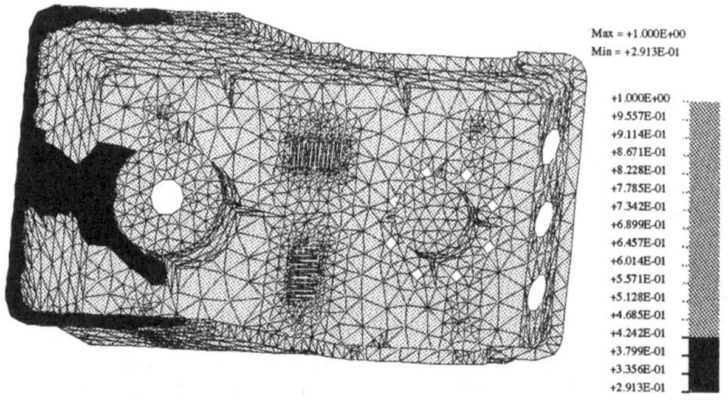

Figure 4. Thickness fraction of skin polymer: Case I

4.1 *Effect of percentage of polymer fill*

The cases 1 and 2 are used here to discuss the effect of percentage of polymer pre-fill. Figures 4 and 5 show the final stage of the gas penetration at the complete filling of the cavity at polymer fill of 85% and 95% respectively. The first case (85% polymer fill) shows that the gas penetrated into the thin section of the part when insufficient polymer was injected into the cavity (the gas penetration is represented by black color in figure 4). In the second case, 95% of the cavity volume is filled by polymer before the gas injection. There is no blowing out of gas into the thin section (see figure 5). but the gas did not penetrate all gas channels because no enough volume was left to be occupied by gas. It can be concluded that the percentage of the polymer pre-fill is a very import parameter for controlling

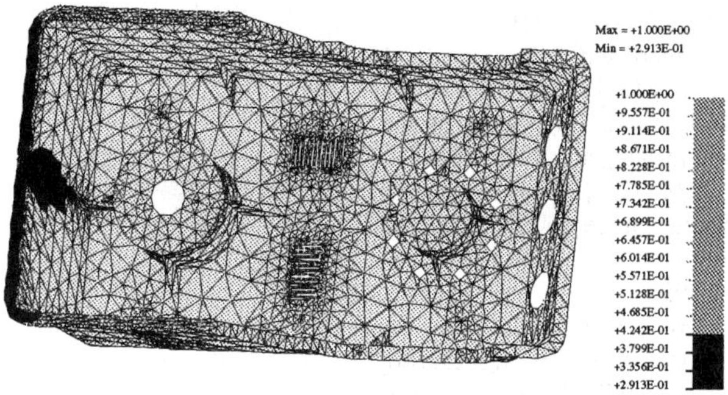

Figure 5. Thickness fraction of skin polymer: Case II

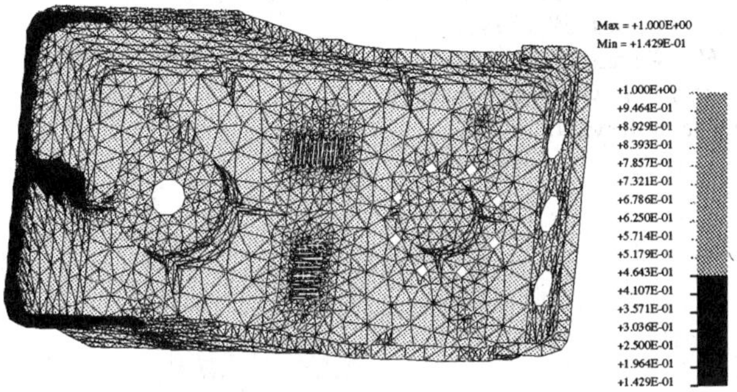

Figure 6. Thickness fraction of skin polymer: Case III

the process. If no enough polymer is injected into the cavity, the gas is blowing out into the thin cavity. A high percentage of polymer pre-fill would not permit the complete penetration of the gas into the gas channels and the advantages of GAIM could not be fully utilized.

4.2 *Effect of dimension of gas channels*

In the case 3, the thickness of the gas channels is doubled to be compared with the case 1. Other conditions remain unchanged. Figure 6 shows the final stage of the gas penetration at the complete filling of the cavity. In this case, the gas succeeded in penetrating into all the gas channels without any blowing through phenomena. The reason is that when the thickness of gas channels increases, the flow resistance in these regions is much smaller than in the rest of the part and the gas flow has much less resistance to go through these preferential channels than thin sections. This comparison demonstrated clearly that the design of gas channels is primordial for a GAIM process and efforts should be made to ensure a good quality of parts.

5 CONCLUSIONS

A numerical model based on a hybrid FEM/Control volume approach is developed for simulating the cavity filling as well as the gas penetration during the GAIM process. The simulation software allows the prediction of the location of polymer fronts, the gas penetration and the thickness fraction of skin polymer during the filling process. The capabilities of the numerical model are demonstrated through the analysis of the filling of a complex part. The influence of different molding parameters such as percentages of polymer fill and different thicknesses of the gas channels are investigated for an experimental mold.

REFERENCES

[1] C.A. Hieber and S.F. Shen, "A Finite Element/Finite Difference Simulation of the Injection Molding Filling Process", J. Non-Newtonian Fluid Mech., 7, (1980)

[2] F. Dupret and L. Vanderschuren, "Calculation of the Temperature Field in Injection Molding",

AICHE J., Vol 34, pp 1959-1972, (1988)
[3] D.M. Gao, K.T. Nguyen and L.P. Hébert, " Numerical Modeling of Mold filling in Injection Molding", Proc. of the 2nd Int. Conf. of the CFD Society of Canada, Toronto, June 1994
[4] K.T. Nguyen and D.M. Gao, "Non-Isothermal Analysis of the Filling Stage in the Injection Molding", The winter annual meeting of ASME, Chicago, November 1994
[5] D.M. Gao, K.T. Nguyen and G. Salloum, "A Numerical Model for Cavity Filling During Gas-Assisted Injection Molding", submitted to NUMIFORM' 95, Ithca, May 1995
[6] C.W. Hirt and B.D. Nichols, "Volume of Fluid(VOF) Method for the Dynamics of Free Boundaries", J. Comp. Physics, 39, (1981)
[7] J.M Floyan and H. Rasmussen, "Numerical Methods for Viscous Flows with Moving Boundaries", Appl. Mech. Rev., 42, (1989)

Simulation of Materials Processing: Theory, Methods and Applications, Shen & Dawson (eds)
© 1995 Balkema, Rotterdam. ISBN 90 5410 553 4

The numerical method of optimum choice of heat transfer conditions on the continuous cast surface

Radosław Grzymkowski & Adam Kapusta
Silesian Technical University, Institute of Mathematics, Poland

ABSTRACT: The parameters determining the process of heat exchange between the cast and surrounding it medium (for example the pouring temperature, the pulling rate, the length of a crystallizer, the flowing velocity and the singular intensity of cooling water in the section of the secondary cooling zone) has the fundamental influence. In the paper it will be presented a certain algorithm for calculation the heat exchange coefficients for the round cast poured on circular machine in the section of the secondary cooling zone.

1 INTRODUCTION

Among many factors influencing the quality of a cast poured continuously, the parameters determining the process of heat exchange between the cast and surrounding it medium has the fundamental influence. These parameters, among others, are the pouring temperature, the pulling rate, the length of a crystallizer, the flowing velocity and the singular intensity of cooling water in the section of the secondary cooling zone. The two last parameters determine the heat exchange coefficients for each section. For example, Müller and Jesschar [1] show, that for the greater part of cast:

$$\alpha = 10\,u + 1000\,(107+0.688\,u)\eta, \tag{1}$$

where α [W/m^2 K] is the coefficient of heat exchange, u [m/s] is the flowing velocity of water from nozzle and η [m^3/m^{2}s] is its singular intensity, when $u \in [11; 32]$, but $\eta \in [0.003; 0.009]$.

By fixing some parameters, it is possible to choose the remaining parameters to obtain the course of the solidification process approached to technological requirements. In the paper it will be presented a certain algorithm for calculation the heat exchange coefficients for the round cast poured on circular machine in the section of the secondary cooling zone, and properly in the presence of (1), parameters of the cooling water assuming the suitable speed of cuticles' growing.

2 THE MATHEMATICAL MODEL

The failure-free work of circular machine causes the generation of pseudo-stationary temperature field, which in an immovable coordinate system, associated with the solidified cast (Fig.1) for the metal solidified in the interval of temperature $[T_S, T_L]$ (T_S - solid temperature, T_L - liquid temperature), is described by the equation:

$$\omega\,\partial_\psi T = a\Big\{\frac{1}{r\,(R+r\cos\varphi)}\partial_r[r\,(R+r\cos\varphi)\partial_r T\,] + \frac{1}{(R+r\cos\varphi)^2}\partial^2_\psi T + \frac{1}{r^2(R+r\cos\varphi)}\partial_\varphi[(R+r\cos\varphi)\partial_\varphi T\,]\Big\}, \tag{2}$$

$$-\frac{\pi}{2} < \varphi \leq \frac{\pi}{2},\quad 0 < \psi \leq \psi^*,\quad 0 < r < D\cos\varphi$$

where r, ϕ and ψ are space coordinates, $T = T(r,\phi,\psi)$ temperature, R radius of curvature of radial continuous casting installation, ω angle velocity of continuous casting, $a = a(T)$ is the substitute thermal conductivity considering a heat of phase change [2], and ψ^* is the end of secondary cooling zone. Because of the small heat conductivity in the direction of the continuous casting pulling [3], the equation (2) can be simplified by dropping out the second component in the right side, in result that we obtain a parabolic equation, in which the coordinate ψ takes a role of time.

Let T^* is the pouring temperature, T^∞ the temperature of environment, λ the thermal conductivity, $\{r_i\ ,\ \phi_i\ ,\ \psi_i\}$, $i=1\ \ldots\ N$, are the

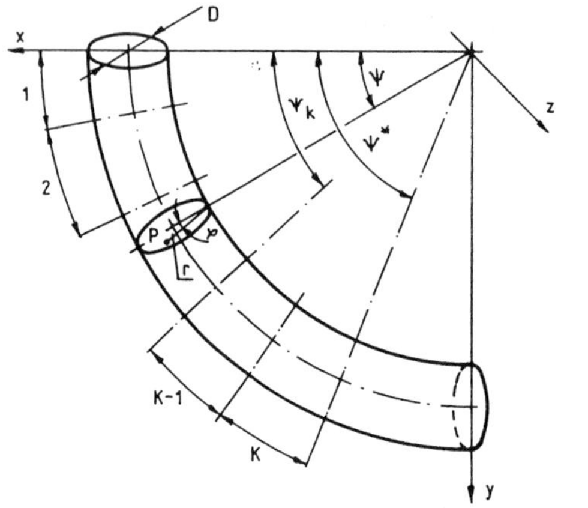

Fig.1. The modelled area.

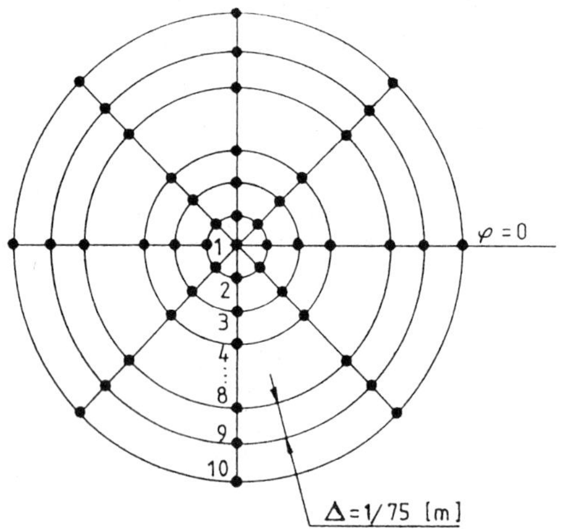

Fig.2. The differential mesh.

isotherm $T=T_S$ accordingly to the technological requirements, and α is the heat exchange coefficient in the secondary cooling zone, where :

$$\alpha = \alpha(\psi) = \sum_{k=1}^{K} [m(\psi-\psi_{k-1}) - m(\psi-\psi_k)]\alpha_k , \qquad m(\xi) = \begin{cases} 0, & \xi \le 0, \\ 1, & \xi > 0. \end{cases} \tag{3}$$

K is the number of considered sections of the secondary cooling zone, α_k is a researched heat exchange coefficient, corresponding to a particular section and ψ_k means the position of surfaces separating the cooling section, where ψ_0 is the pouring surface and $\psi_k=\psi^*$ is the surface closing the secondary cooling zone.

For the assumed foundations, the temperature field is described by the equation:

$$\omega\, \partial_\psi T = a\left\{ \frac{1}{r(R+r\cos\varphi)} \partial_r [r(R+r\cos\varphi)\partial_r T] + \frac{1}{r^2(R+r\cos\varphi)} \partial_\varphi[(R+r\cos\varphi)\partial_\varphi T] \right\}, \tag{4}$$

$$-\frac{\pi}{2} < \varphi \le \frac{\pi}{2}, \quad 0 < \psi \le \psi^*, \quad 0 < r < D\cos\varphi,$$

$$T = T^*, \ \psi = 0, \ -\frac{\pi}{2} < \varphi \le \frac{\pi}{2}, \quad 0 < r < D\cos\varphi,$$

$$-\lambda \partial_{\bar{n}} T = \alpha(T - T^\infty), \ -\frac{\pi}{2} < \varphi \le \frac{\pi}{2}, \quad r = D\cos\varphi, \tag{5}$$

where $\partial_{\bar{n}}$ (.) means the normal derivative to a boundary surface and the parameters α_k , k=1,...,K, determining the function α, should be chosen so, that for each i=1,...,N, $T(r_i,\phi_i,\psi_i) \cong T_s$.

3 THE METHOD OF SOLUTION

The problem formulated above can be solved in several ways. The good precision and effectivness of the algorithm can be obtained using the optimization method. Then, the function of quality (e.g. the criterion of the choice of unknown parameters α_k) should be added to the mathematical model. For example:

$$F(\alpha) = \sum_{i=1}^{N} u_i (T_i - T_s)^2, \tag{6}$$

where u_i , i=1,...,N, mean the weight parameters and $T_i=T(r_i,\phi_i,\psi_i;\alpha)$, i=1,...,N, mean the suitable solution of a boundary problem (4) - (6) with the fixed function α. The minimum of function F is studied by the gradient methods. To this end one should calculate the derivatives of the function F with respect to researched parameters.

$$\nabla_k F = \partial_{\alpha_k} F = 2\sum_{i=1}^{N} u_i (T_i - T_s) w_{ki} , \quad k=1,...,N, \tag{7}$$

where $\quad w_{ki} = \partial_{\alpha_k} T(r_i, \varphi_i, \psi_i; \alpha)$.

The system of equations (4) - (6) is used to calculate unknown functions w_k, $k=1,\dots,K$, wchich appeared in the expression (8). Differentiating (4), (5) with respect to α_k, we obtain K linear boundary problems, for $k=1,\dots,K$, in the form

$$\omega \partial_\psi w_k =$$
$$= a\left\{\frac{1}{r(R+r\cos\varphi)}\partial_r[r(R+r\cos\varphi)\partial_r w_k] + \right.$$
$$\left. + \frac{1}{r^2(R+r\cos\varphi)}\partial_\varphi[r(R+r\cos\varphi)\partial_\varphi w_k]\right\} +$$
$$+ \kappa\left\{\frac{1}{r(R+r\cos\varphi)}\partial_r[r(R+r\cos\varphi)\partial_r T] + \right. \tag{8}$$
$$\left. + \frac{1}{r^2(R+r\cos\varphi)}\partial_\varphi[r(R+r\cos\varphi)\partial_\varphi T]\right\} w_k,$$
$$-\frac{\pi}{2} < \varphi \le \frac{\pi}{2},\; 0 < \psi \le \psi^*,$$
$$0 < r < D\cos\varphi,$$

$$w_k = 0,\; \psi = 0,\; -\frac{\pi}{2} < \varphi \le \frac{\pi}{2}, \tag{9}$$
$$0 < r < D\cos\varphi,$$

$$-\lambda \partial_n w_k = \alpha w_k +$$
$$+ (T - T^\infty)[m(\psi - \psi_{k-1}) - m(\psi - \psi_k)], \tag{10}$$
$$\frac{\pi}{2} < \varphi \le \frac{\pi}{2},\; 0 < \psi \le \psi^*,\; r = D\cos\varphi$$

where $\kappa = da(T)/dT$.

To find a minimum of a function, for example by the simple gradient method [4], one should set the initial approximation of parameters α_k, e.g., $\alpha_k = \alpha_k^{(0)}$. The successive approximations will be found by the formulas :

$$\alpha_k^{(s+1)} = \alpha_k^{(s)} - p_s \nabla_k F^{(s)},\; k=1,\dots,K, \tag{11}$$

where $s=0,1,2,\dots$ is the interpolation number and p_s is the iteration step. The parameters p_s are chosen in such a way, that the following conditions are satisfied:

$$F^{(s)} \quad F^{(s+1)} \ge \gamma p_s[(grad\, F)^{(s)}]^2, \tag{12}$$

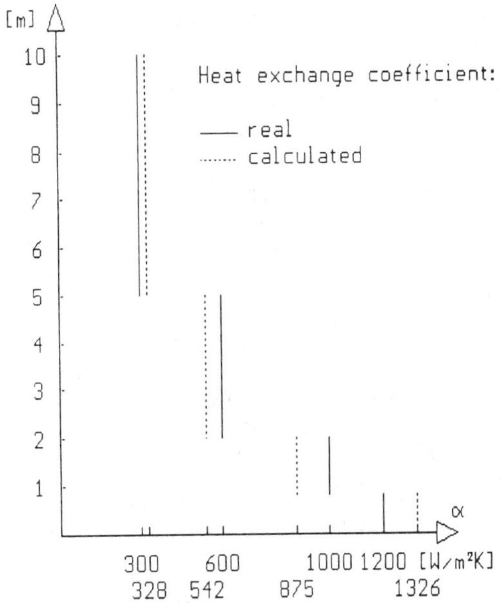

Fig.3. The results of calculations.

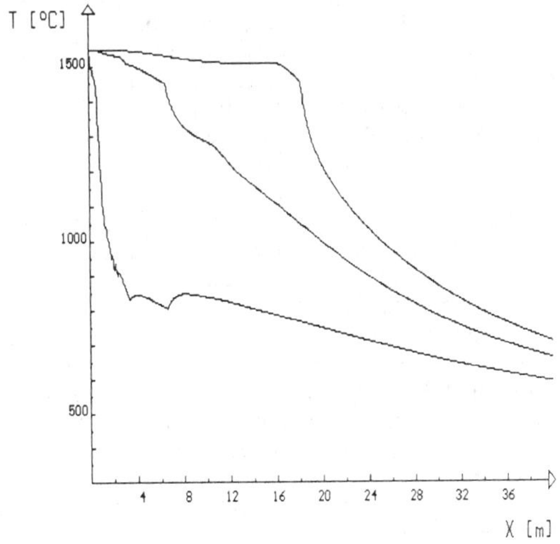

Fig. 4. The cooling curwes along the casting.

for $\gamma \in (0,1)$.

The iteration process is finished if the one of the three conditions is satisfied:

$$|F^{s+1} - F^s| < \epsilon_1, \tag{13}$$

$$\| grad\, F^s \| < \epsilon_2, \tag{14}$$

$$\| \alpha^{s+1} - \alpha^{s} \| \leq \varepsilon_3 \,, \qquad (15)$$

where $\epsilon_1, \epsilon_2, \epsilon_3$ are given small enough real numbers.

4 AN EXAMPLE

The solutions, obtained for the steel cast, are the illustration for researched solutions. It was assumed, that the cast has the diameter D=0.24 [m] and is poured with the velocity w=0.015 [m/s] on the circular machine with the radius R=10 [m]. The considered way of cast's cooling was divided by four sections with the lengths (measured by the circular arc with the radius R): I - 0.8 (crystallizer), II - 1.2, III - 3, IV - 5 [m], and it was assumed, that the pouring temperature is $T^{*}=1550$ [°C]. The thermophisical parameters of the steel were chosen on the base [5].

The respective stages of calculation process require the solution of the boundary problems for the heat conduction equation in the toroidal area. Because of lack of practically useful analytical formulas, the solutions were researched numerically, using the methods of approximate solution for the problems of this kind developed in [5].

The differential mesh was put on the modeled area contained 248 layers in the direction of the cast shift. The knots were located as on the fig.2 in the each layer. The assumption assumed above was used in the way to solve the simple boundary problem for the fixed values of heat exchange coefficients, which were respectively $\alpha_1=1200$, $\alpha_2=1000$, $\alpha_3=600$, $\alpha_4=300$ [W/m^{2}K]. The temperature field obtained in this way was in the assistance for determination of control points position for the inverse problem. It was determining 12 control points, from 2 to 4 in the section, situated on the direction $\phi=0$.

From the calculation carried out it follows that, in spite of the certain interference in data input (the position of the control points), the reproduced values of heat exchange coefficients are slightly different from assumed. The error doesn't exceed 12.5% (Fig.3).

The temperature distribution in the centre, on the surface and in the medium node of continuous casting is shown in Fig. 4.

REFERENCES

[1] Müller H., Jeschar R.: Arch. Eisenhutten, t.44, No.8, 1973, pp. 589-594.

[2] Mochnacki B.: Bull. Pol. Ac. Tech., vol.32, No.34, 1984, pp. 127-143.

[3] Žuravlev W.A., Kitaev E. M.: Teplofizika formirovania neprerivnogo slitka, Mietallurgia, Moskva, 1974.

[4] Ce'a J.: Optimisation, Theorie and Algorithms, Dunod, Paris, 1971.

[5] Kapusta A.: PhD Thesis, Silesian Technical Univ., Gliwice, 1987.

Simulation of Materials Processing: Theory, Methods and Applications, Shen & Dawson (eds)
 ISBN 90 5410 553 4

Three-dimensional finite element simulation of mold filling processes

J.-F. Hétu, Y. Lauzé & A. Garcia-Rejon
Industrial Materials Institute, National Research Council, Boucherville, Que., Canada

ABSTRACT: The numerical simulation of mould filling has become a very important analysis tool since it provides the means to try new part designs, material and processing conditions without having to go through the lengthy trial and error procedure. The coupling of the incompressible Navier-Stokes equations to an hyperbolic surface tracking equation is the most popular method to model free surface flows encountered during the filling of moulds. In this work, the Navier-Stokes equations are first solved with an augmented Lagrangian formulation. Then, a pure transport equation is used to track the flow fronts, which is solved using the Lesaint-Raviart method to ensure accurate representation of the flow fronts. The method is proven to be very cost effective. In order to demonstrate the efficiency of the model , three cases are analyzed: 1) filling of a step cavity; 2) 2D filling of a square cavity; and 3) 3D filling of a car wheel.

1 Introduction

The rapid growth in the use of advanced materials – polymers, metal alloys and composites – in a large number of highly demanding automotive, electronic and consumer goods applications have generated the development of new and more complex material forming processes. A good understanding of the interaction between material and processing conditions is now very important in order to comply to stricter tolerances and demanding service conditions. This is particularly important in the case of forming processes involving the filling of a mould where the coupled phenomena of fluid flow and heat transfer will determine to a large extent the final properties of the moulded part.

The development of numerical simulation tools provides the means to try new part designs, materials and processing conditions without having to go through the lengthy trial and error procedure.

This paper presents a finite element model (FEM) capable of predicting the velocity, pressure fields in the material while filling a mould as well as the location of the flow fronts. In order to demonstrate the effiency of the model three cases are analyzed : i) filling of a step cavity (application to injection molding or high pressure die casting; ii) gravity filling of a square cavity; and iii) 3D filling of a car wheel.

2 Theoretical Background

The main difficulties encountered in the simulation of mould filling can be summarized as follows [1]:

- The computational domain is usually a three dimensional volume having a complex shape.
- The free surface (liquid-air interface) is subject to large deformations and multiple interfaces may come in contact with each other.
- The prediction of the flow boundary layers requires boundary conditions allowing the material to adhere to the cavity walls (no-slip conditions).

Due to these characteristics the model has been developed using a two-step solution algorithm. The first step solves the incompressible Navier-Stokes equations to compute the velocity and pressure fields. Then based on the Volume of Fluid Method (VOF) [5], the position of the flow front is convected using the computed velocity field[6]. The dynamic no-slip boundary conditions are imposed using an augmented Lagrangian formulation [2, 3].

2.1 Flow Equations

The Navier-Stokes equations are solved on the whole computational domain. This implies that the equations are also solved for the air present in the cavity. It is assumed that the air can exit the cavity without restrictions and that its velocity is small compared to the speed of sound. Thus the flow in the empty region of the domain is assumed incompressible. Therefore, in the numerical analysis, the "air" is refered to as a pseudo-fluid. The equations governing the laminar flow of liquid and pseudo-fluid are expressed as

$$\begin{aligned} \rho_l \left(\frac{\partial \vec{u}}{\partial t} + \vec{u} \cdot \nabla \vec{u} \right) &= -\nabla p + \nabla \cdot \sigma_l(\vec{u}) + \rho \vec{g} \\ \nabla \cdot \vec{u} &= 0 \text{ on } \Omega_l \\ \rho_p \left(\frac{\partial \vec{u}}{\partial t} + \vec{u} \cdot \nabla \vec{u} \right) &= -\nabla p + \nabla \cdot \sigma_p(\vec{u}) \\ \nabla \cdot \vec{u} &= 0 \text{ on } \Omega_p \end{aligned} \tag{2.1}$$

with:

$$\sigma(\vec{u}) = 2\mu\dot{\gamma}(\vec{u}) = \mu(\nabla\vec{u} + (\nabla\vec{u})^T) \tag{2.2}$$

In the above equations, t, $\vec{u}$, p, ρ and μ denote time, velocity, pressure, density and viscosity respectively. Subscripts l and p refers to the liquid and pseudo-fluid respectively.

Appropriate boundary conditions complete the statement of the problem:

$$\begin{aligned} \vec{u} &= \vec{u}_0 \quad \text{on} \quad \Gamma_{Dirichlet} \\ \sigma(\vec{u}) \cdot \vec{n} - p\vec{n} &= \vec{t} \quad \text{on} \quad \Gamma_{tractions} \\ \vec{u} &= 0 \quad \text{on} \quad \Gamma_{wall} \end{aligned} \tag{2.3}$$

The boundary condition on Γ_{wall} is imposed only on the filling material.

2.2 Front Tracking

The tracking of the flow front in the mould cavity is modeled using the Volume of Fluid technique (VOF) [5]. This model defines a function $F(\vec{x}, t)$ that is equal to unity at any point filled with fluid or zero elsewhere (the pseudo-fluid). The volume where $F = 1$ thus represents the filled portion of the cavity.

The VOF transport equation can be written as

$$\frac{\partial F}{\partial t} + \vec{u} \cdot \nabla F = 0 \text{ on } \Omega \tag{2.4}$$

with the following initial and boundary conditions:

$$\begin{aligned} F(\vec{x}, t) &= 1 \quad \text{on} \quad \Gamma^- \\ F(\vec{x}, 0) &= 0 \quad \text{at} \quad t = 0 \end{aligned} \tag{2.5}$$

where Γ^- is the inflow region of $\partial\Omega$.

The hyperbolic equation (4) is solved with the element by element technique of Lesaint and Raviart [6, 7].

2.3 No-slip boundary conditions

Since the Navier-Stokes equations are solved on the whole computational domain and because the pseudo-fluid has to exit the cavity freely, boundary conditions imposed on the cavity walls have to change dynamically. Depending on the value of F on the walls, the boundary conditions must satisfy the following conditions

Table 1. Physical properties for the square cavity problem.

Property	Liquid	Pseudo-fluid
Density	1.0	0.01
Viscosity	0.5	0.05
Gravity	−9.81	0.0

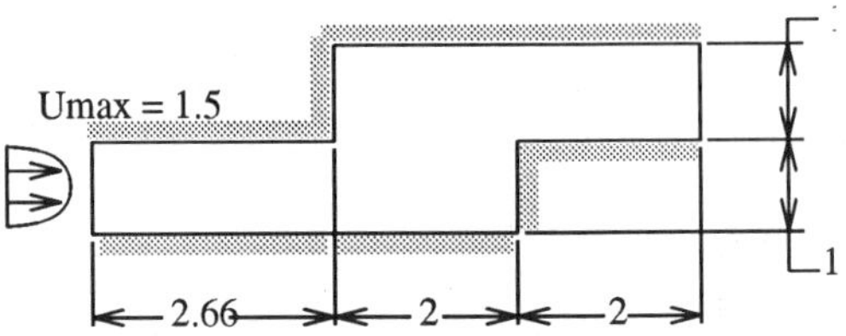

Figure 1. Computational domain for step cavity problem

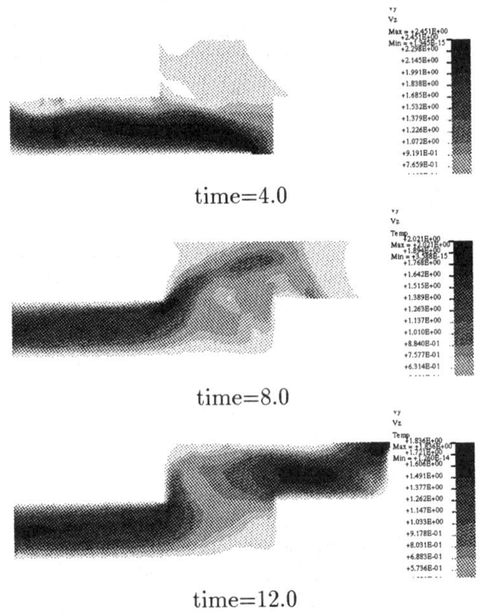

Figure 2. Velocity field at times 0.0, 4.0, 8.0 and 12.0 sec.

$$\begin{aligned} \vec{u} &= 0 \quad \text{when} \quad F = 1 \text{ (Filled)} \\ \vec{\sigma}(\vec{u}) \cdot \vec{n} - p\vec{n} &= 0 \quad \text{when} \quad F = 0 \text{ (Empty)} \end{aligned} \tag{2.6}$$

These boundary conditions are imposed using an augmented Lagrangian technique [3]. For a detailed presentation of this implementation, the reader is refered to [4].

3 Results

In order to prove the validity of the model, three

time=4.0

time=8.0

time=12.0

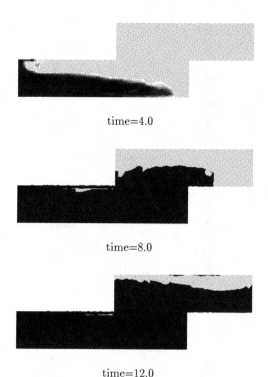

Figure 3. Flow fronts at times 4.0, 8.0 and 12.0 sec.

cases are analyzed:

- filling of a step cavity with a constant flow rate,
- gravity filling of a square cavity,
- filling of a 3D section of a car wheel.

3.1 Filling of a step cavity

This example deals with the filling of a step cavity. It can be representative of the injection moulding of a part having different thickness sections. The computational domain and boundary conditions are shown in Figure 1. The driving force in this case is a fully developed Poiseuille flow imposed at the inlet. This situation corresponds to a Reynolds number at the inlet section equal to 2. Time integration was done using a backward Euler scheme with a timestep of 0.25.

Figure 2 shows the magnitude of the velocity field at various times. Dark areas denote higher velocities. One can see that the no-slip boundary condition is dynamically applied as the liquid touches the walls. It should also be mentioned that the model predicts accurately the decceleration and acceleration the fluid encounters while flowing through the step in the cavity.

Figure 3 shows the position of the free surface at different times. The effect of the gravitational force causes the fluid to race towards the lower right corner before completely filling the inlet region. The combined effect of inertia and gravity forces produced an ondulating free surface which should be minimized in order to avoid moulding problems such as air entrapment and porosity in parts.

3.2 Filling of a square cavity

This case exemplifies the metal forming process known as gravity casting. In this case the flow driving force is the weight of the molten metal. Gravitational forces act on the fluid pushing it to fill the square cavity. The computational domain and boundary conditions are shown in Figure 4. A small pressure of 0.001 is applied at the entrance in order to initiate the movement of the fluid. Dynamic no-slip boundary conditions (i.e. eq. 6) are applied to all the remaining walls of the cavity. All physical properties are summarized in Table 2

Table 2. Physical properties for the square cavity problem.

Property	Liquid	Pseudo-fluid
Density	1.0	0.01
Viscosity	0.2	0.02
Gravity (Y)	−0.98	0.0

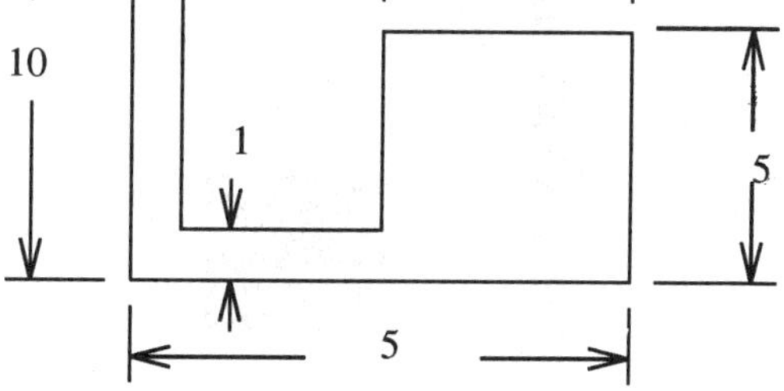

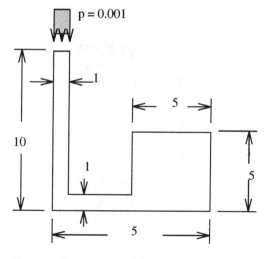

Figure 4. Computational domain for square cavity problem

Figure 5. Flow fronts at times 25, 50, 75, 100 and 125 sec.

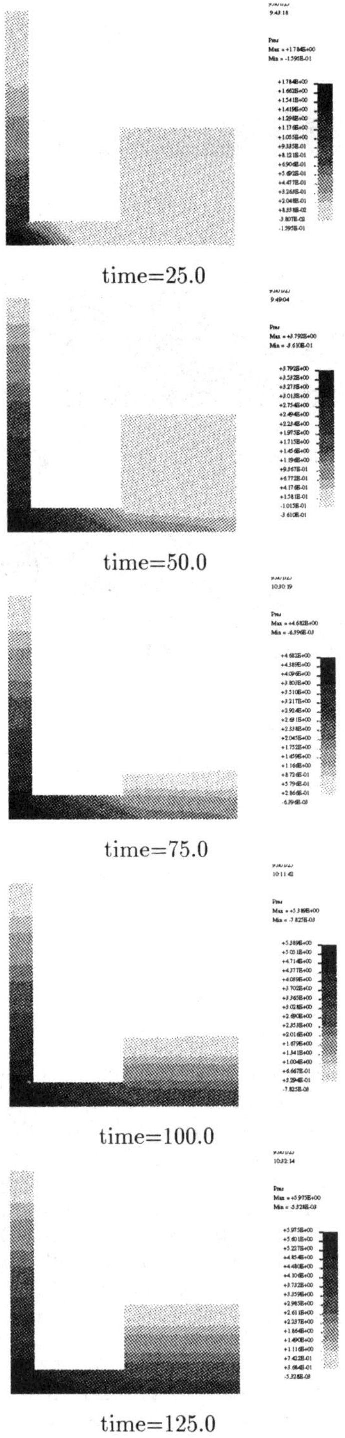

Figure 6. Total pressure at times 25, 50, 75, 100 and 125 sec.

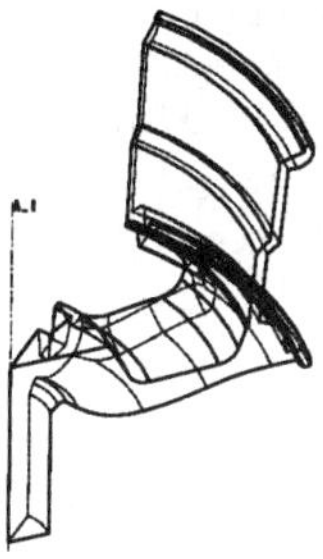

Figure 7. Geometry of a wheel section

Table 3. CPU times (sec.) to compute transient filling of the wheel section.

Step	CPU Time (sec.)
Input phase	18.39
Element equations generation	7330.24
Assembly of system	672.93
LU decomposition	83072.32
Forward/backward subsitution	468.26
Updating for nonlinearity	5.20
Lesaint-Raviart solution	118.43
output phase	28.58
TOTAL TIME	93006.79

Figures 5 and 6 show the location of the flow fronts and the total pressure distribution respectively. The model is capable of predicting the sloshing occurring inside the cavity at the early stages of filling. Once the material leaves the runner and enters the cavity the inertia terms govern the flow behaviour up to the point in which the liquid hits the opposite wall. From this point on the gravitational force tends to stabilize the free surface movement, thus causing a uniform filling pattern.

3.3 3D Filling of a wheel

Figure 7 shows the geometry of a 45° section (1/8) of a car wheel manufactured using a low pressure die casting process. The diameter and height of the wheel are 406mm and 203mm respectively. The total cross section of the spoke is approximately 25 mm thick by 100 mm wide.

The mesh, used for the simulation, has 4599 $P2-P0$ tetrahedral elements. The mould is filled from the bottom of the geometry with a uniform inlet velocity of 102 mm/s. The cross sectional area at the entrance is $307 mm^2$. Under these conditions, the Reynolds number can be estimated as 8.7×10^{-3}. The filling of the wheel was simulated for 20 seconds with a time step of 0.5 sec.

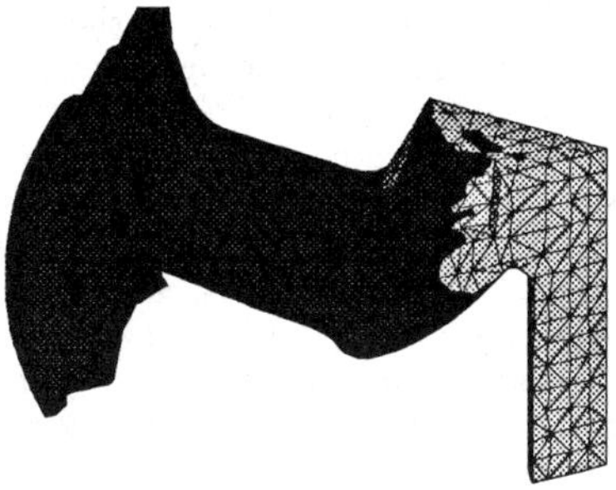

time=10.0

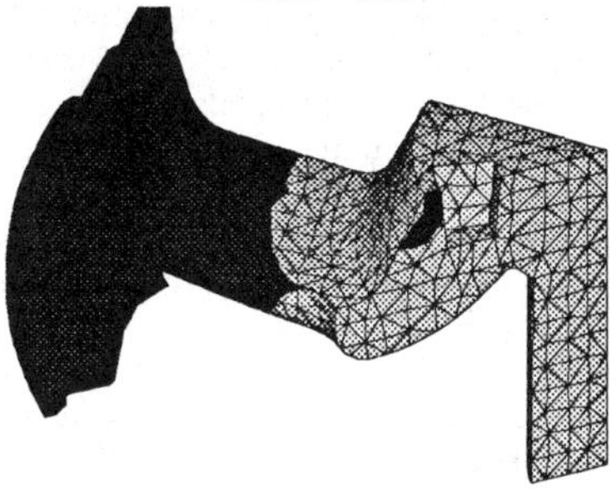

time=20.0

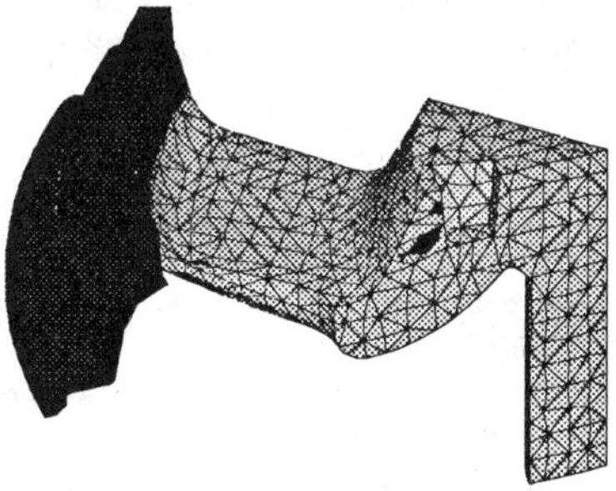

time=30.0

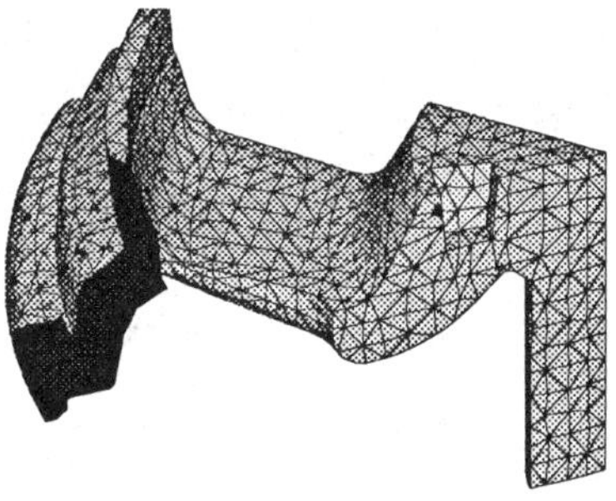

time=40.0

Figure 8. Flow fronts at times 5.0, 10.0, 15.0 and 20.0 sec.

Results of the filling analysis are shown in figure 8. Note that the light gray area represents the filled part of the cavity. CPU times were measured in order to verify the computational efficency of the Lesaint-Raviart method. Table 3 shows the CPU Times measured on an IBM RS6000 model 550. Almost 99% of the time is used to solve the Navier-Stokes equations. The computation time needed to solve the hyperbolic equation with the Lesaint-Raviart method is almost negligeable.

4 Conclusions

The previous results show that the proposed model can predict the position of the free surface and the flow field for complex geometries. The no-slip boundary conditions on the material are satisfied to machine precision. Contrary to the slip boundary condition technique, the imposition of a no-slip boundary condition by an augmented Lagrangian technique allows complete filling of regions where slip is undefined such as corners.

The overall strategy gives good results on the three dimensional case studied. The solution of the hyperbolic transport equation, using the Lesaint-Raviart method, was shown to be very cost efficient. Less than 0.13% of all CPU time is spent solving the transport equation.

References

[1] J. M. Floryan and H. Rasmussen. Numerical methods for viscous flows with moving boundaries. *Appl. Mec. Review*, 42(12):323–345, 1989.

[2] A. Fortin, Y. Demay, and J.-F. Agassant. Computation of stationary interfaces between generalized newtonian fluids. *Revue Européenne des éléments finis*, 1(2):181–196, 1992.

[3] A. Garon and R. Camarero. Imposition of normal and/or tangential boundary conditions by an augmented Lagrangian technique. *Computers and Structures*, 44(1/2):405–408, 1992.

[4] J.-F. Hétu, D.H. Pelletier, and A. Fortin. A finite element method for mold filling simulation. Int. J. Numer. Methods Fluids, 1995.

[5] C.W. Hirt and B.D. Nichols. Volume of fluid (vof) method for the dynamics of free boundaries. *Journal of Computational Physics*, 39:201–225, 1981.

[6] P. Lesaint and P.-A. Raviart. On a finite element method for solving the neutron transport equation. In C de Boor, editor, *Mathematical aspects of finite element in partial differential equations*, pages 89–123, 1974.

[7] R. Touzani. Implementation of the discontinuous finite element method for hyperbolic equations. *Comp. Meth. Appl. Mech. Engng*, 68:115–123, 1988.

Simulation of Materials Processing: Theory, Methods and Applications, Shen & Dawson (eds)
© 1995 Balkema, Rotterdam. ISBN 90 5410 553 4

Slab casting – An asymptotic analysis

R.E.Johnson
The University of North Carolina at Charlotte, N.C., USA

R.E.Smelser
University of Idaho, Engineering in Boise, Idaho, USA

ABSTRACT: A simplified model of a slab or belt caster is developed using asymptotic methods. The model assumes that the slab is thin and the molten metal flow and solid slab deformation are two-dimensional. Three features of the analysis that are of particular interest include: the material flow, the development of the solidification front, and the slab deformation. The resulting expressions are useful in preliminary design estimates. These expressions also provide information needed for code verification.

1 INTRODUCTION

Near net shape manufacturing and concurrent engineering demand sophisticated process models. These process models frequently require large scale numerical codes that are computationally intensive. To achieve solutions in reasonable times often requires large scale parallelism. Many times the engineer is unsure of the physics incorporated in these codes. Consequently, simpler models are valuable for preliminary process design and to verify the more complicated numerical models. Asymptotic methods provide a systematic methodology to generate models to meet these needs.

A simplified model of a slab or belt caster is developed using asymptotic methods. The model assumes that the slab is thin and the molten metal flow and solid slab deformation are two-dimensional. The melt is assumed to behave as a Newtonian fluid, and the rate of deformation of the slab is additively decomposed into elastic, thermal and inelastic components. The asymptotic method leads to simple closed form expressions describing the melt flow and the slab deformation. The energy equation also yields the position of the solidification front.

Three features of the analysis that are of particular interest include: the treatment of the thermal transport, the inclusion of the thermal front development and the explicit calculation of the slab dilatation. The slab shrinks due to thermal contraction in the solidification zone. Contact with the confining belts can be lost. The loss of contact affects the rate at which heat is extracted from the solidifying slab and can lead to local remelting or to a change of metallurgical structure. The asymptotic analysis gives explicit formulae for the shrinkage.

This asymptotic model was developed as an alternative to performing large scale numerical computations. The closed form expressions are useful in preliminary design estimates. The asymptotic model is also suited for real time control applications. The expression for slab dilatation will allow better control of the belt placement for improved thermal contact. Each of the explicit expressions also provides information needed for code verification. A comparison of the asymptotic model with more extensive numerical models can help in understanding what features to incorporate in an improved model. The improved models can be used by designers to plan better control strategies for maintaining belt and slab contact.

2 ASYMPTOTIC SLAB CASTING MODEL

2.1 *Equilibrium equations*

The geometry of the slab casting is shown in Figure 1. A slab of thickness $2\widehat{h}$ solidifies in a caster of length $\widehat{L}$. The caster is inclined at an angle θ to the horizontal. (A superposed caret indicates a dimensional quantity.) The dimensionless momentum equations, (1) and (2), are written in terms of the deviatoric stress, $\hat{s}_{ij} = \hat{\tau}_{ij} + \hat{p}\delta_{ij}$. $\hat{\tau}_{ij}$ are the Cauchy stress components, and $\hat{p} = -\frac{1}{3}\hat{\tau}_{ii}$ is the mean stress.

$$-\frac{\partial p}{\partial x} + \frac{\delta}{\beta}\frac{\partial s_{xx}}{\partial x} + \frac{\partial s_{xy}}{\partial y} = -\rho g \tan\theta \qquad (1)$$

$$-\frac{\partial p}{\partial y} + \frac{\delta}{\beta}\frac{\partial s_{yy}}{\partial y} + \delta^2\frac{\partial s_{xy}}{\partial x} = \delta\rho g \qquad (2)$$

The dimensionless coordinates $x = \widehat{x}/\widehat{L}$ and $y = \widehat{y}/\widehat{h}$ give rise to the thinness parameter $\delta = \widehat{h}/\widehat{L}$. The shear stresses have been made dimensionless by a characteristic shear stress, $\widehat{\tau}_o$, and the normal stresses have been made dimensionless by a characteristic deviatoric stress, $\widehat{s}_o$. These parameters

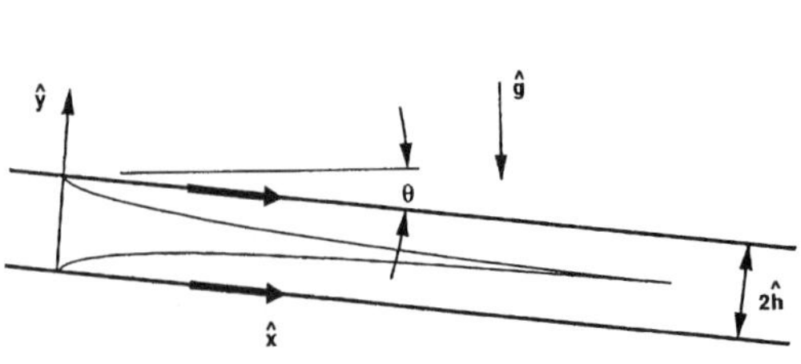

Figure 1: Schematic of the slab caster.

define $\beta = \widehat{\tau}_o/\widehat{s}_o$. The dimensionless gravity parameter is

$$g = \frac{\widehat{h}\widehat{\rho}_l\widehat{g}\cos\theta}{\widehat{\tau}_o}. \tag{3}$$

ρ is a dimensionless density ratio equal to unity in the liquid (l) and $\widehat{\rho}_s/\ \widehat{\rho}_l$ in the solid (s).

2.2 *Constitutive equations*

The total strain rate for the solid shell, $\widehat{D_{ij}}$, is additively decomposed into an elastic part, $\widehat{D^{el}_{ij}}$, an inelastic part, $\widehat{D^{in}_{ij}}$, and a thermal part, $\widehat{D^{th}_{ij}}$. The elastic response of the shell is assumed to be hypoelastic with a shear modulus of $\widehat{G}$ and a bulk modulus $\widehat{\kappa}$. The inelastic response is assumed to be incompressible and is related to the deviatoric stress through an inelastic modulus $\widehat{\mu}_s$. The dimensionless constitutive response in the solid is

$$\begin{aligned} D_{xx} &= \frac{\partial u}{\partial x} \\ &= \frac{\widehat{s}_o\widehat{L}}{2\widehat{\mu}_s\widehat{u}_1}s_{xx} + \frac{\widehat{s}_o}{2\widehat{G}}\ \dot{s}_{xx} - \frac{\widehat{\tau}_o}{\widehat{\kappa}\delta}\ \dot{p} + \Delta\widehat{T}\widehat{\alpha}\ \dot{T}, \end{aligned} \tag{4}$$

$$\begin{aligned} D_{yy} &= \frac{1}{\delta}\frac{\partial v}{\partial y} \\ &= \frac{\widehat{s}_o\widehat{L}}{2\widehat{\mu}_s\widehat{u}_1}s_{yy} + \frac{\widehat{s}_o}{2\widehat{G}}\ \dot{s}_{yy} - \frac{\widehat{\tau}_o}{\widehat{\kappa}\delta}\ \dot{p} + \Delta\widehat{T}\widehat{\alpha}\ \dot{T}, \end{aligned} \tag{5}$$

$$D_{xy} = \frac{1}{2}\left(\frac{1}{\delta}\frac{\partial u}{\partial y} + \frac{\partial v}{\partial x}\right) = \frac{\widehat{\tau}_o\widehat{L}}{2\widehat{\mu}_s\widehat{u}_1}s_{xy} + \frac{\widehat{\tau}_o}{2\widehat{G}}\ \dot{s}_{xy}\ . \tag{6}$$

The deformation is taken to be plane strain. The superposed 'dot' denotes a material time derivative for scalars, and a convective or Jaumann derivative for tensors. The velocity components $\widehat{u}$ and $\widehat{v}$ in the $\widehat{x}$ and $\widehat{y}$ coordinate directions have been made dimensionless by the slab exit velocity $\widehat{u}_1$. (A subscript 0 or 1 denotes the entrance or exit respectively.) A dimensionless time is defined by $t = \left(\widehat{u}_1\widehat{t}\right)/\widehat{L}$, and the dimensionless temperature is

$$T = \frac{\widehat{T} - \widehat{T}_0}{\widehat{T}_1 - \widehat{T}_0} = \frac{\widehat{T} - \widehat{T}_0}{\Delta\widehat{T}}. \tag{7}$$

The liquid behaves as a Newtonian fluid, and laminar flow is assumed. The strain rates are related to the deviatoric stresses through a constant viscosity, $\widehat{\mu}_l$. The relationship is similar to the first terms on the right-hand side of (4) - (6). The laminar flow assumption is consistent with the numerical results of Farouk et al. (1991, 1992) who found that the flow was mostly laminar.

2.3 *Energy equation*

The energy equation for both phases is made dimensionless using the quantities already defined. The dimensionless energy equation for a steady-state process in an Eulerian reference frame is

$$Pe\left(u\delta\frac{\partial T}{\partial x} + v\frac{\partial T}{\partial y}\right) = \delta^2\frac{\partial^2 T}{\partial x^2} + \frac{\partial^2 T}{\partial y^2}. \tag{8}$$

This equation holds in both the solid and liquid phases with the appropriate material parameters for each phase. The Peclet number in (8) is the ratio of the advective thermal transport to the diffusive thermal transport.

$$Pe = \frac{\widehat{\rho}\widehat{c}_v\widehat{u}_1\widehat{h}}{\widehat{k}} \tag{9}$$

Here, $\widehat{\rho}$ is the density, $\widehat{c}_v$ is the specific heat and $\widehat{k}$ is the thermal conductivity.

2.4 *Mass conservation*

The fluid is assumed to be incompressible. The solid slab shrinks as it cools and changes volume elastically in response to the pressure field. The dilatation rate in the slab is

$$\begin{aligned} \text{div}\ \mathbf{u}_s &= D_{xx} + D_{yy} + D_{zz} \\ &= \frac{\partial u_s}{\partial x} + \frac{1}{\delta}\frac{\partial v_s}{\partial y} = 3\left(\Delta\widehat{T}\widehat{\alpha}\ \dot{T} - \frac{\widehat{\tau}_o}{\widehat{\kappa}\delta}\ \dot{p}\right). \end{aligned} \tag{10}$$

(Boldfaced letters denote vectors.) The constitutive equations and the fact that the deformation is plane strain have been used in obtaining (10). The slab dilatation can be related to the evolution of the density and specific volume in a standard way. In addition, overall mass conservation must also be satisfied.

$$\rho = \int_0^{f(x)} \rho u_s(x,y)\,dy + \int_{f(x)}^1 u_l(x,y)\,dy \tag{11}$$

Here the location of the solid-liquid interface is given by $y = f(x)$ and the asymmetry introduced by gravity has been neglected.

2.5 *Interface boundary conditions*

Along the interface between the slab and the belt, a shear stress may exist. The boundary condition is

$$s_{xy} = \mp S_x \quad \text{on } y = 0, 2. \tag{12}$$

where S_x is a prescribed function. Equation (12) indicates the characteristic shear stress, $\widehat{\tau}_o$, is related to the shear stress along the boundary. The normal velocity of the slab and the belt are also the same. This velocity is small compared to the casting speed and is taken to be zero.

Heat transfer occurs along this boundary and is characterized by Newton's law of cooling.

$$\frac{\partial T_s}{\partial y} = \eta\,(T_s - T_w) \qquad \text{on } y = 0, 2. \tag{13}$$

Here $\eta = \widehat{\eta}\widehat{h}/\widehat{k}_s$ is a dimensionless heat transfer coefficient, and T_w is the dimensionless wall temperature.

Along the solid-liquid interface, $y = f(x)$, the shear traction, the tangential velocity and the temperature are continuous. In addition, conservation of mass requires

$$\mathbf{u}_l \cdot \mathbf{n} = -\rho \mathbf{u}_s \cdot \mathbf{n}. \tag{14}$$

In (14), $\mathbf{n}$ is the unit outward normal to the solidifying layer $y = f(x)$. Conservation of energy along the interface is

$$\begin{aligned} \frac{\widehat{\rho}_l\widehat{\mathcal{L}}\widehat{u}_1\widehat{h}}{\widehat{k}_s\Delta\widehat{T}}\mathbf{u}_l \cdot \mathbf{n} &= -\frac{\widehat{\rho}_s\widehat{\mathcal{L}}\widehat{u}_1\widehat{h}}{\widehat{k}_s\Delta\widehat{T}}\mathbf{u}_s \cdot \mathbf{n} \\ &= \left[\text{grad } T_s - \frac{\widehat{k}_l}{\widehat{k}_s}\text{grad } T_l\right]\cdot \mathbf{n}. \end{aligned} \tag{15}$$

The gradient operator denotes the spatial gradient and is dimensionless. $\widehat{\mathcal{L}}$ is the latent heat of fusion.

3 ASYMPTOTIC PROBLEMS

The asymptotic limits are now set by examining the dimensionless quantities in the governing equations and boundary conditions. The dimensionless constitutive equations contain the parameter $\widehat{s}_o$. It is reasonable to assume that the deviatoric stresses are related to the deformation rates. This suggests that

$$\widehat{s}_o = 2\widehat{\mu}_s\frac{\widehat{u}_1}{\widehat{L}}. \tag{16}$$

Additional Parameters can be defined as

$$\gamma = \frac{\widehat{\mu}_s\widehat{u}_1}{\widehat{G}\widehat{L}}, \qquad \alpha = \Delta\widehat{T}\widehat{\alpha}, \qquad \Gamma = \frac{\widehat{\tau}_o}{\widehat{\kappa}\delta\Delta\widehat{T}\widehat{\alpha}}. \tag{17}$$

The parameter Γ is a measure of the strain due to pressure compared to the strain due to thermal contraction. It is taken to be unity since the effects of pressure are the same as those of temperature. Equations (4) and (5) indicate that v is order δ smaller than u and can be rescaled so that $v = \delta V$. If it is assumed that $\gamma \ll \alpha \ll 1$, the dimensionless constitutive expressions suggest asymptotic expansions for the field variables of the form

$$\begin{aligned} u &= u^{(0)} + \alpha u^{(1)} + \cdots \\ V &= V^{(0)} + \alpha V^{(1)} + \cdots \\ &\vdots \end{aligned} \tag{18}$$

These expansions are substituted into the governing equations and boundary conditions and give rise to a sequence of problems in increasing order of α.

3.1 *Asymptotic problem in the solid*

The leading order equilibrium problem in the solid is

$$\frac{\partial p^{(0)}}{\partial x} - \frac{\partial s_{xy}^{(0)}}{\partial y} = \rho g \tan\theta, \tag{19}$$

$$\frac{\partial p^{(0)}}{\partial y} = -\delta\rho g. \tag{20}$$

The effects of gravity are small but are retained. The leading order constitutive equations are

$$\frac{\partial u^{(0)}}{\partial x} = s_{xx}^{(0)}, \qquad \frac{\partial V^{(0)}}{\partial y} = s_{yy}^{(0)}, \tag{21}$$

$$\frac{\partial u^{(0)}}{\partial y} = 0. \tag{22}$$

In arriving at these reductions, variations in the x direction are assumed to be small. This implies $\delta \ll \beta \ll 1$, and terms containing δ/β are taken to be smaller than order α terms. The effects of temperature and pressure will be greater than the effects of normal deviatoric stresses. The leading order boundary conditions are

$$s_{xy}^{(0)} = \mp S_x, \quad V^{(0)} = 0, \qquad \text{on } y = 0, 2. \tag{23}$$

The energy equation contains the Peclet number. An estimate of this parameter is obtained from the values in Table 1, $Pe \simeq 15$. The quantity multiplying the advective terms is δPe. Since the estimate of δ is 0.02 (Section 4.1), this product is 0.3. While this is not strictly a small value, the advective terms will be ignored. This gives

$$\frac{\partial^2 T^{(0)}}{\partial y^2} = 0 \tag{24}$$

for the leading order energy equation. The boundary condition is

$$\frac{\partial T^{(0)}}{\partial y} = \eta\left(T^{(0)} - T_w\right) \qquad \text{on } y = 0, 2. \tag{25}$$

On $y = f(x)$, the temperature is equal to the temperature in the liquid. This will be shown to be the melting temperature, T_m.

3.2 *Leading order solution in the solid*

The temperature field in the solid layer can be obtained from (24) and (25) directly. The result is

$$T_s^{(0)} = \frac{1}{1+\eta f}\left[T_m + \eta f T_w + \eta f\,(T_m - T_w)\,\frac{y}{f}\right]. \tag{26}$$

Table 1: Parameters for slab cating of pure aluminum

$\widehat{\mu}_l$	0.068 gm/cm-s
$\widehat{\rho}_l$	2.357 gm/cm^3
$\widehat{\rho}_s$	2.699 gm/cm^3
$\widehat{k}_s$	2.37 W/cm-K
$\widehat{c}_v^s$	0.90 J/gm-K
$\widehat{T}_m - \widehat{T}_w$	167 K
$\widehat{\mathcal{L}}$	397 J/gm
$\widehat{u}_1$	15.25 cm/s
$2\widehat{h}$	2.0 cm

A similar linear relationship exists for the upper surface. Since gravitational asymmetry has been ignored, attention will be focused on the lower surface.

Equation (22) immediately shows

$$u_s^{(0)} = u_s^{(0)}(x). \tag{27}$$

The pressure in the solid layer is

$$p_s^{(0)}(x, y) = -\delta\rho g y + P_w(x). \tag{28}$$

$P_w(x)$ is an unknown pressure on the belt and remains to be determined. The shear stress in the solid layer is given by

$$s_{xy}^{s(0)} = \left(\frac{dP_w(x)}{dx} - \rho g \tan\theta\right) y + H(x). \tag{29}$$

$H(x)$ is determined from the boundary condition for the shear stress on $y = 0$.

$$H(x) = -S_x \tag{30}$$

The remaining function, $P_w(x)$, is determined by the interface condition along the solid-liquid interface, $y = f(x)$. To leading order in δ, continuity of the shear traction is

$$s_{xy}^l = s_{xy}^s. \tag{31}$$

To complete the solid layer solution requires the solution in the liquid.

3.3 *Asymptotic problem in the liquid*

Equilibrium in the liquid phase is governed by (1) and (2). The dimensionless constitutive equations for the Newtonian fluid are

$$\frac{\partial u}{\partial x} = \frac{\widehat{\mu}_s}{\widehat{\mu}_l} s_{xx}, \qquad \frac{\partial V}{\partial y} = \frac{\widehat{\mu}_s}{\widehat{\mu}_l} s_{yy}, \tag{32}$$

$$\frac{1}{2}\left(\frac{\partial u}{\partial y} + \delta^2 \frac{\partial V}{\partial x}\right) = \frac{\widehat{\mu}_s}{\widehat{\mu}_l} \delta\beta s_{xy}. \tag{33}$$

The dimensionless viscosity, $\mu = \widehat{\mu}_l/\widehat{\mu}_s$, is typically much less than 1. Thus it is reasonable to assume that the longitudinal deviatoric stresses are negligible. The parameter $2\delta\beta/\mu$ in (33) is taken to be of order unity.

The leading order problem for the liquid is given by (19), (20) and (24) with the parameter ρ equal to unity for the liquid. The boundary conditions are set by the conditions along the solid-liquid interface and by symmetry conditions along the centerline of the caster.

3.4 *Leading order solution in the liquid*

Again, the energy equation is integrated directly. The symmetry condition along the slab centerline and the continuity of temperature at the solid-liquid interface imply the temperature in the liquid is

$$T_l^{(0)} = T_m. \tag{34}$$

The pressure in the liquid is

$$p_l(x, y) = p_l(x, 1) + \delta g(1 - y). \tag{35}$$

Continuity of traction along the solid-liquid interface, $y = f(x)$, to $O(\delta/\beta)$ is given by the equality of the pressure in the solid and liquid. Thus, the pressure on the centerline is

$$p_l^{(0)}(x, 1) = p_s^{(0)}(x, f(x)) - \delta g\,(y - f(x)) \tag{36}$$

to leading order in α. This result combined with (28) gives the leading order pressure in the liquid.

The leading order velocity field in the liquid is determined from the equilibrium equation in the x direction, the pressure field in the liquid and the constitutive equation for the shear stress, (33). The velocity in the liquid is

$$u_l^{(0)}(x, y) = u_s^{(0)}(x) - \frac{2\delta\beta}{\mu}\frac{dP}{dx}\left[y\left(1 - \frac{y}{2}\right) - f(x)\left(1 - \frac{f(x)}{2}\right)\right]. \tag{37}$$

The function $P(x)$ is

$$P(x) = P_w(x) + (1 - \rho)\,\delta g f(x) - g\tan\theta\, x. \tag{38}$$

The no-slip condition has been enforced on the tangential velocity in obtaining (37) (Smelser and Johnson, 1995). It remains to determine the two functions $P_w(x)$ and $f(x)$.

3.5 *Global mass conservation*

Global mass conservation is expressed by (11). Since $u_s^{(0)}$ is only a function of x and $u_l^{(0)}$ is know to within two undetermined functions of x, global mass conservation gives

$$[1 - f(1 - \rho)]\, u_s^{(0)}(x) = \rho + \frac{2\delta\beta}{3\mu}\frac{dP}{dx}(1 - f)^3. \tag{39}$$

The function dP/dx is determined from the continuity of shear stress at the solid-liquid interface.

$$\frac{dP}{dx} = S_x + (1-\rho)\left[\delta\frac{df}{dx} - \tan\theta\right] gf \quad (40)$$

The leading order solution for the velocity in the solid is

$$\left[1 - f^{(0)}(1-\rho)\right] u_s^{(0)}(x) = \rho + \frac{2\delta\beta}{3\mu}\left[S_x + (1-\rho)\left(\delta\frac{df^{(0)}}{dx} - \tan\theta\right) gf^{(0)}\right]\left(1 - f^{(0)}\right)^3. \quad (41)$$

3.6 *Deposition of the solid*

The energy balance (15) on $y = f(x)$ combined with (10) determines the remaining unknown, $f(x)$. To $O(\delta)$, (15) gives

$$\mathcal{L}\delta\left(V_s - \frac{df}{dx}u_s\right) = \frac{\widehat{k}_l}{\widehat{k}_s}\left(\frac{\partial T_l}{\partial y}\right) - \frac{\partial T_s}{\partial y} \equiv \Delta\mathcal{T} \quad (42)$$

where $\mathcal{L}$ is the dimensionless latent heat from (15). $\Delta\mathcal{T}$ is evaluated from (26) and (34) and gives

$$\Delta\mathcal{T} = -\frac{\eta\Delta T}{1+\eta f}. \quad (43)$$

Integrating (10) across the solid layer, noting the velocity of the belt is nearly zero, and using (42) one obtains

$$\frac{d}{dx}\left(\overline{u}_s f\right) = -\frac{\Delta\mathcal{T}}{\delta\mathcal{L}} + 3\alpha\int_0^{f(x)}\left(\dot{T} - \Gamma\,\dot{p}\right) dy. \quad (44)$$

The overbar denotes a mean value of the quantity in the layer thickness. To leading order, the solidifying layer thickness is

$$\frac{d}{dx}\left(\overline{u}_s^{(0)} f^{(0)}\right) = \frac{\eta\Delta T}{\delta\mathcal{L}\left[1+\eta f^{(0)}\right]}. \quad (45)$$

Equations (41) and (45) give an ordinary differential equation for the layer thickness, $f^{(0)}(x)$. Solution of this equation requires numerical quadrature. However, in certain cases, analytical expressions can be found.

4 RESULTS

4.1 *Solidifying Layer Thickness*

Equation (45) becomes

$$\left[f^{(0)} - \eta^{-1}\right]\frac{dq}{dX} = 1 \quad (46)$$

using the new variables

$$X = \left(\frac{\Delta T}{\rho\delta\mathcal{L}}\right)x, \qquad q = \frac{\overline{u}_s^{(0)} f^{(0)}}{\rho}. \quad (47)$$

The quantity q can be expressed as

$$q = \frac{f^{(0)}}{1 - af^{(0)}}\left\{1 + \left[S^* + \lambda\left(\tan\theta - \varepsilon f^{(0)}_{,X}\right) f^{(0)}\right]\left(1 - f^{(0)}\right)^3\right\}, \quad (48)$$

where the new parameters

$$S^* = \frac{1}{3}\frac{\widehat{\rho}_l}{\widehat{\rho}_s}\frac{\widehat{\tau}_o S_x \widehat{h}}{\widehat{\mu}_l\widehat{u}_1}, \qquad \varepsilon = \frac{\widehat{\rho}_l}{\widehat{\rho}_s}\frac{\widehat{k}_s\left(\widehat{T}_m - \widehat{T}_w\right)}{\widehat{\rho}_s\widehat{\mathcal{L}}\widehat{u}_1\widehat{h}},$$

$$\lambda = \frac{1}{3}(\rho-1)\cos\theta\frac{\widehat{\rho}_l\widehat{g}\widehat{h}^2}{\widehat{\mu}_l\widehat{u}_1}, \qquad a = \rho - 1,$$

have been defined. S^* characterizes the friction between the slab and the belt, and λ characterizes the ratio of the gravity and viscous forces.

The characteristic length for the solidifying layer can be obtained by taking X to be $O(1)$. This implies

$$\widehat{x} = O\left(\frac{\rho\delta\mathcal{L}}{\Delta T}\widehat{L}\right) = O\left(\varepsilon^{-1}\widehat{h}\right). \quad (49)$$

From the values in Table 1, $\varepsilon \cong 0.02$, and the characteristic length is approximately 0.5 m.

The caster slope is generally 5° or more, and $\tan\theta > 0.1$. Since $\varepsilon \cong 0.02$, the term involving the derivative of f can be neglected in (48). A closed form solution to (46) can be found.

$$X = \left[f + \eta^{-1}\right] q(f) - \int_0^f q(\xi)\, d\xi, \quad (50)$$

where $q(f)$ is given in (48) neglecting the term $\varepsilon f_{,X}$. The superscript (0) has been omitted for convenience.

When the belt friction and caster slope are small, terms involving S^* and $\lambda\tan\theta$ can be neglected in the expression for q. This is equivalent to having a large casting speed or large liquid viscosity. The solidification front is found to be

$$X = a^{-2}\left\{\log(1+af) - \frac{af\left(1 - a\eta^{-1}\right)}{1+af}\right\}. \quad (51)$$

The more general result can be found in Smelser and Johnson (1995). If the density ratio is nearly unity, then $a \cong 0$, and (51) has a very simple form

$$X \simeq f\left[\frac{1}{2}f + \eta^{-1}\right] + O(a) \Rightarrow f \simeq \sqrt{2X + \eta^{-2}} - \eta^{-1}. \quad (52)$$

For perfect thermal contact, η is large and $T_s(y=0) = T_w$. This gives the particularly simple expression

$$f \simeq \sqrt{2X} \quad (53)$$

for the solid layer profile.

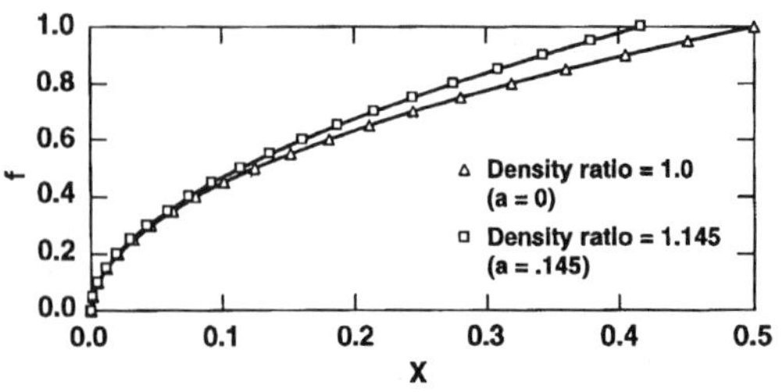

Figure 2: Solid layer profile vs. X for $\eta^{-1} = 0$, vanishing slope and belt friction

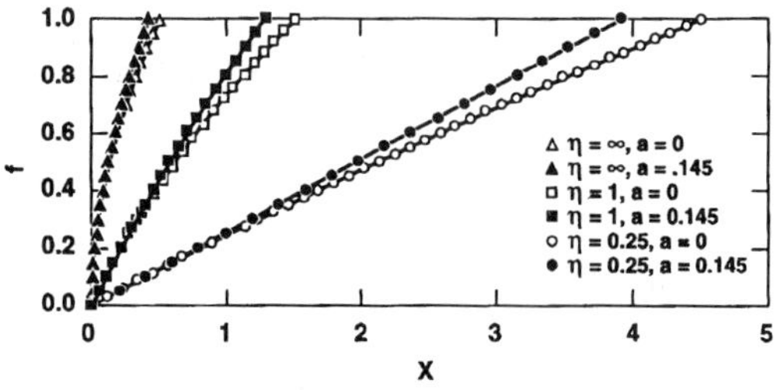

Figure 3: Solid layer profile vs. X for a range of η, two values of a, vanishing slope and belt friction

The solidified layer profile given in (51) and (53) is shown in Figure 2 for perfect thermal contact, $\eta^{-1} = 0$. The distance to full solidification is found when $f = 1$. From (53), $X = 1/2$ or $\widehat{x} = 0.5\varepsilon^{-1}\widehat{h}$. For the density ratio $\rho = 1.145$, the solidification length decreases to $0.417\varepsilon^{-1}\widehat{h}$. The solidification length increases linearly with η^{-1}. Figure 3 shows this effect. As the thermal contact lessens the solidification length increases rapidly.

4.2 *Slab dilatation*

The density of a material particle changes from $\widehat{\rho}_l$ to $\widehat{\rho}_{so}$ as the particle crosses the solidification boundary. Further changes occur in response to the temperature and pressure field in the solid layer. The dilatation in the slab and hence the belt position are related to these changes. The standard relation between density and dilatation integrated along a pathline in the slab gives

$$\log\left(\frac{\widehat{\rho}_s}{\widehat{\rho}_{so}}\right) = -3\alpha\left[(T_s - T_m) - \Gamma\left(p_s - p_f\right)\right]. \quad (54)$$

Here, p_f is the pressure where the particle pathline crosses the solidification front.

If the volume change is assumed to be isotropic, the displacement at the belt surface can be approximated. The specific volume is given by

$$\frac{\Delta\mathcal{V}}{\Delta\mathcal{V}_0} = \frac{\Delta X \Delta Y}{\Delta x \Delta y} = \left(\frac{\Delta Y}{\Delta y}\right)^2 = \left(\frac{\widehat{\rho}_l}{\widehat{\rho}_s}\right)^2 \quad (55)$$

where (X, Y) is a particle position. The displacement of the belt surface can be obtained from (54) and (55) and noting that a is small.

$$Y = f - \int_0^f \sqrt{\frac{\widehat{\rho}_l}{\widehat{\rho}_{so}}}\left\{1 + \frac{3}{2}\alpha\left[(T_s - T_m) - \Gamma\left(p_s - p_f\right)\right] + \cdots\right\} dy \quad (56)$$

An expansion for the exponential function and the binomial expansion have been used in obtaining (56).

Substituting for the temperature in the solid layer, (26), (56) yields

$$Y = \left(1 - \sqrt{\frac{\widehat{\rho}_l}{\widehat{\rho}_{so}}}\right) f + \frac{3}{2}\alpha\sqrt{\frac{\widehat{\rho}_l}{\widehat{\rho}_{so}}} \left\{\frac{1}{2}f\frac{\eta f}{1+\eta f}\left(T_m - T_w\right) + \Gamma\int_0^f \left(p_s - p_f\right) dy\right\}. \quad (57)$$

When the slope of the caster, the belt friction and $1 - \rho$ are small, the integral can be neglected (Smelser and Johnson, 1995). The dimensional displacement is

$$\widehat{Y} = \widehat{h}\left\{1 - \sqrt{\frac{\widehat{\rho}_l}{\widehat{\rho}_{so}}}\left[1 - \frac{3}{4}\widehat{\alpha}\frac{\eta f}{1+\eta f}\left(\widehat{T}_m - \widehat{T}_w\right)\right]\right\} f. \quad (58)$$

The slab dilatation is proportional to the solid layer thickness. The maximum shrinkage can be found by setting f equal to 1.

5 CONCLUSIONS

An asymptotic model for slab casting has been developed. The asymptotic formulation systematically reduces the governing equations. The reduced set of equations was solved analytically. The analytical results give closed form solutions for the temperature field, the solid layer profile and the slab dilatation. These expressions can be used by designers for preliminary design estimates, process control algorithms and verification of numerical simulations.

REFERENCES

Farouk, B., Y. G. Kim & D. Apelian 1991. Numerical study on the solidification processes in a twin-belt caster. *Numer. Heat Transf., Part A* 20:375-393.

Farouk, B., D. Apelian & Y. G. Kim 1992. A numerical and experimental study of the solidification rate in a twin-belt caster. *Metall. Trans. B* 23B:477-492.

Smelser, R. E. & R. E. Johnson 1995. An asymptotic model of slab casting. *Int. J. Mech. Sci.* to appear.

Simulation of Materials Processing: Theory, Methods and Applications, Shen & Dawson (eds)
© 1995 Balkema, Rotterdam. ISBN 90 5410 553 4

Numerical simulation of flow of mushy/semi-solid alloy

Manabu Kiuchi & Suguhiro Fukushima
Institute of Industrial Science, The University of Tokyo, Japan

ABSTRACT

A series of numerical simulations of upsetting and extrusion of mashy (mushy/semi-solid) alloys were performed. The framework of the mathematical model used for this study was proposed by S. Toyoshima. The model was extended and transformed to the FEM program. By this program, deformation and flow features of solid and liquid components in mashy alloys were consistently simulated under various conditions.

The calculated results regarding forming pressure, residual liquid and solid components in product were compared with the experimental measurement and microscopic observation. The both results showed acceptable agreement. However, it was found that the constitutive equation for mashy alloys needs improvement for more precise simulation.

1. INTRODUCTION

In past ten years, mashy (mushy/semi-solid) metal forming technology has made great advance. Various processes, such as mashy extrusion, forging, rolling, die casting, injection moulding as well as rheocasting, thixomoulding and SCR-processing, have been investigated with aims to develop innovative technology for manufacturing metallic products. Some of them are now utilized for actual production. Many kinds of components of automobiles, electrical appliances and industrial machines are successfully manufactured by them. Mashy metal forming is now considered as a highly potential new technology for innovation of metal forming.

In order to promote further development and application of mashy metal forming technology, the study for building up its theoretical base is indispensable. Especially, the method for analysis of flow and deformation of mashy metals is essentially necessary.

In this paper, the mathematical model of mashy alloys (hereinafter, "alloy" is used instead of "metal") and the results of numerical simulation of their flow and deformation are presented.

2. MATHEMATICAL MODELING

In a mashy (mushy/semi-solid) alloy, there exist both of solid phase component and liquid phase component. The solid phase component usually exists as "solid alloy particles" and the liquid phase component exists as "molten alloy" among those particles (See Fig. 1).

Mechanical properties of the mashy alloy change greatly following to the change in its solid fraction as well as they depend upon the shape and size of solid alloy particles. Here, the solid fraction is defined as the weight percentage of the solid alloy particles. The mashy alloys are classified into three groups according to their levels of solid fraction as follows.

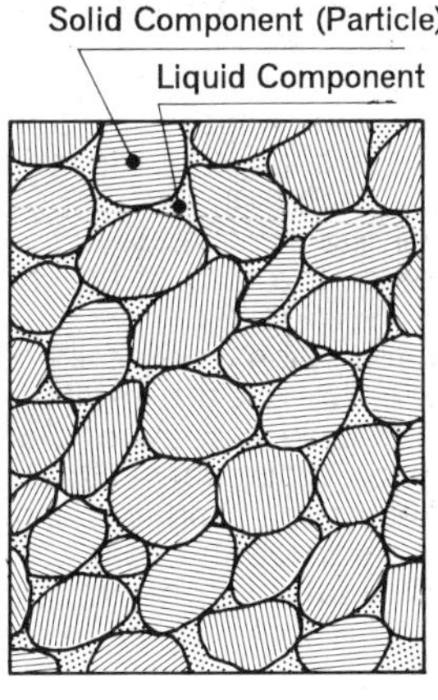

Fig.1 Schematic illustration of mashy (mushy / semi-solid)alloy.

(1) When the solid fraction is lower than roughly 40%, the mashy alloy has very low viscosity and flows like a molten alloy.
(2) When the solid fraction is higher than roughly 85%, the mashy alloy looks like a solid alloy. However, its flow-stress drops remarkably as its solid fraction falls from 100%.
(3) When the solid fraction is in the range from roughly 40% to 85%, the mashy alloy flows and deforms like slurry. It looks like a lump of sherbet or icecream.

In the following discussion, mashy alloys of which solid fraction is higher than 40% will be treated. In this solid fraction range, as mentioned above, they flow like slurry. This means that, in those mashy alloys, solid particles contact with each other and each particle's movement, rotation, slippage and deformation are more of less restricted by others.

The framework of the mathematical model employed for this study was proposed by S. Toyoshima [1]. The main assumptions of the model are described as follows.

(1) The mashy alloy consists of two kinds of components. One is the homogeneous solid porous alloy. Another is the molten alloy which occupies all of the pores included in the solid porous alloy (See Fig. 2).
(2) The solid porous alloy is a skeleton-type continuum, not composed of solid particles.
(3) All of the pores included in the solid porous alloy are linked up with others and compose channels passing through the solid porous alloy.
(4) The molten alloy can flow through those channels and move from one domain to others (See Fig. 3).
(5) The deformation of the solid porous alloy accompanies apparent volume change, that is the change in the volume of the pores (See Fig. 4).
(6) Following to the change in the volume of the pores., the flow of the molten alloy occurs (See Fig. 3).
(7) The deformation of the solid porous alloy and the flow of the molten alloy are dominated by the law of plasticity and the law of fluidity, respectively. However, their deformation and flow can be superposed upon each other. The superposed results represent the flow and deformation of the mashy alloy concerned.

According to the above assumptions, the following equations are introduced.

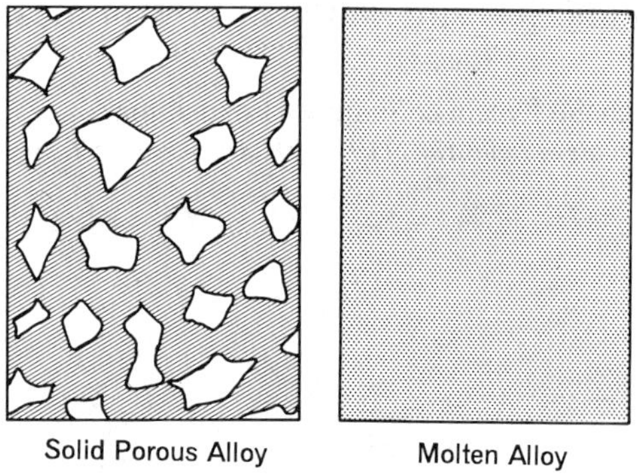

Fig.2 Schematic illustration of assumed solid porous alloy and molten alloy.

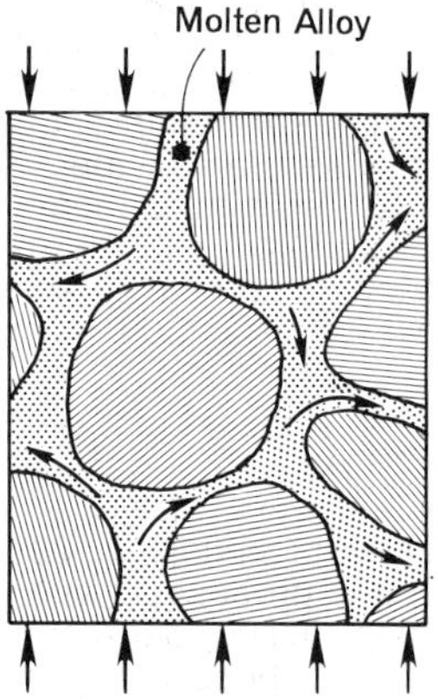

Fig.3 Schematic illustration of flow of molten alloy through channels.

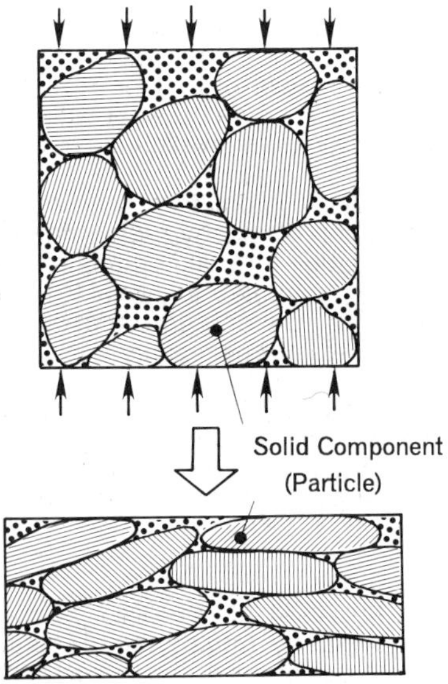

Fig.4 Volume reduction of pores

(1) Total stress acting on the mashy alloy is expressed as follows.

$$\sigma_{Tij} = \sigma_{ij} + \delta_{ij} \bullet P \qquad \text{-----(1)}$$

Here,
σ_{ij} = stress acting on the solid porous alloy,
P = hydrostatic pressure acting on the molten alloy.

(2) As the yield criteria for the solid porous alloy, the yield criteria for general porous materials is introduced. The yield function of the porous alloy is expressed by the next formula.

$$G = 3\,(A \bullet J'_2 + B \bullet J_1{}^2) - C \bullet Y_0{}^2 \qquad \text{-----(2)}$$

Here,
J_1, J_2' = invariants of stress tensor,
Y_0 = uniaxial flowstress of the solid alloy,
A, B, C, = alloy's constants.

(3) From Eq.(2), the following constitutive equation is derived.

$$\sigma_{ij} = 2\bar{\sigma}/3\dot{\bar{\varepsilon}}\ \{ (1/A)\dot{\varepsilon}'_{ij} + \delta_{ij}(1/18B)\dot{\varepsilon}_V \} \qquad \text{-----(3)}$$

Here,
$\bar{\sigma}$ = equivalent stress,
$\dot{\bar{\varepsilon}}$ = equivalent strain rate,
$\dot{\varepsilon}'_{ij}$ = deviatoric strain rate
$\dot{\varepsilon}_V$ = volumetric strain rate.

(4) The flow of the molten alloy in small channels passing through the solid porous alloy is dominated by so-called D'arcy's law. It is expressed by the next equation.

$$V_{xi} \bullet f_l = (K/\mu)\,(\partial P/\partial x_i) \qquad \text{-----(4)}$$

Here,
V_{xi} = relative velocity of the molten alloy to the solid porous alloy,
f_l = fraction of the molten alloy (= liquid fraction),
K = permeability coefficient,
μ = viscosity coefficient of the molten alloy,
x_i = coordinates, (= x, y, z)

(5) The permeability coefficient "K" is expressed by a function of the molten alloy fraction (liquid fraction) "f_l" as below.

$$K = K_0 \bullet (f_l)^2 \qquad \text{-----(5)}$$

Here, K_0 = base value of "K".

(6) As mentioned above, during deformation, the volume of pores changes. According to this, the molten alloy flows from some domains to others. The molten alloy is assumed to be incompressible, therefore, the change in the pores' volume should be equal to the volume of the molten alloy which goes out of or comes into the concerned pores. The balance of their volumes is expressed by the following continuation equation.

$$\partial U/\partial x_i + \partial (f_l \bullet V)/\partial x_i = 0 \qquad \text{-----(6)}$$

Here, U = velocity of the solid porous alloy.

(7) The total stress "σ_{Tij}" satisfies the equilibrium condition.

$$\partial \sigma_{Tij} / \partial x_i = 0 \qquad \text{-----(7)}$$

(8) From above equations, introducing the principle of virtual work and using Galerkin's method, the relationships among discretized stresses, strains, hydrostatic pressure and velocities are obtained. Then, Finite Element Method is introduced and an equation which expresses relationships between nodal velocity, nodal hydrostatic pressure, nodal force and nodal flow rate for one element is formulated. Such equations are superposed upon each other resulting into the following equation which covers all of the elements.

$$\begin{bmatrix} [K1(\hat{U})] & [K2] \\ [\ K3\ ] & [K4] \end{bmatrix} \begin{bmatrix} \{\hat{U}\} \\ \{\hat{P}\} \end{bmatrix} = \begin{bmatrix} \{\hat{F}\} \\ \{\hat{Q}\} \end{bmatrix} \qquad \text{--(8)}$$

Here,
$\{\hat{U}\}$ = velocity of the solid porous alloy at nodal points,
$\{\hat{P}\}$ = hydrostatic pressure acting on the molten alloy at nodal points,
$\{\hat{F}\}$ = force acting on nodal points,
$\{\hat{Q}\}$ = volumetric flow rate at nodal points.

Thus the total FEM equation for the whole domain of the mashy alloy is obtained. By introducing boundary conditions and using Newton-Raphson method, the total FEM equation for the deformation process concerned is formulated and solved. Through the procedure like this, the flow and deformation of the mashy alloy are analyzed.

3. SIMULATION OF UPSETTING

Uniaxial upsetting of cylindrical billets of mashy alloys was simulated by using the above mentioned mathematical model. The tested conditions are as follows.

The diameter and height of billets are φ12mm and 18mm, respectively. The solid fraction of billets was changed in the range from 55 (%) to 100 (%). The strain rate was also changed in the range from 0.5 (sec^{-1}) to 10^4 (sec^{-1}).

Fig. 5 shows some examples of the calculated results concerning the distribution of the solid fraction in the deformed billets when the reduction in height is 25 (%) and 50 (%). From these figures, it is noticed that when the mashy alloy billet deforms, the molten alloy flows out of the solid porous alloy. The molten alloy usually tends to flow from the domain under high hydrostatic pressure to the domain under low hydrostatic pressure. In this sense, under uniaxial upsetting, the molten alloy flows from the center portion of the billet to its side portion. Here, the billet's side surface is a free surface opened to the atmospheric pressure. In the figures, the bold lines show the outer surface configuration of the deformed billet and the contour lines of the solid fraction. The broken line shows the outer surface configuration of the deformed billet of which solid fraction is 100 (%). Therefore, the difference between bold line and broken line corresponds to the volume of the molten alloy which flows out of the billet.

It should be noticed that when the solid fraction of the billet is low and the strain rate, that is,speed of deformation is low, the discrepancy between molten alloy's flow and solid porous alloy's deformation becomes remarkable, and that the flowing out of the molten alloy from the mashy billet is promoted.

Furthermore, from the results, it can be known that even though the solid fraction of the mashy billet is low, if the strain rate is high, the separation of the molten alloy from the solid porous alloy is prevented. This results coincide with the knowledge obtained by experiments. This is a very important characteristic of the flow of the mashy alloy from a view point of application of mashy processings to real manufacturing.

Fig. 6 shows the comparison between predicted flow stress and measured flow stress of mashy billets subjected to uniaxial compression tests. The size of the tested billets is Φ12mm in diameter and 18mm in height. The solid fraction of billets is in the range from 55 (%) to 100(%). The strain rate is 0.5 (sec^{-1}). The abscissa shows the solid fraction of the mashy billet. The ordinate shows the non-dimentional flow stress which is defined as the

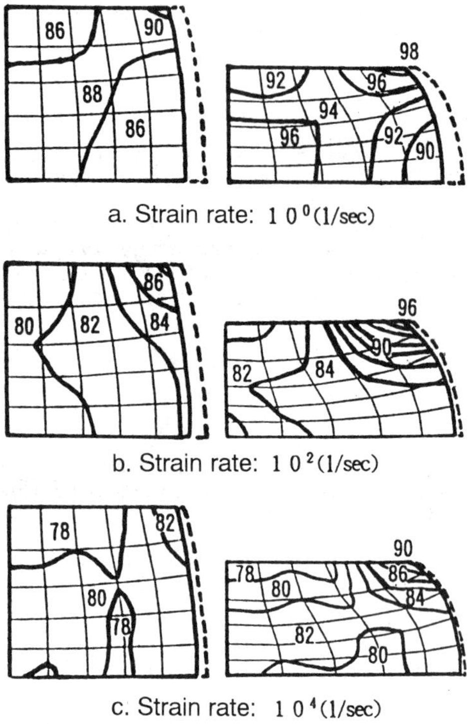

Fig.5 Predicted distribution of solid fraction in deformed billets.

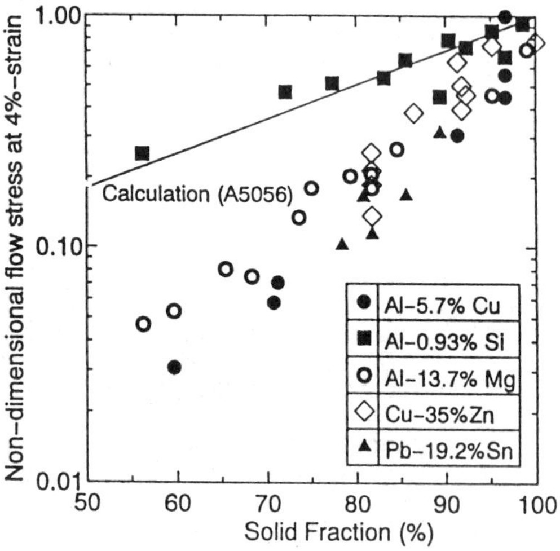

Fig.6 Relationship between flow stress and solid fraction under uniaxial compression.

ratio of the flow stress of the tested mashy billet to that of the billet with 100 (%) solid fraction at its solidus-line temperature.

In the figure, the plotted marks show the measured values of flow stress of several kinds of alloys and a bold line shows the theoretically predicted value. From the figure, the following facts are known.

(1) The flow stress of the mashy alloy decreases remarkably as its solid fraction diminishes.

(2) The decreasing rate of the flow stress of each tested mashy alloy except Al-0.93%S_i is almost same with those of others.

(3) The theoretically predicted flow stress and its decreasing rate agree with those of Al-0.93%S_i. But there is discrepancy between theoretical value and measured results concerning other mashy alloys.

The reason of the above mentioned discrepancy is considered as follows.

(1) In the mashy alloy, the solid component exists as particles, not as porous material. The solid particles can move, rotate and slip to each other rather easily due to existence of the molten alloy among them. As the solid fraction decreases, that means the quantity of the molten alloy among solid alloy particles increases, their movement, rotation and slippage become more and more easy. This results in the drop of the flow stress.

(2) For the employed mathematical model, the solid component of the mashy alloy is assumed to be the porous solid material. This means that all of the solid component in the mashy alloy is connected to each other. This results in that each portion of the solid porous alloy is restricted by others and it can not move, rotate and slip freely. Due to this, the mathematical model can not express so much drop of the flow stress as the real mashy alloy shows.

(3) The reason why the theoretically predicted flow stress shows good agreement with the measured flow stress of Al-0.93%S_i is explained as follows. The tested billets of Al-0.93%Si accidentally had very large grain size. Its grain size was roughly 100 times of others. This means that the solid particles in the mashy billet could not move, rotate and slip freely due to their largeness. They were strongly restricted each other as though they had been connected to each other as the mathematical model assumed.

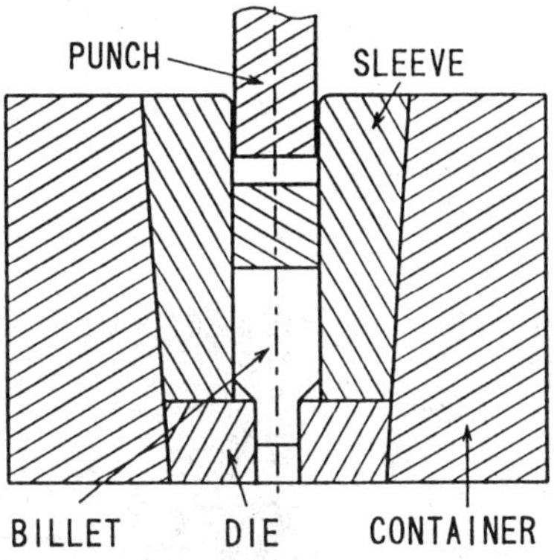

Fig.7 Installation of tool and billet for extrusion test.

4. SIMULATION OF EXTRUSION

The above mentioned simulation method was applied to a mashy extrusion process with aim to know the flow of the molten alloy, the change in the solid fraction of the billet and the distribution of the solid fraction in extruded product.

Fig. 7 shows a schematic illustration of installation of the tool and the billet. The round billets were extruded to round bars. The size of billet was Φ20 (mm) in diameter and 40 (mm) in length. The diameter of extruded bar was Φ10 (mm). The tested alloy was A5056. The punch speed was changed in the range from 5 (mm/sec) to 400 (mm/sec).

Before extrusion, the billet was heated up to the required mashy state and after the extrusion stopped, the extruded bar and the remaining billet were cooled rapidly by water in order to fix the internal structure which were observed afterward.

Fig. 8 shows the internal structures of the extruded bars and the remained billets. In these cases, the initial solid fraction of the billets was 60 (%) and the punch speed was 5 (mm/sec) and 400 (mm/sec) respectively. From the pictures, it can be known that when the punch speed is low, relatively large amount of the molten alloy flows out of the top end of the billet at the initial stage of the extrusion. This results in that the amount of the molten alloy remained in the extruded bar is small.

On the contrary, when the punch speed is high, the molten alloy can hardly flows out of the billet, therefore, almost all of the molten alloy included in the initial mashy billet remains in the extruded bar and the billet.

This is the speed effect on the flow of the molten alloy similar with that pointed out for uniaxial upsetting.

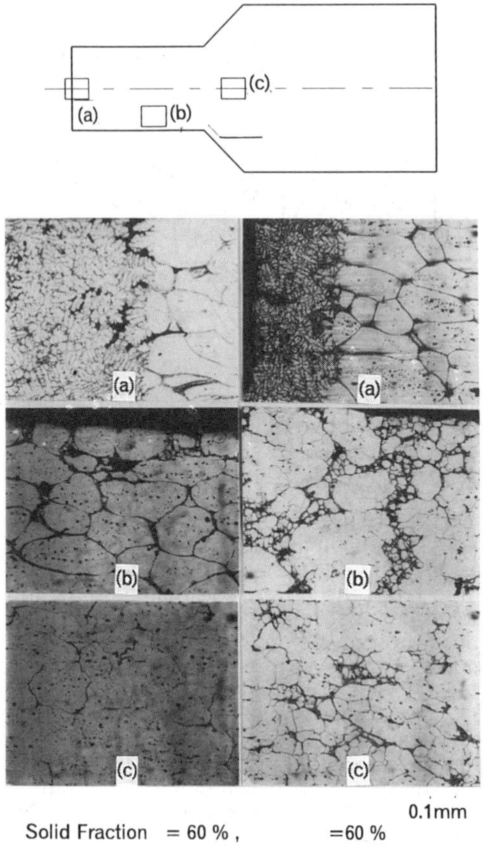

Solid Fraction = 60 % , =60 %

Punch Speed = 5 mm/sec , =400 mm/sec

Fig.8 Internal structures of extruded bar and remained billet.

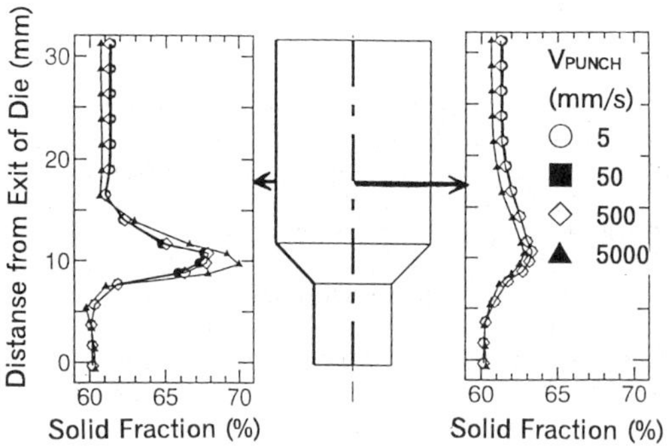

Fig.9 Predicted distribution of solid fraction in a workpiece under extrusion (extruded bar and remained billet)

Fig. 9 shows the calculated distribution of the solid fraction in the extruded bar and the remained billet. The distributions along their common center axis and outer surfaces are shown. The bold line and broken line correspond to different punch speeds. From the curves, the effect of the punch speed similar with those explained referring Fig. 8 are known.

5. CONCLUSION

The mathematical model proposed by S. Toyoshima for the simulation of flow and deformation of mashy alloys was extended and applied to uniaxial upsetting and extrusion. The results are summarized as follows.

(1) The employed mathematical model is usable for analysing general features of flow and deformation of mashy alloys.

(2) In order to analyse the flow and deformation of mashy alloys more precisely, better constitutive equations for solid particles are necessary.

(3) The mathematical model should be formulated referring the experimental measurement and observation.

REFERENCE

[1] Toyoshima, S. : A Numerical Simulation of Forming Processes for Semi-Solid Materials, ISIJ International, vol.31, no.6. (1991), 577.

[2] Masaki, S., Shima, S. : Equations for Deformation of Porous Materials, J. of Japan Society for Technology of Plasticity, vol.18, no.195 (1977), 243.

[3] Lalli, L.A. : A Model for Deformation and Segregation of Solid-Liquid Mixtures, Metall. Trans. A, 16A (1985), 1393.

[4] Kiuchi, M. : Flow Stress of Mashy Metals/Alloys, J. of Japan Society for Technology of Plasticity(J.S.T.P.), vol.17, no.186 (1976), 595.

[5] Kiuchi, M : Extrusion of Mashy Metals/Alloys, J. of J.S.T.P., vol.18, no.199 (1977), 633

Simulation of Materials Processing: Theory, Methods and Applications, Shen & Dawson (eds)

 ISBN 90 5410 553 4

Measurement of heat transfer coefficient in planar flow casting

Joseph Kukura, Kenneth Ford, Amar Singh & Paul H. Steen
Cornell University, Ithaca, N.Y., USA

Tetsuharu Ibaraki
Nippon Steel, Kimitsu, Japan

ABSTRACT: Molten aluminum is spin-cast into a long thin strip using a custom built planar-flow machine. Temperatures in the substrate wheel are measured as a function of time. A numerical solution of the heat balance equations (1D space-time model) is used to best fit the data with the heat transfer coefficient as the free parameter. Measurement technique, numerical model and heat transfer coefficient are reported.

1 BACKGROUND

A schematic of the process is shown in Figure 1. Molten aluminum is forced by overpressure from a nozzle into the gap (1 mm) between the nozzle face and a spinning wheel (195 cm circumference), on which it is quenched. The wheel is a CuBe alloy (0.4% Be) that acts as a heat sink but which, under present operation, is not actively cooled. The duration of a run is short enough that time-dependent thermal effects (e.g. heat up of the wheel) do not dominate the behavior but long enough so that a quasi-steady state of mechanical operation is achieved.

A typical charge of aluminum is 500 gm and is spun to a 10^{-1}mm thickness and 5.8 cm width. This corresponds to 32 m of ribbon. At 10 m/s the duration of a run is 3.2 seconds and for a liquid paquet undergoing solidification, contact with the heat sink within the puddle (length = 2 cm) is only 2 ms. Details of the apparatus, procedure and operability have been reported (Carpenter & Steen 1992, Ibaraki & Steen 1995, Ibaraki 1995).

The contact between ribbon and wheel is relatively poor because of air entrainment. This favors a linear rather than quadratic growth of the solidification front with time. Furthermore, the small superheat (100 K) causes undercooling to be insignificant.

A schematic of the puddle region is shown in Figure 2. The rate of rotation gives a linear circumferencial velocity U that carries the solidified ribbon of thickness Θ off the wheel. The puddle begins upstream at first contact with the wheel and extends until touchdown of the downstream meniscus. The puddle length L is typically 10-20 times the gap distance G between the nozzle face and wheel. The solidification rate can depend on position along the solidification front. Therefore the local rate V* is distinguished from the average rate V. For negligible undercooling, the length of solidification front can be identified with the puddle length.

For steady state operation the mass balance requires,

$$\Theta U = LV. \tag{1}$$

Equation 1 shows that direct measurement of L can be used to determine the solidification rate V since U is prescribed and the thickness (average) is easily measured. This is a 'mechanical' measure of solidification rate. Puddle length has been measured and the solidification rates so obtained have been reported (Ibaraki 1995).

A 'thermal' prediction of the solidification rate is possible, since for long thin puddles, the heat flux (vertical) is orthogonal to the momentum flux (horizontal), on average (Carpenter & Steen 1995). Stated differently, convection effects in the liquid puddle do not influence the heat transfer to a first approximation. Such 'thermal' predictions depend on the heat transfer coefficient for the aluminum/wheel contact. The coefficient includes the effects of conduction and any radiation of heat, as well as influences of non-ideal contacting such as air entrainment. Evidence suggests that coefficients vary with spinning conditions (e.g. air versus vacuum environment) from laboratory to laboratory. Various methods for inferring the heat-transfer coefficient exist. None are direct. They use photocalorimetric or pyrometric measurements of surface temperatures, or microstructural analysis of frozen product (Muhlbach et al. 1987, Davies 1985). Although recent work shows reasonable accuracy (Ludwig & Frommeyer 1992), measurements often have an order of magnitude uncertainty (Davies et al. 1981). Bounds have been proposed (Carpenter & Steen 1990).

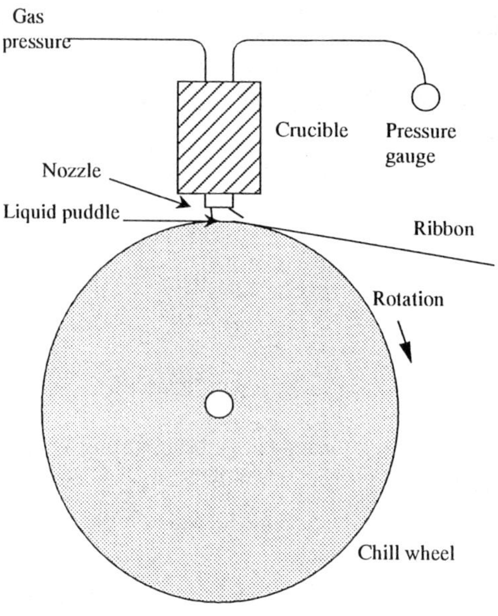

Fig. 1 Process schematic.

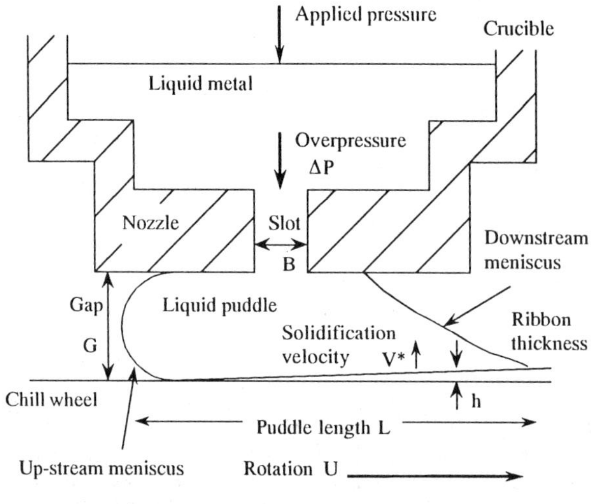

Fig. 2 Schematic of puddle region.

2 EXPERIMENT

Thermocouples have been embedded at various locations in the chill wheel. A sealed mercury-contact slip-ring takes the signals from the wheel to the data acquisition board. We report the temperature along the mid-circumference of the wheel at a depth of 2 mm from the surface of contact. Under present operation, several channels serve to corroborate the one reported signal.

Figure 3 shows the trace of temperature (C) versus time at 300 Hz. The molten metal first contacts the wheel at about 2 seconds. The trace up to that time shows that the measurement is accurate to ± 1 degree. The temperature rises sharply while the thermocouple is under the puddle and then more gradually decreases during the rest of the spin. The charge of metal is exhausted in 12 spins. Note that 4 of the last 5 spins indicate smaller temperature rises. These may be due to shorter contact time (shorter puddle) or poorer contact as can occur near the end of the run as the hydrostatic head is lost. After the last spin, there follows a transient cooling of the wheel. We match a numerical prediction

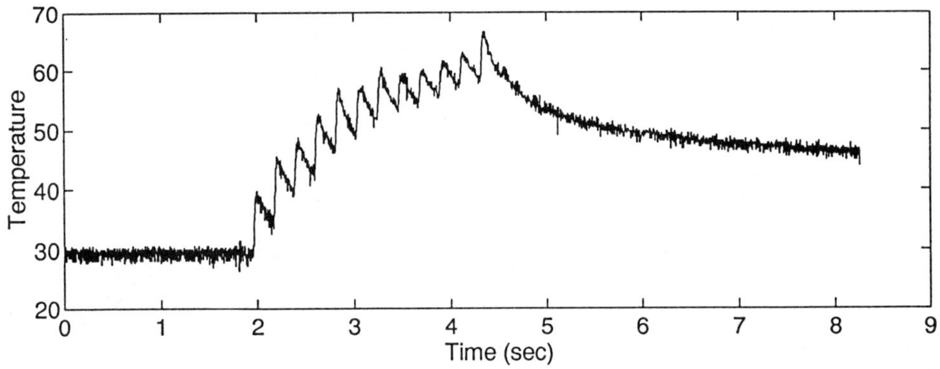

Fig. 3 Temperature (C) versus time(s) at a depth of 2 mm from the wheel contact surface.

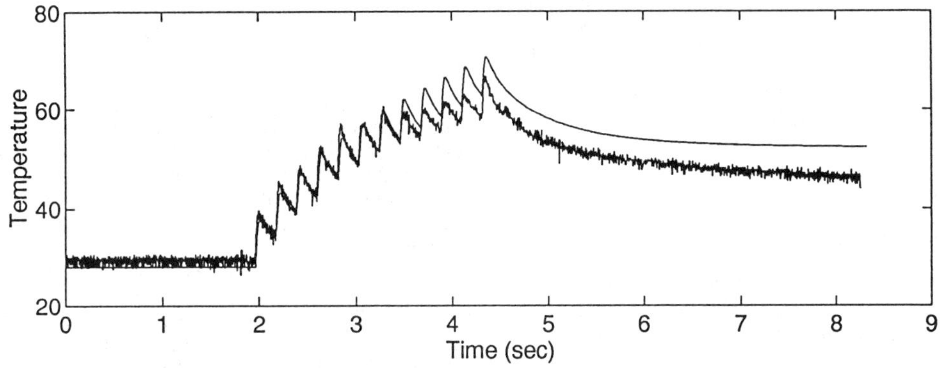

Fig. 4 Comparison of the measurement (Fig. 3) and simulation of temperature at the same depth.

with this signal using the heat transfer coefficient as a parameter.

3 HEAT TRANSFER

For solidification to occur, the superheat of the liquid metal as well as the latent heat of phase change must be conducted through the already solidified ribbon, transferred across the interface of contact, and conducted away into the wheel sink. For our conditions, the superheat represents about 25% of the latent heat. For $H = 3 \times 10^5\ W/m^2K$, the limiting resistance occurs at the interface although for an H ten times as large, the limiting step shifts to the conduction away through the wheel.

4 NUMERICAL MODEL

Heat transfer in the ribbon and wheel is governed by the heat equation. The thinness of the puddle region makes the dominant balance (steady state in the laboratory frame) one between convection in the horizontal direction and conduction in the direction normal to the wheel (y direction). In a frame translating with the wheel and ignoring curvative effects, the equations in the bulk become

$$\partial T / \partial t = \alpha \, \partial^2 T / \partial y^2 \tag{2}$$

where it is convenient to interpret these as *dimensionless* equations where α represents scaled thermal diffusivities. Scales for time and temperature are chosen as $ts \equiv U / L$ and the temperature difference between melting and ambient, $(T_m\text{-}T_a)$, respectively. The length scale δ is the ribbon thickness scale determined by balancing the energy flux due to solidification at a velocity δ/ts with heat conduction across the ribbon held between temperature T_m and T_a.

$$\delta^2 = ts\ k_r (T_m - T_a) / \rho \Delta \eta$$

When the wheel is under the puddle the following boundary conditions are imposed.

$$y = -d/\delta, \quad \partial T/\partial y = h_a(T - T_a)$$

$$y = 0, \quad -(\partial T/\partial y)^+ = B(T^+ - T^-)$$

$$-(\partial T/\partial y)^+ = k(\partial T/\partial y)^-$$

$$y = h(t), \quad T = 1$$

$$\partial T/\partial y = (1 + S)\, dh/dt$$

Here, h(t) is the unknown thickness of the ribbon, $h_a \equiv H_a \delta / k_w$, $B \equiv H\delta / k_w$, $k \equiv k_w/k_r$ and $S \equiv \Delta\eta / C_p(T_h - T_m)$. In these expressions, H_a is convective heat transfer coefficient to the air, k_w and k_r are the conductivity of the wheel and ribbon, respectively, $\Delta\eta$ is the latent heat/mass of solidification, c_p the specific heat of the molten phase; and T_h the temperature of superheat.

Once the wheel emerges from under the puddle (t=1), the ribbon is no longer in contact. For subsequent times equation (2) is solved in the wheel with convection boundary conditions applied at both the inside and contact surfaces. The same coefficient h_a is used for both. The appropriate initial condition is uniform temperature in the wheel (T_a) and no ribbon (h(0)=0).

5 SOLUTION

There have been numerous numerical treatments of heat transfer in planar flow casting (e.g. Ludwig et al. 1990, Wang & Matthys 1991, 1992). Our treatment is distinguished by its simplicity based on a recognition of conditions appropriate to our machine and operation.

The thermal response time of the solidified ribbon is so fast, due to its thinness, that assuming steady conduction in that region is an excellent approximation. With this simplification, the model system is solved using a standard finite-difference scheme. The parameters used are shown in the Table. The first group are measured from the experiment (Figure 3) with the exception of ts which is based on L measured by high-speed cinematography in previous runs under similar conditions. The second group are material properties taken from specification sheets (handbook). δ depends on the first and second groups (see its definition). H_a is estimated based on correlations of heat transfer in turbulent boundary layers on flat plates.

Figure 4 shows the simulation corresponding to H=1.1 x 10^5 W/m^{2}K (solid line). This value is a best fit by eye to the first 1.5 seconds of the run. The corresponding solidification front shows linear growth to a good approximation.

6 DISCUSSION AND SUMMARY

Features of Figure 4 are of note. First, the tail of the simulation has smaller slope and greater temperature than the measurements. Since the slope depends on the convective cooling, it suggests that the H_a used may not be appropriate. The higher temperature may also be due to a simulated heat input (during solidification) that is too high. Clearly the anomalous puddle behavior over the last 4 spins will influence this. On the other hand, the simulation may also be quite sensitive to the value of ts. More accurate values of ts and H_a are within reach.

Table 1. Parameter values.

T_h (C)	760
T_m	660
T_a	29
d (m)	13x10^{-3}
U (m/s)	10.8
ts (s)	2x10^{-3}
ρ (kg/m^3)	2.4x10^3
ρ_w	8.9x10^3
ρ_r	2.7x10^3
c_p (J/kg k)	1.1x10^3
c_{pw}	370
c_{pr}	840
k_w (W/m K)	84
k_r	220
δ (m)	5.4x10^{-4}
Ha (W/m^{2}K)	82
H	1.1x10^5

The 'best-fit' H is obtained by matching the thermal data to a numerical solution of the 'thermal' problem. This can be compared to a 'mechanical' determination using measured puddle length from experiments on the same apparatus. The result is H=1.7 x 10^5 W/m^{2}K (Ibaraki 1995). The agreement is reasonable. We know of no other study where H has been determined using temperature measurements in the substrate. Matching using surface temperatures for a different caster and alloy (FeSi) gives H=1.2 x 10^5 W/m^{2}K (Ludwig & Frommeyer 1992).

This contribution illustrates the measurement of heat-transfer coefficient by numerical solution of a thermal problem. More important than the value of H so-obtained (it is likely peculiar to our machine) is the method. Combining thermal and mechanical measurements are likely to result in measurements of H with greater accuracy than previously possible.

7 ACKNOWLEDGMENTS

The apparatus was designed and built by J.K. Carpenter, with modifications by E.C. Agger and G.M. Tierney. Original construction was supported by NSF grant MSM8711824 and recent experiments by CTS9024461. We thank ALCOA for supplying the aluminum. We thank Wenjie Hu for help with the computations.

REFERENCES

Carpenter, J.K. & P.H. Steen 1992. Planar-flow spin-casting of molten metals: process behavior. *J. Mat Sci.* 27:215.

Ibaraki, T & P.H. Steen (submitted). Behavior of the liquid puddle and operability of planar-flow melt-spinning. *Iron and Steel Institute of Japan. Int. J.*

Ibaraki, T. 1995. Planar-flow melt-spinning: experimental investigation of the liquid puddle and process operability. *Master thesis, Cornell University*.

Carpenter, J.K. & P.H. Steen (accepted). Heat-transfer and solidification in planar-flow melt-spinning. *Int. J. Heat Mass.*

Muhlbach, H., G. Stephani, R. Sellger, & H. Fielder 1987. *Int. J. Rapid Solidification*. 3:83-94.

Davies, H.A. 1985. In S. Steeb & H. Warlimont (eds), *Rapidly quenched metals V*: 101-106. Elsevier Science Publishers B.V., Amsterdam, Holland.

Ludwig, A. & G. Frommeyer 1992. Investigations on the heat transfer during PFC melt-spinning by on-line high-speed temperature measurements. In E.F. Matthys (ed), *Melt-spinning and strip casting: Research and implementation*.: 163-172. TMS.

Davies, H.A., N. Shohodi, & D.H. Warrington 1981. In R. Mehrabian, B.A. Kear & M. Cohen (eds), *Rapid solidification processing: principles and technologies II:* 153-164. Claitor's Publishing Division, Baton Rouge, LA.

Carpenter, J.K. & P.H. Steen 1990. On the heat transfer to the wheel in planar-flow melt spinning. *Met. Trans.* 21B:279-283.

Ludwig, A., G. Frommeyer & L. Granasy 1990. Modelling of crystal growth during the ribbon formation in planar flow casting. *Steel Research*. 61(10):467-471.

Wang, G.X. & E.F. Matthys 1991. Modeling of rapid solidification by melt spinning - effect of heat transfer in cooling substrate. *Mat. Sci. Eng. A* 136:85-97.

Wang, G.X. & E.F. Matthys 1992. Numerical modeling of phase change and heat transfer during rapid solidification: use of control volume integrals with element subdivision. *Int. J. Heat and Mass Transfer* 35(1):141.

Simulation of Materials Processing: Theory, Methods and Applications, Shen & Dawson (eds)
© 1995 Balkema, Rotterdam. ISBN 90 5410 553 4

Three-step finite element approach for centrifugal casting analysis

D. Lochegnies, C. Thibaud & J. Oudin
Mechanical Engineering Laboratory, URA CNRS, University of Valenciennes, France

ABSTRACT: We propose, for the first time, an original approach for a glass centrifugal casting analysis using finite-element viscoplastic models. In this context, three essential steps are proposed to the manufacturer when it is necessary to centrifuge a reference glass : a sensitivity analysis with decision-aid process design curves in order to quantify the capability of the glass to be centrifuged , an optimisation procedure for centrifuging standard geometry products made of this glass and finally a solution research in the case of centrifuging more elaborate products with this glass.

To illustrate our developments, we first study, via numerical analysis, the centrifugal casting of a full tore-shaped gob (this is dropped, with a perfect sticky contact, onto a mould in order to produce a plate) taking primary results on the forming of the NBS-710 glass into account. Then, we show an interesting application of our numerical approach for the centrifuging of a more elaborate product (a bowl) where the use of a distribution finger to drive the molten glass flow is necessary and analysed here.

1 INTRODUCTION

This paper presents a finite-element approach to the centrifugal casting of glass. This automated mass-production glass shaping process is used to form axisymmetric hollow glass items such as dishes, bowls or television picture tubes, where pressing involves, to obtain the same final shape, a larger amount of energy consumption and generates more mechanical stresses in the molten glass.

The present growing need for minimisation of production costs, generated by the increased competition, has lead glass manufacturers to devote special attention in the last few years to the numerical simulation of glass shaping processes: it allows to adapt the mould shape suggested by the designer to the most feasible form.

Henceforth, the behaviour of molten glass at a high temperature (above 500°C) is well known; previous researches were interested in the thermo-dependency of its viscosity [1], assumption limits at high strain rates of its Newtonian flow [2-3] or the homogenising of molten glass in the furnace [4]. Moreover some investigations have been performed to characterise the contact between the glass and a steel mould [5-6]. The interface between the two materials is not only dependent on glass viscosity; the metal surface smoothness and chemical composition are also influent. Next thermal exchanges at the interface were carried out with a time variable heat transfer coefficient which characterise the thermal properties of the gas film created between molten glass and mould micro flaws during the adhesive contact; adhesion of glass, on a heat resistant steel mould, occurred above an initial temperature of about 600°C.

As a result of these previous works on glass properties, numerical approaches, using a finite-element analysis, were more recently developed for blowing and/or pressing operation models [7-8]. Until now, available data for centrifugal casting were rather rare and only few papers refer to general information on the process [9]. In a recent paper [10], we put forward the efficient use of viscoplastic finite-element models for centrifugal casting analysis; primary results were found concerning the adequate range of 800 to 1200°C to centrifuging, the validity of adiabatic conditions for the analysis and the influence of the process parameters on the centrifugal casting success.

In this paper, we propose an original approach for centrifugal casting analysis using finite element viscoplastic models. Three essential steps are proposed to the manufacturer: the sensitivity analysis, the optimisation procedure and finally the solution research.

First of all, the sensitivity analysis permits to quantify, from a final reference disk, the influence of initial glass temperature, mould rotation speed and process time, on the lengthening along the radius, for three different gob heights and radii. In this way, we propose decision aid process design curves for manufacture improvement.

In the second step, these process design curves are used to optimise the generation of a NBS-710 plate 240 mm in diameter and 4.65 mm thick.

Finally, the third step of the approach is linked to difficult casting cases or manufacture problems where the finite element tool constitutes an expert approach to obtain solutions of forming. For instance, using previous information, we obtain the successful centrifuging of a bowl using a distribution finger. In this case, the finger geometry and displacement are defined taking the glass thickness evolution and stresses on the finger during centrifugal casting into account.

2 MODEL FORMULATION

In this paper, all products are made of NBS-710 standard soda-lime glass (this is principally composed of 70.5 wt % of silica, 11.6 wt % of oxide of calcium, 8.7 wt % of oxide of sodium and 7.7 wt % of oxide of potassium; this glass represents 90 % of the world production).

The following description considers standard manufacturing situations where the parison - an unshaped mass of molten glass before it is moulded into its final form - is supposed to fall into the centre of an AISI 316 steel mould (the AISI 316 is heat resistant, used for mould material because it allows to obtain a stable and reproducible forming surface to the glass, even if higher conditioning temperatures up to 600°C are found during casting); in this case, centrifugal casting resembles an axisymmetric process, accomplished in industry via a mould rotation speed of 400 to 800 rpm and a temperature range from 800 to 1100°C.

Centrifuging modelling will be analysed according to NBS-710 soda-lime glass properties. The glass is considered a homogeneous body at high temperatures (above 800°C). It is macroscopically isotropic because of its amorphous state. The NBS-710 molten glass stress/strain-rate relationship is represented by a Newtonian incompressible viscoplastic law in terms of the equivalent Von Mises stress $\overline{\sigma}$ and the equivalent strain rate $\dot{\varepsilon}$ as follows

$$\overline{\sigma} = 3\eta(\theta)\,\dot{\varepsilon} \tag{1}$$

where $\eta(\theta)$ is the viscosity of the NBS-710 soda-lime glass. It is expressed in MPa.s, in the logarithmic form, from the Vogel-Fulcher-Tamman (VFT) equation [1], in the temperature range of 500 to 1400°C, by

$$\log(\eta(\theta)) = -8.655 + \frac{4266}{(\theta - 264.5°C)} \tag{2}$$

For a given temperature, the model is valid if the glass strain rate does not exceed the critical strain-rate value for the onset of non-Newtonian flow (Table 1): this condition is observed for mould rotation speeds and glass temperatures used in standard industrial centrifuging. This means that viscous flow is predominant and thus the elastic effects can be discarded.

Tools are modelled as rigid bodies, on account of their important difference in rigidity with the molten glass.

The penalty function method is used to impose incompressibility in the finite element framework formulation with the updated Lagrangian formulation. 4-node and 3-node axisymmetric isoparametric elements, with an average size within 1 to 2 mm, are used for the mesh. Remeshing takes place during computation when the elements are too distorted.

Table 1. The critical strain rate for the onset of non-Newtonian flow as a function of the glass viscosity.

Temperature - °C -	Viscosity - MPa.s -	Critical strain rate for the onset of non-Newtonian flow - s^{-1} -
900	$1.1\ 10^{-2}$	36.4
990	$1.7\ 10^{-3}$	157.8
1080	$3.7\ 10^{-4}$	494.4

Other characteristics are the density ($\rho = 2.5\ 10^{-6}$ kg.mm^{-3}) of the NBS-710 glass and the acceleration due to gravity ($g = 9.81$ m.s^{-2}).

3 THE MOLTEN GLASS BEHAVIOUR DURING CENTRIFUGING

Before designing an item destined to be produced by centrifugal casting, it is important to quantify the influence of the main parameters such as initial molten glass temperature, mould rotation speed and process time.

3.1 *Description*

With this objective, the glass behaviour in centrifugal casting is analysed from an industrial product of reference: a NBS-710 standard soda-lime plate modelled with a disk 260 mm in diameter and 4 mm thick (these are average dimensions usually observed in commercial items).

Experiments have shown that the initial parison shape (which is the result of dropping a molten glass gob into the mould) is modelled accurately with a "pancake" shape (figure 5, Original mesh) with a width of about three times its height. To take different types of feeders into account, three similar sizes of parison are defined with the respective diameters and thicknesses: (80 mm; 56.82 mm), (100 mm, 30.84 mm) and (120 mm, 20.18 mm) all three are related to the same glass volume of 212.10^3 mm^3 to respect the required final volume, according to the following equation:

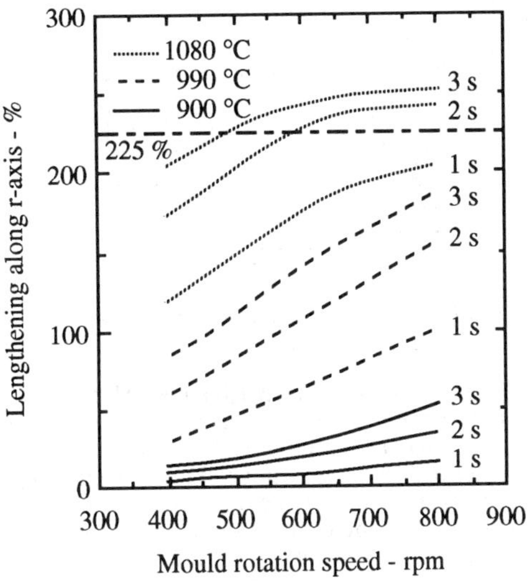

Figure 1. Mould rotation speed and temperature effects on the radius lengthening, from a parison 80 mm in diameter and 56.82 mm thick.

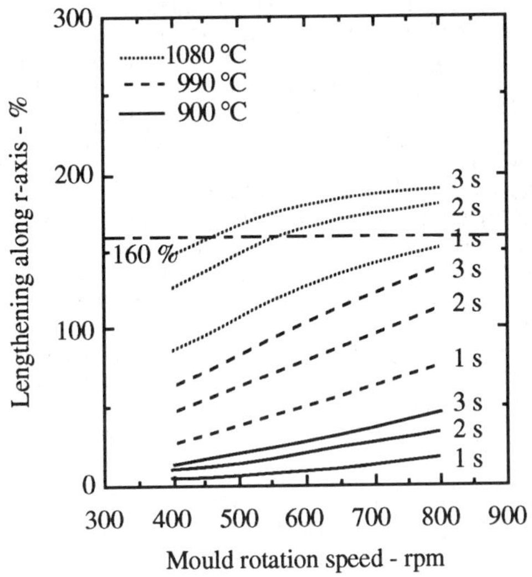

Figure 2. Mould rotation speed and temperature effects on the radius lengthening, from a parison 100 mm in diameter and 30.84 mm thick.

$$\frac{1}{8}\left(\frac{10}{3} - \pi\right)h^3 + \frac{1}{8}(\pi - 4)Dh^2 + \frac{1}{4}D^2h = 67600 \qquad (3)$$

where D is the diameter in mm of the full tore-shaped (i.e. the "pancake" shape) parison and h the height in mm.

Each parison is laid down on a steel mould and adheres perfectly: the mould must impart its rotation speed to the molten glass. According to industrial practices, the centrifugal glass casting process is carried out within a temperature range 800 to 1100°C, a mould rotation speed range from 400 to 800 rpm and a centrifuging time range from 1 to 3 s.

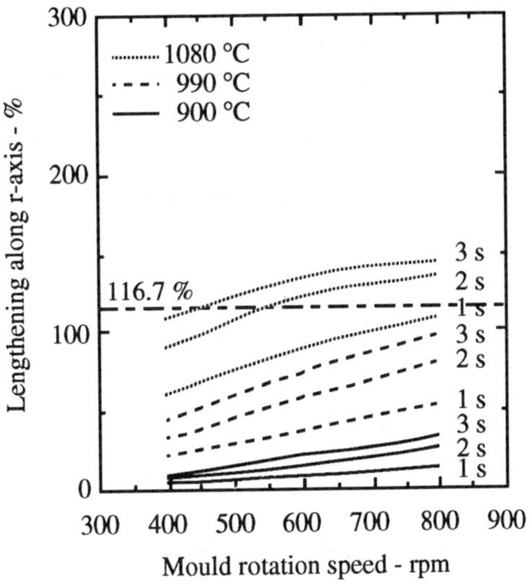

Figure 3. Mould rotation speed and temperature effects on the radius lengthening, from a parison 120 mm in diameter and 20.18 mm thick.

3.2 *Finite element results*

The finite element results obtained with viscoplastic models are represented in a chart with process design curves (figures 1, 2 and 3); beyond a radius of 130 mm (marked on figures 1, 2 and 3 with a discontinuous line) and corresponding to a thickness of 4 mm, a decrease in the lengthening occurs. Indeed, as expressed by the equation of volume conservation of the parison (3), when the diameter becomes important, and with the sticky contact participation, a fixed boundary layer of about 3 to 4 mm thick appears. A finite element thermal analysis, that we carry out on this film, underlines the previous remark: the steel thermal conduction (1.7 $W.m^{-1}.K^{-1}$) and the air thermal convection (30 $W.m^{-2}.K^{-1}$ including the participation of radiation) chill enough glass to harden it. Therefore, centrifugal casting seems to be inappropriate in the production of items thinner than 2.5 mm.

Moreover, process design curves show that to achieve the centrifugal casting of a full tore-shaped parison (to get a centrifuging time of less than 3 s), a minimal NBS-710 glass temperature of 990 °C is necessary, i.e. a viscosity of 1.7 10^{-3} MPa.s. For a desired lengthening and a technically feasible centrifuging time (generally below 3 seconds), the manufacturer will also find, with these curves,

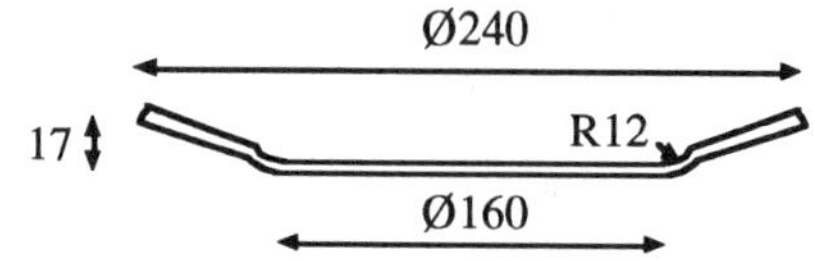

Figure 4. Dimensions of the plate to be centrifuged.

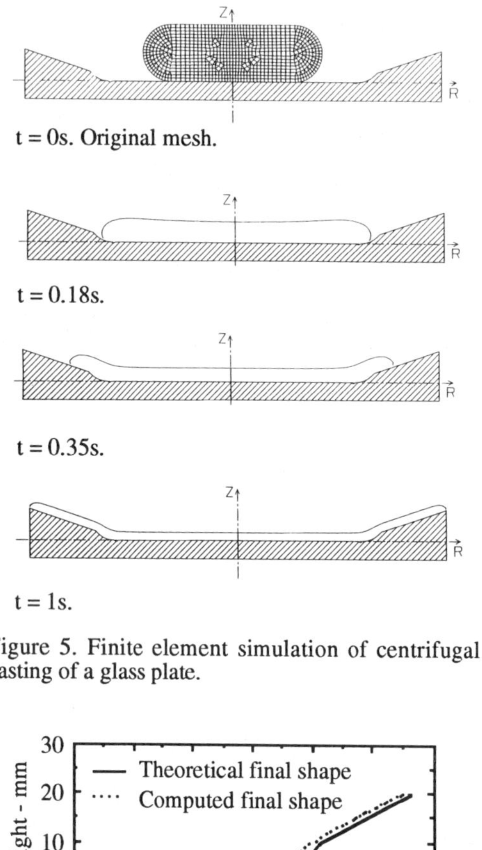

Figure 5. Finite element simulation of centrifugal casting of a glass plate.

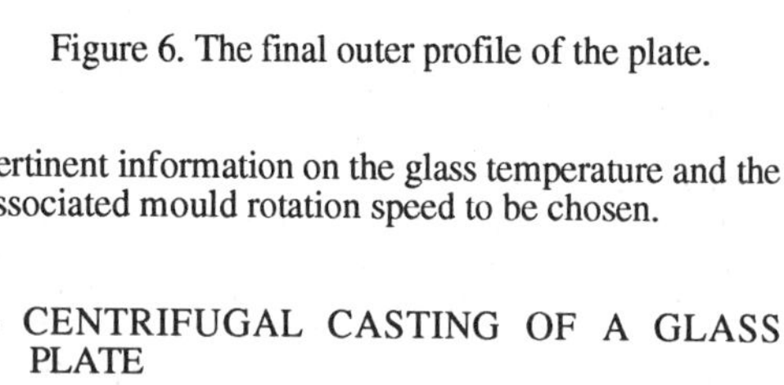

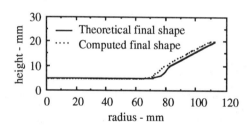

Figure 6. The final outer profile of the plate.

pertinent information on the glass temperature and the associated mould rotation speed to be chosen.

4 CENTRIFUGAL CASTING OF A GLASS PLATE

Finite element analysis of the centrifugal casting process is now applied for the production of a commercial item: a NBS-710 plate 240 mm in

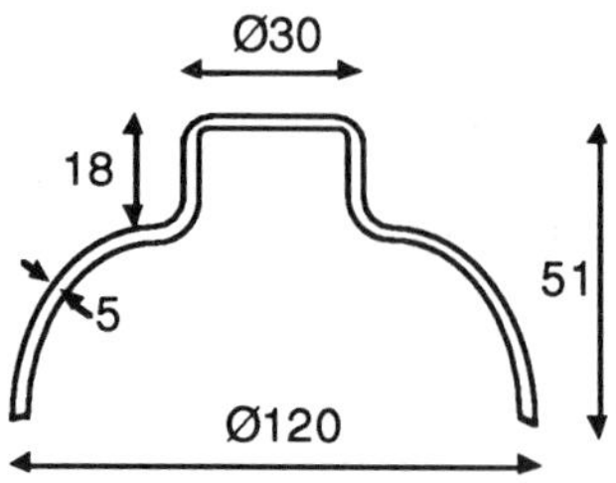

Figure 7. Dimensions of the bowl to be centrifuged.

diameter and 4.65 mm thick (figure 4). This plate has the same volume as the parisons used in part 3 (3).

At the start of the process, a parison 100 mm in diameter and 30.84 mm in height is dropped on the AISI 316 mould with previous sticking boundary conditions. The initial temperature is 1080°C to obtain a viscosity weak enough to print some motifs on plates at the end of the process.

For this parison, with a final lengthening of 140% and an initial temperature of 1080°C, the process design curves (figure 2) propose a process time varying from under 1s to about 2.8 s for a mould rotation speed range within 400 to 800 rpm. A time of 1s is taken to respect production requirements; this corresponds to a mould rotation speed of 700 rpm.

Figure 5 presents four steps in the shaping of the parison; the maximum computed strain rate (61.4 s^{-1}) validates the Newtonian character of the NBS-710 flow during the centrifugal casting (Table 1).

If, at the start of the process, the inner curvature of the glass surface becomes parabolic (figure 5, t = 0.18s) deviations from the parabola are achieved by a reduction in thickness. As shown in figure 6, the centrifugal casting process finally provides a plate with a quasi-homogeneous thickness (included in the margin range of ± 0.3 mm) without any tool to drive the molten glass flow. Deviation from the theoretical profile is predominant in the curvature but it remains within a reasonable margin; to avoid this, a solution would be to increase the radius of the curvature.

5 CENTRIFUGAL FORMING OF A GLASS BOWL WITH THE USE OF A DISTRIBUTION FINGER

If the glass suitability for centrifugal casting has been well mastered for the basic form of plate shaping, the analysis of centrifuging items with a more complex geometry is now conceivable.

In this way, we now analyse the centrifugal casting of a bowl (118 mm in diameter and 51 mm height) (figure 7); because of the great inner curvature of the mould, we must use a distribution finger to form the top of the bowl: this function corresponds to a localised puncher's role.

As previously, a 75.10^3 mm^3 parison, with an initial temperature of 1080°C, is dropped onto the mould with a perfect sticky contact.

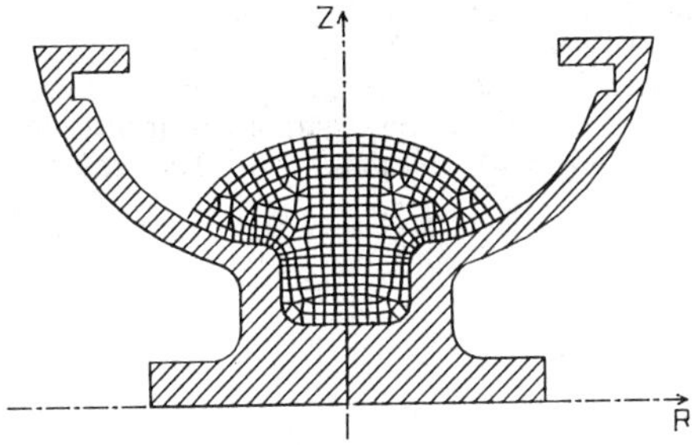

t = 0s.

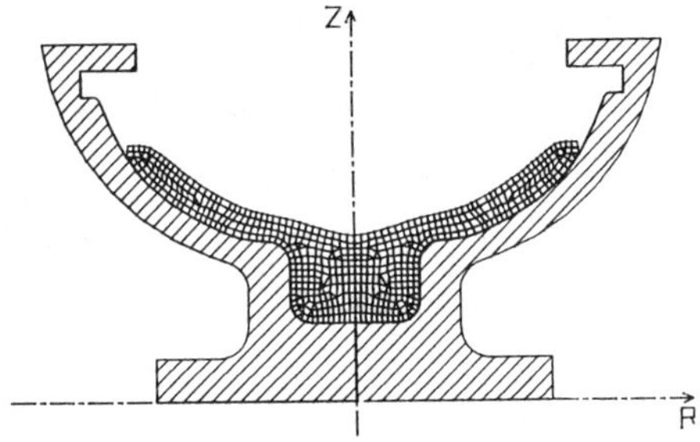

t = 0.82s.

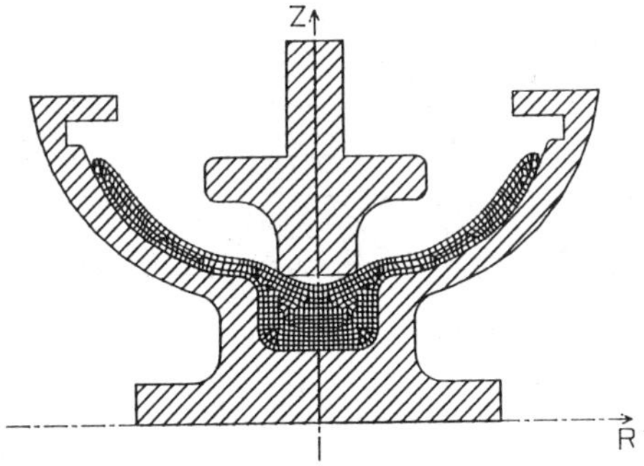

t = 1.44s.

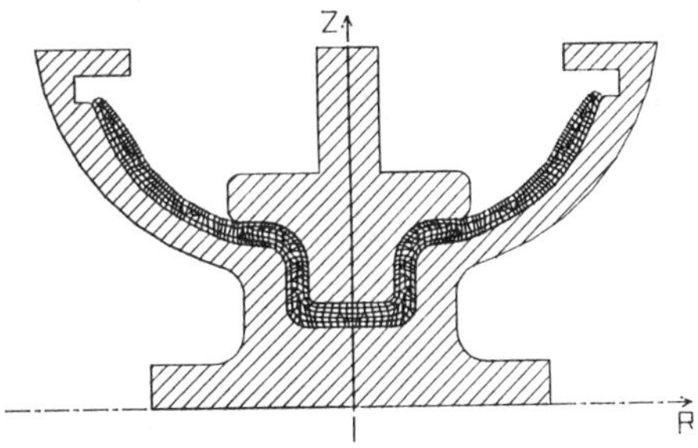

t = 2s.

Figure 8. Finite element simulation of a bowl forming.

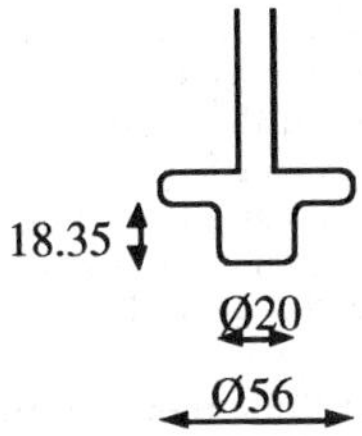

Figure 9. Main dimensions of the distribution finger.

We propose to produce two finite element viscoplastic models of the process:

- The first does not use a distribution finger; it allows to study only the glass flow to evaluate the accurate time when the puncher must act to carry out a successful centrifugal casting: we have to minimise the work of the tool to avoid an increase in stress in the molten glass. Thermal stresses are sufficiently high to try to prevent mechanical ones.
- The second model with the intervention of the distribution finger confirms the time chosen to start the finger forming, considering the final thickness to be obtained. This model also gives also, through the analysis of different geometry and displacements of the puncher, all the process parameters.

In our case, the optimum situation was found, via a finite element analysis:

- a time to start the finger forming equal to 1.44 s,
- with a speed of descent equal to 24 $mm.s^{-1}$,
- and a geometry presented in figure 9 to avoid the formation, near the flange of the puncher, of a weak geometrical excrescence due to the molten glass forced back by pressing loads.

Different steps in the deformation are presented in figure 8: under centrifugal forces the molten glass creeps along the mould surface: gravitational forces and mould temperature allow to maintain the sticky contact.

We observe that the intervention of the puncher in the process holds up the flow, so the speed of spreading of the molten glass on the mould is reduced.

6 CONCLUSION

A new numerical approach is proposed and finite element viscoplastic models are developed for the analysis of the glass centrifugal casting.

(i) The suitability for centrifuging molten glass is studied first via process design curves obtained with the production of a simple reference plate ; these allow the manufacturer to adjust efficiently his future process parameters with regard to the desired final shape according to his prescribed radius lenghtening and/or his thickness.

(ii) Next, the centrifuging of the desired product with a more complex geometry is achieved with finite element analysis, thanks to previous results.

This allows to predict the final form, adjust and fix the casting parameter values according to the final prescribed geometry of the product.

(iii) In the case of centrifuging a more sophisticated glass product, the forming information is also found with numerical analysis, as is the use of a finger distribution to help the glass flow along the turning mould.

In both cases, the finite element tool constitutes an efficient expert approach to obtain the solutions of centrifuging.

7 ACKNOWLEDGEMENTS

The present research work has been supported by the CNRS, the 'Conseil Régional Nord-Pas de Calais' and the 'Ministère de l'Enseignement Supérieur et de la Recherche'; the authors gratefully acknowledge the support of these institutions.

REFERENCES

[1] Fulcher, G.S. 1925. Analysis of recent measurements of the viscosity of glasses. *J. Amer. Ceram. Soc.* 8/6:339-355.

[2] Simmons, J.H. & Simmons, C.J. 1989. Nonlinear viscous flow in glass forming. *Am. Ceram. Soc. Bulll.* 68/11:1949-1955.

[3] Simmons, J.H., Swiler, T.P. & Simmons, C.J. 1993. Studies of non-linear viscous flow in silicate glasses. *Proceedings of the 3rd International Conference on Advances in Fusion and Processing of Glass, Ceram. Trans.* 29:27-34.

[4] Cable, M. 1990. The principles of homogenizing of glass melts. *Glastechn. Ber.* 63K:308-321.

[5] Costa, P. 1972. Untersuchung der thermischen und geometrischen Kontaktverhältnisse zwischen Glas und Metall bei der Formgebung heißen, zähflüssigen Glases. *Glastechn. Ber.* 6:277-287.

[6] Lukishker, EH.M. & Flow, Z.C. 1984. Résistance thermique à l'interface verre-métal et échange thermique lors du formage du verre. *Fizika i Khimiya Stekla.* 10/5:630-634.

[7] Lochegnies, D. & Oudin, J. 1993. Simulations numériques par éléments finis des écoulements newtoniens: application au formage isotherme du verre. *Proceedings of the 11ème Congrès Français de Mécanique of Lille-Villeneuve d'Ascq (France).* 2:29-31.

[8] Williams, J.H., Owen, D.R.J. & Sa, J.M.A.C. 1986. The numerical modelling of glass forming processes. *Proceedings of the XIVth International Congress on Glass of New Delhi. Coll. Papers.* 1:138-145. Calcutta: Centr. Glass & Ceram. Res. Inst.

[9] Trier, W. 1989. Forming processes of hot glass. In O.V. Mazurin (ed.), *A survey of contemporary glass science and technology, XVth International Congress on Glass.* 1:494-515. Leningrad: "Nauka" Leningrad Branch.

[10] Lochegnies, D., Thibaud, C. & Oudin J. 1995. Centrifugal casting of glass plates: a finite-element analysis of process parameter influence. *Glastechn. Ber. Glass Science and Technology.* 68/1.

Simulation of Materials Processing: Theory, Methods and Applications, Shen & Dawson (eds)
© 1995 Balkema, Rotterdam. ISBN 90 5410 553 4

A realistic wall thermal boundary condition for simulating the injection molding of thermoplastics

O. Mal, L. Dheur, P. Pirotte, N. Van Rutten, A. Couniot & F. Dupret
Centre CESAME, Unité de Mécanique Appliquée, Université Catholique de Louvain, Louvain-la-Neuve, Belgium

ABSTRACT: During the filling stage of injection molding, a thermal shock between hot polymer and wall is continuously generated by fountain effect along the moving fluid-wall-atmosphere contact lines. We present a decomposition method that captures this phenomenon, and permits us to accurately simulate the filling of thermoplastic parts. Comparisons with experiments demonstrate the validity of our approach.

1 INTRODUCTION

It is widely accepted that the Hele Shaw approximation is appropriate to model the injection molding of thin parts [1-4]. While a main advantage of this approach is to provide a 2D pressure field, the temperature field remains 3D in non-isothermal situations. First applied to incompressible thermoplastics with a power law viscosity, this method was considerably improved by taking into account the fluid compressibility [4, 5], more realistic viscosity laws [6], a better energy balance, and more complex geometries [7-9]. Its range of applications has been extended to other materials such as thermosets or fiber filled compounds [10].

In the case of thermoplastic parts, thermal calculations remain however very difficult to perform, in view of the high temperature gradients generated by fountain flow in the vicinity of the moving contact lines. This effect takes its origin in the fact that flow fronts are fed by fastly moving hot fluid outgoing from the core of the cavity, while the steel walls are cooled down in order to accelerate the process. This paper presents a decomposition method that permits us to capture this effect, and to obtain very accurate predictions of experimental results.

2 PHYSICAL MODEL

2.1 Basic field equations

For the sake of simplicity, only planar shapes of constant thickness $2h$ will be considered, while thermal boundary conditions will be assumed to be symmetric with respect to the midsurface. The x, y and z symbols will denote the two in-plane and the out-of-plane coordinates, respectively. For a fluid of constant thermal conductivity k , thermal expansion coefficient α , compressibility coefficient γ and specific heat c_p, the mass and energy equations are written as

$$2h\rho_g(\gamma\frac{\partial p}{\partial t} - \alpha\frac{\partial \bar{T}}{\partial t}) = \frac{\partial}{\partial x}(S\frac{\partial p}{\partial x}) + \frac{\partial}{\partial y}(S\frac{\partial p}{\partial y}) , \quad (1)$$

$$\rho_g c_p\frac{DT}{Dt} - \alpha T\frac{Dp}{Dt} = k\frac{\partial^2 T}{\partial z^2} + \eta\dot{\gamma}^2 , \quad (2)$$

where $\frac{D}{Dt}$ denotes material derivation, $\dot{\gamma}$ stands for the shear rate, $\bar{T}$ is the gap-averaged temperature, while the fluidity S depends on the specific mass, the temperature profile, the pressure gradient, the thickness of the cavity and the rheology of the polymer. The specific mass ρ is approximated by a linear expansion

$$\rho = \rho_g(1 - \alpha(T - T_g) + \gamma p) , \quad (3)$$

where T_g and ρ_g are given constants.

2.2 Boundary conditions

Pressure boundary conditions are very simple when the influence of the runner and gate system can be estimated (which we will see to be the case in the present paper). A vanishing pressure is imposed along the flow front. The flow rate is imposed at the gate, and the pressure normal derivative is constrained to vanish along the lateral sides.

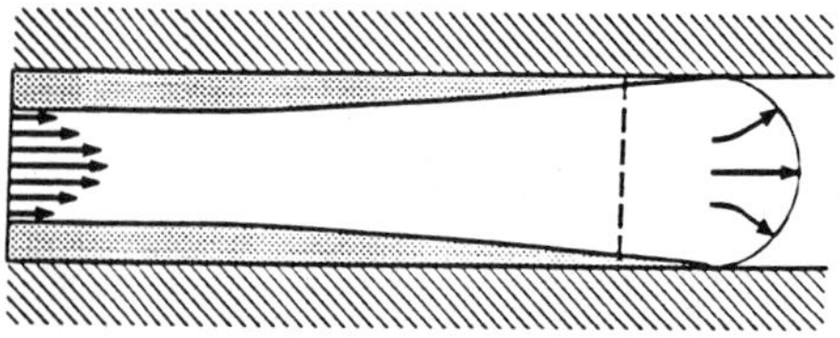

Figure 1 : Schematic view
of the flow in a cavity.

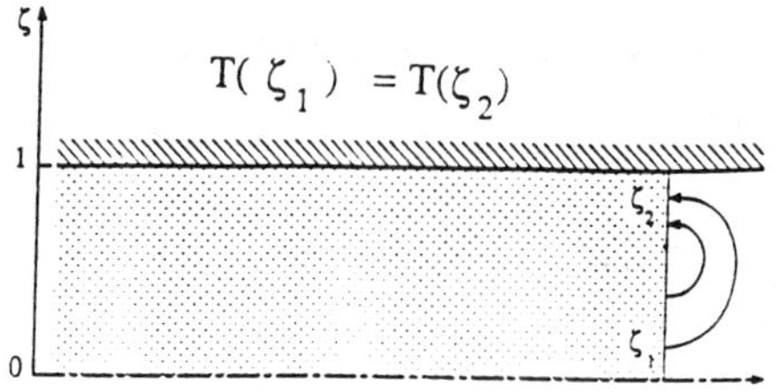

Figure 2 : Simplified model
of fountain flow.

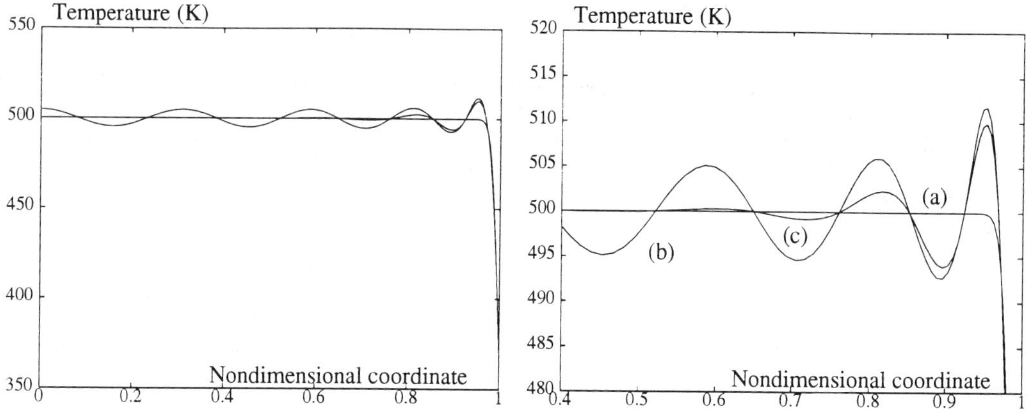

Figure 3 : Approximation of a steep temperature profile (curve a)
by means of Lagrange polynomials (b) and cubic splines (c).
Half thickness (left) and enlarged view (right).

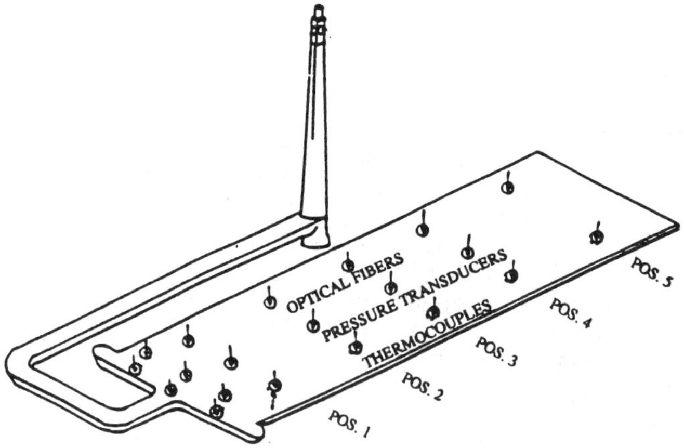

Figure 4 : Sketch of the injected plate
and location of the sensors (from the PEP).

Concerning thermal calculations, different conditions are imposed at the gate, along the flow front, and on the upper and lower walls of the cavity. No condition has to be imposed along the lateral sides of the mold, since in-plane heat conduction is neglected in the Hele Shaw model.

We impose a flat thermal profile at the gate, while temperatures are constrained by fountain effect along the flow front, according to the theory developed by Dupret and Vanderschuren [3]. Basically, the flow domain is divided in a Hele Shaw region and a front zone (Fig. 1). Upstream of the flow front, velocities are essentially parallel to the walls, while material points undergo a strong deflection towards the walls in the vicinity of the front. This effect, which implies a fast advection of heat from the core of the cavity to the walls, is called "fountain flow" [11]. The simplified model of fountain flow which is used in the present paper is illustrated in Fig. 2, where we make use of the nondimensional coordinate $\zeta = z/h$.

Finally, as thermal conduction is not negligible in the z-direction, heat exchanges between polymer and walls must also be modeled. We here introduce the concept of thermal penetration length (e), which represents the progression of heat inside the walls and varies with time. The temperature profile is assumed to be linear inside the walls. In addition, the thermal penetration length is maximized by the distance (e_{cool}) between the polymer-wall interface and the coolant fluid. According to Dheur [12], the equations governing heat transfer between polymer and wall are written as follows :

1. while $e < e_{cool}$:

$$\frac{\partial T_w}{\partial t}\frac{e}{4} + \frac{3}{2}\frac{k_w}{\rho_w c_{pw}}\frac{T_w - T_{cool}}{e} - \frac{1}{\rho_w c_{pw}} q_w = 0 \ , \quad (4)$$

$$\frac{\partial}{\partial t}((T_w - T_{cool})e^2) = 6\frac{k_w}{\rho_w c_{pw}}(T_w - T_{cool}) \ ; \quad (5)$$

2. from the instant $e = e_{cool}$:

$$\frac{\partial T_w}{\partial t}\frac{e_{cool}}{3} + \frac{k_w}{\rho_w c_{pw}}\frac{T_w - T_{cool}}{e_{cool}} - \frac{1}{\rho_w c_{pw}} q_w = 0 \ . \quad (6)$$

In these equations, T_w, T_{cool} and q_w denote the wall and coolant temperatures, and the heat flux entering the wall, while k_w, ρ_w and c_{pw} stand for the wall thermal conductivity, specific mass and specific heat. Both T_w and e are unknown.

3 NUMERICAL ALGORITHM

The algorithm we use to solve Eqs. (1) and (2) is based on a hybrid method combining Galerkin finite elements in the midsurface directions and Chebyshev collocation in the z-direction [3, 5, 9, 12,13]. At any time step, the temperature field prevailing at time t_{n+1} is first integrated with respect to the time on the temporary mesh covering the flow domain at time t_n, by means of a mixed explicit-implicit scheme. The SUPG technique is used in a version adapted to the lubrication model [14]. Next, the flow front is moved and the mesh associated with time step $(n+1)$ is created by means of a sophisticated remeshing technique [13]. The temperature field prevailing at time t_{n+1} is subsequently extrapolated on the new flow domain, and corrected in order to take the effect of fountain flow into account. The pressure field is integrated with respect to time by using an implicit Euler scheme.

Quadratic or bi-quadratic elements are used for approximating the pressure field, while linear or bi-linear elements are used for approximating the temperature field, which is discretized as follows :

$$T(x,y,z,t) = \sum_{i=1}^{N} \tilde{T}_i(z,t)\Phi_i(x,y) \ . \quad (7)$$

Here, the latin subscript i refers to a node of the temporary mesh covering the filled region of the midsurface. Along the z-direction, a cubic spline approximation is used, which is much more efficient than using Lagrange polynomials in order to damp out temperature oscillations in the gapwise direction (Fig.3). Cubic splines are characterized by a relation between temperatures and their second derivatives of the form

$$\sum_{\beta=1}^{m} A_{\alpha\beta}\frac{\partial^2 \tilde{T}_{i\beta}}{\partial \zeta^2} = \sum_{\beta=1}^{m} B_{\alpha\beta}\tilde{T}_{i\beta} + C_{i\alpha} \ , \quad (8)$$

where the greek subscripts refer to the m collocation points. The matrices A and B are tridiagonal, while the vector C depends on the boundary conditions imposed on the walls.

4 ANALYSIS OF THE THERMAL SHOCK AT THE FLOW FRONT

As material points experience fountain effect at the flow front, a thermal shock occurs when the hot core polymer enters in contact with the cold walls. In the injection of thermoplastics, temperature differences as high as 200 K between the polymer melt and the walls are often encountered. Hence, an accurate model of thermal shock is required when one wishes to correctly represent heat transfer in the vicinity of the flow front and the walls.

The thermal shock problem can be solved analytically in the case of an sudden contact oc-

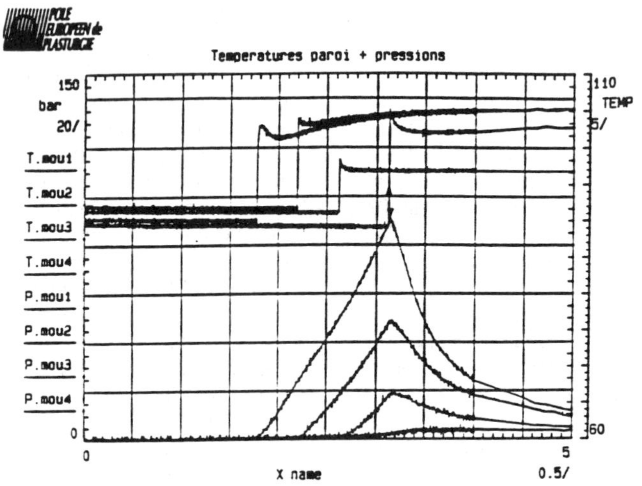

Figure 5 : Evolution of pressures and temperatures, as measured at different sensors of the instrumented plate (from the PEP).

Table 1 : Material properties of PS.

$\eta = \eta_0 e^{-\beta(T-T_0)}\dot{\gamma}^{n-1}$			$\rho = \rho_g(1 - \alpha(T - T_g) + \gamma p)$					
η_0	Pas	$1.2552\ 10^6$	ρ_g	kg/m^3	1024.8	c_p	$J/(kgK)$	2,162.4
β	$1/K$	0.0223	α	$1/K$	$5.8\ 10^{-4}$	k_p	$W/(mK)$	0.17
T_0	K	273	T_g	K	373			
n		0.2931	γ	$1/Pa$	$7.62\ 10^{-10}$			

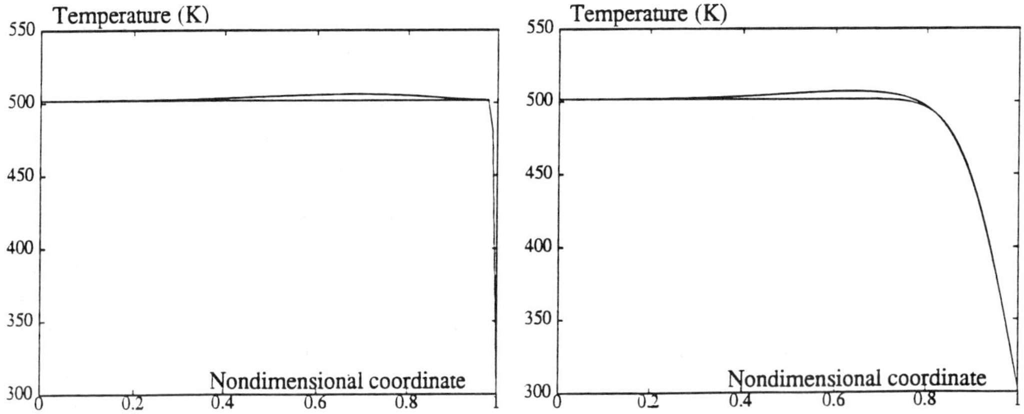

Figure 6 : Temperature profile at a given time and different locations in the part (the lower curves represent the analytical component).

curring between two semi-infinite motionless media (labeled 1 and 2) initially being at different temperatures (T_{10} and T_{20}). The 1D solution can be expressed in terms of the complementary error function $erfc(\xi) = 1 - erf(\xi)$, and depends only upon the thermal properties and initial temperatures of both materials. This solution shows that the interface temperature does not vary with time.

It is easy to find a similar analytical solution when the heat transfer in medium 2 is modeled by the wall thermal model introduced in Eqs. (4) and (5). Again, the interface temperature does not vary with time, and the temperature evolution is expressed as :

$$T_1 = T_{10} + (T_w - T_{10})\, erfc(\frac{(h-z)}{\sqrt{4\frac{k_1}{\rho_1 c_{p1}} t}}) \ , \qquad (9)$$

$$T_2 = T_{20} + (T_w - T_{20})(1 - \frac{z}{h+e}) \ , \qquad (10)$$

$$T_w = \frac{\sqrt{\rho_1 c_{p1} k_1}\sqrt{\frac{1}{\pi}}T_{10} + \sqrt{\rho_2 c_{p2} k_2}\sqrt{\frac{3}{8}}T_{20}}{\sqrt{\rho_1 c_{p1} k_1}\sqrt{\frac{1}{\pi}} + \sqrt{\rho_2 c_{p2} k_2}\sqrt{\frac{3}{8}}} \ , \qquad (11)$$

$$e = \sqrt{6\frac{k_2}{\rho_2 c_{p2}} t} \ , \qquad (12)$$

where T_w denotes the interface temperature.

The key idea of our model is that, certainly during the first instants following fountain flow, but even on a larger time scale (as it will be demonstrated further), the heat transfer between wall and melt is governed by a thermal shock at the flow front. The temperature profile is, therefore, very steep and difficult to capture in the vicinity of the contact line. In practice, unwanted oscillations are generated numerically when classical methods are used. This can easily be explained by the theory of approximation, as shown in Fig. (3) : neither cubic splines nor Lagrange polynomials can approximate the steep imposed temperature profile. Although oscillations are rapidly damped out when cubic splines are used, these are unable to capture in a proper way the temperature profile.

In order to get rid of numerical oscillations, we have developed a decomposition technique. The temperature profile is splitted in two components, the first one (T^a) being given by an analytical shock solution, while the other component (T^*) is unknown and can be calculated by solving the energy equation. The analytical component is the sum of the contributions of two 1D shocks taking place at opposite locations on the walls, and starting when fountain flow occurs. For any given pair of opposite points across the gap, the temperatures T_{10} and T_{20}, and the material data governing Eqs. (9) to (12), are the core melt temperature, the wall temperature before the contact lines reach this pair of positions, and the thermal properties of polymer and walls. The T^* component is governed by heat advection, viscous heating, and weakly by heat conduction. Hence, the gradient of T^* is smooth, which means that this component is accurately approximated by means of cubic splines.

Let us finally observe that refining the collocation points in the vicinity of the walls, in order to capture a singularity which is mainly known, would neither be efficient nor at all be useful. In fact, this is a consequence of the Gibbs effect.

5 COMPARISON WITH EXPERIMENTS

The experiments were performed in the laboratories of the Pôle Européen de Plasturgie (European Polymer Processing Centre, Oyonnax, France). We have used a 420 *tons* hydraulic Billion press. The test part was a 367 × 102 × 4 *mm* plate, fed by a linear gate (Fig. 4). The steel mold was equipped with optical fiber sensors, pressure transducers, and thermocouples linked to a HP data acquisition system. The temperature at the sprue was measured by both an IR thermometer and a thermocouple. All the sensors were placed at different locations on the walls. In view of the expertise of PEP, thermocouples were able to reach a response time as short as 0.2 *ms*, and to record thermal variations of an order of magnitude of 10,000 *K/s*.

Injections of PS (Dow Styron 678E PolyStyrene) and PC (Makrolon 2805 PolyCarbonate) were performed with different operating conditions [15]. Typical short shot results obtained with PC are shown on Fig. 5, where the evolution of temperatures and pressures is plotted at different locations along the mold. One of the thermocouples was defective, as can be observed on the figure. The diagram confirms the sensitivity and precision of the temperature measurements performed by using this apparatus. It also demonstrates the existence of a front thermal shock generated by fountain flow. Indeed, measured wall temperatures (381 *K*) are far below the injection temperature at the sprue (603 *K*), and are very close to the liquid-solid interface temperature T_w predicted by thermal shock theory (Eq. (11)). The temperature increase is sudden, and records accurately the front motion. After thermal shock, the wall temperature varies only slowly with time. Similar results have been observed in the case of PS.

Some experiments have also been simulated using the MOLDSYS software, as improved by means of our thermal shock model. The characterization of PS was performed in the PEP. Material

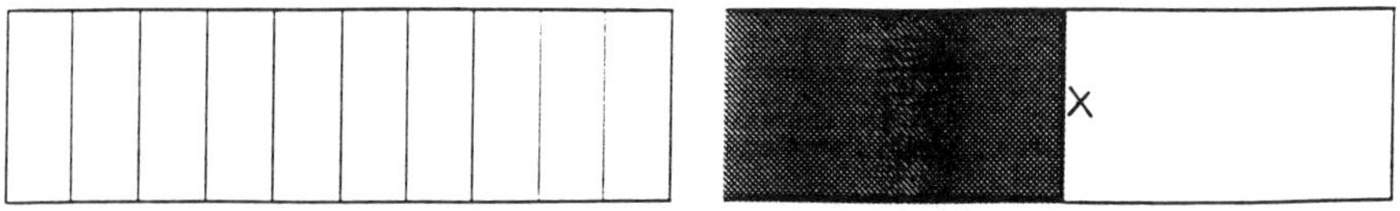

Figure 7 : Short shot experiment : comparison between predicted and experimental (X) last front.

data are detailed in Table 1. A coolant temperature of 294 K and a flow rate of 108.5 cm^3/s were imposed as processing conditions, while the gate temperature was reconstructed from wall experimental measurements by using the theory of thermal shock in an inverse way. This approach, which provides an inlet temperature of 501 K, is very accurate because the core temperature is almost constant along the midsurface during filling. By this way, the effect of the runner system does not need to be taken into account in the simulations, which has allowed us to obtain a very precise validation of our software. Typical temperature profiles calculated either just after thermal shock or some instants later are shown in Fig. 6, together with the T^a component. In the left hand figure, the temperature gradient is very steep near the wall, while no spurious oscillations are generated thanks to our decomposition technique. In the case of PC, the results are basically the same, although the T^* component is more important in view of a higher viscous heating effect.

The same simulation has been used to predict a short shot performed in the same operating conditions, except that the injected load was not sufficient to fill the mold. The final front position is compared in Fig. 7 with its predicted counterpart. The latter was determined by identifying the instant at which simulated pressures reach the peak observed on experimental records when filling stops abruptly (see Fig. 5). Recall indeed that complete filling simulations were performed. By this way, the comparison is highly sensitive to thermal predictions, since pressures are highly dependent on heat transfer during filling. The agreement is excellent, which validates accurately our model.

ACKNOWLEDGMENTS

The authors wish to thank Jean-Pierre Gazonnet, Virginie Durand and Gerard Dechavanne, from the PEP, for their useful advices and friendly collaboration. This paper presents research results of the Belgian Programme on Interuniversity Poles of Attraction, initiated by the Belgian State, Prime Minister's Office for Science, Technology and Culture, and within the frame of a Multimatériaux Project initiated by the Walloon Region of Belgium. The scientific responsibility rests with its authors.

REFERENCES

[1] H. A. Lord and G. Williams. Mold filling studies for the injection molding of thermoplastic materials. part 2 : The transient flow of plastic materials in the cavities of injection-molding dies. *Polym. Eng. Sci.*, 15(8):569–582, 1975.

[2] C. A. Hieber and S. F. Shen. A finite element / finite-difference simulation of the injection-molding filling process. *J. Non-Newt. Fluid Mech.*, 7(1):1–32, 1980.

[3] F. Dupret and L. Vanderschuren. Calculation of the temperature field in injection molding. *AIChE J.*, 34(12):1959–1972, 1988.

[4] H. H. Chiang, C. A. Hieber, and K. K. Wang. A unified simulation of the filling and post-filling stages in the injection-molding. part 1 : Formulation. *Polym. Eng. Sci.*, 31:116–124, 1991.

[5] K. K. Kabanemi and F. Dupret. Analysis of the influence of the packing stage on residual stresses and shrinkage of injection molded parts. In J.-L. Chenot, R. D. Wood, and O. C. Zienkiewicz, editors, *Proc. of the Int. Conf. on Num. Meth. in Ind. Forming. Proc. 1992*, pages 357–363. A. A. Balkema, 1992.

[6] O. Verhoyen, R. Legras, and F. Dupret. A simplified method for introducing the cross viscosity law in the numerical simulation of non-isothermal compressible hele shaw flow. In *Proc. of the PPS European Regional Meeting, Strasbourg, France, august 1994*, pages 175–398, 1994.

[7] S. F. Shen. Simulation of polymeric flows in the injection moulding process. *Int. J. Num. Methods Fluids*, 4:171–183, 1984.

[8] C. A. Hieber, L. S. Socha, S. F. Shen, K. K. Wang, and A. I. Isayev. Filling thin cavities of variable gap thickness : A numerical and experimental investigation. *Polym. Eng. Sci.*, 23(1):20–26, 1983.

[9] A. Couniot, L. Dheur, and F. Dupret. Numerical simulation of injection molding : non-isothermal filling of complex thin parts, including abrupt changes of thickness or bifurcations of the midsurface. In M. Cross, J. F. T. Pittman, and R. D. Wood, editors, *Proc. of the IMA Conf. on Mathematical Modelling for Materials Processing 1991*, pages 381–398. Oxford University Press, 1993.

[10] M. J. Crochet, F. Dupret, and V. Verleye. Injec-

tion molding. In S. G. Advani, editor, *Flow and Rheology in Polymer Composites Manufacturing*, chapter 11, pages 415–463. Elsevier, 1994.
[11] D. J. Coyle, J. W. Blake, and C. W. Macosko. The kinematics of fountain flow in mold-filling. *AIChE J.*, 33(7):1168–1177, 1987.
[12] L. Dheur. *Simulation numérique des échanges thermiques en moulage par injection.* PhD thesis, Université Catholique de Louvain, 1992.
[13] A. Couniot. *Développement d'un algorithme de remaillage automatique pour la simulation du moulage par injection.* PhD thesis, Université Catholique de Louvain, 1991.
[14] L. Vanderschuren. *Simulation numérique du moulage par injection.* PhD thesis, Université Catholique de Louvain, 1989.
[15] P. Pirotte and N. Van Rutten. Simulation numérique du moulage par injection : validation expérimentale. Final work, Université Catholique de Louvain, 1994.

Simulation of Materials Processing: Theory, Methods and Applications, Shen & Dawson (eds)
 ISBN 90 5410 553 4

Determination of cooling temperature history and cooling channel layout in injection molds by using BEM

T. Matsumoto & Masa. Tanaka
Shinshu University, Nagano, Japan

T. Takahashi
Graduate School of Shinshu University, Nagano, Japan

S. Sawada
Toray Industries, Inc.

ABSTRACT: This paper is concerned with the optimum design of cooling lines in injection molds. The temperature or heat-flux history at the boundary of the one-dimensional interval is first determined so that the temperature variation at ejection time become as uniform as possible. In this problem, the final temperature variation at the ejection time is known but the boundary condition is not known. Next, the optimum position and size of the cooling lines that satisfy the cooling history over the cavity surface are determined by using the boundary element method. Some numerical examples for the designs of the cooling lines of two-dimensional model are presented.

1 INTRODUCTION

Injection molding is a standard manufacturing process of products made from polymer. In the process, the molten polymer is first filled in the cavity through the runners. The heat of the polymer is removed by using the slender channels (cooling channels) running near from the cavity surface. The total energy spent for solidification is provided by some coolant flowing through the cooling channels. Thus, an injection mold works as a heat exchanger. From the recent increasing requirements for the production in high quality, further development of the simulation technique, useful to the total design of mold structure and mold temperature control, is strongly expected.

The goal of our study is to develop an approach to determine the optimum layout and size of the cooling channels, the optimum shape of the mold, and the optimum cooling history that can achieve the objective temperature variation of the polymer product at the target ejection time.

In the polymer cooling simulation, the finite difference method has been popularly used (e.g. Manzione 1987) due to the simplicity of its formulation. The time-domain boundary integral method has also been used (Chuang & Szekely 1971; O'Neill 1983; Heinlein, et al. 1986; Vick & Nelson 1993; Zabaras & Mukherjee 1987; Zabaras, et al. 1988), because it can give us accurate results for the heat-flux at the boundary points. The product can be assumed to have a thin geometry, so uni-directional heat conduction is enough to be considered. Although the moving boundaries must be taken into consideration, the time-domain boundary integral method or the boundary element method is easily implemented in one-dimensional problems. The boundary element method is also very effective in the cooling analyses of molds and has been successfully applied to optimum designs of two-dimensional molds (Barone & Caulk 1982), thermal design sensitivity analyses for molds (Matsumoto, et al. 1993a; 1993b).

Several studies on Inverse Stefan Problem (ISP) in theoretical and practical fields has been made so far (e.g. Jochum 1980; Reemtsen & Kirsch 1984; Zabaras, et al. 1988; Bénard & Afshari 1992). Some of them treated the determination of the boundary condition to control the position of the solidification front, others treated the identification of the solid-liquid interface position from the measured data. We have also treated an ISP (Matsumoto 1994) in which the boundary condition at the boundary of the one-dimensional interval is determined so that the temperature variation at ejection time become as uniform as possible.

In this paper, the temperature or heat-flux history at the boundary of the one-dimensional interval is determined first so that the temperature variation at ejection time become as uniform as possible through the inverse analysis based on the minimization of an appropriately defined cost function (Matsumoto 1994). The uni-directional solidification analyses are preformed by using time-domain boundary element method. Next, the optimum layout and size of the

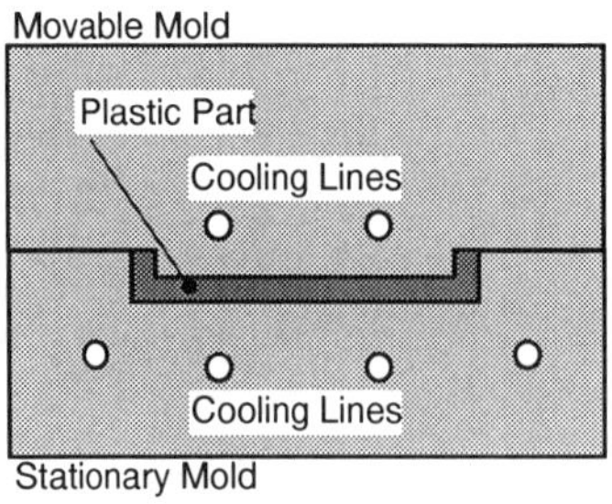

Fig.1 Abstract model of injection mold.

cooling channels and the coolant condition are determined through the cooling analyses based on the boundary element method.

2 PROBLEM DEFINITION

In Fig.1 is shown an illustration of a simplified injection mold. We assume that the plastic part is in molten state and in a uniform temperature when it is injected in the cavity between the molds. Then it is solidified and ejected after it is sufficiently cooled down. The temperature distribution in the plastic part at the ejection time should satisfy a certain condition so that the final ejected part is free of any warpage and residual stress.

In the present study, we first determine the boundary condition over the cavity surface between the molds through solidification analyses so that the temperature distribution of the plastic part at the desired ejection time is as uniform as possible.

Assume that we can neglect the heat conduction in the lateral direction of the polymer product because of its thin geometry. Then, we can treat the heat conduction problem as one-dimensional, governed by the following differential equation:

$$\alpha \frac{\partial^2 T(x,t)}{\partial x^2} - \frac{\partial T(x,t)}{\partial t} = 0, \quad 0 \le x \le b, \ 0 \le t \le \tau \tag{2.1}$$

with the initial condition

$$T(x,0) = T_0 \tag{2.2}$$

where $T(x,t)$ is the temperature at point x and time t, α is the thermal diffusivity, b is the interval size, and τ is the final time when the product is ejected from the mold.

The boundary condition is assumed that the temperature or the heat flux is prescribed by the following relationships:

$$\left. \begin{aligned} T(0,t) &= T_0(t) \\ T(b,t) &= T_b(t) \end{aligned} \right\} \tag{2.3}$$

or

$$\left. \begin{aligned} \frac{\partial T}{\partial x}(0,t) &\equiv q(0,t) = q_0(t) \\ \frac{\partial T}{\partial x}(b,t) &\equiv q(b,t) = q_b(t) \end{aligned} \right\} \tag{2.4}$$

More generally, the heat-flux is related to the temperature as

$$\left. \begin{aligned} \frac{\partial T}{\partial x}(0,t) &\equiv q(0,t) = H\{T(0,t) - \theta_0\} \\ \frac{\partial T}{\partial x}(b,t) &\equiv q(b,t) = H\{T(b,t) - \theta_b\} \end{aligned} \right\} \tag{2.5}$$

where H is the heat transfer coefficient between the mold and polymer, taking different values depending on the state of the phase, θ_0 and θ_b are the mold temperatures at x=0 and b, respectively.

The boundary conditions between the liquid and solid phase are

$$T_s(h(t),t) = T_l(h(t),t) = T_m(t) \tag{2.6}$$

$$k_s \frac{\partial T_s}{\partial x}(h(t),t) - k_l \frac{\partial T_l}{\partial x}(h(t),t) = \rho L \frac{dh(t)}{dt} \tag{2.7}$$

where the subscripts s and l denote the solid phase and liquid phase, respectively, ρ is the density, L is the latent heat, and $h(t)$ is the position of the solid-liquid interface.

Let us assume now that $q(0,t)$ and $q(b,t)$ are expressed with sets of parameters z_0 and z_b, respectively. Then, the present problem, in which $q(0,t)$ and $q(b,t)$ are determined from final ejection time, results in searching the sets of z_0 and z_b that minimize the cost function defined as follows:

$$f = \frac{m \sum_{i=1}^{m} \{T(x_i, \tau; \mathbf{z}_0, \mathbf{z}_b) - T_{\text{target}}(x_i)\}^2}{\sum_{i=1}^{m} \{T_{\text{target}}(x_i)\}^2} \tag{2.8}$$

where x_i is the i-th point of m sampled points uniformly arranged in the interval $[0, b]$.

The steepest descendent method is employed here for the searching algorithm of the minimum point of f. The sensitivities of f with respect to the parameters are calculated approximately by using the finite difference scheme applied to the cost function. In the iterative process of searching the minimum point of f, we must compute the uni-directional solidification problems. The time-domain boundary element method is employed for this purpose (Zabaras & Mukherjee 1988).

After $q(0,t)$ and $q(b,t)$ are determined as the functions of the design parameters z_0 and z_b, they are given as the boundary condition of the cavity surface of the mold and the layout of the cooling channels is next determined. In this second problem, the position of the cooling channels, the cross section size, the flow rate of the coolant, and its temperature can be parameterized with several design variables. Denoting such design variable vector by **b**, the most simplified cost function of the second inverse problem can be defined as

$$g(\mathbf{b}) = \sum_{j=1}^{N_t} \sum_{i=1}^{N_s} \left(T_{ij}(\mathbf{b}) - T_{ij}^{\text{target}}\right)^2 \tag{2.9}$$

where N_s is the total number of nodes over the cavity surface, N_t is the total number of sampling points in time-axis, i refers to a nodal number, j refers to time, $T_{ij}(\mathbf{b})$ is the computed temperature of the cavity surface, T_{ij}^{target} is the target temperature also given by the uni-directional solidification inverse analysis.

In the searching process of the minimum point of the above defined cost function, we are required cooling analyses of the mold. Since injection molds usually consist of complicated structures, the boundary element method is a very convenient analysis tool again. The solidification analysis is not linked to the mold cooling analysis now, therefore, the boundary element method based on the formulation for the governing equation obtained by applying the Laplace transform is utilized.

The basic differential equation for the mold cooling analysis is also the following diffusion equation:

$$\alpha \nabla^2 T(x,t) - \frac{\partial T(x,t)}{\partial t} = 0 \tag{2.10}$$

where x is now the space coordinates.

Taking the Laplace transform to (2.10) gives us

$$\alpha \nabla^2 \tilde{T}(x,s) - s\tilde{T}(x,s) - T_0(x) = 0 \tag{2.11}$$

where $T_0(x)$ is the initial condition for the temperature and $\tilde{T}(x,s)$ is defined by

$$\tilde{T}(x,s) = \int_0^\infty T(x,t) e^{-st} dt \tag{2.12}$$

The integral representation for $\tilde{T}$ is given by

$$\tilde{T}(y) = \alpha \int_\Gamma T^*(x,y)\tilde{q}(x) d\Gamma - \alpha \int_\Gamma q^*(x,y)\tilde{T}(x) d\Gamma + \int_\Omega T^*(x,y) T_0(x) d\Omega \tag{2.13}$$

where Ω is the region, Γ is the boundary of the body, and y is an arbitrary point in Ω. T^* is the fundamental solution of (2.11) and for two-dimensional problems given by

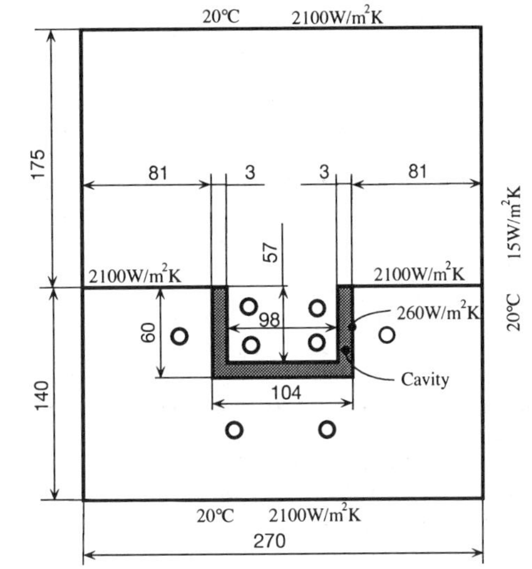

Fig.2 Example computation model.

Table 1. Thermal properties used in boundary integral analysis.

	Liquid	Solid
Thermal conductivity [W/mK]	0.11	0.126
Density [kg/m³]	839	900
Specific heat [J/kgK]	2220	1884
Latent heat [J/kg]	87500	
Melting point [°C]	110	

$$T^*(x,y) = \frac{1}{2\pi\alpha} K_0[kr] \tag{2.14}$$

where K_0 is the modified Bessel function of 0-th order and $k = \sqrt{s/\alpha}$. q^* is the normal derivative of T^*.

Equation (2.13) is equivalent to the following regularized representation:

$$\alpha \int_\Gamma \left\{q^*(x,y) - Q^*(x,y)\right\} d\Gamma \cdot \tilde{T}(y) + \alpha \int_\Gamma q^*(x,y)\left\{\tilde{T}(x) - \tilde{T}(y)\right\} d\Gamma = \alpha \int_\Gamma T^*(x,y)\tilde{q}(x) d\Gamma + \int_\Omega T^*(x,y) T_0(x) d\Omega \tag{2.15}$$

where Q^* is the normal derivative of u^*, which is the particular solution of

$$\alpha \nabla^2 u^*(x,y) + \delta(x-y) = 0 \tag{2.16}$$

Equation (2.15) can be continuously applied to the

point y on the boundary if $\tilde{T}(x)$ satisfies the Hölder continuity at y. Therefore, equation (2.15) can be used as a boundary integral equation.

The advantage of using (2.15) is that it is free of Cauchy principal value integrals. All the strong singularities of the kernel functions are canceled out in advance.

We can solve (2.15) numerically after discretizing Γ and the boundary quantities, and giving the initial condition and the transformed boundary condition. After all the transformed boundary quantities are obtained, time-dependent quantities can be obtained by taking the numerical inverse Laplace transform to them. We use here the Durbin's inversion method (Durbin 1974).

In the searching process of the minimum point of $g(\mathbf{b})$ based on the gradient methods, we need the sensitivity coefficients of the boundary quantities to construct ∇g.

All the quantities in (2.15) are the explicit or implicit functions of design parameters. The sensitivity integral representation for (2.15) is obtained by differentiating (2.15) with respect to an arbitrary design variable, as follows:

$$\begin{aligned}
&\alpha\int_\Gamma \left\{q^*(x,y)-Q^*(x,y)\right\}d\Gamma\cdot\dot{\tilde{T}}(y) \\
&+\alpha\int_\Gamma q^*(x,y)\left\{\dot{\tilde{T}}(x)-\dot{\tilde{T}}(y)\right\}d\Gamma \\
&=\alpha\int_\Gamma T^*(x,y)\dot{\tilde{q}}(x)d\Gamma \\
&-\alpha\int_\Gamma \left\{\dot{q}^*(x,y)-\dot{Q}^*(x,y)\right\}d\Gamma\cdot\tilde{T}(y) \\
&-\alpha\int_\Gamma \left\{\dot{q}^*(x,y)-\dot{Q}^*(x,y)\right\}d\dot{\Gamma}\cdot\tilde{T}(y) \\
&+\alpha\int_\Gamma \dot{q}^*(x,y)\left\{\tilde{T}(x)-\tilde{T}(y)\right\}d\Gamma \\
&+\alpha\int_\Gamma q^*(x,y)\left\{\tilde{T}(x)-\tilde{T}(y)\right\}d\dot{\Gamma} \\
&+\alpha\int_\Gamma \dot{T}^*(x,y)\tilde{q}(x)d\Gamma+\alpha\int_\Gamma T^*(x,y)\tilde{q}(x)d\dot{\Gamma} \\
&+\int_\Omega T^*(x,y)\dot{T}_0(x)d\Omega \\
&+\int_\Omega \dot{T}^*(x,y)T_0(x)d\Omega+\int_\Omega T^*(x,y)T_0(x)d\dot{\Omega}
\end{aligned} \tag{2.16}$$

where an over-scribed dot denote the differentiation with respect to an arbitrary design variable, such as shape design variable, material property of the coolant, boundary condition, and so on. Equation (2.16) can also be discretized in a similar manner as in (2.15) but now we need all the temperature and heat-flux in the transformed space over the boundary to solve it. For the boundary where the boundary quantity is prescribed as the boundary condition, their sensitivities are also known. Therefore, we can solve the discretized form of (2.16) by giving the boundary

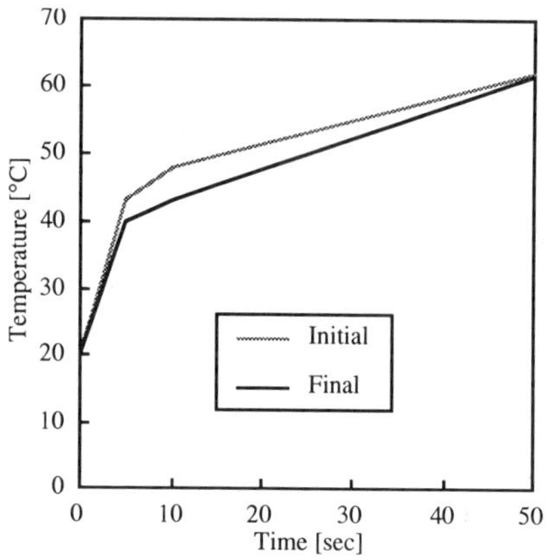

Fig.3 Identified profile of cavity temperature.

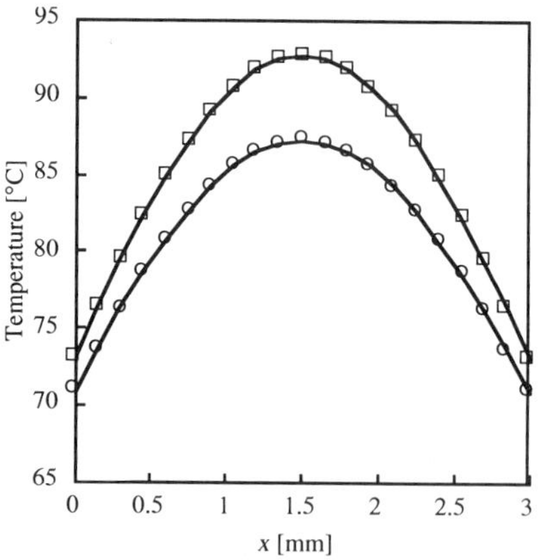

Fig.4 Temperature variation of plastic part at ejection time.

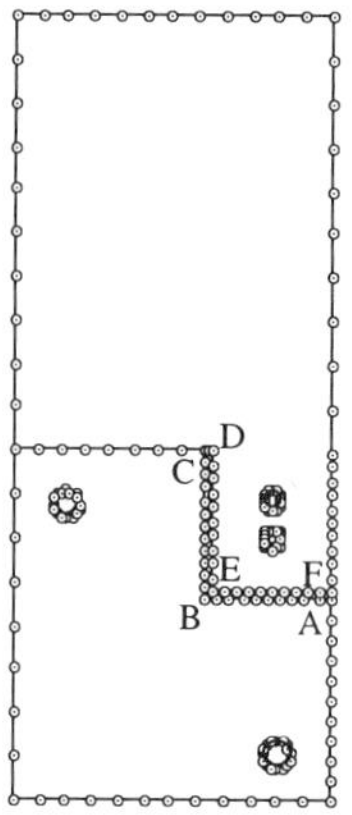

Fig.5 Boundary element discretization of half part of computation model.

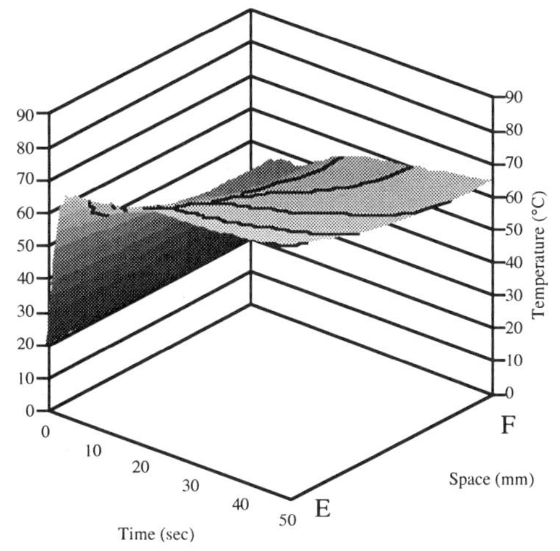

Fig.6 Original temperature profile of EF section of cavity.

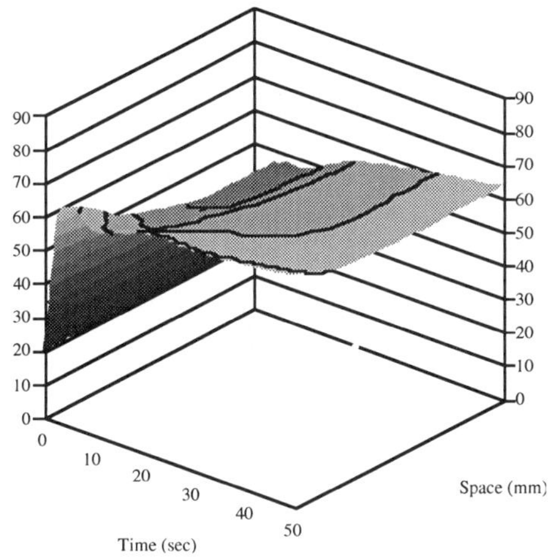

Fig.7 Improved temperature profile of EF section of cavity.

condition for the sensitivity of the boundary quantities.

3 NUMERICAL EXAMPLES

We show in Fig.2 a two-dimensional model of injection mold for a trial computation using the present approach. The upper side is the movable mold and the lower is the stationary one. The molten plastic is first injected into the cavity through the runner that is omitted in the present computation model. The circular holes in Fig.2 represent the initially assumed layout of the cross sections of the cooling lines. The thickness of the plastic part is assumed to be 3[mm]. The thermal properties of the plastic part is listed in Table 1.

First, the cooling condition of the cavity surface is determined through the inverse solidification analysis. The target ejection time is assumed to be 50[sec], and the target temperature of the plastic part at the ejection time is 80°C. The temperature profile of the cavity surface is assumed to be the continuation of the linearly assumed profiles in three intervals. The solid line in Fig.3 is the obtained final profile of the temperature over the cavity surface. Fig.4 shows the temperature variation of the plastic part at ejection time.

Second, the profile of the heat-flux profile from the plastic to the mold, evaluated from the heat-transfer condition is given over the cavity surface as the boundary condition, and the cooling channels are designed through the process described in the above. Considering the symmetry of the computation model, only the half part is discretized with quadratic elements as shown in Fig.5. In Fig.6 and 7 are shown the original and improved temperature profile over EF part of the cavity surface by designing the mold cooling channels. We observe that the temperature variation along EF at ejection time, 50[sec], shown in Fig.7 became more uniform after the optimization analysis than that in the assumed state shown in Fig.6.

CONCLUDING REMARKS

A brief outline of the design process of the cooling channels of injection molds based on the boundary element method is presented. Results of some trial computation for a two-dimensional model is also presented. The results shows that the present approach is effective and promising in the improvement of the initial design.

REFERENCES

Barone M.R. & Caulk D.A. 1982. Optimal arrangement of holes in a two-dimensional heat conductor by a special boundary-integral method. *Int. J. Numer. Methods Eng.* 18: 675-685.

Bénard, C. & Afshari, A. 1992. Inverse Stefan problem: tracking of the interface position from measurements on the solid phase. *Int. J. Numer. Methods Eng.* 35: 835-851.

Burton,T.E. & Rezayat, M. 1987. POLYCOOL 2: A three-dimensional transient mold-cooling simulator. In Bernhardt, E.C. & Manzione L.T. (eds), *Applications of Computer Aided Engineering in Injection Molding:* Chap.9, 269-294. Munich: Hanser.

Chuang, Y.K. & Szekely, J. 1971. On the use of Green's function for solving melting and solidification problems. *Int. J. Heat & Mass Transfer*. 14: 1285-1294.

Crank, J. 1984. *Free and Moving Boundary Problems*. Oxford: Oxford University Press.

Durbin, F. 1974. Numerical inversion of Laplace transform: An efficient improvement to Dubner and Abate's method *Computer J.*, 17: 371-377.

Heinlein, M., Mukherjee, S. & Richmond, O. 1986. A boundary element method analysis of temperature fields and stress during solidification. *Acta Mechanica*. 59: 59-81.

Jochum, P. 1980. The numerical solution of the inverse Stefan problem. *Numer. Math.* 45: 411-429.

O'Neill, K. 1983. Boundary integral equation for moving boundary phase change problems. *Int. J. Numer. Methods Eng.* 19: 1825-1850.

Reemtsen, R. & Kirsch, A. 1984. A method for the numerical solution of the one-dimensional inverse Stefan problem. *Numer. Math.* 45: 253-273.

Matsumoto, T., Tanaka, M. & Miyagawa, M. 1993a. Boundary element system for mold cooling/ heating design. In Brebbia, C.A. & Rencis, J.J. (eds), *Boundary Elements XV, Vol.2, Stress Analysis:* 461-475, Southampton: Computational Mechanics Publications.

Matsumoto, T., Tanaka, M., Miyagawa, M. & Ishii, N. 1993b. Optimum design of cooling lines in injection moulds by using boundary element design sensitivity analysis. *Finite Elements in Analysis and Design*. 14: 177-185.

Matsumoto, T., Tanaka, M., Takahashi, N. 1994. Determination of cooling temperature history in injection molding. In Bui, H.D. & Tanaka, M. (Eds.) *Inverse Problems in Engineering Mechanics*. 295-300.

Vick B. & Nelson D.J. 1993. The boundary element method applied to freezing and melting problems. *Numerical Heat Transfer, Part B*. 24: 263-277.

Zabaras, N. & Mukherjee, S. 1988. An analysis of solidification problems by the boundary element method. *Int. J. Numer. Methods Eng.* 24: 1879-1900.

Zabaras, N., Mukherjee, S. & Richmond, O. 1988. An analysis of inverse heat transfer problems with phase changes using an integral method. *J. Heat Transfer*. 110: 554-561.

Simulation of Materials Processing: Theory, Methods and Applications, Shen & Dawson (eds)
© 1995 Balkema, Rotterdam. ISBN 90 5410 553 4

Finite element simulation of jetting behaviour in metal injection molding

Ken-ichiro Mori, Kozo Osakada & Shinji Takaoka
Department of Mechanical Engineering, Faculty of Engineering Science, Osaka University, Japan

ABSTRACT: The non-isothermal filling into a die cavity in metal injection molding is simulated by the viscoplastic and the heat conduction finite element methods. Complex overlapping between the surfaces of the mixture due to the occurrence of jetting is dealt with by the use of a remeshing scheme. The filling behaviours in metal injection molding into rectangular and ringed dies are two-dimensionally simulated, although the friction and the heat transfer to the die in not only the side surfaces but also the upper and lower surfaces are taken into consideration. The filling behaviour is considerably affected by the die geometry, the gate size, the injection speed and the die temperature. The calculated filling behaviour is in good agreement with the experimental one for a mixture of a steel powder and a binder mainly made from thermoplastic polymers and waxes.

1 INTRODUCTION

In metal injection molding, a metallic powder mixed with a binder made from thermoplastics and waxes is filled into a die cavity. The binder is added to enhance the flowability and the formability in injection molding. After molding, the binder is extracted from the mixture and the compact is subsequently sintered at an elevated temperature. The sintered product has a near 100 percent theoretical density because the diameter of the powder particle is very small, and thus coining and sizing of the sintered product are not required. The metal injection molding is expected to be a net-shape forming process for producing parts with complex geometries.

In metal injection molding, the deformation behaviour of the powder/binder mixture is greatly different from that of thermoplastics because the amount of the binder is not very large, say 40% of the volume. In metal injection molding, the so-called jetting phenomenon is often observed because the mixture hardly exhibits swelling near the die gate (Piccirillo et al. 1991), although the swelling is typical of injection molding of thermoplastics. The small swell is caused by the slip over the die surface and small elastic recovery in the release from the die exit. The deformation behaviour of the mixture is close to a solid rather than a fluid. When the jetting occurs, the mixture is filled into the die cavity with each surface overlapped. Since the overlapping is caused in many portions during injection, it is not easy to treat the overlapping in the computation. Gaspervich (1991) has simulated material flow in metal injection molding under no appearance of jetting by the use of a software package developed for flow analysis of thermoplastics. In commercial software packages for thermoplastics, the jetting cannot be dealt with.

In the present study, material flow and temperature distribution in metal injection molding are simulated by the viscoplastic and heat conduction finite element methods. Complex overlapping due to the occurrence of jetting is dealt with by the use of a remeshing scheme.

2 METHOD OF SIMULATION

2.1 Modelling of metal injection molding

In metal injection molding, the deformation behaviour of the powder/binder mixture is similar to that of metal in hot working. It is observed that the elastic deformation is negligibly small and the volume of the mixture during molding is kept almost constant. Therefore, the mixture during injection is assumed to be an incompressible viscoplastic material.

The material flow and temperature distribution in metal injection molding processes are simulated by the viscoplastic and heat conduction finite element methods, respectively. In the simulation, plane-strain deformation shown in Fig. 1 is assumed because the thickness of the die cavity is constant and in addition small. The mixture is divided into plane-strain quadrilateral elements with four nodal points. Since the deformation behaviour and the temperature distribution are considerably influenced by the upper and lower surfaces of the die, the

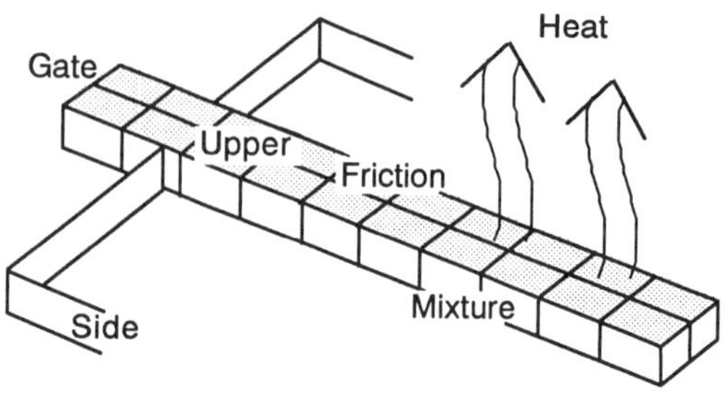

Fig. 1 Modeling of plane–strain metal injection molding.

plane–strain modelling is extended to take this effect into consideration. The friction and the heat transfer to the die in not only the side surfaces but also the upper and lower surfaces are taken into consideration.

2.2 Viscoplastic finite element method

To simulate material flow in metal injection molding, the viscoplastic finite element method based on the plasticity theory for a material with slight compressibility (Osakada 1982, Mori 1982) is used. In the plasticity theory, the material is assumed to obey the yield criterion:

$$\bar{\sigma}^2 = \frac{3}{2}\sigma_{ij}'\,\sigma_{ij}' + g\sigma_m^{\,2}, \qquad (1)$$

where σ_{ij}' is the deviatoric stress, g is a small positive constant (0.01–0.0001), σ_m is the hydrostatic stress and $\bar{\sigma}$ is the equivalent stress. By the pressure sensitivity, the volume of the material is slightly changed during plastic deformation, and thus the stress components can be calculated directly from the strain–rate components. To obtain the exact solution, the following functional Φ for this material is minimized

$$\Phi = \int_A \left[\int_0^{\dot{\bar{\varepsilon}}} \bar{\sigma}\, d\dot{\bar{\varepsilon}}\right] t dA + \int_A \tau_f \Delta v dA + \int_S \tau_f \Delta v t dS, \qquad (2)$$

where $\dot{\bar{\varepsilon}}$ is the equivalent strain–rate, t is the thickness of the die, A is the cross–sectional area, τ_f is the frictional shear stress, Δv is the relative velocity between the mixture and the die and S is the length of the side surface of the mixture. The thickness of the die is equal to that of the mixture during molding. The second term on the right–hand side of equation (2) represents the effect of friction in the upper and lower surfaces of the die and the third term represents that in the side surfaces. On the other hand, the inertial effect is neglected because the injection speed in not very high. The condition of incompressibility is approximately satisfied by a small positive value of g.

The jetting is caused by the small swell of the mixture released from a die gate. The die filling under the jetting phenomenon is treated to be buckling behaviour for a long bar. The gate is slightly skewed to induce asymmetric deformation for the buckling in the die cavity.

2.3 Heat conduction finite element method

Since the material flow in metal injection molding is significantly affected by the distribution of temperature, not only plastic deformation but also the temperature distribution are computed. The effect of the heat transfer to the upper and lower surfaces of the die is included in the two–dimensional differential equation governing non–steady–state heat conduction as follows:

$$\alpha\left(\frac{\partial^2 T}{\partial x^2} + \frac{\partial^2 T}{\partial y^2}\right) - \frac{2h(T-T_d)}{t} - c\rho\frac{\partial T}{\partial t'} + \dot{q} = 0, \qquad (3)$$

where α is the thermal conductivity, T is the temperature, h is the heat transfer coefficient, T_d is the temperature of the die, c is the specific heat, ρ is the density, t' is the time and $\dot{q}$ is the heat generation rate. The temperature change in the thickness direction of the mixture is neglected, and the heat transfer to the upper and lower surfaces is dealt with. The heat transfer to the side surfaces is also taken into consideration. The heat generation due to deformation and friction is obtained from the viscoplastic finite element simulation.

The differential equation of equation (3) is formulated by using the Galerkin method as follows:

$$[A]\{T\} = \{B\}, \qquad (4)$$

where $\{T\}$ is the vector of the nodal temperature. Equation (4) is the linear simultaneous equations of the nodal temperatures. By alternating the deformation and the temperature simulations in each deformation step, the interaction between plastic deformation and temperature during injection molding is included in the simulation.

2.4 Treatment of overlapping between mixture surfaces

Under the occurrence of the jetting, the mixture is filled into the die cavity with each surface overlapped. For the overlapping, no slip between touching surfaces is assumed because of large adhesion of the mixture. The element mesh near the overlapping region is modified as shown in Fig. 2, and this region is treated not to be the surface but the inside of the material after the modification. The change in volume due to the modification is neglected. Weld lines during the molding are obtained by tracing the overlapping. Since cracks are often initiated from the weld lines during sintering, it is important to acquire the information

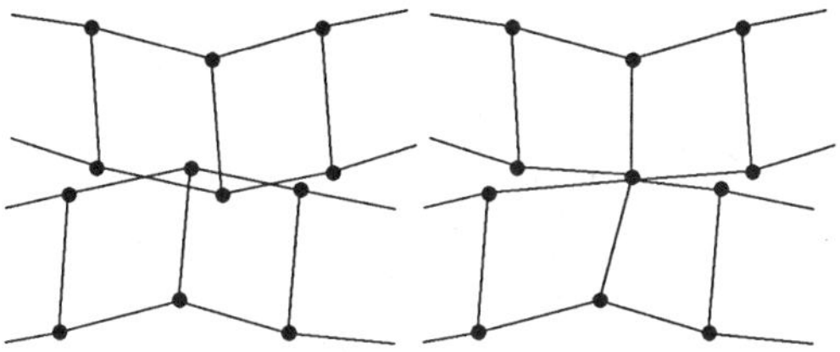

(a) Before modification (b) After modification

Fig. 2 Modification of element mesh near overlapping region.

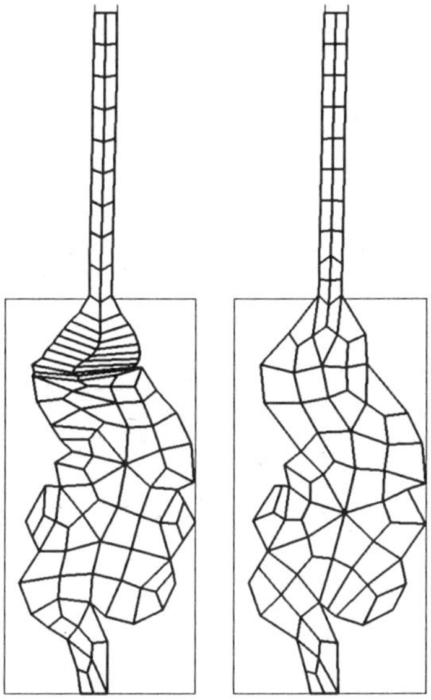

(a) Before remeshing (b) After remeshing

Fig. 3 Remeshing with quadrilateral elements using Delaunay triangulation.

Table 1 Material properties and working conditions used in simulation of plane-strain metal injection molding into rectangular die.

Width of gate w / mm	2, 3, 5
Height of die / mm	50
Width of die / mm	25
Thickness of die t / mm	5
Injection speed v_i / mms^{-1}	42, 127
Flow stress / Pa	$8.91\dot{\bar{\varepsilon}}^{0.57}\exp(3020/T)$
Coefficient of friction	0.16
Temperature of mixture before molding / °C	170

for the weld lines.

When the element mesh is severely distorted, overall remeshing with quadrilateral elements is carried out by the use of the Delaunay triangulation as shown in Fig. 3. The mixture is first divided into triangles, and then the triangles are transformed into quadrilaterals. In the remeshing, the information for temperature distribution and weld lines is transferred from the old mesh to the new one using the shape function.

3 MOLDING INTO RECTANGULAR DIE

3.1 Computational conditions

The material flow in plane-strain metal injection molding into a rectangular die is simulated by the present method. The material properties and the working conditions used in the simulation for the rectangular die are given in Table 1, where T is the temperature of the mixture. The powder is steel and the binder is mainly made from polyethylene and paraffin wax. The flow stress of the mixture is obtained from the capillary viscometer test under different temperatures. The effects of the width of the die gate and the injection speed on the filling behaviour are examined.

For the comparison with the calculated results, the filling behaviour during molding was measured by the use of a high speed video camera. Although the die was not heated in the experiment, the temperature of the die rose from the room temperature because of heating of the mixture before the molding. The temperature distribution of the die measured in the experiment was used as a boundary condition in the calculation.

3.2 Results

The calculated material flows in plane-strain injection molding into the rectangular die for the injection speed v_i=127mm/s and the widths of the gate w=2, 3 and 5mm are shown in Fig. 4. The mixture exhibits considerably complex material flow in the die cavity, and weld lines and internal voids in the molded part are caused by the jetting. The degree of jetting becomes remarkable as the width of the gate decreases, and thus the number of weld lines increases. The weld lines mainly appear in the lower half of the molded parts.

The calculated temperature distribution for v_i=127mm/s and w=3mm is shown in Fig. 5. The mixture is rapidly cooled by the die. The filling behaviour is influenced by the temperature distribution because the mixture has a large temperature sensitivity.

The calculated material flow in plane-strain injection molding into the rectangular die for v_i=42mm/s and w=3mm is shown in Fig. 6. The decrease in injection speed makes the die filling difficult due to the die cooling.

The experimental material flow for v_i=42mm/s

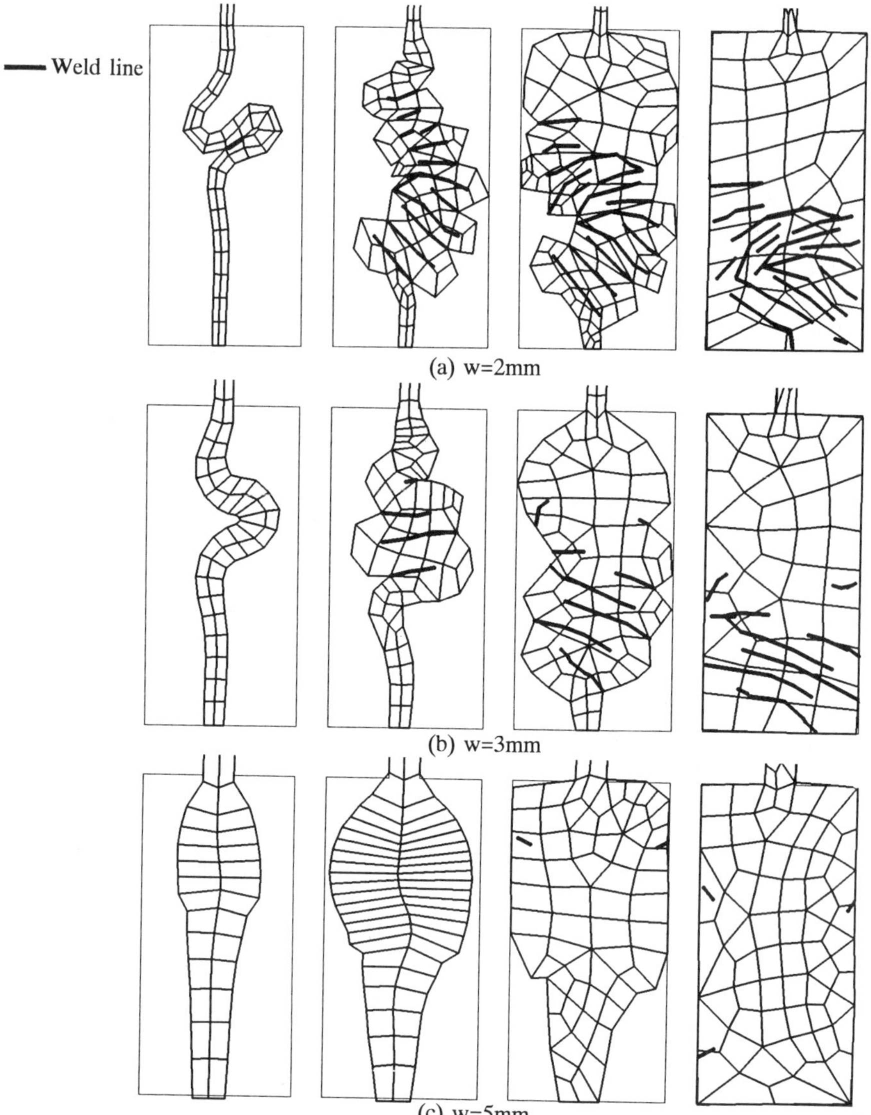

Fig. 4 Calculated material flows in plane–strain injection molding into rectangular die for v_i=127mm/s.

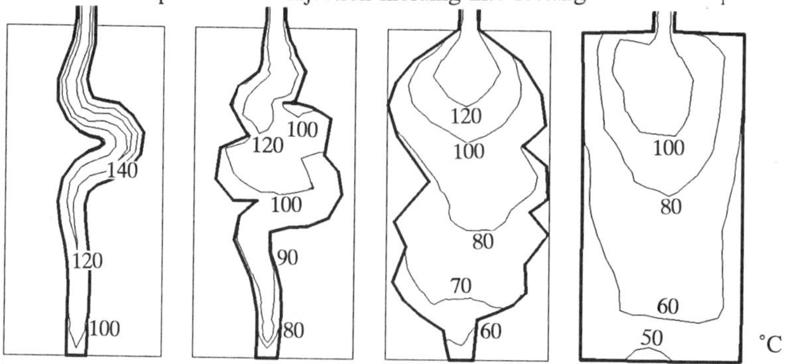

Fig. 5 Calculated temperature distribution for v_i=127mm/s and w=3mm.

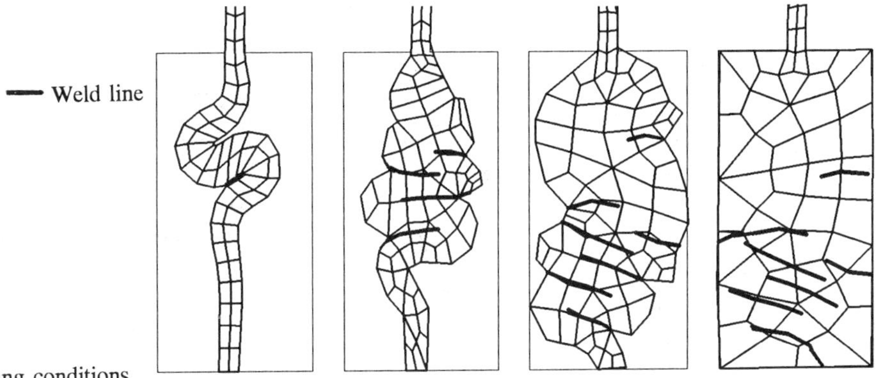

Fig. 6 Calculated material flow in plane–strain injection molding for v_i=42mm/s and w=3mm.

Table 2 Working conditions used in simulation of plane–strain metal injection molding into ringed die.

Width of gate w / mm	2.5
Inner diameter of die / mm	50
Outer diameter of die / mm	20
Thickness of die t / mm	5
Injection speed v_i / mms^{-1}	200
Temperature of die T_d / °C	30, 60

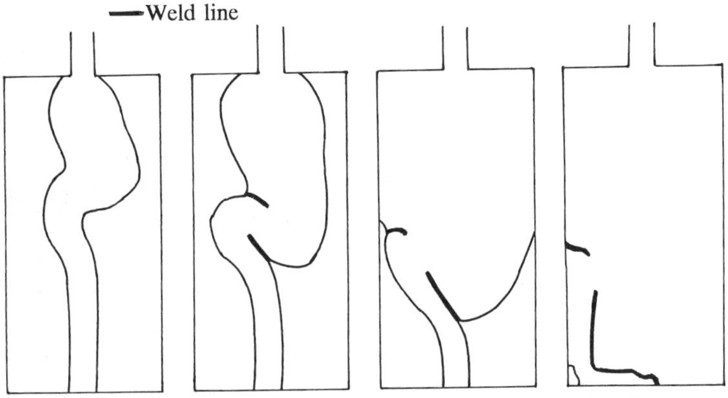

Fig. 7 Experimental material flow for v_i=42mm/s and w=3mm.

and w=3mm is illustrated as a comparison with the calculated result in Fig. 7. Most of the weld lines in the experiment could not be detected by the video camera because of the tight overlapping between the surfaces. The weld lines, however, have a possibility of initiating cracks in the sintering even if the lines are hardly visible to the naked eye. The calculated overall pattern is similar to the experimental one for a steel powder. The present simulation provides useful information for the occurrence of defects such as weld lines.

4 MOLDING INTO RINGED DIE

4.1 Computational conditions

The material flow in plane–strain metal injection molding into a ringed die is simulated by the present method. The working conditions used in the simulation for the ringed die are summarized in Table 2, and the material properties are the same with those in Table 1. Since the filling for the ringed die is more difficult than that for the rectangular die, the number of die gates is increased to two.

4.2 Results

The calculated material flow in plane–strain injection molding into the ringed die for the die temperature T_d=60°C is shown in Fig. 8. Complicated filling occurs due to the collision with the inner surface of the die. The filling behaviour is strongly affected by the geometry of the die.

The calculated material flow for T_d=30°C is illustrated in Fig. 9. The die temperature has a similar tendency to the injection speed.

5 CONCLUSIONS

The metal flow and temperature distribution in metal injection molding are simulated by the viscoplastic and the heat conduction finite element methods, respectively. In the present simulation, the jetting phenomenon during molding is dealt with. The remeshing is the key technique for simulating the

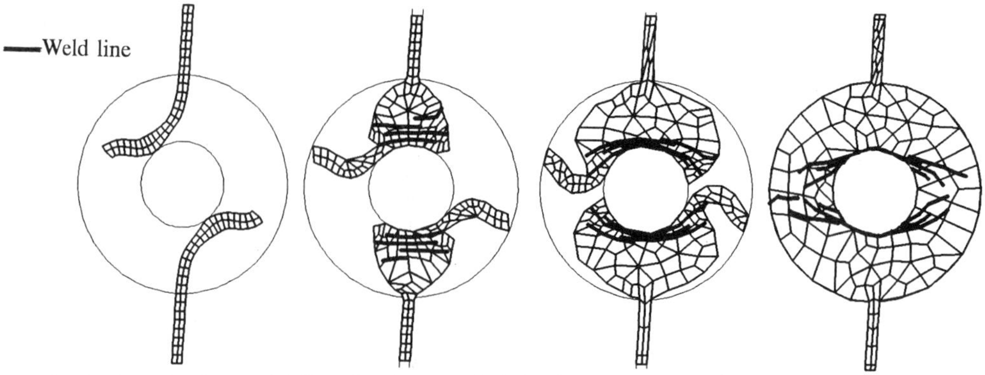

Fig. 8 Calculated material flow in plane–strain injection molding into ringed die for T_d=60°C.

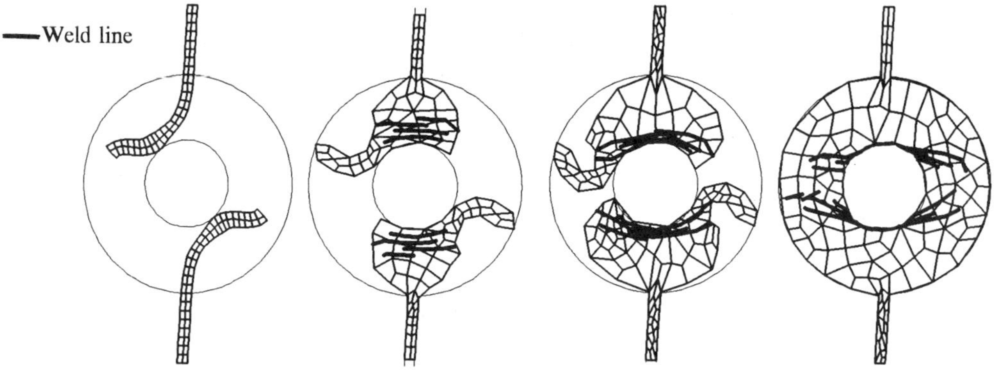

Fig. 9 Calculated material flow for T_d=30°C.

jetting. Although the two–dimensional remeshing schemes have been almost established, existing three–dimensional remeshing schemes do not have enough ability to simulate metal injection molding processes. The three–dimensional jetting is a considerably complicated phenomenon for the simulation, i.e. too much overlapping between the surfaces of the mixture.

The molded parts are sintered to eliminate most of the resulting porosity from a debinding process. The parts greatly shrink during sintering due to the densification. It thus is important to predict the shapes of the sintered products in the process design. The method for simulating nonuniform shrinkage in sintering of powder compacts proposed by Mori et al. (1991, 1992 and 1993) can be utilized for the purpose. The coupling with this method is effective in accomplishing net–shape manufacturing of metal injection molding.

REFERENCES

Gaspervich, J.R. 1991. Practical application of flow analysis in metal injection molding. Int. J. Powder Metallurgy, 27: 133–139.

Mori, K., K. Osakada & T. Oda 1982. Simulation of plane–strain rolling by the rigid–plastic finite element method. Int. J. Mech. Sci., 24: 519–527.

Mori, K., K. Osakada & T. Hirano 1991. Finite element simulation of nonuniform shrinkage during sintering of ceramic products. Hot Isostatic Pressing – Theory and Applications. In M. Koizumi (ed), 29–34. London: Elsevier Applied Science.

Mori, K. 1992. Finite element simulation of nonuniform shrinkage in sintering of ceramic powder compact, Numerical Methods in Industrial Forming Processes, NUMIFORM 92. In J.–L. Chenot et al. (eds), 69–78. Rotterdam: Balkema.

Mori, K., K. Osakada & T. Hirano 1993. Finite element modelling of nonuniform shrinkage in sintering of ceramic powder compact, Computer Aided Innovation of New Materials II. In M. Doyama et al. (eds), 2:1781–1784. Amsterdam: Elsevier Science Publishers.

Osakada, K. J. Nakano & K. Mori 1982. Finite element method for rigid–plastic analysis of metal forming – Formulation for finite deformation. Int. J. Mech. Sci., 24: 459–468.

Piccirillo, N. & D. Lee 1991. Jetting in powder injection molding. Powder Injection Molding, Advances in Powder Metallurgy – 1991. In L.F. Pease & R.. Sansoucy (eds), 2: 119–126.

Simulation of Materials Processing: Theory, Methods and Applications, Shen & Dawson (eds)
© 1995 Balkema, Rotterdam. ISBN 90 5410 553 4

Simulation of grain alignment in mushy-state forging of magnets by distinct element method

Ken-ichiro Mori, Masaaki Otsu, Kozo Osakada & Masanori Shiomi
Department of Mechanical Engineering, Faculty of Engineering Science, Osaka University, Japan

ABSTRACT: A method for simulating the motion of grains in mushy-state forming of magnets is proposed on the basis of the two-dimensional distinct element method. The grains of the magnet are modelled to be a lot of elliptical elements, and the effect of the liquid phase on the motion of the grains is treated as viscous resistance to the movement. The motion of individual grains is obtained by solving the equations of motion for a small time step under the action of the elastic repulsive, frictional and viscous forces. Plastic deformation of a metallic capsule covering the magnet is calculated by the rigid-plastic finite element method, and the obtained motion of the interface between the capsule and magnet is used as a boundary condition in the distinct element simulation. The motion of grains in mushy-state plane-strain upsetting of a rare earth magnet covered with a steel capsule is chosen as an example of the simulation. The degree of grain alignment increases as the reduction in height and the aspect ratio of the grain increase.

1 INTRODUCTION

Although rare earth magnets are generally produced from powder metallurgy, the strength of the sintered magnets is not high. To improve the strength, cast rare earth magnets have been recently developed (Shimoda et al. 1989a and b). The cast magnets are deformed by mushy-state forging or rolling to align the grains mechanically because grains have magnetic anisotropy. This leads to elevation of magnetic properties. In the mushy-state forming, the boundaries of the grains melt and the rotation of the grains becomes easy. The grains cannot be aligned by magnetic forces because the magnet heated to the mushy state loses magnetic properties. Since the magnetic properties are greatly influenced by the degree of grain alignment, it is important to predict the alignment induced by mushy-state forming. It is impossible, however, to simulate the motion of individual grains in mushy-state forming by continuum models such as the finite element method.

To deal with the motion of particles in granular materials, the distinct element method has been developed by Cundall (1979). In this method, a granular material is modelled to an assembly of particles, namely distinct elements, and Newton's equations of motion are solved for individual elements under interactions with neighboring ones. Naturally, the effects of the size, shape and distribution of the particles can be taken into consideration. Although the distinct element method has been mainly used in the field of soil mechanics and geotechnology, Lian et al. (1994) and Tamura et al. (1994) have applied this method to powder forming processes. The distinct element method has a possibility of analysing microscopic behavior in a deforming material.

In the present study, a method for simulating the motion of grains in mushy-state forming of magnets is presented on the basis of the distinct element method. The individual grains in the magnet are modelled to be elliptical elements in the distinct element method.

2 METHOD OF SIMULATION

2.1 Modelling of mushy-state forming

The cast magnets are deformed by mushy-state forming to align the grains mechanically. In mushy-state forming of magnets, the magnet is covered with a metallic capsule not to squeeze the liquid phase out as shown in Fig. 1. In the present study, the distinct element method is used to simulate the motion of the grains in the magnet, and the rigid-plastic finite element method is used for plastic deformation of capsule. For the sake of simplicity, plane-strain deformation is assumed. In the distinct element method, the grains in the magnet are modelled to be a lot of elliptical elements (Ting 1993), and the effect of the liquid phase on the motion of the grains is treated as viscous resistance to the movement. The cross-sectional shapes of the real

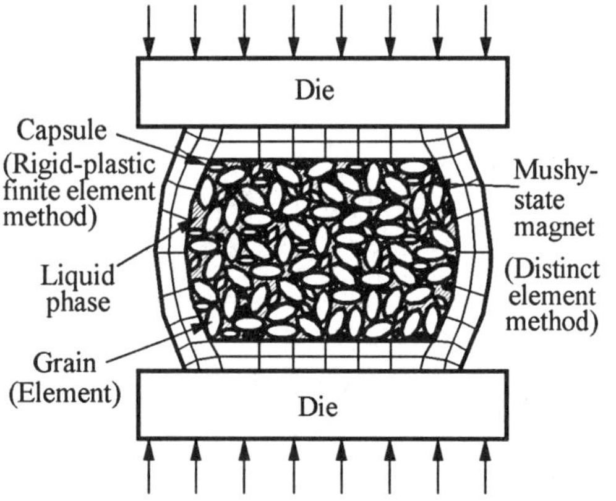

Fig. 1 Model for simulation of mushy-state forming.

grains are not circles but are close to ellipses. The degree of grain alignment induced by the forming is obtained from the motions of grains. The grain alignment results from the use of elliptical elements and not circular ones.

2.2 Distinct element method

Since the grains in the rare earth magnet are intermetallic compounds, the grains do not undergo plastic deformation during the forming. This is convenient to the distinct element simulation because the elements are ordinarily modelled to be rigid.

In the distinct element method, following Newton's equations of motion are solved for individual elements in a small time step as follows:

$$\begin{aligned} m\dot{v} + F_c + F_l &= 0 \\ I\dot{\omega} + M_c &= 0, \end{aligned} \tag{1}$$

where m is the mass of the element, $\dot{v}$ is the acceleration, F_c is the contact force, F_l is the viscous force against the liquid phase, I is the moment of inertia, $\dot{\omega}$ is the angular acceleration and M_c is the contact moment. The contact force and moment are applied to the element only in the case of the contact, whereas the viscous force invariably exists.

The contact force is composed of the elastic repulsive and frictional forces. When two elements touch as shown in Fig. 2, the normal contact force F_n by the elastic repulsion is assumed to be linear with an overlap

$$F_n = \frac{1}{2} K_n \varepsilon, \tag{2}$$

where K_n is the normal spring stiffness. Since the elements are treated to be rigid, the virtual overlap ε between the two elements is introduced to calculate the elastic repulsive force as follows:

$$\varepsilon = AB + A'B' - BB'. \tag{3}$$

The shape of the element is invariable even in the contact. The common normal for the two elements is iteratively determined by solving the following equation by the use of the Newton-Raphson method

$$p\sin\theta - \frac{1}{\alpha} q\cos\theta - \frac{d}{2}\left(1 - \frac{1}{\alpha^2}\right)\sin\theta\cos\theta = 0, \tag{4}$$

where p and q are the x- and y-coordinates at point B', d is the length of the major axis and α is the ratio of the length of the major axis to that of the minor axis, namely the aspect ratio. The treatment of the contact for the elliptical elements is more complex than that for the circular ones mostly used in the distinct element method. The treatment of the contact for the elliptical elements has not be get discussed in detail.

The tangential contact force F_t induced by the elastic repulsion or the friction is expressed by

$$\begin{cases} F_t = \frac{1}{2} K_t \gamma & (K_t \gamma \le \mu F_n) \\ F_t = \mu F_n & (K_t \gamma > \mu F_n), \end{cases} \tag{5}$$

where K_t is the tangential spring stiffness, γ is relative displacement in the tangential direction calculated with the relative movement and rotation between the two elements and μ is the coefficient of friction.

The contact moment M_c is obtained from the normal and tangential contact forces as follows:

$$M_c = \overline{OC} \bullet F_n + \overline{OD} \bullet F_t. \tag{6}$$

From the motion of a circular cylinder in a viscous medium, the viscous force against the liquid phase is approximated by

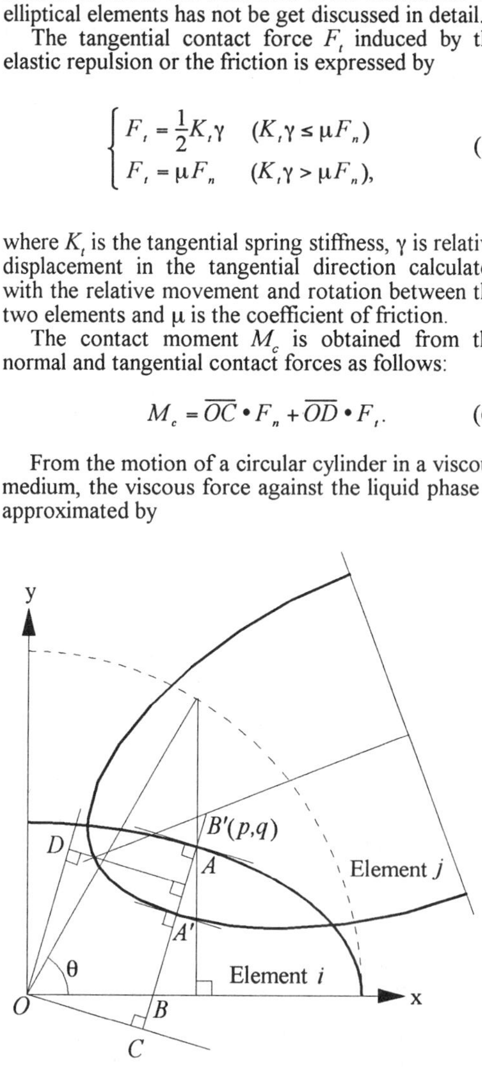

Fig. 2 Contact between two elliptical elements.

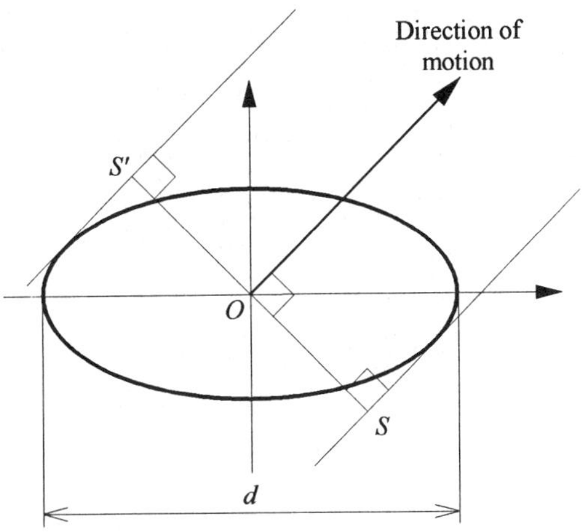

Fig. 3 Projection length perpendicular to motion of element.

$$F_l = \frac{\overline{SS'}}{d} \frac{8\pi\eta_l v}{2\left\{\log\left(\frac{8\eta_l}{\rho_l d v}\right) - \beta\right\} + 1}, \tag{7}$$

where $\overline{SS'}$ is the projection length shown in Fig. 3, η_l is the coefficient of viscosity, v is the velocity, ρ_l is the density of liquid and β is the Euler constant.

2.3 Rigid-plastic finite element method

Plastic deformation of the capsule covering the magnet is calculated by the rigid-plastic finite element method. Although it is desirable that the distinct element method is coupled with the rigid-plastic finite element method, the coupling is complicated. In the distinct element simulation, the pressure at the interface between the capsule and magnet is not accurately calculated because the pressure in the liquid phase is not taken into consideration. Thus, the distinct and the finite element simulations are separately carried out. Plastic deformation of the capsule is first calculated by the rigid-plastic finite element method, and then the distinct element simulation is carried out by using the calculated motion of the interface between the capsule and magnet as a boundary condition.

In the rigid-plastic finite element simulation, the mushy-state magnet is assumed to have a uniform distribution of the macroscopic hydrostatic pressure because the flow stress of the magnet is sufficiently smaller than that of the capsule. The volume constancy of the magnet is introduced into the rigid-plastic finite element formulation by the use of the Lagrange multiplier method (Mori et al. 1991), and the following functional is minimized

$$\Phi = \int_{V_c} \left[\int_0^{\dot{\bar{\varepsilon}}} \bar{\sigma} d\dot{\bar{\varepsilon}}\right] dV + \lambda \dot{\varepsilon}_v V_m, \tag{8}$$

where $\bar{\sigma}$ is the equivalent stress, $\dot{\bar{\varepsilon}}$ is the equivalent strain-rate, λ is the Lagrange multiplier, $\dot{\varepsilon}_v$ is the volumetric strain-rate of the magnet and V_c and V_m are the volumes of the capsule and magnet, respectively. The volumetric strain-rate $\dot{\varepsilon}_v$ of the magnet is expressed to be a function of the nodal velocities at the interface between the capsule and magnet, and only the capsule is divided into the elements. The Lagrange multiplier λ coincides with the macroscopic hydrostatic stress of the magnet in the minimization. The rigid-plastic finite element method for the calculation of capsule is formulated on the basis of the plasticity theory for a material with slight compressibility (Ōsakada et al. 1982 and Mori et al. 1982).

3 SIMULATION OF MUSHY-STATE UPSETTING

3.1 Computational conditions

The motion of grains in mushy-state plane-strain upsetting of a rare earth magnet covered with a mild steel capsule is simulated. The computational conditions used for the finite and the distinct element simulations are given in Tables 1 and 2, respectively. The spring stiffness for the elastic repulsive force is obtained from the elastic finite element simulation of plane-stain side-pressing of an elliptical bar. The relationship between the elastic repulsive force and the displacement is nearly linear.

To set the initial disposition of elements in the magnet before upsetting, the elements are first randomly located under allowing the overlaps between the elements as shown in Fig. 4(a), and then the distinct element simulation is continued without changing the shape of the interface between the capsule and magnet until attaining to a stable state (see Fig. 4(b)). By using this treatment, the initial disposition is almost random.

Table 1 Computational conditions used for rigid-plastic finite element simulation of mushy-state upsetting.

Flow stress (mild steel, 1000°C) /MPa	$\bar{\sigma} = 135\bar{\varepsilon}^{0.1}$
Coefficient of friction between die and capsule	0.25
Ratio of height of magnet to width	1

Table 2 Computational conditions used in distinct element simulation of mushy-state upsetting.

Volume fraction of solid phase ϕ /%	87, 85, 80, 75, 70, 65
Aspect ratio α	2, 2.5, 2.9
Final reduction in height $\Delta h/h$ /%	50
Number of elements	500
Coefficient of friction μ	0.1
Normal spring stiffness K_n /MN mm^{-1}	65
Tangential spring stiffness K_t/MN mm^{-1}	15
Density of solid phase /g mm^{-3}	7.86×10^{-3}
Density of liquid phase ρ_l /g mm^{-3}	6.9×10^{-3}
Coefficient of viscosity η_l /mPa sec	8

3.2 Results

The calculated motion of grains and finite element mesh in mushy-state forging for the volume fraction of solid phase ϕ=80% and the aspect ratio of the grain α=2 are illustrated in Fig. 5. The results obtained by the finite and distinct element simulations are simultaneously shown. The distinct elements are mechanically aligned by the interaction with neighboring ones. The distinct elements are turned to be perpendicular to the upsetting direction as the reduction in height increases. In the case of $\Delta h/h$=50%, the liquid phase is squeezed near the side edge of the magnet.

The calculated motion of grains and finite element mesh in mushy-state forging for ϕ=80% and α=2.9 are shown in Fig. 6. The tendency is similar to that for α=2.

The calculated variation of the average angle of the major axes of elements with the reduction in height for α=2 is shown in Fig. 7. The decreasing rate is relatively large up to 30% reduction. Although the number of elements in the distinct element simulation is not enough to that for the real magnets, the calculated tendency is similar even if the number is increased.

The calculated standard deviation of the angle of the major axes for α=2 is plotted in Fig. 8. Not only the average angle but also the standard deviation are decreased by upsetting.

The calculated distribution of the average angle of the major axes of elements for ϕ=80%, α=2 and $\Delta h/h$=50% is shown in Fig. 9. The degree of grain alignment is large near the center of the magnet.

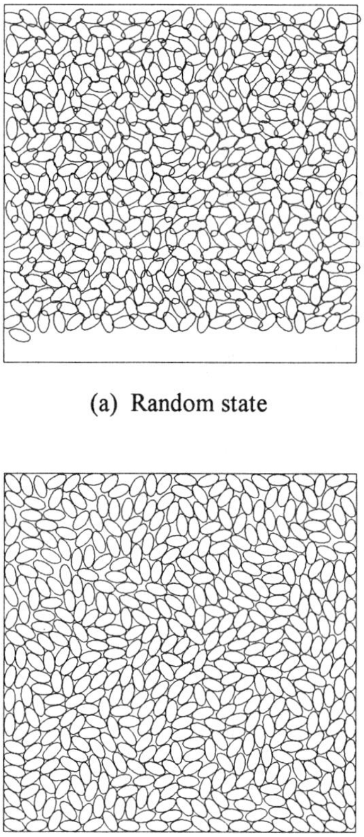

(a) Random state

(b) Stable state

Fig. 4 Initial disposition of elements for ϕ=80% and α=2.

The effect of the volume fraction of solid phase on the degree of grain alignment for α=2 and $\Delta h/h$=50% is plotted in Fig. 10. The average angle of the major axes of elements has a minimum value near 85% volume fraction. This is due to that the rotation of the elements becomes difficult in the case of a high volume fraction.

The effect of the aspect ratio of the grain on the degree of grain alignment for ϕ=80% and $\Delta h/h$=50% is illustrated in Fig. 11. The degree of grain alignment increases as the aspect ratio increases. When the aspect ratio exceeds 3, it is difficult to set the initial disposition of distinct elements, i. e. the solution does not attained to a stable state.

The calculated average angle of the elements is compared with the experimental one for ϕ=80% and α=2.5 in Fig. 12. Since it is not easy to measure the angles of grains for real rare earth magnets, model materials are used in the experiment. The capsule is plasticine and the grains are acrylic resin. Although

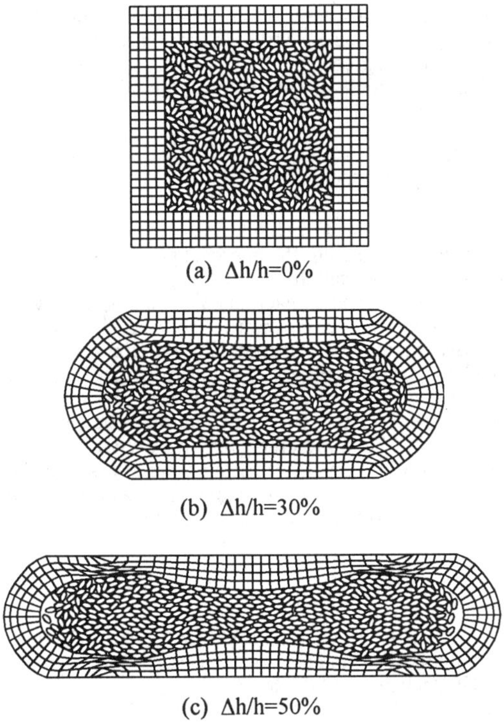

Fig. 5 Calculated motion of grains and finite element mesh for ϕ=80% and α=2.

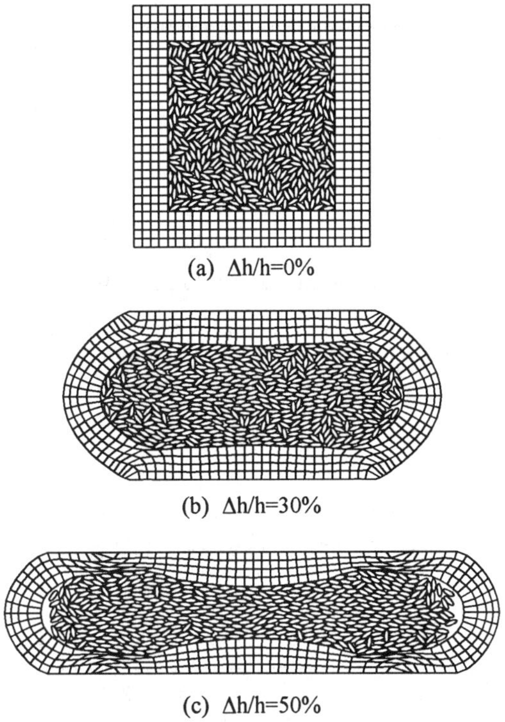

Fig. 6 Calculated motion of grains and finite element mesh for ϕ=80% and α=2.9.

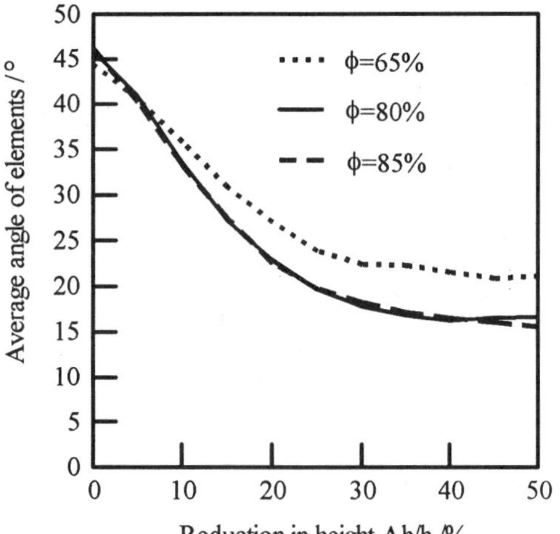

Fig. 7 Variation of average angle of elements with reduction in height for α=2.

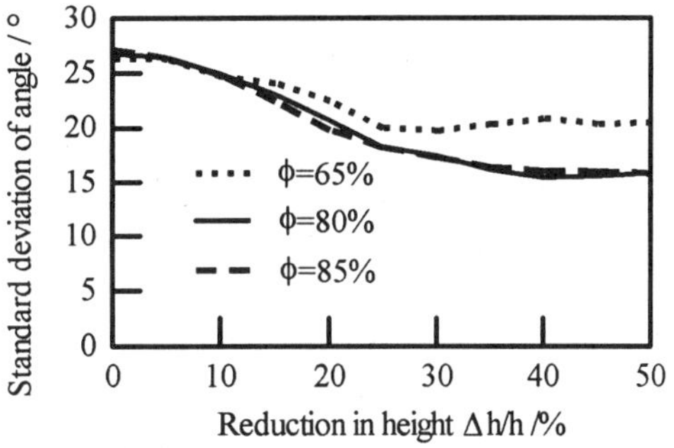

Fig. 8 Variation of standard deviation of angle with reduction in height for α=2.

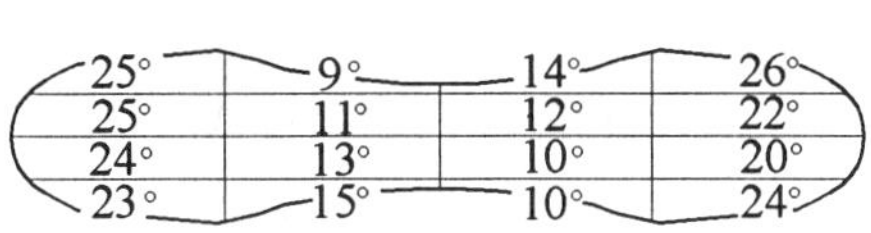

Fig. 9 Distribution of average angle of elements for ϕ=80%, α=2 and Δh/h=50%.

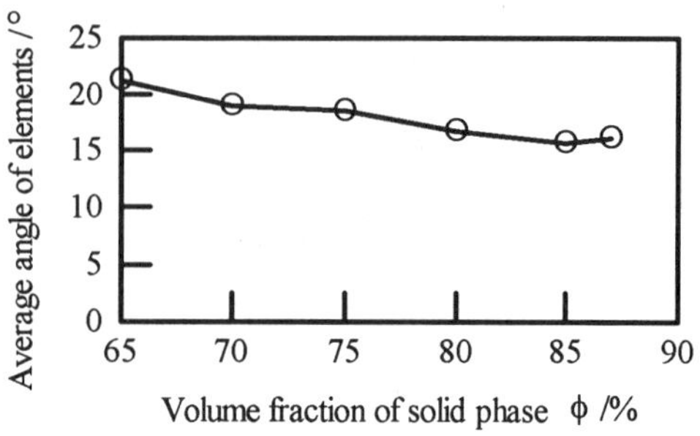

Fig. 10 Relationship between average angle of elements and volume fraction of solid phase for α=2 and $\Delta h/h$=50%.

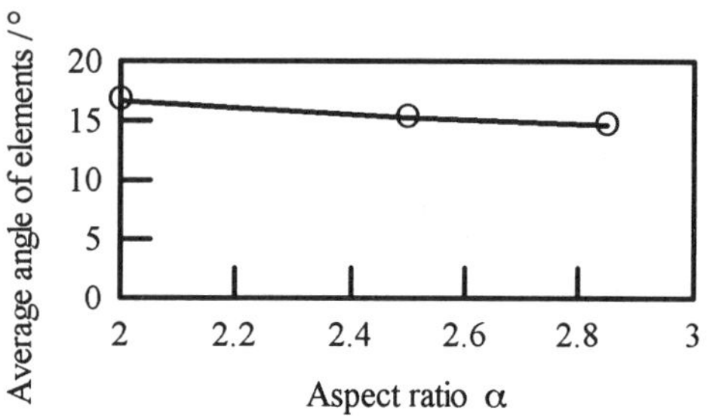

Fig. 11 Relationship between average angle of elements and aspect ratio for ϕ=80% and $\Delta h/h$=50%.

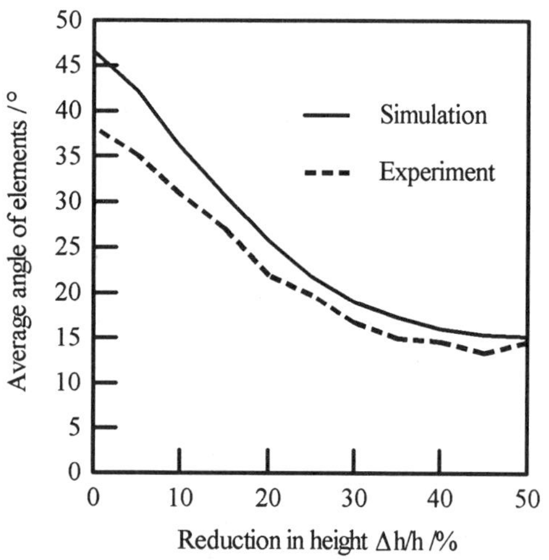

Fig. 12 Comparison between calculated and experimental average angles of elements for ϕ=80% and α=2.5%.

the initial dispositions of grains for the calculation and experiment are a little different, the two results are in good agreement. It is not easy to control the initial dispositions accurately in both calculation and experiment.

4 CONCLUSION

The finite element method does not have enough capability to simulate the microscopic behavior in the deforming material. In the present simulation, both microscopic and macroscopic methods were employed to compensate for each other's disadvantages. The distinct and the finite element methods were employed for portions necessary for microscopic and macroscopic information, respectively. This leads to more efficiency of simulation. Such a method may have wide range of application in the field of forming processes. For the coupling of the microscopic and macroscopic methods, attractive problems still remain.

REFERENCES

Cundall, P.A. & O. D. L. Strack 1979. A discrete numerical model for granular assemblies. Geotechnique 29:47-65.

Lian, J. & S. Shima 1994. Powder assembly simulation by particle dynamics method. Int. J. Numerical Methods in Engineering 37:763-775.

Mori, K., K. Osakada & T. Oda 1982. Simulation of plane-strain rolling by the rigid-plastic finite element method. Int. J. Mech. Sci. 24-9:519-527.

Mori, K., K. Osakada & M. Shiomi 1991. Finite element modelling of forming process of solid metal with liquid phase. J. Materials Processing Tech. 27:111-118.

Osakada, K., J. Nakano & K. Mori 1982. Finite element method for rigid-plastic analysis of metal forming - Formulation for finite deformation. Int. J. Mech. Sci. 24:459.

Shimoda, T., K. Akioka, O. Kobayashi & T. Yamagami 1989a. Hot-working behavior of cast Pr-Fe-B magnets. IEEE Trans. Magnetics 25-5:4099-4104.

Shimoda, T., K. Akioka, O. Kobayashi & T. Yamagami 1989b. Hot-worked anisotropic Pr-Fe-B magnets. 10th Int. Workshop on Rare-Earth Magnets and Their Application 389-398.

Tamura, S., T. Aizawa & J. Kihara 1994. Three-dimensional granular modeling for metallic powder compaction and flow analysis. J. Materials Processing Tech. 42:197-207.

Ting, J. M., M. Khwaja, L. R. Meachum & J. D. Rowell 1993. An ellipse-based discrete element model for granular materials. Int. J. Numerical and Analytical Methods in Geomechanics 17:603-623.

Simulation of Materials Processing: Theory, Methods and Applications, Shen & Dawson (eds)
© 1995 Balkema, Rotterdam. ISBN 90 5410 553 4

Numerical modelling of the injection molding of thermosets

K.T. Nguyen, K.K. Kabanemi & D.M.Gao
Industrial Materials Institute, National Research Council Canada, Boucherville, Que., Canada

ABSTRACT: A general model for predicting the filling of thin cavity by thermosetting polymers was developed. The simulation allows prediction of the flow front and the distribution of the pressure, temperature and the degree of cure throughout the filled domain at any time step. The results were shown for a test case and the influence of the molding conditions on the distribution of the degree of conversion at the end of the filling was given in detail. The numerical approach and algorithm presented here could be useful tool in optimization of industrial forming processes involving complex 3-D calculations.

1. INTRODUCTION

Thermosetting polymers have found a wide range of applications in electronics, aerospace and automotive industry. The increasing popularity of thermosets are due not only to their excellent mechanical and heat resistance properties but also to their greater flexibility in tailoring the ultimate properties by means of composite systems. Injection molding is one of the preferred process for the manufacturing of thermoset parts due to its high production rate and dimensional precision. However, injection molding is a very complicated process, and with thermosets, further complications arise from the interaction between the thermo-mechanical history imposed on the material and the curing reaction since the thermal and rheological properties of the material depend strongly on the degree of reaction. For thermoset materials, the quality of the final products depends largely on the ability to control the reaction. An incomplete reaction results in parts with inferior mechanical properties. A premature curing may give rise to short shots or defects such as porosity, voids or sink marks. A non-uniformly cured part may produce high residual stresses and thus warpage. To achieve the desired properties of the final product and to optimize the process, a strict control of the resin characteristics and a good understanding of the effects of the process history on the material behavior is required. Numerical modelling is a very effective and relatively inexpensive tool to explore various design and processing parameters.

In this work, a model has been developed to simulate the filling of a thin cavity by a thermosetting polymers. The simulation allows the prediction of the flow front position, the distribution of the pressure, temperature and the degree of cure throughout the filled domain at any time step. The influence of the molding conditions on the extent of reaction is demonstrated.

2. MODELLING OF THE REACTION KINETICS

During the injection molding process, the temperature changes with time so that the curing reaction actually occurs under non-isothermal condition. However, during each time increment in the filling stage, it is assumed that the change in temperature is relatively small and therefore the whole process can be considered as a series of quasi-isothermal steps. Thus, the reaction kinetics is assumed to obey the following equation for isothermal curing:

$$\frac{d\alpha}{dt} = Z * e^{-\Delta E/RT} * (1-\alpha)^n \qquad (1)$$

where Z is the rate constant, ΔE is the activation energy, R the gas constant, T the temperature, t the time and n the order of reaction.

The kinetics parameters for the material can be measured using the Differential Scanning Calorimeter (DSC).

3. GOVERNING EQUATIONS

During the filling stage, it is assumed that the thickness of the cavity is very small compared to the other two dimensions. Then the flow in the cavity can be approximated by the usual generalized Hele-Shaw model [Hieber and Shen 1980, Chiang et al. 1991, Shen 1992]. The governing equations for the non-isothermal flow of a generalized Newtonian fluid are given by:

$$\frac{\partial}{\partial x}S\frac{\partial p}{\partial x}+\frac{\partial}{\partial y}S\frac{\partial p}{\partial y}=0 \ , \qquad (2)$$

$$\rho C_p(\frac{\partial T}{\partial t}+u\frac{\partial T}{\partial x}+v\frac{\partial T}{\partial y})=\frac{\partial}{\partial z}k\frac{\partial T}{\partial z}+\eta\dot{\gamma}^2+\dot{Q} \ , \qquad (3)$$

where S is the coefficient of fluidity defined as:

$$S=\int_0^b\frac{z^2}{\eta(T,\dot{\gamma},\alpha)}dz \ , \qquad (4)$$

b is half-thickness of the cavity, ρ is the density, p is the pressure, u and v are the components of the velocity, respectively, in the x and y direction, T is the temperature, C_p is the heat capacity, k is the thermal conductivity, η is the viscosity, $\dot{\gamma}$ is the shear rate, α is the extent of reaction, x and y are the coordinates of the middle plane and z is coordinate in the gapwise direction. $\dot{Q}$ represents the rate of heat generation due to the curing reaction of the resin and is given by:

$$\dot{Q}=\Delta H_t\frac{d\alpha}{dt}, \qquad (5)$$

where ΔH_t is the heat generation of the complete reaction.

In order to take into account the effect of convection on the degree of reaction, the time derivative in equation (1) is replaced by the material derivative, resulting in a transport equation for α:

$$\frac{\partial\alpha}{\partial t}+u\frac{\partial\alpha}{\partial x}+v\frac{\partial\alpha}{\partial y}=Z*e^{-\Delta E/RT}*(1-\alpha)^n \qquad (6)$$

Equations (2), (3) and (6) can be solved simultaneously with appropriate boundary conditions to yield the distribution of pressure, temperature and the degree of cure as functions of time. The coupling between the flow simulation and the curing simulation is through the heat generation in the energy equation and the dependence of viscosity on the degree of reaction. In this work, the solution of equation (6) is of particular interest since it is purely of hyperbolic type and needs special treatment.

4. NUMERICAL IMPLEMENTATION

The pressure equation is solved using the standard Galerkin finite element method [Zienkiewicz 1977] on three-node element mesh. The boundary conditions are as following:

(i) At the gate either the pressure or the flow rate as a function of time is prescribed;

(ii) On the solid wall, the normal velocity vanish; and

(iii) At the flow front, the pressure is set to zero.

A control volume approach is employed to track the flow front advancement [Hirt and Nichols 1981, Floryan and Rasmussen 1989]. A fill factor F, the value of which is between 0 and 1, is associated to each element to indicate the percentage of filling. An F value of 1 represents a completely filled element and 0 an empty element. With the above concept, for every time step, a calculation of pressure equation (1) is performed in order to obtain the velocity distribution in the filled domain. Using the flow rate information based on the velocity field, the fill factor of each element is evaluated by applying the principle of the conservation of mass. Finally, a newly filled domain can be easily defined using the updated fill factor field.

The temperature and curing equations are integrated with known kinematics, combining a finite element method in the middle plane and a collocation procedure in the gapwise direction [Dupret and Vanderschuren 1988]. The hyperbolic nature of these equations is dealt with using a modified SUPG method [Hughes and Brooks 1982]. The time integration of the underlying equations is

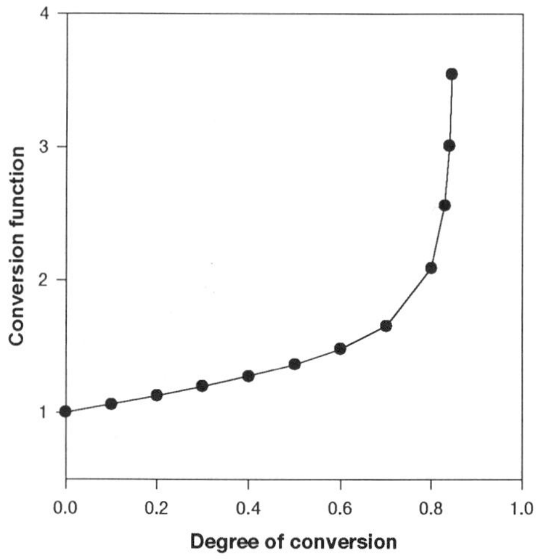

Figure 1: Conversion function for the resin

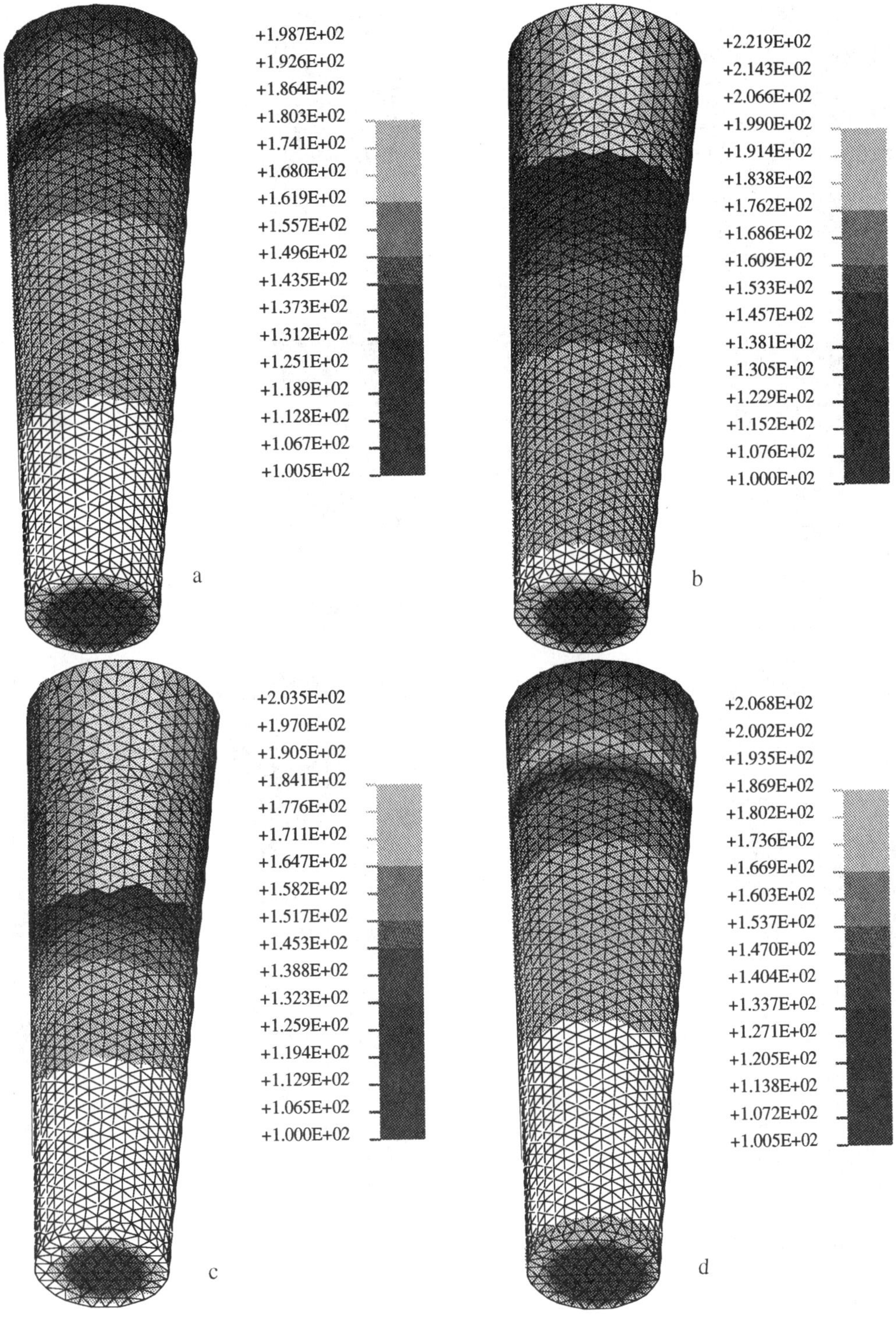

Figure 2: Filling of preform part. Average temperatures (in $^\circ C$) at the end of filling (a and d) and for a premature curing: short shots (b and c).

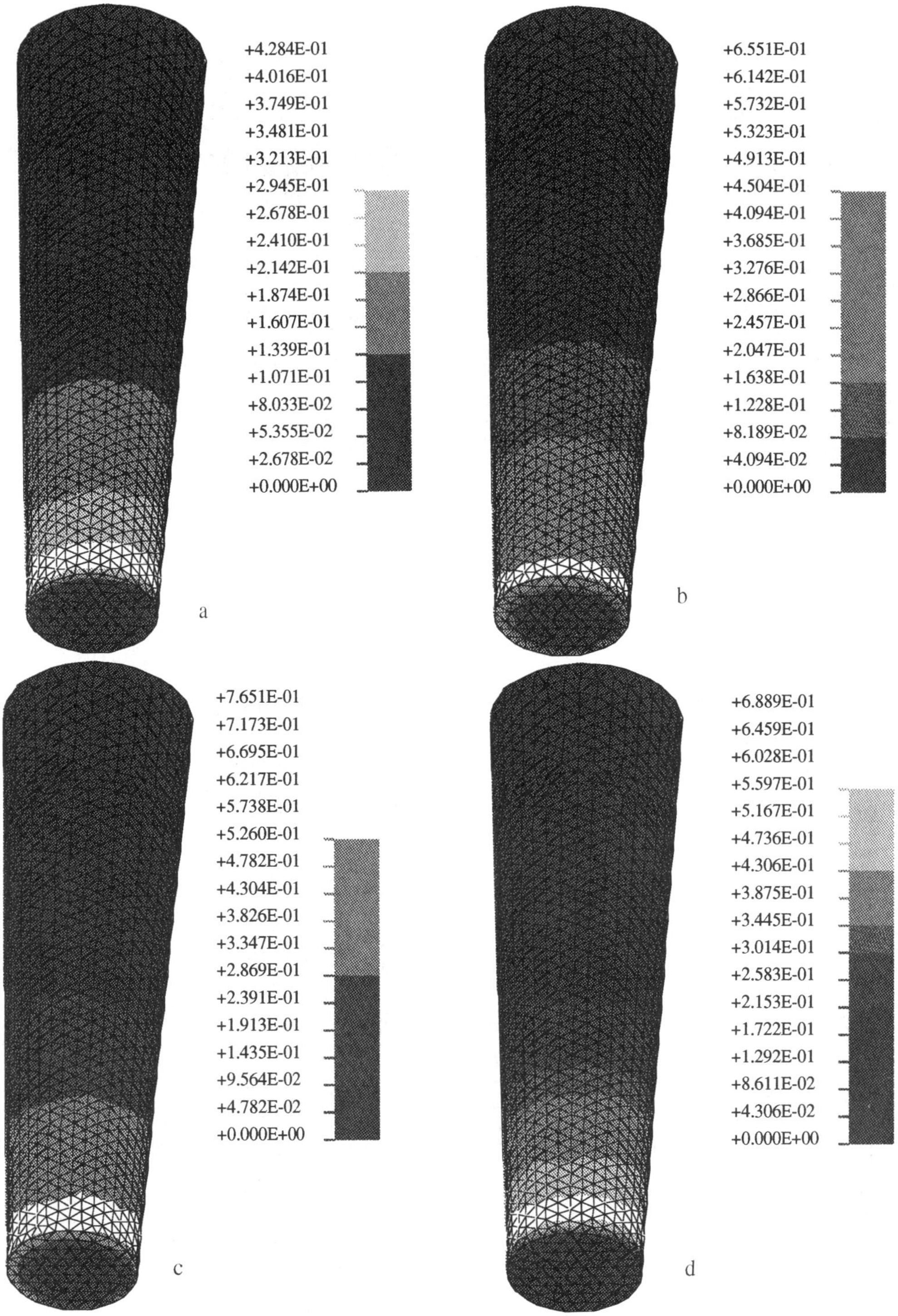

Figure 3: Filling of preform part. Average degree of cure at the end of filling (a and d) and for a premature curing: short shots (b and c).

performed by a first-order semi-implicit Euler method. The classical CFL (Courant-Friedrichs-Lewy) criterion is used for the automatic determination of the time step to guarantee the stability of the scheme. The resulting transport equations are then solved layer by layer in a decoupled fashion. The energy equation is first solved as a advection-diffusion equation, and then the curing equation is integrated with known temperature.

The finite element discretization used for both the temperature field and the degree of cure are those of a 3-node continuous linear triangular element.

The boundary conditions are as following:

(i) At the gate, the temperature and the degree of cure are imposed;

(ii) On the solid wall of the thickness, the temperature is imposed while on the lateral wall the flux vanishes;

(iii) At the front, the inlet temperature and inlet degree of cure are imposed, since the front region is infinitely thin and the residence time of material points in that region is small compared to the characteristic time scale of the filling.

5. RESULTS AND DISCUSSIONS

In this section, the simulation was performed on a non-planar part as depicted in Fig. 2. The part is a preform which is a tapered annular cylinder with one open end and one closed end. The preform has a radius of about 25mm. The part is about 3mm thick and 70mm long. The material employed in this study is an epoxy resin which is commonly used in microchip encapsulation. The viscosity is modelled using the following equation:

$$\eta = A_\eta * e^{\Delta E_\eta / RT} * \dot{\gamma}^{N-1} * \left(\frac{\alpha_g}{\alpha_g - \alpha} \right)^{C_1 + C_2 \alpha} \qquad (7)$$

where A_η, C_1 and C_2 are constants, R is the gas constant, ΔE_η the activation energy, N the power law index and α_g the gel point.

The parameters characterizing the reaction kinetics and the viscosity of the resin are summarized in Table 1 [Laroche et al. 1994]. The molding conditions are given in Table 2.

Figure 1 shows the effect of curing on the viscosity, as represented by the conversion function, i.e., the last term in equation (7). Up to about 50% conversion, the conversion function remains essentially unchanged. At 185ºC, it takes about 11s to reach 50% conversion. Beyond this point, the viscosity begins to increase very fast.

Table 1: Material Properties of the resin system

Material Constants	Resin A
ρ (g/cm^3)	1.83
k (W/m/ºK)	0.507
C_p (J/g/ºK)	0.675
Z (1/s)	3.7095E7
ΔE/R (ºK)	9345.2
n	0.410
ΔH_t (J/g)	45.6
A_η (Pa*s^n)	0.891E-11
ΔE/R (ºK)	13600
α_g	0.85
C_1	0.50
C_2	-0.30
N	0.620

Table 2: Molding Conditions

Case I : Figs 2a, 3a	
Flow Rate (cm^3/s)	1.86
Mold Temperature (ºC	185.
Case II : Figs. 2b, 3b	
Flow Rate (cm^3/s)	1.86
Mold Temperature (ºC)	200.
Case III : Figs. 2c, 3c	
Flow Rate (cm^3/s)	0.74
Mold Temperature (ºC)	185.
Case IV : Figs. 2d, 3d	
Flow Rate (cm^3/s)	1.3
Mold Temperature (ºC)	185.
Melt Temperature (ºC)	100.0

To illustrate the effect of molding conditions on the degree of cure and the pressure requirement, different combinations of flow rates, 1.86cm^3/s, 1.3cm^3/s and 0.74cm^3/s, and mold temperatures, 185ºC and 200ºC, were used. At 1.86cm^3/s, the part is filled in about 10s while at 0.74cm^3/s it takes about 25s.

Representative results are shown in Figures 2 and 3. When the mold temperature is relatively low (185°C) and the flow rate is not too slow (1.86cm^3/s and 1.3cm^3/s) the part is filled completely. The average temperature across the thickness of the part at the end of the filling stage for the above two cases are presented in Figures 2a and 2d. At higher mold temperature (200°C) or at long filling time (0.74cm^3/s) premature curing occurred and the part cannot be filled completely resulting in short shots. Figures 2b and 2c show the average temperature distribution before the flow stops due to premature curing. At the flow front the temperature is low due to the effect of the fountain flow which carries the material from the core to the front. Between the gate and the flow front, the temperature is higher resulted from the heat conduction from the cavity walls, the heat generation due to the curing reaction of the resin and the viscous heating. The influence of the temperature field and the material properties on the degree of cure is highlighted in Figure 3. Since the reaction is faster at higher temperature, the maximum conversion is reached between the gate and the flow front of the sample where a "hot spot" is located while in the front region the conversion remains low. In particular, high degree of cure is reached with low flow rate and high mold temperature conditions (Figures 3b and 3c).

6. CONCLUSIONS

In this work, a general model for predicting the filling of thin cavity by a thermosetting polymers has been presented. The results were shown for a test case and the influence of the molding conditions on the distribution of the degree of conversion at the end of the filling was given in detail. Under extreme conditions, premature curing may occur causing short shots. The non-uniform distribution of the degree of cure may give rise to non-uniform density distribution during the packing phase and hence residual stresses and non-uniform shrinkage causing warpage of the part. In view of numerical results shown above, the numerical approach and algorithm presented here could be useful tool in optimization of industrial forming processes involving complex 3-D calculations.

REFERENCES

Chiang, H.H., Hieber, C.A. and Wang, K.K. 1991. A unified simulation of the filling and postfilling stages in injection molding. Poly. Eng. Sci. 31:116-124.

Dupret, F. and Vanderschuren, L. 1988. Calculation of the temperature field in injection molding. AICHE J. 34:1959-1972.

Floryan, J.M. and Rasmussen, H. 1989. Numerical methods for viscous flows with moving boundaries. Appl. Mech. Rev. 42:323-341.

Hieber, C.A. and Shen, S.F. 1980. A finite element/finite difference simulation of the injection molding filling process. J. Non-Newtonian Fluid Mech. 7:1-32.

Hirt, C.W. and Nichols, B.D., 1981. Volume of fluid (VOF) method for the dynamics of free boundaries. J. Comp. Physics 39:201-225.

Hughes, T.J.R and Brooks A. 1982. A theoretical framework for Petrov-Galerkin methods with discontinuous weighting functions: Application to the streamline-upwind procedure. In *Finite elements in fluids vol. 4*. Wiley: Chichester.

Laroche, D., Nguyen, K.T., Boutin, L. and Bellefleur, E. 1994. Optimization of the transfer molding process in microchip encapsulation. In K. Himasekhar, V. Prasad, T.A. Osswald and G. Batch (eds), Advances in computer aided engineering (CAE) in polymer processing, HTD-Vol 283:209-223.

Shen, S.F. 1992. Grappling with the simulation of non-Newtonian flows in polymer processing. Int. J. Num. Meth. Eng. 34:701-723.

Zienkiewicz, O.C. 1977. *The finite element method, 3rd ed.* McGraw Hill: New York.

Simulation of Materials Processing: Theory, Methods and Applications, Shen & Dawson (eds)
© 1995 Balkema, Rotterdam. ISBN 90 5410 553 4

Process parameter setting for injection molding of plastic parts

K.H.Tan & M.M.F.Yuen
Department of Mechanical Engineering, The Hong Kong University of Science and Technology, Clear Water Bay, Kowloon, Hong Kong

ABSTRACT : The quality criteria of an injection molded product are formulated into a set of fuzzy inequalities by fuzzy set theory. The processing condition for satisfying all these fuzzy inequalities is found by an iterative search algorithm. The search algorithm is implemented by integrating a knowledge based system to a numerical simulation system. An example for molding a flat plate is given to illustrate this method.

1. INTRODUCTION

Optimization of product quality in injection molding process is a difficult task to perform due to its complexity. The quality of an injection molded product is affected by polymer properties, mold geometric parameters and processing condition parameters. The optimization of processing condition is traditionally carried out by a machine operator after the mold is designed and manufactured. Using numerical simulation, it is possible to optimize the injection molding processing condition in the mold design phase. Product quality is evaluated and optimized based on the result of the simulation analysis.

Optimization of the processing condition by process simulation usually relies on the knowledge of a decision maker (DM). Nagarsheth [1] combined flow simulation analysis and statistical techniques to define an optimal region for the critical processing condition. The critical processing condition parameters being used were mold temperature, melt temperature, injection time and injection pressure. Bourdon [2] explained the setting of injection molding machines by means of a process simulation. The procedure involved establishment of a working window, selection of a working point and establishment of parameter correlations at the working point. Pandelidis and Zou [3] broke down the optimization problem into three parts. An approximate feasible molding space (AFMS) was first determined to constraint the search space for the optimization algorithm. Quality was quantified as a function of flow simulation outputs and constituted

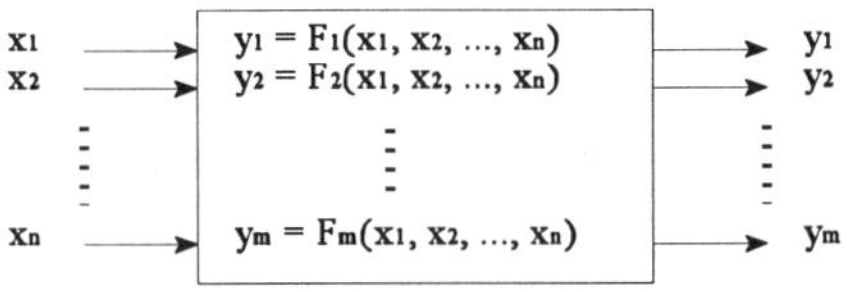

Fig.1. MIMO system representation of a numerical simulation system.

the objective function that must be minimized. The resulting optimization was solved by an iterative search in the constrained space based on conjugate direction method.

The current study is to provide an integrated numerical simulation and knowledge based approach in optimizing the processing condition for the injection molding process. Fuzzy set theory is used to formulate the DM's product quality criteria. These criteria are expressed by a set of fuzzy inequalities. Knowledge on satisfying these fuzzy inequalities are represented by a set of fuzzy rules. These fuzzy rules represent the correlation between each input and output of the system. Fuzzy inference is performed to suggest a processing condition adjustment for the next simulation analysis if any one of the fuzzy inequalities is not satisfied. The procedure repeats itself until all the fuzzy goals are satisfied.

In this paper, the problem formulation will be described first, then the details of the solution scheme will be worked out. An example will also be given for illustration.

2. PROBLEM FORMULATION

A numerical simulation system for injection molding process can be considered as a multi-input/multi-output (MIMO) system, and the system behaviour can be represented by a set of multivariable equations, as shown in figure 1.

The inputs to the system are the processing condition parameters and the outputs are the product quality characteristics. Let the number of inputs be n while that of the outputs be m. Assuming that there are p outputs must be minimized and (m - p) outputs are under constraint, and the n inputs are upper and lower bounded. This optimization problem can be described by a constrained multiobjective optimization problem as shown below,

Minimize

$$y_j = F_j(x_1, x_2, \ldots, x_n) \quad , j = 1 \text{ to } p \qquad (1)$$

Subjected to

$$y_j = F_j(x_1, x_2, \ldots, x_n) \le b_j \quad , j = p \text{ to } m$$

$$x_i^{\min} \le x_i \le x_i^{\max} \quad , i = 1 \text{ to } n$$

As described by Sakawa [4], the problem can be transformed into a set of fuzzy inequalities,

$$y_j = F_j(x_1, x_2, \ldots, x_n) \preceq b_j \quad , j = 1 \text{ to } m$$

$$x_i^{\min} \le x_i \le x_i^{\max} \quad , i = 1 \text{ to } n \qquad (2)$$

where the symbol "$\preceq$" denotes a fuzzy version of the ordinary inequality "$\le$", and b_1, b_2, ... and b_p are aspiration levels of the DM. The objective functions should be essentially smaller than or equal to these values.

Satisfaction of the j-th inequality may be expressed by a fuzzy goal which has a linear membership function,

$$\mu_{satisfied}(y_j) = \begin{cases} 1 & ; y_j \le b_j \\ 1 - \dfrac{y_j - b_j}{d_j} & ; b_j \le y_j \le b_j + d_j \\ 0 & ; b_j + d_j \le y_j \end{cases} \qquad (3)$$

where each d_j is a subjectively chosen constant expressing the limit of the admissible violation of the j-th inequality. It is assumed that the j-th membership function should be 1 if the j-th inequality is well satisfied, 0 if the j-th inequality is violated beyond its limit d_j, and linear from 0 to 1, as shown in figure 2.

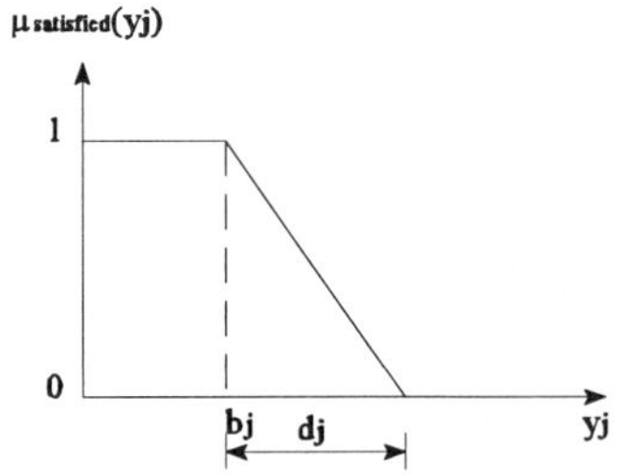

FIG. 2. Linear membership function.

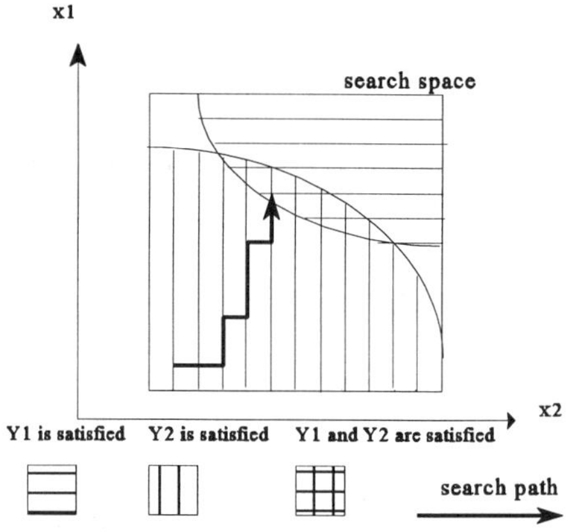

FIG. 3. Search for satisfying all fuzzy goals.

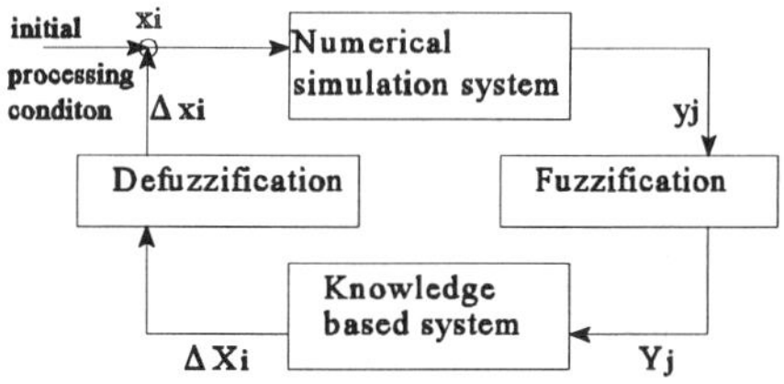

FIG. 4. Optimization approach integrating numerical simulation and knowledge based system.

Based on the above formulation, the problem is equivalent to finding a set of x_i such that all the fuzzy goals are satisfied.

3. SOLUTION SCHEME

This section presents an iterative search algorithm based on process knowledge to search for a set of x_i which satisfies all the fuzzy goals as described in equation 2. The search algorithm is illustrated in figure 3.

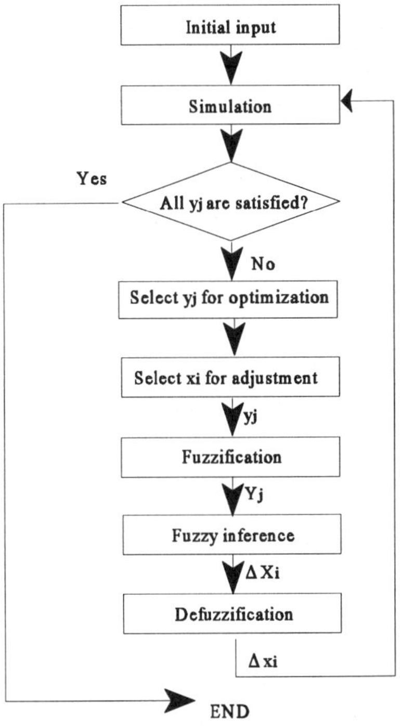

FIG. 5. Iterative search algorithm

The output (y_j) of the simulation analysis will be fuzzified to a linguistic variable [5] (Y_j) according to the associated membership functions. Adjustment of processing condition parameter (ΔX_i), which is also a linguistic variable, will be suggested by the knowledge based system if any one of the fuzzy goals is not satisfied. This suggestion is defuzzified and a numerical value of input adjustment (Δx_i) is obtained for the next simulation analysis. This procedure is repeated until all the fuzzy goals are satisfied. The solution scheme is illustrated in figure 4.

In the following sections, the iterative search algorithm will be first outlined. Then the membership functions of Y_j and ΔX_i, the knowledge representation and the fuzzy inference scheme will be explained in detail. Finally, simple learning algorithms are proposed to improve the performance of the system during the search.

3.1 ITERATIVE SEARCH ALGORITHM

The fuzzy goals will be optimized according to their priority. Fuzzy goals with same priority will be optimized at the same time. Those inputs having the strongest correlation with the outputs will be adjusted. Fuzzy inference is then carried out to find the values of

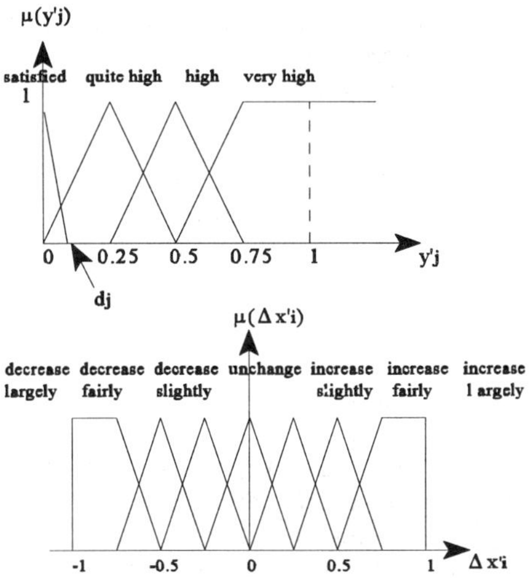

Fig. 6. Membership function of Yj and ΔXi.

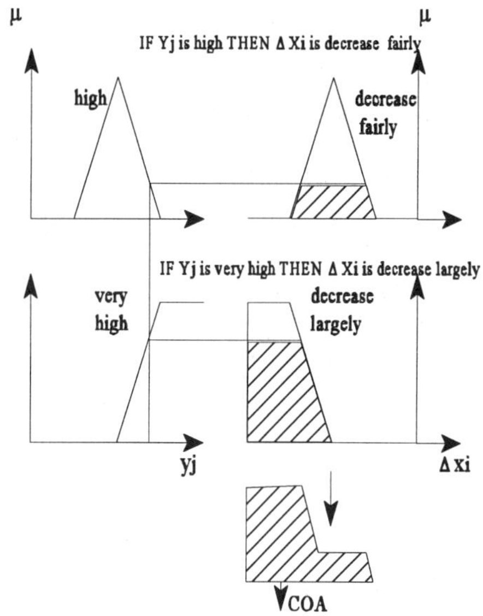

FIG. 7. Fuzzy inference and defuzzification.

input adjustment. Finally, the fuzzy goals will be evaluated by simulation analysis and the procedure repeats until all the fuzzy goals are satisfied. The algorithm is illustrated in figure 5.

3.2 MEMBERSHIP FUNCTON

Assuming that the fuzzy goal is well satisfied if $y_j = y_{satisfied}$, and very unsatisfied if $y_j \geq y_{very\ high}$, and the value of Δx_i is limited from $-\Delta x_{max}$ to $+\Delta x_{max}$. Then the values of y_j and Δxi can be expressed by linguistic variables with membership functions as shown in figure 6. y_j and Δx_i are normalized to y'_j and $\Delta x'_i$ respectively,

$$y'_j = \frac{y_j - y_{satisfied}}{y_{very\ high} - y_{satisfied}} \qquad (4)$$

$$\Delta x'_i = \frac{\Delta x_i}{2\,\Delta x_{max}}$$

3.3 KNOWLEDGE REPRESENTATION

The degree of input adjustment depends on the degree of satisfaction of the fuzzy goal and the correlation of the input and output. This is shown in the following fuzzy matrix for the case of positive correlation.

	unchange	decrease slightly	decrease fairly	decease largely
satisfied	μ_{ij}	0	0	0
quite high	0	μ_{ij}	0	0
high	0	0	μ_{ij}	0
very high	0	0	0	μ_{ij}

Here μ_{ij} is the degree of membership of the correlation. It is a subjective measurement of the strength of the correlation between x_i and y_j. This fuzzy matrix can be represented by a set of fuzzy rules:

IF Yj is satisfied	THEN ΔXi is unchange	τij
IF Yj is quite high	THEN ΔXi is decrease slightly	τij
IF Yj is high	THEN ΔXi is decrease fairly	τij
IF Yj is very high	THEN ΔXi is decrease largely	τij

where τij (=μij) is the truth value of the fuzzy rules.

3.4 FUZZY INFERENCE

The above fuzzy rules are used for fuzzy inference. The rule of inference is the compositional rule [5]. For example, if Y_j is high, then the inference is :

Yj is high
IF Y_j is high THEN ΔX_i is decrease fairly τ_{ij}

ΔX_i is decrease fairly

where $\mu_{decrease\ fairly}(\Delta x_i) = \mu_{high}(y_j) \wedge \tau_{ij}$.

As described by Berenji [6], because of the partial matching attribute of fuzzy rules and the fact that the preconditions of the rules do overlap, usually more than one fuzzy rule can fire at one time. Max-Min

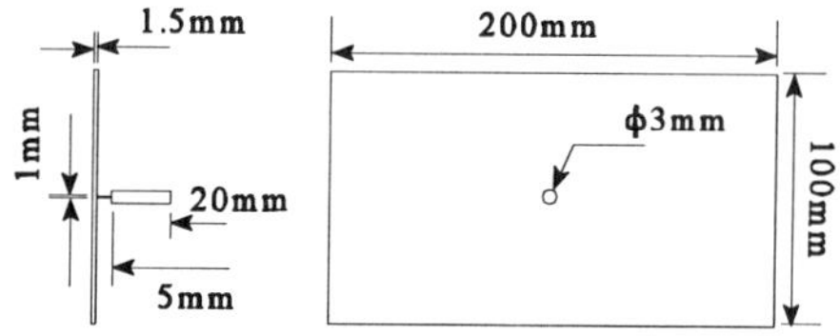

FIG. 8. Plate mold.

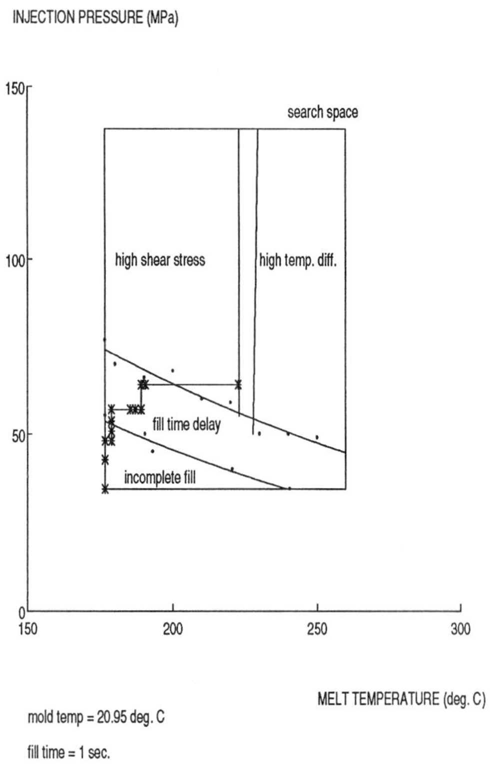

FIG. 9. Search path.

Inference Method [5] is used here for conflict resolution. The result of the fuzzy inference is a membership function and is defuzzified to a crisp value by the Centre of Area Method [6]. The fuzzy inference and defuzzification is illustrated in figure 7.

3.5 LEARNING ALGORITHM

The strength of the correlation between x_i and y_j measured by τ_{ij} (=μ_{ij}) is not necessary a constant during the search. It is also true for the appropriate amount of Δx_{max}. Simple learning algorithms are proposed to improve the system performance by

TABLE 1.

Parameters	Minimum	Maximum
Injection pressure	34.5 MPa	137.9 MPa
Melt temperature	176.57 °C	259.84 °C

TABLE 2.

y_i	Goals	$y_{very\ high}$	Priority
% of Incomplete fill	= 0 %	50 %	First
Actual fill time delay (sec.)	≤ 0 sec.	0.1 sec.	Second
Maximum wall shear stress	≤ 200kPa	724 kPa	Third
Temperature difference	≤ 10 °C	15 °C	Third

TABLE 3.

	Inj. Pressure	Melt temp.
decrease short shot	increase τ=0.76	increase τ=0.76
decrease shear stress	decrease τ=0.13	increase τ=0.87
decrease temp. difference	no effect	decrease τ=1.00

TABLE 4.

		INITIAL	FINAL
Inj. pressure		34.5	64.1
Melt temp.		176.6	222.6
Incomplete fill	y_i	44	0
(%)	Y_i (μ)	very high (1)	satisfied (1)
Fill time delay	y_i	∞	0
(sec.)	Y_i (μ)	very high (1)	satisfied (1)
Max. shear stress	y_i	163.35	254.74
(kPa)	Y_i (μ)	satisfied (1)	satisfied (0.58)
Temp. difference	y_i	16.84	9.35
(°C)	Y_i (μ)	very high (1)	satisfied (1)

acquiring knowledge during the search.

The learning algorithm for updating τ_{ij} is shown below,

$$\tau_{new} = 1 - (1 - \tau_{old})^{\frac{\Delta_{new}}{\Delta_{old}}} \tag{5}$$

where

$$\Delta_{new} = \left(\frac{\Delta y_j}{\Delta x_i}\right)_{new}$$

$$\Delta_{old} = \left(\frac{\Delta y_j}{\Delta x_i}\right)_{old}$$

It has the following characteristics,

1. $\tau_{new} \geq \tau_{old}$ if $\Delta_{new} \geq \Delta_{old}$;
2. τ_{new} tends to 0 if Δ_{new} tends to 0;
3. $\tau_{new} = \tau_{old}$ if $\Delta_{new} = \Delta_{old}$;
4. τ_{new} tends to 1 if Δ_{new} tends to ∞.

The appropriate amount of Δx_{max} should not be too large or too small. It is proposed that a corresponding change for y'_j may be 0.25. Exponential smoothing with a smoothing constant of 0.5 is used to predict the difference ratio between x_i and y'_j, as shown below,

$$\Delta x_{max} = \frac{0.25}{\left(\frac{\Delta y'_j}{\Delta x_i}\right)_{predict}} \tag{6}$$

where

$$\left(\frac{\Delta y'_j}{\Delta x_i}\right)_{predict} = \frac{1}{2}\left[\left(\frac{\Delta y'_j}{\Delta x_i}\right)_{newdata} + \left(\frac{\Delta y'_j}{\Delta x_i}\right)_{lastpredict}\right]$$

4. EXAMPLE

A simple plate mold with pin-point gate at centre is used as shown in figure 8, the material being used is PS. C-MOLD [7] package is used for simulation analysis. The limits of the parameters are summarized in table 1.

For the illustration purpose, only injection pressure and melt temperature will be adjusted during the search, and fill time and mold temperature are kept at 1 second and 20.95 °C respectively. Minimum injection pressure and temperature is used as the initial processing condition. The fuzzy goals are summarized in table 2.

The fuzzy goals are considered to be satisfied if $\mu_{satisfied}(y_i) \geq 0.5$. The knowledge used for the search is summarized in table 3.

The search path is shown in figure 9 and the initial and final (13th) values are given in table 4.

5. CONCLUSION

The quality criteria of an injection molded product were formulated into a set of fuzzy inequalities. A knowledge based system was integrated to a numerical simulation system to search for a solution. An example was given to illustrate this method.

6. REFERENCE

[1] : Nagarsheth, P.S. "Optimization of Certain Processing Parameters for Injection Molding using Flow Simulation Analysis and Statistical Techniques". ANTEC, 1989. 1699-1703.

[2] : Bourdon, K. "The Selection and Setting of Injection Moulding Machines by Means of Process Simulation". ANTEC, 1989. 308-311.

[3] : Pandelidis, I., and Zou, Q. "Optimization of Injection Molding Design. Part II: Molding Conditions Optimization". Polymer Engineering and Science, vol. 30, no. 15, 1990. 883-892.

[4] : Sakawa, M. Fuzzy Sets and Interactive Multiobjective Optimization, 1993. Plenum Press.

[5] : Zimmermann, H.J. Fuzzy Set Theory - and Its Applications. Kluwer Academic Publishers, 1990.

[6] : Berenji, H.R. "Fuzzy Logic Controllers". An Introduction to Fuzzy Logic Applications in Intelligent Systems. Eds Yager, R.R., and Zadeh, L.A. Kluwer Academic Publishers, 1992.

[7] : Advanced CAE Technology, Inc. "C-MOLD - Plastic Process Simulation Software". 1992.

Simulation of Materials Processing: Theory, Methods and Applications, Shen & Dawson (eds)
© 1995 Balkema, Rotterdam. ISBN 90 5410 553 4

Numerical simulation and comparison with water modeling studies of the inertia dominated cavity filling in die casting

K. Venkatesan & R. Shivpuri
Department of Industrial, Welding and Systems Engineering, The Ohio State University, Columbus, Ohio, USA

ABSTRACT

A numerical model based on FLOW-3D, a three dimensional finite difference software based on the SOLA-VOF free surface modeler has been developed, to simulate the transient, inertia dominated and complex cavity filling in high pressure die castings. In order to determine the accuracy of this approach in modeling the complicated flow conditions in die casting, a comparison with experimental filling patterns obtained from water analogy results in the literature was conducted and selected results presented. The comparison of the mold filling patterns from water analogy studies with the results of the numerical simulations, shows that the numerical model appears to simulate the inertia dominated mold filling of die castings very well.

1 *INTRODUCTION*

Die casting processes are high volume production processes which produce geometrically complex parts of nonferrous metals with excellent surface finishes and low scrap rate. Production rates of 200 parts per hour and production batches of 300,000 parts are not uncommon.

The flow of the molten metal from the shot sleeve into the periphery of the casting is accomplished by a runner system, while the entry of the molten metal into the casting is controlled by a gating system. The molten metal is injected into the die cavity under very high pressures and velocities for reduced cycle times. Typical filling times are in the order of milliseconds with typical flow velocities of the order of 40 m/sec. An optimum gating system design creates a preferred filling pattern within the die cavity and consequently produces sound castings. An improper design of the runner gate system or of the die cavity, often results in defects such as lack of fill, cold shut, gas and shrinkage porosity, commonly seen in die castings.

Thus an important requirement of a die casting process model is the adequate simulation of fluid flow during mold filling. Fluid flow during the filling stage of the die cavity is highly transient, inertia dominated and often turbulent with Reynolds numbers in excess of 10000 due to high velocity metal (mold cavities for even large parts are typically filled in a few milliseconds) and rapidly changing geometry in the runner-gate system or in the casting cavity [1]. Within the die cavity, jetting, splashing and liquid droplet and atomized spray formation are quite common [1]. Therefore common simplifications used for modeling less complex flow problems in sandcasting and injection molding, such as the assumption of laminar and viscosity dominated flow (such as Poisseule flow, used for modeling highly viscous flows in injection molding), are not valid for die casting flow modeling.

The solution of the complete Navier Stokes equations (including the momentum terms) to calculate transient velocity and pressure changes, with some degree of turbulence modeling is thus an essential requirement for any fluid flow model used in die casting. The other main requirement is free surface modeling of the fluid flow which is capable of accurately tracking the complicated flow fronts seen in die casting, as well as the formation of jets, droplets and sprays caused by high velocities and splashing. In addition the model should adequately handle the three dimensional geometry of the die cavity and the simultaneous occurrence of fluid flow and heat transfer.

Casting models in the past have considered the phenomena of solidification alone under the premise that flow is not critical in regular sand, or permanent mold castings [2, 3]. Due to the importance of mold filling, a few numerical models have incorporated fluid flow in the modeling of die casting [4-10]. However most of these flow models utilize laminar and viscosity dominated flow assumptions, which are certainly applicable for injection molding of plastics and for sand casting but are not valid for the transient, inertia dominated and turbulent flow conditions prevalent in die castings. Of all the models reviewed [11], FLOW-3D [12], a 3 dimensional CFD code which is an extension of the SOLA [13] and SOLA-VOF [14] family of free surface modelers, appears to have the best potential to simulate the mold filling in high pressure die casting. Its superior free surface modeling capability [11] allows the handling of complicated free surface movement caused by jetting, flow separation, splashing, liquid droplet formation, and turbulence, as well as the transient flow [1] which occurs during the filling of die cavities under high velocities and pressures.

FLOW-3D has traditionally been used to model complex free surface problems in space applications [12]. It was decided to extend FLOW-3D to numerical modeling of mold cavity filling in high pressure die casting. A numerical model based on FLOW-3D was developed, to simulate the transient, inertia dominated and complex cavity filling in high pressure die castings [15-18]. The details of the numerical schemes used and the modeling issues involved in simulating die filling are described in the following section. In order to validate this approach in modeling the complicated flow conditions present in die casting, model predictions were compared with experimental filling patterns obtained from water analogy results reported in literature.

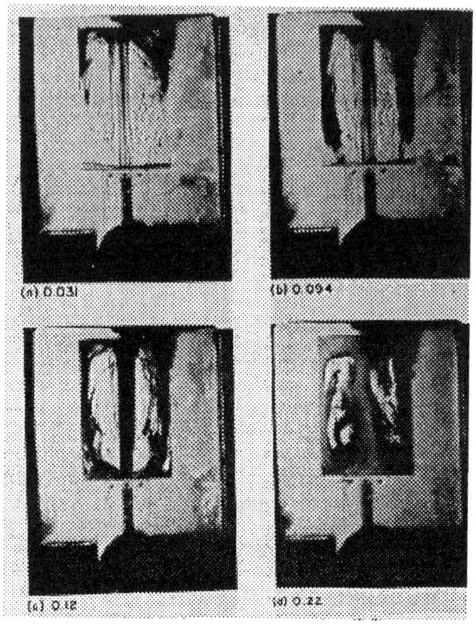

Figure 1a. Cavity fill sequence with a gate velocity of 1800 cm/sec and a 1.25 x 0.052 cm chisel gate as obtained by Smith and Wallace (1963) [1]. Number under each lists time in seconds after start of flow in cavity.

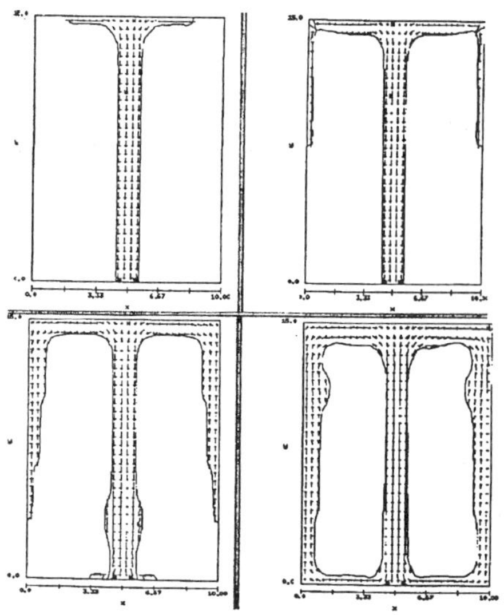

Figure 1b. Fill sequence obtained using the numerical model for the same gate geometry and gate velocity shown in Figure 1a, at approximately similar instants of time.

2 *NUMERICAL APPROACH*

FLOW-3D [12] has evolved from the Marker-and Cell (MAC) [19, 20] and Simplified Marker and Cell (SMAC) [21] finite difference techniques, which use pressure and velocity as primary independent variables. This program calculates dynamics of fluids in three space dimensions and is an extension of the SOLA [13] and SOLA-VOF [14] programs with several additional features. The program utilizes [12]:

the general mass continuity equation, with a turbulent diffusion term and mass source.

$$\frac{\partial}{\partial x}(A_x u)+\frac{\partial}{\partial y}(A_y v)+\frac{\partial}{\partial z}(A_z w)=\frac{1}{\rho}(\mathrm{RDIF}+\mathrm{RSOR}) \qquad 1$$

where A_x, A_y, A_z, are the fractional areas open to flow and u, v, w are the velocities in the x, y and z directions respectively. RDIF is a turbulent diffusion term, RSOR is a mass source term and ρ is the density.

Navier-Stokes equations in three dimensions, with body acceleration, viscous acceleration, wall shear and mass injection source terms. The form of the x-direction momentum balance is achieved by [12]

$$\frac{\partial u}{\partial t}+\frac{1}{V}\left[A_x u\frac{\partial u}{\partial x}+A_y v\frac{\partial u}{\partial y}+A_z w\frac{\partial u}{\partial z}\right]= -\frac{1}{\rho}\frac{\partial p}{\partial x}+G_x+f_x-\frac{\mathrm{RSOR}}{\rho V}u \qquad 2$$

where V is the fractional volume available to flow, p is the fluid pressure, G_x is the body acceleration, f_x is the viscous acceleration, and the last term accounts for the injection of mass at zero velocity. The terms on the LHS account for the momentum effects, and are ignored in most simulations of mold filling in injection molding and other casting operations, as it is assumed that viscosity effects are far more dominant than fluid momentum effects in these operations. This is not an acceptable assumption in die casting, where flow is largely transient and momentum dominated.

For a variable dynamic viscosity μ, the viscous acceleration in the x direction is given by

$$\rho V f_x = ws_x-\left[\frac{\partial}{\partial x}(A_x\tau_{xx})+\frac{\partial}{\partial y}(A_y\tau_{xy})+\frac{\partial}{\partial z}(A_z\tau_{xz})\right] \qquad 3$$

where

$$\tau_{xx}=-2\mu\left[\frac{\partial u}{\partial x}-\frac{1}{3}\left\{\frac{\partial u}{\partial x}+\frac{\partial v}{\partial y}+\frac{\partial w}{\partial z}\right\}\right] \qquad 4a$$

$$\tau_{xy}=-\mu\left[\frac{\partial u}{\partial y}+\frac{\partial v}{\partial x}\right] \qquad 4b$$

$$\tau_{xz}=-\mu\left[\frac{\partial u}{\partial z}+\frac{\partial w}{\partial x}\right] \qquad 4c$$

In the above equations ws_x is the wall shear stress in the x direction. The wall stresses are modeled by assuming a zero tangential velocity on the portion of any area closed to flow. Mesh boundaries are an exception, because they can be assigned a non-zero tangential velocity. Wall shear stresses along mass sources (gates in castings) must be further modified. Mass is injected into the flow from such surfaces with zero tangential momentum. When this mass mixes with the ambient fluid, it slows down the tangential flow much like a viscous stress. Momentum equations in the y and z directions have a similar form.

Fluid energy equation, where the fluid internal energy is assumed a linear function of temperature (specific heat). The equation which contains turbulent diffusion terms, conduction diffusion terms and an energy source term is of the form [12]

$$v\frac{\partial}{\partial t}\rho I + \left(\frac{\partial}{\partial x}(A_x \rho I u) + \frac{\partial}{\partial y}(A_y \rho I v) + \frac{\partial}{\partial z}(A_z \rho I w) = -p\left[\frac{\partial}{\partial x}(A_x u) + \frac{\partial}{\partial y}(A_y v) + \frac{\partial}{\partial z}(A_z w)\right] + RIDIF + TDIF + RISOR \qquad 5$$

Where I is the macroscopic mixture internal energy, which is assumed to be a linear function of temperature, $I = c_V T$, where c_V is the specific heat at constant volume. RIDIF includes the turbulent diffusion effects of I, TDIF is a diffusion term for heat conduction and RISOR is a energy source term.

The main crux of the program is the ability to track fluid interfaces and free surfaces. This is done by defining the fluid configurations in terms of a volume of fluid (VOF) function, F(x,y,z,t). The function represents the volume of fluid per unit volume and satisfies the equation [12]

$$\frac{\partial F}{\partial t} + \frac{1}{V}\left[\frac{\partial}{\partial x}(FA_x u) + \frac{\partial}{\partial y}(FA_y v) + \frac{\partial}{\partial z}(FA_z w)\right] = FDIF + FSOR \qquad 6$$

where FDIF is a diffusion term created due to the turbulent mixing of two fluids, and FSOR is a source term which is the time rate of change of the volume fraction of fluid associated with the mass source for the fluid. F has a value of zero when a computational cell is empty, a value of one when the cell is completely filled with metal and a value between zero and one when the cell is partially filled with liquid (a free surface).

The numerical approximations of the above equations are performed by assuming that the flow region is divided into a mesh of fixed rectangular cells. All variables are located at cell centers except for velocities, which are at cell faces. Most terms in the equations are evaluated explicitly, except the pressures and velocities which are coupled implicitly, by using time advanced pressures in the momentum equations and time advanced velocities in the mass equation.

Explicit approximations of the momentum equations are used to compute the first guess for new time level velocities using the initial conditions or previous time-level values for all advective, pressure and other accelerations. To satisfy the continuity equation, the pressures in each cell are adjusted either implicitly or explicitly and the velocity changes induced by each pressure change are added to the velocities computed in the first step. Finally when there is a free surface or fluid interface, it is updated using Equation 6 to give the new fluid configuration [12]. Repetition of these steps takes place until all full cells satisfy the continuity equation, thereby advancing a solution through the desired time step. The advantage of this algorithm is that pressure and velocity are both included in the iteration and it thus converges faster.

As part of the calculation, boundary conditions are imposed as free surface boundary conditions for the fluid and solid, or wall boundary conditions for dies and pins. Free surface boundary conditions are always imposed as zero velocity shear and specified surface pressure, which could be variable or constant. In this case, the pressure is specified to be a constant atmospheric pressure, which is reasonable if one assumes that there is sufficient venting in the system to prevent pressure build-up. Wall boundary conditions imposed on die walls and pins include zero normal velocity at the wall, and no-slip in the boundary layer, because of the coarse computational meshes.

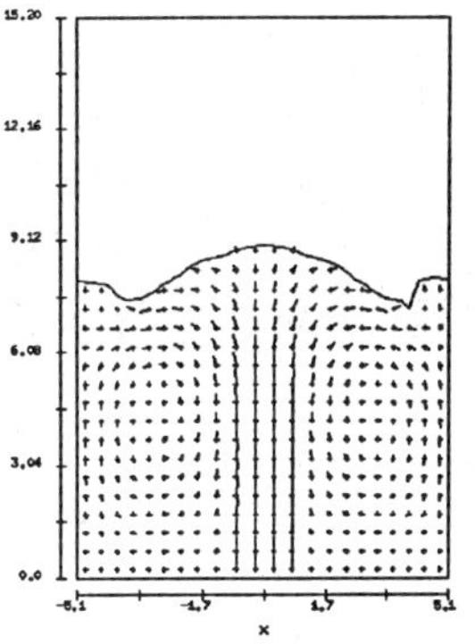

Figure 2. Fill pattern at an instant of time obtained using the numerical model for the same gate geometry shown in Figure 1a, and a very low gate velocity to reduce the effect of fluid momentum.

3 *MODEL VALIDATION: COMPARISON WITH WATER ANALOGY RESULTS*

Before the availability of computational fluid flow techniques, water analogy studies were utilized to visualize mold filling and study the effect of runner and gate geometries. Even though the water analogy method cannot be used to evaluate thermal phenomena, such as the effect of the molten metal temperature and solidification in die casting, the method is useful for visualizing the flow of molten metal, and estimating locations of porosity in the casting cavity, and turbulence in the gating system.

A number of water analogy studies have been conducted over the years by among others, Wallace and his co-workers [1, 22-23], Booth and Allsop [24], Frommer [25], Moorman and Sheptak [26], Mehrabian et al. [27] for different gate, runner and cavity geometries for a range of process conditions. These water analogy studies exhibit the complicated flow conditions such as jetting, splashing and spray formation, which distinguishes the flow in die cavity filling from that of sand casting or injection molding. While the results of these water analogy studies indicate the difficulties in modeling the fluid flow in die casting processes, they also provide an excellent *benchmark* to test the capability of a free surface solver in simulating the mold filling in a die cavity.

Die cavity filling simulations using the developed model, were therefore compared with water analogy studies available in the literature [1, 22-24] for a range of die geometries, gate shapes and sizes, and gate velocities, to test the capability of the numerical model in simulating the mold filling of a die cavity. The gating systems considered include chisel gates, side gates, fan gates, straight tapered gates and tangential tapered gates. The results of some of these comparisons with water analogy results are presented below.

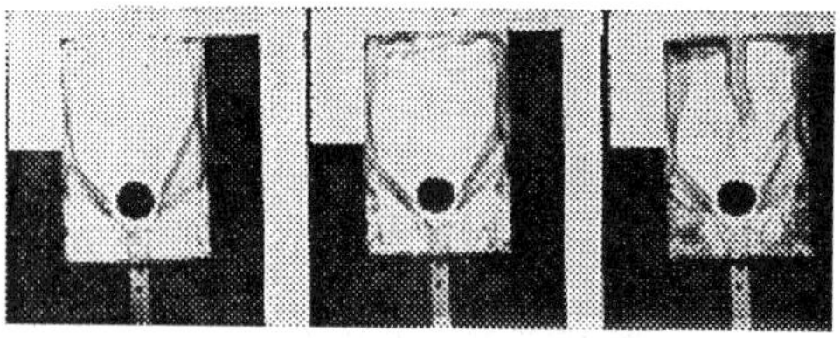

Figure 3a. Flow sequence after stream impingement on 2.5 cm dia core for a gate velocity of 1800 cm/sec with a 1.25 x 0.32 cm gate for the plate die shown in Figure 1a. [23]

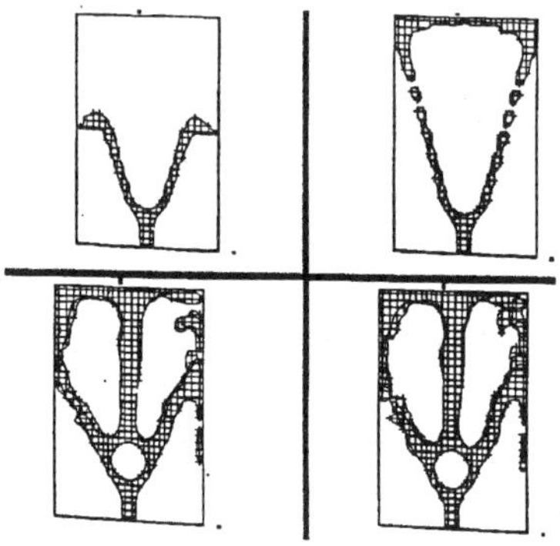

Figure 3b. Flow patterns obtained from the numerical model for the same gate velocities and cavity filling sequence of a cavity with a round core shown in Figure 3a.

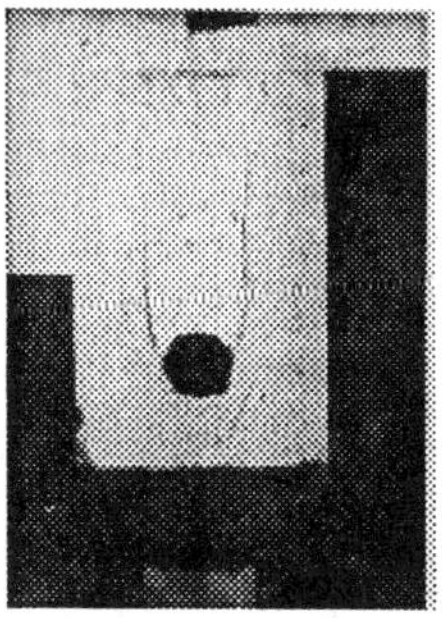

Figure 4a. Flow pattern after stream impingement on a 2.5 cm dia core for a gate velocity of 300 cm/sec with a 10.0 x 0.32 cm gate (continuous fill) for the plate die shown in Figure 1a. [23]

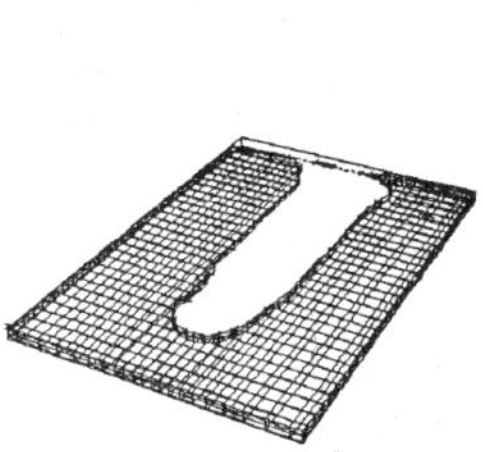

Figure 4b. Flow pattern obtained from the numerical simulations for the cavity, gating system and gate velocity, shown in Figure 4a.

3.1 *Filling of rectangular cavities with chisel gates*

To illustrate the importance of fluid momentum, the filling of a 15 x 10 x 0.3 cm cavity by a 1.25 x 0.32 cm chisel gate with a gate velocity of 1800 cm/s was selected and is shown in Figure 1a [22]. The results of numerical simulations are shown in Figure 1b. The filling patterns predicted by the model are nearly identical to experimental results, except for a small numerical pulse in the jet. The importance of the momentum terms in Navier-Stokes equation, becomes clear during this numerical simulation, as it would not be possible to model the jetting seen here, with simplifying assumptions. A filling pattern at an instant of time, obtained using a very small velocity (50 cm/s) which would reduce the importance of momentum effects, is shown in Figure 2. As can be seen a continuous filling pattern is obtained, which is contrary to the jetting seen in the actual filling pattern (Figure 1a). This hypothetical example clearly shows the dangers of ignoring fluid momentum.

3.2 *Filling of cavities with rectangular and circular cores*

While high fluid velocities are needed in die casting mold filling because of the very low fill times (in the order of milliseconds), the effects of fluid inertia such as jetting are not always desirable. Air is very likely entrapped between the walls and the jet in the filling sequence shown in Figure 1. A commonly used approach to reduce fluid inertia is to gate across the cross-section (fanning the gate), which ensures a solid front fill without air-entrapment. Another means of decreasing inertia is to impact the jet on a die or core surface. The effect exerted on the flow pattern in the die cavity when the entering fluid impinges on cores was investigated under several conditions of jetting and solid front fill, by Stuhrke and Wallace[23]. This was done by placing square and round cores of various sizes.

The effect of impingement from a 1.25 cm wide, 0.32 cm thick gate at a fluid velocity of 1800 cm/sec on a 2.5 cm diameter core and is shown in Figure 3a [23]. As can be seen the jet separates into two streams after impact. These streams are deflected to the side walls of the die cavity and form side runners that flow around the periphery of the die cavity. The side runners meet at the top of the cavity and combine to form a third stream. The resulting stream then flows downwards to the core forming vortices. A similar filling pattern is obtained from the numerical simulations as shown in Figure 3b.

The flow patterns in the die cavity resulting from a solid front flow from a gate the full width of the rectangular cavity around a circular core of 0.625 x 2.5 cm is shown in Figure 4a [23]. The results of the numerical simulation for this case of solid front fill of cavities with a circular core is shown in Figure 4b. The experimental and numerical simulation show that the fluid separates leaving an air pocket behind the core even for the small gate velocity (300 cm/s) employed by Stuhrke and Wallace. For the gate geometries shown in Figures 3 (chisel gate) and 4 (fan gating resulting in continuous fill), a similar match with the experimental results was also obtained [11] for the different core diameters considered by Stuhrke and Wallace [23], but is not reported here for brevity. Thus reducing the stream inertia by fanning the gate does not lower the likelihood of air entrapment as an air pocket is always left behind even for small gate velocities.

The filling sequence using a 5.0 x 2.5 cm rectangular core as obtained by Stuhrke and Wallace [23] is shown in Figure 5a. The numerical sequence of filling as seen in Figure 5b, shows an excellent agreement with the numerical results, including the shape and location of air voids around the rectangular core. Again, continuous filling patterns are not

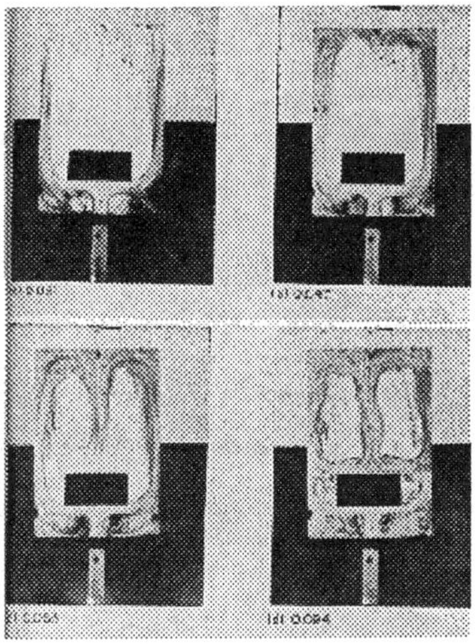

Figure 5a. Flow sequence after stream impingement on a 5 x 2.5 cm rectangular core for a gate velocity of 1800 cm/sec with a 1.25 x 0.32 cm gate for the plate die shown in Figure 1a. [23]

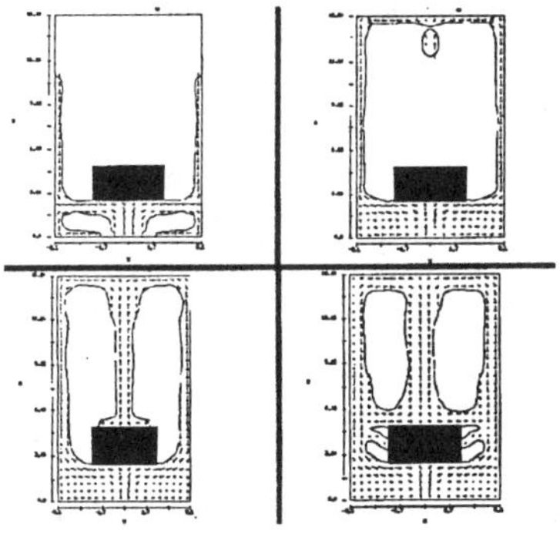

Figure 5b. Flow pattern obtained from the numerical simulations for the cavity with a rectangular core, gating system and gate velocity, shown in Figure 5a.

obtained in experiments and simulations upon impact on a large rectangular core, indicating that placement of cores does not sufficiently reduce fluid inertia in die casting, to promote continuous fill and avoid air-entrapment.

When a jet impinges on a core surface, the pressure on the surface is perpendicular to the surface at every point for an ideal fluid. At the center of the stream in contact with the core surface, the pressure is a maximum and the velocity is a minimum. As the distance from the center of the area of contact increases, the velocity of the fluid in the stream increases and the pressure decreases.

Eventually the pressure of the stream becomes so low that it fails to follow the surface of the round cores and separates. This creates a void of unfilled area behind the core that can lead to entrapped air. This void is thus observed to occur with round cores and rectangular cores during jet fill, and also with round cores during solid front fill both in the experiments and in the numerical simulations

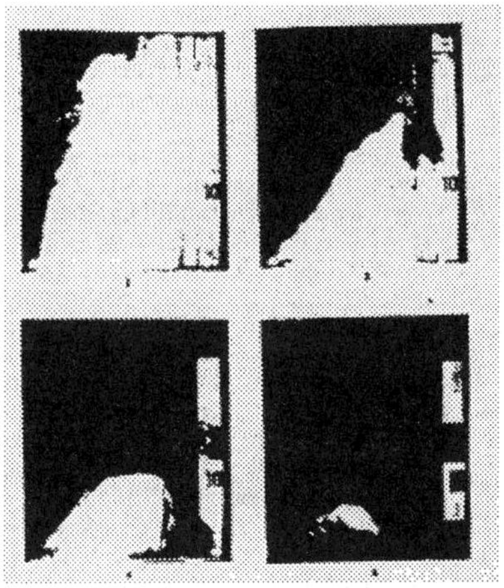

Figure 6a Filling pattern obtained by Booth and Allsop [24], using a parallel tangential runner.

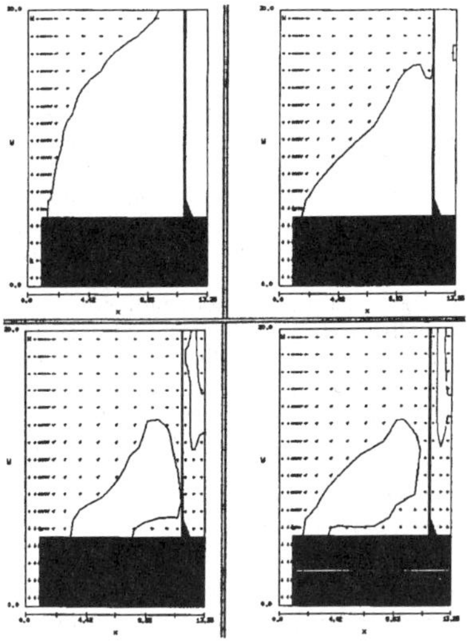

Figure 6b Filling pattern obtained from the numerical simulations for the same geometry as the parallel runner filling sequence shown in Figure 6a.

These results show that cores placed in front of gates will result in the possibility of entrapped air behind the gate for any kind of gate or gate velocity because of the separation of fluid streams. Thus using cores to reduce metal inertia is not a very good idea as it will still create problems of air entrapment, because of likely flow separation.

3.3 *Filling of cavities with tangential gates*

For a comparison with a different gating system, we shall consider the case of tangential runners. Both parallel and tapered tangential runners are widely used by die casters. The trend is however toward the use of tapered runners, which have the advantage of reducing the weight of the runner. Booth and Allsop [24] studied the effects of various designs of tangential runners (parallel and tapered) and cavities on the filling pattern in die cavities using transparent dies and water. The progression of the filling front across the cavity when using a parallel runner (when the gate area is the same as runner area) as obtained by Booth and Allsop [24] is shown in Figure 6a. The numerically obtained filling front progression obtained from the numerical model is shown in Figure 6b. As can be seen from the numerical and experimental results, when the fluid enters the cavity, the flow is almost parallel to the runner. The flow angle between the gate and the direction of flow is increased as soon as the runner is full and the flow is from the entire gate length. When the water reaches the overflow gates, it is deflected along the overflow side of the cavity towards the bottom edge, where it is deflected back towards the gate, thus forming a vortex in the bottom half of the cavity. As can beseen, the numerically obtained filling front progression obtained appears to be a reasonable match for the experimental results and also appears to predict the location of vortex formation. This is not however an ideal filling pattern, as in an actual casting there will be a lot of entrapped air in the region of vortex formation.

4 *CONCLUSIONS*

From these comparisons with water analogy results, it appears that the developed numerical model is capable of handling the complicated free surface profiles, transient flow, jetting, and splashing which occurs during the isothermal filling of die cavities under high velocities and pressures in most situations, although a few situations are not modeled well. Comparisons conducted for several other gating systems and geometries [11], are not reported here for brevity.

Such numerical simulations certainly afford faster and cheaper study of gate and runner designs for the ideal filling of different cavity geometries, as compared with traditionally used water models used to verify gating and part design for complex commercial parts. Numerical models also have the additional capability of non-isothermal flow and solidification modeling, that is certainly not possible with the water models.

5 *ACKNOWLEDGMENTS*

This study was supported by grants from the NSF Engineering Research Center for Net Shape Manufacturing, and by computational resources (Grant number PAS659) provided by the Ohio Supercomputer Center. The authors would also like to thank FlowSciences Inc., for providing a source code of the software FLOW-3D for use in this work.

REFERENCES

1. Stuhrke, W.F. and Wallace, J.F., *AFS Trans*, (1965), vol. 73, p. 569-595.
2 Ohnaka, I., *Numerical Simulation of Casting Solidification in Automotive Applications, ed. Kim, C. and Kim, C-W., TMS*, (1991), p. 35-44.
3 Sahm, P., *Numerical Simulation of Casting Solidification in Automotive Applications, ed. Kim, C. and Kim, C-W., TMS*, (1991), p. 15-34.
4 Wang, K.K., Thurng, L.S. and Wang, S.P., *NADCA Transactions -T91-094,* (1991), p. 281-292.
5 Loong, C., *NADCA Transactions -T91-101*, (1991), p. 299-304.
6 Iwata, Y., Yamamoto, Y., Yonekura, K., *NADCA Transactions -T91-103*, (1991), p. 311-320.
7 Jeurissen, P.C. and van Eldijk, P.C., *Numerical Simulation of Casting Solidification in Automotive Applications, ed. Kim, C. and Kim, C-W., TMS*, (1991), p. 207-220.
8 Lewis, R.W., Usmani, A.S. and Huang, H.C., *Modeling of Casting, Welding and Advanced Solidification Processes-V, TMS*, (1990), p. 3-14.
9 Stoehr, R.A. and Wang, C., *Modeling of Casting, Welding and Advanced Solidification Processes-V, TMS*, (1990), p. 725-732.
10 Anzai, K. and Uchida, T., *Modeling of Casting, Welding and Advanced Solidification Processes-V, TMS*, (1990), p. 741-748.
11 Venkatesan, K. and Shivpuri, R., *Report No. ERC/NSM-C-93-07, Engineering Research Center for Net Shape Manufacturing The Ohio State University*, (1993).
12 FLOW-3D: Computational Modeling Power for Scientists and Engineers, *Flow Sciences Inc., Los Alamos, New Mexico.*
13 Hirt, C.W., Nichols, B.D. and Romers, N.C., *Los Alamos Scientific Laboratory Report*, (1975).
14 Nichols, B.D., Hirt, C.W. and Hotchkiss, R.S., *Los Alamos Scientific Laboratory Report, LA-8355*, (1980).
15 Shivpuri, R., Kuthirakulathu, M. and Mittal, M., *1991 Winter Annual Meeting, ASME Atlanta, GA*, (1991).
16 Venkatesan, K. and Shivpuri, R., *Paper No. Cleveland T93-011, Transactions of the North American Die Casting Association, 17th Int. Die Casting Congress and Expo*, (1993).
17 Venkatesan, K. and Shivpuri, R., *Paper No. MS-39, Transactions of NAMRI/SME, Volume XXII, 1994*, (1994).
18 Venkatesan, K. and Shivpuri, R., *To be published in ASM Journal of Materials Engineering and Performance*.
19 Harlow, F.H. and Welch, J.E., *The Physics of Fluids*, (1965), p.2182.
20 Welch, J.E. and Harlow, F.H., *Los Alamos Scientific Laboratory Report, LA-3425*, (1965).
21 Amsden, A.A. and Harlow, F.H., *Los Alamos Scientific Laboratory Report, LA-4370*, (1970).
22 Smith, W. E. and Wallace, J.F., *AFS Trans*, (1963), vol. 71, p. 325-343.
23 Stuhrke, W. F. and Wallace, J.F., *AFS Trans*, (1964), vol. 72, p. 374-407.
24 Booth, S.E., and Allsop, D.E., *Paper No. G T83-033, Transactions of the North American Die Casting Association, 12th Int. Die Casting Congress and Expo*, (1983).
25 Frommer, L., *J. Springer, Berlin*, (1933).
26 Moorman, J.H and Sheptak, N., *AFS Transactions*, (1962), vol. 70, p. 929.
27 Hong, S., Backman, D.G., Leach, J.L., and Mehrabian, R., *American Die Casting Institute*.

Simulation of Materials Processing: Theory, Methods and Applications, Shen & Dawson (eds)
© 1995 Balkema, Rotterdam. ISBN 90 5410 553 4

Computer simulation of crystallization of PET in injection molding

O. Verhoyen & F. Dupret
Université Catholique de Louvain, CESAME, Unité de Mécanique Appliquée, Louvain-la-Neuve, Belgium

R. Legras
Université Catholique de Louvain, Unité de Physique et de Chimie des Hauts Polymères, Louvain-la-Neuve, Belgium

ABSTRACT : A numerical algorithm is developed in order to predict the crystallization of injection molded polyethylene terephtalate. The scheme is based on a model that describes the behavior of semi-crystalline polymers for arbitrary shear and thermal histories. Both a decoupled and a fully coupled numerical method are investigated. The advantages and drawbacks of these approaches are discussed. Several results will be compared with experiments.

1 INTRODUCTION

The formation of microstructure during the injection molding of semi-crystalline polymers is important with regard to understanding the physical properties of the product. The fast cooling rate and high shear rate experienced by the fluid during the process generate a non-uniform crystallinity degree in the part, whose distribution contributes to the final performance of the product. The aim of our work is to predict by means of numerical simulation the evolution of crystallization during injection molding, and to compare the results with experiments.

2 PHYSICAL MODEL OF CRYSTALLIZATION

Our work is based on the approach developed by Verhoyen *et al.* (1992a, 1995) for predicting the behavior of slowly crystallizing polymers. The model is based on the well known Avrami-Evans equation (Avrami, 1939) :

$$\alpha_v^c(t) = 1 - \exp(-k(T_c)\, t^n) \quad , \qquad (1)$$

where $\alpha_v^c(t)$ stands for the conversion level of crystallization (ranging from 0 to 1) which depends on time (t), k is the kinetic coefficient which depends on crystallization temperature (T_c), and n is the Avrami coefficient. This model has been extended to take into account the influence of several parameters, such as induction time, primary and secondary crystallization and ultimate degree of crystallinity (since crystallization is not complete). This leads to the equation :

$$\frac{\alpha_v(t)}{\alpha_v^\infty} = w_1\left(1 - \exp\left(-k_1(t-t_1)^{n_1}\right)\right) + \left(1 - w_1\right)\left(1 - \exp\left(-k_2(t-t_2)^{n_2}\right)\right) \quad , \qquad (2)$$

where $\alpha_v(t)$ denotes the evolution of the degree of crystallinity in the sample, α_v^∞ is the asymptotic degree of crystallinity when crystallization is terminated, t_1 and t_2 are the induction times related to first and second crystallization, and w_1 defines the relative weight of the first mechanism.

As Eqs. (1) and (2) deal with isothermal crystallization, it is essential to adapt the model in order to account for the highly non-isothermal character of the injection molding process. The transformation we use in the present paper is based on the Nakamura approach (Nakamura *et al.*, 1972) :

$$\alpha_v(T) = \int_{t_0}^{t} \alpha_v^\infty w_1 \frac{\partial}{\partial\tau}\left(1-\exp\left(-\left[\int_{t_1^*}^{\tau} \sqrt[n_1]{k_1}\, ds\right]^{n_1}\right)\right) d\tau + \int_{t_0}^{t} \alpha_v^\infty (1-w_1) \frac{\partial}{\partial\tau}\left(1-\exp\left(-\left[\int_{t_2^*}^{\tau} \sqrt[n_2]{k_2}\, ds\right]^{n_2}\right)\right) d\tau \quad , \qquad (3)$$

where t_0 is the initial time, while t_1^* and t_2^* are the non-isothermal induction times which, following the method proposed by Sifleet *et al.* (1973), are

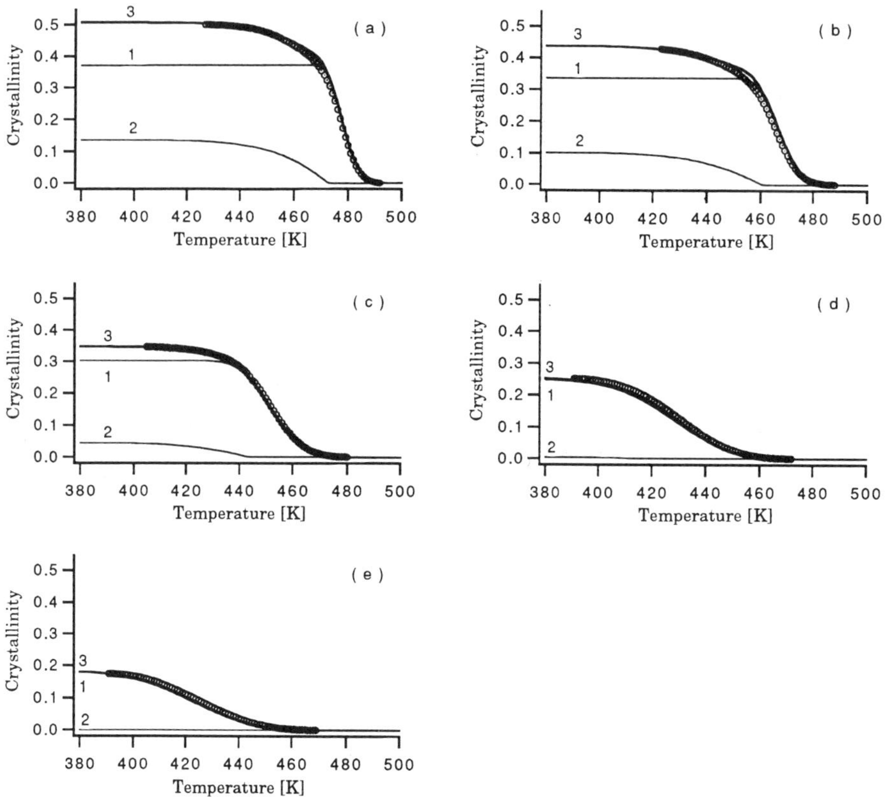

Fig. 1 : Comparison between the experimental and predicted evolutions of crystallinity degree in a DSC apparatus. The experiments have been performed at different constant cooling rates (a. -10 K/min., b. -20 K/min., c. -40 K/min., d. -80 K/min., e. -100 K/min.)

1. Crystallinity degree predicted by the first crystallization mechanism
2. Crystallinity degree predicted by the second crystallization mechanism
3. Total crystallinity degree

o. Experimental results.

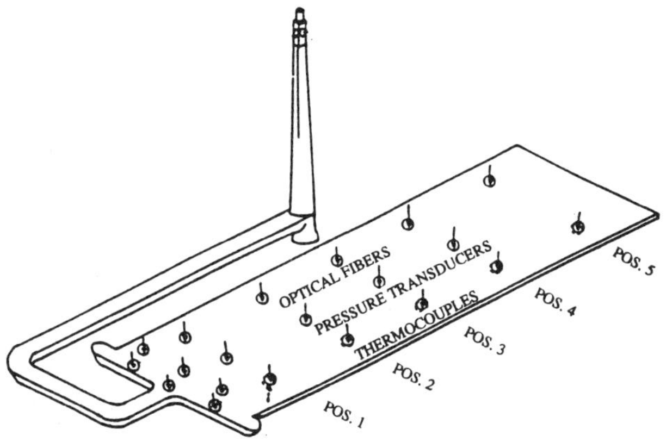

Fig. 2 : Mold geometry. Positions of the optical fibers, pressure transducers and thermocouples.

obtained by integrating a weighted time according to the relation :

$$\int_{t_0}^{t_i^*} \frac{dt}{t_i} = 1 \quad , \quad i = 1 \text{ or } 2 \ . \tag{4}$$

It is important to note that the parameters α_v^∞ , w_1 , k_1, k_2 , t_1 and t_2 in Eqs. (3) and (4) are functions of temperature and shear stress. Hence, the effects of non-isothermal and stress-induced crystallization are taken into account by means of appropriate phenomenological relationships.

In practice, all the parameters and material functions governing Eqs. (3) and (4) can be determined by means of isothermal experiments performed at a constant shear stress. Indeed, for any temperature and shear stress, the coefficients of Eq. (2) can be obtained by fitting the predicted and experimental curves $\alpha_v(t)$. The effect of temperature is determined by performing a set of DSC experiments (at a vanishing shear stress). The effect of shearing is taken into account by performing additional cone and plate rheometer experiments, which provide the evolution of viscosity with respect to time, and can thus be correlated to crystallization.

We have applied this method to characterize the crystallization kinetics of PET. The model was subsequently validated by means of DSC experiments performed at different cooling rates (Fig. 1). The agreement between predicted and experimental results is excellent. We have also characterized the dependence of crystallization kinetics with respect to shear stress. Results obtained by means of cone and plate rheometer experiments will be presented.

3 NUMERICAL SIMULATION SCHEME

Our crystallization model has been introduced in the MOLDSYS software, which is devoted to simulating the filling, packing and cooling stages of injection molding (Dupret and Vanderschuren 1988, Couniot *et al.* 1989, Dupret and Dheur 1992, Kabanemi and Dupret 1992, Couniot *et al.* 1993). The model is simplified by introducing the Hele-Shaw approximation, and the resulting system of equations is two-dimensional for the pressure field and three-dimensional for the temperature field. Briefly, calculations are performed on successive temporary meshes covering the filled part of the mid-surface, which are generated automatically at the successive time steps of the simulation. After Eulerian integration of the 3D temperature field taking into account the thermal shock generated by the fountain effect on the cooled walls (see the companion paper of Mal *et al.* 1995), the fronts are moved. Temperatures are extrapolated on the new flow domain, using a simplified model of fountain flow (Vanderschuren and Dupret 1989) which takes into account the important transverse heat advection which is observed in the front region. The 2D pressure field is subsequently integrated with respect to time using an implicit Euler integration scheme. This model has been improved in order to take into account more realistic viscosity laws (such as the Cross law) without increasing the computational cost (Verhoyen *et al.* 1994).

We will compare two methods for predicting crystallization in injection molded parts.

In the decoupled approach, temperatures, pressures and velocities are first computed without taking the influence of crystallization into account. In a subsequent step, crystallization is evaluated as a function of the thermomechanical history of all the injected material points. Temperature and shear stress histories are obtained by means of a tracking algorithm, which reconstructs the trajectory of any material point in the part by means of an explicit Euler integration scheme (Verhoyen *et al.* 1992b). It is thus possible to recover from previous computations the values of the different field variables (such as pressure, temperature or shear rate) at any time of the process, and to reconstruct their histories for each material point.

The coupled method performs a simultaneous computation of temperature, pressure, velocities and crystallinity distribution during the complete process. It is, therefore, convenient to express our crystallization model in a differential form prior to incorporating it into the system of equations.

We will discuss the advantages and drawbacks of both approaches. The second method is expensive and accurate, since it takes into account the influence of crystallization both on the flow (through its impact on viscosity), and on the heat transfer (through release of crystallization heat). The first method, while being less accurate, has a lower computational cost.

In addition, injection molding experiments have been carried out by filling an instrumented plate with PET at the Pôle Européen de Plasturgie (Oyonnax, France) (Fig. 2). From the molded parts, 2D maps of crystallinity and crystallinity gradient have been measured in the High Polymer Division of UCL, by means of a DSC apparatus and a density gradient column in order to check the model.

Conclusions will be drawn from the comparison between the numerical and experimental approaches.

ACKNOWLEDGEMENTS

The authors wish to thank Jean-Pierre Gazonnet, Virginie Durand and Gerard Dechavanne, from the P.E.P., for their useful advices and friendly collaboration.

This paper presents research results of the Belgian Program on Inter-university Poles of Attraction, initiated by the Belgian State, Prime Minister's Office for Science, Technology and Culture. The scientific responsibility rests with its authors. The work of O. Verhoyen has been supported by a grant from the IRSIA, Belgium (Institut pour la recherche scientifique dans l'industrie et l'agriculture).

REFERENCES

Avrami M., *Kinetics of phase change*, J. Chem. Phys., 7, 1103, (1939); ibid., 8, 212, (1940); ibid., 9, 177, (1941).

Couniot A., Dheur L., Hansen O., Dupret F., *A finite method for simulating injection molding of thermoplastics*, Proc. Numiform'89 conf., Thompson E.G. et al. eds., Balkema, Rotterdam, 235 (1989).

Couniot A., Dheur L., Dupret F., *Numerical Simulation of Injection Molding : Non-Isothermal Filling of Complex Thin Parts, including Abrupt Changes of Thickness or Bifurcations of the Midsurface*, Proc. IMA conf., Bristol, 1991, Cross M. et al. eds., IMA conf. series 42, Clarendon Press, Oxford, 381 (1993).

Dupret F., Vanderschuren L., *Calculation of the Temperature Field in Injection Molding*, AIChE J., vol. 34 n°12, 1959 (1988).

Dupret F., Dheur L., *Modelling and numerical simulation of heat transfer during the filling stage of injection molding*, Proc. Oji Int. Seminar on Adv. Heat Transfer, Tomakomai, Japan, 1990, Tanasawa I. and Lior N. eds., Hemisphere, New-York, 583 (1992).

Kabanemi K., Dupret F., *Analysis of the influence of the packing stage on residual stresses and shrinkage of injection molded parts*, Proc. Numiform'92 conf., Chenot et al. eds., Balkema, Rotterdam, 357 (1992).

Mal O., Dheur L., Pirotte P., Van Rutten N., Couniot A., Dupret F., *A realistic wall thermal boundary condition for simulating the injection molding of thermoplastics*, Proc. Numiform'95 conf., Balkema, Rotterdam, (1995).

Nakamura K., Watanabe T., Katayama K., Amano T., *Some aspects of nonisothermal crystallization of polymers. I. Relationship between crystallization temperature, crystallinity, and cooling conditions*, J. Appl. Polym. Sci., 16, 1077 (1972).

Sifleet W.L., Dinos N., Collier J.R., *Unsteady-State Heat Transfer in a Crystallizing Polymer*, Polym. Eng. Sci., 13, 10 (1973).

Verhoyen O., Dupret F., Legras R., *Application des Cinétiques de Cristallisation à la Simulation Numérique des Procédés de Mise en Oeuvre*, Proc. of the colloque national du GFP, Nov. 1992, Lyon, France, (1992a).

Verhoyen O., Legras R., Dupret F., *Analyse de la Physique de l'Ecoulement en Moulage par Injection par Suivi de Points Matériels*, Proc. of the colloque national du GFP, Nov. 1992, Lyon, France, (1992b).

Verhoyen O., Legras R., Dupret F., *A simplified method for introducing the Cross viscosity law in the numerical simulation of non-isothermal compressible Hele Shaw flow*, Proc. of the PPS European regional meeting , Aug. 29-31 1994, Strasbourg, France, 175 (1994).

Verhoyen O. Legras R., *Mathematical Modelling and Experimental Measurement of Isothermal and Non-isothermal Crystallization Kinetics*, in preparation (1995).

Simulation of Materials Processing: Theory, Methods and Applications, Shen & Dawson (eds)
© 1995 Balkema, Rotterdam. ISBN 90 5410 553 4

Prediction of flow-induced orientation field and mechanical properties of injection molded parts

V. Verleye, G. Lielens, P. Pirotte, F. Dupret & R. Keunings
Centre for Systems Engineering and Applied Mechanics, Université Catholique de Louvain, Louvain-la-Neuve, Belgium

ABSTRACT: On the basis of a detailed numerical simulation of mold filling and flow-induced fiber orientation, we use a micro-mechanical model that allows for the prediction of the anisotropic and non-homogeneous elastic properties of polymer composite molded parts. The micro-mechanical model is implemented within the molding simulation software MOLDSYS developed by Dupret and his group. The emphasis is put on complex, three-dimensional, injection- or compression-molded plastic parts reinforced by long or short fibers. Preliminary results are described.

1 INTRODUCTION

The mechanical properties of injection molded parts made of fiber reinforced polymer are highly dependent on the distribution of fiber orientation that prevails at the end of the molding process. The fiber orientation distribution is itself induced by the kinematics of the flow, but it also influences the rheological behavior of the flowing suspension. Thus, the mechanical properties of the molded part result from an intricate non-linear coupling between process conditions, suspension rheology, flow-induced fiber orientation, and mold geometry.

Much progress has been made over the last decade towards the development of accurate numerical techniques for the simulation of polymer flow in complex geometries (see e.g. the review by Keunings 1989). Applications of this methodology abound in the field of polymer processing (extrusion, blow molding, wire coating). Developments have also taken place in the particular but industrially-important field of molding processes (either by compression or injection). Here, a specific theoretical approach must be adopted in order to tackle the geometrical complexity of the problem (moving boundaries, fountain effect in the vicinity of the flow front, intricate mold geometries involving bifurcations, ribs, etc...). The state-of-the-art in this particular field is described by Crochet et al. (1994). Contributions of the Louvain-la-Neuve group are also described in Couniot et al. (1993) and Dupret et al. (1988, 1992).

The prediction of flow-induced fiber orientation in general flow geometries has been attempted recently on the basis of models valid for dilute suspensions (e.g. Lipscomb et al. 1988, Rosenberg et al. 1990). These approaches have been successfully adapted to molding simulations by Henry de Frahan et al. (1992), and Verleye et al. (1994). However, the industrially-relevant case of concentrated suspensions requires a new approach. While general rheological equations of state have been proposed to describe the behavior of concentrated suspensions (e.g. Dihn and Armstrong 1984, Folgar and Tucker 1984, Sandstrom and Tucker 1993), their use in complex flow simulations has not yet been reported. In the context of molding calculations, simplified equations remain to be developed from the general equations of state. These simplified models should take account of fiber-fiber, fiber-solvent and fiber-wall interactions, while leading to tractable computational tasks.

Finally, the prediction of mechanical properties of the molded part on the basis of the computed distribution of fiber orientation requires an appropriate micro-mechanical model. Several promising approaches have been published over the recent years (e.g. Camacho et al. 1990, Christensen 1979). Here again, reported use of these models are only for simple flow geometries.

We conclude from this brief description of the state-of-the-art that, while many of the basic theo-

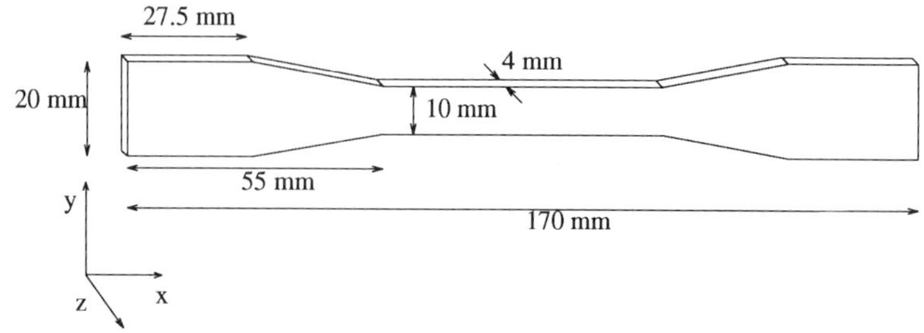

Fig. 1. Part geometry and frame of reference.

a.

b.

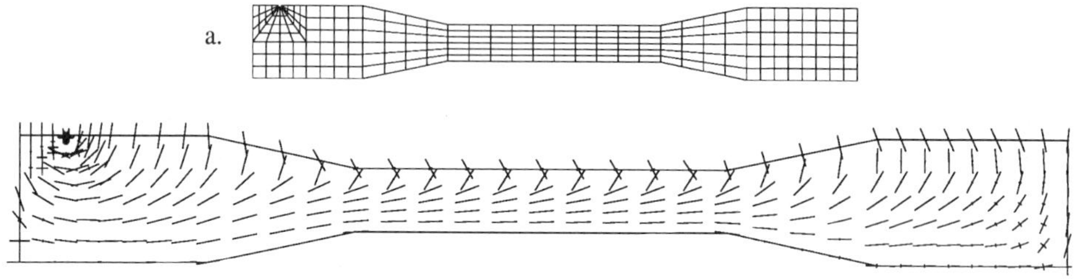

Fig. 2. Part with pinpoint injection gate.
a. Finite element mesh.
b. Orientation field after filling (long fibers).

a.

b.

c.

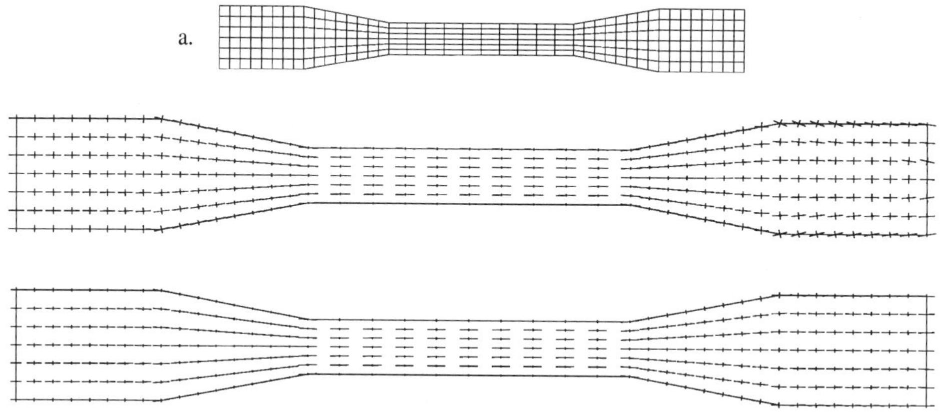

Fig. 3. Part with linear injection gate on the left-hand side.
a. Finite element mesh.
b. Orientation field after filling (long fibers).
c. Orientation field after filling (short fibers).

retical ingredients for the prediction of mechanical properties of molded parts are available, the ability of predicting (and hopefully controlling) the non-linear interplay between process conditions, rheological behavior, flow-induced distribution of fiber orientation, and final mechanical properties of complex composite molded part is currently lacking. The long term objective of the present work is to contribute to fill this gap. Preliminary results are described in the present paper.

2 FLOW AND FIBER ORIENTATION

We wish to calculate the evolution of fiber orientation in a thin molded part, whose thickness is much smaller than the other characteristic dimensions. Under such conditions, it is possible to obtain a good approximation of the flow field with the use of the lubrication approximation (Dupret and Vanderschuren 1988, Dupret and Dheur 1992, Hieber and Shen 1980). Pressure variations through the thickness of the cavity are shown to be negligible, and the pressure field obeys a second-order elliptic equation of the form

$$\underline{\nabla} \cdot (S \underline{\nabla} p) = 0 \ ,$$

where S is the flow conductance factor which depends on the temperature profile, the pressure gradient, the thickness of the cavity and the rheology of the suspension. A power-law viscosity is used in the case of short fibers, while the complex behavior of long fiber suspensions is assumed to give rise to a plug flow where head losses are governed by the phenomenological flow conductance factor S.

Following Advani and Tucker (1987), the fiber orientation field is characterized by a probability distribution function $\psi(\underline{x}, t, \underline{p})$, where $\underline{x}$ is the position vector, t is time, and the vector $\underline{p}$ stands for the unit vector aligned with the fiber. Since calculating ψ directly requires the solution of a six-dimensional problem, it is essential to simplify the model. This is achieved by introducing the orientation tensors $(\underline{\underline{a}}_2, \underline{\underline{a}}_4)$, which are defined respectively as the second and fourth moments of the distribution function ψ. In this paper, we adopt the evolution equation for these moments proposed by Folgar and Tucker (1984) :

$$\overset{\triangleright}{\underline{\underline{a}}_2} = -2\lambda \underline{\underline{a}}_4 : \underline{\underline{D}} + 2C_i\dot{\gamma}(\underline{\underline{I}} - \alpha \underline{\underline{a}}_2) \ ,$$

where $\underline{\underline{D}}$ is the rate of strain tensor, $\overset{\triangleright}{\underline{\underline{a}}_2}$ is a mixed convected time derivative of $\underline{\underline{a}}_2$, λ is a given function of the fiber aspect ratio, C_i is the so-called interaction coefficient, $\dot{\gamma}$ is the magnitude of the strain rate tensor, α equals 2 in 2D and 3 in 3D, and $\underline{\underline{I}}$ stands for the unit tensor.

The drawback of this approach is that the evolution equation for $\underline{\underline{a}}_2$ involves the fouth-order tensor $\underline{\underline{a}}_4$. A closure approximation is thus required in order to directly relate the evolution of $\underline{\underline{a}}_2$ to the velocity field. In the present paper, we have used the natural closure approximation developed by Verleye et al. (1993,1994,1995). Details on the numerical scheme can be found in the cited references of Dupret's group.

3 MECHANICAL PROPERTIES

Our goal is to predict at each point of a molded part the fourth-order elastic stiffness tensor, as well as the second-order thermal dilatation and thermal conductivity tensors, on the basis of the computed flow-induced distribution of fiber orientation. The latter is described by means of the second moment of the orientation distribution function computed in the filling simulation.

In order to predict the mechanical properties at every point of the molded part, we follow a two step method. First, following Camacho et al. (1990), we estimate the properties assuming that the reinforcement is unidirectional (i.e. fully aligned). This is achieved by means of classical micro-mechanical models for discontinuous fiber composites. Second, an orientation averaging procedure is applied that takes into account the actual fiber orientation distribution computed at each point of the molded part.

For the first step, several approaches can be adopted, namely homogeneization using three-dimensional finite element models (e.g. Termonia 1987, Dubois and Keunings 1995), approximate analytical results for idealized composites (Mori and Tanaka 1973, Benveniste 1987, Muneo et al. 1994), or semi-empirical results like the well-known Halpin-Tsaï formulae (Halpin and Kardos 1976). The latter is used in the present paper. We assume that the homogeneized material is tranversally isotropic; its stiffness tensor thus has 5 independent constants B_i, that are computed at each point of the molded part by means of one of the above-mentioned methods.

In the second step, we perform an orientation averaging on the homogeneized material using the predicted orientation tensors. The averaged elastic stiffness tensor is thus obtained as a function of the second and fourth-order orientation tensors, and of the five material constants B_i:

$$\begin{aligned} \overline{C}_{ijkl} &= B_1 a_{ijkl} + B_2(a_{ij}\delta_{kl} + a_{kl}\delta_{ij}) \\ &\quad + B_3(a_{ik}\delta_{jl} + a_{il}\delta_{jk} + a_{jl}\delta_{ik} + a_{jk}\delta_{il}) \\ &\quad + B_4\delta_{ij}\delta_{kl} + B_5(\delta_{ik}\delta_{jl} + \delta_{il}\delta_{jk}) \ , \end{aligned}$$

a.

b.

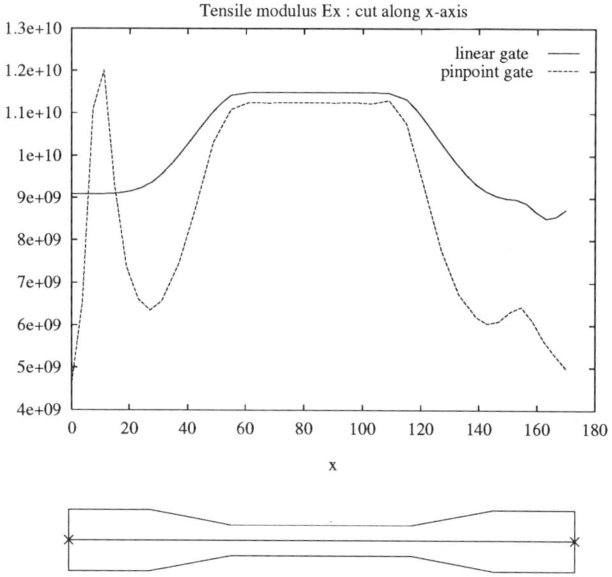

Fig. 4. Comparison of the predicted tensile modulus with two different injection gates (long fibers).
a. Evolution of the tensile modulus E_x [Pa] along a cut.
b. Representation of the cut.

a.

b.

c.

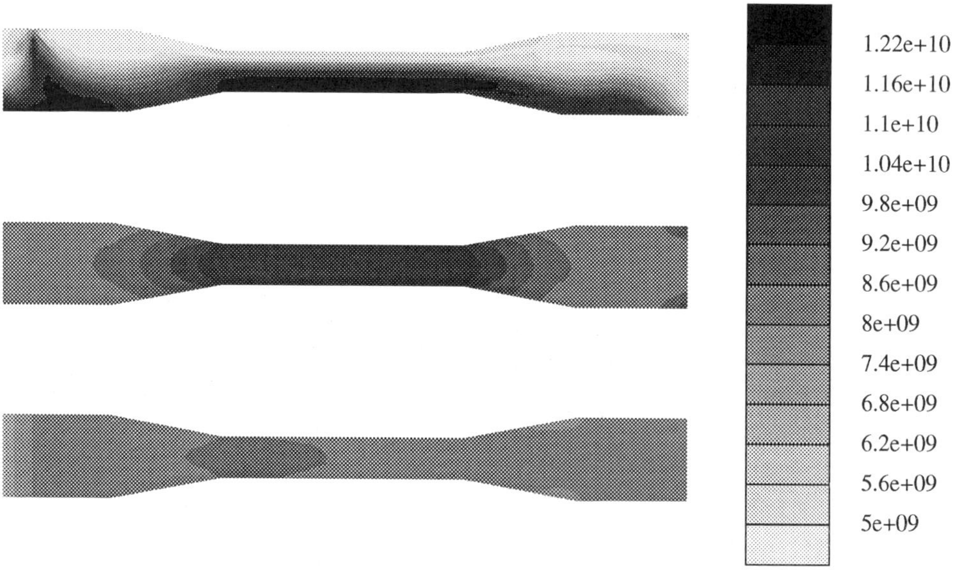

Fig. 5. Isovalues of tensile modulus E_x [Pa] for 30 w% of fibers.
a. Injection of long fibers : pinpoint gate.
b. Injection of long fibers : linear gate.
c. Injection of short fibers : linear gate.

where the components of the fourth-order tensor a_{ijkl} are calculated from a_{ij} using the natural closure approximation (Verleye et al. 1993, 1994, 1995).

Since the molded part is thin, we can use the classical laminate theory (Christensen 1979) in order to average the computed stiffness tensor through the thickness. We thus obtain at the end of the modeling process the laminate membrane and bending stiffness matrices at each point of the part. A structural analysis can then be performed on the molded part, if one so wishes.

4 EXAMPLE AND DISCUSSION

4.1 *Orientation field*

We consider the filling of a 4 mm thick nylon dumbbell-shaped part (Fig. 1). Two types of injection gates are analyzed : a pinpoint gate and a linear one. We specify an isotropic orientation field at the gates. The (fixed) finite element meshes used in both cases are shown in Figs. 2.a and 3.a. The material is Nylon 6-6 reinforced by either long or short glass fibers. The short fibers have a diameter of 0.01 mm, a length of 0.3 mm and an aspect ratio of 30. The long fibers have a diameter of 0.0025 mm. Experimentally, it is found that they rupture as they flow across the feeding system and the gate, so that their average length is reduced from 15 mm to 5 mm inside the mold. Their aspect ratio thus drops to 200. For both long and short fibers, the fiber concentration is 30% by weight. The interaction coefficient is set to 0.005.

Non-isothermal simulations were performed using a 3D orientation model for the short fiber compound and a 2D orientation model for the long fiber compound. In the latter case, the orientation field is assumed constant through the thickness of the part. For short fibers, the orientation tensor $\underline{a}_2$ is computed in several layers parallel to the mid-surface; these results are then averaged through the thickness for later purposes.

The predicted orientation fields are shown in Figs. 2.b, 3.b and 3.c, in terms of the two in-plane eigenvectors of the second-order orientation tensor. The influence of the gate design on the resulting fiber orientation is found to be significant. With the pinpoint gate and long fibers (Fig. 2.b), the diverging flow that occurs at the initial stage of the filling orients the fibers parallel to the flow fronts. Then, the converging flow experienced upstream of the thinner section orients only partially the fibers in the flow direction; the final orientation field is thus spatially non-homogeneous. Very different results are obtained with a linear injection gate (Figs. 3.b and 3.c). With both long and short fibers, it is found that the orientation field is first convected from the gate; then, it is partially oriented in the converging zone. As they converge towards the thinner section of the part, the fibers become oriented in the flow direction; the opposite phenomenon occurs in the diverging zone. As a result, the final orientation field is much more spatially homogeneous than that obtained with the pinpoint gate. At first sight, it would seem that long and short fibers orient in the same manner (Figs. 3.b and 3.c). Actually, for short fibers, we show the thickness-averaged orientation field. In fact, we find that short fibers orient differently in the core region and in the shear zone, which is located closer to the mold walls.

4.2 *Mechanical properties*

On the basis of the above predictions for the orientation field, we have computed the Young modulus E_X in the x-direction (i.e. the main flow direction) at each point of the molded part. With long fibers, it is assumed that there is no variation of fiber orientation through the thickness; the mechanical properties are thus also constant through the thickness. With short fibers, both fiber orientation and mechanical properties vary through the thickness; the results will be expressed as an average E_X modulus obtained by means of the classical laminate plate theory and the properties predicted in the different layers.

The results are shown in Figs. 4 and 5. In all cases, the modulus E_X strongly depends on the flow-induced fiber orientation field. As expected, it is highest when fibers are aligned in the x-direction, and lowest when they are aligned in the y or z directions.

The mechanical properties of the part molded with a pinpoint injection gate are highly anisotropic and spatially non-homogeneous (Figs. 4 and 5.a); this is due to the complex state of orientation obtained at the end of the filling stage. Indeed, it is very difficult to guess the mechanical properties of such a part without the complete simulation of the process; this shows the interest of the global method presented here. The properties of the part obtained with a linear injection gate are easier to guess (Figs 4, 5.b and 5.c). One finds that the part becomes stiffer in the main flow direction over the central region of the part.

5 ACKNOWLEDGEMENTS

The Ph.D. work of G. Lielens and P. Pirotte is supported by the COST 512 European Project, locally financed by the Government of the Walloon Region, Belgium. The paper presents research results of the Belgian Programme on Interuniversity Poles of Attraction, initiated by the Belgian State, Prime Minister's Office for Science, Technology and Culture. The scientific responsability rests with its authors.

6 REFERENCES

Advani, S.G. and C.L. Tucker III 1987. The Use of Tensors to Describe and Predict Fiber Orientation in Short Fiber Composite. *J.Rheology* 31: 751-784.

Benveniste, Y.1987. A New Approach to the Application of Mori-Tanaka's theory in Composite Materials. *Mechanics of Materials* 6: 147-157.

Camacho, C.W., C.L. Tucker III, S. Yalvac, and R.L. McGee 1990. Stiffness and Thermal Expansion Predictions for Hybrid Short Fiber Composites. *Polym. Comp.* 11: 229-239.

Christensen, R.M. 1979. *Mechanics of Composite Materials.* John Wiley and Sons, New York.

Couniot, A., L. Dheur, and F. Dupret 1993. Numerical Simulation of Injection Molding: Non-Isothermal Filling of Complex Thin Parts Including Abrupt Changes in Thickness and Bifurcationof the Mid Surface, in *Mathematical Modeling for Materials Processing*, M. Cross, J.F.T. Pittman, R.D. Wood (Eds), IMA Conf. Series, ClarendonPress, 380-398.

Crochet, M.J., F. Dupret, and V. Verleye 1994. Injection Molding, in *Flow and Rheology in Polymer Composites*, S. Advani (Ed.), Elsevier, 415-463.

Dihn S.M. and R.C. Armstrong 1984. Rheological Equation of State for Semi-Concentrated Fiber Suspensions. *J. Rheol.* 28: 207-227.

Dubois, F. and R. Keunings 1995. Non-Linear Micro-Macro Numerical Analysis of DCB Testing of Thermoplastic Composites, *Proc. 4th Int. Conf. on Computational Plasticity*, Barcelona, Spain, in press.

Dupret, F. and L. Dheur 1992. Modelling and Numerical Simulation of Heat Transfer during the-Filling Stage of Injection Molding. *Proc. of the Oji International Seminar on Advanced Heat Transfer in Manufacturing and Processing of New Materials*, Tomakomai, Japan, Tanasawa I. & Lior N. editors, Hemisphere Publishing Corporation, 583-591.

Dupret, F. and L. Vanderschuren 1988. Calculation of the Temperature Field in Injection Molding. *AIChE Journal* 34 : 1959-1972.

Folgar F. and C.L. Tucker III 1984. Orientation-Behavior of Fibers in Concentrated Suspensions. *J. Reinf. Plast. Comp.* 3: 98-119.

Halpin, J.C and J.L. Kardos 1976. The Halpin-Tsaï equations: a Review. *Polymer Engng. Sci.* 16: 344-352.

Henry de Frahan H., V. Verleye, F. Dupret, and M.J.Crochet 1992. Numerical Prediction of Fiber Orientation in Injection Molding. *Polymer Engng. Sci.* 32: 254-266.

Hieber, C.A. and S.F. Shen 1980. A Finite Element/Finite Difference Simulation of the Injection Molding Filling Process. *J. Non-Newt. Fluid Mech.* 7: 1-32.

Keunings R. 1989. Simulation of Viscoelastic Fluid Flow, in *Fundamentals of Computer Modeling for Polymer Processing*, Ed. C.L. Tucker III, Carl Hanser Verlag, 402-470.

Lipscomb G.G., M.M. Denn, D.U. Hur, and D.V. Boger 1988. The Flow of Fiber Suspensions in Complex Geometries, *J. Non-Newt. Fluid Mech.* 26: 297-325.

Mori, T. and K. Tanaka 1973. Average Stress in Matrix and Average Elastic Energy of Materials with Misfitting Inclusions. *Acta Met.* 21: 571-574.

Muneo, H. and S. Nemat-Nasser 1994. Double-Inclusion Model and Overall Moduli of Multi-Phase Composites. *J. of Engineering Materials and Technology* 116: 305-309.

Rosenberg R., M.M. Denn, and R. Keunings 1990. Simulation of Non-Recirculating Flows of Dilute Fiber Suspensions. *J. Non-Newtonian Fluid Mech.* 37: 317-345.

Sandstrom C.R. and C.L. Tucker III 1993. A Theory for Concentrated Fiber Suspensions with Strong Fiber-Fiber Interactions. *Makromol. Chem.* 68: 291-300.

Termonia, Y. 1987. Computer Model for the Elastic Properties of Short Fibre and Particulate Filled Polymers. *Journal of Materials Science* 22: 1733-1736.

Verleye, V. and F. Dupret 1993. Prediction of Fiber Orientation in Complex Injection Molded Parts. *Proc. of the ME'93 ASME Winter Annual Meeting*, New Orleans, Louisiana (USA), 139-163.

Verleye, V., A. Couniot and F. Dupret 1994. Numerical Prediction of the Orientation Field in Complex Composite Injection Molded Parts. *Proc. of the ME'94 ASME Winter Annual Meeting*, Chicago, Illinois(USA), 265-279.

Verleye, V. and F. Dupret 1995. A Natural Closure Approximation to Simulate the Behavior of Fiber Reinforced Suspensions, submitted.

Simulation of Materials Processing: Theory, Methods and Applications, Shen & Dawson (eds)
© 1995 Balkema, Rotterdam. ISBN 90 5410 553 4

Author index